HANDBOOK OF STATISTICS

VOLUME 26

Handbook of Statistics

VOLUME 26

General Editor

C.R. Rao

ELSEVIER

AMSTERDAM • BOSTON • HEIDELBERG • LONDON • NEW YORK • OXFORD
PARIS • SAN DIEGO • SAN FRANCISCO • SINGAPORE • SYDNEY • TOKYO

Psychometrics

Edited by

C.R. Rao

Center for Multivariate Analysis
Department of Statistics, The Pennsylvania State University
University Park, PA, USA

S. Sinharay

Center for Statistical Theory and Practice
Educational Testing Service
Princeton, NJ, USA

ELSEVIER

AMSTERDAM • BOSTON • HEIDELBERG • LONDON • NEW YORK • OXFORD
PARIS • SAN DIEGO • SAN FRANCISCO • SINGAPORE • SYDNEY • TOKYO

N·H

North-Holland is an imprint of Elsevier
Radarweg 29, PO Box 211, 1000 AE Amsterdam, The Netherlands
The Boulevard, Langford Lane, Kidlington, Oxford OX5 1GB, UK

First edition 2007

Library of Congress Cataloging-in-Publication Data
A catalog record for this book is available from the Library of Congress

British Library Cataloguing in Publication Data
A catalogue record for this book is available from the British Library

ISBN-13: 978-0-444-52103-3
ISBN-10: 0-444-52103-8
ISSN (Series): 0169-7161

For information on all North-Holland publications
visit our website at books.elsevier.com

Transferred to digital print 2007

Printed and bound by CPI Antony Rowe, Eastbourne

Preface

There have been tremendous developments in Psychometric theory, methodology, computation and applications in the past decades; consequently, a vast knowledge base in the form of books, research articles, and conference presentations has accumulated in the field. The *Handbook of Statistics on Psychometrics* highlights recent developments in the theory, methodology, applications, and computational methods. The collection of articles included in this volume are written by the leading experts in Psychometrics, and should serve as a guide to research scholars, reading material for graduate students and source material for consultants and practitioners in universities as well as industries. The broad coverage will provide a general overview of the vast expanses of the field, and will provide the reader with the state-of-the-art of Psychometrics methods that are available as well as the frontiers of research activities in developing new and better methodologies.

The volume begins with a chapter by David Thissen and Lyle Jones, which provides a general overview of Psychometrics. There are several chapters on classical test theory, a primary area in Psychometrics, as well as several chapters on item response theory, another popular area. There are chapters on factor analysis, structural equation modeling, matrix methods, multidimensional scaling, multilevel modeling, latent class analysis, electronic essay grading, and scaling models for preference data, which have found several important applications in the field. The volume includes several chapters intended to provide an in-depth look at operational testing, like those on the National Assessment of Educational Progress, and statistical procedures used in college admission testing. There are also chapters on several topics of special interest, like automatic item generation, causality, computer-adaptive testing, evidence-centered design, value-added modeling, research synthesis, scoring open-ended items, and statistical paradoxes. The last chapter "Future Challenges in Psychometrics" includes short descriptions of potential research problems by several renowned researchers and the Vice-Presidents of the Research and Development divisions of several leading testing companies; this chapter will surely provide guidance to researchers looking for challenging Psychometric problems.

Several people helped us in editing this volume. We express our sincere thanks to all the authors for their excellent contributions and for helping us to publish this volume in a timely manner. We are indebted to Dan Eignor and Shelby Haberman for their help with several reviews and general advice, to Diane Rein for her administrative help, and to Henry Braun, Matthew Johnson, Robert Mislevy, David Thissen, Matthias von Davier, and Neil Dorans for their general advice. We gratefully acknowledge the help of several authors contributing chapters and reviews, and of Russell Almond, Yigal Attali,

Isaac Bejar, Daniel Bolt, Yu Fang, Jean-Paul Fox, Jan de Leeuw, Deping Li, Lou Mariano, Steven Raudenbush, Victoria Savalei, Dipendra Raj Subedi, Bernard Veldkamp, Jack Vevea, Alina von Davier, Hua Wei, Mark Wilson, Wendy Yen, Xueli Xu, Jinming Zhang, and Michael Zieky for their reviews. Our special thanks go to Edith Bomers, Kristi Green, Anita Koch, and Andy Deelen of Elsevier, Amsterdam, for taking a keen interest in this project and for helping us with the final production of this volume. Finally, the volume is a tribute to all individuals who helped in developing this particular area of knowledge.

C.R. Rao
S. Sinharay

Table of contents

Preface v

Table of contents vii

Contributors xvii

Ch. 1. A History and Overview of Psychometrics 1
Lyle V. Jones and David Thissen

1. Introduction 1
2. The origins of psychometrics (circa 1800–1960) 1
3. Psychological scaling 8
4. Psychological measurement (test theory) 10
5. Factor analysis 15
6. Psychological statistics 18
7. Conclusion 21

References 22

Ch. 2. Selected Topics in Classical Test Theory 29
Charles Lewis

1. A modern introduction 31
2. Restriction of range 36
3. Nonlinear classical test theory 39

References 42

Ch. 3. Validity: Foundational Issues and Statistical Methodology 45
Bruno D. Zumbo

1. Introductory remarks 45
2. Foundational issues 47
3. Statistical methods 65
4. Closing remarks 71

References 76

Ch. 4. Reliability Coefficients and Generalizability Theory 81
Noreen M. Webb, Richard J. Shavelson and Edward H. Haertel

1. Introduction 81
2. Reliability Coefficients in Classical Test Theory 81
3. Generalizability theory 93
4. Concluding remarks 120
References 121

Ch. 5. Differential Item Functioning and Item Bias 125
Randall D. Penfield and Gregory Camilli

1. Introduction 125
2. General definition of DIF 127
3. Item response theory approaches for dichotomous items 128
4. Proportion-difference approaches for dichotomous items 132
5. Common odds ratio approaches for dichotomous items 135
6. Logistic regression approaches for dichotomous items 139
7. Classification schemes for dichotomous items 141
8. Item response theory approaches for polytomous items 143
9. Mean-difference approaches for polytomous items 146
10. Multivariate hypergeometric distribution approaches for polytomous items 147
11. Common odds ratio approaches for polytomous items 149
12. Logistic regression approaches for polytomous items 150
13. Differential test functioning and DIF effect variance 152
14. Explaining the sources of DIF 155
15. Steps to conducting DIF analyses: An applied example 157
16. Cautions and limitations 159
References 163

Ch. 6. Equating Test Scores 169
Paul W. Holland, Neil J. Dorans and Nancy S. Petersen

1. Introduction and definitions 169
2. Data collection designs used in test score equating 174
3. Procedures for equating scores 183
4. Best practices and challenges to best practices 197
References 201

Ch. 7. Electronic Essay Grading 205
Shelby J. Haberman

1. Regression analysis 206
2. Composite scales 215
3. Content analysis 218
4. Analysis of discrete responses 226
5. Bayesian analysis 230
6. Conclusions 231
References 232

Ch. 8. Some Matrix Results Useful in Psychometric Research 235
C. Radhakrishna Rao

1. Notation and Basic Matrix Results 235
2. Decomposition of matrices 239
3. Matrix approximations 243
4. Procrustean transformation 247
5. Correspondence analysis 250
6. Metric and multidimensional scaling 252
   Acknowledgements 254
   References 254

Ch. 9. Factor Analysis 257
Haruo Yanai and Masanori Ichikawa

1. Introduction 257
2. Historical development of factor analysis 257
3. The factor analysis model 261
4. Statistical inference in factor analysis 271
5. Factor rotation and estimation of factor scores 281
   Acknowledgements 293
   References 293

Ch. 10. Structural Equation Modeling 297
Ke-Hai Yuan and Peter M. Bentler

1. Introduction 297
2. Model identification 303
3. Estimation and evaluation 305
4. Missing data 327
5. Multiple groups 334
6. Multilevel models with hierarchical data 338
7. Examples 344
   References 348

Ch. 11. Applications of Multidimensional Scaling in Psychometrics 359
Yoshio Takane

1. Introduction 359
2. MDS with the simple Euclidean model 361
3. Individual differences MDS 375
4. Unfolding analysis 385
5. Concluding remarks 395
   References 397

Ch. 12. Multilevel Models in Psychometrics 401
Fiona Steele and Harvey Goldstein

1. Introduction 401
2. Basic Models for Two-level Hierarchical Data Structures 402
3. Models for repeated measures 406
4. Models for multivariate response data 409
5. Models for non-hierarchical structures 412
6. Further extensions 416
7. Estimation procedures and software 417
8. Resources 418

References 418

Ch. 13. Latent Class Analysis in Psychometrics 421
C. Mitchell Dayton and George B. Macready

1. Introduction 421
2. The model for LCA 422
3. Estimation and model fit 423
4. Analyses with score data 428
5. Unconstrained latent class models 432
6. Scaling models 433
7. Models incorporating grouping of respondents 436
8. Covariate latent class models 438
9. Other models 441

References 444

Ch. 14. Random-Effects Models for Preference Data 447
Ulf Böckenholt and Rung-Ching Tsai

1. Introduction 447
2. Thurstonian random utility models 449
3. Identifiability 455
4. Identifying the scale origin 458
5. Estimation 459
6. Applications 460
7. Concluding remarks 466

References 466

Ch. 15. Item Response Theory in a General Framework 469
R. Darrell Bock and Irini Moustaki

1. Introduction 469
2. The general IRT framework 470
3. Item response models 472
4. Estimation of item and group parameters 490
5. Estimation of respondent scores 499
6. Item factor analysis 505
7. Response relations with external variables 508

References 509

Ch. 16. Rasch Models 515
Gerhard H. Fischer

1. Some history of the Rasch model 515
2. Some basic concepts and properties of the RM 517
3. Characterizations and scale properties of the RM 522
4. Item parameter estimation 532
5. Person parameter estimation 546
6. Testing of fit 549
7. The linear logistic test model 564
8. Longitudinal linear logistic models 571
9. Some remarks on applications and extensions of the RM 576
References 578

Ch. 17. Hierarchical Item Response Theory Models 587
Matthew S. Johnson, Sandip Sinharay and Eric T. Bradlow

1. Introduction 587
2. Developing the hierarchical IRT model 589
3. Estimation 592
4. Examples 595
5. Conclusions 604
References 604

Ch. 18. Multidimensional Item Response Theory 607
Mark D. Reckase

1. Introduction 607
2. General forms of MIRT models 609
3. Common forms of MIRT models 612
4. Descriptions of item characteristics 619
5. Descriptions of test characteristics 625
6. The estimation of model parameters 629
7. Applications 631
8. Discussion and conclusions 640
References 641

Ch. 19. Mixture Distribution Item Response Models 643
Matthias von Davier and Jürgen Rost

1. Introduction 643
2. Different perspectives on mixture IRT 644
3. Applications of mixture IRT models 655
4. Mixture IRT for large-scale survey assessments 656
5. Conclusion 658
References 659

Ch. 20. Scoring Open Ended Questions 663
Gunter Maris and Timo Bechger

1. Introduction 663
2. Motivating example 663
3. Theory 665
4. A measurement model for scoring open ended questions 673
5. Motivating example revisited 677
6. Discussion 679
References 680

Ch. 21. Assessing the Fit of Item Response Theory Models 683
Hariharan Swaminathan, Ronald K. Hambleton and H. Jane Rogers

1. Introduction: Models and assumptions 683
2. Checking the assumption of unidimensionality 686
3. Assessing model data fit: Checking model predictions 691
4. Empirical example 707
5. Conclusions 714
Acknowledgements 715
References 715

Ch. 22. Nonparametric Item Response Theory and Special Topics 719
Klaas Sijtsma and Rob R. Meijer

1. Place of nonparametric models in item response theory 719
2. Analyzing test data using nonparametric IRT 727
3. Special topics in NIRT 739
References 743

Ch. 23. Automatic Item Generation and Cognitive Psychology 747
Susan Embretson and Xiangdong Yang

1. Introduction 747
2. Item development and automatic item generation 748
3. Item response models for automatic item generation 753
4. Calibration of the cognitive IRT models 756
5. An application: Cognitive models for algorithmically generated spatial ability items 761
6. Overall discussion 765
References 766

Ch. 24. Statistical Inference for Causal Effects, with Emphasis on Applications in Psychometrics and Education 769
Donald B. Rubin

1. Causal inference primitives 769
2. The assignment mechanism 776
3. Randomization-based modes of causal inference 781
4. Posterior predictive causal inference 786
5. Complications 794

References 796

Ch. 25. Statistical Aspects of Adaptive Testing 801
Wim J. van der Linden and Cees A.W. Glas

1. Introduction 801
2. Response models 802
3. Item calibration 804
4. Ability estimation 811
5. Empirical examples 812
6. Rules for adaptive item selection 814
7. Other statistical issues in adaptive testing 822
8. Concluding comment 833

References 835

Ch. 26. Bayesian Psychometric Modeling From An Evidence-Centered Design Perspective 839
Robert J. Mislevy and Roy Levy

1. Introduction and overview 839
2. An evidentiary perspective 840
3. Evidence-centered design 847
4. Bayesian psychometric modeling 856
5. Discussion 861
6. Concluding remarks 863

References 863

Ch. 27. Value-Added Modeling 867
Henry Braun and Howard Wainer

1. Introduction 867
2. General issues 869
3. Models for estimating teacher effects 875
4. Example 880
5. Current research 884
6. Summing up 888

References 889

Ch. 28. Three Statistical Paradoxes in the Interpretation of Group Differences: Illustrated with Medical School Admission and Licensing Data 893
Howard Wainer and Lisa M. Brown

1. Introduction 893
2. The data 894
3. Simpson's Paradox 895
4. Kelley's Paradox 901
5. Lord's Paradox 910
6. Conclusion 916
Acknowledgements 917
References 918

Ch. 29. Meta-Analysis 919
Larry V. Hedges

1. Introduction 919
2. Effect sizes 920
3. The problem of dependent estimates 923
4. Fixed effects analyses 924
5. Mixed models 942
6. Conclusions 952
References 953

Ch. 30. Vertical Scaling: Statistical Models for Measuring Growth and Achievement 955
Richard J. Patz and Lihua Yao

1. Introduction 955
2. Vertical scaling of achievement tests: Recent and past statistical practice 956
3. Statistical modeling approaches for vertical scaling 960
4. Exploring grade-to-grade growth in achievement: An application of hierarchical modeling 964
5. Discussion 972
References 973

Ch. 31. COGNITIVE DIAGNOSIS

Ch. 31A. Review of Cognitively Diagnostic Assessment and a Summary of Psychometric Models 979
Louis V. DiBello, Louis A. Roussos and William Stout

1. Preliminaries 979
2. Implementation framework for diagnostic assessment 980
3. Reliability, validity, and granularity 993
4. Summary of cognitively diagnostic psychometric models 995
5. Summary and conclusions: Issues and future directions 1023
References 1027

Ch. 31B. Some Notes on Models for Cognitively Based Skills Diagnosis 1031
Shelby J. Haberman and Matthias von Davier

Introduction 1031
1. Retrofitting a test with a hypothesized skill structure 1033
2. Dimensionality 1034
3. Utility and use of skills classifications 1035
4. Latent structure models, latent responses, conjunctive versus compensatory skills 1036
5. Outlook 1037
References 1038

Ch. 32. The Statistical Procedures Used in National Assessment of Educational Progress: Recent Developments and Future Directions 1039
Matthias von Davier, Sandip Sinharay, Andreas Oranje and Albert Beaton

1. Introduction 1039
2. The current NAEP model and estimation method 1041
3. Example: NAEP data and results 1047
4. Alternatives approaches 1049
5. Conclusions 1052
Acknowledgements 1053
Appendix: Sampling students in NAEP 1053
References 1054

Ch. 33. Statistical Procedures Used in College Admissions Testing 1057
Jinghua Liu, Deborah J. Harris and Amy Schmidt

1. Introduction 1057
2. Test design and assembly 1059
3. Item pretesting 1064
4. Item analysis 1066
5. Scaling and equating 1074
6. Reliability and validity 1077
7. Other issues 1086
8. Summary 1088
References 1089

Ch. 34. FUTURE CHALLENGES IN PSYCHOMETRICS

Ch. 34A. Integration of Models 1095
Robert L. Brennan

What constitutes a replication? 1095
What are true scores? 1096
What is error? 1097
No "right" model 1097
References 1097

Ch. 34B. Linking Scores Across Computer and Paper-Based Modes of Test Administration 1099
Daniel R. Eignor

References 1102

Ch. 34C. Linking Cognitively-Based Models and Psychometric Methods 1103
Mark J. Gierl and Jacqueline P. Leighton

Introduction 1103
Developing cognitive models of task performance 1104
Incorporating cognitive models into psychometric methods 1105
Conclusion 1106
References 1106

Ch. 34D. Technical Considerations in Equating Complex Assessments 1107
Ida Lawrence

References 1109

Ch. 34E. Future Challenges to Psychometrics: Validity, Validity, Validity 1111
Neal Kingston

References 1112

Ch. 34F. Testing with and without Computers 1113
Piet Sanders

Ch. 34G. Practical Challenges to Psychometrics Driven by Increased Visibility of Assessment 1117
Cynthia Board Schmeiser

Growth models 1117
Diagnosis, prescription, and instruction 1118
Dual-platform testing 1118

Index 1121

Handbook of Statistics Contents of Previous Volumes 1143

Contributors

Beaton, Albert E., *Lynch School of Education, Boston College, Chestnut Hill, MA 02567*; *e-mail: beatonal@bc.edu* (Ch. 32).

Bechger, Timo, *Cito, Arnhem, The Netherlands*; *e-mail: timo.bechger@citogroep.nl* (Ch. 20).

Bentler, Peter M., *University of California, Los Angeles, CA 90095*; *e-mail: bentler@ucla.edu* (Ch. 10).

Bock, R. Darrell, *Department of Psychology, University of Chicago, Chicago, IL 60637*; *e-mail: rbock@uchicago.edu* (Ch. 15).

Böckenholt, Ulf, *Faculty of Management, McGill University, Montreal, QC, Canada H3A 1G5*; *e-mail: ulf.bockenholt@mcgill.ca* (Ch. 14).

Bradlow, Eric T., *The Wharton School, University of Pennsylvania, Philadelphia, PA 19104*; *e-mail: ebradlow@wharton.upenn.edu* (Ch. 17).

Braun, Henry, *Educational Testing Service, MS-12T, Princeton, NJ 08541*; *e-mail: hbraun@ets.org* (Ch. 27).

Brennan, Robert L., *Center for Advanced Studies in Measurement and Assessment (CASMA), University of Iowa, Iowa City, IA 52242-1529*; *e-mail: robert-brennan@uiowa.edu* (Ch. 34A).

Brown, Lisa M., *Malvern, PA 19355*; *e-mail: lisabrown5@comcast.net* (Ch. 28).

Camilli, Gregory, *Graduate School of Education, New Brunswick, NJ 08901*; *e-mail: camilli@rci.rutgers.edu* (Ch. 5).

Dayton, C. Mitchell, *Department of Measurement & Statistics, University of Maryland, College Park, MD 20742*; *e-mail: cdayton@umd.edu* (Ch. 13).

DiBello, Louis V., *Department of Psychology, University of Illinois at Chicago, Chicago, IL 60607*; *e-mail: ldibello@ets.org* (Ch. 31A).

Dorans, Neil J., *Educational Testing Service, MS 12-T, Princeton, NJ 08541*; *e-mail: ndorans@ets.org* (Ch. 6).

Eignor, Daniel R., *Educational Testing Service, MS 03-T, Princeton, NJ 08541*; *e-mail: deignor@ets.org* (Ch. 34B).

Embretson, Susan, *School of Psychology, Georgia Institute of Technology, Atlanta, GA 30332-0170*; *e-mail: se72@mail.gatech.edu* (Ch. 23).

Fischer, Gerhard H., *University of Vienna, A-1170 Wien (Vienna), Austria*; *e-mail: gh.fischer@univie.ac.at* (Ch. 16).

Gierl, Mark J., *Department of Educational Psychology, University of Alberta, Edmonton, AB, Canada T6G 2G5*; *e-mail: mark.gierl@ualberta.ca* (Ch. 34C).

Glas, Cees A.W., *Faculty of Behavioral Science, Department of Research Methodology, Measurement and Data Analysis, University of Twente, 7500 AE Enschede, The Netherlands*; *e-mail: c.a.w.glas@edte.utwente.nl* (Ch. 25).

Goldstein, Harvey, *Bedford Group, Institute of Education, London WC1H OAL, UK*; *e-mail: harvey@bhgoldstein.co.uk* (Ch. 12).

Haberman, Shelby J., *Educational Testing Service, MS-12T, Princeton, NJ 08541*; *e-mail: shaberman@ets.org* (Chs. 7, 31B).

Hambleton, Ronald K., *University of Massachusetts, Amherst, MA 01003*; *e-mail: rkh@educ.umass.edu* (Ch. 21).

Haretel, Edward H., *School of Education, Stanford, CA 94305-3096*; *e-mail: haertel@stanford.edu* (Ch. 4).

Harris, Deborah J., *ACT Inc., Iowa City, IA 52243*; *e-mail: harris@act.org* (Ch. 33).

Hedges, Lary V., *Northwestern University, Chicago, IL 60637*; *e-mail: l-hedges@northwestern.edu* (Ch. 29).

Holland, Paul W., *Educational Testing Service, MS-12T, Princeton, NJ 08541*; *e-mail: pholland@ets.org* (Ch. 6).

Ichikawa, Masanori, *Tokyo University of Foreign Studies, 3-11-1 Asahi-cho, Fuchu-shi, Tokyo, 183-8534, Japan*; *e-mail: ichikawa.m@tufs.ac.jp* (Ch. 9).

Johnson, Matthew S., *Department of Statistics & CIS, Baruch College, New York, NY 10010*; *e-mail: matthew_johnson@baruch.cuny.edu* (Ch. 17).

Jones, Lyle V., *Department of Psychology, University of North Carolina, Chapel Hill, NC 27599-3270*; *e-mail: lvjones@email.unc.edu* (Ch. 1).

Kingston, Neal, *CTB McGraw-Hill, Monterey, CA*; *e-mail: neal.kingston@gmail.com* (Ch. 34E).

Lawrence, Ida, *Educational Testing Service, MS 19-T, Princeton, NJ 08541*; *e-mail: ilawrence@ets.org* (Ch. 34D).

Leighton, Jacquelline P., *Centre for Research in Applied Measurement and Evaluation, University of Alberta, Edmonton, AB, Canada T6G 2G5*; *e-mail: jacqueline.leighton@ualberta.ca* (Ch. 34C).

Levy, Roy, *Arizona State University, Tempe, AZ 85287*; *e-mail: roy.levy@asu.edu* (Ch. 26).

Lewis, Charles, *Educational Testing Service, MS-12T, Princeton, NJ 08541*; *e-mail: clewis@ets.org* (Ch. 2).

Liu, Jianghua, *Educational Testing Service, MS-15L, Princeton, NJ 08541*; *e-mail: jliu@ets.org* (Ch. 33).

Macready, George B., *Department of Measurement & Statistics, University of Maryland, College Park, MD 20742*; *e-mail: gm2@umail.umd.edu* (Ch. 13).

Maris, Gunter, *CITO, NL-6801 MG, Arnhem, The Netherlands*; *e-mail: gunter.maris@citogroep.nl* (Ch. 20).

Meijer, Rob R., *University of Twente, 7500 AE Enschede, The Netherlands*; *e-mail: r.r.meijer@edte.utwente.nl* (Ch. 22).

Mislevy, Robert J., *Department of Measurement & Statistics, University of Maryland, College Park, MD 20742*; *e-mail: rmislevy@ets.org* (Ch. 26).

Moustaki, Irini, *Department of Statistics, Athens University of Economics and Business, 104 34 Athens, Greece*; *e-mail: moustaki@aueb.gr* (Ch. 15).

Oranje, Andreas, *Educational Testing Service, MS 02-T, Princeton, NJ 08541*; *e-mail: aoranje@ets.org* (Ch. 32).

Patz, Richard, *CTB McGraw-Hill, Monterey, CA*; *e-mail: richard.patz@rjpatz.com* (Ch. 30).

Penfield, Randall D., *University of Miami School of Education, Coral Gables, FL 33124-2040*; *e-mail: penfield@miami.edu* (Ch. 5).

Petersen, Nancy S., *ACT Inc., Iowa City, IA 52243*; *e-mail: nancy.petersen@act.org* (Ch. 6).

Rao, C. Radhakrishna, *Department of Statistics, Pennsylvania State University, University Park, PA 16802*; *e-mail: crr1@psu.edu* (Ch. 8).

Reckase, Mark D., *Michigan State University, East Lansing, MI 48824*; *e-mail: reckase@msu.edu* (Ch. 18).

Rogers, H. Jane, *Neag School of Education, University of Connecticut, Storrs, CT 06269*; *e-mail: jane.rogers@uconn.edu* (Ch. 21).

Rost, Jürgen, *IPN, 24098 Kiel, Germany*; *e-mail: an@j-rost.de* (Ch. 19).

Roussos, Louis A., *Measured Progress, Dover, NH 03820*; *e-mail: lroussos@measuredprogress.org* (Ch. 31A).

Rubin, Donald B., *Department of Statistics, Harvard University, Science Center, Cambridge, MA 02138-2901*; *e-mail: rubin@stat.harvard.edu* (Ch. 24).

Sanders, Piet, *Cito, Arnhem, The Netherlands*; *e-mail: piet.sanders@cito.nl* (Ch. 34F).

Schmeiser, Cyndie, *ACT Inc., Iowa City, IA 52243*; *e-mail: cyndie.schmeiser@act.org* (Ch. 34G).

Schmidt, Amy, *The College Board, New York, NY 10023*; *e-mail: aschmidt@collegeboard.org* (Ch. 33).

Shavelson, Richard J., *School of Education, Stanford, CA 94305-3096*; *e-mail: richs@stanford.edu* (Ch. 4).

Sijtsma, Klaas, *Department of Methodology and Statistics, FSW, Tilburg University, 5000 LE Tilburg, The Netherlands*; *e-mail: k.sijtsma@uvt.nl* (Ch. 22).

Sinharay, Sandip, *Center for Statistical Theory and Practice, Educational Testing Service, Princeton, NJ 08541*; *e-mail: ssinharay@ets.org* (Chs. 17, 32).

Steele, Fiona, *Bedford Group, Institute of Education, London WC1H OAL, UK*; *e-mail: f.steele@ioe.ac.uk* (Ch. 12).

Stout, William, *University of Illinois at Urbana-Champaign, Champaign, IL 61820*; *e-mail: w-stout1@uiuc.edu* (Ch. 31A).

Swaminathan, Hariharan, *University of Miami School of Education, Coral Gables, FL 33124-2040*; *e-mail: hariharan.swaminathan@uconn.edu* (Ch. 21).

Takane, Yoshio, *Department of Psychology, McGill University, Montreal, QC, Canada H3A 1B1*; *e-mail: takane@takane2.psych.mcgill.ca* (Ch. 11).

Thissen, Dave, *University of North Carolina, Chapel Hill, NC 27599-3270*; *e-mail: dthissen@email.unc.edu* (Ch. 1).

Tsai, Rung-Ching, *Department of Mathematics, National Taiwan Normal University, Taipei 116, Taiwan*; *e-mail: rtsai@math.ntnu.edu.tw* (Ch. 14).

van der Linden, Wim J., *Department of Research Methodology, Measurement and Data Analysis, University of Twente, 7500 AE Enschede, The Netherlands*; *e-mail: w.j.vanderlinden@edte.utwente.nl* (Ch. 25).

von Davier, Matthias, *Educational Testing Service, MS 02-T, Princeton, NJ 08541*; *e-mail: mvondavier@ets.org* (Chs. 19, 31B, 32).

Wainer, Howard, *National Board of Medical Examiners, Philadelphia, PA 19104*; *e-mail: hwainer@nbme.org* (Chs. 27, 28).

Webb, Noreen M., *Department of Education, University of California, Los Angeles, CA 90095-1521*; *e-mail: webb@ucla.edu* (Ch. 4).

Yanai, Haruo, *St Luke's College of Nursing, Akashi-cho, Chuo-ku, Tokyo, Japan and National Center for University Entrance Examinations, Komaba, Meguro-ku, Tokyo, Japan*; *e-mail: yanai@rd.dnc.ac.jp* (Ch. 9).

Yang, Xiangdong, *Center for Educational Testing and Evaluation, University of Kansas, Lawrence, KS 66045*; *e-mail: xiangdon@ku.edu* (Ch. 23).

Yao, Lihua, *CTB McGraw-Hill, Monterey, CA*; *e-mail: lihua_yao@ctb.com* (Ch. 30).

Yuan, Ke-Hai, *Department of Psychology, University of Notre Dame, Notre Dame, IN 46556*; *e-mail: kyuan@nd.edu* (Ch. 10).

Zumbo, Bruno D., *Department of ECPS, The University of British Columbia, Vancouver, BC, Canada V6T 1Z4*; *e-mail: bruno.zumbo@ubc.ca* (Ch. 3).

Handbook of Statistics, Vol. 26
ISSN: 0169-7161

DOI: 10.1016/S0169-7161(06)26001-2

1

A History and Overview of Psychometrics

Lyle V. Jones and David Thissen[†]

1. Introduction

Psychometrics is defined well by the subheading that appeared under the title of the journal *Psychometrika* from its inception in 1936 until a cover redesign in 1984, "a journal devoted to the development of psychology as a quantitative rational science." Some definitions are even more inclusive; *WordNet* (developed by George Miller and his colleagues in the cognitive science laboratory at Princeton University; http://wordnet.princeton.edu/) defines *psychometrics* as "any branch of psychology concerned with psychological measurements."

In this chapter, we provide an alternative definition of psychometrics, eschewing the compactness of dictionary definitions, to focus on what researchers in the field of quantitative psychology do, and have done. For that, we begin with a beginning.

2. The origins of psychometrics (circa 1800–1960)

With the active support of graduate students and colleagues of L.L. Thurstone at The University of Chicago, the Psychometric Society was founded in 1935. The Society sponsored the journal *Psychometrika*, of which Volume 1, Number 1 appeared in March, 1936. At nearly the same time, the first edition of J.P. Guilford's (1936) *Psychometric Methods* was published. From one perspective, these events may serve to temporally locate the beginning of a formal sub-discipline of psychometrics. The founding of the Psychometric Society led to the publication in *Science* of Thurstone's presidential address to the meeting of the Society in September of 1936 that presented a strong plea to recognize a mathematical underpinning for psychological research (Thurstone, 1937).

Long before 1935, however, quantitative methods had achieved prominence in psychological research in Europe and in the United States. An early treatise on psychophysics, correlation, and ability testing is that of Brown and Thomson (1921), who had made important contributions to the development of test theory and factor analysis,

[†]We are grateful to Robert C. (Bud) MacCallum, Sandip Sinharay, Howard Wainer, and an anonymous reviewer for very helpful comments on an earlier draft of this chapter. Any errors that remain are, of course, our own.

Table 1
Some participants in various phases of the historical development of psychometrics

"Pioneers in psychometric methods" from Guilford's (1936) *Psychometric Methods*

Ernst Heinrich Weber	Edward Lee Thorndike
Gustav Theodor Fechner	Louis Leon Thurstone
Gerog Elias Müller	Alfred Binet
F.M. Urban	Truman L. Kelley
Sir Francis Galton	Lewis M. Terman
J. McKeen Cattell	Charles E. Spearman

Some of the participants in the WW I Army Alpha/Beta project

Lt. Col. Robert Yerkes	Lt. Arthur S. Otis
Major Lewis M. Terman	Edward L. Thorndike
Major Harold C. Bingham	L.L. Thurstone
Captain Edwin G. Boring	George M. Whipple
Lt. Carl C. Brigham	

Some of the participants in the WW II Army Air Force project[a]

Officers		Enlisted Personnel	
John C. Flanagan	Chester W. Harris	William Angoff	Joseph E. Mooney
Stuart W. Cook	Roger W. Heynes	Robert R. Blake	Albert Pepitone
Meredith P. Crawford	Nicholas R. Hobbs	Urie Bronfenbrenner	Harold M. Proshansky
Frederick B. Davis	Paul Horst	Benjamin Fruchter	Henry W. Riecken
Philip H. Dubois	Lloyd G. Humphreys	Nathaniel L. Gage	Milton Rokeach
Frank J. Dudek	John I. Lacey	Nathan M. Glaser	Elliot Steller
John E. French	Arthur W. Melton	Albert H. Hastorf	Rains B. Wallace
Robert M. Gagne	Neal E. Miller	Harold H. Kelly	Wilse B. Webb
Frank A. Geldard	William G. Molenkopf	Ardie Lubin	George R. Welch
Edwin E. Ghiselli	Donald E. Super	Samuel B. Lyerly	Joseph M. Wepman
James J. Gibson	John W. Thibaut	William McGrath	Wayne S. Zimmerman
J.P. Guilford	Robert L. Thorndike		

Some of the participants in the WW II OSS project[b]

Henry Murray	James G. Miller	Eliot Steller
Urie Bronfenbrenner	Donald W. McKinnon	Percival M. Symonds
Donald W. Fiske	O.H. Mowrer	Edward C. Tolman
John W. Gardner	Theodore M. Newcomb	Robert C. Tryon
David Krech	R. Nevitt Sanford	

[a] A complete listing of about 1500 names associated with the Army Air Force project is provided by DuBois (1947, pp. 377–392).

[b] A complete listing of the staff associated with the OSS project is provided in the Office of Strategic Services Assessment Staff (1948) report.

respectively. Other "pioneers in psychometric methods" were featured in portraits in the frontispiece of the first edition of Guilford's (1936) *Psychometric Methods*; they are listed here in the top panel of Table 1.

Figure 1 places some of the important figures in the early history of psychometrics on a timeline and illustrates some of the linkages among them. It is of interest to trace

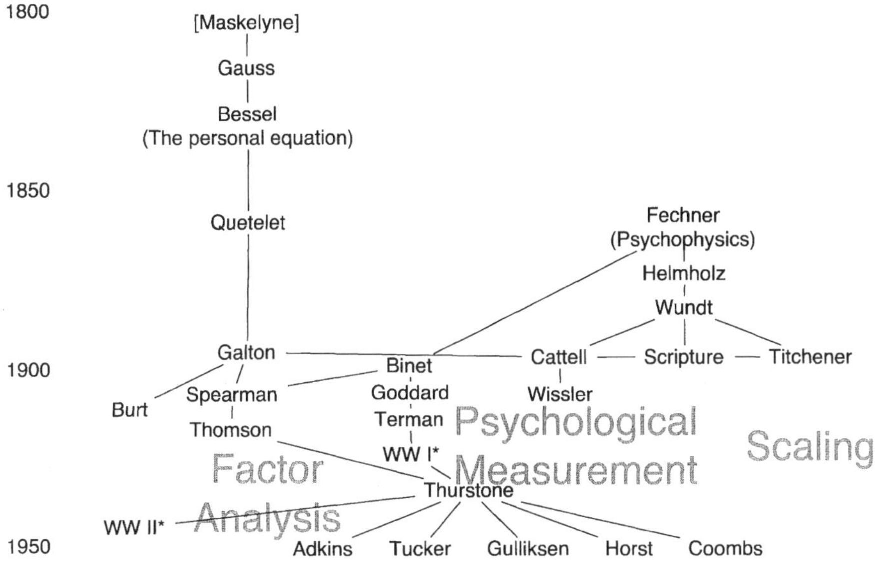

Fig. 1. Some of the important figures in the early history of psychometrics on an approximate timeline. *Some of the participants in the World War I and World War II projects are listed in Table 1.

the influence of some of those scholars and others associated with them, for their works form the roots of contemporary psychometrics.[1]

2.1. *Psychophysics*

The founding of psychometrics might well be thought to correspond with the founding of academic psychology, marked by the publication in 1860 of physicist-philosopher Gustav Fechner's *Elemente der Psychophysik*. Fechner defined psychophysics as "an exact science of the fundamental relations of dependency between body and mind" (Gulliksen in the foreward to Torgerson, 1958, p. v). Boring (1929, p. 286) wrote that Fechner "set experimental quantitative psychology off upon the course which it has followed." Fechner experimentally established the lawful relation between the measure of a physical stimulus (R) and the measure of an observer's sensation thereof (S) as

$$S = k \log R,$$

a formulation that stimulated psychometric research not only in Europe, especially in Germany, but somewhat later in the United States as well.

Hermann von Helmholtz, a physicist–physiologist first at Bonn, then Heidelberg and finally Berlin, contributed to many scientific fields, including the psychology of sensation. He accepted Fechner's proposed logarithmic relation between sensation and stimulus, and he adopted psychophysical methods to support his theories of vision,

[1] The informative review provided by Hilgard (1987) of contributions of individual psychologists has contributed in a variety of ways to the following discussion.

hearing, and tactile sensation. Helmholtz' former assistant was Wilhelm Wundt, whose Leipzig laboratory in 1879 became the first psychology laboratory in the world. Wundt and his students also employed psychophysical methods as primary tools of experimental psychology. In the United States, too, psychophysics was the dominant experimental method for psychology, imported from Germany by psychologists James McKeen Cattell (the first American PhD of Wundt in 1886) at Pennsylvania, Edward Scripture (a Wundt PhD in 1891) at Yale, and Edward Titchener (a Wundt PhD in 1892) at Cornell, among others.

2.2. *Early testing of individual differences*

From another perspective, psychometrics had been founded in the early years of the 19th century by two astronomers, Friedrich Wilhelm Bessel and Carl Friedrich Gauss.

In 1796, Nevil Maskelyne, astronomer royal at the Greenwich Observatory, had dismissed his assistant David Kinnebrook because Kinnebrook reported times of stellar transit nearly a second later than did Maskelyne. The incident was recorded in *Astronomical Observations at Greenwich Observatory*, and was mentioned in a history of the observatory that appeared in 1816 that was noticed by Bessel, director of a new observatory at Königsberg. Bessel acquired and studied the earlier notes of Maskelyne, then undertook a series of independent studies at Königsberg. In 1823 Bessel reported average differences of 1.0 and 1.2 seconds between his observations and those of two other astronomers, a finding that astonished other astronomers who had expected far smaller individual differences.

Bessel also became aware of earlier work by Gauss, an astronomer at Göttingen, who in 1809 had presented the theory of errors of observation employing a statistical distribution now known as the normal or the Gaussian distribution. Bessel then developed a "personal equation" to correct observations for differences among observers. He also discovered that individual differences became smaller with the use of a chronometer with half-second rather than full-second beats. As a result of his studies, the accuracy of astronomic observations improved markedly (see Boring, 1929, pp. 134–142). Bessel's work anticipated the reaction-time experiments that continue to be popular in psychological laboratories.

In Great Britain, Frances Galton, motivated by his interest in human eugenics, was a champion of the measurement of individual differences. Galton was influenced by Adolph Quetelet, Belgian statistician, who had applied the Gaussian law of error to a wide range of human data. Galton designed apparatus to measure a variety of bodily dimensions. Galton's Anthropometric Laboratory opened in London in 1884, and collected data on about 10,000 persons over the ensuing six years. A prominent supporter of Galton's program in the United States was Joseph Jastrow, from 1888 a professor of psychology at the University of Wisconsin. Jastrow had earned a PhD with G. Stanley Hall at Johns Hopkins in 1886. He began a correspondence with Galton in 1887. By 1893, Jastrow (together with anthropologist Franz Boaz) established a version of Galton's Anthropometric Laboratory at the World's Columbian Exposition in Chicago. Galton visited the Laboratory, as did thousands of others whose physical features were measured.

During the late 1880s, James McKeen Cattell spent time in Galton's laboratory and developed an interest in psychometric measures. In 1889, Cattell returned from England to a psychology professorship at the University of Pennsylvania where he established a laboratory of psychology. Largely ignoring the bodily measurements that had dominated Galton's program, he measured his students' hand strength, rate of movement, pain threshold, reaction times, judgments of elapsed time, etc., referring to these as "mental tests" (Cattell, 1890) and expecting them to be related to other evidence of mental ability (Sokal, 1982, p. 329). Cattell moved to Columbia University in 1891, where he was permitted by the university administration to record for all entering freshman students the results from his mental testing program. He promoted his testing program, but without setting forth either a theoretical or an operational justification for it.

The value of the sensory and motor tests developed by Cattell and others was challenged by Clark Wissler, a graduate student at Columbia who studied both with Cattell and with anthropologist Boaz. Wissler found that, for each of Cattell's "mental tests," the correlations with class grades were essentially zero (Wissler, 1901). That finding effectively ended the testing of sensory reactions and motor skills for assessing mental functioning, although the utility of such tests for evaluating sensory disorders in hearing, vision, and tactile sensitivity continues to be recognized.

2.3. *Psychological testing and factor analysis*

At about the same time that anthropometric testing was gaining public attention in Great Britain and the U.S., Alfred Binet and Théophile Simon in France were developing an alternative approach, specifically designed to assess higher cognitive abilities. Binet and Simon (1905) introduced the concept of mental age, determined by earning a test score that is characteristic of children of a given chronological age, and they developed tests suitable for estimating a child's mental age. The intelligence testing movement also was given a boost by the work of Charles Spearman in Great Britain, who analyzed results from a battery of cognitive tests and interpreted the analyses as providing support for a general intelligence factor, g, common to all of the tests (Spearman, 1904).

In 1906, Henry H. Goddard became research director at the Vineland (NJ) Training School for the Feeble-Minded. By 1908 he had visited France, and he provided a translation of the first version of the Binet–Simon scale (Goddard, 1908). Other translations followed; the best known was that of Stanford University psychology professor Lewis M. Terman (1916). Terman adopted the proposal of William Stern, psychologist at the University of Breslau, to construct a quotient of mental age divided by chronological age. Terman multiplied this quotient by 100 to define an IQ score, based upon a child's responses to items on the Stanford–Binet.

In April of 1917, at the initiative of Robert M. Yerkes, president of the American Psychological Association, the APA Council voted "that the president be instructed to appoint committees from the membership of APA to render to the Government of the United States all possible assistance in connection with psychological problems arising in the military emergency" (Yerkes, 1921, p. 8). Almost immediately, this effort was joined by the National Research Council, that proposed to the Surgeon General of the Army a plan for the psychological examination of all Army recruits. By August, 1917, the plan was approved by the Secretary of War.

During the initial planning period in July, 1917, a statistical unit was formed, with Edward L. Thorndike as statistician and Arthur S. Otis and L.L. Thurstone his assistants. Later Major Robert Yerkes acted for the Surgeon General as head of psychological work, assisted by many others, including those listed in the second panel of Table 1. Their efforts culminated in the establishment of a School for Military Psychology at Camp Oglethorpe, Georgia, the development of batteries of psychological tests (Army Alpha and Beta), and the testing during 1918 of nearly two million U.S. Army recruits.

Arthur S. Otis, as a Stanford graduate student, had worked with Terman. For his dissertation Otis devised procedures for scoring multiple-choice items that made feasible group rather than individual testing; these procedures saw their first large-scale application with the administration of the Army tests. Subsequently, Otis's *Group Intelligence Scale*, modeled after the Army Alpha, was the first commercial group-administered mental test. It was published by the World Book Company in 1918. Terman also published a very slightly modified version of Army Alpha as the *Terman Group Test of Mental Ability* in 1920, again with the World Book Company. Otis joined the World Book staff in 1921. In 1923, Terman and colleagues produced the *Stanford Achievement Test*, an objective measure of what students had learned in school.

After noting the impact in the 1920s of Freud's sexual theories on public awareness of psychology, one respected historian noted that "Even more important in popularizing psychology were the Army 'intelligence' tests and the debates they aroused" (Leuchtenburg, 1958, p. 164). Some effects of the testing controversy on educational reform are presented by Cremin (1961, pp. 185–192).

In the early 1920s, Louis Leon Thurstone served as chair of the Department of Applied Psychology at the Carnegie Institute of Technology. In 1923–1924, Thurstone was supported by the Carnegie Corporation of New York to spend a year in Washington, DC to devise improvements to the federal civil service examinations used in the selection of personnel for U.S. government employment. While fulfilling that mission, Thurstone also prepared for the American Council on Education (ACE) a *Psychological Examination for High School Graduates and College Freshmen*. That exam produced two scores, a linguistic score, "L" and a quantitative score, "Q" (Thurstone, 1924). In 1924 Thurstone joined the faculty at The University of Chicago where he and his wife, Thelma Gwinn Thurstone, continued to produce annual versions of that test until 1947. In that year, the Educational Testing Service (ETS) was founded, drawing together the testing programs of ACE, the Carnegie Foundation for the Advancement of Teaching, and the College Entrance Examination Board (CEEB, now known as the College Board). The Scholastic Aptitude Test (SAT) produced by ETS also provided two scores, one for verbal aptitude that correlated higher with college grades in English and history, and one for mathematical aptitude that correlated higher with grades in mathematics and engineering.

During the years in which the Thurstones were developing the ACE tests, L.L. Thurstone also was devising methods for multiple factor analysis (Thurstone, 1931a, 1935). This led to the collection of results from a battery of tests, first from freshmen at the University of Chicago, then from Chicago public high school students. Factor analyses of those data identified seven to nine separate factors of intellect (Thurstone, 1938), results that can be set in opposition to the Spearman one-factor *g* theory. (Later, when he allowed for factors to be correlated, Thurstone granted the possibility that Spearman's *g* might be found as a single second-order factor, the result of analyzing the correlations among the primary factors.)

Applications as well as theoretical developments in factor analysis were stimulated by the appearance of Thurstone's authoritative book *Multiple Factor Analysis* (Thurstone, 1947). Mulaik (1986) provides a review of research literature relating to factor analysis for the 50 years from the 1930s to the 1980s.

In 1941, prior to the entry of the U.S. into World War II, John Flanagan played a role for psychologists similar to that of Yerkes just 24 years earlier. The War Department approved Flanagan's proposal to launch an ambitious testing program within the Army Air Corps to establish the validity of a test battery for selecting and classifying applicants for Air Corps flight crews, initially for pilots, navigators, and bombardiers. A guide to the 19 detailed book-length reports of the program is given by Flanagan (1948). Some 1500 officers and enlisted personnel were engaged in the effort, including many with names that would subsequently become familiar; some are listed in the third panel of Table 1. Scores of the Air Force psychologists became distinguished as contributors to research psychology in post-war years. The Army Air Force's project entailed a unique validity study in which all applicants were accepted in the program, although each had been required to complete a battery of tests to be used potentially for selection and classification. By relating test scores to success in training, a final battery was validated and used later to screen subsequent applicants (DuBois, 1947; Guilford and Lacey, 1947).

After the war, John Flanagan founded the American Institutes for Research (AIR), for the initial sole purpose of providing to commercial airlines a set of selection procedures highly similar to those developed in the Army program. More recently, AIR has engaged in a wide range of research activities, many of which contribute to governmental efforts in support of educational assessment. Other participants in the Army Air Force's program returned to quantitative, experimental, or social psychology programs at a wide range of universities, as well as testing agencies.

In another effort shrouded in a great deal more secrecy during the second World War, a collection of psychologists developed and implemented an extraordinary assessment system for the selection of personnel (otherwise known as spies) for the Office of Strategic Services (OSS). This work is described in a military report, the declassified version of which was published under the titles *Assessment of men: Selection of Personnel for the Office of Strategic Services* (Office of Strategic Services Assessment Staff, 1948) and *Selection of Personnel for Clandestine Operations: Assessment of Men* (Fiske et al., 1993). The staff of "Station S" (for "secret") was largely organized by Henry Murray; some of the dozens of staff members are listed in the final panel of Table 1. The assessment included paper-and-pencil instruments and projective tests as well as a large number of performance and situational exercises that were rated by observers. Extensive data analysis, including the (then new) techniques of multiple correlation and factor analysis, supported the clinical judgment of the staff in recommending candidates. While the extravagance of the multi-day OSS assessment has had few, if any, equals since World War II, the experiences of the OSS staff in that environment were to have extensive consequences for research in personality measurement and personnel selection for a generation of psychologists (Handler, 2001).

Over the decades following World War II, numerous school achievement tests were developed to compete with Terman's Stanford Achievement Test. A number of for-profit and non-profit companies were in the business of producing achievement tests, among them the Psychological Corporation, the Iowa Testing Program, Science Research Associates, Educational Testing Service, CTB/McGraw-Hill, and others. The National Assessment of Educational Progress, a program to monitor school achievement for the nation, was launched by Ralph W. Tyler in the 1960s and periodically has monitored achievement levels of national samples of school children (and state samples, since 1990) to provide "the nation's report card" (see Jones and Olkin, 2004). More recently, almost all states have adopted achievement tests for students in public schools; a mandate for such tests is a feature of the federal program "No Child Left Behind."

2.4. Organizational developments

In 1935, as noted earlier, the Psychometric Society was founded (its charter president was L.L. Thurstone) and the journal *Psychometrika* was launched. In the following year J.P. Guilford (1936) published *Psychometric Methods* that became a standard college text for the study of psychological measurement. The journal *Educational and Psychological Measurement* appeared in 1941. In 1946, the American Psychological Association (APA) created a Division of Evaluation and Measurement, and L.L. Thurstone was its first president. (The name of the division was changed in 1988 to Division of Evaluation, Measurement, and Statistics.) The Society of Multivariate Experimental Psychology was founded in 1960 (according to its website, http://www.smep.org) as an organization of researchers interested in multivariate quantitative methods and their application to substantive problems in psychology; that society publishes the journal *Multivariate Behavioral Research.* In 1964, the National Council on Measurement in Education inaugurated the *Journal of Educational Measurement*, publishing articles on psychometrics as applied to measurement in educational settings. The Society for Mathematical Psychology began to meet in 1968, and publishes the *Journal of Mathematical Psychology*. A number of other journals, including *Applied Psychological Measurement*, *Psychological Methods*, the *Journal of Educational and Behavioral Statistics*, the *British Journal of Mathematical and Statistical Psychology*, *Structural Equation Modeling*, and *Behaviormetrics* have joined the set that publishes psychometric research.

From these historical roots, and with the infrastructure provided by emerging scientific societies, by the second half of the 20th century psychometrics had staked out its intellectual turf: psychological scaling, educational and psychological measurement, and factor analysis. Extensions of those three core topics, along with numerous contributions to applied statistics motivated by the data collection in empirical psychology, have defined the field through the second half of the last century and into the 21st.

3. Psychological scaling

3.1. Thurstone's scaling models and their extensions

S.S. Stevens (1951) wrote that "Measurement is the assignment of numerals to objects or events according to rules." When measurement, thus broadly defined, makes use of behavioral or psychological data, it is called *psychological scaling*, because the assignment of numerals places the objects or events on a *scale*. To make use of behavioral or psychological data, the "rules" for the assignment of numerals are usually based on mathematical or statistical models for those data. Psychological scaling had its origins in psychophysics, with the birth of psychology as a science; but it has expanded to be used in many domains.

From its beginnings in the 19th century, psychophysics had continued to enjoy support in academic psychology during the mid-1920s and early 1930s, a period in which L.L. Thurstone published a remarkable set of papers that proposed adapting psychophysical methods to scale psychological functions that had no clear physical correlates, e.g., human abilities, attitudes, preferences, etc. The Thurstonian scaling methods were generated directly from employing the Gaussian distribution, first to repeated responses from a given individual, then with an extension to the distribution of responses for many individuals. These scaling methods almost immediately were adopted as measurement tools for research in social psychology, and thereby became an important component both of academic and applied psychology. Thurstone considered his development of scaling theory and practice to have been his most important contribution. An integrative summary of Thurstonian scaling procedures is that of Bock and Jones (1968). As noted in subsequent sections of this chapter, Thurstone's scaling methods stimulated later developments in multidimensional scaling and in item response theory. They also influenced the formulation of related scaling methods such as Luce's choice model (Luce, 1959).

In 1950, the Social Science Research Council appointed a Committee on Scaling Theory and Methods to review the status of scaling procedures and their relation to basic research in the social sciences. Committee members were Paul Horst, John Karlin, Paul Lazersfeld, Henry Margenau, Frederick Mosteller, John Volkmann, and Harold Gulliksen, chair. The committee engaged Warren Torgerson as a Research Associate to review and summarize the literature on psychological scaling. The result was the publication of a book that stood for many years as the authoritative summary of scaling procedures (Torgerson, 1958).

3.2. Multidimensional scaling

Various strategies for *multidimensional scaling* (MDS) borrow from Thurstone's scaling models (and their descendents) the idea that similarity data (as might be obtained using various experimental procedures and human judgment) can be represented spatially. However, where Thurstone's scaling models represent objects with real numbers on a single dimension, MDS most often represents objects as points in two- (or higher) dimensional space. In the course of this extension of dimensionality, most procedures for MDS are more data analytic in character, and less based on any psychological process model, than were Thurstone's original procedures. As such, MDS procedures form a bridge between the original methods of psychological scaling and a number of purely data-analytic procedures that are discussed in section 6 of this chapter.

Torgerson (1952) introduced multidimensional scaling using "metric" algorithms that treat the numerical values of the similarity data as if they are on an interval scale.

In the 1960s, simultaneous advances in the theory of statistical optimization and computing made possible the implementation of algorithms for "nonmetric" MDS, which seeks to derive a spatial representation of objects, with coordinates on an interval scale, making use only of the ordinal properties of the original similarity data. A theoretical paper by Luce and Tukey (1964) together with algorithms and computer programs by Shepard (1962a, 1962b), Kruskal (1964a, 1964b), Guttman and Lingoes (Lingoes, 1965), and Young and Torgerson (1967) provided the basis for countless studies in the fields of sensation, perception, and cognition in the 1960s and 1970s (see, for examples, Young and Hamer, 1987). The properties of nonmetric MDS, and additional improvements to methods for its implementation, were among the most prominent topics of psychometric research between the mid-1960s and the mid-1980s.

Tucker and Messick (1963) originated the idea that individual differences could be modeled in MDS, by modeling separate perceptual spaces for individuals having different "viewpoints" about stimulus interrelationships. Additional procedures that combined the idea of individual differences with a spatial representation underlying judgments of similarity were among the new procedures introduced around 1970 (Horan, 1969; Carroll and Chang, 1970; Tucker, 1972; Bloxom, 1974). Torgerson (1986) provided a critical summary of this most active period in the development of procedures for psychological scaling.

While basic MDS has subsequently become a standard data-analytic technique, continuing developments that extend these ideas remain active topics of research, some of which are mentioned in Section 6.

4. Psychological measurement (test theory)

4.1. True score theory

A number of psychological testing programs were initiated in the first half of the 20th century, but were launched with little attention to an underlying theory of testing. When L.L. Thurstone joined the faculty of the University of Chicago, he immediately designed a curriculum for a course in test theory. Later, the material was published under the title, *The Reliability and Validity of Tests* (Thurstone, 1931b). For several decades, this work served to foster numerous contributions to the psychometric literature, largely refinements and embellishments to traditional test theory based on the concept of the "true score."

Test theory is the archetype of a problem unique to psychological research that requires a statistical solution. Considering a test score from a statistical point of view, it is highly desirable to derive an ancillary statement of its precision. In the most basic approach to true score theory, the test score X is considered the sum of a true score T and a random error E,

$$X = T + E.$$

The standard deviation of the errors E is a statement of the (lack of) precision, or standard error, of the test score. That standard error can be estimated using an estimate of the reliability of the score X, where the reliability is the squared correlation between observed scores and the true scores.

Procedures to estimate reliability were developed over the first half of the 20th century, beginning with the seminal papers by Spearman (1910) and Brown (1910) that established the algebraic relation between test length and reliability for (hypothetical) parallel items. The first widely-used internal-consistency estimates of reliability (permitting reliability to be estimated from a single administration of a test, without some arbitrary division into halves) came from Kuder and Richardson (1937) famous paper, in which the equations numbered 20 and 21 came to be the names of statistics estimating reliability (KR-20 and KR-21). Hoyt (1941) provided a more general statement of internal consistency reliability, which was popularized by Cronbach (1951) as "coefficient α." Gulliksen's (1950) landmark text, *Theory of Mental Tests*, provided a summary of the research and developments of the first half of the 20th century, and was the standard theoretical and procedural guide for testing until the publication of the elegant axiomatic treatment of true score theory presented by Lord and Novick (1968). Variants of true score theory, sometimes fairly complex, continue to be used to provide quantitative statements of precision of test scores; Feldt and Brennan (1989) summarize the underlying theory and procedures for tests with a variety of structures.

Cronbach et al. (1972) expanded the concept of reliability to become *generalizability* over a number of possible facets of test administration, for example items or tasks, occasions of measurement, and judges (for responses scored by raters). Generalizability theory has become extremely important for the evaluation of "authentic" or "performance" assessments that often involve extended tasks or exercises administered over a period of time with responses evaluated by trained raters. Brennan (2001) summarizes continuing developments in this area.

During the latter half of the 20th century, as enriched notions of generalizability evolved, there emerged also broadened views of test validity. Cronbach and Meehl (1955) presented the first detailed exposition of "construct validity" as an adjunct to traditional predictive validity. Samuel Messick, in a series of papers culminating with Messick (1980), set forth a further set of considerations that clarify the distinction between construct validity, which he considered *the* validity of a test, and other features such as predictive utility, content coverage, and so on. Messick's formulations have been influential on subsequent revisions of the *Standards for Educational and Psychological Testing* (American Educational Research Association, American Psychological Association, and National Council on Measurement in Education, 1999).

Instruction in test theory is often the only presence of quantitative psychology in the undergraduate psychology curriculum aside from introductory statistics and research methods courses. Influential textbooks on testing include the well-known texts by Anastasi (most recently, Anastasi and Urbina, 1996) and Cronbach (1990) among many others. Summaries of the literature on test theory can be found for the period 1936–1960 in an article by Gulliksen (1961) and for 1961–1985 in a similar article by Lewis (1986).

4.2. *Item response theory (IRT)*

A global shift in graduate instruction and research in test theory accompanied the publication in 1968 of Frederic Lord and Melvin Novick's *Statistical Theories of Mental Test Scores* (Lord and Novick, 1968). In addition to a thorough, modern mathematical and statistical codification of true score theory, that volume included the first comprehensive coverage of IRT. Along with chapters by Lord and Novick on the normal ogive model, additional chapters by Allen Birnbaum (1968) introduced the logistic model, extended an idea that had been suggested by Lord (1953b) to account for guessing in what has come to be called the *three-parameter logistic* model (3PL), and advocated the use of information functions in test development.

IRT focuses the statistical analysis of test data on the responses to items, rather than on the total summed test score which was the basis of true score theory. Most of the models and methods that fall within IRT assume that one or more unobserved (or latent) variables underlie the responses to test items in the sense that variation among individuals on those latent variables explains the observed covariation among item responses (Green, 1954). In IRT models, the relation between the position of individuals on the latent variable(s) and the item responses is described by statistical models that describe the probability of an item response as a function of the latent variable(s). Examples of such models are the aforementioned normal ogive and logistic models; Bock and Moustaki (this volume) provide a detailed overview of IRT, including descriptions of many other models.

Before 1968, research in IRT had proceeded very slowly, with the occasional theoretical contribution, and virtually no application. The seeds of IRT were planted in a seminal paper by L.L. Thurstone (1925) in which he used ideas closely related to what would become the normal ogive model to describe Burt's (1922) data obtained with a British translation of Binet's scale, and another paper by Symonds (1929) in which the idea of the normal ogive model was used to describe data obtained from spelling tests. Guilford's (1936) *Psychometric Methods* described analysis of test data using IRT as though it might soon provide a practical method of item analysis; it actually required more than 40 years for that prediction to be fulfilled.

In a relatively obscure paper published in the *Proceedings of the Royal Society of Edinburgh*, Lawley (1943) described the application of the (then relatively new) method of maximum likelihood to the estimation of ability using the normal ogive model. However, because no practical method existed to produce the requisite item parameters for the model, Lawley's statistical development had no practical use. Tucker (1946) used the normal ogive model as a theoretical device to show that a maximally reliable test is made up of items with difficulty equal to the average ability of the examinees, and Lord (1952, 1953a) also used IRT as a theoretical basis for a description of the relation between ability and summed scores. The work by Lord (1952) and (intellectually) closely related work by the mathematical sociologist Paul Lazarsfeld (1950) clarified the nature of IRT models and emphasized the idea that they were "latent variable" models that explain the observed covariation among item responses as accounted for by an underlying, unobserved variable related to each response. B.F. Green's (1954) chapter in the *Handbook of Social Psychology* provides a lucid explanation of the theoretical structure of IRT models and the related latent class models that assume *local independence*: independence of item responses conditional on the value of the latent variable.

Lord and Novick's (1968) volume placed the normal ogive IRT model on a sound theoretical basis as an integration of a latent variable representing individual differences with the model Thurstone (1927) introduced as the "law of comparative judgment." After 1968, research and development in item response theory increased extraordinarily rapidly. The problem of item parameter estimation was brought into sharp statistical focus by Bock and Lieberman (1970), and a practical method based on maximum likelihood was described by Bock and Aitkin (1981). The latter method is used by most general purpose IRT software. Samejima (1969, 1997) provided the ground-breaking suggestion that IRT could be extended to the analysis of items with more than two ordered responses; Bock's (1972, 1997a) model for more than two nominal responses quickly followed. A number of refinements of, and alternatives to, these models have since been described (see Thissen and Steinberg, 1986). In the 1990s, following a seminal article by Albert (1992), research on applying Markov chain Monte Carlo (MCMC) techniques to estimation in item analysis went forward in several laboratories around the world, opening the way for more complex models (in which IRT plays a part) than are likely feasible using a maximum likelihood approach (Béguin and Glas, 2001; Bradlow et al., 1999; Fox and Glas, 2001, 2003; Patz and Junker, 1999a, 1999b; Wang et al., 2002).

In original work in Denmark that was largely independent of the American and British research that had gone before,[2] Georg Rasch (1960) developed an item response model by analogy with the properties of physical measurement. While Rasch's "one-parameter" logistic model (so named for its only item parameter that reflects difficulty) is closely related to Birnbaum's (1968) two-parameter logistic model, Rasch (1961, 1966, 1977) subsequently developed a philosophical basis for his model that established the Rasch tradition as an alternative to the models of the Thurstone–Lord–Birnbaum traditions. Subsequently, the "Rasch family" (Masters and Wright, 1984) has also expanded to include generalized models for items with more than two response alternatives, and a number of specialized models as well.

Lewis (1986) summarizes the literature on test theory for the period of IRT's explosive growth in the 1970s and early 1980s, and Bock (1997b) provides a more complete history of IRT. Perusal of the contents of this volume makes clear that IRT remains one of the most active areas of research in psychometrics.

4.3. *Equating, test assembly, and computerized adaptive testing*

The problem of *test equating* arises in its most commonly recognized form from the fact that large-scale assessments (such as the SAT) make routine use of alternate forms (certainly over time, and sometimes simultaneously), but all forms must yield comparable reported scores. Numerous statistical algorithms have been proposed for various data

[2] Rasch (1960) refers to Lord (1953b) only once, in a footnote on p. 116, and the normal ogive model only briefly there and on pp. 118 and 122. Rasch also refers, in another footnote, to the choice model (Bradley and Terry, 1952; Luce, 1959); however it appears that these footnotes were added after the fact of Rasch's model development.

collection designs to provide equated scores with various properties, and book-length treatments of the subject describe those methods (Holland and Rubin, 1982; Kolen and Brennan, 2004; von Davier et al., 2003), as do chapters in each edition of *Educational Measurement* (Flanagan, 1951; Angoff, 1971; Petersen et al., 1989; Holland and Dorans, in press). Holland and Dorans (in press) trace the origins of test equating to work by Truman Kelley (1914); Kelley's (1923) *Statistical Methods* included a chapter on "comparable measures" that describes linear equating and equipercentile equating, and refers to alternative proposals by Otis (1916, 1918). For most of the 20th century, test equating was a highly specialized technical process of interest primarily to its practitioners at institutions that created large-scale assessments with alternate forms. While a large psychometric literature accrued on this topic, it was not highly visible.

However, in the 1990s the increasingly salient (and burdensome) use of educational achievement tests in the United States led to requests to link scores on different tests, with the ultimate goal of using the results obtained from one assessment as substitute scores for those that would have been obtained on another test, to reduce the amount of time spent on achievement testing by having the results from one test serve double duty. Test linking then became the subject of acts of Congress, and two committees of the National Research Council published reports describing the many reasons why scores on distinct tests cannot, in general, be made comparable in all respects by any statistical processes (Feuer et al., 1999; Koretz et al., 1999).

Based on a substantial body of work inspired by attempts in the 1990s to extend the reach of test equating beyond its previous scope, Holland and Dorans (in press) use test *linking* as a generic term to encompass (i) *equating*, for producing interchangeable scores, (ii) *scale aligning*, for producing comparable scores (in some senses), and (iii) *predicting*, to make the best prediction of the score on one test from the score on another. For each of those three specific goals, many statistical techniques have been employed, including procedures based on summed scores and on IRT.

Test linking remains an active area of psychometric research as evidenced by the fact that two journals recently published special issues devoted to that area: The *Journal of Educational Measurement*'s special issue "Assessing the population sensitivity of equating functions" appeared in Spring, 2004, and *Applied Psychological Measurement*'s special issue on "Concordance" appeared in July, 2004.

Approaches to test linking may be based on true score theory or on IRT. However, two topics of currently active research in test theory rely primarily on IRT: computerized test assembly and computerized adaptive testing (CAT).The goal of computerized test assembly is usually to create parallel test forms, or at least alternate forms of a test that may be scored on the same scale, from a collection of items (a "pool") that has usually been calibrated using IRT. The scale of the problem can be very large: Hundreds of items may be cross-classified by dozens of content- and format-based attributes as well as by their statistical properties as described by IRT parameters. The numbers of combinations and permutations of those items that are possible if they are to be distributed onto fairly long tests is (literally) astronomical. Human test assembly specialists have assembled test forms for decades using various unspecified heuristic algorithms; computerized test assembly algorithms are intended to assist the human test assemblers (or perhaps replace them) with procedures that produce alternate test forms very quickly (often in ways that will not require subsequent equating of the alternate forms). A special issue of *Applied Psychological Measurement* (1998, volume 22, number 3) includes articles describing several computerized test assembly procedures (see van der Linden, 1998).

Perhaps the most extreme examples of computer-generated tests with linked score scales are those administered by CATsystems, in which each respondent may see a different set of items, sequentially selected to provide optimal information at each stage of the testing process. Volumes edited by Wainer et al. (2000) and van der Linden and Glas (2000) provide accessible entry points for (roughly) three decades of extensive research and development on the subject of CAT. Much of the modern development of IRT in the 1970s and early 1980s was supported by the U.S. Department of Defense in the project to create a CAT version of the Armed Services Vocational Aptitude Battery, or ASVAB. While widespread participation of academic quantitative psychologists in that project has ended and the test is operational, the effects of the project on large-scale testing persist.

5. Factor analysis

5.1. Factor analysis and rotation

Factor analysis is a statistical solution to a psychological problem: Almost as soon as (modern) psychological tests were invented, it was observed that the scores on tests of distinct aspects of ability were correlated across persons. Spearman (1904) sought to explain that correlation by appeal to the idea that the scores on all of the tests reflected, in some measure, variation on an underlying variable he called g (for general intelligence); that article is usually considered the beginning of factor analysis.

While subsequent British research began to consider the idea that there might be more than one factor of ability, it was Thurstone's (1938) *Psychometric Monograph* on the *Primary Mental Abilities* that initiated a line of factor analytic research on the structure of the intellect that would continue for decades. There were problems determining the number of factors, and then of estimating the parameters—the factor loadings, which are the regression coefficients of the observed test scores on the unobserved factors, and the unique (error) variances. Factor analysis was invented long before the theoretical statistics had been developed that would provide methods for the estimation problem for a model with so many parameters, and also long before computational power was available to solve the problem (had suitable algorithms been available). Before the 1960s, creative heuristic methods, largely due to Thurstone, were used.

While others had managed maximum likelihood (ML) estimation for the factor analysis model earlier, it was Jöreskog's (1967) algorithm, and Jennrich and Robinson's (1969) alternative, coinciding with the development of sufficiently fast electronic computers, that gave rise to practical general computer implementations of a statistically optimal procedure. While the earlier heuristic algorithms were also implemented in standard statistical software, the ML procedures permitted consideration of *Factor Analysis as a Statistical Method*, the title of Lawley and Maxwell's (1963, 1971) book.

Another daunting problem with factor analysis came to be called the rotational indeterminacy of factor-analytic solutions. For any factor-analytic model with more than one common factor, there are an infinite number of linear transformations that yield the same fit to the data; in one graphical way of displaying the results, these transformations may be obtained by rotating the axes of the graph (hence the name). Thurstone (1947) proposed the criterion of "simple structure" to select an interpretable version of the solution. For the first decades factor analysts used time-consuming, labor-intensive graphical methods to rotate the solution to simple structure. Before factor analysis could be used routinely, some analytical/computational method for rotation was required. Mulaik (1986, p. 26) wrote that "solving the rotation problem became a kind of Holy Grail for factor analysts to pursue, and many a factor analyst was to make his reputation with a workable analytic scheme of factor rotation." Horst (1941) proposed one of the first analytic methods; Kaiser's (1958) "varimax" method is likely the most successful of the procedures for orthogonal factor solutions. "Direct oblimin" (Jennrich and Sampson, 1966), a modification of a procedure proposed by Carroll (1953), is an example of an analytical rotation algorithm for oblique factor solutions. Browne (2001) provides an excellent overview and review of analytic rotation methods.

Between the 1930s and the 1960s, factor analysis was basically the exclusive domain of quantitative psychologists. By 1976, a textbook in factor analysis was a very large volume indeed (see Harmon, 1976); much had been accomplished. With computer applications prepared to perform the computations involved in the statistical estimation and analytical rotation of the factor loadings, after the 1970s factor analysis became a generally available tool for research psychologists.

John B. Carroll's (1993) 800-page volume *Human Cognitive Abilities: A Survey of Factor Analytic Studies* may represent the culmination of efforts extended toward the original purpose of factor analysis – to draw inferences about the trait structure of human cognition from correlations among test scores. Carroll re-analyzed the data from scores of factor analytic studies from the preceding 50 years, and integrated the results into a single hierarchical structure of human abilities. However, long before that volume was published, factor analysis had mutated into a number of related procedures used for completely different purposes.

5.2. Analysis of covariance structures

In a wide-ranging yet brief review of the history of structural equation modeling, Bentler (1986) attributes to Tucker (1955) the origin of the distinction between "exploratory" and "confirmatory" factor analytic studies. For most of its early history (setting aside the founding article by Spearman that was intended to test his hypothesis about g), factor analysis was an essentially exploratory procedure with an Achilles' heel of rotational indeterminacy.

Bock and Bargmann (1966), in an article entitled *Analysis of Covariance Structures*, described an ML estimation algorithm, and a goodness of fit statistic, for a model formally very much like a factor analysis model but used for a totally different purpose. In matrix notation, the Bock and Bargmann model was identical to the factor analysis model, but some of the matrix elements that are estimated in the factor analysis model were fixed *a priori* to particular values (0s and 1s). The result (in the particular special case described by Bock and Bargmann) was a model for learning, but more general applications soon became apparent.

Jöreskog (1969) described a general approach to confirmatory maximum likelihood factor analysis, in which any of the usual parameters of the model (the factor loadings and/or the unique variances) could be estimated or fixed at a priori values. A first interesting aspect of this advance is that the computer programs based on Jöreskog's algorithm could be used to fit a number of special case models, e.g., Bock and Bargmann's learning curve model, with suitable placement of a priori fixed values. A second aspect of this development was that, if a sufficient number of fixed values are in appropriately chosen locations in the traditional factor analysis model, the long-standing problem of rotational indeterminacy disappears. Thus it became possible to statistically test the goodness of fit of a hypothesized factor-analytic model (including a particular rotation of the axes). Factor analysis thereby graduated from a purely exploratory procedure to become a truly confirmatory, statistical model-fitting activity. Furthermore, the techniques of confirmatory factor analysis can be used for many of the original purposes of exploratory factor analysis if the analyst is willing to specify a priori some (perhaps modest amount) of the structure.

The development of algorithms for ML estimation for (essentially factor-analytic) covariance structure models also became a steppingstone on what then turned out to be a very short crossing to the creation of structural equation modeling.

5.3. Structural equation modeling

Structural equation modeling (SEM) represents a marriage of the factor analysis model to multiple regression, in which regression brought Sewall Wright's (1921) path analysis as her dowry. Given the development of systems for the analysis of covariance structures, or confirmatory factor analysis, it was (in hindsight) a short step to the idea that the (confirmatory) factor analysis model could be used as the "measurement model" for a set of multiple indicators of a latent variable, and those latent variables could simultaneously be involved in multiple regression relations in a "structural model" (Jöreskog and van Thillo, 1973). The structural model describes a hypothesized network of directional and nondirectional relationships among latent variables. Jöreskog (1973), Wiley (1973), and Keesling (1972) developed the idea of SEM nearly simultaneously; Bentler and Weeks (1980) subsequently described an alternative model with essentially the same features.

Software to fit structural equation models is extremely general, and may be used to estimate the parameters, and test the fit, of a wide variety of more specific models. The models described in the previous section for analysis of covariance structures, or confirmatory factor analysis, are subsets of the linear structural equations model, as is the original exploratory factor analysis model. In addition, regression models may be fitted so as to account for measurement error in the observed variables (on either side of the equation) in the sense that the response and predictor variables in the regression(s) are the error-free unobserved, latent variables underlying the observed variables. This feature enabled statistically optimal estimation of the path coefficients that had been desired for path analysis since Wright (1921) first described the procedure. Fitting path models has become one of the most widespread uses of SEM.

The SEM model also subsumed the higher-order factor model, in which correlations among the factor scores of a classical (Thurstone) factor model are, in turn, subjected to factor analysis. Thurstone (1947) had described the concept, and Schmid and Leiman (1957) had developed a multi-stage estimation procedure that involved repeated applications of then-current methods of factor extraction. However, because the factor model is, itself, a regression model, albeit one involving latent variables, the higher-order factor model may be expressed as a regression model involving phantom latent variables (that are not directly related to any observed variables). So expressed, the parameters of the higher-order factor model may be simultaneously estimated using SEM software, allowing the goodness of fit of the model to be evaluated.

In addition, by borrowing the strategy employed by Bock and Bargmann (1966) that placed fixed values in model matrices originally intended to hold estimated parameters, SEM software may be used to fit other models as well. Meredith and Tisak (1990) described the use of the SEM model to fit latent growth models, in which the covariance structure among repeated measurements is analyzed to yield an estimate of the functional form of the unobserved (error-free) growth process. (The analysis of growth involves extension of the SEM model to include the means of the repeated measurements; that extension of SEM has, in turn, had other uses as well.) A similar use of the same strategy permits SEM software to be used to fit variance component models (Jöreskog and Sörbom, 1989). The addition of mechanisms to fit SEMs simultaneously in multiple populations (Jöreskog and Sörbom, 1983) permits SEM software to be used to examine issues of factorial invariance of the kind earlier considered by Meredith (1964).

For the past 30 years, SEM has arguably been the area of most rapid development in quantitative psychology. The generality of the model has permitted its use in countless substantive studies in psychology, as well as in sociology, economics, and other fields. This widespread application has, in turn, motivated extensive methodological work, on topics such as parameter estimation with less restrictive distributional assumptions about the data, and (many) goodness of fit statistics. The story about SEM is not, by any means, written. Still, at this point it can be stated confidently that SEM, offspring of factor analysis that it is, will remain one of the most prominent contributions of quantitative psychology both to research in psychology and to the methods of applied statistics in general.

6. Psychological statistics

From the early part of the 20th century, psychometricians have made contributions to pedagogy in statistics, and to applied statistics more generally. The developments in psychological scaling, test theory, and factor analysis discussed in the preceding sections had as prerequisites the knowledge of fundamental concepts of statistics now taught in the first statistics courses that are universally required in undergraduate and graduate programs in psychology. Psychometricians produced volumes that were among the first texts for such courses; one of the earliest was E.L. Thorndike's (1913) volume *An Introduction to the Theory of Mental and Social Measurements* (2nd edition), which integrated what we would now call a first course in statistics with basic ideas of test theory. (The first edition of Thorndike's book appeared in 1904, but copies of that are now relatively rare.) In the 1920s, Truman Kelley's (1923) *Statistical Methods*, Arthur Otis's (1926) *Statistical Method in Educational Measurement*, and Henry Garrett's (1926) *Statistics in Psychology and Education* appeared. Later texts by Guilford (1942), Edwards (1946) and McNemar (1949) anticipated in their content coverage nearly all subsequent textbooks for the first course in statistics in psychology and education. The everyday work of psychometricians in academic psychology departments has to this day included instruction in applied statistics courses, and countless additional undergraduate and graduate textbooks have been produced.

A driving force for change in instruction and practice in applied statistics over the course of the 20th century was the development of computing machinery. In the early part of that century, publications in statistics (and the practice of statistics) were dominated by the development and implementation of computational strategies, and algorithms for hand computation and the use of nomographs (graphical devices to assist computation) (Appelbaum, 1986). But computing machinery changed that, and psychometricians were again in the forefront. Ledyard Tucker's (1940) description of his modification of the "International Business Machines Corporation scoring machine" to perform matrix multiplication was likely the first publication in *Psychometrika* to describe a non-desk calculator method for statistical computation (Appelbaum, 1986). The modified "scoring machine" used marked multiple-choice answer sheets that had by then become common for large-scale test administration as the input medium, as well as the plug board; it was used in L.L. Thurstone's Psychometric Laboratories at the University of Chicago and at The University of North Carolina.

Psychometricians have been especially interested in providing computational methods for statistical procedures that are broadly useful in psychological research. Appelbaum (1986) notes that perhaps the longest-running topic in the history of *Psychometrika* involved the computation of the tetrachoric correlation that forms the basis of many approaches to item analysis in test theory. Research in psychology yields multivariate data in many experimental and observational contexts; in work that began in the 1960s at the Psychometric Laboratory at the University of North Carolina and continued at the University of Chicago in the 1970s, R. Darrell Bock developed matrix-based algorithms for a number of multivariate linear statistical procedures that were described in the influential text *Multivariate Statistical Methods in Behavioral Research* (Bock, 1975). Algorithms from that volume were implemented in several widely distributed computer programs, and the combination of the text for graduate courses and the computer applications led to widespread use of multivariate methods in psychological research.

Another class of multivariate data analysis techniques on which Bock (1960) also did some work is that of "optimal scaling." Bock's student, Shizuhiko Nishisato, is one of many psychometricians who have developed over the past few decades the theory and computational techniques of these procedures; overlapping sets of these techniques are also known as "dual scaling," "Hayashi's theory of quantification," "*analyse factorielle des correspondences*" ("correspondence analysis"), and "homogeneity analysis" (Nishisato, 1996). These procedures share with psychological scaling and test theory the capacity to quantify categorical data of kinds that appear in research in psychology: categorical item responses, ranks, and paired comparisons (to list a few of the more common). While these procedures are related to the original (metric) multidimensional scaling, their scope has so increased that their heritage is largely obscured.

Closely related work by the Gifi (1990) group at the University of Leiden extends computationally intensive analysis of multivariate data using nonlinear transformations of nominal, ordered, or continuous data to provide nonlinear analogues of such multivariate techniques as regression and canonical correlation, and for cluster analysis. This work and more recent extensions (see Meulman, 2003; Heiser, 2004) often makes use of computer graphics to render visualizations of summaries of high-dimensional multivariate behavioral data that would otherwise be challenging to comprehend.

In work that some might say extends quantitative psychology's interest in multivariate statistics to spaces of infinite dimensionality, Ramsay and Silverman's (1997) *Functional Data Analysis* makes the unit of statistical analysis curves (such as growth curves) or surfaces (such as might describe some response as a function of more than one variable). In this form of analysis as well, the results are often provided graphically.

Indeed, the "results" of many forms of data analysis that have been the special province of psychometrics (psychological scaling, MDS, and various forms of optimal scaling) are essentially spatial representations of some data that are presented graphically, as curves or surfaces. As statistical graphics are at the interface between psychology and statistics (that is, between human perception and the data), quantitative psychologists have often been involved in research and development of statistical graphics even quite aside from the data analytic techniques that may underlie the graphs (Wainer and Thissen, 1993). Leland Wilkinson's (1999) *The Grammar of Graphics* integrates statistical graphics, computer science, and perception in a description of graphics that is essentially psycholinguistic in character (although its language is graphical). Tufte's (1983) volume, *The Visual Display of Quantitative Information*, and Wainer's (2005) *Graphic Discovery: A Trout in the Milk and Other Visual Adventures*, bring ideas about graphical data analysis to a broader audience.

Quantitative psychologists often have collaborated with statisticians to apply statistics to psychological research (e.g., Abelson, Bock, Bush, Green, Hedges, Jones, and Luce in combinations with Bargmann, Mosteller, Olkin, Tukey, and others). And quantitative psychology has also been concerned throughout its history with the topics of experimental design and data analysis. Roger Kirk's (1995) third edition of his classic *Experimental Design: Procedures for the Behavioral Sciences* stands as a monument to the training of graduate students in psychology for nearly four decades. Maxwell and Delaney's (1990) *Designing Experiments and Analyzing Data: A Model Comparison Perspective* integrates modern ideas of research design and data analysis. And in the past two decades, two new topics have become associated with research on experimental design and subsequent data analysis: On the design side, before data are collected, power analysis (Cohen, 1988) has become routine; that process is sufficiently challenging for the complex designs used in behavioral research that it has become its own subspecialty. The same is true for research synthesis (Cooper and Hedges, 1994), also sometimes referred to as meta-analysis (Glass, 1981; Hedges and Olkin, 1985), which seeks to statistically combine the results of collections of experimental studies to describe their common results and the features of variation they exhibit.

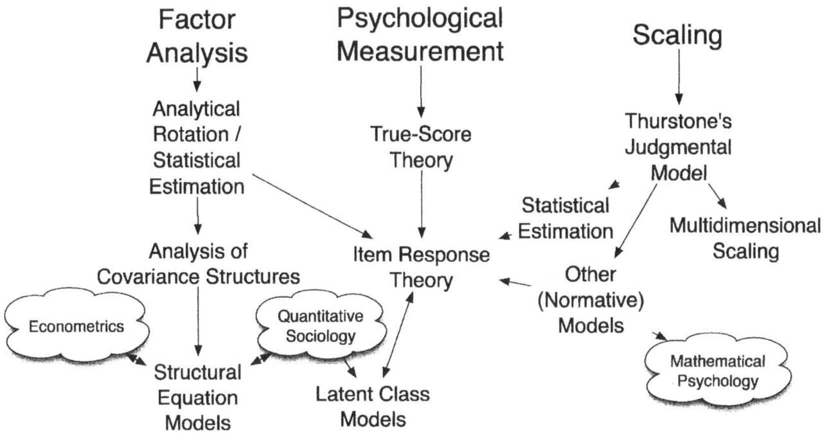

Fig. 2. A graphical description of some of the relations among the traditions of factor analysis, psychological measurement, and scaling and their extensions and developments.

Countless other topics of statistical research could be mentioned that have either captured the interest of psychometricians or been introduced in the literature of quantitative psychology. Suffice it to say that psychometrics has not been merely a consumer of the tools of statistics, but rather it has been one of the contributing fields to applied statistics throughout its long history and is likely to continue to be so.

7. Conclusion

Psychometrics, or quantitative psychology, is the disciplinary home of a set of statistical models and methods that have been developed primarily to summarize, describe, and draw inferences from empirical data collected in psychological research. The set of models and methods that are psychometric may be divided into three major classes. The first of these is psychological scaling, a set of techniques for the assignment of quantitative values to objects or events using data obtained from human judgment. A second class involves a large set of methods and procedures derived from the basic idea of factor analysis – to explain the observed covariation among a set of variables by appeal to the variation on underlying latent (unobserved) random variables that may explain the covariation. A third class of models has produced test theory; while the traditional algebraic true score theory is not a statistical model *per se*, the more recent item response theory is a class of statistical models that combines basic components from psychological scaling and latent explanatory variables from the factor analytic tradition. Figure 2 shows a graphical description of some of the relations among the traditions of factor analysis, psychological measurement, and scaling and their extensions and developments.

In addition, psychometrics has been the source of many other contributions to applied statistics, motivated by new data analytic challenges that have arisen with each new generation of scientific psychology.

We have provided some illustrations of the work of quantitative psychologists in the preceding four sections of this chapter. However, these illustrations have been selective; no attempt has been made to be exhaustive. Indeed, for every reference in this chapter, a dozen or more equally important publications are likely to have been omitted. We have sought to define psychometrics by providing some illustrations of what psychometricians do and have done, not to provide a comprehensive list of either topics or scholars.

While the intellectual history of psychometrics may be traced to roots two centuries in the past, and its existence as an identifiable discipline has been clear for seven decades, quantitative psychology remains a work in progress. Psychometric research has always drawn its motivation from contemporary challenges of psychological and behavioral research, and its tools from statistics. Because both the challenges of psychological research and the available statistical framework change with time, psychometrics continues to mature as a discipline. Subsequent chapters in this volume provide more detailed descriptions of many of the techniques that have been mentioned in this chapter, along with other novel developments.

References

Albert, J.H. (1992). Bayesian estimation of normal ogive item response curves using Gibbs sampling. *Journal of Educational Statistics* **17**, 251–269.

American Educational Research Association, American Psychological Association, and National Council on Measurement in Education (1999). *Standards for Educational and Psychological Testing*. American Psychological Association, Washington, DC.

Anastasi, A., Urbina, S. (1996). *Psychological Testing*, 7th ed. Prentice-Hall, New York.

Angoff, W.H. (1971). Scales, norms and equivalent scores. In: Thorndike, R.L. (Ed.), *Educational Measurement*. 2nd ed. American Council on Education, Washington, DC, pp. 508–600. Reprinted as Angoff, W.H. (1984). Scales, Norms and Equivalent Scores. Educational Testing Service, Princeton, NJ.

Appelbaum, M.I. (1986). Statistics, data analysis, and *Psychometrika*: Major developments. *Psychometrika* **51**, 53–56.

Béguin, A.A., Glas, C.A.W. (2001). MCMC estimation and some model-fit analysis of multidimensional IRT models. *Psychometrika* **66**, 541–561.

Bentler, P. (1986). Structural equation modeling and *Psychometrika*: A historical perspective on growth and achievements. *Psychometrika* **51**, 35–51.

Bentler, P.M., Weeks, D.G. (1980). Linear structural equations with latent variables. *Psychometrika* **45**, 289–307.

Binet, A., Simon, T. (1905). Methodes nouvelles pour le diagnostic du niveau intellectuel des anormaux. *Annee Psychologique* **11**, 191–244.

Birnbaum, A. (1968). Some latent trait models and their use in inferring an examinee's ability. In: Lord, F.M., Novick, M.R. (Eds.), *Statistical Theories of Mental Test Scores*. Addison-Wesley, Reading, MA, pp. 395–479.

Bloxom, B. (1974). An alternative method of fitting a model of individual differences in multidimensional scaling. *Psychometrika* **39**, 365–367.

Bock, R.D. (1960). Methods and applications of optimal scaling. The University of North Carolina Psychometric Laboratory Research Memorandum No. 25.

Bock, R.D. (1972). Estimating item parameters and latent ability when responses are scored in two or more latent categories. *Psychometrika* **37**, 29–51.

Bock, R.D. (1975). *Multivariate Statistical Methods in Behavioral Research*. McGraw-Hill, New York.

Bock, R.D. (1997a). The nominal categories model. In: van der Linden, W., Hambleton, R.K. (Eds.), *Handbook of Modern Item Response Theory*. Springer, New York, pp. 33–50.
Bock, R.D. (1997b). A brief history of item response theory. *Educational Measurement: Issues and Practice* **16**, 21–33.
Bock, R.D., Aitkin, M. (1981). Marginal maximum likelihood estimation of item parameters: an application of the EM algorithm. *Psychometrika* **46**, 443–459.
Bock, R.D., Bargmann, R. (1966). Analysis of covariance structures. *Psychometrika* **46**, 443–449.
Bock, R.D., Jones, L.V. (1968). *The Measurement and Prediction of Judgment and Choice*. Holden-Day, San Francisco, CA.
Bock, R.D., Lieberman, M. (1970). Fitting a response model for n dichotomously scored items. *Psychometrika* **35**, 179–197.
Bock, R.D., Moustaki, I. (this volume), Item response theory in a general framework. Chapter 15 in this volume.
Boring, G.W. (1929). *A History of Experimental Psychology*. Appelton-Century, New York.
Bradley, R.A., Terry, M.E. (1952). Rank analysis of incomplete block designs: I. The method of paired comparisons. *Biometrika* **39**, 324–345.
Bradlow, E.T., Wainer, H., Wang, X. (1999). A Bayesian random effects model for testlets. *Psychometrika* **64**, 153–168.
Brennan, R.L. (2001). *Generalizability Theory*. Springer, New York.
Brown, W. (1910). Some experimental results in the correlation of mental abilities. *British Journal of Psychology* **3**, 296–322.
Brown, W., Thomson, G.H. (1921). *The Essentials of Mental Measurement*. Cambridge University Press, Cambridge.
Browne, M.W. (2001). An overview of analytic rotation in exploratory factor analysis. *Multivariate Behavioral Research* **36**, 111–150.
Burt, C. (1922). *Mental and Scholastic Tests*. P.S. King, London.
Carroll, J.B. (1953). An analytical solution for approximating simple structure in factor analysis. *Psychometrika* **18**, 23–38.
Carroll, J.B. (1993). *Human Cognitive Abilities: A Survey of Factor Analytic Studies*. Cambridge University Press, Cambridge.
Carroll, J.D., Chang, J.J. (1970). Analysis of individual differences in multidimensional scaling via an N-way generalization of "Eckart–Young" decomposition. *Psychometrika* **35**, 283–319.
Cattell, J.M. (1890). Mental tests and measurements. *Mind* **15**, 373–380.
Cohen, J. (1988). *Statistical Power Analysis for the Behavioral Sciences*, 2nd ed. Lawrence Erlbaum Associates, Hillsdale, NJ.
Cooper, H., Hedges, L. (1994). *The Handbook of Research Synthesis*. Russell Sage Foundation, New York.
Cremin, L.A. (1961). *The Transformation of the School: Progressivism in American Education, 1876–1957*. Alfred A. Knopf, New York.
Cronbach, L.J. (1951). Coefficient alpha and the internal structure of tests. *Psychometrika* **16**, 297–334.
Cronbach, L.J. (1990). *Essentials of Psychological Testing*. Harper & Row, New York.
Cronbach, L.J., Gleser, G.C., Nanda, H., Rajaratnam, N. (1972). *The Dependability of Behavioral Measurements: Theory of Generalizability for Scores and Profiles*. Wiley, New York.
Cronbach, L.J., Meehl, P.E. (1955). Construct validity in psychological tests. *Psychological Bulletin* **52**, 281–302.
DuBois P.H. (1947). The classification program. Army Air Forces Aviation Psychology Program Research Reports, Report No. 2. U.S. Government Printing Office, Washington, DC.
Edwards, A.L. (1946). *Statistical Analysis for Students in Psychology and Education*. Rinehart, New York.
Feldt, L.S., Brennan, R.L. (1989). Reliability. In: Linn, R.L. (Ed.), *Educational Measurement*. 3rd ed. American Council on Education/Macmillan, New York, pp. 105–146.
Feuer, M.J., Holland, P.W., Green, B.F., Bertenthal, M.W., Hemphill, F.C. (Eds.) (1999). *Uncommon Measures: Equivalence and Linkage Among Educational Tests*. National Academy Press, Washington, DC.
Fiske, D.W., Hanfmann, E., Mackinnon, D.W., Miller, J.G., Murray, H.A. (1993). *Selection of Personnel for Clandestine Operations: Assessment of Men*. Aegean Park Press, Walnut Creek, CA.

Flanagan, J.C. (1948). The aviation psychology program in the army air forces. Army Air Forces Aviation Psychology Program Research Reports, Report No. 1. U.S. Government Printing Office, Washington, DC.

Flanagan, J.C. (1951). Units, scores, and norms. In: Lindquist, E.F. (Ed.), *Educational Measurement*. American Council on Education, Washington, DC, pp. 695–763.

Fox, J.P., Glas, C.A.W. (2001). Bayesian estimation of a multilevel IRT model using Gibbs sampling. *Psychometrika* **66**, 269–286.

Fox, J.P., Glas, C.A.W. (2003). Bayesian modeling of measurement error in predictor variables using item response theory. *Psychometrika* **68**, 169–191.

Garrett, H.E. (1926). *Statistics in Psychology and Education*. Longmans, Green and Co., New York.

Gifi, A. (1990). *Nonlinear Multivariate Analysis*. Wiley, New York.

Glass, G.V. (1981). *Meta-Analysis in Social Research*. Sage, Beverly Hills, CA.

Goddard, H.H. (1908). The Binet and Simon tests of intellectual capacity. *Training School Bulletin* **5** (10), 3–9.

Green, B.F. Jr. (1954). Attitude measurement. In: Lindzey, G. (Ed.), *Handbook of Social Psychology*, vol. I. Addison-Wesley, Cambridge, MA, pp. 335–369.

Guilford, J.P. (1936). *Psychometric Methods*. McGraw-Hill, New York.

Guilford, J.P. (1942). *Fundamental Statistics in Psychology and Education*. McGraw-Hill, New York.

Guilford, J.P., Lacey, J.L. (Eds.) (1947). Printed classification tests. Army Air Forces Aviation Psychology Program Research Reports, Report No. 5. U.S. Government Printing Office, Washington, DC.

Gulliksen, H.O. (1950). Theory of Mental Tests. Wiley, New York. Reprinted in 1987 by Lawrence Erlbaum Associates, Hillsdale, NJ.

Gulliksen, H. (1961). Measurement of learning and mental abilities. *Psychometrika* **26**, 93–107.

Handler, L. (2001). Assessment of men: Personality assessment goes to war by the Office of Strategic Services assessment staff. *Journal of Personality Assessment* **76**, 558–578.

Harmon, H.H. (1976). *Modern Factor Analysis*, 3rd ed., revised. University of Chicago Press, Chicago.

Hedges, L.V., Olkin, I. (1985). *Statistical Methods for Meta-Analysis*. Academic Press, Orlando, FL.

Heiser, W.J. (2004). Geometric representation of association between categories. *Psychometrika* **69**, 513–545.

Hilgard, E.R. (1987). *Psychology in America: A historical survey*. Harcourt Brace Jovanovich, Orlando, FL.

Holland, P.W., Dorans, N.J. (in press). Linking and equating test scores. In: Brennan, R.L. (Ed.), *Educational Measurement*, 4th ed. In press.

Holland, P.W., Rubin, D.B. (1982). *Test Equating*. Academic Press, New York.

Horan, C.B. (1969). Multidimensional scaling: Combining observations when individuals have different perceptual structures. *Psychometrika* **34**, 139–165.

Horst, P. (1941). A non-graphical method for transforming an arbitrary factor matrix into a simple structure factor matrix. *Psychometrika* **6**, 79–99.

Hoyt, C. (1941). Test reliability estimated by analysis of variance. *Psychometrika* **6**, 153–160.

Jennrich, R.I., Robinson, S.M. (1969). A Newton–Raphson algorithm for maximum likelihood factor analysis. *Psychometrika* **34**, 111–123.

Jennrich, R.I., Sampson, P.F. (1966). Rotation for simple loadings. *Psychometrika* **31**, 313–323.

Jones, L.V., Olkin, I. (2004). *The Nation's Report Card: Evolution and Perspectives*. Phi Delta Kappa International, Bloomington, IN.

Jöreskog, K.G. (1967). Some contributions to maximum likelihood factor analysis. *Psychometrika* **32**, 443–482.

Jöreskog, K.G. (1969). A general approach to confirmatory maximum likelihood factor analysis. *Psychometrika* **34**, 183–202.

Jöreskog, K.G. (1973). A general method for estimating a linear structural equation system. In: Goldberger, A.S., Duncan, O.D. (Eds.), *Structural Equation Models in the Social Sciences*. Academic Press, New York, pp. 85–112.

Jöreskog, K.G., Sörbom, D. (1983). *LISREL User's Guide*. International Educational Services, Chicago.

Jöreskog, K.G., Sörbom, D. (1989). *LISREL 7 User's Reference Guide*. Scientific Software, Chicago.

Jöreskog, K.G., van Thillo, M. (1973). LISREL – A general computer program for estimating a linear structural equation system involving multiple indicators of unmeasured variables. Research Report 73-5. Uppsala University, Department of Statistics, Uppsala.

Kaiser, H.F. (1958). The varimax criterion for analytical rotation in factor analysis. *Educational and Psychological Measurement* **23**, 770–773.

Keesling, J.W. (1972). Maximum likelihood approaches to causal analysis. Unpublished Doctoral Dissertation. University of Chicago.

Kelley, T.L. (1914). Comparable measures. *Journal of Educational Psychology* **5**, 589–595.

Kelley, T.L. (1923). *Statistical Methods*. Macmillan, New York.

Kirk, R.E. (1995). *Experimental Design: Procedures for the Behavioral Science*, 3rd ed. Brooks/Cole, Pacific Grove, CA.

Kolen, M.J., Brennan, R.L. (2004). *Test Equating, Linking, and Scaling: Methods and Practices*, 2nd ed. Springer-Verlag, New York.

Koretz, D.M., Bertenthal, M.W., Green, B.F. (Eds.) (1999). *Embedding Questions: The Pursuit of a Common Measure in Uncommon Tests*. National Academy Press, Washington, DC.

Kruskal, J.B. (1964a). Multidimensional scaling by optimizing goodness of fit to a nonmetric hypothesis. *Psychometrika* **29**, 1–27.

Kruskal, J.B. (1964b). Nonmetric multidimensional scaling: A numerical method. *Psychometrika* **29**, 115–129.

Kuder, G.F., Richardson, M.W. (1937). The theory of the estimation of test reliability. *Psychometrika* **2**, 151–160.

Lawley, D.N. (1943). On problems connected with item selection and test construction. *Proceedings of the Royal Society of Edinburgh* **62-A** (Part I), 74–82.

Lawley, D.N., Maxwell, A.E. (1963). *Factor Analysis as a Statistical Method*. Butterworth, London.

Lawley, D.N., Maxwell, A.E. (1971). *Factor Analysis as a Statistical Method*, 2nd ed. Butterworth, London.

Lazarsfeld, P.F. (1950). The logical and mathematical foundation of latent structure analysis. In: Stouffer, S.A., Guttman, L., Suchman, E.A., Lazarsfeld, P.F., Star, S.A., Clausen, J.A. (Eds.), *Measurement and Prediction*. Wiley, New York, pp. 362–412.

Leuchtenburg, W.E. (1958). *The Perils of Prosperity, 1914–32*. The University of Chicago Press, Chicago.

Lewis, C. (1986). Test theory and *Psychometrika*: The past twenty-five years. *Psychometrika* **51**, 11–22.

Lingoes, J.C. (1965). An IBM program for Guttman–Lingoes smallest space analysis. *Behavioral Science* **10**, 183–184.

Lord, F.M. (1952). A Theory of Test Scores. *Psychometric Monographs*, Whole No. 7.

Lord, F.M. (1953a). The relation of test score to the trait underlying the test. *Educational and Psychological Measurement* **13**, 517–548.

Lord, F.M. (1953b). An application of confidence intervals and of maximum likelihood to the estimation of an examinee's ability. *Psychometrika* **18**, 57–76.

Lord, F.M., Novick, M.R. (1968). *Statistical Theories of Mental Test Scores*. Addison-Wesley, Reading, MA.

Luce, R.D. (1959). *Individual Choice Behavior*. Wiley, New York.

Luce, R.D., Tukey, J.W. (1964). Simultaneous conjoint measurement. *Journal of Mathematical Psychology* **1**, 1–27.

Masters, G.N., Wright, B.D. (1984). The essential process in a family of measurement models. *Psychometrika* **49**, 529–544.

Maxwell, S.E., Delaney, H.D. (1990). *Designing Experiments and Analyzing Data: A model Comparison Perspective*. Wadsworth, Belmont, CA.

Meredith, W. (1964). Notes on factorial invariance. *Psychometrika* **29**, 177–185.

Meredith, W., Tisak, J. (1990). Latent curve analysis. *Psychometrika* **55**, 107–122.

Messick, S. (1980). Test validity and the ethics of assessment. *American Psychologist* **35**, 1012–1027.

Meulman, J.J. (2003). Prediction and classification in nonlinear data analysis: Something old, something new, something borrowed, something blue. *Psychometrika* **68**, 493–517.

McNemar, Q. (1949). *Psychological Statistics*. Wiley, New York.

Mulaik, S.A. (1986). Factor analysis and *Psychometrika*: Major developments. *Psychometrika* **51**, 23–33.

Nishisato, S. (1996). Gleaning in the field of dual scaling. *Psychometrika* **61**, 559–599.

Office of Strategic Services Assessment Staff (1948). *Assessment of Men*. Rinehart, New York.

Otis, A.S. (1916). The reliability of spelling scales, including a 'deviation formula' for correlation. *School and Society* **4**, 96–99.

Otis, A.S. (1918). An absolute point scale for the group measurement of intelligence. *Journal of Educational Psychology* **9**, 239–261, 333–348.
Otis, A.S. (1926). *Statistical Method in Educational Measurement*. World Book Company, Yonkers-on-Hudson, NY.
Patz, R.J., Junker, B.W. (1999a). A straightforward approach to Markov chain Monte Carlo methods for item response models. *Journal of Educational and Behavioral Statistics* **24**, 146–178.
Patz, R.J., Junker, B.W. (1999b). Applications and extensions of MCMC in IRT: Multiple item types, missing data, and rated responses. *Journal of Educational and Behavioral Statistics* **24**, 342–366.
Petersen, N.S., Kolen, M.J., Hoover, H.D. (1989). Scaling, norming and equating. In: Linn, R.L. (Ed.), *Educational Measurement*. 3rd ed. Macmillan, New York, pp. 221–262.
Ramsay, J.O., Silverman, B.W. (1997). *Functional Data Analysis*. Springer, New York.
Rasch, G. (1960). *Probabilistic Models for Some Intelligence and Attainment Tests*. Denmarks Paedagogiske Institut, Copenhagen. Republished in 1980 by the University of Chicago Press.
Rasch, G. (1961). On general laws and the meaning of measurement in psychology. In: *Proceedings of the Fourth Berkeley Symposium on Mathematical Statistics and Probability*. University of California Press, Berkeley, CA, pp. 321–333.
Rasch, G. (1966). An item analysis which takes individual differences into account. *British Journal of Mathematical and Statistical Psychology* **19**, 49–57.
Rasch, G. (1977). On specific objectivity: An attempt at formalizing the request for generality and validity of scientific statements. In: Blegvad, M. (Ed.), *The Danish Yearbook of Philosophy*. Munksgaard, Copenhagen.
Samejima, F. (1969). Estimation of Latent Ability Using a Response Pattern of Graded Scores. *Psychometric Monographs*, Whole No. 17.
Samejima, F. (1997). Graded response model. In: van der Linden, W., Hambleton, R.K. (Eds.), *Handbook of Modern Item Response Theory*. Springer, New York, pp. 85–100.
Schmid, J., Leiman, J.M. (1957). The development of hierarchical factor solutions. *Psychometrika* **22**, 53–61.
Shepard, R.N. (1962a). The analysis of proximities: Multidimensional scaling with an unknown distance function. I. *Psychometrika* **27**, 125–140.
Shepard, R.N. (1962b). The analysis of proximities: Multidimensional scaling with an unknown distance function. II. *Psychometrika* **27**, 219–246.
Sokal, M.M. (1982). James McKeen Cattell and the future of anthropometric mental testing. In: Woodward, W.R., Ash, M.G. (Eds.), *The Problematic Science: Psychology in Nineteenth-Century Thought*. Praeger Publishers, New York, pp. 322–345.
Spearman, C. (1904). "General intelligence", objectively determined and measured. *American Journal of Psychology* **15**, 201–293.
Spearman, C. (1910). Correlation calculated with faulty data. *British Journal of Psychology* **3**, 271–295.
Stevens, S.S. (1951). Mathematics, measurement, and psychophysics. In: Stevens, S.S. (Ed.), *Handbook of Experimental Psychology*. Wiley, New York, pp. 1–49.
Symonds, P.M. (1929). Choice of items for a test on the basis of difficulty. *Journal of Educational Psychology* **20**, 481–493.
Terman, L.M. (1916). *The Measurement of Intelligence*. Houghton Mifflin, Boston.
Thissen, D., Steinberg, L. (1986). A taxonomy of item response models. *Psychometrika* **51**, 567–577.
Thorndike, E.L. (1913). *An Introduction to the Theory of Mental and Social Measurements*. Teachers College, Columbia University, New York.
Thurstone, L.L. (1924). *Psychological Examination for High School Graduates and College Freshmen*. American Council on Education, Washington, DC.
Thurstone, L.L. (1925). A method of scaling psychological and educational tests. *Journal of Educational Psychology* **16**, 433–449.
Thurstone, L.L. (1927). The law of comparative judgment. *Psychological Review* **34**, 278–286.
Thurstone, L.L. (1931a). Multiple factor analysis. *Psychological Review* **38**, 406–427.
Thurstone, L.L. (1931b). *The Reliability and Validity of Tests*. Edwards Brothers, Ann Arbor, MI.
Thurstone, L.L. (1935). *Vectors of the Mind*. University of Chicago Press, Chicago.
Thurstone, L.L. (1937). Psychology as a quantitative rational science. *Science* **85**, 227–232.

Thurstone, L.L. (1938). Primary Mental Abilities. *Psychometric Monographs*, Whole No. 1. University of Chicago Press, Chicago.
Thurstone, L.L. (1947). *Multiple Factor Analysis*. University of Chicago Press, Chicago.
Torgerson, W.S. (1952). Multidimensional scaling: I. Theory and method. *Psychometrika* **17**, 401–419.
Torgerson, W.S. (1958). *Theory and Methods of Scaling*. Wiley, New York.
Torgerson, W.S. (1986). Scaling and *Psychometrika*: Spatial and alternative representations of similarity data. *Psychometrika* **51**, 57–63.
Tucker, L.R. (1940). A matrix multiplier. *Psychometrika* **5**, 289–294.
Tucker, L.R. (1946). Maximum validity of a test with equivalent items. *Psychometrika* **11**, 1–13.
Tucker, L.R. (1955). The objective definition of simple structure in linear factor analysis. *Psychometrika* **20**, 209–225.
Tucker, L.R. (1972). Relations between multidimensional scaling and three-mode factor analysis. *Psychometrika* **37**, 3–27.
Tucker, L.R., Messick, S. (1963). An individual differences model for multidimensional scaling. *Psychometrika* **28**, 333–367.
Tufte, E.R. (1983). *The Visual Display of Quantitative Information*. Graphics Press, Cheshire, CT.
van der Linden, W.J. (1998). Optimal assembly of psychological and educational tests. *Applied Psychological Measurement* **22**, 195–211.
van der Linden, W.J., Glas, C.A.W. (2000). *Computerized Adaptive Testing: Theory and Practice*. Kluwer Academic Publishers, Dordrecht.
von Davier, A., Holland, P.W., Thayer, D. (2003). *The Kernel Method of Test Equating*. Springer, New York.
Wainer, H. (2005). *Graphic Discovery: A Trout in the Milk and Other Visual Adventures*. Princeton University Press, Princeton, NJ.
Wainer, H., Dorans, N., Eignor, D., Flaugher, R., Green, B.F., Mislevy, R.M., Steinberg, L., Thissen, D. (2000). *Computerized Adaptive Testing: A Primer*, 2nd ed. Lawrence Erlbaum Associates, Mahwah, NJ.
Wainer, H., Thissen, D. (1993). Graphical data analysis. In: Keren, G., Lewis, C. (Eds.), *A Handbook for Data Analysis in the Behavioral Sciences: Statistical Issues*. Lawrence Erlbaum Associates, Hillsdale, NJ, pp. 391–457.
Wang, X., Bradlow, E.T., Wainer, H. (2002). A general Bayesian model for testlets: Theory and applications. *Applied Psychological Measurement* **26**, 109–128.
Wiley, D.E. (1973). The identification problem for structural equation models with unmeasured variables. In: Goldberger, A.S., Duncan, O.D. (Eds.), *Structural Equation Models in the Social Sciences*. Academic Press, New York, pp. 69–83.
Wilkinson, L. (1999). *The Grammar of Graphics*. Springer, New York.
Wissler, C. (1901). The correlation of mental and physical tests. *Psychological Review Monograph Supplements* **3** (6).
Wright, S. (1921). Correlation and causation. *Journal of Agricultural Research* **20**, 557–585.
Yerkes, R.M. (1921). Psychological Testing in the US Army. *Memoirs of the National Academy of Sciences*, vol. XV.
Young, F.W., Torgerson, W.S. (1967). TORSCA, a Fortran IV program for nonmetric multidimensional scaling. *Behavioral Science* **13**, 343–344.
Young, F.W., Hamer, R.M. (1987). *Multidimensional Scaling: History, Theory, and Applications*. Lawrence Erlbaum Associates, Hillsdale, NJ.

Handbook of Statistics, Vol. 26
ISSN: 0169-7161

DOI: 10.1016/S0169-7161(06)26002-4

2

Selected Topics in Classical Test Theory

Charles Lewis

It is appropriate to begin this chapter by asking a question: What *is* classical test theory? One possible operational definition would be the contents of Harold Gulliksen's (1950) text, *Theory of Mental Tests*. Gulliksen, to be sure, did not use the label "classical" to describe his subject since, at that time, there was no rival test theory from which to distinguish it. On the other hand, Fred Lord and Mel Novick, writing eighteen years later, needed to make just such a distinction in their *Statistical Theories of Mental Test Scores*.

On pp. 28–29, Lord and Novick (1968) say "The classical test theory model which we shall use throughout the first part of this book is based on a particular, mathematically convenient and conceptually useful, definition of true score and on certain basic assumptions concerning the relationships among true and error scores". The second part of their book, and especially the chapters 17–20 contributed by Allan Birnbaum (1968), marks a clear departure from classical test theory, providing the foundations of item response theory (IRT, or latent trait models, as they were referred to at that time). (See Chapter 15 by Bock and Moustaki in this volume.)

This split suggests another possible definition of classical test theory: anything in test theory other than IRT. However, such a definition might be construed to include such developments as generalizability theory (see Chapter 4 by Webb, Shavelson, and Haertel in this volume) and structural equation modeling (see Chapter 10 by Bentler and Yuan in this volume) that clearly go beyond classical test theory. In addition, the distinction between classical test theory and IRT is not as sharp as is sometimes claimed.

Rod McDonald has long emphasized the essential unity of factor analysis (see Chapter 9 by Yanai and Ichikawa in this volume), classical test theory and item response theory, a position developed fully in his (McDonald, 1999) text *Test Theory: A Unified Treatment*. More recently, Skrondal and Rabe-Hesketh (2004) have introduced generalized linear latent and mixed models (GLLAMMs) to the psychometric world. This family of models includes factor analysis, classical test theory and IRT models as special cases, and covers latent class, multilevel, longitudinal and structural equation models as well. (See Chapter 13 by Dayton and MacReady on latent class analysis and Chapter 12 by Steele and Goldstein on multilevel model, both in this volume.) An argument for the similarities between classical test theory and IRT that follows in these footsteps will be developed in the next section.

For clarity then, let us return to the idea of classical test theory as roughly defined by the topics treated in Gulliksen (1950):

- The first five chapters of the book explore implications of the fundamental concepts of true and error scores, including definitions of parallel tests and test reliability ("the correlation between two parallel forms of the test," p. 13).
- Chapters six through nine explore the effect of increasing the length of a test on its mean, variance, reliability and validity (defined as "the correlation of the test with some criterion," p. 88). In these chapters, it is assumed that "each of the new parts [added to make the test longer] is parallel to the original," p. 69.
- Chapters 10 through 13 consider the effects of group heterogeneity on test reliability and validity. In particular, Gulliksen provides an extensive treatment of the effects of selection, often referred to as restriction of range, based on test scores and other criteria.

This list would seem to provide a good starting point for any review of classical test theory. It is also worth noting that there is no discussion of inferential statistics in Gulliksen's treatment of these topics. In modern statistical terms, all discussion takes place at the level of the test-taker population of interest. This is a point that Lord and Novick (1968) make explicit in their chapters one through six, covering much of the same material.

However, Gulliksen also acknowledges this point in his chapter 14, "A statistical criterion for parallel tests". Here, remarkably, he spells out in some detail the use of likelihood ratio hypothesis testing of several of the implications of the assumption of parallel tests, referring to publications of Wilks (1946) and Votaw (1948) in *The Annals of Mathematical Statistics*. This topic, it should be noted, is not covered by Lord and Novick (1968). (Indeed, they seem to have a strong preference for parameter estimation over hypothesis testing in their use of inferential statistics.) It also seems difficult to justify including this subject in a limited review of classical test theory.

In his chapters 15 and 16, Gulliksen returns to more a traditional test theory topic: the estimation of reliability. Chapter 17 considers speed and power tests, and specifically the effects of speededness on reliability. Chapters 18 and 19 address scoring, standardizing and equating of tests. Chapter 20 discusses the problem of choosing weights to construct composite measures from individual tests. Finally, Chapter 21 gives a detailed treatment of methods for item analysis. Interestingly, instead of "item statistics," Gulliksen refers to "item parameters," a term today associated with IRT models, and discusses the problem of dependence of these classical item parameters on the population of test takers, one of the standard IRT criticisms of classical test theory.

Based on this brief summary of the contents of Gulliksen's text, the following topics will be addressed in this review of classical test theory:

- An introduction to the basic classical theory linear model, including a definition of parallel tests or items, as well as variants introduced by Lord and Novick. This will be done using the statistical framework provided by Skrondal and Rabe-Hesketh (2004) and will address estimation of item and person parameters, as well as a comparison with IRT. In this context, reliability will be defined and the role of test length discussed.

- A discussion of the problem of restriction of range due to explicit selection on a known set of variables, including tests. Here the goal is to estimate predictive validities (as defined by simple, multiple, or partial correlations) for a set of measures relative to specified criteria in an unrestricted population.
- A treatment of classical item analysis in terms of a nonlinear classical test theory model. Specifically, biserial/polyserial and tetrachoric/polychoric correlations will be defined and the relationship of this nonlinear theory to IRT will be described.

Of course, this list does not do full justice to Gulliksen's (400+ page) coverage of the subject. Nonetheless, it is hoped that the present chapter will prove useful to readers seeking to gain an understanding of the basics of classical test theory from a modern point of view. (For additional classical test theory topics, see Chapter 3 by Zumbo on validity, and Chapter 6 by Holland, Petersen and Dorans on scaling and linking, both in this volume.)

1. A modern introduction

Since much has been made of the contrast between classical test theory and item response theory, and since many readers will be more familiar with the latter than the former, this section begins with an introduction to a simple classical test theory model that emphasizes its similarities with a popular IRT model. The framework is that provided by Skrondal and Rabe-Hesketh (2004), and the spirit of the introduction owes much to earlier work by McDonald and others.

Consider two item response models. Using the terminology of generalized linear models, both models have a linear component given by

$$\eta_i = \theta - \beta_i, \quad i = 1, \ldots, n, \tag{1}$$

where θ is a latent random variable and β_i denotes the "difficulty" of item i. For purposes of parameter identification, one may introduce the constraint

$$\sum_{i=1}^{n} \beta_i = 0. \tag{2}$$

The first of the two item response models uses an identity link and independent Normal random components with (homogeneous) variances σ_ε^2 for the n items. The distribution of the random components is conditional on θ. In the language of classical test theory, η_i denotes the true score for item i, the random component is the item's error score, and its variance, σ_ε^2, is referred to as the item error variance, or squared standard error of measurement. This model is a special case of the essentially tau-equivalent (or ETE) model introduced by Lord and Novick (1968, p. 50) and may be appropriate for item (or subtest) scores that are to be treated as continuous. To distinguish it from the more general ETE model, it will be referred to here as the ETE model with homogeneous error variances, and abbreviated as ETEH. (Lord and Novick make no assumption regarding either the error variances or the distribution of error scores in defining the ETE model.)

The second model based on Eqs. (1) and (2) uses a logit link and independent Bernoulli random components (also conditional on θ). It is due to Rasch (1960), and may be the best-known IRT model for binary item responses. (See Chapter 16 by Fisher in this volume.) De Boek and Wilson (2004, pp. 25–26, 44–45) provide a comparison of these two models (Rasch and ETEH) in another recent text emphasizing the unity of linear and nonlinear item response models.

Desirable properties of the Rasch model, sometimes contrasted with characteristics of classical test theory models, can be seen to apply to the ETEH model as well. Let the score of person j on item i be denoted by y_{ij}, for $j = 1, \ldots, N$, and the mean score of person j on the n items be denoted by $\bar{y}_{\cdot j}$. It is well known (and straightforward to show) that the conditional likelihood for the ETEH model given the $\bar{y}_{\cdot j}$ is not a function of the latent variable values θ_j, just as is the case with the Rasch model. However, in contrast to the Rasch model, explicit expressions for the conditional maximum likelihood estimates of the item parameters β_i are available for the ETEH model:

$$\hat{\beta}_i = -(\bar{y}_{i\cdot} - \bar{y}_{\cdot\cdot}). \tag{3}$$

Here $\bar{y}_{i\cdot}$ denotes the mean score on item i, and $\bar{y}_{\cdot\cdot}$ denotes the mean of all scores in the sample. Note that these estimates are "person-free," in the sense that they are unbiased estimates for the β_i, regardless of the values of the θ_j. Moreover, the unbiasedness property continues to hold for the general ETE model of classical test theory, where no assumption is made regarding either the distributional form of the random component or the homogeneity of the item error variances.

Estimation of θ_j in the model considered so far (with the analysis conditional on θ) is likewise straightforward:

$$\hat{\theta}_j = \bar{y}_{\cdot j}, \tag{4}$$

the mean item score for person j, using the restriction that $\bar{\beta}_{\cdot} = 0$. It should be noted that this estimate is not "item free," in the sense that it depends on the particular set of items responded to (and to which the restriction on the item parameters applies). It should also be noted that this point applies to the Rasch model as well.

Now suppose person $N + 1$ is administered a subset S of the original n items, producing responses $y_{i,N+1}$, for $i \in S$. In this case, an unbiased estimator for θ_{N+1} is given by

$$\hat{\theta}_{N+1} = \frac{1}{n_S} \sum_{i \in S} (y_{i,N+1} + \hat{\beta}_i) = \bar{y}_{\cdot(S),N+1} - \bar{y}_{\cdot(S)\cdot} + \bar{y}_{\cdot\cdot}, \tag{5}$$

where n_S denotes the number of items in S, and $\cdot(S)$ as a subscript is used to indicate that a mean has been taken with respect to responses to the items in S. This is an "item-free" estimate in the sense that it takes into account the difficulties of the items that person $N + 1$ responded to. In the context of classical test theory, this estimated value of the latent variable could be thought of as an equated score, where a simple form of linear equating has been applied to match the mean for the original group of N test takers on the new test based on the items in S to the mean of the original test for that same group, namely $\bar{y}_{\cdot\cdot}$. An analogous result regarding the estimation of θ holds for the Rasch model.

Another approach to parameter estimation that is widely used for IRT models is known as marginal maximum likelihood. In this treatment, an additional assumption is made, namely, that the θ_j constitute an independent random sample from some population of interest. For concreteness, assume the θ_j are a random sample from a Normal distribution with mean μ_θ and variance σ_θ^2. The marginal likelihood is produced from the joint likelihood of all quantities by integrating with respect to the θ_j. This function is then maximized with respect to the β_i. For the Rasch model, the resulting marginal maximum likelihood estimates differ from the ones obtained via conditional maximum likelihood, and are no longer "person-free," since they depend on the choice of distribution assumed for the θ_j. For the ETEH model, on the other hand, the two sets of item parameter estimates are identical. The only difference is that the sampling variances for these estimates in the first case are conditional on the θ_j and in the second case (more realistically) they incorporate the assumption that new values of the θ_j are sampled on every replication.

When using marginal maximum likelihood to estimate item parameters in IRT it is customary to modify the estimation of the person parameters as well. So-called EAP (expected a posteriori) and MAP (modal a posteriori) estimates of θ_j are based on the posterior distribution of θ_j given the y_{ij}. In classical test theory, such estimates have a long history. They are thought of as regression estimates (similar to regression estimates for factor scores), where the regression is of the person parameter (or true score) on the item responses (or observed scores). Specifically, the EAP (and MAP) estimate for θ_j is given by

$$\mathrm{E}(\theta_j \mid \bar{y}_{\cdot j}) = \frac{n\sigma_\theta^2 \bar{y}_{\cdot j} + \sigma_\varepsilon^2 \mu_\theta}{n\sigma_\theta^2 + \sigma_\varepsilon^2}. \tag{6}$$

A form of this equation appears in Kelley (1923, p. 214). Here it has been assumed that the parameters μ_θ, σ_θ^2, and σ_ε^2 are known. When appropriate sample estimates are substituted, the resulting estimates of the person parameters are sometimes referred to as empirical Bayes estimates.

A few additional observations regarding this form of the model are in order. First, the variance of the mean item score $\bar{y}_{\cdot j}$ may be expressed as the sum of two components:

$$\mathrm{Var}(\bar{y}_{\cdot j}) = \sigma_\theta^2 + \sigma_\varepsilon^2/n. \tag{7}$$

This is a standard (random effects) analysis of variance decomposition and is described in classical test theory as follows: the observed test score variance is the sum of true score and error variance. Based on this decomposition, it may be appropriate to consider the proportion of observed score variance associated with true scores. This ratio is referred to in classical test theory as the reliability of the observed score $\bar{y}_{\cdot j}$:

$$\rho_n = \frac{n\sigma_\theta^2}{n\sigma_\theta^2 + \sigma_\varepsilon^2}. \tag{8}$$

Kelley's estimate for θ_j may now be rewritten as

$$\mathrm{E}(\theta_j \mid \bar{y}_{\cdot j}) = \rho_n \bar{y}_{\cdot j} + (1 - \rho_n)\mu_\theta. \tag{9}$$

Standard unbiased ANOVA estimates for the variance components are

$$\hat{\sigma}_{\varepsilon}^{2} = \frac{\sum_{i=1}^{n}\sum_{j=1}^{N}(y_{ij} - \bar{y}_{i\cdot} - \bar{y}_{\cdot j} + \bar{y}_{\cdot\cdot})^{2}}{(n-1)(N-1)} \tag{10}$$

and

$$n\hat{\sigma}_{\theta}^{2} = \frac{n\sum_{j=1}^{N}(\bar{y}_{\cdot j} - \bar{y}_{\cdot\cdot})^{2}}{N-1} - \hat{\sigma}_{\varepsilon}^{2}. \tag{11}$$

Substituting these values into the expression for the reliability in (8) gives a consistent estimate of this quantity:

$$\hat{\rho}_{n} = \frac{n\hat{\sigma}_{\theta}^{2}}{n\hat{\sigma}_{\theta}^{2} + \hat{\sigma}_{\varepsilon}^{2}}. \tag{12}$$

It is a standard result in classical test theory (see, for instance, (Hoyt, 1941)) that this estimate may be rewritten in the form

$$\begin{aligned} \hat{\rho}_{n} &= \frac{n}{n-1}\left[1 - \frac{\sum_{i=1}^{n}\sum_{j=1}^{N}(y_{ij} - \bar{y}_{i\cdot})^{2}/(N-1)}{n^{2}\sum_{j=1}^{N}(\bar{y}_{\cdot j} - \bar{y}_{\cdot\cdot})^{2}/(N-1)}\right] \\ &= \frac{n}{n-1}\left[1 - \frac{\sum_{i=1}^{n}s^{2}(y_{ij})}{s^{2}(n\bar{y}_{\cdot j})}\right], \end{aligned} \tag{13}$$

best known today as Cronbach's coefficient alpha (Cronbach, 1951). Here $s^2(y_{ij})$ and $s^2(n\bar{y}_{\cdot j})$ are used to denote sample variances for the scores on item i and the scores on the total test, respectively. It can be shown (see (Novick and Lewis, 1967)) that, for the general ETE model, (13) provides a consistent estimate for test reliability. For still more general classical test theory models, (13) gives a consistent estimate for a lower bound to test reliability.

To further explore properties of reliability, first consider the decomposition of item variance corresponding to the test variance decomposition in (7):

$$\operatorname{Var}(y_{ij}) = \sigma_{\theta}^{2} + \sigma_{\varepsilon}^{2}. \tag{14}$$

Based on (14), item reliability may be defined as

$$\rho_{1} = \frac{\sigma_{\theta}^{2}}{\sigma_{\theta}^{2} + \sigma_{\varepsilon}^{2}}. \tag{15}$$

This expression may be combined with the corresponding one for test reliability in (8) to produce the general form of the Spearman–Brown formula (Spearman, 1910; Brown, 1910):

$$\rho_{n} = \frac{n\rho_{1}}{1 + (n-1)\rho_{1}}. \tag{16}$$

Equation (16) describes the effect of test length on reliability, and is one of the standard results of classical test theory. Note that it only holds for the ETEH model, and not for any more general model. The key restriction is that every item should have the same reliability (and measure the same trait, of course).

Before completing this introduction to some basic concepts of classical test theory, it may be useful to review some of the other models that have been considered. Returning to the initial specification of the ETEH model, it is important to emphasize that, traditionally, no distributional assumptions have been made, either about the random (error) component or about the person parameter θ (or true score). These assumptions have been made in the present development primarily to allow for parametric (likelihood, Bayesian, etc.) inference regarding the model parameters and thus permit a direct comparison with inference for IRT model parameters.

There are at least three features that may be varied to specify different models. Some combinations are well known, while others have received little attention. First, regarding the linear component of the model, the most general form with a single person parameter may be written as

$$\eta_i = \lambda_i \theta - \beta_i, \quad i = 1, \ldots, n, \tag{17}$$

with, say, $\lambda_1 = 1$, as well as restriction (2) on the β_i, chosen for identifiability. Karl Jöreskog (1971) used the label "congeneric" to refer to this specification. Equation (17) may be recognized as the linear component of a factor analysis model with a single factor (θ) and factor "loadings" λ_i. (This linear component is also equivalent to that used in IRT, together with a logit link and binary random component, for the two-parameter logistic model.) When all the λ_i are equal, (17) reduces to the essentially tau-equivalent (ETE) form introduced in (1), and when all the β_i are equal as well, it further reduces to the linear component for what Lord and Novick (1968, p. 47) refer to as tau-equivalent measurements.

The second and third features that are specified in classical test theory models have to do with the variance of the random component for item i. In the development of the ETEH model in this section, the specification was made that

$$\operatorname{Var}(\varepsilon_i|\theta) = \sigma_\varepsilon^2. \tag{18}$$

This may be thought of as a "double-homogeneity" assumption: The error variance is assumed to be constant for all i and for all θ. The most general specification, on the other hand, allows variation both with respect to items and with respect to θ. It might be written as

$$\operatorname{Var}(\varepsilon_i|\theta) = \sigma_{i|\theta}^2. \tag{19}$$

The tau-equivalent, essentially tau-equivalent and congeneric models of classical test theory all (implicitly or explicitly) adopt the form given in (19) for the variances of the random components. That is, these models place no restrictions on the conditional item error variances. The usual name given to $\sigma_{i|\theta}$ is the conditional standard error of measurement (or CSEM) for item i.

The specification for parallel measurements, as given by Lord and Novick (1968, p. 48), adopts the tau-equivalent linear component, together with the restriction that

$$\operatorname{Var}(\varepsilon_i|\theta) = \sigma_{\varepsilon|\theta}^2. \tag{20}$$

In other words, the conditional item error variances are assumed to be constant with respect to items, but may vary as a function of θ. It is important to emphasize that most classical test theory results were originally obtained for parallel measurements. On the other hand, subsequent work showed that, for many of these results, this strong assumption is not required. Also, for most classical results, the conditional item error variance does not play a role. Instead, its expectation with respect to θ is the only quantity that is used:

$$\operatorname{Var}(\varepsilon_i) = \operatorname{E}\big[\operatorname{Var}(\varepsilon_i|\theta)\big] = \sigma_i^2. \tag{21}$$

This allows the rewriting of Eqs. (14) and (15) for the congeneric model as

$$\operatorname{Var}(y_{ij}) = \lambda_i^2\sigma_\theta^2 + \sigma_i^2 \tag{22}$$

for the variance of item i and

$$\rho_{1i} = \frac{\lambda_i^2\sigma_\theta^2}{\lambda_i^2\sigma_\theta^2 + \sigma_i^2} \tag{23}$$

for its reliability.

Now let $\mathbf{w}$ be a vector of item weights and define the weighted composite test score

$$y_{\mathbf{w}j} = \sum_{i=1}^{n} w_i y_{ij}. \tag{24}$$

The variance of this composite may be written as

$$\operatorname{Var}(y_{\mathbf{w}j}) = \left(\sum_{i=1}^{n} w_i\lambda_i\right)^2 \sigma_\theta^2 + \sum_{i=1}^{n} w_i^2\sigma_i^2 \tag{25}$$

and its reliability as

$$\rho_{\mathbf{w}} = \frac{\left(\sum_{i=1}^{n} w_i\lambda_i\right)^2\sigma_\theta^2}{\left(\sum_{i=1}^{n} w_i\lambda_i\right)^2\sigma_\theta^2 + \sum_{i=1}^{n} w_i^2\sigma_i^2}. \tag{26}$$

This last expression is essentially McDonald's (1970) *coefficient omega*: the reliability for a composite made up of congeneric items.

2. Restriction of range

When evaluating the predictive validity of a set of measures used for purposes of selection from an applicant group, a practical problem is that, typically, measures on the criteria of interest are only available for the selected group. In such situations, a common practice is to estimate the relevant (multiple) correlations between predictors and criteria for the applicant group. The model adopted for this purpose was first proposed by Karl Pearson (1903) and developed in its current form by David Lawley (1943). The resulting Pearson–Lawley formulas used to correct correlations for restriction of range are derived in this section, using the notation common in modern multivariate statistics. This may make the current treatment more accessible than that given in Gulliksen (1950, Chapters 11–13) and in Lord and Novick (1968, pp. 140–148).

Suppose there are two vectors of random variables, $\mathbf{x}$ and $\mathbf{y}$, with mean vectors $\boldsymbol{\mu}_{\mathbf{x}}$ and $\boldsymbol{\mu}_{\mathbf{y}}$. Let the joint covariance matrix for these variables be denoted by

$$\boldsymbol{\Sigma} = \begin{bmatrix} \boldsymbol{\Sigma}_{\mathbf{xx}} & \boldsymbol{\Sigma}_{\mathbf{xy}} \\ \boldsymbol{\Sigma}_{\mathbf{yx}} & \boldsymbol{\Sigma}_{\mathbf{yy}} \end{bmatrix}. \tag{27}$$

Now suppose that the conditional mean of $\mathbf{y}$ given $\mathbf{x}$ is a linear function of $\mathbf{x}$, say

$$\mathrm{E}(\mathbf{y}|\mathbf{x}) = \boldsymbol{\beta}_0 + \mathbf{B}\mathbf{x}, \tag{28}$$

and the conditional covariance matrix for $\mathbf{y}$ given $\mathbf{x}$, denoted by $\boldsymbol{\Sigma}_{\mathbf{yy}|\mathbf{x}}$, does not depend on $\mathbf{x}$. These assumptions imply that

$$\boldsymbol{\mu}_{\mathbf{y}} = \boldsymbol{\beta}_0 + \mathbf{B}\boldsymbol{\mu}_{\mathbf{x}}, \tag{29}$$

$$\boldsymbol{\Sigma} = \begin{bmatrix} \boldsymbol{\Sigma}_{\mathbf{xx}} & \boldsymbol{\Sigma}_{\mathbf{xx}}\mathbf{B}' \\ \mathbf{B}\boldsymbol{\Sigma}_{\mathbf{xx}} & \mathbf{B}\boldsymbol{\Sigma}_{\mathbf{xx}}\mathbf{B}' + \boldsymbol{\Sigma}_{\mathbf{yy}|\mathbf{x}} \end{bmatrix}. \tag{30}$$

This is a description of what will be referred to as an "unrestricted" population.

Now introduce a (possibly) random selection variable, say s, with $s = 1$ indicating that an individual in the unrestricted population has been selected and $s = 0$ indicating that the individual has not been selected. Assume that the conditional probability,

$$\Pr(s = 1|\mathbf{x}, \mathbf{y}) = \Pr(s = 1|\mathbf{x}), \tag{31}$$

depends only on $\mathbf{x}$, so that s and $\mathbf{y}$ are conditionally independent, given $\mathbf{x}$. Gulliksen (1950, pp. 130–131) refers to "explicit selection" with respect to $\mathbf{x}$ and "incidental selection" with respect to $\mathbf{y}$.

Next, consider the conditional distribution of $\mathbf{x}$ and $\mathbf{y}$ given $s = 1$. Denote the conditional means by $\boldsymbol{\mu}_{\mathbf{x}|s}$ and $\boldsymbol{\mu}_{\mathbf{y}|s}$ and the conditional covariance matrix by

$$\boldsymbol{\Sigma}_s = \begin{bmatrix} \boldsymbol{\Sigma}_{\mathbf{xx}|s} & \boldsymbol{\Sigma}_{\mathbf{xy}|s} \\ \boldsymbol{\Sigma}_{\mathbf{yx}|s} & \boldsymbol{\Sigma}_{\mathbf{yy}|s} \end{bmatrix}. \tag{32}$$

These are the parameters of the "restricted" population. Because of the conditional independence of s and $\mathbf{y}$, given $\mathbf{x}$,

$$\mathrm{E}(\mathbf{y}|\mathbf{x}) = \mathrm{E}(\mathbf{y}|\mathbf{x}, s = 1), \tag{33}$$

so the regression of $\mathbf{y}$ on $\mathbf{x}$ in the restricted population is the same as in the unrestricted population (Eq. (28)), with

$$\mathbf{B} = \mathbf{B}_s = \boldsymbol{\Sigma}_{\mathbf{yx}|s}\boldsymbol{\Sigma}_{\mathbf{xx}|s}^{-1}, \tag{34}$$

$$\boldsymbol{\beta}_0 = \boldsymbol{\beta}_{0s} = \boldsymbol{\mu}_{\mathbf{y}|s} - \boldsymbol{\Sigma}_{\mathbf{yx}|s}\boldsymbol{\Sigma}_{\mathbf{xx}|s}^{-1}\boldsymbol{\mu}_{\mathbf{x}|s}. \tag{35}$$

Also, the conditional independence of s and $\mathbf{y}$ given $\mathbf{x}$ implies that the residual covariance matrix of $\mathbf{y}$ given $\mathbf{x}$ is the same in the unrestricted and restricted populations:

$$\boldsymbol{\Sigma}_{\mathbf{yy}|\mathbf{x}} = \boldsymbol{\Sigma}_{\mathbf{yy}|\mathbf{x},s} = \boldsymbol{\Sigma}_{\mathbf{yy}|s} - \boldsymbol{\Sigma}_{\mathbf{yx}|s}\boldsymbol{\Sigma}_{\mathbf{xx}|s}^{-1}\boldsymbol{\Sigma}_{\mathbf{xy}|s}. \tag{36}$$

Now suppose the parameters for $\mathbf{x}$ and $\mathbf{y}$ in the restricted population are known, as are the unrestricted population parameters for $\mathbf{x}$, namely, $\boldsymbol{\mu}_{\mathbf{x}}$ and $\boldsymbol{\Sigma}_{\mathbf{xx}}$. It is possible to use this information to reconstruct the remaining unrestricted population parameters as follows:

$$\begin{aligned}\boldsymbol{\mu}_{\mathbf{y}} &= \boldsymbol{\beta}_0 + \mathbf{B}\boldsymbol{\mu}_{\mathbf{x}} = \boldsymbol{\beta}_{0s} + \mathbf{B}_s\boldsymbol{\mu}_{\mathbf{x}} \\ &= \boldsymbol{\mu}_{\mathbf{y}|s} - \boldsymbol{\Sigma}_{\mathbf{yx}|s}\boldsymbol{\Sigma}_{\mathbf{xx}|s}^{-1}\boldsymbol{\mu}_{\mathbf{x}|s} + \boldsymbol{\Sigma}_{\mathbf{yx}|s}\boldsymbol{\Sigma}_{\mathbf{xx}|s}^{-1}\boldsymbol{\mu}_{\mathbf{x}} \\ &= \boldsymbol{\mu}_{\mathbf{y}|s} + \boldsymbol{\Sigma}_{\mathbf{yx}|s}\boldsymbol{\Sigma}_{\mathbf{xx}|s}^{-1}(\boldsymbol{\mu}_{\mathbf{x}} - \boldsymbol{\mu}_{\mathbf{x}|s}),\end{aligned} \tag{37}$$

$$\boldsymbol{\Sigma}_{\mathbf{yx}} = \mathbf{B}\boldsymbol{\Sigma}_{\mathbf{xx}} = \mathbf{B}_s\boldsymbol{\Sigma}_{\mathbf{xx}} = \boldsymbol{\Sigma}_{\mathbf{yx}|s}\boldsymbol{\Sigma}_{\mathbf{xx}|s}^{-1}\boldsymbol{\Sigma}_{\mathbf{xx}}, \tag{38}$$

and

$$\begin{aligned}\boldsymbol{\Sigma}_{\mathbf{yy}} &= \mathbf{B}\boldsymbol{\Sigma}_{\mathbf{xx}}\mathbf{B}' + \boldsymbol{\Sigma}_{\mathbf{yy}|\mathbf{x}} = \mathbf{B}_s\boldsymbol{\Sigma}_{\mathbf{xx}}\mathbf{B}_s' + \boldsymbol{\Sigma}_{\mathbf{yy}|\mathbf{x},s} \\ &= \boldsymbol{\Sigma}_{\mathbf{yx}|s}\boldsymbol{\Sigma}_{\mathbf{xx}|s}^{-1}\boldsymbol{\Sigma}_{\mathbf{xx}}\boldsymbol{\Sigma}_{\mathbf{xx}|s}^{-1}\boldsymbol{\Sigma}_{\mathbf{xy}|s} + \boldsymbol{\Sigma}_{\mathbf{yy}|s} - \boldsymbol{\Sigma}_{\mathbf{yx}|s}\boldsymbol{\Sigma}_{\mathbf{xx}|s}^{-1}\boldsymbol{\Sigma}_{\mathbf{xy}|s} \\ &= \boldsymbol{\Sigma}_{\mathbf{yy}|s} - \boldsymbol{\Sigma}_{\mathbf{yx}|s}\left(\boldsymbol{\Sigma}_{\mathbf{xx}|s}^{-1} - \boldsymbol{\Sigma}_{\mathbf{xx}|s}^{-1}\boldsymbol{\Sigma}_{\mathbf{xx}}\boldsymbol{\Sigma}_{\mathbf{xx}|s}^{-1}\right)\boldsymbol{\Sigma}_{\mathbf{xy}|s}.\end{aligned} \tag{39}$$

Thus the covariance matrix for the unrestricted population may be reconstructed as

$$\boldsymbol{\Sigma} = \begin{bmatrix} \boldsymbol{\Sigma}_{\mathbf{xx}} & \boldsymbol{\Sigma}_{\mathbf{xx}}\boldsymbol{\Sigma}_{\mathbf{xx}|s}^{-1}\boldsymbol{\Sigma}_{\mathbf{xy}|s} \\ \boldsymbol{\Sigma}_{\mathbf{yx}|s}\boldsymbol{\Sigma}_{\mathbf{xx}|s}^{-1}\boldsymbol{\Sigma}_{\mathbf{xx}} & \boldsymbol{\Sigma}_{\mathbf{yy}|s} - \boldsymbol{\Sigma}_{\mathbf{yx}|s}(\boldsymbol{\Sigma}_{\mathbf{xx}|s}^{-1} - \boldsymbol{\Sigma}_{\mathbf{xx}|s}^{-1}\boldsymbol{\Sigma}_{\mathbf{xx}}\boldsymbol{\Sigma}_{\mathbf{xx}|s}^{-1})\boldsymbol{\Sigma}_{\mathbf{xy}|s} \end{bmatrix}. \tag{40}$$

This is Lawley's (1943) result and is also given by Lord and Novick (1968, p. 147).

To see how this formula is used in practice, consider the following special case: a single predictor, x, and a single criterion, y. Equation (40) becomes

$$\begin{bmatrix} \sigma_x^2 & \sigma_{xy} \\ \sigma_{xy} & \sigma_y^2 \end{bmatrix} = \begin{bmatrix} \sigma_x^2 & \dfrac{\sigma_x^2\sigma_{xy|s}}{\sigma_{x|s}^2} \\ \dfrac{\sigma_x^2\sigma_{xy|s}}{\sigma_{x|s}^2} & \sigma_{y|s}^2 - \dfrac{\sigma_{xy|s}^2}{\sigma_{x|s}^2} + \dfrac{\sigma_{xy|s}^2\sigma_x^2}{\sigma_{x|s}^4} \end{bmatrix}, \tag{41}$$

and the correlation between x and y in the unrestricted population is given by

$$\begin{aligned}\rho_{xy} = \frac{\sigma_{xy}}{\sigma_x\sigma_y} &= \frac{\dfrac{\sigma_x^2\sigma_{xy|s}}{\sigma_{x|s}^2}}{\sigma_x\left(\sigma_{y|s}^2 - \dfrac{\sigma_{xy|s}^2}{\sigma_{x|s}^2} + \dfrac{\sigma_{xy|s}^2\sigma_x^2}{\sigma_{x|s}^4}\right)^{1/2}} \\ &= \frac{\sigma_x\sigma_{xy|s}}{\left(\sigma_{x|s}^4\sigma_{y|s}^2 - \sigma_{xy|s}^2\sigma_{x|s}^2 + \sigma_{xy|s}^2\sigma_x^2\right)^{1/2}} \\ &= \frac{\sigma_x\rho_{xy|s}}{\left(\sigma_{x|s}^2 - \rho_{xy|s}^2\sigma_{x|s}^2 + \rho_{xy|s}^2\sigma_x^2\right)^{1/2}}.\end{aligned} \tag{42}$$

This last expression was first given by Pearson (1903) and appears in Gulliksen (1950, p. 137).

More generally, once the covariance matrix given in Eq. (40) has been estimated, any simple, multiple, partial or semipartial correlation involving elements of $\mathbf{x}$ and $\mathbf{y}$ may also be estimated for the unrestricted population. In practice, it is important to include (if possible) all variables used for selection in the vector $\mathbf{x}$. Also, only one correction for restriction of range should be applied in a given setting, regardless of how many correlations will be computed.

For instance, in evaluating the predictive validity of the SAT with regard to performance in a particular college, $\mathbf{x}$ might consist of the three SAT scores (Mathematics, Critical Reading and Writing), together with high school GPA. For simplicity, suppose there is only a single criterion: first year college GPA, making up y. The unrestricted population might be the applicant group for that college in a given year (or the group of all individuals who might have applied to the college). The restricted group might consist of all individuals with a first year GPA at the college in the following year.

To correct for restriction of range in this case, Eq. (40) would be used to estimate the four covariances of the predictors with the criterion in the applicant group, as well as the variance of the criterion in the applicant group, using statistics from the restricted group. These quantities, together with the known sample variances and covariances of the predictors in the applicant group, might then be applied to estimate the simple unrestricted correlations of each predictor with the criterion. Other quantities that might be estimated from the same variances and covariances include the multiple correlation of any combination of the predictors (such as the test scores only) with the criterion, and partial correlations (for instance, of the test scores and criterion, partialling out high school GPA), all for the applicant group. In each case, standard formulas for the correlation of interest would be used, inserting the relevant variances and covariances as though they had all been computed from complete data for the applicant group.

3. Nonlinear classical test theory

Although classical test theory is most closely associated with the application of linear models, such as those described in the previous sections, there is an important exception to this generalization, apparently first introduced by Karl Pearson (1901) and widely used today in contexts ranging from item analysis to structural equation modeling. Specifically, it was clear even in the earliest work on test theory that, for binary test items, a linear model is not usually the most appropriate choice. (That a linear model was sometimes considered to be appropriate for binary items is indicated in the formula for the reliability of a test made up of such items given by Kuder and Richardson, 1937, and known as KR20. This formula, in turn, is a special case of the one for coefficient alpha given in Eq. (13).)

Pearson (1901) proposed two special-purpose correlation indices that he labeled the *biserial* and the *tetrachoric*. The first is intended to indicate the correlation between a binary and a continuous variable, the second to indicate the correlation between two binary variables. The definitions of these two measures of correlation are based on the idea of "latent responses" (see Skrondal and Rabe-Hesketh, 2004, pp. 16, 33, for a modern discussion) that are assumed to underlie observed binary item scores.

Specifically, corresponding to the binary observed score y_i for item i, this approach assumes the existence of a continuous latent item response y_i^*. The two variables are related as follows:

$$y_i = \begin{cases} 1 & \text{if } y_i^* \geqslant y_{i0}^*, \\ 0 & \text{if } y_i^* < y_{i0}^*, \end{cases} \tag{43}$$

with y_{i0}^* representing a threshold for item i. If it is further assumed that y_i^* has a standard Normal distribution, then the threshold parameter is given by

$$\Pr(y_i = 1) = \pi_i = \Pr(y_i^* \geqslant y_{i0}^*) = \Phi(-y_{i0}^*), \tag{44}$$

or

$$y_{i0}^* = -\Phi^{-1}(\pi_i). \tag{45}$$

Gulliksen (1950, p. 368) mentions a linear transformation of (45) as an index of item difficulty for binary item scores. (Interestingly, a modified version of this index, denoted by Δ, is still used for specifying item difficulties for many tests, including the SAT.)

The *biserial* correlation ρ'_{xi} between y_i and a continuous (observable) variable x is simply defined as the "ordinary" (Pearson product-moment) correlation between the latent response y_i^* and x. (The *polyserial* correlation is similarly defined for ordered polytomous y_i.) It is traditionally evaluated based on an additional assumption: the regression of x on y_i^* is linear, or

$$\mathrm{E}(x \mid y_i^*) = \beta_0 + \beta_{1i} y_i^*, \tag{46}$$

with

$$\beta_{1i} = \rho'_{xi} \sigma_x. \tag{47}$$

Based on this assumption it follows that

$$\mathrm{E}(x \mid y_i^* \geqslant y_{i0}^*) = \beta_0 + \beta_{1i} \mathrm{E}(y_i^* \mid y_i^* \geqslant y_{i0}^*), \tag{48}$$

$$\mathrm{E}(x \mid y_i^* < y_{i0}^*) = \beta_0 + \beta_{1i} \mathrm{E}(y_i^* \mid y_i^* < y_{i0}^*). \tag{49}$$

Solving (48) and (49) for β_{1i} gives

$$\beta_{1i} = \frac{\mathrm{E}(x \mid y_i^* \geqslant y_{i0}^*) - \mathrm{E}(x \mid y_i^* < y_{i0}^*)}{\mathrm{E}(y_i^* \mid y_i^* \geqslant y_{i0}^*) - \mathrm{E}(y_i^* \mid y_i^* < y_{i0}^*)}, \tag{50}$$

$$\rho'_{xi} = \frac{\mathrm{E}(x \mid y_i = 1) - \mathrm{E}(x \mid y_i = 0)}{\sigma_x [\mathrm{E}(y_i^* \mid y_i^* \geqslant y_{i0}^*) - \mathrm{E}(y_i^* \mid y_i^* < y_{i0}^*)]}. \tag{51}$$

Since it was assumed that y_i^* has a standard Normal distribution, it follows (using a standard result for the mean of a truncated Normal random variable) that

$$\mathrm{E}(y_i^* \mid y_i^* \geqslant y_{i0}^*) = \frac{\int_{y_{i0}^*}^{+\infty} z\varphi(z)\,\mathrm{d}z}{\int_{y_{i0}^*}^{+\infty} \varphi(z)\,\mathrm{d}z} = \frac{\varphi(y_{i0}^*)}{\pi_i}, \tag{52}$$

$$\mathrm{E}(y_i^*|y_i^* < y_{i0}^*) = \frac{\int_{-\infty}^{y_{i0}^*} z\varphi(z)\,\mathrm{d}z}{\int_{-\infty}^{y_{i0}^*} \varphi(z)\,\mathrm{d}z} = \frac{-\varphi(y_{i0}^*)}{1-\pi_i}, \tag{53}$$

$$\begin{aligned}&\mathrm{E}(y_i^*|y_i^* \geqslant y_{i0}^*) - \mathrm{E}(y_i^*|y_i^* < y_{i0}^*)\\ &\quad= \frac{\varphi(y_{i0}^*)}{\pi_i} - \frac{-\varphi(y_{i0}^*)}{1-\pi_i} = \frac{\varphi(y_{i0}^*)}{\pi_i(1-\pi_i)}.\end{aligned} \tag{54}$$

Substituting in (51) and using (45) gives

$$\rho'_{xi} = \frac{\mathrm{E}(x|y_i=1) - \mathrm{E}(x|y_i=0)}{\sigma_x}\,\frac{\pi_i(1-\pi_i)}{\varphi[-\Phi^{-1}(\pi_i)]}. \tag{55}$$

Traditional estimation of a biserial correlation proceeds by substituting sample quantities for the corresponding population values in (55). Kelley (1923, p. 249) provides an approximate standard error for this estimate ($\hat{\rho}'_{xi}$):

$$\hat{\sigma}(\hat{\rho}'_{xi}) = N^{-1/2}\left\{\frac{[\hat{\pi}_i(1-\hat{\pi}_i)]}{\varphi[-\Phi^{-1}(\hat{\pi}_i)]} - \hat{\rho}'^{2}_{xi}\right\}. \tag{56}$$

It may be worth noting that there is nothing to prevent estimates for the biserial correlation based on (55) from lying outside the interval $[-1, 1]$.

Turning briefly to the *tetrachoric* correlation between two binary item scores, y_i and $y_{i'}$, denoted by $\rho'_{ii'}$, Pearson defined it to be the product-moment correlation between the corresponding latent item response variables y_i^* and $y_{i'}^*$. (The *polychoric* correlation is similarly defined for ordered polytomous y_i and $y_{i'}$.) No simple formulas for $\rho'_{ii'}$ exist, and a variety of tables, nomograms and, more recently, computer algorithms, have been provided to aid in its estimation. (Clearly, the tetrachoric correlation is a function of the joint probability distribution for y_i and $y_{i'}$.) The point of mentioning it in this treatment is that it provides another example of nonlinear models used in classical test theory.

Finally, there is a connection between nonlinear classical test theory and IRT, first given by Tucker (1946), and described by Lord and Novick (1968, pp. 374–378). Consider the two-parameter Normal Ogive IRT model, having a linear component given by (17) and repeated here for an item i:

$$\eta_i = \lambda_i\theta - \beta_i.$$

To complete the model specification, one adopts a probit link and a Bernoulli random component (conditional on θ). This implies that

$$\Pr(y_i = 1|\theta) = \Phi(\lambda_i\theta - \beta_i). \tag{57}$$

To identify the parameters of this model, Tucker (1946) and Lord and Novick (1968, p. 376) set the mean of θ equal to zero and its variance equal to unity (rather than adopting restrictions on the item parameters, as has been done in this chapter). With the additional assumption of a (standard) Normal distribution for θ, the marginal probability of a positive response is given by

$$\Pr(y_i = 1) = \pi_i = \Phi\left[\frac{-\beta_i}{(1+\lambda_i^2)^{1/2}}\right]. \tag{58}$$

Now define a latent response for item i as

$$y_i^* = \lambda_i \theta + \varepsilon_i \tag{59}$$

with

$$\varepsilon_i \sim N(0,1) \tag{60}$$

independently of θ. Then it follows that

$$y_i^* \sim N\big(0, 1 + \lambda_i^2\big). \tag{61}$$

(Note that a latent response so defined no longer has unit variance.) Using (58),

$$\Pr(y_i = 1) = \Pr(y_i^* \geqslant \beta_i), \tag{62}$$

so β_i plays the role of a threshold. Since λ_i is the regression coefficient for the regression of y_i^* on θ, it follows that the biserial correlation between y_i and θ is given by

$$\rho'_{\theta i} = \frac{\lambda_i}{(1 + \lambda_i^2)^{1/2}}, \tag{63}$$

so

$$\lambda_i = \frac{\rho'_{\theta i}}{(1 - \rho'^2_{\theta i})^{1/2}}. \tag{64}$$

Also, from (58),

$$\beta_i = -\Phi^{-1}(\pi_i)\big(1 + \lambda_i^2\big)^{1/2} = \frac{-\Phi^{-1}(\pi_i)}{(1 - \rho'^2_{\theta i})^{1/2}}. \tag{65}$$

To summarize, under certain circumstances the item parameters for the two-parameter Normal Ogive IRT model may be expressed in terms of classical test theory item parameters, with the qualification that a suitable substitute for θ (such as the total test score) must be found to approximate the biserial correlation $\rho'_{\theta i}$.

References

Birnbaum A. (1968). Some latent trait models and their use in inferring an examinee's ability. In: Lord, F.M., Novick, M.R. *Statistical Theories of Mental Test Scores*. Addison-Wesley, Reading, MA, pp. 395–479.

Brown, W. (1910). Some experimental results in the correlation of mental abilities. *British Journal of Psychology* **3**, 296–322.

Cronbach, L.J. (1951). Coefficient alpha and the internal structure of tests. *Psychometrika* **16**, 297–334.

De Boek, P., Wilson, M. (Eds.) (2004). *Explanatory Item Response Models: A Generalized Linear and Nonlinear Approach*. Springer, New York.

Gulliksen, H. (1950). *Theory of Mental Tests*. Wiley, New York.

Hoyt, C. (1941). Test reliability estimated by analysis of variance. *Psychometrika* **6**, 153–160.

Jöreskog, K. (1971). Statistical analysis of sets of congeneric tests. *Psychometrika* **36**, 109–133.

Kelley, T.L. (1923). *Statistical Method*. Macmillan, New York.

Kuder, G.F., Richardson, M.W. (1937). The theory of the estimation of test reliability. *Psychometrika* **2**, 151–160.

Lawley, D. (1943). A note on Karl Pearson's selection formulae. *Royal Society of Edinburgh, Proceedings, Section A* **62**, 28–30.

Lord, F.M., Novick, M.R. (1968). *Statistical Theories of Mental Test Scores*. Addison-Wesley, Reading, MA.

McDonald, R.P. (1970). The theoretical foundations of common factor analysis, principal factor analysis, and alpha factor analysis. *British Journal of Mathematical and Statistical Psychology* **23**, 1–21.

McDonald, R.P. (1999). *Test Theory: A Unified Approach*. Erlbaum, Mahwah, NJ.

Novick, M.R., Lewis, C. (1967). Coefficient alpha and the reliability of composite measurements. *Psychometrika* **32**, 1–13.

Pearson, K. (1901). Mathematical contributions to the theory of evolution. VII. On the inheritance of characters not capable of exact quantitative measurement. *Philosophical Transactions of the Royal Society A* **195**, 79–150.

Pearson, K. (1903). Mathematical contributions to the theory of evolution. XI. On the influence of natural selection on the variability and correlation of organs. *Philosophical Transactions of the Royal Society A* **200**, 1–66.

Rasch, G. (1960). *Probabilistic Models for Some Intelligence and Attainment Tests*. Danish Institute for Educational Research, Copenhagen, Denmark.

Skrondal, A., Rabe-Hesketh, S. (2004). *Generalized Latent Variable Modeling: Multilevel, Longitudinal and Structural Equation Models*. Chapman & Hall/CRC Press, Boca Raton, FL.

Spearman, C. (1910). Correlation calculated with faulty data. *British Journal of Psychology* **3**, 271–295.

Tucker, L.R. (1946). Maximum validity of a test with equivalent items. *Psychometrika* **11**, 1–13.

Votaw Jr., D.F. (1948). Testing compound symmetry in a normal multivariate distribution. *Annals of Mathematical Statistics* **19**, 447–473.

Wilks, S.S. (1946). Sample criteria for testing equality of means, equality of variances, and equality of covariances in a normal multivariate distribution. *Annals of Mathematical Statistics* **17**, 257–281.

Handbook of Statistics, Vol. 26
ISSN: 0169-7161

DOI: 10.1016/S0169-7161(06)26003-6

3

Validity: Foundational Issues and Statistical Methodology

*Bruno D. Zumbo**

1. Introductory remarks

From a statistical point of view, psychometricians are faced solely with an array of numbers whose elements are outcomes of random variables that are most often discrete or ordered categorical in nature. It is the ultimate aim of the psychometrician to use this array to make a variety of meaningful inferences about the examinees and the measurement instrument itself, which should be appreciated as a daunting task.

A long tradition in psychometrics has evolved (as it has in statistics) of concerning ourselves with decomposing observational values into a component that is *deterministic* and a component that is *stochastic* so that relationships between manifest and unobserved variables can be explicitly stated and uncertainty about model parameters can be estimated and used to qualify the inferences that are possible under a given model. In this light, many contemporary measurement inferences are model-based. Contemporary psychometric models involve latent variables, mixture distributions, and hyperparameterized parameter-driven models – all of which are, of course, descendants of a longstanding statistical tradition but are unique because they are located at the *intersection* of examinee and item spaces, *both* of which are typically of interest to measurement specialists (Zimmerman and Zumbo, 2001). As Zumbo and Rupp (2004) remind us, this is central to all major statistical theories of measurement. Specifically, classical test theory generically decomposes the observed score into a deterministic part (i.e., true score) and a stochastic part (i.e., error), generalizability theory further unpacks the stochastic part and redefines part of error as systematic components, and item response theory reformulates the two model components by inducing latent variables into the data structure. Structural equation models and exploratory as well as confirmatory factor analysis models decompose the covariance matrix of multivariate data into deterministic (i.e.,

*I would like to thank the following individuals for feedback and review of an earlier draft of this chapter (in alphabetical order): Greg Camilli, Craig Deville, Brian French, Anne M. Gadermann, Anita M. Hubley, Chris Richardson, Rick Sawatzky, Steve Sireci, and Cornelia Zeisser. Bruno D. Zumbo, Ph.D., is Professor of Measurement and Statistics as well as a member of the Institute of Applied Mathematics and the Department of Psychology at the University of British Columbia. His primary interests are in psychometrics, statistical science, and the mathematical and philosophical foundations of measurement. Send Correspondence to: Professor Bruno D. Zumbo, Department of ECPS, University of British Columbia, Vancouver, B.C., Canada V6T 1Z4, E-mail: bruno.zumbo@ubc.ca, Fax: +1 604 822 3302.

reproduced covariance matrix) and stochastic (i.e., residual matrix) components, which is a model that can be equivalently written as a formulation involving latent variables.

As Zumbo and Rupp (2004) note, even though latent trait indicator values and observed composite scores are typically highly correlated, the injection of a latent continuum into the data matrix has given us the property of item and examinee parameter invariance. For perfect model fit across populations and conditions item and examinee parameter invariance has allowed us, for example, to define conditional standard errors of measurement similar to *generalizability theory*, and has opened up the road for adaptive testing through the use of item and test information functions. For a technical description of invariance and the effects of its violations please see Rupp and Zumbo (2003, 2004, 2006). Still, these advances have not come without a price. Improvements in the level of modeling and in quantifying measurement error have come at the expense of large sample sizes that are typically required for parameter estimation in both frequentist and Bayesian frameworks (see Rupp et al., 2004).

Throughout this chapter, the terms "item" and "task" will be used interchangeably. Furthermore, the terms "test", "measure", "scale", and "assessment" will be used interchangeably even though "tests" are, in common language, used to imply some educational achievement or knowledge test with correct and incorrect responses or partial credit scoring, and "assessment" typically implies some decisions, actions, or recommendations from the test and measurement results and implies a more integrative process (involving multiple sources of information). Furthermore, depending on the context (e.g., structural equation modeling) items or tasks may be referred to as indicators or manifest variables. In the remainder of this chapter, I will contextualize the terminology as much as possible. Finally, in the parlance of day-to-day social and behavioral researchers, clinicians, and policy specialists, tests may be referred to as valid or invalid, but it is widely recognized that such references are, at best, a shorthand for a more complex statement about the validity of inferences made about test scores with a particular sample in a particular context and, more often are inappropriate and potentially misleading.

At this point, the following quotation that is often attributed to the philosopher John Woods may be appropriate "My paper is like a domestic wine, it's not particularly robust or important but I think you may be amused by its pretensions" – with all the various meanings of pretension. With this humorous (yet appropriate) remark in mind, the purpose of this chapter is to shine a spotlight on some foundational and statistical issues in validity theory and validation practice. Due to space limitations, relative to the breadth and scope of the task at hand, for some issues I will provide details whereas in others more general integrative remarks. The chapter is organized as follows. Section 2 discusses several foundational issues focusing on several observations about the current state of affairs in validity theory and practice, introducing a new framework for considering the bounds and limitations of the measurement inferences, and briefly discussing the distinction between measures and indices. Section 3 deals with two statistical methods, variable ordering and latent variable regression, and introduces a methodology for variable-ordering in latent variable regression models in validity research. Section 4 closes the chapter with an overview in the context of remarks around the question "When psychometricians speak of validity what are they really saying?"

Wherever possible, an example will be used throughout this chapter to make matters concrete and motivate several validity questions.

1.1. Example

The domain of validity can be very abstract and philosophic so let me introduce an example using the Center for Epidemiological Studies Depression scale (CES-D) to concretize matters and ground the discussion. The CES-D is useful as a demonstration because it is commonly used in the life, behavioral, health, and social sciences. The CES-D is a 20-item scale introduced originally by Lenore S. Radloff to measure depressive symptoms in the general population (Radloff, 1977). The CES-D prompts the respondent to reflect upon his/her last week and respond to questions, such as "My sleep was restless", using an ordered or Likert response format of "rarely or none of the time (less than 1 day)", "some or a little of the time (1–7 days)", "occasionally or a moderate amount of time (3–4 days)", and "most or all of the time (5–7 days)" during the last week. The items typically are scored from zero (less than 1 day) to three (5–7 days). Composite scores therefore range from 0 to 60, with higher scores indicating higher levels of depressive symptoms.[1] The data presented herein is a sub-sample of a larger data set collected in northern British Columbia, Canada. As part of a larger survey, responses were obtained from a convenience sample of 600 adults in the general population – 290 females ranging in age from 18 to 87 years (mean of 42 years), and 310 males ranging in age from 17 to 82 years (mean of 46 years).

Of course, the composite scale score is not the phenomenon of depression, per se, but rather is related to depression such that a higher composite scale score reflects higher levels of the latent variable depression. Although I will return to this later in the chapter, it is useful to note in the context of the example that there are essential distinctions between (i) the observed items, and their observed score composite, (ii) the latent variable which is defined and estimated in a statistical model, and (iii) the phenomenon, attribute or construct, of depression. Cast in this way, two commonly found questions of test validation are of interest: (a) Given that the items are combined to created one scale score, do they measure just one latent variable? and (b) Are the age and gender of the respondents predictive of the latent variable score on the CES-D? The former question is motivated by psychometric necessities whereas the latter question is motivated by theoretical predictions and investigates known-group gender and age differences on CES-D scores. Although this example is used throughout the chapter, I will return to these two validity questions after discussing some foundational issues in Section 2.

2. Foundational issues

2.1. A series of observations on the current state of affairs in validity theory and practice

There is little question that validity theory and practices have changed over the last century. Angoff (1988), Hubley and Zumbo (1996), Jonson and Plake (1998), and Kane

[1] Some items need to be re-coded to match a larger score being equated to more depressive symptoms.

(2001) provide histories and overviews of validity theory. In brief, the early- to mid-1900s were dominated by the criterion-based model of validity, with some focus on content-based validity models. The early 1950s saw the introduction of, and move toward, the construct model with its emphasis on construct validity; a seminal piece being Cronbach and Meehl (1955). The period post Cronbach and Meehl, mostly the 1970s to date, saw the construct model take root and saw the measurement community delve into a moral foundation to validity and testing by expanding to include the consequences of test use and interpretation (Messick, 1975, 1980, 1988, 1989, 1995, 1998).

Measurement or test score validation is an ongoing process wherein one provides evidence to support the appropriateness, meaningfulness, and usefulness of the specific inferences made from scores about individuals from a given sample and in a given context. The concept, method, and processes of validation are central to constructing and evaluating measures used in the social, behavioral, health, and human sciences because, without validation, any inferences made from a measure are potentially meaningless, inappropriate and of limited usefulness.

The above definition highlights three central features in current thinking about validation. First, it is not the measure per se that is being validated but rather the inferences one makes from a measure. This distinction between the validation of a scale and the validation of the inferences from scores obtained from a scale may appear subtle at first blush but, in fact, it has significant implications for measurement and testing because it highlights that the validity of the inferences one makes from test scores is bounded by place, time, and use of the scores resulting from a measurement operation.

The second central feature in the above definition is the clear statement that inferences made from all empirical measures, irrespective of their apparent objectivity, have a need for validation. That is, it matters not whether one is using an observational checklist, an "objective" educational, economic, or health indicator such as number of students finishing grade 12, or a more psychological measure such as a self-report depression measure, one must be concerned with the validity of the inferences (Zumbo, 1998). It is interesting to note that in my experience in working with health researchers the much exalted medical measures such as blood pressure readings or diagnostic assessments generally have no better psychometric and validity properties than psychosocial or educational measures.

The final central feature in the above definition is that validity depends on the interpretations and uses of the test results and should be focused on establishing the inferential limits (or bounds) of the assessment, test, or measure. In short, invalidity is something that distorts the meaning of test results for some groups of examinees in some contexts for some purposes. Interestingly, this aspect of validity is a slight twist on the ideas of test and item bias (Zumbo and Hubley, 2003). That is, test and item bias aim analyses at establishing the inferential limits of the test – i.e., establishing for whom (and for whom not) the test or item score inferences are valid.

It is instructive at this point to contrast contemporary thinking in validity theory with what is commonly seen in many introductory texts in research methodology in the social, behavioral, health, and human sciences with a series of observations about the state of the field of validity. The purpose here is not to be exhaustive of all aspects of current thinking in validity but rather to shine a light on some points that I believe are key to contemporary validity theory and validation practice.

Observation 1

Validity statements are not dichotomous (valid/invalid) but rather are described on a continuum.

Observation 2

Whereas the quantification of error of measurement aids in the inference from the observed (composite) score to the true score or latent variable, validity theory aids us in the inference from the observed score to the construct of interest, via the latent or unobserved variable. Although there is sometimes confusion even in the technical measurement literature, it is important to note that the construct is not the same as the true score or latent variable, which, in turn in practical settings, is not the same as the observed item or task score. An obvious and popular distortion of these concepts is nearly ubiquitously seen in the use of the term "construct comparability" in, for example, cross-cultural measurement settings. What is often referred to as "construct comparability" is, at best, the equivalence of latent variables. Construct comparability is more than the equivalence of latent variables.

Observation 3

Although it has been controversial, one of the current themes in validity theory is that construct validity is the totality of validity theory and that its demonstration is comprehensive, integrative, and evidence-based. In this sense, construct validity refers to the degree to which inferences can be made legitimately from the observed scores to the theoretical constructs about which these observations are supposed to contain information. In short, construct validity involves generalizing from our behavioral or social observations to the *conceptualization* of our behavioral or social observations in the form of the construct. The practice of validation aims to ascertain the extent to which an interpretation of a test is *conceptually* and *empirically* warranted and should be aimed at making explicit any hidden ethical and social values that overtly or inadvertently influence that process (Messick, 1995).

Observation 4

It is important to highlight that, as Kane (2001) reminds us, there are strong and weak forms of construct validity. The weak form is characterized by any correlation of the test score with another variable being welcomed as evidence for another "validity" of the test. That is, in the weak form, a test has as many "validities" and potential uses as it has correlations with other variables. In contrast to the weak form of construct validity, the strong form is based on a well-articulated theory and well-planned empirical tests of that theory. In short, the strong-form is theory-driven whereas the weak form implies that a correlation with some criterion is sufficient evidence to use the test as a measure of that criterion. In my view (e.g., Zumbo, 2005), the strong form of construct validity should provide an *explanation* for the test scores, in the sense of the theory having explanatory power for the observed variation in test scores – I will return to this point later.

Observation 5

Although it was initially set aside to elevate construct validity, content-based evidence has gained momentum again in part due to the work of Sireci (1998). In addition, as is shown by Sireci, content-based evidence is also growing in influence in the test development process with the use of subject matter experts early in the test development process.

Observation 6

As one can see in Messick (1989) and in Zumbo's (1998) volume on validity, there is a move to consider the *consequences* of inferences from test scores. That is, along with the elevation of construct validity to an overall (unified) validity framework for evaluating test interpretation and use came the consideration of the role of value implications and social consequences as validity evidence contributing to score meaning and use. This movement has been met with some resistance. In the end, Messick (1998) made the point most succinctly when he stated that one should not be concerned simply with the obvious and gross negative consequences of score interpretation, but rather one should consider the more subtle and systemic consequences of "normal" test use. Social consequences and value implications, of course, also feed back into theory development (Hubley and Zumbo, 1996).

It is my interpretation that, by opening measurement and testing to a larger social and cultural interpretation, we are recognizing the social and cultural forces involved in test construction and test use. This, I believe, is a natural consequence of the recognition that test scores are impacted by the situation and living conditions of an examinee and that their test scores are, in part, a result of what the examinee makes of those living conditions. Zumbo and Gelin (in press), for example, build on this idea and expand the realm of test score explanation and validity beyond the cognitive to include psychosocial and cultural variables. In so doing, one embraces Messick's perspective on test consequences and acknowledges the importance of a consequentialist moral tradition such as that of the philosopher John Stuart Mill, who noted that our moral obligation as human beings is to try to make the greatest good – in our case the greatest good from our tests.

The matter and role of consequences still remains controversial today and will regain momentum in the current climate of large-scale test results being used to determine financing and staffing for educational institutions. Likewise, we will see the matter and role of consequences of measurement become prominent as we see health outcome measurement, for example, used to determine financing as well as staffing in public institutions in the United States, Europe, the Asia-Pacific and Canada.

Observation 7

Due to the number of international and cross-cultural efforts in testing and measurement, test adaptation and translation is growing in importance and centrality in the test validation literature and practice (see, for example, Hambleton and Patsula, 1998; Kristjansson et al., 2003). In this context, the matter of measurement invariance is also on the rise in importance and centrality.

Observation 8

There has been some debate as to whether reliability is a necessary but not sufficient condition for validity. It seems to me that this issue is better cast as one of measurement precision so that one strives to have as little measurement error as possible in their inferences. Specifically, reliability is a question of *data quality*, whereas validity is a question of *inferential quality* (Zumbo and Rupp, 2004). Of course, reliability and validity theory are interconnected research arenas, and quantities derived in the former bound or limit the inferences in the latter.

Observation 9

The use of cognitive models as an alternative to traditional test validation has gained a great deal of momentum in the last 15 years. One of the limitations of traditional quantitative test validation practices (e.g., factor-analytic methods, validity coefficients, and multi trait-multi method approaches) is that they are descriptive rather than explanatory. The cognitive models, particularly the work of Embretson (1994), Nichols (1994), and Mislevy (1996), have laid the groundwork to expand the evidential basis for test validation by providing a richer explanation of the processes of responding to tests and hence promoting a richer psychometric theory-building. A similar push for explanatory power has also taken place in the area of differential item functioning where attitudinal, background, and cognitive variables are used to account for differential mediating and moderating variables (Zumbo and Gelin, in press; Zumbo and Hubley, 2003).

Observation 10

As Zumbo and Rupp (2004) and Zumbo and MacMillan (1999) note, it is important to move beyond simple "cold" cognitive models of the sorts found in the cognitive revolution of psychology in the 1960s and 1970s, to more contextualized and social cognitive models that recognize that the examinee functions and lives within a social world that shapes and influences cognition. That is, in many assessment situations, researchers use the word 'cognition' to loosely refer to any process that is somehow grounded in our minds. These psychometric cognitive models seek to explicitly represent the cognitive processes that examinees engage in when responding to items via parameters in mathematical models, which typically consist of augmented IRT models, classification algorithms based on regular IRT models, or Bayesian inference networks that have IRT models as a central component. Developments in cognitive models have often taken place primarily in educational achievement and psychoeducational assessment contexts. An exception was Zumbo et al. (1997) in personality assessment in which they studied the relation of the abstractness and concreteness of items to the psychometric properties of a personality measure. Other advances are being made in the development of simulation-based assessment software that emphasizes a deeper and richer understanding of the cognitive processes required for performing certain tasks in which data are analyzed through Bayesian networks (Mislevy et al., 1999).

Observation 11

The basic idea behind the cognitive models and other explanatory approaches is that, if one could understand why an individual responded a certain way to an item, then that would go a long way toward bridging the inferential gap between test scores (or even latent variable scores) and constructs. Building on this observation, Zumbo (2005) notes it is important to separate the concept of validity from the process of test validation. According to this view, validity per se, is not established until one has an explanatory model of the variation in item responses and the variables mediating, moderating, and otherwise affecting the response outcome. This is a tall hurdle indeed but I believe that it is in the spirit of the earliest work on validity such as that of Cronbach and Meehl (1955) interpreted as a strong form of construct validity; and points to the fact that by focusing on the validation process rather than the concept of validity we have somewhat lost our way as a discipline. This is not to suggest that the activities of the process of validation such as correlations with a criterion or a convergent measure, dimensionality assessment, item response modeling, or differential item or test functioning are irrelevant. Quite the contrary, rather it points to the fact that the information from the validation process needs to be directed toward supporting the concept of validity and not is the end goal itself. My aim is to re-focus our attention on why we are conducting all of these psychometric analyses: to support our claim of the validity of our inferences from a given measure. For example, conducting test and item bias is not just about protecting a test developer or test user against lawsuits; it is also a statistical methodology that ferrets out invalidity that distorts the meaning of test results for some groups of examinees thus establishes the inferential limits of the test.

Observation 12

There should be more discussion about "models" and "modeling" in validity, and their varieties of uses, meanings, and intentions (Zumbo and MacMillan, 1999). As Zumbo and Rupp (2004) note, in validity research, the issue is less about a lack of models for new kinds of test data but rather a lack of awareness in the applied world that these models exist along with a mismatch of assessment instruments and modeling practice. In other words, if test developers are interested in providing examinees and institutions with richer profiles of abilities and developmental progress, the nature of the assessment methods has to change to provide richer data sets from which relevant information can be more meaningfully extracted. What is meant by "more meaningful" will, of course, in the end, depend on the use of the assessment data but, in general, authorities in the field today are beginning to agree that we need more than simple test responses scored 0 and 1 to validate the inferences that are being made from the test data. It has been my view that the key to useful cognitive models is that they need to be explanatory and not just another set of descriptive models in cognitive terms rather than mathematical terms (Zumbo and MacMillan, 1999). It is important to note that not all cognitive models are explanatory. Put differently, a change of terminology in psychometrics from mathematical terminology to cognitive terminology is insufficient to claim true advances in gathering more meaningful and weighty validity evidence.

In this context, it is important to acknowledge the seminal paper by Borsboom et al. (2004). Although, in its core concepts, Borsboom and his colleagues' views share a lot in common with the view of validity I have espoused, I differ from their view on several important philosophical and methodological features. For example, Borsboom and his colleagues argue that a test is valid for measuring an attribute if and only if the attribute exists and variations in the attribute causally produce variations in the outcomes of the measurement procedure.[2] Philosophically this is a very tidy, and as the

[2] A variation on this idea, with a different focus, can be seen in the work of Bollen and Lennox (1991) in their distinction between "measures" as opposed to "indices" in the form of reflective as opposed to formative indicators in the context of structural equation modeling and factor analysis. I discuss these concepts more fully in a section below.

authors' themselves acknowledge, simple idea that has a currency among researchers because it may well be implicit in the thinking of many practicing researchers. From my explanatory-focused view, relying on causality is natural and plausible and provides a clear distinction between understanding why a phenomenon occurs and merely knowing that it does, since it is possible to know that a phenomenon occurs without knowing what caused it. Moreover, their view draws this distinction in a way that makes understanding the variation in observed item and test scores, and hence validity, unmysterious and objective. Validity is not some sort of super-knowledge of the phenomenon we are wishing to measure, such as that embodied in the meta-theoretical views of Messick and others I describe above, but simply more knowledge: knowledge of causes.

I take a different position in the present chapter and elsewhere (e.g., Zumbo, 2005) than Borsboom and his colleagues. My view is that validity is a matter of inference and the weighing of evidence, and that explanatory considerations guide our inferences. I am not fond of the exclusive reliance on "causal" model of explanation of the sort that Borsboom and his colleagues suggest. Their causal notions give us a restricted view of measurement because of the well-known objections to the causal model of explanation – briefly, that we do not have a fully adequate analysis of causation, there are non-causal explanations, and that it is too weak or permissive, that it undermines our explanatory practices. My current leanings are toward inferences to the best explanation. In short, I believe that explanation is key to validity (and supporting the inferences we make from test scores) and that causation as the sole form of explanation is too narrow for the broad scope and context in which measurement and testing is used. In terms of the process of validation (as opposed to validity, itself), the statistical methods, as well as the psychological and more qualitative methods of psychometrics, work to establish and support the inference to the best explanation – i.e., validity itself; so that validity is the explanation, whereas the process of validation involves the myriad methods of psychometrics to establish and support that explanation.

Observation 13

There is one main point I want to make about the various uses of the term "model" as they relate to validity. In working with test developers and psychometricians, I have come to realize that "model" is a favorite word of measurement specialists. We use it in referring to "statistical models", "psychometric models", "IRT models", "cognitive models", etc. Sometimes, unbeknown to us, we use it with various shades of meaning. As Zumbo and MacMillan (1999) note, often times we use it to convey the sense of a: (i) mathematical model, (ii) model in the wider philosophic sense, (iii) explanatory model, (iv) descriptive model, (v) stochastic or random variable model, (vi) logical model, and (vii) computational model, to list but a few. What complicates matters is that these uses of "model" are not mutually exclusive (nor exhaustive) but they do have essential, though subtle, distinctions. Having had measurement specialists point out that they find there to be a confusing array of models, I have become fascinated that a word that is so commonly used and is the source of so much tension among measurement specialists has yet, to my knowledge, to be adequately defined and discussed in the psychometric literature. Even with all of these vague references and multiple uses, I am struck by how many of us are like Pygmalion and fall in love with our models, in good part we fall in love with what we want our model be. In some cases we are very much like the fictional character Pygmalion in that we believe that our particular model of interest is so realistic that we fall in love with it.

Clarity will come from knowing how we are using the word "model" and what we are implying in its use. Zumbo and MacMillan (1999), in very broad strokes, described two different kinds of models as well as their purposes, types, and uses. I found that even this broad discussion of models helps clarify some of the confusion and clear some of the (battle) ground regarding the models used in applications of measurement. For example, it has helped me understand the philosophic basis and practical difference between Rasch and one-parameter IRT by clarifying the Rasch model's epistemological difference. Statistically, the Rasch and one-parameter IRT models are the same but the differences in epistemology between Rasch and IRT modelers leads to rather different validity practices and test data claims.

Observation 14

The function of the psychometric model in validity research is to step in when the data are incomplete. In an important sense, we are going from what we have to what we wish we had. If we had available the complete data or information, then we would know the true score, or theta in IRT models, and no statistics beyond simple summaries would be required. There would be no need for complex models to infer the unobserved score from the observed data and, hence, no need to check the adequacy and appropriateness of such inferences.

The occurrence of complete data or full information, as I describe it, is not commonly encountered, if ever, in the practice of measurement. Naturally, this leads to the common experiences that are the defining characteristics of what we call modeling:

(a) The data you have is never really the data you want or need for your attributions, recommendations or decisions.
(b) No matter how much data you have, it is never enough because without complete information you will always some error of measurement or fallible indicator variable. We get around data and information limitations by augmenting our data with assumptions. In practice, we are, in essence, using the statistical model to creating new data to replace the inadequate data. The most common data augmentation assumption in psychometrics is that the correlation among items is accounted for by an unobserved continuum of variation – of prominence in latent variable and factor analysis models. The key words here are "unobserved" and "continuum", as opposed to observed and categorical.
(c) In contemporary measurement practice, which is heavily-laden with model-based measurement practices, the inferences, in part, come from the model itself. In short, the validity statements rest on the measurement model. In fact, as we noted above, given that one begins with an array of numbers denoting responses to items or tasks for each examinee, it could be argued that the psychometric model "provides" the inferences one can make by being the vehicle for going from what we have to what we wish we had. Therefore, the measurement model is not neutral in validation and, not surprisingly, one's test score interpretations may change depending on the psychometric statistical model being used.

2.2. How the observations on the current state of validity speak to the example

In a broad sense, then, validity is about evaluating the inferences made from a measure. All of the statistical methods discussed in this handbook (e.g., factor analysis, reliability, item analysis, item response modeling, regression, cognitive models and cognitive psychology in testing) are directed at building the evidential basis for establishing valid inferences. In this light, I am bewildered by colleagues who make claims that classical test theory ended with the seminal work of Lord and Novick (1968) or perhaps even earlier with the publication of Gullicksen (1950). A more accurate statement would be that the systematic development and articulation of classical test theory began with these publications – for a description of recent advances in reliability and classical test theory see Zumbo and Rupp (2004).

Taking the CES-D as a case in point to illustrate some of the conceptual features of the current view of validity, gender and age differences are often reported in the depression literature – that is, generally, females score higher than males on depression measures, and depression scores are higher in young adults (although high scores tend to also be found in the elderly). Therefore, any new measure of depression (or the application of a known measure with a new population of respondents) is expected to have, for example, higher depression scores for female respondents. This finding of gender differences is part of what reassures some researchers that they are, in fact, measuring depression. The gender differences are therefore embedded in the network of ideas defining the depression construct hence demonstrating the interplay of empirical findings, theoretical developments, and construct elaboration that is central to test validation. At the same time, the gender and age findings also illustrate the potential for value implications and social consequences. When examining the consequential basis for CES-D score interpretation, one needs evidence of construct validity and an appraisal of the construct label itself, the broader theory, and even broader ideologies (Hubley and Zumbo, 1996). For example, what is the implication of labeling the constellation of listed symptoms of depression in the CES-D as "depression"? What are the implications of our theories of aging for the recognition of depression in the elderly? And vice versa, what are the implications of our empirical results with current measures of depression for theory-building in the area of depression? Likewise, what are the implications of our theories of gender differences and the recognition of women being labeled depressed more often than men? And, for example, given our theories of aging is depression among the elderly viewed as a normal, abnormal, negative, or a realistic state? In addressing these questions, we are bringing the social consequences and value implications to the foreground of the validity discussion and shining a light on the appraisal of the hidden values of inferences made from the CES-D scores. This example of the CES-D vividly demonstrates (a) how one should not be simply concerned with the obvious and gross negative consequences of score interpretation, but rather one should consider the more subtle and systemic consequences of "normal" test use, and (b) how social consequences and value implications feed back into theory development. For now, our discussion has been conceptual and theoretical (after all, validation does involve theorizing), but our focus will turn empirical when we return to the question of age and gender as predictors later in this chapter in the discussion of variable ordering and regression.

2.3. *Different strengths of test inferences: A new framework for measurement inferences*

One must keep in mind the purpose of the testing. The two main purposes for testing are: (a) descriptive: wherein one is assigning numbers to the results of observations for the purpose of obtaining a scale score in scientific or policy research, or (b) decision-making: using the original scale scores to categorize individuals or groups of individuals based on their responses to the test or measure. The latter purpose is the most common use of testing in, for example, the educational testing, diagnostic screening, certification testing and occupational and employment contexts. Two common examples are: (i) individuals are selected or not selected for a job, or (ii) individuals are screened for further testing, interviewing, or training.

The whole purpose of measurement and methodological techniques is to help in establishing the *degree of* validity of the *inferences* made from the scores obtained on a test or assessment battery. The key words in the previous sentence are "degree of" and "inferences". These two features are highlighted in nearly all current views of validity (see, for example, Hubley and Zumbo, 1996, or Zumbo, 1998). That is, as I described above, validity is not an all-or-none decision but rather a matter of degree. This, of course, implies that validation is an on-going process and is not simply something done once and involving only a correlation with a "gold standard". Furthermore, what is being validated are the inferences made from the measurement and not the measurement tool itself. Therefore, one can have a degree of evidential strength for validity and, in fact, as I will describe below, there is a continuum of validity statements that can be articulated in the process of validation.

Below I will introduce a framework that will be useful in the critical analysis of validity evidence.[3] Furthermore, this framework will help answer questions of sampling and representativeness and will aid us in being more explicit about the strength of our validity claims. It is imperative that we be explicit about the limitations or strengths of our validity claims. The following framework is modeled on Draper's (1995) approach to classifying causal claims in the social sciences and, in turn, on Lindley's (1972) and de Finetti's (1974–1975) predictive approach to inference. Although I will rely heavily on Draper's approach (including paraphrasing some of his very clear statements), Draper focused on the issue of causal inferences in quasi-experimental designs whereas I will discuss the matter of measurement validity. In recognition of its bases, I will refer to my new framework as the Draper–Lindley–de Finetti (DLD) measurement validity framework.

The main point to be taken from the DLD framework is the necessity to be explicit about the sorts of inferences one makes and that one is able to make. It is not that some inference is necessarily better than others (because this sort of value judgment needs to take the purpose of the measure into account), but rather that credible and defensible science requires one to be explicit about the sorts of inferences that are made and that can be made in a given context. As Draper states:

[3] An earlier version of this framework was introduced by Zumbo (2001) and then further expanded in Zumbo's invited address to the conference in celebration of the work of Professor Ross Traub upon his retirement, OISE/University of Toronto, 2002.

> *Within this approach, the only inferential elements with objective reality are data values X you have already observed and data values Y you have not yet observed. Inference is then the process of quantifying your uncertainty about future observables Y on the basis of things you know, including the X values and the context in which they were obtained. Informally, one might call X the data you have and Y the data you wish you had; with this terminology, a statistical model supporting your inference is a mathematical story that links these two data sets.* (p. 119)

In the context of validity theory, the data you have and the data you wish you had may differ from each other in two main ways: (1) problems of measurement, and (2) problems of sampling.

1. Problems of measurement

There are two common forms of measurement problems. One form arises when you are trying to quantify something elusive and you are not sure you have got it right. The technical terms for this are construct underidentification or construct misrepresentation. The other form occurs when you may have also included construct(s) irrelevant to the task at hand. The technical term for this is construct irrelevant variance. For example, you are measuring some knowledge domain such as reasoning skills, but the construct "writing ability" could be considered irrelevant because we may be interested in reasoning but it is confounded with writing ability. That is, because the answers are expressed in written form in essays and sometimes good writing masks poor reasoning skills and vice versa.

A useful way of thinking about the tasks or items in an assessment is that they are a sub-set of a (potentially well-defined) finite population of items or tasks from which one or more test forms may be constructed by selection of a sample of items from this population. Obviously, creating all possible items or tasks would be both economically and practically unfeasible. In some situations, one has a clearly articulated domain of items (perhaps even articulated in the form of a test blue-print) so that one would have a very good sense of the exchangeability of the items one has with those that are not in the assembled test. In other cases, one has very little information about the domain being measured and hence has little knowledge of the exchangeability of the items one has with those in the hypothetical item population (or domain). In this case, the data you have is the information on the sampled items or tasks in your assessment and the data you wish you had is the corresponding information on the unsampled items. In essence, this gets at how well one knows the construct or domain being sampled and how well one plans (e.g., task specifications) in the process of constructing the assessment.

Before turning to problems of a sampling nature, I would like to add that measurement error (quantified, for example, by reliability of measurement) is also a related but separate issue here. In the case of measurement error, the data you have are your "measurements" or "observations" which have measurement error or construct irrelevant variance, whereas the data you wish you had are the "true" or error-free scores.

2. Problems of sampling

The second way the data you have and the data you wish you had differ relates to problems of a sampling nature. This sort of problem arises, for example, when there is a finite population of subjects of scientific or policy interest and, due to financial or other constraints, you can only get a subset of the whole population. In this case, the data you have is the information on the sampled individuals and the data you wish you had is the corresponding information on the unsampled people.

The problems of measurement and problems of sampling may, in turn, be thought of as special cases of a *missing data* problem. Also, it is important to note that much of the work in the predictive approach to inference described by Draper is founded on the notion of *exchangeability*. Exchangeability can be thought of as: without prior or additional/supplemental information, the data are similar – and hence, in a mechanical sense, exchangeable.

In part, the DLD framework is a natural extension of ideas in the 1940s and 1950s by Louis Guttman who wrote about generalizing from measures (or tasks) that we have created to more tasks of the same kind. Therefore, this framework is a merger of Guttman's ideas with some Bayesian thinking about inferences. There is also a tidy connection to measurement invariance that I will discuss in more detail below.

Now, when measurement and validation examples are examined from the DLD "predictive point of view", four kinds of inference of varying strength and scope of generality are discernible. Figure 1 depicts these various types of inference graphically. As much as possible I have tried to stay close to Draper's conceptualization, including the terms for the various forms of inference (i.e., calibrative, specific sampling, specific measurement, and general measurement).

Exchangeability of Sampled and Unsampled Items in the Target Construct/Domain (i.e., sampled tasks or items)

Exchangeability of Sampled and Unsampled Units in Target Population (i.e., sampled individuals)	*EXCHANGEABLE*	*NOT EXCHANGEABLE*
EXCHANGEABLE	General Measurement Inference	Specific Sampling Inference
NOT EXCHANGEABLE	Specific Domain Inference	Initial Calibrative Inference

Fig. 1. The various forms of measurement inference.

- Initial Calibrative Inference (calibration inference). This provides the lowest level of inferential strength and generalizability of results. The term "calibrative" is being used here to convey the sense that a measure is simply being applied in an exploratory fashion so that one can observe the measure in action with data. The case could be made that this is not useful measurement at all because item exchangeability is not there and hence not justifiable. However, although I do share this general sentiment, I prefer that the statements that one is limited to be simple calibration statements about whether items or tasks can be comprehended and completed in the most elemental way by some group of individuals.
- Specific sampling inference. This sort of inference shares the domain sampling limitations of calibrative inference except that one has a very good sense of who is being measured – in the sense that the sample is exchangeable to the target population of respondents. Obviously, the statements that one is limited to are specific calibration statements about whether the measure can be used at all for some well-defined population of individuals.
- Specific domain inference. This type of inference is simply of the sort where one can make strong claims (in terms of, for example, reliability and other measurement properties) about what is being measured but the statement is not generalizable because the exchangeability of sampled and unsampled respondents (or examinees) is not justifiable. In fact, this is a very common form of inference in research settings where claims of reliability and validity are limited to the sample at hand.
- General measurement inference. This kind of inference is ideal because it has both exchangeability of items and exchangeability of sampling units. That is, like specific domain inference one has exchangeability of sampled and unsampled items (or tasks) in the target domain *and* one also has the justification for assuming exchangeability of the sampled and unsampled subjects. Therefore, strong validity and measurement claims can be made.

In terms of inferential strength, I conclude that both initial calibrative and specific sampling are less strong than specific domain, which in turn is less than general measurement inference. Clearly, general measurement inference is the strongest. I will return to this later.

Note that Figure 1 depicts the four corners of a matrix, which most certainly has several gradients in between these four corners. Also, the exchangeability of sampled and unsampled items is yet another way of thinking of a limited form of construct validity. As the DLD framework shows, however, construct validity needs to be considered in the light of the exchangeability of sampled and unsampled units (i.e., respondents) in the target population.

What the DLD framework highlights is that you need to consider the sample in two distinct ways. First evidence for the validity of inferences made from task performance needs to be provided with each new sample examined. This is why the process of validation is on going. The inferences you make from your task may be valid for one population (e.g., males) but not another population (e.g., females). Or the inferences you make may be valid for males in one setting (e.g., a large North American city) but not in another setting (e.g., in rural communities in Eastern Europe or perhaps even large cities in Asia). Second, one needs to consider whether the sample with whom you are working (whether in a validation, standard-setting, or normative study) is exchangeable with your population of interest. For example, is the sample of respondents to the CES-D used to validate the inferences representative of target population of interest to users of the CES-D (that is, with respondents from that same population of interest that were not in your sample)?

The objective in this classification and ordering of inferences in the DLD framework is to encourage researchers to be explicit about the types of inferences they can make. That is, we need to be explicit about the information we have about the level of exchangeability of individuals and tasks or items used in our validation study. This explicitness will go a long way toward creating credible scientific measurement evidence. For example, it is sometimes stated in the context of personnel testing that job simulation tasks are used so that one does not have to make any inferences from the test scores; a reliance on a form of what is called face validity. Clearly, from the above framework one is still making inferences in "simulation" tasks because you still do not have the individual doing the job, per se. The simulation may be more realistic, hence shortening the inferential distance from simulation to real job performance, than a job knowledge test, but an inference is still being made from those job simulation tasks to other tasks like them. One should be cautious, however, not to let the DLD framework be interpreted in a weak form, hence allowing the same slippage as is noted above with weak forms of construct validity. This caution also applies to other simulation-like (i.e., performance) tasks such as some language tests.

Note that the DLD framework puts importance on the sampling units (the respondents) and their characteristics, something that this not highlighted enough in the conventional discussions of psychometrics and of validity evidence. Most, but not all, validation studies in the research literature give little time to the exchangeability of the sampled and unsampled units. In addition, there is little discussion in the psychometric literature of matters of complex sampling. Survey data now being collected by many government, health, and social science organizations around the world have increasingly complex structures precipitating a need for ways of incorporating these complex sampling designs into our psychometric models. For example, many surveys entail very complex sampling designs involving stratification and clustering as the components of random sampling. In addition, there can be complexities due to the patterns of nonresponse – either planned or unplanned nonresponse. A common psychometric problem is to compute variances for psychometric models that incorporate both the stochastic modeling as well as the survey design. Advances are being made on item response and latent variable modeling, including item response theory, with complex surveys by, for example, Asparouhov (2005), Asparouhov and Muthen (2005), Cyr and Davies (2005), Kaplan and Ferguson (1999), Mislevy (1991), Muthen and Satorra (1995), Thomas (2001), Thomas and Cyr (2002), and Thomas and Zumbo (2002). It is worth noting that the DLD framework shines a spotlight on the person sampling aspect of measurement, which has mostly held a secondary place to item or domain sampling in psychometrics.

Now, when inferential examples of testing and measurement are examined from the predictive point of view, four kinds of test inferences – of varying strength and scope of generality are discernable. I have found it useful to help one think of the various possibilities from the combinations of the four kinds of test inferences, whether they happen regularly or not. Examining the various combinations of the kinds of test inferences also helps one detail the range of conditions under which generalizations and inferences are expected to hold.

The notions of generalizations and inferences are intimately tied to the notion of invariance in measurement models, and hence to validity. Simply put, measurement invariance allows us model-based generalization and inferences – noting, of course, that model-based inferences are inferences from assumptions in the place of data, as described above. Much has been said in the item response theory literature about invariance, and Rasch specialists make much hay of that model's invariance and specific objectivity properties with, often, implicit and explicit suggestions that one would have greater measurement validity with the Rasch model[4] (Bond and Fox, 2001; Wright, 1997). This suggested supremacy of the Rasch model is an overstatement. In fact, in several applications of the Rasch model one hears the claim that simply fitting the Rasch model gives one measurement item and person parameter invariance, without mention of any bounds to this invariance. The suggestion is that if one fits the Rasch model to data, then one can apply those item parameters to all individuals (imbuing a universality that seems remarkable). Furthermore, if one fits the Rasch model to data, the person parameters (i.e., the theta scores) can be adequately predicted by any items; again, without reference to bounds on this generality. There many advantages to Rasch (or 1-parameter IRT) models in test applications but rarely are they the advantages that advocates of the Rasch model present – one advantage is the ease of IRT test equating in a one-parameter IRT model.

The DLD framework is meant to establish limits, bounds, and conditions for the inferences. It is unfounded to believe that fitting a model to data gives one such unbounded and unbridled universality. Simply put, the main goal of item response modeling should always be to make valid inferences about the examinees. However, inducing latent variable(s) into the data structure, by, for example, fitting a Rasch model to your data cannot mechanically increase the validity of these inferences. In contrast, the DLD framework helps one detail the range of conditions under which invariance is expected to hold. There may be situations (such as those found in a type of calibrative measurement or assessment context) wherein we do not have any sense of a population or sub-population and, in those contexts, we are, in essence, not concerned with invariance. See Rupp and Zumbo (2003, 2004, 2006) for detailed descriptions of invariance in item response theory. In more common contexts, however, the aim is to use a statistical measurement model to draw inferences from calibration samples to the respective populations from which these were drawn. An additional focus is on the range of possible conditions under which invariance is expected to hold.

Simply going about fitting an item response model to data does not necessarily give you measurement invariance. Believing that going about fitting a model to data guarantees you measurement invariance is simply magical thinking! In short, invariance requires that the model be correct (true) in all corners of the data and is an empirical

[4] A similar claim is also often made by advocates of the Rasch model that it has fundamental measurement properties, such as additive item response functions, that make it a superior model that produces "ratio" levels of measurement.

commitment that must be checked. One needs to focus, with the aid of the DLD framework, on the range of possible conditions under which invariance is expected to hold. It depends, then, on the type (or strength) of inferences one wants to draw. Empirically, one goes about investigating the possible conditions and bounds via psychometric methods such as differential item functioning and test level invariance (for recent overviews please see Maller et al., in press; Zumbo and Hubley, 2003). This is a different spin on the use of these psychometric methods; they are not just about fairness issues but rather about empirically investigating the range of possible conditions and hence informing the conclusions (bounds and limitations) of the DLD framework.

It is evident from the DLD framework that invariance is nonsensical without a clear statement about the range of possible conditions of sampled and unsampled items (tasks) and/or sampling units (respondents or examinees). Likewise, invariance is not just a property of item response theory models. Zimmerman and Zumbo (2001) showed that some of the properties and quantities from classical test theory hold for the entire population of individuals, as well as any subpopulation of individuals. Furthermore, many of the examples of invariance we prove in the Zimmerman and Zumbo (2001) paper are of the variety seen in all classical mathematical statistics. For example, we proved that measurements that are parallel in a given population are also parallel in any subpopulation. This was a point also emphasized by Lord and Novick (1968). We provide other new results of the flavor seen in the sentence above about parallel tests. It, of course, should be appreciated that there are quantities in observed score classical test theory that are population specific (and lacking invariance). These include variance of the true scores and, in turn, reliability because it is bound to the variance of the true scores.

It is important to acknowledge that, in some validation settings, there will be an inherent tension that is difficult to balance between the (i) exchangeability of sampled and unsampled items in the target construct/domain (i.e., sampled tasks or items), and (ii) exchangeability of sampled and unsampled units in the target population (i.e., of individuals). This sort of tension is probably less likely in large-scale testing organizations than in research settings with one-time and small-scale measurement. In practice, what should be the trade-off in this tension? Which axis has precedent? In some important senses, the more exchangeable your items or tasks are, perhaps via a longer test, the more difficulty one faces obtaining large numbers of sampling units. Simply put, it is more difficult to obtain large numbers of respondents with very long tests due to the time demands. In this sense, there is usually (but not necessarily always) a trade-off that needs to be balanced between the two forms of exchangeability. In my opinion, the balance between exchangeability of items and respondents should have item exchangeability as the primary concern. There are easily many counter-examples to this advice but, generally, one needs to first know what they are measuring and then find the limits of the inferential bounds on respondents (i.e., sampling units). That is, one should always have at least an adequate amount of item or task exchangeability and then focus on sampling exchangeability. Quite simply, first establish what you are measuring and then try to generalize it. If you do not have some semblance of item and task exchangeability, then you really have nothing to generalize. By this I am suggesting that by establishing item exchangeability one will have to know what they are measuring.

Considering our example with the CES-D, we do not have the inferential strength for general measurement inference because the sample is a convenience sample of respondents. Depending on the exchangeability of the sampled and unsampled items, we are limited to either specific domain or initial calibrative inferences. The question of the exchangeability of the sampled and unsampled items of the CES-D has, to my knowledge, never been directly addressed in the research literature. For example, I know of no content validation studies of the CES-D, which would go a long way in establishing the exchangeability of the items. My sense, however, is that there is moderate support for the item exchangeability and hence our study is limited to specific domain inferences, with the limitations and strengths that this form of inference provides as described above. In addition, although some applied measurement books and articles state otherwise, fitting a Rasch model would not, in and of itself, increase my claims to invariance, generalizability, or validity.

2.4. *Tests/measures and indices: Effect (reflective) versus causal (formative) indicators, respectively, and validity*

In validity research, it is important to distinguish between measures and indices, also referred to as reflective versus formative measures, respectively. For my purposes a measure is defined as a latent variable for which a shift in the value of the latent variable leads to an expected shift in the value of the indicator (i.e., the items are effects indicators). An index is defined as a latent variable for which a shift in the value of the indicator variable (the observed or manifest variable) leads to an expected shift in the value of the latent variable. Consider an example provided by Bollen and Lennox (1991) wherein responses to a series of items on a math test should reflect a student's quantitative ability. An index, on the other hand, has the observed variables (e.g., self-report items) as causal indicators. The typical example, as described by Bollen and Lennox, is the index socio-economic status. They go on to suggest that socio-economic status might be better viewed as formative (an index) but is often treated as reflective.

There are few empirical tests of whether a latent variable is a measure or index, the exception is the vanishing tetrads test of Bollen and Ting (2000). Please note that computing a principal components analysis (PCA) is not sufficient evidence that one has an index, nor does fitting a factor analysis model provide sufficient evidence to claim one has a measure – as you will see below both models could fit the same data. Bollen and Lennox (1991), suggest that a good place to start, and often the only thing available, is a literal thought experiment.

I might add that one can also supplement this thought experiment with a content validation study wherein one asks subject matter experts to consider and rate whether the items (or indicators) are effects or causal; that is, whether the variable is a measure or index, respectively. One can build on the methodologies described for content validity in Sireci (1998) by incorporating questions about whether an item should be considered a causal or effects indicator using methodology in content validity including the coefficients, designs, etc. Also, one could investigate the source of the decision of effects versus causes by talk-aloud protocols and/or by conducting multidimensional scaling of the subject matter experts' judgments. These approaches aid the validity arguments of whether one has effect or causal indicators. What I am suggesting is an extension of Bollen and Lennox's thought experiment to include data from subject matter experts.

My purpose for raising the matter of measures versus indices is that the validity evidence, particularly the empirical validity evidence, may, as Bollen and Lennox suggest, be quite different. The DLD framework, described above, however, will apply to both measures and indices. Therefore, the central matter is that most, if not all, validity research assumes reflective measures; that is, effects indicators. More thought and empirical support as I describe above, needs to be given to deciding on whether one has a measure or index.

Furthermore, it can be shown that the structural equation model for indices (causal indicators) is, in essence, principal components analysis (PCA) whereas the same model for measures (effect indicators) is factor analysis. Factor analysis and PCA are not the same model statistically nor conceptually. Furthermore, PCA is conceptually inappropriate for measures (effects indicators). There may be some statistical reasons for favoring PCA (for example, one may want to avoid the indeterminacy of factor analysis) but in validity research factor analysis is the standard. One should not (mis)use the fact that (a) there are circumstances (e.g., when the variances of the error terms in the factor model are roughly equal or are small) under which PCA is a good approximation to factor analysis, and (b) it is this good approximation that provides the basis for the widespread use of PCA statistics (e.g., number of eigenvalues greater than one, scree plot, or parallel analysis) in the preliminary stages of exploratory factor analysis to choose the number of factors, as reasons for arguing that PCA is the same as factor analysis.

It should be noted, however, that the "causal" perspective of factor analysis should not be interpreted in a narrow sense of the word "cause". The history of factor analysis is replete with debates about the nature of the factors that are determined/uncovered/discovered by factor analysis. I believe it would be too generous of a reading of Bollen and Lennox (1991), as well as the histories of sociology and of factor analysis, to interpret the causal language in a narrow sense. In this light, there is much we can learn about causation from a well-known book on the logic of causal inference by Cook and Campbell (1979) as well as work by Freedman (e.g., Freedman, 1987). Also see Shadish et al. (2002), which has among other things a useful discussion of manipulationist as opposed to non-manipulationist ideas of causation. Also, of course, there is much to be gained from the elegant statistical model for causation proposed by Neyman (1923). In short, one should not interpret the "arrows" in the graphical depictions of structural equation models of factor analysis in a narrow and strict sense of causation.

In the view I am espousing one should instead seek "explanation", and hence factor analysis becomes an explanatory model. As I note above in my critique of Borsboom et al. (2004), I consider the core of validity as one of a quest for explanation for the variation in the observed task or item responses. In this view, factor analysis is a statistical tool to aid in my inference to the best explanation, more broadly defined than just causal explanations.

One should note, as highlighted by Bollen and Lennox, that what we would expect for the composite score for a measure and index to be different in terms of validity. That is, for a measure the indicators or items need to be correlated because it is the correlation that is the source of the measure, in a factor analytic sense. Furthermore, the factor analysis aims to reproduce the covariance among the items therefore there needs to be substantial correlation for the factor analysis to be meaningful. For an index, however, there is no need to expect a correlation among the items or indicators. Also, for a measure one thinks of reliability of measurement as a quality in the internal consistency sense, but for an index that may not make sense in all settings – e.g., whether the indicators of SES are internally consistent is *not* something of concern because it is an index and not a measure. Therefore, the validity evidence (including the dimensionality and error of measurement aspects of validity evidence) could be quite different for an index versus a measure. The point, however, is that this chapter deals with the validity of measures, some of which will be relevant for indices (e.g., the DLD framework) but not all.

Returning to our example of the CES-D, the literature is somewhat mixed. Many, if not most, of the users of the CES-D treat it as a measure. Bollen and Lennox (1991) refer to their work creating some sort of hybrid latent variable with some effects indicators and other causal indicators, but this hybrid model is not, to my knowledge, widely used. I will follow the convention in the depression literature and consider the CES-D as a measure and hence the items as effects; leading us to later use of factor analysis; and not principal components analysis. Furthermore, I fully expect high correlations among the items and to be able to examine the internal consistency of the responses as a way of quantifying error of measurement.

3. Statistical methods

Having dealt with foundational issues, let us now turn to statistical methods for validation studies. There is an extensive and easily accessible literature on the statistical methods of validation research in most introductory and intermediate measurement and assessment texts. Much has been written in the area of educational and achievement testing. An important paper in the context of personality and attitude scales is Thissen et al. (1983).

There is, however, one class of methods that are particularly central to the validation process: structural equation models. These models are particularly important to test validation research because they are a marriage of regression, path analysis, and latent variable modeling (or factor analysis). Given that the use of latent variable structural equation models presents one of the most exciting new developments with implications for validity theory, this section discusses these models in detail. At this point, let me now turn to the two empirical validity questions listed earlier: (a) Given that the items are combined to created one scale score, do they measure just one latent variable?, and (b) Given a set of predictor variables of interest can one order them in terms of importance?

3.1. *Factor analysis: Given that the items are combined to create one scale score, do they measure just one latent variable?*

The question of whether the items measure just one latent variable is addressed by using factor analysis. In the typical confirmatory factor analysis (CFA) model, the score obtained on each item is considered to be a linear function of a latent variable and a stochastic error term. Assuming p items and one latent variable, the linear relationship may be represented in matrix notation as

$$y = \Lambda\eta + \varepsilon, \tag{1}$$

where y is a $(p \times 1)$ column vector of *continuous* scores for person i on the p items, Λ is a $(p \times 1)$ column vector of loadings (i.e., regression coefficients) of the p items on the latent variable, η is the latent variable score for person i, and ε is $(p \times 1)$ column vector of measurement residuals.

In the example with the CES-D, a CFA model with one latent variable was specified and tested using a recent version of the software LISREL. Because the CES-D items are ordinal in nature (i.e., in our case a four-point response scale, and hence not continuous), a polychoric covariance matrix was used as input for the analyses. Using a polychoric matrix is an underlying variable approach to modeling ordinal data (as opposed to an item response theory approach). For a polychoric correlation matrix, an underlying continuum for the polytomous scores is assumed and the observed responses are considered manifestations of respondents exceeding a certain number of latent thresholds on that underlying continuum. Conceptually, the idea is to estimate the latent thresholds and model the observed cross-classification of response categories via the underlying latent continuous variables. Formally, for item j with response categories $c = 0, 1, 2, \dots, C - 1$, define the latent variable y^* such that

$$y_j = c \quad \text{if } \tau_c < y_j^* < \tau_{c+1},$$

where τ_c, τ_{c+1} are the latent thresholds on the underlying latent continuum, which are typically spaced at non-equal intervals and satisfy the constraint $-\infty = \tau_0 < \tau_1 < \cdots < \tau_{C-1} < \tau_C = \infty$. It is worth mentioning at this point that the latent distribution does not necessarily have to be normally distributed, although it is commonly assumed so due to its well understood nature and beneficial mathematical properties, and that one should be willing to believe that this model with an underlying latent dimension is actually realistic for the data at hand.

Suffice it to say that an examination of the model fit indices for our example data with the CES-D, such as the root mean-squared error of approximation (RMSEA = 0.069, 90% confidence interval for RMSEA of 0.063 to 0.074), showed that the one latent variable model was considered adequate. This finding of unidimensionality is important because the single composite score of the CES-D items can, therefore, be easily interpreted as a continuum of depressive symptomology – an increasing score indicating more depression. As well, quantitative indices of measurement error, such as reliability coefficients, are also easily interpreted because of the unidimensionality.

3.2. *Regression with latent variables, MIMIC models in test validation: Given a set of predictor variables of interest can one order them in terms of importance?*

In the context of the CES-D example, are the age and gender of the respondents both important predictors of the latent variable score on the CES-D? This question is often addressed by using ordinary least-squares (OLS) regression to regress the observed composite score of the CES-D onto age and the dummy coded gender variable. The problem with this approach is that the regression results are biased by the measurement error in the observed composite score. Although widely known among statisticians, this bias is unfortunately ignored in most day-to-day validation research.

The more optimal statistical analysis is to use SEM and MIMIC models. MIMIC models, first described by Jöreskog and Goldberger (1975), essentially posit a model stating that a set of possible observed explanatory variables (sometimes called predictors or covariates) affect latent variables that are themselves indicated by other observed variables. In our example of the CES-D, the age and gender variables are predictors of the CES-D latent variable, which itself is indicated by the 20 CES-D items. Our example highlights an important distinction between the original MIMIC models discussed over the last three decades and the most recent developments in MIMIC methodology. In the original MIMIC work, the indicators of the latent variable(s) were all continuous variables. In our case, the indicators for the CES-D latent variables (i.e., the CES-D items) are ordinal or Likert-type variables. This complicates the MIMIC modeling substantially and, until relatively recently, was a major impediment to using MIMIC models in validation research.

The recent MIMIC model for ordinal indicator variables is, in short, an example of the merging of statistical ideas in generalized linear models (e.g., logit and probit models) and structural equation modeling into a generalized linear structural modeling framework (Jöreskog, 2002; Jöreskog and Moustaki, 2001; Muthen, 1985, 1988, 1989). This new framework builds on the correspondence between factor analytic models and item response theory (IRT) models (see, for example, Lu et al., 2005) and is a very general class of models that allows one to estimate group differences, investigate predictors, easily compute IRT with multiple latent variables (i.e., multidimensional IRT), investigate differential item functioning, and easily model complex data structures involving complex item and test formats such as testlets, item bundles, test method effects, or correlated errors with relatively short scales, such as the CES-D.

A recent paper by Moustaki et al. (2004) provides much of the technical detail for the generalized linear structural equation modeling framework discussed in this chapter. Therefore, I will provide only a sketch of the statistical approach to motivate the example with the CES-D. In this light, it should be noted that these models can be fit with either Mplus or PRELIS-LISREL. I chose to use the PRELIS-LISREL software and hence my description of the generalized linear structural equation model will use Jöreskog's notation.

To write a general model allowing for predictors of the observed (manifest) and latent variables, one extends Eq. (1) with a new matrix that contains the predictors x

$$y^* = \Lambda z + Bx + u, \quad \text{where } z = Dw + \delta, \tag{2}$$

u is an error term representing a specific factor and measurement error, and y^* is an unobserved continuous variable *underlying* the observed ordinal variable denoted y, z is a vector of latent variables, w is a vector of fixed predictors (also called covariates), D is a matrix of regression coefficients and δ is a vector of error terms which follows a $N(0, I)$. Recall that, in Eq. (1), the variable being modeled is directly observed (and assumed to be continuous), but in Eq. (2) it is not.

Note that because the PRELIS-LISREL approach does not specify a model for the complete p-dimensional response pattern observed in the data, one needs to estimate the model in Eq. (2) with PRELIS-LISREL in two steps. In the first step (the PRELIS step), one models the univariate and bivariate marginal distributions to estimate the thresholds and the joint covariance matrix of y^*, x, and w and their asymptotic covariance matrix. In the PRELIS step, there is no latent variable imposed on the estimated joint covariance matrix, hence making that matrix an unconstrained covariance matrix that is just like a sample covariance matrix, S, for continuous variables. It can, therefore, be used in LISREL for modeling just as if y^* was directly observed using (robust) maximum likelihood, weighted least-squares estimation methods, etc. Of course, one can use unweighted least-squares and the parameter estimates are consistent, but the standard errors are inaccurate. Thus, the unweighted least-squares is a descriptive, rather than inferential, technique in this context – this will be become important later when we use (finite sample) variable ordering methods based on least-squares types of estimation.

3.3. Variable ordering: Pratt's measure of variable importance

Due to space limitations, I will only provide a sketch of Pratt's variable ordering measure (please see Pratt (1987) and Thomas et al. (1998) for the details). Pratt considered a linear regression of the form

$$y = b_0 + b_1 x_1 + \cdots + b_p x_p + u, \tag{3}$$

where the disturbance term u is uncorrelated with $x_1, x_2, \ldots, x_p$ and is distributed with mean zero and variance σ^2. It is important to note that the linear regression in Eq. (3) is rather generic and hence can apply when the predictors are latent variables as in Eq. (2), the MIMIC model.

The total (standardized) population variance explained by the model in Eq. (3) can be written as

$$R^2 = \sum_j \beta_j \rho_j, \tag{4}$$

where β_j is the usual standardized regression coefficient corresponding to x_j, and ρ_j is the simple correlation between y and x_j. Pratt justified the rule whereby relative importance is equated to variance explained, provided that explained variance attributed to x_j is $\beta_j \rho_j$ – a definition which is widely used in the applied literature (e.g., Green et al., 1978), but as documented by Pratt (1987) and Thomas et al. (1998), it has also been criticized. Pratt justified the measure using an axiomatic approach based largely on symmetry and invariance to linear transformation. He showed that his measure is unique, subject, of course, to his axioms. An additional feature of Pratt's measure is that it allows the importance of a subset of variables to be defined additively, as the sum of their individual importance irrespective of the correlation among the predictors. Other commonly used measures (e.g., the standardized (beta-)weights, the t-values, unstandardized b-weights) do not allow for an additive definition and are problematic with correlated predictor variables.

Thomas et al. (1998) gave a sample interpretation of Pratt's measure based on the geometry of least squares. We considered a sample of N observations fitted to a model of the form (3), so that the observed variables $y, x_1, \ldots, x_p$ comprise vectors in R_N. We assumed, without loss of generality, that the variables are centered to zero,

$$y'1_N = x_1'1_N = \cdots = x_p'1_N = 0,$$

where 1_N is an $N \times 1$ vector of ones. In this case, $\hat{y}$, the fitted value of y, is the projection of y onto the subspace X spanned by the x_i, and has the representation

$$\hat{y} = \hat{b}_1 x_1 + \cdots + \hat{b}_p x_p \tag{5}$$

where the $\hat{b}_j$'s are least squares estimates of the population regression coefficients, b_j, $j = 1, \ldots, p$. We defined the partition of R^2 of x_j, $j = 1, \ldots, p$, to be the signed length of the orthogonal projection of $\hat{b}_j x_j$ onto $\hat{y}$, to the length of $\hat{y}$. By definition, this ratio represents the proportion of R^2 and sums to 1.0. Furthermore, the partitioning is additive so that one could, for example, compute the proportion of R^2 attributable to various subsets of the explanatory variables, irrespective of the correlation among the explanatory variables.

One then can partition the resulting R^2 by applying

$$d_j = \frac{\hat{b}_j r_j}{R^2}, \tag{6}$$

where, as above, $\hat{b}_j$ is the jth standardized regression coefficient (the "beta"), r_j is the simple Pearson correlation between the response and jth explanatory variable in Eqs. (3) and (5).

Several points are noteworthy at this juncture. First, all of the components of the right hand side of Eq. (6) are from a least squares procedure. Second, one can easily investigate "suppressor" variables (also sometimes referred to as "confounders") through the geometric approach described above. By definition, suppressors are variables that are not individually related to the response variable, but do make a significant contribution to the statistical model in combination with one or more of the other explanatory variables. In practice, a suppressor is identified when one has a relatively small value for d_i and a standardized regression coefficient (in the multiple regression model) comparable to the values exhibited by explanatory variables whose d_i are appreciably larger. Third, the index in Eq. (6) can be negative, a result which is counter-intuitive and can be seen as a negative feature of the index. See Thomas et al. (1998) for a discussion of procedures for addressing negative d_i of large magnitude. Fourth, when referring to variable-ordering as a statistical procedure, we are referring to ordering the predictor (explanatory) variables in a regression analysis in terms of *relative importance* as predictors of some dependent (response) variable. Therefore, the relative importance, in our context, is defined for a particular model after having decided on the "best model". Therefore, this is not a procedure for model-building (like stepwise regression). Rather, it is useful in determining the relative importance of predictors after one has decided on the "best model" by the relative proportion of variation accounted for by the predictors. Please see Thomas (1992) and Thomas and Zumbo (1996) for a discussion of variable ordering in the MANOVA and discriminant analysis cases, as well as, Wu et al. (2006) for variable ordering in factor analysis, and Thomas et al. (2006) for variable ordering in logistic regression, which would all be of use in validity studies.

Turning to the CES-D example, the validity researcher may be interested in the question of whether age and gender are predictive of CES-D scale scores. I will conduct the regression two ways: (a) the conventional way with the dependent variable being the observed total score across the 20 items, and (b) using the MIMIC model described above with the latent depressive symptomology variable as the dependent variable. In the latter case, I will use unweighted least-squares[5] estimation in LISREL so as to be able to apply the Pratt index for variable ordering. In both cases, the predictor variables are the respondents' age and gender. Note that the respondents' gender was coded 0 for male and 1 for female respondents. The correlation of age and gender is, as expected from the descriptive statistics of age for each gender, negative.

The ordinary least-squares results, as conventionally performed in validation research, results in an R-squared of 0.013, a value that is small but not unusual in this area of research given that we have only two predictor variables. The standardized b-weights (i.e., the beta weights) for gender and age are 0.064 and −0.084, respectively. The correlations between the observed composite score over the 20 CES-D items are 0.077 and −0.094 for gender and age, respectively. The corresponding Pratt indices computed from Eq. (6) are 0.385 and 0.615 for sex and age, respectively. Therefore, 61.5% of the explained variation (i.e., the R-squared) in the observed CES-D total scale score is attributable to age of the respondents; this makes age the more important of the two predictors. Note that the reliability of the 20 item scale was 0.91 for our sample, as estimated by coefficient alpha. Although introduced herein as a validation research tool for the first time in the research literature, I have already helped implement Pratt indices in an operational validity research program for the General Management Admissions Test (GMAT®) school-based validation reports.

In contrast to the observed score regression, one can estimate the regression of the latent variable (presumably of depressive symptomology) measured by the 20 CES-D items on to age and gender via the MIMIC model in Eq. (2). One can then use the unweighted least-squared parameter estimates (including the ULS model R-squared) and compute the Pratt indices. The Pratt indices can be applied to Eq. (2) because Pratt's original work, and the later geometric work by Thomas et al. (1998) are valid for a very generic regression model described in Eq. (3). The MIMIC model was fit using LISREL, as described above, specifying the standardized solution to get the beta-weights. If one examines the statistical values in the LISREL output, the standardized b-weights are 0.117, and −0.165, for gender and age, respectively. Likewise, the correlations of gender and age with the latent variable are 0.141 and −0.182, respectively, with an R-squared of 0.046. As expected, the MIMIC R-squared is larger than the observed variable R-squared reported above (Lu et al., 2005). Note that this discrepancy in R-squared is event even though our test reliability is 0.91 for our sample, as described above.

[5] Note that the geometry we introduce necessitates (in the finite sample case) a least-squares estimation method, of which weighted least-squares is of course a case. Because, of the limitations described of using weighted least-squares with in our MIMIC model, I do not focus on the standard errors and instead limit the analyses to the consistent parameter estimates and constrain my conclusions to those descriptive in nature, rather formal statistical inference.

It is precisely due to the negative bias of the R-squared for the observed score regression variable model that I recommend using the latent variable MIMIC model in validation studies. Herein, I introduce the Pratt index for the MIMIC model so variable ordering can also be investigated. The corresponding Pratt indices for the MIMIC model are 0.355 and 0.645 for gender and age; these results are similar to those of the observed variable regression above, with age being the more important predictor.

Finally, given the coding for the gender variable, as a group the female respondents scored higher on the latent variable of depression. Likewise, the older respondents tended to have a lower level of depression compared to the younger respondents in this sample, as reflected in the negative regression coefficient. When the predictive relationship of age was modeled in a separate analysis for males and females via this generalized MIMIC model, age was an important (negative) predictor for the female respondents but age was not an important predictor for male respondents. Age is unrelated to depression level for men whereas older women in this sample are less depressed than younger women – there may a nonlinear relationship here. These findings contribute to the empirical underpinnings of CES-D score interpretation. Therefore, this sort of known-groups information is useful to researchers using the CES-D and hence supports, as described at the beginning of this chapter, the inferences made from CES-D test scores and motivates the discussions of value implications and social consequences of CES-D score use.

4. Closing remarks

This chapter had two aims, one fairly general and the other fairly specific. The general aim was to provide some observations and comments on the variety of psychometric models, methods, and theories currently at our disposal, whether for validity and more generally in the wide-sense application of measurement. With this general aim, I intended to go beyond a simple cataloging and description of the variety of strategies, analyses, and statistical applications in validity and draw on my experiences and knowledge in psychometrics, pure and applied mathematics, statistical science, and philosophies of science to shed some light on validity theory and the practice of validation.

The more specific aim of this chapter was to first discuss several foundational issues including presenting a framework to consider the strengths of our measurement inferences depending on the data at hand. Next, I considered statistical methods that are particularly relevant to validation practice. My reading of the vast literature on validity theory and practice dating back to the early parts of the 20th century leaves me with the impression that the history of psychometrics and measurement validity exhibits both a pattern of growing understanding and utility and a series of unending debates on topics of enduring interest. If one spends just a few moments reflecting on the history of many different sciences one will see that this is characteristic of a maturing science.

Integrating and summarizing such a vast domain as validity invites, often rather facile, criticism. Nevertheless, if someone does not attempt to identify similarities among apparently different psychometric, methodological, and philosophic views and to synthesize the results of various theoretical and statistical frameworks, we would probably find ourselves overwhelmed by a mass of independent models and investigations with little hope of communicating with anyone who does not happen to be specializing on "our" problem, techniques, or framework. Hence, in the interest of avoiding the monotony of the latter state of affairs, even thoroughly committed measurement specialists must welcome occasional attempts to compare, contrast, and wrest the kernels of truth from disparate validity positions. However, while we are welcoming such attempts, we must also guard against oversimplifications and confusions, and it is in the interest of the latter responsibility that I write to the more general aim.

4.1. An overview: The melodic line, with some trills and slides

In addition to reviewing the validity literature from a psychometric perspective, this chapter includes three novel methodological contributions: (i) the DLD framework, (ii) the extension of the Bollen and Lennox thought-experiment to content validity studies with subject matter experts, and (iii) a latent variable regression method that allows for variable ordering.

Although a lot of ground has been covered in this chapter, several themes should be evident from the material I have presented above. Let me speak to just a few of these themes. First, there are no widely accepted series of steps that one can follow to establish validity of the inferences one makes from measures in the varied and disparate fields wherein measurement is used. The process of validation, as I see it, involves a weighing and integrating the various bits of information from the whole of psychometric activities from specifying a theory of the phenomenon of interest to test design, scoring and test evaluation, and back to the theory itself. I fall clearly in the camp of validity theorists who see the process of validation as an integrative disciplined activity.

That is, historically, we have moved from a correlation (or a factor analysis to establish "factorial validity") as sufficient evidence for validity to an integrative approach to the process of validation involving the complex weighing of various bodies, sources, and bits of evidence – hence, by nature bringing the validation process squarely into the domain of disciplined inquiry and science. There are many metaphors discussed in the literature for the process of validation: (a) the stamp collection, (b) chains of inference, (c) validation as evaluation, and (d) progressive matrices to name just a few. In my early work on validity I envisioned a "judicial or courtroom" metaphor where all the evidence comes together and is judged, cases are made, evidence (witnesses) come forward and a reasoned body judges the evidence (weighing different aspects) for validity of the inferences made from a test or measure. I am sure I am not alone in the use of the courtroom metaphor. Today I think in less adversarial terms and rather think of it as jazz – as in the musical style. With validation as jazz I want to principally borrow the tenets of sound coming together, but that the coming together is not necessarily scripted. All sorts of notes, chords, melodies, and styles come together (including, of course, improvisation that is particular to that one song or performance) in a creative way to make music.

I was once told by a music teacher that he could not teach me to play jazz. Instead, he could teach me music and some jazz sounds, songs, and styles. It would then be my task to bring it all together, sometimes in an unscripted way, to make jazz music. Perhaps the same applies to the process of validation, there is no one methodology or script that can be applied in all measurement contexts.

There is, however, one framework that is general enough to have broad appeal. Because it was not developed with validation, per se, in mind, this framework has a limitation that needs to be addressed. That is, if one were to insert a component of developing and stating a theory of the phenomenon of interest, the portrayal of the measurement process by Hattie et al. (1999) is a rich and useful framework. A modified version of the Hattie et al. framework would involve a conceptual model (theory) of the phenomenon, leading to conceptual models of the measurement, leading to test and item (task) development, leading to test administration, test use and test evaluation, and back to the conceptual model of the phenomenon; with all the constituent elements and statistical psychometric methods described by Hattie and his colleagues in Figure 1 and the remainder of their chapter. It is noteworthy that there is a feedback back to the conceptual model of the phenomenon. Having said this, however, it is important to note the distinction I make between validity, per se, and the process of validation. I consider validity to be the establishment of an explanation for responses on tasks or items – the emphasis being inference to the best explanation as the governing aspect, and the process of validation informs that explanatory judgment. The modified Hattie et al. framework, therefore, is a useful general description of the process of validation, but is not validity, per se.

Another broad theme in this chapter involves establishing the bounds and studying the limits of our inferences. In this light there are many interconnected ideas I discuss above that deal with the purpose and use of models, as well as a new framework, the DLD framework, to incorporate both the inferences to a domain (or construct) and the inferences to a population of examinees or respondents to your measure.

Throughout this chapter I have highlighted the importance of data modeling and assumptions as empirical commitments. Zumbo and Rupp (2004) remind us that it is the responsibility of mathematically trained psychometricians to inform those who are less versed in the statistical and psychometric theory about the consequences of their statistical and mathematical decisions to ensure that examinees are assessed fairly. As Zumbo and Rupp state, everyone knows that a useful and essential tool such as an automobile, a chainsaw, or a statistical model can be a very dangerous tool if put into the hands of people who do not have sufficient training, handling experience, or lack the willingness to be responsible users.

4.2. *In terms of the psychometrics of validity, when psychometricians speak, what are they really saying?*

With the psychometrician's responsibility in mind, and the general aim of this chapter at hand, let me close by answering a question that I often hear: When psychometricians speak, what are they really saying? Or, put another way, when psychometricians do what they do, what are they doing?

Psychometricians use models to help us go from the data we have, to the data we wish we had, by augmenting our data with assumptions. These assumptions need to be both empirically and conceptually (i.e., from theory) validated. Furthermore, psychometricians use models that are descriptive in nature so that we usually do not know the "processes" involved and hence have, at best, a heuristic purpose. In this light, item response theory is hardly an elaborated process model. This point is important to keep in mind when thinking of alternatives to psychometric models. This lack of explanatory focus has been the root of a long-standing angst among some measurement specialists. The most recent attempt at relieving this angst has been to prevail on cognitive theory to lend an explanatory hand. My only observation on this front is that not all cognitive theories are explanatory so that we need to be careful that, in our quest for explanatory power, we do not inadvertently supplant one heuristic model with another while deluding ourselves that our new model is explanatory. Furthermore, these cognitive theories need to be empirically tested – see, for example, Zumbo et al. (1997) wherein we tested a social-cognition theory of item discrimination.

Psychometricians often have their favorite model. As I noted above, like Pygmalion we sometimes fall in love with our models. It is confusing to hear that "all the models are the same" and "my model is different and you should use it everywhere and all the time". For example, it is confusing for practitioners (and psychometricians) to hear that all models are realizations of factor analysis or generalized linear latent variable models from one quarter, and then hear from another quarter that the Rasch model has special properties that no other model allows, or that generalizability theory is limited because it does not allow invariance of examinees or items. The reason for this confusion is that one can make convincing arguments for the unification of measurement models (see McDonald, 1999; Zumbo and Rupp, 2004; Zimmerman and Zumbo, 2001) and, at the same time, convincing arguments about, for example, the advantages of *generalizability theory* over classical test theory, item response theory over *generalizability theory*, item response theory over classical test theory, and structural equation modeling over *generalizability theory*, classical test theory, and item response theory, and so on (Zumbo and Rupp, 2004). This is not uncommon in mature mathematical disciplines and may well be due, in part, to the distinction between writing a model as a mathematical statement on paper versus the additional conditions that need to be invoked to estimate the parameters in these parameter-driven models. In this light, it is important to note that, in practice, models are not just defined by how they are written on paper but also by the parameterization and estimation with data.

It may come as a surprise to some who advocate the exclusive use of one measurement model in test validation research (e.g., the claim that the Rasch model should be used exclusively to validate all health outcome measures) that it can be shown that IRT (and some forms of factor analysis) is a first-order approximation to generic (classical) test theory model. This is somewhat expected once one recognizes that the generic classical test theory model statement $X = T + E$ is axiomatic to all measurement models (Lord, 1980). The different measurement models arise depending on how one defines T and E; if T is a latent variable, you have IRT or factor analysis, but if T is a complex model statement of the measurement study with facets and domains, one may have a mixed or random effects ANOVA and hence generalizability theory. Depending on the strength of the T and E assumptions one may have weak or strong forms of classical test theory. In fact, one can consider the generic measurement model statement, $X = T + E$, on par with the generic regression model statement in Eq. (3). If you apply the geometry in Zimmerman and Zumbo (2001) one can show that classical test reliability is, in essence, a Pratt index – a partitioning of the explained variation in the test score attributable to the model, just like an R-squared value in regression.

Advocating one measurement model for all occasions is an overstatement. At best, one can say that one particular model might be useful in characterizing and solving several measurement problems or that, depending on the test validation needs, one instantiation of a model with its corresponding parameterization and estimation methods may be more appropriate.

As Zimmerman and Zumbo (2001) note, formally, test data are the realization of a stochastic event defined on a product space $\Omega = \Omega_I \times \Omega_J$ where the orthogonal components, Ω_I and Ω_J, are the probability spaces for items and examinees respectively. The joint product space can be expanded to include other spaces induced by raters or occasions of measurement, a concept that was formalized in *generalizability theory* from an observed-score perspective and the facets approach to measurement from an IRT perspective. Hence, modeling of test data minimally requires sampling assumptions about items and examinees as well the specification of a stochastic process that is supposed to have generated the data – for readers interested in a measure theoretic Hilbert-space approach to the analysis of test data we refer to Zimmerman and Zumbo (2001).

Therefore, as formalized in the DLD framework I described in Section 2.3, two distinct directions of generalizability are typically of interest, which require an understanding of the reliability and validity properties of scores and inferences. As Zumbo and Rupp (2004) note, first, it is of interest to make statements about the functioning of a particular assessment instrument for groups of examinees who share characteristics with those examinees who have already been scored with it. Second, it is of interest to make statements about the functioning of item sets that share characteristics with those items that are already included on a particular test form. For example, it is often of interest to show that the scores and resulting inferences for different examinee groups are comparably reliable and valid if either the same instrument is administered to the different groups, a parallel version of the instrument is administered to the different groups, or selected subsets of items are administered to the different groups. This also specifically implies that researchers should report estimates of reliability coefficients and other parameters for their own data rather than relying on published reports from other data and that validity needs to be continually assessed rather than being taken for granted based on prior assessment calibration(s).

In summary, then, the item (or task) responses created by the interaction of examinees with items (or tasks) on a measure are considered to be *indicators* or markers of unobservable or *latent* variables. I use the term *latent variable* to refer to a random variable that is deliberately constructed or derived from the responses to a set of items and that constitutes the building block of a statistical model (e.g., θ scores in IRT or factor scores in factor analysis). The statistical problem of measuring a latent variable can be characterized as involving two key tasks: (a) to find a set of indicators (items, scales, tasks, performances, or more generally referred to as measurement opportunities) that we believe that the latent variable will imply, and (b) to find a methodology for constructing a summary measure or scalar measure of the latent variable from these indicators. Denoting the set of indicators by $x = (x_1, x_2, \ldots, x_q)$ the second part of the problem is to find a function $\varphi(x)$ so that the numerical value of φ can be regarded as an appropriate scalar measure of the unobserved or latent variable. In this light, it is important to keep in mind that the main goal of modeling test data should always be to make valid inferences about the examinees but inducing latent variables into the data structure cannot mechanically increase the validity of these inferences. No matter how sophisticated the psychometric model, the statement of $\varphi(x)$, and estimation routines have become, a test with poor validity will always remain so.

References

Angoff, W.H. (1988). Validity: An evolving concept. In: Wainer, H., Braun, H.I. (Eds.), *Test Validity*. Lawrence Erlbaum, Hillsdale, NJ.

Asparouhov, T. (2005). Sampling weights in latent variable modeling. *Structural Equation Modeling* **12**, 411–434.

Asparouhov, T., Muthen, B.O. (2005). Multivariate statistical modeling with survey data. Presented at the Conference of the Federal Committee on Statistical Methodology, Office of Management and Budget, Arlington, VA (available via the web at http://www.fcsm.gov/05papers/Asparouhov_Muthen_IIA.pdf).

Bollen, K.A., Lennox, R. (1991). Conventional wisdom on measurement: A structural equation perspective. *Psychological Bulletin* **10**, 305–314.

Bollen, K.A., Ting, K. (2000). A tetrad test for causal indicators. *Psychological Methods* **5**, 3–22.

Bond, T.G., Fox, C.M. (2001). *Applying the Rasch Model: Fundamental Measurement in the Social Sciences*. Lawrence Erlbaum, Hillsdale, NJ.

Borsboom, D., Mellenbergh, G.J., Van Heerden, J. (2004). The concept of validity. *Psychological Review* **111**, 1061–1071.

Cook, T.D., Campbell, D.T. (1979). *Quasi-Experimentation: Design and Analysis Issues for Field Settings*. Houghton Mifflin, Boston.

Cronbach, L.J., Meehl, P. (1955). Construct validity in psychological tests. *Psychological Bulletin* **52** (4), 281–302.

Cyr, A., Davies, A. (2005). Item response theory and latent variable modeling for surveys with complex sample designs: The case of the national longitudinal survey of children and youth in Canada. Presented at the Conference of the Federal Committee on Statistical Methodology, Office of Management and Budget, Arlington, VA (available via the web at http://www.fcsm.gov/05papers/Cyr_Davies_IIIC.pdf).

de Finetti, B. (1974–1975). *Theory of Probability (vols. 1–2)*. Wiley, New York.

Draper, D. (1995). Inference and hierarchical modeling in the social sciences. *Journal of Educational and Behavioral Statistics* **20**, 115–147.

Embretson, S.E. (1994). Applications of cognitive design systems to test development. In: Reynolds, C.R. (Ed.), *Cognitive Assessment: A Multidisciplinary Perspective*. Plenum Press, New York, pp. 107–135.

Freedman, D.A. (1987). As others see us: A case study in path analysis. *Journal of Educational Statistics* **12**, 101–223 (with discussion).

Green, P.E., Carroll, J.D., DeSarbo, W.S. (1978). A new measure of predictor variable importance in multiple regression. *Journal of Marketing Research* **15**, 356–360.

Gullicksen, H. (1950). *Theory of Mental Tests*. Wiley, NY.

Hambleton, R.K., Patsula, L. (1998). Adapting tests for use in multiple languages and cultures. In: Zumbo, B.D. (Ed.), *Validity Theory and the Methods Used in Validation: Perspectives From the Social and Behavioral Sciences*. Kluwer, The Netherlands, pp. 153–171.

Hattie, J., Jaeger, R.M., Bond, L. (1999). Persistent methodological questions in educational testing. *Review of Research in Education* **24**, 393–446.

Hubley, A.M., Zumbo, B.D. (1996). A dialectic on validity: Where we have been and where we are going. *The Journal of General Psychology* **123**, 207–215.

Jonson, J.L., Plake, B.S. (1998). A historical comparison of validity standards and validity practices. *Educational and Psychological Measurement* **58**, 736–753.

Jöreskog, K.G. (2002). Structural equation modeling with ordinal variables using LISREL. Retrieved December 2002 from http://www.ssicentral.com/lisrel/ordinal.htm.

Jöreskog, K.G., Goldberger, A.S. (1975). Estimation of a model with multiple indicators and multiple causes of a single latent variable. *Journal of the American Statistical Association* **10**, 631–639.

Jöreskog, K.G., Moustaki, I. (2001). Factor analysis of ordinal variables: A comparison of three approaches. *Multivariate Behavioral Research* **36**, 341–387.

Kane, M.T. (2001). Current concerns in validity theory. *Journal of Educational Measurement* **38**, 319–342.

Kaplan, D., Ferguson, A.J. (1999). On the utilization of sample weights in latent variable models. *Structural Equation Modeling* **6**, 305–321.

Kristjansson, E.A., Desrochers, A., Zumbo, B.D. (2003). Translating and adapting measurement instruments for cross-cultural research: A guide for practitioners. *Canadian Journal of Nursing Research* **35**, 127–142.

Lindley, D.V. (1972). *Bayesian Statistics, a Review*. Society for Industrial and Applied Mathematics, Philadelphia.

Lord, F.M. (1980). *Applications of Item Response Theory to Practical Testing Problems*. Lawrence Erlbaum, Hillsdale, NJ.

Lord, F.M., Novick, M.R. (1968). *Statistical Theories of Mental Test Scores*. Addison-Wesley, Reading, MA.

Lu, I.R.R., Thomas, D.R., Zumbo, B.D. (2005). Embedding IRT in structural equation models: A comparison with regression based on IRT scores. *Structural Equation Modeling* **12**, 263–277.

Maller, S.J., French, B.F., Zumbo, B.D. (in press). Item and test bias. In: Salkind, N.J. (Ed.), *Encyclopedia of Measurement and Statistics*. Sage Press, Thousand Oaks, CA.

McDonald, R.P. (1999). *Test Theory: A Unified Treatment*. Erlbaum, Mahwah, NJ.

Messick, S. (1975). The standard problem: Meaning and values in measurement and evaluation. *American Psychologist* **30**, 955–966.

Messick, S. (1980). Test validity and the ethics of assessment. *American Psychologist* **35**, 1012–1027.

Messick, S. (1988). The once and future issues of validity: Assessing the meaning and consequences of measurement. In: Wainer, H., Braun, H.I. (Eds.), *Test Validity*. Lawrence Erlbaum Associates, Hillsdale, NJ, pp. 33–45.

Messick, S. (1989). Validity. In: Linn, R.L. (Ed.), *Educational Measurement*, 3rd ed. Macmillan, New York, pp. 13–103.

Messick, S. (1995). Validity of psychological assessment: Validation of inferences from persons' responses and performances as scientific inquiry into score meaning. *American Psychologist* **50**, 741–749.

Messick, S. (1998). Test validity: A matter of consequence. In: Zumbo, B.D. (Ed.), *Validity Theory and the Methods Used in Validation: Perspectives From the Social and Behavioral Sciences*. Kluwer Academic Press, The Netherlands, pp. 35–44.

Mislevy, R.J. (1991). Randomization-based inferences about latent traits from complex samples. *Psychometrika* **56**, 177–196.

Mislevy, R.J. (1996). Test theory reconceived. *Journal of Educational Measurement* **33**, 379–416.

Mislevy, R.J., Steinberg, L.S., Breyer, F.J., Almond, R.G., Johnson, L. (1999). A cognitive task analysis with implications for designing simulation-based performance assessment. *Computers in Human Behavior* **15** (3–4), 335–374.

Moustaki, I., Jöreskog, K.G., Mavridis, D. (2004). Factor models for ordinal variables with covariate effects on the manifest and latent variables: A comparison of LISREL and IRT approaches. *Structural Equation Modeling* **11**, 487–513.

Muthen, B.O. (1985). A method for studying the homogeneity of test items with respect to other relevant variables. *Journal of Educational Statistics* **10**, 121–132.

Muthen, B.O. (1988). Some uses of structural equation modeling in validity studies: Extending IRT to external variables. In: Wainer, H., Braun, H. (Eds.), *Test Validity*. Lawrence Erlbaum, Hillsdale, NJ, pp. 213–238.

Muthen, B.O. (1989). Latent variable modeling in heterogeneous populations. *Psychometrika* **54**, 551–585.

Muthen, B., Satorra, A. (1995). Complex sample data in structural equation modeling. *Sociological Methodology* **25**, 267–316.

Neyman, J. (1923). Sur les applications de la théorie des probabilités aux experiences agricoles: Essai des principes. *Roczniki Nauk Rolniczych* **10**, 1–51, in Polish. English translation by D. Dabrowska and T. Speed (1990). Statistical Science 5, 463–80 (with discussion).

Nichols, P. (1994). A framework for developing cognitively diagnostic assessments. *Review of Educational Research* **64**, 575–603.

Pratt, J.W. (1987). Dividing the indivisible: Using simple symmetry to partition variance explained. In: Pukkila, T., Puntanen, S. (Eds.), *Proceedings of the Second International Tampere Conference in Statistics*. University of Tampere, Tampere, Finland, pp. 245–260.

Radloff, L.S. (1977). The CES-D scale: A self-report depression scale for research in the general population. *Applied Psychological Measurement* **1**, 385–401.

Rupp, A.A., Dey, D.K., Zumbo, B.D. (2004). To Bayes or not to Bayes, from whether to when: Applications of Bayesian methodology to modeling. *Structural Equation Modeling* **11**, 424–451.

Rupp, A.A., Zumbo, B.D. (2003). Which model is best? Robustness properties to justify model choice among unidimensional IRT models under item parameter drift (Theme issue in honor of Ross Traub). *Alberta Journal of Educational Research* **49**, 264–276.

Rupp, A.A., Zumbo, B.D. (2004). A note on how to quantify and report whether invariance holds for IRT models: When Pearson correlations are not enough. *Educational and Psychological Measurement* **64**, 588–599. *Educational and Psychological Measurement* **64** (2004) 991, Errata.

Rupp, A.A., Zumbo, B.D. (2006). Understanding parameter invariance in unidimensional IRT models. *Educational and Psychological Measurement* **66**, 63–84.

Shadish, W.R., Cook, T.D., Campbell, D.T. (2002). *Experimental and Quasi-Experimental Designs for Generalized Causal Inference*. Houghton Mifflin, Boston.

Sireci, S.G. (1998). The construct of content validity. In: Zumbo, B.D. (Ed.), *Validity Theory and the Methods Used in Validation: Perspectives From the Social and Behavioral Sciences*. Kluwer Academic Press, The Netherlands, pp. 83–117.

Thissen, D., Steinberg, L., Pyszczynski, T., Greenberg, J. (1983). An item response theory for personality and attitude scales: Item analysis using restricted factor analysis. *Applied Psychological Measurement* **7**, 211–226.

Thomas, D.R. (1992). Interpreting discriminant functions: A data analytic approach. *Multivariate Behavioral Research* **27**, 335–362.

Thomas, D.R. (2001). Item response theory and its application to the National Longitudinal Survey of Children and Youth. Report prepared for the Survey Research Methods Division, Statistical Society of Canada.

Thomas, D.R., Cyr, A. (2002). Applying item response theory methods to complex survey data. In: *Proceedings of the Statistical Society of Canada, Section on Survey Research Methods*. Statistical Society of Canada, Hamilton, Ontario.

Thomas, D.R., Hughes, E., Zumbo, B.D. (1998). On variable importance in linear regression. *Social Indicators Research: An International and Interdisciplinary Journal for Quality-of-Life Measurement* **45**, 253–275.

Thomas, D.R., Zumbo, B.D. (1996). Using a measure of variable importance to investigate the standardization of discriminant coefficients. *Journal of Educational and Behavioral Statistics* **21**, 110–130.

Thomas, D.R., Zumbo, B.D. (2002). Item response theory and related methods with application to complex sample surveys. Notes for the day long session presented at the Statistics Canada XIX International Symposium "Modeling survey data for social and economic research". Ottawa, Canada.

Thomas, D.R., Zhu, P.C., Zumbo, B.D., Dutta, S. (2006). Variable importance in logistic regression based on partitioning an R-squared measure. In: *Proceedings of the Administrative Sciences Association of Canada (ASAC) Conference*. Banff, AB.

Wright, B.D. (1997). A history of social science measurement. *Educational Measurement: Issues and Practice* **16**, 33–45.

Wu, A.D.L., Zumbo, B.D., Thomas D.R. (2006). Variable and factor ordering in factor analyses: using Pratt's importance measures to help interpret exploratory factor analysis solutions for oblique rotation. Paper presented at the American Educational Research Association Meeting. San Francisco, CA.

Zimmerman, D.W., Zumbo, B.D. (2001). The geometry of probability, statistics, and test theory. *International Journal of Testing* **1**, 283–303.

Zumbo, B.D. (Ed.) (1998). Validity theory and the methods used in validation: Perspectives from the social and behavioral sciences. Special issue of the journal. *Social Indicators Research: An International and Interdisciplinary Journal for Quality-of-Life Measurement* **45** (1–3) 1–359.

Zumbo, B.D. (2001). Methodology and measurement matters in establishing a bona fide occupational requirement for physically demanding occupations. In: Gledhill, N., Bonneau, J., Salmon, A. (Eds.), *Proceedings of the Consensus Forum on Establishing BONA FIDE Requirements for Physically Demanding Occupations*. Toronto, ON, pp. 37–52.

Zumbo, B.D. (2005). Reflections on validity at the intersection of psychometrics, scaling, philosophy of inquiry, and language testing (July 22, 2005). Samuel J. Messick Memorial Award Lecture, LTRC 27th Language Testing Research Colloquium, Ottawa, Canada.

Zumbo, B.D., Gelin, M.N. (in press). A matter of test bias in educational policy research: Bringing the context into picture by investigating sociological/community moderated test and item bias. *Educational Research and Policy Studies*.

Zumbo, B.D., Hubley, A.M. (2003). Item bias. In: Fernández-Ballesteros, R. (Ed.), *Encyclopedia of Psychological Assessment*. Sage Press, Thousand Oaks, CA, pp. 505–509.

Zumbo, B.D., MacMillan, P.D. (1999). An overview and some observations on the psychometric models used in computer-adaptive language testing. In: Chalhoub-Deville, M. (Ed.), *Issues in Computer-Adaptive Testing of Reading Proficiency*. Cambridge University Press, Cambridge, UK, pp. 216–228.

Zumbo, B.D., Pope, G.A., Watson, J.E., Hubley, A.M. (1997). An empirical test of Roskam's conjecture about the interpretation of an ICC parameter in personality inventories. *Educational and Psychological Measurement* **57**, 963–969.

Zumbo, B.D., Rupp, A.A. (2004). Responsible modeling of measurement data for appropriate inferences: Important advances in reliability and validity theory. In: Kaplan, D. (Ed.), *The SAGE Handbook of Quantitative Methodology for the Social Sciences*. Sage Press, Thousand Oaks, CA, pp. 73–92.

Handbook of Statistics, Vol. 26
ISSN: 0169-7161

DOI: 10.1016/S0169-7161(06)26004-8

4

Reliability Coefficients and Generalizability Theory

Noreen M. Webb, Richard J. Shavelson and Edward H. Haertel

1. Introduction

When a person is tested or observed multiple times, such as a student tested for mathematics achievement or a Navy machinist mate observed while operating engine room equipment, scores reflecting his or her performance may or may not agree. Not only may individuals' scores vary from one testing to another, calling into question the defensibility of using only one score for decision-making purposes, but the rankings of individuals may also disagree. The concern of reliability studies is to estimate the consistency of scores across repeated observations. Reliability coefficients quantify the consistency among the multiple measurements on a scale from 0 to 1.

In this chapter we present reliability coefficients as developed in the framework of classical test theory, and describe how the conception and estimation of reliability was broadened in generalizability theory. Section 2 briefly sketches foundations of classical test theory (see the chapter by Lewis for a thorough development of the theory) and focuses on traditional methods of estimating reliability. Section 3 reviews generalizability theory, including applications and recent theoretical contributions.

2. Reliability Coefficients in Classical Test Theory

Classical test theory's reliability coefficients are widely used in behavioral and social research. Each provides an index of measurement consistency ranging from 0 to 1.00 and their interpretation, at first blush, is relatively straightforward: the proportion of observed-score variance attributable to true-scores (stable or nonrandom individual differences) (see Lewis chapter for definitions in Classical Test Theory). Coefficients at or above 0.80 are often considered sufficiently reliable to make decisions about individuals based on their observed scores, although a higher value, perhaps 0.90, is preferred if the decisions have significant consequences. Of course, reliability is never the sole consideration in decisions about the appropriateness of test uses or interpretations.

Coefficient alpha (also known as "Cronbach's alpha") is perhaps the most widely used reliability coefficient. It estimates test-score reliability from a single test administration using information from the relationship among test items. That is, it provides an estimate of reliability based on the covariation among items internal to the test; hence it is also called an internal-consistency coefficient. Cronbach's (2004) (for context, see Shavelson, 2004) reflections on 50 years of coefficient alpha's use foreshadows our treatment of reliability coefficients in this chapter – as a useful but limited tool for practice. More importantly for us, reliability coefficients give insight into limitations of classical test theory and provide background for our treatment of classical test theory's successor, generalizability theory:

> *I no longer regard the alpha formula as the most appropriate way to examine most data. Over the years, my associates and I developed the complex generalizability (G) theory (*Cronbach et al., 1963*;* Cronbach et al., 1972*; see also* Brennan, 2001*;* Shavelson and Webb, 1991*) which can be simplified to deal specifically with a simple two way matrix and produce coefficient alpha* (Cronbach, 2004, p. 403).

In this section, we develop classical test theory's notion of reliability and ways in which it is estimated and applied (see Lewis' chapter for details of this theory). We then show that each traditional estimation method – test-retest, parallel (equivalent) forms, and internal consistency – defines reliability somewhat differently; none is isomorphic with the theoretical definition. We conclude this section by presenting "popular" reliability coefficients and their application before turning to generalizability theory.

2.1. *Theoretical definition of reliability*

Classical test theory (CTT) is often introduced by assuming an infinite population of people, each observed an infinite number of times (see Table 1). The cells of this "infinite matrix" (Cronbach, 2004) contain observed scores – X_{pi} is the observed total score, for example, the total number of mathematics items correct for person p observed under condition i (e.g., person p's total score on occasion i). The average of a person's scores in the infinite matrix (i.e., the average over i of the X_{pi} for person p) is defined as that person's true score – T_p. Note that we denote the true score without specifying the particular condition because of CTT's steady-state assumption – for the period of observation, the person is in a steady state, that is, the person's true score is constant (later on we relax this definition). Because a person's observed scores vary from one condition to the next even though the true score is constant, this variation must be due, in theory, to random measurement error: $e_{pi} = X_{pi} - T_p$. Consequently, a person's observed score is comprised of a true score and error: $X_{pi} = T_p + e_{pi}$. Moreover, the variance of persons' observed scores for condition i equals true-score variance plus error variance, as the covariance (or correlation) of true scores and errors is zero (as is the covariance of errors on two parallel tests): $s^2_{X_i} = s^2_T + s^2_{e_i}$.

We now have sufficient information from CTT to define the reliability coefficient. The reasoning goes like this. If there is little variability in observed scores from one condition to the next, each observed score must be pretty close to the person's true score. Hence, assuming persons vary in their true scores, the correlation between observed and true scores should be quite high. If, however, there is a large bounce in a person's observed score from one condition to the next, the correspondence between observed and true scores must be minimal, error must be substantial, and as a consequence the correlation between observed and true scores must be low.

Table 1
Person × condition score (X_{pi}) "infinite" ("population-universe") matrix

Person	Condition					
	1	2	...	i	...	$k \to \infty$
1	X_{11}	X_{12}	...	X_{1i}	...	X_{1k}
2	X_{21}	X_{22}	...	X_{2i}	...	X_{2k}
...	...	...	...	...	...	...
p	X_{p1}	X_{p2}	...	X_{pi}	...	X_{pk}
...	...	...	...	...	...	...
$n \to \infty$	X_{n1}	X_{n2}	...	X_{ni}	...	X_{nk}

Note: Cronbach, L.J. (2004). My current thoughts on coefficient alpha and successor procedures. *Educational and Psychological Measurement* **64**(3), 401. © 2004 Sage Publications, Inc. Adapted with permission.

Reliability in CTT – the proportion of observed-score variance due to variance among persons' true scores – is defined as the square of the correlation, in the population, between observed and true scores, ρ^2_{XT}:

$$\rho_{XT} = \frac{\sigma_{XT}}{\sigma_X \sigma_T} = \frac{\sigma_{TT}}{\sigma_X \sigma_T} = \frac{\sigma^2_T}{\sigma_X \sigma_T} = \frac{\sigma_T}{\sigma_X} \quad \text{Reliability Index,} \tag{1.1a}$$

$$\rho^2_{XT} = \frac{\sigma^2_T}{\sigma^2_X} \quad \text{Reliability Coefficient.} \tag{1.2b}$$

2.2. *Estimation of the reliability coefficient*

The theoretical reliability coefficient is not practical; we do not know each person's true score. Nevertheless, we can estimate the theoretical coefficient with the sample correlation between scores on two parallel tests. Assume that X and X' are two strictly parallel tests (for simplicity) – that is, tests with equal means, variances, covariances with each other, and equal covariances with any outside measure (see chapter by Lewis). The Pearson product–moment correlation between parallel tests produces an estimate of the theoretical reliability coefficient:

$$r_{XX'} = \frac{s_{XX'}}{s_X s_{X'}} = \frac{s_{(T+e)(T+e')}}{s^2_X} = \frac{s^2_T}{s^2_X}. \tag{2}$$

While we have assumed strictly parallel tests for simplicity in deriving 2, we need not make such strong assumptions. Indeed, reliability coefficients can be obtained with increasingly lenient assumptions (see Lewis' Chapter 2 for details):

- Tau equivalent tests – each person is assumed to have a constant true score over tests but the error variances may vary across tests.
- Essentially tau equivalent tests – each person's true score under one condition (g) differs by an additive constant under a second condition (h): $T_{pg} = T_{ph} + c_{gh}$. (Note:

Table 2a
Schematic sample person × condition score (X_{pi}) matrix

Person	Condition 1	Condition 2	...	Condition i	...	Condition k
1	X_{11}	X_{12}	...	X_{1i}	...	X_{1k}
2	X_{21}	X_{22}	...	X_{2i}	...	X_{2k}
...	...	...	...	...	...	...
p	X_{p1}	X_{p2}	...	X_{pi}	...	X_{pk}
...	...	...	...	...	...	...
N	X_{N1}	X_{N2}	...	X_{Ni}	...	X_{Nk}

Table 2b
Person × condition score matrix with hypothetical data

Person	Condition 1	Condition 2
1	2	1
2	2	3
3	2	3
4	4	3
5	6	5
6	6	5
7	8	6
8	8	6
9	9	8
10	8	9
Mean	5.500	4.900
Variance	7.833	6.100

while the true scores differ by c_{gh} across conditions, true-score variance is constant.) Error variances may differ.

- Congeneric tests – each person's true score may vary across tests but there is an additive constant (c_{gh}) and a positive regression constant (b_{gh}) that relates true-scores across any two tests: $T_{ph} = b_{gh}T_{pg} + c_{gh}$. Neither true-score nor error variances need be equal.

2.3. *Reliability study designs and corresponding reliability coefficients*

To estimate test-score reliability, at a minimum one needs at least two observations (scores) on the same set of persons (Tables 2a and 2b). The correlation between one set of observations with the second, then, provides a reliability coefficient. (Internal consistency reliability estimates follow a slightly more complicated procedure.) From this simple requirement, a wide variety of reliability studies could be designed.

2.3.1. *Test–retest reliability coefficient*

In designing a reliability study to produce two sets of observations, one might give the same test on two occasions, separated (say) by two weeks. In this case, Condition 1 in Table 2b becomes Occasion 1, and Condition 2 becomes Occasion 2. Following Cronbach et al. (1972) and Thorndike (1947), notice that whatever is lasting and general about the test taker (e.g., ability, testwiseness) will contribute to the consistency of the measurement from one occasion to another – i.e., contribute to "wanted" true-score variance and reliability. Something that is lasting but specific about a test taker – e.g., knowledge pertaining to a particular item or set of items on the test – will also contribute to consistency in scores. However, something temporary but general (e.g., disruptions by a lawnmower on a particular test occasion) will contribute to inconsistency from one occasion to another. Something temporary and specific, such as a very recent event (e.g., recent car engine repair creating a negative mental set on a science-test item on ideal gas law illustrated with a piston) might affect performance at time 1 but not at time 2 and hence also give rise to inconsistency in performance. Finally, chance success (or failure!) such as random guessing will also and always contribute to inconsistency from one time to the next. In summary, consistency or reliability with this reliability study design involves lasting general and lasting specific conditions of the test; temporary conditions or events (general and specific) and chance events contribute to error.

The correlation between scores at occasion 1 and occasion 2 produces a test-retest reliability coefficient – a coefficient of stability. From Table 2b, we find the correlation between scores at time 1 and time 2 – i.e., the reliability of the test scores – is 0.908.

2.3.2. Parallel or equivalent forms reliability

A second way to design a reliability study is to create two parallel forms of the test, say Form 1 and Form 2, and give the two forms of the test on the same day. In this case, the correlation between scores on Forms 1 and 2 yields the test's (either form's) reliability, also referred to as a coefficient of equivalence. From Table 2b, considering Condition 1 as Form 1 and Condition 2 as Form 2, the reliability of either form is 0.908.

The parallel forms reliability study defines consistency or reliability somewhat differently than does the retest study. For both retest and parallel forms estimates (a) the lasting-general effect contributes to measurement consistency – i.e., consistent individual differences; (b) the temporary-specific effect – e.g., the car engine example above – contributes to inconsistency across forms; and of course (c) the other random influences contribute to inconsistency. In contrast to retest reliability: (a) the lasting-specific effect – e.g., knowing particular items on one form of the test – will *not* contribute to consistent performance across test forms; and (b) the temporary-general ("lawnmower") effect may contribute to test-score consistency when the two forms of the test are taken at the same time.

2.3.3. Delayed parallel or equivalent forms reliability

It is possible to design a reliability study in which two parallel forms of a test are developed and administered to a sample of people on two different occasions. This produces a design in which the lasting-general effect contributes to consistency but the other effects contribute to error. The correlation between scores on the two forms taken at different times produces a reliability coefficient that we would expect to be less than or equal to either the stability or equivalent forms reliability coefficients above. This may be referred to as a coefficient of stability and equivalence.[1]

2.3.4. *Internal-consistency reliability*

Both retest and parallel-forms reliability are logistically inconvenient. Either the same sample of people must be tested on two occasions or more than one form of a test needs to be developed to find the reliability of a single form. The solution to how to calculate reliability of a single form of a test has a long and illustrious history (e.g., Cronbach, 2004). All solutions involve recognizing that a full length test, given on a single occasion, can be divided into (parallel) parts.

Split-half reliability

Initially tests were divided into half parts and a host of split-half reliability coefficients were derived (for reviews and details, see, for example, Feldt and Brennan, 1989; Haertel, in press). The oldest and probably most widely used split-half coefficient calculates the correlation between scores on two halves of the test (X_1 and X_2) and estimates the correlation – reliability – for a full length test with the Spearman–Brown "prophecy" formula, assuming strictly parallel test halves:

$$_{SB_2}r_{XX'} = \frac{2r_{X_1X_2}}{1 + r_{X_1X_2}}. \tag{3}$$

If we treat Conditions 1 and 2 in Table 2b as corresponding to persons' total scores on two halves of the same test, we can find the reliability of the full length test – i.e., double the length of the half tests – with Spearman–Brown:

$$_{SB_2}r_{XX'} = \frac{2 \cdot 0.908}{1 + 0.908} = 0.952.$$

While all this seems straightforward, it is not and herein lies the controversy. One source of the controversy lies in how to define halves of a test – the first versus the last half? Odd–even halves? Random halves? While the debate goes on, consensus is that the first versus last half method is not a good idea, especially if fatigue and speed apply toward the end of the test, and that content should be stratified in the allocation of items to halves. The debate, in a way, is also over how to define true-score and error variance. Internal consistency coefficients such as the split half method, like parallel-forms coefficients, treat lasting-general (e.g., ability) and temporary-general ("lawnmower") effects as consistency (reliability) in the measurement for most such splits, and lasting- and temporary-specific effects along with random guessing as error.

The second source of controversy revolves around the strictly parallel assumption of the Spearman–Brown formula. To this end, a whole host of split-half reliability coefficients have been developed historically that relax the assumptions (e.g., Flanagan's formula, which requires only the essential tau equivalence assumption). Since, in practice, these coefficients are not as widely used as is a general method, coefficient alpha, we do not provide these formulas (see Feldt and Brennan, 1989; Haertel, in press).

---

[1] For both the equivalent forms and the delayed equivalent forms designs, better estimates of the reliability of forms g and h are given by $\hat{\sigma}_{gh}/\hat{\sigma}_g^2$ and $\hat{\sigma}_{gh}/\hat{\sigma}_h^2$, respectively.

Multi-part reliability and coefficient alpha

Up to this point, we have considered tests split in half to produce two strictly parallel forms to be used to estimate reliability. Scores on one half can be correlated with scores on the other half. If the test is longer than two items, and most tests are, it is possible to split it into thirds, fourths, fifths, and so on until each test item is considered a parallel test itself. This is just the reasoning used in creating general-use internal consistency reliability coefficients, Cronbach's alpha being the most widely known and used.

To see the reasoning behind alpha, we extend the Spearman–Brown application from split-half reliability to multi-part reliability. Rather then adjusting the correlation between scores on test halves, we use the correlation between individual test items – the reliability of a single item test – and adjust this correlation with a generalized Spearman–Brown formula. The Spearman–Brown formula then provides the reliability – internal consistency – of the full-length test with all k items. In making this adjustment using the Spearman–Brown formula, we must assume equal variances and correlations between items. The general Spearman–Brown formula for a test of k items is:

$$_{SB_k}r_{XX'} = \frac{kr_1}{1 + (k-1)r_1}, \tag{3a}$$

where SB_k refers to the Spearman–Brown adjustment for the reliability of a k-item test, k refers to the number of items, and r_1 refers to the inter-item correlation or the reliability of a single-item test.

To see how (3a) works, let's continue the example from Table 2b where Condition 1 refers to a 12-point Likert-type rating on questionnaire item 1 and Condition 2 refers to a similar rating on a strictly parallel questionnaire item 2. In this case, we know that $r_1 = 0.908$ is the reliability of a single item questionnaire. So an estimate of the reliability of the two-item questionnaire would be:

$$_{SB_{K=2}}r_{XX'} = \frac{2 \cdot 0.908}{1 + (2-1) \cdot 0.908} = 0.952.$$

We can also use (3a) to prophesize what the reliability of the test would be with 5 items (or any other number of items):

$$_{SB_{K=5}}r_{XX'} = \frac{5 \cdot 0.908}{1 + (5-1) \cdot 0.908} = 0.980.$$

In general, as parallel items are added to a test, the reliability of the test increases in the manner given by (3a). Conceptually it is helpful to conceive of internal consistency coefficient alpha as the inter-item correlation on a test adjusted to the full length of the test.

In practice, however, the assumption of strictly parallel items is too restrictive. Cronbach (1951) derived an internal consistency coefficient, coefficient alpha, based on a weaker assumption, that of essential tau equivalence; violation of the assumption tends to lower the coefficient.

The general formula for coefficient alpha is typically written as (Cronbach, 2004):

$$\alpha = \frac{k}{k-1}\left(1 - \frac{\sum s_i^2}{s_t^2}\right), \tag{4}$$

where k refers to the number of items on the test, s_i^2 refers to the variance of item i, and s_t^2 refers to the variance of total scores (summed over items) on the test. Again, referring to the data in Table 2b and considering Conditions 1 and 2 to be two questionnaire items, we can calculate the item and total score variance: $s_1^2 = 7.833$, $s_2^2 = 6.100$, and $s_t^2 = 26.489$. Using this information and $k = 2$ in (4) we have:

$$\alpha = \frac{2}{2-1}\left(1 - \frac{7.833 + 6.100}{26.489}\right) = 2\left(1 - \frac{13.933}{26.489}\right) = 0.948.$$

Note that the alpha coefficient gives a lower result than the Spearman–Brown formula. If the sample variances for the two conditions (Table 2b) were identical, the results from the two formulas would be the same.[2]

Coefficient alpha provides a general solution to most all split-half methods that relax the strictly-parallel test assumptions. Moreover, whereas earlier internal consistency formulas such as Kuder–Richardson formulae 20 and 21 dealt with the special case of dichotomously scored items (see Feldt and Brennan, 1989, and Haertel, in press, for details), Cronbach's alpha applies to the more general case of items scored dichotomously or otherwise (e.g., Likert-type scale as in the example above).

Components of variance approach to coefficient alpha

The components of variance approach to alpha begins with the infinite matrix (Table 1) and assumes that persons (rows) and conditions (columns) are sampled from this matrix at random – i.e., the variance-component approach assumes randomly parallel conditions. "The ... assumptions ... are not true in any strict sense, and a naive response would be to say that if the assumptions are violated, the alpha calculations cannot be used. No statistical work would be possible, however, without making assumptions and, as long as the assumptions are not obviously grossly inappropriate to the data, the statistics calculated are used, if only because they can provide a definite result that replaces a hand-waving interpretation" (Cronbach, 2004, p. 404).

In the random model with persons crossed with conditions (for example, persons crossed with items), the observed score for person p on a single item i, X_{pi}, can be divided into four components:

$$X_{pi} = \mu + (\mu_p - \mu) + (\mu_i - \mu) + (X_{pi} - \mu_p - \mu_i + \mu). \tag{5}$$

The first component on the right is the grand mean, μ, constant for all persons. The next term is person p's true score expressed in deviation form $(\mu_p - \mu)$. The metric for the person's true score is the item, where μ_p denotes person p's mean item score over the items in the infinite matrix. (This contrasts with T_p in earlier sections, which was expressed in the total score metric.) The third component is the item effect $(\mu_i - \mu)$ – the average deviation score for item i. And the last term, the residual $(X_{pi} - \mu_p - \mu_i + \mu)$, reflects the departure of observed scores from what would be expected given the first three terms in (5).

[2] If item scores are standardized, the result is what some statistical computer programs call "*standardized alpha*."

Table 3
Expected mean squares and variance component estimates

Expected mean square	Variance component estimators
$E(MS_p) = \sigma_{RES}^2 + n_i\sigma_p^2$	$\hat{\sigma}_p^2 = \frac{MS_p - \hat{\sigma}_{RES}^2}{n_i}$
$E(MS_i) = \sigma_{RES}^2 + n_p\sigma_i^2$	$\hat{\sigma}_i^2 = \frac{MS_i - \hat{\sigma}_{RES}^2}{n_p}$
$E(MS_{RES}) = \sigma_{RES}^2$	$\hat{\sigma}_{RES}^2 = MS_{RES}$

Except for the first component, μ, each of the components of the observed score varies – from one person to another, from one item to another, or from one person-item combination to another. For this simple design, it is readily shown that these last three score components, as defined, must be uncorrelated. Hence, the observed-score variance can be decomposed into three *components of variance* (or simply, variance components):

$$\sigma_{X_{pi}}^2 = \sigma_p^2 + \sigma_i^2 + \sigma_{RES}^2. \tag{6}$$

The first variance component on the right (σ_p^2) is the systematic, error-free variance among persons' score components. The i variance component (σ_i^2) describes the extent to which the item means vary. And the residual variance component (σ_{RES}^2) is what is commonly thought of as variance due to measurement error. (We speak of a residual because in practice, from sample data as in Table 2a, we cannot disentangle the $p \times i$ interaction from random error e.)

Notice that (6) departs from the usual decomposition of observed score variance into true-score variance plus error variance. If we assume strictly parallel tests with equal means, the variance due to conditions (σ_i^2) is zero and drops out. As Cronbach (2004, p. 406) explains, "Spearman started the tradition of ignoring item characteristics because he felt that the person's position on the absolute score scale was of no interest"; condition means were arbitrary. More generally, if our interest is in the performance of a group of examinees relative to one another, and if all of them take the same form of the test, then the score component associated with item difficulties is irrelevant to our intended inferences and should be ignored.

The variance components in (6) can be estimated from sample data (e.g., Tables 2a, 2b) using the random-effects model of the analysis of variance. Variance component estimates can be calculated by setting the observed mean squares in the ANOVA equal to their expectations and solving for the variance components (Table 3). In Table 3, in preparation for the factorial ANOVA approach used in generalizability theory (next section), we use n_i instead of k to represent the number of items and n_p instead of N to represent the number of persons.

With the expected mean squares and estimated variance components in hand, we can write coefficient alpha – proportion of observed-score variance (assuming that $\sigma_i^2 = 0$)

due to true-score variance – as:

$$\alpha = \frac{MS_p - MS_{RES}}{MS_p} = \frac{\hat{\sigma}_p^2}{\hat{\sigma}_p^2 + \frac{\hat{\sigma}_{RES}^2}{n_i}}. \quad (7)$$

As an example, consider the data in Table 2b. Estimating the variance with a repeated-measures ANOVA yields: $MS_{RES} = 0.689$, $MS_i = 1.800$, and $MS_p = 13.244$. Solving for the estimated variance components, we have:

$$\hat{\sigma}_{RES}^2 = MS_{RES} = 0.689,$$

$$\hat{\sigma}_i^2 = (MS_i - MS_{RES})/n_p = (1.800 - 0.689)/10 = 0.111,$$

$$\hat{\sigma}_p^2 = (MS_p - MS_{RES})/n_i = (13.244 - 0.689)/2 = 6.278.$$

Inserting the appropriate estimated variance components into (7) and assuming $n_i = 2$, we have:

$$\alpha = \frac{6.278}{6.278 + \frac{0.689}{2}} = 0.948.$$

2.3.5. Reliability coefficients for special occasions

Up to this point, we have dealt with widely used, standard reliability formulas that apply to total test scores or item scores. Nevertheless, we have not dealt directly with scores on open-ended ("constructed response") tests where "raters" or examiners provide the scores. Moreover, scores other than observed scores, such as difference scores or various kinds of composite scores, are widely used in social, behavioral and educational research. Reliability coefficients have been adapted or developed for such "special" occasions. We review, briefly, a number of "special-occasion" reliability formulas here.

Interrater reliability

Often with essay and other constructed-response tests, two (or more) raters score individuals' performance. Each rater would provide a performance score for each individual (e.g., a total score summed over open-ended test items). Assuming "parallel raters" are created by selection and training, akin to parallel tests, the correlation between the scores assigned by Rater 1 (Condition 1 in Table 2b) and by Rater 2 (Condition 2) to persons can be used to determine interrater reliability – the consistency among raters (in the sense of relative or deviation scores assigned). From the data in Table 2b, the reliability of scores assigned by a single rater (cf. single form) would be 0.908.

Reliability of difference scores

Often interest attaches to an individual's change over time. One way to reflect change is to calculate a difference score – post-test–pretest: $X_{post} - X_{pre}$. In this case:

$$X_{post} = T_{post} + e_{post},$$

$$X_{pre} = T_{pre} + e_{pre}, \quad \text{so that } X_D = (T_{post} - T_{pre}) + (e_{post} - e_{pre}).$$

Assuming that the covariance of the errors on the pre- and post-test is zero, we can write the true-score variance and error variance for the difference score as follows:

$$\sigma^2_{T_D} = \sigma^2_{T_{pre}} + \sigma^2_{T_{post}} - 2\sigma_{T_{pre}T_{post}},$$
$$\sigma^2_{e_D} = \sigma^2_{e_{pre}} + \sigma^2_{e_{post}}.$$

Note that usually the covariance between pre- and post-test scores will be positive, reflecting the positive covariance between true scores on the pre- and post-tests. This implies that the true-score variance of difference scores will be less than the sum of the true-score variances for the pre- and post-tests. However, the error variance of the difference score will be equal to the sum of the pre- and post-test error variances. Consequently, often difference scores have lower reliability than observed scores.

Defining reliability as true (difference)-score variance divided by observed (difference)-score variance, a simple, general computational formula is (Haertel, in press):

$$\rho_{DD'} = 1 - \frac{(1 - \rho_{X_{pre}X'_{pre}})\sigma^2_{X_{pre}} + (1 - \rho_{X_{post}X'_{post}})\sigma^2_{X_{post}}}{\sigma^2_{(X_{post}-X_{pre})}}. \tag{8}$$

If $\sigma^2_{X_{pre}} = \sigma^2_{X_{post}}$, we can write (8) as:

$$\rho_{DD'} = \frac{\bar{\rho}_{XX'} - \rho_{X_{pre}X_{post}}}{1 - \rho_{X_{pre}X_{post}}}, \tag{8a}$$

where $\bar{\rho}_{XX'}$ is the average of the reliabilities of the pre- and post-test scores.

Difference scores have been criticized not only for low reliability but for being problematic in the sense that individuals with high pretest scores are likely to improve less than individuals with low pretest scores (i.e., difference scores are often negatively correlated with pretest scores). Nevertheless, difference scores have been viewed favorably as they provide an unbiased estimator of the true difference for a randomly chosen individual. Moreover, low difference-score reliability does not necessarily mean low statistical power for mean comparisons (see Haertel, in press, for details). Because differences are expressed on a scale with a true zero point (representing no change), the absolute magnitude of the difference is usually of greater interest than the stability of a rank ordering by difference scores. For this reason, the reliability coefficient for the difference score is often the wrong statistic to use; the standard error of the difference score is more informative (see Haertel, in press, for details).

Reliability of composite scores: Weighted and unweighted

The reliability of difference scores is the simplest, most straightforward case of the reliability of composite scores where the difference is a linear combination of two component scores with fixed weights -1 and $+1$: $-1X_{pre} + 1X_{post}$. At the other extreme, a composite score might be formed as a weighted linear combination of a set of tests (components) measuring distinct constructs (e.g., subtests of the Wechsler intelligence test). The weights might be statistically (e.g., multiple regression) determined or defined a priori by expert judges or from theory.

(i) Weighted composites

The composite score, C_p, for person p formed with k component scores, $X_{p1}, \ldots, X_{pk}$ and weights $w_1, \ldots, w_k$ may be written as follows (an additive constant w_0 may also be present):

$$C_p = w_0 + \sum_{i=1}^{k} w_i X_{pi}. \tag{9}$$

To develop a reliability coefficient, we assume that composite scores consist of a true-score component and an error component with errors uncorrelated. Moreover, we conceive of parallel composite scores, C and C', formed from parallel component scores $X_{p1}, \ldots, X_{p2}, \ldots, X_k$ and $X'_{p1}, \ldots, X'_{p2}, \ldots, X'_k$, respectively. With these assumptions, the error variance for the composite score is simply a weighted sum of the error variances of the component scores; the true-score variance, however, is a bit more complicated as it is a weighted sum of the true-component-score variances and covariances. This said, the reliability of composite scores may be written (Haertel, in press):

$$\begin{aligned}\rho_{CC'} &= 1 - \frac{\sigma^2_{E_C}}{\sigma^2_C}\\ &= 1 - \frac{\sum_{i=1}^{k} w_i^2 \sigma^2_{E_i}}{\sigma^2_C} = 1 - \frac{\sum_{i=1}^{k} w_i^2 \sigma^2_{X_i}(1 - \rho_{X_i X'_i})}{\sigma^2_C}.\end{aligned} \tag{10}$$

(ii) Unweighted composites or test battery composite scores

Often a test battery composite will consist of an unweighted composite of subtest mean scores – i.e., simply add up a person's scores on a set of subtests. The reliability of this unweighted composite – often referred to as the reliability of a battery composite score – is a special case of (9) and (10) with all the w_i equal. If the components of a composite are each weighted 1, the formula for weighted composites can be simplified to that for a battery composite score:

$$\rho_{CC'} = 1 - \frac{\sum_{i=1}^{k} \sigma^2_{X_i}(1 - \rho_{X_i X'_i})}{\sigma^2_C}. \tag{11}$$

If the components or subtests in a test battery composite score are scaled to have identical variances, we can simplify (11) to:

$$\rho_{CC'} = \frac{\frac{1}{k-1}\bar{\rho}_{X_i X'_i} + \bar{\rho}_{X_i X_j}}{\frac{1}{k-1} + \bar{\rho}_{X_i X_j}} \quad (i \neq j). \tag{11a}$$

From this formula we see that the battery composite reliability is a function of the number of subtests, the individual subtest reliabilities and the intercorrelation among subtests.

Stratified coefficient alpha

Tests may contain different types of items that can be categorized. For example, reading achievement tests may include both vocabulary and reading comprehension items.

Or, a battery of tests might be categorized into subsets as is done with the Wechsler intelligence test with both verbal (e.g., information, comprehension, arithmetic) and performance (e.g., block design, picture arrangement, matrix reasoning) categories. When components (items or subtests) fall within categories or strata, we might view the composite as a result of stratified random sampling of subtests or items.

When we have such stratification, we would expect that items or subtests within a stratum would correlate more highly with each other than with items or subtests in other strata. Consider a case with k strata, $i = 1, \ldots, k$, with stratum-level total scores X_i and reliabilities α_i. Let X represent the total score obtained as a weighted sum of the X_i. A reliability estimate that takes account this stratification is stratified coefficient alpha:

$$ {}_{stratified}\alpha = 1 - \frac{\sum_{i=1}^{k} \sigma_{X_i}^2 (1 - \alpha_i)}{\sigma_X^2}. \tag{12} $$

When the stratification holds up and correlations within strata are higher than those between strata, coefficient alpha will be smaller than stratified alpha.

2.4. *Concluding comments: The need for more comprehensive reliability theory*

What became evident over 60 years ago, and what we have tried to make clear, is that there are many possible reliability coefficients for any single test. Moreover, different reliability coefficients define true-score and error differently. And these coefficients sometimes produce contradictory estimates of reliability. As Cronbach et al. (1972, p. 6) citing Goodenough (1936) pointed out:

> *[T]he investigator who compares two administrations of the same list of spelling words asks a different question than the investigator who compares performance on two different lists. Inconsistency of observers, inconsistency in the subject's response to different stimulus lists, inconsistency of his response to the same stimulus on different occasions – all of these may be sources of error, but any one comparison will detect some inconsistencies and not others.*

In the next part of this chapter, we present a theory that deals with definitional and interpretative issues of true scores, sources of error, and variation in reliability estimates. This theory is Generalizability theory; it draws on the components of variance approach to reliability.

3. Generalizability theory

As the previous section makes clear, classical test theory treats measurement error as undifferentiated random variation. Cronbach et al. (1972) (see also Brennan, 1997, 2001; Shavelson and Webb, 1981, 1991) introduced the theory of generalizability to make it possible to assess *multiple* sources of measurement error in order to characterize the measurement and improve its design. Generalizability theory pinpoints the sources of measurement error, disentangles them, and estimates each one.

In G theory, a behavioral measurement (e.g., a test score) is conceived of as a sample from a *universe of admissible observations*, which consists of all possible observations on an *object of measurement* (typically a person) that a decision maker considers to be acceptable substitutes for the observation in hand. Each characteristic of the measurement situation (e.g., test form, test item, rater, test occasion) is called a *facet*. The universe of admissible observations is usually defined by the Cartesian product of the levels (called *conditions* in G theory) of the facets.

In order to evaluate the dependability of behavioral measurements, a *generalizability (G) study* is designed to isolate and estimate variation due to the object of measurement and as many facets of measurement error as it is reasonably and economically feasible to examine. A *decision (D) study* uses the information provided by the G-study to design the best possible application of the measurement for a particular purpose. In planning the D-study, the decision maker defines a *universe of generalization*, the set of facets and their levels to which he or she wants to generalize, and specifies the proposed interpretation of the measurement. The decision maker uses the information from the G-study to evaluate the effectiveness of alternative designs for minimizing error and maximizing reliability.

3.1. One-facet designs

Consider a one-facet universe of admissible observations with items (i). The decision maker is interested in the performance of persons drawn from a particular population. The *object of measurement*, then, is persons. Persons are not a source of error, hence not a facet. Assume all persons respond to all items so that persons (p) are crossed with items (i). We denote this crossed design as $p \times i$. With generalization over all admissible items taken from an indefinitely large universe, the decomposition of the observed score is the same as in (5):

$$\begin{aligned} X_{pi} &= \mu && \text{(grand mean)} \\ &\quad + \mu_p - \mu && \text{(person effect)} \\ &\quad + \mu_i - \mu && \text{(item effect)} \\ &\quad + X_{pi} - \mu_p - \mu_i + \mu && \text{(residual effect).} \end{aligned}$$

In G theory, μ_p is called the *universe score* instead of the true score, recognizing that a person may have different universe scores depending on the characterization of the universe.[3] The universe score, then, is defined as the expected value (E) of a person's observed score over the universe of items:

$$\mu_p \equiv \mathrm{E}_i X_{pi}. \tag{13}$$

The population mean for item i is:

$$\mu_i \equiv \mathrm{E}_p X_{pi}. \tag{14}$$

And the mean over both the population and the universe (the "grand mean") is:

$$\mu \equiv \mathrm{E}_p \mathrm{E}_i X_{pi}. \tag{15}$$

[3] Strictly speaking, the universe score is not defined until the objects of measurement are identified in the decision (D) study. A single G-study can give rise to different D studies with different objects of measurement, as noted later in this paper.

Table 4
Estimates of variance components for a random-effects, one-facet $p \times i$ G-study design

Source of variation	Mean square	Expected mean square	Estimated variance component
Person (p)	MS_p	$\sigma^2_{pi,e} + n_i\sigma^2_p$	$\hat{\sigma}^2_p = \frac{MS_p - MS_{pi,e}}{n_i}$
Item (i)	MS_i	$\sigma^2_{pi,e} + n_p\sigma^2_i$	$\hat{\sigma}^2_i = \frac{MS_i - MS_{pi,e}}{n_p}$
pi, e	$MS_{pi,e}$	$\sigma^2_{pi,e}$	$\hat{\sigma}^2_{pi,e} = MS_{pi,e}$

As defined in (6), each score component, except for the grand mean, has a distribution. The distribution of $\mu_p - \mu$ has mean zero and variance $\mathrm{E}_p(\mu_p - \mu)^2 = \sigma^2_p$, which is called the universe-score variance. Similarly, the component for item has mean zero and variance $\mathrm{E}_i(\mu_i - \mu)^2 = \sigma^2_i$. The residual component has mean zero and variance $\mathrm{E}_p\mathrm{E}_i(X_{pi} - \mu_p - \mu_i + \mu)^2 = \sigma^2_{pi,e}$, which indicates the person $\times$ item interaction (pi) confounded with unsystematic or unmeasured error (e). The collection of observed item-level scores, X_{pi}, has a variance $\mathrm{E}_p\mathrm{E}_i(X_{pi} - \mu)^2 = \sigma^2_{X_{pi}}$, which equals the sum of the variance components:

$$\sigma^2_{X_{pi}} = \sigma^2_p + \sigma^2_i + \sigma^2_{pi,e}. \tag{16}$$

Because the nature of the residual component changes depending on the design of the generalizability study, we use the notation pi, e instead of *RES*.

3.1.1. Generalizability and decision studies with a crossed design

The variance components are estimated from an analysis of variance of sample data in the *generalizability (G) study*. For a random-effects $p \times i$ (person by item) design in which a random sample of n_p persons responds to each of n_i randomly sampled items, numerical estimates of the variance components are obtained in the same way as in the components of variance approach to coefficient alpha, that is, by setting the expected mean squares equal to the observed mean squares and solving the set of simultaneous equations that appear in Table 4.

In the *decision (D) study*, decisions usually will be based on the mean over multiple observations rather than on a single observation. The mean score over a sample of n'_i items is denoted as X_{pI} in contrast to a score on a single item, X_{pi}. (Note the switch in notation from k in classical test theory to n'_i in generalizability theory.) A one-facet, crossed D-study design where decisions are to be made on the basis of X_{pI} is, then, denoted as $p \times I$. Note that the number of items used in the D-study, n'_i, may differ from the number of items used in the G-study, n_i. As is seen here, G theory typically uses the mean score metric (e.g., mean of multiple item scores) rather than the total score metric (e.g., sum of multiple item scores).

G theory recognizes that the decision maker might want to make two types of decisions based on a behavioral measurement: relative (norm-referenced) and absolute (criterion- or domain-referenced). A *relative decision* concerns the relative ordering of individuals (e.g., norm-referenced interpretations of test scores). For relative decisions, the error in a random effects $p \times I$ design is defined as:

$$\delta_{pI} \equiv (X_{pI} - \mu_I) - (\mu_p - \mu), \tag{17}$$

where $\mu_p = \mathrm{E}_I X_{pI}$ and $\mu_I = \mathrm{E}_p X_{pI}$. The variance of the errors for relative decisions is:

$$\sigma_\delta^2 = \mathrm{E}_p \mathrm{E}_I \delta_{pI}^2 = \sigma_{pI,e}^2 = \frac{\sigma_{pi,e}^2}{n_i'}. \tag{18}$$

Notice that the main effect of item does not enter into error for relative decisions because all people respond to all items so any difference in items affects all persons and does not change their relative standing. In order to reduce σ_δ^2, n_i' may be increased (analogous to the Spearman–Brown prophesy formula in classical test theory and the standard error of the mean in sampling theory).

Although G theory stresses the importance of variance components and measurement error, it provides a *generalizability coefficient* that is analogous to the reliability coefficient in classical test theory (ratio of universe-score variance to the expected observed-score variance). For the $p \times I$ random-effects design, the generalizability coefficient is

$$\mathrm{E}\rho^2(X_{pI}, \mu_p) = \mathrm{E}\rho^2 = \frac{\mathrm{E}_p(\mu_p - \mu)^2}{\mathrm{E}_I \mathrm{E}_p (X_{pI} - \mu_I)^2} = \frac{\sigma_p^2}{\sigma_p^2 + \sigma_\delta^2}. \tag{19}$$

Sample estimates of the parameters in (19) are used to estimate the generalizability coefficient:

$$\mathrm{E}\hat{\rho}^2 = \frac{\hat{\sigma}_p^2}{\hat{\sigma}_p^2 + \hat{\sigma}_\delta^2}. \tag{20}$$

$\mathrm{E}\hat{\rho}^2$ is a biased but consistent estimator of $\mathrm{E}\rho^2$. When $n_i' = n_i$, $\mathrm{E}\hat{\rho}^2$ is the same as coefficient alpha in (7).

An *absolute decision* focuses on the absolute level of an individual's performance independent of others' performance (cf. criterion- or domain-referenced interpretations). For absolute decisions, the error in a random-effects $p \times I$ design is defined as:

$$\Delta_{pI} \equiv X_{pI} - \mu_p \tag{21}$$

and the variance of the errors is:

$$\sigma_\Delta^2 = \mathrm{E}_p \mathrm{E}_I \Delta_{pI}^2 = \sigma_I^2 + \sigma_{pI,e}^2 = \frac{\sigma_i^2}{n_i'} + \frac{\sigma_{pi,e}^2}{n_i'}. \tag{22}$$

Note that, with absolute decisions, the main effect of items – how difficult an item is – does influence absolute performance and so is included in the definition of measurement error. Also note that $\sigma_\Delta^2 \geqslant \sigma_\delta^2$.

G theory provides an *index of dependability* (Kane and Brennan, 1977) for absolute decisions:

$$\Phi = \frac{\sigma_p^2}{\sigma_p^2 + \sigma_\Delta^2}. \tag{23}$$

For criterion-referenced decisions involving a fixed cutting score (λ), it is possible to define a loss function based on the squared distance from the cut score. In such applications, assuming that λ is a constant that is specified a priori, the error of measurement is:

$$\Delta_{pI} = (X_{pI} - \lambda) - (\mu_p - \lambda) = X_{pI} - \mu_p, \tag{24}$$

and an index of dependability may be defined as:

$$\Phi_\lambda = \frac{\mathrm{E}_p(\mu_p - \lambda)^2}{\mathrm{E}_I \mathrm{E}_p(X_{pI} - \lambda)^2} = \frac{\sigma_p^2 + (\mu - \lambda)^2}{\sigma_p^2 + (\mu - \lambda)^2 + \sigma_\Delta^2}. \tag{25}$$

An unbiased estimator of $(\mu - \lambda)^2$ is $(\bar{X} - \lambda)^2 - \hat{\sigma}^2_{\bar{X}}$ where $\bar{X}$ is the observed grand mean over sampled persons and sampled items in the D-study and $\hat{\sigma}^2_{\bar{X}}$ is the error variance involved in using the observed grand mean $\bar{X}$ as an estimate of the grand mean over the population of persons and the universe of items (μ). For the $p \times I$ random-effects design, $\hat{\sigma}^2_{\bar{X}}$ is:

$$\hat{\sigma}^2_{\bar{X}} = \frac{\hat{\sigma}^2_p}{n'_p} + \frac{\hat{\sigma}^2_i}{n'_i} + \frac{\hat{\sigma}^2_{pi,e}}{n'_p n'_i}. \tag{26}$$

The estimate of Φ_λ is smallest when the cut score λ is equal to the observed grand mean $\bar{X}$; in that case, $\hat{\Phi}_\lambda = \hat{\Phi}$.

Table 5 gives the results of generalizability analyses of the data in Table 2b. For $n'_i = 2$, $\mathrm{E}\hat{\rho}^2 = 0.948$ is the same as coefficient alpha using (7). The error variances for absolute decisions ($\hat{\sigma}^2_\Delta$) and coefficients of dependability ($\hat{\Phi}$) provide information that has no counterpart in classical test theory, that is, consistency of persons' absolute performance independent of others' performance rather than consistency of persons' relative standing. In this example, because the main effect for items ($\hat{\sigma}^2_i = 0.11111$) is small relative to the other estimated variance components, there is little difference between $\mathrm{E}\hat{\rho}^2$ and $\hat{\Phi}$. The results in Table 5 show diminishing returns in $\mathrm{E}\hat{\rho}^2$ and $\hat{\Phi}$ when the length of the test is increased. Estimated levels of generalizability and dependability for a 10-item test ($\mathrm{E}\hat{\rho}^2 = 0.987$ and $\hat{\Phi} = 0.989$) are not appreciably higher than for a 5-item test ($\mathrm{E}\hat{\rho}^2 = 0.979$ and $\hat{\Phi} = 0.975$).

3.1.2. Generalizability and decision studies with a nested design

In a nested one-facet G-study design (items nested within persons, denoted as $i : p$), each person responds to a different set of n_i items. A nested G-study design may be used because the investigator considers the universe of admissible observations to be naturally nested or, when the universe of admissible observations is crossed, for cost or logistical considerations. A person's observed score is decomposed into the following effects:

$$\begin{aligned} X_{pi} = {} & \mu && \text{(grand mean)} \\ & + (\mu_p - \mu) && \text{(person effect)} \\ & + (X_{pi} - \mu_p) && \text{(residual effect).} \end{aligned} \tag{27}$$

Table 5
Generalizability analysis of the data in Table 2b (1-facet, random effects $p \times i$ design)

Source of variation	G-study $\hat{\sigma}^2$	Size of decision study		
		$n'_i = 2$	$n'_i = 5$	$n'_i = 10$
Estimated variance components				
Person (p)	6.27778	6.27778	6.27778	6.27778
Item (i)	0.11111	0.05556	0.02222	0.01111
pi, e	0.68889	0.34444	0.13778	0.06889
Error variances				
$\hat{\sigma}^2_\delta$		0.34444	0.13778	0.06889
$\hat{\sigma}^2_\Delta$		0.40000	0.16000	0.08000
Coefficients				
$\mathrm{E}\hat{\rho}^2$		0.948	0.979	0.989
$\hat{\Phi}$		0.940	0.975	0.987

Table 6
Estimates of variance components for a random-effects one-facet $i : p$ G-study design

Source of variation	Mean square	Expected mean square	Estimated variance component
Person (p)	MS_p	$\sigma^2_{i,pi,e} + n_i\sigma^2_p$	$\hat{\sigma}^2_p = \frac{MS_p - MS_{i,pi,e}}{n_i}$
i, pi, e	$MS_{i,pi,e}$	$\sigma^2_{i,pi,e}$	$\hat{\sigma}^2_{i,pi,e} = MS_{i,pi,e}$

Unlike the one-facet crossed $p \times i$ design, the nested $i : p$ design has no separate term for the item effect. Because μ_i and μ_{pi} are confounded in this design, the item effect is part of the residual term. The variance component for persons, the universe-score variance, is defined in the same way as for crossed designs. The variance component for the residual effect is:

$$\sigma^2_{i,pi,e} = \mathrm{E}_p\mathrm{E}_i(X_{pi} - \mu_p)^2. \tag{28}$$

The variance of the collection of observed scores X_{pi} for all persons and items is:

$$\sigma^2_{X_{pi}} = \sigma^2_p + \sigma^2_{i,pi,e}. \tag{29}$$

Numerical estimates of the variance components are obtained by setting the expected mean squares equal to the observed mean squares and solving the set of simultaneous equations as shown in Table 6.

A one-facet, nested D-study design where decisions are to be made on the basis of X_{pI} is denoted as $I : p$. Because the item (I) effect is confounded with the usual residual term (pI, e), error variance for relative decisions (σ^2_δ) is the same as the error variance for absolute decisions (σ^2_Δ):

$$\sigma^2_\delta = \sigma^2_\Delta = \sigma^2_{I,pI,e} = \frac{\sigma^2_{i,pi,e}}{n'_i}. \tag{30}$$

Consequently, for this design, the generalizability and phi coefficients are the same ($\mathrm{E}\rho^2 = \Phi$).

3.1.3. A crossed generalizability study and a nested decision study

Generalizability theory allows the decision maker to use different designs in the G- and D-studies. Although G-studies should use crossed designs whenever possible to avoid confounding of effects, D-studies may use nested designs for convenience or for increasing sample size, which typically reduces estimated error variance and, hence, increases estimated generalizability. A one-facet, crossed $p \times i$ G-study can be used to estimate variance components, error variance, and generalizability and phi coefficients for a one-facet, nested $I : p$ D-study. The variance component for the effects that are confounded in the nested design (i and pi, e) is the sum of the corresponding variance components in the crossed G-study:

$$\sigma^2_{i,pi,e} = \sigma^2_i + \sigma^2_{pi,e}. \tag{31}$$

The estimated residual variance for use in the D-study, then, is the sum of the following terms from the G-study:

$$\hat{\sigma}^2_{I,pI,e} = \frac{\hat{\sigma}^2_i}{n'_i} + \frac{\hat{\sigma}^2_{pi,e}}{n'_i}. \tag{32}$$

Universe-score variance (σ^2_p) remains the same in the crossed and nested designs.

3.2. Multifacet designs

The procedures for one-facet designs are easily extended to multifacet designs. Consider a two-facet crossed $p \times t \times o$ (person by task by observer) design in which tasks and observers are randomly sampled from infinitely large universes (random-effects design). The observed score for a particular person (X_{pto}) is:

$$\begin{aligned}
X_{pto} = {} & \mu && \text{(grand mean)}\\
& + \mu_p - \mu && \text{(person effect)}\\
& + \mu_t - \mu && \text{(task effect)}\\
& + \mu_o - \mu && \text{(observer effect)}\\
& + \mu_{pt} - \mu_p - \mu_t + \mu && \text{(person} \times \text{task effect)}\\
& + \mu_{po} - \mu_p - \mu_o + \mu && \text{(person} \times \text{observer effect)}\\
& + \mu_{to} - \mu_t - \mu_o + \mu && \text{(task} \times \text{observer effect)}\\
& + X_{pto} - \mu_{pt} - \mu_{po} - \mu_{to} + \mu_p && \\
& + \mu_t + \mu_o - \mu && \text{(residual)},
\end{aligned} \tag{33}$$

Table 7
Expected mean square equations for a random-effects, two-facet, crossed $p \times t \times o$ design

Source of variation	Variance component	Expected mean square equation
Persons (p)	σ_p^2	$EMS_p = \sigma_{pto,e}^2 + n_o\sigma_{pt}^2 + n_t\sigma_{po}^2 + n_t n_o \sigma_p^2$
Tasks (t)	σ_t^2	$EMS_t = \sigma_{pto,e}^2 + n_p\sigma_{to}^2 + n_o\sigma_{pt}^2 + n_p n_o \sigma_t^2$
Observers (o)	σ_o^2	$EMS_o = \sigma_{pto,e}^2 + n_p\sigma_{to}^2 + n_t\sigma_{po}^2 + n_p n_t \sigma_o^2$
pt	σ_{pt}^2	$EMS_{pt} = \sigma_{pto,e}^2 + n_o\sigma_{pt}^2$
po	σ_{po}^2	$EMS_{po} = \sigma_{pto,e}^2 + n_t\sigma_{po}^2$
to	σ_{to}^2	$EMS_{to} = \sigma_{pto,e}^2 + n_p\sigma_{to}^2$
pto, e	$\sigma_{pto,e}^2$	$EMS_{pto,e} = \sigma_{pto,e}^2$

where $\mu = \mathrm{E}_p\mathrm{E}_t\mathrm{E}_o X_{pto}$ and $\mu_p = \mathrm{E}_t\mathrm{E}_o X_{pto}$ and other terms in (33) are defined analogously. The collection of observed scores, X_{pto} has a variance $\mathrm{E}_p\mathrm{E}_t\mathrm{E}_o(X_{pto} - \mu)^2 = \sigma_{X_{pto}}^2$, which equals the sum of the variance components:

$$\sigma_{X_{pto}}^2 = \sigma_p^2 + \sigma_t^2 + \sigma_o^2 + \sigma_{pt}^2 + \sigma_{po}^2 + \sigma_{to}^2 + \sigma_{pto,e}^2. \tag{34}$$

Estimates of the variance components in (34) can be obtained by setting the expected mean squares equal to the observed mean squares and solving the set of simultaneous equations shown in Table 7.

In a fully crossed $p \times T \times O$ decision (D) study, error for relative decisions is defined as:

$$\delta_{pTO} = (X_{pTO} - \mu_{TO}) - (\mu_p - \mu), \tag{35}$$

where $\mu_p = \mathrm{E}_T\mathrm{E}_O X_{pTO}$ and $\mu_{TO} = \mathrm{E}_p X_{pTO}$. The variance of the errors for relative decisions is:

$$\sigma_\delta^2 = \mathrm{E}_p\mathrm{E}_T\mathrm{E}_O\delta_{pTO}^2 = \sigma_{pT}^2 + \sigma_{pO}^2 + \sigma_{pTO,e}^2 = \frac{\sigma_{pt}^2}{n'_t} + \frac{\sigma_{po}^2}{n'_o} + \frac{\sigma_{pto,e}^2}{n'_t n'_o}. \tag{36}$$

The "main effects" of observer and task do not enter into error for relative decisions because, for example, all people are observed on the same tasks so any difference in task difficulty affects all persons and does not change their relative standing.

For absolute decisions, the error in a random-effects $p \times T \times O$ design is defined as:

$$\Delta_{pTO} \equiv X_{pTO} - \mu_p, \tag{37}$$

and the variance of the errors is:

$$\begin{aligned}\sigma_\Delta^2 &= \mathrm{E}_p\mathrm{E}_T\mathrm{E}_O\Delta_{pTO}^2 = \sigma_T^2 + \sigma_O^2 + \sigma_{pT}^2 + \sigma_{pO}^2 + \sigma_{TO}^2 + \sigma_{pTO,e}^2 \\ &= \frac{\sigma_t^2}{n'_t} + \frac{\sigma_o^2}{n'_o} + \frac{\sigma_{pt}^2}{n'_t} + \frac{\sigma_{po}^2}{n'_o} + \frac{\sigma_{to}^2}{n'_t n'_o} + \frac{\sigma_{pto,e}^2}{n'_t n'_o}.\end{aligned} \tag{38}$$

With absolute decisions, the main effects of observers and tasks – the strictness of the observers or the difficulty of the tasks – do influence absolute performance and so are included in the definition of measurement error.

Table 8
Possible random-effects D-study designs from a random-effects, two-facet G-study design

G-study design	D-study design		
$p \times t \times o$	$p \times T \times O$		
	$p \times (T : O)$	or	$p \times (O : T)$
	$(O : p) \times T$	or	$(T : p) \times O$
	$O : (p \times T)$	or	$T : (O \times p)$
	$(O \times T) : p$		
	$O : T : p$	or	$T : O : p$
$p \times (o : t)$	$p \times (O : T)$		
	$O : T : p$		
$p \times (t : o)$	$p \times (T : O)$		
	$T : O : p$		
$o : (p \times t)$	$O : (p \times T)$		
	$O : T : p$		
$t : (p \times o)$	$T : (p \times O)$		
	$T : O : p$		
$(o \times t) : p$	$(O \times T) : p$		
	$O : T : p$	or	$T : O : p$

With these definitions of error variance, the generalizability coefficient ($E\rho^2$) and phi coefficient (Φ) are defined as in (20) and (23).

The results of a two-facet crossed G-study can be used to estimate error variance and generalizability and phi coefficients for a wide range of D-study designs. Any G-study can be used to estimate the effects in a D-study design with the same or more nesting than in the G-study design. Table 8 lists the possible two-facet G- and D-study designs for which p is the object of measurement and is not nested within a facet.

A decision-maker who uses a crossed $p \times t \times o$ generalizability study design may, for example, choose to use a $p \times (T : O)$ design in the decision study for convenience. In this D-study design, each observer rates different tasks and the decision-maker can sample a larger number of tasks in the study than would be the case if tasks and observers were crossed (e.g., if two observers rate performance on 5 tasks, then 10 tasks can be represented). Using the variance components from the $p \times t \times o$ G-study, error variances for relative and absolute decisions for a $p \times (T : O)$ D-study design are:

$$\sigma_\delta^2 = \sigma_{pO}^2 + \sigma_{pT:O}^2 = \frac{\sigma_{po}^2}{n'_o} + \frac{\sigma_{pt,pto,e}^2}{n'_o n'_t}, \tag{39}$$

$$\sigma_\Delta^2 = \sigma_O^2 + \sigma_{pO}^2 + \sigma_{T:O}^2 + \sigma_{pT:O}^2 = \frac{\sigma_o^2}{n'_o} + \frac{\sigma_{po}^2}{n'_o} + \frac{\sigma_{t,to}^2}{n'_t n'_o} + \frac{\sigma_{pt,pto,e}^2}{n'_t n'_o}, \tag{40}$$

where $\sigma_{t,to}^2$ is the sum of σ_t^2 and σ_{to}^2 from the $p \times t \times o$ G-study, and $\sigma_{pt,pto,e}^2$ is the sum of σ_{pt}^2 and $\sigma_{pto,e}^2$. Even if the number of observers and the number of tasks per observer are held constant, nesting tasks within observers will result in improved precision because a larger number of tasks is sampled. Thus, the variance components

σ_t^2 and σ_{pt}^2 are divided by $n_t'n_o'$ instead of n_t', for example, as illustrated in Example 1 below.

EXAMPLE 1 (Generalizability study of job performance measurements of Navy machinist mates). Navy machinist mates ($n_p = 26$) aboard ship were observed by two examiners as they carried out 11 tasks in the engine room (concerning knowledge of gauges, equipment operation, casualty control; Webb et al., 1989). The task score was the proportion of steps in a task (ranging from 12 to 49 steps per task) that were judged to be completed correctly. Two examiners observed each machinist mate on all tasks, so the design of the generalizability study was persons (machinist mates) crossed with tasks and observers ($p \times t \times o$). Tasks and observers were considered random facets.

Table 9 gives the estimated variance components from the G-study, and estimated error variances and generalizability and dependability coefficients for fully crossed, $p \times T \times O$ D-study designs with different numbers of tasks (n_t') and observers (n_o'). Observer was a negligible source of variation. The variance components for the main effect of observer, the interaction between persons and observers, and the interaction between observers and tasks were all zero or near zero. Observers not only rank ordered machinist mates similarly ($\hat{\sigma}_{po}^2 = 0$), but they also gave nearly identical scores, on average, to the machinist mates ($\hat{\sigma}_o^2 = 0$). The task, on the other hand, was a substantial source of variation. The relatively large component for the main effect of task ($\hat{\sigma}_t^2 = 0.00970$) shows that tasks differed in difficulty level. Furthermore, the very large variance component for the interaction between persons and tasks ($\hat{\sigma}_{pt}^2 = 0.02584$; 60% of the total variation) suggests that the relative standing of persons differed across tasks (cf. Shavelson et al., 1993; Shavelson et al., 1999).

The estimated variance components from the G-study suggest that a single observer would probably provide dependable ratings but that multiple tasks are needed to represent job requirements. As can be seen in Table 9, averaging over 11 tasks (the same number as used in the G-study) with a single observer yields moderate estimated generalizability and dependability coefficients ($\mathrm{E}\hat{\rho}^2 = 0.716$; $\hat{\Phi} = 0.649$). Using two observers has no appreciable effect on the results. A very large number of tasks, of questionable feasibility, would be needed to obtain a reasonable level of generalizability (with 17 tasks and one observer, $\mathrm{E}\hat{\rho}^2 = 0.795$; $\hat{\Phi} = 0.742$). Even this level of dependability may be insufficient for individual decision making, however. Using the square root of absolute error variance, the standard error of measurement (SEM) for absolute decisions, to calculate a confidence interval for machinist mates' universe scores shows that observed scores are likely to vary considerably over 17-task samples. For 17 tasks and one observer, $\hat{\sigma}_\Delta = 0.047$; a 95% confidence interval is $X_{pTO} \pm 0.092$. Compared to the possible range of performance scores (0 to 1), the width of this interval is substantial.

For convenience, the decision maker may elect to use a $p \times (T : O)$ decision study design, in which machinist mates perform all tasks, but each examiner is responsible for judging machinist mates on only a subset of tasks. For example, consider a $p \times (T : O)$ decision study in which two examiners each observe machinist mates performing 11

Table 9
Generalizability study of job performance scores (2-facet, random effects $p \times t \times o$ design)

Source of variation	G-study $\hat{\sigma}^2$	Size of decision study		
		$n'_t = 11$ $n'_o = 1$	$n'_t = 11$ $n'_o = 2$	$n'_t = 17$ $n'_o = 1$
Estimated variance components				
Person (p)	0.00626	0.00626	0.00626	0.00626
Task (t)	0.00970	0.00088	0.00088	0.00057
Observer (o)	0.00000[a]	0.00000	0.00000	0.00000
pt	0.02584	0.00235	0.00235	0.00152
po	0.00000[a]	0.00000	0.00000	0.00000
to	0.00003	0.00000	0.00000	0.00000
pto, e	0.00146	0.00013	0.00007	0.00009
Error variances				
$\hat{\sigma}^2_\delta$		0.00248	0.00242	0.00161
$\hat{\sigma}^2_\Delta$		0.00339	0.00330	0.00218
Coefficients				
$\text{E}\hat{\rho}^2$		0.716	0.721	0.795
$\hat{\Phi}$		0.649	0.655	0.742

Note: from Webb, N.M., Shavelson, R.J., Kim, K.S., Chen, Z. (1989). Reliability (generalizability) of job performance measurements: Navy machinists mates. *Military Psychology* **1** (2), 99. © 1989 Lawrence Erlbaum Associates, Inc. Adapted with permission.
[a]Negative estimated variance component set equal to zero.

tasks and the tasks are different for each examiner. The partially nested $p \times (T : O)$ design with $n'_t = 11$ and $n'_o = 2$, then, actually involves using 22 tasks. This design yields $\hat{\sigma}^2_\delta = 0.00124$ and $\hat{\sigma}^2_\Delta = 0.00168$ and $\text{E}\hat{\rho}^2 = 0.835$ and $\hat{\Phi} = 0.788$. Comparing these values to those for the crossed D-study with the same number of observations ($n'_t = 11$ and $n'_o = 2$) shows smaller error variances and larger generalizability coefficients for the partially nested D-study than for the crossed D-study. The error variances in the partially nested design are reduced by virtue of the larger denominators. Because $\hat{\sigma}^2_t$ is confounded with $\hat{\sigma}^2_{to}$ in the partially nested design, it is divided by the number of tasks *and* the number of observers ($n'_t n'_o$). Similarly, $\hat{\sigma}^2_{pt}$ is divided by $n'_t n'_o$ by virtue of its being confounded with $\hat{\sigma}^2_{pto,e}$. The partially nested $p \times (T : O)$ design has an advantage over the fully crossed $p \times T \times O$ design in terms of estimated generalizability as well as convenience.

3.3. Random and fixed facets

G theory is essentially a random effects theory. Typically a random facet is created by randomly sampling levels of a facet (e.g., tasks from a job in observations of job performance). When the levels of a facet have not been sampled randomly from the universe of admissible observations but the intended universe of generalization is infinitely large, the concept of exchangeability may be invoked to consider the facet as random (de Finetti, 1937, 1964). Formally,

> *The random variables $X_1, \ldots, X_n$ are exchangeable if the $n!$ permutations $(X_{k1}, \ldots, X_{kn})$ have the same n-dimensional probability distribution. The variables of an infinite sequence X_n are exchangeable if $X_1, \ldots, X_n$ are exchangeable for each n* (Feller, 1966, p. 225).

Even if conditions of a facet have not been sampled randomly, the facet may be considered to be random if conditions not observed in the G-study are exchangeable with the observed conditions.

A fixed facet (cf. fixed factor in analysis of variance) arises when the decision maker: (a) purposely selects certain conditions and is not interested in generalizing beyond them, (b) finds it unreasonable to generalize beyond the conditions observed, or (c) when the entire universe of conditions is small and all conditions are included in the measurement design. G theory typically treats fixed facets by averaging over the conditions of the fixed facet and examining the generalizability of the average over the random facets (Brennan, 2001; Cronbach et al., 1972). When it does not make conceptual sense to average over the conditions of a fixed facet, a separate G-study may be conducted within each condition of the fixed facet (Shavelson and Webb, 1991) or a full multivariate analysis may be performed with the conditions of the fixed facet comprising a vector of dependent variables (Brennan, 2001; see below).

G theory recognizes that the universe of admissible observations in a G-study may be broader than the universe of generalization of interest in a D-study. The decision maker may reduce the levels of a facet (creating a fixed facet), select (and thereby control) one level of a facet, or ignore a facet. A facet is fixed in a D-study when $n' = N' < \infty$, where n' is the number of levels for a facet in the D-study and N' is the total number of levels for a facet in the universe of generalization. Consider a person by rater by subject matter ($p \times r \times s$) random-effects G-study in which raters (r) observe the behavior of teachers (p) while teaching multiple subject matters (s). The universe of admissible observations is defined by facets r and s of infinite size. Fixing the subject matter facet s in the D-study and averaging over the n_s conditions of facet s in the G-study ($n_s = n'_s$) yields the following universe-score variance:

$$\sigma_\tau^2 = \sigma_p^2 + \sigma_{pS}^2 = \sigma_p^2 + \frac{\sigma_{ps}^2}{n'_s}, \tag{41}$$

where σ_τ^2 denotes universe-score variance in generic terms. When facet s is fixed, the universe score is based on a person's average score over the finite subset of levels of facet s, so the generic universe-score variance in (41) is the variance over persons' mean scores for *just those levels*. Any person effects specific to those particular levels therefore become part of the universe-score variance. Hence, (41) includes σ_{pS}^2 as well as σ_p^2. $\hat{\sigma}_\tau^2$ is an unbiased estimate of universe-score variance for the mixed model only when the same levels of facet s are used in the G and D studies (Brennan, 2001).

The relative and absolute error variances, respectively, are:

$$\sigma_\delta^2 = \sigma_{pR}^2 + \sigma_{pRS}^2 = \frac{\sigma_{pr}^2}{n'_r} + \frac{\sigma_{prs,e}^2}{n'_r n'_s}, \tag{42}$$

$$\sigma_\Delta^2 = \sigma_R^2 + \sigma_{pR}^2 + \sigma_{RS}^2 + \sigma_{pRS}^2 = \frac{\sigma_r^2}{n'_r} + \frac{\sigma_{pr}^2}{n'_r} + \frac{\sigma_{rs}^2}{n'_r n'_s} + \frac{\sigma_{prs,e}^2}{n'_r n'_s}. \tag{43}$$

And the generalizability coefficient and index of dependability, respectively, are:

$$\mathrm{E}\rho^2 = \frac{\sigma_p^2 + \frac{\sigma_{ps}^2}{n'_t}}{\sigma_p^2 + \frac{\sigma_{ps}^2}{n'_s} + \frac{\sigma_{pr}^2}{n'_r} + \frac{\sigma_{prs,e}^2}{n'_r n'_s}}, \tag{44}$$

$$\Phi = \frac{\sigma_p^2 + \frac{\sigma_{ps}^2}{n'_s}}{\sigma_p^2 + \frac{\sigma_{ps}^2}{n'_s} + \frac{\sigma_r^2}{n'_r} + \frac{\sigma_{pr}^2}{n'_r} + \frac{\sigma_{rs}^2}{n'_r n'_s} + \frac{\sigma_{prs,e}^2}{n'_r n'_s}}. \tag{45}$$

EXAMPLE 2 (Generalizability study of teacher behavior). Elementary school teachers were observed by three raters in two subject matters: reading and mathematics (cf. Erlich and Shavelson (1978)). Teachers (p) and raters (r) are considered random. Subject matter (s) is considered fixed because these subject matters were specifically selected to represent standard curriculum areas and it is probably not reasonable to generalize from teachers' behavior (for example, the number of high-level questions teachers ask) when teaching reading and mathematics to their behavior when teaching other subject matters. We consider a crossed design in which three raters observed behavior of all teachers in both reading and mathematics ($p \times r \times s$).

Table 10 gives the estimated variance components from the G-study treating raters (r) and subject matter (s) as random. With subject matter treated as fixed, and substituting values from Table 10 into (41), $\hat{\sigma}_\tau^2 = (1.17262+2.34226/2) = 2.34375$ for the universe score based on a teacher's average score over reading and mathematics. Substituting values from Table 10 into (42)–(45) yields estimated error variances and generalizability and phi coefficients for different numbers of raters in the decision study (n'_r). Results for one and two raters appear in Table 11. When averaging over teacher behavior in reading and mathematics, two raters are needed to reach an estimated level of generalizability of 0.80.

Because teacher behavior varied substantially across reading and mathematics (see, especially, the relatively large $\hat{\sigma}_{ps}^2$ in Table 10, showing that relative standing of teachers differed substantially from reading to mathematics), it may be reasonable to analyze teacher behavior for each subject matter separately, as well as averaging over them. Table 12 gives the results for decision studies of teacher behavior separately for reading and mathematics. When subject matters are analyzed separately, three raters are needed to reach an estimated level of generalizability of 0.80 for reading, but only one rater is needed to reach that level of generalizability for mathematics.

Table 10
Generalizability study of teacher behavior ratings (2-facet $p \times r \times s$ design with all facets treated as random)

Source of variation	$\hat{\sigma}^2$
Teacher (p)	1.17262
Rater (r)	0.01488
Subject matter (s)	0.15774
pr	0.50595
ps	2.34226
rs	0.03869
prs, e	0.87798

Table 11
Decision study error variances and coefficients for study of teacher behavior (2-facet, mixed-effects $p \times r \times s$ design with s fixed)

	Averaging over subject matters	
	$n'_r = 1$	$n'_r = 2$
Universe-score variance		
$\hat{\sigma}^2_\tau$	2.34375	2.34375
Error variances		
$\hat{\sigma}^2_\delta$	0.94494	0.47247
$\hat{\sigma}^2_\Delta$	0.97917	0.48959
Coefficients		
$E\hat{\rho}^2$	0.713	0.832
$\hat{\Phi}$	0.705	0.827

3.4. Symmetry

The purpose of psychological measurement has typically been to differentiate individuals. Recognizing that the focus of measurement may change depending on a particular decision-maker's purpose, Cardinet et al. (1976, 1981) spoke of the *principle of symmetry*: "The principle of symmetry of the data is simply an affirmation that each of the facets of a factorial design can be selected as an object of study, and that the operations defined for one facet can be transposed in the study of another facet" (Cardinet et al., 1981, p. 184). In a persons (p) $\times$ items (i) $\times$ occasions (o) design, while persons may be the focus of measurement for evaluators wishing to make dependable judgments about persons' performance, items may be the focus for curriculum developers wishing to calibrate items for use in item banks (e.g., Wood, 1976). In the latter case, individual differences among persons represent error variation, rather than universe-score variation, in the measurement.

Table 12
Generalizability study of teacher behavior for each subject matter (one-facet, random-effects $p \times r$ design for each subject matter)

Source of variation	Reading			Mathematics	
	$n'_r = 1$	$n'_r = 2$	$n'_r = 3$	$n'_r = 1$	$n'_r = 2$
Estimated variance components					
Person (p)	3.14286	3.14286	3.14286	3.88691	3.88691
Rater (r)	0.15476	0.07738	0.05159	0[a]	0
pr, e	2.26191	1.13096	0.75397	0.50595	0.25298
Error variances					
$\hat{\sigma}^2_\delta$	2.26191	1.13096	0.75397	0.50595	0.25298
$\hat{\sigma}^2_\Delta$	2.41667	1.20834	0.80556	0.50595	0.25298
Coefficients					
$E\hat{\rho}^2$	0.581	0.735	0.807	0.885	0.939
$\hat{\Phi}$	0.565	0.722	0.796	0.885	0.939

Note: from Shavelson, R.J., Webb, N.M. (1991). *Generalizability Theory: A Primer*. Sage, Newbury Park, CA. © 2004 Sage Publications, Inc. Adapted by permission.
[a]Estimated variance component set equal to zero.

The principle of symmetry leads to the possibility of multifaceted populations. (Cardinet et al., 1976, 1981). Educational evaluators may be interested in scholastic achievement of classes, schools, or districts, or in comparisons across years. Or the focus may be on items corresponding to different content units in which the universe-score of interest is that of items (i) nested within content units (c).

Another instance of multifaceted populations is stratified objects of measurement (Brennan, 2001). The population of interest may be persons nested within, say, geographic region, gender, or socio-economic status. If variation due to these attributes of persons is negligible, the decision maker can assume a homogeneous population and so reduce the design of the D-study. If the variation is large, the decision maker may calculate separate variance component estimates for each subgroup.

3.4.1. *Generalizability of group means*

A common area of application of the principle of symmetry is the generalizability of group means (e.g., scholastic achievement of classes, schools, or school districts; student evaluations of teachers; see Kane et al., 1976; Kane and Brennan, 1977, for a detailed explication of the generalizability of class means). In these cases, the group, not the individual student, is the object of study and the sampling of students within groups gives rise to errors of measurement. In a fully random design with persons (p) nested within groups (g) and crossed with items (i), denoted as $(p : g) \times i$, the generalizability coefficient with groups as the object of measurement and when generalizing over persons and items is:

$$\mathrm{E}\rho_g^2 = \frac{\sigma_g^2}{\sigma_g^2 + \frac{\sigma_{p:g}^2}{n'_p} + \frac{\sigma_{gi}^2}{n'_i} + \frac{\sigma_{pi:g,e}^2}{n'_p n'_i}}. \tag{46}$$

In (46), variation among persons within groups ($\sigma_{p:g}^2$) contributes to error variation but not universe-score variation. Designs that fail to recognize explicitly persons as a source of variation (i.e., a groups × items design) would result in misestimation of generalizability due to the confounding of σ_p^2 and σ_{pg}^2.

In some cases, decision makers may be interested in restricted universes of generalization in which one or more sources of variation are fixed. For example, persons may be treated as a fixed facet. In industrial settings where supervisors' ratings of employees (p) in organizations (g) are gathered, employees may be considered as a fixed facet within some period of time. Similarly, in a study of school performance, interest may lie in estimating the performance of the student body during a particular year (Cronbach et al., 1997). For the $(p : g) \times i$ design in which persons are fixed (n_p persons per group) and items are random, the generalizability coefficient for groups is:

$$\mathrm{E}\rho_g^2 \text{ with } p \text{ fixed} = \frac{\sigma_g^2 + \frac{\sigma_{p:g}^2}{n_p}}{\sigma_g^2 + \frac{\sigma_{p:g}^2}{n_p} + \frac{\sigma_{gi}^2}{n'_i} + \frac{\sigma_{pi:g,e}^2}{n_p n'_i}}. \tag{47}$$

Note, however, that Cronbach et al. (1997) caution that in the typical case, when the intended inference concerns the performance of the school with potentially different students, persons (students) should be treated as a random facet, even if all students present in the school are tested.

Alternatively, the focus may be on a fixed set of items. For the $(p : g) \times i$ design in which items are fixed (a fixed set of n_i items) and persons are random, the generalizability coefficient for groups is:

$$\mathrm{E}\rho_g^2 \text{ with } i \text{ fixed} = \frac{\sigma_g^2 + \frac{\sigma_{gi}^2}{n_i}}{\sigma_g^2 + \frac{\sigma_{gi}^2}{n_i} + \frac{\sigma_{p:g}^2}{n_p} + \frac{\sigma_{pi:g,e}^2}{n'_p n_i}}. \tag{48}$$

Cronbach et al. (1997) describe more complex multilevel designs with persons nested within classes, and classes nested within schools. They consider estimation of variance components and school-level standard errors when generalizing to infinite and finite populations. As an alternative to estimating standard errors for school means, Yen (1997) showed how to estimate standard errors for the percent of students in a school above a cutpoint (percent above cutpoint, or PAC) and compared the results for a variety of generalizability study models.

Challenging conventional wisdom that reliability for groups is larger than reliability for persons, Brennan (1995) provides a number of conditions in which the generalizability of group means may be *less* than the generalizability of scores for persons. For the fully random $(p : g) \times i$ design in which *person* is the object of measurement, the generalizability coefficient is:

$$\mathrm{E}\rho_p^2 = \frac{\sigma_g^2 + \sigma_{p:g}^2}{\sigma_g^2 + \sigma_{p:g}^2 + \frac{\sigma_{gi}^2}{n_i'} + \frac{\sigma_{pi:g,e}^2}{n_i'}}. \tag{49}$$

As described by Brennan (1995, p. 393), $\mathrm{E}\rho_g^2$ in (49) will tend to be less than $\mathrm{E}\rho_p^2$ when group means are similar (small σ_g^2), the number of items (n_i') is large, the number of persons within each group (n_p') is small, or variation among persons within groups ($\sigma_{p:g}^2$) is relatively large. Brennan (1995) also discusses conditions under which the generalizability of group means may be less than the generalizability of persons for restricted universes (e.g., persons are fixed and items are random).

3.5. *Multivariate generalizability*

For behavioral measurements involving multiple scores describing individuals' personality, aptitudes, skills, or performance, multivariate generalizability can be used to (a) estimate the reliability of difference scores, observable correlations, or universe-score and error correlations for various D-study designs and sample sizes (Brennan, 2001), (b) estimate the reliability of a profile of scores using multiple regression of universe scores on the observed scores in the profile (Brennan, 2001; Cronbach et al., 1972), or (c) produce a composite of scores with maximum generalizability (Shavelson and Webb, 1981). For all of these purposes, multivariate G theory decomposes both observed variances and covariances into components. Consider a one-facet, crossed $p \times r$ design with teacher behavior divided into two dependent variables – behavior in reading and behavior in mathematics – and the same raters observing teacher behavior in both subject matters. The observed scores for the reading and mathematics variables for person p observed under condition r can be denoted as ${}_1(X_{pr})$ and ${}_2(X_{pr})$, respectively. The variances and covariance of observed scores, $\sigma^2_{1(X_{pr})}$ and $\sigma^2_{2(X_{pr})}$ are decomposed as follows:

$$\begin{aligned}
&\begin{bmatrix} \sigma^2_{1(X_{pr})} & \sigma_{1(X_{pr}),2(X_{pr})} \\ \sigma_{1(X_{pr}),2(X_{pr})} & \sigma^2_{2(X_{pr})} \end{bmatrix} \\
&\quad = \begin{bmatrix} \sigma^2_{1p} & \sigma_{1p,2p} \\ \sigma_{1p,2p} & \sigma^2_{2p} \end{bmatrix} \quad \text{(person)} \\
&\quad + \begin{bmatrix} \sigma^2_{1r} & \sigma_{1r,2r} \\ \sigma_{1r,2r} & \sigma^2_{2r} \end{bmatrix} \quad \text{(rater)} \\
&\quad + \begin{bmatrix} \sigma^2_{1(pr,e)} & \sigma_{1(pr,e),2(pr,e)} \\ \sigma_{1(pr,e),2(pr,e)} & \sigma^2_{2(pr,e)} \end{bmatrix} \quad \text{(residual).}
\end{aligned} \tag{50}$$

In (50) the term $\sigma_{1p,2p}$ is the covariance between universe scores for behavior in reading and behavior in mathematics. Dividing this covariance by the product of the square root of the corresponding variances ($\sigma_{1p} * \sigma_{2p}$) produces the disattenuated correlation between teacher behavior in reading and mathematics. The remaining covariance terms in (50) are error covariance components. The term $\sigma_{1r,2r}$, for example, is the covariance between teacher behavior scores in reading and mathematics due to consistency of rater effects across the two subject matters. Estimates of the covariance components can be obtained by solving the expected mean product equations in analogous fashion to their univariate counterparts.

An important aspect of the development of multivariate G theory is the distinction between linked and unlinked conditions. The expected values of error covariance components are nonzero when levels are linked or jointly sampled (e.g., the same raters are used to produce scores on the two variables in the teacher behavior profile). The expected values of error covariance components are zero when conditions for observing different variables are unlinked, that is, selected independently (e.g., the raters used to obtain scores on one variable in a profile, teaching behavior in reading, are selected independently of the raters used to obtain scores on another variable, teaching behavior in mathematics).

Joe and Woodward (1976) presented a generalizability coefficient for a multivariate composite that maximizes the ratio of universe-score variation to universe-score plus error variation by using statistically derived weights for each dependent variable (e.g., teacher behavior in reading and mathematics). For a one-facet, crossed $p \times r$ design, their multivariate analogue to a univariate generalizability coefficient is:

$$\mathrm{E}\rho^2 = \frac{\boldsymbol{a}'\boldsymbol{V}_p\boldsymbol{a}}{\boldsymbol{a}'\boldsymbol{V}_p\boldsymbol{a} + \frac{\boldsymbol{a}'\boldsymbol{V}_{pr,e}\boldsymbol{a}}{n'_r}}, \tag{51}$$

where $\boldsymbol{V}$ is a matrix of variance and covariance components, n'_r is the number of conditions r in the D-study, and $\boldsymbol{a}$ is the vector of weights that maximizes the ratio of between-person to between-person plus within-person variance component matrices. $\mathrm{E}\hat{\rho}^2$ and $\boldsymbol{a}$ can be obtained by solving the following set of equations:

$$\left[\boldsymbol{V}_p - \rho_k^2(\boldsymbol{V}_p + \boldsymbol{V}_\delta)\right]\boldsymbol{a} = 0, \tag{52}$$

where the subscript k refers to the m ($k = 1, \ldots, m$) roots of (52) and $\boldsymbol{V}_\delta$ is the multivariate analogue to σ_δ^2. For each multivariate generalizability coefficient corresponding to a characteristic root in (52), there is a set of canonical coefficients that defines a composite of the scores. By definition, the first composite is the most reliable while the last composite is the least reliable. Because the data, not the investigators, define the composites of maximum generalizability, this approach would be used in an exploratory, rather than confirmatory, context.

Alternatives to maximizing the generalizability of a composite are to determine variable weights on the basis of expert judgment or use weights derived from a confirmatory factor analysis (Marcoulides, 1994; Short et al., 1986). The most straightforward approach to estimating the generalizability of a composite with weights $\boldsymbol{a}$ is a univariate rather than multivariate analysis. The results of a univariate generalizability analysis would be identical to those of a multivariate generalizability analysis in which the weights of the composite define the $\boldsymbol{a}$ vector in (52).

EXAMPLE 3 (Multivariate generalizability study of teacher behavior). In Example 2, we presented the results of a univariate generalizability study of teacher behavior in

Table 13
Estimated variance and covariance components for multivariate generalizability study of teacher behavior ratings (1-facet, random-effects $p \times r$ design)

Source of variation		(1) Reading	(2) Mathematics
Persons (p)	(1)	3.14286	1.17262
	(2)	1.17262	3.88691
Raters (r)	(1)	0.15476	0.01488
	(2)	0.01488	0[a]
pr, e	(1)	2.26191	0.50595[b]
	(2)	0.50595	0.50595[b]

[a]Negative estimated variance component set equal to zero.
[b]The equal values for these components are correct.

which teachers (p) and raters (r) were considered random and subject matter (s) was considered fixed. Here we present the results of a multivariate generalizability analysis of the same data with teacher behavior in reading and mathematics considered as two dependent variables in a multivariate profile. The design is persons (p) crossed with raters (r).

Table 13 gives the estimated variance and covariance components from the multivariate generalizability study. The estimated variance components (on the main diagonal) are the same as those produced by the univariate analyses of teacher behavior in reading and mathematics (see Table 12). The estimated covariance components provide new information. The positive, nonzero estimated covariance for persons ($\hat{\sigma}_{1p,2p} = 1.17262$) indicates that raters found consistencies in teacher behavior across reading and mathematics. The estimated variance and covariance component for persons can be used to estimate the correlation between teacher behavior in reading and mathematics corrected for attenuation:

$$\frac{\hat{\sigma}_{1p,2p}}{\sqrt{\hat{\sigma}^2_{1p}\hat{\sigma}^2_{2p}}}. \tag{53}$$

Substituting values from Table 13 gives a corrected correlation equal to 0.34, indicating that teachers' universe scores for behavior are modestly correlated in reading and mathematics.

The very small estimated covariance component for raters ($\hat{\sigma}_{1r,2r} = 0.01488$) shows that any disagreements among raters in the overall frequency of teacher behavior observed did not carry over from reading to mathematics. The substantial estimated covariance component for the residual ($\hat{\sigma}_{1(pr,e),2(pr,e)} = 0.50595$) suggests that disagreements among raters about the relative standing of teachers are similar in reading and mathematics; unexplained factors that contribute to the variation of behavior ratings within a subject matter contribute to the covariation between behavior ratings in reading and mathematics, or both.

In this example, a composite that weights mathematics more than reading will have higher estimated generalizability than will a composite that weights the two subject matters equally; compare $E\hat{\rho}^2 = 0.902$ for a composite with weights of 0.25 for reading and 0.75 for math (2 raters in the D-study) with $E\hat{\rho}^2 = 0.832$ for a composite in which reading and mathematics are equally weighted (from Table 11).

3.6. *Additional issues in generalizability theory*

3.6.1. *Variance component estimates*

We treat three issues of concern when estimating variance components: (1) the variability of variance-component estimates, (2) negative estimated variance components, and (3) complexity of estimation in unbalanced designs.

Variability of variance-component estimates

The first concern is that estimates of variance components may be unstable with usual sample sizes (Cronbach et al., 1972). Assuming that mean squares are independent and score effects have a multivariate normal distribution, the sampling variance of an estimated variance component ($\hat{\sigma}^2$) is

$$\operatorname{var}(\hat{\sigma}^2) = \frac{2}{c^2} \sum_q \operatorname{var} MS_q = \frac{2}{c^2} \sum_q \frac{\mathrm{E}MS_q^2}{df_q}, \tag{54}$$

where c is the constant associated with the estimated variance component, $\mathrm{E}MS_q$ is the expected value of the mean square, MS_q, and df_q is the degrees of freedom associated with MS_q (Searle, 1971; Smith, 1978). In (54), the summation is taken over all MS_q that are used to obtain the ANOVA estimate of the variance component.

The variances of estimated variance components will be larger with smaller numbers of observations (reflected in smaller df_q). Moreover, the more mean squares that are involved in estimating variance components, the larger the estimated variances are likely to be. In a two-facet, random-effects $p \times t \times o$ design, for example, σ_p^2 is estimated by $(MS_p - MS_{pt} - MS_{po} + MS_{pto,e})/(n_t n_o)$, c in (54) refers to $n_t n_o$, and MS_q refers to MS_p, MS_{pt}, MS_{po}, and $MS_{pto,e}$. Following Smith (1978), the variance of $\hat{\sigma}_p^2$ is:

$$\begin{aligned}
\operatorname{var}(\hat{\sigma}_p^2) = \frac{2}{n_p - 1}\Bigg[& \left(\sigma_p^2 + \frac{\sigma_{pt}^2}{n_t} + \frac{\sigma_{po}^2}{n_o} + \frac{\sigma_{pto,e}^2}{n_t n_o}\right)^2 \\
& + \frac{1}{n_t - 1}\left(\frac{\sigma_{pt}^2}{n_t} + \frac{\sigma_{pto,e}^2}{n_t n_o}\right)^2 \\
& + \frac{1}{n_o - 1}\left(\frac{\sigma_{po}^2}{n_o} + \frac{\sigma_{pto,e}^2}{n_t n_o}\right)^2 \\
& + \frac{1}{(n_t - 1)(n_o - 1)}\left(\frac{\sigma_{pto,e}^2}{n_t n_o}\right)^2\Bigg].
\end{aligned} \tag{55}$$

In contrast, the variance of $\hat{\sigma}^2_{pto,e}$ is based on only a single mean square:

$$\mathrm{var}\left(\hat{\sigma}^2_{pto,e}\right) = \frac{2}{(n_p - 1)(n_t - 1)(n_o - 1)}\sigma^4_{pto,e}. \tag{56}$$

For the estimated variance components in Table 9 (generalizability study of job performance measurements of Navy machinist mates with a $p \times t \times o$ design), compare estimated standard errors produced by replacing variance components in (55) and (56) with ANOVA estimates of them: $\hat{\sigma}(\hat{\sigma}^2_p) = 0.00246$ (for $\hat{\sigma}^2_p = 0.00626$) and $\hat{\sigma}(\hat{\sigma}^2_{pto,e}) = 0.00013$ (for $\hat{\sigma}^2_{pto,e} = 0.00146$). (The preceding estimates are biased. Unbiased estimates of the standard errors, obtained by replacing df_q with $df_q + 2$ in (54), yield 0.00237 and 0.00013 for $\hat{\sigma}(\hat{\sigma}^2_p)$ and $\hat{\sigma}(\hat{\sigma}^2_{pto,e})$, respectively.)

Smith (1981) also pointed out that the stability of variance component estimates varies across design complexity (the number of facets) and design configurations (the extent of nesting in the design). As an example of the latter, assuming the same number of observations, the sampling variance of $\hat{\sigma}^2_p$ would be smaller in a partially nested $p \times (t : o)$ design than in a fully crossed $p \times t \times o$ design. Compare $\mathrm{var}(\hat{\sigma}^2_p)$ in (57) below for the partially nested $p \times (t : o)$ design to $\mathrm{var}(\hat{\sigma}^2_p)$ in (55) above for the fully crossed $p \times t \times o$ design:

$$\begin{aligned}\mathrm{var}\left(\hat{\sigma}^2_p\right) = \frac{2}{n_p - 1}\Bigg[&\left(\sigma^2_p + \frac{\sigma^2_{po}}{n_o} + \frac{\sigma^2_{pto,e}}{n_t n_o}\right)^2 \\ &+ \frac{1}{(n_o - 1)}\left(\frac{\sigma^2_{po}}{n_o} + \frac{\sigma^2_{pto,e}}{n_t n_o}\right)^2\Bigg].\end{aligned} \tag{57}$$

Smith (1981) used this result to make recommendations for increasing the sampling stability of variance component estimates. He suggested implementing multiple designs in which the number of mean square estimates used to compute variance component estimates is smaller than would be used in a single, more complex design. As an illustration, Smith (1981) considered a situation in which pupils (p) within schools (s) were administered items (i) within test forms (f). A single design configuration that would make it possible to estimate all possible variance components is $(p : s) \times (i : f)$ in which a sample of pupils from each school is administered samples of items from multiple test forms. As an alternative to this comprehensive design, Smith described two designs that require fewer mean squares to compute variance component estimates deemed likely to have the largest errors in the comprehensive design: a $p : [(s \times (i : f)]$ design (different sets of pupils within each school are administered each item in each test form, that is, there is a different set of pupils for each school-item combination) and a $i : [f \times (p : s)]$ design (for each school, samples of items from each test form are administered to a sample of pupils in that school such that there is a different set of items for each form–pupil combination). Together, these alternative designs make it possible to estimate all of the variance components in the single, comprehensive design. Smith showed that estimates of variance components from randomly sampled replications of the designs were more stable for the multiple designs (although less practical) than for the single, comprehensive design (using a fixed number of observations).

Another strategy for increasing the stability of variance component estimates involves the allocation of measurements (e.g., n_i and n_j). In a $p \times i \times j$ design, for example, as $\hat{\sigma}^2_{pij,e}$ increases relative to $\hat{\sigma}^2_{pi}$ and $\hat{\sigma}^2_{pj}$, the optimal solution tends toward $n_i = n_j$; as $\hat{\sigma}^2_{pij,e}$ decreases relative to $\hat{\sigma}^2_{pi}$ and $\hat{\sigma}^2_{pj}$, the optimal solution is to make n_i and n_j proportional to $\hat{\sigma}^2_{pi}/\hat{\sigma}^2_{pj}$ (Smith, 1978).

In contrast to ANOVA-based procedures for estimating standard errors of estimated variance components, approaches for estimating standard errors that do not make any distributional-form assumptions include the jackknife (Tukey, 1958) and bootstrap resampling methods (Efron, 1982; Efron and Tibshirani, 1986; Shao and Tu, 1995). Both methods involve creating multiple samples from a dataset, conducting separate analyses of the samples, and examining the variability of the estimates across the samples. The jackknife procedure creates samples by omitting one data point at a time; the bootstrap procedure samples with replacement. Extending these procedures to obtain estimated standard errors for variance components is not straightforward for multidimensional datasets, however (Brennan, 2001). For example, in a $p \times i$ (persons crossed with items) design it is possible to obtain bootstrap samples by sampling persons with replacement but not items; by sampling items with replacement but not persons; by sampling persons and items with replacement; or by sampling persons, items, and residuals with replacement (which may produce different results; see Brennan, 1994, 2001; see also Wiley, 2000, for guidelines on choosing among bootstrap strategies to obtain the most accurate estimates of standard errors). Furthermore, bootstrap procedures are likely to lead to biased estimates of variance components and standard errors. Wiley (2000) (summarized in Brennan, 2001, p. 188) derived adjustments for the estimates of variance components that eliminate the bias from different bootstrap procedures.

Although exact confidence intervals for variance components are generally unavailable (due to the inability to derive exact distributions for variance component estimates; see Searle, 1971), Satterthwaite (1941, 1946) and Ting et al. (1990) developed procedures for obtaining approximate confidence intervals (see also Burdick and Graybill, 1992). In simulations of $p \times i$ G-study estimated standard errors and confidence intervals with normal data, Brennan (2001) showed that confidence intervals based on the Satterthwaite, Ting et al., and jackknife procedures were fairly accurate, while the results using bootstrap procedures were mixed. Citing Burdick and Graybill (1992), Brennan (2001) noted theoretical reasons to prefer the Ting et al. procedure to the Satterthwaite procedure. For non-normal dichotomous data, the jackknife procedure produced confidence intervals that were more accurate than those obtained with the other methods (Brennan, 1994, 2001).

Negative estimated variance components

Negative estimates of variance components can arise because of sampling errors or because of model misspecification (Searle, 1971). Possible solutions when negative estimates are small in magnitude are to (1) substitute zero for the negative estimate and carry through the zero in other expected mean square equations from the analysis of variance (which produces biased estimates, Cronbach et al., 1972), (2) set any negative estimates of variance components to zero (Brennan, 2001), (3) use a Bayesian approach that sets a lower bound of zero on the estimated variance component (Box and Tiao,

1973; Fyans, 1977), and (4) use maximum likelihood (ML) or restricted maximum likelihood (REML) methods, which preclude negative estimates (Searle, 1971). Although the latter two approaches require a distributional form assumption, usually normality (Brennan, 1994), Marcoulides (1987, 1990) showed in simulations of data using a variety of crossed and nested, balanced and unbalanced designs that REML estimates were more accurate than ANOVA estimates for both non-normal (dichotomous, skewed, and j-shaped) and normal distributions.

Variance component estimation with unbalanced designs

An unbalanced design has unequal numbers of observations in its subclassifications. A nested design may be unbalanced purposively (e.g., achievement test items nested within content categories), dictated by the context itself (e.g., pupils nested within classrooms where class size is not constant), or by unforeseen circumstance (e.g., observers nested within occasions, with an unequal number of observers present at each occasion). Unbalanced situations can also arise in crossed and nested designs due to missing observations.

Estimating variance components in unbalanced designs is not straightforward. Some or all methods have problems of computational complexity, distributional assumptions, biased estimation, require decisions that cannot be justified in the context of generalizability theory, or produce results that are inconclusive (Brennan, 2001). While Henderson's Method 1 produces unbiased estimates for random effects models, two of Henderson's (1953) ANOVA-like methods produce biased estimates with models involving fixed effects (Method 1) or cannot be applied to models with interactions between random and fixed effects (Method 2). Henderson's Method 3 does not have these problems but produces different results depending upon the order in which variance components are estimated (Chiu and Wolfe, 2002), with no obvious basis for choosing among different orders. Maximum likelihood methods assume normality and are computationally complex. MINQUE (minimum norm quadratic unbiased estimation) does not assume normality and does not involve iterative estimation, thus reducing computational complexity, but estimates may be negative and are usually biased. Maximum likelihood procedures and MINQUE are often not viable, however, because sample sizes in G theory tend to be large, necessitating inverting matrices closely related to the number of observations in the data. Moreover, MINQUE requires the investigator to specify a priori weights corresponding to the relative sizes of the variance components (Brennan, 2001). The results may differ for different choices of a priori weights, without a compelling reason to choose among solutions (Searle et al., 1992).

Other approaches to estimating variance components with unbalanced designs involve analyzing one or more balanced subsets of the dataset, the major drawback being that the data analyzed may not be representative of the full dataset. To remedy this problem, Chiu and Wolfe (2002) proposed a method in which the dataset is exhaustively subdivided into all possible fully crossed subsets, into all possible nested subsets, and into all subsets with a modified balanced incomplete block design (*MBIB*, Chiu, 2001). The resulting variance component estimates from the different subsets are then weighted by the subset sample size and averaged. Chiu and Wolfe's (2002) analyses of data sets with two facets and large amounts of missing data (in which they assumed that data were missing at random) produced unbiased and consistent results.

3.6.2. *Hidden facets*

All studies have sources of variance that are not explicitly recognized or represented in the design (Cronbach et al., 1997; Shavelson et al., 1999; Webb et al., 2000). These are called implicit or hidden facets. In some cases, the same condition underlies all observations in the G-study, for example, when all persons are tested on a single occasion. Estimates of generalizability from such data do not take into account the possibility that persons' scores might be different on another occasion and, consequently, overestimate generalizability when interest lies in generalizing over occasions.

In other cases, a facet is linked or confounded with another source of variance (the object of measurement, another facet, or both) such that as the levels of one facet vary, correspondingly the levels of the other source of variance do too. As an example of a hidden facet confounded with the object of measurement, Brennan (2001) describes a piano recital in which students are evaluated by the same panel of judges but each student plays a different musical selection. Here, persons and musical selections are confounded, making it impossible to disentangle the effects of pianist ability and the difficulty of the particular musical selection. When interest lies in generalizing over musical selections, the design with musical selection as a hidden facet may underestimate or overestimate generalizability depending on the magnitudes of the (unknown) effects of musical selection.

As an example of a hidden facet that is confounded with another facet, consider again the notorious hidden facet, occasion. While a test administration might be conceived as a single occasion, an alternative conception of occasion is possible. As a person proceeds through a test or performs a series of tasks, his performance occurs over time. While task is varying, occasion of testing may be considered to vary too. Variability in performance from task to task typically is interpreted as task-sampling variability (Shavelson et al., 1993), but it may also be due to occasion-sampling variability, as pointed out by Cronbach et al. (1997). If occasion is the cause of variability in performance over tasks, then adding tasks to address the task-sampling problem will not improve the dependability of the measurement.

To examine the effects of occasion of testing, Shavelson et al. (1999) re-analyzed Shavelson et al.'s (1993) educational performance assessment data with occasions explicitly recognized as a source of variability in the design. Analysis of the design with persons crossed with tasks, raters, and occasions showed that both the task (person $\times$ task) and occasion (person $\times$ occasion) facets contributed variability, but the lion's share came from task sampling (person $\times$ task), and joint task and occasion sampling (person $\times$ task $\times$ occasion). The person $\times$ task $\times$ occasion interaction was the largest effect by far. In a different study, Webb et al. (2000) reported similar results. These studies showed that explicitly recognizing occasion as a facet of error variance altered the interpretation about the substantial sources of error in the measurement and called into question the interpretation of G studies in which the hidden occasion facet was ignored.

Occasion may be hidden in other respects as well. When a judge scores a particular examinee's performance once (as is usually the case), the occasion of scoring becomes an additional hidden facet (Cronbach et al., 1997), possibly leading to misinterpretation of effects involving judges.

3.6.3. Nonconstant error variance for different true scores

The descriptions of error variance given in previous sections implicitly assume that variance of measurement error is constant for all persons, regardless of universe score (true score in Classical Test Theory). The assumption of constant error variance for different universe scores was first criticized more than half a century ago in the context of Classical Test Theory, including by Lord (1955), who derived a formula for conditional error variance that varies as a function of true score. Lord's approach produced a concave-down quadratic form in which error variances are smaller for very high and very low true scores than for true scores that are closer to the mean. While persons with true scores close to the mean may produce scores that fluctuate from item to item, persons with true scores at the extremes have little opportunity for errors to influence their observed scores.

Brennan (1998) pointed out that Lord's conditional error variance is the conditional error variance for absolute decisions in G theory for a one-facet design with dichotomously-scored items and n_i' equal to n_i. For polytomously-scored items, estimated conditional error variance for absolute decisions for a one-facet $p \times I$ design (Brennan, 2001) is:

$$\hat{\sigma}^2_{\Delta_p} = \frac{\sum_i (X_{pi} - X_{pI})^2}{n_i'(n_i - 1)}. \tag{58}$$

An approximation for estimated conditional error variance for relative decisions for a $p \times I$ one-facet design based on large n_p (Brennan, 2001) is:

$$\hat{\sigma}^2_{\delta_p} = \hat{\sigma}^2_{\Delta_p} + \frac{\hat{\sigma}^2_i}{n_i'} - \frac{2\,\mathrm{cov}(X_{pi}, X_{Pi}|p)}{n_i'}, \tag{59}$$

where

$$\mathrm{cov}(X_{pi}, X_{Pi}|p) = \frac{\sum_i (X_{pi} - X_{pI})(X_{Pi} - X_{PI})}{n_i - 1}. \tag{60}$$

Following Brennan (2001), for a multi-facet random design (e.g., $p \times T \times O$ with persons crossed with tasks and observers), the conditional error variance for absolute decisions is:

$$\sigma^2_{\Delta_p} = \frac{\sigma^2_{t_p}}{n_t'} + \frac{\sigma^2_{o_p}}{n_o'} + \frac{\sigma^2_{to_p}}{n_t'n_o'}, \tag{61}$$

where $\sigma^2_{t_p}$, $\sigma^2_{o_p}$, and $\sigma^2_{to_p}$ are estimated using the $T \times O$ data for person p only. Brennan (1996, 1998) gives complicated formulas for conditional error variances for relative decisions for multifacet random designs. A simplified estimator that assumes very large n_p (Brennan, 2001) is:

$$\hat{\sigma}^2_{\delta_p} = \hat{\sigma}^2_{\Delta_p} + \left(\hat{\sigma}^2_{\Delta} - \hat{\sigma}^2_{\delta}\right). \tag{62}$$

Brennan (2001) also shows procedures for estimating condition relative and absolute error variance for multivariate G studies and for unbalanced designs.

Applying (62) to the generalizability study of job performance measurements of Navy machinist mates (Example 1) for a D-study with 11 tasks and 2 observers gives

$\hat{\sigma}^2_{\Delta_p} = 0.00027$ for a machinist mate with a high observed score (0.96 out of a maximum possible of 1.00) and $\hat{\sigma}^2_{\Delta_p} = 0.00731$ for a machinist mate with an observed score close to the mean (0.75). (As pointed out by Brennan, 2001, estimates of conditional error variances are usually expressed with respect to $\bar{X}_p$ because universe scores are unknown.) These values can be compared to the overall $\hat{\sigma}^2_{\Delta} = 0.00330$ in Table 9. These results show considerable variability in estimated error variance for persons with different estimated universe scores.

3.6.4. Optimizing the decision (D) study design

As shown in previous sections, when designing a decision study that will yield a reasonable level of estimated generalizability, the decision maker may consider both the design of the D-study and the number of observations to be used for each facet in the design. The resources available for the D-study may not accommodate the number of observations needed for a desired level of estimated generalizability, however. In this case, the decision maker may want to know the number of observations per facet that will minimize error (and maximize generalizability) for a fixed number of observations per person (Woodward and Joe, 1973) or for a given set of cost constraints. Marcoulides and Goldstein (1991, 1992), Marcoulides (1993, 1995) (see also Sanders, 1992; Sanders et al., 1991) describe methods for minimizing error variance in decision studies under cost constraints, both total costs and per-facet costs, for univariate and multivariate designs.

In the case of a fixed total number of observations for a multivariate profile or composite in which the observations for each score in the profile are not the same, Brennan (2001, p. 312) describes procedures for allocating observations to minimize error variance of the composite. In Example 3, the multivariate generalizability study of teacher behavior, suppose that the D-study will have six raters who will rate teacher behavior in either reading or mathematics (but not both). If a composite score is desired with weights 0.67 and 0.33 (reading weighted twice as heavily as mathematics), applying Brennan's procedure yields the smallest estimated absolute error variance of the composite ($\hat{\sigma}^2_{\Delta} = 0.27113$) and highest level of dependability ($\hat{\Phi} = 0.885$) for $n'_r = 5$ in reading and $n'_r = 1$ in mathematics.

3.7. Linking generalizability theory and item response theory

Item response theory (IRT), also known as latent trait theory, is a widely used approach to test development that focuses on estimating a person's ability (or other latent trait) and describing properties of items on the basis of assumed mathematical relationships between the latent trait and item responses (Embertson and Reise, 2000; Hambleton et al., 1991; Lord, 1980). A central assumption of IRT is that an individual's probability of responding to an item in a certain way is a function only of the individual's ability. Generally speaking, estimates of a person's ability level do not depend on the group of persons being examined, the particular items used, or other conditions of measurement, such as the occasions of testing or the raters used to judge performance; moreover, estimates of item (or occasion or rater) attributes do not depend on the ability distribution of the examinees. The property of invariance of ability and item (or judge or occasion) parameters makes IRT well suited for addressing issues related to test or scale construction, score equating, computer adaptive testing, and item bias. IRTs focus on items (or other conditions of measurement) as *fixed* entities (Brennan, 2001), however, makes it impossible to estimate sources of error variation that can be used to optimize the design of a future measurement using a different sample of items, raters, and occasions. Hence, Brennan (2001, p. 166) characterizes IRT as principally a scaling model while G theory is truly a measurement model.

Kolen and Harris (1987) and Briggs and Wilson (2004) proposed an approach that brings together the sampling model of G theory with the scaling model of IRT. For example, in a person $\times$ item design in which persons and items are assumed to be randomly drawn from a population and universe of interest, respectively, specifying prior distribution functions for person and item effects makes it possible to define and estimate IRT analogues to variance components used in G theory as well as estimate traditional IRT parameters. Expected, rather than observed, item responses are used to estimate variance components.

Bock et al. (2002) described how to link G theory and IRT in situations where the IRT assumption of local independence – holding constant the abilities influencing test performance, persons' responses to any pair of items are statistically independent – is violated. In the partially nested $p \times (r : i)$ design that is common in large-scale performance assessment, for example, in which different sets of raters evaluate the items, the multiple ratings of each item do not represent distinct items and are not locally independent. Bock et al. (2002) note that treating multiple ratings as if they were separate items in an IRT analysis would produce underestimated standard errors of IRT scale-score and item parameter estimates; they suggest corrections for this downward bias using G theory results.

A number of other studies have used G theory and IRT approaches sequentially, first carrying out generalizability analyses to identify important sources of variation and then applying IRT techniques to diagnose aberrant persons or conditions or combinations of them to help guide test revision and rater training. Bachman et al. (1995), for example, applied generalizability theory and an extension of the Rasch model known as many-facet Rasch measurement (also called multi-faceted Rasch measurement or MFRM) to examine variability in tasks and rater judgments of examinees' foreign language speaking ability (for similar sequences of analyses, see also Lynch and McNamara, 1998; Smith and Kulikowich, 2004; Stahl and Lunz, 1992). Bachman et al.'s generalizability analysis showed a non-negligible examinee $\times$ rater interaction, indicating that raters were not consistent in their rank orderings of examinees. The fit statistics from the many-facet Rasch analysis (e.g., infit, outfit) were used to identify individual raters who were not internally self-consistent in their ratings, that is, who gave unexpectedly high or low ratings compared to their usual rating style, and "bias" statistics (using the FACETS program, Linacre and Wright, 1993), which pointed to specific examinee-rater combinations with scores that deviated from expectation. Consistent with other application studies, Bachman et al. (1995) described the value of diagnostic information from MFRM as feedback to raters to improve their future ratings. Pinpointing discrepant behavior of raters in a G-study may also have ramifications for designing effective training for potential raters (not just those used in the original study), which may reduce variability among future raters, and so reduce the number needed for dependable measurement in a D-study.

In another sequential application of the two theories, Marcoulides and Drezner (2000) developed an alternative model for estimating latent traits such as examinee ability, rater severity, and item difficulty. This model uses a weighted Euclidean distance function to provide both latent trait estimates and two-dimensional graphical representations of conditions or combinations of conditions for use in identifying unusually performing persons, raters, or items (or combinations of them). Applying the model to the teacher behavior data in Example 2, Marcoulides (1999) provided detailed insights into the behavior of raters, showing clearly that raters were more consistent in their ratings of teachers in mathematics than in reading. Marcoulides (1999) also showed that the latent trait estimates produced using the Marcoulides–Drezner model are not sensitive to unusually performing persons or conditions, in contrast to estimation procedures based on the many-facet Rasch model, which makes the assumption of no interactions.

3.8. Computer programs

A number of popular computer packages and programs provide estimates of variance components in generalizability studies. These include SAS (SAS Institute, 1996), SPSS (SPSS, 1997), and S-Plus (MathSoft, 1997). However, only one such program has been developed specifically for generalizability theory: GENOVA (GENeralized analysis Of Variance; Brennan, 2001; Crick and Brennan, 1983). GENOVA handles complete balanced designs (with up to five facets), urGENOVA handles designs that are unbalanced due to nesting and some designs with missing data, and mGENOVA performs multivariate generalizability and decision analyses for a selected set of designs that are balanced or unbalanced due to nesting.

The programs listed above use a variety of methods for estimating variance components, including the ANOVA procedure using expected mean square equations (EMS, Cornfield and Tukey, 1956), Henderson's Method 1 and Method 3 (Henderson, 1953), MINQUE (minimum norm quadratic unbiased estimation) with equal weighting of all variance components (MINQUE(0)) or equal weights for all variance components except the residual (MINQUE(1)), maximum likelihood, and restricted maximum likelihood. GENOVA also provides estimates using Cronbach et al.'s (1972) recommendation to substitute zero for negative estimated variance components wherever they appear in the EMS equations. As noted by Brennan (2001), processing times and memory requirements vary widely, depending on the complexity of the design, the numbers of observations (which can easily exceed 10,000), the estimation method used, and the computational algorithm employed.

4. Concluding remarks

Generalizability theory has sometimes been mischaracterized as "merely" an application of random-effects ANOVA models to item response data. This characterization is unfortunate and misleading (Brennan, 2000). In addition to computational procedures,

G theory provides a conceptual framework and notational system for characterizing alternative intended universes of generalization, distinguishing fixed versus random facets, and designing efficient measurement procedures, among other uses. G theory has also clarified the use and interpretation of coefficients and estimated standard errors from classical test theory. This is of value because, despite the greater sophistication of G theory, and the important applications described in this chapter, classical test theory remains the dominant tool for estimating reliability.

In closing, it bears repeating that reliability is just one consideration in the responsible use and interpretation of tests, and not the most important consideration, at that. The meaning, appropriateness, and consequences of testing must be weighed in any practical situation, and the most reliable measurement procedure may not be the best. That said, it is a commonplace that reliability is prerequisite to validity, and the statistical methods described in this chapter remain critically important for sound measurement practice.

References

Bachman, L.F., Lynch, B.K., Mason, M. (1995). Investigating variability in tasks and rater judgements in a performance test of foreign language speaking. *Language Testing* **12**, 238–257.

Bock, R.D., Brennan, R.L., Muraki, E. (2002). The information in multiple ratings. *Applied Psychological Measurement* **26**, 364–375.

Box, G.E.P., Tiao, G.C. (1973). *Bayesian Inference in Statistical Analysis*. Addison-Wesley, Reading, MA.

Brennan, R.L. (1994). Variance components in generalizability theory. In: Reynolds, C.R. (Ed.), *Cognitive Assessment: A Multidisciplinary Perspective*. Plenum Press, New York, pp. 175–207.

Brennan, R.L. (1995). The conventional wisdom about group mean scores. *Journal of Educational Measurement* **32**, 385–396.

Brennan R.L. (1996). Conditional standard errors of measurement in generalizability theory. Iowa Testing Programs Occasional Paper No. 40. Iowa Testing Programs, University of Iowa, Iowa City, IA.

Brennan, R.L. (1997). A perspective on the history of generalizability theory. *Educational Measurement: Issues and Practice* **16**, 14–20.

Brennan, R.L. (1998). Raw-score conditional standard errors of measurement in generalizability theory. *Applied Psychological Measurement* **22**, 307–331.

Brennan, R.L. (2000). (Mis)conceptions about generalizability theory. *Educational Measurement: Issues and Practice* **19**, 5–10.

Brennan, R.L. (2001). *Generalizability Theory*. Springer-Verlag, New York.

Briggs, D.C. Wilson, M. (2004). Generalizability in item response modeling. Paper presented at the 2004 Annual Meeting of the Psychometric Society, Pacific Grove, CA.

Burdick, R.K., Graybill, F.A. (1992). *Confidence Intervals on Variance Components*. Dekker, New York.

Cardinet, J., Tourneur, Y., Allal, L. (1976). The symmetry of generalizability theory: Applications to educational measurement. *Journal of Educational Measurement* **13**, 119–135.

Cardinet, J., Tourneur, Y., Allal, L. (1981). Extension of generalizability theory and its applications in educational measurement. *Journal of Educational Measurement* **18**, 183–204.

Chiu, C.W.T., Wolfe, E.W. (2002). A method for analyzing sparse data matrices in the generalizability theory framework. *Applied Psychological Measurement* **26**, 321–338.

Chiu, C.W.T. (2001). *Generalizability Theory: Scoring Large-Scale Assessments based on Judgments*. Kluwer Academic, Boston, MA.

Cornfield, J., Tukey, J.W. (1956). Average values of mean squares in factorials. *Annals of Mathematical Statistics* **27**, 907–949.

Crick, J.E., Brennan, R.L. (1983). GENOVA: A generalized analysis of variance system [Computer software and manual]. University of Iowa, Iowa City, IA. Available from http://www.education.uiowa.edu/casma/.

Cronbach, L.J. (1951). Coefficient alpha and the internal structure of tests. *Psychometrika* **16**, 297–334.

Cronbach, L.J. (2004). My current thoughts on coefficient alpha and successor procedures. *Educational and Psychological Measurement* **64**, 391–418.

Cronbach, L.J., Rajaratnman, N., Gleser, G.C. (1963). Theory of generalizability: A liberalization of reliability theory. *British Journal of Statistical Psychology* **16**, 137–163.

Cronbach, L.J., Gleser, G.C., Nanda, H., Rajaratnam, N. (1972). *The Dependability of Behavioral Measurements*. Wiley, New York.

Cronbach, L.J., Linn, R.L., Brennan, R.L., Haertel, E.H. (1997). Generalizability analysis for performance assessments of student achievement or school effectiveness. *Educational and Psychological Measurement* **57**, 373–399.

de Finetti, B.D. (1937). Foresight: Its logical laws, its subjective sources. *Annales de l'Institut Henri Poincare* **7**, 95–158.

de Finetti, B.D. (1964). Foresight: Its logical laws, its subjective sources. In: Kyburg, H.E., Smokler, G.E. (Eds.), *Studies in Subjective Probability*. Wiley, New York, pp. 95–158.

Efron, B. (1982). *The Jackknife, the Bootstrap, and Other Resampling Plans*. SIAM, Philadelphia, PA.

Efron, B., Tibshirani, R. (1986). Bootstrap methods for standard errors, confidence intervals, and other measures of statistical accuracy. *Statistical Science* **1**, 54–77.

Embertson, S.E., Reise, S.P. (2000). *Item Response Theory as Model-Based Measurement*. In: *Item Response Theory for Psychologists*. Erlbaum, Mahwah, NJ.

Erlich, O., Shavelson, R.J. (1978). The search for correlations between measures of teacher behavior and student achievement: Measurement problem, conceptualization problem, or both? *Journal of Educational Measurement* **15**, 77–89.

Feldt, L.S., Brennan, R.L. (1989). Reliability. In: Linn, R.L. (Ed.), *Educational Measurement*. 3rd ed. The American Council on Education, Macmillan, pp. 105–146.

Feller, W. (1966). *An Introduction to Probability Theory and its Applications*. Wiley, New York.

Fyans Jr., L.J. (1977). A new multiple level approach to cross-cultural psychological research. Unpublished doctoral dissertation. University of Illinois at Urbana-Champaign.

Goodenough, F.L. (1936). A critical note on the use of the term 'reliability' in mental measurement. *Journal of Educational Psychology* **27**, 173–178.

Haertel, E.H. (in press). Reliability. In: Brennan, R.L. (Ed.), *Educational Measurement*. 4th ed. Greenwood, Westport, CT.

Hambleton, R.K., Swaminathan, H., Rogers, H.J. (1991). *Fundamentals of Item Response Theory*. Sage Publications, Newbury Park, CA.

Henderson, C.R. (1953). Estimation of variance and covariance components. *Biometrics* **9**, 227–252.

Joe, G.W., Woodward, J.A. (1976). Some developments in multivariate generalizability. *Psychometrika* **41**, 205–217.

Kane, M.T., Brennan, R.L. (1977). The generalizability of class means. *Review of Educational Research* **47**, 267–292.

Kane, M.T., Gillmore, G.M., Crooks, T.J. (1976). Student evaluations of teaching: The generalizability of class means. *Journal of Educational Measurement* **13**, 171–183.

Kolen, M.J., Harris, D.J. (1987). A multivariate test theory model based on item response theory and generalizability theory. Paper presented at the Annual Meeting of the American Educational Research Association, Washington, DC.

Linacre, J.M., Wright, B.D. (1993). *A User's Guide to FACETS: Rasch Model Computer Program, Version 2.4 for PC-Compatible Computers*. MESA Press, Chicago.

Lord, F.M. (1955). Estimating test reliability. *Educational and Psychological Measurement* **16**, 325–336.

Lord, F.M. (1980). *Applications of Item Response Theory to Practical Testing Problems*. Erlbaum, Hillsdale, NJ.

Lynch, B.K., McNamara, T.F. (1998). Using G-theory and many-facet Rasch measurement in the development of performance assessments of the ESL speaking skills of immigrants. *Language Testing* **15**, 158–180.

Marcoulides, G.A., Drezner, Z. (2000). A procedure for detecting pattern clustering in measurement designs. In: Wilson, M., Engelhard, G. (Eds.), *Objective Measurement: Theory into Practice*, vol. 5. Ablex, Stamford, CT, pp. 287–303.

Marcoulides, G.A., Goldstein, Z. (1991). Selecting the number of observations in multivariate measurement studies under budget constraints. *Educational and Psychological Measurement* **51**, 573–584.

Marcoulides, G.A., Goldstein, Z. (1992). The optimization of multivariate generalizability studies with budget constraints. *Educational and Psychological Measurement* **52**, 301–309.

Marcoulides, G.A. (1987). An alternative method for variance component estimation: Applications to generalizability theory. Unpublished doctoral dissertation. University of California, Los Angeles, CA.

Marcoulides, G.A. (1990). An alternate method for estimating variance components in generalizability theory. *Psychological Reports* **66**, 379–386.

Marcoulides, G.A. (1993). Maximizing power in generalizability studies under budget constraints. *Journal of Educational Statistics* **18**, 197–206.

Marcoulides, G.A. (1994). Selecting weighting schemes in multivariate generalizability studies. *Educational and Psychological Measurement* **54**, 3–7.

Marcoulides, G.A. (1995). Designing measurement studies under budget constraints: Controlling error of measurement and power. *Educational and Psychological Measurement* **55**, 423–428.

Marcoulides, G.A. (1999). Generalizability theory: Picking up where the Rasch IRT model leaves off? In: Embretson, S.E., Hershberger, S.L. (Eds.), *The New Rules of Measurement: What Every Psychologist Should Know*. Erlbaum, Mahwah, NJ, pp. 129–152.

MathSoft, Inc. (1997). *S-Plus 4.5 Standard Edition*. Author, Cambridge, MA.

Sanders, P.F. (1992). Alternative solutions for optimizing problems in generalizability theory. *Psychometrika* **57**, 351–356.

Sanders, P.F., Theunissen, T.J.J.M., Baas, S.M. (1991). Maximizing the coefficient of generalizability under the constraint of limited resources. *Psychometrika* **56**, 87–96.

SAS Institute, Inc. (1996). *The SAS System for Windows Release 6.12*. Author, Cary, NC.

Satterthwaite, F.E. (1941). Synthesis of variance. *Psychometrika* **6**, 309–316.

Satterthwaite, F.E. (1946). An approximate distribution of estimates of variance components. *Biometrics* **2**, 110–114.

Searle, S.R., Casella, G., McCulloch, C.E. (1992). *Variance Components*. Wiley, New York.

Searle, S.R. (1971). *Linear Models*. Wiley, New York.

Shao, J., Tu, D. (1995). *The Jackknife and the Bootstrap*. Springer-Verlag, New York.

Shavelson, R.J. (2004). Editor's preface to Lee J. Cronbach's "My current thoughts on coefficient alpha and successor procedures". *Educational and Psychological Measurement* **64**, 389–390.

Shavelson, R.J., Webb, N.M. (1981). Generalizability theory: 1973–1980. *British Journal of Mathematical and Statistical Psychology* **34**, 133–165.

Shavelson, R.J., Webb, N.M. (1991). *Generalizability Theory: A Primer*. Sage Publications, Newbury Park, CA.

Shavelson, R.J., Baxter, G.P., Gao, X. (1993). Sampling variability of performance assessments. *Journal of Educational Measurement* **30**, 215–232.

Shavelson, R.J., Ruiz-Primo, M.A., Wiley, E.W. (1999). Note on sources of sampling variability in science performance assessments. *Journal of Educational Measurement* **36**, 61–71.

Short, L.M., Shavelson, R.J., Webb, M.N. (1986). Issues in multivariate generalizability: Weighting schemes and dimensionality. Paper presented at the Annual Meeting of the American Educational Research Association, San Francisco, CA.

Smith, E.V., Kulikowich, J.M. (2004). An application of generalizability theory and many-facet Rasch measurement using a complex problem-solving skills assessment. *Educational and Psychological Measurement* **64**, 617–639.

Smith, P.L. (1978). Sampling errors of variance components in small sample multifacet generalizability studies. *Journal of Educational Statistics* **3**, 319–346.

Smith, P.L. (1981). Gaining accuracy in generalizability theory: Using multiple designs. *Journal of Educational Measurement* **18**, 147–154.

SPSS, Inc. (1997). *SPlanet. Space Sci. for Windows Release 8.0.0*. Author, Chicago.

Stahl, J.A., Lunz, M.E. (1992). A comparison of generalizability theory and multi-faceted Rasch measurement. Paper presented at the Midwest Objective Measurement Seminar, University of Chicago.

Thorndike, R.L. (1947). Research Problems and Techniques. Report No. 3. AAF Aviation Psychology Program Research Reports. U.S. Government Printing Office, Washington, DC.

Ting, N., Burdick, R.K., Graybill, F.A., Jeyaratnam, S., Lu, T.C. (1990). Confidence intervals on linear combinations of variance components that are unrestricted in sign. *Journal of Statistical Computational Simulation* **35**, 135–143.

Tukey, J.W. (1958). Bias and confidence in not quite large samples. *Annals of Mathematical Statistics* **29**, 614.

Webb, N.M., Schlackman, J., Sugrue, B. (2000). The dependability and interchangeability of assessment methods in science. *Applied Measurement in Education* **13**, 277–301.

Webb, N.M., Shavelson, R.J., Kim, K.S., Chen, Z. (1989). Reliability (generalizability) of job performance measurements: Navy machinists mates. *Military Psychology* **1**, 91–110.

Wiley, E. (2000). Bootstrap strategies for variance component estimation: Theoretical and empirical results. Unpublished doctoral dissertation. Stanford University.

Wood, R. (1976). Trait measurement and item banks. In: DeGruijter, D.N.M., van der Kamp, L.J.T. (Eds.), *Advances in Psychological and Educational Measurement*. Wiley, New York, pp. 247–263.

Woodward, J.A., Joe, G.W. (1973). Maximizing the coefficient of generalizability in multi-facet decision studies. *Psychometrika* **38**, 173–181.

Yen, W.M. (1997). The technical quality of performance assessments: Standard errors or percents of students reaching standards. *Educational Measurement: Issues and Practice* **16**, 5–15.

Handbook of Statistics, Vol. 26
ISSN: 0169-7161

DOI: 10.1016/S0169-7161(06)26005-X

5

Differential Item Functioning and Item Bias

Randall D. Penfield and Gregory Camilli

1. Introduction

The concept of fairness has been a mainstay in educational and psychological testing for the past half century. Within the scope of testing, the term fairness can assume a range of meanings related to the way in which scores on tests or items are used in evaluating examinees for selection and classification decisions. According to the 1999 *Standards for Educational and Psychological Testing* (referred to as the *Standards* hereafter), fairness in testing can be interpreted with respect to bias at the item or test level, equitable treatment within the testing process, and equal opportunity to learn for all groups of examinees (AERA, APA, NCME, 1999). The *Standards* also specify that fairness may be interpreted with respect to equality in test scores (e.g., equal passing rates), although this application is typically restricted to legal requirements that "... other things being equal, a testing alternative that minimizes outcome differences across relevant subgroups should be used" (p. 76).

This chapter is primarily concerned with one aspect of fairness – that of bias. The social concern for bias was born out of the consistent disparity in the mean scores on standardized tests of protected groups relative to their mainstream or prevalent group counterparts. Although the gap in test performance is likely explained by a variety of factors, including educational resources and teacher experience and credential status (Noguera and Akom, 2000; Zwick, 2001), one of the most publicly pervasive factors is the possibility that test items contain language and content to which protected groups have had limited exposure, thus creating an unfairness in the test. This unfairness towards particular groups of examinees has been labeled as bias in the public vernacular of test equity.

The technical testing community, however, has adopted a more narrow definition of bias as *statistical bias*. Cleary (1968) offered the following definition of test bias:

> A test is biased for members of a subgroup of the population if, in the prediction of a criterion for which the test was designed, consistent nonzero errors of prediction are made for members of the subgroup. In other words, the test is biased if the criterion score predicted from the common regression line is consistently too high or too low for members of the subgroup. (p. 115)

Using this unequal prediction definition of bias, the test is unfair for the subgroup of examinees for which the test score underpredicts the criterion. This definition of bias is also echoed by Bond et al. (1996), who stated that bias refers to

> the extent to which a test score and test use are valid for all intended individuals and groups. As such, if an assessment results in scores that systematically underestimate the status of members of a particular group on the construct in question, then the test is biased against members of that group. (p. 119)

However, the mechanism through which bias arises is typically a selection rule based on the predictor. For this reason, the term "selection bias" is often more descriptive than "test bias." In any event, the definition of test bias given above asserts that the measurement of test bias requires the existence of a valid criterion, which is difficult to establish at the time of test construction, and thus hampers practical investigations of test bias prior to the administration of the operational test form. Additional complications encountered in the statistical modeling of test bias (see Camilli and Shepard, 1994, pp. 13–15) have caused the attention of researchers studying bias to turn their attention to the analysis of item-level bias. The extent to which the items of a test are free of bias is generally taken as mild support that the test itself is less likely to be biased.

The development of statistical methods for identifying potentially biased items began in the 1970s, but it wasn't until the mid-to-late 1980s that a general statistical framework was developed that could serve as the foundation for the analysis of item statistical bias. This framework is now referred to as differential item functioning (DIF), a term coined in a seminal chapter on DIF by Holland and Thayer (1988). The following decade bore witness to a tremendous amount of interest and research directed towards the concept of DIF, and its measurement. DIF has become an integral part of test validation, being included among the *Standards* (Standard 7.3), and is now a key component of validity studies in virtually all large-scale assessments. In addition, the study of DIF is being extended to other fields that utilize constructed variables, such as psychology (Bolt et al., 2004; Dodeen and Johanson, 2003; Lange et al., 2002) and the health sciences (Gelin et al., 2004; Gelin and Zumbo, 2003; Iwata et al., 2002; Kahler et al., 2003; Panter and Reeve, 2002). In these areas, the assessment of DIF is not only a concern for selection bias, but also for ensuring the validity of research examining between-group differences on traits measured with constructed variables.

This chapter aims to present a description of many of the commonly employed methods in the detection of item bias. Because much of the statistical detection of item bias makes use of DIF procedures, the majority of this chapter is devoted to a description of statistical methods for the analysis of DIF. DIF detection procedures for dichotomous and polytomous items are presented, as are methods for the categorization of DIF effect in dichotomous items. In addition, several recent innovations in DIF detection are presented, including Bayesian applications, the detection of differential test functioning, and studies examining sources or explanations of DIF. While much of the presentation of DIF procedures for dichotomous items can be found in other sources (e.g., Camilli and Shepard, 1994; Holland and Wainer, 1993), the material related to polytomous DIF detection, Bayesian DIF methods, and differential test functioning are relatively new topics that have been given limited attention in summary presentations.

2. General definition of DIF

Before we set out on a presentation of DIF detection methods, we first define precisely what is meant by DIF. Let us first consider the simple and idealized situation in which the response (Y) to a particular item on the test is determined solely by the target proficiency intended to be measured by the test, denoted here by θ, and random error. We can express the probability distribution of Y conditional on θ by $f(Y \mid \theta)$. Next, suppose that we are concerned with comparing the conditional probability of Y for two groups of examinees, denoted here as the focal and reference groups. Historically, the reference group consists of individuals for which the test was expected to favor (e.g., Whites or males), and the focal group consists of the individuals who were at risk of being disadvantaged by the test (e.g., Blacks, Hispanics, females). If it is specified that the conditional probability of Y is dependent only on θ, and that error distributions for the two groups are identical, then it must be the case that

$$f(Y \mid \theta, G = R) = f(Y \mid \theta, G = F), \tag{2.1}$$

where G corresponds to the grouping variable, R corresponds to the reference group, and F corresponds to the focal group. The situation described in (2.1) is that of no DIF. That is, the absence of DIF is characterized by a conditional probability distribution of Y that is independent of group membership. In the case of a dichotomously scored item (where $Y = 1$ for a correct response and $Y = 0$ for an incorrect response), this suggests that the probability of correct response is identical for individuals belonging to different groups, but sharing the same value of θ. The situation described by (2.1) bodes well for the test developer's assertion that the item is not biased because it states that reference and focal group members with the same level of proficiency have an equal chance of correctly responding to the item, and thus there is no indication of an unfair advantage for one group over the other.

In contrast to the above situation, suppose that the conditional probability of Y is not identical for the reference and focal groups. In this case, individuals having the same level of proficiency, but belonging to different groups, have a different probability distribution of Y. That is, there exists a dependency between group membership and item performance after controlling for θ (again assuming equal error distributions). In the case of a dichotomous item, this conditional dependence implies that reference and focal group members at the same level of θ have a different probability of correct response – the group having the lower conditional probability of correct response is at a disadvantage on that item. It can be said, then, that the item functions differentially for the two groups, or alternatively, DIF is said to exist. All DIF detection statistics, in one way or another, are concerned with either testing the null hypothesis of no DIF, as described by (2.1), or providing a measure of the extent to which the situation described by (2.1) is false.

While DIF is embodied by a conditional dependence of group membership and item performance, the possibility that the nature of the dependence may change across the ability continuum leads to the designation of two general forms of DIF. In the simplest case, the conditional dependence has the same relative magnitude across the θ continuum. That is, the item provides a constant relative advantage for the same group, regardless of the θ level. This situation is referred to as *uniform* DIF because the DIF effect is uniform across the θ continuum. In contrast to uniform DIF, *nonuniform* DIF exists when the conditional dependence changes in magnitude or in direction across the θ continuum. For example, the reference group may have a small relative advantage at low θ levels, but a large relative advantage at high θ levels. A particularly troublesome form of nonuniform DIF, known as *crossing nonuniform* DIF, exists when the reference group has a relative advantage at one end of the θ continuum and the focal group displays a relative advantage at the other end of the θ continuum. The distinction between uniform and nonuniform DIF is of critical importance when conducting and interpreting DIF analyses because different DIF statistics are appropriate depending whether uniform or nonuniform DIF exists.

Whether uniform or nonuniform, the presence of DIF poses a threat to the validity of test scores because it implies that performance on a test item is not only dependent on the target ability, but also on a secondary factor related to group membership. From a multidimensionality perspective, the presence of DIF can be modeled as the presence of a secondary factor influencing item performance for which the distribution differs for the reference and focal groups after conditioning on the target ability (Camilli, 1992; Roussos and Stout, 1996a; Shealy and Stout, 1993b). Between-group differences in the conditional distribution of the secondary factor can lead to between-group differences in the conditional expected value of Y. The group experiencing the lower conditional expected value of Y is at a disadvantage on that item, and this disadvantage may be interpreted as item-level bias.

Although the presence of DIF is a signal that the item may be unfair, statistical bias does not imply that the item is unfair. Rather, the presence of DIF, according to the multidimensionality hypothesis, signals the existence of a secondary factor. Fairness is then established by demonstrating whether or not the secondary factor is intentionally related to the test construct, that is, the objective of the measurement. It is possible, for example, for the secondary factor to be content mandated by test specifications despite being differentially distributed for the reference and focal groups. The establishment that the observed DIF is the result of an unintended secondary dimension is typically pursued through expert review of the item content. If, and only if, the presence of DIF can be attributed to unintended item content or some other unintended item property (e.g., method of administration) is the item said to be unfair. As a consequence, the absence of DIF is a necessary, but not sufficient, condition for item fairness.

3. Item response theory approaches for dichotomous items

Numerous DIF detection procedures have been developed for dichotomous items, and the description of many of them are presented in a variety of sources (Camilli and Shepard, 1994; Hills, 1989; Millsap and Everson, 1993). Despite the variety of available procedures, they all share common theoretical interpretations that are founded in the modeling capabilities of item response theory (IRT). The standard IRT framework consists of a family of parametric models in which the probability of correct response to a particular item of a test is modeled as a function of θ and one or more item-level parameters (Baker, 1992; Lord, 1980). The most general of popular IRT models for dichotomously scored items is the three-parameter logistic model (3PL), for which the probability of correct response conditional on θ is posited using

$$P(Y = 1 \mid \theta) = c + (1 - c)\frac{\exp[Da(\theta - b)]}{1 + \exp[Da(\theta - b)]}, \tag{3.1}$$

where a corresponds to the discriminating power of the item, b corresponds to the difficulty of the item, and c corresponds to the pseudo-guessing factor in the item (the probability of correct response for an individual with asymptotically low proficiency). The function shown in (3.1) describes an S-shaped curve that specifies the probability of correct response as a function of θ. The form of the S-shaped curve for a particular item depends on the values of a, b, and c, and as such the resulting curve is commonly referred to as an item characteristic curve or item response function (IRF). More restricted versions of the 3PL are also employed, including the two-parameter logistic model (2PL) in which c is set equal to zero for the item, and the one-parameter logistic model (1PL) in which c is set equal to zero and a is set equal to unity (or a constant value across items). The constant D is equal to 1.7, and serves the purpose of scaling the logistic models to be very close in form to the normal ogive model.

The IRT framework has proven to be instrumental for the conceptualization and assessment of DIF. The framework of IRT accomplishes this by examining between-group differences in the item's IRF – any between-group difference in an item's IRF signifies a conditional between-group difference in the probability of correct response, and thus DIF. Recalling that an item's IRF is determined by the values of the item-level parameters (i.e., a, b, c), the definition of DIF in the IRT context can be extended to include any between-group differences in the item-level parameters. Notice that the IRT framework affords two different, but compatible, interpretations of DIF: either as a between-group difference in an item's IRF, or as a between-group difference in an item's parameters. Both of these interpretations of DIF have been employed in developing statistics to detect DIF using IRT methods.

Perhaps the most basic application of IRT to the assessment of DIF is the effect size given by the signed area (*SA*) between the IRFs specified for the reference and focal groups, represented by the integral

$$SA = \int_{-\infty}^{\infty} \big[P(Y = 1 \mid \theta, G = R) - P(Y = 1 \mid \theta, G = F)\big]\, d\theta. \tag{3.2}$$

The *SA* effect size index was first proposed by Rudner et al. (1980). The value of *SA* can be expressed as a function of the item parameters for the reference and focal groups (Raju, 1988). Under the assumption that the c-parameter is identical for the reference and focal groups, *SA* is equal to

$$SA = (1 - c)(b_F - b_R). \tag{3.3}$$

A proof of this relationship is provided by Raju. It is of interest to note that, in the case of equal c-parameters for the reference and focal group, the signed area is independent of the a-parameter (even if the a-parameter differs between the groups). Thus, the magnitude of DIF, as measured by *SA*, is dependent only on the between-group difference in the item difficulty parameter (b) scaled by the value of the c-parameter; for a given difference between b_R and b_F, the magnitude of DIF will decrease as c increases. In addition, larger between-group differences in the b-parameter lead to larger conditional between-group differences in the probability of correct response, and thus a larger DIF effect. In the case of the 1PL, for which $c = 0$, *SA* is equal to $b_F - b_R$. When the c-parameter differs for the reference and focal groups, the value of *SA* will be infinite (Raju, 1988), and thus a simple form does not exist for *SA* in the presence of a between-group difference in the c-parameter.

The value of *SA* is informative in the case of uniform DIF, but can be misleading in the case of between-group differences in a-parameter, particularly if the IRFs cross anywhere on the interval of $-3 \leqslant \theta \leqslant 3$ (although an even wider θ interval may be of interest in some applications). This is because the difference between the IRFs will be positive on one side of the point of intersection of the IRFs for the reference and focal groups and negative on the other side, causing the areas on the two sides to partially cancel. As a result, it is possible to have a rather small value of *SA*, despite gross differences in the IRFs of the reference and focal groups. To address this issue, researchers have proposed considering the unsigned area (*UA*) between the reference and focal group IRFs, given by the integral

$$UA = \int_{-\infty}^{\infty} \left| P(Y = 1 \mid \theta, G = R) - P(Y = 1 \mid \theta, G = F) \right| \mathrm{d}\theta. \tag{3.4}$$

When the c-parameter is equal for the reference and focal groups, the value of *UA* has been shown (see Raju, 1988) to be equal to

$$UA = (1 - c) \left| \frac{2(a_F - a_R)}{D a_R a_F} \ln\left[1 + \exp\left(\frac{D a_R a_F (b_F - b_R)}{a_F - a_R} \right) \right] - (b_F - b_R) \right|. \tag{3.5}$$

The form of *UA* shown in (3.5) reduces further when the a-parameter is identical for the reference and focal groups, being equal to

$$UA = (1 - c)|b_F - b_R|. \tag{3.6}$$

When the c-parameter differs between the two groups, an analytic solution to (3.4) does not exist.

We present the *SA* and *UA* indices largely for didactic purposes because, in practice, the *SA* and *UA* indices have several drawbacks that limit their application to measuring DIF magnitude. One drawback is that they do not consider the distribution of examinees across the θ continuum, and thus may yield misleading interpretations of the actual magnitude of DIF observed for a specific population of examinees. A second drawback of *SA* and *UA* is that their direct estimation requires the estimation of the respective item-level parameters for each group, which can require sample sizes that are often difficult to attain, particularly for the focal group. Despite these drawbacks, the *SA* and *UA* indices play an important role in the development of more commonly used DIF statistics, and their interpretation. Specifically, the *SA* and *UA* indices describe important theoretical relationships between the magnitude of DIF and between-group differences in the item parameters, relationships that are often used in interpreting the results of

DIF analyses and simulating responses containing varying levels of DIF in conducting research related to DIF detection statistics. In addition, the *SA* and *UA* indices serve as a theoretical model on which several of the more commonly used DIF detection statistics are based, such as the *SPD* index and SIBTEST, as described below.

While the *SA* and *UA* indices make use of the IRT framework to provide theoretical measures of DIF effect size, they do not offer an efficient method for testing the null hypothesis of no DIF. A more desirable hypothesis test of no DIF can be obtained in terms of the parameters of the IRT models for the reference and focal groups. The simplest such test statistic addresses the null hypothesis that the b-parameters are equal for the reference and focal groups. This test statistic (Lord, 1980) is given by

$$d = \frac{\hat{b}_R - \hat{b}_F}{SE(\hat{b}_R - \hat{b}_F)}, \tag{3.7}$$

where $SE(\hat{b}_R - \hat{b}_F)$ corresponds to the standard error of the difference between the b estimates for the reference and focal groups, given by

$$SE(\hat{b}_R - \hat{b}_F) = \sqrt{\left[SE(\hat{b}_R)\right]^2 + \left[SE(\hat{b}_F)\right]^2}. \tag{3.8}$$

The test statistic, d, is distributed approximately as standard normal, and thus tests of the null hypothesis H_0: $b_R = b_F$ are easily conducted.

If a 1PL model accurately fits the data, then the test statistic d serves as an adequate test of the null hypothesis of no DIF. If, however, a 2PL or 3PL model is used to accurately model the item responses, then only examining the difference between the b-parameters may provide a misleading interpretation of DIF. In this case, Lord's chi-square test of the simultaneous difference between the b- and a-parameters (Lord, 1980) can be computed. This statistic is given by

$$\chi_L^2 = \hat{\mathbf{v}}'\mathbf{S}^{-1}\hat{\mathbf{v}}, \tag{3.9}$$

where $\hat{\mathbf{v}}' = (\hat{a}_R - \hat{a}_F, \hat{b}_R - \hat{b}_F)$, and $\mathbf{S}$ corresponds to the estimated variance-covariance matrix of the between-group difference in the a and b estimates. Lord (1980, p. 223) provides details related to the computation of $\mathbf{S}$. Under the null hypothesis of no DIF, the obtained value of χ_L^2 is distributed as a chi-square variable with two degrees of freedom.

A more recent innovation in IRT-based DIF detection is the likelihood ratio test, which compares the likelihood associated with the studied item's parameters constrained to be equal for the reference and focal groups to that obtained with one or more parameters of the studied item freed to vary between the reference and focal groups. Denoting the likelihood obtained using the compact model for which the studied item parameters are constrained to be equal across groups by $L(\mathrm{C})$, and the likelihood obtained using the augmented model for which the studied item parameters are freed to vary across groups by $L(\mathrm{A})$, a likelihood ratio test can be conducted by computing

$$G^2 = 2\ln\left[\frac{L(\mathrm{A})}{L(\mathrm{C})}\right]. \tag{3.10}$$

The statistic G^2 is distributed approximately as a chi-square variable with degrees of freedom equal to the difference in the number of parameters between the augmented and compact models. Implementation of the likelihood ratio test requires the estimation of the studied item parameters under the augmented and compact model, which can be time intensive if exploratory DIF analyses are conducted for all combinations of parameters for every test item. As a result, nonparametric procedures (such as those discussed in Sections 4 and 5) are often preferred to the IRT likelihood ratio test. Further discussion concerning the theory and application of the likelihood ratio test of DIF, as well as additional IRT approaches to testing DIF, can be found in several sources (Camilli and Shepard, 1994; Thissen et al., 1993).

Each of the IRT DIF detection procedures discussed in this section can be conducted using numerous IRT calibration programs. We recommend using programs that implement marginal maximum likelihood estimation of the item-level parameters due to the stability offered by the marginal maximum likelihood approach. Programs that use marginal maximum likelihood estimation include BILOG-MG 3 (Zimowski et al., 2003), MULTILOG 7 (Thissen et al., 2003), and IRTLRDIF (Thissen, 2001).

4. Proportion-difference approaches for dichotomous items

The IRT approaches described above have theoretical appeal, but their practical implementation can be limited by the requirements of time, group sizes to ensure adequately stable item-level parameter estimates, and reasonable fit of the data to the IRT model being employed. To bypass the requirements of the parametric IRT methods, we can consider the between-group difference in the probability of correct response after conditioning on some observable variable, such as the total test score. Let us represent the total test score by X, such that X has m discrete strata (score levels) represented by $k = 1, 2, \ldots, m$. We can then represent the between-group difference in the probability of correct response conditional on the kth stratum of X by

$$\delta_k = P(Y = 1 \mid X = k, G = R) - P(Y = 1 \mid X = k, G = F). \tag{4.1}$$

Assuming that X is a sufficient statistic for θ, then the null hypothesis of no DIF corresponds to the condition of $\delta_k = 0$, for $k = 1, 2, \ldots, m$ (Zwick, 1990). When X is not sufficient for θ, it may not be the case that $\delta_k = 0$ under the condition of no DIF, and thus DIF detection statistics that incorporate estimates of δ_k may be biased under such conditions. The implication of this for DIF analyses is discussed later in this section.

The population probabilities of correct response for each group are assumed to be unknown, necessitating their estimation from sample data. To describe the estimation of δ_k, let us first organize the responses to item i for the individuals at the kth stratum of X in a contingency table, as shown in Table 1.

In Table 1, A_k corresponds to the frequency of reference group members having a correct response to the item, B_k corresponds to the frequency of reference group members having an incorrect response to the item, C_k corresponds to the frequency of focal group members having a correct response to the item, and D_k corresponds to the frequency of focal group members having an incorrect response to the item. In practice, δ_k

Table 1
Conditional two-way contingency table for a dichotomous item

		Item response (Y)		
		1	0	Total
Group (G)	Reference	A_k	B_k	n_{Rk}
	Focal	C_k	D_k	n_{Fk}
	Total	n_{1k}	n_{0k}	T_k

can be estimated using the conditional between-group difference in the sample proportion correct, given by

$$\hat{\delta}_k = \frac{A_k}{n_{Rk}} - \frac{C_k}{n_{Fk}}. \tag{4.2}$$

Because DIF concerns the conditional between-group difference across all strata, Dorans and Kulick (1986) proposed using a weighted average of the conditional between-group differences in the sample proportion correct as an index of DIF. This weighted average has been termed the standardized p-difference index (SPD), and is given by

$$SPD = \frac{\sum_{k=1}^{m} w_k \hat{\delta}_k}{\sum_{k=1}^{m} w_k}, \tag{4.3}$$

where w_k is the weight assigned to the kth stratum, usually given by n_{Fk}. The values of SPD can range from -1.0 to 1.0; positive values indicate DIF favoring the reference group, and negative values indicate DIF favoring the focal group. The form of SPD given above is a signed index of DIF, because positive and negative values of δ_k can cancel across the m strata. As such, it can be thought of as a nonparametric cousin to the signed IRT area index, SA, described in (3.2). The SPD index, however, is a more versatile index of conditional between-group differences than the SA index because it does not assume a specific parametric form of the IRF for the studied item.

Although a standard error of SPD has been derived (Dorans and Holland, 1993, p. 50), SPD is typically employed as a measure of DIF effect size. The following guidelines have been proposed for interpreting DIF effect size using the SPD index (Dorans and Holland, 1993): $|SPD| < 0.05$ suggests negligible DIF, $0.05 \leqslant |SPD| \leqslant 0.10$ suggests the item is exhibiting a moderate level of DIF and may require further examination; $|SPD| > 0.10$ suggests the item is exhibiting substantial DIF and should be closely examined for the presence of unintended secondary factors.

As was the case for SA, the obtained value of SPD can be misleading if the DIF is crossing in nature, such that the value of δ_k is positive at one end of the ability continuum and negative at the other end of the ability continuum. To examine the unsigned magnitude of the δ_k collapsed across the m strata, an unsigned measure (Camilli and Shepard, 1994, p. 115) can be computed using

$$UPD = \left(\frac{\sum_{k=1}^{m} w_k \hat{\delta}_k^2}{\sum_{k=1}^{m} w_k} \right)^{1/2}. \tag{4.4}$$

The *UPD* index can range from 0 to 1.0, and guidelines similar to those used for *SPD* can be used to interpret the obtained value: $UPD < 0.05$ suggests negligible DIF, $0.05 \leqslant UPD \leqslant 0.10$ suggests moderate DIF, $UPD > 0.10$ suggests substantial DIF.

One of the drawbacks of the *SPD* and *UPD* indices is that the observed total test score is not an ideal measure of θ. This is the case for two reasons. First, the total unweighted test score is not sufficient for the estimate of the latent proficiency metric θ unless the probability of correct response given θ follows the 1PL model (Lord, 1980; Zwick, 1990). Second, the presence of measurement error in X is of particular concern when a between-group difference in the unconditional distribution of θ exists, as this will lead to between-group differences in the distribution of θ conditional on X. As a result, non-zero values of $\hat{\delta}_k$ are expected under the null hypothesis of no DIF, and thus inflated Type I error rates can occur. That the total test score is a sufficient statistic for θ only under the 1PL is a complication that is not isolated to the *SPD* and *UPD* indices, but rather impacts all DIF statistics employing the total test score as the stratifying variable. While this complication will not be given attention in the description of other DIF statistics that stratify on the total test score, we assume it to be understood that its impact applies equally to all such DIF statistics.

Strategies that can compensate for a nonideal X include multivariate matching when additional criteria are available; data reduction techniques such as factor analysis for obtaining a score representing the dominant dimension on the test; or iteratively removing items showing DIF and recomputing X as the stratifying variable. Using such a conditioning strategy, Shealy and Stout (1993a) proposed a measure of the between-group difference in the probability of correct response, an estimate of the signed area DIF effect size measure, *SA*. The resulting statistic, employed in the SIBTEST procedure (Shealy and Stout, 1993a), is computed using

$$\hat{\beta} = \frac{\sum_{k=1}^{m} w_k \left(\bar{Y}_{Rk}^{*} - \bar{Y}_{Fk}^{*}\right)}{\sum_{k=1}^{m} w_k}, \tag{4.5}$$

where w_k is set equal to the number of focal group members at the kth stratum. The values $\bar{Y}_{Rk}^{*}$ and $\bar{Y}_{Fk}^{*}$ correspond to the expected value of the item response (i.e., the probability of correct response) for the reference and focal groups conditional on the true score approximations associated with stratum k for each group. The values $\bar{Y}_{Rk}^{*}$ and $\bar{Y}_{Fk}^{*}$ are computed using

$$\bar{Y}_{Gk}^{*} = \bar{Y}_{Gk} + \hat{M}_{Gk}\left[\hat{V}(k) - \hat{V}_G(k)\right], \tag{4.6}$$

where G represents the group, and

$$\hat{M}_{Gk} = \frac{\bar{Y}_{G,k+1} - \bar{Y}_{G,k-1}}{\hat{V}_G(k+1) - \hat{V}_G(k-1)}, \tag{4.7}$$

$$\hat{V}(k) = \frac{\hat{V}_R(k) + \hat{V}_F(k)}{2}, \tag{4.8}$$

$$V_G(k) = \bar{X}_G + \hat{\rho}(X_k - \bar{X}_G). \tag{4.9}$$

In the above equations $\bar{Y}_{Rk}(= A_k/n_{Rk}$ and $\bar{Y}_{Fk} = C_k/n_{Fk}$ (for dichotomous items). In the final equation, $\bar{X}_G$ represents the mean of the stratifying variable for group G, and $\hat{\rho}$ is an estimate of the reliability of the observed scores, X_k, as given by Cronbach's alpha or some variation thereof (Shealy and Stout, 1993a, p. 192; Zwick et al., 1997, p. 325). A complete derivation of the SIBTEST procedure is presented in Shealy and Stout (1993a). Under the null hypothesis of no DIF, the estimator $\hat{\beta}$ is distributed approximately as a normal variable with a mean of zero, permitting the test statistic

$$z = \frac{\hat{\beta}}{\sqrt{Var(\hat{\beta})}} \tag{4.10}$$

to be used to test the null hypothesis of no DIF, where

$$Var(\hat{\beta}) = \frac{\sum_{k=1}^{m} w_k^2 [Var(\bar{Y}_{Rk}) + Var(\bar{Y}_{Fk})]}{\left(\sum_{k=1}^{m} w_k\right)^2}. \tag{4.11}$$

The theoretical underpinnings of SIBTEST suggest that the procedure should obtain some control over Type I error rates when a between-group difference exists in the distribution of θ conditional on X (as is the case when a between-group difference exists in the unconditional distribution of θ). But, in practice, is SIBTEST robust to between-group differences in the unconditional distribution of θ? The answer is: not entirely – while SIBTEST does maintain a slightly better Type I error rate than other procedures that use an observed stratifying variable (Roussos and Stout, 1996b; Shealy and Stout, 1993a), it still suffers from an inflated Type I error rate when between-group differences in the unconditional distribution of θ exist. As a result, caution should be taken when interpreting the results of any of the proportion-difference approaches in the presence of a between-group difference in the unconditional distribution of θ.

5. Common odds ratio approaches for dichotomous items

An alternative approach to measuring DIF is to consider between-group differences in the odds of correct response after conditioning on the observed test score. Under the null hypothesis of no DIF, the conditional odds of correct response will be equal for the reference and focal groups. This equality can be expressed as

$$\frac{P(Y=1 \mid X=k, G=R)}{1-P(Y=1 \mid X=k, G=R)} = \frac{P(Y=1 \mid X=k, G=F)}{1-P(Y=1 \mid X=k, G=F)}. \tag{5.1}$$

To measure the extent to which equal conditional odds exist, one can consider the conditional odds ratio (ν_k) given by

$$\nu_k = \frac{P(Y=1 \mid X=k, G=R)}{1-P(Y=1 \mid X=k, G=R)} \left[\frac{P(Y=1 \mid X=k, G=F)}{1-P(Y=1 \mid X=k, G=F)}\right]^{-1}. \tag{5.2}$$

Under the null hypothesis of no DIF, $\nu_k = 1$.

In practice, the conditional odds ratio shown in (5.2) is estimated using sample proportions as point estimators of the probability of correct response for the reference and focal group populations. Using the conditional sample proportion correct for the reference group as an estimator of $P(Y = 1 \mid X = k, G = R)$ and the conditional sample proportion correct for the focal group as an estimator of $P(Y = 1 \mid X = k, G = F)$, the conditional odds ratio can be estimated using

$$\hat{\nu}_k = \frac{A_k/n_{Rk}}{B_k/n_{Rk}}\left[\frac{C_k/n_{Fk}}{D_k/n_{Fk}}\right]^{-1} = \frac{A_k D_k}{B_k C_k}. \tag{5.3}$$

The extent to which $\hat{\nu}_k \neq 1.0$ for $k = 1, 2, \ldots, m$ provides evidence against the null hypothesis of no DIF. The value that best characterizes ν_k across the m strata is known as the common odds ratio, denoted here by α. An estimate of α can be obtained by computing a weighted average of the m conditional odds ratios using

$$\hat{\alpha} = \frac{\sum_{k=1}^{m} w_k \hat{\nu}_k}{\sum_{k=1}^{K} w_k}, \tag{5.4}$$

where w_k corresponds to the weight assigned to the kth stratum. Mantel and Haenszel (1959) proposed the weight

$$w_k = \frac{B_k C_k}{T_k}. \tag{5.5}$$

Substituting the above formulation of w_k into (5.4) yields the Mantel–Haenszel common odds ratio, $\hat{\alpha}_{\mathrm{MH}}$, given by

$$\hat{\alpha}_{\mathrm{MH}} = \frac{\sum_{k=1}^{m}\left[\frac{B_k C_k}{T_k}\frac{A_k D_k}{B_k C_k}\right]}{\sum_{k=1}^{m}\frac{B_k C_k}{T_k}} = \frac{\sum_{k=1}^{m}\frac{A_k D_k}{T_k}}{\sum_{k=1}^{m}\frac{B_k C_k}{T_k}}. \tag{5.6}$$

The statistic $\hat{\alpha}_{\mathrm{MH}}$ has received widespread use in the fields of statistics and epidemiology, due to its computational simplicity, consistency, and good efficiency compared to other estimators of α (Breslow, 1981; Donner and Hauck, 1986; Hauck and Donner, 1988; Tarone et al., 1983). In the context of DIF detection, $\hat{\alpha}_{\mathrm{MH}} = 1$ corresponds to the situation of no DIF, $\hat{\alpha}_{\mathrm{MH}} > 1$ corresponds to DIF favoring the reference group, and $\hat{\alpha}_{\mathrm{MH}} < 1$ corresponds to DIF favoring the focal group.

Interpretation of $\hat{\alpha}_{\mathrm{MH}}$ as a DIF effect size estimator is hampered by the fact that $\hat{\alpha}_{\mathrm{MH}}$ is not a symmetric measure of DIF; values corresponding to DIF against the reference group can range from below unity to zero, and values corresponding to DIF against the focal group can range from above unity to infinity. In order to yield a DIF effect size measure that has a symmetric interpretation for the reference and focal groups, it is common to take the natural logarithm of α, represented here by

$$\lambda = \ln(\alpha). \tag{5.7}$$

The value λ is the common log-odds ratio, for which $\lambda = 0$ corresponds to the situation of no DIF, $\lambda < 0$ corresponds to DIF favoring the focal group, and $\lambda > 0$ corresponds to DIF favoring the reference group. The estimate of the common log-odds ratio corresponding to $\hat{\alpha}_{\text{MH}}$ is given by

$$\hat{\lambda}_{\text{MH}} = \ln(\hat{\alpha}_{\text{MH}}). \tag{5.8}$$

The extent to which $\hat{\lambda}_{\text{MH}}$ deviates from zero provides evidence of the presence of DIF; values of $\hat{\lambda}_{\text{MH}}$ less than zero are suggestive of DIF favoring the focal group, and values of $\hat{\lambda}_{\text{MH}}$ greater than zero are suggestive of DIF favoring the reference group.

Standard error estimates of $\hat{\lambda}_{\text{MH}}$ have been proposed by several researchers (Breslow and Liang, 1982; Hauck, 1979; Phillips and Holland, 1987; Robins et al., 1986). Of these, the most widely used in the context of DIF detection is the form proposed by Phillips and Holland (1987) given by

$$Var(\hat{\lambda}_{\text{MH}}) = \frac{\sum_{k=1}^{m} T_k^{-2}(A_k D_k + \hat{\alpha}_{\text{MH}} B_k C_k)(A_k + D_k + \hat{\alpha}_{\text{MH}} B_k + \hat{\alpha}_{\text{MH}} C_k)}{2\left[\sum_{k=1}^{m} \frac{A_k D_k}{T_k}\right]^2}. \tag{5.9}$$

The asymptotic normality of $\hat{\lambda}_{\text{MH}}$ suggests that the null hypotheses of the form H_0: $\lambda = \lambda_0$ can be tested using the standard Wald test given by

$$z = \frac{\hat{\lambda}_{\text{MH}} - \lambda_0}{\sqrt{Var(\hat{\lambda}_{\text{MH}})}}, \tag{5.10}$$

where z is distributed approximately as a standard normal variable.

One of the desirable properties of the estimated common log-odds ratio is that it yields a DIF effect size estimator that can be interpreted in the IRT framework. To illustrate, consider the 2PL, which specifies the probability of correct response using

$$P(Y = 1 \mid \theta) = \frac{\exp[Da(\theta - b)]}{1 + \exp[Da(\theta - b)]}. \tag{5.11}$$

Using (5.11), the conditional odds of correct response can be expressed as

$$\begin{aligned} O(\theta) &= \left(\frac{\exp[Da(\theta - b)]}{1 + \exp[Da(\theta - b)]}\right)\left(1 - \frac{\exp[Da(\theta - b)]}{1 + \exp[Da(\theta - b)]}\right)^{-1} \\ &= \exp\bigl[Da(\theta - b)\bigr]. \end{aligned} \tag{5.12}$$

We can then express the odds ratio conditional on θ as

$$\nu(\theta) = \frac{\exp[Da_R(\theta - b_R)]}{\exp[Da_F(\theta - b_F)]} \tag{5.13}$$

and the conditional log-odds ratio as

$$\ln\bigl[\nu(\theta)\bigr] = D\theta(a_R - a_F) + D(a_F b_F - a_R b_R). \tag{5.14}$$

Under the assumption that the item discrimination parameter is identical for the reference and focal groups ($a_R = a_F = a$), the conditional log-odds ratio reduces to

$$\ln\bigl[\nu(\theta)\bigr] = Da(b_F - b_R). \tag{5.15}$$

The form shown in (5.15) indicates that under the assumption that the reference and focal groups share the same discrimination parameter the conditional log-odds ratio is a constant that is proportional to the between-group difference in the item difficulty parameter. Because the conditional log-odds ratio is constant across θ, and thus is equal to the common log-odds ratio, it follows that the common log-odds ratio can be expressed as

$$\lambda = Da(b_F - b_R). \tag{5.16}$$

Because $\hat{\lambda}_{\text{MH}}$ is an estimator of λ, it follows that $\hat{\lambda}_{\text{MH}}$ can be interpreted as an estimate of $Da(b_F - b_R)$, assuming the item responses follow the 2PL. Furthermore, under the 1PL, for which $a = 1.0$, (5.16) reduces to

$$\lambda = D(b_F - b_R) \tag{5.17}$$

and thus $\hat{\lambda}_{\text{MH}}$ can be interpreted directly with respect the between-group difference in the item difficulty parameter. It is because of this property that $\hat{\lambda}_{\text{MH}}$ has valuable theoretical appeal as a measure of DIF effect size. It should be understood, however, that the relationships described above in (5.16) and (5.17) are only valid under a limited set of circumstances: (a) the stratifying variable used to compute $\hat{\lambda}_{\text{MH}}$ is a sufficient statistic for θ (as is the case in the 1PL), (b) all non-studied items used in constructing the stratifying variable are free of DIF, and (c) the studied item is included in the stratifying variable. These conditions, and their associated rationale, are presented in greater detail by Holland and Thayer (1988) and Zwick (1990). The extent to which these conditions are not met will compromise the interpretation of $\hat{\lambda}_{\text{MH}}$ in terms of a between-group difference in IRT item difficulty parameters.

The drawback of $\hat{\lambda}_{\text{MH}}$ is that it is not a valid measure of DIF when the DIF is nonuniform. As shown in (5.14), if $a_R \neq a_F$, then $\hat{\lambda}_{\text{MH}}$ is not a measure of $b_F - b_R$. The practical implication of this is that, under the condition of $a_R \neq a_F$, it is possible to have a negligible value of $\hat{\lambda}_{\text{MH}}$ despite a substantial difference between b_F and b_R.

The Mantel–Haenszel common log-odds ratio provides a useful tool for assessing DIF, both as a method of conducting hypothesis tests about DIF, and also for answering questions concerning the effect size of the DIF. An alternative method for conducting nondirectional hypothesis tests of uniform DIF is the Mantel–Haenszel chi-square test (Mantel and Haenszel, 1959), given by

$$\chi^2_{\text{MH}} = \frac{\left[\left|\sum_{k=1}^{m}[A_k - E(A_k)]\right| - \frac{1}{2}\right]^2}{\sum_{k=1}^{m} V(A_k)}, \tag{5.18}$$

where

$$E(A_k) = \frac{n_{Rk}n_{1k}}{T_k}, \tag{5.19}$$

$$V(A_k) = \frac{n_{Rk}n_{Fk}n_{1k}n_{0k}}{T_k^2(T_k - 1)}. \tag{5.20}$$

The resulting statistic is distributed approximately as a chi-square variable with one degree of freedom. Assuming the DIF is uniform and the stratifying variable is a sufficient statistic for θ, the chi-square statistic presented in (5.18) provides the uniformly most powerful unbiased test of the null hypothesis of no DIF (Birch, 1964; Holland and Thayer, 1988). Finally, note that the continuity correction in (5.18) brings the chi-square test into closer agreement with an exact conditional test, yet in typical simulation studies the marginal frequencies of the observed-score contingency table have the status of random variables. Thus, researchers conducting simulation studies using χ^2_{MH} should make an explicit determination of whether the correction is consistent with the simulation design.

6. Logistic regression approaches for dichotomous items

The logistic regression approach to studying DIF in dichotomous items was first proposed by Swaminathan and Rogers (1990). In this approach, the dichotomous logistic regression model is used to model the probability of correct response to the studied item as a function of observed test score (X), group membership (G), and the interaction of X and G. The resulting logistic regression model has the form

$$P(Y = 1 \mid X, G) = \frac{\exp(z)}{1 + \exp(z)}, \tag{6.1}$$

where

$$z = \beta_0 + \beta_1 X + \beta_2 G + \beta_3 (XG). \tag{6.2}$$

The coefficient β_2 represents the group effect on performance, and the coefficient β_3 corresponds to the interaction of group membership and ability. The case of no DIF corresponds to $\beta_2 = \beta_3 = 0$, the case of uniform DIF corresponds to $\beta_2 \neq 0$ and $\beta_3 = 0$, and the case of nonuniform DIF corresponds to $\beta_3 \neq 0$.

The values of the regression coefficients, in particular β_2, can serve as useful DIF effect size estimators. Consider the situation in which $\beta_3 = 0$, such that the value of β_2 represents an effect size estimator of uniform DIF. In this situation, if the grouping variable, G, is assigned a value of 0 for the focal group and 1 for the reference group (although the values 1 and 2, or 2 and 3, could have just as easily been used), then the odds of correct response conditional on X for the reference and focal groups are equal to

$$O_R = \exp(\beta_0 + \beta_1 X + \beta_2) \tag{6.3}$$

and

$$O_F = \exp(\beta_0 + \beta_1 X). \tag{6.4}$$

Making use of the odds shown above, the log-odds ratio can be expressed as

$$\ln\left[\frac{O_R}{O_F}\right] = \beta_2 \tag{6.5}$$

indicating that the log-odds ratio is constant across all X. That is, when $\beta_3 = 0$, β_2 is interpreted as a common log-odds ratio, analogous to the Mantel–Haenszel common log-odds ratio, $\hat{\lambda}_{MH}$. This theoretical equivalence between β_2 and $\hat{\lambda}_{MH}$ does not exist in the presence of nonuniform DIF.

To test null hypotheses concerning DIF using logistic regression, likelihood ratio tests can be conducted using a series of three nested models that differ with respect to the DIF-related parameters they contain. The extent to which these three models fit the obtained data sheds light on the form of DIF existing in the responses. The simplest case is a model that does not consider any form of DIF. This model (Model 1) represents the probability of a correct response only as a function of test score, X, such that z is given by

$$z = \beta_0 + \beta_1 X. \tag{6.6}$$

A slightly more complicated model is one in which the probability of correct response is expressed as a function of test score and group, and thus contains a term associated with uniform DIF (β_2). For this model (Model 2), z is given by

$$z = \beta_0 + \beta_1 X + \beta_2 G. \tag{6.7}$$

Finally, we can consider a model that expresses the probability of correct response as a function of test score, group, and the interaction of test score and group. This model (Model 3) contains a term for uniform DIF (β_2) and nonuniform DIF (β_3), such that z is given by

$$z = \beta_0 + \beta_1 X + \beta_2 G + \beta_3 (XG). \tag{6.8}$$

A test for the presence of nonuniform DIF is conducted using a likelihood ratio test by computing

$$\chi^2 = 2\ln\left[\frac{L(\text{Model 3})}{L(\text{Model 2})}\right], \tag{6.9}$$

where L(Model 2) corresponds to the likelihood associated with Model 2, and L(Model 3) corresponds to the likelihood associated with Model 3. The resulting statistic is distributed as a chi-square variable with one degree of freedom. A nonsignificant result leads us to conclude that $\beta_3 = 0$, and thus nonuniform DIF does not exist. In this case, the presence of uniform DIF can be assessed using the likelihood ratio test of

$$\chi^2 = 2\ln\left[\frac{L(\text{Model 2})}{L(\text{Model 1})}\right], \tag{6.10}$$

which is also distributed as chi-square with one degree of freedom. A nonsignificant result leads us to conclude that $\beta_2 = 0$, suggesting that uniform DIF does not exist.

The logistic regression approach holds several potential advantages over the other approaches described to this point (e.g., IRT, conditional proportion-difference, and common odds ratio approaches), and as a result has received substantial use in applied DIF analyses. First, the logistic regression model can inform both uniform and nonuniform DIF, and thus offers a more comprehensive and flexible DIF detection solution than that of the conditional proportion-difference and common odds ratio approaches. Second, because logistic regression uses an observed ability estimate (as opposed to IRT which is based on a latent ability metric), logistic regression can be successfully implemented with group sizes that are substantially smaller than those required for adequate IRT parameter estimation.

Despite the advantages of the logistic regression approach, it is predicated upon the strong assumptions (Camilli and Shepard, 1994, p. 127) that X is a valid representation of ability, and that the logistic regression model is an accurate representation of the relationship between the probability of correct response and observed test score X. It is well known that the observed test score is nonlinearly related to the latent ability metric (Embretson and Reise, 2000; Lord, 1980), and factors such as guessing may preclude the logistic model shown in (6.1) from being an adequate representation of the probability of correct response conditional on ability. That said, in practice they may be realized to an adequate level to warrant the use of the logistic regression model. Given these assumptions, is one better off using the seemingly less restrictive Mantel–Haenszel procedures (common odds ratio and chi-square)? The answer seems to depend on the form of DIF under investigation. If the DIF is uniform, the Mantel–Haenszel approaches have been shown to be slightly more powerful than the logistic regression approach (Hidalgo and López-Pina, 2004). In contrast, if the DIF is nonuniform, then the logistic regression approach will typically display a level of power that exceeds that of the Mantel–Haenszel approaches (Hidalgo and López-Pina, 2004; Rogers and Swaminathan, 1993; Swaminathan and Rogers, 1990).

7. Classification schemes for dichotomous items

Despite the wide variety of statistical procedures available to assess DIF, for any DIF analysis there comes a time when a decision must be made concerning each item. Do the responses to an item display sufficiently little evidence of DIF that the item is labeled as bias-free? Or is there large enough evidence of DIF to warrant additional scrutiny of item content, and the possible removal of the item? Statistical tests of the null hypothesis of no DIF can be used to guide such decisions, but the large sample sizes common to large-scale testing yield so much power that the null hypothesis is often rejected despite extremely small effect sizes. As a result, DIF analysts have gravitated towards decision rules that incorporate significance tests and measures of DIF effect size.

The most widely used DIF decision rule is a classification scheme first developed and used by Educational Testing Service (referred to as ETS hereafter). This scheme classified items into one of three general categories depending on the results of hypothesis tests and the magnitude of a DIF effect size estimator ($\hat{\lambda}_{MH}$). These three categories are:

(A) If either $\hat{\lambda}_{MH}$ is not significantly different from zero, or $|\hat{\lambda}_{MH}| < 0.43$.
(B) If $\hat{\lambda}_{MH}$ is significantly different from zero and $|\hat{\lambda}_{MH}| \geqslant 0.43$, and either: (a) $|\hat{\lambda}_{MH}| < 0.64$, or (b) $|\hat{\lambda}_{MH}|$ is not significantly greater than 0.43.
(C) If $|\hat{\lambda}_{MH}|$ is significantly greater than 0.43 and $|\hat{\lambda}_{MH}| \geqslant 0.64$.

The categories presented above were originally defined not in terms of $\hat{\lambda}_{MH}$ as an effect size estimator, but rather the transformation of $D\text{-}DIF = -2.35\hat{\lambda}_{MH}$, which yields thresholds of 1.0 and 1.5 for the decision rules above (Camilli and Shepard, 1994; Zieky, 1993). Because DIF analyses more frequently employ the $\hat{\lambda}_{\text{MH}}$ as an effect size estimator than *D-DIF*, we specify the categories according to the more widely used $\hat{\lambda}_{\text{MH}}$ index.

The original ETS categorization scheme presented above is deterministic in nature, Platonically labeling items as A, B, or C in a manner that ignores the expected variability in the statistics computed to arrive at the final categorization. That is, a particular item that is categorized as B using one sample may not hold the same categorization for a different sample drawn from the same population as the first sample, solely due to the random variability existing between the two samples. To address this issue, a Bayesian approach to assigning categories to items has been developed (Zwick et al., 1999) that establishes the posterior probabilities that the true level of DIF in the item would lead it to be an A, B, or C item, rather than the deterministic labeling system described above. The posterior probabilities of A, B, and C are obtained by first computing the posterior distribution of λ, using the standard Bayesian form of

$$P(\lambda \mid \hat{\lambda}_{\text{MH}}) = \frac{P(\hat{\lambda}_{\text{MH}} \mid \lambda)P(\lambda)}{\int P(\hat{\lambda}_{\text{MH}} \mid \lambda)P(\lambda)\,\mathrm{d}\lambda}. \tag{7.1}$$

Under the assumption that $\hat{\lambda}_{\text{MH}}$ is normally distributed with mean λ and variance σ^2, and assuming the prior distribution of λ is normal with mean μ and variance τ^2, Bayesian theory asserts (see Lee, 1997) that the posterior distribution is normal with mean and variance equal to

$$E(\lambda \mid \hat{\lambda}_{\text{MH}}) = \frac{\sigma^2\tau^2}{\sigma^2+\tau^2}\left(\frac{\mu}{\tau^2} + \frac{\hat{\lambda}_{\text{MH}}}{\sigma^2}\right), \tag{7.2}$$

$$VAR(\lambda \mid \hat{\lambda}_{\text{MH}}) = \frac{\sigma^2\tau^2}{\sigma^2+\tau^2}. \tag{7.3}$$

In practice μ is estimated using the sample mean across the n values of $\hat{\lambda}_{\text{MH}}$ (where n is the number of items included in the analysis), τ^2 is estimated using the sample variance across the n values of $\hat{\lambda}_{\text{MH}}$, and σ^2 is the variance of $\hat{\lambda}_{\text{MH}}$ estimated using the form presented in (5.9).

Based on the obtained posterior distribution we can obtain the probability that $|\lambda| < 0.43$ (category A), $0.43 \leqslant |\lambda| < 0.64$ (category B), and $|\lambda| \geqslant 0.64$ (category C). This approach does not consider the hypothesis tests included in the original ETS classification scheme presented above, and thus is based purely on the effect size measures. However, the posterior probabilities yield information concerning the confidence one can have that λ lies in the intervals described by the three categories, and thus information concerning the precision of $\hat{\lambda}_{\text{MH}}$ as an estimator of λ is incorporated in the obtained posterior probabilities. It is also of interest to note that the expected value of λ taken across the posterior distribution can be used as a point estimator of λ. The resulting point estimator is a more stable estimate of λ than $\hat{\lambda}_{\text{MH}}$, particularly when group sample sizes are small (Zwick et al., 1999).

A second Bayesian approach to categorizing the magnitude of DIF is to use the posterior-predictive probability (see Zwick et al., 1999), which specifies the future probability of categorization ($|\lambda| < 0.43$, $0.43 \leqslant |\lambda| < 0.64$, and $|\lambda| \geqslant 0.64$) based on the obtained data. In this approach, we consider the probability of observing each possible value of $\hat{\lambda}_{\mathrm{MH}}$ in the future given the observed value of $\hat{\lambda}_{\mathrm{MH}}$ (where a future value of $\hat{\lambda}_{\mathrm{MH}}$ is denoted by $\hat{\lambda}'_{\mathrm{MH}}$). This is obtained by integrating

$$P(\hat{\lambda}'_{\mathrm{MH}} \mid \hat{\lambda}_{\mathrm{MH}}) = \int P(\hat{\lambda}'_{\mathrm{MH}} \mid \lambda) P(\lambda \mid \hat{\lambda}_{\mathrm{MH}})\, \mathrm{d}\lambda \tag{7.4}$$

where it is assumed that future sample sizes are equal to those of the initial sample and that $P(\hat{\lambda}_{\mathrm{MH}} \mid \lambda)$ is specified according to a normal distribution with mean equal to that of the posterior given in (7.2), and variance (see Lee, 1997) equal to

$$VAR(\hat{\lambda}'_{\mathrm{MH}} \mid \hat{\lambda}_{\mathrm{MH}}) = \sigma^2 + \frac{\sigma^2\tau^2}{\sigma^2 + \tau^2}. \tag{7.5}$$

Simulation studies (Zwick et al., 1999) indicate that the Bayesian approaches provide more stable estimates of DIF effect size and DIF classification than the traditional approach of using $\hat{\lambda}_{\mathrm{MH}}$ as a DIF estimator. The improved stability of the DIF estimator is particularly salient under very small group sample sizes (e.g., group sizes of 100 or less). The generation of the posterior distribution and/or the posterior-predictive distribution also provides useful information for test developers in assessing the possibility that the true level of DIF lies in particular ranges of values.

We have limited our discussion of DIF classification schemes to those based on $\hat{\lambda}_{\mathrm{MH}}$ as a DIF effect estimator due to the widespread use of $\hat{\lambda}_{\mathrm{MH}}$ in the testing community. However, a DIF classification scheme for dichotomous items could be constructed based on any of the effect size estimators or significance tests described in Sections 3–6. A simple classification scheme based on the *SPD* index (Dorans and Holland, 1993) was described in Section 4. Other researchers have proposed classification schemes based on the logistic regression approach (Jodoin and Gierl, 2001; Zumbo and Thomas, 1996) and the SIBTEST procedure (Roussos and Stout, 1996b). However, currently the Mantel–Haenszel statistics provide the industry standard.

8. Item response theory approaches for polytomous items

Numerous statistical methods have been proposed for the detection of DIF in polytomous items (Penfield and Lam, 2000; Potenza and Dorans, 1995). While many of these methods are simple extensions of those described for dichotomous items presented in Sections 3–6, the conceptualization of DIF in polytomous items is complicated by the multiple score categories common to polytomous items. That is, for polytomous items a conditional dependence between the item response variable, Y, and grouping variable, G, exists if there is a between-group difference in the conditional probability of any one of the response categories. As was the case with dichotomous items, IRT provides the most comprehensive framework for conceptualizing DIF in polytomous items.

IRT models for polytomous items specify the probability of observing each response option as a function of the latent proficiency θ. Several different polytomous IRT models have been proposed, most of which are variations on two general classes of models: (a) an adjacent categories logistic model, referred to as the generalized partial credit model (GPCM), and (b) a cumulative logistic model, referred to as the graded response model (GRM). The GPCM (Muraki, 1992) presents the probability of attaining response option j given the attainment of response option $j-1$ (represented by η_j) by the 2PL form of

$$\eta_j(\theta) = \frac{\exp[Da(\theta - b_j)]}{1 + \exp[Da(\theta - b_j)]}. \tag{8.1}$$

For an item having r response categories, we denote a particular response category by j, where $j = 0, 1, \ldots, J$, such that $J = r - 1$. There will be J such adjacent category probability functions, one for $j = 1, 2, \ldots, J$. The location, or difficulty, of the transition from category $j - 1$ to category j is reflected in the b_j parameter; the higher the value of b_j, the higher the value of θ required to attain a particular probability of attainment of category j for individuals who had attained category $j - 1$. The a-parameter is associated with the discrimination of the item. Using the adjacent categories formulation, the probability of observing category j can be expressed by the model

$$P(Y = j \mid \theta) = \frac{\exp\left[\sum_{c=0}^{j} Da(\theta - b_c)\right]}{\sum_{c=0}^{J} \exp\left[\sum_{l=0}^{c} Da(\theta - b_l)\right]}. \tag{8.2}$$

Although the scaling factor D has been included in (8.1) and (8.2), it is commonly omitted in specifying the GPCM.

The GRM (Samejima, 1969, 1972, 1997) specifies the conditional cumulative probability of $Y \geqslant j$ for $j > 0$, using

$$\gamma_j(\theta) = \frac{\exp[Da(\theta - b_j)]}{1 + \exp[Da(\theta - b_j)]}. \tag{8.3}$$

As was the case with the GPCM, the GRM posits J such cumulative probability functions. For the jth cumulative probability function the parameter b_j specifies the location of the function; as b_j increases, a higher value of θ is required to attain a specified probability of attaining at least category j. Using this representation, the probability observing category j conditional on θ can be represented by

$$P(Y = j \mid \theta) = \gamma_{j-1}(\theta) - \gamma_j(\theta), \tag{8.4}$$

where it is assumed that $\gamma_{-1}(\theta) = 1$ and $\gamma_J(\theta) = 0$.

Whether the item responses are modeled using the GPCM or the GRM, DIF is said to exist if the parameters b_j, for $j = 1, 2, \ldots, J$, differ for the reference and focal groups. If a between-group difference exists in any one of the b_j, then between-group differences will exist in the resulting conditional probabilities of observing each of the response categories. That said, the conceptualization of DIF effect is substantially more complicated in the case of polytomous items than in dichotomous items for two reasons. First, the interpretation of DIF will vary depending on whether the data follows the GPCM or the GRM. In the case of the GPCM, a between-group difference in b_j indicates a between-group difference in the conditional probability of attaining category j, given the attainment of category $j - 1$. In the case of the GRM, a between-group difference in b_j indicates a between-group difference in the conditional probability of attaining category j or higher.

A second complication arises in the interpretation of the DIF effect because the magnitude of the between-group difference in b_j may vary across the J levels of b_j. Three general categories of patterns of between-group differences in b_j have been proposed (Zwick et al., 1993)

1. Constant DIF, whereby the difference $C_j = b_{jR} - b_{jF}$ is constant across response categories $j = 1, 2, \ldots, J$.
2. Unbalanced DIF, whereby $b_{jR} = b_{jF} + C_j$ for one category, but $b_{jR} = b_{jF}$ for all other categories.
3. Balanced DIF, whereby $b_{jR} = b_{jF} + C_j$ for the lower categories and $b_{jR} = b_{jF} - C_j$ for the higher categories.

Naturally, there are numerous variations on unbalanced and balanced DIF, depending on the number of response categories assumed by the item. What is currently unknown, however, is how the overall magnitude of DIF is related to the pattern of between-group differences in b_j. That is, different patterns of C_j will lead to different manifestations of DIF, and the implication of this on DIF effect size is unknown. In addition, the possibility that between-group differences in the discrimination parameter (a) may exist adds to the complexity of modeling DIF in polytomous items. The implication of between-group differences in the discrimination parameter of polytomous models has not yet been reported in the DIF literature.

The use of IRT in assessing DIF in polytomous items has received only modest attention, likely due to the relatively large sample sizes required for stable parameter estimation of the GPCM and GRM parameters. The most widely used IRT approach to investigating DIF in polytomous items has been the likelihood ratio test of the null hypothesis that $b_{Rj} = b_{Fj}$ for $j = 1, 2, \ldots, J$. The likelihood ratio test assesses the null hypothesis of equal group parameters by considering the difference in the likelihood functions associated with fitting a compact model that uses the same parameters for both the reference and focal groups, L(C), and the likelihood function associated with fitting an augmented model that allows for differences in the step-parameters for the two groups, L(A). Using the ratio of the likelihoods, a likelihood ratio test can be conducted by computing

$$G^2 = 2\ln\left[\frac{L(\mathrm{A})}{L(\mathrm{C})}\right], \tag{8.5}$$

where G^2 is distributed as a chi-square distribution with J degrees of freedom (Kim and Cohen, 1998).

Likelihood ratio tests for the GRM can be implemented using MULTILOG 7 (Thissen et al., 2003). While the likelihood ratio test using the GRM was shown to have greater power than nonparametric methods (such as those discussed in Section 9), it suffers from large Type I error rate inflation when the data are not generated using

Table 2
Conditional contingency table for a polytomous item

		Item response (Y)				
		0	1	...	J	Total
Group (G)	Reference	n_{R0k}	n_{R1k}	...	n_{RJk}	n_{Rk}
	Focal	n_{F0k}	n_{F1k}	...	n_{FJk}	n_{Fk}
	Total	n_{0k}	n_{1k}	...	n_{Jk}	T_k

the GRM (Bolt, 2002). As a consequence, the likelihood ratio test is only appropriate when the DIF analyst is extremely confident that the IRT model employed provides an excellent fit to the response pattern.

9. Mean-difference approaches for polytomous items

As discussed in Section 8, measuring the effect size of DIF in polytomously scored items is complicated by the possibility of between-category differences in the DIF effect. One approach that can be used to address this complication is to assume the restricted interpretation of DIF effect as being the conditional between-group difference in the expected value of the item response variable, Y. If we use an observable test score (X) as the stratifying variable, then the conditional difference at stratum k can be represented by

$$\delta_k = E(Y \mid X = k, G = R) - E(Y \mid X = k, G = F). \tag{9.1}$$

To describe the estimation of δ_k, consider the conditional distribution of responses to the studied item for the reference and focal groups, as shown in Table 2. Using the notation described in Table 2, an estimate of the conditional between-group difference in the expected value of Y is given by

$$\hat{\delta}_k = \frac{R_k}{n_{Rk}} - \frac{F_k}{n_{Fk}}, \tag{9.2}$$

where $R_k = \sum_{j=1}^{J} j n_{Rjk}$ and $F_k = \sum_{j=1}^{J} j n_{Fjk}$. To obtain an index of the DIF effect across all m strata, we can use the standardized mean difference index (Dorans and Schmitt, 1991, 1993), given by

$$SMD = \frac{\sum_{k=1}^{m} w_k \hat{\delta}_k}{\sum_{k=1}^{m} w_k}, \tag{9.3}$$

where w_k is the weight assigned to the kth stratum, usually given by n_{Fk}. Note that the *SMD* index is identical to the *SPD* index generalized to the polytomous item format. The *SMD* statistic is a signed index of DIF, and as such is only appropriate when the same group has the relative advantage across the observed test score continuum.

Two standard error estimators for *SMD* have been proposed (Zwick and Thayer, 1996). Simulation studies of Zwick and Thayer showed that the estimator based on

Mantel's (1963) multivariate hypergeometric framework was the least biased. This variance estimator is given by

$$Var(SMD) = \left(\sum_{k=1}^{m} n_{Fk}\right)^{-2} \sum_{k=1}^{m} n_{Fk}^2 \left(\frac{1}{n_{Fk}} + \frac{1}{n_{Rk}}\right)^2 Var(R_k), \tag{9.4}$$

where

$$Var(R_k) = \frac{n_{Rk} n_{Fk}}{T_k^2 (T_k - 1)} \left[\left(T_k \sum_{j=1}^{J} j^2 n_{jk}\right) - \left(\sum_{j=1}^{J} j n_{jk}\right)^2\right]. \tag{9.5}$$

Using this variance estimator, the test statistic

$$z = \frac{SMD}{\sqrt{Var(SMD)}} \tag{9.6}$$

can be computed, which is distributed approximately as a standard normal variable.

As was the case with *SPD*, *SMD* can suffer from inflated Type I error rates when there is a substantial between-group difference in the mean of the ability distribution. To deal with this, the SIBTEST procedure has been generalized to polytomous items (Chang et al., 1996). The results of a simulation study (Chang et al., 1996) indicate that under conditions of unequal group ability distributions, the SIBTEST procedure displayed a Type I error rate that was closer to the nominal level than that observed for other nonparametric DIF detection procedures.

10. Multivariate hypergeometric distribution approaches for polytomous items

An alternative set of approaches to assessing DIF in polytomous items is based on assuming a multivariate hypergeometric distribution for the conditional bivariate distribution of group membership and item response, as displayed in Table 2. That is, if we treat the marginal values in Table 2 as fixed, then we can test hypotheses concerning the statistic

$$R_k = \sum_{j=1}^{J} j n_{Rjk}. \tag{10.1}$$

Under the null hypothesis of conditional independence of G and Y, the value of R_k will have expected value and variance of

$$E(R_k) = \frac{n_{Rk}}{T_k} \sum_{j=1}^{J} j n_{jk} \tag{10.2}$$

$$Var(R_k) = \frac{n_{Rk} n_{Fk}}{T_k^2 (T_k - 1)} \left[\left(T_k \sum_{j=1}^{J} j^2 n_{jk}\right) - \left(\sum_{j=1}^{J} j n_{jk}\right)^2\right]. \tag{10.3}$$

The derivation of this variance is provided by Mantel (1963, pp. 693–694). Collapsing across the m strata yields the following test statistic for the null hypothesis of no DIF

$$\text{Mantel-}\chi^2 = \frac{\left(\sum_{k=1}^{m} R_k - \sum_{k=1}^{m} E(R_k)\right)^2}{\sum_{k=1}^{m} Var(R_k)}. \tag{10.4}$$

Because the statistic shown in (10.4) was proposed by Mantel (1963), it is denoted by Mantel-χ^2. Under the null hypothesis of no DIF, Mantel-χ^2 is distributed approximately as a chi-square variable with one degree of freedom. It is useful to note that, with the exception of a continuity correction, Mantel-χ^2 reduces to the Mantel–Haesnzel chi-square statistic (χ^2_{MH}) in the case of dichotomous items. It has been proposed (Zwick and Thayer, 1996) that taking the square root of Mantel-χ^2 yields a z-statistic that is distributed approximately as standard normal, and can be used to test directional and nondirectional tests of DIF.

A measure of DIF effect size associated with Mantel-χ^2 is Cox's (1958) noniterative estimator of the noncentrality parameter (ω) of the multivariate hypergeometric distribution, first applied to DIF analysis by Camilli and Congdon (1999). The noncentrality parameter is estimated using

$$\hat{\omega} = \frac{\sum_{k=1}^{m} \sum_{j=1}^{J} j[n_{Rjk} - E(n_{Rjk})]}{\sum_{k=1}^{m} \zeta_k^2}, \tag{10.5}$$

where $E(n_{Rjk})$ is given by

$$E(n_{Rjk}) = \frac{n_{Rk} n_{jk}}{T_k}, \tag{10.6}$$

and ζ_k^2 is computed using

$$\zeta_k^2 = \frac{n_{Rk} n_{Fk}}{T_k(T_k - 1)} \left[\left(\sum_{j=1}^{J} j^2 n_{jk} \right) - \frac{1}{T_k} \left(\sum_{j=1}^{J} j n_{jk} \right)^2 \right]. \tag{10.7}$$

The statistic $\hat{\omega}$ serves as an effect size estimator, such that $\hat{\omega} > 0$ signifies DIF in favor of the reference group, $\hat{\omega} < 0$ signifies DIF favoring the focal group, and $\hat{\omega} = 0$ signifies no DIF. It has been shown that under the condition of constant DIF, the parameter ω is proportional to $b_{jF} - b_{jR}$ when the partial credit model is used to model the probability of item responses (Camilli and Congdon, 1999). As a result, the value of $\hat{\omega}$ can be interpreted in relation to a widely used polytomous IRT model.

The effect size estimator $\hat{\omega}$ can also be used to construct a statistic to test the null hypothesis of no DIF. Using the estimated variance of $\hat{\omega}$ given by Camilli and Congdon (1999) of

$$Var(\hat{\omega}) = \left(\sum_{k=1}^{m} \zeta_k^2 \right)^{-1}, \tag{10.8}$$

the null hypothesis that $\omega = 0$ can be tested using

$$z = \frac{\hat{\omega}}{\sqrt{Var(\hat{\omega})}}. \tag{10.9}$$

The statistic shown in (10.9) is distributed approximately as a standard normal variable. Squaring the statistic shown in (10.9) yields the value of Mantel-χ^2 shown in (10.6), thus demonstrating the relationship between Mantel-χ^2 and $\hat{\omega}$.

The computation of Mantel-χ^2 can be obtained using the DIFAS computer program (Penfield, 2005). Taking the square root of Mantel-χ^2 yields the test statistic displayed in (10.9). This, unfortunately, does not provide direct information concerning the value of $\hat{\omega}$, and thus computation of $\hat{\omega}$ and its associated standard error estimator requires separate programming using a statistical software package such as SPSS or SAS.

11. Common odds ratio approaches for polytomous items

An alternative approach to measuring DIF effect size in a polytomous item is to compute a cumulative common odds ratio. To construct the cumulative common odds ratio for an item having r ordered response categories, we create $J = r - 1$ dichotomizations of Y within each of the m strata of the matching variable. The J dichotomizations are constructed by collapsing the outcomes below category j $(Y < j)$ into one group and the outcomes greater than or equal to category j $(Y \geqslant j)$ into a second group, for $j = 1, 2, \ldots, J$. We denote the cumulative frequency associated with $Y < j$ for the reference and focal groups by n^*_{Rjk} and n^*_{Fjk}. For each of the J dichotomizations, then, we can construct a 2×2 table similar to that displayed in Table 1 for which $A_{jk} = n^*_{Rjk}$, $B_{jk} = n_{Rk} - n^*_{Rjk}$, $C_{jk} = n^*_{Fjk}$, and $D_{jk} = n_{Fk} - n^*_{Fjk}$. Because there will exist J such 2×2 tables at each of the m strata, there will be a total of mJ separate 2×2 tables. If we treat the obtained 2×2 tables at the kth stratum as independent, then an estimator of the common odds ratio for the resulting 2×2 tables is given by

$$\hat{\psi}_{\mathrm{LA}} = \frac{\sum_{k=1}^{m} \sum_{j=1}^{J} A_{jk} D_{jk} / T_k}{\sum_{k=1}^{m} \sum_{j=1}^{J} B_{jk} C_{jk} / T_k}. \tag{11.1}$$

The cumulative common odds ratio estimator shown in (11.1) was first proposed by Liu and Agresti (1996), and was applied to DIF detection in polytomous items by Penfield and Algina (2003). Because $\hat{\psi}_{\mathrm{LA}}$ reduces to the inverse of $\hat{\alpha}_{\mathrm{MH}}$ when $r = 2$, $\hat{\psi}_{\mathrm{LA}}$ can be viewed as a generalization of $\hat{\alpha}_{\mathrm{MH}}$. Transforming $\hat{\psi}_{\mathrm{LA}}$ using

$$\hat{\alpha}_{\mathrm{LA}} = \frac{1}{\hat{\psi}_{\mathrm{LA}}} \tag{11.2}$$

yields an estimator that is consistent in direction with the scale of $\hat{\alpha}_{\mathrm{MH}}$. An estimate of the cumulative common log-odds ratio is given by $\hat{\lambda}_{\mathrm{LA}} = \ln(\hat{\alpha}_{\mathrm{LA}})$, for which $\hat{\lambda}_{\mathrm{LA}} = 0$ suggests no DIF, $\hat{\lambda}_{\mathrm{LA}} > 0$ suggests DIF against the focal group, and $\hat{\lambda}_{\mathrm{LA}} < 0$ suggests DIF against the reference group. The value of $\hat{\lambda}_{\mathrm{LA}}$ has been shown to be proportional to $b_{jF} - b_{jR}$ for the case of constant DIF modeled using the GRM (Penfield and Algina, 2003).

An estimate of the variance of $\hat{\lambda}_{\mathrm{LA}}$ is given by Liu and Agresti (1996), and is described in the context of DIF detection by Penfield and Algina (2003). Assuming asymptotic normality of log-odds ratios, the statistic

$$z_{\mathrm{LA}} = \frac{\hat{\lambda}_{\mathrm{LA}} - C}{\sqrt{Var(\hat{\lambda}_{\mathrm{LA}})}} \tag{11.3}$$

is distributed approximately as standard normal, and can be used to test the null hypothesis that the population cumulative common log-odds ratio equal some constant, C.

In a Monte Carlo study conducted by Penfield and Algina (2003), z_{LA} displayed rejection rates that were similar to those of Mantel-χ^2 and Cox's ω estimator, and $\hat{\lambda}_{\mathrm{LA}}$ displayed expected values that were moderately close to the true effect size simulated in the data when the DIF effect size was constant across the response categories. The quality of $\hat{\lambda}_{\mathrm{LA}}$ as an effect size estimator is expected to less consistent under the condition of unbalanced DIF. Calculation of $\hat{\lambda}_{\mathrm{LA}}$ and z_{LA} can be achieved using the DIFAS computer program (Penfield, 2005).

12. Logistic regression approaches for polytomous items

The logistic regression framework described for dichotomous items (Swaminathan and Rogers, 1990) can be adapted using a variety of methods to accommodate the investigation of DIF in polytomous items. One such method is to recode the r response categories of the polytomous item into $J = r - 1$ coded variables. Each of the J coded variables represents a particular contrast within the r categories, whereby each contrast is established through recoding the original values of the response variable, Y, using a 0/1 coding scheme. The resulting coded variables (denoted by Y^*) are 0/1 dichotomous variables, and thus for each of the J coded variables a logistic regression model of the form

$$P(Y^* = 1 \mid X) = \frac{\exp(\beta_0 + \beta_1 X + \beta_2 G)}{1 + \exp(\beta_0 + \beta_1 X + \beta_2 G)}, \tag{12.1}$$

is fit to the data to examine the presence of DIF within the categories contrasted by the particular coded variable (French and Miller, 1996).

Three different coding schemes have been proposed to obtain the J coded variables; the continuation ratio scheme, the cumulative scheme, and the adjacent categories scheme (Agresti, 1990, pp. 318–322). The continuation ratio scheme creates the dichotomizations using $Y = j$ vs. $Y > j$, for $j = 0, 1, \ldots, J-1$. The cumulative scheme constructs the dichotomizations using $Y \leqslant j$ vs. $Y > j$, for $j = 0, 1, \ldots, J - 1$. The adjacent categories scheme constructs the dichotomizations using $Y = j$ vs. $Y = j+1$, for $j = 0, 1, \ldots, J - 1$. An example of the three coding schemes applied to a polytomous item with four categories is displayed in Table 3 (the contents of Table 3 is based on Table 1 given in French and Miller, 1996).

Within each coding scheme, J logistic regressions can be run such that the null hypothesis H_0: $\beta_2 = 0$ can be assessed through a comparison of the obtained likelihood for the augmented model described in (12.1) and the likelihood for the compact model

Table 3
Coding schemes for logistic regression for a four-category polytomous item

Coding scheme	Dummy variable #	Categories coded as missing	Categories coded as 0	Categories codes as 1
Continuation ratio	1		0	1, 2, 3
	2	0	1	2, 3
	3	0, 1	2	3
Cumulative	1		0	1, 2, 3
	2		0, 1	2, 3
	3		0, 1, 2	3
Adjacent categories	1	2, 3	0	1
	2	0, 3	1	2
	3	0, 1	2	3

given by

$$P(Y^* = 1 \mid X) = \frac{\exp(\beta_0 + \beta_1 X)}{1 + \exp(\beta_0 + \beta_1 X)}, \tag{12.2}$$

which is identical to that described in (12.1) with the exception that it does not contain the group effect term. Using the likelihood obtained for the augmented model, $L(\mathrm{A})$, and that obtained for the compact model, $L(\mathrm{C})$, the following statistic is computed

$$G^2 = 2\ln\left[\frac{L(\mathrm{A})}{L(\mathrm{C})}\right]. \tag{12.3}$$

The statistic shown in (12.3) is distributed approximately as a chi-square variable with one degree of freedom. Notice that the test statistic shown in (12.3) is analogous to that used in the testing of uniform DIF in dichotomous items using logistic regression through a comparison of Model 1 and Model 2 (see Section 6).

An omnibus test of DIF across the J coded variables is only applicable to the continuation ratio scheme because this is the only of the three schemes that yields independent coded variables. To conduct an omnibus test of DIF, the chi-square values obtained using (12.3) for each the J likelihood ratio tests can be added to yield an omnibus chi-square statistic with J degrees of freedom. The results of simulation studies (French and Miller, 1996) indicated that while the results of logistic regression analyses produced detailed descriptions of DIF, its successful implementation with adequate power required relatively large sample sizes (i.e., 2,000). In addition, power tended to be higher using the continuation ratio and cumulative coding schemes than using the adjacent categories, due to the loss of data resulting from the adjacent categories coding scheme.

An alternative approach to using logistic regression for the detection of DIF in polytomous items is to treat the grouping variable as the dependent (or response) variable, and to describe the probability of belonging to the reference group as a function of test score (X), item response (Y), and the interaction of test score and item response, using the logistic function

$$P(G = R \mid X, Y) = \frac{\exp(z)}{1 + \exp(z)}, \tag{12.4}$$

where

$$z = \beta_0 + \beta_1 X + \beta_2 Y + \beta_3 XY. \tag{12.5}$$

The term β_2 corresponds to uniform DIF and term β_3 corresponds to nonuniform DIF. This approach was proposed by Miller and Spray (1993), and is referred to as the logistic discriminant function analysis approach for detecting DIF. Significance tests for the presence of nonuniform and uniform DIF are conducted using the series of likelihood ratio tests as described for logistic regression applied to dichotomous items. Although the logistic discriminant function analysis procedure can be applied to investigate uniform and nonuniform DIF in polytomous items, the power of the this procedure in relation to the other polytomous DIF detection procedures described in this chapter is unknown. Both logistic regression approaches for polytomous items (using J sets of coded variables, or logistic discriminant function analysis) can be conducted using most standard statistical software packages.

13. Differential test functioning and DIF effect variance

The methods described thus far have addressed the concept of fairness and bias at the item level. It is also germane to consider how the individual item-level DIF indices aggregate across all n items of the test, referred to as differential test functioning (DTF). The study of DTF has practical importance because it is possible for a subset of items to contain a level of DIF that is not large enough to warrant flagging at the individual item level, but that aggregates across the subset to result in a substantial impact at the level of the test. In this situation, DTF may signify the possibility of bias despite the lack of conclusive evidence within any single item. This section describes methods used to quantify DTF.

One approach to examining DTF is to consider the between-group differences in the conditional test performance. Shealy and Stout (1993a) proposed generalizing the SIBTEST procedure (presented earlier for detecting DIF in individual items) to permit such an analysis. That is, for some subset of s items, such that the items of the subset are denoted by $i = 1, 2, \ldots, s$, we can consider the weighted average of the conditional between-group differences in the expected sum across the s items of the subset, given by

$$\hat{\beta} = \frac{\sum_{k=1}^{m} w_k (\bar{Y}_{Rk}^{*} - \bar{Y}_{Fk}^{*})}{\sum_{k=1}^{m} w_k}. \tag{13.1}$$

In (13.1) $\bar{Y}_{Rk}^{*}$ and $\bar{Y}_{Fk}^{*}$ correspond to the expected value of the sum of item responses across the s items of the subset for the reference and focal groups conditional on the true score approximations associated with stratum k, and w_k is a weighting factor, typically set equal to the number of focal group members at the kth stratum. Notice that this form is identical that given for SIBTEST applied to single item; SIBTEST applied to single items to detect DIF is simply a restriction of the more general DTF measure where the subset consists of only a single item. Corresponding standard error estimators (identical to those displayed for SIBTEST in DIF detection) exist, and can be employed to create a test statistic that is distributed approximately as standard normal. In addition, SIBTEST as a measure of DTF can accommodate both dichotomous and polytomous item formats.

While the application of SIBTEST has proven useful in some situations (Gierl et al., 2003; Gierl and Khaliq, 2001), its application to more general studies of DTF has two limitations. First, because SIBTEST considers the average between-group difference in the expected subset score, it is a signed measure of DTF. A positive value indicates the reference group members typically outperform focal group members of equal true scores, and a negative value indicates the reverse. The result of this is that opposite DIF effects can cancel one another to yield a net DTF estimate near zero despite substantial levels of DIF at the item level. A second limitation of SIBTEST as applied to the study of DTF is that it is expected to be a rather ineffective measure of DTF when the entire test is used as the subset, and the estimate of ability is based on the scores internal to the test itself. This result is expected due to the ipsitivity inherent in item-level DIF effects (discussed in Section 16), which ensures that the average DIF effect across all items of the test will equal approximately zero (this is a direct result of using the test score itself as an internal measure of ability). Although it was not the intent of Shealy and Stout (1993a) for SIBTEST to be applied as a measure of DTF across the entire test, it is useful to understand this limitation of most measures of differential item functioning using an internal criterion for conditioning.

A second approach for measuring DTF is the DFITS approach first proposed by Raju et al. (1995). The DFITS approach is based on the following logic. First, for an individual having latent proficiency θ, the individual's true score (T) for a test of dichotomously scored items is equal to the expected proportion correct on the test, expressed by

$$T = \sum_{i=1}^{n} P(Y_i = 1 \mid \theta), \tag{13.2}$$

where the subscript i has been inserted to denote the ith item of a test containing n items, and the probability of correct response is specified according to a dichotomous IRT model. If there is no DIF in any of the items of the test, and thus no DTF, it must be the case the true score for an individual with ability θ will be identical regardless of group membership. However, if a DIF effect exists, then between-group differences in item parameters will exist in at least one of the items and, as a result, reference group members with ability θ are expected to have a different true score than focal group members with ability θ. It follows that for a particular individual with ability θ, we can consider the difference between the true score if the individual belongs to the reference group (T_R) and the true score if the individual belongs to the focal group (T_F). The difference between T_R and T_F for a particular individual can be represented by

$$D = T_F - T_R. \tag{13.3}$$

An unsigned index of DTF is then given by the expected value of the squared true score differences integrated over the focal group distribution of θ, given by

$$DTF = \int_{\theta} D^2 f(\theta \mid G = F)\,\mathrm{d}\theta, \tag{13.4}$$

where $f(\theta \mid G = F)$ corresponds to the density of θ for the focal group. The measure of DTF shown in (13.4) can be expressed as

$$DTF = \sigma_D^2 + \mu_D^2. \tag{13.5}$$

The DFITS approach has been extended to tests of polytomous items (Flowers et al., 1999). In this case the same general logic holds, with the difference that the true scores are obtained using polytomous IRT models (e.g., GRM, GPCM) rather than dichotomous models. Implementation of the DFITS paradigm can be accomplished using a computer program written by Raju (1995).

A third approach to measuring DTF is to consider the variance of DIF effect across items with a random effects model. If we define the DIF effect for item i as the common log-odds ratio, λ_i, then the estimate of the DIF effect can be modeled linearly (see Longford et al., 1993) using

$$\hat{\lambda}_i = \mu + \xi_i + \varepsilon_i, \tag{13.6}$$

where μ represents the overall mean DIF effect across items, ξ_i represents the DIF effect of item i, and ε_i represents sampling error. Assuming all test items are scored dichotomously, the DIF effect estimator ($\hat{\lambda}_i$) can be obtained using the Mantel–Haenszel common log-odds ratio as described in Section 5. Using this formulation, the variance of ξ_i (represented by τ^2) provides a measure of global DTF; $\tau^2 = 0$ signifies the absence of DIF, and $\tau^2 > 0$ indicates that between-item differences in DIF effect exist. The situation of $\tau^2 > 0$ can be a result of a few items displaying large DIF effects, or numerous items displaying small DIF effects. Because τ^2 corresponds to the between-item variance component in DIF effect, it is an unsigned index of DTF.

Longford et al. (1993) proposed a maximum likelihood estimator of τ^2. A computationally simpler estimator of τ^2, proposed by Camilli and Penfield (1997), is given by

$$\hat{\tau}^2 = \frac{\sum_{i=1}^{n}(\hat{\lambda}_i - \hat{\mu})^2 - \sum_{i=1}^{n} s_i^2}{n}, \tag{13.7}$$

where $\hat{\lambda}_i$ is equal to the Mantel–Haenszel common log-odds ratio estimator, $\hat{\mu}$ is equal to the mean of $\hat{\lambda}_i$ across the n test items, and s_i^2 is the estimated error variance associated with $\hat{\lambda}_i$ as given by the estimated variance of the Mantel–Haenszel common log-odds ratio in (5.9). Camilli and Penfield also proposed a weighted τ^2 estimator, given by

$$\hat{\tau}_w^2 = \frac{\sum_{i=1}^{n} w_i^2(\hat{\lambda}_i - \hat{\mu})^2 - \sum_{i=1}^{n} w_i}{\sum_{i=1}^{n} w_i^2}, \tag{13.8}$$

where

$$w_i = \frac{1}{s_i^2}. \tag{13.9}$$

While $\hat{\tau}^2$ and $\hat{\tau}_w^2$ were both shown to provide relatively accurate estimates of τ^2 when item responses follow the 1PL and 2PL, $\hat{\tau}_w^2$ was consistently found to be more accurate than $\hat{\tau}^2$ across a variety of conditions (Camilli and Penfield, 1997). In addition, $\hat{\tau}_w^2$ was found to be more accurate than the maximum likelihood estimator of Longford et al. (1993). In the presence of guessing, all estimators were found to underestimate τ^2. As a result, the potential for underestimation must be considered when interpreting τ^2 estimators used in DTF analyses.

The concept of using DIF variance estimators as an index of DTF can be extended to tests consisting exclusively of polytomous items, or to tests consisting of a combination of dichotomous and polytomous items. For example, Penfield and Algina (2005) proposed using the Liu–Agresti cumulative common log-odds ratio as a DIF effect estimator for computing $\hat{\tau}^2$ and $\hat{\tau}_w^2$ in relation to mixed-format tests. Because the Liu–Agresti estimator reduces to the Mantel–Haesnzel estimator in the case of a dichotomous item, the resulting τ^2 estimators can be applied to tests containing dichotomous and polytomous items. Interpretation of the resulting τ^2 estimators is hampered, however, by the highly variable form of DIF observed in polytomous items and the differential weight to overall DTF attributable to dichotomous and polytomous items. Calculation of $\hat{\tau}^2$ and $\hat{\tau}_w^2$ can be conducted in the DIFAS computer program (Penfield, 2005) for tests containing dichotomous and/or polytomous items.

14. Explaining the sources of DIF

While the assessment of DIF is an important aspect to ensuring the fairness and validity of test scores, the statistical assessment of DIF in isolation yields a rather superficial interpretation of item bias. The identification of DIF indicates a conditional between-group difference in item performance, but it provides no information concerning the cause of the DIF. Understanding the cause of DIF is important for two general reasons: (a) it informs whether the presence of DIF indicates an unintended factor determining item responses, and thus whether the DIF signifies a fair or unfair advantage on an item; and (b) it can help test developers better understand item and group properties that are responsible for differential test performance, and consequently may lead to tests that are more fair. This section describes theory and methodology associated with each of these general reasons for understanding the sources of DIF.

Perhaps the most widespread concern for understanding the cause of DIF is to determine whether the observed DIF in an item is actually representative of bias. In this case, items identified as having moderate or large levels of DIF (e.g., using the ETS classification scheme described in Section 7) are subjected to a content review by a panel of experts in an attempt to identify the reason for the observed DIF. If an identified reason corresponds to a construct that was not intended to be measured by the test, then the item is identified as being biased, and would be appropriately revised or removed from the test. If a cause of the DIF is found, but the cause corresponds to a construct that was intended to be measured by the test, then the item is generally retained in the test to preserve the integrity of the content specifications mandated for the test.

A second reason for investigating the sources of observed DIF is to understand item properties that lead to DIF, and possibly unfairness. Several approaches have been proposed for this. One approach is to assess the presence of DIF across predefined subsets of items. For example, if items on a math test can be grouped into subsets according to content, then, using a measure of signed DTF (e.g., SIBTEST), the presence of DIF across all items of the subset can be assessed. In the DIF literature, the subsets of items are termed bundles, and the assessment of aggregated signed DIF across the items of each bundle is referred to as differential bundle functioning (DBF). Let us concern ourselves with the situation in which a particular bundle demonstrates substantial DTF. Because the bundle was defined according to the content peculiar to the items of that bundle, the existence of signed DTF for the bundle provides evidence that the content peculiar to that bundle is responsible for DIF observed in the items of the bundle. Through the examination of which bundles do and do not contain substantial signed DTF, and the respective defining characteristics of the bundles, conclusions can be made concerning item characteristics that are responsible for differential test performance of focal and reference groups. Applications of DBF can be found in numerous studies (Douglas et al., 1996; Gierl et al., 2003; Gierl et al., 2001; Gierl and Khaliq, 2001; Oshima et al., 1998).

In conducting a DBF analysis, the definition of bundles can be constructed a-priori in a confirmatory manner through the use of content experts, or post-hoc in an exploratory manner in which statistical and psychometric methods (e.g., cluster analysis, multidimensional scaling, dual scaling, exploratory factor analysis, etc.) are used to group the items according to two or more dimensions. Gierl et al. (2001) provides an informative overview of issues related to DBF and the associated bundle definition methodology.

While efforts to explain the sources of DIF can occur at the individual item level through an examination of the properties (e.g., content, cognitive processes involved for response) of each item displaying a substantial level of DIF, the application of DBF holds several potential advantages. First, the use of DBF can yield higher power than that associated with item-level DIF analyses because DBF considers aggregated DIF across two or more items. This advantage was previously mentioned in relation to the advantage of DTF over item-level DIF analyses. Second, the bundle may be a more reliable sample of behavior than the individual item (Gierl et al., 2001), and thus may provide a more stable indication of the source of differential performance.

A second approach to identifying sources of DIF is using hierarchical logistic regression (Raudenbush and Bryk, 2002) to model DIF effect as a function of item-level characteristics. In this approach, the logistic regression model described in Section 6 for dichotomous items is applied to each item, and the coefficient representing DIF effect (the uniform DIF effect) in each item is treated as a random variable that can be predicted from item-level characteristics. Using this framework, the proportion of between-item variance in DIF effect that is explained by particular item-level characteristics can be estimated, and thus potential sources of DIF effect can be identified. Because this approach enables the modeling of DIF effect as a function of any number of item-level properties, it offers much greater flexibility than the DBF approach to identifying sources of differential test performance. In addition, robust estimation of the random coefficients provided by the hierarchical approach can provide improved DIF effect estimation over traditional nonnested methods. Swanson et al. (2002) provide a detailed description of the hierarchical logistic regression approach to DIF analysis. Although this approach can be useful, its application is limited by assumptions of model fit, fixed ability estimates, and adequate sample size for parameter estimation. Hierarchical logistic regression analysis can be computed using HLM 5 software (Raudenbush et al., 2000).

A third approach to identifying specific DIF-related factors is to conduct randomized studies to test hypotheses concerning item properties that may cause DIF. This approach begins by developing hypotheses concerning the item properties causing DIF using the results obtained from exploratory DIF analyses conducted on operational test items. Next, two or more versions of a particular item are constructed by varying the level of the item property under investigation. The different versions of the studied item are then administered to random samples of examinees and the amount of DIF observed in the item is compared across the samples. If the samples exhibit different levels of DIF for the studied item, then evidence exists that the manipulated item property is causing DIF. Examples of this approach can be found in several published sources (Bolt, 2000; Scheuneman, 1987; Schmitt et al., 1993).

15. Steps to conducting DIF analyses: An applied example

While much of this chapter has been devoted to describing statistical approaches to measuring DIF, conducting a comprehensive DIF analysis requires of series of steps aimed at measuring DIF and ensuring that the obtained DIF statistics are interpreted appropriately. In this section we outline a set of six steps that practitioners can use in conducting DIF analyses. We then demonstrate these steps using a real dataset.

Step 1. Establish the stratifying variable

If an IRT approach is used for assessing DIF, the stratifying variable is typically of little concern because it is implicit to the IRT analysis itself. In most other cases, however, the DIF analyst must establish an observed score used as the stratifying variable in the analysis. While stratifying variables that are external to the items under investigation can be used, it is often the case that the stratifying variable is the unweighted sum of the responses given to the set of items being examine for DIF (e.g., the observed test score). Although counterintuitive, the studied item should be included in the computation of the stratifying variable because omitting the studied item can induce a bias in the obtained DIF statistics (Zwick, 1990; Zwick et al., 1993).

Step 2. Determine reliability of the stratifying variable

Prior to measuring DIF, the analyst should ensure that the scores comprising the stratifying variable are adequately reliable. Lack of an adequate level of reliability of the stratifying variable can cause DIF statistics to become biased if a between-group difference in the distribution of the target trait exists.

Step 3. Measure between-group differences in the target trait distribution

The presence of a between-group difference in the mean of the target trait can lead to a bias of DIF statistics (Shealy and Stout, 1993a; Zwick, 1990). If a substantial difference exists, then caution should be used in interpreting the obtained measure of DIF because observed DIF may be a result of the between-group difference in target trait. The impact of a between-group difference can become severe when the reliability of the stratifying variable is low, warranting extreme caution when interpreting DIF analyses under these circumstances.

Step 4. Compute DIF statistics

We recommend computing a measure of DIF effect size for each item in concert with a test of the null hypothesis of no DIF. This will ensure that an item is flagged as having substantial DIF only when: (a) a large DIF effect estimate is obtained, and (b) a sufficient level of stability exists in the DIF effect estimate to lead to a rejection of the null hypothesis of no DIF.

Step 5. Conduct content analysis of items flagged as having DIF

For each item flagged as having a substantial level of DIF in Step 4, the content should be examined to identify the secondary trait causing the DIF. If the secondary trait is deemed to be irrelevant to the constructs intended to be measured, then the item should be considered for revision or removal.

Step 6. Remove large DIF items and recomputed DIF statistics

After running the initial DIF analysis and relevant content analysis, all items flagged for revision or removal should be removed from the analysis, and Steps 4 and 5 should be conducted again on the remaining items.

A demonstration of the steps described above will be conducted using the items of the released portion of the mathematics literacy section of the 1995 TIMSS assessment. The section consisted of 20 items (16 dichotomous and 4 polytomous). The reference group consisted of 706 English speaking students administered an English version of the test, and the focal group consisted of 260 French speaking students administered a French version of the test. Both groups resided in Canada. The reliability of the unweighted summated score (estimated using coefficient alpha) was 0.78 and 0.79 for the reference and focal groups, respectively. The mean summated score for the reference and focal groups differed by approximately 0.6 standard deviations (reference group was higher than the focal group) indicating that caution should be used when interpreting the results so that items are not falsely flagged due to the biasing effects of between-group differences on the target trait.

Parametric and nonparametric approaches were used to measure DIF. The nonparametric approaches consisted of two contingency table statistics: (a) $\hat{\lambda}_{\mathrm{LA}}$, and (b) Mantel-$\chi^2$. Note that although $\hat{\lambda}_{\mathrm{LA}}$ and Mantel-χ^2 were introduced for the analysis of DIF in polytomous items, when applied to dichotomous items they reduce to $\hat{\lambda}_{\mathrm{MH}}$ and χ^2_{MH} (with the exception of the continuity correction), respectively. All nonparametric analyses were conducted using the DIFAS computer program (Penfield, 2005). While the meaning (with respect to DIF effect size) of $\hat{\lambda}_{\mathrm{MH}}$ for dichotomous items and $\hat{\lambda}_{\mathrm{LA}}$ for polytomous items is not identical, the respective interpretations are likely parallel enough to permit roughly equivalence comparison of DIF effect across dichotomous and polytomous items (Penfield and Algina, 2003). Parametric approaches consisted of logistic regression analyses conducted using SPSS. For dichotomous items the nonuniform DIF effect of the logistic regression model was obtained by considering the value of $\hat{\beta}_3$ when included in a model containing both $\hat{\beta}_2$ and $\hat{\beta}_3$ (shown in Eq. (6.8)), and the uniform DIF effect was obtained by considering the value of $\hat{\beta}_2$ when using a model that did not include $\hat{\beta}_3$ (shown in Eq. (6.8)). For polytomous items, logistic regression models were conducted using a cumulative dichotomization scheme. Although not displayed in Table 4, all logistic regression analyses were accompanied by the computation of the likelihood ratio statistic for testing the relevant null hypothesis of no DIF.

Table 4 presents the primary results of the DIF analysis. Negative values of $\hat{\lambda}_{LA}$ and $\hat{\beta}$ indicate DIF favoring the focal group and positive values indicate DIF favoring the reference group. Notice the variability in the value of the DIF effect estimator across the 20 items (e.g., $\hat{\lambda}_{LA}$ ranges from -0.716 to 2.474) and the lack of clear delineation between large and small levels of DIF, which is typical of DIF analyses. It is also of interest to note the similarity in the values of $\hat{\lambda}_{LA}$ and $\hat{\beta}_2$ for each dichotomous item (as well as the respective standard errors), which demonstrates the theoretical equivalence between the two estimators under the condition of uniform DIF.

Based on the results shown in Table 4, four items displayed DIF magnitudes that warrant attention: Items 2, 3, 6, and 17. Item 2 displayed a very large uniform DIF effect ($\hat{\lambda}_{LA} = 2.474$, $\hat{\beta}_2 = 2.54$), and the null hypothesis of no DIF was rejected using Mantel-χ^2 as well as the likelihood ratio test of $\hat{\beta}_2 = 0$ ($\chi^2 = 229.165$). The χ^2 value of 229.165 computed for the likelihood ratio test of $\hat{\beta}_2 = 0$ was obtained by taking the difference between the value of "-2 Log likelihood" reported in the SPSS output when $\hat{\beta}_2$ was set equal to zero (1149.68) and when $\hat{\beta}_2$ was free to vary (920.515). The likelihood ratio test of $\hat{\beta}_3 = 0$ also yielded a significant result ($\chi^2 = 6.083$), indicating the possible presence of nonuniform DIF in Item 2. The DIF effects of Items 3, 6, and 17 were also large (although not nearly as large as those of Item 2) and were associated with the rejection of the null hypothesis of no uniform DIF. In the case of Item 6 (a polytomous item), the logistic regression parameter of interest ($\hat{\beta}_2$) significantly differed from zero (determined using likelihood ratio tests) for each of the three steps of the cumulative dichotomization scheme, indicating a relatively constant DIF effect across the steps. The four items discussed above should be subjected to a content review. Any item for which the content review indicates the presence of an unintended secondary trait should be flagged for revision or removal from the test, and the DIF analysis should be conducted again with the flagged items excluded from the stratifying variable.

16. Cautions and limitations

This chapter has outlined just some of the arsenal of statistical methodology developed to detect DIF. But, has the development of this methodology actually served to improve the fairness in test scores? The general conclusion of the DIF community seems to be that the application of DIF methodology has had a modest impact on the fairness of the scores (for a discussion, see Camilli, 1993). It is rare that items are actually removed from a test as a result of a DIF analysis, and the impact of the removal of such items typically has only a negligible impact on test score distributions (Hu and Dorans, 1989).

Table 4
Results of a DIF analysis

Item	Contingency table methods		Logistic regression methods				
	$\hat{\lambda}_{LA}$	Mantel-χ^2	Dichotomous items		Polytomous items		
			$\hat{\beta}_2$ (Uniform)	$\hat{\beta}_3$ (Nonuniform)	$\hat{\beta}_2$ (Step 1)	$\hat{\beta}_2$ (Step 2)	$\hat{\beta}_2$ (Step 3)
1	−0.234	1.201	−0.286	0.116			
	(0.218)		(0.219)	(0.049)			
2	2.474	211.675*	2.554*	0.089*			
	(0.198)		(0.185)	(0.036)			
3	−0.661	12.020*	−0.496*	0.012			
	(0.192)		(0.183)	(0.040)			
4^P	−0.071	0.143			−0.311	0.059	
	(0.185)				(0.234)	(0.196)	
5^P	0.151	0.893			0.528	−0.211	
	(0.168)				(0.194)	(0.217)	
6^P	−0.716	10.850*			−0.850*	−0.614*	−0.746*
	(0.216)				(0.309)	(0.253)	(0.240)
7	−0.157	0.591	−0.102	−0.017			
	(0.205)		(0.197)	(0.040)			
8	−0.541	4.785	−0.553*	−0.080			
	(0.254)		(0.231)	(0.050)			
9	0.142	0.604	0.206	0.020			
	(0.180)		(0.177)	(0.037)			
10	−0.481	3.180	−0.534*	−0.038			
	(0.277)		(0.257)	(0.061)			
11	−0.268	1.859	−0.282	−0.034			
	(0.196)		(0.186)	(0.037)			
12	−0.537	9.054*	−0.561*	0.060			
	(0.180)		(0.184)	(0.039)			
13	0.291	1.512	0.298	−0.057			
	(0.234)		(0.224)	(0.081)			
14	0.140	0.342	0.040	0.012			
	(0.237)		(0.236)	(0.055)			
15	−0.076	0.106	−0.097	−0.030			
	(0.234)		(0.226)	(0.052)			
16	0.198	0.179	0.255	−0.012			
	(0.462)		(0.443)	(0.084)			
17	−0.657	7.798*	−0.544*	−0.045			
	(0.235)		(0.226)	(0.044)			
18	0.151	0.682	0.210	0.036			
	(0.182)		(0.179)	(0.038)			
19	−0.274	1.778	−0.271	0.010			
	(0.204)		(0.199)	(0.043)			
20^P	0.482	7.254*			0.613*	0.399	
	(0.181)				(0.212)	(0.222)	

Note. Values in parentheses correspond to standard errors of the associated statistics. Items denoted by a subscripted P are polytomously scored. The asterisk (*) indicates the rejection of the null hypothesis of no DIF using a Type I error rate of 0.05.

Perhaps the greatest impact that DIF analysis has had on test scores concerns the perception of test fairness; test developers now have statistical methodologies that serve as a robust and objective safety net to guard against bias. In that sense, then, test developers are upholding their duty to generate tests that are as fair as possible. Yet, one cannot escape the reality that DIF provides only a partial solution to the eradication of item and test bias. DIF has several limitations, some statistical and some conceptual, that prohibit it from providing the omnibus panacea for bias. Understanding and respecting these limitations will serve the testing community well in future endeavors of test fairness. The remainder of this section describes a few of the most restrictive limitations inherent in current DIF methodology.

We begin by describing the most pervasive statistical limitation of DIF methodologies – the use of a measure of ability that is internal to the test itself, resulting in measures of DIF that are ipsitive. To describe the ipsitivity problem (also termed the *circularity problem* by Camilli, 1993), consider a test for which there are two items that display a large bias effect against the focal group. The presence of bias against the focal group on the two items yields a situation whereby the test score underestimates ability for numerous focal group members. The biased test scores for the focal group yields a situation whereby reference and focal group members having the same test score will not share the same expected ability level; the focal group members will have a higher expected ability level than the matched reference group members. As a result, the non-biased items of the test will display a small level of DIF favoring the focal group. That is, DIF observed against the reference group is dependent upon the impact of the two biased items on the focal group test score distribution. This between-item dependence in DIF statistics is the ipsitivity problem. The end result of the ipsitivity problem is that the average value of signed DIF indices (e.g., the common log-odds ratio) across all items of a test is approximately zero, and thus the absolute magnitude of DIF indices cannot be used to make statements concerning a pervasive bias across the items of a test. Rather, information concerning item-level bias is based on relative sign and magnitude of the DIF statistics among the items of the test.

While DIF analyses routinely flag items appearing to favor the reference group, such analyses also flag items that favor the focal group. Determining whether the observed DIF in favor of a particular group is attributable to bias in the item, or to bias in other items of the test in favor of the other group can prove difficult. Another potential hurdle posed by the ipsitivity problem is that in the case of pervasive bias where every item is biased against a particular group, none of the items will display substantial DIF. In this case, the conditional underperformance on the items by the disadvantaged group will be offset by the nonvalid matching of reference and focal group members caused by the pervasive bias.

While a complete solution to the ipsitivity problem does not exist, steps can be taken to control some of the problems resulting from stratifying variable contamination. A two-stage screening procedure can be used to purifying the stratifying variable: (a) in the first stage a standard DIF analysis is run and items are screened for removal based on the results of the DIF analysis and expert review, and (b) in the second stage the DIF analysis is rerun using the items retained from the first stage. DIF analysts will admit, however, that there is often such a large range of DIF values across the items of the test that even after an initial round of item removal the DIF statistics obtained in the second stage are highly variable, thus obscuring clear and simple classification of items as being biased.

A second limitation of current DIF detection methodologies concerns the grouping variables commonly used to designate the reference and focal groups. Typical dichotomous grouping variables take the form White vs. Black, White vs. Hispanic, male vs. female, etc. The groupings are, in general, crude and may not consistently distinguish between individuals who have differential exposure to the secondary factor causing item bias. For example, variables such as educational background or socio-economic status may display a stronger relationship with the secondary factor causing the item bias than a gross categorization of White vs. Black. These sentiments were described over a decade ago by Dorans and Holland (1993) by:

> To date, focal groups have been intact, easily defined groups such as Asians, Blacks, Hispanics, and women. References [*sic*] groups have been Whites or men. It could be argued, however, that these intact ethnic groups are merely surrogates for an educational disadvantage attribute that should be used in focal group definition. In fact, within any of these groups, there is probably a considerable degree of variability with respect to educational advantage or disadvantage. Perhaps we should be focusing our group definition efforts toward defining and measuring educational advantage or disadvantage directly. (p. 65)

Other researchers have investigated matching on factors such as educational background (Clauser et al., 1996), and the results appear promising. Further improvements in matching may be obtained by using other nontraditional variables that have been implicated in explaining differential test performance, such as racial identity (Helms, 2004).

Finally, the removal of DIF does not mean that the test is free of bias. The only link we currently have between DIF and item bias is expert judgment specifying properties of items that can be predicted to yield biased test scores. The results of several studies have indicated that expert judges were unable to identify biased items (Reynolds, 1982) and were largely ineffective at flagging the same items as flagged using DIF methodology (Englehard et al., 1990). An alternative use of expert review is that of a sensitivity review, first formally documented by ETS in 1980 (ETS, 1980), and more recently termed a fairness review (ETS, 2003). According to ETS (2003), a sensitivity review makes use of trained expert judges to

> identify and remove invalid aspects of test questions that might hinder people in various groups from performing at levels that allow appropriate inferences about their relevant knowledge and skills. (p. 1)

To this end, the sensitivity review seeks to ensure that: (a) the test is free of content (e.g., wording, figures, situations, etc.) that is sexists, racist, stereotypic, controversial, potentially offensive and upsetting, ethnocentric, specialized to particular settings or occupations (e.g., military topics, legal words, scientific words etc.), or construct-irrelevant stimuli (e.g., maps, charts, etc.); and (b) the items of a test reflect the diversity of gender, race, and ethnicity found in our society (Bond et al., 1996; ETS, 2003; Ramsey, 1993).

A comprehensive sensitivity review could consist of the following five stages: (a) fairness reviewers are trained appropriately, provided access to all pertinent test specifications, and informed of the characteristics of the population taking the test; (b) a preliminary review is conducted of proposed test specifications (particularly content that is potentially controversial or expensive to develop) prior to test construction and pretesting; (c) a review is conducted concurrently with pretest item analyses in which items may be modified or deleted; (d) a review is conducted prior to the administration of operational test forms; and (e) additional reviews of test content are conducted at least once every five years. Variations of the final four stages may be required depending on the specific details guiding test development. A comprehensive discussion of the training of fairness reviewers and the guidelines associated with the review process is beyond the score of this chapter. Detailed descriptions of the sensitivity review process, its rationale, and its implementation, can be found in ACT (2004), ETS (2003) and Ramsey (1993).

References

ACT, Inc. (2004). *Fairness Report for the ACT Assessment Tests*. ACT, Iowa City, IA.

Agresti, A. (1990). *Categorical Data Analysis*. Wiley, New York.

American Educational Research Association (AERA), American Psychological Association (APA), National Council on Measurement in Education (NCME) (1999). *Standards for Educational and Psychological Testing*. APA, Washington, DC.

Baker, F.B. (1992). *Item Response Theory: Parameter Estimation Techniques*. Marcel Dekker, New York.

Birch, M.W. (1964). The detection of partial association, I: The 2 × 2 case. *The Journal of the Royal Statistical Society, Ser. B* **26**, 313–324.

Bolt, D.M. (2000). A SIBTEST approach to testing DIF hypotheses using experimentally designed test items. *Journal of Educational Measurement* **37**, 307–327.

Bolt, D.M. (2002). A Monte Carlo comparison of parametric and nonparametric polytomous DIF detection methods. *Applied Measurement in Education* **15**, 113–141.

Bolt, D.M., Hare, R.D., Vitale, J.E., Newman, J.P. (2004). A multigroup item response theory analysis of the psychopathy checklist – revised. *Psychological Assessment* **16**, 155–168.

Bond, L., Moss, P., Carr, P. (1996). Fairness in large-scale performance assessment. In: Phillips, G.W., Goldstein, A. (Eds.), *Technical Issues in Large-Scale Performance Assessment*. National Center for Education Statistics, Washington, DC, pp. 117–140.

Breslow, N. (1981). Odds ratio estimators when the data are sparse. *Biometrika* **68**, 73–84.

Breslow, N.E., Liang, K.Y. (1982). The variance of the Mantel–Haenszel estimator. *Biometrics* **38**, 943–952.

Camilli, G. (1992). A conceptual analysis of differential item functioning in terms of a multidimensional item response model. *Applied Psychological Measurement* **16**, 129–147.

Camilli, G. (1993). The case against item bias detection techniques based on internal criteria: Do item bias procedures obscure test fairness issues? In: Holland, P.W., Wainer, H. (Eds.), *Differential Item Functioning*. Lawrence Erlbaum, Hillsdale, NJ, pp. 397–413.

Camilli, G., Congdon, P. (1999). Application of a method of estimating DIF for polytomous test items. *Journal of Educational and Behavioral Statistics* **24**, 323–341.

Camilli, G., Penfield, D.A. (1997). Variance estimation for differential test functioning based on Mantel–Haenszel statistics. *Journal of Educational Measurement* **34**, 123–139.

Camilli, G., Shepard, L.A. (1994). *Methods for Identifying Biased Test Items*. Sage, Newbury Park, CA.

Chang, H., Mazzeo, J., Roussos, L. (1996). Detecting DIF for polytomously scored items: An adaptation of the SIBTEST procedure. *Journal of Educational Measurement* **33**, 333–353.

Clauser, B.E., Nungester, R.J., Swaminathan, H. (1996). Improving the matching for DIF analysis by conditioning on both test score and an educational background variable. *Journal of Educational Measurement* **33**, 453–464.

Cleary, T.A. (1968). Test bias: Prediction of grades of Negro and white students in integrated colleges. *Journal of Educational Measurement* **5**, 115–124.

Cox, D.R. (1958). The regression analysis of binary sequences. *Journal of the Royal Statistical Society B* **20**, 215–232.

Dodeen, H., Johanson, G.A. (2003). An analysis of sex-related differential item functioning in attitude assessment. *Assessment and Evaluation in Higher Education* **28**, 129–134.

Donner, A., Hauck, W.W. (1986). The large-sample relative efficiency of the Mantel–Haenszel estimator in the fixed strata case. *Biometrics* **42**, 537–545.

Dorans, N.J., Holland, P.W. (1993). DIF detection and description: Mantel–Haenszel and standardization. In: Holland, P.W., Wainer, H. (Eds.), *Differential Item Functioning*. Lawrence Erlbaum, Hillsdale, NJ, pp. 35–66.

Dorans, N.J., Kulick, E.M. (1986). Demonstrating the utility of the standardization approach to assessing unexpected differential item performance on the Scholastic Aptitude Test. *Journal of Educational Measurement* **23**, 355–368.

Dorans, N.J., Schmitt, A.P. (1991). Constructed response and differential item functioning: A pragmatic approach. ETS Research Report No. 91-47. Educational Testing Service, Princeton, NJ.

Dorans, N.J., Schmitt, A.P. (1993). Constructed response and differential item functioning: A pragmatic approach. In: Bennett, R.E., Ward, W.C. (Eds.), *Construction Versus Choice in Cognitive Assessment: Issues in Constructed Response, Performance Testing, and Portfolio Assessment*. Lawrence Erlbaum, Hillsdale, NJ.

Douglas, J., Roussos, L., Stout, W. (1996). Item bundle DIF hypothesis testing: Identifying suspect bundles and assessing their DIF. *Journal of Educational Measurement* **33**, 465–484.

Educational Testing Service (1980). *An Approach for Identifying and Minimizing Bias in Standardized Tests: A Set of Guidelines*. Author, Princeton, NJ.

Educational Testing Service (2003). *Fairness Review Guidelines*. Author, Princeton, NJ.

Embretson, S.E., Reise, S.P. (2000). *Item Response Theory for Psychologists*. Erlbaum, Mahwah, NJ.

Englehard, G., Hansche, L., Rutledge, K.E. (1990). Accuracy of bias review judges in identifying differential item functioning on teacher certification tests. *Applied Measurement in Education* **3**, 347–360.

Flowers, C.P., Oshima, T.C., Raju, N.S. (1999). A description and demonstration of the polytomous-DFIT framework. *Applied Psychological Measurement* **23**, 309–326.

French, A.W., Miller, T.R. (1996). Logistic regression and its use in detecting differential item functioning in polytomous items. *Journal of Educational Measurement* **33**, 315–332.

Gelin, M.N., Carleton, B.C., Smith, A.A., Zumbo, B.D. (2004). The dimensionality and gender differential item functioning of the mini asthma quality of life questionnaire (MINIAQLQ). *Social Indicators Research* **68**, 91–105.

Gelin, M.N., Zumbo, B.D. (2003). Differential item functioning results may change depending on how an item is scored: An illustration with the center for epidemiologic studies depression scale. *Educational and Psychological Measurement* **63**, 65–74.

Gierl, M., Bisanz, J., Bisanz, G.L., Boughton, K.A. (2003). Identifying content and cognitive skills that produce gender differences in mathematics: A demonstration of the multidimensionality-based DIF analysis paradigm. *Journal of Educational Measurement* **40**, 281–306.

Gierl, M., Bisanz, J., Bisanz, G.L., Boughton, K.A., Khaliq, S.N. (2001). Illustrating the utility of differential bundle functioning analyses to identify and interpret group differences on achievement tests. *Educational Measurement: Issues and Practice* **20** (2), 26–36.

Gierl, M.J., Khaliq, S.N. (2001). Identifying sources of differential item and bundle functioning on translated achievement tests: A confirmatory analysis. *Journal of Educational Measurement* **38**, 164–187.

Hauck, W.W. (1979). The large sample variance of the Mantel–Haenszel estimator of a common odds ratio. *Biometrics* **35**, 817–819.

Hauck, W.W., Donner, A. (1988). The asymptotic relative efficiency of the Mantel–Haenszel estimator in the increasing-number-of-strata case. *Biometrics* **44**, 379–384.

Helms, J.E. (2004). The 2003 Leona Tyler award address: Making race a matter of individual differences within groups. *The Counseling Psychologist* **32**, 473–483.

Hidalgo, M.D., López-Pina, J.P. (2004). Differential item functioning detection and effect size: A comparison between logistic regression and Mantel–Haenszel procedures. *Educational and Psychological Measurement* **64**, 903–915.

Hills, J.R. (1989). Screening for potentially biased items in testing programs. *Educational Measurement: Issues and Practice* **8** (4), 5–11.

Holland, P.W., Thayer, D.T. (1988). Differential item performance and the Mantel–Haenszel procedure. In: Wainer, H., Braun, H.I. (Eds.), *Test Validity*. Erlbaum, Hillsdale, NJ, pp. 129–145.

Holland, P.W., Wainer, H. (1993). *Differential Item Functioning*. Erlbaum, Hillsdale, NJ.

Hu, P.G., Dorans, N.J. (1989, March). The effects of deleting items with extreme differential item functioning on equating functions and reported score distributions. Paper presented at the Annual Meeting of the American Educational Research Association, San Francisco, CA.

Iwata, N., Turner, R.J., Lloyd, D.A. (2002). Race/ethnicity and depressive symptoms in community-dwelling young adults: A differential item functioning analysis. *Psychiatry Research* **110**, 281–289.

Jodoin, M.G., Gierl, M.J. (2001). Evaluating Type I error and power rates using an effect size measure with the logistic regression procedure for DIF detection. *Applied Measurement in Education* **14**, 329–349.

Kahler, C.W., Strong, D.R., Stuart, G.L., Moore, T.M., Ramsey, S.E. (2003). Item functioning of the alcohol dependence scale in a high-risk sample. *Drug and Alcohol Dependence* **72**, 183–192.

Kim, S.-H., Cohen, A.S. (1998). Detection of differential item functioning under the graded response model with the likelihood ratio test. *Applied Psychological Measurement* **22**, 345–355.

Lange, R., Thalbourne, M.A., Houran, J., Lester, D. (2002). Depressive response sets due to gender and culture-based differential item functioning. *Personality and Individual Differences* **33**, 937–954.

Lee, P.M. (1997). *Bayesian Statistics: An Introduction*, 2nd ed. Oxford University Press, New York.

Liu, I.-M., Agresti, A. (1996). Mantel–Haenszel-type inference for cumulative odds ratios with a stratified ordinal response. *Biometrics* **52**, 1223–1234.

Longford, N.T., Holland, P.W., Thayer, D.T. (1993). Stability of the MH D-DIF statistics across populations. In: Holland, P.W., Wainer, H. (Eds.), *Differential Item Functioning*. Lawrence Erlbaum, Hillsdale, NJ, pp. 171–196.

Lord, F.M. (1980). *Applications of Item Response Theory to Practical Testing Problems*. Lawrence Erlbaum, Hillsdale, NJ.

Mantel, N. (1963). Chi-square tests with one degree of freedom: Extension of the Mantel–Haenszel procedure. *Journal of the American Statistical Association* **58**, 690–700.

Mantel, N., Haenszel, W. (1959). Statistical aspects of the analysis of data from retrospective studies of disease. *Journal of the National Cancer Institute* **22**, 719–748.

Miller, T.M., Spray, J.A. (1993). Logistic discriminant function analysis for DIF identification of polytomously scored items. *Journal of Educational Measurement* **30**, 107–122.

Millsap, R.E., Everson, H.T. (1993). Methodology review: Statistical approaches for assessing measurement bias. *Applied Psychological Measurement* **17**, 297–334.

Muraki, R. (1992). A generalized partial credit model: Applications of an EM algorithm. *Applied Psychological Measurement* **16**, 159–176.

Noguera, P.A., Akom, A. (2000). Disparities demystified. *The Nation* (5 June 2000), 29–31.

Oshima, T.C., Raju, N.S., Flowers, C.P., Slinde, J.A. (1998). Differential bundle functioning using the DFIT framework: Procedures for identifying possible sources of differential functioning. *Applied Psychological Measurement* **11**, 353–369.

Panter, A.T., Reeve, B.B. (2002). Assessing tobacco beliefs among youth using item response theory models. *Drug and Alcohol Dependence* **68**, S21–S39.

Penfield, R.D. (2005). DIFAS: Differential item functioning analysis system. *Applied Psychological Measurement* **29**, 150–151.

Penfield, R.D., Algina, J. (2003). Applying the Liu–Agresti estimator of the cumulative common odds ratio to DIF detection in polytomous items. *Journal of Educational Measurement* **40**, 353–370.

Penfield, R.D., Algina, J. (2005). A method for identifying unsigned differential test functioning in tests dichotomous and polytomous item formats. Paper presented at the Annual Meeting of the National Council on Measurement in Education, Montreal, Canada.

Penfield, R.D., Lam, T.C.M. (2000). Assessing differential item functioning in performance assessment: Review and recommendations. *Educational Measurement: Issues and Practice* **19** (3), 5–15.

Phillips, A., Holland, P.W. (1987). Estimators of the variance of the Mantel–Haenszel log-odds-ratio estimate. *Biometrics* **43**, 425–431.

Potenza, M.T., Dorans, N.J. (1995). DIF assessment for polytomously scored items: A framework for classification and evaluation. *Applied Psychological Measurement* **19**, 23–37.

Raju, N. (1995). DFITPU: A FORTRAN program for calculating DIF/DTF [Computer program]. Georgia Institute of Technology, Atlanta, GA.

Raju, N.S. (1988). The area between two item characteristic curves. *Psychometrika* **53**, 495–502.

Raju, N.S., van der Linden, W.J., Fleer, P.F. (1995). IRT-based internal measures of differential functioning of items and tests. *Applied Psychological Measurement* **19**, 353–368.

Ramsey, P.A. (1993). Sensitivity review: The ETS experience as a case study. In: Holland, P.W., Wainer, H. (Eds.), *Differential Item Functioning*. Lawrence Erlbaum, Hillsdale, NJ, pp. 367–388.

Raudenbush, S.W., Bryk, A.S. (2002). *Hierarchical Linear Models: Applications and Data Analysis Methods*, 2nd ed. Sage, Thousand Oaks, CA.

Raudenbush, S.W., Bryk, A.S., Cheong, Y.F., Congdon, R.T. (2000). HLM 5: Hierarchical Linear and Nonlinear Modeling [Computer software]. Scientific Software International, Lincolnwood, IL.

Reynolds, C.R. (1982). The problem of bias in psychological assessment. In: Reynolds, C.R., Gutkin, B. (Eds.), *The Handbook of School Psychology*. Wiley, New York, pp. 178–201.

Robins, J., Breslow, N., Greenland, S. (1986). Estimators of the Mantel–Haenszel variance consistent in both sparse data and large-strata limiting models. *Biometrics* **42**, 311–323.

Rogers, H.J., Swaminathan, H. (1993). A comparison of logistic regression and Mantel–Haenszel procedures for detecting differential item functioning. *Applied Psychological Measurement* **17**, 105–116.

Roussos, L., Stout, W. (1996a). A multidimensionality-based DIF analysis paradigm. *Applied Psychological Measurement* **20**, 355–371.

Roussos, L.A., Stout, W.F. (1996b). Simulation studies of the effects of small sample size and studied item parameters on SIBTEST and Mantel–Haenszel Type I error performance. *Journal of Educational Measurement* **33**, 215–230.

Rudner, L.M., Geston, P.R., Knight, D.L. (1980). Biased item detection techniques. *Journal of Educational Statistics* **5**, 213–233.

Samejima, F. (1969). Estimation of latent ability using a response pattern of graded scores. *Psychometrika Monograph Supplement* **34** (4, Part 2, No. 17).

Samejima, F. (1972). A general model for free response data. *Psychometrika Monograph Supplement* **37** (1, Part 2, No. 18).

Samejima, F. (1997). Graded response model. In: van der Linden, W.V., Hambleton, R.K. (Eds.), *Handbook of Modern Item Response Theory*. Springer, New York, pp. 85–100.

Scheuneman, J.D. (1987). An experimental, exploratory study of causes of bias in test items. *Journal of Educational Measurement* **24**, 97–118.

Schmitt, A.P., Holland, P.W., Dorans, N.J. (1993). Evaluating hypotheses about differential item functioning. In: Holland, P.W., Wainer, H. (Eds.), *Differential Item Functioning*. Lawrence Erlbaum, Hillsdale, NJ, pp. 281–315.

Shealy, R.T., Stout, W.F. (1993a). A model based standardization approach that separates true bias/DIF from group ability differences and detects test bias/DTF as well as item bias/DIF. *Psychometrika* **58**, 197–239.

Shealy, R.T., Stout, W.F. (1993b). An item response theory model for test bias and differential item functioning. In: Holland, P.W., Wainer, H. (Eds.), *Differential Item Functioning*. Lawrence Erlbaum, Hillsdale, NJ, pp. 197–239.

Swaminathan, H., Rogers, H.J. (1990). Detecting differential item functioning using logistic regression procedures. *Journal of Educational Measurement* **27**, 361–370.

Swanson, D.B., Clauser, B.E., Case, S.M., Nungester, R.J., Featherman, C. (2002). Analysis of differential item functioning (DIF) using hierarchical logistic regression models. *Journal of Educational and Behavioral Statistics* **27**, 53–75.

Tarone, R.E., Gart, J.J., Hauck, W.W. (1983). On the asymptotic inefficiency of certain noniterative estimators of a common relative risk or odds ratio. *Biometrika* **70**, 519–522.

Thissen, D. (2001). IRTLRDIF (Version 2) [Computer software]. Retrieved January 17, 2006, from http://www.unc.edu/~dthissen/dl.html.

Thissen, D., Chen, W.-H., Bock, D. (2003). MULTILOG 7 [Computer software]. Scientific Software International, Lincolnwood, IL.

Thissen, D., Steinberg, L., Wainer, H. (1993). Detection of differential item functioning using the parameters of item response models. In: Holland, P.W., Wainer, H. (Eds.), *Differential Item Functioning*. Lawrence Erlbaum, Hillsdale, NJ, pp. 67–113.

Zieky, M. (1993). Practical questions in the use of DIF statistics in item development. In: Holland, P.W., Wainer, H. (Eds.), *Differential Item Functioning*. Lawrence Erlbaum, Hillsdale, NJ, pp. 337–347.

Zimowski, M., Muraki, E., Mislevy, R., Bock, D. (2003). BILOG-MG 3 [Computer software]. Scientific Software International, Lincolnwood, IL.

Zumbo, B.D., Thomas, D.R. (1996). A measure of DIF effect size using logistic regression procedures. Paper presented at the National Board of Medical Examiners, Philadelphia, PA.

Zwick, R. (1990). When do item response function and Mantel–Haenszel definitions of differential item functioning coincide? *Journal of Educational Statistics* **15**, 185–198.

Zwick, R. (2001). What causes the test-score gap in higher education. *Change* **33** (2), 32–37.

Zwick, R., Donoghue, J.R., Grima, A. (1993). Assessment of differential item functioning for performance tasks. *Journal of Educational Measurement* **30**, 233–251.

Zwick, R., Thayer, D.T. (1996). Evaluating the magnitude of differential item functioning in polytomous items. *Journal of Educational and Behavioral Statistics* **21**, 187–201.

Zwick, R., Thayer, D.T., Lewis, C. (1999). An empirical Bayes approach to Mantel–Haenszel DIF analysis. *Journal of Educational Measurement* **36**, 1–28.

Zwick, R., Thayer, D.T., Mazzeo, J. (1997). Descriptive and inferential procedures for assessing differential item functioning in polytomous items. *Applied Measurement in Education* **10**, 321–334.

Handbook of Statistics, Vol. 26
ISSN: 0169-7161

DOI: 10.1016/S0169-7161(06)26006-1

6

Equating Test Scores

Paul W. Holland, Neil J. Dorans and Nancy S. Petersen

1. Introduction and definitions

Test equating is a necessary part of any testing program that continually produces new test forms and for which the uses of these test forms require that the meaning of the score scale be maintained over time. Although different editions or forms of a test are designed to measure the same constructs and are usually built to the same test specifications or test blueprint, they almost always differ somewhat in their statistical properties. For example, one form may be harder than another, so without adjustments, examinees would be expected to receive lower scores on the harder forms. A primary goal of test equating for testing programs is to eliminate the effects on scores of these unintended differences in test form difficulty. For many testing programs, test equating is necessary to be fair to examinees taking different test forms and to provide score-users with scores that mean the same thing, regardless of the test forms taken by examinees.

In high-stakes testing programs, it cannot be overemphasized how important it is that test equating be done carefully and accurately. The released scores are usually the most visible part of a testing program, even though they represent the endpoint of a large test production, administration, and scoring enterprise. An error in the equating function or score conversion function can affect the scores for all examinees. For this reason, the credibility of a testing organization is called into question over test equating problems to a greater degree than when, for example, flawed test questions are discovered in operational tests.

This chapter consists of four main sections. Section 1 provides an introduction to test score equating and gives some important definitions. Section 2 is concerned with the most common data collection designs that are used to equate test scores. In Section 3, we discuss the various methods and procedures for forming equating functions. Section 4 presents best practices in the choice and evaluation of equating designs and functions, and also presents challenges to best practices often encountered in practice.

1.1. Literature on score equating

For two tests, a *link* between their scores is a transformation from a score on one to a score on the other. In the field of educational measurement, there are at least six books that focus on issues in test linking. Holland and Rubin (1982) contains a set of conference papers covering a variety of test equating topics. The most complete coverage of the entire field of test equating and linking is provided by Kolen and Brennan (2004). The book by von Davier et al. (2004b), introduces several new ideas of general use in equating although its focus is on kernel equating. The two book-length reports from the National Research Council, *Uncommon Measures* by Feuer et al. (1999) and *Embedding Questions* by Koretz et al. (1999), are very accessible summaries of informed, professional judgment about the issues involved in linking scores on different educational tests. Finally, Livingston (2004) is a training manual for those who will actually do score equating and it gives a spirited and straightforward description of many of the major issues and procedures encountered in practice.

Readers who are new to test equating will do well to discover the wealth of material that has appeared in the three old editions and the forthcoming edition of *Educational Measurement*. Flanagan (1951), in the first edition, covers some of the test equating techniques that were available when he wrote. He also discusses issues and problems that are still current, which shows how pervasive they are. Angoff (1971), in the second edition, provides a comprehensive introduction to scales, norms and test equating. Petersen et al. (1989) introduced new material developed since the Angoff chapter. Holland and Dorans (2006) provide an historical background for test score linking. In addition to test equating, Holland and Dorans (2006) discuss other ways that scores on different tests are connected or linked together. We will draw heavily from the Holland and Dorans chapter in the material we cover about equating and refer the reader to their treatments of other types of linking, most notably scaling, as appropriate.

1.2. Classes of score linking methods

A *link* between scores on two tests is a transformation from a score on one to a score on the other. Transformations that link the scores on two tests can be of different types. The distinctions between different types of linkages are related to the similarities and differences between the tests as well as to the ways in which the linked scores are intended to be used and interpreted. Several frameworks have been suggested for organizing the different types of links that are used in practice. These are found in Flanagan (1951), Angoff (1971), Mislevy (1992), Linn (1993), Feuer et al. (1999) and Dorans (2000, 2004). Kolen (2004) and Kolen and Brennan (2004) give reviews and syntheses of several frameworks. Holland and Dorans (2006) provide their own framework that builds on and clarifies this prior work.

Holland and Dorans divide linking methods into three basic categories called *predicting*, *scale aligning*, and *equating*. It is important to distinguish between these categories because they are often seen as similar or identical when, in fact, they are not. Testing professionals need to understand these differences, the circumstances when one category is more relevant than another and, when necessary, be able to communicate these distinctions to test users.

1.2.1. Predicting

Predicting is the oldest form of score linking. Predicting has been confused with equating since the earliest days of psychometrics.

The goal of predicting is to predict an examinee's score on a test from other information about that examinee. This other information might be a score on another test or the scores from several other tests. It also might include demographic information (e.g., race/ethnicity), scores on non-cognitive inventories (e.g., motivation), or other types of cognitive information (e.g., grades in selected courses). For this reason, there is always an asymmetry between what is being predicted and what is used to make the prediction, they may be very different in number and character. It has been known since the nineteenth century that, in the simplest case, the usual linear regression function for predicting one test score, **Y**, from another, **X**, is not the inverse of the linear regression function for predicting **X** from **Y** (Galton, 1888). There is an asymmetry between the predictor and the predicted score.

Suffice it to say that asymmetry prevents prediction from meeting one of the fundamental prerequisites of equating, the goal of which is to produce scores that can be used interchangeably. Prediction is often appropriate for other purposes however. An appropriate use of predicting to make a link between two tests occurs when an examinee's score on one test is used to predict how he or she will perform on another test. An example of this is the use of PLAN scores to forecast how an examinee will perform on the ACT a year or so later. For example, periodically a year's worth of ACT data from students who have taken both tests is used to estimate the conditional distribution of ACT scores given the corresponding (English, mathematics, reading, science, and composite) PLAN score (see ACT, 1997). This conditional distribution gives a prediction of the range of likely performance on the ACT given an examinee's PLAN score. If these predictions are applied to examinees who are similar to those in the population from which the prediction equations are derived, then they are likely to be useful. For examinees who are very different from them, these predictions are less likely to be accurate. Holland and Dorans (2006) provide an introduction to prediction and the related procedure called projection for those who wish to explore these approaches.

1.2.2. Scale aligning

The goal of scale aligning is to transform the scores from two different tests onto a common scale. Scaling procedures are about 100 years old, and were initially called methods for creating *comparable scores*.

Scale aligning is the second category in the Holland and Dorans (2006) framework. It has many subcategories, including activities such as battery scaling (Kolen, 2004), anchor scaling (Holland and Dorans, 2006), vertical scaling (Kolen and Brennan, 2004), calibration (Holland and Dorans, 2006), and concordance (Pommerich and Dorans, 2004). Scale aligning or scaling is not discussed much in this chapter. The interested reader is urged to consult Holland and Dorans (2006) and Kolen (2006). The statistical procedures used for scaling can also be used to equate tests. Equating is the focus of this chapter.

1.2.3. Equating

The goal of equating is to establish an effective equivalence between scores on two test forms such that the scores from each test form can be used as if they had come from the same test. To achieve this goal of interchangeable scores, strong requirements must be put on the blueprints for the two tests and on the method used for linking scores. Among other things, the two tests must measure the same construct at the same level of difficulty and with the same accuracy.

All linking frameworks define equating as the strongest form of linking between the scores on two tests. In the Holland and Dorans (2006) framework, equating represents the endpoint of a continuum that begins with methods that make no assumptions about the relationship between the tests being linked (prediction and battery scaling) and proceeds to methods that are appropriate for linking tests that are very similar (concordances and equating). Equating may be viewed as a form of scaling in which very strong requirements are placed on the tests being linked.

1.3. *What constitutes an equating?*

All methods for linking test scores include (a) two or more tests and rules for scoring them, (b) scores on these tests obtained by one or more samples of examinees, and (c) one or more procedures for computing the transformation function. In addition, they all specify either an implicit or explicit population of examinees to which the test linking is to be applied. The thing that distinguishes test equating from other forms of linking is its goal. The goal of score equating is to allow the scores from both tests to be used interchangeably, for any purpose. As noted earlier, this is a very demanding goal and experience has shown that the two tests and the methods used to link them must satisfy very strong requirements to achieve it.

Actually, equating is the first step in a two-step process for putting scores on a new test onto an existing scale used for reporting. The first step in the process is the raw-to-raw equating—the computation of the function that links raw scores on a new test, **X**, to those of an old test, **Y**. In practice, the terms old and new form are used to emphasize that the same construct is being measured by **X** and **Y**. The second step is the conversion of these equated **X**-raw scores to the reporting scale. In practice, there is a scaling function that maps the raw scores of **Y** to the reporting scale, call it $S = \mathrm{s}(y)$, and then the equating function, $y = \mathrm{e}(x)$, is composed with $\mathrm{s}(y)$ to put the raw-scores of **X** onto the reporting scale. Braun and Holland (1982) give an illustrative diagram of this two-step process for a testing program that continually produces and equates new test forms. The composition function, $S = \mathrm{s}(\mathrm{e}(x))$, is called the *score conversion function* for **X**. In this chapter, we will not discuss how scaling functions are obtained, rather we refer the reader to the four editions of *Educational Measurement* (Flanagan, 1951; Angoff, 1971; Petersen et al., 1989; and Kolen, 2006). In this chapter we will concentrate on the linking of raw scores, but it's the final score conversion function that is used in practice to compare the results of different equating functions and to report scores.

1.3.1. *The five requirements for equated scores*

In practice, equating is necessary because of our inability to construct multiple forms of a test that are strictly parallel. Equating is used to fine-tune the test construction process. There are five requirements that are widely viewed as necessary for a linking to be an equating. Those requirements are:

(a) *The equal construct requirement.* The two tests should both be measures of the same construct (latent trait, skill, ability).
(b) *The equal reliability requirement.* The two tests should have the same reliability.
(c) *The symmetry requirement.* The equating transformation for mapping the scores of **Y** to those of **X** should be the *inverse* of the equating transformation for mapping the scores of **X** to those of **Y**.
(d) *The equity requirement.* It should be a matter of indifference to an examinee to be tested by either of the tests that have been equated.
(e) *The population invariance requirement.* The equating function used to link the scores of **X** and **Y** should be the same regardless of the choice of (sub)population from which it is derived.

Both formal and informal statements of subsets of these five requirements appear in a variety of earlier sources including Lord (1950), Angoff (1971), Lord (1980), Petersen et al. (1989), Kolen and Brennan (2004), and Holland and Dorans (2006). Dorans and Holland (2000) explicitly discuss these five requirements and indicate various ways in which the five "... can be criticized as being vague, irrelevant, impractical, trivial or hopelessly stringent", p. 283. Lord (1980) regards (d) as the most fundamental, while Livingston (2004) argues that requirements (d) and (e) are unattainable in practice. Regardless of these differences of opinions, these five requirements have value as criteria for evaluating whether or not two tests can be, or have been, successfully equated. Moreover, they provide a crude and intuitive "theory" of test equating.

In practice, requirements (a) and (b) mean that the tests need to be built to the same content and statistical specifications. Requirement (c) excludes the use of regression methods for test equating. Requirement (d) may be used to explain why both (a) and (b) are needed (Lord, 1980). If the two tests measure different constructs, examinees will prefer the one on which they believe they will score higher. Requirement (d) is, however, hard to evaluate empirically and its use is primarily theoretical (Lord, 1980 and Hanson, 1991). Requirement (e) is easy to apply in practice and can be used to explain why requirements (a) and (b) are needed. If the same blueprint is not used to construct the two tests, then the conversions can differ for different groups of examinees. For example, a conversion table relating scores on a mathematics test to scores on a verbal test developed on data for men would be very different from one developed from data on women, since, in our society, women tend to do less well than men on mathematics tests. Dorans and Holland (2000) use requirement (e), rather than requirement (d) to develop quantitative measures of equitability that indicate the degree to which equating functions depend on the subpopulations used to estimate them.

The difference between prediction and equating has been pointed out repeatedly over the last century, but due to the natural popularity of prediction using regression models and the relative unfamiliarity of test equating, it probably needs continual restatement and clarification. If we examine the five requirements for equating, it is clear that prediction methods need not satisfy any of the five requirements. Requirement (c), in particular, excludes the use of regression methods for test equating. If we are predicting scores on one test from those on another, the tests need not measure the same construct (requirement (a)) nor be equally reliable (requirement (b)) although a more closely related and highly reliable test makes for better a predictor of the scores on another test. Requirement (d) plays no role in prediction. And, prediction may violate requirement (e) as it often makes sense to incorporate population sensitivity directly into the prediction, that is, to include subgroup membership as predictors to improve prediction. See Holland and Dorans (2006) for additional discussion of the difference between prediction and equating.

1.3.2. A crucial concern for all equating methods

A common concern for all of the methods of score equating is how to control for differential examinee ability in the linking process. Suppose, as is often done in practice, that two different tests are given to two different groups of examinees. In examining the distributions of the resulting scores, there are two ubiquitous factors that can make interpreting the results complicated. One is the relative *difficulty* of the two tests (which is what test equating is concerned about), and the other is the relative *ability* of the two groups of examinees on these tests (which is a confounding factor that should be eliminated in the equating process). In test equating we are only interested in adjusting for differences in test characteristics, and we want to control for any differences in examinee ability when making these adjustments.

In practice, there are two distinct procedures for addressing the separation of test difficulty and differential examinee ability. The first is to use a common population of examinees, and the other is to use an anchor measure. Having the same examinees take both tests explicitly controls for differential examinee ability. A variant of the use of a common set of examinees is to use two equivalent samples of examinees from a common population. On the other hand, when it is not possible to have samples of examinees from the same population, their performance on a set of common items or an anchor measure can quantify the ability differences between two distinct, but not necessarily equivalent, samples of examinees. The use of an anchor measure can lead to more flexible data collection designs than the use of common examinees. However, the use of anchor measures requires users to make various assumptions that are not needed when the examinees are either the same or from equivalent samples.

2. Data collection designs used in test score equating

There are numerous data collection designs that have been used for score equating. We will restrict our discussion to those designs that involve complete data in the sense that the examinees are given the opportunity to answer all of the questions on one or both of the tests. Throughout this chapter we will assume that the scoring of omitted or "not reached" items is covered by the rules for scoring the test.

The role of data collection designs to control for differential examinee ability is crucial to successful test equating. This control is accomplished through the use of special data collection designs in which a common group (or equivalent groups) of examinees take each test form or all examinees take a set of common items or an anchor measure. In this section and the next, we give brief summaries of the important data collection designs that have been used over the years. Other good discussions of these designs are given in Lord (1950), Angoff (1971), Petersen et al. (1989), von Davier et al. (2004b) and Kolen and Brennan (2004).

In all of our descriptions, we will identify one or more populations of examinees and one or more samples from these populations. We will assume that all samples are random samples even though in practice this may be only an approximation. See Angoff (1971) and von Davier et al. (2004b) for discussions of the ways that samples are drawn in practice and how they might differ from purely random samples.

While we emphasize the structure of the data collection designs in this section, it should be mentioned here that the *amount* of data collected has a substantial effect on the usefulness of the resulting equatings. Because it is desirable for the statistical uncertainty associated with test equating to be much smaller than the other sources of variation in test results, it is important that the results of test equating be based on samples that are large enough to insure this. This fact should always be kept in mind when selecting a data collection design.

2.1. *The single group (SG) design*

The single group (SG) design is the simplest data collection design. In the single group design, all examinees in a single sample of examinees from population **P** take both tests. The single group design is probably the earliest data collection design used for test linking. It can provide accurate equating results from relatively small sample sizes.

When discussing data collection designs for test linking, it is helpful to use the "design table" notation defined in von Davier et al. (2004b), used earlier in Petersen et al. (1989) and in Braun and Holland (1982). The design table is a schematic representation of the "examinee by test" data matrix, with rows representing populations and samples of examinees and the columns representing the different tests, test sections or test items involved. The design table for the SG design is illustrated in Table 1.

The ✓ mark in the design table indicates that the examinees in the sample for a given row take the tests indicated in the column; otherwise the score data are not collected.

The SG design controls for differential examinee proficiency by having the same examinees take both tests. It has several major uses in the practice of scaling and equating. In using this design, however, it is necessary to assume that an examinee's score on the second test is unaffected by the fact that she or he has previously taken the first form. That is, it must be plausible that practice or other types of order effects can be ignored. In this design, the common population from which the sample is drawn may be clearly identified as, for example, in the concordance between ACT and SAT reported in Dorans et al. (1997). However, the "population" may just be a convenient fiction because there is just a sample of data with no obvious population from which it is a sample.

Table 1
The design table for the SG design

Population	Sample	**X**	**Y**
P	1	✓	✓

Unless we are willing to assume that order effects are negligible, the SG design can only be used if examinees take the items in both tests at exactly the same point in time. One such situation occurs when we have to delete a few questions from a test before reusing it. In this situation, the new form is simply the old form minus the questions that are being deleted. For equating, we use the original sample of examinees that took the old form before those questions were deleted. The SG design can also be used if we need to add a few questions to a test before reusing it. In this situation, the old form is simply the new form without the questions that are being added and we use the data from the sample of examinees that took the test with the new questions included. In both cases, the link from the slightly shorter test to the full test can be viewed as an example of equating using the SG design.

2.2. *The equivalent groups design (EG)*

In most equating situations, it is impossible to arrange for enough testing time for every examinee to take more than one test. The simplest solution is to have two separate samples take each form of the test. In the equivalent groups (EG) design, two equivalent samples are taken from a common population **P**, one is tested with **X** and the other with **Y**. The EG design is often used for equating.

The design table for the EG design, Table 2, clearly shows the pattern of missing data (i.e., the blank cells).

Because examinees take only one test, the issue of order effects does not arise for the EG design. The problem is to select samples that are equivalent in whatever the tests are supposed to measure. In practice, this is done in two ways. First, it is sometimes possible to take two random samples from **P**, and test each with a single test. To reflect this possibility, this design is sometimes called the *random groups design* (Kolen and Brennan, 2004). The two samples are then as "equivalent" as two random samples from the same population can be. Second, the more common situation is to construct the two samples by "spiraling" the test booklets for the two tests. The booklets are alternated in the packaging process so that when the tests are distributed to examinees they are alternated, first **X**, then **Y** and then **X** again, and so on. Certain assumptions must hold in order for spiraling to be feasible. For example, the time limits must be the same for the two tests. When well-executed, spiraled samples are often somewhat more "equivalent" (i.e., less different) than random samples. They are "more equivalent" because they are approximately *stratified* random samples where the strata are the administrative divisions of the tested population (i.e., classrooms, schools, etc.). The effect of stratification on the equivalence of samples, however, is usually small.

Table 2
The design table for the EG design

Population	Sample	**X**	**Y**
P	1	✓	
P	2		✓

The EG design is fairly convenient to administer. It does not require that the two tests have any items in common, but it can be used even if they do. It also has some limitations. It requires large sample sizes to produce accurate equating results. And, it may have some consequences for test security, as in most cases, the old form will have been previously administered. However, when samples sizes are large and forms can be reused, the EG design is usually regarded as a good choice because it avoids the issue of possible order effects that can arise in the SG design where each examinees takes *both* tests.

2.3. The counterbalanced design (CB)

In order to allow for the possibility of order effects in the SG design, the sample is sometimes randomly divided in half and in each sub-sample the two tests are taken in different orders – **X** first and then **Y** or **Y** first and then **X**. The result is the counterbalanced (CB) data collection design.

If we denote a score from **X**, when it is taken first as $\mathbf{X}_1$, and, when it is taken second, as $\mathbf{X}_2$, and similarly for $\mathbf{Y}_1$ and $\mathbf{Y}_2$, then Table 3 describes the CB design.

The CB design contains both the SG and EG designs within it. There are SG designs for both $\mathbf{X}_1$ and $\mathbf{Y}_2$ and $\mathbf{X}_2$ and $\mathbf{Y}_1$. There is an EG design for $\mathbf{X}_1$ and $\mathbf{Y}_1$ and for $\mathbf{X}_2$ and $\mathbf{Y}_2$. Also, if the order is ignored and the counterbalancing is regarded as having eliminated the effect of order, the two samples can be pooled and treated as a simple SG design. This variety of interpretations implies that there are several ways to use the data for the CB design to scale or equate **X** and **Y**. Von Davier et al. (2004b) describe several approaches, and indicate how the naive pooling of the two samples may not be accurate. Their discussion is more complete than that of Angoff (1971). When the SG design is feasible, and the possibility of order effects is substantial, then the CB design is often the preferred option. The main advantage of the CB design is the same as that of the SG design: accurate results from relatively small samples. Its main disadvantage is that it can seldom be fit within an operational administration of a test. Usually, the CB design requires a special study for collecting the data.

2.4. Anchor test or "NEAT" designs

In anchor test designs there are two populations, **P** and **Q**, with a sample of examinees from **P** taking test **X**, and a sample from **Q** taking test **Y**. In addition, both samples take an anchor test, **A**. We follow the terminology of von Davier et al. (2004b) and call this the "Non-Equivalent groups with Anchor Test" (or NEAT) design. Kolen and Brennan

Table 3
The design table for the CB design

Population	Sample	$\mathbf{X}_1$	$\mathbf{X}_2$	$\mathbf{Y}_1$	$\mathbf{Y}_2$
P	1	✓			✓
P	2		✓	✓	

Table 4
The design table for the NEAT design

Population	Sample	X	A	Y
P	1	✓	✓	
Q	2		✓	✓

(2004) and others refer to this as the "common-item nonequivalent groups design" or simply the "common item" or the "anchor test" design.

The use of common items or "anchor tests" arose later than the other data collection designs for test equating and linking. As a way of equating, the earliest versions that we located are in the early 1940s and both scaling to the anchor and scaling on a hypothetical population, described in Holland and Dorans (2006), were used to equate and scale the SAT (Donlon and Angoff, 1971). The NEAT design is used for equating and some forms of scale aligning, as indicated in Holland and Dorans (2006). Table 4 represents the NEAT design.

The role of the anchor test is to quantify the differences between samples from **P** and **Q** that affect their performance on the two tests to be equated, **X** and **Y**. The best kind of an anchor for equating is a test that measures the same constructs that **X** and **Y** measure. The anchor **A** is usually a shorter and less reliable test than the tests to be equated.

Formally, the NEAT design contains two single-group designs within it. This structure underlies both anchor scaling (Holland and Dorans, 2006) and the chain equating methods to be discussed later.

The relationship of **A** to **X** and **Y** can be of two different types. If **A** is a subset of items contained in both **X** and **Y**, then it is called an *internal* anchor test and scores on **A** are used in computing scores on the total tests. If **A** is a separately timed test that each examinee takes in addition to one or the other of the tests being equated, then it is called an *external* anchor test.

The anchor test design improves upon the flexibility of the EG design by allowing the two samples taking **X** and **Y** to be different or "non-equivalent." It also improves upon the SG design by not requiring examinees to take both **X** and **Y**. While the use of anchor tests may *appear to be* a minor variation of the previous data collection designs, the use of common items rather than common examinees involves new assumptions that are not necessary in the use of SG, EG, and CB designs, see Sections 2.1–2.3. Some type of assumption is required in the NEAT design to make up for the fact that **X** is never observed for examinees in **Q** and **Y** is never observed for examinees in **P**. For this reason, there are several distinct methods of scaling and equating tests using the NEAT design. Each of these methods corresponds to making different untestable assumptions about the missing data. Holland and Dorans (2006) discuss these methods and assumptions in more detail. Liou et al. (2001) point out that the NEAT design is a case of missing data that are *missing at random* in the technical sense discussed in Little and Rubin (2002).

There is one case in which the role of the anchor test is different. This is when the two samples are drawn from a common population, $\mathbf{P} = \mathbf{Q}$. This special case of the NEAT design is an "EG design with an anchor test". In this case, the role of the anchor test is the same as that of a covariate in an experimental design. It is used to gain precision in the estimation of the relevant parameters. For this special case, it is not necessary for **A** to measure the same construct that **X** and **Y** do, or even to be a test score. All that matters is for **A** to be correlated with, that is, predictive of, both **X** and **Y** (Lord, 1950). The post stratification procedures described in Section 3.2.1.1 may be viewed as using the anchor test as a covariate. Lord (1955) indicates that Ledyard R Tucker first proposed the EQ design with an anchor test as a way of reaping the benefits of both the EG and the SG designs. The example of an anchor test design given in Liou et al. (2001) mentioned above, is an EG design with both an anchor test and other "anchor information" as well.

One way to think about the difference between the NEAT design and the SG, EG, and CB designs is as the difference between experimental designs versus observational studies (Rosenbaum, 1995). The SG design is like a repeated measures design with a single group and two treatments, the EG design is like a randomized comparison with two treatment groups, while the CB design is like a repeated measures design with a single group and counterbalanced order of treatments. In contrast, the NEAT design is like an observational study where there are two non-randomized study groups that are possibly subject to varying amounts of self-selection.

In Lord (1950) it is shown that the way the anchor test is used in the EG design with anchor test can have dramatic effects on the resulting accuracy of the results. For example, he shows that regardless of the correlation between the anchor measure and the tests, scaling on a hypothetical population produces more stable or less variable estimates of the equating function than does scaling to the anchor. See Holland and Dorans (2006) for more on scaling on a hypothetical population and scaling to the anchor.

When **P** and **Q** are different or "non-equivalent", the statistical role of **A** is to remove bias as well as to increase precision in the estimation of the equating function. When **A** is a miniature version of **X** and **Y** (i.e., a "mini-test" that is shorter and less reliable but otherwise measures the same construct as the two tests to be linked) then it can be expected to do a good job of removing any bias due to the non-equivalence of **P** and **Q**. When **A** is not really a measure of the same construct as **X** and **Y**, or if it is not highly correlated with them, **A** is less useful for removing bias or for increasing precision.

2.5. Practical differences among the SG, EG, CB, and NEAT data collection designs

Each of these data collection designs has advantages and disadvantages that make it more or less useful in practice. For equating, the single group (SG) design requires the smallest sample sizes and the equivalent groups (EG) design requires the largest sample sizes to achieve the same level of accuracy, as measured by the standard error of equating (see Lord, 1950; Holland and Dorans, 2006). The anchor test (i.e., NEAT) designs require sample sizes somewhere in between those of the SG and EG designs depending on how strongly correlated the anchor test is with the two tests to be equated and how similar the two populations are.

We would argue that the ideal design, in theory, is a large equivalent groups (EG) design with an external anchor test. If the anchor test is administered last, only the anchor test can be affected by possible order effects. A comparison of the distributions of the anchor test in the two (equivalent) samples then allows differential order effects to be identified and if they are substantial, the anchor test can be ignored, leaving a simple EG design, where no order effects are possible. If the anchor test is internal to the two tests, then context or order, e.g., item location effects may arise and need to be dealt with.

The primary drawback of the EG design for test equating is that the equated test has to be given at least twice – once to be equated and then again for, at least one, operational use as the old form. In some testing programs, it may be problematic to reuse operational forms. If for test security reasons, special non-operational test administrations are arranged to collect equating data using the EG design, then the issue of examinee motivation arises, as discussed in Holland and Dorans (2006)

The single group (SG) design requires a smaller sample size to achieve the same statistical accuracy as a larger EG design, but it brings with it issues of order effects and it requires twice as much time to administer both tests. A particular problem with the SG design is that there is no way to assess whether order effects exist. The counterbalanced (CB) design, on the other hand, allows order effects to be estimated, but if they are large and different for the two tests, then there may be no option but to ignore the data from the tests given second, and treat the result as an EG design without an anchor test. Because of the greatly reduced sample size, the resulting EG design may produce equating results that are less accurate than was desired. Von Davier et al. (2004b) propose making a formal statistical decision for the CB design to assess the order effects.

2.6. The special problems of the NEAT design

The NEAT design is more flexible than the other designs. For test equating it allows a new test to be used and equated at each successive operational test administration. This is desirable in high stakes situations where test reuse may lead to test security problems.

The use of the NEAT design (Table 4) requires users to make one of several possible sets of untestable, missing-data assumptions in order to interpret the linking results as an equating. In addition, great care must be taken in collecting and analyzing the data in a NEAT design. For example, for both internal and external anchor tests, the statistical properties of the common items should be evaluated to make sure they have not differentially changed from the one test administration to the other. Differential item functioning (DIF) methods may be used to compare the performance of the common items with the two test administrations treated as the reference and focal groups, and the total score on the common items as the matching criterion (see Holland and Wainer, 1993, especially chapter 3). Simple plots of item difficulty values and other statistics may also be used to detect changes in items. Internal common items are susceptible to context effects because they are embedded within different sets of items in the two tests. Changes in widely held knowledge may also lead to changes in performance on anchor test items. For example, a hard question about a new law on a certification exam may become very easy once the law becomes part of the standard training curriculum. There are many examples of this type of "rapid aging" of test questions.

2.6.1. External anchor tests

It is often advised that the anchor test should be a "mini-version" of the two tests being equated (Angoff, 1971). Making the anchor test be a mini-version of the whole test is sometimes in conflict with the need to disguise an external anchor test to make it look like one of the scored sections of the test. For example, to be a mini-version of the test, the anchor test might need to include a variety of item types; whereas, to mirror a specific section of the test, the anchor test might need to include only a limited number of item types. Items in an external anchor test are administered in a separately timed section and usually do not count towards the examinee's score. In many circumstances it is important to disguise the external anchor test so that it appears to be just another section of the test. One reason for this is that some examines may identify the anchor test and, knowing that it does not count towards their final score, skip it or use the time to work on sections that count towards their score (even though they are instructed not to do this). While this type of behavior may appear to benefit them, because of the way that the anchor test is used in equating such behavior may actually result in lowering the scores of all examinees if enough of them do it. This counterintuitive result can be explained as follows. The anchor test is used to compare the performance of the current group of examinees on the anchor test to that of a previous group. If a substantial number of the current examinees under-perform on the anchor test, this will make them appear less able than they really are. As a consequence, the new test will appear to be somewhat easier than it really is relative to the old test. In test equating, a raw score on an easier test is converted to a lower scaled score than that for the same raw score on a harder test. Therefore, the scores reported on the new test will be lower than they would have been had all examinees performed up to their abilities on the anchor test. In practice this effect is likely to be small, but, for those examinees who worked on another section during the anchor test, the effect may be canceled by an increased score on the other section.

Statistical outlier analysis can be used to identify those examinees whose anchor test performance is substantially different from their performance on the operational test, i.e., the scores are so different that both scores cannot be plausible indicators of the examinee's ability. Removing these examinees from the equating sample prevents their unlikely performance from having an undue effect on the resulting equating function.

One major advantage of external NEAT designs is that the extra section may serve multiple purposes, for example, equating, pretesting, and tryout of new item types. This is accomplished by spiraling versions of the test with different content in this "variable" section. This process also can be used to improve test security by limiting the exposure of the anchor test to a relatively small proportion of the total group tested.

2.6.2. Internal anchor tests

Items in internal anchor tests are part of the assessment and count towards each examinee's score. Internal anchor items are usually spread throughout the test. For the observed score equating methods described in this chapter, where the score on the anchor test plays an important role, it is generally accepted that it is desirable for the anchor test to be a mini-version of the two tests. This may be more feasible for internal anchor tests than for external anchor tests.

Because the items in an internal anchor test count towards the score, examinees are unlikely to skip them. On the other hand, once anchor test items have been used in a test administration they may become susceptible to security breaches and known by examinees in future administrations. For anchor items to be effective, they must maintain their statistical properties across the old and new forms. The primary problems with internal anchor tests are context effects and security breaches. Context effects can occur when common items are administered in different locations (e.g., item 10 in one form and item 20 in the other form), under different testing conditions (i.e., paper and pencil versus computer delivered), or when they are adjacent to different kinds of items in the two tests. These effects are well documented (Leary and Dorans, 1985; Brennan, 1992; and Harris and Gao, 2003). Security breaches are an unfortunate reality for many testing programs, and great care is required to prevent them or to recognize them when they occur.

2.6.3. Strengthening the anchor test

When there are only small differences in ability between the two samples of examinees used in the NEAT design, all linear equating methods tend to give similar results, as do all nonlinear equating methods. To the extent that a NEAT design is almost an EG design with an anchor test, the need for the anchor test is minimized.

However, when the two samples are very different in ability, as indicated by their performance on the anchor test, the use of the anchor test information becomes critical, because it is the only means for distinguishing differences in ability between the two groups of examinees from differences in difficulty between the two tests that are being equated. The most important properties of the anchor test are its stability over time (mentioned above) and its correlation with the scores on the two tests being equated. The correlation should be as high as possible. An advantage of internal anchors over external anchors is that their correlations with the tests being equated are usually high because of their part–whole relationship with the total tests.

Anchor test scores should be highly correlated with scores on the tests being equated. Long anchor tests are generally better for equating than short ones, as longer tests are usually more reliable. In practice, it is desirable for both the anchor test and the total tests to have high reliability.

The importance of the test–anchor–test correlation raises the question of whether it makes sense to use supplemental information about the examinees in both groups to increase the correlation. Wright and Dorans (1993) suggest replacing the anchor test with a propensity score (Rosenbaum and Rubin, 1983) that includes both the anchor test and other examinee data that distinguishes the two groups of examinees. Liou et al. (2001) use a missing data model to include surrogate variables along with the anchor test score to adjust for sample differences before equating. This is an important area for further research.

3. Procedures for equating scores

Many procedures have been developed over the years for equating tests. Holland and Dorans (2006) considered three factors when attempting to develop a taxonomy of equating methods:

(1) common population versus common item data collection designs,
(2) observed-score versus true-score procedures, and
(3) linear versus non-linear methods.

Because equating is an empirical procedure, it requires a data collection design, the first factor, and a procedure for transforming scores on one test form to scores on another (the other two factors). Linear methods produce a linear function for mapping the scores from **X** to **Y**, while non-linear methods allow the transformation to be curved. Observed-score procedures directly transform (or equate) the observed scores on **X** to those on **Y**. True-score methods are designed to transform the *true scores* on **X** to the true scores of **Y**. True-score methods employ a statistical model with an examinee's true score defined as their expected observed test score. The statistical models used to date are those of classical test theory and item response theory. Holland and Hoskens (2003) show how these two statistical models may be viewed as aspects of the same model.

In this section, we will limit our discussion to equating methods that use the data collection designs described in Section 2. We will give detailed discussions for only the most common procedures, identifying references for others. First, we will consider procedures used with common population data collection designs and then procedures used with anchor tests or common item designs. Within these two types of data collection designs, we will look at linear and non-linear procedures and observed-score and true-score procedures. For observed-score procedures, we will discuss both linear and non-linear procedures at the same time. For true-score procedures, we will discuss linear and non-linear procedures separately as they are based on different statistical models. In addition, we will be explicit about the population of examinees on which scores are equated. Common to any equating scenario is a population of examinees that we will call the *target population*, **T**, following the usage in von Davier et al. (2004b). In this reference, the target population refers to the source of the samples used to compute the linking function.

3.1. *Procedures for equating scores on complete tests given to a common population*

There are three data collection designs in Section 2 that make use of a common population of examinees: the Single Group (SG), the Equivalent Groups (EG) and the Counterbalanced (CB) designs. They all involve a single population, **P**, which is also the target population. Most of the procedures we discuss in this section apply easily to both the SG and EG designs. To apply them to the CB designs is more complicated. For a discussion of the CB design we refer to von Davier et al. (2004b), Angoff (1971) and Kolen and Brennan (2004).

3.1.1. Observed-score equating methods

These are the oldest and most well studied of the linking methods used for score equating. We will use a single definition of observed-score linking that applies to either linear or non-linear procedures depending on whether additional assumptions are satisfied. This allows us to consider both linear and non-linear observed-score linking methods from a single point of view.

Some notation will be used throughout the rest of this chapter. The *cumulative distribution function*, cdf, of the scores of examinees in the target population, **T**, on test **X** is denoted by $F_{\mathbf{T}}(x)$, and it is defined as the proportion of examinees in **T** who score at or below x on test **X**. More formally, $F_{\mathbf{T}}(x) = P\{\mathbf{X} \leqslant x \mid \mathbf{T}\}$, where $P\{\cdot \mid \mathbf{T}\}$ denotes the population proportion or probability in **T**. Similarly, $G_{\mathbf{T}}(y) = P\{\mathbf{Y} \leqslant y \mid \mathbf{T}\}$, is the cdf of **Y** over **T**. Cumulative distribution functions increase from 0 up to 1 as x (or y) moves from left to right along the horizontal axis. In this notation, x and y may be any real values, not necessarily just the possible scores on the two tests. For distributions of observed scores such as number right or rounded formula scores, the cdfs are step functions that have points of increase at each possible score (Kolen and Brennan, 2004). In Section 3.1.1.2 we address the issue of the discreteness of score distributions in more detail, but will ignore it until then.

3.1.1.1. The equipercentile and linear linking functions The equipercentile definition of *comparable* scores is that x (an **X**-score) and y (a **Y**-score) are *comparable* in **T** if $F_{\mathbf{T}}(x) = G_{\mathbf{T}}(y)$. This means that x and y have the same percentile in the target population, **T**. When the two cdf's are continuous and strictly increasing, the equation $F_{\mathbf{T}}(x) = G_{\mathbf{T}}(y)$ can always be satisfied and can be solved for y in terms of x. Solving for y leads to the *equipercentile function*, $Equi_{\mathbf{YT}}(x)$, that links x to y on **T**, defined by:

$$y = Equi_{\mathbf{YT}}(x) = G_{\mathrm{T}}^{-1}\big(F_{\mathbf{T}}(x)\big). \tag{1}$$

In Eq. (1), $y = G_{\mathbf{T}}^{-1}(p)$ denotes the inverse function of $p = G_{\mathbf{T}}(y)$. We have followed Dorans and Holland (2000), von Davier et al. (2004a), and Holland and Dorans (2006) in explicitly including the target population **T** in the definition of $Equi_{\mathbf{YT}}(x)$. The notation emphasizes that **T** (as well as **X** and **Y**) can influence the form of the equipercentile function.

In general, there is nothing to prevent $Equi_{\mathbf{YT}}(x)$ from varying with the choice of **T**, thereby violating requirement (e) of Section 1.3. The equipercentile function is used for equating, and other kinds of linking. For equating, we expect the influence of **T** to be small or negligible and we call the scores *equivalent*. In other kinds of linking, **T** can have a substantial effect and we call the scores *comparable in* **T**.

If Eq. (1) is satisfied, then $Equi_{\mathbf{YT}}(x)$ will transform the distribution of **X**-scores on **T** so that it is the same as the distribution of **Y**-scores on **T**.

It is sometimes appropriate to assume that the two cdf's, $F_{\mathbf{T}}(x)$ and $G_{\mathbf{T}}(y)$, have the same shape and only differ in their means and standard deviations. To formalize the idea of a common shape, suppose that $F_{\mathbf{T}}(x)$ and $G_{\mathbf{T}}(y)$ both have the form

$$F_{\mathbf{T}}(x) = K\big[(x - \mu_{\mathbf{XT}})/\sigma_{\mathbf{XT}}\big] \quad \text{and} \quad G_{\mathbf{T}}(y) = K\big[(y - \mu_{\mathbf{YT}})/\sigma_{\mathbf{YT}}\big], \tag{2}$$

where K is a cdf with mean zero and standard deviation 1. When Eq. (2) holds, $F_{\mathbf{T}}(x)$ and $G_{\mathbf{T}}(y)$ both have the shape determined by K. In this case, it can be shown that the equipercentile function is the *linear linking function*, $Lin_{\mathbf{YT}}(x)$, defined as

$$Lin_{\mathbf{YT}}(x) = \mu_{\mathbf{YT}} + (\sigma_{\mathbf{YT}}/\sigma_{\mathbf{XT}})(x - \mu_{\mathbf{XT}}). \tag{3}$$

The linear linking function may also be derived as the transformation that gives the **X**-scores the same mean and standard deviation as the **Y**-scores on **T**. Both of the linear and equipercentile functions satisfy the symmetry requirement (c) of Section 1.3.1. This means that $Lin_{\mathbf{XT}}(y) = Lin_{\mathbf{YT}}^{-1}(y)$, and $Equi_{\mathbf{XT}}(y) = Equi_{\mathbf{YT}}^{-1}(y)$, i.e., linking **Y** to **X** is the inverse function for linking **X** to **Y**. In general, the function, $Equi_{\mathbf{YT}}(x)$, curves around the function, $Lin_{\mathbf{YT}}(x)$.

Two important special cases of $Lin_{\mathbf{YT}}(x)$ are the *mean linking function* and the *identity function*. When the two standard deviations in Eq. (3) are equal, then $Lin_{\mathbf{YT}}(x)$ takes on the form $Mean_{\mathbf{YT}}(x) = x + (\mu_{\mathbf{YT}} - \mu_{\mathbf{XT}})$. The mean linking function adjusts the scores of **X** so that they have the same mean as **Y** does on **T**. When both the means and the variances in Eq. (3) are equal, $Lin_{\mathbf{YT}}(x)$ takes on the form $Iden(x) = x$. The identity function makes no adjustment at all to the **X**-scores. It corresponds to assuming that the raw scores on **X** and **Y** are already comparable. Both $Mean_{\mathbf{YT}}(x)$ and $Iden(x)$ are useful when the samples are very small and cannot support accurate estimates of the moments of **X** and **Y** on **T**. They are discussed in more detail in Kolen and Brennan (2004) and Skaggs (2005).

The linear linking function requires estimates of the means and standard deviations of **X**- and **Y**-scores over the target population, **T**. It is easy to obtain these estimates for the SG and EG designs described in Section 2 (see Angoff, 1971, or Kolen and Brennan, 2004). It is less straightforward to obtain estimates for the CB design. Angoff (1971) gives the details of Lord's (1950) linear procedure for the CB design. Von Davier et al. (2004b) describe their method that allows users to control how much the data from the tests that are given second can influence the resulting linear or non-linear kernel equating function.

3.1.1.2. The need to continuize the discrete distributions of scores While there is really only one linear linking function for the SG or EG designs, i.e., the one given by Eq. (3), the equipercentile function can depend on how $F_{\mathbf{T}}(x)$ and $G_{\mathbf{T}}(y)$ are made continuous or "continuized." Test scores are typically integers, such as number-right scores or rounded formula-scores. Because of this, the inverse function, required in Eq. (1), is not well defined – i.e., for many values of p, there is no score, y, for which $p = G_{\mathbf{T}}(y)$. This is not due to the *finiteness of real samples*, but rather to the *discreteness of real test scores*. To get around this, there are two methods of "continuization" of $G_{\mathbf{T}}(y)$ that are in current use.

The first is very old, and has been used at least since Otis (1916) and is described in detail in Kelley (1923). The cdf of test scores is a step function but linear interpolation can make it piecewise linear and continuous. This method is described in Angoff (1971), Petersen et al. (1989) and in great detail in Kolen and Brennan (2004). This form of continuization results in a continuous cdf for which the equation, $p = G_{\mathbf{T}}(y)$, can always be solved for y, using linear methods, to produce $G_{\mathbf{T}}^{-1}(F_{\mathbf{T}}(x))$.

The second approach is to use Gaussian kernel smoothing to continuize the discrete distributions. This alternative method of continuization was introduced by Holland

and Thayer (1989) and is described in detail in von Davier et al. (2004b). The use of Gaussian kernel smoothing results in continuously differentiable cdfs for which the equation, $p = G_{\mathbf{T}}(y)$, can also be solved for y but which requires the use of numerical methods. The older, linear-interpolation method of continuization may also be viewed as kernel smoothing using a uniform kernel on $(-1/2, 1/2)$ rather than a Gaussian kernel. See Holland and Thayer (1989) for more details.

There are two primary differences between these two approaches to continuization. First, the use of linear interpolation results in an equipercentile function that is piecewise linear and continuous. Such functions may have "kinks" that need to be smoothed out by a further smoothing, often called post-smoothing (Fairbank, 1987; Kolen and Brennan, 2004). In contrast, kernel smoothing results in equipercentile functions that are very smooth (i.e., differentiable everywhere) and that do not need further post-smoothing. Second, the equipercentile functions obtained by linear interpolation always map the highest score on **X** into the highest score on **Y** and the same for the lowest scores. The use of kernel smoothing to continuize the cdfs does not necessarily have this property, nor does the use of the linear linking function. While it is sometimes desirable, there are cases where the highest score on an easier test should not be mapped onto the highest score of a harder test. For more discussion of this point, see Petersen et al. (1989), Kolen and Brennan (2004) and von Davier et al. (2004b).

3.1.1.3. Presmoothing score distributions Irregularities in the score distributions cause problems for equipercentile equating. They produce irregularities in the equipercentile equating function that do not generalize to other groups of test-takers. Consequently, it is generally considered advisable to *presmooth* the raw-score frequencies in some way prior to equipercentile equating. The purpose of this step is to eliminate some of the sampling variability present in the raw-score frequencies, in order to produce smoother cdfs for computation of the equipercentile function. If presmoothing is done so as to preserve the essential features of the score frequencies, it will reduce the sampling variability in the estimated frequencies without introducing significant bias. The resulting estimates will be closer to the underlying frequencies in the target population, **T**. When presmoothing is done with a model that does not describe the data well, then the estimated frequencies will be biased estimates of the underlying frequencies in **T**.

A limitation of equipercentile equating is that the equating relationship cannot be computed for any possible scores above the highest observed score or below the lowest observed score. If we could observe the scores of the entire target population, **T**, on both forms of the test, this limitation would not be a problem. Smoothing can help solve this problem because many smoothing methods will produce a smoothed distribution with probabilities (possibly very small) at the highest and lowest score levels, even if no test-takers actually attained those scores.

Kolen and Jarjoura (1987) and Kolen and Brennan (2004) discuss several methods for presmoothing. Von Davier et al. (2004b) describe the use of loglinear models for presmoothing. Their work is based on Holland and Thayer (1989, 2000) and they give examples of presmoothing for the SG, EG, CB and NEAT designs. Livingston (1993) shows that for smallish samples (less than 200) there can be large reductions in equating error when the score frequencies are presmoothed prior to continuization. When the samples are very large, the sampling variability may be small enough that presmoothing does not provide a big improvement, but presmoothing may still be a useful way to remove undesired roughness in the sample frequencies.

The type of data that is available for presmoothing depends on the data collection design. The EG design is the simplest and results in two independent univariate score distributions, one for **X** and one for **Y**. These may be independently presmoothed. The SG design results in a single bivariate score distribution, the joint frequencies for the (**X**, **Y**)-pairs in the sample. For this case, presmoothing should be done on the joint distribution. Presmoothing the **X** and **Y** frequencies separately, as if they were from different samples, ignores the correlation between **X** and **Y** and can lead to incorrect standard errors. The CB design, as well as the NEAT design both result in two independent bivariate score distributions – the $(\mathbf{X}_1, \mathbf{Y}_2)$ and the $(\mathbf{X}_2, \mathbf{Y}_1)$ joint frequencies for the CB design and the (**X**, **A**) and the (**Y**, **A**) joint frequencies for the NEAT design.

In presmoothing data, it is important to achieve a balance between a good representation of the original data and smoothness. Smoothness reduces sampling variability while a good representation of the data reduces the possibility of bias. For example, if a log-linear model is used, it needs to preserve the most important features of the data, such as means, variances and skewnesses, and any other special features. The more parameters that are estimated for the model, the better the model will represent the original data, but the less smooth the fitted model becomes. Using an inadequate presmoothing model can introduce bias into the estimated equating functions. In presmoothing for score linking, it is probably better to fit a few extra parameters than to use a model that does not capture the important regularities in the original data, that is, a little over fitting is usually less harmful than a little under fitting. Holland and Thayer (2000) discuss fitting these models to a variety of data sets that exhibit varying degrees of complexity. Von Davier et al. (2004b) give several additional examples, including a very complicated model for data from a NEAT design.

3.1.1.4. Presmoothing using IRT models IRT observed score equating was proposed by Lord (1980). Basically, it uses an item response theory (IRT) model to presmooth a number-right score distribution prior to equipercentile equating.

Let **X** and **Y** denote number-right scores for two tests with dichotomous test items. Using IRT, the conditional distributions of **X** and **Y** given the common latent ability, θ, may be estimated. This computation exploits the condition of local independence to compute $P\{\mathbf{X} = x \mid \theta\}$ and $P\{\mathbf{Y} = y \mid \theta\}$ as the probabilities of sums of independent but not identically distributed 0/1 variables. See Lord and Wingersky (1984) and Kolen and Brennan (2004) for a detailed example and additional references. The estimated density function for the ability distribution over **T**, $f_{\mathbf{T}}(\theta)$, may be approximated in various ways, and these results combined via numerical integration to obtain $P\{\mathbf{X} = x \mid \mathbf{T}\} = \int P\{\mathbf{X} = x|\theta\} f_{\mathbf{T}}(\theta)\,d\theta$, and $P\{\mathbf{Y} = y \mid \mathbf{T}\} = \int P\{\mathbf{Y} = y \mid \theta\} f_{\mathbf{T}}(\theta)\,d\theta$. The resulting estimated probabilities, $P\{\mathbf{X} = x \mid \mathbf{T}\}$ and $P\{\mathbf{Y} = y \mid \mathbf{T}\}$, are IRT model-based estimates of the original raw-score sample proportions, i.e., the proportion in the sample getting a score of x on **X**, or the proportion getting a score of y on **Y**. Thus, the estimates, $P\{\mathbf{X} = x \mid \mathbf{T}\}$ and $P\{\mathbf{Y} = y \mid \mathbf{T}\}$, may be regarded as a form of presmoothing of the two, unsmoothed sample score distributions. This is a use of an IRT model to presmooth the data, prior to continuization during equipercentile equating. Once the presmoothed estimated score probabilities are in hand, they may be continuized by either linear interpolation or Gaussian kernel smoothing and used to produce a version of the equipercentile function, $Equi_{\mathbf{YT}}(x)$.

IRT observed score equating is seldom used in practice (see Kolen and Brennan, 2004). Compared to other presmoothing methods, it requires various judgments and assumptions and is computationally burdensome. Furthermore, there is no guarantee that these presmoothed estimates of the score distributions will reproduce the types of score frequencies found in real data. For this reason, it is unlikely to be used except in research studies or very special situations with incomplete test data. For situations with data from complete tests, it is neither as efficient nor as flexible as the direct presmoothing of score frequencies using loglinear models, as advocated by Holland and Thayer (2000), von Davier et al. (2004b) and others.

3.1.2. Linear true-score procedures from classical test theory

The use of classical test theory to derive true-score equating procedures is limited to linear methods. While this restricts their usefulness in practice, the ideas of classical test theory give insight into test equating and linking. Our discussion of linear true-score methods for complete tests linked on a common target population will be limited to defining the linear true-score equating function, $TSLin_{\mathbf{YT}}(\tau_{\mathbf{X}})$. See Kolen and Brennan (2004) for a more detailed discussion of true score equating.

Following Holland and Dorans (2006), we use the formulation of classical test theory described in Holland and Hoskens (2003). The true scores, $\tau_{\mathbf{X}}$ and $\tau_{\mathbf{Y}}$, are defined as latent variables underlying each test that have these properties: $\tau_{\mathbf{X}} = \mathrm{E}(\mathbf{X} \mid \tau_{\mathbf{X}}, \mathbf{T})$, and $\tau_{\mathbf{Y}} = \mathrm{E}(\mathbf{Y} \mid \tau_{\mathbf{Y}}, \mathbf{T})$, for any target population, $\mathbf{T}$. From these two assumptions it follows that $\mu_{\mathbf{XT}} = \mathrm{E}(\mathbf{X} \mid \mathbf{T}) = \mathrm{E}(\tau_{\mathbf{X}} \mid \mathbf{T})$, and $\mu_{\mathbf{YT}} = \mathrm{E}(\mathbf{Y} \mid \mathbf{T}) = \mathrm{E}(\tau_{\mathbf{Y}} \mid \mathbf{T})$. To formalize the assertion that $\mathbf{X}$ and $\mathbf{Y}$ "measure the same construct," we assume the true scores are *congeneric*, that is, that they are linearly related by

$$\tau_{\mathbf{Y}} = \alpha\tau_{\mathbf{X}} + \beta, \tag{4}$$

where α and β may depend on the target population, $\mathbf{T}$. The idea behind true-score equating is to estimate α and β and to use Eq. (4) to find the link between the two sets of true scores. Lord (1980) takes the position that only true scores can ever really be equated, but it can be argued that this is a consequence of his very stringent interpretation of the equity condition (d) of Section 1.3.1.

The reliabilities of $\mathbf{X}$ and $\mathbf{Y}$, which we denote by $\rho^2_{\mathbf{XT}}$ and $\rho^2_{\mathbf{YT}}$, are given by the well-known formulas

$$\rho^2_{\mathbf{XT}} = \frac{\sigma^2_{\tau_{\mathbf{X}}\mathbf{T}}}{\sigma^2_{\mathbf{XT}}} \quad \text{and} \quad \rho^2_{\mathbf{YT}} = \frac{\sigma^2_{\tau_{\mathbf{Y}}\mathbf{T}}}{\sigma^2_{\mathbf{YT}}}. \tag{5}$$

Taking the means and variances (over $\mathbf{T}$) of both sides of Eq. (4), we obtain the following equations: $\mathrm{E}(\tau_{\mathbf{Y}} \mid \mathbf{T}) = \alpha\mathrm{E}(\tau_{\mathbf{X}} \mid \mathbf{T}) + \beta$, and $\mathrm{Var}(\tau_{\mathbf{Y}} \mid \mathbf{T}) = \alpha^2 \mathrm{Var}(\tau_{\mathbf{X}} \mid \mathbf{T})$. Rearranging these two equations to find α and β yields

$$\alpha = \frac{\sigma_{\tau_{\mathbf{Y}}\mathbf{T}}}{\sigma_{\tau_{\mathbf{X}}\mathbf{T}}} = \frac{\sigma_{\mathbf{YT}}\,\rho_{\mathbf{YT}}}{\sigma_{\mathbf{XT}}\,\rho_{\mathbf{XT}}} \quad \text{and} \quad \beta = \mu_{\mathbf{YT}} - \alpha\mu_{\mathbf{XT}}. \tag{6}$$

Substituting these values for α and β into Eq. (4), we obtain the *linear true-score equating function*

$$\begin{aligned}\tau_{\mathbf{Y}} &= TSLin_{\mathbf{YT}}(\tau_{\mathbf{X}}) \\ &= \mu_{\mathbf{YT}} + \alpha(\tau_{\mathbf{X}} - \mu_{\mathbf{XT}}) = \mu_{\mathbf{YT}} + \frac{\sigma_{\mathbf{YT}}}{\sigma_{\mathbf{XT}}}\frac{\rho_{\mathbf{YT}}}{\rho_{\mathbf{XT}}}(\tau_{\mathbf{X}} - \mu_{\mathbf{XT}}).\end{aligned} \tag{7}$$

$TSLin_{\mathbf{YT}}(\tau_{\mathbf{X}})$ is very similar to the linear linking function, $Lin_{\mathbf{YT}}(x)$, but its slope involves the ratio of the square roots of the reliabilities, i.e., $\rho_{\mathbf{YT}}/\rho_{\mathbf{XT}}$.

3.1.3. Non-linear true-score procedures from item response theory

Item response theory (IRT) may be used to generalize the linear true-score linking methods of classical test theory to non-linear procedures. Following Holland and Dorans (2006), we adopt the general version of IRT described in more detail in Holland and Hoskens (2003). Assuming a common one-dimensional IRT model with latent variable θ, the true scores of **X** and **Y** may be expressed as

$$\begin{aligned}\tau_{\mathbf{X}} &= \nu_{\mathbf{X}}(\theta) = \mathrm{E}(\mathbf{X} \mid \theta, \mathbf{T}) = \mathrm{E}(\mathbf{X} \mid \theta), \\ \tau_{\mathbf{Y}} &= \nu_{\mathbf{Y}}(\theta) = \mathrm{E}(\mathbf{Y} \mid \theta, \mathbf{T}) = \mathrm{E}(\mathbf{Y} \mid \theta).\end{aligned} \tag{8}$$

The functions, $\nu_{\mathbf{X}}(\theta)$ and $\nu_{\mathbf{Y}}(\theta)$, are the "test characteristic curves" for **X** and **Y**, respectively. They may be estimated using the item parameter estimation techniques of IRT (Yen and Fitzpatrick, 2006). The relationship between the two true scores, $\tau_{\mathbf{X}}$ and $\tau_{\mathbf{Y}}$, in Eq. (8) may be non-linear, and it is found by solving $\tau_{\mathbf{X}} = \nu_{\mathbf{X}}(\theta)$ for θ, and then substituting for θ in $\tau_{\mathbf{Y}} = \nu_{\mathbf{Y}}(\theta)$. This results in

$$\tau_{\mathbf{Y}} = \nu_{\mathbf{Y}}\big(\nu_{\mathbf{X}}^{-1}(\tau_{\mathbf{X}})\big), \tag{9}$$

where $\nu_{\mathbf{X}}^{-1}(\tau_{\mathbf{X}}) = \theta$ is the inverse function of $\tau_{\mathbf{X}} = \nu_{\mathbf{X}}(\theta)$. Eq. (9) generalizes the linear true-score equating function in Eq. (4), which, in turn, may be thought of as a linear approximation to the more general non-linear true-score equating function in Eq. (9). In practice, the function in Eq. (9) is treated as an observed score equating function and applied to observed scores, i.e.,

$$y = e_{\mathbf{Y}}(x) = \nu_{\mathbf{Y}}\big(\nu_{\mathbf{X}}^{-1}(x)\big). \tag{10}$$

The use of $\mathrm{e}_{\mathbf{Y}}(x) = \nu_{\mathbf{Y}}(\nu_{\mathbf{X}}^{-1}(x))$ does produce a symmetric linking relationship as long as the inverse functions of $\nu_{\mathbf{X}}(\theta)$ and $\nu_{\mathbf{Y}}(\theta)$ are well-defined. When the item response functions have non-zero lower asymptotes, see Yen and Fitzpatrick (2006), and the resulting lower asymptotes of $\nu_{\mathbf{X}}(\theta)$ and $\nu_{\mathbf{Y}}(\theta)$ are not the same, the inverse function, $\nu_{\mathbf{X}}^{-1}(x)$, may not be well-defined for the lowest scores on **X**. In these cases, the value of $\nu_{\mathbf{Y}}(\nu_{\mathbf{X}}^{-1}(x))$ is often defined by an arbitrary extrapolation (Lord, 1980; Hambleton and Swaminathan, 1985; Kolen and Brennan, 2004).

3.1.4. Direct IRT scaling procedures

In the framework of item response theory, there is no need to equate raw test scores if scores can be reported on the ability scale (or on a derived scale that is a monotonic transformation of the ability scale). Suppose tests **X** and **Y** are both measures of the same latent ability θ. If IRT item and ability parameters for both tests are estimated simultaneously, then the parameters for both tests will be on the same scale and $\theta_{\mathbf{X}} = \theta_{\mathbf{Y}}$. Thus, when several tests are calibrated concurrently, the need for equating of ability scores on the tests does not arise. If item and ability parameters are estimated separately for each test, the parameter scales for the two tests will be linearly related and $\theta_{\mathbf{X}} = \alpha\theta_{\mathbf{Y}} + \beta$. Once the constants α and β are determined, the equating of ability scores on the two tests is accomplished. This approach bypasses the equating step and directly puts the raw scores from **X** onto the common theta scale. Yen and Fitzpatrick (2006) discuss this extensively.

3.2. Procedures for linking scores on complete tests using common items

The anchor test design is widely used for linking scores because its use of common items to control for differential examinee ability gives it greater operational flexibility than the approaches using common examinees. Examinees need only take one test, and the samples need not be from a common population. However, this flexibility comes with a price. First of all, the target population is less clear-cut for the NEAT design (see Section 2.4) – there are two populations, **P** and **Q**, and either one could serve as the target population. In addition, the use of the NEAT design requires additional assumptions to allow for the missing data – **X** is never observed in **Q** and **Y** is never observed in **P**. We use the term complete test to indicate that everyone in **P** sees all items on **X** and that everyone in **Q** see all items on **Y**. As indicated at the beginning of Section 2, our use of the term "missing data" is restricted to data that are missing by design. The assumptions needed to make allowances for the missing data are not easily tested with the observed data, and they are often unstated. We will discuss two distinct sets of assumptions that may be used to justify the *observed score* procedures that are commonly used with the NEAT design.

Drawing from what they saw being done in practice, Braun and Holland (1982) proposed that the target population for the NEAT design, or what they called the *synthetic population*, be created by weighting **P** and **Q**. They denoted the synthetic population by $\mathbf{T} = w\mathbf{P} + (1 - w)\mathbf{Q}$, by which they meant that distributions (or moments) of **X** or **Y** over **T** are obtained by first computing them over **P** and **Q**, separately, and then averaging them with w and $(1 - w)$ to get the distribution over **T**. When $w = 1, \mathbf{T} = \mathbf{P}$ and when $w = 0$, $\mathbf{T} = \mathbf{Q}$. In practice, w is often taken to be proportional to the two sample sizes from **P** and **Q**. This choice of w is implicit when the data for the anchor test are pooled into a "total group", as done in Angoff (1971) and Petersen et al. (1989). Of course, other choices of w are possible, such as $w = 1/2$ which gives equal weight to **P** and **Q**. There is considerable evidence that the choice of w has a relatively minor influence on equating results, for example, see von Davier et al. (2004b). This insensitivity to w is an example of the population invariance requirement of Section 1.3. The definition of the synthetic population forces the user to confront the need to create distributions (or moments) for **X** on **Q** and **Y** in **P**, where there are no data. In order to do this, assumptions must be made about the missing data.

Equating methods used with the NEAT design can be classified into two major types, according to the way they use the information from the anchor. The first type of missing data assumption commonly employed is of the *post stratification equating* (PSE) type; the second is of the *chain equating* (CE) type. Each of these types of assumptions asserts that an important distributional property that connects scores on **X** or **Y** to scores on the anchor test **A** is the same for any $\mathbf{T} = w\mathbf{P} + (1 - w)\mathbf{Q}$, i.e., is population invariant. Our emphasis here is on the role of such assumptions for observed-score linking because that is where they are the most completely understood at this time, but they are likely to have parallels for true-score linking as well, a topic worthy of future research. In addition to the PSE and CE types of procedures, classical test theory may be used to derive an additional less frequently used *linear observed-score* procedure for the NEAT design – the Levine observed-score equating function (Kolen and Brennan, 2004).

The PSE types of assumptions all have the form that the conditional distribution of **X** given **A** (or of **Y** given **A**) is the same for any synthetic population, $\mathbf{T} = w\mathbf{P} + (1 - -w)\mathbf{Q}$. In this approach, we estimate, for each score on the anchor test, the distribution of scores on the new form and on the old form in **T**. We then use these estimates for equating as if they had actually been observed in **T**. The PSE type of equating assumes that the relationship that generalizes from each equating sample to the target population is a conditional relationship. In terms of the missing data in the NEAT design, this means that conditional on the anchor test score, **A**, the distribution of **X** in **Q** (when it is missing) is the same as in **P** (when it is not missing). In the special case of an EG design with anchor test, $\mathbf{P} = \mathbf{Q}$ and the PSE assumptions hold exactly. When **P** and **Q** are different, the PSE assumptions are not necessarily valid, but there are no data to contradict them.

The CE assumptions all have the form that a linking function from **X** to **A** (or from **Y** to **A**) is the same for any synthetic population, $\mathbf{T} = w\mathbf{P} + (1 - w)\mathbf{Q}$. In this approach, we equate the scores on the new form to scores on the anchor and then equate the scores on the anchor to the scores on the old form. The "chain" formed by these two equatings links the scores on the new form to those on the old form. The CE type of equating approach assumes that the relationship that generalizes from each equating sample to the target population is an equating relationship. It is less clear for the CE assumptions than for the PSE assumptions what is implied about the missing data in the NEAT design.

In the special case of an EG design with anchor test, $\mathbf{P} = \mathbf{Q}$ and the CE assumptions hold exactly. In this special situation, the corresponding methods based on either the PSE or the CE assumptions will produce identical results. When **P** and **Q** are different, the PSE assumptions and CE assumptions can result in equating functions that are different (see von Davier et al., 2004b, for an empirical example) and there are no data to allow us to contradict or choose between either set of assumptions.

We now discuss the consequences of these different assumptions for linking complete tests using common items. We will divide the procedures into observed-score versus true-score procedures and within these into linear versus non-linear methods.

3.2.1. Observed-score linking procedures for the NEAT design

These procedures divide into those of the PSE and those of the CE types, mentioned above. We outline these briefly in turn.

3.2.1.1. The PSE types of linking procedures There are both non-linear and linear PSE procedures. They may be viewed as based on these two assumptions, which we adopt from von Davier et al. (2004b).

PSE1 The conditional distribution of $\mathbf{X}$ given $\mathbf{A}$ over $\mathbf{T}$, $P\{\mathbf{X} = x \mid \mathbf{A} = a, \mathbf{T}\}$ is the same for any $\mathbf{T}$ of the form $\mathbf{T} = w\mathbf{P} + (1 - w)\mathbf{Q}$.

PSE2 The conditional distribution of $\mathbf{Y}$ given $\mathbf{A}$ over $\mathbf{T}$, $P\{\mathbf{Y} = y \mid \mathbf{A} = a, \mathbf{T}\}$ is the same for any $\mathbf{T}$ of the form $\mathbf{T} = w\mathbf{P} + (1 - w)\mathbf{Q}$.

PSE1 and PSE2 are population invariance assumptions because they require that the conditional distributions are the same for any target population of the form $\mathbf{T} = w\mathbf{P} + (1 - w)\mathbf{Q}$.

The clearest examples of procedures of the PSE type are frequency estimation (Angoff, 1971; Petersen et al., 1989, and Kolen and Brennan, 2004), and the PSE version of kernel equating (von Davier et al., 2004b). All PSE procedures estimate the score distribution of $\mathbf{X}$ over $\mathbf{T}$ via the equation:

$$P\{\mathbf{X} = x \mid \mathbf{T}\} = \sum_a P\{\mathbf{X} = x \mid \mathbf{A} = a, \mathbf{P}\} P\{\mathbf{A} = a \mid \mathbf{T}\}. \tag{11}$$

In Eq. (11), assumption PSE1 is used to replace $P\{\mathbf{X} = x \mid \mathbf{A} = a, \mathbf{T}\}$ by $P\{\mathbf{X} = x \mid \mathbf{A} = a, \mathbf{P}\}$. The former should be in the right-side of Eq. (11) to make it a correct statement, but the latter is all that is available from the data on $\mathbf{X}$ and $\mathbf{A}$ in $\mathbf{P}$. The other ingredient of the right-side of Eq. (11), $P\{\mathbf{A} = a \mid \mathbf{T}\}$, is given by

$$P\{\mathbf{A} = a \mid \mathbf{T}\} = wP\{\mathbf{A} = a \mid \mathbf{P}\} + (1 - w)P\{\mathbf{A} = a \mid \mathbf{Q}\}. \tag{12}$$

PSE2 is used to justify a similar equation for $\mathbf{Y}$. These assumptions are discussed extensively in von Davier et al. (2004b).

Once values for the score probabilities, $P\{\mathbf{X} = x \mid \mathbf{T}\}$ and $P\{\mathbf{Y} = y \mid \mathbf{T}\}$, are available, they may be continuized via either linear interpolation or Gaussian kernel smoothing (see Section 3.1.1.2) to obtain estimates of $F_{\mathbf{T}}(x)$ and $G_{\mathbf{T}}(y)$. From $F_{\mathbf{T}}(x)$ and $G_{\mathbf{T}}(y)$ the equipercentile function, $Equi_{\mathbf{YT}}(x)$, is computed via Eq. (1). Frequency estimation and the PSE versions of kernel equating correspond to different ways of continuizing the score probabilities to obtain $F_{\mathbf{T}}(x)$ and $G_{\mathbf{T}}(y)$. Both methods may *presmooth* the joint distributions of $\mathbf{X}$ and $\mathbf{A}$ over $\mathbf{P}$ and $\mathbf{Y}$ and $\mathbf{A}$ over $\mathbf{Q}$, briefly described in Section 3.1.1.3, in order to remove some of the sampling variability that the raw data contain (see Kolen and Brennan, 2004 and von Davier et al., 2004b, for more details).

Linear observed-score PSE equating procedures include (i) Tucker equating (Angoff, 1971; Petersen et al., 1989, and Kolen and Brennan, 2004), (ii) the Braun–Holland method (Braun and Holland, 1982, and Kolen and Brennan, 2004), and (iii) the linear PSE version of kernel equating (von Davier et al., 2004b). The linear PSE version of kernel equating is a way to implement the Braun–Holland procedure and both are directly based on PSE1 and PSE2. Ledyard R Tucker was originally motivated by selection theory in the development of the method that bears his name, Angoff (1971). However, the following versions of PSE1 and PSE2 may also be used to derive Tucker equating with no reference to selection.

TUCK1 (a) The conditional mean of $\mathbf{X}$ given $\mathbf{A}$ over $\mathbf{T}$ is linear in $\mathbf{A}$ and is the same for any $\mathbf{T} = w\mathbf{P} + (1 - w)\mathbf{Q}$, and (b) the conditional variance of $\mathbf{X}$ given $\mathbf{A}$ over $\mathbf{T}$ is constant in $\mathbf{A}$ and is the same for any $\mathbf{T}$.

TUCK2 (a) The conditional mean of $\mathbf{Y}$ given $\mathbf{A}$ over $\mathbf{T}$ is linear in $\mathbf{A}$ and is the same for any $\mathbf{T} = w\mathbf{P} + (1 - w)\mathbf{Q}$, and (b) the conditional variance of $\mathbf{Y}$ given $\mathbf{A}$ over $\mathbf{T}$ is constant in $\mathbf{A}$ and is the same for any $\mathbf{T}$.

TUCK1 and TUCK2 are population invariance assumptions in the same sense that PSE1 and PSE2 are.

The Braun–Holland and linear PSE version of kernel equating do not make the more restrictive assumptions of linear conditional means and constant conditional variances that appear in TUCK1 and TUCK2. For this reason, they may give somewhat different results from the Tucker method when the conditional means are non-linear and/or the conditional variances are not constant.

3.2.1.2. The CE types of linking procedures The idea behind the CE procedures is to first link $\mathbf{X}$ to $\mathbf{A}$ using the data from $\mathbf{P}$, then to link $\mathbf{A}$ to $\mathbf{Y}$ using the data from $\mathbf{Q}$, and finally to combine these two links to link $\mathbf{X}$ to $\mathbf{Y}$ through $\mathbf{A}$. Von Davier et al. (2004b) show that the following two assumptions are sufficient to interpret chain equating as an observed-score equating function for any target population of the synthetic population form.

CE1 The equipercentile function linking $\mathbf{X}$ to $\mathbf{A}$ on $\mathbf{T}$ is the same for any $\mathbf{T}$ of the form $\mathbf{T} = w\mathbf{P} + (1 - w)\mathbf{Q}$.

CE2 The equipercentile function linking $\mathbf{A}$ to $\mathbf{Y}$ on $\mathbf{T}$ is the same for any $\mathbf{T}$ of the form $\mathbf{T} = w\mathbf{P} + (1 - w)\mathbf{Q}$.

CE1 and CE2 are population invariance assumptions because they require that the same linking function be valid for any target population of the form $\mathbf{T} = w\mathbf{P} + (1 - w)\mathbf{Q}$. Like PSE1 and PSE2, these two assumptions are not directly testable with the data that are available in the NEAT design.

Given that it uses the two SG designs within the NEAT design, it is likely that CE was the first equating procedure developed for the NEAT design. However, the earliest reference that we found to what we call the CE procedures was alternate equipercentile equating in Angoff (1971). In addition, it is discussed in Marco et al. (1983) as direct equipercentile equating and Livingston et al. (1990) as chained equipercentile. The chain version of kernel equating, described in von Davier et al. (2004b), differs from the earlier CE methods only in the way that the score distributions are continuized – Gaussian kernel smoothing rather than linear interpolation.

To see that CE procedures based on CE1 and CE2 can be put into the form of Eq. (11) so that they are, in fact, bona fide observed-score equating procedures, apply CE1 to $\mathbf{P}$ and to some other $\mathbf{T}$ of the form $\mathbf{T} = w\mathbf{P} + (1 - w)\mathbf{Q}$. This allows us to conclude that

$$H_{\mathbf{P}}^{-1}\big(F_{\mathbf{P}}(x)\big) = H_{\mathbf{T}}^{-1}\big(F_{\mathbf{T}}(x)\big), \tag{13}$$

where $H_{\mathbf{P}}(a)$ is the cdf of the anchor test $\mathbf{A}$ on $\mathbf{P}$ and $H_{\mathbf{T}}(a)$ is the cdf for $\mathbf{A}$ on $\mathbf{T}$. Then solve for $F_{\mathbf{T}}(x)$ in Eq. (13), so that

$$F_{\mathbf{T}}(x) = H_{\mathbf{T}}\Big(H_{\mathbf{P}}^{-1}\big(F_{\mathbf{P}}(x)\big)\Big). \tag{14}$$

Similarly, applying CE2 with the target population choices of **Q** and the same **T** as above leads to the equation

$$G_{\mathbf{T}}^{-1}(u) = \left(G_{\mathbf{Q}}^{-1}\left(H_{\mathbf{Q}}\left(H_{\mathbf{T}}^{-1}(u)\right)\right)\right). \quad (15)$$

Then, if we form the equipercentile function from **X** to **Y** on **T** we get

$$G_{\mathbf{T}}^{-1}\left(F_{\mathbf{T}}(x)\right) = G_{\mathbf{Q}}^{-1}\left(H_{\mathbf{Q}}\left(H_{\mathbf{P}}^{-1}\left(F_{\mathbf{P}}(x)\right)\right)\right). \quad (16)$$

The right-hand side of Eq. (16) is the chain of links made from the two equipercentile functions (**X** to **A** and **A** to **Y**) that use the data as two SG designs. The left-hand side is the equipercentile function on **T**. Note that the specific **T** does not matter, any $\mathbf{T} = w\mathbf{P} + (1 - w)\mathbf{Q}$, will give the same result. Thus, assuming CE1 and CE2, the CE function given in the right-hand side of Eq. (16) is a bona fide equipercentile equating function on **T**. We do not know how reasonable CE1 and CE2 are, just as we do not know how reasonable PSE1 and PSE2 are, because neither pair of assumptions can be directly tested with the data in the NEAT design.

There are both linear and non-linear versions of CE. Linear observed-score CE equating procedures include (i) the chained *linear linking function* (Angoff, 1971) and (ii) the linear CE version of kernel equating (von Davier et al., 2004b). Because it is derived as a type of equipercentile equating function, the linear CE version of kernel equating is based on assumptions CE1 and CE2. However, we can also show that the chained *linear linking function* is a linear linking function on a target population **T** as defined in Eq. (3) when the linear versions of CE1 and CE2 hold. These are:

- CL1 The linear linking function linking **X** to **A** on **T** is the same for any **T** of the form $\mathbf{T} = w\mathbf{P} + (1 - w)\mathbf{Q}$.
- CL2 The linear linking function linking **A** to **Y** on **T** is the same for any **T** of the form $\mathbf{T} = w\mathbf{P} + (1 - w)\mathbf{Q}$.

Again, CL1 and CL2 are examples of population invariance assumptions.

If we apply CL1 we get the equality $Lin_{\mathbf{AP}}(x) = Lin_{\mathbf{AT}}(x)$ for all x, and any choice of **T** of specified form. This equality can only hold if the slopes and intercepts of the two linear functions are identical, so we must have

$$\frac{\sigma_{\mathbf{AP}}}{\sigma_{\mathbf{XP}}} = \frac{\sigma_{\mathbf{AT}}}{\sigma_{\mathbf{XT}}} \quad \text{and} \quad \mu_{\mathbf{AP}} - \frac{\sigma_{\mathbf{AP}}}{\sigma_{\mathbf{XP}}}\mu_{\mathbf{XP}} = \mu_{\mathbf{AT}} - \frac{\sigma_{\mathbf{AT}}}{\sigma_{\mathbf{XT}}}\mu_{\mathbf{XT}}. \quad (17)$$

From Eq. (17) it follows that CL1 is equivalent to the following assumptions about the mean and standard deviation of **X** in **T**:

$$\sigma_{\mathbf{XT}} = \sigma_{\mathbf{XP}}\frac{\sigma_{\mathbf{AT}}}{\sigma_{\mathbf{AP}}} \quad \text{and} \quad \mu_{\mathbf{XT}} = \mu_{\mathbf{XP}} + \frac{\sigma_{\mathbf{XP}}}{\sigma_{\mathbf{AP}}}(\mu_{\mathbf{AT}} - \mu_{\mathbf{AP}}). \quad (18)$$

Similarly, CL2 is equivalent to the following assumptions about the mean and standard deviation of **Y** in **T**:

$$\sigma_{\mathbf{YT}} = \sigma_{\mathbf{YQ}}\frac{\sigma_{\mathbf{AT}}}{\sigma_{\mathbf{AQ}}} \quad \text{and} \quad \mu_{\mathbf{YT}} = \mu_{\mathbf{YQ}} + \frac{\sigma_{\mathbf{YQ}}}{\sigma_{\mathbf{AQ}}}(\mu_{\mathbf{AT}} - \mu_{\mathbf{AQ}}). \quad (19)$$

Using Eqs. (18) and (19) to find estimates of the means and variances of **X** and **Y** over **T**, we may use the definition of the linear linking function in Eq. (3) to obtain the final result. It can be shown that using the formulas in Eqs. (18) and (19) in the linear linking function, $Lin_{\mathbf{YT}}(x)$, defined in Eq. (3), results in exactly the same linear function as does the chain of the two linear linking functions $Lin_{\mathbf{YQ}}(Lin_{\mathbf{AP}}(x))$. This shows that under CL1 and CL2, the chained linear linking function is a linear observed-score equating function of the form in Eq. (3). A more detailed discussion of these derivations may be found in von Davier et al. (2004b).

3.2.1.3. A linear observed-score equating procedure from classical test theory In addition to the PSE and CE types of procedures, classical test theory may be used to derive an additional *linear observed-score* procedure for the NEAT design – the Levine observed-score equating function, $Lev_{\mathbf{YT}}(x)$ (Kolen and Brennan, 2004). $Lev_{\mathbf{YT}}(x)$ may be derived from two population invariance assumptions that are different from those that we have considered so far and that are based on classical test theory.

Holland and Dorans (2006) use the form of classical test theory discussed in Holland and Hoskens (2003) to derive the Levine observed score equating function from two assumptions. In addition to using the true score definitions for test scores **X** and **Y** found in Section 3.1.2, we define an anchor true score as a latent variable, $\tau_{\mathbf{A}}$, that underlies the observed anchor score **A**, and that satisfies: $\tau_{\mathbf{A}} = \mathrm{E}(\mathbf{A} \mid \tau_{\mathbf{A}}, \mathbf{T})$, for any target population, **T**. From this definition, it follows that: $\mu_{\mathbf{AT}} = \mathrm{E}(\tau_{\mathbf{A}} \mid \mathbf{T})$. To formalize the intuition that **X**, **Y** and **A** all measure the same construct, we assume their true scores are linearly related in a way that holds for all **T**, i.e., the three measures are congeneric. These assumptions are given below to emphasize their similarity to the population invariance assumptions made in Sections 3.2.1.1 and 3.2.1.2.

LL1 $\tau_{\mathbf{X}} = \alpha\tau_{\mathbf{A}} + \beta$, where α and β do not depend on the target population, **T**.
LL2 $\tau_{\mathbf{Y}} = \gamma\tau_{\mathbf{A}} + \delta$, where γ and δ do not depend on the target population, **T**.

These two congeneric assumptions do not necessarily hold for IRT models. IRT models usually imply that the relationships between the true-scores are non-linear (see Section 3.1.3). In this case the equalities in LL1 and LL2 would only be approximate and their approximate validity could very well depend on the choice of **T**.

Holland and Dorans (2006) extract more information out of LL1 and LL2. They show that when the reliabilities for **X**, **Y** and **A** are insensitive to the population over which they are computed, then the formulas for $\sigma_{\mathbf{XT}}$ and $\sigma_{\mathbf{YT}}$ in Levine observed score equating are the same as those derived for linear chain equating.

3.2.2. True-score procedures for the NEAT design

In this section we briefly indicate how the discussion of true-score equating methods of Sections 3.1.2 and 3.1.3 can be applied to the NEAT design. First we consider the linear procedures of classical test theory, and then the non-linear procedures of item response theory.

3.2.2.1. Linear true-score procedures from classical test theory Linear true score equating for the NEAT design was introduced in Levine (1955). It is extensively discussed in Kolen and Brennan (2004). We continue the notation of classical test theory used in Section 3.2.1.3, and continue to make the two congeneric "Levine" assumptions, LL1 and LL2 given there. Holland and Dorans (2006) demonstrate that the two equations relating the true-scores of **X** and **Y** to that of **A** may be written as

$$\tau_{\mathbf{X}} = \mu_{\mathbf{XP}} + \alpha(\tau_{\mathbf{A}} - \mu_{\mathbf{AP}}) = \mathit{TSLin}_{\mathbf{AXP}}(\tau_{\mathbf{A}}), \tag{20}$$

and

$$\tau_{\mathbf{Y}} = \mu_{\mathbf{YQ}} + \gamma(\tau_{\mathbf{A}} - \mu_{\mathbf{AQ}}) = \mathit{TSLin}_{\mathbf{AYQ}}(\tau_{\mathbf{A}}), \tag{21}$$

where, using the usual definitions of reliability (see Section 3.1.2), Holland and Dorans (2006) show that

$$\alpha = \frac{\rho_{\mathbf{XP}}\,\sigma_{\mathbf{XP}}}{\rho_{\mathbf{AP}}\,\sigma_{\mathbf{AP}}}, \tag{22}$$

and

$$\gamma = \frac{\rho_{\mathbf{YQ}}\,\sigma_{\mathbf{YQ}}}{\rho_{\mathbf{AQ}}\,\sigma_{\mathbf{AQ}}}, \tag{23}$$

where $\rho_{\mathbf{XP}}$ denotes the square root of the reliability of **X** over **P**, etc.

Both (20) and (21) are linear true-score equating functions similar to the one given in Eq. (4). Note that in this case, to avoid ambiguity, we indicate the link from $\tau_{\mathbf{A}}$ to $\tau_{\mathbf{X}}$ on **T** by $\mathit{TSLin}_{\mathbf{AXT}}(\tau_{\mathbf{A}})$, etc. The formula in Eq. (20) is the linear true-score equating function linking the true-scores of **A** to those of **X** on population **P**, while the one in Eq. (21) is the linear true-score equating function linking the true-scores of **A** to those of **Y** on population **Q**. The two population invariance assumptions, LL1 and LL2, imply that it does not matter which population is used to compute these linear true-score equating functions so that $\mathit{TSLin}_{\mathbf{AXP}}(\tau_{\mathbf{A}}) = \mathit{TSLin}_{\mathbf{AXT}}(\tau_{\mathbf{A}})$, and $\mathit{TSLin}_{\mathbf{AYQ}}(\tau_{\mathbf{A}}) = \mathit{TSLin}_{\mathbf{AYT}}(\tau_{\mathbf{A}})$, for any **T**. Thus, we can use these two linear true-score equating functions to equate the true-scores of **X** to **Y** on any **T** in exactly the same way as chain equating works for the NEAT design (see Section 3.2.1.2). The linear true-score equating function linking $\tau_{\mathbf{X}}$ to $\tau_{\mathbf{Y}}$ on **T** is given by

$$\begin{aligned}\tau_{\mathbf{Y}} &= \mathit{TSLin}_{\mathbf{AYT}}\big(\mathit{TSLin}^{-1}_{\mathbf{AXT}}(\tau_{\mathbf{X}})\big)\\ &= \mathit{TSLin}_{\mathbf{AYQ}}\big(\mathit{TSLin}^{-1}_{\mathbf{AXP}}(\tau_{\mathbf{X}})\big) = \mathit{TSLin}_{\mathbf{XYT}}(\tau_{\mathbf{X}}).\end{aligned} \tag{24}$$

In order to compute $\mathit{TSLin}_{\mathbf{XYT}}(\tau_{\mathbf{X}})$ for the NEAT design, estimates of α and γ are needed. Kolen and Brennan (2004) give an extensive and careful discussion of this issue.

3.2.2.2. Non-linear true-score procedures from item response theory In a manner similar to that used in Section 3.1.3, item response theory (IRT) can be used to generalize the linear true-score linking methods of classical test theory to non-linear procedures. As in that discussion, we follow Holland and Dorans (2006) and adopt the version of IRT described in more detail in Holland and Hoskens (2003). We assume a one-dimensional IRT model with latent variable θ. There are three IRT assumptions needed to generalize the two conditions LL1 and LL2 of Sections 3.2.1.3 and 3.2.2.1. These three assumptions are again population invariance assumptions and insure that the test characteristic curves for **X**, **Y** and **A** are unchanged across any population of examinees, **T**. (Note this is a weaker assumption than assuming that all item response functions are population invariant.)

IRT1 $\tau_{\mathbf{X}} = \mathrm{E}(\mathbf{X} \mid \theta, \mathbf{T}) = \nu_{\mathbf{XT}}(\theta) = \nu_{\mathbf{X}}(\theta)$, for any $\mathbf{T}$.
IRT2 $\tau_{\mathbf{Y}} = \mathrm{E}(\mathbf{Y} \mid \theta, \mathbf{T}) = \nu_{\mathbf{YT}}(\theta) = \nu_{\mathbf{Y}}(\theta)$, for any $\mathbf{T}$.
IRT3 $\tau_{\mathbf{A}} = \mathrm{E}(\mathbf{A} \mid \theta, \mathbf{T}) = \nu_{\mathbf{AT}}(\theta) = \nu_{\mathbf{A}}(\theta)$, for any $\mathbf{T}$.

If these tests exhibited differential item functioning when the two populations **P** and **Q**, are compared, then we would not expect IRT1–IRT3 to hold.

There are various ways to estimate the test characteristic curves in the NEAT design. We shall outline two of them. If the item parameters of **X** and **A** are estimated on **P** and those of **Y** and **A** are estimated on **Q**, we may then obtain estimates of the true score functions, $\nu_{\mathbf{XP}}(\theta^*)$, $\nu_{\mathbf{AP}}(\theta^*)$, $\nu_{\mathbf{AQ}}(\theta)$, and $\nu_{\mathbf{YQ}}(\theta)$. We have used θ and θ^* to indicate that because they were estimated on different populations the two theta scales may be different. There are several procedures for finding a linear transformation, say $\theta^* = a\theta + b$, to put both tests on the same theta scale. Some methods minimize the difference between $\nu_{\mathbf{AP}}(\alpha\theta + \beta)$ and $\nu_{\mathbf{AQ}}(\theta)$ over the theta scale (Yen and Fitzpatrick, 2006). These approaches force assumption IRT3 to hold as well as possible. Then, $\nu_{\mathbf{XP}}(\theta^*)$ is replaced by $\nu_{\mathbf{XP}}(\alpha\theta + \beta)$ which is then interpreted as $\nu_{\mathbf{XQ}}(\theta) = \tau_{\mathbf{X}}$ from assumption IRT1, above. We then have estimates of the true-score functions for **X** and **Y** on **Q** and they can be combined to obtain the true score equating function on **Q** via

$$\tau_{\mathbf{Y}} = \nu_{\mathbf{YQ}}\big(\nu_{\mathbf{XQ}}^{-1}(\tau_{\mathbf{X}})\big). \tag{25}$$

Under the two assumptions IRT1 and IRT2 above, Eq. (25) will hold for any population **T**.

Another approach is to simultaneously estimate the parameters of all three tests in a concurrent parameter estimation. This puts all three tests on a common theta scale without the need to adjust one of the scales to the other. However, the population on which this holds is the total population that weights **P** and **Q** proportionately to their respective sample sizes, i.e., $\mathbf{T} = w\mathbf{P} + (1 - w)\mathbf{Q}$ where w is proportional to the sample size from **P**. If we then assume both IRT1 and IRT2, we may compute $\tau_{\mathbf{Y}} = \nu_{\mathbf{Y}}(\nu_{\mathbf{X}}^{-1}(\tau_{\mathbf{X}}))$ directly from the estimated true score functions. See Yen and Fitzpatrick (2006) for more on IRT and additional references.

In addition to this form of true score equating, the raw scores of **X** may be put directly onto the θ-scale using the approach mentioned in Section 3.1.4.

4. Best practices and challenges to best practices

In this section we list what we consider to be best equating practices and identify some threats to these practices. We mention possible responses to these threats. Several of these responses are the subject for future research.

4.1. Best practices

Choice of data collection design, quality and similarity of tests to be equated, characteristics of anchor test in relation to total test, sample sizes and examinee characteristics, and choice of analyses all contribute to a successful equating.

4.1.1. Data collection

Data collection is the most important aspect of any equating study. Ideally, the data should come from a large representative sample of motivated examinees that is divided in half either randomly or randomly within strata to achieve equivalent groups. Each half is administered either the new form or the old form of a test. (If timing is generous and examinees are up to the task of taking both tests, a counterbalanced design could be employed in which each half of the sample is broken into halves again and then both the new and old forms are administered to examinees in a counterbalanced order.)

4.1.2. Total tests

The old and new forms are reliable measures of the same construct that are built to the same set of content and statistical specifications. An external anchor is available that is highly related to both total tests; this anchor is used to adjust for any differences in difficulty between test forms and for any differences in ability that may have occurred randomly between groups. The old form, the new form, and the anchor are appropriate in difficulty and content for the population of interest.

4.1.3. Anchor test

It is generally considered good practice to have the anchor test be a mini-version of the total tests being equated. That means it should have the same difficulty, and similar content. For IRT linking methods, it is also desirable to have the anchor test items spread out with respect to difficulty. This spread is also useful for estimation of linear functions used to place item difficulties on a common metric.

When an external anchor is not available, internal anchors are sometimes used. In this case, context effects become a possible issue. To minimize these effects, anchor (or common) items are often placed in the same location within each test.

4.1.4. Equating process

Once the data are available, it is often useful to pre-smooth the sample frequencies, especially when samples are small. Ideally, samples are large enough to make pre-smoothing optional. It is valuable to equate with several different models, including both linear and equipercentile models. In the equivalent groups case, the equipercentile method can be compared to the linear method using the standard error of equating, which describes sampling error, and the difference that matters (DTM), an effect size that can be used to assess whether differences in equating functions have practical significance. Holland and Dorans (2006) describe the DTM, the standard error of equating, and the standard error of the difference in equating or SEED. If the departures from linearity are less than the DTM and less than what would be expected due to sampling error, the linear model is often chosen on the grounds of parsimony because it was not sufficiently falsified by the data. Otherwise, the more general, less falsifiable, equipercentile model is selected.

When an anchor test is used, the items are evaluated via differential item functioning (DIF) procedures to see if they are performing in the same way in both the old and new form samples. All items on both total tests are evaluated to see if they are performing as expected. If they are not, it is often a sign of a quality control problem.

An equating should be checked for its reasonableness. How do we determine reasonableness? We compare the raw-to-scale conversion for the new form to those that have been obtained in the past. Is the new form conversion an outlier? Is it consistent with other difficulty information that may be available for that form and other forms that have been administered in the past? Is the performance of the group taking the new form consistent with the performance of other groups that are expected to be similar to it? For example, in testing programs with large volumes and relatively stable populations, it is reasonable to expect that the new form sample will have a similar scale score distribution to that obtained at the same time the year before. If the test is used to certify mastery, then the pass rates should be relatively stable from year to year, though not necessarily across administrations within a year.

4.2. Challenges to best practices

Large representative motivated samples that result from a random assignment of test forms are not always attainable. Reliability is not always as high as desired. Anchor tests may not be very reliable, especially internal anchors with few items. Anchors, especially external anchors, are not always highly related to the tests being equated. Tests are not always appropriate for the group that takes them. All of these present threats to sound equating.

4.2.1. Data collection issues

Some threats to sound equating are related to the choice of data collection design. Test security is an issue for many high stakes licensure, certification, and admissions tests. To help maintain test security many testing programs want to give a new form of the exam at every administration. Consequently, they do not want to re-administer an old form for equating purposes. Instead, they prefer to use an anchor test or common item design so only a subset of items is re-administered for equating purposes. The non-equivalent group or NEAT design is often used because of the greater flexibility it provides. Statistical procedures are needed to adjust for ability differences between groups when the NEAT design is used. Assumptions need to be made in order to make these adjustments. The assumptions may be flawed.

In fact, more and more testing programs are wanting, because of test security concerns, to move beyond the NEAT design to select the set of common items from multiple forms (from the pool of previously administered items) to further reduce the chances that any candidate has previous exposure to an item on the test. Depending on whether the new form contains any new items, IRT pre- or post-administration equating methods are used. IRT equating is more flexible than observed score methods in that it can be used to create solutions for missing data designs like these. Use of IRT is more complex and requires more assumptions than use of observed score methods. First, there is the choice and appropriateness of the IRT model to the data. Second, there is the actual complexity of the analyses.

To disguise the anchor items in a NEAT design, they are often embedded within sections of scored operational items. Internal anchors or common items may not be located in the same item positions within the old and new forms, making them susceptible to context effects that may diminish their utility as measures of ability. In addition, the common items may be few in number, making the anchor test relatively unreliable and less useful for identifying differences in difficulty between the test forms and differences in ability between the samples.

Small samples lead to weak data for equating. Smoothing may help if the samples are moderate in size. With very small samples, smoothing may be of little help. In these situations, the only option may be to make strong assumptions about the linking function. For example, it may be necessary to assume that it is the identity, $Lin_{\mathbf{YT}}(x) = Iden(x) = x$, or that it differs from the identity by a constant that is estimated by the data, $Lin_{\mathbf{YT}}(x) = Mean_{\mathbf{YT}}(x) = x + (\mu_{\mathbf{YT}} - \mu_{\mathbf{XT}})$.

With the NEAT design, the old and new form sample may perform very differently on the anchor test. In this setting, different equating methods tend to give different answers unless the anchor test is strongly related to both the old and new tests. This divergence of results is indicative of a poor data collection design.

4.2.2. Psychometric properties of the total tests

Characteristics of the test to be equated affect the quality of equating. Pretesting of untried items prior to their operational use produces higher quality exams. The absence of pretesting may result in tests with fewer scorable items than planned. Shorter less reliable tests are hard to equate because a greater portion of score variability is noise. More importantly, unpretested tests can turn out to be different in content and difficulty from the tests to which they are to be equated; these factors increase the difficulty of equating. At the extreme, tests may turn out to be too easy or too difficult for the intended population; this results in data that are not amenable for linking to other tests because the distributions are so skewed, and relationships with other scores are attenuated.

4.2.3. Psychometric properties of the anchor tests

The role of the anchor test is to provide a common score that can be used to adjust for group ability differences before adjusting for test difficulty differences. Scores from short anchor tests tend to have inadequate reliabilities, and consequently less than desirable correlations with the total test scores. Low correlations may also result when the content of the anchor test differs from the total test. Context effects can hurt the comparability of anchor items. Anchors that are too hard or too easy for the target population produce skewed score distributions that are not helpful for equating.

4.2.4. Samples

Unrepresentative or unmotivated samples undermine equating. Special study data collections need to include incentives that ensure that examinees will take the test seriously. Special care should be taken to ensure that only members of the population of interest are included in the samples. If possible, the sample should be representative of the population as well.

Equating can not be done effectively in small samples. The smaller the sample size, the more restricted is the class of plausible equating methods. Smoothing score distributions works in moderate samples, but does not help much with very small samples, especially when it is not clear how representative the sample is of the intended population.

Large ability differences on the anchor test, 0.25 standard deviations or higher, tend to yield situations where equating is unsatisfactory unless the anchor is highly related to both tests to be equated.

4.3. Additional directions for future research

In addition to the avenues for research mentioned earlier, e.g., Sections 2.5 and 2.6.3, there is a need for comprehensive empirical investigations, as well as additional theoretical work. The various challenges discussed in previous portions of Section 4 should be explored via systematic investigations of the appropriateness of different equating procedures in a variety of realistic settings. These empirical investigations have their progenitors, such as the comprehensive studies conducted by Marco et al. (1983), as well as other studies cited in Kolen and Brennan (2004). A variety of factors could be manipulated in a series of studies that examines the robustness of both newer approaches like kernel equating and older linear and nonlinear methods. Foremost among these factors would be the magnitude of ability differences between **P** and **Q** as measured by the anchor items, and the shape of the score distributions. In addition, it would be worthwhile to manipulate difficulty differences between **X**, **Y** and **A**, as well as the reliability of the total score and the anchor score. Correlations of the anchor score with total score, and sample size should also be manipulated and studied. Ideally, real data would be used as the starting point for these studies.

The theory of linear equating has developed to the point where relationships among different linear models are well understood. A theory of non-linear equating is needed. Until that theory is developed, systematic empirical investigations of non-linear equating methods are needed to guide us in the proper use of equating under a variety of circumstances.

References

ACT (1997). *Assessment Technical Manual*. Author, Iowa City, IA.

Angoff, W.H. (1971). Scales, norms and equivalent scores. In: Thorndike, R.L. (Ed.), *Educational Measurement*, 2nd ed. American Council on Education, Washington, DC, pp. 508–600.

Braun, H.I., Holland, P.W. (1982). Observed-score test equating: A mathematical analysis of some ETS equating procedures. In: Holland, P.W., Rubin, D.B. (Eds.), *Test Equating*. Academic Press, New York, pp. 9–49.

Brennan, R.L. (1992). The context of context effects. *Applied Measurement in Education* **5**, 225–264.

Donlon, T.F., Angoff, W.A. (1971). The scholastic aptitude test. In: Angoff, W.A. (Ed.), *The College Board Admissions Testing Program: A Technical Report on Research and Development Activities Relating to the Scholastic Aptitude Test and Achievement Tests*. The College Entrance Examination Board, New York, pp. 15–45.

Dorans, N.J. (2000). Distinctions among classes of linkages. College Board Research Note No. RN-11. The College Board, New York. Retrieved February 7, 2005, from http://www.collegeboard.com/repository/distinctions_among_cl_10503.pdf

Dorans, N.J. (2004). Equating, concordance and expectation. *Applied Psychological Measurement* **28**, 227–246.

Dorans, N.J., Holland, P.W. (2000). Population invariance and the equatability of tests: Basic theory and the linear case. *Journal of Educational Measurement* **37**, 281–306.

Dorans, N.J., Lyu, C.F., Pommerich, M., Houston, W.M. (1997). Concordance between ACT Assessment and recentered SAT I sum scores. *College and University* **73** (2), 24–34.

Fairbank, B.A. (1987). The use of presmoothing and postsmoothing to increase the precision of equipercentile equating. *Applied Psychological Measurement* **11**, 245–262.

Feuer, M.J., Holland, P.W., Green, B.F., Bertenthal, M.W., Hemphill, F.C. (Eds.) (1999). *Uncommon Measures: Equivalence and Linkage Among Educational Tests. Report of The Committee on Equivalency and Linkage of Educational Tests, National Research Council.* National Academy Press, Washington, DC.

Flanagan, J.C. (1951). Units, scores, and norms. In: Lindquist, E.F. (Ed.), *Educational Measurement.* American Council on Education, Washington, DC, pp. 695–763.

Galton, F. (1888). Co-relations and their measurements, chiefly from anthropological data. *Proceedings of the Royal Society of London* **45**, 135–145.

Hambleton, R.K., Swaminathan, H. (1985). *Item Response Theory: Principles and Applications.* Kluwer–Nijhoff, Boston.

Hanson, B.A. (1991). A note on Levine's formula for equating unequally reliable tests using data from the common item nonequivalent groups design. *Journal of Educational Statistics* **16**, 93–100.

Harris, D.J. Gao, X. (2003). A conceptual synthesis of context effect. In: Context Effects: Implications for Pretesting and CBT. Symposium conducted at the 2003 Annual Meeting of the American Educational Research Association, Chicago, IL.

Holland, P.W., Dorans, N.J. (2006). Linking and equating. In: Brennan, R.L. (Ed.), *Educational Measurement*, 4th ed. American Council on Education/Praeger, Westport, CT.

Holland, P.W., Hoskens, M. (2003). Classical test theory as a first-order item response theory: application to true-score prediction from a possibly nonparallel test. *Psychometrika* **68**, 123–149.

Holland, P.W., Rubin, D.B. (1982). *Test Equating.* Academic Press, New York.

Holland, P.W., Thayer, D.T. (1989). *The Kernel Method of Equating Score Distributions.* ETS RR 89-7. Educational Testing Service, Princeton, NJ.

Holland, P.W., Thayer, D.T. (2000). Univariate and bivariate loglinear models for discrete test score distributions. *Journal of Educational and Behavioral Statistics* **25**, 133–183.

Holland, P.W., Wainer, H. (1993). *Differential Item Functioning.* Lawrence Erlbaum Associates, Hillsdale, NJ.

Kelley, T.L. (1923). *Statistical Methods.* Macmillan, New York.

Kolen, M.J. (2004). Linking assessments: Concept and history. *Applied Psychological Measurement* **28**, 219–226.

Kolen, M.J. (2006). Scaling and norming. In: Brennan, R.L. (Ed.), *Educational Measurement*, 4th ed. American Council on Education/Praeger, Westport, CT.

Kolen, M.J., Brennan, R.L. (2004). *Test Equating, Linking, and Scaling: Methods and Practices*, 2nd ed. Springer-Verlag, New York.

Kolen, M.J., Jarjoura, D. (1987). Analytic smoothing for equipercentile equating under the common item nonequivalent populations design. *Psychometrika* **52**, 43–59.

Koretz, D.M., Bertenthal, M.W., Green, B.F. (Eds.) (1999). *Embedding Questions: The Pursuit of a Common Measure in Uncommon Tests. Report of the Committee on Embedding Common Test Items in State and District Assessments, National Research Council.* National Academy Press, Washington, DC.

Leary, L.F., Dorans, N.J. (1985). Implications for altering the context in which test items appear: An historical perspective on an immediate concern. *Review of Educational Research* **55**, 387–413.

Levine, R.S. (1955). *Equating the Score Scales of Alternate Forms Administered to Samples of Different Ability.* ETS RB-55-23. Educational Testing Service, Princeton, NJ.

Linn, R.L. (1993). Linking results of distinct assessments. *Applied Measurement in Education* **6**, 83–102.

Liou, M., Cheng, P.E., Li, M.Y. (2001). Estimating comparable scores using surrogate variables. *Applied Psychological Measurement* **25**, 197–207.

Little, R.J.A., Rubin, D.B. (2002). *Statistical Analysis with Missing Data*, 2nd ed. Wiley, Hoboken, NJ.

Livingston, S.A. (1993). Small-sample equating with log-linear smoothing. *Journal of Educational Measurement* **30**, 23–39.

Livingston, S.A. (2004). *Equating Test Scores (Without IRT).* Educational Testing Service, Princeton, NJ.

Livingston, S.A., Dorans, N.J., Wright, N.K. (1990). What combination of sampling and equating methods works best? *Applied Mesurement in Education* **3** (1), 73–95.

Lord, F.M. (1950). *Notes on Comparable Scales for Test Scores*. ETS RB-50-48. Educational Testing Service, Princeton, NJ.

Lord, F.M. (1955). Equating test scores: A maximum likelihood solution. *Psychometrika* **20**, 193–200.

Lord, F.M. (1980). *Applications of Item Response Theory to Practical Testing Problems*. Erlbaum, Hillsdale, NJ.

Lord, F.M., Wingersky, M.S. (1984). Comparison of IRT true-score and equipercentile observed-score equatings. *Applied Psychological Measurement* **8**, 453–461.

Marco, G.L., Petersen, N.S., Stewart, E.E. (1983). *A Large-Scale Evaluation of Linear and Curvilinear Score Equating Models*. Volumes I and II. ETS RM-83-02. Educational Testing Service, Princeton, NJ.

Mislevy, R.J. (1992). *Linking Educational Assessments: Concepts, Issues, Methods, and Prospects*. ETS Policy Information Center, Princeton, NJ.

Otis, A.S. (1916). The reliability of spelling scales, including a 'deviation formula' for correlation. *School and Society* **4**, 96–99.

Petersen, N.S., Kolen, M.J., Hoover, H.D. (1989). Scaling, norming and equating. In: Linn, R.L. (Ed.), *Educational Measurement*. 3rd ed. Macmillan, New York, pp. 221–262.

Pommerich, M., Dorans, N.J. (Eds.) (2004). Concordance [Special issue]. *Applied Psychological Measurement* **28** (4).

Rosenbaum, P.R. (1995). *Observational Studies*. Springer-Verlag, New York.

Rosenbaum, P.R., Rubin, D.B. (1983). The central role of the propensity score in observational studies for causal effects. *Biometrika* **70**, 41–55.

Skaggs, , G. (2005). Accuracy of random groups equating with very small samples. *Journal of Educational Mesurement* **42** (4), 309–330.

von Davier, A.A., Holland, P.W., Thayer, D.T. (2004a). The chain and post-stratification methods of observed-score equating: Their relationship to population invariance. *Journal of Educational Measurement* **41**, 15–32.

von Davier, A.A., Holland, P.W., Thayer, D.T. (2004b). *The Kernel Method of Test Equating*. Springer, New York.

Wright, N.K., Dorans, N.J. (1993). *Using the Selection Variable for Matching or Equating*. ETS RR-93-4. Educational Testing Service, Princeton, NJ.

Yen, W.M., Fitzpatrick, A.R. (2006). Item response theory. In: Brennan, R.L. (Ed.), *Educational Measurement*, 4th ed. American Council on Education/Praeger, Westport, CT.

Handbook of Statistics, Vol. 26
ISSN: 0169-7161

DOI: 10.1016/S0169-7161(06)26007-3

7

Electronic Essay Grading

Shelby J. Haberman

Electronic essay scoring may at first appear to be an impossible task, for computer programs do not understand essays; however, computers can be employed to calculate statistical summaries of essays, and these summaries can be used to score an essay. Computer scores can be developed so that they predict essay scores provided by human readers, or scores can be produced which correspond to an analytical scale for writing assessment. Software for essay assessment varies considerably in terms of the essay features that are used in the evaluation of an essay and in terms of the measures used to assess the quality of the computer-generated essay score. Basic approaches examined here are regression analysis, composite scales, content analysis, analysis of discrete responses, and Bayesian analysis.

Regression analysis, which is considered in Section 1, typically involves prediction of human holistic essay scores by use of computer-generated essay features. This approach is straightforward, effective, and commonly used. It has the unusual feature that a holistic score is approximated by use of an analytic scale.

Composite scales, the subject of Section 2, do not seek to predict human holistic scores. Instead, they seek to develop analytical scores to measure writing quality which have desired properties such as relatively high reliability coefficients when the same examinee responds to multiple essay prompts. Composite scores need not be entirely computer-generated. They may include both human-generated and computer-generated components. Common methods for development of composite scales are readily applied given some familiarity with multivariate analysis.

As evident in Section 3, content analysis seeks to measure essay quality on the basis of the specific words employed by the writer in response to the prompt. This approach typically relates word choice to human holistic scores. From a methodological standpoint, content analysis is rather difficult due to the very large number of distinct words that may be relevant to a specific prompt. Although regression analysis is an effective tool for content analysis, much work in this area has emphasized techniques originally developed for information retrieval. As a consequence, formal statistical analysis of methods employed in content analysis is quite difficult.

Approaches such as regression analysis do not exploit the fact that typical holistic scores have only a limited number of values. Attempts to rectify this situation by use of methods for discrete data are examined in Section 4. Once again, human holistic scores are predicted, but the approaches are nonlinear and employ maximum likelihood for discrete data rather than linear regression.

In Section 5, an approach is described that applies Bayesian analysis to the prediction of human holistic scores. This approach is related to the analysis of Section 4, but the model used is more complex.

The analysis in this chapter does not treat a number of practical issues that require empirical study. In typical studies, the examinee does not expect to be scored by a computer. It is quite possible that an examinee may modify behavior if he or she expects to be graded by a computer rather than by a person and if the examinee understands the approach employed by the computer software. The analysis also does not address policy issues involving public acceptability of a system for essay scoring that does not involve any real understanding of the essay. There is a further issue that the electronically generated essay features are not necessarily perfect measures. Computer-generated variables such as the rate per word of misspelled words reported need not be the actual rate of misspelled words, and the rate per word of grammatical errors found by a computer program need not be the same as the rate found by an expert reader (Burstein et al., 2004; Haberman, 2004).

Data available do not permit consideration of fairness issues such as whether particular groups of examinees tend to benefit or be harmed by use of computer-generated scoring rather than human scoring.

1. Regression analysis

Regression analysis has been employed in essay scoring since the 1960s (Page, 1966, 1968). The basic approach is straightforward. A sample of $n+1$ essays i, $0 \leqslant i \leqslant n$, is both graded by human graders and processed by computer. Of these essays, essays 1 to n are employed in estimation of the parameters in the regression model. Essay 0 is not actually used in the analysis of data. It is present to discuss the problem of prediction for an observation that has not been employed in estimation of regression parameters. For simplicity, assume that the essays are a simple random sample from a very large population of essays. Let each essay i receive $m \geqslant 1$ numerical scores H_{ij}, $1 \leqslant j \leqslant m$, from different human raters, and for essay feature k, $1 \leqslant k \leqslant p$, let the computer generate a numerical score X_{ik}. For simplicity, assume that the common covariance matrix $\mathrm{Cov}(\mathbf{X})$ of the p-dimensional vector $\mathbf{X}_i$ with coordinates X_{ik}, $1 \leqslant k \leqslant p$, is positive definite. For example, feature 1 might be the square root of the number of characters in the essay, and X_{i1} might be the number of characters in essay i. In the simple case of unidentified raters, one may assume that the score H_{ij} can be decomposed into a true score T_i and error score e_{ij} such that the errors e_{ij} have the same positive variance $\sigma^2(e)$ and are uncorrelated with each other, with the true score T_i, and with the computer-generated variables X_{ik}. Let the best linear predictor of true score T_i by use of the X_{ik} be

$$\hat{T}_i = \alpha + \sum_{k=1}^{p} \beta_k X_{ik},$$

and let the prediction error be

$$r_i = T_i - \hat{T}_i.$$

Assume that r_i has a positive variance $\sigma^2(r)$ and T_i has a positive variance $\sigma^2(T)$. The variance of H_{ij} is then $\sigma^2(H) = \sigma^2(T) + \sigma^2(e)$.

The possibility of effective use of the computer-generated features can be seen through an examination of the reliability coefficient of the human score and the coefficient of determination for linear prediction of the true score by the computer-generated scores of essay features. The reliability coefficient for an individual human score is

$$\rho_{HT}^2 = \frac{\sigma^2(T)}{\sigma^2(H)} = \frac{\sigma^2(T)}{\sigma^2(e) + \sigma^2(T)},$$

and the coefficient of determination for prediction of T_i by the X_{ik} is

$$\rho_{\mathbf{X}T}^2 = 1 - \frac{\sigma^2(r)}{\sigma^2(T)}.$$

It should be noted that ρ_{HT}^2 is also the correlation coefficient of the human scores H_{ij} and $H_{ij'}$ if $j \neq j'$, and the correlation of a human score H_{ij} and a prediction $\hat{T}_i$ of the corresponding true score is $\rho_{HT}\rho_{\mathbf{X}T}$, so that the correlation of the computer-generated prediction $\hat{T}_i$ and a human score H_{ij} exceeds the correlation of the two human scores H_{ij} and $H_{ij'}$ if the multiple correlation coefficient $\rho_{\mathbf{X}T}$ exceeds the correlation ρ_{HT} of a human score H_{ij} and the true score T_i (Holland and Hoskens, 2003).

Were α and the β_k known, then the mean squared error from prediction of the true score T_i by use of $\hat{T}_i$ would be the variance $\sigma^2(r)$ of the residuals r_i. Let

$$\bar{e}_i = m^{-1} \sum_{j=1}^{m} e_{ij}$$

be the mean error for essay i. Because the prediction error from use of $\bar{H}_i$ to predict T_i is $\bar{e}_i$, the mean squared error from prediction of the true score T_i by use of $\bar{H}_i$ is

$$\sigma^2(\bar{e}) = \sigma^2(e)/m,$$

so that the estimate of the true score generated by the computer features is more accurate, in the sense of mean squared error, than is the average human score if $\sigma^2(\bar{e})$ is greater than $\sigma^2(r)$. Given the formulas for reliability and for the coefficient of determination, the computer-based score $\hat{T}_i$ is a more accurate approximation to the true score T_i than is the mean human score $\bar{H}_i$ if

$$1 - \rho_{HT}^2 > \left(1 - \rho_{\mathbf{X}T}^2\right) m \rho_{HT}^2.$$

Thus the computer-based score is increasingly favored as the reliability ρ_{HT}^2 decreases, the number m of human scorers decreases, and the coefficient of determination $\rho_{\mathbf{X}T}^2$ increases.

In practice, use of computer-based scores requires estimation of the parameters α and β_k, $1 \leqslant k \leqslant p$, and accuracy of approximation of the true score is adversely affected, with the effect smaller as the sample size increases. Let the essays i from 1 to n be employed with dependent variables $\bar{H}_i$ and independent variables X_{ik}. Let a be the minimum norm least-squares estimate of α, and let b_k be the minimum norm least-squares estimate of β_k (Rao and Mitra, 1971, p. 51). Then T_i is approximated by

$$T_i^* = a + \sum_{k=1}^{p} b_k X_{ik},$$

and the approximation error is $r_{i*} = T_i - T_i^*$. The approximation T_i^* may be computed both for the essays i from 1 to n used in the estimation of regression parameters and for essay 0 that has not been used in estimation of regression parameters.

The mean squared error V of the approximation T_0^* of T_0 is the basic measure of accuracy for use of the results with new data, for the holistic scores H_{0j} for this essay and the feature scores X_{0k} for this essay have not been employed to find the estimated intercept a and slopes b_k. This mean squared error is obviously greater than $\sigma^2(r)$. To approximate the size B of the increase in mean squared error, let

$$\mathbf{W}_i = (r_i + \bar{e}_i)\mathbf{X}_i,$$

let tr denote the trace of a square matrix, and let $\mathrm{Cov}(\mathbf{W})$ be the covariance matrix of the vector $\mathbf{W}_i$. Minor changes in the standard formula for prediction of a new observation show that nB converges to

$$\sigma^2(r) + m^{-1}\sigma^2(e) + \mathrm{tr}\left(\left[\mathrm{Cov}(\mathbf{X})\right]^{-1}\mathrm{Cov}(\mathbf{W})\right)$$

as the sample size n increases (Draper and Smith, 1998, p. 162), so that for a large sample size n, the effect of estimation of parameters is quite small. This limit of nB does not assume that classical assumptions of regression analysis hold. If c is the maximum conditional expectation given $\mathbf{X}_i$ of the square of $r_i + m^{-1}\bar{e}_i$, then the limit of nB is bounded by $c(p+1)$, so that it is also important that p/n be small. This bound c is the limit of nB and

$$c = \sigma^2(r) + \sigma^2(\bar{e})$$

if $r_i + \bar{e}_i$ and $\mathbf{X}_i$ are independent, as is the case in classical regression analysis; however, this condition is difficult to satisfy in nontrivial cases given the restricted range of human scores H_{ij}.

To assess the value of electronic scoring, estimation of V and $\sigma^2(e)$ is needed. In the case of $\sigma^2(e)$, estimation is quite elementary given standard results from analysis of variance, provided that $m > 1$. One may use

$$s^2(e) = \left[n(m-1)\right]^{-1}\sum_{i=1}^{n}\sum_{j=1}^{m}(e_{ij} - \bar{e}_i)^2$$

(Lord and Novick, 1968, chapter 7). For V, the argument is a bit more difficult. Observe that

$$E\left(\left[\bar{H}_0 - T_0^*\right]^2\right) = V + \sigma^2(e)/m,$$

and $E([\bar{H}_0 - T_0^*]^2)$ may be estimated by use of deleted residuals. It is necessary that $n > p + 1$. For feature score j from 1 to p, let $\bar{X}_j$ be the sample mean of the scores X_{ij}, $1 \leqslant i \leqslant n$. Let $\mathbf{C}$ be the p by p matrix with row j and column k equal to

$$C_{jk} = \sum_{i=1}^{n}(X_{ij} - \bar{X}_j)(X_{ik} - \bar{X}_k),$$

and let C^{jk} be row j and column k of the inverse of $\mathbf{C}$. Let the matrix $\mathbf{R}$ be the n by n matrix with row i and column i' equal to

$$R_{ii'} = n^{-1} + \sum_{j=1}^{p}\sum_{k=1}^{p}(X_{ij} - \bar{X}_j)(X_{i'k} - \bar{X}_k)C^{jk}.$$

This matrix is sometimes termed the hat matrix (Draper and Smith, 1998, pp. 205–207) due to the equation

$$T_i^* = \sum_{j=1}^{p} X_{ij}\bar{H}_j$$

that holds for $1 \leqslant i \leqslant n$. For $1 \leqslant i \leqslant n$, the deleted residual

$$r_{id} = -\frac{\bar{H}_i - T_i^*}{1 - R_{ii}}$$

(Weisberg, 1985) is the error of prediction of $\bar{H}_i$ from X_{ij}, $1 \leqslant j \leqslant p$, if the regression parameters are estimated by use of the $\bar{H}_i'$ and $X_{i'j}$, $1 \leqslant j \leqslant p$, for essays i' between 1 and n such that $i' \neq i$. Thus r_{id} is produced by prediction of $\bar{H}_i$ by deletion of essay i. Then $E([\bar{H}_0 - T_0^*]^2)$ may be estimated by

$$n^{-1}\sum_{i=1}^{n} r_{id}^2,$$

and V may be estimated by

$$\hat{V} = n^{-1}\sum_{i=1}^{n} r_{id}^2 - s^2(\bar{e}).$$

As the sample size n increases, the expectation of $n(\hat{V} - V)$ approaches 0, and the probability approaches 1 that $\hat{V}$ is positive. Thus the desired measure of mean squared error for prediction of a new observation has been obtained by use of results of a conventional regression analysis. In effect, n instances of subsamples of size $n - 1$ have been used, but no explicit computations have been required to accomplish this goal.

Comparison of $s^2(\bar{e})$ and $\hat{V}$ provides a basis for assessment of electronic scoring, for $\hat{V}$ estimates the mean squared error of the approximation T_0^* to the true score T_0 for essay 0, and $s^2(\bar{e})$ estimates the mean squared error $\sigma^2(\bar{e})$ obtained if the true score T_0 for essay 0 is estimated by the mean $\bar{H}_0$ of the m holistic scores H_{0i} for essay 0.

1.1. Example of regression analysis

The challenge in the regression approach is to choose appropriate feature scores X_{ik}. The selection of features has changed over time as computer processing power has increased. Initial efforts involved quite simple features such as number of words, number of semicolons, number of prepositions, average sentence length, and average word length (Page, 1968). Similar selections of essay features continued for some time (Wresch, 1993). More recent efforts with e-rater® have involved more complex features such as rate per word of grammatical errors, rate per word of errors in mechanics, and number of discourse elements in the essay (Burstein et al., 2004). For an illustration of results, consider four prompts previously examined in a study of essay features (Haberman, 2004; Haberman and Qian, 2004). Two prompts are from the TOEFL program and two from the GMAT program. Table 1 provides some basic summary information concerning sample sizes and distributions of holistic scores. To avoid responses that are not really essays, only essays with at least 25 words have been included. In examination of the table, note that each prompt receives a holistic score from two raters, and each score is on a scale of 1 to 6 in which 6 is the highest possible score.

Two kinds of analysis are considered in Table 1. In the first analysis, only $p = 6$ essay features are considered, and none of the features has any specific relationship to the content of the essay. The variables are square root of number of characters (SC), square root of rate per word of errors in grammar and usage detected by the e-rater® Version 2.0 software (SGU), square root of rate per word of spelling errors reported by the software (SSP), square root of rate per word of mechanics errors other than spelling detected by the software (SOM), median Standard Frequency Index (SFI) of words in the essay for which the index is available (Breland et al., 1994; Breland, 1996; Breland and Jenkins, 1997; Carroll, 1970), and the square root (SSI) of Simpson's index for the diversity of content words in the essay (Simpson, 1949). The SFI is a measure of the frequency with which a word appears in English texts. The index has values from 0 to 100, with a lower index corresponding to a less frequent work. The Simpson index is the probability that two randomly selected words from an essay will coincide. The variable transformations are designed to make the predictors more similar in distribution to a normal distribution, to reduce the incidence of outliers, and to improve the prediction of the dependent variable. Intuitively, one would expect that a regression analysis would yield positive coefficients for SC and negative coefficients for SGU, SSP, SOM, SFI, and SSI. These expectations are in fact satisfied. Note that in each case, the estimated value

Table 1
Summary statistics for average holistic score

Program	Prompt	Count	Average	Standard deviation	Skewness	Kurtosis
GMAT	1	5183	3.789	1.118	−0.143	−0.450
GMAT	2	5158	3.842	1.067	−0.119	−0.319
TOEFL	1	4895	4.112	1.039	−0.161	−0.295
TOEFL	2	4884	4.075	1.021	−0.112	−0.238

Table 2
Regression analysis of holistic scores

				$\hat{V}$		
Program	Prompt	No. essays	$s^2(e)$	$p = 6$	$p = 56$	$p = 206$
GMAT	1	5,183	0.356	0.212	0.108	0.070
GMAT	2	5,158	0.346	0.076	0.053	0.043
TOEFL	1	4,895	0.275	0.101	0.098	0.095
TOEFL	2	4,884	0.259	0.123	0.116	0.112

$\hat{V}$ of the mean squared error from estimation of the true score by the regression analysis is less than the estimated mean squared error $s^2(e)$ for estimation of the true score by a single holistic score. Except for the first prompt, the estimated mean squared error $\hat{V}$ for prediction of true score by regression is less than the estimated mean squared error $s^2(\bar{e}) = s^2(e)/2$ from prediction of true score by the average of two human holistic scores.

In the regression analysis with $p = 56$ and $p = 206$, an attempt has been made to introduce some measure of essay content. For this purpose, a stop word is a word that is not customarily used by a search engine. For example, articles such as "the" and "a" are stop words. Various lists are available. The e-rater® software used for many of the computations in this chapter uses a list of about 100 stop words. Words that are not stop words are content words. Let content words W_k be ordered by their total frequency among all essays available for a given prompt, so that W_1 is the most frequent content word for the prompt. Let F_{ik} be the frequency of word W_k in essay i, and let C_i be the total number of content words in essay i. The case of $p = 56$ adds $X_{i(k+6)} = (F_{ik}/C_i)^{1/2}$ for k from 1 to 50, while the case of $p = 206$ adds $X_{i(k+6)} = (F_{ik}/C_i)^{1/2}$ for k from 1 to 200. The square roots are chosen to improve statistical behavior. The ratio F_{ik}/C_i is the fraction of content words that are W_k. The approach to content words used here is rather crude, for plural and singular words are distinguished as well as different verb forms for the same verb. The use of content words has a substantial impact in the GMAT prompts, especially for the first prompt in which $\hat{V}$ goes from 0.212 for $p = 6$ (no content) to 0.070 for $p = 206$ (200 content words). The impact for the TOEFL prompts is very small.

For all three sets of predictors, the regression analysis provides a prediction of the true score that is somewhat more accurate than the prediction provided by one human rater. For the TOEFL examination, each regression provide a smaller mean squared error than would be achieved by two raters. For the GMAT, the second regression for each prompt ($p = 56$) is better in terms of mean squared error than three raters and the third regression ($p = 206$) is better in each case than five raters. These results illustrate the basis of claims made for many years that computers can be employed for essay scoring by use of regression analysis (Page, 1966).

1.2. Mixtures of human and machine scores

Use of scores of human raters or scores generated by computer are not exclusive options. It is also possible to predict the true score of an essay with a combination of $q \leqslant m$ holistic scores and p essay features generated by computer. An approach of this type is found in the GMAT. In GMAT, the score given an essay is normally the average of the score of one human rater and a rounded value of the score provided by the regression analysis. Without use of rounding, one can consider an estimate for T_0 with

$$\bar{H}_{0q} = q^{-1} \sum_{j=1}^{q} H_{0j}$$

and

$$S_{a0} = a\bar{H}_{0q} + (1-a)T_0^*$$

for some constant a between 0 and 1. The GMAT case corresponds to $a = 0.5$ and $q = 1$. For any choice of a, the error is $r_{a0} = S_{a0} - T_0$, and the mean squared error is then

$$V_a = \frac{a^2}{q}\sigma^2(e) + (1-a)^2 V.$$

The minimum value of V_a is achieved for

$$a = \frac{V}{V + \sigma^2(e)/q}.$$

The value of S_{a0} corresponding to this selection of a is

$$S_0^* = \frac{V\bar{H}_{0q} + q^{-1}\sigma^2(e)T_0^*}{V + q^{-1}\sigma^2(e)},$$

and the mean squared error of S_0^* is

$$V^* = \frac{q^{-1}\sigma^2(e)V}{V + q^{-1}\sigma^2(e)}.$$

One may estimate the optimal value of a by

$$\hat{a} = \frac{\hat{V}}{\hat{V} + s^2(e)/q},$$

so that S_0^* is approximated by

$$\hat{S} = \frac{\hat{V}\bar{H}_{0q} + q^{-1}s^2(e)T_0^*}{\hat{V} + q^{-1}s^2(e)}$$

(Haberman and Qian, 2004). The estimated value of the mean squared error of S_0^* is then

$$\hat{V}^* = \frac{q^{-1}s^2(e)\hat{V}}{\hat{V} + q^{-1}s^2(e)}.$$

Table 3
Weighted combination of human and regression scores

Program	Prompt	$V_{0.5}$	$\hat{a}$	$\hat{V}^*$
GMAT	1	0.116	0.232	0.083
GMAT	2	0.100	0.134	0.046
TOEFL	1	0.093	0.263	0.072
TOEFL	2	0.094	0.310	0.080

To illustrate this approach, consider the four prompts under study for $p = 56$ and $q = 1$. Results are shown in Table 3 with $a = 0.5$ and $\hat{a}$.

In terms of mean squared error, the evenly weighted combination of scores is actually less effective than is the regression score for the GMAT prompts. In general, the optimal weighting is somewhat better than is the even weighting.

1.3. *Validity and reliability*

Validity and reliability are further problems for machine scoring of essays. The essay score, whether obtained from human or electronic raters, only applies directly to the specific prompt and testing conditions under which the essay was produced. One must be concerned with applications to other prompts and with attempts to infer more general writing ability or analytical writing ability from the essay score. In the case of regression estimation of holistic scores, validity and reliability research appears to be rather limited. A limited study of 2,595 high school essays that involved 562 students and seven prompts provides a reliability estimate of 0.61 if the regression analysis with $p = 6$ is used (Haberman, 2004). This case should be approached with some caution for several reasons. Relatively few students wrote essays for each prompt, the resolved holistic score available was a single rounded score from 1 to 6 computed from at least two human holistic scores (Burstein et al., 2004; Haberman, 2004), and the reliability was estimated by use of a two-way analysis of variance with examinee and prompt factors. The reliability was based on the estimated variance components for error and for examinee, so that the reliability estimates assume an adjustment for prompt effect has been made. For comparison, the resolved holistic score itself had an estimated reliability of only 0.52.

1.4. *Agreement measures*

The evaluation of regression analysis in this section is somewhat different than has often been seen in the literature on automated essay scoring (Burstein et al., 2004). Much of the discussion has involved rounding of the regression estimate to an integer score with range equal to the range of the original holistic score provided by each rater. Even if a human holistic score is based on more than a single rater, the holistic score used in evaluation and in regression has been an integer holistic score determined from the available holistic scores (Burstein et al., 2004; Haberman, 2004). The criterion used has then been the probability that the human and machine integer scores agree (exact agreement) or the probability that the human and machine scores differ by no more than 1 (adjacent agreement). This procedure has little logical basis given that rounding normally decreases the efficiency of regression analysis and that scores are ordered. In addition, the approach does not consider comparison of prediction by regression analysis with prediction without use of any variables at all.

As an illustration, consider the regression analysis for the first prompt reported in Table 2. Consider the case of $p = 56$. Consider the fraction of integers i and j, $1 \leqslant i \leqslant n$ and $1 \leqslant j \leqslant 2$, such that $|H_{ij} - T_i^*|$ is less than 0.5. In such cases, the rounded value of the regression estimate T_i^* is equal to the holistic score H_{ij}. This fraction with exact agreement is 0.565, so that the estimated probability of error by this criterion is $1 - 0.565 = 0.435$. Similarly, consider the fraction of integers i and j such that $|H_{ij} - T_i^*|$ is less than 1.5, so that the rounded regression estimate differs from the holistic score by no more than 1. This fraction with adjacent agreement is 0.969, so that the probability of error by the criterion of adjacent agreement is 0.031. In study of these fractions, it is important to note that the fraction of integers i and j such that H_{ij} is 4 is 0.333, so that a trivial prediction method for human score that uses no data can achieve an error rate of 0.667. The proportional reduction in error rate due to use of the regression is then a quite modest

$$1 - \frac{0.435}{0.667} = 0.348.$$

Similarly, the fraction of integers i and j such that $|H_{ij} - 4|$ does not exceed 1 is 0.774, so that the error rate by the criterion of adjacent agreement is 0.226. The proportional reduction in error is then a somewhat larger (Dorans and Patsula, 2003)

$$1 - \frac{0.031}{0.226} = 0.863.$$

1.5. Alternative forms of regression analysis

A number of simple variations on regression analysis have been considered. Ridge regression and stepwise regression are natural possibilities, especially in cases in Table 2 in which the number p of predictors is large (Feng et al., 2003). These approaches yield alternative estimates of the true score that are reasonably easily computed with standard statistical software; however, proper estimation of mean squared errors is somewhat more difficult than in the case of ordinary regression. In the examples in Table 2, these alternative approaches appear to have at best only a very limited advantage over ordinary regression analysis.

Regression analysis has the virtue of simplicity, and the methods considered in this section do not require that the dependent variable satisfy linearity, homoscedasticity or normality conditions. The only criterion considered is the extent to which the true holistic score is predicted well. A natural limit on regression analysis is the quality of the holistic scoring system.

2. Composite scales

It is not necessarily true that a computer-generated essay score should be designed to mimic a human holistic score. Given computer-generated essay feature scores X_{ik}, $1 \leqslant k \leqslant p$, with respective means $E(X_k)$ and standard deviations $\sigma(X_k)$ and the average human score $\bar{H}_i$ with mean $E(T) = E(T_i)$ and standard deviation $\sigma(\bar{H}) = [\sigma^2(T) + \sigma^2(\bar{e})]^{1/2}$, consider the standardized features

$$Z_{ik} = \frac{X_{ik} - E(X_k)}{\sigma(X_k)},$$

and

$$G_i = \frac{\bar{H}_i - E(T_i)}{\sigma(\bar{H})}.$$

Composite scores

$$Y_i = \sum_{k=1}^{p} \gamma_k Z_{ik}$$

can be constructed for some constants γ_k, $1 \leqslant k \leqslant p$. In addition, it is possible to consider composite scores

$$U_i = \delta G_i + \sum_{k=1}^{p} \eta_k Z_{ik}$$

based on both human holistic scores and computer-generated essay features. The challenge is to choose appropriate weights γ_k for the components Z_{ik} of Y_i and η_k and δ for the components G_i and Z_{ik} of U_i. Weights can be selected on the basis of judgement concerning relative importance of the various essay features. Weights can also be selected by principal components analysis or by reliability analysis (Gulliksen, 1987, chapter 20). In all methods, some consideration must be given to reasonableness of signs of coefficients. For example, in the case of the analysis reported in Table 2 for $p = 6$, as previously noted, it is normal to expect that an essay is improved as the holistic score average $\bar{H}_i$ increases, the length variable X_{i1} increases (the essay is typically more developed in this case), the square roots X_{i2}, X_{i3}, and X_{i4} of error rates decrease, the median SFI X_{i5} decreases (the essay words are less common), and the square root X_{i6} of Simpson's index of the content word distribution decreases (the vocabulary is more diverse). Thus it is desirable that γ_1 and η_1 are positive, γ_j and η_j are negative for j from 2 to 6, and δ is positive.

2.1. Principal components

In the case of principal component analysis for a composite score Y_0 based exclusively on computer-generated scores, if the correlation matrix of the X_{0k} is known for $1 \leqslant$

$k \leqslant p$, then Y_0 is selected to maximize the variance of Y_0 subject to the constraint that

$$\sum_{k=1}^{p} \gamma_k^2 = 1.$$

In this way, one minimizes the sum of the mean squared errors of the regressions of Z_{ik} on Y_i for $1 \leqslant k \leqslant q$ (Kendall et al., 1983, chapter 43). The variance $\sigma^2(T_0)$ is the largest eigenvalue of the correlation matrix $\mathbf{D}$ of the Z_{ik}, $1 \leqslant k \leqslant p$, and the γ_k are the coordinates of the corresponding eigenvector. The γ_k are uniquely defined by the convention that γ_1 is positive if only one eigenvalue of $\mathbf{D}$ is equal to $\sigma^2(Y_0)$ and if the corresponding eigenvector has a nonzero first coordinate. Similarly, U_0 is selected to maximize the variance of U_0 subject to the constraint that

$$\delta^2 + \sum_{k=1}^{p} \eta_k^2 = 1.$$

In the case in which sampling must be used to estimate the weights, little changes. Let $\bar{X}_k$ and $s(X_k)$ be the respective sample mean and sample standard deviation of the X_{ik}, $1 \leqslant i \leqslant n$, so that the estimated standardized value for feature k is

$$\hat{Z}_i = \frac{X_{ik} - \bar{X}_k}{s(X_k)}.$$

Similarly, let $\bar{H}$ and $s(\bar{H})$ be respective sample mean and sample standard deviation of the $\bar{H}_i$, $1 \leqslant i \leqslant n$, and let

$$\hat{G}_i = \frac{\bar{H}_i - \bar{H}}{s(\bar{H})}.$$

Then

$$\hat{Y}_i = \sum_{k=1}^{p} c_k \hat{Z}_{ik}$$

is selected to maximize the sample variance of $\hat{Y}_i$, $1 \leqslant i \leqslant n$, subject to the constraint that

$$\sum_{k=1}^{p} c_k^2 = 1.$$

Similarly,

$$\hat{U}_i = d\bar{H}_i + \sum_{k=1}^{p} e_k X_{ik}$$

is selected to maximize the sample variance of $\hat{U}_i$, $1 \leqslant i \leqslant n$, subject to the constraint that

$$d^2 + \sum_{k=1}^{p} g_k^2 = 1.$$

Table 4
Principal component composite scale for essay prompts

	GMAT		TOEFL	
Estimate	Prompt 1	Prompt 2	Prompt 1	Prompt 2
c_1	0.400	0.442	0.424	0.440
c_2	−0.472	−0.352	−0.429	−0.364
c_3	−0.534	−0.420	−0.444	−0.355
c_4	−0.443	−0.408	−0.391	−0.379
c_5	−0.367	−0.432	−0.415	−0.466
c_6	0.030	−0.390	−0.338	−0.432
d	0.568	0.558	0.547	0.553
e_1	0.528	0.505	0.471	0.486
e_2	−0.315	−0.234	−0.356	−0.269
e_3	−0.383	−0.306	−0.278	−0.290
e_4	−0.275	−0.271	−0.316	−0.272
e_5	−0.272	−0.321	−0.316	−0.342
e_6	−0.056	−0.331	−0.280	−0.333

In the case of $p = 6$ for the essay prompts reported in Table 2, Table 4 is obtained. The scales are relatively similar for each prompt, except that the weight given to the word diversity measure Z_{i6} is much smaller for the first GMAT prompt and has an inappropriate sign if the average holistic score $\bar{H}_i$ is not used. The reason for the result is not clear.

2.2. *Reliability maximization*

An alternative approach is available if essay portfolios are available for a sample of examinees. In this case, multivariate analysis of variance may be employed to examine the reliability of essay features. With this approach, a composite score can be estimated by canonical analysis to maximize reliability (Haberman, 2004). When applied to the previously described high school essays employed to examine reliability of holistic scores and regression estimates of holistic scores, the coefficients in Table 5 are obtained. Qualitatively, results are fairly similar to those with principal components. With or without the resolved holistic score, the estimated composite scores have reliability of about 0.77. Thus it appears possible to obtain composite scores that are substantially more reliable than human holistic scores or regression approximations to such scores. Whether these composite scores are attractive in terms of validity and in terms of acceptability to users of writing assessments requires further investigation. One point, however, should be emphasized. The individual components of the composite are not notably reliable. The best result, a coefficient of 0.60, was obtained for the essay length variable SC, and a reliability coefficient of 0.56 was obtained for the mechanics variable SOM.

Table 5
Composite scale weights for maximum reliability

Coefficient	Value	Coefficient	Value
c_1	1.081	e_1	0.753
c_2	−0.372	e_2	−0.370
c_3	−0.435	e_3	−0.433
c_4	−0.742	e_4	−0.735
c_5	−0.309	e_5	−0.318
c_6	−0.219	e_6	−0.196
d	0.379		

3. Content analysis

Analysis of content is a relatively difficult problem. It has typically been approached through an examination of the frequencies with which specific words have been used in an essay. This section considers three basic approaches, regression analysis, latent semantic analysis, and content vector analysis.

3.1. Regression analysis

Regression analysis has already been employed to study the effect of content in Section 1 in the cases of $p = 56$ and $p = 206$ predictors. The choice of 50 or 200 words is somewhat arbitrary, and appropriate choices are best explored by examinations of larger classes of prompts than those considered in Table 2. The method used to select words can also be modified, and there are also questions whether it is best to distinguish or not to distinguish closely related words. For example, the analysis reported in Table 2 considers the words "service" and "services" separately. The regression approach is inherently limited in that a meaningful regression can only be performed if the number of different words examined is much smaller than the number of essays examined. It is certainly not evident from results in Table 2 that this limitation has much practical impact, but it does mean that a substantial fraction of the words in the essay are not examined on an individual basis. For instance, in the case of the first GMAT prompt, the 50 most common content words include about 42 percent of all content words found in the collection of essays, while the 200 most common content words include about 63 percent of all words in the essays. There are more than 19,000 distinct content words used in the collection of essays, so that the vast preponderance of words are not considered.

The content measurement in Table 2 is based on relative word frequencies. In the notation used in Section 1, the square root $(F_{ij}/C_i)^{1/2}$ of the relative frequency F_{ik}/C_i of content word W_i was considered. An alternate study of content may be based on the actual word frequencies F_{ik}. This approach, which implicitly includes measurement of essay length, can be remarkably effective at prediction of human holistic scores, even if no other variables are considered. Consider a prediction of the true score T_i in which the predictors are the square roots $F_{ik}^{1/2}$ of the frequencies F_{ik} for content word W_k in essay i, $1 \leqslant k \leqslant 200$, the square root $O_i^{1/2}$ of the frequency $O_i = C_i - \sum_{k=1}^{200} F_{ik}$

of all other content words, and the square root $S_i^{1/2}$ of the frequency S_i in essay i of all stop words. In this case, the estimated mean square errors $\hat{V}$ for the four prompts are 0.069, 0.064, 0.136, and 0.157, respectively. Comparison with Table 2 shows that results are remarkably good for the first GMAT prompt and a bit less successful for the TOEFL prompts. Quite similar results are obtained if the square root of each frequency is replaced by the logarithm of one plus the frequency.

It should be noted that the use of relative frequency without a measure of essay length is much less effective at prediction of holistic score than is use of frequency, and use of the square roots of frequencies plus the measures SGU, SSP, SOM, and SSI used with number of characters in Table 2 is about as effective as the case of $p = 206$ in Table 2. Thus the results here do not provide a true breakthrough in prediction of holistic scores; however, they do indicate that frequencies F_{ik} of individual words can be employed effectively in prediction of holistic score. This issue is important when the method known as latent semantic analysis is explored.

3.2. *Latent semantic analysis*

Latent semantic analysis has been proposed as a technique for electronic essay scoring in which essay content is emphasized (Landauer et al., 1998; Landauer et al., 2003). This method, which is used in the Intelligent Essay Assessor™ (Landauer et al., 2003), differs somewhat from the regression approach in its approach to scoring and in its use of a singular value decomposition. Formal study of its statistical properties appears to be difficult given the very high dimension of many components of the analysis, although modified versions of LSA that restrict the number of words considered can be analyzed in a conventional fashion.

To describe LSA for the n essays under study, let W_k and F_{ik} be defined as in Section 1 for $1 \leqslant k \leqslant p$ for some integer p such that, for $1 \leqslant k \leqslant p$, $F_{ik} > 0$ for at least two essays. Let $F_{+k} = \sum_{i=1}^{n} F_{ik}$ be the total number of uses of word W_k in any of the n essays i, $1 \leqslant i \leqslant n$, and let $f_{ik} = F_{ik}/F_{+k}$ be the fraction of observed uses of word W_k in the n essays that are in essay i. A singular value decomposition (Stewart, 1973, pp. 317–326) is performed on the n by p matrix $\mathbf{S}$ with row i, $1 \leqslant i \leqslant n$, and column k, $1 \leqslant k \leqslant p$, equal to

$$S_{ik} = J_k^{-1} \log(F_{ik} + 1),$$

where J_k is the entropy estimate

$$J_k = -\sum_{i=1}^{n} f_{ik} \log f_{ik}$$

and $0 \log 0$ is 0. Note that S_{ik} is 0 if $F_{ik} = 0$, so that word W_k is not in essay i, and $S_{ik} > 0$ if $F_{ik} > 0$, so that word W_k is found in essay i.

If d is the minimum of n and p, then $\mathbf{S}$ has the singular value decomposition

$$S_{ik} = \sum_{a=1}^{d} \lambda_a t_{ia} u_{ka},$$

where the λ_a are nonnegative and not increasing,

$$\sum_{i=1}^{n} t_{ia}t_{ib} = \begin{cases} 1, & a = b, \\ 0, & a \neq b, \end{cases}$$

and

$$\sum_{k=1}^{p} u_{ka}u_{kb} = \begin{cases} 1, & a = b, \\ 0, & a \neq b. \end{cases}$$

For $0 \leqslant i \leqslant n$, let

$$v_{ia} = \sum_{k=1}^{p} S_{ik}u_{ka}.$$

Then

$$\lambda_a t_{ia} = v_{ia}$$

for $1 \leqslant i \leqslant n$, so that t_{0a}, $1 \leqslant a \leqslant d$, may be defined as

$$v_{0a}/\lambda_a$$

whenever $\lambda_a > 0$. The weighted score S_{ik} for word W_k and essay i, $0 \leqslant i \leqslant n$, $1 \leqslant k \leqslant p$, may be approximated by

$$S_{ikq} = \sum_{a=1}^{q} u_{ka}v_{ia}$$

for each $q \leqslant d$. If $q = d$, then $S_{ikq} = S_{ik}$ for $1 \leqslant i \leqslant n$. Each S_{ikq}, $0 \leqslant i \leqslant n$, is uniquely defined for a given q if the λ_a are all distinct and positive for $1 \leqslant a \leqslant q$ and if $\lambda_q > \lambda_{q+1}$ in the case of $q < d$. Given the choice of q, the new variables S_{ikq}, $1 \leqslant k \leqslant p$, may then used to approximate human holistic scores. Given the linear relationships present between the S_{ikq} and v_{ia}, use of S_{ikq}, $1 \leqslant k \leqslant p$, as predictors of $\bar{H}_i$ is equivalent to use of v_{ia}, $1 \leqslant a \leqslant q$, as predictors of $\bar{H}_i$.

The variables S_{ikq} and v_{ia} are readily criticized as artificial variables with no clear meaning that are dependent on all essays from the sample and not just on essay i. In addition, the S_{ikq} and v_{ia} are rather unstable if the λ_a are similar in value for any $a \leqslant q+1$ (Wilkinson, 1965, chapter 2). The variables have a further limitation that is evident if regression analysis is considered. Consider the case of p fixed and n large. Consider a linear regression analysis with average human holistic score $\bar{H}_i$ and independent variables S_{ikq}, $1 \leqslant k \leqslant p$ for each essay i from 1 to n. Let $X_{ik} = \log(F_{ik} + 1)$ for $1 \leqslant k \leqslant p$. Recall notation in Section 1 for the true score T_0, the estimated true score T_0^*, the residual variance $\sigma^2(r)$, and the mean squared error V. Observe that random variable d_{jk}, $1 \leqslant j \leqslant p$, $1 \leqslant k \leqslant p$, determined by the F_{ik}, $1 \leqslant i \leqslant n$, and $1 \leqslant k \leqslant p$, exist such that

$$S_{ikq} = \sum_{j=1}^{p} d_{jk}X_{ij}.$$

Select a_q and b_{kq}, $1 \leqslant k \leqslant p$, to minimize the residual sum of squares

$$\sum_{i=1}^{n}(\bar{H}_i - G_{iq})^2$$

for

$$G_{iq} = a_q + \sum_{k=1}^{q} b_{kq}S_{kq}, \quad 0 \leqslant i \leqslant n.$$

The mean squared error $V_q = E([T_0 - G_{0q}]^2)$ from the regression analysis based on the S_{ikq} must be at least the variance $\sigma^2(r)$ of the residual r_i from the regression of T_i on the X_{ik} for $1 \leqslant k \leqslant p$. For comparison, consider the mean squared error $V = E([T_0 - T_0^*]^2)$ based on regression analysis in which regression parameters are estimated by observation of $\bar{H}_i$ and X_{ik}, $1 \leqslant k \leqslant p$, for $1 \leqslant i \leqslant n$. Here $n[V - \sigma^2(r)]$ has a finite limit of order $p + 1$, so that $E([T_0 - G_{0q}]^2)$ cannot possibly be less than V by more than a term of order $(p + 1)/n$. Thus for p/n small, SA has very limited potential if used in regression analysis.

It should be noted that as the sample size n increases and p remains constant, the perturbation theory of eigenvalues and eigenvectors may be used to show that $E([T_0 - G_{0q}]^2)$ does approach a limit V_q^* that is $\sigma^2(r)$ for $q = p$ and at least $\sigma^2(r)$ for $q < p$ (Wilkinson, 1965, chapter 2). To find V_q^* for $q < p$, let $\mathbf{D}$ be the p by p matrix with row j and column k equal to $E(X_{ij}X_{ik})$. Let the eigenvalues of $\mathbf{D}$ be μ_a, $1 \leqslant a \leqslant p$, and let the μ_a not increase in a. Let the μ_a be distinct for $a \leqslant q + 1$, and let the eigenvector corresponding to μ_a have coordinates u_{ka}^*, $1 \leqslant k \leqslant p$ and satisfy

$$\sum_{k=1}^{p}\left(u_{ka}^*\right)^2 = 1$$

for $1 \leqslant a \leqslant p$. Let

$$v_{ia}^* = \sum_{k=1}^{p} u_{ka}^* X_{ik}$$

for $1 \leqslant a \leqslant q$, and let

$$S_{ikq}^* = \sum_{a=1}^{q} v_{ia}^*.$$

To simplify analysis, assume that u_{1a}^* is positive for $1 \leqslant a \leqslant q$, and u_{ia} is selected so that u_{1a} is positive if u_{1a} is not 0. Let G_{iq}^* be the best linear predictor of T_i based on v_{ia}^*, $1 \leqslant a \leqslant q$. Then v_{ia} approximates v_{ia}^*, S_{ikq} approximates S_{ikq}^*, and V_q converges to $V_q^* = E([T_0 - G_{0q}^*]^2)$. Obviously V_q^* is at least $\sigma^2(r)$.

A reasonable approach to estimation of V_q^* is to apply the estimation method used for estimation of V in Section 1. The one complication is that one proceeds as if the true independent variables are v_{ia}, $1 \leqslant a \leqslant q$, rather than v_{ia}^*, $1 \leqslant a \leqslant q$. This practice yields an estimate $\hat{V}_q$ of V_q with a bias of order q/n.

Given the approach of least squares, LSA does not necessarily perform notably well. It is important to realize that the v_{i1} are closely related to the logarithm L_i of the number of words in essay i. Consider the first GMAT prompt described in Table 2. Consider use of $p = 200$ content words, and recall that n is 5,183. For the X_{ik}, $1 \leqslant k \leqslant p$, $\hat{V}$ in this case is 0.112, and the corresponding value of $\hat{V}$ based on the single predictor L_i is 0.290. The sample correlation of v_{i1} and L_i is -0.777, but the estimate $\hat{V}_1 = 0.394$ is a bit larger than might be expected. Use of a larger q does have some impact, for the estimate $\hat{V}_{10} = 0.154$ is reasonably small but somewhat greater than $\hat{V}$ for the X_{ik}, $1 \leqslant k \leqslant p$.

If every content word that appears in at least two essays is used rather than just a fixed number of content words, then LSA becomes a formidable task. For example, in the first GMAT prompt, F_{+k} is at least two for more than 9,000 distinct words W_k. A singular value decomposition that involves a matrix of rank more than 5,000 is a difficult undertaking. The statistical properties of such a decomposition are likely to be highly unsatisfactory given that the stability of the u_{ka}, λ_a, and t_{ia} depend strongly on the sizes of the differences between the λ_a. In addition, examination of large-sample properties of LSA is much more difficult if p increases as n increases.

In practice, LSA has normally not employed regression analysis for prediction of T_0. Instead nonparametric regression based on nearest neighbors has been tried. In this approach, nearest neighbors are defined in terms of a cosine similarity measure for the S_{ikq} and S_{0kq}, $1 \leqslant i \leqslant n$ (Landauer et al., 2003). To define the similarity measure, first define the size measure s_{iq} for the S_{ikq}, $1 \leqslant k \leqslant p$, by

$$s_{iq}^2 = \sum_{k=1}^{p} S_{ikq}^2 = \sum_{a=1}^{q} v_{ia}^2.$$

This measure can be regarded as an attempt to assess the amount of content in the essay. It clearly has some relationship to essay length, although a long essay with very few distinct content words would have a relatively small length.

For essays i from 0 to n and j from 0 to n, the size measures s_{iq} and s_{jq} are associated with the cross-product measure

$$d_{ijq} = \sum_{k=1}^{p} S_{ikq} S_{jkq} = \sum_{a=1}^{q} v_{ia} v_{ja}$$

and the cosine measure

$$c_{ijq} = d_{ijq}/(s_{iq} s_{jq}).$$

Note that

$$\sum_{k=1}^{p} (S_{ikq} - a S_{jkq})^2$$

is minimized if $a = d_{ijq}/s_{jq}^2$, and the minimum is $s_{iq}^2(1 - c_{ijq}^2)$.

To define the nonparametric regression estimate, let $j(i)$, $1 \leqslant i \leqslant n$, be defined so that $c_{0j(i)q}$ does not increase in i. For a fixed $N \leqslant n$, the approximation to the true score

T_0 is then

$$H_{0q}^{+} = \frac{\sum_{i=1}^{N} c_{0j(i)q} \bar{H}_{j(i)}}{\sum_{i=1}^{N} c_{0j(i)q}}.$$

The mean squared error V_q^+ is then $V_q^+ = E([T_0 - H_{0q}^+]^2)$. Even for fixed p, the large-sample properties of V_q^+ are complicated due to the use of $j(i)$ and N. Nonetheless, some guidance to reasonable expectations can be obtained from general bounds on the accuracy of nonparametric regression (Stone, 1980). Let the conditional expected value T_i^+ of T_i given v_{ia}^*, $1 \leqslant a \leqslant q$, be a continuously differentiable function of the v_{ia}^*, and let $\sigma^2(r^+)$ be the variance of the residual $T_0 - T_0^+$. Although some complications arise due to the discreteness of the original variables F_{ik} and due to the use of the estimate v_{ia} of v_{ia}^* rather than v_{ia}^*, it is not reasonable to expect that $V_q^+ - \sigma^2(r^+)$ is smaller in order of magnitude than $n^{-2/(2+q)}$. This order of magnitude is based on N of order $n^{2/(2+q)}$. For $q = 10$ and $n = 5,183$, $n^{2/(2+q)}$ is about 4. Both $N = 4$ and $N = 10$ were tried, and the best result was for $N = 10$. With $N = 10$ and $q = 10$, an attempt was made to estimate V_q^+ by the following procedure. For each observation i from 1 to n, $j(i,k)$ was defined for k from 1 to $n-1$ so that $c(i, j(i,k))$ did not increase in k and $j(i,k)$ was not i. The deleted estimate of T_i was then

$$H_{iq}^{+} = \frac{\sum_{k=1}^{N} c_{ij(i,k)q} \bar{H}_{j(i,k)}}{\sum_{k=1}^{N} c_{ij(i,k)q}}.$$

The estimate of V_q^+ was

$$\hat{V}_q^* = n^{-1} \sum_{i=1}^{n} \left(\bar{H}_i - H_{iq}^*\right)^2 - s^2(\bar{e}).$$

This estimate does not recompute the values of c_{ijq} when observation i is deleted; however, the effects of this omission should be modest given the large sample size. The observed value of $\hat{V}_q^*$ is 0.575, a quite large value relative to the value of $\hat{V}$ of 0.112 obtained by an ordinary regression analysis with independent variables X_{ik}, $1 \leqslant k \leqslant p$.

Many alternative approaches to approximation of T_0 have also been considered (Landauer et al., 1998, 2003). Some have used regression analysis with the predictor s_{0q} of T_0 or the two predictors s_{0iq} and H_{0q}^* for $\bar{H}_0$. For the GMAT prompt under study and for $q = 10$, these approaches are not competitive with regression methods based on the v_{ia}, $1 \leqslant a \leqslant q$, although s_{01q} appears to be a more effective predictor of T_0 than is H_{0q}^*. Other approaches have tried scaling methods based on cosine measures for all essays i and j, $1 \leqslant i < j \leqslant n$. Latent semantic analysis has also been used to measure essay coherence through such devices as cosine measures of similarity of sentences within paragraphs.

The weighting in S_{ik} based on entropy can also be modified by replacement of J_k^{-1} by 1 or by replacement of J_k by the *tfidf* weight $\log(n/z_k)$, where z_k is the number of essays i, $1 \leqslant i \leqslant n$, with $F_{ij} > 0$. The *tfidf* weight found in the literature on information retrieval (Salton, 1988, chapters 9–10) has the notable feature that a weight of 0 is given to a word that appears in all essays. The *tfidf* weight has been employed in information retrieval with a cosine measure. It should be noted that application to essays is somewhat different than the original application to information retrieval in which a rather short query is matched to a large collection of documents to find the members of the collection that appear most relevant to the query. In the example under study, weight modifications in LSA do not have major impact on the results of the analysis.

The analysis here has modified latent semantic analysis to permit formal statistical analysis. In traditional latent semantic analysis, the high dimensions involved in the singular value decomposition and the use of nonlinear methods make analysis much more complicated due to the increase of the dimension of the singular value decomposition as the sample size increases. From what can be learned from the feasible formal analysis, latent semantic analysis appears difficult to justify.

3.3. Content vector analysis

Content vector analysis (CVA) is an approach to estimation of holistic scores which is employed by e-rater® (Attali et al., 2003; Burstein et al., 2004). Content vector analysis has some relationship to latent semantic analysis to the extent that similarity measures based on cosines are used and to the extent that conventional application involves very high dimensions. On the other hand, no singular value decomposition is employed in CVA. In addition, CVA differs from LSA in that CVA divides essays into groups by use of holistic scores. In contrast, LSA does not group essays on the basis of holistic score. As is the case for LSA, CVA can be modified so that formal statistical analysis is possible. Because CVA depends on the fractions F_{ik}/C_i rather than on the logarithms $\log(F_{ik}+1)$, prediction of holistic scores by CVA alone is much less effective than is prediction of holistic scores by LSA. This outcome is predictable given what is known from regression analysis. For the first GMAT prompt in Table 2, if the predictors are F_{ik}/C_i, $1 \leqslant k \leqslant p$, for $p = 200$, then $\hat{V}$ is 0.608. If the predictors are $(F_{ik}/C_i)^{1/2}$, $1 \leqslant k \leqslant p$, then $\hat{V}$ is 0.303. Virtually the same result is obtained from predictors $\log[(F_{ik}+1)/(C_i+1)]$. Recall that use of the predictors $\log(F_{ik}+1)$ for $1 \leqslant k \leqslant p$ led to an estimate $\hat{V}$ of 0.112.

To provide a general description of CVA, let the holistic scores H_{ij} have integer values from 1 to $b \geqslant 2$. Let Ξ be the set of all m-dimensional vectors with all coordinates integers from 1 to b, and let A_a, $1 \leqslant a \leqslant b$, be disjoint and nonempty subsets of Ξ such that each member of Ξ is in some A_a. If the vector with coordinates H_{ij}, $1 \leqslant j \leqslant m$, is in A_a, then essay i is in score group $G_i = a$. The score groups are arranged so that the m-dimensional vector with all elements a is in A_a and so that if $H_{ij} \leqslant H_{i'j}$ for $1 \leqslant j \leqslant m$, then $G_i \leqslant G_{i'}$. In the analysis for the first GMAT prompt, G_i is the smallest integer not less than $\bar{H}_i$. This procedure is similar but not identical to the procedure used in e-rater® to obtain a resolved holistic score.

Let I_a, $1 \leqslant a \leqslant b$, be the essays i from 1 to n such that $G_i = a$, and let $w_k = \log(n/z_k)$ be the *tfidf* weighting described at the end of the discussion of latent semantic analysis. Similarity of an essay i to the group of essays I_a is considered by use of a cosine measure similar in nature to cosine measures considered in LSA. For the content words W_k, $1 \leqslant k \leqslant p$, the essay length measure s_i, $0 \leqslant i \leqslant n$, satisfies

$$s_i^2 = \sum_{k=1}^{p} \left[(F_{ik}/C_i) w_k \right]^2.$$

If

$$x_{ak} = \frac{\sum_{i \in A_a} F_{ik}}{\sum_{i \in A_a} C_i}$$

is the fraction of content words in essays in A_a that are equal to W_k, then the length measure u_a for score group a satisfies

$$u_a^2 = \sum_{k=1}^{p} (x_{ak} w_k)^2.$$

The cross-product measure

$$v_{ia} = \sum_{k=1}^{p} (F_{ik}/C_i) x_{ak} w_k^2,$$

and the cosine measure

$$c_{ia} = v_{ia}/(s_i u_a)$$

for $0 \leqslant i \leqslant n$. The estimated holistic score $\hat{H}_{Ci}$ from content vector analysis is the largest integer a such that $c_{ia} \geqslant c_{ia'}$ whenever $a' \neq a$. In e-rater® Version 2.0, the regression analysis employs $\hat{H}_{C0}$ and c_{0b} to predict T_0. Thus the minimum V_C of $E([T_0 - a - b_1 \hat{C}_{C0} - b_2 c_{0b}]^2)$ may be considered. Estimation of this minimum can readily be accomplished as in Section 1. The one detail is that for each observation i from 1 to n, $\hat{H}_{Ci}$ and c_{ib} are replaced by deleted estimates $\hat{H}_{Cdi}$ and c_{idb} obtained by omission of observation i. This deletion does not create a significant computational problem. The resulting estimate of V_C for the first GMAT prompt is 0.417, so that CVA is somewhat less successful at prediction of the true score T_0 than is simple regression analysis based on $(F_{ik}/C_i)^{1/2}$.

In practice, study of CVA involves greater complication due to lack of restriction on the number p of different content words examined. Issues of biased sampling also arise, for in many applications of e-rater®, a stratified sample of 265 essays is used for estimation. Fifty essays are selected for each score group from 2 to 6, but only 15 essays are selected for score group 1. The lower number of scores of 1 reflects the lower frequency of a resolved holistic score of 1 among essays to which e-rater® is customarily applied.

A variation on content vector analysis can also be considered in which an estimated holistic score is obtained by use of a weighted average of human holistic scores assigned to some number of nearest neighbors of the essay under study (Larkey and Croft, 2003). This approach is very similar in nature to the nearest neighbor approach in latent semantic analysis. The basic change is that the cosine measure of similarity of essays i and j

is

$$(s_i s_j)^{-1} \sum_{k=1}^{p} \frac{F_{ik}}{C_i} \frac{F_{jk}}{C_j} w_k^2.$$

This approach suffers from the same basic problems as does the similar approach in latent semantic analysis.

On the whole, regression analysis based on frequencies of content words appears both simpler and more effective than the alternatives that have been proposed to date.

4. Analysis of discrete responses

Because holistic scores provided by individual raters are normally integer measures, it is possible to study the relationship of holistic scores to predicting variables by use of methods for the analysis of ordered discrete responses. A potential gain over regression analysis is the exploitation of restriction on the ranges of the responses.

4.1. Cumulative logit models

One approach that has been considered uses the cumulative logit model, and many alternatives exist (Feng et al., 2003). As an example, consider use of a cumulative logit model to describe the relationship of H_{ij} to X_{ik}, $1 \leqslant j \leqslant m$, $1 \leqslant k \leqslant p$, in Section 1 under the assumption that holistic scores assigned to essays by raters all have integer values from 1 to b. Consider the cumulative logit model that, given the X_{ik}, the H_{ij} are conditionally independent and the probability P_{ih} that $H_{ij} \leqslant h < b$ is equal to

$$\frac{1}{1 + \exp(-\gamma_{ih})},$$

where

$$\gamma_{ih} = \theta_h - \sum_{k=1}^{p} \beta_k X_{ik},$$

(McCullah and Nelder, 1989, chapter 5). In effect, the model postulates that there is an unobserved underlying continuous score

$$C_i = \sum_{k=1}^{p} \beta_k X_{ik}$$

that is the basis of rater behavior. In addition, the model has cut points θ_h for h from 1 to $b - 1$. To each rater j and essay i corresponds an independent score deviation c_{ij} that has a standard logistic distribution, so that $c_{ij} \leqslant x$ with probability $1/(1 + e^{-x})$ for all real x. The underlying rater score is $R_{ij} = C_i + c_{ij}$. The holistic score $H_{ij} = 1$ if $R_{ij} \leqslant \theta_1$. For $1 < h < b$, the score $H_{ij} = h$ if $\theta_{h-1} < R_{ij} \leqslant \theta_h$. If $R_{ij} > \theta_{b-1}$, then $H_{ij} = b$. The restriction that the score deviation has a standard logistic distribution may appear quite strong, but it only serves to identify parameters. The same conditional probabilities P_{ih} are produced if θ_h, β_k, and c_{ij} are all multiplied by the same positive constant.

Maximum likelihood is readily employed to use essays 1 to n to estimate the parameters θ_h and β_k, and the probability P_{ih} is then readily estimated for each essay i from 0 to n. In this fashion, a distribution for H_{ij} is estimated. Let $\hat{P}_{ih}$ denote the resulting estimate of P_{ih}. Note that $\hat{P}_{ih} = P_{ih} = 1$ for $h = b$, for the probability is 1 that $H_{ij} \leqslant b$. Similarly, $\hat{P}_{ih} = P_{ih} = 0$ for $h = 0$, for H_{ij} cannot be less than 1. The conditional mean

$$\sum_{h=1}^{b} h[P_{ih} - P_{i(h-1)}] = b - \sum_{h=1}^{b-1} P_{ih}$$

of H_{ij} given the X_{ik}, $1 \leqslant k \leqslant p$, is then estimated by

$$\hat{H}_i = b - \sum_{h=1}^{b} \hat{P}_{ih},$$

and $\hat{H}_i$ also provides an estimate of the true score T_i.

The estimation procedure considered here may be considered even if the cumulative logit model is not correct, just as the regression analysis in Section 1 may be considered even for an incorrect model (Gilula and Haberman, 1995). Estimates of the parameters θ_h and β_k may be used to estimate T_0 by $\hat{H}_0$, and the mean squared error V is then the expectation of $(T_0 - \hat{H}_0)^2$. To avoid technical problems due to a positive probability of nonexistence of maximum-likelihood estimates requires some procedure for alternative parameter estimates in case maximum-likelihood estimates are unavailable; however, this difficulty can be resolved by a variety of conventions that do not affect large-sample properties. The important issue is to estimate V without the assumption that the model is correct.

The simplest approach to estimation of V is to consider the simple average

$$\bar{M} = n^{-1} \sum_{i=1}^{n} (\bar{H}_i - \hat{H}_i)^2.$$

This average may be used to yield the estimate

$$\bar{V} = \bar{M} - s^2(\bar{e})$$

for V. This estimate has a bias of order n^{-1}. To attempt to adjust for the bias is somewhat more difficult than in Section 1 due to the nonlinear estimation methods used for the cumulative logit model.

One feasible approach is randomly to divide the n essays used in estimation into g approximately equal groups numbered from 1 to g, where g exceeds 1 and n/g is large. Maximum likelihood is applied to the essays not in group f, and one then has estimates $\hat{H}_{if}$ for $\hat{H}_i$ for $1 \leqslant i \leqslant n$. Let

$$c_{if} = (\bar{H}_i - \hat{H}_{if})^2.$$

Table 6
Cumulative logit analysis of holistic scores

			$\hat{V}$			
Program	Prompt	No. essays	$s^2(e)$	$p = 6$	$p = 56$	$p = 206$
GMAT	1	5,183	0.356	0.201	0.103	0.064
GMAT	2	5,158	0.346	0.070	0.048	0.038
TOEFL	1	4,895	0.275	0.0911	0.089	0.086
TOEFL	2	4,884	0.259	0.115	0.108	0.104

Let A_f be the sum of c_{if} for i in group f, and let B_f be the sum of c_{if} for i not in group f. Let A be the sum of the A_f, and let B be the sum of the B_f. Then $C = n^{-1}A - [n(g-1)]^{-1}B$ estimates the difference between the expected value of $(\bar{H}_i - \hat{H}_i)^2$ for i in group f and the expected value of $(\bar{H}_i - \hat{H}_i)^2$ for i not in group f. Because there are approximately $n(g-1)/g$ essays not in group f for each f from 1 to g, the estimate

$$\hat{V} = \bar{V} + C(g-1)/g$$

provides an approximation to V.

In the examples in Table 2, results are rather favorable. Consider Table 6. In the table, estimation for $\hat{V}$ is based on $g = 10$. It is notable that mean squared errors are less than in the case of linear regression.

The analysis in Table 6 does not assume that the model studied is valid, and indeed good reasons exist to believe that no model considered in this section is actually true for the essays under study. A very simple check is to consider the sample correlations of $H_{i1} - \hat{H}_i$ and $H_{i2} - \hat{H}_i$. These correlations should be close to 0 if the model is correct and the sample size is large; however, even for $p = 206$, the sample correlations range from 0.115 and 0.063 for the two GMAT prompts to 0.199 and 0.246 for the TOEFL prompts. Nonetheless, cumulative logit models do appear to perform well relative to regression models.

4.2. Alternative models

Many alternatives to cumulative logit models exist. An obvious alternative is the cumulative probit model in which the score deviation c_{ij} is assumed to have a standard normal distribution, so that the transformation

$$\frac{1}{1 + \exp(-\gamma_{ih})}$$

associated with the logistic distribution function is replaced by the transformation $\Phi(\gamma_{ih})$ associated with the standard normal distribution function Φ. In practice, cumulative probit and cumulative logit models yield very similar results in terms of the predicted true scores, for $\Phi(x)$ and $1/(1+\exp(-1.7x))$ never differ by more than about 0.01 (Lord and Novick, 1968, p. 399).

A number of possibilities exist for use of log-linear models (Haberman, 1979, chapter 6). An analogue to the cumulative logit model is the linear-by-linear log-linear model in which the probability p_{ih} that $H_{ij} = h$ satisfies

$$p_{ih} = \frac{\exp(\gamma_{ih})}{\sum_{a=1}^{b} \exp(\gamma_{ah})}$$

for

$$\gamma_{ih} = \theta_h + \sum_{k=1}^{p} \beta_k \big[h - (b+1)/2\big] X_{ik}$$

for $1 \leqslant h \leqslant b$ and $\sum_{h=1}^{b} \theta_h = 0$. Given maximum-likelihood estimates $\hat{p}_{ih}$, T_i may be estimated by

$$\sum_{h=1}^{b} h \hat{p}_{ih}.$$

A more general model is the multinomial response model that

$$\gamma_{ih} = \theta_h + \sum_{k=1}^{p} \beta_{hk} X_{ik},$$

where $\sum_{h=1}^{b} \beta_{hk} = 0$. Estimation may be accomplished in these models by maximum likelihood without any unusual difficulty.

If the essay features examined are all polytomous, then a very simple log-linear model can be used in which it is assumed that, conditional on $H_{ij} = h$, the features X_{ik}, $1 \leqslant k \leqslant p$, are conditionally independent. This model is a special case of the multinomial response model. If $Q_{hk}(x)$ is the conditional probability that $X_{ik} = x$ given that $H_{ij} = h$ and if $H_{ij} = h$ with probability q_h, then the conditional probability that $H_{ij} = h$ given the X_{ik} is

$$\frac{q_h \prod_{k=1}^{p} Q_{hk}(X_{ik})}{\sum_{a=1}^{b} q_a \prod_{k=1}^{p} Q_{ak}(X_{ik})}.$$

Estimation of q_h and $Q_{hk}(x)$ is straightforward, for q_h can be estimated by the fraction $\hat{q}_h$ of all scores H_{ij} equal to h and $Q_{hk}(x)$ can be estimated by the fraction $\hat{Q}_{hk}(x)$ of scores $H_{ij} = h$ for which $X_{ik} = x$. This approach has been explored in essay assessment for X_{ik} binary (Larkey and Croft, 2003; Rudner and Liang, 2002). Given the estimated conditional probabilities that $H_{ij} = h$ given the X_{ik}, T_i can then be estimated as in the linear-by-linear log-linear model.

The proposed methods have in common a failure to consider model error. This failure can be remedied rather easily in principle, but a substantial computational cost appears to be involved if $b > 1$ raters are used. The simplest case to discuss involves the cumulative probit model. Let the underlying continuous score

$$C_i = \sum_{k=1}^{p} \beta_k X_{ik} + d_i$$

for unknown parameters β_k and for an unobserved normal random variable d_i with mean 0 and variance $\tau^2 < 1$. Again consider unknown cut points θ_h for h from 1 to $b-1$. To each rater j and essay i corresponds a score deviation c_{ij} that has a normal distribution with mean 0 and variance $1-\tau^2$. Assume that the c_{ij} and d_i are mutually independent. The underlying score for rater j for essay i is then

$$R_{ij} = C_i + c_{ij}.$$

Thus the holistic score $H_{ij} = 1$ if $R_{ij} \leqslant \theta_1$. For $1 < h < b$, the score $H_{ij} = h$ if $\theta_{h-1} < R_{ij} \leqslant \theta_h$. If $R_{ij} > \theta_{b-1}$, then $H_{ij} = b$. In this fashion, the conditional probability P_{ih} that $H_{ij} \leqslant h$ is $\Phi(\gamma_{ih})$ for

$$\gamma_{ih} = \theta_h - \sum_{k=1}^{p} \beta_k X_{ik},$$

just as in the regular cumulative probit model. Indeed, for $b = 1$, the new model is indistinguishable from the ordinary cumulative probit model. For $b > 1$, the addition of the parameter τ results in a somewhat more complicated analysis by maximum likelihood with rather uncertain practical benefit given that the model does not change the conditional marginal distributions of the H_{ij} given the X_{ik}.

It is also possible to apply a cumulative logit or cumulative probit model to the average $\bar{H}_i$ rather than to the individual H_{ij}. This possibility was checked in the cumulative logit case for the cases considered in Table 2. Results appear to be very similar to those for the cumulative logit case described in Table 6.

In summary, use of models for discrete data appears appropriate for prediction of holistic scores, and a simple cumulative logit model appears rather satisfactory in practice even if it does not perfectly describe the data.

5. Bayesian analysis

Bayesian analysis of electronic essay scoring has been considered for 473 senior essays from the 1988 National Assessment of Educational Progress for a variant on the cumulative probit model (Johnson, 1996; Page, 1994) in which raters are identified and model error is present. In this analysis, rater j refers to a specific person for each essay. Given the X_{ik} and an unobserved normal random variable d_i with mean 0 and variance $\tau^2 > 0$, the underlying rater score is

$$R_{ij} = \sum_{j=1}^{p} \beta_k X_{ik} + d_i + c_{ij},$$

where the d_i and c_{ij} are independent normal random variables with common mean 0. The standard deviation of c_{ij} is σ_j, so that each judge has an individual degree of variability. In addition, each judge j has distinct cutpoints θ_{hj} for $1 \leqslant h \leqslant b-1$. If $R_{ij} < \theta_{1j}$, then $H_{ij} = 1$, while $H_{ij} = h$ if $\theta_{(h-1)j} < R_{ij} \leqslant \theta_{hj}$. If $R_{ij} > \theta_{(b-1)j}$, then $H_{ij} = b$. This model is sufficiently complex to make estimation somewhat challenging.

To provide estimates, a Bayesian approach is applied in which a prior distribution is applied to the parameters θ_{hj}, σ_j, and β_k. The posterior distribution of the parameters is then approximated by use of Bayesian data augmentation (Tanner and Wong, 1987; Albert and Chib, 1993). This approach is much more computationally intensive than either the regression analysis or cumulative logit or probit analysis previously considered, and the basis for choice of priors is limited in the particular application examined. Nonetheless, a posterior mean of the β_k and of the R_{ij} can certainly be obtained. It is also quite possible to find a posterior mean and variance of the true score T_0, so that results comparable to those obtained in previous sections can be obtained with the Bayesian analysis. The analysis also considers diagnostic procedures to ascertain whether the model has substantial errors, and a residual analysis based on Bayesian inference does demonstrate model deficiencies in the example studied. The difficulty in the analysis is the extent to which the effort is worth while relative to simpler approaches. For example, it is reasonable to ask how much improvement in prediction of true scores is obtained by the Bayesian analysis than would have been obtained by use of linear regression.

6. Conclusions

On the whole, the basic statistical machinery for electronic essay scoring exists and appears to function well. For prediction of holistic scores, regression analysis and cumulative logit analysis are quite effective. Standard methods for construction of composite scales by principal components or by reliability analysis are quite adequate for the construction of composite writing scales. None of the appropriate methodology required is especially complex to use or to evaluate. Nonetheless, room for improvement exists.

Much possible improvement involves human scoring. Better human scoring presumably would improve any machine scoring designed to predict human scoring. Adjustments for rater effects, better methods for human scoring, better raters with better training and better working conditions, randomization of assignment of essays to readers, and better prompts all may affect the quality of the human rater scores used by computers. Research is certainly needed to characterize prompts that elicit responses likely to be scored relatively consistently by human scorers and by machines. One hypothesis is that prompts that are highly structured and provide considerable guidance to the writer are likely to produce particularly good results with both human and machine scoring. Such prompts may be particularly amenable to content analysis. Nonetheless, at this stage, the proposed hypothesis needs to be tested.

The generation of descriptive features of essays is also an area where potential improvements exist. Presumably better scoring is possible if computer software improves the quality of automated analysis of grammar, usage, mechanics, style, word diversity, and vocabulary sophistication. Computer software that can improve the analysis of essay content is particularly desirable.

Classical psychometric issues such as reliability and validity need study in the content of computer grading to ascertain how computer assessment affects the actual purpose of the essay. Given that computer grading does not depend on deep analysis of essay content, one must be concerned about students aware of computer grading producing essays designed for favorable computer scoring rather than for quality as perceived by diligent human scorers. Similar issues arise when essay scoring is used to teach students proper writing practice. If the computer information is not accurate or is limited in scope, then writing development may be distorted.

Computer grading has considerable potential for the analysis of portfolios of student essays. The potential gain in portfolio analysis is the increased data provided. Four essays written by an examinee provide considerable more potential information than just one essay, but some research is required on carefully derived portfolios to determine the manner in which performance varies for different tasks and the extent to which numerical essay features provide reliable information concerning examinees if a portfolio is available. In principle, portfolio analysis may permit meaningful statements about an examinee's grammar, spelling, or word choice that are not available from a single essay.

The effects of computer grading on fairness must be considered. For example, it is not clear how computer scoring differs from human scoring in the evaluation of examinees who are not native speakers of English. Similar questions arise among different groups of examinees who are not native speakers of English. For instance, native speakers of Chinese may be affected by differences between human and machine scoring in a different manner than are native speakers of German. Even among native speakers of English, race and ethnicity may still have effects that require exploration.

References

Albert, J.H., Chib, S. (1993). Bayesian analysis of binary and polytomous response data. *Journal of the American Statistical Association* **88**, 669–679.

Attali, Y., Burstein, J., Andreyev, S. (2003). *E-rater Version 2.0: Combining Writing Analysis Feedback with Automated Essay Scoring*. Educational Testing Service, Princeton, NJ.

Breland, H.M. (1996). Word frequency and word difficulty: A comparison of counts in four corpora. *Psychological Science* **7** (2), 96–99.

Breland, H.M., Jenkins, L.M. (1997). *English Word Frequency Statistics: Analysis of a Selected Corpus of 14 Millions Tokens*. College Entrance Examination Board, New York.

Breland, H.M., Jones, R.J., Jenkins, L. (1994). The college board vocabulary study. Research Report 94-26. Educational Testing Service, Princeton, NJ.

Burstein, J., Chodorow, M., Leacock, C. (2004). Automated essay evaluation: the criterion online writing service. *AI Magazine* **25** (3), 27–36.

Carroll, J.B. (1970). An alternative in Juillard's usage coefficient for lexical frequencies and a proposal for a standard frequency index (S.F.I.). *Computer Studies in the Humanities and Verbal Behavior* **3**, 61–65.

Dorans, N.J., Patsula, L.N. (2003). Using confusion infusion and confusion reduction indices to compare alternative essay scoring rules. Research Report RR-03-09. Educational Testing Service, Princeton, NJ.

Draper, N.R., Smith, H. (1998). *Applied Regression Analysis*, 3rd ed. John Wiley, New York.

Feng, X., Dorans, N.J., Patsula, L.N., Kaplan, B. (2003). Improving the statistical aspects of e-rater ®: Exploring alternative feature reduction and combination rules. Research Report RR-03-15. Educational Testing Service, Princeton, NJ.

Gilula, Z., Haberman, S.J. (1995). Prediction functions for categorical panel data. *The Annals of Statistics* **23**, 1130–1142.

Gulliksen, H. (1987). *Theory of Mental Tests*. Erlbaum, Hillsdale, NJ.

Haberman, S.J. (1979). *Analysis of Qualitative Data, Vol. ii: New Developments*. Academic Press, New York.

Haberman, S.J. (2004). Statistical and measurement properties of features used in essay assessment. Research Report RR-04-21. Educational Testing Service, Princeton, NJ.

Haberman, S.J., Qian, J. (2004). The best linear predictor for true score from a direct estimate and several derived estimates. Research Report RR-04-35. Educational Testing Service, Princeton, NJ.

Holland, P.W., Hoskens, M. (2003). Classical test theory as a first-order item response theory: Application to true-score prediction from a possibly non-parallel test. *Psychometrika* **68**.

Johnson, V.E. (1996). On Bayesian analysis of multirater ordinal data: An application to automated essay grading. *Journal of the American Statistical Association* **91**.

Kendall, M., Stuart, A., Ord, J.K. (1983). *The Advanced Theory of Statistics, vol. 3*, 4th ed. MacMillan, New York.

Landauer, T.K., Foltz, P.W., Laham, D. (1998). Introduction to latent semantic analysis. *Discourse Processes* **25**, 259–284.

Landauer, T.K., Laham, D., Foltz, P.W. (2003). Automated scoring and annotation of essays with the Intelligent Essay Assessor™. In: Shermis, M.D., Burstein, J. (Eds.), *Automated Essay Scoring: A Cross-Disciplinary Perspective*. Lawrence Erlbaum, Mahway, NJ, pp. 87–112.

Larkey, L.S., Croft, W.B. (2003). A text categorization approach to automated essay grading. In: Shermis, M.D., Burstein, J. (Eds.), *Automated Essay Scoring: A Cross-Disciplinary Perspective*. Lawrence Erlbaum, Mahway, NJ, pp. 55–70.

Lord, F.M., Novick, M.R. (1968). *Statistical Theories of Mental Test Scores*. Addison-Wesley, Reading, MA.

McCullah, P., Nelder, J.N. (1989). *Generalized Linear Models*, 2nd ed. Chapman and Hall, Boca Raton, FL.

Page, E.B. (1966). The imminence of grading essays by computer. *Phi Delta Kappan* **47**, 238–243.

Page, E.B. (1968). Analyzing student essays by computer. *International Review of Education* **14**, 210–225.

Page, E.B. (1994). Computer grading of student prose, using modern concepts and software. *Journal of Experimental Education* **62**, 127–142.

Rao, C.R., Mitra, S.K. (1971). *Generalized Inverse of Matrices and Its Applications*. Wiley, New York.

Rudner, L.M., Liang, T. (2002). Automatic essay scoring using Bayes' theorem. *Journal of Technology, Learning, and Assessment* **1** (2).

Salton, G. (1988). *Automatic Text Processing: The Transformation, Analysis, and Retrieval of Information by Computer*. Addison-Wesley, Reading, MA.

Simpson, E.H. (1949). The measurement of diversity. *Nature* **163**, 688.

Stewart, G.W. (1973). *Introduction to Matrix Computation*. Academic Press, New York.

Stone, C.J. (1980). Optimal rates of convergence for nonparametric estimators. *The Annals of Statistics* **8**, 1348–1360.

Tanner, M., Wong, W.H. (1987). The calculation of posterior distributions by data augmentation. *Journal of the American Statistical Association* **82**, 528.

Weisberg, S. (1985). *Applied Linear Regression*, 2nd ed. John Wiley, New York.

Wilkinson, J.H. (1965). *The Algebraic Eigenvalue Problem*. Clarendon, Oxford.

Wresch, W. (1993). The imminence of grading essays by computer – 25 years later. *Computers and Composition* **10**, 45–58.

Handbook of Statistics, Vol. 26
ISSN: 0169-7161

DOI: 10.1016/S0169-7161(06)26008-5

8

Some Matrix Results Useful in Psychometric Research

C. Radhakrishna Rao

1. Notation and Basic Matrix Results

1.1. Matrices

We denote a matrix with p rows and q columns by

$$A(p \times q) = (a_{ij}) = (a_1 \ \cdots \ a_q) = (b_1 \ \cdots \ b_p)', \tag{1.1.1}$$

where a_{ij} is the number in the ith row and jth column, $a_i = (a_{1i}, \ldots, a_{pi})'$ is the ith column p-vector defined as an ordered set of p numbers written vertically, $b_j = (a_{j1}, \ldots, a_{jq})'$ is the jth row q-vector and C' denotes a matrix obtained from C by interchanging rows and columns. Some books on matrix algebra use the notation C^T for C'. A p-vector a is a $p \times 1$ matrix. Two matrices $A = (a_{ij})$ of order $p \times q$ and $B = (b_{rs})$ of order $q \times r$ can be multiplied to yield a matrix of order $p \times r$:

$$AB = \sum_1^q a_{ik} b_{kj}, \quad \text{a } p \times r \text{ matrix.} \tag{1.1.2}$$

Two matrices $A(p \times p)$ and $B(p \times p)$ are said to commute if $AB = BA$. The size of the matrix $p \times q$ in parenthesis is omitted in writing the equations. A p-vector a is a $p \times 1$ matrix. Also $(AB)' = B'A'$ and $(ABC)' = C'B'A'$ and so on. We consider only real numbers as elements of a matrix. For more general treatment of matrices reference may be made to the book by Rao and Rao (1998). In the sequel we refer to this book as RR and quote the Chapter and Section number where a particular result is formulated in more general terms and proved.

A square matrix $A(p \times p)$ is said to be

(i) symmetric if $A = A'$,
(ii) normal if $AA' = A'A$,
(iii) pd (positive definite) if A is symmetric and $a'Aa > 0 \; \forall a \neq 0$,
(iv) psd (positive semidefinite) if A is symmetric and $a'Aa \geqslant 0 \; \forall a$, and
(v) orthogonal (OR) if $AA' = A'A = I_p$, where I_p is an identity matrix of order p. $A(p \times q, q < p)$ is said to be semi-orthogonal (SOR) if $A'A = I_q$. Given an SOR, $A(p \times q)$, we can find a matrix $B(p \times p - q)$ such that $C = (A : B)$ exhibited as a partitioned matrix with columns of B appended to columns of A is OR.

(vi) The trace of a square matrix $A(p \times p)$, denoted by tr A, is defined as the sum of its diagonal elements.

1.2. Rank of a matrix

A set of vectors $\{a_1, \dots, a_r\}$ are said to be independent if $c_1a_1 + \dots + c_ra_r \neq 0$ for numbers $c_1, \dots, c_r$ not all simultaneously zero. The rank of $A(p \times q)$ is the number of independent column (or equivalently) row vectors. When the rank of a matrix is relevant, we use the notation $A(p \times q, r)$ or simply $A(r)$ where $r \leqslant \min(p, q)$ is the rank.

Let $B = \{a_1, \dots, a_s\}$ denote a set of p-vectors. Then

$$S_B = \{c_1a_1 + \dots + c_sa_s\colon \text{ for all real numbers } c_1, \dots, c_s\} \tag{1.2.1}$$

is said to be the vector space generated by the set B. Given S_B, we can find a S_B^C, called a complement of S_B such that $S_B \cap S_B^C = 0$, a vector with all elements zero, and

$$S_B \oplus S_B^C = \{x + y\colon x \in S_B, y \in S_B^C\} = R^p, \tag{1.2.2}$$

where R^p is the set of all p-vectors. S_B^C is not unique. We can also choose the complementary subspace such that the vectors in S_B are orthogonal to all vectors in S_B^C. Such a complement is unique and is represented by $S_B^\perp$.

1.3. Generalized inverse of a matrix

Consider $A(p \times q, r)$, where $r \leqslant \min(p, q)$. Then a generalized inverse of A is defined as a matrix $G(q \times p)$ such that

$$AGA = A. \tag{1.3.1}$$

G satisfying (1.3.1) is not necessarily unique and is usually written as A^- (Rao, 1962; Rao and Rao, 1998, Chapter 8). But any A^- satisfying (1.3.1) provides all possible solutions of a consistent linear equation $Ax = y$ as

$$x = A^-y + (I_q - A^-A)z, \quad \forall z \in R^q, \tag{1.3.2}$$

of which a particular solution is A^-y. If we need a solution with minimum L_2-norm, i.e., a solution with a minimum value for $x'x$, then the solution is $x = Gy = A_m^-y$, where G satisfies the conditions:

$$AGA = A, \qquad GA = (GA)'. \tag{1.3.3}$$

Note that although G satisfying (1.3.3) is not unique, Gy is unique for all such G.

If the equation $Ax = y$ is not consistent, i.e., has no solution, we may need a least squares solution, i.e., an x such that $(Ax - y)'(Ax - y)$ is a minimum. The solution is $x = Gy = A_l^-y$, where G satisfies the conditions:

$$AGA = A, \qquad AG = (AG)'. \tag{1.3.4}$$

If the equation $Ax = y$ is not consistent and the least squares solution is not unique, we may need an x which is a minimum norm least squares solution, i.e., an x with a minimum value of $x'x$ among all values of x which minimize $(Ax - y)'(Ax - y)$. Such a solution is $x = Gy = A^+y$, where G satisfies the conditions:

$$AGA = A, \qquad (AG)' = AG, \qquad (GA)' = GA, \qquad GAG = G. \quad (1.3.5)$$

The G satisfying the conditions (1.3.5) is unique and is called the Moore–Penrose (MP) inverse and is represented by A^+, and so also the minimum L_2 norm least squares solution. A g-inverse satisfying only the conditions

$$AGA = A, \qquad GAG = G \tag{1.3.6}$$

is called a reflexive g-inverse and represented any A_r^-. (The four classes of g-inverses used for different purposes were first introduced in Rao, 1967. A more general definition of a g-inverse called Rao–Yanai g-inverse is given in Rao and Yanai, 1985, and Rao and Rao, 1998, Chapter 8.)

When $A(p \times p)$ is a square matrix of rank p, there is a unique matrix G such that $GA = AG = I_p$. In this case G is written as A^{-1}, the regular inverse of A. Note that $A^{-1}A = AA^{-1} = I_p$, and $(A^{-1})^{-1} = A$ and in general, $(A^+)^+ = A$, a property which holds only for Moore–Penrose g-inverse satisfying (1.3.5).

For a detailed study of generalized inverses and their applications, reference may be made to Rao and Mitra (1971, 1973, 1974).

1.4. The projection operator

Consider the subspace S_A formed by the column vectors of $A(p \times q, r)$ and let S_A^C be any complement of S_A, as defined in (1.2.2), spanned by the columns of a matrix $B(p \times p - r)$. Then any p-vector $x \in R^p$ can be written as

$$x = Aa + Bb = (A \ : \ B)\begin{pmatrix} a \\ b \end{pmatrix}. \tag{1.4.1}$$

Let $\binom{C_1}{C_2}$ be any g-inverse of $(A \ : \ B)$. Then

$$\begin{pmatrix} a \\ b \end{pmatrix} = \begin{pmatrix} C_1 \\ C_2 \end{pmatrix} x \implies Aa = AC_1x = P_Ax, \quad \text{defining } P_A = AC_1, \tag{1.4.2}$$

so that the component of x in S_A is P_Ax, where $P_A = AC_1$ is called the projection operator on S_A along S_B. Similarly $P_B = BC_2$ is called the projection operator on S_B along S_A.

In particular, if $S_B = S_A^\perp$, the orthogonal complement of S_A, then

$$P_A = A(A'A)^- A', \tag{1.4.3}$$

where $(A'A)^-$ is any g-inverse of $A'A$. The projection operator P_A in such a case is unique, and is called an orthogonal projector operator. (Explicit expression for an orthogonal projection operator in the form (1.4.3) was first given in Rao, 1967, and has been found useful in developing a general theory of linear models.)

1.5. Norm of a matrix

There are various ways of defining the norm of a matrix $A(p \times q)$, denoted as $\|A\|$ (Rao and Rao, 1998, Chapter 11). But some expressions used in multivariate analysis are as follows:

$$\begin{aligned}
\|A\| &= \sum\sum |a_{ij}|, && L_1\text{-norm} \\
&= \Big(\sum\sum a_{ij}^2\Big)^{1/2}, && L_2\text{-norm} \\
&= \Big(\sum\sum |a_{ij}|^r\Big)^{1/r}, && L_r\text{-norm } r > 1,\ \text{Minkowski norm} \\
&= (\text{trace } A'MA)^{1/2}, && \text{Mahalanobis norm,}
\end{aligned} \tag{1.5.1}$$

where $M(p \times p)$ is a pd matrix and trace stands for the sum of the diagonal elements of a matrix. Note that $\sum\sum a_{ij}^2 = \text{tr } AA' = \text{tr } A'A$.

The norm or length of a vector a is $\sqrt{a'a}$. The distance between two p-vectors a and b is defined to be $\|a - b\|$, and the angle θ between a and b is $\cos^{-1}(a'b/(a'ab'b)^{1/2})$. A measure of dissimilarity between two $p \times q$ matrices is $\|A - B\|$. In optimization problems where a matrix from a given class has to be fitted to a given matrix as in Procrustes problems we try to minimize a measure of dissimilarity.

1.6. Transformation of vectors

1.6.1. Orthogonal rotation of axes

Consider an orthogonal matrix $A(p \times p)$ and the transformation of two vectors, x_1 and x_2,

$$Ax_1 = y_1, \qquad Ax_2 = y_2, \qquad x_i \in R^p, \quad y_i \in R^p, \quad i = 1, 2. \tag{1.6.1}$$

Then

$$\begin{aligned}
&y_1'y_1 = x_1'A'Ax_1 = x_1'x_1, \qquad y_2'y_2 = x_2'A'Ax_2 = x_2'x_2, \\
&y_1'y_2 = x_1'A'Ax_2 = x_1'x_2
\end{aligned} \tag{1.6.2}$$

so that the norms of vectors and angles between them are preserved by an orthogonal transformation.

Observe that if the components of x are coordinates of a point represented by $x \in R^p$ with reference to an orthogonal set of axes, then the coordinates of x with references to the row vectors of A as axes are the components of Ax. The matrix A provides an orthogonal rotation of the axes.

1.6.2. Oblique rotation of axes

Consider any matrix $A(p \times p)$ with its column vectors normalized to have unit norm, and take them as the new axes. What are the coordinates of x with respect to the new axes?

One version is to consider the *orthogonal projection* of x on each of the column vectors of A as the new coordinates. Thus the ith coordinate is

$$y_i = a_i'x, \quad i = 1, \dots, p \iff y = A'x. \tag{1.6.3}$$

Another version is the *parallel axes system*, i.e., to express x as a linear combination of the column vectors of A

$$x = c_1 a_1 + \cdots + c_p a_p = Ac \implies c = A^{-1}x, \tag{1.6.4}$$

where $c' = (c_1, \ldots, c_p)$ are coordinates referred to new axes. A is referred to as oblique rotation matrix if $A' \neq A^{-1}$. From (1.6.3) and (1.6.4), we have the following relationships between the two versions of coordinates

$$y = A'Ac \quad \text{and} \quad c = (A'A)^{-1}y. \tag{1.6.5}$$

1.7. Transformation of matrices

Let 1_p denote a p-vector with all its elements equal to unity and I_p a $p \times p$ matrix with all its diagonal elements equal to unity and the rest are zeroes and $J_p = I_p - p^{-1}1_p 1'_p$. If we want to center the columns of a matrix $A(p \times q)$, i.e., to make the sum of elements in each column equal to zero by subtracting a suitable number from each element in the column, we need to compute the product $A_c = J_p A$. Similarly to center the rows, we compute the product $A_r = AJ_q$, and to center both the rows and columns we compute $A_{cr} = J_p A J_q$.

Denote by $\Delta(B)$, the diagonal matrix obtained by retaining only the square roots of the diagonal elements in $B'B$. If we want to center the columns as well as normalize them, i.e., to make the sum of squares of elements in each column equal to unity, we need to compute the product

$$_nA_c = A_c \Delta^{-1}(A_c). \tag{1.7.1}$$

Similarly, if we want to center the rows of a matrix and as well as normalize them, we need to compute

$$_nA_r = \Delta^{-1}(A'_r)A_r. \tag{1.7.2}$$

Can we center the rows and columns and also normalize the rows and columns simultaneously? There is no simple way of achieving this. An approximation to this can be obtained by repeatedly doing the operations (1.7.1) and (1.7.2).

2. Decomposition of matrices

Spectral and singular value decompositions of matrices play an important role in optimization problems involving functions of matrices, such as the Procrustes problem in psychometrics (Gower, 2003; Gower and Dijksterhuis, 2004).

2.1. Spectral decomposition of a symmetric matrix

Let $A(p \times p, r)$ be a real symmetric matrix of rank r. Then it has the decomposition (known as spectral decomposition),

$$A = P \wedge P', \qquad PP' = P'P = I_p, \tag{2.1.1}$$

where $\wedge$ is a diagonal matrix with the roots $\lambda_1 \geqslant \cdots \geqslant \lambda_p$ of the determinantal equation

$$|A - \lambda I| = 0 \tag{2.1.2}$$

as diagonal elements. The λ_i, called eigenvalues, are real numbers, some of them may be zero. The expression (2.1.1) can also be exhibited as

$$A = \lambda_1 p_1 p_1' + \cdots + \lambda_r p_r p_r', \tag{2.1.3}$$

where r is the rank of A and $p_1, \ldots, p_r$ are the first r column vectors of P, and p_i is the solution of

$$Ap = \lambda_i p, \quad i = 1, \ldots, r, \tag{2.1.4}$$

and is called the eigenvector associated with the eigenvalue λ_i. Note that when A is of rank $r < p$, the eigenvalues λ_i, $i = r + 1, \ldots, p$ are all zero and the eigenvectors corresponding to zero eigenvalues can be chosen to be any set of orthogonal vectors satisfying the equation $Ap = 0$.

If A is an nnd matrix, $\lambda_i, \ldots, \lambda_r$ are all positive. If A is pd, all the p eigenvalues are positive.

2.2. *Simultaneous decomposition of two symmetric matrices*

Let A and B be symmetric matrices of order p, such that $AB = BA$ (i.e., A and B commute). Then A and B can be simultaneously decomposed as

$$A = P \wedge_1 P' \quad \text{and} \quad B = P \wedge_2 P', \quad PP' = P'P = I, \tag{2.2.1}$$

i.e., with the same eigenvectors but possibly different eigenvalues.

Let A and B be real symmetric matrices of order p of which B is positive-definite, then there exists a matrix $S = (s_1 \ \cdots \ s_p)$ such that

$$\begin{aligned} A &= S \wedge S' = \lambda_1 s_1 s_1' + \cdots + \lambda_p s_p s_p', \\ B &= SS' = s_1 s_1' + \cdots + s_p s_p', \end{aligned} \tag{2.2.2}$$

where $\wedge$ is a diagonal matrix with the diagonal elements $\lambda_1, \ldots, \lambda_p$ as the roots of the determinantal equation $|A - \lambda B| = 0$. (Note that S is not in general an orthogonal matrix.)

2.3. *SVD: Singular value decomposition*

SVD of $A(p \times q)$ plays an important role in matrix optimization problems. We mention some applications of SVD, and also give related matrix decompositions which are useful in psychometric research.

Consider $A(p \times q, r)$. The SVD of A is

$$A = \sigma_1 p_1 q_1' + \cdots + \sigma_r p_r q_r', \tag{2.3.1}$$

where $p_i' p_j = 0 = q_i' q_j$, $i \neq j$ and $p_i' p_i = q_i' q_i = 1$ and $\sigma_1 \geqslant \cdots \geqslant \sigma_r > 0$ are positive numbers called singular values. The expression (2.3.1) can be written as

$$A = P \Delta Q', \qquad P'P = I_r, \qquad Q'Q = I_r, \tag{2.3.2}$$

where $P(p \times r) = (p_1 \ \cdots \ p_r)$ and $Q(q \times r) = (q_1 \ \cdots \ q_r)$ are SOR, and $\Delta(r \times r)$ is a diagonal matrix with $\sigma_1, \ldots, \sigma_r$ in the diagonal. It may be noted that $\sigma_1^2, \ldots, \sigma_r^2$ are eigenvalues of AA' or $A'A$ and p_i, q_i are solutions of

$$AA'p = \sigma_i^2 p \quad \text{and} \quad A'Aq = \sigma_i^2 q. \tag{2.3.3}$$

The expressions (2.3.1) and (2.3.2) for the SVD of $A(p \times q)$ involve r, the rank of A. In general, we can express A without using r as

$$A = P\Delta Q', \qquad P'P = I_p, \qquad Q'Q = I_q \tag{2.3.4}$$

with full orthogonal matrices $P(p \times p)$ and $Q(q \times q)$ and Δ as a $p \times q$ matrix with nonzero elements only in the main diagonal. In the sequel, in proving proportions we use the general form (2.3.4), unless a reference to r is involved. We can also express A as

$$\begin{aligned} A &= P\Delta Q', \qquad P'P = I_q, \qquad Q'Q = I_q, \\ &\Delta(q \times q) \quad \text{if } q = \min\{p, q\}, \\ A &= P\Delta Q', \qquad P'P = I_p, \qquad Q'Q = I_p, \\ &\Delta(p \times p) \quad \text{if } p = \min\{p, q\}. \end{aligned}$$

One particular use of the SVD is in finding the Moore–Penrose g-inverse of a matrix. If $A(p \times q, r)$ has the SVD

$$A = \sigma_1 p_1 q_1' + \cdots + \sigma_r p_r q_r' = P\Delta Q', \tag{2.3.5}$$

then

$$A^+ = \sigma_1^{-1} q_1 p_1' + \cdots + \sigma_r^{-1} q_r p_r' = Q\Delta^{-1}P'. \tag{2.3.6}$$

It is easy to verify that

$$\begin{aligned} &AA^+A = A, \qquad A^+AA^+ = A^+, \\ &(AA^+)' = AA^+, \qquad (A^+A)' = A^+A. \end{aligned}$$

For theory and computation of SVD, reference may be made to Golub and Van Loan (1989), Stewart (1983) and Vaccaro (1991), Datta (1995) and Khattree and Naik (1999).

2.4. MNSVD

Let $M(p \times p)$ and $N(q \times q)$ be pd matrices. Then $A(p \times q, r)$ admits the decomposition

$$A = R\Delta S', \qquad R'MR = I_r, \qquad S'NS = I_r, \tag{2.4.1}$$

where $\Delta(r \times r)$ is a diagonal matrix with singular values $\sigma_1, \ldots, \sigma_r$ in the diagonal.

(For more details on SVD, reference may be made to Rao and Rao, 1998, Chapter 8, and Rao, 2005. The SVD of a matrix can be computed using the subroutine A.13 of Khattree and Naik, 1999, p. 309.)

2.5. *Simultaneous SVD*

Let $A(p \times q)$ and $B(p \times q)$ be such that AB' and $A'B$ are both normal. (A matrix C is said to be normal if $CC' = C'C$.) In such case we can decompose A and B as

$$A = P\Delta_1 Q', \qquad B = P\Delta_2 Q', \tag{2.5.1}$$

with possibly different diagonal matrices Δ_1 and Δ_2.

2.6. *PSVD (Product SVD)*

Let $A(m \times n)$ and $B(p \times n)$ be two matrices such that

$$\begin{aligned}&\dim\{Sp(A') \cap Sp(B')\} = s, \qquad \dim\{Sp(A') \cap Null\,(B)\} = r,\\&\dim\{Null(A) \cap Sp(B')\} = t, \qquad \dim\{Sp(A') \cup Sp(B')\} = k = s + r + t,\\&\dim\{Null(A) \cap Null(B)\} = n - k.\end{aligned}$$

Then we have PSVD of A and B as

$$A = UDX^{-1} \quad \text{and} \quad B' = XJV',$$

where $X(n \times n)$ is nonsingular, $U(m \times m)$ and $V(p \times p)$ are orthogonal matrices, and

$$D = \begin{pmatrix} & s & r & t & n-k \\ s & S & 0 & 0 & 0 \\ r & 0 & I & 0 & 0 \\ w & 0 & 0 & 0 & 0 \end{pmatrix}, \qquad J = \begin{pmatrix} & s & t & p-s-t \\ s & I & 0 & 0 \\ r & 0 & 0 & 0 \\ q & 0 & I & 0 \end{pmatrix},$$

where the bordering letters, $s, r, t, \ldots$ represent the numbers of a columns and rows of submatrices, S = positive-definite diagonal matrix, $w = m - s - r$ and $q = n - k$.

2.7. *QSVD (Quotient SVD)*

Under the conditions as in Section 2.6, we have

$$A = UEX^{-1} \quad \text{and} \quad B = VHX^{-1},$$

where U, V and X are same as in Section 2.6 and E and H are such that

$$E = \begin{pmatrix} & s & r & t & n-k \\ s & T & 0 & 0 & 0 \\ r & 0 & I & 0 & 0 \\ u & 0 & 0 & 0 & 0 \end{pmatrix}, \qquad H = \begin{pmatrix} & s & r & t & n-k \\ s & I & 0 & 0 & 0 \\ t & 0 & 0 & I & 0 \\ v & 0 & 0 & 0 & 0 \end{pmatrix},$$

and T is positive-definite diagonal, $u = m - s - r$ and $v = p - s - t$.

2.8. *QR (rank) factorization*

A matrix $A(p \times q, r)$ can be factorized as

$$A = QR, \tag{2.8.1}$$

where $Q(p \times r)$ has orthogonal columns, i.e., $Q'Q = I_r$ and R is an $r \times q$ upper echelon matrix. A matrix is said to be in upper echelon form if all the elements below the main diagonal are zeroes. When the matrix is square, it is called upper diagonal matrix (Rao and Rao, 1998, Chapter 5). A weaker form of factorization is where R is an $r \times q$ matrix and not of the special type as in (2.8.1).

2.9. *Cholesky decomposition*

Let $A(p \times p, r)$, a square matrix, be nnd. Then it can be factorized as

$$A = KK', \tag{2.9.1}$$

where $K(p \times r)$ is a lower echelon matrix with non-negative elements in the diagonal. If $r = p$, $K(p \times p)$ is a lower triangular matrix with positive elements in the diagonal (Rao and Rao, 1998, Chapter 9).

For further results on matrices of interest in psychometric research, reference may be made to Rao (1955, 1962, 1964, 1967), Takane (2003, 2004) and Yanai (2003).

2.10. *Spectral norm of a matrix*

Consider two vector spaces R^p and R^q with given norms for the vectors denoted by $\|\cdot\|$. Corresponding to any norms in R^p and R^q, we can define an induced norm for a matrix $A(p \times q)$ in the space of $p \times q$ matrices as follows:

$$\|A\| = \max_{0 \neq x \in R^q} \frac{\|Ax\|}{\|x\|},$$

where $Ax \in R^p$. For instance, if the norm of a vector a is defined as $\sqrt{a'a}$ in both R^p and R^q, then

$$\|A\|^2 = \max_{0 \neq x \in R^q} \frac{x'A'Ax}{x'x} = \lambda_1 2 \implies \|A\| = \lambda_1, \tag{2.10.1}$$

where λ_1^2 is the maximum eigenvalue of $A'A$. The norm (2.10.1) is called the spectral norm of A. The usual kinds of norms are defined in (1.5.1).

3. Matrix approximations

We discuss some problems of approximating a given matrix $A(p \times q)$ by a matrix in a specified subclass S. In each case the solution is obtained by minimizing a suitably chosen norm $\|A - B\|$ with respect to $B(p \times q) \in S$. We consider only the L_2-norm (1.5.1) as it is generally used in psychometric research and give reference to papers where more general norms are considered. The L_2-norm of a matrix $C(r \times s)$ is defined as

$$\|C\| = (\operatorname{trace} CC')^{1/2} = (\operatorname{trace} C'C)^{1/2}.$$

The orthogonal projection operator on the space generated by the column vectors of C, as defined in (1.4.3), is denoted by

$$P_c = C(C'C)^- C'.$$

3.1. Regression type approximations

Let $A(p \times q)$, $X(p \times a)$ and $Y(q \times b)$ be given matrices. We study the problem of approximating A by

$$XF + GY' + H$$

under certain conditions on $F(a \times q)$, $G(p \times b)$ and $H(p \times q)$. Let $P_X = X(X'X)^- X'$ denote the projection operator.

3.1.1. Column regression

$$\min_F \|A - XF\| = \|A - P_X A\| \tag{3.1.1}$$

and is attained at $F = (X'X)^- X'A$. The result is well known as in multivariate regression model (Rao, 1973, p. 543). Note that

$$\|A - P_X A + P_X A - XF\|^2 = \|A - P_X A\|^2 + \|P_X A - XF\|^2$$

from which (3.1.1) follows by choosing $P_X A = XF$.

3.1.2. Row regression

$$\min_G \|A - GY'\| = \|A - AP_Y\| \tag{3.1.2}$$

and is attained at $G = AY(Y'Y)^-$.

3.1.3. Column–row regression

$$\min_{F,G} \|A - XF - GY'\| = \|A - P_X A - AP_Y + P_X AP_Y\| \tag{3.1.3}$$

and is attained at

$$XF = P_X A, \qquad GY' = (A - P_X A)P_Y$$
$$\text{or} \quad XF = P_X(A - AP_Y), \qquad GY' = AP_Y.$$

3.1.4. Column–row regression with interaction

The minimum of

$$\|A - XF - GY' - H\| \tag{3.1.4}$$

under the conditions, $\rho(X) = a$, $\rho(Y) = b$ and $\rho(H) = k \leqslant \min(p - a, q - b)$ is attained at

$$XF = P_X A, \qquad GY' = (1 - P_X)AP_Y,$$
$$G = \sigma_1 u_1 v_1' + \cdots + \sigma_k u_k v_k', \tag{3.1.5}$$

where $\sigma_1 u_1 v_1' + \cdots + \sigma_r u_r v_r'$ is the SVD of

$$(I - P_X)A(I - P_Y). \tag{3.1.6}$$

For proofs of results in Sections 3.1.1–3.1.4 under more general conditions, reference may be made to (Rao and Rao, 1998, Chapter 11, Section 11.6).

3.2. Von Neumann theorem

Let $A(p \times q)$ and $B(p \times q)$ have the SVDs of type (2.3.4), $P\Delta_1 Q'$ and $R\Delta_2 S'$, respectively. Then

$$\max_{U'U=I, V'V=I} |\operatorname{tr} AUB'V| = \alpha_1\beta_1 + \cdots + \alpha_r\beta_r, \tag{3.2.1}$$

where α_i are the diagonal elements of Δ_1 and β_j those of Δ_2 and $r = \min(p, q)$. The maximum in (3.2.1) is attained at $U = QS'$ and $V = RP'$.

Von Neumann (1937) stated the result (3.2.1) for square matrices A and B, but the same proof extends to rectangular matrices as shown by Rao (1979, 1980). The result (3.2.1) plays an important role in matrix optimization problems. As special cases we have

$$\max_{U'U=I} |\operatorname{tr} AU| = \operatorname{tr} \Delta_1, \quad \text{when } A \text{ is a square matrix}, \tag{3.2.2}$$

$$\max_{K=K', K^2=K} |\operatorname{tr} A'KA| = \operatorname{tr} \Delta_1^2 = \max_{K'K=I} |\operatorname{tr} A'KA|. \tag{3.2.3}$$

3.3. Eckart–Young theorem

For given $A(p\times q, r)$, the problem is that of minimizing $\|A-B\|$ over matrices $B(p\times q, s \leqslant r)$. Denote the SVD of A by

$$A = P\Delta Q' = \sigma_1 p_1 q_1' + \cdots + \sigma_s p_s q_s' + \cdots + \sigma_r p_r q_r'. \tag{3.3.1}$$

The solution is obtained as follows. $B(p \times q, s \leqslant r)$ can be expressed as KL where $K(p \times s)$ is semi-orthogonal (i.e., $K'K = I_s$), and L is $s \times q$ matrix. Then

$$\begin{aligned} \|A - KL\| &\geqslant \|A - KK'A\| \quad \text{using (3.1.1)}, \\ \|A - KK'A\|^2 &= \operatorname{tr} A'A - \operatorname{tr} KK'AA' \\ &\geqslant \operatorname{tr} A'A - \operatorname{tr} \Delta^2 \quad \text{using (3.2.3)}, \end{aligned} \tag{3.3.2}$$

since KK' is symmetric and $KK'KK' = KK'$. The minimum of $\|A - B\|$ is attained when K and L are as given below:

$$\begin{aligned} K &= (p_1 \cdots p_s), \\ L' &= A'K = \sigma_1 q_1' + \cdots + \sigma_s q_s', \\ B &= KL = \sigma_1 p_1 q_1' + \cdots + \sigma_s p_s q_s' = P_s \Delta_s Q_s'. \end{aligned} \tag{3.3.3}$$

3.4. Reduction of dimensionality

Let the columns of $A(p \times q)$ represent p-measurements taken on q units (individuals or categories). One of the problems in psychometric research is to study the configuration of units in terms of similarities or dissimilarities between them. One of the techniques used is to reduce the p measurements to a few functions of the measurements and study the configuration of units visually by plotting the reduced measurements in 2- or 3-dimensional graphs.

In general, a suitable transformation of the original measurements is done like translating the measurements to a suitable origin, usually the averages of the measurements, scaling the measurements to make them independent of the units of measurement and considering functions of measurements which are uncorrelated. All these can be done by considering a transformation of the type

$$A_* = M^{1/2}(A - \bar{a}1'), \tag{3.4.1}$$

where $\bar{a} = q^{-1}A1_q$ is the vector of the averages and $M(p \times p)$ is a pd matrix. The choice of M as the reciprocal of the variance–covariance matrix of measurements will make the transformed measurements in A_* independent of units of measurements and uncorrelated. However, this information may not be available. In such a case, we may choose M as the identity matrix or a diagonal matrix to correct for differences in units of measurements.

For reduction of dimensionality, we consider transforming the column vectors of A_* to a subspace of dimension s ($\ll p$) spanned by the orthogonal columns of a matrix $K(p \times s)$ and represent the units by the column vectors of KL, where L is an $s \times q$ matrix. We now choose K and L such that $\|A_* - KL\|$ is a minimum for given s. Note that the rank of KL is s and the problem can be solved by using Eckart–Young theorem. Using (3.3.3)

$$K = (a_1 \; \cdots \; a_s), \qquad L = (\sigma_1 b_1 \; \cdots \; \sigma_q b_q), \tag{3.4.2}$$

where a_i, b_i and σ_i are as in the SVD of A_*

$$A_* = \sigma_1 a_1 b_1' + \cdots + \sigma_s a_s b_s' + \cdots + \sigma_r a_r b_r'. \tag{3.4.3}$$

The loss of information in reducing the dimension of measurements to s is usually measured by

$$1 - \frac{\sum_1^s \sigma_i^2}{\sum_1^r \sigma_i^2}. \tag{3.4.4}$$

The coordinates for representing the q units in a s-dimensional Euclidean space are as follows:

Unit	dim 1	dim 2	$\cdots$	dim s
1	$\sigma_1 b_{11}$	$\sigma_2 b_{21}$	$\cdots$	$\sigma_s b_{s1}$
$\vdots$	$\vdots$	$\vdots$	$\ddots$	$\vdots$
q	$\sigma_1 b_{1q}$	$\sigma_2 b_{2q}$	$\cdots$	$\sigma_s b_{sq}$

where $b_{i1}, \ldots, b_{iq}$ are the elements of the vector b_i in (3.4.3). We can display the configuration of the units in the first 2 or 3 dimensions which carry the maximum amount of information. The coordinates in the above table are called canonical coordinates. Reference may be made to Rao (1948, 2005) for various ways of standardizing the measurements and the choice of M matrix.

3.5. Reduced rank regression

Consider a q_2-variate multivariate regression model

$$A_2 = A_1 T + E, \tag{3.5.1}$$

where A_2 is $p \times q_2$ matrix of observations, A_1 is $p \times q_1$ matrix of q_1 covariates, T is $q_1 \times q_2$ matrix of regression coefficients, and E is $p \times q_2$ matrix of errors. In the usual multivariate regression analysis (Rao, 1973, p. 543), T is estimated by

$$\hat{T} = \left(A_1' A_1\right)^{-1} A_1' A_2, \tag{3.5.2}$$

which is unconstrained regression estimate.

The problem is that of estimating T under the condition that rank of T is $s \leqslant q$. For this we need to minimize $\|A_1 T - A_2\|$ under the condition that the rank of T is s. We note that with $\hat{T}$ as in (3.5.2),

$$\begin{aligned}&(A_1 T - A_2)'(A_1 T - A_2)\\&\quad = (A_1 \hat{T} - A_2)'(A_1 \hat{T} - A_2) + (T - \hat{T})' A_1' A_1 (T - \hat{T}),\end{aligned}$$

so that we need only minimize

$$\operatorname{tr}(T - \hat{T})' A_1' A_1 (T - \hat{T}) = \left\| \left(A_1' A_1\right)^{1/2} T - \left(A_1' A_1\right)^{1/2} \hat{T} \right\|^2$$

subject to the condition that the rank of T is s. Using Eckart–Young theorem, the optimum T satisfies the equation

$$\left(A_1' A_1\right)^{1/2} T_* = \sigma_1 a_1 b_1' + \cdots + \sigma_s a_s b_s', \tag{3.5.3}$$

where a_i, b_i and σ_i are as in the SVD

$$\left(A_1' A_1\right)^{1/2} \hat{T} = \sigma_1 a_1 b_1' + \cdots + \sigma_s a_s b_s' + \cdots + a_r b_r'. \tag{3.5.4}$$

For more general results on matrix approximations, reference may be made to Rao (1979, 1980, 1981, 1984, 1989).

4. Procrustean transformation

Let $A_1(p \times q)$ and $A_2(p \times q)$ be two given matrices. In factor analysis, one seeks a transformation matrix $T(q \times q)$ such that A_1 is close to $A_2 T$. Such a transformation is known as the Procrustean transformation. More generally, we consider the problem of finding transformation matrices $T_1(p \times p)$ and $T_2(q \times q)$ belonging to specified classes such that $\|T_1 A_1 - A_2 T_2\|$ is a minimum. We choose the L_2-norm defined as $\|A\| = (\operatorname{tr} AA')^{1/2}$.

4.1. Transformation by orthogonal matrices (OR)

(i) Let $A_1 = P_1 \Delta_1 Q_1'$ and $A_2 = P_2 \Delta_2 Q_2'$ be the SVDs of the type (2.3.4) of A_1 and A_2, respectively, and $T_1(p \times p)$ and $T_2(q \times q) \in OR$. Then

$$\min_{T_1, T_2 \in OR} \|T_1 A_1 - A_2 T_2\| = \left[\operatorname{tr}(\Delta_1 - \Delta_2)'(\Delta_1 - \Delta_2)\right]^{1/2}, \tag{4.1.1}$$

and the minimum is attained when $T_1 = P_2 P_1'$ and $T_2 = Q_2 Q_1'$. Note that

$$\begin{aligned} \|T_1 A_1 - A_2 T_2\|^2 &= \operatorname{tr}(T_1 A_1 - A_2 T_2)'(T_1 A_1 - A_2 T_2) \\ &= \operatorname{tr}\left(A_1' A_1 + A_2' A_2\right) - 2 \operatorname{tr} A_1' T_1' A_2 T_2, \end{aligned} \tag{4.1.2}$$

and by (3.2.1) of von Neumann's theorem

$$\max_{T_1, T_2 \in OR} \operatorname{tr} A_1' T_1' A_2 T_2 = \operatorname{tr} \Delta_1 \Delta_2. \tag{4.1.3}$$

Substituting (4.1.3) in (4.1.2), we have (4.1.1).

(ii) Let $R \Delta S'$ be SVD of $A_1' A_2$ and $T(q \times q)$ be an orthogonal matrix. Then

$$\begin{aligned} \|A_1 - A_2 T\|^2 &= \operatorname{tr}(A_1 - A_2 T)'(A_1 - A_2 T) \\ &= \operatorname{tr}\left(A_1' A_1 + A_2' A_2\right) - 2 \operatorname{tr} A_1' A_2 T, \end{aligned} \tag{4.1.4}$$

$$\max_{T \in OR} \operatorname{tr} A_1' A_2 T = \operatorname{tr} \Delta, \tag{4.1.5}$$

which is attained at $T = S R'$, using (3.2.2) of von Neumann theorem. Hence

$$\min_{T \in OR} \|A_1 - A_2 T\|^2 = \operatorname{tr}\left(A_1' A_1 + A_2' A_2\right) - 2 \operatorname{tr} \Delta. \tag{4.1.6}$$

(iii) In some problems, in addition to an orthogonal transformation a suitable scaling factor may be introduced leading to minimization of $\|A_1 - s A_2 T\|$.

The best estimate of T is $T_* = S R'$ as in (ii). The optimum estimate of s is

$$s_* = \frac{\operatorname{tr} \Delta}{\operatorname{tr} A_2' A_2}. \tag{4.1.7}$$

(iv) Let $P \Delta Q'$ be the SVD of $A(p \times p)$. Then

$$\min_{T \in OR} \|A - T\|^2 = \operatorname{tr}(A' A) + p - 2 \operatorname{tr} \Delta \tag{4.1.8}$$

is attained at $T_* = Q P'$, so that the orthogonal transformation closest to A is T_*.

(v) Consider matrices $A_1(p \times q)$ and $A_2(p \times q)$ and spectral decompositions of

$$A_1 A_1' = P \Delta_1 P', \qquad P_2 = A_2\left(A_2' A_2\right)^{-1} A_2' = Q \Delta_2 Q', \tag{4.1.9}$$

where P_2 is the orthogonal projection operator on S_{A_2}. The problem we consider is that of minimizing

$$\|U A_1 - A_2 T\|, \qquad U \in OR \quad \text{and } T \text{ is arbitrary}. \tag{4.1.10}$$

First, we minimize (4.1.10) w.r.t. to T keeping U fixed, which is a regression problem. Using (3.1.1), the minimum is

$$\|UA_1 - P_2UA_1\|. \tag{4.1.11}$$

Now

$$\begin{aligned}\|UA_1 - P_2UA_1\|^2 &= \operatorname{tr}\big[(I - P_2)UA_1A_1'U'\big]\\ &= \operatorname{tr} A_1A_1' - \operatorname{tr} P_2UA_1A_1'U'.\end{aligned} \tag{4.1.12}$$

By (3.2.1)

$$\operatorname{tr} P_2UA_1A_1'U' \leqslant \operatorname{tr} \Delta_1\Delta_2 \tag{4.1.13}$$

and the maximum is attained when $U = QP'$, so that

$$\min_{U\in OR,T} \|UA_1 - A_2T\| = \operatorname{tr} \Delta_1 - \operatorname{tr} \Delta_1\Delta_2. \tag{4.1.14}$$

4.2. Tucker's problem

Tucker (1951) considered the problem of maximizing $\operatorname{tr} T_2'A_2'A_1T_1$ subject to the constraints $T_i'A_i'A_iT_i = I$, for $i = 1, 2$, where A_1 is $p \times q_1$, A_2 is $p \times q_2$, T_1 is $q_1 \times r$ and T_2 is $q_2 \times r$.

Consider the QR factorizations

$$A_i = Q_iR_i, \qquad Q_i'Q_i = I, \quad i = 1, 2. \tag{4.2.1}$$

The condition $T_i'A_i'A_iT_i = I$, $i = 1, 2$, implies $R_iT_i = L_i$ is orthogonal, $i = 1, 2$, and

$$\begin{aligned}&T_2'A_2'A_1T_1 = L_2'Q_2'Q_1L_1,\\ &\operatorname{tr} L_2'Q_2'Q_1L_1 = \operatorname{tr} Q_2'Q_1L_1L_2' = \operatorname{tr} Q_2'Q_1U,\end{aligned} \tag{4.2.2}$$

where U is an orthogonal matrix. Then

$$\max_{T_1,T_2} \operatorname{tr} T_2'A_2'A_1T_1 = \max_{U\in OR} \operatorname{tr} Q_2'Q_1U = \operatorname{tr} \Delta, \tag{4.2.3}$$

where Δ is the matrix of singular values of $Q_2'Q_1$.

4.3. Meredith's problem

Meredith (1964) considered minimizing the function

$$\sum_{i=1}^{2} \|A_iT_i - Z\|^2, \quad Z'Z = I_r, \tag{4.3.1}$$

which arises in factor analysis as the problem of finding a pair of "oblique rotations" to match A_1 and A_2 to a common matrix Z.

Let $A_i = B_iL_i$, $B_i'B_i = I$, $i = 1, 2$, be the QR factorizations of A_1, A_2. Then (4.3.1) implies

$$\sum_{i=1}^{2} \|B_iS_i - Z\|^2, \quad S_i = L_iT_i, \tag{4.3.2}$$

over S_1, S_2 and Z. For given Z, the minimum of (4.3.2) occurs when

$$S_{i*} = \left(B_i'B_i\right)^{-1}B_i'Z, \quad i = 1, 2. \tag{4.3.3}$$

Then

$$\sum_{i=1}^{2} \|B_i S_{i*} - Z\|^2 = 2r - \operatorname{tr} Z'CZ, \tag{4.3.4}$$

where $C = B_1B_1' + B_2B_2'$. The minimum of (4.3.4) occurs when tr $Z'CZ$ is maximized, which happens when the columns of Z are the first r eigenvectors of C. Then (4.3.4) gives S_{1*} and S_{2*}, and the optimum T_i as $T_{i*} = L_i^{-1}S_{i*}$, $i = 1, 2$.

5. Correspondence analysis

5.1. Two-way contingency table

In Section 3.4, we considered the problem of reducing p-dimensional ratio scale data to a smaller number of functions for graphical representation of individuals in 2 or 3 dimensions. The main tool used is SVD which ensures minimum loss in configuration of individuals in terms of mutual distances between individuals. In psychometric research, we encounter problems where the data may be on a nominal scale. For instance, an observed individual may be classified as belonging to one of p countries and having an attribute which may have q alternatives. The data on n observed individuals can be summarized in a two way $p \times q$ contingency table with the rows representing different attributes and columns, different countries. Let n_{ij} be the number in the ith row and jth column. Define

$$N = (n_{ij}),$$
$$n_{i.} = \sum_{j=1}^{q} n_{ij}, \qquad n_{.j} = \sum_{i=1}^{p} n_{ij}, \qquad n = \sum_{1}^{p}\sum_{1}^{q} n_{ij}. \tag{5.1.1}$$

Let $R(p \times p)$ be the diagonal matrix with $n_{i.}$ as the ith diagonal element and $C(q \times q)$ be the diagonal matrix with $n_{.j}$ as the jth diagonal element. The column (countries) and row (attributes) profiles are as in matrices

$$P = \begin{pmatrix} n_{11}/n_{.1} & \cdots & n_{1q}/n_{.q} \\ \vdots & \ddots & \vdots \\ n_{p1}/n_{.1} & \cdots & n_{pq}/n_{.q} \end{pmatrix} = (c_1 \ \cdots \ c_q),$$
$$Q' = \begin{pmatrix} n_{11}/n_{1.} & \cdots & n_{p1}/n_{p.} \\ \vdots & \ddots & \vdots \\ n_{1q}/n_{1.} & \cdots & n_{pq}/n_{p.} \end{pmatrix} = (r_1 \ \cdots \ r_p), \tag{5.1.2}$$

where c_i is the ith column vector in P and r_i is the ith column vector in Q'.

The problem is to represent the countries (columns of p) as points in a low dimensional Euclidean space (say, E^k of k dimensions) so that dissimilarities between countries appropriately defined in terms of column vectors in P are reflected in Euclidean distances between points in E^k. A similar problem is that of representing attributes as points in E^k, a low dimensional Euclidean space.

5.2. *Representation in lower dimensions (chi-square distance)*

Suppose that we want to represent the countries (the column p-vectors) in a low-dimensional space to study their configuration. The technique developed for this purpose by Benzécri (1973) is known as correspondence analysis. One finds the SVD of

$$R^{-1/2}(P - s1')C^{1/2} = \sigma_1 U_1 V_1' + \cdots + \sigma_r U_r V_r', \tag{5.2.1}$$

where $s = R1$, and chooses the coordinates in E^k as

$$\sigma_1 C^{-1/2} V_1, \quad \ldots, \quad \sigma_k C^{-1/2} V_k, \tag{5.2.2}$$

where the q components of the ith vector, $\sigma_i C^{-1/2} V_i$ are the coordinates of the q countries in the ith dimension. Usually the first two or three dimensions are chosen. Implicit in Benzécri's analysis is the choice of measure of affinity between countries i and j as the chi-square distance

$$d_{ij}^2 = (c_i - c_j)' M^{-1} (c_i - c_j), \tag{5.2.3}$$

where M is the diagonal matrix with the ith diagonal element as $n_{i.}/n$. The squared Euclidean distance between countries i and j in the low-dimensional plot is an approximation to (5.2.3).

NOTE 1. If we want a representation of the different attributes, in a low-dimensional space E^k, we use the same SVD as in (5.2.1) and choose the coordinates in E^k as

$$\sigma_1 R^{-1/2} U_1, \quad \sigma_2 R^{-1/2} U_2, \quad \ldots, \quad \sigma_k R^{-1/2} U_k, \tag{5.2.4}$$

so that no extra computations are needed.

In correspondence analysis, it is customary to plot the coordinates of column profiles (countries) and row profiles (attributes) in the same graph. For practical illustration of correspondence analysis, reference may be made to Rao (1995, 2005).

5.3. *Representation in lower dimensions (Hellinger distance)*

An alternative to chi-square distance, which has some advantages is the Hellinger distance (HD) between ith and jth columns defined by

$$d_{ij}^2 = \left(\sqrt{c_{1i}} - \sqrt{c_{1j}}\right)^2 + \cdots + \left(\sqrt{c_{pi}} - \sqrt{c_{pj}}\right)^2, \tag{5.3.1}$$

where $c_i' = (c_{1i}, \ldots, c_{pi})$, $c_j' = (c_{1j}, \ldots, c_{pj})$, and c_i and c_j are as in (5.1.2) representing the ith and jth column vectors. To obtain coordinates in a low-dimensional space such that the Euclidean distance between points (countries) i and j is an approximation to (5.3.1), we proceed as follows. Consider

$$P = \begin{pmatrix} \sqrt{c_{11}} & \cdots & \sqrt{c_{1q}} \\ \vdots & \ddots & \vdots \\ \sqrt{c_{p1}} & \cdots & \sqrt{c_{pq}} \end{pmatrix}, \qquad G = \begin{pmatrix} n_{.1}/n & \cdots & 0 \\ 0 & \cdots & 0 \\ \vdots & \ddots & \vdots \\ 0 & \cdots & n_{.q}/n \end{pmatrix}, \tag{5.3.2}$$

and compute the SVD

$$(P - p1')G^{1/2} = \lambda_1 U_1 V_1' + \cdots + \lambda_r U_r V_r', \tag{5.3.3}$$

where we may choose $p' = (p_1, \ldots, p_p)$ with

$$p_i = \sqrt{n_{i.}/n} \quad \text{or} \quad p_i = n^{-1}\left(n_{.1}\sqrt{c_{i1}} + \cdots + n_{.q}\sqrt{c_{iq}}\right) \tag{5.3.4}$$

for $i = 1, \ldots, p$. The coordinates in different dimensions of the reduced space are

$$\sigma_1 G^{-1/2} V_1, \quad \sigma_2 G^{-1/2} V_2, \quad \ldots, \quad \sigma_k G^{-1/2} V_k. \tag{5.3.5}$$

If a representation of the column profiles in Q' of (5.1.2) are needed, we start with the matrices

$$Q' = \begin{pmatrix} \sqrt{r_{11}} & \cdots & \sqrt{r_{1p}} \\ \vdots & \ddots & \vdots \\ \sqrt{r_{q1}} & \cdots & \sqrt{r_{qp}} \end{pmatrix}, \qquad H = \begin{pmatrix} n_{1.}/n & \cdots & 0 \\ 0 & \cdots & 0 \\ \vdots & \ddots & \vdots \\ 0 & \cdots & n_{p.}/n \end{pmatrix} \tag{5.3.6}$$

and compute the SVD

$$(Q - \eta 1')H^{1/2} = \mu_1 A_1 B_1' + \cdots + \mu_r A_r B_r', \tag{5.3.7}$$

where we may choose $\eta' = (\eta_1, \ldots, \eta_q)$ with

$$\eta_i = \sqrt{n_{.i}/n}, \quad \text{or} \quad \eta_i = n^{-1}\left(n_{1.}\sqrt{r_{1i}} + \cdots + n_{p.}\sqrt{r_{pi}}\right).$$

The coordinates in different dimensions in the reduced space are

$$\mu_1 H^{-1/2} A_1, \quad \mu_2 H^{-1/2} A_2, \quad \ldots, \quad \mu_k H^{-1/2} A_k. \tag{5.3.8}$$

The advantages in using Hellinger distances are outlined in Rao (1995, 2005). The appropriate software for the two methods using chi-square and Hellinger distances can be found in Khattree and Naik (1999).

6. Metric and multidimensional scaling

In Section 3.4, we considered reduction of dimensionality when the data are available in the form of a matrix $A(p \times q)$ with p measurements taken on each of q individuals (or units). The problem was to represent the individuals as points in an s- (preferably 2- or 3-) dimensional graph without much loss in the configuration (suitably defined) of the points in the full p-dimensional space. This was done by considering the SVD of A and using Eckart–Young theorem.

In some situations, it may not be possible to characterize an individual or a unit in terms of specific measurements, but a direct measure of similarity or dissimilarity between any two individuals may be available. The data will then be in the form of a $q \times q$ matrix, $D = (d_{ij})$, where d_{ij} represents dissimilarity between individuals i and j. Then the problem is to represent the individuals in a 2- or 3-dimensional graph in such a way that the Euclidean distances between points representing the individuals in the low-dimensional space are close to the corresponding measures of dissimilarity.

6.1. Metric scaling

Let us denote the $q \times q$ matrix of dissimilarities by $D = (d_{ij})$. If, in fact, d_{ij}'s are computed from a $p \times q$ matrix of measurements $A = (a_1 \cdots a_q)$, where a_i is the p-vector of measurements on individual i, using the Euclidean distance

$$d_{ij}^2 = (a_i - a_j)'(a_i - a_j), \tag{6.1.1}$$

then, from given D, we can recover A with an appropriate p and a shift of the origin for the measurements (i.e., $A1_q = 0$) as shown below.

Consider the $q \times q$ matrix

$$J = I_q - q^{-1} 1_q 1_q', \tag{6.1.2}$$

where I_q is the identity matrix and 1_q is a vector with each of its q components equal to unity. Then we have the identity

$$A'A = -\frac{1}{2} J (d_{ij}^2) J, \tag{6.1.3}$$

where (d_{ij}^2) is the matrix obtained from D by squaring each element. The next step is to obtain the eigenvalue decomposition

$$-\frac{1}{2} J (d_{ij}^2) J = \sigma_1 b_1 b_1' + \cdots + \sigma_q b_q b_q' \tag{6.1.4}$$

with $\sigma_1 \geqslant \sigma_2 \geqslant \cdots \geqslant \sigma_q$.

If the assumption (6.1.1) is true for some p, then all the eigenvalues $\sigma_1, \ldots, \sigma_q$ will be non-negative, in which case a solution for A' can be obtained from (6.1.3) and (6.1.4) as

$$A' = \left(\sqrt{\sigma_1} b_1 \cdots \sqrt{\sigma_q} b_q\right) = \begin{pmatrix} \sqrt{\sigma_1} b_{11} & \sqrt{\sigma_2} b_{12} & \cdots & \sqrt{\sigma_q} b_{1q} \\ \vdots & \vdots & \ddots & \vdots \\ \sqrt{\sigma_1} b_{q1} & \sqrt{\sigma_2} b_{q2} & \cdots & \sqrt{\sigma_q} b_{qq} \end{pmatrix} \tag{6.1.5}$$

showing that the individuals can be represented in an Euclidean space of dimensions not greater than q, with d_{ij} as the distance between individuals i and j. If an $s < q$ dimensional representation is needed, we choose the coordinates of the q individuals in different dimensions as

Individual	dim 1	dim 2	$\cdots$	dim s
1	$\sqrt{\sigma_1} b_{11}$	$\sqrt{\sigma_2} b_{12}$	$\cdots$	$\sqrt{\sigma_s} b_{1s}$
$\vdots$	$\vdots$	$\vdots$	$\ddots$	$\vdots$
q	$\sqrt{\sigma_1} b_{q1}$	$\sqrt{\sigma_2} b_{q2}$	$\cdots$	$\sqrt{\sigma_s} b_{qs}$

(6.1.6)

We have loss of information in such a representation depending on magnitudes of $\sigma_{s+1}, \ldots, \sigma_q$.

However, the assumption (6.1.1) may not be true, which can be discovered when some of the eigenvalues are negative. If such values are small in magnitude compared to the positive eigenvalues, then we can omit the corresponding terms in the eigen-expansion (6.1.4) and proceed as indicated in steps (6.1.5)–(6.1.6). As a rule one can compute the distances from (6.1.6),

$$\hat{d}_{ij}^2 = \sigma_1(b_{i1} - b_{j1})^2 + \cdots + \sigma_s(b_{is} - b_{js})^2 \tag{6.1.7}$$

for a chosen s and compare the matrices (d_{ij}) and $(\hat{d}_{ij})$ to find the amount of distortion in representing the individuals in a low dimensional Euclidean space. One can also plot the points $(d_{ij}, \hat{d}_{ij})$ in a two-dimensional graph and judge by the monotonicity of the curve to what extent the inequality relationships in $d_{ij}'s$ are reflected in $\hat{d}_{ij}$'s. The method outlined above, called metric scaling, is due to Torgerson (1958) with some clarifications made by Rao (1964) and Gower (1966). For further details, reference may be made to Borg and Groenen (1997) and Shepard et al. (1972).

If metric scaling does not provide a satisfactory fit to observed dissimilarities, one may use the computational algorithm known as multidimensional scaling developed by Kruskal and Wish (1978). This algorithm attempts to obtain a lower-dimensional Euclidean space representation of the individuals in such a way that the inequality relationships of the d_{ij}'s are reflected to the extent possible in the Euclidean distances in the fitted low-dimensional space.

Acknowledgements

I would like to thank the two referees for their useful suggestions which have been incorporated in the revised version.

References

Benzécri, J.P. (1973). *L'analyse de Donées, vols. I, II*. Dunod, Paris.
Borg, I., Groenen, P. (1997). *Modern Multidimensional Scaling*. Springer-Verlag, New York.
Datta, B.N. (1995). *Numerical Linear Algebra and Applications*. Brooks/Cole.
Golub, G.H., Van Loan, C.F. (1989). *Matrix Computations*, 2nd ed. Johns Hopkins University Press.
Gower, J.C. (1966). Some distance properties of latent root and vector methods used in multivariate analysis. *Biometrika* **33**, 325–338.
Gower, J. (2003). A unified approach to oblique Procrustes problems. In: Yanai, H., Okada, A., Shigemasu, K., Kano, Y., Meulman, J.J. (Eds.), *New Developments in Psychometrics*. Springer, Tokyo, pp. 371–380.
Gower, J.C., Dijksterhuis, G.B. (2004). *Procrustes Problems*. Oxford University Press, Oxford.
Khattree, R., Naik, D. (1999). *Applied Multivariate Statistics with SAS Software*, 2nd ed. SAS Institute.
Kruskal, J.B., Wish, M. (1978). *Multidimensional Scaling*. Sage, Beverly Hills, CA.
Meredith, W. (1964). Rotation to achieve factorial invariance. *Psychometrika* **29**, 187–206.
Rao, C.R. (1948). The utilization of multiple measurements in problems of biological classification. *Journal of Royal Statistical Society* **10**, 159–203.
Rao, C.R. (1955). Estimation and tests of significance in factor analysis. *Psychometrika* **20**, 93–111.

Rao, C.R. (1962). A note on generalized inverse of a matrix with applications to problems in mathematical statistics. *Journal of Royal Statistical Society B* **24**, 152–158.

Rao, C.R. (1964). The use and interpretation of principal component analysis in applied research. *Sankhyā A* **26**, 329–358.

Rao, C.R. (1967). Calculus of generalized inverses of matrices: Part I – General theory. *Sankhyā A* **29**, 317–350.

Rao, C.R. (1973). *Linear Statistical Inference and its Applications*. Wiley, New York.

Rao, C.R. (1979). Separation theorems for singular values of matrices and their applications in multivariate analysis. *Journal of Multivariate Analysis* **9**, 362–377.

Rao, C.R. (1980). Matrix approximations and reduction of dimensionality in multivariate statistical analysis. In: Krishnaiah, P.R. (Ed.), *Multivariate Analysis V*. North-Holland, Amsterdam, pp. 3–22.

Rao, C.R. (1981). A lemma on *g*-inverse of a matrix and a computation of correlation coefficients in the singular case. *Commun. Statist. Theor. Meth. A* **10**, 1–10.

Rao, C.R. (1984). Optimization of functions of matrices with applications to statistical problems. In: Poduri, S.R.S. (Ed.), *W.G. Cochran's Impact on Statistics*. Wiley, New York, pp. 191–202.

Rao, C.R. (1989). A lemma on optimization of a matrix function and a review of the unified theory of linear estimation. In: Dodge, Y. (Ed.), *Statistical Data Analysis and Inference*. Elsevier, Amsterdam, pp. 397–418.

Rao, C.R. (1995). A review of canonical coordinates and an alternative to correspondence analysis using Hellinger distance. *Qüestiō* **9**, 23–63.

Rao, C.R. (2005). Canonical variate analysis and related methods for reduction of dimensionality and graphical representation. In: Rao, C.R., Wegman, E.J., Solka, J.L. (Eds.), *Handbook of Statistics*, vol. 24. Elsevier, Amsterdam, pp. 189–211.

Rao, C.R., Mitra, S.K. (1971). *Generalized Inverse of Matrices and its Applications*. Wiley, New York. (Japanese translation, Tokyo, 1973.)

Rao, C.R., Mitra, S.K. (1973). Theory and application of constrained inverse of matrices. *SIAM J. Appl. Math.* **24**, 473–488.

Rao, C.R., Mitra, S.K. (1974). Projections under seminorm and generalized inverse of matrices. *Linear Algebra and its Applications* **9**, 155–167.

Rao, C.R., Yanai, H. (1985). Generalized inverse of linear transformations: A geometric approach. *Linear Algebra and its Applications* **66**, 87–98.

Rao, C.R., Rao, M.B. (1998). *Matrix Algebra and its Applications to Statistics and Econometrics*. World Scientific, Singapore.

Shepard, R.N., Romney, A.K., Nerlov, S.B. (Eds.) (1972). *Multidimensional Scaling: Theory and Applications in Behavioral Sciences*, vol. 1. Seminar Press, New York.

Stewart, G.W. (1983). A method for computing the generalized singular value decompositions. In: Kagstrom, B., Ruhe, A. (Eds.), *Matrix Pencils*. Springer-Verlag, pp. 207–220.

Takane, Y. (2003). Relationships among various kinds of eigenvalue and singular value decompositions. In: Yanai, H., Okada, A., Shigemasu, K., Kano, Y., Meulman, J.J. (Eds.), *New Developments in Psychometrics*. Springer-Verlag, Tokyo, pp. 45–56.

Takane, Y. (2004). Matrices with special reference to applications in psychometrics. *Linear Algebra and its Applications* **388**, 341–361.

Torgerson, W.S. (1958). *Theory and Methods of Scaling*. Wiley, New York.

Tucker, L.R. (1951). A method for synthesis of factor analysis studies. Personal Research Section Report No. 984. Department of the Army, Washington, DC.

Vaccaro, R.J. (1991). *SVD and Signal Processing. II. Algorithms, Analysis and Applications*. Elsevier, Amsterdam.

von Neumann, J. (1937). Some matrix inequalities and metrization of metric spaces. *Tomsk Univ. Rev.* **1**, 286–299.

Yanai, H. (2003). Vectors and matrices in psychometrics with special emphasis on generalized inverse and projection matrices. In: Yanai, H., Okada, A., Shigenmasu, K., Kano, Y., Meulman, J.J. (Eds.), *New Developments in Psychometrics*. Springer-Verlag, Tokyo, pp. 19–32.

Handbook of Statistics, Vol. 26
ISSN: 0169-7161

DOI: 10.1016/S0169-7161(06)26009-7

9

Factor Analysis

Haruo Yanai and Masanori Ichikawa

1. Introduction

Behavioral patterns in various situations are diverse. However, a careful observation shows that certain regularities are common. All of the factors that regulate human behavior are hypothetical. These factors include intelligence and personality traits such as introversion, extroversion, and emotional anxiety; sociological factors such as social status and group cohesiveness; and political attitudes such as conservatism and innovation.

Factor analysis is a multivariate technique designed to analyze correlations among many observed variables and to explore latent factors. Factor analysis was developed by the British psychologist Charles Spearman in the early 20th century as a technique for analyzing intelligence structures. Today, one can find many useful applications of factor analysis in diverse fields such as politics, sociology, education, linguistics, medicine, biology, anthropology, and engineering.

Section 2 gives an overview of the evolution of factor analysis from the early 20th century to the present and a review of applied research in various fields. Section 3 introduces the factor analysis model, and Section 4 deals with statistical inference in factor analysis. Section 5 covers various methods of factor rotation and estimation of factor scores.

2. Historical development of factor analysis

2.1. *The one-factor model*

In his pioneering paper of 1904, Spearman calculated correlation coefficients among six subjects (classics, French, English, mathematics, pitch discrimination, and musical talent) based on the scores of tests administered to children between the ages of 9 and 13. As shown in Table 1, high correlations can be found among the six subjects. One interpretation of a correlation coefficient is that its magnitude reflects the intensity of a common cause of the variables. Let r_{ij} denote the correlation coefficient between the ith and jth tests. We can see from Table 1 an approximate regularity in the nondiagonal

Table 1
Correlations among six subjects

	1	2	3	4	5	6
1 Classics	1.00	0.83	0.78	0.70	0.66	0.63
2 French		1.00	0.67	0.67	0.65	0.57
3 English			1.00	0.64	0.54	0.51
4 Mathematics				1.00	0.45	0.51
5 Pitch discrimination					1.00	0.40
6 Music talent						1.00

elements of the correlation matrix. It is given by

$$\frac{r_{ik}}{r_{jk}} \approx \frac{r_{i\ell}}{r_{j\ell}} \quad (i \neq j, k, \ell;\ j \neq k, \ell;\ k \neq \ell),$$

which can be transformed to the more convenient form

$$r_{ik}r_{j\ell} \approx r_{i\ell}r_{jk}. \tag{2.1}$$

Under Spearman's two-factor theory of intelligence, test scores can be expressed as the sum of a general factor that commonly relates to individual tests and the unique factors that only relate to individual tests. More specifically, let g, x_i, and e_i $(i = 1, \ldots, p)$ denote the general factor, ith test score, and the corresponding unique factor. Then x_i can be written as

$$x_i = \lambda_i g + e_i \quad (i = 1, \ldots, p), \tag{2.2}$$

where λ_i $(|\lambda_i| \leqslant 1)$ is a coefficient representing the intensity of g in x_i. We may assume that all the variables, g, x_i, and e_i, have zero means and that the variances of x_i and g are unity. We also assume that $\mathrm{E}(ge_i) = 0$. Then we have $\mathrm{E}(x_i^2) = \lambda_i^2 + \mathrm{E}(e_i^2)$, where $\mathrm{E}(e_i^2) = 1 - \lambda_i^2$. The unique factor corresponds to an error term in the linear regression model and expresses portions that cannot be explained by the general factor, g. In factor analysis, however, e_i is generally regarded as a factor that has a unique component in the ith variable. Further, we assume $\mathrm{E}(e_i e_j) = 0$ $(i \neq j)$. Then, the correlation coefficient ρ_{ij} between x_i and x_j can be written as

$$\begin{aligned}\rho_{ij} &= \mathrm{E}(x_i x_j)\\ &= \mathrm{E}(\lambda_i g + e_i)(\lambda_j g + e_j)\\ &= \lambda_i \lambda_j,\end{aligned}$$

which explains (2.1). Because λ_i is equal to the correlation coefficient, $\rho_{x_i g} = \mathrm{E}(x_i g)$, between x_i and g, this model implies that $p(p-1)/2$ correlations between the pairs of test scores can be expressed in terms of correlations λ_i $(i = 1, \ldots, p)$ between x_i and the unobservable variable, g.

Influenced by Spearman, British psychologists such as Thomson, Garnet, Burt, and Camp have discussed the utility of the one-factor model. They believed that the correlations among a number of test scores might be explainable in terms of a single factor called the *general ability*. The one-factor model is historically significant as an intermediary to the multiple factor model described below.

2.2. *The multiple factor model*

In the 1930s, the center of factor analysis research moved from Britain to the United States. Thurstone (1935) of North Carolina University expanded Spearman's two-factor theory and advocated the multiple factor theory of intelligence that says that individual tests are explained by more than one factor. The multiple factor model replaces the general factor, g, in (2.2) with $f_1, \dots, f_m$, called the *common factors*, that is,

$$\begin{aligned} x_1 &= \lambda_{11} f_1 + \lambda_{12} f_2 + \cdots + \lambda_{1m} f_m + e_1, \\ x_2 &= \lambda_{21} f_1 + \lambda_{22} f_2 + \cdots + \lambda_{2m} f_m + e_2, \\ &\vdots \\ x_p &= \lambda_{p1} f_1 + \lambda_{p2} f_2 + \cdots + \lambda_{pm} f_m + e_p. \end{aligned} \tag{2.3}$$

In other words, the ith test is expressed as a linear combination of the common factors, $f_1, \dots, f_m$, and the unique factor, e_i. The intensity of the individual factors in the ith test are represented by the coefficients $\lambda_{i1}, \lambda_{i2}, \dots, \lambda_{im}$.

By factor analyzing 57 types of intelligence tests administered to university students, Thurstone (1936) showed that human intelligence can be described by seven factors: memory, word fluency, verbal relation, number, perceptual speed, visualization, and reasoning. Thurstone promoted the multiple factor model by publishing "Multiple Factor Analysis" in 1947.

In 1936, the Psychometric Society was inaugurated and a new journal, Psychometrika, was published. Between the latter part of the 1930 and the beginning of 1950, many results on the following seven points appeared in Psychometrika; (a) estimation of communality, (b) extraction of common factors, (c) determination of the number of common factors, (d) factor rotation, (e) estimation of factor scores, (f) methods to shorten calculation time, and (g) factor indeterminacy.

Spearman and Thurstone did not pay much attention to the problem of factor indeterminacy. Guttman and other researchers continued to investigate this problem untill the late 1970s as an intrinsic problem of factor analysis related to interpretation of factors. Steiger (1979) reviewed the series of studies related to this problem. Comprehensive books by psychometricians such as Horst (1965) and Harman (1976) were published.

2.3. *From exploratory factor analysis to confirmatory factor analysis*

According to Mulaik (1986), the use of factor analysis between 1952 and 1969 was exploratory because researchers did not have any hypotheses on the correlation structure among variables used in the analysis. With the increasing use of factor analysis and accumulation of findings, more attention was paid to verify hypothetical structures behind the observed variables. As a solution to this problem, Guttman (1952) extended the multiple group method that consists of classifying variables into several groups and then generating a factor from each group. The method of Procrustes rotation (Green, 1952;

Schöneman, 1966) was developed to obtain a factor loading matrix that is as close to a hypothetical factor loading matrix as possible. Confirmatory factor analysis (Jöreskog, 1969b) was designed to test hypotheses about the relationships between observed variables and factors. Bock and Bargmann (1966) proposed an analysis of covariance structures. This led to the development of a generalized covariance structure model, the linear structural relations model (LISREL), by Jöreskog and Sörbom (1981).

2.4. *Extension of factor analysis*

Lawley and Maxwell (1971) gave an exposition of statistical theory in factor analysis. In the 1980s, the second edition of a book by Anderson (1984) had an added chapter on factor analysis from the view point of a mathematical statistician. Other books on factor analysis and related methods have been published (Basilevsky, 1994; Bartholomew and Knott, 1999). In the 1990s, independent component analysis attracted attention in the field of engineering (Hyvärinen et al., 2001).

In the factor analysis model, both manifest and latent variables are metrical and thus continuous. According to Bartholomew and Knott (1999), latent variable models can be classified into four types depending on whether the manifest and latent variables are metrical or categorical. Examples are (1) factor analysis of categorical data that uses categorical variables as manifest variables, (2) the latent class model in which both manifest and latent variables are categorical, (3) latent trait analysis in which latent variables are metrical and manifest variables are categorical, and (4) the latent profile model in which manifest variables are metrical and latent variables are categorical.

2.5. *Application of factor analysis*

Applied research that deserves special attention in psychology is the factor analysis of basic abilities of men by Thurstone (1936) mentioned earlier, research by Harrell (1940) in which factors of mechanical abilities were extracted, and research by Carroll (1941) in which factors of linguistic ability were extracted. In personality trait research, Guilford, Allport, and Eysenck developed a personality theory in terms of traits, which is different from the topology theory of personality of Krestchmer, providing a theoretical basis for preparing psychological tests including trait tests. Osgood and Suci (1957) introduced a semantic differential (SD) method that measures several images as objects using pairs of about 40 adjectives, analyzing the structure of semantic dimensions. The five-factor model (FFM) of personality was developed in the 1980s (Goldberg, 1990). Carroll (1993) compiled all available data on intelligence research that used factor analysis between the 1900s and 1980s into a single work.

Today, factor analysis is widely used not only in the field of psychology but also in fields such as politics, literature, biology, and medical science. For example, in anthropology, morphological knowledge has been obtained through factor analysis of correlations between measured traits of human bones and factor analysis of measured traits of animals and plants. According to Gananadesikan and Kettering (1984), who investigated academic journals of the United States in eight main academic areas, factor analysis amounts to 55.3% of multivariate analysis techniques other than regression analysis. It is noteworthy that the percentages of papers that employ factor analysis in the fields of education and psychology are 33.7 and 48.3%. In Japan, the percentages of papers that employ factor analysis in the fields of psychology and educational psychology is increasing every year (Yanai, 2000).

3. The factor analysis model

3.1. *Definition of the model*

Let $\mathbf{x} = (x_1, \ldots, x_p)'$ be an observable random vector with mean vector $\mathrm{E}(\mathbf{x}) = \boldsymbol{\mu}$ and covariance matrix $\mathrm{Var}(\mathbf{x}) = \boldsymbol{\Sigma}$. Then under the factor analysis model, $\mathbf{x}$ can be written in the form

$$\mathbf{x} = \boldsymbol{\mu} + \boldsymbol{\Lambda}\mathbf{f} + \mathbf{e}, \tag{3.1}$$

where $\mathbf{f} = (f_1, \ldots, f_m)'$ and $\mathbf{e} = (e_1, \ldots, e_p)'$ are unobservable random vectors, and $\boldsymbol{\Lambda}$ is a $p \times m$ matrix of regression coefficients. The element λ_{ik} is called the *factor loading* of x_i on f_k, and the entire matrix of factor loadings is sometimes called the *factor pattern*. The elements of $\mathbf{f}$ are called the *common factors* (or simply the *factors*) because each f_k affects all x_is. On the other hand, the elements of $\mathbf{e}$ are called the *unique factors* because each e_i affects only x_i. We assume that

$$\mathrm{E}(\mathbf{f}) = \mathbf{0}, \qquad \mathrm{Var}(\mathbf{f}) = \boldsymbol{\Phi}, \tag{3.2}$$

$$\mathrm{E}(\mathbf{e}) = \mathbf{0}, \qquad \mathrm{Var}(\mathbf{e}) = \boldsymbol{\Psi}, \tag{3.3}$$

where $\boldsymbol{\Phi}$ is a positive-definite matrix and $\boldsymbol{\Psi}$ is a diagonal matrix, of which the diagonal elements ψ_i (>0) are called the *unique variances*. We also assume that

$$\mathrm{Cov}(\mathbf{f}, \mathbf{e}) = \mathbf{0}, \tag{3.4}$$

that is, the common factors and the unique factors are uncorrelated. Assumptions (3.1) to (3.4) imply that $\boldsymbol{\Sigma}$ has a decomposition

$$\boldsymbol{\Sigma} = \boldsymbol{\Lambda}\boldsymbol{\Phi}\boldsymbol{\Lambda}' + \boldsymbol{\Psi} \tag{3.5}$$

and that the covariance matrix between $\mathbf{x}$ and $\mathbf{f}$ is given by

$$\mathrm{Cov}(\mathbf{x}, \mathbf{f}) = \boldsymbol{\Lambda}\boldsymbol{\Phi}, \tag{3.6}$$

of which the elements are called the *factor structures*. The covariance matrix between $\mathbf{x}$ and $\mathbf{e}$ is given by

$$\mathrm{Cov}(\mathbf{x}, \mathbf{e}) = \boldsymbol{\Psi}. \tag{3.7}$$

Eq. (3.1) decomposes each component of $\mathbf{x} - \boldsymbol{\mu}$ as

$$x_i - \mu_i = t_i + e_i \quad (i = 1, \ldots, p), \tag{3.8}$$

where

$$t_i = \lambda_{i1} f_1 + \lambda_{i2} f_2 + \cdots + \lambda_{im} f_m$$

is called the *common part* of x_i and satisfies

$$\mathrm{E}(t_i e_j) = 0 \quad (i, j = 1, \ldots, p)$$

and

$$\mathrm{Cov}(x_i, x_j) = \mathrm{Cov}(t_i, t_j) \quad (i \neq j).$$

There are cases where it may be more appropriate to regard the common factors $f_1, \ldots, f_m$ as nonrandom. But the nonrandom factor model poses a difficulty in inference (Anderson and Rubin, 1956).

3.1.1. *Scale invariance*

In many cases the unit of the measurement of each component of $\mathbf{x}$ is arbitrary. If we change the units of the measurement then

$$\tilde{\mathbf{x}} = \mathbf{D}\mathbf{x}, \qquad \tilde{\boldsymbol{\mu}} = \mathbf{D}\boldsymbol{\mu}, \qquad \tilde{\boldsymbol{\Sigma}} = \mathbf{D}\boldsymbol{\Sigma}\mathbf{D},$$

where $\mathbf{D}$ is a diagonal matrix, of which the ith diagonal element is d_i $(\neq 0)$. The factor analysis model is said to be *scale invariant* because it still holds for $\tilde{\mathbf{x}}$ by replacing $\boldsymbol{\Lambda}$, $\boldsymbol{\Psi}$, and $\mathbf{e}$ with $\tilde{\boldsymbol{\Lambda}} = \mathbf{D}\boldsymbol{\Lambda}$, $\tilde{\boldsymbol{\Psi}} = \mathbf{D}\boldsymbol{\Psi}\mathbf{D}$, and $\tilde{\mathbf{e}} = \mathbf{D}\mathbf{e}$.

If we standardize each of the observed variables by choosing $d_i = \sigma_{ii}^{-1/2}$, then the resulting $\mathbf{D}\boldsymbol{\Sigma}\mathbf{D}$ is a correlation matrix $\mathbf{P} = (\rho_{ij})$. The correlation matrix $\mathbf{P}$ has the decomposition

$$\mathbf{P} = \boldsymbol{\Lambda}^{*}\boldsymbol{\Phi}\boldsymbol{\Lambda}^{*\prime} + \boldsymbol{\Psi}, \tag{3.9}$$

where $\boldsymbol{\Lambda}^{*} = (\lambda_{ik}^{*})$ is a $p \times m$ matrix of *standardized factor loadings* and $\boldsymbol{\Psi}^{*}$ is a diagonal matrix whose diagonal elements, ψ_i^{*}, are the *standardized unique variances*. Obviously, λ_{ik}^{*} and ψ_i^{*} are given by

$$\lambda_{ik}^{*} = \frac{\lambda_{ik}}{\sigma_{ii}^{1/2}}, \qquad \psi_i^{*} = \frac{\psi_i}{\sigma_{ii}}.$$

The standardized unique variance ψ_i^{*} is sometimes called the *uniqueness*. The *communality*, h_i^2, of x_i is the squared multiple correlation between x_i and $f_1, \ldots, f_m$, and hence is the proportion of the variance of x_i that is accounted for by the common factors. Of course, the sum of the communality h_i^2 and the uniqueness ψ_i^{*} is always unity.

3.2. *Indeterminacy of the model*

In exploratory factor analysis, a researcher has no information concerning the values of parameters and it is known that several types of indeterminacy exist. In confirmatory factor analysis, on the other hand, the values of certain parameters are specified in advance, typically by zero factor loadings, on the basis of well-defined hypotheses, to the extent that no indterminacy exists in the parameters.

3.2.1. Indeterminacy of the factor covariance matrix

Suppose we transform $\mathbf{f}$ using a nonsingular matrix $\mathbf{A}$, $\tilde{\mathbf{f}} = \mathbf{A}\mathbf{f}$. Then the covariance matrix of $\tilde{\mathbf{f}}$ is given by $\tilde{\mathbf{\Phi}} = \mathbf{A}\mathbf{\Phi}\mathbf{A}'$. However, the factor analysis model still holds by replacing $\mathbf{\Lambda}$ with $\tilde{\mathbf{\Lambda}} = \mathbf{\Lambda}\mathbf{A}^{-1}$. This implies that the factor covariance matrix $\mathbf{\Phi}$ is indeterminate, and hence we can choose as $\mathbf{\Phi}$ an arbitrary positive-definite matrix. Hence, the factor variances are usually set equal to unity. Then diag $\mathbf{\Phi} = \mathbf{I}_m$, and $\mathbf{\Phi}$ is identical to the correlation matrix. The factor analysis model is called *orthogonal* if $\mathbf{\Phi}$ is taken as $\mathbf{I}_m$ and *oblique* otherwise.

3.2.2. Rotational indeterminacy

Suppose we transform $\mathbf{f}$ as

$$\tilde{\mathbf{f}} = \mathbf{\Phi}^{1/2}\mathbf{T}\mathbf{\Phi}^{-1/2}\mathbf{f},$$

where $\mathbf{T}$ is an orthogonal matrix such that $\mathbf{T}\mathbf{T}' = \mathbf{T}'\mathbf{T} = \mathbf{I}_m$, $\mathbf{\Phi}^{1/2}$ is a square root of $\mathbf{\Phi}$ that satisfies $\mathbf{\Phi} = \mathbf{\Phi}^{1/2}(\mathbf{\Phi}^{1/2})'$, and $\mathbf{\Phi}^{-1/2} = (\mathbf{\Phi}^{1/2})^{-1}$. Then we can easily verify that $\mathrm{Var}(\tilde{\mathbf{f}}) = \mathbf{\Phi}$ and that the factor analysis model still holds by replacing $\mathbf{f}$ and $\mathbf{\Lambda}$ with $\tilde{\mathbf{f}}$ and $\tilde{\mathbf{\Lambda}}$, where $\tilde{\mathbf{\Lambda}} = \mathbf{\Lambda}\mathbf{\Phi}^{1/2}\mathbf{T}'\mathbf{\Phi}^{-1/2}$. This implies that the vector $\mathbf{f}$ of common factors and the factor loading matrix $\mathbf{\Lambda}$ are indeterminate even if we fix $\mathbf{\Phi}$. In the orthogonal model, it follows that the factor loadings are determined only up to permutations and/or changes in sign in the columns of $\mathbf{\Lambda}$. The rotational indeterminacy leads to the process of *factor rotation* for ease of interpretation.

Factor score indeterminacy

We show that, for any set of $\mathbf{\Lambda}$, $\mathbf{\Phi}$, and $\mathbf{\Psi}$, infinitely many different $\tilde{\mathbf{f}}$ and $\tilde{\mathbf{e}}$ exist that satisfy (3.1)–(3.4). Under the factor analysis model, the linear regressions of $\mathbf{f}$ on $\mathbf{x}$ and $\mathbf{e}$ on $\mathbf{x}$ are, in view of (3.6) and (3.7), given by $\mathbf{\Phi}\mathbf{\Lambda}'\mathbf{\Sigma}^{-1}(\mathbf{x} - \boldsymbol{\mu})$ and $\mathbf{\Psi}\mathbf{\Sigma}^{-1}(\mathbf{x} - \boldsymbol{\mu})$. Multiplying $\mathbf{\Psi}^{-1}\mathbf{\Lambda}$ from the right on both sides of (3.5) and letting $\mathbf{\Gamma} = \mathbf{\Lambda}'\mathbf{\Psi}^{-1}\mathbf{\Lambda}$, we have

$$\mathbf{\Sigma}\mathbf{\Psi}^{-1}\mathbf{\Lambda} = \mathbf{\Lambda}(\mathbf{I}_m + \mathbf{\Phi}\mathbf{\Gamma}) = \mathbf{\Lambda}\mathbf{\Phi}\left(\mathbf{\Phi}^{-1} + \mathbf{\Gamma}\right), \tag{3.10}$$

from which we obtain

$$\mathbf{\Sigma}^{-1}\mathbf{\Lambda} = \mathbf{\Psi}^{-1}\mathbf{\Lambda}(\mathbf{I}_m + \mathbf{\Phi}\mathbf{\Gamma})^{-1} = \mathbf{\Psi}^{-1}\mathbf{\Lambda}\left(\mathbf{\Phi}^{-1} + \mathbf{\Gamma}\right)^{-1}\mathbf{\Phi}^{-1}. \tag{3.11}$$

Thus, the covariance matrix, $\mathbf{\Phi}\mathbf{\Lambda}'\mathbf{\Sigma}^{-1}\mathbf{\Lambda}\mathbf{\Phi}$, of the linear regression of $\mathbf{f}$ on $\mathbf{x}$ can be written as

$$\begin{aligned}\mathbf{\Phi}\mathbf{\Lambda}'\mathbf{\Sigma}^{-1}\mathbf{\Lambda}\mathbf{\Phi} &= \mathbf{\Phi}\mathbf{\Gamma}\left(\mathbf{\Phi}^{-1} + \mathbf{\Gamma}\right)^{-1} \\ &= \left\{\mathbf{\Phi}\left(\mathbf{\Phi}^{-1} + \mathbf{\Gamma}\right) - \mathbf{I}_m\right\}\left(\mathbf{\Phi}^{-1} + \mathbf{\Gamma}\right)^{-1} \\ &= \mathbf{\Phi} - \left(\mathbf{\Phi}^{-1} + \mathbf{\Gamma}\right)^{-1},\end{aligned} \tag{3.12}$$

which implies that the covariance matrix of the residual, $\mathbf{f} - \mathbf{\Phi}\mathbf{\Lambda}'\mathbf{\Sigma}^{-1}(\mathbf{x} - \boldsymbol{\mu})$, is given by $(\mathbf{\Phi}^{-1} + \mathbf{\Gamma})^{-1}$.

It can be verified that the factor analysis model given by (3.1) to (3.4) holds by replacing $\mathbf{f}$ and $\mathbf{e}$ with $\tilde{\mathbf{f}}$ and $\tilde{\mathbf{e}}$, where

$$\tilde{\mathbf{f}} = \mathbf{\Phi}\mathbf{\Lambda}'\mathbf{\Sigma}^{-1}(\mathbf{x} - \boldsymbol{\mu}) + \mathbf{u}, \qquad \tilde{\mathbf{e}} = \mathbf{\Psi}\mathbf{\Sigma}^{-1}(\mathbf{x} - \boldsymbol{\mu}) - \mathbf{\Lambda}\mathbf{u},$$

provided that an $m \times 1$ random vector $\mathbf{u}$ is uncorrelated with $\mathbf{x}$ and satisfies $\mathrm{E}(\mathbf{u}) = \mathbf{0}$ and $\mathrm{Var}(\mathbf{u}) = (\boldsymbol{\Phi}^{-1} + \boldsymbol{\Gamma})^{-1}$. Factor score indeterminacy does not exist when the common factors are nonrandom. For the case of a singular $\boldsymbol{\Psi}$, see Krijnen (2002). Factor score indeterminacy has been and is still a source of considerable debate (Maraun, 1996).

3.3. Existence of the decomposition and its uniqueness

3.3.1. Existence of the decomposition $\boldsymbol{\Sigma} = \boldsymbol{\Lambda}\boldsymbol{\Lambda}' + \boldsymbol{\Psi}$

If $\boldsymbol{\Sigma}$ is a covariance matrix of a factor analysis model, then $\boldsymbol{\Sigma}$ has a decomposition of the form given in (3.5). We would like to know under what conditions a positive-definite matrix $\boldsymbol{\Sigma}$ has such a decomposition with m specified. Because of the indeterminacy of the factor covariance matrix $\boldsymbol{\Phi}$, we may assume, without loss of generality, that $\boldsymbol{\Phi} = \mathbf{I}_m$ and consider the decomposition

$$\boldsymbol{\Sigma} = \boldsymbol{\Lambda}\boldsymbol{\Lambda}' + \boldsymbol{\Psi}. \tag{3.13}$$

Then a necessary and sufficient condition for $\boldsymbol{\Sigma}$ to be a covariance matrix of a m-factor model is that a diagonal matrix $\boldsymbol{\Psi}$ exists with positive diagonal elements such that $\boldsymbol{\Sigma} - \boldsymbol{\Psi}$ is positive-semidefinite with $\mathrm{rank}(\boldsymbol{\Sigma} - \boldsymbol{\Psi}) = m$.

First, we consider the case $m = 1$. Then $\mathrm{rank}(\boldsymbol{\Sigma} - \boldsymbol{\Psi}) = 1$, and all the second-order minors of $\boldsymbol{\Sigma} - \boldsymbol{\Psi}$ are zero. Hence, all the second-order minors of $\boldsymbol{\Sigma}$ that do not include a diagonal element, called *tetrads*, must vanish.

PROPOSITION 1 (Theorem 1 of Bekker and de Leeuw, 1987). *If a positive-definite matrix $\boldsymbol{\Sigma}$ cannot be brought, by permutations, into block diagonal form, then the minimum rank of $\boldsymbol{\Sigma} - \boldsymbol{\Psi}$, where $0 \leqslant \psi_i \leqslant \sigma_{ii}$ $(i = 1, \ldots, p)$, is one if and only if, after sign changes of rows and corresponding columns, all its elements are positive such that*

$$\sigma_{ik}\sigma_{j\ell} - \sigma_{i\ell}\sigma_{jk} = 0, \qquad \sigma_{ik}\sigma_{ij} - \sigma_{ii}\sigma_{jk} \leqslant 0$$

for all quadruples $(i \neq j, k, \ell;\ j \neq k, \ell;\ k \neq \ell)$.

Thus, if $\boldsymbol{\Sigma}$ is a covariance matrix of a one-factor model, then ψ_i $(i = 1, \ldots, p)$ is determined by the elements of $\boldsymbol{\Sigma}$ as

$$\psi_i = \sigma_{ii} - \frac{\sigma_{ik}\sigma_{ji}}{\sigma_{jk}} \quad (i \neq j, k;\ j \neq k;\ \sigma_{jk} \neq 0), \tag{3.14}$$

which is obtained by setting a second-order minor of $\boldsymbol{\Sigma} - \boldsymbol{\Psi}$ that includes one diagonal element, $\sigma_{ii} - \psi_i$, equal to zero.

In practice, m is tacitly assumed to be sufficiently smaller than p. In fact, if we take $m = p - 1$, then any positive-definite matrix $\boldsymbol{\Sigma}$ has the decomposition given in (3.13), as shown below. Let θ_p be the smallest eigenvalue of $\boldsymbol{\Sigma}$, and define a positive-definite matrix $\boldsymbol{\Sigma}(\epsilon)$ as

$$\boldsymbol{\Sigma}(\epsilon) = \boldsymbol{\Sigma} - \epsilon\mathbf{I}_p,$$

where $0 < \epsilon < \theta_p$. Let $\boldsymbol{\Sigma}(\epsilon) = \mathbf{L}\mathbf{L}'$, where $\mathbf{L} = (\ell_{ij})$ is a lower diagonal matrix, and let $\boldsymbol{\Lambda}(\epsilon)$ be a $p \times (p - 1)$ matrix consisting of the first $p - 1$ columns of $\mathbf{L}$. Further, let

$\mathbf{\Psi}(\epsilon)$ be a diagonal matrix defined by

$$\mathbf{\Psi}(\epsilon) = \epsilon \mathbf{I}_p + \begin{pmatrix} 0 & \cdots & 0 \\ \vdots & \ddots & \vdots \\ 0 & \cdots & \ell_{pp}^2 \end{pmatrix}.$$

Then we have a decomposition of $\mathbf{\Sigma}$ where

$$\mathbf{\Sigma} = \mathbf{\Lambda}(\epsilon)\mathbf{\Lambda}(\epsilon)' + \mathbf{\Psi}(\epsilon).$$

However, infinitely many such decompositions exist, depending on ϵ.

Then the question naturally arises as to what extent the rank of a positive-definite matrix $\mathbf{\Sigma}$ is reduced by changing its diagonal elements. Ledermann (1937) suggested that the rank of $\mathbf{\Sigma}$ is generally reduced to an integer not smaller than $L(p)$, where $L(p)$ is given by

$$L(p) = \frac{2p + 1 - \sqrt{8p + 1}}{2} \tag{3.15}$$

and is known as the *Ledermann bound*. Guttman (1958) gave a counterexample to $L(p)$ by pointing out that the rank of a tridiagonal matrix cannot be reduced below $p - 1$ by changing its diagonal elements.

The number of nonduplicated elements in $\mathbf{\Sigma}$ is $p(p+1)/2$. The number of parameters in $\mathbf{\Lambda}$ and $\mathbf{\Psi}$ is $p(m+1)$, but $m(m-1)/2$ restrictions are required to eliminate rotational indeterminacy from $\mathbf{\Lambda}$. Hence, one practical meaning of the Ledemann bound is that if m is chosen so that $m > L(p)$, then

$$p(m + 1) - \frac{m(m - 1)}{2} > \frac{p(p + 1)}{2},$$

which implies that the number of the free parameters in the factor analysis model is larger than that in the "saturated" model.

3.3.2. *Uniqueness of the decomposition* $\mathbf{\Sigma} = \mathbf{\Lambda}\mathbf{\Lambda}' + \mathbf{\Psi}$

Next, we consider the uniqueness of the decomposition $\mathbf{\Sigma} = \mathbf{\Lambda}\mathbf{\Lambda}' + \mathbf{\Psi}$ except for the rotational indeterminacy of $\mathbf{\Lambda}$. The uniqueness of the decomposition is closely related to the existence of consistent estimators because if the decomposition is not unique, then a consistent estimator cannot exist.

Suppose that, by a suitable rotation, the mth column, $\boldsymbol{\lambda}_m$, of $\mathbf{\Lambda}$ contains only one nonzero element, λ_{im}. In such a case, the mth factor affects only x_i and is indeed a unique factor. The decomposition can be shown to hold by replacing λ_{im} and ψ_i with $\tilde{\lambda}_{im}$ and $\tilde{\psi}_i$ (> 0) that satisfy

$$\tilde{\lambda}_{im}^2 + \tilde{\psi}_i = \lambda_{im}^2 + \psi_i.$$

Next, suppose that, by a suitable rotation, the mth column, $\boldsymbol{\lambda}_m$, of $\mathbf{\Lambda}$ contains two nonzero elements, that is, $\boldsymbol{\lambda}_m = (0, \ldots, \lambda_{im}, \ldots, \lambda_{jm}, \ldots, 0)'$. In this case, the mth factor affects x_i and x_j and is certainly a common factor, but λ_{im} and λ_{jm} affect only σ_{ij}.

Let $\tilde{\lambda}_{im} = \delta\lambda_{im}$ and $\tilde{\lambda}_{jm} = \delta^{-1}\lambda_{jm}$. It is always possible to find

$$\tilde{\psi}_i = \psi_i + \left(\lambda_{im}^2 - \tilde{\lambda}_{im}^2\right), \qquad \tilde{\psi}_j = \psi_j + \left(\lambda_{jm}^2 - \tilde{\lambda}_{jm}^2\right)$$

that satisfy

$$0 < \tilde{\psi}_i < \sigma_{ii}, \qquad 0 < \tilde{\psi}_j < \sigma_{jj}$$

if we choose δ close to one. Then σ_{ij} is unchanged because $\tilde{\lambda}_{im}\tilde{\lambda}_{jm} = \lambda_{im}\lambda_{jm}$. Hence, the decomposition still can be shown to hold by replacing λ_{im}, λ_{jm}, ψ_i, and ψ_j by $\tilde{\lambda}_{im}$, $\tilde{\lambda}_{jm}$, $\tilde{\psi}_i$, and $\tilde{\psi}_j$.

PROPOSITION 2 (Theorem 5.6 of Anderson and Rubin, 1956). *A necessary condition for uniqueness of the decomposition* $\boldsymbol{\Sigma} = \boldsymbol{\Lambda}\boldsymbol{\Lambda}' + \boldsymbol{\Psi}$ *is that each column of* $\boldsymbol{\Lambda}\mathbf{A}$ *has at least three nonzero elements for every nonsingular* $\mathbf{A}$.

This proposition implies that the condition

$$\operatorname{rank}(\boldsymbol{\Lambda}) = m$$

is necessary for the uniqueness of the decomposition $\boldsymbol{\Sigma} = \boldsymbol{\Lambda}\boldsymbol{\Lambda}' + \boldsymbol{\Psi}$.

PROPOSITION 3 (Theorem 5.1 of Anderson and Rubin, 1956). *A sufficient condition for uniqueness of the decomposition* $\boldsymbol{\Sigma} = \boldsymbol{\Lambda}\boldsymbol{\Lambda}' + \boldsymbol{\Psi}$ *up to multiplication on the right of* $\boldsymbol{\Lambda}$ *by an orthogonal matrix is that if any row of* $\boldsymbol{\Lambda}$ *is deleted, then two disjoint submatrices of rank m remain.*

We shall give a proof of Ihara and Kano (1986), which led them to propose a non-iterative estimator of unique variance. Their proof is simpler than the original one. Assume that a decomposition $\boldsymbol{\Sigma} = \boldsymbol{\Lambda}\boldsymbol{\Lambda}' + \boldsymbol{\Psi}$ exists and that $\boldsymbol{\Lambda}$ satisfies the condition of the theorem. Let $\boldsymbol{\Sigma} = \mathbf{L}\mathbf{L}' + \mathbf{D}$ be another decomposition where $\mathbf{L}$ is a $p \times m$ matrix and $\mathbf{D}$ is a diagonal matrix. To prove the theorem, we show that the diagonal elements of $\mathbf{L}\mathbf{L}'$ are equal to those of $\boldsymbol{\Lambda}\boldsymbol{\Lambda}'$. First, rearrange the rows of $\boldsymbol{\Lambda}$ and $\mathbf{L}$; similarly, rearrange the rows and columns of $\boldsymbol{\Sigma}$. Let

$$\boldsymbol{\Lambda} = \begin{pmatrix} \boldsymbol{\lambda}_1 \\ \boldsymbol{\Lambda}_2 \\ \boldsymbol{\Lambda}_3 \\ \boldsymbol{\Lambda}_4 \end{pmatrix}, \qquad \mathbf{L} = \begin{pmatrix} \boldsymbol{\ell}_1 \\ \mathbf{L}_2 \\ \mathbf{L}_3 \\ \mathbf{L}_4 \end{pmatrix}, \qquad \boldsymbol{\Sigma} = \begin{pmatrix} \sigma_{11} & \boldsymbol{\sigma}_{12} & \boldsymbol{\sigma}_{13} & \boldsymbol{\sigma}_{14} \\ \boldsymbol{\sigma}_{21} & \boldsymbol{\Sigma}_{22} & \boldsymbol{\Sigma}_{23} & \boldsymbol{\Sigma}_{24} \\ \boldsymbol{\sigma}_{31} & \boldsymbol{\Sigma}_{32} & \boldsymbol{\Sigma}_{33} & \boldsymbol{\Sigma}_{34} \\ \boldsymbol{\sigma}_{41} & \boldsymbol{\Sigma}_{42} & \boldsymbol{\Sigma}_{43} & \boldsymbol{\Sigma}_{44} \end{pmatrix},$$

where $\boldsymbol{\lambda}_1$ is the first row of $\boldsymbol{\Lambda}$ and $\boldsymbol{\Lambda}_2$ and $\boldsymbol{\Lambda}_3$ are nonsingular matrices of order $m \times m$. Because $\operatorname{nondiag} \mathbf{L}\mathbf{L}' = \operatorname{nondiag} \boldsymbol{\Lambda}\boldsymbol{\Lambda}'$, we have

$$\boldsymbol{\sigma}_{12} = \boldsymbol{\lambda}_1\boldsymbol{\Lambda}_2' = \boldsymbol{\ell}_1\mathbf{L}_2', \qquad \boldsymbol{\sigma}_{31} = \boldsymbol{\Lambda}_3\boldsymbol{\lambda}_1' = \mathbf{L}_3\boldsymbol{\ell}_1', \qquad \boldsymbol{\Sigma}_{32} = \boldsymbol{\Lambda}_3\boldsymbol{\Lambda}_2' = \mathbf{L}_3\mathbf{L}_2'.$$

Because $\boldsymbol{\Lambda}_2$ and $\boldsymbol{\Lambda}_3$ are nonsingular, so is $\boldsymbol{\Sigma}_{32} = \boldsymbol{\Lambda}_3\boldsymbol{\Lambda}_2' = \mathbf{L}_3\mathbf{L}_2'$, which implies that $\mathbf{L}_2$ and $\mathbf{L}_3$ are nonsingular. Then

$$\boldsymbol{\ell}_1\boldsymbol{\ell}_1' = \boldsymbol{\ell}_1\mathbf{L}_2'\left(\mathbf{L}_2'\right)^{-1}\mathbf{L}_3^{-1}\mathbf{L}_3\boldsymbol{\ell}_1' = \boldsymbol{\ell}_1\mathbf{L}_2'\left(\mathbf{L}_3\mathbf{L}_2'\right)^{-1}\mathbf{L}_3\boldsymbol{\ell}_1'$$

$$= \boldsymbol{\sigma}_{12}\boldsymbol{\Sigma}_{32}^{-1}\boldsymbol{\sigma}_{31}$$
$$= \boldsymbol{\lambda}_1\boldsymbol{\Lambda}_2'(\boldsymbol{\Lambda}_3\boldsymbol{\Lambda}_2')^{-1}\boldsymbol{\Lambda}_3\boldsymbol{\lambda}_1' = \boldsymbol{\lambda}_1\boldsymbol{\Lambda}_2'(\boldsymbol{\Lambda}_2')^{-1}\boldsymbol{\Lambda}_3^{-1}\boldsymbol{\Lambda}_3\boldsymbol{\lambda}_1' = \boldsymbol{\lambda}_1\boldsymbol{\lambda}_1'.$$

In a similar way, we can show that the diagonal elements of $\mathbf{LL}'$ are equal to those of $\boldsymbol{\Lambda}\boldsymbol{\Lambda}'$. This condition implies that $p \geqslant 2m+1$. Anderson and Rubin (1956) showed that the above condition is also necessary for $m = 1$ and $m = 2$.

Ihara and Kano (1986) showed that if $\boldsymbol{\Lambda}$ satisfies the condition of Proposition 3, then each of the unique variances can be written, after rearranging the rows and columns of $\boldsymbol{\Sigma}$, as

$$\psi_1 = \sigma_{11} - \boldsymbol{\sigma}_{12}\boldsymbol{\Sigma}_{32}^{-1}\boldsymbol{\sigma}_{31}, \tag{3.16}$$

of which (3.14) is a special case.

3.4. Properties of the model

3.4.1. Decomposition of $\boldsymbol{\Sigma}^{-1}$

We shall show that if $\boldsymbol{\Sigma}$ is a covariance matrix of a factor analysis model, then $\boldsymbol{\Sigma}^{-1}$ also has a decomposition. In view of (3.7), the linear regression of $\mathbf{e}$ on $\mathbf{x}$ is given by $\boldsymbol{\Psi}\boldsymbol{\Sigma}^{-1}(\mathbf{x}-\boldsymbol{\mu})$, and its covariance matrix by $\boldsymbol{\Psi}\boldsymbol{\Sigma}^{-1}\boldsymbol{\Psi}$. Then

$$\boldsymbol{\Psi}\boldsymbol{\Sigma}^{-1}\boldsymbol{\Psi} = (\boldsymbol{\Sigma} - \boldsymbol{\Lambda}\boldsymbol{\Phi}\boldsymbol{\Lambda}')\boldsymbol{\Sigma}^{-1}(\boldsymbol{\Sigma} - \boldsymbol{\Lambda}\boldsymbol{\Phi}\boldsymbol{\Lambda}')$$
$$= \boldsymbol{\Psi} - \boldsymbol{\Lambda}\boldsymbol{\Phi}\boldsymbol{\Lambda}' + \boldsymbol{\Lambda}\boldsymbol{\Phi}\boldsymbol{\Lambda}'\boldsymbol{\Sigma}^{-1}\boldsymbol{\Lambda}\boldsymbol{\Phi}\boldsymbol{\Lambda}'.$$

By using (3.12), we can rewrite $\boldsymbol{\Lambda}\boldsymbol{\Phi}\boldsymbol{\Lambda}'\boldsymbol{\Sigma}^{-1}\boldsymbol{\Lambda}\boldsymbol{\Phi}\boldsymbol{\Lambda}'$ as

$$\boldsymbol{\Lambda}\boldsymbol{\Phi}\boldsymbol{\Lambda}'\boldsymbol{\Sigma}^{-1}\boldsymbol{\Lambda}\boldsymbol{\Phi}\boldsymbol{\Lambda}' = \boldsymbol{\Lambda}\boldsymbol{\Phi}\boldsymbol{\Lambda}' - \boldsymbol{\Lambda}(\boldsymbol{\Phi}^{-1} + \boldsymbol{\Gamma})^{-1}\boldsymbol{\Lambda}',$$

where $\boldsymbol{\Gamma} = \boldsymbol{\Lambda}'\boldsymbol{\Psi}^{-1}\boldsymbol{\Lambda}$, and we have

$$\boldsymbol{\Psi}\boldsymbol{\Sigma}^{-1}\boldsymbol{\Psi} = \boldsymbol{\Psi} - \boldsymbol{\Lambda}(\boldsymbol{\Phi}^{-1} + \boldsymbol{\Gamma})^{-1}\boldsymbol{\Lambda}'.$$

Multiplying $\boldsymbol{\Psi}^{-1}$ from the left and right on both sides of the above equation, we obtain

$$\boldsymbol{\Sigma}^{-1} = \boldsymbol{\Psi}^{-1} - \boldsymbol{\Psi}^{-1}\boldsymbol{\Lambda}(\boldsymbol{\Phi}^{-1} + \boldsymbol{\Gamma})^{-1}\boldsymbol{\Lambda}'\boldsymbol{\Psi}^{-1}. \tag{3.17}$$

In the case of the orthogonal model, we take $\boldsymbol{\Phi} = \mathbf{I}_m$ and obtain

$$\boldsymbol{\Sigma}^{-1} = \boldsymbol{\Psi}^{-1} - \boldsymbol{\Psi}^{-1}\boldsymbol{\Lambda}(\mathbf{I}_m + \boldsymbol{\Gamma})^{-1}\boldsymbol{\Lambda}'\boldsymbol{\Psi}^{-1}. \tag{3.18}$$

3.4.2. Lower bounds of communalities

Suppose that the common factors are uncorrelated so that $\boldsymbol{\Phi} = \mathbf{I}_m$, and consider the decomposition of the population correlation matrix $\mathbf{P} = (\rho_{ij})$:

$$\mathbf{P} = \boldsymbol{\Lambda}^*\boldsymbol{\Lambda}^{*\prime} + \boldsymbol{\Psi}^*,$$

where $\boldsymbol{\Lambda}^* = (\lambda_{ik}^*)$ is a $p \times m$ matrix of standardized factor loadings and $\boldsymbol{\Psi}^*$ is a diagonal matrix whose diagonal elements ψ_i^* are the standardized unique variances. Then it follows from (3.13) that $\mathbf{P}^{-1} = (\rho^{ij})$ has the decomposition

$$\mathbf{P}^{-1} = (\boldsymbol{\Psi}^*)^{-1} - (\boldsymbol{\Psi}^*)^{-1}\boldsymbol{\Lambda}^*(\mathbf{I} + \boldsymbol{\Lambda}^{*\prime}(\boldsymbol{\Psi}^*)^{-1}\boldsymbol{\Lambda}^*)^{-1}\boldsymbol{\Lambda}^{*\prime}(\boldsymbol{\Psi}^*)^{-1}.$$

Let h_i^2 be the communality of the ith variable and R_i^2 be the squared multiple correlation (SMC) of x_i with the remaining $p-1$ variables. It is known that R_i^2 can be written as

$$R_i^2 = 1 - \frac{1}{\rho^{ii}}.$$

Then we can easily see from the above decomposition of $\mathbf{P}^{-1}$ that $\rho^{ii} \geqslant \psi_i^{*-1} > 0$, and hence

$$R_i^2 = 1 - \frac{1}{\rho^{ii}} \leqslant 1 - \psi_i^* = h_i^2.$$

SMC has long been known as a lower bound to communality (Roff, 1936).

Yanai and Ichikawa (1990) proposed a new lower bound to communality. Suppose that the variables are ordered so that $h_1^2 \geqslant \cdots \geqslant h_p^2$, and let $\theta_1^* \geqslant \cdots \geqslant \theta_p^*$ be the ordered eigenvalues of $\mathbf{P}$. They showed that

$$h_1^2 \geqslant 1 - \theta_p^* \geqslant R_p^2; \tag{3.19}$$

that is, at least for the variable with the largest communality, $1 - \theta_p^*$ is a stronger lower bound of communality. In practice, however, to which variable this lower bound applies is unknown. However, when $m = 1$ and the decomposition of $\mathbf{P}$ is unique, then it can be shown that $h_i^2 \geqslant h_j^2$ and $R_i^2 \geqslant R_j^2$ $(i \neq j)$ are equivalent.

3.4.3. Lower bounds of the number of factors

If $\mathbf{P} = (\rho_{ij})$ is a correlation matrix of a factor analysis model, then it has the decompostion $\mathbf{P} = \mathbf{\Lambda}^*\mathbf{\Lambda}^{*\prime} + \mathbf{\Psi}^*$, and $\mathbf{P} - \mathbf{\Psi}^*$ is a positive-semidefinite matrix with rank $(\mathbf{P} - \mathbf{\Psi}^*) = m$, the number of factors. Hence, m is equal to the number of positive eigenvalues of $\mathbf{P} - \mathbf{\Psi}^*$, of which the diagonal elements are the communalities h_i^2 $(i = 1, \ldots, p)$.

Guttman (1954) gave lower bounds to the number of factors by using the following lemma.

LEMMA 1. *Let* $\mathbf{A}$ *and* $\mathbf{B}$ *be a real symmetric matrix and a positive-semidefinite matrix. Let the* i*th largest eigenvalue of a real symmetric matrix be denoted by* $\mu_i(\cdot)$. *Then*

$$\mu_i(\mathbf{A}) \leqslant \mu_i(\mathbf{A} + \mathbf{B}). \tag{3.20}$$

Let us define $\rho_{i\cdot}^2$ by $\rho_{i\cdot}^2 = \max_{j \neq i} \rho_{ij}^2$, the largest squared correlation among the nondiagonal elements of the ith row of $\mathbf{P}$, and denote by R_i^2 the squared multiple correlation of the ith variable with the remaining $p-1$ variables. Let $\mathbf{P}_1$, $\mathbf{P}_2$, and $\mathbf{P}_3$ be matrices obtained from $\mathbf{P}$ by replacing each of the diagonal elements of $\mathbf{P}$ by 0, $\rho_{i\cdot}^2$, and R_i^2. Moreover, denote the numbers of positive eigenvalues of $\mathbf{P}_1$, $\mathbf{P}_2$, and $\mathbf{P}_3$ by m_1, m_2, and m_3. Because

$$0 \leqslant \rho_{i\cdot}^2 \leqslant R_i^2 \leqslant h_i^2 \quad (i = 1, \ldots, p),$$

it follows from Lemma 1 that

$$m_1 \leqslant m_2 \leqslant m_3 \leqslant m. \tag{3.21}$$

These lower bounds are sometimes used as algebraic criteria for the number of factors. We note that m_1 is equal to the number of the eigenvalues of $\mathbf{P}$ that are greater than unity.

3.5. Related models

3.5.1. Principal component analysis

Principal component analysis was proposed by Hotelling (1933) as a method of analyzing a set of correlated variables. Let $\mathbf{x} = (x_1, \ldots, x_p)'$ be a random vector with $\mathrm{E}(\mathbf{x}) = \boldsymbol{\mu}$ and $\mathrm{Var}(\mathbf{x}) = \boldsymbol{\Sigma}$. The first principal component g_1 is defined as the linear combination

$$g_1 = \boldsymbol{\beta}_1'(\mathbf{x} - \boldsymbol{\mu})$$

that maximizes the variance,

$$\mathrm{Var}(g_1) = \boldsymbol{\beta}_1' \boldsymbol{\Sigma} \boldsymbol{\beta}_1,$$

subject to the constraint that $|\boldsymbol{\beta}_1| = 1$. In general, the kth principal component g_k is defined as the linear combination $g_k = \boldsymbol{\beta}_k'(\mathbf{x} - \boldsymbol{\mu})$ that has the maximum variance, subject to the constraints that $|\boldsymbol{\beta}_k| = 1$ and that g_k is uncorrelated with $g_1, \ldots, g_{k-1}$. Then it can be shown that the variance of g_k is equal to the kth largest eigenvalue, θ_k, of $\boldsymbol{\Sigma}$ and that $\boldsymbol{\beta}_k$ is the corresponding eigenvector of unit length. Hence, $\mathbf{g} = (g_1, \ldots, g_p)$ can be written as

$$\mathbf{g} = \mathbf{B}'(\mathbf{x} - \boldsymbol{\mu}),$$

where $\mathbf{B} = (\boldsymbol{\beta}_1, \ldots, \boldsymbol{\beta}_p)$.

Let us partition $\mathbf{g} = (\mathbf{g}_1', \mathbf{g}_2')'$ and $\mathbf{B} = (\mathbf{B}_1, \mathbf{B}_2)$, where the elements of $\mathbf{g}_1$ and $\mathbf{g}_2$ are the first m and the last $p - m$ principal components and the columns of $\mathbf{B}_1$ $(p \times m)$ and $\mathbf{B}_2$ $(p \times (p - m))$ consist of the corresponding columns of $\mathbf{B}$. Further, let $\boldsymbol{\Theta}_1$ $(m \times m)$ denote a diagonal matrix whose (k, k) element is the kth largest eigenvalue θ_k of $\boldsymbol{\Sigma}$. If $g_1, \ldots, g_m$ account for most of the variation in $\mathbf{x}$, we may write

$$\mathbf{x} = \boldsymbol{\mu} + \tilde{\mathbf{B}}_1 \tilde{\mathbf{g}}_1 + \mathbf{u}, \tag{3.22}$$

where $\tilde{\mathbf{B}}_1 = \mathbf{B}_1 \boldsymbol{\Theta}_1^{1/2}$ is the *principal component loading matrix*, $\tilde{\mathbf{g}}_1 = \boldsymbol{\Theta}_1^{-1/2} \mathbf{g}_1$, and $\mathbf{u} = \mathbf{B}_2'(\mathbf{x} - \boldsymbol{\mu})$. It can be easily verified that $\mathrm{Var}(\tilde{\mathbf{g}}_1) = \mathbf{I}_m$ and $\mathrm{Cov}(\tilde{\mathbf{g}}_1, \mathbf{u}) = \mathbf{0}$. As a substitute for factor analysis, (3.22), which expresses $\mathbf{x}$ in terms of $\tilde{\mathbf{B}}_1$, $\tilde{\mathbf{g}}_1$, and $\mathbf{u}$, may be used. Note, however, that unlike the unique factors of the factor analysis model, u_i and u_j $(i \neq j)$ in (3.22) are correlated in general.

The factor analysis model introduces $p+m$ extra variables to "explain" the observed variables $x_1, \ldots, x_p$ and assumes that each observed variable x_i is a linear combination of the common factors $f_1, \ldots, f_m$ and the unique factor e_i. On the other hand, the principal components are, by definition, linear combinations of the observed variables. Bollen (2002) reviews the use of latent variable models, of which factor analysis is a special case. His third definition of a latent variable is that a latent variable is a nondeterministic function of observed variables. Although principal components are unobservable variables in the sense that they are not present in the data set, they are not latent variables.

In practice, the results of factor analysis and principal component analysis may be close to each other. Bentler and Kano (1990) proved that if a single factor model holds for all p, then the first principal component and the common factor and their corresponding loading vectors become identical as $p \to \infty$. Schneeweiss and Mathes (1995) studied the conditions under which the results of principal component analysis approach those of factor analysis. Jackson (1991) and Jollife (2002) focused on principal component analysis in comparison with factor analysis.

3.5.2. Image theory

Image theory was proposed by Guttman (1953) as an alternative to the factor analysis model. Let $\mathbf{x} = (x_1, \ldots, x_p)'$ be a random vector with $\mathrm{E}(\mathbf{x}) = \mathbf{0}$ and $\mathrm{E}(\mathbf{x}\mathbf{x}') = \boldsymbol{\Sigma}$ ($>\mathbf{0}$), and consider the decomposition

$$\mathbf{x} = \mathbf{x}_I + \mathbf{x}_A,$$

where the ith element, x_{Ii}, of $\mathbf{x}_I$ is the linear regression of x_i on the remaining $p-1$ variables and is called the *image* of x_i. The ith element, x_{Ai}, of $\mathbf{x}_A$, on the other hand, is the residual and is called the *anti-image* of x_i. This is analogous to the decomposition of each x_i into the sum of the common part t_i and the unique factor e_i in the factor analysis model.

We will derive explicit expressions and then investigate the properties of $\mathbf{x}_I$ and $\mathbf{x}_A$. Let us partition $\mathbf{x}$ as $\mathbf{x} = (x_1, \mathbf{x}_2')'$ and accordingly, $\boldsymbol{\Sigma}$ and $\boldsymbol{\Sigma}^{-1}$ as

$$\boldsymbol{\Sigma} = \begin{pmatrix} \sigma_{11} & \boldsymbol{\sigma}_{12} \\ \boldsymbol{\sigma}_{21} & \boldsymbol{\Sigma}_{22} \end{pmatrix}, \qquad \boldsymbol{\Sigma}^{-1} = \begin{pmatrix} \sigma^{11} & \boldsymbol{\sigma}^{12} \\ \boldsymbol{\sigma}^{21} & \boldsymbol{\Sigma}^{22} \end{pmatrix}.$$

From the formulae for the inverse of a partioned matrix (see, for example, Magnus and Neudecker, 1999, p. 11), $\boldsymbol{\sigma}^{12}$ can be written as

$$\boldsymbol{\sigma}^{12} = -\sigma^{11}\boldsymbol{\sigma}_{12}\boldsymbol{\Sigma}_{22}^{-1}.$$

Then the image of x_1 is given by

$$\boldsymbol{\sigma}_{12}\boldsymbol{\Sigma}_{22}^{-1}\mathbf{x}_2 = -\frac{1}{\sigma^{11}}\boldsymbol{\sigma}^{12}\mathbf{x}_2 = x_1 - \frac{1}{\sigma^{11}}(\sigma^{11}, \boldsymbol{\sigma}^{12})\begin{pmatrix} x_1 \\ \mathbf{x}_2 \end{pmatrix}.$$

Thus, we have

$$\mathbf{x}_I = (\mathbf{I}_p - \boldsymbol{\Delta}\boldsymbol{\Sigma}^{-1})\mathbf{x}, \qquad \mathbf{x}_A = \boldsymbol{\Delta}\boldsymbol{\Sigma}^{-1}\mathbf{x}, \tag{3.23}$$

where $\boldsymbol{\Delta} = (\mathrm{diag}\,\boldsymbol{\Sigma}^{-1})^{-1}$. Yanai and Mukherjee (1987) showed that even if $\boldsymbol{\Sigma}$ is singular, then $\mathbf{x}_I$ and $\mathbf{x}_A$ are uniquely determined by replacing $\boldsymbol{\Sigma}^{-1}$ with $\boldsymbol{\Sigma}^{-}$, a generalized inverse of $\boldsymbol{\Sigma}$, where $\boldsymbol{\Sigma}^{-}$ is any matrix that satisfies $\boldsymbol{\Sigma}\boldsymbol{\Sigma}^{-}\boldsymbol{\Sigma} = \boldsymbol{\Sigma}$.

The following results are easily verified.

$$\mathrm{Var}(\mathbf{x}_I) = \boldsymbol{\Sigma} + \boldsymbol{\Delta}\boldsymbol{\Sigma}^{-1}\boldsymbol{\Delta} - 2\boldsymbol{\Delta}, \tag{3.24}$$

$$\mathrm{Var}(\mathbf{x}_A) = \boldsymbol{\Delta}\boldsymbol{\Sigma}^{-1}\boldsymbol{\Delta}, \tag{3.25}$$

$$\mathrm{Cov}(\mathbf{x}_I, \mathbf{x}_A) = \mathbf{\Delta} - \mathbf{\Delta}\mathbf{\Sigma}^{-1}\mathbf{\Delta}. \tag{3.26}$$

Using (3.24)–(3.26), we immediately have

$$\mathrm{Var}(x_j) = \mathrm{Var}(x_{Aj}) + \mathrm{Var}(x_{Ij}),$$
$$\mathrm{Cov}(x_i, x_j) = \mathrm{Cov}(x_{Ii}, x_{Ij}) + \mathrm{Cov}(x_{Ai}, x_{Aj}).$$

It is interesting to note that the vector of anti-image variables of $\mathbf{x}_A$ is equal to $\mathbf{x}$ itself, that is,

$$\begin{aligned}
(\mathbf{x}_A)_A &= \left(\mathrm{diag}\left(\mathrm{Var}(\mathbf{x}_A)^{-1}\right)\right)^{-1} \mathrm{Var}(\mathbf{x}_A)\mathbf{x}_A \\
&= \left(\mathrm{diag}\left((\mathbf{\Delta}\mathbf{\Sigma}^{-1}\mathbf{\Delta})^{-1}\right)\right)^{-1}(\mathbf{\Delta}\mathbf{\Sigma}^{-1}\mathbf{\Delta})^{-1}\mathbf{\Delta}\mathbf{\Sigma}^{-1}\mathbf{x} \\
&= \left(\mathbf{\Delta}\,\mathrm{diag}(\mathbf{\Sigma}^{-1})\mathbf{\Delta}\right)\left(\mathbf{\Delta}^{-1}\mathbf{\Sigma}\mathbf{\Delta}^{-1}\right)\mathbf{\Delta}\mathbf{\Sigma}^{-1}\mathbf{x} \\
&= \mathbf{\Delta}\,\mathrm{diag}(\mathbf{\Sigma}^{-1})\mathbf{x} \\
&= \mathbf{x}.
\end{aligned}$$

In a psychological study, the variables usually represent tests and are regarded as a more or less arbitrary selection of variables from a collection of large number of variables, or the universe of variables. Guttman (1955) showed an interesting relation between the image theory and the factor analysis model: if an m-factor model holds for all p, then

$$t_i \to x_{Ii} \quad (i = 1, \ldots, p) \quad \text{and} \quad \mathbf{\Sigma}^{-1} \to \mathbf{\Psi}^{-1}$$

as $p \to \infty$. This implies that both models are identical and that the common factors are determinate in the infinite universe of variables. This also implies that the corresponding elements of $\mathbf{\Psi}$ and $(\mathrm{diag}\ \mathbf{\Sigma}^{-1})^{-1}$ tend to the same limit. Motivated by this observation, Jöreskog proposed, as an alternative to the standard factor analysis model, an image factor analysis model in which the covariance matrix of unique factors have a special form, $\mathbf{\Psi} = \theta\,(\mathrm{diag}\ \mathbf{\Sigma}^{-1})^{-1}$, where θ is a positive scalar. See, for example, Jöreskog (1969a). Kaiser and Caffrey (1965) made distinction between statistical inference and psychometric inference; the latter makes inference from a selection of varlables to the universe of variables. Then they proposed alpha factor analysis based on a psychometric concept of generalizability.

4. Statistical inference in factor analysis

4.1. Normal theory maximum likelihood method

4.1.1. Likelihood equations

Let $\{\mathbf{x}_1, \ldots, \mathbf{x}_N\}$ be a random sample of size N from a multivariate normal population $\mathrm{N}(\boldsymbol{\mu}, \mathbf{\Sigma})$ for which an m-factor model holds. Because we have no information about parameter values, we may assume, for convenience, that $\mathbf{\Phi} = \mathbf{I}_m$. Let $\mathbf{S}$ be the usual unbiased estimator of $\mathbf{\Sigma}$ defined by

$$\mathbf{S} = \frac{1}{n}\sum_{\alpha=1}^{N}(\mathbf{x}_\alpha - \bar{\mathbf{x}})(\mathbf{x}_\alpha - \bar{\mathbf{x}})',$$

where $n = N - 1$ and $\bar{\mathbf{x}} = N^{-1}\sum_{\alpha=1}^{N}\mathbf{x}_\alpha$. A common practice in factor analysis is to base inference only on $\mathbf{S}$, which follows a Wishart distribution. We can easily see that maximizing the Wishart likelihood function is equivalent to minimizing the following discrepancy function of the maximum likelihood (ML) method:

$$F(\mathbf{\Sigma};\mathbf{S}) = \operatorname{tr}\mathbf{S}\mathbf{\Sigma}^{-1} - \log\left|\mathbf{S}\mathbf{\Sigma}^{-1}\right| - p, \tag{4.1}$$

where $\mathbf{\Sigma} = \mathbf{\Lambda}\mathbf{\Lambda}' + \mathbf{\Psi}$. Then

$$\mathrm{d}F = \operatorname{tr}\mathbf{\Sigma}^{-1}\,\mathrm{d}\mathbf{\Sigma} - \operatorname{tr}\mathbf{\Sigma}^{-1}\mathbf{S}\mathbf{\Sigma}^{-1}\,\mathrm{d}\mathbf{\Sigma},$$

where $\mathrm{d}\mathbf{\Sigma} = \mathbf{\Lambda}(\mathrm{d}\mathbf{\Lambda})' + (\mathrm{d}\mathbf{\Lambda})\mathbf{\Lambda}' + \mathrm{d}\mathbf{\Psi}$. Hence, $\mathrm{d}F$ equals zero for all $\mathrm{d}\mathbf{\Lambda}$ and $\mathrm{d}\mathbf{\Psi}$ if and only if

$$\mathbf{\Sigma}^{-1}(\mathbf{\Sigma} - \mathbf{S})\mathbf{\Sigma}^{-1}\mathbf{\Lambda} = \mathbf{0}, \tag{4.2}$$

$$\operatorname{diag}\mathbf{\Sigma}^{-1}(\mathbf{\Sigma} - \mathbf{S})\mathbf{\Sigma}^{-1} = \mathbf{0}. \tag{4.3}$$

We simplify these equations by using the decomposition of $\mathbf{\Sigma}$ and $\mathbf{\Sigma}^{-1}$. By setting $\mathbf{\Phi} = \mathbf{I}_m$ in (3.11), we have

$$\mathbf{\Sigma}^{-1}\mathbf{\Lambda} = \mathbf{\Psi}^{-1}\mathbf{\Lambda}(\mathbf{I} + \mathbf{\Gamma})^{-1}, \tag{4.4}$$

where $\mathbf{\Gamma} = \mathbf{\Lambda}'\mathbf{\Psi}^{-1}\mathbf{\Lambda}$. Eqs. (4.4) and (4.2) imply that

$$(\mathbf{\Sigma} - \mathbf{S})\mathbf{\Psi}^{-1}\mathbf{\Lambda} = \mathbf{0}. \tag{4.5}$$

Then using the decomposition of $\mathbf{\Sigma}^{-1}$ in (3.18), we have

$$\begin{aligned}(\mathbf{\Sigma} - \mathbf{S})\mathbf{\Sigma}^{-1} &= (\mathbf{\Sigma} - \mathbf{S})\mathbf{\Psi}^{-1} - (\mathbf{\Sigma} - \mathbf{S})\mathbf{\Psi}^{-1}\mathbf{\Lambda}(\mathbf{I}_m + \mathbf{\Gamma})^{-1}\mathbf{\Lambda}'\mathbf{\Psi}^{-1}\\ &= (\mathbf{\Sigma} - \mathbf{S})\mathbf{\Psi}^{-1}.\end{aligned} \tag{4.6}$$

Hence, if $\mathbf{\Lambda}$ satisfies (4.5) then (4.3) can be rewritten as

$$\operatorname{diag}(\mathbf{\Sigma} - \mathbf{S}) = \mathbf{0}. \tag{4.7}$$

In practice, certain restrictions have to be imposed on $\mathbf{\Lambda}$ to eliminate rotational indeterminacy. For convenience, we assume that $\mathbf{\Gamma}$ is a diagonal matrix and that the diagonal elements are distinct and ordered. By setting $\mathbf{\Phi} = \mathbf{I}_m$ in (3.12), we have

$$\mathbf{\Lambda}'\mathbf{\Sigma}^{-1}\mathbf{\Lambda} = \mathbf{\Gamma}(\mathbf{I}_m + \mathbf{\Gamma})^{-1}, \tag{4.8}$$

from which we obtain

$$\mathbf{\Gamma} = \left(\mathbf{I}_m - \mathbf{\Lambda}'\mathbf{\Sigma}^{-1}\mathbf{\Lambda}\right)^{-1} - \mathbf{I}_m. \tag{4.9}$$

The above equations imply that the two restrictions, $\operatorname{nondiag}\mathbf{\Gamma} = \mathbf{0}$ and $\operatorname{nondiag}\mathbf{\Lambda}'\mathbf{\Sigma}^{-1}\mathbf{\Lambda} = \mathbf{0}$, are equivalent. Moreover, if $\mathbf{\Lambda}$ satisfies (4.2), then it follows that

$\boldsymbol{\Sigma}^{-1}\boldsymbol{\Lambda} = \mathbf{S}^{-1}\boldsymbol{\Lambda}$, and hence the following three restrictions are equivalent:

$$\text{nondiag}\,\boldsymbol{\Gamma} = \mathbf{0}, \qquad \text{nondiag}\,\boldsymbol{\Lambda}'\boldsymbol{\Sigma}^{-1}\boldsymbol{\Lambda} = \mathbf{0},$$
$$\text{nondiag}\,\boldsymbol{\Lambda}'\mathbf{S}^{-1}\boldsymbol{\Lambda} = \mathbf{0}. \tag{4.10}$$

Lawley (1940) introduced the maximum likelihood method in factor analysis. Rao (1955) related the maximum likelihood method to canonical correlation analysis, in which the restriction nondiag $\boldsymbol{\Gamma} = \mathbf{0}$ arises naturally. Hence, the factor loadings that satisfy this restriction are sometimes called the canonical loadings. Lawley (1941) studied the maximum likelihood estimation for the case of nonrandom factor model. However, Anderson and Rubin (1956) showed that the maximum likelihood estimators do not exist.

4.1.2. Asymptotic distributions of estimators

Suppose that the maximum likelihood estimators $\hat{\boldsymbol{\Psi}}$ and $\hat{\boldsymbol{\Lambda}}$ can be expanded as

$$\hat{\boldsymbol{\Psi}} = \boldsymbol{\Psi} + \epsilon\boldsymbol{\Psi}^{(1)} + \epsilon^2\boldsymbol{\Psi}^{(2)} + \cdots,$$
$$\hat{\boldsymbol{\Lambda}} = \boldsymbol{\Lambda} + \epsilon\boldsymbol{\Lambda}^{(1)} + \epsilon^2\boldsymbol{\Lambda}^{(2)} + \cdots, \tag{4.11}$$

where $\boldsymbol{\Psi}^{(k)}$ and $\boldsymbol{\Lambda}^{(k)}$ $(k = 1, 2, \ldots)$ are $p \times p$ diagonal matrices and $p \times m$ matrices. Let $\mathbf{U} = n^{1/2}(\mathbf{S} - \boldsymbol{\Sigma})$. By substituting $\mathbf{S} = \boldsymbol{\Sigma} + \epsilon\mathbf{U}$ and the above expansions of $\hat{\boldsymbol{\Lambda}}$ and $\hat{\boldsymbol{\Psi}}$ into (4.5) and (4.7), and then equating the coefficients of ϵ^k $(k = 1, 2)$ on both sides, we obtain

$$\big(\mathbf{U} - \boldsymbol{\Lambda}\boldsymbol{\Lambda}^{(1)\prime} - \boldsymbol{\Lambda}^{(1)}\boldsymbol{\Lambda}' - \boldsymbol{\Psi}^{(1)}\big)\boldsymbol{\Psi}^{-1}\boldsymbol{\Lambda} = \mathbf{0}, \tag{4.12}$$
$$\text{diag}\big(\mathbf{U} - \boldsymbol{\Lambda}\boldsymbol{\Lambda}^{(1)\prime} - \boldsymbol{\Lambda}^{(1)}\boldsymbol{\Lambda}' - \boldsymbol{\Psi}^{(1)}\big) = \mathbf{0}. \tag{4.13}$$

Let us define

$$\boldsymbol{\Phi} = \boldsymbol{\Psi}^{-1} - \boldsymbol{\Psi}^{-1}\boldsymbol{\Lambda}\big(\boldsymbol{\Lambda}'\boldsymbol{\Psi}^{-1}\boldsymbol{\Lambda}\big)^{-1}\boldsymbol{\Lambda}'\boldsymbol{\Psi}^{-1}, \qquad \boldsymbol{\Xi} = \boldsymbol{\Phi} \odot \boldsymbol{\Phi}, \tag{4.14}$$

where the symbol $\odot$ indicates the Hadamard product of matrices. We assume that $\boldsymbol{\Xi}$ is positive-definite and denote the elements of $\boldsymbol{\Xi}^{-1}$ by ξ^{ij}. Anderson and Rubin (1956) showed, by solving (4.12) and (4.13) for $\boldsymbol{\Psi}^{(1)}$, that the limiting distribution of $n^{1/2}(\hat{\psi}_i - \psi_i)$ is normal with zero mean and that

$$n\mathrm{E}(\hat{\psi}_i - \psi_i)(\hat{\psi}_j - \psi_j) = 2\xi^{ij}. \tag{4.15}$$

Because $\hat{\sigma}_{ii} = s_{ii}$ $(i = 1, \ldots, p)$ from (4.7), the maximum likelihood estimator of the standardized unique variance $\psi_i^* = \psi_i/\sigma_{ii}$ is given by $\hat{\psi}_i^* = \hat{\psi}_i/s_{ii}$ and is approximated by

$$\hat{\psi}_i^* - \psi_i^* \approx \sigma_{ii}^{-1}\big\{(\hat{\psi}_i - \psi_i) - \psi_i^*(s_{ii} - \sigma_{ii})\big\}.$$

It can be shown that

$$n\mathrm{E}\big\{(\hat{\psi}_i - \psi_i)(s_{jj} - \sigma_{jj})\big\} = 2\delta_{ij}\psi_i^2,$$

where δ_{ij} is Kronecker's delta with $\delta_{ij} = 1$ if $i = j$ and $\delta_{ij} = 0$ otherwise. Then we obtain

$$n\mathrm{E}(\hat{\psi}_i^* - \psi_i^*)(\hat{\psi}_j^* - \psi_j^*) = 2\sigma_{ii}^{-1}\sigma_{jj}^{-1}\xi^{ij} + 2\psi_i^*\psi_j^*(\rho_{ij}^2 - 2\delta_{ij}\psi_i^*), \quad (4.16)$$

where ρ_{ij} is the correlation coefficient between x_i and x_j and can be written as $\rho_{ij} = \sum_{k=1}^{m} \lambda_{ik}^* \lambda_{jk}^*$ for $i \neq j$. Because the sum of the maximum likelihood estimators of the communality $\hat{h}_i^2$ and standardized unique variance $\hat{\psi}_i^*$ is always unity, the asymptotic covariances of $\hat{h}_i^2$ and $\hat{h}_j^2$ $(i, j = 1, \ldots, p)$ are given by (4.16).

The asymptotic standard errors for the maximum likelihood estimates of the unrotated (canonical) loadings, for both nonstandardized and standardized ones, can be obtained by solving (4.12) and (4.13) for $\mathbf{\Lambda}^{(1)}$ in terms of $\mathbf{U}$. See Lawley and Maxwell (1971) and Jennrich and Thayer (1973). Matrix expressions for the asymptotic covariance matrix of the maximum likelihood estimates of the unrotated factor loadings and unique variances are given by Hayashi and Bentler (2000). In practice, however, factor loadings are usually rotated for ease of interpretation. Methods for obtaining the asymptotic standard errors of rotated factor loadings are described later.

4.1.3. Asymptotic expansions

Approximating the distributions of estimators using the limiting normal distributions may not be very accurate when the sample size is not large enough in relation to the number of observed variables. An approach to coping with this problem is to find asymptotic expansions for the distributions of estimators. We note that the maximum likelihood estimator of a parameter in the factor analysis model can be essentially expressed as a function of the sample covariance matrix $\mathbf{S}$. Let $h(\mathbf{S})$ be a real valued function, not depending on n, of $\mathbf{S}$ and assume that the partial derivatives of $h(\mathbf{S})$ of second order are continuous in a neighborhood of $\mathbf{S} = \mathbf{\Sigma}$. The limiting distribution of

$$Y = n^{1/2}\{h(\mathbf{S}) - h(\mathbf{\Sigma})\}$$

is known to be normal with zero mean as n tends to infinity. Let σ_h^2, m_1, m_3, and m_{11} be defined by

$$\begin{aligned}
&\mathrm{E}\{Y - \mathrm{E}(Y)\}^2 = \sigma_h^2 + O(n^{-1}),\\
&\mathrm{E}(Y) = n^{-1/2}\sigma_h m_1 + O(n^{-3/2}),\\
&\mathrm{E}\{Y - \mathrm{E}(Y)\}^3 = n^{-1/2}\sigma_h^3 m_3 + O(n^{-3/2}),\\
&n\mathrm{E}\left[\frac{\{h(\mathbf{S}) - h(\mathbf{\Sigma})\}(\hat{\sigma}_h^2 - \sigma_h^2)}{\sigma_h^3}\right] = m_{11} + O(n^{-1}).
\end{aligned}$$

In practice, the asymptotic variance σ_h^2 would be unknown, so we consider the Studentized statistic $T = Y/\hat{\sigma}_h$, where $\hat{\sigma}_h^2$ is a consistent estimator of σ_h^2. In general, the distribution function of T is known to have an expansion of the form

$$\begin{aligned}
\mathrm{P}(T \leqslant t) = F\Big(t - n^{-1/2}\big\{\big(m_1 - \tfrac{1}{2}m_{11}\big) + \tfrac{1}{6}(m_3 - 3m_{11})(t^2 - 1)\big\}\Big)&\\
+ O(n^{-1}),& \quad (4.17)
\end{aligned}$$

where $F(\cdot)$ indicates the distribution function of the standard normal distribution.

By substituting $\mathbf{S} = \boldsymbol{\Sigma} + \epsilon\mathbf{U}$ and the expansions given by (4.11) of $\hat{\boldsymbol{\Lambda}}$ and $\hat{\boldsymbol{\Psi}}$ into (4.5) and (4.7), and then equating the coefficients of ϵ^k $(k = 1, 2)$ on both sides, Ichikawa and Konishi (2002) obtained explicit expressions of the elements of $\boldsymbol{\Psi}^{(2)}$ in (4.11). Then m_1, m_3, and m_{11} for $\hat{\psi}_i = h(\mathbf{S})$ are given by

$$m_1 = -\sigma_h^{-1}\psi_i \operatorname{tr}\bigl(\mathbf{I}_m + \boldsymbol{\Gamma}^{-1}\bigr) - 2\sigma_h^{-1} \operatorname{tr} \boldsymbol{\Phi}\boldsymbol{\Theta}_i\boldsymbol{\Phi}\bigl(\mathbf{B} \odot \boldsymbol{\Xi}^{-1}\bigr), \tag{4.18}$$

$$m_3 = 8\sigma_h^{-3} \operatorname{tr}(\boldsymbol{\Phi}\boldsymbol{\Theta}_i)^3 - 24\sigma_h^{-3} \operatorname{tr} \boldsymbol{\Phi}\boldsymbol{\Theta}_i\boldsymbol{\Phi}\boldsymbol{\Theta}_i\mathbf{B}\boldsymbol{\Theta}_i, \tag{4.19}$$

$$m_{11} = 8\sigma_h^{-3} \operatorname{tr}(\boldsymbol{\Phi}\boldsymbol{\Theta}_i)^3 - 16\sigma_h^{-3} \operatorname{tr} \boldsymbol{\Phi}\boldsymbol{\Theta}_i\boldsymbol{\Phi}\boldsymbol{\Theta}_i\mathbf{B}\boldsymbol{\Theta}_i, \tag{4.20}$$

where $\boldsymbol{\Phi}$ and $\boldsymbol{\Xi}$ are defined in (4.14), $\boldsymbol{\Theta}_i$ is a diagonal matrix whose diagonal elements are the ith row (column) of $\boldsymbol{\Xi}^{-1}$, $\mathbf{B} = (\beta_{ij}) = \boldsymbol{\Psi}^{-1}\boldsymbol{\Lambda}\boldsymbol{\Gamma}^{-2}\boldsymbol{\Lambda}'\boldsymbol{\Psi}^{-1}$, and $\boldsymbol{\Pi} = \boldsymbol{\Psi}^{-1}\boldsymbol{\Lambda}\boldsymbol{\Gamma}^{-1}\boldsymbol{\Lambda}'\boldsymbol{\Psi}^{-1}$. The asymptotic variance σ_h^2 of $\hat{\psi}_i$ is given by (4.15). By inverting the expansion in (4.17), we obtain expansions for the quantiles of T, which can be used to construct second-order accurate confidence intervals. Ichikawa and Konishi (2002) gave similar results for the maximum likelihood estimators $\hat{\psi}_i^*$ of the standardized unique variances (uniquenesses) $\psi_i^* = \psi_i/\sigma_{ii}$.

4.1.4. A Newton–Raphson algorithm

In practice, the maximum likelihood estimates $\hat{\boldsymbol{\Lambda}}$ and $\hat{\boldsymbol{\Psi}}$ are calculated by minimizing the discrepancy function given by (4.1) using a numerical method. In the case of exploratory factor analysis, efficient algorithms exist that find the conditional minimum $f(\boldsymbol{\Psi})$ for a given $\boldsymbol{\Psi}$ and then find the overall minimum (Jöreskog, 1967; Jennrich and Robinson, 1969; Clarke, 1970). Let $\boldsymbol{\Lambda}(\boldsymbol{\Psi})$ be the minimizer of $F(\boldsymbol{\Sigma}; \mathbf{S})$ over all $\boldsymbol{\Lambda}$ for a given $\boldsymbol{\Psi}$. Clearly, $\boldsymbol{\Lambda}(\boldsymbol{\Psi})$ is a solution of (4.2), from which we have $\boldsymbol{\Lambda} = \boldsymbol{\Sigma}\mathbf{S}^{-1}\boldsymbol{\Lambda} = (\boldsymbol{\Lambda}\boldsymbol{\Lambda}' + \boldsymbol{\Psi})\mathbf{S}^{-1}\boldsymbol{\Lambda}$, and this can be rewritten as

$$\boldsymbol{\Psi}\mathbf{S}^{-1}\boldsymbol{\Lambda} = \boldsymbol{\Lambda}\bigl(\mathbf{I}_m - \boldsymbol{\Lambda}'\mathbf{S}^{-1}\boldsymbol{\Lambda}\bigr). \tag{4.21}$$

Now, assume that $\boldsymbol{\Lambda}'\mathbf{S}^{-1}\boldsymbol{\Lambda}$ is a diagonal matrix, and let $\mathbf{S} = \mathbf{L}\mathbf{L}'$ and $\mathbf{W} = \mathbf{L}^{-1}$. Further, let $\nu_1 < \cdots < \nu_p$ be the eigenvalues of $\mathbf{A} = \mathbf{W}\boldsymbol{\Psi}\mathbf{W}'$, and let $\boldsymbol{\zeta}_1, \ldots, \boldsymbol{\zeta}_p$ be the corresponding eigenvectors of unit length. Then the conditional minimizer $\boldsymbol{\Lambda}(\boldsymbol{\Psi})$ of F can be shown to be of the form

$$\boldsymbol{\Lambda}(\boldsymbol{\Psi}) = \mathbf{L}\mathbf{Z}_1(\mathbf{I}_m - \mathbf{N}_1)^{1/2}, \tag{4.22}$$

where $\mathbf{N}_1$ is a diagonal matrix whose diagonal elements are the m smallest eigenvalues $\nu_1, \ldots, \nu_m$ of $\mathbf{A}$ and the columns of $\mathbf{Z}_1$ are the corresponding eigenvectors $\boldsymbol{\zeta}_1, \ldots, \boldsymbol{\zeta}_m$. Substituting (4.22) into (4.1), we have

$$f(\boldsymbol{\Psi}) = \operatorname{tr} \mathbf{S}\boldsymbol{\Psi}^{-1} + \log|\boldsymbol{\Psi}| - \sum_{k=1}^{m}\bigl(\nu_k^{-1} + \log \nu_k\bigr) - \log|\mathbf{S}| - (p - m). \tag{4.23}$$

Hence, the minimization is carried out with respect to not the entire set of $p(m + 1)$ parameters in $\boldsymbol{\Lambda}$ and $\boldsymbol{\Psi}$ but only p parameters in $\boldsymbol{\Psi}$.

If $\boldsymbol{\Sigma}$ is decomposed as $\boldsymbol{\Sigma} = \boldsymbol{\Lambda}\boldsymbol{\Lambda}' + \boldsymbol{\Psi}$, then the m smallest eigenvalues of $\boldsymbol{\Sigma}^{-1/2}\boldsymbol{\Psi}(\boldsymbol{\Sigma}^{-1/2})'$ are less than unity and the remaining $p - m$ eigenvalues are equal to unity. Therefore, if the sample size increases, then the $p - m$ largest eigenvalues of $\mathbf{W}\hat{\boldsymbol{\Psi}}\mathbf{W}'$ will all converge to one. This explains the expression of $f(\boldsymbol{\Psi})$ in (4.23) without using the $p - m$ largest eigenvalues.

We assume that the m smallest eigenvalues of $\boldsymbol{\Sigma}^{-1/2}\boldsymbol{\Psi}(\boldsymbol{\Sigma}^{-1/2})'$ are distinct. Then, we have, for $k = 1, \ldots, m$,

$$d\nu_k = \boldsymbol{\zeta}_k'(d\mathbf{A})\boldsymbol{\zeta}_k, \qquad d\boldsymbol{\zeta}_k = (\nu_k\mathbf{I}_p - \mathbf{A})^+(d\mathbf{A})\boldsymbol{\zeta}_k, \qquad d\mathbf{A} = \mathbf{W}(d\boldsymbol{\Psi})\mathbf{W}',$$

where $(\nu_k\mathbf{I}_p - \mathbf{A})^+$ is the Moore–Penrose inverse of $\nu_k\mathbf{I}_p - \mathbf{A}$ and is given by

$$(\nu_k\mathbf{I}_p - \mathbf{A})^+ = \sum_{j \neq k}(\nu_k - \nu_j)^{-1}\boldsymbol{\zeta}_j\boldsymbol{\zeta}_j'.$$

Expressions for the first- and second-order partial derivatives of $f(\boldsymbol{\Psi})$ with respect to the elements of $\boldsymbol{\Psi}$ are given by

$$\frac{\partial f}{\partial \psi_i} = \sum_{j=m+1}^{p} \frac{\nu_k - 1}{\nu_k^2}\tilde{\zeta}_{ik}^2, \tag{4.24}$$

$$\frac{\partial^2 f}{\partial \psi_i \partial \psi_j} = \left(\sum_{r=m+1}^{p} \nu_r^{-2}\tilde{\zeta}_{ir}\tilde{\zeta}_{jr}\right)^2 - \left(\sum_{r=m+1}^{p} \frac{\nu_r - 1}{\nu_r^2}\tilde{\zeta}_{ir}\tilde{\zeta}_{jr}\right)^2 + 2\sum_{k=1}^{m}\sum_{r=m+1}^{p} \frac{1 - \nu_r}{\nu_r^2(\nu_k - \nu_r)}\tilde{\zeta}_{ik}\tilde{\zeta}_{jk}\tilde{\zeta}_{ir}\tilde{\zeta}_{jr}, \tag{4.25}$$

where $\tilde{\boldsymbol{\zeta}}_k = \mathbf{W}'\boldsymbol{\zeta}_k$.

Let $\mathbf{g}$ and $\mathbf{J}$ be the first- and second-order derivatives of $f(\boldsymbol{\Psi})$ with respect to the elements of $\boldsymbol{\Psi}$. The basic Newton–Raphson step for finding the minimum of $f(\boldsymbol{\Psi})$ is to update $\boldsymbol{\psi} = (\psi_1, \ldots, \psi_p)'$ in each iteration by

$$\boldsymbol{\psi}_{\text{new}} = \boldsymbol{\psi}_{\text{old}} + \delta\boldsymbol{\psi},$$

where $\delta\boldsymbol{\psi} = -\mathbf{J}^{-1}\mathbf{g}$. When we are not close enough to the minimum of $f(\boldsymbol{\Psi})$, however, $\mathbf{J}$ is not guaranteed to be positive-definite, and the Newton–Raphson step need not decrease the function. In early iterations, it is sometimes better to replace $\mathbf{J}$ by its expectation, of which the typical element can be expressed as

$$\left(\sum_{r=m+1}^{p} \nu_r^{-1}\tilde{\zeta}_{ir}\tilde{\zeta}_{jr}\right)^2.$$

Using line minimization is also recommended:

$$\boldsymbol{\psi}_{\text{new}} = \boldsymbol{\psi}_{\text{old}} + \alpha(\delta\boldsymbol{\psi}),$$

that is, finding α so that $f(\boldsymbol{\Psi})$ is minimized in the Newton–Raphson direction.

4.1.5. Improper solutions

The maximum likelihood estimates of $\mathbf{\Lambda}$ and $\mathbf{\Psi}$ are calculated using an iterative procedure. However, such a procedure is not guaranteed to give the unique minimizer of the discrepancy function inside of the admissible parameter space. In fact, a diagonal element of $\mathbf{\Psi}$ is often known to approach zero during the iteration and even becomes negative, if allowed to do so. If all the diagonal elements of the final value of $\mathbf{\Psi}$ are positive, then it is called a *proper solution*. On the other hand, if the final value of $\mathbf{\Psi}$ is on the boundary or outside of the admissible parameter space, then it is called an *improper solution* or a *Heywood case*. An iterative procedure is said to be *nonconvergent* if it does not satisfy certain convergence criteria within a specified number of iterations.

The problem of improper solutions and nonconvergence in factor analysis has become widely recognized since Jöreskog (1967) proposed a rapidly convergent method. van Driel (1978) studied the following three causes: (1) sampling fluctuations, (2) nonexistence of a covariance matrix of the factor analysis model that fits the data, and (3) lack of uniqueness of the solution $\mathbf{\Psi}$ of the decomposition $\mathbf{\Sigma} = \mathbf{\Lambda}\mathbf{\Lambda}' + \mathbf{\Psi}$. He proposed examining standard error estimates for $\hat{\psi}_i$ and distinguishing the three causes. We would obtain a proper solution by increasing the sample size if an improper solution is caused by (1). The occurrence of an improper solution or nonconvergence is by no means an exception and tells us explicitly that something is wrong with our data and/or model. Coping with improper solutions and nonconvergence is difficult in practice. Sato (1987) proposed a pragmatic treatment of improper solutions.

4.1.6. Likelihood ratio test

The factor analysis model given in (3.1) to (3.4) cannot be verified directly. However, the hypothesis that $\mathbf{\Sigma}$ is a covariance matrix, as implied by the model, is testable. We shall describe the likelihood ratio test for testing the null hypothesis, $\mathrm{H}_0 : \mathbf{\Sigma} = \mathbf{\Lambda}\mathbf{\Lambda}' + \mathbf{\Psi}$, where $\mathbf{\Lambda}$ is a $p \times m$ matrix with m specified, against the alternative hypothesis, $\mathrm{H}_1 : \mathbf{\Sigma} > \mathbf{0}$, that is, $\mathbf{\Sigma}$ is an arbitrary positive-definite matrix.

Suppose we have proper estimates $\hat{\mathbf{\Psi}}$ and $\hat{\mathbf{\Lambda}}$, and let $\hat{\mathbf{\Sigma}} = \hat{\mathbf{\Lambda}}\hat{\mathbf{\Lambda}}' + \hat{\mathbf{\Psi}}$. Then -2 times the logarithm of the likelihood ratio can be shown to be given by $n\,F(\hat{\mathbf{\Sigma}}; \mathbf{S})$, that is, n times the minimum value of the discrepancy function. As n tends to infinity, the distribution of the likelihood ratio test statistic converges to χ^2 with degrees of freedom, in this case, given by $\frac{1}{2}\{(p-m)^2-(p+m)\}$. Bartlett (1950) suggested replacing n with $n - \frac{1}{6}(2p+5) - \frac{2}{3}m$. It has not yet been shown, however, that this is indeed a special case of the Bartlett adjustment in the true sense of improving the accuracy of the χ^2 approximation to the distribution of the likelihood ratio test statistic.

In usual, m is not specified in advance, and the problem of interest is to select an appropriate number of factors. In such a case, starting with a small values of m, possibly $m = 1$, we test successive values of m until a hypothesis is accepted. Note, however, that the significance level associated with the entire procedure is not known.

The asymptotic robustness of the normal theory methods was extensively investigated in the 1980s. The limiting χ^2 distribution of the likelihood ratio test statistic was shown to hold under conditions much more general than the normality of the observations (Amemiya and Anderson, 1990). See also Browne and Shapiro (1988).

4.1.7. Use of the correlation matrix

If we change the units of measurement, then the resulting covariance matrix is given by $\tilde{\mathbf{S}} = \mathbf{DSD}$, where $\mathbf{D}$ is a nonsingular diagonal matrix. Denote the minimizer of $F(\cdot\,; \tilde{\mathbf{S}})$ by $\tilde{\boldsymbol{\Sigma}} = \tilde{\boldsymbol{\Lambda}}\tilde{\boldsymbol{\Lambda}}' + \tilde{\boldsymbol{\Psi}}$. Because

$$F(\mathbf{D\Sigma D}; \mathbf{DSD}) = F(\boldsymbol{\Sigma}; \mathbf{S}),$$

we have

$$\tilde{\boldsymbol{\Sigma}} = \mathbf{D}\hat{\boldsymbol{\Sigma}}\mathbf{D} = \mathbf{D}\hat{\boldsymbol{\Lambda}}\hat{\boldsymbol{\Lambda}}'\mathbf{D} + \mathbf{D}\hat{\boldsymbol{\Psi}}\mathbf{D}.$$

It follows from (4.7) that $s_{ii} = \hat{\sigma}_{ii}$. Then the maximum likelihood estimates $\hat{\lambda}^*_{ik}$ and $\hat{\psi}^*_i$ of the standardized factor loadings $\lambda^*_{ik} = \lambda_{ik}/\sigma_{ii}^{1/2}$ and the standardized unique variances $\psi^*_i = \psi_i/\sigma_{ii}$ are given by $\hat{\lambda}^*_{ik} = \hat{\lambda}_{ik}/s_{ii}^{1/2}$ and $\hat{\psi}^*_i = \hat{\psi}_i/s_{ii}$. This implies that

$$\hat{\boldsymbol{\Lambda}}^*\hat{\boldsymbol{\Lambda}}^{*\prime} = \mathbf{D}_s\hat{\boldsymbol{\Lambda}}\hat{\boldsymbol{\Lambda}}'\mathbf{D}_s, \qquad \hat{\boldsymbol{\Psi}}^* = \mathbf{D}_s\hat{\boldsymbol{\Psi}}\mathbf{D}_s,$$

where $\mathbf{D}_s$ is a diagonal matrix whose ith diagonal is $s_{ii}^{-1/2}$. Hence, if we use $\mathbf{R} = \mathbf{D}_s\mathbf{S}\mathbf{D}_s$ in place of $\mathbf{S}$ as an input to the numerical procedure for calculating the maximum likelihood estimates, we directly obtain the standardized estimates $\hat{\lambda}^*_{ik}$ and $\hat{\psi}^*_i$. Further, the likelihood ratio test statistic is given by

$$nF(\hat{\boldsymbol{\Sigma}}; \mathbf{S}) = nF\big(\hat{\boldsymbol{\Sigma}}^*; \mathbf{R}\big),$$

where $\hat{\boldsymbol{\Sigma}}^* = \hat{\boldsymbol{\Lambda}}^*\hat{\boldsymbol{\Lambda}}^{*\prime} + \hat{\boldsymbol{\Psi}}^*$.

Finally, we note that while $\hat{\psi}^*_i = s_{ii}^{-1}\hat{\psi}_i$, the standard error (or its estimate) of $\hat{\psi}^*_i$ is not equal to that of $\hat{\psi}_i$ times s_{ii}^{-1}, because s_{ii} itself is a random variable that depends on each sample and is subject to sampling fluctuations.

4.2. Other methods of estimation

4.2.1. Least squares methods

Under non-normal distributions, a class of least squares estimators that minimizes the discrepancy function,

$$F(\boldsymbol{\Sigma}; \mathbf{S}) = \frac{1}{2}\,\mathrm{tr}\big\{(\mathbf{S} - \boldsymbol{\Sigma})\mathbf{W}^{-1}\big\}^2, \tag{4.26}$$

can be used. In (4.26), $\mathbf{W}$ is a weight matrix and $F(\boldsymbol{\Sigma}; \mathbf{S})$ measures the residual $\mathbf{S} - \boldsymbol{\Sigma}$ in the metric of $\mathbf{W}^{-1}$. Common examples are $\mathbf{W} = \mathbf{I}_p$ (ULS; unweighted least squares) and $\mathbf{W} = \mathbf{S}$ (GLS; generalized least squares). The discrepancy function in (4.26), like that of maximum likelihood (ML), satisfies

(i) $F(\boldsymbol{\Sigma}; \mathbf{S}) \geqslant 0$,
(ii) $F(\boldsymbol{\Sigma}; \mathbf{S}) = 0$ if and only if $\boldsymbol{\Sigma} = \mathbf{S}$,
(iii) $F(\boldsymbol{\Sigma}; \mathbf{S})$ is continuous in $\boldsymbol{\Sigma}$ and $\mathbf{S}$,

and the resulting estimators are consistent (Browne, 1982). Both the GLS estimator and the least squares estimator with $\mathbf{W} = \boldsymbol{\Sigma}$ are asymptotically equivalent with the ML estimator under the multivariate normal distribution of observations (Anderson, 1989).

If $\mathbf{W}$ is not a function of $\boldsymbol{\Lambda}$ and $\boldsymbol{\Psi}$, then the minimizer of (4.26) satisfies

$$\mathbf{W}^{-1}(\mathbf{S}-\boldsymbol{\Sigma})\mathbf{W}^{-1}\boldsymbol{\Lambda}=\mathbf{0}, \tag{4.27}$$

$$\operatorname{diag}\mathbf{W}^{-1}(\mathbf{S}-\boldsymbol{\Sigma})\mathbf{W}^{-1}=0. \tag{4.28}$$

Let $\mathbf{W}^{-1}=\mathbf{LL}'$. Then we have, from (4.27), that

$$\mathbf{L}'(\mathbf{S}-\boldsymbol{\Psi})\mathbf{LL}'\boldsymbol{\Lambda}=\mathbf{L}'\boldsymbol{\Lambda}\left(\boldsymbol{\Lambda}'\mathbf{W}^{-1}\boldsymbol{\Lambda}\right),$$

which implies that if $\boldsymbol{\Lambda}'\mathbf{W}^{-1}\boldsymbol{\Lambda}$ is diagonal, then the columns of $\mathbf{L}'\boldsymbol{\Lambda}$ are the eigenvectors of $\mathbf{L}'(\mathbf{S}-\boldsymbol{\Psi})\mathbf{L}$. For a given $\boldsymbol{\Psi}$, the minimizer of F is of the form

$$\boldsymbol{\Lambda}(\boldsymbol{\Psi})=\left(\mathbf{L}^{-1}\right)'\boldsymbol{\Omega}_1\boldsymbol{\Theta}_1^{1/2},$$

where $\boldsymbol{\Theta}_1$ is a diagonal matrix consisting of the m largest eigenvalues of $\mathbf{L}'(\mathbf{S}-\boldsymbol{\Psi})\mathbf{L}$ and the columns of $\boldsymbol{\Omega}_1$ are the corresponding eigenvectors of unit length. This can be used to construct essentially the same algorithm to calculate least squares estimates of $\boldsymbol{\Lambda}$ and $\boldsymbol{\Psi}$ as in ML.

4.2.2. *Principal factor method*

Suppose we take $\mathbf{W}=\mathbf{I}_p$ in (4.26) and wish to estimate $\boldsymbol{\Lambda}^*$ and $\boldsymbol{\Psi}^*$ by using $\mathbf{R}$ instead of $\mathbf{S}$. Then the estimates satisfy the following equations:

$$\left\{\mathbf{R}-\left(\boldsymbol{\Lambda}^*\boldsymbol{\Lambda}^{*\prime}+\boldsymbol{\Psi}^*\right)\right\}\boldsymbol{\Lambda}^*=\mathbf{0}, \tag{4.29}$$

$$\operatorname{diag}\left\{\mathbf{R}-\left(\boldsymbol{\Lambda}^*\boldsymbol{\Lambda}^{*\prime}+\boldsymbol{\Psi}^*\right)\right\}=\mathbf{0}. \tag{4.30}$$

If we have an initial estimate of $\boldsymbol{\Psi}^*$, then (4.26) can be written as

$$\frac{1}{2}\operatorname{tr}\left\{(\mathbf{R}-\boldsymbol{\Psi}^*)-\boldsymbol{\Lambda}^*\boldsymbol{\Lambda}^{*\prime}\right\}^2$$

and the minimizer $\boldsymbol{\Lambda}^*$ that satisfies (4.29) can be calculated from the m largest eigenvalues and the corresponding eigenvectors of $\tilde{\mathbf{R}}=\mathbf{R}-\boldsymbol{\Psi}^*$. This is called the *principal factor* or *principal axis* method. The matrix $\tilde{\mathbf{R}}$ is obtained from $\mathbf{R}$ by replacing each diagonal element of $\mathbf{R}$ by an estimate of communality $h_i^2=1-\psi_i^*$ and is called the *reduced correlation matrix*. An updated estimate of $\boldsymbol{\Psi}^*$ can be obtained from $\mathbf{R}$ and the current estimate $\boldsymbol{\Lambda}^*$ by using (4.30), and this cyclic process can be continued until both (4.29) and (4.30) are satisfied. Such a procedure is called the *iterated principal factor* method. The iterated principal factor method, if convergent, satisfies the same equations as the ULS method. It should be noted, however, that the convergence of such an algorithm can be very slow.

4.2.3. *Criteria for the number of factors*

One of the most commonly used criteria for the number of factors is the Kaiser–Guttman criterion (Kaiser, 1960; Guttman, 1954), which is the number of eigenvalues greater than unity of a sample correlation matrix $\mathbf{R}$. This criterion is motivated by the lower bound, m_1, of the number of factors in (3.21). Although m_3 in (3.21) is a stronger lower bound, Kaiser (1960) remarked that, empirically, m_3 suggested too many factors and did not perform as well as m_1.

The scree test (Cattell, 1966) is based on the plots of the eigenvalues $\hat{\theta}_1^* > \cdots > \hat{\theta}_p^*$ of $\mathbf{R}$ in decreasing order. Typically, the eigenvalues decrease rapidly at first and then form a straight line, with a small slope, to the last one. The scree test takes the beginning of this straight portion as the number of factors.

Horst (1965) proposed taking account of the sampling error of the eigenvalues of $\mathbf{R}$ by generating random data, whch is known as parallel analysis. Let $\hat{\delta}_{1(k)}^* > \cdots > \hat{\delta}_{p(k)}^*$ be the eigenvalues of a correlation matrix based on the kth random sample of size N, which is the same as the original sample, generated from a p-variate normal population, in which the p variates are mutually uncorrelated. Further, let $\bar{\delta}_i^*$ $(i = 1, \ldots, p)$ be the average of $\hat{\delta}_{i(1)}^*, \ldots, \hat{\delta}_{i(K)}^*$. Because the population counterpart of $\hat{\delta}_{i(k)}^*$ is unity, $\bar{\delta}_i^* - 1$ may be used as an estimate of the extent to which $\hat{\theta}_i^*$ is inflated due to sampling error. Thus, we take the number of $\hat{\theta}_i^* - (\bar{\delta}_i^* - 1)$ greater than unity as the number of factors.

4.2.4. Use of the bootstrap methods

Statistical inference in factor analysis is usually based on the assumption of a multivariate normal distribution of observations, which is more or less violated in practice, and the limiting normal distributions of estimators. Formulae for the standard errors of parameter estimates in factor analysis are complicated or may not be expressed in closed forms. One of the advantages of the bootstrap methods introduced by Efron (1979) is that they can be used without analytical derivations. However, caution is needed to use the bootstrap methods in factor analysis. For example, in the case of the bootstrap test for the goodness-of-fit, the covariance structure implied by the null hypothesis should be taken into account in designing a bootstrap simulation (Bollen and Stine, 1992). Another example in exploratory factor analysis is that if we are interested in the factor loading matrix, then we must be aware that it is determined only up to permutations and/or changes in the signs of its columns. Finally, improper solutions may occur frequently in the bootstrap simulation, in particular, when the sample size is not large enough. In exploratory factor analysis, Ichikawa and Konishi (1995) used a Monte Carlo simulation to study the performance of the bootstrap methods in estimating biases and standard errors, constructing confidence intervals, and testing the goodness-of-fit.

4.2.5. Bibliographic notes

Kano and Harada (2000) studied variable selection in exploratory factor analysis, and developed backward elimination and forward selection procedures. Mooijaart (1985) proposed an approach to factor analysis for nonnormal variables. It utilizes not only second order cross products but also third order cross products. Meijer and Mooijaart (1996) studied heteroscedasity of unique variances in one-factor model and compared several types of estimatros by a simulation. Ansari et al. (2002) extended a Bayesian approach to multi level confirmatory factor analysis. Yuan et al. (2002) proposed a unified approach to exploratory factor analysis with nonstandard data such as missing data, nonnormal data, and data with outliers. Their approach is to replace the usual sample covariance matrix $\mathbf{S}$ by another estimate of $\mathbf{\Sigma}$, depending on the nonstandard nature of data. Nonlinear factor analysis is an important extension of the ordinary factor analysis. For recent developments, see Yalcin and Amemiya (2001).

5. Factor rotation and estimation of factor scores

5.1. Factor rotation

5.1.1. Principles of rotation

Under the factor analysis model, the factor correlation matrix $\mathbf{\Phi}$ is indeterminate, and even if we have chosen $\mathbf{\Phi}$, the factor loading matrix $\mathbf{\Lambda}$ is not uniquely determined. An estimate of $\mathbf{\Lambda}$ is usually obtained by choosing $\mathbf{\Phi} = \mathbf{I}_m$ and using a suitable restriction, for example, nondiag $\mathbf{\Lambda}'\mathbf{\Psi}^{-1}\mathbf{\Lambda} = \mathbf{0}$. Such a restriction, of course, is merely for convenience and does not have practical significance. For many applications, the initial factor loading matrix is multiplied from the right by a nonsingular matrix for ease of interpretation. This process is called *factor rotation*. The rotation is said to be *orthogonal rotation* if the common factors after rotation are mutually uncorrelated and is said to be *oblique rotation* otherwise.

Let $\mathbf{f}$, $\mathbf{\Lambda}_f$, and $\mathbf{\Phi}_f$ be the vector of initial factors, corresponding factor loading matrix, and factor correlation matrix. We may assume, without loss of generality, that $\mathbf{\Phi}_f = \mathbf{I}_m$. Suppose we transform $\mathbf{f}$ using a nonsingular matrix $\mathbf{T}$ of order m as $\mathbf{g} = \mathbf{T}\mathbf{f}$. Then the corresponding factor loading matrix $\mathbf{\Lambda}_g$, factor correlation matrix $\mathbf{\Phi}_g$, and factor structure matrix $\mathbf{\Gamma}_g$ are given by

$$\mathbf{\Lambda}_g = \mathbf{\Lambda}_f\mathbf{T}^{-1}, \qquad \mathbf{\Phi}_g = \mathbf{T}\mathbf{T}', \qquad \mathbf{\Gamma}_g = \mathbf{\Lambda}_f\mathbf{T}', \tag{5.1}$$

where

$$\operatorname{diag}(\mathbf{\Phi}_g) = \operatorname{diag}(\mathbf{T}\mathbf{T}') = \mathbf{I}_m. \tag{5.2}$$

In the orthogonal rotation, $\mathbf{T}$ satisfies $\mathbf{T}\mathbf{T}' = \mathbf{I}_m$, so $\mathbf{T}^{-1} = \mathbf{T}'$. Thus,

$$\mathbf{\Lambda}_g = \mathbf{\Lambda}_f\mathbf{T}^{-1} = \mathbf{\Lambda}_f\mathbf{T}', \qquad \mathbf{\Phi}_g = \mathbf{I}_m, \qquad \mathbf{\Gamma}_g = \mathbf{\Lambda}_f\mathbf{T}' = \mathbf{\Lambda}_g. \tag{5.3}$$

In both cases, $\mathbf{\Lambda}_f\mathbf{\Lambda}_f' = \mathbf{\Lambda}_g\mathbf{\Phi}_g\mathbf{\Lambda}_g'$ holds, which implies that the fit of the model remains unchanged after rotation.

What criteria should an interpretable factor loading matrix satisfy? Thurstone (1947) introduced the *simple structure* criterion as a guide to attaining interpretable factors. The three major points are as follows:

1. Any column of the factor loading matrix $\mathbf{\Lambda}$ should have mostly small values as close to zero as possible.
2. Any row of $\mathbf{\Lambda}$ should have only a few elements far from zero.
3. Any two columns of $\mathbf{\Lambda}$ should have some variables with high values in one column and low values in the other columns.

Today, most of the rotation methods involve defining a function of a rotated loading matrix and then rotating the initial loading matrix so that the function is optimized. The function is intended to measure the degree of simple structure. Such a method is called an *analytic rotation*. Browne (2001) reviewed analytic rotation and gives examples.

5.1.2. Orthogonal rotation

Let $\mathbf{A} = (a_{ik})$ and $\mathbf{B} = (b_{ik})$ denote initial and rotated factor loading matrices. The initial factors are orthogonal, and $\mathbf{B} = \mathbf{AT}'$ is obtained from $\mathbf{A}$ by an orthogonal rotation, that is, $\mathbf{TT}' = \mathbf{I}_m$. Kaiser (1958) proposed maximizing

$$\sum_{k=1}^{m} \left\{ \frac{1}{p} \sum_{i=1}^{p} b_{ik}^4 - \frac{w}{p^2} \left(\sum_{i=1}^{p} b_{ik}^2 \right)^2 \right\}, \tag{5.4}$$

which he called the *orthomax criterion*. If $w = 1$, then (5.4) is exactly the sum of the columnwise variance of the squared factor loadings over the m factors and is called the *varimax criterion*, which has been widely used. The expression (5.4) is called the *quartimax criterion* (Carroll, 1953; Neuhaus and Wrigley, 1954) if $w = 0$, the *factor parsimony criterion* (Crawford and Ferguson, 1970) if $w = p$, and the *equamax criterion* if $w = m/2$. If we rotate $\mathbf{A}$ so that the resulting $\mathbf{B}$ maximizes (5.4), then it is sometimes called the *raw orthomax*. Note that if the observed variables are standardized to have unit variances and the factors are orthogonal, then the communality h_i^2 $(i = 1, \ldots, p)$ is the squared length of the ith row of the factor loading matrix. In the normalized orthomax, we first set $a_{ik}^* = a_{ik}/h_i$ to make each row of $\mathbf{A}^* = (a_{ik}^*)$ have unit length. Then we rotate $\mathbf{A}^*$ so that the resulting $\mathbf{B}^*$ maximizes (5.4) and set $b_{ik} = h_i b_{ik}^*$ to make each row of $\mathbf{B} = (b_{ik})$ have the same length as the corresponding row of $\mathbf{A}$.

Now, we express the orthomax criterion (5.4) in terms of vectors and matrices as

$$F(\mathbf{B}) = \frac{1}{p} \operatorname{tr}\{(\mathbf{B} \odot \mathbf{B})'(\mathbf{B} \odot \mathbf{B})\} - \frac{w}{p^2} \operatorname{tr}\{(\mathbf{B} \odot \mathbf{B})'\mathbf{1}_p\mathbf{1}_p'(\mathbf{B} \odot \mathbf{B})\}, \tag{5.5}$$

where the symbol $\odot$ indicates the Hadamard product of matrices and $\mathbf{1}_p = (1, \ldots, 1)'$ is of order $p \times 1$. Then it can be shown that $\mathrm{d}F = \mathbf{F}_d'\mathbf{A}(\mathrm{d}\mathbf{T})'$, where $\mathbf{F}_d$ is given by

$$\mathbf{F}_d = \frac{4}{p}\mathbf{B} \odot \mathbf{B} \odot \mathbf{B} - \frac{4w}{p^2}\mathbf{B} \operatorname{diag}(\mathbf{B}'\mathbf{B}).$$

It can be further shown that $\mathbf{T}$ is the solution of

$$\mathbf{A}'\mathbf{F}_d = \mathbf{T}'\mathbf{L}, \qquad \mathbf{TT}' = \mathbf{I}_m,$$

where $\mathbf{L}$ is a symmetric matrix of Lagrange multipliers, and that $\mathbf{T}$ is expressed as

$$\mathbf{T} = \left(\mathbf{F}_d'\mathbf{AA}'\mathbf{F}_d\right)^{-1/2}\mathbf{F}_d'\mathbf{A},$$

where $(\mathbf{F}_d'\mathbf{AA}'\mathbf{F}_d)^{-1/2}$ is the symmetric square root of $(\mathbf{F}_d'\mathbf{AA}'\mathbf{F}_d)^{-1}$. We note here that $\mathbf{F}_d$ is a function $\mathbf{F}_d(\mathbf{B})$ of the rotated matrix $\mathbf{B} = \mathbf{AT}'$.

In view of the above results, an algorithm for obtaining the rotation matrix $\mathbf{T}$ is to choose any orthogonal matrix $\mathbf{T}_0$ as an initial estimate of $\mathbf{T}$ and set $\mathbf{B}_1 = \mathbf{AT}_0'$. Then update $\mathbf{H}_j$, $\mathbf{T}_j$, and $\mathbf{B}_j$ as

$$\mathbf{H}_j = \mathbf{B}_j'\mathbf{F}_d(\mathbf{B}_j) = \mathbf{P}_j\mathbf{\Delta}_j\mathbf{Q}_j', \qquad \mathbf{T}_j = \mathbf{Q}_j\mathbf{P}_j', \qquad \mathbf{B}_{j+1} = \mathbf{B}_j\mathbf{T}_j',$$

where $\mathbf{P}_j'\mathbf{P}_j = \mathbf{Q}_j'\mathbf{Q}_j = \mathbf{I}_m$ and $\mathbf{\Delta}_j$ is a diagonal matrix whose diagonal elements are the singular-values of $\mathbf{H}_j$ arranged in descending order of magnitude.

If $\mathrm{tr}(\mathbf{T}_j-\mathbf{T}_{j-1})'(\mathbf{T}_j-\mathbf{T}_{j-1})$ approaches zero, we may stop the iteration. For the details of derivation, see Neudecker (1981).

5.1.3. *Oblique rotation*

Factors obtained by an orthogonal rotation method sometimes fail to be interpretable. In such a case, we may obtain more interpretable factors by relaxing the enforced orthogonality restrictions. Let us illustrate with an example. Suppose that factor analysis was applied to four achievement tests English (x_1), social studies (x_2), science (x_3), and mathematics (x_4). The results of orthogonal and oblique rotation are compared in Figure 1. In terms of the simple structure criteria, the oblique factors clearly are more easily interpretable than the orthogonal factors.

If the common factors are orthogonal, then the factor loadings coincide with the factor structures. After oblique rotation, on the other hand, the common factors are correlated and we must distinguish the factor loadings from the factor structures. Figure 2 shows both factor loadings and factor structures represented in terms of projections of a variable x onto the space spanned by two oblique axes, g_1 and g_2. The factor loadings of x on these two factors are given as the oblique projections of x onto g_1 and g_2 by lines parallel to these two axes, while the structure loadings are given as the orthogonal projections onto each of the two axes by lines perpendicular to these axes.

Carroll (1953) proposed minimizing the *oblimin criterion* defined by

$$F(\mathbf{B})=\sum_{k\neq\ell}^{m}\left\{\sum_{i=1}^{p}b_{ik}^2b_{i\ell}^2-\frac{w}{p}\left(\sum_{i=1}^{p}b_{ik}^2\right)\left(\sum_{i=1}^{p}b_{i\ell}^2\right)\right\}. \tag{5.6}$$

This is called the *covarimin criterion* when $w=1$ and the *quartimin criterion* when $w=0$. The covarimin criterion is known to be likely to produce almost uncorrelated factors, while the quartimin criterion is liable to yield highly correlated factors. To compensate for the defects of these two criteria, the *biquatimin criterion*, which is obtained by setting $w=1/2$, was proposed.

Before we describe the minimization of (5.6), we introduce *primary factors* and *reference factors*. Let $g_1,\ldots,g_m$ be common factors with $\mathrm{Var}(g_k)=1$ $(k=1,\ldots,m)$. The reference factors $\tilde{g}_1,\ldots,\tilde{g}_m$ are defined as the linear combinations of $g_1,\ldots,g_m$ such that $\mathrm{Var}(\tilde{g}_k)=1$ $(k=1,\ldots,m)$ and $\mathrm{Cov}(\tilde{g}_k,g_\ell)=0$ $(\ell\neq k)$. The original

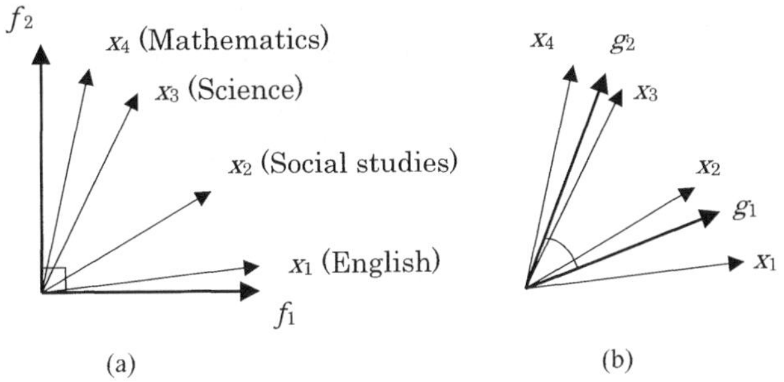

Fig. 1. Comparison of orthogonal and oblique axes. (a) Orthogonal axes (f_1, f_2), (b) oblique axes (g_1, g_2).

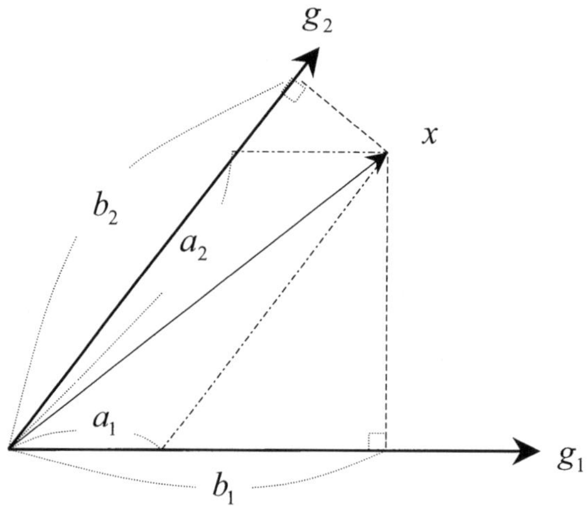

Fig. 2. Factor loading (a_1, a_2) and factor structure (b_1, b_2) of a single variable x upon two oblique axes, g_1 and g_2.

factors that generated the reference factors are called the primary factors. The reference factor $\tilde{g}_k$ is obtained by eliminating from g_k the effect of the other $m - 1$ primary factors using linear regression, and then rescaling the residual to have unit variance. To be more specific, let us denote the vector of primary factors by $\mathbf{g} = (g_1, \ldots, g_m)'$ with $\mathrm{Var}(\mathbf{g}) = \mathbf{\Phi}_g$, and let $\mathbf{\Lambda}_g$ and $\mathbf{\Gamma}_g$ be the corresponding factor loading matrix and factor structure matrix. Further, let us denote the vector of reference factors by $\tilde{\mathbf{g}} = (\tilde{g}_1, \ldots, \tilde{g}_m)'$ with $\mathrm{Var}(\tilde{\mathbf{g}}) = \mathbf{\Phi}_{\tilde{g}}$, and let $\mathbf{\Lambda}_{\tilde{g}}$ and $\mathbf{\Gamma}_{\tilde{g}}$ be the corresponding factor loading matrix and factor structure matrix of the reference factors. Then an explicit expression of $\tilde{\mathbf{g}}$ is

$$\tilde{\mathbf{g}} = \mathbf{D}_g \mathbf{\Phi}_g^{-1} \mathbf{g},$$

where $\mathbf{D}_g = (\mathrm{diag}\, \mathbf{\Phi}_g^{-1})^{-1/2}$. It is easy to verify that

$$\begin{aligned} &\mathrm{Cov}(\tilde{\mathbf{g}}, \mathbf{g}) = \mathbf{D}_g, && \mathbf{\Phi}_{\tilde{g}} = \mathbf{D}_g \mathbf{\Phi}_g^{-1} \mathbf{D}_g, \\ &\mathbf{\Lambda}_{\tilde{g}} = \mathbf{\Gamma}_g \mathbf{D}_g^{-1}, && \mathbf{\Gamma}_{\tilde{g}} = \mathbf{\Lambda}_g \mathbf{D}_g. \end{aligned} \tag{5.7}$$

An interesting relation is that $\tilde{\tilde{\mathbf{g}}} = \mathbf{g}$; that is, the reference factors of reference factors are the original primary factors. It can be seen from the last relation in (5.7) that the columns of the primary factor loading matrix $\mathbf{\Lambda}_g$ and the columns of the reference factor structure matrix $\mathbf{\Gamma}_{\tilde{g}}$ are proportional to each other. Hence, simplification of one implies the simplification of the other.

Now, we are ready to go back to the minimization of (5.6). Let $\mathbf{A}$ be the initial factor loading matrix of orthogonal factors. One approach is to simplify the primary factor loading matrix directly, that is,

$$\text{minimize} \quad F\left(\mathbf{A}\mathbf{T}^{-1}\right) \quad \text{subject to} \quad \mathrm{diag}(\mathbf{T}\mathbf{T}') = \mathbf{I}_m. \tag{5.8}$$

Then the resulting primary factor loading matrix and factor correlation matrix are given by

$$\mathbf{AT}^{-1} \quad \text{and} \quad \mathbf{TT}'. \tag{5.9}$$

This approach is by Jennrich and Sampson (1966) and is called the direct oblimin method. The other approach is to simplify the reference structure matrix, that is,

$$\text{minimize} \quad F(\mathbf{AT}') \quad \text{subject to} \quad \operatorname{diag}(\mathbf{TT}') = \mathbf{I}_m. \tag{5.10}$$

Then the corresponding factor loading matrix and factor correlation matrix of the primary factors are given by

$$\mathbf{AT}'\mathbf{D}^{-1} \quad \text{and} \quad \mathbf{D}(\mathbf{TT}')^{-1}\mathbf{D}, \tag{5.11}$$

where $\mathbf{D} = \{\operatorname{diag}(\mathbf{TT}')^{-1}\}^{-1/2}$. For the detailed algorithm of computation, see, for example, Mulaik (1972) and Takeuchi et al. (1982).

5.1.4. Procrustes and promax rotation

Let $\mathbf{A}$ be an initial factor loading matrix of orthogonal factors. When a hypothetical factor loading matrix $\mathbf{G}$ is available, one can multiply a nonsingular matrix from the right of $\mathbf{A}$ so that the resulting matrix $\mathbf{B}$ is as close to $\mathbf{G}$ as possible to minimize

$$F(\mathbf{B}) = \operatorname{tr}(\mathbf{G} - \mathbf{B})'(\mathbf{G} - \mathbf{B}). \tag{5.12}$$

Such a rotation is known as the *Procrustes rotation* – a term borrowed from Greek mythology.

In the case of the orthogonal Procrustes rotation, $\mathbf{B}$ is of the form $\mathbf{B} = \mathbf{AT}'$, and we minimize (5.12) subject to $\mathbf{TT}' = \mathbf{I}_m$. An expression of $\mathbf{T}$ given by Schöneman (1966) is

$$\mathbf{T} = (\mathbf{G}'\mathbf{AA}'\mathbf{G})^{-1/2}\mathbf{G}'\mathbf{A}, \tag{5.13}$$

where $(\mathbf{G}'\mathbf{AA}'\mathbf{G})^{-1/2}$ is the symmetric square root of $(\mathbf{G}'\mathbf{AA}'\mathbf{G})^{-1}$. Another expression of $\mathbf{T}$ is

$$\mathbf{T} = \mathbf{PQ}', \tag{5.14}$$

where $\mathbf{G}'\mathbf{A} = \mathbf{P}\boldsymbol{\Delta}\mathbf{Q}'$ is the singular-value decomposition of $\mathbf{G}'\mathbf{A}$ with $\mathbf{P}'\mathbf{P} = \mathbf{Q}'\mathbf{Q} = \mathbf{I}_m$. For the derivation, see Ten Berge (1977).

In the case of the oblique Procrustes rotation, we must distinguish two approaches, (5.8) and (5.10), to minimizing (5.12). Here, we show an approximate solution to the second approach (Mulaik, 1972). First, we minimize

$$F(\mathbf{AT}') = \operatorname{tr}(\mathbf{G} - \mathbf{AT}')'(\mathbf{G} - \mathbf{AT}'),$$

without restrictions upon $\mathbf{T}$. The solution is given by

$$\mathbf{T} = \mathbf{G}'\mathbf{A}(\mathbf{A}'\mathbf{A})^{-1}.$$

Because $\mathbf{T}$ is not an admissible transformation matrix to simple reference structure, rescale the rows of $\mathbf{T}$ by $\mathbf{T}^* = \mathbf{D}_T\mathbf{T}$, where $\mathbf{D}_T = (\operatorname{diag}\mathbf{TT}')^{-1/2}$. Note that $\mathbf{T}^*$ does not minimize $F(\cdot)$. Let

$$\mathbf{D}^* = \{\mathrm{diag}(\mathbf{T}^*\mathbf{T}^{*\prime})^{-1}\}^{-1/2} = \mathbf{D}_T\{\mathrm{diag}(\mathbf{T}\mathbf{T}')^{-1}\}^{-1/2} = \mathbf{D}_T\mathbf{D}.$$

Then it follows from (5.11) that the rotated primary factor loading matrix corresponding to $\mathbf{T}^*$ is given by

$$\mathbf{A}\mathbf{T}^{*\prime}\mathbf{D}^{*-1} = \mathbf{A}\mathbf{T}'\mathbf{D}^{-1}.$$

Gower and Dijksterhuis (2004) are concerned with the problem of transforming one set of matrices to represent another set of matrices as closely as possible.

Finding a hypothetical factor loading matrix is sometimes difficult. Hendrickson and White (1964) proposed constructing a target matrix from an orthogonally rotated loading matrix, possibly using the varimax method, and then rotate the initial loading matrix toward the target matrix using the oblique Procrustes rotation described above so as to achieve an oblique simple structure. This method is known as the *promax rotation*.

5.1.5. Variance explained by each factor

If the common factors are orthogonal (uncorrelated), then it follows from the standard theory of regression analysis that the variance of x_i explained by f_k, which we denote V_{ik}, is given by

$$V_{ik} = \mathrm{Var}(x_i)\{\mathrm{Corr}(x_i, f_k)\}^2 = \{\mathrm{Cov}(x_i, f_k)\}^2, \tag{5.15}$$

where $\mathrm{Corr}(x_i, f_k)$ indicates the correlation coefficient between x_i and f_k. The second equality holds because $\mathrm{Var}(f_k) = 1$. By noting that $\mathrm{Cov}(x_i, f_k) = \lambda_{ik}$ and summing over all i, we have the sum of variances explained by the kth factor f_k:

$$V_{\cdot k} = \sum_i \lambda_{ik}^2.$$

We note here that

$$V_{i\cdot} = \sum_k V_{ik} = \sigma_{ii} - \psi_i = \sigma_{ii}h_i^2,$$

which is the variance of the common part t_i in (3.8) and is invariant under the rotation of factors.

If the common factors are oblique (correlated), then no $V_{i\cdot}$ can be decomposed into the sum of m terms. Further, $\mathrm{Cov}(x_i, f_k) = \gamma_{ik}$, where $\boldsymbol{\Gamma} = (\gamma_{ik})$ is the factor structure matrix given by $\boldsymbol{\Gamma} = \boldsymbol{\Lambda}\boldsymbol{\Phi}$. Then the sum of variances explained by the kth factor f_k, ignoring the other factors, is given by

$$V'_{\cdot k} = \sum_i \gamma_{ik}^2.$$

If we eliminate the effect of other $m - 1$ factors from each component of $\mathbf{f} = (f_1, \ldots, f_m)'$ by linear regression and rescale the residual to have unit variance, then the resulting vector $\tilde{\mathbf{f}} = (\tilde{f}_1, \ldots, \tilde{f}_m)'$ consists of the reference factors with

$$\mathrm{Cov}(\mathbf{x}, \tilde{\mathbf{f}}) = \boldsymbol{\Lambda}\mathbf{D}_f,$$

where $\mathbf{D}_f = (\text{diag}\,\Phi^{-1})^{-1/2}$. Then the sum of the variances explained by the kth factor f_k, eliminating the other factors, is the sum of the squared covariances between x_i and $\tilde{f}_k$ and is given by

$$V''_{\cdot k} = \frac{1}{\phi^{kk}} \sum_i \lambda_{ik}^2,$$

where ϕ^{kk} is the kth diagonal element of $\mathbf{\Phi}^{-1}$.

In the case of factor analysis of a correlation matrix, the variables $x_1, \ldots, x_p$ are standardized to have unit variances, and the corresponding expressions for $V_{\cdot k}$, $V'_{\cdot k}$, and $V''_{\cdot k}$, are given by

$$V^*_{\cdot k} = \sum_i \lambda_{ik}^{*2}, \qquad V^{*\prime}_{\cdot k} = \sum_i \gamma_{ik}^{*2}, \qquad V^{*\prime\prime}_{\cdot k} = \frac{1}{\phi^{kk}} \sum_i \lambda_{ik}^{*2},$$

where λ^*_{ik} and γ^*_{ik} are the elements of the standardized factor loading matrix $\mathbf{\Lambda}^*$ and the standardized factor structure matrix $\mathbf{\Gamma}^* = \mathbf{\Lambda}^*\mathbf{\Phi}$.

5.1.6. *Standard errors of rotated factor loadings*

One difficulty in finding standard error formulae for the rotated loadings b_{ik} is that not only the unrotated loadings a_{ik} but also the rotation matrix itself are functions of the data. However, approximate standard errors of the rotated loadings can be obtained by using the delta method from the covariances of the unrotated loadings and the partial derivatives $\partial b_{ik}/\partial a_{j\ell}$ of the rotated loadings with respect to the unrotated loadings. Formulae for the exact derivatives were given, through the use of implicit differentiation, by Archer and Jennrich (1973) for orthogonal rotation, with emphasis on the orthomax family and by Jennrich (1973) for oblique rotation using the generalized Crawford-Ferguson family, which includes the oblimin family. Using a matrix approach, Hayashi and Yung (1999) gave formulae for exact derivatives of orthomax rotated loadings with or without row nomalization.

Because an analytically rotated loading matrix optimizes a corresponding function, it satisfies certain conditions. Hence, the maximum likelihood estimates of the analytically rotated loadings are constrained maximum likelihood estimates and their asymptotic standard errors can be obtained by inverting an augmented information matrix (Jennrich, 1974). See also Ogasawara (1998). Of course, the standard errors of rotated loadings can be evaluated using the bootstrap or jackknife methods. Cudeck and O'Dell (1994) reviewed the use of standard error estimates, in particular those for rotated loadings, in factor analysis.

5.2. *Estimation of factor scores*

It is sometimes useful to "estimate" factor scores of each observation. In the following, we consider the problem of estimating $\mathbf{f}$ by $\hat{\mathbf{f}} = \mathbf{W}'(\mathbf{x} - \boldsymbol{\mu})$, where $\mathbf{W}$ is a $p \times m$ weight matrix. It should be noted, however, that $\hat{\mathbf{f}}$ is not an estimator in the usual sense because $\mathbf{f}$ is not a parameter vector. Rather, the "estimator" is a "predictor" of values for the unobserved random vector $\mathbf{f}$ (Bentler and Yuan, 1997). We assume that the parameters

$\boldsymbol{\mu}$, $\boldsymbol{\Lambda}$, $\boldsymbol{\Phi}$, and $\boldsymbol{\Psi}$ are known. In practice, however, these parameters would be unknown and have to be replaced by corresponding estimates. In the following, we compare four types of commonly used estimators discussed in McDonald and Burr (1967).

The covariance matrix of the joint distribution of $\mathbf{x}$ and $\mathbf{f}$ is

$$\mathrm{Cov}\begin{pmatrix}\mathbf{x}\\ \mathbf{f}\end{pmatrix} = \begin{pmatrix}\boldsymbol{\Sigma} & \boldsymbol{\Lambda\Phi}\\ \boldsymbol{\Phi\Lambda}' & \boldsymbol{\Phi}\end{pmatrix}.$$

The linear regression of $\mathbf{f}$ on $\mathbf{x}$ is given by setting $\mathbf{W} = \boldsymbol{\Sigma}^{-1}\boldsymbol{\Lambda\Phi}$. By using (3.17), the *regression estimator* $\hat{\mathbf{f}}_1$ introduced by Thurstone (1935) can be written as

$$\hat{\mathbf{f}}_1 = \boldsymbol{\Phi\Lambda}'\boldsymbol{\Sigma}^{-1}(\mathbf{x} - \boldsymbol{\mu}) = \left(\boldsymbol{\Phi}^{-1} + \boldsymbol{\Gamma}\right)^{-1}\boldsymbol{\Lambda}'\boldsymbol{\Psi}^{-1}(\mathbf{x} - \boldsymbol{\mu}), \tag{5.16}$$

where $\boldsymbol{\Gamma} = \boldsymbol{\Lambda}'\boldsymbol{\Psi}^{-1}\boldsymbol{\Lambda}$. The conditional mean of $\hat{\mathbf{f}}_1$ given $\mathbf{f}$ is

$$\mathrm{E}(\hat{\mathbf{f}}_1 \mid \mathbf{f}) = \mathrm{E}\left\{\left(\boldsymbol{\Phi}^{-1} + \boldsymbol{\Gamma}\right)^{-1}\boldsymbol{\Lambda}'\boldsymbol{\Psi}^{-1}(\boldsymbol{\Lambda}\mathbf{f} + \mathbf{e}) \mid \mathbf{f}\right\} = \left(\boldsymbol{\Phi}^{-1} + \boldsymbol{\Gamma}\right)^{-1}\boldsymbol{\Gamma}\mathbf{f}.$$

Because $\mathrm{E}(\hat{\mathbf{f}}_1 \mid \mathbf{f}) \neq \mathbf{f}$, the regression estimator $\hat{\mathbf{f}}_1$ is said to be *conditionally biased*. Obviously, $\mathrm{E}(\hat{\mathbf{f}} \mid \mathbf{f}) = \mathbf{W}'\boldsymbol{\Lambda}\mathbf{f}$ holds, and an estimator $\hat{\mathbf{f}}$ is *conditionally unbiased* for all values of $\mathbf{f}$ if and only if $\mathbf{W}'\boldsymbol{\Lambda} = \mathbf{I}_m$. The covariance matrix of $\hat{\mathbf{f}}_1$ is, as given in (3.12),

$$\mathrm{Var}(\hat{\mathbf{f}}_1) = \boldsymbol{\Phi\Lambda}'\boldsymbol{\Sigma}^{-1}\boldsymbol{\Lambda\Phi} = \boldsymbol{\Phi} - \left(\boldsymbol{\Phi}^{-1} + \boldsymbol{\Gamma}\right)^{-1}, \tag{5.17}$$

and is different from $\boldsymbol{\Phi}$.

The ordinary least squares estimator of $\mathbf{f}$ that minimizes

$$\mathrm{E}\left\{\mathrm{tr}(\mathbf{x} - \boldsymbol{\mu} - \boldsymbol{\Lambda}\hat{\mathbf{f}})(\mathbf{x} - \boldsymbol{\mu} - \boldsymbol{\Lambda}\hat{\mathbf{f}})'\right\}$$

is given by

$$\hat{\mathbf{f}}_2 = (\boldsymbol{\Lambda}'\boldsymbol{\Lambda})^{-1}\boldsymbol{\Lambda}'(\mathbf{x} - \boldsymbol{\mu}), \tag{5.18}$$

which was introduced by Horst (1965). It is easily verified that $\hat{\mathbf{f}}_2$ is conditionally unbiased and that the covariance matrix of $\hat{\mathbf{f}}_2$ is given by

$$\mathrm{Var}(\hat{\mathbf{f}}_2) = \boldsymbol{\Phi} + (\boldsymbol{\Lambda}'\boldsymbol{\Lambda})^{-1}\boldsymbol{\Lambda}'\boldsymbol{\Psi}\boldsymbol{\Lambda}(\boldsymbol{\Lambda}'\boldsymbol{\Lambda})^{-1}. \tag{5.19}$$

It should be noted that $\hat{\mathbf{f}}_2$ does not take into account the fact that the unique variances ψ_i of e_i $(i = 1, \ldots, p)$ are unequal.

The weighted least squares estimator of $\mathbf{f}$ that minimizes

$$\mathrm{E}\left\{\mathrm{tr}\,\boldsymbol{\Psi}^{-1/2}(\mathbf{x} - \boldsymbol{\mu} - \boldsymbol{\Lambda}\hat{\mathbf{f}})(\mathbf{x} - \boldsymbol{\mu} - \boldsymbol{\Lambda}\hat{\mathbf{f}})'\boldsymbol{\Psi}^{-1/2}\right\}$$

is given by

$$\hat{\mathbf{f}}_3 = \boldsymbol{\Gamma}^{-1}\boldsymbol{\Lambda}'\boldsymbol{\Psi}^{-1}(\mathbf{x} - \boldsymbol{\mu}). \tag{5.20}$$

This estimator is known as the *Bartlett estimator* (Bartlett, 1937). It is easily verified that the Bartlett estimator $\hat{\mathbf{f}}_3$ is conditionally unbiased, that the covariance matrix of $\hat{\mathbf{f}}_3$ is, in view of (3.10), given by

$$\mathrm{Var}(\hat{\mathbf{f}}_3) = \boldsymbol{\Phi} + \boldsymbol{\Gamma}^{-1}, \tag{5.21}$$

and that

$$\mathrm{E}\,(\hat{\mathbf{f}}_3 - \mathbf{f})(\hat{\mathbf{f}}_3 - \mathbf{f})' = \mathbf{\Gamma}^{-1}. \tag{5.22}$$

The Bartlett estimator $\hat{\mathbf{f}}_3$ can be derived as the conditionally unbiased estimator that minimizes the trace of the mean squared error matrix $\mathrm{E}\,(\hat{\mathbf{f}} - \mathbf{f})(\hat{\mathbf{f}} - \mathbf{f})'$ (Lawley and Maxwell, 1971).

Anderson and Rubin (1956) proposed an estimator $\hat{\mathbf{f}}_4$ of $\mathbf{f}$ that minimizes

$$\mathrm{E}\big\{\mathrm{tr}\,\mathbf{\Psi}^{-1/2}(\mathbf{x} - \boldsymbol{\mu} - \mathbf{\Lambda}\hat{\mathbf{f}})(\mathbf{x} - \boldsymbol{\mu} - \mathbf{\Lambda}\hat{\mathbf{f}})'\mathbf{\Psi}^{-1/2}\big\},$$

subject to the constraint that

$$\mathrm{E}(\hat{\mathbf{f}}\hat{\mathbf{f}}') = \mathbf{W}'\mathbf{\Sigma}\mathbf{W} = \mathbf{\Phi}. \tag{5.23}$$

They showed that the resulting weight matrix is given by

$$\mathbf{W} = \mathbf{\Psi}^{-1}\mathbf{\Lambda}(\mathbf{\Gamma} + \mathbf{M})^{-1},$$

where $\mathbf{M}$ is a symmmetric matrix of Lagrange multipliers and satisfies

$$(\mathbf{\Gamma} + \mathbf{M})\mathbf{\Phi}(\mathbf{\Gamma} + \mathbf{M}) = \mathbf{\Lambda}'\mathbf{\Psi}^{-1}\mathbf{\Sigma}\mathbf{\Psi}^{-1}\mathbf{\Lambda} = \mathbf{\Gamma}\mathbf{\Phi}\mathbf{\Gamma} + \mathbf{\Gamma}.$$

They also showed that if $\mathbf{\Phi} = \mathbf{I}_m$, then $\hat{\mathbf{f}}_4 = (\mathbf{\Gamma}^2 + \mathbf{\Gamma})^{-1/2}\mathbf{\Lambda}'\mathbf{\Psi}^{-1}(\mathbf{x} - \boldsymbol{\mu})$. An explicit expression of $\hat{\mathbf{f}}_4$ given by McDonald (1981) is

$$\hat{\mathbf{f}}_4 = \mathbf{H}\mathbf{V}^{-1}\mathbf{H}'\mathbf{\Lambda}'\mathbf{\Psi}^{-1}(\mathbf{x} - \boldsymbol{\mu}), \tag{5.24}$$

where $\mathbf{\Phi} = \mathbf{H}\mathbf{H}'$ and $\mathbf{V}$ is a symmetric square root of $\mathbf{H}'(\mathbf{\Gamma}\mathbf{\Phi}\mathbf{\Gamma} + \mathbf{\Gamma})\mathbf{H}$. We note here that, as before, the variances of f_k $(k = 1, \dots, m)$ are assumed to be unity, so that $\mathbf{\Phi}$ is identical to the correlation matrix. Hence, an estimator $\hat{\mathbf{f}}$ of $\mathbf{f}$ satisfying the constraint in (5.23) is called a *correlation preserving estimator.*

Krijnen et al. (1996) studied best estimators within each of the classes of linear, linear conditionally unbiased, and linear correlation preserving estimators. Their results on the class of correlation-preserving estimators were extended by Ten Berge et al. (1999).

5.3. *Examples*

We shall give examples of factor analysis by using the data given in Yanai et al. (1990) and the factor analysis program in SAS. The correlations among 12 scales of the New Personality Test (Yanai et al., 1987) based on 200 students at a university in Japan are given in Table 2. The four largest eigenvalues are 3.045, 2.420, 1.733, and 1.010. Because the fourth eigenvalue is close to unity, we chose three as the number of factors.

The factor loading estimates obtained using the maximum likelihood (ML) method are shown in Table 3. The values in the parentheses indicate the associated standard error estimates, ranging from 0.072 to 0.214. The total sum of variances explained by the three common factors is 5.656, which amounts to about 47% of the total variance. In addition, estimated communalities of the 12 scales are shown in the fourth column. For example, "lack of cooperativeness" has the lowest communality: 0.259. This implies that about 74% of the variance of this variable remains unexplained by the three factors.

Table 2
Correlations among 12 scales of the new personality test

	1	2	3	4	5	6	7	8	9	10	11	12
1	1.000	0.525	0.235	0.280	0.136	0.208	0.323	0.049	−0.198	−0.342	−0.189	−0.245
2		1.000	0.311	0.418	0.316	0.377	0.346	0.189	−0.038	−0.453	−0.070	−0.207
3			1.000	0.091	0.336	0.374	0.041	−0.188	−0.262	−0.052	0.165	−0.025
4				1.000	0.070	0.105	0.423	0.322	0.005	−0.232	−0.053	0.003
5					1.000	0.573	0.030	−0.110	−0.052	−0.220	0.046	−0.092
6						1.000	0.118	0.099	−0.002	−0.121	0.197	−0.047
7							1.0000	0.374	0.142	−0.064	0.153	0.130
8								1.000	0.371	0.083	0.248	0.314
9									1.000	0.299	0.175	0.403
10										1.000	0.386	0.518
11											1.000	0.592
12												1.000

1 Social extraversion
2 General activity
3 Empathy
4 Novelty
5 Durability
6 Regularity
7 Self-revelation
8 Aggressiveness
9 Lack of cooperativeness
10 Inferiority feelings
11 Nervousness
12 Depression

Table 3
Maximum likelihood estimates of factor loadings and estimated standard errors

		Factor 1	Factor 2	Factor 3	Communality
1	Extraversion	0.579 (0.080)	0.170 (0.184)	−0.151 (0.080)	0.387
2	Activity	0.702 (0.146)	0.403 (0.214)	−0.110 (0.086)	0.667
3	Empathy	0.312 (0.121)	0.244 (0.142)	0.429 (0.081)	0.341
4	Novelty	0.347 (0.149)	0.360 (0.140)	−0.365 (0.086)	0.384
5	Durability	0.404 (0.124)	0.256 (0.166)	0.488 (0.079)	0.467
6	Regularity	0.385 (0.174)	0.436 (0.156)	0.447 (0.086)	0.538
7	Self-revelation	0.194 (0.193)	0.495 (0.106)	−0.355 (0.093)	0.409
8	Aggressiveness	−0.093 (0.207)	0.531 (0.102)	−0.414 (0.090)	0.461
9	Lack of cooperativeness	−0.334 (0.136)	0.333 (0.127)	−0.192 (0.085)	0.259
10	Inferiority feeling	−0.685 (0.072)	0.167 (0.212)	0.122 (0.074)	0.512
11	Nervousness	−0.406 (0.210)	0.574 (0.143)	0.239 (0.091)	0.551
12	Depression	−0.618 (0.196)	0.545 (0.189)	0.011 (0.086)	0.679
	Variance explained by each factor	2.531	1.928	1.196	5.656

However, it does not necessarily follow that the three-factor model failed to account for a substantial amount of the variances of the 12 scales. Instead, it may be reasonable to guess that scales with low communalities share little in common with the remaining variables.

Although not shown here, we also obtained factor loading estimates using the unweighted least squares (ULS) method. The ML and ULS estimates of the factor loading

Table 4
Varimax rotated factor loadings and their standard error estimates

		Factor 1	Factor 2	Factor 3
1	Extraversion	−0.329 (0.087)	0.473 (0.083)	0.233 (0.077)
2	Activity	−0.262 (0.083)	0.646 (0.070)	0.425 (0.071)
3	Empathy	−0.035 (0.054)	0.021 (0.060)	0.582 (0.062)
4	Novelty	−0.047 (0.066)	0.616 (0.057)	0.043 (0.065)
5	Durability	−0.091 (0.053)	0.034 (0.064)	0.676 (0.056)
6	Regularity	0.040 (0.054)	0.161 (0.063)	0.715 (0.054)
7	Self-revelation	0.156 (0.078)	0.619 (0.057)	0.037 (0.065)
8	Aggressiveness	0.389 (0.092)	0.542 (0.081)	−0.128 (0.070)
9	Lack of cooperativeness	0.454 (0.065)	0.166 (0.085)	−0.159 (0.073)
10	Inferiority feeling	0.630 (0.058)	−0.301 (0.093)	−0.159 (0.068)
11	Nervousness	0.702 (0.047)	0.003 (0.070)	0.241 (0.068)
12	Depression	0.822 (0.038)	0.031 (0.085)	−0.046 (0.060)
	Variance explained by each factor	2.138	1.845	1.676

matrix were determined under different constraints to eliminate rotational indeterminacy. Hence, caution is needed to compare the estimates of the two methods. The ULS estimates of communalities, which are invariant under rotation, are 0.386, 0.669, 0.364, 0.378, 0.426, 0.530, 0.418, 0.508, 0.274, 0.505, 0.529, and 0.671. The maximum difference between the ML and the corresponding ULS estimates of communality is 0.047 in absolute value. The fact that the communalities estimated using the two methods are so close implies that the factor loading matrices estimated using the methods are virtually orthogonal rotations of each other.

Table 4 shows the varimax rotated factor loadings, using those in Table 3 as initial estimates. The 47% of the total sum of variances explained by the three factors remains unchanged and is exactly the same as that obtained from the unrotated loadings. Note that most of the standard error estimates are smaller than those for unrotated loadings, now ranging from 0.038 to 0.093. We recommend taking standard errors of factor loadings into consideration when interpreting factor analysis results. One can easily see that the simple structure criterion is more accentuated in the rotated factor loadings than in the unrotated ones. The 12 scales can be divided into the following three groups depending on which factor is highly loaded in: (1) depression, nervousness, inferiority feeling, and lack of cooperativeness, (2) activity, self-revelation, novelty, aggressiveness, and extraversion, and (3) regularity, durability, and empathy. Referring to the names in the five-factor model (FFM) by Goldberg (1990), mentioned in Section 2, we label the three factors "emotional instability", "extraversion and general activity", and "conscientiousness and agreeableness". According to Dillon and Goldstein (1984), labeling the common factors is more an art than a science, and thus we can safely say that the interpretation of factors largely depends on the researcher's expertise and knowledge of what is really being investigated.

Table 5
Promax rotated factor loadings

		Factor 1	Factor 2	Factor 3
1	Extraversion	−0.298	0.446	0.138
2	Activity	−0.204	0.601	0.318
3	Empathy	0.033	−0.061	0.605
4	Novelty	−0.030	0.630	−0.056
5	Durability	−0.012	−0.062	0.696
6	Regularity	0.128	0.068	0.729
7	Self-revelation	0.176	0.641	−0.043
8	Aggressiveness	0.393	0.593	−0.182
9	Lack of cooperativeness	0.448	0.211	−0.149
10	Inferiority feeling	0.617	−0.267	−0.058
11	Nervousness	0.744	−0.005	0.321
12	Depression	0.833	0.068	0.027
	Variance(1)	2.054	1.694	1.468
	Variance(2)	2.215	2.014	1.905
	Factor correlations	–		
		−0.100	–	
		−0.220	0.288	–

(1) Variance explained by each factor eliminating other factors.

(2) Variance explained by each factor ignoring other factors.

Finally, the factor loadings, rather than the factor structures, obtained using the promax rotation are shown in Table 5. Factor loadings are known to satisfy the simple structure criterion better than do the corresponding factor structures when the variables under factor analysis are positively correlated on the whole. The correlation coefficients among the factors are relatively small, and no significant difference can be found in the factor loadings given in Tables 4 and 5. Thus, we do not need to change the labels given earlier to the varimax rotated factors. However, the varimax rotated factor loadings of activity are −0.262, 0.646, and 0.425, while the promax rotated loadings are −0.204, 0.601, and 0.318. This may suggest that oblique rotation is liable to yield factors that better fulfil the simple structure criterion than orthogonal rotation is.

5.3.1. Statistical software

In exploratory factor analysis, the purpose is to search for a hypothesis by choosing an appropriate number of factors and then obtaining interpretable factors using rotation. Exploratory factor analysis programs in SAS and SPSS cover various methods of estimating factor loadings, of factor rotation, and of estimating factor scores. The coverage of SAS is much wider than that of SPSS. While SAS produces standard error estimates of rotated factor loadings, SPSS does not. The computer program CEFA developed by Browne produces standard error estimates of rotated factor loadings, whether standardized or not. It is available at http://www.quantrm2.psy.ohio-state.edu/browne.

In confirmatory factor analysis, the indeterminacy of the factor analysis model is removed on the basis of a hypothesis about relations between the observed variables and the factors, and the purpose of an analysis is to test such hypothesis. Confirmatory factor analysis can be performed by using SEM software such as AMOS, EQS, and LISREL.

Finally, we note that SEM software can be used in an exploratory context. Consider an orthogonal factor analysis model such that the upper triangular elements of the factor loading matrix are zero. These restrictions are merely for convenience to remove rotational indeterminacy and are not of practical significance. Such a model is on the boundary of the exploratory and confirmatory approaches. Hence, standard error estimates for the unique variances and various fit indices produced by SEM software are correct, though standard error estimates for the factor loadings are not relevant.

Acknowledgements

We deeply thank two anonymous reviewers who gave us useful and constructive comments that helped us improve the manuscript. Special mention must be made of Dr. T.W. Anderson (Stanford University) who read through the manuscript and gave us encouraging comments on revising it. We also acknowledge Dr. Y. Takane (McGill University), Dr. K. Shigemasu (Tokyo University), and Dr. S. Mayekawa (Tokyo Institute of Technology) for their helpful suggestions through personal communications and their publications.

References

Amemiya, Y., Anderson, T.W. (1990). Asymptotic chi-square tests for a large class of factor analysis models. *Annals of Statistics* **18**, 1453–1463.

Anderson, T.W. (1984). *An Introduction to Multivariate Statistical Analysis*, 2nd ed. John Wiley & Sons, New York.

Anderson, T.W. (1989). Linear latent variable models and covariance structures. *Journal of Econometrics* **41**, 91–119.

Anderson, T.W., Rubin, H. (1956). Statistical inference in factor analysis. In: Neyman, J. (Ed.), In: *Proceedings of the Third Berkeley Symposium on Mathematical Statistics and Probability*, vol. 5. University of California Press, Berkeley, CA, pp. 111–150.

Ansari, A., Jedidi, K., Dube, L. (2002). Heterogeneous factor analysis models: A Bayesian approach. *Psychometrika* **67**, 49–78.

Archer, C.O., Jennrich, R.I. (1973). Standard errors for rotated factor loadings. *Psychometrika* **38**, 581–592.

Bartholomew, D.J., Knott, M. (1999). *Latent Variable Models and Factor Analysis*, 2nd ed. Arnold, London.

Bartlett, M.S. (1937). The statistical conception of mental factors. *British Journal of Psychology* **28**, 97–104.

Bartlett, M.S. (1950). Tests of significance in factor analysis. *British Journal of Psychology (Statistics Section)* **3**, 77–85.

Basilevsky, A. (1994). *Statistical Factor Analysis and Related Methods: Theory and Application*. John Wiley & Sons, New York.

Bekker, P.A., de Leeuw, J. (1987). The rank of reduced dispersion matrices. *Psychometrika* **52**, 125–135.

Bentler, P.M., Kano, Y. (1990). On the equivalence of factors and components. *Multivariate Behavioral Research* **25**, 67–74.

Bentler, P.M., Yuan, K.-H. (1997). Optimal conditionally unbiased equivariant factor score estimators. In: Berkane, M. (Ed.), *Latent Variable Modeling with Applications to Causality*. Springer-Verlag, New York, pp. 259–281.

Bock, R.D., Bargmann, R.E. (1966). Analysis of covariance structures. *Psychometrika* **31**, 507–534.

Bollen, K.A. (2002). Latent variables in psychology and the social sciences. *Annual Review of Psychology* **53**, 605–634.

Bollen, K.A., Stine, R. (1992). Bootstrapping goodness-of-fit measures in structural equation models. *Sociological Methods and Research* **21**, 205–229.

Browne, M.W. (1982). Covariance structures. In: Hawkins, D.M. (Ed.), *Topics in Applied Multivariate Analysis*. Cambridge University Press, Cambridge, pp. 72–141.

Browne, M.W. (2001). An overview of analytic rotation in exploratory factor analysis. *Multivariate Behavioral Research* **36**, 111–150.

Browne, M.W., Shapiro, A. (1988). Robustness of normal theory methods in the analysis of linear latent variable models. *British Journal of Mathematical and Statistical Psychology* **41**, 193–208.

Carroll, J.B. (1941). A factor analysis of verbal abilities. *Psychometrika* **6**, 279–307.

Carroll, J.B. (1953). An analytic rotation for approximating simple structure in factor analysis. *Psychometrika* **18**, 23–38.

Carroll, J.B. (1993). *Human Cognitive Abilities, A Survey of Factors – Analytic Studies*. Cambridge University Press, Cambridge.

Cattell, R.B. (1966). The scree test for the number of factors. *Multivariate Behavioral Research* **1**, 245–276.

Clarke, M.R.B. (1970). A rapidly convergent method for maximum-likelihood factor analysis. *British Journal of Mathematical and Statistical Psychology* **23**, 43–52.

Crawford, C.B., Ferguson, G.A. (1970). A general rotation criterion and its use in orthogonal rotation. *Psychometrika* **35**, 321–332.

Cudeck, R., O'Dell, L.L. (1994). Applications of standard error estimates in unrestricted factor analysis: significance tests for factor loadings and correlations. *Psychological Bulletin* **115**, 475–487.

Dillon, W.R., Goldstein, M. (1984). *Multivariate Analysis: Methods and Applications*. John Wiley & Sons, New York.

Efron, B. (1979). Bootstrap methods: another look at the jackknife. *Annals of Statistics* **7**, 1–26.

Gananadesikan, R., Kettering, J.R. (1984). A pragmatic review of multivariate methods in application. In: David, H.A., David, H.T. (Eds.), *Statistics: An Appraisal*. Iowa University Press, pp. 309–337.

Goldberg, L.R. (1990). An alternative "description of personality": The big five factor structure. *Journal of Personalities and Social Psychology* **59**, 1216–1229.

Gower, J.C., Dijksterhuis, G.B. (2004). *Procrustes Problems*. Oxford University Press, New York.

Green, B.F. (1952). The orthogonal approximation of an oblique simple structure in factor analysis. *Psychometrika* **17**, 429–440.

Guttman, L. (1952). Multiple group methods for common-factor analysis: Their basis, computation, and interpretation. *Psychometrika* **17**, 209–222.

Guttman, L. (1953). Image theory for the structure of quantitative variates. *Psychometrika* **18**, 277–296.

Guttman, L. (1954). Some necessary conditions for common-factor analysis. *Psychometrika* **19**, 149–160.

Guttman, L. (1955). The determinacy of factor score matrices with implications for five other basic problems of common-factor theory. *British Journal of Mathematical and Statistical Psychology* **8**, 65–81.

Guttman, L. (1958). To what extent can communalities reduce rank? *Psychometrika* **23**, 297–308.

Harman, H.H. (1976). *Modern Factor Analysis*, 3rd ed. University of Chicago Press, Chicago.

Harrell, W. (1940). A factor analysis of mechanical ability tests. *Psychometrika* **5**, 17–33.

Hayashi, K., Bentler, P.M. (2000). The asymptotic covariance matrix of maximum-likelihood estimates in factor analysis: The case of nearly singular matrix of estimates of unique variances. *Linear Algebra and its Applications* **321**, 153–173.

Hayashi, K., Yung, Y.-F. (1999). Standard errors for the class of orthomax-rotated factor loadings: Some matrix results. *Psychometrika* **64**, 451–460.

Hendrickson, A.E., White, P.O. (1964). PROMAX: a quick method for rotation to oblique simple structure. *British Journal of Mathematical and Statistical Psychology* **17**, 65–70.

Horst, J.L. (1965). A rationale and test for the number of factors in factor analysis. *Psychometrika* **30**, 179–185.

Hotelling, H. (1933). Analysis of a complex of statistical variables into principal components. *Journal of Educational Psychology* **24**, 417–441. 498–520.

Hyvärinen, A., Karhunen, J., Oja, E. (2001). *Independent Component Analysis*. John Wiley & Sons, New York.

Ichikawa, M., Konishi, S. (1995). Application of the bootstrap methods in factor analysis. *Psychometrika* **60**, 77–93.

Ichikawa, M., Konishi, S. (2002). Asymptotic expansions and bootstrap approximations in factor analysis. *Journal of Multivariate Analysis* **81**, 47–66.

Ihara, M., Kano, Y. (1986). A new estimator of the uniqueness in factor analysis. *Psychometrika* **51**, 563–566.

Jackson, J.E. (1991). *A User's Guide to Principal Component Analysis*. Wiley Interscience, New York.

Jennrich, R.I. (1973). Standard errors for obliquely rotated factor loadings. *Psychometrika* **38**, 593–604.

Jennrich, R.I. (1974). Simplified formulae for standard errors in maximum-likelihood factor analysis. *British Journal of Mathematical and Statistical Psychology* **27**, 122–131.

Jennrich, R.I., Robinson, S.M. (1969). A Newton–Raphson algorithm for maximum likelihood factor analysis. *Psychometrika* **34**, 111–123.

Jennrich, R.I., Sampson, P.I. (1966). Rotation for simple loadings. *Psychometrika* **31**, 313–323.

Jennrich, R.I., Thayer, D.T. (1973). A note on Lawley's formulas for standard errors in maximum likelihood factor analysis. *Psychometrika* **38**, 571–580.

Jollife, I.T. (2002). *Principal Component Analysis*, 2nd ed. Springer-Verlag, New York.

Jöreskog, K.G. (1967). Some contributions to maximum likelihood factor analysis. *Psychometrika* **32**, 443–482.

Jöreskog, K.G. (1969a). Efficient estimation in image factor analysis. *Psychometrika* **34**, 51–75.

Jöreskog, K.G. (1969b). A general approach to confirmatory maximum likelihood factor analysis. *Psychometrika* **34**, 183–202.

Jöreskog, K.G., Sörbom, D. (1981). *LISREL V: Analysis of Linear Structural Relationships by Maximum Likelihood and Least Squares*. National Educational Resources Inc., Chicago.

Kaiser, H.F. (1958). The varimax criterion for analytic rotation in factor analysis. *Psychometrika* **23**, 187–240.

Kaiser, H.F. (1960). The application of electronic computers to factor analysis. *Educational and Psychological Measurement* **20**, 141–151.

Kaiser, H.F., Caffrey, J. (1965). Alpha factor analysis. *Psychometrika* **30**, 1–14.

Kano, Y., Harada, A. (2000). Stepwise variable selection in factor analysis. *Psychometrika* **65**, 7–22.

Krijnen, W.P. (2002). On the construction of all factors of the model for factor analysis. *Psychometrika* **67**, 161–172.

Krijnen, W.P., Wansbeek, T.J., ten Berge, J.M.F. (1996). Best linear predictors for factor scores. *Communications in Statistics: Theory and Methods* **25**, 3013–3025.

Lawley, D.N. (1940). The estimation of factor loadings by the method of maximum likelihood. *Proceedings of the Royal Society of Edinburgh, Section A* **60**, 64–82.

Lawley, D.N. (1941). Further investigations in factor estimation. *Proceedings of the Royal Society of Edinburgh, Section A* **61**, 176–185.

Lawley, D.N., Maxwell, A.E. (1971). *Factor Analysis as a Statistical Method*, 2nd ed. Butterworths, London.

Ledermann, W. (1937). On the rank of the reduced correlational matrix in multiple-factor analysis. *Psychometrika* **2**, 85–93.

Magnus, J.R., Neudecker, H. (1999). *Matrix Differential Calculus with Applications in Statistics and Econometrics*, revised ed. John Wiley & Sons, Chichester.

Maraun, M.D. (1996). Metaphor taken as math: Indeterminacy in the factor analysis model. *Multivariate Behavioral Research* **31**, 517–538.

McDonald, R.P. (1981). Constrained least squares estimators of oblique common factors. *Psychometrika* **46**, 337–341.

McDonald, R.P., Burr, E.J. (1967). A comparison of four methods of constructing factor scores. *Psychometrika* **32**, 381–401.

Meijer, E., Mooijaart, A. (1996). Factor analysis with heteroscedastic errors. *British Journal of Mathematical and Statistical Psychology* **49**, 189–202.

Mooijaart, A. (1985). Factor analysis for non-normal variables. *Psychometrika* **50**, 323–342.

Mulaik, S.A. (1972). *The Foundations of Factor Analysis*. McGraw-Hill, New York.

Mulaik, S.A. (1986). Factor analysis and Psychometrika, major development. *Psychometrika* **51**, 23–33.

Neudecker, H. (1981). On the matrix formulation of Kaiser's varimax criterion. *Psychometrika* **46**, 343–345.

Neuhaus, J.O., Wrigley, C. (1954). The quartimax method: An analytical approach to orthogonal simple structure. *British Journal of Mathematical and Statistical Psychology* **7**, 81–91.

Ogasawara, H. (1998). Standard errors for rotation matrices with an application to the promax solution. *British Journal of Mathematical and Statistical Psychology* **51**, 163–178.

Osgood, C.E., Suci, C.J. (1957). *The Measurement of Meaning*. University of Illinois Press.

Rao, C.R. (1955). Estimation and tests of significance in factor analysis. *Psychometrika* **20**, 93–111.

Roff, M. (1936). Some properties of the communality in multiple factor theory. *Psychometrika* **1**, 1–6.

Sato, M. (1987). Pragmatic treatment of improper solutions in factor analysis. *Annals of the Institute of Statistical Mathematics* **39**, 443–455.

Schneeweiss, H., Mathes, H. (1995). Factor analysis and principal components. *Journal of Multivariate Analysis* **55**, 105–124.

Schöneman, P.H. (1966). A general solution of the orthogonal Procrustes problem. *Psychometrika* **31**, 1–10.

Spearman, C. (1904). "General Intelligence", objectively determined and measured. *American Journal of Psychology* **15**, 201–293.

Steiger, J.H. (1979). Factor indeterminacy in the last 1930s and the 1970s: some interesting parallels. *Psychometrika* **44**, 157–167.

Takeuchi, K., Yanai, H., Mukherjee, B.N. (1982). *The Foundation of Multivariate Analysis*. Wiley Eastern, New Delhi.

Ten Berge, J.M.F. (1977). Orthogonal Procrustes rotation of two or more matrices. *Psychometrika* **42**, 267–276.

Ten Berge, J.M.F., Krijnen, W.P., Wansbeek, T., Shapiro, A. (1999). Some new results on correlation-preserving factor scores prediction methods. *Linear Algebra and its Applications* **289**, 311–318.

Thurstone, L.L. (1935). *The Vectors of Mind*. University of Chicago Press, Chicago.

Thurstone, L.L. (1936). The factorial isolation of primary abilities. *Psychometrika* **1**, 175–182.

Thurstone, L.L. (1947). *Multiple Factor Analysis*. University of Chicago Press, Chicago.

van Driel, O.P. (1978). On various causes of improper solutions in maximum likelihood factor analysis. *Psychometrika* **43**, 225–243.

Yalcin, I., Amemiya, Y. (2001). Nonlinear factor analysis as a statistical method. *Statistical Science* **16**, 275–294.

Yanai, H. (2000). Reviews of research trends in educational psychology in Japan during the past year – measurement and evaluation centering on factor analysis. *The Annual Report of Educational Psychology in Japan* **39**, 96–108. In Japanese.

Yanai, H., Ichikawa, M. (1990). New lower and upper bounds for communality in factor analysis. *Psychometrika* **55**, 405–410.

Yanai, H., Kashiwagi, S., Kokusho, R. (1987). Construction of a new personality inventory by means of factor analysis based on Promax rotation. *The Japanese Journal of Psychology* **58**, 158–165. In Japanese.

Yanai, H., Mukherjee, B.N. (1987). A generalized method of image analysis from an intercorrelation matrix which may be singular. *Psychometrika* **52**, 554–564.

Yanai, H., Shigemasu, K., Mayekawa, S., Ichikawa, M. (1990). *Factor Analysis – Theory and Application*. Asakura Publishing Company, Tokyo. In Japanese.

Yuan, K. -H., Marshall, L.L., Bentler, P.M. (2002). A unified approach to exploratory factor analysis with missing data, nonnormal data, and in the presence of outliers. *Psychometrika* **67**, 95–122.

Handbook of Statistics, Vol. 26
ISSN: 0169-7161

DOI: 10.1016/S0169-7161(06)26010-3

10

Structural Equation Modeling[1]

Ke-Hai Yuan and Peter M. Bentler

1. Introduction

In social and behavioral sciences, interesting attributes such as attitudes, IQ, personality traits, political liberalism or conservatism, socio-economic status, etc., cannot be observed directly and are often called latent variables. Although there are several different views about what constitutes a "latent variable" (Bollen, 2002), the influence of such variables can be assessed by multiple indicators that are subject to measurement errors. Due to measurement errors, conventional statistical methodology such as regression and ANOVA/MANOVA cannot be directly used to analyze the relationships among these attributes. By segregating measurement errors from the true scores of attributes, structural equation modeling (SEM) provides a methodology to model the latent variables directly. The methodology of SEM has enjoyed tremendous developments since 1970, and is now widely applied (Hershberger, 2003; MacCallum and Austin, 2000). The idea of multiple indicators for a latent variable is from factor analysis (Spearman, 1904; Thurstone, 1947; Lawley and Maxwell, 1971). SEM is often regarded as an extension of factor analysis in the psychometric literature. This methodology also covers several widely used statistical models in various disciplines. We will present several specific models before introducing the general mean and covariance structures.

Confirmatory factor analysis (CFA) model

Holzinger and Swineford (1939) contains a data set of test scores that comprise the following subtests or variables: Visual Perception, Cubes, Lozenges, Paragraph Comprehension, Sentence Completion, Word Meaning, Addition, Counting Dots, Straight-Curved Capitals. In the original report of Holzinger and Swineford (1939), the first three variables were designed to measure the spatial ability, the next three variables were designed to measure verbal ability, and the last three variables were administered with a limited time and were designed to measure a speed factor in performing the tasks. Let $\mathbf{x} = (x_1, x_2, \ldots, x_9)'$ represent the observed variables, f_1, f_2, and f_3 represent respectively the hypothetical latent spatial, verbal, and speed scores. Then, with

[1]This research was supported by NSF grant DMS04-37167, the James McKeen Cattell Fund, and grants DA01070 and DA00017 from the National Institute on Drug Abuse. We are thankful to two referees for their very constructive comments.

$\mathbf{f} = (f_1, f_2, f_3)'$, Holzinger and Swineford's design can be represented by

$$\mathbf{x} = \boldsymbol{\mu} + \boldsymbol{\Lambda}\mathbf{f} + \mathbf{e}, \tag{1}$$

where $\boldsymbol{\mu} = E(\mathbf{x})$, $E(\mathbf{f}) = \mathbf{0}$, $E(\mathbf{e}) = \mathbf{0}$,

$$\boldsymbol{\Lambda} = \begin{pmatrix} \lambda_{11} & \lambda_{21} & \lambda_{31} & 0 & 0 & 0 & 0 & 0 & 0 \\ 0 & 0 & 0 & \lambda_{42} & \lambda_{52} & \lambda_{62} & 0 & 0 & 0 \\ 0 & 0 & 0 & 0 & 0 & 0 & \lambda_{73} & \lambda_{83} & \lambda_{93} \end{pmatrix}'$$

with λ_{ij} being the regression coefficient of x_i on f_j, and $\mathbf{e} = (e_1, e_2, \ldots, e_9)'$ is the remaining part of $\mathbf{x}$ after partialling out the predictor $\mathbf{f}$. In confirmatory factor analysis, $\mathbf{f}$ is called the latent vector of factors, $\boldsymbol{\Lambda}$ is the factor loading matrix, and $\mathbf{e}$ is the vector of "measurement errors" or "uniquenesses". Conceptually, the three factors should be correlated and we may denote $\boldsymbol{\Phi} = (\phi_{ij}) = \mathrm{Cov}(\mathbf{f})$. Let $\boldsymbol{\Psi} = \mathrm{Cov}(\mathbf{e})$ and assume $\mathbf{f}$ and $\mathbf{e}$ are uncorrelated. Then $\boldsymbol{\Sigma}_0 = \mathrm{Cov}(\mathbf{x})$ can be modeled by

$$\boldsymbol{\Sigma}(\boldsymbol{\theta}) = \boldsymbol{\Lambda}\boldsymbol{\Phi}\boldsymbol{\Lambda}' + \boldsymbol{\Psi}, \tag{2}$$

where $\boldsymbol{\theta}$ contains the unknown elements in $\boldsymbol{\Lambda}$, $\boldsymbol{\Phi}$ and $\boldsymbol{\Psi}$. Ideally, we may want the errors in $\mathbf{e}$ to be uncorrelated, that is, $\boldsymbol{\Psi}$ to be a diagonal matrix. Then the covariances of the nine variables in Holzinger and Swineford's test will just be due to the three factors.

When setting the diagonal matrix of $\boldsymbol{\Phi}$ at 1.0, the off diagonal elements ϕ_{ij} will be the correlations among the latent factors. These correlations cannot be consistently estimated without a model, e.g., not by using the product-moment correlations of the observed variables. Actually, due to measurement errors, the product-moment correlations underestimate the correlations of the latent variables. Unlike the exploratory factor analysis (EFA) model (see Yanai and Ichikawa, Chapter 9, this volume), the zero loadings in a general CFA model are specified a priori based on subject-matter knowledge, and their number and placement guarantee that the model is identified without any rotational constraints (Jennrich, 1974). In the example, each variable has only one nonzero loading, but this is not essential. We can also allow each x_i to load on multiple f_j's, subject to the design or substantive theory as well as to model identification (see below). Although it may be desirable to have $\boldsymbol{\Psi}$ be a diagonal matrix, it is totally legitimate to allow some of the e_i's to correlate. Of course, whether the errors are correlated or each variable only measures one factor has to be and can be tested statistically.

Errors-in-variables model

Let $\mathbf{x}^*$ be a vector, y^* be a variable and the following regression model

$$y^* = \boldsymbol{\beta}'\mathbf{x}^* + e, \tag{3}$$

is of interest, where $\boldsymbol{\beta}$ is the vector of regression coefficients, e is the prediction error. Model (3) is just the standard regression model when $\mathbf{x}^*$ and y^* can be observed directly. Suppose $\mathbf{x}^*$ and y^* are latent variables, and we can only observe $\mathbf{x}$ and y with $\mathbf{x} = \mathbf{x}^* + \boldsymbol{\delta}$ and $y = y^* + \varepsilon$. In order to get efficient and unbiased estimator of $\boldsymbol{\beta}$, we have to consider the following augmented model

$$\begin{pmatrix} y \\ \mathbf{x} \end{pmatrix} = \begin{pmatrix} \boldsymbol{\beta}' \\ \mathbf{I} \end{pmatrix} \mathbf{x}^* + \begin{pmatrix} e + \varepsilon \\ \boldsymbol{\delta} \end{pmatrix}. \tag{4}$$

Model (4) is often called the errors-in-variables model. It has been used in many disciplines including life sciences, econometrics, chemometrics, geology, and others. If we assume $\mathbf{x}^*$ is a random vector with $\mathrm{Cov}(\mathbf{x}^*) = \boldsymbol{\Phi}$, then the covariance structure of (4) is in a similar form as (2). Actually, parameter estimation and model testing of models (2) and (4) are essentially the same, both have to explicitly model the second-order moments of the observed variables. Further development of errors-in-variables models and their applications can be found in Fuller (1987) and Carroll et al. (1995).

Simultaneous equation model

Suppose the ideal demand and supply equations can be expressed as

$$\mathbf{B}\mathbf{y}^* + \boldsymbol{\Gamma}\mathbf{x}^* = \mathbf{0}. \tag{5}$$

The vectors $\mathbf{x}^*$ and $\mathbf{y}^*$ cannot be observed directly, they can only be observed through $\mathbf{x}$ and $\mathbf{y}$ with $\mathbf{y} = \mathbf{y}^* + \boldsymbol{\varepsilon}$ and $\mathbf{x} = \mathbf{x}^* + \boldsymbol{\delta}$. Then (5) becomes

$$\mathbf{B}\mathbf{y} + \boldsymbol{\Gamma}\mathbf{x} = \mathbf{u}, \tag{6}$$

where $\mathbf{u} = \mathbf{B}\boldsymbol{\varepsilon} + \boldsymbol{\Gamma}\boldsymbol{\delta}$ is a vector of random errors. Equations (5) and (6) are standard forms of the simultaneous equation model in econometrics. Assuming $\mathbf{B}$ is nonsingular, (6) can be rewritten as

$$\mathbf{y} = -\mathbf{B}^{-1}\boldsymbol{\Gamma}\mathbf{x} + \mathbf{B}^{-1}\mathbf{u}. \tag{7}$$

Similar to (1) and (4), (7) is also a set of linear equations consisting of observed and latent variables. More in depth discussion on simultaneous equations is provided in Dhrymes (1994). A bibliography emphasizing the closely related field of path analysis is given by Wolfle (2003).

Growth curve model

The factor analysis model (1)–(2) can be extended to also include a mean structure. That is, $\boldsymbol{\mu}$ is a vector of functions of a set of more basic parameters (Sörbom, 1974). One example of a mean structure is the growth curve model. Let $\mathbf{y} = (y_1, y_2, \ldots, y_p)'$ be repeated measures at p time points. The following model

$$\mathbf{y} = \boldsymbol{\Lambda}\mathbf{f} + \mathbf{e}, \tag{8a}$$

has been widely used to study the pattern of change in y during the p time points (Duncan et al., 1999; Meredith and Tisak, 1990; McArdle and Epstein, 1987), where

$$\boldsymbol{\Lambda} = \begin{pmatrix} 1 & 1 & 1 & \cdots & 1 \\ 0 & 1 & \lambda_3 & \cdots & \lambda_p \end{pmatrix}', \tag{8b}$$

$\mathbf{f} = (f_1, f_2)'$ and $\mathbf{e}$ are uncorrelated with $E(\mathbf{f}) = \boldsymbol{\tau} = (\tau_1, \tau_2)'$, $E(\mathbf{e}) = \mathbf{0}$, $\mathrm{Cov}(\mathbf{f}) = \boldsymbol{\Phi}$, and $\mathrm{Cov}(\mathbf{e}) = \boldsymbol{\Psi}$ is a diagonal matrix. Compared to the CFA model in (1), the mean of the latent vector $\mathbf{f}$ is not zero. In addition to the covariance structure given by (2), we also have the following mean structure

$$\boldsymbol{\mu} = \boldsymbol{\Lambda}\boldsymbol{\tau}.$$

Notice that, for each y_i, (8) is just a regression model with the latent intercept f_1 and slope f_2. The rate of growth may be correlated with initial status, which is reflected in the covariance ϕ_{12}. The matrix $\mathbf{\Phi}$ depends on how the λ_j's in (8b) are chosen (Stoel, 2003), different parameterizations lead to different $\mathbf{\Phi}$'s. In many applications, λ_3 to λ_p in (8b) are known constants, and the model represents a growth curve with predetermined or hypothesized growth rates. A special case is the linear growth model with $\lambda_3 = 2, \lambda_4 = 3, \ldots, \lambda_p = p - 1$. A less restricted model is to let λ_3 to λ_p be free parameters. Then (8) represents a growth curve model with an arbitrary growth rate to be determined by $(\lambda_{j+1} - \lambda_j)\tau_2$. In (8b) there are only two growth curve factors, but cubic, quadratic, and other types of more complicated growth structures are possible. Then (8b) would contain more factors with certain fixed factor loadings to represent these more complex growth structures (e.g., Bentler, 2005). For example, $\mathbf{\Lambda}$ may contain orthogonal polynomials.

Factor analytic simultaneous equation model

It is possible to combine the factor analytic model (1) with the simultaneous equation model (6). Effectively this was done by Jöreskog (1973), Keesling (1972) and Wiley (1973), and has become a popular method for structural modeling with the LISREL computer program (Jöreskog and Sörbom, 2001). In this model, the measured variables are partitioned into two sets, and factor analytic models are posited for both sets. Their associated latent factors are related by a simultaneous equation model. We illustrate this model with Neumann's (1994) study on the relationship of alcoholism and psychological symptoms, based on a data set with 10 variables. The two variables in $\mathbf{x} = (x_1, x_2)'$ are respectively Family History for Psychopathology and Family History for Alcoholism, which are indicators for a latent construct of Family History. The 8 variables in $\mathbf{y} = (y_1, \ldots, y_8)'$ are respectively the Age of First Problem with Alcohol, Age of First Detoxification from Alcohol, Alcohol Severity Score, Alcohol Use Inventory, SCL-90 Psychological Inventory, the Sum of the Minnesota Multiphasic Personality Inventory Scores, the Lowest Level of Psychosocial Functioning during the past year, and the Highest Level of Psychosocial Functioning during the past year. With two indicators for each latent construct, these 8 variables are respectively measuring: Age of Onset, Alcohol Symptoms, Psychopathology Symptoms, and Global Functioning. Neumann's (1994) theoretical model for these variables is

$$\mathbf{x} = \boldsymbol{\mu}_x + \mathbf{\Lambda}_x\xi + \boldsymbol{\delta}, \qquad \mathbf{y} = \boldsymbol{\mu}_y + \mathbf{\Lambda}_y\boldsymbol{\eta} + \boldsymbol{\varepsilon}, \tag{9a}$$

and

$$\boldsymbol{\eta} = \mathbf{B}\boldsymbol{\eta} + \mathbf{\Gamma}\xi + \boldsymbol{\zeta}, \tag{9b}$$

where $\boldsymbol{\mu}_x = E(\mathbf{x})$, $\boldsymbol{\mu}_y = E(\mathbf{y})$, $E(\xi) = 0$, $E(\boldsymbol{\eta}) = \mathbf{0}$, $E(\boldsymbol{\varepsilon}) = \mathbf{0}$, $E(\boldsymbol{\zeta}) = \mathbf{0}$,

$$\mathbf{\Lambda}_x = \begin{pmatrix} \lambda_1 \\ \lambda_2 \end{pmatrix}, \qquad \mathbf{\Lambda}_y = \begin{pmatrix} 1 & \lambda_3 & 0 & 0 & 0 & 0 & 0 & 0 \\ 0 & 0 & 1 & \lambda_4 & 0 & 0 & 0 & 0 \\ 0 & 0 & 0 & 0 & 1 & \lambda_5 & 0 & 0 \\ 0 & 0 & 0 & 0 & 0 & 0 & 1 & \lambda_6 \end{pmatrix}',$$

$$\mathbf{B} = \begin{pmatrix} 0 & 0 & 0 & 0 \\ \beta_{21} & 0 & 0 & 0 \\ \beta_{31} & \beta_{32} & 0 & 0 \\ 0 & \beta_{42} & \beta_{43} & 0 \end{pmatrix}, \qquad \mathbf{\Gamma} = \begin{pmatrix} \gamma_{11} \\ 0 \\ 0 \\ 0 \end{pmatrix},$$

and $\boldsymbol{\varepsilon}$, $\boldsymbol{\delta}$, and $\boldsymbol{\zeta}$ are vectors of errors whose elements are all uncorrelated. Similar to the previous model, (9) is a set of equations involving latent and observed variables. Equations (9a) provide the measurement model that relates observed to latent variables, while (9b) provides the simultaneous equation relating the latent variables. Model (9) can be regarded as an extension of (1) where the latent variables are just correlated while in (9) they have a predicative relationship. Model (9) can also be regarded as a multivariate version of (3) and (4).

Linear structural equations model

Another very general formulation was given by Bentler and Weeks (1980). In their approach, the distinction between latent and observed variables is secondary to the distinction between dependent (being predicted by other) variables and independent (not being predicted by other) variables. Any variable in the model (measured, latent, residual, etc.) is either a dependent variable or an independent variable. Even if a variable is a predictor of other variables in the model, as long as it is predicted by one or more variables, it is a dependent variable according to the definition. While all variables may be correlated in the population, only the covariances among the independent variables can be part of the model parameters. The covariances among the dependent variables or between the dependent variables and the independent variables do not vary freely, instead, they are explained by the model through the free parameters. In any application, most correlations among the independent variables will be set as zero due to the hypothesized model structure. For example, **f** and **e** in (1) are independent variables and they do not correlate; the elements of **e** do not correlate either in many applications of the CFA model. Let $\boldsymbol{\xi}$ contain all the independent variables and $\boldsymbol{\eta}$ contain all the dependent variables. The Bentler–Weeks model can be written as

$$\boldsymbol{\eta} = \mathbf{L}\boldsymbol{\eta} + \mathbf{M}\boldsymbol{\xi}, \tag{10a}$$

where **L** and **M** are coefficient matrices. Let

$$\mathbf{B} = \begin{pmatrix} \mathbf{L} & 0 \\ 0 & 0 \end{pmatrix}, \qquad \mathbf{\Gamma} = \begin{pmatrix} \mathbf{M} \\ \mathbf{I} \end{pmatrix}, \qquad \boldsymbol{\nu} = \begin{pmatrix} \boldsymbol{\eta} \\ \boldsymbol{\xi} \end{pmatrix}.$$

Then (10a) is equivalent to

$$\boldsymbol{\nu} = \mathbf{B}\boldsymbol{\nu} + \mathbf{\Gamma}\boldsymbol{\xi}. \tag{10b}$$

Suppose $\mathbf{I} - \mathbf{B}$ is nonsingular. Then all the variables can be expressed as a linear combination of the independent variables, yielding the reduced form

$$\boldsymbol{\nu} = (\mathbf{I} - \mathbf{B})^{-1}\mathbf{\Gamma}\boldsymbol{\xi}.$$

This allows us to relate (10) to data. Some of the variables in $\boldsymbol{\nu}$ are measured variables **x**, i.e., observed data to be modeled, and we obtain those with the selection matrix **G** with elements 0 or 1 such that $\mathbf{x} = \mathbf{G}\boldsymbol{\nu}$. It follows that the observed variables can be expressed as $\mathbf{x} = \mathbf{G}(\mathbf{I} - \mathbf{B})^{-1}\boldsymbol{\Gamma}\boldsymbol{\xi}$. From this we obtain the mean and covariance structure of the observed variables. Denoting $\boldsymbol{\mu}_\xi = E(\boldsymbol{\xi})$ and $\boldsymbol{\Phi} = \mathrm{Cov}(\boldsymbol{\xi})$, we have

$$\boldsymbol{\mu} = E(\mathbf{x}) = \mathbf{G}(\mathbf{I} - \mathbf{B})^{-1}\boldsymbol{\Gamma}\boldsymbol{\mu}_\xi. \tag{11a}$$

The elements of $\boldsymbol{\mu}$ corresponding to independent observed variables are just elements of $\boldsymbol{\mu}_\xi$, while those corresponding to the dependent observed variables are predicted by $\boldsymbol{\mu}_\xi$ according to (11a). The covariance structure of the observed variables is

$$\boldsymbol{\Sigma} = \mathbf{G}(\mathbf{I} - \mathbf{B})^{-1}\boldsymbol{\Gamma}\boldsymbol{\Phi}\boldsymbol{\Gamma}'(\mathbf{I} - \mathbf{B}')^{-1}\mathbf{G}'. \tag{11b}$$

In (11b), the rank of $\boldsymbol{\Sigma}$ is typically less than the rank of $\boldsymbol{\Phi}$, i.e., the number of observed variables in $\mathbf{x}$ is usually less than the number of independent variables in $\boldsymbol{\xi}$ (Bentler, 1982). For example, the number of independent variables in (1) is 12 while the number of observed variables is only 9.

General mean and covariance structures

In view of the variety of models that are available, it is most convenient to describe the statistical theory associated with these models by mean and covariance structures. Let $\boldsymbol{\theta}$ be the vector containing all the free parameters in a model. Then all the models from (1) to (11) can be represented by the mean and covariance structures $\boldsymbol{\mu} = \boldsymbol{\mu}(\boldsymbol{\theta})$ and $\boldsymbol{\Sigma} = \boldsymbol{\Sigma}(\boldsymbol{\theta})$, respectively. Parameter estimates for $\boldsymbol{\theta}$ cannot be obtained by solving the equations involving the latent variables. They have to be obtained by fitting the model implied moments to the corresponding sample moments through minimizing a proper discrepancy function. Thus, SEM is also called mean and covariance structure analysis. When the mean structure is saturated, or when only the covariance structure is of interest as in the CFA model in (1) and (2), the corresponding model is also called a covariance structure model and the method of analysis, covariance structure analysis (CSA).

Our purpose is to provide a systematic treatment for SEM with continuous variables or variables that can be regarded as continuous,[1] where the mean and covariance structures of the observed variables are of interest. In addition to mean and covariance structure models, higher-order moments may also be of interest and can be modeled. Interested readers are referred to Bentler (1983) and Mooijaart (1985) for further discussion on modeling higher-order moments. The models in (1) to (11) relate all the variables through linear equations. There are situations when nonlinear relationships, interactions among independent latent variables in particular, are of interest. These will not be considered here. Interested readers are referred to Lee and Zhu (2000, 2002), Klein and Moosbrugger (2000), Wall and Amemiya (2001), Yang-Wallentin and Jöreskog (2001), Lee et al. (2004), and Marsh et al. (2004) for different ways of model estimation and evaluation. A nonlinear relationship is also necessary when the observed variables cannot be regarded as continuous. One approach assumes that underlying normal variables have been categorized to yield the observed ordered categorical

[1] Dolan (1994) warned that discrete variables with fewer than five categories cannot be regarded as continuous, and treating them as such can lead to bias in estimates (e.g., DiStefano, 2002). When the underlying variables (before categorization) are multivariate normal, Olsson (1979) found that sample skewness created by categorization also plays a substantial role in treating discrete variables as continuous ones.

variables. At a first stage, the thresholds associated with such categorization are estimated. These are used next to obtain estimates of correlations among these underlying variables, called polychoric or polyserial correlations. These correlations are then fit by a CSA model using an appropriate discrepancy function (Flora and Curran, 2004; Muthén, 1984; Jöreskog, 1994; Lee et al., 1992, 1995; Poon and Lee, 1987). Related approaches exist (e.g., Bauer and Curran, 2004; Muthén, 2002). Another approach uses a logistic or normal ogive model to fit the observed categorical data, called item response theory (see, e.g., Maydeu-Olivares, 2001, and related chapters of this volume). These two approaches are equivalent when the normal ogive model is used in the item response theory approach (Maydeu-Olivares, 2005; Takane and de Leeuw, 1987). A recent development for SEM with continuous as well as polytomous data are given by Lee and Song (2003). A generalized approach to various model types is given by Skrondal and Rabe-Hesketh (2004).

Many introductory books on SEM have been written. A recent one by Kline (2005) assumes a minimal level of statistics. The path diagram approach in Loehlin (2004) is unique and stimulates understanding of the material without resorting to equations. Byrne (2006) illustrates the practical aspects of setting up and running models using a variety of real-life examples. A sophisticated overview is Bollen (1989), which covers most of the earlier technical developments in SEM. Several computer programs provide a means of estimating and testing a variety of SEM models. These programs include AMOS, EQS, LISREL, Mplus, Mx, and SAS Proc Calis. The first three of these permit setting up models with easy to understand path diagrams. Many of the recently developed statistics described below are implemented primarily in EQS, but SEM software is constantly updated and the best source for each program is the relevant web site:

- Amos (http://www.spss.com/amos/);
- EQS (http://www.mvsoft.com);
- LISREL (http://ssicentral.com/lisrel/mainlis.htm);
- Mplus (http://statmodel.com);
- Mx (http://vcu.edu/mx/);
- SAS Proc Calis (http://2.sas.com/proceedings/sugi26/p260-26.pdf).

After introducing various models, we will turn to model evaluation and estimation. Section 2 of the chapter will discuss rules for model identification. Section 3 presents statistical development for general mean and covariance structure models, aiming to give an overall picture as well as up-to-date statistical methodology. Sections 4–6 will present the methodology of SEM with missing data and multiple groups as well as multilevel models with nested data. We conclude the chapter with examples based on two real data sets in Section 7.

2. Model identification

Unlike the multiple regression model, where there is only one model for a given set of predictors, there are many different SEM models for a given number of variables. Not only the general model type (see above) needs to be specified, but a particular realization of the given structure has to be specified for the variables at hand. The latent factors, if any, and their indicators have to be specified, as do the potential linear relations among these factors. Any such SEM model will have a set of parameters, and we need to be able to estimate the unknown parameters and evaluate whether the model is consistent with, or significantly different from, data. In order to get a unique set of parameter estimates, the model has to be identified. That is, $\boldsymbol{\mu}(\boldsymbol{\theta}_1) = \boldsymbol{\mu}(\boldsymbol{\theta}_2)$ and $\boldsymbol{\Sigma}(\boldsymbol{\theta}_1) = \boldsymbol{\Sigma}(\boldsymbol{\theta}_2)$ imply $\boldsymbol{\theta}_1 = \boldsymbol{\theta}_2$. A universal sufficient condition for an arbitrary model to be identified is difficult to formulate. It is relatively easy to judge whether a specific model is identified or not. The necessary conditions include:

(I) The scale of each latent variable has to be defined. When the latent variable is an independent variable, we can define its scale by either fixing its variance or fixing the regression coefficient of this latent variable against a variable whose scale is known. For example, in the factor model (1), we can either fix the variance of each factor or one of the loadings of the factor at a prespecified arbitrary value. For the CFA model, many researchers set the factor variances at 1.0 and the coefficients in front of the errors at 1.0, to resemble the EFA models (see Yanai and Ichikawa, Chapter 9, this volume).

(II) With p measured variables, there are p sample means and $p(p+1)/2$ sample variances-covariances. The number of model parameters q in $\boldsymbol{\theta}$ has to be smaller than or equal to $p(p+3)/2 = p + p(p+1)/2$. The model may still be identified when $q = p(p+3)/2$, then the estimated structural model will reproduce the sample means and variances-covariances exactly and the model cannot be tested or rejected. Such a model is equivalent to the model where all the elements in $\boldsymbol{\mu}$ and $\boldsymbol{\Sigma}$ are free, which is also called the saturated or unstructured model.

(III) Although it is possible to see models containing a factor with one indicator in the literature, such models are not identified unless the variable is identical to the factor (i.e., there is no residual error) or the error variance is known. When indicators contain errors, there should be at least two indicators for each latent variable. A model with just two indicators for a factor is also not identified, because there are more parameters than the number of variances and a covariance. But a two variable factor can be identified when it is associated with other factors or variables. If a latent variable is not correlated with, predicts or is predicted by other variables in the model or in the population, then three indicators for the latent variable are necessary to make the overall model identified. When a latent variable with two indicators is associated with other variables in the model but not in the population, then the model is also said to be empirically unidentified because the sample covariances of the two indicators with other variables can be very close to zero. This will make the estimates very unstable.

When a model is not identified, some parameters may still enjoy unique solutions. Then the part of the model formed by these parameters is often called locally identified.

A commonly used strategy for identification is to fix enough parameters (e.g., factor loadings) equal to zero and to leave the errors uncorrelated. In many applications of SEM, it is desirable to let each indicator measure only one factor. This not only makes the model more likely to be identified, but the substantive meaning of the latent variable is clearer (see Anderson and Gerbing, 1988). When several variables measure one and only one factor in a large model, these variables are also called unidimensional. Unidimensionality is strongly encouraged when developing questionnaire or measurement scales.

Model identification is a minimal requirement. The quality of modeling results will depend on many other factors, such as levels of communality of variables and overdetermination of factors (e.g., MacCallum et al., 1999). Furthermore, even though a model is identified, there may exist other models that are equivalent and yield identical global fit (e.g., Raykov and Marcoulides, 2001).

Further discussion on identification can be found in Bekker et al. (1994), Loken (2005), Millsap (2001), Steiger (2002), Wansbeek and Meijer (2000), and Wiley (1973).

3. Estimation and evaluation

We will use $\mathbf{x}$ to represent the p observed variables whose population mean and covariance matrix are $E(\mathbf{x}) = \boldsymbol{\mu}_0$ and $\mathrm{Cov}(\mathbf{x}) = \boldsymbol{\Sigma}_0$. Let $\mathbf{x}_1, \ldots, \mathbf{x}_n$ be a sample and $\bar{\mathbf{x}}$ and $\mathbf{S}$ be the sample mean vector and sample covariance matrix. Ideally, we like to have interesting models $\boldsymbol{\mu}(\boldsymbol{\theta})$ and $\boldsymbol{\Sigma}(\boldsymbol{\theta})$ such that $\boldsymbol{\mu}_0 = \boldsymbol{\mu}(\boldsymbol{\theta}_0)$ and $\boldsymbol{\Sigma}_0 = \boldsymbol{\Sigma}(\boldsymbol{\theta}_0)$. Because $\boldsymbol{\mu}_0$ and $\boldsymbol{\Sigma}_0$ are generally unknown, we have to use $\bar{\mathbf{x}}$ and $\mathbf{S}$ to evaluate the model and to estimate $\boldsymbol{\theta}_0$. When the mean and covariance structure is identified, parameter estimates can be obtained by minimizing a discrepancy function between the data and the model. Several such functions have been well studied, these are the least squares (LS), the normal distribution based maximum likelihood (ML), the normal distribution based generalized LS (GLS), and the GLS based on an arbitrary distribution (AGLS) with finite fourth-order moments. In addition to methods modeling sample means and sample variances–covariances, robust procedures that take into account the distributional shape of the sample have been developed more recently. We will introduce these methods one by one. Because model evaluation is closely related to model estimation, we will discuss them together within each discrepancy function. In the development of SEM methodology, procedures for CSA have been more fully studied. We will make the distinction when necessary to present the results only for CSA.

We need to introduce some notation to facilitate the technical development. For a $p \times p$ symmetric matrix $\mathbf{A}$, let $\mathrm{vec}(\mathbf{A})$ be the operator that stacks the columns of $\mathbf{A}$ and $\mathrm{vech}(\mathbf{A})$ only stacks the elements on and below the diagonal. Then $\mathrm{vec}(\mathbf{A})$ contains duplicated elements of $\mathrm{vech}(\mathbf{A})$. Thus, there exists a duplication matrix $\mathbf{D}_p$ such that $\mathrm{vec}(\mathbf{A}) = \mathbf{D}_p \, \mathrm{vech}(\mathbf{A})$. We denote $\mathbf{s} = \mathrm{vech}(\mathbf{S})$, $\boldsymbol{\sigma} = \mathrm{vech}(\boldsymbol{\Sigma})$, $\mathbf{t}_i = (\mathbf{x}_i', \mathrm{vech}'\{(\mathbf{x}_i - \bar{\mathbf{x}})(\mathbf{x}_i - \bar{\mathbf{x}})'\})'$, $\bar{\mathbf{t}} = (\bar{\mathbf{x}}', \mathbf{s}')'$, $\boldsymbol{\beta} = (\boldsymbol{\mu}', \boldsymbol{\sigma}')'$,

$$\begin{aligned} \mathbf{W}_c &= 2^{-1}\mathbf{D}_p'\left(\boldsymbol{\Sigma}^{-1} \otimes \boldsymbol{\Sigma}^{-1}\right)\mathbf{D}_p, \\ \mathbf{W} &= \mathrm{diag}\left(\boldsymbol{\Sigma}^{-1}, \mathbf{W}_c\right) = \begin{pmatrix} \boldsymbol{\Sigma}^{-1} & \mathbf{0} \\ \mathbf{0} & \mathbf{W}_c \end{pmatrix}. \end{aligned} \tag{12}$$

A dot on top of a function implies derivatives or a Jacobian matrix as in $\dot{\boldsymbol{\beta}}(\boldsymbol{\theta}) = \partial\boldsymbol{\beta}(\boldsymbol{\theta})/\partial\boldsymbol{\theta}'$; double dots on top of a function imply the matrix of second derivatives as in $\ddot{\mu}_i(\boldsymbol{\theta}) = \partial^2 \mu_i(\boldsymbol{\theta})/\partial\boldsymbol{\theta}\partial\boldsymbol{\theta}'$. When the function has more than one set of arguments, a subscript is used to denote the partial derivative. We may omit the argument of the function when it is evaluated at $\boldsymbol{\theta}_0$. We will use $\mathbf{I}$ for the identity matrix and $\mathbf{0}$ for a vector or matrix of zeros, subscripts will be used when their dimensions are not obvious. Convergence in probability and distribution will be denoted by $\xrightarrow{P}$ and $\xrightarrow{\mathcal{L}}$, respectively.

The regularity conditions for SEM are (C1) $\boldsymbol{\theta}_0 \in \boldsymbol{\Theta}$ which is a compact subset of R^q. (C2) $\boldsymbol{\beta}(\boldsymbol{\theta}) = \boldsymbol{\beta}(\boldsymbol{\theta}_0)$ only when $\boldsymbol{\theta} = \boldsymbol{\theta}_0$. (C3) $\boldsymbol{\beta}(\boldsymbol{\theta})$ is twice continuously differentiable. (C4) $\dot{\boldsymbol{\beta}}$ is of full rank. (C5) The vector $(\mathbf{x}_i', \text{vech}'\{(\mathbf{x}_i - \boldsymbol{\mu}_0)(\mathbf{x}_i - \boldsymbol{\mu}_0)'\})'$ has a covariance matrix that is of full rank. Conditions (C1) and (C3) are the standard regularity conditions and are generally satisfied in practice. Condition (C2) implies that the model structure is identified. If the structural model is properly parameterized, condition (C4) will be satisfied. Conditions (C1) and (C2) are for consistency of parameter estimates. Conditions (C3) and (C4) are needed to establish the asymptotic normality of a parameter estimate. (C5) is needed in order for parameter estimates or test statistics for the overall model to have proper asymptotic distributions. (C5) will be satisfied when $\mathbf{x}_i \sim N(\boldsymbol{\mu}_0, \boldsymbol{\Sigma}_0)$ and $\boldsymbol{\Sigma}_0$ is full rank. Further discussions of regularity conditions can be found in Browne (1984), Shapiro (1984), Kano (1986), and Yuan and Bentler (1997a). We will implicitly assume these conditions throughout the chapter.

3.1. Normal theory ML and related procedures

Assuming $\mathbf{x} \sim N_p(\boldsymbol{\mu}, \boldsymbol{\Sigma})$, the log likelihood function of $\boldsymbol{\theta}$ is given by

$$l(\boldsymbol{\theta}) = -\frac{n}{2}\ln\bigl|\boldsymbol{\Sigma}(\boldsymbol{\theta})\bigr| - \frac{n}{2}\operatorname{tr}\bigl\{\boldsymbol{\Sigma}^{-1}(\boldsymbol{\theta})\bigl[\mathbf{S} + \bigl(\bar{\mathbf{x}} - \boldsymbol{\mu}(\boldsymbol{\theta})\bigr)\bigl(\bar{\mathbf{x}} - \boldsymbol{\mu}(\boldsymbol{\theta})\bigr)'\bigr]\bigr\}. \tag{13}$$

Let the ML estimate (MLE) that maximizes $l(\boldsymbol{\theta})$ be $\hat{\boldsymbol{\theta}}$. When all the elements in $\boldsymbol{\mu}$ and $\boldsymbol{\Sigma}$ are free parameters, their MLEs are given, respectively, by $\hat{\boldsymbol{\mu}} = \bar{\mathbf{x}}$ and $\hat{\boldsymbol{\Sigma}} = \mathbf{S}$. When evaluated at $\hat{\boldsymbol{\mu}} = \bar{\mathbf{x}}$ and $\hat{\boldsymbol{\Sigma}} = \mathbf{S}$,

$$l(\bar{\mathbf{x}}, \mathbf{S}) = -\frac{n}{2}\ln|\mathbf{S}| - \frac{np}{2}. \tag{14}$$

It follows from (13) and (14) that

$$2\bigl[l(\bar{\mathbf{x}}, \mathbf{S}) - l(\boldsymbol{\theta})\bigr] = nD_{\text{ML}}\bigl[\bar{\mathbf{x}}, \mathbf{S}, \boldsymbol{\mu}(\boldsymbol{\theta}), \boldsymbol{\Sigma}(\boldsymbol{\theta})\bigr],$$

where

$$D_{\text{ML}}\bigl[\bar{\mathbf{x}}, \mathbf{S}, \boldsymbol{\mu}(\boldsymbol{\theta}), \boldsymbol{\Sigma}(\boldsymbol{\theta})\bigr] = \bigl[\bar{\mathbf{x}} - \boldsymbol{\mu}(\boldsymbol{\theta})\bigr]'\boldsymbol{\Sigma}^{-1}(\boldsymbol{\theta})\bigl[\bar{\mathbf{x}} - \boldsymbol{\mu}(\boldsymbol{\theta})\bigr] + \operatorname{tr}\bigl[\mathbf{S}\boldsymbol{\Sigma}^{-1}(\boldsymbol{\theta})\bigr] - \log\bigl|\mathbf{S}\boldsymbol{\Sigma}^{-1}(\boldsymbol{\theta})\bigr| - p \tag{15}$$

is commonly called the normal theory based ML discrepancy function. Because $l(\bar{\mathbf{x}}, \mathbf{S})$ is a constant, the MLE $\hat{\boldsymbol{\theta}}$ also minimizes $D_{\text{ML}}[\bar{\mathbf{x}}, \mathbf{S}, \boldsymbol{\mu}(\boldsymbol{\theta}), \boldsymbol{\Sigma}(\boldsymbol{\theta})]$. It is obvious that

$$T_{\text{ML}} = nD_{\text{ML}}\bigl[\bar{\mathbf{x}}, \mathbf{S}, \boldsymbol{\mu}(\hat{\boldsymbol{\theta}}), \boldsymbol{\Sigma}(\hat{\boldsymbol{\theta}})\bigr]$$

is the likelihood ratio (LR) statistic. When $\mathbf{x}$ is normally distributed and the model $\boldsymbol{\beta}(\boldsymbol{\theta})$ is correctly specified, standard large sample theory (Ferguson, 1996; Rao, 1973) tells us that

$$T_{\mathrm{ML}} \xrightarrow{\mathcal{L}} \chi^2_{df}, \tag{16}$$

where $df = p(p+3)/2 - q$. Thus, a significantly large T_{ML} when referred to the given chi-square distribution implies that the model structure is most likely misspecified. When $\mathbf{x} \sim N_p(\boldsymbol{\mu}_0, \boldsymbol{\Sigma}_0)$ and $\boldsymbol{\beta}_0 = \boldsymbol{\beta}(\boldsymbol{\theta}_0)$, standard asymptotic theory also implies that $\hat{\boldsymbol{\theta}}$ is consistent and

$$\sqrt{n}(\hat{\boldsymbol{\theta}} - \boldsymbol{\theta}_0) \xrightarrow{\mathcal{L}} N(\mathbf{0}, \boldsymbol{\Omega}_{\mathrm{ML}}), \tag{17}$$

where

$$\boldsymbol{\Omega}_{\mathrm{ML}} = (\dot{\boldsymbol{\beta}}'\mathbf{W}\dot{\boldsymbol{\beta}})^{-1}.$$

To obtain the MLE $\hat{\boldsymbol{\theta}}$, many methods for minimizing nonlinear functions can be used. Our experience indicates that the following iteratively reweighted least squares (IRLS) procedure[2] is very effective for obtaining $\hat{\boldsymbol{\theta}}$ (Green, 1984; Lee and Jennrich, 1979),

$$\boldsymbol{\theta}^{(k+1)} = \boldsymbol{\theta}^{(k)} + \alpha\big(\dot{\boldsymbol{\beta}}^{(k)'}\mathbf{W}^{(k)}\dot{\boldsymbol{\beta}}^{(k)}\big)^{-1}\dot{\boldsymbol{\beta}}^{(k)'}\mathbf{W}^{(k)}\big[\bar{\mathbf{t}}(\boldsymbol{\mu}^{(k)}) - \boldsymbol{\beta}^{(k)}\big], \tag{18}$$

where $\dot{\boldsymbol{\beta}}^{(k)} = \dot{\boldsymbol{\beta}}(\boldsymbol{\theta}^{(k)})$, $\mathbf{W}^{(k)} = \mathbf{W}(\boldsymbol{\theta}^{(k)})$, $\bar{\mathbf{t}}(\boldsymbol{\mu}^{(k)}) = (\bar{\mathbf{x}}', \mathrm{s}' + \mathrm{vech}'[(\bar{\mathbf{x}} - \boldsymbol{\mu}^{(k)})(\bar{\mathbf{x}} - \boldsymbol{\mu}^{(k)})'])'$, and α is fixed at 1.0 or a chosen step-size to allow

$$D_{\mathrm{ML}}\big[\bar{\mathbf{x}}, \mathbf{S}, \boldsymbol{\mu}\big(\boldsymbol{\theta}^{(k+1)}\big), \boldsymbol{\Sigma}\big(\boldsymbol{\theta}^{(k+1)}\big)\big] < D_{\mathrm{ML}}\big[\bar{\mathbf{x}}, \mathbf{S}, \boldsymbol{\mu}\big(\boldsymbol{\theta}^{(k)}\big), \boldsymbol{\Sigma}\big(\boldsymbol{\theta}^{(k)}\big)\big].$$

The matrix $\boldsymbol{\Sigma}(\boldsymbol{\theta}^{(k)})$ in $\mathbf{W}^{(k)}$ can be replaced by the sample covariance matrix $\mathbf{S}$. Then the converged solution corresponds to an estimator $\tilde{\boldsymbol{\theta}}$ that minimizes

$$\begin{aligned} &D_{\mathrm{GLS}}\big[\bar{\mathbf{x}}, \mathbf{S}, \boldsymbol{\mu}(\boldsymbol{\theta}), \boldsymbol{\Sigma}(\boldsymbol{\theta})\big] \\ &\quad = \big[\bar{\mathbf{x}} - \boldsymbol{\mu}(\boldsymbol{\theta})\big]'\mathbf{S}^{-1}\big[\bar{\mathbf{x}} - \boldsymbol{\mu}(\boldsymbol{\theta})\big] + \frac{1}{2}\operatorname{tr}\big\{\big[\boldsymbol{\Sigma}(\boldsymbol{\theta})\mathbf{S}^{-1} - \mathbf{I}\big]^2\big\}. \end{aligned} \tag{19}$$

Equation (19) is the normal theory GLS discrepancy function and

$$T_{\mathrm{GLS}} = nD_{\mathrm{GLS}}\big[\bar{\mathbf{x}}, \mathbf{S}, \boldsymbol{\mu}(\tilde{\boldsymbol{\theta}}), \boldsymbol{\Sigma}(\tilde{\boldsymbol{\theta}})\big]$$

is the corresponding test statistic for the overall model evaluation. The LS procedure corresponds to minimizing

$$D_{\mathrm{LS}}\big[\bar{\mathbf{x}}, \mathbf{S}, \boldsymbol{\mu}(\boldsymbol{\theta}), \boldsymbol{\Sigma}(\boldsymbol{\theta})\big] = \big[\bar{\mathbf{x}} - \boldsymbol{\mu}(\boldsymbol{\theta})\big]'\big[\bar{\mathbf{x}} - \boldsymbol{\mu}(\boldsymbol{\theta})\big] + \frac{1}{2}\operatorname{tr}\big\{\big[\mathbf{S} - \boldsymbol{\Sigma}(\boldsymbol{\theta})\big]^2\big\}$$

for parameter estimates.

When the mean structure $\boldsymbol{\mu}(\boldsymbol{\theta})$ is saturated, (15) becomes

$$D_{\mathrm{ML}c}[\mathbf{S}, \boldsymbol{\Sigma}(\boldsymbol{\theta})] = \operatorname{tr}\big[\mathbf{S}\boldsymbol{\Sigma}^{-1}(\boldsymbol{\theta})\big] - \log\big|\mathbf{S}\boldsymbol{\Sigma}^{-1}(\boldsymbol{\theta})\big| - p \tag{20}$$

---

[2] This procedure will lead to the MLE when $D_{\mathrm{ML}}[\bar{\mathbf{x}}, \mathbf{S}, \boldsymbol{\mu}(\boldsymbol{\theta}), \boldsymbol{\Sigma}(\boldsymbol{\theta})]$ is a convex function of $\boldsymbol{\theta}$. One may use multiple starting values to make sure that the converged solution is not just a local minimizer or use the Hessian matrix to check that it is not just a stationary point (see Chong and Żak, 1996).

and (19) becomes

$$D_{\mathrm{GLS}c}\big[\mathbf{S}, \boldsymbol{\Sigma}(\boldsymbol{\theta})\big] = \frac{1}{2}\operatorname{tr}\big\{\big[\boldsymbol{\Sigma}(\boldsymbol{\theta})\mathbf{S}^{-1} - \mathbf{I}\big]^2\big\}, \tag{21}$$

where the subscript c is used to denote CSA only. We will also denote the corresponding parameter estimates in CSA as $\hat{\boldsymbol{\theta}}_c$ and $\tilde{\boldsymbol{\theta}}_c$ and test statistics as $T_{\mathrm{ML}c}$ and $T_{\mathrm{GLS}c}$, respectively.

For CSA with normally distributed data, Browne (1974) showed that $\tilde{\boldsymbol{\theta}}_c$ has the same asymptotic distribution as $\hat{\boldsymbol{\theta}}_c$ and $T_{\mathrm{GLS}c}$ has the same asymptotic distribution as $T_{\mathrm{ML}c}$. Yuan and Chan (2005) recently showed that, for CSA, the asymptotic equivalence of the ML and GLS procedures does not depend on the distribution of $\mathbf{x}$ but the correctness of the model structure. When model $\boldsymbol{\Sigma}(\boldsymbol{\theta})$ is not correctly specified, ML and GLS are no longer asymptotically equivalent. When modeling the mean and covariance structures simultaneously, it can also be shown that $\tilde{\boldsymbol{\theta}}$ has the same asymptotic distribution as $\hat{\boldsymbol{\theta}}$, and T_{GLS} and T_{ML} follow the same asymptotic distribution. Again, this equivalence depends on the correctness of the model structures but not the distribution of $\mathbf{x}$. Of course, when $\mathbf{x}$ does not follow a multivariate normal distribution, T_{ML} and T_{GLS} will not follow a chi-square distribution in general even for correctly specified models. The distributions of the parameter estimates will no longer be given by (17) either.

When data are not normally distributed, parameter estimates from any of the above procedures are still consistent as long as $\boldsymbol{\beta}(\boldsymbol{\theta})$ is identified and correctly specified. We will mainly focus on the property of $\hat{\boldsymbol{\theta}}$ and the associated statistic T_{ML}. It follows from the central limit theorem that

$$\sqrt{n}(\bar{\mathbf{t}} - \boldsymbol{\beta}_0) \xrightarrow{\mathcal{L}} N(\mathbf{0}, \boldsymbol{\Pi}), \tag{22}$$

where

$$\boldsymbol{\Pi} = \begin{pmatrix} \boldsymbol{\Sigma}_0 & \boldsymbol{\Pi}_{12} \\ \boldsymbol{\Pi}_{21} & \boldsymbol{\Pi}_{22} \end{pmatrix}$$

with $\boldsymbol{\Pi}_{12} = \boldsymbol{\Pi}_{21}' = \mathrm{Cov}(\mathbf{x}, \mathrm{vech}[(\mathbf{x}-\boldsymbol{\mu}_0)(\mathbf{x}-\boldsymbol{\mu}_0)'])$ and $\boldsymbol{\Pi}_{22} = \mathrm{Var}(\mathrm{vech}[(\mathbf{x}-\boldsymbol{\mu}_0)(\mathbf{x}-\boldsymbol{\mu}_0)'])$. In the context of CSA with nonnormally distributed data, $T_{\mathrm{ML}c}$ enjoys some nice properties that are not shared by T_{ML}. We will discuss these properties first and then turn to the limited results for mean and covariance structure analysis. Let q_c be the number of free parameters in $\boldsymbol{\Sigma}(\boldsymbol{\theta})$. Then $df_c = p(p+1)/2 - q_c$ is the degrees of freedom in fitting $\boldsymbol{\Sigma}(\boldsymbol{\theta})$ to $\mathbf{S}$. Let

$$\mathbf{U}_c = \mathbf{W}_c - \mathbf{W}_c\dot{\boldsymbol{\sigma}}(\dot{\boldsymbol{\sigma}}'\mathbf{W}_c\dot{\boldsymbol{\sigma}})^{-1}\dot{\boldsymbol{\sigma}}'\mathbf{W}_c,$$

and κ_j, $j = 1, \ldots, df_c$, be the nonzero eigenvalues of $\mathbf{U}_c\boldsymbol{\Pi}_{22}$. Under the condition of a correctly specified model, by using a Taylor expansion we will obtain (see the appendix of Yuan, Marshall and Bentler, 2002, for detail)

$$T_{\mathrm{ML}c} = \sum_{j=1}^{df_c} \kappa_j z_j^2 + o_p(1), \tag{23}$$

where z_j^2's are independent and each follows a chi-square distribution with 1 degree of freedom, and $o_p(1)$ denotes a random term that approaches zero in probability as $n \to \infty$. Eq. (23) implies that the asymptotic distribution of $T_{\text{ML}c}$ is determined by the fourth-order moments of $\mathbf{x}$ as well as the model structure. Notice that, when $\mathbf{x} \sim N_p(\boldsymbol{\mu}, \boldsymbol{\Sigma})$,

$$\boldsymbol{\Pi}_{22} = \boldsymbol{\Pi}_{\mathcal{N}22} = 2\mathbf{D}_p^+(\boldsymbol{\Sigma} \otimes \boldsymbol{\Sigma})\mathbf{D}_p^{+'} = \mathbf{W}_c^{-1}, \tag{24}$$

where $\mathbf{D}_p^+ = (\mathbf{D}_p'\mathbf{D}_p)^{-1}\mathbf{D}_p'$ is the generalized inverse of $\mathbf{D}_p$. Thus, $\kappa_1 = \kappa_2 = \cdots = \kappa_{df_c} = 1$ and $T_{\text{ML}c}$ asymptotically follows a chi-square distribution. Yuan and Bentler (1999a) called the class of distributions whose fourth-order moments satisfy (24) the pseudo-normal distribution.[3] In addition to the test statistics in covariance structure analysis, other statistics based on analyzing $\mathbf{S}$ will have the same asymptotic distribution within the class of pseudo-normal distributions as well (Yuan and Bentler, 2000a, 2002a). Asymptotic standard errors (SEs) of $\hat{\boldsymbol{\theta}}_c$ based on the normal distribution assumption remain valid within the class of pseudo-normal distributions (Yuan and Bentler, 1999b).

The property that $\hat{\boldsymbol{\theta}}_c$ and/or $T_{\text{ML}c}$ enjoy the same asymptotic distribution for nonnormally distributed data as that for normally distributed data is commonly called asymptotic robustness. Early characterizations of this type of robustness were made by Anderson and Amemiya (1988) and Amemiya and Anderson (1990) for CFA models. Specifically, when $\mathbf{f}$ and $\mathbf{e}$ in a CFA model like (1) are independent and the elements of $\mathbf{e}$ are also independent, Amemiya and Anderson showed that $T_{\text{ML}c}$ still asymptotically follows a chi-square distribution, the SEs for factor loading estimates based on $\mathbf{x} \sim N_p(\boldsymbol{\mu}, \boldsymbol{\Sigma})$ are still asymptotically valid. The results were generalized in various direction by Browne and Shapiro (1988), Kano (1992), Mooijaart and Bentler (1991), Satorra (1992), Satorra and Bentler (1990), Yuan and Bentler (1999a, 1999b).

It follows from (23) that any asymptotic robustness conditions for $T_{\text{ML}c}$ must lead to $\kappa_1 = \kappa_2 = \cdots = \kappa_{df_c} = 1$. When the κ_j's are not equal, $T_{\text{ML}c}$ will not asymptotically follow a chi-square distribution. When $\kappa_1 = \kappa_2 = \cdots = \kappa_{df_c} = \kappa \neq 1$, $T_{\text{ML}c}$ will not approach a chi-square variate either. However, using a consistent estimate $\hat{\kappa}$ for κ, we can rescale the LR statistic to $T_{\text{RML}c} = \hat{\kappa}^{-1}T_{\text{ML}c}$ and

$$T_{\text{RML}c} \xrightarrow{\mathcal{L}} \chi^2_{df_c}. \tag{25}$$

Thus, conditions leading to $\kappa_1 = \kappa_2 = \cdots = \kappa_{df_c} = \kappa$ are also of great interest. Several conditions are necessary to characterize the property of $T_{\text{RML}c}$. For a covariance structure $\boldsymbol{\Sigma}(\boldsymbol{\theta})$, if for any parameter vector $\boldsymbol{\theta}$ and positive constant α there exists a parameter vector $\boldsymbol{\theta}^*$ such that $\boldsymbol{\Sigma}(\boldsymbol{\theta}^*) = \alpha\boldsymbol{\Sigma}(\boldsymbol{\theta})$, then $\boldsymbol{\Sigma}(\boldsymbol{\theta})$ is invariant with respect to a constant scaling factor (ICSF). As noted by Browne (1984), nearly all structural models in current use are ICSF. Let $\mathcal{R}(\dot{\boldsymbol{\sigma}})$ represent the range space spanned by the column vectors of $\dot{\boldsymbol{\sigma}}$. If $\boldsymbol{\sigma} \in \mathcal{R}(\dot{\boldsymbol{\sigma}})$, then the model is said to be quasi-linear (QL). It is easily seen that ICSF implies QL (Satorra and Bentler, 1986). So condition QL is also satisfied by almost all of the models in current use.

[3] Contrasted to the normal distribution whose skewness is zero, a pseudo-normal distribution can have arbitrary skewness.

Under the condition of ICSF, Browne (1984) and Shapiro and Browne (1987) showed that the κ_j's are equal within the class of elliptical symmetric distributions (Fang et al., 1990). Earlier work in this direction by Muirhead and Waternaux (1980) and Tyler (1983) showed that the LR statistics in canonical correlation and several simple covariance structures can be rescaled to asymptotically follow chi-square distributions. The correction factor $\hat{\kappa}$ they used is based on Mardia's (1970) measure of multivariate kurtosis. Satorra and Bentler (1988) observed that $\mathrm{tr}(\mathbf{U}_c\boldsymbol{\Pi}_{22}) = \sum_{j=1}^{df_c}\kappa_j$. They suggested $\hat{\kappa} = \mathrm{tr}(\hat{\mathbf{U}}_c\hat{\boldsymbol{\Pi}}_{22})/df_c$ in constructing $T_{\mathrm{RML}c}$. Under the condition of QL, Yuan and Bentler (1999a) showed that, even when $\hat{\kappa}$ is based on Mardia's multivariate kurtosis, (25) is still valid within the class of pseudo-elliptical distributions.[4] The $\hat{\kappa}$ proposed by Satorra and Bentler makes (25) valid in a even wider class of distributions. In addition to these two correction factors, Kano (1992) constructed an alternative correcting factor by fully utilizing the model structure. The resulting rescaled statistic is also asymptotically robust to a wide class of nonnormal distributions (see Yuan and Bentler, 1999a).

Simulation studies indicate that $T_{\mathrm{RML}c}$ performs quite robustly under a variety of conditions (Curran et al., 1996; Hu et al., 1992; Yuan and Bentler, 1998a). While exceptions have been studied, most data generation in these studies satisfies the asymptotic robustness condition for $T_{\mathrm{RML}c}$ (see Yuan and Bentler, 1999a). In general, however, $T_{\mathrm{RML}c}$ does not approach a variate possessing a chi-square distribution. Instead, it only approaches a variate T with $E(T) = df$. It is likely that the distributional shape of T is far from that of a chi-square. In such cases, $T_{\mathrm{RML}c}$ will not behave like a chi-square. Limited simulation results in Yuan and Bentler (1998a) and Bentler and Yuan (1999) indicate that, when the κ_j's are not equal, $T_{\mathrm{RML}c}$ may behave worse as n increases. When sample size is small, $T_{\mathrm{RML}c}$ as well as $T_{\mathrm{ML}c}$ may not behave like a chi-square for even normally distributed data. It should be noted that most of the conditions for asymptotic robustness are not verifiable in practice. It is misleading to blindly trust that the statistic $T_{\mathrm{ML}c}$ or $T_{\mathrm{RML}c}$ will asymptotically follow a chi-square distribution before the necessary conditions can be verified.

In contrast to CSA, little or no results exist on the asymptotic robustness of T_{ML} when simultaneously modeling mean and covariance structures. Yuan and Bentler (2006a) noted that the asymptotic robustness condition in CSA will not make T_{ML} asymptotically follow a chi-square distribution. But, for correctly specified models, there still exists

$$T_{\mathrm{ML}} = \sum_{j=1}^{df}\kappa_j z_j^2 + o_p(1),$$

where κ_j's are the nonzero eigenvalues of $\mathbf{U}\boldsymbol{\Pi}$ with

$$\mathbf{U} = \mathbf{W} - \mathbf{W}\dot{\boldsymbol{\beta}}(\dot{\boldsymbol{\beta}}'\mathbf{W}\dot{\boldsymbol{\beta}})^{-1}\dot{\boldsymbol{\beta}}'\mathbf{W}, \tag{26}$$

and z_j^2, $j = 1, \ldots, df$ are independent and each follows χ_1^2. The following rescaled statistic with $\hat{\kappa} = \mathrm{tr}(\hat{\mathbf{U}}\hat{\boldsymbol{\Pi}})/df$ has been proposed in different contexts (Satorra, 1992;

[4] Comparing to the class of elliptical distributions that requires $\boldsymbol{\Pi}_{12} = \mathbf{0}$, the class of pseudo-elliptical distributions accommodates a free $\boldsymbol{\Pi}_{12}$ (see Yuan and Bentler, 1999a).

Satorra and Bentler, 1994; Yuan and Bentler, 2000b, 2000b)

$$T_{\mathrm{RML}} = \hat{\kappa}^{-1} T_{\mathrm{ML}},$$

but few or no empirical studies have been conducted to evaluate its performance.

When data are not normally distributed, the asymptotic distribution of $\hat{\boldsymbol{\theta}}$ is characterized by

$$\sqrt{n}(\hat{\boldsymbol{\theta}} - \boldsymbol{\theta}_0) \xrightarrow{\mathcal{L}} N(\mathbf{0}, \boldsymbol{\Omega}_{\mathrm{SW}}), \tag{27}$$

where

$$\boldsymbol{\Omega}_{\mathrm{SW}} = (\dot{\boldsymbol{\beta}}'\mathbf{W}\dot{\boldsymbol{\beta}})^{-1}(\dot{\boldsymbol{\beta}}'\mathbf{W}\boldsymbol{\Pi}\mathbf{W}\dot{\boldsymbol{\beta}})(\dot{\boldsymbol{\beta}}'\mathbf{W}\dot{\boldsymbol{\beta}})^{-1}$$

is commonly called the sandwich-type covariance matrix. Like T_{RML}, $\boldsymbol{\Omega}_{\mathrm{SW}}$ allows us to make an adjustment for distributional violations of the normal theory based ML procedure. When data are normally distributed, $\mathbf{W} = \boldsymbol{\Pi}^{-1}$, so (27) reduces to (17). The sandwich-type covariance matrix was first proposed by Huber (1967). It is now widely used in statistics (White, 1982; Kauermann and Carroll, 2001) and in SEM (Bentler, 1983; Bentler and Dijkstra, 1985; Browne, 1984; Browne and Arminger, 1995; Satorra and Bentler, 1994; Shapiro, 1983; Yuan and Bentler, 1997b, 1997b) under different contexts. Comparing $\boldsymbol{\Omega}_{\mathrm{SW}}$ with $\boldsymbol{\Omega}_{\mathrm{ML}}$, one has to estimate the extra matrix $\boldsymbol{\Pi}$. Simulation results in CSA (Yuan and Bentler, 1997b) indicate that SEs based on (27) match the empirical ones very well for normal as well as nonnormally distributed data. Yuan and Bentler (1997b) recommended using (27) as the default formula for calculating SEs of $\hat{\boldsymbol{\theta}}$.

3.2. *GLS procedures without assuming specific distributions*

With typical nonnormal data in the social and behavioral sciences (Micceri, 1989), the ideal is to have a statistic that approximately follows a chi-square distribution regardless of the underlying distribution of the data. One of the original proposals in this direction was made by Browne (1984). Let $\mathbf{S}_t$ be the sample covariance matrix of $\mathbf{t}_i$. Then $\hat{\boldsymbol{\Pi}} = \mathbf{S}_t$ is consistent for $\boldsymbol{\Pi}$. Thus, an AGLS discrepancy function is given by

$$D_{\mathrm{AGLS}}\big[\bar{\mathbf{x}}, \mathbf{S}, \boldsymbol{\mu}(\boldsymbol{\theta}), \boldsymbol{\Sigma}(\boldsymbol{\theta})\big] = \big[\bar{\mathbf{t}} - \boldsymbol{\beta}(\boldsymbol{\theta})\big]'\hat{\boldsymbol{\Pi}}^{-1}\big[\bar{\mathbf{t}} - \boldsymbol{\beta}(\boldsymbol{\theta})\big]. \tag{28}$$

Note that, when using

$$\hat{\boldsymbol{\Pi}} = \begin{pmatrix} \mathbf{S} & \mathbf{0} \\ \mathbf{0} & 2\mathbf{D}_p^{+}(\mathbf{S} \otimes \mathbf{S})\mathbf{D}_p^{+\prime} \end{pmatrix}$$

in (28), D_{AGLS} reduces to D_{GLS}. Both the GLS and AGLS procedures can be regarded as applications of the method of minimum chi-square (Ferguson, 1958) to mean and covariance structure analysis. Similar to the GLS estimate $\tilde{\boldsymbol{\theta}}$, parameter estimate $\check{\boldsymbol{\theta}}$ that minimizes (28) can be solved by the IRLS procedure as in (18), where the weight matrix $\mathbf{W}^{(j)}$ will be replaced by $\mathbf{S}_t^{-1}$. Because the $\mathbf{S}_t$ is usually of a large dimension, there may exist convergence problems for solving $\check{\boldsymbol{\theta}}$ when n is not large enough (see Hu et al.,

1992; Yuan and Bentler, 1997a). For the asymptotic distribution of $\check{\boldsymbol{\theta}}$, we have

$$\sqrt{n}(\check{\boldsymbol{\theta}} - \boldsymbol{\theta}_0) \xrightarrow{\mathcal{L}} N(\mathbf{0}, \boldsymbol{\Omega}_{\text{AGLS}}), \tag{29}$$

where

$$\boldsymbol{\Omega}_{\text{AGLS}} = (\dot{\boldsymbol{\beta}}' \boldsymbol{\Pi}^{-1} \dot{\boldsymbol{\beta}})^{-1}.$$

It is obvious that, for normally distributed data, $\boldsymbol{\Omega}_{\text{AGLS}} = \boldsymbol{\Omega}_{\text{ML}} = \boldsymbol{\Omega}_{\text{SW}}$. For nonnormally distributed data, $\boldsymbol{\Omega}_{\text{AGLS}} \leqslant \boldsymbol{\Omega}_{\text{SW}}$. However, empirically $\check{\boldsymbol{\theta}}$ can be much less efficient than $\hat{\boldsymbol{\theta}}$ even when data are nonnormally distributed. Also, SEs of $\check{\boldsymbol{\theta}}$ based on (29), with $\hat{\boldsymbol{\Pi}} = \mathbf{S}_t$, are too optimistic compared to those obtained by Monte Carlo. Yuan and Bentler (1997b) proposed

$$\hat{\boldsymbol{\Omega}}_{\text{CAGLS}} = \frac{n-1}{n - p(p+3)/2 - 2} \left(\dot{\boldsymbol{\beta}}'(\check{\boldsymbol{\theta}}) \mathbf{S}_t^{-1} \dot{\boldsymbol{\beta}}(\check{\boldsymbol{\theta}})\right)^{-1}, \tag{30}$$

which is also consistent for $\boldsymbol{\Omega}_{\text{AGLS}}$. Although SEs based on (30) are still smaller than the Monte-Carlo SEs, the biases are much smaller compared to those directly based on (29), especially when p is large.

The corresponding statistic for overall model evaluation is

$$T_{\text{AGLS}} = n D_{\text{AGLS}}\left[\bar{\mathbf{x}}, \mathbf{S}, \boldsymbol{\mu}(\check{\boldsymbol{\theta}}), \boldsymbol{\Sigma}(\check{\boldsymbol{\theta}})\right].$$

In the context of covariance structure analysis, D_{AGLS} reduces to

$$D_{\text{AGLS}c}\left[\mathbf{S}, \boldsymbol{\Sigma}(\boldsymbol{\theta})\right] = \left[\mathbf{s} - \boldsymbol{\sigma}(\boldsymbol{\theta})\right]' \hat{\boldsymbol{\Pi}}_{22}^{-1} \left[\mathbf{s} - \boldsymbol{\sigma}(\boldsymbol{\theta})\right], \tag{31}$$

where $\hat{\boldsymbol{\Pi}}_{22}$ is the sample covariance matrix of vech$[(\mathbf{x}_i - \bar{\mathbf{x}})(\mathbf{x}_i - \bar{\mathbf{x}})']$, $i = 1, 2, \ldots, n$. The procedure of minimizing (31) for parameter estimates $\check{\boldsymbol{\theta}}_c$ and referring $T_{\text{AGLS}c}$ to $\chi^2_{df_c}$ was proposed by Browne (1984). Because $T_{\text{AGLS}c}$ asymptotically follows $\chi^2_{df_c}$ as long as $\boldsymbol{\Pi}_{22}$ exists – without requiring any distributional assumption – the AGLS procedure is commonly referred to as the asymptotically distribution free (ADF) method. The ADF property is desirable. However, the distribution of $T_{\text{AGLS}c}$ may be far from that of $\chi^2_{df_c}$ for typical sample sizes encountered in practice (Hu et al., 1992). Specifically, the mean and variance of $T_{\text{AGLS}c}$ are much greater than those of χ^2_{dfc}. Most correctly specified models are rejected if referring $T_{\text{AGLS}c}$ to $\chi^2_{df_c}$ or T_{AGLS} to χ^2_{df} (Yuan and Bentler, 1997a).

In an effort to find statistics that perform better in rejection rates with smaller n's, Yuan and Bentler (1997a) compared mean and covariance structure analysis with multivariate regression and suggested using

$$\check{\boldsymbol{\Pi}} = \frac{1}{n} \sum_{i=1}^{n} \left(\mathbf{t}_i - \boldsymbol{\beta}(\check{\boldsymbol{\theta}})\right)\left(\mathbf{t}_i - \boldsymbol{\beta}(\check{\boldsymbol{\theta}})\right)'$$

in constructing the AGLS discrepancy function. Because $\check{\boldsymbol{\Pi}} \geqslant \hat{\boldsymbol{\Pi}}$, the greater type I error in T_{AGLS} will be corrected when using $\check{\boldsymbol{\Pi}}^{-1}$ as the weight matrix in formulating an AGLS procedure. Let the corrected AGLS discrepancy function be

$$D_{\mathrm{CAGLS}}[\bar{\mathbf{x}}, \mathbf{S}, \boldsymbol{\mu}(\boldsymbol{\theta}), \boldsymbol{\Sigma}(\boldsymbol{\theta})] = [\bar{\mathbf{t}} - \boldsymbol{\beta}(\boldsymbol{\theta})]' \check{\boldsymbol{\Pi}}^{-1} [\bar{\mathbf{t}} - \boldsymbol{\beta}(\boldsymbol{\theta})]. \tag{32}$$

Because (32) involves $\check{\boldsymbol{\Pi}}$, it seems that one has to obtain $\check{\boldsymbol{\theta}}$ before minimizing (32). Yuan and Bentler (1997a) further showed that $\check{\boldsymbol{\theta}}$ also minimizes (32) and there exists

$$T_{\mathrm{CAGLS}} = n D_{\mathrm{CAGLS}}[\bar{\mathbf{x}}, \mathbf{S}, \boldsymbol{\mu}(\check{\boldsymbol{\theta}}), \boldsymbol{\Sigma}(\check{\boldsymbol{\theta}})] = \frac{T_{\mathrm{AGLS}}}{1 + T_{\mathrm{AGLS}/n}}. \tag{33}$$

Because T_{AGLS}/n approaches zero, T_{CAGLS} asymptotically follows χ^2_{df} as long as $\boldsymbol{\Pi}$ exists. Simulation results in Yuan and Bentler (1997a) imply that the mean of T_{CAGLS} approximately equals *df* for all sample sizes across many distributional conditions. However, at smaller sample sizes T_{CAGLS} tends to over-correct the behavior of T_{AGLS} by having a type I error smaller than the nominal level. Notice that both T_{AGLS} and $T_{\mathrm{AGLS}c}$ are in the form of Hotelling's T^2 statistic. It is well known that T^2 approaches a chi-square distribution as $n \to \infty$, but using an F-distribution will better describe the distribution of T^2. Motivated by this observation, Yuan and Bentler (1999c) further proposed the following F-statistic[5]

$$T_F = \frac{(n - df) T_{\mathrm{AGLS}}}{(n - 1) df} \tag{34}$$

and suggested referring T_F to $F_{df,n-df}$, the F-distribution with *df* and $n - df$ degrees of freedom. It is obvious that T_{CAGLS}, T_{AGLS} and T_F are asymptotically equivalent. Actually, they also perform similarly when n is very large (Yuan and Bentler, 1999c). When n is small, T_F is between T_{CAGLS} and T_{AGLS}, with a slight over-rejection on average but quite satisfactory overall. Both T_F and T_{CAGLS} are asymptotically distribution free and perform quite reliably with finite n. However, both of them use $\check{\boldsymbol{\theta}}$ to start with and thus carry the drawback of the AGLS estimation method with nonconvergences at smaller sample sizes.

In addition to $T_{\mathrm{AGLS}c}$ in the context of CSA, Browne (1984) also proposed a residual-based statistic

$$\begin{aligned} T_{\mathrm{RAGLS}c} = n[\mathbf{s} - \boldsymbol{\sigma}(\hat{\boldsymbol{\theta}})]' \big(\hat{\boldsymbol{\Pi}}_{22}^{-1} - \hat{\boldsymbol{\Pi}}_{22}^{-1} \dot{\boldsymbol{\sigma}}(\hat{\boldsymbol{\theta}}) [\dot{\boldsymbol{\sigma}}'(\hat{\boldsymbol{\theta}}) \hat{\boldsymbol{\Pi}}_{22}^{-1} \dot{\boldsymbol{\sigma}}(\hat{\boldsymbol{\theta}})]^{-1} \dot{\boldsymbol{\sigma}}'(\hat{\boldsymbol{\theta}}) \hat{\boldsymbol{\Pi}}_{22}^{-1}\big) \\ \times [\mathbf{s} - \boldsymbol{\sigma}(\hat{\boldsymbol{\theta}})], \end{aligned} \tag{35}$$

where $\hat{\boldsymbol{\theta}}$ can be the MLE, the GLS or any other consistent estimator that is easier to obtain. $T_{\mathrm{RAGLS}c}$ will approach $\chi^2_{df_c}$ as n tends to infinity. However, $T_{\mathrm{RAGLS}c}$ behaves like $T_{\mathrm{AGLS}c}$ and T_{RAGLS} behaves like T_{AGLS}: all over-reject correct models too often unless n is very large. Parallel to T_{CAGLS}, Yuan and Bentler (1998a) proposed $T_{\mathrm{CRAGLS}c}$ whose performance is almost the same as $T_{\mathrm{CAGLS}c}$, with its empirical mean approximately equal to df_c and under-rejecting the correct model for small sample sizes (Bentler

[5] In the context of multivariate linear model with a saturated covariance matrix, Rao (1951) provided a statistic that can be approximately described by an F-distribution. The degrees of freedom in the approximation obtained by Rao are based on normally distributed data. The T_F in (34) is to test mean and covariance structures simultaneously; the data are not necessarily normally distributed.

and Yuan, 1999; Yuan and Bentler, 1998a). Yuan and Bentler (1998a) also proposed a residual-based F-statistic $T_{\mathrm{RF}c}$ based on $T_{\mathrm{RAGLS}c}$; it also performs similarly to $T_{\mathrm{F}c}$. For mean and covariance structure analysis, statistics T_{CRAGLS} and T_{RF} can be defined parallel to (33) and (34); they are also asymptotically distribution free.

3.3. Robust procedures

When the sample $\mathbf{x}_1, \mathbf{x}_2, \ldots, \mathbf{x}_n$ approximately follows a multivariate normal distribution, then analyzing $\bar{\mathbf{x}}$ and $\mathbf{S}$ by the ML and AGLS procedures will allow us to estimate $\boldsymbol{\theta}_0$ efficiently. Unfortunately, few real data sets follow the normal distribution. This is particularly true for data sets in social and behavioral sciences. Micceri (1989) reported that among 440 large sample achievement and psychometric measures taken from journal articles, research projects, and tests, all were significantly nonnormally distributed. For example, the normalized Mardia's (1970) multivariate kurtosis,

$$M = \left\{\frac{1}{n}\sum_{i=1}^{n}\left[(\mathbf{x}_i - \bar{\mathbf{x}})'\mathbf{S}^{-1}(\mathbf{x}_i - \bar{\mathbf{x}})\right]^2 - p(p+2)\right\}\left\{8p(p+2)/n\right\}^{-1/2},$$

approximately follows $N(0, 1)$ when $\mathbf{x} \sim N(\boldsymbol{\mu}_0, \boldsymbol{\Sigma}_0)$. We have seen many M's that are greater than 100 in students' dissertations and research projects. A huge M may be caused by a few "outliers". In such a situation, removing these outliers followed by modeling $\bar{\mathbf{x}}$ and $\mathbf{S}$ will lead to an efficient parameter estimate $\hat{\boldsymbol{\theta}}$. However, the extra kurtoses in most practical data sets are not just due to a few outliers but to the heavy tails of the underlying population distribution. When a data set possesses moderately heavy tails, $\bar{\mathbf{x}}$ and $\mathbf{S}$ will be inefficient estimates of their population counterparts $\boldsymbol{\mu}_0$ and $\boldsymbol{\Sigma}_0$ (Kano et al., 1993; Tyler, 1983). When the heavy tails are severe, the population 4th-order moment matrix $\boldsymbol{\Pi}_{22}$ may not exist. Then the previously obtained $\hat{\boldsymbol{\theta}}$, $\tilde{\boldsymbol{\theta}}$ and $\check{\boldsymbol{\theta}}$ all have unbounded covariance matrices. In the extreme case, $\boldsymbol{\Sigma}_0$ or even $\boldsymbol{\mu}_0$ may not exist and SEM is questionable. When $\boldsymbol{\Sigma}_0$ exists, it may be possible to use a robust procedure to provide a more efficient estimator of $\boldsymbol{\theta}_0$ than can be obtained by modeling $\bar{\mathbf{x}}$ and $\mathbf{S}$.

Following Huber (1964), many robust statistical methods have been developed (e.g., Huber, 1981; Hampel et al., 1986). Procedures for estimating the population mean vector and covariance matrix are also well studied. Maronna (1976) obtained the properties of M-estimators for the population mean vector and covariance matrix. Lopuhaä (1989, 1991) and Rocke (1996) studied other estimators that can allow the sample to contain nearly 50% contaminated observations. We will mainly use M-estimators for robust SEM procedures, aiming for samples containing a proportion of contaminated but not extreme observations.[6] There exist two main approaches to robust SEM. One is to first get robust estimates for $\boldsymbol{\mu}_0$ and $\boldsymbol{\Sigma}_0$, then treating these estimates as $\bar{\mathbf{x}}$ and $\mathbf{S}$ using ML or AGLS procedures. The other fits the structural model to raw data directly by a robust method without explicitly estimating the saturated model. We will only introduce the first one due to its easier implementation. Readers interested in the direct approach are referred to Yuan and Bentler (2000c) and Yuan, Fung and Reise (2004).

[6] When a large proportion of the sample is contaminated, the sample may come from two or several distributions. Then a mixture model might be more appropriate (Yung, 1997; Arminger et al., 1999; Hoshino, 2001; Muthén, 2001), which is beyond the scope of this chapter.

Standard ML and AGLS procedures are based on modeling $\bar{\mathbf{x}}$ and $\mathbf{S}$ obtained from data in which every $\mathbf{x}_i$ is treated equally. In a robust procedure, cases are differentiated by their distances to the center of the majority of the data, as measured by

$$d(\mathbf{x}, \boldsymbol{\mu}, \boldsymbol{\Sigma}) = \left[(\mathbf{x} - \boldsymbol{\mu})'\boldsymbol{\Sigma}^{-1}(\mathbf{x} - \boldsymbol{\mu})\right]^{1/2}.$$

The farther $\mathbf{x}_i$ is from the center, the less weight it will get in the estimation process. When $\boldsymbol{\mu}$ and $\boldsymbol{\Sigma}$ are saturated, they can be estimated by iteratively solving the following equations

$$\boldsymbol{\mu} = \sum_{i=1}^{n} u_1(d_i)\mathbf{x}_i \Big/ \sum_{i=1}^{n} u_1(d_i), \tag{36a}$$

$$\boldsymbol{\Sigma} = \sum_{i=1}^{n} u_2\left(d_i^2\right)(\mathbf{x}_i - \boldsymbol{\mu})(\mathbf{x}_i - \boldsymbol{\mu})'/n, \tag{36b}$$

where $d_i = d(\mathbf{x}_i, \boldsymbol{\mu}, \boldsymbol{\Sigma})$, u_1 and u_2 are certain weight functions. Two commonly used classes of weight functions for M-estimators are the Huber-type weights and the weights based on the multivariate t-distribution. Let ρ be the percentage of influential cases one wants to control, and r be a constant determined by ρ through $P(\chi_p^2 > r^2) = \rho$. The Huber-type weights are given by

$$u_1(d) = \begin{cases} 1, & \text{if } d \leqslant r, \\ r/d, & \text{if } d > r, \end{cases} \tag{37}$$

and $u_2(d^2) = \{u_1(d)\}^2/\varphi$, where φ is a constant determined by ρ through $E\{\chi_p^2 u_2(\chi_p^2)\} = p$. Let $\hat{\boldsymbol{\mu}}$ and $\hat{\boldsymbol{\Sigma}}$ be the solution to (36). The purpose of φ is to make $\hat{\boldsymbol{\Sigma}}$ unbiased for $\boldsymbol{\Sigma}_0$ when $\mathbf{x} \sim N(\boldsymbol{\mu}_0, \boldsymbol{\Sigma}_0)$. The weights corresponding to a p-variate t-distribution with m degrees of freedom are given by

$$u_1(d_i) = u_2\left(d_i^2\right) = \frac{p+m}{m+d_i^2}. \tag{38}$$

Notice that the only tuning parameter in using (37) is ρ and that in using (38) is m. Motivated by the formula for calculating the sample covariance matrix, Campbell (1980) defined another form of M-estimator by solving

$$\boldsymbol{\mu} = \sum_{i=1}^{n} u(d_i)\mathbf{x}_i \Big/ \sum_{i=1}^{n} u(d_i) \tag{39a}$$

and

$$\boldsymbol{\Sigma} = \sum_{i=1}^{n} u^2(d_i)(\mathbf{x}_i - \boldsymbol{\mu})(\mathbf{x}_i - \boldsymbol{\mu})' \Big/ \left(\sum_{i=1}^{n} u^2(d_i) - 1\right), \tag{39b}$$

where $u(d) = w(d)/d$ with

$$w(d) = \begin{cases} d, & \text{if } d \leqslant d_0 \\ d_0 \exp\left\{-\frac{1}{2}(d - d_0)^2/b_2^2\right\}, & \text{if } d > d_0 \end{cases}, \tag{40}$$

$d_0 = \sqrt{p} + b_1/\sqrt{2}$, b_1 and b_2 are constants. So there are two tuning parameters in (40). Based on empirical experience, Campbell (1980) suggested $b_1 = 2$ and $b_2 = \infty$ or $b_1 = 2$ and $b_2 = 1.25$. When $b_1 = 2$ and $b_2 = \infty$, (39) defines a Huber-type M-estimator; when $b_1 = 2$, $b_2 = 1.25$, (39) defines Hampel-type redescending M-estimator (Hampel, 1974); (39) leads to $\hat{\boldsymbol{\mu}} = \bar{\mathbf{x}}$ and $\hat{\boldsymbol{\Sigma}} = \mathbf{S}$ when choosing $b_1 = \infty$.

Robust M-estimation for $\boldsymbol{\mu}_0$ and $\boldsymbol{\Sigma}_0$ was motivated by ML within the class of elliptical distributions (see Maronna, 1976). However, $\hat{\boldsymbol{\Sigma}}$ generally does not approach $\boldsymbol{\Sigma}_0 = \mathrm{Cov}(\mathbf{x})$ within the class of elliptical distributions. Instead, it approaches $\alpha\boldsymbol{\Sigma}_0$ for a scalar $\alpha > 0$. When the population elliptical distribution is known, one can theoretically calculate the α and rescale $\hat{\boldsymbol{\Sigma}}$ to make it consistent for $\boldsymbol{\Sigma}_0$. Because essentially all the commonly used covariance structure models are ICSF, such an inconsistency will not cause any problems for statistical inference when treating $\alpha\boldsymbol{\Sigma}$ as the covariance matrix. See Yuan and Bentler (1998c) and Yuan, Bentler and Chan (2004) for further discussion on this aspect. So we will still use $\boldsymbol{\Sigma}_0$ instead of $\alpha\boldsymbol{\Sigma}_0$ in the following development.

When replacing $\bar{\mathbf{x}}$ and $\mathbf{S}$ in (15), (19), (20), (21), (28), (31) or (32) by $\hat{\boldsymbol{\mu}}$ and $\hat{\boldsymbol{\Sigma}}$, we will obtain the corresponding robust estimates for $\boldsymbol{\theta}$. Let these robust estimates still be denoted by $\hat{\boldsymbol{\theta}}$, $\tilde{\boldsymbol{\theta}}$ or $\check{\boldsymbol{\theta}}$. Then all the robust properties of $\hat{\boldsymbol{\mu}}$ and $\hat{\boldsymbol{\Sigma}}$ will be inherited by $\hat{\boldsymbol{\theta}}$, $\tilde{\boldsymbol{\theta}}$ or $\check{\boldsymbol{\theta}}$ (see Yuan and Bentler, 1998b, 1998c). In order to have SEs for parameter estimates and test statistics for the overall model evaluation, we need a consistent estimate for the covariance matrix of $\hat{\boldsymbol{\beta}} = (\hat{\boldsymbol{\mu}}', \hat{\boldsymbol{\sigma}}')'$, where $\hat{\boldsymbol{\sigma}} = \mathrm{vech}(\hat{\boldsymbol{\Sigma}})$. Yuan and Bentler (1998c) proposed to use the estimating equation approach to characterize the asymptotic distribution of $\hat{\boldsymbol{\beta}}$ and to estimate its covariance matrix. It is obvious that $\hat{\boldsymbol{\mu}}$ and $\hat{\boldsymbol{\sigma}}$ satisfy

$$\frac{1}{n}\sum_{i=1}^{n} \mathbf{g}_1(\mathbf{x}_i, \hat{\boldsymbol{\mu}}, \hat{\boldsymbol{\Sigma}}) = \mathbf{0}, \qquad \frac{1}{n}\sum_{i=1}^{n} \mathbf{g}_2(\mathbf{x}_i, \hat{\boldsymbol{\mu}}, \hat{\boldsymbol{\Sigma}}) = \mathbf{0},$$

where

$$\mathbf{g}_1(\mathbf{x}, \boldsymbol{\mu}, \boldsymbol{\Sigma}) = u_1\{d(\mathbf{x}, \boldsymbol{\mu}, \boldsymbol{\Sigma})\}(\mathbf{x} - \boldsymbol{\mu})$$

and

$$\mathbf{g}_2(\mathbf{x}, \boldsymbol{\mu}, \boldsymbol{\Sigma}) = u_2\{d^2(\mathbf{x}, \boldsymbol{\mu}, \boldsymbol{\Sigma})\}\,\mathrm{vech}\big[(\mathbf{x} - \boldsymbol{\mu})(\mathbf{x} - \boldsymbol{\mu})'\big] - \boldsymbol{\sigma}$$

corresponding to (36); and

$$\mathbf{g}_1(\mathbf{x}, \boldsymbol{\mu}, \boldsymbol{\Sigma}) = u\{d(\mathbf{x}, \boldsymbol{\mu}, \boldsymbol{\Sigma})\}(\mathbf{x} - \boldsymbol{\mu})$$

and

$$\mathbf{g}_2(\mathbf{x}, \boldsymbol{\mu}, \boldsymbol{\Sigma}) = u^2\{d(\mathbf{x}, \boldsymbol{\mu}, \boldsymbol{\Sigma})\}\,\mathrm{vech}\{(\mathbf{x} - \boldsymbol{\mu})(\mathbf{x} - \boldsymbol{\mu})' - \boldsymbol{\Sigma}\} + \frac{1}{n}\boldsymbol{\sigma}$$

corresponding to (39). Let $\mathbf{g} = (\mathbf{g}_1', \mathbf{g}_2')'$ and $\dot{\mathbf{g}} = \partial\mathbf{g}/\partial\boldsymbol{\beta}'$, then (see Yuan and Bentler, 1998c)

$$\sqrt{n}\begin{pmatrix} \hat{\boldsymbol{\mu}} - \boldsymbol{\mu}_0 \\ \hat{\boldsymbol{\sigma}} - \boldsymbol{\sigma}_0 \end{pmatrix} \xrightarrow{\mathcal{L}} N(\mathbf{0}, \mathbf{V}), \tag{41a}$$

where $\mathbf{V} = \mathbf{H}^{-1}\mathbf{B}\mathbf{H}'^{-1}$ with

$$\mathbf{H} = E\{\dot{\mathbf{g}}(\mathbf{x}, \boldsymbol{\mu}_0, \boldsymbol{\sigma}_0)\}, \qquad \mathbf{B} = E\{\mathbf{g}(\mathbf{x}, \boldsymbol{\mu}_0, \boldsymbol{\sigma}_0)\mathbf{g}'(\mathbf{x}, \boldsymbol{\mu}_0, \boldsymbol{\sigma}_0)\}. \tag{41b}$$

In contrast to (22), we only require $\boldsymbol{\Sigma}_0$ to exist in order for (41) to hold. A consistent estimate of $\mathbf{V}$ can be obtained by using consistent estimates for $\mathbf{H}$ and $\mathbf{B}$; these are given by

$$\hat{\mathbf{H}} = \frac{1}{n}\sum_{i=1}^{n} \dot{\mathbf{g}}(\mathbf{x}_i, \hat{\boldsymbol{\mu}}, \hat{\boldsymbol{\sigma}}), \qquad \hat{\mathbf{B}} = \frac{1}{n}\sum_{i=1}^{n} \mathbf{g}(\mathbf{x}_i, \hat{\boldsymbol{\mu}}, \hat{\boldsymbol{\sigma}})\mathbf{g}'(\mathbf{x}_i, \hat{\boldsymbol{\mu}}, \hat{\boldsymbol{\sigma}}).$$

Now, with $\bar{\mathbf{x}}$, $\mathbf{S}$ and $\hat{\boldsymbol{\Pi}}$ being replaced by $\hat{\boldsymbol{\mu}}$, $\hat{\boldsymbol{\Sigma}}$ and $\hat{\mathbf{V}}$, respectively, all the procedures in the previous two subsections can be applied in a robust manner.

Let $\mathbf{V}_{11}$ and $\mathbf{V}_{22}$ be the submatrices of $\mathbf{V}$ corresponding to asymptotic covariance matrices of $\hat{\boldsymbol{\mu}}$ and $\hat{\boldsymbol{\sigma}}$, respectively. When data follow an elliptical distribution, $\hat{\boldsymbol{\mu}}$ and $\hat{\boldsymbol{\sigma}}$ are asymptotically independent and there exist (Maronna, 1976; Tyler, 1983; Yuan and Bentler, 1998b)

$$\mathbf{V}_{11} = \kappa_1\boldsymbol{\Sigma}_0, \qquad \mathbf{V}_{22} = 2\kappa_2\mathbf{D}_p^{+}(\boldsymbol{\Sigma}_0 \otimes \boldsymbol{\Sigma}_0)\mathbf{D}_p^{+'} + \kappa_3\boldsymbol{\sigma}_0\boldsymbol{\sigma}_0', \tag{42}$$

where the scalars κ_1 to κ_3 are related to the underlying distribution of the data and the weights used in the estimation. A consistent estimator for $\mathbf{V}$ can also be obtained based on (42) when data are truly elliptically distributed. The one based on (41b) is more robust against violation of distributional assumptions (Yuan and Jennrich, 1998). The value of (42) is that, when replacing $\mathbf{S}$ by $\hat{\boldsymbol{\Sigma}}$ and $\hat{\boldsymbol{\Pi}}_{22}$ by $\hat{\mathbf{V}}_{22}$, the rescaled statistic $T_{\mathrm{RML}c}$ in CSA asymptotically follows $\chi^2_{df_c}$ (Yuan, Bentler and Chan, 2004). Parallel to modeling $\bar{\mathbf{x}}$ and $\mathbf{S}$ in mean and covariance structure analysis, the rescaled statistic T_{RML} when modeling $\hat{\boldsymbol{\mu}}$ and $\hat{\boldsymbol{\Sigma}}$ will only approach a distribution whose mean equals df. Of course, we can also obtain the MLE when the specific elliptical distribution form of $\mathbf{x}$ is known (see Fang et al., 1990; Kano et al., 1993). However, in real data analysis, it is unlikely that we know the exact distribution form for a given data set.

All three M-estimators can effectively control the influence of heavy tails or outliers. However, differences exist among the three procedures. Based on empirical experience, Yuan and Bentler (1998b, 1998c) found that, in Huber-type estimators, the effect of abnormal cases is down-weighted but not eliminated. If data are nearly normal, the estimators based on Huber-type weights are still highly efficient. So Huber-type weights are better used for data sets whose distributions are not too far away from normal. By using redescending weights, the effect of outlying cases is minimized, almost equivalent to outlier removal. But the estimators will lose more efficiency than those based on Huber-type weights when data are approximately normal. Yuan and Bentler (1998c) also found that the tuning parameters recommended by Campbell ($b_1 = 2$ and $b_2 = 1.5$) lead to many cases with essentially zero weights, resulting in near singular $\hat{\boldsymbol{\Sigma}}$ and $\hat{\mathbf{V}}$. Thus, b_2 has to be adjusted upwards to have a proper redescending effect. The weights based on a multivariate t-distribution are best used for a data set whose spread can be approximately described by the t-distribution, or for a data set with heavy tails but with no obvious outliers. In addition to these guidelines, statistical procedures have also been proposed for choosing proper weights, mainly in the context of CSA.

For choosing the tuning parameter ρ in using the Huber-type weights of (37), let $u_{2i} = u_2\{d^2(\mathbf{x}_i, \hat{\boldsymbol{\mu}}, \hat{\boldsymbol{\Sigma}})\}$ and

$$\mathbf{x}_i^{(\rho)} = \left\{\sqrt{u_{2i}}(\mathbf{x}_i - \hat{\boldsymbol{\mu}})\right\} \tag{43}$$

at the convergence of (36). Then we can rewrite (36b) as

$$\hat{\boldsymbol{\Sigma}} = \frac{1}{n}\sum_{i=1}^{n} \mathbf{x}_i^{(\rho)}\mathbf{x}_i^{(\rho)\prime},$$

which is just the sample covariance matrix of the $\mathbf{x}_i^{(\rho)}$. Yuan et al. (2000) proposed using (43) as a transformation formula. Working with several practical data sets, they found that the transformed samples $\mathbf{x}_i^{(\rho)}$ are much better approximated by a multivariate normal distribution than the original sample $\mathbf{x}_i$, as measured by Mardia's (1970) skewness and kurtosis. Considering that the sample covariance matrix is most efficient when data are normally distributed, they further proposed using Mardia's kurtosis as a criterion in choosing the tuning parameter ρ, from 0 up by a step of size 0.05 until Mardia's normalized kurtosis for $\mathbf{x}_i^{(\rho)}$ is no longer significant. Using (43), Yuan and Hayashi (2003) proposed using a bootstrap procedure to choose ρ. They discussed a rationale for choosing ρ that makes the statistic $T_{\mathrm{ML}c}$ to approximately follow $\chi^2_{df_c}$. Using real data sets, they showed that $T_{\mathrm{ML}c}$ applied to a properly transformed sample can closely follow $\chi^2_{df_c}$, while the behavior of $T_{\mathrm{AGLS}c}$ and $T_{\mathrm{RML}c}$ remains significantly different from $\chi^2_{df_c}$ when applied to the transformed data. It is obvious that the above criteria for selecting ρ are equally applicable to select m when using the weights based on a multivariate t-distribution. They can also be applied to selecting b_1 and b_2 in applying (39) and (40) by defining

$$u_{2i} = nu_i^2 \Big/ \left(\sum_{i=1}^{n} u_i^2 - 1\right)$$

in (43). Because efficiency of parameter estimates is one of the most important concerns in promoting a statistical methodology, Yuan, Bentler and Chan (2004) proposed to use the empirical efficiency of parameter estimates in selecting a particular transformation. By defining the set of invariant parameters, they found that Huber-type weights in (37) often lead to the most efficient parameter estimates in real data analysis.

Statistical theory for robust estimation has been developed primarily within the class of elliptical distributions (Huber, 1981; Hampel et al., 1986), mainly because analyzing $\hat{\boldsymbol{\Sigma}}$ and $\mathbf{S}$ leads to the same substantive conclusion. In practice, data might contain outliers which will make a true symmetric distribution skewed at the sample level. In such a situation, a robust procedure is definitely preferred (Yuan and Bentler, 2001b). If one believes that the true distribution of $\mathbf{x}$ is skewed, then the results corresponding to analyzing $\hat{\boldsymbol{\mu}}$ and $\hat{\boldsymbol{\Sigma}}$ may not be substantively equivalent to those of analyzing $\bar{\mathbf{x}}$ and $\mathbf{S}$. Hampel et al.'s (1986, p. 401) discussion implies that robust procedures might still be preferred even when $\mathbf{x}$ has a skewed distribution. We recommend that the robust and the classical procedures be compared when having a data set with a highly significant Mardia's multivariate kurtosis.

Techniques for identifying outliers or influential cases are also useful when facing data containing heavy tails or outliers. Berkane and Bentler (1988), Bollen and Arminger (1991), Lee and Wang (1996), Poon and Poon (2002), and Poon and Wong (2004) provided various procedures of influential analysis. With multiple outliers, masking effect may cause difficulty in identifying the true ones (Fung, 1993; Poon et al., 2000; Rousseeuw and van Zomeren, 1990). When true outliers are not obvious, the idea of using empirical efficiency to compare different procedures can also be used to decide whether to keep or remove certain cases or outliers. Evidence in Yuan and Hayashi (2003) and Yuan, Bentler and Chan (2004) implies that outlier removal is not as efficient as a proper robust procedure.

3.4. Misspecified models

SEM involves modeling the relationship among the latent variables as well as that between the observed indicators and the latent variables. For a given number of variables, many models can be formulated. Thus, model misspecification cannot be avoided. Although this also holds in other disciplines of statistical modeling (Box, 1979), it is especially true for SEM. Actually, the perspective that any model is only an approximation to reality dominates the literature in SEM (e.g., MacCallum, 2003). The statistical development for mean and covariance structure models in the previous sections is only for correctly specified models. We will present some statistical results for misspecified models in this section mainly using the ML procedure, partially because most technical developments are based on ML. Most results also hold for AGLS and the robust procedures.

The first concern with a misspecified model is consistency. It is obvious that

$$D_{\mathrm{ML}}\big[\bar{\mathbf{x}}, \mathbf{S}, \boldsymbol{\mu}(\boldsymbol{\theta}), \boldsymbol{\Sigma}(\boldsymbol{\theta})\big] \xrightarrow{P} D_{\mathrm{ML}}\big[\boldsymbol{\mu}_0, \boldsymbol{\Sigma}_0, \boldsymbol{\mu}(\boldsymbol{\theta}), \boldsymbol{\Sigma}(\boldsymbol{\theta})\big]. \tag{44}$$

Let $\boldsymbol{\theta}_*$ be the vector that minimizes $D_{\mathrm{ML}}[\boldsymbol{\mu}_0, \boldsymbol{\Sigma}_0, \boldsymbol{\mu}(\boldsymbol{\theta}), \boldsymbol{\Sigma}(\boldsymbol{\theta})]$. Then, it follows from (44) that $\hat{\boldsymbol{\theta}}$ will converge to $\boldsymbol{\theta}_*$ (see Kano, 1986; Shapiro, 1984). In general, $\boldsymbol{\theta}_*$ does not equal the population $\boldsymbol{\theta}_0$ corresponding to a correctly specified model. Thus, $\hat{\boldsymbol{\theta}}$ is no longer consistent for $\boldsymbol{\theta}_0$. However, this does not mean that all the parameter estimates are inconsistent. Yuan et al. (2003) showed that many parameter estimates in CSA are still consistent even when a model is misspecified. Yuan and Bentler (2006a) further generalized the result to mean and covariance structure analysis. In the following we will present a brief description of the effect of model misspecifications on parameter estimates before discussing their effect on model evaluations.

Let $\boldsymbol{\mu}_*(\boldsymbol{\upsilon})$ and $\boldsymbol{\Sigma}_*(\boldsymbol{\upsilon})$ be the correct mean and covariance structures, thus there exists a vector $\boldsymbol{\upsilon}_0$ such that $\boldsymbol{\mu}_0 = \boldsymbol{\mu}_*(\boldsymbol{\upsilon}_0)$ and $\boldsymbol{\Sigma}_0 = \boldsymbol{\Sigma}_*(\boldsymbol{\upsilon}_0)$. Let the misspecified models be $\boldsymbol{\mu}(\boldsymbol{\theta})$ and $\boldsymbol{\Sigma}(\boldsymbol{\theta})$, due to their missing parameters $\boldsymbol{\vartheta}$ of $\boldsymbol{\upsilon} = (\boldsymbol{\theta}', \boldsymbol{\vartheta}')'$. Then $\Delta\boldsymbol{\theta} = \boldsymbol{\theta}_* - \boldsymbol{\theta}_0$ is the asymptotic bias in $\hat{\boldsymbol{\theta}}$. It is obvious that, if the sample is generated by $\boldsymbol{\mu}^0 = \boldsymbol{\mu}(\boldsymbol{\theta}_0)$ and $\boldsymbol{\Sigma}^0 = \boldsymbol{\Sigma}(\boldsymbol{\theta}_0)$, then $\boldsymbol{\theta}_* = \boldsymbol{\theta}_0$ and $\hat{\boldsymbol{\theta}}$ contains no asymptotic bias. We may regard the true population $(\boldsymbol{\mu}_0, \boldsymbol{\Sigma}_0)$ as a perturbation to $(\boldsymbol{\mu}^0, \boldsymbol{\Sigma}^0)$. Because $\boldsymbol{\theta}_0$ minimizes $D_{\mathrm{ML}}[\boldsymbol{\mu}^0, \boldsymbol{\Sigma}^0, \boldsymbol{\mu}(\boldsymbol{\theta}), \boldsymbol{\Sigma}(\boldsymbol{\theta})]$, under standard regularity conditions, $\boldsymbol{\theta}_0$ satisfies the normal equation $\mathbf{h}(\boldsymbol{\theta}_0, \boldsymbol{\mu}^0, \boldsymbol{\sigma}^0) = \mathbf{0}$, where $\boldsymbol{\sigma}^0 = \mathrm{vech}(\boldsymbol{\Sigma}^0)$ and

$$\mathbf{h}(\boldsymbol{\theta}, \boldsymbol{\mu}, \boldsymbol{\sigma}) = -\dot{\boldsymbol{\mu}}'(\boldsymbol{\theta})\boldsymbol{\Sigma}^{-1}(\boldsymbol{\theta})\big[\boldsymbol{\mu} - \boldsymbol{\mu}(\boldsymbol{\theta})\big] - \dot{\boldsymbol{\sigma}}'(\boldsymbol{\theta})\mathbf{W}_c(\boldsymbol{\theta})\big\{\boldsymbol{\sigma} + \mathrm{vech}\big[\big(\boldsymbol{\mu} - \boldsymbol{\mu}(\boldsymbol{\theta})\big)\big(\boldsymbol{\mu} - \boldsymbol{\mu}(\boldsymbol{\theta})\big)'\big] - \boldsymbol{\sigma}(\boldsymbol{\theta})\big\}.$$

Because

$$\dot{\mathbf{h}}_{\theta}^{-1}(\boldsymbol{\theta}_0, \boldsymbol{\mu}^0, \boldsymbol{\sigma}^0) = (\dot{\boldsymbol{\mu}}'\boldsymbol{\Sigma}^{-1}\dot{\boldsymbol{\mu}} + \dot{\boldsymbol{\sigma}}'\mathbf{W}_c\dot{\boldsymbol{\sigma}})$$

is invertible, under standard regularity conditions (see Rudin, 1976, pp. 224–225), equation $\mathbf{h}(\boldsymbol{\theta}, \boldsymbol{\mu}, \boldsymbol{\sigma}) = \mathbf{0}$ defines $\boldsymbol{\theta}$ as an implicit function of $\boldsymbol{\mu}$ and $\boldsymbol{\sigma}$ in a neighborhood of $(\boldsymbol{\mu}^0, \boldsymbol{\sigma}^0)$. Denote this function as $\boldsymbol{\theta} = \mathbf{g}(\boldsymbol{\mu}, \boldsymbol{\sigma})$. Then $\mathbf{g}$ is continuously differentiable in a neighborhood of $(\boldsymbol{\mu}^0, \boldsymbol{\sigma}^0)$ with

$$\dot{\mathbf{g}}(\boldsymbol{\mu}^0, \boldsymbol{\sigma}^0) = \big(\dot{\boldsymbol{\mu}}'\boldsymbol{\Sigma}^{-1}\dot{\boldsymbol{\mu}} + \dot{\boldsymbol{\sigma}}'\mathbf{W}_c\dot{\boldsymbol{\sigma}}\big)^{-1}\big(\dot{\boldsymbol{\mu}}'\boldsymbol{\Sigma}^{-1} \quad \dot{\boldsymbol{\sigma}}'\mathbf{W}_c\big). \tag{45}$$

It follows from (45) that the perturbation $(\Delta\boldsymbol{\mu}, \Delta\boldsymbol{\sigma})$ causes a perturbation in $\boldsymbol{\theta}$ approximately equal to

$$\Delta\boldsymbol{\theta} \approx \big(\dot{\boldsymbol{\mu}}'\boldsymbol{\Sigma}^{-1}\dot{\boldsymbol{\mu}} + \dot{\boldsymbol{\sigma}}'\mathbf{W}_c\dot{\boldsymbol{\sigma}}\big)^{-1}\big(\dot{\boldsymbol{\mu}}'\boldsymbol{\Sigma}^{-1}\Delta\boldsymbol{\mu} + \dot{\boldsymbol{\sigma}}'\mathbf{W}_c\Delta\boldsymbol{\sigma}\big). \tag{46}$$

Note that the above bias formula is based on a linear approximation. Yuan et al. (2003) and Yuan and Bentler (2004a, 2006a) showed that (46) provides valuable information regarding parameter changes when model is perturbed or misspecified. In the context of CSA,

$$\Delta\boldsymbol{\theta} \approx (\dot{\boldsymbol{\sigma}}'\mathbf{W}_c\dot{\boldsymbol{\sigma}})^{-1}\dot{\boldsymbol{\sigma}}'\mathbf{W}_c\Delta\boldsymbol{\sigma}.$$

If the ijth element of $(\dot{\boldsymbol{\sigma}}'\mathbf{W}_c\dot{\boldsymbol{\sigma}})^{-1}\dot{\boldsymbol{\sigma}}'\mathbf{W}_c$ is zero, then the ith parameter will not be affected by the perturbation on the jth element in $\boldsymbol{\sigma}$. There are also various special cases where a misspecification will not affect the parameter estimates systematically (see Yuan et al., 2003; Yuan and Bentler, 2004a).

Due to biases, positive parameters may have negative estimates (Yuan et al., 2003). In the practice of SEM and EFA, negative estimates for error variances are common and are called improper solutions or Heywood cases (Jöreskog, 1967). In addition to misspecified or underidentified models, large sampling errors due to smaller sample sizes as well as small error variances may result in improper solutions (Anderson and Gerbing, 1984; Boomsma, 1985; Chen et al., 2001; Fornell, 1983; Rindskopf, 1984; van Driel, 1978). Outliers in the sample may also cause improper solutions (Bollen, 1987). When improper solutions are caused by outliers, a robust procedure as in Section 3.3 should resolve the problem. Kano (1998) suggested to fix the corresponding error variances at prespecified positive values when improper solutions are caused by sampling errors and to remove the variables with negative error variance estimates that are caused by model misspecifications. EQS automatically constrains variances so that their estimates are nonnegative. Yuan and Chan (2002) proposed to first estimate submodels that contain the error variances having negative estimates. Because a smaller model is less likely to be misspecified, improper solutions occur less often. These estimates are consistent and can be treated as known in estimating the whole model that involves all the variables. McDonald (2004) reviews various strategies for respecifying improper solutions.

Misspecified models not only affect parameter estimates but also the test statistics. Earlier technical development in CSA was given by Satorra and Saris (1985) and Steiger et al. (1985), who proposed to use the noncentral chi-square to describe the distribution of T_{MLc}. Actually, T_{ML} as well as other LR statistics also asymptotically follow noncentral chi-square distributions under certain assumptions (Wald, 1943). The key assumption behind this development is a sequence of local alternative hypotheses. For the given model structures $\boldsymbol{\mu}(\boldsymbol{\theta})$ and $\boldsymbol{\Sigma}(\boldsymbol{\theta})$, this assumption specifies a sequence of population mean vectors $\boldsymbol{\mu}_0^n$ and covariance matrices $\boldsymbol{\Sigma}_0^n$, indexed by the sample size n. As n increases, $\boldsymbol{\mu}_0^n \to \boldsymbol{\mu}_0$ and $\boldsymbol{\Sigma}_0^n \to \boldsymbol{\Sigma}_0$ at the speed of $1/\sqrt{n}$ each so that $nD_{ML}[\boldsymbol{\mu}_0^n, \boldsymbol{\Sigma}_0^n, \boldsymbol{\mu}(\boldsymbol{\theta}_*^n), \boldsymbol{\Sigma}(\boldsymbol{\theta}_*^n)]$ has a limit that does not depend on n. This limit value is the commonly used noncentrality parameter. In practice, one has a sample of size n whose population means and covariance matrix are unknown but fixed. So the limiting noncentral chi-square distribution obtained based on a sequence of local alternatives may not be applicable in data analysis. Actually, in many contexts, LR statistics have been shown to asymptotically follow normal distributions (Stroud, 1972; Sugiura, 1969; Yanagihara et al., 2005). A very general result was given in Vuong (1989). However, these results do not take the model degrees of freedom into account. Under a fixed alternative, Shapiro (1983) showed that test statistics in CSA generally follow normal distributions and he provided a higher order term in the asymptotic distribution for a class of statistics. By extending the results of Shapiro (1983) and Vuong (1989) to mean and covariance structure models, Yuan et al. (2005) provided a normal distribution description for T_{ML}. Let $\boldsymbol{\mu}_* = \boldsymbol{\mu}(\boldsymbol{\theta}_*)$, $\boldsymbol{\Sigma}_* = \boldsymbol{\Sigma}(\boldsymbol{\theta}_*)$, $\dot{\boldsymbol{\beta}}_* = \dot{\boldsymbol{\beta}}(\boldsymbol{\theta}_*)$, $\ddot{\mu}_{i*} = \ddot{\mu}_i(\boldsymbol{\theta}_*)$, $\ddot{\sigma}_{ij*} = \ddot{\sigma}_{ij}(\boldsymbol{\theta}_*)$,

$$\mathbf{W}_{c*} = 2^{-1}\mathbf{D}_p'\left(\boldsymbol{\Sigma}_*^{-1} \otimes \boldsymbol{\Sigma}_*^{-1}\right)\mathbf{D}_p, \qquad \boldsymbol{\varsigma} = \boldsymbol{\Sigma}_*^{-1}(\boldsymbol{\mu}_0 - \boldsymbol{\mu}_*) = (\varsigma_1, \ldots, \varsigma_p)',$$

$$\mathbf{T} = \boldsymbol{\Sigma}_*^{-1}\left[\boldsymbol{\Sigma}_0 + (\boldsymbol{\mu}_0 - \boldsymbol{\mu}_*)(\boldsymbol{\mu}_0 - \boldsymbol{\mu}_*)'\right]\boldsymbol{\Sigma}_*^{-1}, \qquad \mathbf{K} = (k_{ij}) = \boldsymbol{\Sigma}_*^{-1} - \mathbf{T},$$

$$\mathbf{F} = \begin{pmatrix} \boldsymbol{\Sigma}_*^{-1} & 2^{-1}(\boldsymbol{\varsigma}' \otimes \boldsymbol{\Sigma}_*^{-1})\mathbf{D}_p \\ 2^{-1}\mathbf{D}_p'(\boldsymbol{\varsigma} \otimes \boldsymbol{\Sigma}_*^{-1}) & \mathbf{W}_{c*} \end{pmatrix} \dot{\boldsymbol{\beta}}_*,$$

$$\begin{aligned} \mathbf{Q} = {} & \dot{\boldsymbol{\beta}}_*' \begin{pmatrix} \boldsymbol{\Sigma}_*^{-1} & (\boldsymbol{\varsigma}' \otimes \boldsymbol{\Sigma}_*^{-1})\mathbf{D}_p \\ \mathbf{D}_p'(\boldsymbol{\varsigma} \otimes \boldsymbol{\Sigma}_*^{-1}) & \mathbf{D}_p'\left[(\mathbf{T} - 2^{-1}\boldsymbol{\Sigma}_*^{-1}) \otimes \boldsymbol{\Sigma}_*^{-1}\right]\mathbf{D}_p \end{pmatrix} \dot{\boldsymbol{\beta}}_* \\ & + \frac{1}{2}\sum_{i=1}^p \sum_{j=1}^p k_{ij}\ddot{\sigma}_{ij*} - \sum_{i=1}^p \varsigma_i \ddot{\mu}_{i*}, \end{aligned}$$

$$\mathbf{W}_{*0} = \operatorname{diag}\left(\boldsymbol{\Sigma}_*^{-1}, \mathbf{W}_c\right), \qquad \mathbf{U}_* = \mathbf{W}_{*0} - \mathbf{F}\mathbf{Q}^{-1}\mathbf{F}'.$$

The result of Yuan et al. (2005) can be stated as

$$\sqrt{n}(T_{ML}/n - \mu_{ML}) \xrightarrow{\mathcal{L}} N\left(0, \omega_{ML}^2\right), \tag{47}$$

where

$$\mu_{ML} = D_{ML}(\boldsymbol{\mu}_0, \boldsymbol{\Sigma}_0, \boldsymbol{\mu}_*, \boldsymbol{\Sigma}_*) + \frac{\operatorname{tr}(\mathbf{U}_*\boldsymbol{\Pi})}{n}$$

and

$$\begin{aligned} \omega_{ML}^2 = {} & 4(\boldsymbol{\mu}_0 - \boldsymbol{\mu}_*)'\boldsymbol{\Sigma}_*^{-1}\boldsymbol{\Sigma}_0\boldsymbol{\Sigma}_*^{-1}(\boldsymbol{\mu}_0 - \boldsymbol{\mu}_*) + 2\operatorname{tr}\left[\left(\boldsymbol{\Sigma}_*^{-1}\boldsymbol{\Sigma}_0 - \mathbf{I}_p\right)^2\right] \\ & + \operatorname{tr}\left\{\left[\mathbf{D}_p'\left(\boldsymbol{\Sigma}_*^{-1} - \boldsymbol{\Sigma}_0^{-1}\right) \otimes \left(\boldsymbol{\Sigma}_*^{-1} - \boldsymbol{\Sigma}_0^{-1}\right)\mathbf{D}_p\right](\boldsymbol{\Pi}_{22} - \boldsymbol{\Pi}_{\mathcal{N}22})\right\} \\ & + 4\operatorname{tr}\left(\left\{\left[(\boldsymbol{\mu}_0 - \boldsymbol{\mu}_*)'\boldsymbol{\Sigma}_*^{-1}\right] \otimes \left(\boldsymbol{\Sigma}_*^{-1} - \boldsymbol{\Sigma}_0^{-1}\right)\right\}\mathbf{D}_p\boldsymbol{\Pi}_{21}\right) \end{aligned}$$

with $\mathbf{\Pi}_{\mathcal{N}22}$ being given by (24). Notice that (47) can be applied for both normal and nonnormally distributed data. When data are normally distributed, $\mathbf{\Pi}_{21} = \mathbf{0}$ and $\mathbf{\Pi}_{22} = \mathbf{\Pi}_{\mathcal{N}22}$, thus the formula for ω^2_{ML} is much simpler. When the model is approximately correct, $\mathrm{tr}(\mathbf{U}_*\mathbf{\Pi}) \approx \mathrm{tr}(\mathbf{U}\mathbf{\Pi})$ with $\mathbf{U}$ being given by (26). For normally distributed data, $\mathrm{tr}(\mathbf{U}\mathbf{\Pi}) = df$. So $\mathrm{tr}(\mathbf{U}_*\mathbf{\Pi})$ in (47) accounts for model complexity or the degrees of freedom. Empirical results imply that, for normally distributed data, (47) describes the distribution of T_{ML} better than the corresponding noncentral chi-square distribution unless the null hypothesis is trivially violated. In comparing the means of latent variables with nonnormally distributed data, (47) is better than the noncentral chi-square description even when the effect size is small (see Yuan et al., 2005).

When a model is misspecified, parameter estimates are still asymptotically normally distributed (Vuong, 1989; White, 1982), but the SEs of $\hat{\boldsymbol{\theta}}$ based on $\mathbf{\Omega}_{\mathrm{ML}}$ in (17) or $\mathbf{\Omega}_{\mathrm{SW}}$ in (27) are no longer consistent (see Yuan and Hayashi, in press). Let

$$l_i(\boldsymbol{\theta}) = -\frac{1}{2}\ln\left|\mathbf{\Sigma}(\boldsymbol{\theta})\right| - \frac{1}{2}\left[\mathbf{x}_i - \boldsymbol{\mu}(\boldsymbol{\theta})\right]'\mathbf{\Sigma}^{-1}(\boldsymbol{\theta})\left[\mathbf{x}_i - \boldsymbol{\mu}(\boldsymbol{\theta})\right].$$

Consistent SEs of $\hat{\boldsymbol{\theta}}$ are obtained by the square root of the diagonal elements of

$$\left[\sum_{i=1}^{n}\ddot{l}_i(\hat{\boldsymbol{\theta}})\right]^{-1}\left[\sum_{i=1}^{n}\dot{l}_i(\hat{\boldsymbol{\theta}})\dot{l}_i'(\hat{\boldsymbol{\theta}})\right]\left[\sum_{i=1}^{n}\ddot{l}_i(\hat{\boldsymbol{\theta}})\right]^{-1} \tag{48}$$

or the bootstrap procedure. In summary, misspecified models bring biases to parameter estimates, make the commonly used SEs not consistent, and change the asymptotic distributions of the test statistics from chi-squares to normal distributions.

A further interesting question involves the power of the test statistics. Most existing literature on power is based on the noncentral chi-square distribution (Hancock, 2001; Kim, 2005; MacCallum et al., 1996; Satorra, 2003; Satorra and Saris, 1985), although there are some exceptions (Muthén and Muthén, 2002; Mooijaart, 2003; Yuan and Hayashi, 2003). Noncentral chi-square distributions are valuable when comparing the relative powers of two statistics (see Pitman, 1948), but using them to describe the empirical distributions of overall test statistics at a given alternative is not really valid due to (47) (see also Yuan, 2005a; Yuan and Hayashi, 2003). Another problem with power studies in this field is that none of the existing statistics accurately control type I errors even with normally distributed data when sample size is small (Bentler and Yuan, 1999). For medium to large sample sizes, a bootstrap procedure can be used for the purpose of model testing and power analysis (Beran, 1986; Beran and Srivastava, 1985; Bollen and Stine, 1993; Yuan and Hayashi, 2003; Yung and Bentler, 1996). Because the critical value under the null hypothesis is estimated by simulation rather than from the chi-square distribution, type I errors might be better controlled with a bootstrap procedure. Since heavy tails in the sample will attenuate the power of all the statistics (Yuan and Hayashi, 2003; Yuan, Bentler and Chan, 2004), combining the bootstrap with a robust procedure can increase power in general. Also, in power analysis one has to take into account the nonconverged replications in simulation or bootstrap (see Yuan and Hayashi, 2003). When many nonconvergences occur, bootstrap or simulation approaches to inference and power are problematic. Then improved asymptotics such as

(47) should provide a better approximation to the true power than that based on the non-central chi-square distribution. Further methodological development to evaluate power in SEM is needed.

3.5. *Fit indices*

Referring a statistic $T = nD[\bar{\mathbf{x}}, \mathbf{S}, \boldsymbol{\mu}(\hat{\boldsymbol{\theta}}), \boldsymbol{\Sigma}(\hat{\boldsymbol{\theta}})]$ to χ^2_{df} is to test whether $\boldsymbol{\mu}_0 = \boldsymbol{\mu}(\boldsymbol{\theta}_0)$ and $\boldsymbol{\Sigma}_0 = \boldsymbol{\Sigma}(\boldsymbol{\theta}_0)$. With fixed $\boldsymbol{\mu}_0$ and $\boldsymbol{\Sigma}_0$, unless the models are correctly specified, T generally increases with n according to equations (44) and (47), and is almost surely significant when sample size is large. Alternative criteria, called fit indices, have been developed to measure the approximation between model and data. There are many model fit indices in the literature of SEM. Most of them are defined through a test statistic T, and all aim to get rid of the effect of n. In the context of CSA, some fit indices achieve this by comparing the substantive model to a base model. The commonly used base model for CSA is $\boldsymbol{\Sigma}_b = \mathrm{diag}(\sigma_{11}, \sigma_{22}, \ldots, \sigma_{pp})$. For example, the normed fit index (Bentler and Bonett, 1980) is defined as

$$\mathrm{NFI} = 1 - \frac{T}{T_b} = 1 - \frac{D[\mathbf{S}, \boldsymbol{\Sigma}(\hat{\boldsymbol{\theta}})]}{D[\mathbf{S}, \boldsymbol{\Sigma}_b(\hat{\boldsymbol{\theta}}_b)]}.$$

It is obvious that the effect of n on NFI is much smaller than that on T.

Let H_0 represent the null hypothesis of correct model structure and H_1 represent the alternative hypothesis of a misspecified structure. Under the idealized assumption

$$(T|H_0) \sim \chi^2_{df}, \qquad (T|H_1) \sim \chi^2_{df}(\delta), \tag{49}$$

$\hat{\delta} = T - df$ is an unbiased estimate of the noncentrality parameter $\delta = nD(\boldsymbol{\mu}_0, \boldsymbol{\Sigma}_0, \boldsymbol{\mu}(\boldsymbol{\theta}_*), \boldsymbol{\Sigma}(\boldsymbol{\theta}_*))$ for mean and covariance structures, and $\delta = nD(\boldsymbol{\Sigma}_0, \boldsymbol{\Sigma}(\boldsymbol{\theta}_*))$ for CSA. The larger the δ, the poorer the approximation between the model and the population mean vector and covariance matrix or just the covariance matrix for CSA. Considering CSA again, a model may not be perfect, but it should explain covariances substantially better than the base model does. Thus in CSA, a good model should have a much smaller δ than δ_b that corresponds to the base model. Notice that $\hat{\delta}$ may be negative due to sampling error. Bentler (1990) defined the comparative fit index for CSA as

$$\mathrm{CFI} = 1 - \frac{\max(\hat{\delta}, 0)}{\max(\hat{\delta}, \hat{\delta}_b)}.$$

It is obvious that CFI $\in [0, 1]$ and the effect of n on CFI is much smaller.

A disadvantage of NFI and CFI is that they are defined for CSA only, since there is no natural base model for a mean structure.[7] A fit index that does not refer to a base model is given by Steiger and Lind (1980). This can be used for mean as well as covariance structures. The so called root-mean-square error of approximation (RMSEA) is given by

$$\mathrm{RMSEA} = \sqrt{\max[\hat{\delta}/(ndf), 0]},$$

[7] When there is a mean structure, EQS uses only the covariance structure to define NFI and CFI.

aiming to estimate $\{D_{\mathrm{ML}}[\boldsymbol{\mu}_0, \boldsymbol{\Sigma}_0, \boldsymbol{\mu}(\boldsymbol{\theta}_*), \boldsymbol{\Sigma}(\boldsymbol{\theta}_*)]/df\}^{1/2}$. Notice that RMSEA is solely a function of T. When $T \sim \chi^2_{df}(\delta)$, a confidence interval for δ or $\sqrt{\delta/(ndf)}$ can be obtained using (49) (see Browne and Cudeck, 1993).

Recommended cutoff values for NFI and CFI are 0.95, and that for RMSEA is 0.05. CFI > 0.95 or RMSEA < 0.05 implies a good model.[8] However, useful cutoff values for these indices may depend on many factors (Kenny and McCoach, 2003; Yuan, 2005a). Confidence intervals for RMSEA based on the assumption (49) are also recommended. Unfortunately, (49) does not hold even for T_{ML} with normally distributed data, the confidence interval for RMSEA as printed in SEM programs has little statistical meaning. According to (47), $\hat{\delta}/n$ is generally a biased estimate of $D_{\mathrm{ML}}[\boldsymbol{\mu}_0, \boldsymbol{\Sigma}_0, \boldsymbol{\mu}(\boldsymbol{\theta}_*), \boldsymbol{\Sigma}(\boldsymbol{\theta}_*)]$ even when $T = T_{\mathrm{ML}}$ and data are normally distributed. Because, under $\mathbf{x} \sim N(\boldsymbol{\mu}, \boldsymbol{\Sigma})$, either $D_{\mathrm{ML}}[\boldsymbol{\mu}_0, \boldsymbol{\Sigma}_0, \boldsymbol{\mu}(\boldsymbol{\theta}_*), \boldsymbol{\Sigma}(\boldsymbol{\theta}_*)] > 0$ or $\mathrm{tr}(\mathbf{U}_*\boldsymbol{\Pi}) \neq df$ in (47) is due to model misspecification, it might be more appropriate to abandon the noncentrality parameter and just call $(T - df)/n$ the measure of discrepancy. When data are not normally distributed, $(T - df)/n$ is not a sensible estimate for $D_{\mathrm{ML}}[\boldsymbol{\mu}_0, \boldsymbol{\Sigma}_0, \boldsymbol{\mu}(\boldsymbol{\theta}_*), \boldsymbol{\Sigma}(\boldsymbol{\theta}_*)]$. Raykov (2005) suggests using a bias-corrected bootstrap estimate of the noncentrality parameter. Yuan and Marshall (2004) defined alternative measures for lack of fit that do not depend on (49), they have not yet appeared in SEM programs.

The above three fit indices are available in most SEM software. While there are many other fit indices, they are also among the most widely used fit indices in the applied literature. Further references and discussion on fit indices and their utilities and problems are in Hu and Bentler (1998, 1999), Beauducel and Wittmann (2005), and Yuan (2005a).

3.6. LM and Wald tests

In the specification of a SEM model, some parameters are fixed at known values or are set to be equal; other parameters are free to be estimated. The initial model specifications are usually motivated by a substantive theory or to mimic a previous model in a similar context. Such formulated models may not fit the given data set when evaluated. The Lagrangian multiplier (LM) test allows us to check whether prespecified constraints or fixed parameters are statistically acceptable. The Wald (1943) test evaluates whether free parameters need to be fixed. The LM test was developed in the context of ML estimation with constraint by Aitchison and Silvey (1958). The same test was proposed by Rao (1947), who called it the score test. However, the name LM test is more widely used (Engle, 1984). Lee and Bentler (1980) studied the properties of LM test for CSA within the GLS context. The theory of these tests in SEM was developed by Bentler and Dijkstra (1985), Lee (1985), Bentler (1986), Satorra (1989), and Chou and Bentler (1990) (see also Chou and Bentler, 2002). We will introduce these two tests when $D_{\mathrm{ML}}[\hat{\boldsymbol{\mu}}, \hat{\boldsymbol{\Sigma}}, \boldsymbol{\mu}(\boldsymbol{\theta}), \boldsymbol{\Sigma}(\boldsymbol{\theta})]$ is minimized for parameter estimates $\hat{\boldsymbol{\theta}}$, where $\hat{\boldsymbol{\mu}}$ and $\hat{\boldsymbol{\Sigma}}$ are estimates of $\boldsymbol{\mu}_0$ and $\boldsymbol{\Sigma}_0$, respectively. When $\hat{\boldsymbol{\mu}} = \bar{\mathbf{x}}$ and $\hat{\boldsymbol{\Sigma}} = \mathbf{S}$, $\hat{\boldsymbol{\theta}}$ is the MLE under constraints. We will first present the LM test, followed by the Wald test.

[8] Cutoff values for fit indices vary in the applied literature. Many used CFI > 0.90 to indicate good model.

Suppose not all parameters in $\boldsymbol{\theta}$ are free, they are subject to equality constraints $\mathbf{h}(\boldsymbol{\theta}) = \mathbf{0}$, where $\mathbf{h} = (h_1, h_2, \ldots, h_r)'$ and the h_r's are functionally independent. Then the parameter estimate $\hat{\boldsymbol{\theta}}$ by minimizing $D_{\mathrm{ML}}[\hat{\boldsymbol{\mu}}, \hat{\boldsymbol{\Sigma}}, \boldsymbol{\mu}(\boldsymbol{\theta}), \boldsymbol{\Sigma}(\boldsymbol{\theta})]$ under the constraints satisfies the first order Lagrangian condition

$$\begin{cases} \mathbf{g}(\hat{\boldsymbol{\theta}}) - \dot{\mathbf{h}}'(\hat{\boldsymbol{\theta}})\hat{\boldsymbol{\lambda}} = \mathbf{0}, \\ \mathbf{h}(\hat{\boldsymbol{\theta}}) = \mathbf{0}, \end{cases} \tag{50}$$

where

$$\mathbf{g}(\boldsymbol{\theta}) = \dot{\boldsymbol{\beta}}'(\boldsymbol{\theta})\mathbf{W}(\boldsymbol{\theta}) \begin{pmatrix} \hat{\boldsymbol{\mu}} - \boldsymbol{\mu}(\boldsymbol{\theta}) \\ \hat{\boldsymbol{\sigma}} + \mathrm{vech}\big[\big(\hat{\boldsymbol{\mu}} - \boldsymbol{\mu}(\boldsymbol{\theta})\big)\big(\hat{\boldsymbol{\mu}} - \boldsymbol{\mu}(\boldsymbol{\theta})\big)'\big] - \boldsymbol{\sigma}(\boldsymbol{\theta}) \end{pmatrix}$$

and $\hat{\boldsymbol{\lambda}}$ is a vector of $r \times 1$ Lagrangian multipliers. Notice that when $\boldsymbol{\beta}_0 = \boldsymbol{\beta}(\boldsymbol{\theta}_0)$ and $\mathbf{h}(\boldsymbol{\theta}_0) = \mathbf{0}$, $\mathbf{g}(\boldsymbol{\theta}_0) \xrightarrow{P} \mathbf{0}$; because $\dot{\mathbf{h}}(\boldsymbol{\theta}_0)$ is of full rank, it follows from (50) that $\hat{\boldsymbol{\lambda}} \xrightarrow{P} \boldsymbol{\lambda}_0 = \mathbf{0}$. When $\boldsymbol{\beta}_0 = \boldsymbol{\beta}(\boldsymbol{\theta}_0)$ but $\mathbf{h}(\boldsymbol{\theta}_0) \neq \mathbf{0}$, then (50) implies $\hat{\boldsymbol{\lambda}} \xrightarrow{P} \boldsymbol{\lambda}_0 \neq \mathbf{0}$. Thus, testing $\mathbf{h}(\boldsymbol{\theta}_0) = \mathbf{0}$ is equivalent to testing $\boldsymbol{\lambda}_0 = \mathbf{0}$. Let $\mathbf{B} = \dot{\boldsymbol{\beta}}'\mathbf{W}\boldsymbol{\Pi}\mathbf{W}\dot{\boldsymbol{\beta}}$,

$$\mathbf{A} = \begin{pmatrix} \dot{\boldsymbol{\beta}}'\mathbf{W}\dot{\boldsymbol{\beta}} & \dot{\mathbf{h}}' \\ \dot{\mathbf{h}} & \mathbf{0} \end{pmatrix}, \qquad \mathbf{A}^{-1} = \begin{pmatrix} \mathbf{A}^{11} & \mathbf{A}^{12} \\ \mathbf{A}^{21} & \mathbf{A}^{22} \end{pmatrix}.$$

Then, it follows from the inversion formula for partitioned matrices (Magnus and Neudecker, 1999, p. 11)

$$\begin{aligned} \mathbf{A}^{22} &= -\big[\dot{\mathbf{h}}(\dot{\boldsymbol{\beta}}'\mathbf{W}\dot{\boldsymbol{\beta}})^{-1}\dot{\mathbf{h}}'\big]^{-1}, \\ \mathbf{A}^{11} &= (\dot{\boldsymbol{\beta}}'\mathbf{W}\dot{\boldsymbol{\beta}})^{-1} + (\dot{\boldsymbol{\beta}}'\mathbf{W}\dot{\boldsymbol{\beta}})^{-1}\dot{\mathbf{h}}'\mathbf{A}^{22}\dot{\mathbf{h}}(\dot{\boldsymbol{\beta}}'\mathbf{W}\dot{\boldsymbol{\beta}})^{-1}, \\ \mathbf{A}^{12} &= -(\dot{\boldsymbol{\beta}}'\mathbf{W}\dot{\boldsymbol{\beta}})^{-1}\dot{\mathbf{h}}'\mathbf{A}^{22}, \qquad \mathbf{A}^{21} = \mathbf{A}^{12\,\prime}. \end{aligned}$$

Under the assumptions $\boldsymbol{\beta}_0 = \boldsymbol{\beta}(\boldsymbol{\theta}_0)$ and $\mathbf{h}(\boldsymbol{\theta}_0) = \mathbf{0}$,

$$\sqrt{n}\begin{pmatrix} \hat{\boldsymbol{\theta}} - \boldsymbol{\theta}_0 \\ \hat{\boldsymbol{\lambda}} \end{pmatrix} \xrightarrow{\mathcal{L}} N(\mathbf{0}, \boldsymbol{\Omega}_{\mathrm{LM}}), \tag{51}$$

where

$$\boldsymbol{\Omega}_{\mathrm{LM}} = \begin{pmatrix} \mathbf{A}^{11}\mathbf{B}\mathbf{A}^{11} & \mathbf{A}^{11}\mathbf{B}\mathbf{A}^{12} \\ \mathbf{A}^{21}\mathbf{B}\mathbf{A}^{11} & \mathbf{A}^{22}\mathbf{B}\mathbf{A}^{22} \end{pmatrix}.$$

A consistent estimate for $\boldsymbol{\Omega}_{\mathrm{LM}}$ will be obtained by replacing $\boldsymbol{\theta}$ by $\hat{\boldsymbol{\theta}}$ and $\boldsymbol{\Pi}$ by $\hat{\boldsymbol{\Pi}}$. $\hat{\boldsymbol{\Pi}}$ can be the matrix $\mathbf{S}_t$ in Section 3.1, $\hat{\mathbf{V}}$ in Section 3.3 or $\hat{\boldsymbol{\Omega}}_{\beta_0}$ in the next section, depending on the $\hat{\boldsymbol{\mu}}$ and $\hat{\boldsymbol{\Sigma}}$ used in the estimation process. When $\mathbf{x} \sim N(\boldsymbol{\mu}_0, \boldsymbol{\Sigma}_0)$, $\hat{\boldsymbol{\mu}} = \bar{\mathbf{x}}$ and $\hat{\boldsymbol{\Sigma}} = \mathbf{S}$,

$$\mathbf{A}^{11}\mathbf{B}\mathbf{A}^{11} = \mathbf{A}^{11}, \qquad \mathbf{A}^{11}\mathbf{B}\mathbf{A}^{12} = \mathbf{0}, \qquad \mathbf{A}^{22}\mathbf{B}\mathbf{A}^{22} = -\mathbf{A}^{22}.$$

Let $\tilde{\boldsymbol{\theta}}$ be the estimate of $\boldsymbol{\theta}_0$ by minimizing $D_{\mathrm{ML}}[\hat{\boldsymbol{\mu}}, \hat{\boldsymbol{\Sigma}}, \boldsymbol{\mu}(\boldsymbol{\theta}), \boldsymbol{\Sigma}(\boldsymbol{\theta})]$ without the constraint. When $\mathbf{x} \sim N(\boldsymbol{\mu}_0, \boldsymbol{\Sigma}_0)$, $\boldsymbol{\beta}_0 = \boldsymbol{\beta}(\boldsymbol{\theta}_0)$, $\hat{\boldsymbol{\mu}} = \bar{\mathbf{x}}$ and $\hat{\boldsymbol{\Sigma}} = \mathbf{S}$, the LR statistic for testing the constraint $\mathbf{h}(\boldsymbol{\theta}_0) = \mathbf{0}$ is

$$T_{\mathrm{LR}} = n\big(D_{\mathrm{ML}}\big[\hat{\boldsymbol{\mu}}, \hat{\boldsymbol{\Sigma}}, \boldsymbol{\mu}(\hat{\boldsymbol{\theta}}), \boldsymbol{\Sigma}(\hat{\boldsymbol{\theta}})\big] - D_{\mathrm{ML}}\big[\hat{\boldsymbol{\mu}}, \hat{\boldsymbol{\Sigma}}, \boldsymbol{\mu}(\tilde{\boldsymbol{\theta}}), \boldsymbol{\Sigma}(\tilde{\boldsymbol{\theta}})\big]\big),$$

and $T_{\mathrm{LR}} \xrightarrow{\mathcal{L}} \chi_r^2$ if the constraints hold. The LM statistic is given by

$$T_{\mathrm{LM}} = n\hat{\boldsymbol{\lambda}}'\big(\hat{\mathbf{A}}^{22}\hat{\mathbf{B}}\hat{\mathbf{A}}^{22}\big)^{-1}\hat{\boldsymbol{\lambda}}. \tag{52}$$

Under the null hypothesis $\boldsymbol{\beta}_0 = \boldsymbol{\beta}(\boldsymbol{\theta}_0)$ and $\mathbf{h}(\boldsymbol{\theta}_0) = \mathbf{0}$, T_{LM} and T_{LR} are asymptotically equivalent (see Engle, 1984). Thus, we can refer T_{LM} to χ_r^2 to test whether the constraints are satisfied. It follows from (51) that $T_{\mathrm{LM}} \xrightarrow{\mathcal{L}} \chi_r^2$ even when data are not normally distributed or $\hat{\boldsymbol{\mu}} \neq \bar{\mathbf{x}}$ or $\hat{\boldsymbol{\Sigma}} \neq \mathbf{S}$. When data are normally distributed, both T_{LR} and T_{LM} can be used to test $\mathbf{h}(\boldsymbol{\theta}_0) = \mathbf{0}$. The advantage in T_{LM} is that only the constrained model needs to be estimated, which is more economical. Also, when estimating a model under constraints, one can test any subset of $(h_1, h_2, \ldots, h_r)' = \mathbf{0}$ by focusing on the involved constraints. A significant T_{LM} implies that at least one of the involved constraints is not true in the population.

Equation (50) can be solved by the following IRLS formula[9] for $\boldsymbol{\theta}$ and $\boldsymbol{\lambda}$ (see Aitchison and Silvey, 1958; Browne and Du Toit, 1992)

$$\begin{pmatrix} \boldsymbol{\theta}^{(k+1)} \\ \boldsymbol{\lambda}^{(k+1)} \end{pmatrix} = \begin{pmatrix} \boldsymbol{\theta}^{(k)} \\ \boldsymbol{\lambda}^{(k)} \end{pmatrix} + \alpha \begin{pmatrix} \dot{\boldsymbol{\beta}}^{(k)\prime}\mathbf{W}^{(k)}\dot{\boldsymbol{\beta}}^{(k)} & \dot{\mathbf{h}}^{(k)\prime} \\ \dot{\mathbf{h}}^{(k)} & \mathbf{0} \end{pmatrix}^{-1} \begin{pmatrix} \mathbf{g}^{(k)} - \dot{\mathbf{h}}^{(k)\prime}\boldsymbol{\lambda}^{(k)} \\ -\mathbf{h}^{(k)} \end{pmatrix}, \tag{53}$$

where $\dot{\boldsymbol{\beta}}^{(k)} = \dot{\boldsymbol{\beta}}(\boldsymbol{\theta}^{(k)})$, $\mathbf{W}^{(k)} = \mathbf{W}(\boldsymbol{\theta}^{(k)})$, $\mathbf{h}^{(k)} = \mathbf{h}(\boldsymbol{\theta}^{(k)})$, $\mathbf{g}^{(k)} = \mathbf{g}(\boldsymbol{\theta}^{(k)})$, and α is fixed at 1.0 or chosen to allow

$$\begin{aligned} &D_{\mathrm{ML}}\big[\hat{\boldsymbol{\mu}}, \hat{\boldsymbol{\Sigma}}, \boldsymbol{\mu}\big(\boldsymbol{\theta}^{(k+1)}\big), \boldsymbol{\Sigma}\big(\boldsymbol{\theta}^{(k+1)}\big)\big] + \mathbf{h}^{(k+1)\prime}\mathbf{h}^{(k+1)} \\ &\quad < D_{\mathrm{ML}}\big[\hat{\boldsymbol{\mu}}, \hat{\boldsymbol{\Sigma}}, \boldsymbol{\mu}\big(\boldsymbol{\theta}^{(k)}\big), \boldsymbol{\Sigma}\big(\boldsymbol{\theta}^{(k)}\big)\big] + \mathbf{h}^{(k)\prime}\mathbf{h}^{(k)}. \end{aligned}$$

The Wald test works in the opposite direction from that of the LM test, testing whether $\boldsymbol{\theta}_0$ satisfies the additional constraints

$$\mathbf{c}(\boldsymbol{\theta}_0) = \mathbf{0},$$

where $\mathbf{c} = (c_1, c_2, \ldots, c_s)'$ and the c_j's are functionally independent. Under $\mathbf{c}(\boldsymbol{\theta}_0) = \mathbf{0}$, it follows from (51) that

$$\sqrt{n}\mathbf{c}(\hat{\boldsymbol{\theta}}) \xrightarrow{\mathcal{L}} N(\mathbf{0}, \boldsymbol{\Omega}_W),$$

where

$$\boldsymbol{\Omega}_W = \dot{\mathbf{c}}\mathbf{A}^{11}\mathbf{B}\mathbf{A}^{11}\dot{\mathbf{c}}'.$$

Thus, the Wald statistic for testing $\mathbf{c}(\boldsymbol{\theta}_0) = \mathbf{0}$ is

$$T_W = \mathbf{c}'(\hat{\boldsymbol{\theta}})\hat{\boldsymbol{\Omega}}_W^{-1}\mathbf{c}(\hat{\boldsymbol{\theta}}). \tag{54}$$

When referred to χ_s^2, a nonsignificant T_W implies that all the constraints in $\mathbf{c}(\boldsymbol{\theta}) = \mathbf{0}$ are statistically achievable and a more parsimonious model can be obtained. A special case of T_W is to refer $\hat{\theta}/\mathrm{SE}$ to $N(0, 1)$.

[9] Like in (18), one needs to check conditions to make sure that the converged solution is indeed the constrained MLE.

The LM and Wald tests have been implemented in all SEM software. In EQS, these tests can be used in an a priori fashion based on specific hypotheses, or in an a posteriori fashion to find restrictions to add or free parameters to fix. In LISREL, only the univariate version of each of the tests is available and the LM test is called the model modification index (Sörbom, 1989). When releasing fixed parameters or constraints following the LM tests, an initially poor model can quickly fit the data well. There are concerns that using the LM test in a data-driven manner may capitalize on chance associations in the data (e.g., MacCallum et al., 1992). This is a problem especially in small samples, and some attempts have been made to provide adjusted statistics (e.g., Green et al., 2001; Hancock, 1999). At a minimum, one needs to have an explanation or a new theory when making changes to a model that was originally supported by theory.

The statistics T_{LM} and T_W, as formulated in (52) and (54), are asymptotically distribution free. The statistic T_{LR} also has a rescaled version when data are not normally distributed, which in general does not equal the difference between two rescaled statistics for the nested models (Satorra and Bentler, 2001). We need to note that the asymptotic distributions of the three statistics are obtained based on the assumption that the model $(\boldsymbol{\mu}(\boldsymbol{\theta}), \boldsymbol{\Sigma}(\boldsymbol{\theta}))$, before implementing the constraints $\mathbf{h}(\boldsymbol{\theta}) = \mathbf{0}$, is correctly specified. When $(\boldsymbol{\mu}(\boldsymbol{\theta}), \boldsymbol{\Sigma}(\boldsymbol{\theta}))$ is not correctly specified, these tests may not lead to proper inference about $\mathbf{h}(\boldsymbol{\theta}) = \mathbf{0}$ even when data are normally distributed (see Yuan and Bentler, 2004a).

4. Missing data

Incomplete or missing data exist in almost all areas of empirical research. They are especially common in social and behavioral sciences where surveys or questionnaires are the main sources for acquiring data. Rubin (1976) (see also Little and Rubin, 2002) systematically classified missing data according to the process that causes missing data to occur, he called such a process the missing data mechanism. Missing completely at random (MCAR) is a process in which missingness of the data is independent of both the observed and the missing values; missing at random (MAR) is a process in which missingness of the data is independent of the missing values, but depends on the observed measurements. When the missingness depends on the missing values themselves, the process is not missing at random (NMAR).

Due to its flexibility in modeling survey data and capability of controlling measurement errors, SEM with missing data has been well studied (see Allison, 2003). Most methodology developments focus on the ML procedure by assuming normally distributed data (Allison, 1987; Brown, 1983; Finkbeiner, 1979; Jamshidian and Bentler, 1999; Kim and Bentler, 2002; Lee, 1986; Lee et al., 2003; Muthén et al., 1987; Song and Lee, 2002; Tang and Bentler, 1998). The popularity of ML with missing data is due to several reasons. First, with missing values in the sample, it is not clear how to directly apply the GLS or AGLS procedures. These procedures can be applied when removing all the cases with missing values, but that will result in inconsistent or inefficient parameter estimates. Second, one has to specify a distribution in order to use the EM-algorithm (Dempster et al., 1977) to facilitate parameter estimation. There are many multivariate distributions to choose in principle, without knowing the exact distribution of the population, the normal distribution is a natural and convenient choice. Third, according to Rubin (1976), the property of parameter estimates by any methods is closely related to missing data mechanisms. ML preserves many of its properties when MCAR and MAR mechanisms are ignored while other estimation procedures may not. When missing values are NMAR, ML is still necessary in modeling the missing data mechanism to get consistent parameter estimates (Poon et al., 1997; Tang and Lee, 1998; Lee and Tang, in press). Because the normal theory or any other likelihood function may not be correctly specified, the ML procedure also needs modifications and remedies. We will discuss the ML procedure based on multivariate normal- and t-distribution as well as their extensions for SEM with missing data in this section. Readers interested in ML procedures based on other distributions and the related Markov chain Monte-Carlo method for missing data are referred to Lee et al. (2003) and Song and Lee (2002), and for modeling missing data mechanism that is NMAR are referred to Lee and Tang (in press).

When the ith case contains missing values, let's still use $\mathbf{x}_i$ to denote the $p_i \leqslant p$ observed variables. Let $\boldsymbol{\mu}_i$ be the subvector of $\boldsymbol{\mu}$ and $\boldsymbol{\Sigma}_i$ be the submatrix of $\boldsymbol{\Sigma}$ corresponding to $\mathbf{x}_i$. Then, by ignoring the missing data mechanisms, the log likelihood function based on the normal distribution assumption of $\mathbf{x}_i \sim N_{p_i}(\boldsymbol{\mu}_i, \boldsymbol{\Sigma}_i)$ is given by

$$l(\boldsymbol{\theta}) = \sum_{i=1}^{n} l_i(\boldsymbol{\theta}), \tag{55a}$$

where

$$l_i(\boldsymbol{\theta}) = -\frac{1}{2}\left\{\log\left|\boldsymbol{\Sigma}_i(\boldsymbol{\theta})\right| + \left(\mathbf{x}_i - \boldsymbol{\mu}_i(\boldsymbol{\theta})\right)'\boldsymbol{\Sigma}_i^{-1}(\boldsymbol{\theta})\left(\mathbf{x}_i - \boldsymbol{\mu}_i(\boldsymbol{\theta})\right)\right\}. \tag{55b}$$

The MLE $\hat{\boldsymbol{\theta}}$ can be obtained by maximizing (55) using an iterative procedure as in (18). Since each $\mathbf{x}_i$ may have different marginal variables, the first and second derivatives of the $l(\boldsymbol{\theta})$ in (55) is complicated both analytically and in numerical evaluations. The EM-algorithm facilitates the maximization of (55a) without computing the derivatives of $\boldsymbol{\mu}_i(\boldsymbol{\theta})$ and $\boldsymbol{\Sigma}_i(\boldsymbol{\theta})$ for every missing data pattern. In this algorithm, we do not work with (55) but the likelihood function for the complete data as in (13). There is an E-step and a M-step in EM. The E-step is to calculate the conditional expectation of (13) given the observed data and current parameter values. With missing data being replaced by their expected values, the M-step is to maximize (13) with respect to $\boldsymbol{\theta}$, which can be performed by (18). Let's rewrite (13) as

$$\begin{aligned} l(\boldsymbol{\theta}) = &-\frac{n}{2}\ln\left|\boldsymbol{\Sigma}(\boldsymbol{\theta})\right| \\ &- \frac{n}{2}\operatorname{tr}\left\{\boldsymbol{\Sigma}^{-1}(\boldsymbol{\theta})\left[\frac{1}{n}\sum_{i=1}^{n}\mathbf{x}_i\mathbf{x}_i' - \left(\frac{1}{n}\sum_{i=1}^{n}\mathbf{x}_i\right)\boldsymbol{\mu}'(\boldsymbol{\theta})\right.\right. \\ &\left.\left. - \boldsymbol{\mu}(\boldsymbol{\theta})\left(\frac{1}{n}\sum_{i=1}^{n}\mathbf{x}_i\right)' + \boldsymbol{\mu}(\boldsymbol{\theta})\boldsymbol{\mu}'(\boldsymbol{\theta})\right]\right\}, \end{aligned} \tag{56}$$

and denote by $\mathbf{x}_{io}$ the vector of observed variables for the ith case. The expected value of (56) given the observed data involves $E(\mathbf{x}_i|\mathbf{x}_{io})$ and $E(\mathbf{x}_i\mathbf{x}_i'|\mathbf{x}_{io})$. Suppose the ith case has the pattern $\mathbf{x}_i = (\mathbf{x}_{io}', \mathbf{x}_{im}')'$, where $\mathbf{x}_{im}$ contains the missing variables. Then

$$E(\mathbf{x}_i|\mathbf{x}_{io}) = \begin{pmatrix} \mathbf{x}_{io} \\ E(\mathbf{x}_{im}|\mathbf{x}_{io}) \end{pmatrix},$$

$$E(\mathbf{x}_i\mathbf{x}_i'|\mathbf{x}_{io}) = \begin{pmatrix} \mathbf{x}_{io}\mathbf{x}_{io}' & \mathbf{x}_{io}E(\mathbf{x}_{im}|\mathbf{x}_{io})' \\ E(\mathbf{x}_{im}|\mathbf{x}_{io})\mathbf{x}_{io}' & E(\mathbf{x}_{im}\mathbf{x}_{im}'|\mathbf{x}_{io}) \end{pmatrix}.$$

Corresponding to $\mathbf{x}_i = (\mathbf{x}_{io}', \mathbf{x}_{im}')'$, let

$$\boldsymbol{\mu}^{(c)} = \begin{pmatrix} \boldsymbol{\mu}_o \\ \boldsymbol{\mu}_m \end{pmatrix} \quad \text{and} \quad \boldsymbol{\Sigma}^{(c)} = \begin{pmatrix} \boldsymbol{\Sigma}_{oo} & \boldsymbol{\Sigma}_{om} \\ \boldsymbol{\Sigma}_{mo} & \boldsymbol{\Sigma}_{mm} \end{pmatrix}$$

be the current mean vector and covariance matrix evaluated at $\boldsymbol{\theta}^{(c)}$. It follows from the normal distribution assumption that

$$E(\mathbf{x}_{im}|\mathbf{x}_{io}) = \boldsymbol{\mu}_m + \boldsymbol{\Sigma}_{mo}\boldsymbol{\Sigma}_{oo}^{-1}(\mathbf{x}_{io} - \boldsymbol{\mu}_o),$$
$$E(\mathbf{x}_{im}\mathbf{x}_{im}' \mid \mathbf{x}_{io}) = (\boldsymbol{\Sigma}_{mm} - \boldsymbol{\Sigma}_{mo}\boldsymbol{\Sigma}_{oo}^{-1}\boldsymbol{\Sigma}_{om}) + E(\mathbf{x}_{im}|\mathbf{x}_{io})E(\mathbf{x}_{im}|\mathbf{x}_{io})'.$$

Of course, the pattern of missing values may change from case to case and we will need to put the expected values at the right positions. Once $\sum_{i=1}^n E(\mathbf{x}_i|\mathbf{x}_{io})$ and $\sum_{i=1}^n E(\mathbf{x}_i\mathbf{x}_i'|\mathbf{x}_{io})$ are obtained, the E-step is completed. Let

$$\bar{\mathbf{x}}^{(c)} = \frac{1}{n}\sum_{i=1}^n E(\mathbf{x}_i|\mathbf{x}_{io})$$

$$\mathbf{S}^{(c)} = \frac{1}{n}\sum_{i=1}^n E(\mathbf{x}_i\mathbf{x}_i' \mid \mathbf{x}_{io}) - \bar{\mathbf{x}}^{(c)}\bar{\mathbf{x}}^{(c)\prime}.$$

The M-step is to maximize

$$l(\boldsymbol{\theta}) = -\frac{n}{2}\ln|\boldsymbol{\Sigma}(\boldsymbol{\theta})| - \frac{n}{2}\operatorname{tr}\{\boldsymbol{\Sigma}^{-1}(\boldsymbol{\theta})[\mathbf{S}^{(c)} + (\bar{\mathbf{x}}^{(c)} - \boldsymbol{\mu}(\boldsymbol{\theta}))(\bar{\mathbf{x}}^{(c)} - \boldsymbol{\mu}(\boldsymbol{\theta}))']\} \tag{57}$$

with respect to $\boldsymbol{\theta}$ at $\boldsymbol{\theta}^{(c+1)}$. Note that (57) is just (13) when replacing $\bar{\mathbf{x}}$ by $\bar{\mathbf{x}}^{(c)}$ and $\mathbf{S}$ by $\mathbf{S}^{(c)}$. So the IRLS procedure in (18) can be used to obtain $\boldsymbol{\theta}^{(c+1)}$. This finishes one cycle of the EM-algorithm. One then needs to, in turn, evaluate $\boldsymbol{\mu}^{(c+1)}$ and $\boldsymbol{\Sigma}^{(c+1)}$ at $\boldsymbol{\theta}^{(c+1)}$, complete the E-step by obtaining the $\bar{\mathbf{x}}^{(c+1)}$ and $\mathbf{S}^{(c+1)}$, and proceed to the next M-step by obtaining $\boldsymbol{\theta}^{(c+2)}$. Continue with this process until convergence[10] leads to $\hat{\boldsymbol{\theta}}$. For the saturated model, the M-step, $\boldsymbol{\beta}^{(c+1)} = (\bar{\mathbf{x}}^{(c)\prime}, \operatorname{vech}'(\mathbf{S}^{(c)}))'$, does not need any iteration, which really reflects the beauty of the EM-algorithm.

[10] One has to use multiple starting values and check the Hessian matrix at the convergence to make sure the MLE is indeed obtained.

When $\mathbf{x} \sim N_p(\boldsymbol{\mu}_0, \boldsymbol{\Sigma}_0)$ and all the missing data are either MCAR or MAR, the MLE $\hat{\boldsymbol{\theta}}$ is consistent and asymptotically normally distributed. Let

$$\mathbf{A}(\boldsymbol{\theta}) = -\frac{1}{n}\sum_{i=1}^{n}\ddot{l}_i(\boldsymbol{\theta}),$$

where $l_i(\boldsymbol{\theta})$ is given by (55b). The matrix $\mathbf{A}(\hat{\boldsymbol{\theta}})$ is called the observed information matrix (Efron and Hinkley, 1978). Let $\mathcal{I}(\boldsymbol{\theta}) = E[\mathbf{A}(\boldsymbol{\theta})]$, where the expectation is with respect to the distribution of the observed $\mathbf{x}_i$ and the missing data mechanisms. Suppose $\lim_{n\to\infty}\mathcal{I}(\boldsymbol{\theta}_0)$ exists and we denote it by $\mathcal{I}_{\theta_0}$. Then

$$\sqrt{n}(\hat{\boldsymbol{\theta}} - \boldsymbol{\theta}_0) \xrightarrow{\mathcal{L}} N_q\big(\mathbf{0}, \mathcal{I}_{\theta_0}^{-1}\big). \tag{58}$$

When all the missing data are MCAR, then $\mathcal{I}(\boldsymbol{\theta}_0)$ becomes

$$\mathcal{I}_{\text{MCAR}} = \dot{\boldsymbol{\beta}}'\begin{pmatrix} \frac{1}{n}\sum_{i=1}^{n}\mathbf{G}_i'\boldsymbol{\Sigma}_i^{-1}\mathbf{G}_i & \mathbf{0} \\ \mathbf{0} & \frac{1}{2n}\sum_{i=1}^{n}\mathbf{H}_i'(\boldsymbol{\Sigma}_i^{-1}\otimes\boldsymbol{\Sigma}_i^{-1})\mathbf{H}_i \end{pmatrix}\dot{\boldsymbol{\beta}}, \tag{59}$$

where $\mathbf{G}_i = \partial\boldsymbol{\mu}_i/\partial\boldsymbol{\mu}'$ and $\mathbf{H}_i = \partial\,\mathrm{vec}(\boldsymbol{\Sigma}_i)/\partial\boldsymbol{\sigma}'$. When the model is saturated, $\dot{\boldsymbol{\beta}} = \mathbf{I}$ and (59) becomes a diagonal matrix. The matrix $\mathcal{I}_{\text{MCAR}}$ is often called the expected information matrix, the implicit assumption in taking the expectation is $\mathbf{x}_i \sim N_{p_i}(\boldsymbol{\mu}_i, \boldsymbol{\Sigma}_i)$ or all the missing data are MCAR. When data are MAR, however, $\mathcal{I}_{\theta_0} \neq \mathcal{I}_{\text{MCAR}}$. Kenward and Molenberghs (1998) explained the difference between $\mathcal{I}_{\theta_0}$ and $\mathcal{I}_{\text{MCAR}}$ using simple examples. Specifically, they showed that, for saturated model with $p = 2$, $\mathcal{I}_{\theta_0}$ is not block diagonal. Yuan (2005b) further showed that the structure of $\mathcal{I}_{\theta_0}$ is closely related to the missing data schemes or how the missing values are related to the observed values, the general MAR mechanism alone does not determine the structure of $\mathcal{I}_{\theta_0}$. Because missing data schemes are generally unknown, we cannot analytically calculate $\mathcal{I}_{\theta_0}$. A consistent estimate for $\mathcal{I}_{\theta_0}$ is given by $\mathbf{A}(\hat{\boldsymbol{\theta}})$. So with MAR or MCAR mechanism and $\mathbf{x} \sim N_p(\boldsymbol{\mu}_0, \boldsymbol{\Sigma}_0)$, we can get both the consistent parameter estimates and SEs. Similarly, the likelihood ratio statistic by ignoring MAR or MCAR mechanism also follows a chi-square distribution, that is,

$$T_{\text{ML}} = -2\big[l(\hat{\boldsymbol{\theta}}) - l(\hat{\boldsymbol{\beta}})\big] \xrightarrow{\mathcal{L}} \chi^2_{df}. \tag{60}$$

So, except the need for observed information matrix in computing SEs, when $\mathbf{x} \sim N(\boldsymbol{\mu}, \boldsymbol{\Sigma})$, inference for SEM with missing data that are MAR and MCAR is the same as for complete data. We want to reemphasize that, with MAR data, the observed information has to be used in order to provide consistent SEs for $\hat{\boldsymbol{\theta}}$. Most of the technical development for mean and covariance structure analysis did not make this point clear (e.g., Brown, 1983; Lee, 1986; Allison, 1987; Muthén et al., 1987; Jamshidian and Bentler, 1999).

Although ML with missing data has many nice properties, in modeling real multivariate data, specifying the correct distribution for the likelihood function is always challenging if not impossible. When the data model or distribution specification is incorrect, the ML is commonly called pseudo-ML. Gourieroux et al. (1984) studied the property of pseudo-ML for complete data. Applying the theory of pseudo-ML to mean and covariance structure analysis by ignoring the missing data mechanisms, Arminger and Sobel (1990) provided a sandwich-type covariance matrix to estimate the asymptotic covariance matrix of the pseudo-MLEs. However, Laird (1988) and Rotnitzky and Wypij (1994) implied that, when the distribution form is misspecified, the MLEs are inconsistent unless the missing data mechanism is MCAR. When the distribution is misspecified, the corresponding normal equation is called estimating equation. Liang and Zeger (1986) also implied that, for nonnormally distributed data, MCAR mechanism is necessary for estimators obtained from solving estimating equations to be consistent. Admitting the misspecification of distributions in several normal theory ML based procedures, Yuan and Bentler (2000b) assumed MCAR mechanism to avoid the consistency issues of the MLE. For the same reason, Yuan, Marshall and Bentler (2002) assumed MCAR mechanism in the development for EFA with missing data. However, the limited simulation results in Yuan and Bentler (2000b) indicate that there are no obvious biases in the pseudo-MLEs even when the mechanism is MAR. Actually, none of the above mentioned papers rigorously studied the consistency property of pseudo-MLE with MAR data. Results in Yuan (2005b) imply that, under certain conditions, the normal theory based pseudo-MLE is consistent and asymptotically normally distributed even with MAR data. The asymptotic covariance matrix of the pseudo-MLE is given by a sandwich-type covariance matrix. So we can now assume MAR mechanism in the normal theory ML procedures for nonnormally distributed data.

When data are nonnormally distributed, $\hat{\boldsymbol{\theta}}$ is still consistent when MAR or MCAR mechanisms in the normal theory based likelihood function are being ignored. The asymptotic distribution of $\hat{\boldsymbol{\theta}}$ is given by

$$\sqrt{n}(\hat{\boldsymbol{\theta}} - \boldsymbol{\theta}_0) \xrightarrow{\mathcal{L}} N_q(\mathbf{0}, \boldsymbol{\Omega}_{\theta_0}), \tag{61}$$

where $\boldsymbol{\Omega}_{\theta_0} = \mathcal{I}_{\theta_0}^{-1}\mathbf{B}_{\theta_0}\mathcal{I}_{\theta_0}^{-1}$ with

$$\mathbf{B}_{\theta_0} = \lim_{n\to\infty}\frac{1}{n}\sum_{i=1}^{n} \dot{l}_i \dot{l}_i'.$$

A consistent estimator for $\mathcal{I}_{\theta_0}$ is given by the observed information matrix $\mathbf{A}(\hat{\boldsymbol{\theta}})$ and that for $\mathbf{B}_{\theta_0}$ is given by

$$\hat{\mathbf{B}}_{\theta_0} = \frac{1}{n}\sum_{i=1}^{n} \dot{l}_i(\hat{\boldsymbol{\theta}})\dot{l}_i'(\hat{\boldsymbol{\theta}}),$$

where $l_i(\boldsymbol{\theta})$ is given by (55b). Notice that the $\hat{\boldsymbol{\Omega}}_{\theta_0}$ here has essentially the same analytical expression as (48). The matrix in (48) accounts for misspecifications in both the distribution form and the structural model; $\hat{\boldsymbol{\Omega}}_{\theta_0}$ accounts for misspecification in the distribution form and the lack of modeling missing data mechanisms in (55b).

Similar to the case for complete data, T_{ML} will not approach a chi-square distribution in general with nonnormally distributed data. The AGLS procedure has to be used to obtain a statistic that asymptotically follows a chi-square distribution. Notice that (58) and (61) also apply to the saturated model with

$$\sqrt{n}(\hat{\boldsymbol\beta} - \boldsymbol\beta_0) \xrightarrow{\mathcal{L}} N_q\big(\mathbf{0}, \mathcal{I}_{\beta_0}^{-1}\big)$$

when data are normally distributed, and

$$\sqrt{n}(\hat{\boldsymbol\beta} - \boldsymbol\beta_0) \xrightarrow{\mathcal{L}} N_q(\mathbf{0}, \boldsymbol\Omega_{\beta_0})$$

when data are nonnormally distributed. When data are complete, $\boldsymbol\Omega_{\beta_0}$ reduces to $\boldsymbol\Pi$ in (22). Let the consistent estimator of $\boldsymbol\Omega_{\beta_0}$ be

$$\hat{\boldsymbol\Omega}_{\beta_0} = \begin{pmatrix} \hat{\boldsymbol\Omega}_{\beta_0 11} & \hat{\boldsymbol\Omega}_{\beta_0 12} \\ \hat{\boldsymbol\Omega}_{\beta_0 21} & \hat{\boldsymbol\Omega}_{\beta_0 22} \end{pmatrix}.$$

Then the AGLS procedures can proceed as in Section 2 when replacing $\bar{\mathbf{t}}$ by $\hat{\boldsymbol\beta}$ and $\hat{\boldsymbol\Pi}$ by $\hat{\boldsymbol\Omega}_{\beta_0}$. Especially, the statistic T_{ML} in (60) can be characterized by

$$T_{\mathrm{ML}} = \sum_{j=1}^{df} \kappa_j z_j^2 + o_p(1), \tag{62a}$$

where κ_j are the nonzero eigenvalues of $\mathbf{U}_{\beta_0}\boldsymbol\Omega_{\beta_0}$ with

$$\mathbf{U}_{\beta_0} = \mathcal{I}_{\beta_0} - \mathcal{I}_{\beta_0}\dot{\boldsymbol\beta}(\dot{\boldsymbol\beta}'\mathcal{I}_{\beta_0}\dot{\boldsymbol\beta})^{-1}\dot{\boldsymbol\beta}'\mathcal{I}_{\beta_0}. \tag{62b}$$

Similarly, $\mathcal{I}_{\beta_0}$ can be consistently estimated by

$$\mathbf{A}(\hat{\boldsymbol\beta}) = -\frac{1}{n}\sum_{i=1}^{n} \ddot{l}_i(\hat{\boldsymbol\beta})$$

and $\dot{\boldsymbol\beta}$ by $\dot{\boldsymbol\beta}(\hat{\boldsymbol\theta})$. Let $\hat\kappa = \mathrm{tr}(\hat{\mathbf{U}}_{\beta_0}\hat{\boldsymbol\Omega}_{\beta_0})/df$, then the rescaled statistic is given by

$$T_{\mathrm{RML}} = \hat\kappa^{-1} T_{\mathrm{ML}},$$

which asymptotically follows a distribution with mean df. All other statistics parallel to those in Section 2 are also straightforward to define (see Yuan and Bentler, 2000b). Especially, the AGLS type statistics asymptotically follow χ^2_{df}. However, with missing data, it has not been shown that the rescaled statistics in CSA will asymptotically follow $\chi^2_{df_c}$ even when $\mathbf{x}$ follows an elliptical symmetric distribution. Similarly, a parallel asymptotic robustness condition in CSA with missing data has not been established.

In addition to those procedures parallel to Section 2, there also exists a so called two-stage ML (TML) approach by minimizing

$$\begin{aligned} D_{\mathrm{ML}}\big[\hat{\boldsymbol\mu}, \hat{\boldsymbol\Sigma}, \boldsymbol\mu(\boldsymbol\theta), \boldsymbol\Sigma(\boldsymbol\theta)\big] &= \big[\hat{\boldsymbol\mu} - \boldsymbol\mu(\boldsymbol\theta)\big]'\boldsymbol\Sigma^{-1}(\boldsymbol\theta)\big[\hat{\boldsymbol\mu} - \boldsymbol\mu(\boldsymbol\theta)\big] \\ &\quad + \mathrm{tr}\big[\hat{\boldsymbol\Sigma}\boldsymbol\Sigma^{-1}(\boldsymbol\theta)\big] - \log\big|\hat{\boldsymbol\Sigma}\boldsymbol\Sigma^{-1}(\boldsymbol\theta)\big| - p \end{aligned}$$

for parameter estimate $\breve{\boldsymbol\theta}$, where $\hat{\boldsymbol\mu}$ and $\hat{\boldsymbol\Sigma}$ are the normal theory MLE for the saturated model. Due to missing data, the TML procedure is not equivalent to the ML procedure even when data are normally distributed. Especially,

$$T_{\mathrm{TML}} = n D_{\mathrm{ML}}\big[\hat{\boldsymbol\mu}, \hat{\boldsymbol\Sigma}, \boldsymbol\mu(\breve{\boldsymbol\theta}), \boldsymbol\Sigma(\breve{\boldsymbol\theta})\big]$$

does not approach χ^2_{df} in general even for normally distributed data. But consistent SEs and rescaled statistics can be constructed. Specifically, the TML estimator $\breve{\boldsymbol{\theta}}$ satisfies

$$\sqrt{n}(\breve{\boldsymbol{\theta}} - \boldsymbol{\theta}_0) \xrightarrow{\mathcal{L}} N(\mathbf{0}, \boldsymbol{\Omega}_{\mathrm{TML}}),$$

where

$$\boldsymbol{\Omega}_{\mathrm{TML}} = (\dot{\boldsymbol{\beta}}'\mathbf{W}\dot{\boldsymbol{\beta}})^{-1}(\dot{\boldsymbol{\beta}}'\mathbf{W}\boldsymbol{\Omega}_{\beta_0}\mathbf{W}\dot{\boldsymbol{\beta}})(\dot{\boldsymbol{\beta}}'\mathbf{W}\dot{\boldsymbol{\beta}})^{-1}$$

with $\mathbf{W}$ being defined in (12). The distribution of T_{TML} is characterized by

$$T_{\mathrm{TML}} = \sum_{j=1}^{df} \kappa_j z_j^2 + o_p(1), \tag{63}$$

where κ_j are the nonzero eigenvalues of $\mathbf{U}\boldsymbol{\Omega}_{\beta_0}$ with $\mathbf{U}$ being defined in (26). When a rescaled statistic is computed in the usual way, the concern of Enders and Peugh (2004) that there is no appropriate sample size becomes irrelevant. Comparing (63) to (62), $\mathbf{W}$ is block-diagonal while the $\mathcal{I}_{\beta_0}$ is not block-diagonal in general. When all the missing data are MCAR, $\mathcal{I}_{\beta_0}$ is block-diagonal but it still does not equal $\mathbf{W}$ unless there is no missing data or the proportion of missing data approaches zero (see Yuan and Bentler, 2000b).

Notice that, like the normal theory ML for complete data, the normal theory based ML procedure for missing data is not robust against outliers and will lead to inefficient $\hat{\boldsymbol{\theta}}$ when $\mathbf{x}$ has heavy tails. With missing data, robust procedures for estimating $\boldsymbol{\mu}_0$ and $\boldsymbol{\Sigma}_0$ based on the multivariate t-distribution (Little, 1988) and others (Cheng and Victoria-Feser, 2002; Ruppert, 1992) have been developed. Yuan (2001), Yuan and Bentler (2001a), and Yuan, Marshall and Bentler (2002) studied robust procedures for EFA, SEM and multiple group analysis with missing data. The equations characterizing the distributions of $\hat{\boldsymbol{\theta}}$ in (61) and T_{ML} in (62) also hold when changing the $l_i(\boldsymbol{\theta})$ in (55b) to the likelihood function based on the multivariate t-distribution (see Yuan and Bentler, 2001a; Yuan, Marshall and Bentler, 2002). However, the EM-algorithm has to be based on the multivariate t-distribution, and the expected information matrix will not be given by (59) even when all the missing data are MCAR. When the distribution is misspecified, the normal theory based MLE $\hat{\boldsymbol{\theta}}$ is consistent, while the MLE based on a multivariate t-distribution may not be consistent even for complete data (see Gourieroux et al., 1984).

Before ending this section, we want to emphasize that MCAR, MAR or NMAR is not just an assumption but reflects the truth within the data. Similarly, the true distribution of the data in the normal theory ML does not depend on whether one admits the misspecification of the likelihood function. Whatever the assumption is, once a procedure is chosen, the parameter estimates are the same, whether we call it an MLE or a pseudo-MLE. Although the normal theory MLEs are consistent when the missing data are MCAR or MAR, these procedures still need to be used with caution because the methods themselves cannot tell whether some missing data are NMAR. When data are not normally distributed, procedures that admit misspecification of distributions and make proper adjustments perform better (Enders, 2001; Savalei and Bentler, 2005).

5. Multiple groups

In the practice of data analysis, we often have several samples. For example, we may have samples from males and females, from different ethnic or age groups, or even from different countries. When different groups have the same population distribution, then the samples should be combined for better inference. When the samples are from different populations, it is important to understand how the populations might differ. In the context of SEM, the difference between different groups may be in factor loadings, regression coefficients, or means and variances of factors. Motivated by such practical problems, Jöreskog (1971) developed a maximum likelihood approach to CSA with multiple groups; Lee and Tsui (1982) developed normal theory GLS approach. Sörbom (1974) studied differences in factor means across groups. Considering that practical data may not be normally distributed, Bentler et al. (1987) and Muthén (1989a) proposed GLS approaches to parameter estimation and model testing for multi-group SEM. Satorra (2000) proposed scaled tests in multi-group analysis of moment structures. Because practical data may have missing values as well as heavy tails, Yuan and Bentler (2001a) developed a unified approach to multi-group SEM, which uses more efficient estimates of $\boldsymbol{\mu}_0$ and $\boldsymbol{\Sigma}_0$ than $\bar{\mathbf{x}}$ and $\mathbf{S}$. This approach is a natural extension of the methods in the previous sections and will be described below.

Suppose there are m groups, and for the jth group we have a sample of size n_j whose population mean vector and covariance matrix are $\boldsymbol{\mu}_{j0}$ and $\boldsymbol{\Sigma}_{j0}$, $j = 1, \ldots, m$. When data are normally distributed, we will use the sample mean vectors $\bar{\mathbf{x}}_j$ and sample covariance matrices $\mathbf{S}_j$ to estimate their population counterparts. When samples contain heavy tails or missing data, we may estimate $\boldsymbol{\mu}_{j0}$ and $\boldsymbol{\Sigma}_{j0}$ by the methods in Sections 3.3 or 4. Whatever the estimates are, we denote them by $\hat{\boldsymbol{\mu}}_j$, $\hat{\boldsymbol{\Sigma}}_j$ and let $\hat{\boldsymbol{\beta}}_j = (\hat{\boldsymbol{\mu}}'_j, \hat{\boldsymbol{\sigma}}'_j)'$ with $\hat{\boldsymbol{\sigma}}_j = \text{vech}(\hat{\boldsymbol{\Sigma}}_j)$. Then, it follows from Sections 3 and 4 that

$$\sqrt{n_j}(\hat{\boldsymbol{\beta}}_j - \boldsymbol{\beta}_{j0}) \xrightarrow{\mathcal{L}} N(\mathbf{0}, \boldsymbol{\Pi}_j),$$

where $\boldsymbol{\beta}_{j0} = (\boldsymbol{\mu}'_{j0}, \boldsymbol{\sigma}'_{j0})'$ with $\boldsymbol{\sigma}_{j0} = \text{vech}(\boldsymbol{\Sigma}_{j0})$, and there exists a consistent estimator $\hat{\boldsymbol{\Pi}}_j$ for $\boldsymbol{\Pi}_j$. It is typical that the $\boldsymbol{\mu}_{j0}$ in all the groups are of the same dimension, but we do not need to assume this. When they are not, we let p_j be the dimension of $\boldsymbol{\mu}_{j0}$. Although the p_j's can be different, in practice, the samples in different groups should be comparable. For example, the samples are obtained by administering the same questionnaire to the m groups. The interest is in comparing the mean and covariance structures $\boldsymbol{\beta}_j(\boldsymbol{\theta}_j) = (\boldsymbol{\mu}'_j(\boldsymbol{\theta}_j), \boldsymbol{\sigma}'_j(\boldsymbol{\theta}_j))'$ for $j = 1, \ldots, m$. Similar to the analysis with a single sample, there are ML, GLS, and AGLS procedures for parameter estimation and model evaluation (Bentler et al., 1987). We will only present the procedure based on minimizing D_{ML} for the following reasons: (a) When data are normal, minimizing D_{ML} with data $\bar{\mathbf{x}}_j$ and $\mathbf{S}_j$ generates asymptotically most efficient parameter estimates. (b) For data with influential cases or outliers, using robust estimates $\hat{\boldsymbol{\mu}}_j$ and $\hat{\boldsymbol{\Sigma}}_j$ in D_{ML} still leads to very efficient parameter estimates and statistics approximately following chi-square distributions (Yuan, Bentler and Chan, 2004; Yuan and Hayashi, 2003). (c) The estimation process of minimizing D_{ML} is quite stable, which is very important when modeling several groups simultaneously.

Assuming the m samples are independent and normally distributed, the joint ML discrepancy function parallel to (15) is given by

$$D_{\text{ML}}^{m}(\boldsymbol{\theta}) = \sum_{j=1}^{m} \frac{n_j}{N}\Big(\big[\hat{\boldsymbol{\mu}}_j - \boldsymbol{\mu}_j(\boldsymbol{\theta}_j)\big]'\boldsymbol{\Sigma}_j^{-1}(\boldsymbol{\theta}_j)\big[\hat{\boldsymbol{\mu}}_j - \boldsymbol{\mu}_j(\boldsymbol{\theta}_j)\big] + \text{tr}\big[\hat{\boldsymbol{\Sigma}}_j\boldsymbol{\Sigma}_j^{-1}(\boldsymbol{\theta}_j)\big] - \log\big|\hat{\boldsymbol{\Sigma}}_j\boldsymbol{\Sigma}_j^{-1}(\boldsymbol{\theta}_j)\big| - p\Big), \tag{64}$$

where $N = n_1 + \cdots + n_m$ and $\boldsymbol{\theta} = (\boldsymbol{\theta}_1', \boldsymbol{\theta}_2', \ldots, \boldsymbol{\theta}_m')'$. When all the $\boldsymbol{\theta}_j$ are freely estimated, the $\hat{\boldsymbol{\theta}}_j$, $j = 1, \ldots, m$, obtained by minimizing (15) for the individual groups will simultaneously minimize $D_{\text{ML}}^{m}(\boldsymbol{\theta})$. SEs for each $\hat{\boldsymbol{\theta}}_j$ based on (17) for normal data or (27) for nonnormally distributed data are still asymptotically valid. When all the samples are normally distributed and $\hat{\boldsymbol{\mu}}_j = \bar{\mathbf{x}}_j$ and $\hat{\boldsymbol{\Sigma}}_j = \mathbf{S}_j$, $T_{\text{ML}}^{m} = ND_{\text{ML}}^{m}(\hat{\boldsymbol{\theta}})$ is the LR statistic for testing all the models simultaneously. The rescaled statistic T_{RML}^{m} will be the summation of the individual T_{RML}'s, as are the degrees of freedom. There is no need for extra development when all the $\boldsymbol{\theta}_j$'s are not related. The analysis of multiple groups is interesting only when we put constraints on the separate $\boldsymbol{\theta}_j$'s. When all samples come from the same population, all the $\boldsymbol{\theta}_j$'s may be constrained or equal. In a less restricted setting, only certain parameters may be constrained equal. For example, the factor loadings for the male group may equal those for the female group while factor correlations or error variances are not held to be equal.

Suppose there are r constraints, represented by $\mathbf{h}(\boldsymbol{\theta}) = \mathbf{0}$. Estimation of

$$\boldsymbol{\theta}_0 = \big(\boldsymbol{\theta}_{10}', \boldsymbol{\theta}_{20}', \ldots, \boldsymbol{\theta}_{m0}'\big)'$$

involves minimizing (64) under $\mathbf{h}(\boldsymbol{\theta}) = \mathbf{0}$. So multi-group analysis turns into essentially the same problem as in Section 3.6. Especially, the parameter estimator $\hat{\boldsymbol{\theta}}$ also satisfies the first order Lagrangian condition as in (50), but the $\mathbf{g}$ should be replaced by

$$\mathbf{g}(\boldsymbol{\theta}) = \big(\mathbf{g}_1'(\boldsymbol{\theta}_1), \mathbf{g}_2'(\boldsymbol{\theta}_2), \ldots, \mathbf{g}_m'(\boldsymbol{\theta}_m)\big)'$$

with

$$\mathbf{g}_j(\boldsymbol{\theta}_j) = \frac{n_j}{N}\dot{\boldsymbol{\beta}}_j'(\boldsymbol{\theta}_j)\mathbf{W}_j(\boldsymbol{\theta}_j) \times \begin{pmatrix} \hat{\boldsymbol{\mu}}_j - \boldsymbol{\mu}_j(\boldsymbol{\theta}_j) \\ \hat{\boldsymbol{\sigma}}_j + \text{vech}\big\{\big[\hat{\boldsymbol{\mu}}_j - \boldsymbol{\mu}_j(\boldsymbol{\theta}_j)\big]\big[\hat{\boldsymbol{\mu}}_j - \boldsymbol{\mu}_j(\boldsymbol{\theta}_j)\big]'\big\} - \boldsymbol{\sigma}_j(\boldsymbol{\theta}_j) \end{pmatrix}.$$

The estimate $\hat{\boldsymbol{\theta}}$ can also be obtained by the IRLS procedure[11] as in (53), but the matrix $\dot{\boldsymbol{\beta}}'(\boldsymbol{\theta})\mathbf{W}(\boldsymbol{\theta})\dot{\boldsymbol{\beta}}(\boldsymbol{\theta})$ there should be replaced by

$$\mathbf{H}(\boldsymbol{\theta}) = \text{diag}\big(\mathbf{H}_1(\boldsymbol{\theta}_1), \mathbf{H}_2(\boldsymbol{\theta}_2), \ldots, \mathbf{H}_m(\boldsymbol{\theta}_m)\big)$$
$$\text{with } \mathbf{H}_j(\boldsymbol{\theta}_j) = \frac{n_j}{N}\dot{\boldsymbol{\beta}}_j'(\boldsymbol{\theta}_j)\mathbf{W}_j(\boldsymbol{\theta}_j)\dot{\boldsymbol{\beta}}_j(\boldsymbol{\theta}_j).$$

[11] Again, one needs to check conditions to make sure that the converged solution is indeed the constrained minimum of (64).

To characterize the distribution of $\hat{\boldsymbol{\theta}}$, we need additional notation. Assume $n_j/N \to \gamma_j$ and denote

$$\mathbf{H}_\gamma(\boldsymbol{\theta}) = \mathrm{diag}\big(\mathbf{H}_{\gamma 1}(\boldsymbol{\theta}_1), \mathbf{H}_{\gamma 2}(\boldsymbol{\theta}_2), \ldots, \mathbf{H}_{\gamma m}(\boldsymbol{\theta}_m)\big)$$
$$\text{with } \mathbf{H}_{\gamma j}(\boldsymbol{\theta}_j) = \gamma_j \dot{\boldsymbol{\beta}}_j'(\boldsymbol{\theta}_j)\mathbf{W}_j(\boldsymbol{\theta}_j)\dot{\boldsymbol{\beta}}_j(\boldsymbol{\theta}_j),$$

$$\mathbf{A}_\gamma = \begin{pmatrix} \mathbf{H}_\gamma(\boldsymbol{\theta}_0) & \dot{\mathbf{h}}'(\boldsymbol{\theta}_0) \\ \dot{\mathbf{h}}(\boldsymbol{\theta}_0) & \mathbf{0} \end{pmatrix}, \qquad \mathbf{A}_\gamma^{-1} = \begin{pmatrix} \mathbf{A}_\gamma^{11} & \mathbf{A}_\gamma^{12} \\ \mathbf{A}_\gamma^{21} & \mathbf{A}_\gamma^{22} \end{pmatrix},$$
$$\mathbf{W}_\gamma = \mathrm{diag}\big(\sqrt{\gamma_1}\mathbf{W}_1, \ldots, \sqrt{\gamma_m}\mathbf{W}_m\big),$$
$$\dot{\boldsymbol{\beta}} = \mathrm{diag}(\dot{\boldsymbol{\beta}}_1, \ldots, \dot{\boldsymbol{\beta}}_m), \qquad \boldsymbol{\Pi} = \mathrm{diag}(\boldsymbol{\Pi}_1, \ldots, \boldsymbol{\Pi}_m).$$

Then, under standard regularity conditions, including $\gamma_j > 0$, $j = 1, 2, \ldots, m$,

$$\sqrt{N}(\hat{\boldsymbol{\theta}} - \boldsymbol{\theta}_0) \xrightarrow{\mathcal{L}} N\big(\mathbf{0}, \boldsymbol{\Omega}_{\mathrm{SW}}^m\big),$$

where $\boldsymbol{\Omega}_{\mathrm{SW}}^m = \mathbf{A}_\gamma^{11}\boldsymbol{\Upsilon}_\gamma\mathbf{A}_\gamma^{11}$ with

$$\boldsymbol{\Upsilon}_\gamma = \dot{\boldsymbol{\beta}}'\mathbf{W}_\gamma\boldsymbol{\Pi}\mathbf{W}_\gamma\dot{\boldsymbol{\beta}}.$$

A consistent estimator $\hat{\boldsymbol{\Omega}}_{\mathrm{SW}}^m$ of $\boldsymbol{\Omega}_{\mathrm{SW}}^m$ can be obtained when $\boldsymbol{\theta}$ is replaced by $\hat{\boldsymbol{\theta}}$, γ_j by n_j/N, and $\boldsymbol{\Pi}_j$ by $\hat{\boldsymbol{\Pi}}_j$. SEs of $\hat{\boldsymbol{\theta}}$ follow from square roots of the diagonal elements of $\hat{\boldsymbol{\Omega}}_{\mathrm{SW}}^m$. When data are normally distributed $\hat{\boldsymbol{\beta}}_j = (\bar{\mathbf{x}}_j', \mathrm{vech}'(\mathbf{S}_j))'$, $\boldsymbol{\Pi}_j = \mathbf{W}_j^{-1}$, $\boldsymbol{\Upsilon}_\gamma = \mathbf{H}_\gamma$ and $\boldsymbol{\Omega}_{\mathrm{SW}}^m = \mathbf{A}_\gamma^{11}\boldsymbol{\Upsilon}_\gamma\mathbf{A}_\gamma^{11} = \mathbf{A}_\gamma^{11}$. This corresponds to the standard results obtained when using the normality assumption for all the samples. Similar to Section 3.6, the LM statistic

$$T_{\mathrm{LM}} = N\hat{\boldsymbol{\lambda}}'\big(\hat{\mathbf{A}}_\gamma^{22}\hat{\boldsymbol{\Upsilon}}_\gamma\hat{\mathbf{A}}_\gamma^{22}\big)^{-1}\hat{\boldsymbol{\lambda}}$$

can be used to test the significance of $\mathbf{h}(\boldsymbol{\theta}_0) = \mathbf{0}$.

Let

$$\mathbf{W} = \mathrm{diag}(\mathbf{W}_1, \ldots, \mathbf{W}_m), \qquad \mathbf{U} = \mathbf{W} - \mathbf{W}_\gamma\dot{\boldsymbol{\beta}}\mathbf{A}^{11}\dot{\boldsymbol{\beta}}'\mathbf{W}_\gamma.$$

Then

$$T_{\mathrm{ML}}^m = ND_{\mathrm{ML}}^m(\hat{\boldsymbol{\theta}}) = \sum_{j=1}^{df}\kappa_j z_j^2 + o_p(1), \tag{65}$$

where the κ_j's are the nonzero eigenvalues of $\mathbf{U}\boldsymbol{\Pi}$ and $df = \sum_{j=1}^m p_j(p_j+3)/2 - q + r$. Similar to the one-group case, a simple rescaling on T_{ML}^m can result in a statistic that is better approximated by χ_{df}^2. Let $\hat{\kappa} = \mathrm{tr}(\hat{\mathbf{U}}\hat{\boldsymbol{\Pi}})/df$. Then the rescaled statistic

$$T_{\mathrm{RML}}^m = T_{\mathrm{ML}}^m/\hat{\kappa}$$

approaches a distribution with mean equal to that of χ_{df}^2. A special case is when data are normally distributed and $\hat{\boldsymbol{\beta}}_j = (\bar{\mathbf{x}}_j', \mathrm{vech}'(\mathbf{S}_j))'$. Then $\boldsymbol{\Pi} = \mathbf{W}^{-1}$ and all the κ_j's in (65) are equal to 1. Thus

$$T_{\text{ML}}^{m} \xrightarrow{\mathcal{L}} \chi_{df}^{2},$$

which is the basis for the likelihood ratio statistic in multiple groups.

Notice that, minimizing the $D_{\text{ML}}^{m}(\boldsymbol{\theta})$ in (64) for $\hat{\boldsymbol{\theta}}$ using the MLEs $\hat{\boldsymbol{\beta}}_j$ is a two-stage estimation approach. This is not the only option. With missing data, $\hat{\boldsymbol{\theta}}$ is different from the MLE that is based on $\mathbf{x}_{ij} \sim N_{p_{ij}}(\boldsymbol{\mu}_{ij}, \boldsymbol{\Sigma}_{ij})$, where $\mathbf{x}_{ij}$ is the vector of observed values for the ith case in the jth group, $\boldsymbol{\mu}_{ij}$ is a subvector of $\boldsymbol{\mu}_j$ and $\boldsymbol{\Sigma}_{ij}$ is a submatrix of $\boldsymbol{\Sigma}_j$. The direct MLE of $\boldsymbol{\theta}_0$ is obtained by maximizing

$$l^{m}(\boldsymbol{\theta}) = \sum_{j=1}^{m} \sum_{i=1}^{n_j} l_{ij}(\boldsymbol{\theta}),$$

under the constraint $\mathbf{h}(\boldsymbol{\theta}) = \mathbf{0}$, where

$$l_{ij}(\boldsymbol{\theta}) = -\frac{1}{2}\left\{\log\left|\boldsymbol{\Sigma}_{ij}(\boldsymbol{\theta})\right| + \left(\mathbf{x}_{ij} - \boldsymbol{\mu}_{ij}(\boldsymbol{\theta})\right)'\boldsymbol{\Sigma}_{ij}^{-1}(\boldsymbol{\theta})\left(\mathbf{x}_{ij} - \boldsymbol{\mu}_{ij}(\boldsymbol{\theta})\right)\right\}.$$

The MLE can also be obtained by an IRLS procedure as in (53). The covariance matrix of the MLE is consistently estimated by a sandwich-type matrix in general, and by the inverse of the observed information matrix when all the data are normally distributed. Especially, the likelihood ratio statistic asymptotically follows χ_{df}^{2} when data are normally distributed. Comparing the MLE to the two-stage estimator $\hat{\boldsymbol{\theta}}$ with missing data, the MLE is more efficient than $\hat{\boldsymbol{\theta}}$ when all the samples are normally distributed. When data are not normally distributed, it is not clear which one is more efficient. The two-stage procedure by first obtaining a consistent estimate of the saturated model, as described in this section, is recommended. In the statistical literature, there are more procedures to estimate $\boldsymbol{\mu}_0$ and $\boldsymbol{\Sigma}_0$ than to estimate $\boldsymbol{\theta}_0$ (Cheng and Victoria-Feser, 2002; Little, 1988). When new procedures of estimating $\boldsymbol{\mu}_0$ and $\boldsymbol{\Sigma}_0$ are available, they can be easily adapted into SEM by using (64).

There are important interpretive issues associated with the practice of SEM with multiple groups. As noted, anywhere from no parameters to all parameters may be equal across groups, but some care has to be taken that models are specified correctly for equality to be likely. For example, factor loadings are liable to maintain invariance only when the metric of variables and factors is similar across groups. If factor loadings are not equal across groups, differences in factor covariances or in factor means become difficult to interpret. After all, what is the meaning of a mean difference on a factor when it has different implications in different groups? Meredith (1993), Millsap (1997, 1998) and Lubke et al. (2003) have provided good discussions on the technical issues involved. Unfortunately, in real data analysis, obtaining precise invariance is very difficult, and there is some agreement that small factor loading differences will not seriously impact comparison of results across groups (Byrne et al., 1989; Byrne and Watkins, 2003). A related issue is when T_{ML}^{m} is highly significant before implementing the constraints $\mathbf{h}(\boldsymbol{\theta}) = \mathbf{0}$. Then, a nonsignificant difference in T_{ML}^{m} does not necessarily imply that the model with the constraints is a better model (see Yuan and Bentler, 2004a).

This section only presented multi-group estimation based on the ML discrepancy function. It is straightforward to define a unified approach using the GLS or AGLS discrepancy functions, parallel to those in Bentler et al. (1987) or Muthén (1989a). As in the one-group case, when models are correctly specified, using the GLS approach will lead to parameter estimates and test statistics that are asymptotically equivalent to those given by the ML approach. The AGLS approach will result in estimates having better asymptotic properties, but they tend to be less efficient empirically. The nonconvergence problem of the AGLS discrepancy function in the one-group case will be more severe in the multi-group case because the model size and data dimension in the multi-group case is m times that of the one-group case.

It is well known that one can use a regression approach to deal with the ANOVA model. Similarly, when the interest is in mean comparison with the same covariance structure across all the groups, multi-group analysis also can be modeled as a single sample problem after a proper coding for group membership (Muthén, 1989b). Kano (2001) showed that when all the substantive variables are normally distributed, ML inference still leads to a correct analysis when treating the membership variable as normally distributed.

6. Multilevel models with hierarchical data

Data in social and behavioral sciences are often hierarchically organized. For example, households are nested within neighborhoods, neighborhoods are nested within cities, and cities are further nested within countries; students are nested within classes, classes are nested within schools, and schools are further nested within school districts. Because of the hierarchical structure, cases within a cluster are generally correlated. Analytical methods for such data have to explicitly model these correlations in order to get reliable results. Two main classes of multilevel models that have been well studied are hierarchical linear models (HLM) and multilevel SEM models. Here, we will only present multilevel SEM models and related analytical methods. Interested readers for hierarchical regression models are referred to Raudenbush and Bryk (2002) and Goldstein (2003).

Similar to the conventional SEM model, the multilevel SEM model takes measurement errors into account in the hierarchical data structure (Goldstein and McDonald, 1988; Longford, 1993; McDonald and Goldstein, 1989). Multiple indicators are also necessary in order to model the relationship among the latent constructs. Methodology development for multilevel SEM also started with the normal distribution assumption (du Toit and du Toit, in press; Goldstein and McDonald, 1988; Lee, 1990; Lee and Poon, 1998; Liang and Bentler, 2004; Longford, 1993; McDonald and Goldstein, 1989; Muthén, 1994, 1997). With data sets from many nested units, the normal distribution assumption is more likely to be violated for multilevel data than for a single sample. Analytical methods for multilevel SEM with nonnormally distributed data are also well developed (Lee and Shi, 2001; Lee and Song, 2001; Poon and Lee, 1994; Yuan and Bentler, 2002b, 2003, 2004b, 2005, 2006b). As in conventional SEM, there are GLS and AGLS procedures as well as rescaled statistics for multilevel SEM. We will only present the normal theory based ML procedure and use the sandwich-type covariance matrix and rescaled or residual-based statistics to adjust for violations of the distribution assumption. We will also briefly discuss limited results on asymptotic robustness as well as some special properties when data are elliptically distributed. Although the number of levels can be arbitrary in formulating a multilevel model, we only explicitly consider 2-level models; generalization from a 2-level model to a higher level model is straightforward. As we shall see, the sample size at the highest level plays a key role when inferring parameters at that level as well as when evaluating the overall model structure. Unless there is a large sample size at the highest level, one needs to avoid formulating a model with many levels.

Consider test scores $\mathbf{y}_{ij}$ for student i from school j, $i = 1, 2, \ldots, n_j$, $j = 1, 2, \ldots, J$. Because students from the same school share the same school resource, $\mathbf{y}_{ij}$ with fixed j are generally correlated. Let $\mathbf{z}_j$ be the vector containing school resource variables such as the number of computers, teach/student ratio, etc. Then $\mathbf{z}_j$ only varies at the school level. Liang and Bentler (2004) provided the following formulation of the above data structure

$$\begin{pmatrix} \mathbf{z}_j \\ \mathbf{y}_{ij} \end{pmatrix} = \begin{pmatrix} \mathbf{z}_j \\ \mathbf{v}_j \end{pmatrix} + \begin{pmatrix} \mathbf{0} \\ \mathbf{u}_{ij} \end{pmatrix}, \tag{66}$$

where $\mathbf{u}_{ij}$ represents the level-1 (within-group) components, and $\mathbf{v}_j$ represents the level-2 (between-group) components. It is typically assumed that $(\mathbf{z}'_j, \mathbf{v}'_j)'$ and $\mathbf{u}_{ij}$ are statistically independent, and that the correlations between $\mathbf{y}_{ij}$ and $\mathbf{y}_{i'j}$ are due to $\mathbf{v}_j$. Following Liang and Bentler's setup, we assume $E(\mathbf{u}_{ij}) = \mathbf{0}$. The model implied means and covariances are

$$\boldsymbol{\mu} = E\begin{pmatrix} \mathbf{z}_j \\ \mathbf{v}_j \end{pmatrix} = \begin{pmatrix} \boldsymbol{\mu}_z \\ \boldsymbol{\mu}_v \end{pmatrix},$$

$$\boldsymbol{\Sigma}_b = \mathrm{Cov}\begin{pmatrix} \mathbf{z}_j \\ \mathbf{v}_j \end{pmatrix} = \begin{pmatrix} \boldsymbol{\Sigma}_{zz} & \boldsymbol{\Sigma}_{zv} \\ \boldsymbol{\Sigma}_{vz} & \boldsymbol{\Sigma}_{vv} \end{pmatrix}, \qquad \boldsymbol{\Sigma}_w = \mathrm{Cov}(\mathbf{u}_{ij}).$$

Each of these matrices may be hypothesized to be generated by more basic parameters. As a specific example, with academic achievement data, the $\mathbf{y}_{ij}$ may contain p_1 scores that test quantitative, verbal and spatial abilities. Then we might hypothesize the following CFA model

$$\mathbf{u}_{ij} = \boldsymbol{\Lambda}_w \mathbf{f}_{wij} + \mathbf{e}_{wij}, \qquad \boldsymbol{\Sigma}_w = \boldsymbol{\Lambda}_w \boldsymbol{\Phi}_w \boldsymbol{\Lambda}'_w + \boldsymbol{\Psi}_w, \tag{67a}$$

where $E(\mathbf{f}_{wij}) = \mathbf{0}$, $E(\mathbf{e}_{wij}) = \mathbf{0}$, $\boldsymbol{\Lambda}_w$ is a $p_1 \times 3$ matrix, $\boldsymbol{\Phi}_w$ is a 3×3 correlation matrix and $\boldsymbol{\Psi}_w$ is a $p_1 \times p_1$ diagonal matrix. We may also hypothesize that school resources affect achievement scores through a single dimension latent variable as in

$$\mathbf{v}_j = \boldsymbol{\lambda}_b f_{bj} + \mathbf{e}_{bj}, \tag{67b}$$

and f_{bj} is further predicted by the p_2 variables in $\mathbf{z}_j$ through

$$f_{bj} = \alpha + \boldsymbol{\beta}'\mathbf{z}_j + \epsilon_j, \tag{67c}$$

where $\boldsymbol{\lambda}_b$ is a $p_1 \times 1$ vector, $\boldsymbol{\mu}_z = E(\mathbf{z}_j)$ is a $p_2 \times 1$ vector, $\boldsymbol{\Sigma}_{zz}$ is a $p_2 \times p_2$ free matrix, $E(\mathbf{e}_{bj}) = \mathbf{0}$, $\mathrm{Cov}(\mathbf{e}_{bj}) = \boldsymbol{\Psi}_b$ is a $p_1 \times p_1$ diagonal matrix, $E(\epsilon_j) = 0$, $\psi = \mathrm{Var}(\epsilon_j)$ is a scalar. Thus we have

$$\boldsymbol{\mu}_v = \boldsymbol{\lambda}_b(\alpha + \boldsymbol{\beta}'\boldsymbol{\mu}_z), \qquad \boldsymbol{\Sigma}_{zv} = \boldsymbol{\Sigma}_{zz}\boldsymbol{\beta}\boldsymbol{\lambda}_b',$$
$$\boldsymbol{\Sigma}_{vv} = \boldsymbol{\lambda}_b(\boldsymbol{\beta}'\boldsymbol{\Sigma}_{zz}\boldsymbol{\beta} + \psi)\boldsymbol{\lambda}_b' + \boldsymbol{\Psi}_b. \tag{67d}$$

Equations (67a)–(67d) define an interesting multilevel structure. Let the free parameters in the model be $\boldsymbol{\theta}$, the model structures for (67) and in general can be represented by $\boldsymbol{\mu}(\boldsymbol{\theta})$, $\boldsymbol{\Sigma}_b(\boldsymbol{\theta})$ and $\boldsymbol{\Sigma}_w(\boldsymbol{\theta})$.

Let $\bar{\mathbf{y}}_j = \sum_{i=1}^{n_j} \mathbf{y}_{ij}/n_j$. Then the covariance matrix of $\mathbf{b}_j = (\mathbf{z}_j', \bar{\mathbf{y}}_j')'$ is

$$\boldsymbol{\Sigma}_j = \mathrm{Cov}(\mathbf{b}_j) = \begin{pmatrix} \boldsymbol{\Sigma}_{zz} & \boldsymbol{\Sigma}_{zv} \\ \boldsymbol{\Sigma}_{vz} & \boldsymbol{\Sigma}_{vv} + n_j^{-1}\boldsymbol{\Sigma}_w \end{pmatrix}.$$

Let $\boldsymbol{\sigma}_w = \mathrm{vech}(\boldsymbol{\Sigma}_w)$ and $\boldsymbol{\sigma}_b = \mathrm{vech}(\boldsymbol{\Sigma}_b)$. Then

$$\boldsymbol{\sigma}_j = \mathrm{vech}(\boldsymbol{\Sigma}_j) = \boldsymbol{\sigma}_b + \begin{pmatrix} \mathbf{0}_{p_a \times 1} \\ n_j^{-1}\boldsymbol{\sigma}_w \end{pmatrix} = \boldsymbol{\sigma}_b + n_j^{-1}\mathbf{C}\boldsymbol{\sigma}_w,$$

where $\mathbf{C} = (\mathbf{0}_{p_* \times p_a}, \mathbf{I}_{p_*})'$ with $p_* = p_1(p_1+1)/2$ and $p_a = p_2(2p_1 + p_2 + 1)/2$. If we further assume that the data $\mathbf{y}_{ij}$ and $\mathbf{z}_j$ are normally distributed, the log likelihood function for (66) is (Liang and Bentler, 2004; Yuan and Bentler, 2003)

$$\begin{aligned} l(\boldsymbol{\theta}) = &-\frac{(N-J)}{2}\left\{\log\left|\boldsymbol{\Sigma}_w(\boldsymbol{\theta})\right| + \mathrm{tr}\left[\boldsymbol{\Sigma}_w^{-1}(\boldsymbol{\theta})\mathbf{S}_y\right]\right\} \\ &-\frac{1}{2}\sum_{j=1}^{J}\left\{\log\left|\boldsymbol{\Sigma}_j(\boldsymbol{\theta})\right| + \mathrm{tr}\left[\boldsymbol{\Sigma}_j^{-1}(\boldsymbol{\theta})\mathbf{R}_j(\boldsymbol{\theta})\right]\right\}, \end{aligned} \tag{68}$$

where $N = n_1 + \cdots + n_J$,

$$\mathbf{S}_y = \frac{1}{N-J}\sum_{j=1}^{J} n_j\mathbf{S}_j \quad \text{with } n_j\mathbf{S}_j = \sum_{i=1}^{n_j}(\mathbf{y}_{ij} - \bar{\mathbf{y}}_j)(\mathbf{y}_{ij} - \bar{\mathbf{y}}_j)',$$

and

$$\mathbf{R}_j(\boldsymbol{\theta}) = \left(\mathbf{b}_j - \boldsymbol{\mu}(\boldsymbol{\theta})\right)\left(\mathbf{b}_j - \boldsymbol{\mu}(\boldsymbol{\theta})\right)'.$$

Let $\hat{\boldsymbol{\theta}}$ maximize (68), then it asymptotically follows a normal distribution. To facilitate the technical development we denote

$$\mathbf{W}_w = 2^{-1}\mathbf{D}_{p_1}'\left(\boldsymbol{\Sigma}_w^{-1} \otimes \boldsymbol{\Sigma}_w^{-1}\right)\mathbf{D}_{p_1},$$
$$\mathbf{W}_j = 2^{-1}\mathbf{D}_{(p_1+p_2)}'\left(\boldsymbol{\Sigma}_j^{-1} \otimes \boldsymbol{\Sigma}_j^{-1}\right)\mathbf{D}_{(p_1+p_2)}.$$

Notice that N is the total level-1 sample size and J is the level-2 sample size. Because N is generally much bigger than J, estimates for parameters that appear in the level-1 model will be closer to their population values. We will distinguish level-1 and level-2 parameters in characterizing the distribution of $\hat{\boldsymbol{\theta}}$. Let $\boldsymbol{\theta}_w$ be the parameters that appear in level-1 covariance structure $\boldsymbol{\Sigma}_w(\cdot)$ and the remaining parameters be $\boldsymbol{\theta}_b$, with $\boldsymbol{\theta} =$

$(\boldsymbol{\theta}_b', \boldsymbol{\theta}_w')'$, and

$$\dot{\boldsymbol{\mu}}_b(\boldsymbol{\theta}) = \frac{\partial \boldsymbol{\mu}(\boldsymbol{\theta})}{\partial \boldsymbol{\theta}_b'}, \qquad \dot{\boldsymbol{\mu}}_w(\boldsymbol{\theta}) = \frac{\partial \boldsymbol{\mu}(\boldsymbol{\theta})}{\partial \boldsymbol{\theta}_w'},$$

$$\dot{\boldsymbol{\sigma}}_{bb}(\boldsymbol{\theta}) = \frac{\partial \boldsymbol{\sigma}_b(\boldsymbol{\theta})}{\partial \boldsymbol{\theta}_b'}, \qquad \dot{\boldsymbol{\sigma}}_{bw}(\boldsymbol{\theta}) = \frac{\partial \boldsymbol{\sigma}_b(\boldsymbol{\theta})}{\partial \boldsymbol{\theta}_w'}, \qquad \dot{\boldsymbol{\sigma}}_{ww}(\boldsymbol{\theta}) = \frac{\partial \boldsymbol{\sigma}_w(\boldsymbol{\theta})}{\partial \boldsymbol{\theta}_w'},$$

$$\mathcal{I}_{J11} = \frac{1}{J}\sum_{j=1}^{J}\bigl(\dot{\boldsymbol{\mu}}_b'\boldsymbol{\Sigma}_j^{-1}\dot{\boldsymbol{\mu}}_b + \dot{\boldsymbol{\sigma}}_{bb}'\mathbf{W}_j\dot{\boldsymbol{\sigma}}_{bb}\bigr),$$

$$\mathcal{I}_{J12} = \frac{1}{J}\sum_{j=1}^{J}\bigl(\dot{\boldsymbol{\mu}}_b'\boldsymbol{\Sigma}_j^{-1}\dot{\boldsymbol{\mu}}_w + \dot{\boldsymbol{\sigma}}_{bb}'\mathbf{W}_j\dot{\boldsymbol{\sigma}}_{bw} + n_j^{-1}\dot{\boldsymbol{\sigma}}_{bb}'\mathbf{W}_j\mathbf{C}\dot{\boldsymbol{\sigma}}_{ww}\bigr),$$

$$\mathcal{I}_{21} = \mathcal{I}_{12}',$$

$$\begin{aligned}\mathcal{I}_{J22} = \frac{1}{J}\sum_{j=1}^{J}\bigl(&\dot{\boldsymbol{\mu}}_w'\boldsymbol{\Sigma}_j^{-1}\dot{\boldsymbol{\mu}}_w + \dot{\boldsymbol{\sigma}}_{bw}'\mathbf{W}_j\dot{\boldsymbol{\sigma}}_{bw} + n_j^{-1}\dot{\boldsymbol{\sigma}}_{ww}'\mathbf{C}'\mathbf{W}_j\dot{\boldsymbol{\sigma}}_{bw} \\ &+ n_j^{-1}\dot{\boldsymbol{\sigma}}_{bw}'\mathbf{W}_j\mathbf{C}\dot{\boldsymbol{\sigma}}_{ww}\bigr) \\ &+ \frac{1}{J}\sum_{j=1}^{J} n_j^{-2}\dot{\boldsymbol{\sigma}}_{ww}'\mathbf{C}'\mathbf{W}_j\mathbf{C}\dot{\boldsymbol{\sigma}}_{ww} + \left(\frac{N}{J} - 1\right)\dot{\boldsymbol{\sigma}}_{ww}'\mathbf{W}_w\dot{\boldsymbol{\sigma}}_{ww}.\end{aligned}$$

Then

$$\mathcal{I}_J = \begin{pmatrix} \mathcal{I}_{J11} & \mathcal{I}_{J12} \\ \mathcal{I}_{J21} & \mathcal{I}_{J22} \end{pmatrix}$$

is the information matrix associated with (68). When $\mathcal{I} = \lim_{J\to\infty}\mathcal{I}_J$ exists and all the data are normally distributed,

$$\sqrt{J}(\hat{\boldsymbol{\theta}} - \boldsymbol{\theta}_0) \xrightarrow{\mathcal{L}} N\bigl(\mathbf{0}, \mathcal{I}^{-1}\bigr). \tag{69}$$

Notice that $\mathcal{I}_{J11}$ and $\mathcal{I}_{J12}$ are of magnitude $O(1)$ while $\mathcal{I}_{J22}$ is of $O(N/J)$. So the information for $\boldsymbol{\theta}_w$ is proportional to the average level-1 sample size N/J. Asymptotic covariance matrix for $\hat{\boldsymbol{\theta}}_w$ is proportional to $1/N$ while that for $\hat{\boldsymbol{\theta}}_b$ is proportional to $1/J$ (see Yuan and Bentler, 2002b, 2006b).

Similarly, when all the data are normally distributed, the LR statistic will asymptotically follow a chi-square distribution. Let $\boldsymbol{\beta} = (\boldsymbol{\mu}', \boldsymbol{\sigma}_b', \boldsymbol{\sigma}_w')'$ represent the free parameters in the saturated model and $\hat{\boldsymbol{\beta}}$ be the MLE of $\boldsymbol{\beta}_0$, then

$$T_{\text{ML}} = 2\bigl[l(\hat{\boldsymbol{\beta}}) - l(\hat{\boldsymbol{\theta}})\bigr] \xrightarrow{\mathcal{L}} \chi^2_{df}, \tag{70}$$

where $df = p_1(p_1+1)/2 + (p_1+p_2)(p_1+p_2+3)/2 - q$ with q being the number of free parameters in $\boldsymbol{\theta}$.

When $(\mathbf{z}_j', \mathbf{v}_j')'$ or $\mathbf{u}_{ij}$ are not normally distributed, adjustments need to be made for better inference with the multilevel model. For such a purpose, we need to first characterize the asymptotic distribution of $\hat{\boldsymbol{\beta}}$. We will use the estimating equation approach to study the distribution of $\hat{\boldsymbol{\beta}}$. Let $\mathbf{r}_j = \text{vech}(\mathbf{R}_j)$, $\mathbf{s}_j = \text{vech}(\mathbf{S}_j)$, and $\mathbf{g}_j = (\mathbf{g}'_{j1}, \mathbf{g}'_{j2}, \mathbf{g}'_{j3})'$ with

$$\mathbf{g}_{j1}(\boldsymbol{\beta}) = \boldsymbol{\Sigma}_j^{-1}(\mathbf{b}_j - \boldsymbol{\mu}), \qquad \mathbf{g}_{j2}(\boldsymbol{\beta}) = \mathbf{W}_j(\mathbf{r}_j - \boldsymbol{\sigma}_j),$$
$$\mathbf{g}_{j3}(\boldsymbol{\beta}) = \mathbf{W}_w\big[n_j\mathbf{s}_j - (n_j - 1)\boldsymbol{\sigma}_w\big] + n_j^{-1}\mathbf{C}'\mathbf{W}_j(\mathbf{r}_j - \boldsymbol{\sigma}_j).$$

The normal estimating equation corresponding to parameter $\boldsymbol{\beta}$ is given by

$$\mathbf{g}(\boldsymbol{\beta}) = \frac{1}{J}\sum_{j=1}^{J}\mathbf{g}_j(\boldsymbol{\beta}) = \mathbf{0}.$$

Notice that one cannot obtain $\hat{\boldsymbol{\beta}}$ analytically unless $n_1 = n_2 = \cdots = n_J$. An iterative procedure has to be used, as in (18). Let

$$\mathbf{A}_J(\boldsymbol{\beta}) = -E\big[\dot{\mathbf{g}}(\boldsymbol{\beta})\big], \qquad \mathbf{B}_J(\boldsymbol{\beta}) = \frac{1}{J}\sum_{j=1}^{J}E\big[\mathbf{g}_j(\boldsymbol{\beta})\mathbf{g}'_j(\boldsymbol{\beta})\big].$$

If

$$\mathbf{A} = \lim_{J\to\infty}\mathbf{A}_J(\boldsymbol{\beta}_0), \qquad \mathbf{B} = \lim_{J\to\infty}\mathbf{B}_J(\boldsymbol{\beta}_0)$$

exist, then it follows from the theory for estimating equations that (see Yuan and Jennrich, 1998)

$$\sqrt{J}(\hat{\boldsymbol{\beta}} - \boldsymbol{\beta}_0) \xrightarrow{\mathcal{L}} N(\mathbf{0}, \boldsymbol{\Pi}),$$

where $\boldsymbol{\Pi} = \mathbf{A}^{-1}\mathbf{B}\mathbf{A}^{-1}$. Notice that the estimating equation corresponding to the structural model

$$\boldsymbol{\beta}(\boldsymbol{\theta}) = \big(\boldsymbol{\mu}'(\boldsymbol{\theta}), \boldsymbol{\sigma}'_b(\boldsymbol{\theta}), \boldsymbol{\sigma}'_w(\boldsymbol{\theta})\big)'$$

is given by

$$\mathbf{g}(\boldsymbol{\theta}) = \dot{\boldsymbol{\beta}}'(\boldsymbol{\theta})\mathbf{g}\big[\boldsymbol{\beta}(\boldsymbol{\theta})\big] = \mathbf{0}$$

and there exists

$$-E\big[\dot{\mathbf{g}}(\boldsymbol{\theta})\big] = \dot{\boldsymbol{\beta}}'(\boldsymbol{\theta})\mathbf{A}_J\big[\boldsymbol{\beta}(\boldsymbol{\theta})\big]\dot{\boldsymbol{\beta}}(\boldsymbol{\theta}).$$

It immediately follows that

$$\sqrt{J}(\hat{\boldsymbol{\theta}} - \boldsymbol{\theta}_0) \xrightarrow{\mathcal{L}} N(\mathbf{0}, \boldsymbol{\Omega}_{\text{SW}}), \tag{71}$$

where

$$\boldsymbol{\Omega}_{\text{SW}} = (\dot{\boldsymbol{\beta}}'\mathbf{A}\dot{\boldsymbol{\beta}})^{-1}(\dot{\boldsymbol{\beta}}'\mathbf{A}\boldsymbol{\Pi}\mathbf{A}\dot{\boldsymbol{\beta}})(\dot{\boldsymbol{\beta}}'\mathbf{A}\dot{\boldsymbol{\beta}})^{-1}.$$

It is easy to see that

$$\mathbf{A}_J = \begin{pmatrix} \mathbf{A}_{J11} & \mathbf{0} & \mathbf{0} \\ \mathbf{0} & \mathbf{A}_{J22} & \mathbf{A}_{J23} \\ \mathbf{0} & \mathbf{A}_{J32} & \mathbf{A}_{J33} \end{pmatrix}$$

with

$$\mathbf{A}_{J11} = J^{-1}\sum_{j=1}^{J}\boldsymbol{\Sigma}_j^{-1}, \qquad \mathbf{A}_{J22} = J^{-1}\sum_{j=1}^{J}\mathbf{W}_j,$$

$$\mathbf{A}_{J23} = J^{-1}\sum_{j=1}^{J}n_j^{-1}\mathbf{W}_j\mathbf{C}, \qquad \mathbf{A}_{J32} = \mathbf{A}'_{J23},$$

$$\mathbf{A}_{J33} = \left(\frac{N}{J} - 1\right)\mathbf{W}_w + J^{-1}\sum_{j=1}^{J}n_j^{-2}\mathbf{C}'\mathbf{W}_j\mathbf{C},$$

and $\dot{\boldsymbol{\beta}}'\mathbf{A}\dot{\boldsymbol{\beta}}$ equals the information matrix $\mathcal{I}$ in (69). The form of $\mathbf{B}_J$ depends on the unknown distributions of $(\mathbf{z}'_j, \mathbf{v}'_j)'$ and $\mathbf{u}_{ij}$. When replacing the unknown parameters by $\hat{\boldsymbol{\beta}}$, the resulting $\hat{\mathbf{A}}_J = \mathbf{A}_J(\hat{\boldsymbol{\beta}})$ will be a consistent estimator for $\mathbf{A}$ whether the normality assumption on $\mathbf{z}_j$, $\mathbf{v}_j$ or $\mathbf{u}_{ij}$ holds or not. Similarly, with distributional violations, a consistent estimator of $\mathbf{B}$ is given by

$$\hat{\mathbf{B}}_J = J^{-1}\sum_{j=1}^{J}\mathbf{g}_j(\hat{\boldsymbol{\beta}})\mathbf{g}'_j(\hat{\boldsymbol{\beta}}).$$

Note that the dimension of $\mathbf{A}_J$ or $\mathbf{B}_J$ gets large quickly. In theory, one just needs to have $J > p_1(p_1+1)/2+(p_1+p_2)(p_1+p_2+3)/2$ level-2 units in order for $\hat{\boldsymbol{\Pi}}$ to be positive definite. In practice, one may need a much larger J in order to have a nonsingular $\hat{\boldsymbol{\Pi}}$.

With consistent $\hat{\boldsymbol{\beta}}$ and $\hat{\boldsymbol{\Pi}}$, we can construct AGLS statistics and rescaled statistics. Let

$$\mathbf{U} = \mathbf{A} - \mathbf{A}\dot{\boldsymbol{\beta}}(\dot{\boldsymbol{\beta}}'\mathbf{A}\dot{\boldsymbol{\beta}})^{-1}\dot{\boldsymbol{\beta}}'\mathbf{A}$$

and $\kappa = \operatorname{tr}(\mathbf{U}\boldsymbol{\Pi})/df$. Parallel to that in the conventional SEM, the rescaled statistic is given by

$$T_{\mathrm{RML}} = \frac{T_{\mathrm{ML}}}{\hat{\kappa}}, \tag{72}$$

which also asymptotically follows a distribution whose expectation equals df. The conditions under which T_{RML} in (72) approaches a chi-square distribution were given by Yuan and Bentler (2005). A residual based AGLS statistic that asymptotically follows χ^2_{df} is given by

$$T_{\mathrm{RAGLS}} = J[\hat{\boldsymbol{\beta}} - \boldsymbol{\beta}(\hat{\boldsymbol{\theta}})]'(\hat{\boldsymbol{\Pi}}^{-1} - \hat{\boldsymbol{\Pi}}^{-1}\dot{\boldsymbol{\beta}}(\hat{\boldsymbol{\theta}})[\dot{\boldsymbol{\beta}}'(\hat{\boldsymbol{\theta}})\hat{\boldsymbol{\Pi}}^{-1}\dot{\boldsymbol{\beta}}(\hat{\boldsymbol{\theta}})]^{-1}\dot{\boldsymbol{\beta}}'(\hat{\boldsymbol{\theta}})\hat{\boldsymbol{\Pi}}^{-1})$$
$$\times\,[\hat{\boldsymbol{\beta}} - \boldsymbol{\beta}(\hat{\boldsymbol{\theta}})].$$

Although T_{RAGLS} is asymptotically distribution free, when sample size is not large enough it tends to reject correct models more often than the nominal rate (see Yuan and Bentler, 2003). As noted, J plays the role of sample size. A corrected version of T_{RAGLS}

is

$$T_{\text{CRAGLS}} = \frac{T_{\text{RAGLS}}}{1 + T_{\text{RAGLS}}/J}$$

and a residual based F-statistic is given by

$$T_{\text{RF}} = (J - df)T_{\text{RAGLS}}/[df(J - 1)].$$

As with their counterparts in conventional SEM, both T_{RAGLS} and T_{RF} are asymptotically distribution free and can be referred to χ^2_{df} and $F_{df,J-df}$, respectively.

Limited simulation results in Yuan and Bentler (2003) indicated that T_{CRAGLS} slightly under-rejects the correct models for a smaller sample size J. T_{RF} moderately over-rejects the correct model for smaller J's. The level-1 sample sizes have little or no effect on them. The rescaled statistic T_{RML} in (72) rejects correct models too often for smaller J's, not performing as desirably as its counterpart in the conventional SEM. With nonnormally distributed data, the LR statistic T_{ML} in (70) does not behave like a chi-square variate in general. But there also exist robustness conditions for T_{ML} to approach χ^2_{df} for nonnormally distributed data. Yuan and Bentler (2005) showed that, when the level-1 and level-2 covariance structures do not have overlapping parameters and each level satisfies the asymptotic robustness condition as in the conventional CSA model, T_{ML} is still asymptotically robust. However, as in conventional CSA, the asymptotic robustness conditions cannot be verified. Our recommended statistics are T_{CRADF} and T_{RF}, although these statistics need to be used with caution when J is small. Similarly, SEs of parameter estimates are generally affected by the distributions of $\mathbf{z}_j$, $\mathbf{v}_j$ and $\mathbf{u}_{ij}$ (Yuan and Bentler, 2006b). Under proper conditions, SEs of factor loading estimates at both the between- and within-level are asymptotically robust. Again, the requisite asymptotic robustness conditions are not verifiable. SEs based on (71) are recommended.

This section has mainly concentrated on the MLE $\hat{\boldsymbol{\theta}}$ that maximizes (68), based on model (66). For a two-level model without the $\mathbf{z}_j$ in (66), Muthén (1989b, 1994) proposed a procedure by which parameter estimation and model testing can be proceeded with the two-group option of the previous section, which is now commonly called the MUML procedure. When data are balanced, i.e., $n_1 = \cdots = n_J$, the estimate by MUML is equivalent to the MLE $\hat{\boldsymbol{\theta}}$. The parameter estimate by MUML is still consistent even when data are unbalanced. When data are normally distributed, SEs for parameter estimates and the test statistic for the overall model by MUML may contain biases, depending on the coefficient of variation of $n_1, n_2, \ldots, n_J$ (Yuan and Hayashi, 2005). There has been no development for MUML when data are not normally distributed.

7. Examples

The material presented in the previous sections emphasized the statistical components for various SEM procedures. This section presents two examples using the two data sets introduced in Section 1. Specifically, we will fit the CFA model in (1) and (2) to the data set of Holzinger and Swineford (1939) and the factor analytic simultaneous equation model in (9) to the data set of Neumann (1994). Both examples involve only

CSA models, which are also the most widely used SEM models in practice. We will use EQS 6.1 (Bentler, 2006). Because standard deviations of the 9 variables of Holzinger and Swineford's data differ a lot, we divide them respectively by 6, 4, 8, 3, 4, 7, 23, 20, 36 to keep each marginal standard deviation between 1 and 2. This change of scale has no substantive effect on the model evaluation but makes the iterative procedure to converge faster.

Holzinger and Swineford's data contain $n = 145$ observations. Fitting the sample covariance matrix to the CFA model in (1) and (2), the LR statistic $T_{MLc} = 51.189$, with a p-value $= 0.001$ when referring to χ^2_{24}. So the model does not fit the data well. The normalized Mardia's kurtosis $M = 3.037$ is statistically significant. The large T_{MLc} may also be caused by the heavier tails of the data as compared to a normal distribution. The rescaled statistic $T_{RMLc} = 49.732$, corresponding to a p-value $=$ 0.002, is still significant. The corrected residual-based statistic $T_{CRAGLSc} = 44.549$ and the F-statistic $T_{RFc} = 2.251$, corresponding to larger p-values, but the conclusion remains that the model fits the data poorly. The fit indices based on T_{MLc} are NFI $=$ 0.898, CFI $= 0.942$, RMSEA $= 0.089$. Possibly one could accept this model based on CFI, but the consensus of the test statistics imply that the model contains some serious misspecification. See Browne et al. (2002) for an interesting discussion on resolving conflicting evidence on fit.

The LM statistics indicate that not all the fixed parameters are consistent with the data. In the model, x_9 is not an indicator of f_1 with $\lambda_{91} = 0$. $T_{LM} = 24.370$ for this constraint, implying that a substantial improvement in fit would result if λ_{91} were to be freely estimated. After freeing λ_{91}, $T_{MLc} = 28.099$, $T_{RMLc} = 28.024$, $T_{CRAGLSc} =$ 25.883, $T_{RFc} = 1.161$; and the corresponding p-values are all above 0.2. The fit indices based on T_{MLc} are NFI $= 0.944$, CFI $= 0.989$, RMSEA $= 0.039$, also implying that the model now fits the data well according to the recommended cutoff values by Hu and Bentler (1998). Reexamining the meaning of the variables, we note that a student may need a good spatial judgment in order to perform the test "Straight-Curved Capitals" speedily. It seems reasonable to let x_9 load on f_1.

Those who prefer that each variable loads on only a single factor may consider allowing the e_i's to correlate. The LM test in EQS allows users to evaluate the constraints of either zero factor loadings or zero error covariances. When error covariances are selected, the LM test implies that the constraint $\psi_{78} = 0$ is not supported by data, with $T_{LM} = 24.646$. Freeing the parameter ψ_{78}, the test statistics become $T_{MLc} = 27.842$, $T_{RMLc} = 27.117$, $T_{CRAGLSc} = 25.318$, $T_{RFc} = 1.130$; all are somewhat smaller than their counterparts when freeing λ_{91}. The fit indices based on T_{ML} are NFI $= 0.945$, CFI $= 0.990$, RMSEA $= 0.038$, also slightly more supportive of the correlated error model than the model with a free λ_{91}. From a substantive perspective, "Addition" and "Counting Dots" may share the trait of numerical ability that is not shared by "Straight-Curved Capital", and additional correlation for x_7 and x_8 is needed beyond that accounted for by f_3.

We see that there is not necessarily a unique way to improve a model via the LM test. When λ_{91} is free in the model, the T_{LM} for the constraint $\psi_{78} = 0$ is no longer significant, and $\psi_{23} = 0$ and $\psi_{71} = 0$ are marginally significant with p-values equal to 0.023 and 0.044, respectively. Similarly, when ψ_{78} is free in the model, the T_{LM} for

$\lambda_{91} = 0$ is no longer significant either, and $\lambda_{13} = 0$ is marginally significant with p-value $= 0.031$. Whether these parameters should be further added to the model is subject to judgment. There is a need for a model to be parsimonious and at the same time to replicate data.

Because Holzinger and Swineford's data are not normally distributed, as implied by a significant Mardia's kurtosis, modeling the sample covariance matrix is not the optimal procedure. For this sample, Yuan et al. (2000) suggested using Huber-type weights in (37) with $\rho = 0.05$, because Mardia's kurtosis for $\mathbf{x}_i^{(0.05)}$ in (43) is no longer significant. Yuan and Hayashi (2003) found that $\rho = 0.25$ corresponds to parameter estimates that are empirically most efficient. Also, under the null hypothesis of a correct model, T_{MLc} applying to $\mathbf{x}_i^{(0.25)}$ approximately follows χ^2_{24}. Yuan, Marshall and Weston (2002) also found that $\rho = 0.25$ is supported by cross-validation. Applying the ML procedure to $\mathbf{x}_i^{(0.25)}$ lead to $T_{\text{MLc}} = 48.883$, smaller than the T_{MLc} for the raw data but still significant. The LM test for $\psi_{78} = 0$ is $T_{\text{LM}} = 22.630$, also highly significant when referred to χ^2_1. Adding this parameter to the model results in $T_{\text{MLc}} = 26.557$, whose p-value $= 0.275$ when referred to χ^2_{23}. Similarly, $T_{\text{LM}} = 23.149$ for $\lambda_{91} = 0$ is also highly significant. Freeing λ_{91} in the model results in $T_{\text{MLc}} = 26.453$ and a p-value $= 0.280$ when referred to χ^2_{23}. The T_{MLc}'s for the two modified models are also somewhat smaller than those corresponding to the raw data, reflecting the effect of heavy tails of the raw sample.

Table 1 contains the MLEs $\hat{\boldsymbol{\theta}}$ and their SEs for the original model and modified models, applying to the samples $\mathbf{x}_i$ and $\mathbf{x}_i^{(0.25)}$. The SEs are based on the sandwich-type covariance matrix as in (27). It is obvious that, when the model is modified, both the parameter estimates and their SEs change. The result in Yuan et al. (2003) implies that, when either λ_{91} or ψ_{78} is freed, there will be no systematic change in the estimates for $\lambda_{11}, \lambda_{21}, \lambda_{31}, \lambda_{42}, \lambda_{52}, \lambda_{62}, \phi_{21}, \psi_{11}, \psi_{22}, \psi_{33}, \psi_{44}, \psi_{55}, \psi_{66}$. The differences in the estimates are only reflections of sampling errors.

Comparing those under $\mathbf{x}_i$ with those under the transformed sample $\mathbf{x}_i^{(0.25)}$, the parameter estimates are comparable. Most SEs under $\mathbf{x}_i^{(0.25)}$ are smaller. Yuan, Bentler and Chan (2004) implies that, although parameter estimates using a robust procedures are more efficient, only the SEs for the invariant parameters are comparable. For the parameterization in the CFA model (1) and (2), the invariant parameters are ϕ_{21}, ϕ_{31} and ϕ_{32}. The SEs for the other parameters are not comparable because their population counterparts change when data are transformed.

Notice that once ψ_{78} is free, the estimate for ψ_{99} is not statistically significant. According to the Wald test, we may fix ψ_{99} at zero in a followup analysis. However, $\psi_{99} = \text{Var}(e_9)$ should be positive in theory, and so this restriction does not make sense.

The alcohol and psychological symptom data set of Neumann (1994) has 335 cases. Fitting this data set to the model in (9) leads to $T_{\text{MLc}} = 48.960$, with p-value $= 0.012$ when referring to χ^2_{29}. The fit indices based on T_{MLc} are NFI $= 0.948$, CFI $= 0.966$ and RMSEA $= 0.045$, implying a reasonable model although T_{MLc} is significant at the 0.05 level. The normalized Mardia's multivariate kurtosis for this data set is $M = 14.763$, quite significant when referring to $N(0, 1)$. The rescaled and residual-based AGLS statistics are $T_{\text{RMLc}} = 47.486$ with p-value $= 0.017$, $T_{\text{CRAGLSc}} = 45.018$ with

Table 1
MLEs and their SEs for the CFA model using data from Holzinger and Swineford (1939)

θ	$\mathbf{x}_i$			$\mathbf{x}_i^{(0.25)}$		
	$\hat{\theta}$ (SE)	$\hat{\theta}$ (SE)	$\hat{\theta}$ (SE)	$\hat{\theta}$ (SE)	$\hat{\theta}$ (SE)	$\hat{\theta}$ (SE)
λ_{11}	0.779 (0.116)	0.819 (0.109)	0.820 (0.110)	0.785 (0.094)	0.834 (0.087)	0.832 (0.089)
λ_{21}	0.575 (0.094)	0.544 (0.093)	0.549 (0.094)	0.538 (0.090)	0.503 (0.090)	0.505 (0.090)
λ_{31}	0.722 (0.091)	0.689 (0.088)	0.694 (0.088)	0.693 (0.085)	0.659 (0.081)	0.661 (0.082)
λ_{42}	0.974 (0.084)	0.975 (0.084)	0.975 (0.083)	0.932 (0.076)	0.933 (0.076)	0.934 (0.076)
λ_{52}	0.964 (0.083)	0.964 (0.083)	0.965 (0.083)	0.969 (0.076)	0.968 (0.076)	0.968 (0.076)
λ_{62}	0.938 (0.082)	0.937 (0.083)	0.936 (0.083)	0.895 (0.074)	0.893 (0.074)	0.893 (0.074)
λ_{73}	0.683 (0.080)	0.709 (0.084)	0.440 (0.100)	0.684 (0.078)	0.707 (0.081)	0.428 (0.094)
λ_{83}	0.838 (0.089)	0.899 (0.098)	0.568 (0.105)	0.793 (0.076)	0.835 (0.084)	0.525 (0.092)
λ_{93}	0.718 (0.086)	0.453 (0.093)	1.007 (0.114)	0.668 (0.074)	0.454 (0.084)	0.966 (0.102)
ϕ_{21}	0.541 (0.094)	0.554 (0.091)	0.542 (0.096)	0.577 (0.078)	0.582 (0.075)	0.580 (0.078)
ϕ_{31}	0.520 (0.100)	0.391 (0.112)	0.623 (0.103)	0.476 (0.096)	0.354 (0.103)	0.596 (0.099)
ϕ_{32}	0.334 (0.115)	0.240 (0.117)	0.381 (0.099)	0.340 (0.099)	0.260 (0.103)	0.378 (0.090)
ψ_{11}	0.721 (0.168)	0.657 (0.160)	0.655 (0.164)	0.622 (0.126)	0.544 (0.118)	0.547 (0.123)
ψ_{22}	0.905 (0.140)	0.939 (0.142)	0.934 (0.140)	0.863 (0.114)	0.900 (0.117)	0.898 (0.117)
ψ_{33}	0.560 (0.108)	0.607 (0.097)	0.600 (0.097)	0.584 (0.100)	0.630 (0.092)	0.627 (0.092)
ψ_{44}	0.317 (0.066)	0.315 (0.066)	0.315 (0.066)	0.310 (0.058)	0.307 (0.057)	0.305 (0.057)
ψ_{55}	0.422 (0.072)	0.422 (0.072)	0.421 (0.072)	0.379 (0.059)	0.380 (0.059)	0.381 (0.059)
ψ_{66}	0.409 (0.076)	0.411 (0.077)	0.413 (0.077)	0.390 (0.067)	0.393 (0.067)	0.394 (0.067)
ψ_{77}	0.602 (0.080)	0.567 (0.083)	0.876 (0.090)	0.597 (0.080)	0.565 (0.082)	0.881 (0.093)
ψ_{88}	0.400 (0.112)	0.294 (0.130)	0.780 (0.144)	0.322 (0.079)	0.253 (0.105)	0.675 (0.089)
ψ_{99}	0.542 (0.085)	0.479 (0.076)	0.045 (0.197)	0.515 (0.073)	0.448 (0.067)	0.028 (0.185)
λ_{91}		0.459 (0.103)			0.417 (0.087)	
ψ_{78}			0.388 (0.082)			0.366 (0.080)

p-value $= 0.029$; $T_{\text{RF}c} = 1.643$ with p-value $= 0.022$. Although $T_{\text{CRAGLS}c} = 45.018$ gives the best support for the model, the p-value is still marginal.

Because of the heavier tails of the sample, it is preferable to use a robust procedure. Yuan et al. (2000) found that the transformed sample $\mathbf{x}_i^{(0.1)}$ may have comparable tails to those of the normal distribution, as judged by Mardia's multivariate kurtosis. Yuan and Hayashi (2003) found that $T_{\text{ML}c}$ based on $\mathbf{x}_i^{(0.1)}$ has a shorter right tail than that of χ^2_{29}, and Huber-type weights with $\rho = 0.05$ make $T_{\text{ML}c}$ approximately follow χ^2_{29}. The LR statistic corresponding to $\rho = 0.05$ is $T_{\text{ML}c} = 42.420$ with p-value $= 0.051$; and that corresponding to $\rho = 0.1$ is $T_{\text{ML}c} = 40.985$ with p-value $= 0.069$. Both give more support to the model than the $T_{\text{ML}c}$ based on the raw data. The SEs of $\hat{\boldsymbol{\theta}}$ based on the robust procedure, not reported here, also are smaller than those based on the raw sample.

The LM test identifies $\text{Cov}(\varepsilon_2, \varepsilon_3) = 0$ as the most significant constraint among all the zero error covariances. Freeing this covariance leads to $T_{\text{ML}c} = 38.129$ with p-value $= 0.096$ when using the raw sample; $T_{\text{ML}c} = 33.413$ with p-value $= 0.221$ when using $\mathbf{x}_i^{(0.05)}$; $T_{\text{ML}c} = 32.905$ with p-value $= 0.239$ when using $\mathbf{x}_i^{(0.10)}$. Clearly, adding this error covariance makes the model match the data better statistically, especially when using the robust covariance estimate. The variables involved in the correlation are "Age of First Detoxification from Alcohol" and "Alcohol Severity Score", respectively. Whether it makes sense that these two variables should share additional variance beyond that explained by the model is subject to experts' judgment or substantive theory.

Comparing the two examples, we note that for Holzinger and Swineford's data, the normalized Mardia's kurtosis is $M = 3.037$, and the Huber-type weight with $\rho = 0.25$ allows $T_{\mathrm{ML}c}$ to approximately follow a chi-square distribution. In contrast, Neumann's data are more kurtotic ($M = 14.763$) but require a smaller tuning constant $\rho = 0.05$. It is clear that a more significant sample multivariate kurtosis does not necessarily imply the need for a larger ρ. Compared to the multivariate normal distribution, both the samples only have minor heavier tails. When data are more severely nonnormally distributed, greater difference exists between modeling $\mathbf{S}$ and robust $\hat{\boldsymbol{\Sigma}}$'s.

References

Aitchison, J., Silvey, S.D. (1958). Maximum likelihood estimation of parameters subject to restraints. *Annals of Mathematical Statistics* **29**, 813–828.

Allison, P.D. (1987). Estimation of linear models with incomplete data. *Sociological Methodology* **17**, 71–103.

Allison, P.D. (2003). Missing data techniques for structural equation modeling. *Journal of Abnormal Psychology* **112**, 545–557.

Amemiya, Y., Anderson, T.W. (1990). Asymptotic chi-square tests for a large class of factor analysis models. *Annals of Statistics* **18**, 1453–1463.

Anderson, T.W., Amemiya, Y. (1988). The asymptotic normal distribution of estimators in factor analysis under general conditions. *Annals of Statistics* **16**, 759–771.

Anderson, J.C., Gerbing, D.W. (1984). The effect of sampling error on convergence, improper solutions, and goodness-of-fit indices for maximum likelihood confirmatory factor analysis. *Psychometrika* **49**, 155–173.

Anderson, J.C., Gerbing, D.W. (1988). Structural equation modeling in practice: A review and recommended two-step approach. *Psychological Bulletin* **103**, 411–423.

Arminger, G., Sobel, M.E. (1990). Pseudo-maximum likelihood estimation of mean and covariance structures with missing data. *Journal of the American Statistical Association* **85**, 195–203.

Arminger, G., Stein, P., Wittenberg, J. (1999). Mixtures of conditional mean- and covariance-structure models. *Psychometrika* **64**, 475–494.

Bauer, D.J., Curran, P.J. (2004). The integration of continuous and discrete latent variable models: Potential problems and promising opportunities. *Psychological Methods* **9**, 3–29.

Beauducel, A., Wittmann, W.W. (2005). Simulation study on fit indexes in CFA based on data with slightly distorted simple structure. *Structural Equation Modeling* **12**, 41–75.

Bekker, P.A., Merckens, A., Wansbeek, T.J. (1994). *Identification, Equivalent Models, and Computer Algebra*. Academic Press, Boston.

Bentler, P.M. (1982). Linear systems with multiple levels and types of latent variables. In: Jöreskog, K.G., Wold, H. (Eds.), *Systems Under Indirect Observation: Causality, Structure, Prediction*. North-Holland, Amsterdam, pp. 101–130.

Bentler, P.M. (1983). Some contributions to efficient statistics in structural models: Specification and estimation of moment structures. *Psychometrika* **48**, 493–517.

Bentler, P.M. (1986). *Lagrange Multiplier and Wald tests for EQS and EQS/PC*. BMDP Statistical Software, Los Angeles, CA.

Bentler, P.M. (1990). Comparative fit indexes in structural models. *Psychological Bulletin* **107**, 238–246.

Bentler, P.M. (2005). Latent growth curves. In: Werner, J. (Ed.), *Zeitreihenanalysen*. Logos, Berlin, pp. 13–36.

Bentler, P.M. (2006). *EQS 6 Structural Equations Program Manual*. Multivariate Software, Encino, CA.

Bentler, P.M., Bonett, D.G. (1980). Significance tests and goodness of fit in the analysis of covariance structures. *Psychological Bulletin* **88**, 588–606.

Bentler, P.M., Dijkstra, T. (1985). Efficient estimation via linearization in structural models. In: Krishnaiah, P.R. (Ed.), *Multivariate Analysis VI*. North-Holland, Amsterdam, pp. 9–42.

Bentler, P.M., Lee, S.-Y., Weng, L.-J. (1987). Multiple population covariance structure analysis under arbitrary distribution theory. *Communication in Statistics – Theory and Methods* **16**, 1951–1964.

Bentler, P.M., Weeks, D.G. (1980). Linear structural equations with latent variables. *Psychometrika* **45**, 289–308.

Bentler, P.M., Yuan, K.-H. (1999). Structural equation modeling with small samples: Test statistics. *Multivariate Behavioral Research* **34**, 181–197.

Beran, R. (1986). Simulated power functions. *Annals of Statistics* **14**, 151–173.

Beran, R., Srivastava, M.S. (1985). Bootstrap tests and confidence regions for functions of a covariance matrix. *Annals of Statistics* **13**, 95–115.

Berkane, M., Bentler, P.M. (1988). Estimation of contamination parameters and identification of outliers in multivariate data. *Sociological Methods and Research* **17**, 55–64.

Bollen, K.A. (1987). Outliers and improper solutions: A confirmatory factor analysis example. *Sociological Methods and Research* **15**, 375–384.

Bollen, K.A. (1989). *Structural equations with latent variables*. Wiley, New York.

Bollen, K.A. (2002). Latent variables in psychology and the social sciences. *Annual Review of Psychology* **53**, 605–634.

Bollen, K.A., Arminger, G. (1991). Observational residuals in factor analysis and structural equation models. *Sociological Methodology* **21**, 235–262.

Bollen, K.A., Stine, R. (1993). Bootstrapping goodness of fit measures in structural equation models. In: Bollen, K.A., Long, J.S. (Eds.), *Testing Structural Equation Models*. Sage, Newbury Park, CA, pp. 111–135.

Boomsma, A. (1985). Nonconvergence, improper solutions, and starting values in LISREL maximum likelihood estimation. *Psychometrika* **50**, 229–242.

Box, G.E.P. (1979). Robustness in the strategy of scientific model building. In: Launer, R.L., Wilkinson, G.N. (Eds.), *Robustness in Statistics*. Academic Press, New York, pp. 201–236.

Brown, C.H. (1983). Asymptotic comparison of missing data procedures for estimating factor loadings. *Psychometrika* **48**, 269–291.

Browne, M.W. (1974). Generalized least-squares estimators in the analysis of covariance structures. *South African Statistical Journal* **8**, 1–24.

Browne, M.W. (1984). Asymptotic distribution-free methods for the analysis of covariance structures. *British Journal of Mathematical and Statistical Psychology* **37**, 62–83.

Browne, M.W., Arminger, G. (1995). Specification and estimation of mean and covariance structure models. In: Arminger, G., Clogg, C.C., Sobel, M.E. (Eds.), *Handbook of Statistical Modeling for the Social and Behavioral Sciences*. Plenum, New York, pp. 185–249.

Browne, M.W., Cudeck, R. (1993). Alternative ways of assessing model fit. In: Bollen, K.A., Long, J.S. (Eds.), *Testing Structural Equation Models*. Sage, Newbury Park, CA, pp. 136–162.

Browne, M.W., Du Toit, S.H.C. (1992). Automated fitting of nonstandard models. *Multivariate Behavioral Research* **27**, 269–300.

Browne, M.W., MacCallum, R.C., Kim, C.-T., Andersen, B.L., Glaser, R. (2002). When fit indices and residuals are incompatible. *Psychological Methods* **7**, 403–421.

Browne, M.W., Shapiro, A. (1988). Robustness of normal theory methods in the analysis of linear latent variate models. *British Journal of Mathematical and Statistical Psychology* **41**, 193–208.

Byrne, B.M. (2006). *Structural Equation Modeling With EQS 6*. Erlbaum, Mahwah, NJ.

Byrne, B.M., Shavelson, R.J., Muthén, B. (1989). Testing for the equivalence of factor covariance and mean structures: The issue of partial measurement invariance. *Psychological Bulletin* **105**, 456–466.

Byrne, B.M., Watkins, D. (2003). The issue of measurement invariance revisited. *Journal of Cross-Cultural Psychology* **34**, 155–175.

Campbell, N.A. (1980). Robust procedures in multivariate analysis I: Robust covariance estimation. *Applied Statistics* **29**, 231–237.

Carroll, R.J., Ruppert, D., Stefanski, L.A. (1995). *Measurement Error in Nonlinear Models*. Chapman & Hall, London.

Chen, F., Bollen, K.A., Paxton, P., Curran, P., Kirby, J. (2001). Improper solutions in structural equation models: Causes, consequences, and strategies. *Sociological Methods and Research* **29**, 468–508.

Cheng, T.C., Victoria-Feser, M.P. (2002). High-breakdown estimation of multivariate mean and covariance with missing observations. *British Journal of Mathematical and Statistical Psychology* **55**, 317–335.

Chong, E.K.P., Żak, S.H. (1996). *An Introduction to Optimization*. Wiley, New York.

Chou, C.-P., Bentler, P.M. (1990). Model modification in covariance structure modeling: A comparison among likelihood ratio, Lagrange multiplier, and Wald tests. *Multivariate Behavioral Research* **25**, 115–136.

Chou, C.-P., Bentler, P.M. (2002). Model modification in structural equation modeling by imposing constraints. *Computational Statistics and Data Analysis* **41**, 271–287.

Curran, P.J., West, S.G., Finch, J.F. (1996). The robustness of test statistics to nonnormality and specification error in confirmatory factor analysis. *Psychological Methods* **1**, 16–29.

Dempster, A.P., Laird, N.M., Rubin, D.B. (1977). Maximum likelihood estimation from incomplete data via the EM algorithm (with discussion). *Journal of the Royal Statistical Society B* **39**, 1–38.

DiStefano, C. (2002). The impact of categorization with confirmatory factor analysis. *Structural Equation Modeling* **9**, 327–346.

Dhrymes, P.J. (1994). *Topics in Advanced Econometrics: Linear and Nonlinear Simultaneous Equations*. Springer-Verlag, New York.

Dolan, C.V. (1994). Factor analysis of variables with 2, 3, 5 and 7 response categories: A comparison of categorical variable estimators using simulated data. *British Journal of Mathematical and Statistical Psychology* **47**, 309–326.

Duncan, T.E., Duncan, S.C., Strycker, L.A., Li, F., Alpert, A. (1999). *An Introduction to Latent Variable Growth Curve Modeling: Concepts, Issues, and Applications*. Erlbaum, Mahwah, NJ.

du Toit, S.H.C., du Toit, M. (in press). Multilevel structural equation modeling. In: de Leeuw, J., Kreft, I.G.G. (Eds.), *Handbook of Quantitative Multilevel Analysis*, Kluwer Academic, Boston.

Efron, B., Hinkley, D.V. (1978). The observed versus the expected information. *Biometrika* **65**, 457–487.

Enders, C.K. (2001). The impact of nonnormality on full information maximum likelihood estimation for structural equation models with missing data. *Psychological Methods* **6**, 352–370.

Enders, C.K., Peugh, J.L. (2004). Using an EM covariance matrix to estimate structural equation models with missing data: Choosing an adjusted sample size to improve the accuracy of inferences. *Structural Equation Modeling* **11**, 1–19.

Engle, R. (1984). Wald, likelihood ratio and Lagrange multiplier tests in econometrics. In: Griliches, Z., Intriligator, M.D. (Eds.), *Handbook of Econometrics*, vol. II. North-Holland, Amsterdam, pp. 775–826.

Fang, K.-T., Kotz, S., Ng, K.W. (1990). *Symmetric Multivariate and Related Distributions*. Chapman & Hall, London.

Ferguson, T. (1958). A method of generating best asymptotically normal estimates with application to the estimation of bacterial densities. *Annals of Mathematical Statistics* **29**, 1046–1062.

Ferguson, T. (1996). *A Course in Large Sample Theory*. Chapman & Hall, London.

Finkbeiner, C. (1979). Estimation for the multiple factor model when data are missing. *Psychometrika* **44**, 409–420.

Flora, D.B., Curran, P.J. (2004). An empirical evaluation of alternative methods of estimation for confirmatory factor analysis with ordinal data. *Psychological Methods* **9**, 466–491.

Fornell, C. (1983). Issues in the application of covariance structure analysis: A comment. *Journal of Consumer Research* **9**, 443–448.

Fuller, W.A. (1987). *Measurement Error Models*. Wiley, New York.

Fung, W.K. (1993). Unmasking outliers and leverage points: A confirmation. *Journal of the American Statistical Association* **88**, 515–519.

Goldstein, H. (2003). *Multilevel Statistical Models*, 3nd ed. Arnold, London.

Goldstein, H., McDonald, R.P. (1988). A general model for the analysis of multilevel data. *Psychometrika* **53**, 435–467.

Gourieroux, C., Monfort, A., Trognon, A. (1984). Pseudo-maximum likelihood methods: Theory. *Econometrica* **52**, 681–700.

Green, P.J. (1984). Iteratively reweighted least squares for maximum likelihood estimation, and some robust and resistant alternatives (with discussion). *Journal of the Royal Statistical Society B* **46**, 149–192.

Green, S.B., Thompson, M.S., Poirier, J. (2001). An adjusted Bonferroni method for elimination of parameters in specification addition searches. *Structural Equation Modeling* **8**, 18–39.

Hampel, F.R. (1974). The influence curve and its role in robust estimation. *Journal of the American Statistical Association* **69**, 383–393.

Hampel, F.R., Ronchetti, E.M., Rousseeuw, P.J., Stahel, W.A. (1986). *Robust statistics: The approach based on influence functions*. Wiley, New York.

Hancock, G.R. (1999). A sequential Scheffé-type respecification procedure for controlling Type I error in exploratory structural equation model modification. *Structural Equation Modeling* **6**, 158–168.

Hancock, G.R. (2001). Effect size, power, and sample size determination for structured means modeling and MIMIC approaches to between-groups hypothesis testing of means on a single latent construct. *Psychometrika* **66**, 373–388.

Hershberger, S.L. (2003). The growth of structural equation modeling: 1994–2001. *Structural Equation Modeling* **10**, 35–46.

Holzinger, K.J., Swineford, F. (1939). *A Study in Factor Analysis: The Stability of a Bi-factor Solution. Supplementary Educational Monographs*, vol. 48. University of Chicago, Chicago.

Hoshino, T. (2001). Bayesian inference for finite mixtures in confirmatory factor analysis. *Behaviormetrika* **28**, 37–63.

Hu, L., Bentler, P.M. (1998). Fit indices in covariance structure modeling: Sensitivity to underparameterized model misspecification. *Psychological Methods* **3**, 424–453.

Hu, L.T., Bentler, P.M. (1999). Cutoff criteria for fit indexes in covariance structure analysis: Conventional criteria versus new alternatives. *Structural Equation Modeling* **6**, 1–55.

Hu, L.T., Bentler, P.M., Kano, Y. (1992). Can test statistics in covariance structure analysis be trusted? *Psychological Bulletin* **112**, 351–362.

Huber, P.J. (1964). Robust estimation of a location parameter. *Annals of Mathematical Statistics* **35**, 73–101.

Huber, P.J. (1967). The behavior of maximum likelihood estimates under nonstandard conditions. In: *Proceedings of the Fifth Berkeley Symposium on Mathematical Statistics and Probability*. University of California Press, Berkeley, CA, pp. 221–233.

Huber, P.J. (1981). *Robust Statistics*. Wiley, New York.

Jamshidian, M., Bentler, P.M. (1999). Using complete data routines for ML estimation of mean and covariance structures with missing data. *Journal Educational and Behavioral Statistics* **23**, 21–41.

Jennrich, R.I. (1974). Simplified formulae for standard errors in maximum-likelihood factor analysis. *British Journal of Mathematical and Statistical Psychology* **27**, 122–131.

Jöreskog, K.G. (1967). Some contributions to maximum likelihood factor analysis. *Psychometrika* **32**, 443–482.

Jöreskog, K.G. (1971). Simultaneous factor analysis in several populations. *Psychometrika* **36**, 409–426.

Jöreskog, K.G. (1973). A general method for estimating a linear structural equation system. In: Goldberger, A.S., Duncan, O.D. (Eds.), *Structural Equation Models in the Social Sciences*. Seminar, New York, pp. 85–112.

Jöreskog, K.G. (1994). On the estimation of polychoric correlations and their asymptotic covariance matrix. *Psychometrika* **59**, 381–389.

Jöreskog, K.G., Sörbom, D. (2001). *LISREL 8 Users's Reference Guide*. Lincolnwood, IL, Scientific Software International.

Kano, Y. (1986). Conditions on consistency of estimators in covariance structure model. *Journal of the Japan Statistical Society* **16**, 75–80.

Kano, Y. (1992). Robust statistics for test-of-independence and related structural models. *Statistics and Probability Letters* **15**, 21–26.

Kano, Y. (1998). Improper solutions in exploratory factor analysis: Causes and treatments. In: Rizzi, A., Vichi, M., Bock, H. (Eds.), *Advances in Data Sciences and Classification*. Springer-Verlag, Berlin, pp. 375–382.

Kano, Y. (2001). Structural equation modeling for experimental data. In: Cudeck, R., du Toit, S.H.C., Sörbom, D. (Eds.), *Structural Equation Modeling: Present and Future*. Scientific Software International, Lincolnwood, IL, pp. 381–402.

Kano, Y., Berkane, M., Bentler, P.M. (1993). Statistical inference based on pseudo-maximum likelihood estimators in elliptical populations. *Journal of the American Statistical Association* **88**, 135–143.

Kauermann, G., Carroll, R.J. (2001). A note on the efficiency of sandwich covariance matrix estimation. *Journal of the American Statistical Association* **96**, 1387–1396.

Keesling, J.W. (1972). Maximum likelihood approaches to causal analysis. Ph.D. Thesis. University of Chicago.

Kenny, D.A., McCoach, D.B. (2003). Effect of the number of variables on measures of fit in structural equation modeling. *Structural Equation Modeling* **10**, 333–351.

Kenward, M.G., Molenberghs, G. (1998). Likelihood based frequentist inference when data are missing at random. *Statistical Science* **13**, 236–247.

Kim, K. (2005). The relationship among fit indexes, power, and sample size in structural equation modeling. *Structural Equation Modeling* **12**, 368–390.

Kim, K.H., Bentler, P.M. (2002). Tests of homogeneity of means and covariance matrices for multivariate incomplete data. *Psychometrika* **67**, 609–624.

Klein, A., Moosbrugger, H. (2000). Maximum likelihood estimation of latent interaction effects with the LMS method. *Psychometrika* **65**, 457–474.

Kline, R.B. (2005). *Principles and practice of structural equation modeling*, 2nd ed. Guilford Press, New York.

Laird, N.M. (1988). Missing data in longitudinal studies. *Statistics in Medicine* **7**, 305–315.

Lawley, D.N., Maxwell, A.E. (1971). *Factor analysis as a statistical method*, 2nd ed. Elsevier, New York.

Lee, S.-Y. (1985). On testing functional constraints in structural equation models. *Biometrika* **72**, 125–131.

Lee, S.-Y. (1986). Estimation for structural equation models with missing data. *Psychometrika* **51**, 93–99.

Lee, S.-Y. (1990). Multilevel analysis of structural equation models. *Biometrika* **77**, 763–772.

Lee, S.-Y., Bentler, P.M. (1980). Some asymptotic properties of constrained generalized least squares estimation in covariance structure models. *South African Statistical Journal* **14**, 121–136.

Lee, S.Y., Jennrich, R.I. (1979). A study of algorithms for covariance structure analysis with specific comparisons using factor analysis. *Psychometrika* **44**, 99–113.

Lee, S.Y., Poon, W.Y. (1998). Analysis of two-level structural equation models via EM type algorithms. *Statistica Sinica* **8**, 749–766.

Lee, S.-Y., Poon, W.-Y., Bentler, P.M. (1992). Structural equation models with continuous and polytomous variables. *Psychometrika* **57**, 89–106.

Lee, S.-Y., Poon, W.-Y., Bentler, P.M. (1995). A two-stage estimation of structural equation models with continuous and polytomous variables. *British Journal of Mathematical and Statistical Psychology* **48**, 339–358.

Lee, S.Y., Shi, J.Q. (2001). Maximum likelihood estimation of two-level latent variables model with mixed continuous and polytomous data. *Biometrics* **57**, 787–794.

Lee, S.Y., Song, X.Y. (2001). Hypothesis testing and model comparison in two-level structural equation models. *Multivariate Behavioral Research* **36**, 639–655.

Lee, S.Y., Song, X.Y. (2003). Maximum likelihood estimation and model comparison of nonlinear structural equation models with continuous and polytomous variables. *Computational Statistics and Data Analysis* **44**, 125–142.

Lee, S.-Y., Song, X.-Y., Lee, J.C.K. (2003). Maximum likelihood estimation of nonlinear structural equation models with ignorable missing data. *Journal of Educational and Behavioral Statistics* **28**, 111–124.

Lee, S.Y., Song, X.Y., Poon, W.Y. (2004). Comparison of approaches in estimating interaction and quadratic effects of latent variables. *Multivariate Behavioral Research* **39**, 37–67.

Lee, S.-Y., Tang, N. (in press). Statistical analysis of nonlinear structural equation models with nonignorable missing outcome data. *Psychometrika*.

Lee, S.-Y., Tsui, K.L. (1982). Covariance structure analysis in several populations. *Psychometrika* **47**, 297–308.

Lee, S.Y., Wang, S.J. (1996). Sensitivity analysis of structural equation models. *Psychometrika* **61**, 93–108.

Lee, S.Y., Zhu, H.T. (2000). Statistical analysis of nonlinear structural equation models with continuous and polytomous data. *British Journal of Mathematical and Statistical Psychology* **53**, 209–232.

Lee, S.Y., Zhu, H.T. (2002). Maximum likelihood estimation of nonlinear structural equation models. *Psychometrika* **67**, 189–210.

Liang, J., Bentler, P.M. (2004). A new EM algorithm for fitting two-level structural equation models. *Psychometrika* **69**, 101–122.

Liang, K.Y., Zeger, S.L. (1986). Longitudinal data analysis using generalized linear models. *Biometrika* **73**, 13–22.

Little, R.J.A. (1988). Robust estimation of the mean and covariance matrix from data with missing values. *Applied Statistics* **37**, 23–38.

Little, R.J.A., Rubin, D.B. (2002). *Statistical Analysis With Missing Data*, 2nd ed. Wiley, New York.

Loehlin, J.C. (2004). *Latent variable models: An introduction to factor, path, and structural equation analysis*, 4th ed. Erlbaum, Mahwah, NJ.

Loken, E. (2005). Identification constraints and inference in factor models. *Structural Equation Modeling* **12**, 232–244.

Longford, N.T. (1993). Regression analysis of multilevel data with measurement error. *British Journal of Mathematical and Statistical Psychology* **46**, 301–311.

Lopuhaä, H.P. (1989). On the relation between S-estimators and M-estimators of multivariate location and covariances. *Annals of Statistics* **17**, 1662–1683.

Lopuhaä, H.P. (1991). τ-estimators for location and scatter. *Canadian Journal of Statistics* **19**, 307–321.

Lubke, G.H., Dolan, C.V., Kelderman, H. (2003). Weak measurement invariance with respect to unmeasured variables: An implication of strict factorial invariance. *British Journal of Mathematical and Statistical Psychology* **56**, 231–248.

MacCallum, R.C. (2003). Working with imperfect models. *Multivariate Behavioral Research* **38**, 113–139.

MacCallum, R.C., Austin, J.T. (2000). Applications of structural equation modeling in psychological research. *Annual Review of Psychology* **51**, 201–226.

MacCallum, R.C., Browne, M.W., Sugawara, H.M. (1996). Power analysis and determination of sample size for covariance structure modeling. *Psychological Methods* **1**, 130–149.

MacCallum, R.C., Roznowski, M., Necowitz, L.B. (1992). Model modifications in covariance structure analysis: The problem of capitalization on chance. *Psychological Bulletin* **111**, 490–504.

MacCallum, R.C., Widaman, K.F., Zhang, S., Hong, S. (1999). Sample size in factor analysis. *Psychological Methods* **4**, 84–99.

Magnus, J.R., Neudecker, H. (1999). *Matrix Differential Calculus With Applications in Statistics and Econometrics*. Wiley, New York.

Mardia, K.V. (1970). Measures of multivariate skewness and kurtosis with applications. *Biometrika* **57**, 519–530.

Maronna, R.A. (1976). Robust M-estimators of multivariate location and scatter. *Annals of Statistics* **4**, 51–67.

Marsh, H.W., Wen, Z., Hau, K.T. (2004). Structural equation models of latent interactions: Evaluation of alternative estimation strategies and indicator construction. *Psychological Methods* **9**, 275–300.

Maydeu-Olivares, A. (2001). Multidimensional item response theory modeling of binary data: Large sample properties of NOHARM estimates. *Journal of Educational and Behavioral Statistics* **26**, 51–71.

Maydeu-Olivares, A. (2005). Linear IRT, nonlinear IRT, and factor analysis: A unified framework. In: Maydeu-Olivares, A., McArdle, J.J. (Eds.), *Contemporary Psychometrics*. Erlbaum, Mahwah, NJ.

McArdle, J.J., Epstein, D. (1987). Latent growth curves with developmental structure equation models. *Child Development* **58**, 110–133.

McDonald, R.P. (2004). Respecifying improper structures. *Structural Equation Modeling* **11**, 194–209.

McDonald, R.P., Goldstein, H. (1989). Balanced versus unbalanced designs for linear structural relations in two-level data. *British Journal of Mathematical and Statistical Psychology* **42**, 215–232.

Meredith, W. (1993). Measurement invariance, factor analysis and factorial invariance. *Psychometrika* **58**, 525–543.

Meredith, W., Tisak, J. (1990). Latent curve analysis. *Psychometrika* **55**, 107–122.

Micceri, T. (1989). The unicorn, the normal curve, and other improbable creatures. *Psychological Bulletin* **105**, 156–166.

Millsap, R.E. (1997). Invariance in measurement and prediction: Their relationship in the single-factor case. *Psychological Methods* **2**, 248–260.

Millsap, R.E. (1998). Group differences in regression intercepts: Implications for factorial invariance. *Multivariate Behavioral Research* **33**, 403–424.

Millsap, R.E. (2001). When trivial constraints are not trivial: The choice of uniqueness constraints in confirmatory factor analysis. *Structural Equation Modeling* **8**, 1–17.
Mooijaart, A. (1985). Factor analysis for non-normal variables. *Psychometrika* **50**, 323–342.
Mooijaart, A. (2003). Estimating the statistical power in small samples by empirical distributions. In: Yanai, H., Okada, A., Shigemasu, K., Kano, Y., Meulman, J.J. (Eds.), *New Development in Psychometrics*. Springer-Verlag, Tokyo, pp. 149–156.
Mooijaart, A., Bentler, P.M. (1991). Robustness of normal theory statistics in structural equation models. *Statistica Neerlandica* **45**, 159–171.
Muirhead, R.J., Waternaux, C.M. (1980). Asymptotic distributions in canonical correlation analysis and other multivariate procedures for nonnormal populations. *Biometrika* **67**, 31–43.
Muthén, B. (1984). A general structural equation model with dichotomous, ordered categorical, and continuous latent variable indicators. *Psychometrika* **49**, 115–132.
Muthén, B. (1989a). Multiple group structural modelling with nonnormal continuous variables. *British Journal of Mathematical and Statistical Psychology* **42**, 55–62.
Muthén, B. (1989b). Latent variable modeling in heterogeneous populations. *Psychometrika* **54**, 557–585.
Muthén, B. (1994). Multilevel covariance structure analysis. *Sociological Methods and Research* **22**, 376–398.
Muthén, B. (1997). Latent variable modeling of longitudinal and multilevel data. *Sociological Methodology* **27**, 453–480.
Muthén, B. (2001). Latent variable mixture modeling. In: Marcoulides, G.A., Schumacker, R.E. (Eds.), *New Developments and Techniques in Structural Equation Modeling*. Erlbaum, Mahwah, NJ, pp. 1–34.
Muthén, B. (2002). Beyond SEM: General latent variable modeling. *Behaviormetrika* **29**, 81–117.
Muthén, B., Kaplan, D., Hollis, M. (1987). On structural equation modeling with data that are not missing completely at random. *Psychometrika* **52**, 431–462.
Muthén, L.K., Muthén, B. (2002). How to use a Monte Carlo study to decide on sample size and determine power. *Structural Equation Modeling* **9**, 599–620.
Neumann, C.S. (1994). *Structural equation modeling of symptoms of alcoholism and psychopathology*. Dissertation. University of Kansas, Lawrence, KS.
Olsson, U. (1979). On the robustness of factor analysis against crude classification of the observations. *Multivariate Behavioral Research* **14**, 485–500.
Pitman, E.J.G. (1948). *Lecture Notes on Nonparametric Statistical Inference*. Columbia University Press, New York.
Poon, W.-Y., Lee, S.Y. (1987). Maximum likelihood estimation of multivariate polyserial and polychoric correlation coefficient. *Psychometrika* **52**, 409–430.
Poon, W.-Y., Lee, S.-Y. (1994). A distribution free approach for analysis of two-level structural equation model. *Computational Statistics and Data Analysis* **17**, 265–275.
Poon, W.Y., Lee, S.Y., Tang, M.L. (1997). Analysis of structural equation models with censored data. *British Journal of Mathematical and Statistical Psychology* **50**, 227–241.
Poon, W.Y., Lew, S.F., Poon, Y.S. (2000). A local influence approach to identify multiple multivariate outliers. *British Journal of Mathematical and Statistical Psychology* **53**, 255–273.
Poon, W.Y., Poon, Y.S. (2002). Influential observations in the estimation of mean vector and covariance matrix. *British Journal of Mathematical and Statistical Psychology* **55**, 177–192.
Poon, W.Y., Wong, Y.K. (2004). A forward search procedure to identifying influential observations in the estimation of covariance matrix. *Structural Equation Modeling* **11**, 357–374.
Rao, C.R. (1947). Large sample tests of statistical hypotheses concerning several parameters with applications to problems of estimation. *Proceedings of the Cambridge Philosophical Society* **44**, 50–57.
Rao, C.R. (1951). An asymptotic expansion of the distribution of Wilks' Λ criterion. *Bulletin of the International Statistical Institute* **33** (part II), 177–280.
Rao, C.R. (1973). *Linear Statistical Inference and its Applications*, 2nd ed. Wiley, New York.
Raudenbush, S.W., Bryk, A.S. (2002). *Hierarchical linear models*, 2nd ed. Sage, Newbury Park, CA.
Raykov, T. (2005). Bias-corrected estimation of noncentrality parameters of covariance structure models. *Structural Equation Modeling* **12**, 120–129.
Raykov, T., Marcoulides, G.A. (2001). Can there be infinitely many models equivalent to a given covariance structure model. *Structural Equation Modeling* **8**, 142–149.

Rindskopf, D. (1984). Structural equation models: Empirical identification, Heywood cases, and related problems. *Sociological Methods and Research* **13**, 109–119.

Rocke, D.M. (1996). Robustness properties of S-estimators of multivariate location and shape in high dimension. *Annals of Statistics* **24**, 1327–1345.

Rotnitzky, A., Wypij, D. (1994). A note on the bias of estimators with missing data. *Biometrics* **50**, 1163–1170.

Rousseeuw, P.J., van Zomeren, B.C. (1990). Unmasking multivariate outliers and leverage points (with discussion). *Journal of the American Statistical Association* **85**, 633–651.

Rubin, D.B. (1976). Inference and missing data (with discussions). *Biometrika* **63**, 581–592.

Rudin, W. (1976). *Principles of mathematical analysis*, 3rd ed. McGraw-Hill, New York.

Ruppert, D. (1992). Computing S estimators for regression and multivariate location/dispersion. *Journal of Computational and Graphical Statistics* **1**, 253–270.

Satorra, A. (1989). Alternative test criteria in covariance structure analysis: A unified approach. *Psychometrika* **54**, 131–151.

Satorra, A. (1992). Asymptotic robust inferences in the analysis of mean and covariance structures. *Sociological Methodology* **22**, 249–278.

Satorra, A. (2000). Scaled and adjusted restricted tests in multi-sample analysis of moment structures. In: Heijmans, D.D.H., Pollock, D.S.G., Satorra, A. (Eds.), *Innovations in Multivariate Statistical Analysis: A Festschrift for Heinz Neudecker*. Kluwer Academic, Dordrecht, pp. 233–247.

Satorra, A. (2003). Power of two goodness-of-fit tests in structural equation models: The case of non-normal data. In: Yanai, H., Okada, A., Shigemasu, K., Kano, Y., Meulman, J.J. (Eds.), *New Developments in Psychometrics*. Springer-Verlag, Tokyo, pp. 57–68.

Satorra, A., Bentler, P.M. (1986). Some robustness properties of goodness of fit statistics in covariance structure analysis. In: *American Statistical Association 1986 Proceedings of Business and Economics Sections*. American Statistical Association, pp. 549–554.

Satorra, A., Bentler, P.M. (1988). Scaling corrections for chi-square statistics in covariance structure analysis. In: *American Statistical Association 1988 Proceedings of Business and Economics Sections*. American Statistical Association, pp. 308–313.

Satorra, A., Bentler, P.M. (1990). Model conditions for asymptotic robustness in the analysis of linear relations. *Computational Statistics and Data Analysis* **10**, 235–249.

Satorra, A., Bentler, P.M. (1994). Corrections to test statistics and standard errors in covariance structure analysis. In: von Eye, A., Clogg, C.C. (Eds.), *Latent Variables Analysis: Applications for Developmental Research*. Sage, Thousand Oaks, CA, pp. 399–419.

Satorra, A., Bentler, P.M. (2001). A scaled difference chi-square test statistic for moment structure analysis. *Psychometrika* **66**, 507–514.

Satorra, A., Saris, W. (1985). Power of the likelihood ratio test in covariance structure analysis. *Psychometrika* **50**, 83–90.

Savalei, V., Bentler, P.M. (2005). A statistically justified pairwise ML method for incomplete nonnormal data: A comparison with direct ML and pairwise ADF. *Structural Equation Modeling* **12**, 183–214.

Shapiro, A. (1983). Asymptotic distribution theory in the analysis of covariance structures (a unified approach). *South African Statistical Journal* **17**, 33–81.

Shapiro, A. (1984). A note on the consistency of estimators in the analysis of moment structures. *British Journal of Mathematical and Statistical Psychology* **37**, 84–88.

Shapiro, A., Browne, M. (1987). Analysis of covariance structures under elliptical distributions. *Journal of the American Statistical Association* **82**, 1092–1097.

Skrondal, A., Rabe-Hesketh, S. (2004). *Generalized Latent Variable Modeling: Multilevel, Longitudinal and Structural Equation Models*. Chapman & Hall, London.

Song, X.-Y., Lee, S.-Y. (2002). Analysis of structural equation model with ignorable missing continuous and polytomous data. *Psychometrika* **67**, 261–288.

Sörbom, D. (1974). A general method for studying differences in factor means and factor structures between groups. *British Journal of Mathematical and Statistical Psychology* **27**, 229–239.

Sörbom, D. (1989). Model modification. *Psychometrika* **54**, 371–384.

Spearman, C. (1904). “General intelligence” objectively determined and measured. *American Journal of Psychology* **15**, 201–293.

Steiger, J.H. (2002). When constraints interact: A caution about reference variables, identification constraints, and scale dependencies in structural equation modeling. *Psychological Methods* **7**, 210–227.

Steiger, J.H., Lind, J.M. (1980). Statistically based tests for the number of common factors. Paper presented at the Annual Meeting of the Psychometric Society. Iowa City, IA.

Steiger, J.H., Shapiro, A., Browne, M.W. (1985). On the multivariate asymptotic distribution of sequential chi-square statistics. *Psychometrika* **50**, 253–264.

Stoel, R.D. (2003). *Issues in Growth Curve Modeling*. TT-Publikaties, Amsterdam.

Stroud, T.W.F. (1972). Fixed alternatives and Wald's formulation of the noncentral asymptotic behavior of the likelihood ratio statistic. *Annals of Mathematical Statistics* **43**, 447–454.

Sugiura, N. (1969). Asymptotic expansions of the distributions of the likelihood ratio criteria for covariance matrix. *Annals of Mathematical Statistics* **40**, 2051–2063.

Takane, Y., de Leeuw, J. (1987). On the relationship between item response theory and factor analysis of discretized variables. *Psychometrika* **52**, 393–408.

Tang, M., Bentler, P.M. (1998). Theory and method for constrained estimation in structural equation models with incomplete data. *Computational Statistics and Data Analysis* **27**, 257–270.

Tang, M.L., Lee, S.Y. (1998). Analysis of structural equation model with non-ignorable missing data. *Computational Statistics and Data Analysis* **27**, 33–46.

Thurstone, L.L. (1947). *Multiple Factor Analysis*. University of Chicago Press, Chicago.

Tyler, D.E. (1983). Robustness and efficiency properties of scatter matrices. *Biometrika* **70**, 411–420.

van Driel, O.P. (1978). On various causes of improper solutions in maximum likelihood factor analysis. *Psychometrika* **43**, 225–243.

Vuong, Q.H. (1989). Likelihood ratio tests for model selection and nonnested hypotheses. *Econometrica* **57**, 307–333.

Wald, A. (1943). Tests of statistical hypotheses concerning several parameters when the number of observations is large. *Transactions of the American Mathematical Society* **54**, 426–482.

Wall, M.M., Amemiya, Y. (2001). Generalized appended product indicator procedure for nonlinear structural equation analysis. *Journal of Educational and Behavioral Statistics* **26**, 1–29.

Wansbeek, T., Meijer, E. (2000). *Measurement Error and Latent Variables in Econometrics*. Elsevier, Amsterdam.

White, H. (1982). Maximum likelihood estimation of misspecified models. *Econometrica* **50**, 1–25.

Wiley, D. (1973). The identification problem for structural equation models with unmeasured variables. In: Goldberger, A., Duncan, O. (Eds.), *Structural Equation Models in the Social Sciences*. Academic Press, New York, pp. 69–83.

Wolfle, L.M. (2003). The introduction of path analysis to the social sciences, and some emergent themes: An annotated bibliography. *Structural Equation Modeling* **10**, 1–34.

Yanagihara, H., Tonda, T., Matsumoto, C. (2005). The effects of nonnormality on asymptotic distributions of some likelihood ratio criteria for testing covariance structures under normal assumption. *Journal of Multivariate Analysis* **96**, 237–264.

Yang-Wallentin, F., Marcoulides, G.A. (2001). Robust standard errors and chi-squares for interaction models. In: Schumacker, R.E., Marcoulides, G.A. (Eds.), *New Developments and Techniques in Structural Equation Modeling*. Lawrence Erlbaum, Mahwah, NJ, pp. 159–171.

Yuan, K.-H. (2001). Robust procedures for structural equation modeling with missing data. Invited talk at the International Meeting of the Psychometric Society. Osaka, Japan.

Yuan, K.-H. (2005a). Fit indices versus test statistics. *Multivariate Behavioral Research* **40**, 115–148.

Yuan, K.-H. (2005b). Normal theory ML for missing data with violation of distribution assumptions. Under review.

Yuan, K.-H., Bentler, P.M. (1997a). Mean and covariance structure analysis: Theoretical and practical improvements. *Journal of the American Statistical Association* **92**, 767–774.

Yuan, K.-H., Bentler, P.M. (1997b). Improving parameter tests in covariance structure analysis. *Computational Statistics and Data Analysis* **26**, 177–198.

Yuan, K.-H., Bentler, P.M. (1998a). Normal theory based test statistics in structural equation modeling. *British Journal of Mathematical and Statistical Psychology* **51**, 289–309.

Yuan, K.-H., Bentler, P.M. (1998b). Robust mean and covariance structure analysis. *British Journal of Mathematical and Statistical Psychology* **51**, 63–88.

Yuan, K.-H., Bentler, P.M. (1998c). Structural equation modeling with robust covariances. *Sociological Methodology* **28**, 363–396.

Yuan, K.-H., Bentler, P.M. (1999a). On normal theory and associated test statistics in covariance structure analysis under two classes of nonnormal distributions. *Statistica Sinica* **9**, 831–853.

Yuan, K.-H., Bentler, P.M. (1999b). On asymptotic distributions of normal theory MLE in covariance structure analysis under some nonnormal distributions. *Statistics and Probability Letters* **42**, 107–113.

Yuan, K.-H., Bentler, P.M. (1999c). F-tests for mean and covariance structure analysis. *Journal of Educational and Behavioral Statistics* **24**, 225–243.

Yuan, K.-H., Bentler, P.M. (2000a). Inferences on correlation coefficients in some classes of nonnormal distributions. *Journal of Multivariate Analysis* **72**, 230–248.

Yuan, K.-H., Bentler, P.M. (2000b). Three likelihood-based methods for mean and covariance structure analysis with nonnormal missing data. *Sociological Methodology* **30**, 167–202.

Yuan, K.-H., Bentler, P.M. (2000c). Robust mean and covariance structure analysis through iteratively reweighted least squares. *Psychometrika* **65**, 43–58.

Yuan, K.-H., Bentler, P.M. (2001a). A unified approach to multigroup structural equation modeling with nonstandard samples. In: Marcoulides, G.A., Schumacker, R.E. (Eds.), *Advanced Structural Equation Modeling: New Developments and Techniques*. Lawrence Erlbaum Associates, Mahwah, NJ, pp. 35–56.

Yuan, K.-H., Bentler, P.M. (2001b). Effect of outliers on estimators and tests in covariance structure analysis. *British Journal of Mathematical and Statistical Psychology* **54**, 161–175.

Yuan, K.-H., Bentler, P.M. (2002a). On robustness of the normal-theory based asymptotic distributions of three reliability coefficient estimates. *Psychometrika* **67**, 251–259.

Yuan, K.-H., Bentler, P.M. (2002b). On normal theory based inference for multilevel models with distributional violations. *Psychometrika* **67**, 539–561.

Yuan, K.-H., Bentler, P.M. (2003). Eight test statistics for multilevel structural equation models. *Computational Statistics and Data Analysis* **44**, 89–107.

Yuan, K.-H., Bentler, P.M. (2004a). On chi-square difference and z tests in mean and covariance structure analysis when the base model is misspecified. *Educational and Psychological Measurement* **64**, 737–757.

Yuan, K.-H., Bentler, P.M. (2004b). On the asymptotic distributions of two statistics for two-level covariance structure models within the class of elliptical distributions. *Psychometrika* **69**, 437–457.

Yuan, K.-H., Bentler, P.M. (2005). Asymptotic robustness of the normal theory likelihood ratio statistic for two-level covariance structure models. *Journal of Multivariate Analysis* **94**, 3287–3343.

Yuan, K.-H., Bentler, P.M. (2006a). Mean comparison: Manifest variable versus latent variable. *Psychometrika* **71**, 139–159.

Yuan, K.-H., Bentler, P.M. (2006b). Asymptotic robustness of standard errors in multilevel structural equations models. *Journal of Multivariate Analysis* **97**, 1121–1141.

Yuan, K.-H., Bentler, P.M., Chan, W. (2004). Structural equation modeling with heavy tailed distributions. *Psychometrika* **69**, 421–436.

Yuan, K.-H., Chan, W. (2002). Fitting structural equation models using estimating equations: A model segregation approach. *British Journal of Mathematical and Statistical Psychology* **55**, 41–62.

Yuan, K.-H., Chan, W. (2005). On nonequivalence of several procedures of structural equation modeling. *Psychometrika* **70**, 791–798.

Yuan, K.-H., Chan, W., Bentler, P.M. (2000). Robust transformation with applications to structural equation modeling. *British Journal of Mathematical and Statistical Psychology* **53**, 31–50.

Yuan, K.-H., Fung, W.K., Reise, S. (2004). Three Mahalanobis-distances and their role in assessing unidimensionality. *British Journal of Mathematical and Statistical Psychology* **57**, 151–165.

Yuan, K.-H., Hayashi, K. (2003). Bootstrap approach to inference and power analysis based on three statistics for covariance structure models. *British Journal of Mathematical and Statistical Psychology* **56**, 93–110.

Yuan, K.-H., Hayashi, K. (2005). On Muthén's maximum likelihood for two-level covariance structure models. *Psychometrika* **70**, 147–167.

Yuan, K.-H., Hayashi, K. (in press). Standard errors in covariance structure models: Asymptotics versus bootstrap. *British Journal of Mathematical and Statistical Psychology*.

Yuan, K.-H., Hayashi, K., Bentler, P.M. (2005). Normal theory likelihood ratio statistic for mean and covariance structure analysis under alternative hypotheses. Under review.

Yuan, K.-H., Jennrich, R.I. (1998). Asymptotics of estimating equations under natural conditions. *Journal of Multivariate Analysis* **65**, 245–260.

Yuan, K.-H., Marshall, L.L. (2004). A new measure of misfit for covariance structure models. *Behaviormetrika* **31**, 67–90.

Yuan, K.-H., Marshall, L.L., Bentler, P.M. (2002). A unified approach to exploratory factor analysis with missing data, nonnormal data, and in the presence of outliers. *Psychometrika* **67**, 95–122.

Yuan, K.-H., Marshall, L.L., Bentler, P.M. (2003). Assessing the effect of model misspecifications on parameter estimates in structural equation models. *Sociological Methodology* **33**, 241–265.

Yuan, K.-H., Marshall, L.L., Weston, R. (2002). Cross-validation through downweighting influential cases in structural equation modeling. *British Journal of Mathematical and Statistical Psychology* **55**, 125–143.

Yung, Y.-F. (1997). Finite mixtures in confirmatory factor-analysis models. *Psychometrika* **62**, 297–330.

Yung, Y.-F., Bentler, P.M. (1996). Bootstrapping techniques in analysis of mean and covariance structures. In: Marcoulides, G.A., Schumacker, R.E. (Eds.), *Advanced Structural Equation Modeling: Techniques and Issues*. Lawrence Erlbaum, Hillsdale, NJ, pp. 195–226.

Handbook of Statistics, Vol. 26
ISSN: 0169-7161

DOI: 10.1016/S0169-7161(06)26011-5

11

Applications of Multidimensional Scaling in Psychometrics

Yoshio Takane

1. Introduction

Multidimensional scaling (MDS) is a set of data analysis techniques for analysis of data. Both narrow and broad definitions of MDS exist. Because of space limitation, we take a narrow view of MDS in this paper. According to this view, MDS is a collection of techniques that represent proximity data by spatial distance models. The space is usually Euclidean, although other forms of distance functions (e.g., the city-block distance model) have also been considered in the literature. Again due to space limitation, we focus on the Euclidean case in this paper. See Arabie (1991), and Hubert et al. (1992) for applications of the city-block distance model.

MDS represents a set of stimuli as points in a multidimensional space in such a way that those points corresponding to similar stimuli are located close together, while those corresponding to dissimilar stimuli are located far apart. To illustrate, many road maps have a matrix of inter-city distances. Put simply, MDS recovers a map based on the set of inter-city distances. Given a map it is relatively straightforward to measure the Euclidean distance between the cities. However, the reverse operation, that of recovering a map from inter-city distances (or finding relative locations of the cities so that their mutual distances best agree with a given set of distances) is not as easy. MDS, roughly speaking, is a method to perform this reverse operation.

To illustrate further, let us look at Table 1. This is a matrix of airplane distances between 10 US cities (Kruskal and Wish, 1978). Given a matrix of this sort it would be difficult to find geographic locations of the 10 cities. The story may be totally different, if the cities are identified as in Table 1. Most of us know that the 10 cities should be located as in Figure 1. This is because we already have a fairly good internalized map of North America. It would still be very difficult, if not impossible, for those who are unfamiliar with the geography of North America to construct a map like this based exclusively on the set of inter-city distances. The role of MDS is to construct a map like the one in Figure 1 from a table like Table 1 for those who do not know the "geography" in certain areas. Figure 1 was in fact derived from Table 1 by MDS. In the remaining portions of this paper I would like to elaborate this role of MDS. This will primarily be done by showing various examples of application of MDS.

Table 1
Flying Mileage between 10 US Cities (Kruskal and Wish, 1978)

	Atl	Chi	Den	Hou	LA	Mia	NY	SF	Sea	DC
Atlanta	0	587	1212	701	1936	604	748	2139	2182	543
Chicago	587	0	920	940	1745	1188	713	1858	1737	597
Denver	1212	920	0	879	831	1726	1631	949	1021	1494
Houston	701	940	879	0	1374	968	1420	1645	1891	1220
Los Angeles	1936	1745	831	1374	0	2339	2451	347	959	2300
Miami	604	1188	1726	968	2339	0	1092	2594	2734	923
New York	748	713	1631	1420	2451	1092	0	2571	2408	205
San Franc.	2139	1858	949	1645	347	2594	2571	0	678	2442
Seattle	2182	1737	1021	1891	959	2734	2408	678	0	2329
Wash. DC	543	597	1494	1220	2300	923	205	2442	2329	0

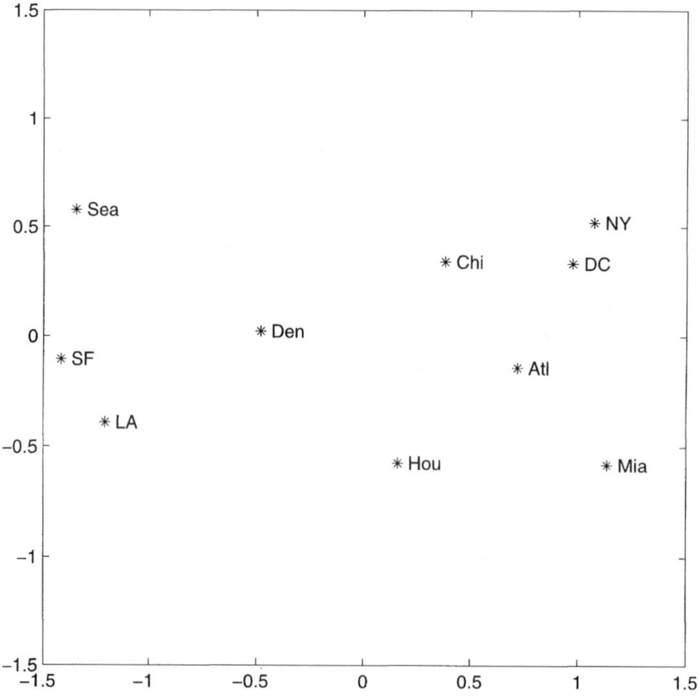

Fig. 1. The map of 10 US cities derived from flying mileage.

This paper consists of three major sections. In the next section (Section 2) we give examples of MDS by the simple Euclidean distance model. In Section 3 we discuss applications of MDS by the weighted Euclidean distance model to capture certain individual differences in proximity judgments. In Section 4 we illustrate applications of unfolding analysis, a special kind of MDS to represent individual differences in preference judgments. Throughout this paper, we keep the amount of technical information minimal. Readers who are interested in a more comprehensive account of MDS are referred to the textbooks listed in the conclusion section of this paper.

2. MDS with the simple Euclidean model

In this section, we present some results from MDS with the simple Euclidean model. Simple MDS is typically applied to a single (dis)similarity matrix. We first present a series of examples of application, followed by a brief account of simple MDS. Most of the results in this section were derived by least squares (LS) MDS. In the final subsection, however, we discuss maximum likelihood (ML) MDS, and illustrate what it can do in the analysis of proximity data.

2.1. Examples of application of simple MDS

EXAMPLE 1 Confusion among Morse Code signals.

Although the example in the previous section involved stimuli (cities) which had an intrinsic geometrical structure, this is not the case in most applications of MDS. Let us first look at Figure 2. This is called confusion matrix (Rothkopf, 1957). Stimuli are 36 Morse Code signals. Two signals were presented in each trial, and subjects were asked to judge whether the two stimuli presented were the "same" or "different". (This is called same–different judgment.) The relative frequencies with which row stimuli were judged the "same" as column stimuli were entered in this table. For example, when the signal indicating letter A was followed by the same signal, 92% of the subjects correctly responded that they were the same. When it was followed by the signal indicating letter K, 22% incorrectly responded that they were identical, and so on. This is an overwhelming array of numbers (36 by 36), and it is almost impossible to find any structure underlying the confusion process by merely inspecting a table like this. MDS is helpful in finding some meaningful structure in the data.

Figure 3 presents the result of MDS of these data (Shepard, 1963). MDS locates the stimulus points in such a way that more confusable (or more similar) stimuli are close together, while less confusable stimuli are far apart. By looking at the figure, it is readily apparent that:

1. A process mediating the confusions between the signals is roughly two-dimensional.
2. One is the total number of components in a signal (i.e., the number of dots plus the number of dashes), and the other is the ratio of the number of dots to the number of dashes in a signal (that is, which type of components is more predominant).

Signals having more components tend to be located toward the top of the figure. Signals with the same number of components are more confusable than those having different numbers of components. Within the same number of components, symbols on the right have more dashes and fewer dots than those on the left. Signals with the same number of dots and the same number of dashes are relatively more confusable, particularly when

	A	B	C	D	E	F	G	H	I	J	K	L	M	N	O	P	Q	R	S	T	U	V	W	X	Y	Z	1	2	3	4	5	6	7	8	9	0	
A	92	04	06	13	03	14	10	13	46	05	22	03	25	34	06	06	09	35	23	06	37	13	17	12	07	03	02	07	05	05	08	06	05	06	02	03	A
B	05	84	37	31	05	28	17	21	05	19	34	40	06	10	12	22	25	16	18	02	18	34	08	84	30	42	12	17	14	40	32	74	43	17	04	04	B
C	04	38	87	17	04	29	13	07	11	19	24	35	14	03	09	51	34	24	14	06	06	11	14	32	82	38	13	15	31	14	10	30	28	24	18	12	C
D	08	62	17	88	07	23	40	36	09	13	81	56	08	07	09	27	09	45	29	06	17	20	27	40	15	33	03	09	06	11	09	19	08	10	05	06	D
E	06	13	14	06	97	02	04	04	17	01	05	06	04	04	05	01	05	10	07	67	03	03	02	05	06	05	04	03	05	03	05	02	04	02	03	03	E
F	04	51	33	19	02	90	10	29	05	33	16	50	07	06	10	42	12	35	14	02	21	27	25	19	27	13	08	16	47	25	26	24	21	05	05	05	F
G	09	18	27	38	01	14	90	06	05	22	33	16	14	13	82	52	23	21	05	03	15	14	32	21	23	39	15	14	05	10	04	10	17	23	20	11	G
H	03	45	23	25	09	32	08	87	10	10	09	29	05	08	08	14	08	17	37	04	36	59	09	33	14	11	03	09	15	43	70	35	17	04	03	03	H
I	64	07	07	13	10	08	06	12	93	03	05	16	13	30	07	03	05	19	35	16	10	05	08	02	05	07	02	05	08	09	06	08	05	02	04	05	I
J	07	09	38	09	02	24	18	05	04	85	22	31	08	03	21	63	47	11	02	07	09	09	09	22	32	28	67	66	33	15	07	11	28	29	26	23	J
K	05	24	38	73	01	17	25	11	05	27	91	33	10	12	31	14	31	22	02	02	23	17	33	63	16	18	05	09	17	08	08	18	14	13	05	06	K
L	02	69	43	45	10	24	12	26	09	30	27	86	06	02	09	37	36	28	12	05	16	19	20	31	25	59	12	13	17	15	26	29	36	16	07	03	L
M	24	12	05	14	07	17	29	08	08	11	23	08	96	62	11	10	15	20	07	09	13	04	21	09	18	08	05	07	06	06	05	07	11	07	10	04	M
N	31	04	13	30	08	12	10	16	13	03	16	08	59	93	05	09	05	28	12	10	16	04	12	04	06	11	05	02	03	04	04	06	02	02	10	02	N
O	07	07	20	06	05	09	76	07	02	39	26	10	04	08	86	37	35	10	03	04	11	14	25	35	27	27	19	17	07	07	06	18	14	11	20	12	O
P	05	22	33	12	05	36	22	12	03	78	14	46	05	06	21	83	43	23	09	04	12	19	19	19	41	30	34	44	24	11	15	17	24	23	25	13	P
Q	08	20	38	11	04	15	10	05	02	27	23	26	07	06	22	51	91	11	02	03	06	14	12	37	50	63	34	32	17	12	09	27	40	58	37	24	Q
R	13	14	16	23	05	34	26	15	07	12	21	37	14	12	12	29	08	87	16	02	23	23	62	14	12	13	07	10	13	04	07	12	07	09	01	02	R
S	17	24	05	30	11	26	05	59	16	03	13	10	05	17	06	06	03	18	96	09	56	24	12	10	06	07	08	02	02	15	28	09	05	05	05	02	S
T	13	10	01	05	46	03	06	06	14	06	14	07	06	05	06	11	04	04	07	96	08	05	04	02	02	06	05	05	03	03	03	08	07	06	14	06	T
U	14	29	12	32	04	32	11	34	21	07	44	32	11	13	06	20	12	40	51	06	93	57	34	17	09	11	06	06	16	34	10	09	09	07	04	03	U
V	05	17	24	16	09	29	06	39	05	11	26	43	04	01	09	17	10	17	11	06	32	92	17	57	35	10	10	14	28	79	44	36	25	10	01	05	V
W	09	21	30	22	09	36	25	15	04	25	29	18	15	06	26	20	25	61	12	04	19	20	86	22	25	22	10	22	19	16	05	09	11	06	03	07	W
X	07	64	45	19	03	28	11	06	01	35	50	42	10	08	24	32	61	10	12	03	12	17	21	91	48	26	12	20	24	27	16	57	29	16	17	06	X
Y	09	23	62	15	04	26	22	09	01	30	12	14	05	06	14	30	52	05	07	04	06	13	21	44	86	23	26	44	40	15	11	26	22	33	23	16	Y
Z	03	46	45	18	02	22	17	10	07	23	21	51	11	02	15	59	72	14	04	03	09	11	12	36	42	87	16	21	27	09	10	25	66	47	15	15	Z
1	02	05	10	03	03	05	13	04	02	29	05	14	09	07	14	30	28	09	04	02	03	12	14	17	19	22	84	63	13	08	10	08	19	32	57	55	1
2	07	14	22	05	04	20	13	03	25	26	09	14	02	03	17	37	28	06	05	03	06	10	11	17	30	13	62	89	54	20	05	14	20	21	16	11	2
3	03	08	21	05	04	32	06	12	02	23	06	13	05	02	05	37	19	09	07	06	04	16	06	22	25	12	18	64	86	31	23	41	16	17	08	10	3
4	06	19	19	12	06	25	14	16	07	21	13	19	03	03	02	17	29	11	09	03	17	55	08	37	24	03	05	26	44	89	42	44	32	10	03	03	4
5	08	45	15	14	02	45	04	67	07	14	04	41	02	00	04	13	07	09	27	02	14	45	07	45	10	10	14	10	30	69	90	42	24	10	06	05	5
6	07	80	30	17	04	23	04	14	02	11	11	27	06	02	07	16	30	11	14	03	12	30	09	58	38	39	15	14	26	24	17	86	69	14	05	14	6
7	06	33	22	14	05	25	06	04	06	24	13	32	07	06	07	36	39	12	06	02	03	13	09	30	30	50	22	29	18	15	12	61	85	70	20	13	7
8	03	23	40	06	03	15	15	06	02	33	10	14	03	06	14	12	45	02	06	04	06	07	05	24	35	50	42	29	16	16	09	30	60	89	61	26	8
9	03	14	23	03	01	06	14	05	02	30	06	07	16	11	10	31	32	05	06	07	06	03	08	11	21	24	57	39	09	12	04	11	42	56	91	78	9
0	09	03	11	02	05	07	14	04	05	30	08	03	02	03	25	21	29	02	03	04	05	03	02	12	15	20	50	26	09	11	05	22	17	52	81	94	0
	A	B	C	D	E	F	G	H	I	J	K	L	M	N	O	P	Q	R	S	T	U	V	W	X	Y	Z	1	2	3	4	5	6	7	8	9	0	

Fig. 2. Confusion data for 36 Morse Code signals (Rothkopf, 1957). (Reproduced from Shepard (1963) with permission from the Humman Factors and Ergonomic Society.)

they are mirror images of one another. Two signals are said to be mirror images when they have exactly the same number of components of each kind, and the components are arranged in exactly reverse orders. In the figure, stimuli that are mirror images of each other are connected by line segments. The configuration is more crowded toward the top. This is because a more variety of signals can be formed with more components. A more variety of signals in turn lead to more confusions.

The role of MDS in this example is rather clear. MDS extracts useful information from an overwhelming array of proximity data in a way easily assimilable to human eyes. This example is perhaps the single most famous example in MDS, and readers may have seen it elsewhere. Buja and his collaborators (Buja and Swayne, 2002; Buja et al., 2001) present further analyses of this data set by XGvis, a very flexible computer program for simple MDS with interactive visualization features.

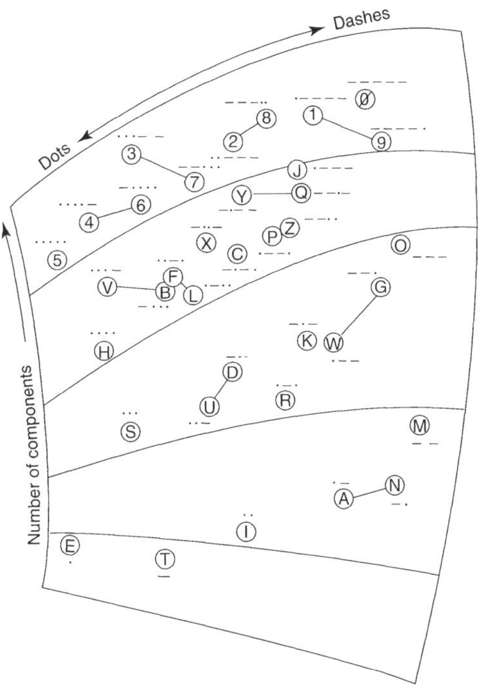

Fig. 3. A stimulus configuration derived from Morse code signal confusion data (Shepard, 1963). (Reproduced with permission from the Humman Factors and Ergonomic Society.)

EXAMPLE 2 Japanese Kana characters.

Figure 4 displays an MDS configuration of 46 Japanese Kana characters (phonetic symbols). Twenty students at the University of Hawaii who did not know Japanese at all were asked to classify these symbols into as many groups as they liked in terms of their similarities in shape (Dunn-Rankin and Leton, 1975). The frequencies with which two symbols were classified into the same groups were used as similarity measures between the symbols for MDS analysis.

Unlike the previous example, this configuration does not seem to permit a straightforward dimensional interpretation, though perhaps the horizontal direction roughly represents the complexity in shape. More complicated symbols tend to be located toward the left hand side of the configuration. The vertical axis is difficult to interpret, however. A cluster analysis, which groups similar stimuli together, was applied to the same set of data, and the groupings of the stimuli obtained by the cluster analysis were superposed on the stimulus configuration derived by MDS. Circular contours enclosing subsets of stimulus points indicate the groupings. Apparently another kind of interpretation (called configural interpretation) is possible of the derived stimulus configuration. (In a dimensional interpretation we seek for meaningful directions in the space, whereas in a configural or neighborhood interpretation we look for meaningful regions in the space.) Our interpretation of the six clusters obtained by the cluster analysis are as fol-

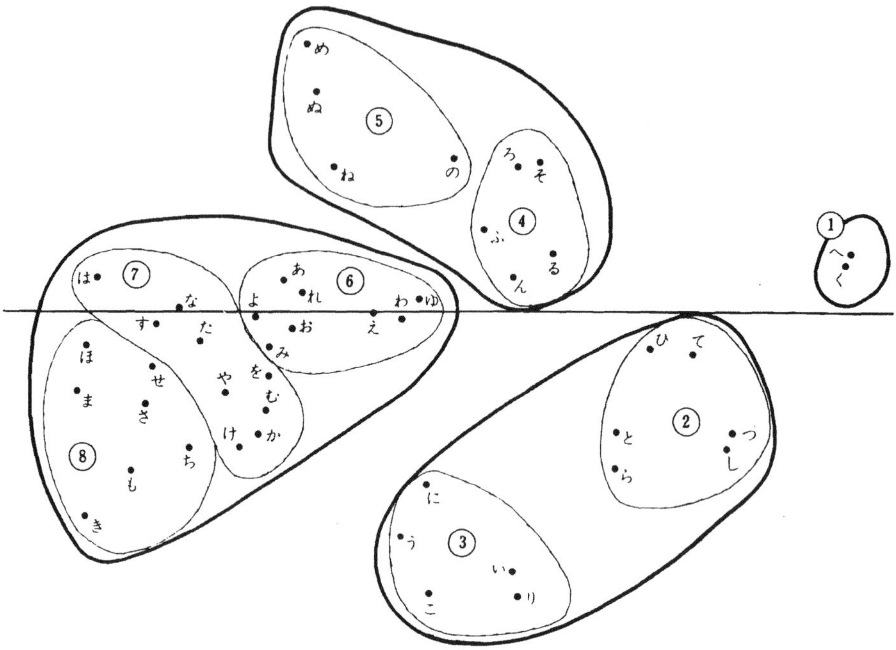

Fig. 4. An MDS configuration of Japanese Kana characters (phonetic symbols).

lows: 1. an angular form, 2. a curved feature, 3. discrete components, 4. a zigzag feature, 5. a round feature, and 6. crossed features. Of the six clusters, the sixth one is the biggest that includes three sub-clusters labelled (6), (7) and (8). Note that at the left end of this cluster, characters having a "double cross" are located. The organizing principle underlying perceived similarities between these symbols seems to be the "distinctive features" that subsets of symbols share in common, and that symbols in other clusters do not.

EXAMPLE 3 Have words.

In what is called stimulus sorting method, subjects are asked to sort a set of stimuli according to the similarity among them. Typically, a group of subjects are given a deck of cards on which stimuli are printed, and are asked to sort the cards into as many piles as they want, so that the stimuli within the same piles are more similar to each other than those grouped into distinct piles. Because of its simplicity, it is a very popular data collection method for similarity data among social scientists. The previous example (Example 2) used the sorting method. We give another example.

Stimuli were 29 Have words used originally by Rapoport and Fillenbaum (1972): 1. accept, 2. beg, 3. belong, 4. borrow, 5. bring, 6. buy, 7. earn, 8. find, 9. gain, 10. get, 11. get rid of, 12. give, 13. have, 14. hold, 15. keep, 16. lack, 17. lend, 18. lose, 19. need, 20. offer, 21. own, 22. receive, 23. return, 24. save, 25. sell, 26. steal, 27. take, 28. use, and 29. want. Ten university students were asked to classify them in terms of their similarity in meaning. A somewhat specialized MDS method (Takane, 1980) was applied to these data to derive the stimulus configuration displayed in Figure 5. (This specialized

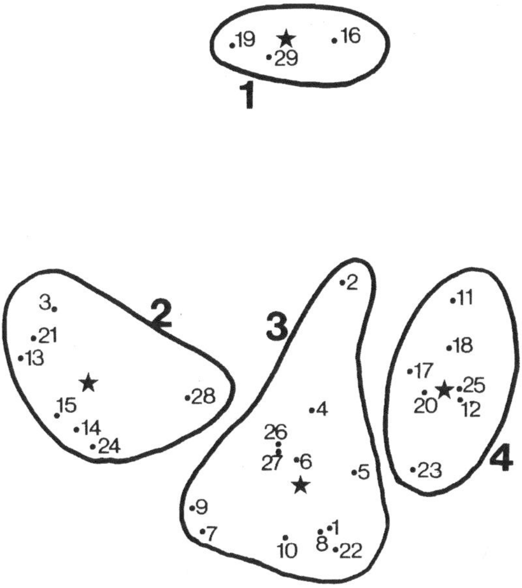

Fig. 5. A stimulus configuration of Have words with four sorting clusters elicited by one of the subjects (Takane, 1980). (Reproduced with permission from the Behaviormetric Society of Japan.)

MDS is analytically similar to dual scaling (Nishisato, 1980) and multiple correspondence analysis (Greenacre, 1984). See (Takane, 1980) for more details.)

This configuration permits a straightforward dimensional interpretation. The horizontal direction contrasts two possible future states, either possession or nonpossession, following the current state of possession. Words such as 13. have, 21. own, and 3. belong are located on the left side, while 12. give, 25. sell, 18. lose, etc. are placed on the opposite side. Thus, the horizontal direction contrasts a stable state of possession (left) with the state of losing possession (right). Similarly, the vertical direction distinguishes two states of current nonpossession, a stable state of nonpossession at the top and an unstable state of nonpossession at the bottom. It seems that the vertical direction represents subtle gradients of sureness of change in the current state. Words such as 16. lack, 19. need, and 29. want are located at the top, which indicate little prospect of change, while 22. receive, 10. get, 8. find, etc. are located at the bottom, which indicate that the change is most probable. Interestingly, 2. beg is located at about the middle, which indicates that some action has been taken to change the state, but the prospect of change is uncertain. Four circular contours in the figure show the four sorting clusters elicited by one of the subjects. The stars indicate the centroids of the four clusters. This subject is most likely to have sorted the stimuli into the four groups by following the dimensional structure described above.

EXAMPLE 4 Color space for the pigeon.

MDS is not restricted to human subjects. The next example shows an instance of MDS with infra-human subjects (Schneider, 1972). Six pigeons were trained to discriminate between colors varying primarily in hue. A pair of colors were presented as two

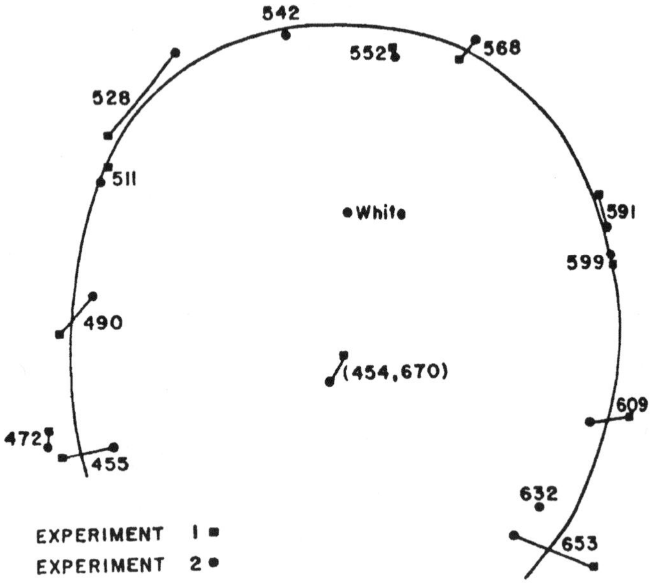

Fig. 6. A two-dimensional color space for the pigeon (Schneider, 1972). (Reproduced with permission from the Psychonomic Society.)

halves of a circle in each trial, and the pigeons were trained to peck the left lever when the two colors were identical, and peck the right lever when they are different. This is a kind of same–different judgment similar to the one used in the Morse Code confusion data. The correct discrimination probabilities between distinct colors were used as dissimilarity measures for MDS analysis. Two experiments were conducted with varying numbers of stimuli (12 in Experiment I, and 15 in Experiment II).

Figure 6 displays the derived stimulus configuration. In this figure, the results of the two experiments were superposed; the stimuli used in Experiment I are indicated by squares, and those in Experiment II by circular dots. Colors are labelled by their wave lengths. A familiar color wheel is observed in pigeon's color space. Starting from purple (associated with the shortest wave length) at the left bottom corner, the configuration goes clockwise through blue, green, and yellow to red (associated with the longest wave length) in a circular manner. For comparison, a color wheel typically found with human subjects is presented in Figure 7 (Ekman, 1954; Shepard, 1964). The two configurations are strikingly similar to each other. This means that pigeons have a color perceiving system similar to that of human subjects.

2.2. *A brief account of simple MDS*

Before proceeding further, a brief account of simple MDS is in order. In simple MDS, we are given a matrix of observed (dis)similarities data on a set of n stimuli. The data may be put in the form of an n by n matrix $\boldsymbol{O}$ whose ijth entry (the element in the ith row and the jth column), o_{ij}, designates the (dis)similarity between stimuli i and j.

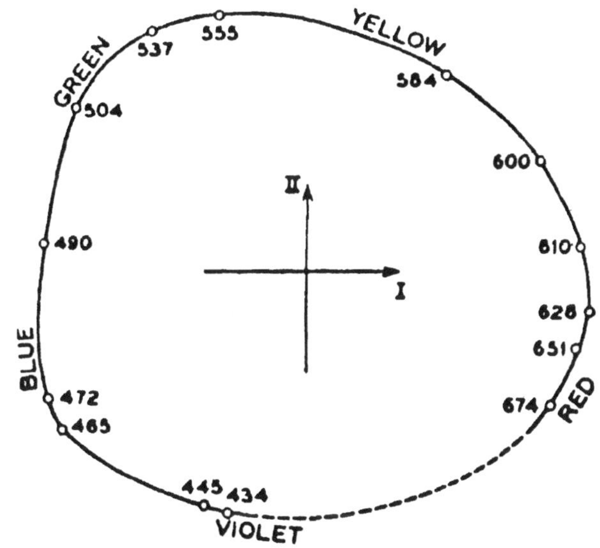

Fig. 7. A two-dimensional color space for humans derived by Shepard (1964) from the data collected by Ekman (1954). (Reproduced from Schneider (1972) with permission from the Psychonomic Society.)

By applying MDS to the data, we obtain an n by A matrix $\boldsymbol{X}$ of stimulus coordinates, where A is the dimensionality of the space in which the stimuli are represented as points. The iath element of $\boldsymbol{X}$, x_{ia}, indicates the coordinate of stimulus i on dimension a. The matrix of stimulus coordinates, $\boldsymbol{X}$, is further converted into a graphic representation (like the ones we saw in the previous subsection) by introducing a Cartesian coordinate system. Since the Euclidean distance is invariant over the shift of origin and the rotation of axes, we may remove the coordinate axes, once stimulus points are located.

Inter-point distances are calculated from $\boldsymbol{X}$. In the simple Euclidean model, the distance between stimuli i and j is defined by

$$d_{ij} = \left\{ \sum_{a=1}^{A} (x_{ia} - x_{ja})^2 \right\}^{1/2}. \tag{1}$$

MDS determines the locations of the stimulus points in such a way that the set of d_{ij} calculated from their coordinates best agree with the set of observed (dis)similarities between the stimuli.

When the observed dissimilarity data are measured on a ratio scale, there exists a method to obtain closed form solutions (Torgerson, 1952). This method first transforms the data into scalar (inner) products (Young and Householder, 1938), which are then subjected to the spectral (eigenvalue and vector) decomposition. However, dissimilarity data collected in psychology are rarely measured on a ratio scale, but only on an ordinal scale. This means that the observed dissimilarity data and inter-point distances are only monotonically related (or in the case of similarity data, inversely monotonically related) with each other. Nonmetric MDS, initiated by Shepard (1962) and Kruskal (1964a,

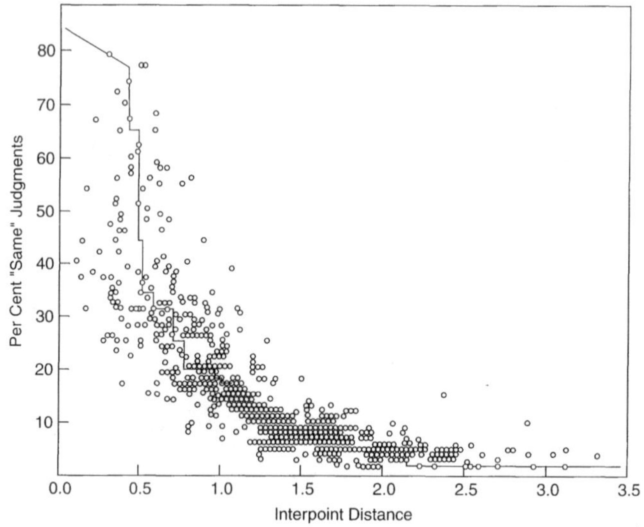

Fig. 8. Shepard's diagram and the best monotonic transformation of the observed confusion probabilities among Morse Code signals (Shepard, 1963). (Reproduced with permission from the Human Factors and Ergonomic Society.)

1964b), monotonically transforms the observed (dis)similarity data (to make them more in line with the underlying distances), and at the same time it finds the best spatial representation of the stimuli. Let g denote the monotonic transformation of the data. (The transformed data are often called "disparities".) Then, nonmetric MDS finds both g and a set of stimulus coordinates x_{ij} in such a way that

$$\phi(g, X) = \sum_{i,j>i} \left(g(o_{ij}) - d_{ij}\right)^2 \tag{2}$$

is minimized under a suitable normalization restriction on distances. All the examples given so far were analyzed by nonmetric MDS, except the one involving airplane distances, in which actual distances were used. When nonmetric MDS was applied to ranked airplane distances, we obtained a configuration virtually indistinguishable from the one derived from the actual distances. (The correlation between the two configurations was 0.999 with the largest displacement of a city of about 70 miles.) This indicates that the ordinal information is often sufficient to derive stable stimulus configurations.

Figure 8 presents the plot of observed confusion probabilities in Morse Code signal confusion data (Example 1) against fitted distances. (This is called Shepard diagram.) The connected line segments in the figure shows the best (inverse) monotonic transformation of the data obtained by a nonmetric MDS procedure. The transformation looks like a negative exponential function of distances. Of course, we had no a priori knowledge of the form of this function before MDS was applied. Nonmetric MDS can be carried out without knowledge of the exact form of the function relating observed (dis)similarity data to underlying distances. The transformation was derived solely on the basis of the (inverse) monotonicity relationship assumed between the data and the distances.

A number of computer programs exist for simple MDS. KYST (Kruskal et al., 1978) is still a good and reliable program for nonmetric MDS. Other available programs are described in Schiffman et al. (1981). See also other textbooks on MDS listed at the end of this article. PROXSCAL (Busing et al., 1997) employs a majorization algorithm for MDS (de Leeuw, 1977, 1988; de Leeuw and Heiser, 1977; Heiser, 1991). PROXSCAL is being incorporated in SPSS.

2.3. *Maximum likelihood MDS*

The next and final example in this section uses maximum likelihood (ML) MDS. This type of MDS allows certain kinds of hypothesis testing as well as assessment of reliability of derived stimulus configurations under some distributional assumptions on observed (dis)similarity data. We briefly discuss ML MDS in preparation for the next example. The exposition is rather terse. For more detailed accounts of ML MDS, see Ramsay (1977, 1978, 1982) and Takane (1978, 1981, 1994), Takane and Carroll (1981), Takane and Sergent (1983), Takane and Shibayama (1986, 1992).

There are three essential ingredients in ML MDS; the representation model, the error model, and the response model. The representation model is the model that represents proximity relations between stimuli. In MDS, this is typically a distance model like (1). Distances are then assumed to be error-perturbed. The error model describes the nature of this error-perturbation process. The error-perturbed distances are further transformed into specific forms of (dis)similarity judgments. The response model prescribes the mechanism of this transformation process. Because a variety of data collection methods exist that elicit a variety of proximity judgments, the response model must be specified for each specific type of data collection method.

Suppose, for simplicity, that we use (1) as the representation model, and that the distance is error-perturbed by

$$\delta_{ijr} = d_{ij} e_{ijr}, \tag{3}$$

where $\ln e_{ijr} \sim \mathrm{N}(0, \sigma^2)$, and r is the index of replication. Suppose further that proximity judgments are obtained on rating scales with a relatively small number of observation categories. We may assume that the dissimilarity between stimuli i and j is rated in category m whenever δ_{ijr} falls between the lower and the upper thresholds of this category, denoted by b_m and b_{m+1}, respectively. Then

$$p_{ijm} \equiv \Pr(b_m < \delta_{ijr} < b_{m+1}). \tag{4}$$

Under the distributional assumption made on δ_{ijr}, this probability can be explicitly evaluated as a function of model parameters ($\{x_{ia}\}$ in the representation model, σ^2 in the error model, and $\{b_m\}$ in the response model). Let f_{ijm} denote the observed frequency with which the dissimilarity between stimuli i and j is rated in category m. Then, the likelihood of the entire set of observed frequencies can be stated as

$$L \propto \prod_{i,j<i,m} p_{ijm}^{f_{ijm}}. \tag{5}$$

Parameters in the model are estimated so as to maximize the likelihood. For response models for other types of proximity judgments, see the papers cited above. We also see another example of response models in the final section of this paper.

The likelihood ratio (LR) statistic can be used to test the significance of the difference between two nested models, one hypothesized under the null hypothesis (H_0) and the other under the alternative hypothesis (H_1). The LR statistic, λ, is defined as

$$\lambda = \frac{L_0^*}{L_1^*}, \tag{6}$$

where L_0^* and L_1^* are the maximum likelihoods obtained under H_0 and H_1, respectively. Under some regularity conditions, $-2\ln\lambda$ asymptotically follows the chi-square distribution with degrees of freedom equal to the difference in the number of parameters between the two models, when the null hypothesis is true. Caution should be exercised, however, when the LR test is used for dimensionality selection. One of the regularity conditions is violated in the comparison between the A-dimensional true model and the $(A+1)$-dimensional model (Takane et al., 2003), and no direct comparison can be made between them. Solutions of specific dimensionality should always be compared against the saturated model. In the categorical rating data described above, the saturated model takes $f_{ijm}/\sum_m f_{ijm}$ as the maximum likelihood estimate of p_{ijm} in (4). The whole testing procedure proceeds as follows: First, the one-dimensional model is compared against the saturated model, and if the difference is significant, the dimensionality is at least two. Then the two-dimensional model is compared against the saturated model, and so on until a solution with no significant difference from the saturated model is found. The first dimensionality with no significant difference from the saturated model is considered the best dimensionality.

Alternatively, AIC may be used to compare goodness of fit (GOF) of various models. The AIC is defined by

$$\mathrm{AIC}_\pi = -2\ln L_\pi^* + n_\pi, \tag{7}$$

where L_π^* is the maximum likelihood of model π, and n_π is the effective number of parameters (Takane, 1994). The model associated with the smallest value of AIC is considered the best fitting model (the minimum AIC criterion). AIC may be used to compare any number of models simultaneously. To use AIC for dimensionality selection, the saturated model should always be included as one of the models in the comparison. In the example we discuss shortly, we mostly use AIC for model selection.

One potential problem with the LR tests or AIC is that it is based on the large sample behavior of the ML estimators. At the moment there is no practical guideline as to what sample size is big enough to rely on the asymptotic properties of the ML estimators. Some small sample behavior of the LR statistic has been investigated by Ramsay (1980).

The Moore–Penrose inverse of the information matrix evaluated at the maximum likelihood estimates of model parameters provides asymptotic variance–covariance estimates of the estimated parameters (Ramsay, 1978). The information matrix, $\boldsymbol{I}(\boldsymbol{\theta})$, is defined as

$$\boldsymbol{I}(\boldsymbol{\theta}) = \mathrm{E}\left[\left(\frac{\partial \ln L}{\partial \boldsymbol{\theta}}\right)\left(\frac{\partial \ln L}{\partial \boldsymbol{\theta}}\right)'\right], \tag{8}$$

Table 2
Summary statistics for MDS analyses of the rectangle data

Model	$-2 \ln L^*$	AIC	n_π
Saturated model	1127.9	2759.9	816
Unconstrained solutions			
3-dimensional	1543.1	1645.1	51
2-dimensional	1571.7	1645.7	37
Constrained solutions			
Area and shape hypothesis	2037.2	2065.2	14
Height and width hypothesis	1910.8	1946.8	18
Schönemann's hypothesis	1707.2	1737.2	15

where $\boldsymbol{\theta}$ is the parameter vector. Under the asymptotic normality of the maximum likelihood estimates, confidence regions may be drawn that indicate how reliably the point locations are estimated. We will see examples of confidence regions in the final section of this paper.

EXAMPLE 5 Similarity of rectangles.

Takane (1981) investigated the psychological dimensions of similarity between rectangles using ML MDS. The rectangles were constructed by factorial combinations of several levels each of the physical area and shape dimensions (Krantz and Tversky, 1975). A single subject made dissimilarity judgments on all pairs of rectangles on seven-point rating scales. This was replicated six times, and frequencies of category responses were cumulated and used as input to the ML MDS procedure.

Table 2 gives a summary of GOF statistics for fitted models. The unconstrained two- and three-dimensional solutions fit about equally well according to the minimum AIC criterion, both of which in turn fit substantially better than the saturated model. For ease of presentation, we display the two-dimensional configuration in Figure 9. (The three-dimensional solution is obtained by slightly bending the two-dimensional configuration along the third dimension.) In this figure, the horizontal direction roughly corresponds with the area dimension, while the vertical direction with the shape dimension. However, some qualifications are needed.

In the figure, rectangles having the same area are connected by line segments. If the horizontal direction indeed corresponds with the area dimension, the connected line segments should all be perpendicular to the horizontal direction. Essentially the same should hold for the vertical direction. We may formally test the validity of the area and shape hypothesis by comparing the GOF of this hypothesis against that of the unconstrained solution. Figure 10 presents the stimulus configuration derived under the area and shape hypothesis. In this figure, rectangles of the same area take equal coordinate values on the horizontal axis, and those with the same shape take equal coordinate values on the vertical axis. Table 2 indicates, however, that the GOF of this configuration is much worse than the unconstrained two-dimensional solution.

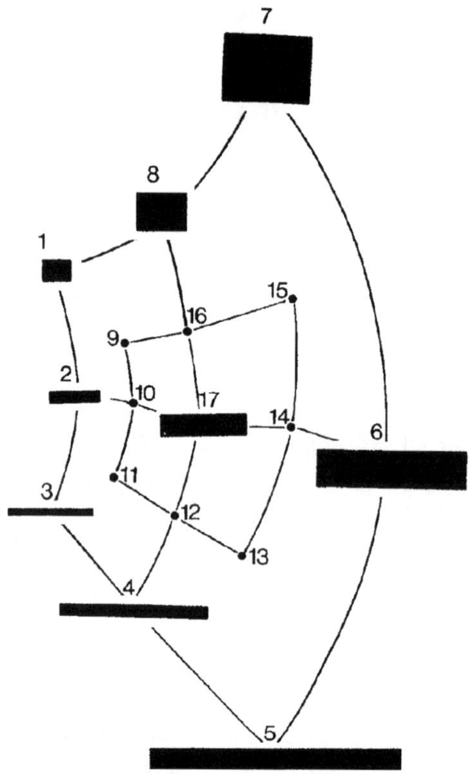

Fig. 9. The unconstrained two-dimensional configuration of rectangles derived by ML MDS (Takane, 1981). (Reproduced with permission from the Psychometric Society.)

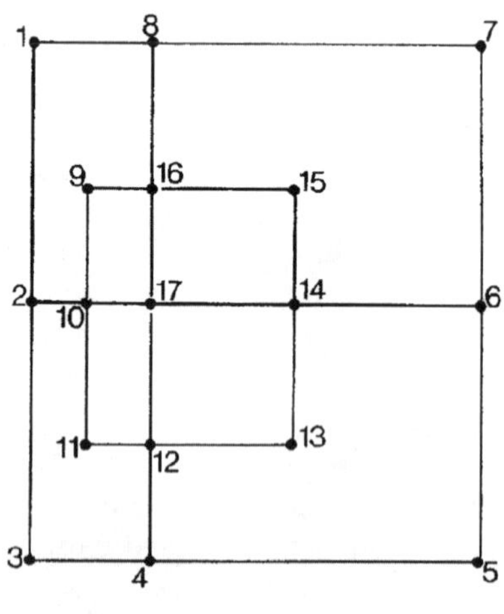

Fig. 10. The stimulus configuration of rectangles derived under the area and shape hypothesis (Takane, 1981). (Reproduced with permission from the Psychometric Society.)

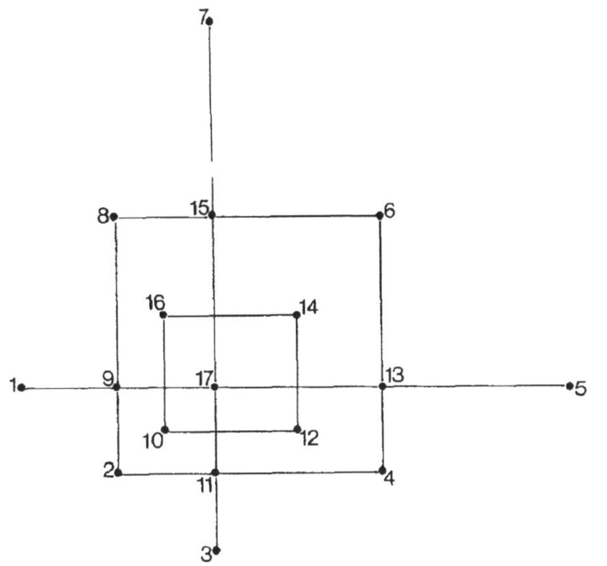

Fig. 11. The stimulus configuration of rectangles derived under the height and width hypothesis (Takane, 1981). (Reproduced with permission from the Psychometric Society.)

Since the area of rectangles is defined as height times width, and the shape is defined as height divided by width, the same set of rectangles can also be characterized by the combination of the (log) height and width dimensions. Again we may formally test how good this characterization is as subjective dimensions of rectangles. Figure 11 displays the stimulus configuration derived under the height and width hypothesis. Table 2 indicates that this hypothesis fits to the data much better than the area and shape hypothesis. It captures one important aspect of the unconstrained solution; the shape difference is not so important when the rectangles are small, but it becomes more important as the rectangles get larger. However, the GOF of the height and width hypothesis is still substantially worse than that of the unconstrained two-dimensional solution.

Schönemann (1977) formalized the hypothesis that the perceived shape difference increases as the area of the rectangles increases. This hypothesis is depicted in Figure 12. This hypothesis was fitted and the GOF of this hypothesis is reported in Table 2. It can be observed that this hypothesis still does not account for the data as well as the unconstrained two-dimensional solution. In Schönemann's hypothesis, area levels were still assumed to be independent of the shape dimension, whereas in the unconstrained two-dimensional configuration, the area dimension is curved along the shape dimension. Curves connecting the rectangles with the same area levels look like arcs drawn from a common focal point, where both the area and the shape of rectangles vanish. Rectangles with the same area levels are equally distant from this focal point. An hypothesis implementing this idea is depicted in Figure 13. Unfortunately, no ML MDS is yet capable of fitting this hypothesis.

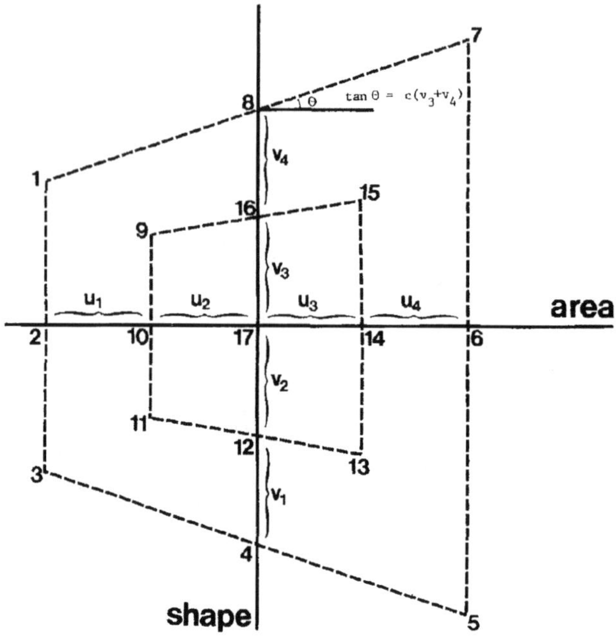

Fig. 12. Schönemann's hypothesis for similarity of rectangles (Takane, 1981). (Reproduced with permission from the Psychometric Society.)

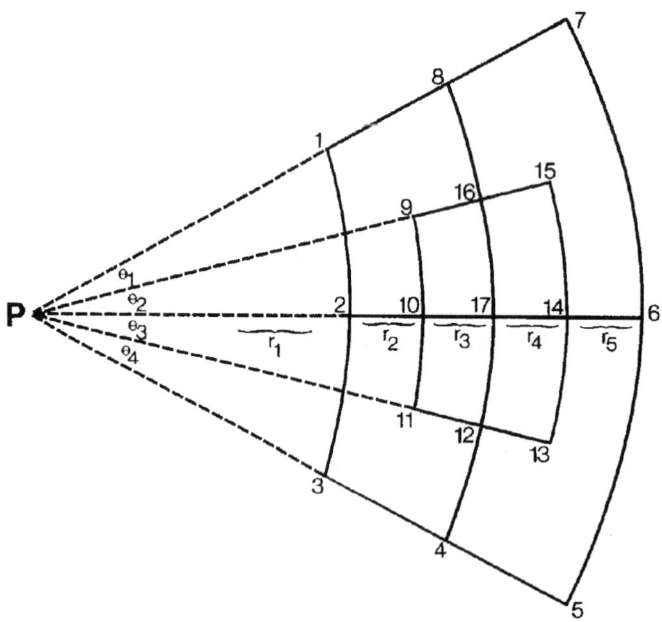

Fig. 13. A focal point hypothesis (Takane, 1981). (Reproduced with permission from the Psychometric Society.)

3. Individual differences MDS

In this section we discuss individual differences (ID) MDS. We primarily focus on a particular kind of ID MDS, called INDSCAL (Carroll and Chang, 1970a), to capture a specific kind of individual differences in (dis)similarity judgments, namely differential weighting of dimensions by different subjects. Again we discuss some examples first, followed by a brief account of the model used to represent the individual differences.

3.1. Examples of individual differences MDS

EXAMPLE 6 Helm's color data.

We have so far been focusing on MDS which derives a single stimulus configuration from a single matrix of observed proximity data. Suppose we have N ($\geqslant 1$) such matrices, each obtained from a different subject. If no systematic individual differences exist, we may analyze them simultaneously and derive a single common stimulus configuration. Alternatively, when some systematic individual differences are suspected, we may apply MDS separately to each single matrix. In this case we obtain N separate stimulus configurations. The question is whether there is a better way to represent differences among matrices of (dis)similarities than applying MDS separately to these data. The answer is "yes", and the technique is called individual differences MDS (Carroll and Chang, 1970a).

Figure 14 helps understand the kind of analysis that can be carried out by the individual differences (ID) MDS. Dissimilarity judgments among 10 colors of varying hue were obtained from 12 subjects (Helm, 1964). Two of the 12 subjects replicated the same experiment twice. The whole data set is given in Borg and Groenen (1997, p. 360). The stimulus configuration displayed in Figure 14 was obtained by applying ID MDS to these data. In this figure, stimuli are labelled as: RP (red–purple), RO (red–orange), Y (yellow), GY1 (green–yellow 1), GY2 (green–yellow 2), Gr (green), Blu (blue), PB (purple–blue), P2 (purple 2), and P1 (purple 1). We can see the familiar color wheel with: 1. Vertical axis roughly contrasting red and green (R–G dimension), and 2. Horizontal axis roughly contrasting yellow and blue (Y–B dimension). Figure 15 represents the weights attached to these two dimensions by different subjects. Subjects N6, for example, put about equal emphasis on both dimensions, while subjects D3 put excessively heavy emphasis on the horizontal axis. This means that N6s judgments of dissimilarities reflect the two dimensions about equally, while D3s judgments were almost exclusively based on the Y–B dimension. In fact, subjects labelled by D are color deficient subjects in the G–R dimension, who are lacking the sensitivity along the G–R dimension. (There was one normal and one color deficient subject who replicated the experiment twice. They are distinguished by symbols a and b in Figure 15.) ID MDS thus attempts to capture individual differences in (dis)similarity judgments by the differential weights attached to different dimensions by different individuals.

The stimulus and weight configurations presented above were derived by a nonmetric ID MDS procedure developed by the author, but they are virtually indistinguishable from those obtained by INDSCAL (Carroll and Chang, 1970b), which assumed a ratio level of measurement. (Stimuli P1 and P2 are somehow interchanged; we followed Borg and Groenen's labelling of the stimuli.)

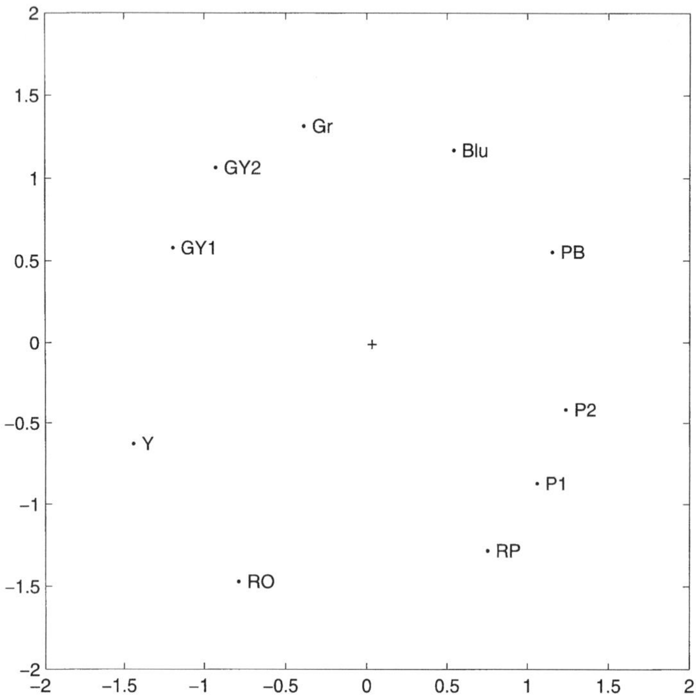

Fig. 14. The two-dimensional common stimulus space derived from Helm's color data.

EXAMPLE 7 Digits.

Not all types of individual differences in (dis)similarity judgments can be represented by differential weighting of dimensions. This example shows an instance of more complicated types of individual differences. The stimuli are 10 digits often used to display time in digital clocks. They were constructed by picking appropriate subsets of seven line segments arranged in the shape of 8 (Figure 16). For example, digit 1 is composed of segments 3 and 6, digit 2 is composed of segments 1, 3, 4, 5, and 7, and so on. In each trial a pair of digits were presented, and subjects were instructed to respond as quickly as possible whether the two presented stimuli were the "same" or "different". In half the trials the same digits were presented, and in the other half, different digits were presented. Reaction times that took the subjects to discriminate two different digits were used as input to ML MDS, especially designed to analyze reaction time data (Takane and Sergent, 1983).

There were two subjects, and the two subjects' data were separately analyzed by simple MDS. Figures 17 and 18 present the results of MDS obtained from the two subjects (Takane, 1994). The two configurations seem completely unrelated. For Subject 1 (Figure 17), the stimuli are organized according to the patterns of presence and absence of line segments. The horizontal axis divides (roughly at the vertical dotted line) the digits having two top (or three) horizontal line segments (on the left) from those lacking them (on the right). The vertical axis divides (roughly at the horizontal dotted line) the digits

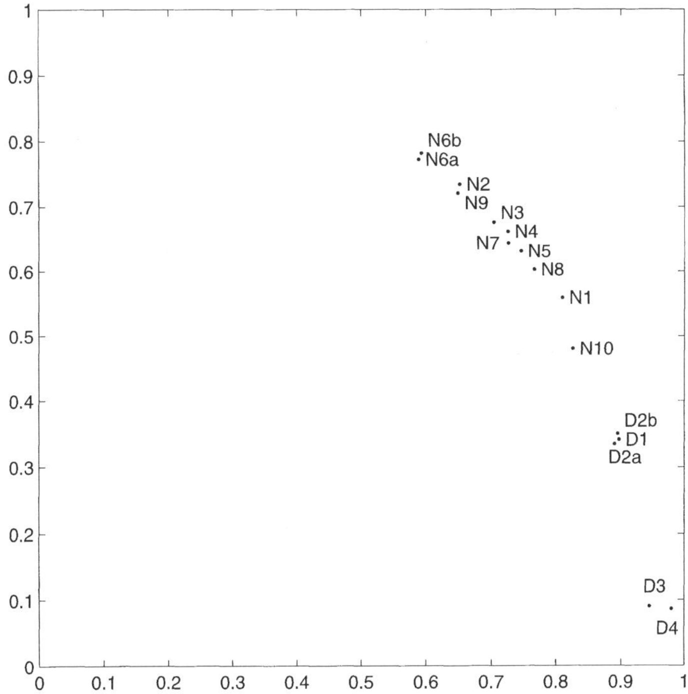

Fig. 15. The plot of individual differences weights attached to the two dimensions of the color space.

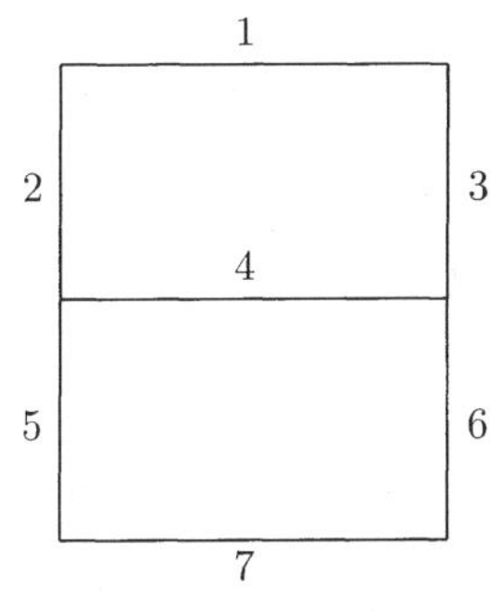

Fig. 16. The construction of segmented numerals (digits) used in Example 7 (Takane, 1994). (Reproduced with permission from Springer.)

having three vertical line segments 2, 3 and 6 (toward the bottom) from those lacking at least one of them. For Subject 2 (Figure 18), on the other hand, the configuration (the unconstrained solution) roughly circles around, starting from 0 and ending with 9. To verify that the configuration is really circular, another solution was obtained under the constraint that all the points lie on a circumference of a circle. The constrained solution was found to fit better according to the minimum AIC criterion. This subject thus seems to perceive the stimuli as numbers (rather than collections of line segments arranged in

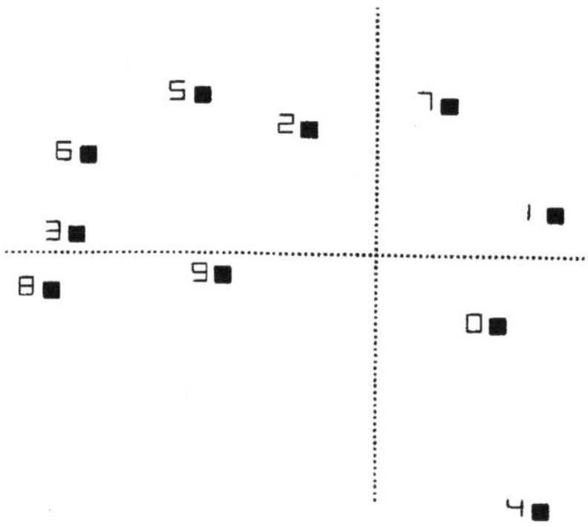

Fig. 17. The configuration of digits for Subject 1 (Takane, 1994). (Reproduced with permission from Springer.)

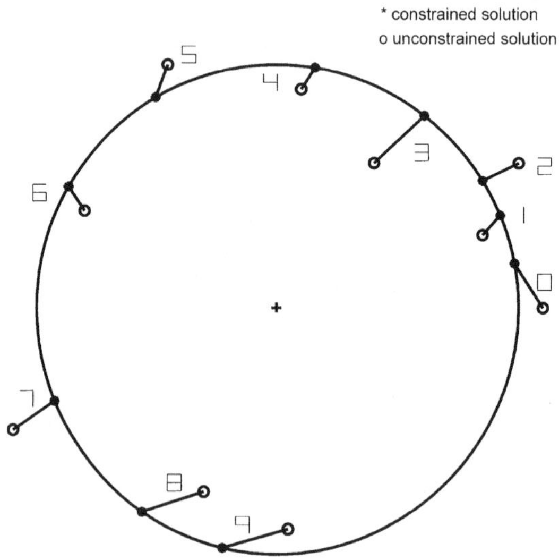

Fig. 18. The configuration of digits for Subject 2 (Takane, 1994). (Reproduced with permission from Springer.)

a specific way), since he took much longer time to discriminate two digits which are closer in numerical value.

The way the same set of stimuli were perceived was entirely different for the two subjects. Clearly, this type of individual differences cannot be captured by the kind of ID MDS described above. Separate analyses by simple MDS are in order, as has been done here. Alternatively, an MDS that allows more general kinds of individual differences may be called for (e.g., Meulman and Verboon, 1993).

EXAMPLE 8 MDS of line drawings of faces.

For multiple sets of (dis)similarity data to be adequately described by the kind of ID MDS described earlier, they must exhibit certain patterns when analyzed separately by

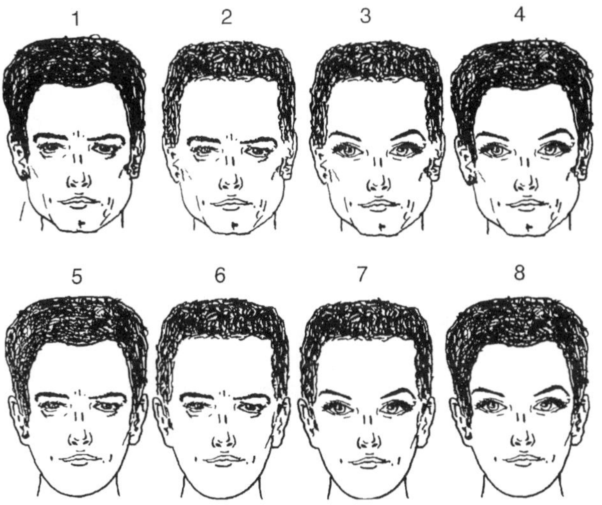

Fig. 19. Stimuli used in Example 8 (Takane and Sergent, 1983) (Reproduced with permission from the Psychometric Society.)

simple MDS. This example shows such patterns. The stimuli are eight faces (Figure 19) constructed by factorially combining two levels each of three features: Hair (long or short), Jaw (smooth or rugged), and Eye (bright or dark). As in Example 7, reaction times were used as input data to ML MDS. Again there were two subjects.

Figures 20 and 21 display derived stimulus configurations for the two subjects. Four dimensions were extracted and interpreted as follows: Dimension 1. Hair, Dimension 2. Jaw, Dimension 3. Eye, and Dimension 4. Sex consistency. The first three dimensions correspond with the three defining features of the stimuli. The fourth dimension is the additional dimension used by the subjects in performing the task. On this dimension (the coordinates of the stimuli on the fourth dimension are given in the right margin of the two figures), the stimuli most consistent with the most typical sex profiles (short hair, rugged jaw and dark eyes for males, and long hair, smooth jaw and light eyes for females) are located at the top, while the remaining stimuli are located downward from the top according to their degrees of inconsistency with these typical sex profiles. The most striking observation is that the two configurations are remarkably similar, in fact almost identical. The two subjects use the same set of dimensions, only their saliency differs across the two subjects. ID MDS discussed earlier is most appropriate to describe this kind of individual differences.

EXAMPLE 9 Facial expression data.

We give a couple of more examples of ID MDS. A set of stimuli employed in the first example are shown in Figure 22. They were constructed by factorially combining two of the most important determinants of facial expressions, namely the curvature of eyes and the curvature of lips (Inukai, 1981). Dissimilarity judgments were obtained on these faces from twenty subjects (10 males and 10 females) using rating scales. Figure 23

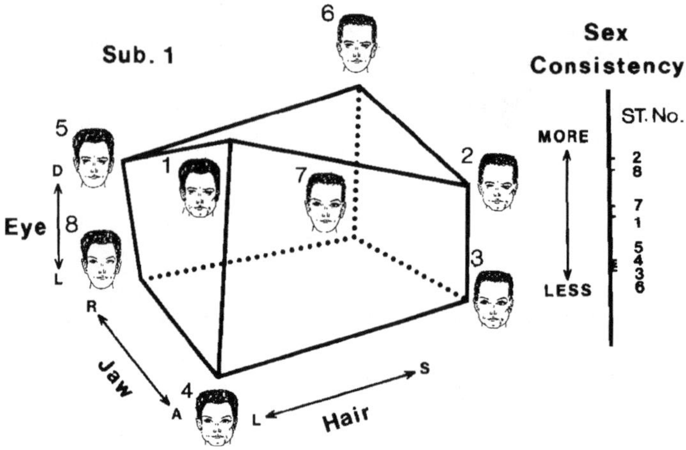

Fig. 20. The four-dimensional configuration of faces for Subject 1 (Takane and Sergent, 1983). (Reproduced with permission from the Psychometric Society.)

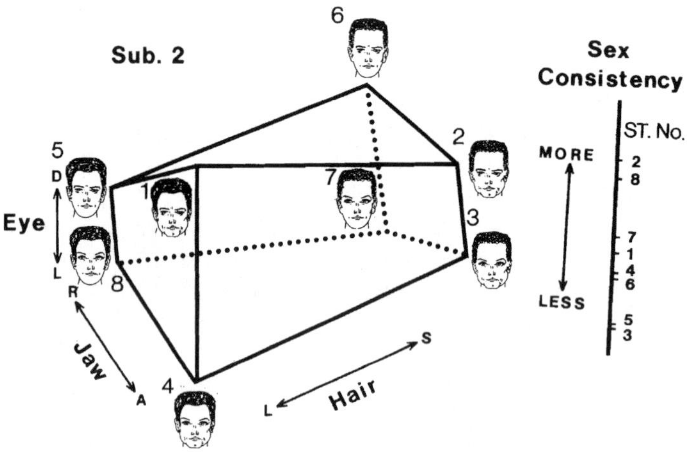

Fig. 21. The four-dimensional configuration of faces for Subject 2 (Takane and Sergent, 1983). (Reproduced with permission from the Psychometric Society.)

shows the two-dimensional common stimulus configuration obtained by INDSCAL (Carroll and Chang, 1970a), assuming that the observed dissimilarity data are measured on a ratio scale. This configuration indicates that subjects' dissimilarity judgments are organized around the two physical attributes (the lips and the eyes dimensions) used to construct the stimuli. Note, however, that the two attributes are not strictly orthogonal in the psychological space. Concave (downward) eyes and convex lips seem to go together, and so do convex eyes and concave lips (lower right corner). The vertical direction roughly corresponds with the eyes dimension and the horizontal direction with the lips dimension.

Fig. 22. Stimuli used in Example 9 (Inukai, 1981).

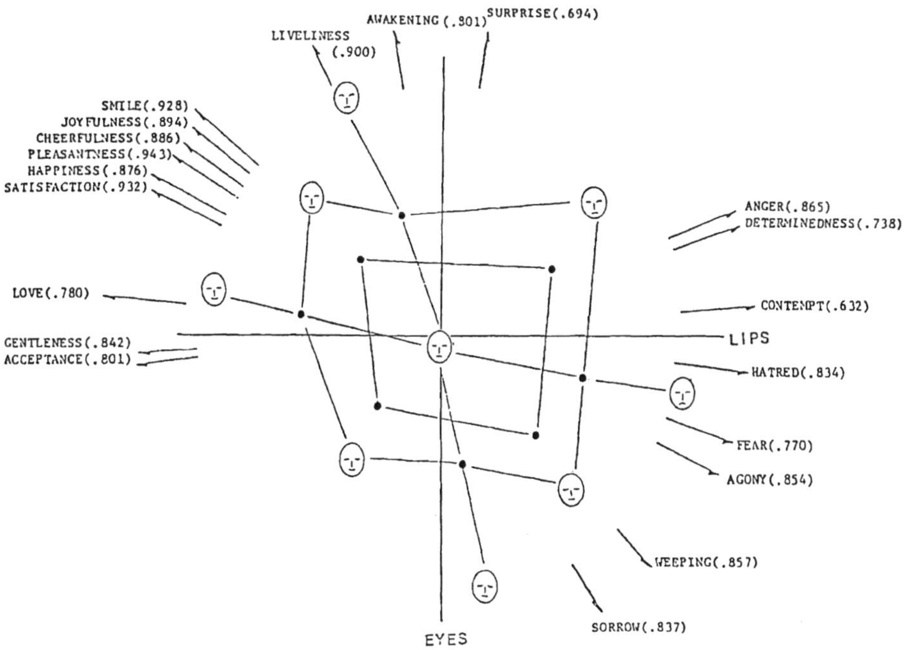

Fig. 23. The two-dimensional common stimulus configuration derived from facial expression data.

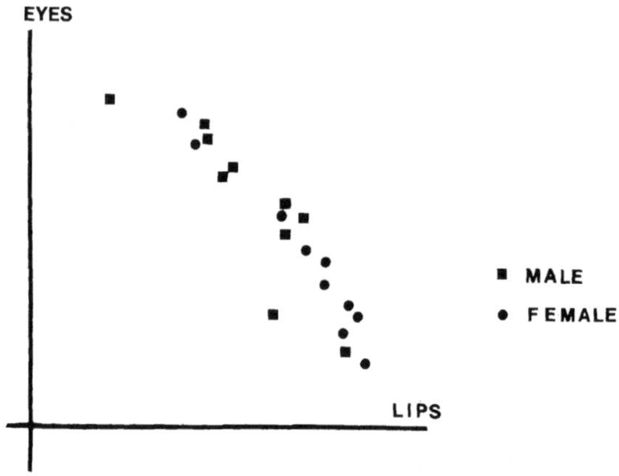

Fig. 24. The two-dimensional individual differences weight space for the facial expression data.

In addition to the dissimilarity data, the experimenter also collected rating data on the same set of stimuli regarding the extent to which they expressed certain emotional dispositions. Directions (designated by arrows in Figure 23) indicate the directions with which certain emotional dispositions are most highly correlated. For example, "smile" is in quadrant 2, and this is the direction in which this emotional disposition is most clearly represented. Similarly, we may characterize other emotional dispositions as: 1. happiness – moderately concave eyes and moderately convex lips, 2. alertness – high concave eyes and straight lips, 3. anger – straight eyes and high concave lips, 4. weeping – moderately convex eyes and moderately convex lips, 5. sleeping – deep convex eyes and straight lips. Numbers in parentheses indicate multiple correlation coefficients.

Interestingly, there seem to be fairly clear sex differences in the evaluation of the two dimensions. Figure 24 shows individual differences weights. Squares represent weights for male subjects and circular dots those for female subjects. Generally speaking, the female subjects tend to put more emphasis on the lips dimension, while the male subjects on the eyes dimension.

EXAMPLE 10 Body-parts data.

The final example in this section involves a developmental change in conceptual relations among words (Takane et al., 1977). Dissimilarity judgments between various body-parts were obtained from a group of adult subjects and a group of 6-year-old children, using the conditional rank-order method. (In this data collection method, one of the stimuli is chosen as the standard stimulus in turn, and the remaining stimuli are rank-ordered in terms of their similarity with the standard.) Figure 25 presents the derived stimulus configuration. The solution is three-dimensional, where the three dimensions are interpreted as: Dimension 1. Contrasts between face terms and limbs terms (both lower and upper limbs) with "body" in between. Dimension 2. Contrasts between upper and lower limbs with "body" and all face terms in the middle. Dimension 3. Represents

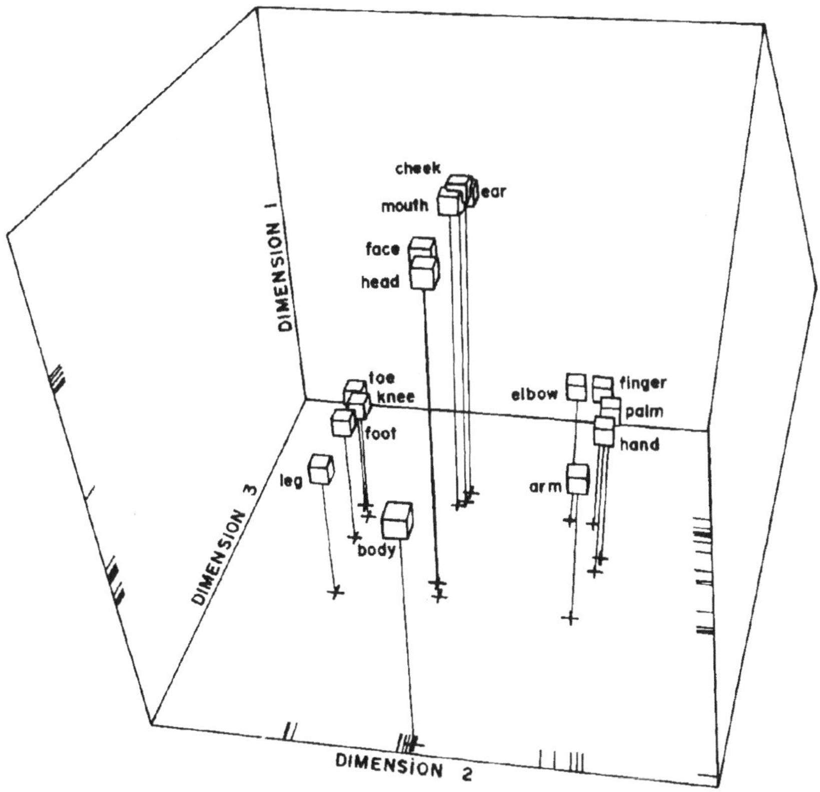

Fig. 25. The three-dimensional common stimulus configuration derived from Jacobowitz' body-parts data (Takane et al., 1977) (Reproduced with permission from the Psychometric Society.)

the whole–part hierarchy. (The term "body" is right in front, and more peripheral parts of the body are located toward the back of the configuration.)

Interestingly, there is a fairly systematic developmental difference in the weighting of these dimensions. Figure 26 depicts individual differences weights. White cubes represent 6-year-old children's weights, while black cubes represent those for adults. As can be seen, 6-year-old children tend to put more emphasis on dimension 2 than dimension 1 or 3, while adults are more heterogeneous. They split into two groups, one group placing more emphasis on dimension 1 and the other on dimension 3, but no adults put most emphasis on dimension 2. The distinction between upper and lower limbs is very important for young children. As they grow older, they come to realize certain parallelism between upper and lower limbs.

3.2. A brief account of ID MDS

In ID MDS, we are given N matrices of proximity data between n stimuli, denoted by n by n matrices, $\boldsymbol{O}_k$ ($k = 1, \ldots, N$), the ijth element, o_{ijk}, of which indicates the (dis)similarity between stimuli i and j obtained from subject k. In ID MDS used in the

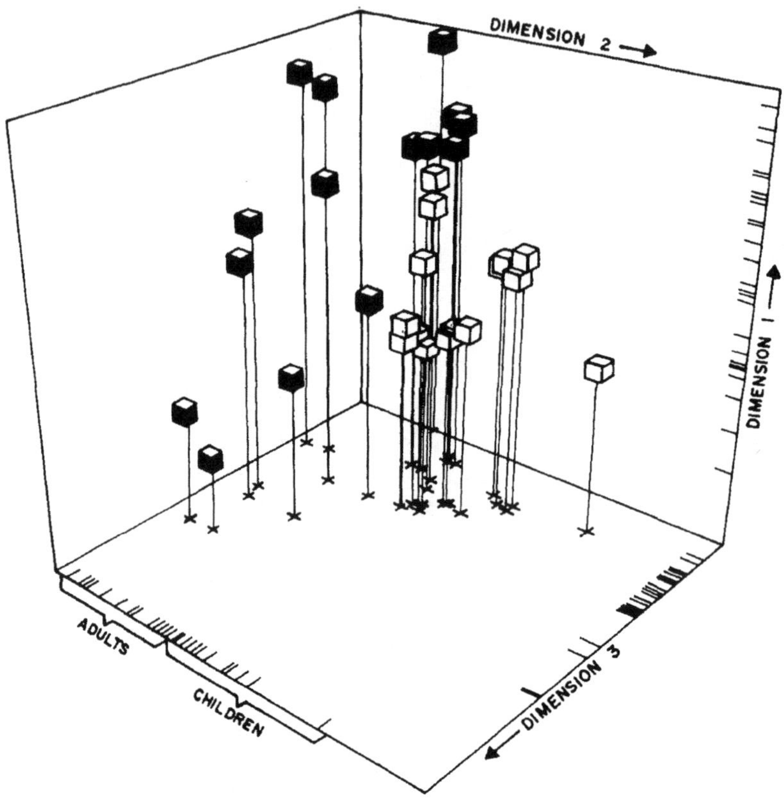

Fig. 26. The three-dimensional individual differences weight space for Jacobowitz' body-parts data (Takane et al., 1977). (Reproduced with permission from the Psychometric Society.)

above examples, we represent o_{ijk} by the weighted Euclidean distance model,

$$d_{ijk} = \left\{ \sum_{a=1}^{A} w_{ka}(x_{ia} - x_{ja})^2 \right\}^{1/2}, \tag{9}$$

where x_{ia} is, as before, the coordinate of stimulus i on dimension a, and w_{ka} is the weight attached to dimension a by subject k. To eliminate scale indeterminacies between the stimulus coordinates and individual differences weights, x_{ia}'s are usually scaled so that $\sum_{i=1}^{n} x_{ia}^2 = n$. By applying ID MDS to the set of $\boldsymbol{O}_k$'s, we obtain the n by A matrix of stimulus coordinates, $\boldsymbol{X}$, and the N by A matrix of individual differences weights, $\boldsymbol{W}$, whose kath element is w_{ka}. These quantities are then plotted in graphical forms.

One important feature of the weighted Euclidean model is that the orientation of coordinate axes is uniquely determined. This means that we interpret the axes derived by the ID MDS, rather than look for some meaningful directions in the space to interpret as in the simple Euclidean distance model (1). This removes the burden of rotating the stimulus configurations in the right orientation as typically required in simple MDS.

Various algorithms have been developed for fitting ID MDS, of which INDSCAL (Carroll and Chang, 1970a) is still the most frequently used algorithm for ID MDS. In this method, observed dissimilarity data were first transformed into inner products (Schönemann, 1972; Young and Householder, 1938), which are then approximated by a weighted inner product model using a special iterative minimization algorithm. (The weighted inner product model is derived from the weighted Euclidean distance model with the same transformation by which the observed dissimilarity data were transformed into inner products.) INDSCAL, however, requires the ratio level of measurement for the observed dissimilarity data. ALSCAL (Takane et al., 1977), on the other hand, allows monotonic transformations of observed (dis)similarity data. However, it fits squared distances to monotonically transformed (dis)similarity data. Consequently, it puts more emphasis on large dissimilarities. PROXSCAL (Busing et al., 1997) is the first workable algorithm that directly fits the weighted Euclidean model to (monotonically transformed) (dis)similarity data.

The weighted distance model poses some difficulty for ML MDS. The number of parameters (individual differences weights) in the weighted distance model increases as more subjects are added to increase the number of observations (unless replicated observations are made within subjects). These are called incidental parameters. Asymptotic properties of the maximum likelihood estimators do not hold in this case. Clarkson and Gonzales (2001) resolved this problem by incorporating the subject weights as random effects, which were then integrated out to define the marginal likelihood. The marginal likelihood is maximized to estimate other (structural) parameters. Takane (1996, 1997) proposed a similar approach in unfolding analysis of multiple-choice questionnaire data.

4. Unfolding analysis

In this section we discuss MDS of preference data. Individual differences are prevalent in preference judgments. Unfolding analysis provides a way of representing individual differences in preference judgments. In unfolding analysis, preference data are regarded as representing proximity relations between subjects' ideal stimuli and actual stimuli.

4.1. Some examples of unfolding analysis

EXAMPLE 11 Family composition preference.

MDS is not restricted to the usual kind of (dis)similarity data. This example pertains to preference data on family compositions, namely how many boys and girls the subjects would like to have in their family. Stimuli are various family compositions constructed by factorially combining several levels of the number of boys (from 0 to 3) and the number of girls (from 0 to 3). Preference rankings were obtained on the 16 family compositions from 82 Belgian students (Delbeke, 1978). The data are presented in Figure 27. The 16 stimuli are numbered so that the index for the number of boys changes more rapidly. Specifically, 1. (0, 0), 2. (1, 0), 3. (2, 0), 4. (3, 0), 5. (0, 1), 6. (1, 1), 7. (2, 1), 8. (3, 1), 9. (0, 2), 10. (1, 2), 11. (2, 2), 12. (3, 2), 13. (0, 3), 14. (1, 3), 15. (2, 3),

No.	1	2	3	4	5	6	7	8	9	10	11	12	13	14	15	16
1	16	6	8	4	7	1	2	12	5	3	10	13	9	11	14	15
2	16	7	4	5	15	1	3	8	10	9	2	6	14	13	11	12
3	16	14	9	7	15	6	4	2	10	5	1	11	8	3	12	13
4	16	15	13	12	14	8	3	5	11	2	1	7	10	4	6	9
5	16	12	10	13	14	8	2	7	11	6	5	1	15	9	3	4
6	16	7	5	6	15	11	1	2	13	10	4	3	14	9	12	8
7	16	12	11	6	15	5	2	3	14	4	1	8	13	7	9	10
8	16	8	6	5	14	9	4	2	15	11	3	1	13	12	7	10
9	14	15	10	12	16	4	2	5	11	3	1	7	13	6	8	9
10	15	16	13	9	14	10	7	8	11	5	4	2	12	6	1	3
11	16	13	4	3	15	1	2	9	11	8	5	7	12	14	10	6
12	16	14	6	3	15	5	2	4	12	8	1	7	13	9	11	10
13	16	14	9	8	15	4	1	11	10	2	5	3	13	12	6	7
14	16	14	10	11	15	9	5	6	13	8	4	2	12	7	3	1
15	15	14	12	16	13	11	9	10	8	7	5	6	2	3	1	4
16	16	6	2	7	8	1	4	10	3	5	9	11	13	14	12	15
17	16	15	13	10	14	9	8	7	12	5	3	6	11	4	1	2
18	16	8	5	2	15	3	1	10	9	4	6	7	14	11	12	13
19	16	14	12	13	15	10	7	6	11	8	1	3	9	5	4	2
20	15	12	10	14	11	6	2	5	9	1	3	8	13	4	7	16
21	16	15	10	11	13	9	8	5	12	7	1	3	14	6	2	4
22	16	14	11	10	15	9	5	6	13	7	1	3	12	8	4	2
23	16	11	6	12	13	4	1	5	14	2	3	7	15	10	9	8
24	16	15	13	12	14	7	8	10	4	1	5	9	2	3	6	11
25	16	14	10	9	15	8	6	5	12	7	1	2	13	11	4	3
26	16	12	11	10	15	5	6	8	14	7	4	2	13	9	3	1
27	14	15	11	9	16	8	1	5	13	6	2	3	12	7	10	4
28	16	13	12	14	10	1	4	15	5	3	2	8	11	6	7	9
29	16	14	8	7	15	10	6	4	13	9	5	1	12	11	3	2
30	1	2	5	7	4	3	6	10	14	12	8	9	16	15	13	11
31	14	15	10	11	16	6	4	7	12	8	1	3	13	9	5	2
32	16	14	11	9	15	10	2	4	13	3	1	6	12	5	7	8
33	16	4	3	8	6	1	2	10	7	5	9	12	13	11	14	15
34	15	7	6	9	4	1	3	12	5	2	10	14	8	11	13	16
35	16	7	6	5	15	8	3	1	14	9	2	4	13	10	11	12
36	16	2	4	5	11	3	1	6	13	7	8	9	15	14	12	10
37	16	7	8	9	6	3	2	12	4	1	10	14	5	11	13	15
38	16	4	9	10	7	3	2	5	11	6	1	13	12	8	14	15
39	16	12	11	10	15	7	5	8	14	6	1	3	13	9	4	2
40	16	2	3	12	4	1	6	10	5	7	8	9	13	11	14	15
41	8	2	3	10	5	1	4	11	6	7	9	12	13	14	15	16
42	16	9	4	7	12	1	2	8	5	3	6	10	11	13	14	15
43	16	15	13	12	14	8	5	7	10	4	1	3	11	6	2	9
44	16	14	6	5	15	9	7	1	13	10	2	3	12	11	8	4
45	16	10	8	9	12	5	4	2	14	11	1	3	15	13	7	6
46	16	13	1	2	15	3	4	6	11	5	7	8	14	12	9	10
47	1	2	5	9	3	4	7	12	6	8	11	14	10	13	16	15
48	10	3	1	5	4	2	7	11	6	8	12	14	9	13	15	16
49	16	13	4	1	15	6	2	3	14	7	5	8	12	10	9	11
50	16	2	7	9	3	1	5	11	4	6	8	13	12	10	14	15
51	16	15	10	13	14	5	7	8	12	6	2	4	11	9	3	1
52	16	15	12	11	14	10	3	4	13	2	1	6	9	8	5	7
53	16	12	11	9	13	10	4	5	14	6	1	2	15	8	3	7
54	16	12	11	9	15	7	5	6	14	10	1	2	13	8	3	4
55	16	15	13	11	14	4	2	6	12	1	3	8	10	5	7	9
56	16	14	13	11	15	6	2	8	12	4	1	3	10	7	5	9
57	16	15	13	10	14	11	6	8	12	5	3	4	9	7	2	1
58	16	15	13	9	14	11	8	5	12	7	3	2	10	4	1	6
59	16	14	8	7	15	12	6	2	13	10	3	1	11	9	4	5
60	12	16	14	10	15	11	8	6	13	7	4	2	9	5	1	3
61	16	15	13	12	14	9	8	6	11	7	2	4	10	5	3	1
62	16	14	10	13	15	9	7	2	11	8	1	4	12	3	5	6
63	16	14	11	13	15	8	6	5	12	7	4	1	10	9	3	2
64	16	12	5	2	15	11	6	1	14	10	7	3	13	9	8	4
65	16	10	11	13	12	9	7	5	14	8	4	2	15	6	3	1
66	16	10	8	3	15	9	5	2	14	11	6	1	13	12	7	4
67	16	10	12	14	11	9	7	3	13	8	1	2	15	6	4	5
68	15	5	3	6	10	2	1	9	8	7	4	11	12	13	14	16
69	16	15	13	10	14	6	8	9	12	5	1	7	11	3	2	4
70	16	12	8	9	15	10	6	4	14	7	5	2	13	11	3	1
71	6	3	5	9	2	1	8	13	4	7	11	15	10	12	14	16
72	16	12	11	10	15	9	5	7	14	6	4	1	13	8	2	3
73	14	13	8	11	12	2	4	15	15	3	1	9	16	7	6	10
74	16	10	11	14	12	9	1	3	13	5	2	4	15	7	6	8
75	16	14	13	15	12	6	10	8	11	5	1	4	9	7	3	2
76	16	12	10	11	14	4	1	3	13	5	2	6	15	7	8	9
77	16	12	4	3	15	9	5	2	14	10	6	1	13	11	8	7
78	16	10	11	13	15	4	2	6	14	3	1	5	12	7	8	9
79	1	2	3	4	9	5	6	7	14	11	10	8	15	16	13	12
80	16	14	11	10	15	9	7	5	13	8	4	2	12	6	3	1
81	16	15	13	11	14	9	8	6	12	7	4	3	10	5	2	1
82	16	12	8	4	15	11	7	3	14	10	6	2	13	9	5	1

Fig. 27. Family composition preference ranking data obtained from 82 Belgium university students (Delbeke, 1978).

and 16. (3, 3), where the first number in the parentheses indicates the number of boys, and the second number indicates the number of girls. In the figure, the smaller numbers indicate more preferred combinations. For example, for the first subject, stimulus 6 (the (1, 1) combination) is the most preferred combination, stimulus 7 (the (2, 1) combination) is the second most preferred combination, and so on. One way to analyze these data is an extension of MDS, called unfolding analysis, in which it is assumed that:

1. Each subject has an ideal point in a multidimensional space corresponding to his or her ideal stimulus.
2. The closer a stimulus point is to one's ideal point, the more preferred that stimulus is by that subject.

Unfolding analysis locates both subjects' ideal points and stimulus points in a joint multidimensional space in such a way that more preferred stimuli are located close to the ideal points, while less preferred stimuli are located far from the ideal.

Figure 28 (Heiser, 1981) shows the result of the unfolding analysis of these data. In this figure, stimulus points are indicated by a pair of numbers (e.g., 1, 3), while subjects' ideal points are indicated by $\times$. There is a strong tendency to prefer large families among the Belgian students. There is also some boy bias (tendency to prefer boys to girls). It is also evident that the difference between two small families (e.g., (1, 0) and (1, 1)) is much larger than the difference between two large families (e.g., (2, 2) and (2, 3)). Adding one girl has more pronounced effects when there is no girl in the family, but it has relatively small impact when there are already two girls in the family. The important point is that these observations, rather obvious in Figure 28, can hardly be made by merely inspecting the data table in Figure 27.

EXAMPLE 12 A marketing research application.

The next example shows importance of unfolding analysis in marketing research (DeSarbo and Rao, 1984). Stimuli are 12 brands of telecommunication devices (e.g., transceivers) characterized by seven descriptor variables such as calling feature (receive only or both send and receive) and style (military or cradle). Preference ratings were obtained from a group of 50 subjects characterized by 11 demographic variables such as age and sex. In the previous example, the stimulus and ideal point configuration was derived based on the observed preference data alone. In contrast, the background information about the stimuli and the subjects was explicitly incorporated in the present analysis, and was utilized in developing useful marketing strategies.

Figure 29 displays the derived stimulus and ideal point configuration. Stimuli are labelled by alphabet (A–I). Ideal points are indicated by circular dots. (White ones indicate *anti-ideal* points, meaning that stimuli further away from the points are more preferred by these subjects.) It can be immediately observed that products A, B, and H are rather unpopular (few ideal points are located close to them), while the others (e.g., I, J, K, and L) are better accepted.

The background information is incorporated by constraining the coordinates of stimulus and ideal points as linear combinations of the background variables. An advantage of this is that the dimensional interpretation becomes easier. Products with high price, walkie-talkie style, and send and receive capability tend to be located toward the top

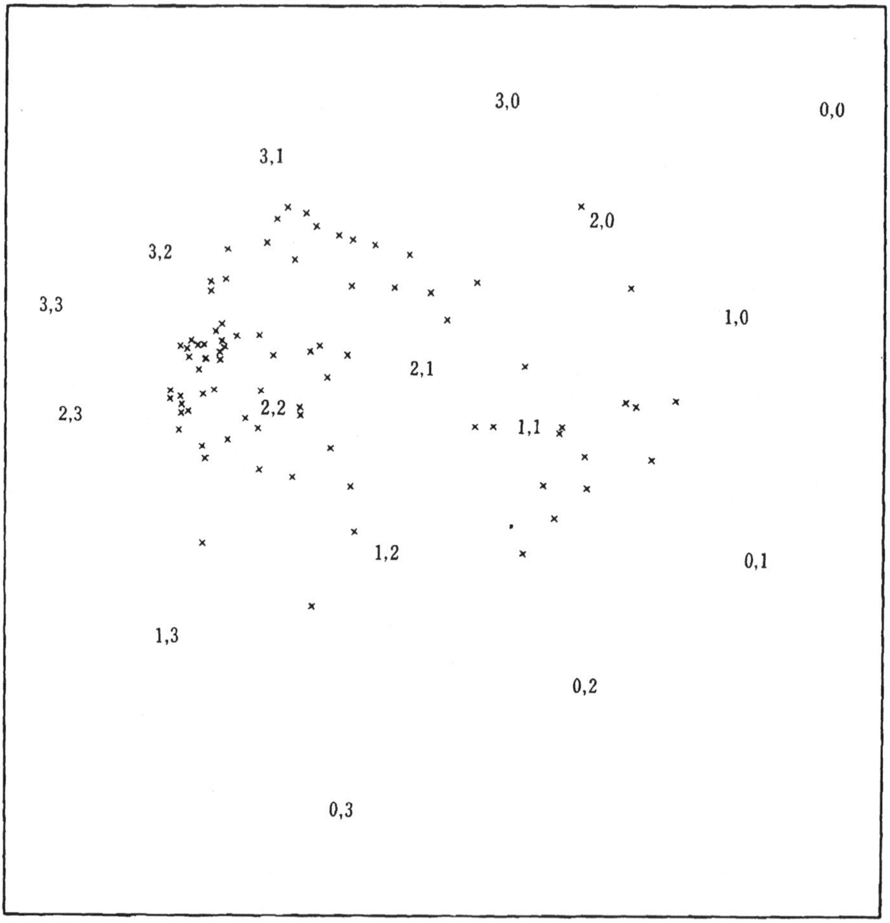

Fig. 28. The joint representation of the stimulus and the ideal points obtained by unfolding analysis of Delbeke's data (Heiser, 1981). (Reproduced with permission from the author.)

of the configuration. Products with send and receive capability, high range, repertory dialing feature, speaker–phone option and stand up set-up tend to be located toward the right. Similarly, subjects who own single dwelling houses with many phone, and are married males having no children living at home, tend to be located toward the top, while older subjects who are educated, own multi-dwelling houses, but are lower in occupational status tend to be located toward the right. Thus, we immediately know what sort of products and subjects are located in which parts of the configuration, and what sort of people prefer, or do not prefer, what sort of products. This provides valuable information regarding which products should be selectively promoted to what sort of people.

Other valuable information can also be obtained by "manipulating" the configuration. Suppose we are interested to know how well new products will sell, if they are

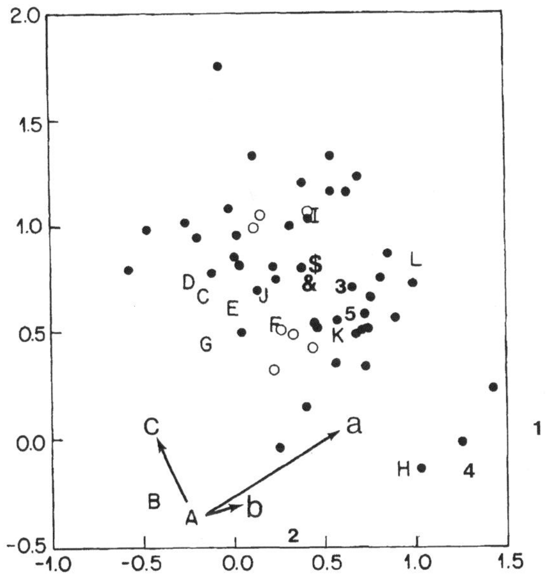

Fig. 29. The joint representation of products and subjects useful in marketing research. (Adapted from DeSarbo and Rao (1984) with permission from the authors.)

put on the market. The prospective products are defined by a set of values that take on the descriptor variables. By applying the weights used to derive the coordinates of the existing products, we can map the new products into the same space in which the existing products as well as subject points are already represented. The location of five new potential products (numbered from 1 to 5) are superposed in the figure. (Product 1 is slightly out of the frame of Figure 29.) It can be seen that products 1, 2 and 4 will not be popular. Products 3 and 5 will be better accepted, although whether they would sell well is rather questionable, since similar products are already on the market.

Essentially the same idea may be utilized for repositioning existing products. For example, product A is currently not very popular. We would like to make it more attractive to more people by changing some of its features. In the current example, product A is modified in three different ways. Symbols a, b, and c in the figure indicate the predicted relocation of product A after these changes: a. Provide send and receive capability for an additional cost of $40; b. Increase the range by additional 200 feet for an additional cost of $20; c. Add repertory dialing and redesign military style at no additional cost. Modifications involved in b and c are almost totally useless to make product A more competitive. a is the best strategy, but it still falls short of attracting many buyers.

A point can be located in the space which will attract the largest number of people. This point is such that the average (squared Euclidean) distance to all subject points is minimal. Thus, it might be called ideal product. This point is indicated by symbol "&" in the figure. This point, although having the minimal average squared distance to all the subjects points, may not be realizable by manipulating the defining features of the product. The realizable ideal product is indicated by symbol "$" in the figure.

This product has send and receive capability, no repertory dialing, range = 1000 ft., speaker–phone option, price = 239.95, lie-down set-up and cradle style. Products I and J come fairly close to this profile, although I is probably better selling than J because the latter has a lot of competitors.

As it turned out, the ideal product is not radically better than some of the existing products. We might give up the idea of developing a product having the best overall popularity, but a product having a strong appeal to a certain group of people, and which does not suffer from strong competitions from other products. For example, a product located in the upper left portion of the configuration will attract people in that region, and will not have major competitors. Thus, it may prove to be a better product to sell. Furthermore, we know what sort of people will like this product. A sales strategy for promoting this product may take this into account.

4.2. *A brief account of unfolding analysis*

In unfolding analysis, we are given an N by n data matrix obtained from N subjects making preference judgments on n stimuli. By subjecting the data matrix to unfolding analysis, we obtain two coordinate matrices, one for subjects' ideal points (an N by A matrix, $\boldsymbol{Y}$), and the other for stimulus points (an n by A matrix, $\boldsymbol{X}$). Let y_{ka} denote the coordinate of subject k's ideal point on dimension a (the kath element of $\boldsymbol{Y}$), and x_{ia} the coordinate of stimulus i on dimension a (the iath element of $\boldsymbol{X}$). The Euclidean distance between ideal point k and stimulus point i is calculated by

$$d_{ki} = \left\{ \sum_{a=1}^{A} (y_{ka} - x_{ia})^2 \right\}^{1/2}. \tag{10}$$

The coordinates of ideal and stimulus points are determined in such a way that observed preference relations between subjects and stimuli are as inversely (monotonically) related to the distances as possible. The distance model used in unfolding analysis is often called ideal point model. In most cases, the coordinates are derived based on observed preference data alone as in Example 11. In some cases, however, additional information about the stimuli and the subjects are available, which may be incorporated in the analysis, as in Example 12. This is often done by constraining the coordinates as linear functions of the background variables.

Unfolding analysis is thus a special kind of MDS. It appeared as if a fitting procedure for unfolding analysis could easily be developed by simple modifications of standard MDS procedures. This expectation, however, turned out to be too optimistic (DeSarbo and Rao, 1984; Kruskal et al., 1978; Takane et al., 1977). The problem is that in many cases, so-called degenerate solutions are obtained. The degenerate solutions are those that fit to the data very well, but are substantively meaningless. Most often, ideal points and stimulus points are completely separated in the configuration, with distances between them being equal or nearly equal.

Kruskal's (1964a, 1964b) original fitting criterion, called stress 1, normalized the (squared) raw stress (like (2)) by the sum of squared distances. This normalization convention does not discourage equal distances. Kruskal and Carroll (1969) proposed to normalize the (squared) raw stress by the sum of squared deviations of distances from the mean to induce more variability among distances. This, however, introduced variability of distances across subjects, while keeping the within-subject distances relatively homogeneous. It was then proposed to assess the fits within subjects, normalized by the within-subject sum of squared deviations, which are then averaged across subjects. This, however, led to another kind of degenerate solution; both ideal points and stimulus points are split into two, and so are the within-subject distances. Other attempts have also been made to prevent degenerate solutions by introducing other kinds of constraints like smoothness in the data transformations (Heiser, 1981, 1989; Kim et al., 1999). More recently, Busing et al. (2005) proposed to penalize the stress function by an inverse of the variation (the ratio of variance to the squared mean) defined on disparities. This strategy, implemented in PREFSCAL, seems to be working. A potential danger (of the penalty method in general) is, of course, that we never know the right amount of nondegeneracy in the solution. For example, by imposing a penalty, we may be getting "too" nondegenerate a solution for a given data set. Some kind of objective criterion such as cross validation is required to choose appropriate values of penalty parameters.

Another interesting idea has been presented recently by Van Deun et al. (2005), who advocate not to try to avoid degenerate solutions, which may occur with good reasons. The ideal point (distance) model to account for individual differences in preference may be adequate only for a subset of subjects. By inspecting some of the degenerate solutions, they have identified that there are at least two other groups of subjects for whom the distance model is not necessarily adequate. For one of these groups, the vector preference model is more appropriate, and for the other group a signed compensatory distance model was more appropriate. In the former, subjects are represented as vectors, and their preferences are predicted by the projections of stimulus points to the vectors. This model is considered as a special case of the ideal point model, where the ideal points are located infinitely far away from the origin (Carroll, 1972). In the signed compensatory distance model, on the other hand, stimulus points are first projected onto the subject vectors, and distances are measured between ideal points and the projected stimulus points. These distances are then assumed to be inversely related to observed preference data. Van Deun et al. proposed to represent the entire data set by a mixture of the three models.

Another possibility is to use the idea of external unfolding (Carroll, 1972). In external unfolding, a stimulus configuration is derived first, and then subjects' ideal points are mapped into the stimulus configuration in such a way that distances between the stimulus points and ideal points best agree with observed preference relations. (This is contrasted with the "internal" unfolding, in which both stimulus and ideal point configurations are derived simultaneously.) A technique developed by Gower (1968) and Carroll (1972) may be used to map the ideal points into the stimulus configuration. A similar approach has been proposed by Meulman (1992), although in her case, the model is PCA (principal component analysis), and the subject points are estimated first rather than the other way round.

This last suggestion entails asymmetric treatment of subjects and stimuli. This may be more natural in unfolding analysis because subjects are typically treated as replications in multivariate analysis. For example, Takane (1996, 1997) proposed models for multidimensional analysis of multiple-choice questionnaire data that combine an item response theory (IRT) model and the ideal point model. It is assumed that both item categories and subjects are represented as points in a multidimensional space, and that the probability of a subject choosing a particular response category is stated as a decreasing function of the distance between them. Subject points are treated as random effects, which are then integrated out for the estimation of the item category points. The subject points may subsequently be mapped into the item category configuration, once the latter is determined. This is akin to external unfolding. A similar idea may be exploited in unfolding analysis in general. The idea is also similar to that of Clarkson and Gonzales (2001) for dealing with individual differences weights in the weighted distance model.

4.3. *Ideal point discriminant analysis*

Ideal point discriminant analysis (IPDA), proposed by Takane (1987), is a method for ML unfolding analysis of two-way contingency tables. Entries in contingency tables are frequencies of joint occurrences of row and column categories, and thus indicate degrees of similarities between them. In IPDA, both row and column categories of contingency tables are represented as points in a multidimensional space. The probabilities of column categories given a specific row category are decreasing functions of the distances between the corresponding row and column points. The row and column points are located in such a way that these probabilities best agree with the observed frequencies of co-occurrences of rows and columns. By incorporating information regarding row categories, a variety of model comparisons are possible, and the best fitting model identified through the minimum AIC criterion. IPDA combines the spatial representation feature of dual scaling (Nishisato, 1980) or correspondence analysis (Greenacre, 1984), and the detailed model evaluation feature of log-linear models (e.g., Bishop et al., 1975) for analysis of contingency tables. (See Takane (1980) and Heiser (1981) for more details of the relationship between unfolding analysis and correspondence analysis. See also Takane et al. (1991) for methods of correspondence analysis with linear constraints.)

Let $\boldsymbol{F} = (f_{ki})$ denote an R by C contingency table, where f_{ki} denotes the observed frequency of row k and column i. Let $\boldsymbol{G}$ denote an R by P matrix of "predictor" variables on rows of the table. When no special structures are assumed on rows of the table, we may set $\boldsymbol{G} = \boldsymbol{I}$. We let $\boldsymbol{Y}$ denote an R by A matrix of the coordinates of the row points, where A indicates, as before, the dimensionality of the representation space. We assume that this $\boldsymbol{Y}$ is obtained by linear combinations of $\boldsymbol{G}$, that is, $\boldsymbol{Y} = \boldsymbol{GB}$, where $\boldsymbol{B}$ is a P by A matrix of weights. Let $\boldsymbol{X}$ denote the C by A matrix of coordinates of column points. We assume that $\boldsymbol{X}$ is obtained by a weighted average of $\boldsymbol{Y}$, namely $\boldsymbol{X} = \boldsymbol{D}_C^{-1}\boldsymbol{F}'\boldsymbol{Y}$, where $\boldsymbol{D}_C$ is a diagonal matrix with column totals of $\boldsymbol{F}$ in its diagonals. (This constraint on $\boldsymbol{X}$ is useful for avoiding degenerate solutions often encountered in unfolding analysis.)

Distances between row and column points can now be calculated by (10), where y_{ka} and x_{ia} are the kath and the iath elements of $\boldsymbol{Y}$ and $\boldsymbol{X}$, respectively. Let $p_{i|k}$ denote the

Table 3
Illustrative data for IPDA. The original data were provided in courtesy of Dr. Cristina Cook, Mental Care Development, Augusta, Maine, and were reported on page 225 of Agresti (1996)

Plotting symbol	Gender	Location	Seat-belt	Seriousness			
				A	B	C	Row total
fun	Female	Urban	No	7462	720	101	8283
fuy			Yes	11713	577	56	12346
frn		Rural	No	3319	710	190	4219
fry			Yes	6228	564	99	6891
mun	Male	Urban	No	10517	566	110	11193
muy			Yes	11052	259	38	11349
mrn		Rural	No	6264	710	233	7207
mry			Yes	6767	353	86	7206
Column total				63322	4459	913	68694

probability of column i given row k. We assume that this probability is given by

$$p_{i|k} = \frac{w_i \exp(-d_{ki}^2)}{\sum_{j=1}^{C} w_j \exp(-d_{kj}^2)}, \tag{11}$$

where w_i ($\geqslant 0$; $\sum_{i=1}^{C} w_i = 1$) indicates the bias for column i. The likelihood for the entire set of observed frequencies in a contingency table is now stated as

$$L \propto \prod_{k=1}^{R} \prod_{i=1}^{C} p_{i|k}^{f_{ki}}. \tag{12}$$

Parameters in the model ($\boldsymbol{B}$ and w_j, $j = 1, \ldots, C$) are estimated in such a way that (12) is maximized. Once the maximum likelihood is obtained, AIC can be calculated by (7) for various model comparisons. The information matrix, defined by (8), can be evaluated at the maximum likelihood estimates of the model parameters. The Moore–Penrose inverse of the information matrix provides variance–covariance estimates of the estimated parameters.

EXAMPLE 13 Seriousness of traffic accidents.

Table 3 (Agresti, 1996) presents observed frequencies of traffic accidents, cross-classified by the seriousness (3 column categories; A. not injured or injured but not transported by emergency medical services, B. injured and transported by emergency medical services but not hospitalized, and C. injured and hospitalized, or injured and died), and 8 accident profiles defined by three binary variables: 1. gender of the driver (male or female), 2. location of accidents (urban or rural), and 3. whether seat-belt was on (yes or no). We apply IPDA to find out how the accident profile variables affect the seriousness of accidents.

Table 4 summarizes the GOF statistics (AIC) for fitted models. The independence model refers to the model in which $p_{i|k} = f_{.i}/f_{..}$, and the saturated model to the model

Table 4
Goodness of fit statistics for various IPDA models

Model	Dimensionality	AIC	# of parameters
Saturated		666.4	(16)
Independence		2594.3	(2)
No constraints on rows	2	664.5	(15)
	1	713.5	(9)
Three main effects	2	656.9*	(7)
	1	715.1	(5)
Gender eliminated	2	1054.8	(4)**
Location eliminated	2	1531.7	(5)
Seat-belt eliminated	2	1495.1	(5)

*Minimum AIC solution.
**Solution almost unidimensional.

in which $p_{i|k} = f_{ki}/f_{k.}$, where $f_{k.}$, $f_{.i}$, and $f_{..}$ are the row total, the column total, and the grand total of $\boldsymbol{F}$, respectively. These non-spatial models serve as bench-mark models against spatial distance models. Table 4 indicates that neither the independence model nor the saturated model fit to the data as well as the best spatial distance model.

Assuming that a spatial representation is adequate, what is the most appropriate dimensionality of the representation space, and is there any simple and meaningful relationship among the row categories? The dimensionality comparison was carried out for both with and without assuming a special relationship among the row categories. Table 4 indicates that in both cases the two-dimensional representation provides a better fit. To see if there is any simple relationship among the row categories, the model without any constraints on rows ($\boldsymbol{G} = \boldsymbol{I}$) and the model with the main effects of the three profile variables as the predictor variables for the row categories were compared. Table 4 indicates that the latter fits to the data better. This means that the derived coordinates of the row categories can be well approximated by linear functions of the main effects of the three profile variables. The model without any constraints on rows, on the other hand, is equivalent to having all seven possible predictor variables that can be formed from the three profile variables (three main effects, three two-way interactions, and one three-way interaction effect). The result thus indicates that there are no significant interaction effects among the profile variables affecting the seriousness of accidents.

Given that no interaction effects are necessary, are all three main effects significant? To see this, one of the three profile variables was removed at a time, and the reduced models were fitted in turn. The results are reported at the bottom portions of Table 4. It is found that none of these reduced models fit as well as the two-dimensional three-main-effect model. This means that the three main effects are all significant. From the size of the increase in the value of AIC, the location variable is the most significant, while the seat-belt variable is a close second.

Figure 30 displays the configuration of the row and column points derived under the best fitting model. Accidents tend to be more serious when they occur in rural areas and when passengers wore no seat belts. Female drivers are slightly more prone to serious accidents than male drivers. Ellipses surrounding the points show 99% confidence

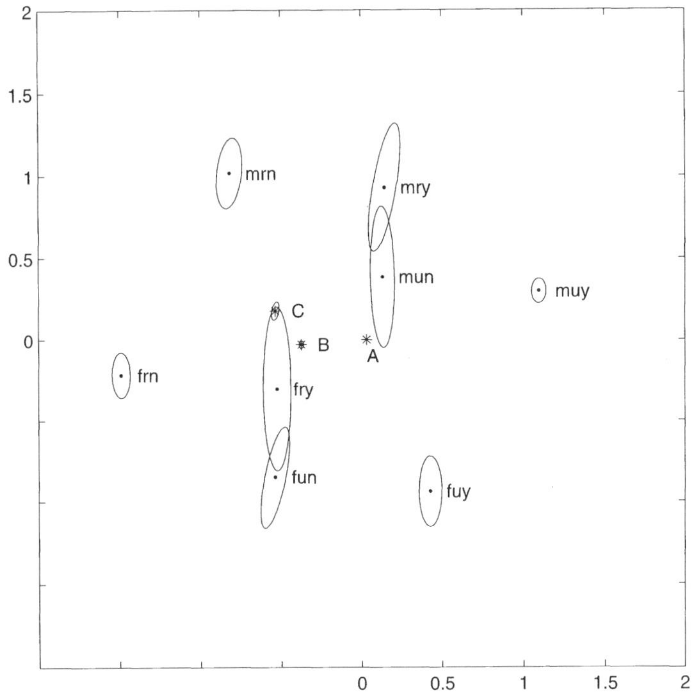

Fig. 30. The best two-dimensional solution with rows constrained by the three main effects: Estimated row and column points and their 99% confidence regions.

regions, indicating how reliably the point locations are estimated. They are all small, indicating that the point locations are very reliably estimated. This, of course, is due to the large sample size. (Note that the confidence region for column category A is so small that we cannot see it in the figure.)

Figure 31 depicts a similar configuration obtained under no constraints on the row categories. Confidence regions are still small, but are nonetheless uniformly larger than the corresponding ones in the previous figure, indicating that point locations are less accurately estimated. This shows the importance of having proper constraints to obtain more reliable estimates of the point locations.

We have quickly run through the kind of analyses that can be performed with IPDA. It was impossible to cover all important aspects of IPDA in this paper. We refer to Takane (1987, 1989, 1998), Takane et al. (1987), van der Heijden et al. (1994) for more detailed accounts of IPDA. De Rooij and Heiser (2005) recently proposed a similar method for analysis of two-way contingency tables.

5. Concluding remarks

Almost a quarter of a century has passed since two definitive articles on MDS (de Leeuw and Heiser, 1982; Wish and Carroll, 1982) appeared earlier in this series.

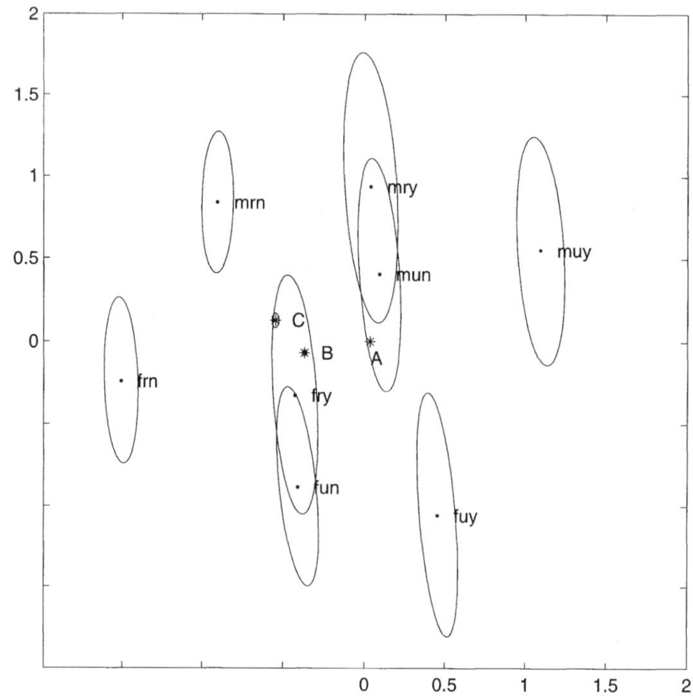

Fig. 31. The unconstrained two-dimensional solution: Estimated row and column points and their 99% confidence regions.

Many significant events took place since then, of which the most significant one was perhaps the so-called computer revolution. The advent of personal computers with massive computational power has had profound impact on the way we think about statistics and data analysis. MDS is no exception. The number of stimuli that can be realistically handled by MDS has been scaled up considerably. There has been a report of MDS with 3,648 stimuli (Buja et al., 2001), which was totally unthinkable 30 years ago. People tended to downplay importance of the local optimum problem. (See Arabie (1973) for some critique.) However, with today's computer it is not at all unrealistic to obtain 1,000 MDS solutions starting from 1,000 random initials to investigate the seriousness of the local optimum problem. Some attempts have also been made to ensure global optimum solutions algorithmically (Groenen and Heiser, 1996; Groenen et al., 1999). XGvis (Buja et al., 2001) allows to investigate, interactively and visually, the nature of locally optimal solutions. Interaction and visualization seem to be the key words for future software developments in MDS.

Nonlinear dimension reduction constitutes another area of computer intensive methods. See Tenenbaum et al. (2000), and Roweis and Saul (2000) for recent developments in this area. Proximity-based multivariate analysis (e.g., kernel methods) typically requires a large amount of computation because of its repetitive operations (inversion, spectral decomposition, etc.) applied to N by N matrices (kernel matrices), where N is the number of cases (subjects) in the data set.

The methodology of MDS is in a maturing stage. Counting only those published after 1990, at least four books (Borg and Groenen, 1997; Cox and Cox, 2001; Everitt and Rabe-Hesketh, 1997; Young and Hamer, 1994), and at least three review papers (Carroll and Arabie, 1998; Nosofsky, 1992; Wood, 1990) have been published. From an application perspective, Kruskal and Wish (1978) is becoming a classic in MDS. With the development of more flexible and more reliable algorithms for MDS and unfolding analysis, it is expected that its applications will grow faster both in number and variety.

References

Agresti, A. (1996). *An Introduction to Categorical Data Analysis*. Wiley, New York.

Arabie, P. (1973). Concerning Monte Carlo evaluation of nonmetric multidimensional scaling algorithms. *Psychometrika* **38**, 607–608.

Arabie, P. (1991). Was Euclid an unnecessarily sophisticated psychologist? *Psychometrika* **56**, 567–587.

Bishop, Y.M.M., Fienberg, S.E., Holland, P.W. (1975). *Discrete Multivariate Analysis: Theory and Practice*. MIT Press, Cambridge, MA.

Borg, I., Groenen, P. (1997). *Modern Multidimensional Scaling: Theory and Applications*. Springer, New York.

Buja, A., Swayne, D.F. (2002). Visualization methodology for multidimensional scaling. *Journal of Classification* **19**, 7–43.

Buja, A., Swayne, D.F., Littman, M.L., Dean, N., Hoffman, H. (2001). XGvis: Interactive data visualization with multidimensional scaling. *Journal of Computational and Graphical Statistics*, in press.

Busing, F.M.T.A., Commandeur, J.J.F., Heiser, W.J. (1997). PROXSCAL: A multidimensional scaling program for individual differences scaling with constraints. In: Bandilla, W., Faulbaum, F. (Eds.), *Softstat '97 Advances in Statistical Software*. Lucius, Stuttgart, Germany, pp. 237–258.

Busing, F.M.T.A., Groenen, P.J.F., Heiser, W. (2005). Avoiding degeneracy in multidimensional unfolding by penalizing on the coefficient of variation. *Psychometrika* **70**, 71–98.

Carroll, J.D. (1972). Individual differences and multidimensional scaling. In: Shepard, R.N., Romney, A.K., Nerlove, S.B. (Eds.), *Multidimensional Scaling. Vol. 1: Theory*. Seminar Press, New York, pp. 105–155.

Carroll, J.D., Arabie, P. (1998). Multidimensional scaling. In: Birnbaum, M.H. (Ed.), *Handbook of Perception and Cognition. Vol. 3: Measurement, Judgment and Decision Making*. Academic Press, San Diego, CA, pp. 179–250.

Carroll, J.D., Chang, J.J. (1970a). Individual differences and multidimensional scaling via an N-way generalization of Eckart–Young decomposition. *Psychometrika* **35**, 282–319.

Carroll, J.D., Chang, J.J. (1970b). Reanalysis of some color data of Helm's by INDSCAL procedure for individual differences multidimensional scaling. In: *Proceedings of the 78th Annual Convention of the American Psychological Association*, pp. 137–138.

Clarkson, D.B., Gonzales, R. (2001). Random effects diagonal metric multidimensional scaling models. *Psychometrika* **66**, 25–43.

Cox, T.F., Cox, M.A.A. (2001). *Multidimensional Scaling*, 2nd ed. Chapman and Hall, Boca Raton, FL.

Delbeke, L. (1978). Enkele analyses op voorkeuoordelen voor gezinssamenstellingen. Centrum voor Mathematishe Psychologie en Psychologische Methodologie. Katholieke Universiteit, Leuven.

de Leeuw, J. (1977). Applications of convex analysis to multidimensional scaling. In: Barra, J.R., Brodeau, F., Romier, G., van Cutsem, B. (Eds.), *Recent Development in Statistics*. North-Holland, Amsterdam, pp. 133–145.

de Leeuw, J. (1988). Convergence of the majorization method for multidimensional scaling. *Journal of Classification* **5**, 163–180.

de Leeuw, J., Heiser, W.J. (1977). Convergence of correction-matrix algorithms for multidimensional scaling. In: Lingoes, J.C. (Ed.), *Geometric Representations of Relational Data*. Mathesis Press, Ann Arbor, MI.

de Leeuw, J., Heiser, W.J. (1982). Theory of multidimensional scaling. In: Krishnaiah, R.P., Kanal, L.N. (Eds.), *Handbook of Statistics 2: Classification, Pattern Recognition, and Reduction of Dimensionality*. North-Holland, Amsterdam, pp. 285–316.

De Rooij, M., Heiser, W.J. (2005). Graphical representations and odds ratios in a distance-association models for the analysis of cross classified data. *Psychometrika* **70**, 99–122.

DeSarbo, W.S., Rao (1984). GENFOLD2: A set of models and algorithms for the GENeral unFOLDing analysis of preference and dominance data. *Journal of Classification* **1**, 147–186.

Dunn-Rankin, P., Leton, D.A. (1975). Differences between physical template matching and subjective similarity estimates of Japanese letters. *Japanese Psychological Research* **15**, 51–58.

Ekman, G. (1954). Dimensions of color vision. *Journal of Psychology* **38**, 467–474.

Everitt, B.S., Rabe-Hesketh, S. (1997). *The Analysis of Proximity Data*. Wiley, New York.

Gower, J.C. (1968). Adding a point to vector diagrams in multivariate analysis. *Biometrika* **55**, 582–585.

Greenacre, M.J. (1984). *Theory and Applications of Correspondence Analysis*. Academic Press, London.

Groenen, P.J.F., Heiser, W.J. (1996). The tunneling method for global optimization in multidimensional scaling. *Psychometrika* **61**, 529–550.

Groenen, P.J.F., Heiser, W.J., Meulman, J.J. (1999). Global optimization in least-squares multidimensional scaling by distance smoothing. *Journal of Classification* **16**, 225–254.

Heiser, W.J. (1981). Unfolding analysis of proximity data. Unpublished doctoral dissertation. Department of Data Theory, Leiden University, Leiden, The Netherlands.

Heiser, W.J. (1989). Order invariant unfolding analysis under smoothness restrictions. In: De Soete, G., Feger, H., Klauer, K.C. (Eds.), *New Developments in Psychological Choice Modeling*. North-Holland, Amsterdam, pp. 3–31.

Heiser, W.J. (1991). A generalized majorization method for least squares multidimensional scaling of pseudodistances that may be negative. *Psychometrika* **55**, 7–27.

Helm, C.E. (1964). Multidimensional ratio scaling analysis of perceived color relations. *Journal of the Optical Society of America* **54**, 256–262.

Hubert, L., Arabie, P., Hesson-Mcinnis, M. (1992). Multidimensional scaling in the city-block metric: A combinatorial approach. *Journal of Classification* **9**, 211–236.

Inukai, Y. (1981). Analysis of perceptual dimensions of schematic facial expressions via three-way multidimensional scaling. *Behaviormetrika* **9**, 1–20.

Kim, C., Rangswamy, A., DeSarbo, W.S. (1999). A quasi-metric approach to multidimensional unfolding for reducing the occurrence of degenerate solutions. *Multivariate Behavioral Research* **34**, 143–180.

Krantz, D.H., Tversky, A. (1975). Similarity of rectangles: An analysis of subjective dimensions. *Journal of Mathematical Psychology* **12**, 4–34.

Kruskal, J.B. (1964a). Multidimensional scaling by optimizing goodness of fit to a nonmetric hypothesis. *Psychometrika* **29**, 1–29.

Kruskal, J.B. (1964b). Nonmetric multidimensional scaling: A numerical method. *Psychometrika* **29**, 115–129.

Kruskal, J.B., Carroll, J.B. (1969). Geometric models and badness of fit functions. In: Krishnaiah, R.P. (Ed.), *Multivariate Analysis II*. Academic Press, New York, pp. 639–671.

Kruskal, J.B., Wish, M. (1978). *Multidimensional Scaling*. Sage Publications, Beverly Hills, CA.

Kruskal, J.B., Young, F.W., Seery, J.B. (1978). How to use KYST, a very flexible program to do multidimensional scaling and unfolding. Technical Report. Bell Laboratories, Murray Hill, NJ.

Meulman, J. (1992). The integration of multidimensional scaling and multivariate analysis with optimal transformations. *Psychometrika* **57**, 539–565.

Meulman, J., Verboon, P. (1993). Points of view analysis revisited: Fitting multidimensional structures to optimal distance components with cluster restrictions on the variables. *Psychometrika* **58**, 7–35.

Nishisato, S. (1980). *Analysis of Categorial Data: Dual Scaling and its Applications*. University of Toronto Press, Toronto.

Nosofsky, R.M. (1992). Similarity scaling and cognitive process models. *Annual Review of Psychology* **43**, 25–53.

Ramsay, J.O. (1977). Maximum likelihood estimation in multidimensional scaling. *Psychometrika* **42**, 337–360.

Ramsay, J.O. (1978). Confidence regions for multidimensional scaling analysis. *Psychometrika* **43**, 145–160.

Ramsay, J.O. (1980). Some small sample results for maximum likelihood estimation in multidimensional scaling. *Psychometrika* **45**, 141–146.
Ramsay, J.O. (1982). Some statistical approaches to multidimensional scaling data. *Journal of the Royal Statistical Society, Series A (General)* **145**, 285–312.
Rapoport, A., Fillenbaum, S. (1972). An experimental study of semantic structures. In: Romney, A.K., Shepard, R.N., Nerlove, S.B. (Eds.), In: *Multidimensional Scaling: Theory and Applications in the Behavioral Sciences*, vol. 2, pp. 93–131.
Rothkopf, E.Z. (1957). A measure of stimulus similarity and errors in some paired associate learning. *Journal of Experimental Psychology* **53**, 94–101.
Roweis, S.T., Saul, L.K. (2000). Nonlinear dimensionality reduction by locally linear embedding. *Science* **290**, 2323–2326.
Schiffman, S.S., Reynolds, M.L., Young, F.W. (1981). *Introduction to Multidimensional Scaling*. Academic Press, New York.
Schneider, B. (1972). Multidimensional scaling of color difference in the pigeon. *Perception and Psychophysics* **12**, 373–378.
Schönemann, P.H. (1972). An algebraic solution for the class of subjective metrics models. *Psychometrika* **37**, 441–451.
Schönemann, P.H. (1977). Similarity of rectangles. *Journal of Mathematical Psychology* **16**, 161–165.
Shepard, R.N. (1962). Analysis of proximities: Multidimensional scaling with an unknown distance function: I and II. *Psychometrika* **27**, 125–140. 219–246.
Shepard, R.N. (1963). Analysis of proximities as a technique for the study of information processing in man. *Human Factors* **5**, 19–34.
Shepard, R.N. (1964). Extracting latent structure from behavioral data. In: *Proceedings of the 1964 Symposium on Digital Computing*. Bell Telephone Laboratories.
Takane, Y. (1978). A maximum likelihood method for nonmetric multidimensional scaling: I. The case in which all empirical pairwise orderings are independent – theory and evaluations. *Japanese Psychological Research* **20**, 7–17. 105–114.
Takane, Y. (1980). Analysis of categorizing behavior by a quantification method. *Behaviormetrika* **8**, 75–86.
Takane, Y. (1981). Multidimensional successive categories scaling: A maximum likelihood method. *Psychometrika* **46**, 9–28.
Takane, Y. (1987). Analysis of contingency tables by ideal point discriminant analysis. *Psychometrika* **52**, 493–513.
Takane, Y. (1989). Ideal point discriminant analysis: Implications for multi-way data analysis. In: Coppi, R., Bolasco, S. (Eds.), *Multiway Data Analysis*. North-Holland, Amsterdam, pp. 287–299.
Takane, Y. (1994). A review of applications of AIC in psychometrics. In: Bozdogan, H. (Ed.), *Proceedings of the First US/Japan Conference on the Frontiers of Statistical Modeling: An Informational Approach*. Kluwer Academic Publisher, Dordrecht, pp. 379–403.
Takane, Y. (1996). An item response model for multidimensional analysis of multiple-choice questionnaire data. *Behaviormetrika* **23**, 153–167.
Takane, Y. (1997). Choice model analysis of the pick any/n type of binary data. *Japanese Psychological Research* **40**, 31–39.
Takane, Y. (1998). Visualization in ideal point discriminant analysis. In: Blasius, J., Greenacre, M.J. (Eds.), *Visualization of Categorical Data*. Academic Press, New York, pp. 441–459.
Takane, Y., Bozdogan, H., Shibayama, T. (1987). Ideal point discriminant analysis. *Psychometrika* **52**, 371–392.
Takane, Y., Carroll, J.D. (1981). Nonmetric maximum likelihood multidimensional scaling from directional rankings of similarities. *Psychometrika* **46**, 389–405.
Takane, Y., Sergent, J. (1983). Multidimensional scaling models for reaction times and same–different judgments. *Psychometrika* **48**, 393–423.
Takane, Y., Shibayama, T. (1986). Comparison of models for the stimulus recognition data. In: de Leeuw, J., Heiser, W.J., Meulman, J., Critchley, F. (Eds.), *Multidimensional Data Analysis*. DSWO Press, Leiden, pp. 119–138.
Takane, Y., Shibayama, T. (1992). Structures in stimulus identification data. In: Ashby, F.G. (Ed.), *Probabilistic Multidimensional Models of Perception and Cognition*. Earlbaum, Hillsdale, NJ, pp. 335–362.

Takane, Y., van der Heijden, P.G.M., Browne, M.W. (2003). On likelihood ratio tests for dimensionality selection. In: Higuchi, T., Iba, Y., Ishiguro, M. (Eds.), *Proceedings of Science of Modeling: The 30th Anniversary Meeting of the Information Criterion (AIC)*. In: *Report on Research and Education*, vol. 17. The Institute of Statistical Mathematics, Tokyo, pp. 348–349.

Takane, Y., Yanai, H., Mayekawa, S. (1991). Relationships among several methods of linearly constrained correspondence analysis. *Psychometrika* **56**, 667–684.

Takane, Y., Young, F.W., de Leeuw, J. (1977). Nonmetric individual differences multidimensional scaling: An alternating least squares method with optimal scaling features. *Psychometrika* **42**, 7–67.

Tenenbaum, J.B., de Silva, V., Langford, J.C. (2000). A global geometric framework for nonlinear dimensionality reduction. *Science* **290**, 2319–2322.

Torgerson, W.S. (1952). Multidimensional scaling: I. Theory and method. *Psychometrika* **17**, 401–409.

van der Heijden, P.G.M., Mooijaart, A., Takane, Y. (1994). In: Greenacre, M.J., Blasius, J. (Eds.), *Correspondence Analysis in the Social Sciences*. Academic Press, New York, pp. 79–111.

Van Deun, K., Groenen, P.J.F., Heiser, W.J., Busing, F.M.T.A., Delbeke, L. (2005). Interpreting degenerate solutions in unfolding by use of the vector model and the compensatory distance model. *Psychometrika* **70**, 45–69.

Wish, M., Carroll, J.D. (1982). Multidimensional scaling and its applications. In: Krishnaiah, R.P., Kanal, L.N. (Eds.), *Handbook of Statistics 2: Classification, Pattern Recognition, and Reduction of Dimensionality*. North-Holland, Amsterdam, pp. 317–345.

Wood, P. (1990). Application of scaling to developmental research. In: von Eye, A. (Ed.), *Statistical Methods in Longitudinal Research, Vol. I: Principles and Structural Change*. Assessment Systems Corporation, St. Paul, MN, pp. 225–256.

Young, F.W., Hamer, R.M. (1994). *Theory and Applications of Multidimensional Scaling*. Erlbaum Associates, Hillsdale, NJ.

Young, G., Householder, A.S. (1938). Discussion of a set of points in terms of their mutual distances. *Psychometrika* **3**, 19–22.

Handbook of Statistics, Vol. 26
ISSN: 0169-7161

DOI: 10.1016/S0169-7161(06)26012-7

12

Multilevel Models in Psychometrics

Fiona Steele and Harvey Goldstein

1. Introduction

Before describing the basic multilevel model, it is useful to reflect on why such models are useful. For many years, social researchers, especially in education, discussed the 'units of analysis' problem, one version of which has also been called the 'ecological fallacy' (see Robinson, 1951). At one extreme, it is possible to study relationships among variables ignoring group structures. At the other extreme we can work solely with group, say school, averages in exploring relationships. Aitkin and Longford (1986) set out the statistical issues associated with various procedures. In an earlier analysis Aitkin et al. (1981) reworked a well-known study on teaching styles which used student level data but ignored school membership (Bennett, 1976). They showed that formerly 'significant' results became non-significant when a multilevel model was used. Woodhouse and Goldstein (1989) showed how the use solely of aggregate level data based upon school means could lead to unstable and misleading conclusions. In addition to the problem of making misleading inferences, failure to model both students and schools simultaneously makes it impossible to study the extent to which school and student characteristics interact to influence the response measurement or measurements. A useful compilation of some early work in the area of psychometrics is the volume edited by Bock (1989).

Multilevel models overcome these problems wherever we have any kind of hierarchical structure, such as individuals grouped within households, themselves grouped within areas (a three-level model). Repeated measures data are an example of a two-level hierarchy with measurement occasions (level 1 units) grouped within individuals (level 2 units). We can readily incorporate multivariate data within this framework. In the simple case of a set of responses on a sample of individuals we think of the set of response variables as forming the lowest level of the data hierarchy; that is, measurements are nested within individuals. Further levels, such as school, can then be added above that of individual. Models for repeated measures and multivariate data are discussed in Sections 3 and 4 and structural equation models are introduced in Section 6.

While the ability to fit models to hierarchically structured data is important, there are many applications where the data structures are more complex. Suppose a student is classified as belonging sequentially to a particular combination of primary (elementary) school and secondary (high) school and we have followed a sample of such students through each school and wish to relate measurements made at the end of secondary school to those made earlier in the primary schools. The students will be identified by a *cross classification* of primary schools and secondary schools. Note that even if we did not have prior measurements, but *did* have identification of the primary and secondary schools we could still carry out a cross classified analysis. Another example is where students are simultaneously classified by the school they attend and the area where they live, both classifications affecting the value of a response variable. Models for such cross classifications will be discussed in Section 5.

A further complication occurs when we cannot assign a lower level unit to a single higher level unit. Suppose that, during secondary schooling many pupils move between schools. If our response is, say, a test score at the end of secondary school, then for such a student we will need to share out the school 'effect' among all the schools attended, using a suitable weighting function. Another example occurs with spatial data where the affect of area will be shared among the area where someone resides and surrounding areas, with weights a function of geographical or other distance. Such models find considerable use in epidemiological studies. These multiple membership models will also be discussed in Section 5.

2. Basic Models for Two-level Hierarchical Data Structures

We begin with a description of simple multilevel models for data from populations with a two-level hierarchical structure. More detailed accounts of these and more general models can be found in Bryk and Raudenbush (2002), Goldstein (2003), Longford (1993) and Snijders and Bosker (1999).

2.1. Random intercept model

For simplicity consider a simple data structure where a response y_{ij} is measured on individual i in school j ($i = 1, \dots, n_j$; $j = 1, \dots, J$), together with a single explanatory variable x_{ij}. For example, the response might be an examination score measured on students at age 16 years and the explanatory variable a test score measured on the same students five years earlier at age 11. Instead of schools we could think of any grouping of individuals, such as households or areas. We wish to model a relationship between the individual response and the explanatory variable, taking into account the effect of school on the mean response. We shall assume in what follows that we are dealing with a continuously distributed response, and for simplicity that this has a Normal distribution. Our data have a simple two-level structure with the schools as higher level units and students as lower level units. The simplest multilevel model that we can fit to such a structure can be written as follows

$$\begin{aligned} y_{ij} &= \beta_{0j} + \beta_1 x_{ij} + e_{ij}, \\ \beta_{0j} &= \beta_0 + u_{0j}, \end{aligned} \tag{2.1}$$

where $\text{var}(e_{ij}) = \sigma_e^2$ and $\text{var}(u_{0j}) = \sigma_{u0}^2$.

The model may also be written as a single equation

$$y_{ij} = \beta_0 + \beta_1 x_{ij} + u_{0j} + e_{ij},$$

where $\beta_0 + \beta_1 x_{ij}$ is referred to as the *fixed part* of the model, and $u_{0j} + e_{ij}$ is the *random part*. The random variable u_{0j} represents the departure of the jth school's intercept from the overall population intercept term β_0. The slope coefficient β_1 is for the present assumed to be the same for all the schools. As mentioned we shall develop the model initially assuming that the random variables have a Normal distribution: $e_{ij} \sim N(0, \sigma_e^2)$, $u_{0j} \sim N(0, \sigma_{u0}^2)$. This model is sometimes called a variance components model, owing to the fact that the residual variance is partitioned into components corresponding to each level in the hierarchy. The variance between schools is σ_{u0}^2, and the variance between students within a given school is σ_e^2.

The similarity between individuals in the same school is measured by the *intra-class correlation* (where, here, 'classes' are schools):

$$\frac{\sigma_{u0}^2}{\sigma_{u0}^2 + \sigma_e^2}.$$

The intra-class correlation (ICC) measures the extent to which the y-values of students in the same school resemble each other as compared to those from individuals in different schools. The ICC may also be interpreted as the proportion of the total residual variation that is due to differences between schools, and is referred to as the *variance partition coefficient* (VPC) as this is the more usual interpretation (see Goldstein, 2003, pp. 16–17).

Having fitted (2.1) we can obtain estimates for the residuals (u_{0j}, e_{ij}) by estimating their expected values, given the data and estimates of the parameters $(\beta_0, \beta_1, \sigma_e^2, \sigma_{u0}^2)$. Of particular interest in our example are estimates of the level 2 residuals u_{0j}, which represent school effects on attainment at age 16, adjusted for prior attainment x_{ij}. In ordinary least squares (OLS) regression residual estimates are obtained simply by subtracting the predicted values of y_{ij} from their observed values, i.e., $r_{ij} = y_{ij} - \hat{y}_{ij}$. In multilevel models, with residuals at more than one level, a more complex procedure is needed. Estimates of u_{0j} are obtained by taking the average of the raw residuals r_{ij} for each school j and multiplying the result by a *shrinkage factor*. This shrinkage factor pulls the estimate of u_{0j} towards zero when the between-school variance σ_{u0}^2 is small relative to the within-school variance σ_e^2, or when the number of students sampled in a school n_j is small.

2.2. *Fixed versus random effects*

An alternative way of allowing for school effects would be to use an analysis of variance (ANOVA) or fixed effects model, which would involve including as explanatory variables a set of dummy variables that indicate the school to which a student belongs. While ANOVA can also be used to compare any number of schools, the random effects approach has a number of advantages over fixed effects models. First, if there are J schools to be compared, then $J-1$ parameters are required to capture school effects and, therefore, if J is large, a large number of parameters need to be estimated. In contrast, in a random effects model only one additional parameter, the between-school variance σ^2_{u0}, is estimated regardless of the number of schools.

Second, the origins of ANOVA lie in experimental design where there are typically a small number of groups under comparison and all groups of interest are sampled. Often we have only a sample of groups (e.g., a sample of schools) and it is the population of groups from which our sample was drawn which is of interest. The ANOVA model does not allow inferences to be made beyond the groups in the sample. The key point about the random variable u_{0j} in the random effects model is that it allows us to treat the samples of units as coming from a universe or population of such units. Thus, the schools (and students) chosen are not typically the principal focus of interest; they are regarded as a random sample from a population of schools and we are concerned with making statements about that population, for example in terms of its mean and variance.

Finally, in a fixed effects model the effects of level 2 explanatory variables cannot be separately estimated. Such variables are confounded with the level 2 effects because any level 2 variable can be expressed as a linear combination of the dummy variables for higher level units. This is a serious drawback of the ANOVA approach as one is often interested in exploring the extent to which the level 2 variation can be explained by observed level 2 characteristics. In the random effects model level 2 variables are not confounded with the level 2 effects u_{0j}, and therefore their effects may be estimated while simultaneously accounting for level 2 variance due to unobserved factors.

2.3. Random slope model

We can elaborate (2.1) by allowing the coefficient β_1 to vary across schools:

$$y_{ij} = \beta_{0j} + \beta_{1j}x_{ij} + e_{ij}, \quad \beta_{0j} = \beta_0 + u_{0j}, \quad \beta_{1j} = \beta_1 + u_{1j}, \tag{2.2}$$

where

$$e_{ij} \sim N(0, \sigma^2_e), \qquad \begin{pmatrix} u_{0j} \\ u_{1j} \end{pmatrix} \sim N(0, \Omega_u), \qquad \Omega_u = \begin{pmatrix} \sigma^2_{u0} & \\ \sigma_{u01} & \sigma^2_{u1} \end{pmatrix}.$$

Model (2.2) is often referred to as a *random coefficient* model by virtue of the fact that the coefficients β_{0j} and β_{1j} in the first equation of (2.2) are random quantities, each having a variance with a covariance between them. Now the slope for school j, β_{1j}, is given by β_1, the average slope across all schools, plus a random departure u_{1j}. The terms u_{0j} and u_{1j} are random departures from β_0 and β_1, or residuals at the school level. As more explanatory variables are introduced into the model, so we can choose to make their coefficients random at the school level thereby introducing further variances and covariances, and this will lead to models with complex covariance structures. One of the aims of multilevel modelling is to explore such potential structures and also to attempt to explain them in terms of further variables.

When coefficients of explanatory variables are permitted to vary randomly across level 2 units, the level 2 variance is no longer constant but depends on those variables with random coefficients. For example, in model (2.2) the school-level variance is

$$\mathrm{var}(u_{0j} + u_{1j}x_{ij}) = \sigma^2_{u0} + 2\sigma_{u01}x_{ij} + \sigma^2_{u1}x^2_{ij}, \tag{2.3}$$

i.e., a quadratic function of x_{ij}.

2.4. *Complex level 1 variation*

We have seen how in a random coefficient model, the level 2 variance is a function of explanatory variables (Eq. (2.3)). We can also allow the level 1 variance to depend on explanatory variables. This leads to a model with *complex level 1 variation* or *heteroskedasticity* at level 1.

Suppose, for example, that we wish to explore whether the between-student variation within schools differs for boys and girls. We introduce a second explanatory variable x_{2ij} which indicates a student's gender (coded 1 for girls and 0 for boys). We then specify a model in which both the mean and variance of y_{ij} depend on gender by including x_{2ij} in the fixed part of the model and allowing separate level 1 residuals for girls and boys. The random intercept version of this model can be written

$$\begin{aligned} y_{ij} &= \beta_{0j} + \beta_1 x_{1ij} + \beta_2 x_{2ij} + e_{0ij}(1 - x_{2ij}) + e_{1ij} x_{2ij}, \\ \beta_{0j} &= \beta_0 + u_{0j}. \end{aligned} \tag{2.4}$$

From (2.4) the student-level variance is $\sigma_{e0}^2(1 - x_{2ij}) + \sigma_{e1}^2 x_{2ij}$ which reduces to σ_{e0}^2 for boys and σ_{e1}^2 for girls.

2.5. *Example*

We now give examples of applications of the models described above using educational data on 4059 London school children from 65 schools (see Rasbash et al., 2005, for further details). The dependent variable is a normalised exam score at age 16. We consider two explanatory variables: a standardised reading test score at age 11 and gender (coded 1 for girls and 0 for boys). Table 1 shows the results from fitting three two-level models to these data. Also presented are the -2 log-likelihood values for the fitted models, which may be used to carry out likelihood ratio tests to compare the fit of nested models.

We begin by considering a simple random intercept model. From this model the ICC (or VPC) is estimated as $0.088/(0.088 + 0.562) = 0.135$. Thus, after accounting for the effects of prior attainment and gender, 13.5% of the remaining variance in age 16 scores is due to differences between schools.

The next model fitted is an elaboration of the random intercept model in which the effect of the age 11 reading test score on attainment at age 16 is allowed to vary across schools, a random slope (or coefficient) model. The difference in -2 log-likelihood value (i.e., deviance change) between this and the random intercept model is 42.6, on 2 degrees of freedom. We therefore conclude that the random slope model is a significantly better fit to the data. In this model, the individual school slopes vary about a mean of 0.553 with estimated variance 0.019. There is a positive covariance between the intercepts and slopes, estimated as 0.015, suggesting that schools with higher intercepts tend to have steeper slopes; on average the effect of prior attainment is stronger in schools with a high mean age 16 score. We note that this conclusion is not invariant under shifts in the values of predictors with random coefficients. In this example, the intercept for a given school is that school's mean attainment at age 16 for boys with a

Table 1
Results from fitting two-level models to educational attainment at age 16

Parameter	Random intercept		Random slope for age 11 score		Complex level 1 variance	
	Est.	(SE)	Est.	(SE)	Est.	(SE)
Fixed						
β_0 (Intercept)	−0.095	(0.043)	−0.112	(0.043)	−0.112	(0.043)
β_1 (Age 11 score)	0.560	(0.012)	0.553	(0.020)	0.553	(0.020)
β_2 (Girl)	0.171	(0.033)	0.176	(0.032)	0.175	(0.032)
Random: Between-school						
σ_{u0}^2 (Intercept)	0.088	(0.017)	0.086	(0.017)	0.086	(0.017)
σ_{u1}^2 (Age 11 score)	–	–	0.015	(0.004)	0.015	(0.004)
σ_{u01} (Intercept/slope covariance)	–	–	0.019	(0.007)	0.019	(0.007)
Random: Within-school						
σ_e^2 (Total)	0.562	(0.013)	0.550	(0.012)	–	–
σ_{e0}^2 (Boys)	–	–	–	–	0.587	(0.021)
σ_{e1}^2 (Girls)	–	–	–	–	0.525	(0.015)
−2 log-likelihood	9330.0		9287.4		9281.4	

mean age 11 score. A different estimate would be obtained for the intercept-slope covariance if age 11 scores were not centred. The positive covariance also implies that the between-school variance increases with prior attainment (see Eq. (2.3)).

The final model allows the within-school between-student residual variance to depend on gender. A likelihood ratio test statistic for a comparison of this model with the constant level 1 variance model is 6 on 1 d.f., so there is strong evidence of heteroskedasticity at the student level. The estimated within-school variance is 0.587 for boys and 0.525 for girls. Thus while on average boys have a lower age 16 score than girls, their scores have greater dispersion than the girls'.

3. Models for repeated measures

The models described in the previous section can also be applied in the analysis of repeated measures where observations over time are the level 1 units and individuals are at level 2. We illustrate multilevel modelling of longitudinal data using a dataset consisting of nine measurements made on 26 boys between the ages of 11 and 13.5 years, approximately three months apart (Goldstein, 1989). Figure 1 shows the mean heights by the mean age at each measurement occasion.

We assume that growth can be represented by a polynomial function with coefficients varying randomly across individuals. Other functions are possible, including fractional

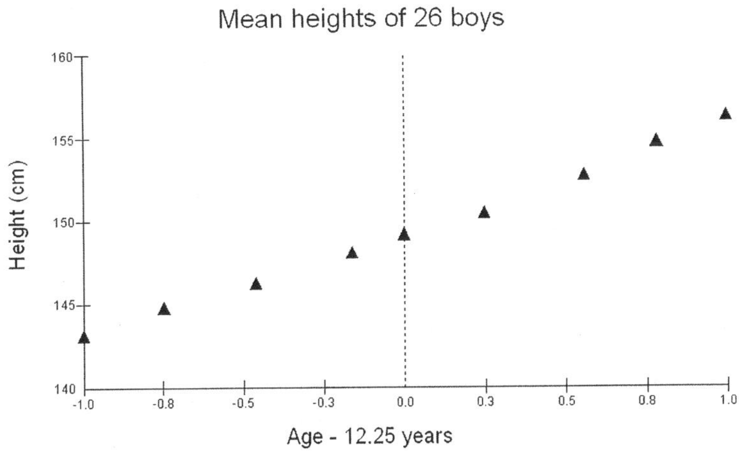

Fig. 1. Mean height by mean age at nine measurement occasions.

polynomials or nonlinear functions, but for simplicity we confine ourselves to examining a fourth order polynomial centred at an origin of 12.25 years. The model we fit can be written as follows:

$$
\begin{aligned}
&y_{ij} = \sum_{h=0}^{4} \beta_{hj} t_{ij}^{h} + e_{ij}, \\
&\beta_{0j} = \beta_0 + u_{0j}, \qquad \beta_{1j} = \beta_1 + u_{1j}, \qquad \beta_{2j} = \beta_2 + u_{2j}, \\
&\beta_{3j} = \beta_3, \qquad \beta_{4j} = \beta_4
\end{aligned}
\tag{3.1}
$$

where

$$
\begin{pmatrix} u_{0j} \\ u_{1j} \\ u_{2j} \end{pmatrix} \sim N(0, \Omega_u), \qquad \Omega_u = \begin{pmatrix} \sigma_{u0}^2 & & \\ \sigma_{u01} & \sigma_{u1}^2 & \\ \sigma_{u02} & \sigma_{u12} & \sigma_{u2}^2 \end{pmatrix},
$$

$$
e_{ij} \sim N\left(0, \sigma_e^2\right).
$$

Eq. (3.1) defines a two-level model with level 1 being 'measurement occasion' and level 2 'individual boy'. Note that we allow only the coefficients up to the second order to vary across individuals; in the present case this provides an acceptable fit. The level 1 residual term e_{ij} represents the unexplained variation within individuals about each individual's growth trajectory. Table 2 shows the maximum likelihood parameter estimates for this model.

For each boy we can also estimate their random effects or 'residuals' (u_{0j}, u_{1j}, u_{2j}), and use these to predict their growth curve at each age. Figure 2 shows these predicted curves. Growth over this period exhibits a seasonal pattern with growth in the summer being about 0.5 cm greater than growth in the winter. Since the period of the growth cycle is a year this is modelled by including a simple cosine term, which could also have a random coefficient.

Table 2
Height modelled as a fourth degree polynomial on age

Fixed effects	Estimate	Standard error	
β_0 (Intercept)	149.0	1.54	
β_1 (t)	6.17	0.35	
β_2 (t^2)	1.13	0.35	
β_3 (t^3)	0.45	0.16	
β_4 (t^4)	−0.38	0.30	
Random effects			
Level 2 (individual) correlation matrix: variances on diagonal			
	u_{0j} (Intercept)	u_{1j} (t)	u_{2j} (t^2)
u_{0j} (Intercept)	61.6		
u_{1j} (t)	0.61	2.8	
u_{2j} (t^2)	0.22	0.66	0.64
Level 1 (occasion) variance = 0.22			
−2 log-likelihood = 625.4			

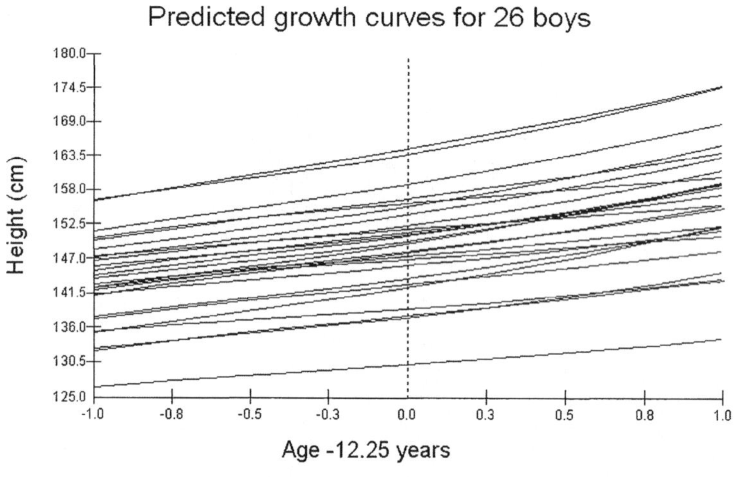

Fig. 2. Predicted growth curves from model with random coefficients for linear and quadratic terms.

In our example we have a set of individuals all of whom have nine measurements. This restriction, however, is not necessary and (3.1) does not require either the same number of occasions per individual nor that measurements are made at equal intervals, since time is modelled as a continuous function. In other words we can combine data from individuals with very different measurement patterns, some of whom may only have been measured once and some who have been measured several times at irregular intervals. This flexibility, first noted by Laird and Ware (1982), means that the multilevel approach to fitting repeated measures data is to be preferred to previous methods based upon a traditional multivariate formulation assuming a common set of fixed occasions.

In these models it is assumed that the level 1 residual terms (e_{ij}) are independently distributed. We may relax this assumption, however, and in the case of repeated measures data this may be necessary, for example where measurements are taken very close together in time. Goldstein et al. (1994) show how to model quite general nonlinear covariance functions and in particular of the form $\text{cov}(e_t, e_{t-s}) = \sigma_e^2 \exp(-g(\alpha, s))$, where s is the time difference between occasions. This allows the correlation between occasions to vary smoothly as a function of their (continuous) time difference. A simple example is where $g = \alpha s$, which in discrete time produces a first-order autoregression, AR(1), model.

For further discussion of multilevel modeling of longitudinal data see Singer and Willett (2003) and Muthén (1997).

4. Models for multivariate response data

Repeated measures data consist of multiple observations on individuals over time. Suppose again that we have more than one observation per individual, but that the observations are for different variables, e.g., responses to a set of items in a questionnaire. In the examples given below we have multivariate responses on students within schools, where y_{rij} denotes the rth response ($r = 1, \ldots, R$) on individual i in school j. In this section we consider two types of multilevel model for multivariate response data. We begin with a multivariate model, which is appropriate when the focus is on estimating the effects of explanatory variables on each response, while accounting for the correlation between responses at the student and school level. We then consider a multilevel factor model which assumes that the pairwise correlations between responses are explained by their mutual dependency on one or more latent variables (factors).

4.1. Multivariate models

A two-level random intercept model for multivariate response data may be written

$$y_{rij} = \beta_{0r} + \beta_{1r} x_{ij} + u_{0rj} + e_{rij}, \tag{4.1}$$

where $u_{0j} = \{u_{0rj}\} \sim N_R(0, \Omega_u)$ and $e_{ij} = \{e_{rij}\} \sim N_R(0, \Omega_e)$.

Eq. (4.1) can be viewed as a 3-level model where the level 1 units are the within-student measurements. The explanatory variables are a set of R dummy variables that indicate the responses (with coefficients β_{0r}), and their interactions with the covariate x_{ij} (with coefficients β_{1r}). The coefficients of the dummy variables are assumed to vary randomly across students (at level 2) and schools (level 3) to obtain the student and school residuals, e_{rij} and u_{0rj}.

There are two main advantages to fitting a multivariate model rather than carrying out a separate analysis for each response. First, we can obtain estimates of the pairwise correlations between responses at each level adjusting for the effects of x_{ij}.

Table 3
Results from fitting a bivariate multilevel model to written paper and coursework exam scores

	Written		Coursework	
	Estimate	(SE)	Estimate	(SE)
Intercept	49.4		69.7	
Female	−2.5	(0.6)	6.8	(0.7)
School-level				
Variance	46.8	(9.2)	75.2	(14.6)
Covariance	24.9	(8.9)	–	–
(Correlation)	*(0.4)*	–		
Student-level	124.6	(4.4)	180.1	(6.2)
Variance				
Covariance	73.0	(4.2)	–	–
(Correlation)	*(0.5)*	–		

Second, individuals with missing data on one or more response can be retained in the analysis; under a 'missing at random' assumption efficient estimates of coefficients and covariance structures at each level are obtained. This relaxation of the requirement for balanced multivariate data is particularly useful for the analysis of data from rotation designs where each respondent answers a random subset from a pool of questions.

We illustrate the use of multivariate models in an analysis of students' scores on two components of a science examination taken by 1905 students in 73 English schools (see Rasbash et al., 2005, for details). The first component is a traditional written paper, and the second is a piece of coursework. A total of 382 responses were missing, but students with only one missing score remain in the analysis sample. The scores on both components have been rescaled so that their maximum is 100, thus enabling comparison of covariate effects across responses. We consider one covariate, the student's gender (coded 1 for female and 0 for male). The results from fitting model (4.1) to these data are presented in Table 3. From the fixed part estimates we conclude that while girls perform worse than boys on the written paper, they do better on the coursework. Turning to the random part of the model, we see that there is greater variation in coursework scores at both the student and school level. The correlations between the coursework and written scores at the student and school level are, respectively, 0.4 and 0.5.

4.2. *Multilevel factor models*

In a multilevel factor model, the student and school level correlations between pairs of responses are assumed to be explained by one or more factors at each level. For simplicity, we assume that there is a single factor at each level. The factor model is an extension of (4.1) and can be written

$$y_{rij} = \beta_{0r} + \beta_{1r}x_{ij} + \lambda_r^{(2)}\eta_j^{(2)} + \lambda_r^{(1)}\eta_{ij}^{(1)} + u_{0rj} + e_{rij}, \tag{4.2}$$

where $\eta_{ij}^{(1)}$ and $\eta_j^{(2)}$ are the factors at the student and school level respectively, and $\lambda_r^{(1)}$ and $\lambda_r^{(2)}$ are the factor loadings. The factors are assumed to be normally distributed. Conditional on the factors and x_{ij}, the responses are assumed to be independent. Thus the residuals u_{0rj} and e_{rij} (often referred to as 'uniquenesses') are now assumed to be uncorrelated across responses. See Goldstein and Browne (2002), Muthén (1994), Skrondal and Rabe-Hesketh (2004) and Steele (2005) for further discussion of multilevel factor analysis.

We illustrate the application of multilevel factor analysis using a dataset of science scores for 2439 students in Hungarian schools (Goldstein, 2003, Chapter 6). The data consist of scores on four test booklets: a core booklet with components in earth science, physics and biology, two biology booklets and one in physics. There are therefore six possible test scores (one earth science, three biology, and two physics). Each student responds to a maximum of five tests, the three tests in the core booklet plus a randomly selected pair of tests from the other booklets. The analysis presented below is based on standardised test scores.

The results from a two-level factor model, with a single factor at each level, are presented in Table 4. The factor variances at the student and school level are estimated as 0.127 (SE = 0.016) and 0.057 (SE = 0.024), respectively. For ease of interpretation, standardised leadings are calculated for each factor as (omitting subscripts) $\lambda_r^{(k)*} = \lambda_r^{(k)}\sqrt{\mathrm{var}(\eta^{(k)})}$, $k = 1, 2$. Because, at each level, the standardised loadings have the same sign across responses, we interpret the factors as student- and school-level measures of overall attainment in science. Biology R3 has the lowest loading. The poor fit for this test is reflected in a relatively high residual variance estimate at both levels. Thus only a small amount of the variance in the scores for this biology test is explained by the student and school level factors, i.e., the test has a low communality.

Table 4
Estimates from two-level factor model with one factor at each level

	Student-level			School-level		
	$\lambda_r^{(1)}$ (SE)	$\lambda_r^{(1)*}$	σ_{er}^2 (SE)	$\lambda_r^{(2)}$ (SE)	$\lambda_r^{(2)*}$	σ_{ur}^2 (SE)
E. Sc. core	1[†]	0.36	0.712 (0.023)	1[†]	0.24	0.098 (0.021)
Biol. core	1.546 (0.113)	0.55	0.484 (0.022)	2.093 (0.593)	0.50	0.015 (0.011)
Biol. R3	0.583 (0.103)	0.21	0.892 (0.039)	0.886 (0.304)	0.21	0.033 (0.017)
Biol. R4	1.110 (0.115)	0.40	0.615 (0.030)	1.498 (0.466)	0.36	0.129 (0.029)
Phys. core	1.665 (0.128)	0.59	0.422 (0.022)	2.054 (0.600)	0.49	0.036 (0.013)
Phys. R2	1.558 (0.133)	0.56	0.526 (0.030)	1.508 (0.453)	0.36	0.057 (0.018)

[†]Parameter constrained for model identification.

5. Models for non-hierarchical structures

5.1. Cross classified models

In the example of children moving from primary to secondary school we have a cross classified structure which can be modelled as follows:

$$y_{i(j_1 j_2)} = (X\beta)_{i(j_1 j_2)} + u_{j_1} + u_{j_2} + e_{i(j_1 j_2)}, \tag{5.1}$$

$(j_1 = 1, \dots, J_1; j_2 = 1, \dots, J_2; i = 1, \dots, n)$, in which the score of student i, belonging to the combination of primary school j_1 and secondary school j_2, is predicted by a linear 'regression' function denoted by $(X\beta)_{i(j_1, j_2)}$. The random part of the model is given by two level 2 residual terms, one for the primary school attended by the student (u_{j_1}) and one for the secondary school attended (u_{j_2}), together with the usual level 1 residual term for each student. We note that the latter may be further modelled to produce complex level 1 variation (see Section 2.4), allowing for example for separate variances for males and females.

As an example consider the analysis carried out by Goldstein (2003, Chapter 11) who fitted cross classified models to 3435 students who attended 148 primary schools and 19 secondary schools in Fife, Scotland. The dependent variable is the overall exam attainment at age 16 and a verbal reasoning score measured on entry to secondary school is included as an explanatory variable. The principal aim was to separate the effect of primary school attended from that of secondary school. Table 5 shows results from two alternative model specifications: a two-level hierarchical model, which ignores the information on primary schools, and the cross classified model.

From the two-level model, the type of model typically used in school effectiveness studies, we would estimate that secondary schools explain $0.28/(0.28 + 4.26) \times 100 = 6.2\%$ of the residual variance in age 16 scores (after accounting for age 12 verbal reasoning). The cross classified model takes into account both secondary and primary school

Table 5
Estimates from analysis of examination scores using hierarchical and cross classified models

	2-level model		Cross classified model	
	Estimate	(SE)	Estimate	(SE)
Fixed				
Intercept	5.99		5.98	
Verbal reasoning	0.16	(0.003)	0.16	(0.003)
Random				
σ^2_{u1} (primary)	–	–	0.27	(0.06)
σ^2_{u2} (secondary)	0.28	(0.06)	0.011	(0.021)
σ^2_e (student)	4.26	(0.10)	4.25	(0.10)
-2 log-likelihood	17172.0		14845.6	

effects on age 16 attainment, and is clearly a much better fit to the data than the two-level hierarchical model (LR statistic = 2326.4, 1 d.f.). From the cross classified model we would conclude that it is primary schools rather than secondary schools that have the strongest effect on attainment; the proportion of variance due to secondary schools is now only 0.24%, compared to 6.0% due to primary schools. The substantive importance of this finding for studies of schooling is that it becomes necessary to take account of achievement during periods of schooling prior to the one immediately being considered (secondary here). However, Goldstein (2003) notes that one reason for the substantially larger primary school variance may be that secondary schools are generally larger than primary schools, so that the sampling variance is smaller.

5.2. *Multiple membership models*

Turning to multiple membership models we consider just the secondary schools from the above example and suppose that we know, for each individual, the weight w_{ij_2} associated with the j_2th secondary school attended by student i with $\sum_{j_2=1}^{J_2} w_{ij_2} = 1$. These weights, for example, may be proportional to the length of time a student is in a particular school during the course of the longitudinal study. Note that we allow the possibility that for some (perhaps most) students only one school is involved, so that one of these weights is one and the remainder are zero. Note that when all level 1 units have a single non-zero weight of 1 we obtain the usual purely hierarchical model. We can write the following model for the case of membership of just two schools (1,2):

$$y_{i(1,2)} = (X\beta)_{i(1,2)} + w_{i1}u_1 + w_{i2}u_2 + e_{i(1,2)}, \tag{5.2}$$

where $w_{i1} + w_{i2} = 1$.

A more general model is

$$y_{i\{j\}} = (X\beta)_{i\{j\}} + \sum_{h\in\{j\}} w_{ih}u_h + e_{i\{j\}}, \tag{5.3}$$

where $\sum_{h\in\{j\}} w_{ih} = 1$, $\mathrm{var}(u_h) = \sigma_u^2$, and $\mathrm{var}(\sum_{h\in\{j\}} w_{ih}u_h) = \sigma_u^2 \sum_{h\in\{j\}} w_{ih}^2$. (The notation $h \in \{j\}$ means for any school h that belong to the set of schools $\{j\}$.) In the particular case of membership of just two schools with equal weights we have $w_{i1} = w_{i2} = 0.5$, $\mathrm{var}(\sum_h w_{ih}u_h) = \sigma_u^2/2$. In other words the contribution to the level 2 variation is just half that for a student who remains in one school, since in the former case the level 2 contribution is averaged over two (random) schools. Note that if we ignore the multiple membership of schools and simply assign students, say, to the final school that they attend, we will underestimate the true extent of between-school variation. This is because, for those students who do attend more than one school, the true level 2 variation is less than that for students who attend a single school. In the model, however, we assume that the level 2 variation for these students is the same as that for those attending a single school, with the result that the overall level 2 variation is underestimated.

A slightly different notation to describe membership relationships is used by Browne et al. (2001). This is particularly useful when we have very complex structures involving mixtures of hierarchical, crossed and multiple membership classifications. Essentially it works by requiring just a single unique identification for each lowest level observation, in the present case a student. Each student then has a relationship with every other type of unit, here primary and secondary schools. The model specifies which classifications are involved and the data structure specifies precisely which schools are involved for each student. Thus our cross classified model would be written as follows

$$y_i = (X\beta)_i + u^{(2)}_{secondary(i)} + u^{(3)}_{primary(i)} + e_i \quad (i = 1, \ldots, n), \tag{5.4}$$

where *primary*(i) and *secondary*(i) refer respectively to the primary and secondary schools attended by student i. The superscripts for the random variables identify the classification; where this is absent, and if there is no ambiguity, it is assumed to be the lowest level, classification (1). Using this notation the multiple membership model would be written as

$$y_i = (X\beta)_i + \sum_{h \in school(i)} w_{i,h} u^{(2)}_h + e_i, \tag{5.5}$$

where $\sum_{h \in school(i)} w_{i,h} = 1$, and $\mathrm{var}(u^{(2)}_h) = \sigma^2_u$.

We can have mixtures of cross classifications and multiple memberships. Thus, for example, pupils may move between schools and also between areas for the cross classification of schools by areas. Such a model can be written in the form

$$y_{i\{j_1\}\{j_2\}} = (X\beta)_{i\{j_1\}\{j_2\}} + \sum_{h \in \{j_1\}} w_{1ih} u_{1h} + \sum_{h \in \{j_2\}} w_{2ih} u_{2h} + e_{i\{j_1\}\{j_2\}}, \tag{5.6}$$

where $\sum_h w_{1ih} = W_1$, $\sum_h w_{2ih} = W_2$, $\mathrm{var}(u_{1h}) = \sigma^2_{u1}$, and $\mathrm{var}(u_{2h}) = \sigma^2_{u2}$.

There are now two sets of higher-level units (schools and areas) which influence the response, each set having a multiple membership structure. Another application of such a model is for household data where individuals move among households and among addresses.

5.3. Representing complex data structures

When we have complex mixtures of hierarchies, cross classifications and multiple memberships, a straightforward way of representing these becomes important. Browne et al. (2001) use simple diagrams for representing such complex structures. Figure 3 represents the cross classified structure described in the example of Section 5.1. The single directional lines indicate a membership relation, and here students are members of just one secondary school and one primary school. Where multiple membership is involved, two parallel lines are used (see Figure 4).

To illustrate the flexibility of these models and their representation using classification diagrams, consider the example of modelling learning groups where the response is modelled at the group level and we have data where each student moves among groups. We can formulate this as a multiple membership model where groups (level 1) 'belong' to individuals (level 2). Suppose, in addition to measuring outcomes at the group level (y_{1j}) we also have a measure of achievement or attitude at the student level (y_{2i}). Recalling that the groups are defined as level 1 units, the group response will have an individual component and this will generally be correlated with the response at the student

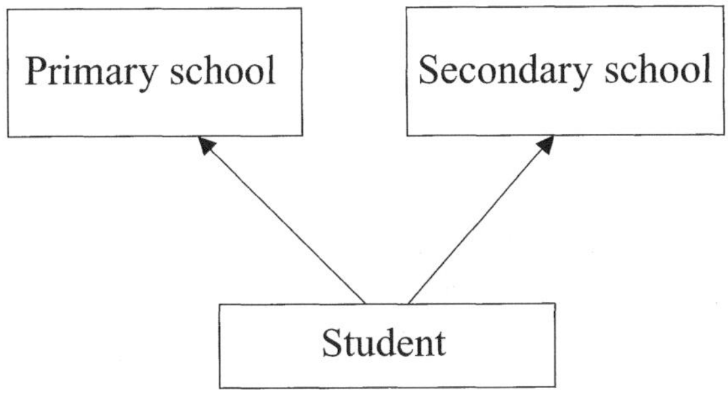

Fig. 3. Classification diagram showing students nested within a cross-classification or primary and secondary schools.

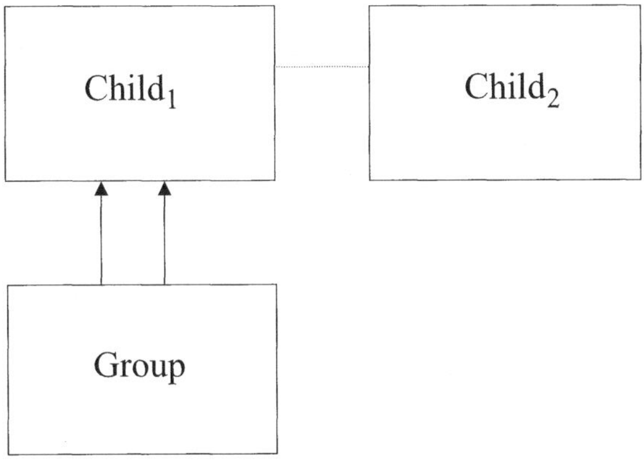

Fig. 4. Classification diagram for a bivariate response with multiple membership structure.

level. We could therefore write such a model as

$$\begin{aligned}
y_{1i} &= (X_1\beta_1)_i + \sum_{j \in group(i)} w_{i,j} u_{1j}^{(2)} + e_i^{(1)}, \qquad \sum_{j \in group(i)} w_{i,j} = 1, \\
y_{2j} &= (X_2\beta_2)_j + u_{2j}^{(2)}, \\
\operatorname{cov}\left(u_{1j}^{(2)}, u_{2j}^{(2)}\right) &\neq 0.
\end{aligned} \tag{5.7}$$

Eq. (5.7) defines a bivariate response model with one response at each level. The first equation refers to a group response and, given suitable data with individuals belonging to different groups, can be used to estimate individual and group effects. The second equation models an individual student response, and from the complete model we can directly estimate the correlation between a student's contribution to the group response and their individual response. Figure 4 shows the relationships using a double arrow for the multiple membership of groups within children and a dotted line joining the two child 'effects' to indicate a bivariate response model.

We can also identify those individuals who may be discrepant, say with low contributions to the group response but high individual effects, and this might be an important diagnostic for learning potential. An alternative formulation for some purposes would be to incorporate the individual level measure as a covariate in the model for the group response. If, however, we had sequential measures on individuals then we might wish to fit trend terms with random coefficients and then the full bivariate response formulation becomes necessary (see Goldstein, 2003, Chapter 5, for such a model in the purely hierarchical case). Further elaborations can be introduced, for example by modelling classes of students (containing learning groups) within teachers and/or schools and so forth.

6. Further extensions

6.1. Further levels

In Sections 2–4 above, models for two-level hierarchical structures were described. All models can be extended to handle further levels of clustering. For example, sampling classes within schools would lead to a three-level structure. If we denote by y_{ijk} the response for student i in class j in school k, the random intercept model in (2.1) generalises to

$$y_{ijk} = \beta_{0jk} + \beta_1 x_{ijk} + e_{ijk}, \quad \beta_{0jk} = \beta_0 + u_{0jk} + v_{0k}, \tag{6.1}$$

where u_{0jk} is the random effect for class j in school k, and v_{0k} is the random effect for school k, both of which are assumed to follow normal distributions. The variance of u_{0jk} represents the between-class within-school variation. Eq. (6.1) may be extended to allow the coefficient of x_{ijk} to vary randomly across classes and/or schools.

6.2. Categorical responses

In the above sections we have described models for continuous responses. These models may be generalised to handle different types of response, including binary, ordered and unordered categorical, count and duration data. For example, a two-level logit model for binary responses may be written

$$\log\left(\frac{\pi_{ij}}{1 - \pi_{ij}}\right) = \beta_{0j} + \beta_1 x_{ij}, \quad \beta_{0j} = \beta_0 + u_{0j} \tag{6.2}$$

where $\pi_{ij} = \Pr(y_{ij} = 1)$. Models for non-normal responses are described in Goldstein (2003, Chapter 4), and further applications are given in Rasbash et al. (2005).

It is also possible to handle mixtures of different response types in a multivariate model. For example, Goldstein (2003, pp. 105–107) describes an application where a binary response indicating whether an individual smokes, is modelled jointly with a continuous response (defined only for smokers) for the number of cigarettes smoked per day. More recently, multilevel factor models have been developed for mixtures of binary, ordinal and continuous items (Goldstein et al., 2006; Skrondal and Rabe-Hesketh, 2004; Steele and Goldstein, 2006).

6.3. Structural equation and item response models

Multilevel factor models may be generalised to structural equation models (SEM). Eq. (4.2) defines the measurement component of a SEM, which describes the relationship between the observed responses and the latent variables, possibly conditional on covariates. In a SEM, a structural component is added in which the latent variables at each level may depend on covariates and other latent variables. Detailed accounts of multilevel SEM include Muthén (1994) and Skrondal and Rabe-Hesketh (2004). A summary is given by Steele (2005).

The traditional (two parameter) item response model (Lord, 1980) is essentially a one-factor model with binary responses which can be written as

$$\operatorname{logit}(\pi_{ri}) = \beta_r + \lambda_r \eta_i + e_{ri}, \quad \pi_{ri} = \Pr(y_{ri} = 1) \tag{6.3}$$

for a set of binary responses (r) for each individual (i). This is a special case of (4.2) but with a logit rather than an identity link function; we can also use a probit link function. In its original formulation each individual's score (or 'latent trait' value), η_i, was treated as a fixed effect but now would generally be formulated as a random effect in (6.3). Model (6.3) is readily extended to the multilevel case, as in (4.2), and to cross classified and multiple membership structures. It can also be extended to handle ordered categorical responses as in 'partial credit' models and a discussion of such models and estimation procedures is given by Fox (2005).

7. Estimation procedures and software

Estimation procedures can conveniently be divided into those based upon maximum likelihood, or approximations such as quasi-likelihood, and those based upon Bayesian Markov Chain Monte Carlo (MCMC) methods. We first look at likelihood-based procedures.

In the case of normally-distributed responses, the two most common procedures are the EM algorithm and the iterative generalized least squares (IGLS) or the related Fisher scoring algorithm. Goldstein (2003, Chapter 2 appendices) gives details of these. These methods are iterative and implemented in major statistics packages including SAS (SAS Institute, 1999), SPlus (Insightful, 2001), SPSS (SPSS, 2003), and Stata (StataCorp, 2005) as well as specialist multilevel modelling packages such as HLM (Raudenbush et al., 2001) and MLwiN (Rasbash et al., 2005). For non-hierarchically structured Normal responses, cross-classified models can be fitted using SAS, SPSS and MLwiN, and the Stata program GLLAMM (Rabe-Hesketh and Skrondal, 2005) while multiple membership models are implemented in MLwiN.

Where responses are discrete, for example binary or count data, we have generalized linear models and estimation tends to become more complicated. Maximum likelihood estimation is commonly carried out using a search procedure whereby the parameter space is explored in a search for the maximum of the likelihood. The computation of the likelihood for a given set of parameter values is commonly carried out using 'quadrature' but this can be very time consuming when there are many random parameters. Thus, for example, Ng et al. (2006) use a simulation-based procedure (see Goldstein,

2003, Appendix 4.3) to compute the likelihood and show that for a large number of parameters this has important timing advantages. Because of this, several approximate procedures are in use. One set of these uses quasi-likelihood estimation based on a Taylor series linearization. Known as marginal or penalized (predictive) quasi-likelihood (MQL, PQL) these will often provide satisfactory estimates but are known to be biased in cases where data are sparse or variances are large. Another procedure that has good properties is a Laplace integral transformation (Raudenbush et al., 2000). Any of these procedures can be used to provide starting values for either full maximum likelihood or MCMC estimation (see below). Another approach is to use an iterated bootstrap which will converge to unbiased estimates but has the disadvantages that it is time-consuming and does not directly provide standard error estimates. We shall not go into details of MCMC estimation procedures here but refer the reader to Browne et al. (2001) and Browne (2004). Multilevel models for discrete response data can be estimated using SAS, Stata or GLLAMM (all of which use quadrature), HLM (PQL and Laplace) and MLwiN (MQL/PQL and MCMC). Very general multilevel models, including those considered in this review, can be fitted using MCMC methods in WinBUGS (Spiegelhalter et al., 2000).

Multilevel factor analysis is implemented in Mplus (Muthén and Muthén, 2004; using two-stage weighted least squares), GLLAMM (adaptive quadrature), and MLwiN (using MCMC). Mplus and GLLAMM can also be used to fit more general structural equation models to any mixture of normal and discrete responses.

Recently published reviews of some of the packages mentioned above are those of De Leeuw and Kreft (2001), Zhou et al. (1999) and Fein and Lissitz (2000). Full reviews of the multilevel modelling capabilities of most mainstream statistical and specialist packages are maintained by the Centre for Multilevel Modelling (http://www.mlwin.com/softrev).

8. Resources

The methodological literature on multilevel modelling is growing rapidly as is the literature on applications. The Centre for Multilevel Modelling endeavours to maintain a selection of the methodological literature and links to other resources such as web sites and training materials. A collection of data sets together with training materials and a version of the MLwiN package that will work with these data sets, is freely available at http://tramss.data-archive.ac.uk. Another useful resource for multilevel modelling is http://www.ats.ucla.edu/stat/mlm/default.htm.

There is a very active email discussion group that can be accessed and joined at http://www.jiscmail.ac.uk/lists/multilevel.html. The group serves as a means of exchanging information and suggestions about data analysis.

References

Aitkin, M., Bonnet, S.N., Hesketh, J. (1981). Teaching styles and pupil progress: a reanalysis. *British Journal of Educational Psychology* **51**, 170–186.

Aitkin, M., Longford, N. (1986). Statistical modelling in school effectiveness studies. *Journal of the Royal Statistical Society A* **149**, 1–43.

Bennett, S.N. (1976). *Teaching Styles and Pupil Progress*. Open Books, London.

Bock, R.D. (Ed.) (1989). *Multilevel Analysis of Educational Data*. Academic Press, San Diego, CA.

Browne, W.J. (2004). *MCMC Estimation in MLwiN*. Institute of Education, London.

Browne, W., Goldstein, H., Rasbash, J. (2001). Multiple membership multiple classification (MMMC) models. *Statistical Modelling* **1**, 103–124.

Bryk, A.S., Raudenbush, S.W. (2002). *Hierarchical Linear Models: Applications and Data Analysis Methods*, 2nd ed. Sage Publications, Newbury Park, CA.

De Leeuw, J., Kreft, I. (2001). Software for multilevel analysis. In: Leyland, A., Goldstein, H. (Eds.), *Multilevel Modelling of Health Statistics*. Wiley, Chichester.

Fein, M., Lissitz, R.W. (2000). Comparison of HLM and MLwiN multilevel analysis software packages: a Monte Carlo investigation into the quality of the estimates. Working Paper. University of Maryland.

Fox, J.P. (2005). Multilevel IRT using dichotomous and polytomous response data. *British Journal of Mathematical and Statistical Psychology* **58**, 145–172.

Goldstein, H. (1989). Flexible models for the analysis of growth data with an application to height prediction. *Rev. Epidem. et Sante Public* **37**, 477–484.

Goldstein, H. (2003). *Multilevel Statistical Models*, 3rd ed. Arnold, London.

Goldstein, H., Browne, W.J. (2002). Multilevel factor analysis modelling using Markov Chain Monte Carlo (MCMC) estimation. In: Marcoulides, G., Moustaki, I. (Eds.), *Latent Variable and Latent Structure Models*. Lawrence Erlbaum Associates, Hillsdale, NJ, pp. 225–243.

Goldstein, H., Healy, M.J.R., Rasbash, J. (1994). Multilevel time series models with applications to repeated measures data. *Statistics in Medicine* **13**, 1643–1655.

Goldstein, H., Bonnet, G., Rocher, T. (2006). Multilevel multidimensional structural equation models for the analysis of comparative data on educational performance. *Journal of Educational and Behavioral Statistics*. In press.

Insightful (2001). *SPlus 6 for Windows Guide to Statistics*. Insightful Corporation, Seattle, WA.

Laird, N.M., Ware, J.H. (1982). Random-effects models for longitudinal data. *Biometrics* **38**, 963–974.

Longford, N.T. (1993). *Random Coefficient Models*. Oxford University Press, New York.

Lord, F.M. (1980). *Applications of Item Response Theory to Practical Testing Problems*. Lawrence Erlbaum Associates, Hillsdale, NJ.

Muthén, B.O. (1994). Multilevel covariance structure analysis. *Sociological Methods and Research* **22**, 376–398.

Muthén, B.O. (1997). Latent variable growth modelling with multilevel data. In: Berkane, M. (Ed.), *Latent Variable Modeling with Applications to Causality*. Springer-Verlag, New York, pp. 3149–3161.

Muthén, L.K., Muthén, B.O. (2004). *Mplus User's Guide, Version 3.0*. Muthén & Muthén, Los Angeles, CA.

Ng, E.S.W., Carpenter, J.R., Goldstein, H., Rasbash, J. (2006). Estimation of generalised linear mixed models with binary outcomes by simulated maximum likelihood. *Statistical Modelling* **6**, 23–42.

Rabe-Hesketh, S., Skrondal, A. (2005). *Multilevel and Longitudinal Modeling using Stata*. Stata Press, College Station, TX.

Rasbash, J., Steele, F., Browne, W., Prosser, B. (2005). *A User's Guide to MLwiN Version 2.0*. University of Bristol.

Raudenbush, S.W., Bryk, A., Cheong, Y.F., Congdon, R. (2001). *HLM 5: Hierarchical Linear and Nonlinear Modeling*. SSI, Lincolnwood.

Raudenbush, S.W., Yang, M., Yosef, M. (2000). Maximum likelihood for generalised linear models with nested random effects via high-order multivariate Laplace approximation. *Journal of Computational and Graphical Statistics* **9**, 141–157.

Robinson, W.S. (1951). Ecological correlations and the behaviour of individuals. *American Sociological Review* **15**, 351–357.

SAS Institute (1999). *SAS/STAT User's Guide, Version 7.1*. SAS Institute, Cary, NC.

Singer, J.D., Willett, J.B. (2003). *Applied Longitudinal Data Analysis*. Oxford University Press, New York.

Skrondal, A., Rabe-Hesketh, S. (2004). *Generalized Latent Variable Modeling: Multilevel, Longitudinal and Structural Equation Models*. Chapman & Hall/CRC, Boca Raton, FL.

Snijders, T.A.B., Bosker, R.J. (1999). *Multilevel Analysis*. Sage Publications, London.

Spiegelhalter, D.J., Thomas, A., Best, N.G. (2000). *WinBUGS Version 1.3 User Manual*. Medical Research Council Biostatistics Unit, Cambridge.

SPSS (2003). *SPSS. Advanced Models 12.0*. SPSS, Chicago.

StataCorp (2005). *Stata 9.0 Base Reference Manual*. Stata Press, College Station, TX.

Steele, F. (2005). Structural equation modeling: multilevel. In: Everitt, B., Howell, D. (Eds.), In: *Encyclopedia of Statistics in Behavioral Science*, vol. 4. John Wiley and Sons, Chichester, pp. 1927–1931.

Steele, F., Goldstein, H. (2006). A multilevel factor model for mixed binary and ordinal indicators of women's status. *Sociological Methods and Research*. In press.

Woodhouse, G., Goldstein, H. (1989). Educational performance indicators and LEA league tables. *Oxford Review of Education* **14**, 301–319.

Zhou, X., Perkins, A.J., Hui, S.L. (1999). Comparisons of software packages for generalized linear multilevel models. *American Statistician* **53**, 282–290.

Handbook of Statistics, Vol. 26
ISSN: 0169-7161

DOI: 10.1016/S0169-7161(06)26013-9

13

Latent Class Analysis in Psychometrics

C. Mitchell Dayton and George B. Macready

1. Introduction

Latent class analysis (LCA) is a statistical method for analyzing and understanding multivariate categorical data. These data arise routinely in the behavioral sciences. In fact, it can be argued that most data collection procedures in the behavioral sciences generate multivariate categorical data. Consider three common examples:

(1) Achievement, aptitude and similar test items generate categorical responses from respondents who are asked to select from prescribed options. These responses are transformed to, say, a categorical, correct/incorrect (i.e., 1/0) scale by comparison with a key. In addition, the 1/0 scores may be transformed to counts of correct responses or modeled using some form of item response theory. Note, however, that the raw data are multivariate categorical.

(2) Surveys may include sets of attitudinal or behavioral items requiring responses in prescribed categories such as "yes" and "no". As in (1), these may be coded 1/0 and summed to create a score or modeled in some way, but the raw data are clearly multivariate categorical.

(3) Opinion surveys and various psychological tests may include items with five or seven response categories such as "strongly agree," "agree," through "strongly disagree." Responses are categorical and it is often reasonable to assume that the categories obey an ordering principle. While these responses may be transformed to derived scores in many different ways, any such conversion requires assumptions that go beyond the nature of the observed data that remain multivariate categorical.

From a mathematical perspective, LCA is closely related to the area of statistics known as discrete mixture models as well as to classical factor analysis. In terms of the former, LCA in its simplest forms is a mixture of product-multinomial distributions whereas, in terms of the latter, LCA may be viewed as a factor analytic model for categorical data. The mathematical model for LCA is summarized in Section 2 while ensuing sections deal with a variety of specific applications and extensions of the underlying model. Some basic references that provide historical and theoretical background for LCA include Lazarsfeld and Henry (1968), Goodman (1974), Haberman (1979) and Heinen (1996). Other overviews of LCA and its applications can be found in Haertel (1984), Bartholomew (1987), Langeheine and Rost (1988), Hagenaars (1990),

Von Eye and Clogg (1994), Rost and Langeheine (1997), Dayton (1999), and Hagenaars and McCutcheon (2002). The web site, http://ourworld.compuserve.com/homepages/jsuebersax/index.htm, maintained by John Uebersax provides many additional references to the latent class literature including citations for interesting applications in education and psychiatry. Throughout this chapter, reference is made to exemplary data sets with special attention to items dealing with attitudes toward abortion issues that have been abstracted from several years of General Social Survey (GSS) data. These abortion data, described in more detail below, are in public-access databases maintained by the National Opinion Research Center (NORC) in Chicago, Illinois at the web site http://webapp.icpsr.umich.edu/GSS/.

2. The model for LCA

For convenience, and without any intended loss of generality, we refer to categorical response variables, unordered or ordered, as "items" and to the individuals, objects, etc., that provide data as "respondents." The basic concept in LCA is that the population of interest is composed of unobserved classes of respondents within which responses to items are stochastically independent. Unlike strata such as sex or socio-economic status, these latent classes do not, in general, have any obvious indicators in the real world. This is not to suggest that classes must be interpreted simply as strata that lack a manifest gold standard. In fact, it is reasonable in some applications, especially those based on repeated observations over time, to interpret latent classes as states that respondents assume with certain probabilities represented by latent class proportions.

The mathematical model for LCA can be represented as follows. Let $\boldsymbol{Y}_i = \{y_{ij}\}$ be the vector-valued response for items $j = 1, \ldots, J$, for the ith respondent from some population. Further, let the response options for the items be defined over a set of distinct, mutually-exclusive values $r = 1, \ldots, R_j$ for the jth item (e.g., for dichotomous responses these values would be $r = 1, 2$). Then, for C distinct latent classes, an unrestricted latent class model is defined as:

$$\Pr(Y_i) = \sum_{c=1}^{C} \theta_c \prod_{j=1}^{J} \prod_{r=1}^{R_j} \alpha_{cjr}^{\delta_{ijr}}. \tag{1}$$

The mixing proportions, θ_c, $c = 1, \ldots, C$, with the usual restriction, $\sum_{c=1}^{C} \theta_c = 1$, are referred to as latent class proportions and represent the sizes of the unobserved latent classes. The conditional probabilities associated with the items, $\alpha_{cj1} \cdots \alpha_{cjR_j}$, represent the probability of response r to item j given membership in the cth latent class. Thus, for each item, there is a vector of R_j conditional probabilities. These conditional probabilities sum to 1 for each item within each latent class: $\sum_{r=1}^{R_j} \alpha_{cjr} = 1$. The Kronecker deltas are defined by

$$\delta_{ijr} = \begin{cases} 1 & \text{iff } y_{ij} = r, \\ 0 & \text{otherwise,} \end{cases} \tag{2}$$

so as to ensure that appropriate conditional probabilities are included in the model (noting that, in general, $a^0 = 1$). The fact that the model is written with products of probabilities associated with item responses is indicative of the assumption that, given knowledge of latent class membership, responses are assumed to be independent (this is sometimes known as the local independence assumption in LCA).

Eq. (1) can be rewritten as

$$\Pr(\boldsymbol{Y}_i) = \sum_{c=1}^{C} \theta_c \Pr(\boldsymbol{Y}_i \mid c), \tag{3}$$

where $\Pr(\boldsymbol{Y}_i \mid c)$ is the conditional probability for response vector, $\boldsymbol{Y}_i$, given latent class c. In this form, the model is recognizable as a finite mixture model with mixing proportions, θ_c, and probability processes defined by $\Pr(\boldsymbol{Y}_i \mid c)$. In effect, the model is a mixture of product-multinomial processes since the distribution for the conditional probabilities, α_{cjr}, for each item within each latent class follows a multinomial density. As the model is written, it is assumed that the sample of respondents arises by a process of simple random sampling. However, it should be noted that this is, typically, not the case for data arising from complex survey designs that incorporate cluster sampling; this situation is treated in later sections (also, see Patterson et al., 2002).

LCA can be viewed as analogous to classical factor analysis (CFA) in the following sense. CFA deals with the observed inter-correlations among a set of continuous response variables. It is postulated that these correlations arise from the fact that the observed variables are themselves correlated with a smaller set of unobserved (i.e., latent), continuous variables called factors. For example, observed inter-correlations among measures of academic achievement in areas such as mathematics, reading, history, etc., might be assumed to arise from a smaller set of factors related to latent abilities such as reasoning, recall, etc. A fundamental notion in CFA is that, if the latent factors are held constant (i.e., partialled out), then the observed correlations vanish. Of course, with real data the correlations will not be completely explained by the factors but this is the ideal. Similarly, in LCA the observed dependencies among a set of categorical response variables (i.e., items) are assumed to vanish if latent class membership is held constant (i.e., partialled out). Of course, this condition will only be approximated with real data.

3. Estimation and model fit

3.1. Estimation

Given the model in Eqs. (1) or (3) for a sample of n independent respondents, the likelihood for the sample is $\lambda = \prod_{i=1}^{n} \sum_{c=1}^{C} \theta_c \Pr(Y_i \mid c)$ and the logarithm of this likelihood is:

$$\Lambda = \sum_{i=1}^{n} \ln \sum_{c=1}^{C} \theta_c \Pr(Y_i \mid c) = \sum_{i=1}^{n} \ln \left\{ \sum_{c=1}^{C} \theta_c \prod_{j=1}^{J} \prod_{r=1}^{R_j} \alpha_{cjr}^{\delta_{ijr}} \right\}. \tag{4}$$

Subject to certain identifiability conditions that are discussed below, maximum likelihood estimates for the distinct parameters in the model can be derived by computing partial derivatives of Λ with respect to the parameters, setting these partial derivatives to 0 and solving simultaneously. The restrictions on the parameters, $\sum_{c=1}^{C} \theta_c = 1$ and $\sum_{r=1}^{R_j} \alpha_{cjr} = 1$, can be imposed on the solution in a variety of ways including the introduction of Lagrange multipliers. If the latent class proportion for the Cth class is rewritten as $\theta_C = 1 - \sum_{c=1}^{C-1} \theta_c$, the typical partial derivative with respect to a latent class proportion is of the form:

$$\frac{\delta \Lambda}{\delta \theta_c} = \sum_{i=1}^{n} \ln \Pr(Y_i \mid c) - \sum_{i=1}^{n} \ln \Pr(Y_i \mid C) = 0. \tag{5}$$

Similarly a partial derivative with respect to a conditional probability for an item is of the form:

$$\frac{\delta \Lambda}{\delta \alpha_{cjr}} = \sum_{i=1}^{n} \ln\left[\theta_c \frac{\delta \Pr(Y_i \mid c)}{\delta \alpha_{cjr}}\right] = 0. \tag{6}$$

Although the partial derivatives are straight-forward to compute, they are non-linear in the parameters and must be solved by iterative procedures. Among the methods available for deriving estimates from real data are the expectation-maximization (EM) algorithm, Newton–Raphson procedures and Bayesian methods. These methods are incorporated into various micro-computer programs such as LEM (Vermunt, 1997) that is based on an EM algorithm, LATENT (Haberman, 1979) that uses Newton methods and Latent Gold (Vermunt and Magidson, 2000) that incorporates Bayesian methods. More details concerning computational approaches can be found in the above references.

One requirement for identification of a latent class model is that the number of free parameters be no greater than one less than the number of distinct response vectors (note that one degree of freedom is lost since the sample size is fixed at n). An unrestricted latent class model with J items and C latent classes requires the estimation of $C - 1$ latent class proportions and $C \sum_{j=1}^{J} (R_j - 1)$ item conditional probabilities given the usual restrictions for probabilities noted above. Thus, for three items with 3, 3 and 4 response options, respectively, and two latent classes, it is necessary to estimate 15 independent parameters. Since there are $3 \times 3 \times 4 = 36$ distinct response patterns, this requirement for identification is met. However, a more fundamental requirement for identification is that the theoretical sampling covariance matrix for the parameter estimates is of full rank. Except for the simplest models, this requirement cannot be established by analytical methods although some programs for LCA do estimate the rank of the empirical sampling covariance matrix and may provide other evidence concerning whether or not identification requirements are met. It should be noted that relying only on the first requirement above can produce misleading conclusions. The classic example in LCA (Goodman, 1974) is for the case of four dichotomous items and three latent classes that represents an unidentified model even though the number of independent parameters (i.e., 14) is two less than the number of distinct response patterns (i.e., 16). In practice, lack of identification may be suspected when estimates for item conditional probabilities go to boundary values of 0 or 1 or when different sets of starting values for parameter estimates does not result in the same solution. Assuming that a latent class model is identified and that ρ_m *and* $\rho_{m'}$ are parameters freely selected from θ_c, α_{cjr}, the asymptotic sampling covariance matrix for the parameter estimates is $V = E\{\partial^2 \Lambda / \partial\rho_m \partial\rho_{m'}\}^{-1}$ evaluated at the point of maximum likelihood. Sample estimates of V are routinely computed by available latent class programs using various approximation methods (e.g., by estimating second derivatives using finite differences).

The log-likelihood function in Eq. (4) is based on the assumption of simple random sampling (SRS) from a defined population. In practice, complex surveys employ stratification and cluster sampling. Thus, survey data are often reported with case-wise sampling weights, w_i, that compensate for stratification that is disproportionate to the size of strata (i.e., some strata may be under- or over-sampled), non-response and related factors (Kish, 1965; Kish and Frankel, 1974; Kalton, 1989). In the context of LCA, Patterson et al. (2002) defined a pseudo-log-likelihood that incorporated sampling weights:

$$\Lambda_w = \sum_{i=1}^{n} w_i \ln \sum_{c=1}^{C} \theta_c \Pr(Y_i \mid c) = \sum_{i=1}^{n} w_i \ln \left\{ \sum_{c=1}^{C} \theta_c \prod_{j=1}^{J} \prod_{r=1}^{R_j} \alpha_{cjr}^{\delta_{ijr}} \right\}. \tag{7}$$

Although little is known about the theoretical properties of pseudo-log-likelihoods, empirical investigations suggest that estimating parameters by maximizing Eq. (7) results in approximately unbiased estimates even though clustering is not explicitly taken into account. However, these same empirical studies indicate that asymptotic estimates for standard errors based on using pseudo-log-likelihoods are negatively biased (i.e., too small). This is not surprising since clustered samples have a reduced effective sample size relative to SRS. In survey research, it is often recommended that this negative bias be corrected by use of a design effect that represents the ratio of the sampling error for the complex design versus comparable data based on SRS. Unfortunately, in the context of LCA, Patterson et al. (2002) found that adjustments based on design effects produced unsatisfactory results and suggested the use of jackknife re-sampling techniques for purposes of estimating standard errors (also, see Wolter, 1985). In all of these analyses, it has been assumed that respondents arise from populations that are sufficiently numerous that finite-correction terms can be ignored.

3.2. *Model fit*

When fitting models to data, it is important to distinguish between confirmatory and exploratory analyses, both of which may be involved in a single research study. The former situation arises when models are specified a priori on the basis of theory or previous findings. In this case, it is appropriate to assess whether or not the data at hand are consistent with these a priori models. Thus, statistical tests that assess goodness-of-fit to the data are based on null hypotheses derived from the theoretical models. For categorical data, significance tests ordinarily entail a comparison between observed frequencies and expected frequencies that are derived by substituting maximum-likelihood estimates for parameters in the theoretical model. The observed frequencies may be viewed as arising from a saturated multinomial distribution. These tests are widely known and are not summarized here but are illustrated and commented upon in connection with applications below. Dayton (1999) provides computational details for the Pearson X^2, likelihood-ratio G^2 and Read–Cressie I^2 goodness-of-fit statistics, the latter being arguably the best choice when small expected frequencies tend to distort the distributions for the first two tests.

It is important to note that classical, frequentist interpretations of significance tests only allow for strong inference when fit of a model is unsatisfactory; that is, when the null hypothesis is rejected. Thus, in a confirmatory analysis, acceptance of the null hypothesis leads to the rather bland conclusion that there is no evidence that the theoretical model is false. But note that this could also be true (and certainly is true) for any number of unspecified models that have not been assessed on the data. Also, for relatively small sample sizes, failure to reject the null may be viewed as simply absence of statistical power. On the other hand, rejection of the null hypothesis does, in fact, provide evidence that the theoretical model is not satisfactory, although this will, in many cases, be opposite to the conclusion of interest to the researcher. However, there certainly are cases where negative findings are of interest as when an independence model is rejected.

In experimental settings, the rubric of significance testing has greater utility since confirming the null hypothesis is ordinarily not of interest. For example, in a two-group experimental design, the null hypothesis for the conventional Student t-test specifies equality of population means. Then, rejection of the null hypothesis provides evidence for the conclusion that is of interest. However, in non-experimental, confirmatory situations where models are fit to data, the use of statistical tests is much less satisfactory since there is no simple binary equality/inequality outcome that covers the range of interests of the researcher.

These same concerns carry over to the case of exploratory analysis. In LCA it is not unusual to fit a series of increasingly more complex latent class models with the aim of locating a model that is a reasonable representation for the data. For nested (i.e., hierarchical) models, with the very important exception noted below, differences in chi-square goodness-of-fit tests can be used to assess the statistical significance of the difference in fit between alternate models (again, see Dayton, 1999, for a summary of these familiar ideas). Thus, it can be decided whether or not the more complex model (that is, the model with more independent parameters being estimated) provides better fit to the data than the simpler model. Note that this is quite different than deciding that either model in any sense represents "good" fit to the data. The important exception to the utilization of difference chi-square tests is for the case of models with different numbers of latent classes. Thus, if one fits unrestricted two-class, three-class and four-class models, say, the differences in chi-square values between these nested models are not, contrary to what one might expect, distributed as chi-square statistics. This is a general condition that pertains to comparing mixture models based on different numbers of components (Titterington et al., 1985). Although the precise mathematical reasons for the failure of these tests is reasonably arcane, a simple example provides some insight into the dilemma. Consider fitting unrestricted two-class and three-class models to data from five dichotomous items. The number of independent parameters in the two-class model is 11 and in the three-class model is 17. Thus, it would seem, from a hierarchic perspective, that 6 restrictions must be placed on the three-class model to reduce it to the two-class model. However, simply setting one of the two independent latent class proportions to 0 accomplishes the same reduction.

3.3. *Model comparison measures*

For the reasons discussed above, the use of statistical significance tests may not be particularly useful in the context of LCA. An alternative approach for comparing models, and for selecting a "best" model, that is gaining currency derives from information theory. Akaike (1973, 1974) advocated the use of his estimate for the Kullback and Leibler (1951) information measure (or, negentropy) as an approach to ordering alternate statistical models based on the same data. Akaike reasoned that Kullback–Leibler information represents a measure of the distance between the "true" model for data and any other model proposed for the same data. His estimate for Kullback–Leibler information, AIC, turns out to be a penalized form of the log-likelihood for the observed data. The penalty depends on the complexity of the model and is simply $2q$, where q is the number of independent parameters estimated in fitting the model by maximum likelihood. In the present notation, $\text{AIC} = -2\ln(\lambda) + 2q = -2\Lambda + 2q$. The Akaike decision-making strategy is to compute the AIC statistic for each model under consideration and to select as the preferred model the one for which AIC is minimized; this is referred to as the min(AIC) strategy. The interpretation of the min(AIC) model is that, among those models being compared, it results in the least loss of information compared to the "true" model which, of course, is never known (or, knowable). In this sense, the model selected by a min(AIC) strategy is the best approximating model *among the models being compared.*

It should be noted that various related measures have been proposed as alternatives to the Akaike AIC. In general, these measures incorporate different penalty terms, usually placing a greater weight on the number of parameters being estimated with the result that simpler models are favored. For example, Schwarz (1978) used Bayesian concepts to derive his measure, BIC, with penalty term $\log_e(n)q$ and Bozdogan (1987) modified Akaike's mathematical derivation to develop CAIC with the penalty term $[\log_e(n)+1]q$. In this chapter, we utilize both the Akaike AIC and Schwarz BIC for model comparisons but suggest that readers consider the merits of alternatives.

From a subjective point of view, a model resulting in lack of fit based on chi-square tests may appear to represent the observed data reasonably well in the sense that discrepancies between observed and expected frequencies are relatively small. Given this, various descriptive measures have been proposed, especially the Index of Dissimilarity defined as:

$$I_D = \sum_{m=1}^{M} \frac{|n_m - \tilde{n}_m|}{2n}, \tag{8}$$

where M is the number of distinct response patterns (e.g., $M = 2^J$ for J dichotomous variables) while n_m and $\tilde{n}_m$ are, respectively, observed and expected frequencies. In practice, values of I_D less than 0.05 may be considered as representing "satisfactory" fit and the measure may be used as the descriptive basis for comparing alternative models. Other descriptive measures include the lack-of-fit index, π^*, proposed by Rudas et al. (1994). This index represents the minimum proportion of respondents that must be deleted from the frequency distribution in order for the proposed model to result in perfect fit (i.e., $G^2 = 0$). Although a highly interpretable and useful index, it has no simple computational approach (see Dayton, 2003, for not-so-simple approaches) and is not available in current latent class programs.

3.4. Computer programs for LCA

The analyses reported in this chapter were generated using the computer programs LEM (Vermunt, 1997), MPlus (Muthén and Muthén, 1998) and Latent Gold (Vermunt and Magidson, 2000). LEM is far more general than other programs with regard to the range of models that can be estimated since, among other things, complex restrictions can be imposed on models using a design matrix. Latent Gold has the advantage of being able to utilize common data bases, such as SPSS, as well as having options for placing prior distributions on the parameters (i.e., Bayesian analyses). For use with data arising from complex sample designs, these programs can accommodate case weights and newer versions of MPlus and Latent Gold have some capacity to deal with cluster sampling.

4. Analyses with score data

For J achievement test items or J dichotomously scored (e.g., agree/disagree) attitude items, a score ranging, in theory, from 0 through J can be created by counting the number of correct (or, agree) responses. Mastery tests, for example, often consist of a set of relatively homogeneous items (e.g., simple arithmetic skills or word-attack decoding skills) for which a minimum "pass" score has been established. Typically, the cut score for passing is established by a panel of judges. A two-class latent class model is a useful supplement (or alternative) for identifying latent subgroups that may (or may not) be identified as masters and non-masters (Macready and Dayton, 1977, 1980). The latent class approach is entirely empirical, does not involve subjective judgments and provides evidence as to whether or not the set of items can reasonably be used to separate respondents into two groups.

4.1. Quiz scores

In this section, our first example uses results from a mastery quiz based on scores from 10 completion items. These items dealt with elementary concepts of matrix algebra and were used over a period of several years in a graduate-level multivariate analysis course taught by the first author of this chapter (actually, the specific items varied from semester to semester but were drawn from the same pool of items and are considered randomly parallel for these analyses). As shown in Figure 1, the distribution of 247 scores is negatively skewed with a heavy "tail" to the left that suggests a mixture of distributions rather than a single distribution. A single binomial distribution with an estimated probability for a correct response equal to 0.77 provides very poor fit to the score distribution ($G^2 = 142.69$ with 9 degrees of freedom; $p < 0.001$). On the other

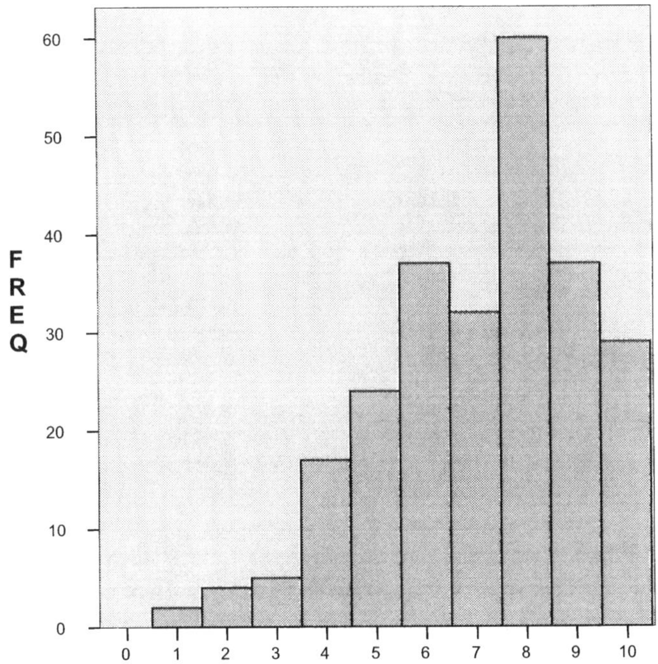

Fig. 1. Matrix algebra scores.

hand, a mixture of two binomial distributions with estimated item conditional probabilities of 0.83 and 0.54 does provide adequate fit at conventional significance levels ($G^2 = 13.37$ with 7 degrees of freedom; $p = 0.064$). It seems reasonable to interpret the first class, with an estimated 60% of the respondents, as "masters" and the second class as "non-masters." In terms of the latent class model in Eq. (1), the conditional probabilities for the Yes responses are constant within classes; i.e., $\alpha_{c1} = \cdots = \alpha_{c10}$ for $c = 1, 2$ where the response subscript has been omitted. It should be noted that for this mixture-of-binomials latent class model, score frequencies are sufficient statistics for maximum likelihood estimation.

Bayes theorem in the form $\Pr(c \mid \boldsymbol{Y}_i) \propto \theta_c \Pr(\boldsymbol{Y}_i \mid c)$ can be used to classify respondents from the score groups, 0 through 10, into the latent classes. This entails substituting maximum likelihood estimates for the unknown parameters in Bayes theorem and classifying respondents into the class for which $\Pr(c \mid \boldsymbol{Y}_i)$ is maximal. As shown in Table 1 for the quiz results, respondents with scores from 0 through 6 are classified into the "non-master" class whereas respondents with scores from 7 through 10 are classified into the "master" class. Note that these classifications involve errors. For example, the 37 respondents with a score of 6 are all classified as "non-masters" although, according to the model being fit to the data, 3.34% of respondents with that score are estimated to arise from the second, or "master" latent class. When these errors of classification are summed, there are about 4.3% of "masters" who are misclassified as "non-masters" and about 9.5% of "non-masters" who are misclassified as "masters,"

Table 1
Bayesian classifications for quiz scores

Score	Frequency	$\theta_1 \Pr(Y_i \mid c = 1)$	$\theta_2 \Pr(Y_i \mid c = 2)$	Classification
0	0	0.0002	0.0000	1
1	2	0.0021	0.0000	1
2	4	0.0111	0.0000	1
3	5	0.0343	0.0002	1
4	17	0.0693	0.0014	1
5	24	0.0960	0.0081	1
6	37	0.0924	0.0334	1
7	32	0.0610	0.0943	2
8	60	0.0264	0.1749	2
9	37	0.0068	0.1923	2
10	29	0.0008	0.0951	2
Total	247	0.4004	0.5997	

resulting in 13.8% estimated total classification error. Letting the proportion of errors of classification be P_e, the estimated proportion of correct classifications is $P_c = 1 - P_e$ or, for the present example, 0.862. However, this proportion may be viewed as somewhat optimistic since simply classifying all respondents in the larger (more generally, largest) latent class represents the baseline rather than 0 (e.g., 60% for the present example). The index, lambda (Goodman and Kruskal, 1979), was proposed as a correction and is of the form $\lambda = \frac{P_c - \theta_{\max}}{1 - \theta_{\max}}$, where $\theta_{\max}$ is the largest estimated latent class proportion. For our example, $\lambda = \frac{0.862 - 0.600}{1 - 0.600} = 0.655$. This can be interpreted to mean that using Bayesian classification based on the latent class model results in about 66% improvement over chance.

4.2. Abortion scores

A different type of situation for which scoring items is useful involves survey items dealing with attitudes toward a defined topic. For example, items dealing with attitudes toward abortion have been included in the GSS for many years and six were selected for analysis. In the GSS, these items are presented with the introduction:

> "Please tell me whether or not *you* think it should be possible for a pregnant woman to obtain a *legal* abortion if ..."

This introduction was followed by specific statements:

- The woman's own health is seriously endangered by the pregnancy;
- She became pregnant as a result of rape;
- There is a strong chance of serious defect in the baby;
- The family has a very low income and cannot afford any more children;
- She is not married and does not want to marry the man;
- She is married and does not want any more children.

Table 2
Counts and percents for responses to abortion items

Item	Yes		No	
	Count	%	Count	%
Woman's health seriously endangered	27187	90.1%	2985	9.9%
Pregnant as result of rape	24691	82.8%	5125	17.2%
Strong chance of serious defect	24487	81.5%	5553	18.5%
Low income – can't afford more children	14487	48.6%	15322	51.4%
Not married	13450	45.2%	16331	54.8%
Married – wants no more children	13204	44.2%	16675	55.8%

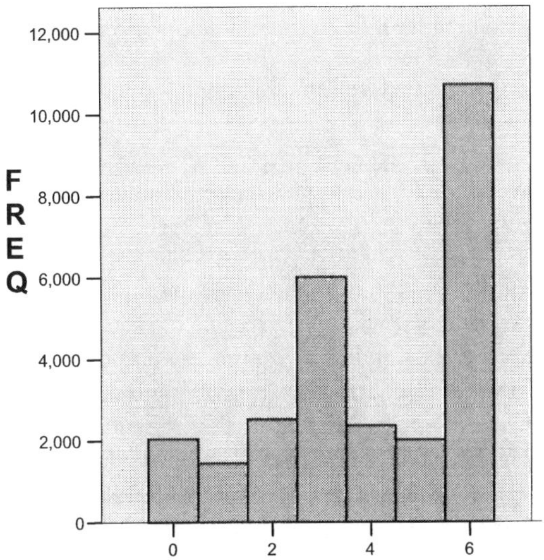

Fig. 2. Abortion agreement score.

Response options comprised Yes, No, Don't Know and No Answer, but for our analyses only the Yes, No responses were considered as non-missing. Counts and percents for the responses for combined data over the years 1972 through 1998 are presented in Table 2. Complete data were available for 27,151 respondents. A score for each respondent with a range of 0 through 6 was created by counting the number of Yes responses. The resulting frequency distribution is shown in Figure 2. It is apparent that these scores do not follow any simple probability model because of the prominent mode at 6 that represents about 28% of respondents who replied Yes to all six items. In addition, there appears to be a local mode near 0 as well as a clearly-defined mode near the center of the score distribution. Thus, these abortion item count data suggest some form of mixture. For the present analyses, it was assumed that the data represent a mixture of Bernoulli processes that generate binomial distributions. As above, we assume that the conditional probabilities for the Yes responses (say) are constant

Table 3
LC models fitted to score data

# of classes	DF for fit	# of independent parameters	Likelihood ratio G^2	AIC	LC proportions	Item conditional probabilities
1	5	1	41,369.82	207,525.68	1.00	0.67
2	3	3	4,370.09	170,529.96	0.45, 0.55	0.98, 0.41
3	1	5	1,023.53	**167,186.34**	0.42, 0.07, 0.51	0.99, 0.01, 0.49

within latent classes; i.e., $\alpha_{c1} = \alpha_{c2} = \alpha_{c3} = \alpha_{c4} = \alpha_{c5} = \alpha_{c6}$ for $c = 1, \dots, C$. The results from fitting one- through three-class models are summarized in Table 3. The degrees of freedom for the chi-square fit statistics are found from the expression: #Cells − #Independent Parameters − 1. Note that the one-class model is not a mixture but is equivalent to estimating and fitting a binomial model for independent items. Also, note that a larger number of classes cannot be fit to these data because degrees of freedom have been exhausted. It is apparent that none of these models provides adequate fit from the perspective of the chi-square statistics although, given the very large sample size (27,151) this result is to be expected (all p-levels in the table are extraordinarily small and not reported). In terms of both information criteria, AIC and BIC (only AIC is displayed in Table 3), the preferred model, among these three, is the three-class model. This model is easy to interpret and appears to be consistent with the characteristics of the frequency distribution displayed in Figure 2. In particular, the first class comprises 42% of the respondents with a conditional probability of 0.99 for responding Yes to an item. This class is, essentially, responsible for the mode at the score of 6. The second class comprising 7% of the respondents has a probability of only about 0.01 for responding Yes to an item and is primarily responsible for the secondary mode near 0. Finally, the third class comprising 51% of the respondents has a probability of 0.49 for responding Yes to an item and generates the bulk of the central range of scores in Figure 2. Referring back to Table 2 in which we note the wide range of positive response rates for the six abortion items, it would seem that three classes each with a constant rate of Yes responses for all six items is unnecessarily simplistic so we explore more complex models in later chapters.

5. Unconstrained latent class models

Exploratory latent class analyses for categorical response data often proceed by fitting unconstrained models with increasing larger numbers of latent classes. The model with a single latent class is referred to as an independence model and is the typical starting point for such analyses. The process of fitting increasing more complex models stops when either degrees of freedom for fit are exhausted or a min(AIC) [min(BIC)] strategy, for example, locates a "best fitting" model. Results for fitting one through seven unrestricted latent classes to frequencies for the $2^6 = 64$ response patterns for the six GSS abortion items are summarized in Table 4. None of the chi-square fit statistics is

Table 4
Unrestricted LC models fitted to abortion item data

# LC	# of parameters	DF	G^2	Pr(G^2)	AIC	BIC
1	6	57	66,146.83	0.000	181,324.29	181,373.54
2	13	50	10,710.17	0.000	125,901.63	126,008.35
3	20	43	704.27	0.000	115,909.73	116,073.91
4	27	36	329.95	0.000	115,549.41	115,771.05
5	34	29	87.65	0.000	115,321.11	**115,600.22**
6	41	22	47.70	0.001	115,295.16	115,631.73
7	48	15	28.87	0.017	**115,290.32**	115,684.37

non-significant but, given the very large sample size of 27,151, this is not unexpected. These analyses illustrate the differential tendencies of AIC which favors more complex models and BIC which favors simpler models due to the latter's greater weight per parameter. For the GSS abortion data, a min(BIC) strategy leads to interpreting the five class model whereas a min(AIC) strategy leads to interpreting the seven class model. For the six and seven class models, problems were encountered in arriving at stable solutions because item conditional probabilities converged to boundaries of 0 or 1 in both cases and because inconsistent solutions arose from different sets of start values. The five class solution was selected for as the best approximating model since it was stable across different sets of start values, consistently showed boundary conditions for the same items and differed only slightly from a Bayesian solution with a relatively uninformative prior distribution (see below for discussion of Bayesian models).

The item profiles for the five class model summarized in Figure 3 provide the basis for a reasonable interpretable of the abortion items. Note that the values in the plot are estimated item conditional probabilities for a Yes response and that latent class proportions are shown to the right of each profile. The two larger classes, estimated to have sizes of 39% and 30%, represent, respectively, respondents who are consistently favorable to abortion for all six items and respondents who seem to distinguish sharply between favoring the first three reasons for abortion and opposing the last three reasons for abortion. The third latent class, estimated to have 17% of respondents, is similar to the second class except that these respondents are less averse to the last three reasons for abortion. The fourth latent class, estimated to have 12% of respondents, is negative toward all six reasons for abortion, but somewhat less so for the first three items. The final class, with only an estimated 1% of respondents may be characterized as "undecided."

6. Scaling models

For psychological scales, Louis Guttman (1947) introduced the notion of locating respondents along a linear (ordinal) dimension by means of a set of ordered items. Assuming three Yes/No items, the items are selected so that a Yes response to item 3 "requires" a Yes response to item 2 and a Yes response to item 2 "requires" a Yes response to item 1. In practice, these requirements may reflect background theory or

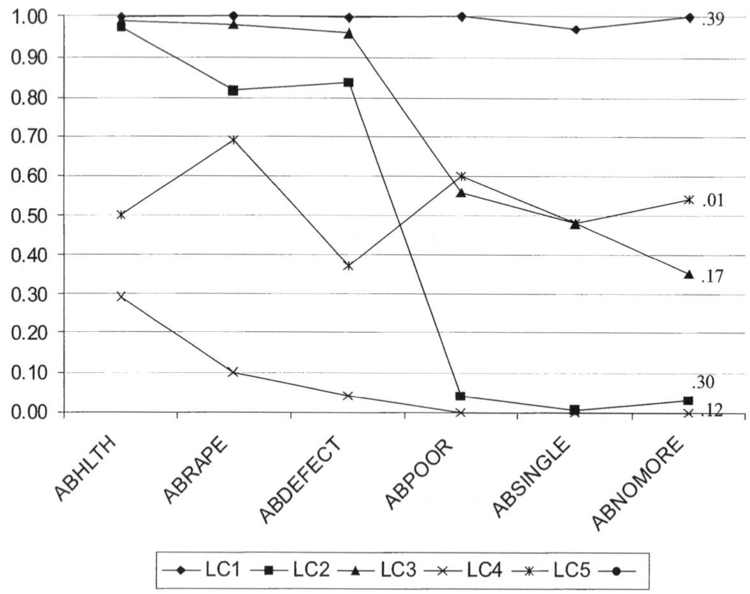

Fig. 3. Profile for five class model.

be based on the test constructor's logical analysis. Thus, if respondents were perfectly scalable, the only observed response patterns for three Yes/No items would be restricted to the ideal patterns NNN, NNY, NYY, and YYY. In practice, even with carefully constructed sets of items, other response patterns such as NYN or YNY are almost always observed. Proctor (1970) presented a probabilistic model for Guttman scaling that incorporated response errors occurring with some constant probability. His model postulated sub-groups of respondents associated with each of the ideal patterns. The remaining response patterns occur because of response errors. Thus, for example, a case that was "truly" NYY might respond NYN if one response error were made or even YNN if three response errors occurred. In effect, every possible, different response pattern can arise from respondents corresponding to any one of the ideal patterns if appropriate errors are assumed to occur. From the perspective of latent class modeling, the ideal patterns correspond to latent classes and response errors represent suitably constrained item conditional probabilities. Dayton and Macready (1976) presented a modified version of Proctor's model in which two types of response errors were assumed: intrusions that occur when a Yes (or, correct) response replaces a No (or, incorrect) in an ideal pattern and omissions that occur when a No response replaces a Yes in an ideal pattern. As for the Proctor model, any observed response can result from a respondent in any latent class by positing appropriate errors of the two types. It is possible to define other types of response-error models. In particular, a constant error rate can be associated with each item or with each ideal pattern (i.e., latent class). For three items and a linear scale as above, the restrictions on the item conditional probabilities in Eq. (1) for each of these types of models would be (where 1 is a Yes, or correct, and 2 is a No, or incorrect, response):

Proctor constant error:

$$\alpha_{111} = \alpha_{121} = \alpha_{131} = \alpha_{221} = \alpha_{231} = \alpha_{331} = \alpha_{212} = \alpha_{312} = \alpha_{322} = \alpha_{412} = \alpha_{422} = \alpha_{432}.$$

Dayton and Macready intrusion–omission errors:

$$\alpha_{111} = \alpha_{121} = \alpha_{131} = \alpha_{221} = \alpha_{231} = \alpha_{331} \quad \text{and} \quad \alpha_{212} = \alpha_{312} = \alpha_{322} = \alpha_{412} = \alpha_{422} = \alpha_{432}.$$

Item specific errors:

$$\alpha_{111} = \alpha_{212} = \alpha_{312} = \alpha_{412} \quad \text{and} \quad \alpha_{121} = \alpha_{221} = \alpha_{322} = \alpha_{422} \quad \text{and} \quad \alpha_{131} = \alpha_{231} = \alpha_{331} = \alpha_{432}.$$

Latent-class specific errors:

$$\alpha_{111} = \alpha_{121} = \alpha_{131} \quad \text{and} \quad \alpha_{221} = \alpha_{231} = \alpha_{212} \quad \text{and} \quad \alpha_{331} = \alpha_{312} = \alpha_{322} \quad \text{and} \quad \alpha_{412} = \alpha_{422} = \alpha_{432}.$$

Note that it is these restrictions on the item conditional probabilities that create the association between the latent classes and the ideal patterns.

Goodman (1975) introduced the novel idea of combining scaling models such as Guttman linear scales with additional unconstrained classes that do not conform to the scaling model. In his original formulation, respondents from the latent classes corresponding to ideal patterns such as NNN, NNY, NYY and YYY respond in complete consistency with expectation (i.e., made no response errors) but respondents from one or more intrinsically unscalable classes respond independently to the items. Thus, all respondents in patterns such as NYN arise from the intrinsically unscalable class(es) whereas respondents in a latent class corresponding to an ideal pattern such as NNY arise both from the associated latent class and from the intrinsically unscalable class(es). Dayton and Macready (1980) generalized the Goodman model by incorporating various types of errors for respondents in the latent classes corresponding to ideal patterns. Thus, in a simple case, for three items there would be a total of five latent classes – four corresponding to the ideal patterns as well as the intrinsically unscalable class. Item conditional probabilities are unconstrained for the intrinsically unscalable class but those associated with the latent classes corresponding to NNN, NNY, NYY and YYY are constrained according to, say, the Proctor model or the item-specific error model. It should be noted that virtually any response data can be adequately represented by a model with a sufficiently large number of intrinsically unscalable classes. Thus, when interpreting the results from a generalized Goodman model it is important to consider the proportion of respondents that correspond to the ideal patterns as opposed to the intrinsically unscalable class(es).

The notion of Guttman scales can be extended to models that incorporate more than a single linear scale underlying the observed responses. With three items, for example, the two linear Guttman scale orderings, NNN, NNY, NYY, YYY, and NNN, NYN, NYY, YYY, constitute a biform scale with five latent classes corresponding to the unique ideal

Table 5
Item error linear scale models fitted to abortion item data

Model[a]	# of parameters	DF	G^2	$\Pr(G^2)$	I_D	AIC	BIC
1	12	51	3723.00	0.000	0.094	3621.00	3202.33
2	19	44	1233.10	0.000	0.045	1145.10	783.89
3	26	37	607.87	0.000	**0.020**	**533.87**	**230.13**

[a]Models are explained in the text.

patterns NNN, NNY, NYN, NYY, and YYY. In effect, items 1 and 2 do not display a prerequisite relationship but a positive response to item 3 does depend on positive responses to both items 1 and 2. In this same manner, triform and more complex scaling models can be conceptualized.

Item responses may represent data at multiple time points as is the case of panel data that arise when a sample of respondents is measured on several occasions. For panel data, an observed response such as NNY could represent a respondent whose attitude toward an issue has shifted from non-agreement to agreement (or, of course, could result from response errors). If only shifts in a positive direction (e.g., No to Yes) are considered, then the ideal patterns for a Guttman scale apply to the panel data. Latent Markov models, which are beyond the scope of this chapter, incorporate the notions of ideal responses along with shifts in attitude that conform to Markov processes (Langeheine, 1994; Collins, 1997).

The six GSS abortion items that have been used illustratively are not well fitted by linear or biform scale models. Among the relatively better fitting (but not well fitting) models are those with item-specific errors. Table 5 displays results for a linear scale in the natural order of the items (see Table 2, above). Model 1 is the linear Guttman scale, Model 2 includes a single unscalable class and Model 3 includes two unscalable classes. Among these models, both AIC and BIC reach their minima for Model 3 that does have a reasonably small value for the index of dissimilarity, I_D. However, the estimated latent class proportions for the two intrinsically unscalable classes are 0.454 and 0.386 suggesting that this more-or-less reasonable fit for a linear scale with item-specific errors can only be achieved by assuming that more than 80% of the respondents are unscalable.

7. Models incorporating grouping of respondents

In general, grouping of respondents in LCA arises in two different ways. First, the design for data collection (e.g., survey design) may involve stratification as a means of ensuring representation across a population of interest. Or, second, the researcher may be interested in comparisons among groups such as sex, socio-economic status, etc.

7.1. *Stratification by design*

Complex survey designs utilize stratification and cluster sampling both of which require modifications to the latent class model presented in Eq. (1). Typically, factors such as stratification and non-response are taken into account in the analysis of survey data by means of case-wise sampling weights (Kish, 1965; Kalton, 1989). In practice, strata are often not sampled proportionally to their representation in the population. This may be done, for example, in order to ensure inclusion of sufficiently large numbers of minority groups to allow for separate analyses for these groups. Even when strata are sampled proportionally to population size, differential response rates and missing data often result in an unbalanced sample. The application of case-wise sampling weights can be used to (approximately) correct for these sources of potential bias and such weights are typically reported with public-use survey sample databases. The use of sampling weights is incorporated into latent class programs such as Latent Gold and MPlus. As noted above, Patterson et al. (2002) proposed obtaining parameter estimates in LCA using a sample-weighted pseudo-log-likelihood in place of Eq. (4) and demonstrated that such estimates were unbiased, at least for the scenarios that were studied.

7.2. *Stratification for group comparisons*

The basic model for LCA can be modified to accommodate stratification variables (Clogg and Goodman, 1984, 1985; Dayton and Macready, 2002). Let an observation for a respondent in stratum s be: $\boldsymbol{Y}_{is} = \{y_{isj}\}$. Then, the unconstrained latent class model in Eq. (1) is rewritten:

$$\Pr(Y_{is}) = \sum_{c=1}^{C} \theta_{c|s} \prod_{j=1}^{J} \prod_{r=1}^{R_j} \alpha_{cjr|s}^{\delta_{isjr}}. \tag{9}$$

Note that the latent class proportion, $\theta_{c|s}$, the item conditional probabilities, $\alpha_{cjr|s}$, as well as the Kronecker deltas, δ_{isjr} include a subscript indicative of membership in a specific stratum. In essence, the model in Eq. (9) is simply the original latent class model written explicitly for the sth stratum. In fact, if parameters for this model are estimated they will be the same as would result from conducting separate latent class analyses for each stratum. For comparison purposes the stratification model in Eq. (9) is referred to as the heterogeneous model. In typical applications, the heterogeneous model is compared with a partially homogeneous as well as to a completely homogeneous model. For a model with S strata, the partially homogeneous model is defined by the restrictions $\alpha_{cjr|s}^{\delta_{isjr}} = \alpha_{cjr}^{\delta_{isjr}}$ for all $s = 1, \dots, S$. For this model, the sizes of the latent classes are allowed to vary across strata but the item conditional probabilities that characterize the nature of the latent classes remain the same across strata so that the interpretation of the classes does not vary from stratum to stratum. On the other hand, the completely homogeneous model is defined by both the restrictions $\alpha_{cjr|s}^{\delta_{isjr}} = \alpha_{cjr}^{\delta_{isjr}}$ for all $s = 1, \dots, S$ and $\theta_{c|s} = \theta_c$ for all $s = 1, \dots, S$. In essence, the stratification is ignored when these restrictions are applied. Statistical comparisons among heterogeneous, partially homogeneous and completely homogeneous models can, in theory, be

Table 6
Example with stratification

Model	DF	Fit G^2	Difference DF	Difference G^2	I_D	AIC	BIC
Heterogeneous	116	153.36	116	153.36	0.005	−78.645	−1024.45
Partially homogeneous	206	313.59	90	160.23	0.021	**−98.414**	**−1778.03**
Homogeneous	218	725.02	12	411.42	0.062	289.003	−1488.43

carried out using conventional difference chi-square tests since these models are nested and do not involve setting latent class proportion to boundary values (i.e., to 0).

Various stratification variables of interest are available for the GSS abortion items. To illustrate these techniques, we selected subjective ratings of social class identification as reported by respondents using the categories lower, working, middle and upper. When the data are divided among these four strata, the resulting frequency distribution contains $4 \cdot 2^6 = 256$ cells and the unconstrained, heterogeneous model with five latent classes requires the estimation of a total of 136 independent parameters (i.e., 16 latent class proportions and 120 item conditional probabilities). In addition, when determining degrees of freedom for model fit, there are four marginal totals representing the numbers of respondents per stratum that are fixed as opposed to only the overall total being fixed. Thus, for this example, the degrees of freedom are 116 and as, shown in Table 6, the heterogeneous model does not fit the data well. The model of partial homogeneity is based on considerably fewer parameters since the item conditional probabilities are constant across strata. From Table 5, it is apparent that none of these models provides particularly good fit in terms of G^2 although the index of dissimilarity for the heterogeneous and partially homogeneous models could be viewed as satisfactory in many applications. Among the models being compared, both the min(AIC) and min(BIC) strategies lead to interpreting the partially homogeneous model. For this model, the item conditional probabilities are constant across strata but the latent class proportions are free to vary. Although not shown here, item conditional probabilities for the partially homogeneous model lead to essentially the same interpretation for the five latent classes as was the case for the non-stratified (homogeneous) model displayed above in Figure 3. However, as shown in Figure 4, there are distinct patterns of differences in latent class proportions associated with the strata. Two apparent trends are for an increasingly larger proportions of respondents estimated in the first latent class (consistent Yes responses) and smaller proportions estimated in the fourth latent class (consistent No responses) when moving from the lower to the upper social classes. On the other hand, the third latent class (Yes for items 1, 2, 3 and intermediate for others) remains relatively stable in size.

8. Covariate latent class models

Differences in latent class structure for population strata can be taken into account as described above although these procedures have practical limitations. First, if the stratification variable is continuous (e.g., age of respondent) it is necessary to create groups

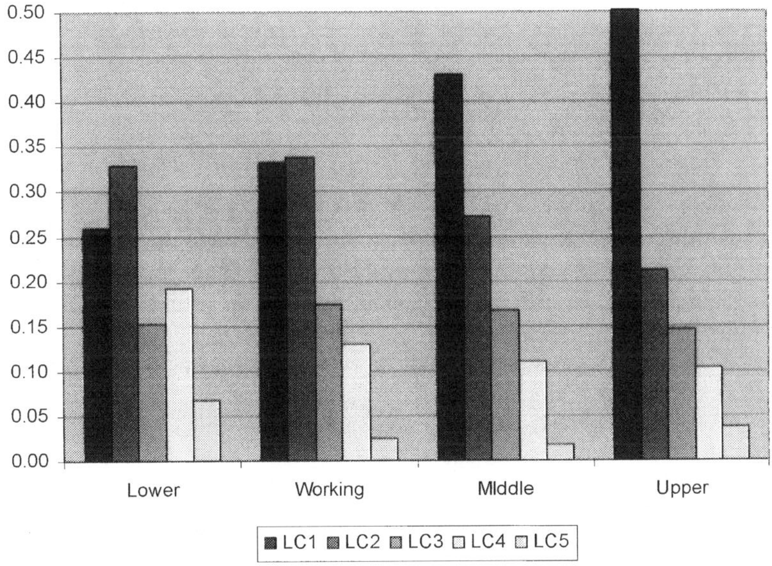

Fig. 4. Latent class proportions for GSS strata.

using some more-or-less arbitrary procedure (e.g., dividing age into quarters of the distribution). Second, it is ordinarily not practical to include more than a small number of stratification variables since, even with a large sample size, small cell sizes tend to occur which results in unstable subgroup analyses. And, third, stratification models can be extremely complex in terms of the number of parameters that are estimated. Dayton and Macready (1988a) proposed a covariate latent class model that incorporates a logistic regression component for the relationship between latent class membership and one or more covariates. This model has the advantage that it incorporates all the modeling possibilities inherent in ordinary multiple regression. That is, covariates can be continuous (e.g., age), categorical using recoding as necessary (e.g., dummy-coded regions of the country) or products of covariates in order to model interactions. In its simplest form, the covariate latent class model is appropriate for two latent classes with a single covariate, Z, but can be generalized as shown below. The basic model posits a functional form for the regression of latent class membership in, say, the first class on the covariate:

$$\theta_{1|Z_i} = g(Z_i \mid \beta) = \frac{e^{\beta_0+\beta_1 Z_i}}{1+e^{\beta_0+\beta_1 Z_i}}. \tag{10}$$

In log-odds form, the model can be written:

$$\ln\left(\frac{\theta_{1|Z_i}}{1-\theta_{1|Z_i}}\right) = \beta_0 + \beta_1 Z_i, \tag{11}$$

which is in the form of a logistic regression model except, of course, that group membership is latent rather than observed. Note that the probability of latent class membership depends upon the specific value, Z_i, of the covariate for a given respondent. Combining the model in Eq. (10) with a conventional latent class model (Eq. (1)) for two latent classes gives:

$$\Pr(Y_i, Z_i) = \sum_{c=1}^{2} \theta_{c|Z_i} \prod_{j=1}^{J} \prod_{r=1}^{R_j} \alpha_{cjr}^{\delta_{ijr}}. \tag{12}$$

Note that latent class membership is conditioned on the covariate but that conditional probabilities for the items are unconditional. This implies that the covariate model, as written, is comparable to a partial homogeneity model in the terminology of stratification models. Thus, the latent structure defined by the item conditional probabilities is assumed to be constant over the range of values for the covariate(s).

The covariate latent class model can accommodate multiple covariates in the obvious manner by expanding the additive model for log-odds:

$$\ln\left(\frac{\theta_{1|Z_i}}{1-\theta_{1|Z_i}}\right) = \beta_0 + \beta_1 Z_{1i} + \cdots + \beta_p Z_{pi}, \tag{13}$$

where Z_i is now a vector of p covariates. Extending the model to more than two latent classes is somewhat less obvious. One approach is to select one of the classes as a reference and to create log-odds models comparing each of the remaining classes with the reference. This topic is treated in more detail in Dayton and Macready (1988a, 1988b).

To illustrate the covariate latent class model, we return to the GSS abortion items but consider fitting a two-class model where the first latent class proportion is regressed on scores representing the respondent's reported strength of religious affiliation. The four options for this item were: strong, not very strong, somewhat strong, and no religion. These categories were encoded 1 through 4 and, in this analysis, were treated as falling along a continuum. A two-class model, while not the preferred model in our earlier analyses, does result in classes corresponding to the two largest latent classes when a five-class model is estimated. That is, the first class in a two-class model represents respondents who are consistently favorable to abortion for all six items and the second class represents respondents who sharply distinguish between favoring the first three reasons for abortion and opposing the last three reasons for abortion. In the five-class solution, these two classes were estimated to comprise 39% and 30% of respondents, respectively. Thus, the model with only two latent classes does capture a substantial proportion of the variation among respondents.

Preliminary study of the functional relationship between religious intensity (Z) and the 0–6 score representing the number of Yes responses to the six abortion items suggested that the log-odds function was best approximated by a cubic equation. Thus, the latent class model was estimated in the form:

$$\ln\left(\frac{\theta_{1|Z_i}}{1-\theta_{1|Z_i}}\right) = \beta_0 + \beta_1 Z_i + \beta_2 Z_i^2 + \beta_3 Z_i^3.$$

The resulting coefficients were $\hat{\beta}_0 = 3.0465$, $\hat{\beta}_1 = -4.2035$, $\hat{\beta}_2 = 1.7827$, $\hat{\beta}_3 = -0.2397$ and the regression for the probability of membership in the first latent class is displayed in Figure 5. The incorporation of a cubic regression component allowed for

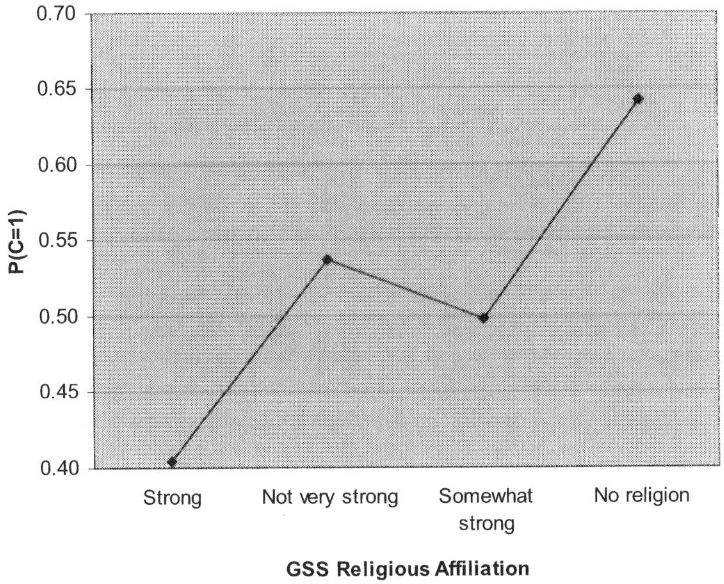

Fig. 5. Regression of yes latent class on intensity of religious affiliation.

the lack of monotonicity for the middle two categories. Indeed, these two categories, while clearly between strong and no religion are quite ambiguous with respect to order. Overall, there is a strong trend relating religious affiliation to predicted membership in the first latent class.

9. Other models

9.1. Adaptive tests

In its simplest form, adaptive testing entails the sequential selection of test items for administration based on prior performance of the respondent (Van der Linden and Glas, 2000; Wainer, 2000). Ordinarily, such testing is administered by computer with a pool of test items that has been pre-tested and calibrated in some way to accommodate the sequential process. Most models for computerized adaptive testing (CAT) in use at this time are based on item response theory (IRT) that assumes an underlying continuous scale (latent trait). However, CAT can be based on latent class models that are conceptually much simpler than IRT in the sense of entailing fewer untestable assumptions. This theory, latent class CAT or LCCAT, was first presented by Macready and Dayton (1992) and was based on earlier work with mastery tests (Macready and Dayton, 1977, 1980). In LCCAT, the goal is to classify the respondent in the appropriate latent class (e.g., master/non-master) with some acceptable, minimum level of error. In brief, there is the usual requirement for a pool of calibrated test items (in this case, items with known latent class characteristics). The sequential testing strategy entails selecting the first item for presentation to a respondent in a more or less arbitrary manner. Then, succeeding items are selected to minimize the misclassification probability for the respondent and testing is complete when this probability falls below a specified criterion. An exemplary application of LCCAT is presented by Macready and Dayton (1992).

9.2. *Two or more latent variables*

The latent class models considered above may be viewed as based on a single underlying latent dimension typified by the structure reflected in the items. It is possible, however, that the items reflect more than a single dimension in much the same manner that factor analytic models incorporate two or more latent variables. For example, items 1, 2 and 3 may reflect variation among two levels of a latent dimension, X, while items 4, 5 and 6 reflect variation among two levels of a latent dimension, Y. Assuming that the latent dimension, X, has C classes and Y has D classes, the basic model for LCA presented in Eq. (1) can be modified to reflect the two-dimensional structure:

$$\Pr(Y_i) = \sum_{c=1}^{C} \sum_{d=1}^{D} \theta_{cd} \prod_{j=1}^{J} \prod_{r=1}^{R_j} \alpha_{cdjr}^{\delta_{ijr}}. \tag{12}$$

If no restrictions are placed on this model, it is equivalent to a one-dimensional latent class model with $C \times D$ latent classes. In practice, various restrictions are imposed to reflect the association of items with latent dimensions and, typically, the latent dimensions are assumed to be orthogonal (i.e., independent). For illustrative purposes, we consider the first three abortion items (Woman's health seriously endangered, Pregnant as result of rape, Strong chance of serious defect) as defining the first latent dimension, X, and the second three abortion items (Low income – can't afford more children, Not married, Married–wants no more children) as defining the second latent dimension, Y. Assuming two latent classes for each latent dimension, X and Y, the latent class proportions are estimated as:

Dimension X/Y	1	2	Sum
1	0.370	0.090	0.460
2	0.434	0.106	0.540
Sum	0.805	0.195	1.000

Note that the ratios of column latent class proportions, 0.370/0.090 and 0.434/0.106, are both 4.12 indicating the independence of the X and Y latent dimensions. Profiles for latent class proportions (not shown) display the association of items with latent classes by the fact that items defining a given latent class have constant conditional probabilities for the other class (e.g., items A, B and C are constant across the levels for X and items D, E and F are constant across the levels for Y).

Latent class models with two or more latent dimensions provide an opportunity for modeling multivariate categorical data that, to a large extent, parallel the techniques of confirmatory factor analysis and structural equation modeling that have been developed for continuous variables. For additional information and examples, good sources are Hagenaars (1993, 1998).

9.3. Linear logistic latent class models

Formann (1982, 1985) proposed a model in which latent class proportions are functionally dependent upon various features of the items themselves. These features may include item content, item structure, etc. that are mapped onto the items in a manner reminiscent of factorial designs. These so-called multi-component latent class models can be combined with covariate models to provide extremely flexible modeling. The main factor limiting the use of such models is the lack of readily available, general purpose software for model estimation.

9.4. Located latent class model

Uebersax (1993) proposed a hybrid latent class model in which the items and the latent classes are both located along a single continuous latent dimension. For example, items A, B and C may have relative scale locations τ_1, τ_2, and τ_3, say, and on the same dimension, two latent classes may have locations β_1 and β_2. The locations of the latent classes are, typically, defined at the median of the cumulative function connecting the latent classes and the items. This function may be the cumulative normal density (or, probit for which Uebersax has a program) or the logistic function (or, logit). The model for a Yes response to item j, given membership in latent class c, with a logistic function would be:

$$\Pr(Y_{ij} \mid c) = \frac{\exp(\alpha(\tau_j - \beta_c))}{1 + \exp(\alpha(\tau_j - \beta_c))},$$

where α is an appropriate scaling constant. The model is a constrained form of a comparable unrestricted latent class model where the cumulative function defines the constraints. More details and examples are provided by Uebersax (1993) who provides a computer program on his web site cited in Section 1 of this chapter.

The located latent class model is similar in concept to the T-class mixture model presented by Lindsay et al. (1991). Both models position latent classes along a discretized underlying continuum with inequality constraints used to define the model. Rather than being locating along a continuum, latent classes may be assumed to be ordered such that membership in latent class $c + 1$ implies a greater likelihood for a Yes response, say, to all items than for latent class c (Croon, 1990). Also, there are restricted latent class models that may be viewed as non-parametric analogues to latent trait models (item response theory) such as Rasch or Birnbaum models. These models are beyond the scope of this chapter but Heinen (1996) presents very insightful comparisons and contrasts between latent class and latent trait models in general.

9.5. Bayesian latent class models

In Bayesian analysis, prior information concerning the parameters is taken into account during estimation. In contrast to frequentist interpretations, Bayesian models incorporate the notion that parameters have distributions representing lack of knowledge about their values prior to collecting empirical data. In LCA all parameters (i.e., latent class proportion and item conditional probabilities) are in the form of probabilities on a 0 to 1 scale. Thus, prior distributions of belief about the values of parameters must be defined on the same scale. Although, in theory, prior distributions could be based on empirical sources, in practice these distributions are most often selected on the basis of their mathematical tractability; that is, so-called natural conjugate prior distributions. Furthermore, since actual prior information is rarely available about the values of parameters, prior distributions are selected that represent little or no prior information (i.e., indifference priors). The mathematics of Bayesian estimation involves combining the prior distribution with the likelihood for the data to arrive at a posterior distribution, the mode of which is taken to be the estimator. The program Latent Gold (Vermunt and Magidson, 2000) allows the specification of prior Dirichelet distributions that are natural conjugate priors for multinomial probability models. The Dirichelet distribution simplifies to the somewhat better known beta distribution if an item is dichotomous or if there are only two latent classes. A practical use for Bayesian models in LCA arises for unidentified models since a Bayesian solution will be unique *given the specified prior distribution*. For identified latent class models, Bayesian estimates tend to be "shrunken" toward the center of the range (e.g., toward 0.5 for probabilities) with the amount of shrinkage dependent upon sample size and the degree of information in the prior distributions. For the GSS abortion items, based on 27,151 respondents, the impact of the virtually non-informative prior that is the default option in Latent Gold has no discernible impact on the resulting estimates when compared to maximum likelihood estimates.

When fitting complex latent class models to data, it may be difficult to achieve stable estimation results. An example of this was encountered when fitting unrestricted latent class models to the six GSS abortion items when the number of latent classes was five or greater. A useful strategy is to use Bayesian methods with non-informative prior distributions (or, nearly so) to estimate the comparable latent class model since it can be expected that, due to including prior information, the Bayesian estimates are stable. Then, it becomes reasonable to accept the conventional solution if it is substantially equivalent to the Bayesian solution. For truly non-identified models, however, there is no "stable" conventional solution although the Bayesian estimates will provide one of an arbitrarily large number of solutions that fit the data equally well.

References

Akaike, H. (1973). Information theory and an extension of the maximum likelihood principle. In: Petrov, B.N., Csake, F. (Eds.), *Second International Symposium on Information Theory*. Akademiai Kiado, Budapest, pp. 267–281.

Akaike, H. (1974). A new look at the statistical model identification. *IEEE Transactions on Automatic Control* **AC-19**, 716–723.

Bartholomew, D.J. (1987). *Latent Variable Models and Factor Analysis*. Charles Griffin, London.

Bozdogan, H. (1987). Model-selection and Akaike's information criterion (AIC): The general theory and its analytical extensions. *Psychometrika* **52**, 345–370.

Clogg, C.C., Goodman, L.A. (1984). Latent structure analysis of a set of multidimensional tables. *Journal of the American Statistical Association* **79**, 762–771.

Clogg, C.C., Goodman, L.A. (1985). Simultaneous latent structure analysis in several groups. In: Tuma, N.B. (Ed.), *Sociological Methodology 1985*. Jossey-Bass, San Francisco, CA.

Collins, L.A. (1997). *Latent Transition Analysis for Windows*. Pennsylvania State University, University Park, PA.

Croon, M.A. (1990). Latent class analysis with ordered latent classes. *British Journal of Mathematical and Statistical Psychology* **43**, 171–192.
Dayton, C.M. (1999). *Latent Class Scaling Analysis*. Sage Publications, Thousand Oaks, CA.
Dayton, C.M. (2003). Applications and computational strategies for the two-point mixture index of fit. *British Journal of Mathematical and Statistical Psychology* **56**, 1–13.
Dayton, C.M., Macready, G.B. (1976). A probabilistic model for validation of behavioral hierarchies. *Psychometrika* **41**, 189–204.
Dayton, C.M., Macready, G.B. (1980). A scaling model with response errors and intrinsically unscalable respondents. *Psychometrika* **45**, 343–356.
Dayton, C.M., Macready, G.B. (1988a). Concomitant-variable latent class models. *Journal of the American Statistical Association* **83**, 173–178.
Dayton, C.M., Macready, G.B. (1988b). A latent class covariate model with applications to criterion-referenced testing. In: Langeheine, R., Rost, J. (Eds.), *Latent Trait and Latent Class Models*. Plenum Press, New York.
Dayton, C.M., Macready, G.B. (2002). Use of categorical and continuous covariates in latent class analysis. In: McCutcheon, A., Hagenaars, J. (Eds.), *Advances in Latent Class Modeling*. Cambridge University Press, Cambridge, UK.
Formann, A.K. (1982). Linear logistic latent class analysis. *Biometrical Journal* **24**, 171–190.
Formann, A.K. (1985). Constrained latent class models: theory and applications. *British Journal of Mathematical and Statistical Psychology* **38**, 87–111.
Goodman, L.A. (1974). Exploratory latent structure analysis using both identifiable and unidentifiable models. *Biometrika* **61**, 215–231.
Goodman, L.A. (1975). A new model for scaling response patterns: An application of the quasi-independence concept. *Journal of the American Statistical Association* **70**, 755–768.
Goodman, L.A., Kruskal, W.H. (1979). *Measures of Association for Cross Classifications*. Springer-Verlag, New York.
Guttman, L. (1947). On Festinger's evaluation of scale analysis. *Psychological Bulletin* **44**, 451–465.
Haberman, S.J. (1979). *Analysis of Quantitative Data (Vol. 2)*. Academic Press, New York.
Haertel, E. (1984). An application of latent class models to assessment data. *Applied Psychological Measurement* **8**, 333–346.
Hagenaars, J.A. (1990). *Categorical Longitudinal Data*. Sage Publications, Newbury Park, CA.
Hagenaars, J.A. (1993). *Loglinear Models with Latent Variables*. Sage Publications, Newbury Park, CA.
Hagenaars, J.A. (1998). Categorical causal modeling: latent class analysis and directed log-linear models with latent variables. *Sociological Methods and Research* **26**, 436–486.
Hagenaars, J.A., McCutcheon, A.L. (Eds.) (2002). *Applied Latent Class Analysis*. Cambridge University Press, Cambridge, UK.
Heinen, T. (1996). *Latent Class and Discrete Trait Models. Advanced Quantitative Techniques in the Social Sciences, Series 6*. Sage Publications, Thousand Oaks, CA.
Kalton, G. (1989). Modeling considerations: discussion from a survey sampling perspective. In: Kasprzyk, D., Duncan, G., Kalton, G., Singh, M.P. (Eds.), *Panel Survey*. John Wiley and Sons, New York, pp. 575–585.
Kish, L. (1965). *Survey Sampling*. John Wiley and Sons, New York.
Kish, L., Frankel, M.P. (1974). Inference from complex samples. *Journal of the Royal Statistical Society, Series B* **36**, 1–37.
Kullback, S., Leibler, R.A. (1951). On information and sufficiency. *Annals of Mathematical Statistics* **22**, 79–86.
Langeheine, R. (1994). Latent variables Markov models. In: Von Eye, A., Clogg, C. (Eds.), *Latent Variables Analysis*. Sage Publications, Thousand Oaks, CA.
Langeheine, R., Rost, J. (Eds.) (1988). *Latent Trait and Latent Class Models*. Plenum Press, New York.
Lazarsfeld, P.F., Henry, N.W. (1968). *Latent Structure Analysis*. Houghton Mifflin, Boston.
Lindsay, B., Clogg, C.C., Grego, J. (1991). Semiparametric estimation in the Rasch model and related models, including a simple latent class model for item analysis. *Journal of the American Statistical Association* **86**, 96–107.
Macready, G.B., Dayton, C.M. (1977). The use of probabilistic models in the assessment of mastery. *Journal of Educational Statistics* **2**, 99–120.

Macready, G.B., Dayton, C.M. (1980). The nature and use of state mastery models. *Applied Psychological Measurement* **4**, 493–516.
Macready, G.B., Dayton, C.M. (1992). The application of latent class models in adaptive testing. *Psychometrika* **57**, 71–88.
Muthén, L.K., Muthén, B.O. (1998). *Mplus: The Comprehensive Modeling Program for Applied Researchers, User's Guide*. Muthén & Muthén, Los Angeles, CA.
Patterson, B., Dayton, C.M., Graubard, B. (2002). Latent class analysis of complex survey data: application to dietary data. *Journal of the American Statistical Association* **97**, 721–729.
Proctor, C.H. (1970). A probabilistic formulation and statistical analysis of Guttman scaling. *Psychometrika* **35**, 73–78.
Rost, J., Langeheine, R. (Eds.) (1997). *Applications of Latent Trait and Latent Class Models in the Social Sciences*. Waxmann, New York.
Rudas, T., Clogg, C.C., Lindsay, B.G. (1994). A new index of fit based on mixture methods for the analysis of contingency tables. *Journal of the Royal Statistical Society, Series B* **56**, 623–639.
Schwarz, G. (1978). Estimating the dimension of a model. *Annals of Statistics* **6**, 461–464.
Titterington, D.M., Smith, A.F.M., Makov, U.E. (1985). *Statistical Analysis of Finite Mixture Models*. John Wiley and Sons, New York.
Uebersax, J.S. (1993). Statistical modeling of expert ratings on medical treatment appropriateness. *Journal of the American Statistical Association* **88**, 421–427.
Van der Linden, W.J., Glas, C.A.W. (Eds.) (2000). *Computerized Adaptive Testing: Theory and Practice*. Springer-Verlag, New York.
Vermunt, J.K. (1997). *The LEM User Manual*. WORC Paper. Tilburg University, The Netherlands.
Vermunt, J.K., Magidson, J. (2000). *Latent Gold: User's Guide*. Statistical Innovations, Belmont, MA.
Von Eye, A., Clogg, C.C. (Eds.) (1994). *Latent Variables Analysis: Applications for Developmental Research*. Sage Publications, Thousand Oaks, CA.
Wainer, H. (Ed.) (2000). *Computerized Adaptive Testing: A Primer*. Lawrence Erlbaum, New Jersey.
Wolter, K.M. (1985). *Introduction to Variance Estimation*. Springer-Verlag, New York.

Handbook of Statistics, Vol. 26
ISSN: 0169-7161

DOI: 10.1016/S0169-7161(06)26014-0

14

Random-Effects Models for Preference Data[1]

Ulf Böckenholt and Rung-Ching Tsai

1. Introduction

Much of human behavior may be viewed as decisions among more or less well-defined options or alternatives. Familiar instances are a citizen deciding whether to vote and if so for whom, a shopper contemplating various brands of a product category, or a physician considering various treatment options. Despite the ubiquity of decision processes, it can be difficult to interpret choice outcomes in an unambiguous way because frequently decisions are subject to such unobserved constraints as time, money, knowledge, or availability of choice options. In contrast, choice data collected in experimental settings where persons are asked directly to state their preferences for a single or multiple sets of options are less prone to these constraints. Both data types, which are referred to as revealed and stated preference data (Louviere et al., 2000), may complement each other. For instance, in an election votes for political candidates represent revealed choice data. Rankings of the same candidates in a survey shortly before the election are an example for stated choice data. Stated preference data may prove useful both for predicting the election outcome and for providing more information about the preference differences among the candidates than would be available from the election results alone. The collection of stated preference data is also useful in situations when it is difficult or impossible to collect revealed preference data because choices are made infrequently or new choice options offered in the studies are yet unavailable on the market. For example, in taste testing studies, consumers may be asked to choose among existing as well as newly developed products.

Most statistical models for the analysis of choice data (or, preference models, in short) have been developed for stated response settings to serve the dual purpose of summarizing choice outcomes and of facilitating the forecasting of choices made by judges facing possibly new or different variants of the choice options (Marshall and Bradlow, 2002). Thus, preference models can provide useful information to identify both option characteristics that influence choices and systematic sources of individual

[1] Ulf Böckenholt was supported by the Social Sciences and Humanities Research Council of Canada. The research of the second author was supported by grant NSC 93-2413-H-003-060 from the National Science Council of Taiwan.

differences in the evaluation of these option characteristics. For example, in an election, voters may base their decision on a set of attributes (e.g., integrity, leadership) but differ in the weights they assign to the attribute values for the different candidates. An application of preference models to voting data may reveal these attributes and provide insights about how voters differ in their attribute-specific assessments of the candidates.

Because preference models focus mainly on choice outcomes, they render little information about the underlying choice processes and the effect of unobserved constraints that frequently are a part of revealed choices. Despite these limitations, preference models can be useful in characterizing a choice process in two important respects provided multiple choices are observed from the same decision maker.

First, they can provide benchmarks for understanding whether decision makers are consistent in making their choices. Consider two pairwise comparisons in which option j is preferred to option k, and option k is preferred to option l. If the judge is consistent, we expect that in a comparison of options j and l, j is preferred to l. If a judge selects option l in the last pairwise comparison then this indicates an intransitive cycle which may be useful in understanding the judgmental process. For instance, in a large-scale investigation with over 4,000 respondents of Zajonc's (1980) proposition that esthetic and cognitive aspects of mentality are separate, Bradbury and Ross (1990) demonstrated that the incidence of intransitive choices for colors declines through childhood from about 50% to 5%. For younger children, the novelty of a choice option plays a decisive role with the result that they tend to prefer the option they have not seen before. The reduction of this effect during childhood and adolescence is an important indicator of the developmental transition from a prelogical to a logical reasoning stage. The diagnostic value of the observed number of intransitive cycles is highest when it is known in advance which option triple will produce transitivity violations (Morrison, 1963). If this information is unavailable, preference models are needed to determine whether intransitivities are systematic or reflective of the stochastic nature of choice behavior.

Second, preference models allow assessing whether decision makers act as utility maximizers. Starting with the work of Thurstone (1927), who invoked the psychophysical concept of a sensory continuum, Thurstone explained the observation that choices by the same person may not be deterministic under seemingly identical conditions. He argued that options should be represented along this sensory continuum by random utility variables which describe the options' effects on a person's sensory apparatus. The choice process is then reduced to selecting the option with the greatest realization. An important implication of this random utility assumption is the regularity condition which states that the probability of choosing an option cannot increase when the choice set is expanded (Block and Marschak, 1960). For example, the probability of choosing option j in a comparison of j and k cannot be lower than the probability of choosing option j in a comparison of j, k and l, say. This assumption is testable and received support in many applications. However, systematic violations of the regularity condition in experimental studies have also been reported (Simonson and Tversky, 1992).

This chapter considers preference models based on Thurstone's random utility approach for the analysis of partial and/or incomplete ranking data. Consider a set of J choice alternatives ($j = 1, \ldots, J$) and n decision makers ($i = 1, \ldots, n$). Incomplete ranking data are obtained when a decision maker considers only a subset of the choice options. For example, in the method of paired comparison, two choice options are presented at a time, and the decision maker is asked to select the more preferred one. In contrast, in a partial ranking task, a decision maker is confronted with all choice options and asked to provide a ranking for a subset of the J options. A special case of an incomplete ranking task is the best–worst method according to which a decision maker is instructed to select the best and worst options out of the offered set of choice options. Both partial and incomplete approaches can be combined by offering multiple distinct subsets of the choice options and obtaining partial or complete rankings for each of them. For instance, in the method of paired comparisons, a judge may be presented with all $\binom{J}{2}$ option pairs sequentially with the instruction to select for each of the pairs the more preferred option (David, 1988).

Thurstone's (1927) random utility paradigm has been highly influential in the development of many preference models, especially in the econometric and transportation literature (for recent reviews, see Marley, 2002; Train, 2003; McFadden, 2001). Numerous applications attest to the usefulness of this approach both for forecasting choices and for providing parsimonious descriptions of how decision makers differ in their preferences at a particular point in time and over time. The next sections outline Thurstonian random utility models for partial and/or incomplete ranking data, and, subsequently, discuss identification and estimation issues in detail. The chapter concludes with two applications.

2. Thurstonian random utility models

According to Thurstone's (1927) random utility approach, a choice of an option j by decision maker i is determined by an unobserved utility assessment, υ_{ij}, that can be decomposed into a systematic and a random part:

$$\upsilon_{ij} = \mu_{ij} + \epsilon_{ij}, \tag{1}$$

where, typically, the person-specific option mean μ_{ij} is assumed to stay the same in repeated evaluations of the option in different choice sets but the random contribution ϵ_{ij} varies from evaluation to evaluation according to some distribution. One may question the use of randomness as a device to represent factors that determine the formation of preferences but are unknown to the observer. However, because, in general, it is not feasible to identify or measure all relevant choice determinants (such as all attributes of the choice options of the person choosing, or of environmental factors), it is not possible to answer conclusively the question of whether the choice process is inherently random or is determined by a multitude of different factors. Fortunately, for the development of statistical models this issue is not critical because either position arrives at the same conclusion that choices are described best in terms of their probabilities of occurrence (Manski and McFadden, 1981).

The choices made by person i for a single choice set can be summarized by an ordering vector $\mathbf{r}_i$. For instance, $\mathbf{r}_i = (h, j, \ldots, l, k)$ indicates that choice option h is judged superior to option j which in turn is judged superior to the remaining options, with the least preferred option being k. The probability of observing this ordering vector for person i can be written as

$$\Pr(\mathbf{r}_i = (h, j, \ldots, l, k)) = \Pr[(\upsilon_{ih} - \upsilon_{ij} > 0) \cap \cdots \cap (\upsilon_{il} - \upsilon_{ik} > 0)]. \quad (2)$$

The distributional assumptions on the ϵ_{ij}'s can then be used to find the probability for each response pattern. Typically, it is assumed that judgments of the J options for judge i are multivariate normal with mean vector $\boldsymbol{\mu}_i$ and covariance matrix $\boldsymbol{\Sigma}_i$. Since it is commonly assumed that judges differ in their mean evaluations $\boldsymbol{\mu}_i$ of the options, but not in how their assigned utilities of the options relate to each other, we set $\boldsymbol{\Sigma}_i = \boldsymbol{\Sigma}$. The assumption that the ϵ_{ij}'s are independently Gumbel or Gamma distributed leads to preference models proposed by Luce (1959) and Stern (1990), respectively. A more general random-effects representation is obtained when the ϵ_{ij}'s are specified semiparametrically using Dirichlet Process priors (Ansari and Iyengar, 2006). This approach provides a robust alternative to the normality assumption. Marden (1995) contains a general discussion of several of these different specifications. In the following, we adopt the popular normality representation but also discuss alternative parametric assumptions when appropriate.

For any given ordering pattern $\mathbf{r}_i$ of J options, a $(J-1) \times J$ contrast matrix, $\mathbf{C}_i$, can be defined to indicate the sign of the differences among its ranked options. For example, for the ordering vectors $\mathbf{r}_i = (j, l, k)$ of $J = 3$, the corresponding contrast matrices take on the form

$$\mathbf{C}_i = \begin{bmatrix} 1 & 0 & -1 \\ 0 & -1 & 1 \end{bmatrix},$$

where the three columns of the contrast matrices correspond to the options j, k, and l, respectively. Equation (2) can then be written as

$$\Pr(\mathbf{r}_i \mid \boldsymbol{\mu}_i, \boldsymbol{\Sigma}) = \Pr(\mathbf{C}_i \boldsymbol{\upsilon}_i > \mathbf{0} \mid \boldsymbol{\mu}_i, \boldsymbol{\Sigma}), \quad (3)$$

where $\boldsymbol{\upsilon}_i = (\upsilon_{i1}, \ldots, \upsilon_{iJ})$ contains the option utility assessments of person i.

Because of the multivariate normality specification of the latent judgments $\boldsymbol{\upsilon}_i$, the distribution of the pairwise differences of the $\boldsymbol{\upsilon}$'s is also multivariate normal. Consequently, the probability of observing the rank order vector $\boldsymbol{r}_i$ can be determined by evaluating a $(J-1)$-variate normal distribution,

$$\Pr(\mathbf{r}_i \mid \boldsymbol{\mu}_i, \boldsymbol{\Sigma}) = \frac{|\boldsymbol{\Gamma}_i|^{-1/2}}{(2\pi)^{(J-1)/2}} \times \int_0^{\infty} \cdots \int_0^{\infty} \exp\left\{-\frac{1}{2}(\mathbf{x} - \boldsymbol{\delta}_i)' \boldsymbol{\Gamma}_i^{-1} (\mathbf{x} - \boldsymbol{\delta}_i)\right\} \mathbf{dx}, \quad (4)$$

where $\boldsymbol{\delta}_i = \mathbf{C}_i \boldsymbol{\mu}_i$ and $\boldsymbol{\Gamma}_i = \mathbf{C}_i \boldsymbol{\Sigma} \mathbf{C}_i'$.

When a decision maker chooses among the options for T choice sets, we obtain a $(J \times T)$ ordering matrix $\mathbf{R}_i = (\mathbf{r}_{i1}, \ldots, \mathbf{r}_{iT})$ containing person's i rankings for each of the choice set. For example, for the rankings of $\mathbf{r}_{i1} = (j, k, l)$, $\mathbf{r}_{i2} = (j, l, m)$, in the two triplet choice sets for options j, k, l, and m, their corresponding contrast matrices respectively take on the forms of $\mathbf{C}_{i1} = \begin{bmatrix} 1 & -1 & 0 & 0 \\ 0 & 1 & -1 & 0 \end{bmatrix}$, and $\mathbf{C}_{i2} = \begin{bmatrix} 1 & 0 & -1 & 0 \\ 0 & 0 & 1 & -1 \end{bmatrix}$.

If we further assume that the judgments obtained in different choice sets are independent given the person-specific parameters $\boldsymbol{\mu}_i$ and $\boldsymbol{\Sigma}$, the probability of observing the rank orders in the two choice sets $\mathbf{R}_i$ can be determined by evaluating a four-variate normal distribution,

$$\Pr(\mathbf{R}_i \mid \boldsymbol{\mu}_i, \boldsymbol{\Sigma}) = \frac{|\boldsymbol{\Gamma}_i|^{-1/2}}{(2\pi)^{4/2}} \times \int_0^{\infty} \cdots \int_0^{\infty} \exp\left\{-\frac{1}{2}(\mathbf{x} - \boldsymbol{\delta}_i)' \boldsymbol{\Gamma}_i^{-1}(\mathbf{x} - \boldsymbol{\delta}_i)\right\} \mathrm{d}\mathbf{x}, \tag{5}$$

where $\boldsymbol{\delta}_i = \mathbf{C}_i \boldsymbol{\mu}_i$ with $\mathbf{C}_i = \begin{bmatrix} \mathbf{C}_{i1} \\ \mathbf{C}_{i2} \end{bmatrix}$ and $\boldsymbol{\Gamma}_i = \begin{bmatrix} \mathbf{C}_{i1}\boldsymbol{\Sigma}\mathbf{C}'_{i1} & \mathbf{0} \\ \mathbf{0} & \mathbf{C}_{i2}\boldsymbol{\Sigma}\mathbf{C}'_{i2} \end{bmatrix}$. More detail on this approach is provided in the next section on preference models for paired comparison data which are employed frequently in experimental research (David, 1988).

2.1. *Paired comparison models*

Because the method of paired comparisons imposes minimal constraints on the response behavior of a judge, it is used in a wide range of applications ranging from sensory testing to investigations of preference and choice behavior. Especially, when differences between options are small, it is desirable to compare them in isolation and to free the judgment process as much as possible from context effects caused by the presence of other options.

Extending (1), we can write the utilities of judge i that underlie the pairwise comparison of options j and k, respectively, as

$$\begin{aligned} \upsilon_{ij(k)} &= \mu_{ij} + \epsilon_{ij(k)}, \\ \upsilon_{ik(j)} &= \mu_{ik} + \epsilon_{ik(j)}. \end{aligned} \tag{6}$$

The terms $\epsilon_{j(k)}$ and $\epsilon_{k(j)}$ capture random variations in the judgmental process, where $\epsilon_{j(k)}$ denotes the random variation of option j when compared to option k. Thus, according to (2), the choice between j and k is determined by the sign of the latent judgment y_{ijk} where

$$y_{ijk} = \upsilon_{ij(k)} - \upsilon_{ik(j)} = \mu_{ij} - \mu_{ik} + \epsilon_{ij(k)} - \epsilon_{ik(j)}. \tag{7}$$

Furthermore, under the assumption that the pairwise judgments are independent,

$$y_{ijk} = \mu_{ij} - \mu_{ik} + \epsilon_{ijk}, \tag{8}$$

where $\epsilon_{ijk} = \epsilon_{ij(k)} - \epsilon_{ik(j)}$ and the variability in the judgments comes solely from the ϵ's which are independently distributed given the person-specific parameters. For example, when the $\boldsymbol{\epsilon}_i$'s are normally distributed, the probabilities of observing the response pattern of judge i, $\mathbf{R}_i = ((j,k),(k,l),(j,l))$ for the option triad (j,k,l) are

$$\Pr\big(\mathbf{R}_i = ((j,k),(k,l),(j,l)) \mid \boldsymbol{\mu}_i, \boldsymbol{\Sigma}\big) = \int_{\mu_{ik}-\mu_{ij}}^{\infty} \int_{\mu_{il}-\mu_{ik}}^{\infty} \int_{\mu_{il}-\mu_{ij}}^{\infty} f(\boldsymbol{\epsilon}_i)\,\mathrm{d}\boldsymbol{\epsilon}_i$$

$$= \int_{\mu_{ik}-\mu_{ij}}^{\infty} \frac{1}{\sqrt{2\pi\psi_{jk}}} \exp\left\{-\frac{\epsilon_{ijk}^2}{2\psi_{jk}}\right\} d\epsilon_{ijk}$$
$$\times \int_{\mu_{il}-\mu_{ik}}^{\infty} \frac{1}{\sqrt{2\pi\psi_{kl}}} \exp\left\{-\frac{\epsilon_{ikl}^2}{2\psi_{kl}}\right\} d\epsilon_{ikl}$$
$$\times \int_{\mu_{il}-\mu_{ij}}^{\infty} \frac{1}{\sqrt{2\pi\psi_{jl}}} \exp\left\{-\frac{\epsilon_{ijl}^2}{2\psi_{jl}}\right\} d\epsilon_{ijl}$$
$$= \Phi\left(\frac{\mu_{ij}-\mu_{ik}}{\sqrt{\psi_{ijk}}}\right) \Phi\left(\frac{\mu_{ik}-\mu_{il}}{\sqrt{\psi_{ikl}}}\right) \Phi\left(\frac{\mu_{ij}-\mu_{il}}{\sqrt{\psi_{ijl}}}\right), \tag{9}$$

where $\psi_{ijk} = \mathbf{a}_{jk}\boldsymbol{\Sigma}\mathbf{a}'_{jk}$ and $\mathbf{a}_{jk} = [1 \quad -1 \quad 0]$ for the (j, k) comparison.

Based on the above formulation, it is necessary to define for each judge a paired comparison matrix $\mathbf{A}_i$ indicating his or her preference on all the pairs. However, when all pairs are compared, it is convenient to define a common design matrix $\mathbf{A}$ which represents the complete paired comparison set and to link the resulting latent judgement to the response outcomes using an indicator function.

For multiple comparisons of J options, the latent pairwise judgments of judge i can be written conveniently as a linear model. For example, in a comparison of the options j, k, l, and m, we obtain:

$$\mathbf{y}_i = \begin{pmatrix} y_{ijk} \\ y_{ijl} \\ y_{ijm} \\ y_{ikl} \\ y_{ikm} \\ y_{ilm} \end{pmatrix} = \begin{pmatrix} 1 & -1 & 0 & 0 \\ 1 & 0 & -1 & 0 \\ 1 & 0 & 0 & -1 \\ 0 & 1 & -1 & 0 \\ 0 & 1 & 0 & -1 \\ 0 & 0 & 1 & -1 \end{pmatrix} \begin{pmatrix} \mu_{ij} \\ \mu_{ik} \\ \mu_{il} \\ \mu_{im} \end{pmatrix} + \begin{pmatrix} \epsilon_{ijk} \\ \epsilon_{ijl} \\ \epsilon_{ijm} \\ \epsilon_{ikl} \\ \epsilon_{ikm} \\ \epsilon_{ilm} \end{pmatrix}$$
$$= \mathbf{A}\boldsymbol{\mu}_i + \boldsymbol{\epsilon}_i, \tag{10}$$

where $\mathbf{A}$ is the paired comparison design matrix. The choice between options j and k is then determined by the sign of y_{ijk} such that

$$w_{ijk} = \begin{cases} 1, & y_{ijk} \geqslant 0, \\ 0, & y_{ijk} < 0. \end{cases} \tag{11}$$

Thus, $w_{ijk} = 1$ indicates that option j is preferred to option k for person i. Moreover, the variance–covariance structure of the $\boldsymbol{\epsilon}_i$ is specified to be a diagonal matrix $\boldsymbol{\Psi}$ (Tsai and Böckenholt, 2002).

Recently, Tsai and Böckenholt (2006) relaxed the conditional independence assumption of the paired judgments given a judge's person-specific parameters. Instead of assuming the difference terms, such as $\epsilon_{ij(k)} - \epsilon_{ik(j)}$ and $\epsilon_{ij(l)} - \epsilon_{il(j)}$, in (7) to be uncorrelated, their dependent-response (DR) model allows the utilities assigned to an option to depend on which option it has been compared to. For example, when an option is compared to two similar options, the utilities in these two comparisons are presumably more correlated than the ones obtained in a comparison with two less similar options. In this case, the difference in $\epsilon_{ij(k)}$ and $\epsilon_{ik(j)}$ cannot be reduced to the pair-specific term ϵ_{ijk}.

The DR model is defined as follows. Let $\boldsymbol{\epsilon} = (\epsilon_{1(2)}, \epsilon_{1(3)}, \ldots, \epsilon_{1(J)}, \epsilon_{2(1)}, \epsilon_{2(3)}, \ldots, \epsilon_{(J)(J-1)})'$ be an $J(J-1)$-dimensional random vector, and $\mathbf{B}$ be an $\binom{J}{2} \times J(J-1)$

matrix where each column of $\mathbf{B}$ corresponds to one of the $\binom{J}{2}$ paired comparisons and each row to one of the $J(J-1)$ terms in $\boldsymbol{\epsilon}_i$. For example, when $J = 3$, $\boldsymbol{\epsilon} = (\epsilon_{1(2)}, \epsilon_{1(3)}, \epsilon_{2(1)}, \epsilon_{2(3)}, \epsilon_{3(1)}, \epsilon_{3(2)})'$ and

$$\mathbf{B} = \begin{pmatrix} 1 & 0 & -1 & 0 & 0 & 0 \\ 0 & 1 & 0 & 0 & -1 & 0 \\ 0 & 0 & 0 & 1 & 0 & -1 \end{pmatrix}. \tag{12}$$

For multiple comparisons with J items, the paired comparison judgments can again be written conveniently as a linear model. For example, in a comparison of the items j, k, l, and m, we obtain:

$$\begin{aligned} \mathbf{y}_i = \begin{pmatrix} y_{ijk} \\ y_{ijl} \\ y_{ijm} \\ y_{ikl} \\ y_{ikm} \\ y_{ilm} \end{pmatrix} &= \begin{pmatrix} 1 & -1 & 0 & 0 \\ 1 & 0 & -1 & 0 \\ 1 & 0 & 0 & -1 \\ 0 & 1 & -1 & 0 \\ 0 & 1 & 0 & -1 \\ 0 & 0 & 1 & -1 \end{pmatrix} \begin{pmatrix} \mu_{ij} \\ \mu_{ik} \\ \mu_{il} \\ \mu_{im} \end{pmatrix} + \begin{pmatrix} \epsilon_{ij(k)} - \epsilon_{ik(j)} \\ \epsilon_{ij(l)} - \epsilon_{il(j)} \\ \epsilon_{ij(m)} - \epsilon_{im(j)} \\ \epsilon_{ik(l)} - \epsilon_{il(k)} \\ \epsilon_{ik(m)} - \epsilon_{im(k)} \\ \epsilon_{il(m)} - \epsilon_{im(l)} \end{pmatrix} \\ &= \mathbf{A}\boldsymbol{\mu}_i + \mathbf{B}\boldsymbol{\epsilon}_i, \end{aligned} \tag{13}$$

where $\boldsymbol{\epsilon}_i = (\epsilon_{ij(k)}, \epsilon_{ij(l)}, \epsilon_{ij(m)}, \epsilon_{ik(j)}, \epsilon_{ik(l)}, \epsilon_{ik(m)}, \ldots, \epsilon_{im(j)}, \epsilon_{im(k)}, \epsilon_{im(l)})'$.

An important feature of the DR model is that it allows testing the conditional independence assumption. However, despite its apparent usefulness, future studies on issues of parameter identifiability and possible decompositions of $\boldsymbol{\Sigma}_\epsilon$ under the DR models are needed. We therefore restrict the discussion of later sections in this chapter to conditional independence models as given by (8).

2.2. *Individual differences*

When preferences for only a single choice set are elicited, the effects of between- and within-judge variability are confounded. For example, if judges are asked to rank J options only once, it is not possible to separately identify the effects of ϵ_{ij} and μ_{ij} in (1). Fortunately, with multiple choice sets, it becomes possible to distinguish explicitly these sources of variability. Although model estimation is simplified considerably if (10) can be reduced to $y_{ijk} = \mu_j - \mu_k + \epsilon_{ijk}$, this disregard for systematic individual differences in the mean utilities is a serious model misspecification that can lead to both incorrect statistical and substantive conclusions (Bock and Jones, 1968; Takane, 1987). Similarly, when using a multinomial logit model to analyze ranking or discrete responses (under the assumption that the ϵ's follow independent Gumbel distributions), the failure to include random variations to account for individual differences results in the unduly restrictive "Independence from Irrelevant Alternative" (IIA) property (Luce, 1959; Skrondal and Rabe-Hesketh, 2003).

In the simplest case, when covariates on judges or options are unavailable, we can represent individual differences as deviations from their mean:

$$\boldsymbol{\mu}_i = \boldsymbol{\mu} + \boldsymbol{\nu}_i, \tag{14}$$

where $\boldsymbol{\mu}$ consists of the overall mean utilities, μ_j, for each option j. The component ν_{ij} captures the degree to which judge i deviates from μ_j and can be specified to be normally distributed with $\boldsymbol{\nu}_i \sim N(\mathbf{0}, \boldsymbol{\Sigma}_\nu)$. When person-specific covariates, such as gender or age of the judges, are available, it is useful to account for individual difference at the between-judge level with

$$\boldsymbol{\mu}_i = \boldsymbol{\mu} + \mathbf{K}\mathbf{x}_i + \boldsymbol{\nu}_i,$$

where the matrix $\mathbf{K}$ contains the regression coefficients. Because of the comparative nature of the data, only the interaction but not the main effects of the covariates can be identified.

In a similar way, the effects of option-specific covariates can be modeled with

$$\boldsymbol{\mu}_i = \mathbf{M}\boldsymbol{\beta}_i + \boldsymbol{\nu}_i,$$

where $\mathbf{M}_{J\times q}$ contains the q $(q < J)$ covariates of the options, $\boldsymbol{\beta}_i$ is a regression vector for person i, $\boldsymbol{\nu}_i$ is an error term representing the effects that are not accounted for by the covariates and $\boldsymbol{\nu}_i \sim N(\mathbf{0}, \boldsymbol{\Sigma}_\nu)$. In addition, $\boldsymbol{\beta}_i$ can be specified to be normally distributed with $\boldsymbol{\beta}_i \sim N(\boldsymbol{\beta}, \boldsymbol{\Sigma}_\beta)$.

Frequently, it can be assumed that there are some unobserved attributes that affect judges in their assessment of the options. This assumption can be tested by using a common-factor model for μ_{ij}:

$$\mu_{ij} = \mu_j + \sum_{r=1}^{R} f_{ir}\lambda_{rj} + \delta_{ij}, \tag{15}$$

where f_{ir} denotes the person-specific weight for attribute r $(r = 1, \ldots, R)$, λ_{rj} denotes the loading of attribute r on option j and δ_{ij} represents the unique characteristic of option j. Assuming that the common factors have the covariance structure of $\mathbf{F}$ and the variabilities in the option evaluations unaccounted for by the common-factors are captured by the variance of δ_{ij}, we have that $\boldsymbol{\Sigma}_\nu = \boldsymbol{\Lambda}\mathbf{F}\boldsymbol{\Lambda}' + \boldsymbol{\Phi}$, where $\boldsymbol{\Phi}$ is a diagonal matrix. Applications of ranking and paired comparison models with common factors are reported by Brady (1989), Chan and Bentler (1998), Maydeu-Olivares and Böckenholt (2005), Takane (1987), Tsai and Böckenholt (2001), and Yu et al. (1998).

It is possible to combine these different sources of individual differences yielding:

$$\boldsymbol{\mu}_i = \boldsymbol{\mu} + \mathbf{K}\mathbf{x}_i + \mathbf{M}\boldsymbol{\beta}_i + \boldsymbol{\Lambda}\mathbf{f}_i + \boldsymbol{\nu}_i,$$

where $\mathbf{x}_i$ contains judge-specific covariates, and $\mathbf{M}$ and $\boldsymbol{\Lambda}$ contain observed and unobserved attributes of the options, respectively. The corresponding regression coefficients are given by $\mathbf{K}$, $\boldsymbol{\beta}_i$, and $\mathbf{f}_i$. Any remaining effects that are not accounted for by the covariates are represented by $\boldsymbol{\nu}_i$.

2.3. Time-dependent choices

When choice data from the same persons are observed over time, a parameter-driven dynamic approach can be used to characterize individual differences (Böckenholt, 2002; Keane, 1997) with

$$\boldsymbol{\mu}_{it} = \boldsymbol{\Upsilon}\boldsymbol{\mu}_{i,t-1} + \boldsymbol{\zeta}_{it}, \tag{16}$$

where $\boldsymbol{\mu}_{it}$ represents the vector mean evaluations for person i at time t, and the $(J \times J)$ coefficient matrix $\boldsymbol{\Upsilon}$ contains the multivariate regression effects between the evaluations of the options at two adjacent time points. The diagonal elements of $\boldsymbol{\Upsilon}$ contain each option's autoregressive term and the off-diagonal elements contain the lagged effects of the $J-1$ remaining random utilities. The J-dimensional random effect $\boldsymbol{\zeta}_{it}$ captures the part of $\boldsymbol{\mu}_{it}$ that cannot be predicted on the basis of past evaluations of the options and is assumed to be normally distributed with mean vector $\mathbf{0}$ and $(J \times J)$ covariance matrix $\boldsymbol{\Sigma}_{\zeta_t}$. Stationarity and stability conditions of (16) are discussed in Reinsel (1997) and Lüthkepol (1993). When the coefficient matrix $\boldsymbol{\Upsilon}$ is diagonal, correlations between the random effects μ_{ijt} decrease exponentially with increasing time lag. However, the size of the time effect may be different for each option j.

2.4. Multi-criterion choices

In many applications, it is useful to collect ranking data with respect to several criteria to understand how they relate to each other. For example, Bradley (1984) reports a multivariate paired comparison experiment in which three vanilla puddings were assessed with respect to color, taste and overall quality. In this application, the perceived dependencies among the criteria could be determined reliably although, on the aggregate level, no significant product differences were found.

In general, when modeling multi-criterion ranking data, one needs to distinguish between two sources of response dependencies: On the one hand, different options are compared with respect to the same criterion (within-criterion dependence), and, on the other hand, the same options are evaluated on different criteria (between-criterion dependence). Böckenholt (1996) suggested the following decomposition of the covariance matrix of the multivariate judgments $(\mathbf{r}_{ia}, \mathbf{r}_{ib}, \ldots, \mathbf{r}_{iq})$ with respect to q criteria:

$$\boldsymbol{\Sigma}_\nu = \boldsymbol{\Sigma}_C \otimes \boldsymbol{\Sigma}_O,$$

where $\boldsymbol{\nu} = (\boldsymbol{\nu}_a, \boldsymbol{\nu}_b, \ldots, \boldsymbol{\nu}_q)$ contains the unobserved utility assessments with respect to each of the q criteria, $\boldsymbol{\Sigma}_C$ represents the associations among the q criteria, and $\boldsymbol{\Sigma}_O$ represents the associations among the J options. For example, in the two-criterion case, when the covariance matrix of the options is represented by an identity matrix, we obtain

$$\boldsymbol{\Sigma}_\nu = (\rho_{ab}\mathbf{I}) \otimes \mathbf{I} = \begin{pmatrix} \mathbf{I} & \rho_{ab}\mathbf{I} \\ \rho_{ab}\mathbf{I} & \mathbf{I} \end{pmatrix},$$

where ρ_{ab} is the correlation between the criteria a and b. This approach is illustrated in the application section of this chapter.

3. Identifiability

According to the random utility framework, choosing the most preferred option is equivalent to selecting the option with the largest utility. Thus, an important feature of the choice process is that it is comparative in nature. A selected option may be the best one out of a set of available options but it may be rejected in favor of other options that are added subsequently to the set of options. Because choices are inherently comparative, the origin of the utility scale cannot be identified on the basis of the choices alone. One option may be preferred to another one but this result does not allow any conclusions about whether either of the options are attractive or unattractive (Guttman, 1946; Mosteller, 1951; Torgerson, 1958). Moreover, because of the discrete-valued structure and comparative nature of preference data, the options' covariance matrix of the unobserved utilities $\mathbf{\Sigma}_\upsilon$ cannot be identified.

For paired comparison data, Arbuckle and Nugent (1973) showed that the standard deviation between each pair is invariant under adding a row/column specific constant to each row and column of the covariance matrix, that is, under any transformation on the covariance matrix of the form $\sigma_{ij} \to \sigma_{ij} + d_i + d_j$. Their results also hold for ranking and paired comparisons under the multiple-judgments framework that $\mathbf{\Sigma}_\upsilon$ and $\mathbf{\Sigma}_\upsilon + \mathbf{d}\mathbf{1}' + \mathbf{1}\mathbf{d}'$ are empirically indistinguishable, given that vector $\mathbf{d}$ is chosen so that the resulting covariance matrix is positive-definite (Tsai, 2000, 2003). In the context of equivalent covariance matrices for multinomial probit models, Bekker and Wansbeek (2000) showed that the set of vectors $\mathbf{d}$ which, with $\mathbf{\Sigma}_\upsilon > 0$, satisfy $\mathbf{\Sigma}_\upsilon + \mathbf{1}\mathbf{d}' + \mathbf{d}\mathbf{1}' > 0$ can be described as $\mathbf{d} = \frac{\mathbf{1}+\mathbf{z}}{1-\mathbf{1}'\mathbf{V}\mathbf{z}}$, where $\mathbf{V} = (\mathbf{\Sigma}_\upsilon + \mathbf{1}\mathbf{1}')^{-1}$ and $\mathbf{z}$ lies within the ellipsoid $\mathbf{z}'(\mathbf{V} + \mathbf{V}\mathbf{1}\mathbf{1}'\mathbf{V})\mathbf{z} < 1 + \frac{1}{\mathbf{1}'\mathbf{V}\mathbf{1}}$. Their results therefore define the set of vectors $\mathbf{d}$ for the equivalent covariance matrices for $\mathbf{\Sigma}_\upsilon$.

The equivalence structures of $\mathbf{\Sigma}_\upsilon$ indicate that not all of its parameters are identified. Thus, constraints need to be imposed to ensure identifiability of the models. The equivalence relation suggests defining the identified parameters based on the utility differences instead of the utilities of the options. Different parameterizations can be obtained by choosing different contrast matrices (Chan and Bentler, 1998; Maydeu-Olivares, 2002; Yao, 1995; Skrondal and Rabe-Hesketh, 2003; Yu, 2000). For any contrast matrix $\mathbf{S}$ of rank $J - 1$, the corresponding identified parameters for the model with mean vector $\boldsymbol{\mu}$ and covariance matrix $\mathbf{\Sigma}_\upsilon$ are

$$\mathbf{D}\mathbf{S}\boldsymbol{\mu} \quad \text{and} \quad d^{-1}\mathbf{S}\mathbf{\Sigma}_\upsilon\mathbf{S}',$$

where $\mathbf{D} = \mathrm{Diag}(\mathbf{S}\mathbf{\Sigma}_\upsilon\mathbf{S}')^{-1/2}$, and d is the $(1, 1)$ element of $\mathbf{S}\mathbf{\Sigma}_\upsilon\mathbf{S}'$. For example, Yao (1995) derived the set of identified parameters under the $(J - 1) \times J$ contrast matrix $\mathbf{C}$ which compares each option to the last one. Denote $\mathbf{\Delta} = \mathbf{C}\boldsymbol{\mu}$ and $\mathbf{\Gamma} = \mathbf{C}\mathbf{\Sigma}_\upsilon\mathbf{C}'$, then the identified parameters are the standardized mean differences, $\frac{\delta_j}{\sqrt{\gamma_{jj}}}$, the ratios of the option differences' variances, $\frac{\gamma_{jj}}{\gamma_{11}}$, and the option differences' correlations, $\frac{\gamma_{jk}}{\sqrt{\gamma_{jj}}\sqrt{\gamma_{kk}}}$ $(j, k = 1, 2, \ldots, J - 1)$ where $\delta_j = \mu_j - \mu_J$, $\gamma_{jj} = \sigma_j^2 + \sigma_J^2 - 2\sigma_{jJ}$, and $\gamma_{jk} = \sigma_{jk} - \sigma_{jJ} - \sigma_{kJ} + \sigma_J^2$. When the identified parameters are formulated using a contrast matrix, it is thus necessary to fix the variance of the difference between two arbitrarily selected options to a constant in order to ensure the identification of the remaining elements in the variance–covariance matrix of the option differences. In addition to identifying the location parameter, the mean utility of one of the options need to be set to a constant. This approach results in a total of $(J - 1) + \frac{J(J-1)}{2} - 1 = \frac{J^2+J-4}{2}$ identified parameters for complete rankings and $\frac{J^2+J-4}{2} + \binom{J}{2} = J^2 - 2$ for paired comparisons $(J > 3)$.

Other identification constraints can be chosen as well that do not involve a contrast matrix. For example, Dansie (1986) proposed to fix the variances of the options' utilities to ones, and the covariance of one option pair to zero. Maydeu-Olivares and Böckenholt (2005) fix all the covariances involving the last utilities to zeros and the variances of the first and last utilities to ones. Thus, the following $\boldsymbol{\Sigma}_\nu^D$ and $\boldsymbol{\Sigma}_\nu^M$ matrices are specified for $J = 4$ in Dansie (1986) and Maydeu-Olivares and Böckenholt (2005), respectively,

$$\boldsymbol{\Sigma}_\nu^D = \begin{bmatrix} 1^\star & \sigma_{21} & \sigma_{31} & 0^\star \\ \sigma_{21} & 1^\star & \sigma_{32} & \sigma_{42} \\ \sigma_{31} & \sigma_{32} & 1^\star & \sigma_{43} \\ 0^\star & \sigma_{42} & \sigma_{43} & 1^\star \end{bmatrix} \quad \text{and} \quad \boldsymbol{\Sigma}_\nu^M = \begin{bmatrix} 1^\star & \sigma_{21} & \sigma_{31} & 0^\star \\ \sigma_{21} & \sigma_2^2 & \sigma_{32} & 0^\star \\ \sigma_{31} & \sigma_{32} & \sigma_3^2 & 0^\star \\ 0^\star & 0^\star & 0^\star & 1^\star \end{bmatrix},$$

where the parameters fixed for identification purposes are marked with an asterisk.

In addition, if a factor-analytic structure (15) is used to account for individual differences in $\boldsymbol{\nu}$, possible identification constraints for $\boldsymbol{\Lambda}$ and $\boldsymbol{\Phi}$ include fixing the uniqueness variance of one option to one, and all the factor loadings involving the same option to zeros. In addition, the upper triangular part of the factor loading matrix needs to be set to zeros to solve the rotational indeterminancy problem (McDonald, 1999). For example, for a three-factor model, we have

$$\boldsymbol{\Lambda} = \begin{bmatrix} \lambda_{11} & 0^\star & 0^\star \\ \lambda_{12} & \lambda_{22} & 0^\star \\ \lambda_{13} & \lambda_{23} & \lambda_{33} \\ \vdots & \vdots & \vdots \\ \lambda_{1,J-1} & \lambda_{2,J-1} & \lambda_{3,J-1} \\ 0^\star & 0^\star & 0^\star \end{bmatrix},$$

$$\boldsymbol{\Phi} = \begin{bmatrix} \phi_1^2 & 0 & \cdots & 0 & 0 \\ 0 & \phi_2^2 & \cdots & 0 & 0 \\ \vdots & \vdots & \ddots & \vdots & \vdots \\ 0 & 0 & \cdots & \phi_{J-1}^2 & 0 \\ 0 & 0 & \cdots & 0 & 1^\star \end{bmatrix}.$$

The above identification constraints on $\boldsymbol{\Sigma}_\nu$ or $\boldsymbol{\Lambda}$ and $\boldsymbol{\Phi}$ also ensure identifiability of all the parameters in $\boldsymbol{\Psi}$ for paired comparisons. An alternative approach which does not require fixing the variance of the utility difference for one option pair is to constrain one of the elements in $\boldsymbol{\Psi}$ to one (say $\psi_{12} = 1$). In this case, the covariance matrix of the reduced covariance matrix of the utility differences is also identified (Tsai, 2003).

All of the above results can be generalized to both incomplete and partial ranking data by appropriately defining the set of contrast matrices for each data type (Böckenholt, 1992). The identified parameters are equal to the ones for complete ranking data when the number of options to be ranked is greater than two.

The existence of equivalence models for preference data complicates the interpretation of the results since preference models that have the same identified parameters may have substantively different interpretations. Consider the following two ranking models, A and B, with random utilities $\boldsymbol{\nu}_A \sim N_3(\boldsymbol{\mu}_A, \boldsymbol{\Sigma}_A)$ and $\boldsymbol{\nu}_B \sim N_3(\boldsymbol{\mu}_B, \boldsymbol{\Sigma}_B)$, where

$$\boldsymbol{\mu}_A = (2 \quad 5 \quad 0)', \qquad \boldsymbol{\mu}_B = (\sqrt{0.8} \quad \sqrt{5} \quad 0)',$$

and

$$\boldsymbol{\Sigma}_A = \begin{bmatrix} 1 & 0 & 0 \\ 0 & 2 & 0 \\ 0 & 0 & 3 \end{bmatrix}, \qquad \boldsymbol{\Sigma}_B = \begin{bmatrix} 1 & 0.7 & 0.6 \\ 0.6 & 1 & 0.5 \\ 0.6 & 0.5 & 1 \end{bmatrix}.$$

These two models yield the same ranking probability distributions of $\{0.038, 0.002, 0.802, 0.148, 0.002, 0.008\}$ for the rank ordering patterns $\{(1,2,3), (1,3,2), (2,1,3), (2,3,1), (3,1,2), (3,2,1)\}$, respectively. The interpretation of these two structures are rather different. Model A assumes different option variances and zero option correlations. In contrast, Model B assumes equal option variances and nonzero correlations between options. Because of the models' indistinguishability, an interpretation in favor of either A or B needs to depend on other factors, such as prior knowledge or some theory about the underlying data generation mechanism. This example also illustrates a frequently overlooked conclusion in the literature: When investigating the model fit of restricted structures for either $\boldsymbol{\mu}$ or $\boldsymbol{\Sigma}_\nu$, it needs to be taken into account that any evidence speaks for an entire class of equivalence models and not just for the restricted model form.

4. Identifying the scale origin

The identifiability issues described in the previous section are caused by the discrete-valued structure of the data and the reduced-rank form of the paired comparison and ranking design matrices. As noted already by Thurstone and Jones (1957), it is possible to overcome the latter limitation under certain conditions. By appropriately modifying the paired comparison or ranking task, we can augment the design matrices (e.g., (10), (13)) to their full-rank form. Böckenholt (2004) discussed different methods that can be used for this purpose. For example, one approach consists of asking respondents to choose between lotteries where in one lottery option j can be won with probability p_s, and nothing otherwise, and in the other lottery option k can be won with probability p_u, and nothing otherwise. These lotteries are written in abbreviated forms as (j, p_s) and (k, p_u), respectively. The status-quo outcome (e.g., "winning nothing") defines the origin of the individual utility scales in a natural way since an option that can be won is a gain relative to the status-quo position.

Let μ_{i0} denote the mean evaluation of the status-quo outcome. Then by setting $\mu_{i0} = 0$, the comparative utility judgment of two lotteries (j, p_s) and (k, p_u) is given by

$$\begin{aligned} y_{ij(p_s)k(p_u)} &= \left[p_s \mu_{ij} + (1 - p_s)\,\mu_{i0}\right] \\ &\quad - \left[p_u \mu_{ik} + (1 - p_u)\,\mu_{i0}\right] + \epsilon_{ij(p_s)k(p_u)} \\ &= p_s \mu_{ij} - p_u \mu_{ik} + \epsilon_{ij(p_s)k(p_u)}, \end{aligned} \tag{17}$$

where $\epsilon_{ij(p_s)k(p_u)}$ denotes the unsystematic response variability in the comparison between the two lotteries. As an illustration, consider the following scenario in which the options j, k, and l (e.g., three different vacation packages) are presented pairwise, first in a non-risky presentation mode and, next, in the form of lotteries:

$$\begin{pmatrix} y_{ijk} \\ y_{ijl} \\ y_{ikl} \\ y_{ij(0.8)k(0.2)} \\ y_{ij(0.8)l(0.2)} \\ y_{ik(0.8)l(0.2)} \\ y_{ij(0.2)k(0.8)} \\ y_{ij(0.2)l(0.8)} \\ y_{ik(0.2)l(0.8)} \end{pmatrix} = \begin{pmatrix} 1 & -1 & 0 \\ 1 & 0 & -1 \\ 0 & 1 & -1 \\ 0.8 & -0.2 & 0 \\ 0.8 & 0 & -0.2 \\ 0 & 0.8 & -0.2 \\ 0.2 & -0.8 & 0 \\ 0.2 & 0 & -0.8 \\ 0 & 0.2 & -0.8 \end{pmatrix} \begin{pmatrix} \mu_{ij} \\ \mu_{ik} \\ \mu_{il} \end{pmatrix} + \begin{pmatrix} \epsilon_{ijk} \\ \epsilon_{ijl} \\ \epsilon_{ikl} \\ \epsilon_{ij(0.8)k(0.2)} \\ \epsilon_{ij(0.8)l(0.2)} \\ \epsilon_{ik(0.8)l(0.2)} \\ \epsilon_{ij(0.2)k(0.8)} \\ \epsilon_{ij(0.2)l(0.8)} \\ \epsilon_{ik(0.2)l(0.8)} \end{pmatrix}. \quad (18)$$

The error terms ϵ_i are specified to be independently distributed with variance terms that may be different for comparisons in the risky and non-risky choice mode. The design matrix in (18) is of full rank. Thus, although only difference judgments are observed, the mean evaluations and their covariance matrix are fully identified. This desirable feature is obtained already when one lottery comparison with $p_s \neq p_u$ is included in the design matrix.

The interpretation of the scale value μ_{ij} as deviation from the status-quo is appealing. Especially when uncertainty plays an integral role in the evaluation of options, such a definition can be implemented easily. Possible fields of applications include choices among (medical) treatment plans, insurance policies, and options with future outcomes.

Methods for identifying an utility scale origin are not only instrumental for avoiding difficulties in the interpretation of the estimated covariance structures, they also provide useful insights about the underlying judgmental process. In particular, by being able to identify the degree of an option's desirability, we can compare preferences for options both in a relative and in an absolute sense across different conditions.

5. Estimation

Model parameters of the models for partial and/or incomplete ranking data can be estimated by standard maximum likelihood methods. Assuming random sampling of judges, we specify the log-likelihood function as

$$\ln L = c + \sum_f n(\mathbf{R}_f) \ln \Pr(\mathbf{R}_f),$$

where $n(\mathbf{R}_f)$ is the frequency of respondents showing the response $\mathbf{R}_f$, and c is a constant. Although the maximization of the likelihood function is straightforward, it requires the computation of the normal probabilities for the response outcomes. For small T and J the multidimensional integrals can be evaluated numerically (deDoncker et al., 1999).

For larger T and J, however, the evaluation of the normal distribution function by numerical integration is not feasible. A number of alternative procedure have been proposed over the years ranging from simulation-based algorithm as the Monte Carlo

Markov-Chain (MCMC) method in a Bayesian framework (Imai and van Dyk, 2005; Yao and Böckenholt, 1999; Yu, 2000), the method of simulated moments (McFadden, 2001; Keane, 1997), to limited-information estimation methods, which use only lower moments of the marginals to estimate the mean and covariance structure of the choice data (Brady, 1989; Chan and Bentler, 1998; Maydeu-Olivares, 2001, 2002). Because of their speed and straightforward implementation, limited-information methods are perhaps most useful for applications of random utility models in routine work. Recently, Maydeu-Olivares and Böckenholt (2005) demonstrated that ranking and paired comparison data can be analyzed readily using the structural equation modeling (SEM) framework implemented in the MPLUS (Muthèn and Muthèn, 2004) package.

Large sample tests of fit are available based on the likelihood ratio (LR) test (G^2) and Pearson's goodness-of-fit test statistic (P^2). Asymptotically, if the statistical model provides an adequate description of the data, then both statistics follow a χ^2-distribution with $(H^T - m - 1)$ degrees of freedom, where m refers to the number of parameters to be estimated and H is number of possible response patterns. The LR test is most useful when $J \leqslant 5$ and $T \leqslant 3$. Otherwise, only a small subset of the possible response patterns may be observed even when the sample size is large. In this case, it is doubtful that the test statistics will follow approximately a χ^2-distribution. However, useful information about a model fit can be obtained by inspecting standardized differences between the observed and fitted probabilities for subsets (e.g., all triplets) of the options. In addition, model selection criteria such as Akaike's (1973) information criterion (AIC) or the Bayesian information criterion, BIC in short, (Schwartz, 1978) can be utilized to compare the considered preference models. Both AIC and BIC are defined based on the penalized likelihoods which adjust the likelihood function for model complexity. In the following applications, we adopt Vaida and Blanchard's (2005) distinction between conditional AIC (cAIC) and marginal AIC (mAIC) with cAIC being the preferred criterion when not only population but also person-specific parameters are of interest.

6. Applications

6.1. *Choosing universities in an exchange program*

In an investigation of how students' choices of foreign universities in a student exchange program depend on the characteristics of the students and the universities, the Vienna University of Economics conducted a survey to establish the ranking of the universities through a paired comparison study (Dittrich et al., 1998). Six universities situated in London (LO), Paris (PA), Milan (MI), Barcelona (BA), St Gall (SG), and in Stockholm (ST) were chosen. Table 1 lists two university-related covariates: location and specialization.

303 students compared the six universities using a trinary response scale that included an indifference category. Since 91 students gave incomplete responses, the following analysis is based on the remaining 212 students who provided a complete set of paired comparisons. The indifference response was replaced by a random draw of the "prefer to" and "not prefer to" response categories. The students can be classified

Table 1
University characteristics (Dittrich et al., 1998)

University	EC	MS	FS	LAT
LO	1	0	0	0
PA	0	1	0	1
MI	1	0	0	1
BA	0	1	0	1
SG	0	0	1	0
ST	0	0	1	0

EC: economics; MS: management science; FS: finance; LAT: Latin countries.

Table 2
List of covariates for the university data of Dittrich et al. (1998)

Variable	Description	Coding
STUD	Main discipline of study	1 = other, 2 = commerce
ENG	Knowledge of English	1 = good, 2 = poor
FRA	Knowledge of French	1 = good, 2 = poor
SPA	Knowledge of Spanish	1 = good, 2 = poor
ITA	Knowledge of Italian	1 = good, 2 = poor
WOR	Full-time employment while studying	1 = no, 2 = yes
DEG	Intention to take an international degree	1 = no, 2 = yes
SEX	Gender	1 = female, 2 = male

STUD: other = economics, business administration, business education.

into four groups based on their gender and main discipline of study (commerce versus others). Other covariates such as knowledge of various languages (English, French, Spanish, and Italian), employment status and intention for getting an international degree are also available. The list of covariates are included in Table 2. The data can be accessed from http://www.blackwellpublishers.co.uk/rss.

Our decision on which person-specific covariates to consider in the models was based on the differences in the dominance index of each school, $d_j = \frac{1}{n}\sum_{j' \neq j}(w_{ijj'} = 1)$, for the subpopulations defined by the covariates. The covariates FRA, SPA, and ITA differentiated the subpopulations the best and were included in the analyses.

Table 3 provides a summary of the models fitted to the data. Models (A) to (E) include either unconstrained or restricted covariance structures for the $\boldsymbol{\mu}_i$'s, but no covariates effects. The single-factor covariance Model (C) gives the best fit. However, the loadings of Model (C) do not seem to reflect the characteristics reported in Table 1.

Next, the three covariates indicating language fluency, FRA, SPA, and ITA are included in Models (F) to (J) to account for individual differences in their evaluations. Based on LR-tests of the nested models in Table 3, models with covariates appear to yield a much better fit than those without covariates effects. To assess the overall fit

Table 3
Summary table of models w/o constrained structures for university data

Model (with $\Psi = \mathbf{I}$)	– log-lik.	K	mAIC	BIC	av. P^2
(A) $\boldsymbol{\Sigma}_\upsilon = \sigma_\upsilon^2\mathbf{I}$,	1563.3	6	3138.6	3158.7	8.3
(B) $\boldsymbol{\Sigma}_\upsilon = \boldsymbol{\Phi}$,	1555.5	11	3133.0	3169.9	6.2
(C) $\boldsymbol{\Sigma}_\upsilon = \boldsymbol{\lambda}\boldsymbol{\lambda}' + \sigma_\delta^2\mathbf{I}$,	1543.1	11	3108.2	3145.1	4.8
(D) $\boldsymbol{\Sigma}_\upsilon = \boldsymbol{\lambda}\boldsymbol{\lambda}' + \boldsymbol{\Phi}$,	1538.8	16	3109.6	3163.3	4.0
(E) $\boldsymbol{\Sigma}_\upsilon$ unconstrained,	1535.7	20	3111.4	3178.5	3.9
(F) (A) with $\boldsymbol{\mu} = \mathbf{X}_3\boldsymbol{\alpha}$	1489.7	21	3021.4	3091.9	8.5
(G) (B) with $\boldsymbol{\mu} = \mathbf{X}_3\boldsymbol{\alpha}$	1481.3	26	3014.6	3101.9	5.9
(H) (C) with $\boldsymbol{\mu} = \mathbf{X}_3\boldsymbol{\alpha}$	1466.3	26	2984.6	3071.9	4.3
(I) (D) with $\boldsymbol{\mu} = \mathbf{X}_3\boldsymbol{\alpha}$	1462.4	31	2986.8	3090.9	3.9
(J) (E) with $\boldsymbol{\mu} = \mathbf{X}_3\boldsymbol{\alpha}$	1460.0	35	2990.0	3107.5	3.8

K: number of parameters;
mAIC = marginal Akaike's information criterion = -2 log-lik. + 2 K;
BIC = Bayesian information criterion = -2 log-lik. + K ln(212).

of the models, we also computed Pearson's χ^2-statistics (P^2) based on the observed and expected frequencies of all possible triplets. In addition, BIC (Schwartz, 1978) and mAIC statistics (Vaida and Blanchard, 2005) are provided in Table 3. We present mAIC instead of cAIC statistics because model inferences are on population rather than person-specific parameters. Among all models, model (H) provides the most parsimonious fit with an average P^2 value of 4.3, and minimum mAIC and BIC values of 2984.6 and 3071.9, respectively.

The parameter estimates and corresponding standard errors of Model (H) are reported in Table 4. First, the intercept effects $\boldsymbol{\alpha}_1' = (2.80\ 1.62\ 1.71\ 0.96\ 1.98\ 0^\star)$ show that on average the colleges can be ranked as London $\succ$ St Gall $\succ$ Paris $\succ$ Milan $\succ$ Barcelona $\succ$ Stockholm, but that the differences between St Gall, Paris, and Milan are not significant. However, it is more meaningful to consider the different subpopulations in Model (H). Table 5 reports the mean utilities of these subpopulations. Inspecting $\hat{\boldsymbol{\alpha}}_2$ and $\hat{\boldsymbol{\alpha}}_4$, we note that good knowledge of French and Italian makes Paris and Milan more attractive, respectively. Interestingly, students who speak Spanish well prefer St Gall to London. This result is unexpected since students who have good knowledge of Spanish do not appear to favor Barcelona. Based on the estimates' standard errors, we conclude that St Gall and Milan are the only universities that appear to be affected by a person's knowledge of Spanish and Italian, respectively.

To interpret the single-factor covariance structure $\boldsymbol{\Sigma}_\upsilon = \boldsymbol{\lambda}\boldsymbol{\lambda}' + \sigma_\delta^2\mathbf{I}$ in model (H), we examine the pattern of the estimated factor loadings

$$\hat{\boldsymbol{\lambda}}' = \begin{pmatrix} \text{LO} & \text{PA} & \text{MI} & \text{BA} & \text{SG} & \text{ST} \\ -0.22 & -0.91 & -1.06 & 0.92 & -0.95 & 0^\star \end{pmatrix}.$$

The loadings indicate that Paris, Milan and St Gall are more similar to each other than the rest of the universities. The university characteristics in Table 1 cannot account for this pattern. Instead, we conclude that individual differences in the comparison of

Table 4
Parameter estimates (standard errors) of model (H) for university data

j	$\hat{\alpha}_{1j}$	$\hat{\alpha}_{2j}$	$\hat{\alpha}_{3j}$	$\hat{\alpha}_{4j}$	$\hat{\lambda}_j$
LO	2.80 (0.31)	0.46 (0.17)	0.18 (0.24)	0.30 (0.23)	−0.22 (0.23)
PA	1.62 (0.30)	1.27 (0.19)	0.25 (0.25)	0.23 (0.23)	−0.91 (0.24)
MI	1.71 (0.32)	0.50 (0.18)	0.08 (0.25)	1.52 (0.26)	−1.06 (0.23)
BA	0.96 (0.29)	0.37 (0.18)	0.22 (0.25)	0.12 (0.23)	0.92 (0.26)
SG	1.98 (0.31)	0.26 (0.18)	1.25 (0.26)	0.20 (0.23)	−0.95 (0.25)
ST	0*	0*	0*	0*	0*

*Fixed for identification purpose; $\hat{\sigma}_\delta^2 = 1.66$ (0.24).
α_1: Intercept; α_2: FRA; α_3: SPA; α_4: ITA.
LO: London; PA: Paris; MI: Milan; BA: Barcelona; SG: St Gall; ST: Stockholm.

Table 5
Estimated mean preferences of subpopulations under model (H)

SPA	good				poor			
ITA	good		poor		good		poor	
FRA	good	poor	good	poor	good	poor	good	poor
LO	3.74	2.83	3.13	2.22	3.38	2.47	2.77	1.86
PA	3.37	0.83	2.91	0.37	2.86	0.32	2.40	−0.13
MI	3.81	2.81	0.76	−0.23	3.64	2.65	0.60	−0.40
BA	1.67	0.94	1.42	0.69	1.23	0.50	0.99	0.25
SG	3.69	3.18	3.28	2.77	1.19	0.67	0.78	0.26
ST	0*	0*	0*	0*	0*	0*	0*	0*

*Fixed for identification purpose.
LO: London; PA: Paris; MI: Milan; BA: Barcelona; SG: St Gall; ST: Stockholm.

universities are related to whether a student is familiar with French, Spanish, or Italian because preferences for Paris, Milan and St Gall are highly influenced by the judges' knowledge of these languages. Finally, the variance component of the unique factors in Model (H), $\hat{\sigma}_\delta^2$, is estimated as 1.66 (0.24), which indicates that the between-judge variance is considerably larger than the within-pair variance ($\sigma^2 = 1$). It accounts for 60–70% of the total variation in each of the pairwise judgments.

Overall, Model (H) provides a satisfactory fit of the data. The average χ^2-statistic P^2 of all triplets is 4.3. Table 6 reports the observed and expected frequencies for pairwise marginals under model (H) and their discrepancies appear to be small.

6.2. Bivariate rankings

In a survey conducted shortly before the 1992 US presidential election, a sample of 81 undergraduate students was asked to order the candidates George Bush [B], Bill Clinton [C], and Ross Perot [P] according to how well each of them represented their

Table 6
Observed (expected) frequencies for pairwise marginals for Model (H) for university data

	LO	PA	MI	BA	SG	ST
LO		140 (139.9)	165 (166.5)	159 (161.0)	131 (133.9)	93 (97.1)
PA	72 (72.1)		161 (167.6)	190 (188.4)	128 (129.3)	129 (130.5)
MI	47 (45.5)	51 (44.4)		138 (137.9)	160 (158.9)	108 (100.0)
SG	53 (51.0)	22 (23.6)	74 (74.1)		107 (102.8)	128 (137.6)
BA	81 (78.1)	84 (82.7)	52 (53.1)	105 (109.2)		142 (141.1)
ST	119 (114.9)	83 (81.5)	104 (112.0)	84 (74.4)	70 (70.9)	

Entry (i, j) indicates the frequency that i is preferred to j.
LO: London; PA: Paris; MI: Milan; BA: Barcelona; SG: St Gall; ST: Stockholm.

views on *education* [ED] and *economy* [EC] issues. The data are reproduced in Table 7. For example, 11 respondents ranked the candidates in the order C, P, B on both criteria.

Our analysis of these data addresses the three questions of (a) whether the respondents distinguished among the political candidates on these two issues, (b) whether individual differences reflect the political orientation of the candidates, and (c) whether political orientation can account for the observed associations between the rankings.

Table 8 contains the goodness-of-fit statistics of four bivariate ranking models and their corresponding likelihood-ratio statistics. In view of the small sample size, the χ^2-approximation of the goodness-of-fit statistics needs to be viewed with caution but it is still useful for the comparison of the different models. Model (A) specifies that the political candidates are perceived as equally similar to each other on both criteria and that the responses to the two criteria are uncorrelated. Individual differences are represented by an identity matrix for both criteria, and the (3×3) covariance matrix $\mathbf{\Sigma}_C$ of the evaluations of the three candidates on the two criteria is set to a null matrix. This model does not provide a satisfactory fit of the data with $G^2 = 59.8$ on 31 degrees of freedom (df). The introduction of equal correlations between the candidates with $\mathbf{\Sigma}_C = \rho\mathbf{I}$ improves significantly the goodness-of-fit statistic. The estimated correlation coefficient $\hat{\rho} = 0.61$ indicates that the respondents judge the three candidates in a similar way on Education and Economy issues.

Further fit improvements are obtained by Models (C) and (D) which allow for a one-factor model to capture individual differences in the evaluation of the candidates. In particular, by letting the correlations between the criteria reflect the one-factor structure with $\mathbf{\Sigma}_C = \rho(\boldsymbol{\lambda}\boldsymbol{\lambda}' + \mathbf{\Phi})$, we obtain an excellent fit of the ranking data. In view of the small sample size, it is difficult to choose between Models (B) and (D), and we therefore present the parameter estimates of both models in Table 9.

We note that the average differences between the political candidates on Education issues are small. Perot and Clinton appear to be more liked than Bush on this topic. Under Model (D), the mean differences between Bush and Perot and Clinton and Perot are estimated as $\hat{\mu}_{BP[ED]} = -0.47$ and $\hat{\mu}_{CP[ED]} = -0.15$, respectively. In contrast, on Economy topics, Clinton appears to be much more liked than either Bush or Perot with a mean difference of about 1 between the other two candidates. The estimated

Table 7
Bivariate ranking data

EC/ED	BCP	BPC	CBP	CPB	PBC	PCB
BCP	1	1	4	2	0	0
BPC	3	4	4	1	1	2
CBP	1	0	6	2	0	0
CPB	2	0	5	11	0	1
PBC	1	4	1	1	2	1
PCB	1	1	4	12	0	2

ED = Education, EC = Economy, B = Bush, C = Clinton, P = Perot.

Table 8
Summary table of bivariate ranking models

Model	G^2	df
(A) $\mathbf{\Sigma}_{\mathrm{O}} = \sigma_{\mathrm{O}}^2\mathbf{I}$, $\mathbf{\Sigma}_{\mathrm{C}} = \mathbf{0}$	59.8	31
(B) $\mathbf{\Sigma}_{\mathrm{O}} = \sigma_{\mathrm{O}}^2\mathbf{I}$, $\mathbf{\Sigma}_{\mathrm{C}} = \rho\mathbf{I}$	26.4	30
(C) $\mathbf{\Sigma}_{\mathrm{O}} = \boldsymbol{\lambda}\boldsymbol{\lambda}' + \mathbf{\Phi}$, $\mathbf{\Sigma}_{\mathrm{C}} = \rho\mathbf{I}$	21.9	28
(D) $\mathbf{\Sigma}_{\mathrm{O}} = \boldsymbol{\lambda}\boldsymbol{\lambda}' + \mathbf{\Phi}$, $\mathbf{\Sigma}_{\mathrm{C}} = \rho\mathbf{\Sigma}_{\mathrm{O}}$	18.7	28

Table 9
Parameter Estimates (Standard Errors) of Models (B) and (D) for bivariate ranking data

Estimates	Model (B)	Model (D)
$\hat{\mu}_{\mathrm{BP[ED]}}$	−0.43 (0.19)	−0.47 (0.20)
$\hat{\mu}_{\mathrm{CP[ED]}}$	−0.14 (0.18)	−0.15 (0.18)
$\hat{\mu}_{\mathrm{BP[EC]}}$	0.08 (0.19)	0.03 (0.19)
$\hat{\mu}_{\mathrm{CP[EC]}}$	1.00 (0.20)	1.06 (0.20)
$\hat{\rho}$	0.61 (0.08)	0.61 (0.08)
$\hat{\lambda}_{\mathrm{B}}$	–	−0.87 (0.20)
$\hat{\lambda}_{\mathrm{C}}$	–	0.74 (0.19)
$\hat{\lambda}_{\mathrm{P}}$	–	0.13⋆

⋆Fixed for identification purposes.

factor loadings under Model (D) for both criteria reflect the political orientation of the candidates. Bush and Clinton are viewed as opposites with estimated loadings $\hat{\lambda}_{\mathrm{B}} = -0.87$ and $\hat{\lambda}_{\mathrm{C}} = 0.74$, respectively. Perot is perceived as somewhat more similar to Clinton than Bush.

7. Concluding remarks

Aided by the many advances in computational technologies, the search for structure in heterogeneous populations has become a fruitful research topic for a wide range of disciplines. Specifically, in the analysis of preference and choice behavior much attention has focused on the extension of existing choice models to consider different forms of variability among respondents. The random-effect models reviewed in this chapter are most useful for providing parsimonious descriptions of how individuals perceive and evaluate choice options. They can be applied effectively in the analysis of stated-preference data because in this case the different sources of variability are readily identifiable. The analysis of revealed choice data is more difficult when little is known about factors that characterize the choice situations. For example, when the set of choice options varies exogenously from person to person in an unobserved way, random-effects choice models cannot provide an unbiased account of individual preference differences. Walker and Ben-Akiva (2002) review methodological approaches designed to overcome some of the complexities introduced by revealed choice data.

In any application it should be kept in mind that preferences may not be well-defined and may thus depend on seemingly irrelevant contextual conditions. For example, different framings of the same choice options may trigger different associations and evaluations with the results that seemingly minor changes in the phrasing of a question or in the presentation format can lead to dramatic changes in the response behavior of a person (Kahneman, 2003). One major conclusion of experimental research on psychological processes that underlie the formation of preferences is that generalizations of scaling results to different choice situations and options require much care and frequently need to be based on additional validation studies.

References

Akaike, H. (1973). Information theory and an extension of the maximum likelihood principle. In: Proceedings of the 2nd International Symposium on Information Theory, Budapest, pp. 267–281.

Ansari, A., Iyengar, R. (2006). Semiparametric Thurstonian models for recurrent choices: A Bayesian analysis. *Psychometrika* **71**, in press.

Arbuckle, J., Nugent, J.H. (1973). A general procedure for parameter estimation for the law of comparative judgment. *British Journal of the Mathematical and Statistical Psychology* **26**, 240–260.

Bekker, P., Wansbeek, T. (2000). Matrix inequality application in econometrics. In: Heijmans, R.D.H., Pollock, D.S.G., Satorra, A. (Eds.), *Innovations in Multivariate Statistical Analysis: A Festschrift for Heinz Neudecker*. Kluwer Academic Publishers, Massachusetts, pp. 51–66.

Block, H., Marschak, J. (1960). Random orderings and stochastic theories of response. In: Olkin, et al. (Eds.), *Contributions to Probability and Statistics*. Stanford University Press, Stanford, pp. 97–132.

Bock, R.D., Jones, L.V. (1968). *The Measurement and Prediction of Judgment and Choice*. Holden-Day, San Francisco, CA.

Böckenholt, U. (1992). Thurstonian models for partial ranking data. *British Journal of Mathematical and Statistical Psychology* **45**, 31–49.

Böckenholt, U. (1996). Analyzing multi-attribute ranking data: Joint and conditional approaches. *British Journal of Mathematical and Statistical Psychology* **49**, 57–78.

Böckenholt, U. (2002). A Thurstonian analysis of preference change. *Journal of Mathematical Psychology* **46**, 300–314.

Böckenholt, U. (2004). Comparative judgments as an alternative to ratings: Identifying the scale origin. *Psychological Methods* **9**, 453–465.

Bradbury, H., Ross, K. (1990). The effects of novelty and choice materials on the intransitivity of preferences of children and adults. *Annals of Operations Research* **23**, 141–159.

Bradley, R.A. (1984). Paired comparisons: Some basic procedures and examples. In: Krishnaiah, P.R., Sen, P.K. (Eds.), *Handbook of Statistics*, vol. 4. North-Holland, Amsterdam, pp. 299–326.

Brady, H.E. (1989). Factor and ideal point analysis for interpersonally incomparable data. *Psychometrika* **54**, 181–202.

Chan, W., Bentler, P.M. (1998). Covariance structure analysis of ordinal ipsative data. *Psychometrika* **63**, 369–399.

Dansie, B.R. (1986). Normal order statistics as permutation probability models. *Applied Statistics* **35**, 269–275.

David, H.A. (1988). *The Method of Paired Comparisons*. Griffin, London.

deDoncker, E., Ciobanu, M., Genz, A. (1999). Parallel computation of multivariate normal probabilities. *Computing Science and Statistics* **31**, 89–93.

Dittrich, R., Hatzinger, R., Katzenbeisser, W. (1998). Modeling the effect of subject-specific covariates in paired comparison studies with an application to university rankings. *Applied Statistics* **47**, 511–525.

Guttman, L. (1946). An approach for quantifying paired comparisons rank order. *Annals of Mathematical Statistics* **17**, 144–163.

Imai, K., van Dyk, D.A. (2005). A Bayesian analysis of the multinomial probit model using marginal data augmentation. *Journal of Econometrics* **124**, 311–334.

Kahneman, D. (2003). Maps of bounded rationality: Psychology for behavioral economics. *American Economic Review* **93**, 1449–1475.

Keane, M.P. (1997). Modeling heterogeneity and state dependence in consumer choice behavior. *Journal of Business and Economic Statistics* **15**, 310–327.

Louviere, J.J., Hensher, D.A., Swait, J.D. (2000). *Stated Choice Methods*. Cambridge University Press, New York.

Luce, R.D. (1959). *Individual Choice Behavior*. Wiley, New York.

Lüthkepol, H. (1993). *Introduction to Multiple Time Series Analysis*. Springer, Berlin.

Manski, C., McFadden, D. (Eds.) (1981). *Structural Analysis of Discrete Data with Econometric Applications*. MIT Press, Cambridge.

Marden, J.I. (1995). *Analyzing and Modeling Rank Data*. Chapman & Hall, London.

Marley, A.A.J. (2002). Random utility models and their applications: recent developments. *Mathematical Social Sciences* **43**, 289–302.

Marshall, P., Bradlow, E.T. (2002). A unified approach to conjoint analysis models. *Journal of the American Statistical Association* **97**, 674–682.

Maydeu-Olivares, A. (2001). Limited information estimation and testing of Thurstonian models for paired comparison data under multiple judgment sampling. *Psychometrika* **66**, 209–228.

Maydeu-Olivares, A. (2002). Limited information estimation and testing of Thurstonian models for preference data. *Mathematical Social Sciences* **43**, 467–483.

Maydeu-Olivares, A., Böckenholt, U. (2005). Structural equation modeling of paired comparison and ranking data. *Psychological Methods* **10**, in press.

McDonald, R.P. (1999). *Test Theory: A Unified Approach*. Lawrence Erlbaum, Mahwah, NJ.

McFadden, D. (2001). Economic choices. *American Economic Review* **91**, 351–378.

Morrison, H.W. (1963). Testable conditions for triads of paired comparison choices. *Psychometrika* **28**, 369–390.

Mosteller, F. (1951). Remarks on the method of paired comparisons: I. The least squares solution assuming equal standard deviations and equal correlations. *Psychometrika* **16**, 3–11.

Muthèn, L., Muthèn, B. (2004). *Mplus 3*. Muthèn & Muthèn, Los Angeles, CA.

Reinsel, G.C. (1997). *Elements of Multivariate Time Series Analysis*. Springer, Berlin.

Schwartz, G. (1978). Estimating the dimension of a model. *Annals of Statistics* **6**, 461–464.

Simonson, I., Tversky, A. (1992). Choice in context: Tradeoff contrast and extremeness aversion. *Journal of Marketing Research* **29**, 281–295.

Skrondal, A., Rabe-Hesketh, S. (2003). Multilevel logistic regression for polytomous data and rankings. *Psychometrika* **68**, 267–287.

Stern, H. (1990). A continuum of paired comparison models. *Biometrika* **77**, 265–273.

Takane, Y. (1987). Analysis of covariance structures and probabilistic binary choice data. *Cognition and Communication* **20**, 45–62.

Thurstone, L.L. (1927). A law of comparative judgment. *Psychological Review* **34**, 273–286.

Thurstone, L.L., Jones, L.V. (1957). The rational origin for measuring subjective values. *Journal of the American Statistical Association* **52**, 458–471.

Torgerson, W.S. (1958). *Theory and Methods of Scaling*. John Wiley & Sons, New York.

Train, K. (2003). *Discrete Choice Methods with Simulations*. MIT Press, Cambridge.

Tsai, R. (2000). Remarks on the identifiability of Thurstonian ranking models: case V, case III, or neither? *Psychometrika* **65**, 233–240.

Tsai, R. (2003). Remarks on the identifiability of Thurstonian paired comparison models under multiple judgment. *Psychometrika* **68**, 361–372.

Tsai, R., Böckenholt, U. (2001). Maximum likelihood estimation of factor and ideal point models for paired comparison data. *Journal of Mathematical Psychology* **45**, 795–811.

Tsai, R., Böckenholt, U. (2002). Two-level linear paired comparison models: Estimation and identifiability issues. *Mathematical Social Sciences* **43**, 429–449.

Tsai, R., Böckenholt, U. (2006). Modeling intransitive preferences: A random-effects approach. *Journal of Mathematical Psychology* **50**, 1–14.

Walker, J., Ben-Akiva, M. (2002). Generalized random utility model. *Mathematical Social Sciences* **43**, 303–343.

Vaida, F., Blanchard, S. (2005). Conditional Akaike information for mixed-effects models. *Biometrika* **92**, 351–370.

Yao, G. (1995). Bayesian estimation of Thurstonian ranking models based on the Gibbs sampler. PhD Dissertation. University of Illinois, Urbana-Champaign, IL.

Yao, G., Böckenholt, U. (1999). Bayesian estimation of Thurstonian ranking models based on the Gibbs sampler. *British Journal of Mathematical and Statistical Psychology* **52**, 79–92.

Yu, P.L.H. (2000). Bayesian analysis of order-statistics models for ranking data. *Psychometrika* **65**, 281–299.

Yu, P.L.H., Lam, K.F., Lo, S.M. (1998). Factor analysis for ranking data. Unpublished manuscript. Department of Statistics, The University of Hong Kong.

Zajonc, R.B. (1980). Feeling and thinking: Preferences need no inferences. *American Psychologist* **15**, 151–175.

Handbook of Statistics, Vol. 26
ISSN: 0169-7161

DOI: 10.1016/S0169-7161(06)26015-2

15

Item Response Theory in a General Framework

R. Darrell Bock and Irini Moustaki

1. Introduction

Item response theory (IRT) deals with the statistical analysis of data in which responses of each of a number of respondents to each of a number of items or trials are assigned to defined mutually exclusive categories. Although its potential applications are much broader, IRT was developed mainly in connection with educational measurement, where the main objective is to measure individual student achievement. Prior to the introduction of IRT, the statistical treatment of achievement data was based entirely on what is now referred to as "classical" test theory. That theory is predicated on the test score (usually the student's number of correct responses to the items presented) as the observation. It assumes that the number of items is sufficiently large to justify treating the test score as if it were a continuous measurement with specified origin and unit. On the further assumption that the items are randomly sampled from a larger item domain, a classical method of estimating measurement error due to item sampling is to score random halves of the test items separately, compute the product-moment correlation of the two scores, and apply the so-called *Spearman–Brown* formula to extend the correlation to that of the full test to obtain its "split-half" reliability. The complement of the reliability coefficient is then proportion of the test variance attributable to measurement error. Various elaborations of the classical theory appear in texts such as Gulliksen (1950), Lindquist (1953), or Lord and Novick (1968).

The results of classical test theory are necessarily limited in application. They are not accurate for short tests, and they ignore sources of error such as variation due to rater effects when responses to the items must be judged subjectively. Although a variance-component extension of classical theory, called *generalizability theory*, treats multiple sources of error (Brennan, 2001), it also assumes that the test scores are continuous measurements. The greatest limitation of the classical theory is, however, its dependence on dichotomous scoring. There is no provision for responses scored in three or more categories, even though this is common practice for performance tests or problem-solving exercises in which differing degrees of student accomplishment are recognized. Classical methods are often applied to this type of item response by arbitrary assignment of numerical values to the categories, a device difficult justify when items are scored in different numbers of categories. Scaling procedures for assigning values to the categories that maximize the ratio of between to within respondent sums of squares are better motivated, but they collapse the data over items without regard to differences in item characteristics (see Nishisato, 1994, for a review).

Item response theory overcomes these limitations by assuming a continuous latent variable representing the student's proficiency in responding to the test items. The probability of a response in any one of two-or-more mutually exclusive categories of an item is assumed to be a function of the student's location on the latent continuum and of certain estimable parameters characteristic of the item. This approach leads to statistical methods of scoring tests, whether with large or small numbers of items, without the assumption that the items are sampled from a defined item domain to which the results generalize. Item response models that yield these probabilities are central to these methods. We discuss in Section 3 the main models in current use.

A further critical assumption in IRT is that item responses, given the respondent's location in the latent space, are statistically independent. Lazarsfeld made use of this principle of "conditional" independence in an analysis of contingency table data as early as 1950, but first stated it clearly in Lazarsfeld (1958). It is important in IRT because it allows straightforward calculation of the likelihood of the model parameters, given the item responses of a sample of respondents. When the number of respondents is large, it leads to efficient, likelihood-based methods for IRT analysis in a wide variety of applications, including not only measurement of individual differences, such as in education or clinical psychology, but also estimation of group effects or population quantiles without explicit calculation of respondent locations or "scores". Moreover, it makes possible new modes of test-item administration, most importantly computerized adaptive testing in which items maximally informative for the individual respondent are selected dynamically during computer-implemented testing sessions (see Wainer, 1990). Extended to multiple latent dimensions, conditional independence allows full-information estimation of factor loadings in the item response models directly from the item response patterns, rather than limited-information factor analysis of inter-item tetrachoric or polychoric correlations. If exogenous quantitative measurements of the respondents are available, IRT can be further extended to include estimation of relationships between the latent variables and external variables. All of these applications are amenable to a likelihood based approach in estimation and hypothesis testing. In the following section, we present a general formulation of item response theory that supports these applications.

2. The general IRT framework

In IRT, we distinguish among observed variables (the respondent's categorical choices), explanatory variables (covariates) and latent variables (unobserved proclivities, proficiencies, or factors). IRT models can be seen as having a measurement part where latent variables together with covariates have an effect on some function of the items and a structural part where covariates affect the latent variables. If a linear structure is adopted for the structural part of the model, that implies a shift in the mean of the latent variable for different values of the covariates. An alternative to the structural part is a multiple group model in which respondents are assigned to mutually exclusive groups and parameters of the corresponding latent distributions are estimated. In some applications group-specific item parameters may also be estimated. It should be noted that multi-group analysis requires a sufficient number of sample members in each group to estimate the model.

Let the groups be indexed in any order by $v = 1, 2, \ldots, V$; let $\mathbf{u}$ be the n-vector of categorical responses to an n-item instrument, with m_j the number of response categories of item j; let $\mathbf{x}$ the r-vector of covariate values. We write $u_j = k$ to mean that u_j belongs to the category k, $k = 1, \ldots, m_j$. The values assigned to the categories are arbitrary but in the case of ordinal manifest variables they must preserve the ordinality property of the variable. Finally, let $\boldsymbol{\theta}$ be the vector of latent variables with elements θ_q, $q = 1, 2, \ldots, p$. Both the item responses and the latent variables are random variables: the latent variables are random in population of respondents; the responses are random both in the population of respondents and in the population of potential responses of a given respondent.

There are $\prod_{j=1}^{n} m_j$ possible response patterns. Let $\mathbf{u}_i = (u_1 = k_1, u_2 = k_2, \ldots, u_n = k_n)$ represent anyone of these. As only $\mathbf{u}$ can be observed any inference must be based on the joint probability of $\mathbf{u}$ for group v written as:

$$f_v(\mathbf{u} \mid \mathbf{x}) = \int_{R(\theta_1)} \cdots \int_{R(\theta_p)} f_v(\mathbf{u} \mid \boldsymbol{\theta}, \mathbf{x}) g_v(\boldsymbol{\theta}) \, \mathrm{d}\boldsymbol{\theta}, \tag{1}$$

where $g_v(\boldsymbol{\theta})$ is the distribution of the latent variables $\boldsymbol{\theta}$ in group v, $f_v(\mathbf{u} \mid \boldsymbol{\theta}, \mathbf{x})$ is the conditional probability of $\mathbf{u}$ given $\boldsymbol{\theta}$ and $\mathbf{x}$ in group v and $R(\theta_q)$ is the range space of θ_q. Note that the density functions $g_v(\boldsymbol{\theta})$ and $f_v(\mathbf{u} \mid \boldsymbol{\theta}, \mathbf{x})$ are not uniquely determined; further restrictions need to be imposed in the selection of those two density functions.

Modelling manifest variables as functions of latent variables and covariates implies that association among the u's can be explained by a set of latent variables and a set of explanatory variables then when all θ's and the x's are accounted for the u's will be independent (conditional or local independence). Therefore, the number of latent variables p and the covariates x must be chosen so that:

$$f_v(\mathbf{u} \mid \boldsymbol{\theta}, \mathbf{x}) = \prod_{j=1}^{n} f_{vj}(u_j \mid \boldsymbol{\theta}, \mathbf{x}). \tag{2}$$

Substituting (2) in (1):

$$f_v(\mathbf{u} \mid \mathbf{x}) = \int_{R(\theta_1)} \cdots \int_{R(\theta_p)} \prod_{j=1}^{n} f_{vj}(u_j \mid \boldsymbol{\theta}, \mathbf{x}) g_v(\boldsymbol{\theta}) \, \mathrm{d}\boldsymbol{\theta}. \tag{3}$$

The latent variables are assumed to be independent with normal distributions, $\theta_q \sim N(\mu_{vq}, \sigma_{vq}^2)$, $q = 1, \ldots, p$. The normality assumption has rotational advantages in the multi-dimensional case. To reduce the complexity of the presentation we limit our discussion in the estimation section to the one-dimensional case.

For a random sample of size N that is classified into V groups the likelihood is written as:

$$\prod_{i=1}^{N} f(\mathbf{u}_i \mid \mathbf{x}_i) = \prod_{v=1}^{V} \prod_{i=1}^{N_v} f_v(\mathbf{u}_i \mid \mathbf{x}_i), \tag{4}$$

where N_v denotes the number of respondents in group v and $f_v(\mathbf{u} \mid \mathbf{x})$ is given in (3).

The log-likelihood is:

$$L = \sum_{i=1}^{N} \log f(\mathbf{u}_i \mid \mathbf{x}_i) = \sum_{v=1}^{V} \sum_{i=1}^{N_v} \log f_v(\mathbf{u}_i \mid \mathbf{x}_i). \tag{5}$$

Note that (5) allows each item to have different parameter values in different groups. However, a primary aim of IRT modelling is to construct items that are measurement invariant. Since measurement invariance (see, e.g., Lord, 1980; Meredith, 1993; Holland and Wainer, 1993; Camilli and Shepard, 1994) requires that items parameters are equal across groups, we therefore constrain all item parameters (e.g., thresholds and slopes) to be equal across groups and allow only the distributions of the latent variables to differ among groups. An exception is when items are suspected of performing differently in the groups – so-called *differential item functioning* (DIF) (Holland and Wainer, 1993). To investigate such effects, the constraints are selectively relaxed for certain items or parameters during parameter estimation.

3. Item response models

The following are commonly employed modes of response modelled in item response theory, with typical examples of each.

Dichotomous. (a) Responses to educational achievement tests are marked "right" or "wrong". (b) Respondents to a public opinion survey are asked to answer "yes" if they endorse a presented statement of opinion and "no" otherwise.

Ordered polytomous. (a) Responses to clinically administered items of an intelligence scale are rated by a clinician in specified categories representing increasing levels of aptness. (b) Respondents to a physical fitness self-report inventory are asked whether they typically engage in some type of physical exercise "once a day", "once a week", "once a month", "almost never".

Nominal polytomous. (a) Students taking a multiple-choice vocabulary test are required to mark one, and only one, of a number of alternative definitions the best describes the meaning of a given word. (b) Respondents in a consumer preference survey are asked which of several flavors of ice cream they most prefer.

Rankings. (a) Members of an association are asked to rank nominated candidates for office in order of preference. (b) Judges of an essay-writing contest are asked to read a certain number of essays and rank them in order of merit.

In this section, we present the best known item response models specifying the probability of response in each category as a function of parameters attributed to items and of a latent value attributed to the respondent. The extension of these models to more than one latent variable and possible respondent-level covariates is deferred to Sections 6 and 7. The present section describes the models and some of their properties; their roles in item parameter estimation and respondent scoring are discussed in subsequent sections.

3.1. *Dichotomous categories*

The most widely used models for dichotomous categories are the normal ogive model and three forms of the logistic model referred to as one-parameter (1-PL), two-parameter (2-PL), and three-parameter (3-PL) logistic.

3.1.1. *Normal ogive model*

The normal ogive model first appeared in connection with item response theory in papers by Lawley (1943) and Lord (1952). Prior to that, the model had a long history in psychophysics going back to Fechner (1860) (who credits his colleague Möbius for the idea), also in psychological scaling by Thurstone (1927) and others, and in bioassay by Bliss (1935), Fisher (Fisher and Yates, 1938) and others (see Finney, 1952). In these fields only one item at a time is studied and the data are frequencies of response aggregated to the group level; the response probabilities are assumed to depend upon a single variable, observable in psychophysics and bioassay and latent in psychological scaling. In item response theory, on the other hand, multiple items are studied jointly, the data are the responses of individuals, and the probabilities depend upon one or more latent variables. In the one-dimensional normal ogive model, respondent i and item j are located respectively at points θ_i and b_j on the latent continuum. The observed response is assumed to be controlled by an unobservable random response-process variable, say,

$$Y_{ij} = a_j(\theta_i - b_j) + \epsilon_{ij}. \tag{6}$$

Either in the population of respondents or in replicate responses of a given respondent, ϵ_{ij} is independent normal with mean 0 and variance 1; a_j is the regression of the response process on θ.

If Y_{ij} is greater than or equal to a threshold value, γ_j, the response falls in category 2 and is assigned a score $u_{ij} = 1$; otherwise, it falls in category 1 and is assigned $u_{ij} = 0$. These so-called *item scores* may be treated as Bernoulli variables, the distribution of which may be expressed in terms of the normal distribution function, $\Phi(y_j)$ parameterized as the so-called *normit*, $y_j = a_j(\theta - b_j)$. Then the item response model is a function of θ,

$$P_j(\theta) = P(u_j = 1 \mid \theta) = \Phi(y_j), \quad -\infty < y_j < +\infty, \tag{7}$$

and

$$Q_j(\theta) = P(u_j = 0 \mid \theta) = 1 - \Phi(y_j).$$

For computational purposes, and also to allow a straight-forward generalization to multiple latent variables, it is convenient to set $c_j = -a_j b_j$ and work with

$$y_j = a_j\theta + c_j. \tag{8}$$

The distribution function of the item score can be expressed concisely in the formula

$$f(u_j \mid \theta) = P_j(\theta)^{u_j} Q_j(\theta)^{1-u_j}. \tag{9}$$

The plot of the response function, $P_j(\theta)$, with respect to θ is an ascending normal ogive with inflection point at $\theta = b_j$. There is a corresponding descending curve for $Q_j(\theta)$ usually plotted along with $P_j(\theta)$. The plot of the functions, called *item characteristic curves* (ICC) or *item response functions*, are mirror images crossing at 0.5 probability, where their slopes are $a_j/\sqrt{2\pi}$ and $-a_j/\sqrt{2\pi}$, respectively. In this context, the slope parameter a_j in the normit is called the item *discriminating power*, b_j the item *location*, and c_j the item *intercept*.

The first derivatives of the response function with respect to the item parameters and latent variable are

$$\partial P_j(\theta)/\partial a_j = \theta\phi(y_j),$$
$$\partial P_j(\theta)/\partial c_j = \phi(y_j),$$
$$\partial P_j(\theta)/\partial \theta = a_j\phi(y_j),$$

where $\phi(y_j)$ is the normal ordinate at y_j.

In connection with maximum likelihood estimation of the respondent's θ (see Section 5), the Fisher information with respect to θ conveyed by the item score is the negative expected value of the second derivative of the log likelihood of θ, given u_{ij}. Assuming the responses to a test consisting of n items are conditionally independent, given θ, then for any dichotomous item response model the first derivative of the log likelihood with respect to θ is

$$\sum_{j=1}^{n} \frac{u_j - P_j(\theta)}{P_j(\theta)Q_j(\theta)} \frac{\partial P_j(\theta)}{\partial \theta}. \tag{10}$$

The second derivative can be expressed as

$$\sum_{j=1}^{n} \frac{(u_j - P_j(\theta)}{P_j(\theta)Q_j(\theta)} \left\{ \frac{\partial^2 P_j(\theta)}{\partial \theta^2} - \frac{(Q_j(\theta) - P_j(\theta))}{P_j(\theta)Q_j(\theta)} \left[\frac{\partial P_j(\theta)}{\partial \theta} \right]^2 \right\}$$
$$- \sum_{j=1}^{n} \frac{1}{P_j(\theta)Q_j(\theta)} \left[\frac{\partial P_j(\theta)}{\partial \theta} \right]^2, \tag{11}$$

(see Bock and Jones, 1968, p. 55). Since expectation of the first term vanishes with increasing n, the *information function* of the maximum likelihood estimator of θ is $\sum_j^n I_j(\theta)$ where

$$I_j(\theta) = I(u_j \mid \theta) = \frac{a_j^2\phi^2(y_j)}{\Phi(y_j)[1 - \Phi(y_j)]} \tag{12}$$

is called the *item* information function. See Lord and Novick (1968, p. 449).

If θ is assumed distributed with finite mean and variance in the respondent population, the indeterminacy of location and scale in the model may be resolved by arbitrary choice of the mean and standard deviation of θ; the default choices in IRT are 0 and 1, respectively. With further assumption that θ is normally distributed, the biserial correlation of Y_j and θ in the population is

$$\rho_j = a_j / \sqrt{1 + a_j^2}. \tag{13}$$

This justifies referring to the a-parameter as "discriminating power". (See Lord and Novick, 1968, Chapter 16, for this and following results for the normal ogive model.)

In classical test theory, the proportion of correct responses, π_j, from respondents in a specified population serves as a measure of item *difficulty*. Although this measure increases with increasing ability and might better be called "facility", the conventional term "difficulty" is firmly established in the testing field. Assuming the normal ogive response model and an $N(0, 1)$ distribution of θ, the difficulty of item j corresponds to the threshold point, γ_j, on the scale of Y_j, and $\pi_j = \Phi(-\gamma_j)$. Then,

$$\gamma_j = a_j b_j / \sqrt{1 + a_j^2}, \tag{14}$$

and $b_j = \Phi^{-1}(\pi_j)/\rho_j$ based on a sample value of π_j is a consistent estimator of the item location. It can be used as a starting value in the iterative estimation of item parameters (see Section 4). For descriptive purposes b_j is preferable to γ_j for characterizing item difficulty because it allows for variation in item discriminating power. In the context of IRT b_j is called "difficulty" and π_j is referred to as the *marginal difficulty* of the item.

A consistent estimator of item discriminating power is available from a one-dimensional factor analysis of the inter-item tetrachoric correlation matrix. From the sample factor loading α_j, the slope parameter is obtained from

$$a_j = \alpha_j / \sqrt{1 - \alpha_j^2}. \tag{15}$$

Computationally, however, this method of obtaining a starting value for iterative estimation of a_j is awkward: a value based on the biserial correlation between the item score and the number-right test score is almost always used in its place.

For numerical work with the normal ogive model, values of the distribution function can be calculated with accuracy 10^{-6} by Hastings' (1955) five-term formula or accuracy 10^{-11} by Cooper's (1968) algorithm.

The inverse function, $y = \Phi^{-1}(\pi)$, is difficult to approximate accurately with a computing formula; it is better obtained by Newton–Raphson iterations

$$y_{t+1} = y_t - \left[\Phi(y_t) - \pi\right]/\phi(y_t), \tag{16}$$

with values from the $\Phi(y)$ approximation. See also Kennedy and Gentle (1980).

3.1.2. 2-PL model

In three US Air Force technical reports, Birnbaum (1957, 1958a, 1958b) obtained numerous results for respondent scoring and classification in which the logistic distribution replaced the normal. (The logistic distribution was first introduced into bioassay by Fisher (Fisher and Yates, 1938), who showed its relationship to his z-transformation of the correlation coefficient.) Birnbaum's reports were not widely known until Lord and Novick (1968) included updated versions in their Chapters 17, 18 and 19. These chapters remain the most thorough analysis of the two- and three-parameter logistic models presently available in the literature. We describe some of the results here.

Birnbaum's two-parameter model is

$$P_j(\theta) = \Psi(z_j) = \frac{e^{z_j}}{1 + e^{z_j}}, \quad -\infty < z_j < +\infty, \tag{17}$$

and

$$Q_j(\theta) = 1 - \Psi(z_j) = \frac{1}{1 + e^{z_j}}.$$

The logit, $z_j = a_j(\theta - b_j) = a_j\theta + c_{jk}$, has the same structure as the normit.

The item characteristic curve of the 2-PL and normal ogive models can be made almost coincident by simple rescaling of the logit. Although the standard deviation of the logistic distribution, $\pi/\sqrt{3}$ (see Gumbel, 1961) would suggest 1.81 as the scale factor, 1.7 gives slightly better fit overall and is standard in IRT. The absolute value of $\Phi(y) - \Psi(1.7y)$ is less than 0.01 anywhere in the range (Haley, 1952). Most of the above results for the normal ogive model apply closely to the 2-PL model: for example, $\Psi^{-1}(\pi_j)/1.7\rho_j$ is very nearly a consistent estimator of b_j.

Unlike the normal distribution function, the logistic function has a simple inverse,

$$z_j = \log\{\Psi(z_j)/[1 - \Psi(z_j)]\}, \tag{18}$$

the "log odds ratio". But more important for IRT is the fact that the derivative of the logistic response function can be expressed in terms of the function values:

$$\partial\Psi(z_j)/\partial z = \psi(z_j) = P_j(\theta)Q_j(\theta). \tag{19}$$

Thus the slope at $\theta - b_j$ is $a_j/4$, and

$$\partial P_j(\theta)/\partial a_j = \theta\psi(z_j), \tag{20}$$

$$\partial P_j(\theta)/\partial c_j = \psi(z_j), \tag{21}$$

$$\partial P_j(\theta)/\partial\theta = a_j\psi(z_j). \tag{22}$$

For a test consisting of n items the likelihood equation for θ is

$$\sum_{j=1}^{n} a_j \left[\frac{u_{ij} - P_j(\theta)}{P_j(\theta)Q_j(\theta)}\right] P_j(\theta)Q_j(\theta) = \Sigma_1^n a_j\left[u_{ij} - \Psi_j(\theta)\right] = 0, \tag{23}$$

the solution of which is the scale score θ_i for respondent i. This shows that $t_i = \Sigma_1^n a_j u_j$ is the sufficient statistic for θ_i.

In application such as personnel selection and qualification, reporting educational achievement test results in population percentiles, setting scholastic proficiency standards, etc., where only the ranking or classification of respondents is at issue, t_i is a straightforward alternative to θ_i. It has the same efficiency as θ_i, is easy to compute, and is simple to explain to respondents by displaying the item weights.

The item information function is

$$I_j(\theta) = \frac{[\partial P_j(\theta)/\partial\theta)]^2}{P_j(\theta)Q_j(\theta)} = a_j^2\psi(z_j). \tag{24}$$

That these results depend upon the cancelling of the quantity $P_j(\theta)Q_j(\theta)$ in the denominator by $\psi(z_j)$ in the numerator, shows this sufficiency property to be unique to logistic item response models. A corresponding result holds for polytomous logistic models (see Section 3.2).

3.1.3. 3-PL model

To allow straightforward machine scoring of item responses, large-scale testing services make use of multiple-choice questions, most commonly with four alternatives. Usually they are administered with instructions to the examinees not to omit items. If in spite of these instructions some examinees fail to mark all items, the omitted items may be implicitly assigned a random response during test scoring. Examinees who do not know the answer to an item and do not omit will either respond blindly or choose indifferently among remaining alternatives after eliminating one or more that they know to be incorrect. Under these conditions of administration, the item response function cannot approach zero at the lowest levels of ability. Interestingly, in toxicological bioassay, a similar problem arises due to "natural mortality of the controls". The correction of the response function used in that field can also be applied to multiple-choice items except that, absent a control group, the probability of correct response on the part of examinees at the lowest extreme of ability must be estimated from the test data. This correction leads to the three-parameter logistic model of Birnbaum (see Lord and Novick, 1968, page 404):

$$P_j(\theta) = \Psi(z_j) + g_j\big[1 - \Psi(z_j)\big] \tag{25}$$

$$= g_j + (1 - g_j)\Psi(z_j), \tag{26}$$

$$Q_j(\theta) = (1 - g_j)\big[1 - \Psi(z_j)\big]. \tag{27}$$

Examinees of ability θ respond correctly with probability $\Psi_j(\theta)$; those remaining have probability g_j of responding correctly by marking blindly or with partial information. The parameter g_j is the lower asymptote of $P_j(\theta)$ and $1-g_j$, that of $Q_j(\theta)$. Raising the lower asymptote to g_j moves the inflection point of the ICC curve to half-way between g_j and 1, where probability is $(g_j + 1)/2$ and the slope is $a_j(1 - g_j)/4$.

The derivatives of the category 2 response function are

$$\partial P_j(\theta)/\partial a_j = (1 - g_j)\theta\psi(z_j), \tag{28}$$

$$\partial P_j(\theta)/\partial c_j = (1 - g_j)\psi(z_j), \tag{29}$$

$$\partial P_j(\theta)/\partial\theta = a_j(1 - g_j)\psi(z_j). \tag{30}$$

The likelihood equation for θ is

$$\sum_{j=1}^{n} a_j\psi(z_j)\frac{u_{ij} - g_j - (1 - g_j)\Psi(z_j)}{g_j[1 - \Psi(z_j)] + (1 - g_j)\Psi(z_j)[1 - \Psi(z_j)]}. \tag{31}$$

The item information function is

$$I_j(\theta) = \frac{a_j^2(1 - g_j)\psi^2(z_j)}{g_j[1 - \Psi(z_j)] + (1 - g_j)\Psi(z_j)[1 - \Psi(z_j)]}. \tag{32}$$

Note that the derivative with respect to θ in the numerator does not cancel $P_j(\theta)Q_j(\theta)$ in the denominator as it did in the 2-PL model: the 3-PL model is not logistic and does not have a sufficient statistic for θ_i.

The normal ogive and the following 1-PL model can be modified similarly to account for the effects of guessing.

The g-parameters are difficult to estimate for easy items that produce very few wrong responses and little if any guessing. Values for parameters of such items must be obtained by conditioned estimation. Zimowski et al. (1996) employ a stochastic constraint based on a beta distribution with an assigned weight of observations to determine the influence of the constraint. Thissen (1991) estimates the logit of the guessing parameter and uses a normal stochastic constraint with an assigned variance to determine the influence. (See du Toit, 2003, in this connection.)

3.1.4. 1-PL Model

Rasch (1960) introduced a version of the logistic model in which the logit is defined as $z_j = \theta - b_j$ and b_j, $j = 1, 2, \ldots, n$, is restricted to sum to zero. This way of scaling θ means that different tests will have different scales, but in practice the scores are adjusted during test development to have the same means and standard deviations in a norming sample from the population of respondents. In applying the model, the test developer must choose from the item pool those items that have the same or nearly the same slope parameters. This requires a preliminary analysis of the item pool using the normal ogive or the 2-PL model, or inspection of classical item statistics such as biserial correlation of the item-by-test number-right score.

When $a = 1$, the sufficient statistic for θ is the number-right score $t_i = \sum_{j=1}^{n} u_{ij}$, and the likelihood equation becomes

$$t_i - \sum_{j=1}^{n} \Psi(z_j) = 0. \tag{33}$$

The item information is equal for all items, and the maximum likelihood estimate of θ is a one-to-one transform of t. This greatly facilitates computation of respondent scale scores, which can be obtained by table lookup. Unlike the scale-scores based on the other models, those of 1-PL cannot produce discrepancies between the rankings of respondents' scale scores and their number-right scores. This feature is sometimes required in achievement-testing programs. Note that in applications that require only ranking of respondents, the model adds no information not already contained in the number-right scores.

3.2. Polytomous categories

The most widely applied models for polytomous item responses are (1) the *graded* models generalized from either the ogive model or the 2-PL model, (2) the *partial credit* models generalized from the 1-PL or 2-PL models, (3) the *nominal* models based on a logistic model for first choices, and (4) the *rating scale* models as special cases of (3).

Models for polytomous items can be expressed in the same manner as those for dichotomous items by the following formal device.

For an item with m_j categories, let $\mathbf{u}_j$ be an m_j-variate vector such that, if the respondent chooses category k of item j, the element $u_{jhk} = 1$ when $h = k$ and 0 when $h \neq k$. The vector is then a (multinomial) Bernoulli variate serving as the item score of the respondent, and a generic polytomous response model can be expressed as

$$P_{jk}(\theta) = P(\mathbf{u}_j \mid \theta) = \prod_{h=1}^{m_j} P_{jh}^{u_{jhk}}(\theta), \tag{34}$$

on condition that

$$\sum_{h=1}^{m_j} P_{jh}(\theta) = 1. \tag{35}$$

This device is merely for consistency of notation; in recording data and in computation, only the item scores $k = 1, 2, \ldots, m_j$ are used.

3.2.1. *Graded categories model*

Samejima (1969) formulated an item response model for graded item scores as a generalization of the normal ogive model or the 2-PL model. In place of a single threshold of the response process, Y_j, has $m - 1$ thresholds, γ_{jk}, dividing the range of the distribution into successive intervals corresponding to ordered categories. In early work of Thurstone (1929) the same concept appeared in a graphical procedure for locating stimulus medians and category boundaries on a latent variable measuring attitude. Bock and Jones (1968) gave a minimum normit chi-square method for estimating these quantities.

In Samejima's model assuming a normally distributed response process, the normit for category k of item j is

$$y_{jk} = a_j(\theta - b_{jk}), \quad k = 1, 2, \ldots, m - 1, \tag{36}$$

where b_{jk} is the location parameter for the upper boundary of category k. Letting $\gamma_{j0} = -\infty$ and $\gamma_{jm} = +\infty$, the response probabilities given θ are

$$\begin{aligned} P_{jk} &= P(\gamma_{j,k-1} \leqslant Y_j < \gamma_{jk} \mid \theta) \\ &= \Phi(y_{jk}) - \Phi(y_{j,k-1}), \quad k = 1, 2, \ldots, m_j - 1 \\ &= 1 - \Phi_{j,m_j-1}, \quad k = m_j, \end{aligned}$$

where $\Phi(y_{j,0}) = 0$.

The b-parameters divide the area under the unit normal curve into proportions $\pi_1, \pi_2, \ldots, \pi_{m_j}$. This means that neighboring categories can be collapsed in the model merely by ignoring the boundary between them.

In practical work, it proves helpful to express the normit in the form

$$y_{jk} = a_j(\theta - b_j + d_{jk}), \qquad \sum_{k=1}^{m_j-1} d_{jk} = 0. \tag{37}$$

This makes b_j an indicator of the overall location of the item on the latent dimension and expresses the category locations as deviations d_{jk}. For computational purposes, however, the parameterization

$$y_{jk} = a_j\theta + c_{jk} \tag{38}$$

with $c_{jk} = -a_j(b_j - d_{jk})$ is more convenient.

In terms of the above Bernoulli variate, the distribution of $\mathbf{u}_j$ is

$$f(\mathbf{u}_j \mid \theta) = \prod_{h=1}^{m_j} P_{jh}^{u_{jhk}}(\theta). \tag{39}$$

The graded model has m_j ICC curves consisting of:

for $k = 1$, a descending ogive centered at b_{j1};
for $k = 2$ through $m - 2$, a single-mode symmetric curve going to zero at $-\infty$ and $+\infty$, with median midway between $b_{j,k-1}$ and b_{jk};
for $k = m$, an ascending ogive centered at $b_{j,m-1}$.

The derivatives of the corresponding curves are:

$$\partial P_{jk}(\theta)/\partial a_j = \theta\big[\phi(y_{jk}) - \phi(y_{j,k-1})\big], \tag{40}$$

$$\partial P_{jk}(\theta)/\partial c_{jk} = \phi(y_{jk}) - \phi(y_{j,k+1}), \tag{41}$$

$$\partial P_{jk}(\theta)/\partial \theta = a_j\big[\phi(y_{jk}) - \phi(y_{j,k-1})\big], \tag{42}$$

where $\phi(y_{j,0}) = 0$ and $\phi(y_{j,m_j}) = 0$.

All category response functions have the same discriminating power and the intervals marked out by the category boundaries determine the partition of the response probability among the categories at a given θ.

The item information function is

$$I_j(\theta) = a_j^2 \sum_{k=1}^{m_j} \frac{[\phi(y_{jk}) - \phi(y_{j,k-1})]^2}{\Phi(y_{jk}) - \Phi(y_{j,k-1})}. \tag{43}$$

However, the information conveyed by response in a given category depends upon the expected frequency of use of the category by respondents of ability θ. The overall information can therefore be partitioned into the separate terms,

$$I_{jk}(\theta) = \frac{a_j^2[\phi(y_{jk}) - \phi(y_{j,k-1})]^2}{\Phi(y_{jk}) - \Phi(y_{j,k-1})}. \tag{44}$$

These results apply *mutatis mutandis* to the logistic version of the graded model. But the derivative of the category response function with respect to θ does not cancel the denominator in the likelihood equation, and there is not a sufficient statistic for θ.

3.2.2. The nominal categories model

When there is no *a priori* order of the response categories, an obvious approach to modelling the choice of one alternative out m_j would be to generalize the response process distribution of the normal ogive model to a multivariate normal distribution. Suppose that associated with the response categories there are processes Y_h, $h = 1, 2, \ldots, m_j$ such that the respondent will choose alternative k if $Y_k > Y_h$, $h \neq k$ (or equivalently $Y_k - Y_h > 0$). Then if the distribution of Y_k and Y_h is, say, $N(\mu_k, \mu_h, \sigma_k, \sigma_h, \rho_{kh})$, the distribution of $Y_k - Y_h$ is normal with, say, mean $\mu_{kh} = \mu_h - \mu_k$ and variance $\sigma_{kh}^2 = \sigma_k^2 + \sigma_h^2 - 2\sigma_k\sigma_h\rho_{kh}$. In terms of the process differences, the problem of evaluating the probability of first choice of one m_j alternatives reduces to evaluating the positive orthant of an $(m_j - 1)$-variate normal distribution.

Consider the case of three alternatives and corresponding Y_h, Y_j, Y_k: the correlation of $Y_k - Y_h$ and $Y_k > Y_j$ is then

$$\rho_{kh,kj} = \sigma_k^2 - \sigma_k\sigma_h\rho_{kh} - \sigma_k\sigma_j\rho_{kj} - \sigma_h\sigma_j\rho_{hj}, \tag{45}$$

and the probability of a first choice of alternative k is

$$\begin{aligned} P(Y_k > Y_h \cup Y_k > Y_j) &= \int_0^\infty \int_0^\infty \phi(\mu_{kh}, \mu_{kj}, \sigma_{kh}, \sigma_{kj}, \rho_{kh,kj})\, \mathrm{d}v\, \mathrm{d}w \\ &= \int_{-\mu_{kh}/\sigma_{kh}}^\infty \int_{-\mu_{kj}/\sigma_{kj}}^\infty \phi(0, 0, 1, 1, \rho_{kh,kj})\, \mathrm{d}x\, \mathrm{d}y. \end{aligned}$$

If the separate response processes are uncorrelated, the correlation of the differences equals 1/2. In that special case the above $(m_j - 1)$-dimensional integrals are not difficult to evaluate numerically (Stuart, 1958), but there is another approach, based on a generalization of the two-category logistic model, that is more attractive for computation and for other reasons.

The two-category model may be thought of arising from a primitive choice model

$$\frac{\pi_1}{\pi_1 + \pi_2}, \tag{46}$$

for positive π_1 and π_2, giving the probability that alternative 1 will be chosen from alternatives 1 and 2 (Bradley and Terry, 1952). Bock (1970) extended this model to the choice of alternative k from among m_j,

$$\frac{\pi_k}{\pi_1 + \pi_2 + \cdots + \pi_{m_j}}. \tag{47}$$

To avoid inequality constraints in estimation, let $\pi_h = e^{z_h}$, $h = 1, 2, \ldots, m_j$ so that

$$\Psi(z_k) = \frac{e^{z_k}}{\sum_{h=1}^{m_j} e^{z_h}}, \tag{48}$$

where the vector $\mathbf{z} = [z_h]$ may be called the *multinomial* logit.

MacFadden (1974) derived the multinomial logit model by assuming independent Weibull error terms.

Gumbel (1961) investigated in some detail the properties of the bivariate logistic distribution expressed as,

$$\Psi(z_1, z_2) = \left(1 + e^{-z_1} + e^{-z_2}\right)^{-1}. \tag{49}$$

In particular, he showed that the marginal distributions are logistic with mean 0 and variance $\pi/3$, and the joint density function is

$$\psi(z_1, z_2) = 2\Psi^3(z_1, z_2)e^{-z_1-z_2}. \tag{50}$$

The expectation of $z_1 z_2$ is $\pi^2/6$, and the correlation is

$$\rho = \frac{\pi^2}{6}\frac{3}{\pi^2} = \frac{1}{2}. \tag{51}$$

The density is asymmetric and has curved regression lines that intersect at 0.5413..., 0.5413...) rather than at (0, 0).

Nevertheless, the volume of the positive quadrant is in good approximation to that of the bivariate normal distribution with correlation 1/2. Bock and Jones (1968) applied both models to prediction of first choices in a food preference study conducted earlier by Thurstone (1959). In choice of three dishes in four different menus, the average absolute difference between the normal and logistic probabilities was 0.025.

In the nominal categories item response model (Bock, 1972), the elements of the multinomial logit are expressed as $z_{jl} = a_{jl}\theta + c_{jl}$, and the probability of response in category k of item j is

$$P_{jk}(\theta) = \Psi(z_{jk}) = \frac{e^{z_{jk}}}{\sum_{\ell=1}^{m_j} e^{z_{j\ell}}}. \tag{52}$$

Because the probabilities are invariant with respect to translation of the logit, the category parameters are subject to the constraints,

$$\sum_{k=1}^{m_j} a_{jk} = 0, \qquad \sum_{k=1}^{m_j} c_{jk} = 0. \tag{53}$$

Alternatively, the parameters may be expressed as functions of $m_j - 1$ linearly independent contrasts, or a value of a_{jk} and c_{jk} may be fixed at zero.

The distribution function of the vector-Bernoulli variate can be expressed as

$$f(\mathbf{u}_j \mid \theta) = \prod_{k=1}^{m_j} \Psi^{u_{jk}}. \tag{54}$$

To obtain the derivatives for category response function h, let D equal the denominator of (52):

$$\frac{\partial P_{jh}(\theta)}{\partial a_{jk}} = \theta\frac{e^{z_{jh}}D - e^{z_{jh}}}{D^2} = \theta\Psi(z_{jh})\big[1 - \Psi(z_{jh})\big], \quad h = k, \tag{55}$$

$$= \frac{(e^{z_{jh}})^2}{D^2} = -\theta\Psi^2(z_{jh}), \quad h \neq k, \tag{56}$$

$$\frac{\partial P_{jh}(\theta)}{\partial c_{jk}} = \Psi(z_{jh})\big[1 - \Psi(z_{jh})\big], \quad h = k, \tag{57}$$

$$= -\Psi^2(z_{jh}), \quad h \neq k, \tag{58}$$

$$\frac{\partial P_{jh}(\theta)}{\partial\theta} = \frac{a_{jh}e^{z_{jh}}D - e^{z_{jh}}\sum_{\ell=1}^{m_j} a_{j\ell}e^{z_{j\ell}}}{D^2} \tag{59}$$

$$= \Psi(z_{jh})\left[a_{jh} - \sum_{\ell=1}^{m_j} a_{j\ell}\Psi(z_{j\ell})\right]. \tag{60}$$

To examine the item characteristic curve for category k it is useful to number in the categories in the order of increasing size of the a-parameters and express the response function as

$$P_{jk}(\theta) = \left[1 + \sum_{\ell=1}^{m_j} e^{z_{j\ell}-z_{jk}}\right]^{-1}. \tag{61}$$

The derivative with respect to θ reveals the shapes of the category characteristic curves. Consider the case of three categories. In category 1,

$$\frac{\partial P_{j1}(\theta)}{\partial\theta} = -\Psi^2(z_{j1})\left[(a_{j2} - a_{j1})e^{z_{j2}-z_{j1}} + (a_{j3} - a_{j1})e^{z_{j3}-z_{j1}}\right], \tag{62}$$

both $a_{j2} - a_{j1}$ and $a_{j3} - a_{j1}$ are positive and $P_{j1}(\theta)$ is strictly decreasing from 1 to 0 with increasing θ.

In the corresponding derivative for category 2, $a_{j1} - a_{j2}$ and $a_{j3} - a_{j2}$ are negative and positive respectively, and the left and right terms are decreasing and increasing in θ. The curve for $P_{j2}(\theta)$ goes to zero at $\pm\infty$ and has a single mode at the point on the θ continuum where

$$\frac{|a_{j3} - a_{j2}|e^{z_{j3}}}{|a_{j3} - a_{j2}|e^{z_{j1}}} = 1;$$

that is, where

$$\theta = \frac{\ln(|a_{j3} - a_{j2}|) - \ln(|a_{j1} - a_{j2}|) + c_{j3} - c_{j1}}{a_{j3} - a_{j2}}.$$

Note that when the a values are equally spaced, the mode is at

$$\theta = \frac{c_{j3} - c_{j1}}{a_{j3} - a_{j2}}.$$

Finally, in the derivative for category 3 both $a_{j1} - a_{j3}$ and $a_{j2} - a_{j3}$ are negative, and $P_{j3}(\theta)$ is strictly increasing from 0 to 1 with increasing θ. A similar pattern is seen with greater numbers of categories: the curves with respect to all intermediate values of the a-parameter are unimodal.

Similar relationships between the category characteristic curves and their item parameters have been noted previously (Samejima, 1972; Andrich, 1978). Wainer et al. (1991) observed that increasing θ implies greater probability of response in a higher category rather than lower if and only if the a-parameter of the higher category is greater than that of the lower. In other words, when $\theta_1 > \theta_2$ and $k > h$, the odds ratio is

$$\frac{P_k(\theta_1)/P_h(\theta_1)}{P_k(\theta_2)/P_h(\theta_2)} > 1,$$

and the log odds simplify to

$$a_k(\theta_1 - \theta_2) - a_h(\theta_1 - \theta_2) > 0,$$

which implies $a_k > a_h$.

Similarly, Andrich (1978) studied the interpretation of the parameters by examining the log odds ratio of successive categories at the points where their characteristic curves cross – that is, where the respondent is indifferent in choosing between those categories:

$$\begin{aligned} &\log \frac{P_{jk}(\theta)}{P_{j,k-1}(\theta)} = (a_{jk} - a_{j,k-1})\theta + c_{jk} - c_{j,k-1} = 0, \\ &\theta = b_{jk} = -\frac{c_{jk} - c_{j,k-1}}{a_{jk} - a_{j,k-1}}. \end{aligned} \tag{63}$$

Although it is plausible that respondents' ability to discriminate between successive categories may differ by category, if discriminating power is equal for all categories, then the differences between the a-parameters will be equal and can be set to a_j, say. In that special case, the b-parameters, which are the inner category boundaries, will correspond to the θ-values that determine the partition of response probabilities among the categories, just as they do in the graded model. In addition, the distance between b_{j1} and b_{j,m_j-1} will be inversely proportional to a_j, and the distances are additive; neighboring categories can then be collapsed just as in the graded model.

More generally, when the spacing of the slopes is unequal, the categories with higher slopes compared to preceding category receive a smaller share of the probability than those with lower slopes.

The likelihood in equation for θ given the item score $\mathbf{u}_{ij}$ of respondent i is

$$\sum_j^n \sum_k^{m_j} \frac{u_{ijk}}{P_{jk}(\theta)} \frac{\partial P_{jk}(\theta)}{\partial \theta} = \sum_j^n \sum_k^{m_j} u_{ijk} a_{jk}\big[1 - \Psi_{jk}(\theta)\big] = 0. \tag{64}$$

The weighted sum of the item scores

$$S_i = \sum_j^n \sum_k^{m_j} a_{jk} u_{ijk} \tag{65}$$

is the sufficient statistic for θ_i just as is in 2-PL and can be similarly used in applications where ranking of the respondents is required.

The item information function is

$$I_j(\theta) = \sum_k^{m_j} a_{jk}^2 \Psi_{jk}(\theta) - \sum_k^{m_j} \sum_h^{m_j} a_{jh} a_{jk} \Psi_{jh}(\theta) \Psi_{jk}(\theta). \tag{66}$$

Other parameterizations of the nominal model that greatly extend its versatility and can be expressed in general terms based on results in Chapter 8 of Bock (1975) for estimation of multinomial response relations. Let the parameterization of the transposed $m_j \times 1$ logit vector be $\mathbf{z}_j$ with elements z_{jk} be

$$\mathbf{z}_j' = \mathbf{K}\mathbf{B}_j\mathbf{A}_j, \tag{67}$$

where

$$\mathbf{K} = [\theta \quad 1],$$

$$\mathbf{B}_j = \begin{bmatrix} a_{j1} & a_{j2} & \cdots & a_{jm_j} \\ c_{j1} & c_{j2} & \cdots & c_{jm_j} \end{bmatrix},$$

and $\mathbf{A}_j$ is the $m_j \times m_j$ projection operator that implements the restriction on the parameters; the diagonal elements of the operator are $1 - 1/m_j$ and the off-diagonal elements are $-1/m_j$. The operator can be replaced by

$$\mathbf{A}_j = \mathbf{S}_j \mathbf{T}_j, \tag{68}$$

where $\mathbf{S}_j$ is an $m_j \times s$, $s \leqslant m_j - 1$ matrix of the coefficients of the linear parametric functions for item j. Then

$$\mathbf{T}_j = (\mathbf{S}_j' \mathbf{S}_j)^{-1} \mathbf{S}_j' \mathbf{A}_j. \tag{69}$$

If s is less than $m_j - 1$ there will be some loss of information in estimating θ, but the loss may be negligible if the parameters are structured favorably.

The quantities estimated after reparameterization are say,

$$\begin{bmatrix} \eta_j \\ \beta_j \end{bmatrix} = \mathbf{B}_j \mathbf{S}_j. \tag{70}$$

The following examples of linear parametric functions are of some interest. They all result in models with no more than $2m_j - 1$ free parameters.

Simple contrasts. The default in most applications is

$$\begin{bmatrix} -1 & -1 & -1 \\ 1 & 0 & 0 \\ 0 & 1 & 0 \\ 0 & 0 & 1 \end{bmatrix}.$$

Parameter trend over categories. Three examination questions pertaining to a reading passage are each scored right-wrong; the number right scorers from 0 to 3 are treated as ordered categories. A degree 2 polynomial trend is observed in the category parameters (see Wainer et al., 1991). The **S**-matrix contains the (un-normalized) orthogonal polynomials:

$$\begin{bmatrix} -3 & 1 \\ -1 & -1 \\ 1 & -1 \\ 3 & 1 \end{bmatrix}.$$

Crossed categories. Responses to an essay question are marked pass-fail and pass-fail for content. The matrix is the basis of a 2×2 analysis-of-variance model, excluding the constant term:

$$\begin{bmatrix} 1 & 1 & 1 \\ 1 & -1 & -1 \\ -1 & 1 & -1 \\ -1 & -1 & 1 \end{bmatrix}.$$

Ordered categories. The categories have a prior ordering. The parametric functions force the ordering of the category parameters; the parameters for category 1 are set to zero:

$$\begin{bmatrix} 0 & 0 & 0 \\ 1 & 1 & 1 \\ 0 & 1 & 1 \\ 0 & 0 & 1 \end{bmatrix}.$$

For further examples, see Thissen and Steinberg (1986). Matrices for parameterizations 1 and 3 are incorporated in the MULTILOG program of Thissen (1991).

The likelihood equation for estimating θ reduces to

$$\sum_{j=1}^{n} \boldsymbol{\beta}'_j \mathbf{T}_j \big[\mathbf{u}_j - \Psi(\mathbf{z}_j)\big]. \tag{71}$$

Note that the sufficient statistic for θ is the sum of the vector-Bernoulli variates premultiplied by $\boldsymbol{\beta}'_j \mathbf{T}_j$. The item information function is

$$I_j(\theta) = \boldsymbol{\beta}'_j \mathbf{T}_j \mathbf{W}_j \mathbf{T}'_j \boldsymbol{\beta}_j, \tag{72}$$

where the $m_j \times m_j$ matrix $\mathbf{W}_j$, with diagonal elements $\Psi(z_{jk})[1 - \Psi(z_{jk})]$ and off-diagonal elements $-\Psi(z_{jh}\Psi(z_{jk}))$, is the covariance matrix of the vector-Bernoulli variate. The second derivative of the log likelihood is exactly the negative of the information function and does not involve the data. In as much as logistic distributions belong to the exponential family, the likelihood surface is everywhere convex and any finite solution of the likelihood equations is unique and can easily be found numerically by Newton–Raphson or EM iterations (see Section 4).

3.2.3. *Nominal multiple-choice model*

An interesting application of the nominal model is recovery of information in wrong responses to multiple-choice items. The item characteristic curves of the model identify the productive wrong alternatives and the range of data values in which they are most effective. The problem arises however, of how to allow for effects of guessing when the test is administered with instructions not to omit items. Samejima (1979) suggested including a *virtual* response category, labelled 0, representing the hypothetical proportion of respondents who do not know the correct answer. She assumed random guessing with probability of correct response $1/m_j$ and $z_{jh} = a_{jh}\theta + c_{jh}$ and wrote the model as, say,

$$P_{jk}(\theta) = \frac{e^{z_{jk}} + m_j^{-1} e^{z_{j0}}}{\sum_{\ell=0}^{m_j} e^{z_{j\ell}}}, \quad k = 1, 2, \ldots, m_j, \tag{73}$$

under the restrictions $\sum_{\ell=0}^{m_j} a_{j\ell} = 0$ and $\sum_{\ell=0}^{m_j} c_{j\ell} = 0$.

But guessing on multiple-choice items is not likely to be entirely random: experience with the 3-PL model indicates that many wrong responders usually rule out a least one of the alternatives before answering. Thissen and Steinberg (1984) therefore substituted a category specific parameter, $g_{jk} \geqslant 0$, $\sum_{k=1}^{m_j} g_{jk} = 1$. To avoid inequality constraints they made a second use of the multinomial logit to express g_{jk} as a function of, say, δ_{jk}:

$$g_{jk} = \frac{e^{\delta_{jk}}}{\sum_{\ell=1}^{m_j} e^{\delta_{j\ell}}}, \tag{74}$$

under the restriction $\sum_{\ell=1}^{m_j} \delta_{j\ell} = 0$.

This generalization of the nominal model has $3m_j - 1$ free parameters.

Setting $D_j = e^{z_{j0}} + \sum_{\ell=1}^{m_j} e^{z_{j\ell}}$ and $E_j = e^{z_{jh}} + g_{jh} e^{z_{j0}}$, the derivatives of category h required in estimation of the item parameters and θ can be expressed as:

$$\frac{\partial P_{jh}(\theta)}{\partial a_{j0}} = \theta e^{z_{j0}} [g_{jh} D_j - E_j] D_j^{-2}, \tag{75}$$

$$\frac{\partial P_{jh}(\theta)}{\partial a_{jk}} = \theta e^{z_{jk}} [D_j - E_j] D_j^{-2}, \quad h = k, \tag{76}$$

$$= -\theta e^{z_{jk}} D_j^{-1}, \quad h \neq k, \tag{77}$$

$$\frac{\partial P_{jh}(\theta)}{\partial c_{j0}} = e^{z_{j0}} [D_j - E_j] D_j^{-2}, \tag{78}$$

$$\frac{\partial P_{jh}(\theta)}{\partial c_{jk}} = e^{z_{jk}} [D_j - E_j] D_j^{-2}, \quad h = k, \tag{79}$$

$$= -e^{z_{jk}} D^{-1}, \quad h \neq k, \tag{80}$$

$$\frac{\partial P_{jh}(\theta)}{\partial \delta} = g_{jh}(1 - g_{jh}) e^{z_{j0}} D_j^{-1}, \quad h = k, \tag{81}$$

$$= -g_{jh} g_{jk} e^{z_{j0}} D_j^{-1}, \quad h \neq k, \tag{82}$$

$$\frac{\partial P_{jh}(\theta)}{\partial \theta} = \left[\left(a_{jk} e^{z_{jh}} + g_{jh} a_{j0} e^{z_{j0}}\right) D_j - E_j \left(a_{j0} e^{z_{j0}} + \sum_{l=1}^{m_j} a_{jl} e^{z_{jl}} \right) \right] D_j^{-2}. \tag{83}$$

These results do not simplify further for substitution in the general expressions in Section 5.1 for the likelihood equation for θ and corresponding information function. As with the 3-PL model, estimation of the g-parameters may require conditioning.

3.2.4. *Partial credit model*

Masters (1982) adapted the rating-scale model proposed by Rasch (1961) and elaborated by Andersen (1977) and Andrich (1978) to apply to ordered categories scored $v_j = 1, 2, \ldots, m_j$ for item j. (In the social science literature this type of category score goes by the name of "Likert scores"; Likert, 1932.) Called the *partial credit* model, Master's version is based on a concept of the respondent processing the item task in successive steps with probability of π_{jk} of success in making the step from k to $k + 1$, where $\pi_{j1} = 1$. Step m_j or the first failure terminates the process; the Likert scores count the steps. Generalizing the 1-PL model by defining the logit as $z_{jk} = \theta - b_{jk}$, with $b_{j1} = 0$, and norming the step probabilities to 1 leads to the following model in $(m_j - 1)$ free item parameters:

$$P_{jk}(\theta) = \Psi(z_{jk}) = \frac{\exp \sum_{\ell=1}^{v_j} z_{j\ell}}{\sum_{h=1}^{m_j} \exp[\sum_{\ell=1}^{h} z_{j\ell}]}, \quad k = v_j. \tag{84}$$

The derivatives for category h of item j are (see Masters, 1982):

$$\frac{\partial P_{jh}(\theta)}{\partial b_{jk}} = -\Psi(z_{jk})\left[1 - \sum_{\ell=k}^{m_j} \Psi(z_{j\ell})\right], \tag{85}$$

$$\frac{\partial P_{jh}(\theta)}{\partial \theta} = -\Psi(z_{jk})\left[1 - \sum_{\ell=k}^{m_j} \ell\Psi(z_{j\ell})\right]. \tag{86}$$

The likelihood equation for estimating, θ_i, the location of respondent i on the latent dimension, given the item scores v_{ij}, is

$$\sum_{j=1}^{n}\left[v_{ijk} - \sum_{\ell=k}^{m_j} \ell\Psi(z_{j\ell})\right] = 0, \tag{87}$$

which shows that the total step count for the respondent to be a sufficient statistic for θ_i. This means that, under the assumptions of the model, the mean Likert score as the same efficiency as the estimated scale score for purposes of ranking respondents.

The corresponding information function is

$$I(\theta) = \sum_{k=1}^{m_j} k^2\Psi(z_{jk}) - \left[\sum_{k=1}^{m_j} k\Psi(z_{jk})\right]^2. \tag{88}$$

As parameterized above, the partial credit model assumes that the unit of scale on the latent dimension is equal to 1. In testing applications, however, it may be more convenient to fix the unit by assuming that the standard deviation of the latent distribution equals 1. In that case the logit may be expressed as $z_{jk} = a(\theta - b_{jk})$, and the common slope, a, can be estimated along with the category parameters.

3.2.5. *Generalized partial credit model*

Muraki (1992) observed that the partial credit model may be generalized to items with varying discriminating powers simply by defining the logit as $z_{jk} = a_j(\theta - b_{jk})$. He also noted that the exponent of the general model can be expressed for category k as

$$ka_j\theta + c_{jk}, \tag{89}$$

where $c_{jk} = \sum_{\ell=1}^{m_j} a_j b_{jk}$, which shows the model to be a special case of the nominal model. The slope parameters in the partial credit model are assigned equally spaced values in the given order of the categories, whereas in Nominal model they are estimated from the sample data. This is also the special case, noted above, of ordered a-parameters with equal differences between successive values that simplifies interpretation of the item parameters and characteristic curves of the nominal model.

The sufficient statistic for θ in the generalized model is the sum of the item step counts weighted by a_j. The model may be formulated in terms of vector-Bernoulli variates rather than Likert scores, such that with the necessary changes to the a-parameters the results for the nominal model apply.

3.2.6. *Rating scale models*

Multiple item rating scales are widely used in the measurement of respondents' attitudes, opinions, personality traits, health status, and similar personal characteristics. All items have the same format – usually a fixed number of ordered categories labelled in the same way throughout the instrument. In modelling rating scale responses it is assumed that a respondent's interpretation of the labels and use of the categories is not influenced by the content of the items. The content itself, however, may vary in the strength of its relationship to the latent variable as reflected in the power of a given item to discriminate among respondents.

These assumptions suggest a response model for, say, m ordered categories with common category parameters and varying discriminating powers. (The assumption that the strength of relationship is the same for all items is not plausible in typical applications of rating scales). Among the item response models described above, these assumptions limit the choice to the graded and generalized partial credit models, with the added requirement that the category parameters are constrained to equality over items.

For the graded model, the normit is therefore,

$$y_{jk} = a_j(\theta - b_j + d_k), \quad k = 1, 2, \ldots, m - 1, \tag{90}$$

under the restriction $\sum_{k=1}^{m-1} d_k = 0$.

For the generalized partial credit model expressed as a special case of the nominal model, the logit is

$$z_{jk} = a_j k\theta - b_j + d_k, \quad k = 1, 2, \ldots, m, \tag{91}$$

under the restriction $\sum_{k=1}^{m} d_k = 0$.

Thus, for an n item scale the rating-scale assumptions reduce the number of free item parameters in these models from nm to $2n + m - 1$. The restriction to common category parameters is easily implemented during estimation of item parameters (see Section 4). The graded and nominal models with varying category parameters may, of course, still be used in preliminary studies to verify the stronger assumptions of the rating-scale model. The graded and generalized partial credit models both in their original and rating-scale form are implemented in the computer program of Muraki and Bock (1991).

3.3. *Ranking model*

Ranking is of particular interest in measurement of preferences for objects of choice – such as political candidates, proposed public policies, occupational goals, or competing consumer goods. When studying preferences there is an important advantage to asking the respondents to rank rather than rate the objects: ranking avoids undesirable effects of respondents' idiosyncratic interpretations of category labels such as "agree moderately", "agree strongly", etc., found in rating forms. In an IRT context the set of objects to be ranked constitute an item, and the objects in the set take the place of categories. For example, in measurement respondents' conservatism, the objects might be proposed alternative solutions to various social problems serving as items. The ranking model can be used to estimate the probability that a given object will be first, second, or any lower choice in the preference rankings. The nominal model is well-suited for this type of analysis because of its computational simplicity.

Böckenholt (2001) used the Luce (1959) principle of independence of first choices from irrelevant alternatives to decompose a rank-ordering of objects into a succession of first choices among successively smaller sets of objects. The principle implies that the probabilities of the first choices are independent and that the probability of the ranking as a whole is the continued product of those probabilities. To express the conditional probability of a respondent's ranking as a function of θ, let H_j be a set containing integers $h_j = 1, 2, \ldots, m_j$ indexing the objects in item j, $j = 1, 2, \ldots, n$. Let K_{ij} be an ordered subset of elements of H_j corresponding to the q objects ranked by respondent i, $q = 1, 2, \ldots, m_j - 1$. Let $k_{ij}^{(\ell)}$, $\ell = 1, 2, \ldots, q$, be an element in K_{ij} corresponding to the object in position ℓ in the rank ordering, and let $\bar{H}_{ij\ell}$ be the subset of $1 + m_j - \ell$ elements of H_j *excluding* elements corresponding to objects above $k_{ij}^{(\ell)}$ in the rank ordering (note that elements of $\bar{H}_{ij\ell}$ not among the elements of K_{ij} need not be ordered).

Then the probability of K_{ij} is

$$P_j(K_{ij} \mid \theta) = \prod_{\ell=1}^{q} \frac{\exp[z(k_{ij\ell})]}{\sum_{h \in \bar{H}_{ij\ell}} \exp z(h)}, \tag{92}$$

where

$$z(h \in \bar{H}_{ij\ell}) = a_{jh}\theta + c_{jh}, \tag{93}$$

is the logit of the nominal model for the probability of each successive first choice.

Provided there is individual variation in the rankings, the item parameters can be estimated in large samples of respondents for even a single item, the characteristic curves corresponding to objects may then be of interest. The multi-stage structure of (92) imposes constraints on the collection of ranking data. Ideally, rankings should be obtained in the form of successive first choices by requiring respondents to rank the objects from most to least preferred. Deviations from this format that are not consistent with the multistage structure of the model may lead to systematic misfits.

4. Estimation of item and group parameters

Although Lord (1953), Birnbaum (1957) and Samejima (1969) made important progress in IRT estimation of respondent scale scores, application of their results was hampered by the absence of a satisfactory method of estimating the item parameters. Lord proposed maximum likelihood estimation jointly with the scale scores, but that ran into "the infinitely many incidental parameters" problem (Neyman and Scott, 1948; Kiefer and Wolfowitz, 1956). Rasch (1960) avoided the problem for the 1-PL model by conditioning the likelihood on the respondent's test score to factor out the latent variable. That approach had limited applicability, however, because it required the highly restrictive assumption that all items have the same discriminating power.

Bock and Lieberman (1970) took the more general approach of maximizing the likelihood in the marginal distribution after integrating numerically over the latent distribution. But their Newton solution of the likelihood equations required inversion of a information matrix of the order of the number of parameters, which for large numbers of items was too computationally intensive for the computing equipment of that time. Finally, Bock and Aitkin (1981) introduced an EM solution that cycled through the items one at a time, was extremely robust, and quite tractable computationally. Bock and Zimowski (1997) extended this procedure to multiple group IRT. These procedures are now in general use. In this section we discuss features of both the Newton and EM solutions in maximum marginal likelihood estimation for the general IRT model.

For ease of exposition we defer the discussion of models with covariate effects to Section 7. Maximization of the marginal log-likelihood given in (5) may be obtained by using an optimization algorithm such as the Newton–Raphson or the E-M algorithm. Alternatively, one can estimate the parameters of item response models using Bayesian estimation methods. A review and implementation of Markov Chain Monte Carlo (MCMC) methods for a variety of IRT models can be found in two papers by Patz and Junker in 1999a, 1999b.

Let us define the item parameters (thresholds and slope) for an item j in group v as the vector $\mathbf{a}_{vj}$ with elements α_{vj} $(c_{vj(0)}, \ldots, c_{v,j,(m_j-1)})$, where $j = 1, \ldots, n$. The parameters related to the latent variable θ are denoted by the vector $\mathbf{b}$ with elements μ_v, σ_v^2, $v = 1, 2, \ldots, V$, where and μ_v and σ_v^2 are the mean and variance respectively of the latent variable in group v. To resolve the indeterminacy of location and scale, the mean and variance of one of the groups may be set arbitrarily. The item parameters are constrained to be equal across groups except in the studies of differential item function mentioned above. In naturally occurring groups and randomized groups receiving different treatments differences between the group parameters are to be expected and will be estimated.

Newton–Raphson

The integration for obtaining the marginal log-likelihood given in (5) is performed numerically by fixed-point quadrature rules or by variable-point adaptive quadrature. These methods have the advantage of not requiring a functional form for the latent distribution: with sufficient number of equally distant points in an open interval, normalized point masses on those points are a suitable representation of the distribution. However, if the latent distribution is assumed normal, Gauss–Hermite quadrature is appropriate. Adaptive quadrature (see Skrondal and Rabe-Hesketh, 2004, pages 165–170) and other methods are discussed below.

Maximization of the numerically integrated likelihood (5) requires the gradient vector and the information matrix with respect to the parameter stack

$$\boldsymbol{\zeta} = \begin{bmatrix} \mathbf{a} \\ \mathbf{b} \end{bmatrix}.$$

The gradient vector is

$$\frac{\partial L}{\partial \boldsymbol{\zeta}} = \sum_{v=1}^{V} \sum_{i=1}^{N_v} \frac{1}{f_v(\mathbf{u}_i)} \frac{\partial f_v(\mathbf{u}_i)}{\partial \boldsymbol{\zeta}}, \tag{94}$$

and the Hessian matrix is

$$\frac{\partial^2 L}{\partial \boldsymbol{\zeta} \partial \boldsymbol{\zeta}'} = \sum_{v=1}^{V} \sum_{i=1}^{N_v} \left[-\frac{1}{f_v(\mathbf{u}_i)^2} \left(\frac{\partial f_v(\mathbf{u}_i)}{\partial \boldsymbol{\zeta}} \right)^2 + \frac{1}{f_v(\mathbf{u}_i)} \frac{\partial^2 f_v(\mathbf{u}_i)}{\partial \boldsymbol{\zeta} \partial \boldsymbol{\zeta}'} \right]. \tag{95}$$

Taking the expected value of the Hessian matrix and under certain regularity conditions, one obtains the information matrix:

$$\mathcal{E}\left[-\frac{\partial^2 L}{\partial \boldsymbol{\zeta} \partial \boldsymbol{\zeta}'} \right] = \mathcal{E}\left[\left(\frac{\partial \log f_v(\mathbf{u}_i)}{\partial \boldsymbol{\zeta}} \right) \left(\frac{\partial \log f_v(\mathbf{u}_i)}{\partial \boldsymbol{\zeta}'} \right) \right]. \tag{96}$$

Let $\boldsymbol{\zeta}^{(0)}$ represent a set of starting values. A Fisher scoring type minimization algorithm generates successive points $\boldsymbol{\zeta}^{(1)}, \boldsymbol{\zeta}^{(2)}, \ldots$, in the parameter such that

$$L[\boldsymbol{\zeta}^{(s+1)}] > L[\boldsymbol{\zeta}^{(s)}].$$

For $s = 0, 1, 2, \ldots$, let $\mathbf{d}^{(s)}$ be the gradient vector and let $\mathbf{B}^{(s)}$ be the inverse of the information matrix at $\boldsymbol{\zeta} = \boldsymbol{\zeta}^{(s)}$. Then the algorithm is

$$\boldsymbol{\zeta}^{(s+1)} = \boldsymbol{\zeta}^{(s)} + \mathbf{B}^{(s)} \mathbf{d}^{(s)}. \tag{97}$$

The Newton–Raphson algorithm uses the inverse of the Hessian matrix whereas the Fisher scoring algorithm (Newton–Gauss) uses the inverse of the empirical information matrix, given the data, which is simply a positively weighted sum of cross-products of the first derivatives of the response functions and easy to compute; if the number of distinct answer patterns in the sample exceed the number of free-parameters and the model is of full rank, the empirical information matrix will be positive-definite. For those item response models that have sufficient statistics (see Section 3), the Hessian matrix is just the negative of the information matrix; otherwise, it includes the term involving the second-derivatives of the response functions and the residuals, and could in principle be indefinite if the sample size is not large. The empirical information matrix is to be distinguished from the theoretical information matrix, which is an expectation of all possible response patterns and is not readily calculable with large numbers of items. Even personal computers are now sufficiently fast for a Newton solution with one-dimensional models involving perhaps as many as 100 items having fewer than 10 response categories. In general, Newton methods require starting values not too far removed from the solution point. For this reason a number of EM steps to obtain improved provisional values of the parameters is a good computing strategy. It is the basis of the IRT implementations of Muraki and Bock (1991) and Zimowski et al. (1996).

EM algorithm

The EM algorithm (Dempster et al., 1977) consists of an expectation step in which the expected score function of the model parameters is computed. The expectation is with respect to the posterior distribution of θ given the observations $f(\theta \mid \mathbf{u})$. In the maximization step updated parameter estimates are obtained.

More specifically, the complete density in group v for an individual i is:

$$f_v(\mathbf{u}_i, \theta_i) = f_v(\mathbf{u}_i \mid \theta_i) g_v(\theta_i). \tag{98}$$

Taking into account the conditional independence assumption, the complete log-likelihood for a random sample of size N is:

$$L = \sum_{i=1}^{N} \log f(\mathbf{u}_i, \theta_i) = \sum_{v=1}^{V} \sum_{i=1}^{N_v} \left[\log \prod_{j=1}^{n} f_{vj}(u_{ij} \mid \theta_i) + \log \phi_v(\theta_i) \right]. \tag{99}$$

The form of $f_{vj}(u_{ij} \mid \theta_i)$ can be different for different types of items. For dichotomous items:

$$f_{vj}(u_{ij} \mid \theta_i) = P_{vj}(\theta)^{u_{ij}} Q_{vj}(\theta)^{1-u_{ij}}, \tag{100}$$

where $P_{vj}(\theta)$ takes different forms depending on the model (see section).

For polytomous items:

$$f_{vj}(u_{ij} \mid \theta_i) = \prod_{h=1}^{m} P_{vjh}^{u_{ijhk}}(\theta), \tag{101}$$

on condition that $\Sigma_{\ell=1}^{m} P_{vj\ell}(\theta) = 1$, where u_{ijhk} takes the value 1 when $h \neq k$ and 0 otherwise.

To resolve the indeterminacy of location and scale, the mean and variance of the latent variable θ in one of the groups may be set arbitrarily. The item parameters (thresholds and slopes) are constrained to be equal across groups but we are interested in estimating group differences for the latent variable.

From (99), we see that the parameter vector $\mathbf{a}_v$ is in the first component of the log-likelihood and the parameter vector $\mathbf{b}_v$ in the second component. Therefore, estimation of $\mathbf{a}_v$ can be done separately from $\mathbf{b}_v$.

The expected score function with respect to the parameter vector $\mathbf{a}_{vj}$ for an individual i and item j takes the form:

$$\mathcal{E} S_i(\mathbf{a}_{vj}) = \int S_i(\mathbf{a}_{vj}) f_v(\theta \mid \mathbf{u}_i) \, \mathrm{d}\theta, \tag{102}$$

where $f_v(\theta \mid \mathbf{u}_i)$ denotes the posterior distribution of the latent variable given what has been observed:

$$f_v(\theta \mid \mathbf{u}_i) = \frac{f_v(\mathbf{u}_i \mid \theta) g_v(\theta)}{f_v(\mathbf{u}_i)}, \tag{103}$$

and

$$S_i(\mathbf{a}_{vj}) = \frac{\partial \log f_v(u_{ij} \mid \boldsymbol{\theta})}{\partial \mathbf{a}_{vj}}.$$

Under the equality constraints of the item parameters the likelihood equations will be summed up with respect to the groups.

There are different ways of approximating the integral in (102) such as Gauss–Hermite fixed quadrature points (Bock and Aitkin, 1981), adaptive quadrature (Schilling and Bock, 2005), Monte Carlo methods (Sammel et al., 1997), and Laplace approximations (Huber et al., 2004).

The Gauss–Hermite method approximates the integral over θ by a weighted summation over a finite number of points giving:

$$\mathcal{E}S_i(\mathbf{a}_{vj}) = \sum_{t=1}^{T} S_i(\mathbf{a}_{vj}) f_v(\theta_t \mid \mathbf{u}_i). \tag{104}$$

The points for the integral approximations are the Gauss–Hermite quadrature points given in Straud and Sechrest (1966). This approximation in effect treats the latent variables as discrete with values $\theta_1, \ldots, \theta_T$ and their corresponding probabilities $\phi(\theta_1), \ldots, \phi(\theta_T)$. To avoid the computational burden as the number of factors increases we need to use a small number of points on each dimension. The implication is that we lose in precision as well as obtaining biased and unstable estimates (see Huber et al., 2004; Schilling and Bock, 2005).

The Monte Carlo approximation generates a large sample of $\theta_1, \ldots, \theta_T$ values from $f^v(\theta \mid \mathbf{u}_i)$ which does not come in a closed form. The expectation becomes

$$\mathcal{E}S_i(\mathbf{a}_{vj}) = \int S_i(\mathbf{a}_{vj}) f_v(\theta \mid \mathbf{u}_i)\,\mathrm{d}\theta = \frac{\int S_i(\mathbf{a}_{vj}) f_v(\mathbf{u}_i \mid \theta) g_v(\theta)\,\mathrm{d}\theta}{\int f_v(\mathbf{u}_i \mid \theta) g_v(\theta)\,\mathrm{d}\theta}. \tag{105}$$

Generating $\theta_1, \ldots, \theta_T$ values from $g_v(\theta)$, (105) is approximated by

$$\mathcal{E}S_i(\mathbf{a}_{vj}) = \frac{\sum_{t=1}^{T} S_i(\mathbf{a}_{vj}) f_v(\mathbf{u}_i \mid \theta_t)}{\sum_{t=1}^{T} f_v(\mathbf{u}_i \mid \theta_t)}. \tag{106}$$

Note that a large number of T is required to obtain precise estimates with the implication of slowing down the convergence of the EM algorithm. Meng and Schilling (1996) proposed to draw directly from the posterior distribution $f^v(\theta \mid \mathbf{u}_i)$ using a Gibbs sampler. They showed that the Gibbs sampler can produce nearly independent samples by retaining every fifth value of the posterior. This is known in the literature as a Monte Carlo EM (MCEM) algorithm.

In adaptive quadrature, points and weights are adapted to the location and dispersion of the likelihood surface for each response pattern during parameter estimation. Schilling and Bock (2005) show that this method of numerical integration can produce fast and accurate solutions using as few as two points on each dimension.

Let us assume that we adopt the Gauss–Hermite quadrature points for the integral approximation. For the dichotomous case, for example, (104) after summing with respect to all sample members becomes:

$$\begin{aligned}\mathcal{E}S(\mathbf{a}_{vj}) &= \sum_{t=1}^{T}\sum_{i=1}^{N_v} S_i(\mathbf{a}_{vj}) f_v(\theta_t \mid \mathbf{u}_i) \\ &= \sum_{t=1}^{T}\sum_{i=1}^{N_v} \frac{(u_{ij} - P_{vj}(\theta_t))}{P_{vj}(\theta_t)(1 - P_{vj}(\theta_t))} \frac{\partial P_{vj}(\theta_t)}{\partial \mathbf{a}_{vj}} f_v(\theta_t \mid \mathbf{u}_i) \\ &= \sum_{t=1}^{T} \frac{\partial P_{vj}(\theta_t)}{\partial \mathbf{a}_{vj}} \frac{r_{vjt} - N_{vt} P_{vj}(\theta_t)}{P_{vj}(\theta_t)(1 - P_{vj}(\theta_t))},\end{aligned} \tag{107}$$

where

$$r_{vjt} = \sum_{i=1}^{N_v} u_{vij} f_v(\theta_t \mid \mathbf{u}_i), \tag{108}$$

and

$$N_{vt} = \sum_{i=1}^{N_v} f_v(\theta_t \mid \mathbf{u}_i). \tag{109}$$

The ML equations become:

$$\mathcal{E}S(\hat{\mathbf{a}}_{vj}) = \sum_{t=1}^{T} \theta^l \big[r_{vjt} - N_{vt} P_{vj}(\theta_t)\big] = 0, \quad l = 0, 1. \tag{110}$$

Similarly, one can obtain the ML equations for the case of nominal and ordinal items.

When items are categorical, the $\mathcal{E}S(\mathbf{a}_{vj}) = 0$ does not provide a closed form solution for the elements of $\mathbf{a}_{vj}$, $j = 1, \ldots, n$. Updating estimates are obtained using one or more Newton or Fisher scoring iterations.

The estimation of the parameters in $\mathbf{b}_v$ depends on the second component of equation (99). The latent variable in each group is assumed to be normally distributed with mean μ_v and variance σ_v^2. We are interested in estimating group means and variances of the latent variable θ.

The expected score function of the parameter vector $\mathbf{b}_v$ where the expectation is taken with respect to $f_v(\theta \mid \mathbf{u}_i)$ is:

$$\mathcal{E}S_i(\mathbf{b}_v) = \int S_i(\mathbf{b}_v) f_v(\theta \mid \mathbf{u}_i)\, d\theta, \quad i = 1, \ldots, N_v, \tag{111}$$

where

$$S_i(\mathbf{b}_v) = \frac{\partial \log g_v(\theta)}{\partial \mathbf{b}_v}, \quad j = 1, \ldots, n,$$

and

$$g_v(\theta) = \frac{1}{\sigma_v^2\sqrt{2\pi}} \exp\big[-(\theta - \mu_v)^2/2\sigma_v^2\big].$$

Summing up with respect to all sample members and approximating the integral using Gauss–Hermite quadrature points we obtain the ML equations:

$$\mathcal{E}S(\mu_v) = \sum_{t=1}^{T}\sum_{i=1}^{N_v} S_i(\mu_v) f_v(\theta_t \mid \mathbf{u}_i)$$
$$= \sum_{t=1}^{T}\sum_{i=1}^{N_v} (\theta - \hat{\mu}_v) f_v(\theta_t \mid \mathbf{u}_i) = 0 \tag{112}$$

and

$$\mathcal{E}S(\sigma_v^2) = \sum_{t=1}^{T}\sum_{i=1}^{N_v} S_i(\sigma_v^2) f_v(\theta_t \mid \mathbf{u}_i)$$
$$= \sum_{t=1}^{T}\sum_{i=1}^{N_v} \left(\hat{\sigma}_v^2 - (\theta - \hat{\mu}_v)^2\right) f_v(\theta_t \mid \mathbf{u}_i) = 0. \tag{113}$$

The steps of the EM algorithm are defined as follows:

Step 1. Choose initial estimates for the model parameters $\mathbf{a}_{vj}$ and $\mathbf{b}_v$, where $j = 1, \ldots, n$.
Step 2. Compute the quantities r_{vjt} and N_{vt} (E-step).
Step 3. Obtain improved estimates for the parameters by solving the ML equations (110), (112) and (113) where the quantities r_{vjt} and N_{vt} as treated as fixed (M-step).
Step 4. Return to Step 2 and continue until convergence is attained.

At the M-step, a Newton or Fisher scoring iteration used to solve the non-linear ML equations in (110). The EM converges very slowly in the neighborhood of the maximum point but it performs well far from it (just the opposite of Newton solutions). This justifies the hybrid approach combining EM and Newton steps mentioned above.

4.1. Sampling properties

The fact that in the EM algorithm there is no need to compute the information matrix does not help to compute standard errors for the estimates. Approximations to the information matrix as well as extensions of the EM have been proposed in the literature.

Following the results from asymptotic theory for ML estimators, sampling variances and covariances for the ML estimates are taken to be the elements of the inverse of the information matrix evaluated at the ML solution. For a set of parameters $\boldsymbol{\zeta}' = (\mathbf{a}, \mathbf{b})$:

$$\left[\operatorname{cov}(\hat{\boldsymbol{\xi}})\right]^{-1} = \mathcal{E}\left[-\frac{\partial^2 L}{\partial \zeta_j \partial \zeta_k}\right]_{\zeta=\hat{\zeta}}, \tag{114}$$

where

$$\frac{\partial^2 L}{\partial \zeta_j \partial \zeta_k} = \sum_{i=1}^{N}\left[\frac{1}{f}\frac{\partial^2 f}{\partial \zeta_j \partial \zeta_k} - \frac{1}{f^2}\frac{\partial f}{\partial \zeta_j}\frac{\partial f}{\partial \zeta_k}\right], \tag{115}$$

and $f \equiv f(\mathbf{u}_i)$. Replacing (115) into (114) and taking the expectations we have:

$$\left[\operatorname{cov}(\hat{\boldsymbol{\zeta}})\right]^{-1} = n\mathcal{E}\left[\frac{1}{f}\frac{\partial f}{\partial \zeta_j}\frac{\partial f}{\partial \zeta_k}\right]_{\zeta=\hat{\zeta}}. \tag{116}$$

The computation of the covariance matrix encounters the problems of having to invert a large matrix as the number of items increases as well as dealing with response patterns $\mathbf{u}_i$ with small probabilities. A full rank parameterization is required; otherwise, a generalized inverse would have to be used. Bock (1989, p. 338) make use of the empirical information matrix:

$$\left[\operatorname{cov}(\hat{\boldsymbol{\zeta}})\right]^{-1} \simeq \sum_{i=1}^{N} \frac{1}{f(\mathbf{u}_i)}\frac{\partial f(\mathbf{u}_i)}{\partial \zeta_j}\frac{\partial f(\mathbf{u}_i)}{\partial \zeta_k}. \tag{117}$$

The dimension of this matrix is equal to the number of parameters estimated in the model. In the case of more than one latent variable, constraints need to be imposed to justify a unique maximum solution before the inversion. Otherwise the information matrix will be singular.

Alternatively, one can use re-sampling or bootstrapping techniques to compute standard errors. However, that requires substantive computational time and will not be feasible for models with more than two or three latent variables.

Louis (1982) proposed a way of computing the observed information matrix within the EM algorithm. His procedure requires the evaluation, in the last iteration of the EM, of the first and second derivatives of functions of the complete-data. He defined the observed information matrix (I_o) as:

$$I_o = -\mathcal{E}\big(B(\boldsymbol{\zeta}, \boldsymbol{\zeta}) \mid \mathbf{u}; \boldsymbol{\zeta}\big) - \mathcal{E}\big(b(\boldsymbol{\zeta})b(\boldsymbol{\zeta})' \mid \mathbf{u}; \boldsymbol{\zeta}\big) + \Delta(\boldsymbol{\zeta})\Delta(\boldsymbol{\zeta})', \tag{118}$$

where $I_o = -\left[\frac{\partial^2 L(\boldsymbol{\zeta}|\mathbf{u})}{\partial\boldsymbol{\zeta}\partial\boldsymbol{\zeta}'}\right]$, $B((\boldsymbol{\zeta}, \boldsymbol{\zeta}) \mid \mathbf{u}; \boldsymbol{\zeta}) = \frac{\partial^2 L(\boldsymbol{\zeta}|\mathbf{u},\mathbf{z})}{\partial\boldsymbol{\zeta}\partial\boldsymbol{\zeta}'}$, $b(\boldsymbol{\zeta}) = \frac{\partial L(\boldsymbol{\zeta}|\mathbf{u},\mathbf{z})}{\partial\boldsymbol{\zeta}}$, $\Delta(\boldsymbol{\zeta}) = \mathcal{E}(b(\boldsymbol{\zeta}) \mid \mathbf{u}; \boldsymbol{\zeta})$. Note that $\Delta(\boldsymbol{\zeta})$ is 0 at the maximum likelihood solution. The terms defined in (118) can be easily computed in the last iteration of the EM algorithm. Verhelst and Glas (1993) implemented Louis's method for the Rasch model with missing data.

4.2. *Goodness-of-fit measures*

As in any type of modelling, in IRT too, is crucial to be able to check the appropriateness of the model fitted using a goodness-of-fit test. We would like to see how well a model fits a particular data set or predicts the associations among a set of items. In IRT modelling, goodness-of-fit can be seen in two different ways. One way is to investigate the overall fit of the model, in other words, how well the model predicts the whole response pattern. The second way is to investigate how well the model fits the one- two- and three-way margins. Those statistics on the margins are considered as measures of fit rather than goodness-of-fit statistics.

We could compare the observed (O) and expected (E) frequencies of the response patterns in total by means of a X^2 Pearson goodness-of-fit or a likelihood ratio test G^2.

$$X^2 = \sum_{r=1}^{R} \frac{(O_r - E_r)^2}{E_r},$$

$$G^2 = 2\sum_{r=1}^{R} O_r \log \frac{O_r}{E_r},$$

where R denotes the total number of response patterns. When N is large, n small and the model is correct the above statistics are asymptotically distributed as χ^2 with degrees of freedom equal to the number of patterns minus the number of independent parameters. As the number of items increases the χ^2 approximation to the distribution of either goodness-of-fit statistic ceases to be valid. However, parameter estimates are still valid but it is difficult to assess the fit of the model. Haberman (1977) derived the conditions for log-linear models under which the asymptotic distribution of the above goodness-of-fit statistics remains the chi-square in the presence of sparseness.

Usually, sparseness is present when the ratio of sample size to the total number of patterns is small (Agresti and Yang, 1987) and there are many response patterns with expected frequency less than 1. Under this condition, tests of fit based these statistics cannot be used. Solutions have been proposed for overcoming the problem of sparseness such as pooling response patterns with small expected frequencies or adding a small number to each response pattern or even considering other distributions such as the normal (Morris, 1975; Koehler and Larntz, 1980) when the number of items increase together with the sample size. Finally, parametric bootstrapping has been also proposed (Bartholomew and Tzamourani, 1999).

Alternatively, we can compute the Pearson chi-square statistic or likelihood ratio statistic or any other measure of discrepancy between observed and expected frequencies (standardized or unstandardized residuals) only for pairs and triplets of responses. The pairwise distribution of any two variables can be displayed as a two-way contingency table, and chi-squared residuals can be constructed in the usual way by comparing the observed and expected frequencies. As a rule of thumb, if we consider the residual in each cell as having a χ^2 distribution with one degree of freedom, then a value of the residual greater than 4 is indicative of poor fit at the 5% significance level. A study of the individual margins provides information about where the model does not fit and suggest ways of improving scales (deleting items or collapsing categories). A detailed discussion on the use of those goodness-of-fit measures for ordinal variables can be found in Jöreskog and Moustaki (2001) and Bartholomew et al. (2002, pp. 213–234). However, for the model with covariate effects the Pearson chi-square statistic or likelihood ratio statistic for pairs and triplets of responses have to be computed for different values of the explanatory variables, which may make the use of these residuals less informative with respect to goodness-of-fit.

In the literature, formal tests have also been developed based on lower order margins by Christoffersson (1975), Muthén (1978), Reiser and VandenBerg (1994), Reiser (1996), Bartholomew and Leung (2002), Maydeu-Olivares and Joe (2005). The hypothesis and tests are formulated as follows:

$$H_0\colon \boldsymbol{\epsilon} = 0$$

where $\boldsymbol{\epsilon}$ can be a vector of first- and second-order marginals with elements:

$$\epsilon_i = \sqrt{n}(p_i - \pi_i) \quad \text{and} \quad \epsilon_{ij} = \sqrt{n}(p_{ij} - \pi_{ij}),$$

where p_i and p_{ij} are the true first- and second-order proportions and π_i and π_{ij} are the ones under the model.

Most tests are based on a statistic defined as:

$$W = \mathbf{e}' \hat{\Sigma}^{-1} \mathbf{e},$$

where $\hat{\Sigma}$ is the estimated covariance matrix of $\boldsymbol{\epsilon}$ and

$$e_i = \sqrt{n}(\hat{p}_i - \hat{\pi}_i) \quad \text{and} \quad e_{ij} = \sqrt{n}(\hat{p}_{ij} - \hat{\pi}_{ij}),$$

where $\hat{p}_i$ and $\hat{p}_{ij}$ are the observed first- and second-order proportions and $\hat{\pi}_i$ and $\hat{\pi}_{ij}$ are the estimated ones under the model. The tests differ with respect to the way the covariance matrix of the residuals is estimated, whether they account both for first- and second-order margins, whether they use standardized or unstandardized residuals and whether the parameters are taken to be known or not.

Finally, instead of testing the goodness-of-fit of a specified model we could use certain statistics for selecting among a set of different models. They provide information about the goodness-of-fit for each model in comparison with other models. That can be useful for the determination of the number of factors or for comparing models with and without covariate effects in addition to latent variables. Sclove (1987) gives a review of some of the model selection criteria used in multivariate analysis such as those due to Akaike, Schwarz and Kashyap. These criteria take into account the value of the likelihood at the maximum likelihood solution adjusted for the number of parameters estimated, the sample size and other factors.

Some of those statistics commonly used are: the Akaike Information Criterion (AIC):

$$\text{AIC} = -2L + 2\nu,$$

where L is the log-likelihood, and ν is the number of estimated model parameters, the Schwarz Bayesian Information Criterion (SBC) also known as Bayesian information criterion BIC:

$$\text{SBC} = -2L + \nu \ln(N),$$

where N is the total number of observations, and the CAIC:

$$\text{CAIC} = -2L + \nu\bigl(1 + \ln(N)\bigr).$$

In comparing different models for the same data, then, one will prefer models with lower values on these statistics since they lead to more parsimonious models.

5. Estimation of respondent scores

An distinct advantage of item response theory over classical test theory is that, given suitable item response models, respondent scores with known optimal properties can be estimated for mixtures of item formats. Following common conventions in the field, IRT scores are referred to here as *scale* scores, to be distinguished from *item* scores (0, 1 for right-wrong scored items and integers $1, 2, \ldots, m_j$ for category scores) and *test* scores (number-right). In this section we discuss scale score estimation in general terms that apply to all of the models in Section 3.

At the present time there are three widely used methods of estimating scale scores: maximum likelihood (ML), Bayes modal, also called MAP (maximum a posteriori), and Bayes, also called EAP (expected a posteriori). All three methods as described here assume that the item parameters are fixed quantities known in advance. This assumption is broadly justified in practical work because item parameter estimation is typically based on very large samples of respondents – often hundreds or thousands. Data from these samples can also be used to estimate, either parametrically or nonparametrically, the latent distributions on which the Bayes methods depend (see Mislevy, 1984).

5.1. ML estimation

Although sufficient statistics may exist for all items in the measuring instrument and can be used in estimating respondent scores by likelihood methods, for greater generality we express the likelihood of θ given the respondent's complete pattern of item scores. We make the usual assumption of conditional independence of item responses given θ (but relax it in certain situations; see Section 5.5) and write the likelihood for an n-item pattern as follows:

Let $u_{ijhk} = 1$ if the response of respondent i is assigned to category k of an m_j-category item j and $h = k$, and 0 otherwise. The likelihood is

$$l_i(\theta) = \prod_j^n \prod_h^{m_j} P_{jh}^{u_{ijhk}}(\theta), \tag{119}$$

where $P_{jh}(\theta)$ is the response function for category h of item j.

The quantity to be maximized is

$$L_i(\theta) = \log l_i(\theta) = \sum_j^n \sum_h^{m_j} u_{ijhk} \log P_{jh}(\theta), \tag{120}$$

and the likelihood equation for θ is, say,

$$G(\theta) = \sum_j^n \sum_h^{m_j} \frac{u_{ijhk}}{P_{jh}(\theta)} \frac{\partial P_{jh}(\theta)}{\partial \theta} = 0. \tag{121}$$

The information with respect to θ conveyed by the response pattern is

$$I(\theta) = \sum_j^n \sum_h^{m_j} \frac{1}{P_{jh}(\theta)} \left[\frac{\partial P_{jh}(\theta)}{\partial \theta}\right]^2. \tag{122}$$

Assuming that the location and scale of the item parameters has been determined during their estimation by specifying the mean and standard deviation of the latent distribution, the maximum likelihood estimate of θ for respondent i can be obtained by Newton–Gauss iterations:

$$\theta_i^{(t+1)} = \theta_i^{(t)} + I^{-1}\big(\theta_i^{(t)}\big)G\big(\theta_i^{(t)}\big). \tag{123}$$

Excluding the forbidden patterns in which responses are all in category 1 or all in category m_j, which have no finite solution, the iterations,

$$I^{-1}(\theta_i^{(t)})G(\theta_i^{(t)}) < 10^{-C},$$

will almost always converge to a precision criterion C from a starting value of $\theta_i = 0$. Convergence is guaranteed for the purely logistic models with their concave likelihood, but at very low values of θ and small n, the 3-PL or nominal multiple-choice model may encounter difficulties. In this situation, and also when the forbidden patterns occur, it is necessary either to assign default values to θ or turn to EAP estimation, which is non-iterative and makes use of information in the latent distribution to determine plausible values of θ. That the pattern information function is the sum of item information functions is useful when developing measuring instrument by selecting items from a larger pool of pre-calibrated items. The plot of the information function (or corresponding standard error function, $I^{-1/2}(\theta)$ shows whether a putative instrument has sufficient precision at cutting points where decisions about respondents will be made. Information in the neighborhood of θ-values at those points should be greater than elsewhere. A general purpose instrument, on the other hand, needs relatively uniform information levels over as wide a range of the latent distribution as possible.

5.2. MAP estimation

The MAP estimator (Birnbaum, 1967) maximizes the posterior probability of θ, given the item scores u_{ijhk},

$$p_i(\theta) = \prod_j^n \prod_h^{m_j} P_{jh}^{u_{ijhk}}(\theta) g(\theta), \tag{124}$$

where $g(\theta)$ is an assumed unimodal proper density function in the population to which respondent i belongs. If this "prior" distribution depends only on location and scale parameters, they are already specified by the arbitrary scaling of the IRT models.

In the corresponding MAP equation,

$$G_i = \sum_j^n \sum_h^{m_j} \left[\frac{u_{ijhk}}{P_{jh}(\theta)} \frac{\partial P_{jh}(\theta)}{\partial \theta} \right] + \frac{\partial \log g(\theta)}{\partial \theta} = 0, \tag{125}$$

the second term is in effect a stochastic constraint that conditions the estimation of θ.

The most common choice of prior is $N(\mu, \sigma)$, in which case the second term of the MAP equation becomes

$$-(\theta - \mu)/\sigma$$

or just $-\theta$ when the standard normal prior is assumed.

The corresponding posterior information function,

$$I(\theta) = \sum_j^n \sum_h^{m_j} \frac{1}{P_{jh}(\theta)} \left[\frac{\partial P_{jh}(\theta)}{\partial \theta} \right]^2 + \frac{1}{\sigma}, \tag{126}$$

includes information conveyed by knowledge of the population to which the respondent belongs. The MAP method has the advantage of providing plausible determinate values of θ for the forbidden patterns. When the data or the response model are unfavorable, convergence can be guaranteed by setting σ to some value less than 1 to increase the influence of the prior.

5.3. EAP estimation

The EAP estimator (Bock and Mislevy, 1982) is the mean of the posterior distribution,

$$\theta_i = \int_{-\infty}^{+\infty} \theta \prod_j^n \prod_h^{m_j} P_{jh}(\theta)^{u_{ijhk}} g(\theta)\, d\theta, \tag{127}$$

and its variance is

$$\sigma_{\theta_i}^2 = \int_{-\infty}^{+\infty} (\theta - \theta_i)^2 \prod_j^n \prod_h^{m_j} P_{jh}(\theta)^{u_{ijhk}} g(\theta)\, d\theta. \tag{128}$$

For the item response models assumed here, these integrals do not have a closed forms, but they are nevertheless easily evaluated numerically – for example, by Gauss–Hermite quadrature when $g(\theta)$ is standard normal. The quadrature formulas are then

$$\theta_i = \sum_{q=1}^{Q} \theta \prod_j^n \prod_h^{m_j} P_{jh}^{u_{ijhk}}(\theta) A(X_q), \tag{129}$$

$$\sigma_{\theta_i} = \sum_{q=1}^{Q} (\theta - \theta_i)^2 \prod_j^n \prod_h^{m_j} P_{jh}^{u_{ijhk}}(\theta) A(X_q), \tag{130}$$

where $A(X_q)$ is the quadrature weight for point X_q (see Straud and Sechrest, 1966). The number of points, Q, required depends upon σ_{θ_i}. The rule of thumb is that at least three points should fall between $\theta_i - \sigma_{\theta_i}$ and $\theta_i + \sigma_{\theta_i}$ over the effective range of the prior.

EAP is the most numerically robust of the three methods and has the advantage of not requiring a functional form for the latent distribution. Any point masses summing to 1 can be employed as weights on equally spaced points. The optimal properties of the Gaussian quadrature formula are lost, but that is of little consequence when the number of quadrature points is reasonably large.

Although the mean of the EAP estimates in the sample is an estimator of the population mean – that is, the mean of the prior – the variance of the estimates is reduced by regression to the mean. In the population, the sum of the score variance and the error variance equals the variance of the latent and distribution; that is,

$$\mathcal{E}(\theta_i - \mu)^2 + \mathcal{E}(\theta_i - \theta)^2. \tag{131}$$

The corresponding mean squares in the sample, say MST and MSE, provide an estimate on of the classical reliability coefficient expressed as the intraclass correlation:

$$R = \frac{\text{MST}}{\text{MST} + \text{MSE}}. \tag{132}$$

5.4. *Multiple matrix sampling*

The most extensive application of item response theory at present is the National Assessment of Educational Progress (NAEP), a nationwide survey of educational outcomes reporting to the states and the general public. The Assessment makes use of specialized scoring procedures exploiting the unique properties of IRT methodology. During each assessment year a probability sample of schools is drawn in the participating states; within each school, a random sample of students is then drawn at each of several grades levels. In group testing sessions, each of these students is randomly assigned a form of a multi-form assessment instrument appropriate to his or her grade level.

The instrument is constructed by assigning items covering several areas of the curriculum to forms in a balanced incomplete block design in which each area of all forms share an equal number items in common. Because of the item overlap between forms, the methods in Section 4 make it possible to estimate item parameters for the entire instrument on single scale in each area. The item formats consist of both machine-scorable multiple-choice items and open-ended items hand-scored in graded categories. Although there are too few items in each area of a form to support reliable reporting of achievement scores to individual students, the aggregation of results gives dependable estimates of average achievement at the state level. Aggregation to the national level allows reports of average performance and also for various demographic and other subsamples of the national population. This method of obtaining item data from individual respondents in more than one area of item content to assess outcomes at the group level is called *multiple matrix sampling*.

The assessment forms also include a certain number of items held over from previous assessment years in order to express assessment results on the same scale from year to year. Estimated parameters for the current year are used in estimating EAP scores and corresponding posterior standard deviations for each student tested. In order to make these scores suitable for conventional least squares analysis of various effects in the data, they are returned to their full variability in the population by the addition of a random deviant drawn from the assumed normal posterior distribution for each student. These augmented scores, referred to as “plausible values”, are used in all subsequent analysis of the results; standard errors are estimated by the jackknife resampling method. For purposes of communicating to the general public in easily understood terms, the reports include the proportions of students whose plausible values fall into arbitrarily defined achievement categories labelled Basic, Proficient, and Advanced. (See Mislevy et al., 1992, for further details.)

5.5. *Failure of conditional independence*

Certain situations arise in educational and psychological testing where the assumption of conditional independence is not tenable. The following are two important examples.

Testlets

Tests of reading comprehension typically consist of a number of text passages containing information that the intended examinees are unlikely to know. Each passages is followed by a number of multiple choice items based on specific facts mentioned in the text. These and similar exercises have been called "testlets" (Wainer and Kiely, 1987). Because an examinee's grasp of the overall meaning of the passage will affect his or her ability to answer any of the following items, it is unlikely that items within the testlet will have the same degree of association as items in different testlets.

The effects of testlets on response association can be investigated by item factor analysis (see Section 6.2). If association within testlets is greater than that attributable to a general factor of reading comprehension, additional specific item-group factors will be found. IRT based item factor analysis provides statistical tests for existence of these "nuisance" factors. The most serious problem they created independence is underestimation of the scale-score standard errors. This occurs because the ignored increased association within the testlet inflates the variance of the test score relative to the measurement error variance. Fortunately, the procedure called item bifactor analysis described in Section 6.2 provides a solution to this problem. On the assumption that the factor structure of a test containing, say, n items divided among m testlets consists of a general factor and m independent item-group factors. This constrained structure has a total of only $2n$ none-zero factor loadings compared to $n(m + 1)$ loadings of an unstructured factor model of the same size. The fit of the bifactor model compared to the full model can also be tested in the IRT analysis. Although the number of additional factors may be large in this type of application, the computational burden is not prohibitive because the bifactor model requires only two-dimensional quadrature (Gibbons and Hedeker, 1992).

The bifactor analysis provides estimates of the item factor loadings and the corresponding intercept parameters. Converting the general factor loadings to item slopes provides the discriminating power parameters required for estimating respondent scores. On the assumption that the posterior distribution of the latent factor variables is multivariate normal, the mean of the general factor marginal distribution estimated by the EAP method is the respondent's proficiency scale score; the corresponding posterior standard deviation then provides a valid measurement-error estimate.

Other approaches to the treatment of testlet effects have been considered by Glas et al. (2000) and Wang et al. (2002).

Multiple ratings

When educational achievement tests contain open ended items (written essays, problem-solving exercises, etc.) that require subjective grading, they are usually graded independently by more than one reader – especially if the results influence student advancement or qualification. As readers are rarely in one-hundred percent agreement, there is additional information in the multiple ratings that should be utilized in score estimation. If the identity of the raters is retained and each rater grades the responses of a sufficient number of examinees, a straightforward approach is to treat the raters of the same item as if they were items of a testlet and to apply the item bifactor analysis described above. Each distinct item-rater set comprises an item-group factor in the model. If the test also contains objectively scored items, they are assigned loadings only on the general factor. This approach makes optimal use of the information in the data while allowing for the considerable association expected in multiple ratings. It also accounts for rater effects, i.e., the tendency of given raters to be consistently lenient or severe.

5.6. Component scores

Bartholomew and Knott (1999) developed the notion of *sufficiency principle* directly related to the dimension reduction in latent variable analysis. One would aim to find, for example, p functions of the data $\mathbf{u}$ named $U_1, \ldots, U_p$ such that the conditional distribution of the latent variables $\boldsymbol{\theta}$ given the functions U_i not to depend on $\boldsymbol{\theta}$. This dimension reduction is achieved under weak regularity conditions when at least $n - p$ of the conditional distributions $f_j(u_j \mid \boldsymbol{\theta})$ are from the exponential family (see Bartholomew and Knott, 1999, p. 20). Writing down the one-parameter exponential family:

$$f_j(u_j \mid \boldsymbol{\theta}) = F_j(u_j)G_j(\psi_j)\exp\bigl[\psi_j d_j(u_j)\bigr], \quad j = 1, \ldots, n, \tag{133}$$

where ψ_j is usually a linear function of the latent variables $\boldsymbol{\theta}$. For example in the two parameter model (single factor), $\psi_i = \alpha_j\theta + c_j$ where α_j are the factor loadings.

All the information about the latent variables in the observed items is found in the posterior distribution:

$$\phi(\boldsymbol{\theta} \mid \mathbf{u}) = \phi(\boldsymbol{\theta}) \prod_{j=1}^{n} f_j(u_j \mid \boldsymbol{\theta}) / f(\mathbf{u}). \tag{134}$$

Substituting (134) into (133) we have:

$$\phi(\boldsymbol{\theta} \mid \mathbf{u}) \propto \phi(\boldsymbol{\theta}) \prod_{j=1}^{n} G_j(\psi_j) \exp \sum_{i=1}^{p} \theta_i U_i, \tag{135}$$

where $U_i = \sum_{j=1}^{n} \alpha_j d_j(u_j)$, $i = 1, \ldots, p$. The U_i are called components and they provide a weighted score of the items for each individual in the sample. The components U_i are considered in the Bayesian sense as minimal sufficient statistic for $\boldsymbol{\theta}$.

More specifically, the components for the two parameter logistic model are

$$U_i = \sum_{j=1}^{n} \alpha_j u_j, \quad i = 1, \ldots, p,$$

and for the nominal logistic model are

$$U_i = \sum_{j=1}^{n} \sum_{h=1}^{m_j} \alpha_{jh} u_{jh}.$$

Component scores do not exist in the case of the probit model and the cumulative models used for modelling ordinal items. In the ordinal case, the parameter ψ_j is not a linear function of the latent variables.

6. Item factor analysis

The most important application of multiple latent variable models discussed in Section 4 is factor analysis of the association among items in the measuring instrument

(Bartholomew, 1980; Bock et al., 1988). Item factor analysis makes the methodology of multiple factor analysis of test scores (Thurstone, 1947; Jöreskog, 1967) applicable at the item level. The estimation methods based on maximum marginal likelihood described in Section 4 provide likelihood ratio tests that serve to answer the fundamental assumption of unidimensional of IRT – namely, that all of the association among item responses be accounted for by a single latent variable. Although conventional goodness-of-fit tests apply when the number of items is small enough to permit counting the frequencies of all patterns and relatively few are zero (Bock and Aitkin, 1981), those tests are of little help with instruments of practical size. Haberman (1977) shows, however, that likelihood ratio tests of nested models for sparse contingency tables are valid for testing the significance of improvement in fit of the item factor model when an additional latent variable is assumed. Thus, only one- and two-dimensional item factor analyses are required to test the assumption of unidimensionality.

Classical methods of item factor analysis apply test-score factoring to the matrix of tetrachoric or polychoric correlations (Jöreskog, 1994). As pointed out below, these methods remain valuable in providing starting values for the iterative estimation procedures of Section 4. Because the classical method works with only pairwise association in the data, it is referred to as "limited information" – as opposed to the "full information" IRT method that works with association of all orders.

In maximum likelihood methods of factor analysis for test scores and other observations treated as continuous, both the observed variables and the latent variables are assumed standard normal. When applied to item responses, this leads to "loadings" in the form of correlations α_{jl} between the $N(0,1)$ response process Y_j of item j and the latent variable θ_l. When the latent variables are assumed uncorrelated, the sum of squares of the loadings over the factors is called the item common factor variance, or *communality*. The remaining residual variance, one minus the communality, is called *uniqueness*. In IRT it is uniqueness that is assumed $N(0,1)$, and the relationships are expressed as regression coefficients, or factor slopes, a_{jl} called factor *discriminating powers*. The relationships between these two quantities are

$$a_{jl} = \frac{\alpha_{jl}}{\sqrt{1 - \sum_{\ell=1}^{p} \alpha_{j\ell}^2}}, \tag{136}$$

and

$$\alpha_{jl} = \frac{a_{jl}}{\sqrt{1 + \sum_{\ell=1}^{p} a_{j\ell}^2}}. \tag{137}$$

Inasmuch as item category thresholds do not exist in multiple factor models, only the factor slopes and item intercepts are estimated when fitting the response models.

Because large numbers of items are typical in item factor analysis, Newton–Gauss solution of the likelihood equations is not deemed practical. The EM solution described in Section 4 is, however, quite feasible with large data sets, but it raises the question of starting values. A good approach is to perform a MINRES (Harman, 1967) principal factor analysis of the matrix of tetrachoric or polychoric correlations while simultaneously estimating the communality's in the diagonal. Highly efficient one-matrix eigen-vector and eigen-value procedures for this purpose are widely available. The eigen-vectors in the descending order of the corresponding eigen-values, when multiplied by the square root of the value, gives the factor loadings in their conventional order. The number of vectors to be calculated equals the assumed dimensionality of the common factor space.

As a result of the rotational indeterminacy of the multiple factor model, there are only $p(p+1)$ free parameters among the factor loadings. Although the EM solution is unaffected by over-parameterization, it is important after each EM step to reorthogonalize the solution to prevent it from wandering away from the original orthogonal principle factors. This can be done by computing the EAP scores of the respondents in the sample, converting them to a correlation matrix, and performing a new eigen-problem to obtain factor loadings for the next EM step (see Schilling and Bock, 2005). At the end of the EM estimation the expected inter-item correlations should be subjected either to another principal factor analysis or to some other rotation to a preferred factor pattern. During the EM steps, the factor loading must be constrained to have communality less than 1. For this purpose Bock et al. (1988) used a stochastic constraint based on a beta distribution.

Provided the matrix of factor loadings is orthogonal with respect to the columns, multiple factor scores for the respondents can computed by the MAP or EAP method generalized to p dimensions. The MAP requires the functional form of the multivariate latent density. If the posterior distribution is assumed multinormal, one-dimensional EAP estimation in the respective margins is possible; otherwise, $(p-1)$-dimensional integration to the desired margin is necessary before EAP estimation. The usual MAP standard errors and EAP posterior standard deviations apply.

6.1. *Multidimensional item response models*

Extending the response models in Section 3 to multiple latent variables is straightforward for those models that have an item discriminating power parameter but do not require parameters that account for guessing. These models include the normal ogive, 2-PL, graded, and generalized partial credit models. Their extension to multiple latent variables requires only the substitution of a, say, p-vector latent variable for the scalar latent variable assumed in Section 3; similarly, a p-vector of discriminating power parameters replaces the scalar parameter.

The multiple latent variable extension of the nominal model is complicated by the presence of the separate discriminating power parameters for the m_j categories. However, on the assumption that only the relative spacing of an item's a-parameters varies and not their magnitude, they could be expressed as the product of an item specific factor and a category specific factor, with the former being a function of the factor slopes. Item factor analysis with the nominal model would then be possible.

6.2. *Item bifactor analysis*

In addition to the general unstructured model with np factor loadings, structured models with certain loadings assumed zero may be of interest. The most important of these is the *bifactor* model (Gibbons and Hedeker, 1992) alluded to in the previous section in connection with item testlets. This model was first proposed for test scores by Holzinger and Swineford (1937) but was soon superseded by the multiple factor model. The version at the item level assumes a general factor representing the respondent's overall proficiency in the item domain, plus, say, s independent item-group factors, where the assignments of items to groups is specified beforehand; that is, items have one loading on the general factor and one, and only one, loading on the mutually exclusive item-group factors.

Stuart (1958), among other special cases for the multivariate normal distribution, showed that, when the covariance matrix is determined by the loadings of the group factors (referred to in the factor analysis literature as the "simple structure" pattern), the probability of the n-vector of item scores, $\mathbf{u}_i = [u_{ijhk}]$, $j = 1, 2, \ldots, n$, can be evaluated in terms of one-dimensional integrals,

$$P(\mathbf{u}_i \mid \boldsymbol{\theta}) = \prod_{l=1}^{s} \int_{R(\theta_l)} \prod_{j=1}^{n} \prod_{h=1}^{m_j} \Phi_{jh}^{w_{jl} u_{ijhk}}(\theta_l) g(\theta_l)\, \mathrm{d}\theta_l, \tag{138}$$

with l indexing the groups, $w_{jl} = 1$ when item j has a non-zero loading on factor l and 0 otherwise. Items that have loadings only on the general factor are considered to belong to group 0.

Gibbons and Hedeker (1992) generalized this to the bifactor model by adding the general factor:

$$P(\mathbf{u}_i \mid \theta_0, \boldsymbol{\theta}) = \int_{R(\theta_0)} \left[\prod_{l=0}^{s} \int_{R(\theta_l)} \left(\prod_{j=1}^{n} \prod_{h=1}^{m_j} \Phi_{jh}^{w_{jl} u_{ijhk}}(\theta_0, \theta_l) \right) g(\theta_l)\, \mathrm{d}\theta_l \right] g(\theta_0)\, \mathrm{d}\theta_0. \tag{139}$$

The two-dimensional numerical integration is very favorable for evaluating the response pattern probabilities: with modern computing facilities, sufficient quadrature points can be used in each dimension to avoid the need for adaptive quadrature. The procedure is readily implemented for both dichotomous and polytomous item responses.

The bifactor model applies wherever respondent scores on the general dimension are of interest and the variation specific to the item groups is included only two satisfy the requirement of multidimensional conditional independence of item responses. Whether the model has accounted for all association in the data can be tested by likelihood ratio with the unstructured model as the alternative. Only the number of free parameters in the constrained and unconstrained models are counted in determining the degrees of freedom.

7. Response relations with external variables

Recently, IRT models have been extended to account for external variables in addition to latent variables (Sammel et al., 1997; Moustaki, 2003). The inclusion of external variables as covariates in IRT modelling can be done in two different ways. One way is to assume that covariates affect directly some function of the response variables and therefore together with the latent variables account for the relationships among the items. In that context, a covariate plays the same role with the latent variable though the covariate is taken to be observed and fixed. Another way is to have covariates affecting items indirectly through the latent variables. For example, when the effect of a categorical covariate is fitted on a latent variable through a simple linear relation is as if we allow for different means of the latent distribution in the groups defined by the covariate. The variability of the latent distribution needs to be fixed for the model to be identified. Alternatively, one could fit the same IRT model within groups defined by the categories of a covariate or by combinations of covariates and allow for measurement invariance (item parameters to be equal across groups) but latent distributions to have different means and variances.

The direct incorporation of covariates has already been defined in (1) via the conditional density ($f_v(\mathbf{u} \mid \boldsymbol{\theta}, \mathbf{x})$). The complete log-likelihood for a random sample of size N becomes:

$$L = \sum_{i=1}^{N} \log f(\mathbf{u}_i, \theta_i \mid \mathbf{x}_i) = \sum_{v=1}^{V} \sum_{i=1}^{N_v} \log f_v(\mathbf{u}_i, \theta_i \mid \mathbf{x}_i). \tag{140}$$

The parameter vector $\mathbf{a}_{vj}$ is now extended to include the regression coefficients: $\mathbf{a}_{vj} = (c^v_{j(0)}, \ldots, c^v_{j,(m_j-1)}, \alpha_{vj}, \beta_{vj1}, \ldots, \beta_{vjr})$, where $j = 1, \ldots, n$. The parameter vector $\mathbf{b}$ remains as before.

The expected score function for the parameter vector $\mathbf{a}_{vj}$ for an individual i and item j is given by

$$ES_i(\mathbf{a}_{vj}) = \int S_i(\mathbf{a}_{vj}) f_v(\theta \mid \mathbf{u}_i, \mathbf{x}_i)\, \mathrm{d}\theta, \tag{141}$$

where $f_v(\theta \mid \mathbf{u}_i, \mathbf{x}_i)$ denotes the posterior distribution of the latent variable given what has been observed in group v,

$$f^v(\theta \mid \mathbf{u}_i, \mathbf{x}_i) = \frac{f_v(\mathbf{u}_i \mid \theta, \mathbf{x}_i) g_v(\theta)}{f_v(\mathbf{u}_i \mid \mathbf{x}_i)},$$

and

$$S_i(\mathbf{a}_{vj}) = \frac{\partial \log f_v(u_{ij} \mid \boldsymbol{\theta}, \mathbf{x}_i)}{\partial \mathbf{a}_{vj}}.$$

The steps of the EM algorithm remain as before.

References

Agresti, A., Yang, M. (1987). An empirical investigation of some effects of sparseness in contingency tables. *Computational Statistics and Data Analysis* **5**, 9–21.

Andersen, E.B. (1977). Sufficient statistics and latent trait models. *Psychometrika* **42**, 69–81.

Andrich, D. (1978). A rating formulation for ordered response categories. *Psychometrika* **43**, 561–573.

Bartholomew, D.J. (1980). Factor analysis for categorical data. *Journal of the Royal Statistical Society, Series B* **42**, 293–321.

Bartholomew, D.J., Knott, M. (1999). *Latent Variable Models and Factor Analysis*, 2nd ed. *Kendall's Library of Statistics*, vol. 7. Arnold, London.

Bartholomew, D.J., Leung, S.O. (2002). A goodness of fit test for sparse 2^p contingency tables. *British Journal of Mathematical and Statistical Psychology* **55**, 1–15.
Bartholomew, D.J., Steele, F., Moustaki, I., Galbraith, J.I. (2002). *The Analysis and Interpretation of Multivariate Data for Social Scientists*. Chapman and Hall/CRC.
Bartholomew, D.J., Tzamourani, P. (1999). The Goodness-of-fit of latent trait models in attitude measurement. *Sociological Methods and Research* **25**, 525–546.
Birnbaum, A. (1957). Efficient design and use of tests of a mental ability for various decision making problems. Series Report No. 58-16. Project No. 7755-23. USAF School of Aviation Medicine, Randolph Air Force Base, Texas.
Birnbaum, A. (1958a). On the estimation of mental ability. Series Report No. 15. Project No. 7755-23. USAF School of Aviation Medicine, Randolph Air Force Base, Texas.
Birnbaum, A. (1958b). Further considerations of efficiency in tests of a mental ability. Technical Report No. 17. Project No. 7755-23. USAF School of Aviation Medicine, Randolph Air Force Base, Texas.
Birnbaum, A. (1967). Statistical theory for logistic mental test models with a prior distribution of ability. Research Bulletin No. 67-12. Educational Testing Service, Princeton, NJ.
Bliss, C.I. (1935). The calculation of the dosage mortality curve. *Annals of Applied Biology* **22**, 134–167. (Appendix by R.A. Fisher).
Bock, R.D. (1970). Estimating multinomial response relations. In: Bose, R.C., et al. (Eds.), *Contributions to Statistics and Probability*. University of North Carolina Press, Chapel Hill, NC, pp. 111–132.
Bock, R.D. (1972). Estimating item parameters and latent ability when responses are scored in two or more nominal categories. *Psychometrika* **37**, 29–51.
Bock, R.D. (1975). *Multivariate Statistical Methods in Behavioral Research*. McGraw-Hill, New York.
Bock, R.O. (1989). Measurement of human variation: A two-stage model. In: Bock, R.D. (Ed.), *Multilevel Analysis of Educational Data*. Academic Press, San Diego, CA, pp. 319–342.
Bock, R.D., Aitkin, M. (1981). Marginal maximum likelihood estimation of item parameters: Application of an EM algorithm. *Psychometrika* **46**, 443–445.
Bock, R.D., Gibbons, R.D., Muraki, E. (1988). Full information item factor analysis. *Applied Psychological Measurement* **12**, 261–280.
Bock, R.D., Jones, L.V. (1968). *The Measurement and Prediction of Judgment and Choice*. Holden-Day, San Francisco, CA.
Bock, R.D., Lieberman, M. (1970). Fitting a response model for n dichotomously scored items. *Psychometrika* **35**, 179–197.
Bock, R.D., Mislevy, R.J. (1982). Adaptive EAP estimation of ability in a microcomputer environment. *Applied Psychological Measurement* **6**, 431–444.
Bock, R.D., Zimowski, M.F. (1997). Multiple group IRT. In: van der Linden, W.J., Hambleton, R.K. (Eds.), *Handbook of Modern Item Response Theory*. Springer-Verlag, New York, pp. 433–448.
Böckenholt, U. (2001). Mixed-effects analyses of rank-ordered data. *Psychometrika* **66**, 45–62.
Bradley, R.A., Terry, M.E. (1952). Rank analysis of incomplete block designs. I. The method of paired comparisons. *Biometrika* **39**, 324–345.
Brennan, R.L. (2001). *Generalizability Theory*. Springer-Verlag, New York.
Camilli, G., Shepard, L.A. (1994). *Methods for Identifying Biased Test Items*, vol. 4. Sage Publications, Thousand Oaks, CA.
Christoffersson, A. (1975). Factor analysis of dichotomized variables. *Psychometrika* **40**, 5–32.
Cooper, B.E. (1968). Algorithm AS2. The normal integral. *Applied Statistics* **17**, 168–187.
Dempster, A.P., Laird, N.M., Rubin, D.B. (1977). Maximum likelihood from incomplete data via the EM algorithm. *Journal of the Royal Statistical Society, Series B* **39**, 1–38.
du Toit, M. (2003). *IRT from SSI*. Scientific Software International, Lincolnwood, IL.
Fechner, G.T. (1860). *Elemente der Psychophysik*. Breitkopf & Hartel, Leipzig.
Finney, D.J. (1952). *Probit Analysis: A Statistical Treatment of the Sigmoid Response Curve*, 2nd ed. Cambridge University Press, London.
Fisher, R.A., Yates, F. (1938). *Statistical Tables for Biological, Agricultural and Medical Research*. Hafner, New York.
Gibbons, R.D., Hedeker, D. (1992). Full-information item bi-factor analysis. *Psychometrika* **57**, 423–436.

Glas, C.A.W., Wainer, H., Bradlow, E.T. (2000). Maximum marginal likelihood and expected a priori estimates for the testlet response model. In: van der Linden, W.J., Glas, C.A.W. (Eds.), *Computerized Adaptive Testing: Theory and Practice*. Kluwer–Nijhoff, Boston.
Gulliksen, H. (1950). *Theory of Mental Tests*. Wiley, New York.
Gumbel, E.J. (1961). Bivariate logistic distributions. *Journal of the American Statistical Association* **56**, 335–349.
Haberman, S.J. (1977). Log-linear models and frequency tables with small expected cell counts. *Annals of Statistics* **5**, 1148–1169.
Haley, D.C. (1952). Estimation of the dosage mortality relationship when the dose is subject to error. Technical Report 15. Applied Mathematics and Statistics Laboratory, Stanford University.
Harman, H.H. (1967). *Modern Factor Analysis*. University of Chicago Press, Chicago.
Hastings Jr., C. (1955). *Approximations for Digital Computers*. Princeton University Press, Princeton, NJ.
Holland, P.W., Wainer, H. (1993). *Differential Item Functioning*. Erlbaum, Hillsdale, NJ.
Holzinger, K.J., Swineford, F. (1937). The bi-factor method. *Psychometrika* **2**, 41–54.
Huber, P., Ronchetti, E., Victoria-Feser, M.-P. (2004). Estimation of generalized linear latent variable models. *Journal of the Royal Statistical Society, Series B* **66**, 893–908.
Jöreskog, K.J. (1967). Some contributions to maximum likelihood factor analysis. *Psychometrika* **32**, 165–178.
Jöreskog, K.J. (1994). On the estimation of polychoric correlations and their asymptotic covariance matrix. *Psychometrika* **59**, 381–389.
Jöreskog, K.G., Moustaki, I. (2001). Factor analysis of ordinal variables: a comparison of three approaches. *Multivariate Behavioral Research* **36**, 347–387.
Kennedy, W.J., Gentle, J.E. (1980). *Statistical Computing*. Marcel Dekker, New York.
Kiefer, J., Wolfowitz, J. (1956). Consistency of the maximum likelihood estimator in the presence of infinitely many incidental parameters. *Annals of Mathematical Statistics* **27**, 887.
Koehler, K., Larntz, K. (1980). An empirical investigation of goodness-of-fit statistics for sparse multinomials. *Journal of the American Statistical Association* **75**, 336–344.
Lazarsfeld, P.F. (1958). Latent structure analysis. In: Koch, S. (Ed.), *Psychology: A Study of a Science*, vol. III. McGraw-Hill, New York.
Lawley, D.N. (1943). On problems connected with item selection and test construction. *Proceedings of the Royal Society of Edinburgh* **61**, 273–287.
Likert, R. (1932). A technique for the measurement of attitudes. *Archives of Psychology*, Monograph 140.
Lindquist, E.F. (1953). *Design and Analysis of Experiments in Psychology and Education*. Hougton Mifflin, Boston.
Lord, F.M. (1952). A theory of test scores. *Psychometric Monograph*, No. 7.
Lord, F.M. (1953). An application of confidence intervals and of maximum likelihood to the estimation of an examinee's ability. *Psychometrika* **18**, 57–76.
Lord, F.M. (1980). *Applications of Item Response Theory to Practical Testing Problems*. Erlbaum, Hillsdale, NJ.
Lord, F.M., Novick, M.R. (1968). *Statistical Theories of Mental Test Scores*. Addison-Wesley, Reading, MA. (With contributions by A. Birnbaum).
Louis, T.A. (1982). Finding the observed information matrix when using the EM algorithm. *Journal of the Royal Statistical Society, Series B* **44**, 226–233.
Luce, R.D. (1959). *Individual Choice Behavior*. Wiley, New York.
MacFadden, D. (1974). Conditional logit analysis of qualitative choice behavior. In: Zarembka, P. (Ed.), *Frontiers of Econometrics*. Academic Press, New York.
Masters, G.N. (1982). A Rasch model for partial credit scoring. *Psychometrika* **47**, 149–174.
Maydeu-Olivares, A., Joe, H. (2005). Limited and full information estimation and testing in 2^n contingency tables: a unified framework. *Journal of the American Statistical Association* **100**, 1009–1020.
Meng, X.L., Schilling, S. (1996). Fitting full-information factors models and an empirical investigation of bridge sampling. *Journal of the American Statistical Association* **91**, 1254–1267.
Meredith, W. (1993). Measurement invariance, factor analysis and factorial invariance. *Psychometrika* **58**, 525–543.
Mislevy, R.J. (1984). Estimating latent distributions. *Psychometrika* **49**, 359–381.

Mislevy, R.J., Johnson, E.G., Muraki, E. (1992). Scaling procedures in NAEP. *Journal of Educational Statistics* **17**, 131–154.
Morris, C. (1975). Central limit theorems for multinomial sums. *Annals of Statistics* **3**, 365–384.
Moustaki, I. (2003). A general class of latent variable models for ordinal manifest variables with covariate effects on the manifest and latent variables. *British Journal of Mathematical and Statistical Psychology* **56**, 337–357.
Muraki, E. (1992). A generalized partial credit model: application of an EM algorithm. *Applied Psychological Measurement* **16**, 159–176.
Muraki, E., Bock, R.D. (1991). *PARSCALE: Parametric Scaling of Rating Data*. Scientific Software, Chicago.
Muthén, B. (1978). Contributions to factor analysis for dichotomous variables. *Psychometrika* **43**, 551–560.
Neyman, J.A., Scott, E.L. (1948). Consistent estimates based on partially consistent observations. *Econometrika* **16**, 1–22.
Nishisato, S. (1994). *Elements of Dual Scaling: An Introduction to Practical Data Analysis*. Erlbaum, Hillsdale, NJ.
Patz, R.J., Junker, B.W. (1999a). A straightforward approach to Markov chain Monte Carlo methods for item response models. *Journal of Educational and Behavioral Statistics* **24**, 146–178.
Patz, R.J., Junker, B.W. (1999b). Applications and extensions of MCMC in IRT: multiple item types, missing data, and rated responses. *Journal of Educational and Behavioral Statistics* **24**, 342–366.
Rasch, G. (1960). *Probabilistic models for some intelligence and attainment tests*. Nielson Lydiche, Copenhagen. (For Denmarks Paedagogiske Institut).
Rasch, G. (1961). On general laws and the meaning of measurement in psychology. In: *Proceedings of the Fourth Berkeley Symposium on Mathematical Statistics and Probability*, vol. 4, pp. 321–324.
Reiser, M. (1996). Analysis of residuals for the multinomial item response model. *Psychometrika* **61**, 509–528.
Reiser, M., VandenBerg, M. (1994). Validity of the chi-square test in dichotomous variable factor analysis when expected frequencies are small. *British Journal of Mathematical and Statistical Psychology* **47**, 85–107.
Samejima, F. (1969). Estimation of latent trait ability using a response pattern of graded scores. *Psychometrika Monograph Supplement* **17**.
Samejima, F. (1972). A general model for free-response data. *Psychometrika Monograph Supplement* **18**.
Samejima, F. (1979). A new family of models for the multiple choice item. Research Report No. 79-4. Department of Psychology, University of Tennessee.
Sammel, R.D., Ryan, L.M., Legler, J.M. (1997). Latent variable models for mixed discrete and continuous outcomes. *Journal of the Royal Statistical Society B* **59**, 667–678.
Schilling, S.G., Bock, R.D. (2005). High-dimensional maximum marginal likelihood item factor analysis by adaptive numerical integration. *Psychometrika* **70**, 533–555.
Sclove, S. (1987). Application of model-selection criteria to some problems of multivariate analysis. *Psychometrika* **52**, 333–343.
Skrondal, A., Rabe-Hesketh, A. (2004). *Generalized Latent Variable Modeling*. Chapman & Hall/CRC, Boca Raton, FL.
Straud, A.H., Sechrest, D. (1966). *Gaussian Quadrature Formulas*. Prentice-Hall, Englewood Cliffs, NJ.
Stuart, A. (1958). Equally correlated variates and the multinormal integral. *Journal of the Royal Statistical Society Series B* **20**, 373–378.
Thissen, D. (1991). *MULTILOG: Multiple Category Item Analysis and Test Scoring Using Item Response Theory*. Scientific Software, Chicago.
Thissen, D., Steinberg, L. (1984). A response model for multiple choice items. *Psychometrika* **49**, 501–519.
Thissen, D., Steinberg, L. (1986). A taxonomy of item response models. *Psychometrika* **51**, 567–577.
Thurstone, L.L. (1927). A law of comparative judgment. *Psychological Review* **34**, 273–286.
Thurstone, L.L. (1929). Theory of attitude measurement. *Psychological Review* **36**, 222–241.
Thurstone, L.L. (1947). *Multiple Factor Analysis*. University of Chicago Press, Chicago.
Thurstone, L.L. (1959). An experiment in the prediction of food preference and the prediction of choice. In: Thurstone, L.L. (Ed.), *The Measurement of Values*. University of Chicago Press, Chicago. Chapter 14.
Verhelst, N.D., Glas, C.A.W. (1993). A dynamic generalization of the Rasch model. *Psychometrika* **58**, 395–415.

Wainer, H., Kiely, G.L. (1987). Item clusters and computerized adaptive testing: A case for testlets. *Journal of Educational Measurement* **24**, 185–201.

Wainer, H. (Ed.) (1990). *Computerized Adaptive Testing: A Primer*. Erlbaum, Hillsdale, NJ.

Wainer, H., Sireci, S.G., Thissen, D. (1991). Differential testlet functioning: Definition and detection. *Journal of Educational Measurement* **28**, 197–219.

Wang, X., Bradlow, E.T., Wainer, H. (2002). A general Bayesian model for testlets: theory and applications. *Applied Psychological Measurement* **26**, 1090–1128.

Zimowski, M.F., Muraki, E., Mislevy, R.J., Bock, R.D. (1996). *BILOG-MG: Multiple-Group IRT Analysis and Test Maintenance*. Scientific Software, Chicago.

Handbook of Statistics, Vol. 26
ISSN: 0169-7161

DOI: 10.1016/S0169-7161(06)26016-4

16

Rasch Models

Gerhard H. Fischer

1. Some history of the Rasch model

Georg Rasch (1901–1980) showed remarkable mathematical talent already at school. In 1919 he decided to study mathematics at the Faculty of Science of the University of Copenhagen and graduated in 1925 with a master's degree in mathematics. In 1930 he defended his dissertation on the application of matrix algebra to difference and differential equations and became Doctor of Science in mathematics. At that time, in Denmark a doctorate was considered a qualification for becoming a full professor. Academic positions for mathematicians, however, were scarce in Denmark in the 1930s, so Rasch had to seek employment as a statistical consultant, amongst others, for the University Department of Hygiene and for the Danish Serum Institute in Copenhagen. To learn about the latest developments in statistics, he 1935/36 spent a year with R.A. Fisher at the Galton Laboratory in London, where he followed the lectures of R.A. Fisher, E. Pearson, and J. Neyman. What impressed him most was Fisher's concept of sufficiency; sufficiency henceforth became a central issue in his thinking.

After World War II, Rasch became involved in psychometrics, both in his capacity as a consultant for projects of the Military Psychology Group at the Danish Ministry of Defense and of the Danish Ministry of Social Welfare. The starting point for the development of his models came in 1952 when he was to analyze follow-up data based on reading tests for adolescents who in elementary school had held a poor reading record. Rasch introduced multiplicative Poisson models where the number of words of a text T_i read within a fixed time by testee (subject) S_v, denoted a_{vi} – or, similarly, the number of reading errors S_v made when reading text T_i – was considered a realization of a Poisson variable. In the case of reading speed, for instance, the parameter λ_{vi} of the Poisson variable was assumed to express the latent reading speed of testee S_v on text T_i. To explain reading performance, however, in terms of both an individual and a text property, Rasch split λ_{vi} into a person factor ξ_v, the reading ability of S_v, and a text factor ϵ_i, the easiness of T_i, so that his model became

$$P(a_{vi}) = \frac{\lambda_{vi}^{a_{vi}}}{a_{vi}!} \exp(-\lambda_{vi}) = \frac{(\xi_v \epsilon_i)^{a_{vi}}}{a_{vi}!} \exp(-\xi_v \epsilon_i). \tag{1}$$

Analogously, if the same person had to read another text T_j under otherwise similar conditions, and if stochastic independence of the outcomes in both tests was assumed, the probability of S_v reading a_{vj} words of text T_j was assumed to be

$$P(a_{vj}) = \frac{(\xi_v \epsilon_j)^{a_{vj}}}{a_{vj}!} \exp(-\xi_v \epsilon_j). \tag{2}$$

As a mathematician, Rasch was of course familiar with the following property of two independent Poisson variables (with parameters λ_{vi} and λ_{vj}, respectively): the probability of jointly observing the outcomes a_{vi} and a_{vj}, given their sum $a_{vi} + a_{vj} =: a_{v.}$,[1] follows a binomial probability function with parameters $a_{v.}$ and $\lambda_{vi}/(\lambda_{vi}+\lambda_{vj}) =: \pi_{vij}$. Inserting $\xi_v \epsilon_i$ for λ_{vi} and $\xi_v \epsilon_j$ for λ_{vj}, respectively, Rasch saw that the person parameter ξ_v canceled, yielding $\pi_{vij} = \epsilon_i/(\epsilon_i + \epsilon_j) =: \pi_{ij}$. Thus, he obtained the remarkable result that the conditional distribution of a_{vi} and a_{vj}, given their sum $a_{v.}$, was binomial with a probability function no longer involving ξ_v,

$$P(a_{vi}, a_{vj} | a_{v.}) = \binom{a_{v.}}{a_{vi}} \left(\frac{\epsilon_i}{\epsilon_i + \epsilon_j}\right)^{a_{vi}} \left(\frac{\epsilon_j}{\epsilon_i + \epsilon_j}\right)^{a_{vj}}. \tag{3}$$

The implications of this somewhat surprising result were not immediately clear to Rasch (cf. Wright, 1980, p. xviii), but eventually he began to realize the significance of his finding with respect to measurement in behavioral science: data from all testees who had read the same two texts T_i and T_j could be pooled to obtain a (maximum likelihood) estimate of π_{ij}, from which in turn an estimate of the quotient of the two text easiness parameters, ϵ_i/ϵ_j, could be derived. This meant a complete *separation* of the text parameters ϵ_i from the person parameters ξ_v and thus enabled a measurement of the *relative* difficulties of the texts that *generalized* over the individual testees. In other words, the result was – albeit in a very specific sense – *independent of the sample of testees*. Symmetrically, testees (i.e., their person parameters) could be compared independently of the sample of texts that happened to be used as test material. Instrumental for this was that each parameter λ_{vi} was conceived as a product of the respective testee's ability and the text's easiness, which had the consequence that a 'comparator' – namely, the conditional probability function (3) – existed depending only on the ϵ_i.

When Rasch in 1952/1953 came to analyze intelligence test data for the Military Psychology Group, he sought to carry the same principles over to the area of item analysis. First, he looked for a suitable *item response function* (IRF) $P(+|S_v, I_i) = f(\xi_v, \epsilon_i)$, where '+' denoted a correct response, ξ_v the testee's 'ability', and ϵ_i the 'easiness' of item I_i. A particularly simple algebraic function mapping the positive reals x on the open interval (0, 1) appeared to be $f = x/(1+x)$, for $x > 0$ (cf. Rasch, 1960, pp. 74–75, or Andersen and Olsen, 2001, p. 15). Starting from this function and inserting for x the product $\xi_v \epsilon_i$, with $\xi_v > 0$ and $\epsilon_i > 0$, Rasch arrived at the model

$$P(+|\xi_v, \epsilon_i) = \frac{\xi_v \epsilon_i}{1 + \xi_v \epsilon_i}, \tag{4}$$

which actually shares with the multiplicative Poisson model the property of separability of the two classes of parameters, item and person parameters (see Sections 2–4 below). This model is now generally denoted the 'Rasch Model' (RM), but is usually

[1] The notation '$y := x$' or '$x =: y$' means that y is defined by x.

presented in reparameterized form, using a logarithmic scale for item and person parameters: $\theta_v := \ln(\xi_v)$ as testee *ability* and $\beta_i := -\ln(\epsilon_i)$ as item *difficulty*.

Generalizing the statistical property of separability of these two classes of parameters, Rasch later formulated the general methodological or epistemic principle that scientific comparisons between *objects* of interest should be carried out independently of the set of *agents* which are instrumental for the comparisons, and vice versa. Comparisons satisfying this demand he denoted 'specifically objective' (Rasch, 1967). During the last two decades of his life, he dedicated all his efforts to the further development and study of *specific objectivity* (SO; Rasch, 1968, 1972, 1977).

For a detailed account of Rasch's life and work, cf. Andersen and Olsen (2001), and on the more general concept of 'Rasch Models', cf. Rost (2001).

2. Some basic concepts and properties of the RM

The RM for dichotomous responses (in the following denoted '+' and '−') is usually written as

$$P(X_{vi} = 1|\theta_v, \beta_i) = \frac{\exp(\theta_v - \beta_i)}{1 + \exp(\theta_v - \beta_i)}, \tag{5}$$

where X_{vi} denotes the binary (Bernoulli) response variable with realization $x_{vi} = 1$ if testee S_v gives response '+' to item I_i, and $x_{vi} = 0$ if S_v's response to I_i is '−'. Parameter θ_v is S_v's latent ability, and β_i the difficulty of item I_i; both parameters are defined on the entire real axis. It is assumed that each response is either '+' or '−', whence it follows that $P(X_{vi} = 0|\theta_v, \beta_i) = 1 - P(X_{vi} = 1|\theta_v, \beta_i)$, or more explicitly,

$$P(X_{vi} = 0|\theta_v, \beta_i) = \frac{1}{1 + \exp(\theta_v - \beta_i)}. \tag{6}$$

The IRFs (5) will be denoted by $f_i(\theta)$ for short where convenient. These IRFs, for $i = 1, \ldots, k$, are parallel curves, they are continuous in both parameters, increasing[2] in θ_v and decreasing in β_i. This corresponds to the intuitive notion that, for any fixed item, the probability of a right response increases with the testee's ability, and for any fixed person, the probability of a right response decreases when the difficulty of the item is increased. Moreover, the definition of the IRFs (5) implies $\lim_{\theta \to -\infty} f_i(\theta) = 0$ and $\lim_{\theta \to \infty} f_i(\theta) = 1$ for all I_i, that is, sufficiently able testees solve any item with a probability close to 1, while guessing of the right response is excluded. It is easy to show that all IRFs have the same maximum slope at $\theta = \beta_i$ (i.e., the same amount of 'discrimination').

For fixed parameters θ_v and ϵ_i, all item responses are assumed to be stochastically independent; therefore, the probability of any response pattern of a testee in a test of length k is the product of k probabilities (5) (for responses '+') or (6) (for responses '−'), respectively. In psychometrics, this kind of independence is denoted *local independence*.

[2] In this chapter, 'increasing' always means 'strictly monotone increasing', 'decreasing' means 'strictly monotone decreasing'.

A consequence of (5) is that the RM defines the person parameters θ_v and the item parameters β_i only up to an arbitrary additive normalization constant c; the latter is usually specified by either setting one item parameter to zero (e.g., $\beta_1 = 0$, thus making item I_1 the anchor or 'normalization' item), or setting $\sum_i \beta_i = 0$, or assuming $E(\Theta) = 0$ for some population of testees. This does not yet fully answer, however, the question of the uniqueness of the scales of θ and β; as will be seen below, this problem is more complicated.

Based on the definition of the RM in (5), a number of useful formulas will now be given, both to illustrate characteristic properties of the RM and to free the derivation and discussion of estimation equations and tests of fit in later sections from too many elementary algebraic manipulations.

Consider the parameters θ_v, $v = 1, \ldots, n$, and β_i, $i = 1, \ldots, k$, to be fixed constants, and assume local independence. Then the likelihood of response pattern $\boldsymbol{x}_v = (x_{v1}, \ldots, x_{vk})$ of testee S_v can be expressed in terms of θ_v and vector $\boldsymbol{\beta} = (\beta_1, \ldots, \beta_k)'$ as

$$L(\boldsymbol{x}_v|\theta_v, \boldsymbol{\beta}) = \prod_i f_i(\theta_v)^{x_{vi}}\left[1 - f_i(\theta_v)\right]^{1-x_{vi}}$$
$$= \frac{\prod_i \exp[x_{vi}(\theta_v - \beta_i)]}{\prod_i [1 + \exp(\theta_v - \beta_i)]} = \frac{\exp(r_v\theta_v - \sum_i x_{vi}\beta_i)}{\prod_i [1 + \exp(\theta_v - \beta_i)]}, \tag{7}$$

where $r_v = \sum_i x_{vi}$ denotes the 'person marginal sum' or 'raw score' of S_v, and $s_i = \sum_v x_{vi}$ the 'item marginal sum' of I_i. The likelihood function of a complete $(n \times k)$ item score matrix $\boldsymbol{X}$ therefore becomes

$$L(\boldsymbol{X}|\boldsymbol{\theta}, \boldsymbol{\beta}) = \prod_v \prod_i \frac{\exp[x_{vi}(\theta_v - \beta_i)]}{1 + \exp(\theta_v - \beta_i)} = \frac{\exp\left(\sum_v r_v\theta_v - \sum_i s_i\beta_i\right)}{\prod_v \prod_i [1 + \exp(\theta_v - \beta_i)]}, \tag{8}$$

with $\boldsymbol{\theta} = (\theta_1, \ldots, \theta_n)'$ the n-vector of person parameters, and the log-likelihood,

$$\ln L(\boldsymbol{X}|\boldsymbol{\theta}, \boldsymbol{\beta}) = \sum_v r_v\theta_v - \sum_i s_i\beta_i - \sum_v \sum_i \ln\left[1 + \exp(\theta_v - \beta_i)\right]. \tag{9}$$

As can be seen from (8) and (9), the likelihood function depends on the data only via the raw scores and item marginal sums, that is, these quantities are jointly *sufficient statistics* for the parameter vectors $\boldsymbol{\theta}$ and $\boldsymbol{\beta}$. This is not only a remarkable property of the RM, it even characterizes the RM (within a framework of locally independent items with increasing IRFs that have upper limits 1 and lower limits 0; see Section 3). In view of the tradition in psychological assessment of utilizing the raw score as a quantification of a testee's test achievement, it is of considerable relevance to ascertain whether the RM fits a given set of data – at least with a sufficient degree of approximation; if this is the case, then the use of the raw score as a summary of an individual's test achievement is grounded on a solid footing. Being able to replace the complete item response vector $(x_{v1}, \ldots, x_{vk})$ by one single number r_v means an enormous data reduction and enables an immediate intuitive comparison of test performances: the greater r_v, the better the achievement.

Another remarkable immediate consequence of the form of likelihood (8), which is independent of the entries in $\boldsymbol{X}$ except for the marginal sum vectors $\boldsymbol{r}$ and $\boldsymbol{s}$, is that all item score matrices with the same marginal sums must have also the same likelihood (probability). Hence the latter is equal to K_{rs}^{-1}, where K_{rs} denotes the number of feasible item score matrices with marginal sum vectors $\boldsymbol{r}$ and $\boldsymbol{s}$.

It has been mentioned above that G. Rasch intended the RM to have the following property: a nontrivial likelihood function should exist which is a function of the item parameters but is independent of the person parameters and is thus serviceable for a specifically objective comparison of the items. Indeed, he showed that this requirement is satisfied by the conditional likelihood of the data given the raw scores of all testees. To see this, consider first the probability that testee S_v obtains raw score r, which is obtained by summing the probabilities (7) over all response patterns $\boldsymbol{y} = (y_1, \ldots, y_k)$ that are compatible with a given raw score $r = \sum_i y_i$,

$$L(r|\boldsymbol{\beta}) = \frac{\exp(r\theta_v)\sum_{\boldsymbol{y}}\prod_i \exp(-y_i\beta_i)}{\prod_i[1+\exp(\theta_v-\beta_i)]}, \tag{10}$$

which can be written more compactly, with $\epsilon_i := \exp(-\beta_i)$ for short, as

$$L(r|\boldsymbol{\beta}) = \frac{\exp(r\theta_v)\gamma_r(\boldsymbol{\epsilon})}{\prod_i[1+\exp(\theta_v-\beta_i)]}, \tag{11}$$

where

$$\gamma_r(\boldsymbol{\epsilon}) = \sum_{\boldsymbol{y}}\prod_i \exp(-y_i\beta_i) = \sum_{\boldsymbol{y}}\prod_i \epsilon_i^{y_i}. \tag{12}$$

This expression is a function of fundamental importance in algebra, the so-called 'elementary symmetric function' of order r of $\epsilon_1, \ldots, \epsilon_k$; the latter parameters are abbreviated as vector $\boldsymbol{\epsilon}$. More explicitly, the elementary symmetric functions are

$$\begin{aligned}
&\gamma_0(\boldsymbol{\epsilon}) = 1,\\
&\gamma_1(\boldsymbol{\epsilon}) = \epsilon_1+\epsilon_2+\cdots+\epsilon_k,\\
&\gamma_2(\boldsymbol{\epsilon}) = \epsilon_1\epsilon_2+\epsilon_1\epsilon_3+\cdots+\epsilon_{k-1}\epsilon_k,\\
&\vdots\\
&\gamma_k(\boldsymbol{\epsilon}) = \epsilon_1\epsilon_2\cdots\epsilon_k.
\end{aligned} \tag{13}$$

For formal completeness, it is useful to define $\gamma_r(\boldsymbol{\epsilon}) = 0$ for $r < 0$ and $r > k$.

Now the conditional likelihood of a response pattern $\boldsymbol{x}_v$, given its raw score r, is obtained by dividing the likelihood function (7) by (11): thereby, both the terms $\exp(r\theta_v)$ and $\prod_i[1+\exp(\theta_v-\beta_i)]$ cancel, leaving

$$L_C(\boldsymbol{x}_v|\boldsymbol{\beta}, r) = \frac{\exp(-\sum_i x_{vi}\beta_i)}{\gamma_r(\boldsymbol{\epsilon})} = \frac{\prod_i \epsilon_i^{x_{vi}}}{\gamma_r(\boldsymbol{\epsilon})}. \tag{14}$$

And the conditional likelihood of the item score matrix $\boldsymbol{X}$, given the vector of all raw scores, $\boldsymbol{r} = (r_1, \ldots, r_n)'$, results from taking the product of (14) over $v = 1, \ldots, n$,

$$L_C(\boldsymbol{X}|\boldsymbol{\beta}, \boldsymbol{r}) = \frac{\prod_v \exp(-\sum_i x_{vi}\beta_i)}{\prod_v \gamma_{r_v}(\boldsymbol{\epsilon})} = \frac{\exp(-\sum_i s_i\beta_i)}{\prod_r \gamma_r(\boldsymbol{\epsilon})^{n_r}}, \tag{15}$$

where n_r denotes the number of testees with raw score r. This is the desired comparator function which does not contain the person parameters and thus can serve as a basis for an estimation of the parameters β_i, in other words, for a 'specifically objective comparison' (or 'measurement') of the item difficulties (see Section 4). Its logarithm is

$$\ln L_C(\boldsymbol{X}|\boldsymbol{\beta}, \boldsymbol{r}) = -\sum_i s_i\beta_i - \sum_r n_r \ln \gamma_r(\boldsymbol{\epsilon}). \tag{16}$$

Within this framework, the following formulae will also be useful. The probability of response $X_i = 1$, given that the raw score is r, is obtained by summing (14) over all response patterns $\boldsymbol{y}^*$ that have $y_i^* = 1$ and are compatible with raw score r,

$$P(X_i = 1|\boldsymbol{\beta}, r) = \sum_{\boldsymbol{y}^*} \frac{\prod_i \epsilon_i^{y_i^*}}{\gamma_r(\boldsymbol{\epsilon})} = \frac{\epsilon_i \gamma_{r-1}^{(i)}(\boldsymbol{\epsilon})}{\gamma_r(\boldsymbol{\epsilon})} =: \pi_{ri}. \tag{17}$$

The summation actually runs over all combinations of responses to items I_l with $l \neq i$ and with raw score $r - 1$. This yields the factor $\gamma_{r-1}^{(i)}(\boldsymbol{\epsilon})$ which denotes the elementary symmetric function of order $r - 1$ of the $k - 1$ parameters ϵ_l with $l \neq i$.

If the number of responses '+' given to item I_i in the subsample of testees with raw score r is denoted by s_{ri}, it follows from (17) that the conditional expectation of S_{ri}, given that the number of testees with raw score r is n_r, is $E(S_{ri}|\boldsymbol{\beta}, n_r) = n_r\pi_{ri}$. Notice that (17) implies

$$\sum_i \pi_{ri} = \frac{\sum_i \epsilon_i \gamma_{r-1}^{(i)}(\boldsymbol{\epsilon})}{\gamma_r(\boldsymbol{\epsilon})} = r, \tag{18}$$

which in turn yields, as a byproduct, that $\sum_i \epsilon_i \gamma_{r-1}^{(i)}(\boldsymbol{\epsilon}) = r\gamma_r(\boldsymbol{\epsilon})$.

Analogously, the conditional probability of S_v having response combination $X_{vi} = 1$ and $X_{vj} = 1$, given that S_v's raw score is r, can be seen to be

$$P(X_{vi}X_{vj} = 1|\boldsymbol{\beta}, r) = \frac{\epsilon_i\epsilon_j \gamma_{r-2}^{(ij)}(\boldsymbol{\epsilon})}{\gamma_r(\boldsymbol{\epsilon})} =: \pi_{rij}, \tag{19}$$

where $\gamma_{r-2}^{(ij)}(\boldsymbol{\epsilon})$ denotes the elementary symmetric function of order $r - 2$ of the ϵ_l with $l \neq i, j$; and the expected number of testees with raw score r having $X_{vi}X_{vj} = 1$, is $n_r\pi_{rij}$.

G. Rasch used to consider the person and item parameters as constants ('fixed effects') and strongly favored the conditional likelihood approach – based on the conditional likelihood function (15) – to estimating the item parameters and to testing of fit of the RM. Another way to look upon the data, however, is to consider the sample of testees a random sample from some population with latent distribution G of the parameters θ. Then the likelihood function of the observed response vector $\boldsymbol{x}_v$, its so-called 'marginal likelihood function', is different from (7): the likelihood of response pattern $\boldsymbol{x}_v$ now is the integral

$$L_M(\boldsymbol{x}_v|\boldsymbol{\beta}, G) = \int_{-\infty}^{\infty} L(\boldsymbol{x}_v|\theta, \boldsymbol{\beta})\,\mathrm{d}G(\theta)$$
$$= \int_{-\infty}^{\infty} \prod_{i=1}^{k} f_i(\theta)^{x_{vi}}\big[1 - f_i(\theta)\big]^{1-x_{vi}}\,\mathrm{d}G(\theta), \quad (20)$$

where the person parameters θ are 'integrated out' of the likelihood function.

Two ways of dealing with $G(\theta)$ are possible: (a) either some parametric form of the distribution is assumed (mostly a normal distribution), or (b) $G(\theta)$ is conceived as a discrete distribution defined on a set of support points θ_l, $l = 1, \ldots, m$ (for details, see Section 4). But even in case (a) with a continuous distribution, the integral $G(\theta)$ has to be approximated, for all practical purposes, by a sum over m quadrature points with weights g_l, for instance, if Gauss–Hermite quadrature is used for the normal distribution (cf. Abramowitz and Stegun, 1974), or if adaptive quadrature (cf. Skrondal and Rabe-Hesketh, 2004) is applied. In this case ('parametric MML'), only a few parameters of the distribution have to be estimated, for instance, σ^2 of the normal distribution G (μ being set to zero for normalization; see Section 4 below). In case (b) (i.e., 'semiparametric MML'), the θ_l and their weights g_l have to be estimated.

In either case, the marginal likelihood of response vector $\boldsymbol{x}_v$ can be written, analogously to (7),

$$L_M(\boldsymbol{x}_v|\boldsymbol{\beta}, G) = \sum_{l=1}^{m} g_l \prod_{i=1}^{k} f_i(\theta_l)^{x_{vi}}\big[1 - f_i(\theta_l)\big]^{1-x_{vi}}$$
$$= \sum_{l=1}^{m} g_l \frac{\exp(r_v\theta_l - \sum_i x_{vi}\beta_i)}{\prod_i[1 + \exp(\theta_l - \beta_i)]}$$
$$= \exp\Big(-\sum_i x_{vi}\beta_i\Big)\sum_{l=1}^{m} g_l \frac{\exp(r_v\theta_l)}{\prod_i[1 + \exp(\theta_l - \beta_i)]}. \quad (21)$$

From this, the marginal probability of observing score r is derived by summing (21) over all score patterns $\boldsymbol{y}$ with the same raw score r,

$$P_M(r|\boldsymbol{\beta}, G) = \sum_{\boldsymbol{y}|r} L_M(\boldsymbol{y}|\boldsymbol{\beta}, G) = \gamma_r(\boldsymbol{\epsilon})\sum_{l=1}^{m} g_l \frac{\exp(r\theta_l)}{\prod_i[1 + \exp(\theta_l - \beta_i)]}. \quad (22)$$

Finally, the marginal log-likelihood of item response matrix $\boldsymbol{X}$ is obtained as the sum over testees of logarithms of terms (21),

$$\ln L_M(\boldsymbol{X}|\boldsymbol{\beta}, G) = -\sum_i s_i\beta_i + \sum_r n_r \ln\Bigg[\sum_l g_l \frac{\exp(r\theta_l)}{\prod_i[1 + \exp(\theta_l - \beta_i)]}\Bigg]$$
$$= -\sum_i s_i\beta_i + \sum_r n_r \ln\bigg[\frac{P_M(r|\boldsymbol{\beta}, G)}{\gamma_r(\boldsymbol{\epsilon})}\bigg]. \quad (23)$$

An interesting relationship between this marginal log-likelihood function and the conditional log-likelihood function (16) is seen when (23) is rewritten as

$$\begin{aligned}\ln L_M(\boldsymbol{X}|\boldsymbol{\beta}, G) &= -\sum_i s_i\beta_i - \sum_r n_r \ln \gamma_r(\boldsymbol{\epsilon}) + \sum_r n_r \ln P_M(r|\boldsymbol{\beta}, G)\\ &= \ln L_C(\boldsymbol{X}|\boldsymbol{\beta}, \boldsymbol{r}) + \ln L_M(\boldsymbol{n}|\boldsymbol{\beta}, G),\end{aligned} \tag{24}$$

where $\boldsymbol{n} = (n_0, n_1, \ldots, n_k)$ denotes the vector of raw score frequencies, and $L_M(\boldsymbol{n}|\boldsymbol{\beta}, G)$ is the multinomial marginal likelihood of $\boldsymbol{n}$. As (24) reveals, the marginal log-likelihood of the item score matrix $\boldsymbol{X}$ can be written as the sum of the conditional log-likelihood (16) and the marginal log-likelihood of the raw score frequencies. This plays an important role for understanding the relationship between conditional and non-parametric marginal maximum likelihood estimation (see Section 4).

It will turn out useful to have formulae analogous to (17) and (19) also under the marginal likelihood framework. The probability that testee S_v attains raw score r and gives response $X_{vi} = 1$ is obtained by multiplying (17) by (22),

$$\begin{aligned}&P_M\big[(X_{vi} = 1) \wedge (R_v = r)|\boldsymbol{\beta}, G\big]\\ &\quad= P(X_{vi} = 1|r, \boldsymbol{\beta}) P_M(r|\boldsymbol{\beta}, G)\\ &\quad= \epsilon_i \gamma_{r-1}^{(i)}(\boldsymbol{\epsilon}) \sum_l g_l \frac{\exp(r\theta_l)}{\prod_i [1 + \exp(\theta_l - \beta_i)]} =: \pi_M(r, i).\end{aligned} \tag{25}$$

Therefore, the expected number of testees being in raw score group r and giving response $X_{vi} = 1$, is $n\pi_M(r, i)$. Similarly, the probability that testee S_v attains raw score r and gives responses $X_{vi} = X_{vj} = 1$ is obtained by multiplying (19) with (22),

$$\begin{aligned}&P_M\big[(X_{vi} X_{vj} = 1) \wedge (R_v = r)|\boldsymbol{\beta}, G\big]\\ &\quad= P(X_{vi} X_{vj} = 1|r, \boldsymbol{\beta}) P_M(r|\boldsymbol{\beta}, G)\\ &\quad= \epsilon_i \epsilon_j \gamma_{r-2}^{(ij)}(\boldsymbol{\epsilon}) \sum_l g_l \frac{\exp(r\theta_l)}{\prod_i [1 + \exp(\theta_l - \beta_i)]} =: \pi_M(r, i, j),\end{aligned} \tag{26}$$

and the expected number of testees with raw score r and $X_{vi} X_{vj} = 1$, is $n\pi_M(r, i, j)$. Notice that (25) and (26), other than (17) and (19), are unconditional (marginal) probabilities.

3. Characterizations and scale properties of the RM

As a next step it is now obvious to ask whether there are any other IRT models that share the remarkable properties of sufficiency of raw scores and item marginals and/or of specific objectivity with the RM. To deal with this question, it is necessary first to define a formal framework within which IRT models are considered. We shall look at models with increasing IRFs with lower limits 0 and upper limits 1. Within that realm, the answer to the above question is 'no': a *family of RMs* follows both from sufficiency of the raw score and from SO. To state this result precisely, however, we have to introduce a notational convention: as long as no specific metric has been established for the latent trait dimension, the person parameters will be denoted by ξ, and the IRFs by $g_i(\xi)$ or $g(\xi, \delta_i)$ with item difficulty parameters δ_i. Notice that in this case the parameters ξ and δ_i can at any time be subjected to increasing scale transformations, entailing changes of the form of the IRFs. If, however, person and item parameters have been transformed such that the IRFs attain the form $f(\phi(\xi)-\psi(\delta_i))$, where f, ϕ, and ψ are suitably chosen increasing functions, the resulting IRFs with subtractive parameters will be denoted as $f_i(\theta-\beta_i)$, with new parameters $\theta := \phi(\xi)$ and $\beta_i := \psi(\delta_i)$. The test length k is always assumed to be $k \geqslant 2$.

Let a unidimensional item response model be assumed for a set of items I_i, satisfying the following assumptions:

ASSUMPTION i. Continuous increasing IRFs $g_i(\xi)$ defined on $\mathbb{R}$, for all I_i;

ASSUMPTION ii. Lower limits $\lim_{\xi\to-\infty} g_i(\xi) = 0$ and upper limits $\lim_{\xi\to+\infty} g_i(\xi) = 1$, for all I_i;

ASSUMPTION iii. Local stochastic independence of the item responses.

These assumptions are rather general and hardly restrict the mathematical form of the IRFs. In fact, the present approach is *nonparametric* because it leaves the IRFs unspecified. The following more specific assumption, however, will be seen to lead to a surprisingly narrow parametric family of item response models:

ASSUMPTION iv. Given a test of length k, the raw score variable $R_v = \sum_{i=1}^{k} X_{vi}$ is a sufficient statistic for parameter ξ_v.

Based on these assumptions the following result can be proved:

THEOREM 1. *Assumptions* i, ii, iii, *and* iv *imply that the item response model must be equivalent to an RM with IRFs* (5).

'Equivalent to an RM' means that there exist an increasing scale transformation $\phi(\xi) =: \theta$ and constants β_i such that all IRFs $f_i(\theta)$ attain form (5). Models equivalent to an RM are empirically indistinguishable from an RM.

A proof of this result, which is fundamental for the RM, can set out from the definition of sufficiency of the raw score R (dropping index v for short): for any response vector $\boldsymbol{x}$ and corresponding raw score r, the conditional likelihood function $L(\boldsymbol{x}|r)$ satisfies

$$L(\boldsymbol{x}|r) = \frac{L(\boldsymbol{x}|\xi)}{L(r|\xi)} = c(\boldsymbol{x}), \tag{27}$$

where c depends on the response vector $\boldsymbol{x}$ but is independent of ξ. By choosing a suitable $\boldsymbol{x}$ and inserting the $g_i(\xi)$ in (27), it can be shown that, for any pair (I_i, I_j),

$$\frac{g_i(\xi)[1-g_j(\xi)]}{[1-g_i(\xi)]g_j(\xi)} = d_{ij}, \tag{28}$$

where d_{ij} is a constant independent of ξ and of $\boldsymbol{x}$ (cf. Fischer, 1995a, pp. 16–18). Since no assumptions about the scale of ξ have been made so far, an increasing scale transformation $\xi \to \theta$ is admissible which carries the hitherto unspecified IRF of item I_i, $g_i(\xi)$, over into the logistic function $f_i(\theta) = \exp(\theta)/[1 + \exp(\theta)]$. Inserting this in (28) and rearranging terms yields $f_j = \exp(\theta - \beta_j)/[1 + \exp(\theta - \beta_j)]$, with $\beta_j := \ln d_{ij}$, for all $j \neq i$. This shows that any model satisfying Assumptions i–iv can be transformed into an equivalent RM with item parameters $\beta_i = 0$ (normalization) and β_j for $j \neq i$; therefore it is empirically indistinguishable from an RM. Another proof was given by Fischer (1968, 1974), employing complete induction. It rested on the obvious property that sufficiency of the raw score $R^{(k)}$ in a test of length k implies sufficiency of the raw score $R^{(k-1)}$ in any subtest of length $k - 1$. (An explicit formal proof of the latter property is found in Fischer, 1995a.)

A result similar to Theorem 1 is due to Birnbaum (1968), who defined 'optimal weights' α_i for right responses such that the weighted score $R = \sum_i \alpha_i X_i$ discriminates optimally at a specified point ξ_0 of the latent continuum. Birnbaum asked: What family of IRT models has the same optimal weights at all $\xi_0 \in \mathbb{R}$? He showed that any such model must be equivalent to a model with IRFs

$$f_i(\theta) = \frac{\exp[\alpha_i(\theta - \beta_i)]}{1 + \exp[\alpha_i(\theta - \beta_i)]}, \quad \alpha_i > 0. \tag{29}$$

This is usually referred to as the 'Birnbaum Model' or 'Two Parameter Logistic Model' (2PL). The parameters α_i, which are proportional to the slopes of the respective IRFs at the points of their maximum steepness, are in psychometrics denoted *discrimination* parameters. If Birnbaum's result is specialized by putting all $\alpha_i = 1$, it is seen that models compatible with the optimality of equal and constant weights of all items at all $\xi_0 \in \mathbb{R}$ are equivalent to RMs.

A different derivation of the RM from sufficiency is due to Andersen (1973a, 1977). He made the following stronger assumption about the IRFs, but a slightly weaker assumption about sufficiency:

ASSUMPTION v. All IRFs are of the form $g_i(\xi) = g(\xi, \delta_i)$, where g is a continuous $\mathbb{R}^2 \to (0, 1)$ function, such that $g(\cdot, \delta_i)$ is a bijective increasing $\mathbb{R} \to (0, 1)$ function, and $g(\xi, \cdot)$ a bijective decreasing $\mathbb{R} \to (0, 1)$ function.

ASSUMPTION vi. There exists a nontrivial sufficient statistic T for ξ that does not depend on the item parameters $\delta_1, \ldots, \delta_k$.

THEOREM 2. *Let Assumptions* iii, v *and* vi *hold for items* $I_1, \ldots, I_k$. *Then there exist increasing scale transformations* $\phi : \xi \to \theta$ *and* $\psi : \delta \to \beta$ *such that the IRFs attain form* (5).

Another variation on the theme of raw score sufficiency is found in Kelderman's (1995) derivation of the RM from the methodological principle of 'measurement interchangeability' (i.e., interchangeability of the item scores within a response vector).

A derivation of the RM similar to that from sufficiency of the raw score R has been given by Roskam and Jansen (1984). Generalizing the notion of *composite ordering* by Ducamp and Falmagne (1969), they defined the *stochastically consistent ordering* of items by means of the following axiom:

ASSUMPTION vii.

$$P(I_i \to I_j) := P(X_i = 0 \wedge X_j = 1 | R_{ij} = 1) = c_{ij},$$

where "$I_i \to I_j$" denotes a dominance relation between I_i and I_j (i.e., indicates that item I_i is more difficult than item I_j), R_{ij} is the raw score of a testee in the subtest of length $k = 2$ composed of items I_i and I_j, and c_{ij} is a constant independent of ξ. This is just a special case of Assumption iv, however, because it expresses that R_{ij} is a sufficient statistic for ξ in that subtest.

A striking feature of the RM is that the ability and the difficulty parameters are concatenated in a particularly simple *subtractive* form, $P(+|\theta, \beta) = f(\theta - \beta)$. (The RM shares this property, however, with other models, like the so-called Normal Ogive Model.) It appears useful to take a look at models with subtractive person and item parameters in general. We therefore start out from the following assumption:

ASSUMPTION viii. There exists a universe of items with IRFs of the subtractive form $f(\theta - \beta)$ for all $\theta, \beta \in \mathbb{R}$, where β is the item difficulty parameter and $f(\cdot)$ a continuous bijective increasing $\mathbb{R} \to (0, 1)$ function.

What class of transformations of the parameters θ and β are compatible with this subtractive form of the IRFs? This is answered by the following

LEMMA 1. *Assumption* viii *implies that, as long as the subtractive form of the IRFs is preserved, the θ and β parameters are unique up to linear transformations $a\theta + b_1$ and $a\beta + b_2$, with $a > 0$ and arbitrary constants b_1 and b_2.*

This result is due to Pfanzagl (1971, p. 171), see also Pfanzagl (1994a, p. 249) and Falmagne (1985, p. 192). The proof utilizes a result on a generalization of the well-known 'functional equation of Cauchy' (Aczél, 1966, p. 142), which requires continuity of $f(\cdot)$ everywhere on its support. The universes of persons and items have therefore to be conceived as 'dense' in the sense that the sets of person and item parameters are compact on $\mathbb{R}$. Under this premise – and using the term 'interval scale' for a scale that is unique up to positive linear transformations – the following can be concluded: Assumption viii implies that testee ability and item difficulty are measurable on two interval scales with a common unit of measurement. This conclusion applied to the RM is in accordance with Colonius (1979), Hamerle (1979), and Fischer (1987, 1995a) as well as with the literature on 'additive conjoint measurement', cf. Krantz et al. (1971). It contradicts, however, many propositions that can be found in psychometric literature where everything has been claimed from a simple ordering of persons and items, on the one hand, to ratio scale properties of person and item parameters, on the other.

Summarizing the results of Theorems 1 and 2 and Lemma 1, we arrive at the following general form of the RM:

$$P(X_{vi} = 1 | S_v, I_i) = \frac{\exp[a(\theta_v - \beta_i) + b]}{1 + \exp[a(\theta_v - \beta_i) + b]}, \tag{30}$$

with arbitrary constants $a > 0$ and $b := b_1 - b_2$, where a is the discrimination common to all items, and b defines the origin of the β-scale relative to that of the θ-scale. We shall denote this a '*family of Rasch models*'. Clearly, for any practical purpose one may set $a = 1$ and $b = 0$, but (30) is useful as a reminder that the scales of θ and β are unique only up to positive linear transformations (and hence can be called 'interval scales') with a common multiplicative constant (or unit of measurement) and arbitrary additive constants (or zero points).

Assumption viii is quite strong, of course, because it assumes subtractivity in terms of the parameters θ and β. It can be shown that, under this premise, a still weaker condition on the existence of a sufficient statistic for θ suffices to deduce the RM.

ASSUMPTION ix. There exists a nontrivial sufficient statistic for θ.

THEOREM 3. *Assumptions* iii, viii, *and* ix *imply that* $f(\cdot)$ *is the logistic function*

$$f(x) = \frac{\exp(ax + b)}{1 + \exp(ax + b)}.$$

In other words, the model must belong to the family of RMs (30). The proof can be found in Pfanzagl (1994a, pp. 257–258). This characterizes the RM as the only model with IRFs of the form $f(\theta - \beta)$ having *some* nontrivial sufficient statistic for θ.

Assumptions about the IRFs made so far are very strong not only because they imply subtractivity of the parameters θ and β, but also because unrestricted continuous variation of both parameters is posited. Test constructors, however, might consider it unrealistic to assume that for any pre-specified constant, c_0, an item can be constructed with difficulty parameter $\beta = c_0$. Therefore, it appears more realistic to assume that the item parameters are certain discrete constants. The person parameters θ, on the other hand, can more easily be imagined to vary continuously, because development of intelligence or effects of cognitive training or treatment are usually conceived as continuous in time or dosage, respectively. Therefore, the following assumption as an alternative to Assumption viii seems to make sense:

ASSUMPTION x. The universe of items consists of a finite number of items, k, with $k \geqslant 3$, where the IRFs are of the form $f(\theta - \beta_i)$, for all $\theta \in \mathbb{R}$ and $i = 1, 2, \ldots, k$, and $f(\cdot)$ is a continuous bijective increasing $\mathbb{R} \to (0, 1)$ function.

Moreover, we formulate the following assumption about the item parameters.

ASSUMPTION xi. There exists a triplet of items, I_i, I_j, I_l, such that the quotient $(\beta_i - \beta_l)/(\beta_j - \beta_l)$ is irrational.

LEMMA 2. *If Assumptions* x *and* xi *hold true, the* θ*-scale has interval scale properties.*

The proof was given by Fischer (1995a) who analyzed also the consequences if, for a given test of k items, all the quotients addressed in Assumption xi are rational: then the interval scale properties hold only for an infinite sequence of equally (narrowly) spaced points on the latent scale, leaving room for nonlinear transformations between these points. Therefore, strictly speaking, the scale level of θ can never be ultimately decided on empirical grounds because there is no empirical way of determining whether quotients $(\beta_j - \beta_i)/(\beta_q - \beta_p)$ are rational or irrational numbers. For all practical purposes, the scale can be considered an interval scale, though.

Another characterization of the RM under the perspective of identifiability of the parameter vector $\boldsymbol{\beta}$ within a semiparametric marginal maximum likelihood (MML) framework is due to Pfanzagl (1994a, 1994b). (On parametric and semiparametric MML estimation, see Section 4 below.) The aspect of identifiability is very important as a necessary prerequisite for consistency. The parameter $\boldsymbol{\beta}$ is denoted 'identifiable from the semiparametric mixture likelihood' $L_M(\boldsymbol{x}|\boldsymbol{\beta}, G)$ if, for all $\boldsymbol{x}$ and any $\boldsymbol{\beta}, G$, $L_M(\boldsymbol{x}|\boldsymbol{\beta}, G) = L_M(\boldsymbol{x}|\boldsymbol{\beta}_0, G_0)$ implies $\boldsymbol{\beta} = \boldsymbol{\beta}_0$.

THEOREM 4. *If Assumptions* iii, x, *and* xi *hold true, the only class of IRT models where* $\boldsymbol{\beta}$ *is identifiable by means of the semiparametric MML approach with probability tending to 1 as* $n \to \infty$, *is the 'family of RMs'* (30).

A completely different approach to deriving the RM is that of Rasch (1967, 1968): he introduced the concept of *specific objectivity* (SO) as a fundamental methodological desideratum. First he defined a *frame of reference* $\langle \mathcal{O}, \mathcal{A}, \mathcal{R} \rangle$, consisting of a set $\mathcal{O} = \{O_v\}$ of *objects*, a set $\mathcal{A} = \{A_i\}$ of *agents* to which the objects are exposed, and a set $\mathcal{R}$ of possible *reactions*, whereby he assumed that for each pair (O_v, A_i) a unique reaction R_{vi} is observable. Comparisons between any two objects should be feasible, according to Rasch, on the basis of the observed reactions of these objects to any $A_i \in \mathcal{A}$, that is, independently of which agent(s) is (are) actually used. In other words, comparisons between objects should be *generalizable* over agents. Similarly, comparisons between any two agents should be feasible when any object O_v is exposed to these agents, and the result should be independent of the particular choice of (one or more) objects for the experiment. Scientific results satisfying this generalizability requirement Rasch denoted *specifically objective*. He called them 'objective' because an arbitrary (or 'subjective') choice of one or more agents (or objects) would not bias the comparison of the objects (or agents). He stressed, however, that this concept of objectivity is specific for a given frame of reference, which fact explains the term 'specific objectivity'.

> Two features seem indispensable in scientific statements: they deal with comparisons, and they must be *objective*. To complete these requirements I have to specify the *kind of comparisons* and the *precise meaning of objectivity*. When doing so I do not feel confident that all sorts of what can justifiably be called 'science' are covered, but certainly a large area of science is. (Rasch, 1967, p. 4.)

Rasch's favorite example of specific objectivity stemmed from classical mechanics: the comparison of masses based on observed accelerations which occur when the masses are exposed to certain forces (as, for instance, in experiments with devices like 'Attwood's machine', used as teaching material in physics). Let the objects of interest be rigid bodies, O_v, each of them characterized uniquely by its respective mass, M_v.

Let the objects be exposed to suitable experimental conditions A_i where forces F_i are applied, producing in each case an observable acceleration A_{vi}. According to the Second Newtonian Axiom (force $=$ mass $\times$ acceleration), acceleration is proportional to force and inversely proportional to mass, so that a comparison between two objects O_v and O_w with respect to their masses can be carried out as follows:

$$\frac{A_{vi}}{A_{wi}} = \left(\frac{F_i}{M_v}\right)\left(\frac{F_i}{M_w}\right)^{-1} = \frac{M_w}{M_v}, \tag{31}$$

which is *independent of* F_i. Therefore, the outcome of the comparison is generalizable over the agents (i.e., forces F_i) that are instrumental for the comparison. The quotient A_{vi}/A_{wi} is the 'comparator (function)' by which the comparison can be carried out.[3]

To apply the same concepts in psychometrics, Rasch (1967) defined the following frame of reference:

ASSUMPTION xii. Let testees S_v be characterized by one parameter $\xi_v \in \mathbb{R}$ each, items I_i by one parameter $\delta_i \in \mathbb{R}$ each, and let a random variable X_{vi} exist for each pair (S_v, I_i), which is characterized by a reaction parameter $\rho_{vi} \in \mathbb{R}$. Moreover, let the reaction parameter be determined by the function $\rho = g(\xi, \delta)$, where g is continuous $\mathbb{R}^2 \to \mathbb{R}$, with $g(\cdot, \delta)$ a bijective increasing $\mathbb{R} \to \mathbb{R}$ function and $g(\xi, \cdot)$ a bijective decreasing $\mathbb{R} \to \mathbb{R}$ function.

Note that, in the present case, the 'unique reaction R' assumed as part of the frame of reference is the distribution of the random variable X_{vi} rather than a single realization (item response) x_{vi}.

The postulate of specific objectivity is the following

ASSUMPTION xiii. A comparator $U(\rho_{vi}, \rho_{wi})$, $\mathbb{R}^2 \to \mathbb{R}$, increasing in ρ_{vi} and decreasing in ρ_{wi}, is assumed to exist for all $\rho_{vi}, \rho_{wi} \in \mathbb{R}$, satisfying the functional equation

$$U\big[g(\xi_v, \delta_i), g(\xi_w, \delta_i)\big] =: V(\xi_v, \xi_w), \tag{32}$$

for all $\xi_v, \xi_w, \delta_i \in \mathbb{R}$, where V is independent of δ_i.

V denotes the comparator function which is assumed to be independent of the item (or agent) parameter δ_i.

A strong result about the structure of function U, about the IRF $g(\xi, \delta)$, and about the comparator V can be derived from Assumptions xii and xiii. The result is due to Rasch (1968, 1972, 1977); however, one of these articles contains no proofs, and two of them give only sketches of a proof, using stronger assumptions than necessary. A more general proof is found in Fischer (1995a), based on results from the theory of functional equations (Aczél, 1966, pp. 310–312). Related earlier work is due to Fischer (1987), Hamerle (1979), Irtel (1987, 1993, 1995), Neuwirth (1987), and Tutz (1989).

[3] Literally speaking, this holds only if the accelerations can be observed without measurement error; in a practical experiment, the results would be only approximately constant across agents. It also deserves mention that acceleration is defined independently of the concepts of mass and force, namely, as the second derivative with respect to time of the distance covered; hence, neither the masses nor the forces need be known in order to carry out the described comparisons.

THEOREM 5. *Assumptions* xii *and* xiii *imply that there exist continuous bijective increasing functions* ϕ, ψ, u, *and* f, *all* $\mathbb{R} \to \mathbb{R}$, *such that*

$$U(\rho_{vi}, \rho_{wi}) = V(\xi_v, \xi_w) = u\big[\phi(\xi_v) - \phi(\xi_w)\big] = u(\theta_v - \theta_w), \tag{33}$$

$$g(\xi_v, \delta_i) = f\big[\phi(\xi_v) - \psi(\delta_i)\big] = f(\theta_v - \beta_i), \tag{34}$$

where $\theta_v := \phi(\xi_v)$ *and* $\beta_i := \psi(\delta_i)$.

This shows that the postulate of SO, together with the technical assumptions of continuity and monotonicity, implies a certain *latent additivity* (Rasch, 1968; or *subtractivity*, in the present notation) of both the IRF $g(\xi, \delta)$ and the comparator U. This property obtains from applying suitable scale transformations $\phi(\xi)$ and $\psi(\delta)$ to the original latent scales, yielding new scales θ and β. Since the latter are unique up to positive linear transformations, so are the functions $\phi(\xi)$ and $\psi(\delta)$; a surprising result, considering that the original scales of ξ and δ were defined only ordinal. This can be summarized by stating that the resulting subtractive representation has interval scale properties.

It has to be emphasized that this result is completely general, it holds for any frame of reference $\langle \mathcal{O}, \mathcal{A}, \mathcal{R} \rangle$ that allows specifically objective comparisons, irrespective of the particular scientific domain. It applies to the case of force and mass in physics (namely, with $\ln A_{vi} = \ln F_i - \ln M_v$) in the same way as to ability and item difficulty in psychology.

Latent subtractivity of ability and difficulty, however, does not yet determine an IRT model. But as far as IRT is concerned, more concrete assumptions about the comparator function U can be added to SO, namely, a specification of the kind of reactions observed plus the assumption that the comparator should be a function of some nontrivial likelihood of the observations: let $A = \{(+\,+), (+\,-), (-\,+), (-\,-)\}$ be the sample space of the possible patterns of responses of two testees S_v and S_w to one item I_i, and let B and C be subsets of this sample space, with $C \subset B \subseteq A$.

ASSUMPTION xiv. For any two subjects S_v and S_w, and any item I_i with difficulty parameter δ_i, there exists a comparator U: $(0, 1) \times (0, 1) \to \mathbb{R}$ of the form $U[g(\xi_v, \delta_i), g(\xi_w, \delta_i)]$, where $g(\xi, \delta_i)$ is the IRF and U satisfies the following conditions:

(a) $U(x, y)$ is continuous increasing in x for all y, and decreasing in y for all x;
(b) $U[g(\xi_v, \delta_i), g(\xi_w, \delta_i)] =: V(\xi_v, \xi_w)$ is independent of δ_i, for all $\xi_v, \xi_w, \delta_i \in \mathbb{R}$ (i.e., the comparison of any S_v and S_w is always possible based on any item I_i and is specifically objective);
(c) Likelihood principle: U is a function of some unconditional likelihood $P(B)$ or of some conditional likelihood $P(C|B)$.

Rasch (1968) showed that under these assumptions the family of RMs (30) is the only solution. His proof was based on an enumeration of all possible sets B and C within the defined sample space A. (The number of cases that had to be discussed was not quite as large as one might think because many cases are symmetric.) The proof is sketched also in Fischer (1995a).

THEOREM 6. *Assumptions* iii, v, *and* xiv *imply that there exist continuous increasing functions* $\phi : \xi \to \theta$ *and* $\psi : \delta \to \beta$, *both* $\mathbb{R} \to \mathbb{R}$, *such that all IRFs assume form* (30). *The* θ *and* β *are unique except for positive linear transformations with a common multiplicative constant.*

Common elements in the formulation of the multiplicative Poisson model and the (dichotomous) RM by G. Rasch were (a) the multiplicative (or, equivalently, additive/subtractive) concatenation of person and item parameters and (b) SO in the form of conditional inference procedures with respect to the item parameters independent of the true values of the person parameters, and vice versa. It suggests itself to ask: What class of psychometric models possesses these properties? For dealing with this question, it has to be specified, though, what kind of reactions are observed. In order to cover such diverse cases as ratings of subjective well-being or of pain of patients, subjects' attitudes toward or aesthetic values of certain objects, the numbers of epileptic or migraine attacks of patients throughout an observation period, the numbers or near-accidents of driving learners observed on given stretches of roads, responses of testees to multiple choice questions or other polytomous test items with ordered categories, or simply testees' responses to dichotomous intelligence test items, Fischer (1995e) made the following assumptions. The general setting is that of the observation of a person S at two time points or under two observational conditions (which may be two items, for instance).

ASSUMPTION xv. Let a person S be observed at two time points (or under two conditions), T_1 and T_2; let the observations be realizations of positive integer-valued random variables H_1 and H_2, both with support $\{0, 1, 2, \ldots\}$, and with probability functions $P_1(j|\theta) > 0$ and $P_2(j|\theta + \eta) > 0$, respectively, for $j = 0, 1, 2, \ldots,$ where θ and $\theta + \eta$ denote S's positions on trait Θ at time points T_1 and T_2, respectively, with $\theta, \eta \in \mathbb{R}$ and $\sum_j P(j|\theta) = \sum_j P(j|\theta + \eta) = 1$.

Parameter η is denoted the 'amount of change' of S between T_1 and T_2. The additive concatenation of θ and η, which has the consequence of interval scale properties of these parameters, is chosen in view of the previous results about SO.

Define local stochastic independence of H_1 and H_2, analogous to (iii), as stochastic independence when parameters θ and η are fixed. The following technical assumptions about the probability functions are also needed:

ASSUMPTION xvi. Let $P_1(j|x)$ and $P_2(j|x)$ be continuous functions of x and let both variables H_1 and H_2 be stochastically increasing in their parameters in the sense that $P_1(j+1|x)/P_1(j|x)$ and $P_2(j+1|x)/P_2(j|x)$ are increasing in x, for all $j = 1, 2, \ldots$ (denoted 'monotone likelihood ratio property'); moreover, let

$$\lim_{x \to -\infty} \frac{P_1(j|x)}{P_1(0|x)} = \lim_{x \to -\infty} \frac{P_2(j|x)}{P_2(0|x)} = 0, \tag{35}$$

for all $j > 0$.

These assumptions assure that H_1 and H_2 are stochastically increasing in θ and that η can be deduced uniquely if at least two probabilities, say $P(0|\theta)$ and $P(1|\theta)$, are known. This will allow application of the results even in cases where the distributions are assumed to be truncated.

The result of Fischer (1995d) can now be formulated as follows:

THEOREM 7. *Let Assumptions* XV *and* XVI *hold true and assume local stochastic independence of* H_1 *and* H_2. *If the conditional probability* $P(h_1, h_2|h_1+h_2 = h)$, *for all* h_1 *and* h_2, *is a function* $V(\eta)$ *independent of* θ, *thus enabling conditional inference on* η, *then* $P_1(j|\theta)$ *and* $P_2(j|\theta+\eta)$ *must be probability functions of power series distributions*

$$P_1(j|\theta) = \frac{m_{1j}\mu^j}{\sum_{l=0}^{\infty} m_{1l}\mu^l} \quad \textit{and} \quad P_2(j|\theta+\eta) = \frac{m_{2j}(\mu\delta)^j}{\sum_{l=0}^{\infty} m_{2l}(\mu\delta)^l}, \tag{36}$$

with $\mu = \exp(c\theta)$, $c > 0$, $\delta = \exp(c\eta)$, *and* $m_{1j}, m_{2j} > 0$ *for all* j.

In statistical literature, the distributions in (36) are called 'power series distributions'. They are exponential families which have been studied in some detail (Noack, 1950), and are well-known to allow conditional inference procedures (cf. Johnson and Kotz, 1969, vol. 1, p. 36). Many elementary distributions, like the binomial or the Poisson, are special cases thereof.

Fischer (1995e) showed that (a) if S is observed at two time points and the variables H_1 and H_2 are truncated above $j = m$, the model is equivalent to the so-called 'Partial Credit Model' (PCM; Masters, 1982) for two polytomous ordered response items with $m+1$ categories; (b) if moreover the probability functions at both time points are assumed to be the same, $P_1(j|x) = P_2(j|x)$ for all j, then the model is equivalent to the so-called 'Rating Scale Model' (RMS; Rasch, 1965; Andrich, 1978); (c) if, as a special case of (b), m is equal 1 (two response categories), the model becomes the RM; and (d) for $P_1(j|x) = P_2(j|x)$ combined with a certain 'ignorability condition' or combined with the assumption that $E(H_1)$ and $E(H_2)$ are proportional to θ and $\theta+\eta$, respectively, the model must be equivalent to the multiplicative Poisson model (cf. Fischer, 1995d). This complements the derivation of the RM from SO in Theorem 6.

Summarizing all this, it can be said that three essentially different approaches to deriving – and hence justifying the choice of – the RM have been taken: (a) postulating sufficient statistics for the person parameter; (b) assuming identifiability of parameters within a marginal maximum likelihood framework; and (c) postulating SO and the feasibility of conditional inference as general methodological principles. The first two of them have a statistical motivation, in other words, practical consequences are brought to the fore; the third is an epistemic principle that seems to be typical for natural sciences. Psychometricians and research workers may have different attitudes or motives for emphasizing or deemphasizing either of these properties; but in any case, these properties are strong arguments for attempting to apply, wherever this appears to be justified, the RM as a measurement model. Yet a warning is indicated already here: the RM is a very restrictive model that is not easily fitted to empirical data. It should therefore serve as an ideal in the development or improvement of test scales rather than as a tool for post hoc analyses of test data. If a researcher has the choice between (a) improving his/her observations (i.e., the test instrument) so that the RM can be fitted and (b) making his/her model more complex in order to fit it to imperfect observations (thus losing the advantageous properties of the RM), the first of these two strategies is recommended. Many 'test scales' with dichotomous items, however, are in use in psychology, education, clinical sciences, etc., where fitting an RM does *not* make sense because some of the assumptions are obviously violated. The RM thus cannot serve as an *omnibus* method for the analysis of arbitrary collections of dichotomous items that happen to be employed with the hope of assessing some hypothetical trait.

4. Item parameter estimation

Since (unidimensional) IRT relates the item responses to an abstract latent variable Θ, it allows to compare testees who have taken different subtests[4] on the same scale. Major recent studies like the 'Program for International Student Assessment' (PISA; see Adams and Wu, 2002) make use of this possibility. Technically, all subsamples of testees who have responded to different subtests can jointly be considered *one incomplete sample of data*. Note, however, that this refers to incompleteness by design. If data are missing for other reasons like, for instance, non-response occurring when testees skip subjectively difficult items or because some items are not reached in a time-limited achievement test, the present derivations are not applicable (on 'ignorable non-response', see Rubin, 1987).

To deal effectively with incomplete data matrices, we generalize the notation by introducing a design matrix $\boldsymbol{B} = ((b_{vi}))$ with elements $b_{vi} = 1$ if testee S_v has responded to item I_i, and $b_{vi} = 0$ otherwise. In order to allow unrestricted algebraic manipulations, it is convenient to complete $\boldsymbol{X}$ by inserting some arbitrary value $x_{vi} = x$ with $0 < x < 1$ whenever $b_{vi} = 0$ (Fischer, 1981; see also Molenaar, 1995). Then the raw scores r_v and the item marginals s_i can be evaluated as

$$r_v = \sum_{i=1}^{k} b_{vi} x_{vi} \tag{37}$$

and

$$s_i = \sum_{v=1}^{n} b_{vi} x_{vi}, \tag{38}$$

and the generalized form of the log-likelihood (9) becomes

$$\ln L(\boldsymbol{X}|\boldsymbol{\theta}, \boldsymbol{\beta}, \boldsymbol{B}) = \sum_{v=1}^{n} r_v \theta_v - \sum_{i=1}^{k} s_i \beta_i - \sum_{v=1}^{n} \sum_{i=1}^{k} b_{vi} \ln\bigl[1 + \exp(\theta_v - \beta_i)\bigr]. \tag{39}$$

[4] Throughout this chapter, the term 'subtest' is used synonymously with 'subsample of items from a given set of items or from an item universe'.

The double sum in the right-hand side of (39) does not depend on the data $\boldsymbol{X}$ except through the sufficient statistics r_v and s_i. Clearly, (39) is an exponential family log-likelihood.

Similarly, the logarithm of the conditional likelihood, analogous to (15), becomes

$$\ln L_C(\boldsymbol{X}|\boldsymbol{\beta}, \boldsymbol{r}, \boldsymbol{B}) = -\sum_{i=1}^{k} s_i\beta_i - \sum_{v=1}^{n} \ln \gamma_{r_v}(\boldsymbol{\epsilon}, \boldsymbol{b}_v), \tag{40}$$

where $\boldsymbol{b}_v = (b_{v1}, \ldots, b_{vk})'$ denotes the design vector for S_v, the elementary symmetric functions γ_{r_v} are defined in terms of the variables $\epsilon_i b_{vi}$ rather than of the ϵ_i alone, and s_i and r_v are defined as in (37) and (38), respectively. Again, the form of the likelihood (40) reveals that the conditional model is an exponential family.

4.1. *Joint maximum likelihood estimation*

In the early days of IRT, a popular method for the estimation of the item parameters was the Joint Maximum Likelihood (JML) method which maximizes (39) as a function of both the β_i and the θ_v (Birnbaum, 1968; Wright and Panchapakesan, 1969; Wingersky et al., 1982). The JML estimation equations are obtained by taking the partial derivatives of (39) with respect to the θ_v and β_i and putting them to zero:

$$\sum_{i=1}^{k} \frac{b_{vi}\exp(\theta_v - \beta_i)}{1 + \exp(\theta_v - \beta_i)} - r_v = 0, \tag{41}$$

$$\sum_{v=1}^{n} \frac{b_{vi}\exp(\theta_v - \beta_i)}{1 + \exp(\theta_v - \beta_i)} - s_i = 0. \tag{42}$$

According to the theory of exponential families (Barndorff-Nielsen, 1978; Andersen, 1980), the ML equations equate the observed values of the sufficient statistics with their expectations; this is obviously the case in (41) and (42).

In the complete data case (all $b_{vi} = 1$), there are n equations (41) and k equations (42). Obviously, (41) cannot be solved whenever $r_v = 0$ or $r_v = k$; therefore, all testees with zero or perfect score have to be removed from the data. Observing that Eqs. (41) with equal raw score r_v are identical, which is a consequence of the sufficiency of R_v, there are at most $k-1$ independent equations (41), for $r_v = 1, \ldots, k-1$. Similarly, if $s_i = 0$ or $s_i = n$, equation (42) cannot be solved; therefore, items with extreme s_i have to be removed. Since one of the Eqs. (42) becomes superfluous when the item parameters are normalized, for instance, by putting $\beta_a = 0$ for some normalization item I_a, there remain at most $k-1$ equations (42). Summarizing this, provided that the data are complete, that all s_i are different, that each raw score $r = 1, \ldots, k-1$ has occurred at least once, and that there are no extreme item or person marginals, the number of independent JML equations is $2(k-1)$. In the incomplete data case, the discussion of the number of estimators and independent equations is in principle analogous, however, there will generally be many more equations (41) because identical raw scores r_v resulting from different design vectors $(b_{v1}, \ldots, b_{vk})$ yield different estimators $\hat{\theta}_v$.

To answer the question of the uniqueness of the normalized JML estimator $(\hat{\boldsymbol{\theta}}, \hat{\boldsymbol{\beta}})$, Fischer (1974; see also 1981) interpreted the RM as an incomplete 'Bradley–Terry–Luce (BTL) Model' for the comparison of testees with items and applied a n.s. condition for the uniqueness of the ML estimator of the BTL which goes back to Zermelo (1929). This condition was much later rediscovered by Ford (1957), see also David (1963) and Harary et al. (1965). The uniqueness result for the RM was independently derived also by Haberman (1977). The crucial condition on the item score matrix $\boldsymbol{X}$ that has to be satisfied for the estimator $(\hat{\boldsymbol{\theta}}, \hat{\boldsymbol{\beta}})$ to be unique, is *well-conditionedness*. It is defined as follows:

ASSUMPTION xvii. $\boldsymbol{X}$ is denoted 'well-conditioned' iff in every possible partition of the items into two nonempty subsets $\mathcal{T}_1$ and $\mathcal{T}_2$, some testee has given response '+' to some item $I_i \in \mathcal{T}_1$ and response '−' to some item $I_j \in \mathcal{T}_2$.

If $\boldsymbol{X}$ fails to be well-conditioned, it is denoted 'ill-conditioned'. Fischer (1981) showed, however, that well-conditionedness is a property of the marginals of $\boldsymbol{X}$ rather than of the matrix itself. A practical method to determine whether a (complete or incomplete) item score matrix $\boldsymbol{X}$ is or is not well-conditioned makes use of a *directed graph* (or 'digraph' for short), D. Assign one-to-one vertices V_i to the items I_i and define the corresponding $k \times k$ adjacency matrix $\boldsymbol{D}$ as follows: for $i \neq j$, put $d_{ij} = 1$ if some testee S_v has given response '+' to item I_i and response '−' to item I_j, and $d_{ij} = 0$ otherwise; and put $d_{ii} = 1$ for all i. The uniqueness of the JML estimator can now be determined by means of digraph D.

LEMMA 3. *The (complete or incomplete) item score matrix $\boldsymbol{X}$ is well-conditioned iff digraph D is strongly connected.*

THEOREM 8. *The JML estimation equations* (41)–(42) *possess a unique finite normalized solution $(\hat{\boldsymbol{\beta}}, \hat{\boldsymbol{\theta}})$ iff the following two conditions hold true: digraph D is strongly connected and no r_v is extreme.*

Very efficient algorithms for determining whether a given digraph is strongly connected are known in graph theory, for instance, taking Boolean powers $\boldsymbol{D}^l$, $l = 1, \ldots, k-1$, of $\boldsymbol{D}$ (cf. Harary et al., 1965, pp. 111–122):

LEMMA 4. *Digraph D is strongly connected iff $\boldsymbol{D}^{k-1}$ contains no zero element.*

For practical purposes, it suffices to consider the powers $\boldsymbol{D}^2, \boldsymbol{D}^4, \ldots, \boldsymbol{D}^{2^t}$, until $2^t \geqslant k - 1$; for further details see Fischer (1981).

The JML equations are easily solved iteratively by an alternating application of the Newton–Raphson method to each of the Eqs. (41) and (42), respectively. It does not pay, however, to go into more technicalities about the JML estimation equations here because the JML approach has more or less been given up. The reason is that, as Andersen (1973b, 1980) has shown, the JML estimates are inconsistent for $n \to \infty$ and k fixed, although consistency does hold for $n \to \infty$ and $k \to \infty$, with $n/k \to \infty$

(Haberman, 1977). Mostly the number of subjects is much larger than that of the items, which is fixed and often even small; hence, the estimator $\hat{\boldsymbol{\beta}}$ is inconsistent. Wright and Panchapakesan (1969) proposed the correction factor $(k-1)/k$ for the β_i which appears to approximately remove the bias, but no rigorous results seem to exist regarding that question. But even if the corrected JML estimates should be acceptable approximations to the parameters, the problem remains that their asymptotic behavior is unknown, hence no asymptotic (e.g., likelihood ratio) tests can be based on them.

4.2. *Conditional maximum likelihood estimation*

In Section 2 it has been shown that the conditional likelihood of $\boldsymbol{X}$, given the raw scores r_v, is independent of the person parameters θ_v and thus can be used for a 'specifically objective' estimation of the item parameters. This approach is called the Conditional Maximum Likelihood (CML) method. The CML estimator $\hat{\boldsymbol{\beta}}$ is obtained by putting the partial derivatives of (40) with respect to β_i, $i = 1, \ldots, k$, to zero. Doing so, two properties of the γ_r-functions in (40) should be taken into account. The function $\gamma_r(\boldsymbol{\epsilon})$ by definition is the sum of products of ϵ_i for all r-tupels of parameters ϵ_i. This is the sum of two terms: the sum over all r-tupels which do *not* contain ϵ_i, denoted $\gamma_r^{(i)}(\boldsymbol{\epsilon})$, plus the sum over all r-tupels containing ϵ_i; the latter sum is ϵ_i times the sum over all $(r-1)$-tupels that do not contain ϵ_i, yielding $\epsilon_i \gamma_{r-1}^{(i)}(\boldsymbol{\epsilon})$. Hence,

$$\gamma_r(\boldsymbol{\epsilon}) = \epsilon_i \gamma_{r-1}^{(i)}(\boldsymbol{\epsilon}) + \gamma_r^{(i)}(\boldsymbol{\epsilon}). \tag{43}$$

Using this, and remembering that $\epsilon_i = \exp(-\beta_i)$, we get

$$\frac{\partial \gamma_r(\boldsymbol{\epsilon})}{\partial \beta_i} = \frac{\partial \gamma_r(\boldsymbol{\epsilon})}{\partial \epsilon_i} \frac{d\epsilon_i}{d\beta_i} = -\epsilon_i \gamma_{r-1}^{(i)}(\boldsymbol{\epsilon}). \tag{44}$$

Now the CML estimation equations are easily obtained by taking the partial derivatives of (40) with respect to the β_i (remembering that now the γ-functions are defined in terms of the arguments $b_{vi}\epsilon_i$, not just of the ϵ_i) and equating them to zero,

$$\sum_{v=1}^{n} \frac{b_{vi}\epsilon_i \gamma_{r_v-1}^{(i)}(\boldsymbol{\epsilon}, \boldsymbol{b}_v)}{\gamma_{r_v}(\boldsymbol{\epsilon}, \boldsymbol{b}_v)} - s_i = 0, \tag{45}$$

for $i = 1, \ldots, k$. Again, these equations could alternatively be obtained from the rule that the observed values of the sufficient statistics, s_i, are equated to their (conditional) expectations, $E(S_i | r_1, \ldots, r_n)$, employing (17).

Looking at the CML equations, two observations can be made. First, if we consider a subgroup G_g of testees with identical design vectors $(b_{v1}, \ldots, b_{vk}) =: \boldsymbol{b}_g$ for all $S_v \in G_g$, the data form a complete item score matrix with respect to the subset of items presented to that group. In other words, the incomplete data can always be viewed as a collection of smaller complete item score matrices for subgroups G_g of testees and subtests $\mathcal{T}_g$. Therefore, the computations can be simplified by suitable grouping of the testees. Second, within such subgroups G_g, the left-hand terms of the CML equations do not depend on the individual testees, but only on their raw scores, so that the testees can be grouped by their raw scores. Hence, the CML equations assume the form, for

$i = 1, \ldots, k$,

$$\sum_g \sum_r n_{gr} \frac{b_{gi} \epsilon_i \gamma_{r-1}^{(i)}(\boldsymbol{\epsilon}_g)}{\gamma_r(\boldsymbol{\epsilon}_g)} - s_i = 0, \tag{46}$$

where $b_{gi} := b_{vi}$ for all $S_v \in G_g$, r runs from 0 to the maximum possible score $\sum_i b_{gi}$ within group G_g, $\boldsymbol{\epsilon}_g$ denotes the vector of those ϵ_i which correspond to items given to all $S_v \in G_g$, and n_{gr} is the number of testees $S_v \in G_g$ who have raw score r. Note that response vectors with extreme raw scores $r_v = 0$ or $r_v = \sum_i b_{vi}$ have no effect on the CML equations and thus need not be removed from the data.

Eqs. (46) are most efficiently solved by applying either the Newton–Raphson method or a suitable quasi-Newton method. For the former approach, the second-order partial derivatives of the log-likelihood function (40) are needed, which involve the second order partial derivatives of the γ-functions, obtained from (44),

$$\frac{\partial^2 \gamma_r(\boldsymbol{\epsilon})}{\partial \beta_i \partial \beta_j} = \epsilon_i \epsilon_j \gamma_{r-2}^{(i,j)}(\boldsymbol{\epsilon}) \tag{47}$$

and

$$\frac{\partial^2 \gamma_r(\boldsymbol{\epsilon})}{\partial \beta_i^2} = \epsilon_i \gamma_{r-1}^{(i)}(\boldsymbol{\epsilon}). \tag{48}$$

Using (47), it can be shown that

$$\frac{\partial^2 \ln L_C(\boldsymbol{X}|\boldsymbol{\beta}, \boldsymbol{r}, \boldsymbol{B})}{\partial \beta_i \partial \beta_j} = -\epsilon_i \epsilon_j \sum_g b_{gi} b_{gj} \sum_r n_{gr} \left[\frac{\gamma_{r-2}^{(i,j)}(\boldsymbol{\epsilon}_g)}{\gamma_r(\boldsymbol{\epsilon}_g)} - \frac{\gamma_{r-1}^{(i)}(\boldsymbol{\epsilon}_g)\gamma_{r-1}^{(j)}(\boldsymbol{\epsilon}_g)}{\gamma_r(\boldsymbol{\epsilon}_g)} \right], \tag{49}$$

for $i \neq j$, and

$$\frac{\partial^2 \ln L_C(\boldsymbol{X}|\boldsymbol{\beta}, \boldsymbol{r}, \boldsymbol{B})}{\partial^2 \beta_i} = -\sum_g b_{gi} \sum_r n_{gr} \left[\frac{\epsilon_i \gamma_{r-1}^{(i)}(\boldsymbol{\epsilon}_g)}{\gamma_r(\boldsymbol{\epsilon}_g)} - \left(\frac{\epsilon_i \gamma_{r-1}^{(i)}(\boldsymbol{\epsilon}_g)}{\gamma_r(\boldsymbol{\epsilon}_g)} \right)^2 \right], \tag{50}$$

for $i = 1, \ldots, k$. Since, for fixed $\boldsymbol{r}$, these derivatives are independent of $\boldsymbol{X}$, they immediately yield the elements of the *information matrix*. Moreover, they allow one to solve the CML equations (45) by means of the Newton–Raphson procedure. (One of the equations, however, is superfluous because of the necessary normalization of the $\hat{\beta}_i$.) But to do so in practice, first a few technical issues need to be addressed.

One problem involved is the evaluation of the elementary symmetric functions. A direct enumeration of all the summands $\prod_i \epsilon_i^{y_i}$ in (12) is impractical, but the γ can be computed by means of serviceable recurrence relations. A reasonably fast and numerically stable method is the following. Suppose that the set of items $\{I_1, \ldots, I_k\}$ is partitioned into two subtests $\mathcal{T}_1 = \{I_1, \ldots, I_{k_1}\}$ and $\mathcal{T}_2 = \{I_{k_1+1}, \ldots, I_k\}$. Then it immediately follows from the definition of γ_r that

$$\gamma_r(\epsilon_1, \ldots, \epsilon_k) = \sum_{l=a}^{b} \gamma_{r-l}(\epsilon_1, \ldots, \epsilon_{k_1})\gamma_l(\epsilon_{k_1+1}, \ldots, \epsilon_k), \tag{51}$$

with $a = \max(0, r - k_1)$ and $b = \min(k - k_1, r)$. (These summation limits immediately follow from the obvious relations $0 \leqslant r - l \leqslant k_1$ and $0 \leqslant l \leqslant k - k_1$.) Recurrence (51) can be used for an efficient computation of the γ_r (Verhelst and Veldhuijzen, 1991), building the γ_r for a test of length k from the γ_l for subtests $\mathcal{T}_1$ and $\mathcal{T}_2$; however, optimizing this approach requires a substantial programming expenditure. A simpler special case is the following: consider a test of length t composed of the items $I_1, \ldots, I_t$, and split it in two subtests $\mathcal{T}_1 = \{I_1, \ldots, I_{t-1}\}$ and $\mathcal{T}_2 = \{I_t\}$. Applying (51),

$$\gamma_r(\epsilon_1, \ldots, \epsilon_t) = \gamma_r(\epsilon_1, \ldots, \epsilon_{t-1}) + \gamma_{r-1}(\epsilon_1, \ldots, \epsilon_{t-1})\epsilon_t, \tag{52}$$

for $0 \leqslant r \leqslant t$ and $t = 1, \ldots, k$, where $\gamma_0 = 1$ and $\gamma_r = 0$ for $r < 0$ or $r > t$. Starting with $t = 2$ and using (52) repeatedly, the following recursive method is established:

$$\begin{aligned} &\gamma_r(\epsilon_1, \epsilon_2) = \gamma_r(\epsilon_1) + \gamma_{r-1}(\epsilon_1)\epsilon_2, \\ &\gamma_r(\epsilon_1, \epsilon_2, \epsilon_3) = \gamma_r(\epsilon_1, \epsilon_2) + \gamma_{r-1}(\epsilon_1, \epsilon_2)\epsilon_3, \\ &\text{etc.} \end{aligned} \tag{53}$$

The partial derivatives $\gamma_{r-1}^{(i)}$ can be obtained by means of the same recurrence relations, rotating the item sequence appropriately. Notice that this procedure is numerically robust because it involves only summations. This method has first been proposed by Andersen (1972) and is sometimes denoted the 'summation method'. A faster method for the $\gamma_{r-1}^{(i)}$ that combines this summation method with an older 'difference method' is described in Fischer and Ponocny (1994); on further details of these methods and on references, cf. Fischer (1995b).

The second order partial derivatives (49) and (50) can be computed by means of the same algorithms, deleting each time one pair of items (I_i, I_j), which, however, requires a considerable amount of computation. To avoid this expense, the software LPCM Win (Fischer and Ponocny-Seliger, 1998) uses the Broyden–Fletcher–Goldfarb–Shanno (BFGS) algorithm which does not need the matrix of second order partial derivatives, yet produces it as a by-product of the iteration process. Each BFGS iteration is therefore much faster than one Newton–Raphson iteration, but the BFGS method requires approximately three to four times more iterations. Another computer program for CML estimation is OPLM (Verhelst et al., 1994).

The properties of the CML estimator are well understood. In finite samples, essentially the same weak condition as in JML estimation suffices for the existence of a unique finite normalized CML estimator (Fischer, 1981):

THEOREM 9. *The normalized CML estimator $\hat{\boldsymbol{\beta}}$ is unique iff digraph D of the item score matrix $\mathbf{X}$ is strongly connected.*

If all item parameters are within the range of the person parameters, strong connectedness of D is almost always attained even in small samples. This shows that, while n increases beyond the point of strong connectedness of D, the additional statistical information contributes to the precision of the estimator, providing favorable asymptotic properties.

THEOREM 10. *The normalized CML estimator $\hat{\boldsymbol{\beta}}$ is consistent and asymptotically normal around $\boldsymbol{\beta}$ with asymptotic covariance matrix equal to the inverse of the information matrix iff the sequence of θ_v satisfies the condition*

$$\sum_{v=1}^{n} \exp(-|\theta_v|) \to \infty, \quad \textit{for } n \to \infty. \tag{54}$$

The consistency and asymptotic normality of the CML estimator was first proved by Andersen (1970, 1973a, 1973b) under somewhat stronger conditions on the distribution of the θ. The n.s. condition (54) is due to Hillgruber (1990), cf. Pfanzagl (1994a, 1994b). It expresses the idea that not too many $|\theta_v|$ should be very large because, in the latter case, almost all raw scores would be zero or perfect, so that not enough statistical information could accumulate as $n \to \infty$.

Therefore, estimates of the asymptotic variances of the estimators β_i are computed by taking, at the point of the CML solution, the inverse of the information matrix with elements obtained by means of (49) and (50); this matrix is approximated iteratively both in the Newton–Raphson and in the BFGS method.

Conditioning on the r_v necessarily leads to some loss of information. Using the concept of 'F-information' (Liang, 1983), a generalization of the Fisher information, Eggen (2000) investigated the loss of efficiency of CML estimation in the RM. He showed, among other results, that this loss of efficiency of CML as compared to MML, assuming a normal latent distribution of θ and a test length of $k \geqslant 20$, is always less than 1%. From a practical point of view, this is negligible. In the case that the latent ability distribution is completely unknown, Pfanzagl (1994a, 1994b) showed that the CML estimators are even asymptotically efficient. Still more important is the fact that several conditional likelihood ratio (CLR) tests for controlling the model and for testing useful null hypotheses can be conveniently based on the CML estimators (see Sections 5 and 7). All these results support the use of CML.

An observation sometimes alarming users of the RM is that estimated standard errors of the $\hat{\beta}_i$ depend on the normalization chosen. The paradox resolves, however, by recalling that the RM does not allow to estimate the β_i as such, but only the differences $\beta_i - \beta_j$. Suppose, for the purpose of illustration, that $k = 3$ items are given to a sample of testees with the same ability parameters and that I_1 is solved only by approximately 1% of the sample, whereas I_2 and I_3 are solved by approximately 50% each. Then the difference $\hat{\beta}_2 - \hat{\beta}_3$ is approximately zero, and if β_2 is put to zero for normalization, the standard error of $\hat{\beta}_3 = \hat{\beta}_3 - \hat{\beta}_2$ is relatively small because the data are very informative about items with a response probability near 0.5. The standard error of $\hat{\beta}_1 = \hat{\beta}_1 - \hat{\beta}_2$, on the other hand, is much larger because little statistical information is available about the position of that item (which has been rarely solved) relative to the position of the other items. Suppose now that alternatively item I_1 is designated as the normalization item, $\beta_1 = 0$. Then the standard error of $\hat{\beta}_1$ technically becomes zero, and both item parameter estimates $\hat{\beta}_2$ and $\hat{\beta}_3$ have large (but quite substantially correlated) standard errors because both differences $\beta_2 - \beta_1$ and $\beta_3 - \beta_1$ are poorly estimated. This shows that the standard errors *must* depend on the normalization. Unfortunately, users of the RM often believe they are estimating item difficulties, while they actually are estimating only difficulty differences between items.

4.3. Marginal maximum likelihood estimation

An approach to estimation that has become very popular in recent years is the Marginal Maximum Likelihood (MML) method (cf., Bock and Aitkin, 1981; Thissen, 1982; Follmann, 1988). It maximizes the marginal log-likelihood function (23). To be more precise, two MML methods have to be distinguished, parametric and semiparametric MML. They differ with respect to the assumptions about and approximation of the latent distribution $G(\theta)$.

The parametric approach introduces assumptions about the latent distribution implying, strictly speaking, an extension of the RM. The enhanced model has, however, the same item parameters. The person parameters are no longer considered fixed effects, rather realizations of a random variable Θ with distribution $G(\theta)$, where G is some parametric family of distributions; most typically, G is chosen to be normal. It has to be emphasized that this approach diverges from Rasch's (1960, 1977) philosophy concerning the separation of item and person parameters. Yet many researchers see an advantage in the consequence that now the data of testees with zero or perfect scores need no longer be ignored; these testees yield serviceable information regarding the latent distribution $G(\theta)$, and finite person parameter estimates are assigned to them.

In the semiparametric MML, the form of the latent distribution is not restricted and is estimated along with the item parameters. Under certain conditions, this approach can even be employed as an alternative technique for obtaining CML estimates of the item parameters, which is of practical relevance in cases of large item samples (say, beyond $k = 200$).

On the other hand, if the assumptions about the latent distribution are wrong (in the parametric case), or if the raw score distribution cannot be fitted perfectly (in the semiparametric case), the item and person parameter estimates may become considerably biased.

We shall deal with the parametric version of MML first, assuming a latent normal distribution $G(\theta)$ with mean zero (which amounts to a normalization of the item parameters) and unknown variance σ^2; the latter will be estimated from the data. Consider the θ_l, for $l = 1, \dots, m$, to be fixed quadrature points for the numerical computation of $G(\theta)$. For formal convenience, however, we transform all θ_l by dividing them through σ, that is, $\theta_l^* := \theta_l \sigma^{-1}$, so that the distribution becomes the $N(0, 1)$, and correspondingly we put $\beta_i^* := \beta_i \sigma^{-1}$. To compensate for this transformation, we have to write the RM in its slightly generalized form (30), with $a = \sigma$ and $b = 0$ (dropping the asterisks for convenience). With these conventions, the marginal log-likelihood, see (23), becomes

$$\ln L_M(\boldsymbol{X}|\boldsymbol{\beta}, G) = -\sigma \sum_i s_i \beta_i + \sum_r n_r \ln\left[\sum_l g_l \frac{\exp(r\sigma\theta_l)}{\prod_i \{1 + \exp[\sigma(\theta_l - \beta_i)]\}}\right]. \tag{55}$$

To obtain estimation equations for the β_i, partial derivatives of (55) with respect to β_i have to be taken and put to zero. This requires some algebra which is not presented in detail here. An expression occurring in these logarithmic derivatives is

$$\frac{\dfrac{g_j \exp(r\sigma\theta_j)}{\prod_i\{1+\exp[\sigma(\theta_j - \beta_i)]\}}}{\sum_l \dfrac{g_l \exp(r\sigma\theta_l)}{\prod_i\{1+\exp[\sigma(\theta_l - \beta_i)]\}}} = \frac{g_j L(\boldsymbol{x}|\theta_j, \boldsymbol{\beta}, \sigma)}{\sum_l g_l L(\boldsymbol{x}|\theta_l, \boldsymbol{\beta}, \sigma)}$$
$$=: P(\theta_j|\boldsymbol{x}, \boldsymbol{\beta}, \sigma) = P(\theta_j|r, \boldsymbol{\beta}, \sigma), \tag{56}$$

which can be interpreted as the posterior probability of S_v having person parameter θ_j (i.e., belonging to a homogeneous class C_j of testees with person parameter θ_j), given S_v's response vector $\boldsymbol{x}_v$; it does not depend on the response vector, though, but only on the raw score r. Other quantities occurring in the derivation are

$$\sum_v P(\theta_j|\boldsymbol{x}_v, \boldsymbol{\beta}, \sigma) =: \tilde{n}_j, \tag{57}$$

the expected numbers of testees with person parameters θ_j (or the expected sizes of classes C_j), for each $j = 1, \ldots, m$. Using (56) and (57), the estimation equations become, after some algebra,

$$\sum_l \tilde{n}_l f_i(\theta_l) - s_i = 0, \tag{58}$$

for $i = 1, \ldots, k$, where $f_i(\theta_l)$ is as defined in (29), with $\alpha_i = \sigma$.

Similarly, the estimation equation for σ is obtained by taking the partial derivative of (55) with respect to σ and equating this to zero. Using the quantities

$$\sum_v P(\theta_j|\boldsymbol{x}_v, \boldsymbol{\beta}, \sigma)x_{vi} =: \tilde{e}_{ji}, \tag{59}$$

the expected numbers of responses '+' to item I_i of all testees with person parameter θ_j (i.e., of all testees in class C_j), for each pair (j, i), the estimation equation for σ becomes, again after some algebra,

$$\sum_i \sum_l \big[\tilde{n}_l f_i(\theta_l) - \tilde{e}_{li}\big](\theta_l - \beta_i) = 0. \tag{60}$$

These $k + 1$ equations (58) and (60) are to be solved for the $k + 1$ parameter estimates $\hat{\beta}_i$ and $\hat{\sigma}$. (Recall the transformations $\theta_l \to \theta_l\sigma^{-1}$, $\beta \to \beta\sigma^{-1}$, and the normalization $E(\Theta) = 0$; the inverse transformations $\hat{\theta}_l \to \hat{\theta}_l\sigma$, $\hat{\beta} \to \hat{\beta}\sigma$ will reproduce the parameters in their original form.) As a practical procedure for solving these equations, one may follow the proposal of Bock and Aitkin (1981) to apply a variant of the so-called expected marginals (EM) algorithm (cf. Dempster et al., 1977). On the basis of some initial values for the $\hat{\beta}_i$ and $\hat{\sigma}$, compute, by means of (56), approximate values for the expectations $\tilde{e}_{li}$ and $\tilde{n}_l$ (this is the so-called E-step of the procedure), and solve Eqs. (58) and (60) numerically for the $\hat{\beta}_i$ and $\hat{\sigma}$ (the so-called M-step), and so forth. The procedure is easy and numerically stable because each $\hat{\beta}_i$ appears in only one of Eqs. (58); these equations can be solved one at a time, so that no problems arise with long tests. The convergence of the EM-algorithm is rather slow, however, in the neighborhood of the solution. MML software was first published by Mislevy and Bock (1986).

Now let the semiparametric approach be considered. The unspecified latent distribution $G(\theta)$ is approximated by a discrete distribution with values (support points) θ_l and weights $g_l := g(\theta_l)$, for $l = 1, \ldots, m$. Both the g_l and the θ_l have to be estimated from the data. The parameter σ can again be set to 1 and thus dropped from all formulae. This approach is equivalent to applying so-called 'Latent Class Analysis (LCA)' (cf. Lazarsfeld and Henry, 1968) where each person is assumed to be member of one latent class C_l, $l = 1, \ldots, m$, with parameters θ_l and class sizes g_l, subject to $\sum_l g_l = 1$, and where the response probabilities for persons $S_v \in C_l$ are the logistic IRFs (5) and (6), with $\theta_v = \theta_l$ (Formann, 1982, 1985, 1995). The log-likelihood (55), with $\sigma = 1$, is maximized as a function of the parameter vectors $\boldsymbol{\beta}$, $\boldsymbol{g}$, and $\boldsymbol{\theta}$, under the constraint $\sum_l g_l = 1$ with $0 < g_l < 1$. In order to satisfy the last inequality, Formann introduced the auxiliary parameters $z_l := \ln[g_l/(1 - g_l)]$.

Employing the method of Lagrange multipliers, the enhanced function $F = \ln L_M(\boldsymbol{X}|\boldsymbol{\beta}, \boldsymbol{\theta}, \boldsymbol{g}) + \lambda \sum_l g_l$ is maximized, where λ is a Lagrange multiplier. Since the second term in F is independent of the β_i, the partial derivatives with respect to the β_i are the same as before, yielding again the estimation Eqs. (58), with $\sigma = 1$. To derive estimation equations for the θ_l, the partial derivative of (55), with $\sigma = 1$, has to be taken with respect to some θ_j. Again, we do not present the full derivation, we just sketch the major steps. First consider the term

$$\begin{aligned} &\frac{\partial}{\partial \theta_j} \sum_l \frac{g_l \exp(r_v \theta_l)}{\prod_i [1 + \exp(\theta_l - \beta_i)]} \\ &\quad = \frac{g_j \exp(r_v \theta_j)}{\prod_i [1 + \exp(\theta_j - \beta_i)]} \big[r_v - E(R|\theta_j, \boldsymbol{\beta})\big], \end{aligned} \tag{61}$$

with $\sum_i f_i(\theta_j) =: E(R|\theta_j, \boldsymbol{\beta})$, the expected raw score for a testee with $\theta_v = \theta_j$. Using (61) and (56), it is found that equating $\partial F/\partial \theta_j$ to zero yields

$$\begin{aligned} \frac{\partial F}{\partial \theta_j} &= \sum_r n_r \frac{\partial}{\partial \theta_j} \ln L_M(\boldsymbol{x}_v|\boldsymbol{\beta}, \boldsymbol{\theta}, \boldsymbol{g}) \\ &= \sum_r n_r P(\theta_j|r, \boldsymbol{\beta}, \boldsymbol{g})\big[r - E(R|\theta_j, \boldsymbol{\beta})\big] = 0, \end{aligned} \tag{62}$$

for $j = 1, \ldots, m$. These are the desired estimation equations for the support points θ_j.

Finally we turn to the estimation of the g_l. To ensure that they are positive quantities, we set $g_l = \exp(z_l)$. Remembering (55) and (56), put to zero the partial derivatives of $F = L_M(\boldsymbol{X}|\boldsymbol{\beta}, \boldsymbol{\theta}, \boldsymbol{g}) + \lambda \sum_l g_l$ with respect to z_l,

$$\frac{\partial F}{\partial z_l} = \frac{\partial}{\partial z_l} \ln L_M(\boldsymbol{X}|\boldsymbol{\beta}, \boldsymbol{\theta}, \boldsymbol{g}) + \lambda \frac{\partial}{\partial z_l} \sum_j \exp(z_j)$$

$$= \sum_r n_r P(\theta_l|r, \boldsymbol{\beta}, \boldsymbol{g}) + \lambda \exp(z_l) = 0. \tag{63}$$

Summing (63) over l, we get $\sum_l \sum_r n_r P(\theta_l|r, \boldsymbol{\beta}, \boldsymbol{g}) + \lambda \sum_l \exp(z_l) = 0$, which in virtue of $\sum_l \exp(z_l) = 1$ yields $\lambda = -\sum_l \sum_r n_r P(\theta_l|r, \boldsymbol{\beta}, \boldsymbol{g}) = n$. Inserting this in (63) and rearranging terms,

$$g_l = n^{-1} \sum_r n_r P(\theta_l|r, \boldsymbol{\beta}, \boldsymbol{g}). \tag{64}$$

These are the estimation equations for the g_l, for $l = 1, \ldots, m$.

So there are three sets of estimation equations: (58) for the β_i, (62) for the θ_l, and (64) for the g_l. Several approaches to solve these equations have been discussed in the literature, for instance, gradient, Newton–Raphson, and quasi-Newton methods. The simplest, albeit not most efficient, way is the following. Begin with trial values for the β_i and θ_l and solve equations (64) iteratively for provisional values for the $\hat{g}_l$; using these values and the provisional β_i, solve equations (62) for improved values for the $\hat{\theta}_l$; and using the current $\hat{g}_l$ and $\hat{\theta}_l$, solve (58) for improved $\hat{\beta}_i$ estimates, and so forth. The convergence of this algorithm is slow, but its relative efficiency becomes the better the greater k. Its most outstanding virtue is that k is practically unlimited.

Since only the first k moments of the latent distribution can be uniquely estimated, even for $n \to \infty$ the distribution G cannot be determined in great detail; the information about G is exhausted if at most $[(k+2)/2]$ support points (or latent classes) have been introduced[5]. For that case, a surprising result about the equivalence of semiparametric MML and CML has been established by De Leeuw and Verhelst (1986), building upon previous results of Tjur (1982) and Cressie and Holland (1983) (see also Follmann, 1988; Lindsay et al., 1991; Pfanzagl, 1993, 1994a, 1994b; and Formann, 1995):

THEOREM 11. *Let $\hat{\boldsymbol{\beta}}_C$ denote the CML estimator of $\boldsymbol{\beta}$. If there exists a distribution G such that $P_M(r|\hat{\boldsymbol{\beta}}_C, \boldsymbol{\theta}, \boldsymbol{g})$, see* (22), *equals the observed relative score frequencies n_r/n, for all $r = 0, \ldots, k$, then $(\hat{\boldsymbol{\beta}}_C, \boldsymbol{\theta}, \boldsymbol{g})$ maximizes also the semiparametric marginal likelihood function $L_M(\boldsymbol{X}|\boldsymbol{\beta}, \boldsymbol{\theta}, \boldsymbol{g})$, making the (normalized) semiparametric MML estimator $\hat{\boldsymbol{\beta}}_M$ identical to the (equally normalized) CML estimator, $\hat{\boldsymbol{\beta}}_M = \hat{\boldsymbol{\beta}}_C$. The solution is unique if the normalized CML estimator is unique.*

This version of the theorem is taken from Lindsay et al. (1991). If the condition stated in the theorem is satisfied, the relative score frequencies and their marginal estimators are said to be 'concordant'. For k items, k independent relative raw score frequencies are observed and are to be matched. If G has m support points, the number of independent parameters of G is $m + (m - 1) = 2m - 1$ (namely, m parameters θ_l and $m - 1$ independent g_l); therefore it is plausible that $2m - 1$ should become $\geqslant k$, whereas a further

[5] Here and in the sequel, the symbol '$[x]$' denotes 'entier (x)'.

increase of m should not improve the fit. If k is odd, concordance – if attainable – should therefore occur with $m = (k+1)/2$, and if k is even, with $m = (k+2)/2$. This is indeed the most that is needed, but there are 'discordant' cases where the condition of the theorem cannot be satisfied with any m. An example is a raw score distribution where $n_r = 0$ for some r; then, obviously, $P_M(r|\hat{\boldsymbol{\beta}}, \boldsymbol{\theta}, \boldsymbol{g})$ cannot fit n_r/n with any finite θ_l and any g_l. On the other hand, there exist cases where concordance is attained even with m smaller than $[(k+2)/2]$. Therefore, n.s. criteria for the concordance condition are not trivial. Such a criterion was first given by De Leeuw and Verhelst (1986). The following more readily applicable results are due to Lindsay et al. (1991). Their criterion is based on the ratio statistics

$$q(r) = \frac{n_r/n}{P_M(r|\hat{\boldsymbol{\beta}}_C, \hat{\theta}_0)}, \tag{65}$$

for $r = 0, \ldots, k$, where $\hat{\theta}_0$ is the estimator of θ_0 resulting from the maximization of the marginal likelihood under the 'homogeneity model' which assumes $m = 1$, that is, where all testees are have the same ability θ_0 (i.e., belong to the same class C_0).

THEOREM 12. *The concordance condition is satisfied if and only if the sequence* $q_0, \ldots, q_k$ *corresponds to the moments of a nonnegative measure.*

A well-known criterion for identifying whether the sequence of q_r statistics is a moment sequence makes use of the determinants of the following matrices (cf., e.g., Karlin and Studden, 1966): for each integer l, $l = 1, \ldots, k$, construct the matrices

$$\boldsymbol{Q}_l = \begin{pmatrix} q(0) & q(1) & \cdots & q(h) \\ q(1) & q(2) & \cdots & q(h+1) \\ \vdots & \vdots & \ddots & \vdots \\ q(h) & q(h+1) & \cdots & q(2h) \end{pmatrix} \quad \text{when } l \text{ is even,} \quad l = 2h,$$

and

$$\boldsymbol{Q}_l = \begin{pmatrix} q(1) & q(2) & \cdots & q(h) \\ q(2) & q(3) & \cdots & q(h+1) \\ \vdots & \vdots & \ddots & \vdots \\ q(h) & q(h+1) & \cdots & q(2h-1) \end{pmatrix} \quad \text{when } l \text{ is odd,} \quad l = 2h-1.$$

COROLLARY 1. *If* $\boldsymbol{Q}_{k-1}$ *and* $\boldsymbol{Q}_k$ *are positive definite, or equivalently, if* $\det(\boldsymbol{Q}_l) > 0$ *for* $l = 0, \ldots, k$, *then there exist infinitely many semiparametric (concordant) MML solutions for G fitting the raw score frequencies perfectly, and hence* $\hat{\boldsymbol{\beta}}_C = \hat{\boldsymbol{\beta}}_M$ *under the same normalization of both estimators.*

The next theorem of Lindsay et al. (1991) indicates that, for fixed k and large n, the concordance condition will usually be satisfied, provided that the population of testees has sufficient heterogeneity.

THEOREM 13. *Let the true latent distribution G have $m \geqslant [(k+2)/2]$ finite support points. Then the concordance condition will be satisfied on a set whose probability tends to 1 for $n \to \infty$, implying that the probability of identity of the (equally normalized) semiparametric MML and CML estimators of $\boldsymbol{\beta}$ tends to* 1.

What these theorems amount to can be summarized as follows: If to a given set of data $\boldsymbol{X}$ the semiparametric MML method is applied with $m = 1, \ldots, [(k+2)/2]$ support points (latent classes) and if the raw score frequencies can be fitted perfectly with one of these models, then the MML estimator $\hat{\boldsymbol{\beta}}_M$ of that model and the CML estimator $\hat{\boldsymbol{\beta}}_C$ are identical (assuming the same normalization in both cases). If this concordance condition is not attained, then it is impossible to fit the raw score distribution perfectly with any m.

These results do not make semiparametric MML and CML 'almost' equivalent for all practical purposes, though. Often researchers want to compare item parameter estimates of, for instance, 'high' and 'low' scoring testees (see Section 6). If testees are selected from the total sample on the basis of their raw scores, however, a subset of the n_r will become zero and hence the semiparametric MML solution will no longer be identical with the CML solution.

4.4. An approximate estimation method

Especially in the early days of the RM, various approximate estimation methods have been in use (e.g., Rasch, 1960). Their disadvantage is that they are not based on a likelihood function and do not have known asymptotic behavior so that (a) no standard errors are estimable and (b) no asymptotic hypothesis tests can be grounded on them. Since, however, exact combinatorial and Monte Carlo tests of fit have become available in the last decade which do not require asymptotic theory and which even appear to be robust with respect to a slight estimation error in the $\hat{\beta}_i$ (see Section 6), approximate methods might again become interesting alternatives to CML and MML. Nevertheless, consistency is a must even for such methods. From this perspective, one promising approximate estimation method will be briefly described.

Fischer (1974) presented a procedure – in the sequel denoted Approximate Method (AM) – based on the numbers n_{ij} of testees with answer '+' to item I_i and '−' to item I_j (implying $n_{ii} = 0$). For a single pair of items (I_i, I_j), it is well known that $n_{ij}/(n_{ij} + n_{ji})$ is a consistent ML estimator of $\epsilon_i/(\epsilon_i + \epsilon_j)$ (cf. Rasch, 1960, p. 172). (This is a conditional approach that does not involve the person parameters and, in the special case where the test consists of just I_i and I_j, even yields the CML estimates.) Hence it follows that n_{ij}/n_{ji} estimates ϵ_i/ϵ_j, and it is easy to show that the random variable $N_{ij}\epsilon_i^{-1} - N_{ji}\epsilon_j^{-1}$ has expectancy zero and variance $(n_{ij} + n_{ji})(\epsilon_i\epsilon_j)^{-1}$. Therefore, consider all pairs of items (I_i, I_j) jointly and minimize the variance-weighted sum of squares,

$$\sum_{i<j} \frac{\left(n_{ij}\epsilon_i^{-1} - n_{ji}\epsilon_j^{-1}\right)^2}{(n_{ij} + n_{ji})(\epsilon_i\epsilon_j)^{-1}} = \sum_{i<j} \frac{(n_{ij}\epsilon_j - n_{ji}\epsilon_i)^2}{(n_{ij} + n_{ji})\epsilon_i\epsilon_j}. \tag{66}$$

The summation in (66) is to be taken with the understanding that pairs (i, j) with $n_{ij} + n_{ji} = 0$ are omitted. This estimation method is not rigorously founded, though, because the terms in the sum of (66) are not independent. The estimation equations resulting from the minimization of (66) are (cf. Fischer, 1974; Molenaar, 1995; Zwinderman, 1995)

$$\epsilon_i^2 = \frac{\sum_{j=1}^{k} y_{ij}\epsilon_j}{\sum_{j=1}^{k} y_{ji}\epsilon_j^{-1}}, \tag{67}$$

for $i = 1, \ldots, k$, with $y_{ij} := n_{ij}^2/(n_{ij} + n_{ji})$ for $n_{ij} + n_{ji} > 0$ and $y_{ij} = 0$ for $n_{ij} + n_{ji} = 0$. Eqs. (67) are easily solved by inserting, on the right-hand sides, trial values for the parameters and obtaining improved values from the left-hand sides, etc. The AM approach has the advantage of (a) being consistent and (b) conditional, that is, independent of the person parameters, (c) requiring very little computational expense both per iteration and in terms of number of iterations, and (d) allowing the data to be incomplete. The disadvantages are (e) that the estimator is not efficient and (f) has unknown asymptotic properties. De Gruijter (1987) reported the method to be extremely fast and accurate as long as the RM holds, but to lack robustness under violations of the model assumptions.

EXAMPLE 1. Item parameter estimation.

To illustrate and compare the results of the application of different item parameter estimation methods, a data sample of $n = 1160$ testees who have taken Gittler's '3DW' cubes test of spatial abilities (Gittler, 1991) is analyzed.[6] The item parameters of the $k = 17$ items are estimated via CML, AM, parametric MML (where the latent distribution of the θ_v is assumed to be normal) and semiparametric MML (with $m = 15 > [(k + 2)/2] = 9$ support points). The item marginals s_i and the raw score frequencies n_r for the total group and three subgroups are given in Table 1, and the results of the item parameter estimation are shown in Table 2.

Table 1
Item and person marginals for Examples 1, 2, 3

i	1	2	3	4	5	6	7	8	9
s_i	786	682	553	476	326	787	502	556	504
i	10	11	12	13	14	15	16	17	
s_i	389	488	695	686	695	589	768	487	
r	0	1	2	3	4	5	6	7	8
n_r	80	33	40	57	52	68	84	73	73
r	9	10	11	12	13	14	15	16	17
n_r	70	90	76	70	73	72	54	50	45

[6] The author thanks G. Gittler for making these data available for the present demonstration.

Table 2
Item parameter estimates by CML, AM, parametric MML, and semiparametric MML for the data of Examples 1, 2, 3

	Total sample				Low scor. group			Med. scor. group			High scor. group		
i	CML	AM	MLn	MLe	CML	MLn	MLe	CML	MLn	MLe	CML	MLn	MLe
1	−1.06	−1.06	−1.06	−1.06	−1.08	−1.09	−1.11	−0.93	−0.99	−0.99	−1.39	−1.40	−1.40
2	−0.48	−0.49	−0.49	−0.48	−0.24	−0.25	−0.25	−0.59	−0.54	−0.54	−0.66	−0.66	−0.66
3	0.17	0.17	0.17	0.17	0.10	0.10	0.10	0.24	0.13	0.13	0.17	0.18	0.17
4	0.56	0.56	0.57	0.56	0.48	0.49	0.49	0.52	0.69	0.69	0.73	0.74	0.74
5	1.38	1.34	1.39	1.38	1.05	1.07	1.08	1.31	1.22	1.22	1.59	1.58	1.59
6	−1.06	−1.08	−1.06	−1.06	−0.91	−0.92	−0.93	−1.13	−1.10	−1.10	−1.47	−1.48	−1.48
7	0.43	0.45	0.43	0.43	0.72	0.73	0.74	0.32	0.40	0.40	0.49	0.50	0.50
8	0.16	0.15	0.16	0.16	0.03	0.02	0.03	0.23	0.36	0.36	0.21	0.22	0.22
9	0.42	0.45	0.42	0.42	0.69	0.70	0.71	0.30	0.38	0.38	0.51	0.51	0.51
10	1.02	1.03	1.03	1.02	1.13	1.15	1.15	1.00	0.92	0.92	1.07	1.07	1.08
11	0.50	0.49	0.50	0.50	0.48	0.49	0.50	0.52	0.68	0.68	0.54	0.55	0.55
12	−0.55	−0.52	−0.56	−0.55	−0.63	−0.64	−0.64	−0.57	−0.59	−0.59	−0.26	−0.26	−0.26
13	−0.50	−0.48	−0.51	−0.50	−0.61	−0.63	−0.63	−0.49	−0.54	−0.54	−0.23	−0.23	−0.24
14	−0.55	−0.56	−0.56	−0.55	−0.42	−0.43	−0.43	−0.53	−0.59	−0.59	−0.92	−0.92	−0.93
15	−0.01	−0.01	−0.01	−0.01	−0.02	−0.03	−0.03	0.13	0.03	0.03	−0.26	−0.26	−0.26
16	−0.95	−0.94	−0.95	−0.95	−1.03	−1.04	−1.06	−0.88	−0.99	−0.99	−0.82	−0.83	−0.83
17	0.51	0.50	0.51	0.51	0.26	0.26	0.26	0.53	0.54	0.54	0.70	0.71	0.71

Legend: 'MLn' denotes 'MML with normal prior', 'MLe' denotes 'MML with empirical prior' (semiparametric MML), and the abbreviations CML and MML are as elsewhere.

The CML estimates were computed by means of LPCM Win (Fischer and Ponocny-Seliger, 1998), the MML estimates using BILOG-MG (Zimowski et al., 1996; Du Toit, 2003). There is an almost perfect concordance between CML and MML results, both in the parametric (normal prior) and semiparametric ('empirical prior') MML, except for the Low Scoring Group: for these data, the BILOG-MG program flagged a warning that the semiparametric MML did not reach convergence (up to a pre-specified degree of accuracy of ± 0.01), but the parameter estimates are nevertheless inconspicuous. The AM estimates, given here only for the Total Sample, are seen to be acceptable approximations to the CML estimates.

5. Person parameter estimation

Once reliable item parameter estimates $\hat{\beta}_i$ are available, the estimation of the person parameters θ can proceed via straightforward maximum likelihood, using equation (41). The estimator $\hat{\theta}$ is denoted the MLE of θ.[7] If the data are complete, only $k - 1$ such equations are to be solved. If a finite MLE exists, it is unique because the sum of probabilities in (41) is increasing in θ. Finding the MLE is therefore easy; a proven numerical

[7] Since person parameter estimation refers to one parameter at a time, index v can be omitted in this section.

procedure that avoids 'spider-web phenomena' is to obtain an approximate solution by means of a simple bisection method and then to improve it via a few Newton iterations.

The MLE has attractive asymptotic properties for $k \to \infty$ but performs poorly in tests of realistic length. Lord (1983) derived the asymptotic (i.e., for $k \to \infty$) variance and bias of the ML estimator:

$$\mathrm{Var}(\mathrm{MLE}) = \frac{1}{\mathcal{I}} + o\big(k^{-1}\big), \tag{68}$$

$$\mathrm{Bias}(\mathrm{MLE}) = -\frac{\mathcal{J}}{2\mathcal{I}^2}, \tag{69}$$

where $\mathcal{I} = \sum_i f_i(\theta)[1 - f_i(\theta)]$ is the Fisher information and $\mathcal{J} = \sum_i \partial^2 f_i(\theta)/\partial\theta^2 = \sum_i [1 - 2f_i(\theta)] f_i(\theta)[1 - f_i(\theta)]$, both evaluated at $\theta = \mathrm{MLE}$. The bias of the MLE is of order $O(k^{-1})$, which is considerable. In addition to this, many researchers view it as a disadvantage of the MLE that no finite estimates exist for extreme raw scores. If, for instance, a study intends to compare test achievements across subpopulations, infinite parameter estimates prevent the application of many standard statistical tools.

Warm (1989) proposed a class of estimators of θ defined as those values of θ which maximize the weighted likelihood

$$L(\boldsymbol{x}|\theta, \boldsymbol{\beta})h(\theta) = \prod_i f_i(\theta)^{x_i}\big[1 - f_i(\theta)\big]^{1-x_i} h(\theta), \tag{70}$$

where $h(\theta)$ is a suitably chosen weight function. Taking the derivative of the logarithm of (70) with respect to θ and setting it to zero yields the estimation equation

$$r - \sum_i f_i(\theta) + \frac{\mathrm{d}\ln h(\theta)}{\mathrm{d}\theta} = 0. \tag{71}$$

If h is constant, the root of (71) is the MLE. If $h(\theta)$ is a prior density function, then (70) is proportional to the posterior density function of θ, so that the value $\hat{\theta}$ maximizing (70) becomes a Bayesian Modal Estimator (BME) of θ, with bias

$$\mathrm{Bias}(\mathrm{BME}) = \mathrm{Bias}(\mathrm{MLE}) + \frac{\mathrm{d}\ln h(\theta)}{\mathrm{d}\theta}\mathcal{I}^{-1}, \tag{72}$$

where $\theta = \mathrm{BME}$; this bias is of order $O(k^{-1})$ (Warm, 1989).

Warm (1989) suggested a new estimator of θ choosing $h(\theta)$ such that the bias (72) is removed. Putting $\mathrm{Bias}(\mathrm{BME}) = 0$ and inserting (69) in (72) gives $\mathrm{d}\ln h(\theta)/\mathrm{d}\theta = \mathcal{J}/(2\mathcal{I})$, which inserted in (71) yields the estimation equation

$$r - \sum_i f_i(\theta) + \frac{\mathcal{J}}{2\mathcal{I}} = 0. \tag{73}$$

Eq. (73) can be solved by means of the same numerical methods as (41). The solution is denoted Warm Likelihood Estimator (WLE). It exists for all r, including extreme scores. Concerning the bias of WLE, Warm (1989) proved the following result:

THEOREM 14. $\text{WLE}(\theta)$ *is unbiased to order* k^{-1}, *that is,* $\text{Bias}(\text{WLE}(\theta)) = 0+o(k^{-1})$. $\text{WLE}(\theta)$ *is asymptotically normally distributed with asymptotic variance equal to the asymptotic variance of* $\text{MLE}(\theta)$, *that is,* $\text{Var}(\text{WLE}(\theta)) \simeq \text{Var}(\text{MLE}(\theta)) \simeq \mathcal{I}^{-1}$.

The asymptotic properties of MLE, BME, and WLE are of limited importance because k in most applications will be small. The smallness of the bias of the WLE and the existence of a finite estimator for $r = 0$ and $r = k$ are often considered criteria by which the WLE should be preferred over the MLE and the BME. It must be mentioned, though, that the application of the estimation equation (73) to extreme scores $r = 0$ or $r = k$ leads to arbitrary estimates $\hat{\theta}$ because Warm's (1989) derivation rests on Lord's (1983) formula (69), which was derived under assumptions excluding the extreme cases $r = 0$ and $r = k$. This is clear also because the bias of an infinite estimator cannot be determined.

Yet another possibility is to use the parameter estimates obtained via MML estimation for computing the mean value of the θ_l from the posterior distribution characterized by the probability function (56), yielding

$$\hat{\theta} = \frac{\sum_l \theta_l P(\theta_l | r, \boldsymbol{\beta}, G)}{\sum_l P(\theta_l | r, \boldsymbol{\beta}, G)}, \tag{74}$$

with MML estimates instead of parameters on the right-hand side of (74). This estimator is denoted Bayesian Expected A Posteriori Estimator (EAP). It has the advantage that finite EAPs result for all values of r. Little seems to be known, however, about their theoretical properties (cf. Hoijtink and Boomsma, 1996).

A basically different approach to assessing the uncertainty (or precision) of $\hat{\theta}$ is interval estimation instead of point estimation. Klauer (1991a) worked this out in detail for the MLE: he constructed randomized Uniformly Most Accurate (UMA) confidence intervals which are formally equivalent to Uniformly Most Powerful Unbiased (UMPU) tests of hypotheses about θ. They are very convenient to test, for instance, the null-hypothesis that an observed $\hat{\theta}$ does not exceed a fixed cutoff or threshold score.

EXAMPLE 2. Person parameter estimation.

Returning to Example 1, we use the CML and MML results of the Total Sample to compute the MLEs, WLEs, and EAPs of θ (shown in Table 3).

The distribution of the MLEs has no finite variance because, for extreme scores $r = 0$ and $r = k$, the MLEs are $-\infty$ and ∞, respectively. Comparing the MLEs and WLEs illustrates that the bias of the MLEs, for $0 < r < k$, becomes the greater the nearer the raw score is to one of the extremes, 0 or k. The EAPs computed under the normal (parametric MML) and empirical (semiparametric MML) priors are so similar that the figures rounded to two figures after the decimal point do not reveal any differences. As was to be expected, the EAPs are biased toward the mean and thus have a smaller variance than the WLEs.

Table 3
MLEs, WLEs, and EAPs of the Person Parameters for Examples 1, 2, 3

r_v	0	1	2	3	4	5	6	7	8
MLE	$-\infty$	−2.98	−2.20	−1.69	−1.31	−0.97	−0.68	−0.40	−0.13
WLE	−3.80	−2.62	−2.01	−1.58	−1.23	−0.93	−0.65	−0.38	−0.13
EAP	−2.20	−1.85	−1.54	−1.28	−1.02	−0.76	−0.54	−0.34	−0.12
r_v	9	10	11	12	13	14	15	16	17
MLE	0.13	0.40	0.67	0.97	1.30	1.69	2.20	2.97	∞
WLE	0.12	0.38	0.64	0.92	1.23	1.58	2.02	2.62	3.81
EAP	0.12	0.34	0.54	0.76	1.01	1.27	1.54	1.85	2.20

6. Testing of fit

The problem of testing of fit in the RM is closely related to the methods of parameter estimation: most of the powerful test methods for assessing fit and/or testing other structural hypotheses about the parameters require consistency of their estimators and knowledge of their asymptotic behavior. Although many approximate tests for item and/or person fit, based on plausibility, have been proposed in the literature, the prevailing methods of current interest are tightly connected to CML or MML estimators and their known asymptotic properties. Some heuristic methods, like Rasch's (1960) graphical control of the model and 'empirical' IRFs, are useful complements to these tests. Tests of fit can be grouped by (a) the type of statistic used (Likelihood Ratio (LR) tests, Pearson-type statistics, Wald-type tests, Lagrange Multiplier (LM) tests, randomization tests, nonparametric methods), (b) the power of the tests with respect to specific alternative hypotheses (violations of sufficiency of the raw score, of increasing parallel IRFs, of specific objectivity, of unidimensionality, of local stochastic independence), and (c) whether they are exact, asymptotic, approximate/heuristic. Space limitations prevent, however, discussing all available methods. So the focus of the following will be on a few rigorously established and sufficiently general methods.

6.1. *Conditional likelihood ratio tests*

G. Rasch generally showed a preference for heuristic graphical methods over significance tests (see, e.g., Rasch, 1960; Andersen and Olsen, 2001). Since SO (i.e., independence of the item parameter estimates of the distribution of the person parameters) is a central feature of the RM, for Rasch an obvious way of controlling the RM was to compare the item parameter estimates per score group (or per score region) to the overall item parameter estimates. Such graphs revealed whether some – and if so, which – of the items violated SO. Andersen (1973c) constructed a conditional likelihood ratio (CLR) test statistic that follows the same rationale. Let the sample of testees be split into score groups G_r, $r = 1, \dots, k-1$, of respective sizes n_r, and estimate the item parameters via CML in each of these groups as well as in the total sample. Let the conditional likelihoods, each evaluated at the respective CML estimate, be denoted $L_C^{(r)}$,

$r = 1, \ldots, k-1$, and L_C, respectively. Under H_0 the same RM holds for all groups, under H_1 a separate RM (with different item parameters) holds in each group.

THEOREM 15. *The test statistic for H_0 against H_1,*

$$-2\ln\lambda = -2\left(\ln L_C - \sum_{r=1}^{k-1} \ln L_C^{(r)}\right), \tag{75}$$

is asymptotically ($n_r \to \infty$ for all r) χ^2-distributed[8] *with $df = (k-1)(k-2)$.*

The degrees of freedom follow the general rule 'number of independent parameters under H_1 minus number of independent parameters under H_0'. The proof was given by Andersen (1973c). The same test was independently suggested by Martin-Löf (1973).

Since an RM has to be fitted in each score group separately, each of them should be 'large'. In practice, often some of the score groups are small, so that 'score regions' have to be formed by collapsing the data from adjoining score groups. The resulting CLR test, with appropriate reduction of the degrees of freedom, is a somewhat conservative test of the above H_0 against the H_1 of heterogeneous item parameters because, within score regions, variances of the item parameters between score groups are ignored. The test moreover remains applicable if the testees are grouped by some external criterion like gender, age, schooling, etc., instead of by raw scores. Note that this type of model test, which has been employed routinely in many applications of the RM since the early 1970s, is equivalent to testing all items simultaneously for DIF. If such a test turns out significant, graphical plots of the item parameter estimates of each of the testee subgroups against the overall estimates and/or graphs of so-called 'empirical' IRFs can give a lead to understanding more clearly possible sources of misfit.

It has to be stressed that the validity of this CLR test hinges on the assumption of model fit under H_1, that is, an RM has to fit in each person group G_g separately. This is not a trivial requirement, often ignored in testing nested models. The danger lies in the possibility that the CLR test statistic is nonsignificant, but that the RM does not fit even under H_1 – which sometimes passes unnoticed: many users seem prone to the erroneous conclusion that the H_0 model fits whenever the CLR test is nonsignificant. Apart from the general principle that null-hypotheses cannot be proved, the problem is that there may be factors causing lack of fit that are not strictly related to the raw scores. The same misunderstanding may happen if the sample is split into two or more subgroups on the basis of, say, type of schooling, gender, age, or the like, if these tests turn out nonsignificant: counting on fit of the RM can go seriously wrong in the presence of factors causing model violations that are unrelated to the splitting criteria. Lack of power of the described CLR tests in such cases has been reported in the literature (cf., e.g., Stelzl, 1979).

Another CLR test aims at detecting departures from the unidimensionality assumption. Suppose the items are partitioned a priori in two subsets (henceforth denoted 'subtests') $\mathcal{T}_1 = \{I_1, \ldots, I_{k_1}\}$ and $\mathcal{T}_2 = \{I_{k_1+1}, \ldots, I_k)$ with test lengths $k_1 \geqslant 2$,

[8] In this chapter, 'χ^2-distributed' always means 'centrally χ^2-distributed'.

$k_2 = k - k_1 \geqslant 2$, and scores r_1 and r_2. Denote the number of testees with score combination (r_1, r_2) by $n_{r_1 r_2}$, of testees with total score $r = r_1 + r_2$ by n_r, and let $L_C^{(1)}$, $L_C^{(2)}$, L_C be the conditional likelihoods for subtest T_1, subtest T_2, and the total test, respectively. The H_0 assumes that one RM fits the total test; under H_1, two separate RMs (i.e., with two independent person parameters per testee) are assumed to fit subtest T_1 and T_2, respectively. Martin-Löf (1973) proved the following result:

THEOREM 16. *The test statistic for H_0 against H_1,*

$$X^2 = -2\left(\sum_r n_r \ln\left(\frac{n_r}{n}\right) - \sum_{r_1}\sum_{r_2} n_{r_1 r_2} \ln\left(\frac{n_{r_1 r_2}}{n}\right) + \ln L_C - \ln L_C^{(1)} - \ln L_C^{(2)}\right), \tag{76}$$

is asymptotically ($n_r \to \infty$ for all r) χ^2-distributed with $df = k_1 k_2 - 1$.

Again it has to be pointed out that unidimensionality cannot be concluded from a nonsignificant χ^2; the H_0 can at best be retained. Note that the two subsets of items must be known beforehand and that in each subset an RM must fit. The test is not admissible as a technique for searching for unidimensional subsets in a given pool of items. To apply the test, a rather large sample is required, otherwise some frequencies n_r might be too small.

EXAMPLE 3. CLR tests of fit.

In order to apply Andersen's CLR test of fit to the 3DW data, an RM should be fitted in each score group G_r, for $r = 1, \ldots, 16$. Obviously, trying to do so would not be successful in the present case, as can be seen from the raw score distribution in Table 1. The sample would have to be app. five to ten times as large to enable parameter estimation in each score group separately. What has widely become practice is to partition the testee sample into just *two* groups, below vs. above the mean (or median) raw score. Obviously, this entails a loss of power to detect departures from the RM. In the present case, splitting the data into groups 'Below' (the mean), $n_1 = 560$ testees with raw scores $r = 0, \ldots, 8$, and 'Above' (the mean), $n_2 = 600$ testees with raw scores $r = 9, \ldots, 17$, leads to the following log-likelihoods: -3573.21 for 'Below' and -3916.91 for 'Above'; and the log-likelihood for the total sample is -7500.95. The CLR test statistic therefore becomes $\chi^2 = -2(-7500.95+3573.21+3916.91) = 21.66$ with df $= 2(17-1)-(17-1) = 16$, with $\chi^2_{16;1-\alpha} = 32.00$ for $\alpha = 0.01$,[9] which is nonsignificant. Similar tests can be carried out splitting the sample by Gender ($n_1 = 861$ males, $n_2 = 299$ females; $\chi^2 = 17.80$), Age ($n_1 = 574$ testees aged 13–16, $n_2 = 586$ testees aged $\geqslant 17$; $\chi^2 = 28.57$) and Education ($n_1 = 902$ high-school or college students, $n_2 = 258$ university students or graduates; $\chi^2 = 16.40$), each with df $= 16$ and $\chi^2_{16;1-\alpha} = 32.00$ for $\alpha = 0.01$. All these tests are nonsignificant. Hence the null-hypothesis of absence of DIF with respect to the splitter variables can be retained.

[9] The critical χ^2-values are based on the Wilson–Hilferty approximation and thus may deviate slightly from the exact values.

Given these results, an applied research worker might be inclined to conclude that "the RM fits the data". Statisticians are usually more reserved: they know that models never fit; models can at best fit to some degree of approximation. If the sample is large and the test powerful enough, the model will always be refuted. This can be nicely demonstrated with the present data by splitting the sample in three subgroups 'Low' ($n_1 = 414$, scores $r = 0, \dots, 6$), 'Mid' ($n_2 = 382$, scores $r = 7 \dots, 11$), and 'High' ($n_3 = 364$, scores $r = 12 \dots, 17$). The CLR test is $\chi^2 = -2(-7500.95 + 2234.46 + 3422.59 + 1816.95) = 53,90$ with $\text{df} = 3(17-1) - (17-1) = 32$ and $\chi^2_{32;1-\alpha} = 53.50$ for $\alpha = 0.01$, which is just significant. Hence, increasing the power of the test by partitioning the sample into narrower raw score regions reveals a significant departure from the assumptions of the model.

Nevertheless, Gittler's 3DW test is a good example of a test that has been constructed to fit the RM: for all practical purposes, the degree of fit is sufficient. This can be demonstrated more convincingly by some graphics in addition to the above CLR tests. A nice tool for the assessment of fit is the comparison of so-called 'empirical IRFs' with the 'theoretical IRFs' that are functions of the CML item parameter estimates. The empirical IRFs are constructed as follows: the relative frequencies of right responses per item and score group are smoothed using a sample-size corrected 'normal kernel smoother' (cf. Hart, 1997) with a bandwidth of three adjacent frequencies. Through the resulting points (ordinates: relative frequencies; abscissae: $\hat{\theta}_r$), a smooth line is drawn employing the `\setquadratic` command of the PiCTeX/LaTeX language which lays local quadratic splines through given points. To the smoothed ordinates, confidence intervals (with $\alpha = 0.01$) are attached so that the deviation of the empirical from the theoretical IRF can be evaluated.[10] Figure 1 shows the results for three selected items (an easy item, I_1 with $\hat{\beta}_1 = -1.06$; an item of medium difficulty, I_9 with $\hat{\beta}_9 = 0.42$; and a difficult item, I_5 with $\hat{\beta}_5 = 1.38$).

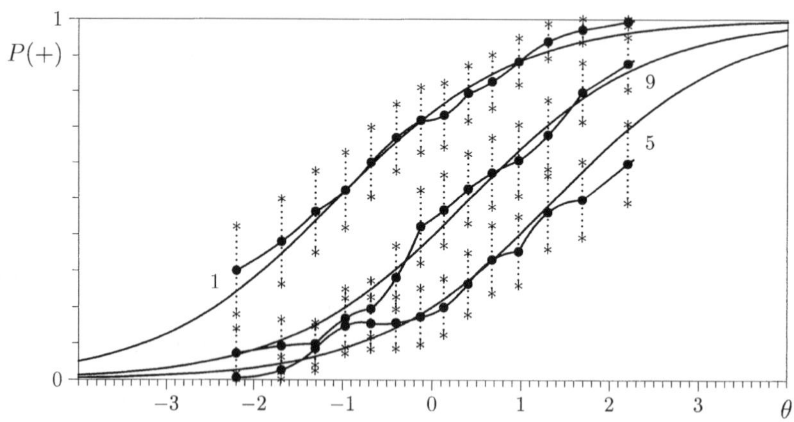

Fig. 1. Empirical and theoretical IRFs for items 1, 5, and 9 of the 3DW test.

[10] A FORTRAN program that automatically produces the LaTeX-file as input for the TeX compiler was written by the author.

The theoretical IRFs lie everywhere within the approximate confidence band of the empirical IRFs. One might speculate whether or not there is an indication of a departure of the empirical from the theoretical IRF of I_1 in low-scoring testees (possibly guessing) or of I_5 in high-scoring testees; but to delineate the lower and upper ranges of the IRFs more precisely, a larger sample with many more low and high scoring testees would be needed. As far as the present data go, the RM can be said to describe the response behavior of the testees sufficiently well.

6.2. Pearson-type tests

When a testee takes a test of k items, the number of feasible response patterns $\boldsymbol{x}$ is $N = 2^k$. In IRT, an observed response pattern $\boldsymbol{x}$ can be considered a realization of a multinomial variable with parameters $\pi_1(\boldsymbol{\phi}), \ldots, \pi_N(\boldsymbol{\phi})$, where $\boldsymbol{\phi}$ is a q-vector of parameters (depending on the type of IRT model). Testing the RM can therefore be carried out within the framework of a general multinomial model. A Pearson-type statistic of the form $X^2 = n\sum_{l=1}^{N}(p_l - \hat{\pi}_l)^2/\hat{\pi}_l$, which is asymptotically χ^2-distributed with df $= N - 1 - q$, should in principle catch all possible deviations from the model assumptions. Unfortunately, such a test is not practical because many observed pattern frequencies would be zero and the corresponding expectations extremely small, even in tests of moderate length and large samples of data. Therefore, Glas (1988) and Glas and Verhelst (1989, 1995), building upon general results about tests of the multinomial distribution (cf. Rao, 1973), developed a class of test statistics composed of certain quadratic forms of deviations between observed and expected frequencies. These statistics are also asymptotically χ^2-distributed, but have the advantage of being applicable in spite of the smallness of the pattern frequencies.

Assume local stochastic independence, let $\boldsymbol{p} = (p_1, \ldots, p_N)'$ and $\boldsymbol{\pi} = (\pi_1, \ldots, \pi_N)'$ denote vectors of the observed and expected relative pattern frequencies, $\boldsymbol{b} = \sqrt{n}(\boldsymbol{p} - \hat{\boldsymbol{\pi}})$ a vector of deviations, and $\boldsymbol{D}_\pi$ a diagonal matrix with diagonal elements π_l, $l = 1, \ldots, N$. With this notation, the Pearson-type statistic can be written as

$$X^2 = \boldsymbol{b}' D_\pi^{-1} \boldsymbol{b}. \tag{77}$$

The class of tests considered is based on a vector $\boldsymbol{d} = \boldsymbol{U}'\boldsymbol{b}$, where $\boldsymbol{U}$ is a (N, m)-matrix of contrasts which condenses the N deviations into a much smaller number m of differences between certain statistics and their expectations. The choice of $\boldsymbol{U}$ should be guided by two practical requirements: m has to be sufficiently small, and the linear combinations of the deviations must reflect the model violations of interest. Moreover, $\boldsymbol{U}$ has to satisfy two formal conditions: (i) the N-vector with all elements equal to 1, denoted $\mathbf{1}$, belongs to the column space of $\boldsymbol{U}$, that is, there exists an m-vector of constants, $\boldsymbol{c}$, such that $\boldsymbol{U}\boldsymbol{c} = \mathbf{1}$, and (ii) the columns of matrix $\boldsymbol{D}_\pi^{-1/2}\partial\boldsymbol{\pi}/\partial\boldsymbol{\phi} =: \boldsymbol{A}$ belong to the column space of $\boldsymbol{D}_\pi^{1/2}\boldsymbol{U}$.

THEOREM 17. *If conditions* (i) *and* (ii) *are satisfied, under* H_0 *the statistic*

$$X^2 = \boldsymbol{b}'\boldsymbol{U}\big(\boldsymbol{U}'\hat{\boldsymbol{D}}_\pi\boldsymbol{U}\big)^{-}\boldsymbol{U}'\boldsymbol{b} = \boldsymbol{d}'\boldsymbol{W}^{-}\boldsymbol{d} \tag{78}$$

is asymptotically χ^2*-distributed with df* $= \operatorname{rank}(\boldsymbol{U}'\boldsymbol{D}_\pi\boldsymbol{U}) - q - 1$.

$\boldsymbol{W}^{-}$ denotes a generalized inverse of $\boldsymbol{U}'\hat{\boldsymbol{D}}_{\pi}\boldsymbol{U} =: \boldsymbol{W}$. For a proof of the theorem, see Glas and Verhelst (1989).

The question is how to find a suitable matrix $\boldsymbol{U}$ of contrasts. Glas and Verhelst (1995) point out a useful strategy: construct $\boldsymbol{U}$ from two building stones, $\boldsymbol{U} = (\boldsymbol{T}|\boldsymbol{Y})$, where $\boldsymbol{Y}$ is a matrix that, via $\boldsymbol{Y}'\boldsymbol{b}$, generates the observed and expected frequencies to be compared, and $\boldsymbol{T}$ satisfies the two conditions (i) and (ii), that is, $\boldsymbol{T}\boldsymbol{c} = \boldsymbol{1}$ and the columns of $\boldsymbol{A}$ belong to the column space of $\boldsymbol{D}_{\pi}^{1/2}\boldsymbol{T}$; moreover, the condition $\boldsymbol{T}'\boldsymbol{b} = \boldsymbol{0}$ should be satisfied. If $\boldsymbol{T}$ satisfies (i) and (ii), so does $\boldsymbol{U}$; and $\boldsymbol{T}'\boldsymbol{b} = \boldsymbol{0}$ implies that the contrasts contained in $\boldsymbol{T}$ have no effect on the test statistic. Glas and Verhelst (1995) describe in more detail how to construct such matrices.

One particular test fitting into this framework is the R_{1C}-test of Glas (1988). The name of this statistic is meant to express that it is a 'first order realization' (referring to response frequencies per item rather than interactions between two or more items) of the construction principles mentioned above and is based on the CML estimator of $\boldsymbol{\beta}$. It was developed to have power against violations of the assumption of parallel increasing IRFs. The focus is on deviations between observed frequencies of responses '+' to items I_i conditional on raw score r, denoted s_{ri}, and their conditional expectations $n_r\pi_{ri}$, obtained from (17).

Let $\boldsymbol{P}_r$ be a $\binom{k}{r} \times k$ matrix whose rows are the response patterns with raw score r; let $\hat{\boldsymbol{D}}_{\pi|r}$ denote the diagonal matrix whose diagonal elements are the conditional pattern probabilities (14) for these response patterns, in the same order, evaluated at the CML estimator $\hat{\boldsymbol{\epsilon}}$; let $\boldsymbol{f}_r$ be the column vector of length $\binom{k}{r}$ whose elements are the observed frequencies of the same patterns; $\boldsymbol{s}_r$ the column vector of length k with elements s_{ri}, the frequencies of right responses to items I_i in score group r; $\boldsymbol{d}_r$ the column vector of length k with elements $n_r^{-1/2}(s_{ri} - n_r\hat{\pi}_{ri})$, where the probabilities π_{ri} are evaluated at $\hat{\boldsymbol{\epsilon}}$. It is easy to verify that $\boldsymbol{P}_r'\boldsymbol{f}_r = \boldsymbol{s}_r$, that is, the vector $\boldsymbol{s}_r$ containing the statistics of interest per score group r is produced from the vector of pattern frequencies, $\boldsymbol{f}_r$, by means of the matrix of contrasts, $\boldsymbol{P}_r$. The H_0 to be tested states that the item parameters generalize over the score groups.

THEOREM 18. *Under H_0, the statistic*

$$R_{1C} = \sum_{r=1}^{k-1} \boldsymbol{d}_r'\big(\boldsymbol{P}_r'\hat{\boldsymbol{D}}_{\pi|r}\boldsymbol{P}_r\big)^{-}\boldsymbol{d}_r = \sum_{r=1}^{k-1} \boldsymbol{d}_r'\hat{\boldsymbol{W}}_r^{-}\boldsymbol{d}_r \tag{79}$$

is asymptotically ($n_r \to \infty$ for $r = 1, \ldots, k-1$) χ^2-distributed with $df = (k-2)(k-1)$.

From a technical point of view, it appears inconvenient that $\hat{\boldsymbol{D}}_{\pi|r}$ and $\boldsymbol{P}_r$ have $\binom{k}{r}$ rows and therefore become unwieldy in long tests. Fortunately, it can be shown that $\boldsymbol{P}_r'\hat{\boldsymbol{D}}_{\pi|r}\boldsymbol{P}_r = \boldsymbol{W}_r$ has elements $\hat{\pi}_{ri}$ in the diagonal and $\hat{\pi}_{rij}$ in the off-diagonal cells. $\hat{\boldsymbol{W}}_r$ can therefore be computed directly using (17) and (19), with $\hat{\boldsymbol{\epsilon}}$ inserted for $\boldsymbol{\epsilon}$.

The same test had already been suggested by Martin-Löf (1973), however, derived under more restrictive assumptions, and it can also be understood as a special case of a family of tests for probabilistic latent trait models proposed by Hamerle (1982). It enables one to test the H_0 that the RM is valid across all score groups, implying that the

IRFs are parallel. If some n_r are too small, adjacent score groups can be collapsed into score regions and the frequencies and expectations added over scores within regions, with appropriate reduction of the degrees of freedom (cf. Glas, 1988). This makes the test somewhat conservative, of course.

To develop this test of fit, the general approach of Glas and Verhelst (1989) to constructing tests within the multinomial distribution would not have been needed, but it proved very helpful for devising several other tests of fit for the RM. Within the MML framework, for instance, a similar statistic, denoted R_{1M} test ('first order frequencies', 'marginal'), was also established. It again compares observed frequencies s_{ri} to their expectations $E(S_{ri}|\boldsymbol{\beta}, G) = n\pi_M(r, i)$, computed by means of (25). The procedure applies both to the parametric and semiparametric MML methods. Let the vector of deviations in score group r be defined as $\boldsymbol{d}_r$ with elements $n^{-1/2}[s_{ri} - n\hat{\pi}_M(r, i)]$, where $\hat{\pi}_M(r, i)$ is the MML estimator of (25), for $r = 1, \ldots, k-1$. Let n_0 and n_k be the numbers of response patterns with raw scores $r = 0$ and $r = k$, respectively, and define $d_0 = n_0 - E(N_0|\hat{\boldsymbol{\beta}}, G)$ and $d_k = n_k - E(N_k|\hat{\boldsymbol{\beta}}, G)$, based on the marginal pattern probabilities (22). Moreover, let $\hat{\boldsymbol{W}}_r$ be the matrix with diagonal elements $\hat{\pi}_M(r, i)$, given in (25), and off-diagonal elements $\hat{\pi}_M(r, i, j)$, given in (26). Finally, suppose that the distribution G is characterized by a q-vector of parameters $\boldsymbol{\omega}$. As in R_{1C}, the H_0 to be tested is the validity of the same RM across score groups, implying parallel IRFs.

THEOREM 19. *Under H_0, the statistic*

$$R_{1M} = \frac{d_0^2}{E(n_0|\hat{\boldsymbol{\beta}}, \hat{\boldsymbol{\omega}})} + \sum_{r=1}^{k-1} \boldsymbol{d}_r' \hat{\boldsymbol{W}}_r^- \boldsymbol{d}_r + \frac{d_k^2}{E(n_k|\hat{\boldsymbol{\beta}}, \hat{\boldsymbol{\omega}})} \tag{80}$$

is asymptotically (i.e., for $n_r \to \infty$) χ^2-distributed with $df = k(k-2) + 2 - q$.

If some n_r are small, adjacent score groups may again be collapsed into score regions and the frequencies and expectations summed over score groups within score regions, with appropriate adjustment of the degrees of freedom (cf. Glas, 1988).

If a testee's position on the latent trait is fixed, the principle of local stochastic independence makes the association between items vanish. Hence, within raw score groups, there will be little association between pairs of items. If the items are multidimensional, the raw score is no longer a sufficient statistic, so that there remains a substantial variance of the latent trait parameters within raw score groups; thus, there will be a considerable association between items. Glas (1988) and Glas and Verhelst (1989) presented asymptotically χ^2-distributed test statistics (namely, R_{2C} and R_{2M} for the CML and MML framework, respectively) for testing the unidimensionality assumption based on the association between pairs of items, again building upon the general principles outlined above.

6.3. Wald-type tests

Glas and Verhelst (1995) described yet another class of tests of fit, the so-called Wald-type tests for testing restrictions on an H_1 model. Suppose, for instance, that the RM is applied independently in two populations of testees, that is, with a total of $p = 2(k-1)$ independent item parameters in the CML framework, and $p = 2(k-1+q)$ parameters in the MML framework. Let the column parameter vectors for the RM in the two populations be denoted $\boldsymbol{\phi}_1$ and $\boldsymbol{\phi}_2$, respectively, and define $\boldsymbol{\phi} = (\boldsymbol{\phi}_1', \boldsymbol{\phi}_2')'$. The null-hypothesis could be, for instance, that some item parameters are equal in both populations, formulated as restrictions on the parameters,

$$h_j(\boldsymbol{\phi}) = \phi_{1j} - \phi_{2j} = 0, \quad \text{for } j = 1, \ldots, m. \tag{81}$$

Let $\boldsymbol{h}(\boldsymbol{\phi}) = (h_1(\boldsymbol{\phi}), \ldots, h_m(\boldsymbol{\phi}))'$, and $\boldsymbol{\Sigma}$ the variance-covariance matrix of the estimator $\hat{\boldsymbol{\phi}} = (\hat{\boldsymbol{\phi}}_1', \hat{\boldsymbol{\phi}}_2')'$, which in this case is a super-diagonal matrix with $\boldsymbol{\Sigma}_1$ and $\boldsymbol{\Sigma}_2$ as diagonal elements; and let $\boldsymbol{T}(\boldsymbol{\phi})$ be the matrix with elements $t_{lj} = \partial h_j(\boldsymbol{\phi})/\partial \phi_l$. Then the Wald statistic is the quadratic form

$$T_W = \boldsymbol{h}'(\hat{\boldsymbol{\phi}})\big[\boldsymbol{T}'(\hat{\boldsymbol{\phi}})\hat{\boldsymbol{\Sigma}}\boldsymbol{T}(\hat{\boldsymbol{\phi}})\big]^{-1}\boldsymbol{h}(\hat{\boldsymbol{\phi}}), \tag{82}$$

which under H_0 is asymptotically χ^2-distributed with df $= m$.

6.4. Lagrange multiplier tests

Finally, Glas and Verhelst (1995) discussed a class of so-called Lagrange multiplier (LM) tests for the RM where the H_0 is obtained from the H_1-model by equating some parameters with certain constants. The LM test of H_0 yields an asymptotic χ^2-statistic with df equal to the number of fixed parameters. A convenient feature of this test is that the model needs to be evaluated only at the ML estimator under the H_0-model. Glas (2001) demonstrated the usefulness of such LM tests for the analysis of differential item functioning (DIF).

Some of the tests mentioned are implemented in the software OPLM (Verhelst et al., 1994). More about the construction of Wald-type and Lagrange multiplier tests for the RM can be found in Hamerle (1982) and Glas and Verhelst (1995).

6.5. Exact tests and approximate Monte Carlo tests

Very powerful approaches to the testing of hypotheses within the RM framework are possible when the item parameters are given. Suppose the item parameters for a set of items (in the following denoted 'item pool') conforming to the RM have been estimated by means of a large calibration sample, so that for practical purposes the parameters can be assumed to be known. Under that assumption, Klauer (1991b) presented a uniformly most powerful unbiased (UMPU) standardized person-fit test of the null-hypothesis that a testee's ability parameter is constant across two subsets of the items ('subtests' $\mathcal{T}_1$ and $\mathcal{T}_2$, respectively, with person parameters θ_1 and θ_2, H_0: $\theta_1 = \theta_2 = 0$). It is important to stress that this is a single-case approach referring to one testee at a time. The method of the test is grounded on the Neyman–Pearson lemma and employs randomization to make the discrete scores continuous. Fischer (1995e, 2001, 2003) suggested also a direct CML estimation of the amount of change between pretest $\mathcal{T}_1$ and posttest $\mathcal{T}_2$, $\eta = \theta_2 - \theta_1$, and extended Klauer's method to a family of generalized RMs (i.e., Partial Credit Models). This approach has some quite attractive properties.

The index v can be dropped because only one testee is being considered. Subtests $\mathcal{T}_1$ and $\mathcal{T}_2$ may be any two samples of items from the pool; they can be disjoint or partially overlapping or even identical subsets, so that there is a great flexibility with respect to applications. Employing again the notation $\epsilon_i = \exp(-\beta_i)$ for the item parameters, let the item parameter vectors of $\mathcal{T}_1$ and $\mathcal{T}_2$ be denoted, w.l.g., $\boldsymbol{\epsilon}_1 = (\epsilon_1, \ldots, \epsilon_{k_1})'$ and $\boldsymbol{\epsilon}_2 = (\epsilon_{k_1+1}, \ldots, \epsilon_k)'$; according to assumptions, some or all of the components of $\boldsymbol{\epsilon}_1$ may be identical with some or all components of $\boldsymbol{\epsilon}_2$. Let θ_1 be denoted by θ, and θ_2 by $\theta + \eta$. The 'change parameter' η, however, can immediately be absorbed into the item parameters, $\beta_i \to \beta_i - \eta$ for all $I_i \in \mathcal{T}_2$, or equivalently, with the notation $\delta = \exp(\eta)$, $\boldsymbol{\epsilon}_2 \to \delta\boldsymbol{\epsilon}_2$. This is convenient because now the person parameter is kept constant, so that it can be conditioned out of the likelihood.

Let r_1 and r_2 be the subtest scores of $\mathcal{T}_1$ and $\mathcal{T}_2$, respectively. The hypothesis test for H_0: $\eta = 0$ (or, equivalently, H_0: $\delta = 1$) is based on the exact conditional distribution of the change score $R_2 - R_1$, given the total score over both subtests, $R_1 + R_2 = r$. Its probability function $P_\delta(r_2 - r_1|r)$ is obtained by summing the likelihoods (7) of all response patterns yielding fixed subtest scores r_1 and r_2 with $r_1 + r_2 = r$, divided by the sum of the likelihoods of all response patterns with total score r. Doing so, it is seen that both the term $\exp(r\theta_v)$ and the product term in the denominator of (7) cancel, so that

$$P_\delta(r_2 - r_1|r) = P_\delta(r_2|r) = \frac{\gamma_{r-r_2}(\boldsymbol{\epsilon}_1)\gamma_{r_2}(\delta\boldsymbol{\epsilon}_2)}{\gamma_r(\boldsymbol{\epsilon}_1, \delta\boldsymbol{\epsilon}_2)}$$
$$= \frac{\gamma_{r-r_2}(\boldsymbol{\epsilon}_1)\gamma_{r_2}(\boldsymbol{\epsilon}_2)\delta^{r_2}}{\sum_{l=a}^{b} \gamma_{r-l}(\boldsymbol{\epsilon}_1)\gamma_l(\boldsymbol{\epsilon}_2)\delta^l}, \tag{83}$$

where the γ are elementary symmetric functions, and a and b are as in (51). A CML estimation equation for δ (and $\eta = \ln\delta$) is immediately obtained by setting the derivative of the logarithm of (83) with respect to δ equal to zero, which yields

$$\sum_{l=a}^{b} P_\delta(l|r)l = \frac{\sum_{l=a}^{b} \gamma_{r-l}(\boldsymbol{\epsilon}_1)\gamma_l(\boldsymbol{\epsilon}_2)\delta^l l}{\sum_{l=a}^{b} \gamma_{r-l}(\boldsymbol{\epsilon}_1)\gamma_l(\boldsymbol{\epsilon}_2)\delta^l} = r_2. \tag{84}$$

It can easily be shown that the discrete distribution defined by (83) is a one-parametric exponential family. Therefore, the CML estimation equation should equate the expected score, $E(R_2|r) = \sum_{l=a}^{b} P_\delta(l|r)l$, with the observed score, r_2, which exactly gives (84). This equation is conveniently solved by means of numerical bisection. Considering only cases with $0 < r < k$ (because for $r = 0$ and $r = k$, the conditional probability is trivially equal to 1 and thus uninformative), the question of the existence and uniqueness of the CML estimator of η is answered as follows (Fischer, 1995e, 2001):

THEOREM 20. *Let $0 < r < k$. Then, for $0 < r_1 < k_1$ and $0 < r_2 < k - k_1$, the CML equation* (84) *has a unique positive root $\hat{\delta} = \exp\hat{\eta}$. For $r_1 = 0$ or $r_2 = k - k_1$, the root $\hat{\delta}$ (and thus also $\hat{\eta}$) diverges to $+\infty$, and for $r_1 = k_1$ or $r_2 = 0$, the root is $\hat{\delta} = 0$ (and thus $\hat{\eta}$ diverges to $-\infty$).*

$P_\delta(r_2|r)$ being a one-parametric exponential family implies that powerful methods for the construction of uniformly most accurate (UMA) confidence intervals for η can be applied, which is formally equivalent to carrying out uniformly most powerful unbiased (UMPU) tests about η (cf., e.g., Lehmann, 1986). Unfortunately, for discrete variables such UMA confidence intervals are not feasible, but randomization of the discrete variable R_2 makes the distribution continuous, thus allowing an optimal solution. Randomization means adding, to the observed score r_2, a random number u drawn from a uniform distribution on the semi-open interval $[0, 1)$, yielding the randomized score $\tilde{r}_2 = r_2 + u$. Consider the hypothesis H_0: $\eta = \eta_0$ (or, equivalently, H_0: $\delta = \delta_0$). A UMPU test of the H_0 is carried out by means of a decision function $\phi(\tilde{r}_2)$,

$$\phi(\tilde{r}_2) = \begin{cases} 1, & \text{for } \tilde{r}_2 \leqslant c_1 \text{ or } \tilde{r}_2 \geqslant c_2, \\ 0, & \text{otherwise,} \end{cases} \tag{85}$$

where c_1 and c_2 are cutoff-scores for the randomized variable $\tilde{R}_2$, $\phi = 1$ means rejecting, and $\phi = 0$ retaining the H_0. These cutoff-scores are determined by the equations

$$\begin{aligned} &E_{\delta_0}\big[\phi(\tilde{R}_2)\big] = \alpha, \\ &E_{\delta_0}\big[R_2\phi(\tilde{R}_2)\big] = \alpha E_{\delta_0}(R_2), \end{aligned} \tag{86}$$

where, for instance, $\eta_0 = 0$ (and $\delta_0 = 1$) in the case of the null-hypothesis of 'no change'. Those values η_* and η^* for which the given $\tilde{r}_2$ falls exactly on the cutoff-scores c_1 and c_2, respectively, are the boundaries of the confidence interval for the unknown η. Eqs. (86) have to be solved by numerical bisection methods. A detailed exposition of such tests and related 'uniformly most accurate' (UMA) confidence intervals is given by Witting (1978), see also Mood et al. (1974); applications to the RM in particular have been carried out by Klauer (1991b) and Fischer (2001, 2003).

This method, elegant as it may be from a theoretical point of view, has two drawbacks: First, the two equations from which the cutoff scores have to be determined are a mix of discrete and continuous quantities and thus can be solved only by numerical methods, requiring the user to have the respective computer software at hands. Second, the outcome depends on the randomization. This implies that, for given subtests T_1 and T_2 and same fixed scores r_1 and r_2, it *may* happen that sometimes H_0 is retained and sometimes rejected, depending on the random number u. This is not easy to explain to the empirical researcher who wants to decide whether a treatment or a training has or has not had an effect on a person's ability parameter θ. Moreover, if the observed score $\tilde{r}_2$ happens to fall short of c_2 and if the researcher is interested to show that there *is* a positive effect η, he/she might be tempted to repeat the randomization until the desired significance occurs.

Therefore Fischer (2003) suggested to apply the so-called 'Mid-P Method' (cf., e.g., Lancaster, 1961) where the random variable U is replaced by the fixed constant $u = 0.5$. The rationale is as follows: the boundaries of the confidence interval, η_* and η^*, are monotone functions of u. The value $u = 0.5$ is the median of the random variable U, therefore the boundaries found in that manner are the medians of all boundaries one would obtain if the randomization were repeated infinitely many times. This seems easy to communicate and probably is exactly what users would like to have. If u is fixed, so are the boundaries for fixed tests T_1 and T_2 and a given H_0, so that the boundaries and the p-values of all feasible raw scores r_2 can be tabulated, for one-sided and two-sided alternatives. The p-value of a score r_2 is the largest α level for which zero is still within the confidence interval. All the user has to do is to look at the appropriate table and to read the p-value corresponding to his/her result. (Only if subtests $\mathcal{T}_1$ or $\mathcal{T}_2$ vary between testees, the researcher would have to resort to computer software for carrying out these tests.)

The described randomized tests can moreover be generalized to groups of testees, for instance, patients under treatment. To test the typical H_0: $\eta = 0$ against a one- or two-sided alternative, it is possible to proceed as follows: determine the p_v-values for all testees S_v and aggregate them by means of a well-known formula by R.A. Fisher,

$$-2\ln(p_1 p_2 \ldots p_n) \sim \chi^2 \quad \text{with df} = 2n, \tag{87}$$

cf. Hedges and Olkin (1985). This yields a test for the H_0 of no change. Formula (87) is exact (i.e., not based on any asymptotic approximation), so that the hypothesis test is also exact and can thus be used in groups of any size $n \geqslant 1$.

A more general approach to testing for person fit is due to Liou and Chang (1992) and Liou (1993). It was generalized and algorithmically improved by Ponocny (2000). Let the RM hold with parameters ϵ_i under H_0, and with $\epsilon_i d_i$ under H_1. The d_i, with $d_i > 0$, are correction coefficients transforming the null-model into H_1. Within the conditional likelihood framework, for any given response vector $\boldsymbol{x}$, a UMPU test of H_0 against a single H_1 can be constructed by application of the Neyman–Pearson lemma. It focuses on the ratio of conditional likelihoods (14),

$$\frac{L_1(\boldsymbol{x}|\boldsymbol{d}, \boldsymbol{\epsilon}, r)}{L_0(\boldsymbol{x}|\boldsymbol{\epsilon}, r)} = \frac{\gamma_r(\boldsymbol{\epsilon}) \prod_i \epsilon_i^{x_i} d_i^{x_i}}{\gamma_r(\boldsymbol{\epsilon}, \boldsymbol{d}) \prod_i \epsilon_i^{x_i}} \propto \prod_i d_i^{x_i}, \tag{88}$$

where $\boldsymbol{d}$ denotes the vector of correction coefficients. All that matters is the right-hand term in (88) since the γ-functions are independent of $\boldsymbol{x}$.

A UMPU test now follows a simple strategy: for all feasible response vectors, compute the CLRs (88) and order the response vectors by decreasing magnitude of (88). This results in a sequence of ordered subsets C_j, $j = 1, \ldots, u$, of response vectors $\boldsymbol{x}$ with identical CLRs. Let the sum of the conditional probabilities for all $\boldsymbol{x} \in C_j$ be denoted P_j. The rejection region of the UMPU test is constructed by accumulating the conditional probabilities under H_0, beginning with those C_j for which the likelihood ratios are highest, and proceeding through the sequence of the C_j as long as the partial sum of the conditional probabilities, $P_l^+ := \sum_{j=1}^{l} P_j$, is $\leqslant \alpha$. If P_l^+ for some l happens to be equal α, the rejection region is $\{C_1, \ldots, C_l\}$. If, however, the desired α-level cannot be realized exactly in this manner, randomization experiments are carried out on the response vectors $\boldsymbol{x} \in C_{l+1}$ as follows: select vectors $\boldsymbol{x} \in C_{l+1}$ with probability $\pi = (\alpha - P_l^+)/P_{l+1}$ and add them to the rejection region. The test procedure is unique for all response vectors except for $\boldsymbol{x} \in C_{l+1}$.

A problem of this person fit test is to generate all possible response vectors. Ponocny (2000) described a procedure that runs on a standard PC up to $k = 30$ items. For $k > 30$, he suggested a Monte Carlo approximation. Both procedures are implemented in the software T-Rasch (Ponocny and Ponocny-Seliger, 1999).

The attractive feature of this approach is that the tests of fit are exact and a fairly wide range of alternative hypotheses is covered. Some examples are: (i) All response vectors have the same probability; put $d_i = \epsilon_i^{-1}$. (ii) Different subgroups of testees, G_g, have different item parameter vectors $\boldsymbol{\epsilon}$; put $d_i = \hat{\epsilon}_i^{(g)}/\hat{\epsilon}_i$, where the $\hat{\epsilon}_i^{(g)}$ are the CML estimates for group G_g, and the $\hat{\epsilon}_i$ are the overall CML estimates. This allows one to test for DIF of several or all items simultaneously. The procedure is applicable even in cases where no alternative item parameter estimates are available. Suppose, for instance, it is hypothesized on the basis of psychological reasoning that items $I_1, \ldots, I_l$ are easier for female than for male testees: put $d_i = d$ with some $d > 1$, $i = 1, \ldots, l$, for the female group. The factor d can be chosen arbitrarily because the resulting UMPU gives the same result for all $d > 0$. (iii) A change of θ occurs between two subtests $\mathcal{T}_1$ and $\mathcal{T}_2$. The change parameter can be absorbed into the item parameters of $\mathcal{T}_2$: for instance, put $d_i = d > 0$ for all items $I_i \in \mathcal{T}_2$, as in (ii). This corresponds to the exact test of $H_0 : \eta = 0$ described above. (iv) Certain patterns of item responses are expected to occur more frequently than predicted under H_0. (v) Certain patterns of responses occur more frequently than predicted under the RM in certain subgroups of testees, G_g. (vi) The item parameters underlie constraints as in the LLTM described in Section 7 below. (On the d_i required in cases (iv) through (vi), see Ponocny, 2000, or Ponocny and Ponocny-Seliger, 1999.)

Another family of exact conditional tests of fit of the RM was proposed by Besag and Clifford (1989), drawing on the so-called ‘switching algorithm’ of Ryser (1963) for the generation of random (0, 1)-matrices with given margins. Later variations of the same approach are due, amongst others, to Rao et al. (1996) and Ponocny (2001). Considerably faster algorithms are the ‘unequal probability sampling algorithm’ by Snijders (1991) and the ‘sequential importance sampling (SIS) algorithm’ of Chen and Small (2005), see also Chen et al. (2005). Software has been published by Ponocny and Ponocny-Seliger (1999). The present exposition follows Ponocny (2001).

These exact conditional tests make use of the fact that in the RM all item score matrices $\boldsymbol{X}$ compatible with given marginal sums $\boldsymbol{r}$ and $\boldsymbol{s}$ have the same probability $K_{\boldsymbol{rs}}^{-1}$. A family of UMPU conditional tests against directed noncomposite alternatives H_1 can be grounded again directly on the Neyman–Pearson lemma. Critical regions are obtained by aggregating item score matrices in the rank order of their quotients of likelihoods under H_1 and H_0, respectively, such that the number of these matrices divided by $K_{\boldsymbol{rs}}$ does not exceed α. (Such nonparametric tests of fit had already been advocated by Rasch, 1960, 1961, 1965, by the way, but at that time the computational problems involved prevented their practical implementation.) Since the H_0 implies equal probability $K_{\boldsymbol{rs}}^{-1}$ of all matrices of interest, the rank order of their likelihood ratios depends only on the likelihoods under H_1. If, moreover, the conditional probability of a matrix $\boldsymbol{X}$ under H_1 increases with a certain test characteristic of interest, T, the procedure for constructing a test of H_0 against H_1 is as follows: generate all matrices with given marginals and count how often the observed value $T = t$ is reached or exceeded. If that relative frequency is $> \alpha$, reject H_0.

An insurmountable obstacle for carrying out such tests in complete form are the computational problems involved: the number of matrices to be generated increases tremendously with n and k. Therefore, Monte Carlo approximations have to be employed. The switching algorithm mentioned above randomly selects a pair (v, w) of rows and a pair (i, j) of columns of $\boldsymbol{X}$, which jointly define a tetrad of elements, and performs the following operation: if $x_{vi} = x_{wj} = 0$ and $x_{vj} = x_{wi} = 1$, change the tetrad to $x_{vi} = x_{wj} = 1$ and $x_{vj} = x_{wi} = 0$; otherwise, leave the tetrad as is and proceed with a new random tetrad, etc. These tetrad operations obviously leave the marginal vectors of $\boldsymbol{X}$ invariant. Besag and Clifford (1989), Rao et al. (1996), and Ponocny (2001) showed how, by means of an infinite sequence of tetrads, each feasible item score matrix with given marginal vectors $\boldsymbol{r}$ and $\boldsymbol{s}$ can be produced with equal probability. Therefore it suffices to count the occurrence of statistics $T \geqslant t$ in a sufficiently long series of such matrices for a consistent estimation of the p-value of a given item score matrix with $T = t$. The software T-Rasch by Ponocny and Ponocny-Seliger (1999) carries out the algorithms necessary for such tests of fit and specifically supports testing the RM against the following alternative hypotheses: (i) violations of local stochastic independence; (ii) increased inter-item association within subscales; (iii) differential item functioning (DIF); (iv) violations of the axiom of sufficiency of the raw score; (v) alternative hypotheses combined of the aforementioned; (vi) increased or decreased item discrimination; (vii) single-peaked IRFs; (viii) satiation or fatigue effects near the end of a test; (ix) cheating. The presently available program version of T-Rasch, however, does not provide an estimate of the Monte Carlo standard error and limits the item score matrices to maximal sizes of 100×30 or 50×50 or 30×100. For larger matrices, the authors recommend to split the sample into subsets and to combine the p-values obtained into a global statistic by means of (87). Chen and Small (2005) demonstrate that their SIS algorithm can handle much larger matrices and estimates both the p-value and its Monte Carlo standard error.

Finally, some nonparametric tests of the monotonicity, unidimensionality, and local independence assumptions of the RM should be mentioned. Scheiblechner (2003) proposed test statistics for 'Isotonic Ordinal Probabilistic' (ISOP) models, which are general unidimensional nonparametric polytomous rating scale models essentially based on the assumption co-monotonicity of the items. A test for violations of the co-monotonicity assumption for item subsets and a procedure for item selection were given, based on the Goodman and Kruskal (1954) 'gamma index' (cf. also Mielke, 1983). Other nonparametric tests of fit were proposed by Junker (2001), Molenaar and Hoijtink (1990), Sijtsma (1998), Sijtsma and Van der Ark (2001), and Stout (1990), to mention a few examples. The advantage of these nonparametric tests of fit is that they can be carried out prior to fitting any parametric IRT model (for instance, the RM) to the data: if the H_0 specified is rejected, so is the parametric model. (A more thorough discussion of the interplay between nonparametric and parametric IRT is beyond the scope of this chapter.)

EXAMPLE 4. Measurement and significance of change.

To illustrate the measurement of and significance testing for change by means of an RM, suppose that two subtests are composed of items from the pool of 60 Standard Progressive Matrices Plus (SPM+) of Raven et al. (2000) and are given to applicants for jobs of bus drivers before and after a strenuous driver training by night (denoted pretest T_1 and posttest T_2, respectively). The aim is to filter out applicants whose test performance changes significantly under conditions of strain. Based on ample experience with the SPM and similar matrices items, it is assumed here (without debate of this admittedly nontrivial question) that the RM fits the SPM+ items. The item parameters β_i, which are estimated on the basis of a sample of $n = 2582$ testees,[11] will henceforth be considered given constants (see Table 4).

Let $\mathcal{T}_1$ be composed of all odd-numbered, and $\mathcal{T}_2$ of all even-numbered items. The estimates of the amounts of change η for all raw score combinations, the corresponding p values, and their exact two-sided confidence intervals for $\alpha = 0.05$ are given here only for total score $r = 30$ (selected arbitrarily for demonstration, see Table 5; the results are

Table 4
Item parameters estimates for the 60 SPM+ items

i	$\hat{\beta}_i$	i	$\hat{\beta}_i$	i	$\hat{\beta}_i$	i	$\hat{\beta}_i$	i	$\hat{\beta}_i$
1	−5.75	2	−5.81	3	−4.32	4	−3.85	5	−2.97
6	−3.29	7	−2.24	8	−1.66	9	−2.77	10	−1.59
11	−0.63	12	−0.44	13	−3.97	14	−3.42	15	−2.61
16	−0.83	17	−1.00	18	−0.34	19	−0.34	20	−0.20
21	−0.37	22	−0.71	23	−0.15	24	0.37	25	−0.79
26	−1.33	27	−0.46	28	−1.33	29	0.10	30	−0.74
31	0.50	32	0.11	33	1.87	34	2.49	35	1.99
36	2.18	37	−0.07	38	0.56	39	0.66	40	1.62
41	2.06	42	1.38	43	1.95	44	1.83	45	1.53
46	2.14	47	1.98	48	2.64	49	0.67	50	1.16
51	0.90	52	1.71	53	2.17	54	2.02	55	2.58
56	3.26	57	2.66	58	3.08	59	2.62	60	3.18

Table 5
CML estimates, two-sided P-values under the H_0 of no change, and two-sided confidence intervals at $\alpha = 0.05$ for the change parameter η, for total score $r = 30$ in Example 4

r_1	r_2	$\hat{\eta}$	p	Conf. Int.	r_1	r_2	$\hat{\eta}$	p	Conf. Int.
30	0	$-\infty$	0.000	$(-\infty, -9.28)$	14	16	0.26	0.713	(−1.14, 1.66)
29	1	−10.39	0.000	$(-\infty, -7.61)$	13	17	0.74	0.306	(−0.66, 2.14)
28	2	−8.74	0.000	(−12.23, −6.48)	12	18	1.21	0.091	(−0.19, 2.63)
27	3	−7.57	0.000	(−10.03, −5.57)	11	19	1.70	0.018	(0.29, 3.15)
26	4	−6.62	0.000	(−8.73, −4.77)	10	20	2.19	0.002	(0.77, 3.68)
25	5	−5.80	0.000	(−7.75, −4.06)	9	21	2.70	0.000	(1.26, 4.24)
24	6	−5.07	0.000	(−6.90, −3.42)	8	22	3.24	0.000	(1.76, 4.82)
23	7	−4.41	0.000	(−6.12, −2.82)	7	23	3.81	0.000	(2.29, 5.46)
22	8	−3.79	0.000	(−5.43, −2.27)	6	24	4.43	0.000	(2.84, 6.22)
21	9	−3.22	0.000	(−4.80, −1.73)	5	25	5.13	0.000	(3.44, 7.05)
20	10	−2.68	0.000	(−4.21, −1.22)	4	26	5.92	0.000	(4.10, 8.05)
19	11	−2.16	0.003	(−3.65, −0.72)	3	27	6.86	0.000	(4.87, 9.32)
18	12	−1.65	0.022	(−3.11, −0.24)	2	28	8.05	0.000	(5.78, 11.53)
17	13	−1.17	0.105	(−2.60, 0.24)	1	29	9.78	0.000	$(6.92, \infty)$
16	14	−0.69	0.338	(−2.10, 0.72)	0	30	∞	0.000	$(8.71, \infty)$
15	15	−0.21	0.765	(−1.61, 1.19)					

[11] The author thanks J. Raven Jr. for making the data available for the present item calibration.

based on the Mid-P Method). Cases with other total scores as well as one-sided intervals could be computed and tabulated in the same manner but have to be omitted for space considerations. The conditions for existence and uniqueness of finite estimators $\hat{\eta}$ were given above; the respective conditions for the boundaries of the confidence intervals are more complicated, they can be found in Klauer (1991a) and Fischer (2001, 2003). An estimate $\hat{\eta}$ is to be interpreted as the distance of migration of the individual on the latent ability dimension between the pretest and posttest time points. The two-sided significance for level $\alpha = 0.05$ can be deduced from the confidence interval: the H_0 of no change is rejected if zero lies outside the confidence interval.

Tabulating all possible results would obviously require much space. The significances can easily be summarized in a single table, though, see Figure 2 (which again is

											1										2										3	
	0	1	2	3	4	5	6	7	8	9	0	1	2	3	4	5	6	7	8	9	0	1	2	3	4	5	6	7	8	9	0	
0					.	s	s	S	S	S	T	T	T	T	T	T	T	T	T	T	T	T	T	T	T	T	T	T	T	T	T	0
1							.	s	s	S	S	T	T	T	T	T	T	T	T	T	T	T	T	T	T	T	T	T	T	T	T	1
2									.	s	s	S	S	T	T	T	T	T	T	T	T	T	T	T	T	T	T	T	T	T	T	2
3	.									.	s	s	S	S	T	T	T	T	T	T	T	T	T	T	T	T	T	T	T	T	T	3
4	s	.									.	s	s	S	S	T	T	T	T	T	T	T	T	T	T	T	T	T	T	T	T	4
5	S	s										.	s	s	S	S	T	T	T	T	T	T	T	T	T	T	T	T	T	T	T	5
6	S	s	s											s	s	S	S	T	T	T	T	T	T	T	T	T	T	T	T	T	T	6
7	T	S	s	.											.	s	S	S	T	T	T	T	T	T	T	T	T	T	T	T	T	7
8	T	S	S	s	.											s	s	S	S	T	T	T	T	T	T	T	T	T	T	T	T	8
9	T	T	S	s	s	.											s	s	S	S	T	T	T	T	T	T	T	T	T	T	T	9
10	T	T	T	S	s	s	.										.	s	s	S	S	T	T	T	T	T	T	T	T	T	T	10
11	T	T	T	S	S	s	s	.										.	s	s	S	S	T	T	T	T	T	T	T	T	T	11
12	T	T	T	T	S	S	s	s	.										.	s	s	S	S	T	T	T	T	T	T	T	T	12
13	T	T	T	T	T	S	S	s	s	.										.	s	s	S	S	T	T	T	T	T	T	T	13
14	T	T	T	T	T	T	S	S	s	s	.										.	s	s	S	S	T	T	T	T	T	T	14
15	T	T	T	T	T	T	T	S	S	s	s	.										.	s	s	S	T	T	T	T	T	T	15
16	T	T	T	T	T	T	T	T	S	S	s	s											.	s	S	S	T	T	T	T	T	16
17	T	T	T	T	T	T	T	T	T	S	S	s	.											.	s	S	S	T	T	T	T	17
18	T	T	T	T	T	T	T	T	T	T	S	S	s	.											.	s	S	S	T	T	T	18
19	T	T	T	T	T	T	T	T	T	T	T	S	S	s	.											s	s	S	T	T	T	19
20	T	T	T	T	T	T	T	T	T	T	T	T	S	S	s	s											s	S	S	T	T	20
21	T	T	T	T	T	T	T	T	T	T	T	T	T	S	S	s	s	.									.	s	S	T	T	21
22	T	T	T	T	T	T	T	T	T	T	T	T	T	T	S	S	s	s	.									.	s	S	T	22
23	T	T	T	T	T	T	T	T	T	T	T	T	T	T	T	S	S	s	s	.									s	S	T	23
24	T	T	T	T	T	T	T	T	T	T	T	T	T	T	T	T	S	S	s	s	.									s	S	24
25	T	T	T	T	T	T	T	T	T	T	T	T	T	T	T	T	T	T	S	S	s	.								.	S	25
26	T	T	T	T	T	T	T	T	T	T	T	T	T	T	T	T	T	T	T	S	S	s	s	.							s	26
27	T	T	T	T	T	T	T	T	T	T	T	T	T	T	T	T	T	T	T	T	S	S	s	s	.						.	27
28	T	T	T	T	T	T	T	T	T	T	T	T	T	T	T	T	T	T	T	T	T	T	S	S	s	.						28
29	T	T	T	T	T	T	T	T	T	T	T	T	T	T	T	T	T	T	T	T	T	T	T	S	S	s	.					29
30	T	T	T	T	T	T	T	T	T	T	T	T	T	T	T	T	T	T	T	T	T	T	T	T	T	S	S	s				30
											1										2										3	
	0	1	2	3	4	5	6	7	8	9	0	1	2	3	4	5	6	7	8	9	0	1	2	3	4	5	6	7	8	9	0	

Fig. 2. Significances of posttest scores r_2 for given pretest score r_1, at $\alpha = 0.05$.
Note: '.' denotes 'significance at $\alpha = 0.10$', 's' denotes '$\alpha = 0.05$', 'S' denotes '$\alpha = 0.01$', and 'T' denotes '$\alpha = 0.001$'. Blank spaces mean 'no significance'.

based on the Mid-P method). For each score combination, it shows the two-sided significances for levels $\alpha = 0.10, 0.05, 0.01, 0.001$. The present author holds that such tables should be part of tests manuals if tests have been found to suffice the requirements of the RM. (For an application, cf. Raven Jr. et al., 2000, Section 3, Appendix 3.)

7. The linear logistic test model

An extension of the RM that has many useful applications is the Linear Logistic Test Model (LLTM). It assumes that the item parameters β_i of the RM can be decomposed into a weighted sum of 'basic parameters', α_l, plus a normalization constant, c. The LLTM has many roots, particularly the linear logistic model of Cox (1970) and related models in biometrics and psychology (Micko, 1969, 1970; Kempf, 1972). One intention was to analyze change under the impact of experimental or testing conditions, treatments given prior to testing, so called 'trend effects', cognitive development, and other factors that may influence response behavior (Fischer, 1972, 1974), another was to describe item difficulty in terms of rules, cognitive operations, or other facets of the item material (Scheiblechner, 1972; Fischer, 1973, 1974).

The LLTM can be conceived as an RM where the item parameters β_i, $i = 1, \dots, k$, are subject to the linear constraints

$$\boldsymbol{\beta} = \boldsymbol{W}\boldsymbol{\alpha} + c\mathbf{1}, \tag{89}$$

where $\boldsymbol{W} = ((w_{il}))$, $i = 1, \dots, k$, $l = 1, \dots, p$, with $p \leqslant k - 1$, is a given matrix of weights, $\boldsymbol{\alpha}$ a column vector of p 'basic' parameters, $\mathbf{1}$ the column vector of k ones, and c the usual normalization constant. The role of c can be explained as follows: if the normalization of $\boldsymbol{\beta}$ in (89) is changed, say, $\beta_i \to \beta_i + c_0$, then this is compensated by the transformation $c \to c + c_0$ on the right-hand side of (89). In other words, changes of the normalization of the underlying RM do not affect the structural part of (89): $\boldsymbol{W}$ and $\boldsymbol{\alpha}$ remain invariant. Applying (89) to two items I_i and I_j, it is seen that the difference of the item difficulty parameters becomes $\beta_i - \beta_j = \sum_l (w_{il} - w_{jl})\alpha_l$, which is independent of c. Suppose the α_l are measures of the difficulty of cognitive operations, O_e, and the weights are $w_{il} = 1$ if operation O_e is needed for item I_i, and $w_{il} = 0$ otherwise. Then the difficulty difference between two items I_i and I_j is explained as the sum of the difficulties of those operations that are needed for item I_i but not needed for I_j, minus the sum of the difficulties of the operations needed for I_j but not needed for I_i.

To understand the LLTM, it is important to realize its relation to the RM: since (89) is a constraint on the β_i, in typical applications the LLTM is more restrictive than the RM in the sense of being harder to fit empirically. But as far as formal considerations are concerned, the LLTM is *more general than the RM*: the RM results from the LLTM by putting, for instance, $p = k - 1$ with $w_{il} = \delta_{il}$, for $i = 1, \dots, k$ and $l = 1, \dots, k - 1$, where δ_{il} is the Kronecker symbol, together with $c = 0$, implying $\beta_k = 0$ as normalization. Thus, the RM is an LLTM with a particular (rectangular) diagonal weight matrix. (This is not the only way of parameterizing an RM as a special case of an LLTM, though.) The LLTM, on the other hand, cannot be formally derived from the RM. This relation will be further illustrated by showing that the uniqueness result of Theorem 9 concerning the RM follows from a more general theorem on uniqueness of the CML estimator in the LLTM (see below).

The most popular estimation methods for the RM – JML, CML, MML – have also been implemented for the LLTM. The most attractive one with respect to the testing of hypotheses, and also the method whose properties are known in all relevant details, is CML. Therefore, and for space considerations, only CML will be treated here.

To simplify the derivation of the CML equations, it is convenient to adopt a *particular* normalization of the item parameters: for a given LLTM with fixed weights w_{ij}, there exists *one* normalization of the item parameters such that c in the right-hand side of (89) vanishes, namely, $\boldsymbol{\beta}^* := \boldsymbol{\beta} - c\mathbf{1} = \boldsymbol{W}\boldsymbol{\alpha}$. Assuming that this normalization has been adopted, one may write w.l.g. (dropping asterisks for simplicity),

$$\boldsymbol{\beta} = \boldsymbol{W}\boldsymbol{\alpha}, \tag{90}$$

keeping in mind, however, that now the β_i in (90) cannot be subjected to any other normalization unless the weight matrix $\boldsymbol{W}$ is changed also.[12]

In view of typical applications, it is necessary to formulate the conditional log-likelihood function and the CML estimation equations for incomplete data. With the notation as in Section 4, the conditional log-likelihood function of the LLTM is immediately obtained from (40) by inserting (90),

$$\ln L_C(\boldsymbol{X}|\boldsymbol{\alpha}, \boldsymbol{W}, \boldsymbol{r}, \boldsymbol{B}) = -\sum_l t_l \alpha_l - \sum_{v=1}^{n} \ln \gamma_{r_v}(\boldsymbol{\epsilon}, \boldsymbol{b}_v), \tag{91}$$

where $t_l = \sum_i \sum_v b_{vi} x_{vi} w_{il} = \sum_i s_i w_{il}$, for $l = 1, \ldots, p$, are the sufficient statistics for the basic parameters α_j. This still is an exponential family, with canonical parameters α_l and sufficient statistics t_l.

As in the case of the CML equations for the RM, let the testees be grouped by their common design vectors $\boldsymbol{b}_v =: \boldsymbol{b}_g$ for all $S_v \in G_g$. Let $\boldsymbol{\epsilon}_g$ be the vector of parameters ϵ_i of items I_i given to all $S_v \in G_g$, n_g the size of group G_g, and n_{gr} the number of testees $S_v \subset G_g$ with raw score r. Then the CML equations for $\boldsymbol{\alpha}$ are, analogous to (46),

$$\sum_i w_{il} \left[\sum_g \sum_r n_{gr} \frac{b_{gi} \epsilon_i \gamma_{r-1}^{(i)}(\boldsymbol{\epsilon}_g)}{\gamma_r(\boldsymbol{\epsilon}_g)} - s_i \right] = 0, \tag{92}$$

for $l = 1, \ldots, p$. They actually are weighted sums of the expressions in (46). For the second-order partial derivatives, one gets

$$\frac{\partial^2 \ln L_C}{\partial \boldsymbol{\alpha} \partial \boldsymbol{\alpha}'} = \boldsymbol{W}' \frac{\partial^2 \ln L_C}{\partial \boldsymbol{\beta} \partial \boldsymbol{\beta}'} \boldsymbol{W}, \tag{93}$$

where the derivatives can be obtained by means of (49) and (50). Note that (92) is independent of the normalization chosen, so that assumption (90) does not imply any loss of generality. From the exponential family property of (91) it follows that any finite solution $\hat{\boldsymbol{\alpha}}$ of (92) must be unique if the model is in canonical (or 'minimal') form (cf.

[12] Some authors define the LLTM in the form of (90), without a normalization constant. This entails complicated algebraic transformations of $\boldsymbol{W}$ whenever the normalization is changed.

Barndorff-Nielsen, 1978; Andersen, 1980). As in the RM, the estimation equations can be solved, for instance, by means of the Newton–Raphson or the BFGS algorithm (see Section 4). A computer program using the latter is LPCM Win by Fischer and Ponocny-Seliger (1998).

A trivial necessary condition for the uniqueness of $\hat{\boldsymbol{\alpha}}$ is that the vector of item parameters in (89), $\boldsymbol{\beta}$, can be decomposed uniquely into $\boldsymbol{\alpha}$ and c. This implies that the matrix of system (89),

$$\mathbf{W}^+ = \begin{pmatrix} w_{11} & \cdots & w_{1p} & 1 \\ \vdots & \ddots & \vdots & \\ w_{k1} & \cdots & w_{kp} & 1 \end{pmatrix}, \tag{94}$$

must have full column rank, that is, rank $p + 1$. If this necessary condition is violated, the LLTM contains too many basic parameters, so that the exponential family is not in minimal form.

A n.s. condition for the existence and uniqueness of a finite CML solution $\hat{\boldsymbol{\alpha}}$, however, is more complicated. It again involves digraph D defined in Section 4. Let the *strong components*[13] of D be C_t, $t = 1, \ldots, u$.

THEOREM 21. *The CML equations* (92) *possess a unique finite solution* $\hat{\boldsymbol{\alpha}}$ *iff the ('alternative') linear system of equations* (95) *with inequality constraints* (96),

$$\sum_{l=1}^{p} w_{il} y_l = d_t, \quad \textit{for all vertices } V_i \in C_t, \textit{ for } t = 1, \ldots, u, \tag{95}$$

where the d_t *are unknown constants satisfying*

$$d_s \leqslant d_t \quad \textit{if a diline } \overrightarrow{V_a V_b} \textit{ exists for some } V_a \in C_s \textit{ and some } V_b \in C_t, \tag{96}$$

has only the trivial solution $\boldsymbol{y} = \mathbf{0}$.

A self-contained proof was given by Fischer (1983), another proof, employing a theorem of Jacobsen (1989), by Fischer (1995b).

System (95)–(96) is denoted the *alternative system*, because either the CML system (92) possesses a unique finite solution $\hat{\boldsymbol{\alpha}}$ and the system (95)–(96) has only the trivial null-solution $\boldsymbol{y} = \mathbf{0}$, or the CML system (92) does not possess a unique finite solution and the system (95)–(96) has some non-null solution $\boldsymbol{y} \neq \mathbf{0}$. Therefore, it suffices to look at the alternative linear system (of equations and inequalities) to decide whether the nonlinear CML equations are uniquely solvable. If this is not the case, either a finite CML solution does not exist, or the solution is not unique; the latter case occurs only if the model is not in minimal (or canonical) form, entailing that there are infinitely many solutions.

A special case of practical importance arises if the (complete or incomplete) data matrix $\boldsymbol{X}$ is well-conditioned, as may be expected in most empirical applications. Well-conditionedness, see Assumption xvii in Section 4, implies that D has only one strong

[13] The strong components are the maximal subsets of strongly connected vertices.

component, which is identical with D. In that case, the alternative system (95)–(96) reduces to the system of linear equations $\sum_l w_{il} y_l = d$, for all i, which has only the trivial solution $\boldsymbol{y} = \boldsymbol{0}$ iff matrix $\boldsymbol{W}^+$ defined in (94) is of full column rank, that is, iff $\text{rank}(\boldsymbol{W}^+) = p + 1$. This is summarized as follows:

COROLLARY 2. *Let the data matrix* $\boldsymbol{X}$ *be well-conditioned. A unique solution* $\hat{\boldsymbol{\alpha}}$ *of the CML equations* (92) *exists iff matrix* $\boldsymbol{W}^+$ *has full column rank,* $\text{rank}(\boldsymbol{W}^+) = p + 1$.

A similar result can be proved also if $u = 2$ (see Fischer, 1995b).

Now it is easy to see that the uniqueness result of Theorem 9 for the RM is a consequence of Theorem 21 and Corollary 2, but the latter results could never be deduced from Theorem 9 alone. This again shows that, from a formal point of view, the RM is a special case of the LLTM.

It is an attractive feature of the CML approach to estimating an LLTM that the asymptotic behavior of the estimator is known: $\hat{\boldsymbol{\alpha}}$ is asymptotically multivariate normal around the true $\boldsymbol{\alpha}$ with an asymptotic covariance matrix equal to the inverse of the information matrix $\boldsymbol{I}$ evaluated at $\hat{\boldsymbol{\alpha}}$. As prerequisite for this to hold it is necessary, though, that (a) the matrix $\boldsymbol{W}^+$ is of full column rank $p + 1$ and (b) condition (54) is satisfied by the sequence of person parameters.

These results can also be applied to a related class of psychometric models: El-Helbawy and Bradley (1977) extended the Bradley–Terry–Luce (BTL) model (Bradley, 1984) by imposing linear constraints on the parameters analogous to (89) and gave a sufficient condition for uniqueness of the ML estimator. Fischer and Tanzer (1994) proved the equivalence of this 'Linear Bradley–Terry–Luce (LBTL) Model' with a special LLTM and thereof derived a weaker n.s. condition for uniqueness of the ML estimator of the LBTL.

7.1. *Testing the fit of an LLTM*

To obtain fit of an LLTM requires that (a) an RM fits the item response data and (b) the item parameters of the RM satisfy the linear constraints (89). Provided that (a) holds true, the H_0 of the validity of (89) can be tested against der H_1 of unrestricted item parameters of an RM by means of a conditional likelihood ratio test: compute $\ln L_C(\boldsymbol{X}|\hat{\boldsymbol{\beta}}, \boldsymbol{r}) =: \ln L_1$, given in (40), under the RM, and $\ln L_C(\boldsymbol{X}|\hat{\boldsymbol{\alpha}}, \boldsymbol{W}, \boldsymbol{r}) =: \ln L_0$, given in (91), under the LLTM. It then follows from general rules for likelihood ratio tests that

$$-2[\ln L_0 - \ln L_1] \overset{\text{as.}}{\sim} \chi^2, \quad \text{with df} = k - p - 1. \tag{97}$$

Alternatives to this CLR test of fit are the LM test suggested by Glas and Verhelst (1995) or Ponocny's (2001) exact combinatorial test, both outlined in Section 6 above.

Estimation and testing of fit with the LLTM (and also some generalizations thereof) can be carried out (a) with specialized software: within the CML framework, by means of LPCM Win (Fischer and Ponocny-Seliger, 1998); within the MML framework, by means of ACER ConQest (Wu et al., 1998) and with a straightforward ML approach, by means of FACETS (Linacre, 1994, 1996); or (b) with general-purpose software (e.g.,

GLIM, S-PLUS): Tuerlinckx et al. (2004) recently gave an overview of the applicability of the latter class of software products for what they call 'Explanatory Item Response Models' (of which the LLTM is one example).

Experience has shown that the CLR test (97) turns out significant in most applications. When this is the case, the question comes into focus whether the LLTM is still useful. Suppose the linear model (89) of item difficulty is at least approximately valid for some universe of items, then it is possible to *predict – at least approximately – the relative item difficulty* for new items of the same universe even for new samples of testees. The LLTM is then still valuable both as a psychological theory of the cognitive complexity of the item material and as a technique for theory-guided design of items with pre-defined difficulty levels. The interest in such technologies has been growing rapidly in recent years (cf. Embretson, 1999; Irvine and Kyllonen, 2002). The LLTM is one tool that appears to be specifically suited for such tasks (cf. Hornke and Habon, 1986).

Extensions of the LLTM have been guided by the aim of improving the fit (especially in cases of partial misspecification of $\boldsymbol{W}$): estimation of some of the weights w_{il} along with the basic parameters α_j (Bechger et al., 2002; Butter et al., 1998; Fischer, 2004), or the inclusion of nonlinearity of the parametric structure (Bechger et al., 2001; Molenberghs and Verbeke, 2004), or the addition of random components (Adams et al., 1997; Mislevy, 1987; Rijmen and De Boeck, 2002; Rijmen et al., 2003), or the modeling of individual differences in solution strategies (Mislevy and Verhelst, 1990), or the extension of the linear structure to the person parameters (Wilson and De Boeck, 2004). Such generalizations are not treated here both for space considerations and because many of them deviate significantly from the basic principles of Rasch models.

7.2. Differential item functioning (DIF)

DIF is closely related to cross-cultural differences in test behavior. If the LLTM is successful as a cognitive model explaining the relative difficulty of items in terms of cognitive operations, it can also be employed to monitor cultural differences of the difficulty or 'availability' of cognitive functions. An exemplary early study of that kind was made by Piswanger (1975) (see also in Fischer and Formann, 1982) who gave the WMT matrices test[14] (Formann and Piswanger, 1979) to Central-European (i.e., Austrian) and African (i.e., Nigerian and Togolese) samples of adolescents. One of the most interesting results was that, in spite of the completely nonverbal nature of the test material, a surprising – yet interpretable – difference occurred with respect to the 'direction' of rules: if a rule had to be applied in a 'horizontal' direction (i.e., from left to right), this made the respective items relatively easier for the Austrian testees; if the direction was 'vertical' or 'horizontal and vertical', this made the respective items relatively easier for the African samples. This result can be understood as the effect of culture: Arabian is written from right to left, and Arabian culture was still quite influential in Africa. Working from right to left, however, was impossible owing to the item format. Therefore, to

[14] 'Wiener Matrizen Test', a matrices test similar to Raven's well-known SPM test, however, designed systematically on the basis of a pre-defined set of operations. In the following it is important to know that in each item the empty space that has to be filled in by the testee is in the lower right-hand corner of the matrix.

detect and apply a rule in the horizontal direction was somewhat less obvious for the African than for the Austrian sample and, as a consequence, vertical or horizontal and vertical directions were *relatively* easier for the Africans. Other similar LLTM studies on 'item bias' or cross-cultural differences are due to Tanzer et al. (1995), Whitely and Schneider (1981), and Van de Vijver (1988).

The LLTM allows one to test for DIF, however, in another way also. Let a test of length k comprise two disjunctive item subsets $\mathcal{T}_1$ and $\mathcal{T}_2$, consisting of k_1 and k_2 items, respectively, with corresponding item parameter vector $\boldsymbol{\beta} = (\boldsymbol{\beta}_1', \boldsymbol{\beta}_2')'$. Suppose that the items $I_i \in \mathcal{T}_2$ are suspected to be susceptible to DIF in testee group G_1 relative to testee group G_2. Under H_0 ('no DIF'), let an RM hold with common item parameter vector $\boldsymbol{\beta}$ for both groups. Under H_1, subtest $\mathcal{T}_1$ has common $\boldsymbol{\beta}_1$ in both groups, but $\mathcal{T}_2$ has different parameter vectors $\boldsymbol{\beta}_2^{(1)}$ and $\boldsymbol{\beta}_2^{(2)}$ in G_1 and G_2, respectively. The H_1 model clearly is an LLTM because it restricts $\boldsymbol{\beta}_1$ to be equal in both groups. The H_0 model is an RM, namely, that special case of the LLTM where $\boldsymbol{\beta}_2^{(1)} = \boldsymbol{\beta}_2^{(2)}$. A CLR test of H_0 against H_1 is readily carried out by computing the CLR statistic

$$-2\big[\ln L_C(\boldsymbol{X}|\hat{\boldsymbol{\beta}}, \boldsymbol{r}) - \ln L_C\big(\boldsymbol{X}_1|\hat{\boldsymbol{\beta}}_1, \hat{\boldsymbol{\beta}}_2^{(1)}, \boldsymbol{r}_1\big) \\ - \ln L_C\big(\boldsymbol{X}_2|\hat{\boldsymbol{\beta}}_1, \hat{\boldsymbol{\beta}}_2^{(2)}, \boldsymbol{r}_2\big)\big] \overset{\text{as.}}{\sim} \chi^2, \tag{98}$$

with df $= k_2$. If the $H_0: \boldsymbol{\beta}_2^{(1)} = \boldsymbol{\beta}_2^{(2)}$ is rejected, one may wish to test another null-hypothesis $H_0: \boldsymbol{\beta}_2^{(1)} = \boldsymbol{\beta}_2^{(2)} + \boldsymbol{\eta}$, with $\boldsymbol{\eta} = (\eta, \ldots, \eta)'$, hypothesizing that the amount of item bias, η, is uniform (i.e., constant across items $I_i \in \mathcal{T}_2$). The test statistic for this H_0 against the former H_1 is analogous to (98), however, with df $= k_2 - 1$.

On the relationship of the Mantel–Haenszel procedure for the detection of DIF and tests of fit in the RM, cf. Holland and Thayer (1988) and Fischer (1995e).

EXAMPLE 5. Measurement and testing of DIF.

To illustrate the use of the LLTM for the study of DIF, we reanalyze a subset of the data of Seliger (1993), taken from Section 2.2.2 of Fischer and Ponocny-Seliger (1998). Seliger used certain construction rules to generate 104 nonverbal letter series; from them, she built several shorter testlets, each of which was presented to a number of school classes (age: 14–17; total sample: $n = 514$). She found that the RM fitted these data reasonably well. On that premise, a subset of $k = 56$ of these items will be used to show how the LLTM can be employed for the assessment of DIF.

In an example of realistic size and, moreover, with an incomplete design like in the present case, the weight matrix $\boldsymbol{W}$ needed for the analysis of DIF is rather large. It is therefore impractical to present it here. The principle of its construction, however, can be demonstrated as follows. Suppose there were only $k = 5$ items, two of which were suspected to be gender-biased, item I_2 presumably favoring males, I_5 favoring females. Under H_1, from a formal point of view, the LLTM has $k = 10$ items, namely, two blocks of five (real) items when given to males and females, respectively, the item parameters of the first, third and fourth item in either block being restricted to be identical. Therefore, the H_1 model has 10 'virtual' (V-) items with parameters β_i^*, $i = 1, \ldots, 10$, which are linear functions of the following basic parameters: $\beta_1, \ldots, \beta_5, \eta_2, \eta_5$, where η_1 and η_5 are the 'effect' parameters of DIF with respect to items I_2 and I_5, respectively. For item I_2 in the male group, we put $\beta_2^* = \beta_2 - \eta_2$, where η_2 is the 'distance' between the difficulties of item I_2 in males vs. females. Analogously, for item I_5 in the female group, we put $\beta_{10}^* = \beta_5 - \eta_5$. We expect positive values of both $\hat{\eta}_2$ and $\hat{\eta}_5$. Under H_0, it is assumed that one RM fits all items in both groups, with common item parameters, that is, $\eta_2 = \eta_5 = 0$. Here, unlike in many typical LLTM applications, the LLTM is the H_1, with $5 + 2 - 1 = 6$ independent parameters, whereas the RM is the H_0, with $5 - 1 = 4$ parameters.

To make the item parameters unique, it is necessary to normalize the β_i, for instance, by putting $\beta_5 = 0$. Then the structure of the weight matrix $\boldsymbol{W}$ results as shown in Table 6 (empty cells are zero).

In terms of the real items, this design is complete because all items are given to all testees. In terms of the V-items, however, the design is incomplete because each testee is presented with only 5 of the 10 V-items. Designs with V-items like the present were the main reason for introducing the incomplete data notation.

Returning now to the empirical example, the test comprises 56 (real) items, so that the number of V-items is $k = 112$. Four items are assumed to favor males, items I_3, I_{13}, I_{18}, I_{20}, and another four to favor females, I_1, I_{16}, I_{30}, I_{37}. Hence, in addition to the item parameters β_i, $i = 1, \ldots, 56$ (with $\beta_{56} = 0$ for normalization), we introduce bias parameters η_i, for $i = 3, 13, 18, 20; 1, 16, 30, 37$; all of them are expected to be positive. The LLTM under H_1 therefore has $56 + 8 - 1 = 63$ independent parameters.

Applying this LLTM with a weight matrix analogous to Table 6 to the Seliger data yields the following estimates and asymptotic standard errors (the latter are given in parentheses; asterisks denote significance at $\alpha = 0.05$): $\hat{\eta}_3 = 0.56^*$ (0.17), $\hat{\eta}_{13} = 0.87^*$ (0.31), $\hat{\eta}_{18} = 0.75^*$ (0.26), $\hat{\eta}_{20} = 0.76$ (0.41); $\hat{\eta}_1 = 0.54^*$ (0.25), $\hat{\eta}_{16} = 0.79^*$

Table 6
Structure of a weight matrix for DIF analysis in Example 5

		β_1	β_2	β_3	β_4	β_5	η_2	η_5
	β_1^*	1						
	β_2^*		1				-1	
G_1	β_3^*			1				
	β_4^*				1			
	β_5^*							
	β_6^*	1						
	β_7^*		1					
G_2	β_8^*			1				
	β_9^*				1			
	β_{10}^*							-1

Note: G_1 and G_2 denote 'Males' and 'Females', respectively. Empty cells are zero.

(0.38), $\hat{\eta}_{30} = 1.12^*$ (0.26), $\hat{\eta}_{37} = 1.05^*$ (0.50). So there is ample evidence of the presence of DIF in these items. An over-all test of the H_0 of no DIF can be made by comparing the log-likelihoods for the RM (under H_0) and the LLTM (under H_1): $\chi^2 = -2[\ln L_C(\text{RM}) - \ln L_C(\text{LLTM})] = -2[-4320.80 + 4304.87] = 31.86$, with df $= 8$ and $\chi^2_{8;1-\alpha} = 15.51$ for $\alpha = 0.05$, which clearly rejects the H_0.

Next, we test a less restrictive H_0 against the same LLTM (H_1 model) as before: to see whether the absolute size of DIF can be assumed to be constant over items ('uniform item bias'), we set all DIF parameters equal to η, collapsing those columns of the weight matrix of the LLTM which correspond to the item bias parameters. (In Table 6, this would have been the last two columns.) This yields another LLTM with $56 + 1 - 1 = 56$ independent parameters. Then the likelihood ratio test gives $\chi^2 = -2[-4305.97 + 4304.87] = 2.20$, with df $= 7$ and $\chi^2_{7;1-\alpha} = 14.07$ for $\alpha = 0.05$, which is nonsignificant. The H_0 of a constant amount of DIF over the items in question, $\hat{\eta} = 0.71^*$ (0.11), can therefore be retained.

8. Longitudinal linear logistic models

The LLTM described in Section 7 enables the customization of Rasch-type models to many designs where experimental or testing conditions, development, learning, treatments, or testing occasions (time points) influence the abilities of the testees. For brevity, all these conditions or influences will be denoted 'treatments'. Suppose there are several treatment groups, G_g, receiving certain treatments B_j, $j = 1, \ldots, m$, or combinations of such treatments. Let the amounts of treatments given to group G_g up to time point T_t be described by a row vector of dosages, $\boldsymbol{q}_{gt} = (q_{gt1}, \ldots, q_{gtm})$, and assign treatment effects η_j to the treatments, collected in a treatment effect (column) vector $\boldsymbol{\eta} = (\eta_1, \ldots, \eta_m)'$; η_j measures the effect of one dosage unit of treatment B_j on the trait parameters θ_v of testees $S_v \in G_g$, corresponding to the transformation $\theta_v \to \theta_v - \eta_j$. The total effect of a treatment combination $\boldsymbol{q}_{gt}$ given to G_g up to T_t is assumed to be $\theta_v \to \theta_v - \boldsymbol{q}_{gt}\boldsymbol{\eta}$. (Treatment interaction terms are excluded here for simplicity; the linear additive structure of the model, however, makes the introduction of interaction effects easy.) A fairly general unidimensional model of change for items conforming to an RM can be formulated in terms of V-item parameters as follows.

8.1. *A unidimensional LLTM of change*

Let there be u different dosage vectors $\boldsymbol{q}_{gt}$; let $\boldsymbol{D}$ denote a supermatrix consisting of u unit matrices $\boldsymbol{I}$ of dimension $(k \times k)$ each, stacked vertically, such that $\boldsymbol{D} = (\boldsymbol{I}, \boldsymbol{I}, \ldots, \boldsymbol{I})'$; let $\boldsymbol{Q}$ consist of the u vectors $\boldsymbol{q}_{gt}$ as rows, where each row is repeated k times, so that the total number of rows becomes ku; let β^*_{gti} be ku V-item parameters measuring the relative difficulty of the test items when given after a cumulative treatment combination $\boldsymbol{q}_{gt}$ in group G_g at time point T_t; and finally let $\mathbf{1}$ be a column vector with all elements equal 1 and length ku. The linear part of the model is then written compactly as

$$\boldsymbol{\beta}^* = \boldsymbol{D}\boldsymbol{\beta} + \boldsymbol{Q}\boldsymbol{\eta} + c\mathbf{1} = \boldsymbol{W}\boldsymbol{\alpha} + c\mathbf{1}. \tag{99}$$

Model (99) is an LLTM where $\boldsymbol{\beta}^*$ denotes the column vector of V-item parameters, $\boldsymbol{\beta}^* = (\beta_{111}, \ldots, \beta_{11k}; \beta_{121}, \ldots, \beta_{12k}; \ldots; \beta_{gt1}, \ldots, \beta_{gtk}; \ldots\ldots)'$, $\boldsymbol{\beta} = (\beta_1, \ldots, \beta_k)'$ the column vector of (real) item parameters, $\boldsymbol{W} = (\boldsymbol{D}, \boldsymbol{Q})$ the weight matrix, and $\boldsymbol{\alpha} = (\boldsymbol{\beta}', \boldsymbol{\eta}')'$ the vector of basic parameters. The latter comprises both the item and effect parameters. Change of the latent trait parameter θ_v under the impact of the treatment(s) is projected into the V-item parameters. This has the advantage that the person parameters θ_v formally remain invariant, implying that the over-all raw score R_v remains a sufficient statistic for θ_v and that the latter can again be eliminated via CML.

All the results and techniques described above for the LLTM apply here too. The CML method again is an effective tool to estimate the basic parameters of the model and to perform tests of hypotheses about the effect parameters. The number of V-items can become quite large, of course, but the model looks much more hideous than it is: it is important to recognize that, for parameter estimation and hypothesis testing, the total number of V-items in the model is *not* a critical quantity; what counts regarding the computability is the number of V-items given to the same testees (i.e., the virtual test length). In a design where all items are given to all testees on o testing occasions, the virtual test length would be $k \times o$; for instance, if the test had 25 items and were given on 6 occasions, the number of V-items per testee would be 150, which would pose no problem at all for CML estimation and for carrying out CLR tests. The criteria for uniqueness of the CML solution given in Theorem 21 and Corollary 2 apply here also, provided that the data are arranged as *one incomplete item score matrix* $\boldsymbol{X}$ of dimension 'testees $\times$ V-items'. Although by conventional standards the present design appears to be complete, from the LLTM perspective the data are structurally incomplete: since groups differ by their treatment combinations and dosages, they also differ with respect to the subsample of V-items presented. A smooth technology for incomplete data is therefore of paramount importance.

Examples of applications can be found in Fischer (1995c), Fischer (1997), and Fischer and Ponocny-Seliger (1998).

8.2. *A multidimensional LLTM of change*

Let the pool of items consist of several subtests T_l, $l = 1, \ldots, h$, where the items $I_i \in T_l$ are unidimensional and conform to one RM each with respective item parameter vectors $\boldsymbol{\beta}_l$, but where the subtests measure different latent traits. Hence, each testee has h person parameters θ_{vl}, $l = 1, \ldots, h$. A model of change can again be written as

$$\boldsymbol{\beta}_l^* = \boldsymbol{D}_l \boldsymbol{\beta}_l + \boldsymbol{Q}\boldsymbol{\eta}_l + c_l \mathbf{1}_l, \tag{100}$$

for $l = 1, \ldots, h$, where the vectors $\boldsymbol{\beta}_l$ and $\boldsymbol{\beta}_l^*$ contain the item and V-item parameters, respectively, of the items $I_i \in T_l$, the vector $\boldsymbol{\eta}_l = (\eta_{1l}, \ldots, \eta_{ml})'$ the treatment effects with respect to the latent dimensions measured by the subtests T_l, and the $\mathbf{1}_l$ are vectors with all elements equal to 1 and appropriate length. There is a one-to-one correspondence of subtests and subdimensions.

If the treatment effects with respect to the h subtests are considered independent parameters, the model in (100) is no more than a collection of h independent models (99), each of which can be estimated separately, and the total conditional likelihood is the product of the h likelihoods. If, on the other hand, some or all of these effect parameters are the same for some or all subtests, (100) becomes a null-model with a smaller number of basic parameters. Note that the basic parameters comprise both the item parameters and the independent effect parameters. The H_0 specifying 'generalizability of (some or all) effects over subtests', can now again be tested against the H_1 of independent effects per subtest by means of a CLR test within the LLTM framework.

Technically, the step from (99) to the multidimensional model (100) is a very simple one: to estimate the basic parameters of (100) and to test hypotheses on them, treat the response vectors of each testee S_v on the subtests $\mathcal{T}_l$, $l = 1, \ldots, h$, as if they stemmed from different virtual testees (V-testees) S^*_{vl}, $l = 1, \ldots, h$, and were responses given to item subsets of *one unidimensional test*, so that the item score matrix becomes structurally incomplete. Eliminate all person parameters θ_{vl}, for $l = 1, \ldots, h$ and $v = 1, \ldots, n$, via CML, conditional on one raw score r_{vl} per V-testee. This is admissible in virtue of the assumption of local stochastic independence and highlights once more the advantages of the CLM approach. Examples of such multidimensional LLTMs of change are given in Fischer and Ponocny-Seliger (1998).

A special kind of treatment effect is the so-called 'trend'. Trend means a migration, between T_1 and T_t, of all testees on one or all latent dimensions, in addition to the transformations of the person parameters caused by the treatments. In the present context, trend can formally be considered a treatment given to all testees with the same dosage, defined as the time interval between T_t and T_1, denoted $T_t - T_1$ for short. This additive effect is assumed to occur in all treatment groups, including untreated control group(s). Formally, however, there is no need to distinguish between treatment and trend effect(s); if there are $m-1$ (real) treatments, trend can simply be defined as an mth treatment with dosages $T_t - T_1$ in all groups. Such trend effect(s) is (are) often explicitly denoted τ (or τ_l for $l = 1, \ldots, h$) in order to emphasize its (their) particular role and to allow for a more explicit formulation of hypotheses.

Typical hypotheses within the framework of LLTMs of change are: (i) 'no change', $\boldsymbol{\eta}_l = \mathbf{0}$ and $\tau_l = 0$ for all l; (ii) 'no treatment effects', $\boldsymbol{\eta}_l = \mathbf{0}$ for some or all l; (iii) 'no trend effects', $\tau_l = 0$ for some or all l; (iv) 'equality of treatment effects', e.g., $\eta_{jl} = \eta_{kl}$ for certain treatments B_j and B_k with respect to some or all subtests $\mathcal{T}_l$; (v) 'generalizability of treatment and/or trend effects over subtest dimensions', e.g., $\boldsymbol{\eta}_a = \boldsymbol{\eta}_b$ for some or all pairs of subtests $(\mathcal{T}_a, \mathcal{T}_b)$; (vi) 'generalizability of treatment effects over subgroups of testees', e.g., $\boldsymbol{\eta}_l^{(g)} = \boldsymbol{\eta}_l^{(f)}$ for some person groups G_g and G_f and some or all $\mathcal{T}_l$. For more details about possible hypothesis tests, cf. Fischer (1995c) or Fischer and Ponocny-Seliger (1998).

Longitudinal designs bear remarkable advantages over designs with only one time point, as far as IRT modeling is concerned: multidimensional extensions of LLTMs for modeling change are easy to realize, and even the elimination of both item and person parameters via CML techniques becomes feasible under certain conditions, so that only the treatment effect parameters have to be estimated; thus the precision of their estimates and the power of the respective CML tests is increased considerably. This conforms ideally with Rasch's principle of freeing 'measurement' – here, of the treatment effect – of influences of all other factors (items, persons). Such favorable cases occur in designs where the testees are tested repeatedly with the same (or parallel) items, as is often done in clinical studies. On a formulation of SO with special reference to treatment effects and on an algebra of such effects, see Fischer (1987).

Suppose, for instance, that the items are symptoms of depression, which certainly do not form a unidimensional scale conforming to the RM: it is well known from clinical psychology that patients may express their depression in the form of individually different patterns of symptoms, such as: sleep disorders; emotional lability; various forms of anxieties; withdrawal; reduction of concentration, abilities, and cognitive functions; impairment of social and occupational abilities; etc.[15] Therefore, it makes sense to formally assume a separate latent trait dimension for each symptom (or item). It has to be stressed, however, that nothing is postulated concerning the probable mutual (linear, nonlinear, or functional) dependence or even the possible independence of these dimensions. Since all person parameters will again be eliminated via CML, the mutual relationship of the latent dimensions may be of any level of complexity, without impairment of the estimation of the effect parameters.

Let patient S_v be characterized by k latent trait parameters θ_{vi}, $i = 1, \ldots, k$, where k is the number of items of a depression inventory. Let the inventory be filled in by each patient at time points T_t, $t = 1, \ldots, s$, let the total amount of treatment(s) given to $S_v \in G_g$ up to time point T_t be described by the dosage vector $\boldsymbol{q}_{gt} = (q_{gt1}, \ldots, q_{gtm})$, and the treatment effects by vector $\boldsymbol{\eta} = (\eta_1, \ldots, \eta_m)'$, possibly including a trend parameter, $\eta_m =: \tau$ with $q_{gtm} = T_t - T_1$. Then the argument of the logistic function of an LLTM for the responses of patient S_v to one item I_i at time point T_t is assumed to be

$$\theta_{vi} - \beta_i - \boldsymbol{q}_{gt}\boldsymbol{\eta} = \theta^*_{vi} - \boldsymbol{q}_{gt}\boldsymbol{\eta}. \tag{101}$$

The item parameter β_i in (101) is not identifiable, but it can immediately be eliminated by amalgamating it with the person parameter, putting $\theta_{vi} - \beta_i =: \theta^*_{vi}$. The latter person parameters – k of them per testee – are denoted 'virtual person (V-person) parameters'. Technically, the model comprises nk V-persons, each of which corresponds to one combination of a (real) testee S_v with a (real) item I_i. Apart from the θ^*_{vi}, (101) contains the terms $\boldsymbol{q}_{gt}\boldsymbol{\eta}$; they formally play the role of V-item parameters. Their number equals the number of person groups by time points, say, k^*. Because of the linear structure of (101) as a function of the effect parameters η_j, the model is a *unidimensional* LLTM for nk V-persons and k^* V-items, with weight matrix $\boldsymbol{W}$ consisting of k^* row vectors $\boldsymbol{q}_{gt}$. Hence, all one has to do is to arrange the item response matrices of (real) testees by (real) items obtained at time points T_t, $t = 1, \ldots, s$, as *one incomplete item score matrix* $\boldsymbol{X}$ of dimension 'V-persons × V-items', and then to apply the LLTM, conditioning on the raw scores of the V-persons. Notice that again the data matrix is *structurally incomplete*.

The nature of the multidimensionality in this model deserves some more comments. Each (real) patient S_v is a point in a k-dimensional parameter space, with coordinates θ_{vi}, $i = 1, \ldots, k$. This enables an appropriate modeling of the responses even for an inventory where all feasible response patterns occur with sizable proportions; person misfit is not possible at a single time point because any response pattern of a testee S_v

[15] In spite of this, clinical psychologists often work with so-called 'depression scales' where the raw score is defined as the number of statements endorsed by the patient. The expression 'depression inventory' is preferable because a collection of questions addressing various symptoms of a disease will in most cases fail to make a unidimensional scale.

can be explained perfectly by means of the parameters $\theta_{v1}, \ldots, \theta_{vk}$. In terms of the V-persons, the model becomes a unidimensional LLTM because the multidimensionality is fully accounted for by the concept of V-persons with independent parameters θ_{vi}. The true dimensionality of the latent space may be smaller than k, which happens if the dimensions assigned to the items are dependent. In principle, the items could even be unidimensional, so that all θ_{vi} lie on one line in k-space. The dependence or independence of the dimensions is not the question of interest, though, since the person parameters are again eliminated via CML. The model focuses on the estimation of and hypotheses testing about the treatment effect parameters η_j as modeled by Eq. (101). Item parameters cannot be identified because they have been amalgamated with the person parameters. Owing to the CML approach, only m basic (or effect) parameters η_j remain to be estimated (including a possible trend effect). The CML estimator is unique provided that the conditions of Theorem 21 or Corollary 2 are met when applied to the incomplete item score matrix $\boldsymbol{X}$ of V-testees $\times$ V-items, and the estimator is asymptotically normal around the true $\boldsymbol{\eta}$. CLR tests concerning the effect and trend parameters are straightforward (see above). Treatment interaction effects can immediately be added, without any change of the model structure, provided that the treatment design permits doing so: if some group(s) G_g had both treatments B_j and B_l, say, with respective dosages q_{gtj} and q_{gtl}, it is possible to introduce an additional treatment-interaction (effect) parameter ρ_{jm} with respective dosages $q_{gtj}q_{gtl}$, and so forth.

It should be stressed that in this model, in spite of its multidimensionality, no identification or rotation problems occur. The effect parameters of interest are estimated consistently if the conditions on the rank of the dosage matrix and on the sequence of latent V-person parameters, analogous to (54), are satisfied for all groups as $n_g \to \infty$. A software focusing on measurement of treatment effects by means of generalized LLTMs is found in Fischer and Ponocny-Seliger (1998).

8.3. The special case of two time points: The LLRA

A particularly simple special case arises if the design comprises just two time points ('pretest' and 'posttest'). Denote the responses of S_v to I_i by x_{vit}, for $t = 1, 2$, and the dosages of the treatments B_j given to S_v between T_1 and T_2, by q_{vj}. Now the V-persons have given only two responses each, and their raw scores are the sums $r_{vi} = x_{vi1} + x_{vi2}$. Conditioning on these r_{vi} implies that the conditional likelihood for the pair of responses (x_{vi1}, x_{vi2}) becomes uninformative whenever r_{vi} is extreme, namely, if $r_{vi} = 0$ or $r_{vi} = 2$; in either case, conditioning on the statistic r_{vi} implies $x_{vi1} = x_{vi2}$. Therefore, the conditional likelihood depends only on the pairs with $r_{vi} = 1$, that is, where the two responses given by S_v to item I_i at time points T_1 and T_2 are different. It can be shown that for this special case the CML estimation equations simplify to

$$\sum_v q_{vl} \sum_i (x_{vi1} - x_{vi2})^2 \left[x_{vi2} - \frac{\exp(\sum_j q_{vj}\eta_j)}{1 + \exp(\sum_j q_{vj}\eta_j)} \right] = 0, \tag{102}$$

for $l = 1, \ldots, m$, where the term $(x_{vi1} - x_{vi2})^2$ acts as a sort of 'filter' for selecting the pairs of responses with $r_{vi} = 1$, ignoring all pairs with $r_{vi} \in \{0, 2\}$. Apart from that filter, the CML equations are ML equations for a simple logit model. Solving them, for instance, by means of the Newton–Raphson method, is straightforward and requires very little computational expense. This model, allowing again for multidimensionality of the item set, was denoted 'Linear Logistic Test Model With Relaxed Assumptions' (Fischer, 1983).

The question of uniqueness of the CML estimator of (102) can once more be answered by means of Theorem 21. Let the design comprise a set G of treatment groups G_g with different dosage vectors $\boldsymbol{q}_g$, $G = \{G_1, \ldots, G_f\}$, such that the dosage matrix $\boldsymbol{Q}_G = ((q_{gj}))$, $g = 1, \ldots, f$; $j = 1, \ldots, m$, has rank$(\boldsymbol{Q}_G) = m$. (The design will typically contain the 'null-treatment' given to an untreated control group, formally represented by a treatment combination with weights $q_{gj} = 0$ for $j \neq m$, that is, except for the trend parameter $\tau = \eta_m$.) Corollary 3 states the uniqueness condition for the CML estimator that follows from an application of Corollary 2.

COROLLARY 3. *Let $\boldsymbol{Q}_G$ have full column rank m. Then the CML equations* (102) *of the LLRA have a unique finite solution $\hat{\boldsymbol{\eta}}$ if, within each treatment group G_g, at least one response sequence* $(0, 1)$ *has been observed on some* (*real*) *item I_{i_g}, and at least one response sequence* $(1, 0)$ *on some* (*real*) *item I_{h_g}.*

Evidently, the uniqueness conditions of Corollary 3 are very weak. In most cases of applications, they are likely to be satisfied even in small groups. Experience has shown that about $n_g \geqslant 30$ testees per treatment group suffice both for parameter estimation and for testing typical hypotheses as mentioned above. This holds in spite of the fact that a large portion of the statistical information in the data is ignored by the 'filter' mentioned. Only tests for generalizability over subgroups of patients, for instance, defined on the basis of age, gender, illness record previous to T_1, etc., require larger treatment groups because the latter have to be split by these external variables and parameter estimates have to be obtained within the subsets of data. These surprising findings regarding the required sample sizes owe to the small number of basic effect parameters that need to be estimated: the statistical information remaining after the 'filtering' is used exclusively to estimate these few effect parameters and to test hypotheses about them. The vast majority of the parameters, namely, the V-person parameters θ_{vi}, have been eliminated beforehand via the CML approach. Applications are mentioned, for instance, in Fischer and Ponocny-Seliger (1998) and Spiel and Glück (1998).

9. Some remarks on applications and extensions of the RM

Applying the RM has recently become quite popular not only in psychology and education, but also in many other scientific domains. It is tempting to use the RM whenever a 'scale' consists of dichotomous observations ('items') and the raw score suggests itself as a useful data reduction or 'measure'. More often than not, such enterprises are futile, however, because the strict limitations of the RM are violated: unidimensionality, no guessing, parallel IRFs (or SO), no DIF with respect to gender, age, education, etc. Unidimensionality of the items requires, on the substantive level, that the items are of very homogeneous content; this often conflicts with psychologists' diagnostic aims.

The requirement of no guessing strictly speaking excludes the popular multiple choice item format; in particular, it is hopeless to fit a RM to personality or attitude questionnaires with two ('yes' vs. 'no') answer categories, because answers determined by the latent trait to be measured are indistinguishable from random responses. Absence of DIF is also a very hard criterion: experience with many intelligence and achievement tests shows that all verbal items – or items having a substantial verbal component – are prone to DIF with respect to the subpopulations mentioned above. Even elementary arithmetic or science problems often show considerable DIF with respect to gender, depending on their content. Therefore, the RM can by no means be considered as an omnibus method for the analysis and scoring of all sorts of tests. Rather, it should be viewed as a guideline for the construction or improvement of tests, as an ideal to which a test should be gradually approximated, so that measurement can profit from the unique properties of the RM.

The difficulties of fitting an RM have strongly and effectively stimulated the development of relaxations and generalizations of the RM. Space only permits to shortly mention some directions of development here.

9.1. *Dichotomous generalizations*

The LLTMs have been discussed sufficiently in Sections 7 and 8. Related to the LLTM are the complete multifactorial RM (Micko, 1969; Kempf, 1972; Fischer, 1974) and the FACETS model of Linacre (1994). Computer programs are LPCM Win 1.0 (Fischer and Ponocny-Seliger, 1998) and FACETS (Linacre, 1996). These models implement linear parametric structures within the RM and can thus be subsumed under the LLTM. Recently, the further development of RMs with linear constraints on the parameters is merging with the mainstream of literature on logistic, loglinear, and generalized linear modeling (cf. De Boeck and Wilson, 2004; Molenberghs and Verbeke, 2005; Skrondal and Rabe-Hesketh, 2004).

Another direction of generalization is the Mixed RM by Rost (1990, 1997), Von Davier (1998), and Rost and Von Davier (1995), which assumes the existence of latent classes between which the item parameters of the RM are allowed to vary. This model appears particularly well suited for cases where there are latent groups of testees with DIF. Fundamental questions concerning the identifiability of mixed models, however, are still unanswered.

One more line of development relaxes the assumption of equal discrimination of all items: the 'One Parameter Logistic Model' (OPLM) by Verhelst et al. (1994), see also Verhelst and Glas (1995), posits that each item belongs to one of a few classes of items with different discrete rational discrimination parameters. Each item is first assigned to one of the classes by means of a heuristic procedure, then its discrimination parameter is considered as fixed and given. The difference between the OPLM and the Two-Parameter Logistic (2PL) Model lies in the nature of the discrimination parameters: in the 2PL, they are free parameters, while in the OPLM they are fixed a priori. Software for the estimation and testing of the OPLM has been published by Verhelst et al. (1994).

Yet another generalization of the RM allows to cope with 'dynamic' processes (e.g., learning) during testtaking (Kempf, 1977; Verhelst and Glas, 1993).

9.2. *Polytomous generalizations*

Andrich (1978) and Masters (1982) have proposed unidimensional polytomous models for items with ordered response categories (like "strongly agree", "rather agree', "rather disagree", "strongly disagree"), namely, the Rating Scale Model (RSM) and the Partial Credit Model (PCM), see also Andersen (1995). The RSM assumes equal response categories for all items, whereas the PCM allows for a different number and/or a different definition of the response categories per item. The PCM contains the RSM, and the RSM contains the RM as special cases. Fischer and Ponocny (1994) have embedded a linear structure in the parameters of these models, analogous to the LLTM, and have developed CML estimation, conditional hypothesis tests, and multidimensional reparameterizations. Glas and Verhelst (1989) have dealt with the same models within an MML framework, and Muraki (1992), with a generalization. These models are particularly well suited for longitudinal and treatment effect studies; a software especially designated to analyzing change is LPCM Win 1.0 (Fischer and Ponocny-Seliger, 1998). Another software for generalized PCMs – named PARSCALE – is due to Muraki and Bock (1991). The individual-centered exact conditional tests of change also remain applicable (cf. Fischer, 2001, 2003).

A still more general class of multidimensional IRT models with embedded linear structures has been developed by Adams et al. (1997) and Meiser (1996), cf. Wilson and De Boeck (2004); the latter authors mainly rely on parametric MML methods (ACER ConQuest by Wu et al., 1998; for an overview of general-purpose software and its use for IRT models, see Tuerlinckx et al., 2004).

Müller (1987), moreover, has proposed an extension of the RM for continuous responses.

References

Abramowitz, M., Stegun, I. (1974). *Handbook of Mathematical Functions*. Dover Publications, New York.

Aczél, J. (1966). *Lectures on Functional Equations and Their Applications*. Academic Press, New York.

Adams, R.J., Wilson, M., Wang, W. (1997). The multidimensional random coefficients multinomial logit model. *Applied Psychological Measurement* **21**, 1–23.

Adams, R.J., Wu, M. (Eds.) (2002). *PISA 2000 Technical Report*. OECD, Organization for Economic Cooperation and Development.

Andersen, E.B. (1970). Asymptotic properties of conditional maximum likelihood estimation. *Journal of the Royal Statistical Society, Series B* **32**, 283–301.

Andersen, E.B. (1972). The numerical solution of a set of conditional estimation equations. *Journal of the Royal Statistical Society, Series B* **34**, 42–54.

Andersen, E.B. (1973a). Conditional inference for multiple-choice questionnaires. *British Journal of Mathematical and Statistical Psychology* **26**, 31–44.

Andersen, E.B. (1973b). *Conditional Inference and Models for Measuring*. Mentalhygiejnisk Forlag, Copenhagen.

Andersen, E.B. (1973c). A goodness of fit test for the Rasch model. *Psychometrika* **38**, 123–140.

Andersen, E.B. (1977). Sufficient statistics and latent trait models. *Psychometrika* **42**, 69–81.

Andersen, E.B. (1980). *Discrete Statistical Models With Social Science Applications*. North-Holland, Amsterdam.

Andersen, E.B. (1995). Polytomous Rasch models and their estimation. In: Fischer, G.H., Molenaar, I.W. (Eds.), *Rasch Models. Foundations, Recent Developments, and Applications*. Springer-Verlag, New York, pp. 271–291.

Andersen, E.B., Olsen, L.W. (2001). The life of Georg Rasch as a mathematician and as a statistician. In: Boomsma, A., van Duijn, M.A.J., Snijders, T.A.B. (Eds.), *Essays on Item Response Theory*. Springer-Verlag, New York, pp. 3–24.

Andrich, D. (1978). A rating formulation for ordered response categories. *Psychometrika* **43**, 561–573.

Barndorff-Nielsen, O. (1978). *Information and Exponential Families in Statistical Theory*. Wiley, New York.

Bechger, T.M., Verhelst, N.D., Verstralen, H.H.F.M. (2001). Identifiability of nonlinear logistic models. *Psychometrika* **66**, 357–372.

Bechger, T.M., Verstralen, H.H.F.M., Verhelst, N.D. (2002). Equivalent linear logistic test models. *Psychometrika* **67**, 123–136.

Besag, J., Clifford, P. (1989). Generalized Monte Carlo significance tests. *Biometrika* **76**, 633–642.

Birnbaum, A. (1968). Some latent trait models and their use in inferring an examinee's ability. In: Lord, F.M., Novick, M.R. (Eds.), *Statistical Theories of Mental Test Scores*. Addison-Wesley, Reading, MA, pp. 395–479.

Bock, R.D., Aitkin, M. (1981). Marginal maximum likelihood estimation of item parameters: An application of an EM-algorithm. *Psychometrika* **46**, 443–459.

Bradley, R.A. (1984). Paired comparisons. Some basic procedures and examples. In: Krishnaiah, P.R., Sen, P.K. (Eds.), *Handbook of Statistics*, vol. IV. Elsevier Science, New York, pp. 299–326.

Butter, R.P., De Boeck, P., Verhelst, N.D. (1998). An item response model with internal restrictions on item difficulty. *Psychometrika* **63**, 47–63.

Chen, Y., Diaconis, P., Holmes, S., Liu, J.S. (2005). Sequential Monte Carlo methods for statistical analysis of tables. *Journal of the American Statistical Association* **100**, 109–120.

Chen, Y., Small, D. (2005). Exact tests for the Rasch model via sequential importance sampling. *Psychometrika* **70**, 1–20.

Colonius, H. (1979). Zur Eindeutigkeit der Parameter im Rasch-Modell (On the uniqueness of parameters in the Rasch model). *Psychologische Beiträge* **21**, 414–416. In German.

Cox, R.D. (1970). *The Analysis of Binary Data*. Methuen, London.

Cressie, N., Holland, P.W. (1983). Characterizing the manifest probabilities of latent trait models. *Psychometrika* **48**, 129–141.

David, H.A. (1963). *The Method of Paired Comparisons*. Charles Griffin, London.

De Boeck, P., Wilson, M. (2004). A framework for item response models. In: De Boeck, P., Wilson, M. (Eds.), *Explanatory Item Response Models. A Generalized Linear and Nonlinear Approach*. Springer, New York, pp. 3–41.

De Gruijter, D.N.M. (1987). On the robustness of the "minimum-chi-square" method for the Rasch model. *Tijdschrift voor Onderwijsresearch* **12**, 185–195.

De Leeuw, J., Verhelst, N. (1986). Maximum likelihood estimation in generalized Rasch models. *Journal of Educational Statistics* **11**, 183–196.

Dempster, A.P., Laird, N.M., Rubin, D.B. (1977). Maximum likelihood from incomplete data via the EM algorithm (with discussion). *Journal of the Royal Statistical Society, Series B* **39**, 1–38.

Ducamp, A., Falmagne, J.C. (1969). Composite measurement. *Journal of Mathematical Psychology* **6**, 359–390.

Du Toit, M. (Ed.) (2003). *IRT from SSI: BILOG-MG, MULTILOG, PARSCALE, TESTFACT*. Scientific Software International, Lincolnwood, IL.

Eggen, T.J.H.M. (2000). On the loss of information in conditional maximum likelihood estimation of item parameters. *Psychometrika* **65**, 337–362.

El-Helbawy, A.T., Bradley, R.A. (1977). Treatment contrasts in paired comparisons: Convergence of a basic iterative scheme for estimation. *Communications in Statistics – Theory and Methods* **6**, 197–207.

Embretson, S.E. (1999). Generating items during testing: Psychometric issues and models. *Psychometrika* **64**, 407–433.

Falmagne, J.-C. (1985). *Elements of Psychophysical Theory*. Clarendon Press, Oxford.

Fischer, G.H. (1968). Neue Entwicklungen in der psychologischen Testtheorie. In: Fischer, G.H. (Ed.), *Psychologische Testtheorie* (Psychological test theory). Verlag Hans Huber, Bern, pp. 15–132. In German.

Fischer, G.H. (1972). A measurement model for the effect of mass-media. *Acta Psychologica* **36**, 207–220.

Fischer, G.H. (1973). The linear logistic test model as an instrument in educational research. *Acta Psychologica* **37**, 359–374.

Fischer, G.H. (1974). *Einführung in die Theorie psychologischer Tests* (Introduction to psychological test theory). Huber, Bern. In German.

Fischer, G.H. (1981). On the existence and uniqueness of maximum-likelihood estimates in the Rasch model. *Psychometrika* **46**, 59–77.

Fischer, G.H. (1983). Logistic latent trait models with linear constraints. *Psychometrika* **48**, 3–26.

Fischer, G.H. (1987). Applying the principles of specific objectivity and generalizability to the measurement of change. *Psychometrika* **52**, 565–587.

Fischer, G.H. (1995a). Derivations of the Rasch model. In: Fischer, G.H., Molenaar, I.W. (Eds.), *Rasch Models. Foundations, Recent Developments, and Applications*. Springer-Verlag, New York, pp. 15–38.

Fischer, G.H. (1995b). The linear logistic test model. In: Fischer, G.H., Molenaar, I.W. (Eds.), *Rasch Models. Foundations, Recent Developments, and Applications*. Springer-Verlag, New York, pp. 131–155.

Fischer, G.H. (1995c). Linear logistic models for change. In: Fischer, G.H., Molenaar, I.W. (Eds.), *Rasch Models. Foundations, Recent Developments, and Applications*. Springer-Verlag, New York, pp. 157–180.

Fischer, G.H. (1995d). Derivations of polytomous Rasch models. In: Fischer, G.H., Molenaar, I.W. (Eds.), *Rasch Models. Foundations, Recent Developments, and Applications*. Springer-Verlag, New York, pp. 293–305.

Fischer, G.H. (1995e). Some neglected problems in IRT. *Psychometrika* **60**, 459–487.

Fischer, G.H. (1997). Unidimensional linear logistic Rasch models. In: Van der Linden, W.J., Hambleton, R.K. (Eds.), *Handbook of Modern Item Response Theory*. Springer-Verlag, New York, pp. 225–243.

Fischer, G.H. (2001). Gain scores revisited under an IRT perspective. In: Boomsma, A., van Duijn, M.A.J., Snijders, T.A.B. (Eds.), *Essays on Item Response Theory*, *Lecture Notes in Statistics*, vol. 157. Springer-Verlag, New York, pp. 43–68.

Fischer, G.H. (2003). The precision of gain scores under an item response theory perspective: A comparison of asymptotic and exact conditional inference about change. *Applied Psychological Measurement* **27**, 3–26.

Fischer, G.H. (2004). Remarks on "Equivalent linear logistic test models" by Bechger et al. (2002). *Psychometrika* **69**, 305–315.

Fischer, G.H., Formann, A.K. (1982). Some applications of logistic latent trait models with linear constraints on the parameters. *Applied Psychological Measurement* **4**, 397–416.

Fischer, G.H., Ponocny, I. (1994). An extension of the partial credit model with an application to the measurement of change. *Psychometrika* **59**, 177–192.

Fischer, G.H., Ponocny-Seliger, E. (1998). *Structural Rasch Modeling. Handbook of the Usage of LPCM-WIN 1.0*. ProGAMMA, Groningen, NL.

Fischer, G.H., Tanzer, N. (1994). Some LLTM and LBTL relationships. In: Fischer, G.H., Laming, D. (Eds.), *Contributions to Mathematical Psychology, Psychometrics, and Methodology*. Springer, New York, pp. 277–303.

Follmann, D. (1988). Consistent estimation in the Rasch model based on nonparametric margins. *Psychometrika* **53**, 553–562.

Ford Jr., L.R. (1957). Solution of a ranking problem from binary comparisons. *American Mathematical Monthly* **64**, 28–33.

Formann, A.K. (1982). Linear logistic latent class analysis. *Biometrical Journal* **24**, 171–190.

Formann, A.K. (1985). Constrained latent class models: Theory and applications. *British Journal of Mathematical and Statistical Psychology* **38**, 87–111.

Formann, A.K. (1995). Linear logistic latent class analysis and the Rasch model. In: Fischer, G.H., Molenaar, I.W. (Eds.), *Rasch Models. Foundations, Recent Developments, and Applications*. Springer-Verlag, New York, pp. 239–255.

Formann, A.K., Piswanger, K. (1979). *Wiener Matrizen-Test. Ein Rasch-skalierter sprachfreier Intelligenztest* (The Viennese Matrices Test. A Rasch-scaled nonverbal intelligence test). Beltz Test Gesselschaft, Weinheim.

Gittler, G. (1991). *Dreidimensionaler Würfeltest (3DW). Ein Rasch-skalierter Test zur Messung des räumlichen Vorstellungsvermögens* (The three-dimensional cubes-test (3 DW). A Rasch-scaled test for spatial ability). Beltz Test Gesellschaft, Weinheim. In German.

Glas, C.A.W. (1988). The derivation of some tests for the Rasch model from the multinomial distribution. *Psychometrika* **53**, 525–546.

Glas, C.A.W., Verhelst, N.D. (1989). Extensions of the partial credit model. *Psychometrika* **54**, 635–659.

Glas, C.A.W., Verhelst, N.D. (1995). Testing the Rasch model. In: Fischer, G.H., Molenaar, I.W. (Eds.), *Rasch Models. Foundations, Recent Developments, and Applications*. Springer-Verlag, New York, pp. 69–95.

Glas, C.A.W. (2001). Differential item functioning depending on general covariates. In: Boomsma, A., van Duijn, M.A.J., Snijders, T.A.B. (Eds.), *Essays on Item Response Theory, Lecture Notes in Statistics*, vol. 157. Springer-Verlag, New York, pp. 131–148.

Goodman, L.A., Kruskal, W.H. (1954). Measures of association for cross classification, Part I. *Journal of the American Statistical Association* **49**, 732–764.

Haberman, S.J. (1977). Maximum likelihood estimates in exponential response models. *The Annals of Statistics* **5**, 815–841.

Hamerle, A. (1979). Über die messtheoretischen Grundlagen von Latent-Trait-Modellen. *Archiv für Psychologie* **132**, 19–39.

Hamerle, A. (1982). *Latent-Trait-Modelle* (Latent trait models). Beltz, Weinheim. In German.

Harary, F., Norman, R.Z., Cartwright, D. (1965). *Structural Models: An Introduction to the Theory of Directed Graphs*. Wiley, New York.

Hart, J.D. (1997). *Nonparametric Smoothing and Lack-of-Fit Tests*. Springer-Verlag, New York.

Hedges, L.V., Olkin, I. (1985). *Statistical Methods for Meta-Analysis*. Academic Press, Orlando, FL.

Hillgruber, G. (1990). Schätzung von Parametern in psychologischen Testmodellen (The estimation of parameters in psychological test models). Diploma Thesis. University of Cologne, Cologne. In German.

Hoijtink, H., Boomsma, A. (1996). Statistical inference based on latent ability estimates. *Psychometrika* **61**, 313–330.

Holland, P.W., Thayer, D.T. (1988). Differential item functioning and the Mantel–Haenszel procedure. In: Wainer, H., Brown, H.I. (Eds.), *Test Validity*. Lawrence Erlabaum, Hillsdale, NJ, pp. 129–145.

Hornke, L.F., Habon, M.W. (1986). Rule-based item bank construction and evaluation within the linear logistic framework. *Applied Psychological Measurement* **10**, 369–380.

Irtel, H. (1987). On specific objectivity as a concept in measurement. In: Roskam, E.E., Suck, R. (Eds.), *Progress in Mathematical Psychology I*. North-Holland, Amsterdam, pp. 35–45.

Irtel, H. (1993). The uniqueness structure of simple latent trait models. In: Fischer, G.H., Laming, D. (Eds.), *Contributions to Mathematical Psychology, Psychometrics, and Methodology*. Springer, New York, pp. 265–275.

Irtel, H. (1995). An extension of the concept of specific objectivity. *Psychometrika* **60**, 115–118.

Irvine, S.H., Kyllonen, P.C. (Eds.) (2002). *Item Generation for Test Development*. Lawrence Erlbaum, Mahwah, NJ.

Jacobsen, M. (1989). Existence and unicity of MLEs in discrete exponential family distributions. *Scandinavian Journal of Statistics* **16**, 335–349.

Johnson, N.L., Kotz, S. (1969). *Distributions in Statistics: Discrete distributions, vol. 1*. Houghton Mifflin, Boston.

Junker, B. (2001). On the interplay between nonparametric and parametric IRT, with some thoughts about the future. In: Boomsma, A., van Duijn, M.A.J., Snijders, T.A.B. (Eds.), *Essays on Item Response Theory, Lecture Notes in Statistics*, vol. 157. Springer-Verlag, New York, pp. 247–276.

Karlin, S., Studden, W.J. (1966). *Tchebycheff Systems: With Applications to Analysis and Statistics*. John Wiley and Sons, New York.

Kelderman, H. (1995). The polytomous Rasch model within the class of generalized linear symmetry models. In: Fischer, G.H., Molenaar, I.W. (Eds.), *Rasch Models. Foundations, Recent Developments, and Applications*. Springer-Verlag, New York, pp. 307–323.

Kempf, W.F. (1972). Probabilistische Modelle experimentalpsychologischer Versuchssituationen (Probabilistic models of designs in experimental psychology). *Psychologische Beiträge* **14**, 16–37. In German.

Kempf, W.F. (1977). Dynamic models for the measurement of 'traits' in social behavior. In: Kempf, W.F., Repp, B. (Eds.), *Mathematical Models for Social Psychology*. Hans Huber, Bern, and John Wiley and Sons, New York, pp. 14–58.

Klauer, K.C. (1991a). An exact and optimal standardized person test for assessing consistency with the Rasch model. *Psychometrika* **56**, 213–228.

Klauer, K.C. (1991b). Exact and best confidence intervals for the ability parameter of the Rasch model. *Psychometrika* **56**, 535–547.

Krantz, D.H., Luce, R.D., Suppes, P., Tversky, A. (1971). *Foundations of Measurement*, vol. I. Academic Press, New York.
Lancaster, H.O. (1961). Significance tests in discrete distributions. *Journal of the American Statistical Association* **56**, 223–234.
Lazarsfeld, P.F., Henry, N.W. (1968). *Latent Structure Analysis*. Houghton Mifflin, Boston.
Lehmann, E. (1986). *Testing Statistical Hypotheses*, 2nd ed. John Wiley and Sons, New York.
Liang, K. (1983). On information and ancillarity in the presence of a nuisance parameter. *Biometrika* **70**, 607–612.
Linacre, J.M. (1994). *Many-Facet Rasch Measurement*. MESA Press, Chicago.
Linacre, J.M. (1996). *A User's Guide to FACETS*. MESA Press, Chicago.
Lindsay, B., Clogg, C.C., Grego, J. (1991). Semiparametric estimation in the Rasch model and related exponential response models, including a simple latent class model for item analysis. *Journal of the American Statistical Association* **86**, 96–107.
Liou, M. (1993). Exact person tests for assessing model-data fit in the Rasch model. *Applied Psychological Measurement* **17**, 187–195.
Liou, M., Chang, C.-H. (1992). Constructing the exact significance level for a person fit statistic. *Psychometrika* **47**, 169–181.
Lord, F.M. (1983). Unbiased estimators of ability parameters, of their variance, and their parallel-forms reliability. *Psychometrika* **48**, 233–245.
Martin-Löf, P. (1973). *Statistiska modeller* (Statistical models). Notes from a seminar held 1969/1970, taken by Rolf Sundberg; reprinted 1973. Institutet för Försäckringsmatematik och Matematisk Statistisk vid Stockholms Universitet, Stockholm. In Norvegian.
Masters, G.N. (1982). A Rasch model for partial credit scoring. *Psychometrika* **47**, 149–174.
Meiser, T. (1996). Loglinear Rasch models for the analysis of stability and change. *Psychometrika* **61**, 629–645.
Micko, H.C. (1969). A psychological scale for reaction time measurement. *Acta Psychologica* **30**, 324.
Micko, H.C. (1970). Eine Verallgemeinerung des Messmodells von Rasch mit einer Anwendung auf die Psychophysik der Reaktionen (A generalization of Rasch's measurement model with an application to the psychophysics of reactions). *Psychologische Beiträge* **12**, 4–22. In German.
Mielke, P.W. Jr. (1983). Goodman–Kruskal tau and gamma. In: Kotz, S., Johnson, N.L. (Eds.), *Encyclopedia of Statistical Sciences*, vol. 3. John Wiley and Sons, New York, pp. 446–449.
Mislevy, R.J. (1987). Exploiting auxiliary information about items in the estimation of Rasch item difficulty parameters. *Applied Psychological Measurement* **12**, 725–737.
Mislevy, R.J., Bock, R.D. (1986). *PC-Bilog: Item Analysis and Test Scoring with Binary Logistic Models*. Scientific Software, Mooresville, IN.
Mislevy, R.J., Verhelst, N.D. (1990). Modeling item responses when different subjects employ different solution strategies. *Psychometrika* **55**, 195–215.
Molenaar, I.W. (1995). Estimation of item parameters. In: Fischer, G.H., Molenaar, I.W. (Eds.), *Rasch Models. Foundations, Recent Developments, and Applications*. Springer-Verlag, New York, pp. 39–51.
Molenaar, I.W., Hoijtink, H. (1990). The many null distributions of person fit indices. *Psychometrika* **55**, 75–106.
Molenberghs, G., Verbeke, G. (2004). An introduction to (generalized non)linear mixed models. In: De Boeck, P., Wilson, M. (Eds.), *Explanatory Item Response Models*. Springer, New York, pp. 111–153.
Molenberghs, G., Verbeke, G. (2005). *Models for Discrete Longitudinal Data*. Springer, New York.
Mood, A.M., Graybill, F.A., Boes, D.C. (1974). *Introduction to the Theory of Statistics*. McGraw-Hill, Auckland.
Müller, H. (1987). A Rasch model for continuous ratings. *Psychometrika* **52**, 165–181.
Muraki, E. (1992). A generalized partial credit model: Application of an EM algorithm. *Applied Psychological Measurement* **16**, 159–176.
Muraki, E., Bock, R.D. (1991). *PARSCALE: Parameter Scaling of Rating Data*. Computer program. Scientific Software, Chicago.
Noack, A. (1950). A class of random variables with discrete distributions. *Annals of Mathematical Statistics* **21**, 127–132.

Neuwirth, E. (1987). *Some Results on Latent Additive Functions for Statistical Models*. Österreichische Akademie der Wissenschaften, Vienna.

Pfanzagl, J. (1971). *Theory of Measurement*. John Wiley and Sons, New York.

Pfanzagl, J. (1993). A case of asymptotic equivalence between conditional and marginal likelihood estimators. *Journal of Statistical Planning and Inference* **35**, 301–307.

Pfanzagl, J. (1994a). On item parameter estimation in certain latent trait models. In: Fischer, G.H., Laming, D. (Eds.), *Contributions to Mathematical Psychology, Psychometrics, and Methodology*. Springer-Verlag, New York, pp. 249–263.

Pfanzagl, J. (1994b). On the identifiability of structural parameters in mixtures: Applications to psychological tests. *Journal of Statistical Planning and Inference* **38**, 309–326.

Piswanger, K. (1975). Interkulturelle Vergleiche mit dem Matrizentest von Formann (Cross-cultural comparisons using the matrices test of Formann). Doctoral dissertation. University of Vienna, Vienna. In German.

Ponocny, I. (2000). Exact person fit indexes for the Rasch model for arbitrary alternatives. *Psychometrika* **65**, 29–42.

Ponocny, I. (2001). Non-parametric goodness-of-fit tests for the Rasch model. *Psychometrika* **66**, 437–460.

Ponocny, I., Ponocny-Seliger, E. (1999). *T-Rasch 1.0*. ProGAMMA, Groningen.

Rao, C.R. (1973). *Linear Statistical Inference and its Applications*, 2nd ed. John Wiley and Sons, New York.

Rao, A.R., Jana, R., Bandyopadhyay, S. (1996). *Sankhyā: The Indian Journal of Statistics, Series A* **58**, 225–242.

Rasch, G. (1960). *Probabilistic Models for Some Intelligence and Attainment Tests*. Pædagogiske Institut, Copenhagen. (Expanded edition 1980[2] with a foreword and afterword by B. Wright. MESA Press, Chicago.)

Rasch, G. (1961). On general laws and the meaning of measurement in psychology. In: Neyman, J. (Ed.), *Proceedings of the IV Berkeley Symposium on Mathematical Statistics and Probability*. Berkeley University Press, Berkeley, CA, pp. 321–333.

Rasch, G. (1965). *Statistiske Seminar* (Statistical Seminar). (Notes taken by J. Stene.) Department of Statistics, University of Copenhagen, Copenhagen. In Danish.

Rasch, G. (1967). An informal report an a theory of objectivity in comparisons. In: van der Kamp, L.J.Th., Vlek, C.A.J. (Eds.), *Measurement Theory. Proceedings of the NUFFIC International Summer Session in Science in "Het Oude Hof"*. The Hague, July 14–28, 1966. University of Leyden, Leyden.

Rasch, G. (1968). A mathematical theory of objectivity and its consequences for model construction. Paper presented at the "European Meeting on Statistics, Econometrics and Management Science, Amsterdam, The Netherlands, September 2–7, 1968".

Rasch, G. (1972). Objectivitet i samfundsvidenskaberne et metodeproblem (Objectivity in the social sciences as a methodological problem). *Nationaløkonomisk Tidsskrift* **110**, 161–196. In Danish.

Rasch, G. (1977). On specific objectivity. An attempt at formalizing the request for generality and validity of scientific statements. In: Blegvad, M. (Ed.), *The Danish Yearbook of Philosophy*. Munksgaard, Copenhagen, pp. 58–94.

Raven Jr., J., Raven, J.C., Court, J.H. (2000). *Standard Progressive Matrices*. Manual for Raven's Progressive Matrices and Vocabulary Scales, Section 3. Harcourt Assessment, San Antonio, TX. Updated 2004.

Rijmen, F., De Boeck, P. (2002). The random weights linear logistic test model. *Applied Psychological Measurement* **26**, 269–283.

Rijmen, F., Tuerlinckx, F., De Boeck, P., Kuppens, P. (2003). A nonlinear mixed model framework for item response theory. *Psychological Methods* **8**, 185–205.

Roskam, E.E., Jansen, P.G.W. (1984). A new derivation of the Rasch model. In: Degreef, E., van Buggenhaut, J. (Eds.), *Trends in Mathematical Psychology*. Elsevier Science, Amsterdam, pp. 293–307.

Rost, J. (1990). Rasch models in latent classes: An integration of two approaches to item analysis. *Applied Psychological Measurement* **3**, 271–282.

Rost, J. (1997). Logistic mixture models. In: Van der Linden, W.J., Hambleton, R.K. (Eds.), *Handbook of Modern Item Response Theory*. Springer-Verlag, New York, pp. 449–463.

Rost, J. (2001). The growing family of Rasch models. In: Boomsma, A., van Duijn, M.A.J., Snijders, T.A.B. (Eds.), *Essays on Item Response Theory, Lecture Notes in Statistics*, vol. 157. Springer-Verlag, New York, pp. 25–42.

Rost, J., Von Davier, M. (1995). Mixture distribution Rasch models. In: Fischer, G.H., Molenaar, I.W. (Eds.), *Rasch Models. Foundations, Recent Developments, and Applications*. Springer-Verlag, New York, pp. 257–268.

Rubin, D.B. (1987). *Multiple Imputation for Nonresponse in Surveys*. Wiley, New York.

Ryser, H.J. (1963). *Combinatorial Mathematics. Carus Mathematical Monographs*. The Mathematical Association of America.

Scheiblechner, H. (1972). Das Lernen und Lösen komplexer Denkaufgaben (The learning and solving of complex reasoning tasks). *Zeitschrift für experimentelle und angewandte Psychologie* **3**, 456–506. In German.

Scheiblechner, H. (2003). Nonparametric IRT: Testing the bi-isotonicity of isotonic probabilistic models (ISOP). *Psychometrika* **68**, 79–96.

Seliger, E. (1993). *Analyse von Buchstabenreihen mittels linear logistischer Testmodelle* (The analysis of letter series by means of linear logistic test models). Master's Thesis. University of Vienna, Vienna. In German.

Sijtsma, K. (1998). Methodology review: Nonparametric IRT approaches to the analysis of dichotomous item scores. *Applied Psychological Measurement* **22**, 3–31.

Sijtsma, K., van der Ark, L.A. (2001). Progress in NIRT analysis of polytomous item scores: Dilemmas and practical solutions. In: Boomsma, A., van Duijn, A.L. (Eds.), *Essays on Item Response Theory*. Springer-Verlag, New York, pp. 297–318.

Skrondal, A., Rabe-Hesketh, S. (2004). *Generalized Latent Variable Modeling*. Chapman & Hall, New York.

Snijders, T.A.B. (1991). Enumeration and simulation methods for 0-1 matrices with given marginals. *Psychometrika* **56**, 397–417.

Spiel, C., Glück, J. (1998). Item response models for assessing change in dichotomous items. *International Journal of Behavior Development* **22**, 517–536.

Stelzl, I. (1979). Ist der Modelltest des Rasch-Modells geeignet, Homogenitätshypothesen zu überprüfen? Ein Bericht über Simulationsstudien mit inhomogenen Daten (Is the test of fit of the Rasch model suited for testing homogeneity hypotheses? A report on simulation studies based on inhomogeneous data). *Zeitschrift für Experimentelle und Angewandte Psychologie* **26**, 652–672.

Stout, W.F. (1990). A new item response theory approach with applications to unidimensionality assessment and ability estimation. *Psychometrika* **55**, 293–325.

Tanzer, N.K., Gittler, G., Ellis, B.B. (1995). Cross-cultural validation of item complexity in an LLTM-calibrated spatial ability test. *European Journal of Psychological Assessment* **11**, 170–183.

Thissen, D. (1982). Marginal maximum likelihood estimation for the one-parameter logistic model. *Psychometrika* **47**, 175–186.

Tjur, T. (1982). A connection between Rasch's item analysis model and a multiplicative Poisson model. *Scandinavian Journal of Statistics* **9**, 23–30.

Tuerlinckx, F., Rijmen, F., Molenberghs, G., Verbeke, G., Briggs, D., Van den Noorgate, W., Meulders, M., De Boeck, P. (2004). Estimation and Software. In: De Boeck, P., Wilson, M. (Eds.), *Explanatory Item Response Models. A Generalized Linear and Nonlinear Approach*. Springer-Verlag, New York, pp. 343–373.

Tutz, G. (1989). *Latent Trait-Modelle für ordinale Beobachtungen* (Latent trait models for ordinal data). *Lehr und Forschungstexte Psychologie*, vol. 30. Springer, Berlin.

Van de Vijver, F.J.R. (1988). Systematizing item content in test design. In: Langeheine, R., Rost, J. (Eds.), *Latent Trait and Latent Class Codels*. Plenum, New York, pp. 291–307.

Verhelst, N.D., Glas, C.A.W. (1993). A dynamic generalization of the Rasch model. *Psychometrika* **58**, 395–415.

Verhelst, N.D., Glas, C.A.W., Verstralen, H.H.F.M. (1994). *OPLM: Computer Program and Manual*. CITO, Arnhem.

Verhelst, N.D., Glas, C.A.W. (1995). The One-Parameter Logistic Model. In: Fischer, G.H., Molenaar, I.W. (Eds.), *Rasch Models. Foundations, Recent Developments, and Applications*. Springer-Verlag, New York, pp. 215–237.

Verhelst, N.D., Veldhuijzen, N.H. (1991). A new algorithm for computing elementary symmetric functions and their first and second derivatives. Measurement and Research Department Reports, 91–1. CITO, Arnhem.

Von Davier, M. (1998). *WINMIRA: A Windows Program for Mixed Rasch Models*. Computer software. University of Kiel, Institute for Science Education, Kiel.

Warm, T.A. (1989). Weighted likelihood estimation of ability in item response models. *Psychometrika* **54**, 427–450.

Whitely, S.E., Schneider, L.M. (1981). Information structure for geometric analogies: A test theory approach. *Applied Psychological Measurement* **5**, 383–397.

Wingersky, M.S., Barton, M.A., Lord, F.M. (1982). *LOGIST User's Guide*. Educational Testing Service, Princeton, NJ.

Wilson, M., De Boeck, P. (2004). Descriptive and explanatory item response models. In: De Boeck, P., Wilson, M. (Eds.), *Explanatory Item Response Models*. Springer, New York, pp. 43–74.

Witting, H. (1978). *Mathematische Statistik* (Mathematical statistics). Teubner, Stuttgart. In German.

Wright, B.D. (1980). Foreword. In: Rasch, G. (Ed.), *Probabilistic Models for Some Intelligence and Attainment Tests*. MESA Press, Chicago, pp. ix–xix.

Wright, B.D., Panchapakesan, N. (1969). A procedure for sample free item analysis. *Educational and Psychological Measurement* **29**, 23–48.

Wu, M.L., Adams, R.J., Wilson, M.R. (1998). *ACER ConQest: Multi-Aspect Software*. Australian Council for Educational Research, Camberwell, Victoria.

Zermelo, E. (1929). Die Berechnung der Turnierergebnisse als ein Maximumproblem der Wahrscheinlichkeitsrechnung (The computation of tournament results as a maximum-probability problem). *Mathematische Zeitschrift* **29**, 436–460. In German.

Zimowski, M., Muraki, E., Mislevy, R.J., Bock, R.D. (1996). *BILOG-MG: Multiple Group Item Analysis and Test Scoring*. Scientific Software, Chicago.

Zwinderman, A.H. (1995). Pairwise parameter estimation in Rasch models. *Applied Psychological Measurement* **19**, 369–375.

Handbook of Statistics, Vol. 26
ISSN: 0169-7161

DOI: 10.1016/S0169-7161(06)26017-6

17

Hierarchical Item Response Theory Models

Matthew S. Johnson, Sandip Sinharay and Eric T. Bradlow

1. Introduction

Item response theory (IRT) models (see Chapter 15 by Bock and Moustaki in this volume) are a class of statistical models used by researchers to describe the response behaviors of individuals to a set of categorically scored items. The most common IRT models can be classified as generalized linear fixed- and/or mixed-effect models. Although IRT models appear most often in the educational testing literature, researchers in other fields have successfully utilized IRT models in a wide variety of applications. For example, Rice (2004) applied an IRT model in a matched case-control study and Bradlow and Zaslavsky (1999) applied IRT models to customer satisfaction survey data. Fienberg et al. (1999) utilized an IRT model to estimate population sizes from multiple non-independent captures in a multiple capture–recapture setting.

To formalize the IRT models, let the random variable X_{ij} denote the ordered categorical response of individual $i \in \{1, \ldots, N\}$ to item $j \in \{1, \ldots, J\}$. Further let $P_{jm}(\theta_i) \equiv \Pr\{X_{ij} = m \mid \theta_i\}$, denote the mth *category response function* for item j, $m \in \{1, \ldots, n_j\}$. When item j is dichotomous the function $P_j(\theta) = P_{j1}(\theta)$ is called the *item response function* (IRF) for item j.

A number of IRT models exist in the statistics and psychometrics literature for the analysis of multiple discrete responses. The models typically rely on the following assumptions:

- *Unidimensionality (U)*: There is a one-dimensional, unknown quantity associated with each respondent in the sample that describes the individual's propensity to endorse the items in the survey (or exam). Let θ_i denote the propensity of individual i.
- *Conditional Independence (CI)*: Given an individual's propensity θ_i, the elements of the item response vector for respondent i, $\boldsymbol{X}_i = (X_{i1}, \ldots, X_{iJ_i})^{\mathrm{t}}$, are independent.
- *Monotonicity (M)*: $\Pr\{X_{ij} > t \mid \theta_i\}$ is a non-decreasing function of an individual's propensity θ_i, for all j and all t. Respondents with high propensities are more likely to endorse items than those with low propensities.

In educational testing psychometricians often refer to the propensity θ_i as the latent ability, or proficiency of individual i.

IRT analyses typically have at least one of the following two goals:[1] (a) to estimate the set of category response functions, denoted $\mathcal{P} = (P_{11}, \ldots, P_{Jk_J})'$; (b) to estimate the propensities $\boldsymbol{\theta} = (\theta_1, \theta_2, \ldots, \theta_N)'$. Simple item response analyses make conditional independence assumptions in order to define the likelihood function. A Bayesian analysis assumes that the item responses are independent conditional on the item response functions $\mathcal{P}$ and the vector of propensities $\boldsymbol{\theta}$. That is,

$$\Pr(\boldsymbol{X} = \boldsymbol{x} \mid \mathcal{P}, \boldsymbol{\theta}) = \prod_i \prod_j \Pr(X_{ij} = x_{ij} \mid \theta_i, \boldsymbol{P}_j) = \prod_i \prod_j P_{jx_{ij}}(\theta_i), \tag{1}$$

where $\boldsymbol{P}_j = (P_{j1}, \ldots, P_{jk_j})'$ denotes the set of category response functions for item j. The Bayesian treats both the response functions and the propensities as random variables, so it makes sense to talk about conditional distributions. Frequentist item response analyses typically treat the response functions as *fixed* but unknown, and treat the propensities as nuisance parameters. Although not required, we focus on the Bayesian formulation of hierarchical IRT models below. In our opinion, the Bayesian approach has the advantage that it allows for exact finite sample inferences.

Under the Bayesian paradigm prior distributions must be placed on the set of model parameters in addition to defining the data model or likelihood in (1). Let $\pi(\boldsymbol{\theta}, \mathcal{P})$ denote the joint prior distribution function on the model parameters. Simple IRT analyses typically assume that individuals are a priori independent and identically distributed (iid), the items are a priori iid, and individuals and items are a priori independent, in which case the joint prior can be written as

$$\pi(\boldsymbol{\theta}, \mathcal{P}) = \prod_i \pi_\Theta(\theta_i) \prod_j \pi_{\mathcal{P}}(\boldsymbol{P}_j). \tag{2}$$

In many real-world problems the iid assumptions placed on the model parameters are unreasonable. Some examples where the assumptions of iid propensities fail include:

- In assessments where students are clustered within school, students from the same school are likely to have related propensities.
- Students with similar background characteristics are going to have similar propensities, e.g., students whose parents finished college will tend to have higher propensities than students whose parents did not finish high school.

Examples of when the assumption of independent response functions fail include:

- When questions are created using templates. The response functions for items created from the same template may be statistically related to each other.
- When multiple raters are used to score items, we can treat each item×rater combination as a pseudo-item. Clearly, it is reasonable to assume that each pseudo-item constructed from the same item (rater) is related. That is, the response functions for the pseudo-items will be clustered by item and by rater.

The prior distribution $\pi(\boldsymbol{\theta}, \mathcal{P})$ must incorporate the structure/context of a given test and we describe a number of these approaches next.

[1] Note that other analyses may not have as their final goal either of these two, but rather a function of either (a) or (b); e.g., the probability that $\theta_i > c$ where c is a passing score.

One way to incorporate dependence into the model is with the use of a hierarchical Bayes model, which decomposes the prior distribution into conditional distributions

$$\pi_1(\boldsymbol{\theta}, \mathcal{P} \mid \eta_1), \quad \pi_2(\eta_1 \mid \eta_2), \quad \ldots, \quad \pi_m(\eta_{m-1} \mid \eta_m),$$

and a marginal distribution $\pi_{m+1}(\eta_m)$ such that

$$\pi(\boldsymbol{\theta}, \mathcal{P}) = \int_{\eta_1 \times \cdots \times \eta_m} \mathrm{d}\pi_1(\boldsymbol{\theta}, \mathcal{P} \mid \eta_1) \cdots \mathrm{d}\pi_m(\eta_{m-1} \mid \eta_m)\, \mathrm{d}\pi_{m+1}(\eta_m).$$

We call η_k the hyperparameters at level k.

Although the decomposition of the prior distribution $\pi(\boldsymbol{\theta}, \mathcal{P})$ described above may be theoretically unnecessary, there may be a number of practical reasons for constructing the prior distribution in this fashion. Reasons for decomposing the prior distribution with a hierarchical model include:

- There is a substantive reason for modeling the parameters in a hierarchical fashion. For example, in the *National Assessment of Educational Progress* (NAEP, see Chapter 32 by von Davier, Sinharay, Beaton, and Oranje in this volume), schools are sampled from primary sampling units and students are sampled within schools. A simple way to incorporate this clustered structure is:

  $$\theta_i \sim N\big(\mu_{s(i)}, \sigma^2\big), \qquad \mu_s \sim N\big(\nu_{p(s)}, \tau^2\big), \qquad \nu_p \sim N\big(0, \omega^2\big),$$

  where $s(i)$ is individual i's school and $p(s)$ is the primary sampling unit containing school s.
- A hierarchical model can be used to borrow information across observational units (individuals or items), thereby making the results more robust. Suppose an examination is made up of several five-option multiple choice items, and that a three-parameter logistic (3PL) model is used for analysis of the data, that is, $P_j(\theta) = c_j + (1 - c_j)\,\mathrm{logit}^{-1}(a_j(\theta - b_j))$. In most applications there is little information about *guessing parameter* c_j as compared to the other parameters. A non-hierarchical analysis might assume a flat normal prior on the logit of the guessing parameter, e.g., $\mathrm{logit}(c_j) \sim N(\mathrm{logit}(1/5), 100)$, while a hierarchical analysis might assume $\mathrm{logit}(c_j) \sim N(\lambda, \zeta^2)$, $\lambda \sim N(\mathrm{logit}(1/5), 100)$, $\zeta^2 \sim \chi^2_1$. Both the hierarchical and non-hierarchical priors have the same prior mean and are relatively flat. However, the hierarchical prior does not allow guessing parameters to vary as far from the population average across items and results in more stable estimates.

Several studies (e.g., Adams et al., 1997; Kamata, 2001; Van den Noorgate et al., 2003) have noted that commonly used IRT models can be formulated as multilevel logistic models.

2. Developing the hierarchical IRT model

Spiegelhalter et al. (1996) identify three steps for the construction of hierarchical Bayesian models. The steps are summarized below:

- The first step defines all model quantities, both observed and unobserved, and describes how those model quantities are related to one another. The relationships are qualitatively defined in terms of the conditional independence structure. This step is called the *structural modeling* step.
- The second step defines the parametric form of the relationships that have been defined in the structural step. This step is called the *probability modeling* step.
- The third step defines the prior distributions for all model parameters. This step is called the *prior modeling* step.

To help clarify the concepts, consider an assessment administered state- or nationwide with respondents having outcome scores $\boldsymbol{X}$. It is quite reasonable to assume that the abilities or outcomes of students from the same school are going to be more closely related than those of students from different schools. That is, students are clustered within school. Describing this clustering would be done in the structural step of the model. If we assume that the distribution of the outcome measures ($\boldsymbol{X}$) is defined by a specific model, e.g., a 3PL model, that would be done as part of the probability modeling step. Finally, the relationship assumed among the model parameters (individual, school-level, etc.) would constitute the prior modeling step. These steps are described below through the use of what are commonly called *directed acyclic graphs* (DAGs; see Anderson, 2003, for a thorough discussion of directed acyclic graphs).

2.1. Structural modeling – The directed acyclic graph

We begin by making the standard IRT assumption that student i's item responses are conditionally independent given the students ability θ_i. However, we must relate the abilities of all students within the same school. One way to model this dependence is to assume the existence of a parameter μ_s (which could, e.g., be a school mean parameter) associated with each school s, such that conditional on μ_s, the students' responses within the school are independent, and that students across different schools are independent; that is, abilities are exchangeable within school. Mathematically we write:

$$\pi(\boldsymbol{\theta} \mid \boldsymbol{\mu}) = \prod_{s=1}^{S} \prod_{i \in \mathcal{I}_s} \pi(\theta_i \mid \mu_s),$$

where S in the number of schools in the sample, $\mathcal{I}_s$ is the index set for examinees in school s, and $\boldsymbol{\mu} = (\mu_1, \ldots, \mu_S)'$.

We could make similar structural assumptions about the data in mathematical form as we have done above for the ability parameters. However, we find it advantageous to describe the structural form of the model graphically. Figure 1 shows the *directed acyclic graph* representing the structural equations for the data. Each model quantity is represented by a node in the graph, and the conditional relationships are defined by the directional edges, which are represented by arrows. For example, the arrow from σ to θ_i in the above figure indicates that σ influences θ_i. Observed quantities (e.g., X_{ij}) are represented by squares and unobserved quantities (e.g., θ_i) are represented by circles.

The class of probability distributions defined by a graph assume that a quantity ν is conditionally independent of all *nondecendent* quantities (i.e., those that can not be

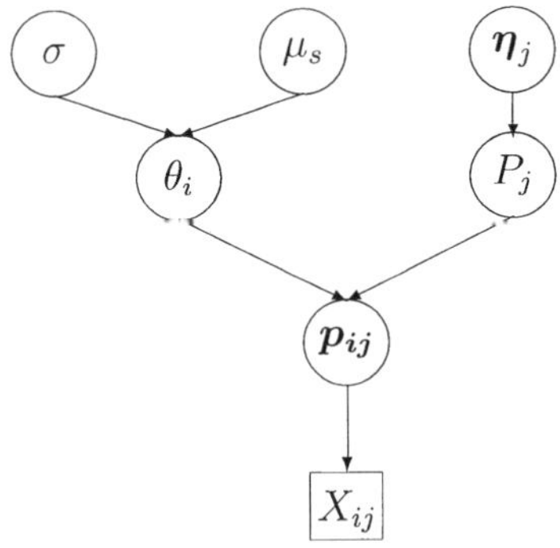

Fig. 1. A directed acyclic graph describing the structural form of the item response data where students (indexed by i) are clustered within schools (indexed by s).

reached by following the directed edges) in the graph given its *parents*, which are defined as all nodes pointing to the node v. For example, the model quantity $\boldsymbol{p}_{ij}$ has two parents, θ_i and P_j. Conditional on these two parent nodes the probability vector $\boldsymbol{p}_{ij}$ is independent of $\boldsymbol{p}_{i'j}$ for all $i' \neq i$, $\boldsymbol{p}_{ij'}$ for all $j' \neq j$, σ, μ_s for all s, and $\boldsymbol{\eta}_j$ for all $j = 1, \ldots, J$.

Figure 1 represents a class of models that assume that the distribution of response X_{ij} of individual i on item j depends only on the probability mass function $\boldsymbol{p}_{ij}$. The model further assumes that the probability mass function is defined by an individual's propensity θ_i and the form of the item response function, denoted P_j. The propensities within a school s are conditionally independent given the school mean μ_s and the within school variance σ^2; the item response function P_j is defined by a parameter vector $\boldsymbol{\eta}_j$.

2.2. *The probability model*

Once all model quantities, and the relationships between these quantities have been described, the second step in constructing a hierarchical model is to define the parametric form of each conditional distribution of each model quantity given its parents.

In the students-within-school example displayed in Figure 1, we must describe the conditional relationships of four child nodes: X_{ij}, p_{ij}, θ_i and P_j given their parents. For the data model we assume that X_{ij} has a discrete distribution defined on the $K_j + 1$ integers $0, \ldots, K_j$; the probability mass function is defined by the vector of probabilities $\boldsymbol{p}_{ij} = (p_{ij0}, \ldots, p_{ijK_j})'$. The vector $\boldsymbol{p}_{ij} = P_j(\theta_i)$, where P_j describes the functional form of an IRT model reasonable for the data set at hand (e.g., the 3PL model for multiple choice data, the GPCM for polytomous data etc.; see, e.g., Chapter 15 by Bock and Moustaki in this book); the vector $\boldsymbol{\eta}_j$ denotes the vector of item parameters for item j. Finally, in order to describe the correlation among abilities within the same school, we could assume that abilities are normally distributed with a mean μ_s which depends on the school for individual i and a variance that is constant across schools; albeit constant variance is not needed and sometimes overly restrictive.

2.3. *The prior model*

The final stage of defining a hierarchical model is where the frequentist and Bayesian formulations of the hierarchical model differ. A frequentist statistician would treat the parentless nodes (e.g., σ, μ_s, $\boldsymbol{\eta}_j$ in Figure 1) as fixed, but unknown parameters; a Bayesian treats the parameters represented by these nodes as random quantities, and therefore must posit a probabilistic model to describe the random nature of the parameters. There is no universally accepted approach for defining prior distributions, but Diaconis and Ylvisaker (1985) discuss three general methodologies for defining prior distribution.

One method, most closely related to the frequentist approach, assumes a *noninformative* prior that should have only a minimal effect on the final results of an analysis. Noninformative priors can be *proper* (i.e., non-negative densities that integrate to unity), or *improper* (i.e., integrate to ∞). Common improper priors include a prior density that is proportional to Lebesgue measure (i.e., $f(\sigma) \propto 1$) and Jeffrey's prior, which forces the posterior to be invariant to one-to-one transformations of any parameter.

The use of improper prior distributions can lead to improper posterior distributions. The only way to be certain that a posterior distribution will be proper is by assuming a proper prior distribution. One popular class of proper prior distributions is the class of *conjugate* priors, which ensure that the posterior (or conditional posterior) distribution of a model parameter will be a member of the same family of distributions as the assumed prior probability model. For example, if we assume that $\theta_i \sim N(\mu_s, \sigma)$, then the conjugate prior distribution on μ_s leads to a normal posterior distribution.

The final way in which prior distributions are specified is by utilizing prior information, or prior knowledge about the problem at hand. For example, researchers may have learned over several years of an assessment that the within-school variance σ^2 is approximately 0.25. A Bayesian analysis of the current assessments data might place a Gamma(0.25, 1) prior on σ^2; the prior mean and variance are both 0.25. Depending on the strength of the prior information, the Gamma distribution could be modified to be either more or less informative.

3. Estimation

3.1. *Maximum marginal likelihood and empirical Bayes*

A non-Bayesian approach for estimating the parameters of an hierarchical Bayes IRT model might leave the prior distributions of some or all of the hyperparameters unspecified. The classical statistician then treats these hyperparameters as fixed and unknown parameters, the same way parameters are treated in non-hierarchical models. For example, assume a hierarchical model with a single hierarchical level describing the variability of the distribution of abilities:

$$X_{ij} \sim \text{Bernoulli}\big(P_j(\theta)\big),$$
$$P_j(\theta) = \big(1 + \exp\{\beta_j - \theta_i\}\big)^{-1},$$

$$\theta_i \sim N(0, \sigma^2).$$

A Bayesian would assume a hyperprior for the level-1 hyperparameter σ^2. The *empirical Bayes* approach treats σ^2 as a fixed but unknown parameter estimated by maximizing the *marginal likelihood* found by integrating the *nuisance parameter* θ_i out of the likelihood. The marginal likelihood for the parentless nodes σ^2 and $\boldsymbol{\beta}$ is

$$L_M(\sigma^2, \boldsymbol{\beta}; \boldsymbol{X}) \propto \prod_{i=1}^{N} \int_{\Theta} \left[\prod_{j=1}^{J} \frac{\exp\{x_{ij}(t - \beta_j)\}}{1 + \exp\{t - \beta_j\}} \right] \exp\{-t^2/2\sigma^2\}\, dt. \tag{3}$$

The marginal likelihood in (3) is maximized with respect to the parentless hyperparameters σ^2 and $\boldsymbol{\beta}$ to derive the *maximum marginal likelihood* (MML) estimates $\hat{\sigma}^2$ and $\hat{\boldsymbol{\beta}}$. However, all descendent parameters (e.g., all θ_i's) have been integrated out of the likelihood, and therefore, cannot be estimated using the likelihood defined in (3).

The empirical Bayes approach takes the MML estimates $\hat{\sigma}^2$ and $\hat{\boldsymbol{\beta}}$ and then treats them as if they were known a priori, thus defining the individual posterior distributions for each observations latent propensity θ_i given $\boldsymbol{\beta} = \hat{\boldsymbol{\beta}}$ and $\sigma^2 = \hat{\sigma}^2$,

$$f(\theta_i \mid \boldsymbol{X} = \boldsymbol{x}_i, \boldsymbol{\beta} = \hat{\boldsymbol{\beta}}, \sigma = \hat{\sigma}) \propto \prod_{j=1}^{J} \frac{\exp\{x_{ij}(\theta_i - \hat{\beta}_j)\}}{1 + \exp\{\theta_i - \hat{\beta}_j\}} \exp\{-\theta_i^2/2\hat{\sigma}^2\}.$$

The empirical Bayes maximum and expected *a posteriori* (MAP, EAP) estimates of θ_i are then defined as the mode and the expected value of the empirical Bayes conditional posterior defined above.

3.2. *Bayesian estimation via Markov chain Monte Carlo*

The goal of Markov chain Monte Carlo (MCMC) methods is to draw a sample of parameter values from the joint posterior distribution. MCMC accomplishes this goal by simulating a random walk process in the parameter space $\boldsymbol{\phi} = (\boldsymbol{\theta}, \mathcal{P}, \eta_1, \eta_2, \ldots, \eta_m)$ that converges to the joint posterior distribution. The following sections review the two most popular MCMC algorithms encountered in the psychometric literature, the *Gibbs sampler* and the *Metropolis–Hastings algorithm*.

3.2.1. *Gibbs sampler* (*Geman and Geman,* 1984; *Gelman et al.,* 2003)

The parameter vector $\boldsymbol{\phi}$ is split into d components, or sub-vectors, $\boldsymbol{\phi} = (\phi_1, \ldots, \phi_d)$. For each iteration of the simulation, the d sub-vectors of $\boldsymbol{\phi}$ are simulated from the posterior distribution of the sub-vector conditioned on the values of the other $d - 1$ sub-vectors. The following pseudo-code describes the process to simulate T steps from the Gibbs sampler:

For $t = 1, 2, \ldots, T$

For $s = 1, \ldots, d$

Sample $\phi_s^{(t)} \sim f(\phi_s \mid \boldsymbol{\phi}_{-s}^{(t-1)}, \boldsymbol{X})$, where $\boldsymbol{\phi}_{-s}^{(t-1)}$ represents all sub-vectors of the parameter vector, except for sub-vector ϕ_s, at their current value.

$$\boldsymbol{\phi}_{-s}^{(t-1)} = \left(\phi_1^{(t)}, \ldots, \phi_{s-1}^{(t)}, \phi_{s+1}^{(t-1)}, \ldots, \phi_d^{(t-1)}\right).$$

Because the distribution of $\boldsymbol{\phi}^{(t+1)}$ given the current state of the parameter vector $\boldsymbol{\phi}^{(t)}$ does not depend further on the history of the sequence $\{\boldsymbol{\phi}^{(1)}, \ldots, \boldsymbol{\phi}^{(t-1)}\}$, the sequence $\{\boldsymbol{\phi}^{(1)}, \boldsymbol{\phi}^{(2)}, \ldots\}$ is a Markov chain. As the length of the chain (T) increases, the distribution of parameter vectors $\boldsymbol{\phi}^{(t)}$ converges to the posterior distribution of $\boldsymbol{\phi}$ given the observed data $\boldsymbol{X}$.

The Gibbs sampler requires a method to draw the parameter vector $\boldsymbol{\phi}_s$ from the conditional posterior $f(\boldsymbol{\phi}_s \mid \boldsymbol{\phi}_{-s}^{(t-1)}, \boldsymbol{X})$. In some cases, this can be done directly. For instances where simulation from the conditional posterior is not straightforward, an alternative method is used. One alternative, discussed here, is the *Metropolis–Hastings* algorithm.

3.2.2. *Metropolis–Hastings algorithm within Gibbs* (*Metropolis and Ulam,* 1949; *Metropolis et al.,* 1953; *Hastings,* 1970; *Chib and Greenberg,* 1995)

Given a target distribution $G(\boldsymbol{\psi} \mid \boldsymbol{X})$ (e.g., the conditional posterior distribution in the Gibbs sampler), which can be computed up to a normalizing constant, a sequence of random values can be simulated whose distribution converges to the target distribution using the following algorithm.

For $t = 1, \ldots, T$

1. Simulate a candidate ψ^* from a candidate distribution that depends on the previous value of the parameter.

$$\psi^* \sim J_t\left(\psi^* \mid \psi^{(t-1)}\right).$$

2. Calculate the acceptance probability

$$\alpha = \frac{g(\psi^* \mid \boldsymbol{X})/J_t(\psi^* \mid \psi^{(t-1)})}{g(\psi^{(t-1)} \mid \boldsymbol{X})/J_t(\psi^{(t-1)} \mid \psi^*)}.$$

3. Set

$$\psi^{(t)} = \begin{cases} \psi^* & \text{with probability } \min(1, \alpha), \\ \psi^{(t-1)} & \text{otherwise.} \end{cases}$$

Note that this particular instance, where a given draw is centered at the previous value, is known as a random-walk Metropolis algorithm. Several authors have utilized MCMC methods for the analysis of item response theory models (see, for example, Albert, 1992; Albert and Chib, 1993; Patz and Junker, 1999; Johnson et al., 1999; Fox and Glas, 2001, 2003). In the sections that follow we discuss three hierarchical models that have been suggested for different situations encountered in testing. Section 4.1 discusses the hierarchical rater model suggested by Patz (1996) and Patz et al. (2002); Section 4.2 reviews the testlet model (Bradlow et al., 1999); and Section 4.3 discusses the related siblings model of Glas and van der Linden (2003) and Johnson and Sinharay (2005).

Although Bayesian analyses and MCMC algorithms are extremely flexible in terms of the types of models that can be estimated, the procedures are not without their difficulties. One of the shortcomings is that MCMC procedures are specialized numerical integration techniques, which can be quite time consuming for large data sets, and require extreme care to make sure that the Markov chain has converged to the posterior distribution of interest (only then are the resulting estimates acceptable). For more on monitoring convergence of MCMC algorithms see, e.g., Sinharay (2004) and the references therein.

4. Examples

4.1. Hierarchical rater model

Consider the multiple ratings problem, where each (subjectively scored) item response by an examinee has been evaluated by more than one rater. Let X_{ijr} denote the score assigned by rater r to individual i's response to item j for $i = 1, \dots, N$, $j = 1, \dots, J$, $r = 1, \dots, R$. For example, consider a writing assessment where students write three short essays ($j = 1, 2, 3$), where each essay is evaluated by four raters ($r = 1, 2, 3, 4$).

One way to treat the problem is to treat each item by rater combination as a separate pseudo-item, and then perform standard item response theory analyses. Let $P_{(jr)k}(\theta_i) \equiv \Pr\{X_{ijr} = k \mid \theta_i\}$, where the subscript (jr) denotes that the item by rater combination is treated as a pseudo-item. Linacre's FACETS model (Linacre, 1994) and the linear logistic test model (LLTM; Fischer, 1973, 1983) both model the response functions $P_{(jr)k}$ with standard IRT models like the partial credit model of Masters (1982). The approach suffers from the assumption of conditional independence of the item–rater combinations given the ability of an examinee. In essence, the pseudo-item approach presumes that an examinee's ability can be perfectly estimated with a single item response rated by an infinite number of raters (Mariano, 2002).

The shortcomings of the FACETS and LLTM approaches can be traced to the fact that the vector of category response functions $\boldsymbol{P}_{(jr)}$ (and therefore item responses) for the item–rater pseudo-items are assumed to be independent within item and across raters (i.e., $\boldsymbol{P}_{(jr)}$ and $\boldsymbol{P}_{(js)}$ are independent) and independent within rater and across items (i.e., $\boldsymbol{P}_{(jr)}$ and $\boldsymbol{P}_{(mr)}$ are independent). Typically raters are given a scoring rubric for each item they rate, and therefore, assuming independence across raters for the same item is unreasonable.

In most situations it is more reasonable to assume that the distribution of the random variable X_{ijr} is determined by three components: (1) the ability of the student θ_i; (2) characteristics of the item $\boldsymbol{\eta}_j$; and (3) the behavior of the rater $\boldsymbol{\lambda}_r$. The directed acyclic graph in Figure 2 demonstrates the structural assumptions of the general hierarchical rater model. Conditional on the examinee's ability θ_i, the scores received by the examinee on two different items rated by two different raters is independent, that is, X_{ijr} is conditionally independent of $X_{ij'r'}$ given θ_i as long as $j' \neq j$ and $r' \neq r$. The ratings of a single rater are conditionally independent across items and examinees given $\boldsymbol{\lambda}_r$.

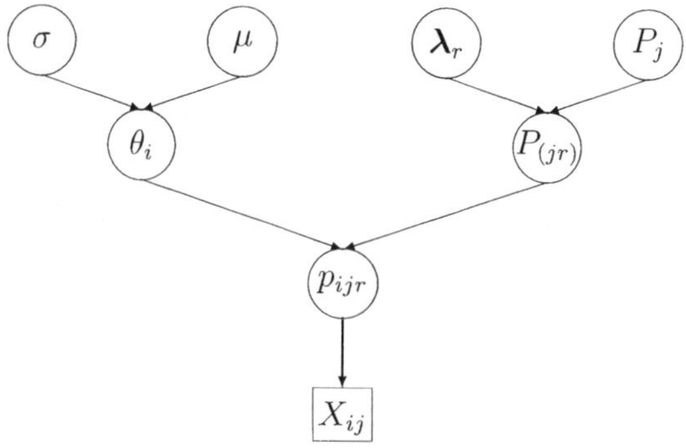

Fig. 2. The directed acyclic graph for the general hierarchical rater model, where examinee $i = 1, \dots, N$ is being rated by rater $r = 1, \dots, R$ on item $j = 1, \dots, J$.

Patz (1996) introduces one such model, which he calls the *hierarchical rater model* (HRM; see also Patz and Junker, 1999; Patz et al., 2002). The HRM is a latent response model (Maris, 1995) that treats the actual responses of examinees to items as latent. The HRM assumes that the latent responses, denoted ξ_{ij}, are conditionally independent across items given the ability θ_i, and models the responses with a standard IRT model. Patz et al. (2002), for example, utilize the partial credit model (Masters, 1982)

$$\operatorname{logit} \Pr\bigl\{\xi_{ij} = \ell \mid \theta_i, \xi_{ij} \in \{\ell, \ell - 1\}\bigr\} = \theta_i - \beta_j - \gamma_{j\ell}. \tag{4}$$

A similar model developed by Verhelst and Verstralen (2001) assumes that the latent responses are continuous random variables.

The HRM goes on to assume that conditional on the latent response ξ_{ij}, the observed ratings X_{ijr} are independent of the abilities θ_i and item characteristics $\boldsymbol{\eta}_j$. The distribution of the observed ratings X_{ijr} is defined in terms of the latent responses ξ_{ij} and rater characteristics $\boldsymbol{\lambda}_r$. Patz et al. (2002) compare the conditional probabilities $\lambda_{\ell kr} \equiv Pr\{X_{ijr} = k \mid \xi_{ij} = \ell, \boldsymbol{\lambda}_r\}$ to those used in discrete signal-detection problems. The conditional probabilities are defined in terms of a matrix of transition probabilities. For example, if a given item is scored on a three-point scale, then we would have the following transition matrix

		Observed rating (k)		
		0	1	2
Ideal	0	λ_{00r}	λ_{01r}	λ_{02r}
rating	1	λ_{10r}	λ_{11r}	λ_{12r}
(ξ)	2	λ_{20r}	λ_{21r}	λ_{22r}

(5)

Each row of the transition matrix in (5) represents a discrete probability distribution. Therefore, $\sum_{k=0}^{2} \lambda_{\xi kr} = 1$ for all ξ and r. Furthermore, in most cases it is safe to assume that the probability distributions in (5) are stochastically ordered by the ideal rating ξ, that is,

$$\Pr\{X_{ijr} \geqslant c \mid \xi_{ij} = t\} \geqslant \Pr\{X_{ijr} \geqslant c \mid \xi_{ij} = s\}$$

for all $t \geqslant s$ and all c. In this setup the transition probabilities are constant across all items, but there could be situations where there is a rater by item interaction, in which case the transition probabilities would also be indexed by the item j.

Patz et al. (2002) model the transition probabilities $\lambda_{\xi kr}$ with the following two parameter model based on the kernel of the normal density:

$$\lambda_{\xi kr} = P[X_{ijr} = k \mid \xi_{ij} = \xi] \propto \exp\left\{-\frac{1}{2\psi_r^2}\left[k - (\xi + \phi_r)\right]^2\right\},$$

$$i = 1, \ldots, N, \quad j = 1, \ldots, J; \quad r = 1, \ldots, R. \tag{6}$$

The rater characteristic $\boldsymbol{\lambda}_r$ is now defined by two parameters ϕ_r and ψ_r. The parameter ϕ_r describes the severity of rater r. The transition probabilities are maximal for the category closest to $\xi + \phi_r$; when the severity parameter is zero ($\phi_r = 0$), the maximal transition probability is $\lambda_{\xi\xi r}$; when $\phi_r < -0.5$, the maximal transition probability is for some category $k \leqslant \xi$, and when $\phi_r > 0.5$ the maximal transition probability is for some category $k \geqslant \xi$. The parameter ψ_r is a measure of how inconsistent the rater is; a rater with a large value of ψ_r is more likely to give two individuals with the same ideal score ξ different scores, than would be a rater with a small inconsistency parameter.

Patz et al. (2002) describe a Markov chain Monte Carlo algorithm for the estimation of the item parameters $\boldsymbol{\beta}$, and $\boldsymbol{\gamma}$, the rater shift parameters $\boldsymbol{\phi}$, rater scale parameters $\boldsymbol{\psi}$, and hyper-parameters of the latent ability distribution. The estimation procedure *augments* (see, for example, Tanner and Wong, 1987) the observed data with the latent response, or "ideal" rating ξ_{ij} of examinee i's response to item j by drawing the ξ_{ij} from their complete conditional posterior distribution. Given the ξ_{ij}, the MCMC estimation procedure for the item parameters $\boldsymbol{\beta}$ and the examinee parameters $\boldsymbol{\theta}$ is straightforward (see, for example, Patz and Junker, 1999). To update the the rater shift and scale parameters ϕ_r and ψ_r, the Patz et al. (2002) algorithm compares the ideal ratings to the ratings given by rater r across all items and subjects rated by the rater. Patz et al. (2002) discuss simulated and real data applications of the model.

4.2. *Testlet model*

Another form of educational data that may violate the conditional independence assumption is a test comprised of *testlets*. A *testlet* is a subset of items generated from a single stimulus. For many testlets, it is difficult to construct examination questions that are conditionally independent given the latent propensity θ_i being measured by the complete test. For example, consider a reading comprehension passage which is the source for five or six different items. Or, for example, consider a mathematical word problem that is followed by a series of questions. In this case, (mis)understanding of the passage or of the text is likely to lead to more (in)correct answers than otherwise expected.

To incorporate this sort of local dependence structure, Bradlow et al. (1999) and Wang et al. (2002) develop the hierarchical Bayes *testlet* model. The structural assumptions for the testlet model are represented by the DAG in Figure 3.

The core assumption to the testlet model is that each testlet d is essentially a separate mini-test constructed to measure the latent propensity $\gamma_{id(j)}$ (item j is a member of testlet $d(j)$); conditional on $\gamma_{id(j)}$, the responses to the items within testlet d are independent. The model further assumes that the testlet propensities $\gamma_{id(j)}$ are conditionally

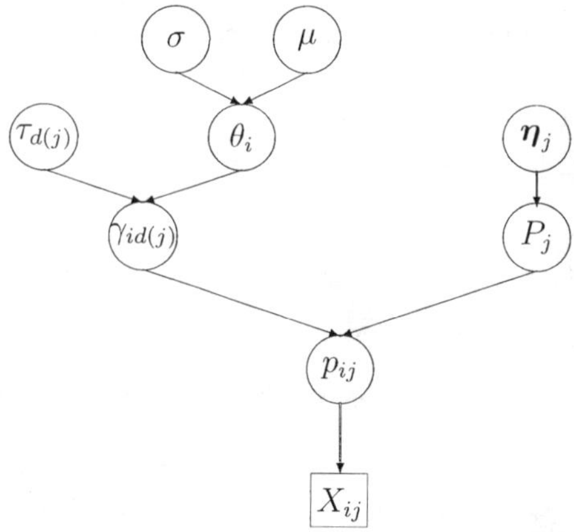

Fig. 3. The directed acyclic graph for the testlet model.

independent across testlets given the overall ability of interest θ_i, which implies that item responses across testlets are conditionally independent given the overall ability θ_i.

In general, any parametric IRT model can be utilized to define the item response function P_j. For example, for multiple choice items we could assume as in Wang et al. (2002), the three-parameter logistic model

$$P(X_{ij} = 1) = c_j + \frac{1 - c_j}{1 + \exp\{-a_j(\gamma_{id(j)} - b_j)\}}. \tag{7}$$

The probability model and prior model assumptions for the Bayesian testlet model have gone through a number of "upgrades" from its initial inception in Bradlow et al. (1999) to current practice. In the initial work, the assumption was that the testlet propensities were normally distributed with mean θ_i and variance τ^2, i.e., $\gamma_{id(j)} \sim N(\theta_i, \tau^2)$. Further, it was assumed that $\theta_i \sim N(\mu, \sigma^2)$. Bradlow et al. (1999) also assumed independent normal priors on $(\log(a_j), b_j, \text{logit}(c_j))'$, and that $\mu = 0$ and $\sigma^2 = 1$ to constrain the location and the scale of the problem. However, this model was found to be too restrictive in three important ways.

First, the assumption that $\tau^2_{d(j)} = \tau^2$ for all $d(j)$ needed to be extended. That is, some testlets exhibit more testlet dependence than others; hence, this was a crucial upgrade to Bradlow et al. (1999) and is first contained in Wainer et al. (2000). Second, assuming independent priors on $(\log(a_j), b_j, \text{logit}(c_j))'$ is also restrictive and was extended in Wang et al. (2004) to allow for a general multivariate normal prior with mean and covariance matrix. The final extension that was done, not to improve model fit but rather to aid in the scientific understanding of testlet models, was to allow for covariates to enter naturally at each stage of the model. That is, Wang et al. (2004) incorporated covariates to explain why some respondents have higher or lower proficiencies, why some items are easier, guessed more often, etc., and most importantly why some testlets have higher within testlet dependence than others.

With this setup, test developers can examine the ratio of the overall ability variation to the total testlet ability variation (i.e., $\sigma^2/(\sigma^2 + \tau^2_{d(j)})$) to determine the severity of local dependence within the testlets. A decision could be made as to whether to correct the items to reduce the amount of local dependence, or to simply use the estimates of the overall abilities (θ_i) from this hierarchical model.

The hierarchical testlet model as given in Bradlow et al. (1999) and Wang et al. (2002) can be implemented via an MCMC algorithm in a number of different ways; however, as in the HRM, the conditional distributions are not easily sampled from. The variety of approaches that can be used include Metropolis–Hastings within Gibbs or data augmentation in which the latent linear predictor $t_{ij} = a_j(\theta_i - b_j - \gamma_{id(j)})$ is sampled within the Markov chain, which then leads to the other conditional distributions mostly being in closed-form.

4.3. *Related siblings model*

4.3.1. *Description of the model*

Item models (e.g., Sinharay et al., 2003) are classes of models from which it is possible to generate/produce items that are equivalent/isomorphic to other items from the same model; they have the potential to greatly reduce costs regarding item-writing. The Graduate Record Examination (GRE), one of the most well-known large-scale tests, recently pretested a number of items automatically generated from item models in its quantitative section. The major goal of the initiative is to be able to calibrate an item model once, and then to use the items generated from it to score examinees on future tests without the need to calibrate the items individually. A simple way to perform that is to assume the same item parameters for all the items from an item model (e.g., Glas and van der Linden, 2003); however, this approach ignores the variance between the item parameters from the same item model. Conceptually, a superior model is the "Related siblings model" (RSM, Glas and van der Linden, 2003; Sinharay et al., 2003; Johnson and Sinharay, 2005), a hierarchical model (the type of which was first suggested by Janssen et al., 2000, although in a different context) whose first component is a simple IRT model, e.g., the 3PL model

$$\Pr(X_{ij} = 1 \mid \theta_i, a_j, b_j, c_j) = c_j + (1 - c_j)\,\mathrm{logit}^{-1}\big(a_j(\theta_i - b_j)\big), \tag{8}$$

where X_{ij} is the score of an examinee with ability θ_i on item j and $\mathrm{logit}^{-1}(x) = \frac{e^x}{1+e^x}$. After making the transformations $\alpha_j \equiv \log(a_j)$, $\beta_j \equiv b_j$, and $\gamma_j \equiv \mathrm{logit}(c_j)$, one uses a normal distribution to relate the item parameters of items within the same item model as

$$(\alpha_j, \beta_j, \gamma_j)' \mid \boldsymbol{\lambda}_{\mathcal{I}(j)}, \boldsymbol{T}_{\mathcal{I}(j)} \sim \mathcal{N}_3(\boldsymbol{\lambda}_{\mathcal{I}(j)}, \boldsymbol{T}_{\mathcal{I}(j)}), \tag{9}$$

where $\mathcal{I}(j)$ is the item model to which item j belongs. The DAG for the model is shown in Figure 4.

The population distribution for the latent abilities is usually assumed to be $\mathcal{N}(0, 1)$. The item model mean vector $\boldsymbol{\lambda}_{\mathcal{I}(j)}$ can be partitioned as $\boldsymbol{\lambda}_{\mathcal{I}(j)} = (\boldsymbol{\lambda}_{\alpha\mathcal{I}(j)}, \boldsymbol{\lambda}_{\beta\mathcal{I}(j)}, \boldsymbol{\lambda}_{\gamma\mathcal{I}(j)})'$, and the diagonal elements of the item model variance $\boldsymbol{T}_{\mathcal{I}(j)}$ will be referred to as $\tau^2_{\alpha\mathcal{I}(j)}$, $\tau^2_{\beta\mathcal{I}(j)}$, and $\tau^2_{\gamma\mathcal{I}(j)}$ respectively.

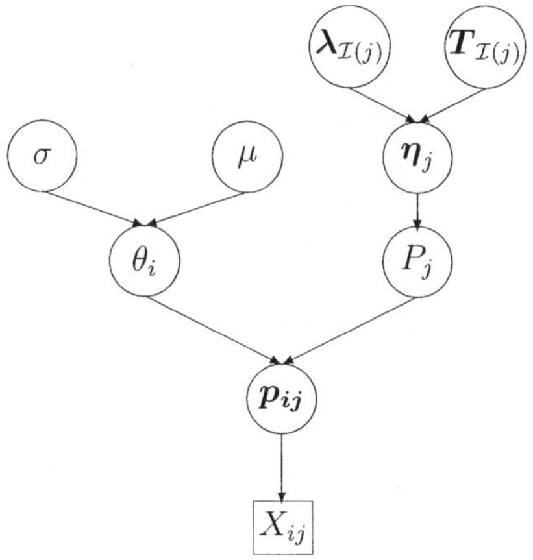

Fig. 4. A directed acyclic graph describing the structural form of the item response data where items (indexed by j) are clustered within item models (indexed by $\mathcal{I}(j)$).

We assume

$$\lambda_{\mathcal{I}(j)} \sim \mathcal{N}(\mu_\lambda, V_\lambda), \quad \mu_\lambda = \big(0, 0, \operatorname{logit}(1/C_{\mathcal{I}})\big)', \quad V_\lambda = 100I, \tag{10}$$

where $C_{\mathcal{I}}$ is the number of multiple choices on item in family $\mathcal{I}$, and I is an identity matrix; independent inverse-Wishart prior distributions (see, for example, Gelman et al., 2003, pp. 574–575) are placed on the item model variances:

$$T^{-1}_{\mathcal{I}(j)} \sim \operatorname{Wishart}_\nu(\mathbf{S}). \tag{11}$$

The prior in (11) implies that the prior mean of $T^{-1}_{\mathcal{I}(j)}$ is $\nu\mathbf{S}$, and that a priori there is information that is equivalent to ν observations of the item parameter vector η_j. Sinharay et al. (2003) use $\nu = 4$, the smallest number to ensure that the density is finite everywhere, and assume $S = \frac{10}{\nu} I_3$. An MCMC algorithm is used to fit the model (details about fitting the RSM using an MCMC algorithm can be found in, e.g., Sinharay et al., 2003).

To graphically summarize the output from an RSM for an item model, Sinharay et al. (2003) suggested the *family expected response function* (FERF) that described the probability that an examinee with ability θ correctly responded to a randomly selected item from the item model. The FERF for item model k is obtained as

$$P(\theta|k) \equiv \int_{\lambda_k, T_k} \int_\eta P(\theta|\alpha, \beta, \gamma)\phi_3(\eta \mid \lambda_k, T_k)\, d\eta\, f(\lambda_k, T_k \mid X)\, d\lambda_k\, dT_k, \tag{12}$$

where $P(\theta \mid \alpha, \beta, \gamma) \equiv \operatorname{logit}^{-1}(\gamma) + (1 - \operatorname{logit}^{-1}(\gamma)) \operatorname{logit}^{-1}(e^\alpha(\theta - \beta))$, $\eta = (\alpha, \beta, \gamma)'$, $\phi_3(\eta|\lambda_k, T_k)$ is the density function of the multivariate normal prior distribution on η and $f(\lambda_k, T_k \mid X)$ is the joint posterior distribution of λ_k and T_k given the response matrix X.

4.3.2. Data example

Let us consider the data from the GRE pretesting mentioned earlier in Section 4.3.1. Sixteen item models were involved; there were four main content areas (remainders or MRE, linear inequality or MLI, quadrilateral perimeter or MQP, and probability or MPR) covered and one item model for each of four difficulty levels (very easy, easy, hard, and very hard) for each content area. The automatic item generation tool generated 10 five-option multiple-choice items from each item model. All of the 160 items were pretested operationally – they were embedded within an operational computer adaptive test, or CAT (taken by 32,921 examinees), to make sure that the examinees were motivated while responding to these items. Each examinee received only four model-based items, one from each content area, and one from each difficulty level so that all four content areas and all four difficulty levels are equally represented. To avoid potential speededness and context effects, the within-section item positions were controlled. The number of examinees receiving any single of these items varied from 663 to 1016.

A simple way to calibrate (i.e., estimating item parameters) in such a situation is to assume the same item response function (the 3PL model for this example) for all items belonging to the same item model. However, this approach ignores any variation between items within a model. If items generated from item models are administered in a future CAT, an examinee receiving a comparatively easier (harder) item from an item model may be getting an unfair advantage (disadvantage). The RSM is more complicated and more difficult to fit, but rightly accounts for the variation between the items in the same item model.

The 3PL model (assuming the same response function for all items in an item model) and the RSM are separately fitted to the data containing the 160 pretested items. As each examinee faced only four pretested items, the usual $\mathcal{N}(0, 1)$ prior distribution would lead to unstable estimates; so the posterior means and standard deviations (SD) from the operational CAT are used as the means and SDs respectively of the normal prior distributions on the proficiency parameters.

Figure 5 shows the estimated FERFs and a 95% prediction interval for each item model, along with the estimated item characteristic curves (ICCs), $P_j(\theta)$, for each item belonging to the item model, all obtained by fitting the RSM. Figure 6, similar to one used in Sinharay et al. (2003), shows the medians of the mean difficulty, the second component of $\boldsymbol{\lambda}_k$ (along the x-axis), and the medians of the within-model variance, the 2nd diagonal element of $\boldsymbol{T}_k$ (along the y-axis), for all item models. For each model, a point (denoted by the model name) shows the mean difficulty versus the within-model variance. A horizontal line around a point denotes an approximate 95% equal-tailed credible interval for mean difficulty; a vertical line around a point denotes a similar credible interval for within-model variance. For example, Figure 6 shows that on an average MLI-IV is the most difficult item model and that model MPR-IV has the greatest within family variation. Different line types are used for different content areas.

Sinharay and Johnson (2005) demonstrate that the simple 3PL model, making a restrictive assumption, does not adequately predict the variance of item difficulties in any of the 16 item models, while the RSM does. Because the variance of item difficulties is an aspect of the data that is of primary interest in this context (one goal of the initiative

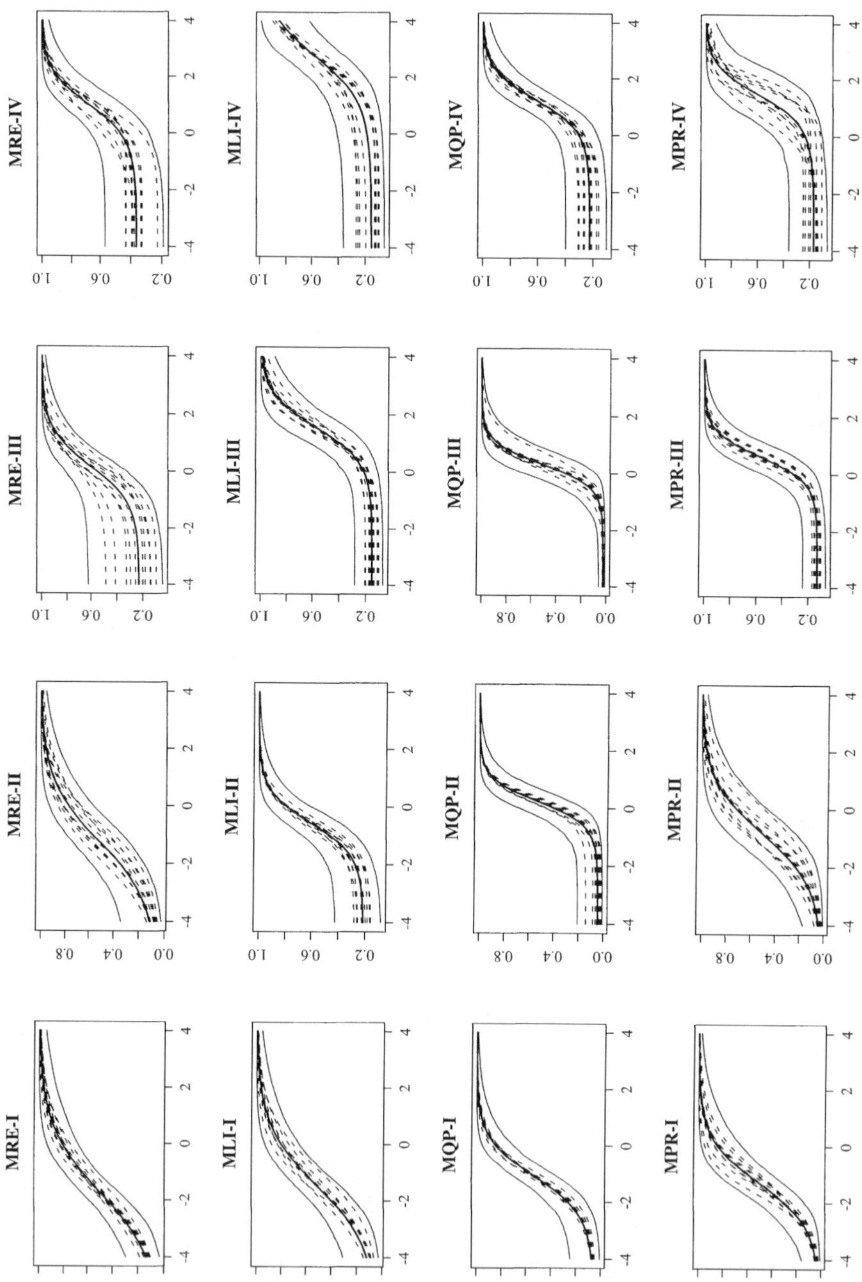

Fig. 5. Estimated FERFs and ICCs for the 16 item models from RSM analysis. The item models are referred to as MRE-I (MRE, very easy), MRE-II (MRE, easy), etc.

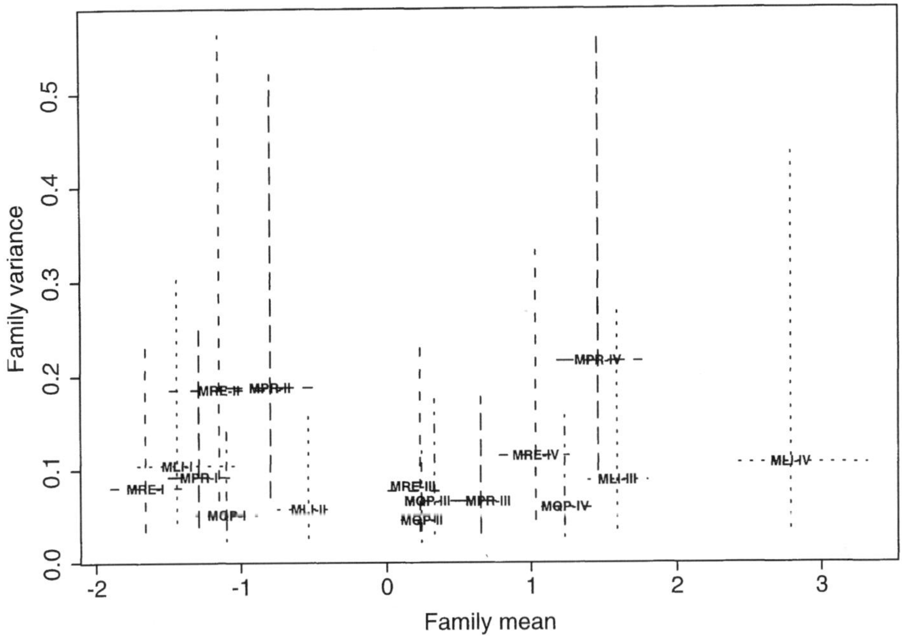

Fig. 6. Posterior summary of the mean and variance of the difficulty parameters for all the item models.

is to create from an item model items whose statistical characteristics vary little), the RSM provides a better description of the data for this example.

4.4. *Other works on hierarchical models*

Mislevy and Bock (1989) laid out a general formulation of hierarchical IRT models. More recently Bolt and Kim (2005) and Pastor (2003) provide (somewhat limited) reviews of hierarchical IRT models. The chapter by von Davier, Sinharay, Oranje, and Beaton on the National Assessment of Educational Progress (NAEP) in this volume (Chapter 32) discusses the NAEP statistical model, which is one among the few hierarchical IRT models that are used operationally in educational testing; the hierarchical part of the NAEP model expresses the person ability parameters using a multivariate linear regression model.

Several other models have been suggested in the IRT literature that are similar to the NAEP model. For example, the three-level hierarchical model of Kamata (2001), which deals with unidimensional abilities in a Rasch-type model (Rasch, 1960), is similar to the NAEP model in the first level of the hierarchy. However, Kamata adds an extra level of hierarchy to account for the fact that examinees may be nested in schools. Johnson and Jenkins (2005) adds yet another level to the NAEP model to account for dependencies within primary sampling units. Cheong and Raudenbush (2000) applied a model similar to that in Kamata (2001) to study the dimensions of a child behavior scale across different demographic groups and different neighborhoods. The hierarchical model of

Fox and Glas (2001) is similar to the NAEP model and the one presented in Kamata (2001); this model employs a two-parameter normal-ogive model in the first level, adds two more levels on the person parameters, uses data augmentation (Albert, 1992), and uses an MCMC algorithm to estimate the model. The model presented in Maier (2001) is similar to that in Fox and Glas (2001) except that the former used a Rasch model (while the latter used a 2-parameter normal ogive model) with a single level of hierarchy on the person level. The two-level model in Adams et al. (1997), as that paper acknowledges, is similar to one used in NAEP. The model in Van den Noorgate et al. (2003) is more general than the above models because it adds hierarchy to a regular IRT model from both person and item side. Fox and Glas (2003) suggested a complicated hierarchical model where proficiency variables at different levels (for example, at the student level or teacher level) can be predictors of a response variable (which can again be a test score, as in their real data example) in a multilevel linear regression model.

5. Conclusions

The examples in this paper demonstrate how hierarchical IRT models can be useful in psychological measurement, especially in situations where simple IRT models fall short. As computers become faster and access to data becomes easier, hierarchical IRT models are sure to find more applications in educational measurement. However, this situation also suggests the use of caution. An application of an MCMC algorithm, the estimation method used most often to fit these models, often does not reveal identifiability problems with the model and unless the investigator is careful, the resulting inference may be inappropriate. Therefore, before using a complicated over-parameterized model, a researcher should always make sure that the model performs considerably better than a simpler alternative, both substantively and statistically, and that the model is well-identified. It is important to remember the following comment from Rubin (1983):

> William Cochran once told me that he was relatively unimpressed with statistical work that produced methods for solving non-existent problems or produced complicated methods which were at best only imperceptibly superior to simple methods already available. Cochran went on to say that he wanted to see statistical methods developed to help solve existing problems which were without currently acceptable solutions.

In the context of hierarchical IRT models, one should check if the model improves over proper application of sampling theory with a simple model. Such attempts have been rare in the literature.

References

Adams, R.J., Wilson, M., Wu, M. (1997). Multilevel item response models: An approach to errors in variables regression. *Journal of Educational and Behavioral Statistics* **22**, 47–76.

Albert, J., Chib, S. (1993). Bayesian analysis of binary and polychotomous response data. *Journal of the American Statistical Association* **88**, 669–679.

Albert, J.H. (1992). Bayesian estimation of normal ogive item response curves using Gibbs sampling. *Journal of Educational Statistics* **17**, 251–269.

Anderson, T. (2003). *An Introduction to Multivariate Statistical Analysis, Patterns of Dependence, Graphical Models*, 3rd ed. John Wiley & Sons, Hoboken, NJ. Chapter 15.

Bolt, D.M., Kim, J.-S. (2005). Hierarchical IRT models. In: *Encyclopedia of Behavioral Sciences*. In press.

Bradlow, E.T., Wainer, H., Wang, X. (1999). A Bayesian random effects model for testlets. *Psychometrika* **64**, 153–168.

Bradlow, E.T., Zaslavsky, A.M. (1999). A hierarchical latent variable model for ordinal data from a customer satisfaction survey with 'No Answer' responses. *Journal of the American Statistical Association* **94**, 43–52.

Cheong, Y.F., Raudenbush, S.W. (2000). Measurement and structural models of children's problem behaviors. *Psychological Methods* **5**, 477–495.

Chib, S., Greenberg, E. (1995). Understanding the Metropolis–Hastings algorithm. *American Statistician* **49**, 327–335.

Diaconis, P., Ylvisaker, D. (1985). Quantifying prior opinion. In: Bernardo, J.M., DeGroot, M.H., Lindley, D.V., Smith, A.F.M. (Eds.), *Bayesian Statistics II*. North-Holland, Amsterdam.

Fienberg, S.E., Johnson, M.S., Junker, B.W. (1999). Classical multilevel and Bayesian approaches to population size estimation using multiple lists. *Journal of the Royal Statistical Society A* **162** (3), 383–405.

Fischer, G.H. (1973). The linear logisit model as an instrument in educational research. *Acta Psychologica* **66**, 269–286.

Fischer, G.H. (1983). Logisit latent trait models with linear constraints. *Psychometrika* **48**, 3–26.

Fox, J.-P., Glas, C.A.W. (2001). Bayesian estimation of a multilevel IRT model using Gibbs sampling. *Psychometrika* **66**, 269–286.

Fox, J.-P., Glas, C.A.W. (2003). Bayesian modeling of measurement error in predictor variable using item response theory. *Psychometrika* **68** (2), 169–191.

Gelman, A., Carlin, J.B., Stern, H.S., Rubin, D.B. (2003). *Bayesian Data Analysis*. Chapman & Hall, New York.

Geman, S., Geman, D. (1984). Stochastic relaxation, Gibbs distributions, and the Bayesian restoration of images. *IEEE Transactions on Pattern Analysis and Machine Intelligence* **6**, 721–741.

Glas, C.A.W., van der Linden, W.J. (2003). Computerized adaptive testing with item cloning. *Applied Psychological Measurement* **27**, 247–261.

Hastings, W. (1970). Monte Carlo sampling methods using Markov chains and their applications. *Biometrika* **57**, 97–109.

Janssen, R., Tuerlinckx, F., Meulders, M., De Boeck, P. (2000). A hierarchical IRT model for criterion-referenced measurement. *Journal of Educational and Behavioral Statistics* **25**, 285–306.

Johnson, M.S., Cohen, W., Junker, B.W. (1999). Measuring appropriability in research and development with item response models. Technical Report 690. Carnegie Mellon University, Department of Statistics, Pittsburgh, PA. http://www.stat.emu.edu/cum-stats/tr/tr690/tr690.html.

Johnson, M.S., Jenkins, F. (2005). A Bayesian hierarchical model for large-scale educational surveys: An application to the national assessment of educational progress. Technical Report RR-04-38. Educational Testing Service.

Johnson, M.S., Sinharay, S. (2005). Calibration of polytomous item families using Bayesian hierarchical modeling. *Applied Psychological Measurement* **29** (5), 369–400.

Kamata, A. (2001). Item analysis by the hierarchical generalized linear model. *Journal of Educational Measurement* **38**, 79–93.

Linacre, J.M. (1994). *Many-Facet Rasch Measurement*. MESA Press, Chicago, IL.

Maier, K.S. (2001). A Rasch hierarchical measurement model. *Journal of Educational and Behavioral Statistics* **26** (3), 307–330.

Mariano, L.T. (2002). Information accumulation, model selection and rater behavior in constructed response student assessments. PhD Thesis. Carnegie Mellon University, Department of Statistics, Pittsburgh, PA. Available for download from http://www.stat.emu.edu/theses/mariano-thesis.pdf.

Maris, E. (1995). Psychometric latent response models. *Psychometrika* **60**, 523–547.

Masters, G.N. (1982). A Rasch model for partial credit scoring. *Psychometrika* **47**, 149–174.

Metropolis, N., Rosenbluth, A.W., Rosenbluth, M.N., Teller, A.H., Teller, E. (1953). Equation of state calculations by fast computing machines. *Journal of Chemical Physics* **21**, 1087–1092.
Metropolis, N., Ulam, S. (1949). The Monte Carlo method. *Journal of the American Statistical Association* **44**, 335–341.
Mislevy, R.J., Bock, R.D. (1989). A hierarchical item–response model for educational testing. In: Bock, R.D. (Ed.), *Multilevel Analysis of Educational Data*. Academic Press, New York.
Pastor, D. (2003). The use of multilevel item response theory modeling in applied research: An illustration. *Applied Measurement in Education* **16**, 223–243.
Patz, R.J. (1996). Markov chain Monte Carlo methods for item response theory models with applications for the national assessment of educational progress. PhD Thesis. Carnegie Mellon University, Pittsburgh, PA.
Patz, R.J., Junker, B.W. (1999). Applications and extensions of MCMC in IRT: Multiple item types, missing data, and rated responses. *Journal of Educational and Behavioral Statistics* **24**, 342–366.
Patz, R.J., Junker, B.W., Johnson, M.S., Mariano, L.T. (2002). The hierarchical rater model for rated test items. *Journal of Educational and Behavioral Statistics* **27** (4), 341–384.
Rasch, G. (1960). *Probabilistic Models for Some Intelligence and Attainment Tests*. Nielsen & Lydiche, Copenhagen.
Rice, K.M. (2004). Equivalence between conditional and mixture approaches to the Rasch model and matched case-control studies. *Journal of the American Statistical Association* **99**, 510–522.
Rubin, D.B. (1983). Some applications of Bayesian statistics to educational data. *The Statistician* **32**, 55–68.
Sinharay, S. (2004). Experiences with Markov chain Monte Carlo convergence assessment in two psychometric examples. *Journal of Educational and Behavioral Statistics* **29**, 463–493.
Sinharay, S., Johnson, M.S. (2005). Analysis of data from an admissions test with item models. Ets rr-05-06.
Sinharay, S., Johnson, M.S., Williamson, D. (2003). Calibrating item families and summarizing the results using family expected response function. *Journal of Educational and Behavioral Statistics* **28**, 295–313.
Spiegelhalter, D.J., Best, N.G., Wilks, W.R., Inskip, H. (1996). Hepatitis B: A case study in MCMC methods. In: Gilks, W.R., Richardson, S., Spiegelhalter, D.J. (Eds.), *Markov Chain Monte Carlo in Practice*. Chapman & Hall/CRC, pp. 21–43.
Tanner, M.A., Wong, W.H. (1987). The calculation of posterior distributions by data augmentation (with discussion). *Journal of the American Statistical Association* **82**, 528–550.
Van den Noorgate, W., De Boeck, P., Meulders, M. (2003). Cross-classification multilevel logistic models in psychometrics. *Journal of Educational and Behavioral Statistics* **28**, 369–386.
Verhelst, N.D., Verstralen, H.H.F.M. (2001). An IRT model for multiple raters. In: Boomsma, A., van Duijn, M.A.J., Snijders, T.A.B. (Eds.), *Essays on Item Response Modeling*. Springer-Verlag, New York, NY, pp. 89–108.
Wainer, H., Bradlow, E., Du, Z. (2000). Testlet response theory: An analog for the 3-PL useful in testlet-based adaptive testing. In: van der Linden, W.J., Glas, C.A.W. (Eds.), *Computerized Adaptive Testing, Theory and Practice*, pp. 245–270.
Wang, X., Bradlow, E.T., Wainer, H. (2002). A general Bayesian model for testlets: Theory and applications. *Applied Psychological Measurement* **26**, 109–128.
Wang, X., Bradlow, E.T., Wainer, H. (2004). User's Guide for SCORIGHT (Version 3.0): A computer program for scoring tests built of testlets including a module for covariate analysis.

Handbook of Statistics, Vol. 26
ISSN: 0169-7161

DOI: 10.1016/S0169-7161(06)26018-8

18

Multidimensional Item Response Theory

Mark D. Reckase

1. Introduction

Psychometric theory deals with the way that information in persons' responses to a series of test tasks, usually called test items, is used to estimate locations of the persons (examinees) on a continuum[1] of performance, aptitude, or other psychological dimension. By continuum in this case is meant an ordering from low to high on the characteristic of interest. One subclass of theories under the general heading of psychometric theory focuses on the relationship between responses to individual items and the locations of persons on the relevant continuum. This subclass is called item response theory (IRT). IRT hypothesizes that a particular mathematical form describes the relationship. A mathematical expression relates the probability of a particular response, or score, for an item to the location on the continuum. Once a mathematical form is hypothesized, the characteristics of the continuum for locating individuals have properties in addition to order – distances between locations become meaningful. Many different mathematical forms have been proposed to represent the connection between the probability of a response to the location on the continuum and it is unclear if each of these mathematical forms constitutes a single theory or if the general approach of using such models is the theory. In general, however, hypothesizing a particular mathematical form to describe the relationship between the probability of a response and the location of the examinee on the continuum is called item response theory.

There is extensive literature on IRT including several excellent textbooks (e.g., Fischer and Molenaar, 1995; Hambleton and Swaminathan, 1985; Hulin et al., 1983; Lord, 1980). Most of the models discussed in the literature on IRT make the assumption that the goal of the measurement task is to locate the examinee on a single continuum. This is usually called the unidimensionality assumption. This means that if the location on this single continuum is known and the hypothesized mathematical form accurately describes the relationship between the probability of a response and the location, then no other information will improve the accuracy of the description of the response process. This kind of assumption is common in many areas of science even though it likely is a simplification of the real state of the world. For example, the speed of a falling object

[1] Psychometric methods also exist for estimating the classification of individuals into discrete classes. These methods are beyond the scope of this chapter.

is usually assumed to be solely a function of the strength of the force of gravity. However, the actual speed of a falling object is influenced by the resistance of the air it falls through, the shape of the object, and other factors. Physics texts are careful to indicate that the equations that describe the speed of a falling object only apply in a vacuum, but the equation for the speed of a falling object is regularly used for practical applications even though it is not absolutely correct for the real life situation.

The unidimensional assumption of most IRT models is similarly a simplification of reality. However, the IRT models might still be useful in the same way that the mathematical equations from Newtonian physics are useful even though they are not perfectly accurate. Test items are actually quite complex tasks that require multiple skills and knowledge to generate a particular response. An example of this complexity is a mathematics test item that presents a practical problem to be solved. Reading skills are required to understand the description of the task. The mathematics task requires problem solving and computational skills as well. Thus, several continua are needed to describe the location of the examinee on the skills required by this item rather than a single one. For these complex cases, the location of the examinee is thought of as a point in a multidimensional space rather then as a point along a single continuum.

Because of the complexity of real test items and the multiple skills and knowledge that examinees use when responding to items, a model based on multiple dimensions is needed. A more realistic mathematical model of the relationship between the probability of response and the location of a person uses a vector to represent the location of the examinee in a multidimensional space. The location of examinee j is represented by the vector $\boldsymbol{\theta}_j = [\theta_{j1}, \theta_{j2}, \ldots, \theta_{jM}]'$, where M is the number of continua (dimensions) on which the examinees need to be ordered to describe the capabilities that come to bear when responding to the test item. The elements of the vector provide coordinates for the location of the examinee in a multidimensional space. Because the location of the individual is given in a multidimensional space, the mathematical models that are used to relate the probability of the response to the location of the examinee are called multidimensional item response theory (MIRT) models. The mathematical expressions for MIRT models will be described in Section 3. A general conceptual framework for the use of MIRT models is given here.

In MIRT, examinees are assumed to be complex entities that differ in many different ways. To totally describe the dimensions on which examinees vary, a large number of dimensions would be needed. A vector whose elements are the coordinates on these dimensions would describe an examinee in this high order space. When an examinee responds to a test item, he or she will draw on the relevant dimensions to determine the response to the item. In theory, if the examinee's location in the full multidimensional space were known, it might be possible to predict the response to a test item with a high degree of certainty. However, the location in the fully specified multidimensional space is never known in practice. At best, some relatively small subset of the dimensions is available for the MIRT model. As a result the relationship between the response and the location in the space is not a deterministic one with a specific form. Instead, the best that can be done is to relate the probability of the response to the estimated location in the multidimensional space and to the estimated test item characteristics. MIRT specifies the form of the mathematical relationship and uses it to estimate the locations of the examinees and the characteristics of the test items.

MIRT was an outgrowth of two other statistics-based methodologies. One has already been mentioned – unidimensional item response theory. The other methodology is factor analysis – a method for identifying the hypothetical dimensions that underlie the relationships among variables in a set of data. MIRT differs from the traditional applications of factor analysis in a number of important ways. First, factor analysis is typically used to analyze a matrix of correlations. Because correlations are a unit free statistic, analyzing a correlation matrix is effectively the same as analyzing variables in z-score form, variables with mean 0 and standard deviation 1. MIRT does not standardize the variables because the differences in item scores provide important information about the characteristics of test items. In a sense, MIRT is a special form of unstandardized factor analysis for special kinds of discrete variables – item scores. McDonald (2000) emphasizes the relationships between the two methodologies.

Second, MIRT was designed for use with variables that are discrete transformations of continuous variables such as the 0 and 1 scores for test items. The use of traditional factor analysis techniques for data of this type has resulted in spurious difficulty factors (Wherry and Gaylord, 1944), especially when the Pearson product-moment correlation is used as the basis for the analysis. The early work of McDonald (1962) in particular focused on devising ways to use factor analytic type methodologies on discrete data. MIRT is an outgrowth of that work.

A third difference between MIRT and the exploratory version of factor analysis is that the goal of exploratory factor analysis is often to identify the minimum number of hypothetical factors that will recover the relationships in the data. Parsimony in description has been a goal of this type of analysis. MIRT procedures have a different goal – the accurate representation of the form of the relationship between the probability of the response and the location in the multidimensional space. The emphasis has not been to minimize the number of dimensions. In fact, Reckase (1997b) analyzed item response data with high numbers of dimensions (25 or more). However, most of the current applications of MIRT have used fewer dimensions because of limitations in estimation programs or because it is useful to graph the results and it is difficult to graph more than three dimensions.

Many of the statistical estimation procedures used for MIRT are very similar to those used for factor analysis. In fact, the commonly used MIRT programs (e.g., NOHARM and TESTFACT) provide results in a form similar to factor analysis programs. In general, however, the results from the MIRT part of the analysis are quite different from those from traditional factor analysis mainly because the variables being analyzed are not standardized to have a mean of zero and a variance of one.

2. General forms of MIRT models

The basic premise of MIRT is that persons vary on many different dimensions. If full knowledge is available on all of the dimensions of variation, a person's characteristics can be represented as a point in a high-dimensional space. The coordinates of the point are given by a vector of real numbers, $\boldsymbol{\theta}_j = [\theta_{j1}, \theta_{j2}, \ldots, \theta_{jM}]'$, where j is the index for the person and M is the number of dimensions in the fully defined space. Differences between persons are indicated by differences in locations in the space. Two assumptions that are often unstated but that are critical to the use of the elements of the $\boldsymbol{\theta}_j$-vector are that the vector indicates a point in a Euclidean space and the elements represent coordinates in a Cartesian coordinate system. The latter point indicates that the elements of the vector represent values on orthogonal axes. These two assumptions allow the use of the standard Euclidean distance formula to determine the distance between the locations of two persons, j and k,

$$\Delta_{jk} = \sqrt{\sum_{g=1}^{M} (\theta_{jg} - \theta_{kg})^2}, \tag{1}$$

where Δ_{jk} is the distance between the two points and the other symbols are previously defined.

Without further assumptions or constraints on the process for estimating the elements of the $\boldsymbol{\theta}_j$-vector, the solution is indeterminate. The dimensions do not have zero points (i.e., no origin) and units of measurement have not been defined. As a result, the outcome of any order-preserving transformation of the points in the space represents the relationships in the data as well as the outcome of any other such transformation. To define the specific characteristics of the space of point estimates of persons' locations, typically called "defining the metric of the space," three constraints are used. The first two constraints are that the mean and variance of estimates of the coordinates for a sample of persons are set to specific values. Any values can be used, but most computer programs set the mean and variance to 0 and 1, respectively. Setting these values fixes the origin and unit of measurement for the estimates of the elements of the $\boldsymbol{\theta}_j$-vector. The unit of measurement and origin could also be set by fixing the parameters of an item, but this is a less commonly used approach.

The third constraint specifies the orientation of the coordinate axes in the multidimensional space. The orientation of the coordinate axes is sometimes fixed by specifying that the direction of best measurement of a test item coincides with the direction specified by an axis. Different items fix the orientation of different axes. Other approaches are to specify a statistical criterion such as maximizing the number of test items that have monotonically increasing probabilities of correct response corresponding to increases in coordinate dimensions.

For realistic situations of modeling the relationship between item responses and persons' locations, the full vector of coordinates is not known. Instead, some smaller number of coordinates is considered, $m < M$, with the value of m based on practical considerations. Often m is 2 or 3 to allow graphic presentation of results. Seldom is m chosen to be more than 5, although there are a few notable examples using m as large as 50 (e.g., Reckase et al., 2000). Because each person is not located in the full M-dimensional proficiency space, it is not possible to develop a functional relationship between the actual response to an item and the location of the person. Instead, MIRT models the probability of response rather then the actual response to reflect that there are likely sources of variation in people that are not included in the model.

Additional constraints that specify the metric of the space are placed on the form of the relationship between the probability of correct response and the location in the m-dimensional coordinate space. Through the specification of the form of the relationship, the specific characteristics of the metric for the space are defined.

The possible forms for the relationship between the probabilities of a correct response to an item and the locations of persons in the space mathematical forms have been proposed. However, for MIRT models for achievement or aptitude test items, the mathematical form is often limited by an additional constraint. The constraint is that the relationship between probabilities of correct response and locations in the space should be monotonically increasing. That is, for an increase in the value of any of the coordinates, the probability of correct response should increase.

MIRT models are typically models for the relationship between the probability of a response to one item and the location of a person in the coordinate space. The model only contains parameters for the characteristics of one item and the location for one person. The item parameters are assumed to be constant. The probability of correct response to the item changes with the location of the person. The model is not for sets of items, although a number of items are analyzed at the same time during estimation of item parameters and tests of fit of the models to item response data. An often unstated assumption is that the model will simultaneously apply to all of the items. It is always possible to fit a model to the data from one item. The simultaneously fitting of the model to multiple items is a stringent test of the usefulness of the model.

Another assumption is often included as part of MIRT. The assumption is that the probability of response to one item conditional on a person's location is independent of the probability of the response to another item conditional on the same location. This is expressed mathematically as

$$P(u_1, u_2, \ldots, u_n|\boldsymbol{\theta}_j) = \prod_{i=1}^{n} P(u_i|\boldsymbol{\theta}_j), \tag{2}$$

where u_i is the response to item i, n is the number of items on the test, and the other symbols are as previously defined. This assumption is important for the estimation of parameters and it is assumed to hold only if $\boldsymbol{\theta}_j$ contains all dimensions needed to respond to all of the items. If too few dimensions are included in $\boldsymbol{\theta}_j$, the assumption of local independence will not likely hold. This provides a means for testing if a sufficient number of dimensions have been included in the modeling of the responses to the items.

The commonly used MIRT models were partly developed as extensions of item response theory (IRT) models that had only a single parameter describing the locations of the persons (see Reckase, 1997a for a brief summary of the development of the models). That is, the persons are assumed to only vary along a single continuum. Accordingly, most MIRT models have these "unidimensional" models as special cases. This feature of the models will be described when specific forms of the models are described.

3. Common forms of MIRT models

Although many types of MIRT models have been proposed, two basic forms appear most frequently in the research literature and in applications of the methodology. The first of these two models is usually called the *compensatory* MIRT model because high values for one set of coordinates for a person's location can offset low values of other coordinates when computing the probability of correct response to an item. That is, persons with quite different locations in the multidimensional space can have the same probability of correct response to an item.

The logistic form of the compensatory MIRT model is given by

$$P(u_i = 1|\boldsymbol{\theta}_j) = \frac{\mathrm{e}^{\mathbf{a}_i'\boldsymbol{\theta}_j + d_i}}{1 + \mathrm{e}^{\mathbf{a}_i'\boldsymbol{\theta}_j + d_i}}, \tag{3}$$

where e is the mathematical constant, 2.718..., that is the base of the natural logarithms, $\mathbf{a}_i$ is an m-element vector of item parameters that indicates the rate the probability of a correct response changes as the location on the corresponding coordinate dimension changes and d_i is a scalar item parameter that is related to the difficulty of the test question. More intuitive meanings for these parameters will be provided in Section 4.

The exponent in Eq. (3) is of the form:

$$a_{i1}\theta_{j1} + a_{i2}\theta_{j2} + \cdots + a_{im}\theta_{jm} + d_i, \tag{4}$$

where the second subscript indicates the coordinate dimension from 1 to m, i is the index for items, and j is the index for persons. From this representation, it is clear that if this sum is equal to a particular value, s, there are many combinations of θ-values that will lead to the same sum, even if the **a**- and d-parameters are held constant. It is this property of the exponent of the model that gives it the label of compensatory. Reducing the value of θ_{jv} can be offset by raising the value of another person coordinate value, θ_{jw}, to yield the same probability of correct response.

The general form of the relationship between the probability of a correct response, $u_i = 1$, and the location of a person in the $\boldsymbol{\theta}$-space can be seen from a plot of the model when $m = 2$. The plot for the case when $a_{i1} = 1.5$, $a_{i2} = .5$, and $d_i = .7$ is presented in Figure 1.

Several important features of the compensatory MIRT model can be noted from the graphs in Figure 1. First, the item response surface increases more quickly along the θ_1-dimension than along the θ_2-dimension. This is a result of the differences in the elements of the **a**-parameter vector. Also, note from the contour plot that the equiprobable contours are straight lines. This is the result of the form of the exponent of e in the model equation. For the example given here, the exponent of e is of the following form:

$$1.5\theta_{j1} + .5\theta_{j2} + .7 = s. \tag{5}$$

If s is equal to 0, then the probability of a correct response is .5 and any combination of θs that result in a value of the expression being 0 define the coordinates of the equiprobable contour of .5. It is clear from Eq. (5) that this is the equation for a

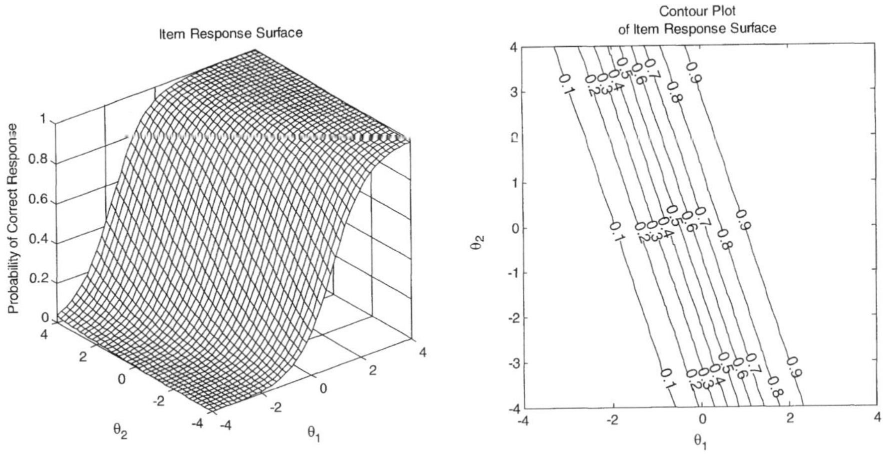

Fig. 1. Graphic representations of the compensatory model – item response surface and equiprobable contours for an item with $a_{i1} = 1.5$, $a_{i2} = .5$, and $d_i = .7$.

straight line. Different equiprobable contours are specified by changing the value of s. The orientation of the equiprobable contours is dependent on the **a**-parameter vector for the item and the distance the lines are from the origin is dependent on the d- and **a**-parameters.

Two variations of the model given in Eq. (3) are often used. One variation includes a parameter that indicates the probability of a correct response for persons who are very low on all of the $\boldsymbol{\theta}$-coordinates. This model is specifically designed for use with multiple-choice items. Such items have a non-zero probability of correct response even when persons taking the items have very little skill or knowledge because it is possible to guess at the answer. The parameter is usually labeled c and the form of the model is given in Eq. (6).

$$P(u_i = 1|\boldsymbol{\theta}_j) = c_i + (1 - c_i)\frac{e^{\mathbf{a}_i'\boldsymbol{\theta}_j + d_i}}{1 + e^{\mathbf{a}_i'\boldsymbol{\theta}_j + d_i}}. \tag{6}$$

For this model, even if the term on the right (which is the same as that in Eq. (3)) is very close to 0, the probability of correct response will still be c_i or slightly greater. The value of c_i is a lower asymptote for the probability of correct response for the item. The c-parameter is a scalar rather than a vector because the process used to respond to items when a person has limited skills or knowledge is not expected to be related to any specific ability. Although the actual psychological processes persons use for responding to items when they have limited ability are not known, they are assumed to be related to guessing. However, the c-parameter estimates are often lower than the chance guessing level for an item. Therefore, the parameter is often called the pseudo-guessing parameter.

The second variation of the model given in Eq. (3) is derived from a family of models that have observable sufficient statistics for the item and person parameters. These models were developed based on the work of Rasch (1960, 1961). A typical characteristic of the Rasch family of models is that the a-parameters for the set of items being analyzed are assumed to have the same value, usually 1. However, if the **a**-parameters in the model presented in Eq. (3) are assumed to be equal, the exponent of e simplifies to

$$a_i(\theta_{j1} + \theta_{j2} + \cdots + \theta_{jm}) + d_i. \tag{7}$$

If the elements of $\mathbf{a}_i$ are equal for all dimensions, there is only an observable sufficient statistic for the sum of the elements of the $\boldsymbol{\theta}$-vector, or alternatively, the sufficient statistic for each of the elements is the same. Therefore, it is not possible to get a unique estimate for each element of the $\boldsymbol{\theta}$-vector. Because unique estimates of the $\boldsymbol{\theta}$s are not available, this approach to creating a simplified version of the model in Eq. (3) is not viable.

The approach that has been used is an extension of a general formulation of a model with observable sufficient statistics that was presented by Rasch (1961). The model can be applied to a wide variety of testing situations, but only the multidimensional version for dichotomous items is presented here. The full version of the model is presented in Adams et al. (1997) as well as other publications. The model using the same symbols as defined above is given by:

$$P(u_i = 1|\boldsymbol{\theta}_j) = \frac{e^{\mathbf{a}_i'\boldsymbol{\theta}_j + \mathbf{w}_i'\mathbf{d}}}{1 + e^{\mathbf{a}_i'\boldsymbol{\theta}_j + \mathbf{w}_i\mathbf{d}}}, \tag{8}$$

where $\mathbf{a}_i$ is a $m \times 1$ vector of scoring weights that are not estimated but are specified in advance by the person using the model; $\mathbf{d}$ is a p-dimensional vector of item characteristics that is constant across all items; and $\mathbf{w}_i$ is a $p \times 1$ set of weights that are set in advance that indicate the way the item characteristics influence the items.

The weight vectors for this model are usually set based on a logical analysis of the content of the items. In many cases, the elements of $\mathbf{a}_i$ and $\mathbf{w}_i$ are 0s or 1s and the vectors serve to select the person dimensions and item characteristics that apply for a specific item. It is the fact that the $\mathbf{a}_i$ and $\mathbf{w}_i$ vectors are set in advance that results in the observable sufficient statistics property of these models. Because the $\mathbf{a}_i$ and $\mathbf{w}_i$ vectors are set in advance, the use of this model has close similarities to confirmatory factor analysis and structural equation modeling. The predicted probabilities of correct response can be compared to probabilities estimated from the observed data to get a fit statistic. That statistic can be used to test hypotheses about the reasonableness of the pre-specified weights. Briggs and Wilson (2003) provide an example of this type of analysis.

The unidimensional logistic model is a special case of this model when θ is a scalar rather than a vector. The model simplifies to $a_i\theta_j + d_i = a_i(\theta_j - b_i)$ when $d_i = -a_ib_i$. The right side of the first equation in the previous sentence is the usual exponent for the unidimensional two-parameter logistic model with difficulty parameter b_i.

To this point, the logistic form of the compensatory MIRT model has been presented. This form of the model is used because it is more mathematically tractable. However, instead of using the cumulative logistic function as the basis of the model, the cumulative normal distribution function could be used. In fact, the early development of these models assumed a cumulative normal form for the test characteristic surface (e.g., McDonald, 1967). The equation for this compensatory MIRT model based on the normal distribution is given by

$$P(u_{ij} = 1|\theta_j) = \int_{-\infty}^{\mathbf{a}_i'\theta_j + d_i} \frac{1}{\sqrt{2\pi}} e^{-\frac{t^2}{2}} \, dt, \tag{9}$$

where all of the symbols have been defined previously.

The normal ogive model in (9) and the logistic model in (3) are very similar in form. Birnbaum (1968, p. 399) indicated that the probability of correct response specified by the two models with the same parameter values will differ by less than .01 over the entire range of $\boldsymbol{\theta}$ if the exponents of e in the logistic model in (3) are multiplied by 1.7. When the constant 1.7 is included in the expression for the logistic model, parameters estimated assuming the model in (9) are often used as estimates for the parameters of the model in (3). The practical implication of this similarity of models is that more computer programs are available for the estimation of model parameters.

The second of the two basic forms of the MIRT models is often called the *non-compensatory* model, although it is more correctly labeled as the *partially compensatory* model because it does not totally remove the compensation effect present in the model in (3). The partially compensatory nature of the model will be described in more detail below. The form of this model was first proposed by Sympson (1978) as an alternative to the model given in (6). The mathematical form of the model is given in Eq. (10).

$$P(u_i = 1|\theta_j) = c_i + (1 - c_i) \prod_{\ell=1}^{m} \frac{e^{1.7a_{i\ell}(\theta_{j\ell} - b_{i\ell})}}{1 + e^{1.7a_{i\ell}(\theta_{j\ell} - b_{i\ell})}}, \tag{10}$$

where ℓ is the index for coordinate dimensions, $b_{i\ell}$ is a parameter that indicates the difficulty of performing the tasks related to dimension ℓ. Note that the initial version of the model included the constant 1.7 so that each term in the product would be similar to a normal ogive model. The item response surface and contour plot for this model are presented in Figure 2 for the two-dimensional case with $a_{i1} = 1.5$, $a_{i2} = .5$, $b_{i1} = -1$, and $b_{i2} = 0$. The c-parameter for the figure has been set to 0 so that it could be more easily compared to Figure 1.

The partially compensatory nature of this model is shown by the curvature of the surface. When the c-parameter is 0, the probability of correct response for an item based on this model can never be higher than the lowest value in the product. An increase in one of the coordinates can improve the overall probability somewhat, but only up to the limit set by the lowest term in the product.

If the terms in the product are represented in this two-dimensional case as p_1 and p_2, it is easy to see that the curves that define the equiprobable contours are hyperbolas in terms of these probability values, $h = p_1 p_2$, where h is the value for any one of the contours, and p_1 and p_2 are the terms in the product in (10). The actual curves plotted in Figure 2 are not hyperbolas because they are mapped through the logistic functions to the θ coordinate system. However, like the hyperbola, they asymptote to a value as the elements of $\boldsymbol{\theta}$ approach infinity.

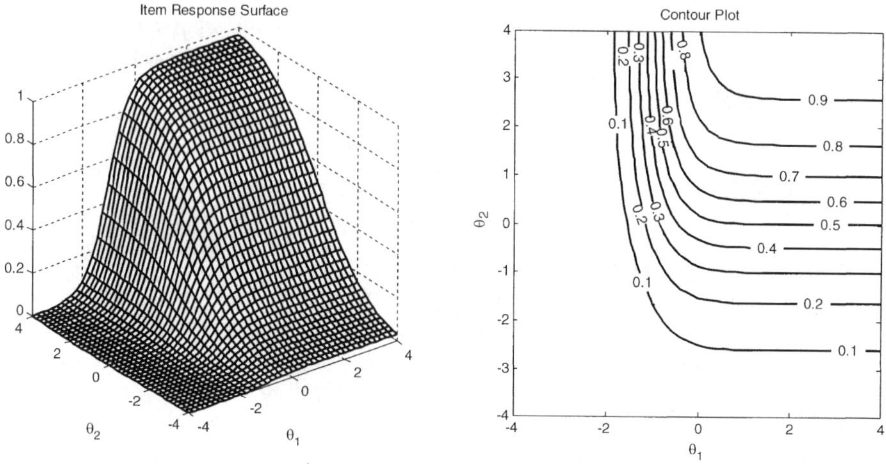

Fig. 2. Graphic representation of the partially compensatory model – item response surface and equiprobable contours for an item with $a_{i1} = 1.5$, $a_{i2} = .5$, $b_{i1} = -1$, $b_{i2} = 0$ and $c_i = 0$.

The model in (10) can be simplified by setting c_i to zero. Then the model parallels the compensatory model given in (3). A further simplification of this model can be obtained by setting all of the $a_{i\ell}$ equal to 1. This model is given by

$$P(u_{ij} = 1|\boldsymbol{\theta}_j) = \prod_{\ell=1}^{m} \frac{e^{(\theta_{j\ell} - b_{i\ell})}}{1 + e^{(\theta_{j\ell} - b_{i\ell})}}. \tag{11}$$

Whitely (1980) suggested this model as a way to describe the interactions among the various cognitive components that enter into the performance on an item. In this case, the terms in the product follow the Rasch model form with only a difficulty parameter for each dimension. Note that if m is 1 in Eqs. (10) and (11), the special cases are the unidimensional three-parameter logistic model and Rasch models, respectively.

The compensatory and partially compensatory forms of MIRT models have quite different properties. For the compensatory model, as any element of $\boldsymbol{\theta}_j$ approaches positive infinity, the value of the exponent approaches positive infinity even if the other elements of the $\boldsymbol{\theta}$-vector are very small negative values. Therefore, as any element of the $\boldsymbol{\theta}$-vector increases, the probability of correct response approaches 1.0. For the partially compensatory model, the product in (10) can never be greater than its lowest value term. If one term of the product is .1 and all other elements of the $\boldsymbol{\theta}$-vector increase so the other terms approach 1.0, the product will only approach .1. Therefore, for the partially compensatory model, the probability of correct response to the item as a whole is limited by the lowest probability of success on a component of the item.

Another difference between the two models is the influence of the number of dimensions in the model on the scaling of the coordinate space. Suppose that the c_i and d_i parameters in (6) are all zero. Also, suppose that a person has elements of the $\boldsymbol{\theta}$-vector that are all zero as well. The probability of correct response in that case is .5. Adding another dimension with a $\theta_{j\ell} = 0$ does not change the probability at all because the exponent remains zero. However, consider the same case for (11) with the b-parameters equal to zero. For one dimension, the probability of correct response is .5, for two dimensions .25, for three dimensions .125, etc. In general, if the probability of correct response for a test item that has a constant probability of mastery for each component of the model, p_c, then probability of correct response will be p_c^m where m is the number of dimensions. If the observed proportion of correct responses for 100 persons at the same point in the θ-space is .5, the values of the item parameters needed to give that proportion correct will change depending on the number of dimensions. That is, the estimates of $\boldsymbol{\theta}$ for a two-dimensional solution will be quite different in magnitude from estimates of $\boldsymbol{\theta}$ for a three-dimensional solution. This complicates the interpretation of the elements of the $\boldsymbol{\theta}$-vector for the partially compensatory model. The range of values that is considered high or low changes depending on the number of dimensions used in the model.

Although the compensatory and partially compensatory MIRT models are quite different in mathematical form and philosophical position on the way that skills and knowledge interact when applied to test items, there is some evidence that the two models fit real test data equally well. Spray et al. (1990) identified item parameters for the two models that gave similar classical item statistics. Although the item response surfaces were very different in some regions of the $\boldsymbol{\theta}$-space, for regions where the density of an assumed bivariate normal distribution was the greatest, the surfaces were similar. Figure 3 shows item response surfaces for the two models that resulted in similar proportions of correct response when a standard bivariate normal distribution of examinees was assumed. The parameters for the two models are given in Table 1.

The parameter values in Table 1 show that the parameters are quite different for the two models even though they appear to have similar functions. The a-parameters for the partially compensatory model are larger than those for the compensatory model. This is an effect of the multiplicative nature of the partially compensatory model.

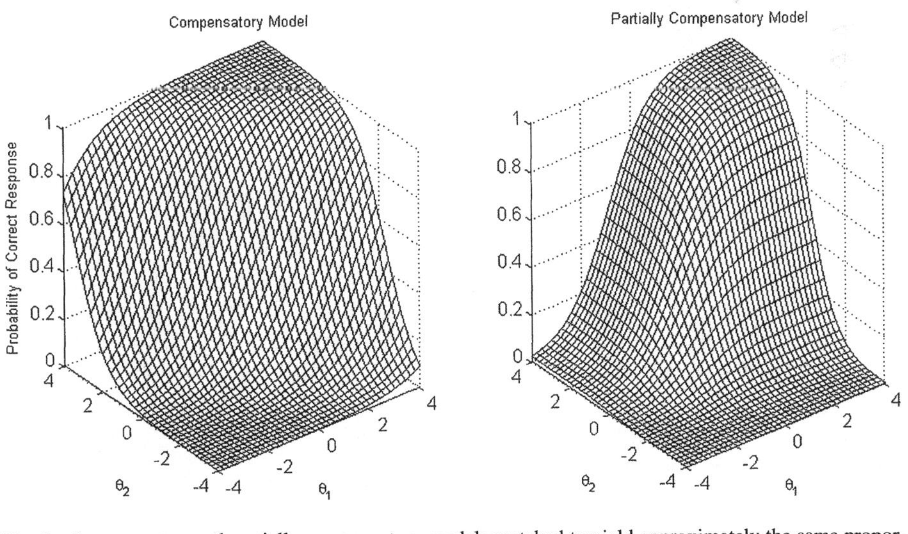

Fig. 3. Compensatory and partially compensatory models matched to yield approximately the same proportion correct assuming a standard bivariate normal distribution with $\rho = 0$.

Table 1
Parameters for matched compensatory and partially compensatory models

Compensatory model		Partially compensatory model	
Parameters	Values	Parameters	Values
a_1	.90	a_1	1.26
a_2	1.31	a_2	1.60
d	−.67	b_1	−.92
		b_2	−.15

Note: The a-parameters for the two models do not have exactly the same meaning and are not comparable. The d-parameter for the compensatory model is on approximately the same scale as $-ab$ for the partially compensatory model.

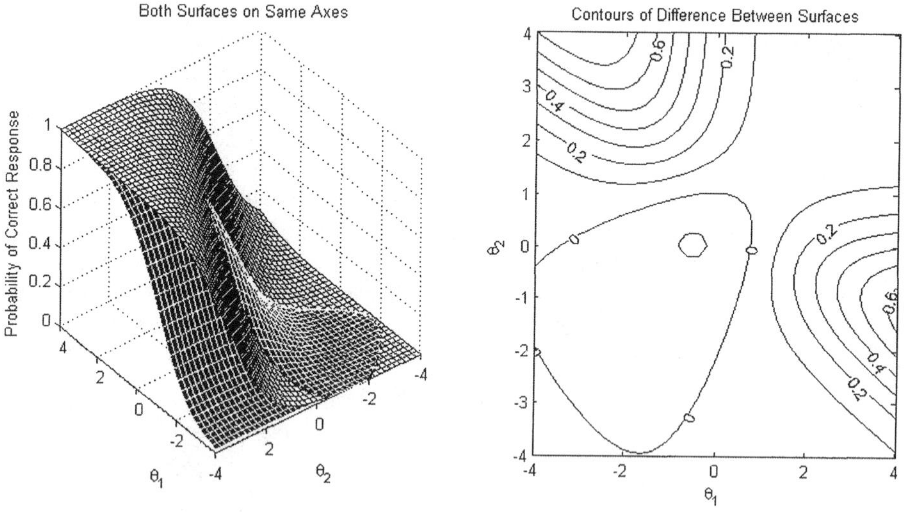

Fig. 4. Compensatory and partially compensatory surfaces on same axes (left panel) and contour plot of the differences between the two surfaces (right panel). Note that the surfaces on the left panel have been rotated to make the intersection of the surfaces more evident.

Figure 4 presents two graphs. The graph on the left shows the surfaces for both models on the same set of axes so that they can be compared. The right pane of the figure shows the contour plot for the difference between the two surfaces. The left panel in Figure 4 shows that the two surfaces intersect in such a way that they are similar near the origin of the $\boldsymbol{\theta}$-space. The surfaces diverge the most when the θ_1 and θ_2 coordinates are opposite in sign and indicate points that are far from the origin. This can be seen clearly in the contour plot in the right pane. In the region around the (0, 0) point in the $\boldsymbol{\theta}$-space, the differences between the two surfaces are close to 0 in probability. However, in the regions around $(-2, 4)$ and $(4, -1)$ the differences in probability are quite large, approximately .7 and .6 respectively.

The form of the contour plot for the difference in the two surfaces shows why the two models yield similar results on classical test statistics (proportion correct of .40 and .38 for the compensatory and partially compensatory, respectively). The large differences in the surfaces occur in regions where the density of the bivariate normal distribution is very small. Further, most skills and knowledge measured by cognitive tests are positively correlated, making persons at locations $(-2, 4)$ even less likely. The similarity of the predicted probabilities for the high density region of the normal distribution and the fact that the partially compensatory model is less mathematically tractable have led to dominance of the compensatory model in the research literature.

4. Descriptions of item characteristics

If the MIRT models provide a reasonable representation of the data obtained from the interaction of persons with a test item, then the characteristics of the test item are described by the surface defined by the model and they are summarized by the parameters of the model. For example, if the compensatory model given in (6) represents the relationship between locations in the $\boldsymbol{\theta}$-space and the proportion of correct responses to an item for a sample of examinees, then the characteristics of the item can be summarized by the **a**-vector, and the scalar c and d parameters.

Unfortunately, the model parameters do not have intuitive meaning. To help in the interpretation of the MIRT models, several statistics have been derived from the parameters to help describe the workings of the models. These statistics indicate the combination of skills best measured by the item, the difficulty of the item, and the usefulness of the item for differentiating between persons at different points in the $\boldsymbol{\theta}$-space.

Generally, the goal of measurement is to differentiate among objects using a particular characteristic of the objects. In this case the objects are people who are located at different places in the $\boldsymbol{\theta}$-space. These locations are represented by vectors of coordinates (e.g., $\boldsymbol{\theta}_j$, $\boldsymbol{\theta}_k$, etc.). Test items are used to gain information about the differences in locations of the persons. Assuming the MIRT model describes the functioning of the test item, the probability computed from the model gives the likelihood of the item score given the persons location in the $\boldsymbol{\theta}$-space. The .5 equiprobable contour is an important feature of an item because the principle of maximum likelihood indicates that if persons respond correctly to an item, they are estimated to be located on the side of the .5 contour that has the higher probability of a correct response. If there is an incorrect response to the item, the person is estimated to be on the side of the .5 contour that has the lower probabilities of correct response. Therefore, the location of the .5 contour is an important characteristic of the item.

Another important characteristic of a test item is how well it differentiates between two persons located at different points in the $\boldsymbol{\theta}$-space. In the context of MIRT models, differentiation is indicated by a difference in the probability of correct response to the item. If the probability of correct response to the item for the locations of two persons is the same, the item provides no information about whether the persons are at the same point or different points. However, if the difference in probability of correct response is large, then it is very likely that the persons are located at different points in the $\boldsymbol{\theta}$-space.

Differences in probability of correct response for an item are largest where the slope of the item response surface is greatest, and when points in the space differ in a way that is perpendicular to the equiprobable contours for the item response surface.

These uses of a test item and the characteristics of the item response surface suggest three descriptive measures for an item: (1) the direction of steepest slope for an item; (2) the distance of the point of steepest slope from the origin of the space; and (3) the magnitude of the slope at the point of steepest slope. These three descriptive measures can easily be derived from the mathematical form of the item response surface.

Measures of direction and distance need to be taken from a particular location. For convenience, the location that is used for these measures is the origin of the space – the $\boldsymbol{\theta}$-vector with all zero elements. Also for mathematical convenience, the development of these measures is done in polar coordinate system so that the direction is easily specified as angles from a particular axis. The compensatory model with the c-parameter set to zero will be used for this development.

The elements of the $\boldsymbol{\theta}_j$ vector in (6), $\theta_{j\ell}$, represent the coordinates of a point in an m-dimensional space. Let θ_j represent the distance from the origin to the point specified by the $\boldsymbol{\theta}_j$ vector. The point and distance are shown in Figure 5 for the two-dimensional case. The angle between the line connecting the point to the origin and coordinate axis ℓ

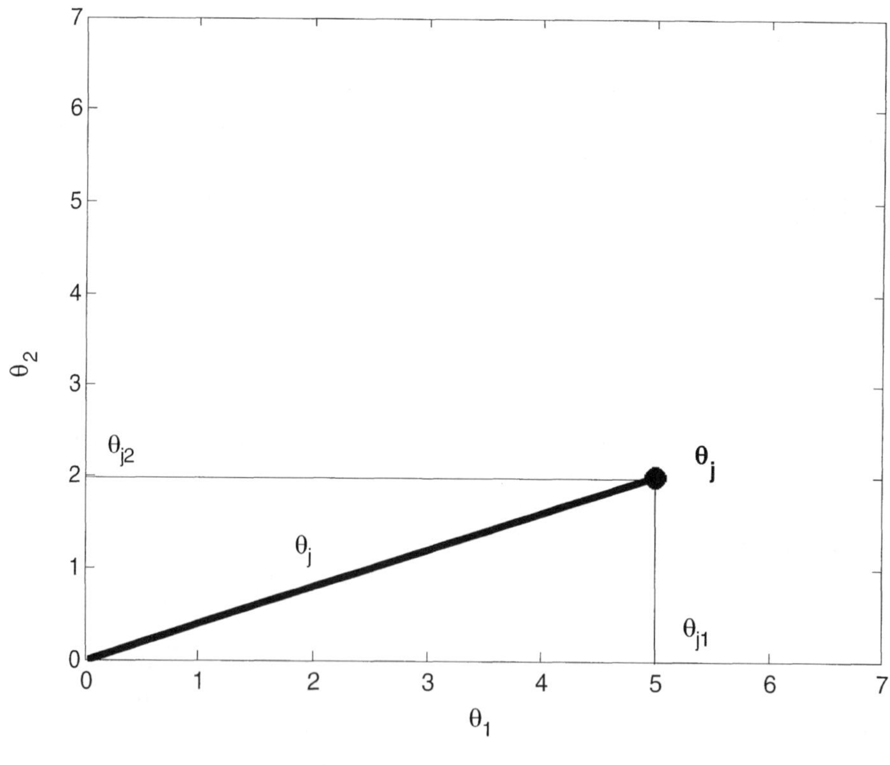

Fig. 5. Polar coordinate representation of person location $\boldsymbol{\theta}_j$.

is given by $\alpha_{j\ell}$. From right triangle trigonometry, the cosine of angle $\alpha_{j\ell} = \theta_{j\ell}/\theta_j$. Solving for $\theta_{j\ell}$ yields $\theta_j \cos\alpha_{j\ell}$. Thus, each element of the coordinate vector can be represented by the distance from the origin times the cosine of the angle between the line to the point and the coordinate axis.

In this two dimensional case, $\alpha_{j2} = 90 - \alpha_{j1}$. More generally, the relationship among the angles between the coordinate axes and the line connecting the origin of the space to the $\boldsymbol{\theta}_j$-point is given by

$$\sum_{\ell=1}^{m} (\cos\alpha_{j\ell})^2 = 1. \tag{12}$$

This relationship is a general property of the relationships of angles with a line represented in an orthogonal coordinate space.

The model in (6) can be reparameterized by substituting a vector composed of elements $[\theta_j \cos\alpha_{j\ell}]$, the scalar value θ_j times the vector of cosines $\cos\boldsymbol{\alpha}_j$, for the vector of coordinates $\boldsymbol{\theta}_j$. After this reparameterization, the MIRT model takes the following form

$$P(u_i = 1|\theta_j, \boldsymbol{\alpha}_j) = \frac{e^{\mathbf{a}_i'\theta_j \cos\boldsymbol{\alpha}_j + d_i}}{1 + e^{\mathbf{a}_i'\theta_j \cos\boldsymbol{\alpha}_j + d_i}}, \tag{13}$$

where θ_j is a scalar parameter that gives the distance of person j from the origin of the space and $\boldsymbol{\alpha}_j$ is a vector of angles between the coordinate axes and the line from the origin to the person's location in the space. This formulation of the model specifies the probability of correct response for any angle from the axes given by $\boldsymbol{\alpha}_g$ and any distance from the origin given by θ_g. The slope in any direction specified by $\boldsymbol{\alpha}_g$ can be determined by taking the first derivative of (13) with respect to θ_g,

$$\frac{\partial P(u_i = 1|\theta_g, \boldsymbol{\alpha}_g)}{\partial\theta_g} = P_{ig}(1 - P_{ig})\mathbf{a}_i' \cos\boldsymbol{\alpha}_g, \tag{14}$$

where $P_{ig} = P(u_i = 1|\theta_g, \boldsymbol{\alpha}_g)$, θ_g is a point on the line from the origin specified by the angles in $\boldsymbol{\alpha}_g$, and g is the index for the specific angle from the origin that is being considered. The subscript for θ and $\boldsymbol{\alpha}$ has been changed from j in (12) to g in (14) because θ_j referred to a specific individual and θ_g refers to a point along a line in direction $\boldsymbol{\alpha}_g$.

The maximum slope in the direction specified by $\boldsymbol{\alpha}_g$ can be determined by taking the second derivative of the item response function and solving for zero. The second derivative is given by

$$\frac{\partial^2 P(u_i = 1|\theta_g, \boldsymbol{\alpha}_g)}{\partial\theta_g^2} = P_{ig}\big(1 - 3P_{ig} + 2P_{ig}^2\big)(\mathbf{a}_i' \cos\boldsymbol{\alpha}_g)^2. \tag{15}$$

There are three solutions when (15) is set to zero, but two of them are for P_{jg} equal to 0 or 1 that lead to values of θ_g that are negative or positive infinity, respectively. The only finite solution for 0 is $P_{jg} = .5$. This means that the slope is steepest along the .5 contour for all directions from the origin. The expression for the steepest slope in direction $\boldsymbol{\alpha}_g$ is simply ¼ $\mathbf{a}_i' \cos\boldsymbol{\alpha}_g$. The slope of the surface along one of the coordinate axes is simply ¼ $a_{i\ell}$, where ℓ indicates the index for the axis, because the cosine is 1 when the angle is 0°and the cosine is 0 when the angle is 90°.

To determine the direction that gives the steepest slope overall, the expression in (14) for the .5 contour is differentiated with respect to $\boldsymbol{\alpha}_g$ and solved for 0. However, before performing the differentiation, the last element in the vector of cosines, cos α_{gm}, is replaced with the following term

$$\cos\alpha_{gm} = \sqrt{1 - \sum_{h=1}^{m-1} \cos^2 \alpha_{gh}}. \tag{16}$$

Adding this constraint insures that the sum of the squared cosines will be equal to 1. This has the effect of making the coordinate axes orthogonal to each other.

Differentiating the expression with the added constraint with respect to each of the angles in the $\boldsymbol{\alpha}_g$-vector, rearranging terms and simplifying results in $m - 1$ expressions of the form

$$a_{ih} - a_{im}\left(\frac{\cos\alpha_{gh}}{\cos\alpha_{gm}}\right), \qquad h = 1, \ldots, m - 1. \tag{17}$$

Setting these expressions equal to zero and solving for each cos α_{gh} yields m expressions of the form

$$\cos\alpha_{ih} = \frac{a_{ih}}{\sqrt{\sum_{h=1}^{m} a_{ih}^2}}. \tag{18}$$

Note that in this equation, the subscript on α has been changed from g to i to indicate that this is now a feature of the item. This expression gives the cosine of the angle between each of the axes and the line that connects the origin to the point of steepest slope on the item response surface. The cosines are called the *directions cosines* for that line. This direction indicates the particular direction between two points in the space that the test item provides the most information about differences in the points.

Another useful piece of information is the distance from the origin to the point of steepest slope in the direction defined by (18). That distance can be determined by substituting the vector of direction cosines from (18) into (14) and then solving for θ_g. The result is the distance from the origin to the point of steepest slope. The equation for the distance is:

$$D_i = \frac{-d_i}{\sqrt{\sum_{h=1}^{m} a_{ih}^2}}. \tag{19}$$

D_i is often called the multidimensional difficulty, or MDIFF, of the item because it is on a scale that is analogous to that used in unidimensional IRT models. In those models, the exponent of e is of the form $a(\theta - b) = a\theta - ab$. The ab term is analogous to the d-parameter in the MIRT model, with the sign reversed. To get the equivalent of the b-parameter, the last term is divided by the discrimination parameter, a.

The slope at the point of steepest slope in the direction specified by (18) can be obtained by substituting (18) into (15) with the probability specified as .5. The result is

$$\text{Slope} = \frac{1}{4}\sqrt{\sum_{h=1}^{m} a_{ih}^2}. \tag{20}$$

This expression without the constant ¼, $\sqrt{\sum_{h=1}^{m} a_{ih}^2}$, is referred to as the multidimensional discrimination (MDISC) for the item. Note that MDISC is in the denominator of (18) and (19). The results in (18), (19), and (20) also apply to the model in (6) with $c_i > 0$. Changing that parameter only raises the item response surface, but does not change the location of the equiprobable contour that has the steepest slope. However, the probability value for that contour is $(c_i + 1)/2$ rather than .5.

The values given in (18), (19), and (20) provide a convenient way to represent items for two or three dimensional cases. Each item can be represented as a vector that points in the direction of increasing probability for the item response surface. The base of the vector is on the equiprobable contour that is the line of steepest slope. The base is a distance MDIFF away from the origin. The length of the vector is a scaled value of MDISC. The direction the vector is pointing is given by (18). An example of this vector representation of an item is given on the contour plot of an item response surface in Figure 6. Further discussion of the development of these measures is given in Reckase (1985) and Reckase and Hirsch (1991).

The vector plot for the item with the parameters specified in the legend for Figure 6 has an MDIFF value of .66, and MDISC value of 1.26, and angles with the two axes of 18° and 72°, respectively. The use of item vectors allows many items to be graphically represented at the same time. It is also possible to show items in three dimensions, something that is not possible with response surfaces or contour plots. Figure 7 shows items from a science test using three dimensions. This graph shows that the items measure in different directions as the MDIFF values for the items change. Relatively easy items measure along θ_1, while more difficult items measure along θ_2 and θ_3.

Another way of representing the way that an item functions is its capability to differentiate between two points in the space. This capability is usually labeled as the *information* provided about the location of the examinee by the item. It is the ratio of the rate of change of the item response surface over the variance of error of estimates at that point in the space. The general equation for information (Lord, 1980) for item i is given by

$$I_i(\theta) = \frac{\left[\frac{\partial P_i(\theta)}{\partial \theta}\right]^2}{P_i(\theta)[1 - P_i(\theta)]}. \tag{21}$$

When $\boldsymbol{\theta}$ is a vector rather than a scalar, the information at the $\boldsymbol{\theta}$-point depends on the direction taken because the slope of the surface depends on the direction, as was shown above. Therefore, the numerator of (21) is changed to include the directional derivative in angle $\boldsymbol{\alpha}$ rather than the standard derivative:

$$I_\alpha(\theta) = \frac{[\nabla_\alpha P_i(\theta)]^2}{P_i(\theta)[1 - P_i(\theta)]}, \tag{22}$$

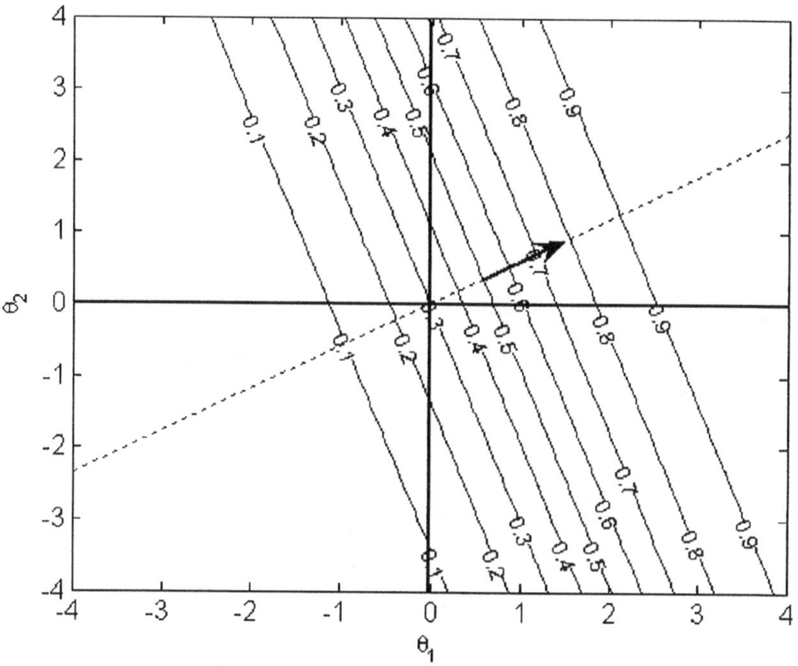

Fig. 6. Item vector representing MDIFF, MDISC, and the direction of best measurement of an item. ($a_1 = 1.2$, $a_2 = .4$, $d = -.84$.)

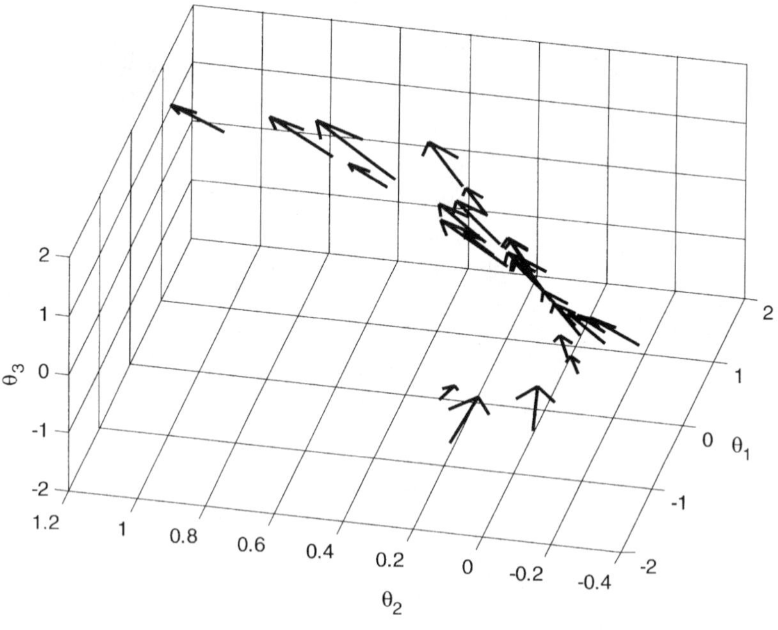

Fig. 7. Vector plot of items in three dimensions.

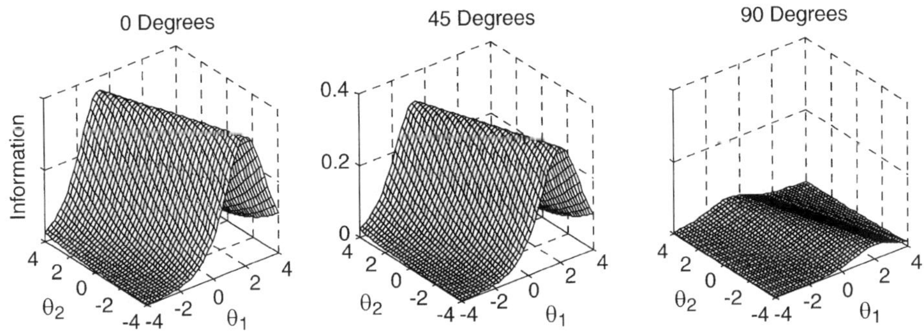

Fig. 8. Information functions for an item in three different directions.

where the directional derivative is defined as follows

$$\nabla_\alpha P_i(\theta) = \frac{\partial P_i(\theta)}{\partial \theta_1} \cos \alpha_1 + \frac{\partial P_i(\theta)}{\partial \theta_2} \cos \alpha_2 + \cdots + \frac{\partial P_i(\theta)}{\partial \theta_m} \cos \alpha_m. \quad (23)$$

Figure 8 shows the information for an item with the same parameters as used in Figure 6 for directions that are 0°, 45°, and 90° from the θ_1 axis. The information functions have the same orientation over the $\boldsymbol{\theta}$-plane, but the heights differ. The orientation has the highest point of the function over the .5-contour of the surface shown in Figure 6. The angle of steepest slope for the surface is 18° which is closest to the 0° direction. Therefore, the information is highest in the left most panel of Figure 8. The 90° direction has substantially lower information than the others because that direction is almost parallel to the equiprobable contours for the surface.

5. Descriptions of test characteristics

IRT provides a number of useful ways to describe the characteristics of a test by combining the characteristics of items. For example, because the probability of correct response to an item is the same as the expected score on an item and because the expected value of a sum is the sum of the expected values, a test characteristic surface can be computed by summing the item characteristic surfaces. The item parameters for 20 items that can be modeled using a two-dimensional coordinate system are given in Table 2. For each of the items on this test, the item characteristic surface was computed. These 20 surfaces were summed to form the test characteristic surface. That test characteristic surface is shown in Figure 9.

The height of the surface represents the estimated true score in the number-correct metric for the test. For examinees who are high on both θ-values, the estimated true score is near the maximum summed score on the test of 20. Those low on both θ-values have estimated true scores near zero. The surface does not have linear equal score contours because of the combination of multiple items. Overall, the surface shows how number-correct scores are expected to increase with increases in the coordinate dimensions.

Table 2
Item parameters for a 20 item test

Item number	a_1	a_2	d	Item number	a_1	a_2	d
1	1.81	.86	1.46	11	.24	1.14	−.95
2	1.22	.02	.17	12	.51	1.21	−1.00
3	1.57	.36	.67	13	.76	.59	−.96
4	.71	.53	.44	14	.01	1.94	−1.92
5	.86	.19	.10	15	.39	1.77	−1.57
6	1.72	.18	.44	16	.76	.99	−1.36
7	1.86	.29	.38	17	.49	1.10	−.81
8	1.33	.34	.69	18	.29	1.10	−.99
9	1.19	1.57	.17	19	.48	1.00	−1.56
10	2.00	.00	.38	20	.42	.75	−1.61

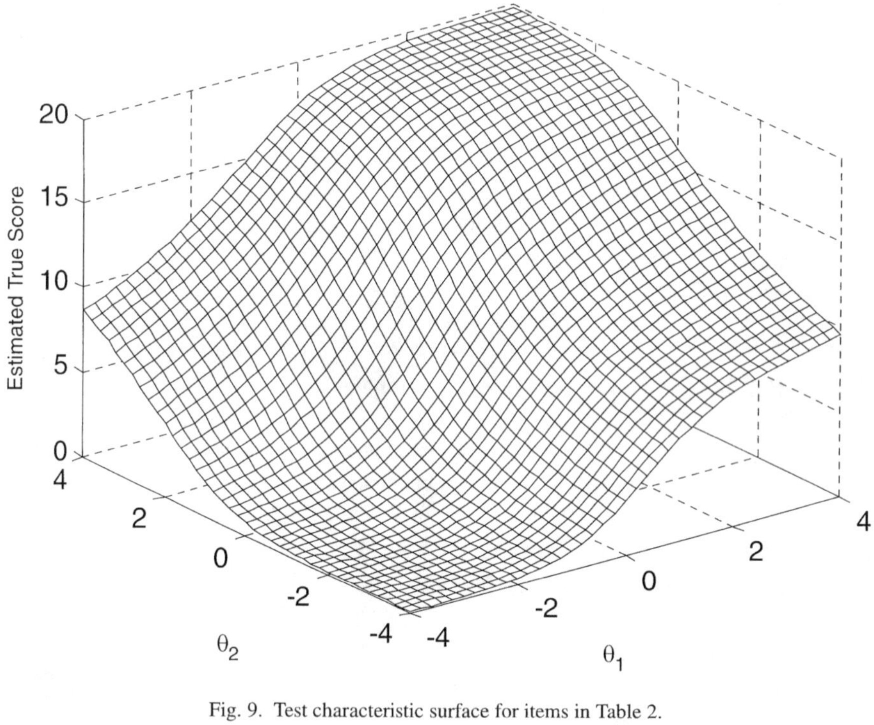

Fig. 9. Test characteristic surface for items in Table 2.

The relationship between item information and the information for the full test is similar to the relationship between the item characteristic surface and the test characteristic surface. However, note that the test information is the sum of the item information values in *a particular direction*. Figure 10 shows the test information for the set of items given in Table 2 in the 0°, 45°, and 90° directions from the θ_1 axis. The different test

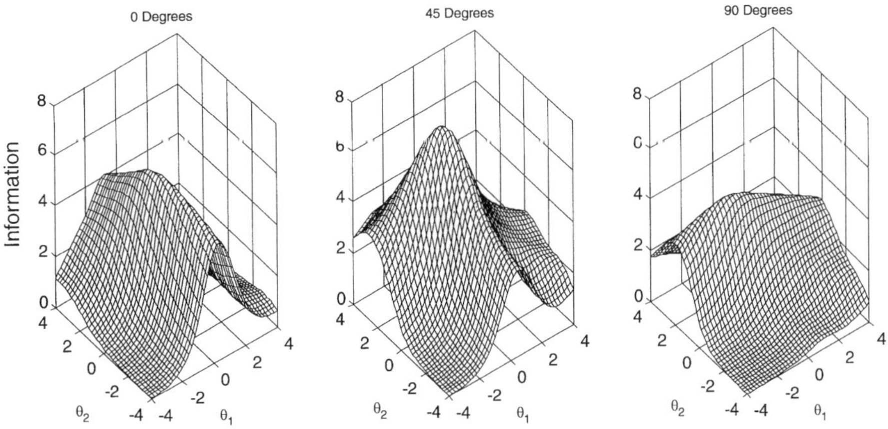

Fig. 10. Test information functions in directions 0°, 45°, and 90° from the θ_1 axis.

information functions show that the form of the information surfaces is not the same for the different directions. The set of items provides more information near the origin of the space in a 45° direction than the other two directions. There are other more subtle differences as well, such as having information higher along the θ_2 dimension when measuring in the 90° direction.

The amount of information provided by a test is related to the direction in the space that the test best differentiates between nearby points. The test shown in Figure 10 is very good at distinguishing between person locations between 0 and 1 on the θ_1 dimension in a direction parallel to that dimension, but is best at distinguishing persons located between 2 and 3 along the θ_2 dimension in the direction parallel to that dimension.

The single direction best measured by this set of items was derived by Wang (1985). She showed that the direction could be determined from the eigenvector of the $\mathbf{a}'\mathbf{a}$ matrix that corresponds to the largest eigenvalue from the eigenvalues/eigenvector decomposition of that matrix. The values of the eigenvector are the equivalent of the elements of the **a**-parameter vector for an item and can be used to find the direction of measurement for a unidimensional calibration of the test. Wang (1985) labeled the line defined by this eigenvector the "reference composite" for the test. In the case of the set of items in Table 2, the reference composite is a line through the origin that has a 37° angle with the θ_1 axis. The reference composite is shown by the long dashed vector in Figure 11.

The concept of a reference composite can be used with sets of items as well as the full test. For example, if the first ten items in the test measure one content area and the second ten items measure a different content area, reference composites can be determined for each set of items and the directions of the unidimensional scale for those two item sets can be determined. Those two reference composites are shown in Figure 12 by the vectors drawn with thick lines. Note that the item vectors for the first ten items, the easier items, are much more tightly grouped around their reference composite than those for the second set of ten items, the harder items. When the item vectors for a set of items are all pointing in the same direction, that set of items can be

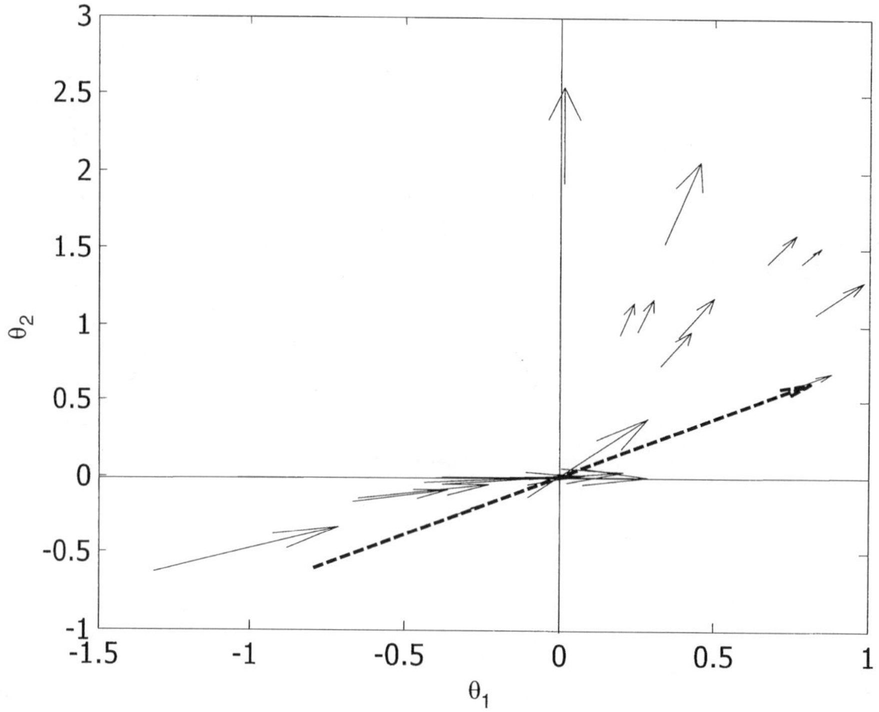

Fig. 11. Item vectors and reference composite vector for the items in Table 2.

represented using a unidimensional model. The first ten items approximately meet that condition.

The second ten items are much more widely dispersed in their directions of best measurement. Although the reference composite for those items indicates the direction of measurement for a unidimensional analysis of that set of items, the null hypothesis that the item responses from this set of items were generated by a unidimensional IRT model would likely be rejected.[2]

The reference composites in Figure 12 show an important relationship between dimensions of sensitivity for test items and the coordinate axes that define the multidimensional space. The coordinate axes for this analysis are a convenient orthogonal frame of reference for representing the item vectors. The reference composites indicate the directions of best measurement for the sets of test items. The dimensions of measurement of sets of items generally do not correspond to the coordinate axes and they are usually not orthogonal. The composites measured by sets of items are often fairly highly correlated. When simulation studies are used to generate data of this type, it is often suggested that the coordinate axes need to be non-orthogonal (i.e., correlated) to

[2] Statistical procedures such as DIMTEST (Stout et al., 1996) are available to test the null hypothesis that the data were generated using a model with a single person parameter. Such procedures were not used for this example because the analysis of a ten item set would not give much power to test the null hypothesis.

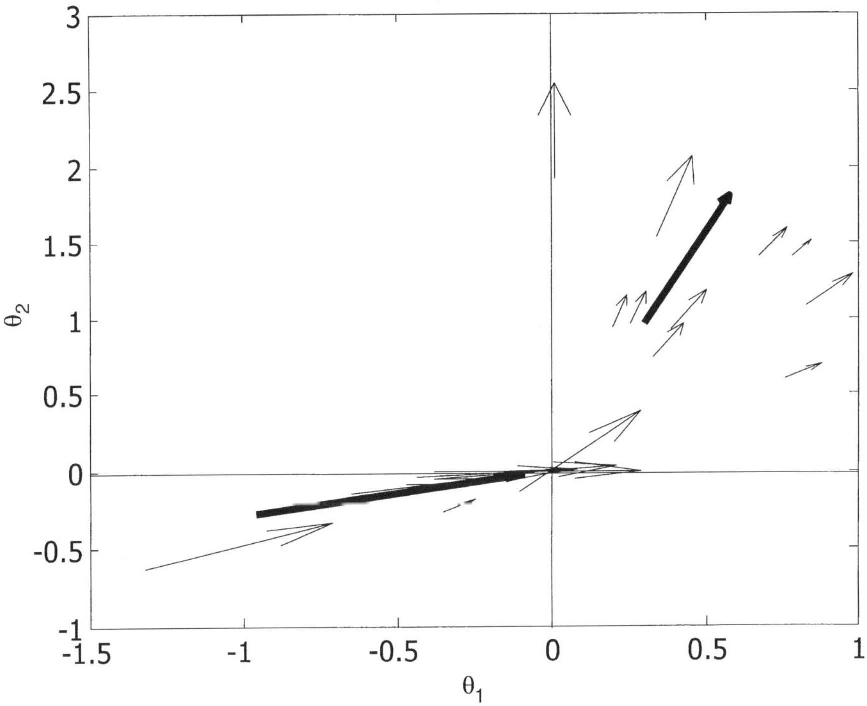

Fig. 12. Reference composites for the first ten and second ten items for the items in Table 2.

represent the correlation between the traits being measured. That is not the case. The dimensions of sensitivity for sets of items in the test can be correlated while the coordinate axes are orthogonal. These two conditions can easily hold at the same time. A common example is a scatter plot between height and weight. The coordinate axes for height and weight are drawn orthogonal to each other. The correlation between height and weight is shown by the form of the scatter plot. The angle between the coordinate axes is not used to represent the correlation between the variables that are represented in the scatter plot.

6. The estimation of model parameters

The most intuitively appealing and creative models are not very useful unless the parameters in the model can be estimated fairly accurately. Estimation procedures for compensatory MIRT models have been available for some time based on the normal ogive version of the compensatory model (see Eq. (9)). Because of the similarity of the logistic model to the normal ogive model, the results of the normal ogive-based programs are often used as estimates of the parameters of the logistic models as well. This is because the increase in estimation error caused by model misspecification is believed to be small relative to other sources of estimation error.

Initial estimation procedures for MIRT were implemented in two computer programs, NOHARM and TESTFACT. NOHARM (Fraser, 1988) is based on the work by McDonald (1967, 1981) to estimate the parameters of the normal ogive model. The method programmed in NOHARM uses a four term polynomial approximation to the normal ogive model. The approximation is used to predict the joint frequency of item scores on pairs of items using least squares polynomial regression. The item response data from the test is tallied to provide the data in the cells of a two-by-two table of frequencies of response combinations for each pair of items (Table 3). In the table, n_{ijuv} gives the frequency of the response pair u, v where u and v can take on the values of 0 and 1 for items i and j. The program uses ordinary least squares to estimate the parameters $\mathbf{a}$ and d that minimizes

$$\sum_{i,j,u,v} (n_{ijuv} - \hat{n}_{ijuv})^2, \tag{24}$$

where $\hat{n}_{ijuv}$ is the value predicted from the polynomial equation. McDonald (1982) and others have found this procedure to be computationally fast and robust. It has been used to estimate parameters for up to 50 dimensions. However, this approach is only used to estimate the item parameters. The $\boldsymbol{\theta}$-vectors are not estimated, but are assumed to be distributed as multivariate normal with a specified variance covariance matrix. The program allows c- and $\mathbf{a}$-parameters to be fixed to specified values to support confirmatory analyses.

The other commonly used program for the estimation of parameters is TESTFACT (du Toit, 2003, Chapter 5). This program uses methodology developed by Bock et al. (1988). Like the methodology used in NOHARM, this program assumes the normal ogive formulation of the MIRT model. However, rather than predicting the elements of the 2×2 table of frequencies, TESTFACT determines the parameter estimates that maximize the probabilities of the observed frequencies of the full response pattern on all of the items for the examinees. Because the full response pattern is used, this methodology has been labeled "full information factor analysis." The maximization is done assuming a multivariate normal distribution of proficiencies. The basic equation for the estimation is given by

$$P(\mathbf{U} = \mathbf{U}_s) = \int_{\theta} L_s(\theta) g(\theta)\, \mathrm{d}\theta, \tag{25}$$

where $\mathbf{U}_s$ is a particular binary response pattern for a set of items, $L_s(\boldsymbol{\theta})$ is the likelihood of response string at a particular point in the θ-space, and $g(\boldsymbol{\theta})$ is the multivariate

Table 3
Frequency of item score pairs for two items

		Item j	
		0	1
Item i	0	n_{ij00}	n_{ij01}
	1	n_{ij10}	n_{ij11}

normal density for the proficiency values. This integral is approximated using m-fold Gauss–Hermite quadrature. Because estimation is done assuming a known marginal distribution of proficiencies, this procedure is called marginal maximum likelihood (MML) estimation. The actual implementation of the estimation procedure uses the EM algorithm (Dempster et al., 1977). The details of the process are given in Bock et al. (1988).

Recently, there has been substantial interest in using Markov chain Monte Carlo (MCMC) methods for the estimation of the parameters of MIRT models. At the time this chapter was written (i.e., 2005), MCMC for MIRT estimation were relatively new, so the properties of those estimates were not well documented. Bolt and Lall (2003) used MCMC methods based on the Metropolis–Hastings algorithm to estimate the parameters of both the compensatory and partially compensatory logistic models presented in Eqs. (8) and (10). Generally, the procedure was able to recover the model parameters in a simulation study, although the recovery was better for the compensatory model than the partially compensatory model. A negative aspect of these methods was that estimation took approximately 40 hours for 31 items and 3000 examinees. This should not be taken as too serious a problem because computers are already substantially faster than those used in 2002. Also, a number of researchers are currently working on the use of this methodology. By the time this chapter is published, there likely will be a number of additional journal articles that update the progress of this methodology. The reader is encouraged to read current publications on these methodologies to get up to date information on their usefulness for MIRT.

7. Applications

There have been relatively few applications of MIRT procedures to practical testing problems, at least as compared to unidimensional IRT procedures. For example, as of 2005, no large scale testing program uses MIRT for test equating while several use unidimensional IRT procedures for that purpose. There is a growing body of research on such applications, but most of the research is on evaluating new procedures or checking methodology through simulation. There are also some confirmatory analyses related to the sensitivity of items to differences along specified dimensions. No doubt, the number of operational applications will increase in the coming years. A few of the areas where applications are actively being developed will be highlighted here. As with any dynamic research area, the current research literature should be checked for the newest developments in this field.

The three areas that have received the most attention to date are (1) the detailed analysis of test content, (2) equating/linking of test calibrations to put them into the same solution space, and (3) computerized adaptive testing. Each of these applications is described in turn.

The MIRT estimation of item parameters and the calculation of the direction of item vectors give a convenient way to study the structure of test content. Item vectors that point in the same direction indicate that the items measure the same composite of the coordinate dimensions. Reckase et al. (1988) showed that when sets of items have item vectors that point in the same direction they can be modeled using a unidimensional

IRT model. Further, Reckase and Stout (1995) determined the general conditions when sets of items that require multiple skills and knowledge to specify a correct response can yield item response data that are well fit by unidimensional models. These findings suggest that if items are clustered into sets with minimal angles between item pairs, the resulting clusters are sets of items that are sensitive to differences on the same composite of skills and knowledge. Miller and Hirsch (1992) demonstrated the process of identifying sets of items sensitive to the same set of skills and knowledge using cluster analysis procedures.

Miller and Hirsch (1992) used NOHARM to estimate the item parameters and then computed the cosine of the angle between two item vectors using the equation

$$\cos\alpha_{ij} = \frac{\mathbf{a}_i'\mathbf{a}_j}{\sqrt{\sum_{k=1}^{m} a_{ik}^2 \sum_{k=1}^{m} a_{jk}^2}} = \sum_{k=1}^{m} \cos\alpha_{ik}\cos\alpha_{jk}. \tag{26}$$

The matrix of cosines was then converted to the angles between the item directions. The angles were used as the dissimilarity measure for a hierarchical cluster analysis using Ward's method (Ward, 1963). The number of clusters is difficult to determine with precision, but considering a combination of the substantive meaning of clusters and the standard errors of the parameter estimates can give reasonable guidance to the number of clusters.

The parameters given in Table 2 are used to demonstrate this procedure. The angles between the items were computed using Eq. (26) and the results were clustered using Ward's method. The results are shown in the dendrogram in Figure 13. In this diagram, the item numbers that are connected together near the horizontal axis have small angles between them. The horizontal bar connecting the items is at the level showing the angular distance on the vertical axis. The dendrogram shows that the 20 items given in Table 2 are generally of two different types. These two types generally are consistent with the two reference composites in Figure 12.

The method of clustering using the angles between items has been used on a wide variety of test data and it has been shown to yield a very fine grained summary of item content (Reckase et al., 1997). The results in Miller and Hirsch (1992) show that interpretable clusters can be identified even when there is a relatively small angle between clusters. Other analyses at ACT, Inc. (reported in Reckase, 1997c) were able to detect item content clusters based on as few as four items. Traditional factor analysis approaches would not detect these item clusters because they did not account for very much variance out of a 60 item test. However, the clustering based on angular distances was quite stable, being replicated on a number of parallel forms of the test.

A second area of the application of MIRT that has received substantial research effort is the linking of MIRT calibrations. The linking of calibrations is important for equating and vertical scaling of test forms and the development of item pools for computerized adaptive testing. The methods for computerized adaptive testing with MIRT also depend on the linking of calibrations for the development of item pools of sufficient size to support sound implementation.

When items are calibrated using any one of the calibration approaches that are available, certain aspects of the calibration are not statistically determined from the data

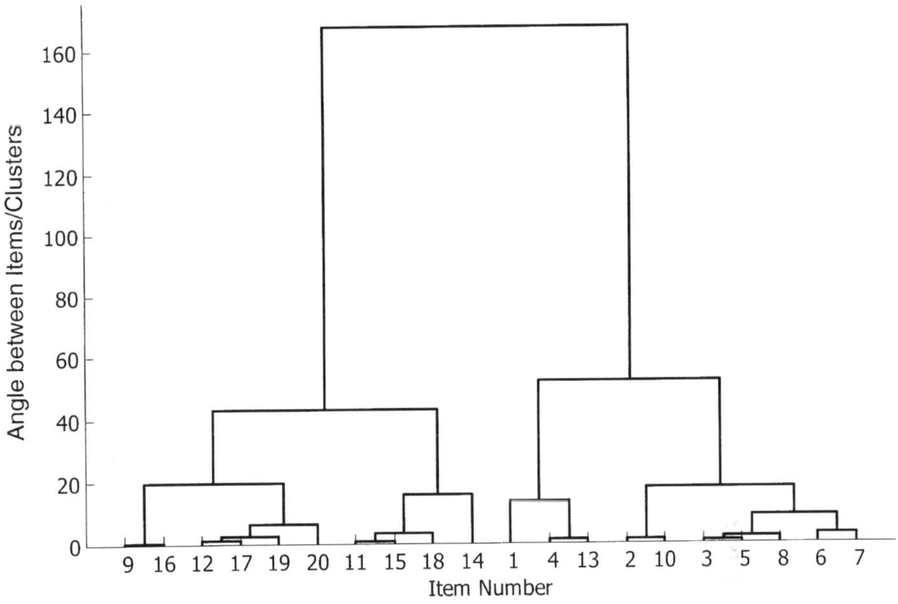

Fig. 13. Dendrogram of item clusters based on item parameters in Table 2.

and these indeterminacies are usually addressed by fixing/constraining certain features of the parameter estimates. Li and Lissitz (2000) lay out these issues and indicate that there are three types of indeterminacy in the solutions. First, the origin of the coordinate system is not defined. This is usually addressed by setting the mean vector of the estimated proficiencies to a vector of zeros. Second, the units along each of the coordinate axes are not determined. This is usually addressed by setting the unit as the standard deviation of the observed proficiencies. These first two types of indeterminacy also exist for unidimensional IRT models so a variety of procedures have been developed to address those issues.

The third type of indeterminacy is the orientation of the axes of the coordinate system in the space defined by the proficiency estimates. NOHARM addresses this type of indeterminacy by fixing the coordinate axes to run parallel to the item vectors for pre-selected items. The solution can later be changed using a variety of rotation procedures to set the orientation of the coordinate axes according to other criteria. It is clear, however, that using a different set of items to anchor the coordinate axes will give a solution that appears different, even though it may be possible to make two solutions look the same after rotation to a common criterion. Other estimation methods use different constraints to set the orientation of the coordinate axes. This means that analyzing the same set of data with two different programs will likely give different "looking" results, different parameter estimates, even though they might be very nearly the same after an appropriate rotation.

Because of these indeterminacies and the different ways that they are addressed by different estimation programs, the estimates of parameters obtained from the programs are to some extent sample specific. That is, if a set of items are calibrated using samples that have different mean vectors and different variance/covariance matrices, the procedures for resolving indeterminacies will cause the parameter estimates to be different when, in fact, they are simple transformations of each other. The goal of MIRT linking/equating procedures is to determine the transformations that will convert a set of item parameter estimates from one sample to estimates of the same parameters from another sample. The transformation determined for items can be used to determine the comparable transformation for abilities. That transformation will keep the person locations in the same place relative to each other, but convert the locations to a different coordinate system.

To make the discussion of the indeterminacy of the solutions concrete, the parameters in Table 2 were used to generate item response data according to the compensatory MIRT model based on two different $\boldsymbol{\theta}$-distributions. In both cases, 2000 simulated student response vectors were generated based on each assumed distribution of $\boldsymbol{\theta}$. The first distribution has a mean vector of $(0, 0)$ and the identity matrix for the variance–covariance matrix. The second distribution has a mean vector of $(.5, -.75)$ and variance–covariance matrix $\left[\begin{smallmatrix} 2.25 & .45 \\ .45 & .5625 \end{smallmatrix}\right]$. The first distribution is consistent with that specified in the TESTFACT estimation program. However, that distribution has a zero intercorrelations between the θ_1 and θ_2 coordinates of the examinees' locations in the space. A zero correlation between coordinates is unlikely in practice because most cognitive proficiencies are related to each other for typical testing populations.

The second distribution specifies a correlation between the θ-coordinates of .4 and a standard deviation for the first set of coordinates that is twice that for the second set of coordinates. Because of the close match to the assumptions built into TESTFACT, the analysis of data generated from the item parameters and the first proficiency distribution should result in parameter estimates that are similar to the generating parameters. The analysis of the data generated from the second distribution should give results that are quite different because the TESTFACT will set the estimated mean vector to a default value of $(0, 0)$ and the default variance covariance matrix to the identity matrix. Because the person parameters have are estimated on a rotated and translated set of coordinates, the item parameters are estimated with a compensating translation and non-orthogonal rotation. For the second set of generated data, the item parameter estimates will not appear similar to the generating parameters. However, the linking of the calibrations to a common set of coordinate axes should show that these two sets of parameter estimates are estimates of the same true parameters.

Table 4 gives the generating parameters and the two sets of estimated parameters side by side so they can be conveniently compared. The estimates from TESTFACT have been multiplied by 1.7 to put them on the logistic model metric because TESTFACT assumes the normal ogive model when estimating parameters. The values are those obtained using the Varimax rotation option from the TESTFACT program. The values labeled Est 1 in the table come from the estimation of the generated data assuming the standard normal distribution of $\boldsymbol{\theta}$s with the identity matrix for the variance covariance matrix. The values labeled Est 2 are for the estimates based on the distribution described above. It is clear from a comparison of the parameter estimates that the Est 1 values are fairly close to the values used to generate the data. The Est 2 values are quite different,

Table 4
Item parameter estimate comparison

Item number	a_1-parameter			a_2-parameter			d-parameter		
	True	Est 1	Est 2	True	Est 1	Est 2	True	Est 1	Est 2
1	1.81	1.51	2.28	0.86	0.86	1.51	1.46	1.27	1.81
2	1.22	1.20	1.36	0.02	0.07	0.74	0.17	0.17	0.77
3	1.57	1.37	1.98	0.36	0.46	1.04	0.67	0.55	1.19
4	0.71	0.60	0.98	0.53	0.63	0.82	0.44	0.44	0.39
5	0.86	0.85	1.01	0.19	0.22	0.56	0.10	0.17	0.44
6	1.72	1.67	2.28	0.18	0.34	1.11	0.44	0.43	1.34
7	1.86	1.66	2.33	0.29	0.36	1.18	0.38	0.43	1.38
8	1.33	1.19	1.81	0.34	0.43	1.00	0.69	0.75	1.20
9	1.19	1.04	1.56	1.57	1.53	1.97	0.17	0.08	−0.33
10	2.00	1.98	2.93	0.00	0.15	0.91	0.38	0.50	1.77
11	0.24	0.18	0.29	1.14	1.07	1.07	−0.95	−0.92	−1.64
12	0.51	0.36	0.69	1.21	1.39	1.33	−1.00	−1.06	−1.64
13	0.76	0.64	0.99	0.59	0.67	0.79	−0.96	−0.99	−0.96
14	0.01	−0.07	0.04	1.94	1.64	1.31	−1.92	−1.81	−3.13
15	0.39	0.25	0.43	1.77	1.55	1.53	−1.57	−1.45	−2.70
16	0.76	0.66	0.92	0.99	1.05	1.14	−1.36	−1.54	−1.63
17	0.49	0.48	0.61	1.10	0.94	1.19	−0.81	−0.76	−1.36
18	0.29	0.15	0.48	1.10	1.08	1.00	−0.99	−1.01	−1.66
19	0.48	0.49	0.54	1.00	0.92	0.99	−1.56	−1.68	−1.97
20	0.42	0.37	0.56	0.75	0.68	0.67	−1.61	−1.66	−2.00
Mean	0.93	0.83	1.20	0.80	0.80	1.09	−0.39	−0.40	−0.44
Standard deviation	0.62	0.59	0.82	0.57	0.48	0.33	0.98	0.97	1.60

with larger mean a-parameters and larger standard deviations for d-parameters. This is not a surprising result because the second data set did not have characteristics that matched the default origin and unit of measurement used in the TESTFACT program. Because TESTFACT constrains the $\boldsymbol{\theta}$-estimates to be uncorrelated, the correlation between the $\boldsymbol{\theta}$-vectors used to generate the data must be accounted for in other ways. The parameter estimates show that adjustment needed to account for the correlation is an increase in the a-parameter estimates and an increase in variance for the d-parameter estimates. However, the parameter estimates from both data sets represent the same correlational structure among the items. To show that, the parameter estimates need to be rotated and translated to be put on the same coordinate system as the generating values.

Approaches to determining transformations for putting all of the item parameters on the same coordinate system have been developed by Li and Lissitz (2000) and Min (2003). Both use Procrustes methods to determine a rotation matrix that rotates the coordinate axes so that the item vectors point in the same directions in the coordinate space. They then solve for translation constants that minimize the difference between the target and estimated d-parameters. The approach presented here is an extension of those methods based on work by Martineau (2004) and Reckase and Martineau (2004). This approach is more general than previously published methods.

The rotation matrix for the coordinate axes is computed using the oblique Procrustes method as specified by Mulaik (1972, p. 297). That method determines the rotation matrix from the following equation

$$\mathbf{Rot} = (\mathbf{a}_a'\mathbf{a}_a)^{-1}\mathbf{a}_a'\mathbf{a}_b, \tag{27}$$

where $\mathbf{a}_b$ is the $n \times m$ matrix of base form discrimination parameters that are the target for the transformation, $\mathbf{a}_a$ is the $n \times m$ matrix of discrimination parameters for the same items on the alternate form, and **Rot** is the $m \times m$ rotation matrix for the discrimination parameters.

For the examples given in Table 4, using the generating **a**-matrix as the target, the rotation matrix for Est 1 is $\left[\begin{smallmatrix} 1.06 & -.09 \\ .07 & 1.09 \end{smallmatrix}\right]$ and that for Est 2 is $\left[\begin{smallmatrix} .75 & -.49 \\ .03 & 1.26 \end{smallmatrix}\right]$. The rotation matrix for Est 1 is very close to an identity matrix. This is because the data were generated according to the same model as is assumed by the TESTFACT program. The rotation matrix for Est 2 is quite different than that for Est 1 because it must account for the differences in variances of the generating θ parameters and the covariance between those parameters. The difference in the diagonal values account for the different variances for the generating θs and non-zero off-diagonal value accounts for the covariance between the θs. The **a**-matrix on the coordinate system for the target set of parameters is given by

$$\widehat{\mathbf{a}}_b = \mathbf{a}_a\mathbf{Rot}, \tag{28}$$

where $\widehat{\mathbf{a}}_b$ is the estimate of the a-parameters on the base form metric. The rotations for Est 1 and Est 2 were applied to the estimated **a**-matrices and the results are presented in Table 5 with the generating parameters. It is clear that the parameters after rotation are much closer than those in Table 4. The means and standard deviations of the a-parameters for the two dimensions after rotation are very similar to statistics for the parameters used to generate the data.

The transformation of the d-parameters from the alternate form to the metric of the base form is given by

$$\mathbf{Trans} = \mathbf{a}_a(\mathbf{a}_a'\mathbf{a}_a)^{-1}\mathbf{a}_a'(\mathbf{d}_b - \mathbf{d}_a), \tag{29}$$

where $\mathbf{d}_b$ is the $n \times 1$ vector of d-parameters for the base form, $\mathbf{d}_a$ is the $n \times 1$ vector of d-parameters for the alternate form, and **Trans** is the $n \times 1$ transformation vector for the d-parameters. The other symbols have been defined above. The estimate of the d-parameters on the base form from those on the alternate form is given by

$$\widehat{\mathbf{d}}_b = \mathbf{d}_a + \mathbf{Trans}. \tag{30}$$

The correlations between the parameter estimates and the true values are also very high. For example, the correlation between the generating d-parameter and Est 2 is .994.

The transformation for the estimates of $\boldsymbol{\theta}$ from the alternate form to the base form metric is given by

$$\widehat{\boldsymbol{\theta}}_b' = \mathbf{Rot}^{-1}\boldsymbol{\theta}_a' + (\mathbf{a}_b'\mathbf{a}_b)^{-1}\mathbf{a}_b'(\mathbf{d}_a - \mathbf{d}_b), \tag{31}$$

Table 5
Item parameter estimate comparison after rotation

Item number	a_1-parameter			a_2-parameter			d-parameter		
	True	Est 1	Est 2	True	Est 1	Est 2	True	Est 1	Est 2
1	1.81	1.66	1.74	0.86	0.80	0.79	1.46	1.29	1.37
2	1.22	1.28	1.04	0.02	−0.03	0.27	0.17	0.18	0.35
3	1.57	1.48	1.50	0.36	0.37	0.35	0.67	0.57	0.55
4	0.71	0.68	0.75	0.53	0.64	0.56	0.44	0.45	0.35
5	0.86	0.92	0.77	0.19	0.16	0.21	0.10	0.18	0.14
6	1.72	1.79	1.73	0.18	0.22	0.29	0.44	0.45	0.53
7	1.86	1.78	1.77	0.29	0.25	0.35	0.38	0.45	0.57
8	1.33	1.29	1.38	0.34	0.37	0.38	0.69	0.76	0.67
9	1.19	1.21	1.22	1.57	1.57	1.72	0.17	0.11	0.24
10	2.00	2.11	2.21	0.00	−0.01	−0.28	0.38	0.52	0.25
11	0.24	0.27	0.24	1.14	1.15	1.20	−0.95	−0.90	−0.88
12	0.51	0.48	0.55	1.21	1.48	1.34	−1.00	−1.04	−0.97
13	0.76	0.72	0.76	0.59	0.67	0.51	−0.96	−0.98	−1.03
14	0.01	0.04	0.07	1.94	1.79	1.64	−1.92	−1.79	−1.95
15	0.39	0.38	0.37	1.77	1.67	1.72	−1.57	−1.43	−1.63
16	0.76	0.77	0.72	0.99	1.08	0.99	−1.36	−1.52	−1.31
17	0.49	0.58	0.49	1.10	0.98	1.20	−0.81	−0.75	−0.76
18	0.29	0.24	0.38	1.10	1.16	1.02	−0.99	−1.00	−1.12
19	0.48	0.58	0.43	1.00	0.95	0.90	−1.56	−1.66	−1.49
20	0.42	0.44	0.44	0.75	0.71	0.58	−1.61	−1.65	−1.82
Mean	0.93	0.94	0.93	0.80	0.80	0.79	−0.39	−.39	−.40
Standard deviation	0.62	0.60	0.61	0.57	0.56	.56	0.98	.97	1.00

where $\boldsymbol{\theta}_a$ is a $1 \times m$ vector of estimates from the alternate form calibration, and $\widehat{\theta}'_b$ is the $1 \times m$ parameter estimate vector after transformation to the coordinate system from the base form.

The transformations presented here are useful for linking calibrations of distinct sets of items to form an item pool for computerized adaptive testing. They can also be used for multidimensional generalizations of horizontal and vertical equating (see Reckase and Martineau, 2004, for an example). In all of these cases, the goals of the analysis can be accomplished by including a common set of items, often called an anchor set, in the tests administered to different groups of examinees. When the groups are at the same educational or grade level, the result is horizontal equating. When the groups differ in educational or grade level, and the non-common items differ in difficulty, the analysis is called vertical scaling. In any case, the set of common items must be sensitive to all of the proficiency dimensions that are of interest. In many cases, the total number of proficiency dimensions is not known. In that case, Reckase and Hirsch (1991) showed that overestimating the number of dimensions was less of a problem than underestimating the number of dimensions. In the latter case, dimensions were projected on top of each other, losing important information. Therefore, calibrations should use enough dimensions to capture all of the meaningful capabilities in all of the tests being administered. This is especially important for vertical scaling because tests of differing levels of difficulty may assess different skills and knowledge. The number of dimensions for analysis should be the union of the dimensions in all of the tests being calibrated.

One of the more interesting applications of MIRT is as the basis for computerized adaptive testing (CAT). CAT is a testing methodology that has strong connections to the sequential design of experiments. In this case, the experiment is the administration of one test item and the result is the response to that test item. The design of a CAT addresses the problem of how to design the following experiments after obtaining the results of prior experiments. The design of subsequent experiments requires the selection of the next test item with the goal of giving results that can be used to determine the location of the examinee in the latent space.

The design of the $n + 1$ experiments is essentially the selection of the item that has characteristics that will reduce the error in estimation of the examinee's location in the latent space by the greatest amount. In order to select the item, that is, to design the next experiment, information must be available about the characteristics of the items that could be used for the experiment. This information is obtained through the calibration of test items using the procedures described above. The calibration results in a data file of prospective test items along with estimates of the parameters that describe the functioning of the item. This data file is typically called an item pool or item bank.

Numerous criteria have been suggested for selecting the next item when CAT is implemented using MIRT models. The most frequently used method for unidimensional versions of CAT is to select the item that provides the most information at the current estimate of $\boldsymbol{\theta}$. In the multidimensional case, the information provided by the item is given by Eq. (22). That formulation of information requires the specification of a direction that is the focus of the assessment. That direction is not easy to specify for a MIRT CAT so other criteria for selecting items have been considered. The method for item selection described here is directly related to the procedure for estimating the $\boldsymbol{\theta}$-vector so that methodology will be described first.

Maximum likelihood estimation is frequently used to obtain the estimate of θ when unidimensional models are used as the basis for CAT. In the unidimensional case, at least one correct and incorrect response is needed before a finite maximum likelihood estimate can be obtained. Before at least one of each response type is obtained, heuristics are used to update the θ-estimates. For MIRT based CAT, the number of types of response strings that do not lead to finite $\boldsymbol{\theta}$-vector estimates is greater than for the unidimensional case, so more elaborate heuristics are required. Acceptable heuristics have not yet been developed. Segall (2000) recommends using Bayesian estimation procedures as an alternative to maximum information item selection and maximum likelihood $\boldsymbol{\theta}$ estimation because it resolves the problem of infinite estimates and it makes the estimation more efficient by using prior information about the distribution of $\boldsymbol{\theta}$. The recommended methods are briefly described below.

The prior distribution for the $\boldsymbol{\theta}$-estimates is assumed to be multivariate normal with a specified mean vector and variance–covariance matrix:

$$f(\theta) = (2\pi)^{-\frac{m}{2}} |\boldsymbol{\Sigma}|^{-\frac{1}{2}} \mathrm{e}^{-\frac{1}{2}(\boldsymbol{\theta}-\boldsymbol{\mu})' \boldsymbol{\Sigma}^{-1}(\boldsymbol{\theta}-\boldsymbol{\mu})}, \tag{32}$$

where $\boldsymbol{\Sigma}$ is the variance/covariance matrix for the $\boldsymbol{\theta}$-values, $\boldsymbol{\mu}$ is a vector of means for the $\boldsymbol{\theta}$-values.

The other variables have been defined earlier. When Segall specifies the prior it is common to use a vector of zeros as the mean vector and to set the variances to 1.0. The covariances are set to values that are expected for the dimensions in the test data. Because most cognitive dimensions are correlated in an unselected population, using a value of .5 or .6 as prior correlations for the variables may be reasonable.

The posterior distribution of the $\boldsymbol{\theta}$-estimate is determined by multiplying the prior distribution by the likelihood of the observed response vector given $\boldsymbol{\theta}$ and then standardizing the distribution to have a total area of 1.0. The likelihood of the observed response vector is given by Eq. (2) above assuming the compensatory MIRT model. When $\boldsymbol{\theta}$ is the parameter of interest and the vector of responses $\mathbf{u}$ is considered fixed, the joint probability is represented by $L(\mathbf{u}|\boldsymbol{\theta})$. The posterior density of $\boldsymbol{\theta}$ given the response vector $\mathbf{u}$ is given by

$$f(\theta|\mathbf{u}) = \frac{L(\mathbf{u}|\theta)f(\theta)}{f(\mathbf{u})}, \tag{33}$$

where $f(\mathbf{u})$ is the marginal probability of $\mathbf{u}$ given by

$$f(\mathbf{u}) = \int_{-\infty}^{\infty} L(\mathbf{u}|\theta)f(\theta)\,\mathrm{d}(\theta). \tag{34}$$

The mode of the posterior density is used as the estimate for $\boldsymbol{\theta}$ because it is more convenient to compute that value than the mean or the median, especially when the number of coordinate dimensions is high. The differences in the various methods for summarizing the posterior distribution are small when a reasonable number of items have been administered.

Segall (2000) recommends using the characteristics of the posterior distribution for $\boldsymbol{\theta}$ to determine the next item to administer. Specifically, the item is selected that will result in the "largest decrement in the size of the posterior credibility region" (p. 62). The details of the mathematical derivation of the practical implementation of this criterion would require more space than is available in this chapter. The interested reader should refer to Segall (2000) for the details. He shows that maximizing the posterior information matrix provides the most decrement in the posterior credibility region. The information is computed for the already administered items, S_{k-1}, and each of the possible items that are candidates for administration, i. The criterion is evaluated at the most recent estimate of the $\boldsymbol{\theta}$ vector. The posterior information matrix is given by

$$I_{i,S_{k-1}}(\theta) = -E\left[\frac{\partial^2}{\partial\theta\partial\theta'}\ln f(\theta|\mathbf{u}_k)\right]. \tag{35}$$

When this value is determined in practice, a number of simplifying assumptions are made. The posterior distribution is assumed to be normally distributed even though the actual posterior distribution is not normal. The mean of the normal approximation is set to the mode of the actual posterior.

When MIRT CAT is implemented in practice, the number of items administered reaches a preset value, or the process continues until the posterior credibility region is smaller than some threshold value. When the stopping criterion has been reached, the estimated $\boldsymbol{\theta}$-vector is assumed to indicate the location of the examinee in the multidimensional space. The elements of the vector can be reported to provide results related to the coordinate axes, or the point can be projected onto reference composites to give estimates of performance on constructs defined by sets of items.

8. Discussion and conclusions

MIRT is a theory that indicates that the interaction between a person and a test item can be represented by probabilistic models that represent the person by a vector of coordinates in a multidimensional space. The application of that theory is through the development and evaluation of the forms of the probabilistic models and the development of statistical methodology for estimating the parameters of the models. If the MIRT model provides a reasonable representation of the item response data from the interaction of persons and test items, then the collection of related statistical methods can be applied to challenging problems in educational and psychological measurement.

The most immediate problem is dealing with the fact that most educational and psychological measurement instruments are not designed to be consistent with unidimensional IRT models. Even those that report single scores and use unidimensional IRT models for reporting scores typically have complex content specifications that imply that multiple skills and knowledge are needed to successfully respond to the test items. These specifications imply that multidimensional models are needed to properly model the item response data. Robustness of the unidimensional models and essential unidimensionality (Stout, 1987) are often used to support the use of the unidimensional models, but there is also an extensive literature checking whether or not unidimensionality assumptions are reasonable for the data from a particular test.

My own perspective on the dimensionality problem is that all matrices of test data need multidimensional models to accurately represent the interaction of persons and test items, but sometimes the multiple dimensions are highly correlated in the examinee sample or the set of items in the test are not very sensitive to differences in many of the dimensions. In the latter cases, the use of unidimensional models may be more efficient ways of getting at the major reference composite for the test, but it is still helpful to understand that other dimensions are in play. That understanding helps put differential item functioning into a formal theoretical context and helps explain the formal requirements for test parallelism.

MIRT is still in its infancy as a psychometric methodology. While there are few applications of MIRT within large scale testing programs as of 2005, there is a growing body of research related to the development of practical methods. At the time of the writing of this chapter, estimation procedures are fairly well developed and the practical issues of linking calibrations and representing analysis results are being addressed. A few applications of MIRT to test design and analysis are beginning to appear.

It is tempting to predict that MIRT procedures will be widely used in large scale testing programs within the next few years. Such predictions are difficult to support. The history of the implementation of computerized adaptive testing shows that very elegant solutions to testing problems do not necessarily get put into practice because of practical issues related to computer availability and test security. It appears that the use of MIRT will improve test design and development, the reporting of test results, and the efficiency of measurement. Whether or not the promise of this theory and methodology becomes a reality will depend largely the level of motivation of researchers in this field. It is hoped that this chapter will provide stimulus for progress in the practical application of this MIRT methodology.

References

Adams, R.J., Wilson, M., Wang, W. (1997). The multidimensional random coefficients multinomial logit model. *Applied Psychological Measurement* **21**, 1–23.

Birnbaum, A. (1968). Some latent trait models and their use in inferring an examinee's ability. In: Lord, F.M., Novick, M.R. (Eds.), *Statistical Theories of Mental Test Scores*. Addison-Wesley, Reading, MA.

Bock, R.D., Gibbons, R., Muraki, E. (1988). Full information item factor analysis. *Applied Psychological Measurement* **12**, 261–280.

Bolt, D.M., Lall, V.F. (2003). Estimation of compensatory and noncompensatory multidimensional item response models using Markov Chain Monte Carlo. *Applied Psychological Measurement* **27**, 395–414.

Briggs, D.C., Wilson, M. (2003). An introduction to multidimensional measurement using Rasch models. *Journal of Applied Measurement* **4**, 87–100.

Dempster, A.P., Laird, N.M., Rubin, D.B. (1977). Maximum likelihood estimation from incomplete data via the EM algorithm (with discussion). *Journal of the Royal Statistical Society, Series B* **39**, 1–38.

du Toit, M. (2003). *IRT from SSI: BILOG-MG, MULTILOG, PARSCALE, TESTFACT*. Scientific Software International, Lincolnwood, IL.

Fischer, G.H., Molenaar, I.W. (1995). *Rasch Models: Foundations, Recent Developments, and Applications*. Springer, New York.

Fraser, C. (1988). NOHARM II: A Fortran program for fitting unidimensional and multidimensional normal ogive models of latent trait theory. The University of New England, Armidale, Australia.

Hambleton, R.K., Swaminathan, H. (1985). *Item Response Theory: Principles and Applications*. Kluwer–Nijhoff, Boston, MA.

Hulin, C.L., Drasgow, F., Parsons, C.K. (1983). *Item Response Theory*. Dow Jones-Irwin, Homewood, IL.

Li, Y.H., Lissitz, R.W. (2000). An evaluation of the accuracy of multidimensional IRT linking. *Applied Psychological Measurement* **24**, 115–138.

Lord, F.M. (1980). *Application of Item Response Theory to Practical Testing Problems*. Lawrence Erlbaum Associates, Hillsdale, NJ.

Martineau, J.A. (2004). The effect of construct shift on growth and accountability models. Unpublished doctoral dissertation, Michigan State University, East Lansing, MI.

McDonald, R.P. (1962). A general approach to nonlinear factor analysis. *Psychometrika* **27**, 397–415.

McDonald, R.P. (1967). *Nonlinear Factor Analysis*. Psychometric Monographs, No. 15.

McDonald, R.P. (1981). The dimensionality of tests and items. *British Journal of Mathematical and Statistical Psychology* **34**, 100–117.

McDonald, R.P. (1982). Some alternative approaches to the improvement of measurement in education and psychology: Fitting latent model. In: Spearritt, D. (Ed.), *The Improvement of Measurement in Education and Psychology: Contributions of Latent Trait Theories*. Australian Council for Educational Research.

McDonald, R.P. (2000). A basis for multidimensional item response theory. *Applied Psychological Measurement* **24**, 99–114.

Miller, T.R., Hirsch, T.M. (1992). Cluster analysis of angular data in applications of multidimensional item-response theory. *Applied Measurement in Education* **5**, 193–211.

Min, K.-S. (2003). The impact of scale dilation on the quality of the linking of multidimensional item response theory calibrations. Unpublished doctoral dissertation, Michigan State University, East Lansing, MI.

Mulaik, S.A. (1972). *The Foundations of Factor Analysis*. McGraw-Hill, New York.

Rasch, G. (1960). *Probabilistic Models for some Intelligence and Attainment Tests*. Danish Institute for Educational Research, Copenhagen.

Rasch, G. (1961). On general laws and the meaning of measurement in psychology. In: *Proceedings of the Fourth Berkeley Symposium on Mathematical Statistics and Probability*, vol. 4, pp. 321–334.

Reckase, M.D. (1985). The difficulty of test items that measure more than one ability. *Applied Psychological Measurement* **9**, 401–412.

Reckase, M.D. (1997a). The past and future of multidimensional item response theory. *Applied Psychological Measurement* **21** (1), 25–36.

Reckase, M.D. (1997b). High dimensional analysis of the contents of an achievement test battery: Are 50 dimensions too many? Paper presented at the meeting of the Society for Multivariate Experimental Psychology, Scottsdale, AZ, October.

Reckase, M.D. (1997c). A linear logistic multidimensional model for dichotomous item response data. In: van der Linden, W.J., Hambleton, R.K. (Eds.), *Handbook of Modern Item Response Theory*. Springer, New York.

Reckase, M.D., Hirsch, T.M. (1991). Interpretation of number correct scores when the true number of dimensions assessed by a test is greater than two. Paper presented at the annual meeting of the National Council on Measurement in Education, Chicago, IL, April.

Reckase, M.D., Martineau, J. (2004). Vertical scaling of science achievement tests. Paper commissioned by the Committee on Test Design for K-12 Science Achievement, Center for Education, National Research Council, Washington, DC, October.

Reckase, M.D., Stout, W. (1995). Conditions under which items that assess multiple abilities will be fit by unidimensional IRT models. Paper presented at the European meeting of the Psychometric Society, Leiden, The Netherlands, July.

Reckase, M.D., Ackerman, T.A., Carlson, J.E. (1988). Building a unidimensional test using multidimensional items. *Journal of Educational Measurement* **25**, 193–203.

Reckase, M.D., Thompson, T., Nering, M. (1997). Identifying similar item content clusters on multiple test forms. Paper presented at the meeting of the Psychometric Society, Gatlinburg, Tennessee, June.

Reckase, M.D., Martineau, J., Kim, J. (2000). A vector approach to determining the number of dimensions needed to represent a set of variables. Paper presented at the annual meeting of the Psychometric Society, Vancouver, Canada, July.

Segall, D.O. (2000). Principles of multidimensional adaptive testing. In: van der Linden, W.J., Glas, C.A.W. (Eds.), *Computerized Adaptive Testing: Theory and Practice*. Kluwer, Dordrecht, The Netherlands.

Spray, J.A., Davey, T.C., Reckase, M.D., Ackerman, T.A., Carlson, J.E. (1990). Comparison of two logistic multidimensional item response theory models. Research Report ONR90-8, ACT, Inc., Iowa City, IA, October.

Stout, W. (1987). A nonparametric approach for assessing latent trait unidimensionality. *Psychometrika* **52**, 589–617.

Stout, W., Habing, B., Douglas, J., Kim, H.R., Roussos, L., Zhang, J. (1996). Conditional covariance-based nonparametric multidimensionality assessment. *Applied Psychological Measurement* **20**, 331–354.

Sympson, J.B. (1978). A model for testing with multidimensional items. In: *Proceedings of the 1977 Computerized Adaptive Testing Conference*, University of Minnesota, Minneapolis.

Wang, M.-m. (1985). Fitting a unidimensional model to multidimensional item response data: The effects of latent space misspecifications on the application of IRT. Unpublished manuscript.

Ward, J.H. (1963). Hierarchical groupings to optimize an objective function. *Journal of the American Statistical Association* **58**, 234–244.

Wherry, R.J., Gaylord, R.H. (1944). Factor pattern of test items and tests as a function of the correlation coefficient: Content, difficulty, and constant error factors. *Psychometrika* **9**, 237–244.

Whitely, S.E. (1980). Multicomponent latent trait models for ability tests. *Psychometrika* **45**, 479–494.

Handbook of Statistics, Vol. 26
ISSN: 0169-7161
2007 Published by Elsevier B.V.
DOI: 10.1016/S0169-7161(06)26019-X

19

Mixture Distribution Item Response Models

Matthias von Davier and Jürgen Rost

Abstract

This chapter provides an overview of research on mixture distribution models for item response data. Mixture distribution models assume that the observed item response data are sampled from a composite population, that is, a population consisting of a number of components or sub-populations. In contrast to multi-group models, mixture distribution models do not assume a known partition of the population but use an un-observed grouping variable to model the item response data. Mixture distribution models do not typically assume that the mixing variable is observed, but offer ways to collect evidence about this variable by means of model assumptions and observed heterogeneity in the data.

The components of a mixture distribution can be distinguished by the differences between the parameters of the assumed distribution model that governs the conditional distribution of the observed data. In the case of item response data, it may be assumed that either different parameter sets hold in different sub-populations or, in even more heterogeneous mixtures, different item response models hold in different sub-populations.

Item response models such as the Rasch model, the 2-parameter and 3-parameter logistic model, and the generalized partial credit model have been extended to mixture distribution IRT models.

This chapter introduces the derivation of mixture distribution models based on different origins and gives an overview of current fields of application in educational measurement, psychology, and other fields.

1. Introduction

This chapter gives an overview of research on mixture distribution models for item response data. Mixture distribution models assume that the observed item response data are sampled from a composite population, that is, a population that consists of a number of components or sub-populations. In contrast to multi-group models, mixture distributions do not assume a known partition of the population but use an un-observed grouping variable to model the item response data. Mixture distribution models do not typically assume that the mixing variable is observed, but offer ways to collect evidence about this variable by means of model assumptions and observed heterogeneity in the data.

The components of a mixture distribution can be distinguished by the differences between the parameters of the assumed distribution model that governs the conditional distribution of the observed data. In the case of item response data, it may be assumed that either different parameter sets hold in different sub-populations or, in even more heterogeneous mixtures, different item response models hold in different sub-populations.

Item response models such as the Rasch model, the 2-parameter and 3-parameter logistic model, and the generalized partial credit model have been extended to mixture distribution IRT models.

This chapter introduces the derivation of mixture distribution models based on different origins and gives an overview of current fields of application in educational measurement, psychology, and other fields.

2. Different perspectives on mixture IRT

Mixture distributions take the form of a weighted sum of conditional distributions (McLachlan and Peel, 2000). Therefore, mixture distributions can be defined in terms of marginal distributions by integrating across mixing components and conditional distributions within the mixing components. The first publication about the analysis of mixture distributions was probably Newcomb (1886), even though Pearson (1894) is often cited as being the first individual to have used mixtures of normal distributions.

In the following, the necessary definitions and assumptions for the models and applications introduced in this chapter will be developed. Assume random variables $x_i \in \{0, \ldots, m_i\}$, for $i = 1, \ldots, I$, a random variable c with values in $\{1, 2, \ldots, C\}$ and a k-dimensional real valued random variable θ, that is $\theta \in \mathbb{R}^k$. The x_i will represent the observed variables, in this context the item response variables, the c variable represents the unobserved mixing variable and may be interpreted as representing the indicator of populations, the θ variable is an (unobserved) latent variable, that often represents ability, proficiency or other constructs in educational and psychological applications of latent variable models such as item response theory. Let the count density of c be $\pi(c)$ and let the conditional density of θ given c be $\phi(\theta \mid c)$. Denote the conditional probability of response x to item i given θ in population c as $p_{ci}(x \mid \theta)$.

With the above definitions and assumptions, and the usual assumption of local independence (LI) of IRT, $p(x_1, \ldots, x_I \mid \theta, c) = \prod_i p_{ci}(x_i \mid \theta)$, the marginal mixture distribution of the observables $x = (x_1, \ldots, x_I)$ is

$$p(x_1, \ldots, x_I) = \sum_{c=1}^{C} \pi(c) \left(\int_\theta \prod_i p_{ci}\big(x_i \mid g_i(\theta)\big) \phi(\theta \mid c) \, \mathrm{d}\theta \right). \tag{1}$$

The expression in parentheses in (1) represents the conditional probability of $(x_1, \ldots, x_I)$ in population c, that is,

$$p(x_1, \ldots, x_I \mid c) = \int_\theta \prod_i p_{ci}\big(x_i \mid g_i(\theta)\big) \phi(\theta \mid c) \, \mathrm{d}\theta, \tag{2}$$

so that the expression in (1) may be rewritten as

$$p(x_1, \ldots, x_I) = \sum_{c-1}^{C} \pi(c) p(x_1, \ldots, x_I \mid c)$$

which conforms to the usual rules for calculations with conditional probabilities.

Applying Bayes' rule allows to determine the posterior probability of class membership as

$$p(c' \mid x_1, \ldots, x_I) = \frac{\pi(c') p(x_1, \ldots, x_I \mid c')}{\sum_{c=1}^{C} \pi(c) p(x_1, \ldots, x_I \mid c)} \tag{3}$$

for latent class c'. The posterior probabilities are important for classifying observations during the estimation of model parameters as well as after the estimation, when observations are assigned to the most likely class (maximum a-posteriori classification) or are assigned multiple imputations of class membership using the posterior distribution as defined in (3).

Note that so far, only local independence (LI) and class dependent response probabilities $p_c(\cdot \mid \theta)$ have been assumed, and no assumptions have been made about the density $\phi(\theta)$ of the ability variable or about the dimensionality of the projection $g_i(\theta)$. For $\phi(\theta)$ it is often assumed to take on a specific convenient form such as the normal density. The $g_i(\theta)$ denote transformations or projections, which may be a uni-dimensional component, i.e., $g_i(\theta) = \theta_k$, of the multivariate $\theta = (\theta_1, \ldots, \theta_k, \ldots, \theta_K)$ for models assuming simple structure.

2.1. *Mixtures of homogeneous sub-populations*

The statistical model behind latent class analysis (LCA; Lazarsfeld and Henry, 1968) is also a discrete mixture distribution model. The model equation for LCA is

$$P(x_1, \ldots, x_I) = \sum_{c=1}^{C} \pi(c) \prod_{i=1}^{I} p_{ci}(x_i)$$

and thus assumes independence of response variables $x_1, \ldots, x_I$ given class c, so that

$$p(x_1, \ldots, x_I \mid c) = \prod_{i=1}^{I} p_{ci}(x_i).$$

The latent class model may be understood as a mixture distribution IRT model with a degenerate conditional ability distribution, i.e., the ability variance can be considered zero within each class, since the $p_{ci}(x_i)$ are defined without an additional (ability) variable θ.

In LCA, the conditional response probabilities are constrained to be the same for all members of a given class (sub-population) c, but are allowed to differ across sub-populations.

It is often convenient to reparameterize the model equation in order to avoid the estimation of bounded probability-based parameters. A useful approach is to choose the logistic transformation and to write the conditional probabilities as

$$p_{ci}(x) = \frac{\exp(\beta_{cix})}{1 + \sum_{y=1}^{M_i} \exp(\beta_{ciy})} = \exp(\beta_{cix} - \delta_{ci}) \tag{4}$$

with $\delta_{ci} = \ln[1 + \sum_{y=1}^{M_i} \exp(\beta_{ciy})]$. This reparameterization does not change the model assumptions, but makes it easier to estimate parameters when additional constraints are to be met. In the form given in (4), the latent class model has obvious similarities to logistic models of item response theory (IRT).

One constrained model derived from unconstrained LCA in (4) is of particular importance for the derivation of mixture IRT models. The central idea is to disentangle item effects and group effects that are confounded in the β_{cix} parameters and therefore require a relatively large set of parameters for each latent class. This means that many parameters have to be estimated in each class, and therefore, the accuracy of these estimates will deteriorate when more classes are added to an existing model. A more parsimonious approach can be taken when using a linear decomposition of the β_{ixc} into item and class parameters such as

$$\beta_{ixc} = \theta_c - \alpha_{ix},$$

for decomposition into one class-level parameter and an item-category parameter, or

$$\beta_{ixc} = x\,(b_i\theta_c - a_i)$$

which yields a model that shares many features with a 2PL IRT model. These linear decompositions greatly reduce the number of necessary parameters. A linear logistic version of LCA was described by Formann (1982, 1985). Haberman (1979) has introduced linear constraints on the parameters and has shown that LCA is equivalent to a log-linear model containing an unobserved variable.

A practical advantage of the linear decomposed version of the latent class model is that each latent class is assigned an "ability level" θ_c and each item has one (set of) parameter(s) (e.g., α_{ix} or (a_i, b_i) in the examples above) that remains the same across latent classes.

2.2. *Mixed Rasch models and mixture IRT models*

The development of mixture IRT models has also been motivated by a number of different phenomena observed when modeling data from potentially non-homogeneous populations. Rost (1990, 1991) developed the mixed Rasch model on the basis of observing different response styles and different test taker strategies. The integration of quantitative ability differences into the modeling of student populations by means of qualitative differences (style, strategy choice) was the goal of developing the mixed Rasch model. The latent class and Rasch models share the local independence assumption, but use quite different variables on which to condition. In the Rasch model, a count of correct responses or credits is sufficient for estimating the ability parameter, whereas the latent class model assumes that an unobserved nominal variable, that is more or less implicitly defined by class profiles, explains all observed dependencies in the data.

2.2.1. The conditional Rasch model

The conditional form of the mixed Rasch model is particularly simple, since it combines the Rasch and latent class models in their most compatible forms. In the Rasch model, the conditional expression for the probability of a response pattern allows one to eliminate the person parameter in the equations necessary to estimate item parameters, thus avoiding problems of inconsistency of the estimators (Neyman and Scott, 1948).

The mixed Rasch model can also adopt the Rasch model in conditional form, so that item parameters can be estimated using this particularly useful approach (von Davier and Rost, 1995). By using the definition of the Rasch model to verify the sufficiency of the total score $r = \sum_i x_i$ we have

$$P(x_1, \ldots, x_I \mid \theta) = \exp(r\theta) \frac{\exp[\sum_{i=1}^{I} \beta_{ix_i}]}{\delta(\beta_{..})}$$

with $\delta(\beta_{..}) = \prod_{i=1}^{I}(1 + \sum_{x=1}^{m_i} \exp(x\theta - \beta_{ix}))$. Conditioning on r by using

$$p(r \mid \theta) = \sum_{\{(x_1,\ldots,x_I):\sum x_i = r\}} P(x_1, \ldots, x_I \mid \theta)$$

yields

$$P(x_1, \ldots, x_I \mid r, \theta) = \frac{\exp[\sum_{i=1}^{I} \beta_{ix_i}]}{\gamma(r)},$$

where the $\gamma(r)$ represents the symmetric function of order r. This function is defined as

$$\gamma(r) = \sum_{\{(x_1,\ldots,x_I):\sum x_i = r\}} \exp\left[\sum_{i=1}^{I} \beta_{ix_i}\right]$$

and represents a sum over terms based on all response patterns that share score r. Only the left-hand side of the above expression depends on θ, so that integrating over θ using $\int_\theta P(x_1, \ldots, x_I \mid r, \theta) p(\theta) \, d\theta = P(x_1, \ldots, x_I \mid r)$ yields

$$P(x_1, \ldots, x_I \mid r) = \frac{\exp[\sum_{i=1}^{I} \beta_{ix_i}]}{\gamma(r)},$$

and finally

$$P(x_1, \ldots, x_I) = \pi_r \frac{\exp[\sum_{i=1}^{I} \beta_{ix_i}]}{\gamma(r)},$$

since r is a function of the x_i. The $\pi_r = P(R = r)$ represent the probability of observing raw score r, which can be estimated from sample counts using $\hat{\pi}_r = \frac{N(R=r)}{N}$. The advantage of using this approach is that the integration over the ability variable θ can be avoided by operating on the basis of the sufficient statistic r for that variable. This approach is viable for models that, like the Rasch model, have sufficient statistics for θ based on observed variables $x_1, \ldots, x_I$ and possibly pre-specified weight constants. If weights are estimated, like in the 2PL, conditional maximum likelihood is infeasible, since the statistic that would be used for conditioning depends on estimated parameters in those cases.

2.2.2. *Combining Rasch model and latent class analysis*

The Rasch model in the above conditional form is finally augmented by including provisions to estimate the model parameters at the sub-population level and to combine these sub-populations in a discrete mixture distribution. Consistent with the above definition of a discrete mixture distribution, we may define

$$P(x_1, \ldots, x_I \mid c) = \pi_{r|c} \frac{\exp[\sum_{i=1}^{I} \beta_{ix_ic}]}{\gamma_c(r)}.$$

For the marginal probability of a response vector, we then have

$$P(x_1, \ldots, x_I) = \sum_{c=1}^{C} \pi_c \pi_{r|c} \frac{\exp[\sum_{i=1}^{I} \beta_{ix_ic}]}{\gamma_c(r)},$$

where, as introduced above, the π_c denote the mixing proportions for $c = 1, \ldots, C$, and the $\pi_{r|c}$ denote the probabilities of the latent score distribution for $r = 0, \ldots, R_{\max}$. For dichotomous items, the maximum score is $R_{\max} = I$ if I denotes the number of items. In the general case, let $x_i \in \{0, \ldots, m_i\}$ be the ith response variable, so that the maximum score for this item is m_i. The maximum raw score is obviously $R_{\max} = \sum_{i=1}^{I} m_i$ in this case, and if estimated without restrictions, the number of parameters for the latent score distribution also is $R_{\max}$ per latent class c.

Figure 1 illustrates how mixtures of IRT models can be understood. Instead of one graph, either several graphs, one for each sub-population need to be drawn. In this example the cloud-shaped figures represent the two different sub-populations.

Within each of the cloud shapes, an IRT model is represented as a graph that connects each observed variable, $X1$ to $X7$, with one latent variable, $T1$. The lack of connection between the two graphs in the figure represents the lack of constraints across the two sub-populations. Hence, the models have the same structure, but need not (and are in no way constrained to have) equal or even similar parameters, which in the case of graphical models would be depicted as letters next to the arrows.

In non-mixture IRT, there would be only one graph, and there is no need to include visual hints to show that separate populations are assumed. The graph as given in Figure 1 does not make a distinction between multi-group IRT and mixture IRT, since graphs usually represent only the structural relationships and variables and do not include information about the availability or missingness of person variables.

2.2.3. *Extensions of the mixed Rasch model*

von Davier and Rost (1995) described the estimation of model parameters for different polytomous mixed Rasch models, among these the mixture version of Andrich's (1988) rating scale model, the equidistance model (Andrich, 1978), Masters' (1982) partial credit model, and a model suggested by Rost (1988) that combines features of the rating scale model and the equidistance model, using conditional maximum likelihood (CML) techniques. As mentioned above, polytomous Rasch models may contain a large number of parameters when the response format allows for many categories. Polytomous models that constrain these parameters efficiently are especially useful in

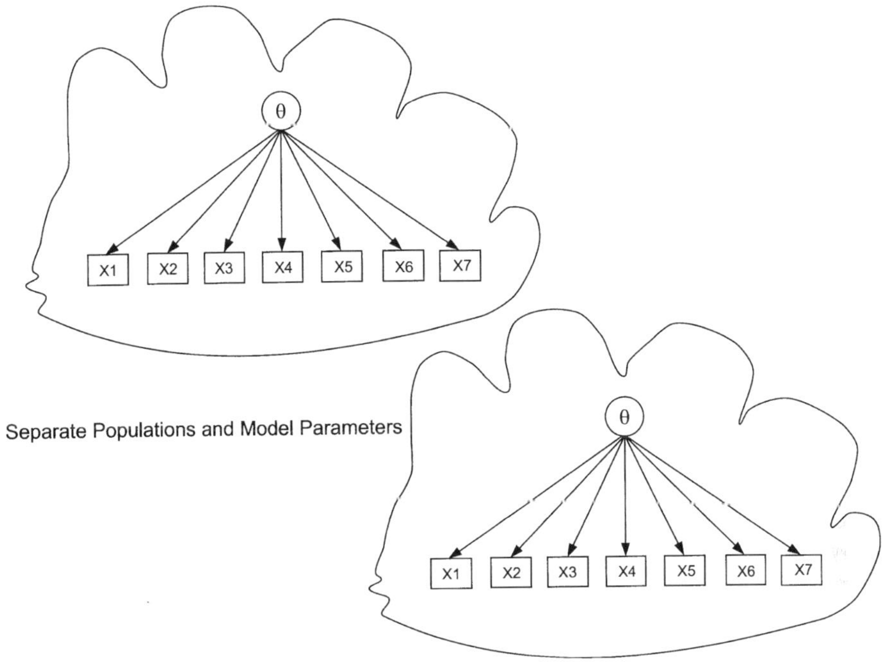

Fig. 1. A mixture IRT model as a graph.

reducing the large number of necessary threshold parameters for such mixture models, while unconstrained models will add a complete set of item parameters for each of the assumed sub-populations.

In addition to the potentially large number of item parameters in polytomous mixed Rasch models, the unobserved class dependent score distributions $\pi_{r|c}$ for $c = 1, \ldots, C$ contribute a considerable number to the total number of parameters to be estimated. von Davier and Rost (1995) use a logistic restriction on the latent score distribution that allows them to parameterize the mixed Rasch model using a discrete score distribution similar to what is used in log-linear pre-smoothing in equating (Holland and Thayer, 2000; von Davier et al., 2003).

2.3. *Log-linear mixture Rasch models*

Kelderman (1984) introduced the log-linear Rasch model by using the raw score as an additional variable making up the $I + 1$ dimensions in a table that is required to specify a model for $i = 1, \ldots, I$ items. The log-linear version of the dichotomous Rasch model can be defined as

$$\log \prod_{i=1}^{I} \frac{P_i(X = x_i \mid r)}{P_i(X = 0 \mid r)} = \left(r\tau_r - \sum_{i=1}^{I} x_i \alpha_i \right), \tag{5}$$

where i is the item index as introduced above and r is the raw score defined as $r = \sum_{i=1}^{I} x_i$. Note that this log-linear model contains structural zeroes, since r is a function of the x_i. The definition above is also suitable for polytomous data, i.e., for $x \in \{0, \dots, m_i\}$, so it also yields a log-linear version of the partial credit Rasch model.

Kelderman and Macready (1990) extend the log-linear Rasch model by introducing a group level index c into (5), which then becomes

$$\log \prod_{i=1}^{I} \frac{P_{ic}(X = x_i \mid r)}{P_{ic}(X = 0 \mid r)} = \left(r\tau_{rc} - \sum_{i=1}^{I} x_i \alpha_{ic} \right)$$

in the general case. Thus, by applying the rationale developed by Haberman (1979), Kelderman and Macready (1990) were able to treat the mixture version of the log-linear Rasch model as a log-linear model with a raw score effect and an unobserved grouping variable c. This model may be subject to additional constraints on the $\tau_{..}$ and $\alpha_{..}$ parameters. For example, the constraint $\tau_{rc} = \tau_r + \tau_c$ decomposes the τ_{rc} parameter into a raw score effect τ_r, which is constant across classes, and an effect τ_c, which is different across classes.

Expressing the Rasch model in the framework of log-linear models allows convenient ways to introduce additional parameters that reflect effects of learning over time or interaction of items with strategy choices. Meiser (1996) and Kreiner and Christensen (2004) developed models that extend the (mixture) log-linear Rasch model in such ways.

2.4. *Hybrid mixtures of IRT models*

Yamamoto (1987, 1989) followed a path in deriving the HYBRID model that is quite different from the typical way of extending the parameter space by employing an unobserved grouping variable. In contrast to common practice in developing discrete mixture distribution models that assumes the same type of model in all mixture components, Yamamoto's (1987, 1989) HYBRID model allows different models in the different components of the mixture distribution. In the most simple HYBRID model, the probability of a response pattern $x = (x_1, \dots, x_I)$ is given by

$$P(x_1, \dots, x_I) = \pi_{\mathrm{LCA}} \prod_{i=1}^{I} \pi_i(x_i) + \pi_{\mathrm{IRT}} \int_{\theta} \left[\prod_{i=1}^{I} p(x_i \mid \theta) \right] \phi(\theta)\, \mathrm{d}\theta$$

with class sizes (or mixing proportions) $\pi_{\mathrm{LCA}} + \pi_{\mathrm{IRT}} = 1$. This model assumes two sub-populations, one in which independence of responses holds (guessing or random responses on all items would be an obvious way to describe this class), and one class in which an IRT model such as the Rasch model or the 2PL model holds.

The HYBRID model can easily be extended to more than two sub-populations. von Davier (1997) presented a HYBRID model for polytomous data that enabled him to use the partial credit model along with more restrictive polytomous Rasch models such as Andrich's (1988) rating scale model and other polytomous, constrained varieties of the Rasch model (Andrich, 1978; Rost, 1988).

Yamamoto and Everson (1995) extended the HYBRID model to a model that can be used for studying speededness effects and that potentially contains a large number of classes. Identifiability and a limit on the number of parameters in this type of HYBRID model are facilitated by using multiple constraints across classes. These restrictions on the parameter space ensure that each sub-population is defined by a switching point $1 < k < I$. For members of the kth class, items located before the switching point k are responded to by using the ability or proficiency according to the assumptions of an IRT model. Items following after the switching point k are responded to in a random fashion, and this sequence of responses can accordingly be modeled by the independence assumption of LCA.

2.5. *Mixture IRT models, with covariates, and partially observed mixtures*

Mislevy and Verhelst (1990) developed a mixture IRT approach for modeling strategy differences in student populations. Mislevy and Verhelst employed marginal maximum likelihood estimation as implemented in the HYBIL software (Yamamoto, 1993) to estimate a mixture IRT model using the 2PL (and constrained versions of the 3PL) for strategy differences in solving mental rotation tasks. Instead of using conditional maximum likelihood and a latent score distribution $(\pi_{r|c})_{r=1..R_{\max}}$ as introduced in the preceding section, multi-parameter IRT mixture models also provide a way to integrate out the ability distribution in each latent class by applying marginal maximum likelihood methods. This can be done either assuming normality with class specific means $\mu_c(\theta)$ and variances $\sigma_c^2(\theta)$ or by using numerical quadrature over a discrete count density.

Smit et al. (1999, 2000) developed a dichotomous mixture IRT model where the posterior probabilities of class membership were calculated using collateral information in conjunction with item responses. von Davier and Yamamoto (2004a) developed a mixture distribution generalized partial credit model (GPCM; Muraki, 1992) and extended the estimation of mixture IRT models by developing a general method to incorporate both observations with known group membership as well as observations with unknown grouping information.

In latent class models, class membership is treated as an unobserved variable, whereas multiple-group models treat class or group membership as observed without error. The approach proposed by von Davier and Yamamoto (2004a) allows one to treat class membership as a variable with partially missing observations, or observed with error. Technically, this is handled by assuming different priors, one that is based on the class sizes π_c for observations with missing class membership and one that is deterministic, i.e., $\pi_c^* = 1$ if the class membership g equals c and $\pi_{c'}^* = 0$ for $c' \neq g$. A more general approach can be taken by assigning each examinee an individual prior distribution for class membership.

2.6. *Diagnostic mixture IRT models*

Mixture distribution models have been extended to models with more than one ability variable. Recently proposed models fall under the term "skill profile models" or "diagnostic models" and assume multiple discrete latent variables, often latent dichotomies that are labeled and interpreted as "mastery/non-mastery" of skills.

von Davier and Yamamoto (2004b) have suggested a general diagnostic model (GDM) that contains the latent class model, the (mixture) Rasch model, and the GPCM, as well as diagnostic models such as multiple classification latent class models (Maris, 1999) and a compensatory version of the fusion (reparameterized unified) model (Hartz et al., 2002), as special cases. This class of models can be viewed as discrete multidimensional mixture IRT models. An estimation algorithm based on marginal maximum likelihood techniques was developed and validated in a parameter recovery study using simulated data (von Davier, 2005).

A central building block of most if not all diagnostic models is a matrix that is often referred to as a Q-matrix (Tatsuoka, 1983), an $I \times K$ matrix that relates the I items to the K skills/attributes/dimensions. The entries q_{ik} are integers in most cases, often $q_{ik} \in \{0, 1\}$. The Q-matrix can be understood as the structural component of the model defining a hypothesis as to which items require which skill or combination of skills.

von Davier (2005) uses a version of the GDM that implements a multidimensional discrete version of the (mixture) generalized partial credit model. This model, pGDM, is therefore suitable for dichotomous and ordinal responses $x \in \{0, 1, 2, \dots, m_i\}$. The model equation for a mixture version of the pGDM is

$$P(X = x \mid \beta_{i\cdot\cdot}, a_{\cdot}, q_{i\cdot}, \gamma_{i\cdot\cdot}, c) = \frac{\exp[\beta_{xic} + \sum_{k=1}^{K} x\gamma_{ikc} q_{ik} a_k]}{1 + \sum_{y=1}^{m_i} \exp[\beta_{yic} + \sum_{k=1}^{K} y\gamma_{ikc} q_{ik} a_k]} \tag{6}$$

with K attributes, or discrete latent traits, $a = (a_1, \dots, a_K)$, latent class c, and a dichotomous design Q-matrix $(q_{ik})_{i=1..I,k=1..K}$. The β_{ixc} are difficulty parameters and the $x\gamma_{ikc}$ term is a discrimination parameter for skill k parameterized like it is done in the GPCM. The a_k are discrete scores determined before estimation and can be chosen by the user. These scores are used to assign real numbers to the skill levels; for example $a(0) = -1.0$ and $a(1) = +1.0$ may be chosen for dichotomous skills. For ordinal skills with s_k levels, the a_k may be defined using $a(x) = x$ for $x = 0, \dots, (s_k - 1)$ or $a(0) = -s_k/2, \dots, a(s_k - 1) = s_k/2$.

Figure 2 illustrates how mixtures of diagnostic models can be understood. Diagnostic models include an hypothesis of which items relate to which attributes or skills. In a graph, such a hypothesis is represented such that certain arrows between latent variables and observed variables are drawn, whereas certain other potential relationships are omitted. In the figure, the skill variable $\theta 1$ is related to the observed variables $X1$ to $X5$, but not to $X6$ and $X7$. The second latent variable, skill $\theta 2$, is related to $X2$, $X4$, $X5$, $X6$, and $X7$ only.

In addition to the diagnostic structure as described above, mixtures of diagnostic models allow that the strength of the hypothesized relations between skills and observed response variables to differ in different sub-populations. This is depicted in Figure 2 as an additional variable denoted as *group?class* in the figure. This variable can be observed, as in multi-group models, or unobserved or latent, as in discrete mixture models.

It can be seen in the graph that the *group?class* variable is not connected to all arrows that go from the skill variables $\theta 1$ and $\theta 2$ to the observed variables. This is an indicator that the skill variables not connected to *group?class* are assumed to be independent

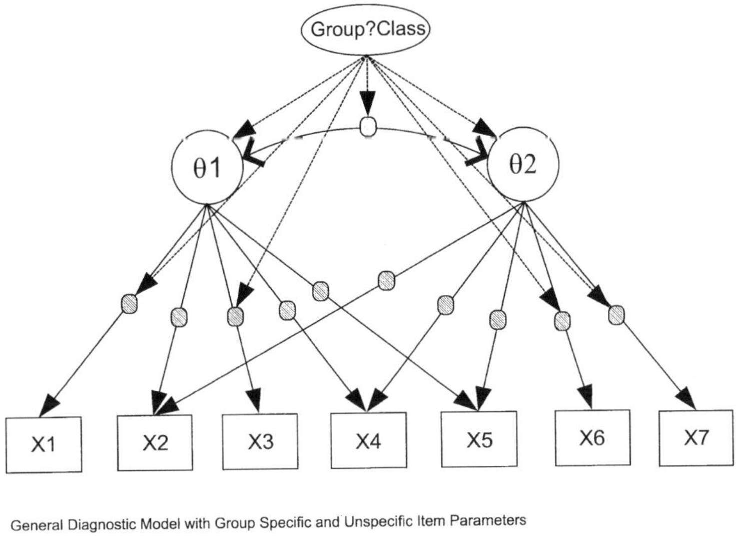

Fig. 2. A mixture general diagnostic model (GDM) with constraints across populations as a graph.

of this variable. What does that mean for the structural item parameters? They are the same across groups or populations for those skill-item arrows that are not connected to *group?class*. In the graph, this is represented by a missing arrow from *group?class* to the connecting arrow. In the mathematical model, this would be indicated by assumptions (constraints) of equality across item parameters.

This pGDM as implemented in the *mdltm* software (von Davier, 2005) has been extended to estimate mixture diagnostic models. *mdltm* allows one to implement models with equality constraints as depicted in the figure. In addition, since many common models for item response data are special cases of the GDM, IRT models such as the Rasch model, the 2PL model, and the GPCM, as well as multidimensional mixture versions of these models, can be estimated with *mdltm*. Using this software, Xu and von Davier (2006) applied different special cases of the pGDM to data from national and international educational survey assessments.

2.7. *Estimation*

For conditional maximum likelihood (CML) and marginal maximum likelihood (MML) estimates, the expectation maximization (EM) algorithm (Bock and Aitkin, 1981) has proven to be very useful in the context of IRT models and mixture IRT models. The mixed Rasch models can be estimated using conditional maximum likelihood techniques with WINMIRA 2001 (von Davier, 2001). Hybrid IRT models can be estimated using HYBIL (Yamamoto, 1993), and hybrid Rasch models can be estimated using WINMIRA 2001 (von Davier, 2001). Log-linear (mixed) Rasch models can be estimated using LOGIMO (Kelderman, 1988) and Lem (Vermunt, 1997). Mixture versions of the general diagnostic model (GDM) can be estimated using marginal maximum likelihood as implemented in the *mdltm* software (von Davier, 2005).

Markov-chain Monte-Carlo (MCMC) estimation of IRT models has been suggested (Junker, 1999; Patz and Junker, 1999 as an alternative to MML techniques in order to allow for the estimation of more complex extensions of IRT models. The greater ease of implementing models with MCMC results from the use of Bayesian estimation techniques that do not require derivatives of likelihood functions or estimation equations. Instead, MCMC techniques iterate through thousands of systematic draws from the model parameter space. This is conducted by following certain algorithmic schemes and requires the specification of prior distributions for each type of parameter. Bolt et al. (2001, 2002) have estimated mixed Rasch models (Rost, 1990) and variants of the HYBRID model (Yamamoto, 1989; Yamamoto and Everson, 1997) using MCMC methods as they are implemented in WinBUGS (Gilks et al., 1996).

2.8. *Testing mixture IRT models*

Global goodness of fit indices used for checking model-data fit that are applicable to LCA may also be used for mixture IRT models. Statistics from the power-divergence family of goodness-of-fit statistics (Cressie and Read, 1984) can be calculated, since mixture IRT models predict counts for each possible response vector. Therefore, fit measures like the likelihood ratio chi-square statistic and the Pearson X^2 are readily computed. One important caveat is that these statistics cannot be tested using the well known χ^2 asymptotics, since the regularity conditions that allow one to use the χ^2 are often not met. Glas and Verhelst (1995) offered alternatives for testing the divergence of observed and expected frequencies globally across all possible response patterns. In this case, observed and expected frequencies of sets of response patterns that share some criteria (e.g., all response patterns that have the same response to some item) are compared.

Alternatives to fit statistics are information criteria like the AIC (Akaike, 1973) or BIC (Schwarz, 1978) that provide a decision rule based on penalization of increases in log-likelihood, done by balancing additional parameters in when increasing model complexity. Increasing model complexity means here, in the case of mixture IRT models, either assuming a less restrictive IRT model within classes or estimating a mixture model with more latent classes.

Mixture IRT models have been shown to provide a powerful test of population invariance of item parameters and ability distribution. Rost and von Davier (1995) suggested the use of the mixed Rasch model here and von Davier (1997) has shown that re-simulation methods can be used to check how many populations are needed to achieve conditional homogeneity and model fit. However, re-simulation based methods are a computationally costly way to assess the accuracy of the model in predicting the observed data. Langeheine et al. (1996) and von Davier (1997) have studied re-simulation methods, more specifically the parametric bootstrap, for generating simulation based null-distributions of goodness-of-fit measures for checking mixture distribution models. For estimating mixture IRT using Bayesian methods, posterior predictive checks are a similar method, which is based on drawing parameters from the stationary posterior distribution of an MCMC chain and generating realizations of the observed variables based on these draws (Sinharay, 2003).

Similar to the generalized Pearson tests, fit diagnostics (Molenaar, 1983) deal with some specific comparison of model prediction and observed data. Fit diagnostics are often focused on single items; in this case they are referred to as item fit measures, or single response patterns, then referred to as person-fit measures. There is a large variety of measures in this category, and many of these are applicable to mixture IRT models, at least as measures of class specific items or person fit. von Davier and Molenaar (2003) described a normally distributed person-fit statistic for the latent class and mixture distribution Rasch models.

3. Applications of mixture IRT models

Research on strategy differences when solving cognitive tasks provided one of the initial applications of mixture IRT models. To name only a selection of research work done along these lines, Mislevy and Verhelst (1990), Rost and von Davier (1993), and recently Rijkjes and Kelderman (2006) fall into this category. The reason for choosing mixture distributions for studying strategy differences is obvious, since the selection of one out of a variety of strategies by different students is usually a choice that does not get recorded. Therefore, strategy usage is in most cases an unobserved variable. Nevertheless, different strategies result in differential item difficulties, since items that are comparably easy under one strategy may prove difficult under an alternative strategy. Rost and von Davier (1993) used the mixed Rasch model to analyze item response data from a test on mental rotation tasks, which depict a cut-out dice as the target, an isomorphic version of the dice cut-out in a different way, along with a couple of non-matching distracters. These tasks can be solved by what is known as either a rotation strategy or an analytic strategy. Depending on strategy use, different items are relatively easier, so that the mixed Rasch model was used in this case to determine two population based on their different profiles of item difficulties that reflect preferred strategies.

Meiser (1996) describes a Mover–Stayer mixed Rasch model for analyzing developmental data containing item responses from multiple assessment waves in longitudinal designs. Movers are students who change their strategy to a more elaborate one, whereas Stayers stick to their initial, often less sophisticated strategy over the assessment cycles. In this application of the mixed Rasch model, there are two latent classes representing two strategies, one more elaborate than the other. Across assessment waves, some individuals may "move up" to the next, more developed strategy, while other individuals in the sample may stay (i.e., may keep using a less developed strategy to solve the items).

Recently, Rijkjes and Kelderman (2006) described a model for strategy shift using the log-linear formulation of mixture Rasch models. Their examples of items that require analysis with mixture distribution item response models are balancing tasks constructed based on developmental theories. Rijkjes and Kelderman based their modeling on theories that describe under which circumstances children may change their solution strategies to maximize performance. In this paper, mixture IRT models are specifically based on developmental theories in the log-linear framework and tested based on empirical data from student performance on cognitive beam-balancing tasks.

Self-reports on attitude measures using rating scales, as well as self-reported behavioral ratings, may vary according to preferred response styles or to the intention to adjust the self-presentation towards an assumed ideal. Rost et al. (1997) presented an application of the mixed Rasch model to the analysis of response styles in large scale personality questionnaire data. The authors used data collected for a national adaptation of a big-five personality inventory.

Eid and Rauber (2000) analyzed response tendencies in organizational surveys. The authors were able to identify individual respondents based on person-fit measures (see von Davier and Molenaar, 2003) that led to model misfit. von Davier and Rost (1995) reanalyzed data from the revised self-monitoring scale (Gangestad and Snyder, 2000) and found that a HYBRID Rasch model in large part confirmed the structure of the construct that stands behind the construction of the self-monitoring scale. Eid and Zickar (2006) gave an overview of the application of mixture IRT models in the analysis of response styles and in the tendency to influence (fake) impressions in non-cognitive tests.

The idea that helped in formulating the HYBRID model is that guessing or random response behavior may be more appropriately modeled on the person level than on the item level. The HYBRID model in its original form (Yamamoto, 1987) included a class of subjects that utilize a random response style on all the items in a test. Later extensions of this model allowed one to model a breaking point at which respondents move from an ability, skill, or proficiency based response mode to a random response mode. Yamamoto and Everson (1997), Boughton and Yamamoto (2006), and Bolt et al. (2002) described applications of the HYBRID speededness model.

Recent extensions of mixture IRT models were aimed at allowing the model to be used with sparse matrix samples of item responses, weighted data, and data that includes prior information on class membership or partial grouping information (Von Davier and Yamamoto, 2004c). These extensions were aimed at questions as to the appropriateness of joint IRT scaling in complex student populations, where some school types are of unknown performance level. The extensions made mixture IRT models suitable for scaling tasks that contain anchor test linkage designs and multiple populations with partially missing data. von Davier (2005) described a class of models for cognitive diagnosis that can also be applied in the mixture distribution framework. Mixture distribution diagnostic models can be estimated using the EM-algorithm as implemented in the *mdltm* software (von Davier, 2005), as can multiple group versions of diagnostic models (Xu and von Davier, 2006).

4. Mixture IRT for large-scale survey assessments

Von Davier and Yamamoto (2004c) presented an application of a mixture distribution generalized partial credit model to large scale survey assessment data from a national version of an international assessment of 15-year-olds. This analysis used a complex sample of students from different school types, consisting of three conventional school types (A, B, and C) that were treated as observed data and five additional school types, treated as missing, since the student proficiency in the school types of less traditional

Table 1
Class proportions for the 3-group IRT model and the 3-group + 3-class model, including the previously unclassified schools

School type	Schools A, B, C only	A, B, C schools plus others	
	$N = 6{,}689$	$N = 9{,}339$	rel. size
C	0.647	0.577	(−)
B	0.281	0.306	(+)
A	0.071	0.116	(+)

type can not be aligned a priori with the traditional schools. The non-traditional school types generally admit a broader range of students. The absolute numbers of these types of schools is quite small both in the sample as well as in the population, and this makes it inevitable that these schools be combined with other types of schools for reporting in many cases, but it is unknown as to which is the correct group of reference.

The scale in this application consists of 65 dichotomous and polytomous scored math items that were suitable for all school types. The assessment was organized in 10 booklets in a matrix design with 14 anchor items. The booklets were administered to approximately 9,400 students from about 180 schools that participated in a federally funded instruction improvement project. About 6,700 students came from the three traditional school types (A, B, and C), and the remaining students came from non-traditional and thus unclassified school types.[1]

The parameters of the 14 anchor items were constrained to be equal across all three groups to establish identifiability and scale linkage across groups. The parameters for the anchor test were taken from the operational calibration, whereas the remaining items were estimated in a 3-class mixture distribution model with partially missing classification information without further constraints. This relates to well-known IRT scale linkage designs with an anchor test that is the same across populations.

The model with three separate school-type based ability distributions was estimated twice, once without the students having missing values on the school type variable (i.e., only students from school types A, B, and C were included), and once including all students (i.e., students from school types A, B, and C, and students from other schools that were assumed to have missing school types and were classified into the known types A, B, and C).

Table 1 shows the class proportions of the school types when computing the 3-group model using only the all-known school type subjects (second column) versus using all subjects, including those from the non-traditional schools (third column). The fourth column indicates the change in size by (−) and (+) signs.

The changes in the class proportions indicate that the students with non-traditional school type information are mainly classified into types A and B schools, as these class

[1] A mixture IRT analysis with school type as an unobserved variable can also be considered, but Von Davier and Yamamoto (2004c) opted for a mixture analysis with partially missing classifications to showcase the applicability of this approach, and more importantly, because the three traditional schools are known to exhibit three largely different low, medium, and high ability distributions.

proportions increase while the proportion in school type C decreases from $\pi_{C,6689} = 0.647$ to $\pi_{C,9339} = 0.577$. Recall that the students with missing school type are classified by means of Bayes' theorem, so that the class sizes play the role of the prior in the classification. School type C dominates the class proportions, so the prior probability is largest for school type C, meaning that school type C should receive 65% of the students if no further information is available. When adding the students with missing school type information, however, the relative proportion of type C decreases, indicating that more students than expected, based on the prior, are classified into the lower performing school types A and B. Von Davier and Yamamoto (2004c) showed that all conditional means decrease when adding the unclassified students to the students from the traditional school types A, B, and C. The conditional standard deviations increase slightly for school types B and C and decrease slightly for school type A. The decrease in all conditional means in the augmented sample corresponds to the observation that the latent distributions were augmented mainly in the lower proficiency levels when adding the unclassified students to the sample of students with known school type.

The largest proportion of the traditional school type sample comes from type C, which represents about 65% of this sample. However the analyses by Von Davier and Yamamoto (2004c) showed that only between 44% and 56% of the students with missing school type are classified into this school type. The authors found that the students from non-traditional schools are classified more according to their proficiency than according to the prior probabilities given by the traditional school type sample. Additional analyses of the data showed that when applying an uninformative prior instead of the prior based on traditional school type proportions for classification, the students from non-traditional schools are classified even more frequently as belonging to the low-achieving school types A and B.

5. Conclusion

Mixture distribution models for item response data range from latent class models to mixture versions of diagnostic models. Common to all such models is the assumption that the observed data stems from a composite population with an unknown number of components. The aim of mixture IRT models is therefore twofold, to identify homogeneous populations where the class specific model assumptions are met, and to unmix the sample in these unobserved homogeneous components of the population.

The increasing level of complexity of models that are available within the mixing component may counteract the necessity of mixtures to some extent. This is, from the perspective of flexibility, desirable, but it may however prove less desirable from the perspective of model selection. While a greater flexibility in choosing between different model assumptions can be viewed as a good thing, the reality is that substantive research seldom provides hypotheses exact enough to derive which model choice is most appropriate. Therefore, many different alternatives may be chosen, estimated, and checked. Among these, several models may provide comparable model-data fit and thus challenge the researcher to make a conscious choice. Parsimony and the requirements of the specific application provide some guidance here, so that the number of models that will actually be taken to the next level of selection is hopefully small enough, and, in the ideal, equals one.

References

Akaike, H. (1973). Information theory and an extension of the maximum likelihood principle. In: Petroy, B.N., Csaki, F. (Eds.), *The Second International Symposium on Information Theory*. Akadémiai Kiado, Budapest, Hungary.

Andrich, D. (1978). A rating formulation for ordered response categories. *Psychometrika* **43**, 561–573.

Andrich, D. (1988). *Rasch Models for Measurement*. Sage, Newbury Park, CA.

Bock, R.D., Aitkin, M. (1981). Marginal maximum likelihood estimation of item parameters: An application of the em algorithm. *Psychometrika* **46**, 443–459.

Bolt, D.M., Cohen, A.S., Wollack, J.A. (2001). A mixture item response model for multiple choice data. *Journal of Educational and Behavioral Statistics* **26**, 381–409.

Bolt, D.M., Cohen, A.S., Wollack, J.A. (2002). Item parameter estimation under conditions of test speededness: Application of a mixture Rasch model with ordinal constraints. *Journal of Educational Measurement* **39** (4), 331–348.

Boughton, K., Yamamoto, K. (2006). A hybrid model for test speededness. In: von Davier, M., Carstensen, C. (Eds.), *Multivariate and Mixture Distribution Rasch Models*. Springer, New York.

Cressie, N., Read, T.R.C. (1984). Multinomial goodness-of-fit tests. *Journal of the Royal Statistical Society, Series B* **46** (3), 440–464.

Eid, M., Rauber, M. (2000). Detecting measurement invariance in organizational surveys. *European Journal of Psychological Assessment* **16**, 20–30.

Eid, M., Zickar, M. (2006). Detecting response styles and faking in personality and organizational assessment by mixed Rasch models. In: von Davier, M., Carstensen, C. (Eds.), *Multivariate and Mixture Distribution Rasch Models*. Springer, New York.

Formann, A. (1982). Linear logistic latent class analysis. *Biometrical Journal* **24**, 171–190.

Formann, A.K. (1985). Constrained latent class models: Theory and applications. *British Journal of Mathematical and Statistical Psychology* **38**, 87–111.

Gangestad, S.W., Snyder, M. (2000). Self-monitoring: Appraisal and reappraisal. *Psychological Bulletin* **126**, 530–555.

Gilks, W.R., Richardson, S., Spiegelhalter, D.J. (1996). *Markov Chain Monte Carlo in Practice*. Chapman and Hall, London.

Glas, C.A.W., Verhelst, N.D. (1995). Testing the Rasch model. In: Fischer, G.H., Molenaar, I.W. (Eds.), *Rasch Models – Foundations, Recent Developments and Applications*. Springer, New York.

Haberman, S.J. (1979). *Qualitative Data Analysis: Vol. 2. New Developments*. Academic Press, New York.

Hartz, S., Roussos, L., Stout, W. (2002). Skills diagnosis: Theory and practice. User Manual for Arpeggio software [Computer software manual], ETS, Princeton, NJ.

Holland, P.W., Thayer, D.T. (2000). Univariate and bivariate loglinear models for discrete test score distributions. *Journal of Educational and Behavioral Statistics* **25**, 133–183.

Junker, B.W. (1999). Some statistical models and computational methods that may be useful for cognitively relevant assessment. Retrieved March 16, 2006, from http://www.stat.cmu.edu/~brian/nrc/cfa/.

Kelderman, H. (1984). Loglinear Rasch model tests. *Psychometrika* **49**, 223–245.

Kelderman, H. (1988). An IRT model for item responses that are subject to omission and/or intrusion errors. Res. Rep. No. 88-16, University of Twente, Enschede, The Netherlands.

Kelderman, H., Macready, G.B. (1990). The use of loglinear models for assessing differential item functioning across manifest and latent examinee groups. *Journal of Educational Measurement* **27**, 307–327.

Kreiner, S., Christensen, K.B. (2004). Analysis of local dependence and multidimensionality in graphical loglinear Rasch models. *Communications in Statistics. Theory and Methods* **33**, 1239–1276.

Langeheine, R., Pannekoek, J., van de Pol, F. (1996). Bootstrapping goodness-of-fit measures in categorical data analysis. *Sociological Methods and Research* **24** (4), 492–516.

Lazarsfeld, P.F., Henry, N.W. (1968). *Latent Structure Analysis*. Houghton Mifflin, Boston.

Maris, E. (1999). Estimating multiple classification latent class models. *Psychometrika* **64** (2), 187–212.

Masters, G.N. (1982). A Rasch model for partial credit scoring. *Psychometrika* **47**, 149–174.

McLachlan, G., Peel, D. (2000). *Finite Mixture Models*. Wiley, New York.

Meiser, T. (1996). Loglinear Rasch models for the analysis of stability and change. *Psychometrika* **61**, 629–645.

Mislevy, R.J., Verhelst, N.D. (1990). Modeling item responses when different subjects employ different solution strategies. *Psychometrika* **55** (2), 195–215.

Molenaar, I.W. (1983). Some improved diagnostics for failure in the Rasch model. *Psychometrika* **48**, 49–72.

Muraki, E. (1992). A generalized partial credit model: Application of an EM algorithm. *Applied Psychological Measurement* **16** (2), 159–176.

Newcomb, S. (1886). A generalized theory of the combination of observations so as obtain the best result. *American Journal of Mathematics* **8**, 343–366.

Neyman, J., Scott, E.L. (1948). Consistent estimates based on partially consistent observations. *Econometrica* **16**, 1–32.

Patz, R., Junker, B. (1999). A straightforward approach to Markov chain Monte Carlo methods for item response models. *Journal of Educational and Behavioral Statistics* **24**, 146–178.

Pearson, K. (1894). Contributions to the mathematical theory of evolution. *Philosophical Transactions A* **185**, 71–110.

Rijkjes, C.P.M., Kelderman, H. (2006). Latent response Rasch models for strategy shifts in problem solving processes. In: von Davier, M., Carstensen, C.H. (Eds.), *Multivariate and Mixture Distribution Rasch Models*. Springer, New York.

Rost, J. (1988). *Quantitative und qualitative probabilistische Testtheorie*. Verlag Hans Huber, Bern, Switzerland.

Rost, J. (1990). Rasch models in latent classes: An integration of two approaches to item analysis. *Applied Psychological Measurement* **14**, 271–282.

Rost, J. (1991). A logistic mixture distribution model for polychotomous item responses. *British Journal of Mathematical and Statistical Psychology* **44**, 75–92.

Rost, J., von Davier, M. (1993). Measuring different traits in different populations with the same items. In: *Psychometric Methodology. Proceedings of the 7th European Meeting of the Psychometric Society*. Gustav Fischer Verlag, Stuttgart, Germany, pp. 446–450.

Rost, J., von Davier, M. (1995). Mixture distribution Rasch models. In: Molenaar, I.W. (Ed.), *Rasch Models – Foundations, Recent Developments and Applications*. Springer, New York, pp. 257–268.

Rost, J., Carstensen, C.H., von Davier, M. (1997). Applying the mixed Rasch model to personality questionnaires. In: Rost, J., Langeheine, R. (Eds.), *Applications of Latent Trait and Latent Class Models in the Social Sciences*. Waxmann Publishing, New York, pp. 324–332.

Schwarz, G. (1978). Estimating the dimensions of a model. *Annals of Statistics* **6**, 461–464.

Sinharay, S. (2003). Practical applications of posterior predictive model checking for assessing fit of common item response theory models. ETS RR-03-33, ETS, Princeton, NJ.

Smit, A., Kelderman, H., van der Flier, H. (1999). Collateral information and mixed Rasch models. *Methods of Psychological Research* **4**, 1–12.

Smit, A., Kelderman, H., van der Flier, H. (2000). The mixed Birnbaum model: Estimation using collateral information. *Methods of Psychological Research Online* **5**, 31–43.

Tatsuoka, K.K. (1983). Rule space: An approach for dealing with misconceptions based on item response theory. *Journal of Educational Measurement* **20**, 345–354.

Vermunt, J.K. (1997). *ℓEM*: A general program for the analysis of categorial data [Computer software]. Retrieved March 17, 2006, from http://www.uvt.nl/faculteiten/fsw/organisatie/departementen/mto/software2.html.

von Davier, M. (1997). *IPN Schriftenreihe: Vol. 157. Methoden zur Prüfung probabilistischer Testmodelle*. IPN, Kiel, Germany.

von Davier, M. (2001). WINMIRA 2001 [Computer software]. Retrieved February 16, 2006, from http://www.winmira.von-davier.de/.

von Davier, M. (2005). A general diagnostic model applied to language testing data, *ETS RR-05-16*, ETS, Princeton, NJ.

von Davier, M., Molenaar, I.W. (2003). A person-fit index for polytomous Rasch models, latent class models, and their mixture generalizations. *Psychometrika* **68**, 213–228.

von Davier, M., Rost, J. (1995). Polytomous mixed Rasch models. In: Fischer, G.H., Molenaar, I.W. (Eds.), *Rasch Models: Foundations, Recent Developments, and Applications*. Springer, New York, pp. 371–379.

von Davier, M., Yamamoto, K. (2004a). A class of models for cognitive diagnosis. Paper presented at the fourth Spearman conference, Philadelphia, PA, October 21.

von Davier, M., Yamamoto, K. (2004b) A class of models for cognitive diagnosis – And some notes on estimation. Paper presented at the ETS Tucker Workshop Seminar, Princeton, NJ, December 2.

von Davier, M., Yamamoto, K. (2004c). Partially observed mixtures of IRT models: An extension of the generalized partial credit model. *Applied Psychological Measurement* **28**, 389–406.

von Davier, A.A., Holland, P.W., Thayer, D.T. (2003). *The Kernel Method of Test Equating*. Springer, New York.

Xu, X., von Davier, M. (2006). *Cognitive Diagnosis for NAEP Proficiency Data*. ETS, Princeton, NJ, in press.

Yamamoto, K. (1987). A model that combines IRT and latent class models. Unpublished doctoral dissertation, University of Illinois at Urbana-Champaign.

Yamamoto, K. (1989). Hybrid model of IRT and latent class models. ETS RR-89-41, ETS, Princeton, NJ.

Yamamoto, K. (1993). HYBIL: A computer program to estimate HYBRID model parameters [Computer software]. ETS, Princeton, NJ.

Yamamoto, K., Everson, H.T. (1995). Modeling the mixture of IRT and pattern responses by a modified HYBRID model. ETS RR-95-16, ETS, Princeton, NJ.

Yamamoto, K., Everson, H.T. (1997). Modeling the effects of test length and test time on parameter estimation using the HYBRID model. In: Rost, J., Langeheine, R. (Eds.), *Applications of Latent Trait and Latent Class Models in the Social Sciences*. Waxmann Publishing, New York, pp. 89–98.

Handbook of Statistics, Vol. 26
ISSN: 0169-7161

DOI: 10.1016/S0169-7161(06)26020-6

20

Scoring Open Ended Questions

Gunter Maris and Timo Bechger

1. Introduction

Scoring open ended questions almost invariably involves the use of human raters. In this chapter we consider how the scores assigned by multiple raters relate to the characteristics of an item, the ability of the student, and the characteristics of the rater(s). The methods introduced in this chapter are an extension of the approach introduced in Maris and Bechger (2003) and differ from earlier methods for the scoring of open ended questions in the literature (e.g., Linacre, 1994; Patz et al., 2000; Verhelst and Verstralen, 2001; Wilson and Hoskens, 2001) mainly in the fact that no attempt is (or needs to be) made to model the behavior of individual raters.

In the first section, an example is introduced to highlight some of the difficulties associated with scoring open ended questions. In the second section, a class of measurement models is specified by a number of assumptions. In the third section, a particular measurement model satisfying the assumptions from section two is developed. In the fourth section we compare this new measurement model with the one that is introduced with the motivating example. Finally, in the last section our results are discussed.

2. Motivating example

We consider an examination consisting of a number of short, longer, and long tasks to which a spoken response is required. The responses of the students are scored on a number of aspects. The aspects include things such as grammatical correctness, fluency, content, vocabulary, ..., and binary scores (correct/incorrect) are used. In our example, the raters are assigned to students in a systematic manner, such that every student is assessed by two raters.

We now consider how this exam is analyzed. To account for the influence of the human raters on the graded responses one could consider a model where the probability that rater r gives a positive rating for the answer to item i by person p, denoted as $X_{pir} = 1$, is the following:

$$P(X_{pir} = 1|\theta_p, \delta_i, a_i, \tau_{pir}) = \frac{\exp[a_i(\theta_p - \delta_i - \tau_{pir})]}{1 + \exp[a_i(\theta_p - \delta_i - \tau_{pir})]}. \tag{1}$$

Table 1
Data layout for the measurement model in Eq. (2)

Raters	Students							X_{p+r}
	1	1	0	1	…	1	0	8
1	2	0	0	1	…	0	1	
	⋮	⋮	⋮	⋮	⋮	⋮	⋮	⋮
	1	1	1	1	…	1	1	15
2	2	1	1	0	1	…	1	
	⋮	⋮	⋮	⋮	⋮	⋮	⋮	⋮
	X_{+i+}		216					

This probability depends on the ability θ_p of the pth person, the difficulty of the ith item δ_i, the discrimination of the ith item a_i, and a person by item by rater interaction parameter τ_{pir}. Obviously, without further assumptions, this model leaves the parameters unidentifiable from $\mathbf{X}$.

In the examination we consider here, it is assumed that the interaction parameter depends on the student and the rater only. That is, there is only an interaction possible between the rater and the person. This gives the following measurement model:

$$P(X_{pir} = 1|\theta_p, \delta_i, a_i, \tau_{pr}) = \frac{\exp[a_i([\theta_p - \tau_{pr}] - \delta_i)]}{1 + \exp[a_i([\theta_p - \tau_{pr}] - \delta_i)]} = \frac{\exp[a_i(\mu_{pr} - \delta_i)]}{1 + \exp[a_i(\mu_{pr} - \delta_i)]}. \tag{2}$$

If the item parameters are known, this measurement model is an exponential family model with the weighted score:

$$X_{p+r} = \sum_i a_i X_{pir} \quad \text{sufficient for } \mu_{pr}.$$

Observe that the unweighted score:

$$X_{+i+} = \sum_p \sum_r X_{pir} \quad \text{is sufficient for } a_i \delta_i.$$

Corresponding to this measurement model, we can depict the data as in Table 1.

With this measurement model, we can still not identify the ability of the student (θ_p) from the observations ($\mathbf{X}$). We now show how ability is estimated in our motivating example. First, the measurement model in Eq. (2) is used to estimate the item difficulties, to evaluate the fit of the measurement model, and to estimate the parameters μ_{pr}. In the second step, it is assumed that:

$$\mu_{pr} = \theta_p - \tau_{pr}. \tag{3}$$

If we substitute the estimate $\hat{\mu}_{pr}$ for μ_{pr}, and assume that the *errors* $\boldsymbol{\tau}_p$ are a random sample from a distribution with mean equal to zero and finite variance we obtain the following estimate for θ_p:

$$\hat{\theta}_p = \frac{1}{R}\sum_r \hat{\mu}_{pr}.$$

This procedure is unsatisfactory for a number of reasons. First, and foremost, we need to consider what happens if the number of raters increases. It is readily found that the item difficulty parameters can be consistently estimated if the number of raters increases. This follows because every rater by person combination is formally treated as an additional student. This is of course a rather bizarre situation. With the responses from only a single student we would be able to estimate to any desired degree of accuracy the item difficulties by choosing an appropriate number of raters. Second, the step that is used to relate the observed sufficient statistics to the estimated ability (see Eq. (3)) is not very well motivated. As an alternative, consider the following procedure: Rather than averaging the estimated parameters μ_{pr} over raters, we may also average the sufficient statistics for these parameters:

$$X_{pi+} = \frac{1}{R}\sum_r a_i X_{pir}$$

to obtain a single score which is then used to estimate the ability. These two procedures may lead to very different results because the relation between the sufficient statistic and the estimated ability is not linear. In Figure 1 this relation is depicted for two hypothetical exams. Estimating the ability of a student corresponds to finding that ability for which the corresponding score is equal to the score of the student.

Third, since raters are assigned to students in a systematic manner, the assumption that the μ_{pr} are a random sample is hardly tenable. This means that the average of the μ_{pr} for a particular student need not be equal to zero. As a consequence, the bias of the raters is *attributed* to the student.

In the following section, a different procedure is developed for analyzing these data. For ease of presentation we restrict attention to binary ratings where every student is assessed by two raters. In the discussion section we consider the extent to which our results generalize to more than two raters and to polytomous ratings.

3. Theory

An important concept that will be used throughout this section is that of a *true* rating. We use X_{pi} (without a subscript r to identify the rater) to denote the true rating for the response of person p to item i. The use of human raters creates a number of problems, the most important ones of which (from a psychometric point of view) are:

1. Raters may not judge different items and/or aspects of the same candidate independently (e.g., the halo-effect);
2. Raters don't always agree: $X_{pi1} \neq X_{pi2}$.

In the following subsection these problems are addressed.

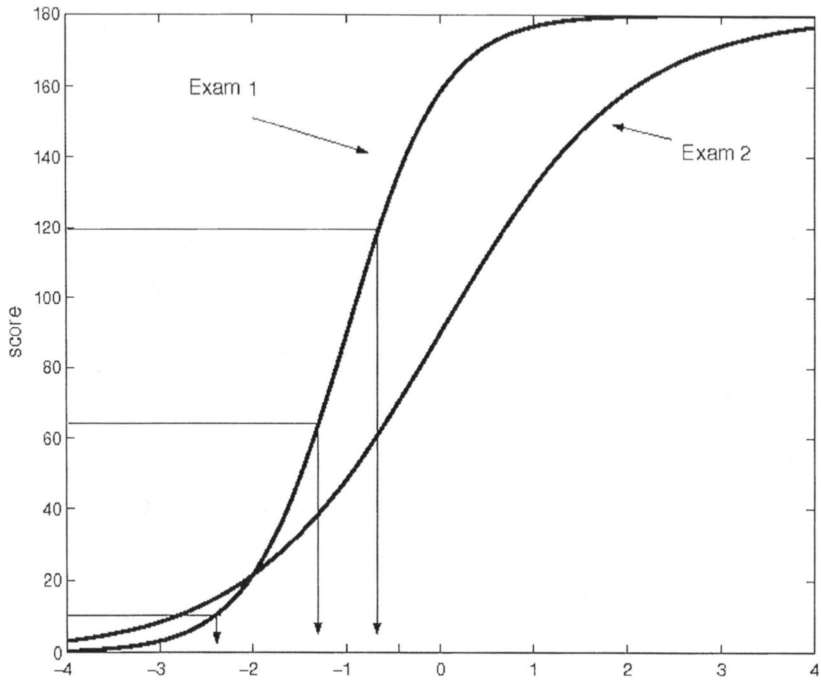

Fig. 1. Relation between score (y-axis) and ability (x-axis) for two hypothetical exams.

3.1. Randomization

In the literature, several methods can be found for the measurement of ability from the responses to open ended questions (Linacre, 1994; Patz et al., 2000; Verhelst and Verstralen, 2001; Wilson and Hoskens, 2001). All these methods have in common that they try to model the behavior of individual raters. This not only complicates the measurement model but also introduces the risk of incorrect inferences about the ability of the students because the behavior of the raters is not adequately modeled. An example of this risk was seen when we introduced the motivating example. We saw that if the assumption that the mean of the rater by person random effect τ_{pr} was not equal to zero, the bias of the raters is attributed to the student.

In this chapter we use a simple way to remove the individual raters from the measurement model. If raters are drawn at random from a (large) list of raters, and assigned at random to person by item combinations, the individual raters become *exchangeable* and hence their individual behavior *need not* be modeled. Formally, we obtain that the ratings are partially exchangeable, which means that

$$
\begin{aligned}
&P(X_{pi1} = y_{pi1}, X_{pi2} = y_{pi2}, X_{pi} = x_{pi} | \theta_p) \\
&\quad = P(X_{pi1} = y_{pi2}, X_{pi2} = y_{pi1}, X_{pi} = x_{pi} | \theta_p).
\end{aligned}
$$

The purpose of assigning raters at random to person by item combinations is to guarantee that the ratings of different responses from the same student are independent.

Without this precaution we run the risk that an overall impression of the rater about the ability of the student, based on the responses to the first few items, influences the ratings of the rest of the responses. This effect is known as the halo effect and it constitutes a violation of the conditional independence assumption that is part of most, if not all, measurement models for scoring open ended questions.

3.2. *More of the same is better*

The main idea that is developed in this section is the following: If two raters agree, there is no immediate reason to doubt their judgment. However, if they disagree there is no immediate reason to prefer the judgment of one over that of the other. Put differently, if raters agree we accept their judgment; if raters disagree we accept neither of the two judgments. This means that disagreement between raters is considered to generate, some form of, missing data. We now give a more formal development of this main idea.

Under the assumption that raters are randomly assigned, the data can, without loss of information, be written as follows:

X_{pi1}	X_{pi2}	$\widehat{X}_{pi}$	C_{pi}
1	1	1	1
1	0	Missing	0
0	1	Missing	0
0	0	0	1

It is seen that $\widehat{X}_{pi}$ contains the grade of the raters when both raters agree. Furthermore, C_{pi} denotes whether or not the raters agreed. In this subsection we show that the *filtered* ratings in $\widehat{X}_{pi}$ contain less observations than either of the original ratings X_{pi1} or X_{pi2}; but this information is of better quality. Formally, this means that the probability with which $\widehat{X}_{pi}$ equals the true rating is larger than the probability with which either of the individual ratings equals the true rating:

$$P(X_{pi1} = X_{pi}|X_{pi1} = X_{pi2}, \theta_p) \geqslant P(X_{pi1} = X_{pi}|\theta_p).$$

In order to obtain this result, we make two assumptions. First, ratings are assumed to be partially exchangeable, and second we assume that $P(X_{ij} = y_i|\theta) \geqslant 0.5$. Partial exchangeability is automatically satisfied if raters are assigned at random. Intuitively, partial exchangeability follows because the probability that X_{i1} refers to the judgment of any rater is equal to the probability that X_{i1} refers to the rating of any other rater. The second assumption is a rather mild one. If the opposite were true, the raters would systematically, and perhaps willingly, provide the wrong rating. Contrary to the first assumption, the truth of which can be guaranteed by assigning raters at random to item by student combinations, the second assumption is difficult to test. For instance, if raters collectively decide to reverse their ratings, there is no way this can be detected.

Under these two assumptions, it can be shown that

$$P(X_{i1} = X_i|X_{i1} = X_{i2}, \theta) \geqslant P(X_{i1} = X_i|\theta).$$

The proof is as follows: First, note that if two raters disagree, one of the ratings equals X_i. Under partial exchangeability, it is equally likely that this is the first rating or the second rating and it follows that

$$P(X_{i1} = X_i | X_{i1} \neq X_{i2}, \theta) = \frac{1}{2}.$$

This implies that

$$\begin{aligned} P(X_{i1} = X_i|\theta) &= P(X_{i1} = X_i | X_{i1} \neq X_{i2}, \theta) P(X_{i1} \neq X_{i2}|\theta) \\ &\quad + P(X_{i1} = X_i | X_{i1} = X_{i2}, \theta) P(X_{i1} = X_{i2}|\theta) \\ &= \frac{1}{2}\big[1 - P(X_{i1} = X_{i2}|\theta)\big] \\ &\quad + P(X_{i1} = X_i | X_{i1} = X_{i2}, \theta) P(X_{i1} = X_{i2}|\theta) \\ &= \frac{1}{2} + \left[P(X_{i1} = X_i | X_{i1} = X_{i2}, \theta) - \frac{1}{2}\right] P(X_{i1} = X_{i2}|\theta). \end{aligned}$$

It follows that

$$P(X_{i1} = X_i|\theta) - \frac{1}{2} = \left[P(X_{i1} = X_i | X_{i1} = X_{i2}, \theta) - \frac{1}{2}\right] P(X_{i1} = X_{i2}|\theta).$$

Hence if $P(X_{i1} = X_i|\theta) \geqslant 0.5$ it follows that:

$$P(X_{i1} = X_i|\theta) - \frac{1}{2} \leqslant P(X_{i1} = X_i | X_{i1} = X_{i2}, \theta) - \frac{1}{2}.$$

This implies, finally, that

$$P(X_{i1} = X_i|\theta) \leqslant P(X_{i1} = X_i | X_{i1} = X_{i2}, \theta)$$

which had to be proven. If the raters were behaving as if they were throwing coins, $P(X_{i1} = X_i|\theta) = P(X_{i1} = X_i | X_{i1} = X_{i2}, \theta) = \frac{1}{2}$.

We have shown that for estimating the true rating, the consistent ratings $\widehat{X}_i$ are better than either of the single ratings. Hence, we consider the consistent ratings $\widehat{X}_i$ as estimates of the true ratings X_i:

$$P(\widehat{X}_i = 1 | C_i = 1, \theta) \approx P(X_i = 1|\theta). \tag{4}$$

Because raters can agree with each other about the wrong thing (e.g., $\widehat{X}_i = 1$ whereas $X_i = 0$), both probabilities in Eq. (4) are not exactly equal. When we introduce a particular model for scoring open ended questions in Section 4 we will give the conditions under which both probabilities in Eq. (4) become exactly equal.

Since it is reasonable to assume that the IRF of the true rating X_i is increasing in θ we assume that:

$$\text{A.} \quad P(\widehat{X}_i = 1 | C_i = 1, \theta) \quad \text{increasing in } \theta. \tag{5}$$

The model specified by assumption A in Eq. (5) entails a loss of information to the extent that the knowledge of whether ratings were consistent or not contains information about ability. The relation between C_i and ability is the subject of the ensuing section. It will be shown that C_i and X_i^* can be analyzed simultaneously at the cost of an additional assumption about the relation between C_i and θ. Notice that any information that remains in the identity of the rater is left unused because it would require a sound understanding of the behavior of individual raters. In general, however, we do not presume to have such understanding.

3.3. *Ability and agreement*

Whether two raters agree depends upon the individual raters, which do not figure in the present exposition, the assignments, and the ability of the respondent. It is reasonable to assume that raters are likely to agree about the performance of extremely incompetent or extremely competent candidates and inconsistencies will occur mainly when raters assess the performances of average candidates. Moreover, the qualifications "incompetent", "average", and "competent" have meaning relative to the difficulty of the assignment. These considerations lead us to model the probability $P(C_i = 1|\theta)$ as a "single-dipped" function of θ that reaches its minimum at θ equal to δ_i, the item difficulty. In Figure 2 an illustration is given. The probability $P(C_i = 1|\theta)$ may be interpreted as a measure of the *assessability* of subjects with ability θ. It achieves its lowest value at $\theta = \delta_i$, and δ_i represents the difficulty of the assignment.

3.4. *Putting it all together*

We now proceed by combining C_i and $\widehat{X}_i$ in a measurement model for the ratings. For each subject, the ratings have as possible outcomes: negative consistent ($\widehat{X}_i = 0$ and $C_i = 1$), inconsistent ($C_i = 0$), and positive consistent ($\widehat{X}_i = 1$ and $C_i = 1$). The

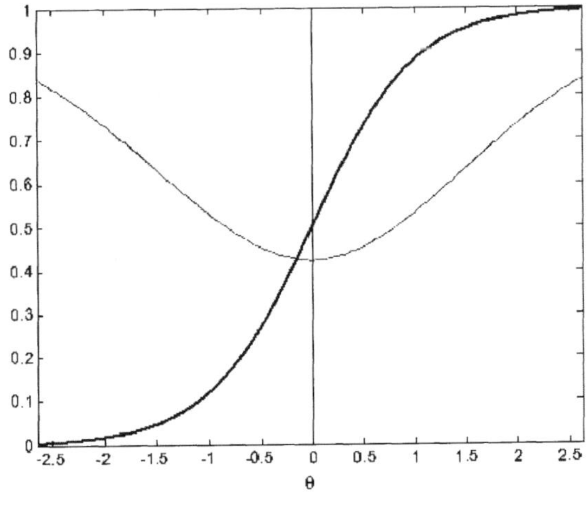

Fig. 2. The thick line is the plot of $P(X_i^* = 1|C_i = 1, \theta)$. The thin line is a plot of $P(C_i = 1|\theta)$. Indicated in the figure is the position of the difficulty of the item, δ_i.

probabilities corresponding to each outcome are:

$$P(\widehat{X}_i = 0, C_i = 1|\theta) = P(\widehat{X}_i = 0|C_i = 1, \theta)P(C_i = 1|\theta),$$
$$P(C_i = 0|\theta),$$
$$P(\widehat{X}_i = 1, C_i = 1|\theta) = P(\widehat{X}_i = 1|C_i = 1, \theta)P(C_i = 1|\theta).$$

If raters are exchangeable, these three outcomes correspond to the different values of the sum of the two ratings $X_{i+} = X_{i1} + X_{i2}$. The relation between X_{i+}, C_i, and $\widehat{X}_i$ is summarized in the following table:

X_{i1}	X_{i2}	$\widehat{X}_i$	C_i	$X_{i1} + X_{i2}$
1	1	1	1	2
1	0	Missing	0	1
0	1	Missing	0	1
0	0	0	1	0

It is seen, for instance, that $X_{i+} = 0$ when the ratings are both negative. Thus, we may proceed by analyzing X_{i+} with any measurement model that satisfies our assumptions:

$$\begin{aligned}&\text{A.}\quad P(\widehat{X}_i = 1|C_i = 1, \theta) = P(X_{i+} = 2|X_{i+} \neq 1, \theta) \quad \text{increasing with } \theta,\\ &\text{B.}\quad P(C_i = 0|\theta) = P(X_{i+} = 1|\theta) \quad \text{is a single-peaked function of } \theta.\end{aligned} \tag{6}$$

We now consider the extent to which our assumptions determine the properties of the resulting measurement model for the summated ratings. Obviously, the probability of a consistent rating is a single-dipped function of ability for any model meeting our assumptions. It is however not clear how the probability of a consistent correct and a consistent incorrect response relate to ability. It is, for instance, not true in general that these probabilities are monotone functions of ability. In the following subsection we further specify the measurement model to obtain more insight in the relation between the probabilities for the different outcomes and ability.

3.5. *To agree or to disagree ...*

It is easily found that any two of the following three probabilities completely determine the distribution of X_{i+}:

$$\begin{aligned}&\text{A.}\quad P(X_{i+} = 2|X_{i+} = 2 \text{ or } X_{i+} = 0, \theta),\\ &\text{C1.}\quad P(X_{i+} = 2|X_{i+} = 2 \text{ or } X_{i+} = 1, \theta),\\ &\text{C2.}\quad P(X_{i+} = 0|X_{i+} = 0 \text{ or } X_{i+} = 1, \theta).\end{aligned} \tag{7}$$

In the earlier sections we assumed that the first assumption (A) in Eq. (7) is a monotone increasing function of ability. Here we consider the remaining two assumptions (C1 and C2) in Eq. (7).

The probability C1 in Eq. (7) can be re-expressed as follows:

$$\begin{aligned} P(X_{i+} = 2 | X_{i+} = 2 \text{ or } X_{i+} = 1, \theta) &= P(X_{i2} = 1 | X_{i1} = 1, \theta) \\ &= P(X_{i1} = 1 | X_{i2} = 1, \theta). \end{aligned}$$

Similarly, we can re-express the probability C2 in Eq. (7) as follows:

$$\begin{aligned} P(X_{i+} = 0 | X_{i+} = 0 \text{ or } X_{i+} = 1, \theta) &= P(X_{i2} = 0 | X_{i1} = 0, \theta) \\ &= P(X_{i1} = 0 | X_{i2} = 0, \theta). \end{aligned}$$

Hence, we see that the probability C1 in Eq. (7) is the probability that the second rater gives a positive rating given that the first rater has given a positive rating. Similarly, the probability C2 in Eq. (7) is the probability that the second rater gives a negative rating given that the first rater has given a negative rating.

We assume that the probability that the second rater gives a positive rating given that the first rater has given a positive rating is an increasing function of ability and that the probability that the second rater gives a negative rating given that the first rater has given a negative rating is a decreasing function of ability. Hence, the class of measurement models we consider is characterized completely by the following three assumptions:

$$\begin{array}{lll} \text{A.} & P(X_{i+} = 2 | X_{i+} = 2 \text{ or } X_{i+} = 0, \theta) & \text{increases with } \theta, \\ \text{C1.} & P(X_{i+} = 2 | X_{i+} = 2 \text{ or } X_{i+} = 1, \theta) & \text{increases with } \theta, \\ \text{C2.} & P(X_{i+} = 0 | X_{i+} = 0 \text{ or } X_{i+} = 1, \theta) & \text{decreases with } \theta. \end{array} \tag{8}$$

As we observed earlier, any two out of these three assumptions are sufficient to determine the distribution of X_{i+}. Hence, we need to check whether these three assumptions are consistent. Furthermore, we need to consider whether these three assumptions are consistent with the assumptions A and B in Eq. (6).

First, we consider whether our assumptions are consistent. We not only show that the assumptions in Eq. (6) are consistent, but also show that C1 and C2 together imply A whereas any other combination (i.e., A and C1, or A and C2) does not necessarily imply the third assumption. In order to simplify the presentation of our results we introduce some shorthand notation. We use $\pi_k(\theta)$ to denote the probability that X_{i+} equals k and $\pi_{k|jk}(\theta)$ to denote the probability that X_{i+} equals k conditionally on the event that X_{i+} equals j or k. Moreover, we use $\rho_k(\theta)$ and $\rho_{k|jk}(\theta)$ to refer to the corresponding odds ratios, defined as follows:

$$\rho_k(\theta) = \frac{\pi_k(\theta)}{1 - \pi_k(\theta)}$$

and

$$\rho_{k|jk}(\theta) = \frac{\pi_{k|jk}(\theta)}{1 - \pi_{k|jk}(\theta)}.$$

Using this shorthand notation we may express our assumptions as follows:

$$\begin{array}{lll} \text{A.} & \pi_{2|02}(\theta) & \text{increases with } \theta, \\ \text{C1.} & \pi_{2|12}(\theta) & \text{increases with } \theta, \\ \text{C2.} & \pi_{0|01}(\theta) & \text{decreases with } \theta. \end{array}$$

Obviously the conditional and unconditional probabilities are related as follows:

$$\pi_{k|jk}(\theta) = \frac{\pi_k(\theta)}{\pi_j(\theta) + \pi_k(\theta)}$$

and

$$\pi_k(\theta) = \frac{\pi_{k|jk}(\theta)}{1 - \pi_{k|jk}(\theta)} \pi_j(\theta) = \rho_{k|jk}(\theta)\pi_j(\theta).$$

Any two of the probabilities in Eq. (7) together with the property that the probabilities of the three possible values for X_{i+} have to sum to one:

$$\pi_0(\theta) + \pi_1(\theta) + \pi_2(\theta) = 1$$

yield a consistent system of three equations in three unknowns. For instance, if we consider the probabilities C1 and C2 in Eq. (7) we find that

$$\begin{aligned} 1 &= \pi_0(\theta) + \pi_1(\theta) + \pi_2(\theta) \\ &= \rho_{0|01}(\theta)\pi_1(\theta) + \pi_1(\theta) + \rho_{2|12}(\theta)\pi_1(\theta) \\ &= \big(\rho_{0|01}(\theta) + 1 + \rho_{2|12}(\theta)\big)\pi_1(\theta) \end{aligned}$$

from which we may conclude that

$$\pi_1(\theta) = \big(\rho_{0|01}(\theta) + 1 + \rho_{2|12}(\theta)\big)^{-1}.$$

Hence we find that

$$\pi_0(\theta) = \frac{\rho_{0|01}(\theta)}{\rho_{0|01}(\theta) + 1 + \rho_{2|12}(\theta)},$$

$$\pi_1(\theta) = \frac{1}{\rho_{0|01}(\theta) + 1 + \rho_{2|12}(\theta)}, \tag{9}$$

and

$$\pi_2(\theta) = \frac{\rho_{2|12}(\theta)}{\rho_{0|01}(\theta) + 1 + \rho_{2|12}(\theta)}.$$

From the distribution of X_{i+} we readily obtain the following expression for $\pi_{2|02}$:

$$\pi_{2|02} = \frac{\rho_{2|12}(\theta)}{\rho_{2|12}(\theta) + \rho_{0|01}(\theta)}$$

and hence

$$\rho_{2|02} = \frac{\rho_{2|12}(\theta)}{\rho_{0|01}(\theta)} = \rho_{2|12}(\theta)\rho_{1|01}(\theta).$$

Obviously, if a probability function is increasing (decreasing) with θ, the corresponding odds ratio is also increasing (decreasing) with θ. Since by assumption C1 $\rho_{2|12}(\theta)$ is an increasing function of θ, and by assumption C2 $\rho_{1|01}(\theta)$ is an increasing function of θ, it follows that also $\rho_{2|02}$ is an increasing function of θ, which means that assumption A follows from assumptions C1 and C2. We have now not only shown that assumption A follows from C1 and C2, but also that the three assumptions in Eq. (8) are consistent. In the following section, we illustrate with an example that assumptions A and C1, or A and C2 in Eq. (8) do not imply the third assumption. Moreover, we also show with an example that assumption B in Eq. (6) and C1 and C2 in Eq. (8) are consistent.

4. A measurement model for scoring open ended questions

Here we consider a particular measurement model that meets the assumptions A, B, C1, and C2. We assume a two parameter logistic model (2PL; Birnbaum, 1968) for C1 and C2 in Eq. (8):

$$P(X_{pi+} = 2|X_{pi+} = 2 \text{ or } X_{pi+} = 1, \theta) = \frac{\exp(a_{i1}\theta_p - \beta_{i1})}{1 + \exp(a_{i1}\theta_p - \beta_{i1})}$$

and

$$P(X_{pi+} = 0|X_{pi+} = 0 \text{ or } X_{pi+} = 1, \theta) = \frac{1}{1 + \exp(a_{i2}\theta_p - \beta_{i2})}.$$

Notice that in order to comply with our assumptions we need to assume that both a_{i1} and a_{i2} are nonnegative.

With these assumptions we readily find that also the conditional probability A in Eq. (8) satisfies a 2PL model:

$$P(X_{pi+} = 2|X_{pi+} = 0 \text{ or } X_{pi+} = 2, \theta) = \frac{\exp([a_{i1} + a_{i2}]\theta_p - [\beta_{i1} + \beta_{i2}])}{1 + \exp([a_{i1} + a_{i2}]\theta_p - [\beta_{i1} + \beta_{i2}])}$$

which is necessarily monotone increasing in θ if both a_{i1} and a_{i2} are nonnegative.

If we allow a_{i2} to be negative, and impose the additional constraint that $a_{i1} + a_{i2} > 0$ we obtain that $P(X_{pi+} = 0|X_{pi+} = 0 \text{ or } X_{pi+} = 1, \theta)$ is an increasing function of ability, contrary to our assumption C2. Similarly, if we allow a_{i1} to be negative, and impose the additional constraint that $a_{i1} + a_{i2} > 0$ we obtain that $P(X_{pi+} = 2|X_{pi+} = 2 \text{ or } X_{pi+} = 1, \theta)$ is a decreasing function of ability, contrary to our assumption C1. Hence, from this example we see that assumptions A and C1 or assumptions A and C2 do not imply the third assumption in Eq. (8).

Some straightforward algebra shows that with our present assumption we obtain the following distribution for X_{i+}:

$$\begin{aligned} P(X_{pi+} = 0|\theta) &= \frac{1}{1 + \exp(a_{i1}\theta_p - \beta_{i1}) + \exp([a_{i1} + a_{i2}]\theta_p - [\beta_{i1} + \beta_{i2}])}, \\ P(X_{pi+} = 1|\theta) &= \frac{\exp(a_{i1}\theta_p - \beta_{i1})}{1 + \exp(a_{i2}\theta_p - \beta_{i2}) + \exp([a_{i1} + a_{i2}]\theta_p - [\beta_{i1} + \beta_{i2}])}, \\ P(X_{pi+} = 2|\theta) &= \frac{\exp([a_{i1} + a_{i2}]\theta_p - [\beta_{i1} + \beta_{i2}])}{1 + \exp(a_{i1}\theta_p - \beta_{i1}) + \exp([a_{i1} + a_{i2}]\theta_p - [\beta_{i1} + \beta_{i2}])} \end{aligned} \tag{10}$$

which is recognized as a special case of the *nominal response model* (NRM; Bock, 1972).

It is readily found that if a_{i1} and a_{i2} are nonnegative, the probability of a consistent response is a single-dipped function of ability. That is, we have now illustrated with an example that assumptions C1 and C2 in Eq. (8) and assumption B in Eq. (6) are consistent.

In the rest of this section we offer an interpretation for the parameters of this measurement model, and show how for this particular model each of the assumptions in Eq. (8) can be tested in isolation from the remaining assumptions.

4.1. Parameter interpretation

First, if the parameters a_{i1} and a_{i2} are known, this is an exponential family model with the weighted sum:

$$X_{p++} = \sum_i \left[a_{i1}(X_{pi+} = 1) + (a_{i1} + a_{i2})(X_{pi+} = 2)\right]$$

sufficient for θ. Put differently, a negative consistent response scores 0 points, an inconsistent response scores a_{i1} points, and a positive consistent response scores $a_{i1} + a_{i2}$ points.

If θ is smaller than β_{i1}/a_{i1}, the probability of a negative consistent response is larger than the probability of an inconsistent response. Similarly, if θ is larger than β_{i2}/a_{i2}, the probability of a positive consistent response is larger than the probability of an inconsistent response. If β_{i1}/a_{i1} is smaller than β_{i2}/a_{i2}, the probability of an inconsistent response is larger than the probability of either of the consistent responses for θ-values larger than β_{i1}/a_{i1} but smaller than β_{i2}/a_{i2}. If β_{i1}/a_{i1} is larger than β_{i2}/a_{i2}, it is true for every value of θ that one of the consistent responses is more likely than an inconsistent response. Clearly, the smaller the number of inconsistent responses, the better. This leads us to propose λ_i:

$$\lambda_i = \beta_{i1}/a_{i1} - \beta_{i2}/a_{i2}$$

as a measure of the item assessability. A negative value indicates that there is an interval of length $-\lambda_i$ over which an inconsistent response is the most likely response; whereas a positive value indicates that it is true for every value of θ that one of the consistent responses is more likely than an inconsistent response.

If θ is smaller than δ_i:

$$\delta_i = \frac{\beta_{i1} + \beta_{i2}}{a_{i1} + a_{i2}}$$

the probability of a negative consistent response is larger than the probability of a positive consistent response; whereas the opposite is true if θ is larger than δ_i. Because δ_i is that value of θ at which the probability of a positive and that of a negative consistent response are equal, we automatically obtain that as λ_i goes to infinity (there are no more inconsistent responses), δ_i becomes the difficulty of the item. This leads us to propose δ_i as a measure of the item difficulty.

Notice that as λ_i goes to infinity, there are no more inconsistent responses, and the raters produce the true rating. In this case we may equate the left- and right-hand side of Eq. (5).

4.2. Checking the model assumptions

Here we show that if the scoring rule is considered to be known and the scores are integer valued, each of the assumptions of the model may be tested in isolation from the remaining assumptions. These additional assumptions are needed in order to make statistical inferences conditionally on the sufficient statistic for the ability parameter valid. As is well known for other exponential family IRT models, such as the Rasch (1960) model, the partial credit model (Andersen, 1977), or the OPLM model (Verhelst and Verstralen, 1997; Verhelst et al., 1984), statistical inferences can be based on the distribution of the responses conditionally on the values of the sufficient statistic for θ (Andersen, 1977). As we show, we can use the fact that our model is an exponential IRT model to motivate a procedure where we first *filter* the data and than analyze the filtered data to evaluate the model assumptions in isolation from the remaining model assumptions.

First, we filter the responses according to one of the assumptions to be tested. For ease of presentation we focus on assumption A. By filtering we mean that we only consider those responses for which the conditioning event ($X_{pi+} \neq 1$) holds. The distribution of these responses is the distribution conditionally on the event that a response is consistent. If we use, as before, C_{pi} to denote whether a response is consistent or not, and $\widehat{X}_{pi}$ to indicate if a consistent response is positive or negative, we may express this distribution as follows:

$$P(\mathbf{X} = \mathbf{x} | \mathbf{C} = \mathbf{c}, \boldsymbol{\theta}, \boldsymbol{\delta}, \boldsymbol{\lambda}; \mathbf{a}) = \prod_p \prod_i \left(\frac{\exp(\hat{x}_{pi}[a_{i1} + a_{i2}][\theta_p - \delta_i])}{1 + \exp([a_{i1} + a_{i2}][\theta_p - \delta_i])} \right)^{c_{pi}}. \quad (11)$$

It is clear that this likelihood does not depend on the item assessability λ_i and depends on the scoring rule only via $a_{i1} + a_{i2}$. Furthermore, observe that neither the truth of assumption C1 or that of assumption C2 has a direct influence on this distribution. That is, if we can evaluate the fit of this conditional distribution, we have effectively tested assumption A in isolation from the remaining assumptions.

Statistical inference based on the conditional distribution in Eq. (11) is hampered by the fact that the person parameters θ_p are incidental. That is, their number increases with the sample size. It is known that in the presence of an increasing number of incidental parameters it is, in general, not possible to estimate the (structural) item parameters consistently (Kiefer and Wolfowitz, 1956; Neyman and Scott, 1948). This problem can be overcome in one of two ways. First, for exponential family models we can base our inferences on the distribution of the data conditionally on the sufficient statistics for the incidental parameters. Obviously, this conditional distribution no longer depends on the incidental parameters. This method is called *conditional likelihood* (CL) inference. Second, if the students can be conceived of as a random sample from a well defined population characterized by an ability distribution G, inferences can be based on the marginal distribution of the data. That is, we integrate the incidental parameters out of the model. Rather than estimating each student's ability, only the parameters of the ability *distribution* need to be estimated. This method is called *marginal likelihood* (ML) inference. Under suitable regularity conditions both methods can be shown to lead to consistent estimates of the item difficulty parameters.

If the scores a_{i1} and $a_{i1} + a_{i2}$ are known constants, our model reduces to an exponential family model. More specifically, the conditional distribution of the response conditionally on the event that it is consistent reduces to the one parameter logistic model (OPLM; Verhelst and Eggen, 1989; Verhelst and Glas, 1995), if the scores are assumed to be integer valued, for which statistical inference both from the marginal and from the conditional likelihood perspective is implemented in the computer program OPLM (Verhelst et al., 1993).

Here, we opt for the first approach and will base our statistical inferences on the distribution of $\mathbf{X}$ conditionally on both $\mathbf{C}$ and the sufficient statistics for the ability parameters. Corresponding to the distribution in Eq. (11) is the following sufficient statistic for θ_p:

$$X_{p++|c} = \sum_{i:c_{pi}=1} [a_{i1} + a_{i2}]X_{pi+}.$$

That is, sufficient statistic is a weighted sum score over the set of items for which a consistent response is obtained. We opt for a conditional approach because in order to base our inferences on the marginal likelihood we need to consider the following likelihood:

$$\begin{aligned} &p(\mathbf{X} = \mathbf{x}|\mathbf{C} = \mathbf{c}, \boldsymbol{\theta}, \boldsymbol{\delta}, \boldsymbol{\lambda}; \mathbf{a}) \\ &\quad = \prod_p \int \prod_i \left(\frac{\exp(\hat{x}_{pi}[a_{i1} + a_{i2}][\theta_p - \delta_i])}{1 + \exp([a_{i1} + a_{i2}][\theta_p - \delta_i])} \right)^{c_{pi}} f(\theta_p|\mathbf{C} = \mathbf{c}, \boldsymbol{\delta}, \boldsymbol{\lambda}; \mathbf{a})\, d\theta. \end{aligned}$$

It is seen that to specify the marginal likelihood correctly we need to specify the distribution of ability conditionally on the consistency of the responses:

$$f(\theta_p|\mathbf{C} = \mathbf{c}, \boldsymbol{\delta}, \boldsymbol{\lambda}; \mathbf{a}) \propto \prod_i P(C_{pi} = c_{pi}|\theta_p, \delta_i, \lambda_i; \mathbf{a}_i) f(\theta),$$

where $f(\theta)$ is the distribution of ability in the population. Clearly, the correct marginal likelihood depends not only on assumption A but also on the remaining model assumptions together with an additional assumption about the distribution of ability in the population. Because we want to evaluate each of our assumptions in isolation we therefore opt for inferences based on the conditional likelihood, which requires no additional assumptions.

A routine but tedious derivation shows that the conditional likelihood can be expressed as follows:

$$\begin{aligned} &P(\mathbf{X} = \mathbf{x}|\mathbf{C} = \mathbf{c}, \mathbf{X}_{++|c} = \mathbf{x}_{++|c}, \boldsymbol{\delta}; \mathbf{a}) \\ &\quad = \prod_p \frac{\prod_i \exp(c_{pi}\hat{x}_{pi}[a_{i1} + a_{i2}]\delta_i)}{\gamma_{x_{p++|c}}(\boldsymbol{\delta}, \mathbf{a})}, \end{aligned} \tag{12}$$

where γ_k is the following generalization of the infamous elementary symmetric function (e.g., Verhelst and Verstralen, 1997; Verhelst et al., 1984):

$$\gamma_k(\boldsymbol{\delta}, \mathbf{a}) = \sum_{\mathbf{x}_p : x_{p++|c}=k} \prod_i \exp\bigl(c_{pi}\hat{x}_{pi}[a_{i1} + a_{i2}]\delta_i\bigr).$$

Having established that we can analyze the filtered data within the conditional likelihood framework, we can use the statistical tests that have been developed (see Glas and Verhelst, 1995, for an introduction) for evaluating the fit of assumption A in Eq. (8) in isolation from the other assumptions. Similarly, the assumptions C1 and C2 can be evaluated.

5. Motivating example revisited

In this section, we compare the use of the model proposed in the previous section (the new method or procedure, for short) for the sum of two ratings with the method that was considered when we introduced our motivating example (the current method or procedure, for short). For ease of presentation, we assume that the item discrimination a_i equals one with the old method, and that a_{i1} and a_{i2} both are equal to one with the new method. Observe, that this reduces the new method to the use of the partial credit model (Andersen, 1977) for the sum of two ratings.

Corresponding to the new method, the data can be depicted as in Table 2. In Table 2, neg. cons, incons., and pos. cons. denote the number of negative consistent, inconsistent, and positive consistent responses for item i, respectively. Any two of these three numbers are the sufficient statistics for the item parameters of the model in Eq. (10). For the estimation of the ability of the student we only need the sufficient statistic X_{p++}.

First, we consider how the item characteristics under the old and the new procedure relate to each other. It is readily observed that the sum of the number of inconsistent responses plus twice the number of positive consistent response equals the sufficient statistic X_{+i+} from the current procedure. This implies that the difficulty in the current procedure is a combination of the item difficulty and assessability from the new procedure. Consider the following situation: with 205 students we obtain the following results:

Table 2
Data layout according to the measurement model in Eq. (10)

Students							X_{p++}
1	2	1	2	...	2	1	23
2	1	0	2	...	1	1	
⋮	⋮	⋮	⋮	⋮	⋮	⋮	⋮
Neg. cons.		200					
Incons.		16					
Pos. cons.		84					

1. The raters disagree about 200 of the students (incons. = 200) and never give a consistent positive rating (pos. cons. = 0); or
2. The raters always agree (incons. = 0) and give a positive consistent rating for 100 students (pos. cons. = 200).

For both these situations, the sufficient statistic X_{+i+} is equal to 200. Put differently, both situations lead to the same estimated item difficulty. However, in the first situation, the item is mainly difficult to judge whereas in the second situation the item is very easy to judge and of moderate difficulty. The conclusion is that difficulty in the old procedure has no clear meaning. In the new procedure, difficulty and assessability are clearly separated which improves the interpretation of the results.

Second, we look at the relation between the estimated ability parameters. Looking at the various scores assigned to students according to the new and the old procedure we see that the sum of the two scores assigned to each student with the old procedure ($X_{p+1} + X_{p+2}$) is the sufficient statistic for the ability with the new procedure (X_{p++}). When we dealt with the item characteristics, we argued that the confounding of difficulty and assessability was a disadvantage of the old procedure. A similar reasoning can be used here. Consider the following situations: With 20 items we obtain the following scores:

1. The raters disagree on 10 responses and always give a negative judgment when they agree; or
2. The raters always agree and give a positive consistent rating for 5 items.

In both situations, the score of the student according to the new procedure is 10. We see that with the new procedure, the score is a combination of the ability and the assessability of the student. This however does not pose any problems. We explicitly assumed that there is a relation between assessability and ability. It is this assumption which merits the confounding of assessability and ability.

At the beginning of this section we considered a variation on the old procedure in which the scores are first averaged and then ability is estimated from this average score. With this alternative procedure, the same statistic is used to estimate ability as with the new procedure. As a consequence, the students are ordered in the same way based on their estimated ability according to the new procedure and according to this variation of the old procedure.

Third, we look at the goodness-of-fit of the two models. A key assumption under the old procedure is that ratings of the answer to a question of the same student are statistically independent. This assumption is difficult to maintain. First, consider the case of perfect raters which always agree with each other (a perfect rater produces the true rating). Such a case clearly violates the statistical independence assumption. In general we expect the different ratings to be dependent because the same behavior is being judged. Moreover, we expect this dependence to be positive in general. Second, consider the case of raters that produce ratings by flipping a coin. Such a case clearly satisfies the conditional independence assumption, and also satisfies the measurement model used in the old procedure. The somewhat bizarre consequence of this is that with the old procedure the model fails for perfect raters but holds if raters assign scores completely at random.

6. Discussion

We formulated a class of models for scoring open ended questions in a number of steps. First, we argued that random assignment of raters to item by person combinations provides an easy method to improve the quality of the data and spread the negative effect of a *poor* rater evenly across the population of students. Second, we showed that under mild assumptions, it holds that consistent ratings are more likely equal to the true rating, compared to using a single rating. This motivated our conception of the consistent rating as an estimate of the true rating. The remaining part of the information (i.e., whether the response was consistent or not) was addressed in the third step where we argued that there is a relation between ability and agreement. Finally, we considered the probability that a second rater judges the response to be (in)correct if the first rater has judged the response as (in)correct.

After a general introduction to the class of measurement models for scoring open ended questions, a particular model from this class was introduced. It was shown that the data could be analyzed according to our assumptions with a NRM for the sum of the individual ratings.

The class of models introduced in this chapter can easily be generalized to more than two judges as well as to polytomous responses. When more judges are employed, the random assignment of raters to item by student combinations still serves both as a method to reduce the risk of violations of conditional independence and to remove the need to model the behavior of individual raters. As a consequence of the random assignment, the ratings are exchangeable, which means that all the relevant information is contained in their sum. A model for the sum of the ratings can be formulated by considering the relevant conditional distributions. Hence, similarly to the case of two raters, we use a filtering approach for formulating the model. The same filtering approach can be used for evaluating the truth of individual assumptions in isolation from the remaining assumptions. To illustrate how a model for multiple raters can be formulated we consider the case with three raters. With three raters, there are four possible outcomes that correspond to the different values for the some of the three ratings:

X_{pi1}	X_{pi2}	X_{pi3}	X_{pi+}
1	1	1	3
1	1	0	2
1	0	1	2
0	1	1	2
1	0	0	1
0	1	0	1
0	0	1	1
0	0	0	0

To formulate a model for three raters, we use the model for two raters together with the law of total probability:

$$
\begin{aligned}
&P(X_{i1} + X_{i2} + X_{i3} = 2|\theta) \\
&\quad = P(X_{i3} = 1|X_{i1} = 1, X_{i2} = 0, \theta)P(X_{i1} = 1, X_{i2} = 0|\theta) \\
&\qquad + P(X_{i3} = 1|X_{i1} = 0, X_{i2} = 1, \theta)P(X_{i1} = 0, X_{i2} = 1|\theta) \\
&\qquad + P(X_{i3} = 0|X_{i1} = 1, X_{i2} = 1, \theta)P(X_{i1} = 1, X_{i2} = 1|\theta).
\end{aligned}
$$

The three terms in the right-hand side of this equation are necessarily equal because with random assignment of raters to person by item pairs it is equally likely that it is the first, the second, or the third rater who disagrees with the other two raters. Also, the first two conditional probabilities, $P(X_{i3} = 1|X_{i1} = 1, X_{i2} = 0, \theta)$ and $P(X_{i3} = 1|X_{i1} = 0, X_{i2} = 1, \theta)$, are equal; as are the marginal probabilities $P(X_{i1} = 1, X_{i2} = 0|\theta)$ and $P(X_{i1} = 0, X_{i2} = 1|\theta)$. This means we may rewrite this expression as follows:

$$
\begin{aligned}
&P(X_{i1} + X_{i2} + X_{i3} = 2|\theta) \\
&\quad = P(X_{i3} = 1|X_{i1} + X_{i2} = 1, \theta)P(X_{i1} + X_{i2} = 1|\theta) \\
&\qquad + P(X_{i3} = 0|X_{i1} + X_{i2} = 2, \theta)P(X_{i1} + X_{i2} = 2|\theta).
\end{aligned}
$$

The first term in the right-hand side of this equation is equal to two times the second term. Hence, we can fully specify the model with an additional assumption regarding either $P(X_{i3} = 1|X_{i1} + X_{i2} = 1, \theta)$ or $P(X_{i3} = 0|X_{i1} + X_{i2} = 2, \theta)$. For instance, a reasonable assumption could be that $P(X_{i3} = 1|X_{i1} + X_{i2} = 2, \theta)$ is an increasing function of ability. Using polytomous responses poses no special problems, the model is formulated by considering a sufficient set of conditional probabilities. The same approach is used as for the case with multiple raters.

References

Andersen, E.B. (1977). Sufficient statistics and latent trait models. *Psychometrika* **42**, 69–81.

Birnbaum, A. (1968). Some latent trait models and their use in inferring an examinee's ability. In: Lord, F.M., Novick, M.R. (Eds.), *Statistical Theories of Mental Test Scores*. Addison-Wesley, Reading, pp. 395–479.

Bock, R.D. (1972). Estimating item parameters and latent ability when responses are scored in two or more nominal categories. *Psychometrika* **37**, 29–51.

Glas, C.A.W., Verhelst, N.D. (1995). Testing the Rasch model. In: Fischer, G.H., Molenaar, I.W. (Eds.), *Rasch Models: Foundations, Recent Developments, and Applications*. Springer-Verlag, New York, pp. 69–95.

Kiefer, J., Wolfowitz, J. (1956). Consistency of the maximum likelihood estimator in the presence of infinitely many nuisance parameters. *Annals of Mathematical Statistics* **27**, 887–906.

Linacre, J.M. (1994). *Many-facetted Rasch Measurement*, 2nd ed. Mesa Press, Chicago.

Maris, G., Bechger, T. (2003). Two methods for the practical analysis of rating data. Tech. Rep. No. 03-01, CITO.

Neyman, J., Scott, E.L. (1948). Consistent estimates based on partially consistent observations. *Econometrica* **16**, 1–32.

Patz, R.J., Junker, B.W., Johnson, M.S. (2000). The hierarchical rater model for rated test items and its application to large scale educational assessment data. Tech. Rep. No. 712, Department of Statistics, Carnegie Mellon University.

Rasch, G. (1960). Probabilistic models for some intelligence and attainment tests. The Danish Institute of Educational Research, Copenhagen. (Expanded edition, 1980, The University of Chicago Press, Chicago).

Verhelst, N.D., Eggen, T.J.H.M. (1989). Psychometrische en statistische aspecten van peilingsonderzoek [Psychometric and statistical aspects of assessment research]. PPON-rapport No. 4, CITO, Arnhem.

Verhelst, N.D., Glas, C.A.W. (1995). The one parameter logistic model: OPLM. In: Fischer, G.H., Molenaar, I.W. (Eds.), *Rasch Models: Foundations, Recent Developments and Applications*. Springer-Verlag, New York, pp. 215–238.

Verhelst, N.D., Verstralen, H.H.F.M. (1997). Modeling sums of binary responses by the partial credit model. Tech. Rep. No. 97-7, CITO, Arnhem.

Verhelst, N.D., Verstralen, H.H.F.M. (2001). IRT models with multiple raters. In: Boomsma, A., Van Duijn, M.A.J., Snijders, T.A.B. (Eds.), *Essays on Item Response Theory*. Springer-Verlag, New York, pp. 89–106.

Verhelst, N.D., Glas, C.A.W., van der Sluis, A. (1984). Estimation problems in the Rasch model: The basic symmetric functions. *Computational Statistics Quarterly* **1** (3), 245–262.

Verhelst, N.D., Glas, C.A.W., Verstralen, H.H.F.M. (1993). OPLM: One parameter logistic model. Computer program and manual, CITO, Arnhem.

Wilson, M., Hoskens, M. (2001). The rater bundle model. *Journal of Educational and Behavioral Statistics* **26** (3), 283–306.

Handbook of Statistics, Vol. 26
ISSN: 0169-7161

DOI: 10.1016/S0169-7161(06)26021-8

21

Assessing the Fit of Item Response Theory Models

Hariharan Swaminathan, Ronald K. Hambleton and H. Jane Rogers

1. Introduction: Models and assumptions

Item response theory provides a framework for modeling and analyzing item response data. The advantages of item response theory over classical test theory for analyzing mental measurement data are well documented (see, for example, Lord, 1980; Hambleton and Swaminathan, 1985; Hambleton et al., 1991). However, item response theory is based on strong mathematical and statistical assumptions, and only when these assumptions are met can the promises and potential of item response theory be realized.

Item response theory postulates that an examinee's performance on a test depends on the set of unobservable "latent traits" that the examinee possesses in common with the other examinees. Once these traits are defined, an examinee's observed score on an *item* is regressed on the latent traits. The resulting regression model, termed an *item response model*, specifies the relationship between the item response and the latent traits, with the coefficients of the model corresponding to parameters that characterize the item. It is this item-level modeling that gives item response theory its advantages over classical test theory.

Typically, an examinee's performance on an item is scored on a discrete scale. The most common scoring scheme is dichotomous, i.e., right/wrong, or (1/0). Polytomous scoring has become popular recently as a result of the emphasis on performance assessment and the desire to assess higher level thinking skills. In this case, the examinee's performance on an item or task is scored on an ordinal scale. In general, if U_i is the response of an examinee to item i, then $U_i = k_i$ where k_i is the category assigned to the response; $k_i = 0$ or 1 for dichotomously scored items, and $k_i = 0, 1, 2, \ldots,$ or $(K_i - 1)$ for polytomously scored items. The number of score categories K_i need not be the same across items. Such mixture scoring has become common in recent years, where some items are scored dichotomously while others are scored polytomously with differing numbers of response categories.

The probability that an examinee responds in category k is specified by the item response model or item response function. The item response function is a monotonically increasing function of the latent traits. In the case of a parametric representation, the item response function is a suitably chosen cumulative density function P that is a function of the underlying set of latent abilities and the characteristics of the item. The common choices for the function $P(x)$ are the normal ogive function $\Phi(x)$ and the logistic function $L(x)$, where

$$\Phi(x) = \int_{-\infty}^{x} \frac{1}{\sqrt{2\pi}} \mathrm{e}^{-\frac{1}{2}t^2}\,\mathrm{d}t,$$

and

$$L(x) = \frac{\mathrm{e}^x}{1+\mathrm{e}^x}.$$

Given the simplicity of the logistic function, it is more commonly employed in item response theory applications. The two functions have close resemblance to each other; in fact, $|L(1.7x) - \Phi(x)| < .01$ for all x. Consequently, the constant 1.7 is frequently included in the logistic formulation of the item response model to maximize the similarity between the two functional forms. If a nonparametric representation is chosen, the item response function needs only to be a monotonically increasing function of the latent trait. For a discussion of nonparametric item response theory, the reader is referred to Sijtsma and Molenaar (2002). We shall only deal with parametric item response theory in this chapter.

In modeling the probability of a response, it is commonly assumed that the *complete* latent space is unidimensional; that is, one underlying latent trait, θ, is sufficient to explain the performance of an examinee on the set of items that comprises the test. Multidimensional item response models have been proposed (Ackerman, 2005; Hattie, 1985; Reckase et al., 1988), but these models have not reached the operational level or practical feasibility of unidimensional models.

For dichotomously scored items, under the assumption that the latent space is unidimensional, the probability of a correct response to item i is given by the three-parameter logistic model as

$$\begin{aligned} P(U_i = 1|\theta, a_i, b_i, c_i) &= c_i + (1-c_i)\frac{\mathrm{e}^{a_i(\theta-b_i)}}{1+\mathrm{e}^{a_i(\theta-b_i)}} \\ &\equiv c_i + (1-c_i)L\big[a_i(\theta-b_i)\big]. \end{aligned} \tag{1.1}$$

The parameters a_i, b_i, c_i are the parameters that characterize the item: b_i is the *difficulty* level of the item, a_i is the *discrimination* parameter, and c_i is the lower asymptote, known as the pseudo chance-level parameter, which is a reflection of the probability that an examinee with a very low level of *proficiency* will respond correctly to the item by chance. The model given in (1.1) is known as the three-parameter model, and is often used with multiple-choice items.

The two-parameter item response model is obtained by setting $c_i = 0$. Thus the two-parameter model has the form

$$P(U_i = 1|\theta, a_i, b_i) = \frac{\mathrm{e}^{a_i(\theta-b_i)}}{1+\mathrm{e}^{a_i(\theta-b_i)}} \equiv L\big[a_i(\theta-b_i)\big]. \tag{1.2}$$

The one-parameter model is obtained by setting $a_i = 1$, $c_i = 0$, and thus has the form

$$P(U_i = 1|\theta, b_i) = \frac{\mathrm{e}^{(\theta-b_i)}}{1+\mathrm{e}^{(\theta-b_i)}} \equiv L\big[(\theta-b_i)\big]. \tag{1.3}$$

While the one-parameter model can be viewed as a special case of the three-parameter or two-parameter models, it was derived independently by Rasch (1960), based on principles of *objective measurement*. Hence the one-parameter model is commonly known as the Rasch model.

In the situation where items are scored on an ordinal scale, the probability that an examinee's score on item i falls in category s ($s = 0, 1, 2, \ldots, K_i - 1$) is given by the model (Samejima, 1969):

$$P(U_i \geqslant s|\theta, a_i, b_{si}) = \frac{e^{a_i(\theta - b_{si})}}{1 + e^{a_i(\theta - b_{si})}}, \tag{1.4}$$

where b_{si} is the boundary for category s. Clearly $P(U_i \geqslant 0|\theta, a_i, b_{si}) = 1$ and $P(U_i > K_i - 1|\theta, a_i, b_{si}) = 0$. It follows that

$$P(U_i = s|\theta, a_i, b_{si}) = P(U_i \geqslant s|\theta, a_i, b_{si}) - P(U_i \geqslant s + 1|\theta, a_i, b_{si}).$$

An alternative model for ordinal response data is the *partial credit model* (Andrich, 1978; Masters, 1982). The partial credit model is a generalization of the Rasch model for ordered responses and can be expressed as

$$P(U_i = s|\theta, b_{ri}) = \frac{e^{\sum_{r=1}^{s}(\theta - b_{ri})}}{1 + \sum_{r=1}^{k_i - 1} e^{\sum_{j=1}^{r}(\theta - b_{ji})}} \quad (s = 0, 1, \ldots, K_i - 1), \tag{1.5}$$

with

$$P(U_i = 0|\theta, b_{ri}) = \frac{1}{1 + e^{\sum_{r=1}^{K_i - 1}(0 - b_{ri})}}.$$

The partial credit model assumes constant discrimination across items. The *generalized partial credit model* (Muraki, 1992) permits varying discrimination parameters across items and is given as

$$P(U_i = s|\theta, a_i, b_{ri}) = \frac{e^{a_i \sum_{r=1}^{s}(\theta - b_{ri})}}{1 + \sum_{r=1}^{k_i - 1} e^{a_i \sum_{j=1}^{r}(\theta - b_{ji})}} \quad (s = 1, \ldots, K_i - 1), \tag{1.6}$$

and for $s = 0$,

$$P(U_i = 0|\theta, a_i, b_{ri}) = \frac{1}{1 + e^{a_i \sum_{r=1}^{K_i - 1}(\theta - b_{ri})}}.$$

More recently, Tutz (1997) and Verhelst et al. (1997) have described alternate models for the analysis of partial credit data. These models can be viewed as the item response theory analog of the continuation ratio model (Agresti, 1990). The models introduced by Tutz (1997) and Verhelst et al. (1997) have the advantage over the partial and the generalized partial credit models in that the item parameters for each category can be interpreted independently of other category parameters. A detailed analysis of the continuation ratio model and related models is provided by Hemker et al. (2001).

While the procedures for fitting the dichotomous and polytomous item response models to data have been well established, the procedures for checking the assumptions and assessing the goodness of fit of the chosen item response model have not received the same level of attention. As pointed out above, item response theory is based on strong mathematical and statistical assumptions, and only when these assumptions are met at least to a reasonable degree can item response theory methods be implemented effectively for analyzing educational and psychological test data and for drawing inferences about properties of the tests and the performance of individuals.

Checking model assumptions and assessing the fit of models to data are routine in statistical endeavors. In regression analysis, numerous procedures are available for checking distributional assumptions, examining outliers, and the fit of the chosen model to data. Some of these procedures have been adapted for use in item response theory. The basic problem in item response theory is that the regressor, θ, is unobservable; this fact introduces a level of complexity that renders a procedure that is straightforward in regression analysis inapplicable in the item response theory context.

Assessing model fit in item response theory requires the following two steps:

(i) Checking the underlying assumptions such as unidimensionality.
(ii) Assessing the agreement between observations and model predictions.

These two steps are described next. The important area of person-fit assessment procedures, however, is not included in this chapter. Interested readers are referred to the review by Meijer and Sijtsma (2001).

2. Checking the assumption of unidimensionality

When unidimensional models are fitted to test data, the validity of the assumption that the *complete* latent space is unidimensional is obviously critical. There are several approaches that can be used for investigating the assumption of unidimensionality. These include linear and nonlinear factor analysis approaches as well as nonparametric approaches that examine the assumption of local independence.

2.1. *Linear factor analysis*

A popular approach to investigating this assumption is to apply a linear factor analysis/principal components procedure. In this approach,

(1) the matrix of inter-item correlations is obtained;
(2) the percent of variance explained by the largest eigenvalue along with the point where a break occurs in the plot of the eigenvalues, or the *scree plot*, are examined;
(3) based on the above considerations a determination is made regarding the dimensionality of the item response data.

In addition to its obvious subjectivity, this approach has several drawbacks. When the item responses are discrete, the inter-item correlations will be small. This is a consequence of the fact that discrete item responses have nonlinear relationships with the underlying ability continuum. This fact is particularly evident when the correlations are computed using phi-coefficients. As a result, even when the model is unidimensional, the percent of variance accounted for by the dominant dimension as indicated by the largest eigenvalue will be relatively small. Tetrachoric correlations provide an improvement over phi-coefficients, but even using tetrachorics, the magnitude of the inter-item correlations will be small. Simulation studies by the authors have shown that in unidimensional item response theory applications, the largest eigenvalue of the matrix of tetrachoric correlations will typically account for only about 25 to 35 percent of the variance. Despite this problem, an investigation of the eigenvalue plot is often a useful first step in the study of test dimensionality.

One possible approach to resolving the problem of the percent of variance accounted for by the dominant eigenvalue is to fit an item response model to the data and then, using the item parameter estimates and a suitably chosen distribution of the trait θ (normal, uniform, or skewed), generate item response data. Once the data are generated, the inter-item correlation matrix is obtained and the percent of variance accounted for by the largest eigenvalue is computed. The purpose of this simulation is to obtain an indicator of the magnitude of the largest eigenvalue to be expected in the unidimensional case. The eigenvalue obtained from analysis of the real item responses can be compared with this "ideal" and a decision made regarding the unidimensionality assumption.

A similar simulation procedure can be used with polytomously scored response data. In this case, the matrix of polyserial correlations is computed with the real and simulated data, the eigenvalues extracted, and a decision made about the assumption of unidimensionality. A more refined approach is to examine the residuals after factor analyzing the simulated and real data using computer software such as LISREL 8 (Joreskog and Sorbom, 2004). The distribution of the residuals for the simulated data will provide a baseline for judging the viability of the unidimensionality assumption for the real data. A problem with this approach is that maximum likelihood procedures will fail if the matrix of tetrachoric correlations is not positive definite.

2.2. *Nonlinear factor analysis and item factor analysis*

The basic problem with the linear factor analysis approach is its linearity. Item response models are inherently nonlinear; in fact, the item response model is essentially a nonlinear factor model, a fact that has long been recognized. The linear factor approach to analyzing item response data can be thought of only as a first-order approximation that is not appropriate for assessing the dimensionality of item response data except in extreme cases where the item response model is linear. In recognition of this fact, Hambleton and Rovinelli (1986) suggested using the nonlinear factor analysis procedure of McDonald (1967, 1982, 1997). The approach by McDonald is to approximate unidimensional as well as multidimensional normal ogive item response functions by Hermite–Chebychev orthogonal polynomials through harmonic analysis. This approach is implemented in the program NOHARM (Fraser, 1988). As statistical procedures for assessing the dimensionality are not given in NOHARM, Gessaroli and De Champlain (1996) suggested a statistic, $X^2_{G/D}$, based on the residual correlations r_{ij} between items i and j. These residuals are transformed using Fisher's transformation, i.e., $z_{ij} = \tanh^{-1} r_{ij}$, and the test statistic is computed as

$$X^2_{G/D} = (N-3)\sum_{i=2}^{n}\sum_{j=1}^{i-1} z^2_{ij},$$

where N is the sample size. This statistic is assumed to have a chi-square distribution with $[1/2n(n-1) - n(1+d)]$ degrees of freedom where n is the test length and d is the number of dimensions. For testing the assumption of unidimensionality, $d = 1$, and the degrees of freedom is $[1/2n(n-5)]$. While it is difficult to justify the distributional assumption on theoretical grounds, Gessaroli and De Champlain (1996) demonstrated through simulations that the statistic maintained the nominal Type I error rate for the unidimensional case, and had good power in detecting two-dimensional data. More recently, Gessaroli et al. (1997) developed a likelihood-ratio type statistic,

$$ALR = \sum_{i=2}^{n}\sum_{j}^{i-1} G^2_{ij}$$

with

$$G^2_{ij} = \sum_{u_i=0}^{1}\sum_{u_j=0}^{1} p_{u_i u_j} \log\left(\frac{\hat{p}_{u_i u_j}}{p_{u_i u_j}}\right),$$

where $p_{u_i u_j}$ is the observed proportion of examinees with dichotomous responses u_i and u_j for items i, j; $\hat{p}_{u_i u_j}$ is the model-predicted joint proportion of examinees with responses u_i and u_j. Gessaroli et al. (1997) have suggested that *ALR* has the same distribution as $X^2_{G/D}$. Maydeu-Olivares (2001) has shown that these statistics do not have the chi-square distributions claimed. However, Finch and Habing (2005) and Tate (2003) examined the behavior of *ALR* and concluded that while this statistic may not have the expected theoretical distribution, it is nevertheless effective in recovering the number of underlying dimensions.

Item factor analysis, developed by Bock et al. (1988) and implemented in the software TESTFACT, allows the factor analysis to be carried out using either the matrix of tetrachoric correlations or a full-information maximum likelihood procedure. De Champlain and Gessaroli (1998) found high Type I error rates with TESTFACT. Tate (2003), on the other hand, found that TESTFACT was able to accurately confirm the hypothesis of unidimensionality when the data were indeed unidimensional, except in cases of extreme item difficulty.

2.3. Examination of local independence

Lord and Novick (1968) have shown that the assumption that the latent space is complete is equivalent to the assumption of local independence. They argued that if the latent space is complete, no additional traits are necessary to account for the responses to the items, and hence the responses to items will be statistically independent when conditioned on the complete set of latent traits. This assertion is analogous to the assertion in linear factor analysis that the correlations among the observed variables will be zero after partialing out or conditioning on the complete set of common factors. Hence, testing the assumption of local independence provides a test of the dimensionality of the test data.

Statistical independence is a strong requirement; a weaker assumption is to require that the covariances among the items be zero when the item responses are conditioned on the complete set of latent traits. In particular, if the test is unidimensional, then conditioning on the latent trait will result in the item covariances being zero. Stout (1987, 1990) used this weaker form of local independence to develop a test for "essential" unidimensionality. Stout argued that a test of length n is essentially unidimensional if the average covariance over all pairs of items is small in magnitude when conditioned on the latent trait. This procedure is implemented in the program DIMTEST (Stout et al., 1991).

The basic procedure in DIMTEST is to: (a) Create an *assessment test* made up of a core subset of m items that is unidimensional using expert judgment or factor analytic methods; (b) create a *partitioning test* using the remaining $n - m$ items to score and group the examinees into subgroups. These subgroups are homogeneous with respect to their test score on the partitioning test and hence are "conditioned" on the latent trait. The score

$$Y_{jk} = \sum_{i=1}^{m} U_{ijk}, \quad j = 1, \ldots, J_k,$$

of each examinee in subgroup k on the assessment test is computed, and the variance of these scores, $\hat{\sigma}_k^2$, is obtained. If local independence holds, then

$$\mathrm{Var}(Y_{jk}) = \sum_{i=1}^{m} \mathrm{Var}(U_{ijk}) = \sum_{i=1}^{m} \hat{p}_{ik}(1 - \hat{p}_{ik}) \equiv \hat{\sigma}_{U_k}^2.$$

Thus, if the assumption of unidimensionality is met, these two quantities are the estimates of the same variance. Stout (1987) demonstrated that

$$E\left(\hat{\sigma}_k^2 - \hat{\sigma}_{Uk}^2 | X\right) \propto \sum \mathrm{Cov}(U_i, U_j | X), \tag{2.1}$$

where X is a score on the partitioning test. Since in the weak form of local independence, the pairwise covariances are zero, a statistic based on $(\hat{\sigma}_k^2 - \hat{\sigma}_{U_k}^2)$ provides a test of essential unidimensionality. Stout (1990) showed that the test statistic can be further refined by correcting for the bias using another subset of unidimensional items, known as *assessment subtest 2*, chosen to be as similar as possible in difficulty to assessment subtest 1.

Hattie et al. (1996) evaluated the performance of DIMTEST using simulated data. They concluded that the Type I error rate was in the acceptable range and that the procedure was sufficiently powerful to reject the null hypothesis when the test data was not unidimensional. The primary drawback with this procedure is the identification of the assessment subtests. This identification is subjective and will affect the test of essential unidimensionality. A further drawback is that the test under investigation must be sufficiently long to support the selection of the assessment subtests. External subtests may be used but these will not always be available. It should be pointed out, too, that

DIMTEST is designed to test the hypothesis that the test is unidimensional – it is not meant for assessing the dimensionality of the test. Finally, Hattie et al. (1996) showed that the test statistic is adversely affected if the tetrachoric correlation matrix is not positive definite, and recommended that the procedure should not be used in this case.

Yen (1984) examined the issue of local independence from the point of view that items based on a common stimulus may violate the local independence assumption. She suggested examining the correlations between residuals $(u_{ia} - P_{ia})$ and $(u_{ja} - P_{ja})$ across pairs of items $i, j = 1, \ldots, n$, $i \neq j$, across examinees $a = 1, \ldots, N$, where u_{ia} is the response of examinee a to item i and P_{ia} is the model-based probability of a correct response. The test statistic is $Q_3 = r_{ij}$, the correlation between the residuals. Yen (1984) indicated that the hypothesis of local independence for pairs of items is equivalent to the hypothesis that the correlation between the residuals is zero. One problem with this approach is that residuals may not be linearly related. A second problem noted by Chen and Thissen (1997) is that the item residuals are not bivariate normally distributed, and hence the distribution of residuals will not have the expected mean and variance. Chen and Thissen (1997) showed the empirical Type I error rates for the Q_3 statistic are higher than the nominal Type I error rates. Glas (2005) and Glas and Suarez Falcon (2003), following the development given by van den Wollenberg (1982) for the Rasch model, provided a statistic for testing local independence based on the difference between observed and expected frequencies

$$d_{ij} = N_{ij} - E(N_{ij}),$$

where N_{ij} is the observed number of examinees responding correctly to items i and j in the group of examinees obtaining a score between $k = 2$ and $k = n - 2$, and

$$E(N_{ij}) = \sum_k N_k P(U_i = 1, U_j = 1 | X = k). \tag{2.1}$$

The joint probability is computed using a recursive algorithm based on that provided by Lord and Wingersky (1984) (see the next section for details of this algorithm). Following the statistic given by van den Wollenberg (1982), Glas and Suarez Falcon suggested the statistic

$$S_{3ij} = d_{ij}^2 \left\{ \frac{1}{E(N_{ij})} + \frac{1}{E(N_{\bar{i}j})} + \frac{1}{E(N_{i\bar{j}})} + \frac{1}{E(N_{\bar{i}\bar{j}})} \right\},$$

where $E(N_{\bar{i}j})$ is defined as in (2.1) with the probability that an examinee with score k responds incorrectly to item i and correctly to item j; $E(N_{i\bar{j}})$ is defined similarly, while $E(N_{\bar{i}\bar{j}})$ is computed with the examinee responding incorrectly to the items. Simulation studies have indicated that the statistic behaves like a chi-square variate with 1 degree of freedom.

Glas and Suarez Falcon (2003) compared the behavior of Yen's Q_3 statistic with the performance of the S_3 statistic and concluded that the Type I error rates for these statistics were below the nominal rate, and that they did not have much power. Their results agreed with previous findings that although local independence was violated, the item response curves were well recovered, and hence robust to violations of the assumption of local independence.

3. Assessing model data fit: Checking model predictions

Assessing whether the test data is unidimensional is only the first step in determining if the assumptions of item response theory are met. Assessment of the fit of the model to data is multi-faceted and must be carried out at the test level as well as the item level. As in any scientific investigation, verification of a theory is most directly carried out by examining the predictions made by the theory. In item response theory, this is accomplished by comparing what the theory predicts with what is observed.

3.1. Assessment of model fit at the test level

Likelihood approaches

The assessment of the fit of model at the test level can be carried out using the likelihood ratio statistic. The probability of a response pattern $U = [u_1 u_2 \dots u_n]$ for the item response data is

$$P(U) = \int \prod_i P_i(\theta)^{u_i} \left(1 - P_i(\theta)^{1-u_i}\right) g(\theta)\,\mathrm{d}\theta \tag{3.1}$$

for the dichotomous case; $P_i(\theta)$ is the probability of a correct response for an examinee on item i given by the item response function specified in Eqs. (1.1) to (1.3), U_i is the response of an examinee to item i, and $g(\theta)$ is the density function of θ. The likelihood function is then given by

$$L = \prod_{\substack{\text{response}\\ \text{patterns}}} P(U)^{f_u}, \tag{3.2}$$

where f_u is the frequency of response pattern U.

In the ideal case, a table that contains all response patterns can be constructed, and based on this table, a likelihood ratio statistic can be calculated. Unfortunately, with n items, there are 2^n response patterns and this number gets large rapidly. Hence it is not feasible to carry out the contingency table approach when the number of items is even moderately large.

When $n \leqslant 10$, and when the number of examinees is large, it is possible to construct a table of frequencies for all response patterns. Let $f_1, f_2, \dots, f_{2^n}$ denote the frequencies for the 2^n response patterns. The likelihood ratio chi-square statistic for the test of fit is

$$G^2 = 2 \sum_{r=1}^{2^n} f_r \log \frac{f_r}{N\widehat{P}(U_r)}, \tag{3.3}$$

where $P(U)$ is given by (3.1) and approximated as

$$\widehat{P}(U) = \sum_{k=1}^{q} P(U|X_k) A(X_k).$$

Here, X is a quadrature point and $A(X)$ is the weight corresponding to the density function $g(X)$. The quantity G^2 defined in (3.3) has an asymptotic chi-square distribution with degrees of freedom equal to $2^n - kn - 1$, where k is the number of item parameters in the model.

When the number of items is large, the procedure given above is not feasible. In this case, traditional likelihood ratio tests are applicable. The fit of the model to the data can be determined by computing $-2\log\lambda$, where λ is the maximum value of the likelihood function given in (3.2) for the fitted model. Under the null hypothesis that the model fits the data against the alternative that the probability of a correct response is random, $-2\log\lambda$ is asymptotically distributed as a chi-square variate with degrees of freedom equal to the number of item parameters estimated.

The likelihood ratio statistic described above can also be used to test the fit of the three-parameter model against the two-parameter model, and the fit of the two-parameter model against the one-parameter model. The likelihood ratio $-2\log(\lambda_A/\lambda_B)$ is computed, where λ_A and λ_B are the maximum values of the likelihood functions (3.1) obtained by fitting the nested models A and B. This ratio is asymptotically distributed as a chi-square variate with degrees of freedom equal to the number of additional parameters estimated in model A in comparison with model B.

In the case of the Rasch model, a likelihood ratio test based on the distribution of item responses conditional on number-right score, the sufficient statistic, is available (Andersen, 1973). This procedure has an advantage over the likelihood ratio tests based on the joint likelihood of item and ability parameters, since in the joint procedure item and ability parameters must be estimated simultaneously, and hence do not yield consistent estimates of item parameters. The marginal maximum likelihood procedure (Bock and Aitkin, 1981) solves this problem and hence there is very little difference between the conditional likelihood ratio procedure and the "marginal" likelihood ratio procedure.

The extension of the likelihood ratio test described above to polytomous item response models is straightforward. The fit of the generalized partial credit model may be compared with the partial credit model (Eqs. (1.5) and (1.6)), for example. The same comparison can be made with the graded response model (Eq. (1.4)) for testing nested hypotheses.

Likelihood ratio tests for testing fit at the item level are also available. These are described in the next section. The likelihood ratio fit statistic at the item level can be added across items to assess model fit at the test level.

Comparison of observed and expected score distributions

The likelihood approach described above is one approach to assessing model data fit. As mentioned earlier, assessing fit is a multifaceted approach and other approaches should be looked at to arrive at a complete picture of model–data fit. One such approach is to compare the observed score distribution with the predicted score distribution. Once the item response model is fitted to the data, it is possible to obtain the predicted number-correct score distribution.

Lord (1980) showed that for a test of length n, the distribution of number-correct score x, $f_n(x|\theta)$ is the compound binomial distribution, and that the relative frequency of a score x can be obtained from the coefficient of t^x in the expansion of the moment generating function $\prod_{i=1}^{n}(P_i t + Q_i)$ where Q_i is the probability of an incorrect response to item i. For example, for a test of length $n = 2$, the number-correct score $x = 0$, 1, or 2. These scores occur with relative frequency

$$f_2(x = 0|\theta) = Q_1 Q_2,$$
$$f_2(x = 1|\theta) = P_1 Q_2 + Q_1 P_2,$$
$$f_2(x = 2|\theta) = P_1 P_2.$$

For long tests the computation of the distribution of number-correct scores using the generating function is extremely tedious, and Lord and Novick (1968) concluded that approximations are inevitable in such situations. However, Lord and Wingersky (1984) developed a simple recursive algorithm for computing the frequency distribution of the number-correct score. The algorithm capitalizes on the fact that if the frequency distribution $f_{n-1}(x|\theta)$ is determined, then when the test is increased by one item, a score $x = k$ is obtained on the n-item test if the examinee obtains a score of $x = k$ on the test with $(n - 1)$ items and responds incorrectly to the nth item, or obtains a score of $(k - 1)$ on the test with $(n - 1)$ items and responds correctly to the nth item, i.e.,

$$f_n[x = k|\theta] = f_{n-1}[x = k|\theta] * Q_n + f_{n-1}[x = k - 1|\theta] * P_n. \tag{3.4}$$

It is relatively straightforward to obtain the conditional distribution of number-correct scores using this algorithm even for very long tests. This algorithm readily extends to the polytomous case.

Once this predicted conditional number-correct score distribution is obtained, the marginal number-correct score distribution is obtained by summing over the examinees at their estimated ability values, $\hat{\theta}_a$:

$$f_n(x) = \sum_{a=1}^{N} f_n(x|\hat{\theta}_a). \tag{3.5}$$

Alternatively, by assuming a density function of θ, $f_n(x)$ is obtained as

$$f_n(x) = \int f_n(x|\theta) g(\theta)\, \mathrm{d}\theta. \tag{3.6}$$

The observed number-correct score distribution is then compared with the model predicted number-correct score distribution and the results are displayed graphically (Hambleton and Traub, 1973). Alternatively, a graphical display of the cumulative distributions may be obtained (see, for example, Hambleton and Han, 2005). A third option is to plot quantiles against each other. The advantage of such Q–Q plots is that if the two distributions are identical, the Q–Q plot will be a straight line. Departures from the straight line are easier to spot than differences in the frequency distributions or the cumulative frequency distributions.

An example of the procedure described above is provided in Figures 1 to 3, where the observed distribution is compared with the frequency distribution obtained after fitting one-, two-, and three-parameter models to an achievement test. The figures indicate that the three-parameter model produces the closest fitting frequency distribution to the observed score distribution. The one-parameter model provides the poorest fit. In these plots, the marginal score distribution is obtained by summing over the estimated

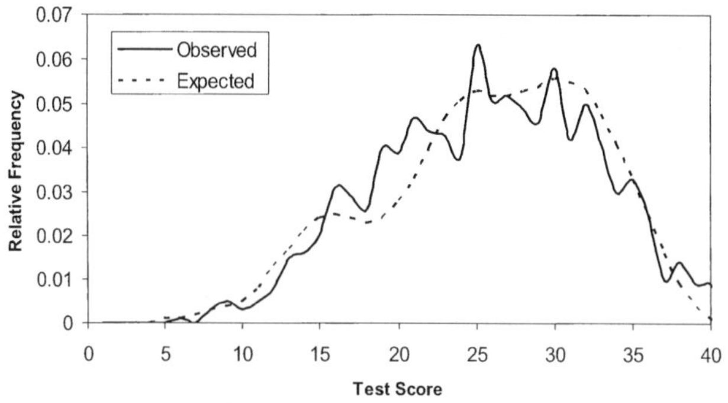

Fig. 1. Observed and expected score distributions based on trait estimates for the one-parameter model.

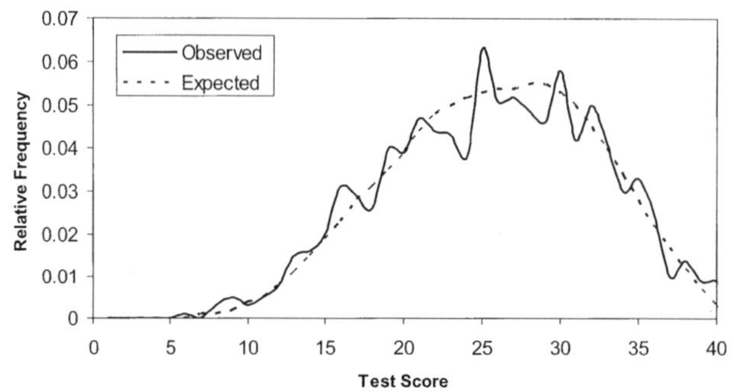

Fig. 2. Observed and expected score distributions based on trait estimates for the two-parameter model.

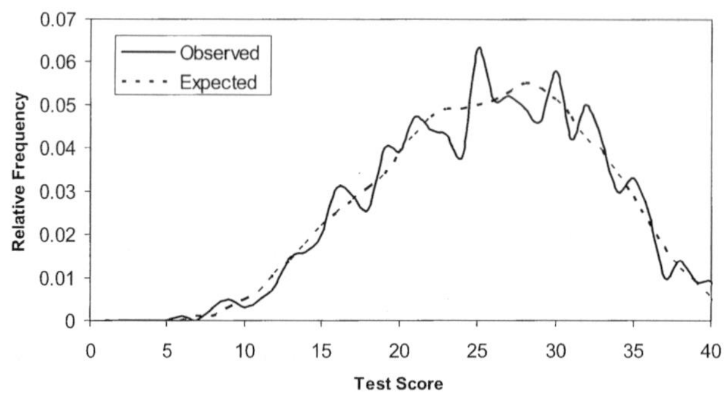

Fig. 3. Observed and expected score distributions based on trait estimates for the three-parameter model.

θ values. The marginal distributions under the assumption that θ is normally distributed are provided in Figures 4 to 6. The differences between the distributions obtained with trait estimates and under the assumption of a normal distribution are small, and the conclusion regarding the fit of the model remains unaltered.

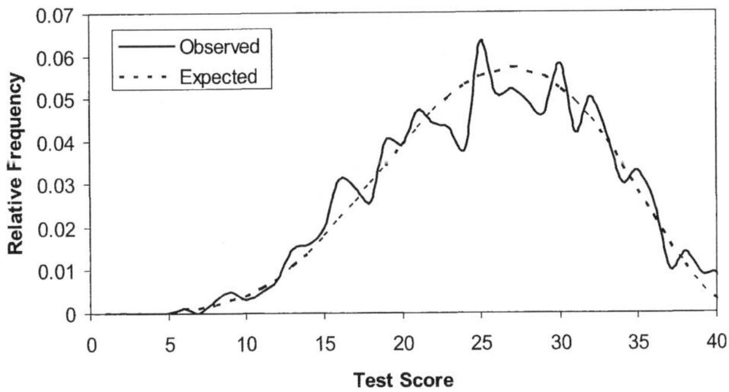

Fig. 4. Observed and expected score distributions assuming a normal trait distribution for the one-parameter model.

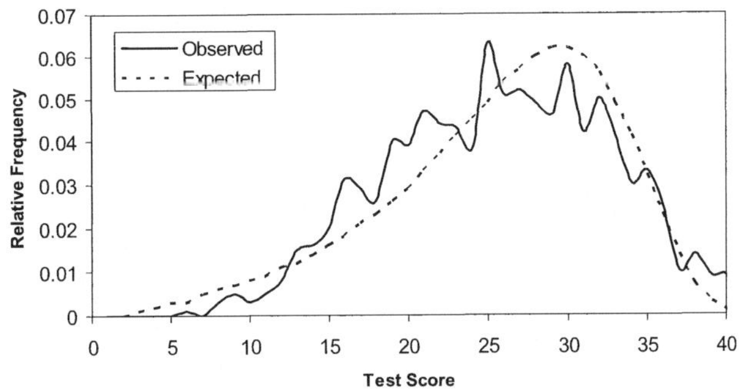

Fig. 5. Observed and expected score distributions assuming a normal trait distribution for the two-parameter model.

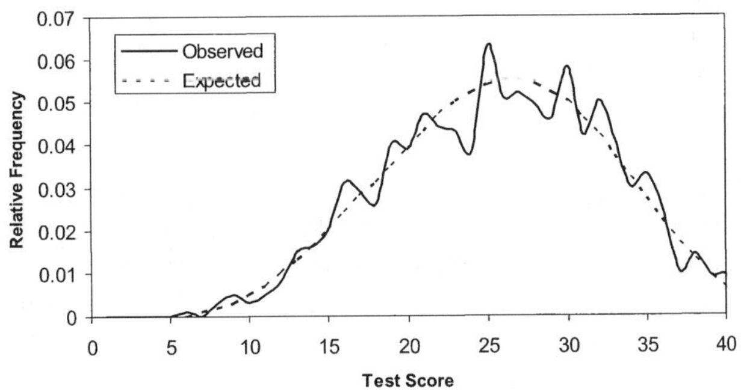

Fig. 6. Observed and expected score distributions assuming a normal trait distribution for the three-parameter model.

The observed and expected distributions may be compared statistically by employing the traditional chi-square test using the statistic

$$X^2 = \sum_{i=1}^{m} \frac{(f_{oi} - f_{ei})^2}{f_{ei}},$$

where f_{oi} and f_{ei} are observed and expected frequencies. The statistic is distributed as a chi-square with $m-1$ degrees of freedom, where m is the number of score groups. When f_{ei} is less than one, it is common to combine score categories to achieve a minimum expected cell frequency of at least one. Alternatively, the Kolmogorov–Smirnoff (K–S) procedure for comparing two cumulative distributions may be used.

3.2. Assessment of model fit at the item level

Assessment of model fit at the item level takes several forms since the theory may be used to predict a variety of outcomes at the item level. These include comparing the observed probability of a correct response to the model-predicted probability of a correct response at various points on the θ continuum or at observed score levels, examining invariance of item parameters across subgroups of examinees, and examining the invariance of ability estimates for partitions of the test into subsets of items. In one sense, assessing fit at the item level is a more stringent test of model fit than assessing fit at the test level, since it can be expected that not all conditions and requirements will be met.

The primary tools for assessing model fit at the item level are discrepancy measures between observations and expectations (Hambleton and Han, 2005; Hambleton et al., 1991; Rogers and Hattie, 1987). These discrepancy measures lend themselves to both graphical displays and statistical tests of significance. While statistical significance tests have been the mainstay in the social and behavioral sciences, they can often be noninformative in the context of item response theory. Graphical displays, on the other hand, while suffering from a certain degree of subjective interpretation, can often provide meaningful insights into the nature of model–data misfit and have been recommended strongly by Hambleton and Rogers (1990), and Hambleton et al. (1991).

Residual analysis

The most frequently used and intuitively appealing approach to assessing IRT model–data fit is to compare predictions based on the model with what was actually observed. Such analysis of residuals is common in model–fitting enterprises. In the IRT context with dichotomous item responses, the model provides the probability of success for an examinee with a given trait value. This probability is computed using the item parameter estimates obtained from the calibration. In practical terms, this probability is the proportion of examinees at that trait value that answers the item correctly. If a sufficient number of examinees with the same trait value were available, this proportion could be compared with the observed proportion correct at that trait value and a residual computed.

Two problems with this approach occur in reality: First, only estimates of trait values are available, and second, few individuals will have identical trait estimates except in the case of the one-parameter model, where there is a one-to-one correspondence between trait value and total score. The most common empirical solution to this problem is to group examinees according to their trait estimates and treat each group as if all examinees were at the same trait value. The probability of a correct response is computed for examinees in that subgroup using a representative trait value (mean or median, for example), or by averaging the probabilities of a correct response for examinees in the interval. This result provides the model-based or expected proportion correct for that interval.

In practice, constructing the trait intervals is not trivial. The requirement of sufficient sample size within intervals to obtain reliable estimates of the observed proportion correct must be balanced against that of sufficiently homogeneous subgroups to make the expected proportion correct a representative value for the interval. At the extremes of the trait continuum, there are few examinees, and hence either the range over which the intervals are constructed must be restricted or the intervals broadened at the extremes.

Of the two criticisms raised, the second does not apply to the Rasch model. In the Rasch model, the number correct score is a sufficient statistic for the trait θ and hence there is a one-to-one correspondence between the number-right score and the trait level. It is therefore possible to compare the observed proportion correct at each score value to the expected proportion correct, i.e., the model-predicted proportion correct score, without resorting to the trait estimates. Residual analysis of this nature was first introduced in the context of the Rasch model by Wright and Panchapakesan (1969) and is a tool that is used commonly for assessing item fit in the Rasch model.

Graphical displays are useful in examining the discrepancy between observed and expected proportions correct. An example of a graphical fit analysis is shown in Figures 7 to 9. In these figures, an item from an achievement test was calibrated using one-, two-, and three-parameter models, and the observed and expected proportions correct calculated in the manner described above. In Figure 7, the plot shows that the model fails to fit the data at the lower end of the trait continuum; the observed proportion correct is considerably higher than the expected proportion correct in the lower intervals. Because of the lower asymptote of zero imposed by the one-parameter model, the model cannot account for examinee guessing and hence does not provide adequate fit. In Figure 8, the fit of the model is improved; this is accomplished in the two-parameter model by lowering the discrimination parameter for the item so that the lower asymptote of zero

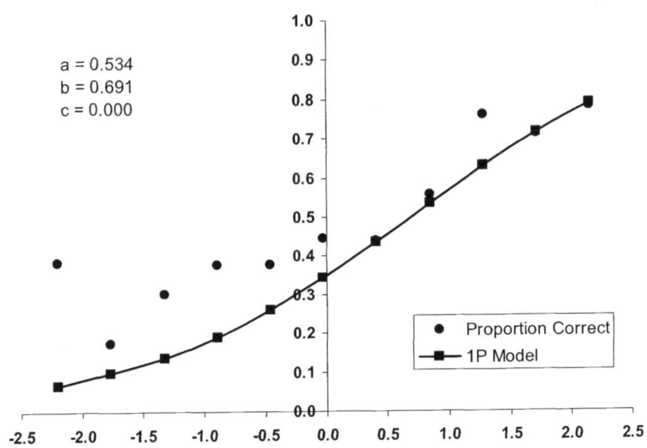

Fig. 7. Fit plot for an item calibrated with the 1P model.

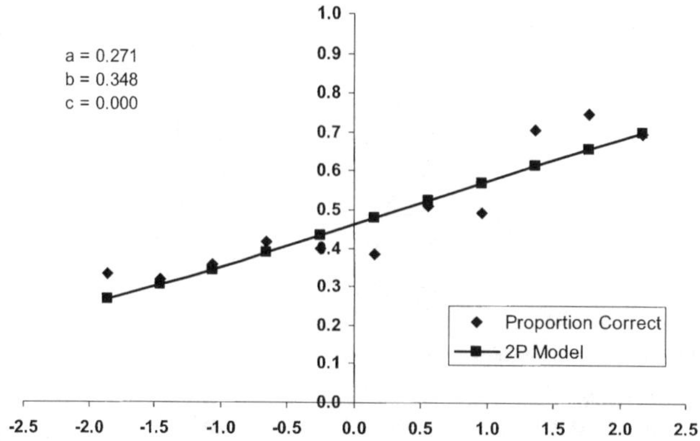

Fig. 8. Fit plot for an item calibrated with the 2P model.

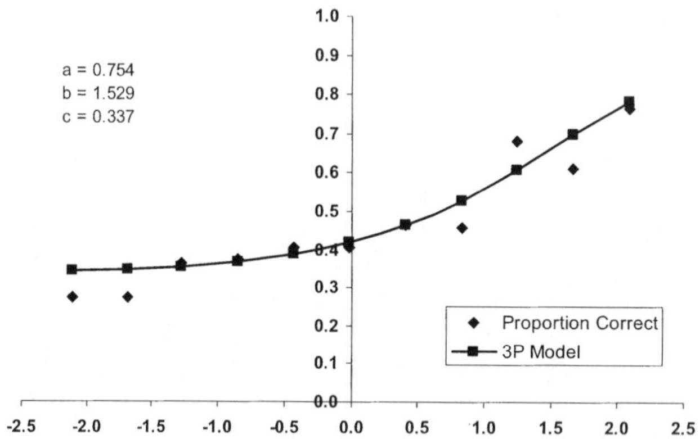

Fig. 9. Fit plot for an item calibrated with the 3P model.

is reached at much lower values of the trait. In Figure 9, where a three-parameter model was fitted, a nonzero lower asymptote is permitted and hence the model is better able to reproduce the observed proportion-correct values. The widely used calibration program BILOG-MG (Zimowski et al., 1996) produces fit plots such as those shown above along with error bands for interpreting departures from the model.

In the case of dichotomous item responses, residuals are obtained by subtracting the expected proportion correct from the observed proportion correct. The residual is generally standardized by dividing by the standard error of the observed proportion correct under the null hypothesis of model–data fit (Hambleton et al., 1991; Wright and Panchapakesan, 1969). Hence, the standardized residual has the form

$$SR_j = \frac{(O_j - E_j)}{\sqrt{\frac{E_j(1-E_j)}{N_j}}}, \tag{3.7}$$

where O_j is the observed *proportion* correct in trait interval j, E_j is the expected *proportion* correct in the interval under the fitted model, and N_j is the number of examinees in the trait interval. In the case of polytomous item responses, residuals can be calculated within each response category. While there is no statistical theory to support the assumption of normality, simulation studies show that the residuals have an approximately normal distribution when the model fits the data (Hambleton et al., 1991).

The distribution of the residuals can be visually inspected to determine the proportion of residuals that are unusually large. The percentage of standardized residuals outside the range $(-2, 2)$ provides an indicator of adequacy of model–data fit. As another interpretational tool, Hambleton et al. (1991) suggested comparing the distribution of standardized residuals to one obtained from analysis of simulated data generated using the item parameter estimates obtained from the calibration of the real data. This distribution provides a baseline for interpreting the extent to which the residual distribution departs from what could be expected when the model fits the data.

Chi-square item fit statistics

Chi-square statistics would seem to be natural tests of the discrepancy between the observed and expected frequencies or proportions computed in a residual analysis. Both Pearson and likelihood ratio statistics have been proposed; these statistics have been standard tools for assessing model fit since the earliest applications of IRT. In its most general form, the Pearson chi-square fit statistic is

$$X^2 = \sum_{j=1}^{J}\sum_{k=1}^{K} N_{jk}\frac{(O_{jk} - E_{jk})^2}{E_{jk}}, \tag{3.8}$$

where j indexes the trait interval and k indexes the response category. In the case of dichotomous item responses, this reduces to

$$X^2 = \sum_{j=1}^{J} N_j\frac{(O_j - E_j)^2}{E_j(1 - E_j)}. \tag{3.9}$$

Variations of this statistic are obtained by varying the numbers of intervals, varying the methods of constructing the intervals (equal frequency or equal width), and varying the methods of obtaining the expected proportion correct (using the mean or median trait value in the interval or averaging the probabilities of response for individuals in the interval). Bock (1972) used J intervals of approximately equal size and used the median trait value in the interval to obtain the expected proportion correct. Yen (1981) specified 10 intervals of equal frequency and used the mean of the probabilities of a correct response as the expected proportion correct. Statistics of this form are generally assumed to have an approximate chi-square distribution with degrees of freedom equal to $J * (K - 1) - m$ degrees of freedom, where m is the number of estimated item parameters in the model. Hence for Yen's statistic (referred to as Q_1), for example, the degrees of freedom are taken as $10 - 3 = 7$ for dichotomous responses under the three-parameter model.

The likelihood ratio statistic has the general form

$$G^2 = 2\sum_{j=1}^{J}\sum_{k=1}^{K} N_j\left(O_{jk}\log\frac{O_{jk}}{E_{jk}}\right),$$

reducing to

$$G^2 = 2\sum_{j=1}^{J} N_j\left(O_{jk}\log\frac{O_j}{E_j} + (1 - O_{jk})\log\frac{1-O_j}{1-E_j}\right) \tag{3.10}$$

in the case of dichotomous item responses. This is the item fit statistic provided in BILOG-MG. McKinley and Mills (1985) assumed the degrees of freedom for the distribution to be $J(K-1) - m$. Mislevy and Bock (1990) argue that since parameter estimation is not based on minimization of the chi-square statistic, no degrees of freedom are lost for the number of item parameters estimated.

A number of problems arise in using chi-square statistics as tests of model–data fit in the IRT context. Principal among them is whether the statistics have the chi-square distribution claimed and if so, whether the degrees of freedom are correctly determined. Glas and Suarez Falcon (2003) note that the standard theory for chi-square statistics does not hold in the IRT context because the observations on which the statistics are based do not have a multinomial or Poisson distribution.

Simulation studies (Yen, 1981; McKinley and Mills, 1985; Orlando and Thissen, 2000, 2003) have shown that the fit statistics in common use do generally appear to have an approximate chi-square distribution; however, the number of degrees of freedom remains at issue. Orlando and Thissen (2000) argued that because the observed proportions correct are based on model-dependent trait estimates, the degrees of freedom may not be as claimed. Stone and Zhang (2003) agreed with the assessment of Orlando and Thissen (2000) and further noted that when the expected frequencies depend on unknown item and ability parameters, and when these are replaced by their estimates, the distribution of the chi-square statistic is adversely affected. Yen (1981), however, concluded that since trait estimates are based on all items, the loss of degrees of freedom due to the use of trait estimates is negligible in the computation of the fit statistic for a single item.

Another obvious problem with chi-square fit statistics computed as described above is that the trait intervals are arbitrary; different choices of intervals will lead to different values of the fit statistics and potentially, to different conclusions about the fit of individual items. A further issue is that of the minimum interval size needed for a chi-square approximation to be valid (disbelief suspended on all other fronts). The usual recommendation is an expected cell frequency of 5.

Given the forgoing, it is not surprising that alternatives to the chi-square statistics described above have been sought. Orlando and Thissen (2000) addressed the issue of the use of trait estimates to establish the intervals in which residuals are computed. They proposed instead that examinees be grouped according to their number correct score, which is not dependent on the model. The recursive formula described in an earlier section for computing the frequency distribution of number correct score given trait value (Eq. (3.4)) is used in obtaining expected proportion correct at each score level. The expected proportion correct of examinees with total score k who get item i correct is:

$$E_{ik} = \frac{\int P_i f(k-1|\theta) g(\theta)\,d\theta}{\int f(k|\theta) g(\theta)\,d\theta}. \tag{3.11}$$

Once the expected proportions are obtained and the observed proportions calculated from the data, a Pearson chi-square fit statistic (referred to as $S - X^2$ by the authors) or a likelihood ratio statistic (referred to as $S - G^2$) is computed in the usual manner as given by Eqs. (3.9) and (3.10). The two statistics are:

$$S - X_i^2 = \sum_{k=1}^{n-1} N_k \frac{(O_{ik} - E_{ik})^2}{E_{ik}(1 - E_{ik})} \tag{3.12}$$

and

$$S - G_i^2 = 2\sum_{k=1}^{n-1} N_k \left(O_{ik} \log \frac{O_{ik}}{E_{ik}} + (1 - O_{ik}) \log \frac{1 - O_{ik}}{1 - E_{ik}} \right). \tag{3.13}$$

Note that O_{ik} and E_{ik} are observed and expected proportions. The degrees of freedom for the statistic are the number of score levels minus the number of estimated item parameters. If all score levels are used in an n-item test, then the degrees of freedom are $(n-1) - m$, where m is the number of parameters in the model. Score levels may be collapsed from the extremes inwards to ensure a desired minimum expected cell frequency. Orlando and Thissen (2000) argued that a minimum expected cell frequency of 1 is acceptable for the chi-square statistic they proposed.

Orlando and Thissen (2000, 2003) compared their fit statistics with that of Yen and with a likelihood ratio statistic and found that the Type I error rates for the Yen Q_1 and likelihood ratio statistics were unacceptably high in short tests, while the Type I error rates for $S - X^2$ remained close to the nominal level. Type I error rates were somewhat higher for $S - G^2$, and the authors recommend use of $S - X^2$ in preference. In longer tests, the Type I error rates improved for Q_1 and were slightly above the nominal values. Glas and Suarez Falcon (2003) found the false alarm rates for $S - G^2$ to be high in short tests with a 10% of misfitting items.

Stone et al. (1994) and Stone (2000), like Orlando and Thissen (2000), point out that the uncertainty in the θ estimates may be responsible for the departures of the chi-square statistics from their expected behavior. They suggested that since the posterior distribution of θ takes into account the uncertainty of θ estimates, using the posterior distribution of θ to develop a fit measure has clear advantages over procedures that use point estimates of θ.

In the E-step of the E–M algorithm for estimating item parameters using a marginal maximum likelihood procedure (Bock and Aitkin, 1981), the expected number of examinees at trait value X_k is

$$\widetilde{N}_{ik} = \sum_{j=1}^{N} P(X_k|\underline{U}_j, \underline{\xi}) = \sum_{j=1}^{N} \frac{L(X_k)A(X_k)}{\sum_{k=1}^{q} L(X_k)A(X_k)}. \tag{3.14}$$

Here, $L(X_k)$ and $A(X_k)$ are the likelihood function and the quadrature weights, respectively, evaluated at quadrature point X_k. Similarly, the expected number of examinees answering the item correctly at trait value X_k is

$$\tilde{r}_{ik} = \sum_{j=1}^{N} u_{ij} P(X_k | \underline{U}_j, \underline{\xi}) = \sum_{j=1}^{N} \frac{u_{ij} L(X_k) A(X_k)}{\sum_{k=1}^{q} L(X_k) A(X_k)}, \tag{3.15}$$

where u_{ij} is the response of examine j to item i. In their approach, Stone et al. (1994) and Stone (2000) use these *pseudocounts*, $\tilde{r}_{ik}$ and $\widetilde{N}_{ik}$, to assess the fit of the item. If P_k is the value of the item response function at quadrature point X_k, then the "observed" frequency is $O_{ik} = \tilde{r}_{ik}$, and the expected frequency is $E_{ik} = \widetilde{N}_{ik} P_{ik}$. Using these values the, the X^2 statistic is computed as

$$X_i^2 = \sum_{k=1}^{K} \frac{(\tilde{r}_{ik} - \widetilde{N}_{ik} P_{ik})^2}{\widetilde{N}_{ik} P_{ik}(1 - P_{ik})}. \tag{3.16}$$

The G^2 statistic is constructed similarly.

Unfortunately, as noted by Donoghue and Hombo (2003), with pseudocounts, the contribution of an examinee is to more than one cell. Hence the use of pseudocounts violates the assumption of independence, and consequently X^2 and G^2 do not have chi-square distributions. Stone et al. (1994) demonstrated through simulation studies that the distribution of these quantities can be approximated by a scaled chi-square distribution. However, the degrees of freedom and the scale factor are not known and must be determined either analytically or through a resampling technique. Stone (2000) used the resampling technique to determine the degrees of freedom and the scale factor. In the resampling technique, replicate data sets are simulated using the item parameter estimates of the original data set. For each replication, the item parameters are estimated, the G^2 (and/or X^2) statistic is computed, and the mean and the variance of the empirical distribution of the statistic are calculated. If the scale factor is denoted by λ and the degrees of freedom by υ, then the mean and the variance of the scaled chi-square variate are $\lambda\upsilon$ and $2\lambda^2\upsilon$, respectively. From the mean and variance of the empirical distribution, the scale factor and the degrees of freedom are determined.

The major advantage of the Stone procedure is that the pseudocounts are the natural by-products of the E–M algorithm, the procedure employed in BILOG-MG, PARSCALE, and MULTILOG. A further advantage is that the procedure extends effortlessly to polytomous item response models. The major drawback is that the procedure is computationally intensive. A computer program, however, is available (Stone, 2004) for carrying out the fit analysis.

Donoghue and Hombo (1999, 2001, 2003) have derived the asymptotic distribution of the statistic described above under the assumption that item parameters are known. They demonstrated that under this restrictive assumption the statistic behaved as expected.

Stone and Zhang (2003) compared the procedure described above with several procedures including that of Orlando and Thissen (2000), and Bock (1972), and that of Donoghue and Hombo (1999). The Orlando–Thissen procedure and the Stone procedures had nominal Type I error rates while the Donoghue–Hombo procedure had unacceptably low Type I error rates. On the other extreme, the Bock procedure had abnormally high Type I error rates and identified almost all items as misfitting. These procedures displayed modest power in detecting misfitting items, with the power increasing with sample size in some instances.

While the procedures developed by Stone et al. (1994), Stone (2000) and Orlando and Thissen (2000) provide improvements over procedures that use estimates of θ in constructing fit statistics, a fundamental problem remains with these procedures. As Sinharay (2005a, 2005b) points out, it has been shown by Chernoff and Lehman (1953) that a chi-square test statistic that depends on the estimates of parameters of a model will not have the limiting chi-square distribution. However, more recently Maydeu-Olivares and Joe (2005) have shown that when the parameter vector is estimated using a consistent and asymptotically normal minimum variance estimator, the statistic based on estimated model parameters has a limiting chi-square distribution.

In the computation of the Orlando–Thissen statistic, item parameter estimates are treated as known. While the uncertainty in the θ estimates does not play a part in the Orlando–Thissen procedure, the uncertainty in item parameter estimates does play a role and this affects the limiting distribution of the statistics. A similar criticism can be leveled against Stone (2000) and the Donoghue and Hombo (1999, 2003) procedure. Glas (1998, 1999) and Glas and Suarez Falcon (2003) have criticized these procedures along the same lines for failing to take into account the stochastic nature of the item parameter estimates.

Glas (1998, 1999) developed a Lagrange Multiplier (LM) procedure that takes into account the uncertainty in the item parameter estimates. The LM procedure of Glas has its roots in the classic procedure for finding extreme values of functions under constraints. Using the notation of Glas and Suarez Falcon (2003), let η_1 denote the vector of parameters of the model, and let η_2 denote a vector of parameters added to this model to obtain a more general model; $h(\eta_1)$ and $h(\eta_2)$ are the first-order derivatives of the log-likelihood function. Under the assumption that the parameters η_1 of the model are estimated by maximum likelihood, i.e., $h(\eta_1) = 0$, Glas and Suarez Falcon (2003) showed that the hypothesis $\eta_2 = 0$ can be tested using the statistic

$$LM = h(\eta_2)'\Sigma^{-1}h(\eta_2), \tag{3.17}$$

where Σ is the covariance matrix of $h(\eta_2)$. Details of the computation of $h(\eta_2)$ and Σ for IRT models using the marginal maximum likelihood are given in Glas (1998, 1999) and for the three-parameter model in Glas and Suarez Falcon (2003). Glas (1998, 1999) showed that the LM-statistic has an asymptotic χ^2-distribution with degrees of freedom equal to the number of parameters in η_2.

Glas (1998, 1999) and Glas and Suarez Falcon (2003) applied the LM procedure to solve a variety of problems in item response theory including assessing model fit and local independence, and assessing person fit. The major advantage of the LM procedure is that it is computationally straightforward, provides a unified approach for examining the assumptions underlying item response theory, and takes into account the uncertainty in the item parameter and ability estimates in constructing the test statistic. Simulation studies by the authors have demonstrated that the LM procedure compares favorably with the Orlando–Thissen procedure, producing slightly lower Type I error rates and improvement in power. However, they found the false positive rates alarmingly high in some cases when compared with that of the Orlando–Thissen procedure. They concluded that while the LM procedure was clearly superior to the Yen Q_1 procedure, the Orlando–Thissen procedure had better overall characteristics than the LM procedure.

A further problem with the Glas and Suarez Falcon approach is the specification of the parameters for the general model. In examining the fit of the three-parameter item response model, they defined a general model in which the difficulty parameters differed among subgroups. While such a parameterization is useful in examining differential item functioning or testing the invariance of item parameters in subgroups, this is only one form of misfit. Another issue is the grouping of examinees into score groups. Different groupings produced different results with respect to both hit rates and false positive rates. Despite these problems, the *LM* method has a sound theoretical basis and offers useful procedures for assessing model fit in the IRT context.

Maydeu-Olivares and Joe (2005) pointed out that assessing goodness of fit in binary latent trait models can be conceptualized in terms of assessing fit in a 2^n contingency table where n is the number of items. Needless to say, as n increases, the number of cells in the contingency table increases exponentially. Since the number of respondents/examinees is usually small in comparison to the number of cells in the table, as n increases the table becomes sparse. The sparseness of the table introduces problems in the estimation of parameters and the assessment of fit. Maydeu-Olivares and Joe (2005) provided a unified framework for assessing fit in a 2^n table. They provided full information procedures that utilize the higher-order joint moments and limited information procedures that use only the first two joint moments for assessing fit in high-dimensional contingency tables. They proposed two statistics, L_r, and M_r where L_r is the statistic based on known model parameters using the first r joint moments (proportions, in this case, and similar to that proposed by Donoghue and Hombo, 2003) while M_r denotes the statistic based on estimated model parameters, again based on the first r joint moments (proportions). Maydeu-Olivares and Joe (2005) show that these statistics, which can be expressed as quadratic forms involving the discrepancies between the observed and expected joint moments weighted by the asymptotic variance–covariance matrix of the sample joint proportions, i.e., Fisher's information matrix, converge in distribution to a chi-square distribution.

The procedure developed by Maydeu-Olivares and Joe (2005) can be viewed as a major advance in the area of testing goodness of fit in 2^n contingency tables and has implications for assessing fit not only in unidimensional item response models but also in multidimensional item response models. The only limitation in the current presentation is that the procedure is applicable only to binary response data.

3.3. Bayesian procedures

In Bayesian analysis, the posterior distribution of a set of parameters contains all the information about the parameters. If ω denotes the vector of parameters, then the posterior distribution of ω given the observations or data is

$$\pi(\omega|y) = \frac{L(y|\omega)\pi(\omega)}{\pi(y)}, \tag{3.18}$$

where y is the data set (the set of item responses of N examinees to n items). In the IRT context, $\omega = [\theta\ \xi]$, where θ is the vector of trait parameters and ξ is the vector of item parameters. The joint posterior distribution of item and trait parameters contains all the information about these two sets of parameters (Swaminathan and Gifford, 1985, 1986) and takes into account the uncertainty in both item and trait parameters.

One of the objectives in Bayesian analysis is to determine if the model can explain the data adequately (Gelman et al., 1996). Using the approach developed by Rubin (1984), Gelman et al. (1996) provided a detailed framework for conducting *Bayesian Posterior Predictive Model Checking* (*PPMC*). Sinharay (2005a, 2005b) and Sinharay and Johnson (2004) applied this procedure for examining model fit in IRT. The model fit approach using *PPMC* as described by Sinharay (2005a, 2005b) requires the determination of the posterior distribution of a replicate data set, y^{rep}, that may be observed if the "mechanism" that generated the observed data y is replicated with the value of ω. Since the values of ω are not known, the posterior distribution of the replicate, y^{rep} can be obtained as

$$\pi\left(y^{\text{rep}}|y\right) = \int \pi\left(y^{\text{rep}}, \omega|y\right) \mathrm{d}\omega = \int \pi\left(y^{\text{rep}}|\omega\right)\pi(\omega|y)\, \mathrm{d}\omega, \tag{3.19}$$

i.e., by averaging the likelihood over the values of ω defined by its posterior distribution as given in Eq. (3.18). Unfortunately, the integral in Eq. (3.19) is multidimensional and cannot be evaluated analytically. Following the suggestion of Rubin (1984), K simulations from the posterior distribution of ω, given in Eq. (3.18), are drawn and combined with a draw from the predictive distribution $\pi(y|\omega)$ evaluated at each draw of ω to yield $y^{\text{rep},k}$, $k = 1, 2, \ldots, K$. Sinharay (2005a) points out that the Markov Chain Monte Carlo (MCMC) algorithm generates draws from the posterior distribution of ω and hence the first step in *PPMC* is accomplished through the *MCMC* algorithm.

Once the replicate data, y^{rep}, is obtained, the fit of the model to the observed data can be obtained by defining a discrepancy measure $D(y^{\text{rep},k})$ between observed and predicted values. Sinharay (2005b) has suggested such discrepancy measures as (i) the tail area probability, (ii) percent correct for items, (iii) observed score distributions, (iv) item-total correlations, (v) unconditional odds ratio, defined for pairs of items as

$$OR_{ij} = \frac{N_{11}N_{00}}{N_{10}N_{01}},$$

where N_{11} is the frequency of examinees responding correctly to both items, with N_{00}, N_{01}, N_{10} defined similarly, and (vi) the conditional odds ratio, conditioned on the number correct score, summed over score categories. Sinharay (2005b) points out that the last discrepancy measure may also serve to detect departures from unidimensionality and weak local independence. The evaluation of these discrepancy measures is carried out primarily graphically.

Through simulation studies as well by applying these procedures to test data, Sinharay (2005a, 2005b) and Sinharay and Johnson (2004) demonstrated that the Posterior Predictive Model Checking procedures are practical, easy to implement although computationally intensive, and provide valuable information regarding model fit when evaluated using a variety of criteria. These procedures readily extend to polytomous models. The *PPMC* is clearly one of the most promising approaches to assessing model fit.

3.4. Assessing model parameter invariance

Invariance of item and ability parameters across subpopulations of examinees is arguably the cornerstone of item response theory. If the item response model fits the data, then the invariance property will be realized. Thus checking the invariance property of item and ability parameters provides a direct means for verifying or validating the theory. Invariance holds if

$$P(u|\theta, G) = P(u|\theta), \tag{3.20}$$

where P is the probability of a response to an item, u is the observed response to the item, θ is the ability level, and G is group membership or a background variable. Thus, the probability of a response should not depend on any other variable but the ability level of the examinee.

Hambleton and Swaminathan (1985) and Hambleton et al. (1991) recommended that invariance of item parameters be examined graphically. They provided numerous examples of plots of item difficulty estimates across subpopulations of examinees that can be used to assess invariance and hence the fit of the model. If invariance holds, the plot of item parameters, such as difficulty and discrimination, will lie on a straight line. In practice, however, estimates and not parameters are available. Because of sampling fluctuations, these estimates will be scattered about the straight line. Checking for outliers will provide an indication of which items do not satisfy the invariance property.

Statistical procedures for assessing invariance of item parameters across subgroups of examinees fall under the class of procedures for examining differential item functioning (DIF). Available statistical procedures for assessing DIF range from model-based procedures to nonparametric procedures (Camilli and Shepard, 1994; Hambleton et al., 1991; Holland and Wainer, 1993). The most directly applicable procedure is to test the equality of the parameters of the item response model across subgroups (Lord, 1980). Holland and Thayer (1988) demonstrated that under the assumption that the Rasch model fits the data, the Mantel–Haenszel procedure provides a test of the equivalence of the difficulty parameters across subgroups. Thus the nonparametric Mantel–Haenszel procedure provides a direct assessment of the invariance of model parameters in the Rasch model and hence, the fit of the Rasch model to the data.

In general, the invariance of item parameters may be tested directly by including in the item response model parameters that indicate group differences. An example of such an approach was provided by Glas (1998, 2001), and Glas and Suarez Falcon (2003) in the context of assessing the fit of the three-parameter model using the Lagrange Multiplier method. The null hypothesis that difficulty parameters are equal across score groups was tested against the alternative that the difficulty parameter differed across the groups by an additive parameter. Similar tests can be developed for other parameters of interest either using such procedures as the Lagrange Multiplier method or the *MCMC* procedure. These statistical tests combined with graphical displays provide powerful methods of examining invariance of parameters across subgroups.

3.5. Role of simulation studies in assessing fit

The notion of using simulated data for assessing model fit was introduced earlier in the context of assessing dimensionality of a set of items. In fact, simulating data and determining the empirical distribution of the test statistic play a critical role in the procedure developed by Stone (2000). In the Bayesian *PPMC* procedure, simulations are carried out to obtain the posterior distribution of the replicate data sets. Hambleton and Rogers (1990) and Hambleton et al. (1991) use simulation to generate an expected distribution of standardized residuals assuming model fit for the purposes of interpreting the observed distribution of standardized residuals.

In essence, simulation approaches can be applied with all the procedures described in this chapter. These simulation studies can provide empirical distributions of the test statistics for which theoretical distributions are not available, or when there is sufficient evidence to throw doubt on the validity of the assumed distribution of the test statistic. For example, in comparing the observed score distribution and the expected score distribution, simulations can be carried out to provide a confidence band for the expected score distribution. These simulated distributions are not unlike those obtained using bootstrap procedures. The combination of simulations and graphical displays provides a powerful tool for assessing model fit (see, for example, Hambleton and Han, 2005).

4. Empirical example

Tables 1 to 5 and Figure 11 contain model fit results of the analyses of data from a large scale assessment. A subset of the data containing the responses of 2043 examinees on 32 multiple-choice questions, four short answer binary-scored questions, and six polytomously-scored items (scored 0 to 4) was analyzed for illustrative purposes. One-, two- and three-parameter models were fitted to the dichotomous items and partial credit, generalized partial credit, and graded response models were fitted to the polytomous items. Figure 10 contains the plot of eigenvalues for the correlation matrix of 42 test items. Table 1 provides the classical item indices and IRT item parameter estimates for each model fitted to the dichotomous items, while Table 2 gives item parameter estimates for each model fitted to the polytomous items. Tables 3 and 4 give chi-square fit statistics for the dichotomous and polytomous items, respectively. Fitting multiple IRT models to the same data set and comparing results is an especially useful strategy for becoming familiar with the data, and comparing the relative advantages of one model over another prior to choosing one to address a measurement problem.

Figure 10 highlights clearly that there is one dominant factor underlying the 42 test items. The plot shows that the first factor (corresponding to the dominant eigenvalue of 11.2) accounts for 26.7% of the score variability, almost nine times that accounted for by the second factor; the second, third, fourth, and subsequent factors account for less than 3% each (corresponding to eigenvalues of 1.3, 1.2, and 1.1, respectively. Taken together, these findings support the unidimensionality assumption.

As pointed out earlier, eigenvalues obtained through simulations can provide a useful baseline for assessing the unidimensionality assumption. To this end, random item response data were simulated for the 2043 examinees on the 42 item test, conditioned on the p-values of the items in the real data. The eigenvalues of the correlation matrix were then obtained from the simulated data set. The largest three eigenvalues obtained from this analysis were 1.27, 1.26, and 1.24. For the actual data, on the other hand, the

Table 1
Classical difficulty values (p), point biserial correlations (r), and IRT item parameter estimates for 1P, 2P, and 3P models for dichotomous items

Item	Classical indices		1P model[1]	2P model		3P model[2]		
	p	r	b	b	a	b	a	c
1	0.91	0.36	−2.24	−1.99	0.87	−2.91	0.76	0.00
2	0.70	0.48	−0.89	−0.86	0.76	−0.52	0.87	0.18
3	0.76	0.40	−1.20	−1.31	0.62	−0.39	0.96	0.38
4	0.84	0.39	−1.68	−1.67	0.71	−1.51	0.70	0.20
5	0.58	0.39	−0.35	−0.41	0.54	0.51	1.35	0.36
6	0.60	0.34	−0.43	−0.60	0.43	0.48	0.90	0.35
7	0.53	0.32	−0.15	−0.21	0.40	0.57	0.64	0.25
8	0.73	0.30	−1.06	−1.51	0.43	0.07	0.77	0.48
9	0.87	0.28	−1.90	−2.38	0.51	−2.27	0.47	0.21
10	0.56	0.55	−0.27	−0.25	0.89	−0.01	1.13	0.13
11	0.67	0.22	−0.78	−1.56	0.28	−0.17	0.37	0.31
12	0.57	0.54	−0.31	−0.28	0.87	−0.06	1.02	0.12
13	0.60	0.54	−0.44	−0.41	0.87	0.06	1.38	0.24
14	0.89	0.43	−2.06	−1.66	1.04	−1.44	1.09	0.24
15	0.88	0.39	−2.03	−1.82	0.86	−1.71	0.81	0.20
16	0.76	0.31	−1.22	−1.66	0.46	−0.01	0.84	0.51
17	0.83	0.49	−1.62	−1.32	1.05	−1.14	1.10	0.16
18	0.73	0.50	−1.03	−0.94	0.84	−0.48	1.10	0.25
19	0.52	0.57	−0.10	−0.10	0.95	0.24	1.79	0.18
20	0.56	0.55	−0.27	−0.25	0.90	−0.08	1.00	0.09
21	0.79	0.37	−1.40	−1.59	0.59	−1.04	0.66	0.27
22	0.71	0.49	−0.93	−0.86	0.80	−0.70	0.84	0.10
23	0.70	0.44	−0.90	−0.95	0.65	−0.66	0.71	0.15
24	0.79	0.42	−1.36	−1.34	0.73	−1.07	0.77	0.17
25	0.84	0.46	−1.69	−1.45	0.93	−1.28	0.93	0.18
26	0.57	0.53	−0.31	−0.29	0.84	0.30	2.26	0.30
27	0.67	0.54	−0.74	−0.65	0.91	−0.38	1.08	0.15
28	0.76	0.44	−1.19	−1.17	0.73	−0.86	0.79	0.17
29	0.70	0.41	−0.89	−0.98	0.61	−0.21	0.87	0.32
30	0.70	0.45	−0.91	−0.93	0.67	0.11	2.28	0.44
31	0.56	0.39	−0.25	−0.30	0.51	0.38	0.87	0.26
32	0.74	0.56	−1.09	−0.90	1.05	−0.47	1.43	0.25
33	0.67	0.48	−0.72	−0.71	0.75	−0.72	0.73	0.00
34	0.67	0.65	−0.70	−0.55	1.31	−0.54	1.29	0.00
35	0.68	0.60	−0.81	−0.65	1.15	−0.64	1.13	0.00
36	0.58	0.61	−0.33	−0.28	1.10	−0.25	1.12	0.00

[1]Common a-parameter = 0.714.

[2]c-parameters for items 33–36 were fixed at zero.

largest three eigenvalues were 11.20, 1.31, and 1.20. Clearly, the evidence to support the assumption of unidimensionality in the actual data is very strong. The second, third, and additional factors in the actual data could not be distinguished from the factors found from the analysis of random error.

Table 2
Partial credit (PC) model, generalized partial credit (GPC) model, and graded response (GR) model item parameters for polytomous items

Item	*b*	*a*	*d*1	*d*2	*d*3	*d*4
PC model						
37	−0.14	0.71	0.95	0.13	−1.02	−0.07
38	−0.28	0.71	0.19	−0.42	0.50	−0.27
39	−0.09	0.71	0.02	0.98	−1.16	0.15
40	−0.07	0.71	0.96	−0.78	−0.17	−0.01
41	0.02	0.71	0.00	0.36	−0.49	0.14
42	−0.44	0.71	0.80	0.24	−0.12	−0.92
GPC model						
37	−0.15	0.72	0.99	0.10	−1.03	−0.06
38	−0.26	0.82	0.30	−0.38	0.39	−0.31
39	−0.08	0.75	0.08	0.92	−1.14	0.14
40	−0.06	0.73	0.99	−0.79	−0.19	−0.01
41	0.03	0.61	−0.13	0.35	−0.55	0.33
42	−0.44	0.77	0.88	0.21	−0.17	−0.92
GR model						
37	−0.11	1.29	1.20	0.23	−0.53	−0.90
38	−0.28	1.52	0.69	0.14	−0.11	−0.72
39	−0.08	1.43	0.84	0.41	−0.48	−0.77
40	−0.03	1.34	1.07	−0.01	−0.33	−0.73
41	0.04	1.30	0.71	0.23	−0.28	−0.65
42	−0.46	1.28	1.15	0.37	−0.31	−1.20

To further check on dimensionality and to investigate the possibility that the polytomous items might be measuring a different dimension from the dichotomous items, an analysis using the DIMTEST program was performed using the polytomous items as the assessment subtest. A t value of 0.118, with a significance level of .547 was obtained, indicating that the polytomous items were measuring the same construct as the dichotomous items.

A review of the classical item statistics for the binary data in Table 1 highlights two things: Items have a wide range of classical item discrimination indices (.27 to .82) and item difficulties (.52 to .91). Even prior to the IRT analysis, the first finding suggests that a discrimination parameter in the IRT model would likely improve model fit. The second suggests that a "guessing" parameter in the IRT model might also improve the model fit, at least for some of the more difficult items. Though we are cautious in placing too much weight on the item fit chi-square significance test results shown in Table 3, it is clear that the chi-square statistic itself can serve as a summary of model-item data departure, and the number of significant chi-square statistics is substantially smaller for the two-parameter model than the one-parameter model and substantially smaller again for the three-parameter model. These findings suggest that the three-parameter model fits the data better than the two-parameter model, and both models fit the data better than the one-parameter model. The PARSCALE chi-square statistic and the Orlando–Thissen

Table 3
PARSCALE (PSL) and Orlando–Thissen (O–T) chi-square fit statistics for 1P, 2P, and 3P IRT models for dichotomous items

Item	1P model		2P model		3P model	
	PSL	O–T	PSL	O–T	PSL	O–T
1	26.40**	33.45	17.00	28.77	98.62**	145.94**
2	19.16	40.76	15.27	41.56	10.57	41.02
3	17.25	52.13	14.34	45.32	10.64	43.79
4	25.66*	50.94	25.39*	44.58	27.91*	52.55*
5	62.09**	107.63**	48.60**	70.36*	17.95	44.54
6	101.71**	172.75**	34.57*	80.47**	16.25	74.09**
7	100.06**	194.01**	31.31	64.65*	14.79	63.70*
8	88.77**	132.75**	31.15*	61.17	22.45	55.99
9	26.22*	53.89*	19.77	25.50	18.17	24.16
10	40.25**	48.45	24.93	28.40	11.74	31.96
11	176.60**	286.19**	23.00	56.95	16.15	52.07
12	23.69	32.80	8.69	23.32	16.23	23.78
13	42.67**	64.51*	32.53*	57.71*	28.90*	52.78
14	37.94**	52.92**	5.82	22.18	8.00	21.48
15	19.20	36.49	16.27	29.38	23.14*	27.92
16	70.98**	112.99**	24.78	59.43	21.39	49.75
17	55.26**	68.90**	15.87	29.39	12.99	32.08
18	17.74	36.77	13.98	33.14	10.74	29.15
19	85.20**	104.41**	60.02**	86.89**	18.08	42.20
20	29.75*	42.49	15.24	27.88	24.65	28.02
21	19.75	59.04*	10.62	54.74	10.27	56.33
22	23.75	54.03	17.76	53.43	20.61	50.31
23	22.41	52.04	15.66	47.28	15.80	50.68
24	19.16	51.98	10.95	51.36	13.27	53.40
25	33.82**	36.87	9.80	23.50	20.59	23.39
26	88.30**	102.39**	84.51**	96.36**	21.63	48.59
27	40.56**	52.81	8.02	40.80	10.77	40.47
28	10.32	48.54	8.51	48.20	12.46	47.66
29	24.56	61.91*	24.08	47.81	14.62	45.28
30	112.29**	150.62**	110.94**	142.48**	25.12*	45.33
31	61.77**	114.71**	36.72**	62.90*	21.20	55.04
32	59.04**	76.23**	13.16	41.35	9.31	26.14
33	21.59	54.44	22.26	50.51	25.08	53.14
34	143.00**	141.62**	12.08	42.39	17.11	39.08
35	87.50**	101.97**	19.17	41.72	20.31	40.15
36	77.09**	97.27**	22.14	52.85*	13.91	50.73

*Number of intervals for PARSCALE fit statistics set at 20 and collapsed to ensure minimum expected cell frequency of 5.

**Number of intervals for Orlando–Thissen fit statistics determined by number of raw score levels and collapsed to ensure minimum expected cell frequency of 5.

statistic show considerable agreement with respect to the items flagged as misfitting, but sufficient discrepancy to highlight the risks of over-interpreting statistically significant fit statistics. For the polytomous items, adding a discrimination parameter had less of

Table 4
PARSCALE (PSL) and Orlando–Thissen (O–T) chi-square fit statistics for the partial credit (PC), generalized partial credit (GPC), and graded response (GR) models for polytomous items

Item	PC model		GPC model		GR model	
	PSL	O–T	PSL	O–T	PSL	O–T
37	67.81	31.97	66.39	20.54	75.09*	31.19
38	70.15*	45.75	50.83	33.84	63.36*	50.30*
39	110.53**	52.55**	84.94**	42.72*	74.71**	50.28*
40	115.37**	39.53*	95.87**	37.41	81.20**	59.91**
41	106.21**	109.75**	89.64**	92.78**	41.10**	72.75*
42	69.53	26.77*	82.00**	16.49	57.59	24.89

*Number of intervals for PARSCALE fit statistics set at 20 and collapsed to ensure minimum expected cell frequency of 5.
**Number of intervals for Orlando–Thissen fit statistics determined by number of raw score levels and collapsed to ensure minimum expected cell frequency of 5.

Table 5
Distributions of standardized residuals for 1P/PC, 2P/GPC, and 3P/GPC IRT models

SR interval	Percent of residuals in interval		
	1P/PC	2P/GPC	3P/GPC
< -3	2.79	0.63	0.72
$(-3, -2)$	5.59	3.42	2.79
$(-2, -1)$	14.23	13.87	12.61
$(-1, 0)$	27.30	33.51	34.59
$(0, 1)$	24.99	28.56	32.07
$(1, 2)$	15.23	14.32	13.51
$(2, 3)$	6.13	4.50	3.15
> 3	3.78	1.17	0.54

an effect on fit, probably because there was not great variation in the discrimination parameters, as Table 2 shows. Looking at the Orlando–Thissen statistics, the generalized partial credit model appeared to fit the item response data a little better overall than did the graded response model. Table 5 shows the distribution of standardized residuals for the 1P, 2P, and 3P models for the dichotomous items in combination with either the partial credit or generalized partial credit model for the polytomous items. These distributions show the clear improvement in overall fit in going from a 1P/PC model to a 2P/GPC model, and further improvement, though not as dramatic, in going from a 2P/GPC model to a 3P/GPC model.

Figure 11 shows plots of item residuals for binary-scored item 19 for the three-parameter, two-parameter, and the one-parameter model, respectively. The gradual improvement in the model fit can be seen as additional parameters are added to the IRT model. Figure 12 shows residual plots for the partial credit and graded response mod-

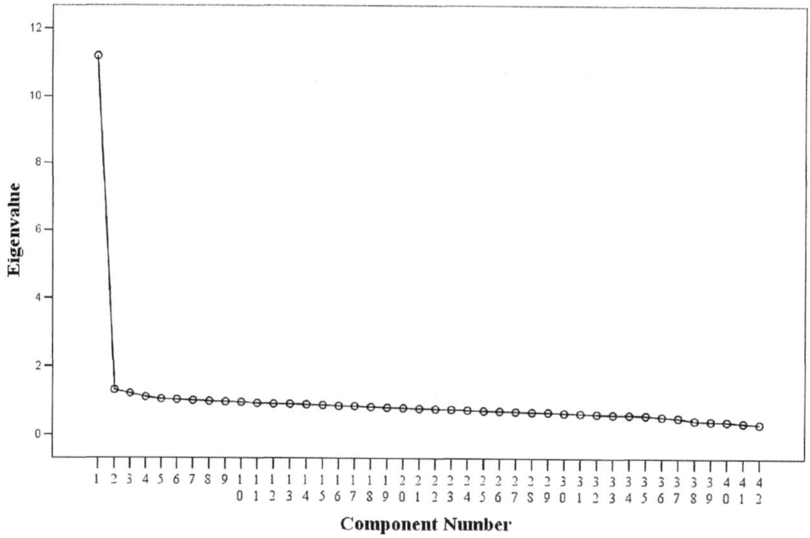

Fig. 10. Plot of the eigenvalues from the item correlation matrix.

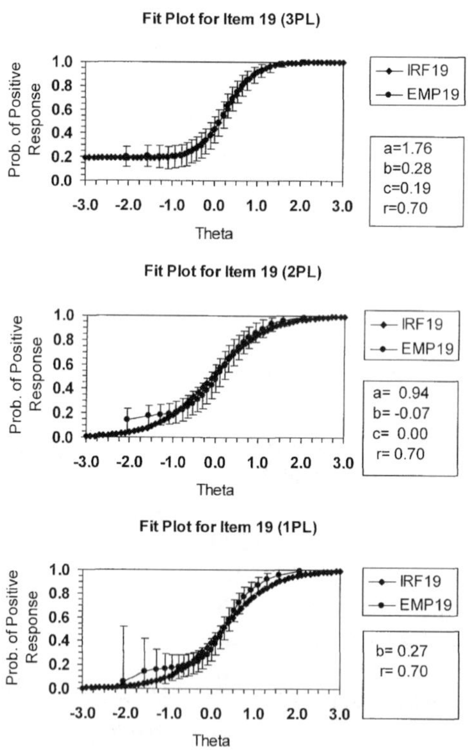

Fig. 11. Residual plots of item 19 (dichotomously-scored item).

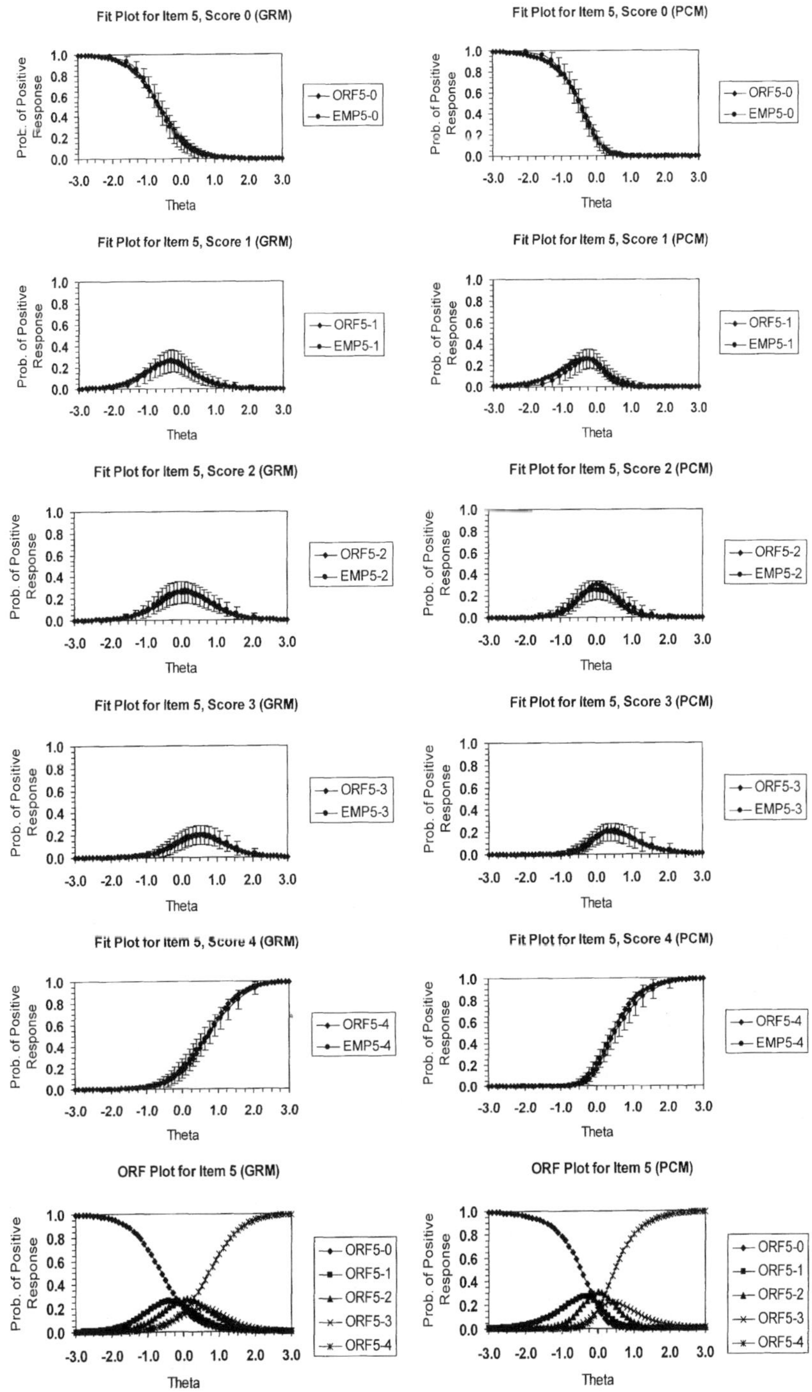

Fig. 12. Residual plots of item 5 (polytomously-scored item).

els for item 41 (item 5 of the polytomously scored items); these plots can be useful in comparing the fit of competing models.

The practical implications of the differences in the model fits would depend on the intended uses of the scores. Figures 1, 2, and 3, and Figures 4, 5, and 6 highlight that the predictions of the test scores are noticeably different from each model, and generally better for models with more item parameters. If the intent was to produce test score norms tables, the results would be very different depending on choice of model, and the most suitable results would come from the more general models. Other empirical investigations could be carried out to investigate model fit and the consequences of model misfit on the intended application such as equating, detecting bias, building tests, and sorting examinees into performance categories.

5. Conclusions

Assessing IRT model fit to item response data is one of the crucial steps before an IRT model can be applied with confidence to estimate proficiency or ability levels of examinees, link tests across administrations, and assess, for example, adequate yearly progress as required by the NCLB legislation. Despite the fundamental importance of this step, assessment of fit receives only the barest of considerations by many test developers and users. Possible reasons for this neglect are the complexity of assessing fit, the lack of understanding of the fit statistics, the absence of comprehensive model fit software, and the fact that the likelihood ratio statistics provided by commercial software often indicate misfit for most of the items with large sample sizes. It is indeed ironic that the highly desirable practice of using large sample sizes to minimize item parameter estimation error results in significant item-fit statistics even when the model fits the data for all practical purposes.

Given the importance of the subject, considerable research has been carried out and advances have been made with respect to fit assessment in item response models, though much of the research is still not being implemented on a wide scale. Some of these model fit procedures are computationally intensive while others are relatively straightforward to implement even without dedicated software, but again, few are being used today.

Hambleton and Han (2005) have suggested a set of activities that practitioners should routinely carry out in the assessment of model fit. These include:

(i) carrying out routine examination of classical indices such as p-values, item-total correlations, the distribution of scores at both the test and item levels, and factor analysis. These analyses may provide an indication of the appropriate item response model to choose as well as identify potential problems in fitting a model (for example, the need to collapse score categories or the suitability of the unidimensionality assumption);

(ii) checking the unidimensionality assumption formally and examining possible violations of local independence that may result from groups of items sharing a common stimuli;

(iii) examining the fit of the model at the test and item levels using graphical as well as statistical procedures;
(iv) checking invariance of item parameters through plots of item parameter estimates and DIF studies (to detect noninvariance across, say, gender, ethnicity, and language groups).

In a final step, Hambleton and Han (2005) suggest examining the consequences of model misfit on outcomes of interest. Since no model can fit a set of observations perfectly, it is important, if possible through simulation studies at least, to examine the consequence of misfit on such important outcomes as determining proficiency or ability scores, assigning examinees to performance categories, equating test scores, optimizing item selection when implementing computer-adaptive tests, and scale drift, to name a few.

As mentioned earlier, assessing model fit is an important part of the test validation process. It is multifaceted, and as in the verification of any scientific theory, it is an ongoing process where only through the accumulation of empirical evidence can one be confident of the appropriateness of item response theory for the solution of a particular measurement problem. The procedures for assessing model–data fit described in this chapter have the potential for addressing the vexing problem of determining if the measurement procedures used are appropriate for addressing the practical measurement problems faced by practitioners.

Acknowledgements

The authors are indebted to Ms. Yue Zhao for assistance in compiling Section 4 of this chapter.

References

Ackerman, T. (2005). Multidimensional item response theory. In: Everitt, B., Howell, D. (Eds.), *Encyclopedia of Statistics in Behavioral Science*. John Wiley & Sons, West Sussex, UK.

Agresti, A. (1990). *Categorical Data Analysis*. Wiley, New York.

Andersen, E. (1973). A goodness of fit test for the Rasch model. *Psychometrika* **38**, 123–140.

Andrich, D. (1978). A rating formulation for ordered response categories. *Psychometrika* **43**, 561–573.

Bock, R.D. (1972). Estimating item parameters and latent ability when responses are scored in two or more nominal categories. *Psychometrika* **37**, 29–51.

Bock, R.D., Aitkin, M. (1981). Marginal maximum likelihood estimation of item parameters: An application of the EM algorithm. *Psychometrika* **46**, 443–449.

Bock, R.D., Gibbons, R.D., Muraki, E. (1988). Full information item factor analysis. *Applied Psychological Measurement* **12**, 261–280.

Camilli, G., Shepard, L.A. (1994). *Methods for Identifying Biased Items*. Sage, Thousand Oaks, CA.

Chen, W., Thissen, D. (1997). Local independence indexes for item pairs using item response theory. *Journal of Educational and Behavioral Statistics* **22**, 265–289.

Chernoff, H., Lehman, E.L. (1953). The use of maximum likelihood estimates in chi-square tests for goodness of fit. *Annals of Mathematical Statistics* **25**, 579–586.

De Champlain, A.F., Gessaroli, M.E. (1998). Assessing the dimensionality of item response matrices with small sample sizes and short test lengths. *Applied Measurement in Education* **11**, 231–253.

Donoghue, J.R., Hombo, C.M. (1999). Some asymptotic results on the distribution of an IRT measure of item fit. Paper presented at the meeting of the Psychometric Society, Lawrence, KS. June.

Donoghue, J.R., Hombo, C.M. (2001). The distribution of an item-fit measure for polytomous items. Paper presented at the meeting of the National Council of Measurement in Education, Seattle, WA. April.

Donoghue, J.R., Hombo, C.M. (2003). A corrected asymptotic distribution of an IRT fit measure that accounts for effects of item parameter estimation. Paper presented at the meeting of the National Council of Measurement in Education, Chicago. April.

Finch, H., Habing, B. (2005). Comparison of NOHARM and DETECT in item cluster recovery: Counting dimensions and allocating items. *Journal of Educational Measurement* **42**, 149–169.

Fraser, C. (1988). NOHARM II: A Fortran program for fitting unidimensional and multidimensional normal ogive models of latent trait theory. University of New England, Centre for Behavioural Studies, Armidale, NSW.

Gelman, A., Meng, X., Stern, H.S. (1996). Posterior predictive assessment of model fitness via realized discrepancies. *Statistica Sinica* **6**, 733–807.

Gessaroli, M.E., De Champlain, A.F. (1996). Using an approximate chi-square statistic to test the number of dimensions underlying the responses to a set of items. *Journal of Educational Measurement* **33**, 157–179.

Gessaroli, M.E., De Champlain, A.F., Folske, J.C. (1997). Assessing dimensionality using likelihood-ratio chi-square test based on a non-linear factor analysis of item response data. Paper presented at the meeting of the National Council on Measurement in Education, Chicago. April.

Glas, C.A.W. (1998). Detection of differential item functioning using Lagrange multiplier tests. *Statistica Sinica* **8** (1), 647–667.

Glas, C.A.W. (1999). Modification indices for the 2-pl and the nominal response model. *Psychometrika* **64**, 273–294.

Glas, C.A.W. (2001). Differential item functioning depending on general covariates. In: Boomsma, A., van Duijn, M.A.J., Snijders, T.A.B. (Eds.), *Essays on Item Response Theory*. Springer, New York, pp. 131–148.

Glas, C.A.W. (2005). Assessing fit to IRT models. In: Everitt, B., Howell, D.C. (Eds.), *Encyclopedia of Behavioral Statistics*. Wiley, West Sussex, UK.

Glas, C.A.W., Suarez Falcon, J.C.S. (2003). A comparison of item-fit statistics for the three-parameter logistic model. *Applied Psychological Measurement* **27** (2), 87–106.

Hambleton, R.K., Han, N. (2005). Assessing the fit of IRT models to educational and psychological test data: A five step plan and several graphical displays. In: Lenderking, W.R., Revicki, D. (Eds.), *Advances in Health Outcomes Research Methods, Measurement, Statistical Analysis, and Clinical Applications*, Degnon Associates, Washington.

Hambleton, R.K., Rogers, H.J. (1990). Using item response models in educational assessments. In: Schreiber, W.H., Ingekamp, K. (Eds.), *International Developments in Large-Scale Assessment*. NFER-Nelson, Windsor, UK, pp. 155–184.

Hambleton, R.K., Rovinelli, R.J. (1986). Assessing the dimensionality of a set of test items. *Applied Psychological Measurement* **10**, 287–302.

Hambleton, R.K., Swaminathan, H. (1985). *Item Response Theory: Principles and Applications*. Kluwer Academic Publishers, Boston, MA.

Hambleton, R.K., Traub, R.E. (1973). Analysis of empirical data using two logistic latent trait models. *British Journal of Mathematical and Statistical Psychology* **24**, 273–281.

Hambleton, R.K., Swaminathan, H., Rogers, H.J. (1991). *Fundamentals of Item Response Theory*. Sage, Newbury Park, CA.

Hattie, J.A. (1985). Methodology review: Assessing unidimensionality of tests and items. *Applied Psychological Measurement* **9**, 139–164.

Hattie, J.A., Krakowski, K., Rogers, H.J., Swaminathan, H. (1996). An assessment of Stout's index of essential unidimensionality. *Applied Psychological Measurement* **20**, 1–14.

Hemker, B.T., van der Ark, L.A., Sijtsma, K. (2001). On measurement properties of continuation ratio models. *Psychometrika* **66**, 487–506.

Holland, P.W., Thayer, D.T. (1988). Differential item functioning and the Mantel–Haenszel procedure. In: Wainer, H., Braun, H.L. (Eds.), *Test Validity*. Lawrence Erlbaum Publishers, Hillsdale, NJ, pp. 129–145.

Holland, P.W., Wainer, H. (Eds.) (1993). *Differential Item Functioning*. Lawrence Erlbaum Publishers, Hillsdale, NJ.

Joreskog, K.G., Sorbom, D. (2004). LISREL 8 [Computer program]. Scientific Software, Chicago.

Lord, F.M. (1980). *Applications of Item Response Theory to Practical Testing Problems*. Lawrence Erlbaum Publishers, Hillsdale, NJ.

Lord, F.M., Novick, M.R. (1968). *Statistical Theories of Mental Test Scores*. Addison-Wesley, Reading, MA.

Lord, F.M., Wingersky, M.S. (1984). Comparison of IRT true-score and equipercentile observed-score "equatings". *Applied Psychological Measurement* **8**, 453–461.

Masters, G.N. (1982). A Rasch model for partial credit scoring. *Psychometrika* **47**, 149–174.

Maydeu-Olivares, A. (2001). Multidimensional item response theory modeling of binary data: Large sample properties of NOHARM estimates. *Journal of Educational and Behavioral Statistics* **26**, 51–71.

Maydeu-Olivares, A., Joe, H. (2005). Limited- and full-information estimation and goodness-of-fit testing in 2^n contingency tables: A unified framework. *Journal of the American Statistical Association* **100**, 1009–1020.

Meijer, R.R., Sijtsma, K. (2001). Methodology review: Evaluating person fit. *Applied Psychological Measurement* **25**, 107–135.

McDonald, R.P. (1967). *Non-Linear Factor Analysis*. Psychometric Monographs, No. 15.

McDonald, R.P. (1982). Linear versus non-linear models in item response theory. *Applied Psychological Measurement* **6**, 379–396.

McDonald, R.P. (1997). Normal ogive multidimensional model. In: Van der Linden, W.J., Hambleton, R.K. (Eds.), *Handbook of Modern Item Response Theory*. Springer, New York, pp. 257–269.

McKinley, R., Mills, C. (1985). A comparison of several goodness-of-fit statistics. *Applied Psychological Measurement* **9**, 49–57.

Mislevy, R.J., Bock, R.D. (1990). BILOG: Item analysis and test scoring with binary logistic models [Computer program]. Scientific Software, Chicago.

Muraki, E. (1992). A generalized partial credit model: Application of an EM algorithm. *Applied Psychological Measurement* **16**, 59–176.

Orlando, M., Thissen, D. (2000). Likelihood-based item-fit indices for dichotomous item response theory models. *Applied Psychological Measurement* **24** (1), 50–64.

Orlando, M., Thissen, D. (2003). Further investigation of the performance of $S - X^2$: An item fit index for use with dichotomous item response theory models. *Applied Psychological Measurement* **27** (4), 289–298.

Rasch, G. (1960). *Probabilistic Models for some Intelligence and Attainment Tests*. Danish Institute for Educational Research, Copenhagen.

Reckase, M.D., Ackerman, T.A., Carlson, J.E. (1988). Building unidimensional tests using multidimensional items. *Journal of Educational Measurement* **25**, 193–203.

Rogers, H., Hattie, J. (1987). A Monte Carlo investigation of several person and item fit statistics for item response models. *Applied Psychological Measurement* **11**, 47–57.

Rubin, D.B. (1984). Bayesianly justifiable and relevant frequency calculations for the applied statistician. *Annals of Statistics* **12**, 1151–1172.

Samejima, F. (1969). *Estimation of Latent Ability Using a Response Pattern of Graded Scores*. Psychometric Monograph, No. 17.

Sijtsma, K., Molenaar, I.W. (2002). *Introduction to Nonparametric Item Response Theory*. Sage, Thousand Oaks, CA.

Sinharay, S. (2005a). Practical applications of posterior predictive model checking for assessing fit of unidimensional item response theory models. *Journal of Educational Measurement* **42**, 375–385.

Sinharay, S. (2005b). Bayesian item fit analysis for unidimensional item response theory models. Paper presented at the meeting of the National Council on Measurement in Education, Montreal. April.

Sinharay, S., Johnson, M.S. (2004). Simulation studies applying posterior predictive model checking for assessing fit of the common item response theory models. Paper presented at the meeting of the National Council on Measurement in Education, San Diego. April.

Stone, C.A. (2000). Monte Carlo based null distribution for an alternative goodness-of-fit test statistic in IRT models. *Journal of Educational Measurement* **37** (1), 58–75.

Stone, C.A. (2004). IRTFIT_RESAMPLE: A computer program for assessing goodness of fit of item response theory models based on posterior expectations. *Applied Psychological Measurement* **28** (2), 143–144.

Stone, C.A., Zhang, B. (2003). Assessing goodness of fit of item response theory models: A comparison of traditional and alternative procedures. *Journal of Educational Measurement* **40** (4), 331–352.

Stone, C.A., Mislevy, R.J., Mazzeo, J. (1994). Classification error and goodness-of-fit in IRT models. Paper presented at the meeting of the American Educational Research Association, New Orleans. April.

Stout, W. (1987). A nonparametric approach for assessing latent trait unidimensionality. *Psychometrika* **52**, 589–617.

Stout, W. (1990). A new item response theory modeling approach with applications to undimensionality assessment and ability estimation. *Psychometrika* **55**, 293–325.

Stout, W.F., Nandakumar, R., Junker, B., Chang, H.H., Steidinger, D. (1991). DIMTEST and TESTSIM [Computer program]. University of Illinois, Department of Statistics, Champaign.

Tate, R. (2003). A comparison of selected empirical methods for assessing the structure of responses to test items. *Applied Psychological Measurement* **27**, 159–203.

Tutz, G. (1997). Sequential models for ordered responses. In: van der Linden, W.J., Hambleton, R.K. (Eds.), *Handbook of Modern Item Response Theory*. Springer, New York, pp. 139–152.

Swaminathan, H., Gifford, J.A. (1985). Bayesian estimation in the two-parameter logistic model. *Psychometrika* **50**, 349–364.

Swaminathan, H., Gifford, J.A. (1986). Bayesian estimation in the three-parameter logistic model. *Psychometrika* **51**, 589–601.

Van den Wollenberg, A.L. (1982). Two new test statistics for the Rasch model. *Psychometrika* **47**, 123–140.

Verhelst, N.D., Glas, C.A.W., de Vries, H.H. (1997). A steps model to analyze partial credit. In: van der Linden, W.J., Hambleton, R.K. (Eds.), *Handbook of Modern Item Response Theory*. Springer, New York, pp. 123–138.

Wright, B.D., Panchapakesan, N. (1969). A procedure for sample free item analysis. *Educational and Psychological Measurement* **29**, 23–48.

Yen, W.M. (1981). Using simulation results to choose a latent trait model. *Applied Psychological Measurement* **5**, 245–262.

Yen, W.M. (1984). Effects of local item dependence on the fit and equating performance of the three-parameter logistic model. *Applied Psychological Measurement* **8** (2), 125–145.

Zimowski, M.F., Muraki, E., Mislevy, R.J., Bock, R.D. (1996). Bilog-MG: Multiple-group IRT analysis and test maintenance for binary items [Computer program]. Scientific Software International, Chicago.

Handbook of Statistics, Vol. 26
ISSN: 0169-7161

DOI: 10.1016/S0169-7161(06)26022-X

22

Nonparametric Item Response Theory and Special Topics

Klaas Sijtsma and Rob R. Meijer

Abstract

This chapter discusses the ideas and assumptions of nonparametric IRT and its merits and drawbacks in analyzing real test and questionnaire data. The place of nonparametric IRT in IRT in general is reviewed and nonparametric IRT is compared to parametric IRT. Procedures for assessing the dimensionality of item response data are explained and compared, and several methods are discussed that evaluate the shape of response functions of items, item steps and response options of items. It is discussed how to investigate whether item response functions are invariantly ordered across the latent variable, and why this is important in practical test usage. Latent class models are discussed as discrete alternatives for continuous latent variable IRT and as model fitting tools in a nonparametric IRT context. The strength of nonparametric IRT is illustrated with recent developments in person-fit analysis using an order-restricted person-response function together with graphical and statistical tools for drawing inferences from such functions about someone's item-response data.

1. Place of nonparametric models in item response theory

Two ways of positioning nonparametric IRT (NIRT) relative to parametric IRT (PIRT) are that NIRT (1) encompasses PIRT by providing a broader theoretical framework from which IRT in general is better understood; and (2) primarily is an exploratory toolkit for the analysis of test and questionnaire data whereas PIRT primarily deals with confirmatory data analysis by fitting specific models to data. This chapter illuminates these two functions of NIRT, and discusses many recent developments. We first introduce some notation, definitions and assumptions.

Let the item score random variable of item j be denoted X_j with realization x_j, $j = 1, \ldots, J$. The item score can be dichotomous with binary scoring, usually for incorrect/correct answers, such that $X_j \in \{0, 1\}$, or ordered polytomous, such that $X_j \in \{0, \ldots, m_j\}$, with $m_j + 1$ being the number of ordered answer categories in, for example, a rating scale. Throughout, we assume that $m_j = m$; thus, each item in the test has the same number of ordered answer categories. Assume that responses to items are driven by one latent variable, denoted θ, or several latent variables, collected in vector $\boldsymbol{\theta} = (\theta_1, \ldots, \theta_Q)$, with $q = 1, \ldots, Q$ indexing latent variables. For simplicity, assume one latent variable unless mentioned otherwise.

Both NIRT and PIRT – together known as item response theory (IRT) – model the probability of having a particular score on item j as a function of θ. For binary item scoring, IRT models probability $P(X_j = 1|\theta) = P_j(\theta)$, which is known as the item response function (IRF). For polytomous items, IRT models probability $P(X_j = x_j|\theta)$, known as the category response function (CRF). Other response functions may be modeled, such as $P(X_j \geqslant x_j|\theta)$. This function is known as the item step response function (ISRF). The ISRFs and the CRFs are related through

$$P(X_j \geqslant x_j|\theta) = \sum_{y=x_j}^{m} P(X_j = y|\theta). \tag{1}$$

Reversely, we have that

$$P(X_j = x_j|\theta) = P(X_j \geqslant x_j|\theta) - P(X_j \geqslant x_j + 1|\theta). \tag{2}$$

Other interesting definitions of response functions exist (e.g., Tutz, 1990; Van der Ark et al., 2002) but are not relevant for this chapter.

The reason for calling one class of IRT models nonparametric and another parametric, is that the former only puts order restrictions on response functions and the latter assumes a specific parametric function, such as the logistic or the normal ogive. For example, several NIRT models typically assume a positive monotone relationship between the binary item score and the latent variable; that is, for all θ pairs, assume that

$$P_j(\theta_a) \leqslant P_j(\theta_b), \quad \text{whenever } \theta_a < \theta_b, \text{ and } j = 1, \ldots, J; \tag{3}$$

whereas a PIRT model such as the 2-parameter logistic model, in addition to monotonicity requires this relationship to be logistic with a location parameter, δ_j, and a slope parameter, α_j, for each item, such that

$$P_j(\theta) = \frac{\exp[\alpha_j(\theta - \delta_j)]}{1 + \exp[\alpha_j(\theta - \delta_j)]}. \tag{4}$$

Another difference, albeit gradual and not as a matter of principle, is that NIRT methods aim primarily at exploring the dimensionality of $\boldsymbol{\theta}$ whereas PIRT models often posit a specific form of dimensionality and then test the fit of the model that hypothesizes this dimensionality to the data. These differences should give the reader some intuition why NIRT (1) is more general than PIRT and (2) as a data analysis tool primarily is exploratory and PIRT confirmatory.

1.1. Assumptions of NIRT and PIRT models

Three classes of assumptions play a central role both in NIRT and PIRT. IRT models are defined by a specific choice from one or more of these classes, for example, with respect to the number of latent variables or the kind of restriction on the response function.

1.1.1. Dimensionality of the data

A distinction can be made between one latent variable and multiple latent variables. Corresponding with this distinction are the assumptions of unidimensionality (UD) and multidimensionality, and strictly unidimensional IRT models and multidimensional IRT models.

1.1.2. Relationship between items

Most IRT models assume that the J-variate distribution of the item scores conditional on $\boldsymbol{\theta}$ – with θ as a special case – is the product of the J marginal conditional distributions of the separate item scores. This assumption is known as local independence (LI). For individual v, define vector $\mathbf{X}_v = (X_{vj})$ with realization $\mathbf{x}_v = (x_{vj})$, $j = 1, \ldots, J$; then, assumption LI is defined as

$$P(\mathbf{X}_v = \mathbf{x}_v|\boldsymbol{\theta}) = \prod_{j=1}^{J} P(X_{vj} = x_{vj}|\boldsymbol{\theta}). \tag{5}$$

An important implication of LI is that the conditional covariance between the scores on items j and k is 0; that is

$$\mathrm{Cov}(X_j, X_k|\boldsymbol{\theta}) = 0, \quad \text{for all pairs } j, k, \text{ and } j \neq k. \tag{6}$$

Reversely, this property, also known as weak LI, does not imply LI (Junker, 1993) because Eq. (6) only deals with bivariate relationships which may be 0 while multivariate relationships are not.

In practical NIRT data analysis, one tries to find the dimensionality of $\boldsymbol{\theta}$ which is adequate for those data, such that LI is satisfied as good as possible. Stout (2002) noted that in real data weak LI is much easier to evaluate than LI, because with weak LI one only needs to evaluate conditional covariances while multivariate relationships between items can be ignored. For realistic J the number of multivariate relationships beyond the $\frac{1}{2}J(J-1)$ conditional covariances is huge and in real data many combinations of scores are not observed (Sijtsma and Junker, 1996). Stout (2002) concluded that, although mathematically not identical, in practice LI and weak LI are so much alike – in fact, it is difficult to imagine that Eq. (6) is satisfied but multivariate relationships deviate significantly from 0 – that the better accessible weak LI property may be used for dimensionality analysis in real data. This topic is discussed extensively in Section 2.1.2.

Assume that vectors with item scores of N individuals are collected in a data matrix $\mathbf{X}_{N\times J} = (\mathbf{X}_1, \ldots, \mathbf{X}_N)^{\mathrm{T}}$, and that the respondents have been sampled i.i.d. (independently and identically distributed) from the cumulative multivariate distribution $G(\boldsymbol{\theta})$, so that

$$P(\mathbf{X}_{N\times J} = \mathbf{x}_{N\times J}|\boldsymbol{\theta}) = \prod_{v=1}^{N}\prod_{j=1}^{J} P(X_{vj} = x_{vj}|\boldsymbol{\theta}). \tag{7}$$

Notice that Eq. (7) is the general likelihood equation of any IRT model that assumes i.i.d. sampling and LI (Eq. (5)). By integrating over $\boldsymbol{\theta}$, the multivariate distribution of

$\mathbf{X}_{N\times J}$ is obtained; that is, for $G(\boldsymbol{\theta})$ we have that

$$P(\mathbf{X}_{N\times J} = \mathbf{x}_{N\times J}) = \prod_{v=1}^{N} \int_{\theta_1} \cdots \int_{\theta_Q} \prod_{j=1}^{J} P(X_{vj} = x_{vj} | \boldsymbol{\theta}) \, \mathrm{d}G(\boldsymbol{\theta}). \tag{8}$$

Modeling the item response data collected in a sample of N respondents involves restricting the response functions on the right-hand sides of Eqs. (5) and (8) and the latent variable distribution $G(\boldsymbol{\theta})$ in Eq. (8).

1.1.3. Relationship of item score and latent trait

The most important distinction is between a nonparametric restriction on the response function as in Eq. (3) and a parametric definition as in Eq. (4). The model that is based on assumptions UD, LI and Eq. (3), which is also known as assumption M, is known as the monotone homogeneity model (Mokken, 1971; also, see Sijtsma and Molenaar, 2002). The model based on UD, LI and Eq. (4) is the 2-parameter logistic model (e.g., Baker and Kim, 2004). It may be noted that, compared with the IRFs of the 2-parameter logistic model the nonparametric IRFs (1) need not be S-shaped, and need not even be smooth; (2) can have several inflexion points, and may be constant over several ranges of θ; and (3) can have lower asymptotes or minimum values greater than 0 and upper asymptotes or maximum values smaller than 1.

Inserting in the likelihood (Eq. (7)) the conditional probabilities, $P_j(\theta_v)$ from the 2-parameter logistic model (Eq. (4)) if $x_{vj} = 1$ was observed in cell (v, j) of $\mathbf{X}_{N\times J}$, and $Q_j(\theta_v) = 1 - P_j(\theta_v)$ if $x_{vj} = 0$ was observed; and solving for the item parameters, α_j and δ_j, $j, \ldots, J$, and the latent variable values θ_v, $v = 1, \ldots, N$, amounts to joint maximum likelihood (ML) estimation of the model parameters. Insertion of these conditional probabilities in Eq. (8), choosing a distribution for $G(\boldsymbol{\theta})$, such as the multivariate normal, integrating $\boldsymbol{\theta}$ out of the equation, and solving for the item parameters, is known as marginal maximum likelihood estimation. Inserting the marginal ML estimates of the item parameters thus obtained in the likelihood in Eq. (7) and solving for θ yields ML estimates of the person parameters. Other estimation methods exist in addition to joint ML and marginal ML (e.g., Baker and Kim, 2004), but are not discussed here.

Restricting response probabilities in Eq. (8) by, for example, assumption M as reflected in Eq. (3) does not lead to ML estimates of latent variable and item parameters as in Eq. (4), simply because these parameters do not appear in the likelihood of the model. However, these order restrictions are sufficient to restrict the data, $\mathbf{X}_{N\times J}$, such that observable consequences can be derived that can be used to evaluate the fit of the model to the data. Moreover, together with UD and LI, assumption M can be shown to imply an ordinal scale for persons which can be estimated from the data and, if a restriction on the ordering of the IRFs is added as a fourth assumption, also implies a scale for items. Another approach to evaluating NIRT models comes from order-restricted latent class analysis (LCA) modeling. This approach replaces the continuous latent variable by an ordered, discrete latent class variable, and restricts the ordering of response probabilities conditional on these latent classes. Bayesian methods (Hoijtink and Molenaar, 1997; Van Onna, 2002) and ML methods (Vermunt, 2001) are used to evaluate the fit of the model to the data, and allocate respondents to one of the ordered latent classes of the fitting model.

1.1.4. Other assumptions in NIRT

Eq. (8) defines a latent structure that underlies the multivariate distribution of the item-response data in $\mathbf{X}_{N \times J}$, but Suppes and Zanotti (1981) showed that it is always possible, for finite J, to construct a unidimensional random variable, ξ, which is not necessarily latent, such that the item score variables in $\mathbf{X}_{N \times J}$ satisfy LI (Eq. (5)). Thus, any data set can be "forced" to satisfy Eq. (8) meaning, reversely, that Eq. (8) does not restrict the data and in that sense does not define a practically useful IRT model. Other results exploring the boundaries of what is minimally required to have an identifiable IRT model are presented by Junker (1993) and Ellis and Junker (1997); also see Junker and Ellis (1997). Ignoring the many subtleties in their work, what matters here is to note that if one restricts the dimensionality of the data and assumes LI, an assumption like M (Eq. (3)) or a parametric restriction like Eq. (4) is needed to have an IRT model that restricts the data.

In general, it is true that PIRT models restrict the data in $\mathbf{X}_{N \times J}$ more than NIRT models. For example, the IRF of the 2-parameter logistic model (Eq. (4)) is a special case of the order-restricted IRF of the monotone homogeneity model (Eq. (3)). Given that both models assume UD and LI, it follows that the 2-parameter logistic model is a special case of the monotone homogeneity model. Many comparable relationships can be found within pairs of PIRT and NIRT models, but within some pairs the hierarchy is not unequivocal (e.g., De Koning et al., 2002) or may even be reversed (e.g., Sijtsma and Hemker, 1998). Thus, while it is true in general that NIRT is less restrictive than PIRT it is not true that each NIRT model is less restrictive than each one of the PIRT models.

One of the themes of NIRT is in what degree assumptions may be relaxed such that particular measurement properties still are implied while no parametric restrictions are needed. Stout (1987, 1990) and Junker (1991, 1993) studied the class of essential unidimensional (EU) IRT models. EU models are based on the following assumptions:

- Essential unidimensionality (EU); one dominant latent variable, θ, and Q nuisance variables, collected in $\boldsymbol{\theta} = (\theta, \theta_1, \ldots, \theta_Q)$, are assumed much as in factor analysis where a distinction is sometimes made between common factors and unique factors;
- Essential independence (EI); for infinitely many items the mean of the inter-item covariances conditional on $\boldsymbol{\theta}$ equals 0; thus,

$$\frac{2}{J(J-1)} \sum_{j=1}^{J-1} \sum_{k=j+1}^{J} \left|\mathrm{Cov}(X_j, X_k|\boldsymbol{\theta})\right| \to 0, \quad \text{for } J \to \infty. \tag{9}$$

- Weak monotonicity (WM); weak monotonicity replaces the stronger assumption M, by assuming that the mean of J IRFs, also known as the test response function, is monotone in each of the θ_qs from $\boldsymbol{\theta}$. That is,

$$J^{-1} \sum_{j=1}^{J} P_j(\boldsymbol{\theta}_a) \leqslant J^{-1} \sum_{j=1}^{J} P_j(\boldsymbol{\theta}_b),$$

$$\text{whenever } \boldsymbol{\theta}_a \leqslant \boldsymbol{\theta}_b, \text{ coordinatewise.} \tag{10}$$

WM does not restrict the individual IRFs, as long as their mean is monotone nondecreasing. This means that assumption M is dropped as a requirement for individual IRFs but that it is active for the test response function under the name of WM.

Stout (1990, Theorem 3.2) showed that for EU models based on EI and WM, that total score

$$X_+ = \sum_{j=1}^{J} X_j, \tag{11}$$

is a consistent ordinal estimator of the dominant θ, and Junker (1991) generalized this result to polytomous EU models, based on EI and ISRFs defined as $P(X_j \geqslant x_j|\theta)$. The replacement of assumptions UD, LI, and M by assumptions EU, EI, and WM, respectively, implies a more general nonparametric model. As a result, properties implied by UD, LI, and M with respect to stochastically ordering persons on θ by means of X_+ and investigating assumption M in real data are lost (these properties are discussed in Sections 1.2.1 and 2.2.1, respectively). However, the more general EU model is more flexible than, for example, the monotone homogeneity model and thus accommodates more data sets while maintaining attractive measurement properties; see Stout (2002) for a summary.

1.2. Differences between NIRT and PIRT

In this section we further explore the two important characteristics of NIRT, which are that NIRT (1) encompasses PIRT by providing a broader theoretical framework from which IRT in general is better understood; and (2) primarily is an exploratory toolkit for the analysis of test and questionnaire data whereas PIRT primarily deals with confirmatory data analysis by fitting specific models to data. An example of the first characteristic is the scale of measurement in NIRT and an example of the second characteristic is the estimation of response functions and the way these estimates are used in practical data analysis.

1.2.1. Measurement of individuals

The metric of the θ scale of the 2-parameter logistic model can be derived as follows. Define the odds of a success on item j as $\Omega_j = \frac{P_j(\theta)}{Q_j(\theta)}$, then $\ln \Omega_j = \alpha_j(\theta - \delta_j)$. For two respondents, v and w, the log-odds ratio equals

$$\ln \frac{\Omega_{vj}}{\Omega_{wj}} = \alpha_j(\theta_v - \theta_w). \tag{12}$$

Because NIRT models typically do not parameterize the response function, a θ metric for differences as in Eq. (12) is not available. Alternatively, NIRT models for dichotomous items based on the assumptions of UD, LI and M imply an ordinal scale in the sense that latent variable θ is stochastically ordered by total score X_+; that is, for any value t of θ and any pair of total scores, $x_{+a} < x_{+b}$, we have that

$$P(\theta \geqslant t|X_{+a} = x_{+a}) \leqslant P(\theta \geqslant t|X_{+b} = x_{+b}). \tag{13}$$

This is known as stochastic ordering of the latent variable by the observed sum score (SOL; Hemker et al., 1997; also see Grayson, 1988; Hemker et al., 1996; Huynh, 1994, for a discussion of the more general monotone likelihood ratio property). An implication of Eq. (13) is that

$$E(\theta|X_{+a} = x_{+a}) \leqslant E(\theta|X_{+b} = x_{+b}). \tag{14}$$

Thus, in practice any IRT model based on UD, LI, and M implies the ordering of respondents on latent variable θ by means of the ordering on the observable sum score X_+, except for random error. Note that this applies to all NIRT and PIRT models that assume monotonicity for the IRFs in addition to UD and LI. The parametric 2-parameter logistic model (based on UD, LI, and Eq. (4)) and the nonparametric monotone homogeneity model (based on UD, LI, and Eq. (3)) are examples. Other examples of parametric models based on UD and LI and with monotone IRFs are the much used Rasch model or 1-parameter logistic model, which defines IRFs as logistic with $\alpha_j = 1$ for all j (thus, in Eq. (4), $\alpha_j = 1$); and the 3-parameter logistic model, which adds a lower asymptote to the 2-parameter logistic model, denoted λ_j (with $0 \leqslant \lambda_j \leqslant 1$, $j = 1, \ldots, J$). Hemker et al. (1997) discussed failure of SOL in many polytomous-item IRT models, and Van der Ark (2005) provided many reassuring robustness results.

The importance of the SOL result is that for ordering individuals on the θ scale it suffices to use an IRT model that assumes property M in addition to UD and LI, but not necessarily an increasing logistic or normal ogive IRF. This result has positive consequences, not only for NIRT models but also for PIRT models.

First note that the relatively weak assumptions of NIRT models imply an ordinal scale. Moreover, as a kind of bonus for not asking for an abundance of structure in the data, as a rule NIRT models allow relatively many items in a scale. Even though these items would allow for reliable measurement, several would not be tolerated by the 2-parameter logistic model because their IRFs are not logistic. Among these rejected items would be items with IRFs that are asymmetrical, constant over some ranges of θ but steep over others, and have minimum values greater than 0 and maximum values smaller than 1 (see, e.g., Meijer and Baneke, 2004). However, such items contribute to reliable measurement provided that their IRFs are relatively steep locally on θ. It is important to note that such contributions do not depend on the IRF being logistic, normal ogive, or otherwise parametric, but on assumption M. Ramsay (2000) discusses a method for estimating the item and test information functions in an NIRT context, and Meijer et al. (1995) discuss methods for estimating the reliability of single items based on methods for estimating total score (X_+) reliability in an NIRT context (Sijtsma and Molenaar, 1987). Also note that among the items not well described by PIRT models may be items that add to the coverage of the variable represented by θ.

Second, when PIRT models fit to the data at the expense of perhaps some of the items that would be fitted by an NIRT model their numerical θ scale is a highly convenient tool for applications such as equating and adaptive testing. However, the SOL property also has advantages for PIRT models because it justifies the use of the number-correct X_+ score for communicating test results to teachers, parents, students, and also researchers as an intuitively appealing alternative for the inaccessible θ scale.

An interesting suggestion for an alternative scale that may replace the arbitrary and difficult to interpret θ scale was done by Ramsay (1996); also see Rossi et al. (2002). Basically, they propose to use a scale based on response probabilities, $P_j(\theta)$, because, unlike θ or monotone transformations of θ, a probability like $P_{12}(\theta_{\text{Mary}}) = 0.74$ represents a property of Mary relative to Item 12 which is invariant with respect to any scale of measurement. Thus, it lacks the arbitrariness of other scales, like θ. A curve is proposed which results from plotting the J response probabilities against one another as a function of θ. Starting from the scale's zero point, say, θ_0, the position of person v on this curve can be determined by means of the *arc length* measure $s(\theta_v)$ as

$$s(\theta_v) = \int_{\theta_0}^{\theta_v} \left[\sum_{j=1}^{J} \left(\frac{\mathrm{d}P_j(\theta)}{\mathrm{d}\theta} \right)^2 \right]^{1/2} \mathrm{d}\theta. \tag{15}$$

For more details, see Ramsay (1996) and Rossi et al. (2002).

We conclude by noting that NIRT has considerably contributed to defining scales that are either easy to obtain and understand (X_+) or firmly rooted in empirical data [$s(\theta)$], and circumvent the complexity and the arbitrariness of the θ scale. Clearly, such scales are also advantageous to PIRT.

1.2.2. *Model diagnostics*

From the perspective of parsimony it is convenient to assume that IRFs are, for example, logistic, and that items can be described using no more than a few parameters. Many tests have been constructed and item banks assembled that generate data that may be approximated by means of unidimensional or multidimensional PIRT models, but in itself there is no compelling reason why real data must exhibit the often hypothesized parametric curvature (Meredith, 1965). Detailed data analysis often reveals IRFs that deviate from this regular shape, with deviations ranging from small and irrelevant oscillations to deep dips and decreases along parts of the θ scale. It is interesting to know how PIRT and NIRT models react to such deviations from assumption M. See Meijer and Baneke (2004) for a confrontation of PIRT and NIRT models and an application to data from the Depression content scale of the Minnesota Multiphasic Personality Inventory-2 (Butcher et al., 1989).

Estimating a PIRT model is like stretching a grid over the data which is flexible only in some directions but not in others, and then summarizing the curvature and location of the grid as far as its flexibility allows it. This summary consists of one or a few parameters, which are valid for the data to the degree in which the hypothesized IRFs match the IRFs identified from the data. Obviously, peculiarities in the data may not be caught by logistic or other parametric curves but, if properly revealed, could help the researcher to make useful decisions about the items. For example, assume that the true IRF increases at the low and middle θs, but decreases at the higher θs. The logistic curve that is fitted to the data generated by means of this nonmonotone IRF accommodates this local decrease by a necessarily monotone curve which increases with a relatively flat slope (Junker and Sijtsma, 2001a). Thus, the item is identified as a weakly discriminating item, and may be a candidate for removal from the test.

An interesting question is, whether for given distribution $G(\theta)$ different models for the IRFs can yield the same distribution of the data, $P(\mathbf{X}_{N\times J} = \mathbf{x}_{N\times J})$ (Eq. (8)). Douglas (2001) showed that there is only one correct IRT model, and that nonparametric methods can consistently estimate the IRFs. This means that if an IRF is found in data analysis of which the shape does not match that predicted by a particular PIRT model, that this PIRT model is incorrect as a model for the data. Based on this result, Douglas and Cohen (2001) used nonparametric estimation of IRFs to study the fit of PIRT models. This was done by finding, given the distribution of θ that is used for estimating a particular PIRT model (or that results from the estimation procedure), the IRF from that PIRT model that matches the nonparametrically estimated IRF the closest. The discrepancy between the two functions indexes model misfit, and this was evaluated by means of re-sampling.

In general, NIRT approaches use kernel smoothing or another nonparametric regression method to estimate the IRF or other response functions from the data, and thus catch the monotone and nonmonotone parts of the curve. The value of knowing the whole curve mentioned in the previous example is that not only the dip at the higher θs is identified, but that it is also revealed that this item may actually discriminate well at the lower and middle regions of the θ distribution. Even though it fails assumption M, such an item may be included in the test when the researcher has only few items available, most of which discriminate well at the high θs but not at the low θs. Thus, here the item's strength may outbalance its weakness and the researcher may decide to include the item in the test so as to improve measurement precision at the lower θs. In another context, such as marketing, if many IRFs are found to exhibit this bell shape, the conclusion could be that another model is needed to analyze the data; that is, one that accommodates measurement of preferences by means of unimodal IRFs (e.g., Johnson and Junker, 2003; Post, 1992).

2. Analyzing test data using nonparametric IRT

The relatively weak sets of assumptions on which they are based lend NIRT models a flexibility which provides excellent opportunities for studying item and test properties. Thus, the outcomes on an NIRT analysis may give many clues of how to improve a test or questionnaire. We distinguish outcomes with respect to the dimensionality of the data and outcomes with respect to the fit a unidimensional NIRT model to the data. A sensible data analysis may start with a dimensionality analysis, identifying item subsets that predominantly are driven by one θ, followed by fitting an NIRT model to each of the dimensionally distinct item subsets so as to ascertain the measurement properties for each. Next, we review some methods for dimensionality analysis and model fitting.

2.1. *Dimensionality analysis*

At the basis of all NIRT methods for dimensionality analysis is conditional association (CA; Holland and Rosenbaum, 1986). Let vector $\mathbf{X}$ contain ordered polytomous items scores. Split $\mathbf{X}$ into two disjoint part vectors: $\mathbf{X} = (\mathbf{Y}, \mathbf{Z})$. Define f_1 and f_2 to be nondecreasing functions in the item scores from $\mathbf{Y}$, which means that f_1 and f_2 are nondecreasing coordinatewise in $\mathbf{Y}$; and define g to be some function of the item scores in $\mathbf{Z}$. Then UD, LI, and M imply CA; that is,

$$\operatorname{Cov}\bigl[f_1(\mathbf{Y}), f_2(\mathbf{Y}) | g(\mathbf{Z}) = z\bigr] \geqslant 0, \quad \text{for all } z. \tag{16}$$

CA is a powerful tool for investigating the fit of NIRT models to data because there are so many ways in which item subsets and functions of item scores can be defined. We only mention two examples that are of direct relevance to dimensionality analysis under NIRT.

2.1.1. *Dimensionality analysis based on nonnegative inter-item covariance*

Ignore item subset $\mathbf{Z}$, and define $f_1(\mathbf{Y}) = X_j$ and $f_2(\mathbf{Y}) = X_k$, then CA reduces to nonnegative inter-item covariance within the whole group; that is,

$$\operatorname{Cov}(X_j, X_k) \geqslant 0, \quad \text{all pairs } j, k;\ j \neq k. \tag{17}$$

That is, in the whole group all $\frac{1}{2}J(J-1)$ inter-item covariances must be nonnegative. A negative covariance indicates misfit of the model of monotone homogeneity – defined by assumptions UD, LI, and M – for at least one of the items involved.

How is Eq. (17) related to dimensionality analysis? Before we go into this topic, we first define coefficient H for the scalability of a set of J items and coefficient H_j for the scalability of item j with respect to the other $J - 1$ items in the set. Let the maximum possible covariance of items j and k, given the marginal distributions of the $(m + 1) \times (m + 1)$ frequency table, be denoted $\operatorname{Cov}(X_j, X_k)_{\max}$; then coefficient H (Loevinger, 1948; Mokken, 1971) is defined as

$$H = \frac{\sum_{j=1}^{J-1} \sum_{k=j+1}^{J} \operatorname{Cov}(X_j, X_k)}{\sum_{j=1}^{J-1} \sum_{k=j+1}^{J} \operatorname{Cov}(X_j, X_k)_{\max}}, \tag{18}$$

and the item coefficient H_j is defined as

$$H_j = \frac{\sum_{k \neq j} \operatorname{Cov}(X_j, X_k)}{\sum_{k \neq j} \operatorname{Cov}(X_j, X_k)_{\max}}, \quad j = 1, \ldots, J. \tag{19}$$

Note that $\operatorname{Cov}(X_j, X_k)_{\max} > 0$, for all (j, k) pairs, implies that the denominators of H and H_j are positive; and that nonnegative inter-item covariance (Eq. (17)) implies that the numerators are nonnegative; thus, $H \geqslant 0$ and $H_j \geqslant 0$, for $j = 1, \ldots J$. Further, combining this result with $\operatorname{Cov}(X_j, X_k) \leqslant \operatorname{Cov}(X_j, X_k)_{\max}$, for all (j, k) pairs, we obtain that $0 \leqslant H \leqslant 1$ and $0 \leqslant H_j \leqslant 1$, for all $j = 1, \ldots, J$. Mokken (1971, p. 152) showed that coefficient H is a positively weighted average of the H_js.

Mokken (1971) (see Mokken and Lewis, 1982; Molenaar, 1997; Sijtsma and Molenaar, 2002, Chapter 5) proposed a bottom-up, sequential item selection procedure that clusters items on the basis of their relationship with the θ they predominantly measure, using coefficient H as the procedure's loss function. Noting first that concepts like IRF slope are not well defined in an NIRT context, Mokken et al. (1986) then argued that, keeping the distribution of θ and the J item locations constant, both H and H_j are increasing functions of the slopes of logistic IRFs, $P_j(\theta)$ (Eq. (4)). Their line of reasoning may be generalized to non-smooth IRFs that are roughly S-shaped, and that are typical of an NIRT context (e.g., see Emons et al., 2002 and Hessen, 2003, Chapter 3), and to ISRFs, $P(X_j \geqslant x_j|\theta)$ (Van Onna, 2004). Now, assume that all items in the set from which the procedure selects measure the same θ. The procedure selects items one-by-one into a cluster such that, in each selection step, H is maximized given all possible choices from the unselected items at that moment (i.e., at that selection step). Thus, in each step items are selected that have response functions with relatively steep slopes. These are items that hang together well due to the θ that they have in common. The procedure stops if in the next step none of the unselected items is selected due to an H_j with the selected items that is lower than some user-specified positive lowerbound, c.

Next, assume that different subsets of items in the set from which the procedure selects measure different θs. Then, from the items not selected into the first cluster the procedure tries to select a second cluster, a third, and so on, until no items are left to be clustered. Thus, assumption M forms the basis of the dimensionality investigation: Items that measure the same θ have steep regressions (IRFs or ISRFs) on that θ and are singled out from the larger item set and selected into the same cluster. The item selection procedure is implemented in the computer program Mokken Scale analysis for Polytomous items (MSP; Molenaar and Sijtsma, 2000).

2.1.2. *Dimensionality analysis based on nonnegative conditional inter-item covariance*

For dichotomously scored items define a rest score based on $\mathbf{X}$ as,

$$R_{(-j,-k)} = \sum_{h \neq j,k} X_h, \tag{20}$$

and let $g(\mathbf{Z}) = R_{(-j,-k)}$; and define $f_1(\mathbf{Y}) = X_j$ and $f_2(\mathbf{Y}) = X_k$. Then CA implies that,

$$\begin{aligned} &\operatorname{Cov}[X_j, X_k | R_{(-j,-k)} = r] \geqslant 0, \\ &\quad \text{all } r = 0, 1, \ldots, J-2; \text{ and all } j, k;\ j \neq k. \end{aligned} \tag{21}$$

That is, in the subgroup of respondents that have the same restscore r, the covariance between items j and k must be nonnegative. This special case of CA proves to be useful for dimensionality analysis that is based on investigating weak LI, as we discuss next (see Nandakumar et al., 1998, for generalizations to polytomous items).

Stout (1987, 1990) and his coworkers (Stout et al., 1996; Zhang and Stout, 1999a) proposed methods for investigating weak LI; see Eq. (6). Justified by consistency results derived by Stout (1987) for dichotomous items and Junker (1991) for polytomous items, they estimated $\operatorname{Cov}(X_j, X_k|\theta)$ twice by inserting not only $\hat{\theta} = R_{(-j,-k)}$ but also $\hat{\theta} = X_+$ and then taking the mean of both estimates, thus obtaining smaller bias than had only one of these θ estimates been used. Estimated conditional covariances equal or close to zero were taken to indicate weak LI. Since these results are based on asymptotics, two realistic situations involving finite numbers of items, one pertaining to unidimensionality and the other to multidimensionality, may be distinguished.

First, assume that all J items measure the same θ, what can one say about $\mathrm{Cov}(X_j, X_k|\hat{\theta})$ with either $\hat{\theta} = R_{(-j,-k)}$ or $\hat{\theta} = X_+$ inserted for θ? For $\hat{\theta} = R_{(-j,-k)}$ or other estimates that do not involve X_j and X_k we have by CA that $\mathrm{Cov}(X_j, X_k|\hat{\theta}) \geqslant 0$ (Eq. (21)); and for $\hat{\theta} = X_+$ it has been shown that $\mathrm{Cov}(X_j, X_k|\hat{\theta}) < 0$ both for items that agree with the Rasch model (Junker, 1993, p. 1370) but also for i.i.d. coin flips. However, given only the general conditions of UD, LI, and M little is known about results for the sign let alone the magnitude of the conditional covariance, thus rendering this covariance somewhat problematic for investigating weak LI.

Second, when the items measure Q latent variables to a different degree (i.e., some items predominantly measure one latent variable, other items measure another, and so on), then we may assume that θ_α is a linear combination of these latent variables. The performance on the Q latent variables is estimated by means of scalars $R_{(-j,-k)}$ and X_+, that both summarize test performance but ignore multidimensionality. However, Zhang and Stout (1999b) showed that the sign of $\mathrm{Cov}(X_j, X_k|\theta_\alpha)$ [and the estimator $\mathrm{Cov}(X_j, X_k|\hat{\theta}_\alpha)$] provides useful information about the dimensionality of the data. Roughly speaking, the covariance is positive if the two items measure approximately the same latent variable and negative if they clearly measure different latent variables. This sign behavior is the basis of an algorithm called DETECT (Zhang and Stout, 1999a) that splits a given set of items into clusters that together approach estimated weak LI as good as possible given all possible cluster composites.

Assume that several of the latent variables measured by the items are dominant and that others are nuisance variables. Consider an arbitrary partitioning, denoted $\mathcal{P}$, of the set of items into disjoint clusters; and let $d_{jk}(\mathcal{P}) = 1$ if the items j and k are in the same cluster, and $d_{jk}(\mathcal{P}) = -1$ if they are not. The loss function of DETECT, denoted $D_\alpha(\mathcal{P})$, is defined as

$$D_\alpha(\mathcal{P}) = \frac{2}{J(J-1)} \sum_{j=1}^{J-1} \sum_{k=j+1}^{J} d_{jk}(\mathcal{P}) E\big[\mathrm{Cov}(X_j, X_k|\theta_\alpha)\big]. \tag{22}$$

The DETECT procedure tries to find the partitioning of the items, denoted $\mathcal{P}^*$, that maximizes Eq. (22). This is the item clustering that best approximates weak LI, and it is taken as the best description of the dimensionality of the data.

2.1.3. *Comparative research*

Van Abswoude et al. (2004a) compared MSP and DETECT with respect to their effectiveness in retrieving the true dimensionality from simulated data sets. They concluded that, in general, DETECT was better capable of retrieving a simulated dimensionality structure than MSP. However, DETECT required greater sample sizes to arrive at reliable results and for smaller sample sizes appeared to be more vulnerable to chance capitalization than MSP. It was recommended to use different procedures next to one another and use their strengths – for DETECT its sensitivity to LI and for MSP its sensitivity to IRF steepness – to arrive at better results than are possible using only one procedure.

Other cluster procedures based on nonnegative inter-item covariance and H coefficients (Van Abswoude et al., 2004b) and nonnegative conditional inter-item covariance and weak LI (Roussos et al., 1998) exist, but are not considered in further detail. Van Abswoude (2004) suggested a cluster approach that combines the unique orientations of DETECT towards dimensionality evaluation and Mokken's procedure towards evaluation of the response function's slope into one algorithm, such that for a minimally required quality level (or higher) in terms of item discrimination the optimal clustering in terms of dimensionality is obtained. This research is in progress.

2.1.4. Conditional covariance functions

Douglas et al. (1998) suggested estimating the conditional covariance function, $\text{Cov}(X_j, X_k|\theta)$, and using its specific form to find out whether only one θ drives item performance or whether other latent variables or particular characteristics of the test procedure are also influential. Let $P_{jk}(\theta) = P(X_j = 1, X_k = 1|\theta)$, and estimate the conditional covariance as

$$\widehat{\text{Cov}}(X_j, X_k|\hat{\theta}) = \widehat{P}_{jk}(\hat{\theta}) - \widehat{P}_j(\hat{\theta})\widehat{P}_k(\hat{\theta}). \tag{23}$$

In Eq. (23), estimate $\hat{\theta}$ is the percentile rank of rest score, $R_{(-j,-k)}$ (Eq. (20)). The conditional probabilities are estimated using kernel smoothing (e.g., Douglas, 1997; Ramsay, 1991). The minimum and maximum values of $\text{Cov}(X_j, X_k|\theta)$ depend on the IRFs, $P_j(\theta)$ and $P_k(\theta)$, and are expected to vary across θ. To eliminate this effect, instead of using the odds ratio to express the relationship between items j and k, a function of the odds ratio, denoted $f[\alpha(\theta)]$, is used (for details, see Douglas et al., 1998) which has several desirable properties, two of which are that (1) $f[\alpha(\theta)] = 0$ indicates independence, $f[\alpha(\theta)] = -1$ maximum negative dependence, and $f[\alpha(\theta)] = 1$ maximum positive dependence; and (2) $\text{sign}(f[\alpha(\theta)]) = \text{sign}[\text{Cov}(X_j, X_k|\theta)]$.

Function $f[\alpha(\theta)]$ is used for studying multidimensionality phenomena at the level of item pairs such as caused by a second common latent variable, denoted θ_0, when it can be shown that while $\text{Cov}(X_j, X_k|\theta, \theta_0) = 0$ by LI, ignoring θ_0 is revealed by $\text{Cov}(X_j, X_k|\theta) > 0$. Indeed, for an item pair that was suspected of measuring a second latent variable it was found that the estimated function was consistently positive across the whole θ range. Another application concerned the investigation of effects of speededness (also, see Habing, 2001), which was argued to be manifest as a violation of pairwise weak LI in areas of θ where finishing all items was not accomplished by all respondents.

2.2. Fitting NIRT models to dimensionally distinct item clusters

Once the item set has been divided into dimensionally distinct clusters, each of which represent one dominant latent variable, each cluster is subjected to an item analysis that serves to establish the measurement quality of the scale defined by the items in a cluster. This part of the analysis concentrates on the response functions. Two properties are of particular interest. One is whether the response functions are monotone functions, so that SOL (Eq. (13)) is established, and the other is whether the response functions intersect. This is relevant for establishing an item ordering, as is explained in Section 2.2.2.

2.2.1. Diagnosing monotonicity and other response function shapes
Like restscore $R_{(-j,-k)}$ (Eq. (20)), restscore $R_{(-j)}$, defined as

$$R_{(-j)} = \sum_{k \neq j} X_k, \tag{24}$$

estimates θ, a result which is justified on the basis of the same stochastic ordering and consistency arguments. Based on this result, for dichotomous items $P[X_j = 1|R_{(-j)}]$ is used to estimate $P(X_j = 1|\theta)$. Junker (1993) showed that UD, LI, and M together imply manifest monotonicity (MM),

$$P[X_j = 1|R_{(-j)}] \quad \text{nondecreasing in } R_{(-j)};\ R_{(-j)} = 0, \ldots, J-1. \tag{25}$$

Contrary to intuition, a monotonicity result as in Eq. (25) is not obtained when $R_{(-j)}$ is replaced by X_+ (Junker and Sijtsma, 2000).

The conditional probabilities $P[X_j = 1|R_{(-j)} = r]$, for all r, can be used to estimate the IRF by means of nonparametric regression. This is done by estimating, for each value of r, the proportion of the population that have item j correct, plotting these proportions as a function of r, and then checking visually for MM. This approach, which is known as binning, yields a limited number – at most J – of discrete points of the IRF. Each sample deviation,

$$\widehat{P}[X_j = 1|R_{(-j)} = r] > \widehat{P}[X_j = 1|R_{(-j)} = s], \quad r < s, \tag{26}$$

is tested for significance by means of a normal approximation to the binomial test (Molenaar, 1973; Molenaar and Sijtsma, 2000). Karabatsos and Sheu (2004) proposed a Bayesian approach using Markov Chain Monte Carlo simulation to evaluating assumption M for J items simultaneously. This procedure also gives information about item fit.

Alternatively, kernel smoothing methods may be used to obtain a continuous estimate of the IRF, the ISRFs, and the so-called option response curves, $P(X_j = x_j|\theta)$, with X_j nominal, representing the options of a multiple-choice item (Ramsay, 1991, 1997). Jack-knife procedures may be used to estimate confidence envelopes. The program MSP (Molenaar and Sijtsma, 2000) produces discrete estimates of IRFs and ISRFs based on binning and the program TestGraf (Ramsay, 2000) produces quasi-continuous estimates of IRFs, ISRFs, and option response curves.

Rossi et al. (2002) proposed a methodology that uses ML estimation based on EM to obtain the logit transformation of the IRF, denoted $\lambda_j(\theta) = \ln \Omega_j$, by means of a linear combination of polynomials,

$$\lambda_j(\theta) = \sum_{k=1}^{K} \boldsymbol{\beta}_j^{\mathrm{T}} \boldsymbol{\phi}(\theta), \tag{27}$$

in which $\boldsymbol{\phi}(\theta) = (\phi_1(\theta), \ldots, \phi_K(\theta))$ are basis functions, so-called B-splines, chosen by the researcher and used to approximate adjacent segments of $\lambda_j(\theta)$, each weighed by the corresponding coefficient in $\boldsymbol{\beta}_j = (\beta_1, \ldots, \beta_K)$. These weights are estimated from the data given the choice of basis functions, and they control the smoothness of $\hat{\lambda}_j(\theta)$. As with kernel smoothing, too much irregularity in the curve cannot be separated from sampling error, but too much smoothness polishes away systematic and interesting phenomena that are useful to diagnose the curve.

Another counterintuitive result is that Eq. (25) is invalid for polytomous items. Substituting X_+ for $R_{(-j)}$ does not solve this problem (result due to Hemker; cited in Junker and Sijtsma, 2000). Thus, probabilities, $P[X_j \geqslant x_j | R_{(-j)} = r]$, in which $R_{(-j)}$ is the sum of $J - 1$ polytomous variables, may be estimated, plotted, checked, and tested for deviations of monotonicity, but only heuristically. An alternative may be the following (Junker, 1996; Sijtsma and Van der Ark, 2001). Let D_j be a dichotomization of the polytomous item score, X_j, at score x, such that

$$D_j = 0 \quad \text{if } X_j \leqslant x, \quad \text{and} \quad D_j = 1 \quad \text{if } X_j > x,\ j = 1, \ldots, J; \tag{28}$$

and let

$$D_{(-j)} = \sum_{k \neq j} D_k. \tag{29}$$

Using $D_{(-j)}$, the following "manifest monotonicity" result is true for polytomous items,

$$P[X_j \geqslant x | D_{(-j)}] \quad \text{nondecreasing in } D_{(-j)};\ D_{(-j)} = 0, \ldots, J - 1;$$
$$x = 1, \ldots, m. \tag{30}$$

Eq. (30) is also true if different items are dichotomized at different values of x_j. Thus, assumption M may be investigated using a multitude of "manifest monotonicity" results, first, by varying score x_j in $P[X_j \geqslant x_j | D_{(-j)}]$ and, second, by trying different versions of $D_{(-j)}$. How to make sensible decisions to reduce the abundance of results and draw one conclusion about assumption M with respect to item j is a topic of our present research.

2.2.2. *Investigating invariant item ordering*

Some applications require that the ordering of the J items in the test by difficulty is the same for each respondent taking the test. A set of J items – dichotomous or polytomous, with $m + 1$ ordered scores for each item – has an invariant item ordering (IIO; Sijtsma and Junker, 1996; Sijtsma and Hemker, 1998) if the items can be ordered and numbered accordingly, such that

$$E(X_1|\theta) \leqslant E(X_2|\theta) \leqslant \cdots \leqslant E(X_J|\theta), \quad \text{for each } \theta. \tag{31}$$

In this ordering ties may occur for some values or intervals of θ. Integration of the conditional expectation in Eq. (31) across $G(\theta)$ yields an ordering of item means, denoted μ_{X_j}, for $j = 1, \ldots, J$,

$$\mu_{X_1} \leqslant \mu_{X_2} \leqslant \cdots \leqslant \mu_{X_J}. \tag{32}$$

If an IIO has been established by one of the methods that is discussed in this section, the ordering of the item mean scores estimated from the sample can be used to estimate the item ordering as it applies to each individual θ and each subgroup from the population of interest. That is, use $\hat{\mu}_{X_j} = N^{-1} \sum_i X_{ij}$, for $j = 1, \ldots, J$, to estimate the sample mean for each item, order these sample means from small to large, and consider the result to be the item ordering that is valid for each θ. For dichotomous items, the nonparametric model defined by assumptions UD, LI, M, and IIO is the double monotonicity model (Mokken, 1971).

An example of a situation that requires an IIO is one in which the items are presented to the respondents in a fixed order, usually in ascending difficulty (opposite to the ordering in Eq. (31)), to make respondents feel at ease by presenting the easiest items first. This is common practice in intelligence testing (e.g., Bleichrodt et al., 1984), and the IIO property is applied here both at the individual and at the age-group level. For example, John, who belongs to the second-youngest age group, skips the first five items because these are of trivial difficulty to his age group, and then continues trying items of ascending difficulty until he fails, say, three consecutive items. Because the next items are more difficult, John is expected to fail almost all items of greater difficulty, meaning that the item level no longer matches his ability level. The higher someone's ability level, the longer it takes to fail three consecutive items. A similar procedure is followed for the other age groups, with each next age group skipping more of the easier items and starting at a higher item difficulty level.

This procedure with starting rules, tailored to the group level, and stopping rules, tailored to the individual level, may be seen as a kind of adaptive testing using a small "item bank". Within an age group, each respondent starts at the same difficulty level and is administered the same sequence of items until he/she fails at a string of items of which the difficulty exceeds his/her ability level by far. Metaphorically, one might say that this way of testing "scans" a wide neighborhood on the θ scale in which a respondent's θ value is located, while "true" adaptive testing tries to target his/her θ value with each new item narrowing the neighborhood as quickly as possible, given the present knowledge of θ.

Adaptive testing in general uses fewer items to obtain the same measurement accuracy, but requires huge item banks to attain this goal. The vast majority of test applications in psychology and elsewhere do not have access to huge item banks because (1) they are much too expensive if the application is not large-scale and commercially productive, and (2) the construction of items in tests and questionnaires is more difficult than in an educational environment that is often concerned with the reproduction of knowledge. For example, it is easier to construct items for arithmetic than for the personality domain of *Dysphoria*, because with arithmetic each new problem poses a new challenge to the student while with *Dysphoria* each new item is likely to repeat basic questions like "I often feel unhappy", "I take a gloomy view of things", and "I am often down in the dumps". Variations on these basic questions as would be required in an item bank, would yield little more than a reshuffling of words and phrases without providing a better coverage of the latent variable. Thus, we support the point of view that for personality assessment short inventories consisting of the same high-quality items that are presented to everyone are more appropriate than computerized adaptive testing procedures based on item banks (see, e.g., Reise and Henson, 2000). Also, if such short tests have an IIO the quality of the assessment could be improved even further.

Sijtsma and Junker (1996) (also see Sijtsma and Molenaar, 2002, Chapter 6) presented a survey of methods for investigating IIO for dichotomous items. Several methods evaluate pairs of discrete estimates of IRFs. To understand their general principle,

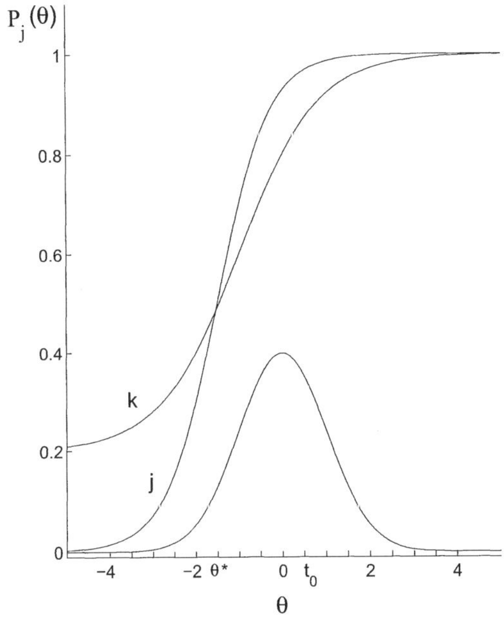

Fig. 1. Two item response functions that intersect at θ^* and a normal-shaped distribution of θ that is separated in two disjoint parts by $\theta = t_0$.

consider the two IRFs in Figure 1, and notice that they intersect at θ^*. Now, suppose that this true situation is unknown and that possible intersection is evaluated in the subgroups with $\theta \leqslant t_0$ and $\theta > t_0$ – assuming that it would be possible to make this sharp distinction with respect to θ. This evaluation typically takes the form of comparing the sign of the difference of the two item means in each of these subgroups. If the signs are the same, this is taken as evidence of nonintersection of the IRFs; and if they are different, this is taken as evidence of intersection. The latter conclusion would be justified but the former can be highly misleading, as the next example shows.

Let $f(\theta)$ be the density of θ. In the subgroup with $\theta \leqslant t_0$, which is the subgroup with the relatively low (L) θ values, denote the item means by μ_{jL} and μ_{kL}; then we have that

$$\mu_{jL} - \mu_{kL} = \frac{\int_{-\infty}^{t_0} [P_j(\theta) - P_k(\theta)] f(\theta)\, \mathrm{d}\theta}{\int_{-\infty}^{t_0} f(\theta)\, \mathrm{d}\theta}. \tag{33}$$

The sign of this difference is positive because in Figure 1 most of the density mass of $f(\theta)$ is where the difference between the IRFs is positive; that is, to the right of θ^*. For the relatively high (H) θ subgroup we have, similarly, that

$$\mu_{jH} - \mu_{kH} = \frac{\int_{t_0}^{\infty} [P_j(\theta) - P_k(\theta)] f(\theta)\, \mathrm{d}\theta}{\int_{t_0}^{\infty} f(\theta)\, \mathrm{d}\theta}, \tag{34}$$

which also is positive because the difference of the IRFs is positive for each θ in the interval of integration. Thus, a subgrouping based on t_0 masks the intersection. Obviously, the most effective subdivision would be obtained for $t_0 = \theta^*$, but other subdivisions in the neighborhood of θ^* would also produce opposite item mean orderings in subgroups. Also, the θ continuum may be divided into a greater number of disjoint and exhaustive intervals, using boundary values, $t_{01}, t_{02}, \ldots$.

In practice, subdivisions are necessarily based on estimates of θ and thus cannot make a "clean cut" as with t_0 in the example. The following trade-off is relevant here. If only two or a small number of subgroups are identified, the subgroup size is large and the method has much power but often in the "wrong" neighborhood on the θ scale (Figure 1); that is, far from θ^* (see the example in Figure 1). If a large number of subgroups is identified, in principle intersections are more easily spotted but may remain unseen because of little power due to small subsample sizes. Let us now look at the proposed methods.

Let $A_{(-j,-k)}$ denote a set of F items not including items j and k, with $1 \leqslant F \leqslant J-2$; let $A_{(-j,-k)}$ be defined *ad hoc*; and let

$$S_{(-j,-k)} = \sum_{h \in A_{(-j,-k)}} X_h \tag{35}$$

be an unweighed sum score based on the items in $A_{(-j,-k)}$, with realization s. Sum score $S_{(-j,-k)}$ is used to define intervals on the θ scale. Suppose $A_{(-j,-k)} = \{X_h\}$, so that $S_{(-j,-k)} = X_h$, with $X_h = 0, 1$. Thus, $S_{(-j,-k)}$ defines two subgroups with densities of θ conditional on X_h, denoted $f_0(\theta)$ (for $X_h = 0$) and $f_1(\theta)$ (for $X_h = 1$), which overlap partly, and have means that are ordered,

$$E_0(\theta) \leqslant E_1(\theta) \tag{36}$$

(this ordering follows from the SOL property in Eq. (13) and its implication for expected values in Eq. (14), with X_h substituted for X_+). Thus, X_h and other sum scores may be used to define subgroups that correspond with intervals on θ in a stochastic ordering sense. Then inspection of pairs of discrete IRF estimates boils down to checking whether

$$P[X_j = 1 | S_{(-j,-k)} = s] \leqslant P[X_k = 1 | S_{(-j,-k)} = s], \quad \text{for all } s, \tag{37}$$

and this is done for all $\frac{1}{2}J(J-1)$ item pairs in the test. Three possibilities for $S_{(-j,-k)}$ have been proposed:

- *Restscore method.* Let $A_{(-j,-k)}$ contain all $J-2$ items. Rosenbaum (1987) suggested using $S_{(-j,-k)} = \sum_{h \neq j,k} X_h$, usually denoted restscore $R_{(-j,-k)}$ (Eq. (20)), with realizations $r = 0, \ldots, J-2$. Success probabilities are estimated in each restscore group, and the ordering must be the same across r, as in Eq. (37) but with r substituted for s. Note that for moderate to large test length J, restscore $R_{(-j,-k)}$ is a relatively accurate estimate of θ and defines many intervals for evaluating IIO, but that subgroup sizes may be small. More accuracy can be gained by joining small adjacent subgroups.
- *Item-splitting method.* Define $S_{(-j,-k)} = X_h$, with $X_h = 0, 1$, for $h \neq j, k$, as before. For item pair (j, k), Eq. (37) is evaluated once for $X_h = 0$ and once for $X_h = 1$. Obviously, item-splitting uses a crude division of the θ scale, but compensates for that

by doing this for each item $h \neq j, k$. As item h is more difficult, the "fuzzy" boundary between the two groups – due to error – shifts to the right so that the item-splitting method uses several sub-groupings and is able to pick up most intersections of the IRFs. This method is mathematically equivalent with Mokken's (1971, pp. 132–133) $\mathbf{P}(+,+)$/$\mathbf{P}(-,-)$ matrices method.

- *Restscore-splitting method.* Sijtsma and Junker (1996) suggested using $R_{(-j,-k)} \leqslant r$ and $R_{(-j,-k)} > r$, with $r = 0, \ldots, J-3$, and evaluating Eq. (37) for each value of r (which is substituted for s). This method combines the use of the more reliable sum score, $R_{(-j,-k)}$, for subdivision with the greater group size produced by splitting on the basis of item score X_h.

While each of these methods evaluates only item pairs and produces many results for the researcher to combine, coefficient H^{T} (Sijtsma and Meijer, 1992) provides an overall impression of the degree in which J IRFs intersect. H^{T} is formally identical to coefficient H in Eq. (18), but reverses the roles of persons and items; thus, it is H applied to the transposed data matrix $\mathbf{X}^{\mathrm{T}} = \mathbf{X}_{J \times N}$; hence coefficient H^{T}. Its use in real data analysis was explained by Sijtsma and Meijer (1992). Te Marvelde et al. (2005) generalized H^{T} to ordered, polytomous item scores. Other work on investigating IIO for polytomous items is scarce, but research is in progress (e.g., Karabatsos and Sheu, 2004; Van der Ark and Bergsma, 2004; also see Sijtsma and Hemker, 1998).

2.3. *Ordered latent class approaches to NIRT*

Let ζ denote a discrete version of a latent variable, with values $w = 1, \ldots, W$. LCA models assume that a division of the group of respondents into a limited number of W subgroups or classes, each characterized by an unobserved latent variable value, $\zeta = w$, produces LI,

$$P(\mathbf{X}_v = \mathbf{x}_v | \zeta = w) = \prod_{j=1}^{J} P(X_j = x_j | \zeta = w). \tag{38}$$

Because the classes are not defined a priori, but estimated from the data, they are considered to be latent; hence, latent class analysis.

Using the property of conditional probabilities for independent events, A and B, that $P(A \wedge B) = P(B)P(A)$, and applying LI to $P(A)$, we may write the LCA model as (Goodman, 2002; Heinen, 1996, p. 44; McCutcheon, 2002),

$$P(\mathbf{X}_v = \mathbf{x}_v \wedge \zeta = w) = P(\zeta = w) \prod_{j=1}^{J} P(X_{vj} = x_{vj} | \zeta = w). \tag{39}$$

The probability that a randomly chosen respondent produces item-score vector $\mathbf{X} = \mathbf{x}$, is

$$P(\mathbf{X} = \mathbf{x}) = \sum_{w=1}^{W} P(\zeta = w) \prod_{j=1}^{J} P(X_j = x_j | \zeta = w). \tag{40}$$

This equation shows how LCA models the J-variate distribution of the item scores in terms of latent class probabilities, $P(\zeta = w)$, and probabilities of having particular item scores $X_j = x_j$ ($j = 1, \ldots, J$) given class membership, $P(X_j = x_j|\zeta = w)$.

The class probabilities and the conditional probabilities can be estimated from the data for several choices of the number of latent classes, W. In practical data analysis, W often varies between 1 and 5. The parameter estimates for the best-fitting model are used to estimate the discrete distribution of ζ, $P(\zeta = w)$, with $w = 1, \ldots, W$. This distribution can be used together with the conditional probabilities, $P(X_j = x_j|\zeta = w)$, to assign people to latent classes. For respondent v, this is done using probabilities $P[\zeta = w|\mathbf{X}_v]$, for $w = 1, \ldots, W$, on the basis of which he/she is assigned to the class that has the greatest subjective probability.

The likelihood of the unrestricted LCA model is

$$\begin{aligned} &P(\mathbf{X}_{N\times J} = \mathbf{x}_{N\times J}|\zeta = w) \\ &\quad = \prod_{v=1}^{N}\sum_{w=1}^{W} P(\zeta = w)\prod_{j=1}^{J} P(X_{vj} = x_{vj}|\zeta = w). \end{aligned} \tag{41}$$

This likelihood is unidentified; also, see Eq. (8) for continuous latent variable θ. Hoijtink and Molenaar (1997) adapted work by Croon (1991) for binary-item, unidimensional, order-restricted LCA models based on assumptions M and IIO to an LCA version of the model of monotone homogeneity – only M, but no IIO – with a restricted number of discrete latent variables, for example, $\boldsymbol{\zeta} = (\zeta_1, \zeta_2)$. The resulting response probabilities, $P_j(\boldsymbol{\zeta})$ are order-restricted; that is,

$$\begin{aligned} &P_j(\zeta_1 = 1, \zeta_2 = w_2) \leqslant \cdots \leqslant P_j(\zeta_1 = W_1, \zeta_2 = w_2), \\ &\quad w_2 = 1, \ldots, W_2, \end{aligned} \tag{42}$$

and

$$\begin{aligned} &P_j(\zeta_1 = w_1, \zeta_2 = 1) \leqslant \cdots \leqslant P_j(\zeta_1 = w_1, \zeta_2 = W_2), \\ &\quad w_1 = 1, \ldots, W_1. \end{aligned} \tag{43}$$

These and other restrictions are used to estimate the model in a Bayesian framework, using Gibbs sampling to simulate the sampling distribution of several test diagnostics for properties based on monotonicity and conditional association, and to reproduce the distribution of X_+ as good as possible so as to determine the appropriate number of latent classes. This work was generalized by Van Onna (2002) to polytomous items and ordered ISRFs (Eq. (2)), for the LCA version of assumption M in the monotone homogeneity model,

$$\begin{aligned} &P(X_j \geqslant x_j|\zeta_a) \leqslant P(X_j \geqslant x_j|\zeta_b), \\ &\quad \text{whenever } \zeta_a < \zeta_b;\ \text{for } x_j = 1, \ldots, m;\ \text{for } j = 1, \ldots, J, \end{aligned} \tag{44}$$

and two versions of the double monotonicity model (all three models were described in more detail by Sijtsma and Hemker, 1998, for continuous θ).

Vermunt (2001) used a logit formulation of three basic types of IRT models for polytomous items, the cumulative probability model, the adjacent category model, and the continuation ratio model (also, see Hemker et al., 2001, and Mellenbergh, 1995; Van der Ark et al., 2002 provide the hierarchical relationships between the nonparametric versions of these models). For example, for the cumulative probability model, also known as the monotone homogeneity model (Molenaar, 1997) or the nonparametric graded response model (Hemker et al., 1997) – here with discrete ISRF defined in Eq. (44) – define the log odds as,

$$\log \Omega^{j}_{\zeta,x} = \log \frac{P(X_j \leqslant x - 1|\zeta)}{P(X_j \geqslant x|\zeta)}. \tag{45}$$

For ζ_a and ζ_b the difference between the log odds is,

$$\log \Omega^{j}_{\zeta_a,x} - \log \Omega^{j}_{\zeta_b,x} \geqslant 0, \quad \text{whenever } \zeta_a < \zeta_b, \tag{46}$$

which is readily checked to be identical to Eq. (44). Thus, Eq. (46) provides an alternative formulation of assumption M for the polytomous-item monotone homogeneity model. Another restriction on the the log odds for items j and k, is that,

$$\log \Omega^{j}_{\zeta,x} - \log \Omega^{k}_{\zeta,x} \geqslant 0, \quad \mu_j \leqslant \mu_k. \tag{47}$$

Eq. (47) restricts the ordering of the response probabilities, $P(X_j \leqslant x|\theta)$, across j to be the same, except for possible ties, for each x; and for each ζ. For the cumulative probability formulation, this restriction assures an IIO (Scheiblechner, 1995; Sijtsma and Hemker, 1998). Models with inequality constraints are estimated using specialized procedures based on ML. Extensions of the models include multidimensionality, local dependence, and covariates.

3. Special topics in NIRT

3.1. Overview

NIRT researchers have contributed to several special topics in psychometrics. Examples are skills assessment (DiBello et al., 1995; McDonnell Hartz, 2002; Junker and Sijtsma, 2001b), unfolding models (Johnson and Junker, 2003; Post, 1992), and a generalization of the H coefficient (Eq. (18)) to a multilevel context (Snijders, 2001; also see Reise et al., 2006). Reviews of NIRT have been provided by Sijtsma (1998), Junker (2001) and Stout (2002). NIRT models have been used on numerous occasions for constructing scales. Sijtsma and Molenaar (2002, pp. 149–150) list many examples of abilities, traits and attitudes for which European psychologists, sociologists, political science researchers, and marketing researchers successfully constructed scales by fitting specific NIRT models to the data. North American NIRT specialists have concentrated more on modeling educational test data, for example, by means of response curve fitting (Ramsay, 1991), dimensionality assessment (Bolt, 2001; Stout et al., 1996), and differential item functioning (DIF) analysis (Shealy and Stout, 1993; Stout, 2002), but Santor et al. (1994) and Santor and Ramsay (1998) used NIRT for personality assessment. Bouwmeester and Sijtsma (2004) employed NIRT measurement data analysis tools from both sides of the Atlantic for constructing a scale for the Piagetian developmental concept of transitive reasoning.

Many new lines of research suggest themselves as soon as one starts to think about these topics. For example, DIF could be combined with IIO in the study of unidirectional DIF, as defined by Hessen (2003, pp. 22–23), following Shealy and Stout (1993). Consider two meaningful subgroups from the group of interest, and denote one as reference (R) group and the other as focal (F) group. Exactly what it means to identify such groups has been defined mathematically by Ellis and Van den Wollenberg (1993) by means of their local homogeneity concept. Usually, the focal group is suspected to be at a disadvantage when taking the test, often due to language deficiencies or cultural differences when such differences are expected not to play a role in test performance but in fact do. Absence of DIF is defined as

$$P_j^R(\theta) = P_j^F(\theta), \quad \text{for all } \theta; \text{ and for } j = 1, \ldots, J. \tag{48}$$

This definition implies that an IRF that is not exactly the same in groups R and F reflects some kind of DIF. Unidirectional DIF means that $P_j^R(\theta)$ and $P_j^F(\theta)$ do not intersect although they may touch at one or more θs or intervals of θ; that is, technically these IRFs have an IIO. The methods discussed earlier may be used to investigate IIO for pairs of IRFs (stemming from the same item) because samples from subgroups R and F do not share observations and the IRFs can be treated as they were generated by two different items. Then, after an IIO has been established a measure such as the area between the IRFs can be used to express the amount of unidirectional DIF (also, see Raju, 1988); that is, for

$$P_j^R(\theta) \geqslant P_j^F(\theta), \quad \text{for all } \theta, \tag{49}$$

we determine

$$AREA = \int_\theta \left[P_j^R(\theta) - P_j^F(\theta)\right] f(\theta)\,\mathrm{d}\theta. \tag{50}$$

The IRFs can be estimated by means of binning or kernel smoothing, or another method that yields smooth estimates, and *AREA* can be calculated using numerical integration. The bulk of the work in nonparametric DIF analysis has been done by Stout (2002) and his co-workers, and what we did here was tie DIF together with IIO in an attempt to suggest future lines of research.

3.2. *Person-fit analysis using the person response function*

We finish this chapter by discussing a recent example from our own work in NIRT, in particular person-fit analysis (PFA; e.g., Emons et al., 2005; Meijer, 2003). We believe that this example nicely illustrates the potential of NIRT as a flexible and rich toolkit, by introducing individual diagnosis using the person response function (PRF), using kernel smoothing and jackknifing to obtain estimates of this PRF and corresponding confidence bands, respectively, and testing for deviation from the expected decreasing trend – performance deteriorates as items grow more difficult – by means of logistic regression.

PFA studies individuals' vectors of J item scores, collected in $\mathbf{X}_v$. Given the probabilities of having item j correct given θ_v, and the monotonicity assumption of most IRT models, item-score vectors are more unlikely the larger the number of item-score pairs in which the relatively easy item was failed and the more difficult item was succeeded. Aberrant item-score vectors may be due, for example, to a lack of concentration in high-ability examinees when trying the easiest items and to answer copying to the most difficult items from other examinees by low-ability examinees. Similar to DIF, PFA attempts to uncover multidimensionality underlying test performance. This helps to diagnose test performance by taking advantage of the information available in the J item scores.

Many statistics have been proposed that summarize for each item-score vector, $\mathbf{X}_v$ ($v = 1, \ldots, N$), the degree in which an IRT model fits that vector or the degree in which the vector agrees with the vectors produced by a group of people (see Meijer and Sijtsma, 2001, for a review; and Karabatsos, 2003, for a comparison of 36 PFA statistics by means of simulated data). A recent development is the comprehensive PFA methodology proposed by Emons et al. (2005). It consists of three stages:

- *Global person-fit test.* Van der Flier's (1982) global person-fit statistic $ZU3$ – one of the most powerful person-fit statistics identified by Karabatsos (2003) – is used to identify fitting and misfitting item-score vectors. $ZU3$ is a monotone function of the likelihood of the data given an IRT model based on UD, LI, M, and IIO.
- *Graphical person-response function analysis.* The PRF (Sijtsma and Meijer, 2001) provides detailed information about the misfitting item-score vectors that were flagged by $ZU3$. For a respondent v, the PRF – which is the "mirror image" of the IRF – gives the probability of a correct response (scored 1) as a function of item difficulty. In this context, item difficulty is defined as the proportion-incorrect on item j,

$$\beta_j = 1 - \int_\theta P_j(\theta)\,\mathrm{d}G(\theta). \tag{51}$$

For person-response variable, S_v, which replaces item score X_j, and a hypothetical item domain on which the continuous item difficulty, β, is defined, which replaces latent variable θ, the PRF is defined as

$$P_v(\beta) \equiv P(S_v = 1|\beta). \tag{52}$$

This function is nonincreasing when the J IRFs do not intersect (Sijtsma and Meijer, 2001). Eq. (52) is estimated from J item scores by means of kernel smoothing. This is done by rank ordering the items according to increasing $\hat{\beta} = N^{-1}\sum_{v=1}^{N}(1 - X_{vj})$; and letting each item be focal point once, denoted β_0. The estimate of $P(S_v = 1|\beta_0)$ is obtained by taking a weighted average of the scores, X_{vj}, of person v for the focal item and the neighboring items. Weights are defined by the kernel function, $\omega(\beta_j)$, which may be based on the Gaussian or some truncated function, such that,

$$\widehat{P}_v(\beta_0) = \sum_{j=1}^{J} \omega(\beta_j) x_{vj}. \tag{53}$$

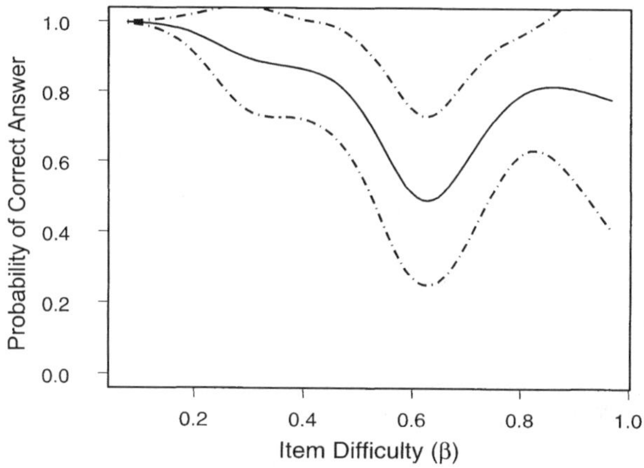

Fig. 2. Estimated person response function and 90% confidence bands, based on intelligence data (from: Emons et al., 2005; Figure 9F).

More details are provided by Emons et al. (2004). Suppose that a misfitting item-score vector that was flagged by $ZU3$ has an estimated U-shaped PRF; see Figure 2 for a real-data example based on intelligence test data (Emons et al., 2005). In this example, the local increase occurs at relatively difficult items ($\hat{\beta} \geqslant 0.63$, approximately), which suggest that the difficult items were easier for this person than the items of moderate difficulty. Confidence bands – here, 90% bands – for this function may be obtained by means of a jackknife procedure (see Emons et al., 2004, for details). It can be seen that for approximately $\hat{\beta} = 0.75$ and higher the estimated PRF exceeds the upper confidence band for the lowest value of the PRF (at approximately $\hat{\beta} = 0.63$). Thus, success probabilities for values $\hat{\beta} > 0.75$ are significantly larger than the minimum success probability at $\hat{\beta} = 0.63$, which is approximately $P_v(\hat{\beta}) = 0.5$. We conclude that Figure 2 gives evidence of person-misfit.

- *Local PRF analysis.* Sijtsma and Meijer (2001) and Emons (2003) discuss conservative local statistical tests for evaluating increases in the PRF that take only the part of the PRF into account where the increase occurs. Alternatively, Emons et al. (2004) use logistic regression to model the U-shape (e.g., Figure 2) and other shapes suggested by the estimated PRF. Denote the PRF for respondent w by $P_w(\beta)$, and assume that the items have been ordered from easy to difficult and numbered accordingly. Then, for a U-shaped PRF the logistic regression model is fitted to the item rank numbers; that is

$$\ln\left(\frac{P_w(\beta)}{1 - P_w(\beta)}\right) = \gamma_0 + \gamma_1 j + \gamma_2 j^2. \tag{54}$$

This quadratic model may be fitted against a linear model ($\gamma_2 = 0$; in addition $\gamma_1 = 0$ may be assumed). Emons et al. (2004) discuss the interpretation of results and also other logistic models that accommodate differently shaped PRFs.

The combination of global testing, graphical inspection of the PRF for misfitting-item-score vectors, and local testing of increases found in the PRF and fitting logistic regression models together help to better diagnose the misfit indicated by $ZU3$. Auxiliary information from earlier testing, personality characteristics, personal history and social-economic background may further enhance the interpretation of test performance.

References

Baker, F.B., Kim, S.-H. (2004). *Item Response Theory. Parameter Estimation Techniques*, 2nd ed., revised and expanded. Marcel Dekker, New York.

Bleichrodt, N., Drenth, P.J.D., Zaal, J.N., Resing, W.C.M. (1984). *Revisie Amsterdamse Kinder Intelligentie Test* (Revision of Amsterdam Child Intelligence Test). Swets & Zeitlinger, Lisse, the Netherlands.

Bolt, D.M. (2001). Conditional covariance-based representation of multidimensional test structure. *Applied Psychological Measurement* **25**, 244–257.

Bouwmeester, S., Sijtsma, K. (2004). Measuring the ability of transitive reasoning, using product and strategy information. *Psychometrika* **69**, 123–146.

Butcher, J.N., Dahlstrom, W.G., Graham, J.R., Tellegen, A., Kaemmer, B. (1989). *MMPI-2: Manual for Adminstration and Scoring*. University of Minnesota Press, Minneapolis.

Croon, M.A. (1991). Investigating Mokken scalability of dichotomous items by means of ordinal latent class analysis. *British Journal of Mathematical and Statistical Psychology* **44**, 315–331.

De Koning, E., Sijtsma, K., Hamers, J.H.M. (2002). Comparison of four IRT models when analyzing two tests for inductive reasoning. *Applied Psychological Measurement* **26**, 302–320.

DiBello, L.V., Stout, W.F., Roussos, L.A. (1995). Unified cognitive/psychometric diagnostic assessment likelihood-based classification techniques. In: Nichols, P.D., Chipman, S.F., Brennan, R.L. (Eds.), *Cognitively Diagnostic Assessment*. Erlbaum, Hillsdale, NJ, pp. 361–389.

Douglas, J.A. (1997). Joint consistency of nonparametric item characteristic curve and ability estimation. *Psychometrika* **62**, 7–28.

Douglas, J.A. (2001). Asymptotic identifiability of nonparametric item response models. *Psychometrika* **66**, 531–540.

Douglas, J., Cohen, A. (2001). Nonparametric item response function estimation for assessing parametric model fit. *Applied Psychological Measurement* **25**, 234–243.

Douglas, J., Kim, H., Habing, B., Gao, F. (1998). Investigating local dependence with conditional covariance functions. *Journal of Educational and Behavioral Statistics* **23**, 129–151.

Ellis, J.L., Junker, B.W. (1997). Tail-measurability in monotone latent variable models. *Psychometrika* **62**, 495–523.

Ellis, J.L., Van den Wollenberg, A.L. (1993). Local homogeneity in latent trait models. A characterization of the homogeneous monotone IRT model. *Psychometrika* **58**, 417–429.

Emons, W.H.M. (2003). Investigating the local fit of item-score vectors. In: Yanai, H., Okada, A., Shigemasu, K., Kano, Y., Meulman, J.J. (Eds.), *New Developments in Psychometrics*. Springer, Tokyo, pp. 289–296.

Emons, W.H.M., Meijer, R.R., Sijtsma, K. (2002). Comparing the empirical and the theoretical sampling distributions of the U3 person-fit statistic. *Applied Psychological Measurement* **26**, 88–108.

Emons, W.H.M., Sijtsma, K., Meijer, R.R. (2004). Testing hypotheses about the person-response function in person-fit analysis. *Multivariate Behavioral Research* **39**, 1–35.

Emons, W.H.M., Sijtsma, K., Meijer, R.R. (2005). Global, local and graphical person-fit analysis using person response functions. *Psychological Methods* **10**, 101–119.

Goodman, L.A. (2002). Latent class analysis. The empirical study of latent types, latent variables, and latent structures. In: Hagenaars, J.A., McCutcheon, A.L. (Eds.), *Applied Latent Class Analysis*. Cambridge University Press, Cambridge, UK, pp. 3–55.

Grayson, D.A. (1988). Two-group classification in latent trait theory: Scores with monotone likelihood ratio. *Psychometrika* **53**, 383–392.

Habing, B. (2001). Nonparametric regression and the parametric bootstrap for local dependence assessment. *Applied Psychological Measurement* **25**, 221–233.

Heinen, T. (1996). *Latent Class and Discrete Latent Trait Model. Similarities and Differences.* Sage, Thousand Oaks, CA.

Hemker, B.T., Sijtsma, K., Molenaar, I.W., Junker, B.W. (1996). Polytomous IRT models and monotone likelihood ratio of the total score. *Psychometrika* **61**, 679–693.

Hemker, B.T., Sijtsma, K., Molenaar, I.W., Junker, B.W. (1997). Stochastic ordering using the latent trait and the sum score in polytomous IRT models. *Psychometrika* **62**, 331–347.

Hemker, B.T., Van der Ark, L.A., Sijtsma, K. (2001). On measurement properties of continuation ratio models. *Psychometrika* **66**, 487–506.

Hessen, D.J. (2003). Differential item functioning. Types of DIF and observed score based detection methods. Unpublished Ph.D. dissertation, University of Amsterdam, The Netherlands.

Hoijtink, H., Molenaar, I.W. (1997). A multidimensional item response model: Constrained latent class analysis using the Gibbs sampler and posterior predictive checks. *Psychometrika* **62**, 171–189.

Holland, P.W., Rosenbaum, P.R. (1986). Conditional association and unidimensionality in monotone latent variable models. *The Annals of Statistics* **14**, 1523–1543.

Huynh, H. (1994). A new proof for monotone likelihood ratio for the sum of independent Bernoulli random variables. *Psychometrika* **59**, 77–79.

Johnson, M.S., Junker, B.W. (2003). Using data augmentation and Markov Chain Monte Carlo for the estimation of unfolding response models. *Journal of Educational and Behavioral Statistics* **28**, 195–230.

Junker, B.W. (1991). Essential independence and likelihood-based ability estimation for polytomous items. *Psychometrika* **56**, 255–278.

Junker, B.W. (1993). Conditional association, essential independence and monotone unidimensional item response models. *The Annals of Statistics* **21**, 1359–1378.

Junker, B.W. (1996). Examining monotonicity in polytomous item response data. Paper presented at the Annual Meeting of the Psychometric Society, June 27–30, 1996, Banff, Alberta, Canada.

Junker, B.W. (2001). On the interplay between nonparametric and parametric IRT, with some thoughts about the future. In: Boomsma, A., van Duijn, M.A.J., Snijders, T.A.B. (Eds.), *Essays on Item Response Theory.* Springer-Verlag, New York, pp. 247–276.

Junker, B.W., Ellis, J.L. (1997). A characterization of monotone unidimensional latent variable models. *The Annals of Statistics* **25**, 1327–1343.

Junker, B.W., Sijtsma, K. (2000). Latent and manifest monotonicity in item response models. *Applied Psychological Measurement* **24**, 65–81.

Junker, B.W., Sijtsma, K. (2001a). Nonparametric item response theory in action: An overview of the special issue. *Applied Psychological Measurement* **25**, 211–220.

Junker, B.W., Sijtsma, K. (2001b). Cognitive assessment models with few assumptions, and connections with nonparametric item response theory. *Applied Psychological Measurement* **25**, 258–272.

Karabatsos, G. (2003). Comparing the aberrant response detection performance of thirty-six person-fit statistics. *Applied Measurement in Education* **16**, 277–298.

Karabatsos, G., Sheu, C.-F. (2004). Order-constrained Bayes inference for dichotomous models of unidimensional nonparametric IRT. *Applied Psychological Measurement* **28**, 110–125.

Loevinger, J. (1948). The technique of homogeneous tests compared with some aspects of "scale analysis" and factor analysis. *Psychological Bulletin* **45**, 507–530.

McDonnell Hartz, S. (2002). A Bayesian framework for the unified model for assessing cognitive abilities: Blending theory with practicality. Unpublished Ph.D. dissertation, University of Illinois at Urbana-Champaign, IL, USA.

McCutcheon, A.L. (2002). Basic concepts and procedures in single- and multiple-group latent class analysis. In: Hagenaars, J.A., McCutcheon, A.L. (Eds.), *Applied Latent Class Analysis.* Cambridge University Press, Cambridge, UK, pp. 56–85.

Meijer, R.R. (2003). Diagnosing item score patterns on a test using item response theory-based person-fit statistics. *Psychological Methods* **8**, 72–87.

Meijer, R.R., Baneke, J.J. (2004). Analyzing psychopathology items: A case for nonparametric item response theory modeling. *Psychological Methods* **9**, 354–368.

Meijer, R.R., Sijtsma, K. (2001). Methodology review: Evaluating person fit. *Applied Psychological Measurement* **25**, 107–135.

Meijer, R.R., Sijtsma, K., Molenaar, I.W. (1995). Reliability estimation for single dichotomous items based on Mokken's IRT model. *Applied Psychological Measurement* **19**, 323–335.

Mellenbergh, G.J. (1995). Conceptual notes on models for discrete polytomous item responses. *Applied Psychological Measurement* **19**, 91–100.

Meredith, W. (1965). Some results based on a general stochastic model for mental tests. *Psychometrika* **30**, 419–440.

Mokken, R.J. (1971). *A Theory and Procedure of Scale Analysis*. De Gruyter, Berlin.

Mokken, R.J., Lewis, C. (1982). A nonparametric approach to the analysis of dichotomous item responses. *Applied Psychological Measurement* **6**, 417–430.

Mokken, R.J., Lewis, C., Sijtsma, K. (1986). Rejoinder to the Mokken scale: A critical discussion. *Applied Psychological Measurement* **10**, 279–285.

Molenaar, I.W. (1973). Simple approximations to the Poisson, Binomial and Hypergeometric distributions. *Biometrics* **29**, 403–407.

Molenaar, I.W. (1997). Nonparametric model for polytomous items. In: van der Linden, W.J., Hambleton, R.K. (Eds.), *Handbook of Modern Item Response Theory*. Springer, New York, pp. 369–380.

Molenaar, I.W., Sijtsma, K. (2000). User's manual MSP5 for Windows. A program for Mokken scale analysis for polytomous items. ProGAMMA, Groningen, The Netherlands.

Nandakumar, R., Yu, F., Li, H.-H., Stout, W.F. (1998). Assessing unidimensionality of polytomous data. *Applied Psychological Measurement* **22**, 99–115.

Post, W.J. (1992). *Nonparametric Unfolding Models: A Latent Structure Approach*. DSWO Press, Leiden, The Netherlands.

Raju, N.S. (1988). The area between two item characteristic curves. *Psychometrika* **53**, 495–502.

Ramsay, J.O. (1991). Kernel smoothing approaches to nonparametric item characteristic curve estimation. *Psychometrika* **56**, 611–630.

Ramsay, J.O. (1996). A geometrical approach to item response theory. *Behaviormetrika* **23**, 3–16.

Ramsay, J.O. (1997). A functional approach to modeling test data. In: van der Linden, W.J., Hambleton, R.K. (Eds.), *Handbook of Modern Item Response Theory*. Springer, New York, pp. 381–394.

Ramsay, J.O. (2000). A program for the graphical analysis of multiple choice test and questionnaire data. Department of Psychology, McGill University, Montreal, Canada.

Reise, S.P., Henson, J.M. (2000). Computerization and adaptive administration of the NEO-PI-R. *Assessment* **7**, 347–364.

Reise, S.P., Meijer, R.R., Ainsworth, A.T., Morales, L.S., Hays, A.D. (2006). Application of group-level item response models in the evaluation of consumer reports about health plan quality. *Multivariate Behavioral Research*, in press.

Rosenbaum, P.R. (1987). Comparing item characteristic curves. *Psychometrika* **52**, 217–233.

Rossi, N., Wang, X., Ramsay, J.O. (2002). Nonparametric item response function estimates with the EM algorithm. *Journal of Educational and Behavioral Statistics* **27**, 291–317.

Roussos, L.A., Stout, W.F., Marden, J. (1998). Using new proximity measures with hierarchical cluster analysis to detect multidimensionality. *Journal of Educational Measurement* **35**, 1–30.

Santor, D.A., Ramsay, J.O. (1998). Progress in the technology of measurement: Applications of item response models. *Psychological Assessment* **10**, 345–359.

Santor, D.A., Ramsay, J.O., Zuroff, D.C. (1994). Nonparametric item analysis of the Beck depression inventory: Evaluating gender item bias and response option weights. *Psychological Assessment* **6**, 255–270.

Scheiblechner, H. (1995). Isotonic ordinal probabilistic models (ISOP). *Psychometrika* **60**, 281–304.

Shealy, R.T., Stout, W.F. (1993). An item response theory model for test bias and differential item functioning. In: Holland, P.W., Wainer, H. (Eds.), *Differential Item Functioning*. Erlbaum, Hillsdale, NJ, pp. 197–239.

Sijtsma, K. (1998). Methodology review: Nonparametric IRT approaches to the analysis of dichotomous item scores. *Applied Psychological Measurement* **22**, 3–31.

Sijtsma, K., Hemker, B.T. (1998). Nonparametric polytomous IRT models for invariant item ordering, with results for parametric models. *Psychometrika* **63**, 183–200.

Sijtsma, K., Junker, B.W. (1996). A survey of theory and methods of invariant item ordering. *British Journal of Mathematical and Statistical Psychology* **49**, 79–105.

Sijtsma, K., Meijer, R.R. (1992). A method for investigating intersection of item response functions in Mokken's nonparametric IRT model. *Applied Psychological Measurement* **16**, 149–157.

Sijtsma, K., Meijer, R.R. (2001). The person response function as a tool in person-fit research. *Psychometrika* **66**, 191–207.

Sijtsma, K., Molenaar, I.W. (1987). Reliability of test scores in nonparametric item response theory. *Psychometrika* **52**, 79–97.

Sijtsma, K., Molenaar, I.W. (2002). *Introduction to Nonparametric Item Response Theory*. Sage, Thousand Oaks, CA.

Sijtsma, K., Van der Ark, L.A. (2001). Progress in NIRT analysis of polytomous item scores: Dilemma's and practical solutions. In: Boomsma, A., van Duijn, M.A.J., Snijders, T.A.B. (Eds.), *Essays on Item Response Theory*. Springer, New York, pp. 297–318.

Snijders, T.A.B. (2001). Two-level non-parametric scaling for dichotomous data. In: Boomsma, A., van Duijn, M.A.J., Snijders, T.A.B. (Eds.), *Essays on Item Response Theory*. Springer, New York, pp. 319–338.

Stout, W.F. (1987). A nonparametric approach for assessing latent trait unidimensionality. *Psychometrika* **52**, 589–617.

Stout, W.F. (1990). A new item response theory modeling approach with applications to unidimensionality assessment and ability estimation. *Psychometrika* **55**, 293–325.

Stout, W.F. (2002). Psychometrics: From practice to theory and back. *Psychometrika* **67**, 485–518.

Stout, W.F., Habing, B., Douglas, J., Kim, H., Roussos, L., Zhang, J. (1996). Conditional covariance based nonparametric multidimensionality assessment. *Applied Psychological Measurement* **20**, 331–354.

Suppes, P., Zanotti, M. (1981). When are probabilistic explanations possible? *Synthese* **48**, 191–199.

Te Marvelde, J.M., Van der Ark, L.A., Sijtsma, K. (2005). A method to investigate invariant item ordering for polytomous items. Internal Report, Department of Methodology and Statistics FSW, Tilburg University.

Tutz, G. (1990). Sequential item response models with an ordered response. *British Journal of Mathematical and Statistical Psychology* **43**, 39–55.

Van Abswoude, A.A.H. (2004). Dimensionality assessment under nonparametric IRT models. Unpublished Ph.D. dissertation, Tilburg University, The Netherlands.

Van Abswoude, A.A.H., Van der Ark, L.A., Sijtsma, K. (2004a). A comparative study of test data dimensionality assessment procedures under nonparametric IRT models. *Applied Psychological Measurement* **28**, 3–24.

Van Abswoude, A.A.H., Vermunt, J.K., Hemker, B.T., Van der Ark, L.A. (2004b). Mokken scale analysis using hierarchical clustering procedures. *Applied Psychological Measurement* **28**, 332–354.

Van der Ark, L.A. (2005). Practical consequences of stochastic ordering of the latent trait under various polytomous IRT models. *Psychometrika* **70**, 283–304.

Van der Ark, L.A., Bergsma W.P. (2004). Investigation of invariant item ordering of polytomous items. Submitted for publication.

Van der Ark, L.A., Hemker, B.T., Sijtsma, K. (2002). Hierarchically related nonparametric IRT models, and practical data analysis methods. In: Marcoulides, G., Moustaki, I. (Eds.), *Latent Variable and Latent Structure Modeling*. Erlbaum, Mahwah, NJ, pp. 40–62.

Van der Flier, H. (1982). Deviant response patterns and comparability of test scores. *Journal of Cross-Cultural Psychology* **13**, 267–298.

Van Onna, M.J.H. (2002). Bayesian estimation and model selection in ordered latent class models for polytomous items. *Psychometrika* **67**, 519–538.

Van Onna, M.J.H. (2004). Estimates of the sampling distribution of scalability coefficient H. *Applied Psychological Measurement* **28**, 427–449.

Vermunt, J.K. (2001). The use of restricted latent class models for defining and testing nonparametric and parametric item response theory models. *Applied Psychological Measurement* **25**, 283–294.

Zhang, J., Stout, W.F. (1999a). The theoretical DETECT index of dimensionality and its application to approximate simple structure. *Psychometrika* **64**, 213–249.

Zhang, J., Stout, W.F. (1999b). Conditional covariance structure of generalized compensatory multidimensional items. *Psychometrika* **64**, 129–152.

Handbook of Statistics, Vol. 26
ISSN: 0169-7161

DOI: 10.1016/S0169-7161(06)26023-1

23

Automatic Item Generation and Cognitive Psychology

Susan Embretson and Xiangdong Yang

1. Introduction

In automatic item generation, optimally informative items for an examinee are developed during the course of testing. That is, prior to the test, the items do not exist. Instead, the items are generated on the fly. Since the generated items are new, they have not been subjected to empirical tryout. Consequently, no calibrated parameters are available. Instead, item parameters are predicted from a psychometric model based on the relationship of item stimulus features or family membership to item parameters. Automatic item generation may seem futuristic, but in fact several projects have already demonstrated feasibility (Bejar, 1996; Bejar et al., 2003; Embretson, 1999, 2002a, 2002b; Deane et al., 2005).

Automatic item generation has been an interesting possibility, in part, for its potential to solve some pressing practical problems in test development. That is, large numbers of new items are continually needed for many testing programs. For high-stakes testing, such as tests for selection, placement and certification, continual supplies of new items are needed to counter threats to test validity. Item familiarity may change what is measured by the test. Problems with item overexposure or security are commonplace in high-stakes testing. For computerized adaptive testing (CAT), which has become the state of the art in psychological and educational measurement, especially large item banks are often required. For example, for many tests, examinees must be measured with comparable levels of precision over a broad range of ability. Sufficient representation of items at the various trait levels is required, which consequently leads to large item banks.

Automatic item generation requires both appropriate psychometric models and substantive research to replace calibrated item parameters with predictions. Appropriate psychometric models are available as generalized item response theory (IRT) models. Some models permit item parameters to be predicted from scored features or from class memberships of individual items.

The substantive research foundation for automatic item generation rests heavily on cognitive psychology. Irvine and Kyllonen's (2002) edited volume from an international conference includes item generation research for items involving both simple and complex cognitive processes. Two methods are currently available for automatic item generation; the item model approach and the item structure approach. The nature and strength of impact of cognitive theory depends on which approach is used.

It should be noted that cognitive psychology has influenced many aspects of ability and achievement testing in addition to automatic item generation. That is, cognitive psychology has been a basis for defining new measurement constructs (e.g., Kyllonen and Christal, 1990; Das and Naglieri, 1997), providing a basis for diagnostic assessment (e.g., Sheehan, 1997; Tatsuoka, 1985) and providing a basis for automated scoring (e.g., Bennett and Bejar, 1998). Embretson and Gorin (2001) present a more detailed elaboration of these other areas of impact.

The focus of this chapter is item generation and the role of IRT models that permit cognitive variables to predict item parameters. The chapter is organized into several sections. Section 2 presents an overview of the methods of item generation and the research requirements for application. Section 3 presents some appropriate psychometric models that are based on item response theory (IRT). Section 4 presents some estimation procedures for the psychometric models. Section 5 presents an example of automatic item generation to spatial ability. Finally, Section 6 presents a summary.

2. Item development and automatic item generation

In this section, three methods of item development are described: traditional, item models and cognitive design systems. Then, the foundations required for successful automatic item generation are reviewed.

Methods of item development

Traditional item development

For most tests, items are carefully handcrafted by human item writers. The process can be more appropriately described as an art, rather than as a science. Item specifications guide the construction of items; however, the specifications do not rigorously prescribe item content. That is, a wide variety of items may fit into the same general content considerations (e.g., Topic Area) or vaguely described processing levels (e.g., Abstract vs. Concrete). Once the items are produced, reviews by committees are undertaken to assure the justifiability of the keyed answer and to examine item content for various equity issues. Then, items are subjected to empirical tryout to determine item quality. Items that are construct-relevant have certain expected patterns of correlations with other items or subtests. Items that do not meet the expected pattern are eliminated. High levels of item attrition are fairly typical. Then, of course, item parameters are estimated using an appropriate psychometric model, given that sufficient data are available. Finally, the items can be entered into the item bank.

Although traditional item development procedures have yielded the majority of available tests, unfortunately these procedures have been unable to keep pace with the huge demand for new items. Adding new items to the item bank typically entails considerable time, often one or more years, and considerable expense.

Table 1
An example of the item model approach

Mary flew on an airplane 5,000 miles at 500 miles per hour. How many hours did she travel?
a. 2
b. 5
c. 10
d. 100
Substitutions
5,000 = [500, 1,000, 1,500, etc.]
500 = [10, 50, 100, 200, 300, etc.]
airplane = [sailboat, car, sled, skateboard, motorcycle, space ship, etc.]
miles = [kilometers, lengths, feet, etc.]
flew = [rode, sailed, etc.]
Mary [Agnes, Indio, Jamal, Robin, etc.]

Item models

In the item model approach (e.g., Bejar et al., 2003), an existing item with adequate psychometric qualities serves as a model for new items. That is, one or more features become variables by permitting substitutions. The features that are substituted are viewed as irrelevant to the problem-solving process. Thus, a family of new items is created. Ability can then be estimated from the new items, without empirical item tryout, if the properties of the item model are inheritable to each new item, as hypothesized.

Singley and Bennet (2002) describe the Math Test Creation Assistant which allows various combinations of substitutions for mathematical problem-solving items. Table 1 presents a hypothetical item model for problem solving. The terms with the bold underlines, 500 and 5,000, may be substituted for other combinations of numerical values and the alternatives adjusted accordingly. Although many new items could be generated, the stem would vary little. Further substitutions could involve the character, the mode of transportation and the type of rate. However, notice that with some substitutions, such as the mode of transportation, other features of the problem must change as well. For example, if "boat" is substituted for "airplane" than the verb must change from "flew" to "sailed". Also, the rate probably should change, to be realistic.

Cognitive psychology is involved in two ways in the item model approach. First, the range of permissible substitutions in the construct-relevant variables (e.g., the numbers) must be constrained so as to not change processing substantially. For example, introducing combinations of rate and distance that involve more laborious calculations would increase difficulty quantitatively. Research on processing involved in arithmetic operations is relevant to the specified range. Second, psycholinguistic networks and inheritances are needed to determine what combinations of substitutions make sense, such as the mode of transportation and the appropriate verb for traveling. Recently, Deane et al. (2005) have made progress with a natural language item generator for quantitative reasoning items based on psycholinguistic principles to enter appropriate constraints.

Cognitive design system approach

In the cognitive design system approach (Embretson, 1998), item structures are based directly on an information-processing theory of the item type. A cognitive theory speci-

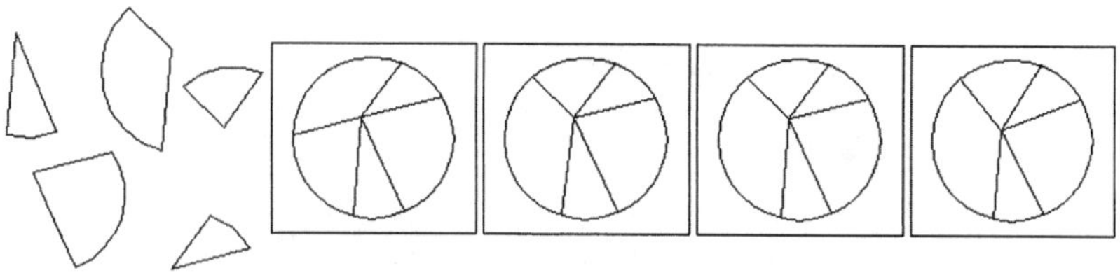

Fig. 1. A spatial ability item.

fies processes in item solution, the impact of processes on performance and the impact of stimulus features on processes. Thus, the item stimulus properties are related to processing difficulties, which in turn, are related to item difficulty and other item psychometric properties. Although the primary psychometric property is item difficulty, item discrimination may be influenced by the processes as well. For example, peripheral processes that are not central to item difficulty may lower item discrimination.

An item structure is defined as a unique combination of the processing-related features. Consider the spatial ability item shown in Figure 1. The task for the examinee is to determine which of four constructed figures can be obtained by rearranging and rotating the pieces in the stem. According to an empirically-supported theory (Embretson, 2000; Embretson and Gorin, 2001), processing involves encoding, falsifying perceptually inappropriate alternatives and confirming the key. In turn, stimulus features such as the number of pieces in the stem, the number of alternatives with perceptually inappropriate pieces, rotation and displacement of the pieces, and so forth determine the difficulty of the processes. Other features, such as the overall shape of the construct object (i.e., circle), are hypothesized to have little or no impact on processing. Thus, an item structure is defined by combinations of the processing-relevant variables. Given a range of outer shapes, centerpoint and radiant positions and so forth, infinitely many items can be constructed for the same structure.

Steps to automatic item generation

The item model approach requires existing items with established validity, while the cognitive design system approach may be applied either to new measures or existing measures. For existing measures, adequate validity is already established while for new measures, an extra stage to establish validity is required. However, for new measures, algorithmic item generation can occur earlier in the process. In algorithmic item generation, a human item writer generates items to a rigorous set of specifications (see Embretson, 1998 for a more detailed treatment in the context of the cognitive design system approach). Automatic item generation goes a step further; a computer program writes the items according to specifications.

Understanding sources of item variation

In the initial stages of research, the goal is to develop an understanding of sources of item variation. For the item model approach, the impact of stimulus variations on psychometric properties must be studied. Significant sources of variation between similar items must be controlled. For the cognitive design system approach, a plausible cognitive processing model for the existing ability test items must be developed. The goal is to understand sources of item variation in the item bank as a whole. For either approach, studies typically begin with a literature review to examine sources of differences in item difficulty. For the cognitive design system approach, a cognitive model of the underlying processing components and the stimuli that determine their difficulty is sought. However, cognitive tasks that are studied in the laboratory typically differ from items that appear on tests; laboratory tasks are often quite easy and presented in a simple verification format rather than a multiple choice format. Thus, a more complex model may need to be postulated to adequately represent processing for solving ability test items.

The next step is to empirically support the variables identified as impacting item variation. For the item model approach, item variations in the identified variables can be correlated with psychometric properties to determine impact. For the cognitive design system approach, data on two primary dependent variables on the ability test items must be obtained: item difficulty and item response time. Although item difficulty is the primary interest, response time modeling is needed also to support a processing model. The dependent variables are mathematically modeled from item stimulus features that are postulated to impact processing. The cognitive design system approach may involve several related studies for several reasons. First, item stimulus features on existing tests are usually incidentally correlated due to existing item development practices. This produces multicollinearity among the independent variables, which leads to complexity in assessing the relative importance of a variable. Thus, additional studies are needed to unconfound the correlated features. Second, since any one method of developing a theory has limitations, converging operations for supporting the model are needed. For example, eyetracker studies and simulation studies also provide information about the plausibility of the processing model.

Algorithmic item generation and revised cognitive model

The next stage directly concerns test design. For the item model approach, the issue is to determine if variations from the model lead to item differences in psychometric properties. Substitutions to the item model are created and submitted to empirical tryout. The greatest success is obtained when the item variations lead to no significant changes in item psychometric properties. For the cognitive design system approach, the issue is to determine if manipulated stimulus features impact processing difficulty and item performance. The item features that are manipulated are based on the cognitive model variables from the preceding stage. Previously correlated features are manipulated independently by crossing the various levels of the stimulus features. For example, in the spatial ability item in Figure 1, the rotation of the pieces in the key (binary) can be crossed with the number of perceptually-falsifiable distractors.

The newly constructed items are then studied empirically to determine the impact of the stimulus design features on item performance. Although new items can be calibrated in a tryout, the main focus is on impact of the manipulations. For the item model approach, items that are derived from the same item model should have highly similar empirical properties. For the cognitive model approach, the design features should be sufficiently predictive of item difficulty and other psychometric indicators, as well as response time.

Item generation by algorithms
As noted above, a computer program must be developed for item generation. Considerable effort may be required to develop an item generator program and an item administration program linked to the generator, as described below. However, even more crucial are the substantive features to be embedded in the item generator. For the item model approach, the features that can be substituted must be identified and a list of permissible substitutions must be developed. Further, constraints among permissible substitutions must be specified, as illustrated above in the problem-solving example. For the cognitive design system approach, item structures must be defined and sampled stimulus content must be specified. An item structure is defined by a unique combination of the values for all variables in the cognitive model. Thus, items from the same structure carry the same sources and levels of cognitive complexity, and consequently the same predicted psychometric properties. Content features not included in the cognitive model may be varied. For example, shape of the assembled object, position of the centerpoint and section for the attachment of the radiant may be varied for the spatial ability items.

Empirical tryout of item generation
The final stage involves actual tryout of generated items. Online testing is essential because continuous data are needed to evaluate the quality of the design principles, as new combinations of features emerge. New psychometric issues arise with adaptive item generation, as well. First, new diagnostic measures are needed to evaluate the design features, and their various combinations, for yielding items of predicted psychometric quality. Since the design features probably will be contained in IRT models, as noted above, perhaps current indices for item fit can be extended to assess the design features. For the item model approach, hierarchical IRT models provide estimates of variability among the item isomorphs. High quality items have been automatically generated from the item models and pretested in large-scale assessments (Sinharay and Johnson, 2005). Second, further research is needed on how uncertainty in the item parameters (i.e., because they are predicted from design features) impacts ability estimates. Several studies (Bejar et al., 2003; Embretson, 1999; Mislevy et al., 1993) have found that measurement error increases modestly when item parameters are predicted rather than calibrated. These studies further suggest that the impact of item uncertainty can be readily countered by administering a few more items. However, in the context of online testing, it may be possible to monitor individual examinees for the impact of item uncertainty. Research on indices to diagnose online problems is also needed.

Computer programs for adaptive item generation
Adaptive item generation requires two types of computer programs: (1) an item generator program that actually creates the items and (2) an adaptive testing program that

can be interfaced with the generator. For the item model approach, the generator program produces item variants from the item model with lists of substitutions. For the cognitive design system approach, the item generator program produces items to target levels and sources of cognitive complexity in the cognitive design system approach. Item structures, which are blueprints for the items, are essential for item production. The structures carry the specific sources of cognitive complexity and predicted item difficulty.

The adaptive testing administration program must then be interfaced with the item generator. For the item model approach, a fully compatible integration of a generator and a test administration program has been accomplished (see Bejar et al., 2003). For the cognitive design system approach, although both the test administration program and a generator for several item types have been developed, they have not yet been interfaced (Embretson, 2000).

3. Item response models for automatic item generation

Special psychometric models are now available to model items that are generated from the automatic item generation approach. In the cognitive design system approach, since item stimulus features are incorporated in item generation to affect the cognitive processes, strategies, and knowledge structures, the psychometric model to be used must be capable of incorporating such cognitive variables in the model. Many item response theory (IRT) models have been developed for this purpose (Adams et al., 1997; Embretson, 1984, 1985, 1999; Fischer, 1973; Maris, 1995). In general, cognitive IRT models provide at least two important contributions to automatic item generation: (1) the goodness of fit of the resulting model can be used to evaluate the cognitive theory underlying the item generation (Embretson, 1998), and (2) parameters for the set of cognitive variables provide an approach to quantify the sources of item difficulty on which subsequent item banking can be relied.

In this section, we present three cognitive IRT models that are relevant to automatic item generation. Two models are especially useful for the cognitive design system approach while the last model is especially useful for the item model approach. More extended treatments of cognitive IRT models are available in textbooks such as De Boeck and Wilson (2004).

Linear logistic test model

The linear logistic test model (LLTM; Fischer, 1973) is an extension to the unidimensional Rasch (1960) model to incorporate item stimulus features into the prediction of item success. Let X_{ij} be the response to item j, $j = 1, 2, \ldots, J$, by the examinee i, $i = 1, 2, \ldots, N$, with $X_{ij} = 1$ for a correct response and 0 otherwise, the Rasch model is given as

$$P(X_{ij} = 1|\theta_i, b_j) = \frac{\exp(\theta_i - b_j)}{1 + \exp(\theta_i - b_j)}, \tag{1}$$

where b_j is the difficulty of item j and θ_i is the latent ability of examinee i. In Eq. (1), b_j is an item-specific parameter and doesn't connect to cognitive variables involved in the item. To incorporate item stimulus features into the model, the LLTM decomposed the item difficulty parameter b_j into a linear combination of K "basic" parameters $(K \leqslant J)$. The LLTM is given as

$$P(X_{ij} = 1|\theta_i, \mathbf{q}_j, \boldsymbol{\eta}) = \frac{\exp(\theta_i - \sum_{k=1}^{K} q_{jk}\eta_k)}{1 + \exp(\theta_i - \sum_{k=1}^{K} q_{jk}\eta_k)}, \tag{2}$$

where $\mathbf{q}_j$ is the row vector of prespecified weights of the K variables involved in item j, $\mathbf{q}_j = \{q_{j1}, q_{j2}, \dots, q_{jK}\}$, $\boldsymbol{\eta}$ is the vector of the basic parameters of the K variables or the "effects" of the K variables on the psychological complexity of item j, $\boldsymbol{\eta}' = \{\eta_1, \eta_2, \dots, \eta_K\}$, where $\boldsymbol{\eta}'$ is the transpose of vector $\boldsymbol{\eta}$. It is assumed that q_{j1} is set to be 1 so that η_1 is an intercept.

In cognitive tasks, the K variables are interpreted as K different cognitive operations or processes in an item, q_{jk} can be referred to as the number of times the kth operation is required to solve the item. η_k can be referred to as the difficulty of the kth cognitive operation or process. The tabular arrangement between the set of items (usually in rows) and the set of cognitive components (usually in columns) is called the Q matrix, in which q_{jk} specifies the value of item j taking on the kth cognitive variable. Q is a (J, K) matrix with rank K.

In general, b_j will not equal $\sum_{k=1}^{K} q_{jk}\eta_k$ as the prediction will not be perfect in practice. Since b_j is item-specific and $K \leqslant J$, the LLTM is in fact a special case of the Rasch model. The Rasch model is the saturated model of the former. As such, one way to assess the fit of the LLTM to response data from automatic generated items is to compare its fit to that of the Rasch model via standard likelihood-ratio test. The degree of freedom of such test is the difference of the number of parameters between the two models.

The LLTM also allows interactions between the K variables. A new variable can be created as the product of the two variables that interact and added to the Q matrix.

2PL-constrained model

Research has shown that, not only item difficulties, but also item discriminations, can be predictable from the item design features (Embretson, 1999). Although LLTM estimates the impact of item design features on item difficulty, it does not allow varying item discriminations across items since it belongs to the Rasch model family. To resolve this problem, Embretson (1999) proposed an IRT model that is a generalization of the two parameter logistic (2PL) model. This model, referred to as 2PL-constrained model, is capable of incorporating item design features for both item difficulties and discriminations. Let a_j be the item discrimination parameter for item j, the 2PL model is given as

$$P(X_{ij} = 1|\theta_i, a_j, b_j) = \frac{\exp[a_j(\theta_i - b_j)]}{1 + \exp[a_j(\theta_i - b_j)]}. \tag{3}$$

Similar to LLTM, a linear constraint is imposed on a_j and the resulting 2PL-constrained model can then be written as

$$P(X_{ij} = 1|\theta_i, \mathbf{q}_j, \boldsymbol{\eta}, \boldsymbol{\tau}) = \frac{\exp\left[\sum_{k=1}^{K} q_{jk}\tau_k(\theta_i - \sum_{k=1}^{K} q_{jk}\eta_k)\right]}{1 + \exp\left[\sum_{k=1}^{K} q_{jk}\tau_k(\theta_i - \sum_{k=1}^{K} q_{jk}\eta_k)\right]}, \tag{4}$$

where $\boldsymbol{\tau}$ is the vector of the basic parameters of the K variables on the discrimination of item j, $\boldsymbol{\tau}' = \{\tau_1, \tau_2, \ldots, \tau_K\}$. Similarly, τ_k can be interpreted as the "effect" of the kth variable on the discrimination power of the item j. As for LLTM, it is assumed that q_{j1} is set to be 1 so that τ_1 and η_1 are intercepts.

The 2PL-constrained model is a special case of the 2PL model with the latter as the saturated model of the former. Again, likelihood ratio test can be conducted to assess the relative fit of the 2PL-constrained model to the 2PL model. In fact, the 2PL-constrained model is also a generalization of the LLTM. As such, sequential model comparisons can be conducted in practice to facilitate model selection.

Hierarchical IRT model for item structure

In automatic item generation, items are generated from a set of developed item structures, each of which is a specified combination of the item stimulus features. Since item stimulus features are mapped onto the cognitive variables underlying the item solution and calibrated through psychometric models as mentioned above, item properties (both cognitive and psychometric) have been specified in each item structure. Other approaches to item generation, such as generating families of item isomorphs from a set of "item templates", share the similar rationale although the linkage between item isomorphs and the templates is established at a relatively shallower level. Empirical studies have been conducted to address the important question of whether items that are generated from the same item template or structure have comparable psychometric characteristics (Embretson, 1998; Enright et al., 2002). While items within structure show similar psychometric characteristics in general, the variability among those items is still noticeable to some degrees.

Glas and van der Linden (2003) proposed a two-level IRT model to incorporate the variability of within-structure item parameters by assuming that the within-structure items are randomly sampled from the structure. Let j_p, $j = 1, 2, \ldots, J_p$, $p = 1, 2, \ldots, P$, be the jth item within the pth item structure, the first-level model is the common three parameter logistic (3PL) model, which is given as

$$P(X_{ij_p} = 1|\theta_i, a_{j_p}, b_{j_p}, c_{j_p}) = c_{j_p} + (1 - c_{j_p})\frac{\exp[a_{j_p}(\theta_i - b_{j_p})]}{1 + \exp[a_{j_p}(\theta_i - b_{j_p})]}, \tag{5}$$

where X_{ij_p} be the response to the jth item within the pth item structure by examinee i and $a_{j_p}, b_{j_p}, c_{j_p}$ are the corresponding item discrimination, difficulty and guessing parameters for the item, respectively. Let $\boldsymbol{\xi}_{j_p} = \{\log a_{j_p}, b_{j_p}, \operatorname{logit} c_{j_p}\}$ be the transformed item parameter vector for the item j_p, the second-level model assumes that $\boldsymbol{\xi}_{j_p}$ has a multivariate normal distribution

$$\boldsymbol{\xi}_{j_p} \sim N(\boldsymbol{\mu}_p, \boldsymbol{\Sigma}_p), \tag{6}$$

where $\boldsymbol{\mu}_p$ is the vectors that contain the mean values of the item parameters in the pth item structure and $\boldsymbol{\Sigma}_p$ is the corresponding covariance matrix. The transformations of the item discrimination and guessing parameters were proposed such that the assumption of multivariate normality can hold.

Although it is not obvious, the two-level IRT model is in fact a generalization of the 2PL-constrained model. Since each item structure represents a particular combination of the item stimulus features, a particular item structure p, $p = 1, 2, \ldots, P$, can then be symbolically represented by a row vector $\mathbf{d}_p = \{d_{p1}, d_{p2}, \ldots, d_{pK}\}$, where d_{pk} specifies the value of item structure p taking on the kth cognitive variable. The collection of $\mathbf{d}_p$, $p = 1, 2, \ldots, P$, can be represented as $\mathbf{D}_{P\times K}$. Let $\mathbf{S}_{J\times P}$ be the matrix whose element s_{jp} specifies whether item j belongs to the item structure p or not, with $s_{jp} = 1$ for a positive response and 0 otherwise. Further, let $\boldsymbol{\Lambda}_{K\times M}$ be the matrix whose element λ_{km} is the basic parameter of the kth cognitive variable on the mth item parameter. In a 3PL model, $M = 3$, and, without loss of generality, let the 1st, 2nd and 3rd parameter be item discrimination, difficulty and guessing parameter, respectively. Then, it is easy to show that

$$\mathbf{Q}_{J\times K} = \mathbf{S}_{J\times P}\mathbf{D}_{P\times K}, \tag{7}$$

where $\mathbf{Q}_{J\times K}$ is the Q matrix specified in both LLTM and the 2PL-constrained model and

$$\boldsymbol{\xi}_{j_p} = \boldsymbol{\Lambda}'_{K\times M}\mathbf{d}'_p + \boldsymbol{\varepsilon}_{j_p}, \tag{8}$$

where $\boldsymbol{\varepsilon}_{j_p} \sim N(\mathbf{0}_p, \boldsymbol{\Sigma}_p)$. Several IRT models can be derived by imposing different constraints in the two-level IRT model as above. For example, by constraining the guessing parameter to be 0 for each item, a random-effect 2PL-constrained model can be obtained. If further item discrimination parameter is constrained to be 1 for each item, a random-effect LLTM is obtained. In both cases, within-structure variability can be modeled as structure-specific, as shown in Eq. (8), or as common across structures. Therefore, the two-level IRT model generalizes the 2PL-constrained model in two aspects. First, the two-level IRT model assumes that, not only item discrimination and difficulty parameters, but also the guessing parameter, can be predicted from the set of item design features. Second, the two-level IRT model recognizes that the prediction from the set of the design features is not perfect, even if they explain most of the variation in the item parameters, and incorporates an item-specific random error term to improve the model fit. This formulation highlights a random sampling interpretation of the items with the same item structure, which is consistent with the rationale of the automatic item generation approach.

4. Calibration of the cognitive IRT models

Calibration of the parameters in the cognitive IRT models plays an essential role in their applications. Commonly used methods include the joint maximum likelihood estimation (JMLE; Birnbaum, 1968), the conditional maximum likelihood estimation (CMLE; Anderson, 1970, 1972) and the marginal maximum likelihood estimation (MMLE;

Bock and Lieberman, 1970). In JMLE, both the items and ability parameters in the models are simultaneously estimated. However, parameter estimates from JMLE may suffer the issue of inconsistency since the number of parameters to be estimated increases as the number of observations (examinees and/or items) increases. This is the problem that were discussed by Neyman and Scott (1948) using the term *structural parameters* and *incidental parameters*. In the present context, the item parameters are structural parameters and the examinees' abilities are the incidental parameters. Alternatively, in CMLE, the minimal sufficient statistics of the incidental parameters are substituted for the incidental parameters themselves to resolve this problem. However, in IRT models, sufficient statistics for the examinees' abilities are available only for Rasch model and its extensions such as the LLTM. Therefore, alternative methods have to be developed for other IRT models.

The MMLE provides a general approach to estimate the structural parameters by integrating over the incidental parameter distributions. Since it is applicable to any type of IRT models, this section presents calibration of the aforementioned IRT models using MMLE. Prior to De Boeck and Wilson (2004) developments, MMLE was available only in specialized IRT programs for each cognitive model. Mostly, these programs were not generally available and those that were available did not include the whole family of cognitive models. Recently, De Boeck and Wilson (2004) linked the family of explanatory item response models to generalized linear and nonlinear models, which thus permitted readily accessible estimation methods from widely available statistical software (e.g., SAS). Empirical estimation procedures, such as MCMC simulation (Spiegelhalter et al., 2003) can be applied to most cognitive IRT models. However, they are not widely available and are rather slow to converge. In this section we review the methods that can be applied to generalized linear and nonlinear models.

Let $\mathbf{x}_i$ be the response pattern of examinee i to the set of J items, $\mathbf{x}_i = \{X_{i1}, X_{i2}, \ldots, X_{iJ}\}, i = 1, 2, \ldots, N$, the conditional probability of observing $\mathbf{x}_i$ can the generally denoted as $P(\mathbf{x}_i|\theta_i; \mathbf{Q}, \boldsymbol{\Lambda}, \mathbf{E})$, where θ_i, $\mathbf{Q}$, and $\boldsymbol{\Lambda}$ are defined as before. $\mathbf{E}$ is a stacked vector of the random error ε_{mj_p} which is associated with the mth parameter of the jth item with item structure p. $\mathbf{E}$ has MJ rows with M being the number of columns in $\boldsymbol{\Lambda}$.

Specifically, for LLTM, $M = 1$, $\boldsymbol{\Lambda} = [\boldsymbol{\eta}]$, and $\mathbf{E}$ is a zero vector. On the usual local independence assumption,

$$P(\mathbf{x}_i|\theta_i; \mathbf{Q}, \boldsymbol{\Lambda}, \mathbf{E}) = \prod_{j=1}^{J} P(X_{ij} = 1|\theta_i, \mathbf{q}_j, \boldsymbol{\eta})^{x_{ij}} \big(1 - P(X_{ij} = 1|\theta_i, \mathbf{q}_j, \boldsymbol{\eta})\big)^{1-x_{ij}}, \tag{9}$$

where $P(X_{ij} = 1|\theta_i, \mathbf{q}_j, \boldsymbol{\eta})$ is given by Eq. (2). Similarly, for the 2PL-constrained model, $M = 2$, $\boldsymbol{\Lambda} = [\boldsymbol{\tau}, \boldsymbol{\eta}]$, and $\mathbf{E}$ is a zero vector as well. $P(X_{ij} = 1|\theta_i, \mathbf{q}_j, \boldsymbol{\eta})$ in Eq. (9) is replaced by $P(X_{ij} = 1|\theta_i, \mathbf{q}_j, \boldsymbol{\eta}, \boldsymbol{\tau})$, which is given by Eq. (4).

For the two-level hierarchical IRT model, $M = 3$, $\boldsymbol{\Lambda}$ is defined as before, and, for all items j belonging to the item structure p, $\boldsymbol{\varepsilon}_{j_p} \sim N(\mathbf{0}_p, \boldsymbol{\Sigma}_p)$. Then, $P(X_{ij} = 1|\theta_i, \mathbf{q}_j, \boldsymbol{\eta})$ in Eq. (9) is replaced by $P(X_{ij_p} = 1|\theta_i, \mathbf{q}_{j_p}, \boldsymbol{\Lambda}, \boldsymbol{\varepsilon}_{j_p})$, which is given by combining Eqs. (5) and (8).

Under the two-level hierarchical IRT model, the marginal probability of observing response pattern $\mathbf{x}_i$ is given by

$$P(\mathbf{x}_i|\mathbf{Q}, \boldsymbol{\Lambda}, \boldsymbol{\Sigma}) = \int \cdots \int P(\mathbf{x}_i|\theta_i; \mathbf{Q}, \boldsymbol{\Lambda}, \boldsymbol{\Sigma}, \mathbf{E}) P(\theta_i) P(\mathbf{E}|\boldsymbol{\Sigma})\, \mathrm{d}\theta_i\, \mathrm{d}\mathbf{E}, \quad (10)$$

where $\boldsymbol{\Sigma}$ is a block diagonal matrix with order MP. Each block of $\boldsymbol{\Sigma}$ is $\boldsymbol{\Sigma}_p$. To resolve the indeterminacy of the scales in the item and ability parameters, the scale for θ_i is conventionally set as $N(0, 1)$, although other distributions or scales are possible. In MMLE, θ_i is viewed as the latent ability of a random examinee sampled from a population, the subscript i is then dropped in subsequent expressions.

From Eq. (10), the marginal likelihood function for the sample is given as

$$L = \prod_{i=1}^{N} P(\mathbf{x}_i|\mathbf{Q}, \boldsymbol{\Lambda}, \boldsymbol{\Sigma}). \quad (11)$$

For an item parameter u_t, the tth element of the stacking vector of the parameters in $\boldsymbol{\Lambda}$ and $\boldsymbol{\Sigma}$, the marginal log-likelihood equation can be written as

$$\begin{aligned} \frac{\partial \log L}{\partial u_t} = 0 &= \sum_{i=1}^{N} \frac{\partial}{\partial u_t}\left[\log P(\mathbf{x}_i|\mathbf{Q}, \boldsymbol{\Lambda}, \boldsymbol{\Sigma})\right] \\ &= \sum_{i=1}^{N} \left[P(\mathbf{x}_i|\mathbf{Q}, \boldsymbol{\Lambda}, \boldsymbol{\Sigma})\right]^{-1} \frac{\partial}{\partial u_t} \\ &\quad \times \left[\int \cdots \int P(\mathbf{x}_i|\theta; \mathbf{Q}, \boldsymbol{\Lambda}, \boldsymbol{\Sigma}, \mathbf{E}) P(\theta) P(\mathbf{E}|\boldsymbol{\Sigma})\, \mathrm{d}\theta\, \mathrm{d}\mathbf{E}\right]. \end{aligned} \quad (12)$$

For LLTM and the 2PL-constrained model, Eq. (12) can be simplified as

$$\frac{\partial \log L}{\partial u_t} = \sum_{i=1}^{N} \left[P(\mathbf{x}_i|\mathbf{Q}, \boldsymbol{\Lambda})\right]^{-1} \frac{\partial}{\partial u_t} \int P(\mathbf{x}_i|\theta; \mathbf{Q}, \boldsymbol{\Lambda}) P(\theta)\, \mathrm{d}\theta = 0. \quad (13)$$

In implementation, the set of unknown parameters in $\boldsymbol{\Lambda}$ and/or $\boldsymbol{\Sigma}$ is estimated by solving Eqs. (12) and (13).

Two important issues have to be resolved in order to solve Eqs. (12) and (13). First, the integrals appeared in both equations do not have closed-form solutions and approximation methods may have to be applied. Second, both equations can not be solved analytically even after the approximation to the integration has been resolved and numerical optimization techniques are required to estimate the parameters.

Many approaches to approximate the marginal log-likelihood have been proposed (Pinheiro and Bates, 1995). One common approach is the Gauss–Hermite quadrature approximation (Abramowitz and Stegun, 1964). In this approach, the integral involved in the marginal loglikelihood equation is approximated numerically by a weighted average of the integrand evaluated at a set of predetermined quadrature points (nodes). For example, the Gaussian quadrature approximation to the term that involves integral in

Eq. (13) is given as

$$\sum_{r=1}^{R} P(\mathbf{x}_i | X_r; \mathbf{Q}, \boldsymbol{\Lambda}) W(X_r), \tag{14}$$

where X_r is the rth quadrature point (node) and $W(X_r)$ is the corresponding weight. For Eq. (12) in which multiple integrals are involved, a Cartesian product rule can be used to approximate the marginal loglikelihood through multidimensional Gaussian quadrature points (Naylor and Smith, 1982). Adequate approximation can usually be obtained with an adequate number of quadrature points with appropriate centering and scaling.

To illustrate the approaches of centering and scaling of quadrature points, Eq. (13) is manipulated by applying the following relation

$$\frac{\partial}{\partial u_t} P(\mathbf{x}_i | \theta; \mathbf{Q}, \boldsymbol{\Lambda}) = P(\mathbf{x}_i | \theta; \mathbf{Q}, \boldsymbol{\Lambda}) \frac{\partial}{\partial u_t} \bigl[\log P(\mathbf{x}_i | \theta; \mathbf{Q}, \boldsymbol{\Lambda})\bigr], \tag{15}$$

which leads to

$$\begin{aligned} \frac{\partial \log L}{\partial u_t} &= \sum_{i=1}^{N} \int \frac{\partial}{\partial u_t} \bigl[\log P(\mathbf{x}_i | \theta; \mathbf{Q}, \boldsymbol{\Lambda})\bigr] \left[\frac{P(\mathbf{x}_i | \theta; \mathbf{Q}, \boldsymbol{\Lambda}) P(\theta)}{P(\mathbf{x}_i | \mathbf{Q}, \boldsymbol{\Lambda})} \right] \mathrm{d}\theta \\ &= \sum_{i=1}^{N} \int \frac{\partial}{\partial u_t} \bigl[\log P(\mathbf{x}_i | \theta; \mathbf{Q}, \boldsymbol{\Lambda})\bigr] P(\theta | \mathbf{x}_i, \mathbf{Q}, \boldsymbol{\Lambda}) \, \mathrm{d}\theta, \end{aligned} \tag{16}$$

where $P(\theta | \mathbf{x}_i, \mathbf{Q}, \boldsymbol{\Lambda})$ is the posterior distribution of the latent ability in the model, conditional on the response data, the design matrix and the current estimates of the parameters. One way to centering and scaling the quadrature points is based on the first two moments of this posterior distribution of θ. In this case, only the first and second moments of the posterior distribution need to be estimated and the centering and scaling of the quadrature points is identical to each examinee. An alternative approach, called adaptive Gaussian quadrature by Pinheiro and Bates (1995), is to center and scale the set of quadrature points individually. In this approach, the empirical Bayes estimate of θ_i, denoted as $\hat{\theta}_i$, and its associated asymptotic variance are estimated for examinee i and the set of quadrature points is recentered at $\hat{\theta}_i$ and rescaled by the estimated variance of $\hat{\theta}_i$.

An alternative to approximate the marginal loglikelihood function is to approximate the integrand, instead of the integral, such that the integration of the resulting expression from the approximation to the integrand has a closed form. One approach is the Laplace's method for integral (Tierney and Kadane, 1986). In this approach, the integrand of the marginal likelihood function is reformulated in an exponential form and is approximated by a second-order Taylor series expansion to the integrand around the current estimate of the unknown parameter that maximizes the log-likelihood function. The resulting expression is proportional to a normal distribution whose integral can be solved explicitly. Pinheiro and Bates (1995) showed that the Laplace's method of approximation is a special case of the adaptive Gaussian quadrature approximation, where only one node is used. Another related approach is to use first-order Taylor series expansion to linearize the nonlinear function in the

IRT model and then apply the estimation routine developed for linear mixed models to estimate the unknown parameters in the model. In this approach, two closely related approximate methods are called penalized quasi-likelihood (PQL) method and marginal quasi-likelihood (MQL) method, respectively (Breslow and Clayton, 1993; Wolfinger and O'Connell, 1993). In PQL, approximation to the nonlinear function is made using first-order Taylor expansion around the current estimates for both the random and fixed parameters whereas, in MQL, the approximation is around the estimates for fixed parameters and 0 for random effects.

Once the issue of integration involved in the marginal log-likelihood equation has been resolved, both Eqs. (12) and (13) can be solved numerically through standard optimization algorithms. Commonly used algorithms for IRT models include the Newton–Raphson algorithm and the Expectation–Maximization (EM) algorithm (Dempster et al., 1977). Let $\hat{\mathbf{u}}^{s+1}$ be the estimated values of the item parameters $\mathbf{u}$ at the $(s+1)$th iteration, where $\mathbf{u}$ is the stacking vector of the parameters in $\boldsymbol{\Lambda}$ and $\boldsymbol{\Sigma}$, the Newton–Raphson algorithm proceeds as the following

$$\hat{\mathbf{u}}^{s+1} = \hat{\mathbf{u}}^{s} - \left[\frac{\partial^2 \log L}{\partial \mathbf{u} \partial \mathbf{u}'}\right]^{-1} \left[\frac{\partial \log L}{\partial \mathbf{u}}\right]_{\mathbf{u}=\hat{\mathbf{u}}^{s}} \tag{17}$$

until the differences between parameter estimates in successive iterations are smaller than some prespecified criterion for convergence. For most IRT models, however, the Newton–Raphson algorithm is not computationally efficient since it involves the inverse of the Hessian matrix with the order of MJ, evaluated at the estimates of the current iteration (Bock and Lieberman, 1970). To resolve this problem, Bock and Aitkin (1981) adapted the EM algorithm to estimate the IRT model parameters. In the EM algorithm, the set of random effects $(\theta, \mathbf{E})$ is considered as missing data and the data matrix $\mathbf{X}$ is considered as observed (incomplete) data. Then, $P(\mathbf{X}|\theta, \mathbf{Q}, \boldsymbol{\Lambda}, \mathbf{E}, \boldsymbol{\Sigma})$ is the probability density function of the incomplete data whereas $P(\mathbf{X}, \theta, \mathbf{E}|\mathbf{Q}, \boldsymbol{\Lambda}, \boldsymbol{\Sigma})$ is the joint probability density function of the unobserved (complete) data. Instead of maximizing the marginal log-likelihood equation given by Eq. (12) and/or (13), the expectation of the "complete data" log-likelihood, denoted as $E[\log P(\mathbf{X}, \theta, \mathbf{E}|\mathbf{Q}, \boldsymbol{\Lambda}, \boldsymbol{\Sigma})|\mathbf{X}, \mathbf{Q}, \boldsymbol{\Lambda}, \boldsymbol{\Sigma}]$ is optimized. Within each iteration, the EM algorithm consists of two steps which are called the E-step and the M-step, respectively. In the E-step, $E[\log P(\mathbf{X}, \theta, \mathbf{E}|\mathbf{Q}, \boldsymbol{\Lambda}, \boldsymbol{\Sigma})|\mathbf{X}, \mathbf{Q}, \boldsymbol{\Lambda}, \boldsymbol{\Sigma}]$ is computed conditional on the current estimates of $\boldsymbol{\Lambda}$ and $\boldsymbol{\Sigma}$. In the M-step, $E[\log P(\mathbf{X}, \theta, \mathbf{E}|\mathbf{Q}, \boldsymbol{\Lambda}, \boldsymbol{\Sigma})|\mathbf{X}, \mathbf{Q}, \boldsymbol{\Lambda}, \boldsymbol{\Sigma}]$ is maximized so that updated estimates of $\boldsymbol{\Lambda}$ and $\boldsymbol{\Sigma}$ are obtained. These two steps alternate until the optimization process is converged. For the two-level hierarchical IRT model, $\log P(\mathbf{X}, \theta, \mathbf{E}|\mathbf{Q}, \boldsymbol{\Lambda}, \boldsymbol{\Sigma})$ is given as

$$\sum_{i=1}^{N} \log P(\mathbf{x}_i|\theta, \mathbf{E}, \mathbf{Q}, \boldsymbol{\Lambda}, \boldsymbol{\Sigma}) + \log P(\mathbf{E}|\boldsymbol{\Sigma}) + \log P(\theta), \tag{18}$$

and the expectation is with respect to the conditional posterior density for the random effects (incidental parameters) in θ and $\mathbf{E}$. In implementation, the EM algorithm for IRT models can be greatly simplified because of the assumption of local independence across items, across examinees, and the independence between examinee and items.

5. An application: Cognitive models for algorithmically generated spatial ability items

To illustrate the cognitive models, data were obtained from a project on the generation of spatial ability items (Embretson, 2000) for the Armed Services Vocational Aptitude Battery (ASVAB). Spatial ability items, such as shown in Figure 1, appear on the Assembling Objects (AO) Test of the ASVAB. All except the final stage in automatic item generation for these items was completed in the project.

Using the cognitive design system approach, a processing model with five stages was developed for existing AO items. The five stages were (1) Encoding, bringing the pieces in the stem to a meaningful representation, (2) Falsification, searching for perceptually implausible distractors, (3) Confirmation I, comparing the stem to the key, (4) Confirmation II, comparing the stem to nonfalsifiable distractors and (5) Decision. Eight variables represented the difficulty of these processes, as shown in Table 2. For Encoding, processing difficulty was hypothesized to increase with the number of pieces in the stem but to decrease if the pieces had obvious verbal labels (e.g., right triangle). For Falsification, processing difficulty was expected to decrease with the number of distractors with obvious perceptual mismatches to the stem pieces. For Confirmation I, processing difficulty was expected to increase if the pieces in the stem were rotated or displaced with respect to their position in the key. For Confirmation II, processing difficulty was expected to increase if nonfalsifiable distractors had pieces that mismatched by small angular disparities or if the expected number of comparisons to find the mismatched pieces was large. The cognitive model based on these variables achieved moderately high levels of prediction of item difficulty and response

Table 2
Estimates for LLTM and 2PL-Constrained models

	LLTM			2PL-Constrained					
Item predictor	η	se	t	η	se	t	τ	se	t
Constant	−3.693	0.175		−4.204	0.353		2.083	0.600	
Encoding									
Number of pieces	0.118	0.041	2.87	0.477	0.069	6.91	0.611	0.108	5.67
Verbal labels	0.141	0.025	5.74	−0.064	0.041	−1.56	−0.348	0.057	−6.11
Falsification									
Number of falsifiable	−0.198	0.040	−4.96	−0.207	0.050	−4.10	0.099	0.107	.93
Confirmation I									
Rotation	0.634	0.054	11.74	0.486	0.076	6.37	−0.358	0.136	−2.64
Displacement	0.274	0.055	4.96	0.424	0.081	5.22	0.438	0.148	2.95
Confirmation II									
Angular disparities	0.187	0.101	1.85	−0.305	0.146	−2.08	−0.816	0.287	−2.84
Comparison cycles	0.395	0.060	6.61	0.101	0.066	1.53	−0.725	0.198	−3.66
Decision									
Target distance	0.379	0.023	16.26	0.316	0.033	9.64	−0.077	0.057	−1.34

time (i.e., correlations in the .70's) for both the ASVAB items (Embretson, 2000; Embretson and Gorin, 2001) and for items on the Revised Minnesota Paper Form Board (Embretson, 2000).

In the example presented below, the cognitive model variables provided a basis for algorithmic generation of new items. Item specifications were developed to generate items with known sources of processing complexity and predictable levels of item difficulty. An earlier application of cognitive IRT models to a restricted set of items (Embretson and Yang, 2002) showed the same sources of cognitive complexity and levels of predictions as had been obtained for the original ASVAB items.

Method

Item structures were defined by crossing four sources of processing difficulty in a Pieces (3) X Falsification (4) X Rotation (2) X Displacement (2) design, yielding a total of 48 item structures. Other features, such as outer shape of the key, centerpoint position and number of mismatched pieces were counterbalanced across the item structures. Some features are related to other variables in the cognitive model while others (e.g., outer shape) were not. A total of 128 items were generated.

To estimate the impact of independently manipulating the cognitive model variables, the generated items were given empirical tryout. A sample of 321 undergraduates were randomly assigned to test forms of 46 items each, including a core set to link forms.

Results

Initial data analysis revealed that 30 items had insufficient discriminations. These items were eliminated from further analysis. Table 3 presents the number of parameter estimates, log likelihoods, AIC index, model comparisons and fit statistics for several cognitive IRT models. All models were estimated with the SAS NLMIXED procedure, using marginal maximum likelihood with adaptive quadrature points approximation approach. Since the EM algorithm is not available in SAS NLMIXED, Quasi-Newton method was used to optimize the likelihood equation instead. Shown are estimates for a null model (item difficulties and item slopes are constrained to a common value), the LLTM, the 2PL-Constrained, the Rasch model and the 2PL model. The AIC index was computed for all models and a log likelihood fit index for the models (see Embretson, 1997) was computed for three models. The log likelihood fit index yields estimates that are similar in magnitude to a multiple correlation coefficient, using the null model as a baseline (i.e., no differential item information) and the 2PL as the saturated model. For nested models, −2 times the log likelihoods is approximately distributed as chi square, with the degrees of freedom equal to the change in the number of parameter estimates between models. These values are shown on Table 3 as χ^2/df for various comparisons.

Table 3 shows that the LLTM provided significant prediction of the item parameters, as compared to the null model ($\chi^2/\text{df} = 110.34$). The 2PL-Constrained model was indicated as better than the LLTM, as the AIC index was smaller and the nested model comparison ($\chi^2/\text{df} = 9.33$, $p < .01$) was statistically significant. The fit index shows that both the LLTM and the 2PL-Constrained model provide moderately high prediction of the item parameters from relatively few parameters. The remaining models shown

Table 3
Log likelihoods and fit statistics for cognitive IRT models

Model	Number estimates	$-2 \ln L$	AIC	Model χ^2/df	Fit index $\Delta^{1/2}$
Null	2	10,821.00	10,823.00	–	–
LLTM	9	9,827.90	9,845.90	110.34[a]	.735
2PL-Constrained	18	9,743.90	9,779.9	9.33[b]	.755
Rasch	98	9,445.26	9,641.26	1.22[c]	.854
2PL	196	9,325.92	9,717.92	2.35[d]	–

[a]Null versus LLTM.
[b]LLTM versus 2PL-Constrained.
[c]Rasch versus 2PL.
[d]2PL-Constrained versus 2PL.

on Table 3, the Rasch model and the 2PL model, are not cognitive IRT models, as they contain no predictive information from the item stimuli. However, interestingly, the 2PL model was not indicated as better than the Rasch model because the AIC is larger and the nested model comparison ($\chi^2/\text{df} = 1.22$, $p > .05$) was not statistically significant.

Table 2 shows the parameter estimates, standard errors and t-tests obtained for the LLTM and the 2PL-Constrained model. For LLTM, all parameter estimates were significant at the .05 level except for angular disparity ($t = 1.85$). All relationships were in the hypothesized direction, except for Verbal Labels. For the 2PL-Constrained model, significant parameter estimates for item difficulty at the .05 level for all variables except Verbal Labels, Angular Disparities and Comparison Cycles. These variables had strong negative parameter estimates for slope, indicating that items were less discriminating when these variables are high. Significant parameter estimates for slope were found for all variables except Number of Falsifiable Distractors and Target Distance.

To illustrate hierarchical IRT models for automatic item generation, Table 4 presents parameter estimates, standard errors and t-tests obtained from several hierarchical IRT models. All models in Table 4 were estimated using SAS GLIMMIX using PQL method. LLTM was also included in Table 4 in order to compare parameter estimates obtained from different programs. It can be seen that all item predictors have similar magnitude of parameter estimates with identical directions, which suggests somewhat comparable results from the two different estimation methods. Two different hierarchical IRT models were estimated. The hierarchical IRT model with single item structure assumes that all of the items are generated from a single item structure and variability among item parameters is modeled through a random-effect term, which is assumed to have a normal distribution with mean 0 and variance σ^2_{item}. This model has no item predictor except for a common item difficulty. As can be seen from Table 4, the common item difficulty is estimated as -1.486, which suggests that on average the items are relatively easy for the examinees. The within-structure parameter variability, with no item predictor included, is estimated as .812. The Hierarchical IRT model with item predictor, which is a random-effect LLTM with common within-structure variance, re-

Table 4
Estimates for hierarchical IRT models from SAS GLIMMIX

	LLTM			Hierarchical IRT model with item predictors			Hierarchical IRT model with single item structure	
Parameter	Estimate	se	t	Estimate	se	t	Estimate	se
Fixed effects								
Constant	−3.577	0.162		−3.716	0.449		−1.486	0.113
Encoding								
Number of pieces	0.114	0.038	3.008	0.210	0.104	2.014	–	–
Verbal labels	0.137	0.023	5.952	0.070	0.068	1.038	–	–
Falsification								
Number of falsifiable	−0.193	0.037	−5.211	−0.220	0.103	−2.136	–	–
Confirmation I								
Rotation	0.615	0.050	12.306	0.649	0.142	4.575	–	–
Displacement	0.268	0.052	5.144	0.316	0.143	2.208	–	–
Confirmation II								
Angular disparities	0.183	0.095	1.926	0.114	0.234	0.487	–	–
Comparison cycles	0.384	0.055	6.989	0.386	0.145	2.662	–	–
Decision								
Targeted distance	0.369	0.022	16.750	0.289	0.063	4.587	–	–
Random effects								
σ^2_{item}	–	–		0.332	0.070		0.812	0.139
$\sigma^2_{\text{examinee}}$	0.837	0.086		0.906	0.092		0.905	0.092

duces such within-structure variability substantially through the inclusion of the item predictors, which is estimated as .332. Although theoretically possible, a random-effect LLTM with structure-specific error terms was not estimated for the AO items due to limited number of items within each structure (98 items with 49 different structures). Similar magnitude and identical directions were obtained from the hierarchical IRT model for these item predictors, while the corresponding standard errors obtained from the hierarchical IRT models were much larger. Since model fit information obtained from PQL or MQL methods cannot be used for model testing (Tuerlinckx et al., 2004), model fit index calculated as in Table 3 cannot be done for such models.

Discussion

The results clearly support the validity of the cognitive model for the algorithmically generated items. Item difficulty was strongly predicted by both the LLTM and the 2PL-Constrained model. Further, most variables in the model had the hypothesized pattern of relationship with item difficulty. Combined with other data reported (i.e., response time models) in Embretson (2000), the estimated models describe well both the level and source of cognitive complexity in the generated items.

An important question is the utility of the predictions for automatic item generation. The levels of prediction achieved were comparable to multiple correlations in the .70's. Other research has indicated that while measurement error is increased by using predicted item parameters in place of calibrated values (Mislevy et al., 1993; Embretson, 1999), the increase is not large. In fact, at the obtained prediction levels, the increased measurement error can be readily reduced by administering a couple of additional items.

Although the data in the example were not generated by a computer program, such a program was developed for AO items shortly after the study was completed. To be suitable for automatic item generation, all cognitive model variables must be anticipated directly from the generating item specifications. In the study described above, the number of falsifiable distractors was scored by raters. A study was undertaken to determine how to predict falsifiability from the size and degree of object differences. The results indicated that a combination of degree of angular disparity and object size could be quantified to predict falsifiability (reported in Embretson, 2002a, 2002b). Finally, generated items were tried empirically with operational AO items. The results indicated that reasonably good prediction of their psychometric properties was obtained using the cognitive model. Although the item generation program has not yet been attached to a CAT administration program, the research foundation indicates good potential for automatic generation of AO items.

6. Overall discussion

A substantial foundation for automatic item generation is being developed. In this chapter, both the cognitive foundations and the psychometric foundations are reviewed. Both the item model approach and the cognitive design system approach to item generation were reviewed. The item model approach has the advantage of being applicable to item generation relatively quickly as it requires a lesser cognitive foundation. The cognitive design approach has the advantages of explicating construct validity at the item level because the level and the source of cognitive complexity in an item are quantified. Further, the cognitive design system approach produces a greater diversity of items, thus limiting practice effects when item structures are repeated.

The requirements for automatic generation of items were reviewed for both the item model and the cognitive design system approach. The several stages of research effort include: (1) understanding the sources of item variation, (2) algorithmic item generation and a revised understanding of item variation, (3) item generation by algorithms, (4) empirical tryout of generated items and (5) computer programs for adaptive item generation (i.e., the item generator program plus an adaptive testing program). Although seemingly a long process, in fact adaptive item generation may be feasible for some item types in the near future.

The psychometric foundation for automatic item generation was also reviewed. A class of appropriate IRT models was reviewed and estimation procedures for the parameters were presented. The models reviewed included the linear logistic test model (LLTM), the 2PL-constrained model and the hierarchical IRT model. The latter was shown to produce a broad family of models appropriate for item generation with certain constraints applied. Then, estimation methods were reviewed. It was shown that marginal maximum likelihood estimates for the family of cognitive IRT models are available in widely-distributed software. Finally, the application of the models was illustrated with data on the algorithmic generation of spatial ability items, as appears on a large volume test. The spatial ability test has now most of the requisite research foundation for automatic item generation. Adequate prediction of item parameters for newly generated items using both LLTM and 2PL-constrained has been found. Random effects versions of these models were also illustrated, to examine within item structure variability.

In conclusion, although automatic item generation is not currently employed for large volume tests, actual implementation may occur in the near future. Efforts in many locations are leading to feasible systems. It will be interesting to observe how other conceptualizations of testing may be influenced by the emergence of operating systems of automatic item generation.

References

Abramowitz, M., Stegun, I.A. (1964). *Handbook of Mathematical Functions with Formulas, Graphs, and Mathematical Tables*. Dover, New York.

Adams, R.J., Wilson, M., Wang, W.C. (1997). The multidimensional random coefficients multinomial logit model. *Applied Psychological Measurement* **21**, 1–23.

Anderson, E.B. (1970). Asymptotic properties of conditional maximum likelihood estimators. *Journal of the Royal Statistical Society, Series B* **32**, 283–301.

Anderson, E.B. (1972). The numerical solution of a set of conditional estimation equations. *Journal of the Royal Statistical Society, Series B* **34**, 42–54.

Bejar, I.I. (1996). Generative response modeling: Leveraging the computer as a test delivery medium. RR-96-13, Educational Testing Service, Princeton, NJ.

Bejar, I.I., Lawless, R., Morley, M.E., Wagner, M.E., Bennett, R.E., Revuelta, J. (2003). A feasibility study of on-the-fly item generation in adaptive testing. *The Journal of Technology, Learning, and Assessment* **2** (3).

Bennett, R.E., Bejar, I.I. (1998). Validity and automated scoring: It's not only the scoring. *Educational Measurement: Issues and Practice* **17**, 9–17.

Birnbaum, A. (1968). Some latent trait models and their use in inferring an examinee's ability. In: Lord, F.M., Novick, M.R. (Eds.), *Statistical Theories of Mental Test Scores*. Addison-Wesley, Reading, MA, pp. 397–479.

Bock, R.D., Aitkin, M. (1981). Marginal maximum likelihood estimation of item parameters: Application of an EM algorithm. *Psychometrika* **46**, 443–459.

Bock, R.D., Lieberman, M. (1970). Fitting a response model for *n* dichotomously scored items. *Psychometrika* **35**, 179–197.

Breslow, N.E., Clayton, D.G. (1993). Approximate inference in generalized linear mixed models. *Journal of the American Statistical Association* **88**, 9–25.

Das, J.P., Naglieri, J.A. (1997). Intelligence revised: The planning, attention, simultaneous, successive (PASS) cognitive processing theory. In: Dillon, R.F. (Ed.), *Handbook on Testing*. Greenwood Press, Westport, CT, pp. 136–163.

Deane, P., Graf, E., Higgins, D., Futagi, Y., Lawless, R. (2005). Model analysis and model creation: Capturing the task model structure of quantitative item domains. ETS Research Report, Educational Testing Service, Princeton, NJ.

De Boeck, P., Wilson, M. (2004). *Explanatory Item Response Models: A Generalized Linear and Nonlinear Approach*. Springer-Verlag, New York, NY.
Dempster, A.P., Laird, N.M., Rubin, D.B. (1977). Maximum likelihood from incomplete data via the EM algorithm. *Journal of the Royal Statistical Association, Series B* **39**, 1–38.
Embretson, S.E. (1984). A general latent trait model for response processes. *Psychometrika* **49**, 175–186.
Embretson, S.E. (1985). Multicomponent latent trait models for test design. In: Embretson, S. (Ed.), *Test Design: Developments in Psychology and Psychometrics*. Academic Press, New York, pp. 195–218.
Embretson, S.E. (1997). Multicomponent latent trait models. In: van der Linden, W., Hambleton, R. (Eds.), *Handbook of Modern Item Response Theory*. Springer-Verlag, New York, pp. 305–322.
Embretson, S.E. (1998). A cognitive design system approach to generating valid tests: Application to abstract reasoning. *Psychological Methods* **3**, 380–396.
Embretson, S.E. (1999). Generating items during testing: Psychometric issues and models. *Psychometrika* **64**, 407–434.
Embretson, S.E. (2000). Generating assembling objects items from cognitive specifications. HUMRRO Report No. SubPR98-11, Washington, DC. November.
Embretson, S.E. (2002a). Generating abstract reasoning items with cognitive theory. In: Irvine, S., Kyllonen, P. (Eds.), *Generating Items for Cognitive Tests: Theory and Practice*. Erlbaum, Mahwah, NJ.
Embretson, S.E. (2002b). Measuring human intelligence with artificial intelligence. In: Sternberg, R.J., Pretz, J. (Eds.), *Cognition and Intelligence*. Cambridge University Press.
Embretson, S.E., Gorin, J. (2001). Improving construct validity with cognitive psychology principles. *Journal of Educational Measurement* **38** (4), 343–368.
Embretson, S.E., Yang, X. (2002). Modeling item parameters from cognitive design features using non-linear mixed models. Symposium paper presented at the annual meeting of the American Educational Research Association, New Orleans, LA. April.
Enright, M.K., Morley, M., Sheehan, K.M. (2002). Items by design: The impact of systematic feature variation on item statistical properties. *Applied Measurement in Education* **15**, 49–74.
Fischer, G.H. (1973). The linear logistic test model as an instrument in educational research. *Acta Psychologica* **37**, 359–374.
Glas, C.A.W., van der Linden, W. (2003). Computerized adaptive testing with item cloning. *Applied Psychological Measurement* **27**, 247–261.
Irvine, S., Kyllonen, P. (2002). *Item Generation for Test Development*. Erlbaum Publishers, Mahwah, NJ.
Kyllonen, P., Christal, R. (1990). Reasoning ability is (little more than) working memory capacity? *Intelligence* **14**, 389–434.
Mislevy, R.J., Sheehan, K.M., Wingersky, M. (1993). How to equate tests with little or no data. *Journal of Educational Measurement* **30**, 55–76.
Maris, E. (1995). Psychometric latent response models. *Psychometrika* **60**, 523–547.
Naylor, J.C., Smith, A.F.M. (1982). Applications of a method for the efficient computation of posterior distributions. *Applied Statistics* **31**, 214–225.
Neyman, J., Scott, E.L. (1948). Consistent estimates based on partially consistent observations. *Econometrica* **16**, 1–32.
Pinheiro, J.C., Bates, D.M. (1995). Approximation to the log-likelihood function in the nonlinear mixed-effects model. *Journal of Computational and Graphical Statistics* **4**, 12–35.
Rasch, G. (1960). *Probabilistic Models for Some Intelligence and Attainment Tests*. Paedagogiske Institut, Copenhagen.
Sheehan, K.M. (1997). A tree-based approach to proficiency scaling and diagnostic assessment. *Journal of Educational Measurement* **34**, 333–354.
Sinharay, S., Johnson, M. (2005). Analysis of data from an admissions test with item models. ETS Research Report RR-05-06, ETS, Princeton, NJ.
Singley, M.K., Bennet, R.E. (2002). Item generation and beyond: Applications of schema theory to mathematics assessment. In: Irvine, S.H., Kyllonen, P.C. (Eds.), *Item Generation for Test Development*. Erlbaum Publishers, Mahwah, NJ.
Spiegelhalter, D., Thomas, A., Best, N., Lunn, D. (2003). BUGS: Bayesian inference using Gibbs sampling. MRC Biostatistics Unit, Cambridge, England.

Tatsuoka, K.K. (1985). A probabilistic model for diagnosing misconceptions in the pattern classification approach. *Journal of Educational Statistics* **12**, 55–73.

Tierney, L., Kadane, J.B. (1986). Accurate approximations for posterior moments and marginal densities. *Journal of the American Statistical Association* **81**, 82–86.

Tuerlinckx, F., Rijmen, F., Molenberghs, G., Verbeke, G., Briggs, D., Van den Noortgate, W., Meulders, M., De Boeck, P. (2004). Estimation and software. In: De Boeck, P., Wilson, M. (Eds.), *Explanatory Item Response Models: A Generalized Linear and Nonlinear Approach*. Springer-Verlag, New York.

Wolfinger, R., O'Connell, M. (1993). Generalized linear mixed models: A pseudo-likelihood approach. *Journal of Statistical Computation and Simulation* **48**, 233–243.

Handbook of Statistics, Vol. 26
ISSN: 0169-7161

DOI: 10.1016/S0169-7161(06)26024-3

Statistical Inference for Causal Effects, with Emphasis on Applications in Psychometrics and Education[1]

Donald B. Rubin

Abstract
A central problem in statistics, psychometrics and education, is how to draw inferences about the causal effects of treatments (i.e., interventions) from randomized and nonrandomized data. For example, does the new job-training program really improve the quality of jobs for those trained relative to the jobs they would get without training, or does exposure to that chemical in drinking water increase cancer rates relative to drinking water without that chemical, or, of particular relevance in this volume, does that educational intervention improve the test scores of students relative to the standard program? This presentation provides an overview of the approach to the estimation of such causal effects based on the concept of potential outcomes, including a relatively detailed presentation of the Bayesian approach.

1. Causal inference primitives

This chapter begins with the description of the first part of a framework for causal inference that is now commonly referred to as "Rubin's Causal Model" (RCM; Holland, 1986), for a series of articles developing this framework (Rubin, 1974, 1975, 1976a, 1977, 1978, 1980). A second major part of that framework for causal inference, the assignment mechanism, is described in Section 2. The classical use of the framework in randomized experiments, due to Neyman (1923) and Fisher (1925), is described in Section 3, and its use in Bayesian inference for causal effects generally, due to Rubin (1978), is presented in Section 4. The remaining section discusses some extensions and complications. Other approaches to causal inference, such as graphical ones (e.g., Pearl, 2000), I find conceptually less satisfying, for reasons discussed, for instance, in Rubin (2004b, 2005). The presentation here is essentially a very shortened and compact version of the perspective presented in Imbens and Rubin (2006a), and a greatly expanded version is given in Imbens and Rubin (2006b).

[1]This chapter is a modified and substantially expanded version of the first chapter in volume 25 of the Handbook of Statistics, Bayesian Thinking: Modelling and Computation, edited by D.K. Dey and C.R. Rao (2005).

1.1. Units, treatments, potential outcomes

For causal inference, there are several primitives – concepts that are basic and on which we must build. A "unit" is a physical object, e.g., a person, at a particular point in time. A "treatment" is an action that can be applied or withheld from that unit. We focus on the case of two treatments, although the extension to more than two treatments is simple in principle although not necessarily so with real data.

Associated with each unit are two "potential outcomes": the value of an outcome variable Y at a point in time when the active treatment is applied and the value of that outcome variable at the *same* point in time when the active treatment is withheld. The objective is to learn about the causal effect of the application of the active treatment relative to the control (its being withheld) on Y.

For example, the unit could be "you now" with your headache, the active treatment could be taking aspirin for your headache, and the control could be not taking aspirin. The outcome Y could be the intensity of your headache pain in two hours, with the potential outcomes being the headache intensity if you take aspirin and if you do not take aspirin.

Notationally, let W indicate which treatment the unit, you, received: $W = 1$ the active treatment, $W = 0$ the control treatment. Also let $Y(1)$ be the value of the potential outcome if the unit received the active version, and $Y(0)$ the value if the unit received the control version. The causal effect of the active treatment relative to its control version is the comparison of $Y(1)$ and $Y(0)$ – typically the difference, $Y(1) - Y(0)$, or perhaps the difference in logs, $\log[Y(1)] - \log[Y(0)]$, or some other comparison, possibly the ratio.

We can observe only one or the other of $Y(1)$ and $Y(0)$ as indicated by W. The key problem for causal inference is that, for any individual unit, we observe the value of the potential outcome under only one of the possible treatments, namely the treatment actually assigned, and the potential outcome under the other treatment is missing. Thus, inference for causal effects is a missing-data problem – the "other" value is missing. For example, a gain score measures a change in time, in particular, a change from pre-test to post-test on one unit, and so is not a causal effect.

1.2. Relating this definition of causal effect to common usage

The definition of a causal effect provided so far may appear a bit formal and the discussion a bit ponderous, but the presentation is simply intended to capture the way we use the concept in everyday life. Also this definition of causal effect as the comparison of potential outcomes is frequently used in contemporary culture, for example, as revealed by movies. Let us consider some movie plots to illustrate this point.

Most of us have probably seen parts of "It's a Wonderful Life" with Jimmy Stewart as George Bailey more times than we can remember (around Christmas time). In this movie, at one point in George's life, he becomes very depressed and sincerely wishes he had never been born. At the appropriate moment, a wingless angel named "Clarence" shows him exactly what the world he knows would be like if, contrary to fact, he had not been born.

The actual world is the real observed potential outcome, but Clarence shows George the other potential outcome, the counterfactual unobserved one, and George experiences this other world as a real phenomenon, just as real as his actual world. Not only are there obvious consequences, like his own children not being born, but there are many other untoward events. For example, his younger brother, Harry, who was, in the actual world, a World War II hero, in the counterfactual world drowned in a skating accident at age eight, because George was never born, and thus was not there to save Harry as he did in the actual world. And there was the pharmacist, Mr. Gower, who filled the wrong prescription and was convicted of manslaughter because George was not there to catch the error as he did in the actual world.

The causal effect of George not being born is the comparison of (a) the entire stream of events in the actual world with him in it, to (b) the entire stream of events in the counterfactual world without him in it. Fortunately for George, he has Clarence to show him the counterfactual potential outcome, and George regrets ever having wished he'd never been born, and he returns to the actual world.

Another movie often shown around Christmas time is based on Dickens' classic novel "A Christmas Carol". Here, the wealthy and miserly Ebenezer Scrooge is fortunate enough to have a visit from a trio of ghosts. Although the first two ghosts make Scrooge feel guilty about things he has done in the distant and recent past, the third ghost, the most frightening of all, the Ghost of Christmas Future, is the most effective. The ghost reveals to Ebenezer the potential outcome that will occur if Ebenezer continues his current mean-spirited ways. Not surprisingly, Scrooge does not want to live this potential outcome, and because it is the future, can reject it in favor of the other potential outcome, by altering his behavior. Neither outcome was counterfactual at the time of the third ghost's visit, but one was by the end of the movie.

These and modern movies with similar themes clearly indicate that the causal effects of actions or treatments are the comparison of potential outcomes under each action. A current list includes, among others, "Sliding Doors", "The Family Man", and "Mr. Destiny". Sometimes one of the potential outcomes is clearly counterfactual, as in "It's a Wonderful Life", sometimes the revealed potential outcomes are not yet counterfactual because they are in the future, as in "A Christmas Carol". Sometimes the movie is not clear about whether one of the potential outcomes is counterfactual or not. But, invariably, the causal effect of actions involves the comparison of potential outcomes, which are the stream of events at the same times but after different actions.

1.3. Relationship to the "but-for" concept in legal damages settings

Probably the setting outside science where causal effects are most seriously considered is in law. Suppose you committed an action that you should not have, *and as a result of that action*, someone else suffered damages. That is, the causal effect of your action relative to the absence of that action is the comparison of potential outcomes, the first the actual, the second the counterfactual. For example, let's suppose that because you were driving and talking on a cell phone, you ran a stop sign, hitting another car. "But for" your negligence, the other car would not have been involved in any accident. The causal effect of your negligence is the amount of damage in this accident.

One of the more interesting and expensive recent examples of trying to assess causal effects in a legal damages setting has revolved around the alleged misconduct of the tobacco industry (i.e., lying about the health risks of smoking) in the United States and other countries. Here, the world with the alleged misconduct is the actual world, with all its cigarette smoking and health-care expenditures. The other potential outcome, the counterfactual one, is the world without the alleged misconduct and presumably with reduced cigarette smoking, and different amounts and kinds of health-care expenditures. One cannot simply assume the complete absence of cigarette smoking in a counterfactual world without the alleged misconduct of the tobacco industry, and so only a fraction of all expenditures that may be attributable to smoking are the causal effect of the alleged misconduct.

This point has been made repeatedly in the tobacco litigation by both plaintiffs' experts and defendants' experts (including me) in both their reports and their testimony. For example, the well-known economist, Franklin Fisher, in an expert witness report for the plaintiffs, (1999, p. 2, item #6) wrote:

> It is necessary to generate a stream of damages taking into account that not all smoking-related expenditures result from the alleged behavior of the defendants. Thus, the smoking-related expenditures estimated by Glenn Harrison and Wendy Max, and Henry Miller [experts for the plaintiffs] need to be adjusted by what Jeffrey Harris [expert for the plaintiffs, Harris, 1999] has calculated to be the proportion of total smoking-attributable expenditures caused by defendants' improper conduct. The monetary amount of damage resulting from the defendants' alleged behavior in each past and future year is thus calculated by multiplying the annual smoking related expenditures by the proportion caused by defendants' improper conduct.

Also, the September 22, 1999 news conference held to announce the United States filing of its lawsuit against the tobacco industry, Assistant Attorney General Ogden (1999) stated:

> The number that's in the complaint is not a number that reflects a particular demand for payment. What we've alleged is that each year the federal government expends in excess of 20 billion on *tobacco* related medical costs. What we would actually recover would be our portion of that annual toll that is the result of the illegal conduct that we allege occurred, and it simply will be a matter or proof for the court, which will be developed through the course of discovery, what that amount will be. So, we have not put out a specific figure and we'll simply have to develop that as the case goes forward.

These positions are supported by the Federal Judicial Center's "Reference Manual on Scientific Evidence" (2000, p. 284):

> The first step in a damages study is the translation of the legal theory of the harmful event into an analysis of the economic impact of that event. In most cases, the analysis considers the difference between the plaintiff's economic position if the harmful event had not occurred and the plaintiff's actual economic position. The damages study restates the plaintiff's position "but for" the harmful event; this part is often called the *but-for analysis*. Damages are the difference between the but-for value an the actual value.

The "but-for" analysis compares the observed actual world potential outcome with the alleged misconduct to the counterfactual world potential outcome without the alleged misconduct. The difference between the monetary values in these worlds is the basis for calculating damages. This is not necessarily an easy quantity to estimate in the tobacco litigation, but it is the relevant causal estimand (i.e., the causal quantity to be estimated).

1.4. Learning about causal effects: Replication and the Stable Unit Treatment Value Assumption – SUTVA

How do we learn about causal effects? The answer is replication, more units. The way we personally learn from our own experience is replication involving the same physical object (me or you) with more units in time. That is, if I want to learn about the effect of taking aspirin on headaches for me, I learn from replications in time when I do and do not take aspirin to relieve my headache, thereby having some observations of $Y(0)$ and some of $Y(1)$. When we want to generalize to units other than ourselves, we typically use more objects; that is what is done in medical experiments and social science experiments, for example, involving students and possible educational interventions, such as "Value Added Assessment" (e.g., Rubin et al., 2004).

Suppose instead of only one unit we have two. Now in general we have at least four potential outcomes for each unit: the outcome for unit 1 if both unit 1 and unit 2 received control, $Y_1(0, 0)$; the outcome for unit 1 if both units received the active treatment, $Y_1(1, 1)$; the outcome for unit 1 if unit 1 received control and unit 2 active, $Y_1(0, 1)$, and the outcome for unit 1 if unit 1 received active and unit 2 received control, $Y_1(1, 0)$; and analogously for unit 2 with values $Y_2(0, 0)$, etc. In fact, there are even more potential outcomes because there have to be at least two "doses" of the active treatment available to contemplate all assignments, and it could make a difference which one was taken. For example, in the aspirin case, one tablet may be very effective and the other quite ineffective.

Clearly, replication does not help unless we can restrict the explosion of potential outcomes. As in all theoretical work with applied value, simplifying assumptions are crucial. The most straightforward assumption to make is the "stable unit treatment value assumption" (SUTVA – Rubin, 1980, 1990a) under which the potential outcomes for the ith unit just depend on the treatment the ith unit received. That is, there is "no interference between units" and there are "no versions of treatments". Then, all potential outcomes for N units with two possible treatments can be represented by an array with N rows and two columns, the ith unit having a row with two potential outcomes, $Y_i(0)$ and $Y_i(1)$.

Obviously, SUTVA is a major assumption. But there is no assumption-free causal inference, and nothing is wrong with this. It is the quality of the assumptions that matters, not their existence or even their absolute correctness. Good researchers attempt to make such assumptions plausible by the design of their studies. For example, SUTVA becomes more plausible when units are isolated from each other, as when using, for the units, intact schools rather than students in the schools when studying an educational intervention, such as a smoking prevention program (e.g., see Peterson et al., 2000).

The stability assumption (SUTVA) is very commonly made, even though it is not always appropriate. For example, consider a study of the effect of vaccination on a contagious disease. The greater the proportion of the population that gets vaccinated, the less any unit's chance of contracting the disease, even if not vaccinated, an example of interference. Throughout this chapter, we assume SUTVA, although there are other assumptions that could be made to restrict the exploding number of potential outcomes with replication.

In general, the N units may receive neither the active treatment $W_i = 1$ or the control treatment $W_i = 0$. For example, some of the units may be in the future, as when we want to generalize to a future population. Then formally W_i must take on a third value, $W_i = *$ representing neither 1 nor 0; we often avoid this extra notation here.

1.5. Covariates

In addition to (1) the vector indicator of treatments for each unit in the study, $W = \{W_i\}$, (2) the array of potential outcomes when exposed to the treatment, $Y(1) = \{Y_i(1)\}$, and (3) the array of potential outcomes when not exposed, $Y(0) = \{Y_i(0)\}$, we have (4) an array of covariates $X = \{X_i\}$, which are, by definition, unaffected by treatment, such as pretest scores. All causal estimands involve comparisons of $Y_i(0)$ and $Y_i(1)$ on either all N units, or a common subset of units; for example, the average causal effect across all units that are female as indicated by their X_i, or the median causal effect for units with X_i indicating male and $Y_i(0)$ indicated failure on the post-test.

Thus, under SUTVA, all causal estimands can be calculated from the matrix of "scientific values" with ith row: $(X_i, Y_i(1), Y_i(1))$. By definition, all relevant information is encoded in X_i, $Y_i(0)$, $Y_i(1)$ and so the labelling of the N rows is a random permutation of $1, \ldots, N$. In other words, the N-row array

$$\bigl(X, Y(0), Y(1)\bigr) = \begin{bmatrix} X_1 & Y_1(0) & Y_1(1) \\ & \vdots & \\ X_i & Y_i(0) & Y_i(1) \\ & \vdots & \\ X_N & Y_N(0) & Y_N(1) \end{bmatrix}$$

is row exchangeable.

Covariates (such as age, race and sex) play a particularly important role in observational studies for causal effects where they are variously known as possible "confounders" or "risk factors". In some studies, the units exposed to the active treatment differ on their distribution of covariates in important ways from the units not exposed. To see how this issue influences our formal framework, we must define the "assignment mechanism", the probabilistic mechanism that determines which units get the active version of the treatment and which units get the control version. The assignment mechanism is the topic of Section 2.

1.6. A brief history of the potential outcomes framework

The basic idea that causal effects are the comparisons of potential outcomes seems so direct that it must have ancient roots, and we can find elements of this definition of causal effects among both experimenters and philosophers. See, for example, the philosopher John Stuart Mill, who, when discussing Hume's views, offers (Mill, 1973, p. 327):

> If a person eats of a particular dish, and dies in consequence, that is, would not have died if he had not eaten of it, people would be apt to say that eating of that dish was the source of his death.

And Fisher (1918, p. 214) wrote:

> If we say, "This boy has grown tall because he has been well fed," we are not merely tracing out the cause and effect in an individual instance; we are suggesting that he might quite probably have been worse fed, and that in this case he would have been shorter.

Despite the insights evident in these quotations, apparently there was no formal notation for potential outcomes until Neyman (1923), which appears to have been the first place where a mathematical analysis is written for a randomized experiment with explicit notation for the potential outcomes. This notation became standard for work in randomized experiments from the randomization-based perspective (e.g., Pitman, 1937; Welch, 1937; McCarthy, 1939; Anscombe, 1948; Kempthorne, 1952; Cox, 1958; Hodges and Lehmann, 1970, Section 9.4; Brillinger et al., 1978). Neyman's formalism was a major advance because it allowed explicit frequentist probabilistic causal inferences to be drawn from data obtained by a randomized experiment, an approach discussed in Section 3.

Independently and nearly simultaneously, Fisher (1925) created a somewhat different method of inference for randomized experiments, also based on the special class of randomized assignment mechanisms, discussed in Section 3. The notions of the central role of randomized experiments seems to have been "in the air" in the 1920's, but Fisher was apparently the first to recommend the actual physical randomization of treatments to units and then use this randomization to justify theoretically an analysis of the resultant data.

Despite the almost immediate acceptance in the late 1920's of Fisher's proposal for randomized experiments, and Neyman's notation for potential outcomes in randomized experiments, this same framework was not used outside randomized experiments for a half century thereafter, apparently not until Rubin (1974), and these insights therefore were entirely limited to randomization-based frequency inference.

The approach used in nonrandomized settings during the half century following the introduction of Neyman's seminal notation for randomized experiments was based on mathematical models relating the observed value of the outcome variable $Y_{\text{obs}} = \{Y_{\text{obs},i}\}$ to covariates and indicators for the treatment received, and then to define causal effects as parameters in these models; here, $Y_{\text{obs},i} = Y_i(1)W_i + Y_i(0)(1 - W_i)$. This approach is illustrated by the Lord's Paradox example of Section 2.2. The same statistician would simultaneously use Neyman's potential outcomes to define causal effects in randomized experiments and the observed outcome setup in observational studies. This led to substantial confusion because the role of randomization cannot even be stated using observed outcome notation.

The framework that we describe here that uses potential outcomes to define causal effects, generally has been called the "Rubin Causal Model" – RCM by Holland (1986) for work initiated in the 1970's (Rubin, 1974, 1977, 1978). This perspective conceives of all problems of statistical inference for causal effects as missing data problems with a mechanism for creating missing data in the potential outcomes (Rubin, 1976a). Of course, there were seeds of the RCM before 1974. Already mentioned is Neyman (1923), but in economics there is Tinbergen (1930), Haavelmo (1944) and Hurwicz (1962).

The potential outcome framework seems to have been basically accepted and adopted by most workers by the end of the twentieth century. Sometimes the move was made explicitly, as with Pratt and Schlaifer (1984) who moved from the "observed outcome" to the potential outcomes framework in Pratt and Schlaifer (1988). Sometimes it was made less explicitly as with those who were still trying to make a version of the observed outcome notation work in the late 1980's (e.g., see Heckman and Hotz, 1989), before fully accepting the RCM in subsequent work (e.g., Heckman, 1989, after discussion by Holland, 1989).

The movement to use potential outcomes to define causal inference problems seems to be the dominant one at the start of the 21st century. See, for example, Baker (1998), Dempster (1990), Efron and Feldman (1991), Gelman and King (1991), Greenland and Poole (1988), Greenland et al. (1999), Holland (1988a, 1988b), Kadane and Seidenfeld (1990), Robins (1989), Rosenbaum (1987, 2002), Smith and Sugden (1988), Sobel (1990, 1995, 1996), Sugden (1988), and their references. A recent article exploring whether the full potential outcomes framework can be avoided when conducting causal inference is Dawid (2000) with discussion. Also see Cox (1992). For more discussion of my views on this transition, see Rubin (2005).

2. The assignment mechanism

Even with SUTVA, inference for causal effects requires the specification of an assignment mechanism: a probabilistic model for how some units were selected to receive the active treatment and how other units were selected to receive the control treatment. We first illustrate this model in two trivial artificial examples, and then present formal notation for this model.

2.1. Illustrating the criticality of the assignment mechanism

Consider a teacher who is considering one of two treatments to apply to each of his eight students, a standard and a new one. This teacher is a great teacher: he chooses the treatment that is best for each student! When they are equally effective, he tosses a coin. Table 1 gives the hypothetical potential outcomes in test scores under each treatment for these students, and so also gives their individual causal effects. The asterisks indicate the chosen treatment for each student and their observed potential outcomes. The column labelled "W" shows which treatment each student received, $W_i = 0$ or $W_i = 1$ for the ith student.

Notice that the eight individual causal effects, displayed in the rightmost column, indicate that the typical student will do better with the standard treatment: the average causal effect is two points in favor of the standard. But the teacher, who is conducting ideal educational practice for the benefit of his students, reaches the opposite conclusion from an examination of the observed data: the students assigned the new treatment do,

Table 1
Perfect teacher

Potential outcomes			Observed data		
	$Y(0)$	$Y(1)$	W	$Y(0)$	$Y(1)$
	13	14	1	?	14
	6	0	0	6	?
	4	1	0	4	?
	5	2	0	5	?
	6	3	0	6	?
	6	1	0	6	?
	8	10	1	?	10
	8	9	1	?	9
True averages	7	5	Observed averages	5.4	11

on average, twice as well as the students assigned the control, with absolutely no overlap in their distributions! Moreover, if the teacher now applies the new treatment to all students in a population of students like the eight in the study, he will be disappointed: the average test score will be closer to five points under the new treatment rather than the eleven points seen in this earlier study.

What is wrong? The simple comparison of observed results assumes that treatments were *randomly assigned*, rather than as they were, to provide maximal benefit to the students. We will have more to say about randomized experiments, but the point here is simply that the assignment mechanism is crucial to valid inference about causal effects, and the teacher used a "nonignorable" assignment mechanism (defined shortly). With a posited assignment mechanism, it is possible to draw causal inferences; without one, it is impossible. It is in this sense that when drawing inferences, a model for the assignment mechanism is more fundamental than a model for the potential outcomes: Without positing an assignment mechanism, we basically cannot draw causal inferences.

More precisely, notice that the teacher, by comparing observed means, is using the three observed values of $Y_i(1)$ to represent the five missing values of $Y_i(1)$, effectively imputing or filling in $\bar{y}_1$ for the five $Y_i(1)$ question marks, and analogously effectively filling in $\bar{y}_0$ for the three $Y_i(0)$ question marks. This process makes sense for point estimation if the three observed values of $Y_i(1)$ were randomly chosen from the eight values of $Y_i(1)$, and the five observed values of $Y_i(0)$ were randomly chose from the eight values of $Y_i(0)$. But under the actual assignment mechanism, it does not. It would obviously make much more sense under the actual assignment mechanism to impute the missing potential outcome for each student to be less than or equal to that student's observed potential outcome.

2.2. *Lord's paradox*

We now consider a "paradox" in causal inference that is easily resolved with the simple ideas we have already presented, despite the controversy that it engendered in some literatures. This example illustrates how important it is to keep this perspective clearly in mind when thinking about causal effects of interventions and when reading the remainder of this chapter. Lord (1967) proposed the following example:

> A large university is interested in investigating the effects on the students of the diet provided in the university dining halls and any sex differences in these effects. Various types of data are gathered. In particular, the weight of each student at the time of arrival in September and the following June are recorded.

The result of the study for the males is that their average weight is identical at the end of the school year to what it was at the beginning; in fact, the whole distribution of weights is unchanged, although some males lost weight and some males gained weight – the gains and losses exactly balance. The same thing is true for the females. The only difference is that the females started and ended the year lighter on average than the males. On average, there is no weight gain or weight loss for either males or females. From Lord's description of the problem quoted above, the quantity to be estimated, the estimand, is the difference between the causal effect of the university diet on males and the causal effect of the university diet on females. That is, the causal estimand is the difference between the causal effects for males and females, the "differential" causal effect.

The paradox is generated by considering the contradictory conclusions of two statisticians asked to comment on the data. Statistician 1 *observes* that there are no differences between the September and June weight distributions for either males or females. Thus, Statistician 1 concludes that

> . . . as far as these data are concerned, there is no evidence of any interesting effect of diet (or of anything else) on student weight. In particular, there is no evidence of any differential effect on the two sexes, since neither group shows any systematic change. (p. 305)

Statistician 2 looks at the data in a more "sophisticated" way. Effectively, he examines males and females with about the same initial weight in September, say a subgroup of "overweight" females (meaning simply above-average-weight females) and a subgroup of "underweight" males (analogously defined). He notices that these males tended to gain weight on average and these females tended to lose weight on average. He also notices that this result is true no matter what group of initial weights he focuses on. (Actually, Lord's Statistician 2 used a technique known as covariance adjustment or regression adjustment.) His conclusion, therefore, is that after "controlling for" initial weight, the diet has a differential positive effect on males relative to females because for males and females with the same initial weight, on average the males gain more than the females.

Who's right? Statistician 1 or Statistician 2? Notice the focus of both statisticians on gain scores and recall that gain scores are not causal effects because they do not compare potential outcomes. If both statisticians confined their comments to *describing* the data, both would be correct, but for causal inference, both are wrong because these data cannot support any causal conclusions about the effect of the diet without making some very strong assumptions.

Back to the basics. The units are obviously the students, and the time of application of treatment (the university diet) is clearly September and the time of the recording of the outcome Y is clearly June; accept the stability assumption. Now, what are the potential outcomes and what is the assignment mechanism? Notice that Lord's statement of the problem has reverted to the already criticized observed variable notation, Y_{obs}, rather than the potential outcome notation being advocated.

The potential outcomes are June weight under the university diet $Y_i(1)$ and under the "control" diet $Y_i(0)$. The covariates are sex of students, male vs. female, and September weight. But the assignment mechanism has assigned everyone to the new treatment! There is no one, male or female, who is assigned to the control treatment. Hence, there is absolutely no purely empirical basis on which to compare the effects, either raw or differential, of the university diet with the control diet. By making the problem complicated with the introduction of the covariates "male/female" and "initial weight", Lord has created partial confusion. For more statistical details of the resolution of this paradox, see Holland and Rubin (1983), and for earlier related discussion, see for example, Lindley and Novick (1981), or Cox and McCullagh (1982). But the point here is that the "paradox" is immediately resolved through the explicit use of potential outcomes. Either answer could be correct for causal inference depending on what we are willing to assume about the control diet.

2.3. *Unconfounded and strongly ignorable assignment mechanisms*

We have seen that a model for the assignment mechanism is needed for all forms of statistical inference for causal effects. The assignment mechanism gives the conditional probability of each vector of assignments given the covariates and potential outcomes:

$$\Pr\big(W \mid X, Y(0), Y(1)\big). \tag{1}$$

Here W is a N by 1 vector and X, $Y(1)$ and $Y(0)$ are all matrices with N rows. A specific example of an assignment mechanism is a completely randomized experiment with N units, where $n < N$ are assigned to the active treatment, and $N - n$ to the control treatment.

$$\Pr\big(W \mid X, Y(0), Y(1)\big) = \begin{cases} 1/C_n^N & \text{if } \sum W_i = n, \\ 0 & \text{otherwise.} \end{cases} \tag{2}$$

An "unconfounded assignment mechanism (Rubin, 1990b)" is free of dependence on either $Y(0)$ or $Y(1)$:

$$\Pr\big(W \mid X, Y(0), Y(1)\big) = \Pr(W \mid X). \tag{3}$$

With an unconfounded assignment mechanism, at each set of values of X_i that has a distinct probability of $W_i = 1$, there is effectively a completely randomized experiment. That is, if X_i indicates sex, with males having probability 0.2 of receiving the active treatment and females having probability 0.5 of receiving the active treatment, then essentially one randomized experiment is prescribed for males and another for females.

The assignment mechanism is "probabilistic" if each unit has a positive probability of receiving either treatment:

$$0 < \Pr\big(W_i = 1 \mid X, Y(0), Y(1)\big) < 1. \tag{4}$$

The assignment mechanism is fundamental to causal inference because it tells us how we got to see what we saw. Because causal inference is basically a missing data problem with at least half of the potential outcomes missing, when we have no understanding of the process that creates missing data, we have no hope of inferring anything about the missing values. That is, without a stochastic model for how treatments are assigned to individuals, formal causal inference, at least using probabilistic statements, is impossible. This statement does not mean that we need to know the assignment mechanism, but rather that without positing one, we cannot make any statistical claims about causal effects, such as unbiased estimation or the coverage of confidence intervals or the coverage of Bayesian posterior intervals, all defined in Sections 3 and 4.

Unconfounded probabilistic assignment mechanisms, also called "strongly ignorable" in Rosenbaum and Rubin (1983a), often allow particularly straightforward estimation of causal effects from all perspectives, as we see shortly. Therefore, these assignment mechanisms form the basis for inference for causal effects in more complicated situations, such as when assignment probabilities depend on covariates in unknown ways, or when there is noncompliance with the assigned treatment. Probabilistic unconfounded assignment mechanisms, which essentially are collections of separate completely randomized experiments at each distinct value of X_i, form the basis for the analysis of observational nonrandomized studies, as we see later.

2.4. *Confounded and ignorable assignment mechanisms*

A confounded assignment mechanism is one that depends on the potential outcomes:

$$\Pr\bigl(W|X, Y(0), Y(1)\bigr) \neq \Pr(W|X). \tag{5}$$

A special class of possibly confounded assignment mechanisms are particularly important to Bayesian inference: ignorable assignment mechanisms (Rubin, 1978). Ignorable assignment mechanisms are defined by their freedom from dependence on any missing potential outcomes:

$$\begin{aligned} &\Pr\bigl(W|X, Y(0), Y(1)\bigr) = \Pr(W|X, Y_{\text{obs}}),\\ &\quad \text{where} \quad Y_{\text{obs}} = \{Y_{\text{obs},i}\}\\ &\quad \text{with} \quad Y_{\text{obs},i} = W_i Y_i(1) + (1 - W_i) Y_i(0). \end{aligned} \tag{6}$$

Ignorable confounded assignment mechanisms arise in practice, especially in sequential experiments. Here, the next unit's probability of being exposed to the active treatment depends on the success rate of those previously exposed to the active treatment versus the success rate of those exposed to the control treatment, as in "play-the-winner" designs (e.g., see Efron, 1971): expose the next patient with higher probability to whichever treatment appears to be more successful.

All unconfounded assignment mechanisms are ignorable, but not all ignorable assignment mechanisms are unconfounded (e.g., play-the-winner designs). Seeing why ignorable assignment mechanisms play a critical role in causal inference is easily seen in the two trivial examples that started this section, the first, in Section 2.1, involving a nonignorable treatment assignment mechanisms, and the second, in Section 2.2, involving an unconfounded but nonprobabilistic assignment mechanism.

3. Randomization-based modes of causal inference

Fundamentally, there are three formal statistical modes of causal inference; one is Bayesian, discussed in Section 4, which treats the potential outcomes as random variables, and two are based only on the assignment mechanism, which treat the potential outcomes as fixed but unknown quantities. Rubin (1990b) describes these three as well as a combination, which is fundamentally not as conceptually tight, and so is not discussed here as a distinct mode. Of the two distinct forms of randomization-based inference, one is due to Neyman (1923) and the other is due to Fisher (1925). Both will first be described in the absence of covariates, X. The randomization-based modes can also be called assignment-based modes because they extend classical randomization-based methods in randomized experiments.

3.1. Fisherian randomization-based inference

Fisher's approach is the more direct conceptually and is therefore introduced first. It is closely related to the mathematical idea of proof by contradiction. It basically is a "stochastic proof by contradiction" giving the significance level (or p-value) – really, the plausibility – of the "null hypothesis", which often is that there is absolutely no treatment effect whatsoever. Fisher's method only works for the set of units with $W_i = 1$ or 0, and not for units with $W_i = *$, so in this subsection, we assume that all units are exposed to either the active treatment or the control treatment.

The first element in Fisher's mode of inference is the null hypothesis, which is usually that $Y_i(1) \equiv Y_i(0)$ for all units: the treatments have absolutely no effect on the potential outcomes. Under this null hypothesis, all potential outcomes are known from the observed values of the potential outcomes, Y_{obs}, because $Y(1) \equiv Y(0) \equiv Y_{\text{obs}}$. It follows that, under this null hypothesis, the value of any statistic, S, such as the difference of the observed averages for units exposed to treatment 1 and units exposed to treatment 0, $\bar{y}_1 - \bar{y}_0$, is known, not only for the observed assignment, but for all possible assignments W.

Suppose we choose a statistic, S, such as $\bar{y}_1 - \bar{y}_0$, and calculate its value under each possible assignment (assuming the null hypothesis) and also calculate the probability of each assignment under the randomized assignment mechanism. In many classical experiments, these probabilities are either zero or a common value for all possible assignments. For example, in a completely randomized experiment with $N = 2n$ units, n are randomly chosen to receive treatment 1 and n to receive treatment 0. Then any assignment W that has n 1's and n 0's has probability $1/C_n^N$, and all other W's have zero probability. Knowing the value of S for each W and its probability, we can then calculate the probability (under the assignment mechanism and the null hypothesis) that we would observe a value of S as "unusual" as, or more unusual than, the observed value of S, S_{obs}. "Unusual" is defined a priori, typically by how discrepant S_{obs} is from the typical values of S under the null hypothesis. This probability is the plausibility (p-value or significance level) of the observed value of the statistic S under the null hypothesis: the probability of a result (represented by the value S_{obs} of the statistic, S) as rare, or more rare, than the actual observed value if the null hypothesis were true, where the probability is over the distribution induced by the assignment mechanism.

This form of inference is elegant: Unless the data suggest that the null hypothesis of absolutely no treatment effect is false (for an appropriate choice of statistic, S), it is not easy to claim evidence for differing efficacies of the treatments.

Fisher's approach can be extended to other "sharp" null hypotheses, that is, a null hypothesis such that from knowledge of Y_{obs}, $Y(1)$ and $Y(0)$ are known; e.g., an additive null, which asserts that for each unit, $Y(1) - Y(0)$ is a specified constant, e.g., 3. The collection of such null hypotheses that do not lead to an extreme p-value can be used to create interval estimates of the causal effect assuming additivity. Extensions to other statistics and other fully specified assignment mechanisms, including unconfounded and even nonignorable ones, are immediate, because all potential outcomes are known from Y_{obs}, and thus the probabilities of any assignment are known; Fisher, however, never discussed such extensions. Notice that Fisher's perspective provides no ability to generalize beyond the units in the experiment, nor to consider "nuisance" null hypotheses when there are multiple treatments, as in factorial designs (e.g., Cochran and Cox, 1957). These limitations are not present in Neyman's approach.

3.2. *Neymanian randomization-based inference*

Neyman's form of randomization-based inference can be viewed as drawing inferences by evaluating the expectations of statistics over the distribution induced by the assignment mechanism in order to calculate a confidence interval for the typical causal effect. The essential idea is the same as in Neyman's (1934) classic article on randomization-based (now often called "designed-based") inference in surveys. Typically, an unbiased estimator of the causal estimand (the typical causal effect, e.g., the average) is created, and an unbiased, or upwardly biased, estimator of the variance of that unbiased estimator is found (bias and variance both defined with respect to the randomization distribution). Then, an appeal is made to the central limit theorem for the normality of the estimator over its randomization distribution, whence a confidence interval for the causal estimand is obtained.

To be more explicit, the causal estimand is typically the average causal effect $\overline{Y(1)} - \overline{Y(0)}$, where the averages are over all units in the population being studied, and the traditional statistic for estimating this effect is the difference in sample averages for the two groups, $\bar{y}_1 - \bar{y}_0$, which can be shown to be unbiased for $\overline{Y(1)} - \overline{Y(0)}$ in a completely randomized design. A common choice for estimating the variance of $\bar{y}_1 - \bar{y}_0$ over its randomization distribution, in completely randomized experiments with $N = n_1 + n_2$ units, is $se^2 = s_1^2/n_1 + s_0^2/n_0$, where s_1^2, s_0^2, n_1, and n_0 are the observed sample variances and sample sizes in the two treatment groups. Neyman (1923) showed that se^2 overestimates the actual variance of $\bar{y}_1 - \bar{y}_0$, unless additivity holds (i.e., unless all individual causal effects are constant), in which case se^2 is unbiased for the variance of $\bar{y}_1 - \bar{y}_0$. The standard 95% confidence interval for $\overline{Y(1)} - \overline{Y(0)}$ is $\bar{y}_1 - \bar{y}_0 \pm 1.96se$, which, in large enough samples, includes $\overline{Y(1)} - \overline{Y(0)}$ in at least 95% of the possible random assignments.

Neyman's form of inference is less direct than Fisher's. It is really aimed at evaluations of procedures: in repeated applications, how often does the interval $\bar{y}_1 - \bar{y}_0 \pm 1.96se$ include $\overline{Y(1)} - \overline{Y(0)}$? Nevertheless, it forms the theoretical foundation for much of what is done in important areas of application, including in social science. However, Neyman's approach is not prescriptive in the sense of telling us what to do to create an inferential procedure, but rather it tells us how to evaluate a proposed procedure for drawing causal inferences. Thus, it really is not well-suited to deal with complicated problems except in the sense of evaluating proposed answers that are obtained by insight or another method. Fisher's approach also suffers from this disadvantage of lack of prescription, in fact, more so, because there is little guidance in Fisher's approach for which test statistics to use or how to define "more unusual".

3.3. *The role for covariates in randomized experiments*

As stated earlier, covariates are variables whose values are not affected by the treatment assignment, for example, variables that are recorded before randomization into treatment groups (e.g., year of birth, pre-test scores in an educational evaluation). In classical randomized experiments, if a covariate is used in the assignment mechanism, as with a blocking variable in a randomized block design, that covariate must be reflected in the analysis because it affects the randomization distribution induced by the assignment mechanism. Also, covariates can be used to increase efficiency of estimation, even when not used in the assignment mechanism.

The point about efficiency gains can be seen in the context of a completely randomized experiment in education with X = pre-test and Y = post-test. From either the Fisherian or Neymanian perspectives, we can use covariates to define a new statistic to estimate causal estimands. For example, one can use the difference in average observed gain scores, $(\bar{y}_1 - \bar{x}_1) - (\bar{y}_0 - \bar{y}_0)$ – where $\bar{x}_1$ and $\bar{x}_0$ are average observed pre-test scores for those exposed to $W = 1$ and $W = 0$, respectively – rather than the difference in average post-test scores, $\bar{y}_1 - \bar{y}_0$, to estimate $\overline{Y(1)} - \overline{Y(0)}$. Suppose X and Y are correlated, which is to be expected for pre-tests and post-tests. From the Neymanian perspective, the variance of the difference in average gain scores should be less than the variance of the difference in average post-test scores, which translates into smaller estimated variances and therefore shorter confidence intervals. From the Fisherian perspective, this reduced variance translates into more powerful tests of the null hypothesis: under an alternative hypothesis, smaller real deviations from the null hypothesis are more likely to lead to more significant p-values.

This point is easily seen in examples. Suppose as an extreme case, that the new treatment adds essentially 10 points to everyone's pre-test score, whereas the old treatment does nothing. The observed gain scores have essentially zero variance in each treatment group, whereas the post-test scores have the same variances as the pre-test scores. This result means that the Neymanian confidence interval for the treatment effect based on the average treated minus control difference in gain scores is much shorter than the corresponding interval based on post-test scores. Also, the observed value of the difference of gain scores is the most extreme value that can be observed under Fisher's null hypothesis, and so the observed result with gain scores is as significant as possible, which is not true for the difference in post-test scores.

3.4. Propensity scores

Suppose that the assignment mechanism is unconfounded:

$$\Pr\big(W|X(1), Y(0)\big) = \Pr(W|X)$$

(e.g., older males have probability 0.8 of being assigned the new treatment; younger males, 0.6; older females, 0.5; and younger females, 0.2). With large N, because of the random indexing of units, by appealing to de Finetti's (1963) theorem, we can write $\Pr(W|X)$ as

$$\Pr(W|X) \propto \int \prod_{1}^{N} e(X_i|\phi)\pi(\phi)\,\mathrm{d}\phi, \tag{7}$$

where the function $e(X_i|\phi)$ gives the probability that a unit with value X_i of the covariate has $W_i = 1$ as a function of the parameter ϕ with prior (or marginal) probability density function $\pi(\phi)$.

Assignment mechanisms for which the representation in (7) is true and for which all $0 < e(X_i|\phi) < 1$ (for all X_i, ϕ) are called "regular", whether or not the functions $e(\cdot|\phi)$ and $\pi(\phi)$ are known. The unit level assignment probabilities, $e_i = e(X_i|\phi)$, are called propensity scores (Rosenbaum and Rubin, 1983a). Regular designs, which essentially imply strongly ignorable treatment assignment (Rosenbaum and Rubin, 1983a), are the major template for the analysis of observational, nonrandomized studies, and propensity scores are the key ingredients of regular designs. That is, even with an observational data set, we try to structure the problem so that we can conceptualize the data as having arisen from an underlying regular assignment mechanism. Four situations need to be distinguished: (1) when the propensity scores are known, that is, when the function $\pi(\phi)$ and the parameter ϕ are known; (2) when the functional form $e(X_i|\phi)$ is known, but ϕ is not; (3) when the function form, $e(X_i|\phi)$, is unknown, but it is known to be a stochastic function of X_i, and (4) when the assignment mechanism is not even known to be unconfounded. The techniques of analysis are similar across settings, however.

3.5. Known propensity scores

When the propensity scores are known, the assignment mechanism is essentially known. As a result, simple generalizations of Fisherian and Neymanian modes of inference can be applied, for example, by considering the number of treated $\sum W_i$ and number of controls $\sum(1 - W_i)$ to be fixed by design. In particular, Horvitz–Thompson (1952) estimation (or ratio-adjusted versions, see Cochran, 1963), where observations are weighted by the inverse probabilities of their being observed, play a key role for both randomization-based modes of inference because these estimates are unbiased for average treatment effects over the randomization distribution with no modeling assumptions. As the overlap in propensity scores in the treatment and control groups becomes more limited (i.e., as propensity scores approach zero or one), the Neymanian variance of the estimators for the average causal effect increases, with the result that confidence intervals become very wide, and the Fisherian randomization distribution has more of its probability mass on the observed randomization, with the result that it becomes very difficult to get a "significant" p-value. If there is no, or little, overlap with the propensity scores in the treatment groups, no sharp causal inference is possible using the basic Fisherian or Neymanian perspectives. This is a critical issue that researchers must appreciate, no matter what their field.

In general, with the assignment-based modes of inference and known propensity scores that take many values, it is usually acceptable to create several (e.g., 5–10) subclasses of propensity scores to recreate a randomized block experiment (i.e., a series of completely randomized experiments with different propensities across them). This conclusion is based on Cochran's (1968) advice. Alternatively, pairs of treatment-control units can be created that are matched on the propensity scores, thereby recreating a paired comparison experiment.

3.6. Unknown propensity scores, but regular design

When the propensity scores are unknown, but the function $e(X_i|\phi)$ is known, the obvious first step is to estimate them, i.e., that is, estimate ϕ, and thereby e_i, typically using maximum likelihood estimates of ϕ. When the function $e(X_i|\phi)$ is not known, various methods can be used to estimate propensity scores (e.g., discriminant analysis, logistic regression, probit regression). In either case, typically estimated propensity scores are used as if they were known, and often this leads to more precision than using true propensity scores (Rubin and Thomas, 1992a).

Generally, with a design that is known to be regular, the issues that arise with estimated propensity scores are the same as with known ones, and the reduction to a paired-comparison or randomized-block design is acceptable when there is enough overlap in the estimated propensity scores. A common technique is the use of matched sampling to create treated-control pairs whose values of X are "close". Typically, each treated unit is matched to one "close" control unit, and the unused control units are discarded. Various definitions of "close" can be formulated: there are techniques for scalar X (Rubin, 1973); ones for multivariate X summarized by the scalar (estimated) propensity score or best linear discriminant (Cochran and Rubin, 1973; Rosenbaum and Rubin, 1983a), or using multivariate metrics, such as the Mahalanobis metric (Rubin, 1976b, 1976c, 1979); or methods can be combined, such as Mahalanobis metric matching within propensity score calipers (Rosenbaum and Rubin, 1985; Rubin and Thomas, 1996; also see Rubin, 2006a).

The closely related technique of subclassification is also commonly used when no control units are to be discarded, as discussed years ago in Cochran (1968). Subclassification is used to reconstruct an underlying, hypothetical randomized block design.

Some theoretical results, for instance, concerning "Equal Percent Bias Reducing" (EPBR, Rubin, 1976b, 1976c) matching methods are important, as are extensions involving affinely invariant matching methods with ellipsoidal distributions (Rubin and Thomas, 1992b) and further extensions involving discriminant mixtures of ellipsoidal distributions (Rubin and Stuart, 2006). Much work is currently taking place on theoretical, practical, and computational aspects of matching algorithms (see page 1 of Rubin, 2006b for some recent references).

3.7. *Observational studies – possibly confounded*

In order to draw statistical inferences in observational studies, a model for the assignment mechanism is needed, and this defines the template into which we can map the data from an observational study. That is, we need to posit a particular form for the assignment mechanism, and the major template that we try to use is the class of complicated randomized experiments, i.e., regular designs. Although designs that are known to be regular but that have unknown propensity scores are not that common in practice (because of the need to know all covariates used in the assignment mechanism), they are the most critical template for inference for causal effects from observational data. That is, we attempt to assemble data with enough covariates that it becomes plausible (or initially arguable) that the unknown assignment mechanism is unconfounded given these covariates. Then an observational study can be analyzed using the techniques for a regular design with unknown propensity scores. The resulting causal inferences will be valid under the assumption of strong ignorability.

That is, we begin by estimating propensity scores. If there is little or no overlap in the distributions of the estimated propensity scores in the treatment and control groups, there is no hope of drawing valid causal inferences from these data without making strong external assumptions. The message that sometimes a data set cannot support a decent causal inference is very important for all researchers to understand and accept. The desirability of discarding irrelevant units from the control group is also an important point to recognize. For example, in the tobacco litigation (Rubin, 2002), never smokers who have characteristics not matching any smokers were discarded. Sometimes it may even be necessary to discard some treated units as "unmatchable", and then their characteristics should be carefully described because of the limitations on generalization of results. Sometimes subclassification can be used. The example in Rubin (1997) on treatments for breast cancer compares results from randomized experiments and an observational study based on subclassification.

A key idea is that, like good experiments, good observational studies are designed, not simply "found". When designing an experiment, we do not have any outcome data, but we plan the collection, organization and analysis of the data to improve our chances of obtaining valid, reliable and precise causal answers. The same exercise should be done in an observational study: Even if outcome data are available at the design stage, they should be put aside.

Because observational studies are rarely known to be unconfounded, we are concerned with sensitivity of answers to unobserved covariates. Because in my view this and other complications are better dealt with from the model-based perspective, these are addressed after discussing Bayesian methods, although methods described by Rosenbaum (2002) are appropriate from the randomization-based perspective.

4. Posterior predictive causal inference

Bayesian causal inference for causal effects requires a model for the underlying data, $\Pr(X, Y(0), Y(1))$, and this is where "science" enters. A great virtue of the RCM framework we are presenting is that it separates science – a model for the underlying data, from what we do to learn about science – the assignment mechanism, $\Pr(W|X, Y(0), Y(1))$. Notice that together, these two models specify a joint distribution for all observables, an approach commonly called Bayesian.

4.1. *The posterior predictive distribution of causal effects*

Bayesian inference for causal effects directly and explicitly confronts the missing potential outcomes, $Y_{\text{mis}} = \{Y_{\text{mis},i}\}$ where $Y_{\text{mis},i} = W_i Y_i(0) + (1 - W_i)Y_i(1)$. The perspective takes the specification for the assignment mechanism and the specification for the underlying data, and derives the posterior predictive distribution of Y_{mis}, that is, the distribution of Y_{mis} given all observed values

$$\Pr(Y_{\text{mis}}|X, Y_{\text{obs}}, W). \tag{8}$$

This distribution is posterior because it is conditional on all observed values (X, Y_{obs}, W) and predictive because it predicts (stochastically) the missing potential outcomes. From this distribution and the observed values of the potential outcomes, Y_{obs}, the observed assignments, W, and observed covariates, X, the posterior distribution of any causal effect can, in principle, be calculated.

This conclusion is immediate if we view the posterior predictive distribution in (8) as specifying how to take a random draw of Y_{mis}. Once a value of Y_{mis} is drawn, any causal effect can be directly calculated from the drawn value of Y_{mis} and the observed values of X and Y_{obs}, e.g., the median causal effect for males: $\text{med}\{Y_i(1) - Y_i(0)|X_i \text{ indicate males}\}$. Repeatedly drawing values of Y_{mis} and calculating the causal effect for each draw generates the posterior distribution of the desired causal effect. Thus, we can view causal inference entirely as a missing data problem, where we multiply-impute (Rubin, 1987, 2004a) the missing potential outcomes to generate a posterior distribution for the causal effects. We now describe how to generate these imputations. A great advantage of this general approach is that we can model the data on one scale (e.g., log (income) is normal), impute on that scale, but transform the imputations before drawing inferences on another scale (e.g., raw dollars).

4.2. *The posterior predictive distribution of Y_{mis} under ignorable treatment assignment*

First consider how to create the posterior predictive distribution of Y_{mis} when the treatment assignment mechanism is ignorable (i.e., when (6) holds). In general:

$$\Pr(Y_{\text{mis}}|X, Y_{\text{obs}}, W) = \frac{\Pr(X, Y(0), Y(1)) \Pr(W|X, Y(0), Y(1))}{\int \Pr(X, Y(0), Y(1)) \Pr(W|X, Y(0), Y(1))\, \mathrm{d}Y_{\text{mis}}}. \tag{9}$$

With ignorable treatment assignment, Eq. (9) becomes:

$$\Pr(Y_{\text{mis}}|X, Y_{\text{obs}}, W) = \frac{\Pr(X, Y(0), Y(1))}{\int \Pr(X, Y(0), X(1))\, \mathrm{d}Y_{\text{mis}}}. \tag{10}$$

Eq. (10) reveals that, under ignorability, all that we need model is the science $\Pr(X, Y(0), Y(1))$. We can ignore an ignorable assignment mechanism.

4.3. de Finetti's theorem applied to model the data

Because all information is in the underlying data, the unit labels are effectively just random numbers, and hence the array $(X, Y(0), Y(1))$ is row exchangeable. With essentially no loss of generality, therefore, by de Finetti's (1963) theorem, we have that the distribution of $(X, Y(0), Y(1))$ may be taken to be i.i.d. (independent and identically distributed) given some parameter θ, with prior distribution $p(\theta)$:

$$\Pr\bigl(X, Y(0), Y(1)\bigr) = \int \left[\prod_{i=1}^{N} f\bigl(X_i, Y_i(0), Y_i(1)|\theta\bigr) \right] p(\theta)\,\mathrm{d}(\theta). \tag{11}$$

Eq. (11) provides the bridge between fundamental theory and the common practice of using i.i.d. models. Of course, there remains the critical point that the functions $f(\cdot|\theta)$ and $p(\theta)$ are rarely, if ever known, and this limitation will haunt this form of inference despite its great flexibility. We nevertheless proceed with this approach with completely general $f(\cdot|\theta)$ and $p(\theta)$.

4.4. Assumption: Parametric irrelevance of marginal distribution of X

Without loss of generality, we can factor $f(X_i, Y_i(0), Y_i(1)|\theta)$ into:

$$f\bigl(Y_i(0), Y_i(1)|X_i, \theta_{y\cdot x}\bigr) f(X_i|\theta_x),$$

where $\theta_{y\cdot x} = \theta_{y\cdot x}(\theta)$ is the parameter governing the conditional distribution of $Y_i(0), Y_i(1)$ given X_i, and analogously, $\theta_x = \theta_x(\theta)$ is the parameter governing the marginal distribution of X. The reason for doing this factorization is that we are assuming X is fully observed, and so we wish to predict the missing potential outcomes Y_{mis} from X and the observed potential outcomes, Y_{obs}, and therefore must use $f(Y_i(0), Y_i(1)|X_i, \theta_{y\cdot x})$.

To do this, we factor $f(Y_i(0), Y_i(1)|X_i, \theta_{y\cdot x})$ into either

$$f\bigl(Y_i(0)|X_i, Y_i(1), \theta_{0\cdot x1}\bigr) f\bigl(Y_i(1)|X_i, \theta_{1\cdot x}\bigr)$$

when $Y_i(0)$ is missing, or

$$f\bigl(Y_i(1)|X_i, Y_i(0), \theta_{1\cdot x0}\bigr) f\bigl(Y_i(0)|X_i, \theta_{0\cdot x}\bigr)$$

when $Y_i(1)$ is missing; here the various subscripted θ's are all functions of θ governing the appropriate distributions in an obvious notation.

These factorizations allow us to write (11) as

$$\int \prod_{i\in S_1} f\bigl(Y_i(0)|X_i, Y_i(1), \theta_{0\cdot x1}\bigr) \prod_{i\in S_1} f\bigl(Y_i(1)|X_i, \theta_{1\cdot x}\bigr) \tag{12a}$$

$$\times \prod_{i\in S_0} f\bigl(Y_i(1)|X_i, Y_i(0), \theta_{1\cdot x0}\bigr) \prod_{i\in S_0} f\bigl(Y_i(0)|X_i, \theta_{0\cdot x}\bigr) \tag{12b}$$

$$\times \prod_{i\in S_*} f\bigl(Y_i(0), Y_i(1)|X_i, \theta_{y\cdot x}\bigr) \tag{12c}$$

$$\times \prod_{i=1}^{N} f(X_i|\theta_x) p(\theta)\,\mathrm{d}\theta, \tag{12d}$$

where $S_0 = \{i|W_i = 0\}$, $S_1 = \{i|W_i = 1\}$ and $S_* = \{i|W_i = *\}$.

Notice that the first factor in (12a), times the first factor in (12b), times (12c) is proportional to the posterior predictive distribution of Y_{mis} given θ (i.e., given X, Y_{obs} and θ), $\Pr(Y_{\text{mis}}|X, Y_{\text{obs}}, \theta)$. Also notice that the remaining factors in (12), that is the second factor in (12a) times the second factor in (12b) times (12d) is proportional to the posterior distribution of θ, $\Pr(\theta|X, Y_{\text{obs}})$, which is equal to the likelihood of θ, $\mathcal{L}(\theta|X, Y_{\text{obs}})$, times the prior distribution of θ, $p(\theta)$.

Let us now assume that $\theta_{y\cdot x}$ and θ_x are a priori independent:

$$p(\theta) = p(\theta_{y\cdot x}) p(\theta_x). \tag{13}$$

This assumption is not innocuous, but it is useful and is standard in many prediction environments. For an example of a situation where it might not be reasonable, suppose X includes students' scores on a standardized test going back many years; the relationships among previous X tests may provide useful information for predicting $Y(0)$ (i.e., Y without intervention), from X, using for example, a time-series model (e.g., Box and Jenkins, 1970).

For simplicity in the presentation here, we make assumption (13), although, as with all such assumptions, it should be carefully considered. Then the integral over θ_x in (12) passes through all the products in (12a), (12b) and (12c), and we are left with the integral over (12d); (12d) after this integration is proportional to $p(\theta_{y\cdot x})\,\mathrm{d}\theta_{y\cdot x}$.

As a consequence, (12) becomes

$$\int \Pr(Y_{\text{mis}}|X, Y_{\text{obs}}, \theta_{y\cdot x}) \Pr(\theta_{y\cdot x}|X, Y_{\text{obs}})\,\mathrm{d}\theta_{y\cdot x}, \tag{14}$$

where the second factor in (14) is proportional to the product of the second factors in (12a) and (12b), and the first factor in (14) is, as before, proportional to the product of the first factors of (12a) and (12b) times (12c).

4.5. Assumption: No contamination of imputations across treatments

We now make an assumption that is sometimes implicitly made and sometimes explicitly not made, which is the case discussed in Section 3. Specifically, we now assume that entirely separate activities are to be used to impute the missing $Y_i(0)$ and the missing $Y_i(1)$. This is accomplished with two formal assumptions:

$$f\big(Y_i(0), Y_i(1)|X_i, \theta_{y\cdot x}\big) = f\big(Y_i(0)|X_i, \theta_{0\cdot x}\big) f\big(Y_i(1)|X_i, \theta_{1\cdot x}\big) \tag{15}$$

and

$$p(\theta_{y\cdot x}) = p(\theta_{0\cdot x}) p(\theta_{1\cdot x}). \tag{16}$$

Thus, in (15) we assume $Y_i(0)$ and $Y_i(1)$ are conditionally independent given X_i and $\theta_{y\cdot x}$, and in (16), that the parameters governing these conditional distributions are a priori independent. Consequently, $f(Y_i(0)|X_i, Y_i(1), \theta_{0\cdot x1}) = f(Y_i(0)|X_i, \theta_{0\cdot x})$, and

$f(Y_i(1)|X_i, Y_i(0), \theta_{1\cdot 0x}) = f(Y_i(1)|X_i, \theta_{1\cdot x})$. Thus, (12) or (14) can be written in four distinct parts with associated activities as follows.

1. Using the control units, obtain the posterior distribution of $\theta_{0\cdot x}$:

$$p(\theta_{0\cdot x}|X, Y_{\text{obs}}) \propto \mathcal{L}(\theta_{0\cdot x}|X, Y_{\text{obs}})p(\theta_{0\cdot x}) \\ \propto \prod_{i\in S_0} p\big(Y_i(0)|X_i, \theta_{0\cdot x}\big)p(\theta_{0\cdot x}).$$

2. Using $\theta_{0\cdot x}$, obtain the conditional posterior predictive distribution of the missing $Y_i(0)$:

$$\prod_{i\in S_1\cup S_*} \Pr\big(Y_i(0)|X_i, \theta_{0\cdot x}\big).$$

3. Using the treated units, obtain the posterior distribution of $\theta_{1\cdot x}$:

$$\Pr(\theta_{1\cdot x}|X, Y_{\text{obs}}) \propto \mathcal{L}(\theta_{1\cdot x}|X, Y_{\text{obs}})p(\theta_{1\cdot x}) \\ \propto \prod_{i\in S_1} \Pr\big(Y_i(1)|X_i, \theta_{1\cdot x}\big)p(\theta_{1\cdot x}).$$

4. Using $\theta_{1\cdot x}$, obtain the conditional posterior predictive distribution of the missing $Y_i(1)$:

$$\prod_{i\in S_0\cup S_*} \Pr\big(Y_i(1)|X_i, \theta_{1\cdot x}\big).$$

For simulation, perform steps 1–4 repeatedly with random draws, thereby multiply imputing Y_{mis} (Rubin, 1987, 2004a).

4.6. *Simple normal example illustrating the four steps*

To illustrate the idea of imputing the missing potential outcomes, suppose there is no X and $f(\cdot|\theta)$ is normal with means (μ_1, μ_0), variances (σ_1^2, σ_0^2) and zero correlation.

The units with $W_i = 1$ (i.e., $i \in S_1$) have $Y_i(1)$ observed and are missing $Y_i(0)$, and so their $Y_i(0)$ values need to be imputed. To impute $Y_i(0)$ values for them, intuitively we need to find units with $Y_i(0)$ observed who are exchangeable with the $W_i = 1$ units, but these units are the units with $W_i = 0$ (i.e., $i \in S_0$). Therefore, we estimate (in a Bayesian way) the distribution of $Y_i(0)$ from the units with $W_i = 0$, and use this estimated distribution to impute $Y_i(0)$ for the units missing $Y_i(0)$.

Because the n_0 observed values of $Y_i(0)$ are a simple random sample of the N values of $Y(0)$, and are normally distributed with mean μ_0 and variance σ_0^2, with the standard independent noninformative prior distributions on (μ_0, σ_0^2), we have for the posterior distribution of σ_0^2:

$$\sigma_0^2/s_0^2 \sim \text{inverted } \chi^2_{n_0-1}/(n_0 - 1);$$

and for the posterior distribution of μ_0 given σ_0:

$$\mu_0 \sim N\big(\bar{y}_0, s_0^2/n_0\big);$$

Table 2
Observed data in artificial example

X	$\bar{y}_0$	n_0	s_0	$\bar{y}_1$	n_1	s_1
HI	300	90	60	400	10	60
LO	100	10	60	200	90	60

Table 3
Causal inferences for example in Table 2

	$X = HI$	$X = LO$	Population $= \frac{1}{2}HI + \frac{1}{2}LO$
$E(\bar{Y}_1 - \bar{Y}_0 \mid X, Y_{\text{obs}}, W)$	100	100	100
$V(\bar{Y}_1 - \bar{Y}_0 \mid X, Y_{\text{obs}}, W)^{1/2}$	20	20	$10\sqrt{2}$

and for the missing $Y_i(0)$ given μ_0 and σ_0:

$$Y_i(0) \ni W_i \neq 0 \overset{\text{i.i.d.}}{\sim} N(\mu_0, s_0^2).$$

The missing values of $Y_i(1)$ are analogously imputed using the observed values of $Y_i(1)$.

When there are covariates observed, these are used to help predict the missing potential outcomes using, for example, one regression model for the observed $Y_i(1)$ given the covariates, and another regression model for the observed $Y_i(0)$ given the covariates.

4.7. Simple normal example with covariate – numerical example

For a specific example with a covariate, suppose we have a large population of students with a covariate X_i indicating parental education, which is dichotomous, *HI* versus *LO*, with a 50%/50% split in the population. Suppose that a random sample of 100 with $X_0 = HI$ is taken, and 10 are randomly assigned to the active treatment, and 90 are randomly assigned to the control treatment. Further suppose that a random sample of 100 with $X_i = LO$ is taken, and 90 are randomly assigned to the active treatment and 10 are assigned to the control treatment. The outcome Y is a test score a year after randomization, with $Y_{i,\text{obs}}$ and X_i observed for all 200 units; X_i is effectively observed in the population because we know the proportions of X_i that are *HI* and *LO* in the full population.

Suppose the hypothetical observed data are as displayed in Table 2. Then the inferences based on the normal-model are as shown in Table 3. The obvious conclusion in this artificial example is that the treatment increases test scores for both those with *HI* and *LO* parental education by about 100 points, and thus for the population, which is a 50%/50% mixture of these two subpopulations. In this sort of situation, the final inference is insensitive to the assumed normality of $Y_i(1)$ given X_i and of $Y_i(0)$ given X_i; see Pratt (1965) or Rubin (1987, 2004a, Section 2.5) for the argument. But, in general, this is not so.

4.8. *Dangers of model-based extrapolation with some regular designs*

A great strength of the model-based approach is that it allows us to conduct causal inference by predicting all of the missing potential outcomes from observed values. The problem with this approach is the need to specify the distribution $f(\cdot|\theta)$ and $p(\theta)$, which sometimes can implicitly involve extrapolations that are extremely unreliable. This situation can be easily conveyed by a simple example based on the one in Section 4.7.

Suppose that the units with *LO* parental education are split equally into *NONE* and *LITTLE*, and further suppose that the 10 with *LO* assigned the control treatment all are *LITTLE*; the 90 with *LO* assigned the active treatment are 50 *NONE* and 40 *LITTLE*. Now, although the comparison of treatment versus control for $X_i = HI$ is unaffected, the comparison of treatment versus control for the $X = NONE$ group is entirely dependent on our model specifications. That is, there are no $X = NONE$ units in the control condition, and so to impute $Y_i(0)$ values for the $X_i = NONE$ treated units, we must rely entirely on some external information.

For example, suppose we associate *NONE* with $X_i = 0$, *LITTLE* with $X_i = 1$, and *HI* with $X_i = 2$, and claim that $Y_i(0)$ is linearly related to X_i (given θ). In the treatment group, we can then impute the missing $Y_i(0)$ for units with $X_i = NONE$ even though there are no units with $X_i = NONE$ and $Y_i(0)$ observed, based on the assumed linear relationship between $Y_i(0)$ and X_i estimated from the 10 control units with $X_i = 1$ and the 90 with $X_i = 2$. Moreover, as the sample sizes get bigger and bigger, the posterior variance of the estimated average causal effect shrinks to zero, so we are certain of our answer for the $X_i = NONE$ causal effect, even though it is reliant on an assumption that may be only implicitly recognized: the linear relationships between the control potential outcomes and the covariate, which allows the extrapolation to take place.

Because of the issues of model-based extrapolation, propensity score methods (i.e., based on matching or subclassification) are highly relevant to the application of Bayesian methods to causal inference. For example, it is always a good idea to examine the overlap in multivariate X distributions between treatment and control groups, and, by design, to create treated and control units with very similar distributions of X. Then, formal model-based imputations that rely on interpolation can be made with these samples, thereby avoiding extrapolation. This approach is illustrated in a psychological/medical study in Reinisch et al. (1995), where matching was used to select a subset of controls, and linear model adjustments were made in the resulting samples.

4.9. *Nonignorable treatment assignment mechanisms*

With nonignorable treatment assignment, the simplifications in previous sections, which follow from ignoring the specification for $\Pr(W|X, Y(0), Y(1))$, do not follow in general, and the analysis typically becomes far more difficult and uncertain. As a simple illustration, take the example in Section 4.7 and assume that everything is the same except that only Y_{obs} is recorded, so that we do not know whether X_i is *HI* or *LO* for anyone. The actual assignment mechanism is now

$$\Pr\bigl(W|Y(0), Y(1)\bigr) = \int \Pr\bigl(W|X, Y(0), Y(1)\bigr)\,\mathrm{d}P(X)$$

because X itself is missing, and so treatment assignment depends explicitly on the potential outcomes, both observed and missing, because both are generally correlated with the missing X_i.

Inference for causal effects, assuming the identical model for the science, now depends on the implied normal mixture model for the observed Y data within each treatment arm, because the population is a 50%/50% mixture of those with *LO* and *HI* parental education, and these subpopulations have different probabilities of treatment assignment. Here the inference for causal effects is sensitive the propriety of the assumed normality and/or the assumption of a 50%/50% mixture, as well as to the prior distributions on μ_1, μ_0, σ_1 and σ_0.

If we mistakenly ignore the nonignorable treatment assignment and simply compare the sample means of all treated with all controls, for example, using the simple model of Section 4.6, we have $\bar{y}_0 = 0.9(300) + 0.1(100) = 280$ versus $\bar{y}_1 = 0.1(400) + 0.9(200) = 220$; doing so, we reach the incorrect conclusion that the active treatment hurts the test score in the population. This sort of example is known as "Simpson's Paradox" (Simpson, 1951) and can easily arise with incorrect analyses of nonignorable treatment assignment mechanisms, and thus indicates why such assignment mechanisms are to be avoided whenever possible. Randomized experiments are the most direct way of avoiding nonignorable treatment assignments. Other alternatives are ignorable designs with nonprobabilistic features so that all units with some specific value of covariates are assigned the same treatment, like the extreme example in Section 4.8. In practice, the analyses of observational studies proceeds as if they were ignorable, as discussed previously. Then, to assess the consequences of this assumption, sensitivity analyses can be conducted under various hypothetical situations.

Typically sensitivity analyses utilize the idea of a fully missing covariate, U, such that treatment assignment is ignorable given U but not given the observed data. The relationships between U and W, and between U, $Y(0)$, and $Y(1)$, all given X, are then varied. See, for example, Rosenbaum and Rubin (1983b) and Cornfield et al. (1959). Extreme versions of sensitivity analyses examine bounds (e.g., see Imbens and Manski, 2004). Bounds on point estimates can be viewed as extreme versions of sensitivity analyses where we obtain estimates that might be obtained over extreme distributions of the unobserved covariate U. Although often very broad, such bounds can play an important role in informing us about the sources of sharpness in inferences. Some relevant references for this approach include Manski et al. (1992), Manski and Nagin (1998), and Horowitz and Manski (2000).

Related techniques for assessing nonignorable designs include the formal role of a second control group (Rosenbaum, 1987) and tests of unconfoundedness (Rosenbaum, 1984). Generally I prefer to use evidence to produce better estimates rather than to test assumptions. That is, if there is evidence in the data available to test an assumption, then there is evidence for how to generalize the questionable assumption, and thereby improve the estimation of causal effects.

5. Complications

There are many complications that occur in real world studies for causal effects, many of which can be handled much more flexibly with the Bayesian approach than with randomization-based methods. Of course, the models involved, including associated prior distributions, can be very demanding to formulate in a practically reliable manner. Also, Neymanian evaluations are still important. Here I simply list some of these complications with some admittedly idiosyncratically personal references to current work from the Bayesian perspective.

5.1. Multiple treatments

When there are more than two treatments, the notation becomes more complex but is still straightforward under SUTVA. Without SUTVA, however, both the notation and the analysis can become very involved. The exploding number of potential outcomes can become especially serious in studies where the units are exposed to a sequence of repeated treatments in time, each distinct sequence corresponding to a possibly distinct treatment. Much of the field of classical experiment design is devoted to issues that arise with more than two treatment conditions (e.g., Kempthorne, 1952; Cochran and Cox, 1957).

5.2. Unintended missing data

Missing data, due perhaps to unit dropout or machine failure, can complicated analyses more than one would expect based on a cursory examination of the problem. Fortunately, Bayesian/likelihood tools for addressing missing data such as multiple imputation (Rubin, 1987, 2004a) or the EM algorithm (Dempster et al., 1977) and its relatives, including data augmentation (Tanner and Wong, 1987) and the Gibbs sampler (Geman and Geman, 1984) are fully compatible with the Bayesian approach to causal inference outlined in Section 4. Gelman et al. (2003, Parts III and IV) provide guidance on many of these issues from the Bayesian perspective.

5.3. Noncompliance with assigned treatment

Another complication, common when the units are people, is noncompliance. For example, some of the subjects assigned to take the active treatment take the control treatment instead, and some assigned to take the control manage to take the active treatment. A nice example of this is given in Sommer and Zeger (1991). Initial interest focuses on the effect of the treatment for the subset of people who would comply with their treatment assignment under either assignment. Early work related to this issue can be found in economics (e.g., Tinbergen, 1930; Haavelmo, 1944) and elsewhere (e.g., Zelen, 1979; Bloom, 1984). Much progress has been made in recent years on this topic (e.g., Baker, 1998; Baker and Lindeman, 1994; Goetghebeur and Molenberghs, 1996; Angrist et al., 1996; Imbens and Rubin, 1997; Little and Yau, 1998; Hirano et al., 2000). In this case, sensitivity of inference to prior assumptions can be severe, and the Bayesian approach is well-suited to not only revealing this sensitivity but also to formulating reasonable prior restrictions.

5.4. *Truncation of outcomes due to death*

In other cases, the unit may "die" before the final outcome can be measured. For example, in an experiment with new fertilizers, a plant may die before the crops are harvested and interest may focus on both the effect of the fertilizer on plant survival and the effect of the fertilizer on plant yield when the plant survives. Or with an educational intervention designed to improve final high school test scores, students who drop out of school before taking the test effectively have their data "truncated due to death". This problem is far more subtle than it may at first appear to be, and valid approaches to it have only recently been formulated (Rubin, 2000, 2006c; Zhang and Rubin, 2003). Surprisingly, the models also have applications in economics (Zhang et al., 2006) and the evaluation of job-training programs.

5.5. *Direct and indirect causal effects*

Another topic that is far more subtle than it first appears to be is the one involving direct and indirect causal effects. For example, the separation of the "direct" effect of a vaccination on disease from the "indirect" effect of the vaccination that is due solely to its effect on blood antibodies and the "direct" effect of the antibodies on disease. This language turns out to be too imprecise to be useful within our formal causal effect framework. This problem is ripe for Bayesian modelling as briefly outlined in Rubin (2004b, 2005).

5.6. *Principal stratification*

All the examples in Sections 5.3–5.5 can be viewed as special cases of "principal stratification" (Frangakis and Rubin, 2002), where the principal strata are defined by partially unobserved intermediate potential outcomes, namely in our examples: compliance behavior under both treatment assignments, survival under both treatment assignments, and antibody level under both treatment assignments. This appears to be an extremely fertile area for research and application of Bayesian methods for causal inference, especially using modern simulation methods such as MCMC (Markov Chain Monte Carlo); see, for example, Gilks et al. (1995).

5.7. *Combinations of complications*

In the real world, such complications typically do not appear simply one at a time. For example, a randomized experiment in education evaluating "school choice" suffered from missing data in both covariates and longitudinal outcomes; also, the outcome was multicomponent as each point in time; in addition, it suffered from noncompliance that took several levels because of the years of school. Some of these combinations of complications are discussed in Barnard et al. (2003) in the context of the school choice example, and in Mealli and Rubin (2002) in the context of a medical experiment.

Despite the fact that Bayesian analysis is quite difficult when confronted with these combinations of complications, I believe that it is still a far more satisfactory attack on the real scientific problems of causal inference than the vast majority of ad hoc frequentist approaches commonly in use today.

5.8. More missing data

The problem of missing data, both in covariates and outcomes, is very common in practice. Standard methods (e.g., as in Little and Rubin, 2002) are highly valuable here, and special methods, for instance, for dealing with missing covariates in propensity score analyses (D'Agostino and Rubin, 1999) are also relevant. Outcomes that are censored, e.g., survival data, can be viewed as a special but very important case of missing or coarsened data (Heitjan and Rubin, 1991). Moreover, dealing with combined complications, such as missing outcomes with noncompliance (Frangakis and Rubin, 1999), is important, as is clustering in design issues (Frangakis et al., 2002).

References

Angrist, J.D., Imbens, G.W., Rubin, D.B. (1996). Identification of causal effects using instrumental variables. *Journal of the American Statistical Association* **91**, 444–472. As Applications Invited Discussion Article with discussion and rejoinder.

Anscombe, F.J. (1948). The validity of comparative experiments. *Journal of the Royal Statistical Society, Ser. A* **61**, 181–211.

Baker, S.G. (1998). Analysis of survival data from a randomized trial with all-or-none compliance: Estimating the cost-effectiveness of a cancer screening program. *Journal of the American Statistical Association* **93**, 929–934.

Baker, S.G., Lindeman, K.S. (1994). The paired availability design: A proposal for evaluating epidural analgesia during labor. *Statistics in Medicine* **13**, 2269–2278.

Barnard, J., Hill, J., Frangakis, C., Rubin, D. (2003). School choice in NY city: A Bayesian analysis of an imperfect randomized experiment. In: Gatsonis, C., Carlin, B., Carriquiry, A. (Eds.), In: *Case Studies in Bayesian Statistics*, vol. V. Springer-Verlag, New York, pp. 3–97. With discussion and rejoinder.

Bloom, H.S. (1984). Accounting for no-shows in experimental evaluation designs. *Evaluation Review* **8**, 225–246.

Box, G.E.P., Jenkins, G.M. (1970). *Time Series Analysis: Forecasting and Controls*. Holden-Day, San Francisco.

Brillinger, D.R., Jones, L.V., Tukey, J.W. (1978). Report of the statistical task force for the weather modification advisory board. In: *The Management of Western Resources*, vol. II. *The Role of Statistics on Weather Resources Management*. Government Printing Office, Washington, DC. Stock No. 003-018-00091-1.

Cochran, W.G. (1963). *Sampling Techniques*, 2nd ed. Wiley, New York.

Cochran, W.G. (1968). The effectiveness of adjustment by subclassification in removing bias in observational studies. *Biometrics* **24**, 295–313.

Cochran, W.G., Cox, G.M. (1957). *Experimental Designs*, 2nd ed. Wiley, New York. Reprinted as a *Wiley Classic* (1992).

Cochran, W.G., Rubin, D.B. (1973). Controlling bias in observational studies: A review. *Sankhya-A* **35**, 417–446.

Cornfield, J., et al. (1959). Smoking and lung cancer; recent evidence and a discussion of some questions. *Journal of the National Cancer Institute* **22**, 173–203.

Cox, D.R. (1958). *The Planning of Experiments*. Wiley, New York.

Cox, D.R. (1992). Causality: Some statistical aspects. *Journal of the Royal Statistical Society, Ser. A* **155**, 291–301.

Cox, D.R., McCullagh, P. (1982). Some aspects of covariance (with discussion). *Biometrics* **38**, 541–561.

D'Agostino, R. Jr., Rubin, D.B. (1999). Estimation and use of propensity scores with incomplete data. *Journal of the American Statistical Association* **95**, 749–759.

Dawid, A.P. (2000). Causal inference without counterfactuals. *Journal of the American Statistical Association* **95**, 407–424. With discussion.

de Finetti, B. (1963). Foresight: Its logical laws, its subjective sources. In: Kyburg, H.E., Smokler, H.E. (Eds.), *Studies in Subjective Probability*. Wiley, New York.

Dempster, A.P. (1990). Causality and statistics. *Journal of Statistical Planning and Inference* **25**, 261–278.

Dempster, A.P., Laird, N., Rubin, D.B. (1977). Maximum likelihood from incomplete data via the EM algorithm. *Journal of the Royal Statistical Society, Ser. B* **39**, 1–38. With discussion and reply.

Efron, B. (1971). Forcing a sequential experiment to be balanced. *Biometrika* **58**, 403–417.

Efron, B., Feldman, D. (1991). Compliance as an explanatory variable in clinical trials. *Journal of the American Statistical Association* **86**, 9–17.

Federal Judicial Center (2000). *Reference Manual on Scientific Evidence*, 2nd ed. Federal Judicial Center, Washington, DC.

Fisher, R.A. (1918). The causes of human variability. *Eugenics Review* **10**, 213–220.

Fisher, R.A. (1925). *Statistical Methods for Research Workers*, 1st ed. Oliver and Boyd, Edinburgh.

Fisher, F. (1999). Expert Witness report for the plaintiffs. U.S. vs. Tobacco Industry.

Frangakis, C., Rubin, D.B. (1999). Addressing complications of intention-to-treat analysis in the combined presence of all-or-none treatment-noncompliance and subsequent missing outcomes. *Biometrika* **86**, 366–379.

Frangakis, C.E., Rubin, D.B. (2002). Principal stratification in causal inference. *Biometrics* **58**, 21–29.

Frangakis, C., Rubin, D.B., Zhou, X.-H. (2002). Clustered encouragement designs with individual noncompliance: Bayesian inference with randomization, and application to advance directive forms. *Biostatistics* **3**, 147–177. With discussion and rejoinder.

Gelman, A., King, G. (1991). Estimating incumbency advantage without bias. *American Journal of Political Science* **34**, 1142–1164.

Gelman, A., Carlin, J., Stern, H., Rubin, D. (2003). *Bayesian Data Analysis*, 2nd ed. CRC Press, New York.

Geman, S., Geman, D. (1984). Stochastic relaxation. Gibbs Distributions, and the Bayesian Restoration of Images. *IEEE Transactions on Pattern Analysis and Machine Intelligence* **6**, 721–741. November.

Gilks, W.R., Richardson, S., Spiegelhalter, D.J. (1995). *Markov Chain Monte Carlo in Practice*. CRC Press, New York.

Goetghebeur, E., Molenberghs, G. (1996). Causal inference in a placebo-controlled clinical trial with binary outcome and ordered compliance. *Journal of the American Statistical Association* **91**, 928–934.

Greenland, S., Poole, C. (1988). Invariants and noninvariants in the concept of interdependent effects. *Scandinavian Journal of Work and Environmental Health* **14**, 125–129.

Greenland, S., Robins, J.M., Pearl, J. (1999). Confounding and collapsibility in causal inference. *Statistical Science* **14**, 29–46.

Haavelmo, T. (1944). The probability approach in econometrics. *Econometrica* **15**, 413–419.

Heckman, J.J. (1989). Causal inference and nonrandom samples. *Journal of Educational Statistics* **14**, 159–168.

Heckman, J.J., Hotz, J. (1989). Alternative methods for evaluating the impact of training programs. *Journal of the American Statistical Association* **84**, 862–874. With discussion.

Heitjan, D., Rubin, D.B. (1991). Ignorability and coarse data. *The Annals of Statistics* **19**, 2244–2253.

Hirano, K., Imbens, G., Rubin, D.B., Zhou, X. (2000). Assessing the effect of an influenza vaccine in an encouragement design. *Biostatistics* **1**, 69–88.

Hodges, J.L., Lehmann, E. (1970). *Basic Concepts of Probability and Statistics*, 2nd ed. Holden-Day, San Francisco.

Holland, P.W. (1986). Statistics and causal inference. *Journal of the American Statistical Association* **81**, 945–970.

Holland, P.W. (1988a). Causal inference, path analysis, and recursive structural equation models. *Sociological Methodology*, 449–484.

Holland, P.W. (1988b). Comment on "Employment discrimination and statistical science" by A.P. Dempster. *Statistical Science* **3**, 186–188.

Holland, P.W. (1989). It's very clear. Comment on "Choosing among alternative nonexperimental methods for estimating the impact of social programs: The case of manpower training" by J. Heckman and V. Hotz. *Journal of the American Statistical Association* **84**, 875–877.

Holland, P.W., Rubin, D.B. (1983). On Lord's paradox. In: Wainer, H., Messick, S. (Eds.), *Principals of Modern Psychological Measurement: A Festschrift for Frederic M. Lord*. Earlbaum, Hillsdale, NJ, pp. 3–25.

Horowitz, J.L., Manski, C.F. (2000). Nonparametric analysis of randomized experiments with missing covariates and outcome data. *Journal of the American Statistical Association* **95**, 77–84.
Horvitz, D.G., Thompson, D.J. (1952). A generalization of sampling without replacement from a finite population. *Journal of the American Statistical Association* **47**, 663–685.
Hurwicz, L. (1962). On the structural form of interdependent systems. In: Nagel, E., Suppes, P., Tarski, A. (Eds.), *Logic, Methodology, and Philosophy of Science*. In: *Proceedings of the 1960 International Congress*. Stanford University Press, Stanford, CA.
Imbens, G.W., Manski, C.F. (2004). Confidence intervals for partially identified parameters. *Econometrica* **72**, 1845–1857.
Imbens, G.W., Rubin, D.B. (1997). Bayesian inference for causal effects in randomized experiments with noncompliance. *The Annals of Statistics* **25**, 305–327.
Imbens, G., Rubin, D.B. (2006a). Rubin causal model. In: *The New Palgrave Dictionary of Economics*, 2nd ed. Palgrave McMillan, New York. In press.
Imbens, G.W., Rubin, D.B. (2006b). *Causal Inference in Statistics and the Medical and Social Sciences*. Cambridge University Press, Cambridge, UK.
Kadane, J.B., Seidenfeld, T. (1990). Randomization in a Bayesian perspective. *Journal of Statistical Planning and Inference* **25**, 329–346.
Kempthorne, O. (1952). *The Design and Analysis of Experiments*. Wiley, New York.
Lindley, D.V., Novick, M.R. (1981). The role of exchangeabiltiy in inference. *Annals of Statistics* **9**, 45–58.
Little, R.J.A., Rubin, D.B. (2002). *Statistical Analysis With Missing Data*, 2nd ed. Wiley, New York.
Little, R.J., Yau, L. (1998). Statistical techniques for analyzing data from prevention trials: Treatment of no-shows using Rubin's causal model. *Psychological Methods* **3**, 147–159.
Lord, F.M. (1967). A paradox in the interpretation of group comparisons. *Psychological Bulletin* **68**, 304–305.
Manski, C.F., Nagin, D.S. (1998). Bounding disagreements about treatment effects: A study of sentencing and recidivism. *Sociological Methodology* **28**, 99–137.
Manski, C., et al. (1992). Alternative estimates of the effect of family structure during adolescence on high school. *Journal of the American Statistical Association* **87**, 25–37.
McCarthy, M.D. (1939). On the application of the z-test to randomized blocks. *Annals of Mathematical Statistics* **10**, 337.
Mealli, F., Rubin, D.B. (2002). Assumptions when analyzing randomized experiments with noncompliance and missing outcomes. *Health Services Outcome Research Methodology* **3**, 225–232.
Mill, J.S. (1973). *A System of Logic. Collected Works of John Stuart Mill*, vol. 7. University of Toronto Press, Toronto.
Neyman, J. (1923). On the application of probability theory to agricultural experiments: Essay on principles, Section 9. Translated in *Statistical Science* **5** (1990), 465–480.
Neyman, J. (1934). On the two different aspects of the representative method: The method of stratified sampling and the method of purposive selection. *Journal of the Royal Statistical Society, Ser. A* **97**, 558–606.
Pearl, J. (2000). *Causality: Models, Reasoning and Inference*. Cambridge University Press, Cambridge, UK.
Peterson, A.V., et al. (2000). Hutchinson smoking prevention project: Long-term randomized trial in school-based tobacco use prevention – results on smoking. *Journal of the National Cancer Institute* **92**, 1979–1991.
Pitman, E.J.G. (1937). Significance tests which can be applied to samples from any population. III. The analysis of variance test. *Biometrika* **29**, 322–335.
Pratt, J.W. (1965). Bayesian interpretation of standard inference statements. *Journal of the Royal Statistical Society, Ser. B* **27**, 169–203. With discussion.
Pratt, J.W., Schlaifer, R. (1984). On the nature and discovery of structure. *Journal of the American Statistical Association* **79**, 9–33. With discussion.
Pratt, J.W., Schlaifer, R. (1988). On the interpretation and observation of laws. *Journal of Econometrics* **39**, 23–52.
Reinisch, J., et al. (1995). In utero exposure to phenobarbital and intelligence deficits in adult men. *Journal of the American Medical Association* **274**, 1518–1525.
Robins, J.M. (1989). The control of confounding by intermediate variables. *Statistics in Medicine* **8**, 679–701.
Rosenbaum, P.R. (1984). From association to causation in observational studies: The role of tests of strongly ignorable treatment assignment. *Journal of the American Statistical Association* **79**, 41–48.

Rosenbaum, P.R. (1987). The role of a second control group in an observational study. *Statistical Science* **2**, 292–316. With discussion.

Rosenbaum, P.R. (2002). *Observational Studies*, 2nd ed. Springer-Verlag, New York.

Rosenbaum, P.R., Rubin, D.B. (1983a). The central role of the propensity score in observational studies for causal effects. *Biometrika* **70**, 41–55.

Rosenbaum, P.R., Rubin, D.B. (1983b). Assessing sensitivity to an unobserved binary covariate in an observational study with binary outcome. *Journal of the Royal Statistical Society, Ser. B* **45**, 212–218.

Rosenbaum, P.R., Rubin, D.B. (1985). Constructing a control group using multivariate matched sampling incorporating the propensity score. *The American Statistician* **39**, 33–38.

Rubin, D.B. (1973). Matching to remove bias in observational studies. *Biometrics* **29**, 159–183. Printer's correction note 30, p. 728.

Rubin, D.B. (1974). Estimating causal effects of treatments in randomized and nonrandomized studies. *Journal of Educational Psychology* **66**, 688–701.

Rubin, D.B. (1975). Bayesian inference for causality: The importance of randomization. In: *The Proceedings of the Social Statistics Section of the American Statistical Association*, pp. 233–239.

Rubin, D.B. (1976a). Inference and missing data. *Biometrika* **63**, 581–592.

Rubin, D.B. (1976b). Multivariate matching methods that are equal percent bias reducing, I: Some examples. *Biometrics* **32**, 109–120. Printer's correction note p. 955.

Rubin, D.B. (1976c). Multivariate matching methods that are equal percent bias reducing, II: Maximums on bias reduction for fixed sample sizes. *Biometrics* **32**, 121–132. Printer's correction note p. 955.

Rubin, D.B. (1977). Assignment of treatment group on the basis of a covariate. *Journal of Educational Statistics* **2**, 1–26.

Rubin, D.B. (1978). Bayesian inference for causal effects: The role of randomization. *The Annals of Statistics* **7**, 34–58.

Rubin, D.B. (1979). Using multivariate matched sampling and regression adjustment to control bias in observational studies. *Journal of the American Statistical Association* **74**, 318–328.

Rubin, D.B. (1980). Comment on "Randomization analysis of experimental data: The Fisher randomization test" by D. Basu. *Journal of the American Statistical Association* **75**, 591–593.

Rubin, D.B. (1987). *Multiple Imputation for Nonresponse in Surveys*. Wiley, New York.

Rubin, D.B. (1990a). Comment: Neyman (1923) and causal inference in experiments and observational studies. *Statistical Science* **5**, 472–480.

Rubin, D.B. (1990b). Formal modes of statistical inference for causal effects. *Journal of Statistical Planning and Inference* **25**, 279–292.

Rubin, D.B. (1997). Estimating causal effects from large data sets using propensity scores. *Annals of Internal Medicine* **127**, 757–763.

Rubin, D.B. (2000). The utility of counterfactuals for causal inference. Comment on A.P. Dawid, 'Causal inference without counterfactuals'. *Journal of the American Statistical Association* **95**, 435–438.

Rubin, D.B. (2002). Using propensity scores to help design observational studies: Application to the tobacco litigation. *Health Services & Outcomes Research Methodology* **2**, 169–188.

Rubin, D.B. (2004a). *Multiple Imputation for Nonresponse in Surveys*. Wiley, New York. Reprinted with new appendices as a *Wiley Classic*.

Rubin, D.B. (2004b). Direct and indirect causal effects via potential outcomes. *The Scandinavian Journal of Statistics* **31**, 161–170; 195–198, with discussion and reply.

Rubin, D.B. (2005). Causal inference using potential outcomes: Design, modeling, decisions. 2004 Fisher Lecture. *Journal of the American Statistical Association* **100**, 322–331.

Rubin, D.B. (2006a). The design versus the analysis of observational studies for causal effects: Parallels with the design of randomized trials. *Statistics in Medicine*, 25th anniversary issue, in press.

Rubin, D.B. (2006b). *Matched Sampling for Causal Effects*. Cambridge University Press.

Rubin, D.B. (2006c). Causal inference through potential outcomes and principal stratification: Applications to studies with 'censoring' due to death. In: *Statistical Science*. With discussion and rejoinder. In press.

Rubin, D.B., Stuart, E.A. (2006). Affinely invariant matching methods with discriminant mixtures of ellipsoidally symmetric distributions. *Annals of Statistics*. In press.

Rubin, D.B., Thomas, N. (1992a). Characterizing the effect of matching using linear propensity score methods with normal covariates. *Biometrika* **79**, 797–809.

Rubin, D.B., Thomas, N. (1992b). Affinely invariant matching methods with ellipsoidal distributions. *Annals of Statistics* **20**, 1079–1093.
Rubin, D.B., Thomas, N. (1996). Matching using estimated propensity scores: Relating theory to practice. *Biometrics* **52**, 249–264.
Rubin, D.B., Stuart, E.A., Zanutto, E.L. (2004). A potential outcomes view of value-added assessment in education. *Journal of Educational and Behavioral Statistics* **29**, 103–116.
Simpson, E.H. (1951). The interpretation of interaction in contingency tables. *Journal of the Royal Statistical Society, Ser. B* **13**, 238–241.
Smith, T.M.F., Sugden, R.A. (1988). Sampling and assignment mechanisms in experiments, surveys and observational studies. *International Statistical Review* **56**, 165–180.
Sobel, M.E. (1990). Effect analysis and causation in linear structural equation models. *Psychometrika* **55**, 495–515.
Sobel, M.E. (1995). Causal inference in the social and behavioral sciences. In: Arminger, Clogg, Sobel (Eds.), *Handbook of Statistical Modeling for the Social and Behavioral Sciences*. Plenum, New York, pp. 1–38.
Sobel, M.E. (1996). An introduction to causal inference. *Sociological Methods & Research* **24**, 353–379.
Sommer, A., Zeger, S. (1991). On estimating efficacy from clinical trials. *Statistics in Medicine* **10**, 45–52.
Sugden, R.A. (1988). The 22 table in observational studies. In: Bernardo, DeGroot, Lindley, Smith (Eds.), *Bayesian Statistics 3*. Oxford University Press, New York, pp. 785–790.
Tanner, M.A., Wong, W.H. (1987). The calculation of posterior distributions by data augmentation. *Journal of the American Statistical Association* **82**, 528–550. With discussion.
Tinbergen, J. (1930). Determination and interpretation of supply curves: An example. *Zeitschrift fur Nationalokonomie*. Reprinted in Henry, Morgan (Eds.), *The Foundations of Economics*.
Ogden, D.W. (Acting Assistant Attorney General) (1999). News Conference (Sept. 22, 1999). http://www.usdoj.gov/archive/ag/speeches/1999/tobaccopc92299.htm.
Welch, B.L. (1937). On the z test in randomized blocks and Latin squares. *Biometrika* **29**, 21–52.
Zelen, M. (1979). A new design for randomized clinical trials. *New England Journal of Medicine* **300**, 1242–1245.
Zhang, J., Rubin, D.B. (2003). Estimation of causal effects via principal stratification when some outcomes are truncated by 'death'. *Journal of Educational and Behavioral Statistics* **28**, 353–368.
Zhang, J., Rubin, D., Mealli, F. (2006). Evaluating the effects of training programs on wages through principal stratification. *Advances in Economics* **21**. Submitted for publication.

Handbook of Statistics, Vol. 26
ISSN: 0169-7161

DOI: 10.1016/S0169-7161(06)26025-5

Statistical Aspects of Adaptive Testing

Wim J. van der Linden and Cees A.W. Glas

1. Introduction

In psychometrics, it is customary to view the responses of subjects $n = 1, \ldots, N$ to the items $i = 1, \ldots, I$ in an educational or psychological test as realizations u_{ni} of random response variables U_{ni}. The distribution of these variables is usually represented by a response model. A large variety of alternative response models has been developed in item response theory (van der Linden and Hambleton, 1997).

Since the responses reveal as much about the properties of the items as the abilities of the subjects that take them, we have to treat the variables U_{ni} as nonidentically distributed across items and subjects. It usually makes sense to assume independence for the variables for the same subject (within-subject or "local" independence). Provided there is no cooperation between them, the variables can also be assumed to be independent between subjects. Statistically, the analysis of response data on tests therefore deals with an instance of nonidentical but independently distributed variables U_{ni}.

Suppose the ability of subject n is represented by a (possibly vector-valued) parameter $\boldsymbol{\theta}_n$ and the properties of item i by a parameter vector $\boldsymbol{\xi}_i$. Typically, the interest is in estimating the parameters $\boldsymbol{\theta}_n$ and less in $\boldsymbol{\xi}_i$. The usual approach is to estimate the item parameters with a likelihood marginalized over the subject parameters or in a Bayesian fashion. This step is known as item calibration. Once the item parameters can be treated as known, the response model can be used as a model for measuring future subjects, i.e., estimating their ability parameters $\boldsymbol{\theta}_n$ from realizations $(u_{n1}, \ldots, u_{nI})$.

But even if the parameters $\boldsymbol{\xi}_i$ can be treated as known, the fact that they are item dependent requires special attention when assembling a test. As shown below, collectively the item parameters determine the accuracy of the estimator of $\boldsymbol{\theta}_n$. Inasmuch as the response models are typically nonlinear, the accuracy of these estimators also depends on the unknown parameters $\boldsymbol{\theta}_n$ themselves. The combination of these features creates a problem of optimal design in test theory that is much more complicated than the standard optimal design problem in linear regression with known predictors (van der Linden, 2007, Section 4.7).

For a fixed test, a practical way out of this problem is to specify a target function for the accuracy of the estimator $\hat{\boldsymbol{\theta}}$ of the ability parameter $\boldsymbol{\theta}$ over its space and use the known item parameters $\boldsymbol{\xi}_i$, $i = 1, \ldots, I$, to select a test from the inventory of calibrated items that approximates the target best in some sense. A more efficient way, however, is an *adaptive* test design. In adaptive testing, the test is administered by a computer that has a well-chosen pool of calibrated items stored in its memory. The estimator of $\boldsymbol{\theta}_n$ is updated after an item is completed and the next item is selected to be most efficient at the last update. The test is stopped if a fixed number of items has been taken or a predetermined level of accuracy for $\hat{\boldsymbol{\theta}}$ is realized. Extensive computer simulations have shown that, typically, adaptive testing realizes the same accuracy for $\hat{\boldsymbol{\theta}}$ as a well-designed fixed test with some 50% of the items.

The intuitive idea underlying the first attempts at adaptive testing was that a test had better measurement properties if the difficulties of the items matched the ability of the examinee. Items that are too easy or difficult have predictable responses and cannot provide much information about the ability of the examinee. In fact, this intuition was the same as that of a good examiner in an oral exam who avoids asking questions for which he knows in advance that they are answered correctly or incorrectly by the examinee.

For two reasons it took a long time before adaptive testing could be implemented in large-scale testing programs. First, realistic response models for the distribution of the variables U_{ni} had to be developed before the notion of adaptation could be formalized. Second, estimation of $\boldsymbol{\theta}_n$ is generally computationally intensive, and we had to wait for computers to become more powerful and affordable before estimation in real time was feasible. The first large-scale applications of adaptive testing were introduced in the early 1990s in testing programs for personnel selection, college admission, and licensure and certification testing. Adaptive testing is now also common in such applications as course placement in education and diagnostics in psychology.

The introduction of adaptive testing has led to a host of practical problems that all have serious statistical aspects. In this chapter, we first introduce a few response models that are used in adaptive testing and discuss estimation of their item and subject parameters. We then discuss adaptive estimation of the subject parameters and treat the statistical aspects of several of these more practical problems.

2. Response models

A basic unidimensional model for dichotomously scored responses is the three-parameter logistic (3PL) model (Birnbaum, 1968). In this model, the probability of a correct response $U_{ni} = 1$ of subject n on item i is given by

$$\begin{aligned} p_i(\theta_n) \equiv \Pr(U_{ni} = 1 \mid \theta_n) &\equiv c_i + (1 - c_i)\Psi\big[a_i(\theta_n - b_i)\big] \\ &\equiv c_i + (1 - c_i)\frac{\exp[a_i(\theta_n - b_i)]}{1 + \exp[a_i(\theta_n - b_i)]}, \end{aligned} \tag{1}$$

where θ_n is a scalar parameter for the ability of subject n, and $\boldsymbol{\xi}_i = (a_i, b_i, c_i)$ contains the parameters for the discriminating power, difficulty, and height of the lower asymptote of the response probability of item i, respectively. The last item parameter has been introduced to represent instances of random guessing on test items. In some applications, which will be discussed below, it proves practical to replace the logistic function $\Psi(\cdot)$ by the normal cumulative distribution function $\Phi(\cdot)$. Then, the probability of a correct response becomes

$$p_i(\theta_n) \equiv c_i + (1 - c_i)\Phi\big[a_i(\theta_n - b_i)\big]. \tag{2}$$

This model is known as the three-parameter normal ogive (3PNO) model.

In practice, it is usually assumed that there is one (dominant) ability θ that explains the test performances. But empirical model fit analyses may show that a unidimensional response model does not adequately describe the data, for example, when the test items consist of worded mathematical problems whose solution require a verbal in addition to a mathematical skill (Reckase, 1985). In such cases, a response model with a multidimensional ability parameter is more appropriate (McDonald, 1967; Andersen, 1985; Adams et al., 1997).

In a multidimensional response model, the logistic function in (1) and the normal ogive function in (2) are redefined as $\Psi(\mathbf{a}_i^t\boldsymbol{\theta}_n - b_i)$ and $\Phi(\mathbf{a}_i^t\boldsymbol{\theta}_n - b_i)$, respectively, where $\mathbf{a}_i$ is a Q-dimensional vector of discrimination parameters $(a_{i1}, \ldots, a_{iq}, \ldots, a_{iQ})$, and $\boldsymbol{\theta}_n$ is a vector of ability parameters $(\theta_{n1}, \ldots, \theta_{nq}, \ldots, \theta_{nQ})$ of the same dimension.

Under the additional assumption that the subjects' ability parameters can be taken to be a simple random sample from a Q-variate normal distribution, that is,

$$\boldsymbol{\theta}_n \sim \text{MVN}(\boldsymbol{\mu}_{\mathcal{P}}, \boldsymbol{\Sigma}_{\mathcal{P}}), \tag{3}$$

the model is equivalent with a full-information factor analysis model (Takane and de Leeuw, 1987). For this reason, the discrimination parameters are sometimes referred to as factor loadings and the ability parameters as factor scores.

In multiple-group IRT, different ability distributions are assumed for different populations.

Note that (3) can be used as a common prior distribution for the ability parameters of the subjects. In addition, it often is useful to introduce a distribution for the items parameters,

$$\boldsymbol{\xi}_i = (a_i, b_i, c_i) \sim \text{MVN}(\boldsymbol{\mu}_{\mathcal{I}}, \boldsymbol{\Sigma}_{\mathcal{I}}), \tag{4}$$

as well. This distribution can also be used in two different ways. First, as a (common) prior distribution to support parameter estimation when the data matrix is ill-conditioned, i.e., contains too little information and can be appropriately described by different sets of parameter values. Mislevy (1986) considers a number of Bayesian approaches that involve prior distributions for the item parameters. Second, the distribution can be used as an empirical distribution for the domain of possible items when modeling hierarchical relations between them. An example is the application to item-cloning techniques, in which a computer algorithm is used to generate families of items from so-called parent items.

Glas and van der Linden (2003, 2006) and Sinharay et al. (2003) present a multilevel response model for this application in which it is assumed that the parameters of each item are randomly sampled from a multivariate normal distribution associated with a parent item. Analogous to multiple-group IRT, this model allows for normal distributions with different means and covariance matrices for each family.

3. Item calibration

Adaptive testing is based on the availability of a large pool of calibrated test items. The calibration process can have two different designs:

1. A pretesting design in which fixed subsets of items are administered to fixed subsets of subjects in a series of pretest sessions, and the response model is fitted to the data to obtain item-parameter estimates.
2. An adaptive design in which new items are seeded into an operational adaptive test and the responses are used to estimate their item parameters along with the subjects' ability parameters.

Of course, it is also possible to use combinations of these sampling designs, for example, in a setup with initial item parameter estimation through a fixed design and further improvement of the estimates during operational adaptive testing. In both types of designs, the subjects never respond to all available items. Either type of design can be captured by a design vector $\mathbf{d}_n$ with elements d_{ni}, $i = 1, \ldots, I$, that are equal to one if n responded to item i and zero otherwise.

Two different estimation procedures for the item parameters are in use. The first is a marginal maximum likelihood (MML) procedure that builds on the fact that in item calibration only the item parameters are of interest. Consequently, the ability parameters are treated as nuisance parameters, which are integrated out of the likelihood function assuming a population distribution. The resulting marginal likelihood is maximized with respect to the item and population parameters. The second procedure is Bayesian with prior distributions for all parameters. This approach has become feasible through the use of Markov chain Monte Carlo (MCMC) computational methods. These methods circumvent the cumbersome direct evaluation of high-dimensional integrals in MML estimation for more complicated models. The two approaches to item parameter estimation are further explained in the next sections.

3.1. MML estimation

Currently, MML estimation (Bock and Aitkin, 1981) is the most frequently used technique for item calibration. One reason for this is its presence in widely disseminated software packages, for example, *Bilog-MG* (Zimowski et al., 2002) for unidimensional models for dichotomously scored items and *Multilog* (Thissen et al., 2003) or *Parscale* (Muraki and Bock, 2002) for polytomously scored items. MML estimation for multidimensional IRT models (Bock et al., 1988) has been implemented in the software packages *Testfact* (Wood et al., 2002) and *ConQuest* (Wu et al., 1997). All these packages compute MML estimates of the structural parameters in the model in a simultaneous fashion; this approach will be outlined below. But MML estimates can also be obtained through a two-step procedure, for example, that implemented in the *Mplus* software (Muthén and Muthén, 2003).

To obtain generality, the MML estimation equations will be derived for the multidimensional version of the 3PL model. The extension of MML estimation to a Bayesian version of it with prior distributions of the item parameters will be addressed at the end of this section.

Let $\mathbf{U}_n$ be the response vector of subject n and $\boldsymbol{\xi} = (\boldsymbol{\xi}_1, \ldots, \boldsymbol{\xi}_I)$. Using the assumption of local independence, the probability of responses $\mathbf{u}_n$ given $\boldsymbol{\theta}_n$ is

$$\Pr(\mathbf{u}_n \mid \boldsymbol{\theta}_n, \boldsymbol{\xi}) = \prod_{i=1}^{I} \Pr(U_{ni} = u_{ni} \mid \boldsymbol{\theta}_n, \boldsymbol{\xi}_i)^{d_{ni}}. \tag{5}$$

It is assumed that u_{ni} takes a dummy value if $d_{ni} = 0$. We will use the convention $u_{ni} = 9$ and, for mathematical treatment, define $\Pr(U_{ni} = 9 \mid \boldsymbol{\theta}_n, \boldsymbol{\xi}_i) = 1$. Since $\mathbf{u}_n$ determines the values of d_{ni}, the latter do not appear on the left side of (5).

Direct maximization of the likelihood function in (5) with respect to $\boldsymbol{\theta}$ does generally not produce consistent estimators because the number of subject parameters goes to infinity with the sample size. Therefore, Bock and Aitkin (1981) suggested to treat the $\boldsymbol{\theta}$s as unobserved realizations of stochastic variables that have the population distribution in (3). Assume that the response vectors of N subjects have been collected in an $N \times I$ matrix, $\mathbf{U}$. The marginal likelihood function is defined as

$$L(\boldsymbol{\xi}, \boldsymbol{\mu}_{\mathcal{P}}, \boldsymbol{\Sigma}_{\mathcal{P}}; \mathbf{U}) = \prod_{n=1}^{N} \Pr(\mathbf{u}_n \mid \boldsymbol{\xi}, \boldsymbol{\mu}_{\mathcal{P}}, \boldsymbol{\Sigma}_{\mathcal{P}}), \tag{6}$$

with

$$\Pr(\mathbf{u}_n \mid \boldsymbol{\xi}, \boldsymbol{\mu}_{\mathcal{P}}, \boldsymbol{\Sigma}_{\mathcal{P}}) = \int \cdots \int \Pr(\mathbf{u}_n \mid \boldsymbol{\theta}_n, \boldsymbol{\xi}) g(\boldsymbol{\theta}_n \mid \boldsymbol{\mu}_{\mathcal{P}}, \boldsymbol{\Sigma}_{\mathcal{P}}) \, \mathrm{d}\boldsymbol{\theta}_n. \tag{7}$$

Glas (1992, 1999) showed how to derive the likelihood equations for (6) from Fisher's identity (Efron, 1977). Let $\boldsymbol{\eta}$ denote all structural parameters in the model, that is, $(\boldsymbol{\xi}, \boldsymbol{\mu}_{\mathcal{P}}, \boldsymbol{\Sigma}_{\mathcal{P}})$, where we choose the diagonal and lower-diagonal elements of $\boldsymbol{\Sigma}_{\mathcal{P}}$ as free parameters. The first-order derivatives with respect to $\boldsymbol{\eta}$ can be written as

$$\mathbf{h}(\boldsymbol{\eta}) = \frac{\partial}{\partial \boldsymbol{\eta}} \log L(\boldsymbol{\eta}; \mathbf{U}) = \sum_{n} \mathcal{E}\big(\boldsymbol{\omega}_n(\boldsymbol{\eta}) \mid \mathbf{u}_n, \boldsymbol{\eta}\big), \tag{8}$$

with

$$\boldsymbol{\omega}_n(\boldsymbol{\eta}) = \frac{\partial}{\partial \boldsymbol{\eta}} \log p(\mathbf{u}_n, \boldsymbol{\theta}_n \mid \boldsymbol{\eta}), \tag{9}$$

and where the expectation is with respect to the posterior distribution

$$p(\boldsymbol{\theta}_n \mid \mathbf{u}_n, \boldsymbol{\eta}) = \frac{\Pr(\mathbf{u}_n \mid \boldsymbol{\theta}_n, \boldsymbol{\xi}) p(\boldsymbol{\theta}_n \mid \boldsymbol{\mu}_{\mathcal{P}}, \boldsymbol{\Sigma}_{\mathcal{P}})}{\int \cdots \int \Pr(\mathbf{u}_n \mid \boldsymbol{\theta}_n, \boldsymbol{\xi}) p(\boldsymbol{\theta}_n \mid \boldsymbol{\mu}_{\mathcal{P}}, \boldsymbol{\Sigma}_{\mathcal{P}}) \, \mathrm{d}\boldsymbol{\theta}_n}. \tag{10}$$

The equality in (8) is closely related to the EM algorithm (Dempster et al., 1977) for finding the maximum of a likelihood. The EM algorithm is applicable in situations where direct inference based on the marginal likelihood is complicated but the complete-data likelihood equations, i.e., equations based on $\boldsymbol{\omega}_n(\boldsymbol{\eta})$, are easier to solve. The EM algorithm does so by iteratively improving on estimates $\boldsymbol{\eta}^{(t-1)}$, solving $\sum_n \mathcal{E}(\boldsymbol{\omega}_n(\boldsymbol{\eta}^{(t)}) \mid \mathbf{u}_n, \boldsymbol{\eta}^{(t-1)}) = 0$ with respect to $\boldsymbol{\eta}^{(t)}$ until satisfactory convergence is reached.

We illustrate the application of Fisher's identity for the estimation of the population parameters for the qth component of $\boldsymbol{\theta}$. If the subject parameters $\boldsymbol{\theta}_n$ had been observed, the mean μ_q would be equal to

$$\mu_q = \frac{1}{N}\sum_{n=1}^{N}\theta_{nq}. \tag{11}$$

Since they are not observed, we take the posterior expectation of (11) with respect to $\boldsymbol{\theta}_n$ to obtain

$$\mu_q = \frac{1}{N}\sum_{n=1}^{N}\mathcal{E}(\theta_{nq} \mid \mathbf{u}_n, \boldsymbol{\xi}, \boldsymbol{\mu}_{\mathcal{P}}, \boldsymbol{\Sigma}_{\mathcal{P}}). \tag{12}$$

Observe that (12) is a function of the response vector $\mathbf{u}_n$, the item parameters $\boldsymbol{\xi}$, as well as the population parameters $\boldsymbol{\mu}_{\mathcal{P}}$ and $\boldsymbol{\Sigma}_{\mathcal{P}}$.

Analogously, the equations for the elements of $\boldsymbol{\Sigma}_{\mathcal{P}}$ are

$$\sigma_q^2 = \frac{1}{N}\left[\sum_{n=1}^{N}\mathcal{E}\big(\theta_{nq}^2 \mid \mathbf{u}_n, \boldsymbol{\xi}, \boldsymbol{\mu}_{\mathcal{P}}, \boldsymbol{\Sigma}_{\mathcal{P}}\big) - N\mu_q^2\right], \tag{13}$$

and

$$\sigma_{qq'} = \frac{1}{N}\left[\sum_{n=1}^{N}\mathcal{E}(\theta_{nq}\theta_{nq'} \mid \mathbf{u}_n, \boldsymbol{\xi}, \boldsymbol{\mu}_{\mathcal{P}}, \boldsymbol{\Sigma}_{\mathcal{P}}) - N\mu_q\mu_{q'}\right]. \tag{14}$$

Thus, the population mean of the subjects' abilities is estimated as the mean of the posterior expected values of the subjects' parameters. Likewise, the population variances and covariances of these abilities are estimated as the posterior expected variances and covariances between the subjects' parameters. The estimation equations for the item parameters are found in a similar way; for their equations as well as the standard errors of all estimates, refer to Glas (2000). Further, for the model to become identifiable, a number of restrictions on the parameters is required (see Béguin and Glas, 2001).

The multiple integrals above can be evaluated using Gauss–Hermite quadrature. A critical factor in using Gauss–Hermite quadrature is the number of variables that can be integrated simultaneously. Wood et al. (2002) indicate that the maximum number of variables is 10 with adaptive quadrature, 5 with nonadaptive quadrature, and 15 with Monte Carlo integration.

The consequences of the adding of prior distributions for the item parameters to the MML procedure are not too complicated. If (4) is used as a common prior distribution for the item parameters, the likelihood defined by (6) should be multiplied by the prior density to obtain

$$L(\boldsymbol{\xi}, \boldsymbol{\mu}_{\mathcal{P}}, \boldsymbol{\Sigma}_{\mathcal{P}}, \boldsymbol{\mu}_{\mathcal{I}}, \boldsymbol{\Sigma}_{\mathcal{I}} \mid \mathbf{U}) = L(\boldsymbol{\xi}, \boldsymbol{\mu}_{\mathcal{P}}, \boldsymbol{\Sigma}_{\mathcal{P}}; \mathbf{U})\,p(\boldsymbol{\xi} \mid \boldsymbol{\mu}_{\mathcal{I}}, \boldsymbol{\Sigma}_{\mathcal{I}}). \tag{15}$$

Mislevy (1986) uses the marginal posterior distributions for this likelihood to obtain Bayes modal estimates of the parameters, that is, he estimates these parameters as the mode of their posterior distributions. In one of his approaches, $p(\boldsymbol{\xi} \mid \boldsymbol{\mu}_{\mathcal{I}}, \boldsymbol{\Sigma}_{\mathcal{I}})$ is treated as a subjective prior with known hyperparameters $\boldsymbol{\mu}_{\mathcal{I}}$ and $\boldsymbol{\Sigma}_{\mathcal{I}}$. In another approach, labeled empirical Bayes approach, the parameters of $p(\boldsymbol{\xi} \mid \boldsymbol{\mu}_{\mathcal{I}}, \boldsymbol{\Sigma}_{\mathcal{I}})$ are estimated along with the ability parameters. In that case, $\boldsymbol{\mu}_{\mathcal{I}}$ and $\boldsymbol{\Sigma}_{\mathcal{I}}$ have their own (known) prior distributions, say, $p(\boldsymbol{\mu}_{\mathcal{I}}, \boldsymbol{\Sigma}_{\mathcal{I}} \mid \boldsymbol{\mu}_{\mathcal{I}0}, \boldsymbol{\Sigma}_{\mathcal{I}0})$.

Glas and van der Linden (2003) argue that for some applications (e.g., item cloning), the item parameters should be treated as nuisance parameters as well and consider a likelihood that is marginalized not only with respect to $\boldsymbol{\theta}_n$ but also with respect to $\boldsymbol{\xi}_i$.

3.2. Evaluation of model fit in an MML framework

The reliability and validity of adaptive testing is based on the assumption that the response model properly fits the data. Orlando and Thissen (2000) and Glas and Suárez-Falcon (2003) review the available statistical tests of the various assumptions underlying response models. Most statistics are based on the difference between the observed and expected response frequencies.

As an example, consider a statistic of differential item functioning, i.e., differences between the response probabilities of subgroups of subjects that are not related to the abilities that are tested. The particular concern is that these differences disadvantage one of the subgroups. A comparable problem in adaptive testing is item-parameter drift that is, a change in an item parameter between pretesting and operational testing or during operational testing.

Suppose we have two subgroups, which we label as the focal group and the reference group. A test based on the differences between the observed and expected correct scores would proceed as follows: First, MML estimates of the item parameters are obtained using the data of the two groups simultaneously, typically with the assumption that their subjects are sampled from different ability distributions. Then, the statistic

$$\sum_{n=1}^{N} y_n d_{ni} u_{ni} - \sum_{n=1}^{N} y_n d_{ni} \mathcal{E}\big(P_i(\boldsymbol{\theta}_n) \mid \mathbf{u}_n, \boldsymbol{\xi}, \boldsymbol{\mu}_{\mathcal{P}}, \boldsymbol{\Sigma}_{\mathcal{P}}\big) \tag{16}$$

is calculated, where $y_n = 1$ if subject n belongs to the focal group and zero otherwise. Observe that $\mathcal{E}(P_i(\boldsymbol{\theta}_n) \mid \mathbf{u}_n, \boldsymbol{\xi}, \boldsymbol{\mu}_{\mathcal{P}}, \boldsymbol{\Sigma}_{\mathcal{P}})$ is the posterior expectation of a correct response of subject n on item i. Glas (1998) shows that if the square of (16) is weighted by the estimated variance of $\sum_{n=1}^{N} y_n d_{ni} u_{ni}$, the resulting statistic has an asymptotic χ^2-distribution with one degree of freedom.

This test of differential item functioning belongs to a general class of so-called Rao score tests or Lagrange multiplier tests for item response models (Glas, 1999; Glas and Suárez-Falcon, 2003), which also contains tests of the shape the response function $P_i(\boldsymbol{\theta}_n)$ and the assumption of local independence between item responses.

3.3. Bayesian estimation

In Bayesian inference, all parameters are considered as random variables. Samples from their posterior distribution can be generated using an MCMC procedure. These samples can be used to approximate the functionals over the posterior distributions that are of interest. Albert (1992) shows how to apply Gibbs sampling (Gelfand and Smith, 1990) for the 2PNO model (that is, the model in (2) with $c_i = 0$), and Johnson and Albert (1999) have generalized the procedure to the 3PNO model. We work with the 3PNO instead of the 3PL model because the Gibbs sampler is easier to implement for it. For all practical purposes, the results from the two models are indistinguishable.

To implement a Gibbs sampler, the parameter vector is divided into a number of components. Iterative sampling from the conditional posterior distributions of each component given the previously sampled values for all other components generates a Markov chain. The chain is continued after it has stabilized to generate a sample from the joint posterior distribution of the parameters.

Johnson and Albert (1999, also see Béguin and Glas, 2001) introduced a data-augmentation scheme for the 3PNO that results in tractable conditional posterior distributions. The scheme is based on the following interpretation of the 3PNO. Suppose that the examinee either knows the correct answer with probability $\Phi(\lambda_{ni})$, $\lambda_{ni} = \mathbf{a}_i^t\boldsymbol{\theta}_n - b_i$, or does not know the correct answer with probability $1 - \Phi(\lambda_{ni})$. In the first case, a correct response is given with probability one, and in the second case the examinee guesses the correct response with probability c_i. Thus, the marginal probability of a correct response is equal to $\Phi(\lambda_{ni}) + c_i(1 - \Phi(\lambda_{ni}))$.

It proves to be convenient to introduce a vector of binary variables W_{ni} such that

$$W_{ni} = \begin{cases} 1 & \text{if subject } n \text{ knows the correct answer to item } i, \\ 0 & \text{otherwise.} \end{cases} \tag{17}$$

So, if $W_{ni} = 0$, subject n guesses the response to item i, and if $W_{ni} = 1$, subject n knows the right answer and gives a correct response. Consequently, the conditional probabilities of $W_{ni} = w_{ni}$ given $U_{ni} = u_{ni}$ are

$$\begin{aligned} &P(W_{ni}{=}1 \mid U_{ni} = 1, \lambda_{ni}, c_i) \propto \Phi(\lambda_{ni}), \\ &P(W_{ni}{=}0 \mid U_{ni} = 1, \lambda_{ni}, c_i) \propto c_i\bigl(1 - \Phi(\lambda_{ni})\bigr), \\ &P(W_{ni}{=}1 \mid U_{ni} = 0, \lambda_{ni}, c_i) = 0, \\ &P(W_{ni}{=}0 \mid U_{ni} = 0, \lambda_{ni}, c_i) = 1. \end{aligned} \tag{18}$$

The observed responses $\mathbf{u}$ of subjects are augmented with latent data $\mathbf{w} = (w_{11}, \ldots, w_{NI})^{\mathrm{T}}$. In addition, following Albert (1992), the data are also augmented with latent data $\mathbf{z} = (z_{11}, \ldots, z_{NI})^{\mathrm{T}}$, which are realizations of independent normally distributed random variables Z_{ni} with mean λ_{ni} and standard deviation equal to one. For the variables w and z, it holds that $z_{ni} > 0$ if $w_{ni} = 1$ and $z_{ni} \leqslant 0$ if $w_{ni} = 0$. Hence,

$$\begin{aligned} p(z_{ni} \mid w_{ni}, \lambda_{ni}) &\propto \phi(z_{ni}; \lambda_{ni}, 1) \\ &\quad \times \bigl[\mathbf{I}(z_{ni} > 0)\mathbf{I}(w_{ni} = 1) + \mathbf{I}(z_{ni} \leqslant 0)\mathbf{I}(w_{ni} = 0)\bigr], \end{aligned} \tag{19}$$

where $\phi(\cdot; \lambda_{ni}, 1)$ is the normal density with mean λ_{ni} and standard deviation equal to one and $\mathbf{I}(\cdot)$ is an indicator function that takes the value one if its argument is true and zero otherwise.

We introduce normal prior distributions for a and b, that is, $a \sim N(\mu_a, \sigma_a)$ and $b \sim N(\mu_b, \sigma_b)$, whereas the conjugate distribution $Beta(\alpha, \beta)$ is chosen as the prior distribution for the guessing probability, c. A noninformative prior is obtained by taking $Beta(1, 1)$. As an alternative, we could use (4) as the prior distribution of the item parameters. Finally, the joint prior distribution of $(\boldsymbol{\mu}_{\mathcal{P}}, \boldsymbol{\Sigma}_{\mathcal{P}})$ is specified as the product of the normal and inverse-Wishart densities $p(\boldsymbol{\mu}_{\mathcal{P}} \mid \boldsymbol{\Sigma}_{\mathcal{P}})$ and $p(\boldsymbol{\Sigma}_{\mathcal{P}})$, respectively (see, for instance, Box and Tiao, 1973).

The joint posterior of $\boldsymbol{\eta} = (\boldsymbol{\xi}, \boldsymbol{\theta}, \boldsymbol{z}, \boldsymbol{w}, \boldsymbol{\mu}_{\mathcal{P}}, \boldsymbol{\Sigma}_{\mathcal{P}})$ has density

$$\begin{aligned} p(\boldsymbol{\eta} \mid \mathbf{U}) &= p(\mathbf{z}, \mathbf{w} \mid \mathbf{U}; \boldsymbol{\xi}, \boldsymbol{\theta},)p(\boldsymbol{\theta} \mid \boldsymbol{\mu}_{\mathcal{P}}, \boldsymbol{\Sigma}_{\mathcal{P}})p(\boldsymbol{\mu}_{\mathcal{P}} \mid \boldsymbol{\Sigma}_{\mathcal{P}})p(\boldsymbol{\Sigma}_{\mathcal{P}})p(\boldsymbol{\xi}) \\ &= \prod_{n=1}^{N}\prod_{i=1}^{I} p(z_{ni} \mid w_{ni}; \lambda_{ni})p(w_{ni} \mid u_{ni}; \lambda_{ni}, c_i)p(\boldsymbol{\theta}_n \mid \boldsymbol{\mu}_{\mathcal{P}}, \boldsymbol{\Sigma}_{\mathcal{P}}) \\ &\quad \times p(\boldsymbol{\mu}_{\mathcal{P}} \mid \boldsymbol{\Sigma}_{\mathcal{P}})p(\boldsymbol{\Sigma}_{\mathcal{P}})p(\mathbf{a}, \mathbf{b})p(\mathbf{c}), \end{aligned} \tag{20}$$

where $p(w_{ni} \mid u_{ni}, \lambda_{ni}, c_i)$ and $p(z_{ni} \mid w_{ni}, \lambda_{ni})$ follow from (18) and (19), respectively. Although the distribution has an intractable form, the conditional distributions of the components $\boldsymbol{\xi}, \boldsymbol{\theta}, z, w, \mu, \boldsymbol{\Sigma}_{\mathcal{P}}$ given all others are tractable and easy to sample from.

Therefore, the Gibbs sampler consists of five steps:

1. Draw $\boldsymbol{\mu}_{\mathcal{P}}$ and $\boldsymbol{\Sigma}_{\mathcal{P}}$ conditionally on $\boldsymbol{\theta}$,
2. Draw $\mathbf{w}$ and $\mathbf{z}$ conditional on $\boldsymbol{\theta}$, $\boldsymbol{\xi}$, and $\mathbf{U}$,
3. Draw $\boldsymbol{\theta}$ conditional on $\mathbf{z}$, $\mathbf{a}$, $\mathbf{b}$, $\boldsymbol{\Sigma}_{\mathcal{P}}$, and $\boldsymbol{\mu}$,
4. Draw $\mathbf{a}$, $\mathbf{b}$ conditional on $\mathbf{z}$ and $\boldsymbol{\theta}$,
5. Draw $\mathbf{c}$ conditional on $\mathbf{w}$ and $\mathbf{U}$.

Details on these steps are given in Béguin and Glas (2001). Steps 3–5 are straightforward: Steps 3 and 4 boil down to sampling from the normal posterior distribution that is associated with a normal regression model and a normal prior distribution. Step 5 involves the well-known case of a binomial model with a beta as prior distribution, which leads to sampling from a beta posterior distribution.

The first two steps are as follows.

Step 1. The prior distribution for $(\boldsymbol{\mu}_{\mathcal{P}}, \boldsymbol{\Sigma}_{\mathcal{P}})$ we have chosen is a normal-inverse-Wishart. Their families are conjugate with the multivariate normal (3). Let $\boldsymbol{\Lambda}_0$ be the scale matrix and v_0 the degrees of freedom for the prior of $\boldsymbol{\Sigma}_{\mathcal{P}}$. Further, let $\boldsymbol{\mu}_0$ be the prior mean and κ_0 a weight for the prior of $\boldsymbol{\Sigma}_{\mathcal{P}}$, which can be interpreted as the number of prior measurements on the $\boldsymbol{\Sigma}_{\mathcal{P}}$ scale (Gelman et al., 1995). Then, the posterior distribution is normal-inverse-Wishart with parameters

$$\begin{aligned} \boldsymbol{\mu}_N &= \frac{\kappa_0 N}{\kappa_0 + N}\boldsymbol{\mu}_0 + \frac{N}{\kappa_0 + N}\bar{\boldsymbol{\theta}}, \\ v_N &= v_0 + N, \\ \kappa_N &= \kappa_0 + N, \\ \boldsymbol{\Lambda}_N &= \boldsymbol{\Lambda}_0 + \mathbf{S} + \frac{\kappa_0 N}{\kappa_0 + N}(\bar{\boldsymbol{\theta}} - \boldsymbol{\mu}_0)(\bar{\boldsymbol{\theta}} - \boldsymbol{\mu}_0)^{\mathrm{T}}, \end{aligned}$$

where $\mathbf{S}$ is a matrix with entries $S_{pq} = \sum_{n=1}^{N}(\theta_{np} - \bar{\theta}_p)(\theta_{nq} - \bar{\theta}_q)$ and $\bar{\theta}_p$ and $\bar{\theta}_q$ are the means of the proficiency parameters of dimension p and q, respectively.

Samples from the joint posterior distribution of $\boldsymbol{\mu}_{\mathcal{P}}$ and $\boldsymbol{\Sigma}_{\mathcal{P}}$ are obtained through the following procedure:

1. Draw $\boldsymbol{\Sigma}_{\mathcal{P}} \mid \boldsymbol{\theta}$ from the Inverse-Wishart$_{v_N}(\boldsymbol{\Lambda}_N^{-1})$.
2. Draw $\boldsymbol{\mu}_{\mathcal{P}} \mid \boldsymbol{\Sigma}_{\mathcal{P}}$ from $N(\boldsymbol{\mu}_N, \boldsymbol{\Sigma}_{\mathcal{P}}/\kappa_N)$.

The hyperparameters κ_0, v_0, $\boldsymbol{\mu}_0$ and $\boldsymbol{\Lambda}_0$ lead to a noninformative prior distribution if $\kappa_0 \to 0$, $v_0 \to -1$ and $|\boldsymbol{\Lambda}_0| \to 0$. This limiting case, which is a multivariate version of Jeffrey's prior density, leads to the posterior distribution

$$\begin{aligned} \boldsymbol{\Sigma}_{\mathcal{P}} \mid \boldsymbol{\theta} &\sim \text{Inverse-Wishart}_{N-1}(\mathbf{S}^{-1}), \\ \boldsymbol{\mu}_{\mathcal{P}} \mid \boldsymbol{\Sigma}_{\mathcal{P}} &\sim N(\bar{\boldsymbol{\theta}}, \boldsymbol{\Sigma}_{\mathcal{P}}/N). \end{aligned} \tag{21}$$

Step 2. The posterior $p(\mathbf{z}, \mathbf{w} \mid \mathbf{U}; \boldsymbol{\xi}, \boldsymbol{\theta})$ is factored as $p(\mathbf{z} \mid \mathbf{U}; \mathbf{w}, \boldsymbol{\xi}, \boldsymbol{\theta})p(\mathbf{w} \mid \mathbf{U}; \boldsymbol{\xi}, \boldsymbol{\theta})$, and values of $\mathbf{w}$ and $\mathbf{z}$ are obtained through the following procedure:

1. Draw w_{ni} from the distribution of W_{ni} given the data $\mathbf{U}$ and $\boldsymbol{\xi}, \boldsymbol{\theta}$ in (18).
2. Draw z_{ni} from the conditional distribution of Z_{ni} given $\mathbf{w}$, $\boldsymbol{\theta}$ and $\boldsymbol{\xi}$, which follows from (18) as

$$Z_{ni} \mid \mathbf{w}, \boldsymbol{\theta}, \boldsymbol{\xi} \sim \begin{cases} N(\eta_{ni}, 1) \text{ truncated at the left at } 0 & \text{if } w_{ni} = 1, \\ N(\eta_{ni}, 1) \text{ truncated at the right at } 0 & \text{if } w_{ni} = 0. \end{cases}$$

As start values for the parameters, we suggest using the MML estimates by one of the computer programs listed earlier. The number of burn-in iterations for the Markov chain to stabilize should be found empirically. One helpful tool is to start multiple MCMC chains from different points and evaluate their convergence by comparing the between- and within-chain variances (see, for instance, Robert and Casella, 1999, p. 366).

3.4. *Evaluation of model fit in a Bayesian framework*

Posterior predictive checks (Meng, 1994) are one of the standard tools for the evaluation of model fit in a Bayesian framework with MCMC computation. They can readily be used to evaluate the fit of IRT models (Hoijtink, 2001; Glas and Meijer, 2003; Sinharay, 2005).

Bayesian analogues of the likelihood-based test statistics above can be calculated as a by-product of the Markov chain. Once it has stabilized, at every iteration the sampled values for the parameters $\boldsymbol{\eta}$ and $\boldsymbol{\theta}$ are used to generate replicate data $\mathbf{U}^{\text{rep}}$ under the response model. The check on the fit of the model consists of a comparison between the same test statistic computed from the original data $\mathbf{U}$ and the generated data $\mathbf{U}^{\text{rep}}$ for the sampled parameter values. Particularly, we compute the proportion of times the former has a smaller value than the latter, i.e., the Bayesian p-value.

As an example, a Bayesian analogue of the statistic for differential item function in (16) is given. For the original data, the statistic is calculated as

$$T(\mathbf{U}, \boldsymbol{\eta}) = \sum_{n=1}^{N} y_n d_{ni} u_{ni} - \sum_{n=1}^{N} y_n d_{ni} P_i(\boldsymbol{\theta}_n), \tag{22}$$

where $P_i(\boldsymbol{\theta}_n)$ is evaluated at the current draws of the item and subjects' parameters and y_n is still the indicator variable used to denote membership of the focal and reference group. The calculation is repeated to obtain the version of the same statistic for the generated data, $T(\mathbf{U}^{\text{rep}}, \boldsymbol{\eta})$. Formally, the Bayesian p-value is defined as

$$\Pr\{T(\mathbf{U}, \eta) \leqslant T(\mathbf{U}^{\text{rep}}, \eta) \mid \mathbf{U} = \mathbf{u}\}. \tag{23}$$

Possible other model violations, for instance, with respect to shape of the response functions $P_i(\boldsymbol{\theta}_n)$, the assumption of local independence, and the behavior of individual subjects, can be detected by replacing (22) by another appropriate Pearson-type statistic.

4. Ability estimation

If the item parameters have been estimated with sufficient precision, they can be treated as the true parameters when estimating the subject parameter, $\boldsymbol{\theta}$. This practice simplifies real-time ability estimation in operational adaptive testing considerably. Again, likelihood-based and Bayesian methods can be used for estimating $\boldsymbol{\theta}$. As we will show in the next sections, both classes entail different rules for adapting the item selection to the estimates of $\boldsymbol{\theta}$.

In the first approach, the likelihood associated with (5),

$$L(\boldsymbol{\theta}; \mathbf{u}_n) = \prod_{i=1}^{I} P(U_{ni} = u_{ni} \mid \boldsymbol{\theta}, \boldsymbol{\xi}_i)^{d_{ni}}, \tag{24}$$

is maximized with respect to $\boldsymbol{\theta}$. The first-order and second-order partial derivatives needed to compute the MLEs and their standard errors can be found in Segall (1996); see also Segall (2000).

For a unidimensional θ, an attractive alternative to the regular MLE is Warm's (1989) weighted likelihood estimator (WLE). The estimator is the maximizer of the likelihood in (24) weighted by a function $w(\theta)$. That is, it maximizes

$$L(\theta; \mathbf{u}_n) = w(\theta) L(\theta \mid \mathbf{u}_n), \tag{25}$$

where the weight function is defined to satisfy

$$\frac{\partial w(\theta)}{\partial \theta^2} = \frac{H(\theta)}{2I(\theta)}, \tag{26}$$

with

$$H(\theta) = \sum_{i=1}^{I} d_{ni} \frac{[p_i'(\theta)][p_i''(\theta)]}{p_i(\theta)[1 - p_i(\theta)]}, \tag{27}$$

$$I(\theta) = \sum_{i=1}^{I} d_{ni} \frac{[p_i'(\theta)]^2}{p_i(\theta)[1 - p_i(\theta)]} \tag{28}$$

and $p_i'(\theta)$ and $p_i''(\theta)$ are the first- and second-order derivatives of $p_i(\theta)$. For a linear test, the WLE is attractive because of its unbiasedness to the order I^{-1}.

In a Bayesian framework, estimation is based on the posterior distribution of θ given $\mathbf{u}_n$ in (10). We either use the full posterior distribution or a point estimate derived from it. Examples of point estimates are the posterior mode (Bayes Modal or BM estimation; also known as the Maximum A Posteriori or MAP estimation, Lord, 1980) and its expected value (Expected A Posteriori or EAP estimate, Mislevy, 1986) Use of the EAP estimator in adaptive testing is discussed extensively in Bock and Mislevy (1988).

The choice between likelihood-based and Bayesian estimators in adaptive testing has many aspects. First of all, as is well known, for a noninformative prior, the ML and BM estimators are identical. For informative priors, the choice between an ML and BM estimator is generally one between favoring a smaller bias or a smaller mean-squared error (MSE). However, the WLE in (25) has been shown to have smaller bias without sacrificing its MSE. On the other hand, the use of empirical prior distributions estimated from the individual subject's response times on the items tends to lead to a much more favorable MSE for the Bayesian estimates (see below). Second, in the beginning of the test, when the posterior distribution tends to be considerably skewed, the EAP estimator behaves differently from the BM estimator. Interestingly, the impact of these differences have not yet been researched. Third, the item-selection rule in use for the test interacts with the estimation errors. It is therefore generally impossible to derive the behavior of estimators analytically. The only feasible way to study them is using computer simulation. In fact, more substantial remaining errors in the item parameters may also interact with the selection rule and, for instance, lead to a capitalization on chance (van der Linden and Glas, 2000a). Fourth, a fully Bayesian approach has the advantage of allowing for estimation error in θ when selecting the items. This feature leads to much more robust item selection in the beginning of the test when point estimates have a large likelihood of being off target and, hence, easily lead to suboptimal item selection. The best choice of estimator might be an eclectic one in which we combine different estimators for different stages in the test (van der Linden and Pashley, 2000, Section 2.3).

5. Empirical examples

Much research has been done on the properties of MML and Bayesian estimators for fixed tests. For adaptive test, however, the properties of the estimators for the item parameters depend on such factors as the calibration design and the distribution of the latent abilities, whereas the estimators of the subject parameters depend on additional factors such as the composition of the item pool as well as the presence of content constraints on the test, collateral information, and item-exposure control. Several of these factors are explained in Sections 5 and 6. Without any claim of generality, we present the results from two small studies, one on item calibration and the other on ability estimation, to illustrate some of the differences between MML and Bayesian estimation.

Table 1 shows the results from a comparison between MML item calibration and calibration using a fully Bayesian approach with MCMC computation. We drew 2,000 ability parameters from a standard normal distribution. For these parameters and the 15 sets of item parameters shown in the first columns of Table 1, item responses vectors were generated according to the 3PNO model. For the Bayesian method, guessing

Table 1
Results from item parameter recovery study: Posterior standard deviations for Bayesian estimates and standard errors for MML estimates between brackets ($N = 2{,}000$)

Item	True item parameters			Bayesian estimates			MML estimates		
	a_i	b_i	c_i	a_i	b_i	c_i	a_i	b_i	c_i
1	0.50	−1.00	0.20	0.53	−0.99	0.25	0.58	−1.99	0.20
				(0.07)	(0.11)	(0.10)	(0.08)	(0.29)	(0.09)
2	0.50	−0.50	0.20	0.61	−0.48	0.25	0.66	−1.08	0.19
				(0.08)	(0.11)	(0.08)	(0.08)	(0.22)	(0.08)
3	0.50	0.00	0.20	0.44	0.04	0.22	0.42	−0.03	0.22
				(0.07)	(0.14)	(0.07)	(0.07)	(0.37)	(0.08)
4	0.50	0.50	0.20	0.45	0.43	0.18	0.40	1.12	0.23
				(0.08)	(0.16)	(0.05)	(0.09)	(0.37)	(0.07)
5	0.50	1.00	0.20	0.52	0.84	0.14	0.58	1.56	0.18
				(0.10)	(0.18)	(0.04)	(0.15)	(0.22)	(0.04)
6	1.00	−1.00	0.20	1.08	−0.90	0.24	1.10	−1.14	0.19
				(0.15)	(0.11)	(0.08)	(0.13)	(0.15)	(0.08)
7	1.00	−0.50	0.20	1.08	−0.42	0.21	0.99	−0.60	0.19
				(0.12)	(0.11)	(0.06)	(0.13)	(0.16)	(0.07)
8	1.00	0.00	0.20	1.19	0.04	0.22	0.95	−0.22	0.16
				(0.15)	(0.13)	(0.04)	(0.13)	(0.14)	(0.06)
9	1.00	0.50	0.20	1.17	0.64	0.20	0.95	0.33	0.17
				(0.17)	(0.16)	(0.03)	(0.16)	(0.13)	(0.05)
10	1.00	1.00	0.20	0.90	0.84	0.15	0.71	0.83	0.15
				(0.12)	(0.15)	(0.02)	(0.14)	(0.15)	(0.05)
11	1.50	−1.00	0.20	1.85	−1.18	0.23	1.68	−0.97	0.18
				(0.28)	(0.13)	(0.06)	(0.25)	(0.11)	(0.08)
12	1.50	−0.50	0.20	1.64	−0.43	0.21	1.62	−0.38	0.24
				(0.24)	(0.11)	(0.04)	(0.29)	(0.11)	(0.05)
13	1.50	0.00	0.20	2.24	0.33	0.27	1.72	−0.06	0.25
				(0.36)	(0.17)	(0.02)	(0.34)	(0.10)	(0.04)
14	1.50	0.50	0.20	1.72	0.56	0.19	1.45	0.12	0.15
				(0.25)	(0.16)	(0.02)	(0.24)	(0.09)	(0.03)
15	1.50	1.00	0.20	1.42	0.98	0.21	1.16	0.52	0.16
				(0.21)	(0.19)	(0.02)	(0.22)	(0.09)	(0.03)

parameters c_i had a common beta (5, 17) prior distribution; for the two other item parameters, we chose noninformative prior distributions. The Gibbs sampler was run for 12,000 iterations; the first 2,000 were burn-in iterations. As shown in Table 1, the parameter estimates were generally within their estimated confidence band of two standard deviations. Further, we found no systematic differences between the MML and the Bayesian estimates.

To illustrate the difference between EAP and WML estimation of the ability parameter, we generated a pool of 200 items with responses following the 2-parameter logistic model. The item difficulties were drawn from a standard normal distribution and the item discrimination parameter from the distribution of $Y = .5 + .25X$, with for X a χ^2-distribution with one degree of freedom. We then simulated adaptive tests of 10, 20 or 40 items using the maximum-information rule in (34) below. For each test length,

Table 2
Results from ability parameter recovery study (10,000 replications per condition)

Mean estimate

Estimation method	Number of items	Normal sample	True ability values				
			2.00	1.00	0.00	−1.00	−2.00
EAP	10	−0.00	2.20	1.14	0.04	−1.23	−2.12
	20	0.00	2.10	1.07	−0.00	−1.09	−2.15
	40	−0.00	2.05	1.03	−0.00	−1.03	−2.05
WML	10	−0.03	2.01	1.00	0.00	−1.00	−1.99
	20	0.00	2.00	1.00	0.00	−1.00	−1.99
	40	0.01	2.00	0.99	0.00	−0.99	−2.00

Mean absolute error

Estimation method	Number of items	Normal sample	True ability values				
			2.00	1.00	0.00	−1.00	−2.00
EAP	10	0.38	0.60	0.46	0.32	0.47	0.57
	20	0.32	0.45	0.34	0.27	0.35	0.54
	40	0.25	0.35	0.26	0.21	0.25	0.36
WML	10	0.33	0.53	0.35	0.26	0.33	0.52
	20	0.25	0.37	0.25	0.20	0.27	0.41
	40	0.24	0.32	0.25	0.20	0.24	0.33

10,000 response vectors were generated for ability parameters drawn from a standard normal distribution as well as for each of the fixed values $\theta = 2.0, 1.0, \ldots, 2.0$. The mean estimate and the mean absolute error of the ability estimates are given in Table 2. Note that the EAP estimates had a slight outward bias, while the WML estimates were virtually unbiased. Further, the mean absolute errors of the WML estimates were nearly almost smaller than those of the EAP estimates. So, in the present study, the WML estimator performed better.

6. Rules for adaptive item selection

We follow the same distinction between a likelihood-based and a Bayesian approach as above, and discuss item selection from both perspectives. The following notation will be used. The items in the pool are still denoted as $i = 1, \ldots, I$, but, in addition, we use $k = 1, \ldots, m$ to denote the items in the adaptive test. Thus, i_k is the index of the item in the pool administered as the kth item in the test. For convenience, we suppress subject index n in the rest of this chapter.

Assume the kth item has to be selected. The preceding $k - 1$ items form the set $S_k = \{i_i, \ldots, i_{k-1}\}$. The set of items in the pool remaining after $k-1$ items have been selected is $R_k = \{1, \ldots, I\} \backslash S_{k-1}$. The subjects have responses $\mathbf{U}_{k-1} = (U_{i_1}, \ldots, U_{i_{k-1}})$ with realizations $\mathbf{u}_{k-1} = (u_{i_1}, \ldots, u_{i_{k-1}})$ on the items in S_k. The point estimate of θ based on these responses is denoted as $\hat{\theta}_{\mathbf{u}_{k-1}} = \hat{\theta}_{k-1}$. The posterior distribution of θ has density $p(\theta \mid \mathbf{u}_{k-1})$. Item k is selected from set R_k to be optimal at $\hat{\theta}_{k-1}$ or (some functional on) $p(\theta \mid \mathbf{u}_{k-1})$.

The item pool is assumed to be calibrated by the three-parameter logistic model in (1). Item-selection rules for adaptive testing with polytomous items are discussed, for instance, in Dodd et al. (1995) and Koch and Dodd (1989). For rules for adaptive testing with multidimensional items, see Segall (1996, 2000) or Veldkamp and van der Linden (2002).

6.1. Likelihood-based selection

The second-order derivative in (24) reflects the curvature of the observed likelihood function of θ relative to its scale. Its negative is the observed information measure

$$J_{\mathbf{u}_{k-1}}(\theta) = -\frac{\partial}{\partial\theta^2} \ln L(\theta; \mathbf{u}_{k-1}). \tag{29}$$

The expected valued of this measure over the response variables is the Fisher information:

$$I_{\mathbf{U}_{k-1}}(\theta) = E\big[J_{\mathbf{U}_{k-1}}(\theta)\big]. \tag{30}$$

For the 3PL model, the observed and expected information measures are identical. For this model, (30) was already given in (28); using our current notation and ignoring the missing data indicator d_{ni}, it can be written as

$$I_{\mathbf{U}_{k-1}}(\theta) = \sum_{j=1}^{k-1} \frac{[p'_{i_j}(\theta)]^2}{p_{i_j}(\theta)[1 - p_{i_j}(\theta)]}, \tag{31}$$

with

$$p'_{i_j}(\theta) \equiv \frac{\partial}{\partial\theta} p_{i_j}(\theta). \tag{32}$$

For several other response models, (29) and (30) are different.

For a fixed test, with increasing test length, the variance of the MLE of θ is known to converge to the reciprocal of (31). In addition, (31) is easy to calculate and additive in the items. For this reason, Birnbaum (1968) introduced Fisher's information as the main criterion for fixed-test assembly, where it is typically treated as function of θ and is more commonly known as the test information function.

A popular choice in adaptive testing is to select the kth item to maximize (31) at $\theta = \hat{\theta}_{k-1}$. Formally, the rule can be presented as

$$i_k \equiv \arg\max_j \big\{ I_{\mathbf{U}_{k-1}, U_j}(\hat{\theta}_{k-1});\ j \in R_k \big\}. \tag{33}$$

Because of the additivity of the information function, the rule boils down to

$$i_k \equiv \arg\max_j \big\{ I_{U_j}(\hat{\theta}_{k-1});\ j \in R_k \big\}. \tag{34}$$

This rule is known as the maximum-information rule in adaptive testing.

For an asymptotic motivation of maximum-information selection, see Chang and Ying (2006). They show that the standard result of the MLE of θ converging to its true value with sampling variance equal to the reciprocal of (31) still holds, provided the values for the discrimination and guessing parameters in the item pool are bounded away from 0 and ∞ and 1, respectively, but item selection is otherwise unconstrained. They also present a slight modification of the likelihood equation for the 3PL model that is sufficient to prevent multiple roots.

An alternative to (31) is the Kullback–Leibler information (Lehmann and Casella, 1998, Section 1.7) defined as the distance between the distribution of response U_{i_j} on candidate item j at the true ability θ_0 and the current ability estimate $\hat{\theta}_{k-1}$:

$$K_{i_j}(\hat{\theta}_{k-1}, \theta_0) \equiv \mathcal{E}\left[\ln \frac{L(\theta_0; U_{i_j})}{L(\hat{\theta}_{k-1}; U_{i_j})}\right], \tag{35}$$

where the expectation is over U_{i_j} at θ_0. For the 3PL model,

$$K_{i_j}(\hat{\theta}_{k-1}, \theta_0) = p_{i_j}(\theta_0) \ln \frac{p_{i_j}(\theta_0)}{p_{i_j}(\hat{\theta}_{k-1})} + \left[1 - p_{i_j}(\theta_0)\right] \ln \frac{1 - p_{i_j}(\theta_0)}{1 - p_{i_j}(\hat{\theta}_{k-1})}. \tag{36}$$

Because of local independence, (35) is additive in the items and we can focus on the information by the candidate items. To deal with the fact that θ_0 is unknown, Chang and Ying (1996) propose to integrate (35) over an interval about the current ability estimate, $[\hat{\theta}_{k-1} - \delta_k, \hat{\theta}_{k-1} + \delta_k]$, where δ_k is an arbitrary constant decreasing with k. This suggestion leads to the item-selection rule

$$i_k \equiv \arg\max_j \left\{ \int_{\hat{\theta}_{k-1}-\delta_k}^{\hat{\theta}_{k-1}+\delta_k} K_j(\hat{\theta}_{k-1}, \theta)\, d\theta;\ j \in R_k \right\}. \tag{37}$$

If the constants δ_k are well chosen, (37) can be expected to be robust with respect to the estimation errors in $\hat{\theta}_{k-1}$ in the beginning of the test. As already observed, these errors can be extremely large and it may take some time for the test to recover from initial suboptimal item selection due to large errors, particularly if the subject happens to produce a few unlikely response at this stage.

6.1.1. *Likelihood-weighted selection*

The same wish of more robust item selection underlies the rule proposed by Veerkamp and Berger (1997). They take the likelihood function associated with the responses $\mathbf{u}_{k-1}$ to express the plausibility of the various values of θ given the data. Hence, their suggestion of a likelihood-weighted information criterion:

$$i_k \equiv \arg\max_j \left\{ \int_{-\infty}^{\infty} L(\theta; \mathbf{u}_{k-1}) I_{i_k}(\theta)\, d\theta;\ j \in R_k \right\}. \tag{38}$$

In the beginning of the test, the likelihood function is flat and values away from the MLE receive substantial weight. Toward the end of the test the likelihood function becomes more peaked, and nearly all of the weight will go to the values close to the MLE.

Although the idea was presented for Fisher's information measure, it can easily be extended to include the Kullback–Leibler information and then becomes an alternative to (37). Also, it is interesting to reflect on the differences between (38) and the weighted MLE in (25), where, the other way around, the likelihood function is enhanced with an information-related weight to produce a better point estimate of θ.

6.2. Bayesian selection

Adaptive testing is a natural area for the application of sequential Bayesian methods, with the posterior distribution of θ after the preceding item serving as the prior distribution for the selection of the next item. If exchangeability of the subjects is assumed and the prior distribution for the first item is taken to be an estimate of the population distribution of θ, such as that in (3), the method becomes fully empirical Bayes. Another motivation for the use of a Bayesian method is further improved robustness of the item selection during the first stage of the test relative to (37) and (38).

The posterior distribution after $k-1$ items has density $p(\theta \mid \mathbf{u}_{k-1})$. The basic idea is to select the kth item such that the posterior distribution

$$p(\theta \mid \mathbf{u}_{k-1}, U_{i_k}) \propto p(\theta \mid \mathbf{u}_{k-1}) p(U_{i_k} = u_{i_k} \mid \theta), \tag{39}$$

or a functional defined on it, becomes optimal in some sense. This general idea is known as preposterior analysis.

6.2.1. Owen's approximate procedure

Owen (1969; see also 1975) pioneered the application of Bayesian methods to adaptive testing. Due to the computational restrictions of his time, he had to find an approximation to the full preposterior analysis that he actually wanted. To derive his approximation, he choose to work with the 3PNO model in (2).

Owen's procedure started with a normal prior distribution for the selection of the first item. Next, the rule for choosing items $k = 2, 3, \ldots$ was

$$\left| b_{i_k} - \mathcal{E}(\theta \mid \mathbf{u}_{k-1}) \right| < \delta \tag{40}$$

for a small value of $\delta \geqslant 0$. His suggestion was to stop when the posterior variance became smaller than a predetermined tolerance. Observe that (40) selects an item that is closer than δ to the current EAP estimate of θ.

Since the normal prior is not conjugate with the binomial distribution of the response variables, the result is an intractable posterior distribution. Much of Owen's work was devoted to finding closed-form approximations to the true posterior variance and mean. The approximation for the mean that he suggested was motivated by its convergence to the true value of θ in mean square for $k \to \infty$ (Owen, 1975, Theorem 2).

6.2.2. Preposterior selection

The rules in this section illustrate full preposterior item selection. We show a few rules that predict the responses on the items $i \in R_k$ and select the kth item with an optimal expected update of the posterior quantity we are interested in. The prediction of the response on item i is provided by the predictive posterior distribution, which has probability function

$$p(u_i \mid \mathbf{u}_{k-1}) = \int p(u_i \mid \theta)p(\theta \mid \mathbf{u}_{k-1})\,\mathrm{d}\theta. \tag{41}$$

An obvious application is to select the item with the minimum expected posterior variance,

$$i_k \equiv \arg\min_j \left\{ \sum_{u_j=0}^{1} \mathrm{Var}(\theta \mid \mathbf{u}_{k-1}, U_j = u_j)p(u_j \mid \mathbf{u}_{k-1});\ j \in R_k \right\}. \tag{42}$$

The expression for the variance in this criterion follows immediately from (39). The criterion is equivalent to minimization of the preposterior risk under a quadratic loss function for the EAP estimator.

From a Bayesian perspective, the rule in (38) can be generalized by replacing the likelihood function by the posterior density of θ as weight:

$$i_k \equiv \arg\max_j \left\{ \int I_{U_j}(\theta)p(\theta \mid \mathbf{u}_{k-1})\,\mathrm{d}\theta;\ j \in R_k \right\}. \tag{43}$$

A further improvement on (43) can be based on a preposterior argument. We would then predict the posterior weighted information in the responses on the next item and calculate its expectation over the predictions:

$$\begin{aligned} i_k \equiv \arg\max_j \Bigg\{ \int & J_{\mathbf{u}_{k-1}, U_j = u_j}(\theta)g(\theta \mid \mathbf{u}_{k-1}, U_j = u_j)\,\mathrm{d}\theta \\ & \times p(u_j \mid \mathbf{u}_{k-1});\ j \in R_k \Bigg\}. \end{aligned} \tag{44}$$

The last two rules can also be formulated with the Kullback–Leibler information and are then alternatives to (37). Also, we could predict more than one item ahead. For a review of these and other Bayesian item-selection rules for the 3PL model as well as empirical comparisons between some of them, see van der Linden (1998) or van der Linden and Pashley (2000). Applications to polytomous models are presented in Penfield (2006).

6.2.3. *Collateral information*

Item selection can be further improved by using collateral information on the subjects, such as earlier test scores, biographical data, etc. If the information is used in a Bayesian framework, it should be incorporated in the prior distribution of θ. As a result, it allows us to start the test from an individual empirical prior instead of a common (subjective) prior distribution.

Procedures for adaptive testing with a prior distribution regressed on background variables X_p, $p = 0, \ldots, P$, are described in van der Linden (1999). They are based on linear regression of θ on the predictor variables:

$$\theta = \beta_0 + \beta_1 X_1 + \cdots + \beta_P X_P + \varepsilon, \tag{45}$$

with

$$\varepsilon \sim N\big(0, \sigma^2\big). \tag{46}$$

If the items have been calibrated, substitution of (46) into the response model leads to a logistic regression model with unknown parameters $(\beta_0, \beta_1, \ldots, \beta_P, \sigma)$ and unknown errors ε_n. It is a standard application in logistic regression to estimate the parameters using an EM algorithm with the ε_n as missing data. This should be done along with the regular item calibration. Once the parameters have been estimated, an obvious prior distribution for subject n is

$$p(\theta_n) = N(\hat{\theta}_{0n}, \sigma) \tag{47}$$

with

$$\hat{\theta}_{0n} = \beta_0 + \beta_1 x_{1n} + \cdots + \beta_P x_{Pn}. \tag{48}$$

In adaptive testing, the response times on the items form a unique source of collateral information. They are automatically recorded. When a new item has to be selected, the time spent on the previous item can be used to retrofit the previous prior distribution. This retrofitting is in addition to the regular update of the likelihood using the response on the previous item. If the speed and ability at which the population of subjects works are dependent, the retrofitted prior quickly becomes an informative individual empirical prior for the estimation of ability parameter.

To explore the benefits of response times as item-level collateral information, van der Linden (2007) (see also van der Linden, 2006a) used a lognormal model for the response time $T_i = t_i$ of a subject on item i operating at speed $\tau \in (-\infty, \infty)$ during the test,

$$p(t_i; \tau, \alpha_i, \beta_i) = \frac{\alpha_i}{t_i\sqrt{2\pi}} \exp\left\{-\frac{1}{2}\big[\alpha_i\big(\ln t_i - (\beta_i - \tau)\big)\big]^2\right\}, \tag{49}$$

where $\beta_i \in (-\infty, \infty)$ and $\alpha_i \in (0, \infty)$ are parameters for the time intensity and discriminating power of item i. To import the information in t_i on θ, the model was extended with a second-level distribution

$$(\theta, \tau) \sim \text{MVN}(\boldsymbol{\mu}_{\mathcal{P}}, \boldsymbol{\Sigma}_{\mathcal{P}}), \tag{50}$$

which relates θ to the speed τ for the population of subjects. It is assumed that the free parameters in $\boldsymbol{\mu}_{\mathcal{P}}$ and $\boldsymbol{\Sigma}_{\mathcal{P}}$ have been estimated during item calibration. (A few of the parameters have to be fixed to make the model identifiable.)

After the response times on $k-1$ items have been recorded, the posterior density of the subject's speed parameter given these times,

$$p(\tau \mid \mathbf{t}_{k-1}) \propto p(\tau) \prod_{j=1}^{k-1} p(t_{i_j}; \tau), \tag{51}$$

is calculated from (49), adopting a prior distribution $p(\tau)$. Along with the conditional density of the ability given speed, $p(\theta \mid \tau)$, which follows from the known distribution in (50), the posterior distribution allows us to calculate the posterior predictive density of θ given the response times $\mathbf{t}_{k-1}$ as

$$p(\theta \mid \mathbf{t}_{k-1}) = \int p(\theta \mid \tau) p(\tau \mid \mathbf{t}_{k-1})\, \mathrm{d}\tau. \tag{52}$$

When selecting the kth item, this predictive density is used to retrofit the prior in the posterior distribution of θ after the previous $k - 1$ items. The selection of the kth item is then based on the following prior distribution:

$$p(\theta \mid \mathbf{u}_{k-1}, \mathbf{t}_{k-1}) \propto p(\theta \mid \mathbf{t}_{k-1}) \prod_{j=1}^{k-1} P(u_{i_j} \mid \boldsymbol{\theta}). \tag{53}$$

If the times are measured on a logarithmic scale, both factors in the integrand in (51) are normal, and $p(\theta \mid \mathbf{t}_{k-1})$ is normal with known mean and variance (for their expressions, see van der Linden, 2007). The procedure is therefore easy to implement.

Figure 1 shows the results from a computer simulation study of adaptive tests of 10 and 20 items. It appears that, for a moderate correlation of $\rho_{\theta\tau} = .20$, a 10-item adaptive test with the use of the response times is approximately equally accurate as a 20-item test without it.

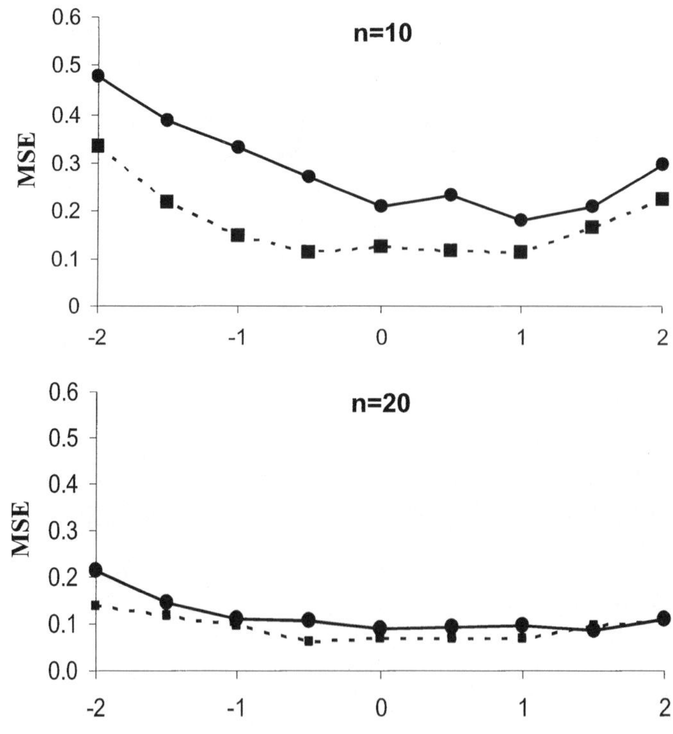

Fig. 1. MSE functions for an adaptive test without and with the use of response times for items selection ($n = 10$ and 20).

6.2.4. Item cloning

Bayesian hierarchical modeling is also convenient for item selection from a pool with families of cloned items. As indicated earlier, an appropriate model for such a pool is a response model extended with a separate second-level distribution of the item parameters for each family.

Let $f = 1, \dots, F$ denote the families in the pool. For notational convenience, we assume that the test consists of one random item from each of a selected subset of families and use f_k to denote the family in the pool that provides the kth item in the test. For the 3PL model, an obvious generalization of (4) is to choose

$$\boldsymbol{\xi}_f = (a_f, b_f, c_f) \sim \mathrm{MVN}(\boldsymbol{\mu}_f, \boldsymbol{\Sigma}_f), \quad f = 1, \dots, F, \tag{54}$$

as the second-level distributions. The means $\boldsymbol{\mu}_f$ and covariance matrices $\boldsymbol{\Sigma}_f$ are assumed to be estimated from samples of items from the families in the pool during a previous calibration study.

Item selection now has to follow a two-stage process: first, a family is selected and then an item is picked from the family. The selection of the family can still be adapted to the current estimate of θ but the selection of the item has to be random.

Suppose the kth family is to be selected. The responses on the items from the $k-1$ previously selected families are denoted as $\mathbf{u}_{k-1} = (u_{f_1}, \dots, u_{f_{k-1}})$. The update of the posterior distribution of θ after these responses is

$$p(\theta \mid \mathbf{u}_{k-1}) \propto p(\theta) \prod_{j=1}^{k-1} \int \cdots \int p(u_{f_j} \mid \theta, \boldsymbol{\xi}_j) p(\boldsymbol{\xi}_j \mid \boldsymbol{\mu}_{k-1}, \boldsymbol{\Sigma}_j) \, \mathrm{d}\boldsymbol{\xi}_j. \tag{55}$$

The selection has to be optimal with respect to this posterior distribution.

Analogous to (42), the family could be selected as

$$f_k \equiv \arg\min_j \left\{ \sum_{u_j=0}^{1} \mathrm{Var}(\theta \mid \mathbf{u}_{k-1}, U_j = u_j) p(u_j \mid \mathbf{u}_{k-1});\ j \in R_k \right\}, \tag{56}$$

where R_k now denotes the set of candidate families available for selection and $u_j = 0, 1$ are the possible responses to a random item from candidate $j \in R_k$.

If the interest is in a criterion based on an information measure, we could define the expected information in a random item from family f as

$$I_f(\theta) = \int \cdots \int \left[-\frac{\partial}{\partial^2 \theta} \ln L(\theta; u_f, \boldsymbol{\xi}_f) \right] p(\boldsymbol{\xi}_f \mid \boldsymbol{\mu}_f, \boldsymbol{\Sigma}_f) \, \mathrm{d}\boldsymbol{\xi}_f, \tag{57}$$

and, for example, use posterior weighting to select the kth family as

$$f_k = \arg\max_j \left\{ \int I_j(\theta) p(\theta \mid \mathbf{u}_{k-1}) \, \mathrm{d}\theta;\ j \in R_k \right\}. \tag{58}$$

The loss of information due to the sampling of the items from their families depends on the relation between the within-family and between-family item variation. For an empirical study of these effects for the rule in (56), see Glas and van der Linden (2003).

7. Other statistical issues in adaptive testing

When the first large-scale testing programs moved to an adaptive format in the early 1990s, it quickly became clear that calibrating an item pool and adopting an item-selection algorithm were not enough. In fact, several new problems with interesting statistical aspects presented themselves. In this section, we discuss a few of them.

First, just like paper-and-pencil tests, to maintain their content validity (and, therefore, their fit to the response model), adaptive tests have to meet an extensive set of content specifications. This set has to be imposed while maintaining the adaptive nature of the item selection. As we will see below, the consequence is a complicated constrained sequential optimization problem, typically with hundreds of constraints, which has to be solved in real time.

Second, an equally urgent new problem was item-pool security. By its very nature, an item-selection rule in adaptive testing picks the best items in the areas on the ability scale where the subjects' estimates are. Typically, a small set of highly discriminating items tends to dominant the whole scale. For an unconstrained test, this set need not be larger than 5% of the entire item pool or so. Since the pool is in continuous use for some time, it is relatively easy for test takers to find out what the dominant items are and share them with future test takers. As explained below, an effective reaction by the testing agency is to build a random experiment into the test. The experiment has to create more evenly distributed exposure rates for the item without sacrificing too much of the other desired features of the test.

Third, fixed tests are usually assembled with an eye on the time limit in force for them. In order to schedule the subjects efficiently, most real-world adaptive tests also have a fixed length and time limit. But different subjects get different sets of items and the items in the pool typically vary considerably in their time intensity. As a result, some of the subjects may run out of time whereas others have more than enough time to finish their test. This phenomenon is known as differential speededness of the test. To make adaptive tests fair, they should not suffer from it. As we will show, a solution to this problem entails the necessity to calibrate the items with respect to their time properties as well and use the parameter estimates in the item-selection rule.

Finally, for security reasons, it is not feasible to provide test takers with copies of the test they have taken. In order to given them the opportunity to interpret their scores relative to the content of typical test items, it is standard practice to release a fixed test assembled to the same specifications and equate the score on the adaptive test to a number-correct score on this reference test. The problem of how to equate an adaptive test to a fixed test is not trivial, though.

In the next section, we discuss a solution to adaptive item selection that can not only be used to solve the problem of imposing identical content specifications on the test for all subjects but offers a more general framework that resolves the other problems as well.

7.1. Content specifications

The necessity to impose a set of content specifications on an adaptive test introduces a fundamental dilemma in item selection: To optimize the accuracy of the estimator of θ, the items have to be selected sequentially, with an update of the estimator after each item. However, to realize a set of content specification, we have to select them simultaneously; otherwise, during a test, we might soon observe that some types of items tend to be under- or overrepresented but that attempts to correct for this would lead to the violation of some of the other specifications or the selection of much less favorable items for the accuracy of the ability estimator.

A more general solution of this problem is possible if we treat it as an instance of constrained combinatorial optimization, that is, the choice of a combination of items from the pool that satisfies a set of constraints representing the content specifications and is optimal with respect to the ability estimator. This approach would also allow us to deal with any requirement for the test that can be formulated as an explicit constraint on the selection of its items.

To implement the idea, we use a selection process with the following steps:

1. Begin the adaptive test with the selection of a full-size fixed test that is optimal at the initial estimate of θ and meets all constraints.
2. Administer one item from the test that is optimal at the estimate.
3. Update the estimate of θ.
4. Reassemble the full-size test to be optimal at the updated estimate of θ while still meeting all constraints and fixing the items that have already been administered.
5. Repeat Steps 2–4 until the adaptive test is completed.

The full-size tests that are assembled, which the subject does not see, are known as shadow tests. Because each of the shadow tests meets all constraints, the adaptive test does. Similarly, since all shadow tests are assembled to be optimal at the current estimate of θ and the best item at this estimate is administered, the adaptive test is our best choice given all the constraints (van der Linden and Reese, 1998; van der Linden, 2000).

Instead of as a sequence of m constrained combinatorial optimization problems, we can also view the shadow-test approach is as an application of a projection method: Because the item-selection algorithm is not allowed to look back and undo any earlier choices of items that appear to lead to constraint violation later in the test, the only option left is to look forward and project the rest of the test. The shadow-test approach does so by calculating a projection that is both feasible with respect to the constraints and optimal with respect to the ability estimator, and then picks the best item from the projection.

The only problem left is that of how to calculate the shadow tests in real time. To do so, we formalize the test-assembly problem more explicitly as a combinatorial optimization problem with binary variables $x_i, i = 1, \dots, I$, that take the value one if the item is selected for the shadow test and zero otherwise. The variables are used to model the content specifications as a set of constraints, which typically fall into three different classes: (i) constraints on categorical item attributes, such as content categories, item format (e.g., open-ended or multiple-choice format), and problem type (e.g., knowledge, analysis, or application), (ii) constraints on quantitative attributes, such as statistical parameters, word counts, and expected response time, and (iii) logical constraints to deal with conditional item selection, such as the requirement that one item cannot be selected if any from a given set is.

Suppose the test is of size of m. As an example of a categorical constraints, we use the set V_c for the indices of the items in the pool that belong to category c. Likewise, q_i is the value of item i for a quantitative attribute that we want to constrain. To illustrate the use of logical constraints, we assume that the items with the indices in set V_e cannot be selected for the same test. Finally, the items are selected using the maximum-information rule in (34).

The shadow-test model for the selection of the kth item in the adaptive test is

$$\text{maximize} \sum_{i=1}^{I} I_i(\hat{\theta}_{k-1})x_i \quad \text{(test information)} \tag{59}$$

subject to

$$\sum_{i=1}^{I} x_i = m \quad \text{(test length)}, \tag{60}$$

$$\sum_{i \in V_c} x_i \gtreqless m_c, \ \text{for all } c \quad \text{(categorical attributes)}, \tag{61}$$

$$\sum_{i=1}^{I} q_i x_i \gtreqless b_q \quad \text{(quantitative attributes)}, \tag{62}$$

$$\sum_{i \in S_k} x_i = k - 1 \quad \text{(previous items)}, \tag{63}$$

$$\sum_{i \in V_e} \leqslant 1, \ \text{for all } e \quad \text{(item exclusion)}, \tag{64}$$

$$x_i \in \{0, 1\}, \ \text{for all} i \quad \text{(range of variables)}. \tag{65}$$

In (61) and (62), m_c and b_q are bounds in the equalities or inequalities on the number of items from category c and the sum of the quantitative attributes (e.g., sum of the expected response times) in the test, respectively. The constraint in (63) serves a double purpose: It fixes all $k - 1$ items that have already been administered to the subject. At the same time, its presence automatically accounts for the attributes of these fixed items in all other constraints in the model.

The only dynamic part of the model is the objective function in (59) and the constraint on the previous items in (63). They contain quantities that have to be updated before the next shadow test can be calculated ($I_i(\hat{\theta}_{k-1})$, S_k, and k).

The model is entirely linear in the variables. A solution to it can be found through a call to a standard integer solver prior to the selection of an item. Since the solution of later shadow-test models are always within the feasible space of a previous model, the next shadow test can be found using a "hot start" of the solver. For a modern solver with optimized settings, the running times for a real-world adaptive test are negligible.

The model in (59)–(65) can easily be used with any of the other item-selection rule in Section 5. It can also be extended with constraints at other levels in the tests (e.g., item sets with common stimuli and subtests), constraints for more complicated logical selection, etc. For these and other implementation details, see van der Linden (2007).

Figure 2 shows the bias and MSE functions for a 50-item adaptive version of the Law School Admission Test (LSAT) from a pool of 753 items with 100 simulated subjects at $\theta = -2.0, -1.5, \ldots, 2.0$. The test had over 400 constraints that were automatically satisfied for each simulated subject. Except for the lower end of the scale (the item pool

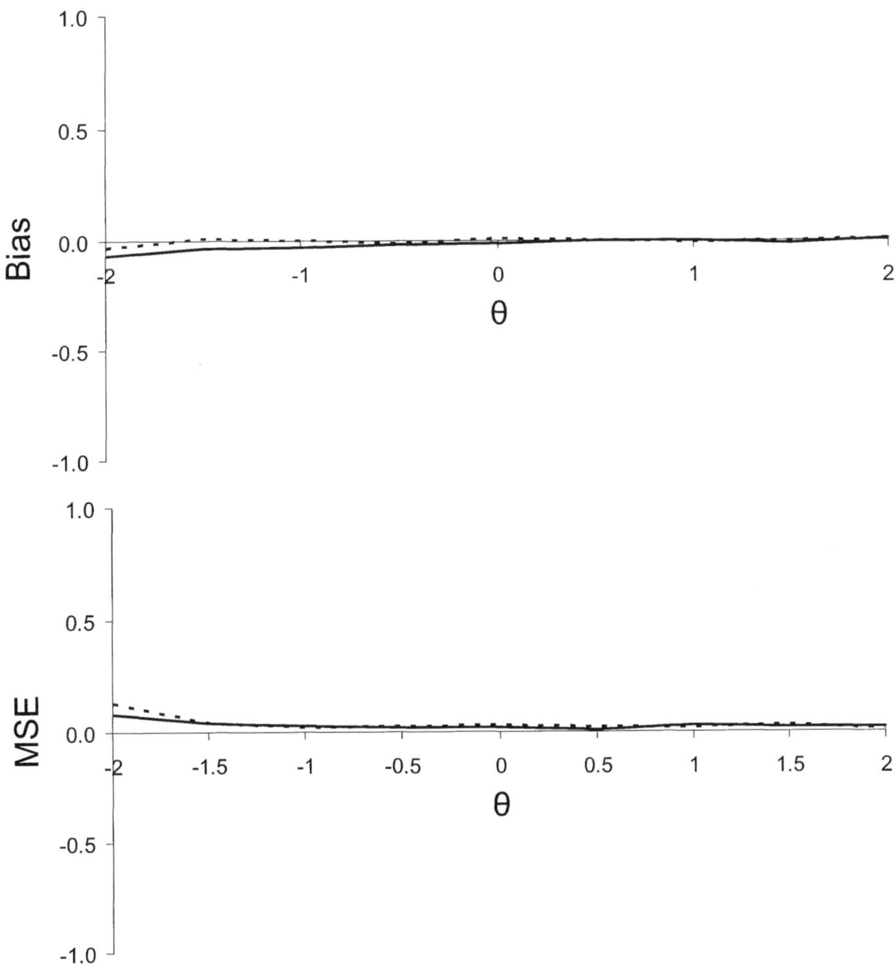

Fig. 2. Bias and MSE functions for 50-item adaptive version of the LSAT without constraints (solid curves) and with constraints (dashed curves).

was known to be on the difficult side), the functions for the conditions with and without these constraints are hardly distinguishable.

7.2. *Item-exposure control*

The exposure rates of the items in the pool can be controlled by running a probability experiment before or after a subject begins the test. The first experiments introduced in the literature use the former option. The most popular one is the Sympson–Hetter experiment (1985) (see also Hetter and Sympson, 1997). The experiment is conducted before a newly selected item is administered; the outcome is the decision either to administer the item or to pass and select the next best item at the current estimate of θ. If an item is passed, it is removed from the pool for the subject.

The probability of selecting an item is fixed by the setup of the test (item-selection rule; composition of the item pool; ability estimator; etc.). The only way to control the exposure rates (= probabilities of item administration) for this type of experiment is through manipulation of the conditional probabilities of administering an item given its selection. Because their joint effect on the distribution of the exposure rates is the result of a complicated interaction between them, these conditional probabilities have to be set through an iterative process of simulated adaptive test administrations. The process is generally cumbersome; for a realistic adaptive test it is not unusual to have to run some 150 simulations before a set of probabilities that guarantee a maximum exposure rate below say, .30, at selected ability values is found. For an evaluation and alternative implementations of Sympson–Hetter item-exposure control, see van der Linden (2003).

The other class of methods of random control is based on a probability experiment conducted before a subject begins the test. The experiment is to determine which items are eligible for the subject and which are not. If an item is eligible, it remains in the pool for the subject; otherwise it is removed.

To model the experiment, we consider two different kinds of events: the events of an item being eligible (E_i) and an item being administered (A_i). It holds that

$$A_i \subset E_i \tag{66}$$

for all i. We want to control the conditional exposure rates given θ, $\Pr(A_i \mid \theta)$, because otherwise subjects of comparable ability levels might collude and discover items with dangerously high local exposure rates (Stocking and Lewis, 1998). From (66),

$$\Pr(A_i \mid \theta) = \Pr(A_i, E_i \mid \theta) = \Pr(A_i \mid E_i, \theta) \Pr(E_i \mid \theta) \tag{67}$$

for all possible values of θ.

Imposing a maximum exposure rate, $r^{\max}$,

$$\Pr(A_i \mid \theta) = \Pr(A_i \mid E_i, \theta) \Pr(E_i \mid \theta) \leqslant r^{\max}, \tag{68}$$

or

$$\Pr(E_i \mid \theta) \leqslant \frac{r^{\max}}{\Pr(A_i \mid E_i, \theta)}, \quad \Pr(A_i \mid E_i, \theta) > 0. \tag{69}$$

From (67), it follows that

$$\Pr(E_i \mid \theta) \leqslant \frac{r^{\max} \Pr(E_i \mid \theta)}{\Pr(A_i \mid \theta)}, \quad \Pr(A_i \mid \theta) > 0. \tag{70}$$

The result in (70) is used to form a recurrence relation. Suppose n subjects have already taken the test, and we want to know the probabilities of eligibility for subject $n+1$ at selected points θ_g, $g = 1, \ldots, G$. The probabilities are

$$\Pr^{(n+1)}(E_i \mid \theta_g) = \min\left\{\frac{r^{\max} \Pr^{(n)}(E_i \mid \theta_g)}{\Pr^{(n)}(A_i \mid \theta_g)}, 1\right\}. \tag{71}$$

By substitution, it is easy to show that the probabilities remain equal, increase, or decrease if they are equal, smaller, or greater than $r^{\max}$ for the previous subject. This feature of self-adjustment enables us to just begin the test, without the necessity of any prior study to find optimal setting for the control parameters.

To implement the method, we conceive of G different versions of the item pool, one at each θ_g, $g = 1, \ldots, G$, and record the following information: (i) the number of earlier subjects who visited item pool g and took item i and (ii) the number of earlier subjects who visited item pool g when item i was eligible. Let α_{nig} and ε_{nig} denote these numbers. The probabilities for $n+1$ are estimated as

$$\Pr(E_i \mid \theta_g) = \min\left\{\frac{r^{\max} \varepsilon_{nig}}{\alpha_{nig}}, 1\right\}. \tag{72}$$

During the test, we ignore the differences between the subject's true and estimated ability and use the probabilities at the point θ_g closest to the estimated ability. For a choice of $G = 10$ or so, the impact of the estimation error on the actual exposure rates has shown to be negligible for all practical purposes.

In order to get smooth behavior of the probabilities of eligibility, it is recommended to update the numbers α_{nig} and ε_{nig} using the technique of fading, which was developed for the update of posterior probabilities in Bayesian networks (e.g., Jensen, 2001, Section 3.3.2). This type of update involves a factor w slightly smaller than one, which is used to weigh the earlier events relative to current event. For example, α_{nig} is then updated as $\alpha^*_{(n+1)ig} = w\alpha^*_{nig} + 1$ when item i is administered to subject n and as $\alpha^*_{(n+1)ig} = w\alpha^*_{nig}$ when it is not. Updating with a fading factor w produces estimates of the probabilities of eligibility based on a effective sample of the last $1/(1-w)$ subjects, with a weight equal to one for subject n and weights approaching zero for the subjects $n-1, n-2, \ldots$. A practical value is $w = .999$, which amounts to an effective sample size of 1,000.

In practice, we should use this method of item-exposure control in combination with the shadow-test approach in the preceding section to allow for all other specifications that have to be imposed on the test. The only thing that has to be done is to introduce the constraints

$$x_i = 0 \tag{73}$$

for the items that are ineligible in the model for the shadow test in (59)–(65). The update of the probabilities in (71) then requires a slight modification to account for the possibility of overconstraining. (If infeasibility should happen, all items remain in the pool;

the algorithm automatically adjusts the future probabilities of eligibility for an extra exposure.) For these and other details of item–exposure control using random ineligibility constraints, see van der Linden and Veldkamp (2004, 2006).

An example of the results for the method for a 25-item section of the LSAT is given in Figure 3. The results are based on 10,000 simulated subjects. The exposure rates were controlled at the nine values $\theta_g = -2.0, -1.5, \ldots, 2.0$. We show the resulting rates for the cases of no exposure control and control with $r^{\max}$ equal to .15 and .10. The rates are shown at five different values θ_g; the rates at for the other values were entirely comparable.

7.3. *Differential speededness*

Effective constraints for making a test equally speeded for all subjects can only be formulated when the item pool is calibrated with respect to the time properties of the items as well. Suppose that the lognormal model for response times in (49) has been used. The properties of the items are then captured by (known) parameters β_i and α_i for their time intensity and discriminating power, respectively. In addition, we assume that the response times of the subject are used to update the estimate of the test taker's speed parameters τ during the test.

From a measurement point of view, the item parameters are nuisance parameters for which we have to control. We will select the items such that their impact on the total amount of time that is required for the test does not vary across the subjects. A moment's reflection reveals that, provided their values are not unduly large, the parameters α_i do not have much impact on the distribution of the total time. The total logtime has variance $\sum a_i^{-2}/m^2$. Since the time on an item does usually not exceed two minutes or so, the variability of the total time on a 25-item test is already ignorable for all practical purposes. Therefore, for real-world tests, it is typically unnecessary to control for the impact of α_i.

The answer to the question of whether we should treat the speed parameter as a nuisance or an intentional parameter depends on the goal of the test. If the test is a pure power test, τ should be treated as a nuisance parameter. In this case, we should constrain the selection of the items to give each subject enough time to relative to the time limit. If it is intended to have a speed aspect as well, τ should be treated as an intentional parameter, and the selection of the items should not be constrained with respect to τ. (In fact, it would then be correct to score the test using the vector $(\hat{\theta}, \hat{\tau})$ but this is not common yet.)

We first treat both the item parameters and the speed parameter as nuisance parameters and constrain the time spent on the test with respect to their impact. Since τ is unknown, it has to be estimated during the test. In order to account properly for the estimation error, we adopt the posterior distribution in (51) as the estimate of τ. This posterior distribution can used be to calculate the predictive distributions on all remaining items in the pool for the subject. If $k - 1$ items have been selected, the predictive densities are

$$p(t_i \mid \mathbf{t}_{k-1}) = \int p(t_i \mid \tau) p(\tau \mid \mathbf{t}_{k-1})\, \mathrm{d}\tau, \tag{74}$$

for all $i \in R_k$.

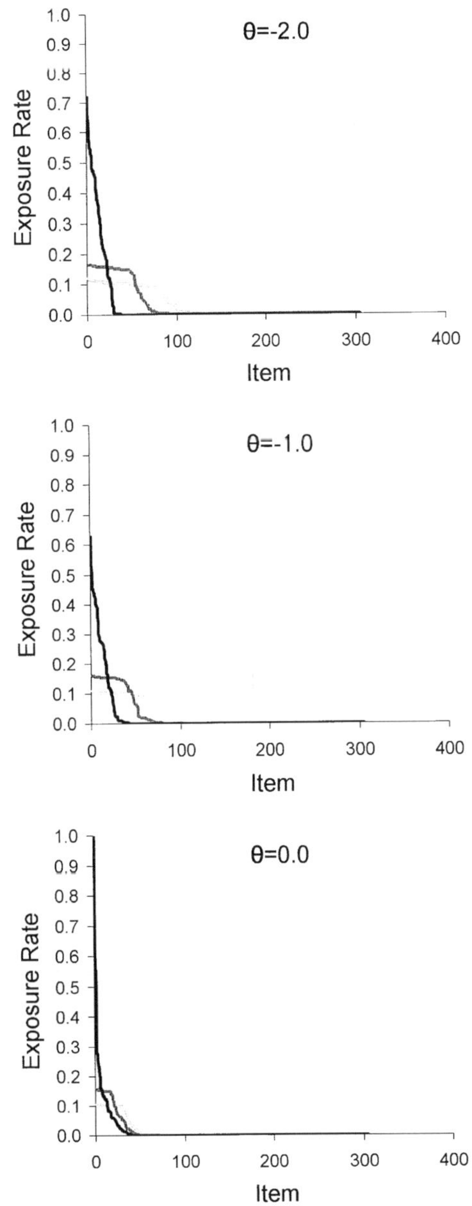

Fig. 3. Conditional item–exposure rates at $\theta = -2.0\ (1)\ 2.0$ without control (dark curves), control with $r^{\max} = .15$ (gray curves), and control with $r^{\max} = .10$ (light gray curves).

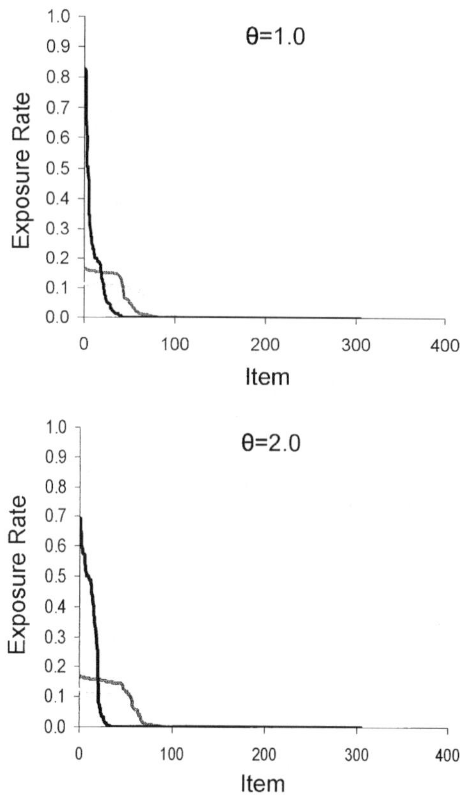

Fig. 3. *(Continued.)*

Let t_i^π be the πth quantile in the predictive distribution for item i; that is, it holds that

$$\int^{t_i^\pi} p(t_i \mid \mathbf{t}_{k-1})\, \mathrm{d}t_i = \pi. \tag{75}$$

If $t^{\max}$ is the time limit for the test, we impose the following constraint on the shadow test for item k:

$$\sum_{i \in S_{k-1}} t_i + \sum_{i \in R_k} t_i^\pi x_i \leqslant t^{\max} \tag{76}$$

(van der Linden et al., 1999). The first term in (76) is the actual total time recorded for the first $k-1$ items in the test; the second term is the sum of the πth quantiles in the predicted distributions on the remaining $n-k$ items. The constraint requires the items to be selected such that the projected time on the entire test not be larger than $t^{\max}$.

Observe that (76) is linear in the variables x_i. Also, if logtimes are used, the densities in (74) are normal, and so is $p(t_j \mid \mathbf{t}_{k-1})$. The quantiles in (75) are then easy to calculate.

To make the procedure conservative, it makes sense to begin the test with $\pi = .50$ but let π increase toward the end of it.

If τ is an intentional parameter, we control the selection of the items only for the impact of the item parameters. As already indicated, we ignore α_i and focus on β_i. A convenient way to identify the model in (49) is to set μ_τ in (50) equal to zero. Under this assumption, for a subject taking the test at the average speed in the population, the expected time on item i on a logarithmic scale is equal to β_i. Thus, on the natural scale, the expected time is equal to $\exp(\beta_i)$.

The expected total time on the test can now be controlled by including the following constraint in the model for the shadow test:

$$\sum_{i=1}^{I} \exp(\beta_i) x_i \leqslant t^{\max} - \delta, \tag{77}$$

where δ is a speededness parameter to be selected by the testing agency (the larger δ, the less speeded the test). No matter the value of δ chosen, (77) makes the test equally speeded for all subjects.

Observe that (77) is independent of k; the constraint on the previous items in (63) guarantees that its left-hand sum automatically contains the values of β_i for the $k - 1$ items already administered. Also, the constraints in (76) and (77) are formulated at the level of the total test. If stricter control, in the sense of more homogeneous speededness throughout the test, is required, we can divide the total testing time into equally long periods and impose the constraints at the level of subsequent blocks of items for the periods.

The following example is derived from van der Linden (2007, Section 9.5). For one of the adaptive tests from the Armed Services Vocational Aptitude Battery (ASVAB), we calibrated the items using a constrained version of the lognormal model in (49) and simulated adaptive test administrations for subjects in the range of values for the speed parameter τ we found in the sample of subjects. The first plot in Figure 4 shows the total time used by the simulated subjects in two of the ability groups. At the lower speed levels, the subjects in these two groups ran out of time. The second plot is for the replication of the same simulation but now with the constraint in (76) in the shadow-test model. The two ability groups no longer ran out of time.

7.4. Observed-score reporting

Classical methods of score equating that have been used in adaptive testing are equipercentile equating with randomly equivalent groups of subjects and true-score equating using the test characteristic function (for a description of these methods, see Kolen and Brennan, 2004). The first method requires an equating study in which the population distributions of the scores on the two tests are to be estimated. The equating transformation is defined as the transformation that equates the quantiles in the two distributions. These equating studies are time consuming and expensive. They are particularly cumbersome for the equating of an adaptive test to a fixed reference test, since they have to be repeated each time an item pool is changed. The second method is more efficient

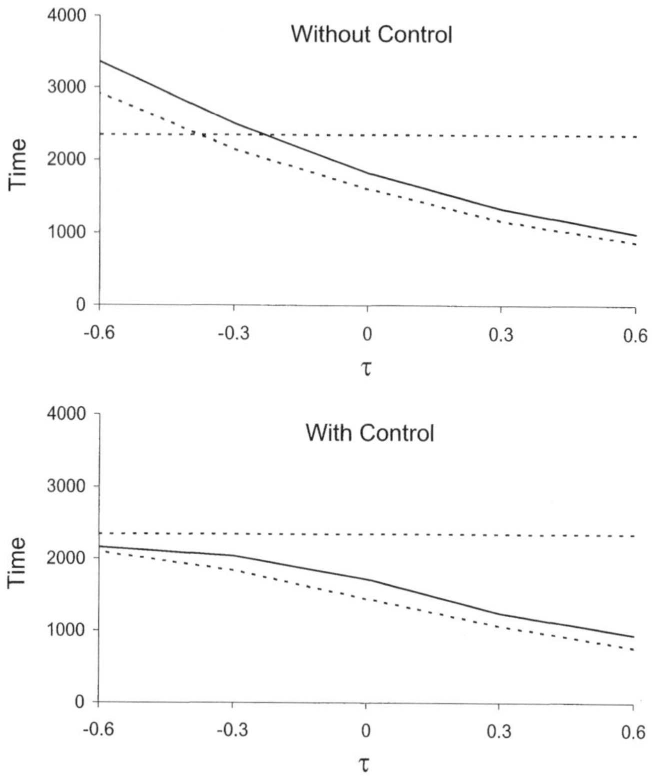

Fig. 4. Mean time used in adaptive test by subjects in ability groups $\theta = 2.0$ (solid curves) and $\theta = 1.5$ (dashed curves) without and with control for speededness (time limit: 2340 secs = 39 mins).

but confounds the estimated true scores on the reference test with its observed number-correct scores. More importantly, both methods have been shown to be seriously biased because the single equating transformation that they use has to compromise between the observed-score distributions of the subjects at widely different ability levels (van der Linden, 2006b).

A more efficient way of equating the scores on an adaptive test to a linear test is to select the items to yield a number-correct scale for the adaptive test that is identical to that of the reference test for each subject. This can be done using a simple set of linear constraints on the items (van der Linden, 2001). From a practical point of view, the alternative is attractive because it avoids the necessity of an equating study prior to the operational use of the adaptive test. From a statistical point, it is attractive because it equates the scores on the two tests locally at the estimate of θ and thereby avoids much of the bias inherent in the classical equating methods.

The set of linear constraints follows from a condition on two tests to yield identical observed number-correct scores conditional on θ derived in van der Linden and Luecht (1998). Suppose we have two tests of n items with number-correct scores X and Y. For a subject with ability θ, the probabilities of a correct response on the two tests are denoted as $p_i(\theta)$ and $p_j(\theta)$, $i, j = 1, \ldots, m$. The rth power of the probability of item i is denoted as $p_i^r(\theta)$. The two tests have identical conditional distributions of X and Y given θ if and only if

$$\sum_{i=1}^{m} p_i^r(\theta) = \sum_{j=1}^{m} p_j^r(\theta), \quad r = 1, \ldots, m, \tag{78}$$

that is, if the powers of their success probabilities up to the mth order are equated. In addition, the importance of these conditions vanishes with the order of the power. In fact, if the test length increases, the conditions for $r \geqslant 2$ become superfluous. For the typical test lengths in educational and psychological testing, the conditions have to be satisfied only for the first 2 or 3 powers.

Observe that the conditions in (78) are linear in the items. It is thus a simple step to incorporate them in the shadow-test model. The model then automatically produces a test for with the same observed-score distribution as the reference test for each of the subjects.

Let $j = 1, \ldots, n$ denote the items in the reference test and $\mathcal{T}_{r(k-1)} = \mathcal{T}_r(\hat{\theta}_{k-1})$ the right-hand sum of (78) for the rth power. These target values can be calculated easily from the response model. The following constraints impose the target values on the shadow tests for item k:

$$\sum_{i=1}^{I} p_i^r(\hat{\theta}_{k-1})x_i \leqslant \mathcal{T}_{r(k-1)} + \varepsilon, \quad \text{for all } r \leqslant R, \tag{79}$$

$$\sum_{i=1}^{I} p_i^r(\hat{\theta}_{k-1})x_i \geqslant \mathcal{T}_{r(k-1)} - \varepsilon, \quad \text{for all} r \leqslant R, \tag{80}$$

where $R = 2, 3$ and ε is a small tolerance needed to avoid infeasibility due to an equality constraint. In the beginning of the test, when $\hat{\theta}_{k-1}$ has a tendency to larger estimation errors, it may be useful to add similar constraints at the values $\hat{\theta}_{k-1} + \gamma_k$ and $\hat{\theta}_{k-1} - \gamma_k$, where γ_k should be chosen to decrease with k. Alternatively, we could constraint the sums of powers of the response probabilities at fixed set of values θ_g, $g = 1, 2, 3$.

The best choice of constraints for adaptive testing still has to be researched. We therefore show an example of the use of (78) for the equating of the conditional number-correct-score distributions on two fixed versions of the LSAT. One version served a 101-item reference test; the other was assembled from a pool of over 600 items subject to the constraints in (79) and (80) at the fixed values $\theta_g = -1.2$, .0, and 1.2 for $R = 2$. Their conditional observed-score distributions given $\theta = -1.5, -0.5, 0.5$, and 1.5 are compared in Figure 5. The distributions appear to be identical for all practical purposes.

8. Concluding comment

The use of adaptive tests is a relatively new phenomenon in educational and psychological testing but over the last decade its technology has quickly matured. In this chapter,

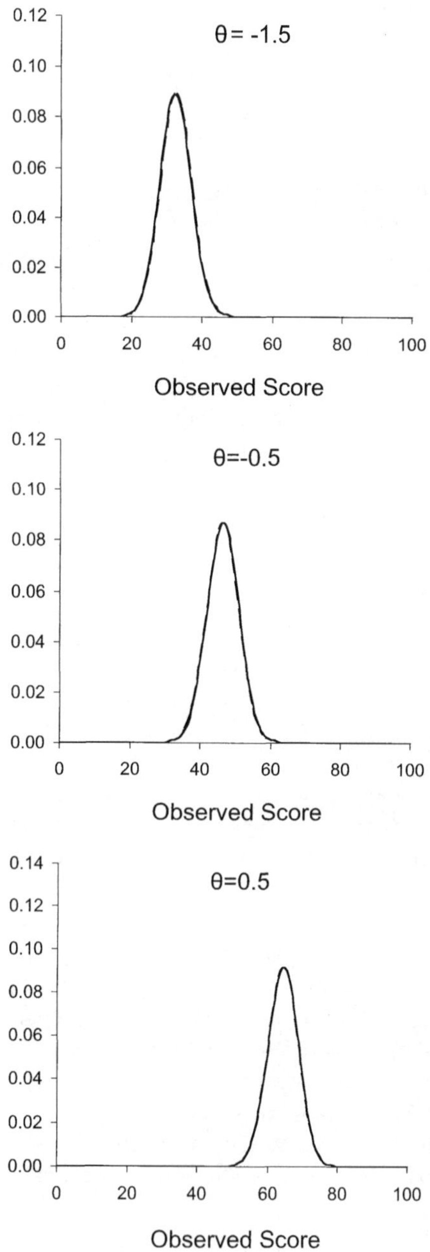

Fig. 5. Distributions of number-correct score on adaptive test (dashed curves) and fixed reference test (solid curves) for subject at $\theta = -1.5, (1)1.5$. (Note: differences between the curves are hardly discernible.)

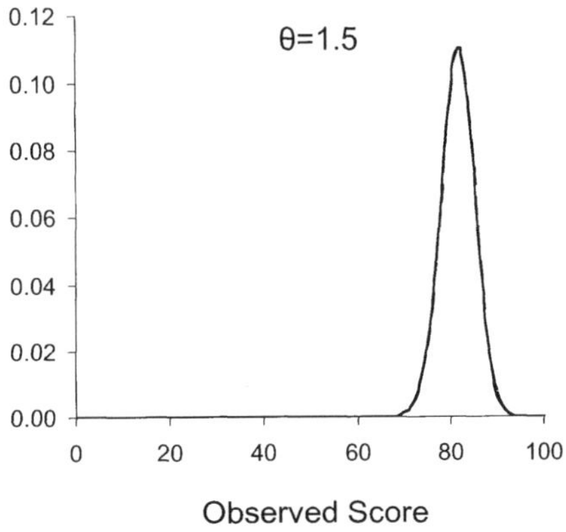

Fig. 5. (*Continued.*)

we reviewed some of its statistical aspects but had to leave others untouched. For example, we did not discuss any of the aspects of adaptive testing from an item pool with polytomously scored items or items that require a response model with a multidimensional ability parameter. The latter has additional complexity because we have to adapt the item selection to the individual components of $\boldsymbol{\theta}$ or a combination of them. Another area that we did not discuss is the use of statistical procedures for checking response vectors for possible aberrances due to design errors in the test or deviant behavior by subjects. A review of Bayesian procedures for fixed tests is given in Glas and Meijer (2003). But these procedures may not have much power for adaptive tests, where the response probabilities for the model in (1) quickly approximate a value close to .50 if the adaptation is successful and wild response patterns become likely. Additional information, however, can be found in the pattern of the response times on the items. Finally, we have not treated such important areas as item-pool design for adaptive testing, testing with adaptation at a higher level in the test than the individual items (e.g., multistage testing), and newer developments in multidimensional adaptive testing. For these and other aspects, see Chang (2004), van der Linden and Glas (2000b, 2007), or Wainer (2000).

References

Adams, R.J., Wilson, M.R., Wang, W.C. (1997). The multidimensional random coefficients multinomial logit model. *Applied Psychological Measurement* **21**, 1–23.

Albert, J.H. (1992). Bayesian estimation of normal ogive item response functions using Gibbs sampling. *Journal of Educational Statistics* **17**, 251–269.

Andersen, E.B. (1985). Estimating latent correlations between repeated testings. *Psychometrika* **50**, 3–16.

Béguin, A.A., Glas, C.A.W. (2001). MCMC estimation and some fit analysis of multidimensional IRT models. *Psychometrika* **66**, 541–562.

Birnbaum, A. (1968). Some latent trait models and their use in inferring an examinee's ability. In: Lord, F.M., Novick, M.R. (Eds.), *Statistical Theories of Mental Test Scores*. Addison-Wesley, Reading, MA, pp. 397–479.

Bock, R.D., Aitkin, M. (1981). Marginal maximum likelihood estimation of item parameters: An application of an EM-algorithm. *Psychometrika* **46**, 443–459.

Bock, R.D., Mislevy, R.J. (1988). Adaptive EAP estimation of ability in a microcomputer environment. *Applied Psychological Measurement* **6**, 431–444.

Bock, R.D., Gibbons, R.D., Muraki, E. (1988). Full-information factor analysis. *Applied Psychological Measurement* **12**, 261–280.

Box, G., Tiao, G. (1973). *Bayesian Inference in Statistical Analysis*. Addison-Wesley Publishing Company, Reading, MA.

Chang, H.-H. (2004). Understanding computerized adaptive testing: From Robbins–Monro to Lord and beyond. In: Kaplan, D. (Ed.), *Handbook of Quantitative Methods for the Social Sciences*. Sage, Thousand Oaks, CA, pp. 117–133.

Chang, H.-H., Ying, Z. (1996). A global information approach to computerized adaptive testing. *Applied Psychological Measurement* **20**, 213–229.

Chang, H.-H., Ying, Z. (2006). Nonlinear designs for logistic item response models with application in computerized adaptive tests. *The Annals of Statistics*, in press.

Dempster, A.P., Laird, N.M., Rubin, D.B. (1977). Maximum likelihood from incomplete data via the EM algorithm (with discussion). *J. Roy. Statist. Soc. B* **39**, 1–38.

Dodd, B.G., DeAyala, W.R., Koch, W.R. (1995). Computerized adaptive testing with polytomous items. *Applied Psychological Measurement* **19**, 5–22.

Efron, B. (1977). Discussion on maximum likelihood from incomplete data via the EM algorithm (by A. Dempster, N. Liard, and D. Rubin). *J. Roy. Statist. Soc. B* **39**, 29.

Gelfand, A.E., Smith, A.F.M. (1990). Sampling-based approaches to calculating marginal densities. *Journal of the American Statistical Association* **85**, 398–409.

Gelman, A., Carlin, J.B., Stern, H.S., Rubin, D.B. (1995). *Bayesian Data Analysis*. Chapman and Hall, London.

Glas, C.A.W. (1992). A Rasch model with a multivariate distribution of ability. In: Wilson, M. (Ed.), *Objective Measurement: Theory into Practice*, vol. 1. Ablex Publishing Corporation, New Jersey, NJ, pp. 236–258.

Glas, C.A.W. (1998). Detection of differential item functioning using Lagrange multiplier tests. *Statistica Sinica* **8** (1), 647–667.

Glas, C.A.W. (1999). Modification indices for the 2-pl and the nominal response model. *Psychometrika* **64**, 273–294.

Glas, C.A.W. (2000). Item calibration and parameter drift. In: van der Linden, W.J., Glas, C.A.W. (Eds.), *Computerized Adaptive Testing: Theory and Practice*. Kluwer–Nijhoff Publishing, Boston, MA, pp. 183–200.

Glas, C.A.W., Meijer, R.R. (2003). A Bayesian approach to person fit analysis in item response theory models. *Applied Psychological Measurement* **27**, 217–233.

Glas, C.A.W., Suarez Falcon, J.C. (2003). A comparison of item-fit statistics for the three-parameter logistic model. *Applied Psychological Measurement* **27**, 87–106.

Glas, C.A.W., van der Linden, W.J. (2003). Computerized adaptive testing with item cloning. *Applied Psychological Measurement* **27**, 247–261.

Glas, C.A.W., van der Linden, W.J. (2006). Modeling item parameter variability in item response models. *Psychometrika*, in press.

Hetter, R.D., Sympson, J.B. (1997). Item exposure control in CAT-ASVAB. In: Sands, W.A., Waters, B.K., McBride, J.R. (Eds.), *Computerized Adaptive Testing: From Inquiry to Operation*. American Psychological Association, Washington, DC, pp. 141–144.

Hoijtink, H. (2001). Conditional independence and differential item functioning in the two-parameter logistic model. In: Boomsma, A., van Duijn, M.A.J., Snijders, T.A.B. (Eds.), *Essays on Item Response Theory*. Springer, New York, pp. 109–130.

Jensen, F.V. (2001). *Bayesian Networks and Graphs*. Springer, New York.

Johnson, V.E., Albert, J.H. (1999). *Ordinal Data Modeling*. Springer, New York.

Koch, W.R., Dodd, B.G. (1989). An investigation of procedures for computerized adaptive testing using partial credit scoring. *Applied Measurement in Education* **2**, 333–357.

Kolen, M.J., Brennan, R.L. (2004). *Test Equating, Scaling, and Linking: Methods and Practices*. Springer, New York.

Lehman, E.L., Casella, G. (1998). *Theory of Point Estimation*, 2nd ed. Springer, New York.
Lord, F.M. (1980). *Applications of Item Response Theory to Practical Testing Problems*. Erlbaum, Hillsdale, NJ.
McDonald, R.P. (1967). *Nonlinear Factor Analysis*. Psychometric Monographs, No. 15.
Meng, X.L. (1994). Posterior predictive p-values. *Ann. Statist.* **22**, 1142–1160.
Mislevy, R.J. (1986). Bayes modal estimation in item response models. *Psychometrika* **51**, 177–195.
Muthén, L.K., Muthén, B.O. (2003). *Mplus version 2.14*. Los Angeles, CA.
Muraki, E., Bock, R.D. (2002). *PARSCALE: Parameter Scaling of Rating Data*. Scientific Software, Inc., Chicago, IL.
Orlando, M., Thissen, D. (2000). Likelihood-based item-fit indices for dichotomous item response theory models. *Applied Psychological Measurement* **24**, 50–64.
Owen, R.J. (1969). A Bayesian approach to tailored testing. Research Report 69-92, Educational Testing Service, Princeton, NJ.
Owen, R.J. (1975). A Bayesian sequential procedure for quantal response in the context of adaptive mental testing. *Journal of the American Statistical Association* **70**, 351–356.
Penfield, R.D. (2006). Applying Bayesian item selection approaches to adaptive tests using polytomous items. *Applied Measurement in Education* **19**, 1–20.
Reckase, M.D. (1985). The difficulty of test items that measure more than one ability. *Applied Psychological Measurement* **9**, 401–412.
Robert, C.P., Casella, G. (1999). *Monte Carlo Statistical Methods*. Springer, New York.
Segall, D.O. (1996). Multidimensional adaptive testing. *Psychometrika* **61**, 331–354.
Segall, D.O. (2000). Principles of multidimensional adaptive testing. In: van der Linden, W.J., Glas, C.A.W. (Eds.), *Computerized Adaptive Testing: Theory and Practice*. Kluwer Academic Publishers, Boston, MA, pp. 53–73.
Sinharay, S. (2005). Assessing fit of unidimensional item response theory models using a Bayesian approach. *Journal of Educational Measurement* **42**, 375–394.
Sinharay, S., Johnson, M.S., Williamson, D.M. (2003). Calibrating item families and summarizing the results using family expected response functions. *Journal of Educational and Behavioral Statistics* **28**, 295–313.
Stocking, M.L., Lewis, C. (1998). Controlling item exposure conditional on ability in computerized adaptive testing. *Journal of Educational and Behavioral Statistics* **23**, 57–75.
Sympson, J.B., Hetter, R.D. (1985). Controlling item-exposure rates in computerized adaptive testing. In: *Proceedings of the 27th Annual Meeting of the Military Testing Association*, Navy Personnel Research and Development Center, San Diego, CA, pp. 973–977.
Takane, Y., de Leeuw, J. (1987). On the relationship between item response theory and factor analysis of discretized variables. *Psychometrika* **52**, 393–408.
Thissen, D., Chen, W.-H., Bock, R.D. (2003). *Multilog*. Scientific Software International, Lincolnwood, IL.
van der Linden, W.J. (1998). Bayesian item-selection criteria for adaptive testing. *Psychometrika* **62**, 201–216.
van der Linden, W.J. (1999). A procedure for empirical initialization of the ability estimator in adaptive testing. *Applied Psychological Measurement* **23**, 21–29.
van der Linden, W.J. (2000). Constrained adaptive testing with shadow tests. In: van der Linden, W.J., Glas, C.A.W. (Eds.), *Computerized Adaptive Testing: Theory and Practice*. Kluwer Academic Publishers, Boston, MA, pp. 27–52.
van der Linden, W.J. (2001). Computerized adaptive testing with equated number-correct scoring. *Applied Psychological Measurement* **25**, 343–355.
van der Linden, W.J. (2003). Some alternatives to Sympson–Hetter item–exposure control in computerized adaptive testing. *Journal of Educational and Behavioral Statistics* **28**, 249–265.
van der Linden, W.J. (2006a). A lognormal model for response times on test items. *Journal of Educational and Behavioral Statistics* **31**, 181–204.
van der Linden, W.J. (2006b). Equating scores from adaptive to linear tests. *Applied Psychological Measurement* **30**, in press.
van der Linden, W.J. (2007). Using response times for item selection in adaptive testing. *Journal of Educational and Behavioral Statistics* **32**, in press.
van der Linden, W.J., Glas, C.A.W. (2000a). Capitalization on item calibration error in adaptive testing. *Applied Measurement in Education* **13**, 35–53.

van der Linden, W.J., Glas, C.A.W. (Eds.) (2000b). *Computerized Adaptive Testing: Theory and Practice.* Kluwer Academic Publishers, Boston, MA.

van der Linden, W.J., Glas, C.A.W. (Eds.) (2007). *Elements of Adaptive Testing*. Springer, New York, in press.

van der Linden, W.J., Hambleton, R.K. (Eds.) (1997). *Handbook of Modern Item Response Theory*. Springer, New York.

van der Linden, W.J., Luecht, R.M. (1998). Observed-score equating as a test assembly problem. *Psychometrika* **63**, 401–418.

van der Linden, W.J., Pashley, P.J. (2000). Item selection and ability estimation in adaptive testing. In: van der Linden, W.J., Glas, C.A.W. (Eds.), *Computerized Adaptive Testing: Theory and Practice*. Kluwer Academic Publishers, Boston, MA, pp. 1–25.

van der Linden, W.J., Reese, L.M. (1998). A model for optimal constrained adaptive testing. *Applied Psychological Measurement* **22**, 259–270.

van der Linden, W.J., Veldkamp, B.P. (2004). Constraining item-exposure rates in computerized adaptive testing with shadow tests. *Journal of Educational and Behavioral Statistics* **29**, 273–291.

van der Linden, W.J., Veldkamp, B.P. (2006). Conditional item-exposure control in adaptive testing using item-ineligibility probabilities. *Journal of Educational and Behavioral Statistics* **31**, in press.

van der Linden, W.J., Scrams, D.J., Schnipke, D.L. (1999). Using response-time constraints to control for differential speededness in computerized adaptive testing. *Applied Psychological Measurement* **23**, 195–210.

Veerkamp, W.J.J., Berger, M.P.F. (1997). Some new item-selection criteria for adaptive testing. *Journal of Educational and Behavioral Statistics* **22**, 203–226.

Veldkamp, B.P., van der Linden, W.J. (2002). Multidimensional adaptive testing with constraints on test content. *Psychometrika* **67**, 575–588.

Wainer, H. (Ed.) (2000). *Computerized Adaptive Testing: A Primer*. Lawrence Erlbaum Associates, Hilsdale, NJ.

Warm, T.A. (1989). Weighted likelihood estimation of ability in item response theory with tests of finite length. *Psychometrika* **54**, 427–450.

Wood, R., Wilson, D.T., Gibbons, R.D., Schilling, S.G., Muraki, E., Bock, R.D. (2002). *TESTFACT: Test Scoring, Item Statistics, and Item Factor Analysis*. Scientific Software International, Inc., Chicago, IL.

Wu, M.L., Adams, R.J., Wilson, M.R. (1997). *ConQuest: Generalized Item Response Modeling Software*. Australian Counsil for Educational Research.

Zimowski, M.F., Muraki, E., Mislevy, R.J., Bock, R.D. (2002). *Bilog-MG*. Scientific Software International, Lincolnwood, IL.

Handbook of Statistics, Vol. 26
ISSN: 0169-7161

DOI: 10.1016/S0169-7161(06)26026-7

26

Bayesian Psychometric Modeling From An Evidence-Centered Design Perspective

Robert J. Mislevy and Roy Levy

1. Introduction and overview

An educational assessment consists of an evidentiary argument for drawing inferences about what students know or can do from a set of limited observations, and mechanisms and processes that instantiate the argument. Psychometrics is an ever-growing domain rich with nuance. Though models may vary in a multitude of ways, psychometrics is built upon the general principle of inference from observations – behaviors and occurrences that manifest themselves in the world – to claims – abstract statements and expressions about examinees or subjects more broadly conceived. Modern psychometrics is characterized by the use of probabilistic reasoning in the form of statistical models to instantiate this principle and facilitate the desired inference in the face of uncertainty.

For the construction of a statistical model, Gelman et al. (1995) stressed that "The model should be consistent with knowledge about the underlying scientific problem and the data collection process" (p. 3). At a rudimentary level, this means that all variables believed to be relevant should be included, the effects of the sampling design should be accounted for, etc. More conceptually, the model employed should be built so that it is connected to all that is known about the problem at hand. Accordingly, we ought to construct and construe jointly the observational setting, theory, and statistical model to be employed in an assessment. In what follows, we describe a principled approach to assessment design that unifies ideas from fields of cognitive science, statistical modeling, and educational assessment, all within the framework of evidentiary reasoning.

This chapter is aimed at making explicit the groundwork necessary for assessment design and use. Though they may differ widely in terms of their surface features, all assessments may be framed as exercises in reasoning through evidentiary arguments. This chapter describes the processes involved in reasoning through assessment arguments. The result is not a recipe for making a perfect model for use in an assessment. Indeed, no model is perfect and model criticism is crucial for learning about both the domain and the assessment. Rather, by highlighting the foundations of model construction, the approach advocated here provides a framework that is grounded in first principles of assessment and inference.

2. An evidentiary perspective

2.1. *Evidentiary and inferential language in assessment*

Evolving complexities and emerging practices in the development, administration, scoring, and use of assessments have rendered popular assessment terms ambiguous or limited in breadth at best and misleading or irrelevant at worst. Expressions such as "item," "answer," and "score," which on the surface appear to be quite general, are, in fact, limited in their application; their shortcomings become painfully apparent in designing or describing complex, innovative assessments (Behrens et al., 2004).

Innovative assessments, particularly while under development, are better served by a richer, more general terminology. One such language is provided in the "evidence-centered" design (ECD) framework described by Mislevy et al. (2003). In this section, foundational terms that are indispensable for an evidence-based approach are briefly described. The balance of this chapter characterizes common and emerging assessment practices in terms of ECD, with a particular focus on those parts of the argumentation that are instantiated in the statistical model.

A *claim* is a declarative statement about what an examinee knows or can do. Claims vary in their specificity (compare "the examinee can subtract" to "the examinee can subtract negative improper fractions") as determined by the purpose of the assessment. Formative assessments often involve highly specific claims; summative assessments tend to have broader claims. Claims may be structured (e.g., "the examinee can subtract negative improper fractions" is a sub-claim of "the examinee can subtract"). Multiple, structured claims are often relevant when the assessment serves multiple purposes (e.g., summative rankings, content specific feedback).

Claims are hypotheses about the examinee; addressing these hypotheses is the goal of the assessment. As discussed in greater detail below, information pertinent to addressing the claim(s) is expressed in terms of student model variables, typically latent variables representing the underlying construct of interest.

In order to gain information regarding the student model variables and ultimately the claims to which they are relevant, *observations* or *observable outcome variables* (observables, for short) are collected to serve as evidence. To obtain observations, examinees are presented with tasks; the result of these interactions is a *work product*. In the traditional language of assessment, work products are termed "answers." However, the term "answer" implies the presence of a question whereas in assessment it may be that there is no question, but a task to be attempted (e.g., recite a speech, perform a dance, fix a malfunctioning computer network). Evaluations of work products, possibly complex and possibly multivariate, possibly automated and possibly by human raters, produce values for observables. These values constitute evidence to update beliefs about student model variables, through relationships suggested by the substantive considerations and reflected in the statistical model.

2.2. *Assessments as evidentiary arguments*

What all assessments share is a reasoning structure (Mislevy et al., 2003). In particular, an assessment is a system for reasoning from what examinees do, produce, or say to more broad aspects of what they know or are able to do. The thrust of the assessment is that the inferences are broader than the particular observations on which they are founded. Sound assessment practices justify the inferences. Upon accepting that the name of the game is validity, and that validity concerns "the degree to which empirical evidence and theoretical rationales support the adequacy and appropriateness of inferences and actions based on test scores or other modes of assessment" (Messick, 1989, p. 13), an evidentiary argument structure for assessments is natural (Mislevy et al., 2003). In an evidentiary framework, assessments are a mechanism for placing the assessor in position to obtain evidence, obtaining the evidence, managing and processing the evidence, and drawing inferences and decisions during and after the assessment.

Following Schum (1987, 1994) (see Mislevy, 1994), inference is characterized as reasoning from what is known to what is unknown in the form of explanations, conclusions, or predictions. Inferences are based on observations, which become *evidence* only when they are linked to a particular claim or desired inference (Schum, 1987). Understanding what counts as evidence therefore comes to understanding what counts as relevant to a particular claim. This in turn comes to understanding the relationships of the various entities in the domain, which in assessment is how potentially observable data are related to broader hypotheses about what examinees know or can do.

These relations are often effectively conveyed through a narrative logic. Toulmin (1958) diagrams (Figure 1) are convenient knowledge representations for portraying all the actors in the narrative and their influence on the outcome. Ultimately, the inference will flow from the data to the claim, in assessment, from the observed behaviors and statements from examinees to hypotheses about their knowledge, skills, and abilities more broadly conceived. The justification for the inference comes from the warrant, which states the relation between the observation and the claim, often in the opposite direction as the inferential path. A warrant expresses that under some constellation of conditions (e.g., the claim is true) certain outcomes (i.e., the observed data) are to be expected as a result. The conflict between the directionality of the inference (from observations to claims) and the directionality of the warrant (from claims to observations; hence the use of a double-headed arrow in Figure 1) has important implications for explicit modeling, which are further complicated by the presence of uncertainty, as discussed below. Warrants are not observations, per se, but rather paths along which inferences may travel. They require justification, which comes in the form of a backing, which may be the result of theory, prior empirical research, or, ideally a combination of both.

Warrants have varying degrees of strength; formal syllogisms represent powerful warrants (Jaynes, 2003; Mislevy et al., 2003). A syllogistic warrant expressing the relation of the data (D) to the claim (C) takes the form

$$C \rightarrow D, \tag{1}$$

which may be read "If C is true (or occurs), D is true (or occurs)." This warrant, in conjunction with the information that C is true (occurs), logically implies D is true (occurs).

Note the directionality of the warrant is from (C) to (D). However, the direction of inference is the reverse. As a result, observing (D) does *not* justify an inference to (C).

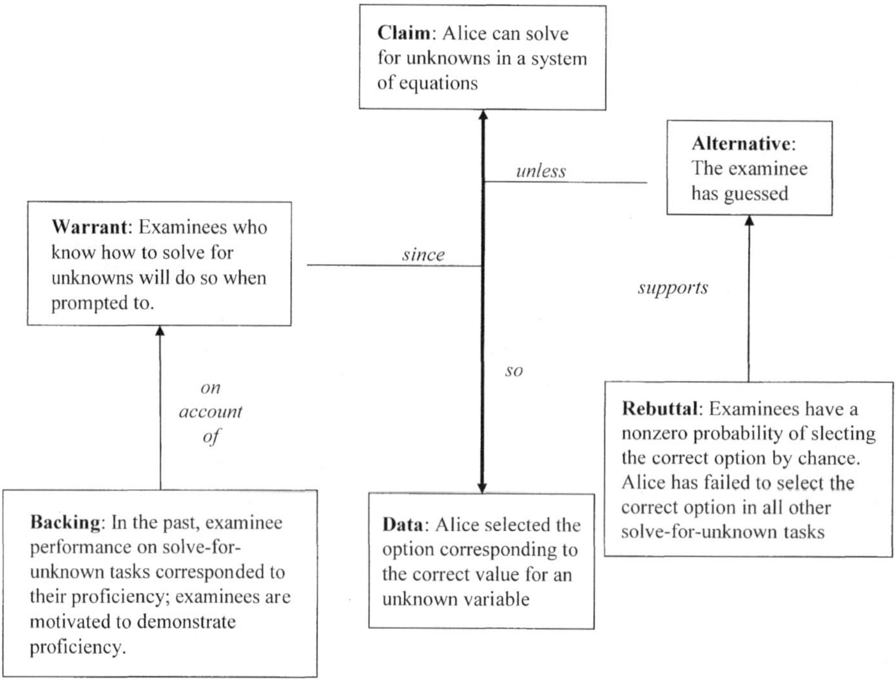

Fig. 1. A Toulmin diagram for the argument structure in educational assessment.

What is needed is reason to believe that the denial of (C), say $(\neg C)$, will *not* lead to (D). For if $(\neg C)$ leads to (D), then observing (D) is insufficient to draw a conclusion of (C) or $(\neg C)$. In psychometric terms, this corresponds to (D) failing to discriminate between (C) and $(\neg C)$.

In the schema of Toulmin diagrams, the possibility of observing data when a claim does not hold is captured by alternative explanations. Alternative hypotheses, like those represented in the claims, require empirical support from rebuttal (with respect to the inference to the truth of the claim) observations.

In assessment, the flow of warrants is *from* examinees' states or levels of knowledge, strategies, or proficiencies, *to* characteristics of their actions in particular, relevant, circumstances. The recognition of a need to consider alternative explanations in assessment is not new. Rasch (1960, p. 73) stated:

> Even if we know a person to be very capable, we cannot be sure that he will solve a difficult problem, nor even a much easier one. There is always a possibility that he fails – he may be tired or his attention is led astray, or some other excuse may be given.

Lord and Novick (1968, p. 30) reported:

> Most students taking college entrance examinations are convinced that how they do on a particular day depends to some extent on "how they feel that day." A student who receives scores which he considers surprisingly low often attributes this unfortunate circumstance

> to a physical or psychological indisposition or to some more serious *temporary* state of affairs not related to the fact that he is taking the test that day.

In justifying the use of probabilistic models for representing examinee behavior (a point which we will return to in some detail) Samejima (1983, p. 159) wrote:

> There may be an enormous number of factors eliciting his or her specific overt reactions to a stimulus, and, therefore, it is suitable, even necessary, to handle the situation in terms of the probabilistic relationship between the two.

Holland (1990, p. 579) cautioned us:

> It is important to remember that performance on a test can depend on many factors in addition to the actual test questions – for example, the conditions under which the test was given, the conditions of motivation or pressure impinging on the examinee at the time of testing, and his or her previous experiences with tests....

Deterministic rules for deductive and inductive inference, while powerful when applicable, are inappropriate in light of the continued possibilities of alternative explanations. Probabilistic relations offer greater flexibility for characterizing relationships between claims and observations and hence, between student model and observable variables in the face of uncertainty (Samejima, 1983; as quoted above).

Recognizing assessment as an evidentiary argument entails two requirements for the design and conduct of assessment. The first is to determine situations in which students at different states of proficiency are likely to act in different ways. This is the problem of task design, and it is addressed by an understanding of the nature of the proficiencies at issue and the ways they are manifest. Note that the flow of reasoning needed to construct tasks is down through the warrant: IF a student's knowledge were such and such, THEN she would probably act in such and such manner.

Given suitable observational situations, the second requirement is for mechanisms to reverse the flow of reasoning, now from data about what examinees say, do, or produce to claims about their knowledge and proficiencies. This kind of reasoning can be carried out, in sophisticated and nuanced ways, without statistical models (e.g., see Moss, 1994, on hermeneutic approaches to inference in educational assessment). In order to manage the intricacies of the evidentiary structures in assessment, with the possibility of supporting and refuting sources of evidence, psychometric models harness the powerful tools of probability based reasoning to structure, manage, and integrate information in a systematic and explicit manner. Bayesian modeling in particular provides the necessary mechanisms to support reverse reasoning in the probabilistic framework. The balance of this chapter describes principles of Bayesian reasoning in complex environments and ECD as a schema that recognizes and takes advantage of the evidentiary nature of assessment arguments. As a consequence, ECD and Bayesian inference constitute a normative framework for the design of assessments that support the desired reverse reasoning.

2.3. *Bayesian inference*

Drawing from Schum (1987, 1994) and Jaynes (2003) we maintain that probability based reasoning can play a central role in all forms of inference, including inference in educational measurement and related fields (Mislevy, 1994). Beliefs and uncertainty regarding variables are captured by probability distributions. An inferential process is thus the characterization and evaluation of probability distributions in light of evidence. Once some evidence is observed, Bayesian inference is a framework for the incorporation and propagation of evidence to arrive at the posterior distribution for unknown variables. Bayes' theorem for the posterior distribution of θ given some evidence X may be written as

$$P(\theta \mid X) = \frac{P(\theta) \times P(X \mid \theta)}{P(X)} \propto P(\theta) \times P(X \mid \theta), \tag{2}$$

where $P(\theta)$ is the prior probability distribution for θ, $P(X)$ is the marginal probability of X, and $P(X \mid \theta)$ is the conditional probability of X given θ. Conditioning on X (i.e., treating X as known) $P(X \mid \theta)$ is treated as a likelihood function for θ. The change in our beliefs about θ from the time before X is observed ($P(\theta)$) to the time after X is observed ($P(\theta \mid X)$) is a function of this likelihood expression. These likelihoods – the relative probabilities of the observed value of X given the possible values of θ that may have produced the observed value – allow for the deductive reasoning about the possible values of X, given the value of θ. And as just shown, these likelihoods also allow, via Bayes' theorem, for inductive reasoning about the possible values of θ, once some datum, X, is observed (Jensen, 1996; Mislevy, 1994). In assessment, the (possibly vector-valued) θs are student model variables, representing potential states, levels, or configurations of proficiency. The Xs are observables obtained from the interaction of examinees and the assessment system. The conditional probabilities, namely $P(X \mid \theta)$, effect a quantitative aspect of the warrant: Given a known value of θ, what is the distribution of aspects of action that might be anticipated in a given task situation?

A large number of psychometric models have been developed over the past century, each with their own particular forms of θs, Xs, and $P(X \mid \theta)$s. Classical test theory, item response theory (IRT), latent class analysis (LCA), and factor analysis are familiar examples. In each case, terminology and analytic procedures have been developed to estimate the parameters of conditional probability distributions and of population distributions of θs, as well as support posterior inferences for θs given Xs as in Eq. (2). In the following sections, we will use an IRT example to illustrate ideas. More recent work has capitalized on developments in Bayesian modeling that encompass prior work, but allow further for more flexible and modular model construction and estimation. The following section summarizes one such development, Bayesian networks.

2.4. *Bayesian networks*

When the number of variables in a problem increases, as is the case in complex assessments, application of Bayes' theorem in its form given in Eq. (2) becomes computationally intractable as the calculation of $P(X)$ becomes cumbersome, involving integration over the distributions of continuous variables. As discussed below, this computation is made easier by (a) the use of discrete variables and (b) graphical models that structure the computations. More efficient techniques to represent variables and apply Bayes' theorem across a large system of variables have been developed in the form of Bayesian

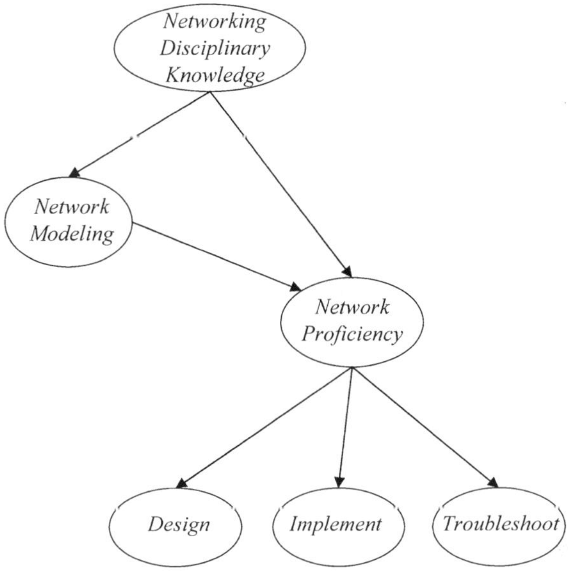

Fig. 2. The NetPASS proficiency model.

networks (BNs; Jensen, 1996, 2001; Lauritzen and Spiegelhalter, 1988; Pearl, 1988; Spiegelhalter et al., 1993). BNs support probability-based reasoning as a means of transmitting complex observational evidence throughout a network of interrelated variables. The relationships of variables in a BN constitute the reasoning structures of the network. As in the preceding discussion of Bayes' theorem, the likelihoods within the network that define the deductive reasoning structures support subsequent inductive reasoning (Mislevy, 1994).

A BN is a graphical model (of which Figure 2 is an example) of a joint probability distribution over a set of random variables, and consists of the following elements (Jensen, 1996):

- A set of variables (represented by ellipses or boxes and referred to as *nodes*) with a set of *directed edges* (represented by arrows) between nodes indicating the statistical dependence between variables. Nodes at the source of a directed edge are referred to as *parents* of nodes at the destination of the directed edge, their *children*.
- For discrete variables, each has a set of exhaustive and mutually exclusive states.
- The variables and the directed edges together form an acyclic directed graph (Brooks, 1998; Jensen, 1996, 2001; Pearl, 1988). These graphs are directed in that the edges follow a "flow" of dependence in a single direction (i.e., the arrows are always uni-directional rather than bi-directional). The graphs are acyclic in that following the directional flow of directed edges from any node it is impossible to return to the node of origin.
- For each endogenous variable (i.e., one with parents), there is an associated set of conditional probability distributions corresponding to each possible pattern of values of the parents.

- For each exogenous variable (i.e., one without parents), there is an associated unconditional probability table or distribution.

The structure of the graph conveys the dependence and conditional independence relationships in the model (see Pearl, 1988, for rules on conditional independence relations in graphs). We also note that BNs are networks of discrete variables. Network representations of models with continuous variables belong to a broader class of graphical models (Almond and Mislevy, 1999). Models with continuous variables share the above features save that continuous variables have associated (conditional) probability distributions rather than a finite set of states. The key difference is that models with continuous variables require integration to obtain $P(X)$ whereas BNs are less computationally cumbersome, requiring summation over the discrete states.

2.5. *Illustrative examples*

We pause to introduce two examples that will be revisited throughout the chapter. Consideration of both examples reveals how a common framework underlying assessment arguments supports assessments that vary widely in terms of their surface features.

The first example is a mathematics assessment to be used in admissions for higher education. The hypothesis to be addressed pertains to how well prepared the student is for collegiate, quantitative work in terms of the necessary proficiencies in the domain of mathematics. This is a broad claim, and necessarily so due to the wide variety in higher education generally and with respect to the role of mathematics in various curricula.

Owing to the coarse grainsize of this claim, the evidence should be broad and inclusive. Examinees should demonstrate declarative knowledge and low-level procedural skills in the area of mathematics. As the goal is supporting inference about an intentionally broad hypothesis of interest, evidence should be collected from the entirety of the domain. This shows up as content balancing, and restrictions on task sets that ensure that all chief aspects of the domain have been addressed. Appeals to efficiency of the assessment argument and the need to re-instantiate that argument justify a standardized administration in which the examinees are familiar with the format and response process (Mislevy et al., 2003). To this we add that standardization and familiarity with the administration process and response format serve to rule out alternative hypotheses for poor performance on the tasks related to unfamiliarity.

Our second example comes from the domain of computer networking, a rich domain with multiple, interrelated aspects of proficiency. Accordingly, claims regarding the student's proficiency in areas of network design, implementation, and troubleshooting were formulated. A key feature of the domain of computer networking is the role of procedures. Experts are not characterized by how much they know, but what they do and how they do it.

An immediate implication is that assessing the declarative knowledge of students simply does not suffice. Evidence necessary for addressing sophisticated claims in the domain comes from the manner in which students work through problems. Interest lies in the student's abilities to model the computer network and address the necessary features (e.g., create a network to satisfy the client's requirements, troubleshoot faults) in a manner that is efficient and accurate. The need for complex forms of evidence to address multiple, related claims led to the development of NetPASS (Behrens et al., 2004) a simulation-based assessment of computer networking skills.

3. Evidence-centered design

In discussing the virtues of different approaches to assessment design, Messick (1994, p. 16) stated:

> A construct-centered approach would begin by asking what complex of knowledge, skills, or other attributes should be assessed, presumably because they are tied to explicit or implicit objectives of instruction or are otherwise valued by society. Next, what behaviors or performances should reveal those constructs, and what tasks or situations should elicit those behaviors?

ECD elaborates and extends on the key points in Messick's guiding message. Thinking more broadly and characterizing assessment as a situation in which an inference is desired, the relevant questions become: What is (are) the target inference(s)? What information should be collected to inform upon what is unknown in order to facilitate those inferences? What would we need to see in order to be able to make the desired inferences? How do we put ourselves in a position of obtain that information?

ECD is a prescriptive framework for structuring assessment systems by connecting the desired inferences to the tasks administered to examinees. The focus is, as implied by the name, on evidence. What counts as evidence? What needs to be done to collect evidence? Upon obtaining evidence, how do we evaluate the merit of the evidence? How is evidence incorporated and managed?

3.1. The conceptual assessment framework

The heart of an assessment is the Conceptual Assessment Framework (CAF), which consists of models that capture the evidentiary argument of an assessment. Substantive, domain-specific work in domain analysis, domain modeling, task templates, etc. all lead to the models in the CAF. In turn, the CAF dictates the flow of the assessment in terms of how tasks will be presented to examinees, the implications for scoring and accumulating responses, and how the assessment yields inferences about examinees.

The *proficiency model* contains variables that represents features of the examinees that are relevant to the domain and the desired claims. *Task models* describe the structure of the situations needed to obtain evidence. *Evidence models* dictate the ways in which an examinee's performance is rendered into evidence and incorporated to inform about examinees. The proficiency model is instantiated for each examinee, yielding the *student models*, which contain the same variables but are localized to each examinee. Similarly, a task model may be used to author multiple tasks. Likewise, there may be multiple instantiations of an evidence model, known as *link models*. Lastly, the *assembly model* describes how the student, task, and evidence models work in tandem in light of restrictions and specifications of the assessment. These main models of the CAF relevant to evidentiary reasoning in assessment are described in detail below.

3.2. *The proficiency and student models*

The student models are instantiations of the population-level proficiency model. In Messick's terms, the proficiency model represents the structure of knowledge, skills, and attributes to be assessed. In a more evidentiary framework, the student model for an examinee summarizes the evidential impact of the observables for addressing claims regarding the examinee. Inferences and decisions regarding an examinee (e.g., rankings, placement, certification, feedback, as depends on the purpose of assessment) are based on the state of the student model after the assessment. In other settings, the state of the student model influences the manner in which the assessment unfolds, as in the case of adaptive testing or intelligent tutoring systems.

As any examinee's set of knowledge, skills, and abilities are typically unknown, the student model variables are latent variables. This reflects an epistemological, rather than ontological stance to student model variables in the sense that the variables and the distribution of those variables are represent not the student, but rather *our thinking* about the student (Mislevy et al., 2003).

Student model variables and the distributions of those variables are tools for us to express our beliefs about students. Framing our beliefs about students in terms of distributions of variables links the substantive inferences we would like to make to the statistical machinery in the evidence models used to incorporate and synthesize evidence in the form of observed data. More specifically, the student model is the joint distribution of the variables in the student model. Examinees interact with an assessment system so that distributions of and beliefs about student model variables can be updated.

Mathematics admissions assessment

The proficiency model for the mathematics admissions assessment contains a single variable that serves to summarize information obtained regarding the examinee's overall mathematics proficiency. As there may be theoretically infinitely many levels of proficiency, the latent variable is modeled as continuous. Such a choice highlights the epistemological stance in the student model. The choice of single, continuous latent variable does not corresponding to the assertion that there is a single underlying dimension that is "mathematics proficiency" upon which people differ. Rather, it is a purposive choice for summarizing *our beliefs* based on information obtained in the assessment with the explicit purpose of addressing, in this case, a quite general claim.

NetPASS assessment

The proficiency model for NetPASS contains six latent variables that correspond to examinee proficiencies in terms of Networking Disciplinary Knowledge, Networking Proficiency, Network Modeling, Designing networks, Implementing networks, and Troubleshooting networks. Each proficiency model variable was posited to be an ordered discrete variable taking on any of five levels corresponding to locations in terms of the Cisco Networking Academy Program sequence of courses (Novice, Semester 1, Semester 2, Semester 3, Semester 4).

Figure 2 contains the graphical model for the BN for the proficiency model, which reflects the flow of dependence and conditional independence relationships among the variables. The specification of the joint distribution came to specifying the distribution of each variable conditional on its parent(s). Details are provided in Levy and Mislevy (2004). What is of interest here is the manner in which the proficiency model reflects the substantive beliefs about the domain. For example, the model was constructed with the restriction that Network Modeling cannot take on a higher value than Network Disciplinary Knowledge. This reflects the belief that declarative knowledge is a prerequisite for visualization and modeling proficiency. Statistically, the use of a BN affords flexibility in the specification of the proficiency model to incorporate such a constraint. Substantively, this is exemplifies the incorporation of knowledge about the domain and students (i.e., how certain skills relate) into the statistical model, as Gelman et al. (as quoted above) advised when they stressed that the model should be consistent with substantive knowledge.

As our examples illustrate, few restrictions are placed on the proficiency model in terms of the nature and type of variables. The proficiency model may be unidimensional or multidimensional, the variables may be continuous or discrete (ordered or otherwise), and may consist of both observed and latent variables, though typically the latent variables are of more interest. Further, multiple proficiency models (and hence, multiple student models for each examinee) may be constructed for a given assessment, such as in situations in which the intended purposes vary (e.g., summary score reporting and substantively relevant feedback from the same assessment).

The principle use of student model variables in many assessments involves applying a set of *scoring rules* to get test scores or reports. Scoring rules are functions of the posterior distribution of a student model variable. A desire to summarize the posterior distribution into a more manageable quantity leads to the use of point estimates. Popular choices include the posterior mean and mode, the expected a posteriori (EAP) and maximum a posteriori (MAP) values, respectively (Thissen and Wainer, 2001). In assessments with multiple student model variables (e.g., NetPASS) scores may be obtained by computing functions of the multivariate posterior distribution or by computing functions on each variable separately to produce a profile with a collection of scores.

3.3. *Task models*

Task models define the variables that characterize the tasks in terms of the interactions the examinee will have with the tasks. In Messick's terms, they detail the situations examinees will encounter so as to elicit behavior to serve as evidence relevant to the claims of the assessment.

Task models specify materials and instructions presented to examinees, situational variables related to the context of the task, and the work products to be produced. From one task model, many tasks can be created. Tasks that are instantiations of the same task model share the same variables but differ on the values of the variables.

From an assessment design perspective, using task models creates efficiencies for creating new tasks, in particular, for automated item generation (Bejar, 2002; Irvine and Kyllonen, 2002; Mislevy et al., 2002b, 2003). Which features of tasks to vary, and what the effect of varying those features will have on how examinees interact with tasks is a direct result of modeling the domain, including the cognitive activities, knowledge representations, and behaviors involved in becoming and being an expert in the domain. In terms of the development of a statistical model, the connection between task features and task functioning is receiving an increasing amount of attention in the psychometric community (e.g., Sheehan and Mislevy, 1990) and is another key instance where assessment design and development relies on the intersection of psychometrics, cognitive science, and substantive knowledge of the domain.

Mathematics admissions assessment

The general claim of interest implies that the evidence necessary to support the claim ought to be broad. To ensure adequate representation of the domain, multiple task models may need to be developed, corresponding to the distinct aspects of the domain (algebra, geometry, data analysis, etc.). Task model variables for an algebra task may include the number of unknowns to solve for, the number of equations, the complexity of the equations, the location of the unknowns in the equations, features of distractors, etc. Task model variables for a geometry task may include the presence or complexity of a coordinate graph or figure, the number of unknowns, features or distractors, etc. Tasks may be created by varying the values of these variables in manner guided by links from tasks back to the domain (Sheehan and Mislevy, 1990). Though the task models in this assessment differ on what variables are included, they all share the same form of the work product, namely, an indicated response from among a set of five choices.

NetPASS assessment

Tasks were constructed to provide evidence for claims regarding the student's proficiencies in designing, implementing, and troubleshooting computer networks. That is, the goal is to obtain evidence of the Design, Implement, and Troubleshoot variables in the student models. For each area, three tasks were constructed to provide evidence regarding the student's accuracy and efficiency. Task model variables include specifications of the scenario, instructions, the tools made available, and the forms of the resulting work products.

Figure 3 provides a screenshot of the description of the scenario and instructions for one of the Design tasks. The student is presented with a network diagramming tool with a blank tablet and a set of devices; the student then creates a diagram by dragging and dropping icons as they see fit (Figure 4). The network diagramming tool is used to capture the student's mental model of the topology of the network. Within the diagramming tool, the student selects a device to configure. A window pops up which allows for the specification of the device (Figure 5). As discussed in the next section, tasks that yield complex work products that reflect multiple aspects of proficiency require complex evidence models for modeling performance.

3.4. Evidence and link models

In Messick's terms, evidence models address the question of what behaviors would need to be been in order to make inferences about the constructs. In an evidentiary framework, evidence models characterize what and how information counts as evidence. Following Schum (1987, 1994), data only count as evidence when it becomes relevant

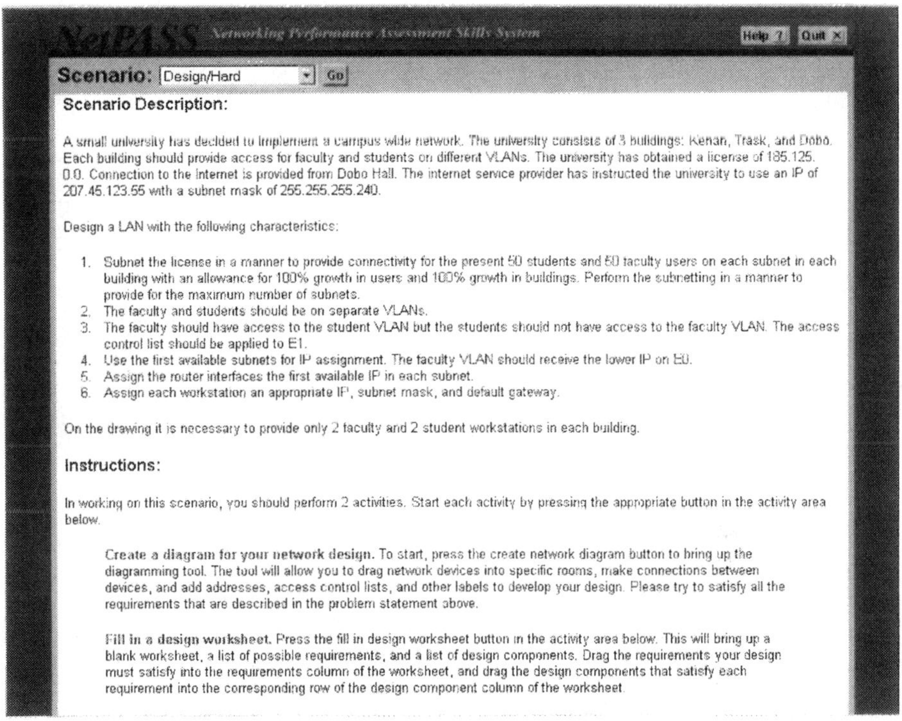

Fig. 3. Scenario description and instructions for a Design task.

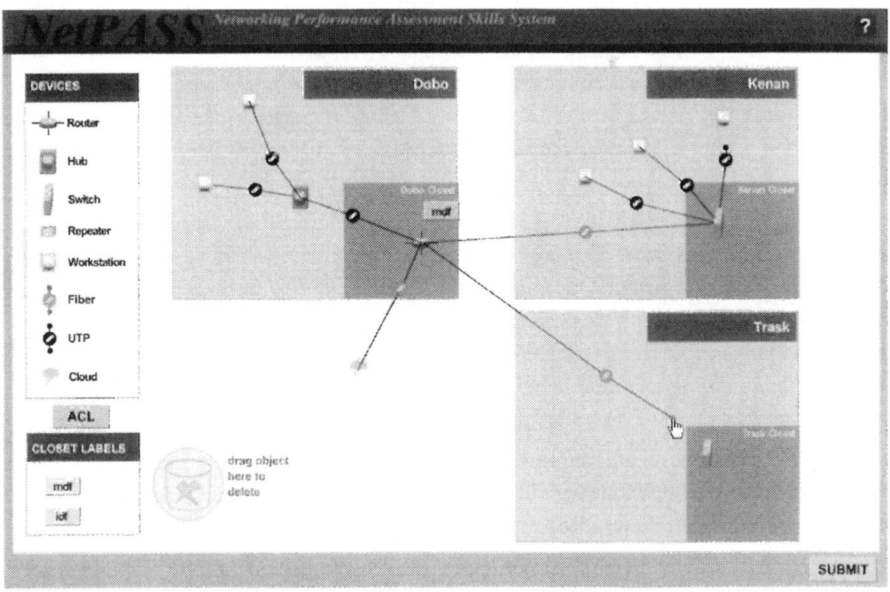

Fig. 4. Use of the network diagramming tool to map a network.

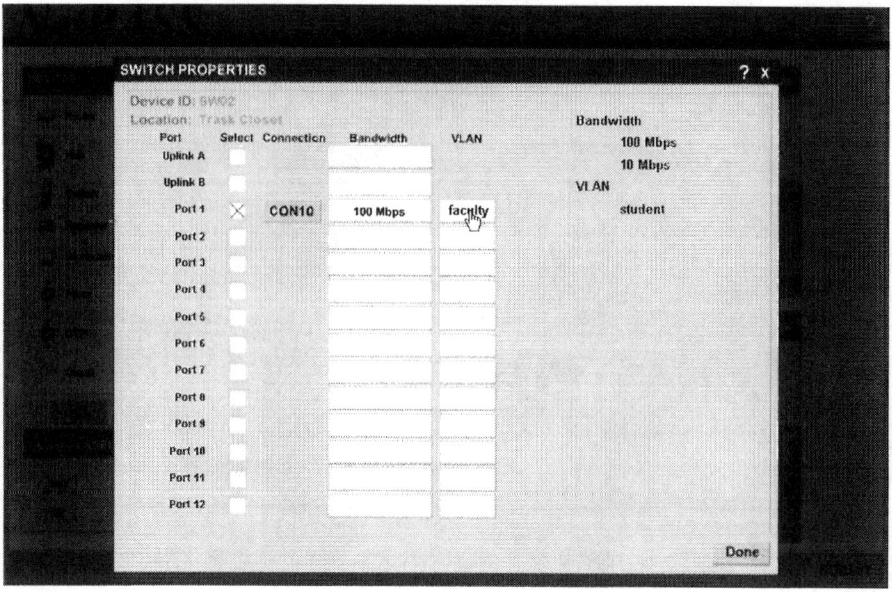

Fig. 5. Configuring a switch in NetPASS.

to some claim. The student model is comprised of examinee variables that are relevant for the claim(s) of the assessment. Therefore, observable data count as evidence when they become linked to the student model variables. Evidence models consist of two components that detail the connection between the raw, observable features of what an examinee does or produces to the student model variables.

Evidence identification is the mechanism by which the work product is characterized along relevant dimensions, which in turn are captured by values for the observable variables. Our two examples show the levels of complexity evidence identification processes can take.

Mathematics admissions assessment

The tasks in this assessment result in work products of the same form. The evidence identification rules evaluate the salient features of each work product. As correctness of response is the only feature of interest, there is one observable created for each task. The evidence identification rule for creating the observables for the mathematics assessment is

$$X_{ij} = \begin{cases} 1, & \text{if the Work Product for examinee } i \text{, task } j \text{ indicates a correct response,} \\ 0, & \text{otherwise.} \end{cases}$$

More complex evidence identification rules may be built for graded response/partial credit scoring as would be appropriate if one of the incorrect responses is indicative of some knowledge.

NetPASS assessment
Multiple evidence identification rules characterize the work product along several dimensions. In NetPASS, evidence identification consists of evaluating the features of the complex work products. To begin, the submitted network diagram is converted to an XML format text representation which is then scored according to detailed, complex rules. Different tasks require different evidence identification processes. In network Design tasks, the work products include the log file of all the commands an examinee enters in manipulating the network. One evidence identification rule for the log file evaluates the state of the network once the examinee finishes, to assess the correctness of the procedure. Another evidence identification rule evaluates the sequence of actions taken (all of which are captured in the log file) to assess the efficiency with which the examinee proceeded through the task and the quality of their rationale.

The result is a characterization of their work in terms of observables, which are the data used to update the student model variables. In NetPASS, all observables are coded as an ordered, discrete variable taking on any of three levels.

The computer networking example highlights an additional feature of evidentiary reasoning in assessment. A single work product may contain multiple evidentiary thrusts. In these cases, when a single work product or set of work products from a single task is used to form multiple observables, there exist dependencies among the values of the observables due to the common origin of the work product(s). The proper modeling of these situations will be discussed below.

The formulation of principled, useful (in the sense that use of the resulting observables results in accurate inferences and decisions regarding examinees) evidence identification rules may seem straightforward in familiar assessments such as dichotomously scored multiple choice items. Things are far from trivial in complex, innovative assessments. Constructing evidence identification rules requires the confluence of expertise from subject matter experts, psychometricians, information systems experts, and database managers. Lessons learned from proper analysis and modeling of the domain (Williamson et al., 2004) and empirical studies of distinctions between experts and novices (DeMark and Behrens, 2004) must be leveraged with restrictions in the management and processing of work products.

For any assessment, the resulting observables capture the relevant features of the work product(s). The second component of the evidence model, the *evidence accumulation* process, consists of a psychometric model that connects the observables to the student model variables. Typically, this is a statistical model that characterizes the conditional distribution of the observables given the student model variables or the joint distribution of the student model and observable variables.

Mathematics admissions assessment
The evidence accumulation component of the evidence model requires a probabilistic model for linking values on the dichotomous observables to the student model variable. The three parameter logistic model (3-PLM; Birnbaum, 1968; Hambleton and Swaminathan, 1985) may be written as

$$P(X_{ij} = 1 \mid \theta_i, b_j, a_j, c_j) = c_j + (1 - c_j)\frac{e^{a_j(\theta_i - b_j)}}{1 + e^{a_j(\theta_i - b_j)}}, \tag{3}$$

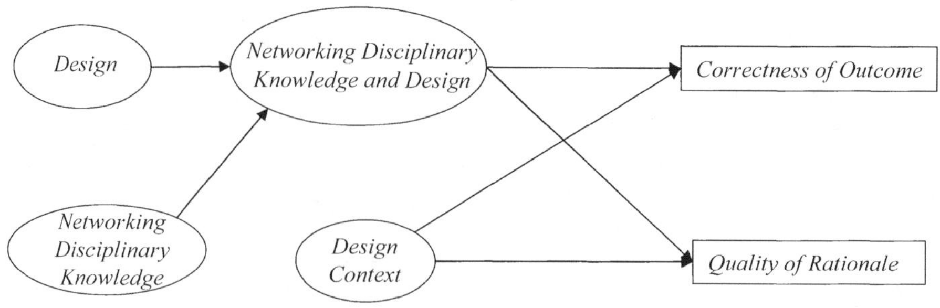

Fig. 6. An evidence model BN for a Design task.

where θ_i is the single student model variable for examinee i, and b_j, a_j, and c_j are parameters for task difficulty, task discrimination, and a lower-asymptote, respectively, for task j.

The model is justified through principled reasoning of the evidentiary possibilities. Substantively, location (b_j) and slope (a_j) parameters account for the recognition that tasks may differ, by construction, in difficulty and discrimination power. A lower asymptote parameter (c_j) accounts for the possibility of responding correctly when the examinee has the lowest levels of proficiency.

Statistically, a lower asymptote allows for better fit of the model. More importantly, it serves to accommodate a potential alternative hypothesis, namely guessing (see Figure 1). That is, a warrant is constructed that implies that success on the task is indicative of high levels of proficiency. The assessment argument proceeds by reverse reasoning from the observation of a correct response to beliefs about proficiency. Guessing is an alternative explanation that may also have led to a correct response. A measurement model that includes a lower asymptote need not attempt to address definitively which of these two (of possibly many) explanations is actually correct. Rather, the inclusion of a lower asymptote is recognition of the possibility of this alternative hypothesis, and a mechanism to temper the reverse reasoning from the observed data to the student model variable.

NetPASS assessment

The statistical models are represented by evidence model BNs, which reflect the dependence of the observables on the student model variables of interest. Figure 6 depicts the DAG for an evidence model BN for a Design task. The model connects the student model variables at the left (see Figure 2) to the observables at the right. Several points are of interest here. First, recall that multiple observables were obtained from the same task. A hallmark of innovative, computer-based assessments is the possibility for obtaining multiple forms of evidence from the administration of a single task. The observables are modeled as dependent on Network Disciplinary Knowledge and Design, a variable introduced for modeling convenience that reflects the combination of Network Disciplinary Knowledge and Design from the student model. Since both observables were obtained from the presentation of a single task, there may be a dependency between them due to their common origin. A context variable is used to account for this possible dependency and induce conditional independence between the observables.

The evidence models for the Implement and Troubleshooting tasks mimic the Design evidence model in that the (a) evidence identification process leads to multiple observables per task, (b) BNs reflect the dependence of the observables on the student model variables, and (c) task-specific context variables used to induce conditional independence among the variables obtained from the same task. There are three instantiations of each evidence model, one for each of the three tasks in each area of the domain. More details regarding the statistical modeling of NetPASS can be found in Levy and Mislevy (2004).

Evidence accumulation takes place when values for observables are conditioned on and the posterior distribution is evaluated. This may take place in a piece-wise fashion, in which evidence model BNs are docked into the student model one at a time. The observables are then conditioned on to obtain the posterior distribution of the student model variables. Once the evidence is propagated throughout the network the evidence model BN is discarded and a new evidence model BN is docked into the student model (Almond and Mislevy, 1999).

On one side of the assessment argument, the student model contains the variables of inferential interest for an examinee. On the other side of the assessment argument, the task models specify the raw work products to be produced by the examinees. The evidence models serve as adapters between these two, processing the salient features of the work products into the values of the observables and updating the student model variables in light of the values of the observables. This highlights the interactivity of the models and the primary role of evidence in the reasoning structure underlying educational assessments.

3.5. *The assembly model*

The assembly model governs how the student, evidence, and task models interact to conform to requirements motivated by the purpose of the assessment. Examples include the balancing of task content, number of tasks to be presented, and the precision with which the student model variables are to be measured.

Mathematics admissions assessment

As the claim refers to proficiency in mathematics broadly conceived, the assembly model insures that tasks targeting the various aspects of the domain (e.g., algebra, geometry, data analysis, etc.) are included. If multiple linear (e.g., paper and pencil) forms are to be created, the assembly model contains specifications to insure comparability across forms. In adaptive assessment in which tasks are selected for each examinee on the fly, the assembly model dynamically creates a form for each examinee in an attempt to maximize information.

NetPASS assessment

All tasks are assembled and available online for all examinees. As an examinee-directed diagnostic assessment, NetPASS allows the user to select the task(s) on the fly. This departs from more popular adaptive assessments in which the assessment system assembles the tasks. Allowing the examinee to control which tasks are administers cedes control of the assessment and the claims addressed to the examinee (Levy et al., 2006).

4. Bayesian psychometric modeling

4.1. The probability model

A student model contains potentially unobservable variables characterizing examinee proficiency on the knowledge, skills, and abilities of interest. Let

$$\boldsymbol{\theta}_i = (\theta_{i1}, \ldots, \theta_{iP}) \tag{4}$$

be the vector of P student model variables for examinee i, $i = 1, \ldots, N$. The collection of student model variables for all examinees is denoted $\boldsymbol{\theta}$.

Task models define those characteristics of a task that need to be specified. Such characteristics are expressed by task model variables; for each task j, $j = 1, \ldots, J$. Let these variables be denoted by the vector

$$\mathbf{Y}_j = (Y_{j1}, \ldots, Y_{jL_f}), \tag{5}$$

where L_f is the number of task model variables for a task from task model f. The full collection of task model variables is denoted $\mathbf{Y}$.

The evidence identification process of the evidence models defines how to extract relevant features from an examinee's work product(s) to yield the values of observables. Let

$$\mathbf{X}_j = (X_{j1}, \ldots, X_{jM_j}) \tag{6}$$

be the vector of M_j potentially observable variables for task j. X_{ijm} is then the value of observable m from the administration of task j to examinee i. $\mathbf{X}_i$ denotes the collection of values of observables (i.e., the response pattern) from examinee i. The complete collection of the values of the observables, that is, the values for all observables from all tasks for all examinees is denoted as $\mathbf{X}$.

The BN for each student model is a probability distribution for $\boldsymbol{\theta}_i$. An assumption of exchangeability (Lindley and Smith, 1972) entails a common prior distribution (i.e., before any responses to tasks are observed the student model is in the same state for all examinees). Beliefs about the expected values and associations among the student model variables are expressed through the structure of the model and higher level hyperparameters $\boldsymbol{\lambda}$. Thus, for all examinees,

$$\boldsymbol{\theta}_i \sim P(\boldsymbol{\theta}_i \mid \boldsymbol{\lambda}). \tag{7}$$

As illustrated in examples below, the higher level parameters $\boldsymbol{\lambda}$ can serve the varied purposes of (a) identifying the location, scale, and (in the case of multidimensional models) orientation of latent axes, (b) defining features of the population, or (c) specifying prior expectations, especially about relationships.

The evidence models define how the observables are dependent on the state of the student model. Let π_{jmk} be (proportional to) the probability for responding to observable m from task j with a value of k. The parenthetical qualifier is necessary in the case where the observable is modeled as a continuous variable. $\boldsymbol{\pi}_{jm}$ is then the probability structure associated with observable m from task j. The distribution of the value for observable m from task j for examinee i is then

$$X_{ijm} \sim P(X_{ijm} \mid \boldsymbol{\theta}_i, \boldsymbol{\pi}_{jm}). \tag{8}$$

In short, for any examinee, the distribution for the observables is defined by the values of the student model variables and the conditional distributions of observables given student model variables. Thus if we knew both the values of the student model variables and the conditional distribution of observables given student model variables, we would know the distribution of the observables. Of course in practice, the situation with the student model variables and the observables is reversed: we have values for the observables but not the student model variables; hence the use of Bayes' theorem to reason from observations to student model variables (Mislevy, 1994).

When there are a large number of levels of student model variables and/or of the observables (as is certainly the case of continuous variables), there are a very large number of π_{jmk}s. It may be the case that further structure exists for modeling the $\boldsymbol{\pi}_{jm}$s, which may be expressed via

$$\boldsymbol{\pi}_{jm} \sim P(\boldsymbol{\pi}_{jm} \mid \boldsymbol{\eta}_{jm}), \tag{9}$$

where $\boldsymbol{\eta}_{jm}$ are higher level hyperparameters for observable m (e.g., characteristics of the appropriate evidence model and the task j from which m is obtained); prior beliefs about such parameters are expressed through higher level distributions, $P(\boldsymbol{\eta}_{jm})$. As illustrated below, in psychometric models it is often the case that the conditional distributions depend on $\boldsymbol{\eta}_{jm}$ in a deterministic rather than a stochastic manner. The complete set of conditional probability distributions for all evidence models for all observables is denoted $\boldsymbol{\pi}$; the complete set of parameters that define those distributions is denoted $\boldsymbol{\eta}$.

The joint probability of all parameters can be expressed as a recursive set of conditional distributions

$$\begin{aligned} P(\boldsymbol{\lambda}, \boldsymbol{\eta}, \boldsymbol{\theta}, \boldsymbol{\pi}, \mathbf{X}) = {} & P(\boldsymbol{\lambda}) \times P(\boldsymbol{\eta} \mid \boldsymbol{\lambda}) \times P(\boldsymbol{\theta} \mid \boldsymbol{\lambda}, \boldsymbol{\eta}) \times P(\boldsymbol{\pi} \mid \boldsymbol{\lambda}, \boldsymbol{\eta}, \boldsymbol{\theta}) \\ & \times P(\mathbf{X} \mid \boldsymbol{\lambda}, \boldsymbol{\eta}, \boldsymbol{\theta}, \boldsymbol{\pi}). \end{aligned} \tag{10}$$

This expression can be simplified in light of additional knowledge and assumptions we bring to the assessment context by taking advantage of the conditional independence relationships implied in Eqs. (7)–(9), yielding:

$$P(\boldsymbol{\lambda}, \boldsymbol{\eta}, \boldsymbol{\theta}, \boldsymbol{\pi}, \mathbf{X}) = P(\boldsymbol{\lambda}) \times P(\boldsymbol{\theta} \mid \boldsymbol{\lambda}) \times P(\boldsymbol{\eta}) \times P(\boldsymbol{\pi} \mid \boldsymbol{\eta}) \times P(\mathbf{X} \mid \boldsymbol{\theta}, \boldsymbol{\pi}). \tag{11}$$

Following Gelman et al. (1995, as quoted above), the goal then becomes to specify the forms of the various terms in Eq. (11).

Once data are collected, values for observables are conditioned on to arrive at the posterior distribution $P(\boldsymbol{\theta}, \boldsymbol{\pi}, \boldsymbol{\eta}, \boldsymbol{\lambda} \mid \mathbf{X})$. In operational assessments conditional probabilities, $\boldsymbol{\pi}$, and the parameters that govern them, $\boldsymbol{\eta}$, are treated as known and the focus is on estimation of student model variables, $\boldsymbol{\theta}$, or higher level parameters $\boldsymbol{\lambda}$ that characterize the examinee population. Estimation is then of the posterior distribution of $\boldsymbol{\theta}$ and $\boldsymbol{\lambda}$ conditional on $\boldsymbol{\pi}$ in addition to $\mathbf{X}$. In such cases, $\boldsymbol{\eta}$ need not be conditioned on, as $\boldsymbol{\eta}$ only contributes to the posterior distribution via $\boldsymbol{\pi}$. Once $\boldsymbol{\pi}$ is known, knowledge regarding $\boldsymbol{\eta}$ does not alter the distribution of the remaining parameters (for more on this concept, d-separation, see, e.g., Jensen, 1996, 2001). In cases where $\boldsymbol{\pi}$ is deterministically rather than stochastically related to $\boldsymbol{\eta}$, conditioning on $\boldsymbol{\eta}$ is equivalent to conditioning on $\boldsymbol{\pi}$, and may lead to conveniences in modeling and interpreting results (e.g., IRT, as discussed below).

In calibration settings, the focus is on estimating the parameters defining and governing the conditional distributions, $\boldsymbol{\pi}$ and $\boldsymbol{\eta}$. If student model variables are known, estimation of $\boldsymbol{\pi}$ and $\boldsymbol{\eta}$ comes to evaluating the posterior distribution conditional on $\boldsymbol{\theta}$ in addition to $\mathbf{X}$. $\boldsymbol{\lambda}$ need not be conditioned on, as $\boldsymbol{\theta}$ d-separates (Jensen, 1996, 2001) $\boldsymbol{\lambda}$ from the rest of the parameters. Such a practice is common in ongoing assessment systems, which employ responses to tasks with parameters treated as known to estimate student model variables and then condition on estimates of student model variables to estimate the parameters for piloted tasks.

In emerging assessments without calibrated tasks, joint estimation of student model variables and evidence model parameters is required, and only the data are conditioned on. Calibration studies for innovative assessments have been demonstrated for both student model and evidence model parameters in complex assessments (Levy and Mislevy, 2004; Mislevy et al., 2002a).

4.2. Estimation

In a Bayesian analysis, model fitting comes to obtaining or estimating the posterior distribution. Three techniques for representing a distribution are (a) calculating an analytical result, (b) drawing independent samples, and (c) drawing dependent samples. Obtaining an analytical solution is ideal yet, with only a few simple problems as exceptions, impractical or impossible due to the computational requirements necessary to evaluate the marginal distribution of the data in the denominator of Eq. (2). Drawing independent samples from the distribution offers an empirical approximation to the distribution. In many cases, it is not possible to draw independent samples from the posterior distribution, especially if it is not of common form as is often the case in complex models. Although analytical computation or independent sampling from the posterior distribution of the model parameters may be possible in simple problems, such computations are intractable in complex models. Drawing dependent samples is sufficient if the samples can be drawn throughout the support of the distribution of interest and in the correct proportions (Gilks et al., 1996a).

Markov chain Monte Carlo (MCMC; Brooks, 1998; Gilks et al., 1996b; Smith and Roberts, 1993; Spiegelhalter et al., 2003; Tierney, 1994) estimation consists of drawing possibly dependent samples from the distribution of interest and as such provides an appropriate framework for computation in Bayesian analyses (Gelman et al., 1995). A complete treatment and description of MCMC estimation is beyond the scope and intent of this chapter; suffice it to say that MCMC estimation consists of drawing from a series of distributions that is in the limit equal to drawing from the true posterior distribution (Gilks et al., 1996a). That is, to empirically sample from the posterior distribution, it is sufficient to construct a Markov chain that has the posterior distribution as its stationary distribution. An arbitrarily large number of iterations may be performed resulting in values of the unknown parameters that form an empirical approximation for the posterior distribution.

In some applications, full exploration of the posterior distribution is not desired; rather, point estimates of model parameters are preferred. In such situations, variants of the EM algorithm (Dempster et al., 1977) may be employed. EM approaches estimation by judiciously choosing to treat or introduce unknown entities in the model as missing data. Estimation proceeds by providing starting points for the unknown model parameters and then iteratively computing (1) provisional values of the missing data given the model parameters and (2) maximizing functions of the now complete data with respect to the model parameters. Typically, the provisional values of the missing data are obtained by taking their expected values. A flexible extension is stochastic EM (Diebolt and Ip, 1996), where the provisional values are obtained by taking a draw from the conditional distribution of the missing data given the observed data and the current values of the model parameters.

4.3. The examples revisited

Mathematics admissions assessment

There is one student model variable, $\boldsymbol{\theta}_i = (\theta_i)$ and only one observable per task, which permits us to drop the subscript m from our notation. An assumption of exchangeability for examinees results in the use of a common prior. Using a normal prior for all examinees

$$P(\boldsymbol{\theta} \mid \boldsymbol{\lambda}) = \prod_i P(\theta_i \mid \boldsymbol{\lambda}) = \prod_i N(\theta_i \mid \mu, \sigma^2), \tag{12}$$

where $\boldsymbol{\lambda} = (\mu, \sigma^2)$ are the variables that govern the prior distribution of the student model variable. Exploiting the conditional independence assumptions, the conditional probability of the complete data factorizes as

$$P(\mathbf{X} \mid \boldsymbol{\theta}, \boldsymbol{\pi}) = \prod_i \prod_j P(X_{ij} \mid \theta_i, \boldsymbol{\pi}_j) = \prod_i \prod_j P(X_{ij} \mid \theta_i, \boldsymbol{\eta}_j). \tag{13}$$

The latter equality in Eq. (13) holds because, in IRT models, the conditional probabilities $\boldsymbol{\pi}$ are deterministically rather than stochastically related to the parameters in $\boldsymbol{\eta}$. That is, we may rewrite the three parameter logistic model (Eq. (3)) as

$$\begin{aligned} P(X_{ij} = 1 \mid \theta_i, \boldsymbol{\eta}_j) &= P(X_{ij} = 1 \mid \theta_i, b_j, a_j, c_j) \\ &= c_j + (1 - c_j)\frac{e^{a_j(\theta_{i1} - b_j)}}{1 + e^{a_j(\theta_{i1} - b_j)}}, \end{aligned} \tag{14}$$

where $\boldsymbol{\eta}_j = (b_j, a_j, c_j)$ are parameters for task difficulty, task discrimination, and a lower-asymptote, respectively.

To complete the model specification, the prior distributions for $\boldsymbol{\lambda}$ and $\boldsymbol{\eta}$ must be specified. Often, the parameters in $\boldsymbol{\lambda}$ are set to 0 and 1, respectively, to identify the location and scale of the latent dimension. An exchangeability assumption with regard to the tasks permits the prior for $\boldsymbol{\eta}$ to be factored as

$$P(\boldsymbol{\eta}) = \prod_j P(\boldsymbol{\eta}_j). \tag{15}$$

Furthermore, an assumption of independence *a priori* permits the prior distribution of parameters for any one observable to be factored as

$$P(\boldsymbol{\eta}_j) = P(b_j, a_j, c_j) = P(b_j) \times P(a_j) \times P(c_j). \tag{16}$$

It should be noted that specifying parameters as independent in the prior does not restrict them to be independent in the posterior (Gelman, 1996).

An assumption of exchangeability leads to the use of a common prior for the b_j; likewise for the a_j and the c_j. Typical choices for the priors include normal distributions for b_j, lognormal distributions for a_j, and beta distributions for the c_j. These choices reflect choices consistent with theory in that difficulty parameters can range over the real line, discrimination parameters should be positive, and lower asymptote parameters should be bounded by zero and one. What's more, these distributions are quite flexible and offer natural interpretations (e.g., use of the Beta(5, 17) distribution for c_j corresponds to the interpretation that the prior belief is akin to having observed four out of twenty people guess correctly). Other choices for prior distributions have been employed in IRT; see Rupp et al. (2004) for a recent review.

NetPASS assessment

In NetPASS there are six student model variables, $\boldsymbol{\theta}_i = (\theta_{i1}, \theta_{i2}, \theta_{i3}, \theta_{i4}, \theta_{i5}, \theta_{i6})$. An exchangeability assumption allows the joint distribution of the student model variables over all examinees to be factored into the product of distributions for each examinee (see the first equality in Eq. (12)). Taking advantage of the conditional independence assumptions implicit in the graph for student model (Figure 2), the distribution for the student model becomes

$$\begin{aligned} P(\boldsymbol{\theta}_i \mid \boldsymbol{\lambda}) = {} & P(\theta_{i1} \mid \boldsymbol{\lambda}) \times P(\theta_{i2} \mid \theta_{i1}, \boldsymbol{\lambda}) \times P(\theta_{i3} \mid \theta_{i1}, \theta_{i2}, \boldsymbol{\lambda}) \\ & \times P(\theta_{i4} \mid \theta_{i3}, \boldsymbol{\lambda}) \times P(\theta_{i5} \mid \theta_{i3}, \boldsymbol{\lambda}) \times P(\theta_{i6} \mid \theta_{i3}, \boldsymbol{\lambda}), \end{aligned} \tag{17}$$

where $\boldsymbol{\lambda}$ are parameters that define the conditional distributions of the variables.

Exploiting the conditional independence assumptions, the conditional probability of the complete data factorizes as

$$\begin{aligned} P(\mathbf{X} \mid \boldsymbol{\theta}, \boldsymbol{\pi}) &= \prod_i \prod_j \prod_m P(X_{ijm} \mid \boldsymbol{\theta}_i, \boldsymbol{\pi}_{jm}) \\ &= \prod_i \prod_j \prod_m P\big(X_{ijm} \mid \boldsymbol{\theta}_i^{jm}, \boldsymbol{\eta}_{jm}\big), \end{aligned} \tag{18}$$

where $\boldsymbol{\theta}_i^{jm}$ are the relevant student model variables for observable m of task j. As in Eq. (13) the latter equality in Eq. (18) holds because the conditional probabilities

π are deterministically rather than stochastically related to the parameters in η. All the variables in the NetPASS model are ordered and discrete. As such, the model is a complex, hierarchical latent class model. The conditional probabilities for the latent classes are restricted via an IRT model (e.g., Formann, 1985) that allows for modeling ordered polytomous responses (Samejima, 1969). Normal and truncated normal priors were used for the parameters specifying the associations between variables. Complete details are given by Levy and Mislevy (2004).

In closing this discussion of NetPASS, we remark that, unlike the case of the mathematics admissions assessment, an *a priori* assumption of exchangeability regarding the tasks was *not* warranted. The three tasks from each task model were constructed to be of different difficulties. Following the advice of Gelman et al. (1995, as quoted above), this knowledge was incorporated by specifying different priors for observables that came from tasks that were expected to differ in difficulty. An assumption of *conditional exchangeability* allowed for the use of the same prior for observables from tasks expected to be of the same difficulty (Levy and Mislevy, 2004). This again highlights the epistemological stance of the modeler. Prior beliefs were modeled to be consistent with expert expectations. Pilot test data were used to update and refine those beliefs in the form of posterior distributions.

5. Discussion

5.1. Bayesian modeling in psychometrics

Bayesian modeling and estimation strategies have their roots in early applications to classical test theory (Novick et al., 1971) and factor analysis (Martin and McDonald, 1975). Bayesian techniques are receiving an increasing amount of attention in IRT (Albert, 1992; Albert and Chib, 1993; Almond and Mislevy, 1999; Béguin and Glas, 2001; Bolt and Lall, 2003; Mislevy, 1986; Patz and Junker, 1999a, 1999b; Rupp et al., 2004; Swaminathan and Gifford, 1985, 1986), LCA (Hoijtink, 1998; Hoijtink and Molenaar, 1997), factor analysis (Lee, 1981; Rowe, 2003), structural equation modeling (Arminger and Muthén, 1998; Scheines et al., 1999), multilevel modeling (Fox and Glas, 2001; Seltzer et al., 1996), cognitive diagnoses (Martin and VanLehn, 1995), and the psychometric modeling community generally (Mislevy, 1994).

Bayesian techniques have proven useful in modeling assessment settings such as adaptive testing (Almond and Mislevy, 1999) and accounting for psychometric phenomena such as testlet effects (Bradlow et al., 1999). The emergence of research specialized to particular aspects of Bayesian modeling such as investigations in MCMC convergence assessment (Sinharay, 2004) and model fit (Sinharay and Stern, 2003) as applied to psychometric situations constitutes another step in the maturation of Bayesian modeling as an approach to psychometric problems.

5.2. Evidence-centered design

Since Spearman, the evidence accumulation process has been the domain of psychometricians. Given data, select a mathematical model that represents the hypothesized data generating process and estimate the parameters that govern the model. Questions concerning differential item functioning (DIF), changes in item or test functioning over time (e.g., drift, equating), or suspicious responses (e.g., cheating, item exposure and breach in security) were all part of system maintenance. Similarly, features of common assessment systems such as the use of (imperfect) raters or a single stimulus (e.g., reading passage) on which to base multiple items were considered nuisances.

As a principled approach to assessment design, ECD offers a contrasting take on psychometric modeling. This is not to say that the psychometric models commonly used and investigating the issues listed above have no place in ECD. The contrast is in genesis of the model. Rather than try to tackle the issue of what model to employ, given the data, the model is developed in conjunction with the creation of the remaining features of the assessment, in particular the student models, task models, and the evidence identification process. For example, the use of a single stimulus for multiple tasks or, similarly, the development of multiple observables from a single work product may induce dependencies among the observables due to their common origin. Instead of attempting to understand and account for these dependencies post hoc, or worse, neglecting them altogether, ECD incorporates them at the initial model construction phase, as the tasks, evidence identification rules, and the psychometric models are developed jointly. To the question "Which has primacy, the data or the model?" the appropriate answer is that the question leads to a false dichotomy. Neither the data nor the model is much use without the other; neither enjoys a primary role and neither is relegated to be a supporting actor. A principled, joint development with connections to the student models, task models, and evidence identification rules leads to a statistical model and an assessment that is set up for success in terms of leading to accurate, trustworthy inferences regarding the claims of interest.

The use of ECD does not make the resulting assessment immune to threats to model adequacy (e.g., DIF). Data-model fit assessment and model criticism remain necessary and important. However, the use of ECD has two main advantages with respect to model evaluation. First, the development of the psychometric model jointly with the proficiency model, task models, and evidence identification rules leads to a model, grounded in substantive theory, that is more likely to lead to solid inferences than one that is developed by starting with data and then trying to sort out and create a model post hoc. Second, by linking psychometric models with the domain-specific features, alternative, competing models must also be linked to these features. Alternative models should not come from the imagination, but from competing theoretically viable hypotheses about the domain, student model, task models, etc. Researchers have recently been calling for renewed attention to the role of statistical modeling of assessment data for generating feedback to task writers and modelers or gaining substance knowledge of phenomenon such as DIF (e.g., Ackerman, 1996; Stout et al., 1996; Swaminathan and Rogers, 1990). ECD features these connections, both in terms of the feedback after a model has been created and employed, as discussed above, in the construction of the model prior to data collection.

6. Concluding remarks

A Bayesian approach offers great flexibility and power in psychometric modeling. With that flexibility comes the potential for incorporating information regarding the substantive domain and the observational setting. ECD offers a principled approach to assessment design and statistical model construction and serves a prescriptive approach for analysts to take advantage of this power in terms of assessment design, model estimation, and substantive inference.

References

Ackerman, T.A. (1996). Graphical representation of multidimensional item response theory analysis. *Applied Psychological Measurement* **20**, 311–329.

Albert, J.H. (1992). Bayesian estimation of normal ogive item response curves using Gibbs sampling. *Journal of Educational Statistics* **17**, 251–269.

Albert, J.H., Chib, S. (1993). Bayesian analysis of binary and polychotomous response data. *Journal of the American Statistical Association* **88**, 669–679.

Almond, R.G., Mislevy, R.J. (1999). Graphical models and computerized adaptive testing. *Applied Psychological Measurement* **23**, 223–237.

Arminger, G., Muthén, B.O. (1998). A Bayesian approach to nonlinear latent variable models using the Gibbs sampler and the Metropolis–Hastings algorithm. *Psychometrika* **63**, 271–300.

Béguin, A.A., Glas, C.A.W. (2001). MCMC estimation and some model-fit analysis of multidimensional IRT models. *Psychometrika* **66**, 541–562.

Behrens, J.T., Mislevy, R.J., Bauer, M., Williamson, D.W., Levy, R. (2004). Introduction to Evidence Centered Design and lessons learned from its application in a global e-learning program. *International Journal of Testing* **4**, 295–301.

Bejar, I.I. (2002). Generative testing: From conception to implementation. In: Irvine, S.H., Kyllonen, P.C. (Eds.), *Item Generation for Test Development*. Erlbaum, Hillsdale, NJ, pp. 199–217.

Birnbaum, A. (1968). Some latent trait models and their use in inferring an examinee's ability. In: Lord, F.M., Novick, M.R. (Eds.), *Statistical Theories of Mental Test Scores*. Addison-Wesley, Reading, MA, pp. 397–479.

Bolt, D.M., Lall, V.F. (2003). Estimation of compensatory and noncompensatory multidimensional item response models using Markov chain Monte Carlo. *Applied Psychological Measurement* **27**, 395–414.

Bradlow, E.T., Wainer, H., Wang, X. (1999). A Bayesian random effects model for testlets. *Psychometrika* **64**, 153–168.

Brooks, S.P. (1998). Markov chain Monte Carlo method and its application. *The Statistician* **47**, 69–100.

DeMark, S.D., Behrens, J.T. (2004). Using statistical natural language processing for understanding complex responses to free-response tasks. *International Journal of Testing* **4**, 371–390.

Dempster, A.P., Laird, N.M., Rubin, D.B. (1977). Maximum likelihood from incomplete data via the EM algorithm. *Journal of the Royal Statistical Society, B* **39**, 1–38.

Diebolt, J., Ip, E.H.S. (1996). Stochastic EM: Method and application. In: Gilks, W.R., Richardson, S., Spiegelhalter, D.J. (Eds.), *Markov Chain Monte Carlo in Practice*. Chapman and Hall, London, pp. 259–273.

Formann, A.K. (1985). Constrained latent class models: Theory and applications. *The British Journal of Mathematical and Statistical Psychology* **38**, 87–111.

Fox, J.P., Glas, C.A.W. (2001). Bayesian estimation of a multilevel IRT model using Gibbs sampling. *Psychometrika* **66**, 271–288.

Gelman, A. (1996). Inference and monitoring convergence. In: Gilks, W.R., Richardson, S., Spiegelhalter, D.J. (Eds.), *Markov Chain Monte Carlo in Practice*. Chapman and Hall, London, pp. 131–143.

Gelman, A., Carlin, J.B., Stern, H.S., Rubin, D.B. (1995). *Bayesian Data Analysis*. Chapman & Hall, London.

Gilks, W.R., Richardson, S., Spiegelhalter, D.J. (1996a). Introducing Markov Chain Monte Carlo. In: Gilks, W.R., Richardson, S., Spiegelhalter, D.J. (Eds.), *Markov Chain Monte Carlo in Practice*. Chapman and Hall, London, pp. 1–19.

Gilks, W.R., Richardson, S., Spiegelhalter, D.J. (Eds.) (1996b). *Markov Chain Monte Carlo in Practice*. Chapman and Hall, London.

Hambleton, R.K., Swaminathan, H. (1985). *Item Response Theory: Principles and Applications*. Kluwer, Boston.

Hoijtink, H. (1998). Constrained latent class analysis using the Gibbs sampler and posterior predictive p-values: Applications to educational testing. *Statistica Sinica* **8**, 691–711.

Hoijtink, H., Molenaar, I.W. (1997). A multidimensional item response model: Constrained latent class analysis using the Gibbs sampler and posterior predictive checks. *Psychometrika* **62**, 171–189.

Holland, P.W. (1990). On the sampling theory foundations of item response theory models. *Psychometrika* **55**, 577–601.

Irvine, S.H., Kyllonen, P.C. (Eds.) (2002). *Item Generation for Test Development*. Erlbaum, Hillsdale, NJ.

Jaynes, E.T. (2003). *Probability Theory: The Logic of Science*. Cambridge University Press, New York.

Jensen, F.V. (1996). *An Introduction to Bayesian Networks*. Springer-Verlag, New York.

Jensen, F.V. (2001). *Bayesian Networks and Decision Graphs*. Springer-Verlag, New York.

Lauritzen, S.L., Spiegelhalter, D.J. (1988). Local computations with probabilities on graphical structures and their application to expert systems (with discussion). *Journal of the Royal Statistical Society, B* **50**, 157–224.

Lee, S.-Y. (1981). A Bayesian approach to confirmatory factor analysis. *Psychometrika* **46**, 153–160.

Levy, R., Mislevy, R.J. (2004). Specifying and refining a measurement model for a computer based interactive assessment. *International Journal of Testing* **4**, 333–369.

Levy, R., Behrens, J.T., Mislevy, R.J. (2006). Variations in adaptive testing and their on-line leverage points. In: Williams, D.D., Howell, S.L., Hricko, M. (Eds.), *Online Assessment, Measurement and Evaluation: Emerging Practices*. Information Science Publishing, Hershey, PA, pp. 180–202.

Lindley, D.V., Smith, A.F.M. (1972). Bayes estimates for the linear model. *Journal of the Royal Statistical Society, B* **34**, 1–42.

Lord, F.M., Novick, M.R. (1968). *Statistical Theories of Mental Test Scores*. Addison-Wesley, Reading, MA.

Martin, J.K., McDonald, R.P. (1975). Bayesian estimation in unrestricted factor analysis: A treatment for Heywood cases. *Psychometrika* **40**, 505–517.

Martin, J.D., VanLehn, K. (1995). A Bayesian approach to cognitive assessment. In: Nichols, P., Chipman, S., Brennan, R. (Eds.), *Cognitively Diagnostic Assessment*. Erlbaum, Hillsdale, NJ, pp. 141–165.

Messick, S. (1989). Validity. In: Linn, R.L. (Ed.), *Educational Measurement*. 3rd ed. Macmillan, New York, pp. 13–103.

Messick, S. (1994). The interplay of evidence and consequences in the validation of performance assessments. *Educational Researcher* **23**, 13–23.

Mislevy, R.J. (1986). Bayes modal estimation in item response models. *Psychometrika* **51**, 177–195.

Mislevy, R.J. (1994). Evidence and inference in educational assessment. *Psychometrika* **59**, 439–483.

Mislevy, R.J., Senturk, D., Almond, R.G., Dibello, L.V., Jenkins, F., Steinberg, L.S., Yan, D. (2002a). Modeling conditional probabilities in complex educational assessments. CSE Technical Report 580, National Center for Research on Evaluation, Standards, and Student Testing (CRESST), Los Angeles.

Mislevy, R.J., Steinberg, L.S., Almond, R.G. (2002b). On the roles of task model variables in assessment design. In: Irvine, S., Kyllonen, P. (Eds.), *Generating Items for Cognitive Tests: Theory and Practice*. Erlbaum, Hillsdale, NJ, pp. 97–128.

Mislevy, R.J., Steinberg, L.S., Almond, R.G. (2003). On the structure of educational assessments. *Measurement: Interdisciplinary Research and Perspectives* **1**, 3–62.

Moss, P. (1994). Can there be validity without reliability?. *Educational Researcher* **23** (2), 5–12.

Novick, M.R., Jackson, P.H., Thayer, D.T. (1971). Bayesian inference and the classical test theory model: Reliability and true scores. *Psychometrika* **36**, 261–288.

Patz, R.J., Junker, B.W. (1999a). A straightforward approach to Markov chain Monte Carlo methods for item response models. *Journal of Educational and Behavioral Statistics* **24**, 146–178.

Patz, R.J., Junker, B.W. (1999b). Applications and extensions of MCMC for IRT: Multiple item types, missing data, and rated responses. *Journal of Educational and Behavioral Statistics* **24**, 342–366.

Pearl, J. (1988). *Probabilistic Reasoning in Intelligent Systems: Networks of Plausible Inference*. Kaufmann, San Mateo, CA.

Rasch, G. (1960). *Probabilistic Models for some Intelligence and Attainment Tests*. University of Chicago Press, Chicago. Reprinted 1980.

Rowe, D.B. (2003). *Multivariate Bayesian Statistics: Models for Source Separation and Signal Unmixing*. CRC Press, Boca Raton, FL.

Rupp, A.A., Dey, D.K., Zumbo, B.D. (2004). To Bayes or not to Bayes, from whether to when: Applications of Bayesian methodology to modeling. *Structural Equation Modeling* **11**, 424–451.

Samejima, F. (1969). Estimation of latent ability using a response pattern of graded scores. *Psychometrika Monograph*, No. 17 (4, Pt. 2).

Samejima, F. (1983). Some methods and approaches of estimating the operating characteristics of discrete item responses. In: Wainer, H., Messick, S. (Eds.), *Principals (sic) of Modern Psychological Measurement*. Lawrence Erlbaum Associates, Hillsdale, NJ, pp. 154–182.

Scheines, R., Hoijtink, H., Boomsma, A. (1999). Bayesian estimation and testing of structural equation models. *Psychometrika* **64**, 37–52.

Schum, D.A. (1987). *Evidence and Inference for the Intelligence Analyst*. University Press of America, Lanham, MD.

Schum, D.A. (1994). *The Evidential Foundations of Probabilistic Reasoning*. Wiley, New York.

Seltzer, M.H., Wong, W.H., Bryk, A.S. (1996). Bayesian analysis in applications of hierarchical models: Issues and methods. *Journal of Educational and Behavioral Statistics* **21**, 131–167.

Sheehan, K., Mislevy, R.J. (1990). Integrating cognitive and psychometric models to measure document literacy. *Journal of Educational Measurement* **27**, 255–272.

Sinharay, S. (2004). Experiences with Markov chain Monte Carlo convergence assessment in two psychometric examples. *Journal of Educational and Behavioral Statistics* **29**, 461–488.

Sinharay, S., Stern, H.S. (2003). Posterior predictive model checking in hierarchical models. *Journal of Statistical Planning and Inference* **111**, 209–221.

Smith, A.F.M., Roberts, G.O. (1993). Bayesian computation via the Gibbs sampler and related Markov chain Monte Carlo methods. *Journal of the Royal Statistical Society, B* **55**, 3–23.

Spiegelhalter, D.J., Dawid, A.P., Lauritzen, S.L., Cowell, R.G. (1993). Bayesian analysis in expert systems. *Statistical Science* **8**, 219–247.

Spiegelhalter, D.J., Thomas, A., Best, N.G., Lunn, D. (2003). WinBUGS version 1.4: User manual. Cambridge Medical Research Council Biostatistics Unit. http://www.mrc-bsu.cam.ac.uk/bugs/.

Stout, W., Habing, B., Douglas, J., Kim, H.R., Roussos, L., Zhang, J. (1996). Conditional covariance-based nonparametric multidimensionality assessment. *Applied Psychological Measurement* **20**, 331–354.

Swaminathan, H., Gifford, J.A. (1985). Bayesian estimation in the two-parameter logistic model. *Psychometrika* **50**, 349–364.

Swaminathan, H., Gifford, J.A. (1986). Bayesian estimation in the three-parameter logistic model. *Psychometrika* **51**, 589–601.

Swaminathan, H., Rogers, H.J. (1990). Detecting differential item functioning using logistic regression procedures. *Journal of Educational Measurement* **27**, 361–370.

Thissen, D., Wainer, H. (2001). *Test Scoring*. Lawrence Erlbaum Associates, Mahwah, NJ.

Tierney, L. (1994). Markov chains for exploring posterior distributions. *Annals of Statistics* **22**, 1701–1728.

Toulmin, S.E. (1958). *The Uses of Argument*. Cambridge University Press, Cambridge.

Williamson, D.W., Bauer, M., Steinberg, L.S., Mislevy, R.J., Behrens, J.T., DeMark, S.F. (2004). Design rationale for a complex performance assessment. *International Journal of Testing* **4**, 303–332.

Handbook of Statistics, Vol. 26
ISSN: 0169-7161
2007 Published by Elsevier B.V.
DOI: 10.1016/S0169-7161(06)26027-9

Value-Added Modeling

Henry Braun and Howard Wainer

1. Introduction

The term value-added models (VAM) refers to a family of statistical models that are employed to make inferences about the effectiveness of educational units, usually schools and/or teachers. They are characterized by their focus on patterns in student score gains over time, rather than on student status at one point in time. In particular, they attempt to extract from the data on score trajectories estimates of the contributions of schools or teachers to student learning.

Interest in such models has been fueled by an increasing emphasis in the United States on holding the public education system accountable for student learning. In 2001, the U.S. Congress passed a law, the No Child Left Behind Act (NCLB, 2002) that requires states receiving federal funds to establish standards for proficiency in reading and mathematics in grades 3–8 based on performance on standardized assessments. Moreover, states must set yearly targets for the proportions of students in each school achieving those standards.[1] NCLB marks a decisive shift away from evaluating districts and schools on the basis of inputs (e.g., per-pupil expenditures) to judging them on the basis of outcomes (e.g., student performance).

The increasingly stringent targets (for each grade) that a school must meet are termed "adequate yearly progress" or AYP. Schools that fail repeatedly to meet their AYP requirements are subject to sanctions. There have been a number of concerns raised with respect to the technical aspects of AYP (Linn, 2004, 2005). One is that AYP involves a comparison in the proportions proficient between different cohorts (e.g., students in grade 4 in the current academic year in comparison to students in grade 4 in the previous year). Accordingly, the apparent change in effectiveness of the school is confounded with intrinsic differences between the cohorts.

Perhaps the most serious objection, however, is that judging schools on the basis of student status at the end of an academic year without regard to their status at the beginning of the year can be unfair to schools serving student populations either with substantial deficits upon entry or with high migration rates, or both. Although some of these schools may be helping their students make excellent progress, too few students

[1] For more information on this act, consult Bourque (2005). The expectation is that essentially all students will achieve proficiency by the 2013–14 academic year.

reach the proficient level, causing the school to be sanctioned. By contrast, some schools serving advantaged student populations may meet the requirements of AYP despite the fact that the students are learning at a slow pace. Lastly, schools may have little or nothing to do with the observed progress that students make.[2]

To remedy this situation, a number of policy makers and measurement specialists have argued that it would be both fairer and more helpful if schools were judged on the basis of how much growth their students achieved. The critical point is that to evaluate growth properly, each student must be tested twice – once at the beginning of the academic year and once at the end. More realistically, students are likely to be tested at the end of each academic year.[3] Recall that the testing provisions of the NCLB Act mandate such a regular testing regime in reading and mathematics in grades 3 through 8. As a result, states are building databases containing longitudinal student records – precisely what is required for the application of VAMs. At first blush, the prospect of evaluating schools on the basis of an appropriate, aggregate measure of their students' growth is appealing and it has advantages over the current AYP regulations. Nonetheless, difficulties remain and these are treated more fully below. Some districts in Tennessee and the city of Dallas, Texas are using value-added approaches to evaluate schools. Other states are planning to follow suit.

The NCLB Act also addresses the issue of teacher quality. Interestingly, it does so in a more traditional manner by focusing on such teacher characteristics as training and certification. At present, teacher evaluation in most schools and districts is a notoriously subjective enterprise. In principle, the introduction of a quantitative and, arguably, more objective component into the process should be welcomed. Indeed, this was the motivation behind the development of the so-called layered model and its adoption by the state of Tennessee in 1993.[4] Around the same time, a two-stage covariance model was developed and implemented in Dallas, Texas (Webster and Mendro, 1997). The stated intent of utilizing a value-added strategy is to "isolate" the contribution a teacher makes to his or her students. Early implementations have employed the results of VAM to inform decisions about teacher professional development. It is likely, however, that in the near future some states will seek to employ these results to make administrative decisions about their teachers.

This possibility has raised many questions. Some of the issues parallel those raised in the context of school accountability, while others are specific to teacher accountability. Despite many technical concerns, there is a great deal of interest among policy-makers in moving forward. This is due, in large part, to the intuitive appeal of the value-added approach; namely, "that if good teaching is critical to student learning, then can't evidence of student learning (or its absence) tell us something about the quality of teaching"? (Braun, 2005a, p. 21).

[2] To take an extreme example, if we observe children's stature every year, some schools may have students who grow much more than the students at other schools, but few would argue that schools had much to do with the outcomes.

[3] Note that end-of-year testing means that student gains during the academic year are confounded with gains (or losses) during the summer.

[4] See Sanders et al. (1997) for an accessible description of the layered model.

There are a number of sources for further information on value-added approaches to evaluation. The monograph by McCaffrey et al. (2003) provides an historical perspective, technical reviews of extant models, as well as a thoughtful discussion of research issues. More recent compilations of VAM-related research can be found in a special issue of the *Journal of Educational and Behavioral Statistics* (Wainer, 2004) and in a volume of conference proceedings (Lissitz, 2005). Non-technical descriptions of VAM with discussion of policy implications are provided in Braun (2005b), Choi et al. (2005), and National Association of State Boards of Education (2005).

Outline

Before turning to the specifics of different VAMs, the next section addresses a number of general issues that pertain to all such models. Each issue addresses a particular aspect of the use of VAMs in educational settings and leads to caution in the interpretation and use of VAM results. Section 3 then presents two VAMs currently used operationally and two VAMs that have appeared in the research literature. Section 4 presents an example of the results yielded by an analysis employing a particular VAM, while Section 5 describes some current research. The final Section 6 discusses the future of this technology and policy implications of its use.

2. General issues

Causal inference

A fundamental issue concerns the problem of causality.[5] The goal in almost all applications of VAM is the estimation of the (relative) effectiveness of each teacher in a defined group of teachers. That is, the purpose is to make a causal inference with respect to the contribution that a teacher makes to the academic achievement of the students in his or her class in comparison to the contributions made by other teachers. It is crucial to make clear the distinction between statistical description and causal inference.

Suppose the average gain of students in Mr. Jones' class is 8 points, while the average gain of students in Mr. Smith's class is 12 points. That is description. However, as a result of the application of a VAM we obtain estimated "teacher effects", which we are invited to treat as indicators of relative teacher effectiveness. For example, suppose the effect associated with Mr. Jones is 2, while the effect associated with Mr. Smith is 6.[6] The causal interpretation is that if the students in Mr. Jones' class had been taught instead by Mr. Smith, their average gain would have been $6 - 2 = 4$ points greater.

Of course, the students in Mr. Jones' class were not taught by Mr. Smith. So the causal interpretation involves (as it almost always does) an unobservable or counterfactual conditional. It is the use of the counterfactual that distinguishes causal inference from simple description – and makes it so much more problematic. (For more on counterfactuals, see Holland, 2005.) Indeed, the domain of causal inference is broad and

[5] For ease of exposition, most of the following discussion is framed with reference to teachers. The issues are essentially the same for schools.

[6] The estimated teacher effect will typically be different numerically from the average gain in the class.

deep, encompassing philosophical, scientific and methodological considerations. See, for example, Rubin (1974, 1978), Rubin et al. (2004), Holland (1986), Rosenbaum (1984) and Shadish et al. (2002), as well as the references therein.

Under what circumstances, then, can causal inferences justified? In effect, when the data collection strategy, the statistical model employed, and the assumptions invoked yield a plausible substitute for the counterfactual. It is well-known, for example, that causal inferences are most credible when the data have been obtained from a large, well-designed randomized experiment. In such a case, the groups of units (e.g., students) exposed to the different treatments (e.g., teachers) are sufficiently similar in the aggregate so that the comparison of the outcomes between groups can be taken as an estimate of what would have been observed had a single group somehow been exposed to both treatments.

When, instead, the data are drawn from an observational study, more elaborate models and stronger assumptions must be invoked before the estimated effects can be reasonably interpreted as causal effects. Typically, these models involve both matching on observed covariates and making assumptions about how the covariates describe those aspects of the experimental and control conditions that were different. These assumptions cannot be tested directly, although, in a particular case, indirect support might be adduced on substantive grounds, from the data, or both. This elaborate machinery (i.e., the covariates and assumptions) is required to compensate for the fact that the groups of units exposed to each treatment are not known to be – and usually are not – equivalent. Moreover, the lack of equivalence may be directly related to the outcomes of interest. Ignoring the lack of equivalence can result in biased estimates and lead to incorrect inferences. (This is often referred to as "selection bias".) The models used to analyze observational data, of which VAMs are a special case, are all intended to "adjust" for prior differences in the groups so as to make the counterfactual substitution more plausible.

Randomized experiments are relatively rare in the social sciences. Typically, data are drawn from observational studies, with school district data a prime example: Students and teachers ordinarily do not find themselves in a particular school – or in the same class in that school – by a truly random mechanism. For example, students' families can decide to live in a specific district because of the school system. They may also exert some influence on the school and class to which students are assigned. Teachers, by dint of seniority, can also have some say about their assignment. Moreover, principals' decisions can also be based on criteria that depart from randomness. There are many processes at work in determining the composition of classes and no one has yet modeled them successfully.

In the absence of statistical randomization, one cannot discount the possibility that the estimated school and/or teacher effects are confounded with prior differences among students across the different classes. To the extent that is the case, the estimated teacher effects can be biased indicators of teachers' contributions to student learning. Now, each VAM attempts to reduce the bias by making various adjustments. For some models, the adjustments are based primarily on students' demographic characteristics. In others, they are based on students' test score histories. In yet others, a combination of demographics and test scores may be employed. The fundamental concern, however, is that in making causal inferences, no statistical model, however complex, and no method of analysis, however sophisticated, can fully compensate for the lack of randomization. In Holland's (1986, p. 959) words, there can be "no causation without manipulation".

Whatever the approach used, the success of the adjustment in producing an appropriate counterfactual substitution is difficult to ascertain. The use of simple gain scores usually cannot be recommended (Holland, 2005; McCaffrey et al., 2005) and even regression models can lead one astray (Petersen and Novick, 1976; Ladd and Walsh, 2002). For example, Ladd and Walsh (2002) analyze data drawn from the accountability systems employed by South Carolina and North Carolina. They show that measurement error in prior year test scores, which are used as predictors in a regression model, results in biased estimates of school effectiveness. More complex models may be more successful but there are still many pitfalls. The extensive literature on "Lord's Paradox" (Lord, 1967; Holland and Rubin, 1983; Wainer, 1991; Wainer and Brown, 2004) testifies to the subtleties involved in making regression adjustments. In this context, covariance adjustments that account for differences in student characteristics are not a panacea. For example, if the student characteristics are correlated with class effects, then covariance models can "overadjust", resulting in estimated teacher effects that are under-dispersed.

There is an additional difficulty in making causal inferences with respect to teachers. Strictly speaking, the output of a VAM should be considered to be a set of estimated "classroom effects", since student data are aggregated on the basis of which students were in the same classroom during the academic year. Treating these classroom effects as indicators of teacher effectiveness requires making two linked assumptions: (1) Each statistical estimate must be interpreted as the causal effect of the corresponding classroom, and (2) That causal effect must be attributed entirely to the teacher in that classroom.[7]

As this discussion makes clear, a causal interpretation of a statistical estimate can be problematic in this setting. With regard to the second assumption, it is likely that a student's learning during the year is a function not only of the pedagogical (and other) strategies of the teacher but also of the dynamics within the class, the personal history of the student, events and the general environment of the school, and so on.[8] To be sure, some of these are due to the teacher, some can only be influenced by the teacher, while others are essentially beyond the teacher's control. Here again, we need to make some effort to disentangle the various factors that contribute to student learning. However, as Raudenbush (2004) and others have pointed out, this is a daunting task given the kinds of studies that are now undertaken.

Missing data

Another potential source of selection bias is missing data. In many district data bases, it is not uncommon for as much as 30–40 percent of the data to be missing; that is,

[7] Paraphrasing Kupermintz (2003), it is the identification of the classroom effect with teacher effectiveness that makes VAM useful to policy makers, school administrators and teachers.

[8] For example, schools can differ with respect to the leadership and policies, availability of resources and differential treatment by the district, and so on. To the extent that such characteristics are associated with student learning, comparisons among teachers at different schools will be confounded with differences among schools in these factors.

students will be missing one or more test scores and/or the link to a specific teacher in one or more grades. All VAMs assume that missing data are missing at random (MAR), which means that, conditional on the observed student characteristics and test scores included in the model, the process generating the pattern of missingness and the test score outcomes are independent (Rubin, 1976; Little and Rubin, 1987). In other words, the conditional distribution of the missing scores equals the conditional distribution of observed scores.

There are many ways in which the MAR assumption can fail. It fails, for example, if (all other things being equal) weaker students are more likely to be absent on the day of testing. They may be absent because they choose to do so, they may be encouraged to do so, or they may no longer be living in the district. It is usually impossible to directly check the MAR assumption and, moreover, there is good reason to believe that it does not hold in educational settings. Simplistic approaches to the problem cannot be recommended. For example, utilizing only those students with complete records will lead to biased inferences. More sophisticated strategies can ameliorate the problem somewhat.

Precision

Even if it were feasible to collect data free of selection bias, there is a problem related to the precision of VAM estimates of teacher effects. In experimental terms, each teacher plays the role of a treatment and the amount of information available to estimate the teacher (treatment) effect is relatively small, consisting of the data from just a few classes. This is in contrast with most experiments in which the number of treatments is small and the number of observations per treatment is large.

With a small number of students contributing to the estimated effect for a particular teacher, the "averaging out" power of randomization can't work for all teachers in a given year. Suppose, for example, that there are a small number of truly disruptive students in a cohort. While all teachers may have an equal chance of finding one (or more) of those students in their class, each year only a few actually will – with potentially deleterious impact on the academic growth of the class in that year.[9] The bottom line is that even if teachers and students come together in more or less random ways, estimated teacher effects can be too variable from year to year to be of value in making consequential decisions.

Some approaches to VAM attempt to address this problem by incorporating data for each student from many years and over many subjects. Indeed, that can be helpful but it brings along with it other assumptions that must be validated. Estimated teacher effects can also be obtained by averaging results over successive cohorts of students.

Even at the school level, modest sample sizes can result in year-to-year volatility when simple gain score models are employed (Kane and Steiger, 2002). A subsequent critique by Rogosa (2003) exposes some of the difficulties with such simple models.

[9] The issue is that if the teacher has to allocate a disproportionate amount of effort to classroom management, then there is a concomitant reduction in the time to devoted to instruction. Another instance arises if the class includes many transient students. In that case, the teacher's work with those students may go unrecognized either because they have left the class before testing begins or because they have not been in the class long enough to count toward the teacher's record.

The literature has little to offer on the stability of estimated school effects based on more complex models.

Test data as the basis for inference
From a practical point of view, evaluating teachers on the basis of their students' academic growth means relying on scores obtained from standardized tests. The quality of these tests can vary across jurisdictions, as well as by subject and grade within jurisdictions. Consequently, the validity of the test scores for this use must be established. Both substantive and technical issues arise. Chief among the substantive issues are: (1) The degree of articulation between the actual test content and the test standards and (2) The shifts in content across grades. Chief among the technical issues are: (1) The construction and maintenance of the score scale over time and (2) The measurement properties of the scale.

To the extent that a test does not adequately measure the content standards established by the jurisdiction, it is said to suffer from construct under-representation (Messick, 1989). Severe construct under-representation may preclude use of the test to make decisions about both students and teachers. In general, a test that is more narrowly focused can encourage "teaching to the test", a tactic in which students are drilled on specific tasks that are likely to appear. This can result in test score gains that do not generalize to gains in the broader domain. Conversely, students of teachers who do try to teach the full curriculum may show less improvement than expected because the test is not sensitive to the full range of their learning gains. In both cases, estimated teacher effects can be misleading.

Standardized test instruments are the end result of a lengthy process of design and development, shaped by a multitude of goals and constraints (Braun, 2000). Two tests developed for the same grade can differ both in the emphasis placed on different content areas and the weights used to combine subscores into an overall score. Such choices by test developers can have implications for the relative effectiveness of teachers who vary in their curricular emphases.

One of the final steps in test analysis is the conversion of raw scores into scale scores. When different forms are used each year (for a particular subject and grade), the conversion includes an equating step, which introduces some error. Of greater concern is the practice of linking the score scales of the different tests (in a given subject) administered to a sequence of grades. This is called vertical scaling and also introduces error into the system (Doran and Cohen, 2005).

A problematic aspect of vertical scaling is that different procedures can yield patterns of increasing, decreasing or stable variances in test score distributions from lower to higher grades. There is no absolute basis for preferring one pattern to another and the choice will influence the distributions of gains at different points of the score scale (Dorans, 2005). This, in turn, will impact estimates of teacher effects.

It is also possible that vertical scaling can induce a form of construct under-representation, again leading to biased estimates of teacher effects (Reckase, 2004; Schmidt et al., 2005). Specifically, Schmidt et al. (2005) studied the Michigan mathematics curriculum from the 2nd to the 8th grades. The display below is an adaptation of Figure 4 of Schmidt et al. (2005).

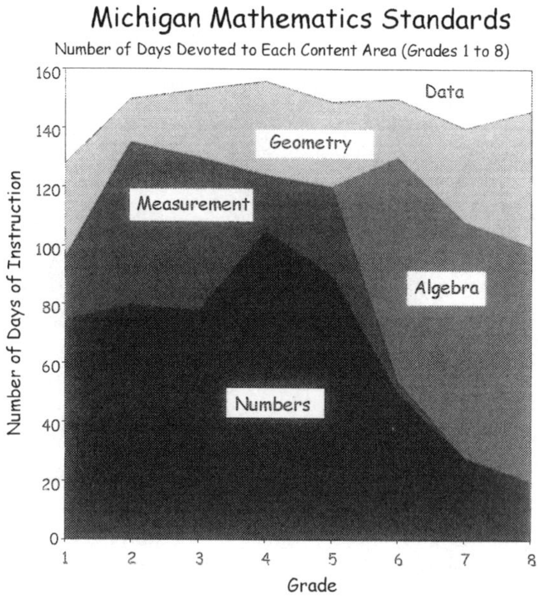

As is evident, there is a shift in focus from number concepts and measurement in the early/middle grades to algebra and geometry in the later grades. Even within the domain of number concepts, emphasis moves from whole number manipulation in the early grades to more complex numbers (fractions, decimals) later on. The authors conclude that "math is NOT math". That is, simply labeling course content as mathematics conceals enormous differences in what is being taught.

What then does the unidimensional scale over 7 grades that is generated by the vertical scaling procedure represent? Schmidt et al. speculate that it may reflect more students' general proficiency in mathematics and less those aspects of achievement that are sensitive to pedagogical intervention. To the extent that this interpretation is correct, a portion of the useful information in the test data will have effectively been discarded. That is, those items that do not load heavily on the dominant dimension will not contribute much to the measured growth of the student. This loss has clear implications for measuring teacher effects.

Many of the VAMs that have been proposed do not actually demand that the test scores be vertically scaled. They simply require that scores in successive grades be approximately linearly related. In most cases, that is a reasonable assumption. However, the test scores that are made available often have already been vertically scaled (for other reasons) and so the issues discussed above are still relevant.

VAMs assume that the score scale can be treated as an interval scale (i.e., equal differences have equal meaning at every point along the scale). While this assumption is not likely to hold exactly for any mental measurement scale, it can often be taken as a reasonable approximation for a well-constructed test scored with a suitable test scoring model. However, when the scale has been derived through a vertical scaling across several grades, the assumption is much less plausible in view of the substantial changes in test content. This problem is related to the question of whether the test score scale is linearly related to teacher effectiveness. One implication is that, at most, teacher effects should only be compared for teachers in the same grade.

A final consideration concerns the problem of measurement error. Test scores are fallible measures of achievement, hence measurement error constitutes a non-trivial fraction of the total variation in scores. Further, measurement error is not constant along the score scale, with the pattern of variation depending on the particular measurement model employed. Extant VAMs do not account for this heteroscedasticity. McCaffrey et al. (2003) speculate on its impact on estimated teacher effects.

3. Models for estimating teacher effects

Introduction

There are a number of choices to be made in developing a VAM. Eleven among them are:

- The structure of the database
- The number of cohorts
- The number of years of data for each cohort
- The number of subjects for each cohort
- The structure of the model
- The number of levels
- The number of estimation stages
- The use of covariates
- Treatment of teacher or school effects as fixed or random
- Treatment of teacher or school effects as permanent or subject to decay
- Treatment of missing data

Each choice has implications for the operating characteristics of the resulting system No one set of choices is optimal under all circumstances. Primary goals and local constraints will determine which combination of features is likely to prove most efficient and practical, although there is still much uncertainty about the relative advantages of different models in a particular setting. McCaffrey et al. (2005) and Webster (2005) provide some information on such model comparisons.

Clearly, the most straightforward way to estimate the contribution of the teacher is to obtain a simple gain score for each student and compute the average for the class. Teachers could then be ranked on the basis of these averages. There are a number of difficulties with this approach, the most obvious being that gains can only be computed for those students with scores in adjacent grades. Restricting attention to those students can result in selection bias. Measurement error in the prior year scores can also introduce bias. For more details, see Ladd and Walsh (2002).

In view of the above discussion, more complex methods of analysis are called for. Recall that, from a theoretical perspective, the goal is to generate a plausible counterfactual that enhances the credibility of the desired causal inferences. With that in mind, we first present two VAMs that have been used operationally to estimate teacher effects. We then discuss two more VAMs that have been used for research purposes.

Dallas value-added accountability system

The VAM employed in Dallas can be described as a two-stage, three-level model for estimating school effects. A full explication is contained in Webster (2005). The model employs student scores from two successive years, referred to as the pre-test and the post-test scores. Each subject is analyzed separately. The database is constructed from the records of all students enrolled in district schools for the specified grades. In advance of the first stage, data are compiled for four student demographic characteristics (race/ethnicity, gender, limited English proficiency status and eligibility for free or reduced price lunch), as well as three census tract variables that together serve as proxies for student socio-economic status.

In the first stage, the pre-test and post-test scores are regressed separately on the student characteristics, the census tract variables as well as selected two-factor and three-factor interactions. The residualized pre-test and residualized post-test scores serve as inputs for the second stage. The intent of the first stage is to remove, as much as possible, the influence of student characteristics on the results of the subsequent analyses. The intent is to provide a fairer basis on which to make school evaluations.

Specifically, let

$$y_{ij} = b_0 + b_1 X_{1ij} + \cdots + b_p X_{pij} + \varepsilon_{ij}. \tag{1}$$

Here, i indexes students within schools j, y denotes a current or prior test score outcome, $\{X\}$ denotes a set of student characteristics and selected interactions, $\{b\}$ denotes a set of regression coefficients, ε_{ij} denotes independent, normally distributed deviations with a common variance for all students.

The coefficients of Eq. (1) are estimated for each possible choice of y. Typically, ordinary least squares is employed. Interest, however, focuses not on the estimated coefficients, but on the residuals from the regression. For each fitted regression, the residuals are standardized.

The second stage comprises a two-level hierarchical linear model or HLM (Raudenbush and Bryk, 2002). The first level consists of a family of linear regressions, one for each school. A student's residualized post-test score is regressed on one or more residualized pre-test scores.[10] At the second level, the intercept and regression coefficients from the first level model for each school are regressed on a set of school characteristics.

Suppose we use a $\sim$ to denote a standardized residual. Then level 1 takes the following form:

$$\widetilde{Z}_{ij} = c_{0j} + c_{1j}\widetilde{P}_{ij,1} + c_{2j}\widetilde{P}_{ij,2} + \delta_{ij} \tag{2}$$

and level 2 takes the form:

$$c_{0j} = G_{00} + \sum_{k=1}^{m} G_{0k} W_{kj} + u_{0j},$$

[10] Note that more than one end-of-year test may be given in a single subject.

$$c_{1j} = G_{10} + \sum_{k=1}^{m} G_{1k} W_{kj},$$

$$c_{2j} = G_{20} + \sum_{k=1}^{m} G_{2k} W_{kj} \quad (3)$$

In level 1, i indexes students within schools j, $\widetilde{Z}_{ij}$ denotes a student's adjusted current test score, $\widetilde{P}_{ij,1}$ and $\widetilde{P}_{ij,2}$ denote a student's adjusted prior test scores, $\{c\}$ denote a set of regression coefficients, δ_{ij} denotes independent, normally distributed deviations with a common variance for all students.

Note that the term "adjustment" refers to the results of carrying out the stage 1 analysis. In principle, more than two prior measures of prior achievement could be employed. In level 2, $\{W\}$ denotes a set of m school characteristics, including various indicators of the demographic composition of the school, multiple indicators of the SES of the school community, school mobility and school crowding, $\{G\}$ denotes a matrix of regression coefficients, u_{0j} denotes a school-specific deviation of its intercept in the level 1 equation from the general linear regression relating school intercepts to school characteristics.

The level 2 model is fit using multi-level software. The estimated school effect is a reliability-adjusted estimate of u_{0j}. This is sometimes called an empirical Bayes estimate because it is equal to the estimate of u_{0j} obtained from a least squares regression for that school alone shrunk toward the estimated regression plane, with the amount of shrinkage inversely proportional to the relative precision of that estimate. (See Braun, 2006, for an introduction to empirical Bayes methodology.)

Finally, for each student, a residualized gain score is computed by taking the difference between the criterion in the level one model (i.e. the student's residualized post-test score) and the prediction of that criterion using the overall district regression coefficients. The residualized gains are then standardized for each grade to have a mean of 50 and a standard deviation of 10 over the district. The teacher effectiveness index is the average of the standardized, residualized gains for the students in the class. As has been pointed out by Ladd and Walsh (2002), and others, the Dallas model may result in estimated teacher effects that are underdispersed. See also Clotfelter and Ladd (1996).

Educational value-added assessment system

Another approach, based on mixed model methodology, was proposed by Sanders et al. (1997) and is known as the Educational Value-added Assessment System (EVAAS). Among VAMs, the EVAAS and its variants are the most widely used and discussed. We present a simplified version of the model for estimating teacher effects, in which we employ the notation of Ballou et al. (2004):

$$y_t^k = b_t^k + u_t^k + e_t^k, \quad (4)$$

$$y_{t+1}^{k+1} = b_{t+1}^{k+1} + u_t^k + u_{t+1}^{k+1} + e_{t+1}^{k+1}, \quad (5)$$

where y_t^k = student score in grade k, year t, b_t^k = district mean score in grade k, year t, u_t^k = contribution of the teacher in grade k, year t, e_t^k = unexplained variation in student score in grade k, year t.

Eqs. (4) and (5) are treated as a mixed model, with the b coefficients estimated as fixed effects and the u coefficients estimated as random effects. Models for subsequent years follow the same pattern. In the aggregate, such a set of equations is referred to as a "layered model" or a "persistence model". As indicated above, a critical issue revolves around how to fairly label the terms in the model. In the standard presentation of the EVAAS model, the parameter u_t^k is defined as the contribution of the teacher in grade k, year t to the student's score. Now, since this term is common to all students who were in the same class in that grade and year, it would be more accurate to label it as a "class effect." The class effect parameters are introduced to see how much of the variance in student gains can be accounted for by the different classes in which the students were enrolled. With this nomenclature, it is clear that directly interpreting a class effect as a measure of teacher effectiveness requires a transition from statistical description to a particular causal attribution.

Sanders argues that the structure of the layered model precludes the need to include student characteristics in the model. That is, since each student acts as her own control, the contribution of her gain in a particular year to her teacher's estimated effect is appropriately adjusted for the student's previous (and future!) gains. Moreover, the correlations of raw gains with student characteristics are much weaker than those between current test scores and student characteristics.

A key assumption of the persistence model is that the class effect in a given year persists in future years. That is, the term u_k^t which first appears in the equation for year t, also appears in the equation for years $t+1$ and so on. The persistence model can be expanded to include longitudinal data from other subjects. The covariance structures can be quite complex; see Sanders et al. (1997) for details. Employing a multivariate, longitudinal database in the estimation of teacher effects brings more information to bear – but at the cost of making more assumptions about the processes generating the data. Sanders asserts that this approach mitigates the effects of missing data not being MAR. As of yet, this claim has not been verified empirically.

Cross-classified model

This is a special case of a hierarchical linear model (Raudenbush and Bryk, 2002). It is one of the models fit in a study by Rowan et al. (2002) and employs longitudinal student records. Test scores are modeled as a polynomial in time with random coefficients. Specifically, the test score, y_{ijt}, for the jth student in school i at time (grade) t can be represented as

$$\begin{aligned} y_{ijt} &= \alpha + \beta t + \delta t^2 + \alpha_i + \beta_i t + \alpha_{ij} + \beta_{ij} t + \gamma' z_{ijt} + \theta_{1(ij)} \\ &\quad + \cdots + \theta_{t(ij)} + e_{ijt}. \end{aligned}$$

Here α, β, and δ are the parameters of a common quadratic function of time, $\{\alpha_i\}$ and $\{\beta_i\}$ are school-specific random intercepts and slopes, $\{\alpha_{ij}\}$ and $\{\beta_{ij}\}$ are teacher-specific random intercepts and slopes. Thus, both the random intercept and the random slope have school and teacher components. Further, z_{ijt} denotes a vector of student characteristics (some of which may vary over time), γ is a common vector of regression coefficients. The θ's represent class (or teacher) residual effects, which persist over time, and are assumed to be independent of one another and of the residual term e_{ijt}. McCaffrey et al. (2003) demonstrate that, for this model, only the current year teacher contributes to the gain score for a student. This is analogous to the situation with the layered model. Discussion of several HLM-related approaches to estimating school effects can be found in Choi and Seltzer (2005).

Variable persistence model

A more general mixed model, called the variable persistence model, has been proposed by McCaffrey et al. (2004). In this model, the class effect is allowed to decay over time.

$$y_t^k = b_t^k + u_t^k + e_t^k, \tag{6}$$

$$y_{t+1}^{k+1} = b_{t+1}^{k+1} + \alpha_{k+1,k} u_t^k + u_{t+1}^{k+1} + e_{t+1}^{k+1}. \tag{7}$$

Eq. (6) repeats Eq. (4) but Eq. (7) differs from Eq. (5) by inclusion of a multiplier $\alpha_{k+1,k}$, which captures the degree to which the class effect in the previous year persists in the following year. As new years are added to the two years represented here, new persistence parameters are introduced. As with the persistence model, the variable persistence model can be generalized to accommodate situations in which a student has more than one teacher in a given year. It can also include random school effects, if desired.

Values of the multiplier less than one represent decay of the class effect. The estimation of the decay parameter greatly complicates the estimation process and procedures that can accommodate large data sets for the layered model are inadequate. Lockwood et al. (2006) have developed a Bayesian formulation that can yield parameter estimates for the variable persistence model.

McCaffrey et al. (2004) point out that the other VAMs presented here are special cases of the variable persistence model in its most general form. Unfortunately, that does not automatically imply that estimated teacher effects from this model are superior to those from (say) the persistence model. The reason is that the latter model may provide better protection against model misspecification due to omitted variables (i.e., student characteristics). On the other hand, if teacher effects do dampen rapidly, the variable persistence model may be preferred, even in the presence of omitted variables.

At this point, there is no general rule for choosing among models, although knowledge of the specific context application, as well as the uses to which the estimated effects will be put, may offer more guidance. Note that the situation here is different than in standard regression problems where various goodness-of-fit measures can provide clear direction. While some can be used here, the key concern is sensitivity to critical assumptions which are almost certainly false.

In the few instances where both models have been fit to the same data, the estimated α's are different from zero but much less than one. With such low persistence parameters, the estimated teacher effects are similar to those that would be obtained by modeling student attainment scores – in contrast to the persistence model for which the estimated teacher effects are similar to those that would be obtained by modeling student gains (D. McCaffrey, personal communication, August 16, 2005). Not surprisingly, the estimated teacher effects from the two models differ substantially. In one case, the correlations ranged from .73 to .87 for mathematics and from .77 to .84 for reading. Moreover, teacher effects from the variable persistence model are more dispersed, so that more teachers are identified as meaningfully different from the district average (McCaffrey et al., 2005).

4. Example

To illustrate the application of VAM, we present the results of applying the EVAAS to data extracted from the Early Childhood Longitudinal Study – Kindergarten Cohort, usually denoted as the ECLS-K (see, for example, NCES, 2005a, 2005b). This study has tracked a cohort of students from the beginning of kindergarten through the end of the fifth grade.[11] Students have been administered tests in reading and mathematics on six occasions: Kindergarten-fall, kindergarten-spring, first grade-fall, first grade-spring, third grade-spring, and fifth grade-spring. Since only a thirty percent subsample was tested on the third occasion, we do not use those scores. Fifth grade scores were not available at the time of the analysis. With the data from the remaining four occasions, we are able to estimate a model over three transitions. Scores for each subject were reported on a single cross-grade scale constructed through application of item response theory. For further details, consult Rock and Pollack (2002).

The results reported are taken from a larger study (Braun et al., 2006), which employed software provided by the RAND Corporation (Lockwood et al., 2006). The software uses a Bayesian implementation and the estimated school effect is taken to be the median of the posterior distribution of the school's contribution to the average student. It should be emphasized that the estimated school effects presented are based on a single subject and only one cohort; ordinarily EVAAS reports estimated school effects based on at least two subjects and averaged over three cohorts. Therefore, the presented effects will be subject to greater error than would be typically the case. This data set comprises students attending 337 different schools. In order to be included in the analysis, students must have been enrolled in the same school on all four testing occasions.

A single run of the model, yields estimated school effects for three transitions. Table 1 displays summary statistics for the sets of estimated school effects for reading and math for transition three (i.e. grade one – spring to grade three – spring) and Figure 1 displays the corresponding frequency plots. For each subject, the estimated effects

Table 1
Summary statistics for estimated school effects for transition from grade one-spring to grade three-spring

	Mean	Standard deviation	Minimum	Maximum
Reading	−.005	3.39	−9.4	10.5
Math	−.003	2.79	−7.7	7.6

[11] Students will be tracked through the end of the twelfth grade.

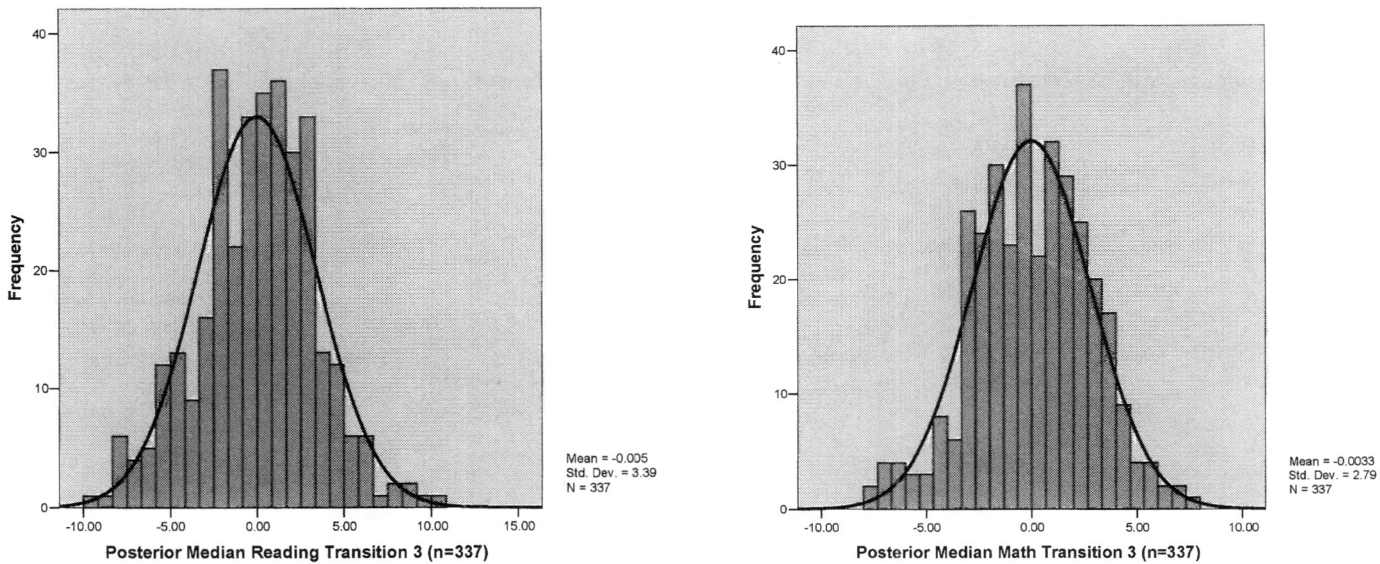

Fig. 1. Histograms of estimated school effects for reading and math for transition from grade one-spring to grade three-spring.

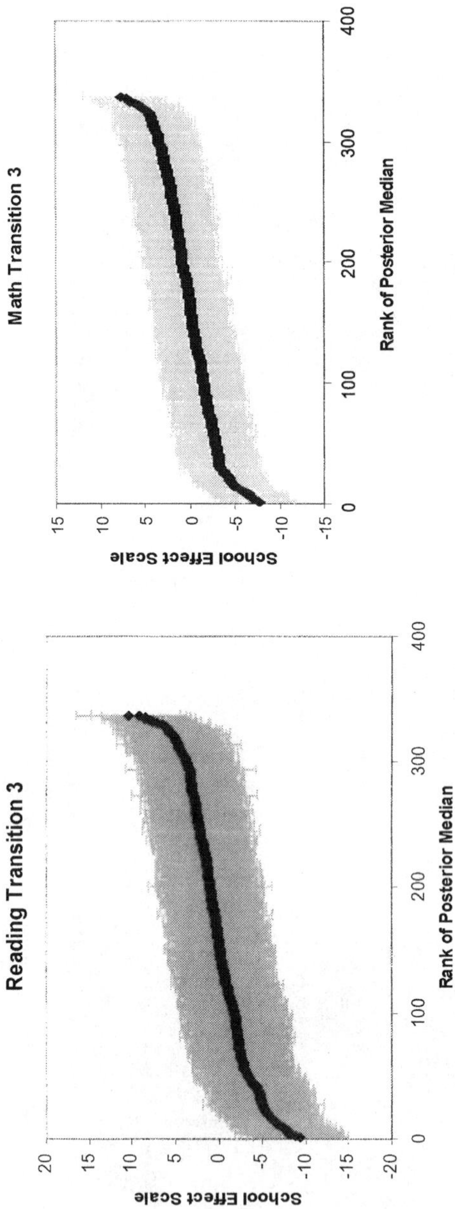

Fig. 2. Caterpillar plots for estimated school effects for reading and math for transition from grade one-spring to grade three-spring. Bars are 4 standard errors long.

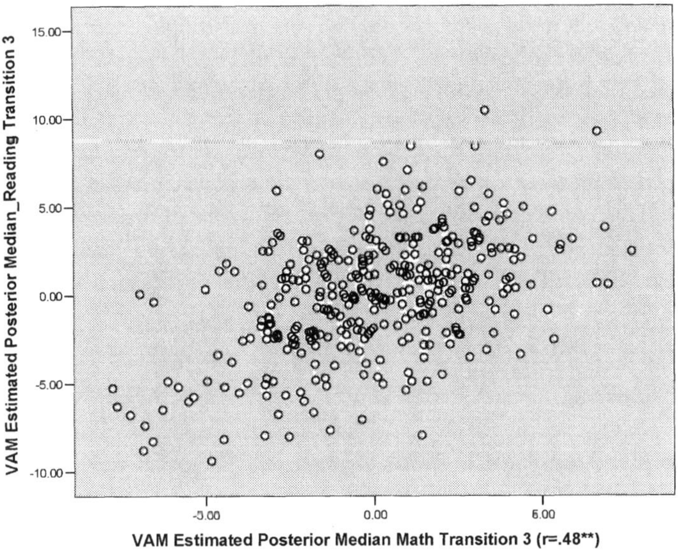

Fig. 3. Plot of estimated reading effects against estimated math effects for 337 schools for transition from grade 1-spring to grade 3-spring.

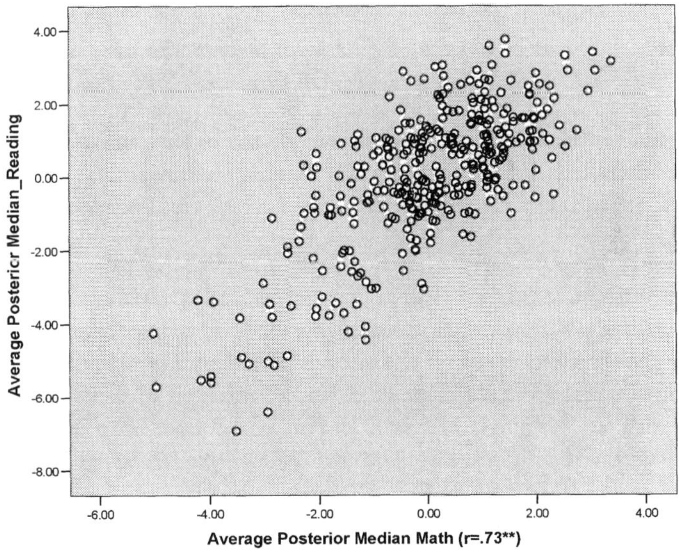

Fig. 4. Plot of average over three transitions of estimated reading effects against average estimated math effects for 337 schools.

are centered at zero, approximately symmetric and somewhat longer tailed than would be expected under normality. Results for the other transitions are qualitatively similar.

Figure 2 displays caterpillar plots for the two sets of estimates. Only the most extreme estimated effects can be considered significantly different from zero. Of course, with more data the 95% confidence intervals would be shortened and it is likely that more of the estimated effects would be found significant.

Figure 3 displays a scatterplot of the estimated effects for reading and for math. As one might expect there is a reasonably strong positive association, with a correlation of 0.48. Since the estimated effects for both subjects for all three transitions for each school are available, it is possible to calculate an average school effect for each subject. Figure 4 displays a scatterplot of the average effects for reading and math. There is a stronger positive association, with a correlation of 0.73.

5. Current research

Sensitivity of estimated effects to model assumptions

(i) Model structure

The previous section described the many dimensions along which model structures for value-added analysis may vary. McCaffrey et al. (2005) present a somewhat different taxonomy comprising seven different approaches to estimating teacher effects. Comparisons among models can be made by means of theoretical calculations, empirical analysis and simulations.

With respect to the first approach, McCaffrey et al. (2004) demonstrate that essentially all VAMs that have been proposed can be treated as special cases of the variable persistence model introduced in that article. On that basis, they explore the differences among the models with particular attention to their sensitivity to various assumptions about the nature and distribution of omitted covariates.[12] The overall conclusion is that no one model dominates the others under all plausible circumstances. The optimal strategy depends on how the covariates are distributed across classes, schools and groups of schools.[13]

Unfortunately, by definition such information is unavailable and *ad hoc* adjustments may have problematic results. That is, even if observed student characteristics are strongly associated with outcomes, adjusting for those covariates cannot always be recommended. The difficulty is that if student covariates are correlated with true teacher effects, then such adjustments can result in biased estimates of the teacher effects. For example, if students with high SES tend to enroll in classes with teachers who truly superior, adjusting for SES will underestimate the relative advantage of their teachers. McCaffrey et al. (2005) present an analysis of the mean squared errors (MSEs) of seven different models under a variety of assumptions about the relative size of the variances at different levels and the strength of the correlation between outcomes in successive years. While they are able to make some general conclusions, much depends on the true "state of nature."

[12] Omitted covariates are a concern in observational studies because they are a principal source of selection bias in the estimation of model parameters.

[13] This last phrase refers to the notion of schools in a district forming "strata" with little mixing of students across strata over time.

Empirical analysis involves comparing the estimated effects generated by different models, typically by computing correlations (occasionally supplemented by scatterplots). Ballou et al. (2004) compare the basic persistence model with three variants representing different approaches to incorporating covariates into the model. Data were drawn from a large, diverse district over five years. Test scores in five subjects for grades 3 through 8 were analyzed, although only estimated teacher effects for math, reading and science (grades 4 through 8) were reported. With a few exceptions, correlations were greater than 0.90.

McCaffrey et al. (2005) compare the persistence and variable persistence model employing data from a large urban school district. The data set comprises math and reading scores for a single cohort of students (9295 students) followed from grade 1 through grade 5. They find that the estimated persistence parameters range from about 0.07 to 0.20 for math and from about 0.10 to 0.23 for reading. That is, the persistence parameters are significantly different from one (the value assumed by the persistence model) but different from zero, as well. Nonetheless, they report that correlations between estimated teacher effects range from 0.73 to 0.87 for math and from 0.77 to 0.84 for reading.[14]

The advantage of a simulation is that because the true values of teacher effects are known, model estimates can be compared directly to their estimands. The disadvantage, of course, is that it is difficult to ascertain how closely the simulation captures all the salient features of real-world settings. McCaffrey et al. (2004) report the result of a small simulation intended to investigate the impact of model misspecification on estimated teacher effects. In particular, they fit the layered model to data generated from a persistence model with a range of alphas less than one. Using the correlation between true and estimated effects as the criterion, they find that misspecification does not materially degrade the quality of the estimates.

Tekwe et al. (2004) report the results of a simulation comparing estimated school effects for four models: A simple change score fixed effects model, two versions of a HLM (with and without adjustments for student covariates) and the layered model. Data for grades 3, 4 and 5 were obtained from a Florida school district with 22 elementary schools. With the exception of the HLM with covariate adjustments, correlations among the model estimates typically exceeded 0.90. Correlations with the estimates from the HLM with covariance adjustments typically exceed 0.70. Tekwe et al. (2004) conclude that there does not appear to be a substantial advantage to using more complex models rather than simpler ones. As the authors note, there are many limitations to the study and its generalizability to other settings is not clear.

In response to Tekwe et al. (2004), Wright (2004) carried out a simulation employing a factorial design for the different parameters: number of students, gain patterns and degree of departure from MAR. He compared a simple gain score model with two more complex, longitudinal models. Using a MSE criterion, he concludes that the more complex models are to be preferred in view of their lower MSE in those cells of the design that are more likely to represent real world data.

[14] The authors note that the estimated teacher variance components from the variable persistence model are typically much larger than those from the layered model, which results in the former declaring a greater number of estimated teacher effects significantly different from the average.

Test score scale The discussion in Section 2 argued that, in view of the changing character of what is taught in mathematics, the underlying scale of longitudinal measurement is not fixed. If that is the case in mathematics, perhaps the most linearly hierarchical of academic subjects, it is more likely to be the case in domains such as history or literature, in which there is greater variation in both the nature and sequencing of the content over grades. A plausible argument can be made that summarizing student performance by reporting scores on a vertically linked scale can reduce the instructional sensitivity of the results that serve as the input for a value-added analysis. This would run counter to the purpose of the process.

Unfortunately, there has been very little empirical work in this area. For example, no one has studied differences in estimated teacher effects by carrying out analyses based on scores scaled within year to scores scaled (using a variety of models) across years. The interval scale assumption has also not been much examined, with the exception of a study by Braun et al. (2006). They used data from the Early Childhood Longitudinal Study – Kindergarten Cohort (ECLS-K) to examine the robustness of estimated teacher effects to both the interval scale assumption and model structure.

In their analysis of the ECLS-K database, Rock and Pollack (2002) built eight-step ordered developmental scales for math and reading, suitable for kindergarten through grade 3. Based on their response patterns, students were assigned to a step at each administration of the test battery. Thus, it was possible to construct a sequence of transition matrices for each school and subject that displayed the conditional distribution of the level at the later administration given the level at the previous administration. Braun et al. (2006) developed a measure of school efficacy that capitalized on the difference between the school-specific transition matrix and an aggregate transition matrix obtained by pooling data across schools.

Based on a conventional IRT analysis, students were also assigned a scale score reported on a single, cross-grade scale. Accordingly, Braun et al. (2006) were able to obtain estimated teacher effects by applying the persistence model to the scale scores. Estimates could be compared by subject and transition, yielding $3 \times 2 = 6$ comparisons. Correlations were around 0.8.

(ii) Missing data

"What we don't know won't hurt us, its what we do know that ain't"
Will Rogers

Model approaches that require longitudinal data are more likely to confront problems of missing data than are models that require only cross-sectional data. VAMs assume that the data are missing at random (MAR) so that the parameter estimates are unbiased. Unfortunately, with respect to educational databases, such assumptions are both untestable and heroic. It is all the more important, therefore, to study the fragility of the inferences to the MAR assumption.

There have been at least three such sensitivity studies done so far: Wright (2004), McCaffrey et al. (2005) and Shirley et al. (2006). These studies vary both with respect to the structure of the simulation and how sensitivity was characterized. All three studies focused on estimated teacher effects. Briefly, they found that although the rank ordering of estimated effects was reasonably robust, there was enough instability to make it inadvisable to use such estimates – on purely statistical grounds – as the sole or primary indicator of teacher performance.

Utility of VAM-based evaluations

(i) Volatility of estimated effects
The utility of estimated teacher effects depends both on their statistical properties and the credibility of their interpretation as causal effects. With respect to the former, Lockwood et al. (2002) examined the stability of the rank order of estimated effects based on a simulation employing a relatively simple two-stage VAM. They found that estimating extreme ranks or percentiles is very difficult under conditions likely to obtain in practice; i.e. moderate heterogeneity among teachers and substantial sampling variance. The consequence is that the 95% confidence interval for an extreme percentile is likely to cover more than half the feasible range. They also make the case that, despite the larger sample sizes, the situation is not likely to be much better for ranking schools.

(ii) External validation of estimated effects
In view of the extended use of VAMs in many districts in Tennessee and in Dallas, Texas, one might expect that there would be published reports examining the relationship between (say) estimated teacher effects and independently arrived at judgments of teacher efficacy. However, only three such reports have come to light.

Gallagher (2004) describes a small study carried out at a single innovative school in California. Using HLM to estimate value-added teacher effects, she found positive but moderate correlations in literacy and positive but weak correlations in mathematics with teacher evaluation scores. Milanowski (2004) correlated teacher evaluations obtained through an elaborate, theoretically based system with value-added estimates also derived from an HLM. The first stage of the HLM obtains an estimated classroom effect adjusted for students' prior test scores and five student characteristics. Using data from Cincinnati public schools, he found modest correlations for science, reading and mathematics, with those for mathematics being the highest. Finally, Dee and Keys (2004), analyzing data from the Tennessee class size experiment, looked at the relationship between teacher evaluations based on the state's career ladder evaluation system and student test scores, adjusted for student demographics and class characteristics (but not prior year test scores). They found a modest relationship between student performance and teacher position on the career ladder.

(iii) Causal attribution
High-stakes use of VAM results rest on the supposition (often not explicitly stated) that the estimated effect associated with a teacher is an adequate indicator of that teacher's efficacy relative to an appropriate comparison group of teachers. Section 2 describes the pitfalls in making the transition from statistical description to causal attribution in observational studies subject to selection bias. One can imagine undertaking sensitivity studies to ascertain the robustness of estimated teacher effects to unobserved heterogeneity as has been done in medicine (Rosenbaum and Rubin, 1983) and in demography

(Montgomery et al., 1986).[15] On the one hand, the longitudinal nature of the data could make this approach quite informative; on the other hand, the complexity of the constellation of selection bias processes makes the modeling challenge quite daunting.

(iv) Beyond VAM estimates
Rubin et al. (2004) examined the problem of making causal inferences on the basis of a value-added analysis and concluded that it would require assumptions of "heroic proportions." They suggested, instead, that it might be appropriate – and feasible – to compare those districts that had implemented VAM with those that had not, in order to estimate the causal effect of incorporating VAM into a district's accountability system. As in a quasi-experimental design, this inference could be made by carefully matching districts on a complex mixture of background variables. Serendipitously, at about the same time that they made this suggestion the state of Pennsylvania instituted a pilot VAM program involving 32 districts from across the state. A study is currently underway comparing those districts (on a number of dependent variables) to a matched set of districts that had not adopted VAM. The results should be available in 2006.

(v) Future developments
The testing programs that support VAM generally don't yield information fine-grained enough to guide individual instruction. Typically, they are not designed to do so and, moreover, the scores are not available before the end of the school year. It is possible, however, that addressing these issues may also help to resolve one of the difficult challenges facing the use of VAMs, specifically how to construct meaningful change scores when the content of the courses change over time (e.g., "math is not math").

Suppose that a student's overall score in a subject is supplemented by a series of sub-scores, each focusing on a different dimension or content area, resulting in a vector of changes scores that could offer some pedagogical guidance. (This would require modifications to some VAMs, but not EVAAS, for example, if there were interest in estimating teacher effects from such data.) Ordinarily, a test that provides reliable estimates of proficiency for a set of sub-skills is likely to impose an unacceptable burden on both schools and students. However, a combination of adaptive testing (Wainer, 2000) and empirical Bayes estimation (Thissen and Wainer, 2001, Chapter 9; Braun, 2006) can yield sub-scores sufficiently reliable for instructional uses, even when based on a relatively small number of items. Of course, adaptive testing would require computerized administration which, in turn, would facilitate timely feed-back. Certainly, practical implementation of such a scenario would require considerable effort, but the goal is well worth striving for.

6. Summing up

VAM is a methodology that has attracted a great deal of interest on the part of policy makers because it promises to provide information with respect to efficacy of schools

[15] Note that both these studies involved binary response data.

and teachers through the use of longitudinal student records. However, there are a number of technical concerns, which lead many measurement specialists to counsel caution. These concerns are not just academic niceties – they are central to the proposed uses of VAM, especially those related to accountability. Indeed, given the complexity of educational settings, we may never be satisfied that VAM can be used to appropriately partition the causal effects of the teacher, the school and the student on measured changes in standardized test scores. Although the results of more sophisticated VAMs can fall short of our desires, statistics based on simple gains can be even more problematic. While the latter are simpler to use for regulatory purposes, they should never be interpreted as estimating school or teacher value-added. Nonetheless, there is likely to be continuing confusion between indicators based on simple summaries of student gains and those based on the outcomes of a true value-added analysis.

One can envision VAM results serving as a component of a multidimensional system of evaluation; however, this possibility would have to be evaluated in light of the specifics of such a system and the context of use. In the meanwhile, one can support the use of VAM results as a first screen in a process to identify schools in need of assistance or teachers in need of targeted professional development. Subsequent stages should involve interviews and/or direct observation since statistical analyses cannot identify the strategies and practices employed by educators. In some districts, VAM results have been used successfully in this manner by principals and teachers in devising individualized improvement strategies.

Value-added modeling is an area of active research and one can expect a continuing flow of results that will refine our understanding of the different models and their operating characteristics in a variety of settings. Such research should be encouraged and it is hoped that the fruits of this work will inform future policy development.

References

Ballou, D., Sanders, W., Wright, P. (2004). Controlling for student background in Value-Added assessment of teachers. *Journal of Educational and Behavioral Statistics* **29** (1), 37–65.

Bourque, M.L. (2005). The history of no child left behind. In: Phelps, R.P. (Ed.), *Defending Standardized Testing*. Lawrence Erlbaum Associates, Hillsdale, NJ, pp. 227–254.

Braun, H.I. (2000). A post-modern view of the problem of language assessment. In: Kunnan, A.J. (Ed.), *Studies in Language Testing 9: Fairness and Validation in Language Assessment*. Selected papers from the 19th Language Testing Research Colloquium, University of Cambridge, Local Examinations Syndicate, Cambridge, UK, pp. 263–272.

Braun, H.I. (2005a). Value-added modeling: What does due diligence require? In: Lissitz, R. (Ed.), *Value Added Models in Education: Theory and Applications*. JAM Press, Maple Grove, MN, pp. 19–39.

Braun, H.I. (2005b). Using student progress to evaluate teachers: A primer on value-added models. Educational Testing Service, Princeton, NJ.

Braun, H.I. (2006). Empirical Bayes. In: Green, J., Camilli, G., Elmore, P. (Eds.), *Complementary Methods for Research in Education*. American Educational Research Association, Washington, DC, pp. 243–258.

Braun, H.I., Qu, Y., Trapani, C.S. (2006). Robustness of statistical inferences from value-added models. In preparation.

Choi, K., Seltzer, M. (2005). Modeling heterogeneity in relationships between initial status and rates of change: Latent variable regression in a three-level hierarchical model. CSE Report 647, National Center for Research on Evaluation, Standards and Student Testing, UCLA, Los Angeles, CA.

Choi, K., Goldschmidt, P., Yamahiro, K. (2005). Exploring models of school performance: From theory to practice. In: Herman, J., Haertel, E. (Eds.), *National Society for the Study of Education*, vol. 104. Blackwell, Malden, MA.

Clotfelter, C., Ladd, H.F. (1996). Recognizing and rewarding success in public schools. In: Ladd, H. (Ed.), *Holding Schools Accountable: Performance-based Reform in Education*. Brookings Institution, Washington, DC.

Dee, T.S., Keys, B.J. (2004). Does merit pay reward good teachers? Evidence from a randomized experiment. *Journal of Policy Analysis and Management* **23** (3), 471–488.

Doran, H.C., Cohen, J. (2005). The confounding effects of linking bias on gains estimated from value-added models. In: Lissitz, R. (Ed.), *Value Added Models in Education: Theory and Applications*. JAM Press, Maple Grove, MN, pp. 80–100.

Dorans, N.J. (2005). Vertical linking: Issues relevant to assessing growth. ETS Research Report, Educational Testing Service, Princeton, NJ, in press.

Gallagher, H.A. (2004). Vaughn Elementary's innovative teacher evaluation system: Are teacher evaluation scores related to growth in student achievement? *Peabody Journal of Education* **79** (4), 79–107.

Holland, P.W. (1986). Statistics and causal inference. *Journal of the American Statistical Association* **81**, 945–970.

Holland, P.W. (2005). Counterfactuals. In: Everitt, B.S., Howell, D.C. (Eds.), *Encyclopedia of Statistics in Behavioral Sciences*, vol. 2. Wiley, Chichester.

Holland, P.W., Rubin, D.B. (1983). On Lord's paradox. In: Wainer, H., Messick, S. (Eds.), *Principals of Modern Psychological Measurement*. Lawrence Erlbaum Associates, Hillsdale, NJ, pp. 3–25.

Kane, T.J., Steiger, D.O. (2002). Volatility in school test scores: Implications for test-based accountability systems. In: Ravitch, D. (Ed.), *Brookings Papers on Education Policy*. Brookings Institution, Washington, DC, pp. 235–260.

Kupermintz, H. (2003). Teacher effects and teacher effectiveness: A validity investigation of the Tennessee Value Added Assessment System. *Educational Evaluation and Policy Analysis* **25** (3), 287–298.

Ladd, H.F., Walsh, R.P. (2002). Implementing value-added measures of school effectiveness: Getting the incentives right. *Economics of Education Review* **21**, 1–17.

Linn, R.L. (2004). Rethinking the No Child Left Behind accountability system. Paper presented at the Center for Education Policy Forum, Washington, DC. Available online at http://www.ctredpol.org.

Linn, R.L. (2005). Conflicting demands of "No Child Left Behind" and state systems: Mixed messages about school performance. *Education Policy Analysis Archives* **13** (33). Retrieved May 15, 2006 from http://epaa.asu.edu/epaa/v13n33/. June 28.

Lissitz, R. (Ed.) (2005). *Value Added Models in Education: Theory and Applications*. JAM Press, Maple Grove, MN.

Little, R.J.A., Rubin, D.B. (1987). *Statistical Analysis with Missing Data*. Wiley, New York.

Lockwood, J.R., Louis, T.A., McCaffrey, D.F. (2002). Uncertainty in rank estimation: Implications for value-added modeling accountability systems. *Journal of Educational and Behavioral Statistics* **27** (3), 255–270.

Lockwood, J.R., McCaffrey, D.F., Mariano, L.T., Setodji, C. (2006). Bayesian methods for scalable multi-subject value-added assessment. *Journal of Educational and Behavioral Statistics*, submitted for publication.

Lord, F.M. (1967). A paradox in the interpretation of group comparisons. *Psychological Bulletin* **68**, 304–305.

McCaffrey, D.F., Lockwood, J.R., Koretz, D.M., Hamilton, L.S. (2003). *Evaluating Value-Added Models for Teacher Accountability*. The RAND Corporation, Santa Monica, CA.

McCaffrey, D.F., Lockwood, J.R., Koretz, D., Louis, T.A., Hamilton, L. (2004). Models for value-added modeling of teacher effects. *Journal of Educational and Behavioral Statistics* **29** (1), 67–101.

McCaffrey, D.F., Lockwood, J.R., Mariano, L.T., Setodji, C. (2005). Challenges for value-added assessment of teacher effects. In: Lissitz, R. (Ed.), *Value Added models in Education: Theory and Practice*. JAM Press, Maple Grove, MN, pp. 111–144.

Messick, S. (1989). Validity. In: Linn, R. (Ed.), *Educational Measurement*. American Council on Education, Washington, DC.

Milanowski, A. (2004). The relationship between teacher performance evaluation scores and student achievement: Evidence from Cincinnati. *Peabody Journal of Education* **79** (4), 33–53.

Montgomery, M., Richards, T., Braun, H.I. (1986). Child health, breastfeeding and survival in Malaysia: A random effects logit approach. *Journal of the American Statistical Association* **81**, 297–309.

National Association of State Boards of Education. (2005). *Evaluating Value-Added: Findings and Recommendations from the NASBE Study Group on Value-Added Assessments*. Association of State Boards of Education, Alexandria, VA.

National Center for Education Statistics. (2005a). Early Childhood Longitudinal Study, Kindergarten Class of 1998–99 (ECLS-K): Third Grade Methodology Report. Retrieved February 1, 2006 from http://nces.ed.gov/pubsearch/pubsinfo.asp?pubid=2005018.

National Center for Education Statistics. (2005b). Early Childhood Longitudinal Study, Kindergarten Class of 1998–99 (ECLS-K): Third Grade Psychometric Report. Retrieved February 1, 2006 from http://nces.ed.gov/pubsearch/pubsinfo.asp?pubid=2005062.

No Child Left Behind Act of 2001, Pub. L. No. 107-110, 115 Stat. 1425 (2002).

Petersen, N.S., Novick, M.R. (1976). An evaluation of some models for culture-fair selection. *Journal of Educational Measurement* **13** (1), 3–29.

Raudenbush, S.W. (2004). Schooling, statistics, and poverty: Can we measure school improvement? The ninth annual William H. Angoff Memorial Lecture, Educational Testing Service, Princeton, NJ.

Raudenbush, S., Bryk, A. (2002). *Hierarchical Linear Models: Applications and Data Analysis Methods*, 2nd ed. Sage Publications, Newbury Park, CA.

Reckase, M.D. (2004). The real world is more complicated than we would like. *Journal of Educational and Behavioral Statistics* **29** (1), 117–120.

Rock, D.A., Pollack, J.M., (2002). A model-based approach to measuring cognitive growth in pre-reading and reading skills during the kindergarten year. ETS Research Report RR-02-18, Educational Testing Service, Princeton, NJ.

Rogosa, D.R. (2003). Confusions about consistency in improvement. Retrieved September 8, 2004 from http://www.stat.stanford.edu/~rag/api/consist.pdf.

Rosenbaum, P.R. (1984). From association to causation in observational studies: The role of tests of strongly ignorable treatment assignment. *Journal of the American Statistical Association* **79**, 41–48.

Rosenbaum, P.R., Rubin, D.B. (1983). Assessing sensitivity to an unobserved binary covariate in an observational study with binary outcome. *Journal of the Royal Statistical Society, Series B* **45**, 212–218.

Rowan, B., Correnti, R., Miller, R.J. (2002). What large-scale survey research tells us about teacher effects on student achievement: Insights from the *Prospects* study of elementary schools. *Teacher College Record* **104**, 1525–1567.

Rubin, D.B. (1974). Estimating causal effects of treatments in randomized and non-randomized studies. *Journal of Educational Psychology* **66**, 688–701.

Rubin, D.B. (1976). Inference and missing data. *Biometrika* **63**, 581–592.

Rubin, D.B. (1978). Bayesian inference for causal effects: The role of randomization. *The Annals of Statistics* **7**, 34–58.

Rubin, D.B., Stuart, E.A., Zanutto, E.L. (2004). A potential outcomes view of value-added assessment in education. *Journal of Educational and Behavioral Statistics* **29** (1), 103–116.

Sanders, W., Saxton, A., Horn, B. (1997). The Tennessee value-added assessment system: A quantitative outcomes-based approach to educational assessment. In: Millman, J. (Ed.), *Grading Teachers, Grading Schools: Is Student Achievement a Valid Evaluational Measure?* Corwin Press, Inc., Thousand Oaks, CA, pp. 137–162.

Schmidt, W.H., Houang, R.T., McKnight, C.C. (2005). Value-added research: Right idea but wrong solution? In: Lissitz, R. (Ed.), *Value Added Models in Education: Theory and Applications*. JAM Press, Maple Grove, MN, pp. 145–165.

Shadish, W.R., Cook, T.D., Campbell, D.T. (2002). *Experimental and Quasi-Experimental Designs for Generalized Causal Inference*. Houghton Mifflin Company, Boston, MA.

Shirley, K., Braun, H.I., Wainer, H.I., Lockwood, J.R., McCaffrey, D.F. (2006). The sensitivity to missing data of statistical inferences from value-added models. In preparation.

Tekwe, C.D., Carter, R.L., Ma, C., Algina, J., et al. (2004). An empirical comparison of statistical models for value-added assessment of school performance. *Journal of Educational and Behavioral Statistics* **29** (1), 11–35.

Thissen, D., Wainer, H. (2001). *Test Scoring*. Lawrence Erlbaum Associates, Hillsdale, NJ.

Wainer, H. (1991). Adjusting for differential base-rates: Lord's Paradox again. *Psychological Bulletin* **109**, 147–151.

Wainer, H. (2000). *Computerized Adaptive Testing: A Primer*. Lawrence Erlbaum Associates, Hillsdale, NJ. First published in 1990.

Wainer, H. (2004). Introduction to a special issue of the Journal of Educational and Behavioral Statistics on Value-Added Assessment. *Journal of Educational and Behavioral Statistics* **29** (1), 1–2.

Wainer, H., Brown, L. (2004). Two statistical paradoxes in the interpretation of group differences: Illustrated with medical school admission and licensing data. *The American Statistician* **58**, 117–123.

Webster, W.J. (2005). The Dallas school-level accountability model: The marriage of status and value-added approaches. In: Lissitz, R. (Ed.), *Value Added Models in Education: Theory and Applications*. JAM Press, Maple Grove, MN, pp. 233–271.

Webster, W., Mendro, R. (1997). The Dallas value-added accountability system. In: Millman, J. (Ed.), *Grading Teachers, Grading Schools: Is Student Achievement a Valid Evaluation Measure?* Corwin Press, Inc., Thousand Oaks, CA, pp. 81–99.

Wright, S.P. (2004). Advantages of multivariate longitudinal approach to educational value-added assessment without imputation. Paper presented at the National Evaluation Institute, Colorado Springs, CO. July.

Handbook of Statistics, Vol. 26
ISSN: 0169-7161
2007 Published by Elsevier B.V.
DOI: 10.1016/S0169-7161(06)26028-0

28

Three Statistical Paradoxes in the Interpretation of Group Differences: Illustrated with Medical School Admission and Licensing Data

Howard Wainer and Lisa M. Brown

Abstract

Interpreting group differences observed in aggregated data is a practice that must be done with enormous care. Often the truth underlying such data is quite different than a naïve first look would indicate. The confusions that can arise are so perplexing that some of the more frequently occurring ones have been dubbed paradoxes. In this chapter we describe three of the best known of these paradoxes – Simpson's Paradox, Kelley's Paradox, and Lord's Paradox – and illustrate them in a single data set. The data set contains the score distributions, separated by race, on the biological sciences component of the Medical College Admission Test (MCAT) and Step 1 of the United States Medical Licensing Examination™ (USMLE). Our goal in examining these data was to move toward a greater understanding of race differences in admissions policies in medical schools. As we demonstrate, the path toward this goal is hindered by differences in the score distributions which gives rise to these three paradoxes. The ease with which we were able to illustrate all of these paradoxes within a single data set is indicative of how wide spread they are likely to be in practice.

"To count is modern practice, the ancient method was to guess"
Samuel Johnson

"Evidence may not buy happiness, but it sure does steady the nerves"
Paraphrasing Satchel Paige's comment about money

1. Introduction

Modern policy decisions involving group differences are both based on, and evaluated by, empirical evidence. But the understanding and interpretation of the data that comprise such evidence must be done carefully, for many traps await the unwary. In this chapter we explore three statistical paradoxes that can potentially mislead us and illustrate these paradoxes with data used in the admission of candidates to medical school, and one measure of the success of those admissions.

The first is known as Simpson's Paradox (Yule, 1903; Simpson, 1951) and appears when we look at the aggregate medical school application rates by ethnicity. The second is Kelley's Paradox (Wainer, 2000), which shows its subtle effect when we examine the success rates of minority medical students on Step 1 of the U.S. Medical Licensing Exam (USMLE-1). And, finally, the third paradox, which was first described by Lord (1967), emerges when we try to estimate the effect size of medical school training on students.

The balance of this chapter is laid out as follows: in Section 2 we describe the data that form the basis of our investigation and provide some summarizations; in Section 3 we describe Simpson's Paradox and demonstrate its existence within our data and show how to ameliorate its effects through the method of standardization; in Section 4 we describe Kelley's Paradox and use our data to illustrate its existence; in Section 5 we demonstrate Lord's paradox and describe how its puzzling result can be understood by embedding the analysis within Rubin's model for causal inference. We conclude in Section 6 with a discussion of these findings.

2. The data

There are many steps on the path toward becoming a physician. Two important ones that occur early on are tests. The first test, the Medical College Admission Test (MCAT), is usually taken during the junior or senior year of college and is one important element in gaining admission to medical school. The second test is Step 1 of the United States Medical Licensing Exam (USMLE). Step 1 is the first of a three-part exam a physician must pass to become licensed in the United States. This test is usually taken after the second year of medical school and measures the extent to which an examinee understands and can apply important concepts of the basic biomedical sciences. For the purposes of this investigation we examined the performance of all black and white examinees who's most recent MCAT was taken during the three-year period between 1993 and 1995. Two samples of examinees testing during this time were used in the analyses. The first sample of approximately 83,000 scores comprises all black and white examinees whose most recent MCAT was taken during this time. This sample includes all examinees rather than being limited to only those applying to medical school. Additionally, because the sample reflects performance of examinees who had taken the MCAT after repeated attempts, the initial scores from low scoring examinees who repeated the examination to improve their performance were not included. This makes these average scores somewhat higher than those reported elsewhere (http://www.aamc.org).

The funnel of medical school matriculation continued with about 48,000 (58%) of those who took the MCAT actually applying to medical school; of these about 24,000 (51%) were actually accepted. And finally, our second sample of approximately 22,000 (89%) of the candidates who were accepted to allopathic medical schools, sat for Step 1 three years after their last MCAT attempt. By limiting our sample to those who entered medical school the year after taking the MCAT and took Step 1 two years later, we have excluded those who progressed through these steps in less typical amounts of time. But this seems like a plausible way to begin, and the conclusions we reach using this assumption should not be very far from the truth.

In Table 1 we present the distributions of MCAT-Biological Sciences[1] scores for two racial groups along with selected conditional probabilities. The first column in the upper portion of Table 1 shows the MCAT scores; we grouped some adjacent extreme score categories together because the sample sizes in the separate categories were too small in one or the other of the two groups to allow reliable inferences. The first section of the table shows the distributions of MCAT scores by race for black and white candidates whose most recent attempt was between 1993 and 1995. The second and third sections present the number of examinees from each group who applied to allopathic medical schools the following year and the respective acceptance rates. The final section shows the distribution of MCAT scores among those in our sample who matriculated to medical school and took Step 1 of the USMLE three years after their last MCAT attempt.

The bottom portion of Table 1 presents selected conditional probabilities at each level of MCAT score that were derived from the frequencies on the top portion in the indicated fashion.

For the purposes of this discussion there are three important characteristics of Table 1: (i) the higher the MCAT score the greater the likelihood of applying to medical school, being selected, and eventually taking Step 1, (ii) at every MCAT score level the proportion of black MCAT takers taking Step 1 is higher than for white applicants, and (iii) despite this, the Step 1 rates for whites overall was higher than for blacks. If we have not made any errors in our calculations how do we account for this remarkable result? Are black students sitting for the licensing exam with greater likelihood than whites? Or with lesser? This is an example of Simpson's Paradox and in the next section we discuss how it occurs and show how we can ameliorate its effects.

3. Simpson's Paradox

The seeming anomaly in Table 1 is not rare. It shows up frequently when data are aggregated. Indeed we see it also in the probabilities of applying to medical school. Let us examine a few other examples to help us understand both how it occurs and what we ought to do to allow us to make sensible inferences from such results.

On September 2, 1998 the *New York Times* reported evidence of high school grade inflation. They showed that a greater proportion of high school students were getting top grades while at the same time their SAT-Math scores had declined (see Table 2). Indeed, when we look at their table, the data seem to support this claim; at every grade level SAT scores have declined by 2 to 4 points over the decade of interest. Yet the

[1] MCAT is a test that consists of four parts – Verbal Reasoning, Physical Sciences, Biological Sciences and a *Writing Sample*. The Biological Sciences score is the one that correlates most highly with subsequent performance on Step 1 of the USMLE, and so we used it as the stratifying variable throughout our study. None of our conclusions would be changed if we used an amalgam of all parts of the test, but the interpretations could get more complex. Therefore henceforth when we use the term "MCAT" we mean "MCAT *Biological Sciences*."

Table 1
Selected medical school application and licensing statistics

Frequencies												
Last MCAT score All MCAT takers 1993–1995				Applied to medical school 1994–1996			Accepted at medical school 1994–1996			USMLE Step 1 test volumes 1996–1998		
MCAT-BS score	Black	White	Total	Black	White	Total	Black	White	Total	Black	White	Total
3 or less	1,308	1,168	2,476	404	238	642	8	1	9	6	1	7
4	1,215	2,094	3,309	482	557	1,039	52	10	62	39	10	49
5	1,219	3,547	4,766	582	1,114	1,696	202	45	247	116	36	152
6	1,269	5,289	6,558	752	1,983	2,735	417	163	580	256	140	396
7	1,091	6,969	8,060	748	3,316	4,064	518	636	1,154	338	589	927
8	1,234	11,949	13,183	868	6,698	7,566	705	2,284	2,989	537	2,167	2,704
9	702	13,445	14,147	544	8,628	9,172	476	4,253	4,729	340	4,003	4,343
10 or more	660	29,752	30,412	511	20,485	20,996	475	14,244	14,719	334	12,786	13,120
Total	**8,698**	**74,213**	**82,911**	**4,891**	**43,019**	**47,910**	**2,853**	**21,636**	**24,489**	**1,966**	**19,732**	**21,698**

Table 1
(*Continued*)

Selected conditional probabilities												
Probability of MCAT taker applying to medical school				Probability of MCAT taker being accepted to medical school			Probability of MCAT taker Taking USMLE Step 1			Probability of med school acceptee Taking USMLE Step 1		
MCAT-BS score	Black	White	Total	Black	White	Total	Black	White	Total	Black	White	Total
3 or less	0.31	0.20	0.26	0.01	0.00	0.00	0.00	0.00	0.00	0.75	0.78	
4	0.40	0.27	0.31	0.04	0.00	0.02	0.03	0.00	0.01	0.75	1.00	0.79
5	0.48	0.31	0.36	0.17	0.01	0.05	0.10	0.01	0.03	0.57	0.80	0.62
6	0.59	0.37	0.42	0.33	0.03	0.09	0.20	0.03	0.06	0.61	0.86	0.68
7	0.69	0.48	0.50	0.47	0.09	0.14	0.31	0.08	0.12	0.65	0.93	0.80
8	0.70	0.56	0.57	0.57	0.19	0.23	0.44	0.18	0.21	0.76	0.95	0.90
9	0.77	0.64	0.65	0.68	0.32	0.33	0.48	0.30	0.31	0.71	0.94	0.92
10 or more	0.77	0.69	0.69	0.72	0.48	0.48	0.51	0.43	0.43	0.70	0.90	0.89
Total	**0.56**	**0.58**	**0.58**	**0.33**	**0.29**	**0.30**	**0.23**	**0.27**	**0.26**	**0.69**	**0.91**	**0.89**

Table 2
A decade of high school grades and SAT scores: are students getting better or worse?

Grade average	Percentage of students getting grades		Average SAT Math scores		
	1988	1998	1988	1998	Change
A+	4%	7%	632	629	−3
A	11%	15%	586	582	−4
A−	13%	16%	556	554	−2
B	53%	48%	490	487	−3
C	19%	14%	431	428	−3
Overall average			**504**	**514**	**10**

From NY Times September 2, 1998.

article also reported that over the same time period (1988–1998) SAT-Math scores had in fact gone up by ten points.

How can the average SAT score increase by 10 points from 1988 to 1998, while at the same time decrease at every grade level? The key is the change in the percentages of children receiving each of the grades. Thus, although it is true that SAT-Math scores declined from 632 to 629 for A+ students, there are nearly twice as many A+ students in 1998. Thus in calculating the average score we weight the 629 by 7% in 1998 rather than by only 4%. The calculation of the average SAT score in a year partitioned by grade level requires both high school grades and SAT scores for students with those grades. As we will demonstrate shortly, we can make the anomaly disappear by holding the proportional mix fixed across the two groups.

As a third example, consider the results from the National Assessment of Educational Progress shown in Table 3. We see that 8th grade students in Nebraska scored 6 points higher in mathematics than their counterparts in New Jersey. Yet we also see that both white and black students do better in New Jersey. Indeed, all other students do better in New Jersey as well. How is this possible? Once again it is an example of Simpson's Paradox. Because a much greater percentage of Nebraska's 8th grade students (87%) are from the higher scoring white population than in New Jersey (66%), their scores contribute more to the total.

Given these results, we could ask, "Is ranking states on such an overall score sensible?" It depends on the question that these scores are being used to answer. If the question is something like "I want to open a business. In which state will I find a higher proportion of high-scoring math students to hire?" this unadjusted score is sensible. If, however, the question of interest is "I want to enroll my children in school. In which state are they likely to do better in math?" a different answer is required. If your children have a race (it doesn't matter what race), they are likely to do better in New Jersey. If questions of this latter type are the ones that are asked more frequently, it makes sense to adjust the total to reflect the correct answer. One way to do this is through the method of standardization, in which we calculate what each state's score would be if it were based upon a common demographic mixture. In this instance one sensi-

Table 3
NAEP 1992 8th grade Math scores

Other					
	State	White	Black	Non white	**Standardized**
Nebraska	**277**	281	236	259	**271**
New Jersey	**271**	283	242	260	**273**
Proportion of population					
Nebraska		87%	5%	8%	
New Jersey		66%	15%	19%	
Nation		69%	16%	15%	

ble mixture to use is that of the nation overall. Thus, after standardization the result obtained is the score we would expect each state to have if it had the same demographic mix as the nation. To create the standardized score for New Jersey we multiple the average score for each subgroup by their respective percentages in the nation, e.g., $(283 \times 0.69) + (242 \times 0.16) + (260 \times 0.15) = 273$. Because New Jersey's demographic mix is not very different from the national mix, its score is not affected much (273 instead of 271), whereas because of Nebraska's largely white population its score shrinks substantially (271 instead of 277).

Simpson's Paradox is illuminated through a clever graphic developed by Jeon et al. (1987) (and independently reinvented by Baker and Kramer, 2001). In Figure 1 the solid line represents what Nebraska's average score would be with any proportion of white students. The solid point at "87% white" shows what the score was with the actual percentage. Similarly, the dashed line shows what New Jersey's average score would be for any percentage of whites, with the unshaded point showing the actual percentage. We can readily see how Nebraska's average point is higher than New Jersey's. The unshaded rectangle represents what both states' averages would be with a hypothetical population of 69% white – the standardization mix. This plot shows that what particular mixture is chosen for standardization is irrelevant to the two state's relative positions, since the two states' lines are parallel.

The use of standardization is not limited to comparing different states with one another. Indeed it may be even more useful comparing a state with itself over time. If there is a change in educational policy (e.g., per pupil expenditure) standardization to the demographic structure of the state at some fixed point in time allows us to estimate the effect of the policy change uncontaminated by demographic shifts.

Now we can return to the data about MCAT examinees in Table 1 with greater understanding. Why is it that the overall rate for taking Step 1 is lower for blacks than for white examinees, when we see that the rate is higher for blacks (often markedly higher) at each MCAT score level? The overall rate of 23% for black students is caused by a combination of two factors: policy and performance. For many policy purposes it would be well if we could disentangle these effects. As demonstrated in the prior example, one path toward clarity lies in standardization. If we wish to compare the rates for black and

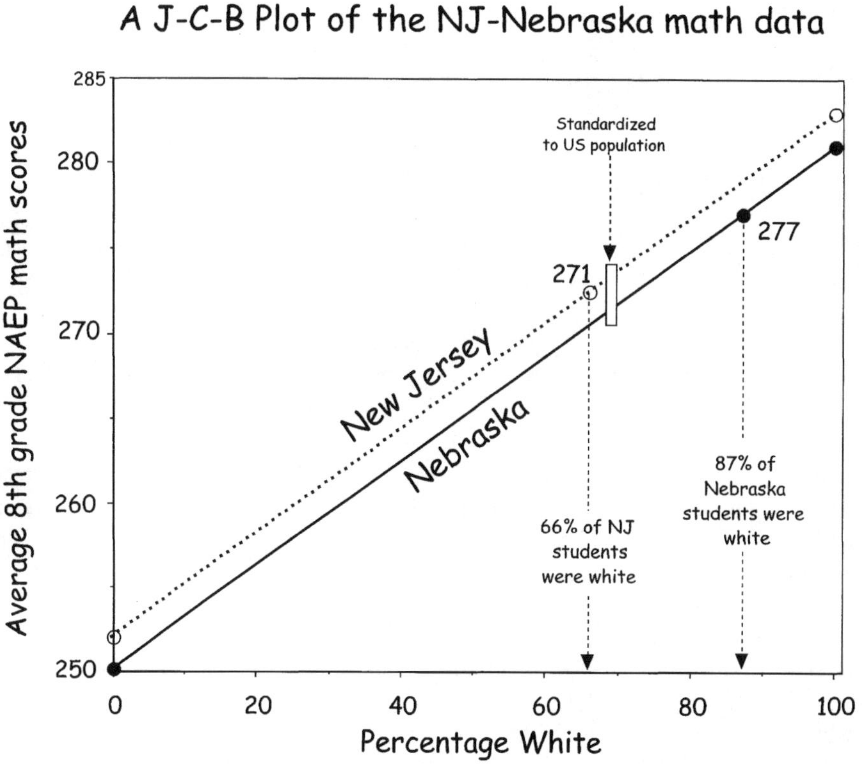

Fig. 1. A graph developed by Jeon et al. (1987) that illuminates the conditions for Simpson's Paradox as well as how standardization ameliorates it.

white students that current policy generates we must rid the summary of the effects of differential performance and estimate standardized rates. Standardized rates can be obtained by multiplying the Step 1 rates of each stratum of both black and white students by the score distribution of white students. Multiplying the two columns of Step 1 rates in Table 4 by the score distribution of whites (in bold) yields the two final columns, which when summed are the standardized rates; standardized to the white score distribution. Of course the white summary stays the same 27%, but the standardized Step 1 rate for black students is 41%. We can use this information to answer to the question:

> If black students scored the same on the MCAT as white students what proportion would go on to take Step 1?

Comparing the total Step 1 rates for blacks and whites after standardization reveals that if black and white candidates performed equally well on the MCAT, blacks would take Step 1 at a rate 54% higher than whites. The standardization process also allows for another comparison of interest. The difference between the standardized rate of 41% for blacks and the actual rate of 23% provides us with the effect of MCAT performance on Step 1 rates of black students. This occurs because white students are more heavily

Table 4
Distributions of Step 1 rates by ethnicity with standardized totals

MCAT score	Step 1 rates		Percentage of		Standardized Step 1 rates	
	Black	White	Black	White	Black	White
3 or less	1%	0%	15%	**2%**	0%	0%
4	3%	1%	14%	**3%**	0%	0%
5	10%	1%	14%	**5%**	1%	0%
6	20%	3%	15%	**7%**	1%	0%
7	31%	9%	13%	**9%**	3%	1%
8	44%	18%	14%	**16%**	7%	3%
9	48%	30%	8%	**18%**	9%	5%
10 or more	51%	43%	8%	**40%**	20%	17%
Total	23%	27%	23%		41%	27%

concentrated at high MCAT scores, which have a higher rate of taking Step 1. Standardization tells us that if black students had that same MCAT distribution their rate of taking Step 1 would almost double.

4. Kelley's Paradox

We now turn to a second statistical paradox that appears in the same data set. The score distributions of the two ethnic groups under consideration are shown in Figure 2. The score distributions are about one standard deviation apart. This result matches rather closely what is seen in most other standardized tests (e.g., SAT, NAEP, LSAT, GRE).[2]

The distribution of MCAT scores in the medical school population is shown as Figure 3. The means of these distributions are a little closer together, but because the individuals that make up these distributions are highly selected, they have somewhat smaller standard deviations (they are leptokurtic). The difference between them, in standard deviation units, is 18% larger.

As shown by the data that make up Table 1, a black candidate with the same MCAT score as a white candidate has a greater likelihood of admission to medical school. Often much greater; at MCAT scores of 5 or 6 a black candidate's probability of admission is about twelve times that of a white candidate. There are many justifications behind a policy that boosts the chances of being selected for medical school for black applicants. One of these is the expectation that students from less privileged social backgrounds who do well must be very talented indeed and hence will do better still when given the opportunity. This point of view was more fully expressed in the August 31, 1999 issue of the *Wall Street Journal*. In it an article appeared about a research project done

[2] For further general documentation see Bowen and Bok (1998). Details on NAEP are in Johnson and Zwick (1988); on SAT see http://www.collegeboard.com/prod_downloads/about/news_info/cbsenior/yr2003/pdf/table_3c.pdf; on GRE "*Sex, race, ethnicity and performance on the GRE General Test: 2003–2004*" (Author, 2003).

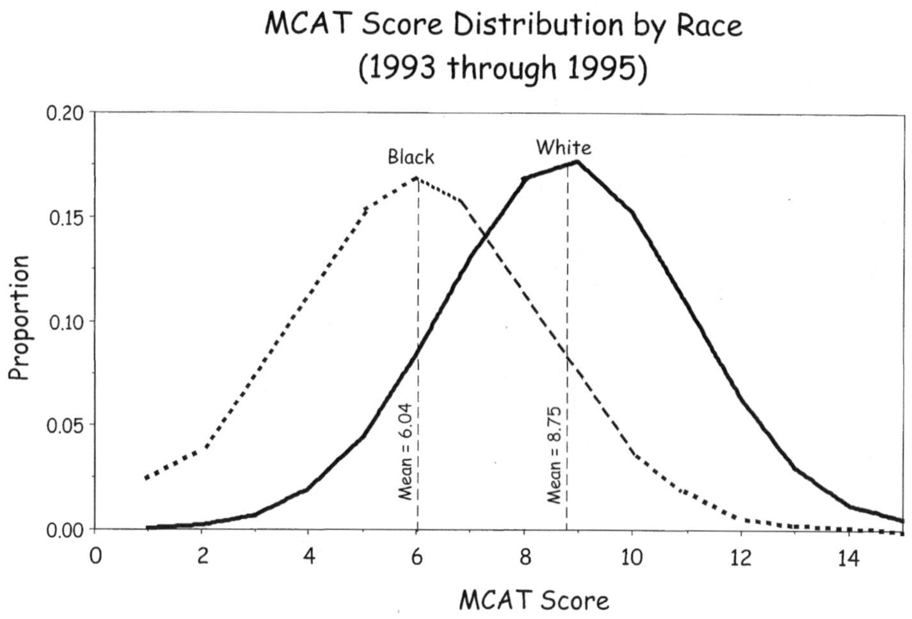

Fig. 2. MCAT score distributions by race aggregated for all examinees taking the exam for the last time between 1993 and 1995.

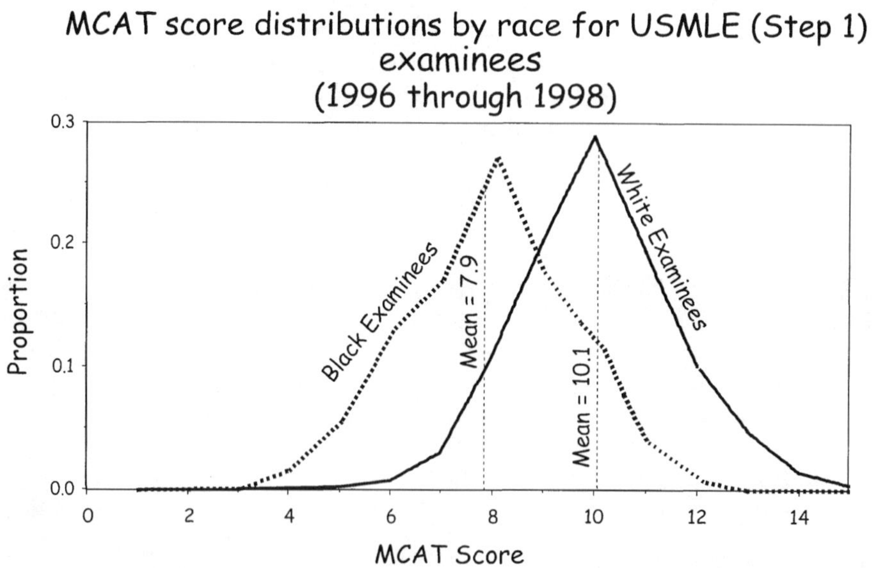

Fig. 3. MCAT score distributions by race aggregated for all medical students who took the USMLE Step 1 from 1995 through 1998.

under the auspices of the Educational Testing Service called "Strivers." The goal of "Strivers" was to aid colleges in identifying applicants (usually minority applicants) who have a better chance of succeeding in college than their test scores and high school grades might otherwise suggest. The basic idea was to predict a student's SAT score from a set of background variables (e.g., ethnicity, SES, mother's education, etc.) and characterize those students who do much better than their predicted value as "Strivers". These students might then become special targets for college admission's officers. In the newspaper interview the project's director, Anthony Carnevale said, "When you look at a Striver who gets a score of 1000, you're looking at someone who really performs at 1200". Harvard emeritus professor Nathan Glazer, in an article on Strivers in the September 27, 1999 *New Republic* indicated that he shares this point of view when he said (p. 28) "It stands to reason that a student from a materially and educationally impoverished environment who does fairly well on the SAT and better than other students who come from a similar environment is probably stronger than the unadjusted score indicates."

But before this intuitively appealing idea can be accepted it has a rather high theoretical hurdle to clear. To understand this hurdle and be able to attach its relevance to medical licensing, it is worthwhile to drop back a century and trace the origins of the fundamental issues underlying this proposal.

When we try to predict one event from another we always find that the variation in the prediction is smaller than that found in the predictor. Francis Galton (1889) pointed out that this always occurred whenever measurements were taken with imperfect precision and was what he called "regression toward the mean." This effect is seen in some historical father-child height data shown in Figure 4. Note how the father's heights vary over a 44-centimeter range (from 152 to 196 centimeters) while the prediction of their children's heights, shown by the dark dashed line, varies over only a 30-centimeter range (from 158 to 188 centimeters). What this means is that fathers that are especially tall are predicted to sire children that are tall but not as tall, and fathers who are short are predicted to have children that are short but not as short as their fathers. In this instance it is the imperfect relationship between father's and son's heights, rather than the imperfect precision of measurement, that gives rise to the regression effect.

While regression has been well understood by mathematical statisticians for more than a century, the terminology among appliers of statistical methods suggests that they either thought of it as a description of a statistical method or as only applying to biological processes. The economic statistician Frederick C. Mills (1924) wrote "the original meaning has no significance in most of its applications" (p. 394).

Stephen Stigler (1997, p. 112) pointed out that this was "a trap waiting for the unwary, who were legion." The trap has been sprung many times. One spectacular instance of a statistician getting caught was "in 1933, when a Northwestern University professor named Horace Secrist unwittingly wrote a whole book on the subject, *The Triumph of Mediocrity in Business*. In over 200 charts and tables, Secrist 'demonstrated' what he took to be an important economic phenomenon, one that likely lay at the root of the great depression: a tendency for firms to grow more mediocre over time." Secrist (1933) showed that the firms with the highest earnings a decade earlier were currently performing only a little better than average; moreover, a collection of the more poorly

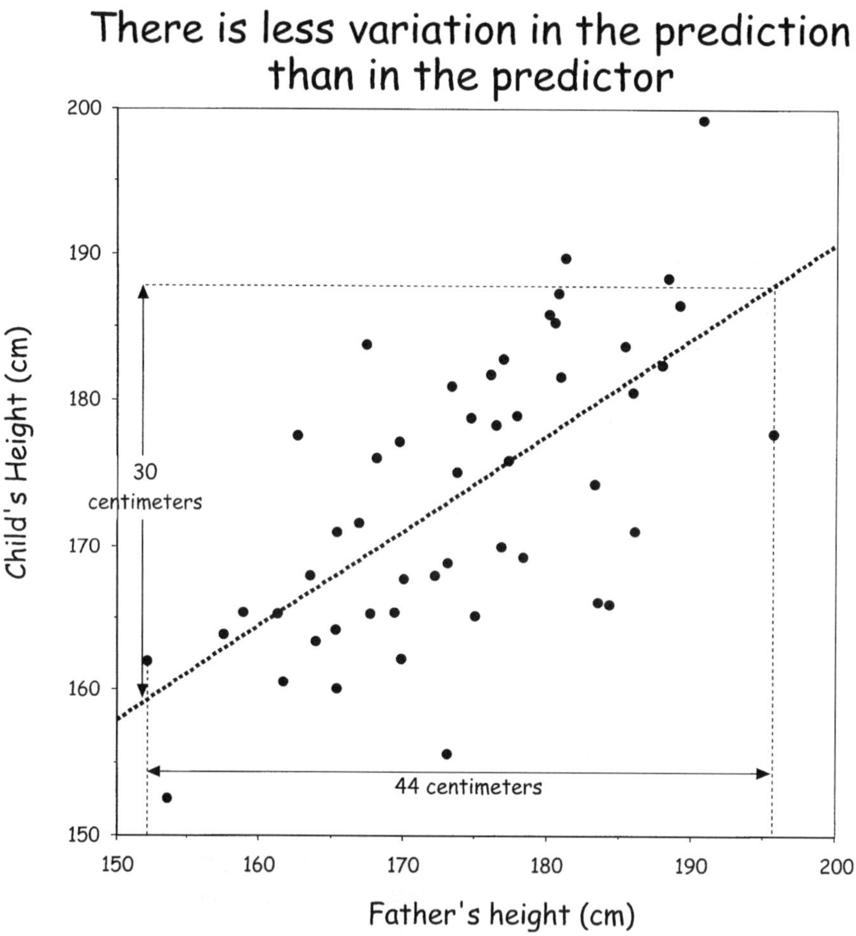

Fig. 4. The height of a child is roughly predictable from the height of its parent. Yet the distance between the shortest and tallest adults is greater than distance between the predicted statures of the most extreme children.

performing firms had improved to only slightly below average. These results formed the evidence supporting the title of the book. Harold Hotelling (1933), in a devastating review, pointed out that the seeming convergence Secrist obtained was a "statistical fallacy, resulting from the method of grouping." He concluded that Secrist's results "prove nothing more than that the ratios in question have a tendency to wander about." He then demonstrated that the firms that had the highest earnings now were, on average, only slightly above average ten years earlier. And firms with the lowest earnings now were only slightly worse, on average, than the average ten years previous. Thus showing the reverse of what Secrist claimed. Hotelling's argument was not original with him. Galton not only showed that tall fathers had sons who were, on average shorter than they were, but he also showed that very tall sons had fathers who were shorter than they were.

It is remarkable, especially considering how old and well-known regression effects are, how often these effects are mistaken for something substantive. Although Secrist himself was a professor of statistics, Willford I. King (1934) wrote a glowing review of Secrist's book, was president of the American Statistical Association! This error was repeated in a book by W.F. Sharpe (1985), a Nobel laureate in economics (p. 430) who ascribed the same regression effect Secrist described to economic forces. His explanation of the convergence, between 1966 and 1980, of the most profitable and least profitable companies was that "ultimately economic forces will force the convergence of profitability and growth rates of different firms." This continuing misunderstanding led Milton Friedman (1992), yet another Nobel laureate in economics, to try to set his colleagues straight. He discussed one of the principal theses of a book by Baumol et al. (1989), and its review by Williamson (1991); that the rates of growth of various countries tend to converge. Friedman agreed with their thesis but not their explanation, which, he pointed out, was statistical, not economic.

Truman Kelley (1927) described a specific instance of a regression formula of great importance in many fields, although it was proposed for use in educational testing. It shows how you can estimate an examinee's true score from his/her observed score on a test. "True score" is psychometric shorthand for the average of the person's observed scores if they took essentially identical tests[3] over and over again forever. Kelley's equation relates the estimated true score ($\hat{\tau}$) to the observed score (x). It tells us that the best estimate is obtained by regressing the observed score in the direction of the mean score (μ) of the group that the examinee came from. The amount of the regression is determined by the reliability (ρ) of the test. Kelley's equation is

$$\hat{\tau} = \rho(x) + (1 - \rho)\mu. \tag{1}$$

Note how Kelley's equation works. If a test is completely unreliable ($\rho = 0$), as would be the case if each examinee's score was just a random number, the observed score would not count at all and the estimated true score is merely the group mean. If the test scores were perfectly reliable ($\rho = 1$) there would be no regression effect at all and the true score would be the same as the observed score. The reliability of virtually all tests lies between these two extremes and so the estimated true score will be somewhere between the observed score and the mean.

A diagram aids intuition about how Kelley's equation works when there are multiple groups. Shown in Figure 5 are the distributions of scores for two groups of individuals, here called Group 1 (lower scoring group) and Group 2 (higher scoring group). If we observed a score x, midway between the means of the two groups the best estimate of the true score of the individual who generated that score depends on which group that person belonged to. If that person came from Group 1 we should regress the score downward; if from Group 2 we should regress it upward.

[3] "Essentially identical tests" is shorthand for what psychometricians' call "parallel forms" of the test. This means tests that are constructed of different questions but span the same areas of knowledge, are equally difficult, and are equally well put together. In fact, as one part of the formal definition is the notion that if two tests were truly parallel a potential examinee would be completely indifferent as to which form was actually presented.

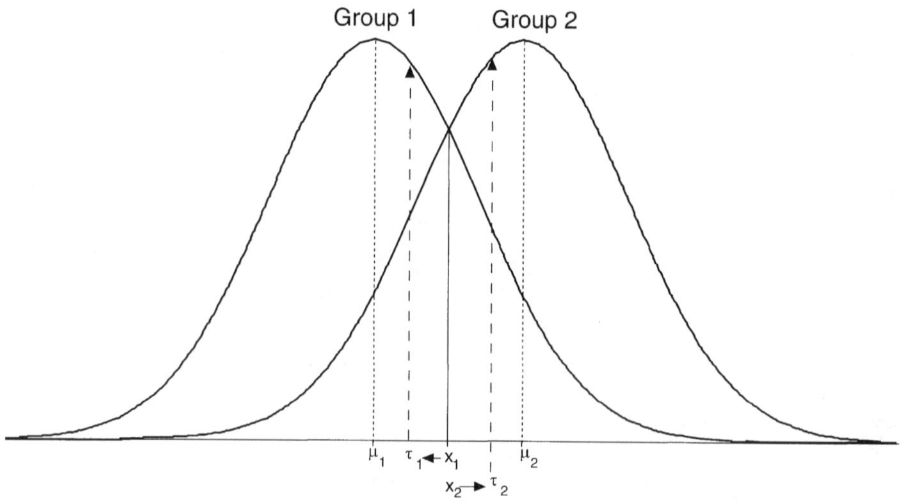

Fig. 5. A graphical depiction of Kelley's equation for two groups. The two distributions and their means are shown. Also indicated is how the true scores are regressed when two identical observed scores come from each of the two different score distributions.

The regression effect occurs because there is some error in the observed score. The average error is defined to be zero, and so some errors will be positive and some negative. Thus if someone from a low scoring group has a high score we can believe that to some extent that person is the recipient of some positive error, which is not likely to reappear upon retesting, and so we regress their score downward. Similarly, if someone from a high scoring group has an unusually low score, we regress that score upward.

So far this is merely an equation – a mathematical tautology. What is the paradox? Webster defines a paradox as a statement that is opposed to common sense and yet is true. So long as Kelley's equation deals solely with abstract groups named 1 and 2, no paradox emerges. But consider the similarity of Figure 5 with Figures 2 and 3 and consider the logic of "Strivers" which suggests explicitly that if we find someone from Group 1 with a high score, despite their coming from an environment of intellectual and material deprivation we strongly suspect their true ability ought to be considered as being somewhat higher. Similarly, someone who comes from a more privileged background, but who scores low, leads us to suspect a lack of talent and hence ought to be rated lower still. The underlying notion of "Strivers" points in the opposite direction of what would be expected through the application of Kelley's equation. This is the source of the paradox.

Harvard statistician, Alan Zaslavsky (2000) in a letter that appeared in the statistics magazine *Chance* tried to salvage the strivers idea with a more statistical argument. He did not question the validity of Kelley's equation, when it matched the situation. But he suggested that we must determine empirically what is the correct distribution toward which we regress the observed score. Does a person selected from the lower distribution remain a part of that group after being selected? Or does the very act of being selected obviate past associations? He described a metaphorical race in which

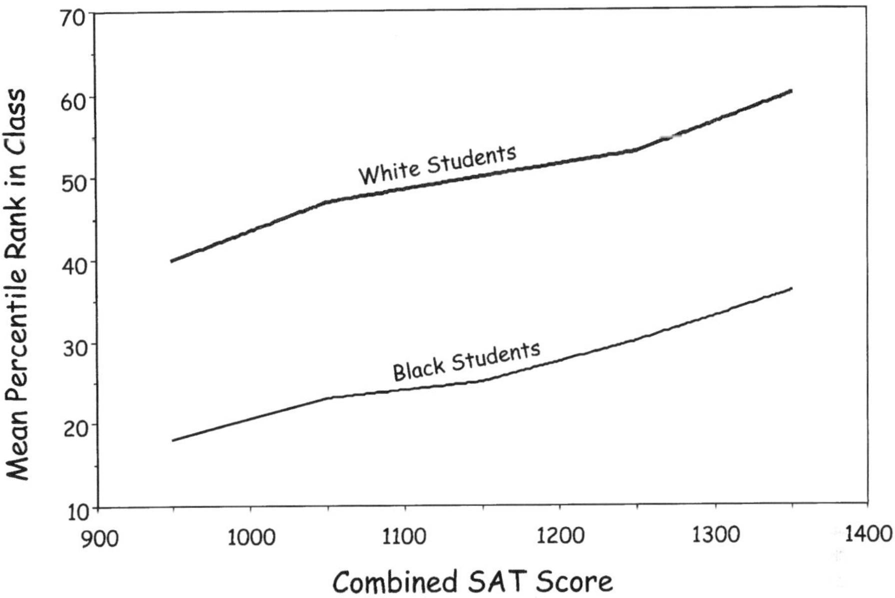

Fig. 6. An accurate revision of Figure 3.10 from Bowen and Bok (1998) showing that at all levels of SAT score Black students performance in college courses are much lower than white students with matched SAT scores. This bears out the prediction made by Kelley's equation.

one set of participants was forced to carry heavy weights. But after they were selected, they rid themselves of the weights and then ran much faster than other participants who had the same initial speed but who had not been carrying weights.

Thus the issue, as Zaslavsky views it, is an empirical one. Is social background like a weight that can be shed with proper help? Or is it more like height, a characteristic one is pretty much stuck with?

One empirical test of Kelley's equation within this context was carried out by William Bowen and Derek Bok (1998, Figure 3.10) in their exhaustive study of the value of affirmative action. They prepared a figure (analogous to Figure 6) that shows that black students' rank in college class is approximately 25 percentile points lower than white students with the same SAT scores. In this metric, the effect is essentially constant over all SAT scores, even the highest. This confirms, at least in direction, the prediction described in Figure 5; the performance in college of students with the same SAT score is different in a way predicted by their group membership.

In a more thorough analysis Ramist et al. (1994) used data from more than 46,000 students, gathered at 38 different colleges and universities to build a prediction model of college performance based on precollegiate information. One analysis (there were many others) predicted first year college grade point average from SAT score, from High School Grade Point Average (HS-GPA), and from both combined. They then recorded the extent to which each ethnic group was over or under predicted by the model. An extract of their results (from their Table 8) is shown in Table 5. The metric reported is grade points, so that a one-point difference corresponds to one grade level (a B to

Table 5
A summary of Ramist et al. (1994) results showing the size and direction of the errors in predictions for various models

Predictor	Ethnic group Asian American	White	Hispanic	Black
HS-GPA	0.02	0.03	−0.24	−0.35
SAT	0.08	0.01	−0.13	−0.23
HS-GPA & SAT	0.04	0.01	−0.13	−0.16
Sample sizes	3,848	36,743	1,599	2,475

a C, for example). The entries in this table indicate the extent to which students were over-predicted by the model based on the variables indicated. Thus the entry "0.02" for Asian Americans for a prediction based on just their high school grades means that Asian Americans actually did very slightly (.02 of a grade level) better than their high school grades predicted. And the "−0.35" for Black Americans means that they did about a third of a point worse than their grades predicted. Note that while SAT scores also over-predict minority college performance, the extent of the error they make is somewhat smaller.

The results from these studies are very clear; the common regression model over-predicts non Asian minority populations. This result matches the regression phenomenon we described as Kelley's paradox. And it matches it in amount as well as direction. Moreover, these results are not the outcome of some recent social change, but have been observed for decades (Linn, 1982; Reilly, 1973).

The licensing of physicians

If Kelley's equation holds for medical students we should expect that the scores for black medical students on Step 1 should be lower than those for white students who had the same MCAT score. Figure 7 carries the principal results of our study. The error bars around each point are one standard error of the difference of the means in each direction (adjusted by the Bonferroni inequality to deal with the eight comparisons being made). Hence if the bars do not overlap the difference in the associated means is statistically significant well beyond the nominal levels.

As is evident, at each MCAT score white medical students score higher on Step 1 than matched black medical students. The shaded horizontal line stretching across the figure represents the range of passing scores used during the time period 1996 through 1998. The unusual data point for white medical students with MCAT scores of "3 or less" is not an error. It represents a single student who switched into undergraduate science courses late in her collegiate career but graduated from Wellesley with a 3.6 GPA. Her medical school gambled that despite her low MCAT score she would be a good bet. Obviously the bet paid off.

The data shown in Figure 7 match the direction predicted by Kelley's equation, but what about the amount? Figure 8 is similar to Figure 7 except that we have ungrouped

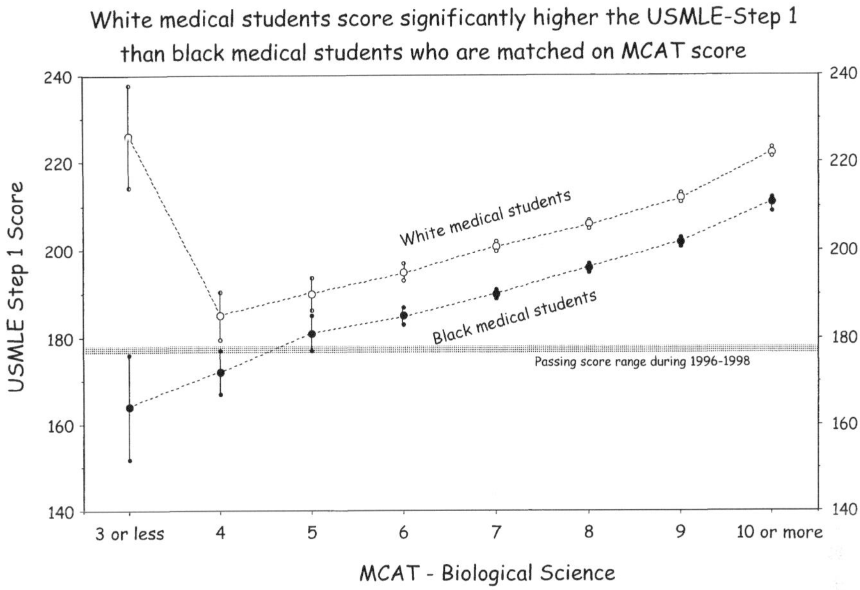

Fig. 7. Analysis of USMLE showing that at all MCAT score levels Black medical students performance on Step 1 are much lower than white medical students with matched MCAT scores. This bears out the prediction made by Kelley's equation.

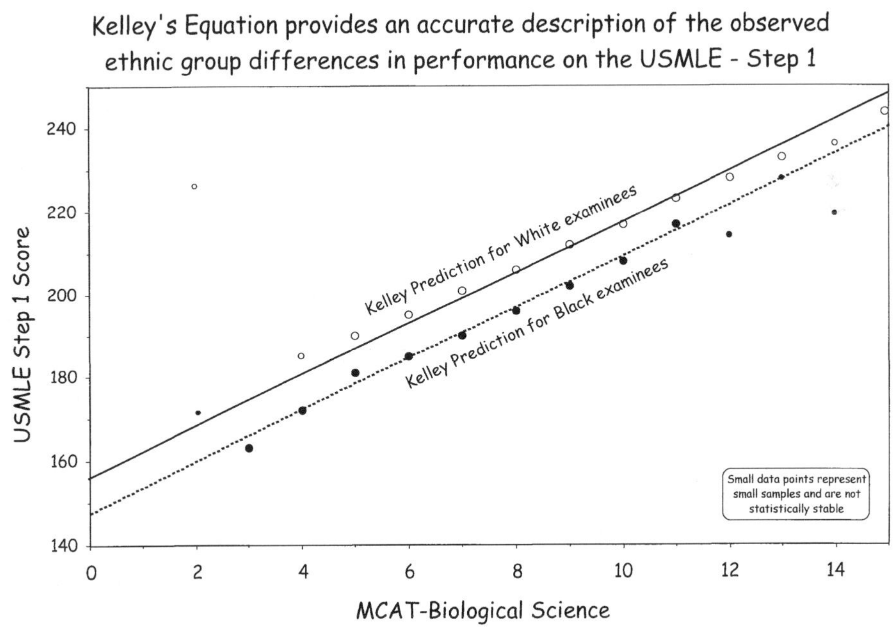

Fig. 8. A repeat of the results shown in Figure 7 with the Kelley predictions superimposed. This bears out the prediction made by Kelley's equation in both direction and size.

the data points at the ends of the distribution (some of the data points thus depicted are extremely unstable, being based on very few individuals) to present as unvarnished a view as possible. We have reduced the physical size of the data points that represent very few individuals. The lines drawn on the plot are the predictions made by substituting group means and the correlation between MCAT and USMLE into Kelley's equation. This yields two lines about ten points apart. We view this as relatively conclusive evidence that ethnicity, at least in this instance, is not like Zaslavsky's metaphorical weights, but rather it is a characteristic that is not apparently cast off when an individual enters medical school.

Unfortunately, academic ability in the instances we have examined is more like height than Zaslavsky's metaphorical weights. Neither Mr. Carnevale nor Mr. Glazer is correct. Thus when you look at a Striver who gets a score of 1000, you're probably looking at someone who really performs at 950. And, alas, a Striver is probably *weaker* than the unadjusted score indicates.

This result should be distressing for those who argue that standard admission test scores are unfair to students coming from groups whose performance on such admission tests is considerably lower than average (Freedle, 2003; Mathews, 2003) and that they under-predict the subsequent performance of such students. Exactly the opposite of that is, in fact true, for there is ample evidence that students from groups who are admitted with lower than usual credentials on average, do worse than expected.[4] The extent to which they perform more poorly is almost entirely predictable from Kelley's equation.

5. Lord's Paradox

We have observed that the performance of the two groups on the outcome variable, the USMLE Step 1 score, depends on both performance on the predictor variable, the MCAT score, and on group membership. Faced with this observation it is natural to ask:

How much does group membership matter in measuring the effect of medical school?

What does this question mean? One plausible interpretation would be to examine an individual's rank among an incoming class of medical students, and then examine her rank after receiving a major portion of her medical education. If her rank did not change we could conclude that the effect of medical school was the same for that individual as it was for the typical medical student. If we wish to measure the effect of medical school on any group we might compare the average change in ranks for that group with

[4] Due to the nested structure of the data, we also used a multi-level modeling approach to determine whether the regression effect was due to, or lessened by, differences between medical schools. We found that 10% of the variation in Step 1 scores was due to differences between medical schools and the MCAT-Biological Sciences score explained 76% of this between-school variation. Although we found significant variation in the regression effect for blacks across schools the fixed effect closely approximated the regression effect reported in this chapter. Furthermore, holding MCAT score constant, black students were predicted to have lower Step 1 scores in all 138 schools. The percentage of blacks in the sample from each school explained 18% of the variance in this effect and indicated that schools with more blacks have a smaller (albeit substantively trivial) regression effect.

another. But this is not the only plausible approach. Alternatively we might use the pre-medical school ranks as a covariate and examine the differences between the groups' average medical school rank after adjusting for the pre-medical school rank. How we might do this and how we interpret the results is the subject of this section.[5]

We begin the investigation by:

(a) Drawing a random sample from the USMLE Step 1 takers of 200 white examinees and 200 black examinees.
(b) Then we rank these 400 examinees on both their MCAT scores and their Step 1 scores.
(c) Next we subtract each examinee's rank on the Step 1 from that person's rank on the MCAT, and
(d) Calculate the average difference for white and for black examinees.

We found that white examinees' ranks improved, on average, about 19 places. This was, of course, balanced by a decline of 19 places in rank among black examinees, or a total differential effect of 38.

But, as we mentioned before, taking the difference in ranks is not the only way to estimate this effect. Alternatively we could use the MCAT rank as a covariate and look at the ranks of the individuals on the adjusted USMLE Step 1 (the residuals on Step 1 ranks after a linear adjustment for MCAT score). When we did exactly this we found that white examinees' Step 1 ranks, after adjusting for MCAT scores, improved, on average, about 9 places, with black examinees' ranks declining the same 9 places, for a total differential effect of 18.

The results of these two analyses were substantially different. Which is the right answer? This question was posed previously by Fred Lord (1967) in a two-page paper that clearly laid out what has since become known as Lord's paradox. He did not explain it. The problem appears to be that the analysis of covariance cannot be relied upon to properly adjust for uncontrolled preexisting differences between naturally occurring groups. A full explanation of the paradox first appeared fully sixteen years later (Holland and Rubin, 1983) and relies heavily on Rubin's model for causal inference (Rubin, 1974).

The paradox, as Lord described it, was based on the following hypothetical situation:

> A large university is interested in investigating the effects on the students of the diet provided in the university dining halls .... Various types of data are gathered. In particular, the weight of each student at the time of his arrival in September and his weight the following June are recorded. (p. 304)

Lord framed his paradox in terms of the analyses of two hypothetical statisticians who come to quite different conclusions from the data in this example.

The first statistician calculated the difference between each student's weight in June and in September, and found that the average weight gain in each dining room was zero. This result is depicted graphically in Figure 9 with the bivariate dispersion within each

[5] We use ranks rather than test scores to circumvent the problems generated by the two different tests being scored on different scales and having very different reliabilities. It is not that such an alternative path could not be taken, but we felt that for this illustration it would be cleaner, simpler and more robust to assumptions if we stuck with analyses based on the order statistics.

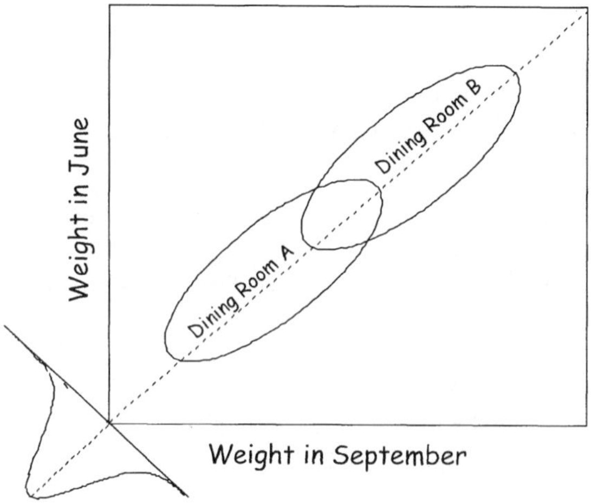

Fig. 9. A graphical depiction of Lord's Paradox showing the bivariate distribution of weights in two dining rooms at the beginning and end of each year augmented by the 45° line (the principal axis).

dining hall shown as an oval. Note how the distribution of differences is symmetric around the 45° line (the principal axis for both groups) that is shown graphically by the distribution curve reflecting the statistician's findings of no differential effect of dining room.

The second statistician covaried out each student's weight in September from his/her weight in June and discovered that the average weight gain was greater in Dining Room B than Dining Room A. This result is depicted graphically in Figure 10. In this figure the two drawn-in lines represent the regression lines associated with each dining hall. They are not the same as the principal axes because the relationship between September and June is not perfect. Note how the distribution of adjusted weights in June is symmetric around each of the two different regression lines. From this result the second statistician concluded that there was a differential effect of dining room, and that the average size of the effect was the distance between the two regression lines.

So, the first statistician concluded that there was no effect of dining room on weight gain and the second concluded there was. Who was right? Should we use change scores or an analysis of covariance? To decide which of Lord's two statistician's had the correct answer requires that we make clear exactly what was the question being asked. The most plausible question is causal, "What was the causal effect of eating in Dining Room B?" But causal questions are always comparative[6] and the decision of how to estimate the standard of comparison is what differentiates Lord's two statisticians. Each statistician made an untestable assumption about the subjunctive situation of what would have been a student's weight in June had that student not been in the dining room of interest. This devolves directly from the notion of a causal effect being the difference between what happened under the treatment condition vs. what happened under the control condition.

[6] The comedian Henny Youngman's signature joke about causal inference grew from his reply to "How's your wife?" He would then quip, "Compared to what?"

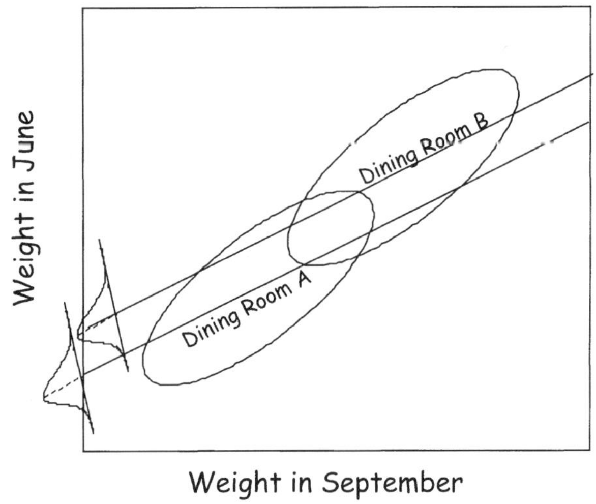

Fig. 10. A graphical depiction of Lord's Paradox showing the bivariate distribution of weights in two dining rooms at the beginning and end of each year augmented by the regression lines for each group.

The fundamental difficulty with causal inference is that we can never observe both situations. Thus we must make some sort of assumption about what would have happened had the person been in the other group. In practice we get hints of what such a number would be through averaging and random assignment. This allows us to safely assume that, on average, the experimental and control groups are the same.

In Lord's set-up the explication is reasonably complex. To draw his conclusion the first statistician makes the implicit assumption that a student's control diet (whatever that might be) would have left the student with the same weight in June as he had in September. This is entirely untestable. The second statistician's conclusions are dependent on an allied, but different, untestable assumption. This assumption is that the student's weight in June, under the unadministered control condition, is a linear function of his weight in September. Further, that the same linear function must apply to all students in the same dining room.

How does this approach help us to untangle the conflicting estimates for the relative value of medical school for the two racial groups? To do this requires a little notation and some algebra.

The elements of the model[7] are:

1. A population of units, P.
2. An "experimental manipulation," with levels T and C and its associated indicator variable, S.
3. A subpopulation indicator, G.
4. An outcome variable, Y.
5. A concomitant variable, X.

[7] This section borrows heavily from Holland and Rubin (1983, pp. 5–8) and uses their words as well as their ideas.

The purpose of the model is to allow an explicit description of the quantities that arise in three types of studies:

(a) Descriptive studies.
(b) Uncontrolled causal studies.
(c) Controlled causal studies.

A *descriptive study* has no experimental manipulation so there is only one version of Y and X and no treatment indicator variable S.

Controlled and *uncontrolled causal studies* both have experimental manipulations and differ only in the degree of control that the experimenter has over the treatment indicator, S. In a controlled causal study, the values of S are determined by the experimenter and can depend on numerous aspects of each unit (e.g., subpopulation membership, values of covariates) but not on the value of Y, since that is observed after the values of S are determined by the experimenter. In an uncontrolled causal study the values of S are determined by factors that are beyond the experimenter's control. Critical here is the fact that in a controlled study S can be made to be statistically independent of Y_C and Y_T whereas in an uncontrolled causal study this is not true.

The causal effect of T on Y (relative to C) for each unit in P is given by the difference $Y_T - Y_C$. The average causal effect of T versus C on Y in P is $E(Y_T - Y_C)$, which equals $E(Y_T) - E(Y_C)$. This shows us how the unconditional means of Y_T and Y_C over P have direct causal interpretations. But since T and C are usually not observable on the same unit, $E(Y_T)$ and $E(Y_C)$ are not typically observable.

In a causal study, the value of Y that is observed on each unit is Y_S, so that when $S = T$, Y_T is observed and when $S = C$, Y_C is observed. The expected value of Y for the "treatment group" is $E(Y_T|S = T)$ and for the "control group" is $E(Y_C|S = C)$. Without random assignment, there is no reason to believe that $E(Y_T)$ should equal $E(Y_T|S = T)$, or that $E(Y_C)$ should equal $E(Y_C|S = C)$. Hence neither $E(Y_T|S = T)$ nor $E(Y_C|S = C)$ have direct causal interpretation.

Consider that $E(Y_T|S = T)$ and $E(Y_T)$ are related through

$$E(Y_T) = E(Y_T|S = T)P(S = T) + E(Y_T|S = C)P(S = C). \tag{2}$$

There is the obvious parallel version connecting $E(Y_C|S = C)$ with $E(Y_C)$. The second term of (2) is not observable. This makes explicit the basis of our earlier assertion about the shortcomings of $E(Y_T|S = T)$ and $E(Y_C|S = C)$ for making direct causal interpretations.

Note that Eq. (2) involves the average value of Y_T, among those units exposed to C. But $E(Y_T|S = C)$ and its parallel $E(Y_C|S = T)$ can never be directly measured except when Y_T and Y_C can both be observed on all units. This is what Holland and Rubin (1983, p. 9) term "the fundamental problem of causal inference."

With this model laid out, let us return to the problem of measuring the differential effect of medical school.

Study design

P: 400 medical students in the years specified.
T: Went to medical school.
C: Unknown.

Variables measured

G: Student race ($W = 1$, $B = 2$).
X: The rank of a student on the MCAT.
Y: The rank of a student on Step 1 of the USMLE.

This layout makes clear that the control condition was undefined – no one was exposed to C ($S = T$ for all students) – and so any causal analysis must make untestable assumptions. As is perhaps obvious now, the two different answers we got to the same question must have meant that we made two different untestable assumptions. This will become visible by making the inference explicit.

The causal effect of medical school for black and white students is

$$D_i = E(Y_T - Y_C | G = i), \quad i = 1, 2, \tag{3}$$

and so the difference of average causal effects is

$$D = D_1 - D_2. \tag{4}$$

This can be expressed in terms of individual subpopulation averages,

$$D = \left[E(Y_T|G = 1) - E(Y_C|G = 1)\right] - \left[E(Y_T|G = 2) - E(Y_C|G = 2)\right]. \tag{5}$$

We can profitably re-arrange this to separate the observed Y_T from the unobserved Y_C

$$D = \left[E(Y_T|G = 1) - E(Y_T|G = 2)\right] - \left[E(Y_C|G = 1) - E(Y_C|G = 2)\right]. \tag{6}$$

The first approach estimated the effect of medical school by just looking at the difference in the ranks on MCAT and Step 1. Doing so made the (entirely untestable) assumption that an individual's response to the control condition, whatever that might be, is given by his/her rank on the MCAT

$$Y_C = X \tag{7}$$

yielding,

$$E(Y_C|G = i) = E(X|G = i). \tag{8}$$

The second approach estimated the effect of medical school by using the students' rank on the MCAT as a covariance adjustment, which corresponds to the following two conditional expectations:

$$E(Y_T|X, G = i), \quad i = 1, 2, \tag{9}$$

and the mean, conditional, improvement in rank in group i at X is

$$U_i(X) = E(Y_T - X|X, G = i), \quad i = 1, 2. \tag{10}$$

Hence, the difference in these conditional ranks at X is

$$U(X) = U_1(X) - U_2(X). \tag{11}$$

The second analysis assumes that the conditional expectations in (9) are linear and parallel. Thus we can write

$$E(Y_T|X, G = i) = a_i + bX, \quad i = 1, 2. \tag{12}$$

Substituting into (10) yields

$$U_i(X) = a_i + (b - 1)X, \quad i = 1, 2. \tag{13}$$

And hence (11) simplifies to

$$U(X) = a_1 - a_2. \tag{14}$$

The second approach correctly interprets $U(X)$ as the average amount that a white student's ($G = 1$) rank will improve over a black student ($G = 2$) of equal MCAT score. This is descriptively correct, but has no direct causal interpretation since U is not directly related to D. To make such a connection we need to make the untestable assumption, related to (7) that

$$Y_C = a + bX. \tag{15}$$

Where b is the common slope of the two within-groups regression lines in (12). This allows the interpretation of $U(X)$ as the difference in the causal effects D in Eq. (4).

Both of these assumptions seem to stretch the bounds of credulity, but (15) seems marginally more plausible. However deciding this issue was not our goal. Instead we wished to show how subtle an argument is required to unravel this last paradox in the investigation of group differences. The interested reader is referred to Holland and Rubin (1983) or Wainer (1991) for a fuller description of how Rubin's Model of causal inferences helps us to understand this subtle paradox.

6. Conclusion

"What we don't know won't hurt us,
it's what we do know that ain't"

Will Rogers

This chapter, and the research behind it, has two goals. The first is to publicize more broadly the pitfalls that await those who try to draw inferences from observed group differences. The second is to provide analytic tools to allow the construction of bridges over those pitfalls.

Group differences must be examined if we wish to evaluate empirically the efficacy of modifications in policy. But such comparisons, made naively, are very likely to lead us astray.

Ridding ourselves of Simpson's Paradox through the use of standardization is straightforward. But we must always remember that there may be another, unnoticed, variable that could reverse things again. Inferences must be made carefully. The only reasonably certain way to be sure that stratification by some unknown variable will not reverse your inference is to have random assignment to groups. When assignment is not random the possibility of Simpson's Paradox is always lurking in the background.[8]

Kelley's Paradox is not so much a summary statistic pointing in the wrong direction, as it is an indication that our intuition has been improperly trained. The fact that economists fall into this trap more often than others may reflect on their training, but why this should be the case is a phenomenon that we will not try to explain.

Lord's Paradox is the newest of this triad. It occurs when data analysts use their favorite method to assess group differences without careful thought about the question they are asking. It is, by far, the most difficult paradox to disentangle and requires clear thinking. It also emphasizes how the assessment of group differences often entails making untestable assumptions. This too should give us pause when we try to draw strong conclusions.

Acknowledgements

We are grateful to the Association of American Medical Colleges for allowing us to use their MCAT data. More specifically we would like to thank Ellen Julian for her help in providing us with the information we required. In addition, this entire project was prompted by Don Melnick, who not only suggested that we do it but who has supported our inquiry. We are not unaware of the sensitive nature of some of the questions we are asking and fully appreciate Don's support and helpful comments on an earlier draft of this paper. David Swanson and Douglas Ripkey generously provided us with a file of the USMLE results paired with MCAT scores; obviously a key element in this investigation. An earlier version of this chapter was read and commented on by our colleagues Ron Nungester and Brian Clauser; we thank them for helping us excise errors and fuzzy thinking and a shortened version appeared in the *American Statistician* **58**(2), 117–123 (Wainer and Brown, 2004). We are also grateful to the sharp eyes of an anonymous referee who found several errors that had escaped our notice. This research was paid for by NBME and we thank our employer for providing such fascinating opportunities. Of course, the opinions expressed here are ours, as are any errors we may have made along the way. This work is collaborative in every respect and the order of authorship is random.

[8] Benjamin Disraeli (1804–1881) was twice prime minister of England (1868, 1874–1880). At an earlier time in his career he was an outspoken critic of Sir Robert Peel's (1788–1850) free-trade policies, and to support his criticism he offered data defending the Corn Laws (1845). Peel offered counter data that justified his desire to repeal them. The two sets of data seemed contradictory, and Disraeli, not knowing about Simpson's Paradox (or the use of standardization to correct it), exclaimed out of frustration, "Sir, there are lies, damn lies and statistics."

References

Author (2003). *Sex, Race, Ethnicity and Performance on the GRE General Test: 2003–2004*. Educational Testing Service, Princeton, NJ.

Baker, S.G., Kramer, B.S. (2001). Good for women, good for men, bad for people: Simpson's paradox and the importance of sex-specific analysis in observational studies. *Journal of Women's Health and Gender-Based Medicine* **10**, 867–872.

Baumol, W.J., Blackman, S.A.B., Wolff, E.N. (1989). *Productivity and American Leadership: The Long View*. MIT Press, Cambridge and London.

Bowen, W.G., Bok, D. (1998). *The Shape of the River*. Princeton University Press, Princeton, NJ.

Freedle, R.O. (2003). Correcting the SAT's ethnic and social-class bias: A method for reestimating SAT scores. *Harvard Educational Review* **73** (1), 1–43.

Friedman, M. (1992). Do old fallacies ever die? *Journal of Economic Literature* **30**, 2129–2132.

Galton, F. (1889). *Natural Inheritance*. Macmillan, London.

Holland, P.W., Rubin, D.B. (1983). On Lord's paradox. In: Wainer, H., Messick, S. (Eds.), *Principals of Modern Psychological Measurement*. Lawrence Erlbaum Associates, Hillsdale, NJ, pp. 3–25.

Hotelling, H. (1933). Review of *The triumph of mediocrity in business* by Secrist H. *Journal of the American Statistical Association* **28**, 463–465.

Jeon, J.W., Chung, H.Y., Bae, J.S. (1987). Chances of Simpson's paradox. *Journal of the Korean Statistical Society* **16**, 117–125.

Johnson, E.G., Zwick, R.J. (1988). The NAEP Technical Report, Educational Testing Service, Princeton, NJ.

Kelley, T.L. (1927). *The Interpretation of Educational Measurements*. World Book, New York.

King, W.I. (1934). Review of *The triumph of mediocrity in business* by Secrist H. *Journal of Political Economy* **42**, 398–400.

Linn, R. (1982). Ability testing: Individual differences and differential prediction. In: Wigdor, A.K., Garner, W.R. (Eds.), *Ability Testing: Uses, Consequences and Controversies, Part II*. In: *Report of the National Academy of Sciences Committee on Ability Testing*. National Academy Press, Washington, DC.

Lord, F.M. (1967). A paradox in the interpretation of group comparisons. *Psychological Bulletin* **68**, 304–305.

Mathews, J. (2003). The bias question. *The Atlantic Monthly*, 130–140.

Mills, F.C. (1924). *Statistical Methods. Applied to Economics and Business*. Henry Holt, New York.

Ramist, L., Lewis, C., McCamley-Jenkins, L. (1994). *Student Group Differences in Predicting College Grades: Sex, Language and Ethnic Group*. The College Board, New York.

Reilly, R.R. (1973). A note on minority group bias studies. *Psychological Bulletin* **80**, 130–133.

Rubin, D.B. (1974). Estimating causal effects of treatments in randomized and nonrandomized studies. *Journal of Educational Psychology* **66**, 688–701.

Secrist, H. (1933). *The Triumph of Mediocrity in Business*. Bureau of Business Research, Northwestern University, Evanston, IL.

Sharpe, W.F. (1985). *Investments*, 3rd ed. Prentice-Hall, Englewood Cliffs, NJ.

Simpson, E.H. (1951). The interpretation of interacting contingency tables. *Journal of the Royal Statistical Society B* **13**, 238–241.

Stigler, S. (1997). Regression toward the mean, historically considered. *Statistical Methods in Medical Research* **6**, 103–114.

Wainer, H. (1991). Adjusting for differential base-rates: Lord's paradox again. *Psychological Bulletin* **109**, 147–151.

Wainer, H. (2000). Kelley's paradox. *Chance* **13** (1), 47–48.

Wainer, H., Brown, L. (2004). Two statistical paradoxes in the interpretation of group differences: Illustrated with medical school admission and licensing data. *The American Statistician* **58**, 117–123.

Williamson, J.G. (1991). Productivity and American leadership: A review article. *Journal of Economic Literature* **29** (1), 51–68.

Yule, G.U. (1903). Notes on the theory of association of attributes of statistics. *Biometrika* **2**, 121–134.

Zaslavsky, A. (2000). On Kelly's paradox. *Chance* **13** (3), 3.

Handbook of Statistics, Vol. 26
ISSN: 0169-7161

DOI: 10.1016/S0169-7161(06)26029-2

29

Meta-Analysis

Larry V. Hedges

1. Introduction

The growth of the research enterprise in many areas of science has led to a large body of related research studies. The sheer volume of research related to many topics of scientific or policy interest poses a problem of how to organize and summarize these findings in order to identify and exploit what is known and focus research on promising areas (see Garvey and Griffith, 1971). This problem has arisen in fields as diverse as physics, chemistry, experimental biology, medicine, and public health. In each of these fields, as in the social sciences, accumulation of quantitative research evidence has led to the development of systematic methods for the quantitative synthesis of research (see Cooper and Hedges, 1994). Although the term meta-analysis was coined to describe these methods in the social sciences (Glass, 1976), the methods used in other fields are remarkably similar to those in the social sciences (Cooper and Hedges, 1994; Hedges, 1987).

Meta-analysis refers to an analysis of the results of several studies for the purposes of drawing general conclusions. Meta-analysis involves describing the results of each study via a numerical index of effect size (such as a correlation coefficient, a standardized mean difference, or an odds ratio) and then combining these estimates across studies to obtain a summary. The specific analytic techniques involved will depend on the question the meta-analytic summary is intended to address. Sometimes the question of interest concerns the typical or average study result. For example in studies that measure the effect of some treatment or intervention, the average effect of the treatment is often of interest (see, e.g., Smith and Glass, 1977). When the average effect is of interest, effect size estimates are typically combined across studies to obtain an estimate of the average effect size.

In other cases the degree of variation in results across studies will be of primary interest. When the variability of treatment effect sizes is of interest, it is often characterized via a test for heterogeneity of effects or an estimate of the variance of effect parameters across studies. In yet other cases, the primary interest is in the factors that are related to study results. For example, meta-analysis is often used to identify the contexts in which a treatment or intervention is most successful or has the largest effect (see, e.g., Cooper, 1989). In such cases analyses typically examine whether study-level covariates (moderator variables reflecting study context) are related to effect size via meta-analytic models that are analogues to the analysis of variance or regression analysis.

In any of these analysis situations, two different families of statistical methods (each corresponding to a somewhat different inference model) are available. The fixed effects statistical methods correspond to a model of inference to the studies that have been observed. The mixed effects statistical methods correspond to a model of inference to a hypothetical population of studies from which the observed studies are a random sample.

The term meta-analysis is sometimes used to connote the entire process of quantitative research synthesis. More recently, it has begun to be used specifically for the statistical component of research synthesis. This chapter deals exclusively with that narrower usage of the term to describe statistical methods only. However it is crucial to understand that in research synthesis, as in any research, statistical methods are only one part of the enterprise. Statistical methods cannot remedy the problem of data that are of poor quality. Excellent treatments of the non-statistical aspects of research synthesis are available in Cooper (1989), Cooper and Hedges (1994), and Lipsey and Wilson (2001).

2. Effect sizes

Effect sizes are quantitative indexes that are used to express the results of a study in meta-analysis. That is, effect sizes reflect the magnitude of the association between variables of interest in each study. There are many different effect sizes and the effect size used in a meta-analysis should be chosen so that it represents the results of a study in a way that is easily interpretable and is comparable across studies. In a sense, effect sizes should put the results of all studies "on a common scale" so that they can be readily interpreted, compared, and combined.

It is important to distinguish the effect size estimate in a study from the effect size parameter (the true effect size) in that study. In principle, the effect size estimate will vary somewhat from sample to sample that might be obtained in a particular study. The effect size parameter is in principle fixed. One might think of the effect size parameter as the estimate that would be obtained if the study had a very large (essentially infinite) sample size, so that the sampling variation is negligible.

The choice of an effect size index will depend on the design of the studies, the way in which the outcome is measured, and the statistical analysis used in each study. Most of the effect size indexes used in the social sciences will fall into one of three families of effect sizes: the standardized mean difference family, the odds ratio family, and the correlation coefficient family.

2.1. The standardized mean difference

In many studies of the effects of a treatment or intervention that measure the outcome on a continuous scale, a natural effect size is the standardized mean difference. The standardized mean difference is the difference between the mean outcome in the treatment group and the mean outcome in the control group divided by the within group standard deviation. That is the standardized mean difference parameter is

$$\delta = \frac{\mu^T - \mu^C}{\sigma},$$

where μ^T is the population mean in the treatment group, μ^C is the population mean outcome in the control group, and σ is the population within-group standard deviation of the outcome. This effect size is easy to interpret since it is just the treatment effect in standard deviation units. The usual sample estimate of δ is

$$d = \frac{\overline{Y}^T - \overline{Y}^C}{S},$$

where $\overline{Y}^T$ is the sample mean of the outcome in the treatment group, $\overline{Y}^C$ is the sample mean of the outcome in the control group, and S is the within-group standard deviation of the outcome. The standardized mean difference d is known to be approximately normally distributed with variance

$$v = \frac{n^T + n^C}{n^T n^C} + \frac{d^2}{2(n^T + n^C)},$$

where n^T and n^C are the treatment and control group sample sizes, respectively (see Hedges and Olkin, 1985). Note that this variance can be computed from a single observation of the effect size if the sample sizes of the two groups within a study are known. Several variations of the standardized mean difference are also sometimes used as effect sizes (see Rosenthal, 1994).

2.2. *The log odds ratio*

In many studies of the effects of a treatment or intervention that measure the outcome on a dichotomous scale, a natural effect size is the log odds ratio. The log odds ratio is just the log of the ratio of the odds of a particular one of the two outcomes (the target outcome) in the treatment group to the odds of that particular outcome in the control group. That is, the log odds ratio parameter is

$$\omega = \log\left(\frac{\pi^T/(1-\pi^T)}{\pi^C/(1-\pi^C)}\right) = \log\left(\frac{\pi^T(1-\pi^C)}{\pi^C(1-\pi^T)}\right),$$

where π^T and π^C are the population proportions in the treatment and control groups, respectively, that have the target outcome. The corresponding sample estimate of ω is

$$\hat{\omega} = \log(OR) = \log\left(\frac{p^T/(1-p^T)}{p^C/(1-p^C)}\right) = \log\left(\frac{p^T(1-p^C)}{p^C(1-p^T)}\right),$$

where p^T and p^C are the proportion of the treatment and control groups, respectively that have the target outcome. The log odds ratio is widely used in the analysis of data that have dichotomous outcomes and is readily interpretable by researchers who frequently encounter this kind of data. It also has the same meaning across studies so it is suitable for combining (see Fleiss, 1994). As in the case of the standardized mean difference, the log odds ratio is approximately normally distributed in large samples and its variance is approximately

$$v = \frac{1}{n^T p^T} + \frac{1}{n^T(1-p^T)} + \frac{1}{n^C p^C} + \frac{1}{n^C(1-p^C)},$$

where n^T and n^C are the treatment and control group sample sizes, respectively.

There are several other indexes in the odds ratio family, including the risk ratio (the ratio of proportion having the target outcome in the treatment group to that in the control group or p^T/p^C) and the risk difference (the difference between the proportion having a particular one of the two outcomes in the treatment group and that in the control group or $p^T - p^C$). For a discussion of effect size measures for studies with dichotomous outcomes, including the odds ratio family of effect sizes, see Fleiss (1994).

2.3. *The correlation coefficient*

In many studies of the relation between two continuous variables, the correlation coefficient ρ is a natural measure of effect size. Often this correlation is transformed via the Fisher z-transform

$$\zeta = \frac{1}{2}\log\left(\frac{1+\rho}{1-\rho}\right)$$

in carrying out statistical analyses. The corresponding sample estimate of the correlation parameter is r, and the sample estimate of ζ is the z-transform of r given by

$$z = \frac{1}{2}\log\left(\frac{1+r}{1-r}\right),$$

which has a approximately a normal distribution with variance

$$v = \frac{1}{n-3},$$

where n is the sample size of the study, and it is used in the same way as are the variances of the standardized mean difference and log odds ratio to obtain confidence intervals.

The statistical methods for meta-analysis are quite similar, regardless of the effect size measure used. Therefore, in the rest of this chapter we do not describe statistical methods that are specific to a particular effect size index, but describe them in terms of a generic effect size measure T_i. We assume that the T_i are normally distributed about the corresponding θ_i with known variance v_i. That is, we assume that

$$T_i - N(\theta_i, v_i), \quad i = 1, \ldots, k.$$

This assumption is very nearly true for effect sizes such as the Fisher z-transformed correlation coefficient and standardized mean differences. However for effect sizes such as the untransformed correlation coefficient, or the log-odds ratio, the results are not exact, but remain true as large sample approximations. For a discussion of effect size measures for studies with continuous outcomes, see Rosenthal (1994) and for a treatment of effect size measures for studies with categorical outcomes see Fleiss (1994).

3. The problem of dependent estimates

It often occurs that several effect size estimates are associated with a study and these estimates are not statistically independent. There are three common situations in which this occurs. First, each individual may be measured using several different measures of outcome (e.g., different measures of mathematics achievement), and an effect size may be computed using data on each of the several measures. Second, the same individuals may be measured at several different points in time (e.g., just after treatment and then at various followup periods) using the same measuring instruments, and an effect size may be computed at each timepoint. Third, in studies with multiple treatment groups, an effect size may be computed by comparing each of the several treatment groups with the same control group, and because the estimates so computed share a common control group, they are not independent.

In each of the cases mentioned, effect sizes are usually sorted into groups so that there is only one effect size per study and thus all of the effect size estimates in a particular group are independent. For example, groups might be effect sizes based on the same measure (e.g., mathematics computation), the same followup interval (e.g., approximately 6 months after treatment), or the same particular treatment (e.g., the standard [not enhanced] variation of treatment versus control). The summary of independent effect size estimates across groups can be accomplished via standard meta-analytic methods (see, e.g., Cooper and Hedges, 1994).

However, reviewers often want to carry out analyses that involve combining data across groups – analyses that involve effect size estimates that are not independent. Two types of combined analyses are the most common. One analysis involves estimating a mean effect size across all types of outcomes or treatment variations (e.g., across mathematics computation and problem solving, across several different followup intervals, or across several variations of the treatment). Such analyses are often desired to answer the most global questions about whether the treatment had an impact. A second kind of analysis involves estimating differential effects of treatment. This type of analysis is often desired to answer questions about whether the treatment has a bigger effect on some outcomes than others, at some followup intervals than others or whether certain variations of treatment have bigger effects.

Three strategies are commonly used for handling effect size data involving nonindependence of this type. The first strategy is to explicitly model the correlations among the effect sizes using multivariate methods (see, e.g., Hedges and Olkin, 1985; Olkin and Gleser, 1994; Kalaian and Raudenbush, 1996). This strategy is the most elegant, providing the most efficient estimates of effects and accurate results of significance tests. Unfortunately, because it requires knowledge of the covariance structure of the effect size estimates (which in turn requires knowledge of the dependence structure of the raw data in each study), the information needed to implement this strategy is rarely available and therefore this strategy is rarely used.

The second strategy is to first compute a within-study summary from nonindependent effect size estimates and then summarize the (independent) summaries across studies. For example, to compute the overall mean effect size (across effect sizes of all different types from all different studies) one might first compute the mean effect size within each study as a kind of synthetic effect size for the study. The effect sizes within a study will not be independent, but the synthetic effect sizes (the study-average effect sizes) from different studies will be independent.

Similarly, one might compute the difference between the average effects of two types (different outcome measures, different followup intervals, or different treatment types) by first computing the difference between the two effects within each study as another type of synthetic effect size. The effect sizes within a study are not independent, but the synthetic effect sizes (the difference between effect sizes of the two types) from different studies will be independent.

In either case, because the synthetic effect sizes are independent, they can be combined across studies using conventional methods for meta-analysis. There is, however, one problem in using conventional meta-analytic methods to combine the synthetic effect sizes: the standard errors of these synthetic effect sizes depend on the correlation structure of the within-study effect sizes from which they are computed, and thus are typically not known.

The third strategy is to ignore the fact that some of the effect size estimates are not independent, and use the same meta-analytic procedures that would have been used of the effect size data were independent. While this might seem to be misguided, it may not be too misleading if relatively few studies report more than one effect size. Moreover, we will show in this chapter that it may lead to conservative results for tests of the difference between average effects of different types.

Although the first strategy is the most elegant, the second and third strategies are by far the most widely used strategies for dealing with nonindependent effect size estimates in meta-analysis. The purpose of this chapter is to examine these two strategies and provide some guidance for their use. We derive some properties of estimates and statistical tests that may be useful for conducting analyses. Finally we suggest how to carry out sensitivity analyses that may be useful in interpreting results.

4. Fixed effects analyses

We begin by examining procedures for dealing with nonindependent effect sizes in fixed effects statistical procedures.

4.1. Model and notation

Begin by supposing that a total of K studies, each with m outcomes, are possible. Denote the effect size estimate in the jth outcome in the ith study by T_{ij} and the corresponding effect size parameter by θ_{ij}. Then the data layout could be as simple as

$$T_{ij}, \quad i = 1, \ldots, K; \; j = 1, \ldots, m.$$

Unfortunately, practical meta-analytic problems seldom lead to data layouts this simple. In most cases, not all studies measure all outcomes, leading to complex patterns of missing data. Often, however, there is a subset of studies, all of which estimate effect sizes on the same subset of outcomes (and perhaps some other outcomes as well).

While many statisticians would advocate that meta-analyses combining or comparing across outcomes should be confined to the subset of studies that estimate effect sizes on the same set of outcomes, there are sometimes compelling reasons to include as many studies or outcomes as possible. Consequently, we consider a data layout in this chapter that includes the possibility that not all studies measure every outcome.

Suppose that effect sizes are estimated on N different outcomes and that k_j studies provide estimates of effect size on the jth outcome and let $k = k_1 + k_2 + \cdots + k_N$ be the total number of effect sizes. Assume however that there is a subset of m outcomes that are of primary interest (e.g., the most frequently measured outcomes in the studies at hand) and a subset $p \leqslant \min\{k_1, k_2, \ldots, k_m\}$ of the studies provide effect size estimates on all m outcomes and that these estimates are not independent. We lose no generality in assuming that these are the first p studies and the first m outcomes. Note that this data layout assumes that there is a core of p studies that provide effect size data on all outcomes, and that other studies provide effect size data on only a single outcome. Of course, if all studies provide effect size data on all outcomes, then $N = m$ and $p = k_1 = \cdots = k_m = K$.

The idealized data layout with a total of K studies and N outcomes (without missing data) is illustrated below, along with the observed number of effects in each study and the observed number of studies having each effect on each outcome (k_j for the jth outcome) and the total number of effects for each study (n_i for the ith study):

Studies	Possible effects							Observed number	Unweighted means
1	T_{11}	T_{12}	...	T_{1m}	...	T_{1N}		n_1	$\overline{T}_1$
2	T_{21}	T_{22}	...	T_{2m}	...	T_{2N}		n_2	$\overline{T}_2$
$\vdots$	$\vdots$	$\vdots$	$\vdots$	$\vdots$	$\vdots$	$\vdots$		$\vdots$	$\vdots$
p	T_{p1}	T_{p2}	...	T_{pm}	...	T_{pN}		n_p	$\overline{T}_p$
$\vdots$	$\vdots$	$\vdots$	$\vdots$	$\vdots$	$\vdots$	$\vdots$		$\vdots$	$\vdots$
K	T_{K1}	T_{K2}	...	T_{Km}	...	T_{KN}		n_K	$\overline{T}_K$
Observed number	k_1	k_2	...	k_m	...	k_N		$\sum n_i = \sum k_j$	
Weighted means	$\overline{T}_{\bullet 1}$	$\overline{T}_{\bullet 2}$	...	$\overline{T}_{\bullet m}$	...	$\overline{T}_{\bullet N}$		$\overline{T}_{\bullet\bullet}$	

EXAMPLE. An example of a real meta-analysis with 13 studies and 2 outcomes was reported by Becker (1990). In this example, the studies were randomized experiments on the effects of coaching for the Scholastic Aptitude Test (SAT), and the outcomes were effect sizes expressed as standardized mean change scores on the mathematics and verbal scales. The data layout is given below. In this example, there are effects on two outcomes, so $m = 2$. The first three studies have effect size estimates for both of the two outcomes of primary interest, so $p = 3$. Study 4 also measures an effect on outcome one, but all of the other studies (studies 5 through 13) measure effects only on the second outcome. Thus $n_1 = 2$, and $n_2 = 2$, and $n_3 = 2$, but $n_4 = \cdots = n_{13} = 1$. Similarly, there are 4 studies that measure effects on outcome one and 12 studies that measure effects on outcome two, so $k_1 = 4$ and $k_2 = 12$. The analysis might focus on only the first three studies that have effects on both outcomes, involving a total of 6 effect size estimates. It is also possible that the analyst wishes to consider the average effect size on all of the outcomes on which effects are measured (a total of 16 estimates). Similarly, it is possible that the analyst wishes to consider differences between the mean effect sizes on the first two outcomes using data from all studies that estimated effects on either of those outcomes (a total of 16 effect estimates).

Studies	Effects		Observed number	Unweighted means
1	T_{11}	T_{12}	$n_1 = 2$	$\overline{T}_1$
2	T_{21}	T_{22}	$n_2 = 2$	$\overline{T}_2$
3	T_{31}	T_{32}	$n_3 = 2$	$\overline{T}_3$
4	T_{41}	–	$n_4 = 1$	$\overline{T}_4$
5	–	T_{52}	$n_5 = 1$	$\overline{T}_5$
6	–	T_{62}	$n_6 = 1$	$\overline{T}_6$
7	–	T_{72}	$n_7 = 1$	$\overline{T}_7$
8	–	T_{82}	$n_8 = 1$	$\overline{T}_8$
9	–	T_{92}	$n_9 = 1$	$\overline{T}_9$
10	–	$T_{10,2}$	$n_{10} = 1$	$\overline{T}_{10}$
11	–	$T_{11,2}$	$n_{11} = 1$	$\overline{T}_{11}$
12	–	$T_{12,2}$	$n_{12} = 1$	$\overline{T}_{12}$
13	–	$T_{13,2}$	$n_{13} = 1$	$\overline{T}_{13}$
Observed number	$k_1 = 4$	$k_2 = 12$	$\sum n_i = \sum k_j = 16$	
Weighted means	$\overline{T}_{\bullet 1}$	$\overline{T}_{\bullet 2}$	$\overline{T}_{\bullet\bullet}$	

We describe the sampling distribution of the T_{ij} in two parts. Let the standard error of T_{ij} be σ_{ij}. For the ith study, we assume that all of the observed effect size estimates associated with that study have the multivariate normal distribution, so that the mean of T_{ij} is θ_{ij}, the variance of T_{ij} is σ_{ij}^2, and the covariance of T_{ia} and T_{ib} is $\rho_{iab}\sigma_{ia}\sigma_{ib}$. However, for the first p studies and the first m outcomes, the correlation structure can be described more comprehensively. For the T_{ij} where $i \leqslant p$ and $j \leqslant m$, the T_{ij} are not statistically independent of the $T_{ij'}$. Describe the dependence among these estimates by saying that the vector $\boldsymbol{T} = (T_{11}, \ldots, T_{p1}, T_{12}, \ldots, T_{p2}, \ldots, T_{pm})'$ has a multivariate normal distribution with mean vector $\boldsymbol{\theta} = (\theta_{11}, \ldots, \theta_{p1}, \theta_{12}, \ldots, \theta_{pm})'$ and covariance matrix

$$\boldsymbol{\Sigma} = \begin{pmatrix} \boldsymbol{\Sigma}_{11} & \boldsymbol{\Sigma}_{12} & \cdots & \boldsymbol{\Sigma}_{1m} \\ \boldsymbol{\Sigma}_{12} & \boldsymbol{\Sigma}_{22} & \cdots & \boldsymbol{\Sigma}_{2m} \\ \vdots & \vdots & \ddots & \vdots \\ \boldsymbol{\Sigma}_{1m} & \boldsymbol{\Sigma}_{2m} & \cdots & \boldsymbol{\Sigma}_{mm} \end{pmatrix}, \tag{1}$$

where $\boldsymbol{\Sigma}_{\boldsymbol{aa}}$ is a $p \times p$ diagonal matrix given by $\boldsymbol{\Sigma}_{\boldsymbol{aa}} = \text{diag}(\sigma_{a1}^2, \ldots, \sigma_{ap}^2)$ each $\boldsymbol{\Sigma}_{\boldsymbol{ab}}$ is a $p \times p$ diagonal matrix given by $\boldsymbol{\Sigma}_{\boldsymbol{ab}} = \text{diag}(\sigma_{a1}\sigma_{b1}\rho_{1ab}, \ldots, \sigma_{ap}\sigma_{bp}\rho_{pab})$, where ρ_{iab} is the correlation between T_{ia} and T_{ib}.

This distributional assumption is the large sample approximation typically used to develop the theory of statistical tests used in meta-analysis. In univariate meta-analytic situations this corresponds to assuming that the (independent) studies have effect size estimates that are normally distributed with known variances. For example the standardized mean difference and z-transformed correlation coefficient (and their multivariate analogues) meet this assumption reasonably well.

4.2. *Analyses involving synthetic effect sizes*

Consider the strategy of combining effect size estimates within each study into averages or differences within studies (synthetic effect sizes) before combining them across studies to estimate an overall effect size or a difference between effect sizes. This strategy has the advantage that the synthetic effect size estimates from different studies are independent. The difficulty presented in using such synthetic effect size estimates is that, unless the covariances of the estimates combined into the synthetic estimate are known, it is usually difficult to compute an accurate variance of the composite. Consequently, analysts usually use a crude estimate of the variance of the synthetic effect size estimate (typically a conservative over estimate of its variance). The combined analysis of such estimates using standard meta-analytic methods (and crude approximations of the variances) typically involves weighted combinations with weights that are crude approximations to the inverse of the actual variance of the estimate. In this section we derive some theorems on the behavior of such estimates and in the next sections apply the theorem to two commonly used types of synthetic effect sizes.

Let $\widehat{T}_1, \ldots, \widehat{T}_K$ be a set of synthetic estimates of the effect size parameters $\theta_1, \ldots, \theta_K$, let $\sigma_1^2, \ldots, \sigma_K^2$ be the actual variances of the estimates, and let $\hat{\sigma}_1^2, \ldots, \hat{\sigma}_K^2$ be the estimates of the variances (which may be substantially biased). If the effect size parameters are homogeneous, that is if $\theta_1 = \cdots = \theta_K = \theta$, the standard fixed effects estimate of θ

$$\widehat{T}_{\bullet} = \frac{\sum_{i=1}^K w_i \widehat{T}_i}{\sum_{i=1}^K w_i}, \tag{2}$$

where $w_i = 1/\hat{\sigma}_i^2$, will be an unbiased estimate of θ (see, e.g., Shadish and Haddock, 1994). However the usual estimate of the variance (the square of the standard error) computed in the usual analysis

$$v_{\bullet} = \left(\sum_{i=1}^K w_i \right)^{-1}, \tag{3}$$

will be incorrect. In this case the exact variance of the weighted mean $\widehat{T}_\bullet$ is stated in the fact below.

FACT 1. The variance of $\widehat{T}_\bullet$ defined in (2) is

$$V(\widehat{T}_\bullet) = \frac{\sum_{i=1}^{K} w_i^2 \sigma_i^2}{(\sum_{i=1}^{K} w_i)^2} = v_\bullet \left(\frac{\sum_{i=1}^{K} w_i \left(\frac{\sigma_i^2}{\hat{\sigma}_i^2}\right)}{\sum_{i=1}^{K} w_i} \right), \tag{4}$$

where $w_i = 1/\hat{\sigma}_i^2$.

Thus the ratio

$$R = \frac{V(\widehat{T}_\bullet)}{v_\bullet} = \left(\sum_{i=1}^{K} w_i a_i \right) \Big/ \sum_{i=1}^{K} w_i \tag{5}$$

of the true variance $V(\widehat{T}_\bullet)$ to $v_\bullet$ is a weighted mean of the $a_i = \sigma_i^2/\hat{\sigma}_i^2$. Note that if $\hat{\sigma}_i^2$ is chosen so that it is larger than σ_i^2, the true variance of $\widehat{T}_\bullet$ is less than $v_\bullet$, and the amount by which it is less is the (weighted) average of degree to which each individual variance is smaller than the actual variance.

A test for heterogeneity of effects across studies is a test of the hypothesis

$$H_0\colon \theta_1 = \cdots = \theta_K.$$

The usual test for heterogeneity of effects across studies, when applied to the composite effect size estimates, uses the statistic

$$Q = \sum_{i=1}^{K} w_i (\widehat{T}_i - \widehat{T}_\bullet)^2, \tag{6}$$

where in this case, $w_i = 1/\hat{\sigma}_i^2$. The test rejects the null hypothesis when Q exceeds the $100(1-\alpha)$ percent point of a chi-square distribution with $(K-1)$ degrees of freedom. When approximate variances are used to compute weights (and $\widehat{T}_\bullet$), Q will not have the usual chi-squared distribution with $(K-1)$ degrees of freedom when the null hypothesis is true. The general form of the distribution of Q is given in the following theorem, which can be used to study the effects on homogeneity tests of using weights that are not the reciprocals of the variance of the estimates.

THEOREM 1. *The sampling distribution of Q defined in* (6) *is given approximately by $Q/g \sim \chi_h^2$. In other words Q is distributed approximately as g times a chi-square with h degrees of freedom, where $g = K_2/2K_1$, $h = 2K_1^2/K_2$, and K_1 and K_2 are the first two cumulants of Q given by*

$$K_1 = \sum_{i=1}^{K} w_i \sigma_i^2 - \frac{(\sum_{i=1}^{K} w_i^2 \sigma_i^2)}{\sum_{i=1}^{K} w_i}$$

and

$$K_2 = 2\sum_{i=1}^{K} w_i^2\sigma_i^4 - \frac{4[\sum_{i=1}^{K} w_i^3\sigma_i^4]}{\sum_{i=1}^{K} w_i} + \frac{2(\sum_{i=1}^{K} w_i^2\sigma_i^2)^2}{(\sum_{i=1}^{K} w_i)^2},$$

where $w_i = 1/\hat{\sigma}_i^2$.

PROOF. To obtain the approximate sampling distribution of Q, we use a result of Box (1954), which gives the approximate sampling distribution of quadratic forms in dependent normal variables. Write $Q = \boldsymbol{T}'\boldsymbol{M}\boldsymbol{T}$, for a $K \times K$ matrix $\boldsymbol{M}$, and $\mathbf{T}' = (\widehat{\boldsymbol{T}}_1, \dots, \widehat{\boldsymbol{T}}_K)$. Note that $\boldsymbol{T}$ has a K-dimensional multivariate normal distribution with covariance matrix $\boldsymbol{V} = \mathrm{diag}(\sigma_1^2, \dots, \sigma_K^2)$. Theorem 3.1 in Box (1954) says that Q is distributed as approximately a constant g times chi-square with h degrees of freedom, where g and h are functions of the first two cumulants of Q (K_1 and K_2, respectively), specifically $g = K_2/2K_1$ and $h = 2K_1^2/K_2$. A standard result on the distribution of quadratic forms gives the first two cumulants of Q as $K_1 = \mathrm{tr}(\boldsymbol{V}\boldsymbol{M})$ and $K_2 = 2\,\mathrm{tr}(\boldsymbol{V}\boldsymbol{M}\boldsymbol{V}\boldsymbol{M})$, where $\mathrm{tr}(\boldsymbol{X})$ is the trace of the square matrix $\boldsymbol{X}$ (see, e.g., Rao, 1973). Writing $\boldsymbol{M} = \boldsymbol{D} - c_w\boldsymbol{w}\boldsymbol{w}'$ with $\boldsymbol{D} = \mathrm{diag}(w_1, \dots, w_K)$, $\boldsymbol{w} = (w_1, \dots, w_K)'$, and $c_w = 1/(w_1 + \dots + w_K)$, we obtain the cumulants as $K_1 = \mathrm{tr}(\boldsymbol{V}\boldsymbol{D}) - c_w\,\mathrm{tr}(\boldsymbol{w}'\boldsymbol{V}\boldsymbol{w})$ and $K_2 = 2\,\mathrm{tr}(\boldsymbol{V}\boldsymbol{D}\boldsymbol{V}\boldsymbol{D}) - 4c_w\,\mathrm{tr}(\boldsymbol{w}'\boldsymbol{V}\boldsymbol{D}\boldsymbol{V}\boldsymbol{w}) + 2c_w^2\,\mathrm{tr}(\boldsymbol{w}'\boldsymbol{V}\boldsymbol{w}\boldsymbol{w}'\boldsymbol{V}\boldsymbol{w})$, which, after simplification gives the result. □

4.2.1. *Estimating the overall mean effect size*

It often occurs that a study will measure outcomes on multiple related constructs. While it is possible and often desirable to synthesize results on measures of each outcome separately, reviewers sometimes wish to summarize the results across all outcomes. One way to do so is to compute an effect size from data on a composite measure of the higher level construct (such as an average or total score on a test battery) for each study.

Combining effect estimates across different outcomes is usually motivated by an interest in a higher level construct (such as school achievement) that includes all of the lower level constructs reflected in particular outcomes (such as reading, and mathematics achievement). When an analyst has such an interest, it will almost always be true that the effects on the different outcomes combined will be highly correlated (such as reading and mathematics achievement, which typically have correlations in excess of 0.8).

Unfortunately, measures of the composite are often not available in studies, but effect sizes on the components of the composite are reported. In such cases, reviewers may wish to use data from effect sizes on each component of the construct in a synthesis as an alternative to the direct analysis of effects on the composite. If the components of the composite are measured on the same individuals, and the measures are positively correlated, then the effect sizes on different components in the same study will not be independent.

REMARK. If the correlation structure of the component variables (which implies the correlation structure of the effect sizes computed from them) is known, then fully efficient analyses are possible using multivariate methods. Methods for estimation and testing hypotheses about a composite effect size from components and for testing homogeneity across components when the covariance structure of the effect sizes is known were given in Hedges and Olkin (1985).

Often the correlation structure of component outcomes is not reported. One alternative is to impute values for the correlations and proceed as if these imputed values were known using multivariate methods. This approach has two disadvantages. First multivariate methods are cumbersome to use. Moreover, when the correlation between components is relatively high (as it must be to justify the arguments that the components are measures of the same construct), multivariate methods provide only a small gain in efficiency over simpler alternatives (such as picking a single outcome in each study to include in the synthesis and ignoring the rest). Therefore the gain in efficiency is often not viewed as worth the effort of a multivariate analysis, especially when most or all of the data on correlation structure must be imputed.

A widely used approach to dependence in meta-analysis is to use only a single estimate from each study. If the ith study has $n_i \leqslant N$ effect size estimates, the composite estimates are computed for a total of K studies. Often, the composite within-studies mean estimate for the ith study

$$\overline{T}_i = (T_1 + \cdots + T_{n_i})/n_i \tag{7}$$

is used as an estimate of the mean within-study effect size parameter

$$\bar{\theta}_i = (\theta_1 + \cdots + \theta_{n_i})/n_i. \tag{8}$$

If the effect sizes are homogeneous across studies, that is if all the $\theta_{ij} = \theta$, then each $\overline{T}_i$ is an unbiased estimate of θ.

REMARK. The unweighted average is typically used within studies because estimates of different outcomes from the same study often have very similar standard errors, and thus weights for different estimates would be essentially equal. When estimates are correlated, the optimal weights (weights that minimize the variance of the composite) depend on the correlation structure of the estimates. However when estimates are highly correlated, as they are likely to be in situations where one wishes to combine across them, the unweighted mean is nearly as efficient as the optimally weighted composite (see Hedges and Olkin, 1985).

The problem with using $\overline{T}_i$ as the (single) composite effect size for the ith study is the estimation of its variance, which (through the weights used in combining effect size estimates across studies) is used to compute the standard error of the combined effect size estimate. If the individual study weights are inaccurate, so is the standard error of the weighted average effect size estimate across studies.

The exact variance of $\overline{T}_i$ is

$$\tilde{\sigma}_i^2 = \frac{\bar{\sigma}_i^2}{n_i} + \frac{1}{n_i^2}\sum_{a \neq b}^{n_i} \sigma_{ia}\sigma_{ib}\rho_{iab}, \tag{9}$$

where $\bar{\sigma}_i^2 = (\sigma_{i1}^2 + \cdots + \sigma_{in_i}^2)/n_i$ is the unweighted average of the variances of the effect sizes in the ith study and the sum runs over all unequal values of a and b from 1 to n_i. The value of $\tilde{\sigma}_i^2$ can be calculated if all of the intercorrelations of the effect size estimates are known, but if they are not, it can be useful to carry out sensitivity analyses to see how various possible values of the intercorrelations affect the variance of each $\overline{T}_i$ or the combined variance of the weighted average of the $\overline{T}_i$ across studies.

The formulation of $\tilde{\sigma}_i^2$ in (9) makes it clear that $\tilde{\sigma}_i^2 = \bar{\sigma}_i^2/n_i$ when the components of the composite are independent. It is less clear in this formulation that $\tilde{\sigma}_i^2$ is often much closer to $\bar{\sigma}_i^2$ in situations of practical interest. In most situations in which reviewers wish to combine component effects into a composite, the correlations among components will be relatively high. If the effect size measures are standardized mean differences, the variances are also typically nearly identical (since they depend most strongly on sample sizes). If all of the variances are identical to $\bar{\sigma}_i^2$, all the $n_i = n$, and the intercorrelations have the same value ρ, then

$$\tilde{\sigma}_i^2 = \bar{\sigma}_i^2\left[1 + (n-1)\rho\right]/n.$$

Thus when ρ is high, $\tilde{\sigma}_i^2$ is close to $\bar{\sigma}_i^2$. For example, when $n = 2$ and $\rho = 0.8$, $\tilde{\sigma}_i^2 = 0.9\bar{\sigma}_i^2$ and even if $n = 2$ and $\rho = 0.6$, $\tilde{\sigma}_i^2 = 0.8\bar{\sigma}_i^2$. It is clear that if the $\rho_{iab} > 0$, $\tilde{\sigma}_i^2$, is always smaller than $\bar{\sigma}_i^2$, so that $\bar{\sigma}_i^2$ is a conservative estimate of $\tilde{\sigma}_i^2$ (that is $\bar{\sigma}_i^2$ overestimates $\tilde{\sigma}_i^2$). Therefore a sensible (but conservative) value to use for the variance of the within-study mean effect size $\overline{T}_i$ is $\bar{\sigma}_i^2$, the average of the variances of the effect sizes combined into $\overline{T}_i$.

If $\bar{\sigma}_i^2$ is used as an estimate of the variance of $\overline{T}_i$, weights computed as $w_i = 1/\bar{\sigma}_i^2$ would be used to compute the estimate of θ. The weighted mean

$$\overline{T}_{\bullet} = \frac{\sum_{i=1}^{k} \overline{T}_i/\bar{\boldsymbol{\sigma}}_i^2}{\sum_{i=1}^{k} 1/\bar{\boldsymbol{\sigma}}_i^2} \tag{10}$$

will be an unbiased estimate of θ. However Fact 1 implies that the standard error

$$v_{\bullet} = \left(\sum_{i=1}^{k} 1/\bar{\boldsymbol{\sigma}}_i^2\right)^{-1} \tag{11}$$

will not be correct. In this case the exact variance of the weighted mean $\overline{T}_{\bullet}$ is

$$V(\overline{T}_{\bullet}) = v_{\bullet}\left[\left(\sum_{i=1}^{K} w_i a_i\right) \Big/ \sum_{i=1}^{K} w_i\right] = v_{\bullet}R, \tag{12}$$

where $a_i = \tilde{\sigma}_i^2/\bar{\sigma}_i^2$ and $w_i = 1/\bar{\sigma}_i^2$. Since each of the a_i is less than or equal to 1, the true variance of $\overline{T}_{\bullet}$ is less than $v_{\bullet}$, and the amount by which it is less is the (weighted) average of degree to which each individual variance is larger than the actual variance. If all of the variances are identical to $\bar{\sigma}^2$, all the $n_i = n$, and the intercorrelations have the same value ρ, then

$$V(\overline{T}_{\bullet}) = v_{\bullet}\left[1 + (n-1)\rho\right]/n,$$

which for large values of ρ will be reasonably close to (but smaller than) $v_\bullet$

The usual test of the hypothesis H$_0$: $\theta = 0$, rejects if the statistic

$$Z = \left|\overline{T}_\bullet/\sqrt{v_\bullet}\right| > c_{\alpha/2},$$

where c_α is the $100(1-\alpha)$ percent point of the standard normal distribution. Because $v_\bullet$ overestimates the true variance of $\overline{T}_\bullet$, the probability that the test based on Z will reject is less than α when the null hypothesis is true. In particular, the probability that the test based on Z rejects when the null hypothesis is true is

$$1 + \Phi\left(-c_\alpha/\sqrt{R}\right) - \Phi\left(c_\alpha/\sqrt{R}\right),$$

where R is given in (12) above. This implies that the test of the hypothesis that the mean effect size is zero is conservative (has a rejection rate less than the nominal significance level). The test is not only conservative, but it may be quite conservative. For example, if $k = 10$, $m = 2$ and $\rho = .5$, then R computed from (12) would be $R = 1.5/2.0 = 0.75$. Consequently the true rejection rate of the test based on Z at the $\alpha = 0.05$ significance level would be

$$1 + \Phi\left(-1.96/\sqrt{0.75}\right) - \Phi\left(1.96/\sqrt{0.75}\right) = 0.024.$$

It may be of interest to conduct sensitivity analyses to see how the standard error of the combined effect size estimate changes as a function of the intercorrelations among the effect sizes. This can be done by inserting values in (9) above, but it will usually be more intuitive to simplify the structure somewhat, for example by assuming that the intercorrelation has the same value ρ for all studies.

To test the hypothesis

$$\text{H}_0\text{: } \bar{\theta}_1 = \cdots = \bar{\theta}_K$$

of heterogeneity of synthetic effects across studies uses the statistic

$$Q_M = \sum_{i=1}^{K} w_i\left(\overline{T}_i - \overline{T}_\bullet\right)^2, \tag{13}$$

where in this case, $w_i = 1/\bar{\sigma}^2$. Theorem 1 can be used to obtain the distribution of Q_M. When all of the effect estimates are independent Q_M has the chi-square distribution with $(K-1)$ degrees of freedom. When the effects are not independent Theorem 1 implies that Q_M has a distribution that is a constant g times a chi-square with h degrees of freedom.

The conservative bias in estimation of the variance of the individual composites (that is, $\bar{\sigma}_i^2$ used to estimate the exact variance is generally larger than $\tilde{\sigma}_i^2$) naturally leads to smaller values of Q_M, since each term is divided by a number that is larger than the actual variance of $\overline{T}_i$.

When all of the studies have effects in all m groups and the intercorrelations are identical, then $g = [1 + (m-1)\rho]/m$, and $h = (K-1)$. For example, if $K = 10$, $m = 2$ and $\rho = 0.5$, then Q would be distributed as $1.5/2.0 = 0.75$ times a chi-square with $10 - 1 = 9$ degrees of freedom, and when the null hypothesis of homogeneity is

Table 1
Effect size estimates and variances from 13 studies of the effects of coaching on SAT verbal and mathematics scores

	Data						Derived quantities			
Study	n^T	n^C	T_{i1}	v_{i1}	T_{i2}	v_{i2}	$\bar{T}_i$	$\bar{\sigma}_i^2$	D_i	$2\bar{\sigma}_i^2$
1	145	129	0.12	0.0042	0.13	0.0042	0.125	0.0042	−0.01	0.00837
2	13	14	0.08	0.0368	0.00	0.0368	0.04	0.0368	0.08	0.07364
3	16	17	0.51	0.0449	0.14	0.0449	0.325	0.0449	0.37	0.0899
4	188	122	0.12	0.0033			0.12	0.0033		
5	28	22			0.22	0.0195	0.22	0.0195		
6	39	40			0.09	0.0122	0.09	0.0122		
7	22	17			0.14	0.0250	0.14	0.0250		
8	48	43			0.14	0.0106	0.14	0.0106		
9	25	74			−0.01	0.0128	−0.01	0.0128		
10	37	35			0.14	0.0133	0.14	0.0133		
11	24	70			0.18	0.0134	0.18	0.0134		
12	16	19			0.01	0.0276	0.01	0.0276		
13	154	111			0.17	0.0037	0.17	0.0037		

Note: These data are from Becker (1990).

true, the probability of getting a value of Q_M larger than 16.92 (the $\alpha = 0.05$ critical value of the chi-square with 9 degrees of freedom) is 0.007, much smaller than the nominal 0.05.

If the correlations among effect sizes are not known, but it is known (or hypothesized for sensitivity analysis) that all of the $\bar{\sigma}_i^2$ overestimate $\tilde{\sigma}_i^2$ by the same fraction a (that is, $\tilde{\sigma}_i^2 = a\bar{\sigma}_i^2$), then Q_M has a distribution that is a times a chi-square with $(K - 1)$ degrees of freedom. Thus the actual critical value of the test at level α (corrected for conservative estimation of the variance) is c_α/a, where c_α is the nominal critical value. For example if $K = 10$, then the nominal $\alpha = 0.05$ critical value of Q is 16.92, but the corrected critical value is $(16.92)(0.75) = 12.69$.

If an exact test of homogeneity is desired, there are two possibilities. If all of the correlations among effect sizes were known, then the exact variances of the composites could be computed. If these exact variances were substituted for the $\bar{\sigma}_i^2$ in Q_M, then the test statistic would have the correct distribution and an exact test could be done. Alternatively, the exact values of the variances would permit calculation of the constants g and h using Theorem 1 to obtain the distribution of Q_M (when it was computed using the $\bar{\sigma}_i^2$).

EXAMPLE. Return to the example of the SAT coaching data. The effect size estimates and their variances from the $K = 13$ studies, as reported by from Becker (1990), are given in Table 1, along with the unweighted average effect size $\bar{\boldsymbol{T}}_i$ and the average variance $\bar{\boldsymbol{\sigma}}_i^2$ for each study. The weighted mean (10) is $\bar{\boldsymbol{T}}_\bullet = 0.130$ with an estimated variance from (11) of $v_\bullet = 0.0007179$, corresponding to a standard error of 0.0268. Thus we compute a 95 percent confidence interval for θ as $0.077 \leqslant \theta \leqslant 0.183$. One advantage of this dataset is that the correlation parameter between effect sizes on outcomes one and two (SAT mathematics and SAT verbal scores, respectively) is actually known to be $\rho = 0.66$ (see Becker, 1990). This value makes it possible to compute the exact variance of each of the $\overline{\boldsymbol{T}}_i$, and consequently the exact variance of $\overline{\boldsymbol{T}}_\bullet$, which we compute as $\boldsymbol{V}(\overline{\boldsymbol{T}}_\bullet) = 0.0006926$, corresponding to an exact standard error of 0.02632. We see that the ratio of $v_\bullet$ to the exact variance of $\overline{\boldsymbol{T}}_\bullet$ is $R = 0.98$, so that the approximation to the variance had only a slight effect on the standard error of the weighted mean effect size and the confidence interval computed from this exact standard error ($0.078 \leqslant \theta \leqslant 0.181$) is very close to that based on the approximation. Turning to the homogeneity test, the value of the test statistic is $Q_M = 4.334$. If all of the effect size estimates were independent, this statistic would have a chi-square distribution with 12 degrees of freedom if the null hypothesis is true, yielding a p-value of $p = 0.977$. Using the exact variances of the $\overline{\boldsymbol{T}}_i$ and Theorem 1, we obtain the cumulants $K_1 = 11.525$ and $K_2 = 22.253$, yielding $g = 0.965$ and $h = 11.938$, so that $Q_M/0.965$ has a chi-squared distribution with 11.938 degrees of freedom. Compared with this later distribution the p-value of the Q_M statistic is $p = 0.972$, very close to the original value based on ignoring dependence. Dependence had little effect in part because few of the total studies (only 3 of 13) actually had dependent data.

4.2.2. Aggregating within-study contrasts between correlated effect sizes

Reviewers often wish to estimate the differential effect of treatments, that is, the difference between the effects of a treatment on different outcomes, at different time points, or the differential effects of different versions of the same treatment. If the different effects are measured on the same individuals, then the effect size estimates will not be statistically independent.

REMARK. Omnibus tests of the difference between groups of effect sizes permit a flexible analysis to examine the difference between average effect sizes in groups of studies in cases when not every study has a measure in every group. When possible, it may be desirable to compute contrasts of study results between groups within studies, and then aggregate these within-study contrasts across groups. Such a comparison is logically superior to contrasts of across study average effect sizes since each within-study contrast controls for characteristics of the study that might affect results.

Note that when the variance of all of the effect size estimates in a study are approximately the same, the fixed effects estimate of the average of the differences will give approximately the same results as the difference of the average. This need not be so for random effects analysis however, because the variance component of the differences may be quite different that of the individual components, which would affect the weighting, possibly substantially. For example, it is possible that the effect sizes of the individual components have substantial variation across studies associated with differences in study design or procedures, while the differences are not affected by such variation because both components of the difference are measured in the same study (and therefore share the study design and procedures).

A widely used approach in meta-analysis is to compute a within-study difference between effect size estimates, and aggregate these synthetic effect size estimates across studies. Denote the within-study difference between effect size estimates by

$$D_i = \overline{T}_{ia} - \overline{T}_{ib}, \tag{14}$$

which is an estimate of the difference between within-study effect size parameters

$$\Delta_i = \theta_{ia} - \theta_{ib}. \tag{15}$$

If there are p differences and there is homogeneity of the effect size differences, that is if $\Delta_1 = \cdots = \Delta_p = \Delta$, then each D_i is an unbiased estimator of Δ. As in the case of the composite effect size parameter, the two effect size estimates involved in the difference are statistically dependent and this raises the problem of how to estimate the variance of D_i. Since the exact variance of D_i is

$$\widehat{\sigma}_i^2 = \sigma_{ia}^2 + \sigma_{ib}^2 - 2\sigma_{ia}\sigma_{ib}\rho_{iab}, \tag{16}$$

it follows that if $\rho > 0$, a conservative estimate of $\widehat{\sigma}_i^2$ is given by assuming that $\rho_{iab} = 0$, that is by estimating $\widehat{\sigma}_i^2$ by $\sigma_{ia}^2 + \sigma_{ib}^2 = 2\bar{\sigma}_i^2$. If the variances are identical to $\bar{\sigma}_i^2$, then the variance of D_i reduces to

$$\widehat{\sigma}_i^2 = 2\bar{\sigma}_i^2(1 - \rho_{iab}).$$

Note that this formulation is actually quite conservative if ρ is large. When the variances are equal, the ratio of the exact variance $\widehat{\sigma}_i^2$ to the conservative estimate ($2\bar{\sigma}_i^2$) is $(1 - \rho_{12i})$, so that when $\rho = 0.8$, the ratio is 0.2 and even if $\rho = 0.6$, the ratio is 0.4. If it is known that $\rho > 0.5$ (which may be plausible for highly reliable tests of closely related achievement constructs) then a much less conservative estimate of the variance of D_i is $\bar{\sigma}_i^2$.

Since the D_i, $i = 1, \ldots, p$, are independent, they can be combined to obtain an average effect and its standard error via the usual fixed effects methods (see, e.g., Shadish and Haddock, 1994). Specifically, the weighted mean will be

$$\overline{D}_{\bullet} = \frac{\sum_{i=1}^{p} D_i/\bar{\sigma}_i^2}{\sum_{i=1}^{p} 1/\bar{\sigma}_i^2}, \tag{17}$$

where $w_i = 1/\bar{\sigma}_i^2$, which will be an unbiased estimate of Δ. However the variance (the square of the standard error) computed in the usual analysis will be

$$v_{\bullet} = 2\left(\sum_{i=1}^{p} 1/\bar{\sigma}_i^2\right)^{-1} \tag{18}$$

(or $v_{\bullet}/2$ if the estimate $\bar{\sigma}_i^2$ is used in place of $2\bar{\sigma}_i^2$ as the variance of D_i), which will not be correct.

From Fact 1, the exact variance of the weighted mean $\overline{D}_{\bullet}$ is given by

$$V(\overline{D}_{\bullet}) = v_{\bullet}\left[\left(\sum_{i=1}^{p} w_i \frac{\widehat{\sigma}_i^2}{\bar{\sigma}_i^2}\right) \Big/ \sum_{i=1}^{p} w_i\right] = Rv_{\bullet}. \tag{19}$$

If all of the variances are identical to $\bar{\sigma}^2$ and the intercorrelations have the same value ρ, then

$$V(\overline{D}_\bullet) = \left(2\bar{\sigma}^2/k\right)(1-\rho) = v_\bullet(1-\rho),$$

which will be smaller than $v_\bullet$, and much smaller when ρ is large.

A test for heterogeneity of effects across studies is a test of the hypothesis H_0: $\Delta_1 = \cdots = \Delta_k$. The usual test for heterogeneity of effects across studies, when applied to the composite effect size estimates and the estimated variance given above uses the statistic

$$Q_D = \sum_{i=1}^{p} (D_i - \overline{D}_\bullet)^2/2\bar{\sigma}_i^2. \tag{20}$$

The usual test rejects the null hypothesis when Q_D exceeds the $100(1-\alpha)$ percent point of a chi-square distribution with $(p-1)$ degrees of freedom. When the effects are not independent Theorem 1 implies that Q_D will not generally have the usual chi-squared distribution with $(p-1)$ degrees of freedom, but a distribution that is a constant g times a chi-square with h degrees of freedom.

Note that the conservative bias in estimation of the variance of the individual composites (that is, $2\bar{\sigma}_i^2$ used to estimate the exact variance $\widehat{\sigma}_i^2$ is generally larger than $\widehat{\sigma}_i^2$) naturally leads to smaller values of Q_D, since each term is divided by a number that is larger than the actual variance of $\overline{T}_i$. In other words the test for heterogeneity based on Q_D will tend to reject less often than expected given the nominal significance level. If the correlations between effect sizes are large, then the rejection rate of the test based on Q_D will be *much* less than the nominal rejection rate.

For example, when all of the studies have effects in all m groups and the intercorrelations are identical, then $g = (1-\rho)$, and $h = (k-1)$. For example, if $p = 10$, and $\rho = 0.5$, then Q_D would be distributed as $(1-0.5) = 0.5$ times a chi-square with $10-1 = 9$ degrees of freedom. The actual rejection rate for the test at nominal significance level $\alpha = 0.05$ when the null hypothesis is true would be less than 0.0001.

If a test of homogeneity is desired, there are two possibilities. If the correlations between the effect sizes composing the D_i were known, then the exact variances of the difference could be computed. If these exact variances were substituted for $2\bar{\sigma}_i^2$, then this modified version of the test statistic Q_D would have the correct distribution and an exact test could be done. Alternatively, the exact values of the variances would permit calculation of the constants g and h using Theorem 1 to obtain the distribution of Q_D (when it is computed using $w_i = \bar{\sigma}_i^2$).

If the correlations among effect sizes are not known, but it is known (or hypothesized for sensitivity analysis) that all of the correlations between effects have the same value ρ, then $2\bar{\sigma}_i^2$ overestimates $\widehat{\sigma}_i^2$ by the same fraction $1/(1-\rho)$ for all i, that is, $\widehat{\sigma}_i^2 = 2\bar{\sigma}_i^2/(1-\rho)$, so that Q_D has a distribution that is $(1-\rho)$, times a chi-square with $(k-1)$ degrees of freedom. Thus the actual critical value of the test at level α (corrected for conservative estimation of the variance) is $(1-\rho)c_\alpha$, where c_α is the nominal critical value. For example if $p = 10$, and $\rho = 0.5$, then the nominal $\alpha = 0.05$ critical value of Q is 16.92, but the corrected $\alpha = 0.05$ critical value is $(16.92)(0.5) = 8.46$.

The usual test of the hypothesis H_0: $\Delta = 0$, rejects if the statistic

$$Z = \left|\overline{D}_\bullet/\sqrt{v_\bullet}\right| > c_{\alpha/2},$$

where c_α is the $100(1-\alpha)$ percent point of the standard normal distribution. Because $v_\bullet$ overestimates the true variance of $\overline{D}_\bullet$, the probability that the test based on Z will reject is less than α when the null hypothesis is true. In particular, the probability that the test based on Z rejects when the null hypothesis is true is

$$1 + \Phi\left(-c_\alpha/\sqrt{R}\right) - \Phi\left(c_\alpha/\sqrt{R}\right),$$

where R is given in (19) above. This implies that the test of the hypothesis that the mean effect size is zero is conservative (has a rejection rate less than the nominal significance level). The test is not only conservative, it may be quite conservative. For example, if $p = 10, m = 2$ and $\rho = 0.5$, then R computed from (19) would be $R = 1.5/2.0 = 0.75$. Consequently, when the null hypothesis that $\Delta = 0$ is true, the actual rejection rate of the test based on Z at the $\alpha = 0.05$ nominal significance level would be

$$1 + \Phi\left(-1.96/\sqrt{0.75}\right) - \Phi\left(1.96/\sqrt{0.75}\right) = 0.024.$$

It may be of interest to conduct sensitivity analyses to see how the standard error of the combined effect size estimate changes as a function of the intercorrelations among the effect sizes. This can be done by inserting values in (16) above, but it will usually be more intuitive to simplify the structure somewhat, for example by assuming that the intercorrelation has the same value ρ for all studies.

EXAMPLE. Return to the example of the SAT coaching data. The effect size estimates and their variances from the $K = 13$ studies, as reported by from Becker (1990), are given in Table 1, along with the difference between effect sizes on SAT mathematics and SAT verbal for $p = 3$ studies, and the average variance $\bar{\sigma}_i^2$ for each study. The weighted mean (17) is $\overline{\boldsymbol{D}}_\bullet = 0.028$ with an estimated variance from (18) of $v_\bullet = 0.006937$, corresponding to a standard error of 0.0833. Thus we compute a 95 percent confidence interval for θ as $-0.135 \leqslant \theta \leqslant 0.191$. One advantage of this dataset is that the correlation parameter between effect sizes on outcomes one and two (SAT mathematics and SAT verbal) is actually known to be $\rho = 0.66$ (see Becker, 1990). This value makes it possible to compute the exact variance of each of the $\overline{\boldsymbol{T}}_i$, and consequently the exact variance of $\overline{\boldsymbol{D}}_\bullet$, which we compute as $\boldsymbol{V}(\overline{\boldsymbol{D}}_\bullet) = 0.002358$, corresponding to an exact standard error of 0.0486. We see that the ratio of $v_\bullet$ to the exact variance of $\overline{\boldsymbol{D}}_\bullet$ is $R = 0.340$, so that the approximation to the variance had a very large effect on the standard error of the weighted mean effect size. Note however that the effect is, as expected, conservative. That is, the estimated variance $v_\bullet$ is much larger than the true variance. Note here that if we had used $\bar{\sigma}_i^2$ instead of $2\bar{\sigma}_i^2$ as an estimate of the variance of D_i, the ratio R of the true variance to the estimated variance would have been $R = 0.68$, still far from 1.0 (indicating that the variance was overestimated), but much closer to 1.0.

Turning to the homogeneity test, the value of the test statistic is $Q_D = 1.510$. If all of the effect size estimates were independent, this statistic would have a chi-square distribution with 2 degrees of freedom if the null hypothesis is true, yielding a p-value of $p = 0.470$. Using the exact variances of the D_i and Theorem 1, we obtain the cumulants $K_1 = 0.680$ and $K_2 = 0.4624$, yielding $g = 0.340$ and $h = 2.000$, so that $Q_D/0.340$ has a chi-squared distribution with 2.000 degrees of freedom. Thus the statistic $1.51/.34 = 4.44$ is the adjusted homogeneity test statistic that has a chi-square distribution under the null hypothesis. Compared with this later distribution the p-value of 4.44 is $p = 0.108$, very different from the original value based on ignoring dependence. Dependence had a large effect because all of the studies involved in computing D_i values had dependent data and our estimates $2\bar{\sigma}_i^2$ for the variance of the D_i was quite conservative (that is, they were large overestimates).

4.3. *Estimation and testing of mean effect sizes for different outcomes ignoring dependence entirely*

Because the effect sizes from different studies that estimate a given outcome are independent, the usual estimates of the group mean effect sizes and their standard errors have the expected properties. That is, the standard errors of the weighted group mean effect sizes for each outcome separately are correct and the usual significance tests for these means individually have the nominal significance levels. However tests for differences across outcomes in average effect size and estimates of the grand mean effect size (across outcomes) are affected by dependence and the associated tests and estimates do not have the distributions that would be expected under independence. In this section we determine the effects of dependence on these statistics.

4.3.1. *Testing for between-group differences in average effect size ignoring dependence entirely*

The test for between–outcome differences in meta-analysis uses an analogue of the analysis of variance (Hedges, 1981). Specifically, the test of the hypothesis

$$\mathrm{H}_0\colon\ \bar{\theta}_{\bullet 1} = \bar{\theta}_{\bullet 2} = \cdots = \bar{\theta}_{\bullet m} = \theta$$

uses the test statistic

$$Q_B = \sum_{i=1}^{m} w_{\bullet j}(\overline{T}_{\bullet j} - \overline{T}_{\bullet\bullet})^2, \tag{21}$$

where $\overline{T}_{\bullet j}$ is the weighted mean of the effect size estimates for the jth outcome given by

$$\overline{T}_{\bullet j} = \frac{\sum_{i=1}^{k_j} w_{ij} T_{ij}}{\sum_{i=1}^{k_j} w_{ij}}, \tag{22}$$

$\overline{T}_{\bullet\bullet}$ is the weighted grand mean given by

$$\overline{T}_{\bullet\bullet} = \frac{\sum_{j=1}^{m}\sum_{i=1}^{k_j} w_{ij} T_{ij}}{\sum_{j=1}^{m}\sum_{i=1}^{k_i} w_{ij}} = \frac{\sum_{j=1}^{m} w_{\bullet j}\overline{T}_{\bullet j}}{\sum_{j=1}^{m} w_{\bullet j}}, \tag{23}$$

and $w_{\bullet j}$ is given by

$$w_{\bullet j} = \sum_{i-1}^{kj} w_{ij} = 1/v_{\bullet j}. \tag{24}$$

When the effect sizes are independent, this statistic has the chi-square distribution with $(m-1)$ degrees of freedom under the null hypothesis. When the effect sizes are not independent, this test statistic has a sampling distribution that depends of the covariance matrix $\boldsymbol{V}$ of the vector $\overline{\boldsymbol{T}} = (\overline{T}_{\bullet 1}, \ldots, \overline{T}_{\bullet m})'$ of the group mean effect size estimates. The diagonal elements of $\boldsymbol{V}$ are just the variances of the $\overline{T}_{\bullet j}$, for example the jth diagonal element of $\boldsymbol{V}$ is $v_{jj} = 1/w_{\bullet j}$. The off-diagonal elements of $\boldsymbol{V}$ are the covariance of $\overline{T}_{\bullet a}$ and $\overline{T}_{\bullet b}$ given by $v_{ij} = \rho_{\bullet ab}/\sqrt{w_{\bullet a} w_{\bullet b}}$ where $\rho_{\bullet ab}$ is the correlation between $\overline{T}_{\bullet a}$ and $\overline{T}_{\bullet b}$ given by

$$\rho_{\bullet ab} = \frac{\sum_{i=1}^{p} \sqrt{w_{ia} w_{ib}} \rho_{iab}}{\sqrt{(\sum_{i-1}^{k_a} w_{ia})(\sum_{i-1}^{k_b} w_{ib})}}, \tag{25}$$

which is a kind of weighted average of the study-specific correlations between groups a and b. Note that if $p < k_a$ or $p < k_b$ (e.g., if studies do not have effect size estimates in every group), the consequence for the correlation $\rho_{\bullet ab}$ is exactly the same as averaging in a correlation of $\rho_{iab} = 0$ for that study. Because it is rarely the case that all studies have measures in each group, the correlations between group mean effect sizes will often be rather small even when the study-specific correlations that exist are rather high. Theorem 2 gives the sampling distribution of Q_B.

THEOREM 2. *The sampling distribution of Q_B is approximately by $Q_B/g \sim \chi_h^2$, in other words Q_B is distributed approximately as g times a chi-square with h degrees of freedom, where $g = K_2/2K_1$, $h = 2K_1^2/K_2$, and K_1 and K_2 are the first two cumulants of Q_B given by*

$$K_1 = m - \frac{(\sum_{j=1}^{m} \sum_{i=1}^{m} w_{\bullet i} w_{\bullet j} v_{ij})}{\sum_{i=1}^{m} w_{\bullet i}}$$

and

$$K_2 = 2\left[\sum_{j=1}^{m} \sum_{i=1}^{m} w_{\bullet i} w_{\bullet j} v_{ij}^2 - \frac{2[\sum_{j=1}^{m} w_{\bullet j} (\sum_{i=1}^{m} w_{\bullet i} v_{ij})^2]}{\sum_{i=1}^{m} w_{\bullet i}} + \frac{(\sum_{j=1}^{m} \sum_{i=1}^{m} w_{\bullet i} w_{\bullet j} v_{ij})^2}{(\sum_{i=1}^{m} w_{\bullet i})^2}\right],$$

$v_{jj} = 1/w_{\bullet j}$, *and* $w_{\bullet j}$ *is given in* (24), $v_{ij} = \rho_{\bullet ab}/\sqrt{w_{\bullet a} w_{\bullet b}}$, *and* $\rho_{\bullet ab}$ *is given in* (25).

PROOF. As in the proof of Theorem 1, write $Q_B = \boldsymbol{T}'\boldsymbol{M}\boldsymbol{T}$, for a $m \times m$ matrix $\boldsymbol{M}$, and $\mathbf{T}' = (\overline{\boldsymbol{T}}_{\bullet 1}, \ldots, \overline{\boldsymbol{T}}_{\bullet m})$. Note that $\boldsymbol{T}$ has a m-dimensional multivariate normal distribution with covariance matrix $\boldsymbol{V}$ given above (which is not diagonal in this case). Theorem 3.1 in Box (1954) says that Q_B is distributed as approximately a constant

g times chi-square with h degrees of freedom, where g and h are functions of the first two cumulants of Q_B (K_1 and K_2, respectively), specifically $g = K_2/2K_1$ and $h = 2K_1^2/K_2$. A standard result gives the first two cumulants of Q_B as $K_1 = \mathrm{tr}(\boldsymbol{VM})$ and $K_2 = 2\,\mathrm{tr}(\boldsymbol{VMVM})$, where $\mathrm{tr}(\boldsymbol{X})$ is the trace of the square matrix $\boldsymbol{X}$ (see, e.g., Rao, 1973). Writing $\boldsymbol{M} = \boldsymbol{D} - c\boldsymbol{w}\boldsymbol{w}'$ with $\boldsymbol{D} = \mathrm{diag}(w_1, \ldots, w_m)$, $\boldsymbol{w} = (w_1, \ldots, w_m)'$, and $c_w = 1/(w_1 + \cdots + w_m)$, we obtain the cumulants as $K_1 = \mathrm{tr}(\boldsymbol{VD}) - c_w\,\mathrm{tr}(\boldsymbol{w}'\boldsymbol{V}\boldsymbol{w})$ and $K_2 = 2\,\mathrm{tr}(\boldsymbol{VDVD}) - 4c_w\,\mathrm{tr}(\boldsymbol{w}'\boldsymbol{VDV}\boldsymbol{w}) + 2c_w^2\,\mathrm{tr}(\boldsymbol{w}'\boldsymbol{V}\boldsymbol{w}\boldsymbol{w}'\boldsymbol{V}\boldsymbol{w})$, which, after simplification gives the result. □

Note that when the effect sizes are independent, that is when $v_{ij} = 0$ if $i \neq j$, then g reduces to $g = 1$ and h reduces to $h = m - 1$, as expected. Also note that $h \geqslant 1$ in any event. Also note that the degrees of freedom are fractional, but the probability statements involving the distribution remain valid.

Several results follow immediately from this theorem.

COROLLARY 1. *When $m = 2$, $h = 1$ and $g \leqslant 1$, thus the probability that Q_B exceeds the level α critical value of the chi-square distribution with 1 degree of freedom is less than α. That is carrying out the test using Q_B with critical values assuming that effect sizes estimates are independent leads to a conservative test.*

COROLLARY 2. *When $m > 2$, but the variances of the $\overline{T}_{\bullet j}$'s are equal, $h = m - 1$, but $g < 1$, thus the probability that Q_B exceeds the level α critical value of the chi-square distribution with $(m - 1)$ degrees of freedom is less than α. That is carrying out the test using Q_B with critical values assuming that effect sizes estimates are independent leads to a conservative test.*

COROLLARY 3. *When $m > 2$ and the variances of the $\overline{T}_{\bullet j}$ are not too unequal (within an order of magnitude or so), $h < m - 1$, but $g < 1$, thus the thus the probability that Q_B exceeds the level α critical value of the chi-square distribution with $(m - 1)$ degrees of freedom is less than α. That is carrying out the test using Q_B with critical values assuming that effect sizes estimates are independent leads to a conservative test.*

COROLLARY 4. *When the correlations between the $\overline{T}_{\bullet a}$ and $\overline{T}_{\bullet b}$ are small, the test is only slightly conservative, but when the correlations are large, it can be* very *conservative.*

COROLLARY 5. *If the correlation structure between the $\overline{T}_{\bullet a}$ and $\overline{T}_{\bullet b}$ is known, a less conservative test can be carried out using critical values obtained from the approximate distribution given above. Alternatively, sensitivity analyses can be carried out to determine the p-value that would have been obtained under different assumptions about the correlation structure.*

4.4. Estimating the grand mean effect size ignoring dependence entirely

The weighted grand mean effect size estimate $\overline{T}_{\bullet\bullet}$, which is the usual estimate of the effect size θ when the hypothesis of homogeneity of effects is true, is an estimate of θ, but the usual computation of the

$$v_{\bullet\bullet} = \left(\sum_{j=1}^{m} w_{\bullet j}\right)^{-1} = \left(\sum_{j-1}^{m}\sum_{i-1}^{k_j} w_{ij}\right)^{-1} \tag{26}$$

will underestimate the true standard error of $\overline{T}_{\bullet\bullet}$ when data from different outcomes are positively correlated. The variance of $\overline{T}_{\bullet\bullet}$ is given by

$$V(\overline{T}_{\bullet\bullet}) = \frac{\sum_{a=1}^{m}\sum_{b=1}^{m} w_{\bullet a} w_{\bullet b} v_{ab}}{(\sum_{j=1}^{m} w_{\bullet j})^2} = v_{\bullet\bullet} + \frac{\sum_{a\neq b}^{m} \sqrt{w_{\bullet a} w_{\bullet b}}\rho_{\bullet ab}}{(\sum_{j=1}^{m} w_{\bullet j})^2}. \tag{27}$$

Because the second term on the right-hand side of (27) is a sum of quantities greater than or equal to 0 some of which are greater than 0 (since some of the $\rho_{\bullet ab}$ are larger than 0), it must be greater than 0 and so the actual variance of $\overline{T}_{\bullet\bullet}$ is greater than $v_{\bullet\bullet}$. In other words $v_{\bullet\bullet}$ underestimates the true variance of $\overline{T}_{\bullet\bullet}$.

The usual test of the hypothesis H_0: $\theta = 0$, which rejects if the statistic

$$Z = \left|\overline{T}_{\bullet\bullet}/\sqrt{v_{\bullet\bullet}}\right| > c_{\alpha/2},$$

where c_α is the $100(1-\alpha)$ percent point of the standard normal distribution. Because $v_{\bullet\bullet}$ underestimates the true variance of $\overline{T}_{\bullet\bullet}$, the probability that the test based on Z will reject is greater than α when the null hypothesis is true. In particular, the probability that the test based on Z rejects when the null hypothesis is true is

$$1 + \Phi\left(-c_\alpha/\sqrt{R}\right) - \Phi\left(c_\alpha/\sqrt{R}\right),$$

where $R = V(\overline{T}_{\bullet\bullet})/v_{\bullet\bullet}$. Similarly, confidence intervals will be too short by a factor of $\sqrt{R}$.

It may be of interest to conduct sensitivity analyses to see how the standard error of the combined effect size estimate or confidence intervals based upon it change as a function of the intercorrelations among the effect sizes. This can be done by inserting values in (27) above, but it will usually be more intuitive to simplify the structure somewhat, for example by assuming that the intercorrelation has the same value ρ for all studies.

EXAMPLE. Return to the example of the SAT coaching data. Using the data given in Table 1 we use (22) to compute value of $\overline{\boldsymbol{T}}_{\bullet 1} = 0.133$ and $\overline{\boldsymbol{T}}_{\bullet 2} = 0.129$, and we use (24) to compute $v_{\bullet 1} = 0.001687$ and $v_{\bullet 2} = 0.000919$. We compute the value of Q_B as $Q_B = 0.00446$. Comparing this value with the chi-squared distribution with 1 degree of freedom (the null sampling distribution of Q_B when effect sizes are independent), we obtain a p-value of $p = 0.947$. Using the exact correlation between effect sizes on outcome one and outcome two supplied by Becker (1990), we compute the exact correlation between $\overline{\boldsymbol{T}}_{\bullet 1}$ and $\overline{\boldsymbol{T}}_{\bullet 2}$ as $\rho_{\bullet 12} = 0.237$. We then apply Theorem 2 with these values of $v_{\bullet 1} = 0.001687$, $v_{\bullet 2} = 0.000919$, and $\rho_{\bullet 12} = 0.237$ to obtain the cumulants $K_1 = 0.774$ and $K_2 = 1.197$, which yield $g = 0.774$ and $h = 1.000$. Thus the actual distribution of Q_B can be defined by saying that $Q_B/0.774$ has a chi-squared distribution with 1 degree of freedom. Comparing the value 0.00446/0.774 with the chi-squared distribution with 1 degree of freedom yields a p-value of $p = 0.939$.

We compute the weighted grand mean (23) as $\overline{\boldsymbol{T}}_{\bullet\bullet} = 0.131$ with an estimated variance from (26) of $v_{\bullet\bullet} = 0.000595$, corresponding to a standard error of 0.0244. Thus we compute a 95 percent confidence interval for θ as $0.083 \leqslant \theta \leqslant 0.178$. Using the known correlation parameter between effect sizes on outcomes one and two (SAT mathematics and SAT verbal) from Becker (1990), $\rho = 0.66$, we compute the exact variance of $\overline{\boldsymbol{T}}_{\bullet\bullet}$ as $\boldsymbol{V}(\boldsymbol{T}_{\bullet\bullet}) = 0.000729$, corresponding to an exact standard error of 0.0270. We see that the ratio of $v_{\bullet\bullet}$ to the exact variance of $\overline{\boldsymbol{T}}_{\bullet\bullet}$ is $R = 1.225$, so that the approximation to the variance had a substantial effect on the standard error of the weighted mean effect size. Note however that the effect is, as expected, anti-conservative. That is, the estimated variance $v_{\bullet\bullet}$ is smaller than the true variance.

5. Mixed models

We begin by examining procedures for dealing with non-independent effect sizes in mixed effects statistical procedures. When there is only a single fixed effect (the mean) such procedures are often called random effects models in meta-analysis.

5.1. Model and notation

Suppose that the data layout is the same as in the fixed effects case. Under the fixed effects model, the θ_{ij} were fixed, but unknown, constants. Under this assumption the variance of T_{ij} is simply v_{ij}. In the mixed model, the θ_{ij} are not fixed but sampled from a universe of θ_{ij} values (corresponding to a universe of study effects for jth outcome and the ith study). Therefore it is necessary to distinguish between the variance of T_{ij} assuming fixed θ_{ij} and the variance of T_{ij} that incorporates the variance of θ_{ij} as well. The former is the conditional sampling variance of T_{ij} and the later is the unconditional sampling variance of T_{ij}.

We describe the sampling distribution of the T_{ij} in two parts. Let the standard error of T_{ij} given θ_{ij} be σ_{ij}. Assume that

$$T_{ij} = \theta_{ij} + \varepsilon_{ij} = \mu_j + \xi_{ij} + \varepsilon_{ij},$$

where ε_{ij} is the sampling error of T_{ij} as an estimate of θ_{ij} and θ_{ij} itself can be decomposed into the mean μ_j of the population of effects from which the effects of the jth outcome are sampled and the error ξ_{ij} of θ_{ij} as an estimate of μ_j. In this decomposition only μ_j is fixed, and we assume that ξ_{ij} and ε_{ij} are random and independently distributed with mean 0. The variance of ε_{ij} is σ_{ij}^2, the conditional sampling error variance of T_{ij}, which is known. The variance of the population from which the ξ_{ij} are sampled is τ^2, assumed to be the same across groups. Equivalently, we might say that τ^2 is the within-groups (of studies) variance of the population effect size parameters, which is why τ^2 is often called the between-studies variance component.

Because the effect size θ_{ij} is a value obtained by sampling from a distribution of potential values, the unconditional sampling variance of T_{ij} involves τ^2. In particular, the unconditional sampling variance of T_{ij} is

$$\sigma_{ij}^{2*} = \sigma_{ij}^2 + \tau^2. \tag{28}$$

For the T_{ij} where $i \leqslant p$ and $j \leqslant m$, we can describe the dependence succinctly by saying that the vector $\boldsymbol{T} = (T_{11}, \ldots, T_{p1}, T_{21}, \ldots, T_{pm})'$ has a multivariate normal distribution with mean vector $\boldsymbol{\mu} = (\mathbf{1}'\mu_1, \ldots, \mathbf{1}'\mu_2, \ldots, \mathbf{1}'\mu_m)'$ (where $\mathbf{1}'$ is a p-dimensional column vector of ones) and covariance matrix

$$\boldsymbol{\Sigma}^* = \begin{pmatrix} \boldsymbol{\Sigma}^*_{11} & \boldsymbol{\Sigma}^*_{12} & \cdots & \boldsymbol{\Sigma}^*_{1m} \\ \boldsymbol{\Sigma}^*_{12} & \boldsymbol{\Sigma}^*_{22} & \cdots & \boldsymbol{\Sigma}^*_{2m} \\ \vdots & \vdots & \ddots & \vdots \\ \boldsymbol{\Sigma}^*_{1m} & \boldsymbol{\Sigma}^*_{2m} & \cdots & \boldsymbol{\Sigma}^*_{mm} \end{pmatrix},$$

where $\boldsymbol{\Sigma}^*_{\boldsymbol{aa}}$ is a $p \times p$ diagonal matrix given by $\boldsymbol{\Sigma}^*_{\boldsymbol{aa}} = \mathrm{diag}(\sigma^2_{a1} + \tau^2, \ldots, \sigma^2_{ap} + \tau^2)$, $\boldsymbol{\Sigma}^*_{\boldsymbol{ab}}$ is a $p \times p$ diagonal matrix given by $\boldsymbol{\Sigma}^*_{\boldsymbol{ab}} = \mathrm{diag}(\sigma_{a1}\sigma_{b1}\rho_{1ab} + \alpha\tau^2, \ldots, \sigma_{ap}\sigma_{bp}\rho_{pab} + \alpha\tau^2)$, where ρ_{iab} is the conditional correlation between T_{ia} and T_{ib} given θ_{ia} and θ_{ib}, the correlation induced by correlation of the within-study sampling errors (the ε_{ij}'s), and α is the within-study correlation of the random effects (the correlation of the ξ_{ia}'s and ξ_{ib}'s). Thus we could write $\boldsymbol{\Sigma}^*_{\boldsymbol{aa}}$ and $\boldsymbol{\Sigma}^*_{\boldsymbol{ab}}$ as

$$\boldsymbol{\Sigma}^*_{\boldsymbol{aa}} = \boldsymbol{\Sigma}_{\boldsymbol{aa}} + \tau^2\mathbf{I}$$

and

$$\boldsymbol{\Sigma}^*_{\boldsymbol{ab}} = \boldsymbol{\Sigma}_{\boldsymbol{ab}} + \alpha\tau^2\mathbf{I}.$$

Note that this model makes two assumptions. The first is that the within-outcome variance component is the same in each outcome (since there is no group index on the τ^2). Second, we assume that the study-specific random effects for each study have a common correlation α. The latter assumption requires further discussion.

REMARK. Note that the model, as stated here, allows for the possibility that there are correlations between the random effects contained in the correlated effect sizes. Specifically, it allows the possibility that the correlation between ξ_{ij} and $\xi_{ij'}$ is α, and this value does not differ across studies or between different outcomes. Such correlations between the random effects (multiplied by the appropriate standard deviations) are called covariance components. Such correlations are often plausible.

For example, suppose the groups consist of different outcome variables such as reading and mathematics achievement. It is plausible that studies that find a larger than average effect on reading (a positive ξ_{i1}) also find a larger than average effect on mathematics (a positive ξ_{i2}) because the particular implementation of the treatment in that study was more generally effective. Alternatively, one might expect that studies that find a larger than average effect on reading (a positive ξ_{i1}) find a smaller than average effect on mathematics (a positive ξ_{i2}) because the particular implementation of the treatment in that study concentrated more on reading than mathematics.

Models that can account for such correlations among the random effects are called covariance components models. Estimation of covariance components typically requires a considerable amount of data–data that is usually not available in the meta-analytic context and is beyond the scope of this chapter. For a discussion of covariance component models in meta-analysis see Becker and Schram (1994).

5.2. Analyses involving synthetic effect sizes

Consider the strategy of combining effect sizes into averages or differences within studies (synthetic effect sizes) before combining them across studies (using random effects procedures) to estimate an overall effect size or a difference between effect sizes.

5.2.1. Estimating the overall mean effect size

Suppose that reviewer wishes to use only a single composite estimate from each study, and uses the mean $\overline{T}_i$ of the estimates in the study given in Eq. (7). If the effect sizes are such that all the $\mu_j = \mu$, then each $\overline{T}_i$ is an unbiased estimate of μ.

The problem with using $\overline{T}_i$ as the (single) composite effect size for the ith study is the estimation of its (conditional) variance, which is used to compute the standard error of the combined effect size estimate. If the individual study variances are inaccurate, so is the standard error of the weighted average effect size estimate across studies. Inaccurate variances of the $\overline{T}_i$ contribute to inaccuracy in two ways: directly by inaccurate estimates of the sampling error variances and indirectly through their effect on estimation of the between-studies variance component.

A sensible suggestion made in connection with the fixed effects analysis is to use $\bar{\sigma}_i^2$, the average of the variances of the estimates averaged to obtain $\overline{T}_i$. This variance is smaller than the exact variance of $\overline{T}_i$, which will lead to an underestimation of the variance component. However the overall estimate of the variance of $\overline{T}_i$ will often be quite close to the true value.

A direct computation shows that the variance of $\overline{T}_i$ under the random effects model is $\tilde{\sigma}_i^2 + \gamma_i \tau^2$, where

$$\gamma_i = \left[1 + (n_i - 1)\alpha\right]/n_i. \tag{29}$$

Note that the γ_i are in general a function of both n_i, the number of effect sizes in each study, and α, the correlation of the random effects within studies. Note also that if $\alpha = 1$ so that there is a single study-specific random effect, then $\gamma_i = 1$.

Since the $\overline{T}_i, i = 1, \ldots, K$, are independent, they can be combined using standard random effects methods to obtain an average effect and its standard error (see, e.g., Shadish and Haddock, 1994). However, because the variance of the random effect associated with each $\overline{T}_i$ is not necessarily the same (unless all the n_i are the same or unless $\alpha = 1$), the procedure for estimating τ^2 needs to be modified slightly. Specifically, τ^2 can be estimated by

$$\hat{\tau}_M^2 = \begin{cases} \frac{Q_M-(K-1)}{a_M} & \text{if } Q_M \geqslant (K-1), \\ 0 & \text{if } Q_M < (K-1), \end{cases} \tag{30}$$

where a is given by

$$a_M = \sum_{i=1}^{K} w_i \gamma_i - \frac{\sum_{i=1}^{K} w_i^2 \gamma_i}{\sum_{i=1}^{K} w_i}, \tag{31}$$

$w_i = 1/\bar{\sigma}_i^2$ and Q_M is the homogeneity statistic given in (13) in connection with the fixed effects analysis. Note that the fact that $\bar{\sigma}_i^2$ used for the variance of $\overline{T}_i$ in Q_M is an overestimate of its actual variance leads to bias in the form of consistent underestimation of τ^2, since each term of Q_M is an underestimate of the ratio of the ith squared deviation to the ith variance. An expression for the bias of $\hat{\tau}^2$ is given in the following theorem.

THEOREM 3. *The bias* $E\{\hat{\tau}_M^2\} - \tau_M^2$ *in the estimate of* τ^2 *given by* (30) *is*

$$\text{Bias}\{\hat{\tau}_M^2\} = \frac{\sum_{i=1}^K \frac{\tilde{\sigma}_i^2}{\bar{\sigma}_i^2} - (\sum_{i=1}^K w_i)^{-1}(\sum_{i=1}^K w_i(\frac{\tilde{\sigma}_i^2}{\bar{\sigma}_i^2})) - (K-1)}{a},$$

where $w_i = 1/\bar{\sigma}_i^2$ *and* a *is the constant given in* (31) *used in the estimate of* τ^2.

PROOF. Writing Q_M as a quadratic form in $\boldsymbol{T}' = (\overline{\boldsymbol{T}}_1, \ldots, \overline{\boldsymbol{T}}_K)$, which has covariance matrix $\boldsymbol{V} = \text{diag}(\tilde{\sigma}_1^2+\gamma_1\tau^2, \ldots, \tilde{\sigma}_K^2+\gamma_K\tau^2)$ under the mixed model. Writing $\boldsymbol{M} = \boldsymbol{D} - c\boldsymbol{w}\boldsymbol{w}'$ with $\boldsymbol{D} = \text{diag}(w_1, \ldots, w_K)$, $\boldsymbol{w} = (w_1, \ldots, w_K)'$, and $c_w = 1/(w_1 + \cdots + w_K)$, the expected value of Q_M is $\mathrm{E}\{Q_M\} = \text{tr}(\boldsymbol{V}\boldsymbol{M}) = \text{tr}(\boldsymbol{V}\boldsymbol{D}) - c_w\,\text{tr}(\boldsymbol{w}'\boldsymbol{V}\boldsymbol{w})$, which, in turn, leads to $\mathrm{E}\{\hat{\boldsymbol{\tau}}_M^2\}$ and the bias. □

Note that the bias does not depend on τ^2, but only on the ratio $\tilde{\sigma}_i^2/\bar{\sigma}_i^2$ of the exact to estimated conditional variances of the $\overline{T}_i$. Specifically, the bias depends on the degree to which the average ratio and the weighted average ratio (with weights $w_i = 1/\bar{\sigma}_i^2$), respectively, depart from 1.

The random effects weighted mean will be

$$\overline{T}_\bullet^* = \frac{\sum_{i=1}^K \overline{T}_i/(\bar{\boldsymbol{\sigma}}_i^2 + \gamma_i\hat{\tau}_M^2)}{\sum_{i=1}^K 1/(\bar{\boldsymbol{\sigma}}_i^2 + \gamma_i\hat{\tau}_M^2)}. \tag{32}$$

The variance (the square of the standard error) computed in the usual analysis will be

$$v_\bullet^* = \left(\sum_{i=1}^K 1/(\bar{\boldsymbol{\sigma}}_i^2 + \gamma_i\hat{\tau}_M^2)\right)^{-1}, \tag{33}$$

which will not be correct. Using Fact 1, the true variance $V(\overline{T}_\bullet^*)$ is

$$V(\overline{T}_\bullet^*) = v_\bullet^* \left(\sum_{i=1}^K w_i^* \frac{\tilde{\sigma}_i^2 + \gamma_i\tau_M^2}{\bar{\sigma}_i^2 + \gamma_i\hat{\tau}_M^2}\right) \Big/ \sum_{i=1}^K w_i^*. \tag{34}$$

Note that the true variance is $v_\bullet^*$ times a weighted mean of the $a_i = (\tilde{\sigma}_i^2 + \gamma_i\tau_M^2)/(\bar{\sigma}_i^2 + \gamma_i\hat{\tau}_M^2)$, the ratio of the true unconditional variance to the estimated unconditional variance (using the weights $w_i^* = 1/(\bar{\sigma}_i^2 + \gamma_i\hat{\tau}_M^2)$).

REMARK. Note, that the biases in the two quantities in the denominator of each term of $v_\bullet^*$ go in opposite directions. That is, $\bar{\sigma}_i^2$ overestimates $\tilde{\sigma}_i^2$, but $\hat{\tau}^2$ underestimates τ^2, so the biases tend to cancel. In fact, if all of the variances are identical to $\bar{\sigma}^2$, then $\overline{T}_\bullet^*$ is the unweighted sample mean of $\overline{T}_1, \ldots, \overline{T}_K$ and $v_\bullet^*$ is exactly equal to the usual sample estimate of the variance of the mean of $\overline{T}_1, \ldots, \overline{T}_K$, namely $S_{\overline{T}}^2/K$. Thus we would expect $v_\bullet^*$ to be much closer to $V(\overline{T}_\bullet^*)$ than $v_\bullet$ is to $V(\overline{T}_\bullet)$.

The usual test of the hypothesis H_0: $\theta = 0$, rejects if the statistic

$$Z^* = \left|\overline{T}_\bullet^*/\sqrt{v_\bullet^*}\right| > c_{\alpha/2},$$

where c_α is the $100(1-\alpha)$ percent point of the standard normal distribution. The rejection rate of this statistic is often remarkably close to nominal. In fact, if all the variances are identical, Z^* is exactly the usual t-test statistics computed from the $\overline{T}_i$.

5.2.2. Aggregating within-study contrasts between correlated effect sizes

Reviewers often wish to estimate the differential effect of treatments, that is, the difference between the effects of a treatment on different outcomes, at different time points, or the differential effects of different versions of the same treatment. If the different effects are measured on the same individuals, then the effect size estimates will not be statistically independent.

Suppose the analyst wishes to estimate the average within-study difference between effect size estimates using D_i, defined in (14) which is an estimate of Δ_i defined in (15). In this case the object of the estimation is not a constant value of the treatment effect size, but the average, μ_Δ of the Δ_i values in the universe of studies from which the Δ_i have been sampled.

As in the case of the fixed effects model, the two effect size estimates involved in the difference are statistically dependent and this raises the problem of how to estimate the variance of D_i. Under the random effects model, the exact variance of D_i is

$$\widetilde{\sigma}_i^{2*} = \sigma_{i1}^2 + \sigma_{i2}^2 - 2\sigma_{i1}\sigma_{i2}\rho_{i12} + 2(1-\alpha)\tau^2 = \widetilde{\sigma}_i^2 + 2(1-\alpha)\tau^2.$$

It follows that if $\rho > 0$, a conservative estimate of $\widetilde{\sigma}_i^2$ is given by assuming that $\rho_{i12} = 0$, that is by estimating $\widetilde{\sigma}_i^{2*}$ by $\sigma_{i1}^2 + \sigma_{i2}^2 + 2(1-\alpha)\tau^2$. If the variances are identical to $\bar{\sigma}_i^2$, then the variance of D_i reduces to

$$\widetilde{\sigma}_i^2 = 2\bar{\sigma}_i^2(1 - \rho_{12i}) + 2(1-\alpha)\tau^2.$$

Note that this formulation is actually quite conservative if ρ and α are large. If it is known that $\rho > 0.5$ and $\alpha > 0.5$ (which may be plausible for highly reliable tests of closely related achievement constructs) then $\bar{\sigma}_i^2 + 2(1-\alpha)\tau^2$ will usually be a much less conservative estimate of the variance of D_i and it has much to recommend it. In such cases, substituting $\bar{\sigma}_i^2 + 2(1-\alpha)\tau^2$ for $2\bar{\sigma}_i^2 + 2(1-\alpha)\tau^2$ in analyses is desirable.

Since the D_i, $i = 1, \ldots, p$, are independent, they can be combined using standard random effects methods to obtain an average effect and its standard error (see, e.g., Shadish and Haddock, 1994). However the procedure for estimating τ^2 needs to be modified slightly. In this case τ^2 can be estimated by

$$\hat{\tau}_D^2 = \begin{cases} \frac{Q_D-(p-1)}{2a(1-\alpha)} & \text{if } Q_D \geqslant (p-1), \\ 0 & \text{if } Q_D < (p-1), \end{cases} \tag{35}$$

where a is given by

$$a = \sum_{i=1}^{p} w_i - \frac{\sum_{i=1}^{p} w_i^2}{\sum_{i=1}^{p} w_i}, \tag{36}$$

where the $w_i = 1/2\bar{\sigma}_i^2$ and Q_D is the heterogeneity statistic given in (20). Note that the fact that $\bar{\sigma}_i^2$ used for the variance of $\overline{T}_i$ is an overestimate of its actual variance leads to bias in the form of consistent underestimation of τ^2, since each term of Q is an underestimate of the ratio of the ith squared deviation to the ith variance. An expression for the bias of $\hat{\tau}_D^2$ is given in the following theorem.

THEOREM 4. *The bias* $E\{\hat{\tau}_D^2\} - \tau_D^2$ *is given by*

$$\text{Bias}\{\hat{\tau}_D^2\} = \frac{\sum_{i=1}^{p} \frac{\widehat{\sigma}_i^2}{\bar{\sigma}_i^2} - (\sum_{i=1}^{p} w_i)^{-1} (\sum_{i=1}^{p} w_i (\frac{\widehat{\sigma}_i^2}{\bar{\sigma}_i^2})) - (p-1)}{2a(1 \quad \alpha)},$$

where $w_i = 1/2\bar{\sigma}_i^2$ *and a is the constant given in* (36) *used in the estimate of* τ^2.

PROOF. The proof is essentially identical to that of Theorem 3. □

The random effects weighted mean will be

$$\overline{D}_{\bullet}^{*} = \frac{\sum_{i=1}^{p} \overline{D}_i/(2\bar{\boldsymbol{\sigma}}_i^2 + 2(1-\alpha)\hat{\tau}_D^2)}{\sum_{i=1}^{p} 1/(2\bar{\boldsymbol{\sigma}}_i^2 + 2(1-\alpha)\hat{\tau}_D^2)}. \tag{37}$$

The variance (the square of the standard error) computed will be

$$v_{\bullet}^{*} = \left(\sum_{i=1}^{p} 1/(2\bar{\boldsymbol{\sigma}}_i^2 + 2(1-\alpha)\hat{\tau}_D^2)\right)^{-1}, \tag{38}$$

which will not be exactly correct. The ratio of the true variance $V(\overline{D}_{\bullet}^{*})$ to $v_{\bullet}^{*}$, the variance computed in the analysis of the composites

$$\frac{V(\overline{D}_{\bullet}^{*})}{v_{\bullet}^{*}} = \left(\sum_{i=1}^{p} w_i^{*} \frac{\tilde{\sigma}_i^2 + 2(1-\alpha)\tau^2}{2\bar{\sigma}_i^2 + 2(1-\alpha)\hat{\tau}_D^2}\right) \Big/ \sum_{i=1}^{p} w_i^{*} \tag{39}$$

is a weighted mean of the $(\tilde{\sigma}_i^2 + 2(1-\alpha)\tau^2)/(2\bar{\sigma}_i^2 + 2(1-\alpha)\hat{\tau}_D^2)$, the ratio of the true unconditional variance to the estimated unconditional variance (using the weights $w_i^{*} = 1/(2\bar{\sigma}_i^2 + 2(1-\alpha)\hat{\tau}_D^2)$).

Note however, that the biases in the two quantities in the denominator of each term of $v_{\bullet}^{*}$ go in opposite directions. That is, $2\bar{\sigma}_i^2$ overestimates $\widehat{\sigma}_i^2$, but $\hat{\tau}^2$ underestimates τ^2, so the biases tend to cancel. For example, if all of the variances are identical to $\bar{\sigma}^2$, then $\overline{D}_{\bullet}^{*}$ is the unweighted sample mean of $\overline{D}_1, \ldots, \overline{D}_p$ and $v_{\bullet}^{*}$ is exactly equal to the sample estimate of the variance of the mean of $\overline{D}_1, \ldots, \overline{D}_p$, namely $S_{\overline{D}}^2/K$.

The usual test of the hypothesis H_0: $\theta = 0$, rejects if the statistic

$$Z^* = \left|\overline{D}^*_{\bullet}/\sqrt{v^*_{\bullet}}\right| > c_{\alpha/2},$$

where c_α is the $100(1-\alpha)$ percent point of the standard normal distribution. The rejection rate of this statistic is often remarkably close to nominal. In fact, if all the variances are identical, Z^* is exactly the usual t-test statistics computed from the $\overline{D}_i$.

5.3. *Testing for between-outcome differences in average effect size ignoring dependence entirely*

The mixed model test for between-outcome differences in meta-analysis uses an analogue of mixed model analysis of variance. Specifically, the test of the hypothesis

$$H_0\colon\ \mu_{\bullet 1} = \mu_{\bullet 2} = \cdots = \mu_{\bullet m} = \mu$$

uses the test statistic

$$Q_B^* = \sum_{j=1}^{m} w^*_{\bullet j}\left(\overline{T}^*_{\bullet j} - \overline{T}^*_{\bullet\bullet}\right)^2, \tag{40}$$

where $\overline{T}^*_{\bullet j}$ is the weighted mean of the jth group given by

$$\overline{T}^*_{\bullet j} = \frac{\sum_{i=1}^{k_j} w^*_{ij} T_{ij}}{\sum_{i=1}^{k_j} w^*_{ij}}, \tag{41}$$

$\overline{T}^*_{\bullet\bullet}$ is the weighted grand mean given by

$$\overline{T}^*_{\bullet\bullet} = \frac{\sum_{j=1}^{m}\sum_{i=1}^{k_j} w^*_{ij} T_{ij}}{\sum_{j=1}^{m}\sum_{i=1}^{k_j} w^*_{ij}} = \frac{\sum_{j=1}^{m} w^*_{\bullet j}\overline{T}^*_{\bullet j}}{\sum_{j=1}^{m} w^*_{\bullet j}}, \tag{42}$$

and $w^*_{\bullet j}$ is given by

$$w^*_{\bullet j} = \sum_{i=1}^{k_j} w^*_{ij}, \tag{43}$$

and $w^*_{ij} = 1/(v_{ij} + \hat{\tau}^2)$. In the case of the mixed model, τ^2 is typically estimated by

$$\hat{\tau}^2 = \begin{cases} \frac{Q_W - m(p-1)}{a} & \text{if } Q_W \geqslant m(p-1), \\ 0 & \text{if } Q_W < m(p-1), \end{cases} \tag{44}$$

where Q_W is pooled within outcomes homogeneity statistic

$$Q_W = \sum_{i=1}^{m}\sum_{j=1}^{k_i} w_{ij}\left(T_{ij} - \overline{T}_{\bullet j}\right)^2, \tag{45}$$

where $w_{ij} = 1/\sigma_{ij}^2$ are the fixed effects weights, a is given by $a = a_1 + \cdots + a_m$, and a_j is given by

$$a_j = \sum_{i=1}^{k_j} w_{ij} - \frac{\sum_{i=1}^{k_j} w_{ij}^2}{\sum_{i=1}^{k_j} w_{ij}}.$$

Note that Q_W can be written as $Q_W = Q_{W1} + \cdots + Q_{Wm}$, where Q_{Wi} is the test statistic for homogeneity of effect sizes within the ith group. Thus the estimate of τ^2 given in (44) can be seen as obtained by pooled (across outcomes) the *within*-outcomes heterogeneity statistics. Although the Q_{Wj} are not independent, the estimate of τ^2 derived from pooling across outcomes is still unbiased.

Note that the effect size estimates that are combined to yield $\overline{T}^*_{\bullet j}$ are independent, so the variance of $\overline{T}^*_{\bullet j}$ is approximately $1/w^*_{\bullet j}$. However, the $\overline{T}^*_{\bullet j}$ are *not* independent of one another in the model examined in this chapter. Thus test statistics such as Q_B^* that were designed for comparing independent means will have not have their nominal distributions under the model in this chapter.

When the effect sizes are independent, the test statistic Q_B^* has the chi-square distribution with $(m - 1)$ degrees of freedom under the null hypothesis. When the effect sizes are not independent, this test statistic has a sampling distribution that depends of the covariance matrix $\boldsymbol{V}^*$ of the vector $\overline{\boldsymbol{T}}^* = (\overline{T}^*_{\bullet 1}, \ldots, \overline{T}^*_{\bullet m})'$. The diagonal elements of $\boldsymbol{V}^*$ are just the variances of the $\overline{T}^*_{i\bullet}$, for example the ith diagonal element of $\boldsymbol{V}^*$ is $v_{ii} = 1/v^*_{\bullet j}$, the variance of $\overline{T}^*_{\bullet j}$ computed under the random effects analysis. The off-diagonal elements of $\boldsymbol{V}^*$ are the covariance of $\overline{T}^*_{\bullet a}$ and $\overline{T}^*_{\bullet b}$ given by $v_{ij} = \rho_{\bullet ab}/\sqrt{w^*_{\bullet a} w^*_{\bullet b}}$ where $\rho_{\bullet ab}$ is the correlation between $\overline{T}^*_{\bullet a}$ and $\overline{T}^*_{\bullet b}$ given by

$$\rho^*_{\bullet ab} = \frac{\sum_{i=1}^{p} \frac{\sqrt{w_{ia} w_{ib}}\rho_{iab} + \alpha\tau^2}{w^*_{ia} w^*_{ib}}}{\sqrt{(\sum_{i=1}^{k_a} w^*_{ia})(\sum_{i=1}^{k_b} w^*_{ib})}}, \tag{46}$$

which is a kind of random effects weighted average of the study-specific correlations between groups a and b. The random effects average correlation $\rho^*_{\bullet ab}$ is often less than the fixed effects average $\rho_{\bullet ab}$ since, in the model considered here, the random effects increase the variance of effect size estimates but have no effect on the covariances between them. The sampling distribution of Q_B^* under the model in this paper is given in the following theorem.

THEOREM 5. *The sampling distribution of Q_B^* is approximately by $Q_B^*/g \sim \chi_h^2$, in other words Q_B^* is distributed approximately as g times a chi-square with h degrees of freedom, where $g = K_2/2K_1$, $h = 2K_1^2/K_2$, and K_1 and K_2 are the first two cumulants of Q_B^* given by*

$$K_1 = \sum_{i=1}^{m} w^*_{\bullet i} v^*_{ii} - \frac{(\sum_{j=1}^{m}\sum_{i=1}^{m} w^*_{\bullet i} w^*_{\bullet j} v^*_{ij})}{\sum_{i=1}^{m} w^*_{\bullet i}}$$

and

$$K_2 = 2\sum_{j=1}^{m}\sum_{i=1}^{m} w_{\bullet i}^* w_{\bullet j}^* (v_{ij}^*)^2 - \frac{4[\sum_{j=1}^{m} w_{\bullet j}^* (\sum_{i=1}^{m} w_{\bullet i}^* v_{ij}^*)^2]}{\sum_{i=1}^{m} w_{\bullet i}^*} + \frac{2(\sum_{j=1}^{m}\sum_{i=1}^{m} w_{\bullet i}^* w_{\bullet j}^* v_{ij}^*)^2}{(\sum_{i=1}^{m} w_{\bullet i}^*)^2},$$

$v_{jj}^* = 1/w_{\bullet j}^*$ *is given in* (43), $v_{ij}^* = \rho_{\bullet ab}^*/\sqrt{w_{\bullet a}^* w_{\bullet b}^*}$, *and* $\rho_{\bullet ab}^*$ *is given in* (46).

PROOF. As in the proof of Theorem 2, write $Q_B^* = \boldsymbol{T}'\boldsymbol{M}\boldsymbol{T}$, for a $m \times m$ matrix $\boldsymbol{M}$, and $\mathbf{T}' = (\overline{\boldsymbol{T}}_{\bullet 1}^*, \ldots, \overline{\boldsymbol{T}}_{\bullet m}^*)$. Note that $\boldsymbol{T}$ has a m-dimensional multivariate normal distribution with covariance matrix $\boldsymbol{V}^*$ given above. Following the same procedure as in the proof of Theorem 2 yields the result. □

Note that when the effect sizes are independent, that is when $v_{ij} = 0$ if $i \neq j$, and if errors of estimation of τ^2 in $w_{\bullet j}^*$ can be ignored, then g reduces to $g = 1$ and h reduces to $h = m - 1$, as expected. Also note that $h \geqslant 1$ in any event. Also note that the degrees of freedom are fractional, but the probability statements involving the distribution remain valid.

Several results follow immediately from this theorem.

COROLLARY 1. *When* $m = 2$, $h = 1$ *and* $g \leqslant 1$, *thus the probability that* Q_B^* *exceeds the level* α *critical value of the chi-square distribution with* 1 *degree of freedom is less than* α. *That is carrying out the test using* Q_B *with critical values assuming that effect sizes estimates are independent leads to a conservative test.*

COROLLARY 2. *When* $m > 2$, *but the variances of the* $\overline{T}_{\bullet j}^*$*'s are equal*, $h = m - 1$, *but* $g < 1$, *thus the probability that* Q_B^* *exceeds the level* α *critical value of the chi-square distribution with* $(m - 1)$ *degrees of freedom is less than* α. *That is carrying out the test using* Q_B^* *with critical values assuming that effect sizes estimates are independent leads to a conservative test.*

COROLLARY 3. *When* $m > 2$ *and the variances of the* $\overline{T}_{\bullet j}^*$ *are not too unequal (within an order of magnitude or so)*, $h < m - 1$, *but* $g < 1$, *thus the probability that* Q_B^* *exceeds the level* α *critical value of the chi-square distribution with* $(m - 1)$ *degrees of freedom is less than* α. *That is carrying out the test using* Q_B^* *with critical values assuming that effect sizes estimates are independent leads to a conservative test.*

COROLLARY 4. *When the correlations between the* $\overline{T}_{\bullet a}^*$ *and* $\overline{T}_{\bullet b}^*$ *are small, the test is only slightly conservative, but when the correlations are large, it can be* very *conservative.*

COROLLARY 5. *If the correlation structure between the* $\overline{T}_{\bullet a}^*$ *and* $\overline{T}_{\bullet b}^*$ *is known, a less conservative test can be carried out using critical values obtained from the approximate* *distribution given above. Alternatively, sensitivity analyses can be carried out to determine the p-value that would have been obtained under different assumptions about the correlation structure.*

5.4. *Estimating the grand mean effect size ignoring dependence entirely*

The weighted grand mean effect size estimate $\overline{T}^*_{\bullet\bullet}$ is the usual estimate of the effect size μ under the mixed model when the hypothesis of homogeneity of effect sizes across outcomes is true. In the case of dependence across outcome groups, $\overline{T}^*_{\bullet\bullet}$ is still an estimate of μ, but the usual computation of the variance

$$v^*_{\bullet\bullet} = \left(\sum_{j=1}^{m} w^*_{\bullet j}\right)^{-1} = \left(\sum_{j=1}^{m}\sum_{i=1}^{k_j} w^*_{ij}\right)^{-1} \tag{47}$$

will underestimate the true standard error of $\overline{T}^*_{\bullet\bullet}$ when data from different outcomes are positively correlated. The variance of $\overline{T}^*_{\bullet\bullet}$ is given by

$$V(\overline{T}^*_{\bullet\bullet}) = \frac{\sum_{a=1}^{m}\sum_{b=1}^{m} w^*_{\bullet a} w^*_{\bullet b} v^*_{ab}}{(\sum_{j=1}^{m} w^*_{\bullet j})^2} = v^*_{\bullet\bullet} + \frac{\sum_{a\neq b}^{m} \lambda_{ab}\rho^*_{\bullet ab}}{(\sum_{j=1}^{m} w^*_{\bullet j})^2}, \tag{48}$$

where

$$\lambda_{ab} = \frac{\sqrt{V(\overline{T}^*_{\bullet a})V(\overline{T}^*_{\bullet b})}}{v^*_{\bullet a} v^*_{\bullet b}}.$$

Note that the exact variances in the numerator of λ_{ab} involve the exact value of τ^2, while the weights in the denominator of λ_{ab} involve the *estimated* values of τ^2. Thus except for estimation error, $v_{\bullet a} = 1/V(\overline{T}^*_{\bullet a})$ and $\lambda_{ab} = \sqrt{w^*_{\bullet a} w^*_{\bullet b}}$. Because the second term on the right-hand side of (48) is a sum of quantities greater than or equal to 0 some of which are greater than 0 (since some of the $\rho^*_{\bullet ab}$ are larger than 0), it must be greater than 0 and so the actual variance of $\overline{T}^*_{\bullet\bullet}$ is greater than $v^*_{\bullet\bullet}$. In other words $v^*_{\bullet\bullet}$ underestimates the true variance of $\overline{T}^*_{\bullet\bullet}$.

The usual test of the hypothesis H_0: $\mu = 0$, which rejects if the statistic

$$Z^* = \left|\overline{T}_{\bullet\bullet}/\sqrt{v^*_{\bullet\bullet}}\right| > c_{\alpha/2},$$

where c_α is the $100(1-\alpha)$ percent point of the standard normal distribution. Because $v^*_{\bullet\bullet}$ underestimates the true variance of $\overline{T}^*_{\bullet\bullet}$, the probability that the test based on Z^* rejects is greater than α when the null hypothesis is true. In particular, the probability that the test based on Z^* rejects when the null hypothesis is true is

$$1 + \Phi\left(-c_\alpha/\sqrt{R^*}\right) - \Phi\left(c_\alpha/\sqrt{R^*}\right),$$

where $R^* = V(\overline{T}^*_{\bullet\bullet})/v^*_{\bullet\bullet}$. Similarly, confidence intervals will be too short by a factor of R^*.

It may be of interest to conduct sensitivity analyses to see how the standard error of the combined effect size estimate or confidence intervals based upon it change as a function of the intercorrelations among the effect sizes. This can be done by inserting values into (48) above, but it will usually be more intuitive to simplify the structure somewhat, for example by assuming that the intercorrelation has the same value ρ for all studies.

6. Conclusions

Analysts often desire to combine effect size data that are not independent. Although multivariate methods have been developed that provide optimally efficient estimates, valid standard errors, and statistical test with valid significance levels, these multivariate methods are seldom used because they require access to data on the correlation structure of the estimates that is seldom available and cannot be often be usefully estimated from the effect size data available. Consequently, analysts often use approximate methods that do not require knowledge of the correlation structure.

One approach involves creating synthetic effect sizes from the dependent effect sizes in each study, and then combining these (independent) synthetic estimates of effect size across studies. These methods provide unbiased estimates of average effect size parameters, but conservative (that is, overestimates) of standard errors. Thus they lead to confidence intervals for average effects that are too wide, and significance tests on the mean that reject less often than nominal. The amount by which they are conservative depends on the correlation structure and the amount of the effect size data that actually involves correlated estimates. Under some circumstances (e.g., few studies that actually provide multiple, correlated effect size estimates) they may be only slightly conservative. Tests of heterogeneity of the synthetic effect sizes also reject less often than nominal, that is they suggest heterogeneity less often than they should.

A second approach is to simply ignore the dependence among positively correlated effect size estimates. This approach also leads to unbiased estimates of the grand mean effect size parameter, but conservative (that is overestimates) of standard errors. Thus they lead to confidence intervals for the grand mean effect that are too wide, and significance tests on the grand mean that reject less often than nominal. Omnibus tests for differences among group mean effect sizes have more complex behavior, but they are generally also conservative, rejecting less often than nominal. This leads to the remarkable situation in which ignoring dependence is actually conservative. However this is only the case if the dependent effect size estimates are positively correlated. Positive correlations are plausible in cases where the dependent effect sizes are based on measures of closely related constructs. However, negative correlations are possible, particularly when constructs are not closely related, and in such cases ignoring dependence may be anticonservative.

One interesting finding is that random effects methods involving synthetic effect sizes generally perform much better in terms of standard errors than do fixed effects procedures. This happens because random effects procedures compute the variance of each effect size from two sources, one theoretical (which is selected to be an underestimate) and one empirical which tends to overestimate variance exactly when the theoretical component underestimates it. Thus these two biases tend to cancel each other. Indeed, they do so exactly when all studies have the same conditional variances. This suggests a new reason to prefer random effects methods in meta-analysis.

References

Becker, B.J. (1990). Coaching for the scholastic aptitude test: Further synthesis and appraisal. *Review of Educational Research* **60**, 373–417.

Becker, B.J., Schram, C. (1994). Examining explanatory models through research synthesis. In: Cooper, H., Hedges, L.V. (Eds.), *The Handbook of Research Synthesis*. The Russell Sage Foundation, New York.

Box, G.E.P. (1954). Some theorems on quadratic forms applied to the study of analysis of variance problems, I. Effect of inequality of variance in the one-way classification. *Annals of Mathematical Statistics* **25**, 290–302.

Cooper, H. (1989). *Integrating Research*. Sage Publications, Thousand Oaks, CA.

Cooper, H., Hedges, L.V. (1994). *The Handbook of Research Synthesis*. The Russell Sage Foundation, New York.

Fleiss, J. (1994). Measures of effect size for categorical data. In: Cooper, H., Hedges, L.V. (Eds.), *The Handbook of Research Synthesis*. The Russell Sage Foundation, New York.

Garvey, W.D., Griffith, B.C. (1971). Scientific communication: Its role in the conduct of research and the creation of knowledge. *American Psychologist* **26**, 349–361.

Glass, G.V. (1976). Primary, secondary, and meta-analysis of research. *Educational Researcher* **5**, 3–8.

Hedges, L.V. (1981). Distribution theory for Glass's estimator of effect size and related estimators. *Journal of Educational Statistics* **6**, 107–128.

Hedges, L.V. (1987). How hard is hard science, how soft is soft science?: The empirical cumulativeness of research. *American Psychologist* **42**, 443–455.

Hedges, L.V., Olkin, I. (1985). *Statistical Methods for Meta-analysis*. Academic Press, New York.

Kalaian, H., Raudenbush, S.W. (1996). A multivariate mixed linear model for meta-analysis. *Psychological Methods* **1**, 227–235.

Lipsey, M., Wilson, D. (2001). *Practical Meta-Analysis*. Sage Publications, Thousand Oaks, CA.

Olkin, I., Gleser, L. (1994). Stochastically dependent effect sizes. In: Cooper, H., Hedges, L.V. (Eds.), *The Handbook of Research Synthesis*. The Russell Sage Foundation, New York.

Rao, C.R. (1973). *Linear Statistical Inference and its Applications*. John Wiley, New York.

Rosenthal, R. (1994). Parametric measures of effect size. In: Cooper, H., Hedges, L.V. (Eds.), *The Handbook of Research Synthesis*. The Russell Sage Foundation, New York.

Shadish, W., Haddock, K. (1994). Combining estimates of effect size. In: Cooper, H., Hedges, L.V. (Eds.), *The Handbook of Research Synthesis*. The Russell Sage Foundation, New York.

Smith, M.L., Glass, G.V. (1977). The meta-analysis of psychotherapy outcome studies. *American Psychologist* **32**, 752–760.

Handbook of Statistics, Vol. 26
ISSN: 0169-7161

DOI: 10.1016/S0169-7161(06)26030-9

30

Vertical Scaling: Statistical Models for Measuring Growth and Achievement

Richard J. Patz and Lihua Yao

Abstract

This chapter examines psychometric modeling approaches to vertically scaled educational assessments. We discuss historical approaches, and we examine model assumptions widely used in current practice. We identify a weakness in existing approaches that leaves users overly vulnerable to implausible (e.g., disordinal) grade-by-grade results, and we develop and demonstrate the general feasibility of an alternative approach grounded in hierarchial modeling. The approach we introduce and explore is a hierarchical multi-group item response theory model that allows explicit estimation of the functional form of the grade-to-grade growth patterns. The modeling framework we introduce is very general and incorporates widely used item response theory approaches to vertical scaling as special cases. We explore properties of the model and the estimation algorithms using simulated and real data from achievement tests.

1. Introduction

Test forms may be scaled in order to support valid interpretations regarding growth and change. Scaling may be used in conjunction with linking and equating to facilitate comparisons of results arising from the use of different test forms. Vertical scaling refers to the calibration of test forms (i.e., "levels") intended to be used at an ordinal set of points in the development of examinee proficiency in a domain of knowledge (Patz and Hanson, 2002). Prominent examples of vertical scales include those underlying the achievement tests commonly used for measuring achievement in elementary and secondary schools. Many of these testing programs employ a set of unique test forms of sequentially greater difficulty for testing students in progressive grade levels of schooling. Such tests may report "scale scores" on a common scale that spans the range of tested grades.

When differences in population proficiency at adjacent levels are modest in comparison to differences among examinees within levels, and when the expectations or standards against which examinees are to be measured overlap extensively, then linking the adjacent test levels to a common scale will make sense and help to provide meaningful information. These conditions are generally well satisfied in the case of achievement test batteries that measure development of proficiency in broad domains such as reading and mathematics, for example, although it is certainly possible to configure standards and test blueprints for such tests with minimal grade-to-grade overlap.

Tests that are vertically scaled are intended to support valid inferences regarding growth over time, since a student may take forms that become intentionally more difficult as the student progresses through years of schooling. Vertical scaling may also support the use of 'out-of-level' testing, where a student of one grade level may take a test form built for a different grade level, as may be a prescribed accommodation for some students.

The inferences that vertical scaling of test forms can support have important limitations. Because adjacent test levels are not parallel forms, vertical scaling does not constitute an equating of forms in the strict sense, but instead represents a form of test linking. Furthermore, the validity of inferences based on such a linking will depend on the strength of the link, and this strength will generally diminish with the distance between levels under most vertical scaling data collection designs. That is, although a set of tests may measure achievement in writing on a scale that spans grades 3–7, comparisons between scores based on the third and seventh grade forms will typically not be well supported, although comparisons based on adjacent forms may be very well supported.

In this chapter we examine vertical scaling issues in their statistical modeling context. In Section 2 we discuss recent and historical statistical practice for creating vertical scales. We discuss some historical approaches to vertical scaling, and we point out some limitations in widely employed current practice. We discuss in Section 3 general modeling issues that must be considered in evaluating approaches to vertical scaling. In Section 4 we develop and implement a hierarchical model with a layer that explicitly models the functional form of grade-to-grade growth in mean proficiency. We conclude with a general discussion of the various modeling approaches.

2. Vertical scaling of achievement tests: Recent and past statistical practice

Tests and assessments that measure growth across large grade spans on a common scale have a long history. For example, editions of the *California Achievement Tests* published before 1980 (e.g., CTB/McGraw-Hill, 1979) used Thurstone (1928) scaling, as described in Gulliksen (1950, pp. 284–286). Under this version of Thurstone scaling, raw scores (i.e., number correct scores) for equivalent groups of examinees are normalized and linearly equated.

The Iowa Tests of Basic Skills have been vertically scaled using a "scaling test" design, wherein a special test composed of items selected from each test level is administered to special samples of students. Within- and between-grade variability and growth is determined on the basis of the scaling test results, and grade level specific forms are linearly equated to the scale (Hieronymus et al., 1982).

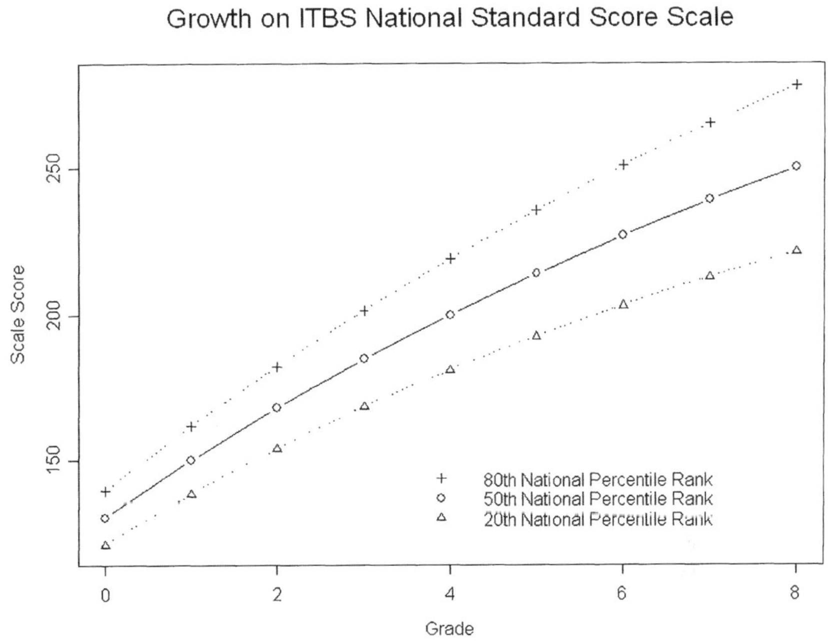

Fig. 1. Grade-to-grade growth on the ITBS national standard score scale, derived from *ITBS Guide to Research and Development*, 2003.

The Iowa Tests report student test results in terms of grade equivalent scores and normalized standard scores. A student's grade equivalent score is defined as the grade level for which the student's performance is the national median. Figure 1 depicts grade level growth on Iowa's developmental standard scale, as can be inferred from information presented in the *ITBS Guide to Research and Development* (Riverside Publishing, 2003). Figure 1 shows grade on the horizontal axis (with Kindergarden as grade 0), and scale score for the national 50th, 20th, and 80th percentile ranks. A grade equivalent score associated with any scale score may be identified by finding the grade (on the horizontal axis) associated with any scale score (from the vertical axis) using the line for the 50th national percentile rank.

We note as an aside that Figure 1 displays a pattern of increasing within-grade variability as grade increases. Other authors have found decreasing or stable within-grade variability depending on the method for creating the vertical scale, and the topic has been a matter of some controversy (see, for example, Camilli et al., 1993).

It is generally recognized that grade equivalent scores must be interpreted with some caution. Although it is possible, for example, to report grade-equivalent scores that differ (significantly) from a given student's grade level, such a score should not be taken as evidence that the student could advance several grades. A more subtle misinterpretation of grade equivalent scores can arise when growth across grades is nonlinear. For example, a pattern of decelerating growth is visible in the ITBS standard score scale and is also visible in empirically derived vertical scale results for other published tests.

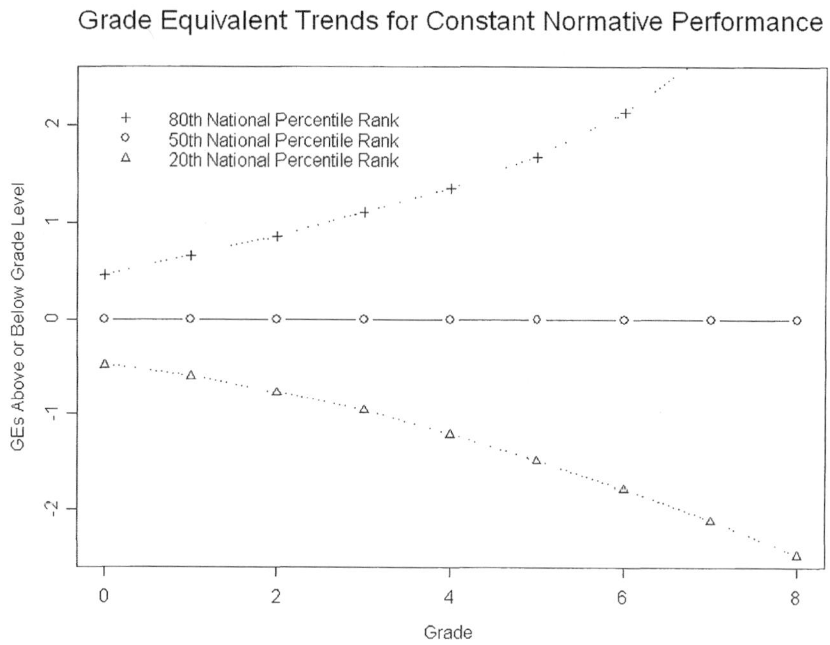

Fig. 2. Examinees or schools with consistent normative performance may appear otherwise when grade equivalent scores are used to characterize nonlinear growth.

Figure 2 displays, for students at the 20th, 50th, and 80th national percentile ranks, the number of grades that each is above or below grade level. A student (or group of students) scoring consistently at the 20th national percentile rank would appear to be falling farther and farther behind over time. Similarly, students scoring consistently at the 80th national percentile rank would appear to be pulling farther ahead. In each case consistent normative performance may be misinterpreted when grade equivalent scores are used.

Note that grade equivalent scores are uniquely defined only when the pattern of growth across grade levels is monotonic. Some empirical evidence (e.g., Heynes, 1978) suggests that students may fall back during a summer away from school, and achievement test batteries standardized for both fall and spring norms (e.g., CTB/McGraw-Hill's Comprehensive Test of Basic Skills, Fifth Edition), manage observed "summer slumps" by creating separate sets of grade equivalent scores for fall and spring. Finally, we note that grade equivalent scores may be derived from a vertical scale, but that grade equivalents per se do not constitute a vertical scale (since, for example, the amount of growth from one grade to the next is defined to be one grade equivalent unit).

2.1. *Applications of item response theory to vertical scaling*

Item response theory (IRT; Lord, 1980) models describe the probability of responding correctly (or to a degrees of partial correctness) to each of a set of items, for students with proficiency characterized by values on a continuous scale. Examples include the one-parameter (Rasch, 1960), two-parameter, or three-parameter logistic (3PL; Birnbaum, 1968) models, typically used for dichotomously scored multiple-choice items, and (two-parameter or generalized) partial credit models (e.g., Masters, 1982; Muraki, 1992; Yen, 1993) for polytomously scored constructed response items. When any of these IRT models are estimated based on examinee response data, the resulting "latent variable" proficiency estimate for any examinee is statistically identified as that variable that explains the statistical dependence of responses to multiple items by individual examinees. Conditional independence of item responses given proficiency is a fundamental IRT modeling assumption, and it has direct implications not only for model building and estimation, but also for interpretation of reported scale scores. Scale scores thus have immediate interpretation not only as the general proficiency on the domain defined by the items on the exam, but also in terms of likely success for an examinee on particular items. This observation is exploited, for example, in standard setting procedures that order items from easiest to most difficult and elicit judgments regarding required levels of mastery on the ordered item set (e.g., bookmark standard setting as introduced by Lewis et al., 1996; see also Mitzel et al., 2001).

Scales based on IRT models have become the most widely used in vertically scaled achievement tests in the United States. The Rasch (1960) IRT model has been widely used for many decades. This model is the basis for scaling the widely used Stanford Achievement Tests and Metropolitan Achievement Tests, for example, as well as a number of state testing programs (e.g., California's STAR testing program). The application of the Rasch model to vertical scaling applications received significant attention and scrutiny (e.g., Slinde and Linn, 1979; Gustafsson, 1979; Loyd and Hoover, 1980; Holmes, 1982). This literature highlights some difficulties in interpreting results when linking carried out using one set of examinees is used to support interpretations regarding the performance of another set of examinees, especially when the different examinee groups have different ability distributions.

The application of other IRT models (e.g., 3PL) in vertical scaling contexts has been widely employed more recently. These models are used in the vertical scaling of TerraNova series of tests (including the CTBS/5 and CAT/6 tests) published by CTB/McGraw-Hill, as well as in state testing programs such as those in Mississippi and Tennessee. Harris and Hoover (1987), and Harris (1991) studied the application of the 3PL model to vertical scaling, and found reasons to be cautious about interpreting the results for largely the same reasons identified for the Rasch model studies (i.e., the ability level of the examinees involved in the linking of test forms has an impact on the resulting scores and interpretations).

Comparisons of IRT and Thurstone scales in a variety of contexts may be found in Yen (1986), Yen and Burket (1997), and Williams et al. (1998). A particularly interesting discussion of historical approaches to the development of vertical scales for achievement test batteries, with an emphasis on the advantages and disadvantages of grade-equivalent scores, may be found in the 1984 edition of *Educational Measurement: Issues and Practice* (Hoover, 1984a, 1984b; Burket, 1984). Because of its widespread use, we focus on IRT approaches to vertical scaling in the remainder of this chapter.

Several compelling features are provided by achievement tests when they are vertically scaled. They facilitate the estimation and tracking of growth over time, as we may obtain repeated measures on individual students using different, age- and grade-appropriate test forms. This would help determine how much growth has occurred over time and in different regions of the proficiency range. Second, it would appear that vertically scaled achievement tests would allow more robust comparisons from single, cross-grade administrations. In particular, vertical scales allow comparisons of one grade level to another, and one cohort of students to another at any point in time.

Vertical scaling of test forms also enables important comparisons regarding test items. Vertical scaling can lead to more efficient field testing of new content, as items targeted for one grade might be found to be of more appropriate difficulty for an adjacent grade. Final form selection for a target grade can then identify appropriate items from a larger pool, when each item in the pool has parameters on a common scale. In addition, standard setting judgments can be made more precisely in theory, since a richer set of items (from adjacent levels of the test) may be ordered and the scale may thus be more finely segmented as the density of items increases.

All of these advantages that vertical scaling promises to deliver rest on the information provided by the vertical scaling analysis. In the case of IRT models, the information comes in the form of model parameters, and valid inference depends on the appropriateness and fit of the model, and the accuracy with which model parameters can be estimated. For these reasons, careful examination of modeling approaches that support vertical scaling is warranted.

3. Statistical modeling approaches for vertical scaling

3.1. Modeling complex educational assessment phenomena

Inference problems that are complex may sometimes be broken down into a series of simpler problems to be tackled in sequence. Educational assessment programs regularly employ this "divide-and-conquer" strategy. For example, the National Assessment of Educational Progress (NAEP) has the complicated task of measuring and tracking over time the achievement of students in the United States as a whole and in selected subgroups of the nation. Divide-and-conquer features of the NAEP analyses include: (1) errors in the rating process for constructed-response items are monitored during the scoring, but then assumed to not exist in subsequent analysis; (2) IRT item parameters are estimated using a simple $N(0, 1)$ model for proficiency, and then assumed to be fixed and known at their estimated values in subsequent analysis (Patz and Junker, 1999b).

Similarly, operational scoring of standardized tests typically proceeds under an assumption that the test's characteristics (e.g., item parameters, raw-score-to-scale-score conversions), are known and fixed at values based on an earlier analysis. Although it would be possible to re-estimate test characteristics for the population being scored, that would not be practical in most cases. Furthermore, for certain inferences regarding test scores (e.g., comparisons with a norm group), it would not be desirable to re-estimate the item and test characteristics.

Under a divide-and-conquer approach, the results of each sub-analysis will generally be easy to interpret, and this may be especially important for validity in assessment contexts. Furthermore, sources of uncertainty or model misspecification may be easier to identify and easier to isolate from other analyses. For example, if an IRT model does not fit response data well, this fact may be easier to identify in an analysis that focuses on the IRT model and does not include parameters for rater accuracy or parameters for subpopulation effects.

Although divide-and-conquer approaches may have these advantages and may also be computationally tractable where more unified analyses may not, a clear disadvantage of a divide-and-conquer analysis is the difficulty associated with propagating errors or uncertainty from one part of the analysis to another. For example, standard errors associated with mean achievement levels in NAEP do not reflect any uncertainty attributable to errors in the human rating process.

Inferences based on vertical scaling data may be amenable to either divide-and-conquer or unified analyses. Unified analysis may take the form of concurrent estimation of a single unidimensional IRT model that spans the grade levels and models the population proficiency distribution for each grade (Bock and Zimowski, 1996; Zimowski et al., 1996). Divide-and-conquer approaches involve separate analysis of data from each grade level (or possibly pairs of levels), followed by an analysis focused on the relationships between the grade-by-grade results.

The relative performance of divide-and-conquer versus unified analysis (i.e., separate versus concurrent IRT estimation) in the case of horizontal equating (i.e., equating forms intended to have equal difficulty) has been studied (Hanson and Béguin, 2002; Kim and Cohen, 1998; Petersen et al., 1983). There is some evidence that when the model is correctly specified concurrent calibration produces more accurate results than separate calibration. Hanson and Béguin (2002) provided some evidence to indicate that this is at least partly due to the fact that there is one set of parameter estimates for the common items, which is based on more data than the two separate sets of parameter estimates for the common items in separate calibration.

Several variations of divide-and-conquer are currently in use for vertical scaling, and these are discussed and compared to a unified (i.e., concurrent calibration) approach by Karkee et al. (2003) and Hoskens et al. (2003). These analyses using real assessment data fail to indicate any advantage to the unified approach, and instead suggest that separate calibration by grade level is superior. Patz and Hanson (2002) conjectured that whereas a concurrent calibration may produce more accurate results when the model is correctly specified, it may be that separate estimation or fixed estimation is more robust to model misspecification than concurrent estimation.

A noticeable feature of existing statistical approaches (unified and divide-and-conquer) to vertical scaling is that no assumptions are made regarding the progression of proficiency across the grade levels. Instead, assumptions are made regarding the test items (e.g., that items appearing on two test levels have common technical characteristics on each), and inferences regarding progression of proficiency across grade levels are made by examining the resulting distributions of estimated student proficiencies. When those distributions have implausible characteristics, the validity of the scaling is rightly called into question.

For example, in the early years of California's STAR testing program, the state's ninth grade population, year after year, was reported to be less proficient in reading on average than its eighth grade population (see, for example, Shrag, 1999). In part because this phenomenon (the "ninth grade dip") showed up in other states using the same standardized test battery, it was generally believed that the problem was likely a problem with the vertical scaling of the standardized test or the appropriateness of the vertical scale for newly developed forms of the standardized test.

Now that California and most other states have moved away from nationally standardized tests in favor of custom state tests, they face the prospect of vertically scaling their NCLB reading and math tests with some caution. In many cases, states have developed their standards and tests for different grade levels at different points in time and with different stakeholders. The resulting system of standards and test blueprints may lack coherent cross grade articulation of progressively more challenging content that publishers of nationally standardized tests build with great care. Furthermore, the methods widely used in these vertical scaling analyses (for state tests and nationally standardized tests) depend on strong assumptions regarding the behavior of relatively few item parameters and not at all on reasonable assumptions regarding how very large numbers of examinees might be expected to behave on average.

As a general statistical matter, assumptions (e.g., ordinality, smoothness) regarding the progression of proficiency across grade levels may be articulated as constraints on appropriate model parameters, resulting in a fitted model in which both items and examinees may be expected to display technical characteristics that conform to reasonable expectations. In Section 4 we demonstrate the feasibility of such a statistical modeling approach to establishing a vertical scale. First we lay some groundwork by describing a general modeling framework.

3.2. *A general IRT modeling framework for vertical scaling*

In order to develop and explore comprehensive and unified modeling approaches it is necessary to introduce a formal statistical model. We find Patz and Junker (1999b) description useful here. Item response data consists of a set $\{X_{ij}\colon i = 1, 2, \ldots, N;\ j = 1, 2, \ldots, J\}$ of J discrete observations on each of N individual examinees. IRT models quantify examinee i's propensity to answer item j correctly based on the examinee's location θ_i on a latent scale, and on characteristics of the items captured by parameters β_j. Covariates $Y = (Y_1, Y_2, \ldots, Y_N)$ on examinees (e.g., demographic or grade level information) and covariates $Z = (Z_1, Z_2, \ldots, Z_J)$ on items may be specified. We let $\lambda = (\lambda_1, \lambda_2, \ldots, \lambda_L)$ represent the parameters of the distribution of θ in the population; for example, if θ is assumed to be normally distributed, then perhaps $\lambda = (\mu, \sigma)$, the usual normal mean and standard deviation parameters. We assume independence of item responses within examinees given proficiency (i.e., local independence), and independence of item responses across subjects given θ and the relevant covariates. These assumptions imply that

$$
\begin{aligned}
p(X|\beta, \lambda, Y, Z) &= \int p(X|\theta, \beta, \lambda, Y, Z)\, p(\theta|\beta, \lambda, Y, Z)\, \mathrm{d}\theta \\
&= \int p(X|\theta, \beta, Y, Z)\, p(\theta|\lambda, Y, Z)\, \mathrm{d}\theta
\end{aligned}
$$

$$= \int p(X|\theta, \beta, Z) p(\theta|\lambda, Y)\, \mathrm{d}\theta$$

$$= \prod_{i=1}^{N} \int \prod_{j=1}^{J} p(X_{ij}|\theta_i, \beta_j, Z_j) p(\theta_i|\lambda, Y_i)\, \mathrm{d}\theta_i, \tag{1}$$

where X_{ij} is the response of examinee i to item j (Patz and Junker, 1999b). Most standard IRT models and estimation methods may be characterized in relation to this general framework.

In a fully Bayesian analysis, a prior distribution $p(\beta, \lambda|Y, Z)$ for the parameters λ and β is specified. It is usually most convenient to assume that all parameters are a-priori independent, so that $p(\beta, \lambda|Y, Z) = p(\beta|Z)p(\lambda|Y) = \prod_1^J p_j(\beta_j|Z) \prod_1^L p_\ell(\lambda_\ell|Y)$. We are interested in features of the joint posterior distribution $p(\theta, \beta, \lambda|X, Y, Z)$, which under the assumptions above we may write as

$$\begin{aligned} &p(\theta, \beta, \lambda|X, Y, Z) \\ &\quad \propto p(X|\theta, \beta, \lambda, Y, Z) p(\theta|\beta, \lambda, Y, Z) p(\beta|\lambda, Y, Z) p(\lambda|Y, Z) \\ &\quad = p(X|\theta, \beta, Z) p(\theta|\lambda, Y) p(\beta|Z) p(\lambda|Y). \end{aligned} \tag{2}$$

This joint posterior distribution presents a comprehensive and unified view of all unknown quantities of interest in the model, including their expected or most probable values, their statistical dependence on one another, and the degree of precision with which they may be characterized. Estimation and examination of this distribution will support unified analyses as generally described above. If the joint posterior distribution $p(\theta, \beta, \lambda|X, Y, Z)$ could not be estimated easily, we might resort to a divide-and-conquer approach involving a sequence of analyses, each of which focuses on one part of the model and makes simplifying assumptions regarding others.

The standard IRT models commonly estimated using marginal maximum likelihood via the E–M algorithm (Bock and Aitken, 1981) may be seen as a special case of the general model described above. In this case $p(X|\theta, \beta)$ is a standard unidimensional IRT model (e.g., 3PL), and characteristics λ of the proficiency distribution are fixed so that $p(\theta|\lambda) = N(0, 1)$. Separate calibration approaches to the vertical scaling problem may be viewed as repeated estimation of the simple IRT model (once for each grade), followed by a linear transformation of results based on separate analysis of parameters of items common to two or more grades. (See, for example, Hanson and Béguin, 2002.)

The standard unified analysis of vertical scaling data, concurrent estimation of a unidimensional, multiple group, IRT model, may also be seen as a special case of the general framework. Here again, $p(X|\theta, \beta)$ is a unidimensional IRT model, the covariate Y of interest is the grade level of the student, and features of the proficiency distribution λ are $\{\mu_g, \sigma_g\}_{g=1}^G$, the mean and standard deviation of the population at each grade g.

In the remainder of this chapter we introduce a hierarchical model for $p(\theta|\lambda, Y)$ that includes a functional form for mean growth across grades. This level will allow us to explicitly estimate a mean growth trajectory, and to do so under assumptions that such a trajectory should have some degree of regularity. More generally, the approach is an example of how estimation for one "small area" (i.e., one grade level) may be informed by estimation of other areas of the model (i.e., adjacent grade levels).

An alternative exploration using this framework may be found in Patz and Yao (2005), where proficiency θ is allowed to be multidimensional and $p(X|\theta, \beta)$ is a multidimensional IRT model. Whereas multidimensional IRT analyses can explain how the definition of proficiency changes over grades, unidimensional IRT analyses are likely to remain central in the actual building of assessments and the reporting of their results.

In order to implement the approaches and to retain flexibility to easily examine modeling alternatives, we develop Markov chain Monte Carlo (MCMC) algorithms. MCMC may be implemented in a straightforward manner for fitting item response models (Patz and Junker, 1999a), and variations of standard IRT models may be developed and examined with relative ease using MCMC (e.g., Patz and Junker, 1999b; Béguin and Glas, 2001).

4. Exploring grade-to-grade growth in achievement: An application of hierarchical modeling

In this section we develop and implement a version of (1) that includes a hierarchical model for $p(\theta|\lambda, Y)$ with a functional form for mean growth across grades. Patz and Hanson (2002) suggested such an approach for vertical scaling problems but did not implement it. Principles of hierarchical modeling and Bayesian data analysis would suggest that such an approach might result in more regular patterns of mean growth across grade levels, as estimates for any grade level mean will "borrow strength" from the information available at neighboring grades.

In this chapter we are able to demonstrate plausibility of the approach and reasonableness of parameter estimates using a simple simulation study and an analysis of real vertical scaling data. We introduce one functional form (quadratic growth across grades) and estimate the model using a specific set of prior distributions. We note at the start that significant further investigation (including, for example, sensitivity of results to prior specification) will be required before the approach we recommend is adopted in practice.

4.1. Model specification

In K-12 achievement tests, the mean growth curves often appear to follow an approximately quadratic form, with the growth rate noticeably steeper at the lower grades and leveling off as the grades increase. We will examine the hierarchical modeling approach assuming a general quadratic form (i.e., one that would accommodate growth that accelerates over grades just as easily as growth that decelerates). We will assume standard item response theory models for dichotomous and polytomous items.

For a dichotomous item j we will assume that the probability of a correct response for an examinee with ability θ_i follows the three-parameter logistic (3PL) model:

$$P_{i,j} = P(X_{i,j} = 1|\theta_i, \beta_j) = \beta_{3,j} + \frac{1 - \beta_{3,j}}{1 + e^{(-\beta_{2,j}\theta_i + \beta_{1,j})}}. \tag{3}$$

In this case $\beta_j = (\beta_{1,j}, \beta_{2,j}, \beta_{3,j})$ are the parameters for the jth item.

Let

$$P_{i,j}(X_{i,j}|\theta_i, \beta_j) = P_{i,j}^{X_{i,j}}(1 - P_{i,j})^{1-X_{i,j}}.$$

For a polytomous item j, we assume the probability of a response k for an examinee with ability θ_i is given by the generalized or two parameter partial credit model (2PPC) as:

$$P_{i,j,k} = P(X_{i,j} = k|\theta_i, \beta_j) = \frac{e^{(k-1)\beta_{2,j}\theta_i - \sum_{t=1}^{k}\beta_{\delta_t,j}}}{\sum_{m=1}^{K_j} e^{[(m-1)\beta_{2,j}\theta_i - \sum_{t=1}^{m}\beta_{\delta_t,j}]}}, \tag{4}$$

where

$$\beta_j = (\beta_{\delta_2,j}, \ldots, \beta_{\delta_{K_j},j}, \beta_{2,j})$$

are the parameters for jth item, $\beta_{\delta_1,j} = 0$, and $k = 1, 2, \ldots, K_j$, where K_j is the number of response categories for the jth item. The latent trait space can be unidimensional or multidimensional. In this section, the latent trait space is unidimensional.

Let

$$P_{i,j}(X_{i,j}|\theta_i, \beta_j) = \prod_{k=1}^{K_j} P_{i,j,k}^{1_{(X_{i,j}=k)}},$$

where

$$1_{(X_{i,j}=k)} = \begin{cases} 1 & \text{if } X_{i,j} = k, \\ 0 & \text{otherwise.} \end{cases}$$

The item parameters for all items on the test are

$$\beta = (\beta_1, \ldots, \beta_j, \ldots, \beta_J),$$

where

$$\beta_j = (\beta_{1,j}, \beta_{2,j}, \beta_{3,j})$$

for 3PL items and

$$\beta_j = (\beta_{\delta_2,j}, \ldots, \beta_{\delta_{K_j},j}, \beta_{2,j})$$

for 2PPC items. The ability parameters for examinees are

$$\theta = (\theta_1, \ldots, \theta_N).$$

Thus the likelihood takes the form

$$P(X|\theta, \beta) = \prod_{i=1}^{N} P(X_i|\theta_i, \beta) = \prod_{i=1}^{N}\prod_{j=1}^{J} P(X_{i,j}|\theta_i, \beta_j). \tag{5}$$

A two-level model for proficiencies can be specified by re-writing one term in Eq. (2):

$$p(\lambda|Y) = p(\lambda|Y, \tau)p(\tau|Y).$$

Here we allow τ to characterize the distribution of population parameters.

We assume that for students in grade g, with population mean and standard deviation μ_g and σ_g, we assume

$$P(\theta|\mu_g, \sigma_g) \sim N(\mu_g, \sigma_g).$$

The population parameters λ are thus

$$\lambda = (\mu_1, \sigma_1, \mu_2, \sigma_2, \ldots, \mu_G, \sigma_G),$$

where each μ_g, σ_g is the mean and standard deviation of the population distribution for grade $g = 1, 2, \ldots, G$.

Thus,

$$P(\lambda|Y) = \prod_{g=1}^{G} P(\mu_g, \sigma_g|Y, \tau)P(\tau|Y) \tag{6}$$

$$= \prod_{g=1}^{G} P(\mu_g|\tau)P(\sigma_g)P(\tau|Y). \tag{7}$$

Here we will use τ to specify a quadratic function for mean proficiencies over grades. That is $\tau = (a, b, c)$ and

$$P(\mu_g|\tau) = P(\mu_g|a, b, c) \sim N\big(ag^2 + bg + c, \sigma_\mu = 0.1\big), \tag{8}$$

$$P(\sigma_g) \sim N(\mu_\sigma = 1, \sigma_\sigma = 0.1). \tag{9}$$

Finally, we assume a flat prior for quadratic parameters

$$p(\tau) = P(a, b, c) = 1. \tag{10}$$

4.2. Estimation method and software

Markov Chain Monte Carlo (MCMC) methods sample from the posterior distribution as in Eq. (2), which were obtained from Eq. (5) (the likelihood), (7), (8) and (9). Markov Chain Monte Carlo algorithms are well suited to IRT models (see, for example, Patz and Junker, 1999a, 1999b). The algorithm has been implemented in the software program BMIRT (Yao, 2003). The BMIRT microcomputer software program was used for calibrations of multiple groups. Identifiability of the model is achieved by constraining the mean and standard deviation of the person ability distribution equal to 0 and 1, respectively, for one of the groups. Means and standard deviations for all other groups may then be estimated along with the other model parameters. Additional discussion of identifiability may be found in Patz and Hanson (2002).

For comparison, Pardux (Burket, 2002), a microcomputer software program was used for all calibrations and equatings in the present study. Pardux can estimate dichotomous and polytomous IRT models for both single groups and for multiple groups using the E–M algorithm. In the single group case Pardux constrains the mean and standard deviation of the person ability distribution equal to 0 and 1, respectively, during the item parameter estimation process to obtain model identification. In the multiple group case Pardux constrains the mean and standard deviation of the person ability distribution equal to 0 and 1, respectively, for one of the groups and estimates the means and standard deviations for all other groups during the item parameter estimation process, in order to obtain model identification. Marginal maximum likelihood is used to estimate item parameters. Although Pardux estimates means and standard deviations for each grade level, no prior distributions are specified for these parameters and no functional form is assumed or estimated. The Pardux program can also be used for equating and linking. Pardux uses the Stocking–Lord (1983) test characteristic curve method for equating/linking.

We specify prior distributions for item parameters. Proper prior distributions can be helpful in obtaining stable parameter estimates, particularly true for the 3PL model (see Patz and Junker, 1999b). For multiple choice items (3PL model), we assume the following priors:

$$\begin{aligned}
&\beta_{1,j} \sim N\left(\mu_{\beta_{1,j}}, \sigma^2_{\beta_{1,j}}\right),\\
&\log(\beta_{2,j}) \sim N\left(\log(\mu_{\beta_{2,j}}), \sigma^2_{\beta_{2,j}}\right),\\
&\beta_{3,j} \sim beta(a, b).
\end{aligned}$$

For constructed response items (2PPC model), the item parameters are

$$\beta_j = (\beta_{\delta_2,j}, \ldots, \beta_{\delta_{K_j},j}, \beta_{2,j}).$$

The priors are taken to be lognormal for $\beta_{2,j}$ and normal for $\beta_{\delta_k,j}$. And the means and standard deviations of the prior distributions are $\mu_{\beta_{2,j}} = 0$, $\mu_{\beta_{\delta_k,j}} = 1$, $\sigma_{\beta_{\delta_k,j}} = 1$, $\sigma_{\beta_{2,j}} = 1$, where $k = 1, \ldots, K_j$.

4.3. *Examining model estimation with simulated data*

4.3.1. *Simulated data*

We examine characteristics of the model estimation using a simulated data set with characteristics similar to what was observed in a vertical scaling analysis using real assessment data. In particular we simulate 1000 examinees in each of 5 grade levels. The ability distribution for each grade was taken to be normal. The means across grades were taken to fall along a quadratic function, namely $(-1.65, -0.65, 0, 0.40, 0.60)$. The standard deviations are specified to be sd $= (0.906, 0.965, 1, 1.110, 1.006)$. The number of common items between grades are $(18, 18, 18, 19)$, and they are all MC items. For each grade, there are 13 constructed responce items: seven two-level items, one three-level item, and five five-level items. The total number of items for each grade 3–7 are 53, 53, 54, 55, and 51.

Figure 3 displays the test characteristic curves (TCCs) for Grade 1 to Grade 5 from the true parameter files that we used to simulate the data. Also visible in the graph is a vertical line indicating the mean of the examinee proficiency for each grade.

4.3.2. *Estimation results*

We fit the hierarchical BMIRT model in 15000 iterations of the MCMC algorithm, using the first 5000 to reach stationarity ("burn-in"), and estimating parameters using

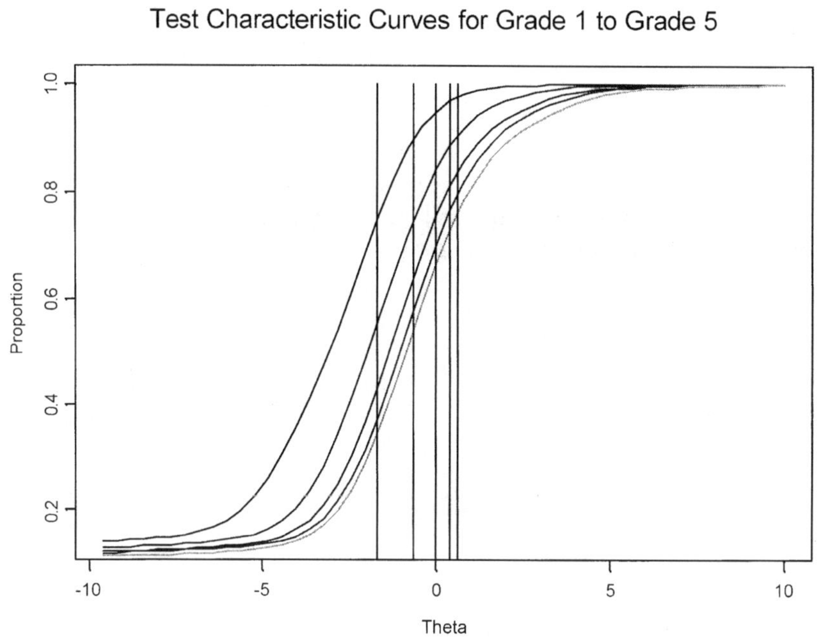

Fig. 3. Test characteristic curves for simulated vertical scaling test forms. Vertical lines show population means by grade.

Table 1
BMIRT EAP population parameter estimates (posterior standard deviations in parentheses) from a simulated data set, with (nonhierarchical) Pardux estimates for comparison. Parameters a, b, and c indicate the shape of the quadratic growth specified for the prior distribution of grade-level means

Variable	True	Pardux	BMIRT (SE)
a	−0.13	–	−0.16 (0.08)
b	1.35	–	1.48 (0.48)
c	−2.85	–	−2.94 (0.58)
μ_1	−1.65	−1.82	−1.66 (0.11)
μ_2	−0.65	−0.71	−0.63 (0.06)
μ_3	0.00	0.00	0.00 (0.03)
μ_4	0.40	0.41	0.37 (0.07)
μ_5	0.60	0.58	0.50 (0.08)
σ_1	0.91	1.05	0.90 (0.11)
σ_2	0.97	1.08	1.01 (0.10)
σ_3	1.00	1.01	1.00 (0.05)
σ_4	1.11	1.16	1.20 (0.11)
σ_5	1.01	0.97	0.93 (0.10)

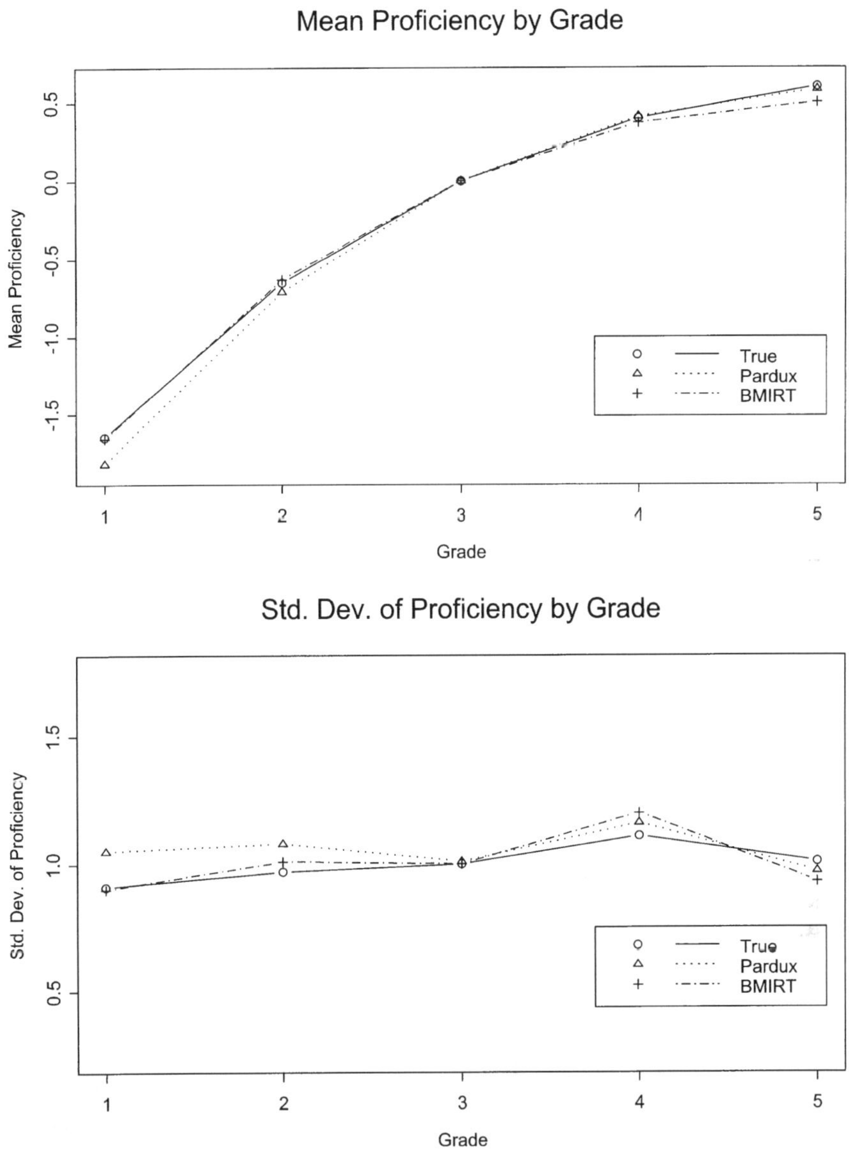

Fig. 4. True and estimated proficiency distribution means and standard deviations across grades from simulation.

the last 10000 iterations. Quick inspection, sufficient for our exploratory purposes here, suggests that the Markov chain reached stationarity.

Results for population parameters are presented in Table 1, and graphically in Figure 4. The BMIRT software appears to do a plausible job of capturing population parameters. There appears to be some divergence in the extreme grades, with BMIRT

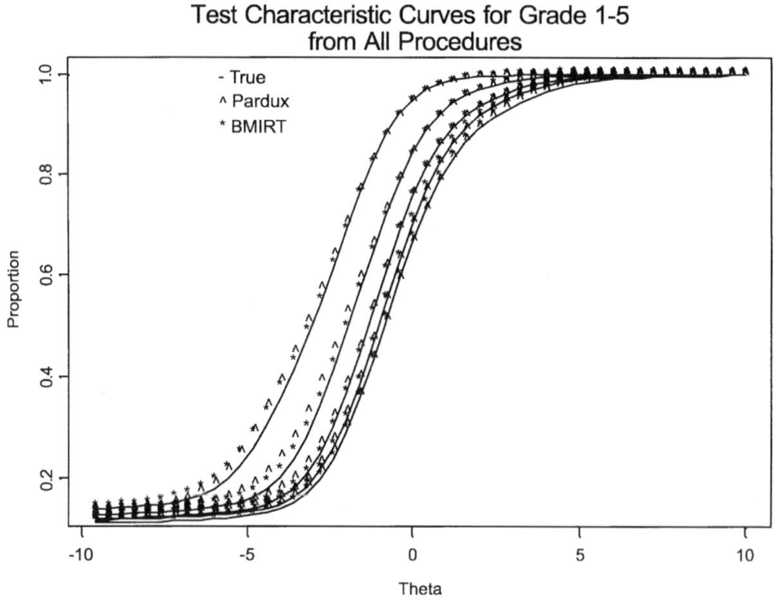

Fig. 5. True and estimated (by Pardux and BMIRT) test characteristic curves for each grade.

doing a better job at the lowest grade and Pardux doing a better job at the higher grade. More testing of the program is warranted.

Results for item parameters are summarized graphically as the corresponding test characteristic curves are displayed in Figures 5 and 6. The BMIRT software appears to do a plausible job of capturing item parameters.

It is worth emphasizing again that the purpose of this simulation and model estimation exercise was to examine behavior of the modeling approach and software program in an example for which the BMIRT model fits exactly. It appears to behave adequately in this case. One might expect that BMIRT fitted growth curves would be close to the true curves, given the quadratic form of the prior on grade means, and this appears visible to some extent. It will be important in future work to examine the impact of making an incorrect assumption regarding the functional form of the growth curves. Comparisons with Pardux estimates simply provide some additional context for examining the results, but no conclusion regarding the relative performance of the two programs is suggested or supported by these analyses.

4.4. *Vertically scaling real achievement test data*

We next examine the behavior of the hierarchical model when fit to real achievement test data. For this purpose we use data collected as a part of the national standardization studies conducted for the *Comprehensive Test of Basic Skills, Fifth Edition* (*CTBS/5*; CTB/McGraw-Hill, 1998). We examine the mathematics results for grades 2–8, utilizing nationally representative samples of 1700 examinees per grade. We compare the resulting grade level means to those published by CTB/McGraw-Hill (1998). Published

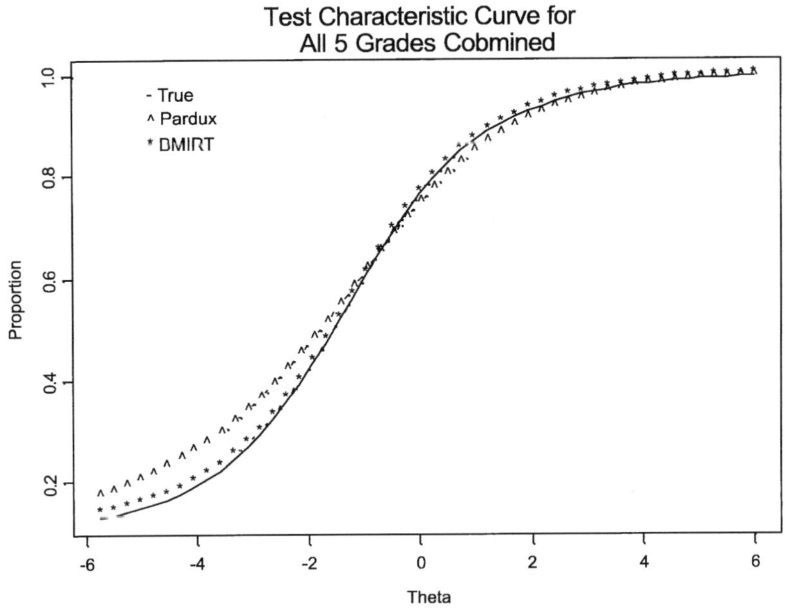

Fig. 6. True and estimated (by Pardux and BMIRT) test characteristic curves for a test comprising all items across all grade levels.

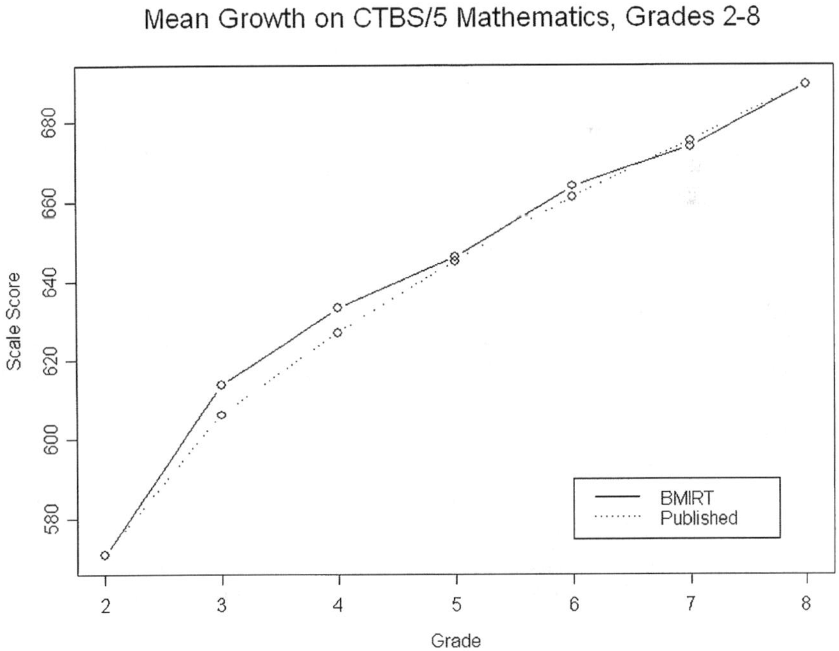

Fig. 7. Growth on CTBS/5 Mathematics between grades 2 and 8.

results are based on simultaneous analysis of fall and spring testing data, and are subject to additional smoothing procedures. To align the scales, we linearly transform our results so that the grade 2 mean and grade 8 mean agree, allowing us to examine how our estimates of where this growth occurs compares to the published results. Figure 7 displays the results. We can see that our results display a generally similar trend to the published results. This analysis suggests that the approach may plausibly be applied to real data sets, with reasonable grade-to-grade trends resulting. More extensive development and testing is warranted for this particular model, its variations, and this particular estimation approach.

5. Discussion

In this chapter we have examined general issues in modeling educational assessment data for which vertical scaling may be appropriate. We have described the divide-and-conquer and unified approaches currently in practice, and we have developed and explored a unified approach that might overcome significant limitations of existing approaches. In particular, we developed an approach that might be most useful when there is reason to believe that the growth in proficiency might be relatively smooth, and where other information regarding the relative proficiencies of grade levels might be strengthened.

The hierarchical approach is a more general version of concurrent estimation of the unidimensional IRT model. That is, the marginal Bayes or maximum likelihood estimation of the multi-group IRT model is consistent with the hierarchical model with flat priors for parameters characterizing the grade-by-grade proficiency distributions. When proper priors are specified for these components of the hierarchical model, the hierarchial model may be viewed as a constrained version of the multi-group IRT model. To the extent that the prior distributions capture real features of growth patterns, they promise to enable more reliable estimation in vertical scaling problems.

As in any fully Bayesian analysis, information contained in the prior distributions is combined with information contained in the data (through the likelihood function), so that the final results derived from the posterior distribution reflect both prior and data. In particular, specifying regular (i.e., quadratic in our example) growth in the prior does not force regular growth in the posterior estimates of grade level means. Although we obtain posterior estimates of parameters characterizing grade-to-grade growth (i.e., quadratic parameters a, b, and c in our example), posterior estimates μ_g for each grade g will deviate from this estimated curve. The approach will have an effect of smoothing grade-to-grade growth, and the degree of smoothness is controlled by the strength of the prior distributions (e.g., by the size of σ_μ). More work remains to be done to identify appropriate choices of these parameters for particular analyses.

Patterns of growth for individuals and populations is of direct interest, and having a model that captures salient features of growth trajectories would provide an important tool for examining these features. In this chapter we have successfully fit a model with a quadratic form for grade-to-grade growth, and we have shown by straightforward simulation that the estimation approach can capture model parameters. More work is needed to make the estimation algorithm more robust, and to identify those conditions under which the approach would provide superior or inferior results. Furthermore, the reasonableness of the quadratic growth assumption should be evaluated, and alternatives to quadratic growth (e.g., splines) should be investigated. A useful feature of the MCMC fitting algorithm is that changes to distributional assumptions and functional relationships are relatively easy to implement, so these investigations should be practical in the near future.

The results of these analyses suggest that (1) more comprehensive modeling approaches are possible; (2) patterns of growth, which will remain of interest, can be modeled parametrically and better understood and controlled; and (3) that we have some additional work to do before we are able to supplant existing divide-and-conquer approaches and reap the benefits in practice that more comprehensive modeling might bring.

Finally, we note that the approaches we have proposed and explored in this chapter relate to a single administration of a cross-grade assessment. This type of data is of significant interest, and it is generally the only type of data that may be available when a vertical scale must be established. The growth that is modeled and analyzed captures changes in the distribution of proficiency from one grade level to the next. Perhaps the most powerful information available to inform us about growth over grade levels, however, will come from longitudinal data collected as students progress over years of schooling. Such data is just now becoming available in large volume as states have established assessment programs and data warehouses that capture information every year for all students in a range of grades. This data will allow us to examine individual growth trajectories, as advocated by Willett (1994), Singer (1998), and Singer and Willet (2003), for example, and to re-examine the definition of vertical scales for achievement tests. We expect the tools we have developed here for hierarchical analysis to facilitate this promising area of study.

References

Béguin, A.A., Glas, C.A. (2001). MCMC estimation and some model-fit analysis of multidimensional IRT models. *Psychometrika* **66**, 541–561.

Birnbaum, A. (1968). Some latent trait models and their use in inferring an examinee's ability. In: Lord, F.M., Novick, M.R. (Eds.), *Statistical Theories of Mental Test Scores*. Addison-Wesley, Reading, MA.

Bock, R.D., Aitken, M. (1981). Marginal maximum likelihood estimation of item parameters: Application of an EM algorithm. *Psychometrika* **46**, 443–459.

Bock, R.D., Zimowski, M.F. (1996). Multiple group IRT. In: van der Linden, W.J., Hambleton, R.K. (Eds.), *Handbook of Modern Item Response Theory*. Springer-Verlag, New York.

Burket, G.R. (1984). Response to Hoover. *Educational Measurement: Issues and Practice* **3**, 15–16.

Burket, G. (2002). *PARDUX* [Computer Program]. CTB/McGraw-Hill.

Camilli, G., Yamamoto, K., Wang, M.M. (1993). Scale shrinkage in vertical equating. *Applied Psychological Measurement* **17**, 379–388.

CTB/McGraw-Hill (1998). TerraNova Technical Report. Monterey, CA.

Gulliksen, H. (1950). *Theory of Mental Tests*. John Wiley & Sons, New York.

Gustafsson, J.E. (1979). The Rasch model in vertical equating of tests: A critique of Slinde and Linn. *Journal of Educational Measurement* **16**, 153–158.

Hanson, B.A., Béguin, A.A. (2002). Obtaining a common scale for IRT item parameters using separate versus concurrent estimation in the common item nonequivalent groups equating design. *Applied Psychological Measurement* **26**, 3–24.

Harris, D.J. (1991). A comparison of Ango's Design I and Design II for vertical equating using traditional and IRT methodology. *Journal of Educational Measurement* **28**, 221–235.

Harris, D.J., Hoover, H.D. (1987). An application of the three-parameter IRT model to vertical equating. *Applied Psychological Measurement* **11**, 151–159.

Heynes, B. (1978). *Summer Learning and the Effects of Schooling*. Academic Press, New York.

Hieronymus, A.N., Linquist, E.F., Hoover, H.D. (1982). *Iowa Test of Basic Skills: Manual for School Administrators*. Riverside Publishing Company, Chicago.

Hoskens, M., Lewis, D.M., Patz, R.J. (2003). Maintaining vertical scales using a common item design. Paper presented at the annual meetings of the National Council on Measurement in Education, Chicago, IL.

Holmes, S.E. (1982). Unidimensionality and vertical equating with the Rasch model. *Journal of Educational Measurement* **19**, 139–147.

Hoover, H.D. (1984a). The most appropriate scores for measuring educational development in elementary schools: GE's. *Educational Measurement: Issues and Practice* **3**, 8–14.

Hoover, H.D. (1984b). Rejoinder to Burket. *Educational Measurement: Issues and Practice* **3**, 16–18.

Karkee, T., Lewis, D.M., Hoskens, M., Yao, L. (2003). Separate versus concurrent calibration methods in vertical scaling. Paper presented at the annual meetings of the National Council on Measurement in Education, Chicago, IL.

Kim, S., Cohen, A.S. (1998). A comparison of linking and concurrent calibration under item response theory. *Applied Psychological Measurement* **22**, 131–143.

Lewis, D.M., Mitzel, H., Green, D.R. (1996). Standard setting: A bookmark approach. In: Green, D.R. (Chair), ITR-based Standard-Setting Procedures Utilizing Behavioral Anchoring. Symposium presented at the 1996 Council of Chief State School Officers National Conference on Large-Scale Assessment, Phoenix, AZ.

Lord, F.M. (1980). *Applications of Item Response Theory to Practical Testing Problems*. Lawrence Erlbaum Associates, Hillsdale, NJ.

Loyd, B.H., Hoover, H.D. (1980). Vertical equating using the Rasch model. *Journal of Educational Measurement* **17**, 179–193.

Masters, G.N. (1982). A Rasch model for partial credit scoring. *Psychometrika* **47**, 149–174.

Mitzel, H.C., Lewis, D.M., Patz, R.J., Green, D.R. (2001). The Bookmark procedure: Psychological perspectives. In: Cizek, G.J. (Ed.), *Setting Performance Standards: Concepts, Methods, and Perspectives*. Lawrence Earlbaum, Mahwah, NJ, pp. 249–281.

Muraki, E. (1992). A generalized partial credit model: Application of an EM algorithm. *Applied Psychological Measurement* **16**, 159–176.

Patz, R.J., Hanson B. (2002). Psychometric issues in vertical scaling. Paper presented at the annual meetings of the National Council on Measurement in Education, New Orleans, LA.

Patz, R.J., Junker, B.W. (1999a). A straightforward approach to Markov chain Monte Carlo methods for item response models. *Journal of Educational and Behavioral Statistics* **24**, 146–178.

Patz, R.J., Junker, B.W. (1999b). Applications and extensions of MCMC in IRT: Multiple item types, missing data, and rated responses. *Journal of Educational and Behavioral Statistics* **24**, 342–346.

Patz R.J., Yao L. (2005). Methods and models for vertical scaling. Paper presented at the ETS Conference, Linking and Aligning Scores and Scales: A Conference in Honor of Ledyard R. Tucker's Approach to Theory and Practice, Educational Testing Service, Princeton, NJ, June 24.

Petersen, N.S., Cook, L.L., Stocking, M.L. (1983). IRT versus conventional equating methods: A comparative study of scale stability. *Journal of Educational Statistics* **8**, 137–156.

Rasch, G. (1960). *Probabilistic Models for Some Intelligence and Attainment Tests*. Danish Institute for Educational Research, Copenhagen.

Shrag, P. (1999). Phonics, Testing goofs and the ninth-grade dip. *The Sacramento Bee*. July 7.

Singer, J.D. (1998). Using SAS PROC MIXED to fit multilevel models, hierarchical models, and Individual Growth Models. *Journal of Educational and Behavioral Statistics* **24**, 323–355.

Singer, J.D., Willet, J.D. (2003). *Applied Longitudinal Data Analysis: Modeling Change and Event Occurence*. Oxford University Press, London.

Slinde, J.A., Linn, R.L. (1979). A note on vertical equating via the Rasch model for groups of quite different ability and tests of quite different difficulty. *Journal of Educational Measurement* **16**, 159–165.

Stocking, M.L., Lord, F.M. (1983). Developing a common metric in item response theory. *Applied Psychological Measurement* **7**, 201–210.

Thurstone, L.L. (1928). The absolute zero in intelligence measurement. *Psychological Review* **35**, 175–197.

Willett, J.B. (1994). Measurement of change. In: Husen, T., Postlethwaite, T.N. (Eds.), *International Encyclopedia of Education*. 2nd ed. Elsevier Science Press, Oxford, UK, pp. 671–678.

Williams, V.S.L., Pommerich, M., Thissen, D. (1998). A comparison of developmental scales based on Thurstone methods and item response theory. *Journal of Educational Measurement* **35**, 93–107.

Yao, L. (2003). BMIRT. Bayesian Multivariate Item Response Theory estimation [Computer software]. CTB/McGraw-Hill, Monterey, CA.

Yen, W.M. (1986). The choice of scale for educational measurement: An IRT perspective. *Journal of Educational Measurement* **23**, 299–325.

Yen, W.M. (1993). Scaling performance assessments: Strategies for managing local item independence. *Journal of Educational Measurement* **30**, 187–213.

Yen, W.M., Burket, G.R. (1997). Comparison of item response theory and Thurstone methods of vertical scaling. *Journal of Educational Measurement* **34**, 293–313.

Zimowski, M.F., Muraki, E., Mislevy, R.J., Bock, R.D. (1996). BILOG–MG: Multiple-group IRT analysis and test maintenance for binary items [Computer Program]. Scientific Software, Inc, Chicago, IL.

31

Cognitive Diagnosis

Ch. 31A. Review of Cognitively Diagnostic Assessment and a Summary of Psychometric Models
Louis V. DiBello, Louis A. Roussos and William Stout

Ch. 31B. Some Notes on Models for Cognitively Based Skills Diagnosis
Shelby J. Haberman and Matthias von Davier

Editors' note

Cognitive diagnosis or profile scoring is an emerging area – so there are different schools of thought – hence this chapter consists of two papers. The first paper, by DiBello, Roussos and Stout, provides a broad overview of how users can conduct cognitive diagnostic assessments and includes a description of psychometric models used. The second paper, by Haberman and von Davier, points to some challenges that face users of cognitively diagnostic models.

Handbook of Statistics, Vol. 26
ISSN: 0169-7161

DOI: 10.1016/S0169-7161(06)26031-0

31A

Review of Cognitively Diagnostic Assessment and a Summary of Psychometric Models[1,2]

Louis V. DiBello, Louis A. Roussos and William Stout

Abstract
This paper is divided into two main sections. The first half of the paper focuses on the intent and practice of diagnostic assessment, providing a general organizing scheme for a diagnostic assessment implementation process, from design to scoring. The discussion includes specific concrete examples throughout, as well as summaries of data studies as appropriate.

The second half of the paper focuses on one critical component of the implementation process – the specification of an appropriate psychometric model. It includes the presentation of a general form for the models as an interaction of knowledge structure with item structure, a review of each of a variety of selected models, separate detailed summaries of knowledge structure modeling and item structure modeling, and lastly some summarizing and concluding remarks.

To make the scope manageable, the paper is restricted to models for dichotomously scored items. Throughout the paper, practical advice is given about how to apply and implement the ideas and principles discussed.

1. Preliminaries

As a context for describing existing approaches to cognitive diagnostic modeling, this paper describes an assessment implementation framework, including assessment design and subsequent development, model specification, estimation, and examinee score reporting. In the second half, the paper presents a survey of psychometric models, focusing primarily on several current diagnostic models and approaches. The paper promotes further research along two streams: (1) research in the various disciplines required for

[1]DiBello and Roussos are co-first authors.

[2]The second half of this paper represents a focused updating of an earlier report by Roussos (1994). In addition we acknowledge the important review of statistical approaches to cognitively relevant assessment by Junker (1999), especially for the idea of grounding a discussion of psychometric models within the broader framework of assessment design. We follow that approach here.

effective diagnostic assessment design, field test trials, and evaluation in operational educational testing settings that may benefit from skills diagnostic scoring, and (2) further statistical and psychometric investigations of these models, methods and applications. The terms "skills" and "attributes" are used interchangeably throughout.

2. Implementation framework for diagnostic assessment

As noted by Junker (1999), the challenges of designing a diagnostic assessment are: "how one wants to frame inferences about students, what data one needs to see, how one arranges situations to get the pertinent data, and how one justifies reasoning from the data to inferences about the student." This recognition of assessment design as a process of reasoning from evidence was summarized in the NRC Report, *Knowing What Students Know* (Pellegrino et al., 2001), as the "assessment triangle" consisting of three main components: cognition, observation, and interpretation. The evidence centered design (ECD) paradigm developed by Mislevy, Almond, and their colleagues (see, e.g., Almond et al., 2003; Mislevy et al., 2003, 2004) frames that recognition into a systematic and practical approach to assessment design. In this section we adapt and augment the ECD paradigm to describe a framework for the entire implementation process for diagnostic assessment, elaborating on practical aspects of cognitively based assessment design and also going beyond design to illuminate practical issues in regard to estimation and score reporting.

We conceptualize the diagnostic assessment implementation process as involving the following main components:

(1) Description of assessment purpose;
(2) Description of a model for the latent skills of diagnostic interest (the skills space);
(3) Development and analysis of the assessment tasks (e.g., test items);
(4) Specification of a psychometric model linking performance to latent skills;
(5) Selection of statistical methods for model estimation and evaluating the results; and
(6) Development of systems for reporting assessment results to examinees, teachers, and others.

As we will detail next, the components of a successful implementation process are necessarily nonlinear, requiring considerable interaction and feedback between the components and demanding close collaboration between users, test designers, cognitive psychologists, and psychometricians.

2.1. *Assessment purpose*

The purpose of the assessment should be clearly delineated, and this purpose has strong implications for the description of the latent attribute space. For example, if the purpose is to diagnose human competency on a user-selected set of multiple skills, this would seem to require an explication of what it means to be competent in each of the skills and would seem to point toward a small number of discrete levels of classification on each skill, e.g., the dichotomy of competent versus not competent on each skill. However, if the purpose of the assessment is a more traditional one of ranking examinees along some broad general competency scale, then an appropriate skill space selection would seem to be a single continuous scale.

In particular, the purpose of the assessment will have significant impact on whether the targeted latent attribute or skill space will be modeled with one or more than one variable, and whether the variables will be discrete for classification purposes or continuous for scaling purposes. It is essential to note that targeted skill inferences can be supported by the user-selected tasks only if proper attention is given to skill information provided by the tasks. Hence, the determination of purpose in the first step interacts with choosing tasks that provide appropriate information about skills.

Already, one can see the nonlinear nature of diagnostic assessment implementation in that components (2) and (3) above are necessarily partly initiated at step (1). However, it is still convenient to think of the components as separate, even though successful implementation requires tight interaction and integration among the six components.

There are two kinds of practical skills diagnostic settings considered: (1) analysis of existing assessment data using more complex skills-based models in hopes of extracting richer information than provided by unidimensional scaling or other existing analyses, and (2) designing a test from the beginning for a skills-based diagnostic purpose. In designing an effective diagnostic assessment, satisfying the assessment purpose is the prime motivation, so delineating the assessment purpose should be accomplished with care. However, many of the applications with complex diagnostic models that are actually found in the literature are cases in which the skills diagnosis is conducted as a post-hoc analysis, called retrofitting by some, sometimes as a demonstration of a new statistical model and/or method, and sometimes as an attempt to extract richer information than the assessment was designed for. One good example of where the consideration of assessment purpose preceded and motivated the discussion of the development of a skills diagnosis implementation procedure is that of Klein et al. (1981), who investigated the diagnosis of student errors in math problems involving the addition and subtraction of fractions. Another example is the LanguEdge study of Jang (2005, 2006) and Roussos et al. (2006b). The Jang study is an example of an existing test in the sense that the LanguEdge English Language Learning (ELL) assessment items already existed. Jang defined a new classroom purpose for this test and subsequently built a successful skills framework to extract diagnostic information that would be useful for teachers and learners in an ELL class. We cite both of these examples (along with a few others, as needed) throughout this paper to illustrate important points about diagnostic assessment design and implementation.

2.2. *Description of the attribute space*

As mentioned above, the assessment purpose leads naturally to the question of what is to be assessed about the test takers – what proficiencies are involved and what types of inferences are desired. The second component in our implementation process then requires a detailed formulation of the skills and other attributes that will be measured in order to accomplish the test purpose. In this step, a detailed representation of the skills or attributes space is developed in light of the purpose and based on cognitive science, educational psychology, measurement, and relevant substantive literature. Often such literature will not lead to only one particular attribute or skills representation, and in such cases multiple representations should be considered before a selection is made. Jang (2005, 2006) and Roussos et al. (2006b) provide an excellent example of such a detailed literature investigation as Jang considered how to represent reading comprehension in second-language learning. Jang reported great controversy as to whether reading comprehension was decomposable into skills; and that, among the many researchers who did believe in a skills representation, there was still further controversy about what exactly those skills should be. Jang based her final selection of the skills space representation on careful consideration of the several possibilities given in the literature and also on task analyses of component (3). Since in this example the tasks already existed, Jang carefully analyzed the tasks before settling on a final selection of the skills representation.

In general, although substantive cognitive theory is essential for successful design and implementation, we underscore here that the arbiter of which representation is selected for the attribute space is not solely or even primarily which representation is most cognitively or psychologically correct. The substantive theory and knowledge must always be considered within the context of the purpose of the assessment. If the purpose is mainly to rank examinees along a single competency scale, a unidimensional continuous ability model may be the best solution even though most cognitive psychologists would not seriously believe that such a model captures much of any depth about real cognition. Further, the number of skills cannot be overly large and must be controlled by issues of statistical tractability.

2.3. *Development and analysis of the assessment tasks*

Logically, considering assessments as systems for reasoning about mental capabilities according to evidence from tasks administered to examinees, the choice of tasks should be based primarily on a consideration of the type and amount of evidence needed to support desired inferences about examinee attributes. Ideally, test developers should consider a wide variety of possible tasks, choosing feasible ones that best match the purpose of the assessment. As we demonstrate throughout the survey of the psychometric models in Section 4, task developers should not avoid tasks that require combinations of multiple skills per task.

In cases where the assessment instrument already exists, the question becomes how much information can be provided about the set of desired attributes by the test. In the preferred case when new tasks are explicitly designed for a new diagnostic assessment, the task development phase should also include a detailed analysis of the tasks to understand how many and what kinds of skills or attributes are involved, at what level of difficulty, and in what form of interaction. For example, Jang (2005, 2006) and Roussos et al. (2006b), in analyzing an already existing test for use as a diagnostic assessment, identified possible examinee solution strategies for each of the tasks (the test items), conducted verbal think-aloud protocols with appropriate subjects, consulted the test specification codes from the original test developers, analyzed the task textual features (such as number of words in the task stimulus, difficulty of the vocabulary in the questions asked, etc.), and conducted a nonparametric latent dimensionality analysis using data that had been previously collected from administration of the test forms. Through the use of all these analyses together, Jang was able to develop a final skills space for her assessment design that simultaneously satisfied the new teaching and learning purpose of her assessment and was theoretically defensible from the viewpoint of existing linguistic literature. In reporting her results, she stressed that it was important that the number of skills not be so large as to be statistically unsupportable. In particular, she initially identified through substantive analyses 32 possible processes or features that could have been used as skills or attributes, a number she knew was not statistically supportable. She then reduced this to 16 and finally to 9, a number she found was both statistically and substantively supportable. She noted that further reductions by combining or eliminating skills, while possibly resulting in still increased reliability, would have not been supported by substantive theory in that either skills that are too distinct would have had to have been combined or skills would have been eliminated such that some items would have had no assigned skills. In general, the number of skills should not be reduced too much by deleting or combining so as to result in the remaining skills lacking diagnostic utility or interpretability. Making this compromise between the initially proposed set of skills and a set of skills that can be successfully statistically assessed is a vital aspect of carrying out a successful skills diagnostic analysis.

In developing the skills space, it is also important to keep in mind how the skills interact with each other both in terms of correlational relationships among skills and the nature of the interactions among the multiple skills required by individual tasks. One of the more important distinctions for within-task interaction is one that we refer to as conjunctive versus compensatory. By "conjunctive" interaction of skills on a task, we mean a task for which successful application of all the required skills seems necessary for successful performance on the task as a whole – lack of competency on any one required skill represents a severe obstacle to successful performance on the task as a whole. By "compensatory" (or "fully compensatory") interaction of skills on a task, we mean that a high enough level of competence on one skill can compensate for a low level of competence on another skill to result in successful task performance, even to the extreme "disjunctive" case in which the probability of successful task performance is high so long as competency occurs on any one or more of the required skills. For example, a disjunctive model is appropriate when the various attributes really represent alternative strategies for solving the item. In that case, successful performance of the item only requires that one of the possible strategies be successfully applied.

The choice of mode of attribute interaction clearly depends on the diagnostic setting, including the purpose of the assessment, and how the skills or attributes are defined. Jang (2005, 2006) and Roussos et al. (2006b) noted that depending on the skill representation she chose, Jang could have been led to either a conjunctive, compensatory, or a mixture of the two models. She purposely chose a conjunctive skill representation for her particular setting because based on her task analyses and her reading of the literature, most of the skill interactions seemed cognitively conjunctive in nature, and the few that were not were amenable to being combined into a single skill. Understanding of this type of attribute interaction within tasks is important because it will help determine which psychometric model is most appropriate and interpretable for the intended diagnostic assessment.

Another important part of understanding skill interactions is that there may be important relationships between the skills, such as, certain skills tending to co-occur on tasks, an ordering of the difficulty of the skills, or mastery of one skill only occurring when another skill has been mastered. Also, skill pairs tend to be positively correlated in general so that choosing models that presume this correlational structure can be useful. Including such information increases the accuracy of a skills diagnostic analysis if the added complexity can be reasonably well fit by the data.

Still another important outcome of the task analysis process is the development of specific task-skill coding rules that are used for an initial substantive assignment of skills to tasks so that independent coders will reliably agree on these assignments. The matrix that relates skill assignment to tasks is referred to as the Q matrix – the rows are the tasks, the columns are the skills or attributes, and the entries are 1's and 0's indicating, respectively, whether a specific skill is intended or not intended to be measured by a particular task. One can consider a proposed Q matrix as a hypothesized linking of skills and tasks, based on theoretical and other substantive considerations. Subsequent statistical analyses can result in the initial Q matrix being modified to improve statistical performance. An example of such a Q matrix modification performed by Jang (2005, 2006) is briefly discussed later in this paper. A successful skills diagnosis critically depends on high quality Q matrix development.

Once diagnostic assessment data have been analyzed, one useful internal consistency check of the hypothesized Q matrix is the consideration of whether there are any mismatches between observed or logically inferred task difficulties (proportion of test takers successfully performing each task) and the difficulty of the skills or attributes that are assigned to the tasks. For the dichotomous case of mastery/nonmastery of skills, we speak of the estimated population proportion of masters of an attribute as the difficulty of that attribute. For example, in the case of a conjunctive model, a Q matrix entry should not assign a purportedly difficult skill to an easy task.

2.4. *Psychometric model specification*

The model discussions in the second half of this paper focus on component (4), the selection of the psychometric model. We focus primarily on psychometric models that explicitly contain multiple examinee proficiency variables corresponding to the skills or attributes that are to be diagnosed. Some researchers (other than the authors) claim the use of residual methods to derive diagnostic benefits from unidimensional item response theory (IRT) models can suffice. It should be noted that Tatsuoka's rule space method (Tatsuoka, 1983, 1984, 1990) based upon unidimensional IRT ability and person fit statistics, can be considered a residual approach that does employ a Q matrix. In the interests of space, we restrict our discussion mostly to models with an explicit Q matrix or equivalent. The only exceptions are the standard unidimensional IRT models, which are included as a base of comparison because of their widespread use in educational measurement.

One important distinction between the models is that the proficiency variables may be modeled as discrete or continuous depending on the purpose of the assessment as described above. Another important distinction, also mentioned above, will be whether the required skills for a specific task interact in a conjunctive or compensatory manner. The psychometric model itself is a mathematical function specifying the probability of a particular item response in terms of examinee skills and item characteristics (the item parameters). For a given set of skill variables and a type of skill interaction (conjunctive versus compensatory), a variety of particular models with varying simplifying assumptions may be compared and selected for suitability to the specific testing purpose at hand. For example, one might assume that the item success probability is the same for all examinees who are nonmasters of any one or more required skills, regardless of how many or which required skills are nonmastered. Another simplifying assumption is that items that involve the same skills have the same values for their item parameters and hence the same success probability for each examinee. These types of assumptions reduce the number of item parameters to be estimated, thus reducing standard errors of estimation. This can be especially useful when the total number of items measuring a given attribute is small or if the number of examinees in the sample is small. But these kinds of parameter reductions may also introduce unwanted bias if the assumptions are not warranted.

2.5. *Model calibration and evaluation*

In the next step of the implementation process, the selected model must be calibrated (estimation of both item parameters and examinee population parameters), the calibration must be evaluated, and then examinees classified on the skills (e.g., for the dichotomous classification case, estimation of mastery versus nonmastery of each skill for each examinee). Models with large numbers of skills and/or large average numbers of skills per item are in particular danger of being nonidentifiable (see further discussion below). Moreover, even if identifiability holds, test designs must provide sufficient numbers of items per skill, and not too many skills per item in order for accurate estimation to be possible.

The reality of much diagnostic testing is that for tests of reasonable length the number of items that contribute to the measurement of each skill will be relatively small compared to the numbers of items that typically contribute to the estimation of unidimensional scale values. For this reason, it is necessary to keep models as simple as possible while satisfying the constraints imposed by the diagnostic purpose. Even the simplest models that are able to satisfy the test purpose often will exhibit significant parametric complexity, at least more than is encountered in typical unidimensional models. Careful consideration of both parameter identifiability and systematic model evaluation help achieve effective calibration.

The discussion in this section is broadly divided into two parts: (1) estimation and computation and (2) evaluation of the estimation results.

2.5.1. *Commonly used estimation and computational methods*

Commonly used estimation methods In principle, models can be non-Bayesian or enhanced in a partially Bayesian or fully Bayesian manner. That is, model parameters can be posited in one of three ways: all parameters are estimated without prior distributions, a portion of the parameters can have prior distributions imposed (such as for the latent skills mastery/nonmastery vectors), or all of the parameters can be assigned a joint prior distribution.

Related to this selection of the level of Bayesian involvement in the model, three general estimation approaches have come to be discussed, some heavily used and others having the potential to be used. These are joint maximum likelihood estimation (JMLE), which seems natural for the non-Bayesian models, maximum marginal likelihood estimation (MMLE), which has been widely used for the partially Bayesian models that place a prior just on the population parameters for the latent examinee skills vector space, and Bayesian Expected *a posteriori* (Bayes EAP) estimation for fully Bayesian models that place priors on all model parameters. Bayes maximum *a posteriori* (Bayes MAP) estimation is possible also. Although conditional maximum likelihood is possible (see, for example, Junker, 1999, p. 67), we are unaware of any serious applications in skills level assessment. In many real data examples with fully Bayesian models, instead of analytically computing the Bayes EAP, Markov Chain Monte Carlo (MCMC) estimation is used to simulate the Bayesian posterior distributions of parameters. Expectation–maximization (EM) computational approaches for fully or partially Bayes models are sometimes used too.

Commonly used computational methods Because the models are usually parametrically complex and the data sets are often large (e.g., 25 or more items and 2000 or more examinees), the choice of computational method becomes an important practical issue. First, we note that a direct application of JMLE seems largely unused currently in the skills diagnostic setting, partially because of the lack of good theoretical asymptotic consistency results, which is true even for parametrically simple scaling models like the two-parameter logistic model. However, the ever increasing speed of low-cost computing seems to allow for the practical use in fitting skills diagnosis models of the JMLE exhaustive search and divide-and-conquer approach. In such an approach, one fixes item parameters while estimating examinee parameters, then proceeds vice versa, and iterates until some convergence criterion is satisfied.

As discussed above, an alternative estimation approach to JMLE is MMLE, which can be implemented by providing an examinee skills vector distribution and integrating over that distribution. The EM computational procedure can be used for this purpose. In the empirical Bayes tradition, the examinee skills distribution could be and sometimes is updated using the observed data, perhaps by using a hierarchical Bayes approach. Once the model has been thus calibrated, a straightforward likelihood approach can be used to do examinee skills classification.

MCMC is a probabilistically based computational method that is used to generate Markov chains of simulated posterior distributions for all the model parameters given the data (see Gelman et al., 1995). Largely due to the influence of the pair of Patz and Junker (1999a, 1999b) articles, MCMC has become popular in many skills diagnostic applications; in particular for the fully Bayesian Fusion model (Roussos et al., 2006a) and the NIDA and DINA models (de la Torre and Douglas, 2004).

It is interesting to note that MCMC could be used from either a partially Bayesian or non-Bayesian perspective. However, the most common application seems to be for fully Bayesian models, often with hierarchical structure to give it an appropriate empirical

Bayes flavor. The primary inferential outcome is the joint posterior distribution of the item and examinee parameters. Many MCMC references exist, and the Patz and Junker (1999a, 1999b) papers and the Gelman et al. (1995) book are especially recommended to readers considering parametrically complex models such as the models surveyed here.

EM algorithms have also been used in estimating skills diagnostic models, for example, by von Davier in calibrating his General Diagnostic Model (GDM) (von Davier et al., 2006; von Davier, 2005; von Davier and Yamamoto, 2004). The EM algorithms are often much more computationally efficient relative to MCMC algorithms. On the other hand, the EM algorithm tends to be more difficult to extend to new models or model variants compared to MCMC. As Patz and Junker (1999a, p. 14) point out, EM algorithms are not as "straightforward" (see also Junker, 1999, p. 70) to apply for parametrically complex models. Further, MCMC provides a joint estimated posterior distribution of both the test's item parameters and the examinee skills parameters, which may provide better understanding of the true standard errors involved (as argued by Patz and Junker, 1999a, 1999b). Moreover, MCMC routines adapt easily to produce posterior predictive model diagnostics. MCMC also provides a ready capability for comparing model parameter prior and posterior distributions as a measure of parameter identifiability (Sinharay, 2006).

2.5.2. Broad-based evaluation of estimation results and connections with validity

When a complex diagnostic model is applied to data, the results of the estimation process must be carefully scrutinized. In this section, we deliberately broaden the typical model checking process to include evaluation of statistical convergence, interpretation of parameter estimates for the items and the ability distribution, traditional model fit statistics, reliability, internal validity, and external validity. In particular, we emphasize the intertwining of psychometrically oriented validity assessment along with essential statistical model fit procedures in evaluating the appropriateness and effectiveness of the statistically calibrated model.

Evaluation of statistical convergence No matter which of the above estimation methods is used, all such estimation methods require checking for convergence – either convergence to within some specified tolerance as in a JMLE or EM algorithm or convergence to the desired posterior distribution in the case of MCMC. Because the statistical information one obtains from MCMC estimation (a full posterior distribution) is richer than that obtained from an EM algorithm (an estimate and its standard error), the evaluation of whether convergence has occurred is more difficult in the MCMC case.

If nonconvergence occurs, one can revisit the model building steps and reconsider the Q matrix and the selected model to see where changes may be warranted. For example, if a skill is assigned to items having a large range of difficulty, then the MCMC algorithm may not be able to converge to a single level of difficulty for the skill. There is much more to say about MCMC convergence. The reader is referred to Cowles and Carlin (1996), Gelman et al. (1995), and Sinharay (2004) for excellent and thorough discussions.

Interpretation of model parameter estimates Given that the parameter estimation procedure has converged, the estimates for the ability distribution and item parameters should be evaluated for internal consistency, reasonability, and concurrence with substantive expectations. For example, a key issue for mastery/nonmastery diagnostic models is whether the proportion of examinees estimated as masters on each skill is relatively congruent with substantive theoretical expectations. If a skill turned out much harder or easier than expected (e.g., from a standard-setting perspective), the Q matrix should be revisited and the item difficulty levels investigated for the items to which the skill has been assigned. In addition, the choice of tasks for that skill can be revisited to see if other more appropriate tasks can be found (e.g., if the proportion of masters for a skill is too low, one could try replacing the harder tasks for that skill with easier ones). Ultimately, the definition of the skill may need to be adjusted, for example, by suitable modification of Q or in a more basic way leading to a new set of tasks.

In Jang (2005, 2006) and Roussos et al. (2006b), one of the skills was found by statistical analysis to be easier than expected whereas the other eight skill difficulties were ordered as expected. In this case, Jang did not feel it necessary to conduct any further modifications for purposes of her study, but she plans future work to investigate this particular skill.

Next, the item parameters should be inspected in detail because they play a key role in determining the success of the diagnosis. For every model, the item parameters indicate how well the items performed for diagnostic purposes. The item parameters may indicate, for example, how well each item discriminates for each skill between masters and nonmasters. In the analysis of Jang (2005, 2006) and Roussos et al. (2006a, 2006b), the estimated item parameters led Jang to eliminate about 9% of her Q matrix entries because certain items did not discriminate well on some of the skills that had been assigned to them. In general, such a reduction in number of parameters may be beneficial simply because fewer parameters can be better estimated with the same observed data; but, more importantly, eliminating nonsignificant Q matrix entries reduces the very real possibility of confusion that can arise from examinees labeled nonmasters performing nearly as well as those labeled masters on the items corresponding to the Q entries being eliminated.

Model checking statistics Standard definitions of model checking appropriately focus on assessing the degree of fit between the estimated model and the observed data. For skills diagnostic implementation success and, hence, for this review, we expand this focus to include some aspects of consequential validity. This broader perspective is intended to subsume, refocus, and broaden traditional statistical model-checking when specialized to skills diagnostic assessment. More precisely, as mediated by the assessment purpose, enacted by the assessment design, and controlled by the skills classification setting (particularly testing constraints), the goal is that the choice of the model and the subsequent model estimation and examinee skills classification provide accurate enough calibration to allow the practitioner's skills-level assessment goals to be met. Much the same argument can be made for any type of assessment. The satisfaction of appropriate validity standards is essential to achieve success. Because of the greater complexity of diagnostic assessment, it is especially critical that validity concerns are explicitly addressed from the initial phases of determination of test purpose and assessment design through the entire implementation process.

A concrete example is a study by Zhang et al. (2006) in which four skills are measured in an English language learning test of 40 items. In this particular case the standard two-parameter logistic (2PL) IRT model did a better job of predicting total score. In part, the reason was that students who were classified as lacking all skills or having mastered all skills were not broken down as finely as they were on the estimated continuous IRT ability scale. Consequently, in a comparison of classical model fit statistics between the more complex skills-diagnostics-focused model versus the unidimensional scale-focused model with fewer parameters, the unidimensional model performed better than the complex model at predicting total score. But the unidimensional model did not provide a direct basis for inferring which of the specific skills were mastered and not mastered. This particular total-score-based fit statistic was not germane to the skills classification purpose.

The development of appropriate fit statistics that are sensitive to the skills diagnostic purpose of assessment is an important area for further research. Model fit evaluation must be sensitive to the primary diagnostic purpose of the assessment. For example, in the case of posterior predictive model checking mentioned later, the discrepancy measures investigated should be sensitive to aspects of the data consequential for diagnostic performance.

Once a model has been calibrated from data, one relatively straightforward and easily interpretable approach to evaluating model fit is to compare observed and model-predicted statistics. For example, the Bayesian RUM (Reparameterized Unified Model) approach (Henson et al., 2004, 2005) checks both item difficulties and item pair correlations. When MCMC has been used to produce posterior distributions for the model parameters, one can simulate from these posterior distributions and estimate the predicted distribution of a statistic or discrepancy measure based on the simulated data, and then compare the observed value of the statistic with the predicted distribution. This particular approach is called posterior predictive model checking (see Gelman and Meng, 1996; Gelman et al., 1995; Sinharay, 2005; and especially Sinharay, 2006, which is a skills diagnosis application). From our perspective of consequential validity used to augment classical model fitting for diagnostic assessments, the challenge in applying posterior predictive model checking and other common fit techniques is selecting test statistics that reflect aspects of data that are important for the primary goal of diagnostic assessment.

As a specific example, in the application of Jang (2005) and Roussos et al. (2006a, 2006b), the mean absolute difference (MAD) between predicted and observed item proportion-correct scores was 0.002, and the MAD for the item-pair correlations was 0.049, supporting the claim of good fit. One can also compare the fit between the observed and predicted score distributions. Shown in Figure 1 is a comparison between the observed and predicted score distributions from the real data analysis of Jang (2005, 2006) and Roussos et al. (2006a, 2006b). The misfit at the very lowest and highest parts of the distribution were expected as the mastery/nonmastery examinee model overestimated the scores of the lowest scoring examinees and underestimated the scores of the highest scoring examinees. Because the goal of the analysis was to estimate mastery/nonmastery rather than to scale examinees, this misfit actually had no effect on the mastery/nonmastery classification.

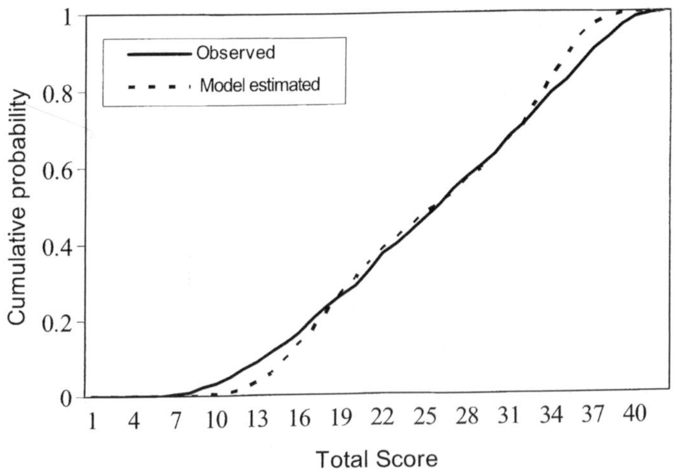

Fig. 1. Comparison of observed and model-estimated score distributions (from Jang, 2005, 2006 and Roussos et al., 2006a, 2006b).

Other standard fit statistics have also been used in real data analyses. Sinharay (2006) examined several model diagnostic approaches applied to a real data analysis of the mixed-number subtraction data of Tatsuoka (1984) and Tatsuoka et al. (1988), using a DINA-like Bayes-net model due to Mislevy (1995). Sinharay's methods demonstrated several instances of poor model fit, as well as severe nonidentifiability of model parameters, as determined by comparing prior and posterior distributions of item parameters. Although the focus in the Sinharay (2006) article was Bayes nets, the model fit diagnostics and approaches are broadly applicable to cognitive diagnosis models such as those discussed here.

Another standard fit statistic is the log-likelihood statistic that is typically used to compare models, especially nested models. This statistic with others was employed by von Davier (2005) in an analysis of data Educational Testing Service *Test of English as a Foreign Language* in which he compared the predicted log likelihood of the manifest distribution for a compensatory mastery/nonmastery diagnostic model to that of a unidimensional 2PL model. It was interesting to note the 2PL model had the larger log likelihood. This was a case in which the Q matrix was nearly simple structure (four skills with most items having only one skill), and the estimated posterior probabilities of mastery were highly correlated with 2PL scale score. In such a case as this, one could make the argument that most of the available compensatory-model-based classification information is captured by the scale score, and so the lower-reliability, less-well-fitting classification information may not add substantial new information to the scaled score. This is an example where caution should be exercised in applying a multidimensional skills diagnosis model in a situation where a unidimensional model better fits the data.

Reliability estimation Standard reliability coefficients as estimated for assessments modeled with a continuous unidimensional latent trait do not translate directly to discrete latent space modeled cognitive diagnostic tests. We note, however, that conceptions of reliability from first principles do still apply. Diagnostic skills classification reliability can be conceptualized in terms of the twin notions of, on the one hand the correspondence between inferred and true attribute state, and on the other hand the consistency of classification if the same assessment were independently administered to the same examinee twice. The reader is referred to Henson et al. (2006).

For example, to estimate classification reliability, Henson et al. (2004) use the calibrated model to generate parallel sets of simulated data, estimate mastery/nonmastery for each simulated examinee on each set, and calculate the proportion of times that an examinee is classified correctly according to the known true attribute state (thus producing an estimate of correct classification rate for the attribute/skill) and proportion of times an examinee is classified the same for that skill on two parallel tests (thus producing an estimated test-retest consistency rate). These agreement rate calculations based on simulated data provide reasonable estimates of actual rates. The calculated rates also can be adjusted for agreement by chance, for example, with the Cohen kappa statistics. Zhang et al. (2006) applied this procedure in an analysis of large scale English language assessment data (about 33 items per form, 2 forms, 4 skills, approximately 1.3 skills per item and 11 items per skill) where they typically found skills classification test–retest consistency indices of 0.8 with the Cohen's Kappa value being typically 0.6. These levels of skills classification reliability could be judged sufficient for low-stakes uses.

Internal validity checks Several measures of internal validity are important to evaluate. The examinee "scores" of interest are the proficiency estimates resulting from using the psychometric model. One kind of internal validity check is to measure the differences in observed behavior between examinees who are classified differently. We call these statistics "internal validity checks" because they use the test data itself to help verify the authenticity of the model. As one illustration of such a procedure that has been developed in the case of a conjunctive model (Hartz and Roussos, 2005; Roussos et al., 2006a), one defines an examinee to be a master of a particular item (an "item master") if the examinee is classified as a master of each of the skills required by that item. For each item, this separates the set of examinees into two subsets: the set of all item masters of that item, and the set of all examinees who are not masters of that item. A simple comparison used in the "IMstats" procedure of Hartz and Roussos (2005) is to note the average observed proportion of examinees answering the item correctly among the item masters versus the item nonmasters. If these two proportions are close, then the inferred skill classification has little effect on performance of that item. That indicates that the item, the Q matrix and the skills coding of the item should be investigated. A similar measure, called EMstats, has been developed for examinee performance by Hartz and Roussos (2005).

As an example of the use of the IMstats internal validity check in a real data analysis, we present in Figure 2 the IMstats results from Jang (2005, 2006) and Roussos et al. (2006a, 2006b) for one of the test forms she analyzed. These results indicate a high degree of internal validity for the skills diagnosis as the mean score differences between item masters and item nonmasters are quite large for the vast majority of the items. The results also point to certain problematic (for example, Item 1 in Figure 2) items. Jang investigated these items and discovered that they tended to be either extremely easy or

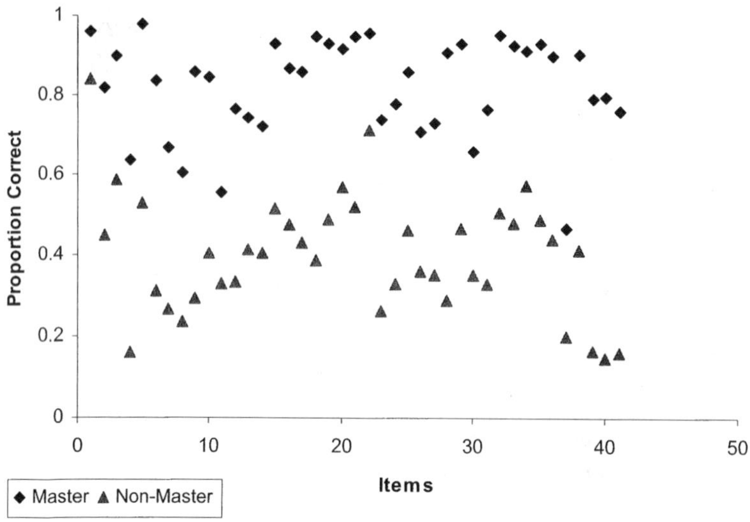

Fig. 2. Internal validity check comparing performance difference between item masters and item masters and item-masters (from Jang, 2005, 2006 and Roussos et al., 2006a, 2006b).

Table 1
Effect of diagnosis-based remediation on seriousness of student errors based on pretest and posttest data. Results reproduced from Table 5 in Tatsuoka and Tatsuoka, 1997

Pretest	Posttest		
	Serious	Nonserious	Total
Serious	21	93	114 (51%)
Nonserious	1	107	108 (49%)
Total	22 (10%)	200 (90%)	222

extremely hard. As noted by Jang, the test she used had been originally intended as a norm-referenced test with the purpose of placing examinees on a continuous scale, thus requiring items of a wide range of difficulty. Jang quite rightly notes that such a test is not necessarily the best one for doing skills diagnosis, and the identification of these problematic items through the internal validity analysis drove home this point.

External validity checks We use "external validity" to refer to criterion-related validity – statistics that relate test-derived ability estimates to some other ability criterion external to the test. One example of such research is a study in which Tatsuoka and Tatsuoka (1997) conducted skills diagnosis in a classroom setting (more than 300 students in all) performing diagnosis at a pretest stage, providing remediation instruction based on the diagnosis, and then evaluating the effectiveness of the diagnosis-based instruction with a posttest diagnosis. The results indicated that the diagnosis was very effective. For example, as shown in Table 1 (Table 5 in Tatsuoka and Tatsuoka, 1997), for the

114 students diagnosed with serious errors at pretest, the diagnosis-based remediation resulted in 93 of them improving to having only nonserious errors, while only one of the 108 students who had nonserious errors at pretest was assessed as having a serious error at posttest. Jang (2005) and Roussos et al. (2006a, 2006b) also conducted skills diagnosis in a pretest–posttest setting, but on a much smaller sample (27 students). Another example is a protocol study by Gierl (1997) comparing diagnostic classification results to observed student performance on algebra items from a large scale standardized test. There is a need for more validity studies like these. It would also be useful to have validity studies that include control groups or alternate treatment groups, which were not used in the above studies.

2.6. *Score reporting*

The last step in our framework for a diagnostic assessment implementation process is the development of the score reports, which, for example, would go to students, teachers, and parents in an educational diagnostic assessment setting. Such score reports must be easy to read and interpret and, most importantly, must fit into the use environment. For example, the information on the score report must be useful to teachers and students. From our perspective of consequential validity, the score reports may include action advice such as prescribed instructional or learning activities for a given classification outcome. One example of a diagnostic score report in current practice is the College Board's "Score Report *Plus*™" that is sent to each student who takes the PSAT/NMSQT™ (Preliminary SAT/National Merit Scholarship Qualifying Test). This is the first nationally standardized test to give some limited diagnostic skills-based feedback (other than the use of subscores or proportions correct on subsets of items). The information provided for the PSAT/NMSQT is based on the diagnostic methodology of Tatsuoka (1983). Interested readers are strongly encouraged to visit the College Board website (http://www.collegeboard.com/student/testing/psat/scores/using.html) that shows what the score report looks like and gives links for further information. Jang (2005) and Roussos et al. (2006a, 2006b) developed score reports that give more extensive information and is the best example we know of a complete user-friendly diagnostic report.

3. Reliability, validity, and granularity

Determining the appropriate level of reliability and validity necessary for successful educational applications of cognitive diagnostic assessment is not well researched. Measures of success depend as much on user needs, user environments, and market acceptance as on technical aspects of reliability and validity. But it seems clear that more work is needed for determining standards of sufficient statistical quality for various types of diagnostic assessment.

Within the IRT psychometric paradigm there is a clear trade-off between increasing model complexity to more closely align the psychometric model with skills-based psychological models for test takers at the expense of increased estimation error and even possible nonidentifiability, as opposed to improving reliability that can be achieved with simpler, low-dimensional continuous proficiency representations, at the possible expense of low congruence with cognitive theory. This trade-off can be thought of as a version of the well-known statistician's bias-variance trade-off. In fact the pursuit of cognitive diagnosis within the IRT probability modeling framework enforces strong feasibility constraints on modeling complexity. Because of the combined effects of non-identifiability along with the danger of having a large number of parameters, the best we can expect to do is to modestly expand model complexity from a single ability variable to a few skill proficiencies – on the order of ten skills or so for a standard length test. A 40-item mathematics test, for example, could not possibly statistically support diagnosis of 50 skills, even if there are compelling substantive rationales for the 50 skills. We believe that advancing psychometric and test development capabilities so that 40-item mathematics tests can be used to effectively diagnose, say, even as few as 5 or 6 specified skills, will provide significant teaching and learning benefits.

Two strong guidelines are presented for practitioners wishing to carry out successful, psychometrically rigorous skills diagnosis. First it is necessary for reasons of feasibility to keep the models as parametrically simple as possible within the diagnostic purpose of the test. Second, the question of how simple can a diagnostic model be and still accomplish the diagnostic purpose is not as straightforward as it may sound and requires close collaboration between statisticians and subject matter experts. In our view, parsimony is important, provided it is correctly applied. The aim is not to achieve the simplest model that fits the data, but rather the simplest model that accomplishes the diagnostic purpose and fits the data reasonably well. Thus, one could choose a more complex model, even if some traditional fit measures were lower, provided substantive argument supports the model.

Let us consider how these principles apply to a few real assessment examples, moving from unidimensional assessments to multidimensional examples. One prominent application setting for skills diagnosis models is that of standardized educational achievement tests, which are designed to measure a single dominant dimension. The dominant dimension, such as math or verbal proficiency, may be decomposed into a set of skills, but they are generally highly correlated with each other. In other settings, however, such as classroom applications of specially designed instruments for diagnosing math errors as analyzed by Tatsuoka and Tatsuoka (1997) or in application of instruments developed for diagnosing the presence of psychological disorders as analyzed by Templin and Henson (2006), the assessment instruments are designed for cases that are more strongly multidimensional.

In cases where a single dimension tends to dominate, low-dimensional or even unidimensional skills diagnosis models may be appropriate. (Here, the number of skills to be diagnosed can be larger than the number of latent dimensions, which may even be one.) In such cases the Q-matrix-based skills diagnosis model might be combined with a standard unidimensional modeling assumption to enrich the unidimensional score scale with interpretable skill information, such as identifying a correspondence between specific skill masteries and thresholds on the score scale.

In the more strongly multidimensional cases, the skills diagnosis model provides still further information beyond that which a single unidimensional scale can provide.

For example, Tatsuoka and Tatsuoka (1997) provide a tree diagram for nine knowledge states where the different states of mastery cannot be ordered by a unidimensional scale. Templin and Henson (2006) provide an example in a psychological testing setting where data were analyzed from an instrument used to diagnose pathological gambling based on 10 DSM (Diagnostic and Statistical Manual of the American Psychological Association) criteria. They forcefully demonstrated that the inferred latent criteria were strongly multidimensional (e.g., criteria 3, 6, and 7 were practically uncorrelated with the other seven criteria), and they emphasized that the criteria-profile they estimated for each individual has the potential to lead to better prevention and treatment programs than would result from merely estimating whether or not each individual is a pathological gambler.

It should be stressed here that careful and systematic design for cognitive diagnosis is necessary but not sufficient. That is, given a diagnostic test purpose, an identified skill space structure, and a selection of tasks that experts believe will measure the designated skills, it is still necessary through data analysis and various model evaluation steps to confirm that the intended and expected diagnostic performance did indeed occur. At the pilot data analysis stage, it may be helpful to replace one psychometric model with a simpler model to achieve improved statistical estimation and diagnostic performance, for example, by imposing further constraints to reduce the number of item parameters. The key consideration from the point of view of diagnostic assessment is that the model simplifications be made within the framework of the diagnostic purpose. If a diagnostic test is desired to assess examinee skill levels, for example, on four user-defined skills, then it is necessary to verify that the proposed four-skill diagnostic instrument is statistically sound, in addition to being valid from a design perspective. Simpler four-skill models may result in improved diagnostic performance by reducing estimation variability. Even the originally proposed skills space and task designs may need to be thoroughly revisited to determine if they can be improved.

Such statistical supportability procedures are no different in spirit from the rigorous statistical analyses done to support a unidimensional test model. Finally, if there appears no reasonable way to statistically support diagnosis of four distinct skills with the test, a more fundamental redesign from first principles is necessary, starting by reviewing the purpose and proceeding through the components of the implementation process.

To engage with the test client and the design and development teams to discover why a particular four-skill design did not achieve statistical supportability will provide a valuable service toward clarifying and satisfying the ultimate diagnostic testing needs. For example, some modification of the original four skills model may indeed achieve statistical supportability and thus satisfy the user's assessment purpose. The purpose-driven statistician's role is essential within the implementation process to help the client find a way to satisfy their diagnostic assessment needs, provide hard scientific information when their goal has been reached, and determine the degree of statistical quality.

4. Summary of cognitively diagnostic psychometric models

In the last two decades psychometricians have proposed a number of models apropos cognitive diagnosis. This section reviews and compares these models within an overarching framework that allows practitioners to see clearly the differences and similarities in the models. Although the current treatise is limited in scope to models for dichotomously scored items, the general model for dichotomous items can be easily generalized to polytomously scored items, and the organizing structure is equally applicable to polytomous-item models.

The framework for discussing the models consists of the following sections:

Section 4.1: General model for item-knowledge interaction.
Section 4.2: Specific models for item-knowledge interaction.
Section 4.3: Knowledge structure modeling assumptions.
Section 4.4: Item structure modeling assumptions.
Section 4.5: Summary table of the models.

Section 4.1 presents a general model that encompasses all the specific cognitive diagnosis models reviewed in this paper. It reviews the deterministic cognitive diagnosis model underlying this general model, introduces relaxations of the deterministic model along with psychological explanations for these relaxations, and presents a generalized model equation. We also briefly discuss two other general model frameworks that readers may find useful to further investigate.

Section 4.2 then reviews each of the specific models included in the paper describing how each represents the interaction of item structure with knowledge structure. Unidimensional item response theory (IRT) models are included even though they are usually not used as, nor intended be, cognitive diagnosis models. They are included as a basis for comparison because they are well known for modeling the performance of examinees on tests and, as such, help motivate modifications needed to produce effective psychometric models for skills diagnosis. We also include a brief description of Fischer's (1973, 1983) LLTM even though it does not model the ability variable as a multidimensional vector of multiple cognitive skills. As we describe below, we see LLTM's use of a Q matrix as an historically pivotal development that lies mid-way between unidimensional models such as the Rasch model and fully diagnostic models such as those on which we focus here.

Table 2 presents a list of the models included in this review along with pertinent references and with the abbreviations that will be used to refer to these models.

We should note that Adams et al. (1997) introduced a model that is actually more general than the MIRT-C model of Reckase and McKinley (1991) and includes some of the other models. Similarly, von Davier (2005), von Davier et al. (2006), von Davier and Yamamoto (2004) presented a generalized form of the compensatory model of Maris (1999), within a much more general modeling framework. Our paper focuses on specific instantiations of these models that have been presented in the literature. These more general modeling approaches of von Davier and of Adams et al. are discussed in Section 4.1.3 below.

Because our summary generally focuses on models that include an explicit specification of skills or attributes of interest as delineated by a Q matrix (or equivalent), we do not present general latent class models or loglinear models (see, e.g., Hagenaars, 1993; Rost, 1990; or Kelderman and Macready, 1990) or even skill mastery models that do not involve Q in their model structure (e.g., Macready and Dayton, 1977, 1980).

Table 2
List of reviewed cognitive diagnosis models

Model abbreviation	Model name	Reference(s)
1PL	One-parameter logistic	Rasch (1961)
2PL	Two-parameter logistic	Birnbaum (1968)
3PL	Three-parameter logistic	Birnbaum (1968)
DINO	Deterministic-Input Noisy-Or	Templin and Henson (2006)
GLTM	General component latent trait	Embretson (1985, 1997)
HYBRID	HYBRID	Gitomer and Yamamoto (1991)
LLTM	Linear logistic test	Fischer (1983)
MCLCM-C	Compensatory MCLCM (multiple classification latent class)	Maris (1999)
MCLCM-D	Disjunctive MCLCM	Maris (1999)
MIRT-C	Compensatory multidimensional IRT	Reckase and McKinley (1991)
MIRT-NC	Noncompensatory MIRT	Sympson (1978)
MLTM	Multicomponent latent trait	Whitely (1980), Embretson (1997)
RLCM	Restricted latent class	Haertel (1984, 1990)
RUM	Reparameterized unified cognitive/psychometric	DiBello et al. (1995) Hartz (2002), Hartz and Roussos (2005)

After describing the above models in detail in Section 4.2, Sections 4.3 and 4.4 separately focus on the knowledge structure and the item structure, respectively, comparing and summarizing how the different models represent each of these components.

For simplicity in the current paper, we will generally use the term "skills" to refer to the latent traits or, more generally, cognitive attributes or components being measured on examinees or test takers. We recognize that the latent examinee characteristics of interest do not always fit well with the use of this term (for example, in diagnosing the amount of knowledge an examinee has or in diagnosing an illness or disease). We will also use the term "ability parameter" as a general way of referring to any variable used to model the level of examinee proficiency or standing on a skill or attribute.

The goal of a model is to represent the performance of an examinee on an item in terms of the skills required in responding to the item and the proficiency of the examinee on these skills. The general idea behind all the models is that the higher the proficiency of an examinee on the skills required for the item, the higher the probability the examinee will get the item right.

4.1. General model for item-knowledge interaction

The general model form that is presented here is a slight generalization of the form described in DiBello et al. (1995). The psychological justification/explanation for the model is also taken from DiBello et al. (1995).

4.1.1. Deterministic cognitive diagnosis model

To arrive at the general model form, we first start with a description of the deterministic psychological model for cognitive diagnosis. We begin, as in Tatsuoka (Tatsuoka, 1984,

1985, 1990; Tatsuoka and Tatsuoka, 1987), with a test of I items, and a list of K skills of interest for diagnosis on the test. Let Q be the $I \times K$ skills-by-items incidence matrix that specifies which skills must be mastered in order for an examinee to give a correct answer on each item:

$$q_{ik} = \begin{cases} 1 & \text{if item } i \text{ requires skill } k, \\ 0 & \text{otherwise.} \end{cases}$$

Implicit in the selection of these skills is a particular solution strategy. An alternative solution strategy would likely result in a different list of skills, though a large degree of overlap would typically be expected.

Next, we define for the jth examinee a vector $\underline{\alpha}_j = (\alpha_{1j}, \alpha_{2j}, \ldots, \alpha_{Kj})$ denoting the state of mastery of the examinee on each of the skills. In this deterministic psychological model of cognitive diagnosis, mastery is modeled as a dichotomous variable (though, in general, examinee mastery could be modeled with either dichotomous, polytomous or continuous variables):

$$\alpha_{jk} = \begin{cases} 1 & \text{if examinee } j \text{ has mastered skill } k, \\ 0 & \text{otherwise.} \end{cases}$$

Finally, we assume that the cognitive requirement for the multiple skills within an item is conjunctive – that is, answering the item correctly requires mastery of all the skills required by that item. Then the probability an examinee correctly responds to an item is determined from the following deterministic model:

$$P(X_{ij} = 1 | \underline{\alpha}_j) = \begin{cases} 1 & \text{if } \underline{\alpha}_j \text{ contains 1s for all skills required for item } i, \\ 0 & \text{otherwise.} \end{cases}$$

Thus, a computationally simpler form would be:

$$P(X_{ij} = 1 | \underline{\alpha}_j) = \prod_{k=1}^{K} (\alpha_{jk})^{q_{ik}}.$$

4.1.2. Relaxations of the single-Q deterministic model

As described above in this deterministic model, knowing an examinee's state of mastery on the skills of an item determines whether an examinee gives a correct response on the item. Though not explicitly mentioned above, this deterministic assumption holds also for application of the skills. In other words, if an examinee has mastered a skill, then the deterministic model assumes the examinee has a probability of 1 of successfully executing it on an item that requires it; and, conversely, if an examinee has not mastered a skill, then the examinee has a probability of 0 of successfully executing it on an item that requires it. In reality, however, other factors play a role in whether an examinee gets an item right and DiBello et al. list four that seem the most salient:

STRATEGY: The examinee may use a different strategy from that presumed by the skills listed in Q.

COMPLETENESS: An item may require skills in addition to those listed in Q. If so, then Q is said to be not complete for that item.

POSITIVITY: An examinee who has "mastered" a skill will sometimes fail to execute it correctly; and, conversely, an examinee who has *not* "mastered" a skill will sometimes execute it correctly. This lack of perfect positivity can be explained by how, in practice, skills are defined. From a practical point of view, skills cannot be overly atomistic (i.e., indivisible). Otherwise there would be so many fine grained skills that it would be difficult to estimate the mastery of all of them separately. Because the defined skills are divisible, any possible definition of mastery of a skill must always allow some lack of perfect positivity. Some models deal with positivity by representing skill mastery level with continuous variables and employing item response functions (IRFs) that are sensitive to the continuous scale of the skill mastery variables. Other models maintain the representation of skill mastery as dichotomous variables and deal with positivity by relaxing the probability of 1 given mastery to allow a probability somewhat less than 1 and/or by relaxing the probability of 0 given nonmastery by allowing a probability somewhat greater than zero.[1]

SLIPS: The student may commit a random error. This is a "catch-all" category for all the remaining nonsystematic error in an examinee's responses. One example of how this could be manifested is when an examinee accidentally bubbles in the wrong response category on a multiple-choice test.

Presented next is the general form of the cognitive diagnosis model – the interaction of item structure and knowledge structure that is based upon the above psychological reasoning.

First, some additional notation needs to be defined:

ν_i = the number of strategies for item i.
S_{li} = [Examinee applies lth strategy of the ν_i strategies for item i].
$\underline{\psi}_j$ = skill mastery proficiency vector for examinee j.
C_{li} = [Examinee correctly executes all the cognitive components for strategy l on item i].
p_s = probability of a random slip.

The general form of the cognitive diagnosis model is then given by:

$$P(X_{ij} = 1|\underline{\psi}_j) = (1 - p_s)\sum_{l=1}^{\nu_i}\bigl[P(S_{li}|\underline{\psi}_j)P(C_{li}|S_{li}, \underline{\psi}_j)P(X_{ij} = 1|S_{li}, C_{li}, \underline{\psi}_j)\bigr]. \quad (1)$$

The equation can be most easily understood as mimicking an abstract summary of the cognitive progression of an examinee as the examinee solves the item. The entire equation is based on the examinee not committing a random slip, so it begins with the probability of not slipping, which is $1 - p_s$. Given that no slipping occurs, the examinee first chooses a particular strategy S_{li}. This is represented in the model by the first term in the product within the summation, the probability that an examinee with skill mastery vector $\underline{\psi}_j$ uses the lth strategy for item i, namely $P(S_{li}|\underline{\psi}_j)$. Then, given that particular

[1] It is not difficult to extend the dichotomous mastery models to allow for multiple discrete categories.

strategy choice, the examinee must successfully execute the skills for the item that the strategy calls for. This is represented in the model by the probability that an examinee of given mastery levels on the skills and given strategy successfully executes the skills as specified by Q for the item, namely $P(C_{li}|S_{li}, \underline{\psi}_j)$. Finally, given a particular strategy selection and given successful execution of the skills, the examinee must still execute whatever other skills are necessary to solve the item that may have been left out of Q. This is represented by the last probability in the product within the summation, the probability of a correct response, given skills mastery levels, strategy selection, and successful execution of the skills, namely $P(X_{ij} = 1|S_{li}, C_{li}, \underline{\psi}_j)$.

4.1.3. Other general diagnostic model frameworks

GDM Von Davier's General Diagnostic Model (GDM) (von Davier et al., 2006; von Davier, 2005; von Davier and Yamamoto, 2004) is defined in a very general manner. Part of the general form of the model is provided by a link function. By setting this link function in various ways, a large class of models can be generated as special cases of the GDM, including uni- and multidimensional Rasch models, 2PL model, the generalized partial credit model, the FACETS model (Linacre, 1989), and Maris' (1999) compensatory MCLCM, as well as more general partial noncompensatory models, including conjunctive diagnostic models. Analyses of real and simulated data have been published on the compensatory discrete MIRT version of the models. The estimation software is EM-algorithm based MML and has demonstrated good computational efficiency. The reader is referred to the references for further detail.

MRCMLM The random coefficient multinomial logit model (RCMLM) was proposed by Adams and Wilson (1996) and extended to a multidimensional model (MRCMLM) (Adams et al., 1997). The functional form of this model is quite flexible and generates a wide range of unidimensional and multidimensional models, including unidimensional and multidimensional Rasch models, Fischer's (1973, 1983) LLTM, Master's (1982) partial credit model, the FACETS model (Linacre, 1989), Wilson's ordered partition model (Wilson, 1992), Whitely's (1980) MLTM, Andersen's (1985) multidimensional Rasch model for repeated testing, and Embretson's (1991) multidimensional model for learning and change. In addition to a design matrix similar to the Q matrix that specifies the contribution of each skill to item difficulty, there is a second design matrix called a scoring matrix. The scoring matrix extends the flexibility of the model and provides the capability for handling testlets (called item bundles in Wilson and Adams, 1995) and even allows a conjunctive version. Examples are provided in which the Q matrix is not simple structure – i.e. some items require more than one skill. The person population parameters are assumed to be random variables and used mainly for the purpose of parameter estimation. They are decomposed into a linear combination of multiple parameters. The general model allows polytomous items, linear constraints on item parameters and flexible scoring rules. Estimation methods attempted include MML estimation that was successful up to three or four dimensions, and CML estimation in the case of Rasch versions of the model. The reader is referred to the references for further detail (Adams et al., 1997; Wilson and Adams, 1995; Adams and Wilson, 1996).

The next section will now review each of the specific models from Table 2, describing how each represents the interaction of the knowledge structure and item structure. The following two sections after that will separately review the knowledge structure and item structure, respectively, in general terms, summarizing key points from the detailed reviews of the individual models. The knowledge structure review will summarize the variety of ways that skill mastery level $\underline{\psi}_j$ has been seen to be modeled in Section 4.2; and the item structure review will summarize the variety of approaches seen in modeling $P(C_{li}|S_{li}, \underline{\psi}_j)$, $P(X_{ij} = 1|S_{li}, C_{li}, \underline{\psi}_j)$, and $P(S_{li}|\underline{\psi}_j)$. Thus, Section 4.4 on item structure will be divided into three subsections – one describing item structures for modeling performance on the cognitive components in Q $(P(C_{li}|S_{li}, \underline{\psi}_j))$, another describing item structures for modeling performance on abilities missing from Q $(P(X_{ij} = 1|S_{li}, C_{li}, \underline{\psi}_j))$, and another describing item structures for modeling strategy selection $(P(S_{li}|\underline{\psi}_j))$.

4.2. *Specific models for item-knowledge interaction*

This detailed discussion of specific models will generally progress from the simplest models to the more complex models; however, there is no strict ordering in terms of complexity for some of the models in the middle of the spectrum. For convenience, we will mostly use the acronyms for each model without naming it. The reader is referred to the table for name-acronym pairs. The Item-Knowledge interaction modeling of the cognitive diagnosis models will now be reviewed for each model, explicitly connecting the structure of each model to the different components of the general model. It is in this section that the various modeling approaches of Table 2 are made explicit.

4.2.1. *Unidimensional logistic models: 1PL, 2PL, & 3PL* (Rasch, 1961; Birnbaum, 1968)*; and LLTM* (Fischer, 1973, 1983)

The IRF for the 3PL model is as follows:

$$P(X_{ij} = 1|\theta_j) = c_i + \frac{1 - c_i}{1 + \exp[-1.7a_i(\theta_j - b_i)]},$$

where θ_j = level of proficiency of examinee j on the presumed unidimensional construct that the item is measuring; a_i = the discrimination parameter for item i and is proportional to the maximum slope of the IRF, it is "the degree to which item response varies with ability level" (Lord, 1980); c_i = the so-called "guessing parameter," which is the probability of a correct response for an examinee of very low proficiency, the lower asymptote of the IRF; b_i = the difficulty level of item i which is the level of proficiency at which an examinee has a probability of a correct response that is halfway between 1.0 and c_i.

When $c_i = 0$, the 2PL model results. When $a_i = 1$ and $c_i = 0$, the 1PL model results. Since the 1PL, 2PL, and 3PL models are unidimensional, they do not have the capability to be directly connected to a set of multiple skills. So, these models are not generally regarded as directly applicable to modeling multiple skills in item performance. Usually two or more skills are conjectured to have an effect on many, if not most, of the items and these skills are usually regarded as distinct dimensions that should appear in the model. (Multidimensional IRT models are discussed below.)

In terms of the general model (Eq. (1)), there is only one possible strategy and the number of skills for any item is one, and indeed the same skill for all items. Successful performance on the skill is the same as successful performance on the item, i.e., the modeled skill is assumed to capture all the skills involved in solving the item (completeness of Q). Lack of perfect positivity in the skill is dealt with by modeling mastery level as a continuous variable and employing a logistic IRF to determine probability of a correct response conditioned on mastery level. The probability of a random slip is assumed equal to zero. Thus,

$$\begin{aligned} p_s &= 0, & P(S_{1i}|\underline{\psi}_j) &= 1, \\ \underline{\psi}_j &= \theta_j, & P(C_{1i}|S_{1i}, \underline{\psi}_j). &= \text{the 3PL IRF}, \\ \nu_i &= 1, & P(X_{ij} = 1|C_{1i}, S_{1i}, \underline{\psi}_j) &= 1. \end{aligned}$$

We include here a brief description of LLTM (Fischer, 1973, 1983) even though it does not model the ability variable as a vector of multiple cognitive skills. As conceived by Fischer, the LLTM does not provide, and was not intended to provide, the capability for cognitive skills diagnosis of individual students. We nonetheless include LLTM in this discussion because, in our interpretation, the analysis of item difficulty into factors of difficulty, which can represent "cognitive operations" as noted by Fischer, was a key step historically in the evolution of cognitive diagnostic models from unidimensional IRT models. LLTM includes a Q matrix specifying from a given list of types of "cognitive operations" for each item, the number of each such operations required by the item. Consequently, even though LLTM ability is unidimensional, we see LLTM's use of a Q matrix as an historically pivotal development that lies mid-way between unidimensional models such as the Rasch model and fully diagnostic models such as those on which we focus below.

The IRF for this model was described by Fischer as a restricted form of the Rasch or 1PL, in the sense that the linear expression for item difficulty in terms of cognitive operations can be thought of as a constraint on item difficulty. The 1PL modeling framework was extended to include "cognitive operations" within the model. These cognitive operations were viewed by Fischer as attributes of items, and not as skills of examinees. Once cognitive determinants of item difficulty have been identified, represented by the LLTM Q matrix, a natural next step is to partition overall ability measured by the test into factors of ability, one for each cognitive determinant of item difficulty. As was done by Tatsuoka, each cognitive item difficulty factor was mapped in a one-to-one fashion to the student's ability to perform on items that involve that factor. Even though LLTM does not directly diagnose examinee mastery on each skill, it is included in our discussion here because it was the first to employ a Q matrix representation and was, thus, an important stepping stone toward the more complex models and, as such, provides important insights in terms of understanding the more complex models.

Specifically, the LLTM IRF has the same basic form as that for the 1PL model with the only item parameter being b_i, but in LLTM the b_i parameter is further explicated as:

$$b_i = \sum_{k=1}^{K} q_{ik}\eta_k + c,$$

where q_{ik} = the frequency with which skill k is a factor in the solution of item i, η_k = the level of difficulty attributed to skill k, and c = arbitrary constant that fixes the origin of the scale.

The q_{ik} variable is almost always dichotomous, indicating whether or not the skill influences item difficulty.

In terms of the general model, there is only one possible strategy. Even though the model allows for any number of skills per item, the performance on the separate skills is not modeled. By using a single latent trait, the model effectively employs a projection of all the skills mastery levels onto a unidimensional representation that can be thought of as some implicit combination of the skills. Thus, θ can be thought of as a weighted average of the skills mastery levels of an examinee. With such an interpretation the model can be seen to be a special case of a compensatory model. In Section 4.3 there is a discussion of when this assumption is appropriate. Also note that in the case of q_{ik} being dichotomous, all items that measure the same set of skills are assumed to have the same difficulty level. Successful performance on the skills is modeled as being the same as successful performance on the item, i.e., Q is considered to be complete. Lack of perfect positivity is dealt with in the same way as with the logistic models above: overall skill mastery level is modeled as continuous and the logistic function determines a probability of a correct response for a given level of overall skill mastery level.

4.2.2. *Compensatory multidimensional IRT model: MIRT-C* (Reckase and McKinley, 1991)

The IRF for the MIRT-C model can be written as follows:

$$P(X_{ij} = 1|\underline{\theta}_j) = c_i + (1 - c_i)\left[1 + \mathrm{e}^{-1.7\|\underline{a}_i\|\left(\frac{\underline{a}_i^t\underline{\theta}_j}{\|\underline{a}_i\|} - b_i\right)}\right]^{-1},$$

where $\underline{\theta}_j$ = mastery level vector for examinee j, indicating level of mastery of examinee j on each of the skills required by the item, $\underline{a}_i$ = vector of discrimination parameters (a_{ik}, $k = 1, \ldots, K$), for item i indicating the amount of discrimination that item i has for each of the skills that the item is sensitive to, c_i = the so-called "guessing parameter," which is the probability of a correct response for an examinee who has very low level of mastery on all the skills required by the item, b_i = the difficulty parameter for item i, and $\|\underline{a}_i\| = (\sum_{k=1}^{K} a_{ik}^2)^{1/2}$.

This model was first introduced by Reckase and McKinley (1991) and has been popularized for use in cognitive diagnosis by Adams et al. (1997), who employ a Rasch-model based version in which all the a_i parameters are set equal to 1, c_i is set to 0, and a Q matrix determines which skills correspond to which items. The compensatory nature of the item can be seen more clearly if we look at the IRF for the case of two dimensions with $c_i = 0$, $b_i = 0$, and $a_1 = a_2 = 1$. The IRF for this special case is:

$$P(X_{ij} = 1|\theta_1, \theta_2) = \frac{1}{1 + \mathrm{e}^{-1.7(\theta_1+\theta_2)}}.$$

If we let θ_1 take on a very low value, say $\theta_1 = -2$, and let θ_2 compensate for this low value of θ_1 by taking on a very high value, say $\theta_2 = +2$, we get $P(X_{ij} = 1|\theta_1, \theta_2) = 0.5$, the same value we would get if we had $\theta_1 = \theta_2 = 0$. Clearly in this model a high level of mastery on one skill can compensate for a low level on another to give the effect of a medium level on both. For cognitive diagnosis, this model is generally an improvement over the unidimensional IRT models in that multidimensionality is allowed. (See Section 4.3 below for discussion of this issue.) However, the compensatory nature of the interaction of the skills does not match cognitive assumptions in many diagnostic settings. In particular, cognitive analysis of item performance often requires successful performance (i.e., high level of mastery) on all the required skills for an item in order to have success on the item as a whole. In these circumstances, at least from a cognitive perspective, a conjunctive model appears to be more appropriate for the cognitive assumptions.

In terms of the general model, there is only one possible strategy. The model allows for a multitude of skills to influence an item through a multidimensional examinee skill mastery parameter. Successful performance on the skills is modeled as being the same as successful performance on the item, i.e., Q is assumed to be complete. Lack of perfect positivity in the cognitive components is dealt with by modeling ability as continuous variables and employing a logistic IRF. The probability of a random slip is assumed equal to zero. Thus,

$$\begin{aligned} &p_s = 0, && P(S_{1i}|\underline{\psi}_j) = 1, \\ &\underline{\psi}_j = \underline{\theta}_j, && P(C_{1i}|S_{1i}, \underline{\psi}_j) = \text{the MIRT-C IRF}, \\ &\nu_i = 1, && P(X_{ij} = 1|C_{1i}, S_{1i}, \underline{\psi}_j) = 1. \end{aligned}$$

4.2.3. Noncompensatory MIRT model: MIRT-NC (Sympson, 1978)
The item response function (IRF) for the MIRT-NC model is as follows:

$$P(X_{ij} = 1|\underline{\theta}_j) = \prod_{k=1}^{K} \frac{1}{1 + \mathrm{e}^{-1.7a_{ik}(\theta_{jk}-b_{ik})}},$$

where θ_{jk} = level of mastery of examinee j on skill k, a_{ik} = discrimination parameter of item i on skill k, and b_{ik} = the difficulty level of item i on skill k.

The knowledge structure of this model is multidimensional, as is generally desired; and the interaction of the skills with item performance is *noncompensatory*, in the sense that its multiplicative form results in a low probability of a correct response on the item if an examinee has very low proficiency on any one skill, regardless of how high a proficiency the examinee has on any other skill. More precisely, the model is not fully compensatory – we use the term "noncompensatory" for convenience. Note that the term *conjunctive* is also sometimes used instead of noncompensatory. A reformulation of the model helps to clarify the conjunctive/noncompensatory nature. Note that each term of the product is a 2PL IRF. Let

$$Y_{ikj} = \begin{cases} 1 & \text{if examinee } j \text{ successfully executes skill } k \text{ on item } i, \\ 0 & \text{otherwise.} \end{cases}$$

Then,

$$P(X_{ij} = 1|\underline{\theta}_j) = \prod_{k=1}^{K} P(Y_{ikj} = 1|\theta_{jk}).$$

A justification for the use of this product is the assumption that the examinee must successfully execute all the required skills and that these operations are independent of each other for a given set of skill levels. By independence, the joint probability of successfully executing all the skills is the product of the conditional probabilities of success for the individual skills.

In terms of the general model, there is only one possible strategy. The model allows for a multitude of skills to influence an item through a multidimensional examinee skill mastery parameter space. Successful performance on the skills is modeled as being the same as successful performance on the item, meaning that Q is assumed to be complete. Lack of perfect positivity is dealt with by modeling skill mastery as a vector of continuous variables and employing logistic functions for probability of success on executing each skill. The probability of a random slip is assumed equal to zero. Thus,

$$\begin{aligned} p_s &= 0, & P(S_{1i}|\underline{\psi}_j) &= 1, \\ \underline{\psi}_j &= \underline{\theta}_j, & P(C_{1i}|S_{1i}, \underline{\psi}_j) &= \text{the MIRT-NC IRF}, \\ \nu_i &= 1, & P(X_{ij} = 1|C_{1i}, S_{1i}, \underline{\psi}_j) &= 1. \end{aligned}$$

4.2.4. Multicomponent latent trait model: MLTM (Embretson, 1985, 1997; Whitely, 1980)

This model can be viewed as an extension of MIRT-NC. Like MIRT-NC, it employs unidimensional IRFs to model the probability of successful examinee performance on each skill and uses the product of these IRFs to model the probability of successful examinee performance on all the skills of an item simultaneously. In addition, MLTM models the probability of successful examinee performance on the item as a whole, conditioned on whether or not the examinee successfully executed all the skills. The modeling of examinee performance on the item (given successful execution of the skills) amounts to including the modeling of examinee performance on the non-Q skills.

First, let Y_{ikj} be defined as above as a dichotomous indicator of the success of examinee j in executing skill k on item i. Then, the IRF for MLTM can be written as follows:

$$P(X_{ij} = 1|\underline{\theta}_j) = a \prod_{k=1}^{K} P(Y_{ikj} = 1|\theta_{jk}) + g \left[1 - \prod_{k=1}^{K} P(Y_{ikj} = 1|\theta_{jk}) \right],$$

where θ_{jk} = level of mastery of examinee j on skill k, $a = P(X_{ij} = 1|\prod_{k=1}^{K} Y_{ikj} = 1)$, $g = P(X_{ij} = 1|\prod_{k=1}^{K} Y_{ikj} = 0)$, $P(Y_{ikj} = 1|\theta_{jk}) = \frac{1}{1+e^{-(\theta_{jk}-b_{ik})}}$ (a one-parameter logistic function), and b_{ik} = the difficulty level of item i on skill k.

In terms of the general model, this model can be viewed as having two strategies with each strategy linked to examinee performance on the skills. The first strategy is that the examinee responds to the item by trying to successfully execute the required skills; and the second is that the examinee, having *un*successfully executed one of the skills, "guesses." The probability of the first strategy is defined to be the probability of successfully executing all the required skills for the item. The probability an examinee applies the second strategy (guessing) is simply the probability that an examinee does not successfully execute the required skills for the item.

Successful performance on the skills is *not* modeled as being the same as successful performance on the item, which essentially means that Q is allowed to be *not* complete. Only when $a = 1$ is Q modeled as complete for the item. If $a = 1$ and $g = 0$ this model reduces to MIRT-NC.Lack of perfect positivity in the skills is dealt with by modeling skill mastery as a vector of continuous variables and employing logistic functions to determine success probability on each skill. The probability of a random slip is assumed equal to zero. Thus, $p_s = 0$, $\underline{\psi}_j = \underline{\theta}_j$, $\nu_i = 2$,

$$P(S_{1i}|\underline{\psi}_j) = \prod_{k=1}^{K} P(Y_{ikj} = 1|\theta_{jk}),$$

$$P(S_{2i}|\underline{\psi}_j) = 1 - \prod_{k=1}^{K} P(Y_{ikj} = 1|\theta_{jk}),$$

$$P(C_{1i}|S_{1i}, \underline{\psi}_j) = 1, ^{4}$$

$$P(C_{2i}|S_{2i}, \underline{\psi}_j) = g \quad \text{(the skill in this case, guessing, is not very cognitive)},$$

$$P(X_{ij} = 1|C_{1i}, S_{1i}, \underline{\psi}_j) = a, \quad \text{and}$$

$$P(X_{ij} = 1|C_{2i}, S_{2i}, \underline{\psi}_j) = 1.$$

MLTM also includes a version that allows for multiple cognitive strategies, which we now describe. In this version of the model, the strategies have an order of application such that the probability that each strategy is applied is equated to the probability that the previous strategies in the order were not successfully applied. Additionally all examinees are assumed to use the same order of strategy application. Specifying the order of strategy execution and the skills for each strategy is crucial to the model. The skill mastery vector, $\underline{\theta}_j$, now includes skills for all the strategies. Note that this model for multiple cognitive strategies fits the two-strategy MLTM case above where the probability of applying the second strategy, guessing, was equal to the probability of nonsuccess on the first strategy, i.e., the probability of nonsuccess in the execution of the skills for the single-strategy MLTM case.

[4] This probability is determined given S_{1i} whose probability is equal to $P(C_{1i}|\underline{\psi}_j)$. That is why this is 1.

Let S_{li} = [Examinee applies strategy l on item i], C_{li} = [successful execution of the skills for strategy l on item i], K_l = the set of skills for strategy l.

Then, the probability that strategy L is applied on item i is given by:

$$P(S_{Li}|\underline{\theta}_j) = \prod_{l=1}^{L-1}\left[1 - P(C_{li}|S_{li}, \underline{\theta}_j)\right],$$

where

$$P(C_{li}|S_{li}, \underline{\theta}_j) = \prod_{k \in K_l} P(Y_{ikj} = 1|\theta_{jk}).$$

As in the case of a single cognitive strategy, MLTM in the case of multiple cognitive strategies does not model successful performance on the skills for a given strategy as being exactly the same as successful performance on the item. That is, Q is allowed to be incomplete for each strategy. The probability of successful performance on item i for strategy l is given by:

$$P(X_{ij} = 1|\underline{\theta}_j) = P(S_{li}|\underline{\theta}_j)P(C_{li}|S_{li}, \underline{\theta}_j)\,a_l,$$

where a_l = the probability of successful performance on the item given successful performance on the cognitive components of strategy l. This probability is a constant over all examinees and items but may vary with strategy.

In terms of the general model:

$p_s = 0,$

$\underline{\psi}_j = \underline{\theta}_j,$

ν_i = the number of strategies for item i,

$$P(S_{Li}|\underline{\psi}_j) = \prod_{l=1}^{L-1}\left[1 - P(C_{li}|S_{li}, \underline{\theta}_j)\right],$$

$$P(C_{li}|S_{li}, \underline{\psi}_j) = \prod_{k \in K_l} P(Y_{ikj} = 1|\theta_{jk}),$$

$$P(X_{ij} = 1|C_{li}, S_{li}, \underline{\psi}_j) = a_l.$$

4.2.5. *General component latent trait model: GLTM* (Embretson, 1985, 1997)

This model is a mixture of the MLTM for a single cognitive strategy and the LLTM.

As with MLTM, Y_{ikj} is defined as before. Then the item response function (IRF), as with MLTM, is written as follows:

$$P(X_{ij} = 1|\underline{\theta}_j) = a\prod_{k=1}^{K} P(Y_{ikj} = 1|\theta_{jk}) + g\left[1 - \prod_{k=1}^{K} P(Y_{ikj} = 1|\theta_{jk})\right],$$

where, as with MLTM, θ_{jk} = level of mastery of examinee j on skill k, $a = P(X_{ij} = 1|\prod_{k=1}^{K} Y_{ikj} = 1)$, $g = P(X_{ij} = 1|\prod_{k=1}^{K} Y_{ikj} \neq 1)$, $P(Y_{ikj} = 1|\theta_{jk}) = \frac{1}{1+e^{-(\theta_{jk}-b_{ik})}}$, and b_{ik} = the difficulty level of item i on cognitive component k.

But now, as with LLTM, the b_{ik} difficulty parameter is broken down as follows:

$$b_{ik} = \sum_{f=1}^{p} \eta_{fk} q_{ikf} + \beta_k,$$

where q_{ikf} = complexity (usually 1 or 0) of difficulty factor f in skill k on item i, η_{fk} = weight of difficulty factor f in overall difficulty of skill k, and β_k = normalization constant.

In LLTM each η_{fk} stands for the effect on item difficulty resulting from the item response requiring the application of required skill k, but in MLTM and GLTM the effect on the item response of the required skills is more directly represented through the θ_{jk} examinee parameters. In GLTM, each η_{fk} is interpreted as a difficulty factor that comes from breaking down the kth skill into the stimuli that are its building blocks.

In terms of the general model, GLTM has the exact same representation as the single-strategy form of MLTM (except for the decomposition of the difficulty parameter described above). Therefore, the comparison with the general model will not be repeated here for GLTM.

4.2.6. Restricted latent class model: RLCM (Haertel, 1984, 1990)
RLCM is termed a *restricted* latent class model because the number and types of deterministic latent response vectors allowed in the model are restricted by Q.

With this model, unlike the models discussed above, the level of mastery of an examinee is *not* represented as a continuous variable on the dimensions defined by the skills. Instead, examinee ability is modeled by a K-dimensional vector $\underline{\alpha}$ of 0's and 1's. This vector indicates for each skill whether an examinee is a master (1) or nonmaster (0) of the skill. RLCM takes into account lack of perfect positivity by introducing: π_i = probability an examinee get item i right, given the examinee has mastered all the skills for item i, r_i = probability an examinee gets item i right, given that the examinee has *not* mastered at least one of the skills for item i.

The IRF for RLCM is then defined by:

$$P(X_{ij} = 1|\underline{\alpha}_j) = \pi_i^{\prod_{k=1}^K \alpha_{jk}^{q_{ik}}} r_i^{1-\prod_{k=1}^K \alpha_{jk}^{q_{ik}}}.$$

Cognitively the model simply says that if the examinee has mastered all the skills specified as required for an item, the examinee has one (likely high) probability of getting the item right. Whereas, if the examinee has not mastered all the skills of an item, the examinee has another (likely small) probability of getting the item right. Note that with RLCM if an examinee has not mastered one or more required skills for an item, it does not matter how many nor which particular ones have not been mastered. This is an example of the deliberate introduction of modeling parsimony to improve statistical tractability. This type of skill interaction is often referred to as "conjunctive" (analogous to noncompensatory) because the model intends for a high probability of correct response to require mastery of all the skills required for the item.

Historically, RLCM can be viewed as an extension of a model proposed by Macready and Dayton (1977) that used a similar IRF but assumed all items are measuring the same skills and, thus, classifies examinees into one of two categories, either having mastered all the skills or having nonmastered at least one. The RLCM model has been more recently referred to as "DINA" (Junker and Sijtsma, 2001), which stands for "Deterministic-Input, Noisy-And." The acronym highlights two features of the model: the α vector by itself that can be used to deterministically determine which items an examinee will get right and wrong (see description of deterministic model in Section 4.1.1), but also the fact the item response is probabilistic and is intended to require mastery of all the skills (the "And" part) in order to result in a high probability of a correct response. For a more detailed explanation, readers should consult Junker and Sijtsma (2001).

In terms of the general model, there is only one possible strategy. The model allows for a multitude of skills to influence an item. Successful performance on the skills is modeled as being the same as successful performance on the item, which means that Q is assumed to be complete. Lack of perfect positivity in the skills is dealt with by allowing π_i to be less than 1 and allowing r_i to be greater than 0. The probability of a random slip is assumed equal to zero. Thus,

$$p_s = 0, \qquad P(S_{1i} | \underline{\psi}_j) = 1,$$

$$\underline{\psi}_j = \underline{\alpha}_j = (\alpha_{1j}, \alpha_{2j}, \ldots, \alpha_{Kj}), \qquad P(C_{1i} | S_{1i}, \underline{\psi}_j) = \text{ the RLCM IRF},$$

$$\nu_i = 1, \qquad P(X_{ij} = 1 | C_{1i}, S_{1i}, \underline{\psi}_j) = 1.$$

4.2.7. HYBRID model (Gitomer and Yamamoto, 1991)

This model, from one perspective, is very similar to RLCM, indeed RLCM is included within HYBRID. In addition to the latent classes postulated by RLCM, HYBRID adds one more class of examinees whose item responses follow a unidimensional IRT model.

Thus, based on an examinee's observed response vector, the examinee is either classified into one of the $\underline{\alpha}$ latent classes or is simply assigned an estimate of unidimensional ability from a unidimensional IRT model (such as 1PL, 2PL, or 3PL). The IRF is given by:

$$P(X_{ij} = 1 | \text{examinee } j) = \begin{cases} \text{RLCM IRF} & \text{if examinee } j \text{ is estimated to belong to a latent class,} \\ P_i(\theta_j) & \text{if examinee } j \text{ cannot be classified into any latent class,} \end{cases}$$

where θ_j = unidimensional overall mastery level of an examinee who cannot be classified into a latent class, $P_i(\theta_j)$ = either the 1PL, 2PL, or 3PL unidimensional IRF.

In terms of the general model, this model can be viewed as having two strategies, even though the originators of the model probably did not intend for their model to be viewed this way. Examinees who use the first strategy are sensitive to the varied skills that have been specified as being required for the items, i.e., they are using the Q-based strategy. For these examinees, skill mastery is modeled as membership in a latent class. Examinees who use the second strategy are not using the Q-based skills and are modeled as though they are only sensitive to a unidimensional combination of all the skills, as though the items were measuring only a single skill. For these examinees, skill mastery is modeled as a continuous unidimensional latent trait. The probability of the first strategy is simply the probability that an examinee is in one of the latent classes. The probability of the second strategy is simply $1 - P$ (first strategy). HYBRID assumes that the probability a randomly chosen examinee uses a particular strategy does not vary across items and that an examinee uses the same strategy for all items on the test. This latter assumption concerns the test as a whole and is, thus, not captured in our general model which is for a single item. The comparison with the general model is the same as for RLCM, except for the addition of the second strategy. Thus,

$$
\begin{array}{ll}
p_s = 0, & P(S_{1i}|\underline{\psi}_j) = P(S_1) = \text{probability an examinee is} \\
 & \text{classified into one of the latent classes,} \\
\underline{\psi}_j = \underline{\alpha}_j \text{ or } \theta_j, & P(S_{2i}|\underline{\psi}_j) = 1 - P(S_1), \\
\nu_i = 2, & P(C_{1i}|S_{1i}, \underline{\psi}_j) = \text{RLCM IRF}, \\
 & P(C_{2i}|S_{2i}, \underline{\psi}_j) = P_i(\theta_j) \quad \text{(1PL, 2PL, or 3PL IRF)}, \\
 & P(X_{ij} = 1|C_{1i}, S_{1i}, \underline{\psi}_j) = 1, \\
 & P(X_{ij} = 1|C_{2i}, S_{2i}, \underline{\psi}_j) = 1.
\end{array}
$$

4.2.8. Deterministic-Input, Noisy-Or: DINO (Templin and Henson, 2006)
This model is very similar to RLCM (also known as DINA), except instead of being *conjunctive* (the “And” in DINA), it is *disjunctive* (the “Or” in DINO). By disjunctive we mean that this model intends for a high probability of a correct response on an item to occur whenever *any* of the skills are mastered (the “Or” part). Thus, the item and ability parameters are nearly the same as for RLCM, but the IRF takes on a disjunctive form.

Examinee ability is modeled the same as with RLCM using a K-dimensional vector $\underline{\alpha}$ of 0’s and 1’s. DINO takes into account the lack of perfect positivity by introducing: π_i = probability examinees get item i right, given they have mastered *at least one of* the skills for item i, r_i = probability examinees get item i right, given they have *not* mastered *any* of the skills for item i.

The IRF for DINO is then defined by:

$$P(X_{ij} = 1|\underline{\alpha}_j) = \pi_i^{[1-\prod_{k=1}^{K}(1-\alpha_{jk})^{q_{ik}}]} r_i^{\prod_{k=1}^{K}(1-\alpha_{jk})^{q_{ik}}}.$$

With DINO, if an examinee has mastered one or more required skills for an item, it does not matter how many have been mastered nor which particular ones have been mastered. This is another example of the deliberate introduction of modeling parsimony to improve statistical tractability. As mentioned above, this type of skill interaction is referred to as “disjunctive” (analogous to compensatory) because the model intends for a high probability of an *incorrect* response on the item to require *nonmastery of all* the skills required for the item.

In terms of the general model, the only difference from RLCM is $P(C_{1i}|S_{1i}, \underline{\psi}_j) =$ the DINO IRF.

4.2.9. Reparameterized Unified Model: RUM (DiBello et al., 1995; Hartz, 2002)
The original Unified Model of DiBello et al. (1995), although conceptually attractive, had nonidentifiable item parameters and, hence, was statistically inestimable without further constraints on the parameters. There was also a version that allowed for multiple strategies, which will not be discussed here. Hartz (2002) parsimoniously reparameterized the single-strategy Unified Model, and this is the model that has been extensively

researched and applied, and which will be presented here. Hartz also added a Bayesian hierarchical structure for estimation purposes and called the enhanced model the Fusion Model. Our focus here is on the RUM IRF, although readers interested in the Bayesian component are advised to consult Hartz (2002) or Hartz and Roussos (2005).

Like HYBRID, RUM can be viewed as an extension of RLCM. Note that HYBRID does not change the part of RLCM that determines the classification of the examinees into cognitively based (i.e., Q-based) latent classes. Instead, HYBRID merely places hard-to-classify examinees under the umbrella of a unidimensional IRT model, thus moderating the knowledge structure assumptions of RLCM. RUM, on the other hand, faithfully maintains a knowledge structure based on cognitively meaningful latent classes and, instead, brings more elaborated cognitive modeling into the item structure in an attempt to improve the classification process.

The authors of RUM intended for the model to be a unifying bridge between cognitively over-simplified unidimensional IRT models and intractable cognitive-based expert systems with thousands of nodes. In addition, within the context of cognitive diagnosis models alone, RUM can be viewed as bridging the gap between the more cognitively complex MLTM (single strategy) and GLTM and the statistically more tractable but cognitively more constrained RLCM and HYBRID.

Like MLTM and GLTM, RUM models examinee performance on the individual skills. However, RUM replaces the continuous skill level parameters with discrete ones. RUM uses RLCM-like item parameters (focused on dichotomous skill mastery) instead of logistic functions (focused on continuous skill mastery) to model examinee performance on the skills. (Research has been completed with a version of RUM in which skill levels can be any finite number of ordered categories. That work will not be discussed here.)

Thus, in terms of model complexity RUM is effectively in between (MLTM, GLTM) and (RLCM, HYBRID). As described for RLCM, RUM models the ability of examinee j by a vector $\underline{\alpha}_j$ of 0's and 1's, indicating mastery (1) or nonmastery (0) on each of the K skills in Q. As described for MLTM and MIRT-NC, RUM models the joint probability of successfully executing all the skills required for an item as the product of the success probabilities for the individual skills, an assumption of skill execution independence.

Thus, as in MIRT-NC and MLTM, let

$$Y_{ikj} = \begin{cases} 1 & \text{if examinee } j \text{ successfully executes skill } k \text{ on item } i \\ & \text{(or if examinee } j \text{ is not required to execute skill } k \text{ on item } i\text{),} \\ 0 & \text{otherwise.} \end{cases}$$

Assuming skill execution independence,

$$P\left(\prod_{k=1}^{K} Y_{ikj} = 1|\underline{\alpha}_j\right) = \prod_{k=1}^{K} P(Y_{ikj} = 1|\underline{\alpha}_j).$$

In MLTM, positivity is dealt with by modeling $P(Y_{ikj} = 1|\underline{\alpha}_j)$ as a logistic function with structural parameters specific to skill k on item i. In RLCM, positivity is dealt with by modeling $P(\prod_{k=1}^{K} Y_{ikj} = 1|\underline{\alpha}_j)$ as two constants, one (high) value for when $\prod_{k=1}^{K} \alpha_{jk}^{q_{ik}} = 1$, and another (low) value for when $\prod_{k=1}^{K} \alpha_{jk}^{q_{ik}} = 0$.

The original Unified Model maintained structural parameters specific to skill k on item i, but simplified the model by using RLCM-like constants. Specifically, in the original Unified Model (UM),

$$P(Y_{ikj} = 1|\underline{\alpha}_j) = \pi_{ik}^{\alpha_{jk}} r_{ik}^{(1-\alpha_{jk})},$$

where $\pi_{ik} = P(Y_{ikj} = 1|\alpha_{jk} = 1)$ and $r_{ik} = P(Y_{ikj} = 1|\alpha_{jk} = 0)$.

Thus, $P(\prod_{k=1}^{K} Y_{ikj} = 1|\underline{\alpha}_j) = \prod_{k=1}^{K} \pi_{ik}^{\alpha_{jk}q_{ik}} r_{ik}^{(1-\alpha_{jk})q_{ik}}$.

With UM, if an examinee has not mastered a required skill for an item, the probability of a correct response depends on which skills have not been mastered. The positivity of the nonmastered skill determines how likely the examinee will successfully execute it given nonmastery on it. If all the skills for an item have sufficiently high positivity, then the probability that an examinee who has not mastered one or more skills will successfully execute all the skills required for an item, will be approximately the same (very close to 0) value no matter how many or which skills the examinee has not mastered. If such a strong assumption is justified then this part of UM can be effectively replaced by the simpler representation used in RLCM.

The above parameterization was not identifiable, as also pointed out by Maris (1999). To ameliorate the nonidentifiability, Hartz (2002) reparameterized UM as follows:

$$P\left(\prod_{k=1}^{K} Y_{ikj} = 1|\underline{\alpha}_j\right) = \pi_i^* \prod_{k=1}^{K} r_{ik}^{*(1-\alpha_{jk})q_{ik}},$$

where $\pi_i^* = \prod_{k=1}^{K} \pi_{ik}^{q_{ik}}$ and $r_{ik}^* = \frac{r_{ik}}{\pi_{ik}}$.

The parameter π_i^* is the probability that an examinee having mastered all the Q required skills for item i will correctly apply all the skills. For an examinee who has not mastered a required skill, the item response probability will be multiplicatively reduced by an r_{ik}^* for each nonmastered skill, where $0 < r_{ik}^* < 1$. The more strongly the item depends on mastery of a skill, the lower the item response probability should be for a nonmaster of the skill, which translates to a lower r_{ik}^* for that skill on that item. Thus, r_{ik}^* functions like a reverse indicator of the strength of evidence provided by item i about mastery of skill k. The closer r_{ik}^* is to 0, the more discriminating item i is for skill k.

Like MLTM (and unlike RLCM), RUM also allows for Q to be not complete and, thus, does *not* equate successfully executing the Q-specified skills with successfully executing the item. RUM includes a model for the ability of the examinee to successfully execute skills that are necessary for getting the item right but that have been left out of the Q-specified skills for an item. These skills are modeled using a unidimensional IRT model with a continuous latent trait, η_j. Thus, the full IRF for RUM can be written as follows:

$$P(X_{ij} = 1|\underline{\alpha}_j, \eta_j) = P\left(\prod_{k=1}^{K} Y_{ikj} = 1|\underline{\alpha}_j\right) P_{c_i}(\eta_j),$$

where $P_{c_i}(\eta_j) = \frac{1}{1+\exp\{-1.7[\eta_j-(-c_i)]\}}$, a 1PL model with difficulty parameter $-c_i$ and ability parameter η_j.

The η_j parameter functions as a unidimensional projection or combination of the levels of skill mastery for an examinee on the non-Q skills. The c_i parameter is referred to by DiBello et al. and by Hartz as the "completeness parameter" and is interpreted as an indicator of the degree to which Q is complete in its coverage of the skills. Hartz constrained c_i to be greater than 0, with larger values indicating greater completeness for Q. When c_i is 3 or more, this term will be very close to 1 for most values of η_j, and when c_i is near 0, the η_j variation will have its maximum influence on the RUM IRF.

The $P_{c_i}(\eta_j)$ part of the model simply recognizes that it is usually neither desirable nor possible to list in Q every possible skill that goes into solving an item for a given strategy. Indeed, this part of the model might sometimes be appropriately viewed as representing higher-order skills, such as self-monitoring behavior or the ability to assimilate the separate cognitive components into a coherent whole, which are less amenable to being modeled as discrete cognitive components (as suggested by Samejima, 1995).

If all examinees with the same latent skill class $\underline{\alpha}$ have the same η, then the completeness term could be eliminated by setting c_i equal to 10 (or any large number) and reducing the other item parameters by appropriate multiplicative constants. Also, as described above, if the positivity of all the cognitive components for an item were sufficiently high, then this part of RUM could further reduce to the corresponding part of RLCM.

In terms of the general model,

$$
\begin{aligned}
p_s &= 0, & P(S_{1i}|\underline{\psi}_j) &= 1,\\
\underline{\psi}_j &= (\underline{\alpha}_j, \eta_j), & P(C_{1i}|S_{1i}, \underline{\psi}_j) &= \pi_i^* \textstyle\prod_{k=1}^{K} r_{ik}^{*(1-\alpha_{jk})q_{ik}},\\
\nu_i &= 1, & P(X_{ij} = 1|C_{1i}, S_{1i}, \underline{\psi}_j) &= P_{c_i}(\eta_j).
\end{aligned}
$$

4.2.10. Disjunctive MCLCM (Maris, 1999)

Maris (1999) proposed several Multiple Classification Latent Class Models, including a disjunctive model which we refer to here as MCLCM-D. Let Y_{ijk} be the same as defined above as for MLTM, MIRT-NC, and RUM, all of which assumed skill execution independence such that

$$P\left(\prod_{k=1}^{K} Y_{ikj} = 1|\underline{\alpha}_j\right) = \prod_{k=1}^{K} P(Y_{ikj} = 1|\underline{\alpha}_j).$$

If Q is complete, then

$$P(X_{ij} = 1|\underline{\alpha}_j) = \prod_{k=1}^{K} P(Y_{ikj} = 1|\underline{\alpha}_j).$$

The RUM IRF takes on this form when Q is complete. Maris (1999) proposed a model of this form using the same parameterization of UM, thus rediscovering the Q-complete version of the original Unified Model. He called it a "conjunctive" MCLC model because the conjunction of successful skill execution on all the skills is required for success on the item.

Maris also proposed what he called a "disjunctive" model as follows:

$$P(X_{ij} = 1|\underline{\alpha}_j) = 1 - \prod_{k=1}^{K} [1 - P(Y_{ikj} = 1|\underline{\alpha}_j)],$$

where he again used the same parameterization as UM for $P(Y_{ikj} = 1|\underline{\alpha}_j = 1)$ and $P(Y_{ikj} = 1|\underline{\alpha}_j = 0)$, namely, $\pi_{ik} = P(Y_{ikj} = 1|\alpha_{jk} = 1)$ and $r_{ik} = P(Y_{ikj} = 1|\alpha_{jk} = 0)$.

Thus, the MCLCM-D IRF can be written as

$$P(X_{ij} = 1|\underline{\alpha}_j) = 1 - \prod_{k=1}^{K} (1 - \pi_{ik})^{\alpha_{jk} q_{ik}} (1 - r_{ik})^{(1-\alpha_{jk}) q_{ik}}.$$

As in UM, these item parameters are not identifiable, but a reparameterization similar to that of Hartz (2002) could be implemented. The interpretation of this disjunctive model is that a high probability of success at the item level will occur so long as examinee j has a high probability of successfully executing at least one of the required cognitive components for the item (i.e., the examinee is a master of at least one of the required components). Conversely, a low probability of success will occur only if the examinee is a nonmaster on all the cognitive components. This disjunctive model might be termed the maximum or extreme compensatory model, as it allows mastery on any one component to compensate for nonmastery on all the others. This model may also be viewed as a more complex version of DINO. In both DINO and MCLCM-D, a high probability of success on an item is intended to result from mastery of *any* of the skills; but, in MCLCM-D this probability varies depending on which skills have been mastered, whereas in DINO this probability is the same no matter which skills or how many have been mastered (as long as at least one has been mastered).

In terms of the general model,

$$\begin{aligned} &p_s = 0, && P(S_{1i}|\underline{\psi}_j) = 1, \\ &\underline{\psi}_j = \underline{\alpha}_j, && P(C_{1i}|S_{1i}, \underline{\psi}_j) = \text{MCLCM-D IRF}, \\ &\nu_i = 1, && P(X_{ij} = 1|C_{1i}, S_{1i}, \underline{\psi}_j) = 1. \end{aligned}$$

4.2.11. Compensatory MCLCM (Maris, 1999)

Among several Multiple Classification Latent Class Models proposed by Maris (1999), see also von Davier et al. (2006), von Davier (2005), von Davier and Yamamoto (2004), was a compensatory model in the form of MIRT-C, but using the dichotomous $(\alpha_{jk}, k = 1, \dots, K)$ ability variables instead of the continuous $\underline{\theta}_j$ variables of the MIRT-C model.

Thus, the form of the IRF is as follows:

$$P(X_{ij} = 1|\underline{\alpha}_j) = \frac{\exp[\sum_{k=1}^{K} a_{ik}\alpha_{jk} - d_i]}{1 + \exp[\sum_{k=1}^{K} a_{ik}\alpha_{jk} - d_i]},$$

where a_{ik} is the amount of increase in probability on the log-odds scale when α_{jk} goes from 0 to 1, and d_i is a threshold parameter.

This model has the attractive feature of allowing practitioners to maintain discrete (mastery, nonmastery) skills while implementing cognitive diagnosis in a situation for which a compensatory model is more appropriate than a noncompensatory model. One could imagine using this model in comparison with the complete-Q RUM to see whether a conjunctive or compensatory model is more appropriate for a given psychological situation. It is important also to note a model sensitivity issue here. Even in cases in which a conjunctive model is better aligned with cognitive theory, a comparison of both conjunctive and compensatory discrete models can indicate if the model outcomes are sensitive to the model differences. Research is currently being conducted to investigate this issue.

In terms of the general model,

$$p_s = 0, \qquad P(S_{1i}|\underline{\psi}_j) = 1,$$

$$\underline{\psi}_j = \underline{\alpha}_j, \qquad P(C_{1i}|S_{1i}, \underline{\psi}_j) = \text{MCLCM-C IRF},$$

$$\nu_i = 1, \qquad P(X_{ij} = 1|C_{1i}, S_{1i}, \underline{\psi}_j) = 1.$$

4.3. *Knowledge structure modeling assumptions*

This section of the paper, in somewhat broad strokes to aid clarity of presentation, reviews and, in particular, compares and contrasts the nature of the knowledge structures for the models detailed in Section 4.2. The two main distinctions to be made are: (1) Dimensionality (unidimensional vs. multidimensional) and (2) Skill mastery scale (continuous vs. discrete). Table 3 summarizes the knowledge structure of the models with respect to these two dimensions. Readers are advised to consult Section 4.2 and Table 2 as the model abbreviations will be used throughout this section and the detailed model descriptions will sometimes be referred to.

Table 3
Knowledge structure of reviewed cognitive diagnosis models. U = Unidimensional, M = Multidimensional, C = Continuous, D = Discrete

Skill mastery scale	Dimensionality		
	U	M	U/M
C	1-2-3PL LLTM	MIRT-C MIRT-NC MLTM GLTM	
D		DINO MCLCM-C MCLCM-D RLCM RUM (Q complete)	
C/D		RUM	HYBRID

4.3.1. Dimensionality

The knowledge structure is explicitly unidimensional for some models (LLTM, 1PL, 2PL, and 3PL) and is explicitly multidimensional for the others (MIRT-C, MIRT-NC, MCLCM-C, MCLCM-D, MLTM, GLTM, RLCM, DINO, and RUM). One model, HYBRID, is a mix of unidimensional and multidimensional modeling. For HYBRID, most of the examinees can be assigned a multidimensional vector of skill mastery levels, but examinees who do not fit the multidimensional model are assigned a unidimensional overall mastery level estimated from a unidimensional IRT model (1PL, 2PL, or 3PL).

Given that cognitive diagnosis models are meant to be applied in situations where individual items are assumed to be measuring a number of skills, standard unidimensional IRT models (1PL, 2PL, and 3PL) lack any direct link between the individual skills and their unidimensional ability variable. LLTM does include an explicit link between skills and items through the item component of the IRF, even though its ability is unidimensional.

In LLTM the unidimensional ability can be interpreted as some sort of composite of the multiple skills. Thus, according to the model, two examinees who have the same average level of skill mastery (not necessarily the same set of mastered and not-mastered skills) are expected to have the same probability of a correct response on any item. For this assumption to hold at least roughly, LLTM implicitly requires the multiple skills be highly related. This representation could hold true in some situations. For example, consider a classroom of students learning a subject for which all start off with zero knowledge. All students receive the same instruction and thus learn the multiple skills (to the best of their ability) in the same order. Since skills are usually highly related in such a setting, the degree to which a child learns one skill could be the same for all skills. And since all skills are learned in the same order, the unidimensional knowledge structure model would hold approximately true. Thus, unidimensional models can be useful in some cognitive diagnosis settings. However, in situations where students with the same average ability over the multiple skills vary significantly on their levels of mastery on the individual skills, multidimensional cognitive diagnosis models would be necessary in order to effectively assess skill mastery levels. Multidimensional models vary widely, of course. One important distinction is whether the knowledge structure is continuous or discrete, or a mixture of the two. This distinction is discussed next with respect to all the models. Beyond this distinction, models with the same knowledge structure may also differ in their item structures which will be discussed in Section 4.4.

4.3.2. Skill mastery scale

The second distinction in knowledge structures to be highlighted is that of continuous skill mastery scale vs. discrete skill mastery scale. The models that have unidimensional knowledge structure (LLTM, 1PL, 2PL, and 3PL) all have continuous representations for knowledge via a single real-valued number of a continuous scale, which along with the item parameters completely determines the examinees probability of getting the item right.

Of the models that allow the required skills for an item to be multidimensional, MLTM, GLTM, MIRT-C, and MIRT-NC represent knowledge as continuous latent traits with the number of latent traits for an item being the number of skills of the knowledge structure involved in responding to the item.

On the other hand, of the remaining multidimensional models, RLCM, DINO, MCLCM-C, MCLCM-D, and RUM (with Q complete) all use a discrete representation of the knowledge structure, namely a *latent class* defined by a vector of dichotomous skill mastery parameters, $\underline{\alpha}_j = (\alpha_{j1}, \alpha_{j2}, \ldots, \alpha_{jK})^t$, where, as described earlier in Sections 4.1 and 4.2, $\alpha_{jk} = 1$ indicates examinee j has mastered skill k, and $\alpha_{jk} = 0$ indicates nonmastery.

We ignore here the question of whether the models have mathematically equivalent discrete representations. For example, 1PL is mathematically equivalent to a model for which the ability variable is discrete with $I + 1$ ordered levels, where I is the number of items. For purposes of this paper, we consider the continuity or discreteness as explicitly defined. In the examples surveyed here, the number of discrete ordered levels for ability variables is two. In published cases using more than two discrete ordered levels, the number of such levels is usually much lower than the number of items.

The remaining models that allow for multidimensional knowledge structure, employ a mixture of continuous and discrete representations, though in very different ways. The HYBRID model uses virtually the same knowledge structure as RLCM except that, in addition to the discrete latent classes, HYBRID also models some examinees to have their knowledge characterized instead by one of the unidimensional IRT models (1PL, 2PL, or 3PL), which, as mentioned above, employ a continuous representation of skill mastery level. The RUM knowledge structure representation is similar to that of RLCM and HYBRID. All three have the same latent class knowledge representation for skill mastery, a vector of dichotomous skill mastery level parameters. The major difference between RUM and the other two models is that RUM not only includes model parameters for mastery of the Q-specified skills, but also includes a continuous unidimensional parameter to provide an approximate model for any skills that may be needed to get the item right and which are not included in the Q-specified skills. Note that this use of IRT modeling to augment the latent class knowledge structure based on Q is much different from the use of the unidimensional IRT model in HYBRID. In HYBRID the knowledge structure of an examinee is either one of the restricted latent classes or an IRT ability estimate. One or the other case is assumed to hold for all items. By contrast, with RUM all examinees are posited both a latent class $\underline{\alpha}$ (for their Q-specified skills) and a continuous IRT ability estimate (summarizing their non-Q skills). Of all the cognitive diagnosis models, RUM seems to be the only one that attempts to model examinee differences in mastery on skills not included in Q. This flexibility may turn out to be of consequence in many modeling situations, though the continuous term as currently parameterized has often been difficult to work with.

A comparison of the above knowledge structure representations with the knowledge structure representation of the purely deterministic cognitive diagnosis model and with the relaxations of knowledge structure described in Section 4.1 can provide some guidance to deciding what type of knowledge structure is preferred. In individual cases, such preference would depend on how different the modeled situation is from the deterministic situation. Specifically, it depends on the positivity of the skills, the completeness of Q, and the purpose of the cognitive diagnosis. If Q is approximately complete or if examinees of equal mastery on the Q skills do not vary significantly on their non-Q skills, then a model with no representation of the non-Q skills may perform very well. If the testing situation is truly concerned only with determination of mastery vs. nonmastery (even in the presence of low positivity of the skills) or if the skills have a high degree of positivity, then the discrete (for example, the mastery/nonmastery dichotomy) representation of mastery on the skills would seem to be preferred since the estimation will be more efficient than for models with continuous mastery variables. Other trade-offs in complexity between knowledge structure and item structure can also be considered as described in Section 4.2.

4.4. *Item structure modeling assumptions*

In this section the item structure modeling assumptions will be discussed in general terms to demonstrate the broad variety of such assumptions present in cognitive diagnosis models. The models have many differences in the details of the implementations of these structures, and the reader is referred to Section 4.2 for a more detailed description of the assumptions for each model. This section is divided into three subsections: Item structure for Q skills, Item structure for non-Q skills, and Item structure for multiple strategies. All of the models are included in the first subsection, but only three of the models reviewed in this paper have item structure for non-Q skills and only two have item structure for multiple cognitive strategies. All of the models assume that responses to individual items are statistically independent conditioned on ability. This is commonly referred to as the assumption of Local Independence or Conditional Independence.

4.4.1. *Item structure for Q skills*

The two main distinctions to be made among the item structure representations for the Q skills are: (1) Number of skills (single vs. multiple) and (2) Skill interaction (compensatory/disjunctive vs. noncompensatory/conjunctive). By *compensatory/disjunctive* is meant that no matter how low a level of mastery an examinee has on one skill, it is possible in the mathematical model of the IRF that a high level of mastery on another required skill can compensate (in terms of probability of a correct response) for the low level of mastery on the first skill *to any desired degree*. Suppose an item depends on only two skills. Then even if an examinee is extremely weak in mastery of one skill, by having high enough mastery of the second skill the examinee can still have a high probability of a correct response. By *noncompensatory/conjunctive* is meant that an examinee must be of sufficiently high ability on all the required skills of an item in order to have a high probability of getting the item right. Weakness on one skill may be partly compensated by strength on another but it cannot be compensated for to an arbitrarily high degree. Partial compensation is allowed in all the noncompensatory models as one of the relaxations (positivity) of the deterministic cognitive diagnosis model described in Section 4.1. Indeed, any item response model that demonstrates monotonicity in each of the ability variables must demonstrate some amount of partial compensation (van der Linden, personal communication).

Table 4
Item structure for Q skills. S = Single, M = Multiple, C = Compensatory-Disjunctive, NC = Noncompensatory-Conjunctive, N.A. = Not applicable

Skill interaction	Number skill components		
	S	M	S/M
C		LLTM MIRT-C MCLCM-D MCLCM-C	
NC		MIRT-NC GLTM MLTM RLCM RUM DINO	
N.A.	1-2-3PL		
NC/N.A.			HYBRID

Table 4 summarizes the item structure of the models with respect to these two dimensions. Note that when an item structure models only a single skill, then the skill interaction dimension is not applicable.

Number of skills Given that cognitive diagnosis models are meant to be applied in situations where individual items are assumed to be measuring a number of skills, the standard unidimensional IRT models (1PL, 2PL, and 3PL) are seen, as discussed above, to not provide direct links between multiple skill demands of items and multiple skill levels of examinees. All the remaining models allow for the influence of multiple skills in their IRFs. In HYBRID, as mentioned in the previous section, most of the examinees are modeled with a vector of dichotomous variables which indicate mastery or nonmastery of the multiple skills modeled by the test items; but a few examinees are modeled with a unidimensional ability. Thus, HYBRID is listed as a model that mixes a single-skill representation with a multiple-skill representation.

Skill interaction Type of skill interaction assumed by a model is most important when items require multiple skills, but, as we have seen above, does not strictly require a multidimensional representation of skill mastery. Specifically, LLTM does include the effects of multiple skills in the item structure portion of its IRF, but skill mastery is essentially represented as some sort of unspecified average of mastery over all the skills, a unidimensional representation. Although the representation of multiple skills is not explicit in LLTM, as we discussed earlier, it can be thought of as equivalent to a compensatory skill mastery interaction model. Consequently, we have so labeled it in our corresponding tables. Clearly, LLTM was not intended for diagnosis of examinee mastery on individual skills, and we include it here as an historically important bridge in the evolution from nondiagnostic unidimensional models to multidimensional diagnostic models. MIRT-C and MCLCM-C each use a weighted sum of mastery level variables on the skills in their IRFs, thus making them also compensatory models. MCLCM-D is referred to by Maris (1999) as a "disjunctive" model because a high probability of a correct response will occur so long as at least one of the required skills has been mastered, which could be viewed as the maximum possible compensation.

All of the remaining models that have item structures that allow for multiple skills assume that the mastery levels on the skills required by the item interact in a noncompensatory or "conjunctive" manner in the IRF.

From a psychological perspective, the interaction of skills in examinee item response behavior may sometimes be better modeled as compensatory and other times better modeled as conjunctive. From the classical model fit perspective, however, it may be that either type may fit approximately equally well in practice. For example, even though the compensatory models in principle allow compensation to an arbitrarily high degree among the skills, in practice the skills required for a given item may be highly enough correlated so as to preclude large amounts of compensation for a given examinee. In situations where this is true, a compensatory model might sometimes be favored over an equivalent noncompensatory model if the parameters of the compensatory model are fewer in number or easier to estimate for a particular setting. However, as will be described below, the available noncompensatory models may also contain other psychologically valuable modeling components – such as item structure for non-Q skills (Section 4.4.2) and multiple cognitive strategies (Section 4.4.3) – that are not found in the compensatory models thus making it difficult to find an equivalent compensatory model. Perhaps the highlighting of such modeling gaps will result in new models to fill these gaps as needed.

4.4.2. Item structure for non-Q skills

Only three models have item structure for non-Q skills: MLTM, GLTM, and RUM. As mentioned in Section 4.3, among all the models only RUM explicitly models differences in examinee mastery on non-Q skills. MLTM and GLTM essentially assume that all examinees have the same level of mastery on the non-Q skills. MLTM and GLTM have exactly the same item structure for modeling performance on the non-Q skills. They both use a constant probability that is independent of both examinees and items (indeed, this constant could be interpreted as the probability of "not slipping," although that was not the intended interpretation in Embretson, 1985, 1997). RUM takes a different approach, using a 1PL model with a completeness item parameter that ranges from 0 to 3 (the negative of the 1PL difficulty parameter). Thus, RUM allows different degrees of Q completeness from item to item, and uses a unidimensional parameter to approximately model the non-Q skills.

As discussed in the Knowledge structure section (Section 4.3 above), the modeling of the non-Q skills, as in RUM, is not always necessary. Even though completeness and positivity are theoretically separate issues, the two can become confounded. For example, as discussed earlier, if examinees of the same level of mastery on the skills are also of the same level of mastery on the non-Q skills, performance on the non-Q skills cannot be distinguished from performance on the Q skills, and Q incompleteness is totally accounted for within the positivity terms. Of course there is also the trivial case where modeling performance on non-Q skills is not necessary because Q is complete, or very nearly so. Though a high degree of completeness is unlikely on theoretical grounds, we believe that when the number of skills is at least moderately large, the skills in Q may "explain" (much like in a principal components analysis) most of the observed systematic variation. In such cases, we can eliminate the completeness term from the model with little penalty to effective skills mastery assessment.

4.4.3. Multiple cognitive strategies

The term *strategy* as we use it here refers to a hypothesis about how examinees respond to one or more test items. A strategy is made concrete by the specification of the skills that are required for each item, i.e., the Q matrix. Only MLTM allows for alternate strategies in the item response behavior of examinees. Even though HYBRID can also be viewed as modeling two specific strategies, only one of the strategies is cognitively based. In MLTM the alternative strategies are explicitly modeled in the skills. In HYBRID, even though only a single set of skills is modeled, the alternative unidimensional model within HYBRID can be viewed as modeling a single skill for all the items. As mentioned above, this single skill could be viewed as a unidimensional composite of the multiple skills with the composite being the same for all the items.

Although the modeling of multiple strategies in parametric IRT skills diagnosis has been implemented to only a limited degree, the integration of multiple strategies into unidimensional and latent class IRT models has been studied more extensively, including, for example, Mislevy and Verhelst (1990), Kelderman and Macready (1990), and Rost (1990). Thus, the modeling of multiple strategies in parametric IRT skills diagnosis is an area where advancements in the near future can be expected.

4.5. Summary table of the models

The above review of cognitive diagnosis models can be summarized with a table highlighting important differentiating characteristics that the above review revealed. Four features seem especially relevant in assessing the degree of complexity of a model:

(1) *Multidimensionality*. The skills can be modeled as separate dimensions and more than one can be allowed to occur in a given item. We note here the term multidimensional is used to refer to the indicated dimensions. Even if a four-skill model were shown to have lower dimensionality for a particular dataset, the model itself mediates the true statistical dimensionality for this dataset and the indicated dimensionality that may be based on cognitive or pedagogical theory.
(2) *Conjunctive vs. compensatory item response modeling*. The psychological nature of how the cognitive components interact in an item should help determine which model is more appropriate. Conjunctive models are better aligned with cognitive theory in some cases in which it is strongly believed that high levels of ability on all required skills for an item are required for high likelihood of correct response to that item.
(3) *Incompleteness of Q*. The degree to which Q may be considered complete or nearly complete should be considered in choosing a model. When the number of specified skills is large, Q will usually be approximately complete, by analogy with principal

Table 5
Classification of cognitive diagnosis models

Model	Multidimensional	Noncompensatory	Q incomplete	Multiple strategies
1PL	N	N	N	N
2PL	N	N	N	N
3PL	N	N	N	N
LLTM	N	N	N	N
MIRT-C	Y	N	N	N
MIRT-NC	Y	Y	N	N
GLTM	Y	Y	Y	N
MLTM	Y	Y	Y	Y
MCLCM-C	Y	N	N	N
MCLCM-D	Y	N	N	N
DINO	Y	N	N	N
RLCM	Y	Y	N	N
HYBRID	Y	Y	N	Y
RUM	Y	Y	Y	N

components analysis. But when the specified number of skills is small, this part of the modeling process could be important.

(4) *Multiple cognitive strategies*. There will often exist more than one way to solve an item correctly and this can result in more than one set of skills available for success in solving the item.

A fifth important feature is that a model allow for the influence of multiple skills on examinee item performance. As noted above, all the models, except 1PL, 2PL, and 3PL, allow for such an influence, usually through multidimensional modeling of examinee proficiency. Even though this is an important characteristic of a cognitive diagnosis model, since it is shared by all the models meant for cognitive diagnosis, it is not included as a separate entry in this list. Table 5 summarizes the cognitive diagnosis models reviewed in this paper by noting whether or not (Y for Yes, N for No) each model contains the above characteristics. The first four models in the table are the unidimensional models; the next four models are the ones with multidimensional continuous skill mastery representations; and the last five models are the models with multidimensional discrete skill mastery representations. Within each of these sets the models are placed in order according to increasing cognitive complexity as determined by the headings of the table, though the ordering is only partial in some cases since there are ties.

Finally, it is important to point out that models having a time component were purposely left out of this review to maintain a reasonable scope. These models include the multidimensional Rasch model for learning and change by Embretson (1991), the LLTM with relaxed assumptions by Fischer (1989), the tutoring systems of Anderson et al. (1995) and of VanLehn et al. (1998), and the stochastic learning theory model for a knowledge space by Falmagne (1989). A separate summary of how learning over time is modeled should be addressed in the future.

5. Summary and conclusions: Issues and future directions

The purpose of this paper is to provide for psychometricians and statisticians a broad survey of the basic ideas, models, and statistical tools that constitute the emerging field of skills diagnostic assessment, to provide a focused survey of the field's rapidly growing literature, and to lay out for researchers and practitioners the steps needed to effectively conduct such assessments. A secondary goal is to illuminate some of the unresolved issues and controversies of the skills diagnostic assessment field, as a guide for future research. This statistical subfield promises exciting research opportunities and promises to play a vital societal role as assessments developed specifically to support teaching and learning become increasingly important in schools and other instructional settings.

Skills diagnostic assessment research to date has made considerable and important progress. In particular, a rich, varied, and useful collection of parametrically complex and potentially useful stochastic models have been developed, which were carefully and systematically surveyed in the second portion of the paper. To situate the model survey within a proper context of assessment design, the first portion of the paper presented a practitioner-oriented framework for the broad basic steps of the process required to conduct a skills diagnostic assessment, including modeling and statistical analysis.

Scientific disagreements and controversies exist in both the methods and goals of the approach we advocate, and in the interpretations of progress that has occurred to date. We provide our point of view here, and the reader is encouraged also to read the literature for alternative points of view. Well established and largely agreed-upon recommendations concerning which combinations of model, inference method, computational approach, and evaluation of assessment effectiveness that should work best for practitioners in various settings do not yet exist and are still subject to controversy. In spite of the lack of consensus concerning various aspects, the foundations and basic building blocks of the field are well developed, numerous models and several statistical approaches are available for application and study, and various appropriate computational software packages exist. Thus, skills diagnostic assessment studies can currently be carried out to pursue the questions that remain.

Alternative statistical computation methods such as MCMC and EM algorithms can be compared along several important dimensions. For example, the EM algorithm demonstrates advantages over MCMC such as greater computational efficiency and direct focus on estimation of particular parameters. EM disadvantages include greater difficulty in developing the analytical likelihood functions necessary to implement new models, especially complex models, and less straightforward determination of standard errors (Junker, 1999; Patz and Junker, 1999a, 1999b).

MCMC advantages over EM include relatively greater ease in setting up new models, and the generation of the estimated posterior distribution of all the parameters. The estimated posterior distribution directly aides determination of multiple modes and better control of standard errors. These advantages come at certain costs, including longer computation time, and greater difficulty of determining convergence since the convergence is not that of individual parameters, but convergence of a joint posterior with many variables (all the model parameters). These issues, as they relate specifically to cognitive diagnostic models, should be a target of future setting-specific research.

As we argue in the paper, development of effective diagnostic assessment can be attempted now, whether for experiment or for trial and/or operational use within a program. Better diagnostic performance can be expected if the data come from a test carefully designed for the purposes of skills diagnosis. The selection of an appropriate cognitive diagnostic model based on an analysis of the cognitive interaction between the skills and the items on the test, almost always requires consultation of the literature and close collaboration among psychometric and substantive experts, in addition to empirical checking and confirmation. As emphasized in the paper, one should plan a broad-based evaluation of the effectiveness of one's model-driven skills assessment from a variety of perspectives, including statistical convergence of the estimation procedure, interpretation of model parameter estimates for reasonableness and internal consistency, evaluation of model diagnostics, reliability estimation, and gathering multiple pieces of evidence in support of internal and external validity. Comparison of fit statistics based on a cognitive diagnostic model with fit statistics based on a standard unidimensional IRT model may have scientific or academic interest, but, as we note in the text of this paper, whether or not a unidimensional model has better statistical fit does not determine, in and of itself, the appropriateness or usefulness of the diagnostic model. For the fit study to be most useful, the fit measures used to vet the model must be sensitive to the diagnostic purpose of the test.

Given the available foundations of cognitive diagnostic modeling, the field is at a stage where conducting pilot skills diagnostic studies involving real test data and for various test design settings (including tests already designed and originally intended for continuous scaling of examinees rather than diagnosing them) is a natural and necessary next step in the evolution of the field. The resulting feedback from numerous such real-data-based pilot studies should sharpen the focus of future model building research and associated statistical methodology development, should contribute to the foundations of diagnostic assessment design and task development, and should resolve many of the outstanding open issues that remain. Such diagnostic research will make clearer which combinations of models, procedures, and computational approaches are to be preferred, as a function of varying instructional settings and assessment purposes.

One fundamental requirement for the approach we advocate here is that the skills and the Q matrix of the cognitive diagnostic model represent what substantive experts believe to be the essential skills and their item/skill interactions. That is, to optimize effectiveness, a skills diagnostic analysis of data must accept *ab ovo* the approximate validity of the proposed cognitive diagnostic model. Then, as long as the diagnostic model with its expert-chosen Q matrix fits the data reasonably well, based on fit measures that are relevant to the skills diagnostic analysis, it will be appropriate to use the model to draw skills diagnostic conclusions, even if the diagnostic model fits slightly less well in a classical sense than a unidimensional model, say. As we note in the body of the paper, parsimony is vitally important, but only what we would term "parsimony well conceived": The simplest solution is preferred among all possible solutions that satisfy the diagnostic purpose.

We also note that high correlations among the skills and/or low individual skill reliabilities have the potential to cause measurement problems. In general, it is helpful if the diagnostic model and estimation method used are equipped to model and estimate inter-skill correlations. We note that a most extreme case may occur in which five skills, say, are defined with perfectly unidimensional data. In such a case the ordered individual skill classifications would be completely determined by four thresholds on the unidimensional scale. As long as the individual skills are clearly defined substantively in terms of a Q matrix that connects cognitive skills and items, it may be an extremely useful service to teachers and learners to know that each particular scale score denotes a particular profile of skill proficiencies. The point of the example is that even in the most extreme case that data are perfectly unidimensional, there may be statistically supportable, useful diagnostic information that can be given to teachers and learners, stated in terms of the cognitive diagnostic skills rather than an abstract scale score.

The typical formative assessment student profiles that educators and students require for their teaching and learning purposes are almost always more substantively complex and detailed than what a single score can provide. For example, each examinee might be assigned a nine component discrete mastery/nonmastery algebra skills diagnostic profile as a result of an algebra skills diagnostic assessment. This sort of additional parametric complexity includes a real danger that selected models may be nonidentifiable or, more generally, may demonstrate unstable estimation or computation. The careful user needs to watch for this possibility.

On one hand the use of cognitive diagnostic models presents some very real dangers due to model complexity, and on the other hand, the use of overly simple models fails to provide the diagnostic benefit that is required. Thus, one of the key emerging challenges facing both researchers and practitioners is the finding of a comfortable middle ground of moderate model complexity such that rigorous statistical thinking supports the use of the selected models and the consequent statistical methodology chosen, but also where there is just enough well-estimated parametric complexity to provide student profiles with enough cognitive skill information to be useful for student and instructor.

It is important to note that most test data that have been currently available for skills diagnostic analysis have been from tests intended for continuous unidimensional scaling. By contrast, future skills diagnostic assessments ideally will be generated from tests specifically designed to assess discrete levels of mastery of a targeted set of skills. Such designed tests may display much stronger latent multidimensionality as will be determined by future research and applications.

The DSM pathological gambling diagnostic example (Templin and Henson, 2006) of Section 3 provides an instance where the profile information resulting from the multiple latent components of the complex multidimensional diagnostic model improves on the practitioner's capacity to draw conclusions about gambling profiles. The multiple dimensions that were found were so strongly dimensionally distinct from each other that, *a fortiori*, information was provided that could not have been made available from a standard unidimensional model based analysis. The further consequential validity question of whether this additional information obtained can in practice actually improve the diagnosis or treatment of gambling addiction is a subject of interesting future research. Expanding the store of such real data examples, some more explicitly defended by statistical analyses and especially some involving consequential validity, is another important goal for future research.

Some would argue that relatively simple and perhaps even approximately unidimensional IRT models, accompanied by targeted sophisticated statistical analyses, can produce as much diagnostic information as can be reasonably expected for test lengths appropriate for classroom use. We disagree that this assertion should be uniformly applied to all (or even most) skills diagnostic settings and welcome the opportunity to address the issue in future studies and applications. Of course, the resolution of this controversy may have differing general conclusions for the different types of student test data. In particular, for example, there may be one answer for assessments that are strongly unidimensional and for which the diagnostic question is translated into providing better skills-based information about the score scale. A different answer may be expected for tests that are designed for the explicit purpose of providing more highly multidimensional, discrete skill profile information.

Reliability has always been a conceptual cornerstone of psychometrically based educational testing, and remains so for diagnostic testing. Perhaps the most salient question for skills diagnostic testing is understanding and establishing standards for this new notion of skills classification reliability. The paper summarized one practical method of defining and estimating an index of reliability. Given estimates of diagnostic classification reliability, the literal transfer of standardized test reliability standards to diagnostic testing may not be appropriate. If, for example, a Cronbach alpha or KR20 of at least 0.9, say, represents high reliability for a high stakes college admission test, it does not necessarily follow that the same level of 0.9 estimated correct classification reliability or skills classification test–retest consistency is required for a successful diagnostic test intended for classroom use. The needed standards of reliability for these diagnostic classification reliability indices must be determined in reference to the changed interpretation of the new scale and the changed diagnostic testing setting and purposes for the test, and must evolve independently through research and experience over time.

A multitude of interesting and important research questions remain concerning skills diagnostic assessment, especially concerning the statistical aspects of such diagnoses. These questions include foundational questions as well as applications questions, and comprise psychometric and statistical aspects as well as applications work that combines the statistics with substantive areas of cognitive science, pedagogy, and learning theory. The benefits of widely providing effective skills diagnostic assessment in the near future, designed to improve teaching and learning in instructional settings, are potentially very large. Moreover, as this paper argues, based on the current state of the art, practitioners and researchers alike can currently carry out skills diagnostic assessments in multiple ways with tools to evaluate performance.

We hope this concluding discussion spurs increased activities along two different and vital streams of research. The first is to assess how wide and deep the value of skills diagnosis assessment is when applied in real instructional settings. This is partly a question of how much value is added by the diagnostic information from carefully designed and analyzed diagnostic tests. The consequential validity question is whether the added information directly and demonstrably improves teaching and learning.

The second stream of research is statistical and psychometric, namely to continue to refine various diagnostic approaches, to introduce new ones, and to compare the performances of the major approaches with the goal of setting-specific optimization of the retrieval of skills diagnostic information. Some models, statistical methods, computational approaches, and evaluation approaches will fade away but the needs are so rich and varied that the authors predict that many will survive and prosper.

References

Adams, R.J., Wilson, M. (1996). Formulating the Rasch model as a mixed coefficients multinomial logit. In: Engelhard, G., Wilson, M. (Eds.), *Objective Measurement III: Theory in Practice*. Ablex, Norwood, NJ.

Adams, R.J., Wilson, M., Wang, W.-C. (1997). The multidimensional random coefficients multinomial logit model. *Applied Psychological Measurement* **21**, 1–23.

Almond, R.G., Steinberg, L.S., Mislevy, R.J. (2003). A framework for reusing assessment components. In: Yanai, H., Okada, A., Shigemasu, K., Kano, Y., Meulman, J.J. (Eds.), *New Developments in Psychometrics*. Springer, Tokyo, pp. 281–288.

Andersen, E.B. (1985). Estimating latent correlations between repeated testings. *Psychometrika* **50**, 3–16.

Anderson, J.R., Corbett, A.T., Koedinger, K.R., Pelletier, R. (1995). Cognitive tutors: Lessons learned. *The Journal of the Learning Sciences* **4**, 167–207.

Birnbaum, A. (1968). Some latent trait models and their use in inferring an examinee's ability. In: Lord, F.M., Novick, M.R. (Eds.), *Statistical Theories of Mental Test Scores*. Addison-Wesley, Reading, MA.

Cowles, M.K., Carlin, B.P. (1996). Markov chain Monte Carlo convergence diagnostics: A comparative review. *Journal of the American Statistical Association* **91**, 883–904.

de la Torre, J., Douglas, J.A. (2004). Higher-order latent trait models for cognitive diagnosis. *Psychometrika* **69**, 333–353.

DiBello, L.V., Stout, W.F., Roussos, L.A. (1995). Unified cognitive/psychometric diagnostic assessment likelihood-based classification techniques. In: Nichols, P.D., Chipman, S.F., Brennan, R.L. (Eds.), *Cognitively Diagnostic Assessment*. Erlbaum, Mahwah, NJ, pp. 361–389.

Embretson, S.E. (1985). Multicomponent latent trait models for test design. In: Embretson, S.E. (Ed.), *Test Design: Developments in Psychology and Psychometrics*. Academic Press, New York, pp. 195–218.

Embretson, S.E. (1991). A multidimensional latent trait model for measuring learning and change. *Psychometrika* **56**, 495–515.

Embretson, S.E. (1997). Multicomponent response models. In: van der Linden, W.J., Hambleton, R.L. (Eds.), *Handbook of Modern Item Response Theory*. Springer, New York, pp. 305–321.

Falmagne, J.-C. (1989). A latent trait theory via a stochastic learning theory for a knowledge space. *Psychometrika* **54**, 283–303.

Fischer, G.H. (1973). The linear logistic model as an instrument in educational research. *Acta Psychologica* **37**, 359–374.

Fischer, G.H. (1983). Logistic latent trait models with linear constraints. *Psychometrika* **48**, 3–26.

Fischer, G.H. (1989). An IRT-based model for dichotomous longitudinal data. *Psychometrika* **54**, 599–624.

Gelman, A., Meng, X.-L. (1996). Model checking and model improvement. In: Gilks, W.R., Richardson, S., Spiegelhalter, D.J. (Eds.), *Markov Chain Monte Carlo in Practice*. Chapman and Hall/CRC, Boca Raton, FL. Chapter 11.

Gelman, A., Carlin, J.B., Stern, H.S., Rubin, D.R. (1995). *Bayesian Data Analysis*. Chapman & Hall, London.

Gierl, M.J. (1997). An investigation of the cognitive foundation underlying the rule-space model. Doctoral dissertation, University of Illinois at Urbana-Champaign, 1996. *Dissertation Abstracts International* **57** (08), 5351-B. University Microfilms No. AAC 97-02524.

Gitomer, D.H., Yamamoto, K. (1991). Performance modeling that integrates latent trait and class theory. *Journal of Educational Measurement* **28**, 173–189.

Haertel, E.H. (1984). An application of latent class models to assessment data. *Applied Psychological Measurement* **8**, 333–346.

Haertel, E.H. (1990). Continuous and discrete latent structure models of item response data. *Psychometrika* **55**, 477–494.

Hagenaars, J. (1993). *Loglinear Models with Latent Variables*. Sage Publications, Newbury Park, CA.

Hartz, S.M. (2002). A Bayesian framework for the Unified Model for assessing cognitive abilities: Blending theory with practicality. Unpublished doctoral dissertation, University of Illinois, Champaign, IL.

Hartz, S.M., Roussos, L.A. (2005). The Fusion Model for skills diagnosis: Blending theory with practice. ETS Research Report, Educational Testing Service, Princeton, NJ. In press.

Henson, R.A., Roussos, L., Templin, J.L. (2004) Cognitive diagnostic "fit" indices. Unpublished ETS Project Report, Princeton, NJ.

Henson, R.A., Roussos, L., Templin, J.L. (2005). Fusion model "fit" indices. Unpublished ETS Project Report, Princeton, NJ.

Henson, R.A., He, X., Roussos, L.A. (2006). Cognitive diagnosis attribute level discrimination indices. *Applied Psychological Measurement*. In press.

Jang, E.E. (2005). A validity narrative: Effects of reading skills diagnosis on teaching and learning in the context of NG TOEFL. Unpublished doctoral dissertation, University of Illinois, Champaign, IL.

Jang, E.E. (2006). Pedagogical implications of cognitive skills diagnostic assessment for teaching and learning. Paper presented at the annual meeting of the American Educational Research Association, San Francisco, CA, April.

Junker, B. (1999). Some statistical models and computational methods that may be useful for cognitively-relevant assessment. Prepared for the Committee on the Foundations of Assessment, National Research Council, November 30, 1999.

Junker, B., Sijtsma, K. (2001). Cognitive assessment models with few assumptions, and connections with nonparametric item response theory. *Applied Psychological Measurement* **25**, 258–272.

Kelderman, H., Macready, G.B. (1990). The use of loglinear models for assessing differential item functioning across manifest and latent examinee groups. *Journal of Educational Measurement* **27**, 307–327.

Klein, M., Birenbaum, M., Standiford, S., Tatsuoka, K.K. (1981). Logical error analysis and construction of tests to diagnose student 'bugs' in addition and subtraction of fractions. Tech. Rep. 81- 6-NIE, University of Illinois, CERL, Urbana.

Linacre, J.M. (1989). *Many-Facet Rasch Measurement*. MESA Press, Chicago.

Lord, F.M. (1980). *Application of Item Response Theory to Practical Testing Problems*. Lawrence Erlbaum Associates, Hillsdale, NJ.

Macready, G.B., Dayton, C.M. (1977). The use of probabilistic models in the assessment of mastery. *Journal of Educational Statistics* **2**, 99–120.

Macready, G.B., Dayton, C.M. (1980). The nature and use of state mastery models. *Applied Psychological Measurement* **4**, 493–516.

Maris, E. (1999). Estimating multiple classification latent class models. *Psychometrika* **64**, 187–212.

Masters, G.N. (1982). A Rasch model for partial credit scoring. *Psychometrika* **47**, 149–174.

Mislevy, R.J. (1995). Probability-based inference in cognitive diagnosis. In: Nichols, P., Chipman, S., Brennan, R. (Eds.), *Cognitively Diagnostic Assessment*. Erlbaum, Hillsdale, NJ, pp. 43–71.

Mislevy, R.J., Verhelst, H. (1990). Modeling item responses when different subjects imply different solution strategies. *Psychometrika* **55**, 195–215.

Mislevy, R.J., Steinberg, L.S., Almond, R.G. (2003). On the structure of educational assessments. *Measurement: Interdisciplinary Research and Perspectives* **1**, 3–67.

Mislevy, R.J., Almond, R.G., Lukas, J.F. (2004). A brief introduction to Evidence-Centered Design. CSE Technical Report 632, The National Center for Research on Evaluation, Standards, and Student Testing (CRESST), Center for the Study of Evaluation (CSE), UCLA, Los Angeles, CA.

Patz, R.J., Junker, B. (1999a). A straightforward approach to Markov chain Monte Carlo methods for item response models. *Journal of Educational and Behavioral Statistics* **24**, 146–178.

Patz, R.J., Junker, B. (1999b). Applications and extensions of MCMC in IRT: Multiple item types, missing data, and rated responses. *Journal of Educational and Behavioral Statistics* **24**, 342–366.

Pellegrino, J.W., Chudowsky, N., Glaser, R. (2001). *Knowing what Students Know: The Science and Design of Educational Assessment*. National Academy Press, Washington, DC.

Rasch, G. (1961). On general laws and the meaning of measurement in psychology. In: Neyman, J. (Ed.), In: *Proceedings of the Fourth Berkeley Symposium on Mathematical Statistics and Probability*, vol. 4. University of California Press, Berkeley, CA.

Reckase, M.D., McKinley, R.L. (1991). The discriminating power of items that measure more than one dimension. *Applied Psychological Measurement* **15**, 361–373.

Rost, J. (1990). Rasch models in latent classes: An integration of two approaches to item analysis. *Applied Psychological Measurement* **14**, 271–282.

Roussos, L.A. (1994). Summary and review of cognitive diagnosis models. Unpublished manuscript, University of Illinois.

Roussos, L., DiBello, L., Stout, W. (2006a). Diagnostic skills-based testing using the Fusion-Model-Based Arpeggio system. In: *Cognitively Diagnostic Assessment in Education: Theory and Practice*. Cambridge University Press, Cambridge.

Roussos, L.A., DiBello, L.V., Henson, R.A., Jang, E.E., Templin, J.L. (2006b). Skills diagnosis for education and psychology with IRT-based parametric latent class models. In: Embretson, S., Roberts, J. (Eds.), *New Directions in Psychological Measurement with Model-based Approaches*. American Psychological Association, Washington, DC.

Samejima, F. (1995). A cognitive diagnosis method using latent trait models: Competency space approach and its relationship with DiBello and Stout's unified cognitive-psychometric diagnosis model. In: Nichols, P.D., Chipman, S.F., Brennan, R.L. (Eds.), *Cognitively Diagnostic Assessment*. Erlbaum, Mahwah, NJ, pp. 391–410.

Sinharay, S. (2004). Experiences with Markov chain Monte Carlo convergence assessment in two psychometric examples. *Journal of Educational and Behavioral Statistics* **29**, 461–488.

Sinharay, S. (2005). Assessing fit of unidimensional IRT models using a Bayesian approach. *Journal of Educational Measurement* **42**, 375–394.

Sinharay, S. (2006). Model diagnostics for Bayesian networks. *Journal of Educational and Behavioral Statistics* **31**, 1–33.

Sympson, J.B. (1977). A model for testing with multidimensional items. In: Weiss, D.J. (Ed.), *Proceedings of the 1977 Computerized Adaptive Testing Conference*. University of Minnesota, Department of Psychology, Psychometric Methods Program, Minneapolis, pp. 82–88.

Tatsuoka, K.K. (1983). Rule space: An approach for dealing with misconceptions based on item response theory. *Journal of Educational Measurement* **20**, 345–354.

Tatsuoka, K.K. (1984). Caution indices based on item response theory. *Psychometrika* **49**, 95–110.

Tatsuoka, K.K. (1985). A probabilistic model for diagnosing misconceptions in the pattern classification approach. *Journal of Educational Statistics* **12**, 55–73.

Tatsuoka, K.K. (1990). Toward an integration of item-response theory and cognitive error diagnoses. In: Frederiksen, N., Glaser, R.L., Lesgold, A.M., Shafto, M.G. (Eds.), *Diagnostic Monitoring of Skill and Knowledge Acquisition*. Erlbaum, Hillsdale, NJ.

Tatsuoka, K.K., Tatsuoka, M.M. (1987). Bug distribution and pattern classification. *Psychometrika* **52**, 193–206.

Tatsuoka, K.K., Tatsuoka, M.M. (1997). Computerized cognitive diagnostic adaptive testing: Effect on remedial instruction as empirical validation. *Journal of Educational Measurement* **34**, 3–20.

Tatsuoka, K.K., Linn, R.L., Yamamoto, K. (1988). Differential item functioning resulting from use of different solution strategies. *Journal of Educational Measurement* **25**, 301–319.

Templin, J.L., Henson, R.A. (2006). Measurement of psychological disorders using cognitive diagnosis models. *Psychological Methods*. In press.

VanLehn, K., Niu, Z., Siler, S., Gertner, A. (1998). Student modeling from conventional test data: A Bayesian approach without priors. In: Goetl, B., et al. (Eds.), *Proceedings of the Intelligent Tutoring Systems Fourth International Conference ITS 98*. Springer-Verlag, Berlin, pp. 434–443.

von Davier, M. (2005). A general diagnostic model applied to language testing data. ETS Research Report: RR-05-16, Educational Testing Service, Princeton, NJ.

von Davier, M., Yamamoto, K. (2004). A class of models for cognitive diagnosis. Paper presented at the ETS Spearmann Conference, The Inn at Penn, Philadelphia, PA, Invitational Conference, October.

von Davier, M., DiBello, L., Yamamoto, K. (2006). A shortlist of models for cognitive diagnosis. In: Klieme, E., Leutner, D. (Eds.), *Assessment of Competencies in Educational Contexts*. Hogrefe & Huber Publishers, Bern. In press.

Whitely, S.E. (1980). Multicomponent latent trait models for ability tests. *Psychometrika* **45**, 479–494. Name is now Embretson, S.E.

Wilson, M. (1992). The ordered partition model: An extension of the partial credit model. *Applied Psychological Measurement* **16**, 309–325.

Wilson, M., Adams, R.J. (1995). Rasch models for item bundles. *Psychometrika* **60**, 181–198.

Zhang, Y., Puhan, G., DiBello, L.V., Henson, R.A., Templin, J.L. (2006). Estimating reliability of student profile scores for Next-Generation TOEFL. ETS Research Report series, Educational Testing Service, Princeton, NJ. Submitted for publication.

Handbook of Statistics, Vol. 26
ISSN: 0169-7161

DOI: 10.1016/S0169-7161(06)26040-1

31B

Some Notes on Models for Cognitively Based Skills Diagnosis

Shelby J. Haberman and Matthias von Davier

Cognitive diagnostic models have been proposed in recent years. Although the explicit intention has been to employ such models in assessments designed for cognitive diagnosis, these models have commonly been employed to produce cognitive diagnosis from assessments not designed for this purpose. This strategy raises a number of questions. First, what really happens when existing, well established assessment instruments are retrofitted using cognitive diagnosis models? Are the hypothesized multiple skills distinguishable from overall performance on the test? Second, is there evidence for the presence of the number and the structure of the hypothesized multiple skills, or can the observed data appropriately be described using fewer ability variables or using a simpler skill structure? Third, what is the utility or added value of reports that contain skill profiles? How do educators and other consumers of these reports actually use such student data? Fourth, what distinguishes models for skills diagnosis from other models with multiple latent variables such as multidimensional IRT or multiple classification latent class models?

Introduction

Cognitive diagnostic assessment traditionally has been performed through the use of specialized tests designed for this purpose. The Woodcock–Johnson III™ Tests of Cognitive Abilities are a well-known example. These tests are traditionally analyzed by straightforward statistical methods. However, within psychometrics, recently a much different picture of cognitive diagnosis has emerged in which statistical models are employed to obtain cognitive diagnosis information from conventional assessments not designed for cognitive diagnosis purposes. It is appropriate to ask whether use of these statistical models accomplishes its purpose. Presumably, if the application to cognition is really successful, then validity studies should relate results of analyses based on cognitive diagnostic models to results that are obtained from use of conventional tests of cognition. As far as we are aware, such a validity study has not yet been attempted, although limited validity studies of other kinds have been conducted to evaluate the skill assessments produced by cognitive diagnostic models (see, for example, the discussion on validity in DiBello, Roussos and Stout in Chapter 31A).

Without a validity study, examination of the value of cognitive diagnostic models must be performed on a much more limited basis. The models in effect seek to find multivariate proficiency measures to explain the observed patterns of responses to items in conventional tests. These multivariate measures are not necessarily difficult to find. In the old SAT® I examination, it is a straightforward matter to consider a combined measure of quantitative proficiency based on the Math examination and a measure of verbal proficiency based on the Verbal examination. To be sure, the correlations of the proficiencies appear to be rather high, say more than 0.7, but a claim that the complete SAT measures more than one proficiency appears reasonable enough.

The challenge to cognitive diagnostic models arises when a single test designed to measure a single proficiency is analyzed in an attempt to extract more information. For instance, the SAT I Verbal examination was designed to provide a highly reliable test of verbal proficiency. Such design leads to a test relatively well described by use of conventional univariate item response theory. To be sure, such an item response model will not provide a perfect description of the data, but the room for improvement appears to be modest. In addition, some causes of model variation are much more prosaic than are the causes considered by cognitive diagnostic models. Multiple responses are based on items that refer to a common reading passage, so that testlet effects may appear. Occasional examinee blunders arise in which gridding errors are not caught, pages are omitted for no obvious reason, or speededness may have an effect. Nonetheless, such an examination is relatively unfavorable for models that envision multivariate proficiencies. The burden of proof lies on the developer of a multivariate item response model to demonstrate that multivariate parameters can be reliably estimated, that the multivariate model approximates the observed pattern of responses better than the univariate model, and that the multivariate parameters cannot be more accurately estimated by a simple regression of the multivariate parameter estimates on the conventional univariate estimate (Haberman, 2005).

This approach that models should not have more parameters than necessary can be regarded as contradicting a market demand for systems to diagnose skills without use of a custom-designed test. Clients of testing agencies want to report more data about examinees and expect that these richer reports can be used by teaching professionals to modify their instruction. Nonetheless, a claim concerning market demand does not ensure either that reporting multiple skills for a given assessment is actually warranted or that such a multidimensional report has a higher utility for the user of such data than a report that contains only one normed or equated scale score, a percentile, or a single proficiency level (e.g., "proficient" or "above basic"). This note addresses some of the questions that arise when models for cognitive diagnosis are considered for reporting student performance on cognitive tasks. Section 1 considers problems associated with an existing test that is provided with a more complex cognitive structure than was envisioned when the assessment was originally designed. Section 2 addresses a related question, the issue of dimensionality, that arises even if an instrument was constructed based on the assumption of multiple skills. Section 3 talks about actual use of skill profiles, an issue that needs further research. Section 4 addresses the relationship between models for skills diagnosis and other statistical models with multiple latent variables.

1. Retrofitting a test with a hypothesized skill structure

Traditionally, theory or hypothesis driven test construction aimed at developing a reliable measure for an operationalization of one construct at a time. An assessment may contain more than one of these unidimensionally defined and constructed measures. More exploratory test development strategies use statistical tools such as principal component analysis that enable one to derive subsets of items reflecting empirically defined and distinguishable entities. Once a structure of an assessment has been established, protective measures take place during subsequent administrations, which aim at preserving the original structure of the assessment.

In this note, we discuss a slightly different approach that has been taken in order to allow the use of skill diagnosis tools with existing test instruments. This approach is often referred to as retrofitting. It assumes that a given one- or low-dimensional assessment can be augmented by an expertly generated Q-matrix that relates the assessment to a more complex theory than the one originally applied to develop the instrument. The Q-matrix is utilized to derive a multidimensional skill structure for the retrofitted assessment. An examination of basic tools of classical test theory indicates the problematic nature of retrofitting existing unidimensional assessments with a multidimensional skill structure. Formulas commonly used to assess reliability are also being used to select items, often with the effect that items most representative of the total score are retained while less representative items are often removed from a scale. In addition, tasks that function differently in different subpopulations are also considered undesirable for classical applications of tests, which usually aim at providing only one (or very few) score(s) that should ideally be comparable and equally reliable and valid across different subgroups of the target population. Testing organizations have highly sophisticated procedures to ensure that tasks entering operational use are of this type, so that operational versions of assessments may be less than ideal material for finding deviations from unidimensionality. Unidimensionality in turn, implies that only one score or ability estimate is necessary (see von Davier, 2005) to predict observed variability of responses across students, so that skill diagnostic models that assume more than one dimension will likely fail to provide evidence that more than one skill is required to perform successfully on the assessment.

Retrofitting an existing assessment with a cognitive diagnosis structure also implies that some form of arbitrary choice has been made. The original test construction has taken place under one paradigm, whereas the cognitive diagnosis retrofit assumes another paradigm. Often the difference lies in the assumption of one (or a few) continuous latent variables versus a multitude of discrete (mastery/nonmastery) latent variables, for which the intercorrelations have to be estimated. In the case of an originally unidimensional assessment, the resulting correlations will be rather high, so that the choice of the Q-matrix design, more specifically the number and allocation of skills becomes arbitrary, since the highly correlated skills found for unidimensional assessment data are almost interchangeable. So far, most examples of applications of cognitive diagnosis have made use of retrofitted data. However, we will assume in the following that assessment data are based on a test that was constructed based on the assumption of multiple, empirically distinguishable skills.

2. Dimensionality

Even if an assessment was actually designed according to a rationale that encompasses more than one skill domain, so that different subsets of the tasks representing the assessment tap into more than one skill or student attribute, research has to establish that the hypothesized structure can be identified reliably for the assessment in the targeted population. Based on extensions of classical test theory, Haberman (2005) has shown that in cases where skills correlate considerably, predictions of subscores based on a total test score can be more reliable than the subscores themselves. Apart from the skills correlations and other structural variables, this result depends also on the length of the subscales, but obviously bears relevancy for many applications of diagnostic skill assessments where dichotomous skill mastery/nonmastery assessments are based on as few as three or four, often dichotomous, items.

There are examples of large-scale assessments where multidimensional data were analyzed with lower-dimensional models and comparable fit was achieved using a model with fewer dimensions. As just one example, Cartsensen et al. (2004) analyzed large scale assessment data from a national reading assessment option embedded in a large scale international assessment. The analysis with the multidimensional random coefficients multinomial logit model (Adams et al., 1997) using the hypothesized 8-factor structure crossing content domains and cognitive dimensions resulted in a similar fit compared to a 3-dimensional model with a mixed domain/cognitive structure of the design matrix. De la Torre and Douglas (2004) analyzed simulated test data with multiple skills and found that the simulated positively correlated skills could be modeled using a higher-order unidimensional latent trait. If such a data reduction is possible without noticeable loss of model-data fit, the question of why we should opt for a model with more latent variables seems appropriate, especially if the hypothesized skill variables represent constructs that are not observables. If there are directly observed variables, it is easier to argue for maintaining a multitude of correlated entities, even though some may argue that aggregating correlated observed variables is a good idea.

We would like to advocate the principle of parsimony to be applied for cognitive diagnosis models and skills assessment systems. Apart from an abstract and somewhat sterile principle, there is evidence that parsimony should be a guiding principle in terms of utility. Thissen-Roe et al. (2004) have shown in the area of physics education that out of a large number of hypothesized misconceptions in student learners, only very few misconceptions could be found empirically. A physics-didactics proverb would be that students physics skills are not as complex as teachers and other physics education experts wish they were. Even if additional misconceptions may exist, they may be extremely rare and therefore found only occasionally even in large scale assessments. If desired, they could potentially be identified by specifically targeted tests, which would then contain several tasks intended to measure a single (unidimensional) variable, namely the absence or presence of a comparably rare math misconception.

Another perspective is brought to bear from the philosophy of science, for example, Hitchcock and Sober (2004) discuss the balance between prediction and accommodation, in which prediction is termed as the component of model data fit that is generalizable and withstands cross-validation in additional samples whereas accommodation represents the always present over-fitting when estimating the parameters of a statistical model. Since random variation and nonpersistent effects of administration will be present in any data collection, accommodation of that noise is inevitable. However, because each estimate of the unobserved entities is based on relatively fewer observed entities, models with a larger amount of entities to be estimated per observation are more prone to tip the balance towards accommodation. For models with fewer unobserved entities, the nonpersistent effects and random variations can be expected to be more balanced, for each entity to be estimated is based on averaging over a relatively larger base of observed quantities.

3. Utility and use of skills classifications

A related issue to the dimension of the unobserved latent structure necessary to fit the observed data is the utility and actual use of the reported skills based on a multidimensional skill structure. If the main use is description, one common way to describe the observations is with respect to some order relation. However, an order relation is difficult to establish and visualize in multiple dimensions. Common ways to develop a measures by which to order are either to ignore the multidimensional subscore results and to use the unidimensional results, or to construct a composite based on (weighted) subscores.

If an order relation is not sought, profiles of skills are a useful way to compare observations in somewhat more qualitative terms. If many skills, say more than 10, are involved, the number of different skill profiles is already quite large. As mentioned above, this comparably large set of potential skill profiles may again be subjected to data-reduction techniques like latent class analysis, log-linear modeling, item response theory, etc. (see de la Torre and Douglas, 2004). Other considerations also may result in the need to eliminate or combine certain skills. One important issue is the reliability and the differential validity of the different skills in the sense of meriting separate reporting. If there are many skills that are virtually perfectly correlated, then reporting these skills separately provides no real information not already provided by a single composite score based on the joint collection of tasks that previously represented separate skills. Another extreme is that the skills reported do not possess sufficient reliability in the sense that skill classifications are to a large extent random. In that case, skill profiles that are different with respect to such unreliable skills are not distinguishable in statistical terms. Besides from a lack of reliability, one cannot make valid distinctions between skill profiles in such cases.

Multiple skills may be aggregated or modeled using higher-order models (de la Torre and Douglas, 2004). However, if higher-order variables are the target of inference, they are the ones being reported and used by clients employing the assessment outcomes for their decisions. In that case, the skills diagnosis system becomes an intermediate processing step and will be an internal variable, much in the same way as are internal layers of a neural network. They are necessary to some extent but are subject to scrutiny, and attempts to prune randomly activated or redundant nodes seem inevitable. A related issue is whether a model with an intermediate layer is necessary in such cases. If higher-order processing is considered necessary, skills in the intermediate layer may be replaceable by direct links between the observations and the variables actually reported and used. Models of this kind with multiple latent variables have been available for some time.

4. Latent structure models, latent responses, conjunctive versus compensatory skills

Models for cognitive diagnosis reinvent ideas from – and are essentially an application of – models for categorical data with more than one unobserved variable. Cognitive diagnosis models have a lot in common with multiple classification latent class models (Maris, 1999) and with log-linear models with unobserved variables (Haberman, 1979). In addition, multidimensional item response theory (M-IRT) is based on ideas some of which are found nowadays in diagnostic modeling. This similarity to other models can be used to learn about the specific issues researchers face when using diagnostic models.

One long-standing issue in use of models with multidimensional latent variables is assumptions concerning interactions of the latent skill variables. The joint effect of a specific skill profile on the conditional probability of successfully answering an item has to be specified in any model that contains more than one latent skill variable. Are effects incremental, or are all skills as specified in the Q-matrix assumed to be present simultaneously as a necessary condition to move from a low probability of solving an item to a high probability? In the literature on M-IRT, such issues are discussed in terms of compensatory versus noncompensatory models (Reckase, 1985, 1997), whereas diagnostic modeling, since latent variables tend to be discrete, discuss similar matters more in terms of Boolean logic, for example, the NIDA (noisy input, deterministic AND) model (Maris, 1999; Junker and Sijtsma, 2001) assumes that mastery/nonmastery levels of multiple skill are mapped through Boolean AND operation on some discrete space, whereas the DINA model assumes a deterministic input, and a noisy Boolean AND operation. Obviously, additional models are easily derived based on other Boolean operators such as OR or XOR, or by using Boolean AND for some skills and Boolean OR operators for some other skills.

Other models allow more general types of unobserved skill variables, such as, for example, ordinal skill levels, and base the joint effect of multiple skills on linear combinations of the skill variables (see von Davier, 2005; von Davier and Rost, 2007, Chapter 19 in this volume), which is quite similar to a latent confirmatory factor analysis with discrete latent factors.

Without going into details, decisions about how multiple skills interact are important in determining the functional arguments for model based conditional probabilities given skill profiles, but in most cases such decisions are not directly determined by the theory or hypothesis the diagnostic model tries to reflect. Hypotheses about cognitive variables such as skills required to perform on math problems may provide information concerning skill interaction, but the decision whether the AND is noisy or the input of the AND is noisy are only so specific in rare cases.

What is known or at least can be studied, however, is how different assumptions perform in studies involving misspecified models, where the model is specified in one way and the data generation is based on a different model. De la Torre and Douglas (2004) analyzed data generated from a conjunctive diagnostic model with a compensatory model in one condition, and data generated from a compensatory model with a noncompensatory model in another. Their conclusion was that compensatory models can fit data generated from conjunctive models relatively better than the other way around. This is not surprising when one takes into account that conjunctive models assume all required skills need to be present in order to increase the likelihood of a success to the maximum, but are not as strict in constraining the likelihood of success if only one out of several required skills is missing.

In contrast to certain conjunctive models and compensatory models, noncompensatory models produce fewer conditional probability levels given different skill patterns. Even for a simple example with only two required skills for an item, for the four different possible skill combinations $(0, 0)$ $(0, 1)$ $(1, 0)$ and $(1, 1)$ there are two levels of conditional probabilities for a noncompensatory model instead of four levels for a compensatory model. This holds because noncompensatory models assume that only the presence of all required skills for an item will raise the success probability from the nonmastery level to a mastery level. Conjunctive models are not as strict, since here the terms determining the conditional success probability are joint multiplicative, so that a smaller numeric value due to a lack of one skill cannot be compensated totally, but alleviated somewhat (unless the term equals zero) if all remaining skills are mastered. Conjunctive models of the multiplicative type could therefore be labeled semi-compensatory in the system of terms commonly used in M-IRT.

The decision in favor or against a specific modus operandi on how and if skills interact can be tested in terms of model-data fit. Noncompensatory models may often perform worse than the more flexible conjunctive (multiplicative connection of skill terms) and compensatory models (additive connection of skill based terms) when compared with respect to model-data fit. The decision between a compensatory model and a conjunctive model may be based on the question whether each of the skills when lacking should be capable of reducing the probability of success, in some cases down to zero, even if all other required skills are present for the item(s).

5. Outlook

The variety of available models for cognitive diagnosis adds complexity to the responsibilities of researchers. Models of the diagnostic type allow one to derive sophisticated skill profile estimates for student reports to a level that may be beyond what a decision maker can consume for larger groups of students. In addition, the differences between the models may be minimal in terms of model-data fit, so that a decision for one specific model based on the reproducibility of observed data alone will not suffice. Checking model-data fit is necessary and should focus on how well the observed data are predicted by the different types of statistics assumed in the models considered, for example, with resimulations methods like the parametric bootstrap (Langeheine et al., 1996) or posterior predictive checks (Sinharay, 2005). However, even less computationally costly checks such as residual analyses may prove useful for comparisons between more or less parsimonious ways to model the data. In addition to such internal model-fit checks, cross-validation with additional samples seems advisable to ensure that the model chosen based on some measure of fit in one data set is also an appropriate choice in at least a second data set. Moreover, validation studies that target the utility of the outcome statistics derived from the different models are needed. This allows one to compare alternative models with respect to predicting behavioral data collected outside of the assessment used to select the models.

References

Adams, R.J., Wilson, M., Wang, W.-C. (1997). The multidimensional random coefficients multinomial logit model. *Applied Psychological Measurement* **21**, 1–23.

Cartsensen, C.H., Voss, A., Bos, W. (2004). Reading comprehension of primary school students: One unified cognitive process or several interacting component processes? Paper presented at the 69th Meeting of the Psychometric Society, Monterey, CA.

de la Torre, J., Douglas, J.A. (2004). Higher-order latent trait models for cognitive diagnosis. *Psychometrika* **69**, 333–353.

Haberman, S.J. (1979). *Analysis of Qualitative Data: New Developments*. Academic Press.

Haberman, S.J. (2005). When can subscores have value? ETS Research Report RR-05-08, Educational Testing Service, Princeton, NJ.

Hitchcock, C., Sober, E. (2004). Prediction versus accommodation and the risk of overfitting. *The British Journal for the Philisophy of Science* **55** (1), 1–34.

Junker, B., Sijtsma, K. (2001). Cognitive assessment models with few assumptions, and connections with nonparametric item response theory. *Applied Psychological Measurement* **25**, 258–272.

Langeheine, R., Pannekoek, J., van de Pol, F. (1996). Bootstrapping goodness-of-fit measures in categorical data analysis. *Sociological Methods and Research* **24** (4), 492–516.

Maris, E. (1999). Estimating multiple classification latent class models. *Psychometrika* **64**, 187–212.

Reckase, M.D. (1985). The difficulty of test items that measure more than one ability. *Applied Psychological Measurement* **9**, 401–412.

Reckase, M.D. (1997). The past and future of multidimensional item response theory. *Applied Psychological Measurement* **21** (1), 25–36.

Sinharay, S. (2005). Assessing fit of unidimensional IRT models using a Bayesian approach. *Journal of Educational Measurement* **42**, 375–394.

Thissen-Roe, A., Hunt, E., Minstrell, I. (2004). The DIAGNOSER project. Combining assessment and learning. *Behavioral Research Methods, Instruments, and Computers* **36** (2), 234–240.

von Davier, M. (2005). A general diagnostic model applied to language testing data. ETS Research Report RR-05-16, Educational Testing Service, Princeton, NJ.

von Davier, M., Rost, J. (2007). Mixture distribution item response models. In: Rao, C.R., Sinharay, S. (Eds.), *Handbook of Statistics*, vol. 26. Elsevier, Amsterdam. Chapter 19 in this volume.

Handbook of Statistics, Vol. 26
ISSN: 0169-7161

DOI: 10.1016/S0169-7161(06)26032-2

32

The Statistical Procedures Used in National Assessment of Educational Progress: Recent Developments and Future Directions

Matthias von Davier, Sandip Sinharay, Andreas Oranje and Albert Beaton

Abstract

The National Assessment of Educational Progress (NAEP) is an ongoing survey of the performance of the school students in the U.S. in a number of subject areas, including reading, writing, and mathematics. This chapter discusses in detail the statistical model, the current estimation technique, possible alternatives, and future directions of psychometric research in NAEP.

The reporting method used in NAEP relies on a complex latent regression model and sampling weights with some of the assessments involving more than a hundred thousand examinees. Estimating the model parameters, therefore, is not straightforward for NAEP data, and neither is reporting of the results. For example, in order to generate maximum (marginal) likelihood estimates of the parameters of the latent regression model, multivariate (of up to five dimensions) integrals must be evaluated. Also, computing standard errors is not straightforward because of the complicated nature of the model and the complex sampling scheme. Still, the current estimation method used in NAEP performs respectably. However, the method has often been criticized recently. A number of researchers have suggested alternatives to the current statistical techniques in NAEP.

1. Introduction

The National Assessment of Educational Progress (NAEP) is the only regularly administered and congressionally mandated national assessment program. NAEP is an ongoing survey of the academic achievement of school students in the United States in a number of subject areas like reading, writing, and mathematics. NAEP policies are determined by the National Assessment Governing Board (NAGB). NAEP is carried out by the National Center for Education Statistics (NCES), a part of the U.S. Department of Education, and is administered to samples of students (see Appendix) in grades 4, 8, and 12.

A document called the Nation's Report Card[1] reports the results of NAEP for a number of academic subjects on a regular basis. Comparisons are provided to the previous assessments in each subject area. NAEP results are reported in terms of an average scale score in the subject for all students at a grade level in the U.S. and also as the percentage of students in the grade who attained predefined levels of proficiency in the subject. The levels are defined by the NAGB and indicate what students should know and be able to do at each grade level. In addition to producing these numbers for the nation as a whole, NAEP reports the same kind of results for states, urban districts, etc., as well as for different subpopulations (based on such characteristics as gender, race/ethnicity, school type) of the student population.

The principal dilemma NAEP faces is the conflicting interests between a relatively short testing time for students and the reliable and valid assessment of a broad framework. A short testing time (e.g., less than 1 hour) is particularly motivated by concerns about students motivation and performance, given the fact that NAEP is a low stakes test, and given that fatigue effects affect performance in most lengthy tests in fundamental ways. As a result, NAEP only assesses each student in a portion of the contents of the framework, but assesses the total framework across students by use of a balanced incomplete block (BIB) design. In a BIB design, blocks of items are paired systematically across booklets such that each block appears together with each other block in at least one booklet. Also, blocks appear an equal number of times in each of the block positions (either two or three). There are situations where these objectives conflict in which case preference is given to a balanced position over balanced pairing by replicating several pairings, thus, adding booklets.

The immediate result is that the scores of different students are not comparable. However, NAEP is not required nor allowed to disseminate individual results. Instead, results are reported by subgroup (e.g., males and females) and, subsequently, a subgroup based analysis approach is conducted to overcome the incomparability of students. Effectively, the testing dilemma has shifted to an analysis problem.

A simple solution to this analysis problem would be to assess the relative difficulty for each test booklet that is given, and computing simple booklet scores adjusted for difficulty and aggregating for subgroups. Instead, response behavior of test takers are modeled using item response theory (IRT; see Chapter 15 by Bock and Moustaki in this volume), analyzing response patterns rather than scores and taking into account uncertainty as a function of proficiency. In addition, because blocks of items are systematically linked across booklets in the BIB design, every item can be linked to every other item using equality constraints.

Yet, with as few as 2 items (e.g., NAEP Writing assessment) and usually no more than 20 items (e.g., NAEP Reading and Mathematics assessments) per scale in a booklet, the uncertainty for an individual student might still be too high to make substantive claims about his or her performance. Instead, NAEP considers groups of students rather than individuals, and utilizes the fact that enough information is available for groups to warrant a point estimate of proficiency for that group. Because the proficiency distribution is latent, a latent regression (Mislevy, 1984, 1985) is carried out with group

[1] A version is available on the website: http://nces.ed.gov/nationsreportcard/.

indicators as independent variables and the effect of belonging to a certain group can be estimated. In summary, the NAEP method exploits the many advantages of item response theory to overcome sparse testing to produce reliable results for subgroups.

A number of extensions and alternative estimation methods have been suggested to improve or replace the current operational method. The Markov Chain Monte Carlo (MCMC) estimation approach (Johnson and Jenkins, 2004; Johnson, 2002) focuses on implementing a fully Bayesian method for joint estimation of IRT and regression parameters. Direct estimation for normal populations (Cohen and Jiang, 1999) tries to replace the multiple regression by a model that is essentially equivalent to a multigroup IRT model with various additional constraints. The YGROUP approach (von Davier and Yu, 2003) implements seemingly unrelated regressions (Zellner, 1962) and offers generalized least-squares (GLS) solutions. Researchers von Davier and Sinharay (2004) apply a stochastic expectation–maximization (EM) algorithm to perform the estimation.

The following section gives an overview of the NAEP model and the estimation method it currently uses. Section 3 discusses an example. Section 4 discusses some model extensions that have been suggested. Section 5 lays out conclusions and future directions.

2. The current NAEP model and estimation method

2.1. *The model*

In the NAEP model, the unique p-dimensional latent proficiency vector for examinee i is $\boldsymbol{\theta}_i = (\theta_{i1}, \theta_{i2}, \dots, \theta_{ip})'$. For example, for a mathematics assessment, $\boldsymbol{\theta}_i$ denotes the multidimensional mathematics skill vector of examinee i, and its components relate to mathematics sub-skills like algebra, geometry, data analysis, etc. In NAEP, p could be between 1, 2, 3, 4 and 5. The response vector to the examination items for examinee i is $\boldsymbol{y}_i = (\boldsymbol{y}_{i1}, \boldsymbol{y}_{i2}, \dots, \boldsymbol{y}_{ip})$, where $\boldsymbol{y}_{ik}$, a vector of responses, contributes information about θ_{ik}. The likelihood for an examinee is given by

$$f(\boldsymbol{y}_i|\boldsymbol{\theta}_i) = \prod_{q=1}^{p} f_1(\boldsymbol{y}_{iq}|\theta_{iq}) \equiv L(\boldsymbol{\theta}_i; \boldsymbol{y}_i). \tag{1}$$

The terms $f_1(\boldsymbol{y}_{iq}|\theta_{iq})$ above follow a univariate IRT model, usually with 3PL (Lord and Novick, 1968) or generalized partial credit model (GPCM; Muraki, 1992) likelihood terms. Though most NAEP assessments measure multiple sub-skills, each item in a NAEP assessment is constructed to measure only one sub-skill; this is often known as "between items multidimensionality" (e.g., Wang et al., 1997) as opposed to "within items multidimensionality" where an item may measure more than one sub-skill. The approach used in NAEP to deal with the latent variables can also be used to handle "within items multidimensionality." To simplify notation, the dependence of $f(\boldsymbol{y}_i|\boldsymbol{\theta}_i)$ on item parameters is suppressed.

Suppose $\mathbf{x}_i = (x_{i1}, x_{i2}, \dots, x_{im})$ are m demographic and educational characteristics for the examinee. Conditional on $\boldsymbol{x}_i$, the examinee proficiency vector $\boldsymbol{\theta}_i$ is assumed to follow a multivariate normal prior distribution, i.e., $\boldsymbol{\theta}_i|\mathbf{x}_i \sim N(\Gamma'\mathbf{x}_i, \Sigma)$. The mean parameter matrix Γ and the variance matrix Σ are assumed to be the same for all examinees.

In practice, NAEP collects information on hundreds of variables $\mathbf{z}_i$s on each examinee, runs a principal component analysis on them and uses a number of principal components ($\mathbf{x}_i$s) that explain 90% of the variation in the $\mathbf{z}_i$s for further analysis.

Under this model, $L(\Gamma, \Sigma|\boldsymbol{Y}, \boldsymbol{X})$, the (marginal) likelihood function for (Γ, Σ) based on the data $(\boldsymbol{X}, \boldsymbol{Y})$, is given by

$$\prod_{i=1}^{n} \int f_1(y_{i1}|\theta_{i1}) \dots f_1(y_{ip}|\theta_{ip}) \phi(\boldsymbol{\theta}_i|\Gamma' \boldsymbol{x}_i, \Sigma) \, \mathrm{d}\boldsymbol{\theta}_i, \tag{2}$$

where n is the number of examinees, and $\phi(\cdot|\cdot, \cdot)$ is the multivariate normal density function.

The complete NAEP model obtained by combining IRT and a latent regression in two levels is referred to either as a conditioning model, a multilevel/hierarchical IRT model (Adams et al., 1997; Chapter 17 by Johnson et al. in this volume), or a latent regression model.

Besides reliable reporting of most subgroups, there are other advantages associated with this approach. First, the underlying item response model grants the opportunity to use advanced equating techniques across years based on common items and/or randomly equivalent populations. Also, this model provides a straightforward approach to multi-dimensional scales. While the subscales have been determined before conditioning, patterns of corresponding responses on items of different subscales can be translated into scale covariances at the sample or subgroup level. In addition, this overcomes some of the disadvantages of multidimensional item response theory, such as indeterminacy of parameters. Finally, the uncertainty in the parameters of the latent regression can be used to quantify measurement error. Assuming that the group membership information is precise, the uncertainty of the regression parameters can contribute to the uncertainty reflected in the latent distribution. Based on this, an imputation model can be constructed, imputing proficiency for all students and deriving *plausible values* (explained later in Section 2.3) for each individual as part of their vector of group memberships. The variance between these plausible values can be regarded measurement error.

2.2. NAEP estimation and the MGROUP programs

NAEP uses a three-stage estimation process to fit the above model.

- The first stage, *scaling*, fits an IRT model (consisting of 3PL and GPCM terms) to the examinee response data, and estimates the item parameters.
- The second stage, *conditioning*, assumes that the item parameters are fixed at the estimates found in scaling, and fits the latent regression model to the data, i.e., estimates Γ and Σ in its first part. In the second part of the conditioning stage, *plausible values* (multiple imputations) for all examinees are obtained as described in Section 2.3. The plausible values are used to estimate examinee subgroup averages.

- The third stage, *variance estimation*, estimates the variances corresponding to the examinee subgroup averages, using multiple imputation (e.g., Rubin, 1987) in combination with a jackknife approach (see, e.g., Johnson, 1989; Johnson and Rust, 1992).

In the conditioning step, the goal is to fit the model given by Eq. (2) to the data. If the $\boldsymbol{\theta}_i$'s were known, the maximum likelihood estimators of Γ and Σ would be

$$\begin{aligned} \widehat{\Gamma} &= (\boldsymbol{X}'\boldsymbol{X})^{-1}\boldsymbol{X}'(\boldsymbol{\theta}_1, \boldsymbol{\theta}_2, \ldots, \boldsymbol{\theta}_n)', \\ \widehat{\Sigma} &= \frac{1}{n}\sum_i(\boldsymbol{\theta}_i - \Gamma'\boldsymbol{x}_i)(\boldsymbol{\theta}_i - \Gamma'\boldsymbol{x}_i)'. \end{aligned} \tag{3}$$

However, the $\boldsymbol{\theta}_i$'s are latent variables and hence unknown. Mislevy (1984, 1985) shows that the maximum likelihood estimates of Γ and Σ under unknown $\boldsymbol{\theta}_i$'s can be obtained using an expectation–maximization (EM) algorithm (Dempster et al., 1977). The EM algorithm iterates through a number of expectation steps (E-steps) and maximization steps (M-steps). The expression for $(\Gamma_{t+1}, \Sigma_{t+1})$, the updated value of the parameters in the $(t + 1)$th M-step, is obtained as

$$\Gamma_{t+1} = (\boldsymbol{X}'\boldsymbol{X})^{-1}\boldsymbol{X}'\big(\tilde{\boldsymbol{\theta}}_{1t}, \tilde{\boldsymbol{\theta}}_{2t}, \ldots, \tilde{\boldsymbol{\theta}}_{nt}\big), \tag{4}$$

$$\Sigma_{t+1} = \frac{1}{n}\sum_i\big[\mathrm{Var}(\boldsymbol{\theta}_i|\boldsymbol{X}, \boldsymbol{Y}, \Gamma_t, \Sigma_t) + (\tilde{\boldsymbol{\theta}}_{it} - \Gamma'_{t+1}\boldsymbol{x}_i)(\tilde{\boldsymbol{\theta}}_{it} - \Gamma'_{t+1}\boldsymbol{x}_i)'\big], \tag{5}$$

where $\tilde{\boldsymbol{\theta}}_{it} = E(\boldsymbol{\theta}_i|\boldsymbol{X}, \boldsymbol{Y}, \Gamma_t, \Sigma_t)$ is the posterior mean for examinee i given the preliminary parameter estimates of iteration t. The process is repeated until the estimates Γ and Σ converge. The MGROUP set of programs at Educational Testing Service consists of various implementations of this EM algorithm.

Formulas (4) and (5) require the values of the posterior means $E(\boldsymbol{\theta}_i|\boldsymbol{X}, \boldsymbol{Y}, \Gamma_t, \Sigma_t)$ and the posterior variances $\mathrm{Var}(\boldsymbol{\theta}_i|\boldsymbol{X}, \boldsymbol{Y}, \Gamma_t, \Sigma_t)$ for the examinees. Correspondingly, the $(t + 1)$th E-step computes the two required quantities for all the examinees.

Different versions of the MGROUP program are used, depending on the statistical method used to perform the E-step. BGROUP uses numerical integration, NGROUP (e.g., Beaton, 1988) uses Bayesian normal theory, and CGROUP (Thomas, 1993) uses Laplace approximation.

The CGROUP and BGROUP programs are used operationally in NAEP. When the dimension of $\boldsymbol{\theta}_i$ is less than or equal to two, BGROUP, which applies numerical quadrature to approximate the integral, is used. When the dimension of $\boldsymbol{\theta}_i$ is larger than two, NGROUP or CGROUP may be used. NGROUP is not used for operational purposes in NAEP because it assumes normality of both the likelihood and the prior distribution and may produce biased results if these assumptions are not met (Thomas, 1993).

CGROUP, the approach currently used operationally in NAEP and several other large-scale survey assessments, uses the Laplace approximation of the posterior mean and variance to approximate the means and covariances of θ_js as

$$\begin{aligned} &E(\theta_j|\boldsymbol{X}, \boldsymbol{Y}, \Gamma_t, \Sigma_t) \\ &\quad\approx \hat{\theta}_{j,\mathrm{mode}} - \frac{1}{2}\sum_{r=1}^{p} G_{jr}G_{rr}\hat{h}_r^{(3)}, \quad j = 1, 2, \ldots p, \end{aligned} \tag{6}$$

$$
\begin{aligned}
&\operatorname{cov}(\theta_j, \theta_k | \boldsymbol{X}, \boldsymbol{Y}, \Gamma_t, \Sigma_t) \\
&\quad \approx G_{jk} - \frac{1}{2} \sum_{r=1}^{p} (G_{jr} G_{kr} G_{rr}) \hat{h}_r^{(4)} \\
&\qquad + \frac{1}{2} \sum_{r=1}^{p} \sum_{s=1}^{r} \biggl\{ \biggl[1 - \frac{1}{2} I(r = s) \biggr] \hat{h}_r^{(3)} \hat{h}_s^{(3)} G_{rs} \\
&\qquad \times (G_{js} G_{ks} G_{rr} + G_{js} G_{kr} G_{rs} + G_{jr} G_{ks} G_{rs} + G_{jr} G_{kr} G_{ss}) \biggr\}, \\
&\quad j, k = 1, 2, \ldots p,
\end{aligned} \tag{7}
$$

where

$$
\begin{aligned}
\hat{\theta}_{j,\mathrm{mode}} &= j\text{th dimension of the posterior mode of } \boldsymbol{\theta}, \\
h &= -\log\bigl[f(y|\boldsymbol{\theta}) \phi(\boldsymbol{\theta} | \Gamma_t' x, \Sigma_t) \bigr], \\
G_{jk} &= \left(\frac{\partial h}{\partial \theta_r \partial \theta_s} \bigg|_{\hat{\theta}_{\mathrm{mode}}} \right)_{jk}^{-1}, \\
\hat{h}_r^n &= n\text{th pure partial derivative of } h \text{ with respect to } \theta_r, \\
&\quad \text{evaluated at } \hat{\theta}_{\mathrm{mode}}.
\end{aligned} \tag{8}
$$

Details about formulas (6) and (7) can be found in Thomas (1993, pp. 316–317).

The Laplace method does not provide an unbiased estimate of the integral it is approximating and may provide inaccurate results if p is large, and the number of items measuring each skill is small, which is not rare in NAEP. Further, if the posterior distribution of $\boldsymbol{\theta}_i$'s is multi-modal (which can occur, especially for small number of items), the method may perform poorly. Therefore the CGROUP method is not entirely satisfactory. Figure 2 in Thomas (1993), plotting the posterior variance estimates of 500 randomly selected examinees using the Laplace method and exact numerical integration for two-dimensional $\boldsymbol{\theta}_i$, shows that the Laplace method provides inflated variance estimates for examinee proficiency estimates with large posterior variance (see also von Davier and Sinharay, 2004). The departure may be more severe for $\boldsymbol{\theta}_i$'s for higher dimensions.

2.3. *Estimating group results with plausible values*

After completing the EM algorithm in the conditioning step, multiple imputations – in this context called plausible values – for each examinee's proficiency are drawn from an approximation of the conditional posterior distribution of proficiency, using the following algorithm:

1. Draw $\Gamma^{(m)} \sim N(\widehat{\Gamma}, V(\widehat{\Gamma}))$, where $V(\widehat{\Gamma})$ is the estimated variance of the maximum likelihood estimate $\widehat{\Gamma}$ obtained from the EM algorithm.

2. Conditional on the generated value $\Gamma^{(m)}$ and fixed variance $\Sigma = \hat{\Sigma}$ (obtained from the EM algorithm), calculate the mean $\bar{\boldsymbol{\theta}}_i$ and variance Σ_i for the conditional posterior of each examinee i in the sample.
3. Draw a single $\boldsymbol{\theta}_i \sim N(\bar{\boldsymbol{\theta}}_i, \Sigma_i)$ independently for each examinee $i = 1, 2, \ldots, N$.

The above three steps are repeated M times, producing M sets of imputations (plausible values) for each examinee in the sample. The plausible values are used to approximate posterior expected values as well as other reporting statistics, such as percentiles and percentages above achievement levels.

One of the major goals of NAEP is to estimate the performance of examinee subgroups by computing the average value of some function $g(\boldsymbol{\theta})$ over all individuals in that group. Some examples of $g(\boldsymbol{\theta})$'s that are of usual interest are: (i) $g(\boldsymbol{\theta}) = \theta_j$, when the interest is in the average of one subscale, (ii) $g(\boldsymbol{\theta}) = \sum_j w_j\theta_j$, when the interest is in average of an overall/composite ability (where the weights w_j's are usually determined by the test developers), (iii) $g(\boldsymbol{\theta}) = I_{\sum_j w_j\theta_j > K}$, when the interest is in the proportion of examinees with overall/composite ability above a cutpoint.

To estimate the average value of $g(\cdot)$ for all individuals in group G, the NAEP operational analysis calculates the posterior expected value of

$$\hat{g}_G(\boldsymbol{\theta}) = \frac{1}{n_G} \sum_{i \in G_s} w_i g(\boldsymbol{\theta}_i),$$

with $n_G = \sum_{i \in G_s} w_i$, and G_s denoting the members of group G who are in the sample s. Then the above expected value is estimated, using the plausible values, as

$$\tilde{g}_G = \frac{1}{M} \sum_{m=1}^{M} \frac{1}{n_G} \sum_{i \in G_s} w_i g\big(\boldsymbol{\theta}_i^{(m)}\big).$$

2.4. *Variance estimation*

The third stage in the three-stage NAEP operational analysis is to find the variance of the group estimators $\tilde{g}_G$. There are two primary sources of uncertainty associated with the estimates. First, NAEP samples only a portion of the entire population of students. Second, the true values of the sampled students' proficiencies $\boldsymbol{\theta}_i$s are not directly observed; examinees simply provide multiple discrete indicators (item responses) of their proficiency by responding to the assessment items.

The operational analysis, therefore, splits the total variation into two additive pieces: the variation due to sampling, and the variance due to latency of proficiencies. NAEP uses a jackknife repeated-sampling estimate of the sampling variance (Efron, 1982) and the multiple imputation or plausible-value methodology (Rubin, 1987) to estimate the variance due to latency.

2.4.1. *Variance due to latency of the proficiency $\boldsymbol{\theta}_i$'s*

The variance due to the latency of $\boldsymbol{\theta}_i$, or measurement error of $\boldsymbol{\theta}_i$, is the variance of the latent proficiency given the item responses. If the abilities $\boldsymbol{\theta}_i$ were observed directly, then the measurement error would be zero. However, because we estimate the proficiencies based on a finite number of item responses for each examinee, the measurement error will be nonzero.

The operational analysis estimates the variance of $\tilde{g}_G$ due to latency of $\boldsymbol{\theta}_i$ for subgroup G, denoted U_G, by calculating the variance of the posterior expected values over the plausible values. That is,

$$U_G = \frac{1}{M-1}\sum_{m=1}^{M}\left(\hat{g}_G^{(m)} - \tilde{g}_G\right)^2.$$

2.4.2. Variance due to sampling of students

The variance due to sampling of students arises from the fact that NAEP uses only a sample of students from the total student population and that different samples would lead to different estimates. The NAEP estimation procedure uses a jackknife estimator (Efron, 1982) to estimate the variance due to sampling.

For example, in NAEP 1998, thirty-six primary sampling unit (PSU; described in the appendix) strata were formed from the 72 PSUs sampled with noncertainty. The strata were composed of PSU's from adjacent regions so that PSU strata were relatively similar in socioeconomic makeup. The schools from the 22 PSUs sampled with certainty were distributed to 26 pseudo-PSU strata based on information about the schools. Combining the 36 noncertainty PSU strata with the 26 certainty PSU strata yielded 62 PSU strata. (These numbers vary in different assessment years due to differences in assessment design and participation.) In turn, a stratum was chosen from the 62 PSU strata, a single PSU was dropped from the stratum, and the estimated mean $\tilde{g}_G$ was recalculated (after increasing the weight on the remaining PSU in the stratum chosen to compensate for the missing PSU). In most assessments after 2001, schools instead of PSUs were directly sampled and assigned to the 62 strata (i.e. 124 units). In noncertainty schools, schools were assigned to one of 124 units while in certainty schools students were individually assigned to those 124 units.

Let $\hat{g}_G^{(l)}$ denote the estimate obtained when a PSU is dropped from lth PSU stratum. The NAEP jackknife estimate of sampling variance, denoted V_G, is defined as

$$V_G = \sum_{l=1}^{62}\left(\tilde{g}_G - \hat{g}_G^{(l)}\right)^2.$$

In practice the NAEP operational analyses use only the first plausible value to calculate the group estimate $\hat{g}_G^{(p)}$ for the replicate set p, and this may have an impact on the estimates of the standard errors.

2.4.3. Other approaches for variance estimation

Alternative approaches to estimating the sampling variance of the group statistics have been explored by Longford (1995) and Raudenbush et al. (1999). In both cases the authors use superpopulation methods (Potthoff et al., 1992) to derive estimates of sampling variance; Longford (1995) employs a classical approach and Raudenbush et al. (1999) apply Markov Chain Monte Carlo (MCMC; Gilks et al., 1996) techniques to perform a

Bayesian analysis. However, like the current NAEP estimation procedures, these methods are based on plausible values, assume that the item parameters are known and fixed at estimates found in scaling.

3. Example: NAEP data and results

From 1990 to 2003, NAEP assessments in Mathematics and Reading in grades 4 and 8 were analyzed using the NAEP model described above – the CGROUP program was used. The Mathematics assessment in grades 4 and 8 contain five subscales. Reading in grade 4 is assessed in two subscales and in grade 8 in three subscales. Rounded sample sizes for the national results, including private schools, are displayed in Table 1, showing an increase in 2002 reflecting the use of State samples being combined into a National sample as opposed to a separate National sample before 2002. Also, the table shows which assessments took place by year. Several results will be discussed to illustrate what kind of information yielded from these models could be of interest to school administrators, policy makers, educators, and researchers. All results are taken from the publicly available online analysis tool accessible via http://nces.ed.gov/nationsreportcard.

An important part of NAEPs results is characterized by trends showing whether the status of education is improving or not. This is a fundamental outcome of the year to year linking based on item response theory, common items, and randomly equivalent samples. For example, 4th grade students performed better in Reading in both 2003 and 2002 compared to 2000, 1998, and 1994, but they have reached the same level students had in 1992. In grade 8, students have not improved in 2003, 2002, and 1998, but have in all those years improved over 1994 and 1992. In Math, 4th graders have shown continuously improvement over the years 1990, 1992, 1996, 2000, and 2003, albeit that a temporary stagnation showed between 1996 and 2000. The same continuously improving pattern shows in 8th graders, though a stagnation occurs between 1992 and 1996.

Table 1
Approximate sample sizes in NAEP

Population	Mathematics		Reading	
	Grade 4	Grade 8	Grade 4	Grade 8
1990	3400	3400		
1992	7200	7700	6300	9500
1994			7400	10100
1996	6900	7100		
1998			7800	11200
2000	13900	15900	8100	
2002			140500	115200
2003	190100	153200	187600	155200

Note. Source: NAEP Data Tool at http://www.nces.ed.gov/nationsreportcard

For state legislators it might be of interest to see whether their state follows the same trends as the national results, or whether their state may out- or under-perform. Additionally, it might be of interest to compare state results with other similar states to see whether they do better or worse. One implication could be to investigate whether certain policies or good practices could be adopted from better performing states to improve education further. An important factor in American demographics is the regional differences in population composition and, hence, a similar state could be defined in terms of race/ethnicity composition. States that are fairly similar are, for example, Maryland, Delaware, and North Carolina or Vermont, Maine, and New Hampshire, or Arizona and New Mexico. Subsequently, it can be derived from NAEP that in 2003 grade 8 Mathematics Maryland and Delaware performed similar to the nation, but that North Carolina outperformed both states and the nation. Also, all four jurisdictions showed an increase between 1992 and 2003, while North Carolina showed the largest improvement.

The choice of race/ethnicity composition to define similar states can be further supported by looking at differences between racial groups in terms of scale scores. For example, in 2003 grade 8 Reading, White and Asian American students perform similar and higher than Black, Hispanic, and Native American students, who in turn perform similar. Furthermore, between 1992 and 2003 White, Black, and Hispanic students improved their performance while Asian American students showed no change in performance between those years. This triggers another important policy issue, which is the difference between subgroups coined as gaps, and especially, closing of the achievement gap between racial groups. From the grade 4 Mathematics national assessment it can be concluded that the White–Black and the White–Hispanic achievement gap is narrowing between 2000 and 2003. However, in 2003 these gaps are still 27 and 22 scale score points, which amounts to more than .65 of a standard deviation unit. In grade 12 Reading between 1998 and 2002 those two gaps did not change significantly. More specifically, all three race/ethnicity groups performed lower in 2002 compared to 1998 by the same amount. Yet, the gaps are 25 to 20 points, an effect size of approximately .65. These results can also be derived from NAEP for specific states, for example, to find out which state is more successful in closing the achievement gap.

Another possible area of interest to educators, legislators, and teachers might be the framework of NAEP and the differentiation between knowledge and skills areas, translated into subscales. At the heart of the population characteristics estimation and the development of subsequent software (e.g., CGROUP in addition to BGROUP) is the multi-dimensional character of the assessment. For example, the Reading assessment in grades 8 and 12 covers three distinct, though related, areas: reading for literary experience, reading for information, and reading to perform a task, which are combined to a composite scale in most reports. However, these areas can also be analyzed separately. While 12th graders overall Reading performance declined between 1998 and 2002, this was not true for the reading to perform a task subscale, which remained level.

Finally, there are many research questions that are too specific, or sensitive from a privacy and individual disclosure risk standpoint, to be made publicly available. For those cases, a restricted use secondary users file is offered that includes the imputations generated from CGROUP. These imputed values can be applied to a wide variety of modeling techniques, such as Hierarchical Linear Modeling and Structural Equation

Modeling, enabling the researcher to obtain reasonable complex sample and measurement standard error estimates.

4. Alternatives approaches

A number of modifications of the current NAEP methodology have been suggested in the literature. These evolved out of criticisms of (i) the complex nature of the NAEP model, and (ii) the approximations made at different stages of the NAEP estimation process.

4.1. Application of seemingly unrelated regressions

Seemingly unrelated regressions (SUR; Zellner, 1962; Greene, 2000) are a special case of multiple, multivariate regressions, that is applicable if all dependent variables are paired with exactly the same set of predictors.

More formally, if the dependent variable is directly observed and has k dimensions $\boldsymbol{\theta} = (\theta_1, \ldots, \theta_k)^{\mathrm{T}}$ and there is an m-dimensional predictor variable $\boldsymbol{x} = (x_1, \ldots, x_m)^{\mathrm{T}}$ with $\boldsymbol{\theta} = \boldsymbol{x}\Gamma^{\mathrm{T}} + \varepsilon$ where Γ is a $m \times k$ matrix of regression coefficients, where all regression equations contain the same set of predictors, it can be shown that the separate ordinary least squares (OLS) estimators for the columns Γ_k are equivalent to the simultaneous generalized least-square estimators (see Greene, 2000, p. 616, Section 15.4.2). Cohen and Jiang (1999) conjecture without a formal proof that this property may also apply in the case of indirectly observed dependent variables, i.e., in cases where $\boldsymbol{\theta}$'s are estimated and not observed, as in NAEP.

Comparison of YGROUP (von Davier and Yu, 2003) that implements SUR in NAEP estimation and CGROUP have so far shown that YGROUP provides estimates of Γ close but not identical to those of CGROUP. Noticeable differences may be seen in cases where CGROUP's multidimensional approach can borrow information based on covariances between dimensions. This is especially true if the dimensions are highly correlated and are measured with varying precision across dimensions. YGROUP is faster than CGROUP and requires fewer iterations in some cases. This suggests that YGROUP may be used to generate starting values for fully multidimensional approaches like CGROUP or stochastic EM version of MGROUP (von Davier and Sinharay, 2004) in order to reduce computational costs.

4.2. Application of an stochastic EM method

Researchers von Davier and Sinharay (2004) approximate the posterior expectation and variance of the examinee proficiencies $\boldsymbol{\theta}_i$ in (4) and (5) using importance sampling (e.g., Gelman et al., 2003).

The posterior distribution of $\boldsymbol{\theta}_i$, denoted as $p(\boldsymbol{\theta}_i|\boldsymbol{X}, \boldsymbol{Y}, \Gamma_t, \Sigma_t)$, is given by

$$p(\boldsymbol{\theta}_i|\boldsymbol{X}, \boldsymbol{Y}, \Gamma_t, \Sigma_t) \propto f(\boldsymbol{y}_{i1}|\theta_{i1}) \ldots f(\boldsymbol{y}_{ip}|\theta_{ip}) \phi(\boldsymbol{\theta}|\Gamma_t' \boldsymbol{x}_i, \Sigma_t) \tag{9}$$

using (2). The proportionality constant in (9) is a function of $\boldsymbol{y}_i$, Γ_t, and Σ_t. Let us denote

$$q(\boldsymbol{\theta}_i|\boldsymbol{X},\boldsymbol{Y},\Gamma_t,\Sigma_t) \equiv f(\boldsymbol{y}_{i1}|\theta_{i1})\ldots f(\boldsymbol{y}_{ip}|\theta_{ip})\phi(\boldsymbol{\theta}|\Gamma_t'\boldsymbol{x}_i,\Sigma_t). \tag{10}$$

We drop the subscript i for convenience for the rest of the section and let $\boldsymbol{\theta}$ denote the proficiency of an examinee.

We have to compute the conditional mean and variance of $\boldsymbol{\theta}$ with the density $p(\boldsymbol{\theta}|\boldsymbol{X},\boldsymbol{Y},\Gamma_t,\Sigma_t)$. If we can generate a random sample $\boldsymbol{\theta}^1, \boldsymbol{\theta}^2, \ldots, \boldsymbol{\theta}^n$ from a distribution $h(\boldsymbol{\theta})$ approximating $p(\boldsymbol{\theta}|\boldsymbol{X},\boldsymbol{Y},\Gamma_t,\Sigma_t)$ reasonably, we can approximate the conditional mean by A/B, where

$$A = \frac{1}{n}\sum_{j=1}^{n}\frac{\boldsymbol{\theta}^j q(\boldsymbol{\theta}^j|\boldsymbol{X},\boldsymbol{Y},\Gamma_t,\Sigma_t)}{h(\boldsymbol{\theta}^j)}; \qquad B = \frac{1}{n}\sum_{j=1}^{n}\frac{q(\boldsymbol{\theta}^j|\boldsymbol{X},\boldsymbol{Y},\Gamma_t,\Sigma_t)}{h(\boldsymbol{\theta}^j)}.$$

Similarly, we can approximate the variance as the ratio C/B, where

$$C = \frac{1}{n}\sum_{i=1}^{n}\frac{(\boldsymbol{\theta}^i - E(\boldsymbol{\theta}^i|\boldsymbol{X},\boldsymbol{Y},\Gamma_t,\Sigma_t))(\boldsymbol{\theta}^i - E(\boldsymbol{\theta}^i|\boldsymbol{X},\boldsymbol{Y},\Gamma_t,\Sigma_t))'q(\boldsymbol{\theta}^i|\boldsymbol{X},\boldsymbol{Y},\Gamma_t,\Sigma_t)}{h(\tilde{\boldsymbol{\theta}}^i)}.$$

Researchers von Davier and Sinharay (2004) use a multivariate t distribution with four degrees of freedom as the importance-sampling density. The mean and variance of the importance-sampling density are the same as the mode and curvature of the posterior distribution of θ. Because the posterior mode and posterior curvature (information) is already computed in the MGROUP program, this is a convenient choice. Simulated data and real data analyses indicate that the stochastic EM method provides a viable alternative to CGROUP for fitting the NAEP model.

4.3. *Multilevel IRT using Markov Chain Monte Carlo methods*

Johnson and Jenkins (2004) and Johnson (2002) suggest a MCMC estimation method (e.g., Gelman et al., 2003, Gilks et al., 1996) that can be adapted to combine the three steps (scaling, conditioning, and variance estimation) of the MGROUP program. They use a Bayesian hierarchical model that includes the IRT component and the linear regression component of the NAEP model discussed earlier; in addition, their model accommodates clustering of the examinees within schools and PSUs, using a linear mixed effects (LME) model.

The model accommodates clustering of the examinees within schools and PSUs with the help of a LME model of the form

$$\begin{aligned}
&\boldsymbol{\theta}_i|\nu_{s(i)},\Gamma,\mathbf{x}_i,\Sigma \sim N(\nu_{s(i)}+\Gamma'\mathbf{x}_i,\Sigma),\\
&\nu_s|\eta_{p(s)},T \sim N(\eta_{p(s)},T),\\
&\eta_{p(s)}|\Omega \sim N(0,\Omega),
\end{aligned}$$

where $s(i)$ is the school that examinee i attends and $p(s)$ is the PSU (school district, state, etc.) that school s belongs to. This idea is similar to that proposed by Longford (1995) and Raudenbush et al. (1999).

Prior distributions are assumed where required, a determination currently made empirically. The joint posterior distribution of the item and ability parameters and the mean and variance parameters in the regression model are computed, and a Gibbs sampler samples from the joint posterior distribution. However, Johnson and Jenkins (2004) deal only with unidimensional student proficiency. Johnson (2002) handles data with three-dimensional student proficiency, but incorporates only three background variables.

4.4. Estimation using generalized least squares

von Davier and Yon (2004) applies generalized least-squares (GLS) methods to the conditioning model used in NAEP, employing an individual variance term derived from the IRT measurement model. A quasi-Newton estimation algorithm using the Broyden–Fletcher–Goldfarb–Shanno (BFGS) algorithm (Dennis and Schnabel, 1983) was developed and implemented. This method eliminates some basic limitations of classical approaches to regression model estimation. von Davier and Yon (2004) discuss similarities with feasible GLS methods and the White robust covariance estimator.

The proposed estimator can be viewed as a GLS estimator that uses the available information about the uncertainty associated with the indirectly observed dependent variable $\boldsymbol{\theta}$. The GLS estimator is defined as

$$\Gamma = \left(X^{\mathrm{T}}\Omega X\right)^{-1} X^{\mathrm{T}}\Omega\Theta$$

and includes a $N \times N$ matrix

$$\Omega = \begin{pmatrix} \sigma_1^2 & \ldots & \sigma_{1N} \\ \ldots & \sigma_i^2 & \ldots \\ \sigma_{N1} & \ldots & \sigma_N^2 \end{pmatrix}$$

of error variances and covariances. The classical OLS approach is to assume uncorrelated errors (i.e., all $\sigma_{ij} = 0$) and homoscedasticity (i.e., $\sigma_i^2 = \sigma_j^2$) for all $i, j \in \{1, \ldots, N\}$. von Davier and Yon's (2004) approach retains the assumption of uncorrelated errors but relaxes the homoscedasticity assumption using

$$\hat{\sigma}_i^2 = \hat{s}^2(\tilde{\theta}|\vec{x}) + \hat{s}^2(\theta|\vec{y}_i, \vec{x}_i),$$

which assumes that the individual residual variance is the sum of the averaged squared residual

$$\hat{s}^2(\tilde{\theta}|\vec{x}) = \frac{N}{N-K}\sum_{i=1}^{N}(\tilde{\theta}_i - \vec{x}_i\Gamma)^2$$

and the individual posterior variance, given data and background variables.

Analysis of real and simulated data have shown that the approach does not deviate from homoscedastic estimation if data are balanced and the variance in precision across subjects is small. As expected, for unbalanced designs and large variation across booklets, the results from weighted estimation differ moderately from those from the homoscedastic approach.

4.5. Other modifications

The other important works on modification of the current NAEP methodology include Bock (2002) and Thomas (2002).

5. Conclusions

The methodology currently used in NAEP to estimate population distributions is the focus of ongoing research. Suggestions either to ignore auxiliary information when estimating distributions or to incorporate this information in a different way are discussed in the literature. The suggestions aimed at limiting the number of background variables (e.g., Cohen and Jiang, 1999) stand to some extent in contrast to current developments in the estimation of statistics from complex surveys (see the literature on small area estimates, or model-based estimates; Rao, 2003). Research that focuses on extending the model or making the current model less restrictive (Thomas, 2002) may be viewed as the more promising approach when the goal is to make the model more appropriate for increasingly complex samples.

The complexity of NAEP's operational model, as well as its integration into a large operational project that must produce results in a reliable way that enables stable comparisons, makes trials with different model extensions a nontrivial matter. Therefore, an evolution of increasingly flexible models seems preferable to a revolution replacing the current model by some even more complex alternative. At the same time, alternative ways of modeling the data must be explored as well, as a way to study the impact on reporting results of different ways of modeling the data. In addition, the convergence between IRT and hierarchical linear models and other schools of modeling data from complex samples should be studied. If possible, desirable features of these models should be incorporated into the operational marginal estimation model of NAEP.

Research that aims at weakening or removing some technical assumptions and approximations seems promising. The increased availability of powerful computers makes some approximations that were necessary in the past obsolete. Among these seemingly minor modifications are methods that aim at producing more general estimation algorithms. This is a valuable approach since it maintains continuity, but at the same time allows greater flexibility. It may not have the same appeal as a totally new method, but it has the advantage that introducing generalizations gradually keeps fallback positions and well-established solutions for common issues in estimation and reporting at hand.

More substantial modifications that aim at replacing the operational approach need to provide answers to important questions regarding NAEP's charge to provide comparable measures across assessment cycles over time. The current method is based on maximum likelihood estimates using methods from IRT. This method has a long history of research, and linking across time points is a standard process in this approach. Alternative approaches will probably only find consideration if they can perform this task equally well.

Acknowledgements

The authors would like to thank Dan Eignor, Robert Mislevy, John Donoghue, Shelby Haberman, and John Mazzeo for valuable input and helpful suggestions and comments on previous versions of this chapter.

Appendix: Sampling students in NAEP

NAEP is prohibited by law from reporting results for individual students (Jones and Olkin, 2004), schools or school districts, and is designed to obtain optimal estimates of subpopulation characteristics rather than optimal point estimates of individual performance. To assess national performance in a valid way, NAEP must sample a wide and diverse body of student knowledge. To avoid the burden involved with presenting each item to every examinee, NAEP selects students randomly from designated grade and age population (First, a sample of schools are selected according to a detailed stratified sampling plan, as mentioned in, e.g., Beaton and Zwick, 1992, and then students are sampled within the schools) and administers to each student one of a number of booklets that contains only a subset of items. This process is sometimes referred to as "matrix sampling of items." For example, the year 2000 fourth grade NAEP in Mathematics contained 173 items split in 26 booklets. Each item is developed to measure one of five subscales: (i) Number and operations, (ii) Measurements, (iii) Geometry, (iv) Data Analysis, (v) Algebra. An item can be a multiple-choice item or a constructed response item. Background information is collected on the students through questionnaires that are completed by students, teachers, and school administrators. These background variables are used in various ways in the NAEP assessment. For one, background information is used to subset the student population for purposes of reporting into policy relevant subgroups like "limited English proficiency YES/NO". Another purpose background variables are used for is to provide auxiliary information to model the proficiency distribution in subgroups as a composite of distributions that are specified by the outcome on these background variables. For example, the background variables used to modeling the composite population distribution in the 2000 fourth grade NAEP Mathematics assessment was providing 381 numbers for each student. The above description clearly shows that NAEP's design and implementation are fundamentally different from those of a large-scale testing program aimed at individual scores.

In addition to NAEPs content sampling scheme that ensures a broad coverage of the assessment domain, NAEP samples geographic areas to ensure a representative picture of the nations educational progress. The basic primary sampling unit (PSU) sampling design is a stratified probability sample with a single PSU drawn in each stratum. A single PSU consists of a consolidated metropolitan statistical area (CMSA), a metropolitan statistical area (MSA), and a New England county metropolitan area (NECMA), a county, or group of contiguous counties in the United States.

The sampling frame consists of 1027 PSU's; 290 of the PSU's are MSA's. The largest 22 PSU's are included in the sample with certainty. The remaining PSU's are first stratified according to region of the country and MSA status of the PSU. Crossing the region of the country (Northeast, Southeast, Central, West) with MSA status (MSA, non-MSA) defines the eight major strata. Within each of the major stratum, further stratification is achieved by ordering PSU's on additional socioeconomic characteristics, yielding 72 strata with approximately equal populations according to the 1990 Census. Within each of the 72 strata, a single PSU is drawn with probability proportional to the population of the PSU. Combining the 22 certainty PSU's with the 72 noncertainty PSU's gives a total of 94 primary sampling units from which schools are sampled.

Sampling of Schools: Schools are sampled within PSU's with a sampling probability proportional to a measure of school size with two adjustments. In order to increase the precision of estimates for the group means of minority and private school students, schools designated high minority (HM; the Black and Hispanic students account for more than ten percent of the total student population) and private schools were oversampled. The sampling probability for high minority schools was double that of nonhigh minority schools; and the sampling probability was tripled for private schools. A total of 483 schools participated in the NAEP 1998 eighth grade reading assessment.

Sampling of Students: Students are sampled randomly from a list of all enrolled students in the specified grade. In public schools with low minority enrollment, Black and Hispanic students are oversampled. The oversampling is accomplished by taking all nonselected Black and Hispanic students and sampling from this list the same number of Black and Hispanic students in the original student sample from the school. A similar method was used to oversample students with disabilities (SD) and students of limited English proficiency (LEP).

To account for the differential probabilities of selection, and to allow for adjustments for nonresponse, each student is assigned a sampling weight. Let w_i denote the sampling weight for examinee i. The interested reader can find more information on the derivation of the sampling weights in NAEPs technical reports available online.

References

Adams, R.J., Wilson, M., Wu, M. (1997). Multilevel item response models: An approach to errors in variables regression. *Journal of Educational and Behavioral Statistics* **22** (1), 47–76.

Beaton, A.E. (1988). Expanding the new design. The NAEP 1985–86 Technical Report, Educational Testing Service, Princeton, NJ.

Beaton, A.E., Zwick, R. (1992). Overview of the National Assessment of Educational Progress. *Journal of Educational Statistics* **17** (2), 95–109.

Bock, R.D. (2002). *Issues and Recommendations on NAEP Data Analysis*. NAEP Validity Studies Panel, American Institutes for Research, Palo Alto, CA.

Cohen, J.D., Jiang, T. (1999). Comparison of partially measured latent traits across normal populations. *Journal of the American Statistical Association* **94** (448), 1035–1044.

Dempster, A.P., Laird, N.M., Rubin, D.B. (1977). Maximum likelihood from incomplete data via the EM algorithm. *Journal of the Royal Statistical Society, B* **39**, 1–38.

Dennis, J.E., Schnabel, R.B. (1983). *Numerical Methods for Unconstrained Optimization and Nonlinear Equations*. Prentice-Hall, Englewood Cliffs, NJ.

Efron, B. (1982). The Jackknife, the Bootstrap and Other Resampling Plans. *CBMS-NSF Regional Conference Series in Applied Mathematics*. SIAM, Philadelphia. Monograph 38.

Gelman, A., Carlin, J.B., Stern, H.S., Rubin, D.B. (2003). *Bayesian Data Analysis*. Chapman and Hall, New York.

Gilks, W.R., Richardson, S., Spiegelhalter, D.J. (1996). *Markov Chain Monte Carlo in Practice*. Chapman and Hall, New York.

Greene, W.H. (2000). *Econometric Analysis*, 4th ed. Macmillan, New York.

Johnson, E.G. (1989). Considerations and techniques for the analysis of NAEP data. *Journal of Education Statistics* **14**, 303–334.

Johnson, E.G., Rust, K.F. (1992). Population inferences and variance-estimation for NAEP data. *Journal of Educational Statistics* **17** (2), 175–190.

Johnson, M.S. (2002). A Bayesian hierarchical model for multidimensional performance assessments. Paper presented at the annual meeting of the National Council on Measurement in Education, New Orleans, LA.

Johnson, M.S., Jenkins, F. (2004). A Bayesian hierarchical model for large-scale educational surveys: An application to the national assessment of educational progress. ETS Research Report, Educational Testing Service, Princeton, NJ, in preparation.

Jones, L.V., Olkin, I. (Eds.) (2004). *The Nation's Report Card: Evolution and Perspectives*. Phi Delta Kappa International, Bloomington, IN.

Longford, N.T. (1995). Model-based methods for analysis of data from the 1990 NAEP trial state assessment. NCES 95-696, National Center for Education Statistics, Washington, DC.

Lord, F.M., Novick, M.R. (1968). *Statistical Theories of Mental Test Scores*. Addison-Wesley, Reading, MA.

Mislevy, R. (1984). Estimating latent distributions. *Psychometrika* **49** (3), 359–381.

Mislevy, R.J. (1985). Estimation of latent group effects. *Journal of the American Statistical Association* **80** (392), 993–997.

Muraki, E. (1992). A generalized partial credit model: application of an EM algorithm. *Applied Psychological Measurement* **16** (2), 159–177.

Potthoff, R.F., Woodbury, M.A., Manton, K.G. (1992). Equivalent sample size and "equivalent degrees of freedom" refinements for inference using survey weights under superpopulation models. *Journal of the American Statistical Association* **87** (418), 383–396.

Rao, J.N.K. (2003). *Small Area Estimation*. Wiley-Interscience, New York.

Raudenbush, S.W., Fotiu, R.P., Cheong, Y.F. (1999). Synthesizing results from the trial state assessment. *Journal of Educational and Behavioral Statistics* **24**, 413–438.

Rubin, D.B. (1987). *Multiple Imputation for Nonresponse in Surveys*. John Wiley & Sons Inc., New York.

Thomas, N. (1993). Asymptotic corrections for multivariate posterior moments with factored likelihood functions. *Journal of Computational and Graphical Statistics* **2**, 309–322.

Thomas, N. (2002). The role of secondary covariates when estimating latent trait population distributions. *Psychometrika* **67** (1), 33–48.

von Davier, M., Sinharay, S. (2004). Application of the Stochastic EM method to latent regression models. ETS Research Report RR-04-34, Educational Testing Service, Princeton, NJ.

von Davier, M., Yon, H. (2004). A conditioning model with relaxed assumptions. Paper presented at the annual meeting of the National Council on Measurement in Education, San Diego, CA.

von Davier, M., Yu, H.T. (2003). Recovery of population characteristics from sparse matrix samples of simulated item responses. Paper presented at the annual meeting of the National Council on Measurement in Education, Chicago, IL.

Wang, W., Wilson, M., Adams, R.J. (1997). Rasch models for multidimensionality between and within items. In: Wilson, M., Engelhard, G. (Eds.), *Objective Measurement: Theory into Practice*, vol. IV. Ablex, Norwood, NJ.

Zellner, A. (1962). An efficient method of estimating seemingly unrelated regressions and tests for aggregation bias. *Journal of the American Statistical Association* **57** (298), 348–368.

Handbook of Statistics, Vol. 26
ISSN: 0169-7161

DOI: 10.1016/S0169-7161(06)26033-4

Statistical Procedures Used in College Admissions Testing

Jinghua Liu, Deborah J. Harris and Amy Schmidt

1. Introduction

Large-scale college admissions testing, as we know it today, began in June of 1926 when the Scholastic Aptitude Test was administered to slightly over 8,000 young men who had aspirations to go to college. The test was administered by the College Entrance Examination Board, an organization created in 1900 and originally made up of a select group of colleges and universities that wanted to instill some uniformity into a college admissions process wherein each higher education institution had a unique set of requirements and examinations. The original Scholastic Aptitude Test consisted of nine subtests: definitions, arithmetical problems, classification, artificial language, antonyms, number series, analogies, logical inference, and paragraph reading (Lawrence et al., 2003). Since then, the SAT Reasoning Test™, or just simply the SAT, has evolved to measure developed verbal and mathematical reasoning skills. The College Board, the membership organization that sponsors the SAT, has stated that the SAT is a "demanding test that examines fundamental math and reading abilities that are crucial to success in college and adult life" (College Board, 2001, p. 4). In March 2005 the College Board introduced a new version of the SAT. The new version consists of a critical reading section, a math section, and a writing section, which includes an essay and multiple-choice items. See Lawrence et al. (2003) for more detail on the history of the SAT Reasoning Test.

In the late 1950s, a new test, the ACT®, joined the SAT, and these two assessments have dominated the landscape in large-scale college admissions testing ever since. The ACT Assessment is owned by ACT, Inc., a not-for-profit organization dedicated to "measurement and research in support of individuals making educational and career transitions and decisions" (ACT, 1997, p. 1). The American College Testing Company, as it was known then, was established by E.L. Lindquist in 1959. Lindquist, who was a member of the Board of Trustees for the College Board before he founded ACT, was interested in moving college achievement testing from an aptitude testing model to more of an achievement testing model. Lindquist believed that a college admissions test should measure a student's readiness for college and the extent to which a student is prepared to profit from college experience (Peterson, 1983;

Schmeiser, 2004). The current version of the ACT Assessment Program includes an interest inventory and other components, as well as a battery of four mandatory multiple-choice tests: English, Mathematics, Reading, and Science. In February 2005, an optional writing component that includes a student-produced essay was added to the battery.

Approximately 1.4 million college-bound seniors now take the SAT each year, and more than 80 percent of four-year colleges use college admissions tests such as the SAT as part of their admission process when evaluating a test-taker's readiness for college-level work. The ACT Assessment Program (hereinafter called the ACT) was taken by 1.2 million students in the 2004 graduating class; nearly 3,000 postsecondary institutions (including scholarship agencies, state educational systems, individual public and private universities, four-year colleges, junior and community colleges, nursing schools, and technical schools) require or recommend that applicants submit ACT Assessment results (ACT, 1997, 2004a).

Although the ACT and the SAT have somewhat different underlying philosophies (i.e., a focus on achievement versus a focus on developed reasoning ability) and somewhat different structures, the two assessments function quite similarly in their ability to predict first-year grade point average in college. In one concordance study the ACT composite and sum scores (English, Mathematics, Reading, and Science) correlated 0.92 with SAT sum scores (verbal plus math) for 103,525 students who took both the ACT and the SAT, and used both scores in applying for admission during the 1996–1997 period (Dorans, 1999).

College admissions testing programs have a unique set of advantages over other large-scale assessments but they also face a unique set of challenges. Students who take college admissions tests are generally very motivated to do well, and are willing to pay the fees associated with taking the test and reporting results to colleges to which they are applying. The large, relatively steady volume of test-takers allows the test sponsors to create very high-quality assessments at a reasonable price. These are clear advantages, but because of the high-stakes nature of these assessments, sponsors of college admissions tests must also cope with a great deal of criticism from a wide variety of constituents – high school counselors, college admissions officers, parents, students, researchers, and the media, to name a few – and a great deal of effort goes into producing research to support, improve, and defend the use of the tests.

In terms of psychometric considerations, college admissions testing programs must emphasize certain aspects of technical quality that may or may not be as salient in other testing programs. For example, while licensure and certification tests require a great deal of measurement precision around the pass/fail point of the scale, college admissions tests usually require good measurement precision all along the scale continuum. College admissions tests are administered multiple times per year, and test scores from each administration must be comparable across different versions of the test and multiple testing years; therefore, the ability to place the tests on an appropriate scale and equate different test forms to one another becomes of paramount importance. Because of the high-stakes nature of college admissions, it is vitally important that college admission test scores exhibit a high degree of reliability, and validity evidence must be gathered and evaluated on a regular basis. Arguably one of the most important characteristics of college admissions tests in this age of increasing diversity in the test-taking population is that they are as fair as possible. Fairness is one of the thorniest issues that test sponsors and developers face, especially as there is no single technical definition for it (AERA et al., 1999). However, content development procedures and statistical analyses have been developed to ensure that items appearing on college admissions tests are as fair as possible to all test-takers.

In this chapter we review each of these topics. The concepts of reliability, validity, equating and linking are topics that are covered in greater depth in a more general context in other chapters, but we will discuss these concepts so that the reader can see their specific applications in the context of college admissions testing. For example, we focus on equating plans and methods because the frequent administration of college admissions tests requires that greater emphasis be paid to score comparability. In terms of validity, we concentrate primarily on predictive evidence, because although it has become more important in recent years to amass validity evidence of a variety of sorts to support the use of college admissions tests, the primary type of validity evidence that has been collected historically has been predictive in nature. In addition, we discuss topics that are not covered in other chapters, such as test design and assembly, item pretesting models, item analysis, and so on. While these latter topics are certainly ones that are important in other types of testing programs, the large-scale nature of college admissions testing ensures that the testing organizations that create these tests are very cognizant of the "state of the art" techniques in each of these areas. We hope that the reader comes away with a deeper understanding of the care that goes into the process of creating a high-quality, reliable, and valid college admissions test.

This chapter is organized in such a way that follows the chronological order of a testing cycle: creating a test, administering and scoring it, and making use of the results. Section 2 outlines the procedures and guidelines for test design and assembly. Section 3 describes item pretesting: the trying out of items before they are operationally administered. Section 4 discusses item analysis, such as assessing item difficulty, item discrimination, and differential item functioning. Section 5 presents test equating and scaling. Section 6 describes test reliability, standard errors of measurement, and test validity. Section 7 discusses other important issues.

2. Test design and assembly

In this section we will briefly discuss the process of test design and test assembly, from the development of test specifications to the assembly of a final test form. Before test developers begin to build a test, they should be able to answer such questions as: what is/are the intended purpose(s) of the test? What is the target population of test-takers? And what constructs should be measured by the test? The answers to these questions will influence the development of both the content and statistical test specifications.

2.1. *Defining the purpose(s) of the test*

The first step in designing a test is to ascertain the purpose of the test. How is the information derived from the test intended to be used? Is it a test of general knowledge or is it a test of minimum competence? Is it a placement test? Every aspect of the development of a test grows out of the test's purpose. The more specific a test developer can be at this point, the better able s/he is to design a successful test.

The purpose of the SAT, for example, is to give college admission professionals meaningful information to help guide their decisions about how likely students are to succeed at their institutions (College Board, 2004a). Providing information for college admissions is also one of the primary purposes of the ACT, which is designed to help high school students develop postsecondary educational plans and to help postsecondary educational institutions meet the needs of their students (ACT, 1997). It should be noted that although a test is designed with a particular purpose in mind, its use may evolve.

2.2. *Defining the target population of the test*

In conjunction with defining the intended purpose(s) of the test, test developers must know the population(s) to be served by the test. Test developers should be able to describe the characteristics of intended test-takers, including the major subgroups in the population. For example, the SAT and ACT are both designed for test-takers who are seeking entrance to colleges, most of whom are high school students in the United States. This population includes examinees from diverse geographical regions and well as varied academic and demographic backgrounds.

2.3. *Defining the construct(s) to be measured*

Test developers must know what underlying construct the test is intended to measure. The ACT and SAT have been designed to measure general areas that have been determined to be relevant to virtually all college and university programs. This ensures the tests have broad applicability. The SAT measures verbal and mathematical reasoning abilities that students develop over time, both in and out of school. It emphasizes critical thinking and problem-solving skills that are essential for college-level work. The ACT contains four tests of educational achievement: English, Mathematics, Reading and Science. Underlying the ACT Assessment tests of educational achievement is the belief that students' preparation for college is best assessed by measuring, as directly as possible, the academic skills that they will need to perform successfully at the college level.

2.4. *Defining the test specifications*

Once the purpose, population, and construct of the test are determined, test developers need to define the desired attributes of the test in detailed specifications to ensure the purpose of the testing is achieved. Test specifications are a detailed blueprint of all the important dimensions of the test. The purpose of the test specifications is to help ensure that the test does indeed measure the construct intended at the appropriate level for the target population, and that multiple forms of a test are built to the same blueprint, to ensure they are measuring the same thing. There are two major types of test specifications: content specifications and statistical specifications.

Content specifications

Content specifications may include various types of information, at various levels of detail. Key information contained in the content specifications includes: item content and skills to be measured; item formats and item types; scoring methods and rubrics; and the overall make up of a form, including the numbers of each item format/type and the relative weight to be given to each part of the domain that is to be measured (ETS, 2002).

The SAT and the ACT content specifications are developed through similar processes. Both the College Board and the ACT conduct extensive and frequent surveys of high school and college faculty to determine what is being taught in the U.S. schools, to identify college success skills, and to ensure the content appropriateness of the tests (College Board, 2004a; ACT, 1992, 1998, 2000, 2003). A series of consultant panels or review committees, made up of national content experts, college faculty, and high school teachers, are convened, at which the experts reach consensus regarding the important knowledge and skills to be on the tests given current and expected curricular trends. This cycle of reviewing test specifications, curriculum guides and national standards, administering curriculum surveys, and meeting with content experts is an ongoing process for both the ACT and the SAT.

Statistical specifications

Statistical specifications usually involve the *difficulty* index and the *discrimination* index of individual items on the test. The simplest measure of item difficulty is the proportion correct: the proportion of test-takers who answer the item correctly among who attempt to answer the item (i.e., p-value). ACT uses proportion correct as a difficulty index; ETS uses the delta index for the SAT, which is derived from the p-value (see Section 4.1 for more details).

When test developers set the difficulty specifications for a test, they frequently set a target average difficulty for the entire test, and either a standard deviation and/or a distribution of individual item difficulties. Item difficulty specifications may interact with content specifications to ensure, for example, that the most difficult items do not all come from a single content area.

Difficulty specifications need to reflect the purpose of the test. For example, college admissions test specifications may display a wide range of item difficulty, with the largest concentration of items at the average ability level of the test taking population. In general, test developers want to target the item difficulties to the ability level at which the finest discrimination among test-takers is desired. It should be kept in mind that the observed p-value and delta value are group dependent, and they should not be compared directly across different administrations.

Both the ACT and SAT are constructed to be centered at middle difficulty. The target mean item difficulty for the ACT is about 0.58, with a range of difficulties from about 0.20 to 0.89 (ACT, 1997). The distribution of item difficulties is set so that the test will effectively differentiate among test-takers who vary widely in their level of achievement. The mean delta for the SAT is about 11.4 (SAT-V) and 12.0 (SAT-M), which roughly corresponds to p-values of 0.66 and 0.60, respectively. The delta distri-

Table 1
Difficulty distributions and mean discrimination indices for ACT assessment test items, 2003–2004

Difficulty range	Observed difficulty distributions (frequencies)			
	English	Mathematics	Reading	Science
0.00–0.09	0	0	0	0
0.10–0.19	4	19	0	2
0.20–0.29	3	39	5	14
0.30–0.39	22	54	26	35
0.40–0.49	40	44	48	33
0.50–0.59	81	60	54	45
0.60–0.69	124	56	61	38
0.70–0.79	110	60	28	42
0.80–0.89	65	26	18	28
0.90–1.00	1	2	0	3
No. of items[a]	450	360	240	240
Mean difficulty[b]	0.64	0.53	0.57	0.58
Mean discrimination[c]	0.51	0.52	0.5	0.49
Avg. completion rate[d]	93	92	94	95

Note. Taken from ACT Technical Manual, ACT, in press.
[a]Five forms consisting of the following number of items per test: English 75, Mathematics 60, Reading 40, Science 40.
[b]Difficulty is the proportion of examinees correctly answering the item.
[c]Discrimination is the item-total score biserial correlation coefficient.
[d]Mean proportion of examinees who answered each of the last five items.

bution for the SAT is a unimodal distribution with more middle difficulty items (that is, items answered correctly by one-third to two-thirds of test-takers), and fewer very easy (answered correctly by at least two-thirds of test-takers) or very difficult (answered correctly by less than one-third of test-takers) items.

Another aspect of statistical specifications is item discriminating power. Each item in a test should be able to distinguish higher ability test-takers from lower ability test-takers with respect to the construct being tested. A common measure is the biserial correlation, which is the correlation between test-takers' performance on a single item (correct/incorrect for a multiple-choice item) and his or her performance on the total test (see Section 4.2 for more details). Test developers may set a different criterion for these correlations depending on the purpose of the test. For a general test such as ACT and SAT, the goal is to try to discriminate across the entire range of ability. Items on the ACT have a minimum biserial correlation of 0.20 (ACT, 1997); on the SAT, the criterion of biserial correlation for unacceptable items varies between from below 0.2 to below 0.4, depending on the item difficulty and item type (ETS, 2005). Examples of the distributions of item difficulties, the mean difficulty and discriminations are given in Table 1 for the four ACT tests.

2.5. *Content and fairness review*

In addition to content specifications and statistical specifications, there are other issues that need to be considered in assembling a test, such as fairness issues. Items on the SAT and the ACT are written by content specialists in the disciplines measured by the tests. These content experts represent diverse racial and ethnic groups, in order to help ensure that no items are offensive or inappropriate toward any culture. After items are written, they undergo a series of content and fairness reviews to make certain they are fair in content and tone for all test-takers.

In developing the SAT, for example, ETS employs two procedures that constitute the process known as "fairness review". The first is the sensitivity review, a process in which all ETS products and services – including individual test questions, tests, instructional materials, publications, and other products – are evaluated for their demonstration of awareness to the contributions of various groups to society in the United States. Items in the SAT, as well as those in other ETS products, are also constructed to avoid the use of stereotyping and language, symbols, words, phrases, or examples that are sexist, racist, or otherwise potentially offensive, inappropriate, or negative toward any group (ETS, 2002). The second procedure, known as Differential Item Functioning (DIF), is conducted to detect when people from different groups who have approximately equal knowledge and skill perform in substantially different ways on a test question (see Section 4.4). After the internal reviews are completed, external committees review preliminary versions of the test forms, and appropriate changes are made based on the committees' comments. On the SAT, DIF is conducted using both pretest data and operational data (see Section 4.4).

ACT test forms are subjected to several reviews as well. First, items are checked by ACT staff for content accuracy and conformity to ACT style. Then the overall test forms are reviewed again by different reviewers, to ensure the forms conform to ACT style and to check for "cuing"; that is, clues in other items that could allow students to correctly answer a particular item without having the requisite skill or knowledge. The forms are then submitted for review by external content and fairness experts. Two panels, one for content and one for fairness, are then convened to discuss with ACT staff the experts' comments. Finally, appropriate changes are made (if the changes are significant, another review is conducted), and forms are ready for operational administration. The total development cycle can take two and a half years, with at least 20 independent reviews done on each item (ACT, 1997). In all, the reviews are performed to help ensure that each test-taker's level of achievement is accurately and fairly evaluated. Additional reviews are conducted after the forms are administered.

2.6. *Test assembly*

Items judged to be acceptable after the content and fairness reviews are then pretested and analyzed (see Section 3 for item pretesting and Section 4 for item analysis). Items that are statistically sound are placed in an item pool, ready to be selected and assembled into intact test forms. Item selection is usually subject to various rules, based on the test specifications, to constrain the selection of items for test forms. Preliminary forms are constructed by selecting items from the item pool that match the constraints.

The SAT program currently employs test creation software that uses an automated item selection (AIS) algorithm to assemble tests. This method requires a variety of constraints with different weights, and these constraints are used as rules to select items. The model attempts to satisfy the target test properties by minimizing failures to satisfy properties at the aggregate level, and attempts to provide some control over allowable failures by weighting each constraint (Stocking and Swanson, 1992). Consequently, by using AIS, the quality of the assembled test is usually assured. Preliminary forms of the ACT are also constructed by using software that selects items from the item pool that match the content, cognitive and statistical specifications of each test (ACT, 1997).

3. Item pretesting

Pretesting allows for the collection of examinee responses to test items before the items are used operationally (e.g., before the items count towards the test-takers' scores). Items may be pretested using various designs. They may be pretested at regular test administrations, or under special arrangements separate from regular test administrations; they may be either embedded in a separately timed section of a test, or mixed with operational items in an intact test book. In this part of the chapter, we discuss the goals of pretesting in more detail and present several designs for collecting pretesting data.

3.1. Purpose of pretesting

One of the major purposes of item pretesting is to obtain item statistics before the items are used operationally and contribute to an examinee's score. Accurate item statistics are necessary for building tests to the specifications discussed earlier, which are employed to ensure the items will be at the appropriate difficulty level for the targeted examinee population; to distinguish high from low ability test-takers, and not to advantage or to disadvantage one subgroup over another. The second major purpose of pretesting items is to detect flaws in items. There are many possible ways in which items may be flawed. For example, a multiple-choice item may have an incorrect response (a distractor) that appeals to the more able test-takers, suggesting that the intended correct answer (the key) may not be the best answer, or that the item is written in a confusing manner that distracts test-takers inappropriately, or an item may have content flaws. A flawed item can be rejected or revised and re-pretested. The two major purposes of pretesting are related. Often item flaws are signaled by unusual item statistics.

In order to have good item statistics on which to assemble test forms, care must be taken in obtaining the pretest data. Pretest data should be obtained from participants who come from and are representative of the intended test-taking population, and these participants should be sufficiently motivated to perform as well as they can on the pretest items. Various designs that address these issues are taken up in the following section.

3.2. Pretesting data collection design

There are certain key issues that must be considered in pretest data collection, such as the characteristics of the sample on which the data is to be gathered, whether the data collection design is part of an operational administration, where items are positioned in pretesting, and whether pretest items can be distinguished from operational items. There are advantages and disadvantages associated with each type of data collection design, which range from security concerns to obtaining optimal item statistics, to context effects and the impact of these effects on the examinees' operational scores. A good pretesting design seeks to balance these issues.

Pretest sample

Items that are to be pretested should, if possible, be given to a sample from the same test-taking population for whom the test is intended. A major advantage of sample representativeness is that the test-takers who take the pretest items are from the same test-taking population who will later be administered the test form the pretest items may appear on, and hence item statistics are usually more accurate and stable (other things, such as examinee motivation, being equal) from pretest to operational use. On the other hand, this design may be associated with some security risks, in that the same test-takers may see these pretest items again in the future (though this may be controlled to some extent by the lag in time between pretesting and operational use of an item), and are thus more motivated to copy or memorize the items.

To avoid the security risks associated with pretesting in the same population, pretesting in a different population (e.g., pretest items designed for high school test-takers using college students) may be considered, but the resulting item statistics are then based on the pretest population, and not on the test-taking population. The statistics can be adjusted for differences in populations, but the method is far from ideal.

Pretest administration

Pretest items can be given in a regular test administration along with operational items. The SAT and ACT both pretest items during regular test administrations. This design ensures that data are obtained from a motivated and representative test-taking population, which in turn ensures that the pretest statistics accurately reflect the operational statistics to be calculated later on. The disadvantage is that in order to collect pretest data under operational conditions, test administration time must be extended.

Pretest items may also be given during a specially-arranged administration, separate from the regular test administrations, in which the examinees take pretest items only. This design allows more pretest items to be administered, but context effects (i.e., item position, item order, test length, etc.) may be an issue, depending on how closely the pretest forms mimic the operational test forms. SAT Subject Tests (special content-based tests that augment the SAT) employ this design. Separate pretests are sent to volunteer high school and college classes, which administer the pretests under different conditions than those associated with an actual administration. The initial forms of the ACT Writing Test were pretested in a special study, because forms had to be pretested prior to introducing the test operationally in February 2005. An advantage of completely separate pretests is shortened operational testing time. A major disadvantage is that this pretesting method typically results in a sample of convenience that is not necessarily representative of the test-taking population. Item statistics are not as stable from pretest to operational use as those obtained from a representative sample under operational test taking conditions. Test-takers probably are not as motivated as they would be in a real test administration, which may lead to poor pretest statistics. Another disadvantage may be an increased security risk.

Pretest item location
Pretest items can be embedded within operational sections so that examinees are not sure if they are responding to an operational item or a pretest item. This method typically provides optimal item statistics for the pretest items; however, this must be balanced by the effect that pretesting this way may have on the examinee's operational score, where pretest items require time and energy that may be diverted from the items that contribute to an examinee's score. Even if the items are appended at the end of the operational questions, there is no guarantee that examinees will answer all the questions in the order given, as many examinees tend to "skip around" a test section. There is the additional concern that fatigue and lack of time will affect the pretest statistics if the items are all at the end of the test.

Pretesting may also occur in a separately timed section of the test packaged along with the operational test and given at the same administration. The ACT and SAT can both pretest in this way because the two tests contain several test sections. For example, the SAT test book consists of operational sections containing operational items, and a variable section containing either equating or pretest items. The variable section can appear in any of the sections of the test, so that test-takers do not know which section contains pretest items (or equating items) and which sections contain operational items. This means that the test-takers work just as hard on the pretest items as on the rest of the test. Use of a separate section for pretesting allows the SAT program to obtain pretest data from a representative test-taking population under realistic testing conditions, yielding highly accurate item statistics. Because every test administration includes pretesting, the item pool is constantly replenished. Thus, the program remains viable even with regularly scheduled mandated test form disclosures (for operational items only). Operationally, this is a very inexpensive way to collect data (Liu and Walker, 2002).

4. Item analysis

Item analysis provides a statistical description of how a particular item functions on a test for a specific group of test-takers. There are a variety of ways in which item functioning can be assessed when conducting an item analysis, such as assessing how difficult an item is; how well an item distinguishes high ability from low ability test-takers; how the distractors on a multiple-choice item behave; how an item performs across different subgroups of examinees; and whether an item has a high non-response rate. In the following sections, we will examine each of these aspects.

4.1. Assessing item difficulty

As discussed earlier, the simplest measure of item difficulty for a given group of test-takers is the p-value: the proportion of the test-takers who attempted to answer the item and who answered the item correctly. This value is routinely computed in all item analyses.

For the SAT Program, p-values are converted onto a standard scale called delta index:

$$Delta = 13 + 4 \times z, \tag{1}$$

where $(1 - p)$ is first converted to a normalized z-score and then linearly transformed to a scale with a mean of 13 and a standard deviation of 4. Suppose an item has a p-value of 0.25. We draw a normal curve and find the z-score below which 75% of the area below the curve falls. This z-score is 0.67. We multiply the z-score by 4 and plus the result from 13, and then get the delta value, 15.68. Deltas are inversely related to p-values: the lower the p-value, the higher the delta, and the more difficult the item.

The conversion of p-values provides raw delta values that reflect the difficulty of the items from the particular test-takers from a particular administration. This measure of item difficulty then must be adjusted to correct for differences in the ability of different test-taking populations. Delta equating is a statistical procedure used to convert raw delta values to equated delta values. This procedure involves administering some old items with known equated delta values, along with new items. Each old item now has two difficulty measures: the equated delta, which is on the scale to be employed, and the observed delta from the current group of test-takers. The linear relationship between the pairs of observed and equated deltas on the old items is used to determine scaled values for each of the new items. Delta equating is essential because the groups taking a particular test may differ substantially in ability from one administration to another. Through delta equating, item difficulties can be compared directly to each other.

Examining item difficulty, either through p-values or deltas, is done to ensure that the final test form meets the test specifications. In general, items that are too easy or too difficult are not very informative, because either all examinees will get them right or will miss them. Item difficulty for some non-multiple choice items, such as essays written to a prompt, is often assessed by looking at the score distributions.

4.2. Assessing item discrimination

Another important characteristic of an item is item discrimination. Each item in a test should be able to distinguish higher ability test-takers from lower ability test-takers with respect to the construct being tested. An item is considered discriminating if proportionately more test-takers who are high in the ability being measured, generally defined by the total score on the test, answer the item correctly than do test-takers low in the ability measured. Note that item difficulty can constrain item discrimination, in that if most or very few examinees are responding correctly to an item, the discrimination is restricted.

There are a number of indices used in assessing the discriminating power of an item. The index currently used on both the SAT and the ACT is the biserial correlation coefficient (r_{bis}), which measures the strength of the relationship (correlation) between examinees' performance on a single item and some criterion, usually defined as performance on the total test. A very low or negative correlation indicates that the item does not measure what the rest of the items on the test are measuring (and should perhaps be removed from the test), while a very high correlation (close to $+1$) suggests that the information provided by the item is probably redundant with the information provided by the other items.

Items with extreme biserials should be examined carefully. Although items should rarely be discarded on the basis of statistics alone, an extreme r_{bis} serves as a flag to draw attention to the item, which should lead to further investigation for possible defects such as key error, ambiguity in the stem, more than one correct answer, and so on. In addition to examining the biserial, ACT divides examinees into low-, medium-, and high-performing groups by the individuals' scores on the test as a whole, using cutoff scores of the 27th and the 73rd percentile points in the distribution of those scores. These points are chosen to maximize the critical ratio of the difference between the mean scores of the upper and lower groups, under the assumption that the standard errors of measurement in each group is the same and that the scores for the entire examinee population are normally distributed (Millman and Greene, 1989).

The r_{bis} is appropriate for all dichotomous (right/wrong) scored items. When assessing discrimination for some non multiple-choice items, such as the essay, other methods can be used. For example, for the essay in the ACT Writing Test, ACT examines the distribution of prompt scores conditioned on an external variable, such as English achievement as measured by the ACT English Test.

4.3. Distractor analysis

When we conduct item analysis on a multiple-choice question, we not only evaluate how the key (the correct response) functions, but we also examine how the distractors behave. This can be done by examining the percent of examinees in both high and low scoring groups who choose each distractor (ACT's method) or by drawing an empirical item response curve (SAT's method).

Figure 1 shows an example containing multiple empirical item response curves, one for each option. Figure 1 illustrates a psychometrically sound 5-choice multiple-choice item. The total score on the test appears on the horizontal axis and the percentage of people at each possible score level on the test who choose each response appears on the vertical axis. The curves are drawn by computing the percentage of people choosing each response and fitting a smoothed function to the points. "A" is the key. The line associated with choosing "A" keeps rising from the bottom left corner and climbing higher towards the top right corner. This indicates that test-takers who do better on the total test have a greater tendency to answer this item correctly. Distractors B, C, D and E show decreasing frequencies from left to right, which means they only attract those test-takers at lower ability levels. The higher the test-takers' scores, the less likely they are to select any of these distractors. Thus all the distractors appear to be good ones.

Figure 2 illustrates a possibly flawed item. B is the key, but one would not easily know that from looking at the plot. At all score levels, more test-takers either choose C, or omit the question. It shows that the correlation between this item and the total score is low. The large numbers of test-takers omitting the item, and/or choosing distractor C, may be a sign that the item is ambiguous, and that test-takers are unable to determine the correct answer. This is an item that would likely not be included in a final form of a test.

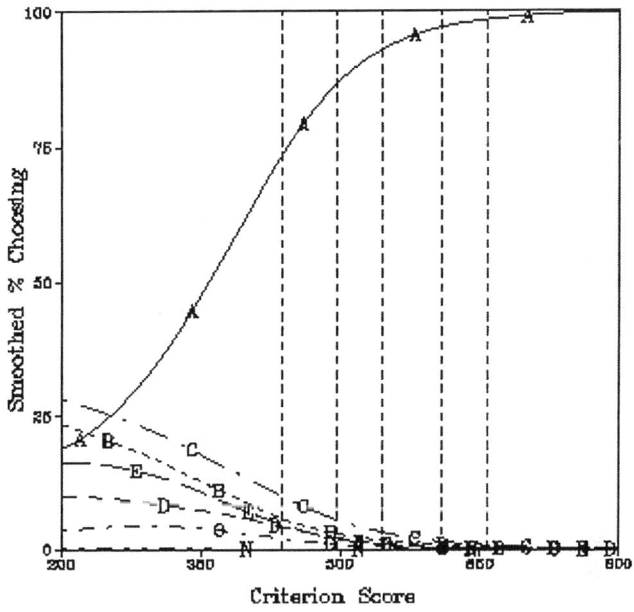

Fig. 1. Empirical item response curve of a good item.

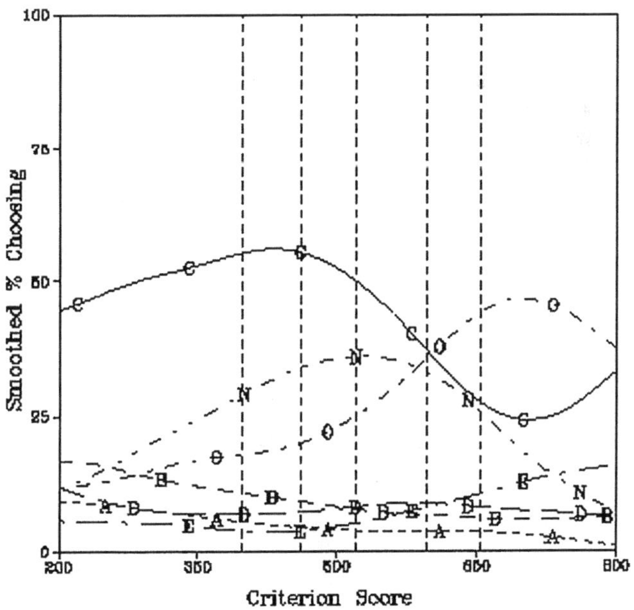

Fig. 2. Empirical item response curve of an item that is possibly flawed.

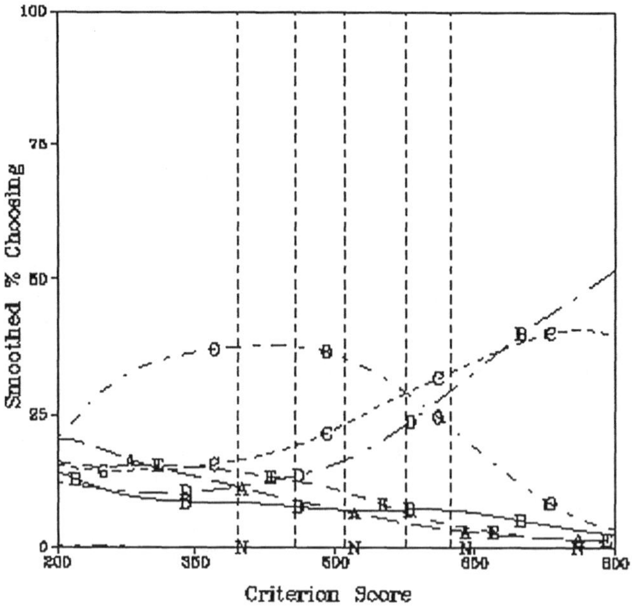

Fig. 3. Empirical item response curve of items with two possible keys.

Figure 3 presents an item that appears to have two possible keys. Both C and D function like a key, with frequencies rising from left to right. Across most of the score range (say, below 700), proportionally more test-takers tend to choose C rather than D even though D is the key. It might be the case that either the stem is ambiguous, or both C and D are acceptable. This item would be flagged for further scrutiny by content experts to verify the key.

Generally speaking, one is looking for the pattern where the high ability examinees choose the key, and the low ability examinees are spread pretty evenly across all the distractors. Distractors that are not chosen by any examinee are not functioning well; distractors that draw a large number of high ability examinees may indicate a wrong key or ambiguity in the item.

4.4. Differential Item Functioning (DIF) analysis

Differential Item Functioning, or DIF, is another item analysis procedure that is typically performed. DIF indicates "a difference in item performance between two comparable groups of examinees; that is, the groups that are matched with respect to the construct being measured by the test" (Dorans and Holland, 1993, p. 35). Theoretically, if test-takers from two different groups have the same ability level, they should have the same probability of getting an item correct. The two groups are referred to as the focal group and the reference group, where the focal group is the focus of analysis and the reference group is the basis for comparison. There are a variety of statistical methods that can be used to identify DIF. Two commonly used methods for DIF assessment are the Mantel–Haenszel approach (Holland and Thayer, 1988), used for DIF detection, and the standardization approach (Dorans and Kulick, 1986; Dorans and Holland, 1993), used for DIF description. Both methods are designed for use with multiple-choice items, and require data from significant numbers of examinees to provide reliable results. Both the ACT and SAT use these two procedures to examine DIF.

The Mantel–Haenszel (MH) procedure

Holland and Thayer (1988) adapted the procedure for the study of matched groups developed by Mantel and Haenszel (1959) and used it as a DIF detection device. The MH method is based on a 2 (groups) × 2 (item scores) table at each level of a matching variable m, where m is usually the total test score:

Group	Item score		
	Right	Wrong	Total
Focal group	R_{fm}	W_{fm}	N_{fm}
Reference group	R_{rm}	W_{rm}	N_{rm}
Total group	R_{tm}	W_{tm}	N_{tm}

The null hypothesis for the MH method is that "the odds of getting the item correct at a given level of the matching variable is the same in both the focal group and the reference group, across all m levels of the matching variable" (Dorans and Holland, 1993, p. 39). With the MH approach, the constant odds ratio, α_{MH}, is defined as:

$$\alpha_{\text{MH}} = \frac{\left[\sum_m \frac{R_{rm} W_{fm}}{N_{tm}}\right]}{\left[\sum_m \frac{R_{fm} W_{rm}}{N_{tm}}\right]}, \tag{2}$$

where m = score levels $= 1, \ldots, m$. Holland and Thayer (1985) converted α_{MH} into a difference in deltas so that it is more convenient for test developers to use:

$$\text{MH D-DIF} = -2.35 \ln[\alpha_{\text{MH}}]. \tag{3}$$

Positive values of MH D-DIF favor the focal group, while negative values favor the reference group. For a detailed discussion of the MH D-DIF method, see Dorans and Holland (1993).

Standardization approach

Standardization is a flexible method to describe DIF in that this approach is based on all available appropriate data to estimate conditional item performance given ability matching and avoids contamination caused by model misfit (Dorans and Holland, 1993). The index of standardization that is currently used in the ACT and SAT programs is the standardized p-difference (STD P-DIF), originally developed by Dorans and Kulick

(1983). The standardized p-difference is defined as:

$$\text{STD P-DIF} = \sum_m \left[\frac{W_m (P_{fm} - P_{rm})}{\sum_m W_m} \right] = \sum_m \left[\frac{W_m D_m}{\sum_m W_m} \right], \tag{4}$$

where $P_{fm} = \frac{R_{fm}}{N_{fm}}$ and $P_{rm} = \frac{R_{rm}}{N_{rm}}$ are the proportions correct in the focal and reference groups at score level m, and D_m is the difference between these two proportions. What is critical about this standardization approach is the use of common weighting factor $(W_m / \sum_m W_m)$ with both P_{fm} and P_{rm}. In practice, $W_m = N_{fm}$ in that it "gives the greatest weight to differences in P_{fm} and P_{rm} at those score levels most frequently attained by the focal group under study" (Dorans and Holland, 1993, p. 49).

At ETS, the current practice for the SAT is to run DIF analyses in which Whites are the reference group, while African Americans, Hispanics, Asian Americans and Native Americans each separately serves as the focal groups; and to run an analysis in which males are the reference group and females are the focal group. Both the standardization approach and the Mantel–Haenszel method are used for item screening for the SAT. On the basis of the MH D-DIF statistic, and/or STD P-DIF values, a classification scheme was developed at ETS to put items into one of the three DIF categories: negligible DIF (labeled A), intermediate DIF (labeled B), or large DIF (labeled C). Any C-DIF item will either be dropped after pretesting or will be revised and then re-pretested. The operational item pool of SAT does not contain C-DIF items.

Similarly, the procedures currently used for DIF analysis for the ACT are the standardized difference in proportion-correct (STD) procedure, which is similar as STD P-DIF used for the SAT, and the Mantel–Haenszel common odds-ratio (MH) procedure, which is the same as α_{MH} defined by formula (2). Items with STD or MH values exceeding pre-established criteria are flagged. Flagged items are reviewed by the content specialists for possible explanations of the unusual results, and actions are taken as necessary to eliminate or minimize the influence of the problem item.

4.5. *Test speededness: Not reaching rates*

Item analysis also includes the calculation of the percentages of examinees not reaching items. Large numbers of examinees not reaching an item could be an indication that the test is speeded, suggesting speed rather than lack of knowledge or ability was the reason examinees did not get the item correct. Both ACT and SAT examine the item omission rates and not reaching rates (see Table 1 for an example of the completion rates for ACT tests). Table 1 provides an example that includes the completion rate for ACT tests, computed as the average proportion of examinees who answered each of the last five items on a test.

Speededness is also assessed from the perspective of fairness. The SAT uses a measure of differential speededness to examine whether or not a test unfairly favors or disfavors certain group(s). Differential speededness refers to the "existence of differential response rates between focal group members and matched reference group members to items appearing at the end of a section" (Dorans and Holland, 1993, p. 56). This standardized not-reached (NR) difference is expressed as:

$$\text{STD P-DIF (NR)} = \sum_m W_m\left(P_{fm}(NR) - P_{rm}(NR)\right) \Big/ \sum_m W_m. \tag{5}$$

The same criteria are used here as with STD P-DIF to put items into three DIF classifications. The data show, for example, that across the testing year 2002–2003, all the SAT verbal (SAT-V) items demonstrate negligible STD P-DIF (NR) values for the male/female comparison, with STD P-DIF (NR) near zero. For the SAT math (SAT-M), two items were classified as intermediate, or B-DIF (NR) items, but the degree of STD P-DIF (NR) was not great enough to cause concern.

4.6. Three stages of item analysis

Item analysis is usually performed several times at different stages in the "life span" of an item. In chronological order, we discuss the three stages of item analysis as: pretest item analysis, preliminary item analysis, and final item analysis.

Pretest item analysis is conducted after the new items are first tried out in pretest form. As discussed in the pretesting section, one of the major purposes of pretesting items is to obtain reliable item statistics, which serve as indicators of how the items will perform when used as operational items in a test form. These item statistics help test developers assemble tests to meet test specifications. Another major purpose of item pretesting is to detect flawed items. A flawed item is either discarded, or revised and re-pretested. Therefore, pretest item analysis ensures that items on the final test form have sound statistics.

Once items are pretested and determined to be acceptable they are assembled into a final form of a test. Once the test form is administered, but before all the answer sheets are scored, another item analysis is conducted. Because this analysis precedes the completion of scoring, this stage is called preliminary item analysis. The purpose of preliminary item analysis is to determine whether the key for each item is in fact the one being used for scoring, whether one of the distractors is arguably the correct response, or whether the item is otherwise seriously flawed (Hecht and Swineford, 1981; Crone and Kubiak, 1997). Preliminary item analysis serves as a further check on errors that may have slipped through the pretest item analysis or on errors that may have been made between the time of test assembly and the time of test scoring; for example, a wrong key may have been entered into the computer system, or an item flaw went undetected in the pretest analysis. For those testing programs that do not have the luxury of pretesting all the items to appear in final forms, preliminary item analysis is especially critical.

Final item analysis is performed after all answer sheets from a test administration have been received and scored. The primary purposes of final item analysis are to assist in the overall evaluation of the entire test and to further evaluate individual items on the full population. The overall evaluation provides information on test performance. It focuses on the main aspects of the tests, such as the test difficulty, reliability and speededness. It is used to determine how well the statistical specifications were satisfied by the test items, and how well the test has in fact performed its intended function. In addition, final item analysis also yields item statistics to be used in testing programs that are able to re-use items. In general, final item statistics are more appropriate than pretest statistics in assembling future final forms. Even for the ACT, which does not reuse items, final item statistics are important as feedback on the quality of the test development process.

Because of the care that goes into the building of the ACT and SAT forms, it is very rare to find a flawed item on an operational test form. In addition to the many content and statistical reviews, there is also the actual review done by the test examinees themselves. After every test administration, examinees may submit "item challenges" regarding any item they feel is questionable. Once a challenge is raised, content specialists review the item. "In the event that a problem is found with an item, actions are taken to eliminate or minimize the influence of the problem item as necessary. In all cases, the person who challenges an item is sent a letter indicating the results of the review" (ACT, 1997).

5. Scaling and equating

Because of the high stakes nature of admissions testing, both the ACT and SAT administer multiple forms throughout the school year. In order to provide scores that are comparable across test administrations, scale/d scores (note that ACT uses "scale" and SAT uses "scaled") are reported instead of raw scores, and equating is conducted to ensure that the reported scores have the same meaning across different forms and across years. In this section we discuss these concepts as they pertain to college admissions tests.

5.1. Scaling

Both the ACT and SAT report scaled scores, rather than raw scores. Scaled scores are designed to retain their meaning across different forms (i.e., a scale score of 25 on the ACT Science Test should indicate the same level of achievement, respectively, regardless of the particular ACT form they were obtained on), whereas raw scores can vary in meaning depending on which form they were obtained (i.e., a raw score of 40 on a somewhat easy form is not the same as a raw score of 40 on a somewhat more difficult form). According to the *Standards for Educational and Psychological Testing* (AERA et al., 1999), detailed information on how a scale is developed should be provided to test score users in order to facilitate score interpretation.

As discussed in Petersen et al. (1989), there are a variety of methods one can use when setting a score scale. The SAT and ACT scales were set in very different ways, yet both sets of resulting reported scores are useful for college admissions. Both tests have been in existence for many years, yet circumstances required the original scales be adjusted in the 1990s. The SAT scales were recentered in 1995 because of concern about the changes in the test-taker population and the resulting effects of continued use of the original scales. There was interest in being able to "repopulate the top end of the score scale" (Dorans, 2002, p. 3). The current ACT scale was set in 1989 when the content of the ACT tests underwent major revisions, and it was determined that the scores on the pre-1989 and post-1989 tests were not directly comparable (ACT, 1997).

Until 1995, SAT scores were reported on a 200-to-800 scale that was based on a 1941 Reference Group of slightly more than 10,000 people. The scaled score mean was set to 500, and the standard deviation was set to 100. Since 1995, the College Board 200-to-800 scale used for the SAT has derived its meaning from a 1990 Reference Group of slightly more than one million test-takers. In redefining the SAT scale, Dorans adhered to seven properties of a well-aligned scale: (1) the average score for the reference group should be near the midpoint of the score scale; (2) the distribution of scores in the reference group should be unimodal; (3) the distribution of scores in the reference group should be symmetric; (4) the shape of the distribution of scores in the reference group should follow a recognized distribution, such as the normal distribution; (5) the working range of scores should extend enough beyond the reported range of scores to permit shifts in the population away from the midpoint of the scale without causing problems with the endpoints of the scale; (6) the number of scaled score points should not exceed the number of raw score points, and (7) the reporting scale should be viewed as an infrastructure in need of periodic repair (Dorans, 2002). Dorans imparted these properties to the SAT verbal and math scales by normalizing the test scores in the 1990 Reference Group (to achieve a symmetric, unimodal distribution); setting the scaled score mean to 500 (the center of the 200-to-800 scale); and setting the scaled score standard deviation to 110 (creating a working scaled score range of roughly 170 to 830). In this way, the marginal distributions of both verbal and math scores were identical in the 1990 Reference Group.

The ACT scales range from 1 to 36, and target means for each of the four tests were set at 18, using data from college-bound students at the beginning of twelfth grade in a nationwide sample in 1988. The scales were set such that the average standard error of measurement is approximately 2 points for each test score, and the conditional standard errors of measurement are approximately equal. The scaling process for the ACT Assessment consisted of three steps. First, weighted raw score distributions for both national and college-bound groups of examinees were computed, with the weighting based on the sample design. Second, the weighted raw score distributions were smoothed in accordance with a four-parameter beta compound binomial model (Lord, 1965; Kolen, 1991; Kolen and Hanson, 1989). Finally, the smoothed raw score distributions for twelfth-grade college-bound examinees were used to produce initial scaled scores, which were rounded to integers ranging from 1 to 36 for the tests, with some adjustments made in attempting to meet the specified mean and standard error of measurement, to avoid score "gaps" and to avoid having too many raw scores converting to a single scale score. This process resulted in the final raw-to-scale score conversions. The ACT Composite score is not scale, but is calculated by rounding the average of the four test scores.

The new ACT writing test has lead to the development of two additional scales. The Direct Writing subscore is based on the scoring rubric and ranges from 2 to 12. The Combined English/Writing score scale was created by standardizing the English and Direct Writing scores, weighting them by 2/3 and 1/3, respectively, and using a linear transformation to map these combined scores onto a scale that ranged from 1 to 36. These transformed scores were then rounded to integers to form the reported score scale.

5.2. Equating

Once score scales for reporting purposes have been established they are maintained by equating. Therefore, scores from different forms are interchangeable. A variety of equating designs and methods exist. There are two types of equating designs employed in the current SAT program: the nonequivalent groups anchor test design (*NEAT*); and the random/equivalent groups (*EG*) design. At each administration of a new SAT form, the new form is equated to multiple old SAT forms through a *NEAT* design. The old forms were either administered to a group with a similar ability level, and/or they were administered at core administrations of the SAT that contribute large numbers of scores to the SAT cohort. This design has produced stable equatings because it directly acknowledges the important role that the old form equating plays in placing a new form on scale (Dorans et al., 2004). At each SAT administration with two new forms, the first new form is equated through the *NEAT* design, while the second new form is equated to the first one through an *EG* design. The spiraling procedure used in the administration and the large numbers of examinees who take each form usually ensures equivalent groups in the same administration. Typical SAT equatings employ a variety of equating models, linear as well as nonlinear models, observed scores as well as true score models, and equatings that employ poststratification.

The ACT equating design uses a carefully selected sample of examinees from one of the national test dates, and administers a spiraled set of forms to that sample. One of the forms, an anchor form, in the set has already been equated, and serves as the link to the score scale; the other forms are new forms. The use of randomly equivalent groups is an important feature of the equating procedure and provides a basis for confidence in the continuity of scales. Scores on the alternate forms are equated to the score scale through the anchor form, which is already on scale, using equipercentile equating methodology. In equipercentile equating, a score on Form X of a test and a score on Form Y are considered to be equivalent if they have the same percentile rank in a given group of examinees. The equipercentile equating results are subsequently smoothed using an analytic method described by Kolen (1984) to establish a smooth curve, and the equivalents are rounded to integers. The conversion tables that result from this process are used to transform raw scores on the new forms to scaled scores.

Different equating designs have advantages and disadvantages, but for large scale programs like the ACT and SAT, which have carefully developed test forms and the availability of large, stable examinee groups, it is likely that most methods would work satisfactorily. Stability studies (McHale and Ninneman, 1994; Modu and Stern, 1975; Harris and Kolen, 1994; Hanson et al., 1997) and careful monitoring suggest that both the SAT and ACT equating designs are working satisfactorily in terms of providing stable scale or scaled scores.

One way of evaluating an equating is to examine whether the equating function is invariant across subpopulations, such as gender or ethnic groups, as it is in the total population. That is, is the equating function derived using subgroup data the same as the equating function derived using the data from the total population (Dorans and Holland, 2000)? Research indicates population invariance was achieved for SAT-V and SAT-M equatings across gender groups (Dorans and Holland, 2000; Liu et al., 2006). ACT has examined the stability of equating results defined by examinee ability (Harris and Kolen,

1986; Yi et al., 2004) and found that the conversions obtained from different subgroups were similar to the conversions obtained by using the total group.

5.3. *Norms/ranks*

In addition to providing information on how a score scale is established and maintained over time, test sponsors typically provide a wealth of interpretive information. College admissions test scores are generally interpreted by comparing an examinee's scores to those of other examinees. The SAT provides percentile ranks of SAT-V, SAT-M, and composite scores (SAT V+M) to aid in score interpretation. The percentile ranks range from 1 to 99, indicating the percent of examinees scoring below a given score. The national percentile ranks for SAT-V and SAT-M scores from a testing year are based on scores from the most recent college-bound seniors cohort. For example, the percentile ranks for the SAT verbal and math scores for the 2004–2005 test administrations are based on scores from the 2004 college-bound seniors cohort.

The ACT provides accumulative percents (the percent of examinees scoring at or below a given score) for students for each test score, subscore, and the composite score, which are based on the most recent scores of the seniors who graduated during the previous three years and who tested as 10th, 11th, or 12th graders under standard conditions, either on a national test date or as part of a state testing program. In addition, the ACT periodically collects national norm information. Both the ACT and SAT report additional sets of ranks, such as state ranks, when appropriate.

The SAT score scale allows for the comparison of scores across the math and verbal tests because the average of each is about 500, although the user is cautioned that because there is always some inaccuracies in scores, an examinee could score slightly higher on one test and still have the same level of ability in both skills. Therefore there must be at least a 60-point difference between a student's verbal and math scores to consider one score better than the other.

Because each of the four ACT tests were scaled separately, one should not directly compare scores; instead, one should compare percentile ranks. However, the question of whether an examinee is stronger in reading than in mathematics, as assessed by the ACT, can only be answered relative to the group of examinees making up the reference group, and whether the answer is "yes" or "no" may vary, depending on the group.

6. Reliability and validity

6.1. *Reliability*

The potential for error, or uncertainty, exists in all cognitive measurements, and it is therefore important for the user to have information that will help determine how much confidence to have in a test score. "Error" is often interpreted as inconsistency: how likely is the examinee to obtain the same score if he/she were tested over and over and over? Scores that fluctuate widely would be of little use to a college admissions officer trying to make enrollment decisions – the decision would become in large part a process based on chance or luck, similar to rolling a die, rather than one based on whether a student appears to have the requisite skills needed to succeed at a particular college.

Several factors can impact how consistent scores will be, including test factors (e.g., test specifications, number of items, score scale), examinee factors (e.g., propensity to guess, knowledge level), and situational factors (e.g., the amount of fatigue an examinee has on a particular occasion, the particular set of items administered on a particular occasion). Reliability indices are used to indicate the expected consistency in scores for a particular test. Because it is generally impractical to administer tests over and over to the same examinees, several different types of reliability coefficients have been developed to estimate score consistency.

One type of reliability coefficient used is internal consistency. What is typically termed "test–retest" reliability attempts to get at the consistency of scores for an examinee that is administered the same test form more than once. Because there are problems with obtaining this type of data (such as an examinee remembering how they responded to a question the first time it was administered), it is typically estimated using a single administration of a particular test form by computing an internal consistency measure. Both the ACT and SAT use the KR-20 formula to compute the internal consistency estimate; however, because the SAT uses formula scoring, a modification of the formula is used (Dressel, 1940):

$$\mathit{reliability} = \frac{n}{n-1}\left[1 - \frac{\sum_{i=1}^{n} p_i q_i + \sum_{i=1}^{n} k^2 p_i' q_i' + 2\sum_{i=1}^{n} k p_i p_i'}{\sigma_t^2}\right], \tag{6}$$

where p_i = proportion of total sample responding correctly to item i, p_i' = proportion of total sample responding incorrectly to item i, $q_i = 1 - p_i$, $q_i' = 1 - p_i'$, k = correction factor for formula scoring (0.250 for 5-choice, 0.333 for 4-choice, and 0 for student produced response items), n = total number of items in the subscore test section, and σ_t^2 = variance of total formula scores for the item type or test section.

This estimate is identical to coefficient alpha.

Internal consistency estimates are computed on raw scores. However, the scores on which decisions are made are scale/d scores, which are the ones reported to users, so it is important to also compute estimates of reliability on these scale/d scores. Alternate forms reliability is usually estimated using either repeater data or data collected during a special study. There are drawbacks to both approaches, in that naturally occurring repeater examinees are self-selected and tend to consist of examinees who did not score as high as they would have liked on the original testing. In addition, they often engage in activities designed to improve their skill level during the second testing (e.g., they take an additional semester of high school mathematics). Special study examinees often lack the motivation to perform at their best on two forms of an assessment. In a special analysis conducted using two sets of repeater data and realistic simulation data, where both a test/retest and an alternate forms situation were mimicked, the observed reliability as estimated by the correlation between two reported test scores ranged from 0.86 to 0.93 for the ACT English test. The reliability of scaled scores can also be estimated by using scores from a single administration of a test and applying statistical formulas.

In addition to scaled score reliability, the scaled score standard errors of measurement (SEM) is reported. While reliability looks at score consistency over a group of examinees, and is reported on a scale from zero (no consistency) to one (perfect consistency), the SEM is reported on the reported score metric, and is an estimate of how much we expect scores to typically vary across testings. If the normal distribution can be used to approximate the distribution of measurement error, about two-thirds of the examinees can be expected to be mismeasured by less than 1 SEM. For example, a test-taker got 500 on his/her SAT-V, and the SEM on the SAT-V is 30. It means if the test-taker takes the test again, the chance that his/her second score will fall in the range of 470–530 (500 ± 30) is 68%. According to the *Standards* (AERA et al., 1999, p. 29), "The standard errors of measurement is generally more relevant than the reliability coefficient once a measurement procedure has been adopted and interpretation of the scores has become the user's primary concern."

The SEM of a test is not constant across all levels of ability (Lord, 1980). Rather, the SEM varies depending on the ability level and the psychometric properties of the items comprising the test. These SEMs are referred to as conditional standard errors of measurement (CSEMs). The CSEM examines the SEM conditioned on score level, allowing a user to examine whether, for example, low scoring examinees and high scoring examinees are measured with the same precision.

Plots of the CSEM are provided in Figure 4 for the four ACT tests. The minimum scale scores plotted were chosen such that only an extremely small proportion of examinees are expected to have a true scale score lower than the minimum plotted score for each test for each administration. The ACT tests were scaled to have a relatively constant SEM across most of the true score range (see ACT, 1997). Scale score average standard errors of measurement were estimated using a four-parameter beta compound binomial model as described in Kolen et al. (1992). The estimated scale score reliability for test i (REL_i) was calculated as:

$$REL_i = 1 - \frac{SEM_i^2}{\sigma_i^2}, \tag{7}$$

where SEM_i is the estimated scale score average standard errors of measurement and σ_i^2 is the observed scale score variance for test i. Table 2 shows the range of scale score reliabilities and average SEMs for six recent ACT forms.

Figure 5 presents the SAT CSEM plots for 13 SAT Verbal forms administered between 1999 and 2001. Although there are some variations across individual forms, the trend is the same: greater measurement precision (or smaller CSEMs) is gained for scores between about 300 and 700, where the majority of the scores are located. Dorans (1984) computed the conditional standard error of measurement (CSEM) based on the three-parameter IRT model:

$$p_i(\theta_j) = c_i + \frac{1 - c_i}{1 + \mathrm{e}^{-1.7a_i(\theta_j - b_i)}}, \tag{8}$$

where $p_i(\theta_j)$ is the probability of answering item i correctly for examinee j, given the three item parameters (a_i: the discriminating power of the item; b_i: item difficulty; and

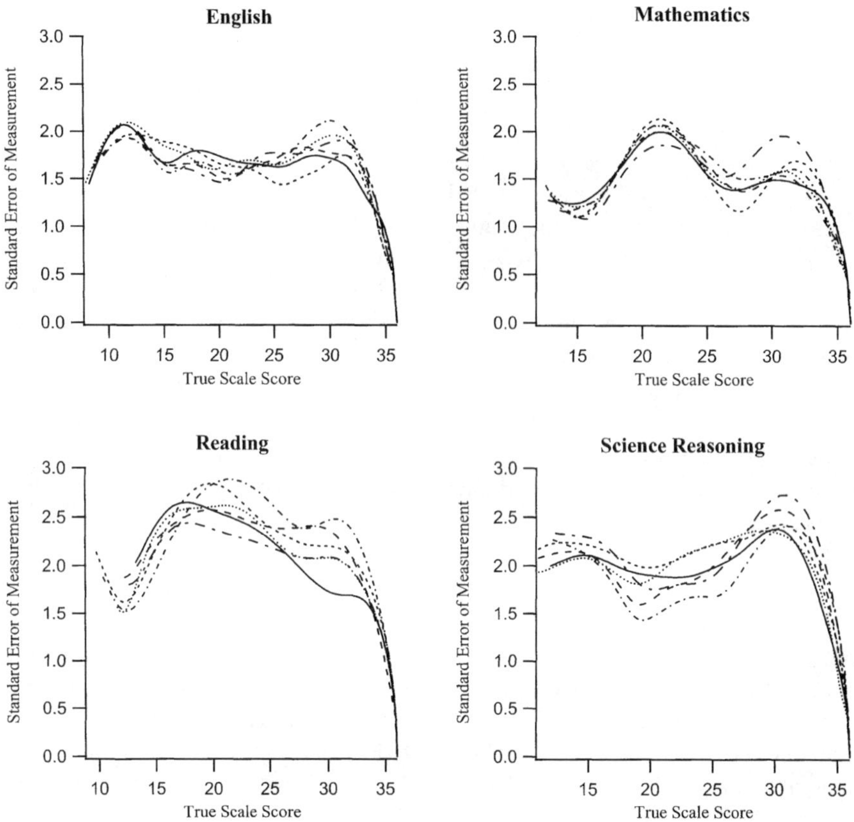

Fig. 4. Estimated conditional standard errors of measurement of the four tests of the six national ACT assessment administrations in 2002–2003.

c_i: the guessing parameter of the item). The CSEM for a given number right score, is:

$$CSEM(RS) = \sqrt{\sum_{i=1}^{N} P_i(\theta_j)Q_i(\theta_j)}, \tag{9}$$

where $Q_i(\theta_j) = 1 - P_i(\theta_j)$, and N is the total number of items on the test. When formula scoring is used, the CSEM may be computed by:

$$CSEM(FS) = \sqrt{\sum_{i=1}^{N}\left[\frac{k_i}{k_i - 1} * CSEM(RS)\right]^2}, \tag{10}$$

where k_i is the number of options associated with item i (i.e., 5-choice). An overall SEM is calculated based on CSEMs and the number of test-takers:

$$SEM = \sqrt{\frac{\sum_{m=0}^{N}(CSEM_m^2 \cdot J_m)}{J}}, \tag{11}$$

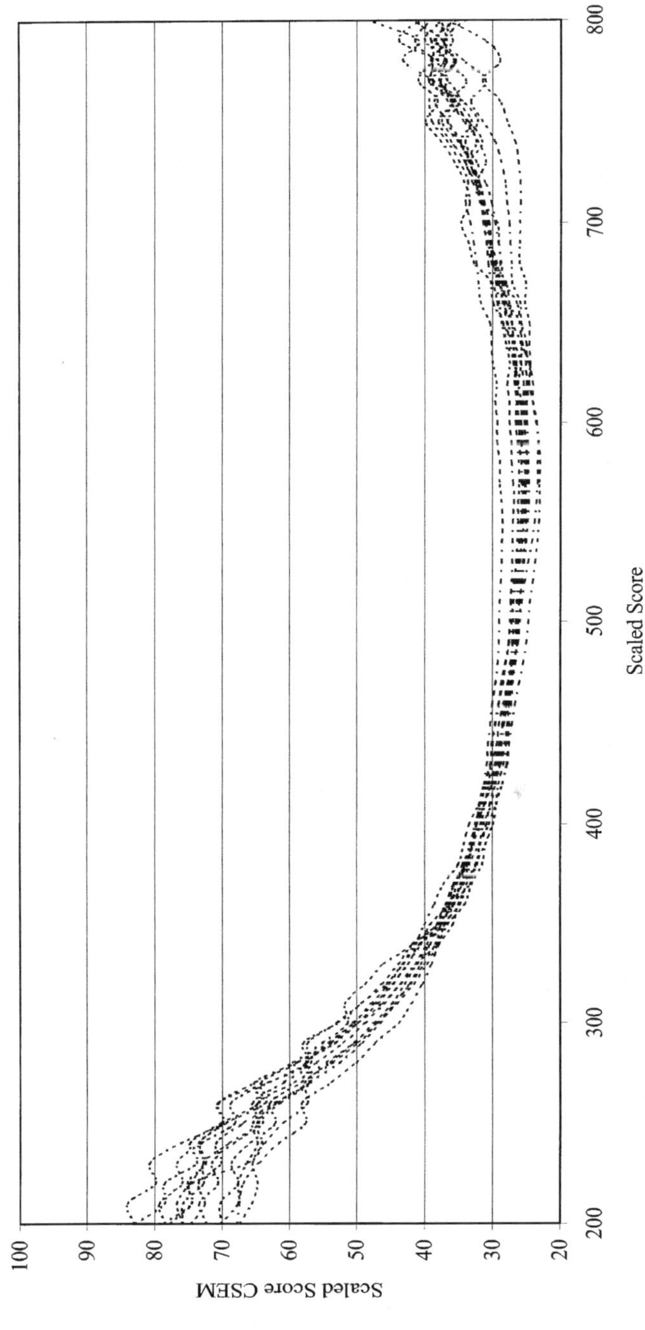

Fig. 5. The scaled score CSEMs of 13 SAT-V administered from 1999–2001.

Table 2
Scale score reliability and average standard errors of measurement summary statistics for the six national ACT assessment administrations in 2002–2003

Test/subtest	Scale score reliability			Average SEM		
	Median	Min	Max	Median	Min	Max
English	0.91	0.90	0.91	1.68	1.67	1.74
Mathematics	0.89	0.89	0.91	1.58	1.57	1.65
Reading	0.84	0.82	0.85	2.36	2.22	2.50
Science	0.79	0.77	0.84	1.96	1.72	2.12
Composite	0.96	0.96	0.96	0.89	0.87	0.92

Table 3
Reliability estimates and standard errors of measurement (SEM) of scaled scores for SAT based on 13 forms administered from 1999–2001

Test/subtest	Scaled score reliability			SEM		
	Average	Min	Max	Average	Min	Max
Verbal	0.91	0.90	0.93	30	29	32
Mathematics	0.92	0.90	0.93	31	28	33

where J_m is the number of examinees obtaining observed score m, and J is the total number of examinees. IRT scaled score reliability is calculated from the SEM using formula (7). The average and the range of SAT scaled score reliability and SEM for thirteen SAT forms are presented in Table 3. As can be seen, both SAT-V and SAT-M are highly reliable, with reliability estimates around 0.92.

Both the SAT and ACT have added writing components to their assessments in the spring of 2005. The ACT added an optional writing component, while the SAT added a mandatory writing component. Once the writing assessments are introduced, additional reliability coefficients will be estimated, specifically rater consistency and reliability index making use of generalizability theory variance components. Initial data from a preliminary study estimates the inter-rater reliability of the ACT essay as 0.92 to 0.94, which indicates the percentage of times two independent raters assigned an essay an identical score. The variance component for examinees was 63% of the total variance, indicating that most of the variability in scores was due to differences in examinee ability, rather than prompts or raters. The estimated reliability of the combined English/Writing score is 0.91, with a SEM of 1.67.

The various reliability statistics reported for the ACT and the SAT indicate that admissions staff, or other test score users, can have confidence that the reported scores indicate examinees' level of achievement and not random fluctuations.

6.2. Validity

The topic of validity is covered in depth in another chapter in this volume; in this section we concentrate on the kinds of validity evidence that are typically collected to support the inferences made from college admissions test scores. One of the major purposes of both of the ACT and the SAT is to predict success in college; therefore, historically, predictive validity evidence has been the primary type of evidence collected to support the use of these scores. Typically, predictive validity models examine the degree to which SAT or ACT scores, by themselves and in combination with high school grades, predict important college outcomes, such as first-year grade point average, retention, and graduation. ACT scores are often used for making course placement decisions, and so placement validity evidence (that is, evidence that demonstrates the ability of ACT scores to accurately predict student success in particular college courses), is also collected. In recent years it has also become important to amass different types of validity evidence, as both the College Board and ACT are asked more frequently than ever before, perhaps as a result of the standards movement, to explain exactly what the SAT and the ACT assessments measure. Other uses that are made of college admissions test scores that are explicitly or implicitly sanctioned by the testing organizations, such as the use of scores as a piece of evidence to decide whether to award a college scholarship or a high school diploma to a student, also need to be validated by the collection of appropriate evidence. While the above-mentioned types of validity evidence are the primary types that are collected, both testing organizations, in keeping with the *Standards* (AERA et al., 1999), are keenly aware of the need to collect validity evidence on an on-going basis to support the use of their assessments and to educate users regarding their responsibility to collect validity evidence when test scores are used for purposes for which the tests were not designed. In the next section we briefly review a small portion of the literature that exists that supports the use of ACT and SAT scores for admissions and placement. The reader should note that there are literally hundreds of studies examining the predictive validity of college admissions test scores. The studies presented here are primarily those conducted or sponsored by the testing organizations, and are only meant to be a small sampling even of these.

As mentioned previously, the most common college outcome variable used in predictive validity studies is first-year grade point average in college (FYGPA), which is used most often not only because it can be collected relatively easily when compared to four-year grade point average and graduation, but also because it is considered an important measure in itself in that it reflects the judgment of faculty from several different, important disciplines during a year when all of the students are typically taking a similar set of introductory courses (Bridgeman et al., 2000). Typically, in a predictive validity analysis, a multiple linear regression model is specified with FYGPA as the dependent variable and test scores and a measure of high school achievement (usually high school grades or class rank) included as independent variables. Correlation coefficients that reflect the strength of relationship between the predictors and FYGPA are computed and reported. Results of many studies that have been conducted over the years indicate that there is a significant, positive relationship between both ACT and SAT test scores and FYGPA, and that when a measure of high school success is added, the relationship is even stronger. In its technical manual, ACT reports that the median correlation coefficient between ACT scores and FYGPA is 0.42, and when high school grades are added, it rises to 0.53 (ACT, 2006). These results are based on data obtained from 291 post-secondary institutions from students who entered college in 1997 (all correlation coefficients are uncorrected in restriction of range). Results from studies sponsored by the College Board are very similar. The College Board reports that the mean weighted correlation coefficient between SAT scores and FYGPA is 0.42, and rises to 0.55 when high school grades are added (Ramist, 1984). (These results are based on data obtained from 685 post-secondary institutions for students who entered college between the mid-1960s and the early 1980s.) A more recent College Board study (Bridgeman et al., 2000), which was conducted after a new SAT was introduced in 1994 and recentered in 1995, showed slightly lower correlations between the predictors and FYGPA (.35 for SAT scores alone and 0.44 for SAT scores and high school grades, based on data from 23 post-secondary institutions for students who entered college in 1994 and 1995), but this discrepancy was explained by the fact that these coefficients were uncorrected for restriction of range (Burton and Ramist, 2001). Once the coefficients from both the older and newer studies were corrected, the differences between them were negligible. (For a more complete discussion of the problem of range restriction, see Gulliksen (1950), and for a good discussion of that and other technical issues in conducting predictive validity studies, see Burton and Ramist (2001).)

The studies that we have discussed so far involve calculating correlation coefficients across entire populations of students, but many test score users and policy makers are also concerned with whether test scores function the same way for women and men and for members of various racial or ethnic subgroups. The types of studies that address these questions are often referred to as predictive differential analysis (PDA). These studies have generally found that both ACT and SAT scores predict FYGPA about equally well across different racial/ethnic subgroups; however, test scores and high school grades tend to over-predict college performance for African American and Hispanic students and for males in general, and under-predict college performance for White and Asian students and for females in general (ACT, 1997; Bridgeman et al., 2000; Burton and Ramist, 2001; Noble, 2003; Ramist et al., 1993). Although estimates vary across studies, in general, the over- and under-prediction phenomena is reduced when both test scores and high school grades are used together to predict first-year grade point average. Some researchers have suggested that most, if not all, of the differential prediction results from differential academic preparation in high school (Noble et al., 1992), while others have been able to reduce the effect when taking into account such variables as different course-taking patterns and grading policies in college (Ramist et al., 1993) or the level of selectivity of the post-secondary institution (Bridgeman et al., 2000).

Even though FYGPA remains the criteria most often used in predictive validity studies, studies that have examined the ability of test scores to predict college outcomes such as four-year GPA and graduation have also been conducted. Similar to what has been found with studies that predict FYGPA, both high school grades and test scores are significant predictors of both four-year GPA and graduation, although the strength of the relationship is somewhat lower for four-year grades than for FYGPA (Burton and Ramist, 2001; Bridgeman et al., 2004).

While the use of multiple linear regression models remains a popular way to evaluate predictive validity, the results of such models can be difficult for lay people to interpret, and testing organizations have been often criticized because of the seemingly small incremental validity that test scores provide over high school grades in predicting college performance. Other methodologies have been used to provide information that may be more intuitive for the layperson than correlation coefficients. For example, Noble and Sawyer (2002) specified logistic regression models using various levels of FYGPA as a dependent measure, and they found that both high school grades and ACT scores were effective in predicting FYGPA levels of 2.0, 2.5 and 3.0, while ACT scores were more useful in predicting higher levels of FYGPA. Bridgeman et al. (2004) took an even simpler approach and presented a series of tables with the percentages of students that fit into each category of high school grades, SAT score, and academic intensity of high school coursework, for each of two levels of FYGPA and four-year GPA (2.5 and 3.5), and for each of four levels of college selectivity. The results indicate that there are substantial differences in the percentage of students who succeed at the end of one year or four years of college by SAT score level, even when the intensity of the high school curriculum and high school grades are taken into account.

In addition to using test scores for admission purposes, post-secondary institutions may also be interested in using these scores, along with high school performance information, in making course placement decisions for entering freshman. While the College Board's SAT program offers a different set of assessments for this purpose, (i.e., the SAT Subject Tests), scores from the ACT Assessment can be used both for admission and placement (ACT, 1994, 1997; Allen and Sconing, 2006; Noble and Sawyer, 1997). ACT bases their placement validity argument on both content and statistical evidence, pointing to the fact that the ACT is developed to reflect the academic skills and knowledge that are acquired in typical college-preparatory high school classes, although they recommend that college faculty who are considering using ACT scores for placement should carefully evaluate the test content to assess the degree of alignment between what is tested on the ACT and the skills that are required for success in specific courses at specific institutions (ACT, 1997). ACT also collects statistical evidence to support the use of ACT scores in placement. In its technical manual, ACT reports that, after the new ACT Assessment was introduced in 1989, ACT collected course grades across a variety of entry-level course types from over 90 post-secondary institutions in order to establish the placement validity of ACT scores (ACT, 1997). Logistic regression models were specified to estimate the probability of success for each course at each level of ACT score, with success defined as attaining a B or higher at the end of the course. Results indicate that ACT scores allow institutions to make more accurate placement decisions across a variety of entry-level courses than if they had not used the scores. Using a hierarchical logistic regression model, ACT extended this research to establish "benchmark" ACT scores that indicate readiness for three common first-year credit-bearing college courses: English composition, college algebra, and biology (Allen and Sconing, 2006), allowing high schools to assess how well they are preparing students for the rigors of college courses.

Many more studies exist that support the use of college admissions test scores, both from a content and statistical perspective, for a variety of purposes; this select review was intended only to give the reader a general sense of the kinds of studies that are typically done on an on-going basis. As was previously mentioned, the need to continually collect validity evidence is a never-ending process, and both major testing organizations offer free and easy-to-use predictive validity services that make it simple for colleges and universities to participate. Because a revised SAT, which includes a writing section, was introduced in March 2005, the College Board is planning on replicating and expanding the work done by Bridgeman and his colleagues (2000) to support the continued use of the SAT in college admissions. The study will be expanded to provide placement validity evidence for the new writing section, as well as an estimate of the incremental predictive validity provided by the writing section for use in admissions. ACT will also be carefully evaluating its new essay component, introduced in February 2005, and will continue to provide updated validity information that supports the use of ACT scores for admission and placement. Preliminary validity data from 2005 found the median accuracy rate for placement into a standard English composition course for the combined English and Writing test scores was 74% for the institutions in the sample studied (ACT, 2005).

7. Other issues

One of the primary goals of creating a large-scale, standardized college admissions test is to produce scores for individual test-takers that are comparable across forms, administration dates, and years. Test-score users depend on the test developers to provide this comparability so that candidates for admissions can be compared fairly, and so that trends across time can be maintained. There are several seemingly-disparate topics that influence the issue of score comparability that do not fit neatly into any of the previous sections but will be addressed in this section: the effects of coaching on test scores; testing special populations; and the detection of cheating.

The effect of coaching on test scores

There are many companies and individuals that offer various services for helping students prepare for college admissions tests, and a major criticism that has been leveled against testing companies is that affluent students have a strong advantage over their less-wealthy peers because of their access to these sometimes very expensive services. The larger test preparation companies often claim that their services will result in extremely large test score increases, and if their claims were true, it would pose a serious threat to the validity of test scores. Both ACT and the College Board advise students to have a basic familiarity with their tests in terms of content, item types, testing time limits, strategies for guessing, etc., and, in keeping with the *Standards* (AERA et al., 1999), they both provide free copies of test forms and answer keys so that students can become familiar with the test in advance of taking the test. Both organizations report data annually demonstrating that students who take more college-preparatory courses in high school also obtain higher test scores, and one can draw the inference that test scores are positively related to a high-quality, long-term educational experience (ACT, 2004b;

College Board, 2004b). The large test preparation firms, in contrast, offer very intensive short-term instruction and practice on the content that is being tested, as well as information on test-taking tips and "tricks". Research conducted by ACT (1997), ETS on behalf of the College Board (Powers and Rock, 1998), and an external researcher not affiliated with any of these organizations (Briggs, 2004) all reach the same conclusion – the effects of test preparation on test score change is minimal at best. ACT reports that students who engaged in a test preparation course showed no gain in ACT composite score over students who did not take a prep course, but that students who used practice tests and workbooks showed a modest gain of 0.2 points on the ACT composite, which ranges from 1 to 36 (ACT, 1997). Powers and Rock (1998) found that students who received coaching gained, on average, 8 points on the verbal section and 18 points on the math section, on a scale of 200 to 800, and they noted that a full third of the students who received coaching either did not experience any score change, or experienced a score decline. Briggs's (2004) results are similar. He found that the effect of coaching was about 30 points for verbal and math scores combined, with math scores gaining more from coaching than verbal scores. While these results are persuasive to some researchers, many members of the public still believe that expensive coaching classes can significantly increase test scores, and effectively communicating that this is simply not so remains an on-going challenge.

Testing special population

Large-scale college admissions testing programs must adhere to the applicable laws and to the *Standards* (AERA et al., 1999) pertaining to the testing of students with disabilities. Both major testing organizations offer a variety of accommodations to students who qualify, such as large print and Braille versions of test forms, or tests administered by a reader, or through the use of an audiocassette. However, the overwhelming number of students who take the SAT and the ACT with accommodations, many of whom have a learning disability, simply request extra time, typically either 50% or 100% extra time. Both the College Board and ACT maintain guidelines and procedures for determining student eligibility for the various accommodations that are provided, and all accommodations are made with the intent of 'leveling the playing field'; that is, of providing test scores for students with disabilities that are comparable to test scores achieved by students without disabilities who take the test without accommodations. Testing organizations should provide evidence of score comparability to the extent that it is practically possible; however, even though a fairly large number of students take college admissions tests with accommodations, because of the differences between students in terms of their individual disabilities, it is difficult to get enough data on large enough groups of students with disabilities who are similar enough to one another be able to conduct such analyses. Studies have been conducted that compare the performance of students with disabilities to students without disabilities under regular time and extended time conditions (Camara et al., 1998; Bridgeman et al., 2003), and in general, they have found that students without disabilities do not exhibit significant score gains when provided with extra time while students with disabilities do. These findings provide some evidence of comparability, but more research in this area is needed.

Security procedures and detection: Prevention of unfair advantage
Another threat to the ability to compare scores across students occurs when students fail to follow directions and attempt to gain unfair advantages. Much time, effort and money is spent in developing appropriate security procedures for the delivery of college admissions tests. Both the ACT and the College Board provide extensive manuals to test center staff with strict guidelines about how test materials should be handled, exactly how the tests should be administered and collected, and what kinds of student behaviors are prohibited, etc. (ACT, 2004b; College Board, 2004b). Toll-free numbers are provided to the center staff to report serious security breaches and pre-printed forms are provided to facilitate reporting of minor disturbances or breaches of procedures. Scores for individual students who have been reported can be delayed while a security investigation takes place. In addition, when students retake a test and exhibit a much larger than expected score differences between the two tests, their scores will be flagged for review. If the review resulted in substantial evidence of score invalidity, scores may not be reported.

Concordance tables relating SAT and ACT scores
Most four-year colleges and universities that require test scores for admissions will accept either SAT or ACT scores, and because of this, admission offices need a way to compare the performance of students who submit scores from one assessment to the performance of students who submit scores from the other. Because of this need on the part of the test score users, ACT, the College Board and ETS cooperated to analyze data from students who took both tests in order to create tables that provide the correspondence between ACT and SAT scores (Dorans, 1999; Dorans et al., 1997). Because of the differences in test specifications and test-taking populations, SAT and ACT scores cannot be considered interchangeable; however, because scores on the SAT sum (verbal plus math) and the ACT composite or sum correlate at 0.92, the production of the concordance tables is justified (Dorans, 1999; Dorans et al., 1997). While concordance tables are provided, and will be updated due to the introduction of the revised SAT and the ACT with writing, caution is recommended in using such tables, because SAT scores and ACT scores are not interchangeable, and concordance results are highly dependent on the samples used to produce them.

8. Summary

In this chapter, we review the statistical procedures used in college admissions testing. Topics discussed above include test design, item pretesting, item analysis, test equating and scaling, and test reliability and validity. Information contained in other chapters has been augmented by the unique information provided in this chapter, such as test design and item analysis, to give the reader an appreciation for the full role statistics play in large-scale operational testing programs, and a deeper understanding of the care that goes into the process of creating high-quality, reliable and valid college admissions tests.

References

ACT (1992). Content validity of ACT's Educational Achievement Tests. American College Testing (ACT), Iowa City, IA.

ACT (1994). ACT assessment course placement service: Interpretive guide. ACT, Iowa City, IA.

ACT (1997). ACT Assessment technical manual. ACT, Iowa City, IA.

ACT (1998). Content validity of ACT's Educational Achievement Tests. American College Testing (ACT), Iowa City, IA.

ACT (2000). Content validity evidence in support of ACT's Educational Achievement Tests: ACT's 1998–1999 National Curriculum Study. American College Testing (ACT), Iowa City, IA.

ACT (2003). Content validity evidence in support of ACT's Educational Achievement Tests: ACT National Curriculum Survey 2002–2003. American College Testing (ACT), Iowa City, IA.

ACT (2004a). 2004 National Score Report. ACT, Iowa City, IA.

ACT (2004b). Supervisor's manual of instructions: National testing 2004–2005 (September, October and December 2004 only). ACT, Iowa City, IA.

ACT (2005). ACT Writing Test: Preliminary technical report. ACT, Iowa City, IA.

ACT (2006). ACT Assessment technical manual. ACT, Iowa City, IA. In press.

Allen, J., Sconing, J. (2006). Using ACT assessment scores to set benchmarks for college readiness. ACT, Iowa City, IA. In press.

American Educational Research Association, American Psychological Association, National Council on Measurement in Education (1999). *Standards for Educational and Psychological Testing*. AERA, Washington, DC.

Bridgeman, B., McCamley-Jenkins, L., Ervin, N. (2000). Predictions of freshman grade-point average from the revised and recentered SAT I: Reasoning Test. College Board Report No. 2001-1; ETS Report No. 00-1, College Entrance Examination Board, New York, NY.

Bridgeman, B., Trapani, C., Curley, W.E. (2003). Effect of fewer questions per section on SAT I scores. College Board Report No. 2003-2, College Entrance Examination Board, New York, NY.

Bridgeman, B., Pollack, J., Burton, N.W. (2004). Understanding what SAT reasoning test scores add to high school grades: A straightforward approach. College Board Report No. 2004-4, College Entrance Examination Board, New York, NY.

Briggs, D.C. (2004). Evaluating SAT coaching gains, effects and self selection. In: Zwick, R. (Ed.), *Rethinking the SAT: The Future of Standardized Testing in University Admissions*. RoutledgeFalmer, New York, pp. 217–233.

Burton, N.W., Ramist, L. (2001). Predicting success in college: SAT studies of classes graduating since 1980. College Board Report No. 2001-2, College Entrance Examination Board, New York, NY.

Camara, W.J., Copeland, T., Rothschild, B. (1998). Effects of extended time on the SAT I: Reasoning Test: Score growth for students with learning disabilities. College Board Report No. 98-7, College Entrance Examination Board, New York, NY.

College Board (2001). What does the SAT measure and why does it matter? College Entrance Examination Board, New York, NY.

College Board (2004a). How the SAT is made: What goes into the nation's most widely used college admissions test. College Entrance Examination Board, New York, NY.

College Board (2004b). SAT program supervisor's manual: Instructions and scripts for administering the SAT Reasoning Test and the SAT Subject Tests (October, November, December 2004 and January 2005 only). College Entrance Examination Board, New York, NY.

Crone, C., Kubiak, A. (1997). Introduction to the statistics used in test assembly. Unpublished Statistical Report, Educational Testing Service, Princeton, NJ.

Dorans, N.J. (1984). Approximate IRT formula score and scaled score standard errors of measurement at different score levels. Statistical Report SR 84-188, Educational Testing Service, Princeton, NJ.

Dorans, N.J. (1999). Correspondences between ACT™ and SAT I® scores. College Board Report No. 99-1, ETS Report No. 99-2, College Entrance Examination Board, New York, NY.

Dorans, N.J. (2002). The recentering of SAT scales and its effect on score distributions and score interpretations. College Board Report No. 2002-11, College Entrance Examination Board, New York, NY.

Dorans, N.J., Holland, P.W. (1993). DIF detection and description: Mantel–Haenszel and Standardization. In: Holland, P.W., Wainer, H. (Eds.), *Differential Item Functioning*. Lawrence Erlbaum Associates, Hillsdale, NJ, pp. 35–66.

Dorans, N.J., Holland, P.W. (2000). Population invariance and the equatability of tests: Basic theory and the linear case. *Journal of Educational Measurement* **37** (4), 281–306.

Dorans, N.J., Kulick, E. (1983). Assessing unexpected differential item performance of female candidates on SAT and TSWE forms administered in December 1977: An application of the standardization approach. ETS Report 83-9, Educational Testing Service, Princeton, NJ.

Dorans, N.J., Kulick, E. (1986). Demonstrating the utility of the standardization approach to assessing unexpected differential item performance on the Scholarship Aptitude Test. *Journal of Educational Measurement* **23**, 355–368.

Dorans, N.J., Lyu, C.F., Pommerich, M., Houston, W.M. (1997). Concordance between ACT Assessment and recentered SAT I sum scores. *College and University* **73** (2), 24–34.

Dorans, N.J., Liu, J., Hammond, S. (2004). The role of the anchor test in achieving population invariance across subpopulations and test administrations. Paper presented at the Annual Meeting of the National Council of Measurement in Education, San Diego, CA.

Dressel, P.L. (1940). Some remarks on the Kuder–Richardson reliability coefficient. *Psychometrika* **5**, 305–310.

Educational Testing Service (2002). ETS standards for quality and fairness. Educational Testing Service, Princeton, NJ.

Educational Testing Service (2005). College Board SAT Program: Spring 2005 administrations of SAT Reasoning Test Statistical Procedures. Educational Testing Service, Princeton, NJ.

Gulliksen, H. (1950). *Theory of Mental Tests*. Wiley, New York.

Hanson, B.A., Harris, D.J., Kolen, M.J. (1997). A comparison of single-and multiple linking in equipercentile equating with random groups. Paper presented at the Annual Meeting of the American Educational Research Association, Chicago, IL.

Harris, D.J., Kolen, M.J. (1986). Effect of examinee group on equating relationships. *Applied Psychological Measurement* **10**, 35–43.

Harris, D.J., Kolen, M.J. (1994). Stability checks in random groups equating. Paper presented at the Annual Meeting of the American Educational Research Association, New Orleans, LA.

Hecht, L.W., Swineford, F. (1981). Item analysis at educational testing service. Unpublished Statistical Report, Educational Testing Service, Princeton, NJ.

Holland, P.W., Thayer, D.T. (1985). An alternative definition of the ETS delta scale of item difficulty. ETS Report number 85-43, Educational Testing Service, Princeton, NJ.

Holland, P.W., Thayer, D.T. (1988). Differential item performance and the Mantel–Haenszel procedure. In: Wainer, H., Braun, H.I. (Eds.), *Test Validity*. Lawrence Erlbaum Associates, Hillsdale, NJ, pp. 129–145.

Kolen, M.J. (1984). Effectiveness of analytic smoothing in equipercentile equating. *Journal of Educational Statistics* **9**, 25–44.

Kolen, M.J. (1991). Smoothing methods for estimating test score distributions. *Journal of Educational Measurement* **28**, 257–282.

Kolen, M.J., Hanson, B.A. (1989). Scaling the ACT Assessment. In: Brennan, R.L. (Ed.), Methodology Used in Scaling the ACT Assessment and the P-ACT+. American College Testing Program, Iowa City, IA.

Kolen, M.J., Hansen, B.A., Brennan, R.L. (1992). Conditional standard errors of measurement for scaled scores. *Journal of Educational Measurement* **29**, 285–307.

Lawrence, I.M., Rigol, G.W., Van Essen, T., Jackson, C.A. (2003). A historical perspective on the content of the SAT. College Board Report No. 2003-3, College Entrance Examination Board, New York, NY.

Liu, J., Walker, M. (2002). Will the new admission test include a variable section? Unpublished Statistical Report, Educational Testing Service, Princeton, NJ.

Liu, J., Cahn, M.F., Dorans, N.J. (2006). An application of score equity assessment: Invariance of linkage of new SAT® to old SAT across gender groups. *Journal of Educational Measurement* **43** (2), 113–129.

Lord, F.M. (1965). A strong true score theory, with applications. *Psychometrika* **30**, 239–270.

Lord, F.M. (1980). *Applications of Item Response Theory to Practical Testing Problems*. Lawrence Erlbaum, Hillsdale, NJ.

Mantel, N., Haenszel, W.M. (1959). Statistical aspects of the analysis of data from retrospective studies of disease. *Journal of the National Cancer Institute* **22**, 719–748.

McHale, F.J., Ninneman, A.M. (1994). The stability of the score scale for the scholastic aptitude test from 1973 to 1984. ETS Statistical Report SR-94-24, Educational Testing Service, Princeton, NJ.

Millman, J., Greene, J. (1989). The specification and development of tests of achievement and ability. In: Linn, R.L. (Ed.), *Educational Measurement*, 3rd ed. American Council on Education and Macmillan, New York, pp. 335–366.

Modu, C.C., Stern, J. (1975). The stability of the SAT score scale. ETS Research Bulletin RB-75-9, Educational Testing Service, Princeton, NJ.

Noble, J. (2003). The effects of using ACT composite score and high school average on college admission decisions for racial/ethnic groups. ACT Research Report 2003-1, ACT, Iowa City, IA.

Noble, J., Sawyer, R. (1997). Alternative methods for validating admissions and course placement criteria. AIR Professional File Number 6, Association for Institutional Research, Tallahassee, FL.

Noble, J., Sawyer, R. (2002). Predicting different levels of academic success in college using high school GPA and ACT composite scores. ACT Research Report 2002-4, ACT, Iowa City, IA.

Noble, J., Crouse, J., Sawyer, R., Gillespie, M. (1992). Ethnic/gender bias and the differential performance hypothesis: Implications for performance on the ACT assessment. Paper presented at the annual meeting of the American Educational Research Association, San Francisco.

Peterson, J.J. (1983). *The Iowa Testing Programs: The First Fifty Years*. University of Iowa Press, Iowa City.

Petersen, N.S., Kolen, M.J., Hoover, H.D. (1989). Scaling, norming, and equating. In: Linn, R.L. (Ed.), *Educational Measurement*, 3rd ed. American Council on Education and Macmillan, New York, pp. 221–262.

Powers, D.E., Rock, D.A. (1998). Effects of coaching on SAT I: Reasoning Test scores. College Board Report No. 98-6, College Entrance Examination Board, New York.

Ramist, L. (1984). Validity of the ATP tests. In: Donlon, T. (Ed.), *The College Board technical handbook for the Scholastic Aptitude Test and the Achievement Tests*, College Entrance Examination Board, New York.

Ramist, L., Lewis, C., McCamley-Jenkins, J. (1993). Student group differences in predicting college grades: Sex, language, and ethnic groups. College Board Report No. 93-1, College Entrance Examination Board, New York.

Schmeiser, C.B. (2004). Reaffirming our raison d'etre: The ACT assessment. Paper presented at the Annual Meeting of the American Psychological Association, Honolulu, HI.

Stocking, M.L., Swanson, L. (1992). A method for severely constrained item selection in adaptive testing. Research Report, RR-92-37, Educational Testing Service, Princeton, NJ.

Yi, Q., Harris, D.J., Gao, X. (2004). Invariance of IRT equating across different sub-populations taking a science achievement test. Paper presented at the Annual Meeting of the National Council on Measurement in Education, San Diego, CA.

34

Future Challenges in Psychometrics

Ch. 34A. Integration of Models
Robert L. Brennan

Ch. 34B. Linking Scores Across Computer and Paper-Based Modes of Test Administration
Daniel R. Eignor

Ch. 34C. Linking Cognitively-Based Models and Psychometric Methods
Mark J. Gierl and Jacqueline P. Leighton

Ch. 34D. Technical Considerations in Equating Complex Assessments
Ida Lawrence

Ch. 34E. Future Challenges to Psychometrics: Validity, Validity, Validity
Neal Kingston

Ch. 34F. Testing with and without Computers
Piet Sanders

Ch. 34G. Practical Challenges to Psychometrics Driven by Increased Visibility of Assessment
Cynthia Board Schmeiser

Editors' note

We requested short articles on potential research problems from several authors, including experts in the field and leaders of the research and development divisions of several leading testing companies. This chapter includes those articles and is intended to help advance the field. Note that even though these articles appear in one chapter, no collaboration was involved.

Handbook of Statistics, Vol. 26
ISSN: 0169-7161

DOI: 10.1016/S0169-7161(06)26034-6

34A

Integration of Models[1]

Robert L. Brennan

Our psychometric models are quite well developed, but there are occasional misconceptions and paradoxes (see Brennan, 1998a, 2001c) and, in some respects, the models are not well integrated. In considering relationships among models, it is important to attend to both mathematical and conceptual issues. Mathematics can address syntactical similarities, but semantics and inference require attention to conceptual issues. Both conceptually and mathematically, it can be argued that classical test theory is largely a special case of generalizability theory. Also, it can be argued mathematically that classical test theory is a special case of item response theory (IRT) (Holland and Hoskens, 2003). However, from a conceptual perspective, IRT is essentially a scaling model, whereas classical test theory and generalizability theory are measurement models. The essential difference, as I see it, is that a measurement model has a built-in, explicit consideration of error. There are also differences among these models in what constitutes replications, definitions of true scores, and conceptualizations of error.

What constitutes a replication?

Whether reliability is quantified as a coefficient or a standard error of measurement (SEM), it involves quantifying the consistencies and/or inconsistencies in examinee scores over replications. In this sense, the concept of reliability and its estimates involves grappling with the question: "What constitutes a replication of a measurement procedure?" (See Brennan, 2001a.) In generalizability theory the notion of replications is operationalized in part by specifying which sets of conditions of measurement (i.e., facets) are *fixed* for all replications and which are *random* (i.e., variable) over replications.

Careful thought about replications requires that an investigator have clear answers to two questions: (1) what are the intended (possibly idealized) replications of the measurement procedure, and (2) what are the characteristics of the data actually available, or to be collected, to estimate reliability? It is particularly important to note that if a facet is intended to be random (Question 1), but it is effectively fixed (e.g., only one instance) in a particular set of data (Question 2), then a reliability coefficient computed using

[1] Much of this part is an abbreviated version of a section in Brennan (2006).

the data will likely overestimate reliability, and error variance will be underestimated. For example, Brennan (2001b, pp. 127–129) shows how differences in the magnitudes of traditional coefficients of reliability are partly explainable in terms of fixed versus random facets.

Historically, in IRT, the terms "fixed" and "random" have not been used widely. However, these notions play a role in the theory. Specifically, in typical treatments of IRT, the n items are *fixed*, or, more correctly, the parameters of the items are fixed. That is, a replication would involve a set of n items with *identically* the same item parameters. In effect, there is no assumed sampling of content in the usual IRT models, whereas other models permit content sampling. One clear implication is that, all other things being equal, standard errors of ability estimates in IRT will tend to be smaller than SEMs in classical test theory or generalizability theory (see Lee et al., 2000, pp. 14–16).

In theory, one possible route to relaxing the assumption of fixed item parameters in IRT models is to employ Bayesian procedures with prior distributions for the item parameters. Doing so might lead to greater integration of IRT and generalizability theory, but much research remains to be done. (See Kolen and Harris, 1987, for a different ad hoc approach to integrating generalizability theory and item response theory.)

What are true scores?

Theories of measurement make repeated reference to true scores, or scores on some latent proficiency. Since true scores are unobservable, they must be defined for these theories to have any utility. Importantly, "true score" does not mean the same thing in all of these models.

Classical test theory and generalizability theory employ an expected-value notion of true score. By contrast, when IRT is used with dichotomously-scored items, some of the arguments among proponents of the one-parameter logistic (1PL) and two-parameter logistic (2PL) models vis-à-vis the three-parameter logistic (3PL) model are essentially arguments about what shall be considered true score. The 3PL model with its non-zero lower asymptote is reasonably consistent with defining true score as an expected value, because it acknowledges that a low-ability examinee has a positive probability of a correct response. By contrast, the 1PL and 2PL models require that low ability examinees have a probability-of-correct response approaching zero. In this sense, it appears that these latter models are based on defining true score in the more platonic sense of "knowing" the answer to an item, as opposed to getting it correct.

Recall, as well, that in traditional treatments of IRT, the n items in an analysis are fixed, which means that true scores given by the test characteristic curve are for the fixed set of items. By contrast in classical test theory and generalizability theory, true score (or universe score) is defined as the expected value of observed scores over forms that are "similar" in some sense. These differences, which are often unacknowledged, have important theoretical and practical implications.

What is error?

In classical theory and generalizability theory, error does not mean mistake, and it does not mean model misfit. Rather, important aspects of error are defined directly or indirectly by the investigator. A principal difference between classical theory and generalizability theory is that the latter permits an investigator to assess directly the influence of different facets on investigator-defined error variance.

There is no error term per se in IRT, but there are different notions of error that are often discussed in conjunction with the model. For example, the extent to which the model does not fit the data is a type of error. Rarely, however, is model misfit reported as a quantified amount; rather, various methodologies are employed to assess model (mis)fit. Also, conditional standard errors of ability estimates in IRT are usually used in much the same way that conditional SEMs are used in classical test theory or generalizability theory. Sometimes, however, this can be misleading.

For example, it is often stated that, under IRT assumptions, conditional standard errors of estimates are *larger* at the extremes of the score scale than in the middle. By contrast, in classical test theory and generalizability theory, almost always conditional SEMs are *smaller* at the extremes than in the middle. This difference is almost always an artifact arising from the choice of the θ scale (see Brennan, 1998b, pp. 326–328), and there is no a priori theoretical virtue to any particular choice of scale (Lord, 1980, pp. 84–85).

Also, in traditional IRT, standard errors cannot distinguish among multiple sources of error. Recognizing this, Bock et al. (2002) have suggested an ad hoc modification of information functions based on results from a generalizability analysis. The net effect is an adjustment of the IRT standard errors so that they incorporate error attributable to multiple facets; that is, in a sense, the procedure induces more random error into the model.

No "right" model

The 20th century (particularly the second half) witnessed the development of superb measurement and scaling models, but they are not yet particularly well integrated, and there is no "right" model. I submit that integrating these models is one of the principal psychometric challenges at the beginning of the 21st century.

References

Bock, R.D., Brennan, R.L., Muraki, E. (2002). The information in multiple ratings. *Applied Psychological Measurement* **26**, 364–375.

Brennan, R.L. (1998a). Misconceptions at the intersection of measurement theory and practice. *Educational Measurement: Issues and Practice* **17** (1), 5–9. 30.

Brennan, R.L. (1998b). Raw-score conditional standard errors of measurement in generalizability theory. *Applied Psychological Measurement* **22**, 307–331.

Brennan, R.L. (2001a). An essay on the history and future of reliability from the perspective of replications. *Journal of Educational Measurement* **38**, 295–317.

Brennan, R.L. (2001b). *Generalizability Theory*. Springer-Verlag, New York.

Brennan, R.L. (2001c). Some problems, pitfalls, and paradoxes in educational measurement. *Educational Measurement: Issues and Practice* **20** (4), 6–18.

Brennan, R.L. (2006). Perspectives on the evolution and future of educational measurement. In: Brennan, R.L. (Ed.), *Educational Measurement*, 4th ed. American Council on Education/Praeger, Westport, CT.

Holland, P.W., Hoskens, M. (2003). Classical test theory as a first-order item response theory: Application to true-score prediction from a possibly nonparallel test. *Psychometrika* **68**, 123–149.

Kolen, M.J., Harris, D.J. (1987). A multivariate test theory model based on item response theory and generalizability theory. A paper presented at the Annual Meeting of the American Educational Research Association, Washington, DC, April.

Lee, W., Brennan, R.L., Kolen, M.J. (2000). Estimators of conditional scale-score standard errors of measurement: A simulation study. *Journal of Educational Measurement* **37**, 1–20.

Lord, F.M. (1980). *Applications of Item Response Theory to Practical Testing Problems*. Erlbaum, Hillsdale, NJ.

Handbook of Statistics, Vol. 26
ISSN: 0169-7161

DOI: 10.1016/S0169-7161(06)26041-3

34B

Linking Scores Across Computer and Paper-Based Modes of Test Administration

Daniel R. Eignor

The past five to ten years have witnessed an increased level of interest in the computerization of existing large-scale paper-and-pencil testing programs. The availability of hardware to support this sort of testing has been a major contributor to this level of increased interest. For innovative computerized testing procedures, such as computer adaptive testing (CAT), the availability of software to support such processes has played a major role (see Wainer, 2000). While in the case of transitioning paper-and-pencil testing programs to CAT there have been some unexpected wrinkles, such as the unanticipated need for extremely large pools of items for secure CAT administration (see Wainer and Eignor, 2000), one issue has remained pervasive across most existing testing programs considering a switch from paper to computer mode of administration: the need to link scores across the two modes of administration. For existing testing programs, undertaking such a linking has proven to be a most challenging enterprise in that the results have often been treated in an inappropriate fashion, with related repercussions. The issue remains as a future challenge to the psychometric community because at present there is a lack of clarity as to how to best undertake such a linking and how best to share the results with score users in operational contexts. In what follows, an attempt is made to provide some of this needed clarity.

The major reason for needing to link computer- and paper-based scores in testing programs is that both modes of testing are likely to occur together, at least for some transition period. For instance, computer-based facilities may not initially be available at all testing sites, and paper-based forms of the test will need to be used at these sites. This will be particularly true for large-scale testing programs that test at an international level. However, even if it were the case that paper-and-pencil testing could be immediately phased out when computer-based testing is introduced, scores from the paper version would continue to be reported for some period of time until they are officially no longer accepted. The ideal situation in this context would be one in which the computer- and paper-based scores can be reported on a common reporting scale. Depending on the nature of the computer-based test, this may or may not be possible.

Holland and Dorans (2006) have developed general definitions or descriptions for a wide variety of score linking methods, three of which are important in the context of linking computer- and paper-based test scores: equating, calibration, and concordance.

Holland et al. (Chapter 6 in this volume) refer to the latter two methods as procedures for aligning scales.

Equating is used to refer to the linking of scores on test forms that measure the same construct at approximately the same level of difficulty and with the same level of reliability. Equated scores can be treated as being interchangeable and can be reported on the same score reporting scale. When an equating has taken place, the resulting equated scores are subgroup invariant; this is, the equating transformation will be the same for all relevant subgroups of the test taking population.

Calibration is used to refer to the linking of scores on test forms that measure the same construct, but in many cases at different levels of difficulty, and in all cases with different levels of reliability. Calibrated scores are not subgroup invariant, and the linking transformation will be different for different subgroups of the test taking population. Because of this, calibration is best carried out using samples that are representative of the whole population. Calibrated scores from different test forms have often been treated as though they were interchangeable, and scores reported on a common reported score scale. In certain limited instances this can be viewed as being for the most part appropriate, while in most other instances, this clearly should not be done. For instance, it would be viewed as appropriate to report scores on 95 and 100 item paper-and-pencil test forms, both built to the same level of difficulty and measuring the same construct, on the same scale as an outcome of the calibration process, because the forms will differ only slightly in reliability. (In effect, the calibrated scores are treated as if they were equated scores.) On the other hand, scores on 50 item and 100 item paper-and-pencil test forms measuring the same construct and built to the same level of difficulty should be reported on separate scales because the difference in reliability of the scores across the two forms is substantial.

Concordance refers to the linking of scores on test forms that measure somewhat similar constructs, often at different levels of difficulty and reliability. Scores that have been concorded can not be treated as if they were interchangeable, and separate reporting scales need to be set up for the tests. A concordance table then relates scores reported on the two different score reporting scales. Like calibrated scores, concorded scores will vary across different subgroups from the relevant population, and hence, it is important that samples that are representative of the population be employed. Proper use of a concordance table depends to a great extent on how it is presented and explained to the score user. Without appropriate explanation, a score user might use a concordance table to directly translate a score on one scale to a score on another scale, in essence treating the scores as if they were interchangeable.

The simplest form of a computer-based test occurs when a paper-and-pencil form of a test is put on the computer in a linear fashion, i.e., the items are presented in the same order they would have been presented in the corresponding paper-and-pencil test form, and all that differs between the two tests is the mode of administration. In this situation, linked scores from different forms, one given in paper mode and the other in computer mode, can frequently be viewed as being the result of an equating, and scores from both modes can be reported on the same scale and appropriately treated as being interchangeable. Holland and Dorans (2006) discuss procedures for determining whether a particular linking of scores qualifies as an equating of the scores, and these procedures should definitely be applied in this context. In many instances, the same form that already exists in paper-and-pencil is put on the computer in a linear fashion. It should not be assumed that scores here are automatically interchangeable; in many instances the computer mode of administration may affect the level of difficulty of the items (see Mazzeo and Harvey, 1988). However, in most instances, after appropriate application of the equating process, the resulting scores can be treated as interchangeable.

Many of the large-scale paper-and-pencil tests that recently have been computerized have also taken advantage of the machinery that allows CAT to be possible and have created adaptive tests. In this case, test forms are built that are tailored to the ability level of each examinee. Wang and Kolen (2001) discuss reasons why a linking of such scores to scores on a paper-and-pencil form should not be treated as an equating. Instead the linking should be considered to be a calibration. Two fairly obvious reasons for this is that it is unlikely that the CAT administered to a particular examinee will be at the same level of difficulty as the paper form, which typically would be centered on middle difficulty. Also, for typical CATs, which are fixed length in nature, i.e., all examinees are administrated CATs of the same length, the actual CAT test length will be determined so that the CATs will provide approximately the same overall level of reliability as the paper test in a representative group of examinees. While this should ensure reasonably comparable overall standard errors of measurement, it in no way ensures that the conditional standard errors of measurement, which are the relevant measures, are equal. This situation violates one of the assumptions of equating, the equity assumption (see Kolen and Brennan, 2004).

It should be noted that recent CAT/paper-and-pencil comparability studies that have been conducted have considered the derived linking relationship to constitute an equating, and scores from the two modes have been reported on the same scale rather than more appropriately reported on separate scales. (See Eignor, 2005, for a review of these studies.) Only when the groups used in the calibration process are representative samples from the relevant population of examinees and these samples of examinees are both computer literate and familiar with CAT procedures is it reasonable to even consider that the effects of the inappropriate treatment of the calibrated scores as if they were equated scores will be minimal. Unfortunately, most studies have not been conducted under such conditions, and scores should have been reported on separate scales. If a relationship between scores is needed, the relationship should be treated as if it is the outcome of a concordance study, to be described next.

For instances where the new computer-based test has been purposely constructed to differ from the paper-and-pencil test in the construct being measured, perhaps though the use of innovative computer-based item types, then linked scores from the two modes can not be considered to be interchangeable in the sense of an equating, and scores should be reported on different scales. Typically a concordance table will be created that relates reported scores across the two reporting scales. Experience has shown however (see Eignor, 2005) that great care needs to be taken in how the concordance table is constructed and in describing exactly how the table should be used. The table should not be used to directly translate a score on one scale to a score on the other scale. The table might be used to tentatively transfer a cut-point on the paper-and pencil scale to the computer-based scale, but the new tentative cut-point would then need to be validated through actual application. One big mistake the producers of concordance tables have made is to present the concordance relationship between the two scales on a point by point basis. This encourages users to use the concordance table as if it were an equating table and substitute a "comparable" paper-based score for an individual's computer-based score so that all scores can be processed through the existing paper-based system. In this situation, users do not want to employ two sets of reporting scales and systems, one for the computer test and one for the paper-and-pencil test, and hence will inappropriately use the concordance table to place all scores on a common scale, sometimes even if warned not to do so.

To summarize, the move of a number of existing large scale paper-based programs to the computer has created and will continue to create some serious challenges for the process of score reporting. Only when paper-and-pencil forms are directly put up on computer can the subsequent linking possibly be considered an equating and the resulting scores used interchangeably. In all other situations, scores from the computer-based test should be reported on a different scale than scores on the paper-based test. This is particularly true for CAT and paper-and-pencil scores that have been linked via the calibration process. However, this is a situation that is typically not tolerated well by score users, and if a point-by-point concordance table is produced, it is highly likely that it will be used inappropriately. Clearly, working out the details of how best to present and use the results of a linking of scores on paper-and-pencil and computer forms of tests remains a challenge for the future.

References

Eignor, D.R. (2005). Linking scores derived under different modes of test administration. Paper presented at the ETS Scaling and Linking Conference, Princeton, NJ, June.

Holland, P.W., Dorans, N.J. (2006). Linking and equating. In: Brennan, R.L. (Ed.), *Educational Measurement*, 4th ed. American Council on Education, Washington, DC, in press.

Kolen, M.J., Brennan, R.L. (2004). *Test Equating, Scaling, and Linking: Methods and Practices*, 2nd ed. Springer, New York.

Mazzeo, J., Harvey, A.L. (1988). The equivalence of scores from automated and conventional educational and psychological tests. College Board Report 88-08, College Entrance Examination Board, New York.

Wainer, H. (Ed.) (2000). *Computerized Adaptive Testing: A Primer*. 2nd ed. Erlbaum, Hillsdale, NJ.

Wainer, H., Eignor, D.R. (2000). Some caveats, pitfalls, and unexpected consequences of implementing large-scale computerized testing. In: Wainer, H. (Ed.), *Computerized Adaptive Testing: A Primer*. 2nd ed. Erlbaum, Hillsdale, NJ.

Wang, T., Kolen, M.J. (2001). Evaluating comparability in computerized adaptive testing: Issues, criterion, and an example. *Journal of Educational Measurement* **38**, 19–49.

Handbook of Statistics, Vol. 26
ISSN: 0169-7161

DOI: 10.1016/S0169-7161(06)26042-5

34C

Linking Cognitively-Based Models and Psychometric Methods

Mark J. Gierl and Jacqueline P. Leighton

Introduction

A cognitive diagnostic assessment (CDA) is designed to model examinees' cognitive performances and yield specific information about their strengths and weaknesses. While most psychometric models are based on latent trait theories (see section on IRT MODELS), CDA requires a cognitive information processing approach to model the psychology of test performance because the score inference is specifically targeted to the examinees' cognitive skills. Latent trait theories posit that a small number of stable underlying characteristics or traits can be used to explain test performance. Individual differences on these traits account for variation in performance over a range of test-taking situations. Because these traits are specified at a large grain size and are deemed to be stable, trait performance is often used to classify or rank examinees. Cognitive information processing theories require a much deeper understanding of trait performance where the psychological features of how a trait can produce a performance become the focus of inquiry. With a cognitive approach, problem solving is assumed to require the careful processing of complex information using a relevant sequence of operations.[1] Examinees are expected to differ in the knowledge they possess and the processes they apply thereby producing response variability in each test-taking situation. Because these knowledge structures, cognitive components, and processing skills are specified at a small grain size and are expected to vary among examinees within any testing situation, cognitive theories can be used to understand and evaluate specific information processing skills that affect test performance. To make inferences about these cognitive skills, cognitive *models* are required.

A cognitive model in educational measurement refers to a simplified description of human problem solving on standardized tasks at some convenient grain size or level of detail, which helps characterize academic knowledge structures and skill processes, so as to facilitate explanation and prediction of students' performance, including their

[1] Although we describe information processing as following a sequential trajectory in this chapter, we acknowledge that some cognitive processes may be executed in parallel. The reader is referred to Logan (2002) for a review of serial versus parallel processing in complex cognition.

strengths and weaknesses (Leighton and Gierl, 2006). These models provide an interpretative framework that can guide test construction so performance can be linked to specific cognitive inferences about performance. These models also provide the means for connecting cognitive principles with measurement practices, as Snow and Lohman (1989) explain:

> As a substantive focus for cognitive psychology then, 'ability,' the latent trait θ in EPM (educational and psychometric measurement) models, is not considered univocal, except as a convenient summary of amount correct regardless of how obtained. Rather, a score reflects a complex combination of processing skills, strategies, and knowledge components, both procedural and declarative and both controlled and automatic, some of which are variant and some invariant across persons, or tasks, or stages of practice, in any given sample of persons or tasks. In other samples of persons or situations, different combinations and different variants and invariants might come into play. Cognitive psychology's contribution is to analyze these complexes. (pp. 267–268)

Developing cognitive models of task performance

A cognitive model of task performance provides a description of the processing skills, strategies, and knowledge components use by examinees to solve test items. Cognitive processes occur as a sequence of internal events where information in short- and long-term memory interacts. Short-term memory is seen as a storage area associated with limited capacity, fast access, and conscious awareness. Long-term memory is seen as a storage area associated with unlimited capacity, slow access, and unconscious awareness. Verbal reports are assumed to provide an accurate representation of the examinees' thought processes when the reported information enters short-term memory.[2] Given these assumptions, cognitive models of task performance can be generated by conducting empirical investigations of the processes, including the knowledge and skills, students use *in vivo* to respond to items on tests. One approach for generating a model is to have examinees think-aloud (Ericsson and Simon, 1993; Leighton, 2004) as they respond to items and to record the information requirements and cognitive processes required to solve the task. The model created using information gleaned from this method can be used to generate a simplified description of the thinking processes for a group of students on a particular task. The model is then evaluated by comparing its fit and predictive utility to competing models using examinee response data from tests as a way of substantiating its components and structure. After extensive evaluation, scrutiny, and revision, the model can be generalized to other groups of examinees and different problem-solving tasks.

These models support cognitively-based test score interpretations because test performance is directly connected to examinees' cognitive skills. Strong inferences about examinees' cognitive skills can be made because the small grain size in these models help illuminate the knowledge and skills required to perform competently on testing

[2] Information in long-term memory can be consciously experienced if it is transferred into short-term or "working" memory. However, until this information is accessed and attended to, it will not be consciously experienced.

tasks. Specific diagnostic inferences can also be generated when items are developed to measure different components and processes in the model. Unfortunately, little is known about the information processing steps and cognitive structures that characterize student performance in most testing situations. As a result, relatively few models exist. Even when these models are available, they rarely guide psychometric analyses because they are usually restricted to a narrow academic domain (or, even, tasks within a domain); they are expensive to develop, initially, and to refine, over time, because they require extensive studies of problem solving on specific tasks; and they require cognitive measurement expertise, which is uncommon. Thus, future research efforts should focus on the development of cognitive models of task performance, particularly for CDAs where the target of inference is information on examinees' cognitive strengths and weaknesses.

Incorporating cognitive models into psychometric methods

Because cognitive models are rare, few assessments are designed or analyzed using these models, although some notable exceptions exist. For example, tests generated from Embretson's cognitive design system (Embretson, 1998), Mislevy's evidence center design principles (Mislevy et al., 2003), and Tatsuoka's rule-space model (cf. Tatsuoka et al., 2004) are typically based on cognitive models of task performance. To illustrate the explicit modeling of cognitive performance using an information processing approach, we describe the attribute hierarchy method (AHM; Leighton et al., 2004). The AHM is a psychometric method for classifying examinees' test item responses into a set of structured attribute patterns associated with different components specified in a cognitive model of task performance. An attribute is a description of the procedural or declarative knowledge needed to perform a task in a specific domain. The cognitive model is presented as a hierarchy of structured attributes (which is operationalized using a binary adjacency matrix; see Leighton et al., 2004, p. 211). The hierarchy defines the psychological ordering among the attributes required to solve a test problem and, thus, serves as the cognitive model of task performance. By using the attribute hierarchy to develop test items, maximum control is achieved over the specific attributes measured by each item. By using the attribute hierarchy to interpret test performance, scores and inferences about performance can be linked directly to the cognitive model, thereby identifying processes and skills within the model where examinees' have particular strengths and weaknesses. Consequently, the attribute hierarchy has a foundational role in the AHM because it represents the construct and, by extension, the cognitive competencies that underlie test performance. Few psychometric methods are guided by a cognitive model of task performance. And the methods that do exist, such as the AHM, are in their early stages of development. Thus, future research should also focus on the development of new psychometric methods that include cognitive models of test performance thereby providing a psychologically-based assessment framework for developing, scoring, and interpreting tests.

Conclusion

If the challenges in linking cognitively-based models and psychometric methods are to be realized, then a more productive relationship between cognitive psychologists and psychometricians is also required. This serves as our most important challenge. This relationship may take the form of a *cognitive measurement specialist* who understands the fields of cognitive psychology and psychometrics and who promotes this new form of theory-based assessment (e.g., Leighton, 2004; Snow and Lohman, 1989). Regardless of the form, this inter-disciplinary relationship, which was initially viewed with great promise and optimism, has not been as fruitful or productive as once anticipated. The challenges of integrating cognitive principles into measurement practices has proven to be much more complex than expected. Added to this fact are the professional pressures of maintaining a disciplinary focus which, at times, makes it difficult to combine knowledge across disciplines and promote broader research objectives. However, the integration of cognitive psychology and psychometrics is important and these barriers must be overcome, particularly as stakeholders are beginning to demand that tests yield more useful information about examinees' cognitive skills. If ranking examinees on a trait of interest is primary to the test score inference, then existing latent trait theories are adequate. But if understanding and evaluating examinees' cognitive skills is primary to the test score inference, then new procedures – which integrate cognitive models with psychometric methods – are imperative.

References

Embretson, S.E. (1998). A cognitive design system approach to generating valid tests: Application to abstract reasoning. *Psychological Methods* **3**, 380–396.

Ericsson, K.A., Simon, H.A. (1993). *Protocol Analysis: Verbal Reports as Data*. MIT Press, Cambridge, MA.

Leighton, J.P. (2004). Avoiding misconceptions, misuse, and missed opportunities: The collection of verbal reports in educational achievement testing. *Educational Measurement: Issues and Practice* **23**, 6–15.

Leighton, J.P., Gierl, M.J. (2006). Identifying and evaluating cognitive models in educational measurement. Submitted for publication.

Leighton, J.P., Gierl, M.J., Hunka, S. (2004). The attribute hierarchy model: An approach for integrating cognitive theory with assessment practice. *Journal of Educational Measurement* **41**, 205–236.

Logan, G.D. (2002). Parallel and serial procesing. In: Pashler, H., Wixted, J. (Eds.), *Steven's Handbook of Experimental Psychology, Vol. 4: Methodology in Experimental Psychology*. 3rd ed. Wiley, New York, pp. 271–300.

Mislevy, R.J., Steinberg, L.S., Almond, R.G. (2003). On the structure of educational assessments. *Measurement: Interdisciplinary Research and Perspectives* **1**, 3–62.

Snow, R.E., Lohman, D.F. (1989). Implications of cognitive psychology for educational measurement. In: Linn, R.L. (Ed.), *Educational Measurement*. 3rd ed. American Council on Education, Macmillan, New York, pp. 263–331.

Tatsuoka, K.K., Corter, J.E., Tatsuoka, C. (2004). Patterns of diagnosed mathematical content and process skills in TIMSS-R Across a sample of 20 countries. *American Educational Research Journal* **41**, 901–926.

Handbook of Statistics, Vol. 26
ISSN: 0169-7161

DOI: 10.1016/S0169-7161(06)26043-7

34D

Technical Considerations in Equating Complex Assessments

Ida Lawrence

In recent years, assessment designers, developers, and psychometricians have been faced with demands for the inclusion of increasingly sophisticated and complex assessment tasks within large scale tests designed for making high stakes decisions. These complex assessments often contain mixtures of multiple choice and constructed response questions or contain only the latter. In those situations where we wish to rely on traditional equating procedures (Kolen and Brennan, 2004) to ensure production of comparable scores across multiple test editions given to nonequivalent groups of test takers, we need to recognize and address the challenges of equating these complex assessments.

These challenges are manifested in several ways. When assessments contain a mixture of constructed response and multiple choice questions, practitioners must decide whether or not to include constructed response questions in the anchor set used for equating. In making this decision, psychometric and operational factors must be considered. For example, due to security issues it may prove infeasible or inadvisable to reuse constructed response questions to allow for traditional equating. Some kinds of constructed response questions, for instance, one-sentence essay prompts, pose particular security challenges because they are easily memorized by test takers and thus may no longer be considered secure once administered. The practical constraint of not repeating constructed response questions across administrations means that such questions cannot be used as anchor items for equating multiple test forms across administrations. This in turn means that equating will be conducted based solely on multiple choice items, which may not contain adequate representation of the construct being measured by the full assessment. What data are needed to understand the impact on equating results of using only multiple choice items in the anchor set? Is it possible to perform an adequate equating under these conditions?

A different challenge arises when constructed response questions can be included in the anchor set. An important assumption when using an anchor test to adjust for ability differences between nonequivalent groups of test takers is that items in the anchor set perform similarly across administrations. While various statistical techniques can be used to screen for and eliminate multiple choice items in the anchor set that behave dissimilarly across old form and new form samples, the situation is more complicated when constructed response items are included in the anchor set. When answers to constructed response questions are scored by judges, rather than by machines, data are needed to ensure that the standards used to score the questions are the same across the pools of judges who scored the constructed response question when it was part of the old form and when it was part of the new form. Research focused on mechanisms to control for differences in scoring at different administrations is relevant in this regard. These mechanisms include working closely with subject matter experts to provide the clearest possible scoring rubrics and adequate training of judges. When test questions are scored by humans, instead of by machines, research is needed to evaluate whether the mechanisms for ensuring that scoring is done using consistent standards have been successful. A variety of methods for monitoring rater consistency across test administrations can be used, and these methods each have strengths and weaknesses. (Bejar et al., 2006, review relevant literature on human scoring and approaches to monitoring the quality of scoring by humans; Williamson et al., 2004, discuss monitoring of automated scoring.) But the greatest challenge lies in how best to use the information. If data indicate that scoring is not stable over time, options for using the information include dropping the question(s) as a linking item(s), rescoring the question with additional training of raters, or using a statistical procedure for adjusting anchor test scores for rater leniency or stringency. Obviously, these issues are even more salient in assessments that are wholly made up of constructed response questions. These assessments typically contain relatively few items, so a change in scoring rigor on one question can have major implications for score comparability.

The aforementioned practical issues surrounding the use of constructed response items in assessments used for making high stakes decisions raise the following research questions:

1. What quality control and statistical procedures are needed to ensure form-to-form comparability of tests containing constructed response items?
2. Can constructed response questions be used effectively in anchor tests for equating across nonequivalent groups of test takers?
3. If so, how can constructed response items be most effectively used as part of an anchor test?
4. What are best practices for scoring constructed response questions and carrying out statistical monitoring of scoring to ensure high quality in terms of accuracy and validity?
5. What quality control steps are needed to ensure stability of scores when tests containing constructed response items are reused across administrations?
6. When assessments containing constructed response questions are reused across administrations, what kind of evidence is needed to ensure that scores on the reused assessments can be reported without a check on the stability of the scoring of the constructed response portions?

Another set of challenges occurs in the situation in which form-to-form comparability across nonequivalent groups of test takers needs to be established but it is simply not possible to collect anchor test information. Research is needed on innovative approaches to ensuring score comparability in those situations where repeating items is not possible. For example, the licensing test for architects (Kenney, 1997) faced just such a situation where neither pretesting nor anchor equating was feasible for security reasons. Kenney developed an innovative assessment where forms are constructed such that the difficulty of the items is held constant by design such that the forms are "randomly equivalent." Research is needed on best practices for establishing form-to-form comparability in suboptimal equating situations where the use of an anchor test and the use of other alternatives, like field testing, spiraling old and new forms, or pre-equating, are not feasible. For example, when anchor test information is not available, it may be possible to collect and use demographic information on test takers. Research is needed on the practicality of using demographic information to match score distributions against a synthetic reference population and obtaining score conversions for new forms of the tests in these demographically matched samples. Nonstatistical approaches are also worthy of investigation. For example, research is needed on test development approaches that might be used where traditional equating is impossible. Examples of such test development approaches are evidence-centered design (Mislevy et al., 2003), use of item models (Bejar et al., 2003), and use of item shells, where the aim is to control variability of item difficulty through specification of item features and understanding of factors that influence item difficulty. While none of these alternative approaches can replace traditional equating methods for ensuring comparability of scores across multiple forms, operational constraints may require creative use of these alternatives, if nothing else, to prevent the creation of forms that are too different to be equated successfully. However, the strength of these approaches will depend on a research base to demonstrate their effectiveness at achieving score equity.

References

Bejar, I., Lawless, R., Morley, M., Wagner, M., Bennett, R., Revuelta, J. (2003). Feasibility study of on-the-fly item generation in adaptive testing. *Journal of Technology, Learning, and Assessment* **2** (3). Available from http://www.jtla.org.

Bejar, I.I., Mislevy, R.J., Williamson, D.M. (2006). Human scoring. In: Williamson, D.M., Mislevy, R.J., Bejar, I.I. (Eds.), *Automated Scoring for Complex Constructed Response Tasks in Computer Based Testing*. Lawrence Erlbaum Associates, Mahwah, NJ, in press.

Kenney, J.F. (1997). New testing methodologies for the architect registration examination. *CLEAR Exam Review* **8** (2), 23–29.

Kolen, M.J., Brennan, R.L. (2004). *Test Equating, Scaling, and Linking: Methods and Practices*, 2nd ed. Springer, New York.

Mislevy, R.J., Almond, R.G., Lukas, J.F. (2003). A brief introduction to evidence-centered design. ETS Research Report RR-03-16, Educational Testing Service, Princeton, NJ, July.

Williamson, D.M., Bejar, I.I., Hone, A.S. (2004). Automated tools for subject matter expert evaluation of automated scoring. *Applied Measurement in Education* **17** (4), 323–357.

Handbook of Statistics, Vol. 26
ISSN: 0169-7161

DOI: 10.1016/S0169-7161(06)26044-9

34E

Future Challenges to Psychometrics: Validity, Validity, Validity

Neal Kingston

Any realtor will tell you the three most important factors contributing to the value of a property are location, location, and location. Similarly, the three most important factors contributing to the value of a testing program are validity, validity, and validity. Unfortunately, it is *relatively* easy to develop sophisticated models to help us reduce error of estimation by a few percent, so the vast majority of recent psychometric research has focused on more accurately modeling and assessing the same constructs we have always measured rather than breaking new ground for better assessments.

What is needed is more attention to the confluence of psychometrics, cognitive psychology, developmental psychology, learning theory, curriculum, and other theoretical bases of our assessment blueprints. It is not enough to develop parsimonious models – we need models that support appropriate actions. For example, currently most truly diagnostic tests require individual administration – we need strong psychometrics to support group administered tests that not only provide accurate diagnoses, but also provide valid prescription leading to demonstrable educational improvement.

Snow and Lohman (1989) presented the implications of cognitive psychology for educational measurement, but there has been little groundswell in the last decade and a half to address the issues they raise. A review of the table of contents of Volume 69 (2004) of *Psychometrika* showed one pertinent article (De La Torre and Douglas, 2004) out of 34. A similar review of Volume 42 (2005) of the *Journal of Educational Measurement* showed one pertinent article (Gorin, 2005) out of 17. Susan Embretson and Kikumi Tatsuoka (to name but two) have done important bodies of research in this area, but much more is needed.

Steven J. Gould (1989) posited that the Burgess Shale demonstrated much greater diversity of life in the Cambrian Period than exists today and that many families and species of this period were evolutionary dead ends that have no living progeny. Although a controversial theory within the field of evolutionary biology, there is an easier to demonstrate parallel with item types. In the early part of the 20th century there were far more item types used in large scale assessment than are used currently. The need for efficient scoring (originally by hand using stencils but eventually by high speed optical scanning machines) led to an evolutionary dead end for many item types that might effectively measured many constructs of interest. Although there has been a resurgence of the use of constructed response items in large-scale assessment in the last decade (including but not limited to direct measures of writing), most tests today look very much like tests from a half century ago.

The use of computers allows for the efficient administration and scoring of a large variety of item types – including those that previously were seen as evolutionary dead ends and new item types that are much more dynamic in nature. One noteworthy example is the work of Bejar and Braun (1999) on the simulation of architectural practice and the knotty problems associated with scoring such novel item types. Much work in this area has also taken place in the field of medical testing, for example, Clauser et al. (1997).

One challenge we face in all of this stems from the depth and breadth of knowledge required for a quantum advance in the field of educational testing. It is sufficiently challenging for most of us to keep up with one narrow field let alone with many. Collaboration within interdisciplinary teams is another, perhaps more reasonable, approach. Regardless of the challenges, the potential positive impact from advances in these areas is immense.

References

Bejar, I.I., Braun, H.I. (1999). Architectural simulations: From research to implementation: Final report to the National Council of Architectural Registration Boards (RM-99-2). Educational Testing Service, Princeton, NJ.

Clauser, B.E., Margolis, M.J., Clyman, S.G., Ross, L.P. (1997). Development of automated scoring algorithms for complex performance assessments: A comparison of two approaches. *Journal of Educational Measurement* **34**, 141–161.

De La Torre, J., Douglas, J.A. (2004). Higher order latent trait models for cognitive diagnosis. *Psychometrika* **69** (3), 333–353.

Gorin, J.S. (2005). Manipulating processing difficulty of reading comprehension questions: The feasibility of verbal item generation. *Journal of Educational Measurement* **42** (4), 351–373.

Gould, S.J. (1989). *Wonderful Life: The Burgess Shale and the Nature of History*. Norton and Company, New York.

Snow, R.E., Lohman, D.F. (1989). Implications of cognitive psychology for educational measurement. In: Linn, R.L. (Ed.), *Educational Measurement*. 3rd ed. American Council on Education, Macmillan, New York, pp. 263–331.

Handbook of Statistics, Vol. 26
ISSN: 0169-7161

DOI: 10.1016/S0169-7161(06)26045-0

34F

Testing with and without Computers

Piet Sanders

In the coming years, psychometrics will have to address four major challenges.

The first challenge is maximizing the increasing potential of the computer for several aspects of test development: item construction, test assembly, models for data analysis, and reporting of scores.

Item construction is the aspect of test development that has been, and will continue to be, most affected by computer technology. The computer has already made it possible to develop and administer items or tasks which could not be administered previously owing to the costs or to the risks involved in carrying out tasks in real conditions. The introduction of more advanced interactive multimedia capability will undoubtedly increase the realism of the items or tasks presented to those taking the tests. The computer has also started to change the way items are being constructed. While the construction of items is still mainly done by content experts, automatic item generation techniques will eventually take over a large part of the work of these content experts. The ultimate challenge here is the further development of item generation programs that can construct items or tasks according to the qualitative and quantitative specifications of test constructors or that will automatically generate new items dependent upon the responses given by students during a test administered by computer.

The analysis of item responses can be considered the essence of psychometrics. The administration of tests via computer or Internet will influence this aspect of test development in several ways. The computer and the Internet has made it possible to collect responses from many more students to larger numbers of items than was the case in the past. This increase in the number of observations will necessitate further investigation of the different fit statistics that are used to determine the adequacy of the item response models that are used to analyze students' responses. However, the computer can collect not only more but also different responses. The computer can record the time it takes to answer an item, and this information can be used to obtain a more accurate or more efficient estimate of a person's ability. Response formats in which students can express their belief in the likelihood that an answer may be correct, therefore giving credit for partial knowledge, are easy to administer by computer. Until now, the application of current item response models for time and confidence measurement has been limited. More powerful computers, however, will make other response formats possible, and new item response models for these new response formats will have to be developed.

One aspect of test development which is an excellent example of how the computer has changed test development is that of automatic test assembly. The introduction of linear programming in psychometrics has resulted in algorithms and software that allow test constructors to assemble tests automatically according to their specifications. Much progress has also been made in computerized adaptive testing, which can be considered a special version of automatic test assembly. Since the presence of item banks is an essential prerequisite for automatic test assembly and computerized adaptive testing, the design, management, and protection of item banks will continue to be an important area of psychometric research.

Another aspect of test development where computers can make a major contribution is that of providing information to students, teachers, schools, and other interested parties. A good example is the student-monitoring system for primary education that has been developed by Cito, the Dutch National Institute for Educational Measurement. It is a system for longitudinal assessment that enables teachers and schools to monitor their students' progress for most primary school subjects throughout primary education. At the moment, this low-stakes assessment system consists of paper-based tests and computer-based (adaptive) tests. The reports provided to the students and the teachers allow not only norm-referenced but also criterion-referenced score interpretations. Reports on the performance of a school in comparison to other schools are also available. Results of the analyses done with item response models are the basis for these reports. Integrating cognitive theory into these models in the near future, however, would substantially enhance the usefulness of these reports for the students and the teachers. The computer is able to detect particular patterns of responses that can, for example, help to identify content misconceptions, reasoning errors, or learning difficulties.

In the national examination systems of many countries, raters are used to score the responses of the students to open-ended questions or tasks the students have to perform. Despite the training and scoring rubrics provided to raters, they continue to give different scores, sometimes extremely different scores, to the constructed responses of the same students. Although automated scoring systems such as those developed by Educational Testing Service in Princeton can help correct for rater differences with certain response formats, these systems cannot be used for what are called performance tests. The second challenge for psychometrics would be to improve the quality of performance tests by developing the appropriate psychometric models and tools for analyzing these tests. The same holds for portfolios, which consist of a number of tests with different response formats. Questions regarding the number of tests needed to make a valid and reliable assessment of a student or the efficient administration of these tests require more attention from measurement professionals.

The third challenge for psychometrics is offered by international comparison studies like PISA. The quality of the solutions for a number of problems that have been the object of study in psychometrics for many years, including differential item functioning, horizontal and vertical equating, and standard-setting procedures, can undoubtedly benefit from applying them in international comparison studies. Psychometrics or educational measurement could make a major contribution by developing international proficiency scales and worldwide standards. In Europe, a start has been made by relating language examinations from different countries to the proficiency standards formulated in the Common European Framework.

The fourth challenge for psychometrics is attracting more students to the measurement profession. There is currently a worldwide lack of people who have expertise in psychometrics, educational measurement, or assessment. It seems that there are at least two ways to solve this problem. Firstly, testing organizations should cooperate with university departments of education, psychology, econometrics, and statistics to attract talented students. Testing organizations should offer students scholarships for dissertations on measurement topics, traineeships, and possibly the guarantee of a job after completion of graduate studies. Professional measurement organizations could also offer certification programs for different levels of expertise in assessment. These certification programs should be directed to individuals, in particular teachers, who have experience developing assessment procedures. Certification of the expertise of these individuals should be based on their theoretical knowledge of measurement concepts as well as their practical experience in developing and administering assessment procedures. For these certification programs to become successful, employers or school boards should take measurement expertise into account in the hiring and promotion of individuals.

Handbook of Statistics, Vol. 26
ISSN: 0169-7161

DOI: 10.1016/S0169-7161(06)26046-2

34G

Practical Challenges to Psychometrics Driven by Increased Visibility of Assessment

Cynthia Board Schmeiser

Educational assessment is in the midst of an unprecedented period of visibility at all levels – international, national, state, and local. One of the most important contributors to this visibility has been the passage of the No Child Left Behind (NCLB) Act in 2001, which gave assessment a greatly expanded role in holding states and schools accountable for educating our children.

With increased visibility comes new challenges for psychometrics. These challenges involve both finding solutions to problems arising out of the practical application of assessment in education today and finding these solutions quickly. Although there are virtually hundreds of issues needing resolution, I will focus on only three: growth models; diagnosis, prescription, and instruction; and dual-platform testing.

Growth models

Because of increased school accountability requirements under NCLB, states are confronted with the need to identify viable models for measuring growth in student achievement over time. Since the passage of NCLB, there has been greater interest among states in the use of growth models for school accountability. While growth models have been used for decades in research and program evaluation, policymakers and educational leaders are now seeking assistance in using growth models to provide useful information in states' accountability systems.

The psychometric issues associated with measuring growth are not new. What are new are the practical realities of the context within which growth is to be measured, including: (1) student attrition and mobility, (2) increased use of multiple measures not only within grades but between grades, (3) different student cohorts who may be assessed at different times, (4) wide variation in input variables among both students and schools, and (5) subgroups that have small sample sizes and missing data. We need psychometric models that are sufficiently flexible to accommodate these realities while providing growth estimates that are technically defensible and informative.

Diagnosis, prescription, and instruction

A second challenge to psychometrics arises out of a need to help children who are falling behind in their achievement. Students who are failing to meet state standards may be lacking one or more of the foundational skills that are necessary prerequisites for meeting the standard. True to the purposes of NCLB, it is important to develop a more efficient and effective diagnostic testing model that provides technically sound information about student achievement on state standards and the foundational skills that undergird these standards and that directly links that information to instructional next steps. We need psychometric models that allow us to offer detailed diagnoses of student achievement that: (1) are obtained efficiently, (2) report accurate scores directly tied to the curriculum, (3) link directly with state standards and the prerequisite knowledge and skills for each standard, (4) connect individual achievement to instructional prescriptions that define what the student needs to do next to remediate differences and strengthen achievement, and (5) facilitate differential instruction to individual students or groups of students needing to strengthen varying subsets of knowledge and skills.

While diagnostic testing is not new, the context within which diagnostic testing is needed is now different. More apparent than ever before are the sizable number of students who are falling behind and the fact that these students are being identified far too late in their K-12 education. Diagnostic testing needs to occur earlier, must be part of a student's ongoing record of progress, must become more efficient (i.e., use a shorter testing time across broad domains), must measure achievement at both the state standard level and a more detailed level focused on the prerequisites underlying each standard, and must be tied directly to instructional next steps.

Dual-platform testing

Increased visibility of assessment has also brought to the forefront the issue of articulating scores derived from an assessment program delivered through more than one platform – for example, in paper-and-pencil form and on computer. As more states are considering how computer-based testing can be delivered in their states, they inevitably confront inequities in the availability of computers across schools. Schools having computers want to use them to make the assessment process more efficient, while schools that lack computers cannot do this. The challenges here are to enable scores arising from the administration of the same test on different platforms to have comparable meaning and to report these scores on a common scale that can be easily understood and used by the public across schools and districts. At the same time, mode effects arising out of differences in test scores attributable to different delivery platforms cannot be ignored. Embedded in these challenges are issues associated with score comparability, equating of test forms and pools, and scaling. Up to now, there has been very little published, peer-reviewed research on this issue at the K-12 level.

* * *

By raising these three challenges, I do not intend to minimize the significant progress that has been made in the field, progress that this volume demonstrates in abundance. My examples are intended only to illustrate some of the new demands placed on psychometrics as a result of the increased visibility of assessment and a growing demand by the public for transparency of assessment results to evaluate school effectiveness.

The context surrounding assessment is changing. Practical requirements are driving changes in the way assessment results are being derived, reported, and used. The burden of many of these changes falls on psychometrics. We certainly have much to do, but I am confident we can find practical solutions to the new challenges we face.

Subject Index

2PNO model 808

A Christmas Carol 771
ability 516, 517, 529, 747–749, 751, 752, 757, 758, 766
– latent 754, 758, 759
– spatial 748, 750, 751, 761, 766
absolute decision 96
accommodation 956
ACER ConQuest 578
achievement test batteries 956
achievement tests 956, 959, 973
ACT 1057, 1058
ACT Assessment Program 1058
active treatment 770
adaptation 802, 835
adaptive design 804
adaptive item generation 752, 765
adaptive quadrature 494
adaptive test 441, 734, 802, 804, 813, 815, 817–820, 822, 823, 831, 833, 835, 888
additivity 525, 529
adequate yearly progress 867
adjacency matrix 534
adjacent category model 738
affinely invariant matching methods 785
affirmative action 907
agents 517, 527
AGLS procedures 311
agreement measures 213
AIC 370, 371, 377, 392–394
Akaike Information Criterion (AIC) 427, 499
algorithmic item generation 750, 751, 765
alleged misconduct 772
allopathic medical schools 895
ALSCAL 385
alternative system 566
alternative system of equations 566
American Council on Education 6
American Institutes for Research 7
American Psychological Association 995
analysis of discrete responses 205, 226
analysis of item responses 1113
analysis of variance 33, 403, 404, 938, 948
anchor set 1107, 1108
anchor test 1107–1109
anchor test or "NEAT" design 177–182
angles between the items 632
ANOVA 34
anti-ideal points 387
approximate method for RM (AM) 544
Army Air Corps 7
Army Alpha 6
aspirin 770
assessability 669
assessment 979, 1027
– design 979–983, 985, 988, 994, 995, 1024–1026
– diagnostic 979, 1027
– – skills 979, 1027
– formative, *see also* assessment, diagnostic 1022, 1025, 1026
– implementation framework 979–993, 995
– implementation process, *see* assessment, implementation framework
– tasks, *see also* skills diagnosis, Q matrix 980, 981, 983, 984, 988
– – dichotomous items 979, 996
– – polytomous items 996, 1000
– – task analysis 980, 982–984
– – task design, *see* task development
– – task development 980, 982–984, 995, 1024
– – task modeling, *see also* skills diagnosis, models; skills diagnosis, Q matrix; skills diagnosis, model fit 979–1026
– – task stimulus 982
– – think-aloud protocols 982, 993
assignment mechanism 769, 774, 776, 787
Assistant Attorney General Ogden 772
association between items 555, 561
asymptotic covariance matrix 538
asymptotic expansions 274
asymptotic normality
– of CML estimator 538

asymptotic robustness 309, 310, 332, 339, 344
asymptotically distribution free 312–314, 327, 343, 344
Attwood's machine 527
automated scoring 748, 1108
automated scoring systems 1114
automatic item generation 747, 748, 750, 753, 755, 756, 765, 766
– item model (approach) 748, 753, 765
– item structure (approach) 748, 750, 752, 753, 755, 757, 762, 766
automatic item generation techniques 1113
automatic test assembly 1114
autoregressive model 409
average causal effect 782
averaging results 872
AYP 867

balanced incomplete block 1040
Bartlett adjustment 277
basic parameters 575
– of LLTM 564, 566, 570, 572
Bayes model estimates 806, 812
Bayes theorem 429, 658
Bayesian 781, 787
Bayesian analysis 205, 206, 230, 704, 705
Bayesian approach 769, 794
Bayesian data analysis 964
Bayesian data augmentation 231
Bayesian estimate 813
Bayesian estimation 638, 807, 812
Bayesian estimator 812
Bayesian expected a posteriori estimator (EAP) 548
Bayesian inference 769, 843, 857
Bayesian information criterion (BIC) 499
Bayesian methods 786
Bayesian modal estimator (BME) 547
Bayesian modelling 795
Bayesian network 827, 844, 856
Bayesian p-value 811
Bayesian procedures 1096
Bayesian/likelihood tools 794
Bernoulli variate vector 479
best practices 197, 199
bias 125, 126, 155, 161, 535, 875, 947
– BME of person parameter 547
– MLE of person parameter 547
– WLE of person parameter 547
biased estimates 871
biased indicators 870
bifactor analysis 504
BILOG-MG 698, 700, 702
binomial distribution 531
Birnbaum model 524
bisection method 547
biserial correlation 31, 39–42, 1067
bivariate distribution 913
black 898
blocking variable 783
BM estimation 812
BM estimator 812
BMIRT 966
bookmark standard setting 959
bootstrap 654
bootstrap methods 280
bootstrapping 497, 498
boundaries of confidence intervals 558
boundary conditions 433
boundary values 425, 438
bounds 793
Bradley–Terry–Luce (BTL) model 534
Broyden–Fletcher–Goldfarb–Shanno (BFGS) algorithm 537, 566
Broyden–Fletcher–Goldfarb–Shanno (BFGS) method 538
but-for concept 771

calibration 955, 1099–1101
California Achievement Tests 956
California's STAR testing 959, 961
canonical analysis 217
Cartesian coordinate system 610
CAT, *see also* computerized adaptive testing 14, 15, 638, 639
categorical constraints 824
categorical response data 416
category response function 588, 720
causal attribution 887
causal effect 769, 770, 888, 912, 914, 916
causal inference 787, 869, 870, 871, 911, 913, 916
causal interpretation 869
causal questions 912
causality 869
CE linking procedures 191–195
certification 868
certification programs 1115
challenges to equating 199, 201
change
– amount of 557
– effect parameter 558
– hypothesis of no change 558, 559
– parameter 556
change score 557
characterizations of Rasch model 522
cheating 561

chi-square 654
choice model 481
Cholesky decomposition 243
Cito 1114
city-block distance 359
Clarence 770
class membership 645
classical experiment design 794
classical reliability coefficient 502
classical test theory 29–42, 45, 55, 62, 74, 81–93, 469, 499, 1095–1097
– error 82
– observed score 82
– parallel tests 82, 83
– true score 82
classical theory 1097
classification diagrams 414, 415
classroom effects 871
cluster analysis 632
cluster sampling 423, 425, 428, 437
clustering 796
CML 648
co-monotonicity of items 561
coefficient alpha, Cronbach's 34, 39
coefficient H_j for the scalability of item j 728
coefficient H^T 737
coefficient H for the scalability of a set of J items 728
coefficient of determination 207
coefficient omega, McDonald's 36
cognitive 50–53, 55, 74
– process 747
– psychology 747, 749, 1111
– variable 748, 753, 756
cognitive components 616
cognitive design system 748, 753, 761, 765
cognitive diagnosis 979, 1027
cognitive diagnostic assessment 1103
cognitive diagnostic models 1031, 1032, 1036
cognitive model 1103–1105
cognitive operations 564, 568
Cohen's Kappa 991
collateral information 818, 819
college admissions testing 1057
College Board 6, 993
College Entrance Examination Board 6
column regression 244
column–row regression 244
column–row regression with interaction 244
combinatorial optimization problem 823
Common European Framework 1115
common factors 261
common item equating 190–197
common log-odds ratio 136–139, 154
common odds ratio 136
common population equating 183, 190
common scale 955, 956
communality 262, 506
comparator 528
comparator function 516, 520, 528, 529
comparison
– specifically objective 527, 528
comparisons, scientific 517, 527
compensatory MIRT model 612
compensatory model 616, 617, 619, 620, 1037
complete designs 570
completely randomized experiment 780, 781, 782
complex data structures 412, 416
– notation 414, 415
complex level-1 variation 405
complex survey 423, 425
complex survey designs 437
components of variance 88
components score 505
composite scales 205
composite score 215, 217
Comprehensive Test of Basic Skills 958, 970
computer adaptive testing 1099, 1101
computer-based testing 1118
computerized adaptive testing (CAT), *see also* CAT 13, 14, 441, 470, 601, 638, 747, 1114
computerized test assembly 14, 15
computers
– use of in schools 1118
conceptual assessment framework 847
concordance 1099–1102
– nonparametric MML 542–544
concurrent estimation 961, 963, 972
conditional
– distribution of change scores 557
– estimator 545
conditional AIC 460
conditional association 727
conditional covariance 721
conditional covariance functions 731
conditional distribution 872
conditional expectations 915, 916
conditional independence, *see* local independence 470, 471, 493, 500, 503, 959
conditional inference 531
conditional likelihood inference 675
conditional likelihood ratio (CLR) 559
conditional likelihood ratio (CLR) tests 569, 571
conditional maximum likelihood (CML) 566, 572, 647, 648, 653

– uniqueness of a CML solution 566
conditional maximum likelihood (CML) equations
– for basic parameters 565
– of LLTM 566
conditional maximum likelihood (CML) estimator 566
conditional maximum likelihood (CML) method 565, 575
conditional posterior distribution 1044
conditional posterior predictive distribution 790
conditional ranks 916
conditional standard errors of measurement (CSEMs) 1079
conditioning 1041
confidence interval 782
– MLE 548
– uniformly most accurate (UMA) 557
confidence region 371, 395
configural interpretation 363
confirmatory factor analysis 297, 614
confounded assignment mechanisms 780
confounders 774
congeneric model 35, 36
congeneric tests 84
conjoint measurement
– additive 525
conjunctive models 1037
consistency 534
– of CML estimator 538
– of estimator 544, 545
– of MML estimator 527
consistent ordinal estimator
– total score relative to latent trait 724
constrained combinatorial optimization 823
constrained sequential optimization problem 822
constraint 610, 823, 825, 830–832
constraints on parameters 564
construct under-representation 873
construct validity 11, 48–50, 52, 55, 59, 60
construct(s) 1060
content analysis 205, 218
content experts 1113
content misconceptions 1114
content sampling 1096
content specification 822, 823, 1061
content validity 63, 72
content vector analysis 224, 225
continuation ratio model 739
continuizing discrete distributions 185, 186
continuous responses 578
contour plot 612, 615, 618
contrasts
– linear 553, 554
control 921–923
control condition 912, 913
control group 914
control treatment 770
controlled and uncontrolled causal studies 914
controlled causal studies 914
coordinates 609
correlation 910
correlation coefficient 207
correspondence analysis 19, 250
cosine similarity measure 222
counterbalanced design (CB) 177, 179, 180
counterfactual conditional 869
counterfactual substitution 870
counterfactuals 869
covariance adjustment 778, 871, 915
covariance components models 943
covariance structures 344, 878
covariate latent class model 439, 440
covariates 470, 471, 498, 499, 508, 509, 774, 783, 870, 875, 884, 962
criterion-referenced score interpretations 1114
cross classification 402
cross classified models 412, 413
cross-cultural research 568
cross-sectional data 886
crossing nonuniform DIF 128
CSA, *see* covariance structures
CTB/McGraw-Hill 8, 959, 970
cumulative common log-odds ratio 149, 155
cumulative common odds ratio 149
cumulative logit analysis 228, 231
cumulative logit model 226, 227, 230
cumulative probability model 738
cumulative probit 228
cumulative probit analysis 231
cumulative probit model 230

damn lies 917
dampen 879
data augmentation 794, 808
data collection 174–182, 198, 200, 956
data collection strategy 870
decision (D) study
– crossed design 95–105
– fixed facet 104
– nested design 97
– random facet 103
– unbalanced design, vs. balanced 115
decomposition of item parameters 564
degenerate solutions 390–392

deleted residuals 209
delta equating 1067
delta index 1067
demographic composition 877
demographic mixture 898
dendrogram 632
depression
– inventory 574
derivatives of the response function
– 1-PL 478
– 2-PL 476
– 3-PL 477
– graded categories 480
– nominal categories 484
– normal categories 482
– normal ogive 474
derivatives of the response model
– nominal multiple-choice 487
– partial credit 488
descriptive study 914
design
– of treatments 576
design effect 425
design matrix 532
design of item banks 1114
design principle 752
designed-based inference 782
DETECT
– item selection procedure
– – computer program 730
diagnosis 1118
diagnostic assessment 748
diagnostic mixture IRT models 651
diagnostic models 651, 652
diagnostic structure 652
dichotomous 472, 493, 494
DIF 127
difference judgment 459
difference method 537
differential bundle functioning 156
differential item function 491
differential item functioning (DIF) 51, 52, 62, 67, 126, 472, 550, 556, 560, 561, 568, 569, 576, 577, 739, 807, 1063, 1070, 1114
– uniform item bias 571
differential speededness 822, 828
differential test functioning 152
difficulty
– of an item 517, 529
difficulty parameter 616
difficulty specifications 1061
digraph 534
dimensional interpretation 363, 387
dimensionality 687, 993–995, 1004, 1015, 1016, 1025, 1026
– multidimensional 990, 994–996, 1000, 1003, 1004, 1016, 1019, 1021, 1022, 1025, 1026
– unidimensional 981, 982, 984, 985, 989, 990, 994–996, 1000–1004, 1016, 1022–1026
dimensionality analysis 727
dimensionality assessment 739
DIMTEST 689, 690
DINA 1036
direct and indirect causal effects 795
direct causal interpretation 914
directed acyclic graph (DAG) 590, 591, 597–600
directed line (diline) 566
direction 622
direction of steepest slope 620
directional derivative 625
directions cosines 622
directions of best measurement 628
Dirichlet distributions 444
discarding irrelevant units 786
discrepancy function 272
discrete mixture models 421
discriminant mixtures of ellipsoidal distributions 785
discriminating power 474, 491, 506
discrimination
– of items 517
discrimination parameter 524, 577, 636
dissimilarity 361, 367–369, 375, 376, 383, 385
dissimilarity measure 632
distance between two points 610
distractor analysis 1068
distribution
– logistic 475
distribution function
– bivariate logistic 481
– normal 475
– vector-Bernoulli 482
divide-and-conquer 960
dosage
– matrix 576
– of a treatment 571
– unit 571
dosage vector 574
double monotonicity model 734, 738
DSM 995, 1025
dual scaling 19
dual-platform testing 1118
– and comparability of scores 1118
– and equating 1118
– and mode effects 1118
– and scaling 1118

Dutch National Institute for Educational Measurement 1114

e-rater® 210, 211, 224, 225
E-step of EM algorithm 541
EAP estimates 814
EAP estimator 502, 812, 818
EAP (expected a posteriori) 500, 501, 813
easiness
– of a text 515, 516
– of an item 516
Ebenezer Scrooge 771
Eckart–Young theorem 245, 252
ecological fallacy 401
education 769
educational testing 980, 984, 990, 993, 994, 1025, 1026
Educational Testing Service 6, 8, 903, 1114
effect
– parameters 572, 574, 575
effect size 919, 952
efficiency
– of CML estimation 538
– of estimator 545
efficiency gains 783
elementary symmetric functions 519, 520, 533, 535, 536, 557
– recurrence relations 536, 537
eligibility 827
eligible 826, 827
ellipsoidal distributions 785
empirical Bayes 593, 817
empirical Bayes approach 807
empirical Bayes estimate 877
empirical Bayes estimation 888
empirical evidence 893
empirical information matrix 497
empirical IRFs 549, 552
empirically indistinguishable from RM 523, 524
entropy 219, 223
EPBR 785
EQS 303, 327, 345
Equal Percent Bias Reducing 785
equality constraints 494
equating 13, 14, 632, 637, 649, 873, 955, 956, 1076, 1099, 1100
equating definition 170–174
equating process 198, 199
equidistance model 648
equipercentile equating functions 184, 193, 195
equiprobable contour 612, 613, 619
equity assumption 1101
equivalence 870
– to RM 524
equivalence models 457
equivalence of MML and CML 544
equivalence structures 456
equivalent groups 956
equivalent groups (EG) design 176, 177, 179, 180
error 1097
– conceptualization 1095
– sources of 1097
error score 29–31
error variance 31–33, 35, 36, 1096, 1097
errors-in-variables 298
essential independence 723
essential unidimensionality 640, 723
essentially tau-equivalent model 31, 35
estimated school effects 880
estimated teacher effects 869, 885, 886
estimated true score 625, 905
estimation 748, 757, 760, 763, 765, 766, 769, 959
– conditional maximum likelihood estimation (CMLE) 756, 757
– joint maximum likelihood estimation (JMLE) 756, 757
– marginal maximum likelihood estimation (MMLE) 651, 756–758
estimation equations
– approximate method for RM (AM) 544, 545
– CML 535
– for item parameters 540, 541
– for person parameters 541
– for probability function 542
– JML 534
– of latent variance in parametric MML 540
– parametric MML 540
– Warm estimator (WLE) 547
estimation of model parameters 629
estimation of person parameters 546
estimation of teacher effects 878
estimator
– CML 535, 537, 575
Euclidean distance 253, 359, 360, 367, 384, 385
– simple 360, 361, 367, 385
– weighted 360, 384, 385
Euclidean distance formula 610
Euclidean space 610
evidence model 847, 850
evidence-centered design 840, 847, 861, 1109
exact conditional likelihood ratio (CLR) tests 560

exact tests 556, 559
exchangeability 666
exchangeability, exchangeable 57–60, 62, 63
expectation–maximization 1043
Expectation–Maximization (EM) algorithm 417, 491–494, 496, 497, 509, 631, 653, 656, 760, 794, 805, 819, 963, 966, 1043
Expected A Posteriori or EAP estimate 812
expected marginals (EM) algorithm 541
expected value 914
experimental manipulations 914
explanation, explanatory (model) 49–53, 64, 67, 69, 73, 74
exponential family 505, 531, 533, 664
– canonical (minimal) form 566
– one-parametric 557
exponential family model 674
exposure control 828
exposure rate 822, 826, 828
external anchor tests 181, 182
external unfolding 391, 392
external validation 887

F-information 538
F-statistic 313
FACETS 1000
facets 1095
– multiple 1097
FACETS model 577
factor analysis 15–18, 21, 29, 35, 45, 51, 52, 54, 55, 63–65, 67, 69, 72, 74, 75, 421, 423, 442, 609
factor loading 261, 475
factor pattern 261
factor rotation 281
factor score indeterminacy 263
factor slopes 506
factor structures 261
fairness 125, 128, 155, 159, 161, 1063
families 821
families of cloned items 821
families of items 803
family of RMs 522, 526
fatigue effect 561
field testing 960
filter
– for uninformative data 575
filtering 675
final item analysis 1073
finite mixture model 423
Fisher scoring 492, 495, 496, 500
Fisher scoring algorithm 417
Fisherian randomization distribution 784
Fisherian randomization-based inference 781
Fisher's approach 782, 783
Fisher's identity 805, 806
Fisher's information 474, 815
Fisher's information measure 817
fit diagnostics 655
fit indices 323, 324, 345, 346
fit of the model 960
fit statistics 614, 1113
fixed 875, 1095, 1096
fixed effect 539, 920, 924, 935
fixed effects model 403, 404
focal group 1070, 1072
force 527, 529
forecasting 447
form-to-form comparability 1108, 1109
frame of reference 527, 529
Franklin Fisher 772
frequentist approaches 795
full information factor analysis 630
fully missing covariate 793
functional equation
– of associativity 528
– of Cauchy 525
fundamental problem of causal inference 914
fusion model 986, 1011

gain patterns 885
gain scores 783
gamma index of ordinal association 561
Gauss–Hermite 494, 496
Gauss–Hermite quadrature 502, 806
Gauss–Hermite quadrature approximation 758
Gaussian distribution 4, 9
general diagnostic model (GDM) 652
general intelligence factor, *g* 5
general linear regression 877
general proficiency 874
General Social Survey 422
generalizability 11, 527
– of effects 573
– of treatment effects 576
– over agents 528
– over samples of persons 516
– over subtests 573
generalizability, multivariate 109–112
generalizability (G) coefficient 96
– group means 107
generalizability (G) study 94
– multifacet designs 99–105
– one-facet designs 94–99
– vs. decision (D) study 94
generalizability theory 93–120, 469, 1095–1097

– error, absolute 96
– error, relative 96
– error variance 96
– estimated variance components 95
– facet, hidden 116
– universe of admissible observations 93
– universe of generalization 94
– universe score 94
– variance components 95
generalized inverse 236, 270, 497
generalized least squares 1051
generalized linear models 31–36, 577
generalized partial credit model 144, 685, 692, 959
generalized RMs 556
George Bailey 770
Gibbs sampler 494, 593, 594, 794, 808, 809, 813
GLS, *see* covariance structures; GLS procedures
GLS procedures 311
GOF (goodness of fit) 370, 371, 373, 393
Goodman model 435
goodness-of-fit 497–499, 654
goodness-of-fit measures 879
GPA 39, 908
grade-equivalent scores 957–959
grade-to-grade growth 964, 972
graded response model 144
graded responses 663
gradient method 542
gradient vector 492
graphical approach 769
graphical control of RM 549
graphical depiction 913
graphical display 693, 696, 697, 706, 707
GRE 901
group differences 916
growth 868, 955, 956, 958, 960
growth curve model 299, 406, 409
growth curves 964
growth models 1117
growth patterns 972
growth trajectory 963
guessing 576
Guttman linear scales 435
Guttman scale 435, 436
Guttman scale orderings 435
Guttman scaling 434

halo effect 667
hat matrix 209
headache 770
health-care expenditures 772
Hellinger distance 251
Hessian matrix 492
heteroscedasticity 875
heteroskedasticity 405
Heywood case 277
hierarchical cluster analysis 632
hierarchical model 587, 589, 964, 972
hierarchical rater model 595, 596
hierarchical structure 401, 418
High School Grade Point Average (HS-GPA) 907
history 774
HLM 885
horizontal equating 961, 1114
Horvitz–Thompson estimation 784
human scoring 1108
HYBIL 653
HYBRID model 650
hyperparameter 593
hypothetical population 899

ICC curves
– nominal 474
ideal point discriminant analysis, *see also* IPDA 392
ideal (subject) point 387, 392
identifiability 424, 527, 531
identification 424
identified 444
ignorability condition 531
ignorable designs 793
ignorable treatment assignment 787
ignoring nonignorable treatment assignment 793
image theory 270
improper solutions 277
imputation model 1041
imputations 1045
impute 789
incidental parameter 675
incomplete data 532, 535, 545, 565, 574
incomplete designs 570
indeterminacy 633, 634
index of dependability 96
index of dissimilarity 427, 436, 438
indicators 869
indifference category 460
individual differences weights 382, 384, 385, 392
INDSCAL 380, 385
ineligibility constraints 828
ineligible 827
inferences 956, 961
information 623, 625, 627, 638

information function 501, 815
– 2-PL 477
– 3-PL 478
– graded categories 480
– multiple category 500
– nominal categories 484, 486
– normal ogive 474
– partial credit 488
information functions 625, 1097
information matrix 370, 393, 492, 496, 536, 538, 567
information-processing theory 749
instruction 1118
Intelligent Essay Assessor™ 219
intentional parameter 828, 831
intentionally more difficult 956
interaction parameter 664
interactive multimedia capability 1113
internal anchor tests 181, 182
internal unfolding 391
international comparison studies 1114
international proficiency scales 1114
interpretation 959
interpreting group differences 893
interval scale 525, 526, 529
interventions 769
intraclass correlation 403, 405
intransitive cycle 448
intrinsically unscalable classes 435
invariance 46, 50, 58, 61–63, 68, 74, 696, 706
invariant item ordering 733
inverse-Wishart 809, 810
Iowa Testing Program 8
Iowa Tests of Basic Skills 956
IPDA 392, 393, 395
IQ 5
irrational quotient of item parameter differences 526
IRT, *see also* item response theory 12, 15, 29–33, 35, 41, 42, 607, 962, 984, 989, 993, 994, 996, 997, 1002–1004, 1009, 1011, 1012, 1016, 1017, 1019, 1021, 1024, 1026
IRT equating 196, 197
isotonic ordinal probabilistic (ISOP) models 561
ITBS 957
item
– difficulty 1061
– discrimination 1061
– marginal sum 518
– parameters 518
item analysis 1066
item assessability 674, 677
item bank 747, 748, 751, 753, 1114
item bias, *see* DIF
– uniform 569
item characteristic surface 625, 626
item cloning 807, 821
item construction 1113
item content 748
item design 568
item development 748, 751
item difficulty 475, 674, 677, 1066
– prediction of 568
item discrimination 561, 1067
item factor analysis 505, 688
– EAP scores 507
– full information 506
– limited information 506
– MINRES 506
item generation programs 1113
item information 626
item intercept 474
item location 474
item parameter 747, 748, 752, 755–766, 960
– difficulty 754, 756
– discrimination 754, 756
– guessing 755, 756
item pool 556, 557
item response function (IRF) 516, 522
– increasing 523
item response model 417, 492, 643, 1113
item response models for confidence measurement 1113
item response models for time 1113
item response surface 615, 617
item response theory, *see also* IRT 9, 12, 21, 29, 31, 45, 51, 53, 54, 60–62, 66, 67, 74, 75, 118, 128, 469, 587, 604, 607, 683, 684, 686, 691, 706, 801, 958, 984, 1035, 1095, 1096
item response theory (IRT) model 748, 752, 757, 760, 765
– 2PL-constrained model 754–758, 761–764, 766
– cognitive 753, 756, 757, 762, 763, 766
– generalized 747
– hierarchical 752, 755, 757, 758, 760, 763, 764, 766
– linear logistic test model (LLTM) 753–764, 766
– Rasch model 754, 757, 762, 763
– three parameter logistic (3PL) model 755, 756
– two parameter logistic (2PL) model 754, 755, 762, 763
item scores 473

item specification 748, 762, 765
item step response function 720
item type 749, 753, 765, 1111, 1112
item vectors 623, 627
item-cloning techniques 803
item-exposure control 812, 826, 827
item-parameter drift 807
item-pool design 835
item-selection algorithm 823
item-selection rule 815, 816, 818, 822, 825
item-splitting method 736
iterated bootstrap 418
iterative generalized least squares 417
It's a Wonderful Life 770

J-C-B Plot 900
jackknife 425
Jeffrey's prior density 810
Jimmy Stewart 770
job-training programs 795
John Stuart Mill 774
joint maximum likelihood (JML) 533, 565

Kelley predictions 909
Kelley's equation 905, 906, 908–910
Kelley's estimate 33
Kelley's Paradox 893, 894, 901, 908, 917
Kullback–Leibler information 816–818
KYST 369

L_r-norm 238
lack of randomization 871
Lagrange multiplier (LM)
– tests 549, 556, 807
– method of 541
LanguEdge 981
Laplace 494
Laplace approximation 1043, 1044
Laplace integral transformation 418
latent additivity 529
latent class analysis (LCA) 541, 645, 1035
latent class models 1031
latent class proportion 422, 424, 427, 430, 433, 436–440, 442, 443
latent classes 577
latent distribution 539, 541, 544
latent distribution of person parameters 539
latent gold 424, 428, 437, 444
latent growth models 18
latent Markov models 436
latent regression 1040, 1042
latent score distribution 649
latent semantic analysis 219, 221–225
latent space 684, 686, 688
latent trait 683, 965
latent trajectory models 406, 409
latent variables 297–299, 301, 302, 304, 319, 322, 339, 470, 959
law of comparative judgment 13
layered model 868, 878
LCA 646
learning difficulties 1114
least squares methods 278
least squares polynomial regression 630
least squares regression 877
least squares solution 236
Ledermann bound 265
legal damages 772
LEM 424, 428, 653
less privileged social backgrounds 901
Levine equating 191, 195, 196
likelihood 675
– conditional 523
– of item score matrix 518
– of response pattern 518
likelihood equation 758, 760, 762, 805, 816
– multiple category 500
likelihood function 691, 692, 758, 759
likelihood principle 529
likelihood ratio 688, 691, 692, 699–701
– monotone 530
likelihood ratio (LR) tests 549
– conditional 549
likelihood ratio statistic 498
– conditional 567
likelihood ratio test 140, 277, 405
likelihood ratio test G^2 497
likelihood-weighted information criterion 816
limits of IRFs 523
linear Bradley–Terry–Luce (LBTL) model 567
linear computer test 1100
linear constraints
– on parameters 564
linear equating functions 184, 185, 188, 189, 191–193, 195, 196
linear factor analysis 686, 688
linear logistic 646
linear logistic model with relaxed assumptions (LLRA) 575
linear logistic test model (LLTM) 560, 564, 571, 595
– testing of fit 567
linear programming 1114
linear regression 206
linear relationship 792
linear structural equations model 301
linear transformation

– of parameters 529
linearly equated 956
linking 14, 873, 955, 956
linking of MIRT calibrations 632
linking of the calibrations 634
LISREL 300, 303
local independence 423, 471, 611, 645, 688–690, 703, 705, 721, 801, 805, 807, 811, 816, 962, 1018
local stochastic independence 523, 555, 561, 573
located latent class model 443
log-likelihood
– conditional 521
– – of item score matrix 533
– marginal 521, 522, 539
– multinomial 522
– of item score matrix 532
log-linear model 229, 650, 1036
log-linear modeling 1035
log-linear pre-smoothing 649
log-linear Rasch model 649, 650
logical constraints 823, 824
LOGIMO 653
logistic distribution 226, 228
logistic function 476, 614, 803
logistic model 615
logistic regression 139, 140, 143, 150
logistic restriction 649
logit model 575
lognormal model 831
lognormal model for response times 828
longitudinal data 406, 413, 878, 886, 973
longitudinal outcomes 795
longitudinal student records 868, 878
Lord's Paradox 775, 777, 893, 894, 910, 911, 913, 917
lower asymptote 613
lower bounds of communalities 267
lower order margins 498
LPCM Win 1.0 578
LR (likelihood ratio) 370
– statistic 370
– test 370
LSAT 901

M-step of EM algorithm 541
machine scoring, *see* automated scoring
Mahalanobis metric 785
Mahalanobis metric matching within propensity score calipers 785
Mahalanobis norm 238
management of item banks 1114
manifest monotonicity 732, 733
Mantel–Haenszel 569, 1071
MAP equation 501
MAP estimator 501
MAP (maximum a posteriori) 500
MAR 872, 878, 885, 886
marginal AIC 460
marginal difficulty 475
marginal distribution 491
marginal likelihood 593, 805, 1041
marginal likelihood function 805
marginal likelihood inference 676
marginal maximum likelihood (MML) 565, 631, 653, 692, 701, 804, 963, 967
– nonparametric 541
marginal maximum likelihood (MML) method 539
– parametric 539
– semiparametric 539
marginal quasi-likelihood (MQL) 760, 764
Markov chain 810
Markov chain Monte Carlo (MCMC) 417, 491, 593–595, 597, 600, 604, 631, 654, 795, 804, 807, 810, 812, 858, 964, 966, 973
Martin-Löf test 550
mass 527, 529
mastery tests 428, 441
matched on the propensity scores 785
matched sampling 785
matching 792, 870
Maximum A Posteriori or MAP estimation 812
maximum likelihood 12, 15, 206, 227, 229, 230
– conditional 32, 33
– marginal 33, 527, 539
maximum likelihood estimation 638
– marginal
– – parametric 521
– – semiparametric 521
maximum likelihood method 271
maximum marginal likelihood 506
maximum slope 621
maximum-information rule 815, 824
maximum-information selection 816
MCAT 894, 895, 899–901, 910, 911, 915, 916
MCAT score 908, 911
MCAT-Biological Sciences 895
MCEM 494
MDIFF 622, 624
MDISC 623, 624
mdltm software 653
MDS, *see also* multidimensional scaling 9, 10, 20, 371, 375, 376, 378, 385, 387, 390, 395–397
– ID (individual differences) 375, 378, 379, 383–385

– ML (maximum likelihood) 361, 369, 371–373, 376, 379, 385
– nonmetric 367–369
– simple 361, 362, 366, 369, 376, 378, 379, 385
mean and covariance structures 297, 302, 334
mean squared error 207, 208, 211–213, 221, 223
measure growth 956
measurement 748, 752, 765, 979, 1027
– construct 748
– educational 747, 766
– psychological 766
measurement error 871, 875, 1045
measurement invariance 472
measurement model 664
measurement profession 1115
mechanism for creating missing data 775
Medical College Admission Test (MCAT) 893, 894
mental age 5
mental tests 5
Meredith's problem 249
meta-analysis 20, 919, 952
method of standardization 898
metric
– interval 522
metric scaling 253
Metropolis–Hastings algorithm 593, 594, 631
Metropolitan Achievement Tests 959
Mid-P Method 558, 563, 564
min(AIC) 432, 433, 438
min(AIC) model 427
min(AIC) strategy 427
min(BIC) 432, 433, 438
minimum expected posterior variance 818
Minkowski norm 238
MIRT 608, 609, 640
missing at random 872, 886
missing covariates 796
missing data 327, 328, 330–334, 337, 410, 794–796, 872, 875, 886
missing data problem 787
misspecified models 319, 320, 322
mixed effects 920, 942
mixed model 948, 951
mixed Rasch model (RM) 577, 646
mixture distribution 650
mixture distribution models 643
mixtures of normal distributions 644
ML, *see* normal theory ML
ML estimator of person parameter (MLE) 546
MLE 811, 815, 817
MML 812, 813
MML estimates 804, 807, 810
MML estimation 804
MML procedure 806
MNSVD 241
model 45, 53, 54
model fit 675
model fit analyses 803
model misfit 1097
model selection criteria 499
model-based extrapolation 792
models for data analysis 1113
modern simulation methods 795
Mokken's item selection procedure 728
Mokken's $\mathbf{P}(+,+)/\mathbf{P}(-,-)$ matrices method 737
moment sequence 543
monotone homogeneity model 722, 724, 725, 738
monotone likelihood ratio 530
monotonic transformation 368, 385
monotonicity assumption 611, 722
monotonicity of IRFs 517
Monte Carlo 494
Monte Carlo approximation 559, 560
Monte Carlo method
– test of fit 544
Monte Carlo tests 556
Moore–Penrose g-inverse 237, 241
MPlus 428, 437
MSP, Mokken Scale analysis for Polytomous items
– computer program 729
multi-component latent class models 443
multi-group 472, 491, 493
multi-group IRT model 972
multidimensional
– LLTM of change 574
multidimensional ability parameter 803
multidimensional difficulty 622
multidimensional discrimination 623
multidimensional IRT 964
multidimensional IRT models 578, 607, 608, 804, 1036
multidimensional items 555
multidimensional response model 803
multidimensional scaling 254, 359, 369
multidimensional space 608
multidimensionality
– of LLTM 572
multifactorial RM 577
multilevel factor analysis 410, 411
multilevel model 338, 339, 341, 401, 418
– categorical responses 416
– estimation 417, 418

– fixed part 403
– random part 403
– software 417, 418
multilevel residuals 403, 409
– shrinkage 403
multilevel response model 803
multilevel time series model 409
multilevel/hierarchical IRT 1041, 1050
multinomial distribution 555
multinomial logit model 453, 481, 1034
multinomial model 553
multinomial response model 229
multiple correlation coefficient 207
multiple factor model 259
multiple group 334, 335, 337
multiple group model 471
multiple imputation 794
multiple matrix sampling 503
multiple membership models 402, 413, 414
multiple ratings 504
multiple treatments 794
multiple-indicators–multiple-causes (MIMIC) model 66–68, 70, 71
multiplicative Poisson model 531
multiply-impute 787
multivariate analysis of variance 217
multivariate data 401
multivariate item response model 1032
multivariate response data 409, 411
multivariate response model 409, 410
multivariate X 785

National Assessment of Educational Progress (NAEP) 8, 503, 589, 603, 604, 898, 901, 960, 1039
national examination systems 1114
national median 957
NCLB 867, 868, 962
nearest neighbor 225
Nebraska 898, 899
NetPASS 847, 848, 850, 853–855, 860
never smokers 786
New Jersey 898, 899
New Republic 903
Newton method 547
Newton scoring 495, 496
Newton–Gauss iterations 500
Newton–Raphson 275, 491, 492, 538, 542, 566, 576, 760
Newtonian axiom, second 528
Neyman–Pearson lemma 556, 559, 560
Neymanian evaluations 794
Neymanian randomization-based inference 782
Neymanian variance 784
Neyman's approach 783
NIDA 1036
No Child Left Behind (NCLB) act 867, 1117, 1118
Nobel laureate 905
NOHARM 609, 630, 633, 687
nominal 472
nominal multiple-choice model 501
non-hierarchical data structures 412, 416
non-response
– ignorable 532
non-risky mode 459
noncompensatory model 615, 1037
noncompliance 794, 796
nonconvergent 277
nonequivalent groups anchor test design (NEAT) 1076
nonignorable assignment mechanism 777
nonignorable treatment assignment mechanisms 792
nonignorable treatment assignments 793
nonlinear classical test theory 39–42
nonlinear factor analysis 687
nonmonotone item response function 726
nonnormal 310, 314
nonnormal data 308, 309, 311, 312, 322, 331, 335, 338, 344, 348
nonparametric
– item response theory (IRT) 523, 561, 719
– tests of fit 561
nonparametric regression 223
nonparametric tests 549
nonstatistical approaches 1109
nonuniform DIF 128, 140
norm-referenced score interpretations 1114
normal cumulative distribution function 803
normal distribution function 473, 614
normal example 790
normal example with covariate 791
normal ogive function 803
normal ogive model 41, 525, 615, 630
normal theory ML 331, 333
normal-inverse-Wishart 809
normalization
– in LLTM 565
– standard error of estimators 538
normalization constant 564
normalization of item parameters 565
normalization of parameters 518, 539
normit 473
norms 958
Northwestern University 903

nuisance 828
nuisance parameter 807, 828
null hypothesis 781, 783
number of dimensions 616, 637, 638
numerical quadrature 651

objectivity, specific 517
objects 517, 527
oblimin criterion 283
oblique Procrustes 636
oblique rotation 281
observational data 870
observational nonrandomized studies 780
observational study 786, 870, 884
observed information matrix 497
observed score 905, 906
observed-score equating methods 184–193, 195
observed-score reporting 831
odds ratio 919, 921
Office of Strategic Services 7
omitted covariates 884
omitted variables 879
one-factor model 257
one-parameter 959
one-parameter logistic (1PL) model 491, 577, 675
one-parameter model 684, 692
open-ended questions 1114
optimal design 801
optimal scaling 19, 20
optimally weighted composite 930
ordered latent class approaches to NIRT 737
ordering
– composite 524
– stochastically consistent 524
ordinal 472
ordinal scale 367, 529
ordinary least squares 876
origin of the space 620
orthogonal axes 610
orthogonal coordinate space 621
orthogonal projection 238
orthogonal rotation 281
orthomax criterion 282
Otis's group intelligence scale 6
out-of-level testing 956
Owen's approximate procedure 817

p-value 781, 1066
paired comparison 449, 451, 452, 454, 455, 457
paired comparison experiment 785
paradoxes 893
parallel forms 956
parallel IRFs 554, 555, 576
parallel tests or items 30, 35, 36
parameter 747, 748, 752, 760, 762, 766
– basic 754
– incidental 757
– structural 757
parametric
– IRT 523
parametric bootstrap 1038
parametric irrelevance 788
parametric MML 555
Pardux 966
partial credit model (PCM) 531, 556, 578, 648, 675, 685, 692
partially compensatory model 615–617
Pearson chi-square 498, 654, 699, 701
Pearson-type statistic 549, 811
Pearson-type tests 553
pedagogical strategies 871
penalized quasi-likelihood (PQL) 760, 763, 764
performance tests 1114
persistence 885
persistence model 878, 879
person
– marginal sum 518
– parameters 518
person fit 574
person fit test 556, 559
person response function 740
person-fit analysis 740
person-fit statistic 655
personal equation 4
personality assessment 734
physical randomization 775
PISA 1114
platonic 1096
plausible values 1041, 1044, 1046
play-the-winner designs 780
Poisson distribution 531
Poisson model, multiplicative 515
polychoric correlation 31, 41
polyserial correlation 31, 40
polytomous 493
polytomous generalizations of the RM 578
polytomous mixed Rasch models 648
pool of items 572
portfolios 232, 1114
posited assignment mechanism 777
posterior distribution 639, 759, 790, 963, 972
posterior information function 501
posterior mean 1043
posterior predictive causal inference 786

posterior predictive checks 810
posterior predictive distribution 789
posterior predictive distribution of causal effects 787
posterior variance 1043
posttest 556
potential outcomes 770, 775
potential outcomes framework 774
power analysis 20
power series distribution 531
power test 828
powerful tests 783
precision 747
precision of VAM estimates 872
predicting 170–174
prediction of first choices 482
predictive differential analysis (PDA) 1084
preference 359, 361, 385, 387, 390, 391
PREFSCAL 391
preliminary item analysis 1073
preposterior analysis 817
preposterior argument 818
preposterior item selection 817
prescription 1118
presmoothing score distributions 186–188
pretest 556
pretest administration 1065
pretest item analysis 1073
pretest item location 1066
pretest sample 1065
pretesting 1064
pretesting data collection design 1064
primary factors 283
primary sampling unit 1046
primitives 770
Princeton 1114
principal component analysis 269
principal components 215
principal factor analysis 506
principal factor method 279
principal stratification 795
prior differences 870
prior distribution 638, 963, 967, 972
prior modeling 590
probability modeling 590
processing difficulty 751, 761, 762
Procrustean transformation 247
Procrustes problem 238, 239
Procrustes rotation 285
Proctor model 435
proficiency model 847, 848
proficiency standards 1115
projection operator 237
promax rotation 286
propagating errors 961
propensity scores 784, 796
proportional reduction in error 214
protection of item banks 1114
proximity 359–362, 369, 370, 375, 383, 385, 396
PROXSCAL 369, 385
PSAT/NMSQT 993
PSE linking procedures 191–193
pseudo-guessing parameter 613
pseudo-log-likelihood 425, 437
PSVD (product SVD) 242
Psychological Corporation 8
psychological scaling 8, 9, 18, 20, 21
psychometric model 747, 748, 753, 755
Psychometric Society 1, 8
psychometrics 769
– practical challenges to 1117, 1119
Psychometrika 19
psychophysics 3, 9
purpose of pretesting 1064

Q matrix, *see* skills diagnosis, Q matrix
QR (rank) factorization 242
QSVD (quotient SVD) 242
quadratic forms 553, 556
quadrature
– Gauss–Hermite 521
– points 521
quadrature formulas 502
quadrature weights 541
qualitative specifications 1113
quantitative specifications 1113
quasi-likelihood
– marginal (MQL) 418
– penalized or predicted (PQL) 418
quasi-Newton method 542, 762

random 875, 1095, 1096
random assignment 914, 917
random assignment of raters 679
random coefficient model 404
random coefficients 878
random effects 878, 943–947, 949, 952
random effects model 401, 418
random intercept model 405
random sample 911
random slope model 404
random utility 448, 449, 460
random/equivalent groups (EG) design 1076
randomization
– of discrete scores 556
– of scores 558

– randomized tests 559
randomization distribution 782
randomization tests 549
randomization-based 794
randomization-based modes of causal inference 781
randomization-based perspective 786
randomized and nonrandomized data 769
randomized assignment mechanisms 775
randomized block design 783
randomized block experiment 785
randomized experiments 769, 775, 777, 781, 793, 870
range, restriction of 30, 36–39
ranking 454, 472
ranking data 449, 457, 459
Rao score tests 807
Rao–Yanai *g*-inverse 237
Rasch 54, 61, 63, 74
Rasch family of models 614
Rasch model (RM) 32, 33, 497, 516, 517, 616, 646, 675, 685, 690, 692, 697, 706, 725
– equivalence with 523
– extension of 564
– family of RMs 526
Rasch model to vertical scaling 959
raters 663, 1114
rating 663
– filtered 667
– inconsistent 669
– negative consistent 669
– positive consistent 669
rating scale model (RSM) 531, 578, 648
rating scales 656
ratio scale 367, 380
ratio statistics
– nonparametric MML 543
raw score 518, 1074
RCM 769, 775, 776, 786
reaction 527
reading ability 515
reading speed 515
realism 1113
reasoning errors 1114
reduced rank regression 247
reduction of dimensionality 245
reference composite 627, 628, 640
reference factors 283
reference group 1070, 1072
Reference Manual on Scientific Evidence 772
reference test 822, 831, 832
regression 46, 55, 65–72, 74, 876, 912
regression adjustment 778
regression analysis 205, 206, 208, 210, 211, 213, 214, 218, 221–223, 225, 227, 231
regression effect 905, 910
regression estimation 213
regression lines 912, 913
regression model 871
regression toward the mean 903
regular assignment mechanism 784
regular designs 784, 786, 792
regularity condition 448
related siblings model 594, 599, 600
relative decision 95
relative teacher effectiveness 869
reliability 11, 30, 33–36, 39, 50, 51, 55, 57, 59, 62, 65, 66, 70, 74, 75, 82, 83, 213, 217, 231, 905, 1077, 1095, 1096
– overestimate 1096
reliability analysis 215
reliability coefficient 83–93, 207
– coefficient alpha 87
– composite scores 91
– difference scores 90
– internal consistency 86
– interrater 90
– parallel or equivalent forms 85
– split-half 86
– stratified coefficient alpha 92
– test–retest 84
reliability, *see* skills diagnosis, reliability
remediation 1118
repeated measures 401, 406, 409, 960
replication 773, 1095
reporting of scores 1113
requirements for equated scores 172–174
resampling 497
rescaled statistic 310, 317, 327, 332, 333, 335, 336, 338, 343–345
research synthesis 20, 920
residual 207, 498, 499, 690, 698, 699, 876
residual analysis 696, 697
residual-based statistic 313, 314, 339, 343–346
residualized gain score 877
residualized gains 877
residualized post-test 876
residualized pre-test 876
respondent scores 499
response formats 1113
response function 500
response model 801, 802, 807
– 1-PL 478
– 2-PL 475
– 3-PL 477
– bifactor 507
– generalized partial credit 488

– graded categories 479
– multidimensional 507
– nominal categories 480
– normal ogive model 473
– partial credit 487
– ranking 489
– rating scale 489
response modes
– nominal multiple-choice 486
response pattern 471
response time 819, 823, 835
response variable 517
responses at different levels 415
restrictions
– on parameters 556
restscore method 736
restscore-splitting method 737
revealed choice 466
revealed choice data 447
revealed preference data 447
ridge regression 214
risk factors 774
risky mode 459
robust procedure 305, 314, 315, 318–320, 322, 333, 346, 347
robustness
– of estimator 545
rotation 15–17, 633, 635, 636
rotational indeterminacy 263, 272
row exchangeable 774, 788
row regression 244
Rubin's causal model 769, 775
Rubin's model 911, 916
Rubin's model for causal inference 894

sampling error 1046
sampling properties 496
sampling weights 425, 437
SAS GLIMMIX 763, 764
SAS NLMIXED 762
SAT 39, 40, 901, 925, 933, 934, 1057
SAT Reasoning Test™ 1057
SAT score 898, 907, 908
SAT-Math 895, 898
satiation effect 561
scale
– of items 576
– transformation
– – linear 525
scale aligning 170–174
scale alignment 1100
scale properties
– of RM 522
scale scores 499, 955, 959
scale transformation
– increasing 522
– isotonic 523
– linear 529
– monotone 529
scale transformations
– linear 526
scaled scores 1074
scaling 955, 1074, 1095, 1097
scaling test 956
scatterplot 884
Scholastic Aptitude test 6, 925
school characteristics 877
school choice 795
school intercepts 877
Schwarz BIC 427
science 787
Science Research Associates 8
score comparability 1108, 1109
score distributions 906
score linking 1099, 1102
Score Report Plus 993
score scale 1074
score trajectories 867
scoring 365
scoring rubrics 1114
scoring rules 849
scree test 280
second-order margins 499
security issues 1107
seemingly unrelated regressions 1049
selection bias 872
SEM, *see also* structural equation modeling 17, 18
SEM software 303
semiparametric 555
sensitive 793
sensitivity 879, 886
sensitivity analysis 793, 932, 933, 936, 937, 940, 941, 951
sensitivity of inference to prior assumptions 794
sensitivity review 162, 163
separability of classes of parameters 516
separate 961, 963
separation of parameters 516, 539
sequential Bayesian methods 817
sequential experiments 780
sequential importance sampling (SIS) algorithm 560
SES 877

shadow test 823, 825, 830, 833
shadow-test approach 823, 827
shadow-test model 824, 833
sharp null hypotheses 782
Shepard diagram 368
SIBTEST procedure 134, 143, 147
signed area 129
signed compensatory distance 391
significance level 781
similarity 363, 364
simple gain score 875
simple gain score model 872, 885
simple structure 281
simple two-stage VAM 887
Simpson's index 210
Simpson's Paradox 793, 893–895, 898–900, 917
simulation 886
simultaneous equation model 299
simultaneous SVD 242
single group (SG) design 175, 176, 179, 180
single-case approach 556
single-peaked IRFs 561
singular value decomposition 219, 222, 224
skill profile models 651
skills assessment 739
skills diagnosis 979, 1027
– ability model 980, 982, 983, 989, 996–998, 1000–1002, 1008–1018, 1021
– – cognitive attributes 980–984, 997, 1002
– – cognitive operations 1002
– – latent class 996, 997, 1008–1011, 1013, 1014, 1017
– – latent trait 990, 997, 1003, 1005, 1007, 1009, 1012, 1016
– – mastery 984, 986, 988–992, 994, 995, 998, 999, 1001–1009, 1011, 1013, 1015–1020, 1022, 1025
– estimation 979, 980, 985–991, 993, 995, 999, 1000, 1011, 1018, 1020, 1023–1025
– – Bayes 985–987, 989, 990, 1011
– – CML 986, 1000
– – convergence 986, 987, 1023, 1024
– – correlation, effect of 984, 1025
– – EAP 986
– – EM 986, 987, 1000, 1023
– – identifiability 985, 987, 990, 994, 1010, 1012, 1014, 1025
– – JMLE 986, 987
– – MAP 986
– – MCMC 986, 987, 989, 1023, 1024
– – MMLE 986
– – posterior 986, 987, 989, 990, 1023, 1024
– – prior 985, 986, 990
– model fit 987–990, 994, 1016, 1020, 1024
– – posterior predictive check 987, 989
– models
– – Bayes nets 990
– – compensatory 983–985, 990, 996, 997, 1000, 1003, 1004, 1010, 1014–1021
– – completeness 998, 1002–1007, 1009, 1012, 1013, 1015, 1017, 1018, 1020–1022
– – conjunctive 983–985, 991, 998, 1000, 1004, 1005, 1008, 1010, 1013, 1015, 1018–1021
– – deterministic 996–999, 1008–1010, 1017, 1018
– – DINA 986, 990, 1008–1010
– – DINO 997, 1010, 1014–1017, 1019, 1022
– – disjunctive 983, 997, 1010, 1013, 1014, 1018–1020
– – GDM 987, 1000
– – GLTM 997, 1007, 1008, 1011, 1015, 1016, 1019, 1020, 1022
– – HYBRID 997, 1009–1011, 1015–1017, 1019, 1021, 1022
– – identifiability 985, 987, 990, 994, 1010, 1012, 1014, 1025
– – item parameters 985–988, 990, 995, 1000–1007, 1010–1014, 1020
– – item structure 979, 996, 997, 999, 1001, 1016, 1018–1021
– – item-knowledge interaction 979, 996–1018, 1024
– – knowledge structure 979, 996, 997, 999, 1001, 1015–1018
– – LLTM 996, 997, 1000–1003, 1007, 1008, 1015, 1016, 1019, 1020, 1022
– – MCLCM 997, 1000, 1013–1017, 1019, 1020, 1022
– – MIRT 996, 997, 1000, 1003–1006, 1011, 1013–1016, 1019, 1020, 1022
– – MLTM 997, 1000, 1005–1008, 1011–1013, 1015, 1016, 1019–1022
– – MRCMLM 1000
– – NIDA 986
– – noncompensatory 997, 1000, 1004, 1005, 1008, 1015, 1018–1020, 1022
– – parsimony 994, 1008, 1010, 1024
– – positivity 999, 1002–1004, 1006, 1008–1012, 1017, 1018, 1020

– – Rasch model 996, 1000, 1002, 1003, 1022
– – RLCM 997, 1008–1013, 1015–1017, 1019, 1022
– – RUM 989, 997, 1010–1013, 1015–1017, 1019, 1020, 1022
– – slips 999, 1002, 1004–1006, 1009, 1020
– – strategy 983, 998–1022
– person parameter 1000
– Q matrix 984, 987, 988, 990, 991, 994, 996, 998–1022, 1024, 1025
– reliability 983, 990, 991, 993–995, 1024, 1026
– scoring 980, 991, 993, 994, 1000, 1025, 1026
– – classification 980, 981, 985, 986, 989–991, 993
– – mastery 985, 989, 991, 995, 998, 1016–1018, 1025
– – profile scores 1025
– – subscores 993
– skills space 988
– models 1000
slope parameter 475
small area 963
smoking 772
solution strategies 568
sorting
– method 364
sparseness 498
Spearman–Brown formula 34
specific objectivity 529
specific objectivity (SO) 519, 527–529, 531, 549, 576
spectral decomposition 239
spectral norm of a matrix 243
speed 819, 831
speed parameter 819, 828
speededness model 656
speededness parameter 831
spider-web phenomenon 547
squared multiple correlation 268
stable unit treatment value assumption (SUTVA) 773, 776, 794
standard error of measurement 31, 1095
– conditional 35
standard errors 496, 497
– ability estimates 1096
standard errors of measurement (SEM) 1078
standard frequency index 210
standard setting 959, 960
standard-setting procedures 1114
standardization 899, 900, 917
standardization approach 1071
standardization mix 899
standardized mean difference 919–921, 931
standardized mean difference index 146
standardized p-difference index 133
standardized rate 900
standardized residual 498, 499, 698, 699, 707
standardized tests 873
standards 956
Stanford Achievement Tests 6, 959
state standards
– student performance on 1118
stated choice data 447
stated preference 466
stated preference data 447
statistic 781
statistical description 869
statistical inference 769
statistical model 870
statistical randomization 870
statistical specifications 1061
Step 1 scores 911
stepwise regression 214
stimulus (item) feature 747, 750, 751, 753, 756
stochastic EM 1049
stochastic ordering
– latent variable by total score 724
stochastic proof by contradiction 781
strategy differences 651, 655
Strivers 903, 906, 910
strong components of digraph 566
strong connectedness 534
strongly ignorable assignment mechanisms 779, 780
strongly ignorable treatment assignment 784
structural equation model (modeling) 45, 46, 52, 64, 65, 67, 74
structural equation modeling 17, 18, 442, 614
structural equation models 401, 417
structural incompleteness 574
structural modeling 590
structure of test content 631
student covariates 884
student model 847, 848, 856
student-monitoring system 1114
students' demographic characteristics 870
students' test score histories 870
subclasses of propensity scores 785
subclassification 785, 786
subtests
– multidimensional 572
subtractivity 525, 526, 529
subtractivity of parameters 523

sufficiency 524, 531
– of person and item marginal sums 518
sufficiency of the raw score 561
sufficiency principle 505
sufficient statistic 492, 500, 505, 523, 565, 613, 614, 647, 664, 674, 676, 677
– nontrivial 526
summation method 537
summer slumps 958
SVD: singular value decomposition 239, 240
switching algorithm 560
symmetry 106
Sympson–Hetter experiment 826
Sympson–Hetter item-exposure control 826
symptoms
– of depression 574
– patterns of 574
synthetic effect size 924, 927, 934, 944, 952

T-class mixture model 443
tables
– significance of change 558
target population 1060
task model 847, 849
tasks 1114
tau equivalent tests 83
tau-equivalent model 35
teacher characteristics 868
teacher effects 872, 873
teacher evaluation 868
teacher quality 868
teaching to the test 873
technique of fading 827
template for inference for causal effects from observational data 786
Terman group test of mental ability 6
TerraNova series of tests 959
test assembly 1113
test blueprints 956
test characteristic curves 967
test characteristic function 831
test characteristic surface 625, 626
test development 747, 1113
test information 626
test information function 815
test length
– virtual 572
test response function 723
test score distributions 873
test specifications 1060
test speededness 1072
test theory 10, 11, 13, 18, 21
TESTFACT 609, 630, 634–636, 688
TestGraf
– computer program 732
testlet 503, 597
tests of unconfoundedness 793
tetrachoric correlation 31, 39, 41
tfidf 223, 224
The Triumph of Mediocrity in Business 903
theory of errors 4
three-parameter logistic (3PL) model 501, 589, 598, 601, 616, 725, 802, 808, 815, 816, 818, 853, 859, 964
three-parameter model 684, 692
three-parameter normal ogive (3PNO) model 803, 808, 812, 817
Thurstone scales 959
Thurstone scaling 956
time intensity 819, 822, 828
Toulmin diagram 841, 842
training 868
transformation, *see* scale transformation 522, 634, 636
transition matrix 886
transitivity violations 448
translation 635
treatment 921–923, 934, 946
– effects 571
– groups 571
treatment condition 912
treatment effect 559, 573, 575
treatment group 914
treatments 571, 769, 770
trend effect 573
true ability 906
true rating 665
true score 29, 31, 33–35, 905, 1095, 1096
true score theory 10, 11
true-score equating 188–190, 195, 197, 831
truncation of outcomes due to death 795
Tucker's problem 249
two parameter partial credit model 965
two-parameter 959
two-parameter logistic (2PL) model (or Birnbaum model) 505, 524, 577, 673, 720, 725
two-parameter model 684, 692
two-stage covariance model 868
two-way analysis of variance 213

UMPU test
– about person parameter 548
unbiased estimator 782
unconditional maximum likelihood 533
unconditional means 914
unconfounded assignment 779
unconfounded assignment mechanism 779
unconfounded probabilistic assignment mechanisms 780
uncontrolled causal studies 914
underdispersed 877
unequal probability sampling algorithm 560
unfolding analysis 360, 385, 387, 388, 390–392, 397
unfolding models 739
unidimensional IRT 609, 972
unidimensional model 628
unidimensional scale 627
unidimensionality 576, 686–689, 705
– LLTM 571
unidimensionality assumption 555, 607
unidimensionality of items 550, 551
unidirectional DIF 740
unified analyses 961
uniform DIF 128, 140
uniformly most powerful unbiased (UMPU) tests 556, 558–560
uninformative data 575
unintended missing data 794
unique factors 261
unique variances 261
uniqueness 262, 506
– of CML estimator 567
– of CML estimator (LLRA) 576
– of CML estimator (RM) 537, 566
– of JML estimator (RM) 534
– of ML estimator (LBTL) 567
– of scales 518
uniqueness of CML estimator 557
– of change parameter 557
unit level assignment probabilities 784
United States Medical Licensing Exam (USMLE) 894
United States Medical Licensing Examination 893
units 770
universe of items 525
– dense 525
universe score 1096
unknown propensity scores 785
unobserved intermediate potential outcomes 795
unsigned area 130
unstandardized residuals 498, 499
untestable assumption 916
USMLE 893, 895, 910
USMLE Step 1 910, 911
utility maximizers 448
utility models 449

vaccination 795
valid inferences 956
valid interpretations 955
validity 30, 36, 39, 45–76, 213, 217, 231, 750, 764, 956, 961, 987–1026, 1083, 1111
– construct 48, 50, 52, 55, 59, 60, 765
– content 63, 72
– external 992, 993
– internal 991, 992
– test 747
value added assessment 773
value-added modeling (VAM) 868, 869, 871, 874, 876, 880, 886, 888, 889
variable persistence model 879, 885
variance 783
variance components 213
variance components model 403, 405
variance estimation 1045
variance partition coefficient 403, 405
varimax 16
varimax criterion 282
varimax rotation 634
vector plot 623, 624
vector preference model 391
vertical equating 1114
vertical scale 958, 973
vertical scaling 632, 873, 955, 956, 958, 960, 961, 972
vertically scaled 874, 956
vertices 534
virtual
– item parameters 574
– items 569, 572
– person parameters 574
– persons 575
– test length 572
virtual (V-) persons 575
volatility of estimated effects 887
Von Neumann theorem 245

Wald-type tests 549, 555, 556
Wall Street Journal 901
Ward's method 632
Warm estimator (WLE) 547
weak local independence 721
weak monotonicity 723
weight matrix 574

weighted likelihood estimator (WLE) 811
weighted MLE 817
well-conditioned data 534, 567
well-conditioned item score matrices 566
Wellesley 908
white 898
WINMIRA 2001 653
within school variance 591

WLE 811
WML estimates 814
WML estimation 813
WML estimator 814
worldwide standards 1114

X^2 Pearson 497
XGvis 362, 396

Handbook of Statistics
Contents of Previous Volumes

Volume 1. Analysis of Variance
Edited by P.R. Krishnaiah
1980 xviii + 1002 pp.

1. Estimation of Variance Components by C.R. Rao and J. Kleffe
2. Multivariate Analysis of Variance of Repeated Measurements by N.H. Timm
3. Growth Curve Analysis by S. Geisser
4. Bayesian Inference in MANOVA by S.J. Press
5. Graphical Methods for Internal Comparisons in ANOVA and MANOVA by R. Gnanadesikan
6. Monotonicity and Unbiasedness Properties of ANOVA and MANOVA Tests by S. Das Gupta
7. Robustness of ANOVA and MANOVA Test Procedures by P.K. Ito
8. Analysis of Variance and Problems under Time Series Models by D.R. Brillinger
9. Tests of Univariate and Multivariate Normality by K.V. Mardia
10. Transformations to Normality by G. Kaskey, B. Kolman, P.R. Krishnaiah and L. Steinberg
11. ANOVA and MANOVA: Models for Categorical Data by V.P. Bhapkar
12. Inference and the Structural Model for ANOVA and MANOVA by D.A.S. Fraser
13. Inference Based on Conditionally Specified ANOVA Models Incorporating Preliminary Testing by T.A. Bancroft and C.-P. Han
14. Quadratic Forms in Normal Variables by C.G. Khatri
15. Generalized Inverse of Matrices and Applications to Linear Models by S.K. Mitra
16. Likelihood Ratio Tests for Mean Vectors and Covariance Matrices by P.R. Krishnaiah and J.C. Lee
17. Assessing Dimensionality in Multivariate Regression by A.J. Izenman
18. Parameter Estimation in Nonlinear Regression Models by H. Bunke
19. Early History of Multiple Comparison Tests by H.L. Harter
20. Representations of Simultaneous Pairwise Comparisons by A.R. Sampson
21. Simultaneous Test Procedures for Mean Vectors and Covariance Matrices by P.R. Krishnaiah, G.S. Mudholkar and P. Subbaiah
22. Nonparametric Simultaneous Inference for Some MANOVA Models by P.K. Sen

23. Comparison of Some Computer Programs for Univariate and Multivariate Analysis of Variance by R.D. Bock and D. Brandt
24. Computations of Some Multivariate Distributions by P.R. Krishnaiah
25. Inference on the Structure of Interaction Two-Way Classification Model by P.R. Krishnaiah and M. Yochmowitz

Volume 2. Classification, Pattern Recognition and Reduction of Dimensionality
Edited by P.R. Krishnaiah and L.N. Kanal
1982 xxii + 903 pp.

1. Discriminant Analysis for Time Series by R.H. Shumway
2. Optimum Rules for Classification into Two Multivariate Normal Populations with the Same Covariance Matrix by S. Das Gupta
3. Large Sample Approximations and Asymptotic Expansions of Classification Statistics by M. Siotani
4. Bayesian Discrimination by S. Geisser
5. Classification of Growth Curves by J.C. Lee
6. Nonparametric Classification by J.D. Broffitt
7. Logistic Discrimination by J.A. Anderson
8. Nearest Neighbor Methods in Discrimination by L. Devroye and T.J. Wagner
9. The Classification and Mixture Maximum Likelihood Approaches to Cluster Analysis by G.J. McLachlan
10. Graphical Techniques for Multivariate Data and for Clustering by J.M. Chambers and B. Kleiner
11. Cluster Analysis Software by R.K. Blashfield, M.S. Aldenderfer and L.C. Morey
12. Single-link Clustering Algorithms by F.J. Rohlf
13. Theory of Multidimensional Scaling by J. de Leeuw and W. Heiser
14. Multidimensional Scaling and its Application by M. Wish and J.D. Carroll
15. Intrinsic Dimensionality Extraction by K. Fukunaga
16. Structural Methods in Image Analysis and Recognition by L.N. Kanal, B.A. Lambird and D. Lavine
17. Image Models by N. Ahuja and A. Rosenfield
18. Image Texture Survey by R.M. Haralick
19. Applications of Stochastic Languages by K.S. Fu
20. A Unifying Viewpoint on Pattern Recognition by J.C. Simon, E. Backer and J. Sallentin
21. Logical Functions in the Problems of Empirical Prediction by G.S. Lbov
22. Inference and Data Tables and Missing Values by N.G. Zagoruiko and V.N. Yolkina
23. Recognition of Electrocardiographic Patterns by J.H. van Bemmel
24. Waveform Parsing Systems by G.C. Stockman

25. Continuous Speech Recognition: Statistical Methods by F. Jelinek, R.L. Mercer and L.R. Bahl
26. Applications of Pattern Recognition in Radar by A.A. Grometstein and W.H. Schoendorf
27. White Blood Cell Recognition by F.S. Gelsema and G.H. Landweerd
28. Pattern Recognition Techniques for Remote Sensing Applications by P.H. Swain
29. Optical Character Recognition – Theory and Practice by G. Nagy
30. Computer and Statistical Considerations for Oil Spill Identification by Y.T. Chien and T.J. Killeen
31. Pattern Recognition in Chemistry by B.R. Kowalski and S. Wold
32. Covariance Matrix Representation and Object-Predicate Symmetry by T. Kaminuma, S. Tomita and S. Watanabe
33. Multivariate Morphometrics by R.A. Reyment
34. Multivariate Analysis with Latent Variables by P.M. Bentler and D.G. Weeks
35. Use of Distance Measures, Information Measures and Error Bounds in Feature Evaluation by M. Ben-Bassat
36. Topics in Measurement Selection by J.M. Van Campenhout
37. Selection of Variables Under Univariate Regression Models by P.R. Krishnaiah
38. On the Selection of Variables Under Regression Models Using Krishnaiah's Finite Intersection Tests by J.L. Schmidhammer
39. Dimensionality and Sample Size Considerations in Pattern Recognition Practice by A.K. Jain and B. Chandrasekaran
40. Selecting Variables in Discriminant Analysis for Improving upon Classical Procedures by W. Schaafsma
41. Selection of Variables in Discriminant Analysis by P.R. Krishnaiah

Volume 3. Time Series in the Frequency Domain
Edited by D.R. Brillinger and P.R. Krishnaiah
1983 xiv + 485 pp.

1. Wiener Filtering (with emphasis on frequency-domain approaches) by R.J. Bhansali and D. Karavellas
2. The Finite Fourier Transform of a Stationary Process by D.R. Brillinger
3. Seasonal and Calendar Adjustment by W.S. Cleveland
4. Optimal Inference in the Frequency Domain by R.B. Davies
5. Applications of Spectral Analysis in Econometrics by C.W.J. Granger and R. Engle
6. Signal Estimation by E.J. Hannan
7. Complex Demodulation: Some Theory and Applications by T. Hasan
8. Estimating the Gain of a Linear Filter from Noisy Data by M.J. Hinich
9. A Spectral Analysis Primer by L.H. Koopmans
10. Robust-Resistant Spectral Analysis by R.D. Martin
11. Autoregressive Spectral Estimation by E. Parzen

12. Threshold Autoregression and Some Frequency-Domain Characteristics by J. Pemberton and H. Tong
13. The Frequency-Domain Approach to the Analysis of Closed-Loop Systems by M.B. Priestley
14. The Bispectral Analysis of Nonlinear Stationary Time Series with Reference to Bilinear Time-Series Models by T. Subba Rao
15. Frequency-Domain Analysis of Multidimensional Time-Series Data by E.A. Robinson
16. Review of Various Approaches to Power Spectrum Estimation by P.M. Robinson
17. Cumulants and Cumulant Spectra by M. Rosenblatt
18. Replicated Time-Series Regression: An Approach to Signal Estimation and Detection by R.H. Shumway
19. Computer Programming of Spectrum Estimation by T. Thrall
20. Likelihood Ratio Tests on Covariance Matrices and Mean Vectors of Complex Multivariate Normal Populations and their Applications in Time Series by P.R. Krishnaiah, J.C. Lee and T.C. Chang

Volume 4. Nonparametric Methods
Edited by P.R. Krishnaiah and P.K. Sen
1984 xx + 968 pp.

1. Randomization Procedures by C.B. Bell and P.K. Sen
2. Univariate and Multivariate Multisample Location and Scale Tests by V.P. Bhapkar
3. Hypothesis of Symmetry by M. Hušková
4. Measures of Dependence by K. Joag-Dev
5. Tests of Randomness against Trend or Serial Correlations by G.K. Bhattacharyya
6. Combination of Independent Tests by J.L. Folks
7. Combinatorics by L. Takács
8. Rank Statistics and Limit Theorems by M. Ghosh
9. Asymptotic Comparison of Tests – A Review by K. Singh
10. Nonparametric Methods in Two-Way Layouts by D. Quade
11. Rank Tests in Linear Models by J.N. Adichie
12. On the Use of Rank Tests and Estimates in the Linear Model by J.C. Aubuchon and T.P. Hettmansperger
13. Nonparametric Preliminary Test Inference by A.K.Md.E. Saleh and P.K. Sen
14. Paired Comparisons: Some Basic Procedures and Examples by R.A. Bradley
15. Restricted Alternatives by S.K. Chatterjee
16. Adaptive Methods by M. Hušková
17. Order Statistics by J. Galambos
18. Induced Order Statistics: Theory and Applications by P.K. Bhattacharya
19. Empirical Distribution Function by F. Csáki
20. Invariance Principles for Empirical Processes by M. Csörgő

21. M-, L- and R-estimators by J. Jurečková
22. Nonparametric Sequential Estimation by P.K. Sen
23. Stochastic Approximation by V. Dupač
24. Density Estimation by P. Révész
25. Censored Data by A.P. Basu
26. Tests for Exponentiality by K.A. Doksum and B.S. Yandell
27. Nonparametric Concepts and Methods in Reliability by M. Hollander and F. Proschan
28. Sequential Nonparametric Tests by U. Müller-Funk
29. Nonparametric Procedures for some Miscellaneous Problems by P.K. Sen
30. Minimum Distance Procedures by R. Beran
31. Nonparametric Methods in Directional Data Analysis by S.R. Jammalamadaka
32. Application of Nonparametric Statistics to Cancer Data by H.S. Wieand
33. Nonparametric Frequentist Proposals for Monitoring Comparative Survival Studies by M. Gail
34. Meteorological Applications of Permutation Techniques Based on Distance Functions by P.W. Mielke Jr
35. Categorical Data Problems Using Information Theoretic Approach by S. Kullback and J.C. Keegel
36. Tables for Order Statistics by P.R. Krishnaiah and P.K. Sen
37. Selected Tables for Nonparametric Statistics by P.K. Sen and P.R. Krishnaiah

Volume 5. Time Series in the Time Domain
Edited by E.J. Hannan, P.R. Krishnaiah and M.M. Rao
1985 xiv + 490 pp.

1. Nonstationary Autoregressive Time Series by W.A. Fuller
2. Non-Linear Time Series Models and Dynamical Systems by T. Ozaki
3. Autoregressive Moving Average Models, Intervention Problems and Outlier Detection in Time Series by G.C. Tiao
4. Robustness in Time Series and Estimating ARMA Models by R.D. Martin and V.J. Yohai
5. Time Series Analysis with Unequally Spaced Data by R.H. Jones
6. Various Model Selection Techniques in Time Series Analysis by R. Shibata
7. Estimation of Parameters in Dynamical Systems by L. Ljung
8. Recursive Identification, Estimation and Control by P. Young
9. General Structure and Parametrization of ARMA and State-Space Systems and its Relation to Statistical Problems by M. Deistler
10. Harmonizable, Cramér, and Karhunen Classes of Processes by M.M. Rao
11. On Non-Stationary Time Series by C.S.K. Bhagavan
12. Harmonizable Filtering and Sampling of Time Series by D.K. Chang
13. Sampling Designs for Time Series by S. Cambanis

14. Measuring Attenuation by M.A. Cameron and P.J. Thomson
15. Speech Recognition Using LPC Distance Measures by P.J. Thomson and P. de Souza
16. Varying Coefficient Regression by D.F. Nicholls and A.R. Pagan
17. Small Samples and Large Equations Systems by H. Theil and D.G. Fiebig

Volume 6. Sampling
Edited by P.R. Krishnaiah and C.R. Rao
1988 xvi + 594 pp.

1. A Brief History of Random Sampling Methods by D.R. Bellhouse
2. A First Course in Survey Sampling by T. Dalenius
3. Optimality of Sampling Strategies by A. Chaudhuri
4. Simple Random Sampling by P.K. Pathak
5. On Single Stage Unequal Probability Sampling by V.P. Godambe and M.E. Thompson
6. Systematic Sampling by D.R. Bellhouse
7. Systematic Sampling with Illustrative Examples by M.N. Murthy and T.J. Rao
8. Sampling in Time by D.A. Binder and M.A. Hidiroglou
9. Bayesian Inference in Finite Populations by W.A. Ericson
10. Inference Based on Data from Complex Sample Designs by G. Nathan
11. Inference for Finite Population Quantiles by J. Sedransk and P.J. Smith
12. Asymptotics in Finite Population Sampling by P.K. Sen
13. The Technique of Replicated or Interpenetrating Samples by J.C. Koop
14. On the Use of Models in Sampling from Finite Populations by I. Thomsen and D. Tesfu
15. The Prediction Approach to Sampling Theory by R.M. Royall
16. Sample Survey Analysis: Analysis of Variance and Contingency Tables by D.H. Freeman Jr
17. Variance Estimation in Sample Surveys by J.N.K. Rao
18. Ratio and Regression Estimators by P.S.R.S. Rao
19. Role and Use of Composite Sampling and Capture-Recapture Sampling in Ecological Studies by M.T. Boswell, K.P. Burnham and G.P. Patil
20. Data-based Sampling and Model-based Estimation for Environmental Resources by G.P. Patil, G.J. Babu, R.C. Hennemuth, W.L. Meyers, M.B. Rajarshi and C. Taillie
21. On Transect Sampling to Assess Wildlife Populations and Marine Resources by F.L. Ramsey, C.E. Gates, G.P. Patil and C. Taillie
22. A Review of Current Survey Sampling Methods in Marketing Research (Telephone, Mall Intercept and Panel Surveys) by R. Velu and G.M. Naidu
23. Observational Errors in Behavioural Traits of Man and their Implications for Genetics by P.V. Sukhatme
24. Designs in Survey Sampling Avoiding Contiguous Units by A.S. Hedayat, C.R. Rao and J. Stufken

Volume 7. Quality Control and Reliability
Edited by P.R. Krishnaiah and C.R. Rao
1988 xiv + 503 pp.

1. Transformation of Western Style of Management by W. Edwards Deming
2. Software Reliability by F.B. Bastani and C.V. Ramamoorthy
3. Stress–Strength Models for Reliability by R.A. Johnson
4. Approximate Computation of Power Generating System Reliability Indexes by M. Mazumdar
5. Software Reliability Models by T.A. Mazzuchi and N.D. Singpurwalla
6. Dependence Notions in Reliability Theory by N.R. Chaganty and K. Joagdev
7. Application of Goodness-of-Fit Tests in Reliability by B.W. Woodruff and A.H. Moore
8. Multivariate Nonparametric Classes in Reliability by H.W. Block and T.H. Savits
9. Selection and Ranking Procedures in Reliability Models by S.S. Gupta and S. Panchapakesan
10. The Impact of Reliability Theory on Some Branches of Mathematics and Statistics by P.J. Boland and F. Proschan
11. Reliability Ideas and Applications in Economics and Social Sciences by M.C. Bhattacharjee
12. Mean Residual Life: Theory and Applications by F. Guess and F. Proschan
13. Life Distribution Models and Incomplete Data by R.E. Barlow and F. Proschan
14. Piecewise Geometric Estimation of a Survival Function by G.M. Mimmack and F. Proschan
15. Applications of Pattern Recognition in Failure Diagnosis and Quality Control by L.F. Pau
16. Nonparametric Estimation of Density and Hazard Rate Functions when Samples are Censored by W.J. Padgett
17. Multivariate Process Control by F.B. Alt and N.D. Smith
18. QMP/USP – A Modern Approach to Statistical Quality Auditing by B. Hoadley
19. Review About Estimation of Change Points by P.R. Krishnaiah and B.Q. Miao
20. Nonparametric Methods for Changepoint Problems by M. Csörgő and L. Horváth
21. Optimal Allocation of Multistate Components by E. El-Neweihi, F. Proschan and J. Sethuraman
22. Weibull, Log-Weibull and Gamma Order Statistics by H.L. Herter
23. Multivariate Exponential Distributions and their Applications in Reliability by A.P. Basu
24. Recent Developments in the Inverse Gaussian Distribution by S. Iyengar and G. Patwardhan

Volume 8. Statistical Methods in Biological and Medical Sciences
Edited by C.R. Rao and R. Chakraborty
1991 xvi + 554 pp.

1. Methods for the Inheritance of Qualitative Traits by J. Rice, R. Neuman and S.O. Moldin
2. Ascertainment Biases and their Resolution in Biological Surveys by W.J. Ewens
3. Statistical Considerations in Applications of Path Analytical in Genetic Epidemiology by D.C. Rao
4. Statistical Methods for Linkage Analysis by G.M. Lathrop and J.M. Lalouel
5. Statistical Design and Analysis of Epidemiologic Studies: Some Directions of Current Research by N. Breslow
6. Robust Classification Procedures and their Applications to Anthropometry by N. Balakrishnan and R.S. Ambagaspitiya
7. Analysis of Population Structure: A Comparative Analysis of Different Estimators of Wright's Fixation Indices by R. Chakraborty and H. Danker-Hopfe
8. Estimation of Relationships from Genetic Data by E.A. Thompson
9. Measurement of Genetic Variation for Evolutionary Studies by R. Chakraborty and C.R. Rao
10. Statistical Methods for Phylogenetic Tree Reconstruction by N. Saitou
11. Statistical Models for Sex-Ratio Evolution by S. Lessard
12. Stochastic Models of Carcinogenesis by S.H. Moolgavkar
13. An Application of Score Methodology: Confidence Intervals and Tests of Fit for One-Hit-Curves by J.J. Gart
14. Kidney-Survival Analysis of IgA Nephropathy Patients: A Case Study by O.J.W.F. Kardaun
15. Confidence Bands and the Relation with Decision Analysis: Theory by O.J.W.F. Kardaun
16. Sample Size Determination in Clinical Research by J. Bock and H. Toutenburg

Volume 9. Computational Statistics
Edited by C.R. Rao
1993 xix + 1045 pp.

1. Algorithms by B. Kalyanasundaram
2. Steady State Analysis of Stochastic Systems by K. Kant
3. Parallel Computer Architectures by R. Krishnamurti and B. Narahari
4. Database Systems by S. Lanka and S. Pal
5. Programming Languages and Systems by S. Purushothaman and J. Seaman
6. Algorithms and Complexity for Markov Processes by R. Varadarajan
7. Mathematical Programming: A Computational Perspective by W.W. Hager, R. Horst and P.M. Pardalos

8. Integer Programming by P.M. Pardalos and Y. Li
9. Numerical Aspects of Solving Linear Least Squares Problems by J.L. Barlow
10. The Total Least Squares Problem by S. van Huffel and H. Zha
11. Construction of Reliable Maximum-Likelihood-Algorithms with Applications to Logistic and Cox Regression by D. Böhning
12. Nonparametric Function Estimation by T. Gasser, J. Engel and B. Seifert
13. Computation Using the OR Decomposition by C.R. Goodall
14. The EM Algorithm by N. Laird
15. Analysis of Ordered Categorial Data through Appropriate Scaling by C.R. Rao and P.M. Caligiuri
16. Statistical Applications of Artificial Intelligence by W.A. Gale, D.J. Hand and A.E. Kelly
17. Some Aspects of Natural Language Processes by A.K. Joshi
18. Gibbs Sampling by S.F. Arnold
19. Bootstrap Methodology by G.J. Babu and C.R. Rao
20. The Art of Computer Generation of Random Variables by M.T. Boswell, S.D. Gore, G.P. Patil and C. Taillie
21. Jackknife Variance Estimation and Bias Reduction by S. Das Peddada
22. Designing Effective Statistical Graphs by D.A. Burn
23. Graphical Methods for Linear Models by A.S. Hadi
24. Graphics for Time Series Analysis by H.J. Newton
25. Graphics as Visual Language by T. Selkar and A. Appel
26. Statistical Graphics and Visualization by E.J. Wegman and D.B. Carr
27. Multivariate Statistical Visualization by F.W. Young, R.A. Faldowski and M.M. McFarlane
28. Graphical Methods for Process Control by T.L. Ziemer

Volume 10. Signal Processing and its Applications
Edited by N.K. Bose and C.R. Rao
1993 xvii + 992 pp.

1. Signal Processing for Linear Instrumental Systems with Noise: A General Theory with Illustrations from Optical Imaging and Light Scattering Problems by M. Bertero and E.R. Pike
2. Boundary Implication Results in Parameter Space by N.K. Bose
3. Sampling of Bandlimited Signals: Fundamental Results and Some Extensions by J.L. Brown Jr
4. Localization of Sources in a Sector: Algorithms and Statistical Analysis by K. Buckley and X.-L. Xu
5. The Signal Subspace Direction-of-Arrival Algorithm by J.A. Cadzow
6. Digital Differentiators by S.C. Dutta Roy and B. Kumar
7. Orthogonal Decompositions of 2D Random Fields and their Applications for 2D Spectral Estimation by J.M. Francos

8. VLSI in Signal Processing by A. Ghouse
9. Constrained Beamforming and Adaptive Algorithms by L.C. Godara
10. Bispectral Speckle Interferometry to Reconstruct Extended Objects from Turbulence-Degraded Telescope Images by D.M. Goodman, T.W. Lawrence, E.M. Johansson and J.P. Fitch
11. Multi-Dimensional Signal Processing by K. Hirano and T. Nomura
12. On the Assessment of Visual Communication by F.O. Huck, C.L. Fales, R. Alter-Gartenberg and Z. Rahman
13. VLSI Implementations of Number Theoretic Concepts with Applications in Signal Processing by G.A. Jullien, N.M. Wigley and J. Reilly
14. Decision-level Neural Net Sensor Fusion by R.Y. Levine and T.S. Khuon
15. Statistical Algorithms for Noncausal Gauss Markov Fields by J.M.F. Moura and N. Balram
16. Subspace Methods for Directions-of-Arrival Estimation by A. Paulraj, B. Ottersten, R. Roy, A. Swindlehurst, G. Xu and T. Kailath
17. Closed Form Solution to the Estimates of Directions of Arrival Using Data from an Array of Sensors by C.R. Rao and B. Zhou
18. High-Resolution Direction Finding by S.V. Schell and W.A. Gardner
19. Multiscale Signal Processing Techniques: A Review by A.H. Tewfik, M. Kim and M. Deriche
20. Sampling Theorems and Wavelets by G.G. Walter
21. Image and Video Coding Research by J.W. Woods
22. Fast Algorithms for Structured Matrices in Signal Processing by A.E. Yagle

Volume 11. Econometrics
Edited by G.S. Maddala, C.R. Rao and H.D. Vinod
1993 xx + 783 pp.

1. Estimation from Endogenously Stratified Samples by S.R. Cosslett
2. Semiparametric and Nonparametric Estimation of Quantal Response Models by J.L. Horowitz
3. The Selection Problem in Econometrics and Statistics by C.F. Manski
4. General Nonparametric Regression Estimation and Testing in Econometrics by A. Ullah and H.D. Vinod
5. Simultaneous Microeconometric Models with Censored or Qualitative Dependent Variables by R. Blundell and R.J. Smith
6. Multivariate Tobit Models in Econometrics by L.-F. Lee
7. Estimation of Limited Dependent Variable Models under Rational Expectations by G.S. Maddala
8. Nonlinear Time Series and Macroeconometrics by W.A. Brock and S.M. Potter
9. Estimation, Inference and Forecasting of Time Series Subject to Changes in Time by J.D. Hamilton

10. Structural Time Series Models by A.C. Harvey and N. Shephard
11. Bayesian Testing and Testing Bayesians by J.-P. Florens and M. Mouchart
12. Pseudo-Likelihood Methods by C. Gourieroux and A. Monfort
13. Rao's Score Test: Recent Asymptotic Results by R. Mukerjee
14. On the Strong Consistency of M-Estimates in Linear Models under a General Discrepancy Function by Z.D. Bai, Z.J. Liu and C.R. Rao
15. Some Aspects of Generalized Method of Moments Estimation by A. Hall
16. Efficient Estimation of Models with Conditional Moment Restrictions by W.K. Newey
17. Generalized Method of Moments: Econometric Applications by M. Ogaki
18. Testing for Heteroscedasticity by A.R. Pagan and Y. Pak
19. Simulation Estimation Methods for Limited Dependent Variable Models by V.A. Hajivassiliou
20. Simulation Estimation for Panel Data Models with Limited Dependent Variable by M.P. Keane
21. A Perspective Application of Bootstrap Methods in Econometrics by J. Jeong and G.S. Maddala
22. Stochastic Simulations for Inference in Nonlinear Errors-in-Variables Models by R.S. Mariano and B.W. Brown
23. Bootstrap Methods: Applications in Econometrics by H.D. Vinod
24. Identifying Outliers and Influential Observations in Econometric Models by S.G. Donald and G.S. Maddala
25. Statistical Aspects of Calibration in Macroeconomics by A.W. Gregory and G.W. Smith
26. Panel Data Models with Rational Expectations by K. Lahiri
27. Continuous Time Financial Models: Statistical Applications of Stochastic Processes by K.R. Sawyer

Volume 12. Environmental Statistics
Edited by G.P. Patil and C.R. Rao
1994 xix + 927 pp.

1. Environmetrics: An Emerging Science by J.S. Hunter
2. A National Center for Statistical Ecology and Environmental Statistics: A Center Without Walls by G.P. Patil
3. Replicate Measurements for Data Quality and Environmental Modeling by W. Liggett
4. Design and Analysis of Composite Sampling Procedures: A Review by G. Lovison, S.D. Gore and G.P. Patil
5. Ranked Set Sampling by G.P. Patil, A.K. Sinha and C. Taillie
6. Environmental Adaptive Sampling by G.A.F. Seber and S.K. Thompson
7. Statistical Analysis of Censored Environmental Data by M. Akritas, T. Ruscitti and G.P. Patil

8. Biological Monitoring: Statistical Issues and Models by E.P. Smith
9. Environmental Sampling and Monitoring by S.V. Stehman and W. Scott Overton
10. Ecological Statistics by B.F.J. Manly
11. Forest Biometrics by H.E. Burkhart and T.G. Gregoire
12. Ecological Diversity and Forest Management by J.H. Gove, G.P. Patil, B.F. Swindel and C. Taillie
13. Ornithological Statistics by P.M. North
14. Statistical Methods in Developmental Toxicology by P.J. Catalano and L.M. Ryan
15. Environmental Biometry: Assessing Impacts of Environmental Stimuli Via Animal and Microbial Laboratory Studies by W.W. Piegorsch
16. Stochasticity in Deterministic Models by J.J.M. Bedaux and S.A.L.M. Kooijman
17. Compartmental Models of Ecological and Environmental Systems by J.H. Matis and T.E. Wehrly
18. Environmental Remote Sensing and Geographic Information Systems-Based Modeling by W.L. Myers
19. Regression Analysis of Spatially Correlated Data: The Kanawha County Health Study by C.A. Donnelly, J.H. Ware and N.M. Laird
20. Methods for Estimating Heterogeneous Spatial Covariance Functions with Environmental Applications by P. Guttorp and P.D. Sampson
21. Meta-analysis in Environmental Statistics by V. Hasselblad
22. Statistical Methods in Atmospheric Science by A.R. Solow
23. Statistics with Agricultural Pests and Environmental Impacts by L.J. Young and J.H. Young
24. A Crystal Cube for Coastal and Estuarine Degradation: Selection of Endpoints and Development of Indices for Use in Decision Making by M.T. Boswell, J.S. O'Connor and G.P. Patil
25. How Does Scientific Information in General and Statistical Information in Particular Input to the Environmental Regulatory Process? by C.R. Cothern
26. Environmental Regulatory Statistics by C.B. Davis
27. An Overview of Statistical Issues Related to Environmental Cleanup by R. Gilbert
28. Environmental Risk Estimation and Policy Decisions by H. Lacayo Jr

Volume 13. Design and Analysis of Experiments
Edited by S. Ghosh and C.R. Rao
1996 xviii + 1230 pp.

1. The Design and Analysis of Clinical Trials by P. Armitage
2. Clinical Trials in Drug Development: Some Statistical Issues by H.I. Patel
3. Optimal Crossover Designs by J. Stufken
4. Design and Analysis of Experiments: Nonparametric Methods with Applications to Clinical Trials by P.K. Sen

5. Adaptive Designs for Parametric Models by S. Zacks
6. Observational Studies and Nonrandomized Experiments by P.R. Rosenbaum
7. Robust Design: Experiments for Improving Quality by D.M. Steinberg
8. Analysis of Location and Dispersion Effects from Factorial Experiments with a Circular Response by C.M. Anderson
9. Computer Experiments by J.R. Koehler and A.B. Owen
10. A Critique of Some Aspects of Experimental Design by J.N. Srivastava
11. Response Surface Designs by N.R. Draper and D.K.J. Lin
12. Multiresponse Surface Methodology by A.I. Khuri
13. Sequential Assembly of Fractions in Factorial Experiments by S. Ghosh
14. Designs for Nonlinear and Generalized Linear Models by A.C. Atkinson and L.M. Haines
15. Spatial Experimental Design by R.J. Martin
16. Design of Spatial Experiments: Model Fitting and Prediction by V.V. Fedorov
17. Design of Experiments with Selection and Ranking Goals by S.S. Gupta and S. Panchapakesan
18. Multiple Comparisons by A.C. Tamhane
19. Nonparametric Methods in Design and Analysis of Experiments by E. Brunner and M.L. Puri
20. Nonparametric Analysis of Experiments by A.M. Dean and D.A. Wolfe
21. Block and Other Designs in Agriculture by D.J. Street
22. Block Designs: Their Combinatorial and Statistical Properties by T. Calinski and S. Kageyama
23. Developments in Incomplete Block Designs for Parallel Line Bioassays by S. Gupta and R. Mukerjee
24. Row-Column Designs by K.R. Shah and B.K. Sinha
25. Nested Designs by J.P. Morgan
26. Optimal Design: Exact Theory by C.S. Cheng
27. Optimal and Efficient Treatment – Control Designs by D. Majumdar
28. Model Robust Designs by Y.-J. Chang and W.I. Notz
29. Review of Optimal Bayes Designs by A. DasGupta
30. Approximate Designs for Polynomial Regression: Invariance, Admissibility, and Optimality by N. Gaffke and B. Heiligers

Volume 14. Statistical Methods in Finance
Edited by G.S. Maddala and C.R. Rao
1996 xvi + 733 pp.

1. Econometric Evaluation of Asset Pricing Models by W.E. Person and R. Jegannathan
2. Instrumental Variables Estimation of Conditional Beta Pricing Models by C.R. Harvey and C.M. Kirby

3. Semiparametric Methods for Asset Pricing Models by B.N. Lehmann
4. Modeling the Term Structure by A.R. Pagan, A.D. Hall and V. Martin
5. Stochastic Volatility by E. Ghysels, A.C. Harvey and E. Renault
6. Stock Price Volatility by S.F. LeRoy
7. GARCH Models of Volatility by F.C. Palm
8. Forecast Evaluation and Combination by F.X. Diebold and J.A. Lopez
9. Predictable Components in Stock Returns by G. Kaul
10. Interset Rate Spreads as Predictors of Business Cycles by K. Lahiri and J.G. Wang
11. Nonlinear Time Series, Complexity Theory, and Finance by W.A. Brock and P.J.F. deLima
12. Count Data Models for Financial Data by A.C. Cameron and P.K. Trivedi
13. Financial Applications of Stable Distributions by J.H. McCulloch
14. Probability Distributions for Financial Models by J.B. McDonald
15. Bootstrap Based Tests in Financial Models by G.S. Maddala and H. Li
16. Principal Component and Factor Analyses by C.R. Rao
17. Errors in Variables Problems in Finance by G.S. Maddala and M. Nimalendran
18. Financial Applications of Artificial Neural Networks by M. Qi
19. Applications of Limited Dependent Variable Models in Finance by G.S. Maddala
20. Testing Option Pricing Models by D.S. Bates
21. Peso Problems: Their Theoretical and Empirical Implications by M.D.D. Evans
22. Modeling Market Microstructure Time Series by J. Hasbrouck
23. Statistical Methods in Tests of Portfolio Efficiency: A Synthesis by J. Shanken

Volume 15. Robust Inference
Edited by G.S. Maddala and C.R. Rao
1997 xviii + 698 pp.

1. Robust Inference in Multivariate Linear Regression Using Difference of Two Convex Functions as the Discrepancy Measure by Z.D. Bai, C.R. Rao and Y.H. Wu
2. Minimum Distance Estimation: The Approach Using Density-Based Distances by A. Basu, I.R. Harris and S. Basu
3. Robust Inference: The Approach Based on Influence Functions by M. Markatou and E. Ronchetti
4. Practical Applications of Bounded-Influence Tests by S. Heritier and M.-P. Victoria-Feser
5. Introduction to Positive-Breakdown Methods by P.J. Rousseeuw
6. Outlier Identification and Robust Methods by U. Gather and C. Becker
7. Rank-Based Analysis of Linear Models by T.P. Hettmansperger, J.W. McKean and S.J. Sheather
8. Rank Tests for Linear Models by R. Koenker
9. Some Extensions in the Robust Estimation of Parameters of Exponential and Double Exponential Distributions in the Presence of Multiple Outliers by A. Childs and N. Balakrishnan

10. Outliers, Unit Roots and Robust Estimation of Nonstationary Time Series by G.S. Maddala and Y. Yin
11. Autocorrelation-Robust Inference by P.M. Robinson and C. Velasco
12. A Practitioner's Guide to Robust Covariance Matrix Estimation by W.J. den Haan and A. Levin
13. Approaches to the Robust Estimation of Mixed Models by A.H. Welsh and A.M. Richardson
14. Nonparametric Maximum Likelihood Methods by S.R. Cosslett
15. A Guide to Censored Quantile Regressions by B. Fitzenberger
16. What Can Be Learned About Population Parameters When the Data Are Contaminated by J.L. Horowitz and C.F. Manski
17. Asymptotic Representations and Interrelations of Robust Estimators and Their Applications by J. Jurečková and P.K. Sen
18. Small Sample Asymptotics: Applications in Robustness by C.A. Field and M.A. Tingley
19. On the Fundamentals of Data Robustness by G. Maguluri and K. Singh
20. Statistical Analysis With Incomplete Data: A Selective Review by M.G. Akritas and M.P. La Valley
21. On Contamination Level and Sensitivity of Robust Tests by J.Á. Vissek
22. Finite Sample Robustness of Tests: An Overview by T. Kariya and P. Kim
23. Future Directions by G.S. Maddala and C.R. Rao

Volume 16. Order Statistics – Theory and Methods
Edited by N. Balakrishnan and C.R. Rao
1997 xix + 688 pp.

1. Order Statistics: An Introduction by N. Balakrishnan and C.R. Rao
2. Order Statistics: A Historical Perspective by H. Leon Harter and N. Balakrishnan
3. Computer Simulation of Order Statistics by Pandu R. Tadikamalla and N. Balakrishnan
4. Lorenz Ordering of Order Statistics and Record Values by Barry C. Arnold and Jose A. Villasenor
5. Stochastic Ordering of Order Statistics by Philip J. Boland, Moshe Shaked and J. George Shanthikumar
6. Bounds for Expectations of L-Estimates by T. Rychlik
7. Recurrence Relations and Identities for Moments of Order Statistics by N. Balakrishnan and K.S. Sultan
8. Recent Approaches to Characterizations Based on Order Statistics and Record Values by C.R. Rao and D.N. Shanbhag
9. Characterizations of Distributions via Identically Distributed Functions of Order Statistics by Ursula Gather, Udo Kamps and Nicole Schweitzer
10. Characterizations of Distributions by Recurrence Relations and Identities for Moments of Order Statistics by Udo Kamps

11. Univariate Extreme Value Theory and Applications by Janos Galambos
12. Order Statistics: Asymptotics in Applications by Pranab Kumar Sen
13. Zero-One Laws for Large Order Statistics by R.J. Tomkins and Hong Wang
14. Some Exact Properties of Cook's D_I by D.R. Jensen and D.E. Ramirez
15. Generalized Recurrence Relations for Moments of Order Statistics from Non-Identical Pareto and Truncated Pareto Random Variables with Applications to Robustness by Aaron Childs and N. Balakrishnan
16. A Semiparametric Bootstrap for Simulating Extreme Order Statistics by Robert L. Strawderman and Daniel Zelterman
17. Approximations to Distributions of Sample Quantiles by Chunsheng Ma and John Robinson
18. Concomitants of Order Statistics by H.A. David and H.N. Nagaraja
19. A Record of Records by Valery B. Nevzorov and N. Balakrishnan
20. Weighted Sequential Empirical Type Processes with Applications to Change-Point Problems by Barbara Szyszkowicz
21. Sequential Quantile and Bahadur–Kiefer Processes by Miklós Csörgő and Barbara Szyszkowicz

Volume 17. Order Statistics: Applications
Edited by N. Balakrishnan and C.R. Rao
1998 xviii + 712 pp.

1. Order Statistics in Exponential Distribution by Asit P. Basu and Bahadur Singh
2. Higher Order Moments of Order Statistics from Exponential and Right-truncated Exponential Distributions and Applications to Life-testing Problems by N. Balakrishnan and Shanti S. Gupta
3. Log-gamma Order Statistics and Linear Estimation of Parameters by N. Balakrishnan and P.S. Chan
4. Recurrence Relations for Single and Product Moments of Order Statistics from a Generalized Logistic Distribution with Applications to Inference and Generalizations to Double Truncation by N. Balakrishnan and Rita Aggarwala
5. Order Statistics from the Type III Generalized Logistic Distribution and Applications by N. Balakrishnan and S.K. Lee
6. Estimation of Scale Parameter Based on a Fixed Set of Order Statistics by Sanat K. Sarkar and Wenjin Wang
7. Optimal Linear Inference Using Selected Order Statistics in Location-Scale Models by M. Masoom Ali and Dale Umbach
8. L-Estimation by J.R.M. Hosking
9. On Some L-estimation in Linear Regression Models by Soroush Alimoradi and A.K.Md. Ehsanes Saleh
10. The Role of Order Statistics in Estimating Threshold Parameters by A. Clifford Cohen

11. Parameter Estimation under Multiply Type-II Censoring by Fanhui Kong
12. On Some Aspects of Ranked Set Sampling in Parametric Estimation by Nora Ni Chuiv and Bimal K. Sinha
13. Some Uses of Order Statistics in Bayesian Analysis by Seymour Geisser
14. Inverse Sampling Procedures to Test for Homogeneity in a Multinomial Distribution by S. Panchapakesan, Aaron Childs, B.H. Humphrey and N. Balakrishnan
15. Prediction of Order Statistics by Kenneth S. Kaminsky and Paul I. Nelson
16. The Probability Plot: Tests of Fit Based on the Correlation Coefficient by R.A. Lockhart and M.A. Stephens
17. Distribution Assessment by Samuel Shapiro
18. Application of Order Statistics to Sampling Plans for Inspection by Variables by Helmut Schneider and Frances Barbera
19. Linear Combinations of Ordered Symmetric Observations with Applications to Visual Acuity by Marios Viana
20. Order-Statistic Filtering and Smoothing of Time-Series: Part I by Gonzalo R. Arce, Yeong-Taeg Kim and Kenneth E. Barner
21. Order-Statistic Filtering and Smoothing of Time-Series: Part II by Kenneth E. Barner and Gonzalo R. Arce
22. Order Statistics in Image Processing by Scott T. Acton and Alan C. Bovik
23. Order Statistics Application to CFAR Radar Target Detection by R. Viswanathan

Volume 18. Bioenvironmental and Public Health Statistics
Edited by P.K. Sen and C.R. Rao
2000 xxiv + 1105 pp.

1. Bioenvironment and Public Health: Statistical Perspectives by Pranab K. Sen
2. Some Examples of Random Process Environmental Data Analysis by David R. Brillinger
3. Modeling Infectious Diseases – Aids by L. Billard
4. On Some Multiplicity Problems and Multiple Comparison Procedures in Biostatistics by Yosef Hochberg and Peter H. Westfall
5. Analysis of Longitudinal Data by Julio M. Singer and Dalton F. Andrade
6. Regression Models for Survival Data by Richard A. Johnson and John P. Klein
7. Generalised Linear Models for Independent and Dependent Responses by Bahjat F. Qaqish and John S. Preisser
8. Hierarchial and Empirical Bayes Methods for Environmental Risk Assessment by Gauri Datta, Malay Ghosh and Lance A. Waller
9. Non-parametrics in Bioenvironmental and Public Health Statistics by Pranab Kumar Sen
10. Estimation and Comparison of Growth and Dose-Response Curves in the Presence of Purposeful Censoring by Paul W. Stewart

11. Spatial Statistical Methods for Environmental Epidemiology by Andrew B. Lawson and Noel Cressie
12. Evaluating Diagnostic Tests in Public Health by Margaret Pepe, Wendy Leisenring and Carolyn Rutter
13. Statistical Issues in Inhalation Toxicology by E. Weller, L. Ryan and D. Dockery
14. Quantitative Potency Estimation to Measure Risk with Bioenvironmental Hazards by A. John Bailer and Walter W. Piegorsch
15. The Analysis of Case-Control Data: Epidemiologic Studies of Familial Aggregation by Nan M. Laird, Garrett M. Fitzmaurice and Ann G. Schwartz
16. Cochran–Mantel–Haenszel Techniques: Applications Involving Epidemiologic Survey Data by Daniel B. Hall, Robert F. Woolson, William R. Clarke and Martha F. Jones
17. Measurement Error Models for Environmental and Occupational Health Applications by Robert H. Lyles and Lawrence L. Kupper
18. Statistical Perspectives in Clinical Epidemiology by Shrikant I. Bangdiwala and Sergio R. Muñoz
19. ANOVA and ANOCOVA for Two-Period Crossover Trial Data: New vs. Standard by Subir Ghosh and Lisa D. Fairchild
20. Statistical Methods for Crossover Designs in Bioenvironmental and Public Health Studies by Gail E. Tudor, Gary G. Koch and Diane Catellier
21. Statistical Models for Human Reproduction by C.M. Suchindran and Helen P. Koo
22. Statistical Methods for Reproductive Risk Assessment by Sati Mazumdar, Yikang Xu, Donald R. Mattison, Nancy B. Sussman and Vincent C. Arena
23. Selection Biases of Samples and their Resolutions by Ranajit Chakraborty and C. Radhakrishna Rao
24. Genomic Sequences and Quasi-Multivariate CATANOVA by Hildete Prisco Pinheiro, Françoise Seillier-Moiseiwitsch, Pranab Kumar Sen and Joseph Eron Jr
25. Statistical Methods for Multivariate Failure Time Data and Competing Risks by Ralph A. DeMasi
26. Bounds on Joint Survival Probabilities with Positively Dependent Competing Risks by Sanat K. Sarkar and Kalyan Ghosh
27. Modeling Multivariate Failure Time Data by Limin X. Clegg, Jianwen Cai and Pranab K. Sen
28. The Cost–Effectiveness Ratio in the Analysis of Health Care Programs by Joseph C. Gardiner, Cathy J. Bradley and Marianne Huebner
29. Quality-of-Life: Statistical Validation and Analysis An Example from a Clinical Trial by Balakrishna Hosmane, Clement Maurath and Richard Manski
30. Carcinogenic Potency: Statistical Perspectives by Anup Dewanji
31. Statistical Applications in Cardiovascular Disease by Elizabeth R. DeLong and David M. DeLong
32. Medical Informatics and Health Care Systems: Biostatistical and Epidemiologic Perspectives by J. Zvárová
33. Methods of Establishing In Vitro–In Vivo Relationships for Modified Release Drug Products by David T. Mauger and Vernon M. Chinchilli

34. Statistics in Psychiatric Research by Sati Mazumdar, Patricia R. Houck and Charles F. Reynolds III
35. Bridging the Biostatistics–Epidemiology Gap by Lloyd J. Edwards
36. Biodiversity – Measurement and Analysis by S.P. Mukherjee

Volume 19. Stochastic Processes: Theory and Methods
Edited by D.N. Shanbhag and C.R. Rao
2001 xiv + 967 pp.

1. Pareto Processes by Barry C. Arnold
2. Branching Processes by K.B. Athreya and A.N. Vidyashankar
3. Inference in Stochastic Processes by I.V. Basawa
4. Topics in Poisson Approximation by A.D. Barbour
5. Some Elements on Lévy Processes by Jean Bertoin
6. Iterated Random Maps and Some Classes of Markov Processes by Rabi Bhattacharya and Edward C. Waymire
7. Random Walk and Fluctuation Theory by N.H. Bingham
8. A Semigroup Representation and Asymptotic Behavior of Certain Statistics of the Fisher–Wright–Moran Coalescent by Adam Bobrowski, Marek Kimmel, Ovide Arino and Ranajit Chakraborty
9. Continuous-Time ARMA Processes by P.J. Brockwell
10. Record Sequences and their Applications by John Bunge and Charles M. Goldie
11. Stochastic Networks with Product Form Equilibrium by Hans Daduna
12. Stochastic Processes in Insurance and Finance by Paul Embrechts, Rüdiger Frey and Hansjörg Furrer
13. Renewal Theory by D.R. Grey
14. The Kolmogorov Isomorphism Theorem and Extensions to some Nonstationary Processes by Yûichirô Kakihara
15. Stochastic Processes in Reliability by Masaaki Kijima, Haijun Li and Moshe Shaked
16. On the supports of Stochastic Processes of Multiplicity One by A. Kłopotowski and M.G. Nadkarni
17. Gaussian Processes: Inequalities, Small Ball Probabilities and Applications by W.V. Li and Q.-M. Shao
18. Point Processes and Some Related Processes by Robin K. Milne
19. Characterization and Identifiability for Stochastic Processes by B.L.S. Prakasa Rao
20. Associated Sequences and Related Inference Problems by B.L.S. Prakasa Rao and Isha Dewan
21. Exchangeability, Functional Equations, and Characterizations by C.R. Rao and D.N. Shanbhag
22. Martingales and Some Applications by M.M. Rao
23. Markov Chains: Structure and Applications by R.L. Tweedie

24. Diffusion Processes by S.R.S. Varadhan
25. Itô's Stochastic Calculus and Its Applications by S. Watanabe

Volume 20. Advances in Reliability
Edited by N. Balakrishnan and C.R. Rao
2001 xxii + 860 pp.

1. Basic Probabilistic Models in Reliability by N. Balakrishnan, N. Limnios and C. Papadopoulos
2. The Weibull Nonhomogeneous Poisson Process by A.P. Basu and S.E. Rigdon
3. Bathtub-Shaped Failure Rate Life Distributions by C.D. Lai, M. Xie and D.N.P. Murthy
4. Equilibrium Distribution – its Role in Reliability Theory by A. Chatterjee and S.P. Mukherjee
5. Reliability and Hazard Based on Finite Mixture Models by E.K. Al-Hussaini and K.S. Sultan
6. Mixtures and Monotonicity of Failure Rate Functions by M. Shaked and F. Spizzichino
7. Hazard Measure and Mean Residual Life Orderings: A Unified Approach by M. Asadi and D.N. Shanbhag
8. Some Comparison Results of the Reliability Functions of Some Coherent Systems by J. Mi
9. On the Reliability of Hierarchical Structures by L.B. Klebanov and G.J. Szekely
10. Consecutive k-out-of-n Systems by N.A. Mokhlis
11. Exact Reliability and Lifetime of Consecutive Systems by S. Aki
12. Sequential k-out-of-n Systems by E. Cramer and U. Kamps
13. Progressive Censoring: A Review by R. Aggarwala
14. Point and Interval Estimation for Parameters of the Logistic Distribution Based on Progressively Type-II Censored Samples by N. Balakrishnan and N. Kannan
15. Progressively Censored Variables-Sampling Plans for Life Testing by U. Balasooriya
16. Graphical Techniques for Analysis of Data From Repairable Systems by P.A. Akersten, B. Klefsjö and B. Bergman
17. A Bayes Approach to the Problem of Making Repairs by G.C. McDonald
18. Statistical Analysis for Masked Data by B.J. Flehinger†, B. Reiser and E. Yashchin
19. Analysis of Masked Failure Data under Competing Risks by A. Sen, S. Basu and M. Banerjee
20. Warranty and Reliability by D.N.P. Murthy and W.R. Blischke
21. Statistical Analysis of Reliability Warranty Data by K. Suzuki, Md. Rezaul Karim and L. Wang
22. Prediction of Field Reliability of Units, Each under Differing Dynamic Stresses, from Accelerated Test Data by W. Nelson

23. Step-Stress Accelerated Life Test by E. Gouno and N. Balakrishnan
24. Estimation of Correlation under Destructive Testing by R. Johnson and W. Lu
25. System-Based Component Test Plans for Reliability Demonstration: A Review and Survey of the State-of-the-Art by J. Rajgopal and M. Mazumdar
26. Life-Test Planning for Preliminary Screening of Materials: A Case Study by J. Stein and N. Doganaksoy
27. Analysis of Reliability Data from In-House Audit Laboratory Testing by R. Agrawal and N. Doganaksoy
28. Software Reliability Modeling, Estimation and Analysis by M. Xie and G.Y. Hong
29. Bayesian Analysis for Software Reliability Data by J.A. Achcar
30. Direct Graphical Estimation for the Parameters in a Three-Parameter Weibull Distribution by P.R. Nelson and K.B. Kulasekera
31. Bayesian and Frequentist Methods in Change-Point Problems by N. Ebrahimi and S.K. Ghosh
32. The Operating Characteristics of Sequential Procedures in Reliability by S. Zacks
33. Simultaneous Selection of Extreme Populations from a Set of Two-Parameter Exponential Populations by K. Hussein and S. Panchapakesan

Volume 21. Stochastic Processes: Modelling and Simulation
Edited by D.N. Shanbhag and C.R. Rao
2003 xxviii + 1002 pp.

1. Modelling and Numerical Methods in Manufacturing System Using Control Theory by E.K. Boukas and Z.K. Liu
2. Models of Random Graphs and their Applications by C. Cannings and D.B. Penman
3. Locally Self-Similar Processes and their Wavelet Analysis by J.E. Cavanaugh, Y. Wang and J.W. Davis
4. Stochastic Models for DNA Replication by R. Cowan
5. An Empirical Process with Applications to Testing the Exponential and Geometric Models by J.A. Ferreira
6. Patterns in Sequences of Random Events by J. Gani
7. Stochastic Models in Telecommunications for Optimal Design, Control and Performance Evaluation by N. Gautam
8. Stochastic Processes in Epidemic Modelling and Simulation by D. Greenhalgh
9. Empirical Estimators Based on MCMC Data by P.E. Greenwood and W. Wefelmeyer
10. Fractals and the Modelling of Self-Similarity by B.M. Hambly
11. Numerical Methods in Queueing Theory by D. Heyman
12. Applications of Markov Chains to the Distribution Theory of Runs and Patterns by M.V. Koutras
13. Modelling Image Analysis Problems Using Markov Random Fields by S.Z. Li
14. An Introduction to Semi-Markov Processes with Application to Reliability by N. Limnios and G. Oprişan

15. Departures and Related Characteristics in Queueing Models by M. Manoharan, M.H. Alamatsaz and D.N. Shanbhag
16. Discrete Variate Time Series by E. McKenzie
17. Extreme Value Theory, Models and Simulation by S. Nadarajah
18. Biological Applications of Branching Processes by A.G. Pakes
19. Markov Chain Approaches to Damage Models by C.R. Rao, M. Albassam, M.B. Rao and D.N. Shanbhag
20. Point Processes in Astronomy: Exciting Events in the Universe by J.D. Scargle and G.J. Babu
21. On the Theory of Discrete and Continuous Bilinear Time Series Models by T. Subba Rao and Gy. Terdik
22. Nonlinear and Non-Gaussian State-Space Modeling with Monte Carlo Techniques: A Survey and Comparative Study by H. Tanizaki
23. Markov Modelling of Burst Behaviour in Ion Channels by G.F. Yeo, R.K. Milne, B.W. Madsen, Y. Li and R.O. Edeson

Volume 22. Statistics in Industry
Edited by R. Khattree and C.R. Rao
2003 xxi + 1150 pp.

1. Guidelines for Selecting Factors and Factor Levels for an Industrial Designed Experiment by V. Czitrom
2. Industrial Experimentation for Screening by D.K.J. Lin
3. The Planning and Analysis of Industrial Selection and Screening Experiments by G. Pan, T.J. Santner and D.M. Goldsman
4. Uniform Experimental Designs and their Applications in Industry by K.-T. Fang and D.K.J. Lin
5. Mixed Models and Repeated Measures: Some Illustrative Industrial Examples by G.A. Milliken
6. Current Modeling and Design Issues in Response Surface Methodology: GLMs and Models with Block Effects by A.I. Khuri
7. A Review of Design and Modeling in Computer Experiments by V.C.P. Chen, K.-L. Tsui, R.R. Barton and J.K. Allen
8. Quality Improvement and Robustness via Design of Experiments by B.E. Ankenman and A.M. Dean
9. Software to Support Manufacturing Experiments by J.E. Reece
10. Statistics in the Semiconductor Industry by V. Czitrom
11. PREDICT: A New Approach to Product Development and Lifetime Assessment Using Information Integration Technology by J.M. Booker, T.R. Bement, M.A. Meyer and W.J. Kerscher III
12. The Promise and Challenge of Mining Web Transaction Data by S.R. Dalal, D. Egan, Y. Ho and M. Rosenstein

13. Control Chart Schemes for Monitoring the Mean and Variance of Processes Subject to Sustained Shifts and Drifts by Z.G. Stoumbos, M.R. Reynolds Jr and W.H. Woodall
14. Multivariate Control Charts: Hotelling T^2, Data Depth and Beyond by R.Y. Liu
15. Effective Sample Sizes for T^2 Control Charts by R.L. Mason, Y.-M. Chou and J.C. Young
16. Multidimensional Scaling in Process Control by T.F. Cox
17. Quantifying the Capability of Industrial Processes by A.M. Polansky and S.N.U.A. Kirmani
18. Taguchi's Approach to On-line Control Procedure by M.S. Srivastava and Y. Wu
19. Dead-Band Adjustment Schemes for On-line Feedback Quality Control by A. Luceño
20. Statistical Calibration and Measurements by H. Iyer
21. Subsampling Designs in Industry: Statistical Inference for Variance Components by R. Khattree
22. Repeatability, Reproducibility and Interlaboratory Studies by R. Khattree
23. Tolerancing – Approaches and Related Issues in Industry by T.S. Arthanari
24. Goodness-of-fit Tests for Univariate and Multivariate Normal Models by D.K. Srivastava and G.S. Mudholkar
25. Normal Theory Methods and their Simple Robust Analogs for Univariate and Multivariate Linear Models by D.K. Srivastava and G.S. Mudholkar
26. Diagnostic Methods for Univariate and Multivariate Normal Data by D.N. Naik
27. Dimension Reduction Methods Used in Industry by G. Merola and B. Abraham
28. Growth and Wear Curves by A.M. Kshirsagar
29. Time Series in Industry and Business by B. Abraham and N. Balakrishna
30. Stochastic Process Models for Reliability in Dynamic Environments by N.D. Singpurwalla, T.A. Mazzuchi, S. Özekici and R. Soyer
31. Bayesian Inference for the Number of Undetected Errors by S. Basu

Volume 23. Advances in Survival Analysis
Edited by N. Balakrishnan and C.R. Rao
2003 xxv + 795 pp.

1. Evaluation of the Performance of Survival Analysis Models: Discrimination and Calibration Measures by R.B. D'Agostino and B.-H. Nam
2. Discretizing a Continuous Covariate in Survival Studies by J.P. Klein and J.-T. Wu
3. On Comparison of Two Classification Methods with Survival Endpoints by Y. Lu, H. Jin and J. Mi
4. Time-Varying Effects in Survival Analysis by T.H. Scheike
5. Kaplan–Meier Integrals by W. Stute
6. Statistical Analysis of Doubly Interval-Censored Failure Time Data by J. Sun
7. The Missing Censoring-Indicator Model of Random Censorship by S. Subramanian
8. Estimation of the Bivariate Survival Function with Generalized Bivariate Right Censored Data Structures by S. Keleş, M.J. van der Laan and J.M. Robins

9. Estimation of Semi-Markov Models with Right-Censored Data by O. Pons
10. Nonparametric Bivariate Estimation with Randomly Truncated Observations by Ü. Gürler
11. Lower Bounds for Estimating a Hazard by C. Huber and B. MacGibbon
12. Non-Parametric Hazard Rate Estimation under Progressive Type-II Censoring by N. Balakrishnan and L. Bordes
13. Statistical Tests of the Equality of Survival Curves: Reconsidering the Options by G.P. Suciu, S. Lemeshow and M. Moeschberger
14. Testing Equality of Survival Functions with Bivariate Censored Data: A Review by P.V. Rao
15. Statistical Methods for the Comparison of Crossing Survival Curves by C.T. Le
16. Inference for Competing Risks by J.P. Klein and R. Bajorunaite
17. Analysis of Cause-Specific Events in Competing Risks Survival Data by J. Dignam, J. Bryant and H.S. Wieand
18. Analysis of Progressively Censored Competing Risks Data by D. Kundu, N. Kannan and N. Balakrishnan
19. Marginal Analysis of Point Processes with Competing Risks by R.J. Cook, B. Chen and P. Major
20. Categorical Auxiliary Data in the Discrete Time Proportional Hazards Model by P. Slasor and N. Laird
21. Hosmer and Lemeshow type Goodness-of-Fit Statistics for the Cox Proportional Hazards Model by S. May and D.W. Hosmer
22. The Effects of Misspecifying Cox's Regression Model on Randomized Treatment Group Comparisons by A.G. DiRienzo and S.W. Lagakos
23. Statistical Modeling in Survival Analysis and Its Influence on the Duration Analysis by V. Bagdonavičius and M. Nikulin
24. Accelerated Hazards Model: Method, Theory and Applications by Y.Q. Chen, N.P. Jewell and J. Yang
25. Diagnostics for the Accelerated Life Time Model of Survival Data by D. Zelterman and H. Lin
26. Cumulative Damage Approaches Leading to Inverse Gaussian Accelerated Test Models by A. Onar and W.J. Padgett
27. On Estimating the Gamma Accelerated Failure-Time Models by K.M. Koti
28. Frailty Model and its Application to Seizure Data by N. Ebrahimi, X. Zhang, A. Berg and S. Shinnar
29. State Space Models for Survival Analysis by W.Y. Tan and W. Ke
30. First Hitting Time Models for Lifetime Data by M.-L.T. Lee and G.A. Whitmore
31. An Increasing Hazard Cure Model by Y. Peng and K.B.G. Dear
32. Marginal Analyses of Multistage Data by G.A. Satten and S. Datta
33. The Matrix-Valued Counting Process Model with Proportional Hazards for Sequential Survival Data by K.L. Kesler and P.K. Sen
34. Analysis of Recurrent Event Data by J. Cai and D.E. Schaubel
35. Current Status Data: Review, Recent Developments and Open Problems by N.P. Jewell and M. van der Laan

36. Appraisal of Models for the Study of Disease Progression in Psoriatic Arthritis by R. Aguirre-Hernández and V.T. Farewell
37. Survival Analysis with Gene Expression Arrays by D.K. Pauler, J. Hardin, J.R. Faulkner, M. LeBlanc and J.J. Crowley
38. Joint Analysis of Longitudinal Quality of Life and Survival Processes by M. Mesbah, J.-F. Dupuy, N. Heutte and L. Awad
39. Modelling Survival Data using Flowgraph Models by A.V. Huzurbazar
40. Nonparametric Methods for Repair Models by M. Hollander and J. Sethuraman

Volume 24. Data Mining and Data Visualization
Edited by C.R. Rao, E.J. Wegman and J.L. Solka
2005 xiv + 643 pp.

1. Statistical Data Mining by E.J. Wegman and J.L. Solka
2. From Data Mining to Knowledge Mining by K.A. Kaufman and R.S. Michalski
3. Mining Computer Security Data by D.J. Marchette
4. Data Mining of Text Files by A.R. Martinez
5. Text Data Mining with Minimal Spanning Trees by J.L. Solka, A.C. Bryant and E.J. Wegman
6. Information Hiding: Steganography and Steganalysis by Z. Duric, M. Jacobs and S. Jajodia
7. Canonical Variate Analysis and Related Methods for Reduction of Dimensionality and Graphical Representation by C.R. Rao
8. Pattern Recognition by D.J. Hand
9. Multidimensional Density Estimation by D.W. Scott and S.R. Sain
10. Multivariate Outlier Detection and Robustness by M. Hubert, P.J. Rousseeuw and S. Van Aelst
11. Classification and Regression Trees, Bagging, and Boosting by C.D. Sutton
12. Fast Algorithms for Classification Using Class Cover Catch Digraphs by D.J. Marchette, E.J. Wegman and C.E. Priebe
13. On Genetic Algorithms and their Applications by Y.H. Said
14. Computational Methods for High-Dimensional Rotations in Data Visualization by A. Buja, D. Cook, D. Asimov and C. Hurley
15. Some Recent Graphics Templates and Software for Showing Statistical Summaries by D.B. Carr
16. Interactive Statistical Graphics: the Paradigm of Linked Views by A. Wilhelm
17. Data Visualization and Virtual Reality by J.X. Chen

Volume 25. Bayesian Thinking: Modeling and Computation
Edited by D.K. Dey and C.R. Rao
2005 xx + 1041 pp.

1. Bayesian Inference for Causal Effects by D.B. Rubin
2. Reference Analysis by J.M. Bernardo
3. Probability Matching Priors by G.S. Datta and T.J. Sweeting
4. Model Selection and Hypothesis Testing based on Objective Probabilities and Bayes Factors by L.R. Pericchi
5. Role of P-values and other Measures of Evidence in Bayesian Analysis by J. Ghosh, S. Purkayastha and T. Samanta
6. Bayesian Model Checking and Model Diagnostics by H.S. Stern and S. Sinharay
7. The Elimination of Nuisance Parameters by B. Liseo
8. Bayesian Estimation of Multivariate Location Parameters by A.C. Brandwein and W.E. Strawderman
9. Bayesian Nonparametric Modeling and Data Analysis: An Introduction by T.E. Hanson, A.J. Branscum and W.O. Johnson
10. Some Bayesian Nonparametric Models by P. Damien
11. Bayesian Modeling in the Wavelet Domain by F. Ruggeri and B. Vidakovic
12. Bayesian Nonparametric Inference by S. Walker
13. Bayesian Methods for Function Estimation by N. Choudhuri, S. Ghosal and A. Roy
14. MCMC Methods to Estimate Bayesian Parametric Models by A. Mira
15. Bayesian Computation: From Posterior Densities to Bayes Factors, Marginal Likelihoods, and Posterior Model Probabilities by M.-H. Chen
16. Bayesian Modelling and Inference on Mixtures of Distributions by J.-M. Marin, K. Mengersen and C.P. Robert
17. Simulation Based Optimal Design by P. Müller
18. Variable Selection and Covariance Selection in Multivariate Regression Models by E. Cripps, C. Carter and R. Kohn
19. Dynamic Models by H.S. Migon, D. Gamerman, H.F. Lopes and M.A.R. Ferreira
20. Bayesian Thinking in Spatial Statistics by L.A. Waller
21. Robust Bayesian Analysis by F. Ruggeri, D. Ríos Insua and Jacinto Martín
22. Elliptical Measurement Error Models – A Bayesian Approach by H. Bolfarine and R.B. Arellano-Valle
23. Bayesian Sensitivity Analysis in Skew-elliptical Models by I. Vidal, P. Iglesias and M.D. Branco
24. Bayesian Methods for DNA Microarray Data Analysis by V. Baladandayuthapani, S. Ray and B.K. Mallick
25. Bayesian Biostatistics by D.B. Dunson
26. Innovative Bayesian Methods for Biostatistics and Epidemiology by P. Gustafson, S. Hossain and L. McCandless
27. Bayesian Analysis of Case-Control Studies by B. Mukherjee, S. Sinha and M. Ghosh
28. Bayesian Analysis of ROC Data by V.E. Johnson and T.D. Johnson

29. Modeling and Analysis for Categorical Response Data by S. Chib
30. Bayesian Methods and Simulation-Based Computation for Contingency Tables by J.H. Albert
31. Multiple Events Time Data: A Bayesian Recourse by D. Sinha and S.K. Ghosh
32. Bayesian Survival Analysis for Discrete Data with Left-Truncation and Interval Censoring by C.Z. He and D. Sun
33. Software Reliability by L. Kuo
34. Bayesian Aspects of Small Area Estimation by T. Maiti
35. Teaching Bayesian Thought to Nonstatisticians by D.K. Stangl

Edwards Brothers Malloy
Ann Arbor MI. USA
May 7, 2013